Principles of Conservation Biology

THIRD EDITION

Principles *of* Conservation Biology

THIRD EDITION

Martha J. Groom
University of Washington
Bothell and Seattle

Gary K. Meffe
Department of Wildlife Ecology and Conservation
University of Florida

C. Ronald Carroll
Director for Science in the River Basin Science and Policy Center
University of Georgia

Sinauer Associates, Inc. • Publishers
Sunderland, Massachusetts U.S.A.

About the Cover

Mountains from Zhangjiajie National Forest Park, Hunan Province, China. Zhangjiajie was the first national forest park created in China in 1982, and protects 4810 hectares of forests, mountains, and many rare species from the pressures of population growth. Similar areas of high mountains have long been revered in Chinese culture, with protections beginning 4200 B.P. as Imperial or temple gardens. An IUCN Category V protected area, the park is one of over 2000 sites in China's protected area network. (Photograph © John Wong/RubberBall Productions/PictureQuest.)

Principles of Conservation Biology, Third Edition

Copyright © 2006 by Sinauer Associates, Inc.

For information, address
Sinauer Associates, 23 Plumtree Road, Sunderland, MA 01375 USA
FAX: 413-549-1118
E-mail: publish@sinauer.com
Internet : www.sinauer.com

Library of Congress Cataloging-in-Publication Data

Groom, Martha J., 1962-
    Principles of conservation biology / Martha J. Groom, Gary K. Meffe, C. Ronald Carroll.— 3rd ed.
      p. cm.
    Rev. ed. of: Principles of conservation biology / Gary K. Meffe. 2nd ed. 1997.
    ISBN-10: 87893-518-5 (alk. paper)
    ISBN-13: 978-0-87893-518-5
    1. Conservation biology. I. Meffe, Gary K. II. Carroll, C. Ronald (Carl Ronald), 1941- III. Meffe, Gary K. Principles of conservation biology. IV.Title.

QH75.M386 2005
333.95'16—dc22   2005016894

Printed in U.S.A.

To the students of conservation biology,
in whose collective hands the future of biodiversity rests,
and to the pioneers of the field, upon whose shoulders we stand.

Brief Contents

Contents

UNIT I
Conceptual Foundations for Conservation Biology 1

UNIT II
Focus on Primary Threats to Biodiversity 171

CHAPTER 8

Overexploitation 253

John D. Reynolds and Carlos A. Peres

CHAPTER 9

Species Invasions 293

Marjorie Wonham

CHAPTER 10

Biological Impacts of Climate Change 333

Camille Parmesan and John Matthews

CHAPTER 11

Conservation Genetics: The Use and Importance of Genetic Information 375

Kim T. Scribner, Gary K. Meffe, and Martha J. Groom

UNIT *III*

Approaches to Solving Conservation Problems   417

CHAPTER 14

Protected Areas: Goals, Limitations, and Design 509

Hugh P. Possingham, Kerrie A. Wilson, Sandy J. Andelman, and Carly H. Vynne

CHAPTER 15

Restoration of Damaged Ecosystems and Endangered Populations 553

Peggy L. Fiedler and Martha J. Groom

Contributing Chapter Authors

Sandy J. Andelman, Tropical Ecology, Assessment and Monitoring Initiative, Conservation International

Deborah M. Brosnan, Sustainable Ecosystems Institute, Portland, Oregon

J. Baird Callicott, Department of Philosophy and Religion Studies, University of North Texas

Blair Csuti, Department of Biology, Portland State University

John B. Dunning, Jr., Department of Forestry and Natural Resources, Purdue University

Gareth Edwards-Jones, School of Agricultural and Forest Sciences, University of Wales, Bangor

Peggy L. Fiedler, Blasland, Bouck, & Lee Sciences, Walnut Creek, CA

John H. Matthews, Section of Integrative Biology, University of Texas, Austin

Reed F. Noss, Department of Biology, University of Central Florida

Gordan H. Orians, Department of Biology, University of Washington

Camille Parmesan, Section of Integrative Biology, University of Texas, Austin

Carlos A. Peres, School of Environmental Sciences, University of East Anglia, United Kingdom

Hugh P. Possingham, The Ecology Centre, The University of Queensland, Australia

H. Ronald Pulliam, Institute of Ecology, University of Georgia

John D. Reynolds, School of Biological Sciences, Simon Fraser University, Burnaby, Canada

Kim T. Scribner, Department of Fisheries and Wildlife and Department of Zoology, Michigan State University

Carly H. Vynne, Department of Biology, University of Washington

Kerrie A. Wilson, The Ecology Centre, The University of Queensland, St. Lucia, Australia

Marjorie J. Wonham, Centre for Mathematical Biology, University of Alberta, Edmonton, Canada

Preface

A great deal has changed in the world since the Second Edition of this book in 1997 and our approach to writing and structuring the book has changed as well. Most importantly, Martha Groom joined us in 2000 and took on the task of managing a major revision of the text. To guide the revision, Martha surveyed over 60 instructors on their needs for teaching conservation biology. Martha also conducted focus groups of graduate and undergraduate students at the University of Washington, who taught us much about their perceptions and concerns regarding conservation biology. This input made it clear that a major reorganization of the text was appropriate. We are grateful for all the suggestions we received—we only wish it were possible to integrate more of these insights.

In this edition, we increased our coverage of marine conservation issues by adding the chapters on overexploitation and species invasion, both of which are particularly large problems in the marine realm. We also increased our coverage of issues in countries other than the U.S., although the book has its primary focus on U.S. and global conservation issues. In response to our surveys, we increased representation of modeling approaches and discussion of models in conservation planning. We have incorporated a greater spectrum of guest essays and case studies, which are featured in every chapter. We worked to keep the flavor of multiple voices of experience and opinion that characterized the first two editions. Conservation is necessarily an enterprise of shared work, and we are fortunate to have many colleagues willing to describe their efforts openly to help others learn.

This book is structured to introduce essential concepts in conservation biology, beginning with an overview of the history of the field, biodiversity, and the major threats to biodiversity. We then explore ethical and economic constructs that affect our attitudes toward biodiversity and natural resources, and how these might be structured to better motivate conservation of biodiversity. With this background, we review in depth some of the most prominent drivers of biodiversity loss: habitat degradation and loss, habitat fragmentation, overexploitation, species invasion, and global climate change. After this grounding, we focus the rest of the book on

mechanisms for conserving biodiversity. We begin with a transitional chapter on conservation genetics that describes both the problems of genetic losses and the promise of genetic tools for conservation. Our focus then increases in scale to encompass species and landscape-level approaches, and then ecosystem-level approaches to conservation in a broad sense. We explore how protected areas, restoration and reintroduction programs, and sustainable development serve as conservation tools. We conclude by discussing how conservation biologists can become better integrated into policy processes, and look toward the future of conservation biology as we search for means to protect biodiversity and improve human lives as our demands on this Earth grow.

Biodiversity conservation and an end to human suffering are entwined goals, on both ethical and pragmatic grounds. Unfortunately, they are not so entwined that solutions that will improve the status of each will also compliment each other. Indeed, we may best relieve human suffering at a much lower level of conservation than we might prefer, and we could conserve biodiversity in many areas without helping people escape poverty to a significant degree. Thus, the ethical dilemmas in biodiversity conservation are profound. How do we weigh the needs of poor people against those of nonhuman species? How do we protect the interests of future generations of wild nature and meet the needs of humanity now? Working through these issues to develop solutions that meet the needs of human populations and conserve the broadest spectrum of biodiversity is the challenge of our generation. We hope that each person chooses to make both personal and professional efforts to help us all find better solutions.

Supplements

Companion Website (www.sinauer.com/groom)

The *Principles of Conservation Biology*, Third Edition companion website contains a variety of study materials and supplemental resources to accompany the textbook. Study questions, suggested readings, and Web links for each chapter help the student to master the material presented in the textbook and provide direction for further

study. In addition, supplemental essays, case studies and boxes expand on the book's coverage of selected topics.

Instructor's Resource CD (ISBN 0-87893-295-X)

Available to qualified adopters, the Third Edition Instructor's Resource CD contains all of the illustrations and tables from the textbook, for use in lecture presentations and other course documents. All figures are provided as JPEG files and are also included in ready-to-use PowerPoint® presentations. In addition, the IRCD includes a set of suggested exercises—class, lab, computer, and field-based—for instructors to use with their classes.

Acknowledgements

Martha Groom

Typically, people acknowledge their intellectual roots first, and personal ties last. In this case, I'm blessed to be married to an inspirational mathematical ecologist and oceanographer, Danny Grünbaum, who is simultaneously a tremendous intellectual and emotional support. Not a day goes by when I do not love and appreciate him more than the day before. I would not have been able to complete this book without his direct support in caring for me and our children, and his thoughtfulness in discussing many issues that are covered in this book. I owe my greatest thanks to Danny for this, and much, much more.

Our parents, Anne and Len Groom and Zdenka and Branko Grünbaum have been fundamental supports. For their help in this effort, and all they have given us and our children throughout our lives—thank you! I also thank my sister, Lisa Firke, for her unflagging support.

Late in 2001 our sons Maks and Sam were born. They have affected how I have approached this book in several ways. Pragmatically, I slowed down the pace of my work, which delayed publication by years. But more importantly, Sam and Maks sharpened my focus on a major conundrum in conservation biology—how can we balance the needs of people and the needs of all the rest of biodiversity? As a conservationist, my guiding passion is to enhance protection of biodiversity. But now, as a mother too, the fierceness of my wish to protect my children from want and suffering is overpowering. Having children has intensified my set of motivating beliefs: that biodiversity has intrinsic value, and that we must protect it from undue harm as we provide for ourselves; that all people should have the ability to provide for their families, and that we have an obligation to make this a reality; and that education has the power to transform and inspire people. This has translated to a greater attention throughout this book to integrating human issues into conservation science, and to providing opportunities for students to wrestle with the moral, practical, and scientific issues of conservation.

Writing and editing a book gives one a welcome excuse to thank people who have poured their time and talent into helping you in the past. Peter Kareiva was my Ph.D. advisor and role model for critical thinking and the use of mathematical analysis to provide insight into applied conservation practice. A first-rate teacher, he inspired me to learn how to convey ideas simply and provocatively—a skill I still work on year after year. He offered the gifts of his mentorship with great dedication, and I continue to benefit from his many lessons and encouragement. Peter Feinsinger advised me in my M.S. work, and left his indelible mark in sparking my interest in plant-animal interactions and their role in conservation, as well as passing along his penchant for the muddle that is real-world conservation and the teaching of young children. I benefited greatly from being a part of the Tropical Conservation and Development group at University of Florida, and the mentorship of John Robinson and Kent Redford. Doug Schemske, Bob Paine, Dee Boersma, Jane Brockman, John Terborgh, and Hal Feiveson all sharpened my ecological and evolutionary understanding, and encouraged my interests in conservation and teaching. At each institution, the web of people with whom I've worked continues to be a source of great inspiration. Thank you all!

I was partially supported by the University of Washington, Bothell in Winter 2005, and my department chair, JoLynn Edwards, has been particularly helpful throughout. I worked on this book at the Helen R. Whitely Center in the summers of 2003 and 2005, and while on sabbatical at Princeton University in Autumn 2004.

Gary Meffe

As one ages, one tends to increasingly understand and appreciate the great and lasting influence that others have had upon their lives. Now some 15 years after inception of the First Edition, these influences, for me, are in clearer focus. So I gratefully thank those mentors instrumental in my development as an ecologist and conservation biologist, in particular Ted Stiles, Bob Vrijenhoek, Jim Collins, and the late W. L. Minckley. Each in his own way had profound influences on my life and my career, and for that I am in their debt. I hope that debt is partially repaid by passing on their wisdom and energies to the next generation of professionals. There is no more important gift in this life than good and solid parents to support you and send you off in the right direction, and Edward and Mary Meffe did that and more. I only wish they were still here to experience the rewards for their efforts. Finally, my wife Nancy Meffe continues to support, inspire, challenge, and to love me despite my

shortcomings. Others have shown me how to be a better scientist; she has shown me how to be a better person. I am still learning.

Ron Carroll

In the Second Edition I acknowledged the critically important, but generally unrecognized, contributions to conservation that are being made by small grassroots and regional organizations. The intervening years have only deepened my respect for these organizations. They are often the first to call attention to environmental problems and to engage the academic community in finding solutions. Of special importance now, these organizations work tirelessly to gain the support of local communities and to encourage good environmental legislation.

At a personal level, I want to acknowledge my mentor, Dan Janzen, for opening my eyes to the wonders of evolution and tropical ecology, my parents for encouraging my passion for studying decidedly non-charismatic organisms like bugs, my students who always impress me that the future of conservation biology is in good hands, and most importantly, my wife Carol for her constant support and for challenging my ecological pronouncements when the logic is faulty.

From all of us

A host of people participated in discussions about the book, and provided critiques of chapter drafts, including J. Ruesink, A. Edwards, J. Payne, T. Hass, J. Ginsberg, N. Dolšak, A Peterson, R. Irwin, K. Howe, S. Vignieri, F. Bonier, E. Holmes, L. Crozier, A. Solomon, W. Palen, U. Valdez, P. Townsend, C. Vynne, D. Froelich, E. Buhle, E. Skewes, S. Graham, P. Feinsinger, K. Almasi, N. Hahn, A. Clark, L. Addis, T. Billo, A. van Buren, F. Oryazun, F. Prado, K. Braun, E. Jones, C. Frey, L. Payne, D. Hahn, B. Semmens, and many undergraduate and graduate students from conservation courses at the University of Washington, Seattle and Bothell, and the University of Georgia. We owe special thanks to Michael Gillespie for his work in providing some new contexts and contributions to Chapter 4. Thank you for sharing your ideas—the book is certainly better for your work.

Finally, the entire staff at Sinauer has been remarkable in working with us through all the starts and stops of trying to do a major revision of a textbook while balancing families and full-time jobs. Andy Sinauer has been extremely supportive of this book in all its editions. Kathaleen Emerson helped manage production in the late stages, and Marie Scavotto organized the marketing aspects. David McIntryre did wonders in finding appropriate photos, Christopher Small oversaw all of the production, Joanne Delphia and The Format Group created beautiful figures, Joan Gemme did a lovely and meticulous job paging and designing elements of the book, and Jefferson Johnson did a fantastic job creating the overall design of the book. Jennifer Garrett and Mara Silver expertly copyedited the manuscript; a difficult task due to the many divergent writing styles of our contributors. But it is to Sydney Carroll, our production editor, that we owe the lion's share of our thanks. She has seen this book through all the intricacies of its production, was endlessly patient, and also kept Martha sane through to the end. This is a terrific group of people to work with, and we look forward to another edition in the future.

UNIT I

Conceptual Foundations for Conservation Biology

1

What Is Conservation Biology?

Gary K. Meffe, C. Ronald Carroll, and Martha J. Groom

When the last individual of a race of living things breathes no more, another heaven and another earth must pass before such a one can be again.

William Beebe, 1906

Expanding Human Demands on Earth

The natural world is a far different place now than it was 10,000 years ago, or even 100 years ago. Every natural ecosystem on the planet has been altered by humanity, some to the point of collapse. Many species have gone prematurely extinct, natural hydrologic and chemical cycles have been disrupted, billions of tons of topsoil have been lost, genetic diversity has eroded, and the very climate of the planet may have been disrupted significantly. What is the cause of such vast environmental change? Very simply, the cumulative impacts of 6.4 billion people (Figure 1.1), have stressed the many ecological support systems of the planet. Although it took hundreds of years for the human population to reach 1 billion people, we increased to six times that size in a little more than a century. As a consequence, biological diversity (**biodiversity**, for short), the grand result of evolutionary processes and events tracing back several billion years, is itself at stake and rapidly declining. One of the many species suffering the consequences of ecological destruction is *Homo sapiens*, the perpetrator of it all.

All people should recognize the degree to which human impacts affect the natural world, and in turn, diminish our abilities to prosper. Our population explosion over the past century is not yet over, as annually the world's population increases by 77 million each year (the equivalent of adding the population of the United States every 3.8 years). Fortunately, global population growth rate finally slowed beginning in the 1990s. Worldwide, human populations should reach nearly 9 billion by 2050 (8.92 billion, range 7.4–10.6 billion; World Population Prospects 2002; United Nations 2004), and our population is unlikely to stabilize at a size much below 9–11 billion (United Nations 2004) (Figure 1.2). But it is not solely how many of us that is the problem, but how damaging our use of resources has been.

Figure 1.1 Estimated global human population size from the last Ice Age to the present, illustrating the exponential nature of human population growth since the Industrial Revolution. Notice that we reached 6 billion people in 1999 and now number more than 6.4 billion.

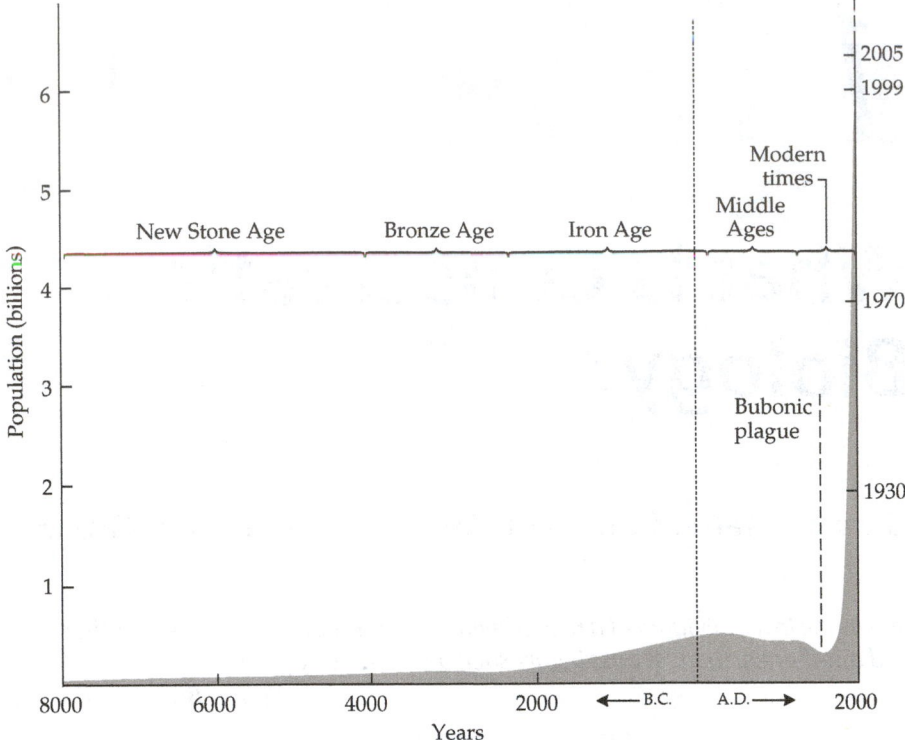

The fraction of the world's natural resources consumed by humans is staggering. For example, 35% of productivity from the ocean shelf (Pauly and Christensen 1995), and 60% of freshwater runoff are claimed for our use (Postel et al. 1996). **Net primary productivity (NPP)**—the energy from the sun that is transformed into plant biomass, and that is the base of all food webs—is co-opted to an alarming extent by our species, with estimates ranging from 20% (Imhoff et al. 2004) to 31%–32% globally (these higher estimates include loss of productivity due to land clearing; Vitousek et al 1986; Rojstaczer et al. 2001). Per capita consumption world-

wide has increased 3% per year for the past 30 years (Hawken et al. 1999), and will continue to increase in the future.

Importantly, the level of human appropriation of NPP is highly unequal throughout the world, with cities in industrialized nations of North America and Western Europe and the large populations in Southeast Asia consuming up to 60%–80% of their regional NPP (Imhoff et al. 2004). Consumption levels in the U.S. particularly, are unsustainable and vastly higher than those in other countries. For example, with about 4% of the world's population, the U.S. alone accounts for 30% of the

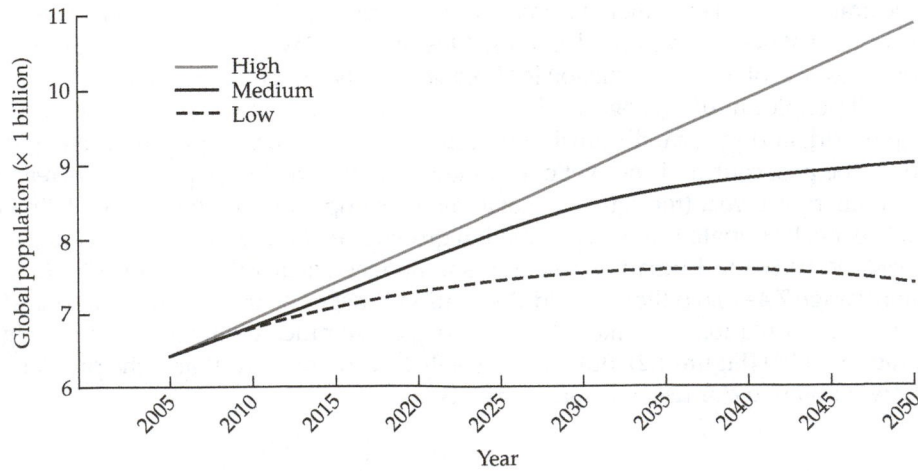

Figure 1.2 United Nations projections for human population growth to 2050. (Data from UN World Population Division, World Population Prospects 2002.)

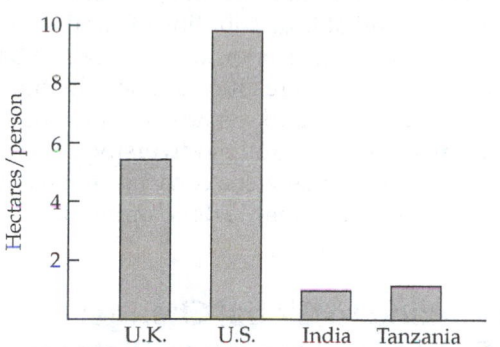

Figure 1.3 Number of global hectares per person needed to support current lifestyles in the U.K., U.S., India, and Tanzania. (Modified from World Wildlife Fund 2003.)

world's daily oil consumption (U.S. Department of Energy 2002), or 250 times as much as India, which has about a sixth of the world's population (Myers 1987). As some less industrialized, but more densely populated countries, such as China, adopt cultural consumptive habits more similar to the U.S., the increasing demands will be enormous and unsustainable.

To help place these impacts in perspective, several studies estimate the "ecological footprint" of our impacts on the globe (e.g., Wackernagel et al. 2002). An ecological footprint calculates how much land and water resources we consume to grow our food, support our lifestyles, and assimilate our wastes. The aggregate portrait is sobering—beginning in the mid-1970's our consumption patterns exceeded Earth's annual production capacity, and our demands continue to grow due to ever-increasing demands for energy, food, and forests (Figure 1.3). A number of footprint calculators are available for individuals to get a rough estimate of their own impacts, and compare their consumption rates with those elsewhere in the world (e.g., at www.myfootprint.org). Many countries and local communities are using these indices for public education and regional planning to encourage reductions in resource use. These indices indicate that it would take four Earths to support the world's population at the level of consumption typical of the U.S. (Wilson 2002).

A final compelling portrait of our "human footprint" was recently produced that shows that more than 83% of Earth's surface bears the imprint of our activities (Sanderson et al. 2002). By overlaying datasets depicting human population density, land use data showing degree of transformation from natural habitats to built up areas, road densities and other means of access to natural areas, and networks of electrical power across the globe, Sanderson and his colleagues were able to provide a more tangible map of our impacts that is neither aggregated as the ecological footprint calculations are, nor restricted to impacts on NPP. The map shows that 83% of the land surface is influenced by one or more of the following: human population density greater than $10/km^2$, agricultural land use, built up areas, access within 15 km of a road, major river or coast, and nighttime light bright enough to be picked up by satellite sensors (Figure 1.4). Further, fertile lands that can grow wheat, rice, or maize are almost entirely transformed (98%; Sanderson et al. 2002).

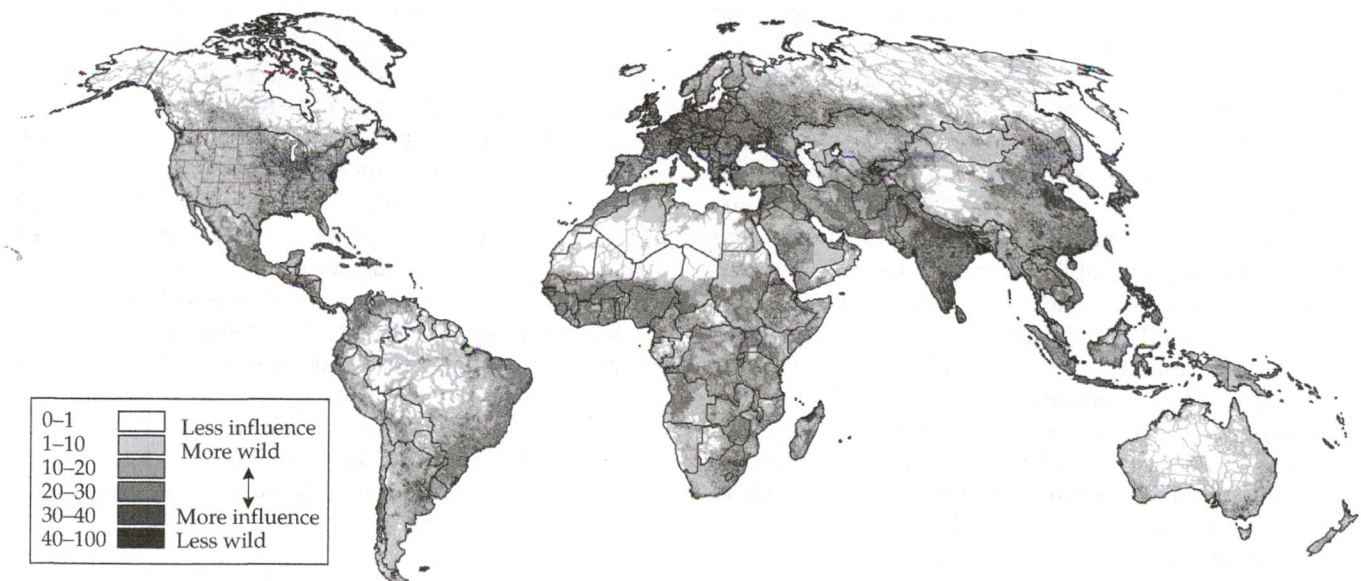

Figure 1.4 Map of the human footprint. Darker regions have borne a greater impact from human development than lighter regions. (From Sanderson et al. 2002 © American Institute of Biological Sciences.)

In summary, we have arrived at the age where human influences predominate across the globe—what some have termed the "Anthropocene" era of Earth (Steffen and Tyson 2001). Although the enormity of this realization and the seeming inevitability of human consumption patterns and population growth outstripping our planet's resources can easily lead to a feeling of helplessness and apathy in the face of so much destruction, there are reasons for optimism.

First, world population growth rates have slowed substantially in the last three decades, and a number of countries were able to significantly lower their population growth rates in a short period of time. Examples include Costa Rica, Cuba, Mexico, Venezuela, and Thailand. Most European countries, such as Hungary and West Germany, have even had periods of negative growth in the past few decades. Birth rates are high where family survival depends on being successful in an unskilled and uneducated labor pool, that is, where there are strong economic incentives for large families. The corollary is that education and the appropriate kinds of economic development can greatly reduce population growth rates. This insight should give us hope that equitable and sustainable development will reduce population growth while relieving intense poverty, and also help preserve biodiversity.

Second, in addition to encouraging signs of slowing human population growth, we all have the potential to change how much and what we consume. In developing countries the expansion of highly commercialized agriculture and forestry has displaced the rural poor into city slums or onto steep hillsides and other ecologically fragile areas. In the industrialized world, the wealthiest nations consume a disproportionate share of global resources, and produce the most waste. These patterns are reversible, given enough will.

For example, the U.S. is currently the largest consumer of energy and producer of greenhouse gases. Politicians and citizens of the U.S. are often worried that any significant efforts to reduce consumption of energy and greenhouse gas emissions would cause a sharp decrease in our "standard of living," and thus are cautious about reversing our consumptive habits. Germany, despite its current economic problems, provides an interesting benchmark. Germans enjoy a high standard of living, but use only about half the amount of energy as U.S. citizens. Further, Germany plans to meet half of its energy needs with low-emission renewable sources by 2050 and to reduce greenhouse gas emissions 21% by 2012. Thus, by following practices similar to Germany's, the U.S. should be able to reduce its consumption of energy and emission of greenhouse gases substantially, without compromising the general standard of living.

The many ways that human population growth rates and consumption patterns can be humanely reduced have several features in common: gender equity, access to education, equitable distribution of rural income, and development of rural economies that are not based on exploitation of natural resources. The take-home message is that we must think broadly about conservation. The **stewardship** of natural biodiversity requires that a strong link be forged between conservation biology and environmentally **sustainable development**.

Responding to Global Change: The Field of Conservation Biology

The field of **conservation biology** is a response by the scientific community to this biodiversity crisis. It is a relatively recent, synthetic field that applies the principles of ecology, biogeography, population genetics, economics, sociology, anthropology, philosophy, and other theoretically based disciplines to the maintenance of biological diversity throughout the world. It is recent in that it is a product of the 1980s, although its roots go back centuries. It is synthetic in that it unites traditionally academic disciplines such as population biology and genetics with the applied traditions of wildlife, fisheries, and land management and allied fields. It is most of all challenging and imperative, in that it is motivated by human-caused global changes that have resulted in the greatest episode of mass extinction since the loss of the dinosaurs 65 million years ago.

Environmentally, we are at the most critical point in the history of humanity, and the current population of students and professionals has a unique place in that history: Of the hundreds of thousands of human generations that ever existed, no previous generation has had to respond to possible annihilation of a large percentage of the species diversity on the planet by humans. Unless humanity acts quickly and in a significant way, the next generation may not have this opportunity. The necessity to act now, not later, makes conservation biology, in every sense of the word, a "crisis discipline" (Soulé 1985); but as Redford and Sanjayan (2003) argue, crises are not excuses for weak science. One of the major developments needed in conservation is a shift from a reactive analysis of each crisis to a proactive science that permits us to anticipate developing crises and to prepare scientifically grounded contingency plans. Beyond this, we also must find the means to lead in the formation of solutions (Redford and Sanjayan 2003).

Many would ask, "What's so new about conservation biology? People have been doing conservation for decades, even centuries." This is true, but conservation biology as a field of endeavor differs in at least three ways. First, it now includes, and has been partially led by, major contributions from theoretically oriented academi-

cians, whose ecological, genetic, and social models are being applied to real-world situations. The unfortunate and false dichotomy of "pure" and "applied" research is finally breaking down, as academic researchers, policy makers, and resource managers have joined intellects, professional experience, and perspectives to address local to global conservation problems.

Second, much of traditional conservation was rooted in an economic, **utilitarian** philosophy whose primary motivation was to maintain high yields of selected species for harvest. Nature was seen as providing benefits to people, mostly from Western nations, through highly visible, selected components such as deer, trout, minerals, or timber, and was managed for maximization of a single or a few species, a small subset of the huge diversity of nature. Conservation biology views all of nature's diversity as important and having inherent value. With this perspective, management has been directed primarily toward stewardship of the world's biodiversity and natural ecosystems, rather than toward management of single species only for our benefit. Four detailed perspectives on the field of conservation biology are offered in essays in this chapter from academic (Essay 1.1 by Erica Fleishman), government agency (Essay 1.2 by Jamie M. Clark), nongovernmental organization, or NGO (Essay 1.3 by Kathryn S. Fuller), and private landowner (Essay 1.4 by Bill McDonald) viewpoints.

Conservation biologists recognize that diverse and functioning ecosystems are critical not only to the maintenance of the few species we harvest, but also to perpetuation of the nearly limitless variety of life forms of which we know little or nothing. The conservationist realizes that intact and functioning ecosystems are also important as life-support systems for the planet, and are critical to our own continued survival and well-being as a species (Odum 1989; Daily 1997).

Third, conservation biology fully recognizes and embraces the contributions that need to be made by nonbiologists to the conservation of biodiversity. In particular, the social sciences, economics, and political sciences may ultimately have more impact on real advances or losses in conservation than the biological sciences. Unless major changes can be made in the way that humanity does business with the natural world, and in humanity's destructive patterns of population growth and resource consumption, it would appear that much of our biological knowledge of conservation will be rendered useless under the sheer weight of the human presence.

A goal of conservation biology is to understand natural ecological systems well enough to maintain their diversity in the face of an exploding human population that has fragmented, simplified, homogenized, and destroyed many ecosystems. Thus, conservation biology tries to provide the basis for intelligent and informed management of highly disrupted ecosystems.

In 1965, the ecologist G. Evelyn Hutchinson described the natural world as an "ecological theater" serving as a stage for an "evolutionary play." Perhaps no better metaphor sums up the mission of conservation biology: to retain the actors in that evolutionary play and the ecological stage on which it is performed. Conservation biology strives to maintain the diversity of genes, populations, species, habitats, ecosystems, and landscapes, and the processes normally carried out by them, such as natural selection, biogeochemical cycling, photosynthesis, energy transfer, and hydrologic cycles. It is a dynamic play, with players and action on many different spatial and temporal scales, old actors disappearing and new ones arriving. But the play ultimately comes down to one thing: dynamic evolutionary processes in a changing ecological background. Conservation biology attempts to keep those normal evolutionary processes working within a functioning ecological setting.

A Brief History of Conservation Biology

The global effort to conserve and protect the natural environment is a recent phenomenon, though efforts to conserve economically important natural resources have a long history. Although we may think of environmental destruction as a product of recent times—and certainly the scale of contemporary destruction is unprecedented—significant environmental degradation has always accompanied humankind (Chapters 3 and 6). Prehistoric humans caused extinctions through overexploitation, habitat modification, and species introductions, and often changed ecosystems drastically through burning, clearing, and cultivation (Chapter 3).

In the classical Greek period, Aristotle commented on the widespread destruction of forests in the Baltic region. At the same time in southern Asia, forests were felled to meet the growing need for timber to build trading ships to serve expanding mercantile centers such as Constantinople (now Istanbul). The barren landscapes that we associate with much of Turkey, Syria, Iraq, and Iran are unnatural deserts resulting from massive exploitation of fragile woodlands. Indeed, this part of Asia had been known in earlier times as the "land of perpetual shade." The Mediterranean region of Italy and Greece was likewise heavily wooded before human settlement.

Diamond (1992) argues that virtually wherever humans have settled, environmental destruction has been the rule; he and others (e.g., Redford 1992) largely debunk the notion of the "noble savage," primitive but wise peoples who had great concern for natural resources. In the humid Tropics, early agrarian societies dealt with declining resources by moving when yields

ESSAY 1.1

A Perspective on the Role of Academia in Conservation Biology
Permeable Walls in the Ivory Tower

Erica Fleishman, *Stanford University*

■ Academia esteems the growth and exchange of ideas and information. By sanctioning the pursuit of knowledge for its own sake, academic institutions allow conservation science to evolve and flourish. Yet academic conventions also can hinder application of conservation science. Accordingly, many action-oriented conservation biologists use academia as a base camp rather than a fixed residence. We value the autonomy academia grants us to formulate and evaluate theory, but recognize that theory alone cannot solve most conservation challenges.

Like it or not, conservation and land use decisions are not based strictly on science. Trade-offs between ecological, social, and economic criteria are inevitable and often necessary. Any science that enters the decision-making process, however, must be objective and reliable, with uncertainties articulated fully. High standards of scientific quality increase the probability that a conservation strategy will be successful once implemented. High standards also enhance the credibility of science in the eyes of managers, politicians, and the public. Consequently, I believe that conservation biologists in academia have three primary responsibilities. First, we must ensure that our scientific assumptions, methods, and inferences are clear and justified. Second, we must direct our science toward the needs of practitioners and decision-makers. Third, we must advocate—by deeds, not dictates—the relevance of science for real-world conservation.

Many disciplines are germane to conservation efforts in the twenty-first century. Academia provides an open forum for interaction among biological, physical, and social scientists, historians, legal scholars, and experts in other fields. Multidisciplinary collaborations unite depth with breadth, and frequently generate creative solutions to contemporary management challenges. The emergence of predictive approaches to conservation planning, for example, is largely due to partnerships among ecologists, geographers, and statisticians.

Scientists today have ample field and analytic techniques to identify and explain observed distributions of species and their habitats. Planning for the future, however, demands that we consider how current patterns may respond to different types of environmental and social change. We also must evaluate whether and how humans realistically can affect those alternative outcomes. Emerging predictive approaches are statistically robust and practical, with flexibility to consider economic and social priorities. Academia's characteristic intellectual freedom encourages development of scientific tools that incorporate both ecological and human dimensions of conservation.

In academia, research and teaching are intertwined. Like many of my colleagues, I can testify to the influence of mentors throughout my education who encouraged my interests and stood as examples of dedication to science and service. Because conservation biology emerged as a distinct field in the late 1980s, just one generation has been trained formally as conservation biologists. But as universities and colleges develop synthetic programs in conservation biology, students still need grounding in the conceptual and empirical traditions from which conservation biology emerged. Rapid responses to many environmental crises are possible only because we can draw from decades and even centuries of careful research in disciplines such as evolution, landscape ecology, and population biology. We must bear in mind that overlooking past contributions is not only disingenuous, but wastes time that we frankly don't have.

Because most students are unlikely to pursue careers in resource sciences, conservation biologists in academia have a pivotal opportunity to educate the general public about biological diversity and its relationship to land use. We also can cultivate scientific literacy within society at large by illustrating how critical thinking—in essence, confronting assumptions or hypotheses with data—can be applied to any situation that benefits from informed assessment and reaction.

Moreover, academia increasingly is appreciating the importance of outreach. As a result, colleges and universities are making concerted efforts to train both students and professors to connect more openly with the media. As communication improves, conservation biologists are realizing that the challenges of balancing objectivity and opinion, and of capturing an audience without compromising the facts, are inherent to both science and journalism. Several formal programs (some based at universities, some in partnership with museums, aquaria, or other educational institutions) have been founded on the principle that the academic voice is a vital component to societal debates about the future of all levels of biological diversity. Such programs provide rigorous training to both early and later-career conservation biologists in the hope that all will become more effective communicators whether they are interacting with corporations or with reporters.

To inform or influence management and policy, it is essential to engage in active dialogue with those making decisions on the ground. Conservation science is far more likely to be integrated into the management process when academics make the effort to understand the everyday opportunities and constraints of practitioners. Publication is an essential part of the scientific process. Publication furnishes access to information, at least some degree of quality assurance, and validation by a knowledgeable community of scientific peers. But because relatively few managers have time or desire to read journal articles, appending "conservation recommendations" or "conservation implications" to a manuscript has little real impact. Instead, academics must work directly with practitioners, emphasizing that cooperative research will be truly applicable to management only when practitioners define their objectives and management alternatives honestly and explicitly. For example, academics need to be far more vocal in explaining that even the most exacting monitoring plan has limited

value if data are not used to guide work on the ground. Academics must work closely with agencies and stakeholders to evaluate what environmental changes will trigger changes in management, and what those management changes might be.

Within our home institutions, academics need to advocate modification of the standard reward structure to facilitate communication with practitioners and, by extension, to build conservation capacity. Academia typically recognizes publication as a researcher's greatest achievement—and preferably publications with high "impact," as measured by number of citations in other publications. This metric of accomplishment is unambiguous and equitable, but its correlation with conservation action arguably is weak for the reasons described earlier. Thus, in addition to publication and teaching, academia should reward meaningful partnerships with resource professionals in the public and private sectors. How can we affect such changes? Junior researchers can work to include nontraditional measures of success in their employment contracts and performance reviews. Senior researchers,

meanwhile, can exert their influence as department heads or as leaders of new environmental initiatives.

From an academic base, conservation biologists have considerable latitude to collaborate with other academics, managers, and local communities. Especially before they have become well established, some academics fear they will lose professional integrity by communicating freely with industry representatives or grassroots environmental organizations. Their concern is not entirely without justification. Personally, I think the potential benefits of accepting any good-faith invitation to participate in tangible conservation efforts far outweigh the potential risks. Conservation biologists cannot expect agencies and the public to heed science if we perpetuate barriers between academia and the policy process. This certainly does not mean that we should compromise our scientific principles. Conservation biologists must resist pressure to misrepresent science even if doing so might promote certain conservation objectives. Likewise, misrepresentations of science by others must be corrected promptly and, if necessary, publicly. In 2004, the Society for Con-

servation Biology developed a code of ethics to better equip its members around the world to grapple with ethical dilemmas. Among other things, the code encourages all conservation professionals to volunteer their services for the public good at a level appropriate to their financial abilities. The code also emphasizes responsibilities to human welfare and social equity.

Academia can be the best of both worlds. Many conservation biologists chose their career because they are fascinated by the natural world and want to protect its integrity. Thanks to laboratory facilities, field stations, sabbaticals, library networks, and other advantages, academics typically can spend a portion of their time conducting research and communicating their enthusiasm to the next generation of conservation professionals. At the same time, we can reach out to colleagues and communities whose management concerns are literal and immediate. With the freedom of academia comes the responsibility to ensure that, as the paradigms of conservation biology shift to meet new conservation challenges, so do the walls of the ivory tower. ■

---

began to drop and local game became scarce, an apparently sustainable strategy given enough land base. Yet, even these shifting cultivators drove species to extinction, and changed the character of natural communities. Many, if not most societies have had some lasting, destructive impact on the natural world.

However, some societies have certainly minimized their environmental influences and lived in a more sustainable fashion than most. Some shifting cultivators practiced, and some still practice, forms of conservation management. In many tropical regions, complex tree gardens helped stabilize land use (see Carroll 1990 for examples), and some shifting cultivators practiced a kind of management of natural succession. Today, in "Dammar" agroforestry in Sumatra, for example, natural forest plots are converted over a period of 10–20 years into complex modified forests based primarily on dammar (*Shorea javanica*), a tree that is tapped for resins, and other economically important native trees (Mary and Michon 1987). The plots are structurally similar to natural successional plots and likely help support regional biodiversity. In terms of financial returns, Dammar agroforestry outperforms rubber plantations, cinnamon/coffee polyculture, and rubber agroforestry. Compared to these systems, Dammar agroforestry also

sequesters more carbon and shelters more biodiversity. Indeed, Dammar agroforestry contains about half of the bird and plant species found in primary forest (Ginoga et al. 2002). Although we may think of conservation management as a modern Western notion, management of natural resources has been practiced in many other cultures, often for much longer periods (Figure 1.5).

We would be remiss, however, if we failed to point out the fragility of these traditional systems in the modern, interconnected global marketplace. While shifting cultivation may be sustainable over a large area, it is not when people are confined to small indigenous reserves. As smaller indigenous cultures become connected with modern societies, their choices and practices change, often toward less sustainable practices. To continue with the Dammar example, the practice is disappearing in one region for two unexpected reasons. First, the establishment of Burkit National Park appropriated a major portion of Dammar forestry land and put severe constraints on the use of the remaining land. In particular, the long fallow period needed became increasingly difficult to accommodate. Second, a growing urban market created great demand for rice and, to a lesser extent, coffee and cloves. In response to these two factors, Dammar agroforestry around Burkit National Park has

Figure 1.5 Highly diverse agroforestry systems, such as the Dammar system from Indonesia, can be found in many tropical regions. This photograph shows a similar agroforestry system from southeastern Mexico, locally known as a "huerto," or tree-garden. These traditional agroforestry systems of mixed, cultivated perennials are structurally similar to old, second-growth natural forests, and may contain nearly as many tree species per hectare. (Photograph by C. R. Carroll.)

been largely replaced by dryland rice and coffee cultivation.

Nevertheless, Dammar agroforestry remains a viable agroforestry system when the land base is adequate. The contribution of Dammar agroforestry to both human welfare and biodiversity conservation has been recognized, and the practices are now regarded as a significant contributor to the United Nations Millennium Development Goals (Garrity 2004).

For many centuries, societies recognized and worked to counteract some of the harm caused by overexploitation of species and lands. In Europe, where most land was held by royalty or the very wealthy, early conservation efforts took the form of private game management and maintenance of royal preserves and private manor lands. Yet, until the eighteenth and nineteenth centuries, little notice was given to problems of the **commons**, the public lands. As a consequence, exploitation of these common-use resources led to the deforestation of most of Europe by the early eighteenth century. This occurred even earlier in Great Britain, where many of the native forests were destroyed by the twelfth century (McKibben 1989); the demand for charcoal to supply home heating and industrial needs led to virtual elimination of the remaining public forests by the late eighteenth century. Similarly, in Asia, conservation efforts were game-oriented and largely restricted to the private lands of the privileged. An artist's early rendition of a forest and pastoral scene in China juxtaposed against a later photograph of the same place, which depicted an eroded and barren landscape, is said to have been the telling argument made to the Theodore Roosevelt administration by

forester Gifford Pinchot in his successful campaign to establish the U.S. Forest Service in 1905.

Since the end of the nineteenth and throughout the twentieth century, conservation began to become an important goal for many nations, and broadened from efforts to safeguard important game species to those intended to protect all of biodiversity both within a nation, and across borders. To illustrate this progression, we will describe changes in the nature of conservation in the U.S.

Conservation in the United States

Europeans colonizing America found a landscape that, by comparison with a highly exploited Europe, must have seemed pristine. Aboriginal peoples had exploited natural resources and driven some species to extinction, but their low population densities and lack of technologies for widespread devastation prevented wholesale destruction. Native Americans apparently made extensive use of fire to manage lands for both agriculture and game. Some historians argue that Atlantic coastal lands cleared by Native Americans became important colonization sites for European settlers and helped them survive their first winters (Russell 1976).

During the colonial period, North American forests were extensively exploited for lumber, ship masts, naval stores (gum and turpentine), and charcoal for heating. Huge tracts were cleared for agriculture. Demand for forest products in Europe and domestic demand by a rapidly growing population were eagerly met by exploiting the seemingly endless forests. Later, forests were again called upon to provide lumber for vast railroad

networks and building construction as the nation expanded westward. In coastal areas, salt marshes were harvested for salt hay (*Spartina*) to feed cattle before the opening of the prairies to grain farming.

The value of forests as an economic resource was not the only philosophical perspective held by the colonists, however. Religious attitudes of some groups, especially the Puritans, held that the forest was the abode of the devil. This is perhaps not an unfamiliar attitude even today, for many children's stories place witches, trolls, and goblins in deep, dark forests, and many otherwise reasonable adults are more frightened in a forest than in the heart of a large city with high crime rates.

Thus, the forests were beset by increasing economic demands and were perceived to be endless and vaguely evil—hardly a nourishing environment for conservation. Conservation did, of course, develop in North America, but it required several centuries after initial European colonization to become firmly established. Perhaps it was necessary first to develop a significant population whose livelihood was not intimately tied to forest exploitation.

American conservation efforts can be traced to three philosophical movements, two of the nineteenth century and one of the twentieth (Callicott 1990). The **Romantic-Transcendental Conservation Ethic** was derived from the writings of Ralph Waldo Emerson and Henry David Thoreau in the East, and John Muir in the West. Emerson and Thoreau were the first prominent North American writers to argue, in the mid-1800s, that nature has uses other than human economic gain. Specifically, they spoke of nature in a quasi-religious sense, as a temple in which to commune with and appreciate the works of God. Nature was seen as a place to cleanse and refresh the human soul, away from the tarnishings of civilization. This was the philosophical and aesthetic position that Muir used as he argued for a national movement to preserve nature in its wild and pristine state, and condemned its destruction for material and economic gain. John Muir's movement flourishes today in the form of many citizen conservation groups; his direct organizational legacy is the Sierra Club.

This noneconomic view was countered by the so-called **Resource Conservation Ethic**, made popular by the forester Gifford Pinchot at the turn of the twentieth century. His was an approach to nature based in the popular utilitarian philosophy of John Stuart Mill and his followers. Pinchot saw only "natural resources" in nature and adopted the motto, "the greatest good of the greatest number for the longest time" (Pinchot 1947). Nature, to Pinchot, was an assortment of components that were either useful, useless, or noxious to people. Note the **anthropocentric** valuing of nature, not because it is part of "God's design" (as per the Romantic-Transcendentalists), but because natural resources feed the economic machine and contribute to the material quality of life. Pinchot

(1947) once stated that "the first great fact about conservation is that it stands for development."

Pinchot's approach to conservation stressed equity—a fair distribution of resources among consumers, both present and future—and efficiency, or lack of waste. This led to adoption of the **multiple-use** concept for the nation's lands and waters, which remains the mandate of the U.S. Forest Service and Bureau of Land Management. Under multiple use, many different uses of the land are attempted simultaneously, such as logging, grazing, wilderness preservation, recreation, and watershed protection. Because a market economy may or may not be efficient and has little to do with equity, government regulation or outright public ownership of resources was deemed necessary to develop and enforce conservation policy.

These two movements thus created a schism, with the preservationists (Muir, Emerson, Thoreau) advocating pure wilderness and a spiritual appreciation for nature, and conservationists (Pinchot) adopting a resource-based, utilitarian view of the world. A third movement, born of this century, emerged with the development of evolutionary ecology. This **Evolutionary-Ecological Land Ethic** was developed by Aldo Leopold in his classic essays, published shortly after his death in *A Sand County Almanac* (1949), and in other writings. Leopold was educated in the Pinchot tradition of resource-based conservation, but later saw it as inadequate and scientifically inaccurate. The development of ecology and evolution as scholarly disciplines conclusively demonstrated that nature was not a simple collection of independent parts, some useful and others to be discarded, but a complicated and integrated system of interdependent processes and components, something like a fine Swiss watch. There are really only a few parts of a watch that appear to be of direct utility to its owner, namely, the hour, second, and minute hands (back when watches had hands). However, proper functioning of these parts depends on dozens of unseen components that must all function well and together. Leopold saw ecosystems in this context, and this is the context in which modern ecology first developed. This **equilibrium** view was subsequently replaced by a dynamic, **nonequilibrium** ecological perspective, discussed later in this chapter. Nevertheless, the Leopold land ethic remains as the philosophical foundation for conservation biology.

Much of modern conservation is based on various mixtures of these three philosophies. The Resource Conservation Ethic of the late nineteenth century is still a dominant paradigm followed by public resource agencies such as the U.S. Forest Service, under which U.S. forest tracts are seen as economic resources to be managed for multiple human uses. The Romantic-Transcendental Conservation Ethic, though more typically without the overt religious rationale of its early proponents, is the

basis for activism by many private conservation organizations throughout the world, whose goals are to save natural areas in a pristine state for their inherent value. This difference has resulted in repeated confrontations among so-called "special interest groups."

Leopold's Evolutionary-Ecological Land Ethic is the most biologically sensible and comprehensive of any approach to nature and should serve as the philosophical basis for most decisions affecting biodiversity. It is the only system that can provide even moderately useful predictions about our effects on the natural world, but it is still only part of the total decision-making process; the economic, spiritual, and social needs of people must also be met. It is curious that management decisions concerning natural areas can be made without recourse to evolutionary ecology, yet this still routinely happens in many resource agencies. Similarly, it would be a fruitless, counterproductive, and ethically suspect exercise to base comprehensive land use decisions solely on evolutionary ecology without regard to the people who will be affected.

Most natural areas today are remnant patches of formerly contiguous habitats in landscapes dominated by human economic endeavors (Figure 1.6). The biological activity within any one of these natural areas is strongly dependent on what happens outside its boundaries. Any long-term security for a natural area will come about only when it is accepted as an integral and contributing part of broader economic and development planning. Just as the Evolutionary-Ecological Land Ethic grew out of traditional disciplines to meet the emerging crises in biodiversity, so also are the traditional disciplines of re-source economics and anthropology giving rise to new interdisciplinary views, sometimes called "ecological economics" and "ecological anthropology," views that stress long-term environmental sustainability.

Modern conservation biology: A synthesis

Modern conservation biology has sought to replace both the extreme Romantic Preservationist and the exploitative utilitarian philosophies of the nineteenth century with a balanced approach that looks to an ethic of stewardship for philosophical guidance, and a melding of natural and social sciences for theory and practice. This interdisciplinary context is necessary for conservation biology to flourish and make contributions to a sustainable biosphere.

By the 1960s and into the 1980s, it was becoming painfully obvious to many ecologists that prime ecosystems throughout the world, including their favorite study sites, were disappearing rapidly. Biodiversity, the outcome of millions of years of the evolutionary process, was being carelessly discarded, and, in some cases, willfully destroyed. Previous conservation efforts, while focusing on important components of nature such as large vertebrates, soils, or water, still had not embraced the intricacies of complex ecosystem function and the importance of all the "minor," less charismatic, biotic components such as insects, nematodes, fungi, and bacteria. It was time to change this attitude, and many people began writing on these subjects (e.g., Dasmann 1959, Ehrenfeld 1970, Soulé and Wilcox 1980, Frankel and Soulé 1981, and Schonewald-Cox et al. 1983). These books helped to lay the groundwork for today's conser-

Figure 1.6   An aerial photograph showing a mixed natural and human-dominated landscape in South Carolina. Lighter areas are agricultural fields and human housing developments, darker patches are forests, fields, and streams. (Photograph courtesy of Savannah River Ecological Laboratory.)

ESSAY 1.2

Working with U.S. Government Agencies in Biodiversity Conservation

Jamie Rappaport Clark, *Executive Vice President, Defenders of Wildlife, and Former Director, U.S. Fish and Wildlife Service*

■ Over the years, biology has furthered our understanding of the interconnectedness and interdependencies that keep ecosystems functioning. This increased understanding has validated the inherent value of nature and supported the realization that a vibrant economy ultimately depends on a healthy environment. Public awareness and support of environmental protection led to the passage of important environmental laws such as the Endangered Species Act, the National Environmental Policy Act, the Clean Water Act, and the Clean Air Act.

Protecting the environment remains a tricky balancing act. Most people want economic development to be compatible with environmental protection and they expect government to reconcile conflicts between the two.

In many situations, however, government actually causes this conflict through contradictory mandates of their environmental and land management agencies. The U.S. Forest Service conserves our nation's forest resources, but also allows for the harvesting of timber. The National Marine Fisheries Service conserves living marine resources off America's shores, but is also charged with maintaining viable commercial fisheries. The National Park Service's mission includes not only preserving the ecological and historical integrity of parks, but also offering quality public recreational opportunities. How do you balance the beauty and diversity of these public lands while at the same time ensuring a safe and memorable experience for the nearly three hundred million visitors to our national parks?

To cope with the duality of these mandates, many government agencies have explored the practical implementation of concepts like sustainability and ecosystem management. Both of these concepts, however, are still only vaguely defined. From the resource manager's perspective, they need to be able to say authoritatively with the support of sound science that they have identified the limit on how much human interaction an ecosystem can sustain. To date, such limits still have not been defined for most cases.

Conservation biology can help us identify those thresholds. Presently, academicians in this fast evolving field are discussing what constitutes a sustainable level of human resource interaction and whether we can recognize it and manage for it. Similar discussions are taking place with respect to ecosystem management. Meanwhile, natural resource managers, through the process of adaptive management, are doing their best to put these ideas into practice. While doing so, they are confronted with real-world, real-life constraints. Limited funding, staffing, and resources are some of the most challenging obstacles affecting all natural resource management agencies. The politicization of natural resources decision-making is causing increasing conflicts at all levels of government. Additionally, each agency is also faced with a set of constraints unique to the resources under its care.

The U.S. Fish and Wildlife Service's (USFWS) mission is to manage fish, wildlife, and plants and their habitats. To carry out this mission, the agency is often guided by statutory deadlines. For example, under the Endangered Species Act, the USFWS must make decisions (listing determinations, consultation decisions) based on the best *available* science. For practical and statutory compliance reasons, the USFWS does not have the luxury of waiting for better science or more science to become available. There is often inherent internal conflict regarding what constitutes the best available science and whether it is complete enough to make informed decisions.

Wildlife trade issues, for instance, require the USFWS to examine not only the status of a species like the African elephant (*Loxodonta africana*), its habitat, and its interaction with other species, but also to determine who wants to trade ivory, and the social and economic situations that drive the ivory trade. Decisions to list a species as threatened or endangered take into account not only population size, but also the degree of threat, based on factors like disease, habitat loss, and commercial use. We need to understand all these factors if we are to make smart choices about how to best conserve species and their habitats.

To achieve conservation success, natural resource managers must forge partnerships with all segments of society that are in play in a given region. This point was clearly brought home in the Pacific Northwest with the Northern Spotted Owl controversy of the early 1990s. A landscape challenge as complex as conserving resources in forested lands across the Pacific Northwest can be accomplished only with the participation of local communities, industry, private groups, and other government agencies at all levels, including internationally.

Throughout the country, the USFWS is exploring innovative ways to fulfill their conservation mission. They partner with the timber industry to provide habitat for Red-cockaded woodpeckers (*Picoides borealis*) in the southeast, work with ranchers and farmers to restore natural habitat while ensuring continued economic viability for landowners, and they strategize with expanding municipalities like San Diego to make sure their development plans remain consistent with the conservation of imperiled species. In forging such partnerships, the USFWS faces two real-world constraints: (1) the incomplete knowledge of ecosystem functions and processes, and (2) the ever increasing societal demands on the landscape that cannot be ignored.

Enter the conservation biologist, who will have a significant role in the future direction and management of our nation's natural heritage. To make the best decisions possible, resource managers require solid, dependable information and sound, science-based approaches to guide them on decisions on everything from local planning and zoning initiatives to state and national environmental laws and compliance. As government agencies experiment more with approaches to ecosystem management, they will need more

options to explore. Managers will need a better understanding of metapopulation dynamics. They will require an understanding of the scientific principles and practical concepts for designing future refuges. To connect protected lands, resource managers will need to better understand the theories and application of linking fragmented habitats. All of these challenges are the very issues that the science of conservation biology seeks to address.

Many of the resource agencies will use this evolving science to help in prioritizing research needs. Conservation biologists can help decide where to focus limited resources by identifying the circumstances and the species that most need research initiatives such as genetic or behavioral studies, or population viability analyses. Such tactical research is essential to everyday decision making. Vital to this is the link between ongoing activities, important conservation initiatives, and specific research needs. Conservation biologists can and should play a pivotal role in creating that link.

Providing resource managers the information to make sound decisions is only the beginning. The task of conserving biodiversity is too large and important for just one agency. That is why partnerships are crucial, especially with private landowners. Because private lands are increasingly important for maintenance of biological diversity, we need to do a better job of engaging landowners in conservation issues and opportunities. We need to identify all the stakeholders and invite them to help find a solution that can protect both biodiversity and a sustainable economy. We need to provide our partners with flexibility and certainty in the face of uncertainty about biological processes and human activities.

With this in mind, the USFWS developed a suite of management tools to accomplish this task, including Habitat Conservation Plans, Candidate Conservation Agreements, and Safe Harbor Agreements. In each case, an agreement takes some of the risk of proactive conservation efforts away from the landowner and offers them some degree of security and certainty. Central to all is the concept of adaptive management, which acknowledges the evolving nature of scientific knowledge.

The USFWS often relies on conservation biologists to identify the areas of uncertainty and to help address them by devising a range of actions along with a feedback loop to monitor progress. As more public–private partnerships take wing, conservation biologists will need to help natural resource managers further develop and implement adaptive management principles, and to devise other means to achieve conservation goals. Collaborative, comprehensive efforts, with a focus on the concepts of basic ecology, landscape conservation, and restoration, are the only way we are going to successfully tackle the challenges facing us in this new century.

Biodiversity conservation requires everyone to assume some of the responsibility. The overwhelming challenge of invasive alien species is a case in point. According to a Cornell University study, invasive species inflict damages of $138 billion annually on the U.S. Further, they are contributing to the decline of 35 percent of threatened and endangered species. These biological invaders have infested more than 100 million acres of the U.S. and are spreading across the nation at a rate of 3 million acres per year.

To address this challenge, we need to broaden awareness of the ecological consequences of foreign species. We need to enlist a broader constituency. Boaters can help prevent the spread of zebra mussels (*Dreissena polymorpha*) from one body of water to another by checking for biological hitchhikers whenever they take their vessels out of the water. To stem nonnative species invasions, increased efforts toward prevention, eradication, and control mechanisms are essential. We are currently relying mostly on biological controls and pesticides, both of which can be problematic. The next generation of conservation biologists must help find better ways. Further, as global trade increases, we need sound science-based risk analysis to identify the potential problems caused by new species invading our borders.

Conservation biologists are also finding themselves increasingly involved with captive propagation. This wildlife recovery method has proven useful as an emergency tool where the original threats to a species in the wild are also being mitigated.

The USFWS has worked with partners to restore populations of California Condors, Mexican gray wolves, Puerto Rican Parrots, and black-footed ferrets.

For some species, wild populations have fallen so low that the remaining animals are in danger of becoming extremely inbred, a real challenge in addressing the potential for genetic bottlenecks. For the Florida panther (*Puma concolor coryi*), the population had been so reduced that the rate of inbreeding has increased significantly over natural levels. The closest related subspecies, cougars from Texas (*P. c. stanleyana*), were introduced into south Florida to inject diversity into the population's gene pool, with careful attention given to avoid swamping the locally adapted Florida panther genome. This is an area requiring further study and conservation biology is the science best suited to deliver the information wildlife managers need in making the right call in these difficult situations.

The twenty-first century presents formidable challenges. Some scientists believe that the pressures put on ecosystems by society are already causing a new wave of mass extinctions. In many ways we stand at a crossroads. In the final chapter of *Silent Spring*, Rachel Carson wrote "The road we have been traveling is deceptively easy, a smooth superhighway on which we progress with great speed, but at its end lies disaster. The other fork on the road . . . offers our last, our only chance to reach a destination that assures the preservation of our earth." She penned those words in 1962. I believe we still have a chance to change the path we are on. Conservation biologists will play a significant role in determining our success along that path.

As this new century gets underway, conservation biologists have the opportunity to help natural resource managers make a strong case for the preservation of biological diversity and to provide policy makers with the scientific basis and the tools to change the course we're on. I believe we will ultimately be successful and that history will remember this time not for a catastrophic loss of biodiversity but for a heroic choice made by humankind: to value what we leave of the land more than what we build on it. ■

Figure 1.7 The first issue of the journal *Conservation Biology*, published in May 1987. (Photograph courtesy of E. P. Pister.)

vation biology by melding good evolutionary ecology with human resource use, and providing a vision of where modern conservation should go, and they motivated a large cadre of scientists to put conservation at the forefront of their research and personal agendas.

In 1985, the Society for Conservation Biology was formed, a large membership rapidly grew, and a new journal, *Conservation Biology* (Figure 1.7), was developed to complement existing journals such as *Biological Conservation* and *The Journal of Wildlife Management*. More recently, the Society for Conservation Biology launched *Conservation in Practice* in 1999, with the goal of putting conservation science into practice, and conservation practice into science. Thus, in the past two decades, the thrust and outlook of conservation dramatically changed, and continues to change as conservation science matures. From this point forward the field has continued to develop rapidly. Many textbooks and many more professional books on diverse aspects of conservation biology have been produced. The Society for Conservation Biology has grown from a fledgling organization with a small journal to a major international scientific society with multiple publications. And graduate and undergraduate programs in conservation biology have developed in colleges and universities throughout the world.

Students of conservation biology today should be excited to know that the science of conservation is still developing, and needs many bright minds to determine its future di-

rections. Anyone who thinks that much of the work has already been done, and that there is little room left for contributions, does not yet understand the many challenges of conservation biology; hopefully, the following chapters will set that record straight.

A large number of international conservation organizations are active today, as well as numerous national and local organizations, and government organizations at all levels. These conservation groups are responsible for directing public and government attention to particular places and species in need of our help, and developing major approaches to effecting conservation across the globe. Beyond these organizations, groups of citizens are increasingly taking on conservation issues and working to improve the future for their children and grandchildren.

Conservation is undertaken at all levels: by citizens working to restore degraded lands or stop overexploitation of wild species, by local governments regulating land and water uses, by NGOs working toward opportunities for conservation that support local communities, by state and national governments enacting legislation to prevent extinction, and by international agreements designed to curb climate change and the loss of biodiversity worldwide. By knowing more about these activities, we can help push for new actions at levels higher than ourselves, while working within our own communities for positive change.

Guiding Principles for Conservation Biology

Three principles or themes that serve as working **paradigms** for conservation biology will appear repeatedly throughout this book (Table 1.1). A paradigm is "the world view shared by a scientific discipline or community" (Kuhn 1972), or "the family of theories that undergird a discipline" (Pickett et al. 1992). A paradigm underlies, in a very basic way, the approach taken to a discipline, and guides the practitioners of that discipline. We believe these three principles are so basic to conservation practice that they should permeate all aspects of

TABLE 1.1 *Three Guiding Principles of Conservation Biology*

Principle 1:	Evolution is the basic axiom that unites all of biology. (The evolutionary play)
Principle 2:	The ecological world is dynamic and largely nonequilibrial. (The ecological theater)
Principle 3:	Human presence must be included in conservation planning. (Humans are part of the play)

ESSAY 1.3

The Role of Science in Defining Conservation Priorities for Nongovernmental Organizations (NGOs)

Kathryn S. Fuller, *World Wildlife Fund*

■ The proposition that science should play a key role in setting conservation priorities seems self-evident: After all, where would conservation *be* without the sciences of biology and ecology? Isn't science the foundation of the environmental movement?

Science indeed lies at the heart of conservation, but the relationship is complex. Understanding how science contributes to conservation requires us to examine a range of disciplines that would once have seemed completely alien to it. Today conservation science increasingly incorporates economics, social science, geography, and knowledge management into the planning process.

A crucial part of that process has been the emergence of the field of conservation biology. The professional membership organization, the Society for Conservation Biology, is now 20 years old, with over 5,000 active members, and the competition to publish in its journal has grown dramatically.

How has science, through the maturing study of conservation biology, influenced the setting of conservation priorities, as practiced by nongovernmental institutions, which are increasingly responsible for moving conservation initiatives forward?

Biological science no longer exclusively sets the boundaries of conservation. This is due in part to the uniquely multidisciplinary nature of modern conservation, which is the product of years of evolving philosophy and practice. My own organization, World Wildlife Fund (WWF), is a useful case study in this evolution. When we began in 1961, we concentrated our efforts on individual species, animals like the Arabian oryx, the rhinoceros, and the giant panda, our organization's symbol. We emphasized scientific research and hands-on fieldwork. The application of the broader principles of conservation biology have forced us to look beyond individual species' requirements, to incorporate ecological processes, environmental change, and most importantly, to protect viable and representative areas of all natural habitats.

The focus on habitats, in turn, led us toward the humans who interact with those habitats and the connection between human poverty and resource destruction. Now, every day, WWF addresses itself to what is perhaps conservation's bitterest irony: Some of the world's poorest people struggle to survive alongside the world's greatest natural treasures. Beyond the borders of parks live people desperate for cropland and firewood. Adjacent to herds of wildlife in Africa are villagers without an adequate source of protein. And around the world is a vastly increasing new category of refugees, fleeing not tyrants but a deteriorating environment.

Clearly, unless we consider economic and social realities we will fail in our efforts to preserve biodiversity in the long term. So WWF seeks ways to marry the preservation of biological diversity with environmentally sound economic development. This transition from "pure" conservation to one that considers both conservation and development means we can no longer closet ourselves behind laboratory doors. We must delve into areas unfamiliar to conservationists, like anthropology, sociology, economics, and political science. And recognizing that the best-designed projects will fail without ongoing funding, we must take on the role of conservation financiers, brokering debt-for-nature swaps and creating new financial mechanisms to leverage our limited resources into lasting change. What we do know from experience in places as different as Nepal, Bhutan, New Guinea, the Guianas, and South Africa is that it is simply a myth that only developed nations can afford to undertake, or are interested in promoting, biodiversity conservation.

Given all this, it might be easy to go on and say that science has less of a claim on today's conservation agenda—fighting for attention as it is with the fields of economics and politics. But that would be a mistake. Because not only does science lie at the heart of conservation, it is now more critical than ever. Science determines our conservation priorities and pro-

vides the blueprint. We use various tools—sustainable development, conservation finance, and yet more science—to create strong and enduring on-the-ground conservation networks.

We can only guess at the number of species on this planet. Some estimates put the number at 30 million or more, but with millions still to be identified, most of this is highly educated guesswork. What we do know is that we are losing species at an almost unimaginable rate. The renowned biologist E.O. Wilson says we are on the brink of a catastrophic extinction of species—of a kind unseen since the demise of dinosaurs 65 million years ago.

When confronted with mass extinctions on this scale, the inevitable temptation is to throw up one's hands and ask, "Where to begin?" Science can tell us where to start a path in a rational and comprehensive manner, and equally important, it can help correct our path even as we forge it. Science also provides the kind of foresight that every conservation organization desperately needs—the ability to look ten and twenty years into the future and envision where we need to be.

Of course, setting conservation priorities for our planet will never be simple or straightforward. As a start, we know that tropical moist forests contain at least half of all Earth's species. Tropical forests, especially on islands, are in fact the crucible of modern conservation. Knowing this only takes us so far, however, since it still leaves us with a range of other habitats to incorporate into our planning: tropical dry forests, temperate, Mediterranean, and boreal forests, grasslands, and of course, marine and freshwater systems. But scientists at WWF and elsewhere are working to identify key natural areas featuring exceptional concentrations of endemic species and unique ecological phenomena that are representative of the full measure of biodiversity. By concentrating efforts in these areas where the potential payoffs are greatest, conservationists can respond in a more informed and systematic way to the challenge of preserving biodiversity.

We have learned about the importance of examining ecosystems at large spatial and temporal scales, and the more we have learned about ecological systems, the more we have learned how interconnected they are. We have also learned about the importance of events occurring at large time scales, including El Niño and other natural large-scale disturbances such as fire. Conservation biology has forced us to examine food webs and the far-reaching effects of large predators in ecological systems, and in the case of marine systems, the removal of top predators and the resulting effects on other trophic levels. The combined effect has caused most conservation organizations to think at larger scales beyond the boundaries of individual protected areas—the scales at which most ecological systems operate.

Science can and must contribute to the fruitful mélange of ideas currently circulating in the conservation field.

Without the help of science, we cannot hope to tackle the truly foreboding problems facing our planet today—problems that in fact were first identified by scientists: climate change, fragmentation and degradation of habitat, and their result, the loss of biological diversity.

Already, we are seeing exciting and promising new sustainable-use techniques at work: sustainable harvesting of nontimber products like fruits, seeds, medicinal plants, and wild game; agroforestry methods that combine traditional crops with multiple-purpose trees; restoration ecology and watershed protection. Scientific methodologies such as geographic information systems (GIS), remote sensing, decision support systems, and high technology solutions to tracking the movements and dispersal of endangered species continue to revolutionize the field of conservation biology.

Science anchors the economic and political exigencies of modern conservation in intellectual bedrock. Conservation biology has forced us to expand our temporal and spatial reach to fully incorporate today's conservation challenges. Although foundations and endowments encourage scientists to think in small and discrete terms, the problems confronting the world are so massive that scientists must scale their thinking accordingly. The need for solid science to inform decisive action by regional, national, and international nongovernmental organizations and other groups has never been so great. If conservation biologists fail to come up with a long-term vision for what success looks like in all of the biologically important places in the world, surely others who wish to exploit the resources of those regions for short-term gain have their own blueprint for development. Doing good conservation science has never been a more urgent occupation. ■

---

conservation efforts and should be a presence in any endeavor in the field.

PRINCIPLE 1: EVOLUTIONARY CHANGE   The population geneticist Theodosius Dobzhansky once said, "Nothing in biology makes sense except in the light of evolution." Evolution is indeed the single principle that unites all of biology; it is the common tie across all areas of biological thought. Evolution is the only reasonable mechanism able to explain the patterns of biodiversity that we see in the world today; it offers a historical perspective on the dynamics of life. The processes of evolutionary change are the "ground rules" for how the living world operates.

Conservationists would do well to repeatedly recall Hutchinson's metaphor, "the ecological theater and the evolutionary play," discussed earlier. Because conservation issues all lie within the biological arena, evolution should guide their solution. Answers to conservation problems must be developed within an evolutionary framework; to do otherwise would be to fight natural laws (Meffe 1993), a foolish approach that could eventually destroy the endeavor.

The genetic composition of most populations is likely to change over time, whether due to drift in small populations, immigration from other populations, or natural selection (discussed in Chapter 11). From the perspective of conservation biology, the goal is not to stop genetic (and thus evolutionary) change, not to try and conserve the *status quo*, but rather to ensure that populations may continue to respond to environmental change in an adaptive manner.

PRINCIPLE 2: DYNAMIC ECOLOGY   The ecological world, the "theater" of evolution, is dynamic and largely nonequilibrial. The classic paradigm in ecology for many years was the "equilibrium paradigm," the idea that ecological systems are in equilibrium, with a definable stable point such as a "climax community." This paradigm implies closed systems with self-regulating structure and function, and embraces the popular "balance of nature" concept. Conservation under this paradigm would be relatively easy: Select pieces of nature for protection, leave them undisturbed, and they will retain their species composition and function indefinitely and in balance. Would that it were so simple!

The past several decades of ecological research have taught us that nature is dynamic (Pickett et al. 1992). The "balance of nature" concept may be aesthetically pleasing, but it is inaccurate and misleading; ecosystems or populations or gene frequencies may appear constant and balanced on some temporal and spatial scales, but other scales soon reveal their dynamic character. This principle applies to ecological structure, such as the number of species in a community, as well as to evolutionary structure, such as characteristics of a particular species. Conservation actions based on a static view of ecology or evolution will misrepresent nature and be less effective than those based on a more dynamic perspective.

The contemporary dominant paradigm in ecology recognizes that ecological systems are generally not in dynamic equilibrium, at least not indefinitely, and have no long-term stable points (Botkin 1990). Regulation of

ecological structure and function is often not internally generated; external processes, in the form of natural disturbances such as fires, floods, droughts, storms, earth movement, and outbreaks of diseases or parasites are frequently of overriding importance. Indeed, we now know that biodiversity in ecosystems as different as prairies, temperate and tropical forests, and the intertidal zone are maintained by nonequilibrial processes (Figure 1.8). Ecosystems consist of patches and mosaics of habitat types, not of uniform and clearly categorized communities.

It is important to understand that our emphasis on nonequilibrial processes does not imply that species interactions are ephemeral or unpredictable, and therefore unimportant. Communities are not chaotic assemblages of species; they do have structure. Embedded within all communities are clusters of species that have strong interactions, and in many cases, these interactions have a long evolutionary legacy. Nevertheless, this does not mean that community structure is invariant and that species composition does not change at some scale of space and time. Change at some scale is a universal feature of ecological communities.

Conservation within this paradigm focuses on dynamic processes and physical contexts. An important research goal for conservation biologists is to understand how the interplay between nonequilibrial processes and the hierarchy of species interactions determines commu-

(A)

(B)

(C)

(D)

Figure 1.8 Nonequilibrial processes play a major role in most ecosystems. Surface disturbances by bison create openings or "wallows" in prairies (A). Hurricanes and other storms open gaps in both temperate (B) and tropical (C) forests. Wave action (D) and tidal changes on rocky shorelines open up disturbance patches. (A, photograph courtesy of J. Wolfe; B, Congaree Swamp, South Carolina after Hurricane Hugo, 1989, by R. Sharitz; C, lower montane forest in Costa Rica, by C. R. Carroll; D, coral rock in the Dominican Republic, Caribbean Sea, by M. C. Newman.)

nity structure and biodiversity. Ecosystems are open systems with fluxes of species, materials, and energy, and must be understood in the context of their surroundings (Pickett et al. 1992). A further implication is that conservation reserves cannot be treated in isolation, but must be part of larger conservation plans whose design recognizes and accounts for spatial and temporal change (Petraitis et al. 1989; Pickett and Ostfeld 1995).

PRINCIPLE 3: HUMAN PRESENCE   Humans are and will continue to be a part of both natural and degraded ecological systems, and their presence must be included in conservation planning. Conservation efforts that attempt to wall off nature and safeguard it from humans will ultimately fail. As discussed in principle 2, ecosystems are open to the exchange of materials and species, and to the flux of energy. Because protected areas are typically surrounded by lands and waters intensively used by humans, it will be impossible to isolate them completely from these outside influences. There is simply no way to "protect" nature from human influences, and those influences must be taken into account in planning efforts. Indeed, isolating protected areas may carry its own liability in terms of increased extinction probabilities for many species.

On the positive side, there are benefits to be gained by explicitly integrating humans into the equation for conservation. First, people who have been longtime residents in the region of a protected area often know a great deal about local natural history. This "indigenous knowledge" can be useful in developing protected area management plans, and local residents can play important roles as staff (for example, as guards or environmental educators). Second, protected areas should be "user friendly" to build public support. Two ways to achieve this are through zoning that allows limited public access to portions of the protected areas with established nature trails, and through bringing ecological knowledge about the protected area into formal and informal educational programs. Most cultures take pride in their natural heritage, and a critical mission for all conservation biologists is to build upon that pride through public education. If people do not perceive that the protected area has any value to them, they will not support it.

Finally, native human cultures are a historical part of the ecological landscape and have an ethical right to the areas where they live. Aboriginal and tribal peoples from alpine to tropical regions have existed for millennia in their local systems, and to displace them in the name of conservation is simply unethical. Furthermore, they themselves add other types of diversity—cultural and linguistic diversity—which Earth is rapidly losing. Impoverishment of indigenous human cultures and languages is as large a problem as is impoverishment of

other levels of biological diversity. What's more, some of these cultures have developed sustainable methods of existence that can serve as models for modern sustainable development.

We must equally recognize that indigenous cultures have the right to control their destiny. We would be hopelessly naive to imagine that indigenous cultures can remain unchanged and unaffected by outside influences. What we can do is understand their internal systems of values and their knowledge of local natural resources, and then try to work with them toward the twin goals of conservation of biodiversity and sustainable economic development.

We must also incorporate problems of modern cultures into conservation, for they will have the largest influences on resource use. Many conservationists feel that the only realistic path to conservation in the long term is to ensure a reasonable standard of living for all people. Of course, this involves achieving greater equity among peoples, with less disparity between the "haves" and the "have-nots." In part, this will involve convincing some to accept lower standards of living so that others may climb out of desperate poverty, with the result that all will have lesser impact on biodiversity. This will not be an easy task. It will also involve attention to a number of other issues, including birth control, revised concepts of land ownership and use, education, health care, and empowerment of women.

Some postulates of conservation biology

Of course, the foundation of conservation biology is much broader than these three principles. For example, Michael Soulé, a cofounder of the Society for Conservation Biology, lists four postulates and their corollaries that characterize value statements relevant to conservation biology (Soulé 1985). Like the principles listed above, these postulates help to define the ethical and philosophical foundations for this field. Soulé's first postulate is that *diversity of organisms is good*. Humans seem to inherently enjoy diversity of life forms (called **biophilia** by E. O. Wilson [1984]), and seem to understand that natural diversity is good for our well-being and that of nature. A corollary of this postulate is that untimely extinction (that is, extinction caused by human activities) is bad. His second postulate, *ecological complexity is good*, is an extension of the first, and "expresses a preference for nature over artifice, for wilderness over gardens" (Soulé 1985). It also carries the corollary that simplification of ecosystems by humans is bad. The third postulate, *evolution is good*, has already been discussed above, and carries the corollary that interference with evolutionary patterns is bad. The final postulate is that *biotic diversity has intrinsic value*, regardless of its utilitarian value. This postulate recognizes inherent value in non-

human life, regardless of its utility to humans, and carries the corollary that destruction of diversity by humans is bad. This is perhaps the most fundamental motivation for conservation of biodiversity.

These postulates can be, and have been debated, as can any philosophical position that by definition cannot be founded on an entirely objective, scientific basis. Nevertheless, they are explicitly or implicitly accepted by many, both in and out of the conservation profession. Aspects of these arguments will be pursued further in Chapter 4.

Pervasive Aspects of Conservation Biology Efforts

Conservation biologists seek solutions to a daunting problem: how to preserve the evolutionary potential and ecological viability of a vast array of biodiversity, preserving the complexity, dynamics, and interrelationships of natural systems, in the face of humankind's propensity to try to control, simplify, and conquer those systems. To accomplish this, conservation biology has evolved into a complex multidisciplinary field that is united by the need to respond swiftly to the unfolding biodiversity crisis despite considerable uncertainty.

A discipline responding to an immense crisis

In crises, action must often be taken without complete knowledge, because to wait to collect the necessary data could mean inaction that would destroy the effort at hand. Such immediate action requires working with available information with the best intuition and creativity one can muster, while tolerating a great deal of uncertainty. This, of course, runs counter to the way that scientists are trained, but is nonetheless necessary given the practical matters at hand.

Conservation biologists are often asked for advice and input by government and private agencies regarding such issues as design of nature reserves, potential effects of introduced species, propagation of rare and endangered species, or ecological effects of development. These decisions are usually politically and economically charged and cannot wait for detailed studies that take months or even years. The "expert" is expected to provide quick, clear, and unambiguous answers (which is, of course, generally impossible), and is looked upon askance if such answers are not there, or seem counterproductive to short-term economic gain. This is a major challenge for conservation biologists, who must walk a fine line between strict scientific credibility, and thus conservatism and possibly inaction, versus taking action and providing advice based on general and perhaps incomplete knowledge, thereby risking their scientific reputations.

A multidisciplinary science

No single field of study prepares one to be a conservation biologist, and the field does not focus on input from any single area of expertise. It is an eclectic, broad discipline, to which contributions are needed from fields as different as molecular genetics, biogeography, philosophy, landscape ecology, policy development, sociology, population biology, and anthropology. This multidisciplinary nature is illustrated in Figure 1.9, in which the overlapping fields of natural and social sciences contribute to the special interdisciplinary identity of conservation biology.

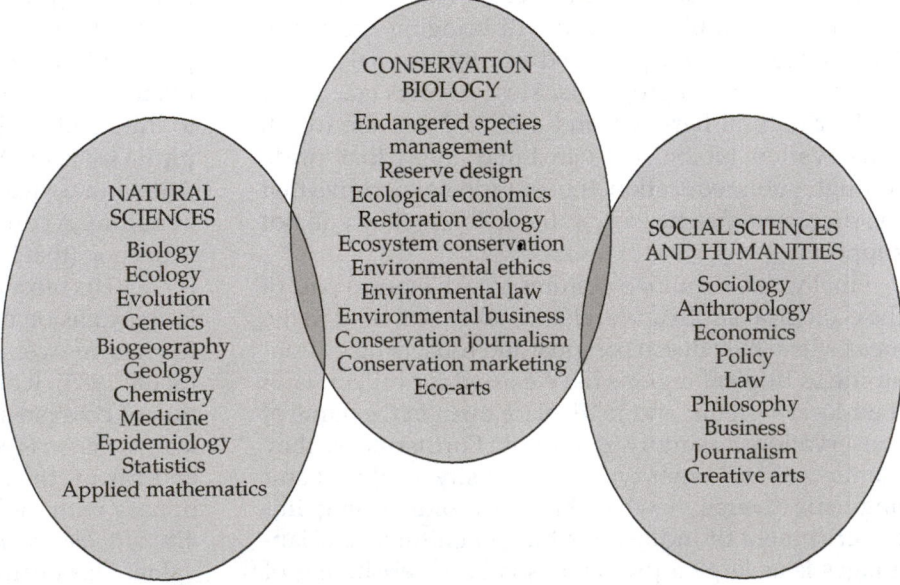

Figure 1.9 The interdisciplinary nature of conservation biology merges many traditional fields of natural and social sciences, and the humanities. This list of relevant subdisciplines and interactions is not exhaustive. Many connections have been made, and more could productively be created in the future.

CONSERVATION BIOLOGY
Endangered species management
Reserve design
Ecological economics
Restoration ecology
Ecosystem conservation
Environmental ethics
Environmental law
Environmental business
Conservation journalism
Conservation marketing
Eco-arts

NATURAL SCIENCES
Biology
Ecology
Evolution
Genetics
Biogeography
Geology
Chemistry
Medicine
Epidemiology
Statistics
Applied mathematics

SOCIAL SCIENCES AND HUMANITIES
Sociology
Anthropology
Economics
Policy
Law
Philosophy
Business
Journalism
Creative arts

ESSAY 1.4

A Private Landowner's Perspective
Conservation Biology and the Rural Landowner

Bill McDonald, *Malpai Borderlands Group*

To this rural private landowner, who leases public land for livestock grazing, the emerging discipline of conservation biology embodies both my greatest hopes for the future and my worst fears. Hope—that the best scientific minds will work with the best managerial minds to help us come to grips with the fallout from the remarkable changes of this past century, and help chart a sustainable course to the future. Fear—that the tendency to use big government, in the mistaken belief that government alone can tackle massive issues such as biodiversity loss, will add conservation biology to the growing list of buzzwords abhorred by many rural landowners, and thus make it an impediment to the very effort it represents.

The complexity of our ecosystems, on whatever scale you wish to define the term, simply defies our complete comprehension. Yet, as human beings, we are the only species with the intellectual capacity to recognize the consequences of our collective actions and consciously attempt change for the better. As the dominant species on Earth, to strive to do better is both our responsibility and our hope for survival. It is not easy work. A popular way to attempt to effect positive change is through government edict. In some very clear black and white cases (direct pollution of waters, for instance), this can be a successful approach. When we get to more complex situations, however, this approach results in partial success at best, and often in complete failure. This is particularly true when those who will be most directly affected by the "chosen course of action" are not involved in determining and implementing that course.

I am involved in a different approach. The Malpai Borderlands is a term used to describe a million-acre region in southeastern Arizona and southwestern New Mexico. The region is open space, mountains, and valleys, and its use by people is almost exclusively for cattle grazing. My family has maintained our ranch here for 98 years. Of the families who live here, many like mine are descended from the area's original homesteaders. This region is habitat to many species of plants and animals, some considered rare or endangered.

The Malpai Borderlands Group is composed of area landowners, scientists, and other stakeholders, the latter defined as anyone who has an interest in the future of the place and is willing to work to make it happen. At our invitation, federal and state land agency personnel are included in our effort; federal and state land makes up 47% of the ownership. The goal statement of our group reads as follows:

> Our goal is to restore and maintain the natural processes that create and protect a healthy, unfragmented landscape to support a diverse, flourishing community of human, plant and animal life in our Borderlands Region. Together, we will accomplish this by working to encourage profitable ranching and other traditional livelihoods that will sustain the open space nature of our lands for generations to come.

Early on, we identified two major threats to the natural diversity and health of our lands. First, is the historical suppression of fire, which is leading to a landscape dominated by woody shrub species at the expense of grasses. Second, is the threat of commercial and residential development. Both are also threats to the future of ranching livelihoods, which require both open space and healthy grasses.

While acknowledging that mistakes have been made in the past, and that there is still much to be learned about the effects of grazing on semiarid grasslands, we believe that ranching livelihoods—which depend directly on this large open space resource for its survival—are the best hope for future sustainability of that resource. To date, after ten years of existence, our group has some impressive results to show for our efforts, not the least of which is improved coordination and communication between government agencies and private landowners and between the different agencies themselves.

We have completed four prescribed burns, the first in the history of the area. The burn plans have involved wilderness study areas, two states, multiple private landowners, offices of five different government agencies in both states, coordination with Mexico, and adherence to regulations of the National Environmental Policy Act and the National Antiquities Act. Most challenging was addressing the issue of how fire would impact endangered species. While the burns were successful, with some 70,000 acres improved, the effort required to plan and implement them, one burn at a time, was expensive and exhausting. We therefore embarked on a search for a more comprehensive way to allow fire to beneficially affect the landscape. Working with the appropriate agencies, we came up with a plan that addresses all the issues in advance and we are now on the verge of getting approval from the U.S. Fish and Wildlife Service, whose office of Ecological Services has enforcement responsibility for endangered species, the source of the most contentious issues.

Our group has supported a cattle ranching family in their effort to protect a population of Chiricahua leopard frogs (*Rana chiracahuaensis*), a federally listed species that resides in stock tanks on their ranch. This effort blossomed into a joint commitment with the state wildlife department that has resulted in improved habitat for the frog and an enhanced cattle operation for the rancher. The effort eventually expanded to include the public schools of the nearby community of Douglas, Arizona, where interest in the leopard frog from teachers and students alike resulted in the construction of ponds that have become a temporary sanctuary for the frogs until they can be dispersed to natural habitats on ranches. The recent signing of a Safe Harbor Agreement between the Malpai Borderlands Group and the Fish and Wildlife Service will facilitate expansion of leop-

ard frog habitats, while protecting rights of the private property owners.

We initiated a unique program of grassbanking, where ranchers have access to grass on another ranch in exchange for conservation action of value equal to the value of the grass. For the first users of the grassbank, this meant conveyance of conservation easements to the Malpai Borderlands Group. That means that the private lands on those ranches will never be subdivided. The group now holds conservation easements on a dozen area ranches containing over 75,000 acres of private land permanently protected from commercial or residential development.

There have been a number of other actions taken or facilitated by the group that, while perhaps not as dramatic, have nudged the land a little closer to a long lasting, healthy, and sustainable open space future. Most important of all, we are working together, creating as we go a structure of support for actions that promote the biological diversity of our area and the long-term viability of our ranching livelihoods.

This grassroots alternative to traditional land management approaches is based on the voluntary actions of individuals. Our approach does not, and will never, involve coercion or the force of law. Our approach has been embraced by government agencies, politicians from both major parties, and by most of the news media. It is not, however, completely without its critics. A few landowners remain suspicious of an effort that welcomes the involvement of agency personnel and other stakeholders, particularly The Nature Conservancy. There are also those in the environmental community who do not believe that cattle grazing and healthy semiarid grasslands can coexist, period. We find ourselves between these two poles, in what we call the "radical center." We believe our approach is the one that brings results.

Where does conservation biology fit into such an effort? The role of conservation biology should be informational, certainly. Sound scientific information is crucial to helping us understand what actions to take that will be beneficial to biological diversity, and to be able to analyze the effects of actions already taken. Equally important, conservation biology's role must be supportive. It is important to champion and communicate to others those efforts that are showing results.

Will results come fast enough? Conservation biology has been called a crisis science, which certainly suggests an urgency for its application. The question of how fast, however, becomes irrelevant when we are struggling for something that works at all. The idea that you can artificially speed up a process and then inflict that approach upon all the relevant habitats of the world will ensure failure by changing the very dynamics that made the process initially successful. The continued failure of grand schemes is the real threat to the future diversity of the planet, not the pace or scope of the truly successful efforts. As our effort in the Malpai Borderlands shows, it takes time and hard work to build the trust relationships necessary to achieve real success, and it takes time and hard work to maintain them. This crisis does not call for a few broad strokes, but for millions of little ones. ■

Several features of this conceptualization of conservation biology are of note. First is the melding of the formerly "pure" fields of population biology and ecology with the "applied" fields that encompass natural resource management. The historical distinction between these disciplines is beginning to blur, and practitioners in these areas are working together toward a common goal. Second is the need for a strong philosophical basis and input from the social sciences. Because the need for conservation in the first place is the direct result of human intervention into natural systems, concern for humanistic viewpoints is vital for reducing present and future confrontations between human expansion and the natural world. Finally, this conceptualization illustrates that conservation biology is a holistic field because protection involves entire ecosystems, and multidisciplinary approaches and cooperation among disparate groups will be the most successful approach.

A strong cross-disciplinary perspective is desirable and necessary for success in conservation. The interests of natural resource agencies for their conservation employees have been expressed as being less in narrow, disciplinary skills than in "real-world" problem-solving abilities. These include "(1) cross-disciplinary breadth as well as disciplinary depth; (2) field experience; (3) language and communications skills; and (4) leadership skills, especially a mix of diplomacy and humility" (Jacobson 1990). Cannon et al. (1996) also indicated the strong need for development of human interaction skills in conservation biologists. A broad, liberal education and an ability to communicate across disciplines, combined with strength within a specialized area, is probably an ideal combination for success and real contributions in conservation biology.

An inexact science

Ecological systems are complex, often individualistic, and currently unpredictable beyond limited generalities. The public, and even other scientists, often do not appreciate this and cannot understand why conservation biologists rarely provide a simple answer to an environmental problem. The reason is, of course, that there usually *is* no simple answer. Ecological systems are complex, their dynamics are expressed in probabilities, **stochastic** influences may be strong, and many significant processes are nonlinear. *Uncertainty is inherently part of ecology and conservation, and probabilistic, rather than prescriptive answers to problems are the norm.*

Conservation biologists increasingly employ modeling techniques and statistical analyses (particularly likelihood and Bayesian approaches) to define an envelope of likely scenarios that may answer a given question.

Thus, a critical area for conservation biologists to become familiar with are such quantitative approaches to problem solving and definition. It is beyond the scope of this book to teach these approaches, but you will see examples of their use in most chapters. Used in efforts to understand the workings of a relatively undisturbed ecosystem, the effect of a threat to that ecosystem, or the effect of a management intervention, these statistical and mathematical approaches can help us to find where the answers are most likely to lie, or at the least, to better define where we need more information to find the answers. While we often cannot know a single answer to a problem, we often may be able to define which answers are most likely to be wrong or right, and work within the range of probable answers.

Thus, the conservation biologist often faces a credibility gap, not because he or she is incompetent, or because the field is poorly developed, but because even the simplest of ecosystems is far more complicated than the most complex of human inventions, and most people have not the slightest notion that this is the case. This gap can easily be exploited by representatives of special interest groups, such as lawyers, engineers, and developers, all of whom are used to dealing with concrete situations that can be easily quantified, and for which a "bottom line" can be extracted. There is never an easy bottom line in ecology, and we can only hope to educate others to that fact, rather than be forced to develop meaningless and dangerous answers that have no basis in reality. The conservation biologist must think "probabilistically" and understand the nature of scientific uncertainty. Consequently, conservationists should include safety margins in the design of management and recovery strategies, as does an engineer in the design of a bridge or an aircraft.

A primary "safety net" that conservation biologists advocate is the adoption of the **precautionary principle**. The environmental equivalent of the Hippocratic oath, "First, do no harm," the precautionary principle exhorts us to avoid practices that could lead to irrevocable harm or serious environmental degradation in the absence of scientific certainty about whether such harm will occur. If an ongoing practice is suspect, then it should be suspended until and unless it is shown not to be harmful. Beyond this, it also calls on people to search for alternatives to potentially damaging practices. Essentially, this is the ultimate safety margin that prevents us from taking potentially damaging actions unless and until we are reasonably sure they will cause no serious harm.

Not only conservationists hold fast to the precautionary principle; many politicians see the wisdom of acting with care when the environmental or human welfare stakes are high. The precautionary principle is a core part of the environmental policies of the European Union. Although less prevalent in U.S. politics, many have pushed for its widespread adoption to protect human health, as well as biodiversity. Former U.S. Environmental Protection Agency director Christine Todd Whitman held the position that "policymakers need to take a precautionary approach to environmental protection. . . . We must acknowledge that uncertainty is inherent in managing natural resources, recognize it is usually easier to prevent environmental damage than to repair it later, and shift the burden of proof away from those advocating protection toward those proposing an action that may be harmful."

One of the strongest statements suggesting extensive use of the precautionary principle is the Wingspread Agreement, formulated at an international meeting of government officials, scientists, lawyers, and environmental and labor activists (Box 1.1), although this wording has not been adopted as policy by any government. Although not explicitly written into many U.S. laws, precaution is implicit in the U.S. **Endangered Species Act**, as well as a number of other important pieces of environmental legislation.

The precautionary principle is imbedded in many laws and international agreements. One of the most important for biodiversity conservation comes in the preamble to the 1992 Convention on Biological Diversity and the Framework for Convention on Climate Change: "Where there is a threat of significant reduction or loss of biological diversity, lack of full, scientific certainty should not be used as a reason for postponing cost-effective measures to avoid or minimize such a threat."

Peru, Costa Rica, and Australia all have recently enacted legislation to protect biodiversity that envokes the precautionary principle. Increasingly, conservation biologists are working to define when and how the precautionary principle should be included in strategies to protect biodiversity (Cooney 2004).

A value-laden science

Science is supposed to be value-free. It is presumably completely objective and free from such human frailties as opinions, goals, and desires. Because science is done by humans, however, it is never value-free, but is influenced by the experiences and goals of the scientists, although they often will not admit that. "Too many teachers, managers, and researchers are trapped by the Western positivist image of science as value-free; . . . Biologists must realize that science, like everything else, is shot through with values. Sorting out the norms behind positions is the initial step of critical thinking" (Grumbine 1992).

Unlike many other areas of science, conservation biology is "mission-oriented" (Soulé 1986). The goal is clearly to conserve natural ecosystems and biological processes, which are held as intrinsically valuable by conservation biologists.

BOX 1.1 Wingspread Statement on the Precautionary Principle

Recognizing the need for guidance on environmental policy, an international group of scientists, government officials, lawyers and environmentalists met January 23–25, 1998 at the Wingspread Center in Racine, Wisconsin. Following two days of discussion, the group issued the following concensus statement, which has served to guide environmental policy planning.

Statement

The release and use of toxic substances, the exploitation of resources, and physical alterations of the environment have had substantial unintended consequences affecting human health and the environment. Some of these concerns are high rates of learning deficiencies, asthma, cancer, birth defects, and species extinctions, along with global climate change, stratospheric ozone depletion, and worldwide contamination with toxic substances and nuclear materials.

We believe existing environmental regulations and other decisions, particularly those based on risk assessment, have failed to adequately protect human health and the environment, the larger system of which humans are but a part.

We believe there is compelling evidence that damage to humans and the worldwide environment is of such magnitude and seriousness that new principles for conducting human activities are necessary.

While we realize that human activities may involve hazards, people must proceed more carefully than has been the case in recent history. Corporations, government entities, organizations, communities, scientists, and other individuals must adopt a precautionary approach to all human endeavors.

Therefore, it is necessary to implement the Precautionary Principle: When an activity raises threats of harm to human health or the environment, precautionary measures should be taken even if some cause and effect relationships are not fully established scientifically. In this context the proponent of an activity, rather than the public, should bear the burden of proof.

The process of applying the Precautionary Principle must be open, informed, and democratic, and must include potentially affected parties. It must also involve an examination of the full range of alternatives, including no action.

The question of values and advocacy in conservation science have been debated in conservation journals and within various scientific societies (e. g. Barry and Oelschlaeger 1996 and associated responses; Meffe 1996). Whether and how conservation scientists should become involved in policy development is a major issue (see Chapter 17). An emerging, consensus answer seems to be that scientists have a clear responsibility to society to lend their knowledge and expertise toward the value-laden goal of biodiversity preservation, but that good, objective science must serve as a foundation for reaching that goal. Objectivity in how science is conducted cannot be compromised to reach predetermined goals, for then all scientific credibility is lost.

A science with an evolutionary time scale

In contrast to traditional resource management, whose currency includes maximum sustained yields, economic feasibility, and immediate public satisfaction with a product, the currency of conservation biology is long-term viability of ecosystems and preservation of biodiversity *in perpetuity*. A conservation biology program is successful not when more deer are harvested this year, or even when more natural areas are protected, but when a system retains the diversity of its structure and function over long time periods, and when the processes of evolutionary adaptation and ecological change are permitted to continue. If there is a common thread running throughout conservation biology, it is the recognition that evolution is *the* central concept in biology, and has played and should continue to play *the* central role in nature.

A science of eternal vigilance

The price of ecosystem protection is eternal vigilance. Even "protected" areas may be destroyed in the future if they contain resources that are deemed desirable enough by powerful groups or individuals. A case in point is the United States' Arctic National Wildlife Refuge, an area set aside for its ecological significance, but repeatedly under pressure (and again as of this writing) to be opened up for oil extraction as world political affairs affect the price and availability of oil. What appears secure today may well be exploited tomorrow for transitory resource use, and the conservation biologist must continually be protective of natural areas and must stay on top of policy developments that affect conservation. Natural ecosystems can easily be destroyed, but they cannot be created, and at best only partially restored.

A Final Word

Ecological systems are complex, and situations are often unique. What makes sense in one system or circumstance will be inapplicable in another. Idiosyncrasies abound, as do conflicting demands. Conservation sce-

narios need to be defined and pursued individually, not by "prescription." Conservation biology is not easy, but it is not hopelessly complicated either, and much research and application remains to be done, as we emphasize throughout this book. Above all, it can provide exciting and unparalleled career opportunities for people interested in solving real-world problems. The world's biodiversity desperately needs bright, energetic, and imaginative people who will dedicate their work to making a difference. And they certainly can, and must.

Summary

1. Exponential human population growth and consumptive habits in the last few centuries have affected the natural world to the extent that massive alteration of habitats and associated biological changes threaten the existence of thousands of species and basic ecosystem processes. The field of conservation biology developed over the last 40 years as a response of the scientific community to this crisis. Conservation biology differs from traditional resource conservation in being motivated not by utilitarian, single-species issues, but by the need for conservation of entire systems and all their biological components and processes.

2. Conservation practices have a varied history around the world, but generally have focused on human use of resources. In the U.S., two value systems dominated resource conservation early in the twentieth century. The Romantic-Transcendental Conservation Ethic of Emerson, Thoreau, and Muir recognized that nature has inherent value and should not simply be used for human gain. The Resource Conservation Ethic of Pinchot was based on a utilitarian philosophy of the greatest good for the greatest number of people; many resource agencies in the U.S. and elsewhere follow this view. Aldo Leopold's Evolutionary-Ecological Land Ethic developed later, and is the most biologically relevant perspective, recognizing the importance of ecological and evolutionary processes in producing and controlling the natural resources we use. Much of modern conservation biology has grown from and is guided by Leopold's land ethic.

3. Three overriding principles guide all of conservation biology. First, evolution is the basis for understanding all of biology, and should be a central focus of conservation action. Second, ecological systems are dynamic and nonequilibrial; change must be a part of conservation. Finally, humans are a part of the natural world and must be included in conservation actions.

4. Conservation biology has some unusual characteristics not always found in other sciences. It is a crisis discipline that requires multidisciplinary approaches. It is an inexact science that operates on an evolutionary time scale. It is a value-laden science that requires long-term vigilance to succeed. It also requires of its practitioners innovation, flexibility, multiple talents, and an understanding of the idiosyncrasies of ecological systems, but offers outstanding career challenges and rewards.

Please refer to the website www.sinauer.com/groom for Suggested Readings, Web links, additional questions, and supplementary resources.

Questions for Discussion

1. How would you explain the significance of human population growth and the human ecological footprint for biodiversity conservation?

2. What would conservation practice be like if we primarily followed the principles of Pinchot's Resource Conservation Ethic? How would it differ from the present day focus of Conservation Biology? Is the present focus of conservation biology preferable to the Romantic-Transcendental Conservation Ethic of Emerson, Thoreau and Muir? How?

3. How would you answer a conservation skeptic who asserted that because ecological processes are nonequilibrial, conservation of current communities is misguided since they are destined to change?

4. Why are multiple disciplinary perspectives important in conservation biology?

5. How are the views of conservation biology and conservation practice similar among the four essayists (Erica Fleishman, Jamie Rappaport Clark, Kathryn Fuller, and Bill McDonald) featured in this chapter? How are they different?

2

Global Biodiversity

Patterns and Processes

Gordon H. Orians and Martha J. Groom

The most wonderful mystery of life may well be the means by which it created so much diversity from so little physical matter. The biosphere, all organisms combined, makes up only about one part in ten billion of the earth's mass. It is sparsely distributed through a kilometer-thick layer of soil, water, and air stretched over a half billion square kilometers of surface.

E. O. Wilson, 1992

What Is Biodiversity and Why Is it Important?

Biological diversity, or **biodiversity**, is the sum total of all living things—the immense richness and variation of the living world. Biodiversity can be considered at many levels of biological variation, ranging from genetic variability within a species, to the biota of some selected region of the globe, to the number of evolutionary lineages and the degree of distinctness among them, to the diversity of ecosystems and biomes on Earth. There is no one "correct" level at which to measure biodiversity because different scientific issues and practical problems find their focus at different levels.

Few people are aware of the full spectrum of biodiversity, because our own experience focuses on interactions with people and those species that interact directly with people, or a few others that attract our attention. Yet if we are to understand why it is important to preserve biodiversity, we must appreciate the richness that lies at each level—from genes to biomes.

The various levels of biodiversity are best understood from a hierarchical perspective, from genes, through populations and species, to communities, ecosystems, and landscapes. Further, we also can describe biodiversity in terms of its variation in composition, structure, and functioning. This value of a hierarchical perspective is described and illustrated by Reed Noss in Essay 2.1. The following discussion proceeds from lower to higher levels in the hierarchy of biodiversity.

ESSAY 2.1

Hierarchical Indicators for Monitoring Changes in Biodiversity

Reed F. Noss, *University of Central Florida*

■ Biodiversity and how to save it is the subject matter of conservation biology. If you have read this far in this book, or even skimmed its pages, two things about biodiversity should be clear: (1) it is complex, and (2) it is always changing. How on earth can a conservation biologist or land manager deal with this mess?

First, we need to make some sense of the complexity of nature. We can dissect the biodiversity concept into meaningful components, and yet retain some idea of how they all fit together, by appealing to hierarchy theory. There are several kinds of hierarchies in nature, including the familiar levels of biological organization (such as genes, populations, species, communities), hierarchies of space and time, and hierarchies of rates. There are also ethical hierarchies; for instance, many people care about the suffering of individual animals, but some also care about the loss of species, ecosystems, and biomes. All of these hierarchies are nested; that is, higher levels enclose lower levels and, to a great extent, constrain their behavior. A tree is part of a forest stand, the stand is part of a landscape, the landscape is part of a physiographic region, and so on. If the physiographic region is inundated by a volcanic flow, everything nested within it also goes.

Biodiversity is not just species diversity. A comprehensive approach to biodiversity conservation must address multiple levels of organization and many different spatial and temporal scales. Most definitions of biodiversity recognize its hierarchical structure, with the genetic, population–species, community–ecosystem, and landscape levels considered most often. Each of these levels can be further divided into compositional, structural, and functional components. Composition includes the genetic constitution of populations, the identity and relative abundances of species in a natural community, and the kinds of habitats and communities distributed across the landscape. Structure includes the sequence of pools and riffles in a stream, downed logs and snags in a forest, and the vertical layering and horizontal patchiness of vegetation. Function includes the climatic, geologic, hydrologic, ecological, and evolutionary processes that generate biodiversity and keep it forever changing.

Change is universal, but some kinds of change threaten biodiversity. Changes in climate, changes in disturbance regime (such as fire suppression, or conversely, increases in ignitions), introductions of novel chemicals into the environment, and species introductions or deletions are all changes likely to degrade native biodiversity. These kinds of changes happen naturally, but often occur faster and are of greater magnitude with human activity. In order to have any chance of protecting biodiversity against the onslaught of these factors, we must have early warning of change; hence the need for monitoring. Because biodiversity is multifaceted and hierarchical, those indicators that we select as targets for monitoring should represent all of this complexity. Otherwise, something might fall through the cracks.

Land managers are familiar with the use of *indicator species,* often selected to represent a suite of species with similar habitat requirements. As a well-known example, the Northern Spotted Owl

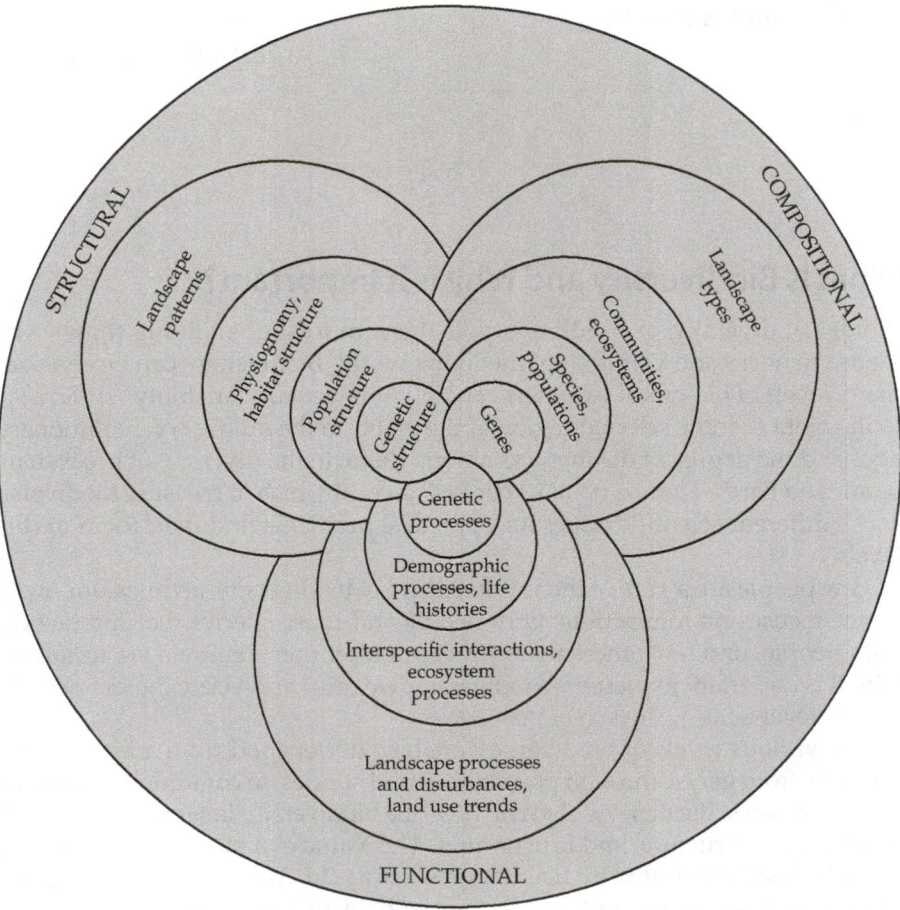

Figure A   Compositional, structural, and functional attributes of biodiversity at four levels of organization. (From Noss 1990.)

(*Strix occidentalis caurina*) was selected by managers of national forests in the Pacific Northwest as a surrogate for all other species associated with old-growth forests. However, the use of indicator species has encountered problems, including biased selection criteria, false assumptions about species–habitat relationships, unwarranted extrapolations from one species to others, and flawed design of monitoring programs. But species will continue to be useful as indicators, particularly if we focus on those most sensitive to human activities and those that play pivotal roles in their ecosystems. For example, change in the abundance of woodpeckers may warn us of possible changes in populations of other species that use woodpecker cavities. But we should not carry our extrapolations too far. The idea of one species representing all others that share a similar habitat is not ecologically realistic.

Indicators for monitoring biodiversity must consist of much more than a set of indicator species. Because biodiversity is distributed hierarchically, so too should indicators be. A framework for selecting biodiversity indicators might follow a nested hierarchy of compositional, structural, and functional elements (Figure A). A monitoring program should select a broad range of indicators that correspond to critical management questions, such as: Are populations of rare species being maintained in sizes and distributions that assure long-term demographic and genetic viability? Are the natural structure and species composition of the community being maintained? Is the configuration of the landscape adequate to permit normal movements of organisms? Using the example of a managed forest landscape, some measurable indicators that might help answer such questions are listed in Table A. Only with such a comprehensive approach will conservation biologists be able to track changes in biodiversity and obtain the information they need to preserve it. ■

TABLE A *Hierarchical Indicators for Monitoring Biodiversity*

GENETIC

Composition
- Allelic diversity
- Presence/absence of rare alleles

Structure
- Heterozygosity
- Phenotypic polymorphism

Function
- Symptoms of inbreeding depression or genetic drift (reduced survivorship or fertility, abnormal sperm, reduced resistance to disease, morphological abnormalities or asymmetries)
- Inbreeding/outbreeding rate
- Rate of genetic interchange between populations (measured by rate of dispersal and subsequent reproduction of migrants)

POPULATION–SPECIES

Composition
- Absolute and relative abundance, density, basal area, cover, importance value for various species

Structure
- Sex ratio, age distribution, and other aspects of population structure for sensitive species, keystone species, and other special interest species
- Distribution and dispersion of special interest species across the region

Function
- Population growth and fluctuation trends of special interest species
- Fertility, fecundity, recruitment rate, survivorship, mortality rate, individual growth rate, and other individual and population health parameters
- Trends in habitat components for special interest species (varies by species)
- Trends in threats to special interest species (depends on life history and sensitivity of species in relation to land use practices and other influences)

COMMUNITY–ECOSYSTEM

Composition
- Identity, relative abundance, frequency, richness, and evenness of species and guilds (in various habitats)
- Diversity of tree ages or sizes in community (stand)
- Ratio of exotic species to native species in community (species richness, cover, and biomass)
- Proportions of endemic, threatened, and endangered species

Structure
- Frequency distribution of seral stages (age classes) for each forest type and across all types
- Average and range of tree ages within defined seral stages
- Ratio of area of natural forest of all ages to area in clear-cuts and plantations
- Abundance and density of snags, downed logs, and other defined structural elements in various size and decay classes
- Spatial dispersion of structural elements and patches

- Foliage density and layering (profiles), and horizontal diversity of foliage profiles in stand
- Canopy density and size, dispersion of canopy openings
- Areal extent of each disturbance event (e.g., fires)

Function
- Frequency, intensity, return interval, or rotation period of fires and other natural and anthropogenic disturbances
- Cycling rates for various key nutrients (e.g., N, P)
- Intensity or severity of disturbance events
- Seasonality or periodicity of disturbances
- Predictability or variability of disturbances
- Human intrusion rates and intensities

LANDSCAPE

Composition
- Identity, distribution, richness, and proportions of patch types (such as forest types and seral stages) across the landscape
- Total amount of late successional forest interior habitat
- Total amount of forest patch perimeter and edge zone

Structure
- Patch size frequency distribution for each seral stage and forest type, and across all stages and types
- Patch size diversity index
- Size frequency distribution of late successional interior forest patches (minus defined edge zone, usually 100–200 m)
- Forest patch perimeter:area ratio
- Edge zone:interior zone ratio
- Fractal dimension
- Patch shape indices
- Patch density
- Fragmentation indices
- Interpatch distance (mean, median, range) for all forest patches and for late successional forest patches
- Juxtaposition measures (percentage of area within a defined distance from patch occupied by different habitat types, length of patch border adjacent to different habitat types)
- Structural contrast (magnitude of difference between adjacent habitats, measured for various structural attributes)
- Road density (mi/mi² or km/km²) for different classes of road and all road classes combined

Function
- Disturbance indicators (see above)
- Rates of nutrient, energy, and biological transfer between different communities and patches in the landscape

Components of Biodiversity

Genetic diversity

Genetic variability is the ultimate source of biodiversity at all levels. The number of genes found in different species ranges over more than three orders of magnitude (from about 500 in a parasitic bacterium to more than 20,000 in a mouse or a mustard plant), although not all of those genes code for products. In addition, most genes exist in several forms, or **alleles**, so the possible number of combinations of genes is enormous, much larger than the number of individuals present in any species. This genetic variability is the material upon which the agents of evolution act.

Recent advances in molecular biology provide the tools we need to measure the amount of genetic variability present in organisms. Measures of variability within local populations provide important clues that help us understand the nature of forces acting on genetic variation. Such knowledge has important practical applications in, for example, designing captive breeding programs for rare species to reduce deleterious effects of inbreeding, or determining the best sources of individuals for reestablishing populations in areas from which they have been exterminated. Measurements of the amount of genetic differences among species enable us to determine rates of evolution and establish phylogenetic relationships among organisms.

Genes are the fundamental unit of biodiversity, without which no evolutionary change occurs. Specific genes or combinations of genes allow individuals to tolerate polluted conditions, exploit resources more efficiently, or compete better with other species. Abilities to tolerate temperature changes, or disperse great distances, which are at least in part genetically based, may be crucial to the persistence of species in the face of a rapidly changing climate. Preservation of genetic diversity is thus an important goal of conservation biology.

Preservation of genetic diversity in crops and livestock breeds is also important. Distinct varieties of domesticated species are adapted to a wide range of environmental conditions. These varieties are especially important for human welfare, particularly in the poorest areas of the world where other supporting institutions are weak or absent. Further, greater agricultural efficiency means that less land must be cultivated or serve as rangeland to meet the needs of a given population, leaving more land available for preservation of biodiversity.

Population-level diversity

An extremely important component of biodiversity is the variation found within members of a species or population. A population is a group of individuals of a particular species living in a specific area at the same time. Genetic differences among individuals within a population, and among different populations of a species reflect the evolutionary history of that population, and shape its future potential for adaptation. It is in local populations where responses to environmental challenges occur, where adaptations arise, and where genetic diversity is maintained and reshuffled each generation. We are concerned here primarily with genetic differences, because only heritable differences can be passed on to future generations. The potential rate of evolutionary change of a population is proportional to the amount of available genetic variability. Variation among populations of a species is the result primarily of adaptations to local ecological conditions. Locally adapted populations of a widespread species may have particular genes and gene combinations critical for survival and reproduction in those particular areas. The nature and extent of between-population genetic variability reveals much about the evolutionary history of populations. The theoretical and practical importance of population-level genetic diversity is treated in greater detail in Chapter 11.

Individual populations may have great conservation value. Many wide-ranging species consist of a number of genetically isolated or semi-isolated populations, each of which is adapted to a different environment. For example, populations of some plant species have evolved high levels of tolerance to metals in the soil (Antonovics et al. 1971). If we were attempting to re-establish a population of a plant on metal-contaminated soil, we would be successful only if we planted genotypes that could tolerate those conditions.

Guppies (*Poecilia reticulata*) in Trinidad streams have very different genetically based color patterns, clutch sizes, and offspring sizes depending on whether they occur in streams with or without predatory fishes (Reznick and Bryga 1987; Reznick et al. 1990). If we were trying to reintroduce an extirpated population of these guppies into a stream with predators, we would need to use individuals having the traits that promote coexistence with predators. Some plant species have markedly different genetically based growth forms in low- and high-elevation ecosystems (Clausen et al. 1940). Such genetically based **plasticity** in life history characters, which is probably a pervasive feature of most organisms, is under-appreciated by conservation biologists.

The ecological functioning of each population and the services it provides may be largely independent of other populations of the species. For example, if a population of a bee is exterminated from a river basin due to pesticides, the flowers that depend on that bee for pollination have lost that service, even if the "species" still exists elsewhere, unless some other species can assume that role. Thus, the loss of local populations reduces the ability of ecosystems to provide goods and services to people and other species living there. For this reason, the massive declines in abundances and ranges of many species are of great conservation concern (Gaston and Spicer 2004).

Human cultural diversity

The variety of human cultures is an important component of population-level diversity. Cultural diversity embodies a reservoir of human knowledge, skills, values, and management traditions that have evolved for thousands of years. Different cultures interact with nature in many ways that reciprocally influence the development of human societies and their impacts on nature. Cultural practices represent "solutions" to the challenges of finding or growing food and defeating infections and disease in a given habitat. In Essay 2.2, Mark Plotkin and Adrian Forsyth describe some of the sophisticated solutions indigenous cultures devised to maintain productive agricultural systems in difficult environmental conditions.

There are 6526 distinct spoken languages, with the greatest diversity concentrated in tropical regions, with a few exceptions (e.g., Australia, United States) (Figure 2.1). To the extent that language reflects cultural differences, this provides at least a lower bounds on the total number of distinct cultures worldwide. Cultural diversity can also be measured by counting the number of localized, indigenous human populations, which are most numerous in tropical moist forest habitats (see Figure 2.1).

The importance of cultural diversity is widely unappreciated. Essentially, cultural diversity encompasses many of the mechanisms that allow people to adapt to changing conditions in their environment.

Diversity of species

Most people think of diversity of species when they hear and use the term "biodiversity." Species are fundamental units of evolution, and species are the primary targets of some of the most powerful pieces of conservation legislation, such as the U.S. Endangered Species Act (ESA) and the Convention on International Trade in Species (CITES). Therefore we will emphasize species diversity in this chapter and throughout the book. Yet we should keep in mind that population-level diversity can be as or more important for conservation, and that communities, ecosystems, landscapes, or ecoregions may be the most appropriate level for some conservation activities.

A species is a group of actually or potentially interbreeding natural populations that are reproductively isolated from other such groups. Deciding whether two populations constitute different species may be difficult because speciation is often a grad-

ual process. Once a species becomes separated into two or more populations, the daughter populations may evolve independently for a long time before they become reproductively incompatible, or they may become reproductively incompatible before they evolve noticeable morphological, physiological or behavioral differences. Thus, separate populations within a species may exist at various stages in the process of becoming separate species. Conservation biologists are often concerned with preserving evolving populations that have not yet become full species. We do this because the genetic differences evolving populations have accumulated may enable them to survive in different future environments. We seek to preserve the differing evolutionary lineages represented by each subspecies, incipient species or full species, as well as the potential for variable evolutionary outcomes that will unfold as species or subspecies persist separately or interact.

Although species are generally defined in terms of reproductive isolation, such information is often not avail-

(A)

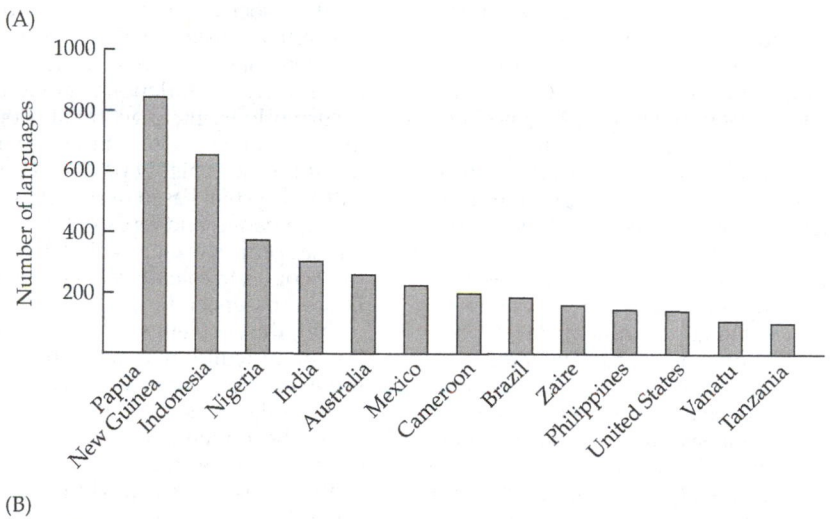

(B)

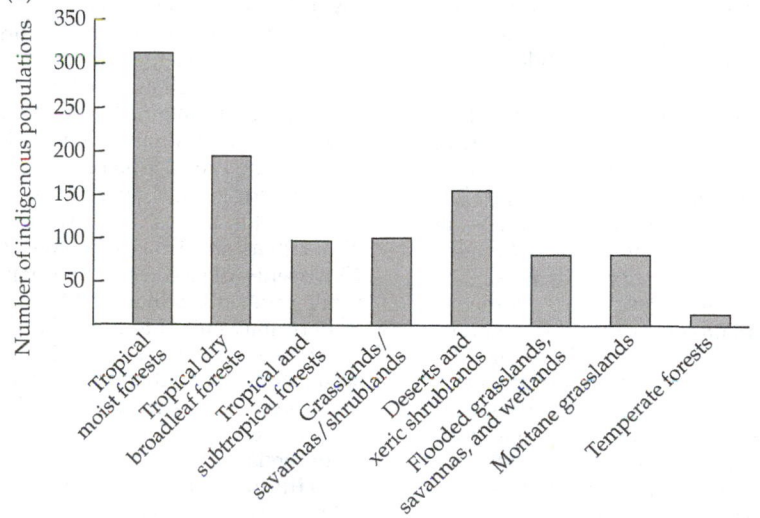

Figure 2.1   Linguistic diversity (A) and numbers of indigenous cultures (B) across the world. (Modified from Mulder and Coppolillo 2005.)

ESSAY 2.2

The Importance of Indigenous Knowledge Systems

Mark J. Plotkin, *Amazon Conservation Team,* **and Adrian Forsyth,** *Gordon and Betty Moore Foundation*

■ The philosopher-naturalist Laurens Van der Post remarked that the most damning legacy of colonialism has been its relentless tendency to separate tropical peoples from their land and culture. "The great mistake people make is in thinking about the aboriginal races in South Africa—the Bushmen and Hottentots and the like—as primitive peoples. They were actually very sophisticated societies and cultures with immense awareness and very important values."

The sophistication of many indigenous cultures is most evident in their agricultural systems. Over 200 years ago, a Jesuit priest in lowland Bolivia, Francisco Eder, noted that an extraordinary irrigation system in the local savannas, which had been devised in pre-Columbian times, still supported an enormous population of Mojeno Indians. Thanks to the influence of the padre and his cohorts, as well as introduced diseases, this Amerindian agricultural system was abandoned and the savannas that once fed the teeming local populace today are covered with a scrub vegetation supporting a few mangy cows.

The chinampa system, developed in pre-Columbian Mexico, remains in use, albeit in a much more restricted range than it once covered. Usually constructed in swampy or lacustrine environments, chinampas consist of garden plots created by building small islands using layers of vegetation and mud. The surrounding waters seep in and provide the necessary moisture, while mud scooped up from canals around the plot is periodically added as fertilizer to the garden on top of the island. According to Dr. Jim Nations of Conservation International, chinampas not only produce food and other useful crops year-round, they do not deplete the soil or require artificial fertilizers or pesticides.

A similar system was rediscovered in the altiplano of northwestern Bolivia. Dr. Alan Kolata of the University of Chicago had long been intrigued by a series of ridges and depressions in this remote region that seemed to indicate some form of agricultural system that involved raised planting beds surrounded by canals. The local Aymara Indians, however, employed no such system. Kolata and his colleagues worked with the Aymaras to dig a series of raised beds surrounded by ditches and the results were a crop yield seven times greater than the local average.

Here we see three different fates for three independently evolved agricultural systems: The Mojeno system has been lost, the chinampa system still exists, and the altiplano system would have remained lost if it had not been for an intrepid anthropologist and his indigenous colleagues. How many other equally valid and possibly even more productive systems have completely disappeared without a trace? The superiority complex that drives outsiders from Western societies to want to replace "primitive" systems with an ill-suited one developed on foreign soils continue to this day: The Indonesian government is currently trying to get tribal peoples in highland New Guinea to forgo their traditional and highly productive agricultural system based on sweet potatoes or sago palms, and replace it with rice, an inappropriate crop. This colonial attitude comes from cultural biases and a lack of understanding.

Yet, the problem we face is not just loss of agricultural systems themselves, but extinction of cultivar diversity as well. Indigenous agricultural systems typically contain many different varieties of a single crop, much as a farmer in the industrialized world may grow several different types of corn: one to feed his family, another to feed his animals, still another to make popcorn, and another to sell as a cash crop. The Amerindian farmer in the Amazon usually cultivates distinct varieties of cassava for the production of bread, beer, meal, porridge, and whatever other end products he desires. The Tirio Indians of northeastern Amazonia have at least 15 varieties of cassava in their gardens while other tribes like the Machiguenga of Peru may cultivate several dozen.

These crop varieties often harbor adaptations that enhance yields or increase resistance to pests and diseases. Recently, crosses of indigenous and commercial varieties have led to tremendous improvements. A barley plant from Ethiopia was crossbred with barley in California, providing resistance to the lethal yellow dwarf virus, which threatened an industry worth $160 million per year. The yield of cassava in Africa was increased ten-fold because of disease resistance provided by cassava from the Amazon. And scientists have recently found that a variety of sunflower being cultivated by the Havasupai Indians in the American southwest offers resistance to a blight attacking sunflower crops in the Old World.

The late economic botanist Edgar Anderson once stumbled across an Indian garden in Guatemala that initially seemed more of a rat's nest than a productive agricultural plot. It was only after careful study that he realized how much more sophisticated the botanist was the farmer:

> In terms of our American and European equivalents the garden was a vegetable garden, an orchard, a medicinal plant garden, a rubbish heap, a compost heap, and a bee-yard. There was no problem of erosion though it was at the top of a steep slope; the soil surface was practically all covered and apparently would be during most of the year. Humidity would be kept up during the dry season and plants of the same sort were so isolated from one another by intervening vegetation that pests and diseases could not readily spread from plant to plant. . . . I suspect that if one were to make a careful study of such an American Indian garden, one would find it more productive than ours in terms of pounds of vegetables and fruit per man-hour per square foot of ground.

At a time when there are ever more mouths to feed on this planet, Anderson's admonition to look more closely at the form and function of indigenous systems seems like advice worth following. Modern management techniques often overlook and disparage these indigenous systems, which are based on centuries of in situ sustainable existence, in favor of high-tech but often inappropriate and expensive systems that fail. We can learn a great deal about environmental management (and humility) from such cultures. ■

able. For example, the concept cannot be applied to organisms that reproduce asexually. Also, fossils, the source of most information about organisms that lived in the past, rarely provide clues about breeding behavior. Finally, information about reproductive behavior is not available for most species that live today. Therefore, many species are recognized on the basis of their morphology. For example, in efforts to complete an inventory of the biota of Costa Rica, biologists have relied heavily on distinguishing **morphospecies**, particularly among arthropods (INBIO 1995). The morphospecies concept is useful, because it can be applied when other data are unavailable. Moreover, given that evolution is, to use Darwin's words, "descent with modification," organisms that look alike generally share many alleles that code for their body structures. This is why most species designations based on morphology have been supported by subsequent genetic data.

As closely related species continue to evolve in genetic isolation from one another, they diverge and accumulate genetic differences. Thus, the number of unique genes, and the morphological and physiological traits they encode, increases in a lineage of organisms over time. These unique traits are important components of the rich fabric of life on Earth. To determine the degree and significance of these differences, biologists attempt to reconstruct **phylogenies** or **cladograms**. A phylogeny is a hypothesis that describes the history of descent of a group of organisms from its common ancestor. A lineage is represented as a branching "tree," each node of which represents a speciation event. That is, the phylogeny shows the order in which lineages are hypothesized to have split. Systematists use a wide variety of information—morphological, developmental, physiological, behavioral, and molecular—to reconstruct phylogenies. Computer programs are used to infer the most likely evolutionary relationships among organisms in a lineage (e.g., PHYLIP, Felsenstein 2004). Conservation biologists use phylogenies for many purposes, as we will describe in Chapter 11.

The biological classification system in use today was first proposed by the Swedish biologist, Carolus Linnaeus in 1758. Linnaeus gave each species two names, one identifying the species itself, and the other the genus (a group of closely related species) to which it belongs in the Linnean system. Species and genera are further grouped into a hierarchical system of higher taxonomic categories—families, orders, classes, phyla, kingdoms, and domains. This system provides unique names for each species and is an aid to memory and communication. As much as possible, taxonomists classify organisms hierarchically in ways that reflect their evolutionary history, constructing classifications in which the higher taxonomic categories contain all of the descendants of their most recent ancestor, but no other species; that is, categories that are monophyletic.

How Many Species Are There?

Approximately 1.75 million living and 300,000 fossil species have been described and given scientific names (Table 2.1). On average, about 300 new species are formally described and named every day, and no slowdown is in sight. New animal phyla, Cycliophora and Loricifera, whose members live in interstitial spaces in ocean sediments, were first described within the past 25 years! Not surprisingly, because they are based on incomplete and indirect evidence, estimates of the actual number of living species vary widely. Current estimates of the total number of living species range from 10 million to as high as 50 million or more (May 1988; Wilson 1992; Gaston and Spicer 2004). In other words we do not know within an

TABLE 2.1   *Number of Living Species in Major Phyla[a]*

Domain/ kingdom	Phylum	Number of described species	Number of estimated species	Percent described
Viruses	—	5,000	500,000	1
Bacteria	—	4,760	1,000,000	0.5
Archaea	—	259	Unknown	
Eukarya				
Protista[b]	—	80,000	500,000	16
Fungi	Eumycota	80,000	1,500,000	5
Plantae	Bryophyta	15,000	30,000	50
	Trachaeophyta	272,655	500,000	55
Animalia	Porifera	15,000	—	—
	Cnidaria	10,000	—	—
	Platyhelminthes	25,000	—	—
	Rotifera	1,800	—	—
	Bryozoa	5,000	—	—
	Nematoda	25,000	400,000	6
	Arthropoda	1,065,000	9,000,000[c]	12
	Annelida	15,000	—	—
	Mollusca	70,000	200,000	35
	Echinodermata	7,000	—	—
	Chordata	57,739	60,000	96

Source: Adapted from data in Margulis and Schwartz 2000; Lecointre et al. 2001; IUCN Red list 2004; Gaston and Spicer 2004; Maddison and Schultz 2004.

[a]Major phyla listed have >1000 described species.

[b]Protista lumps 5–7 distinct kingdoms within the domain Eukaryota.

[c]Estimates range from 6–30 million.

order of magnitude the number of living species. Thus, a large fraction of species likely to be exterminated during the twenty-first century will disappear before they have been named, much less understood ecologically.

The immense richness of viruses, bacteria, archaea, protists, and unicellular algae is still largely uncatalogued. Within the domain Archaea, new phylum-level groups are discovered every year (Furhman and Campbell 1998), and the richness of species within these groups is unknown. There are few studies of viruses except those that attack people, our domesticated plants and animals, and the organisms we study scientifically. How many types attack noncrop plants and insects is unknown, as is the number of marine forms. Similarly, bacteria and protists living in soils or that attack invertebrates have scarcely been examined. Terrestrial algae, especially those living on bark and rocks, have been little studied. All estimates of the number of species in these groups are crude guesses at best (Torsvik et al. 2002).

About 80,000 species of fungi have been described, but the total number living today likely exceeds 1 million. For example, the 12,000 fungal species described in Britain is about six times the number of native vascular plants described for the same area. If there were, on average, six fungal species for each vascular plant species worldwide (the ratio found in Britain), the number of fungal species would be approximately 1.6 million (Hawksworth 1991). The global ratio could be much higher or lower than the British one, but we have little else to base estimates upon.

Taxonomists believe that the number of species of nematodes is very large because millions of individuals may be present in 1 square kilometer of soil or mud; more than 200 species have been reported in samples of just a few cubic centimeters of coastal mud (Poinar 1983). Yet, almost nothing is known about species ranges and rates of species turnover geographically, so global estimates are very uncertain, although these data make a compelling case for a vast diversity of nematode species.

Approximately 30,000 species of mites have been described but, because knowledge of tropical mite faunas is extremely poor, the actual number of living species could easily exceed 1 million. Nearly 1 million species of insects, the world's most speciose group of organisms, have been described, but this is certainly a small fraction of the total. Most (>55%) of the insects collected by fogging the canopies of tropical trees, for example, are members of undescribed species (Erwin 1991), and samples taken within 70 km of one another at four sites in Manaus, Brazil had only 1% of their species in common (Erwin 1983). These samples hint at a vast unexplored diversity of arthropods, just in the canopies of tropical trees.

Earth is a relatively unexplored planet biologically. Not only are most living species still undescribed, but very little is known about the life history and ecological relations of most of the species that have been named.

The rate of description of new species today is higher than it ever has been, but the rate is nonetheless inadequate to accomplish a reasonable inventory prior to the likely extinction of many of them.

Diversity of Higher Taxa

Recent advances in molecular evolution have painted a radically different portrait of taxonomic diversity than existed even 20 years ago. Until recently, taxonomists classified all living organisms into five kingdoms: Animalia, Plantae, Fungi, Protista, and Monera (all prokaryotes). Today, we recognize that there are deep evolutionary divisions among the prokaryotes. They are now divided into two large domains, the Archaea and the Bacteria (Woese et al. 1990; Figure 2.2). The remaining four kingdoms are grouped together in the third domain, the Eukarya (see Figure 2.2). The genetic diversity among the Bacteria (which contains all familiar prokaryotes such as gram-positive bacteria, cyanobacteria, and green sulfur bacteria) and Archaea (which are unfamiliar to most, but include organisms adapted to life in extreme environments such as deep sea hydrothermal vents, hot springs, the guts of animals, and hypersaline environments) is as great as that among the long-recognized kingdoms among the Eukaryotes. Two new kingdoms are now recognized among the Archaea (Crenarchaeota and Euryarchaeota) and at least six more are recognized among the Bacteria. Similarly, the protists, which now are known to be vastly more diverse than previously recognized, have been divided into seven more kingdoms. These discoveries are exciting because they force us to consider the importance of organisms whose size make them generally invisible to us, yet are evolutionarily equal to the plant and animal kingdoms that have a greater hold on our consciousness. To appreciate this in another way, the diversity among all plants, fungi, and animals in small subunit RNA, the molecule used to create the modern phylogeny of all life on Earth, is only 10% of the diversity spread across all the rest of the Eukaryotes, Archaea, and Bacteria (Olsen and Woese 1996). Freeman (2002) describes their importance well:

> Bacteria and archaea are ancient, abundant, ubiquitous and diverse… their abundance is well-documented. A teaspoon of good-quality soil contains billions of microbes. In sheer numbers, the bacterium *Prochlorococcus*—found in the plankton of the world's oceans—may be the dominant life-form on the planet… a drop of seawater contains a population equivalent to a large city. Yet *Prochlorococcus* was first described and named only recently—in 1998.
>
> Bacteria and archaea are also found in almost every conceivable habitat. On land, they live in environments as unusual as oxygen-free mud, hot springs, salt flats, the roots of plants, and the guts of animals. They have been discovered living in bedrock to a

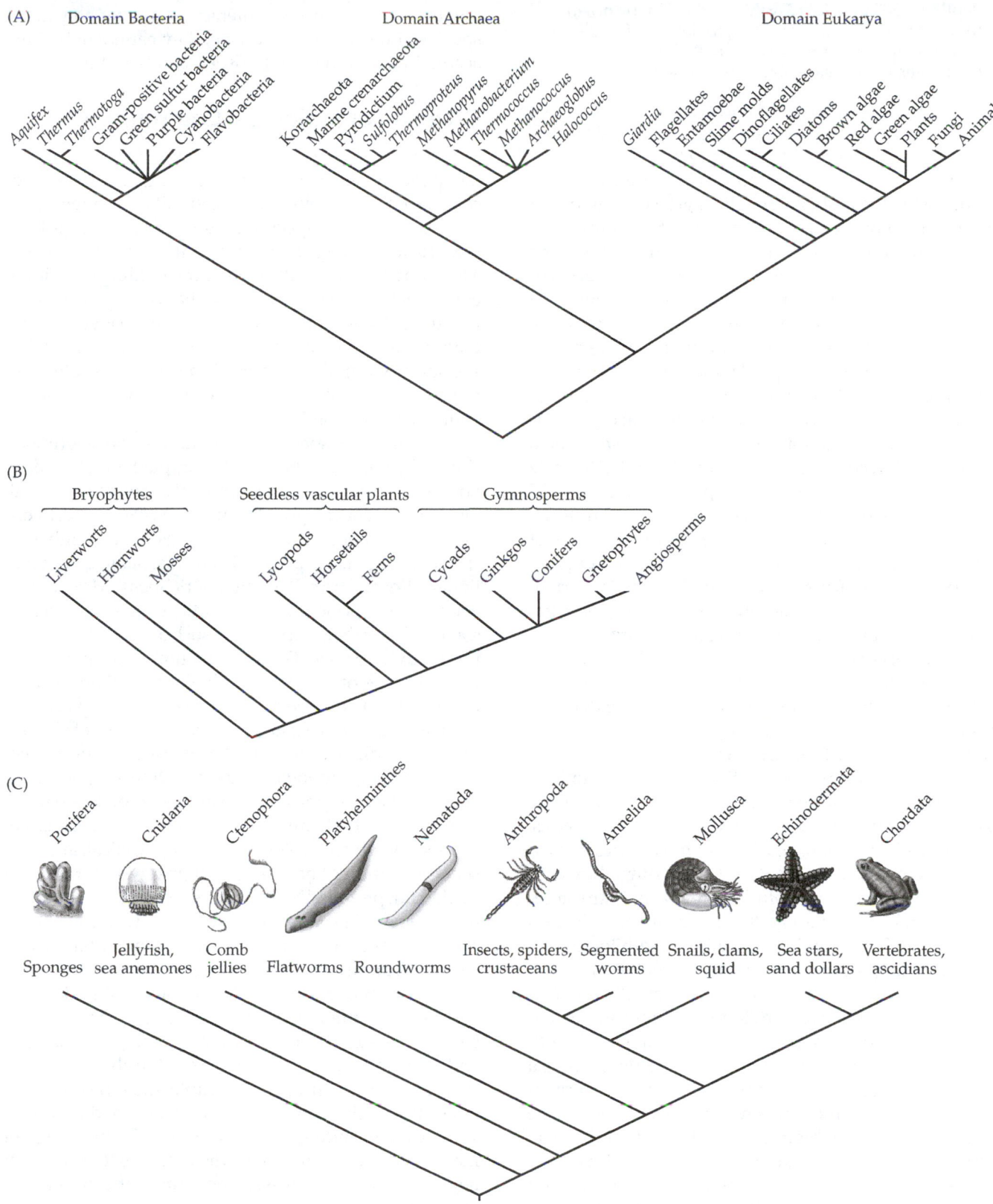

Figure 2.2 (A) Domains of biodiversity: Bacteria, Archaea, and Eukarya. Note that plants, fungi, and animals, evolved relatively recently within the Eukarya. (B) Major groups of plants. (C) Major groups of animals. (Modified from Freeman 2002.)

depth of 1500 m below Earth's surface. In the ocean they are found from the surface to depths of 10,000 m, and at temperatures ranging from 0°C in Antarctic sea ice to over 110°C near submarine volcanos.

Organisms so widespread and adaptable may not be in danger of extinction. Nonetheless we should be concerned that such incredible diversity be maintained because it may be of critical functional importance.

The distinctness of evolving lineages is an important component of biodiversity. The higher taxonomic categories of the Linnaean biological classification system provide rough measures of distinctness of lineages. For example, phyla in the animal kingdom are distinguished by their developmental patterns and their adult body plans (e.g., having two or three embryonic tissue types). Measured by diversity of phyla, marine biodiversity is much higher than terrestrial biodiversity even though far fewer species of marine than of terrestrial organisms have been described. Of the 96 phyla (most of them prokaryotes) recognized by Margulis and Schwartz (1998), about 69 have marine representatives and 55 have terrestrial ones. Of the 35 extant phyla of multicellular animals, 34 have marine representatives, and 16 of these are exclusively marine. From this perspective, preservation of marine biodiversity is more important than might be suggested simply by comparing the numbers of species in marine and terrestrial environments.

The preservation of evolutionarily distinct lineages above the level of species is important for a number of reasons. First, the evolutionary potential of life depends upon the distinctness of evolving lineages, not just the number of species. Lineages that have been evolving separately for long periods of time have many unique genes and gene combinations that would be lost were those lineages to become extinct. Closely related species, on the other hand, share nearly all of their alleles. For example, we share an estimated 99% of our DNA with chimpanzees. Second, evolutionary lineages are storehouses of information about the history of life. Scientists can read and interpret this information with increasing accuracy using modern phylogenetic methods. Third, the integrated functioning of ecosystems depends, in part, on the variety of species in them. For example, microorganisms have more diverse mechanisms for obtaining energy from the environment and decompose a wider variety of substances than multicellular organisms. Without microorganisms, ecosystems would be unable to provide, at high rates, such goods and services to humankind as absorbing and breaking down pollutants, storing and cycling nutrients, forming soils, and maintaining soil fertility. The numbers and kinds of fuels, construction materials, and medicines we obtain from nature depend upon the evolutionary distinctness of species. Finally, the aesthetic benefits we receive from nature are strongly correlated with the variety of living organisms with which we interact. No matter how many species of beetles there may be, they cannot substitute aesthetically for mammals, fishes, corals, or butterflies.

Diversity of Biological Communities

The distinctness of biological communities is defined by the species within them and their interactions. The species composition of ecological communities changes over space because each species has unique adaptations to its physical and biological environment. As we move across a landscape, species will be present according to their tolerance for the physical conditions that vary spatially (temperature, salinity, light levels, soil chemistry, etc.), and if they are able to survive their interactions with other species. Community membership is never absolute, although suites of species may be commonly found together under similar conditions.

The number of species in a location can be determined simply by counting them, but a simple tabulation does not serve all purposes. The commonly used measures fall into three major categories that emphasize, respectively, numbers, evenness, and differences. The number of species of organisms present in an area, habitat, or evolutionary lineage, is called species richness. Measures of evenness, or species diversity indices, weigh species by some index of their importance, such as their abundance, productivity, or size. For a given number of entities, the highest value of evenness is obtained if all are equally abundant. The degree of difference among species (or populations, or biotas) is measured by indices of similarity. These indices are also used to assess the degree of genetic similarity among evolutionary lineages, or the diversity of habitat types across landscapes or ecosystems.

Indices of both diversity and species richness are commonly used in ecological, biogeographical, and conservation studies because each type gives useful information not provided by the others. Ecologists commonly use diversity measures to assess the adverse effects of pollution and other types of environmental disturbances. Typically a stressed ecological community experiences losses of species and increases in abundance—and hence dominance—of a few species. Multimetric biological indexes quantify diverse biological changes, and an examination of the nature of the changes yields clues as to their causes (for example, the Index of Biotic Integrity, widely used to evaluate stream condition; Karr 1991). Understanding causes is essential for the design of management plans that have the potential to counteract negative effects of stresses and restore the systems to their former states.

Conservation biologists usually use unweighted measures of species richness because the many rare species that characterize most biotas are often of greater conservation interest than the more common ones that dominate

weighted indices of diversity. In addition, because accurate estimates of population densities on geographic scales seldom are available, species lists are the only available information for most areas.

However, species richness is a poor indicator of differences among biological communities in the degree to which they retain significant species interactions. Neither the abundance nor the functional roles in a community are indicated through a total species count, and thus we do not have an indication of which species play significant roles or those that play lesser ones. Further, species richness does not differentiate between native and nonnative species, and the species richness of degraded communities may be increased initially by an influx of nonnative species. Finally, we may most wish to know the role and importance of variability in communities, the dynamical responses of communities to disturbance, or several other factors that influence both the current status of a community or its ability to recover following disturbances. The extent to which communities can recover from disturbance on their own, or may require intervention such as ecological restoration, is important for conservation biologists to understand. Yet, other measures of species diversity, though they might indicate differences in the degree of dominance of the community by one or more species, also can not distinguish among these ecologically important differences. Unfortunately, at present we do not have many good indicators of the functioning of these more complex aspects of biological communities (see Chapter 18).

Biological communities are of conservation interest largely because all species are influenced by their interactions with other species. Evolutionary change takes place in the context of ecological communities, and thus alterations to communities will change the course of evolution. More immediately, changes in the abundance or presence of individual species can have strong effects on the remaining species in a community, which can cause species extinctions or other disruption (see Chapter 3). Thus, conservation of biological communities may be a necessary means to conserving species and evolutionary processes.

Ecosystem and Biome Diversity, and the World's Ecoregions

Classification of terrestrial ecological systems typically has been based on the shapes and life-forms of the plants that dominate the structure of those communities (von Humboldt 1806; Raunkaier 1934; Dansereau 1957; Halle et al. 1978). Köppen (1884) even used plant life-form distribution to define climates. Holdridge's (1967) widely used life zone system, on the other hand, is based entirely upon climatic variables. The diversity of schemes reflects the varied goals of biologists who classify large ecological units.

Although ecological communities typically gradually grade into one another, recognition of major divisions is useful for analysis. Based on the composition of the biotas of different regions, Earth has been divided into eight major biogeographic realms: the Palearctic, Afrotropic, Indo–Malay, Australasia, Oceania, Nearctic, Neotropic, and Antarctic (see Plate 1). Each of these biogeographic realms contains a number of **biomes**, large ecological units that are identified on land on the basis of the dominant type of vegetation (see Plate 1), and, in the sea, on ocean currents and spatial patterns of primary productivity (Figure 2.3).

Terrestrial biomes change along environmental (precipitation, elevation or temperature, and latitudinal) gradients (Figure 2.4). At temperate latitudes, commonly recognized biomes change along a precipitation gradient from mesophytic forest, to woodland, tallgrass prairie, shortgrass prairie, and finally, desert. With increasing latitude, mesophytic forests change into conifer-dominated boreal forests, and eventually tundra. In tropical regions, biomes change along precipitation gradients (rainforest, evergreen seasonal forests, dry forests, thorn woodland, desert scrub, and desert) and elevational gradients (rainforest, montane rainforest, cloud forest, elfin woodland, páramo). Finer divisions of biomes into ecosystem types are based on drainage, soil type, slope, and species composition. Classifications of ecosystems and biomes are inevitably somewhat arbitrary, but they identify ecological units that are useful for a variety of purposes. Moreover, the richness of human experience depends upon the richness of biomes as well as the richness of species.

Recently, a group of scientists led by the Conservation Science Program at the WWF-US reclassified Earth's biomes into 867 terrestrial **ecoregions** (see Plate 2; Olson et al. 2001). An ecoregion is a relatively large area containing a distinct assemblage of natural communities and ecological conditions, characterized by a dominant and widespread assemblage of species. With contributions from over 1000 scientists, the WWF map of ecoregions has become the most comprehensive description of higher-level landscape diversity available, and is a widely accepted base map for conservation planning.

Preservation of representatives of all habitat types in all ecoregions is necessary for preservation of species, because species will not survive without sufficient quantities of their natural habitats. Captive propagation can, and does, serve a role in keeping species alive for short periods until they can be reintroduced into the wild (see Chapter 15). But captive propagation is of little ultimate use if there are no suitable habitats into which to reintroduce the species. Managing species in zoos and botanical gardens is expensive, and an animal in a cage or a plant in a garden lives isolated from its natural physical and biological environment.

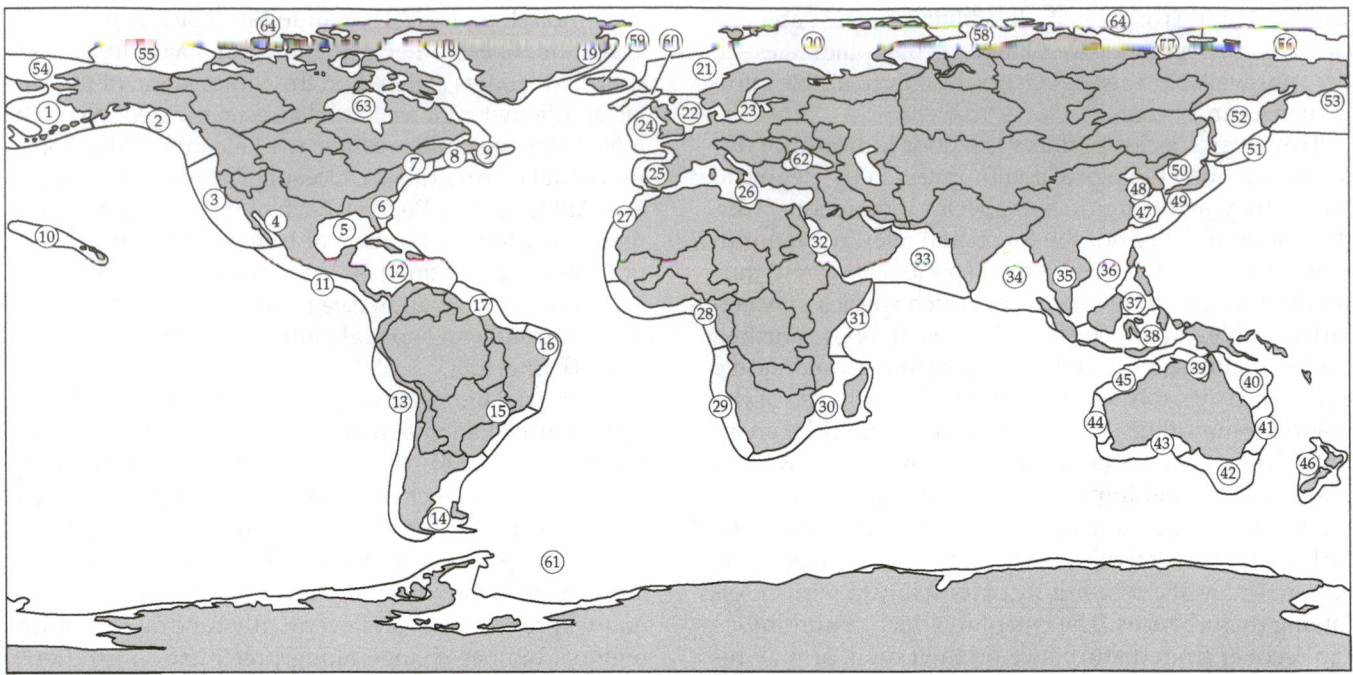

1. East Bering Sea	14. Patagonian Shelf	27. Canary Current	40. Northeast Australia	53. West Bering Sea
2. Gulf of Alaska	15. South Brazil Shelf	28. Guinea Current	41. East-Central Australia	54. Chukchi Sea
3. California Current	16. East Brazil Shelf	29. Benguela Current	42. Southeast Australia	55. Beaufort Sea
4. Gulf of California	17. North Brazil Shelf	30. Agulhas Current	43. Southwest Australia	56. East Siberian Sea
5. Gulf of Mexico	18. West Greenland Shelf	31. Somali Coastal Current	44. West-Central Australia	57. Laptev Sea
6. Southeast U.S. Continental Shelf	19. East Greenland Shelf	32. Arabian Sea	45. Northwest Australia	58. Kara Sea
7. Northeast U.S. Continental Shelf	20. Barents Sea	33. Red Sea	46. New Zealand Shelf	59. Iceland Shelf
8. Scotian Shelf	21. Norwegian Sea	34. Bay of Bengal	47. East China Sea	60. Faroe Plateau
9. Newfoundland-Labrador Shelf	22. North Sea	35. Gulf of Tailand	48. Yellow Sea	61. Antartic
10. Insular Pacific-Hawaiian	23. Baltic Sea	36. South China Sea	49. Kuroshio Current	62. Black Sea
11. Pacific Central-American	24. Celtic-Biscay Shelf	37. Sulu-Celebes Sea	50. Sea of Japan	63. Hudson Bay
12. Caribbean Sea	25. Iberian Coastal	38. Indonesian Sea	51. Oyashio Current	64. Artic Ocean
13. Humboldt Current	26. Mediterranean	39. North Australia	52. Sea of Okhotsk	

Figure 2.3   Large marine ecosystems of the world and their associated major watersheds. These ecosystems account for the majority of ocean productivity due to their proximity to land masses. However, this view neglects other important marine biomes, particularly pelagic zones making up the majority of the ocean basins. (Modified from map created by NOAA and the University of Rhode Island, 2002.)

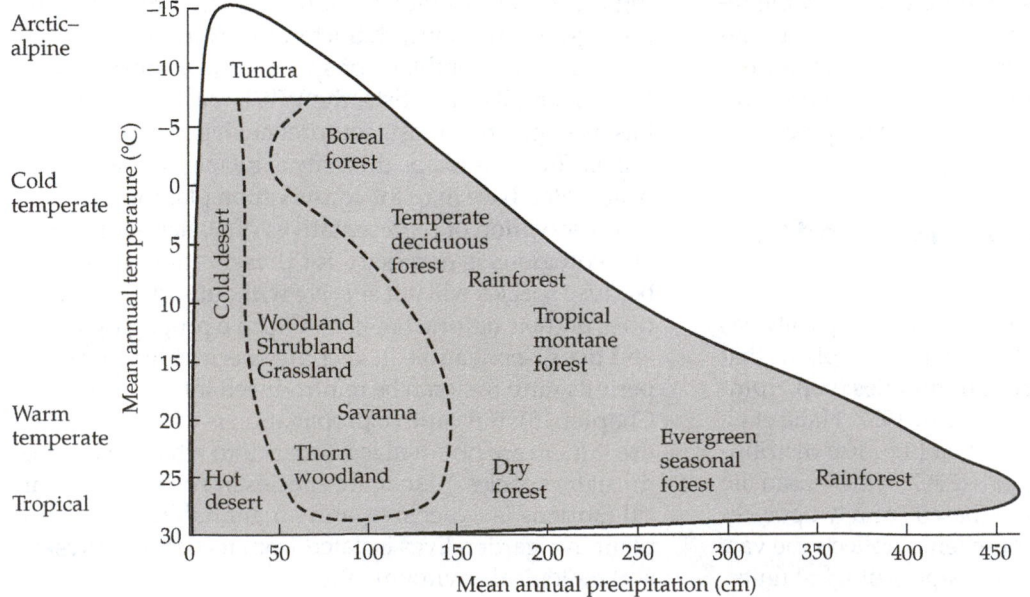

Figure 2.4   Biomes and climate. Distributions of the major biomes are plotted on axes of mean annual precipitation. Within the region bounded by the dashed line, factors such as seasonality of drought, fire, and grazing strongly affect which type of vegetation is present (Modified from Whittaker 1970.)

Ecosystems are often a conservation target because critical functions that support human populations are maintained by this level of biodiversity. As ecosystems are degraded by human activities, their ability to provide critical ecosystem services and to support a broad array of biological functions is compromised, as we discuss in detail later in this chapter. Biomes across the world are being modified at different rates, and by different forces (see Chapter 6). Our focus at the biome level is often to understand the drivers of change in these areas, and how our options for mitigating these changes may be used for conservation (Millennium Ecosystem Assessment 2005).

Species Richness over Geological Time

The number of species on Earth at any moment in time is the result of the difference between the rate of produc-

tion of new species and the rate of loss of existing species. The rate of species formation depends both on events that generate barriers to gene flow (**vicariance** events), and on the number of species—the more species, the more ranges there are to be separated, the more species available to cross existing barriers, and the more opportunities for divergence. Rates of loss of species depend on the kinds and severity of disturbances, and the strength of factors that permit species to coexist. Many such factors exist but their operation depends fundamentally on the existence of trade-offs in the abilities of species to deal with the factors that influence their abundance and distributions.

The fossil record, though very incomplete, provides a rough measure of trends in species richness during the history of life on Earth (Table 2.2). Cellular life in the form of bacteria evolved about 3.8 billion years ago (bya); eu-

TABLE 2.2 *Earth's Geological History*

Era	Period	Onset	Major physical changes on Earth	Major evolutionary events
Cenozoic	Quaternary	2 mya	Cold/dry climate; repeated glaciations	Earliest homonids, ~3.5 mya
	Tertiary	65 mya	Continents near current positions; climate cools	Diversification of angiosperms and insects
Mesozoic	Cretaceous	144 mya	Northern continents attached; Gondwana begins to drift apart; meteorite strikes Yucatán Peninsula	Dinosaurs extinct at end of period
	Jurassic	206 mya	Two large continents form: Laurasia (north) and Gondwana (south); climate warms	First birds appear
Triassic		251 mya	Pangaea slowly begins to drift apart; hot/wet climate	First dinosaurs, mammals appear
Paleozoic	Permian	296 mya	Continents aggregate into Pangaea; large glaciers form; dry climates form in interior of Pangaea	Mass extinction of marine forms
	Carboniferous	360 mya	Climate cools; marked latitudinal climate gradients	First reptiles appear
	Devonian	409 mya	Continents collide at end of period; asteroid probably collides with Earth	First insects, ferns, vascular and seed plants, and amphibians appear
	Silurian	439 mya	Sea levels rise; two large continents form; hot/wet climate	First land plants, jawed fishes appear
	Ordovician	510 mya	Gondwana moves over South Pole; massive glaciation, sea level drops 50 m	Further diversification of marine invertebrates; first vertebrates appear
	Cambrian	542 mya	O_2 levels approach current levels	Animal taxa diversify rapidly; most phyla appear
Precambrian		1.5 bya		Origin of multicellular eukaryotes
		2.0 bya		Oldest eukaryotic fossils
		2.5 bya	O_2 level at >1% of current level	
		3.6 bya	O_2 first appears in atmosphere	
		3.8 bya		Oldest prokaryotic fossils
		~4 bya		Origin of life
		4.5 bya	Earth forms	

karyotic organisms probably evolved about 2 bya. Although many species must have lived in ancient times, species richness appears to have been low during the first 2 billion years of Earth's existence. During the late Precambrian, the richer Ediacaran fauna, consisting of strange frond- and disc-shaped, soft-bodied animals and some forms that appear to be arthropods and echinoderms, evolved. The first explosion of biodiversity took place during the early Cambrian period. Some Cambrian species may be members of phyla that left no surviving descendants. As measured by the number of phyla, animal life may have been more diverse during the Cambrian period than at any time since (Gould 1989).

The fossil record for marine invertebrates with hard skeletons is good enough to provide a general picture of the number of evolutionary lineages present at different times in the past. The Cambrian explosion was followed, about 60 million years later, by the extensive radiation of the Paleozoic fauna. Following the Permian mass extinction, the modern fauna evolved, and overall family richness has increased steadily throughout the Mesozoic and Cenozoic eras to a maximum today (Figure 2.5). Ap-

proximately 40,000 species of marine invertebrates have been described from the Paleozoic and Mesozoic eras, a number that increased to about 250,000 in the late Cenozoic era. The fossil record of terrestrial animals is much poorer, especially among such species-rich groups as insects, which fossilize poorly. The fossil record of vertebrates, particularly mammals, is much better. It indicates that richness, as measured by the number of orders, is slightly higher today than earlier in the Cenozoic.

Terrestrial vascular plants appeared by the early Silurian and their richness increased rapidly during the Devonian when seed-bearing plants first appeared. Species richness has continued to increase overall but the numbers of species of ferns and gymnosperms has decreased, whereas the number of species of angiosperms has increased dramatically (Figure 2.6).

Rates of species formation

Speciation rates have varied greatly over evolutionary time. Large numbers of new evolutionary lineages originated three times during the history of life. The first event, known as the Cambrian explosion, took place about half a billion years ago. The second, about 60 million years later, resulted in the Paleozoic fauna. Biodiversity was greatly reduced 300 million years later by the great Permian mass extinction, which was followed by the Triassic explosion that led to our modern biota. Although all three of these explosions resulted in many new species, they were qualitatively very different. Virtually all the major

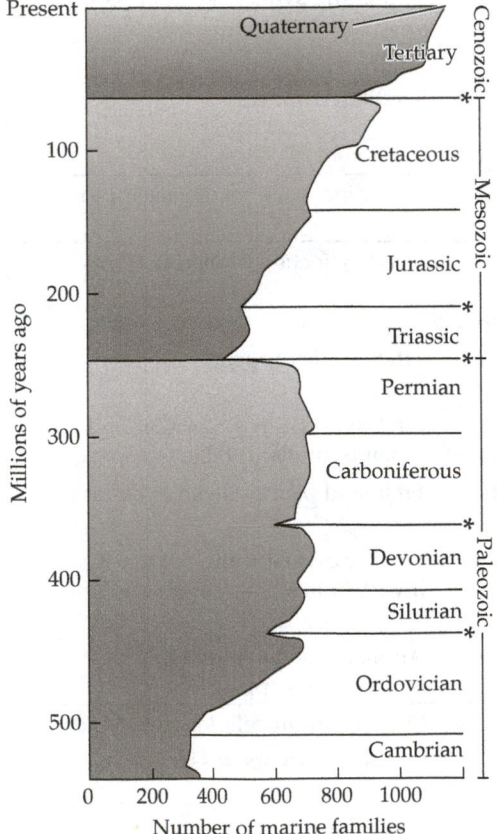

Figure 2.5 Diversity of marine families from the Cambrian to the present. The asterisks mark the five major mass extinction events.

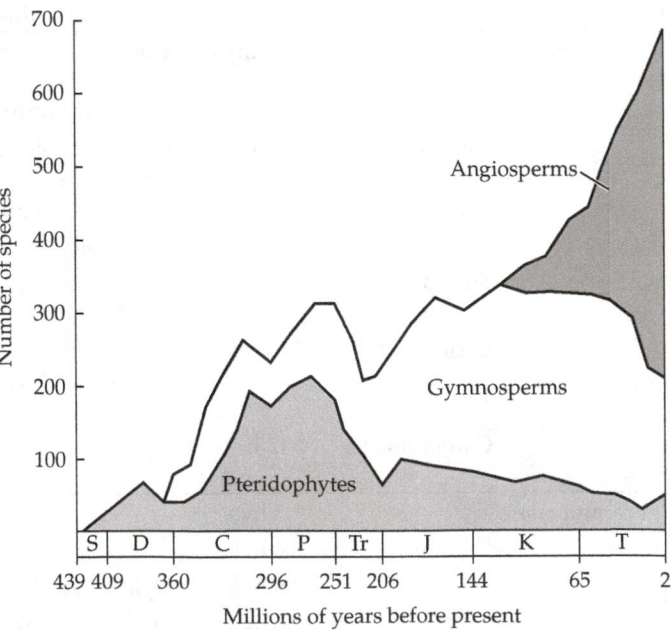

Figure 2.6 Terrestrial plant species richness. Ferns, gymnosperms, and angiosperms have, in turn, dominated the world's flora. (Modified from Signor 1990.)

groups of living organisms appeared in the Cambrian period, along with some phyla that subsequently became extinct. The Paleozoic and Triassic explosions greatly increased the number of families, genera, and species, but no new phyla of multicellular organisms evolved.

High rates of speciation have been favored by three factors: mass extinctions, increasing separation of land masses across Earth, and evolution of new life forms and types of species interactions. Mass extinctions tend to be followed by increases in rates of speciation because those lineages that survive inherit a biologically depauperate Earth, an ecological setting favorable for the evolution of many different life-forms and body plans. Increasing provinciality, such as occurred during the breakup of Pangaea and, later, Gondwana, stimulated speciation, because following their separation, evolution on each of the new continents produced many new endemic species. Much of the Cenozoic era increase in numbers of species appears to be due to such provincialization. South America, Africa, and Southeast Asia are all rich in species, but they share few species or genera because they have been separated from one another for many millions of years. Increasing complexity of ecological interactions also has stimulated rates of speciation. On land, the richness of vascular plants increased dramatically with the evolution of angiosperms and their complicated interactions with animals during reproduction and dispersal of seeds. The evolution of flight undoubtedly also contributed to the great diversity of both insects and birds by allowing better exploitation of the third dimension on land. In the oceans, diversity was stimulated by evolution of organisms' ability to burrow into sediments and to swim in open water. The number of species living together also increased as a result of finer adaptations to particular environmental conditions and ways of exploiting the environment made possible by minor variations in morphology, physiology, and behavior.

Rates of extinction

Extinctions have occurred at all times throughout the history of life, but rates have changed dramatically. Paleontologists distinguish between "normal" or "background" extinction rates and the much higher rates associated with mass extinctions (Sepkoski and Raup 1986; Figure 2.7). The first of six mass extinctions, at the end of the Cambrian period, destroyed about half of the known animal species. About 75% of species became extinct in the second event at the end of the Devonian period. The most extreme of the mass extinctions, at the close of the Permian period eliminated about 95% of both marine and terrestrial organisms. In the oceans, trilobites declined to extinction and brachiopods almost became extinct. On land, the trees that formed the great coal forests

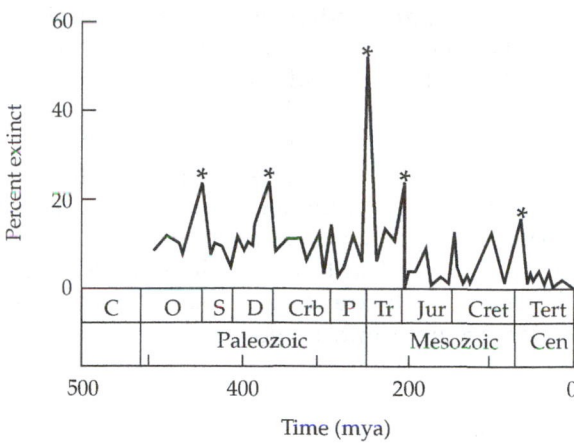

Figure 2.7 Extinctions of families through geologic time. The five historical mass extinction events are marked with an asterisk.

became extinct, as did most lineages of amphibians. At the end of the Triassic period, nearly all ammonites and approximately 80% of reptile species vanished, resulting in an overall loss of about 65% of species. The fifth extinction event at the end of the Cretaceous is most famous because dinosaurs and other large reptiles became extinct during that period, along with most marine lineages and about 75% of all species. Discussed in greater detail in Chapter 3, the current and sixth mass extinction event, promulgated by human expansion over the planet, initially exterminated large mammals and island species, but if current trends continue, organisms of all sizes and lineages will be seriously affected.

Following each of the previous mass extinctions, biodiversity expanded, but species richness did not reach its previous value until after an average lag of about 10 million years, and in some cases much longer (Kirchner and Weil 2000; see Figure 2.5). However, even with such long lags, species richness has not been directly affected by mass extinction events during most of geological history. Their major effect has been to eliminate some lineages, thereby making ecological room for other lineages that proliferated (over geological time intervals) following episodes of mass extinction.

Current patterns of species richness

Earth is not uniform and neither is the distribution of organisms across its surface. Some important general patterns in the geographical distribution of species richness have been discovered, but much remains unknown, in large part because the inventory of living organisms is so incomplete. Also, the distributions of species are best known for temperate regions where most taxonomists and ecologists live and work. Tropical regions, where

most of the world's species live, are poorly known bio-logically. Nonetheless, some places on Earth clearly have exceptionally high biodiversity (tropical rainforests, coral reefs) whereas others are virtually devoid of life (extremely dry tropical and polar deserts) (Figure 2.8). These patterns are discussed in greater detail in the later section, "Latitudinal Gradients in Species Richness."

Much of the interest in describing and interpreting current patterns of distribution of biodiversity has been stimulated by a growing concern about the future of bio-diversity on Earth (Gaston 2000). If efforts to preserve Earth's biodiversity are to be successful, they must be based on accurate information on where species are found, why they are there, and what is threatening their continued existence. Recently, there has been a revolu-tion in the quantity of new data on distributions and in the development of new technical tools, such as distri-bution-mapping schemes (i.e., **geographic information systems** or **GIS**) and remote-sensing technology, with which to analyze complex data sets.

To describe spatial patterns of biodiversity, ecologists and biogeographers have found it useful to divide species richness into three major components: alpha (α) -richness, beta (β) -richness, and gamma (γ) -richness (Fig-ure 2.9). **Alpha-richness** refers to the number of species found in a small, homogeneous area; β-**richness** refers to

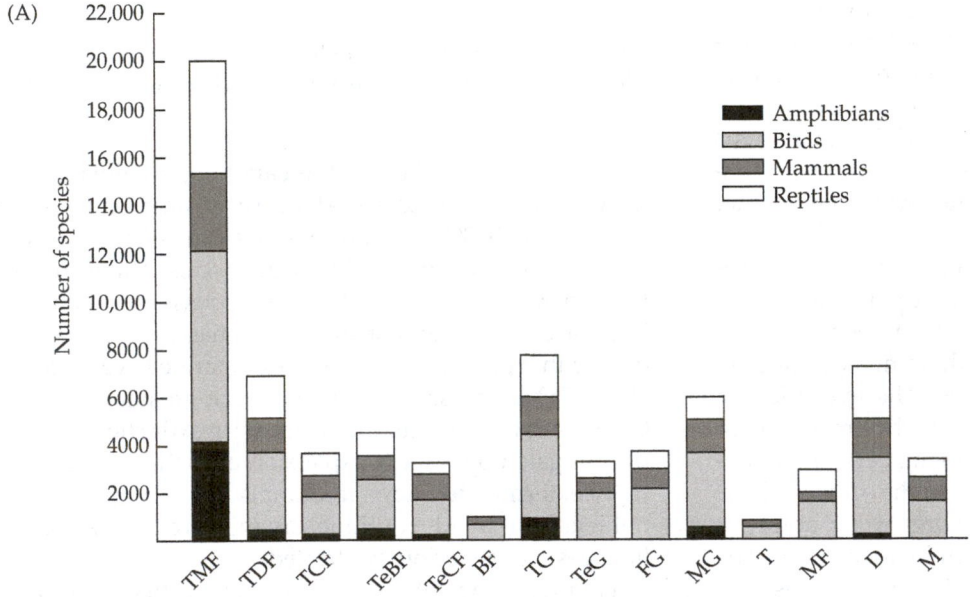

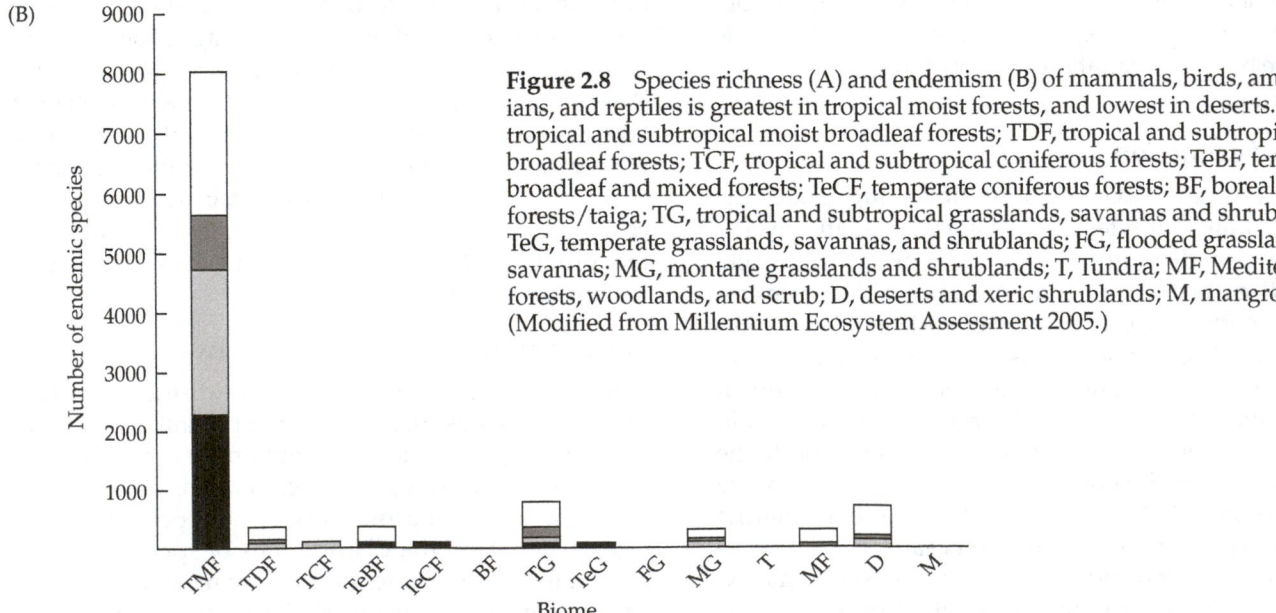

Figure 2.8 Species richness (A) and endemism (B) of mammals, birds, amphib-ians, and reptiles is greatest in tropical moist forests, and lowest in deserts. TMF, tropical and subtropical moist broadleaf forests; TDF, tropical and subtropical dry broadleaf forests; TCF, tropical and subtropical coniferous forests; TeBF, temperate broadleaf and mixed forests; TeCF, temperate coniferous forests; BF, boreal forests/taiga; TG, tropical and subtropical grasslands, savannas and shrublands; TeG, temperate grasslands, savannas, and shrublands; FG, flooded grasslands and savannas; MG, montane grasslands and shrublands; T, Tundra; MF, Mediterranean forests, woodlands, and scrub; D, deserts and xeric shrublands; M, mangroves. (Modified from Millennium Ecosystem Assessment 2005.)

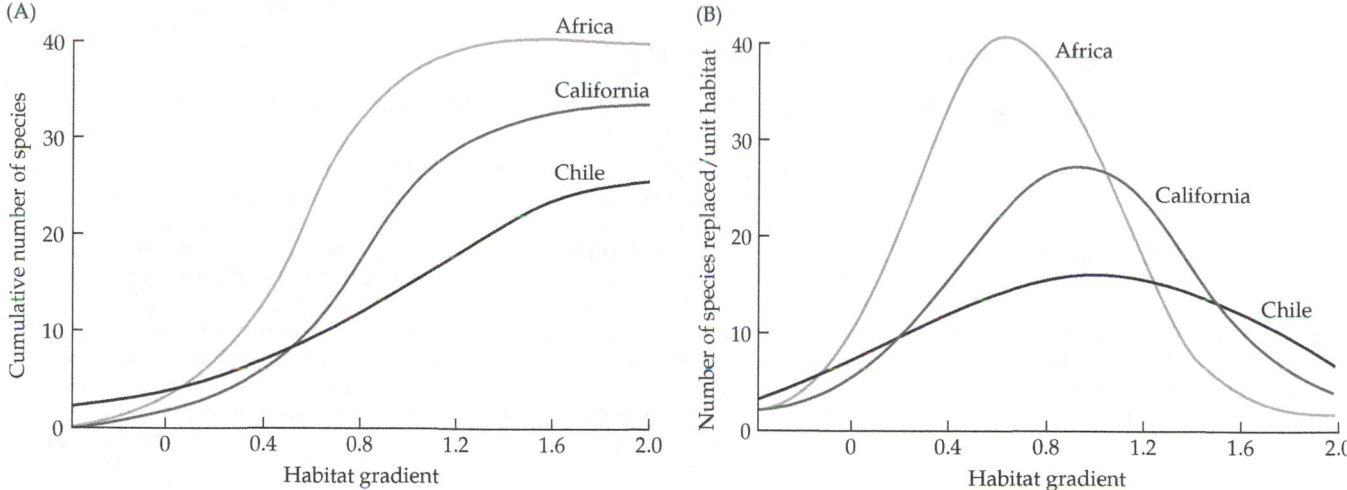

Figure 2.9 Species turnovers along habitat gradients. (A) Species accumulation curves for birds across habitat gradients in Mediterranean vegetation (from dry scrub to woodland) in southern Africa, California, and Chile. (B) The differentials of the curves in (A) give the rate of species turnover across the gradient. (Modified from Cody 1975.)

the rate of change in species composition across habitats or among communities; and **γ-richness** refers to changes across larger landscape gradients. A high β-richness means that the cumulative number of species recorded rapidly increases as additional areas are censused along some environmental gradient. Species may also drop out rapidly along such gradients, resulting in a high rate of species turnover. Susan Harrison and Jim Quinn discuss some of the ways in which studies of β-richness enhance our understanding of regional ecology, and inform conservation planning in Box 2.3.

The rate at which the species composition of communities changes across environmental gradients is determined by the sizes of species ranges and the degree to which species are habitat specialists. The ranges of tropical species are much less known than the ranges of their temperate counterparts, but on average terrestrial tropical species have smaller ranges than species of higher latitudes. The altitudinal ranges of species on the slopes of tropical mountains appear to be narrower than ranges of species on temperate mountains, but this impression may be an artifact of inadequate sampling of rich tropical biotas (Colwell and Hurtt 1994).

An inevitable and important corollary of patterns of species richness is that areas with high α-richness inevitably have many rare species. A tropical wet forest in South America or Southeast Asia, for example, may harbor between 300 and 400 species of trees per square kilometer, whereas a temperate forest harbors an order of magnitude less. However, the number of trees per hectare is roughly the same in tropical and temperate forests. It follows that most of the tree species in tropical forests

must be present at very low densities. Many of those species are more abundant elsewhere, but some species evidently are present only at low densities throughout their ranges.

The following sections describe some major patterns in the distribution of species and discuss mechanisms that may generate these patterns. An understanding of both the patterns and the mechanisms that generate and maintain them can be used to develop strategies to protect or recover biodiversity.

Patterns of Endemism

A species that is found in a particular region but nowhere else is said to be **endemic** to that region. However, what is an appropriate spatial scale for assessing endemism varies greatly. All species are, as far as we know, endemic to Earth. At the opposite extreme, some species are restricted to single desert springs, small islands, or isolated mountain tops. Regions with many endemic species are the result of one or more major events that caused the ranges of many taxa to separate at approximately the same place. Causes of such geographical isolation include continental drift, mountain building, climate change, and sea level rises. Following such isolation, many taxa may undergo evolutionary radiations in the same general area. Vicariance due to continental drift has been extremely important in generating the high degrees of endemism found in the biotas of Madagascar, Australia, New Guinea, and New Caledonia. Owing to their isolation, islands often have high proportions of endemic species, but, given that islands often have relatively impoverished biotas, high

BOX 2.1 The Importance of β-Diversity

Susan Harrison and Jim Quinn, *Dept of Environmental Science and Policy, UC Davis*

Ecologists who study biological diversity, as well as conservation biologists looking for ways to preserve it, often find it useful to follow Robert Whittaker's (1960) approach of separating diversity, as measured in species richness, into geographic components. Local (alpha, α) richness refers to the number of species coexisting in a single site, such as one sampling plot or one patch of habitat. Regional (gamma, γ) richness is the total number of species in a large collection of sites that represent the whole region of interest. Turnover or differentiation (beta, β) richness links local and regional diversity; β-richness = γ-richness/(average α-richness), in Whittaker's formulation. Beta-richness is a measure of the spatial variation in species composition among sites within a region; it is high when a region contains many distinct habitats with different species, and low when the same species are found throughout a region. Beta-richness is inversely proportional to average geographic range size, for a group of species in a given geographic area.

Why it may be useful to partition diversity in this way can be demonstrated with an example of the tropical-temperate diversity gradient. Ecologists have argued for many years about why there are so many more species in the tropics than the temperate zone. Some very approximate figures for ants are provided in Table A.

Ants are much more diverse in Costa Rica than California, and in California than Alaska, whether one considers the whole region (γ) or the one-hectare level (α). But ants are also more diverse farther south in terms of the *ratio* of total to local diversity, reflecting higher differentiation or β-richness. (Note that because these calculations do not take into account the large differences in area between Alaska, California, and Costa Rica, they considerably underestimate the differences in β-richness among the three regions.) This cannot be explained simply by some of the standard hypotheses for the latitudinal diversity gradient, such as the greater area or higher energy input of the tropics. However, the elevated β-richness of

tropical ants is consistent with Janzen's (1967) argument that because seasonal variation in temperature is small in the Tropics, tropical species tend to have narrow climatic tolerances, which in turn gives them narrower geographic ranges, and thus produces higher β-richness in tropical faunas than occurs in temperate faunas.

Another example concerns grassland ecology and management. Grazing by large animals is said to be important in maintaining diversity in many grassland ecosystems, because it may open up space for more species to coexist. Yet it is important to consider how a disturbance such as grazing may affect diversity at different scales (Olff and Ritchie 1999; Stohlgren et al. 1999; Harrison et al. 2003). While grazing may promote the coexistence of more species within small study plots, it can also reduce regional diversity by eliminating grazing-intolerant species. Of course, the highest diversity of all might be found in a region containing a mix of both grazed and ungrazed areas.

Both the statistics and the terminology for components of diversity vary widely among authors. Beta-richness metrics have been developed that are similar to Whittaker's in concept, but are more robust to variation in sampling properties such as the number and completeness of samples (Colwell and Coddington 1994; Lande 1996). Some authors have distinguished two kinds of β-richness: turnover of species in relation to environmental variation, such as the gradients of elevation and moisture studied by Whittaker (1960), and turnover due to the existence of different species in similar but separated habitats, called ecological equivalency by Shmida and Wilson (1985) and γ-richness by Cody (1993).

Habitat patchiness or fragmentation could actually promote β-richness, for example if competing species can more easily coexist on separate patches of habitat. Alternatively, fragmentation could reduce β-richness if only a certain subset of species such as good dispersers and habitat generalists ("weeds") are able to survive on fragments. In a grassland experiment, fragmentation led to lower α-richness but higher β- and γ-richness of plants (Quinn and Robinson 1987). In a naturally patchy system of serpentine outcrops, sampling sites on small isolated outcrops supported lower α-, higher β-, and equal γ-richness of plants restricted to serpentine, compared with sampling sites on large continuous outcrops (Harrison 1997, 1999). Both studies suggest that patchy habitats can have positive as well as negative effects on diversity.

Beta-richness is an important aspect of devising geographic strategies for conservation. A common dilemma is whether to invest in acquiring new reserves or in enlarging existing ones. While a larger reserve will nearly always harbor more species than a smaller one (higher α), a large number of well-chosen small reserves could best capture the total diversity of a region if each reserve tends to support different species (higher β). The best tradeoff between number and sizes of reserves is an empirical question, one that depends on details of each case. Quinn and Harrison (1988) found that in 24 of 25 park and island data sets, the β-richness enhancing benefits of multiple parks or islands outweighed the α-richness enhancing effects of large park or island size, in terms of obtaining the greatest number of species in a given total area.

TABLE A *Approximate Number of Ant Species*

	Area (km^2)	Total number of species	Number of species/ha	Ratio
Alaska	225,540	15	5	3
California	63,000	240	30	8
Costa Rica	51,100	1,500	150	10

One way to combine the concepts of α- and β-richness is the **rarity-weighted richness index (RWRI)**, designed to rate geographic areas according to their unique biodiversity. This index uses data sets in which species occurrences are represented on maps consisting of equal-area grid cells. The geographic range of each species is the set of grid cells on which it occurs. The rarity-weighted richness index of each grid cell is

$$RWRI = \sum_{i=1}^{N} \frac{1}{h^i}$$

where h^i is the number of grid cells occupied by species i, and N is the number of species found in that particular cell. Using this index, geographic areas that contain the largest number of narrowly distributed species are rated as the most diverse. For the U.S., these hotspots include Hawaii, coastal California, Death Valley, the Florida panhandle, and the southern Appalachian Mountains, all regions in which geographic isolation has produced an exceptional number of unique, localized species (Figure A; Stein et al. 2000).

Conservation organizations such as The Nature Conservancy stress the importance of creating well-diversified "portfolios" of protected sites, analogous to diverse investment portfolios (Groves et al. 2002). The goal of protecting as much biodiversity as possible within a region by protecting multiple sites is increasingly being approached using tools for systematic conservation planning that consider geographic patterns in biodiversity. The general strategy is to condition future land acquisitions or other conservation actions on the properties of the network of sites that is already protected. New additions to the network should favor sites with species or habitats that are not already protected, as this will tend to maximize overall richness in the network (Scott et al. 1993; Margules and Pressey 2000).

One example of an algorithm for finding a nearly optimal network of sites to protect maximum diversity within a region is to begin by protecting the locations with the highest α-richness (see Chapter 14). New sites are then sequentially added in order of their β-richness, or more specifically, how many species they add to the list of those protected by all previous acquisitions combined. This method is not perfect in the sense that it may not find the absolutely optimal network (this turns out to be a surprisingly difficult computational problem). But it has the advantage of simplicity, being easy for policymakers and the public to understand and accept, and it reflects the incremental nature of real-world policymaking (Margules and Pressey 2000).

Conservation organizations that use these systematic analyses acknowledge that their actual conservation actions are never "ideal" for protecting biodiversity because they are influenced by many other goals and constraints. Nevertheless, such analyses are valuable because they inform stakeholders how well they are doing compared with what could be achieved.

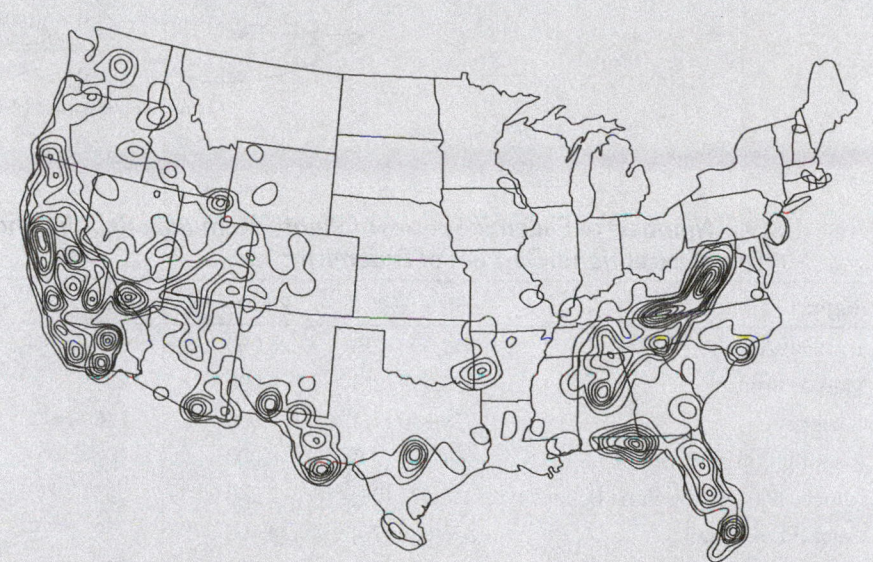

Figure A   Hot spots of rarity and species richness in the lower 48 United States. Read as a topographic map with concentric circles showing higher values of the rarity-weighted species richness index (RWRI). Hotspots are found in CA, the Death Valley region of Nevada, the Appalachian Mountains, and the Florida panhandle and Everglades. Many other regions of higher diversity are found in other parts of the U.S., and the Hawaiian islands (not shown) have the greatest concentration of range-restricted species by far. To achieve a high RWRI both α- and β-diversity must be high. (Modified from Stein et al. 2000.)

endemism often is not associated with high species richness. Coral reefs associated with islands, however, are an exception. The island-studded western Pacific Ocean is an example of a region of maximum species richness and endemism for corals, reef fishes, snails, and lobsters (Figure 2.10; Briggs 1996; Roberts et al. 2002). Finally, continental areas in the Tropics often contain both high species richness and high degrees of endemism (see Figure 2.8).

Patterns of endemism differ greatly among taxa (Table 2.3). Both the Cape region of South Africa and southwestern Australia have extremely high numbers of endemic plant species but very few endemic mammals or birds (Cowling 1992; Myers et al. 2000). These differences arise because plants can speciate via polyploidy in much smaller areas than can vertebrates, among which polyploidy is rare. However, there are strong correla-

Figure 2.10 The Indo-West Pacific is a marine diversity hotspot. The distribution of species richness for damselfish (Pomacentridae) shown here is typical of many marine taxa in the region. (From Briggs 1996.)

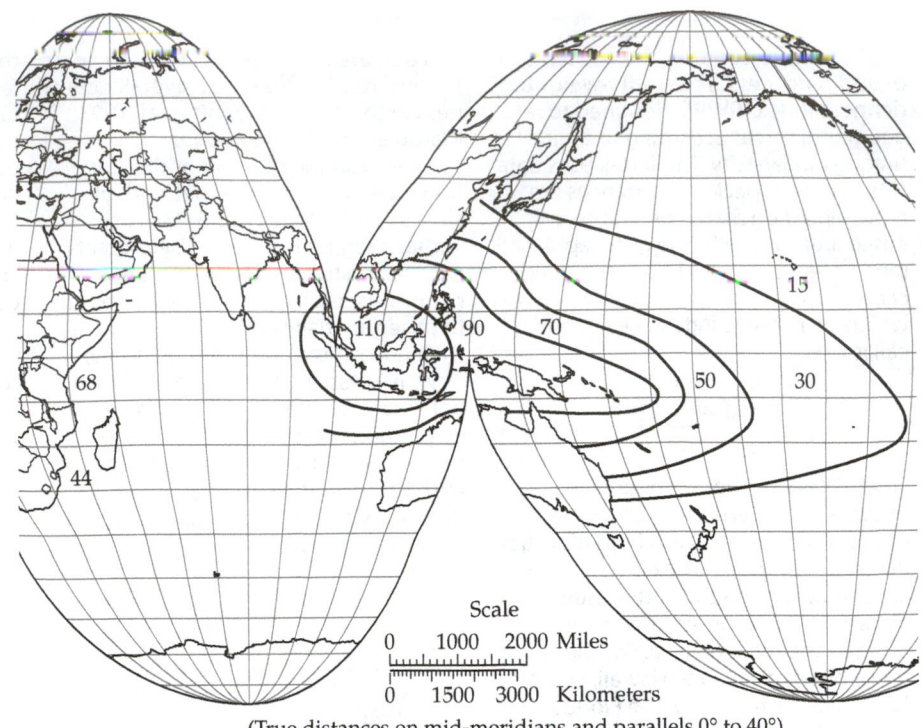

Scale
0   1000   2000 Miles
0   1500   3000 Kilometers
(True distances on mid-meridians and parallels 0° to 40°)

TABLE 2.3   *Numbers of Endemic Species of Plants, Mammals, Reptiles and Amphibians in Regions of Exceptional Degrees of Endemism[a]*

Region	Size (%)[b]	Plants	Birds	Mammals	Reptiles	Amphibians
Tropical Andes	314,500 (25)	20,000	677	68	218	604
Mesoamerica	231,999 (20)	5,000	251	210	391	307
Caribbean	29,840 (11)	7,000	148	49	418	164
Atlantic Forest of Brazil	91,930 (7.5)	8,000	181	73	60	253
Choco/Darien/Western Ecuador	16,471 (24)	2,250	85	60	63	210
Cerrado of Brazil	356,630 (20)	4,400	29	19	24	45
Madagascar	59,038 (10)	9,704	199	84	301	187
West African Forests	126,500 (10)	2,250	90	45	46	89
Cape Floristic Province	18,000 (24)	5,682	6	9	19	19
Mediterranean Basin	110,000 (5)	13,000	47	46	110	32
Sundaland	125,000 (8)	15,000	139	115	268	179
Wallacea	52,020 (15)	1,500	249	123	122	35
Philippines	9,023 (3)	5,832	183	111	159	65
Indo-Burma	100,000 (5)	7,000	140	73	201	114
South-central China	64,000 (8)	3,500	36	75	16	51
Western Ghats/Sri Lanka	12,450 (7)	2,180	40	38	161	116
Southwest Australia	33,336 (11)	4,331	19	7	50	24
Polynesia/Miconesia	10,024 (22)	3,334	174	9	37	3

Source: Data from Meyers et al. 2000.

[a]1% or greater of global total for endemic plants or vertebrates.

[b]Size (%) refers to remaining primary vegetation in km² and to percent remaining of original extent.

tions among patterns of endemism in mammals, birds, and reptiles, all of which require relatively large areas for geographic speciation.

Latitudinal Gradients in Species Richness

In both marine and terrestrial environments tropical regions harbor many more species in most higher taxonomic categories than higher-latitude regions. For example, Arctic waters have about 100 species of tunicates, 400 species are known from temperate waters, and more than 600 species inhabit tropical seas. The richness of species, genera, and families of bivalve mollusks peaks in tropical regions and declines rapidly with increasing latitude (Figure 2.11). A similar pattern is found among the fauna living on hard substrates (Thorson 1957) and benthic living and fossil foraminiferans (Buzas and Gibson 1969; Stehli et al. 1969). The number of ant species found in local regions increases from about 10 at 60°N latitude, to as many as 2000 species in equatorial regions. Greenland hosts 56 species of breeding birds, New York, 105, Guatemala 469, and Colombia 1395. Latitudinal gradients in species richness of birds and mammals in North and Central America are illustrated in Figure 2.12.

There are relatively few exceptions to these latitudinal patterns, in which highest richness occurs in mid- or high latitudes, or there is no correlation with latitude. However, most of these cases occur when the scale of analysis is smaller (<20° latitude), or a lower taxonomic category is examined (e.g., orders). But parasites, ichneumonid parasitoid wasps, and aquatic plants (reviewed in Willig et al. 2003) all appear to be exceptions even at large scales of analysis. Seaweeds show no consistent pattern in relation to latitude (Bolton 1994). Also, the richness of marine birds and mammals is greater at high latitudes (Willig et al. 2003).

Although latitudinal gradients in species richness have been known for a long time, determining their causes has proven difficult. For a starter, latitude or any other

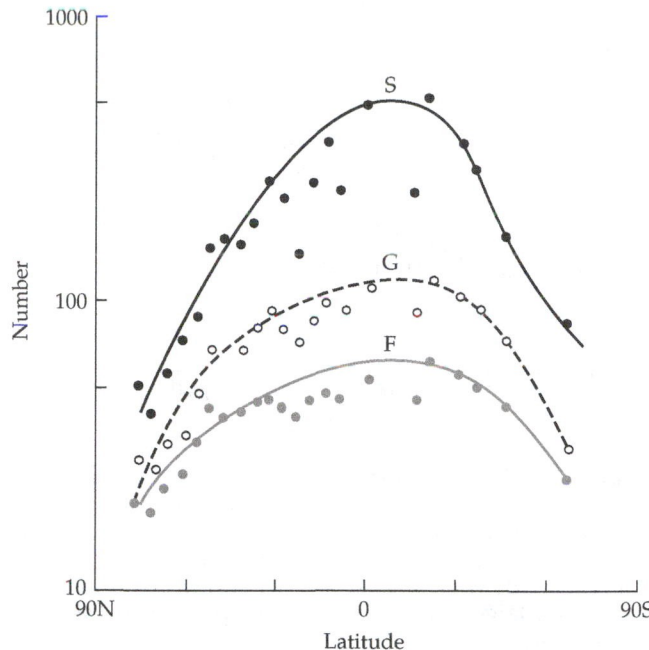

Figure 2.11 Latitudinal species richness in bivalve mollusks. Points are average numbers of species (S), genera (G), and families (F). (Modified from Stehli et al. 1969.)

positional variable, cannot by itself be a determinant of species richness patterns. Because of their differing requirements, it seems likely that different combinations of factors may determine the distributions of bacteria, fungi, vascular plants, amphibians, and mammals, yet some suite of factors that co-varies with latitude must operate to increase species richness at low relative to high latitudes. Differences in species richness must be generated by differences in the rates of speciation, extinction, immigration and/or emigration among locations. Because latitude cannot directly influence speciation, extinction, or movement rates, it must be a surrogate for some combination of variables that can directly influence species

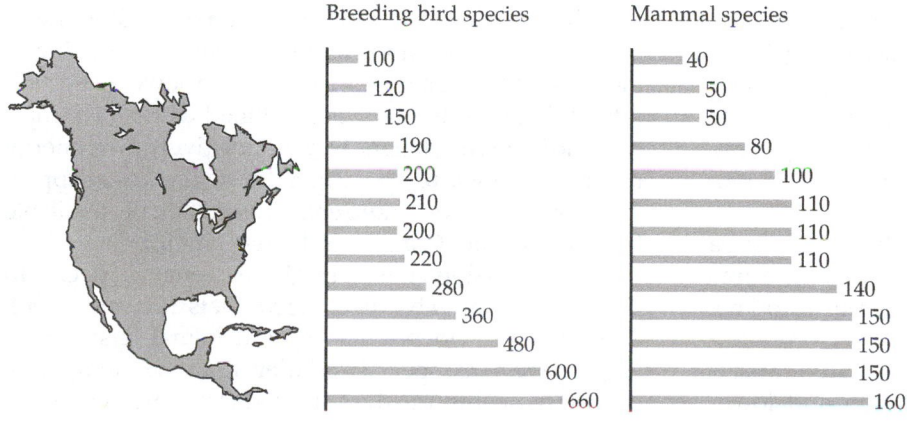

Figure 2.12 Latitudinal gradients of species richness of birds and mammals in North and Central America. Species richness corresponds to latitude map at left. The numbers for birds are from breeding species only. (Modified from Briggs 1996.)

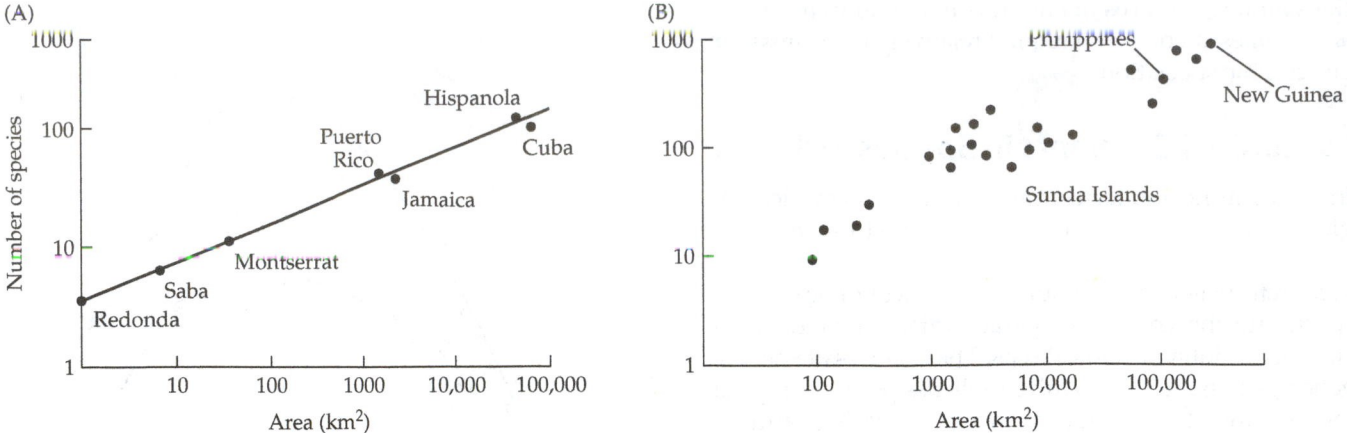

Figure 2.13   Relationship between area and number of species on islands of various sizes. (A) The number of species of amphibians and reptiles found on selected islands in the West Indies. (B) The number of land and freshwater bird species on the Sunda Islands, the Philippines, and New Guinea. These islands are close to the Asian continent, and many were connected to the mainland during glacial periods. Therefore, many of the larger islands are relatively species-rich. (Modified from MacArthur and Wilson 1967.)

richness and that are correlated with position. One likely correlated factor is area, which can influence both speciation and extinction rates.

One of the first ecological relationships to be established empirically was the relationship between area and number of species (Arrhenius 1921). This relationship, which is true for both continental and island biotas (Rosenzweig 1995), is commonly expressed using a power function of the form

$$S = cA^z$$

where S is the number of species, A is the area, and c and z are constants fitted to the data. On a logarithmic scale, this relationship plots as a straight line where c is the y-intercept and z is the slope of the line (Figure 2.13). Analyses of species–area relationships in many groups of organisms revealed that values of z range from 0.15 for continents, to between 0.25 and 0.45 for islands associated with a continent, to 0.19 for comparisons among continents or biogeographic provinces (Rosenzweig 1995). The regular relationship between the size of an area and the number of species it supports, which is most readily observed using island data, was a key empirical generalization in the development of the theory of island biogeography (MacArthur and Wilson 1967).

The Tropics have a larger climatically similar area than any other ecological zone because the surface area of latitudinal bands decreases toward the poles and because temperatures change relatively slowly between the equator and 20°N and 20°S latitudes. Larger area is correlated with higher rates of speciation and lower rates of extinction, and thus, also is correlated with high species richness. Large areas can facilitate the long-term survival of populations, which in turn, allows the accumulation of the genetic differences that may prevent hybridization should the populations come together at some time in the future. This favors greater speciation rates. Populations of vertebrates in lower latitudes show greater genetic differentiation, which is consistent with greater capacity for speciation in the Tropics (Martin and McKay 2004). Therefore, area is probably a contributor to high species richness in the Tropics, but so are productivity and available energy, as discussed in the next section.

Species Richness–Energy Relationships

Available energy affects local species richness because the more energy that is available, the more biomass per unit area that can be supported. More biomass enables more individuals to coexist in an area, potentially resulting in more individuals per species and, hence, lower extinction rates. If high productivity is combined with moist tropical conditions, organisms expend relatively little energy maintaining their temperatures and moistures at appropriate levels, so even more energy is potentially available for reproduction (Connell and Orias 1964).

More available energy might also enable species to persist on relatively specialized diets (Gaston 2000). There is some evidence that tropical animals may, on average, have more specialized diets than their temperate counterparts, but much more research is needed to es-

tablish the degree to which this is true and to identify the dimensions along which dietary specialization has evolved (Beaver 1979; Marquis and Braker 1994).

Primary production, the major source of available energy, is determined primarily by temperature, moisture, and nutrient availability. Highest terrestrial production is found in regions with high rainfall and year-round warm temperatures. Production drops with elevation because of lowered temperatures. In many areas of the world, the amount of precipitation determines total production. In arid and semiarid regions, annual production can be predicted fairly accurately from the amount and distribution of precipitation.

Marine production is also limited by temperature, but more often by shortage of nutrients in surface waters (Longhurst 1998). Organisms and nutrients sink in the water column. Therefore, processes that keep organisms close to the surface where photosynthesis is possible, and those that bring nutrients and organisms back to the surface, are major determinants of marine productivity. Productivity is highest in areas where vertical turbulence (upwelling) is produced by friction between tidal streams and the seabed and at the confluence of two major ocean currents; it is lowest in the open ocean (Koblentz-Mishke et al.1970; Bunt 1975; Longhurst 1998). Primary productivity also is higher near continental margins due to the runoff of nutrients from the land. Finally, primary production decreases with depth because light levels decrease rapidly with depth even in clear water.

The hypothesis that available energy contributes to species richness is supported by a correlation between productivity and species richness along both latitudinal and altitudinal gradients. A strong correlation exists between realized annual evapotranspiration, a measure of available energy to plants, and tree species richness in North America (Currie and Paquin 1987). Similarly, Gentry (1988) demonstrated a strong correlation between plant species richness in Neotropical forests and absolute annual precipitation, a variable strongly positively correlated with productivity.

However, the relationship between productivity and species richness is not simple. Many of the world's most productive systems, such as estuaries, seagrass beds, and hot springs, are species-poor. Conversely, plant species richness is higher in semiarid regions with nutrient-deficient soils than in similar areas with richer soils. The remarkably rich plant communities of the Cape region of South Africa (fynbos) and the extreme southwest corner of Australia (Figure 2.14) are found on highly infertile soils (Kruger and Taylor 1979; Bond 1983; Rice and Westoby 1983). Nitrogen levels in those soils are an order of magnitude lower than in lowland California and Chile (Specht and Moll 1983), where species richness is lower. Evidence suggests that soil infertility itself is a major contributor to the inverse correlation between soil fertility and plant species richness because low fertility favors abilities to exploit slightly different microhabitats more efficiently (Tilman 1982, 1985; Cowling et al. 1992; Willis et al. 1996). Fynbos plants are remarkably variable in leaf morphology (Cody 1986) and growth phenologies (Kruger 1981; Cowling 1992). These differences are probably correlated with a corresponding diversity of carbon-fixing strategies. In addition, the low palatability of leaves of plants growing on nutrient-poor soils results in a rapid accumulation of flammable biomass, leading to high frequencies of fires in these summer-dry climates.

Soil infertility also may favor high species richness indirectly via interactions with seed dispersers. Compared

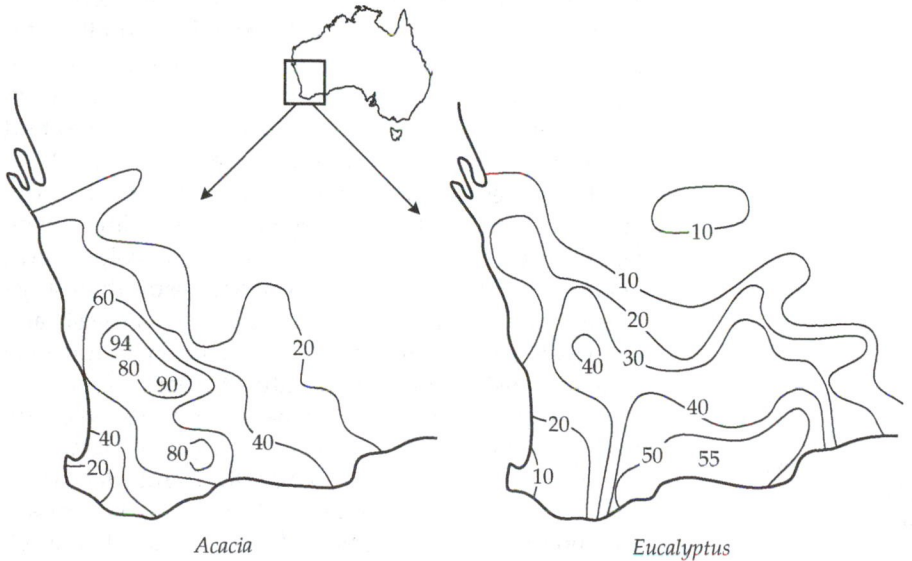

Acacia   *Eucalyptus*

Figure 2.14 Numbers of species of *Eucalyptus* and *Acacia* in southwestern Australia. Notice how many species in these two genera can be found living in proximity. Similar patterns are found among many other genera of plants in southwestern Australia. (Modified from Lamont et al. 1984.)

to birds, ants move seeds relatively short distances, but they usually bury them. Also, ants pick up and move seeds that offer much smaller rewards than those that attract birds. Australia, a continent with notoriously poor soils, has the highest proportion of plants with ant-dispersed seeds of any continent (Berg 1975; Westoby et al. 1991) and it has unusually high ant species richness (Andersen 1983; Greenslade and Greenslade 1984). Ant-dispersed seeds are also common in South African fynbos (Milewski and Bond 1982). In combination, these factors could result in very high α-species richness and unusually high β-species richness. Species richness on scales of one hectare is higher in southwest Australia than in rainforests even though the latter accumulate more species at slightly larger scales (Figure 2.15).

There is a consistent unimodal relationship between species richness and depth for many marine taxa. Maximum richness of most groups is found at depths between 2000 and 4000 meters (Rex 1973, 1983). In contrast, coral species richness peaks at depths between 15 and 30 meters because corals, which obtain much of their energy from photosynthetic algae embedded in their tissues, are confined to the photic zone (Huston 1994).

In addition, in many ecosystems, species richness increases with primary production at low production levels, but declines at high levels of primary production (Huston 1994). This phenomenon has been called "the paradox of enrichment" (Rosenzweig 1971) because addition of fertilizer to aquatic and terrestrial plant communities often results in sharp decreases in species richness (Huston 1980).

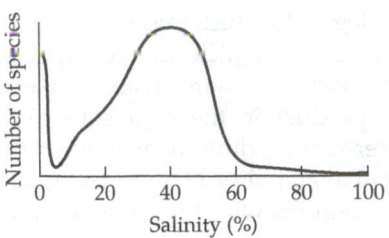

Figure 2.16 Estimated species richness of aquatic invertebrates living in waters of different salinity throughout the world. The two peaks correspond to fresh water (<2% salinity) and seawater (30%–40% salinity). (Modified from Kinne 1971.)

Most highly productive but species-poor systems are distributed as relatively small, fragmented patches whose physical environments differ strikingly from those of the surrounding, more extensive ecosystems. Evolutionary biologists believe that the combination of major physical environmental differences and isolation of the patches results in fewer species of organisms evolving adaptations to those unusual environments. For the same reason, species adapted to the more common surrounding environments are less likely to survive well in those rare habitat types. Both fresh water and seawater are widely distributed, but waters of intermediate salinity and waters that are more saline than the oceans are rare. The species richness of aquatic invertebrates throughout the world parallels this pattern (Figure 2.16).

Available energy may influence species richness indirectly via vegetation structure. The structural complexity of vegetation is strongly correlated with potential evapotranspiration and, hence, primary productivity. In most terrestrial environments, plants provide most of the physical structure within which the activities of all other organisms are carried out. Coral reefs create complex structures in tropical marine environments that influence fish species richness (Holbrook et al. 2002). Structurally complex communities have a greater variety of microclimates, a greater variety of resources, more ways in which to exploit those resources, and more places in which to find shelter from predators and the physical environment.

The richness of species in most taxa is positively correlated with structural complexity. Structurally simple habitats, such as the open ocean, grasslands, and cold deserts, generally support fewer species of organisms than structurally more complex communities, such as forests and coral reefs. Animal groups that exploit the environment in three dimensions are most sensitive to plant community structure. A positive correlation between foliage height diversity and bird species richness exists in many plant communities on all continents (MacArthur and MacArthur 1961; MacArthur 1964). Similarly, the richness of species of web-building spiders is positively correlated with the het-

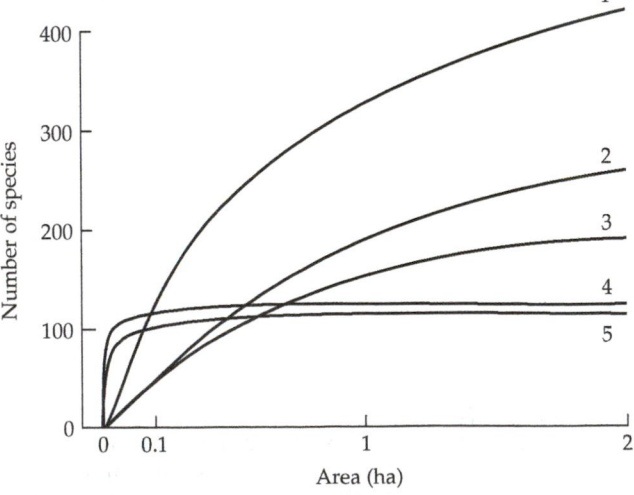

Figure 2.15 Plant species richness in Borneo and in southwestern Australia. Curves 1, 2, and 3 represent trees greater than 10 cm in diameter in mixed forests in Borneo. Curves 4 and 5 are for plots in southwestern Australia. (Modified from Lamont et al. 1984.)

erogeneity in heights of the tips of vegetation to which such spiders attach their webs (Greenstone 1984).

In contrast, there is no consistent relationship between vegetation structure and lizard species richness in hot deserts in North and South America, Africa, and Australia (Pianka 1986). The number of lizard species in similar habitats in African deserts averages twice that in North America; Australian deserts are about twice as rich as African ones. The main differences are due to the presence in Australia of non-lizardlike lizards and nocturnal species. Only one nocturnal lizard is found at the North American sites, whereas there are four nocturnal species at the African and eight at the Australian desert sites. Mammal-like species (the monitors) and worm-like species add to the richness of Australian desert lizard faunas. These differences relate primarily to the long-term evolutionary history of deserts on the three continents, not to differences in today's vegetation (Pianka 1986).

Disturbance and Species Richness

One explanation for the latitudinal patterns of species richness comes from considerations of the extreme disturbances caused by climate change over Earth's history (Dynesius and Jansson 2000). During the Pleistocene and before, periods of rapid climatic change (influenced by orbital oscillations of Earth), which were more intense toward the poles, clearly had enormous influence on worldwide patterns of species richness. Rapid climate fluctuations caused increases in extinction rate as species were unable to adapt to new conditions, or migrate to more hospitable ones, and the more extreme fluctuations toward the poles resulted in more extinctions at high latitudes. Importantly, these fluctuations also should have changed patterns of speciation, favoring characteristics that lead to lower speciation rates among the survivors

(less specialization, greater vagility, larger ranges), particularly in higher latitudes that were most affected by climate change (Dynesius and Jansson 2000). This combination of selective factors in response to climate cycles could contribute to the latitudinal gradient of species richness we see today (Figure 2.17). At a finer spatial scale, physical and biological disturbances clearly influence the number of species present in many ecological communities. Small disturbances occur much more frequently than large disturbances. Physical perturbations include heavy rains, strong winds, landslides, earthquakes, and fires. Biological disturbances, most of which are small scale, include activities of predators and parasites, tree-falls, activities of competitors, and trampling.

Physical disturbances influence species richness by destroying habitat structure, selectively killing individuals of different species, and sterilizing soils. Physical disturbances typically also have important indirect effects on biodiversity by altering the biological interactions that affect species richness. For example, fire, one of the most important terrestrial physical disturbances, kills plants and animals, destroys soil organic matter, and redistributes nutrients, resulting in a new ecological community with a dramatically different physical structure and altered biological interactions.

Theoretical and empirical studies both suggest that disturbances do not have consistent effects on species richness. The hypothesis that lack of disturbances might favor high species richness was originally advanced by Sanders (1968), and Sanders and Hessler (1969), who noted that productive estuaries and continental shelves in most latitudes have few species of benthic animals, but the cold, dark, unproductive floor of the deep sea is very species-rich. Sanders suggested that the deep sea supported many species because its environments had relatively constant levels of temperature, salinity, and

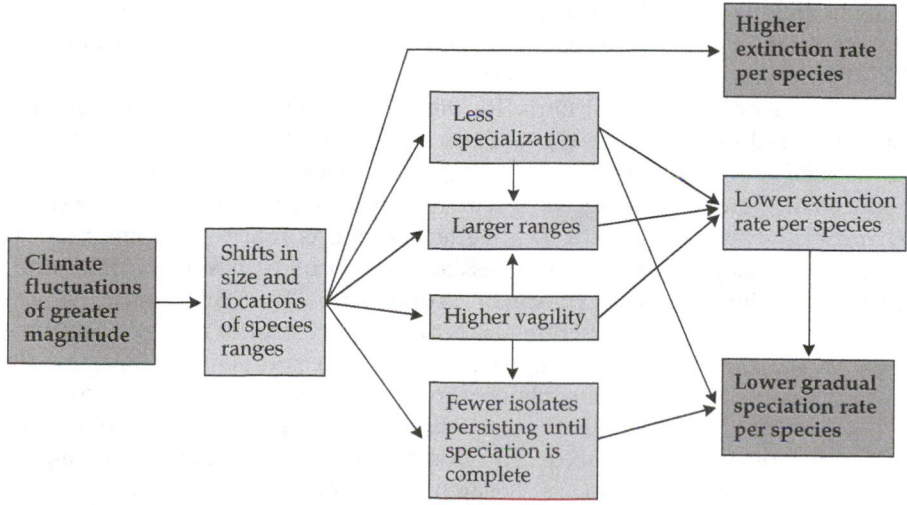

Figure 2.17 Sudden climate shifts influence species richness by changing speciation and extinction rates. Selective pressure for species to be less specialized, more vagile, and have larger range sizes result in slower speciation rates. Direct effects of climate change increase extinction rates. The net result of these forces is a decrease in species richness. (Modified from Dynesius and Jansson 2000.)

oxygen concentration and had been so for millennia. However, food resources in the deep sea are patchily distributed, and local disturbances are caused by the feeding, burrowing, and mound-building activities of animals (Grassle 1989, 1991). These activities create a patchiness similar to that produced by tree-falls in terrestrial environments. They should create conditions suitable for many different species.

Biological interactions can also influence species richness. For example, competition could reduce α-richness if some species exclude others from ecosystems. However, competition could increase α-richness by favoring finer habitat segregation among species (Rosenzweig 1995). Predation can increase species richness if predators prey preferentially upon competitive dominants, thereby preventing competitive exclusion. A well-known example is the increase in species richness in rocky intertidal communities of the Pacific Coast of North America. This increase is a result of selective predation by the sea star *Pisaster ochraceus* on the competitively dominant mussel *Mytilus californianus* (Paine 1974). On the other hand, predation can reduce species richness by preventing vulnerable species from living in an area. Experimental evidence exists for all of these outcomes, but how often and where these forces exert their influence are unknown (Orians and Kunin 1991). Analyzers of the literature have reached different conclusions, in part because they used different criteria for including studies in their samples (Connell 1975, 1983; Schoener 1983).

Some researchers have hypothesized that many organisms are held at sufficiently low population densities such that competition rarely occurs (Connell 1975). Yet other researchers have suggested that competition occurs primarily during unusually hard times, "competitive crunches," when resources are scarce (Wiens 1977). Which of these two scenarios applies depends upon whether low population densities are caused by harsh physical conditions or by scarcity of consumable resources.

Often species richness is highest at intermediate levels of disturbance (Figure 2.18). The **Intermediate Disturbance Hypothesis** states that physical disturbances or predation should augment species richness where other sources of disturbance are few, but should reduce species richness when exogenous disturbance is high (Connell 1978; Abugov 1982). If disturbances are rare, species richness may decline owing to competitive exclusion. If disturbances are common, many species may be unable to complete their life cycles during the short intervals between disturbances.

Interactions between local and regional species richness

The structure of local ecological communities may be determined by local interactions among species and their environments, and by processes operating at larger spatial

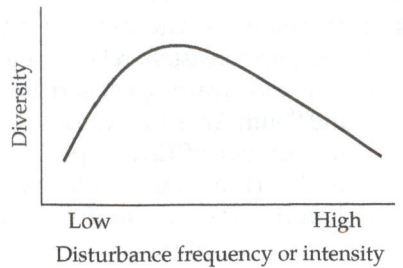

Figure 2.18 Model of the intermediate disturbance hypothesis. Species diversity is lowest at high and low levels (frequencies or intensities) of disturbance, and highest at an intermediate level. (Modified from Connell 1978.)

and temporal scales (MacArthur 1972; Ricklefs 1987). In other words, to understand why a particular set of species exists in a specific place, ecologists must study more than local competition, predation, and mutualistic interactions. Larger-scale processes, such as dispersal, speciation, and historical biogeography cannot be ignored.

The ecological interactions that appear to influence current patterns of species richness, in combination with physical environmental factors, could set absolute limits to species richness. In other words, we can ask if there are limits to the number of species that can live in a given ecological community. If so, are those limits regularly reached? Is there a critical limit to overlap in resource use that prevents more species from being accommodated unless dramatic changes occur in the nature of available resources? Does the chemical warfare between plants and herbivores set limits to species richness or does it offer possibilities for more species? Are there limits to the size of mimicry systems, and do mimicry systems allow more species to persist in ecosystems than would be possible without them? Have the richness-generating interactions between plants and their pollinators and between plants and their seed dispersers been exhausted? To answer these questions research must be carried out at many different spatial scales.

For many diverse taxa, including gall wasps (Cornell 1985), birds (Ricklefs 1987), tiger beetles (Pearson and Juliano 1993), fishes (Griffiths 1997), and primates (Eeley and Lawes 1999), local species richness appears to increase along with regional species richness, but at a slower rate. This pattern suggests that rigid limits to local species richness are not set by competition, predation, and parasitism (Cornell 1999). Thus, although ecological interactions are typically very intense in local communities they apparently influence abundances more strongly than numbers of species. Moreover, their influence on the numbers of coexisting species appears to be weaker than the influence of regional processes. A major challenge in ecological research is to determine how regional processes act to determine local species richness.

The importance of biodiversity

We asserted in the first chapter that a normative postulate of conservation biology is that biodiversity is worthy of being conserved for its own sake. Yet, many people do not find this normative argument compelling, and seek anthropocentric justifications, primarily economic ones, for the conservation of biodiversity, as discussed in Chapters 4 and 5. In the past 15 years, conservationists have focused on describing the nature of the links between biodiversity and ecosystem processing (Chapin et al. 1997). Ecosystems provide a large array of goods and services vital to human existence, including water and air purification, cycling of critical elements and water, development and retention of rich soils, regulation of water flow, carbon sequestration, decomposition, and primary and secondary production (Table 2.4). Claire Kremen discusses the importance of one of these services, pollination, in Box 2.2.

If clean air and water, fertile soils, pollination of crops, and other "services" depend upon the maintenance of a broad diversity of native species, this becomes a powerful motivator for their conservation. For example, in an experiment manipulating species richness in a prairie ecosystem in Minnesota, Tilman and Downing (1994) showed that drought resistance was greater in species-rich plots than in species-poor plots. Many observational, experimental and theoretical studies have found that species-rich communities have higher values of some metric of ecosystem functioning than species-poor communities. If generally true, then preserving as many species as possible should be a primary goal.

Yet, ecologists still know little about how attributes of communities—including species richness, relative abundances, and functional traits of individual species (such as ability to fix nitrogen or seed dispersal effectiveness)—

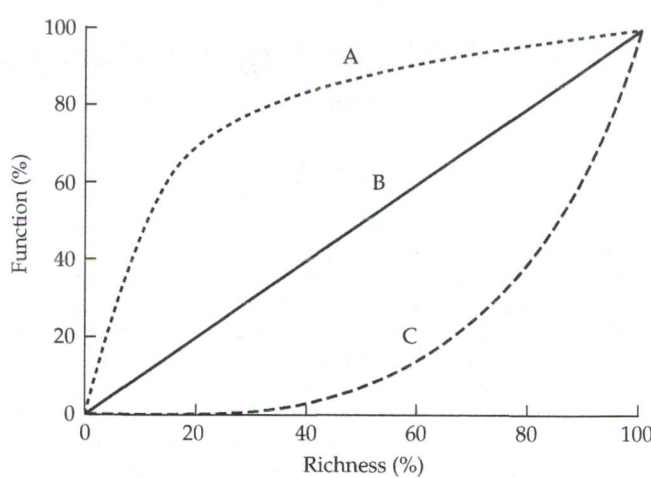

Figure 2.19   Ecosystem function could increase quickly with species richness (A), linearly (B), or as an accelerating function (C). Increases in ecosystem function with species richness could occur due to increases in ecological functionality either because of changes in the number, abundance, or functional traits held by the species in the community. (Modified from Kremen 2005.)

interact to influence the magnitude and stability of ecosystem processes (Figure 2.19). For example, are relatively few or many species required for high values of some ecosystem process (curve A versus C in Figure 2.19)? Or is the relative abundance of key species, or the diversity of functional groups, most strongly related to ecosystem processes?

Considerable debate has sprung up regarding the interpretation of experimental investigations of relationships between species richness and ecosystem processes. The importance of rare species to ecosystem functioning is poorly known, and many studies show that a particular ecosystem process, such as primary productivity, may reach a plateau at only modest species richness (e.g., 20%–30% of maximum richness) (Schwartz et al. 2000). Only a few studies show a linear increase of ecosystem functioning with species richness (e.g., Naeem and Li 1997), although several show that a more diverse community yields the least variability in biomass (e.g., Mc-Naughton 1977; Tilman 1996). Further, most experimental investigations to date used synthetic communities, where less than 40 species are brought together in different mixtures in lab or field plots at small scales. The results of these studies may not be good predictors of processes and patterns in natural ecosystems where the assembled species have long histories of interaction. Yet, these experiments point to interesting principles to test in the field, and the scale of some experiments is appropriate for the process of interest. For example, microcosm experiments of soil communities are complex enough to show that functional diversity of species (i.e., differences in the roles they play in

TABLE 2.4   Examples of Ecosystem Services

Supporting services
Soil formation, nutrient cycling, primary production
Provisioning services
Food, fresh water, fuelwood, fiber, biochemicals, genetic resources
Regulating services
Climate regulation, disturbance regulation, regulation of hydrologic flows, water purification, air purification, disease regulation, erosion control, biological control, pollination
Cultural services
Spirituality, religious inspiration, aesthetic(s), inspiration, education, recreation, sense of place, cultural heritage

Source: Costanza et al. 1997 and Millenium Ecosystem Assessment 2005.

BOX 2.2 The Ecology and Management of Crop Pollination Services

Claire Kremen, *Princeton University*

Animal pollinators, generally bees, are required to produce fruits and seeds for one or more cultivars of 70% of the world's 1300 crops (Roubik 1995). Less than a dozen bee species are managed for crop pollination worldwide, but hundreds to thousands more species (out of 20,000 existing bee species) visit and likely pollinate crop plants (Parker et al. 1987; Free 1993). In fact, most farmers rely on a single managed pollinator, *Apis mellifera*, a species that is increasingly difficult to manage due to diseases and genetic introgression from an aggressive subspecies (Delaplane and Mayer 2000). Wild (unmanaged) bees also contribute to crop pollination (Parker et al. 1987; Kevan et al. 1990; Free 1993); these contributions may become increasingly important, or even critical, for guaranteeing the world's food supply if current declines in management of *Apis mellifera* continue (Wardell et al. 1998).

Locally, up to 40 wild bee species visit and pollinate individual crops in California, Costa Rica, and Indonesia (Kremen et al. 2002a; Klein et al. 2003a; Ricketts 2004); generally members of these guilds visit and pollinate multiple crops as well as native and alien species. In these systems, diversity, abundance, and composition of these bee communities on farms strongly depend on their distance to natural habitat and on site-level environmental characteristics. In both tropical and temperate settings, the most diverse communities are also the most abundant (see also Steffan-Dewenter and Tscharntke 2001). In watermelon fields in California, as agricultural land use has intensified, a number of bee species have become locally extinct. Those that were most extinction-prone were also among the largest and most efficient pollinators (Larsen et al. 2004), reducing crop pollination via three interrelated losses: species richness, aggregate abundance, and most efficient pollinating species. Farms with the least intensive use received sufficient pollination services from wild bee communities alone, while farms at the high end had to import managed honeybee colonies to fields in order to receive sufficient pollination (Kremen et al. 2002b). In the Indonesian system, the farms with the greatest bee diversity included rare, highly efficient pollinator species that contributed services disproportionately to their abundance; a species-rich pollinator assemblage may also be more capable of providing consistent services across the spectrum of weather conditions and spatial and temporal availability of coffee bloom because of complementarity among species (Klein et al. 2003a).

A principal landscape factor influencing pollination services in all three systems appears to be the proximity and proportional area of nearby natural habitat (Klein et al. 2003a; Kremen et al. 2004; Ricketts et al. 2004). Pollination function provided by wild bees is not only greater, but is also more stable and predictable, as the proportion of natural habitat increases (Kremen et al. 2004; Figure A).

Protecting habitat for wild bees within the agro-natural landscape or restoring habitat in small patches on farms should increase the services that wild pollinators provide to agriculture, although alterations of farming practices may also encourage pollination services (Klein et al. 2003b). In coffee, an important cash crop, the value of pollination services stemming from maintaining natural forests rivals alternative high value uses of that land (Ricketts et al. 2004), while in lower value crop systems that currently rely on managed bees, maintaining natural habitat both reduces rental costs for bees and provides an insurance policy in the event of further decline, or even total loss, of managed honeybee stocks (Kremen et al. 2002b). The relationship between pollination function and natural habitat can be used to establish targets for conservation and restoration to provide the service by native bee communities at a desired level (Kremen et al. 2004). Studies of the crop, alien, and wild plant resources required by different pollinators and their occurrence across the landscape (Steffan-Dewenter and Tscharntke 2001; Kremen et al. 2002a; Steffan-Dewenter and Kuhn 2003; Westphal et al. 2003) allow the development of land management protocols that take into account spatial and temporal availability of floral resources in both conserved and farmed portions of the landscape. Models that include the relationships between foraging scale and resource availability, versus productivity, dispersal, and population persistence of populations of pollinating species, can then be used to determine how the agro-natural landscape should be managed for pollination function.

Figure A The relationship between the amount and variability of pollen reaching watermelon flowers provided by native bees and the proportion of natural, upland habitat within 2.4 km of the farm site. (I) Mean estimated pollen deposition over 10 minutes (natural log-transformed) against the proportion of upland habitat. Black circles, organic farms; gray circles, conventional farms. (II) Variability in pollen reaching watermelon flowers (natural-log transformed coefficient of variation) on organic farms against the proportion of upland habitat. (Modified from Kremen et al. 2004.)

the ecosystem), and not species richness, influences the rate of decomposition (Heemsbergen et al. 2004).

Increasingly researchers are using large-scale removal experiments and comparative studies of communities with varying species membership and abundances to gauge the role of individual species as well as diversity on ecosystem functioning (e.g., Kremen et al. 2002a,b; Diaz et al. 2003). A number of biologists believe that ecological differences among species should create the conditions for an increase in ecosystem process with diversity, while others believe that only the attributes of a few dominant species will control that relationship (Figure 2.20). A growing consensus focuses on the importance of differences in the ecological roles of species, and the joint contributions these differences make (Diaz and Cabido 2001; Loreau et al. 2001). However, measuring the significance of a species for a particular ecosystem process is much harder than counting numbers of species or estimating population densities. Full understanding of how biodiversity affects the ability of an ecosystem to provide goods and services over time awaits much further work.

The Future of Biodiversity Studies

The incomplete state of our knowledge of the identities, taxonomic relationships, and distributions of the vast majority of the world's organisms means that the primary work of cataloging biodiversity is yet to be done. Today relatively few scientists are being trained as taxonomists. Therefore, increasing the cadre of competent taxonomists, particularly in tropical nations, is an important goal (Mikkelsen and Cracraft 2001; Gotelli 2004). How this inventory should be carried out is the subject of much debate. Some biologists have committed themselves to an intense global survey, aimed at the discovery and classification of all species (e.g., the Species 2000 Project). Others, pointing to the shortage of people, funds, and time, believe that the only realistic hope lies in the rapid recognition and preservation of those threatened habitats that contain the largest number of endemic species (Conservation International strategy), or to conserve representatives of the habitats of every ecoregion (World Wildlife Fund and The Nature Conservancy strategies). They give the inventory task a lower immediate priority.

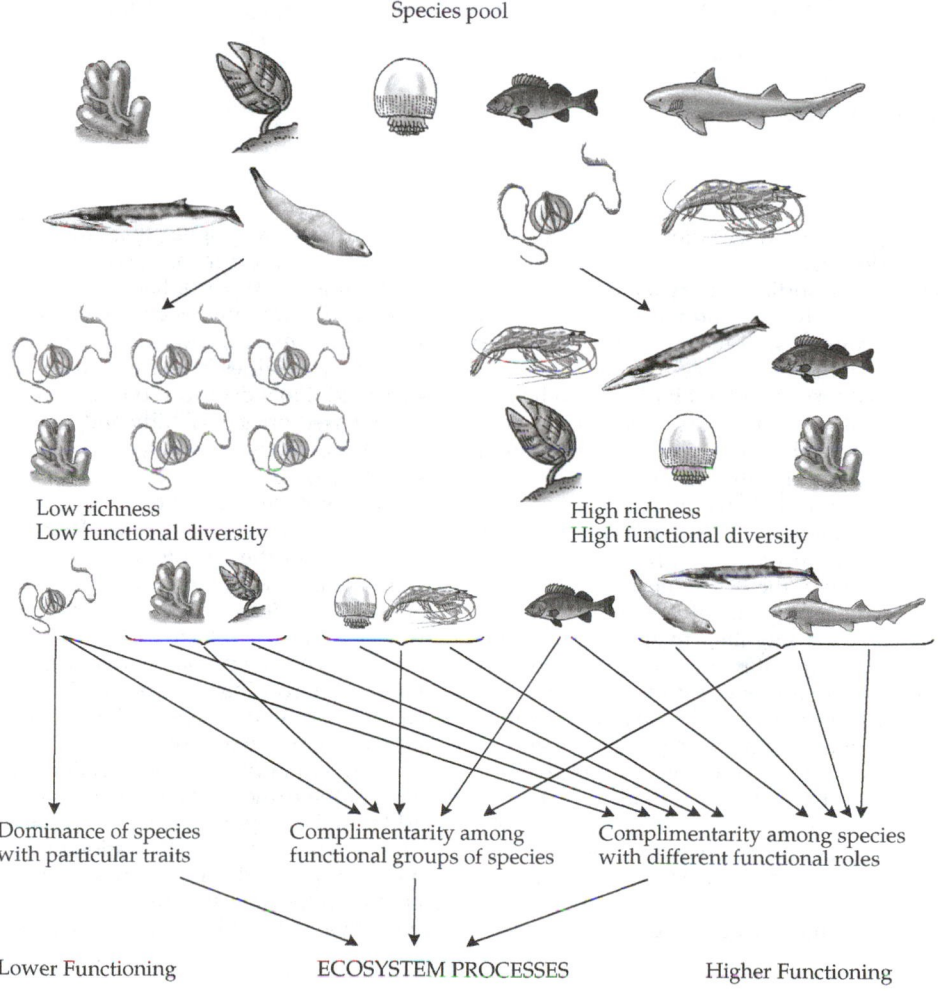

Figure 2.20 Ecosystem function can be influenced by the functional diversity among species in a community via several mechanisms. Representation within a community of more than a single or few functional groups, or of many species with distinct functional roles should lead to higher functioning of various ecosystem processes. (Modified from Loreau et al. 2001.)

Species pool

Low richness
Low functional diversity

High richness
High functional diversity

Dominance of species with particular traits

Complimentarity among functional groups of species

Complimentarity among species with different functional roles

Lower Functioning   ECOSYSTEM PROCESSES   Higher Functioning

E. O. Wilson (1992) strongly advocated a strategy that combines global surveys designed to achieve a complete biodiversity inventory in 50 years with special attention to areas of unusually high species richness and endemism. His strategy has three components. The first is a Rapid Assessment Program (RAP) that would investigate, within a few years, poorly known ecosystems. RAP teams would be formed, consisting of experts on groups such as flowering plants, reptiles, mammals, birds, fishes, and butterflies that are well enough known to be inventoried quickly and accurately. These groups would then serve as proxies for the entire biota (see Essay 2.3 by Lily Rodriguez; for a discussion of applications to priority setting in biodiversity conservation see Chapter 14). The next stage would be to establish research stations in areas believed to be major hotspots of diversity. Inventories and ecological studies would then be carried out there and in surrounding regions. The third stage, with a time frame of 50 years, would combine the inventories from RAP and the intensive studies at a small number of research stations with monographic studies of many groups of organisms to provide a more complete picture of global biodiversity and its distribution.

While a reasonably complete inventory of the biota of a region, or significant components of it, is being achieved, a monitoring program also needs to be established. Such a program should be designed to detect trends in biodiversity and to identify impending problems to which attention should be directed (National Resource Council 2000). Without a monitoring program society cannot know if its efforts to preserve biodiversity are succeeding or why and where they are failing.

Human societies will need to greatly increase their investments in biodiversity studies during future decades if we are to understand the patterns of biodiversity more fully. The scope of the untapped wealth residing in biodiversity, as well as the importance of inventorying biodiversity and understanding ecological relationships among species, is unappreciated. It is doubtful that humans can devise a sustainable future without a more complete knowledge of biodiversity.

The relationship between area and species richness has major practical implications for the location, design, and management of parks and reserves that are established and maintained to preserve biodiversity. There is increasing evidence that even the largest parks and re-

ESSAY 2.3

Rapid Inventories for Conservation

Lily O. Rodríguez, *CIMA, Cordillera Azul*

■ Developed under the Rapid Assessment Program (RAP, of Conservation International, CI) or as Rapid Biological Inventories (RBI, by The Field Museum of Chicago), rapid inventories are a relatively new conservation tool designed to assess the biological importance of priority sites for conservation. Using short field trips, usually of no more than four weeks, these inventories evaluate the biodiversity values of sites of global conservation interest, providing scientifically based recommendations for prompt establishment of protected areas and conservation action plans. Identifying conservation targets and their threats, and generating recommendations for immediate conservation action are the main goals of these inventories.

RAP or RBI results rely highly on the expertise of the team undertaking the inventory, as there is no single methodology that is or can be used in all places with all taxa. Here, I provide some insights and recommendations based on my experience after partici-

pating in inventories over the last decade.

Multidisciplinary teams are formed for each expedition, usually with a core group of well-recognized scientists (taxonomists and ecologists), combined with resident field biologists, who are both specialists and specialists-in-training. Each specialist selects the methods and groups that will be the focus of its work, based on abundance, rarity or uniqueness, and capacity to document the group in a short period of time. None of these inventories are meant to yield complete information on the site.

Planning the trip and identifying 2–4 campsites are crucial steps to ensure exciting results. Satellite imagery is used first to assess the general area and locate points of main interest. The preliminary selection of sites to inventory is made based on this first step, combined with a review of any previous research in the area or nearby locations, as well as local knowledge. Overflights to the area are then made by a few team members.

During those overflights, team members select points for easy access to as many different types of habitats as possible, verify any unique characteristics seen in the images, and assess impacts of human activities in the area to be surveyed. Final decisions on sites are made based on accessibility and transportation. Although relatively expensive, helicopters are the best way to reach remote areas, and to relocate the team between main campsites. To maximize time for the inventory, camps and trails should be prepared prior to arrival of the team. While preparing the trails (usually loop trails), satellite campsites can be prepared and used by the team according to distances, intensity of work in different habitats, taxa, and technique requirements.

Different approaches are needed for each of the principle taxonomic groups addressed in rapid inventories. Plants provide the best characterization of habitat types and communities associated with the area and are therefore a must in any rapid inventory. Transects,

quadrat plots, collections, and photographs are some of the techniques that can be used to characterize the flora, providing quantitative information, field observations and general collections of the diversity of plants. Focal groups, such as ferns or understory vegetation, are important as they provide better resolution of habitat differences than do trees. Groups of special value such as orchids may also add to the richness of the results.

Birds are one of the best-documented groups around the world. Because of their high species numbers, and the fact that the range, distribution, and biology is reasonably well known for most species or groups of species, birds are used as habitat indicators. Avian data can be used to provide quick comparisons among sites, and to guide considerations such as connectivity, size of the protected area, and conservation status and perturbation of the site. Most commonly used methods involve auditory surveys along trails, including song recording and point counts. Assessing the status of local populations of hunted species is an important step for management recommendations.

Fishes are the main community associated with aquatic habitats. Often easy to collect, they can be captured with small nets that are for scientific use only. A high percentage of the fish community can thus be surveyed. Interviewing local people with photographs in hand provides rich information on consumed species and seasonal variations. Usually fishes contribute to

the inventory with rare records or new species and the presence of suites of species can help relate basins and aquatic communities (Figure A).

Insects, because of their immense species richness, can rarely be used as a group for rapid inventories. The best choice is to take experts on specific taxonomic groups that are best known and easily seen in few days. Butterflies, beetles, dung beetles, and spiders are some candidates for rapid inventories, but, as always, this will depend on the availability of experts and the scope of knowledge of the group.

Mammals (especially large mammals) experience heavy human pressures and are important to document, providing a primary indicator of the conservation quality of the area. Tracks and visual encounters are used most frequently to document mammalian species. However, small mammals such as rodents and bats are speciose groups that if specifically surveyed can provide a more detailed evaluation of the site. A trade-off between time and value is usually applied here, with bats (which are easily censused with mist nets) being one of the best choices.

Because of their sensitivity to disturbance, frogs are usually included in rapid inventories. Easily located by their calls, auditory transects and visual encounters are the most common techniques employed to survey for frogs. Snakes, because of their low population densities and secretive habits, are difficult to include; however, they can add value as indicators of

regional faunas. Because aquatic and terrestrial turtles as well as crocodilians are frequently hunted they are always included and evaluated as target species for conservation actions. Tissue collection of amphibians (especially in mountain ecosystems, where populations declines due to disease appear to occur more often) is now recommended to be able to evaluate possible causes of decline in amphibian populations.

Geology, soil diversity, freshwater characteristics, and substratum are surveyed to understand and document the uniqueness of the area and to indicate species distributions and plant community types. An additional but extremely important aspect is to include human dimensions in the rapid inventory through rapid social asset inventories. This consists of visiting the communities surrounding the site surveyed, and is achieved through a diversity of techniques (interviews of local leaders, authorities, key people of the community, and focal groups; meetings with the whole community; and visits to their crop fields), the team explores the history, demography, economy, social organization and structure, and the use of natural resources. While identifying patterns of social organization and opportunities for capacity building, these inventories contribute to hone the recommendations, and to engage local participation for future conservation action.

Rapid inventories should not be taken as the sole source of information on biodiversity. While they are a quick and effective way to document biodiversity, they do not provide complete species lists, and abundance information is limited and depends on the areas visited and the seasons in which inventory occurs. Furthermore, it is highly expensive, and requires a core of highly trained taxonomists who can quickly recognize what is new to science and relevant to the area, and who are able to compare findings with the widest possible range of other sites.

It is also important that local people, as well as local scientists participate in the inventory. Being part of the fieldwork will give the residents a first-hand insight of the global importance and will increase their support and interest in the area, as well as contribute to developing pride in the site. Recognizing species that are of no use to them but that can be of high scientific value (rare or unique species)

Figure A Members of a RAP team examining fish specimens collected that day. (Photograph courtesy of L. Rodríguez.)

usually gives them a new perspective on the value of protected areas and their potential to attract new and compatible activities such as research and tourism.

Student participation is usually limited because the time devoted in the field to the learning process is expensive. However, it is a wonderful opportunity for training and increasing local capacities.

Presenting results to local, regional, and national key players is another important issue. Immediately after completing fieldwork and before leaving the area, a meeting with local communities to present preliminary results will inform them of the value of their area and enhance relationships. After some processing and analysis, a more elaborate report should be given to regional authorities, academics, and conservation, tourism, and other interests. A similar presentation in the capital will enable the results to be highlighted to proper decision makers. But the most important step is to involve any local institutions that will be in charge of the next steps for conservation management and oversight of the protected area.

In 1992, I was the herpetologist for the RAP to Tambopata, a large expanse of Andean forest in southeastern Peru, from lowland rainforest at 200 m to cloud forest at 2400 m (Figure B). Importantly, Tambopata also included the only piece of lowland Amazonian grasslands (pantanal) in Peru. Because Roy McDiarmid and others had studied amphibians in Tambopata in the lowlands, I assumed any new species I might find would be in the highlands. As expected, I added new species and occurrence records at a high elevation site, yet I also found a new species at

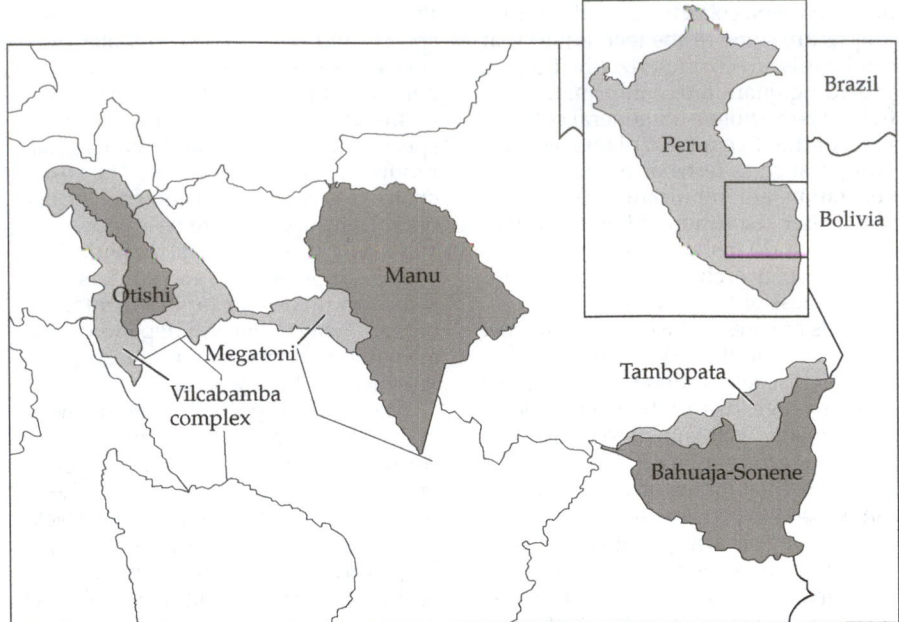

Figure B   Locations of RAP expeditions to Tambopata and Megantoni (light gray), areas adjacent to globally important national parks (Manu Biosphere Reserve, Otishi, and Bahuaja-Sonene National Parks are in gray) in southeastern Peru.

only 500 m—*Epipedobates simulans*, a poison arrow frog. Thus, RAP inventories can add new species, even in areas that have been well studied previously.

Eleven species of reptiles and amphibians, including a new snake, three lizards, and seven frogs were discovered in a more recent inventory (May 2004) of the Megantoni Reserved Zone, the strip of land connecting Manu Biosphere Reserve and the Vilcabamba complex. Data in this case supported the categorization of this area as a national sanctuary and has provided very valuable recommendations for management and zoning of the area.

Rapid biological inventories are an efficient tool for conservation. They provide quick results and recommendations that can be used immediately by decision makers. For example, the Tambopata reserved zone was created adjacent to Bahuaja-Sonene National Park, and the boundaries of this park were expanded based on the Tambopata inventory. Despite the fact that RAP/RBI is expensive and reliant on the expertise of the team, this modern tool is a good solution to fill gaps of information needed to complete protected area systems, and to enable local governments to take further steps for effective conservation. ■

serves are too small to maintain viable populations of those species with the largest aerial requirements (discussed in Chapters 7 and 14). Many of the national parks of the United States have already lost their largest mammal species, and the trend continues (Newmark 1987). Therefore, additional studies of the species–area relationship and its causes are essential to inform management of protected areas. Meanwhile, existing knowledge of species–area relationships is being used to guide establishment of protected areas. A notable example is the megareserve system of Costa Rica, which is designed to preserve about 80% of the biodiversity of the country

over the long term (Figure 2.21). Each megareserve includes natural areas and areas managed for economically valuable products. Some conservation areas remain the homes of indigenous people who continue to use the environment in their traditional ways. The largest of the Costa Rican reserves, La Amistad Biosphere Reserve, is a mosaic of more than 500,000 ha and includes three national or international parks, a large biological reserve, five Indian reservations, and two large forest reserves.

On a theoretical level, biodiversity studies continue to be needed to resolve the many uncertainties surrounding the historical and present-day ecological processes

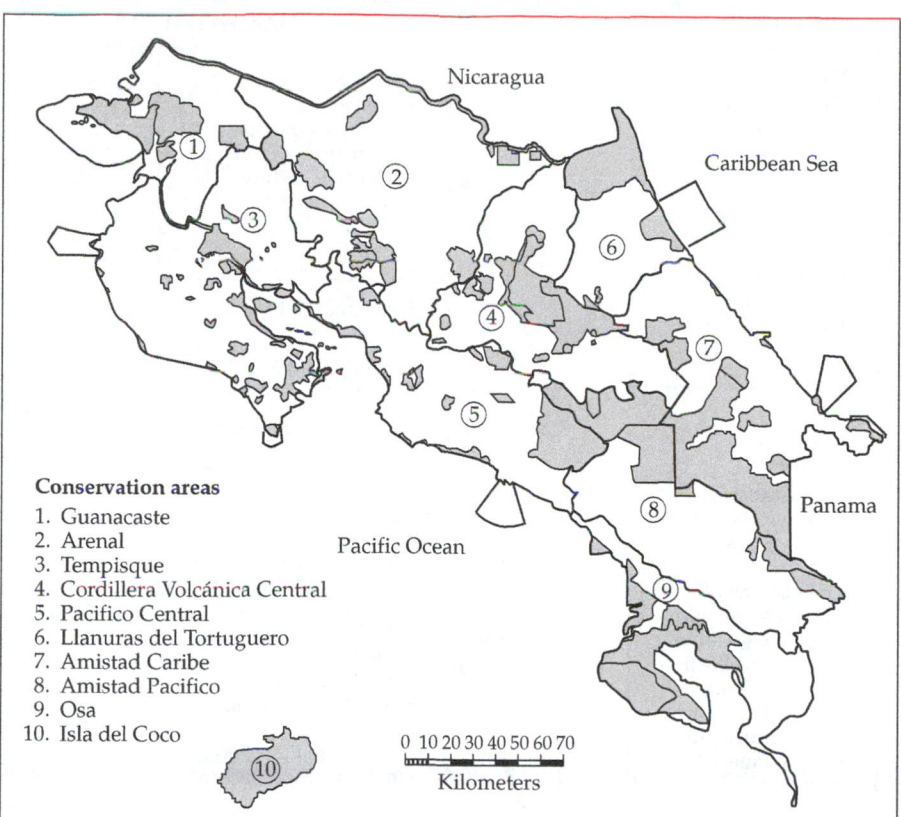

Figure 2.21 The extent of Costa Rica's conservation areas as of 1996. Shaded areas indicate protected areas, which are organized into 10 distinct conservation areas.

Conservation areas
1. Guanacaste
2. Arenal
3. Tempisque
4. Cordillera Volcánica Central
5. Pacifico Central
6. Llanuras del Tortuguero
7. Amistad Caribe
8. Amistad Pacifico
9. Osa
10. Isla del Coco

that determine today's patterns of biodiversity. Modern phylogenetic techniques that enable systematists to develop soundly based phylogenies are being combined with biogeographical studies to provide a more complete picture of the history of the distribution of life on Earth. Processes operating over ecological time frames are increasingly being studied using manipulative experiments in which some or all species are removed from restricted areas, or species are introduced. Many "natural experiments," are being studied to gain ecological insights from them—volcanic eruptions that eliminate the biotas of islands, the massive inadvertent movement of species around the world by human travel; deliberate introductions for agricultural, aesthetic, or pest-control purposes; and habitat fragmentation by conversion of natural landscapes to ones dominated by highly modified communities that are managed to channel most of their productivity to human uses. Such scenarios provide opportunities to examine the results of manipulations over longer time frames than is possible with investigator-initiated experiments.

Studies of the influences of human activities on species distributions and species richness are adding rapidly to our understanding of the roles of the varied processes that interact to cause the patterns in the distribution of biodi-

versity on Earth that so fascinate us today. This knowledge is also being used to attempt to reduce the rates of species extinction and to restore landscapes so that they can continue to support the array of species originally found in them and the evolutionary processes that generate new species. For example, to promote survival of existing species and the evolutionary processes that generate new species, devoting large areas to biodiversity preservation will be critical.

Recently, the Millennium Ecosystem Assessment (MA), a large coalition of international development and conservation organizations, governments, and scientists has come together to assess the status of Earth's ecosystems, the goods and services they provide, and the likely effects of potential pathways of human economic development on the future provisioning of these services and human well-being (Figure 2.22). The MA focuses both globally, and on subglobal regions of particular concern due to the difficulty of human existence or potential for serious declines in human welfare in these regions. Published in 2005, this is the first comprehensive assessment of the condition and status of global ecosystem services. The information summarized in the MA will be used to guide development policy both regionally and globally. In addition, the MA will help focus re-

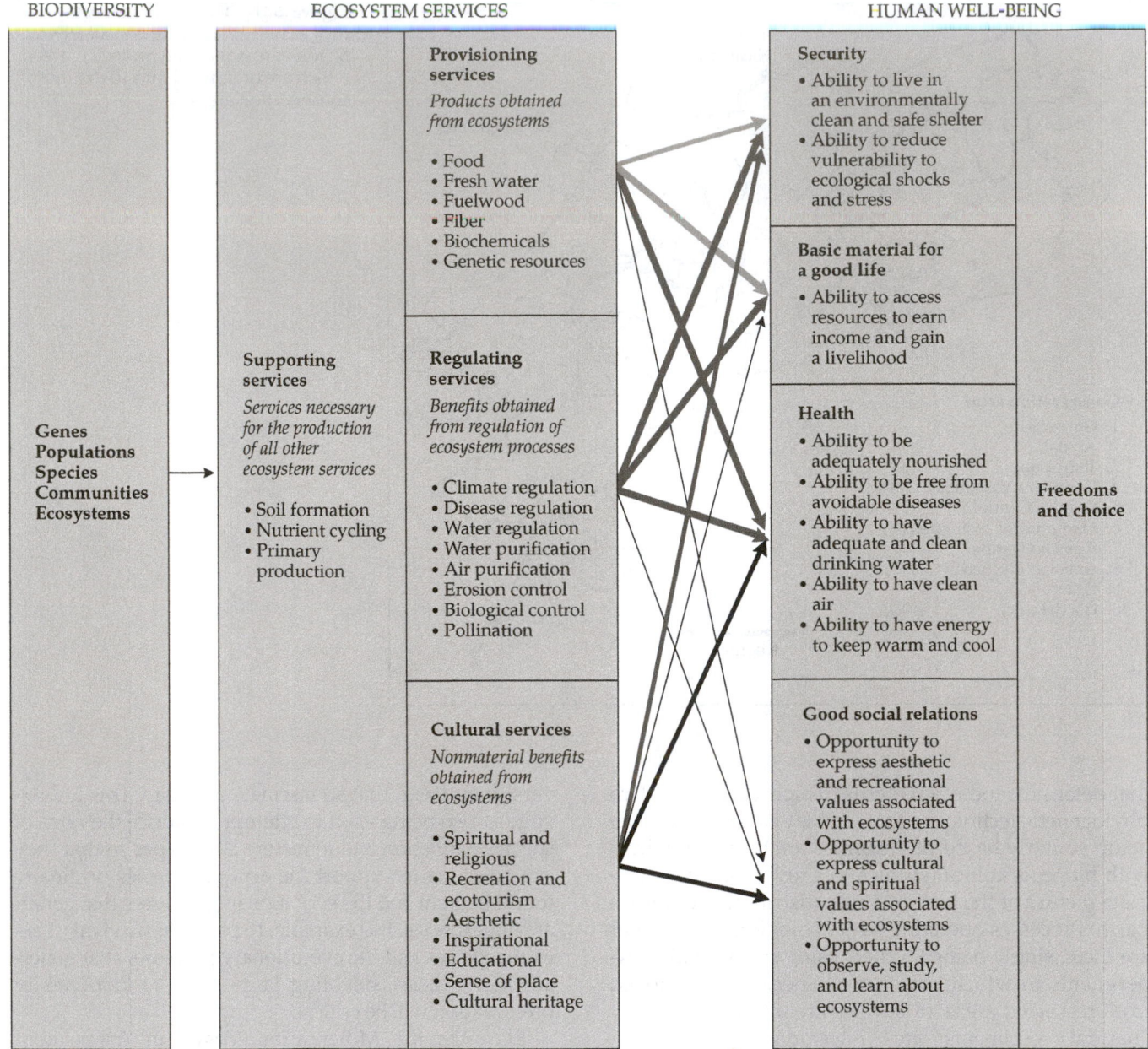

BIODIVERSITY   ECOSYSTEM SERVICES   HUMAN WELL-BEING

Figure 2.22 Guiding framework for the Millennium Ecosystem Assessment, showing linkages and feedback among biodiversity, ecosystem processes, and human welfare. Thicker arrows indicate a stronger linkage between ecosystem services and constituents of human well-being. (Modified from Millennium Ecosystem Assessment 2003.)

search on the connections between the status of biodiversity and ecosystem services, as well as the reciprocal influences of biodiversity and human welfare.

Summary

1. Biodiversity is a term that refers to the variety of living organisms, their genetic diversity, the diversity of evolutionary lineages, and the types of ecological communities into which they are assem-

bled. Different measures of biodiversity emphasize, respectively, number, evenness, and difference. The number of species—species richness—found in different areas is the measure of biodiversity most commonly used in conservation biology, both because accurate information on abundances is typically not available and because rare species are often of great conservation interest. Ecologists commonly use measures of biodiversity weighted by abundance, biomass, or productivity because more

common species typically dominate ecosystem processes.

2. About 1.75 million living and 300,000 fossil species have been described, but estimates of the total number of living species range from 10 million to as high as 50 million or more. Although species richness is higher on land than in the oceans, 34 of the 35 extant animal phyla have marine representatives, and 16 are exclusively marine.

3. Cellular life in the form of bacteria evolved about 3.8 billion years ago; eukaryotic organisms evolved about 2 billion years ago. The first major explosion of biodiversity took place in the early Cambrian period, and except during times of mass extinctions, the number of species has increased since then. More species are probably alive today than at any other time in the history of life, even though some taxa had more species in the past than they do today. The number of species present at any moment is the result of the difference in the rate of formation of new species and the rate of extinction of existing species. High rates of speciation have been favored by mass extinctions, the breakup of the continents, and the evolution of more diverse body plans that enabled animals to burrow, swim, and fly.

4. Several broad patterns characterize the current distribution of species. Areas that have experienced long geographical isolation and that have great topographic relief often support many endemic species. Tropical regions often have both high species richness and endemism, and islands have high endemism of both marine and terrestrial species.

5. Among most taxa, more species live in tropical regions than at higher latitudes. The large land area of the Tropics contributes to this pattern. Species richness is also positively correlated with available energy, which enables more individuals to be supported per unit area, and with structural complexity, which is positively correlated with available energy. On land, structure is provided primarily by vascular plants, whereas in many marine communities, animals such as corals generate most structure. There is a positive correlation between productivity and species richness at low levels of productivity, but species richness typically declines at higher productivities. Some highly productive systems, such as salt marshes, seagrass beds, and hot springs are species-poor. Plant species richness is extremely high in some unproductive semiarid regions with poor soils. Island communities are poorer in species than comparable mainland communities at all latitudes.

6. Climatic oscillations producing swift, radical changes in climate increased extinction rates and likely decreased speciation rates toward the poles, thus acting as a strong driver of latitudinal patterns in species richness.

7. Both physical (heavy rains, strong winds, landslides, earthquakes, and fires) and biological disturbances (tree-falls and activities of predators and competitors) influence species richness but their effects are highly varied. Most disturbances are small and directly affect only small areas, but they may influence more distant regions via indirect effects. Strong competition and predation may eliminate species from particular areas but intermediate levels of predation, by preventing competitive exclusion, may increase local species richness.

8. Despite the demonstrated importance of local biological interactions, local species richness in many taxa appears to increase without apparent limit along with increasing regional species richness. This surprising result suggests either that local interactions exert strong effects on abundances of species but not on their numbers, or that the influence of regional processes on numbers of species in local communities is strong enough to override local influences.

9. The state of knowledge of Earth's biodiversity is so poor that the primary work of cataloging biodiversity is yet to be done. Resources currently devoted to this task are inadequate, especially given the rate at which species are becoming extinct. Better information on biodiversity, its distribution, and its causes is needed for wise management of Earth's biotic resources. Further, research that allows us to understand the contributions of biodiversity to human welfare, and the effects of development on ecosystem services is needed to help us preserve the richness of biodiversity and human existence.

> Please refer to the website www.sinauer.com/groom for Suggested Readings, Web links, additional questions, and supplementary resources.

Questions for Discussion

1. The history of life has been punctuated by five episodes during which extinction rates were very high. If extinction is a normal process, and if life has rediversified after each mass extinction, why should we be worried about the prospects of high extinction rates during this century? How does the current extinction spasm differ from previous ones?

2. Given that millions of species are yet to be described and named, how should the limited human and financial resources available for taxonomic research be allocated? Should attention be concentrated on poorly known taxa? Should efforts be directed toward areas threatened with habitat destruction so that species can be collected before they are eliminated? Should major efforts be directed to obtain complete "all taxa" surveys of selected areas? How and by whom should these decisions be made?

3. Indices of species diversity that are weighted by abundance, biomass, or productivity are used frequently by ecologists, but seldom by conservation biologists. For what purposes might conservation biologists wish to use weighted indices instead of simple lists of species?

4. Many conservation efforts are directed at particular local areas harboring rare species or having high species richness. Why is concentrating only on local problems insufficient as an effective conservation strategy?

5. For which animal taxa would you expect species richness to be most positively correlated with plant community structure? Mammals? Amphibians? Insects? Why?

3

Threats to Biodiversity

Martha J. Groom

Extinction is the most irreversible and tragic of all environmental calamities. With each plant and animal species that disappears, a precious part of creation is callously erased.

Michael Soulé, 2004

Human impacts are now a pervasive facet of life on Earth. All realms—terrestrial, marine, and freshwater—bear our imprint; our pollution spans the globe, our fisheries extend throughout the world's oceans, and our feet tread across almost every surface on Earth. By many estimates, we use substantial and increasing fractions of Earth's primary productivity (Vitousek et al. 1986; Pauly and Christensen 1995; Postel et al. 1996), and our total ecological impact may already extend beyond Earth's capacity to provide resources and absorb our wastes (Wackernagel and Rees 1996).

As humans became a widespread and numerous species, our agricultural expansion forever changed vast landscapes; our hunting and our transport of invasive, commensal species drove numerous aquatic and terrestrial species extinct. When highly organized societies began to settle and grow throughout the globe, the pace of transformation of terrestrial and aquatic habitats sharply increased, and our use of natural resources began to dramatically outstrip natural rates of replacement. Thus, humans have had enormous impacts on the form and diversity of ecosystems. Ultimately, we have set in motion the sixth great mass extinction event in the history of the Earth—and the only one caused by a living species.

Human population and consumption pressures are the root threat to biodiversity (see Chapter 1; Figure 3.1). Increasing numbers of humans, and most importantly, increasing levels of consumption by humans create the conditions that endanger the existence of many species and ecosystems: habitat degradation and loss, habitat fragmentation, overexploitation, spread of invasive species, pollution, and global climate change. Species extinction, endangerment, and ecosystem degradation are not the aims of human societies, but are the unfortunate by-product of human activities. Because our practices are unsustainable, we strongly erode the natural capital that we have used to flourish, thus endangering our and our descendents' future.

Figure 3.1 Major forces that threaten biological diversity. All arise from increases in human population and consumption levels, often mediated through our activities on the land and sea. Extinction and severe ecosystem degradation generally result from multiple impacts and from synergistic interactions among these threats.

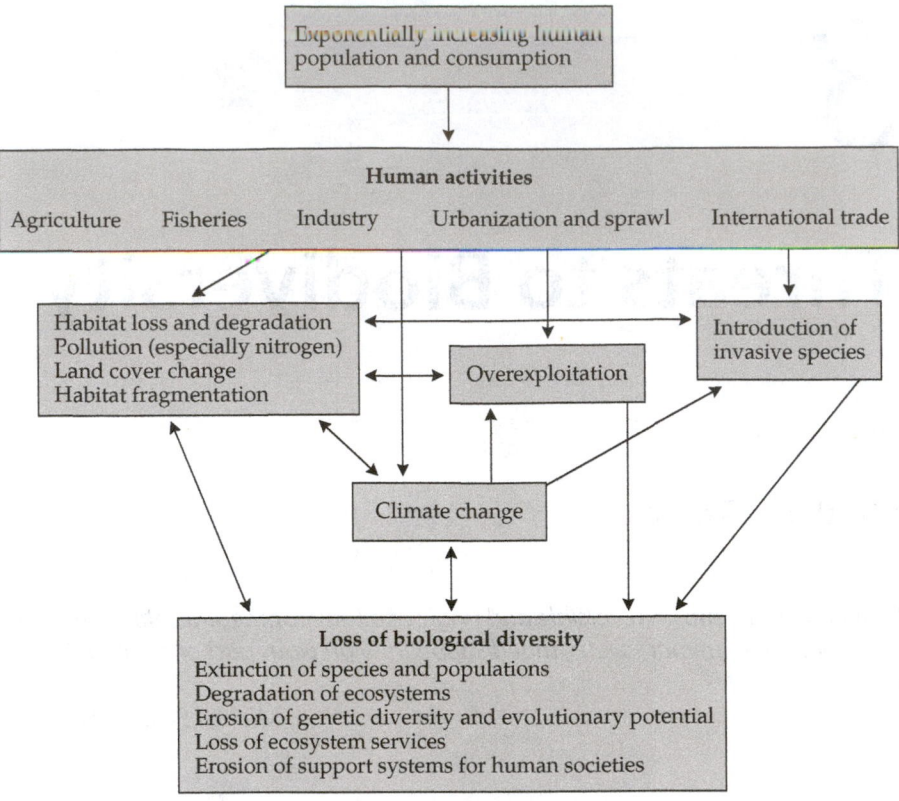

Major Threats to Biodiversity and Their Interaction

In this chapter, I provide an overview of patterns of extinction and species endangerment, and describe efforts employed to slow and reverse these trends. The first step is to review the major types of threats to biodiversity, while laying the groundwork for the more in-depth coverage of these topics later in the book.

Habitat degradation includes the spectrum of total conversion from a usable to an unusable habitat type (or "habitat loss"), severe degradation and pollution that makes a habitat more dangerous or difficult for an organism to live in, and fragmentation that can reduce population viability. Habitat degradation can be caused by a host of human activities including industry, agriculture, forestry, aquaculture, fishing, mining, sediment and groundwater extraction, infrastructure development, and habitat modification as a result of species introductions, changes in native species abundance, or changes in fire or other natural disturbance regimes. In addition, most forms of pollution affect biodiversity via their degradation of ecosystems. Chapter 6 contains a full discussion of the impacts of various forms of habitat loss and degradation, while the phenomenon and effects of habitat fragmentation are detailed in Chapter 7.

Overexploitation, including hunting, collecting, fisheries and fisheries by-catch, and the impacts of trade in species and species' parts, constitutes a major threat to biodiversity. Most obviously, a direct impact of overexploitation is the global or local extinction of species or populations. Less obvious, the decrease in population sizes with exploitation can lead to a cascade of effects that may alter the composition and functionality of entire ecosystems (Estes et al. 1989; Redford 1992; Pauly et al. 1998). Overexploitation is discussed in detail in Chapter 8.

The spread of **invasive species**, species that invade or are introduced to an area or habitat where they do not naturally occur, is also a significant threat to biodiversity. Invasive species can compromise native species through direct interactions (e.g., predation, parasitism, disease, competition, or hybridization), and also through indirect paths (e.g., disruption of mutualisms, changing abundances or dynamics of native species, or modifying habitat to reduce habitat quality). The process and impacts of species invasions are described in detail in Chapter 9.

Anthropogenic climate change is perhaps the most ominous threat to biodiversity of the present era. Climate change appears to have caused mass extinctions seen in the geologic record, and because the pace of climate change is predicted to be at least as fast and extreme as the most severe shifts in climate in the geologic

record, the effect on biodiversity is expected to be enormous. Coupled with the extensive transformation of Earth's ecosystems, widespread overexploitation of populations, and introductions of species to new areas in the globe, we can expect the effects of future shifts in climate to usher in an extremely severe mass extinction event. The probable biological impacts of the climate changes underway today are examined in full in Chapter 10.

As we are becoming more aware of the global impacts of climate change, we also are learning more about the global extent of pollution. We now recognize that in addition to direct discharge of chemicals into the environment, many are circulated atmospherically. Toxic compounds, such as lead, mercury, and other heavy metals, are found in trace amounts even in remote areas of Antarctica and the Arctic (Bargagli 2000; Clarke and Harris 2003), and are transported from industrial sources through the atmosphere. Importantly, many compounds can have subtle yet profound impacts on the endocrine systems of wild animals (see Chapter 6). Since the 1960s we have been aware of the dangerous consequences from many noxious chemicals, particularly those that **bioaccumulate** or magnify in

the food chain (Figure 3.2). In Essay 3.1, Peter Ross describes potential impacts of one particularly noxious class of pollutants—persistent organic pollutants (POPs)—on killer whales (*Orcinus orca*). POPs include the infamous DDT, which caused the decline of many raptor populations via eggshell thinning.

Diseases are also becoming more widespread, and our recognition of their impacts is increasing. Among the most noticeable and worrisome are outbreaks of disease that decimate coral reefs; as corals die, the complex community of fishes, algae, and invertebrate species is compromised (Harvell et al. 1999). Andy Dobson describes several examples of how diseases threaten populations and indirectly play an enormous role in altering ecosystems (Essay 3.2). Often, diseases become more dangerous as a result of interactions with the stresses caused by pollutants.

Typically, species and ecosystems face multiple threats. Importantly, the joint effects of several threats may be the ultimate cause of biodiversity losses. For example, the Dodo (*Raphus cucullatus*) went extinct on Mauritius in 1681 due to human overexploitation com-

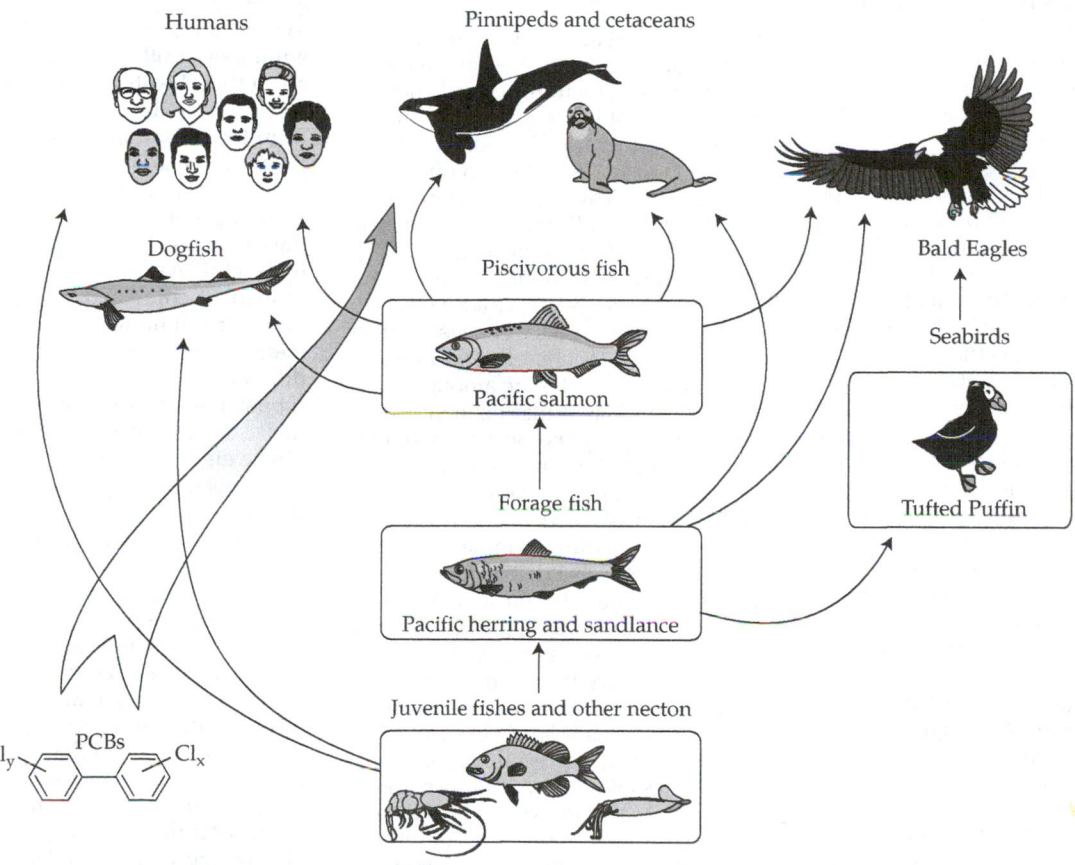

Figure 3.2 Toxic chemicals that accumulate in fatty tissues, such as PCBs and dioxins, concentrate in the tissues of organisms at the top of the food chain. (Modified from Ross and Birnbam 2003.)

ESSAY 3.1

Killer Whales as Sentinels of Global Pollution

Peter S. Ross, *Institute of Ocean Sciences, Canada*

■ Persistent organic pollutants (POPs) comprise a large number of industrial and agricultural chemicals and by-products. While these chemicals have widely varying applications, they share three key features: They are *persistent, bioaccumulative,* and *toxic* (PBT). The physico-chemical characteristics of these chemicals dictate the degree to which they break down in the environment (persistence), the degree to which they are metabolically broken down or to which they accumulate in organisms (bioaccumulative potential), and the degree to which they bind to certain cellular receptors or mimic natural (endogenous) hormones in vertebrates (toxicity).

Marine mammals are often considered vulnerable to the accumulation of high concentrations of POPs as a result of their high position in aquatic food chains, their long life spans, and their relative inability to eliminate these contaminants. Because POPs are oily (lipophilic), they are easily incorporated into organic matter and the fatty cell membranes of bacteria, phytoplankton, and invertebrates at the bottom of the food chain. As these components are grazed upon by small fishes and other organisms at low trophic levels, both the lipids and the POPs are consumed. In turn, these small fishes are consumed by larger fishes, seabirds, and marine mammals that occupy higher positions in aquatic food chains. However, lipids are burned off at each trophic level and are utilized for maintenance, growth, and development, while the POPs are left largely intact. This leads to biomagnification, with increasing concentrations of POPs found at each trophic level. In this way, fish-eating mammals and birds are often exposed to high levels of POPs, even in remote parts of the world.

The killer whale (*Orcinus orca*) is one of the most widely distributed mammals on the planet. Although elusive and poorly studied in many parts of the world, these large dolphins have been the subject of ongoing study in the coastal waters of British Columbia, Canada, and Washington State. A long-standing photo-identification catalogue based on unique markings has facilitated the study of populations in this

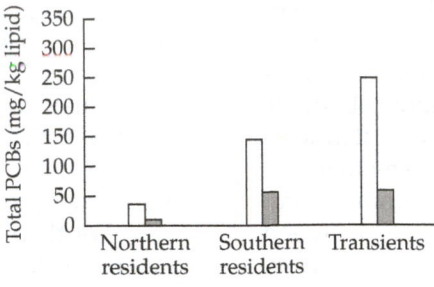

Figure A   Transient killer whales represent the most PCB-contaminated marine mammal on the planet, reflecting their high trophic level (they consume marine mammals), long life span, and relative inability to eliminate these contaminants from their bodies. Females (in gray) are less contaminated than males (in white) because they transfer these contaminants to their offspring via fat-rich milk. (Modified from Ross et al. 2000.)

region (Ford et al. 2000). Several communities, or ecotypes, frequent these coastal waters, including the salmon-eating resident killer whales, the marine mammal-eating transient killer whales, and the poorly characterized offshore killer whales. There are two communities of resident killer whales: the northern residents that ply the waters of northern British Columbia, and the southern residents that straddle the international boundary between British Columbia and Washington. Our discovery that killer whales are among the most contaminated marine mammals in the world highlights concerns about the way in which POPs move great distances around the planet with relative ease (Ross et al. 2000; Figure A).

Studying marine mammal toxicology is not easy. Because deceased individuals are not generally considered representative of the free-ranging population, biopsy samplings of blubber from live individuals have increasingly been used to generate high-quality information about contaminant levels in their tissues. However, the resulting data are meaningless if nothing is known about the individual from which the biopsy sample originates. Age, sex, condition, and dietary preferences all represent important "confounding" or "natural" factors that affect the concentration of contaminants in the animal's tissue. In

the case of killer whales, the photo-identification catalogue provides such a backdrop, with each sample originating from an individual of known age, sex, and community (hence dietary preference). Consequently, we were able to document that males became increasingly contaminated as they aged, while females became less contaminated as they transferred the majority of their contaminant burden to their nursing calves via fat-rich milk.

Many POPs, including the polychlorinated biphenyls or PCBs and the pesticide DDT, are highly toxic. Laboratory animal studies have conclusively demonstrated that such chemicals are endocrine disrupting, with effects noted on reproduction, the immune system, and normal growth and development. Studies of wildlife are more challenging, as free-ranging populations are exposed to thousands of different chemicals, and a number of other natural factors can affect their health. Captive-feeding studies have demonstrated that herring from the contaminated Baltic Sea affect the immune and endocrine systems of harbor seals (Ross et al. 1996). Studies of wild populations provide clues about the impact of POPs on marine mammals. However, as is the case with humans, a combined "weight of evidence" from numerous lines of experimental and observational evidence in different species provides the most robust assessment of health risks in animals such as killer whales (Ross 2000). This weight of evidence is based on inter-species extrapolation, and depends upon the conserved nature of many organ, endocrine, and immunological systems among vertebrates.

While regulations have resulted in many improvements for certain POPs, new chemicals are designed each year. Killer whales represent a warning about those chemicals that are unintentionally released, that may travel great distances through the air, and that end up at high concentrations at the top of food chains. Given their tremendous habitat requirements and that of their prey (salmon), Northeastern Pacific killer whales are serving as sentinels of global pollution, and a reminder that we indeed inhabit a "global village." ■

ESSAY 3.2

Infectious Disease and the Conservation of Biodiversity

Andy Dobson, *Princeton University*

Dissect any vertebrate, invertebrate, plant, or fungus and you will find a whole community of organisms living within their tissues; the richness of this community of parasites will increase as you examine the host's tissues at finer scales of resolution. Unfortunately, ecologists and conservation biologists often overlook the huge variety of biodiversity that lives in, upon, and often at the expense of free-living species. Parasites and microorganisms are a major component of biodiversity, perhaps making up as much as 50% of all living species (Price 1980; Toft 1991). Ironically, whereas few people worry about their long-term conservation (Sprent 1992), it is important not to ignore parasites and pathogens, as many have profound effects that fundamentally influence the evolution of populations and the functioning of ecosystems.

The Zen of Parasitism

Parasites become problems when natural systems are perturbed, or when species escape regulation by their parasites; an important illustration of this occurs when invasive species escape from their natural pathogens. Green crabs (*Carcinus maenus*) host a significant diversity of parasites in their natural range along the Atlantic coast of Europe (Torchin et al. 2002). This parasite diversity is considerably reduced in the many areas of the world where green crabs have been accidentally introduced and have established as invading species. This absence of pathogens may make a significant contribution to the crab's ability to invade, as it considerably reduces the energy each crab puts into resisting the invasion of its body by a diversity of parasitic species. Indeed, in areas where crabs have successfully invaded they may grow to five times the size of the largest crabs found in their native range.

This effect appears to be an important general result; in detailed comparative studies of the most successful invasive animal species from a variety of taxa, Torchin et al. (2003) showed that parasite diversity is considerably reduced in areas where the species has invaded. Similar results were found in the fungal and viral pathogens of

plants that have invaded the United States (Mitchell and Power 2003). All of this suggests that the parasitic, under-observed half of biodiversity plays a major role in regulating the abundance of the more familiar free-living species.

Parasites can have dramatic effects at ecosystem levels. Rinderpest virus (RPV), a morbillivirus that causes widespread mortality in ungulates, was first introduced into East Africa at the end of last century with cattle (Plowright 1982). RPV spread throughout sub-Saharan Africa, producing mortality as high as 90% in some species. Travelers through the region report that in some places the ground was littered with carcasses and the vultures were so satiated they could not take off (Simon 1962). Even today, the observed geographic ranges of some species are thought to reflect the impact of the great rinderpest pandemic. Vaccination of cattle produced a remarkable and unforseen effect; the incidence of RPV in wildebeest and buffalo declined rapidly and calf survival in these and other wild ungulates increased significantly (Talbot and Talbot 1963; Plowright 1982). This led to a rapid increase in the density of these species; in the Serengeti, wildebeest numbers increased from 250,000 to over a million between 1962 and 1976, and buffalo numbers nearly doubled over the same period and expanded their range. This increase in herbivore density produced a significant increase in some carnivore species, particularly lions and hyenas.

A significant threat to endangered species may be pathogens acquired from species with large populations that sustain continued infections. In this case, the pathogen is present in one host species and invades another, and two things can happen: The combined population densities of the potential host species may be insufficient to sustain the pathogen and it dies out, or the parasite sustains itself in the new community of hosts.

Pathogens with Multiple Hosts

When pathogens infect multiple host species it is likely that some hosts are more resistant to the pathogen than others; West Nile virus provides an impor-

tant example. Crow species are highly susceptible to the disease and die within a week of infection; in contrast, House Sparrows (*Passer domesticus*) seem more able to withstand infection (interestingly, House Sparrows are an invading species in the U.S.; their native range overlaps that of West Nile virus, suggesting the ghost of past natural selection for resistance [Campbell et al. 2002]).

At a further extreme, Nipah and Hendra virus have caused deadly outbreaks of disease in humans and domestic livestock in Australia, Malaysia, and Bangladesh. The main hosts of these viruses are fruit bats (*Pteropus hypomelanus*, *P. vampyrus*, and several other species), where the high levels of prevalence imply that they exist as a relatively benign pathogen. In undisturbed habitats, there is little contact between humans, their livestock, and bats. However, massive habitat conversion in Australia and Malaysia has compressed the range of fruit bats so that the only trees left for them to roost in are those associated with intensive agricultural areas. This increases rates of contact between bats and livestock or humans. Pathogen transmission usually occurs when fruits that the bats have been feeding on drop from trees and are consumed by pigs or horses. These partially infected fruits can trigger the first case of a disease outbreak, which spreads through the livestock into the agricultural workers and on to their families and friends. When this occurred in Malaysia, several million pigs had to be culled, and over a hundred humans were infected, more than 65% of whom died.

Pathogens that use multiple hosts create a double-edged problem for ecologists and conservation biologists. Pathogens like Nipah virus represent one extreme, where habitat conversion increases human exposure to novel pathogens. There is essentially no way of predicting when similar novel pathogens will emerge; all we can say is that the frequency of these events will increase as humans increase their rates of contact with novel environments and their potential hosts.

At the other extreme are multi-host pathogens that are vector transmitted and use mosquitoes, ticks, and fleas for

transmission between wildlife reservoirs and humans and their domestic live-stock. The specificity of these pathogens is determined by the feeding choice of their vectors. In some cases this choice will be very specific and the pathogen will only occasionally spread from the reservoir into a novel host. This seems to be the case for West Nile virus, which is less of a problem in Europe, where it is transmitted predominantly by mosquitoes that specialize in birds. In the U.S., West Nile virus is transmitted by a hybrid mosquito that feeds on both birds and mammals. However, when vectors have a choice of hosts, the diversity of species present in a habitat will have a major buffering impact on the scale of an epidemic outbreak. Ostfeld and colleagues have shown that this is the case for Lyme disease in the U.S., where the diversity of hosts available for ticks leads to a significant reduction in the rate of attack on hosts that are susceptible to the disease (Ostfeld and Keesing 2000; LoGuidice et al. 2003). This buffering effect is even stronger for pathogens where the abundance of the vectors is independent of that of the hosts (Dobson 2004); this will be the case for mosquito-transmitted diseases such as malaria, Dengue fever, and West Nile virus. This creates an important incentive to conserve biodiversity, particularly in the event of future climate warming that will allow the classic tropical pathogens to spread to the temperate zone.

Most discussion of the response of vectored pathogens to climate change has focused on examining how increased temperature allow pathogens and mosquitoes to successfully complete their life cycle development in the temperate zone. Once the pathogen can develop in a shorter time period than the mosquitoes' life expectancy, the pathogen can establish. However, vectored pathogens will also be moving down a biodiversity gradient as they spread from the tropics to the temperate zone. There will be less choice for the mosquitoes, so they can focus their infective bites on the most common species; in many places this

will be *Homo sapiens* and their commensal domestic species.

Role of Predators in Buffering Outbreaks and Keeping Herds Healthy

Predators and scavengers may provide an unsuspected ecosystem service by preventing infectious disease outbreaks. Work by Packer et al. (2003) suggests that when predators selectively remove infected individuals from populations of prey species they will significantly reduce the burden of disease within the prey population; this may even lead to increases in the abundance of prey in the presence of predators! Evidence in support of this is provided in populations of game species where culling of predators has led to increases in infectious diseases and parasites as the host becomes more abundant. In studies of game birds in northern Britain the abundance of parasitic worms varies inversely with gamekeeper abundance (Hudson et al. 1992; Dobson and Hudson 1994). One of the gamekeeper's traditional jobs is to remove foxes and birds of prey; however these predators differentially remove heavily parasitized birds from the bird populations. Removing the foxes leads to a general increase in parasitic worm burdens that have a major impact on bird numbers.

In a similar fashion, the control of scavengers such as wolves, coyotes, and vultures may have permitted the emergence of prion diseases such as scrapie and chronic wasting disease in Europe and the U.S. (Prusiner 1994; Westaway et al. 1995). The natural foci of prion diseases seem to be areas characterized by very poor soils such as chalk grasslands. Ungulates that graze in these habitats are extremely nutrient stressed; they are particularly deficient in phosphorus and this leads them to chew on the carcasses of individuals who have failed to survive harsh winters. When scavenging canids are removed there are considerably more carcasses to chew on, and prion diseases that are present at very low levels in the population can slowly become

endemic at higher prevalences. Once they get into domestic livestock, they can produce devastating economic impacts (Anderson et al. 1996). They provide an important example of how previously obscure and little-known pathogens can quickly become quite significant when we perturb natural ecosystems.

To summarize, parasites and pathogens remain an important consideration in the management of captive and free-living populations of threatened and endangered species. Epidemiological theory suggests that pathogens shared among several species present a larger threat to the viability of endangered species than do specific pathogens. However, there is a way parasites and pathogens may be used to conservation advantage: Pathogens could be effectively employed as biological control agents to reduce the densities of introduced rats, cats, and goats that are a major threat to many endangered island species. Obviously, caution has to be exercised when considering introduction of any pathogen into the wild, so this method of pest control should be restricted to isolated oceanic islands (Dobson 1988). However, the majority of extinctions recorded to date in wild populations have occurred on oceanic islands (Diamond 1989).

Clearly, parasites and diseases are emerging as important considerations in conservation biology. The enormous expansion of our ecological understanding of parasites and their hosts in the last fifteen years means that ecologists now see a predictable structure in conditions that foster disease outbreaks. Epidemics can no longer be considered purely stochastic events that occur as random catastrophes. We now have a significant mathematical framework that delineates the general conditions under which a disease outbreak will occur (Anderson and May 1986, 1991; Grenfell and Dobson 1995). A major challenge for conservation biologists is to apply and extend this framework so it can minimize the disease risk to endangered species of plants and animals. ■

---

bined with nest predation by introduced cats, dogs, pigs, and rats. Of great concern is the likelihood that some threats may be **synergistic**, whereby the total impact of two or more threats is greater than what you would expect from their independent impacts (Myers 1987). Corals often are stressed physiologically by increases in

temperature, but may also be more susceptible to fungal pathogens when stressed (Harvell et al. 1999). This synergism suggests that the combined effects of global warming and increasing transport of disease organisms among coral reefs could precipitate catastrophic declines. At times, synergisms develop through the inter-

action of a threat and population density. In an experimental study, Linke-Gamenick and her colleagues (1999) showed that a toxic chemical reduced survivorship of a capitellid polycheate worm, but at high concentrations its impacts grew more severe with increasing density.

Further, many threats can intensify as they progress, a process known as "snowballing." Invasion of plant communities in western Australia by the alien root pathogen *Phytophthora* causes death of woody species and an increase in herbaceous cover, which can suppress germination and early growth of woody seedlings (Richardson et al. 1996; Figure 3.3). The changes in the plant community in turn decrease the suitability of the community for many animal species, which may further reduce the capacity of the animal community to foster the development of woody cover. Thus, the initial invasion pushes the entire system into a new balance that makes recovery of the original system difficult.

Finally, many species are threatened by interactions between large impacts, such as direct mortality from harvest, and more subtle impacts on their biology and population dynamics. For example, a variety of changes in the genetic structure of populations can enhance their risk of extinction. Species may lose functional genetic diversity due to prolonged isolation in small populations, or the loss of entire populations, which may leave them less able to cope with stresses of habitat degradation or climate change. These genetic threats, as well as many genetic tools for conservation are discussed in Chapter 11. Intrinsic demographic factors, such as rarity, low reproductive rates, or low dispersal rates can further predispose a population or species to extinction, as discussed later in this

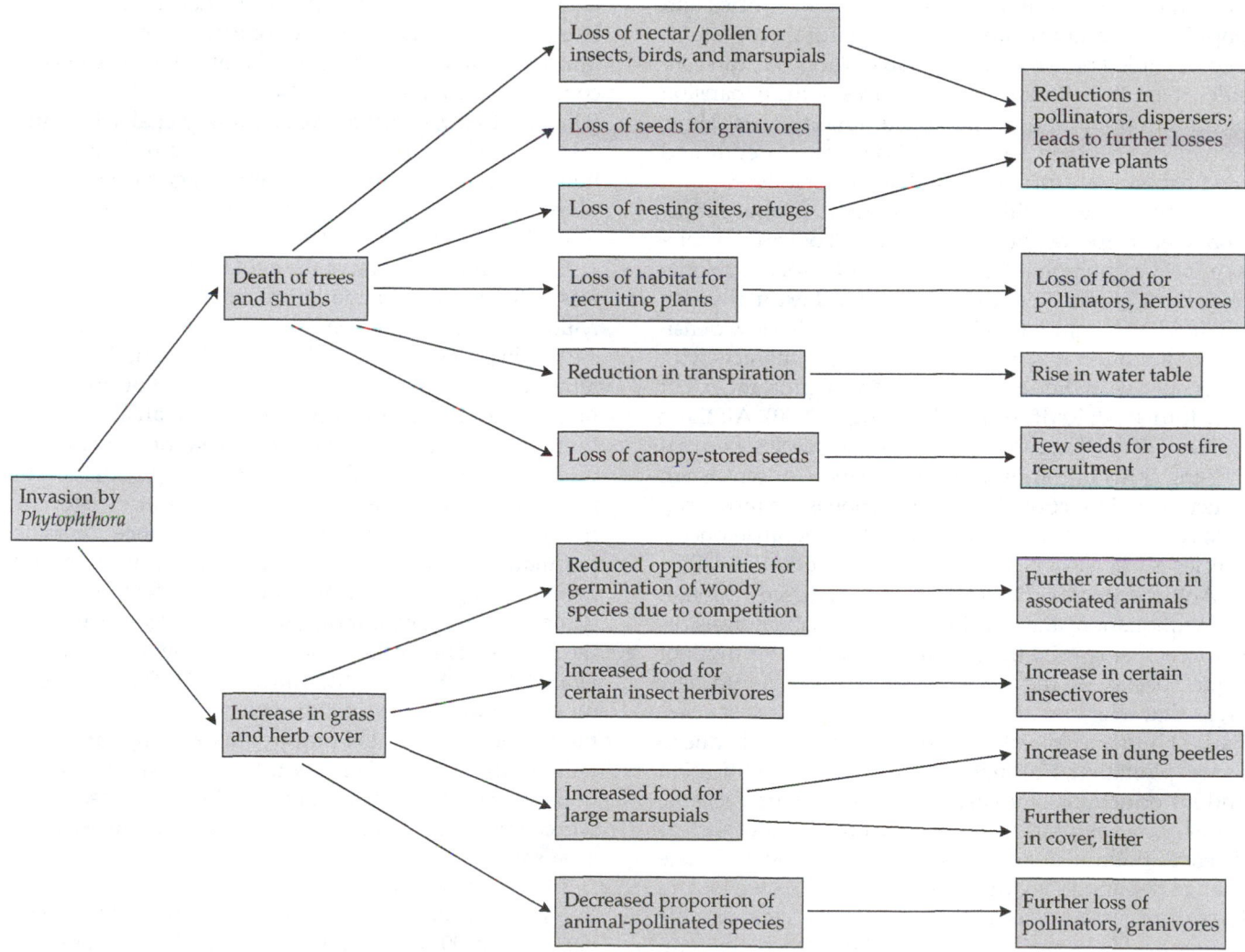

Figure 3.3   "Snowballing" effect of invasion of the alien root pathogen *Phytophthora cinnamomi* into shrublands and woodlands of western Australia. (Modified from Richardson et al. 1996.)

chapter and in Chapter 12. Importantly, species may vary in their responses to different threats based on their biology (Owens and Bennett 2000; Isaac and Cowlishaw 2004). Such demographic predispositions to risk can be exaggerated via reduced reproduction or increased mortality due to anthropogenic factors. Typically, both genetic and demographic threats interact with habitat and climate change, overexploitation, or species invasion to cause species loss, and ultimately changes in community and ecosystem function. Thus, although rarely the primary cause of extinction, genetic and demographic factors are important contributors to endangerment that must be considered in most conservation situations.

Anthropogenic Extinctions and Their Community and Ecosystem Impacts

The most obvious and extreme unwanted effects of human development are the extinction of species and populations, and transformation of natural ecosystems into degraded or even uninhabitable places. We can consider these the ultimate consequences of the expansion of human populations and consumption levels. Here, we will discuss patterns of extinction in some depth, and treat habitat transformation in detail in Chapter 6.

Extinction can be either global or local. **Global extinction** refers to the loss of a species from all of Earth, whereas a **local extinction** refers to the loss of a species in only one site or region. In addition, **ecological extinction** can occur when a population is reduced to such a low density that although it is present, it no longer interacts with other species in the community to any significant extent (Redford 1992; Redford and Feinsinger 2000). All these forms of extinction can affect remaining species, perhaps causing shifts in community composition, or ecosystem structure and function. Global extinction is the most tragic loss resulting from human activities, because once a species is lost entirely, it cannot be recreated.

Anthropogenic extinctions are caused directly through overexploitation, and also indirectly via habitat transformations that restrict the population size and growth of some species, or the introduction of species into new areas that over-consume or outcompete native species. Earliest human- caused extinctions were probably due to overexploitation, but increasingly habitat modification and introductions of invasive species to islands became primary causes. Only recently have other factors such as disease, pollution, and anthropogenic global climate change begun to play major roles as well. As we look to the future, synergisms among these factors and climate change are likely to accelerate extinction rates in the coming century (Myers 1987; Myers and Knoll 2001).

Because of the inherent spottiness of the fossil record, it is difficult to discern extinction events prior to recorded history, or to document the nature of changes to ecological communities. Our knowledge of prehistoric effects of humans on biodiversity is thus limited to a few cases where the fossil record can provide a clear trail, and to fairly gross-scale changes in ecosystems. Similarly, our incomplete knowledge of living taxa also makes this task difficult even after ecological records were kept in detail. Yet, some patterns are traceable.

Most notably, the Pleistocene extinction of megafauna (mammals, birds, and reptiles over 44 kg in body size) and other vertebrates speaks of widespread impacts of humans that have forever changed ecological communities. The demise of between 72% and 88% of the genera of large mammals in Australia, North America, Mesoamerica, and South America coincided with the arrival of humans in each continental region (44,000–72,000 years ago in Australia and 10,000–15,000 years ago in North and South America; Figure 3.4). Certainly the coincidence of pulses of human colonization or population growth and the loss of taxa is suggestive of a strong role of "Earth's most ingenious predator" in the loss of these creatures (Steadman and Martin 2003). However, rapid climate change and concomitant vegetation change also took place in most cases (Diamond 1989; Guthrie 2003; Barnosky et al. 2004), and loss of many taxa in Alaska or Northern Asia where human populations were never large suggests that climate played a major, or in a few cases the only role in megafaunal extinctions (Barnosky et al. 2004).

Careful consideration of the evidence suggests that the loss of the megafauna may have resulted from a combination of range contraction due to climate change and decreases in population sizes due to hunting and habitat alteration by humans (see Figure 3.4). For example, mammoths often survived longest in areas without human populations. Although glaciations eventually caused extinctions of all mammoths, fragmentation of mammoth ranges among larger groups of Pleistocene human hunters may have tipped the balance for some populations (Barnosky et al. 2004). Both hunting and habitat change (e.g., burning of savannahs to improve game and forage conditions) seem the largest drivers of the demise of many large mammals in the conterminous United States (Martin 2001; Miller et al. 1999), resulting in more pronounced extinction events among these taxa than could be accounted for by climate change alone. Finally, humans were likely contributors to animal extinctions in Africa, although because human presence in Africa is so ancient, it is much more difficult to establish a causal link.

Dramatic extinction events occurred among birds as Polynesians colonized Pacific Islands 1000–3000 years ago. Over 2000 species (particularly flightless rails), and over 8000 populations were driven extinct by overexploitation, habitat alteration, and the introduction of rats, pigs, and other commensal mammals carried by the

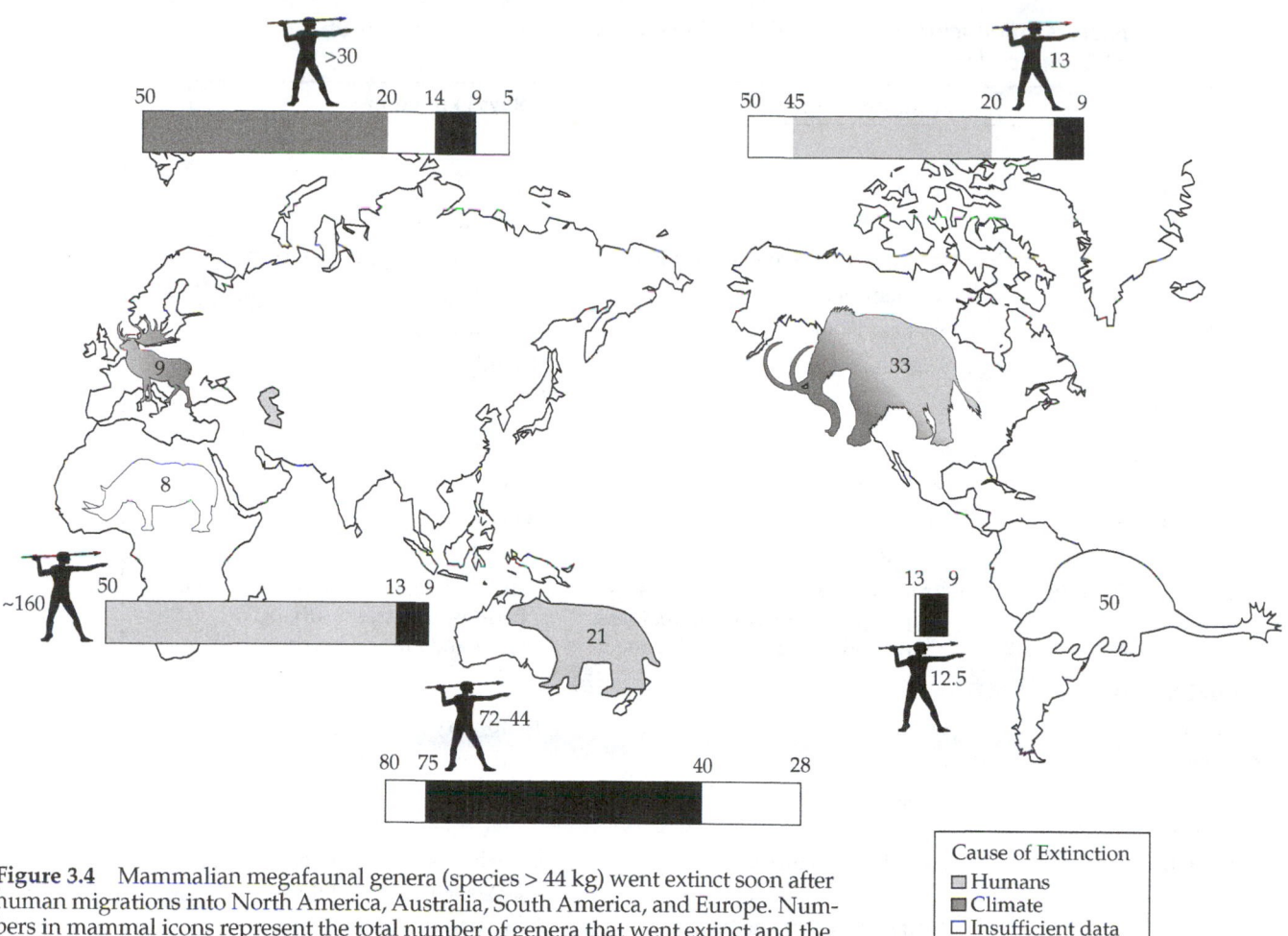

Figure 3.4 Mammalian megafaunal genera (species > 44 kg) went extinct soon after human migrations into North America, Australia, South America, and Europe. Numbers in mammal icons represent the total number of genera that went extinct and the shading indicates cause of extinction; see inset legend. Bars indicate the period of extinction (in kya), and shading indicates the magnitude of extinctions during that time; Black = many, dark gray = some, light gray = few, white = one or none. Numbers next to human icons indicate when humans arrived to the continent. Rapid climate warming (occurring from 14–10 kya) contributed to extinction in many cases, particularly in Eurasia and South America. In South America climate change and the arrival of humans entirely coincided with an extinction spasm of 50 genera. (Modified from Barnosky et al. 2004.)

Cause of Extinction
☐ Humans
☐ Climate
☐ Insufficient data

Polynesians (Steadman 1995). In some cases, extinction occurred within the first 100 years after an island was colonized (Steadman 1995), while in others, species may have persisted for a thousand years, or to the present day (Steadman and Martin 2003).

A useful lesson emerges from study of the patterns of extinction among birds in the Pacific islands. Certain factors are correlated with rapid extinction of taxa, and others with their persistence or at least delayed extinction (Table 3.1). Most importantly, where humans introduced many species of predators (particularly rats), consumers, and certain weeds, and began cultivation, extinction was the predominant outcome among certain avian groups (Steadman and Martin 2003). In other words, where invasive species and habitat degradation were combined, extinctions were most common. Other abiotic and biotic

factors, such as the size and shape of islands and their species richness, were also predictors of extinction risk.

Somewhat surprisingly, patterns of extinction in more recent times are often obscure because despite the existence of historical records, few species were studied well enough for the causes of extinction to be understood. Further, many species particularly sensitive to human activities undoubtedly were lost before they were recorded, particularly across Europe, parts of China, and Africa or other locations with a long history of human occupation (Balmford 1996). Often the cases that are well understood are examples of rapid overexploitation, such as the extinction of Stellar's sea cow (*Hydrodamalis gigas*) and the Great Auk (*Pinguinus impennis*).

Birds are the only taxa for which causes of extinction can be ascribed in the majority of cases. Since 1500, we

TABLE 3.1 *Factors Influencing Vertebrate Extinction Following Human Colonization of Oceanic Islands*

	Promotes Extinction	Delays Extinction
Abiotic factors		
Island size	Small	Large
Topography	Flat, low	Steep, rugged
Bedrock	Sandy, noncalcareous, sedimentary	Limestone, steep, volcanics
Soils	Nutrient rich	Nutrient poor
Isolation	No near islands	Many near islands
Climate	Seasonally dry	Reliably wet
Biotic factors		
Plant diversity	Depauperate	Species-rich
Animal diversity	Depauperate	Species-rich
Marine diversity	Depauperate	Species-rich
Terrestrial mammals	Absent	Present
Species-specific traits	Ground-dwelling, flightless, large, tame, palatable, colorful feathers, long and straight bones for tools	Canopy-dwelling, volant, small, wary, bad-tasting, drab feathers, short and curved bones
Cultural factors		
Occupation	Permanent	Temporary
Settlement pattern	Island-wide	Restricted (coastal)
Population growth and density	Rapid; high	Slow; low
Subsistence	Includes agriculture	Only hunting, fishing, gathering (especially in marine zone)
Introduced plants	Many species, invasive	Few species, noninvasive
Introduced animals	Many species, feral populations	Few or no species, no feral populations

Source: Modified from Steadman and Martin 2003.

know that at least 129 species went extinct (IUCN 2004). Habitat loss and degradation was a factor in most cases, particularly for species with narrow ranges. The introduction of alien invasive species, such as the black rat, was especially influential in the extinction of island endemics. Finally, overexploitation for food, feathers, or the pet trade, contributed to the demise of many species, including spectacular examples such as the massive overharvest of the Passenger Pigeon (*Ectopistes migratorius*) and Carolina Parakeet (*Conuropsis carolinensis*), both of which once had populations in the millions. As discussed above, extinction often was caused by a combination of two or more factors.

Indirect impacts of extinctions on animal and plant communities

As dramatic as these prior human-mediated extinction events have been, an equally dramatic adjustment of our concepts of "undisturbed" or "pristine" communities or ecosystems is now necessary. The extinction of large-bodied species, as well as untold numbers of more poorly fossilized species, is likely to have caused significant changes in the composition, character, and extent of ecological communities. Further, ongoing extinctions of species or populations, and even the reduction of some species to low population sizes, are changing present-day ecosystems.

Where species depend on their interactions with other species, extinction can have ripple effects as these interactions are disrupted. Thus, the loss of key species can spark a suite of indirect effects—a **cascade effect** of subsequent, or **secondary extinctions**, and substantive changes to biological communities. Secondary extinctions, those caused indirectly by an earlier extinction, are most likely to occur when species rely on a single or a few species as prey or as critical mutualists. For example, a plant that relies on a single bat species for pollination will not be able to reproduce should that bat go extinct. Similarly, a carnivore specializing on two species of insects may be unable to maintain itself if one of its prey species went extinct.

BOX 3.1 Cascade Effects Resulting from Loss of a Critical Species or Taxon, or from Species Introductions

A wide variety of studies have shown that losses of a single or group of critical species can have a cascade effect on biological communities, with implications for the functioning of ecosystems. The loss of top predators is most commonly cited as causing cascade effects. Reduction of top predators typically results in increases in prey, and thus enhanced populations of medium-sized predators (mesopredators), which in turn leads to strong declines of their prey species (especially birds and mammals) (e.g., Palomares et al. 1995, Crooks and Soulé 1999, and Terborgh et al. 1999, 2001). Herbivore release from the loss of top predators can reduce plant populations and plant community diversity, which in turn reduces diversity of other herbivores (e.g., Estes et al. 1989, Leigh et al. 1993, and Ostfeld et al. 1996). Cascade effects can lead to the scavenger community as well (Berger 1999; Terborgh et al. 1999). Similar cascade effects have also been seen in food webs with insect top predators (Rosenheim et al. 1993; Letourneau et al. 2004), although these are much less studied, so little is known about the commonness of these effects.

Studies of lake ecosystems has repeatedly shown that loss of piscivo-rous fishes release fish that graze on zooplankton, leading to a reduction in zooplankton, increases in phytoplankton, and broad rearrangements in community composition (e.g., Carpenter et al. 1985, Vanni et al. 1990, Carpenter et al. 2001, and Lazzaro et al. 2003). Effects can include changes in water clarity and large scale shifts in macroinvertebrate abundance and diversity (Nicholls 1999). Importantly, cascades do not always occur when piscivorous fish are removed, but only under specific conditions (Benndorf et al. 2002). Both ecosystem and community level effects can result from the loss of anadromous fishes. In the northwest of North America reduction in, or loss of, salmon populations can cause decreases in the input of nutrients to inland streams (Schindler et al. 2003), and loss of food for grizzly bears, bald eagles, killer whales, and predaceous fishes (e.g., Francis 1997 and Willson et al. 1998).

The loss of large-bodied species can often cause ripple effects through a community. Many studies in tropical forest have documented that the loss of many large-bodied species preferred by hunters leads to release of their prey, and loss of any services to other community members (Dirzo and Miranda 1991; Redford 1992; Redford and Feinsinger 2001). Reduction of plant diversity can result through enhanced granivory and herbivory (Terborgh and Wright 1994; Ganzhorn et al. 1999), or through loss of seed dispersal, and sometimes pollination, services (e.g., Chapman and Onderdonk 1998, Andersen 1999, Hamann and Curio 1999, and Wright et al. 2000).

Many cascade effects can be caused by changes in the abundance of ecosystem engineers. Loss of ecosystem engineers results in large structural changes in ecosystem, such as loss of pools created by beavers (Naiman et al. 1988), or shifts in plant diversity with loss of grazing by bison (Knapp et al. 1999). Loss of detritivores in streams can also cause large-scale changes (Flecker 1996).

Finally, cascade effects can result from species introductions. Introduction of species to marine estuaries or to lake systems has been shown to result in massive reduction in native algae, loss of native crayfish, molluscs, and other invertebrates, and even in a reduction in waterfowl, fishes, and amphibians (e.g., Olsen et al. 1991, Hill and Lodge 1999, and Nyström et al. 2001).

Cascade effects are probable in any community where strong interactions among species occur, be they predator–prey, mutualistic, or competitive in character, and thus are likely to have occurred prior to recorded history as well as in the few recorded cases from recent times (Box 3.1). One of the best-known examples of a cascade effect occurred when local extinction of sea otters (*Enhydra lutris*), which were aggressively hunted for their pelts, led to a transformation of marine kelp forest communities off the Pacific coast of North America (Estes et al. 1989). Sea otters are heavy consumers of sea urchins, and their absence led to an urchin population explosion, which in turn leads to overgrazing of kelp and other algae by the urchins, creating "urchin barrens" (Estes et al. 1989). Thus, because kelp forests are a haven for a broad variety of fishes and other species (Dayton et al. 1998), the local extinction of sea otters leads to the local extinction of many other species, and a radical change in the nature of the structure and composition of the community.

Cascade effects are difficult to demonstrate in the hyper-diverse tropics, but are likely to occur there. Tropical forests in which species have been driven locally extinct ("empty forests," Redford 1992) or depleted ("half-empty forests," Redford and Feinsinger 2001) now may lack effective populations of key interactors. The loss of critical seed dispersers could eventually result in the extinction of disperser-dependent tree species (Janzen 1986; see Essay 12.3). Depletion of top predators may cause the **ecological release** of prey species that may in turn drive down populations of their prey, as occurred with sea otters (Terborgh et al.1999; see Case Study 7.3). Thus, many tropical communities may appear healthy on the surface, but are in fact destined to decline in diversity due to disruption of critical interactions through local extinction of pollinators, seed dispersers, and top predators.

Overexploitation of great whales from 1700 to the mid-1900s had enormous impacts on marine ecosystems

(Roberts 2003). The removal of such large consumers released their prey, causing changes throughout the food chain. Industrial-scale fishing of large-bodied fish species that occupy higher places on the food chain also has initiated dramatic cascading effects throughout marine ecosystems (Dayton et al. 1995; Parsons 1996). The net result may be marine communities that bear little resemblance to the structure and abundances that would have been typical before humans began whaling and fishing on extensive scales (Pauly et al. 1998). Presumably, the loss of megafauna in the Pleistocene had similar community-level impacts to those that we can describe from more contemporary events.

Although it would be helpful in conservation to know which species are most critical to communities, in practice it is not simple to identify which species have the largest impacts, or are involved in the greatest number of strong interactions (Berlow et al. 1999). **Dominant species** are those that are very common and that also have strong effects on other members of the community (Figure 3.5). Examples include reef-building corals, forest trees, and large herbivores, such as deer. **Ecosystem engineers**, those species such as beavers or elephants that strongly modify habitat, are also ones whose absence or presence will change communities (Naiman et al. 1986; see Case Study 9.1). Generally, both community dominants and ecosystem engineers are relatively easy to identify.

A **keystone species** is a species that has a greater impact on its community than would be expected by the contribution of its overall numbers or biomass (Paine 1969; Power et al. 1996; see Figure 3.5). If a bat species is necessary for pollinating many species, and no other

species can serve its role, then it would be considered a keystone species. Similarly, large carnivores frequently act as keystone species through their impacts on other predators, a wide variety of prey species, and the competitors of their prey (Crooks and Soulé 1999). Ecosystem engineers are usually also keystone species, as defined above. Unfortunately, unlike dominant species whose impacts are easily discernable, often a keystone species is not so easily recognized and is only discovered after its numbers have been reduced and the impacts become obvious.

Current Patterns of Global Endangerment

As we consider the present era, the challenges to biodiversity have intensified in many respects. Where prehistoric impacts were dominated by overexploitation, moderate habitat modification, and introduction of human commensals, in recorded history we add the problems of pollution and human infrastructure development, and ultimately human-mediated climate change. In our era, vastly larger human populations and greater consumptive habits ensure that each primary threat has increased in magnitude and extent. In Chapter 6 we will examine how these threats have resulted in habitat degradation across the globe, but here we will focus on effects on species. To help us direct our efforts, and to help motivate social and political will to act on this biodiversity crisis, it may help to review what is known about global patterns of species endangerment, as well as those of selected countries.

Figure 3.5 Keystone and dominant species can have large impacts on biological communities. Keystone species by definition make up only a small proportion of the biomass of a community, yet have a large impact, whereas dominant species have impacts that are more proportional to their biomass or abundance. Reductions in the biomass or extinction of keystone or dominant species can be expected to cause cascade effects in communities. Many rare and common species have low impacts, and changes in their abundance may not have noticeable effects on other species. (Modified from Power et al. 1996.)

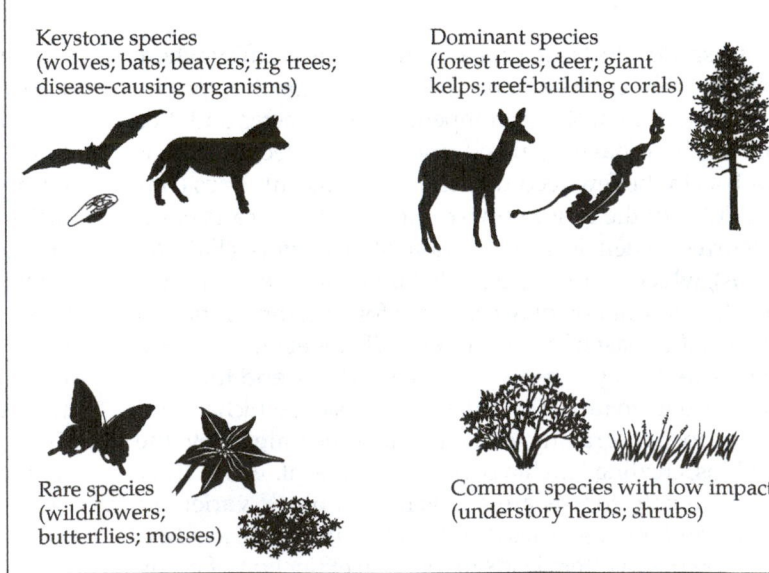

Globally threatened species

The best data on global endangerment are collated in the IUCN Red List of Threatened Species (www.redlist.org). The Red List classifies all species reviewed into one of nine categories (Box 3.2), with three primary categories of endangerment, in order of the risk of extinction: Critically Endangered (CR), Endangered (EN), and Vulnerable (VU). Very specific rules have been adopted to standardize rankings of each species, and to allow use of the list to index changes in the status of biodiversity over time (see Box 3.1, Table A; IUCN 2001). Taxonomic experts, conservationists, and other biologists work together in teams to conduct the reviews, and thus assure that the best available information is used in each case, although large uncertainties due to incomplete knowledge often make judgments difficult. Thus, the Red List is seen as a work in progress, undergoing constant revision both to document true changes in status, and to reflect updates in our knowledge. Through these efforts, the IUCN Red List has become the most complete database on global status of species available.

Complete evaluations have been undertaken for bird and amphibian species, and are nearly complete for mammals, and gymnosperms (conifers, cycads, and ginkgos) among plants. However, only a small percentage of all other taxa have been reviewed (about 6% of reptiles and fishes; 0%–3% of invertebrates, and 1%–5% of other plant groups; Table 3.2). Overall only 2.5% of described species have been evaluated. Among those species that have been evaluated, many are so poorly known that they cannot be categorized, and are given a Data Deficient ranking (9.5%; see Box 3.1).

Of the 38,046 species evaluated as of November 2004, 41% are endangered (CR, EN, or VU), 10% are Near Threatened or Conservation Dependent (meaning they may become endangered in the coming decades), while 38% are designated as Least Concern (indicating a very low risk of extinction for the foreseeable future). The Red List also includes a conservative tally of extinction, recording a total of 784 extinctions, plus 60 extinctions in the wild (where the only living individuals are in captivity or cultivation) (Baillie et al. 2004). Globally, 317 ma-

TABLE 3.2 *Number of Globally Threatened Species by Taxon*

	Described species	Evaluated species	Threatened species	Percent of described species threatened	Percent of evaluated species threatened
Vertebrates					
Mammals	5416	4853	1101	20	23
Birds	9917	9917	1213	12	12
Reptiles	8163	499	304	4	61
Amphibians	5743	5743	1856	32	32
Fish	28,600	1721	800	3	46
Invertebrates					
Insects	950,000	771	559	0.1	73
Molluscs	70,000	2163	974	1	45
Crustaceans	40,000	498	429	1	86
Others	130,200	55	30	0.02	55
Plants					
Mosses	15,000	93	80	0.5	86
Ferns	13,025	210	140	1	67
Gymnosperms	980	907	305	31	34
Dicotyledons	199,350	9473	7025	4	74
Monocotyledons	59,300	1141	771	1	68
Lichens	10,000	2	2	0.02	100
Total	1,545,594	38,046	15,503	1	41

Note: A "threatened species" includes any species designated as CR, EN, or VU by the IUCN Red List.
Source: Modified from IUCN 2004.

BOX 3.2 The IUCN Red List System

The IUCN Red List System, a systematic listing of species in threat of extinction, was initiated in 1963 to be used in conservation planning efforts around the globe. Over time, hundreds of scientists have worked to create listing criteria that have been carefully defined to be maximally useful as a diagnostic tool to help establish extinction risk over all taxa. Species are assigned to one of nine categories, which indicate their threat status or their status in the review process (Figure A). These categories are defined as follows:

Extinct (EX)

A taxon is Extinct when there is no reasonable doubt that the last individual has died. A taxon is presumed extinct when exhaustive surveys in known and expected habitat, at appropriate times (diurnal, seasonal, annual) to the taxon's life cycle and life form, throughout its historic range have failed to record an individual.

Extinct in the Wild (EW)

A taxon is Extinct in the Wild when it is known only to survive in cultivation, in captivity, or as a naturalized population (or populations) well outside the past range, and there is no reasonable doubt that the last individual in the wild has died, as outlined under EX.

Critically Endangered (CR)

A taxon is Critically Endangered when the best available evidence indicates that it meets any of the criteria A–E in Table A for Critically Endangered species, and is therefore facing an extremely high risk of extinction in the wild.

Endangered (EN)

A taxon is Endangered when the best available evidence indicates that it meets any of the criteria A–E for Endangered (see Table A) and is therefore facing a very high risk of extinction in the wild.

Vulnerable (VU)

A taxon is Vulnerable when the best available evidence indicates that it meets any of the criteria A–E for Vulnerable (see Table A) and is therefore facing a high risk of extinction in the wild.

Near Threatened (NT)

A taxon is Near Threatened when it has been evaluated against the criteria but does not qualify for Critically Endangered, Endangered, or Vulnerable now, but is close to qualifying for or is likely to qualify for a threatened category in the near future.

Least Concern (LC)

A taxon is deemed Least Concern when it has been evaluated against the criteria and it neither qualifies for the previously described designations (Critically Endangered, Endangered, Vulnerable, or Near Threatened), nor is it likely to qualify in the near future.

Widespread and abundant taxa are included in this category.

Data Deficient (DD)

A taxon is Data Deficient when there is inadequate information to make a direct or indirect assessment of its risk of extinction based on its distribution, population status, or both. Every effort is made to use this category as a last resort, as this is not a category of threat, but only indicates more information is needed to make a status determination.

Not Evaluated (NE)

A taxon is Not Evaluated if it is has not yet been evaluated against the criteria.

Assignment to one of the three threatened categories (CR, EN, or VU) is made on the basis of a suite of quantitative standards adopted in 1994 that relate abundance or geographic range indicators to extinction risk (see Table A). The different criteria and their quantitative values (A–E) were chosen through extensive scientific review, and are aimed at detecting risk factors across the broad diversity of species that must be considered (IUCN 2001).Qualification under any of the criteria A–E is sufficient for listing; however, evaluations are always made as completely as possible for use in evaluating changes in status over time, and for conservation planning purposes. Thus, the status of a taxon will be evaluated according to most of these criteria, as

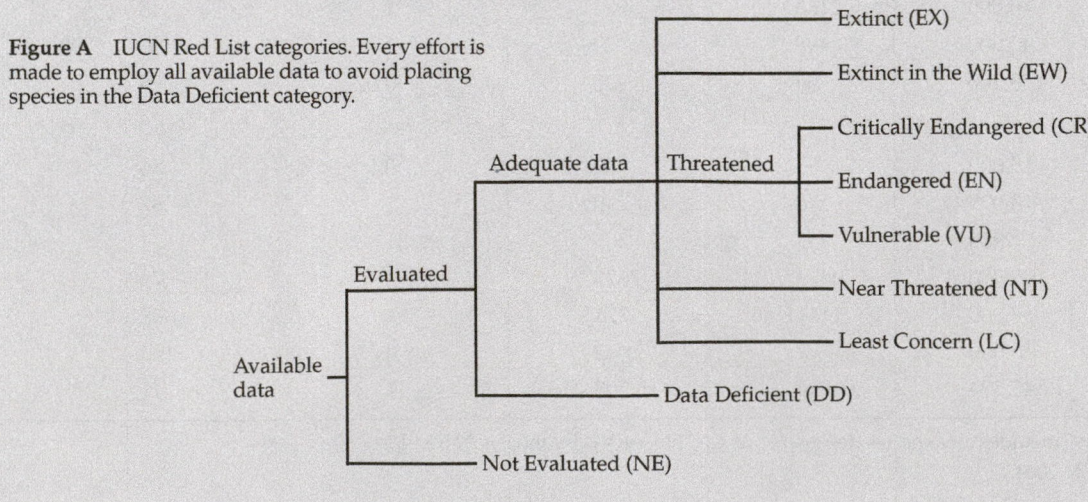

Figure A   IUCN Red List categories. Every effort is made to employ all available data to avoid placing species in the Data Deficient category.

far as is possible given current knowledge.

A major advance in risk evaluation, the Red List criteria require efforts to place quantitative bounds on our knowledge, and explicitly allow for uncertainty. The assignments to category are not assignments of priority, but rather a reflection of our current best judgment of how great the risk of extinction is for this species, given the best available information at present. All species on the list must be reevaluated at least once every ten years.

In addition to quantifying risk of extinction, the Red List compiles data on the nature of the threats to the species. These evaluations are useful for initial efforts to conserve the threatened species, and in aggregate can guide efforts to reduce threatening processes.

TABLE A *Overview of Criteria (A–E) for Classifying Species as CR, EN, or VU in IUCN Red List*

Criterion	Critically Endangered (CR)	Endangered (EN)	Vulnerable (VU)	Qualifiers
A.1 Reduction in population size	>90%	>70%	>50%	Over 10 years or 3 generations in the past where causes are reversible, understood, and have ceased
A.2–4 Reduction in population size	>80%	>50%	>30%	Over 10 years or 3 generations in the past, future, or combination, where causes are not reversible, not understood, or ongoing
B.1 Small range (extent of occurrence)	<100 km^2	<5000 km^2	<20,000 km^2	Plus two of (a) severe fragmentation or few occurrences (CR = 1, EN = 2–5, VU = 6–10), (b) continuing decline, (c) extreme fluctuation
B.2 Small range (area of occupancy)	<10 km^2	<500 km^2	<2000 km^2	
C Small and declining population	<250	<2500	<10,000	Mature individuals, plus continuing decline either over a specific rate in short time periods, or with specific population structure or extreme fluctuations
D.1 Very small population	<50	<250	<1000	Mature individuals
D.2 Very small range	—	—	<20 km^2 or <5 locations	Capable of becoming CR or EX within a very short time
E Quantitative analysis	>10% in 100 years or 3 generations	>20% in 20 years or 5 generations	>50% in 100 years	Estimated extinction risk using quantitative models, e.g., population viability analyses

Source: IUCN 2001.

rine, 2981 freshwater, and 13,657 terrestrial species are considered endangered, and recorded extinctions are similarly apportioned.

For the most complete evaluated taxa, amphibians and gymnosperms stand out as particularly threatened (see Table 3.2). Cycads, an ancient group of gymnosperms, are especially vulnerable, with 52% endangered. The true level of threat is undoubtedly higher than these estimates due to the large number of Data Deficient rankings: 1290 amphibians (23%), 360 mammals, 78 birds and 77 gymnosperms all are too poorly known to be ranked, but certainly some of these are endangered.

Among mammals, ungulates, carnivores, and primates are particularly at risk mostly due to habitat degradation and overexploitation (Baillie et al. 2004). Albatrosses, cranes, parrots, pheasants, and pigeons are particularly threatened among the birds due to bycatch, habitat loss, the pet trade, and direct exploitation (Birdlife International 2004). Amphibians appear to be at greatest risk of extinction, with a high fraction of these species listed as critically endangered (21%; IUCN 2004). A wide variety of threats affect amphibian populations throughout the world, many of which exert synergistic effects on these sensitive animals; these threats

are described by Joe Pechmann and David Wake in Case Study 3.1.

While levels of endangerment among the remaining groups reflect a tendency for worrisome cases to be put forward for evaluation ahead of general analyses (for example, see crustaceans and mosses in Table 3.2), it seems likely that high levels of threat may occur in many other groups. Turtles and tortoises have been more completely assessed among the reptiles, and 42% are endangered (Baillie et al. 2004). Marine and freshwater fishes are placed on the list in increasing numbers, suggesting a serious level of threat, whatever the exact percentage. Two messages from these statistics are clear: First, a large fraction of vertebrate and gymnosperm diversity is in great danger of extinction over the next century, and second, we know very little about the status of most other taxa.

Globally threatened processes

Not only are species at risk of extinction, but some processes that undergird ecosystem functions, or that are glorious in and of themselves, are put at risk from human activities. Lincoln Brower discusses the concept of an "endangered biological phenomenon," in which a species is likely to survive, but some spectacular aspect of its life history, such as the mass annual migration of monarch butterflies between Mexico and the United States, is in jeopardy of disappearing (Essay 3.3). The mass migrations of springbok in southern Africa have already been eliminated but the seasonal migrations of vast herds of wildebeest and zebra in the Serengeti still exist. Not only is this mass migration an amazing spectacle, but the Serengeti grasslands are adapted to the impacts of high densities of these grazers, as well as other ungulates. The loss of wildebeest migra-

tions would be tragic, even though wildebeest would survive in many places.

What factors are most threatening to biodiversity globally?

As species are evaluated, all threats faced at present or in the past that led to endangerment are coded into the Red List database. Our knowledge about the threats faced by species varies tremendously as certain types of threats are easier to document (e.g., forest conversion versus competition from invasive species), and for some species it is only possible to say that they are declining and endangered, but not to diagnose why. Most species face multiple threats. Finally, the red list is dominated by evaluations primarily for vertebrates and some vascular plants, and for areas where many biologists already work, perhaps reflecting biases in interest among biologists more than intrinsic levels of threat (Burgman 2002). Thus, we cannot be sure that the patterns that may hold for birds will serve for mollusks and other species groups (and indeed, we should expect that they will not in many cases). Nonetheless, to a limited degree, we can use these data to give us a sense of the pervasiveness of different types of threats.

Habitat loss and degradation is the most pervasive and serious threat to mammals, birds, amphibians, and gymnosperms (Figure 3.6). Overexploitation is the most pervasive threat to fishes, and a predominant one for mammals and birds as well, whereas invasive species pose particular risks for birds on islands (Baillie et al. 2004). Amphibians are more challenged than birds and mammals by pollution that directly kills individuals, and by changes in the abundance of native species, particularly diseases. Intrinsic factors, such as a limited reproductive capacity or limited

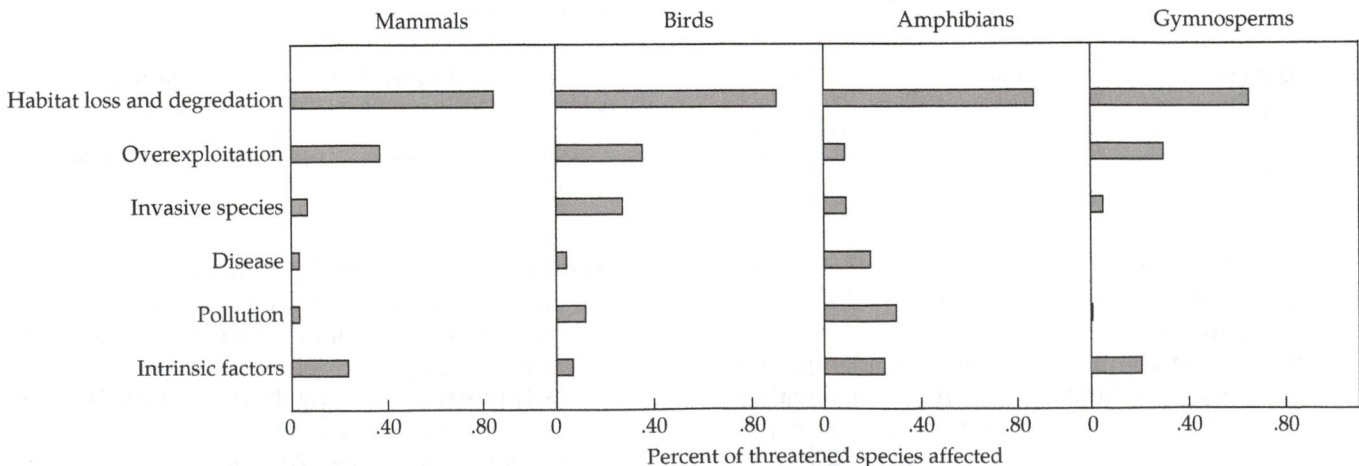

Figure 3.6 Habitat loss and degradation is the greatest threat to global biodiversity among mammals, birds, amphibians, and gymnosperms. Because not all threats are documented, this figure underestimates threat levels among Red Listed species. Overexploitation includes both direct mortality and by-catch. (Modified from IUCN 2004.)

ESSAY 3.3

An Endangered Biological Phenomenon

Lincoln P. Brower, *Sweet Briar College*

■ Much of conservation research focuses on describing diminishing species diversity and on understanding the processes that lead species to small populations, and thence to extinction. Here I discuss endangered biological phenomena, defined as a spectacular aspect of the life history of an animal or plant species, involving a large number of individuals, that is threatened with impoverishment or demise. The species per se need not be in peril; rather, the phenomenon it exhibits is at stake (Brower and Malcolm 1991).

Examples of endangered biological phenomena include the ecological diversity associated with naturally flooding rivers, the vast herds of bison of the North American prairie ecosystems, the synchronous flowering cycles of bamboo in India, the 17-year and 13-year cicada emergence events in eastern North America, and scores of current animal migrations. Instances of the latter include seasonal migrations of the African wildebeest and North American caribou, the wet and dry season movements of Costa Rican sphingid moths, the billion individuals of 120 songbird species that migrate from Canada to Neotropical overwintering areas, and the highly disrupted migrations of numerous whale species.

There are two principal reasons why animal migrations are endangered by human activities. First, migrant species move through a sequence of ecologically distinct areas, any one of which could become an Achilles' heel. Second, aggregation of the migrants can occur, making the animals especially vulnerable. The major impact on migratory species is due to accelerating habitat modification throughout the world. Even when problems are recognized, mitigation is difficult because of varying policies and enforcement abilities in the different countries the animals occupy during the different phases of their migration cycles. The extraordinary migration and overwintering behaviors of the monarch butterfly in North America well exemplify the concept of endangered phenomena.

The monarch butterfly (*Danaus plexippus*) is a member of the tropical subfamily Danainae, which contains 157 known species. It is alone in its subfam-

ily for having evolved extraordinary spring and fall migrations (Figure A) that allow it to exploit the abundant *Asclepias* (milkweed) food supply across the North American continent, becoming one of the most abundant butterflies in the world. Remarkably, and in contrast to vertebrate migrations, the monarch's orientation and navigation to its overwintering sites is carried out by descendants three or more generations removed from their migrant forebears. Its fall migration, therefore, is completely inherited, with no opportunity for learned behavior. This, together with the vastness of the migration and overwintering aggregations, constitutes a unique biological phenomenon.

Two migratory populations of the monarch occur in North America. The

larger one occurs east of the Rocky Mountains and undoubtedly represents the stock from which the smaller, western North American migration evolved. Both migrations are threatened because the aggregation behavior during winter concentrates the species into several tiny and vulnerable geographic areas.

By late summer, the monarch population in eastern North America builds to an estimated 0.5–3 billion individuals over an enormous area east of the Rocky Mountains. Beginning in late August, the adult butterflies migrate to central Mexico, where they overwinter for more than five months in high-elevation fir forests, about 90 km west of Mexico City (see Figure A). Here the butterflies coalesce by the hundreds of millions into dense and stunningly spectacular aggregations that festoon

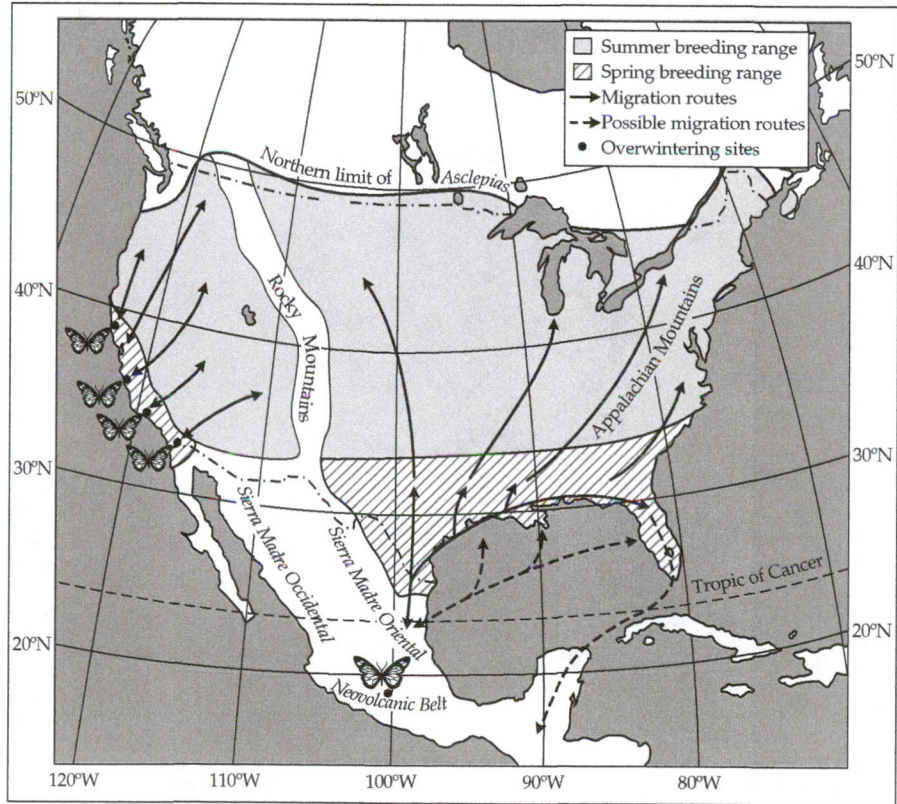

Figure A   Fall and spring migrations of the eastern and western populations of the monarch butterfly in North America. (Modified from Brower 1995.)

0.1 to 5 hectares of forest. The butterflies, effectively in cold storage, remain sexually inactive until the approach of the vernal equinox.

Survivors from the Mexican overwintering colonies begin migrating northward in late March to lay their eggs on sprouting milkweed plants (*Asclepias* spp.) (see Figure A). These 8-month-old remigrants then die and their offspring, produced in late April and early May, continue the migration northward to Canada. Over the summer, two or three more generations are produced as each previous generation dies. Toward the end of August, butterflies of the last summer generation enter reproductive diapause and the cycle begins anew as these monarchs migrate instinctively southward to the overwintering grounds in Mexico.

To date, about 30 overwintering areas have been discovered on 12 isolated mountain ranges in central Mexico at elevations ranging from 2900 to 3400 m. This elevational band coincides with a summer fog belt where boreal-like oryamel fir (*Abies religiosa*) forests occur that probably are a relict ecosystem from the Pleistocene. The five largest and least disturbed butterfly forests occur in an astoundingly small area of 800 km². By clustering on the trees in the cool and moist environment, individual monarchs are able to survive in a state of reproductive inactivity until the following spring.

Until recently, human impact on the high-elevation fir forests has been less than that on other forest ecosystems in Mexico. However, negative developments began to occur in the 1970s. Commercial harvest of trees, including thinning and clear-cutting has increased illegal removal of logs and firewood, and local charcoal manufacturing in pits dug within the fir forest. Villages have expanded, reaching locations up the mountainsides, and this expansion has lead to an increased frequency of fires associated with forest clearing for planting corn and oats, and also with killing young trees for local home construction. Forest lepidopteran pests have invaded some areas of the fir ecosystem, probably due to stress caused by thinning and deforestation at lower elevations. As a result of increasing lepidopteran pests, spraying of the organic pesticide, *Bacillus thuringiensis*,

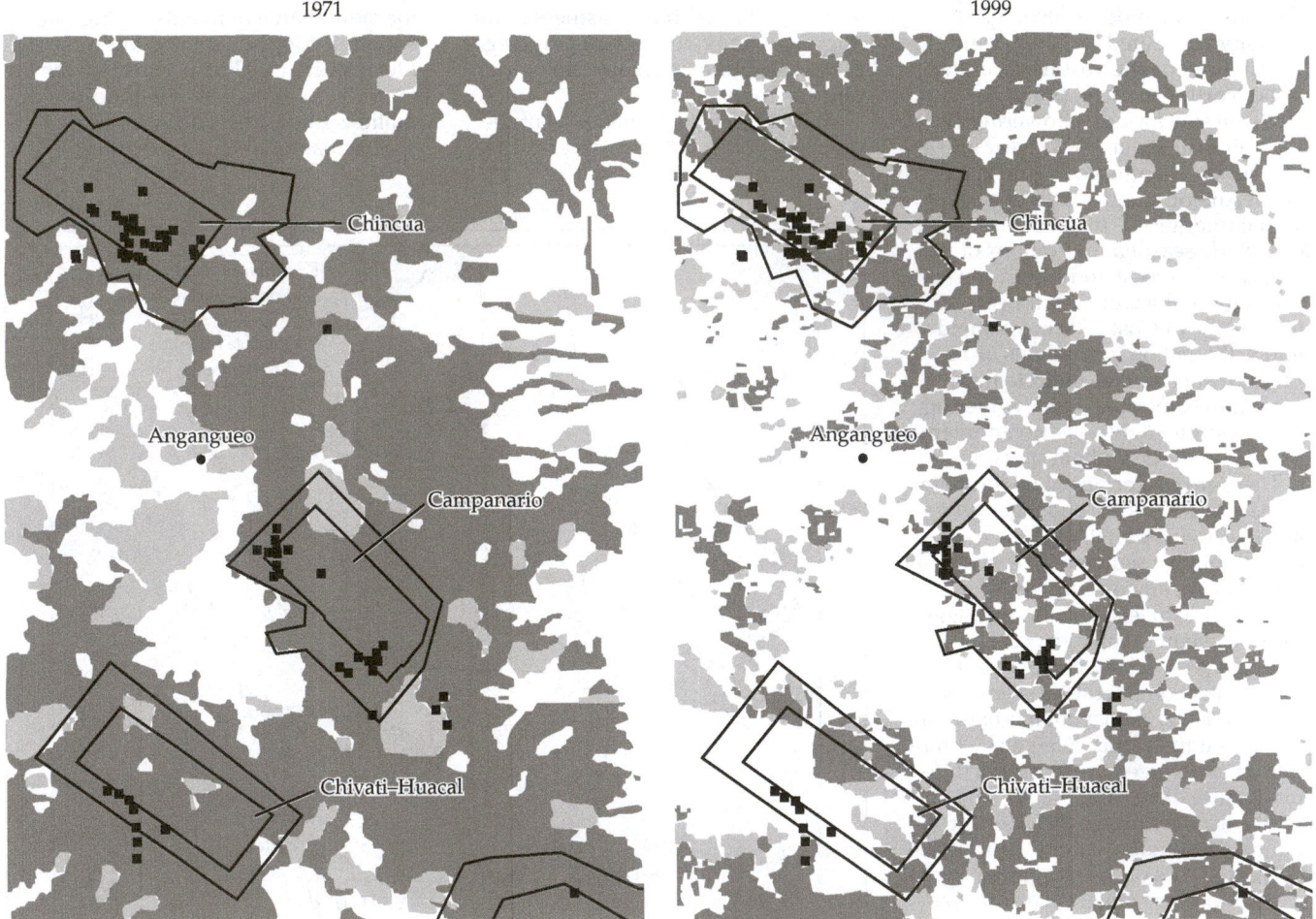

Figure B   Forest cover has dropped dramatically within, and adjacent to, the three major monarch butterfly reserves from 1971 to 1999. Forest is shown in dark gray, forest habitat that is somewhat degraded is shown in light gray, and highly modified areas are in white. Squares denote any butterfly overwintering sites that have been recorded, but not all those shown persist today. For each reserve, the inner line denotes core areas and the outer line buffer zones. (Modified from Brower et al. 2002.)

was initiated and considered for widespread use without adequate knowledge of its effect on monarchs. Most recently, heavy ecotourism is trampling the infrastructure and generating severe dust precipitation on vegetation along paths through the colonies; because of the income generated, there is demand to open all colonies to tourists.

Encroachment on existing monarch overwintering sites, even those in reserved status, has been extensive in the past two decades. A comparison of photos from 1971 and 1999 show shocking reductions in forests both outside and inside some of the largest and most important reserves (Figure B; Brower et al. 2002). The remaining fragments of forests experience a disrupted microclimate that will not support the butter-

flies once fragments become too small or perforated via selective logging.

As a result of documenting these drastic habitat losses, in November 2000, then President Ernesto Zedillo accepted recommendations to revise and increase the area protected to a total of 56,259 ha and to redefine the area as the "Monarch Butterfly Biosphere Reserve" (Brower et al. 2002). Importantly, this new decree was associated with a trust fund to compensate local inhabitants for the profits they would have made through logging. Notwithstanding this new decree, illegal logging has accelerated and is severely degrading critical areas of the reserve (Galindo and Honey-Roses 2004). Conservation of this migratory phenomenon will not succeed without enforcement of logging bans together

with appropriate compensation to affected people. Public education about the values of protecting these forests and the remarkable spectacle of nature that the monarch butterfly migrations represent is desperately needed.

If we fail to conserve these overwintering sites, we will not lose the monarch butterfly as a species, because numerous nonmigratory populations will persist in its tropical range. However, the monarch's spectacular North American migrations will soon be destroyed if extensive overwintering habitat protection and management are not implemented on a grand scale. The impending fate of this remarkable insect is an omen, warning us that we must incorporate the concept of endangered biological phenomena into our plans for conserving biodiversity. ∎

---

dispersal ability, are critical factors in endangerment for a large number of species as well (see Figure 3.6).

Where are species most at risk worldwide?

Some biomes contain greater fractions of threatened species, particularly biomes that are already species rich (Figure 3.7). Tropical and subtropical moist and dry forests, grasslands, savannas, shrublands, montane grasslands, and xeric biomes all have substantial numbers of threatened mammals, birds, and amphibians. As global climate change intensifies, many worry that montane habitats in particular will become unsuitable for many species, greatly increasing biodiversity losses.

Across the globe, there are particularly high numbers of threatened species in South America, Southeast Asia, sub-Saharan Africa, Oceania, and North America (IUCN 2004). To a large degree, this distribution reflects the vastly greater numbers of species in the tropics, as well as intensifying pressures, particularly in Southeast Asia and sub-Saharan Africa. The greater numbers in North America reflect both high diversity and levels of threat in the Hawaiian Islands, California, and Florida, as well as extensive habitat modification, particularly for agriculture, and substantial efforts to identify species at risk of extinction. Lower numbers of endangered species in some regions reflect low species diversity due to severe environments and low human densities (Antarctica, Saharan Africa, and to a lesser extent Northern Asia).

Endangered species in the United States

Among the first countries to address the problems of endangered species seriously, the United States has listed 1264 species as threatened or endangered as of Feb-

ruary 2005 (USFWS 2005), and 1143 are on the Red List (IUCN 2004). In addition, more detailed data on the locations and status of at risk species, populations, and ecosystems has been collected on a state-by-state basis through Natural Heritage Programs (now expanded to include Canada and Latin America, and managed by NatureServe). The total list of at risk populations in the U.S. and Canada now includes over 15,500 occurrences (NatureServe 2005). Overall, a third of U.S. species are considered to be at risk of extinction (Stein et al. 2000), and the U.S. is second only to Ecuador in the number of species thought to be at risk of extinction globally (IUCN 2004).

Although a greater number of threatened species in the U.S. are plants (>5000), freshwater species (mussels, crayfish, stoneflies, and fishes) are threatened in higher proportions than any other group (Figure 3.8). Freshwater mussel species appear to be particularly diverse in the U.S., so the fact that 70% are imperiled is of global importance. Neither biodiversity nor the risk of extinction is distributed uniformly across the U.S. California is the state with the greatest species richness, as well as the greatest number of endemic species, but Hawaii has the greatest number of species at risk and the greatest number of species that have become extinct (Stein et al. 2002).

Using the expanded list of imperiled species existing in the natural heritage database as well as federally listed species as of 1996, Wilcove et al. (1998) classified threats for 1880 species and populations for which sufficient information was available. Over 85% of these were threatened by habitat degradation, while 49% were affected by invasive species, 24% by pollution, 17% by overexploitation and 3% by disease (Table 3.3). Vertebrates, freshwa-

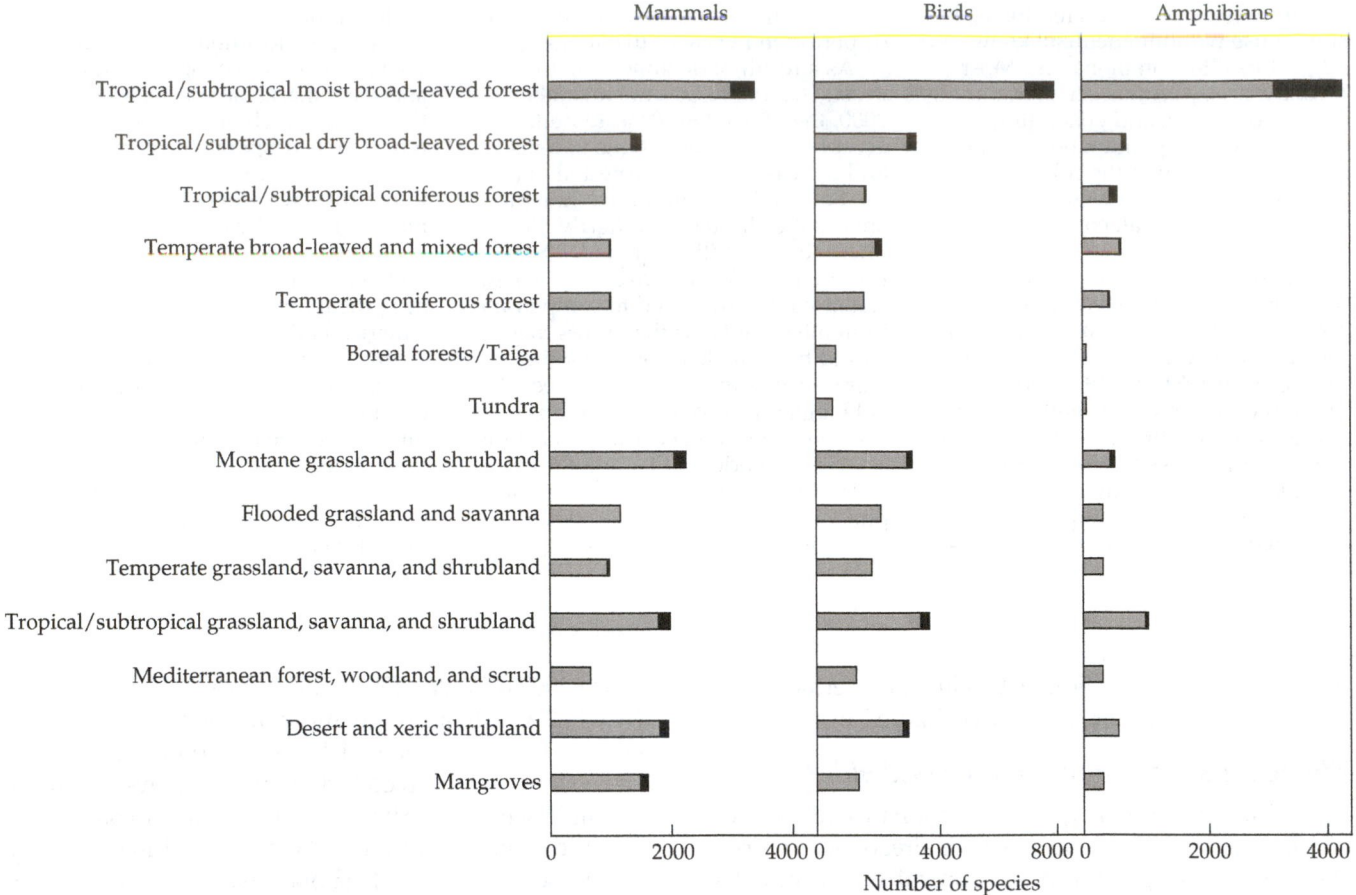

Figure 3.7 Species richness of mammal, bird, and amphibian species in the major biomes of the world. Black portions of the bars indicate number of threatened species. (Modified from Baillie et al. 2004.)

TABLE 3.3 *Percent of U.S. Threatened and Endangered Species Affected by Five Types of Threat*[a]

	Mammals	Birds	Reptiles	Amphibians	Fishes	Freshwater mussels	Crayfish	Other invertebrates	Plants
Threat	(85)	(98)	(35)	(60)	(213)	(102)	(67)	(143)	(1055)
Habitat loss and degradation	89	90	97	87	94	97	52	96	81
Overexploitation	45	33	66	17	13	15	0	43	10
Invasive species	27	69	37	27	53	17	4	46	57
Pollution	19	22	53	45	66	90	28	20	7
Disease	8	37	8	5	1	0	0	0	1

Note: Numbers in parentheses are the number of threatened species with threat data available.

Source: Wilcove et al. 1998.

[a]Listed by taxonomic group.

Figure 3.8 Proportion of species threatened with extinction by plant and animal groups in the United States. (From Stein et al. 2000.)

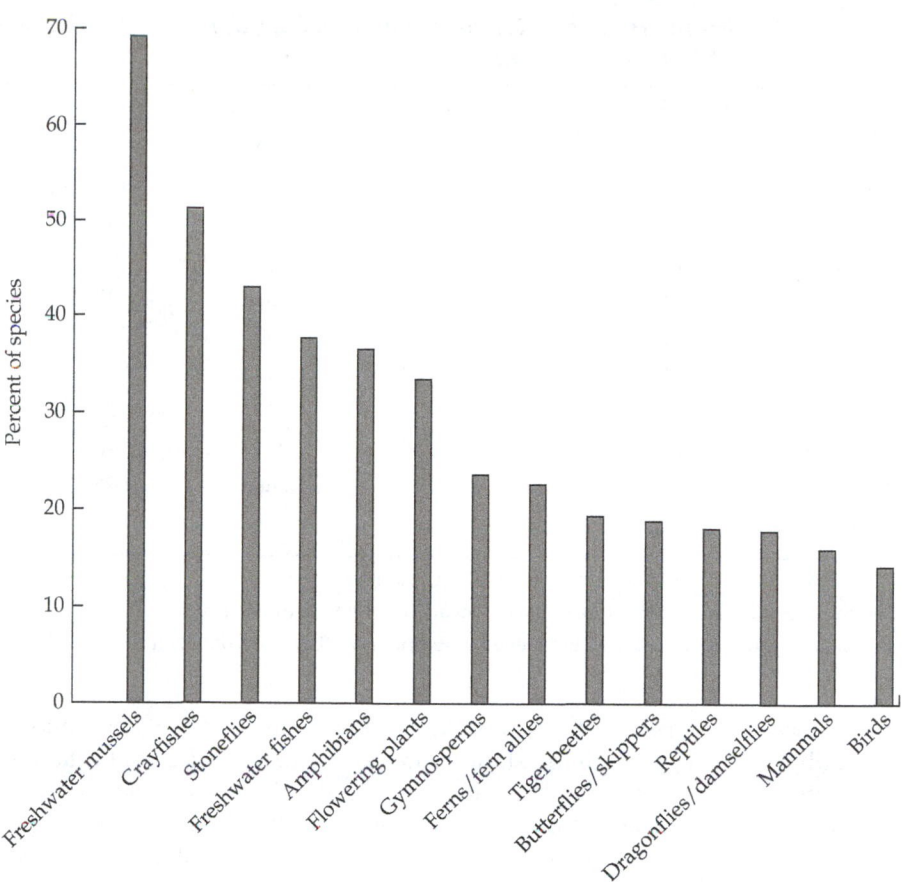

ter mussels, butterflies, and other invertebrates were threatened particularly by habitat destruction, while pollution and siltation impacts were most severe for aquatic species. Disease has limited distributions of many populations, and may have contributed to extinction of birds in the Hawaiian Islands (Van Riper et al. 1986). Invasive species pose threats to a greater proportion of plant and bird species in the Hawaiian Islands than in the continental U.S., and also to fish populations more than to other taxa.

Czech et al. (2000) examined the associations among 18 causes of endangerment for 877 species that were listed in the U.S. and Puerto Rico as of 1994. Urbanization was most frequently associated with other threats, as it involves both destruction of habitat directly and depletion of resources to support urban populations. Urbanization is also associated strongly with pollution, nonnative species, and the development of roads. Roads are not directly threatening to many species, but facilitate many forms of development that are threatening, such as urbanization, logging, industrial development, mining, and agriculture. Agricultural development was the most widespread threat, and associated most strongly with urbanization, reservoirs and pollution.

Threatened species in other countries

Although many countries maintain lists of threatened species, few have been analyzed in detail. Based on Red List data, Ecuador and the U.S. have the greatest number of endangered species (Table 3.4). Ecuador has a much higher number of listed plants than any other country. This has both a biological basis and a fortuitous one. Ecuador is a center of plant diversity where development, particularly in the Andean slopes has been extremely rapid recently, and thus a substantial fraction of evaluated plant species are threatened. Also, due to substantial efforts by concerned botanists, more than 2000 species have been evaluated—about half of those known from Ecuador. Such concerted efforts at reviewing threatened status have not been made for other centers of plant diversity, nor have they been undertaken for other taxa (for example, invertebrates) within the same country. Similarly, high numbers of species threatened in the U.S. reflects both true threat levels and much greater activity by its large cadre of biologists.

Although many countries do not harbor as long a list of globally threatened species, many have particularly high fractions of their species at risk. For example, over 80% of all plants and 30% of all vertebrates evaluated in

TABLE 3.4   *Numbers of Threatened Species (CR, EN, VU) in 12 Countries with the Greatest Overall Numbers of Red Listed Species*

	Mammals	Birds	Reptiles	Amphibians	Fishes	Invertebrates	Plants (%)[a]	Total
Ecuador	34	69	10	163	12	48	1815 (81)	2151
United States	40	71	27	50	154	561	240 (63)	1143
Malaysia	50	40	21	45	34	19	683 (60)	892
Indonesia	146	121	28	33	91	31	383 (60)	833
China	80	82	31	86	47	4	443 (73)	773
Mexico	72	57	21	190	106	41	261 (69)	748
Brazil	74	120	22	24[b]	42	34	381 (68)	697
Australia	63	60	38	47	74	283	56 (33)	621
Colombia	39	86	15	208	23	0	222 (69)	593
India	85	79	25	66	28	23	246 (71)	552
Madagascar	49	34	18	55	66	32	276 (80)	530

Source: IUCN 2004 Red List Summary tables 5, 6a, and 6b.

[a]Percent refers to the percent of those species evaluated that are threatened.

[b]A complete revision of the list of threatened amphibians in Brazil has not been completed.

Madagascar are at risk of extinction. Madagascar has been subject to very high deforestation rates and other pressures, and also has a very high number of endemic organisms. Thus, the island nation is a focus of considerable conservation interest (Case Study 3.2).

Australia lists 1554 threatened species, the majority of which are vascular plants, followed by birds and mammals (Australian Government Department of Environment and Heritage 2005), while only 621 of these are also Red Listed and few of these are plants. Habitat degradation and invasive species are the most common factors in endangerment. Australia's landmass is roughly equal to that of the United States, but it has a very high degree of endemism, with roughly 85% of flowering plants, mammals, and freshwater fishes unique to this continent. Thus, most threatened species in Australia represent ones at risk globally.

Recently, Li and Wilcove (2005) summarized threats to vertebrates in China. Overexploitation is overwhelmingly the most important threat, contributing to endangerment of 78% of 437 species, and habitat degradation was also a pressing concern for 70% of these imperiled vertebrates (Figure 3.9). Nearly all endangered reptiles in China are overexploited for food and medicines. This is not surprising given the importance of traditional medicine for China's population, and strong dependence on wild species for protein (see Chapter 8). Pollution is a cause of endangerment in 20% of the cases, but as in other parts of the world, freshwater fishes are particularly susceptible with over 40% threatened by industrial pollutants and pesticides. Dams do not yet appear as a major problem, perhaps because they are still rela-

tively new and unstudied. Invasive species appear much less important, although this could be because they are little studied, or because global trade has a very long history and the species most vulnerable to invasives have already gone extinct (Li and Wilcove 2005).

Overall, understanding patterns of global endangerment, or even threatened status at a national level is fraught with uncertainty. These difficulties have led to some controversy concerning the relative rate of extinction currently compared with those occurring in geologic history (Box 3.3). Although, these issues are not entirely resolved, the consensus is that current rates of extinction already exceed "background" rates of extinction, and are at least approaching those of the mass extinctions.

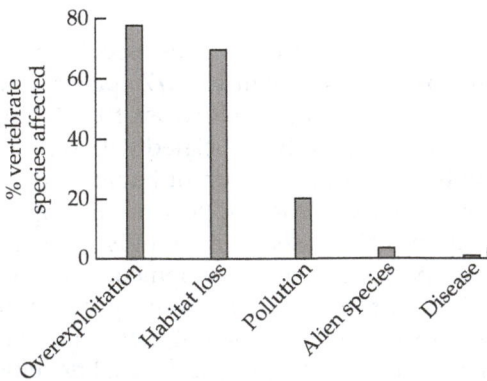

Figure 3.9 Contribution of major threats to endangerment among vertebrate species in China. (From Li and Wilcove 2005.)

BOX 3.3 Have We Set in Motion the Sixth Mass Extinction Event?

For a number of decades conservation biologists have called attention to the current rate of global extinction in extreme terms. Estimates of extinction rates vary, with most ranging from 50–10,000 times background extinction rates seen in the geologic record (Wilson 1992; May et al. 1995; Millenium Ecosystem Assessment 2005). To put our influences on biodiversity in context, many have tried to compare the current rate of extinction to the rates discernable from the fossil record. It is helpful to understand how these estimates are made, to provide the most defensible estimates for public discussion.

Before humans were dominant on Earth, the "background" rate of extinctions is estimated to have been about 0.1–10 species per year (May et al. 1995). During mass extinction events (see Figure 2.7), the rate of extinctions was 1–2 orders of magnitude higher (Raup 1994). Because neither the number of species on Earth during the mass extinctions is known (nor precisely how rapidly these extinctions may have occurred during a given 10,000–1,000,000 year interval), this estimate is highly imprecise, and thus difficult to compare with current estimates of extinction rates. Further, 95% of fossils are of marine invertebrates, while about 85% of extant species are terrestrial (May et al. 1995). How extinctions of marine invertebrates compared to extinctions in terrestrial and freshwater habitats is unknown.

Estimating current extinction rates is surprisingly difficult as well. This comes partly from the fact that it is much harder to prove that something does not exist, than that it does. To prove a species is extinct one needs to search exhaustively throughout a species' range, during appropriate seasons, and sometimes for many years. The recent rediscovery of the Ivory-billed Woodpecker (*Campephilis principalis*; Fitzpatrick et al. 2005) long believed to be extinct, illustrates this difficulty. As this is obviously impossible to accomplish for all species on Earth, conservation biologists primarily use two methods to try to estimate current extinction rates.

First, we can derive extinction rate estimates from species–area curves. As described in Chapter 2, species richness increases with area. Species richness therefore also declines with decreases in area; thus as habitats are severely degraded, we can expect to see a concomitant loss of species. Conservation biologists thus use estimates of habitat change combined with those for species richness to project species losses (Figure A). The proportional reduction in species is then calculated as:

$$\Delta S = z \, \Delta A$$

where S is the number of species, z is the slope of the species–area curve for a given location, and A is the area of that location (May et al. 1995). The difficulty is that while increasingly we can estimate changes in forest cover with accuracy (ΔA), we do not know the relationship between extinction and reduction in area (z). We also don't know the total number of species that may inhabit a given location, and thus we further extrapolate when we translate proportional reduction to an absolute number of losses per year. For example, tropical forest cover is being lost at a rate of 0.5%/year (Achard et al.

2002), and if the value of z is 0.2 (as is commonly assumed), then we can expect to lose 0.125% of these species each year. If there are roughly 5 million species in tropical forests, then we are losing 5000 species per year. Of course, if the total number of species is smaller or larger, or z is higher or lower, we would get a different total (e.g., if z = 0.15 and only 1 million species exist in tropical forests, then the loss is 750 species per year). Despite the crudeness of these estimates, reasonable values of the parameters yield substantially more than 10 species per year (a high estimate of background extinction rates).

Second, extinction rate estimates are often based on numbers of directly observed extinctions (i.e., recorded in the IUCN Red List, or comparable national databases). The Red List documents 844 extinctions since 1500, a loss of 2.2% of all evaluated species and 0.04% of all described species. Certainly, this is an underestimate of global extinctions (and even more of local extinctions) because only a few taxa and locations have been documented well enough to allow a defini-

Continued on next page

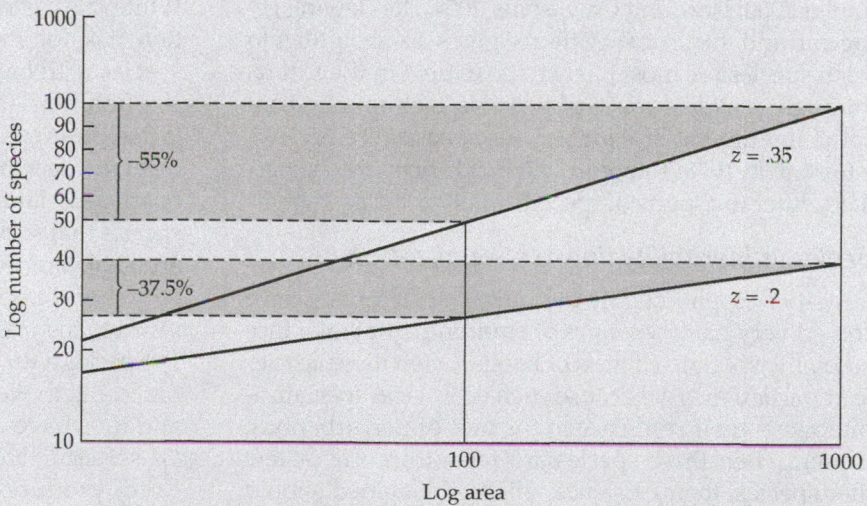

Figure A   Species–area relationships based on the equation $S = cA^z$, with $c = 10$ and $z = 0.2$ or 0.35. Note that a 90% decrease in area, from 1000 to 100 ha, would result in a predicted loss of 37.5% to 55% of the species, for a z value of 0.2 and 0.35 respectively. Greater z values (steeper species–area relationships) imply greater species losses per unit area. Most z values calculated from data fall between 0.15 and 0.35.

Box 3.3 *continued*

tive determination of extinction. Because we know the Red List total is too low, attempts are made to correct for this by extrapolating the rates for the best-known group to all species.

For example, roughly 100 birds and mammals have gone extinct between 1900 and 2000. To compare this to the fossil record, we need to calculate the loss as a percentage of all birds and mammals (currently 15,333 species). This translates to the loss of 0.65% of all birds and mammals each century, and 1 bird or mammal species per year. Average extinction rates of birds and mammals in the fossil record are on the order of 0.003 species per year (McKinney 1997). Thus, the current extinction rate is 333 times greater than the background rate of extinction.

By these and related methods of estimation, we can conclude that the current rate of extinction already exceeds background rates of extinction, and either already or will soon equal or exceed those observed in the other five mass extinction events of Earth's history. Even losses less extreme than those seen in the mass extinction events are abnormal; a loss of 30% of the species on Earth is roughly equivalent to what occurs on average once in 10 million years (Raup 1994). Yet, more than 30% of amphibians, and over 40% of all species evaluated by the IUCN are threatened with extinction today.

Perhaps more important than the fact that species are going extinct at a great rate, should be the reflection that the time needed to recover levels of species diversity similar to those prior the extinction event is exceedingly long—e.g., 5–10 million years for coral reefs (Jablonski 1995). Our actions are causing biodiversity losses that cannot be recovered in our children's lifetime, or even that of our grandchildren or great grandchildren.

What Types of Species Are Most Vulnerable to Extinction?

Although we do not know the absolute rate of extinction of described and undescribed species today (see Box 3.3), we can describe with greater confidence which types of species are most likely to go extinct. Many attributes can make a species more vulnerable to extinction, including rarity, narrow habitat range, large area requirements, low reproductive rates, extreme specialization or coevolutionary dependencies. Further, species may be vulnerable to different threats, depending on their traits (Owens and Bennett 2000; Isaac and Cowlishaw 2004). Reviewing geological and historical patterns gives us insight into which species are most likely to go extinct in the future. Although a wide variety of possible factors may contribute to vulnerability for various species (see reviews by Ehrenfeld 1970; Terborgh 1974; McKinney 1997), most relate either to aspects of specialization or rarity.

Species vulnerability due to specialization

Many species, particularly in tropical moist forests, have evolved very narrow ranges of environmental tolerance and highly specialized diets or habits. Often these species are at particular risk because such ecological specialization lowers their resilience in the face of perturbations. Further, when these species are reliant on one or few other species, their existence will be threatened should these other populations decline or go extinct, as discussed earlier for the case of plant species dependent on one or few species of pollinators or seed dispersers.

Top carnivores, with low densities, large body size, high tropic position, and large area requirements are often cited as being particularly vulnerable to extinction, particularly through habitat degradation, and also overexploitation (direct or indirectly of their prey) (Purvis et al. 2000). While this suite of factors certainly seems associated with greater extinction risk, large body size does not seem to have led to added risk for marine invertebrates (McKinney 1997), and other scientists have found stronger correlations between extinction risk and other life history traits, such as low rates of reproduction.

Because so many large-bodied species went extinct in the Pleistocene, it has long been assumed that large body size per se carries higher risks of extinction—in part because larger species were more valuable to hunters. While this has validity, close analysis reveals that extinction risk for Pleistocene megafauna was greatest for species with low reproductive rates, regardless of body size (Johnson 2002). Thus, what seems to have happened is that those species that could not replace themselves to meet the pace of human hunting went extinct. Further, nearly all large-bodied, slow-reproducing mammal species that survived to the present day were nocturnal, arboreal, alpine or deep-forest dwellers, suggesting that only those species that could elude human hunters survived to the present day (Johnson 2002).

Species with low reproductive rates typically can rely on adults to weather minor environmental variations, and thus have adapted to follow what is often called a "K-selected" life history strategy. These species are long lived, produce few young (which in some cases are cared for extensively by their parents), have high juvenile survivorship, and abandon reproduction when environmental conditions are poor to maintain high adult survivorship. Overall, these life history characteristics may make them unable to cope with the rapid changes occurring today. Because the evolutionary response to

stresses is to reduce reproduction, such species will decline in abundance, eventually to extinction.

Certainly, the decline and extinction of marine mammal species follows this pattern. Larger species were hunted initially, but these were also the species with the slowest maturity and lowest reproductive rates, and thus unable to replace themselves under industrial-scale exploitation (Roberts 2003). Many species of sharks, rays, and bony fishes today that have lower reproductive rates also show signs of greater vulnerability to exploitation, and thus greater extinction risks (Jennings et al. 1998; Dulvy et al. 2000). Often the most vulnerable species have later maturation, which also increases their susceptibility, as harvest before individuals have reproduced is thus more likely.

Vulnerability of rare species

A species may be rare for many reasons: because of a highly restricted geographic range, because of high habitat specificity, or because of low local population densities, as well as due to a combination of these characteristics (Rabinowitz et al. 1986; Gaston 1994; Table 3.5). Often a species may have attributes that seem to confer abundance, combined with ones that create rarity, and thus it is considered to be rare overall. For example, a species that occurs across an entire ocean basic (broad geographic range) and in high abundance, but is only found in a few isolated habitats such as deep-sea vents (high habitat specificity), is considered a rare marine species. Similarly, some species are never abundant anywhere, although their range may be quite extensive across broad geographic areas, as well as habitat types. Higher extinction rates are correlated particularly with species with re-

stricted ranges and low density (Jablonski 1991; Gaston 1994; Rosenzweig 1995; Purvis et al. 2000).

Species with highly restricted habitat preferences, such as a plant growing only within a rare soil type or seabirds that can only nest on cliffs with the correct prevailing winds, are particularly vulnerable to habitat degradation, as once their habitat is destroyed they are unable to adjust to another location. Common species that aggregate only in very particular locations, such as many bat species that roost in caves, are not only vulnerable to habitat degradation, but can be overexploited with ease. Often, recognizing habitat restrictions challenges us to transcend our experience to understand what environmental factors are critical to a marine invertebrate, an insect, or a fern.

Many of the highest estimates of future extinction rates are derived from our observations of extremely narrow ranges, or extreme endemism among many species in the tropics—particularly plants and insects, most of which are undescribed species (see Box 3.3). We rarely know why these species are so narrowly distributed, although many may be extreme habitat specialists or simply may not have dispersed to other places where they could persist, and thus have a restricted range for historical reasons. However, many analyses have shown that a restricted geographic range is the strongest predictor of extinction risk (e.g., carnivores and primates, Purvis et al. 2000; birds, Manne et al. 1999).

To better understand how extreme endemism can imperil species, consider an event documented in the mid-1980s. A group of scientists were doing RAP inventories in the western Andes of Ecuador, finding rapid turnover

TABLE 3.5 *Seven Forms of Species Rarity, Based on Three Distributional Traits*

Population size		Geographic range			
		Large		**Small**	
Somewhere large		Common	Locally abundant over a large range in a specific habitat	Locally abundant in several habitats but restricted geographically	Locally abundant in a specific habitat but restricted geographically
Everywhere small		Constantly sparse over a large range and in several habitats	Constantly sparse in a specific habitat but over a large range	Constantly sparse and geographically restricted in several habitats	Constantly sparse and geographically restricted in a specific habitat
		Broad	**Restricted**	**Broad**	**Restricted**
			Habitat specificity		

of plant species, with large numbers of endemics. On a single ridge, Centinela, they found 90 endemic plant species. Just after their work was completed, the entire ridge was cleared for agriculture, and the 90 plant species were forever destroyed (Dodson and Gentry 1991). Although it is possible that some of these species may be found in other locations, many are likely now globally extinct, as well as the numerous associated animal species (especially insects) we can presume specialized on these endemic plants. In the words of E. O. Wilson (1992),

> [Centinela] deserves to be synonymous with silent hemorrhaging of biological diversity. When the forest on the ridge was cut, a large number of rare species were extinguished. They went just like that, from full healthy populations to nothing, in a few months. Around the world such anonymous extinctions—call them "centinelan extinctions"—are occurring, not open wounds for all to see and rush to stanch but unfelt internal events, leakages from vital tissue out of sight.

Island communities are typically rich in endemics, but also poorer in species than comparable mainland communities. Low species richness on islands is attributed to a combination of low colonization rates of new species (the functional equivalent of a low rate of speciation), and high extinction rates (because populations are usually small and subject to decimation by local catastrophes and stochastic variation) (MacArthur and Wilson 1967). Also, island communities have experienced extremely high extinction rates of species during recent centuries, primarily due to anthropogenic introductions of mammalian predators (mammals other than bats disperse poorly across ocean barriers) and mainland diseases.

Many groups are more endangered on island systems than on mainlands. In the new world, amphibians are much more endangered among the islands of the Caribbean (84% are at risk of extinction) compared with the adjacent continental landmasses of Mesoamerica (54%), South America (31%), and North America (21%; Young et al. 2004). As discussed earlier, avian extinctions have been much more common on islands, particularly due to combination of pressure from rapid habitat change, overexploitation, and invasive species that may act in an ecological role not represented on the island (e.g., introduced mammalian predators), and thus represent a challenge outside the evolutionary experience of many island species. However, carnivores and primates are no more likely to be endangered on islands than on continents (Purvis et al. 2000).

Small populations are generally more vulnerable to extinction than large ones. This has been seen in several comparisons of faunal surveys taken 50–100 years apart. Nearly 40% of small populations of birds (with less than 10 breeding pairs) went locally extinct in the California Channel Islands over an 80-year period compared with only 10% of those populations with 10–100 pairs, only 1 population with 100–1000 pairs, and none of those with greater than 1000 pairs (Jones and Diamond 1976).

In the Baltic Sea, intense harvests of skates have caused the smallest populations to disappear, while larger ones are still extant (Dulvy et al. 2000). Many species with small populations may belong to species with large body size, which is often correlated with slower reproductive rates and other life history attributes that increase vulnerability to extinction.

Widely dispersed species that are always rare may be vulnerable to local extinction events, although they are less likely to become globally extinct. Some marine invertebrates may fall into this category, as well as populations of top carnivores. Although such species may not become globally extinct, local extinctions are potentially very damaging to communities and ecosystems, as discussed earlier.

We may need to worry most about **artificial rarity**, where a once widespread or abundant species has been reduced to a very small population size through human activities. Species that are artificially rare seem more likely to go extinct than species that are evolutionarily adapted to rarity (Kunin and Gaston 1997). Thus, small population size alone may be less predictive than a sudden reduction to small population size (which is one rationale for the criteria for listing species in the Red List; see Box 3.2).

"Bad luck": Extrinsic causes of extinction due to human activities

In addition to these categories, we might also add one for species who simply have "bad luck" (Raup 1991). These species are not intrinsically vulnerable due to their traits, but rather have the misfortune to be in the wrong place at the wrong time, or of being particularly palatable to humans. That is, living in areas of greatest human impact, such as on arable soils, in river systems that are heavily used for commerce and subject to substantial pollution, or along coasts where human populations congregate, will thrust species that share those locations directly into harm's way. For example, freshwater fishes in the most extensively altered ecosystems may all share an increased likelihood of extinction, despite having a wide variety of traits (Duncan and Lockwood 2001). Purvis et al. (2000) estimated that roughly half of the variation in extinction risk among carnivores and primates related to their exposure to anthropogenic disturbances, particularly to high rates of habitat loss or intense commercial or bushmeat exploitation.

Environmental changes may occur very rapidly, as they appear to have in the mass extinction events of geologic history, and then vulnerability to extinction may be widely shared among species (Raup 1991). The intensification of human pressures on ecosystems may have the same effect as the environmental changes of the past.

Our need to grow enough food to feed the increasing human population, particularly if large proportions eat meat, requires greater conversion of land for agriculture (see Chapter 6). Increasing rates of exploitation may hit those species with low reproductive rates hardest, but to some degree all species that are marketable may be endangered as local demand for bushmeat, for example, explodes with local human population density (see Case Study 8.3), or as commercial fisheries strive to meet world demand for protein (see Chapter 8). Global trade is increasing transport of species (see Chapter 9), and therefore increasing the chance that invasive species will take hold in some new place. Finally, our consumption of fossil fuels and other practices that drastically increase greenhouse gases has begun to change our climate (see Chapter 10)—which may be the worst luck of all.

An awareness of factors that enhance vulnerability to extinction can help us formulate plans to counteract these factors to some degree. Understanding which traits increase vulnerability also may help us predict which species will most need protections as human pressures increase in areas where they have been slight. However, it is also possible that increasingly species will best be categorized as simply having the bad luck to be in the path of human development.

Economic and social contexts of endangerment

Underlying threats caused by human development are powerful economic and social drivers. Perhaps most of all, economic growth and rising affluence do the most to increase the pace of habitat conversion, overexploitation of marine species for food, pollution of waterways, and transformation of our atmosphere that is changing our climate. In the U.S., causes of endangerment are those primarily associated with rapid economic growth—urbanization and agricultural expansion, infrastructure development (roads, reservoirs, wetland modification), and the byproducts of these activities (pollution, habitat degradation) (Czech et al. 2000). Moreover, geographic centers of endangerment currently occur where high diversity meets rapidly expanding economic activity in southern California, east-central Texas, and southern Florida (Dobson et al. 1997; Czech et al. 2000). As Czech et al. (2000) put it, "The list of endangered species is growing because the scale of the integrated economy, and therefore the causal network of species endangerment, is increasing."

At the other end of the economic continuum, the majority of people struggle in poverty (UN Millennium Project 2005). Extreme poverty afflicts over 1 billion people, who live on less than $1/day, are chronically hungry, lack clean water and sanitation, are burdened with preventable diseases, and often lack shelter and sufficient clothes. An additional 2.7 billion live on scarcely $2/day, and are just able to meet their basic needs, but no more. Thus, it is

no wonder that unsustainable levels of burning, small-scale agriculture, grazing, and bushmeat hunting occur wherever these practices help the poorest people to survive. Biodiversity losses are clearly associated with this human tragedy, but even more serious are the signs that erosion of ecosystem services increasingly undercut the ability of human populations to escape extreme and moderate poverty (Millennium Ecosystem Assessment 2005).

Further, globalization spreads the dangers of economic expansion as countries prioritize international trade in natural resources to enhance their economies. This can exacerbate poverty where trade is not tied to sustainable practices that improve the lives of local people, but rather provide cheap goods that fuel the economies of developed nations. Often, pressures to overexploit and transform land come principally from decisions of large multinational corporations, and consumer demand in far-off countries. In addition, globalization also has radically accelerated the spread of nonnative species. Rapid economic expansion in China and other densely populated countries has raised concern that pressures on developing countries will intensify sharply in the near term (Millennium Ecosystem Assessment 2005).

Responses to the Biodiversity Crisis

Clearly, we live in a time of crisis for biodiversity. While many fine points can be debated, and some uncertainties remain, it is certain that humans have wrought radical change throughout the globe. Moreover, as our populations grow and consume more, ecosystems will continue to degrade, more species will go extinct, and human suffering on a staggering scale will continue. How can we respond to crises of this magnitude? Conservation biologists and promoters of human development agree that solutions will need to come from a mixture of (1) scientific analysis of and communication about the drivers of change in biodiversity and human welfare, (2) technological improvements, (3) legal and institutional instruments, (4) economic incentives and plans, and (5) social interventions. The mechanisms for achieving solutions will include preventative measures, such as establishment of protected areas (see Chapter 14), targeted interventions at genetic, species, and ecosystem levels that integrate ecological understanding with community-based problem solving (see Chapters 11–13), restoration of damaged ecosystems or endangered populations (see Chapter 15), and creation of truly sustainable forms of development (see Chapter 16).

Most approaches to conservation focus on species, ecological communities or ecosystems, or landscapes, although some also focus more broadly (and vaguely) on biodiversity overall (Redford et al. 2003). Because biodiversity conceptually spans the range from genes to ecosystems, and spatial scales from single ecological

communities to landscapes and larger, effective conservation is most likely to emerge from multiple approaches undertaken at a range of spatial scales, directed toward different conservation targets (species, ecosystems or landscapes). Conservation approaches focus from single species in a single site to all species across the globe, including species or ecosystem targets, and often explicitly including human communities and development (Redford et al. 2003).

Conservationists recognize that species approaches alone will not be sufficient to conserve biodiversity, and that actions on larger spatial scales targeting ecosystems and landscapes may be more efficient and effective as ecological processes will be more likely to remain intact when conservation is undertaken at these scales (Redford et al. 2003). However, conservation is usually achieved by many smaller actions, undertaken at smaller spatial scales. Overall, to address the numerous threats to biodiversity, conservationists need methods to prioritize conservation activities, and plans for how to achieve conservation at various scales. As Redford et al. (2003) put it, conservationists' generally are asking "where" questions to set geographical priorities, and "how" questions about developing and implementing strategies to conserve conservation targets at priority places.

The remainder of this section focuses on just two issues in promoting solutions—the use of national and international agreements to address our biodiversity crisis, and the issues involved in providing clear indicators of problems and of progress. The focus here will remain more on actions that are centered on species, whereas a discussion of approaches aimed more at habitat protection is included in Chapter 6, and longer treatments of approaches to solving the biodiversity crisis are included in Chapters 11–18.

Laws and international agreements that address biodiversity loss

Foremost among efforts to conserve biodiversity are national laws and international agreements that limit human impacts to threatened species and ecosystems, and establish a mandate for governmental and citizen action. Major international agreements, the landmark U.S. Endangered Species Act (ESA), and other key U.S. laws that protect biodiversity are described by Daniel Rohlf in Case Study 3.3. These institutional policy instruments are most effective when they reflect ecological and evolutionary understandings (as in the ESA, where protecting distinct population segments was given importance for future evolution and local community interaction), are well linked to human activities that can be amended, and have adequate enforcement

Many of the laws to protect biodiversity do so via lists of threatened species, as described earlier. In a broad

sense, these lists are used to set priorities for species protection, although of course many other factors—including economic and political ones—determine which species are targeted for conservation actions (Czech and Kaufman 2001). Laws and international agreements more effectively achieve conservation goals when they reduce the tendency for economic and political pressures to be paramount in decision-making, instead requiring a fair examination of potential harms and benefits to biodiversity. This may be accomplished via effective enforcement (e.g., the provision for citizen lawsuits in the ESA, or for severe trade restrictions to accompany failure to abide by animal trade prohibitions under CITES; see Case Study 3.3), and by mechanisms that promote economic incentives and disincentives (as are being developed under the Convention on Biological Diversity), or that increase public involvement in solution creation (e.g., Safe Harbor agreements under ESA).

Some conservationists have cautioned against using threatened species lists as too literal a guide for priorities. Because so many groups are poorly known, there is under-representation for many taxonomic groups (e.g., nearly all groups of invertebrates and nonvascular plants) and in areas of the world (most of the developing world, and most marine habitats). Prudence suggests we should be very cognizant that the biases in our knowledge could lead us to ignore taxa, or indeed other levels of biodiversity besides that of species that may be critical conservation targets (Possingham et al. 2000; Burgman 2002). Thus, often policies that are directed explicitly toward broader protections for habitat rather than species, or toward enhancing sustainable practices may be preferred to avoid such biases. However, usually conservation is not an either-or venture, but rather requires a suite of complementary approaches, some of which include admittedly imperfect policy instruments that motivate and guide action at many levels.

International agreements are meant to serve as motivators to action. At the 2002 Johannesburg World Summit on Sustainable Development, 190 countries committed to the reduction of biodiversity losses significantly by 2010. This in turn motivated efforts to develop more specific policies for conservation on global, national, and local levels. Further, this has motivated an examination of how we can track both problems and progress.

Identifying driving factors and trends in species endangerment

To create an effective plan to recover declining species, or to address biodiversity loss more broadly in some region of the world, a necessary step is to use rigorous scientific methods to determine which factors most influence population decline, and address our conservation actions to remove or ameliorate these threats (Caughley and Gunn

1996). While to some degree this is obvious, Caughley and Gunn found numerous examples where assumptions have led well-meaning conservationists astray. Deducing which causal factors drive declines should involve clearly delineating a plan of analysis to separate potential threats; this plan should include setting up testable hypotheses and predictions that follow from these hypotheses, and reviewing all evidence to evaluate each hypothesis carefully. To do this it is necessary to learn enough about the species' natural history to develop reasonable hypotheses. Commonly, the agent or agents of decline are not obvious, so conservation scientists have to become good detectives.

Beyond assessing the drivers of endangerment in particular cases, it is critical to track status trends to set priorities for intervention, and also to know when we are succeeding in protecting species. The IUCN has kept data on species endangerment for many years, and can now track the progress of the best-studied groups. The Red List Index (RLI) tallies changes in status due to either a deterioration or improvement of all threatened and near-threatened species since 1988 (Butchart et al. 2004). As applied to birds, RLI analyses show a portrait of steady deterioration of status (e.g., moving from vulnerable to endangered) (Figure 3.10). The RLI for birds has decreased by nearly 7% since 1988, roughly equivalent to 10% of all species in NT, VU, EN, and CR moving up into the next higher category of threat. For albatrosses and petrels the decline has been particularly steep, equivalent to a 25% drop in RLI since 1995, indicating a very perilous situation. A retrospective analysis for amphibians also indicates steep declines of roughly 15% in RLI (Baillie et al. 2004).

Most species groups are too poorly known to adequately evaluate trends, but a variety of indices help bridge this gap. These range from the index of biotic integrity that is used to assess stream health (Karr 1991), to composite indices that attempt to summarize global changes for the general public, business community, and government leaders. One of these composite indices, the Living Planet Index (LPI), summarizes change over time in populations of over 1100 terrestrial, freshwater, and marine vertebrate species (Loh et al. 2005). Because they are much better known, most of the data are drawn from bird and mammal species, on land in temperate climes. At present, these are the best data available, but it may be possible to weight the data to compensate for these biases as more data are collected on currently under-represented and little known groups (Loh et al. 2005). The LPI shows that terrestrial vertebrates have declined by 25% between 1970 and 2000 (Figure 3.11A). Further, freshwater species have declined more sharply than either terrestrial or marine vertebrates (see Figure 3.11B). Both these trends are disturbing. Importantly, the results of the LPI are discussed internationally, and are succeeding in conveying a straightforward and compelling message about overall trends.

Increasingly it is helpful to examine indicators of future change based on economic and social parameters (see also

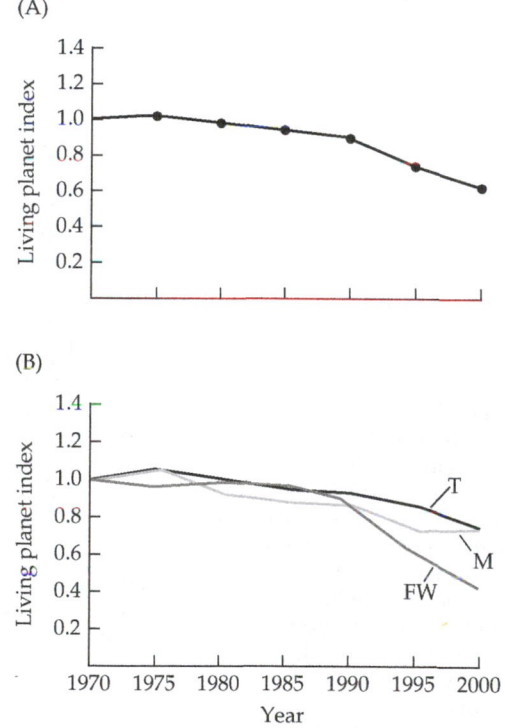

(A)

(B)

Figure 3.11   The Living Planet Index (LPI) tracks population trends for over 1100 terrestrial, freshwater, and marine vertebrates. (A) Terrestrial vertebrates have declined 25% from 1970 to 2000. (B) Freshwater (FW) vertebrates have declined more than either terrestrial (T) or marine (M) vertebrates. (Modified from Loh et al. 2005.)

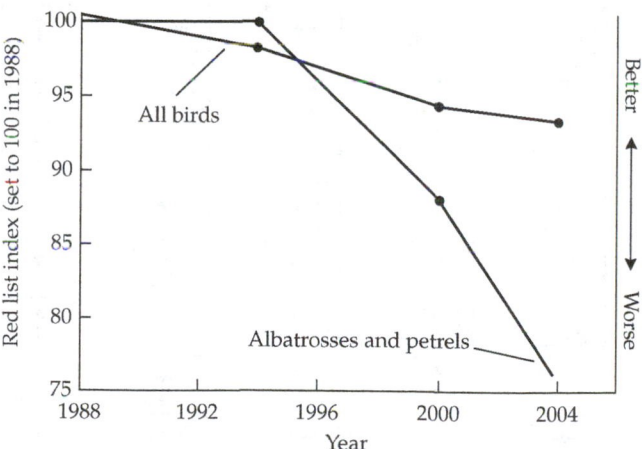

Figure 3.10   The Red List Index (RLI) shows a worsening of threatened status between 1988 and 2004 for all Red Listed bird species and a precipitous decline for albatrosses and petrels. (Modified from Butchart et al. 2004.)

Chapter 18). McKee et al. (2003) developed a stepwise regression model that predicted 87.9% of the variance in the density of threatened mammal and bird species based on human population densities and species richness in 114 countries. Because the model showed no bias, they used it to project potential impacts of increases in human population density on future levels of threat. If human populations grow as predicted by the United Nations and the correlations between human density and density of threatened mammals and birds do not change, then the number of threatened mammals and birds should increase by 7% by 2020 and by 14.7% by 2050 (McKee et al. 2003). While it does not direct specific action, the model can be used to help convince decision makers that interventions are needed now, and that delay has serious consequences.

The most comprehensive survey of the current status of biodiversity as it supports human life was made by a consortium of thousands of scientists in the Millennium Ecosystem Assessment (MA), completed in March 2005. Their conclusions are perhaps the most sobering of all. Although we have substantially increased food production worldwide, nearly all other indicators of ecosystem services have declined sharply since 1950, as human population grew from 2.5 to nearly 6.5 billion, and economic growth worldwide increased nine-fold (and in the U.S., twenty-five-fold). Threatening processes are ex-

pected to intensify, particularly habitat loss and degradation via pollution, species invasion, and climate change (Figure 3.12). Inland and coastal waters and grasslands have already born the greatest impacts, and are expected to continue to be subject to intense pressures. The results of the MA should serve as a particularly strong wakeup call that, unless we wish to leave a substantially impoverished and dangerous world to our children, we need to make far-reaching changes in our global, national, and local behavior now.

To achieve our goals for conserving biodiversity, we will be well served by working to understand better the factors that drive human behavior, and the means that can effectively change destructive behaviors. Jeffrey Sachs, leader of the UN Millennium Development Project recently noted that the world's poorest people do not lack the *will* to follow more sustainable practices, but rather they and their governments lack the *means* (Sachs 2005). Human values, needs, and desires are powerfully, and quite imperfectly, translated into economic constructs that guide public policy and individual actions. One key to a long-term solution to the biodiversity and human poverty crises is to find ways to translate our ethics and the importance of biodiversity into our economics, and thus more integrally into our decision making.

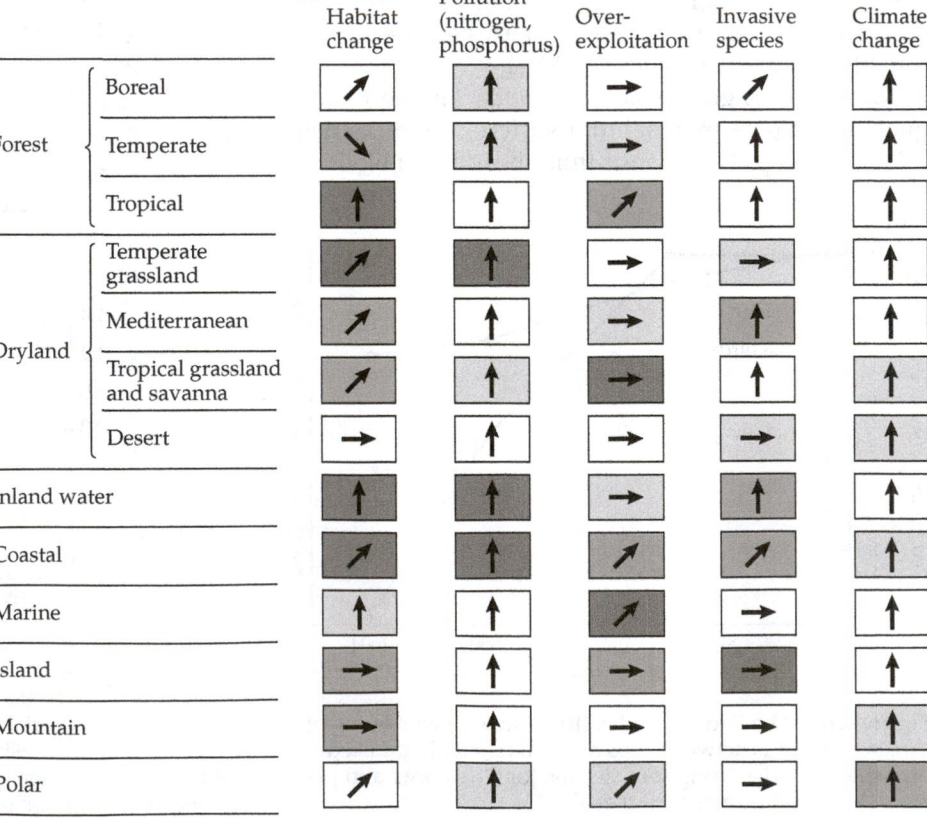

Figure 3.12 Projected trends in threatening processes in different habitat types in the coming decades. Shading indicates the intensity of each process' impacts on biodiversity over the past century, with dark gray indicating the most impact and white the least. Straight upward arrows indicate very rapid increase in intensity of this threat at present; diagonally upward facing arrows indicate increases in the threat; sideways arrows indicate continuing levels of the threat; and downward facing arrows indicate declines in threat intensity. (Modified from Millennium Ecosystem Assessment 2005.)

CASE STUDY 3.1

Enigmatic Declines and Disappearances of Amphibian Populations

Joseph H. K. Pechmann, Western Carolina University, and David B. Wake, University of California, Berkeley

The First World Congress of Herpetology in 1989 and a National Research Council workshop in 1990 brought the accelerating losses of amphibian biodiversity to the forefront of conservation concerns. It long had been obvious that habitat destruction and alteration, introduction of nonnative species, pollution, and other human activities exact an increasingly heavy toll upon amphibians as well as on other taxa. By the time of these meetings, however, herpetologists realized that some of the declines and disappearances of amphibian populations and species were unusual in one respect: The causes of their loss were undetermined. These unexplained losses occurred in isolated areas relatively protected from most human impacts, particularly in montane regions in tropical/subtropical Australia, the western United States, Costa Rica, and other parts of Latin America. Especially noteworthy, detailed retrospective studies of declines in amphibians in large national parks in California (Drost and Fellers 1996; Fellers and Drost 1993) and in the Monteverde Cloud Forest Preserve in Costa Rica (Pounds et al., 1997) presented the first conclusive evidence of community-wide declines of amphibians.

Population declines and range contractions have continued. For example, during the latter 1990's 35 of 55 species disappeared from a study area in the Fortuna Forest Reserve in Panama (Lips 1999; Young et al. 2001). However, many of these cases are complicated by the fact that not all sympatric amphibian species appeared to be affected.

A general problem with these reports was the lack of perspective, with a focus on what may be special cases. To rectify this situation, about 500 amphibian biologists were enlisted to participate in a Global Amphibian Assessment (GAA) conducted from 2000 to 2004 (Stuart et al. 2004). The assessment attempted to evaluate every species of amphibian across the planet, a challenging task when one considers the fact that the vast majority of species are tropical and most of these have been little studied. The GAA documented that the reports were representative of a widespread phenomenon, and that the problem is more acute than generally had been thought; 32.5% of the known species of amphibians are "globally threatened" (Figure A). The assessment also showed that community-wide amphibian declines were more likely to be encountered in the tropics. Many species showed no evidence of declines, however.

The causes of many documented declines and disappearances remain poorly understood despite a burgeoning literature on the subject. Nearly half of the 435 amphibian species classified as rapidly declining by the GAA are threatened primarily by "enigmatic" (unidentified) processes (Stuart et al. 2004). In this case study we briefly review the taxonomic and spatial patterns of these enigmatic losses and their major hypothesized causes. Although the enigmatic declines are our focus, well-documented threats such as habitat loss and overharvesting that are the primary cause of most losses of amphibian biodiversity (including over half of the rapid declines, Stuart et al. 2004) require attention as well.

Although amphibians are more threatened than some other groups, their plight is not unique (Gibbons et al. 2000). The case study of amphibian declines illustrates several themes applicable to studies of biodiversity losses in all taxa:

1. Biodiversity loss has many causes that are not mutually exclusive; several factors may act simultaneously on one population.

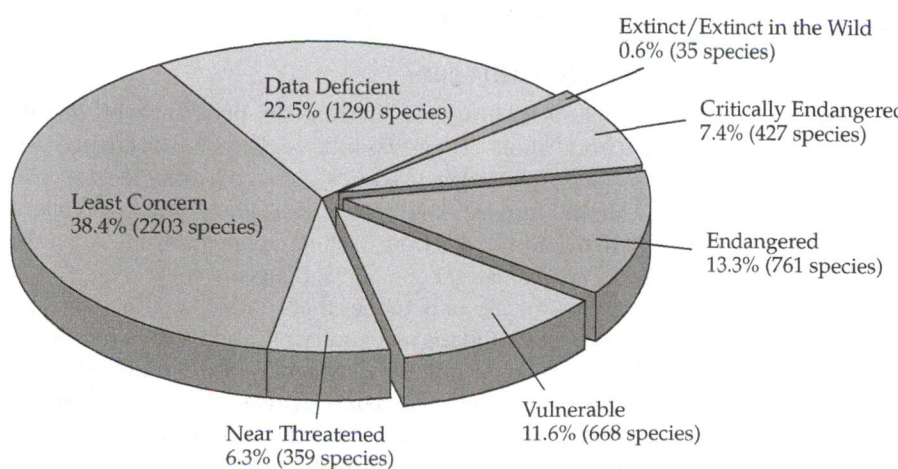

Extinct/Extinct in the Wild
0.6% (35 species)

Data Deficient
22.5% (1290 species)

Critically Endangered
7.4% (427 species)

Least Concern
38.4% (2203 species)

Endangered
13.3% (761 species)

Near Threatened
6.3% (359 species)

Vulnerable
11.6% (668 species)

Figure A   Over one-third of amphibian species have a threatened designation under the IUCN Red List. (From Global Amphibian Assessment 2004.)

2. Species, and populations within species, vary with respect to the importance of different threats.

3. The combined effect of several stressors can be worse than the sum of their individual effects. These synergistic interactions are especially difficult to study.

4. Experimental demonstration that a factor may affect a population is not proof that it is responsible for observed population declines.

5. No area on Earth is protected from human activities.

Species declining for enigmatic reasons are concentrated in the tropics, especially in Mesoamerica, the northern Andes, Puerto Rico, southern Brazil, and eastern Australia, although the phenomenon may be underestimated in poorly-monitored regions such as Africa (Stuart et al. 2004). Most species (82%) having unexplained declines are frogs and toads in the families Bufonidae, Hylidae, and Leptodactylidae. Ecologically, enigmatic declines are associated with species found in forests, streams, and tropical montane habitats (Stuart et al. 2004). A variety of hypotheses as to the underlying causes of these enigmatic declines have been proposed and studied by a host of conservation scientists.

Ultraviolet-B Radiation

Increased ultraviolet-B (UV-B) radiation is a possible explanation for declines and losses of amphibian populations in comparatively undisturbed areas. A large body of experimental evidence indicates that ambient levels of UV-B are harmful to some amphibian species at some locations, although effects may vary among populations, species, and geographic regions and with ecological factors such as elevation, water chemistry, and the presence of other stressors (reviewed by Blaustein and Kiesecker 2002; Blaustein et al. 2003a). For example, egg mortality in the western toad and Cascades frog was higher when eggs were exposed to ambient levels of sunlight in natural ponds in Oregon than when 100% of UV-B was blocked (Blaustein et al. 1994a; Kiesecker and Blaustein 1995). UV-B had no effect on the survival of Pacific treefrog eggs in the same experiments. The Pacific treefrog was found to have higher levels of an enzyme, photolyase, that facilitates the repair of DNA damaged by UV-B (Blaustein et al. 1994a). Ambient UV-B did not affect survival of western toad eggs in a field experiment conducted in Colorado (Corn 1998). Differences between the Oregon and Colorado results could be due to differences in genes, ecological conditions, or experimental design. UV-B had a greater effect on survival of long-toed salamander larvae from low-elevation populations than those from high-elevation populations under identical conditions in the lab (Belden and Blaustein 2002).

There is also much experimental evidence that UV-B has a greater effect on amphibians when combined with other stressors (Blaustein and Kiesecker 2002; Blaustein et al. 2003a). For example, UV-B decreased the survival of *Rana pipiens* eggs in acidic water but not at near-neutral pH (Long et al. 1995). In some experiments western toad and Cascades frog eggs died as a result of UV-B increasing their susceptibility to a pathogenic fungus, *Saprolegnia* (Kiesecker and Blaustein 1995; Kiesecker et al. 2001a).

This body of experiments helps us understand some of the impacts of UV-B on amphibians, but does not show that UV-B is responsible for population declines and disappearances. It remains unknown whether the affected populations were exposed to increased UV-B, or that the effects of UV-B extend from individuals to the population level.

Ozone depletion is one factor that may have increased exposure of amphibians to UV-B, even in some locations at low latitudes (Middleton et al. 2001). Many other factors also affect the UV-B dose experienced by amphibians, including the attenuation of UV-B by water and by dissolved organic matter in water, breeding phenology, and behavior (e.g., Corn and Muths 2002, Palen et al. 2002). Climate warming and anthropogenic acidification may reduce concentrations of dissolved organic matter and thereby increase the depth to which UV-B penetrates aquatic habitats (Schindler et al. 1996; Yan et al. 1996).

Ecologists disagree on interpretation of research on the extent to which natural levels of dissolved organic matter currently protect amphibians from harmful levels of UV-B (Blaustein et al. 2004; Palen et al. 2004). Global climate change can affect precipitation, and consequently water depths and UV-B exposure (Kiesecker et al. 2001a). In dry years eggs will be closer to the surface of the water and thus more exposed to UV-B. A complicating factor is that in some regions, such as the Rocky Mountains, montane snow melt and amphibian breeding occur earlier in the season in dry years, when UV-B radiation is lower (Corn and Muths 2002). These issues illustrate why it is difficult to evaluate possible relationships between global climate change and UV-B exposure (Blaustein et al. 2004; Corn and Muths 2004). Integration of information on lethal and sublethal effects of UV-B throughout an amphibian's life cycle with demographic models will bring the phenomenon into a population dynamics framework (Biek et al. 2002; Vonesh and De la Cruz 2002).

Disease Pathogens

Pathogens may cause declines and disappearances of amphibian populations. There was little evidence for this hypothesis until a pathogen new to science, *Batrachochytrium dendrobatidis* (a chytrid fungus), was found in dead and dying frogs collected in the mid-1990's from declining populations in Australia and Panama (Berger et al. 1998; Longcore et al. 1999). *B. dendrobatidis* attacks only tissues that contain keratin, which include the mouthparts in anuran (frog and toad) tadpoles and the skin in salamanders and metamorphosed anurans (Berger et al. 1998; Bradley et al. 2002; Davidson et al. 2003). Research to date suggests that chytrid infections cause little or no mortality of anuran tadpoles or salamanders, although it can decrease the growth rates of the former, making them more susceptible to predators or other stressors (e.g., Davidson et al.

2003; Parris and Beaudoin 2004). *B. dendrobatidis* can kill metamorphosed frogs, either by producing a toxin or by interfering with skin functions such as respiration and osmoregulation (e.g., Berger et al. 1998; Rachowicz and Vredenburg 2004).

Batrachochytrium dendrobatidis has been isolated from other frog populations undergoing mass mortality events associated with population declines and disappearances, including *Rana yavapaiensis, R. chiricahuensis,* and *H. arenicolor* in Arizona (Bradley et al. 2002), the midwife toad (*Alytes obstetricans*) in Peñalara Natural Park, Spain (Bosch et al. 2001; Martinez-Solano et al. 2003b), and (retrospectively, using preserved specimens) several species in Las Tablas, Costa Rica (Lips et al. 2003a). It has also been isolated from populations not known to be in decline, including tiger salamanders in Arizona (Davidson et al. 2003), *Litoria wilcoxii* or *jungguy* (taxonomy uncertain) in Australia (Retallick et al. 2004), and *Xenopus laevis* in southern Africa (Weldon et al. 2004). Many individuals in these populations had light infections and appeared healthy. Antimicrobial peptides located in the skin may provide some species with natural defenses against *B. dendrobatidis* (Rollins-Smith et al. 2002a, b). Environmental conditions in some locations may not be favorable for the fungus. For example, *B. dendrobatidis* can grow and reproduce at temperatures of 4°C–25°C, whereas growth ceases at 28°C and 50% mortality occurs at 30°C (Piotrowski et al. 2004). This may explain why amphibian population declines and disappearances associated with chytrids have been observed in cool tropical montane areas (Berger et al. 1998; Lips et al. 2003a), but not in tropical lowlands where temperatures are often above 30°C.

Several scenarios for the association between chytrid fungus infection and population declines and disappearances are possible: (1) amphibians have long coexisted with *B. dendrobatidis,* and observed population changes are cyclical phenomenon that previous researchers may not have noticed; (2) amphibians have long coexisted with *B. dendrobatidis,* but environmental change or stress has made them more susceptible to the fungus or the fungus more pathogenic to the amphibian; and (3) *B. dendrobatidis* is a novel disease recently introduced to susceptible populations around the world through human activities such as the pet trade.

Scenario 1 would follow that of many wildlife diseases, including ranaviruses in North American tiger salamanders and in United Kingdom common frogs (Daszak et al. 2003). This scenario is considered unlikely for amphibian chytridiomycosis because of the large magnitude of the population changes and the lack of recovery in many cases (Daszak et al. 2003).

There are many ways in which an established coexistence between chytrids and amphibians may have changed over time, as in scenario 2. For example, increases in UV-B (Kiesecker and Blaustein 1995), exposure to pesticides (Taylor et al. 1999; Gilbertson et al. 2003), or other stresses can result in immunosuppression and disease emergence. Climate change can induce droughts, causing amphibians to aggregate around water bodies and increasing their exposure to waterborne diseases such as *B. dendrobatidis* (Pounds et al. 1999; Burrowes et al. 2004). Immunosuppression would be likely to increase the prevalence of many diseases, however, not just chytridiomycosis.

Several pieces of data are consistent with the hypothesis that *B. dendrobatidis* is a novel pathogen that has recently been spread around the world with the assistance of humans. For four cases where *B. dendrobatidis* was associated with population declines and disappearances, chytrids could not be detected in museum samples collected prior to the population crash (Berger et al. 1998; Fellers et al. 2001; Lips et al. 2003a). Disease would have to have been very prevalent to allow detection from the small number of samples that were available, however (Lips et al. 2003a). Little genetic variation has been found so far in *B. dendrobatidis* collected around the world, although additional work is needed (Morehouse et al. 2003). Two amphibians that have been transported all over the world, the American bullfrog (*Rana catesbeiana*) and the African clawed frog (*Xenopus laevis*), are carriers of *B. dendrobatidis* (Daszak et al. 2004; Hanselmann et al. 2004; Weldon et al. 2004). The sudden, catastrophic nature of some declines and disappearances also suggests an introduced pathogen (Daszak et al. 2003).

If the chytrid is novel to most areas, where might it have originated? Some workers suggest it is endemic in populations of *Xenopus* in Africa (Weldon et al. 2004). The earliest documented case of chytridiomycosis was in a *X. laevis* collected in South Africa in 1938 (Weldon et al. 2004). The chytrid seems to have a benign relationship to this species, and *X. laevis* is a popular laboratory animal exported worldwide beginning with its use in pregnancy tests in the 1930s.

Introduced Species

Predation and competition from introduced species other than pathogens may have caused declines of some amphibian populations in isolated, seemingly protected areas. For example, the introduction of trout for sport-fishing is thought to have been an important factor in some disappearances of frogs in the Sierra Nevada of California. All but 20 of the 4131 mountain lakes of the state were fishless in the 1830s, as were most high-elevation streams in the Sierra (Knapp 1996). Stocking in Yosemite National Park reached a peak of a million fish each year in the 1930s and 1940s (Drost and Fellers 1996). Stocking has recently been reduced in the Sierra Nevada and discontinued in the region's national parks, in part because of concern about its effects on amphibians (Carey et al. 2003).

The best-documented effects of introduced fishes are for the mountain yellow-legged frog (*Rana muscosa*). It was once a common frog in high-elevation lakes, ponds, and streams of the Sierra Nevada (Grinnell and Storer 1924), but disappeared from over 85% of historical sites in the Sierra (Bradford et al. 1994b; Drost and Fellers 1996; Vredenburg 2004). *Rana muscosa* has a larval period of 1–4 years in the Sierra, and therefore requires permanent bodies of water, as do fishes (Wright and Wright 1949; Knapp and Matthews 2000). There is a strong

negative association between the presence of fish and of *R. muscosa*, even when habitat type and isolation are taken into account (e.g., Bradford 1989; Knapp and Matthews 2000). Densities of *R. muscosa* are low in the few sites where it does co-occur with fishes (Knapp and Matthews 2000; Knapp et al. 2001; Vredenburg 2004). *Rana muscosa* recolonizes lakes from which fishes have disappeared (Knapp et al. 2001) or have been removed experimentally (Vredenburg 2004; Figure B).

Although introduced fishes can account for the disappearance of *R. muscosa* from some sites, *R. muscosa* also disappeared from lakes without fishes in Sequoia, Kings Canyon, and Yosemite National Parks (Bradford 1991; Bradford et al. 1994b; Drost and Fellers 1996). Furthermore, some disappearances occurred during or after the late 1970s, after fish stocking was on the wane. One school of thought is that these other disappearances are an indirect result of fish introductions (Bradford 1991; Bradford et al. 1993; Knapp and Matthews 2000; Vredenburg 2004). According to this hypothesis, *R. muscosa* exists in metapopulations in which populations sometimes go extinct due to natural stochastic processes such as winterkill, drought, disease outbreaks, and predation. Sites are then recolonized by individuals migrating along streams. Stocking of fishes in lakes and streams has reduced the number, size, connectivity, and average habitat quality of *R. muscosa* populations, however. Thus, populations are now more likely to go extinct, and less likely to be recolonized if they do. The time lag between fish stocking and the disappearances of some populations is the

"extinction debt" predicted by metapopulation theory (Hanski 1998; Hanski and Ovaskainen 2002).

Introduced fishes and bullfrogs (*Rana catesbeiana*) are thought to have contributed to widespread range reductions of lowland yellow-legged frogs, *Rana boylii* (Hayes and Jennings 1986; Kupferberg 1997) and red-legged frogs, *Rana aurora* (Adams 2000; Hayes and Jennings 1986; Kiesecker et al. 2001b) in the western U.S. Their role is difficult to evaluate, however, because fish introductions, bullfrog introductions, and habitat alterations have occurred concomitantly across the landscape (Hayes and Jennings 1986). Several experiments have sought to disentangle these factors. Kupferberg (1997) found that bullfrog tadpoles outcompeted *R. boylii* tadpoles, and documented that breeding populations of *R. boylii* were greatly reduced in a stretch of stream that was unaltered except for the presence of *R. catesbeiana*. Experiments with *R. aurora* and native Pacific treefrog (*Pseudacris regilla*) tadpoles concluded that direct effects of the introduced species were less important than the widespread conversion of temporary ponds to permanent ponds, which provide better habitat for the introduced than the native species (Adams 2000). Kiesecker et al. (2001b) found that these habitat alterations intensified competition between *R. aurora* and *R. catesbeiana* tadpoles.

Negative relationships between the distribution of introduced fishes and the distribution and abundance of other amphibians have also been found, including the Pacific treefrog (*Pseudacris regilla*) in the Sierra Nevada (Matthews et al. 2001) and most native species in the mountains of northern Spain (Brana et al. 1996; Martinez-Solano et al. 2003a). The effects of fishes on *P. regilla* and other species are apparently localized, perhaps because they use some habitats that fish cannot, such as ephemeral ponds and terrestrial areas, which may ameliorate landscape-level effects of fish stocking (Bradford 1989; Matthews et al. 2001). *Bufo boreas* and *B. canorus* frequently breed in fishless temporary ponds and produce toxins that fish avoid (Bradford 1989; Drost and Fellers 1996) and these fishes may have little affected their declines. Because fishes have been introduced into nearly all montane systems on the planet, the possibility of impacts exists for many amphibian species, most of which are as yet unstudied.

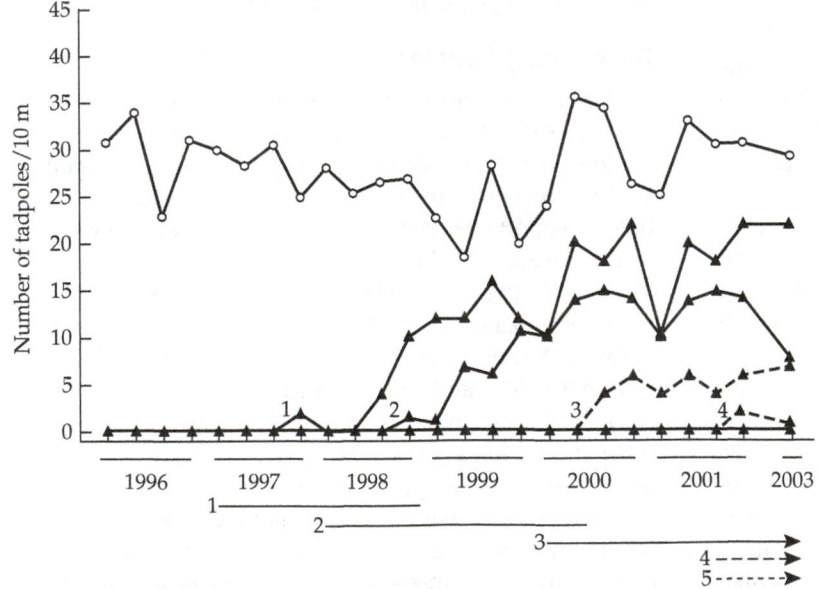

Figure B Density of larval *Rana mucosa* in 21 lakes in the Sierra Nevada, California, from 1996 to 2003. Filled triangles represent trout removal lakes (*n* = 5), and numbers correspond to individual lakes. Open circles indicate fishless control lakes (*n* = 8). The horizontal bars beneath the graph indicate the time period over which trout were removed. *R. mucosa* had not yet recolonized lake 5 after trout were removed in 2001. Control lakes with trout (*n* = 8) never had any frogs present during the duration of the study. No tadpole counts were made in 2002. (Modifed from Vredenburg 2004.)

Chemical Pollutants

Pesticides, herbicides, heavy metals, and other chemical pollutants can have lethal, sublethal, and indirect effects on all organisms, including amphibians (Sparling et al. 2000; Blaustein et al. 2003a; Linder et al. 2003). Chemical pollution might seem to be a minor threat in isolated, protected areas; however, chemical contaminants may be transported long distances through at-

mospheric processes. Further, extremely low, supposedly safe doses of chemicals can harm biota (e.g., Marco et al. 1999; Relyea and Mills 2001). For example, exposure to 0.1 ppb of the herbicide atrazine causes gonadal deformities in northern leopard frogs (*Rana pipiens*) and African clawed frogs (*Xenopus laevis*), whereas the U.S. drinking water standard for atrazine is 3 ppb (Hayes et al. 2003, Hayes 2004).

The best-documented connection between long-distance transport of chemical pollutants and enigmatic changes in amphibian populations is in California. Prevailing winds transport pesticides and other contaminants from areas of intensive agriculture in the Central Valley to national parks, forests, and wilderness areas in the Sierra Nevada (Fellers et al. 2004; LeNoir et al. 1999; Sparling et al. 2001). Disappearances of *Rana aurora draytonii*, *R. boylii*, *R. cascadae*, and *R. muscosa* populations in California are correlated with the amount of agricultural land use and pesticide use upwind (Davidson et al. 2001; Davidson 2004). *Rana muscosa* that were translocated to an area of Sequoia National Park from which they had disappeared (despite the absence of fish) developed higher tissue concentrations of chlordanes and a DDT metabolite than were found in persisting *R. muscosa* populations 30 km away (Fellers et al. 2004).

These studies provide correlative evidence that pesticides contributed to declines and disappearances of amphibian populations in the Sierra Nevada. Acceptance of this hypothesis will require additional evidence connecting the presence of pesticides and their effects on individuals to effects on populations. This connection cannot be assumed. For example, leopard frogs remain abundant in many areas highly contaminated with atrazine in spite of this herbicide's negative effects on reproductive organs (Hayes et al. 2003). Interactions between pesticides transported long distances and other factors have been suggested as a cause of amphibian declines and disappearances in other protected areas, including Monteverde and Las Tablas, Costa Rica (Pounds and Crump 1994; Lips 1998) and The Reserva Forestal Fortuna, Panama (Lips 1999), but remain little investigated at these sites.

Acid rain is another type of pollution that could affect amphibian populations in isolated areas (Harte and Hoffman 1989). Chemical analyses of water samples in several regions, however, suggested that acid precipitation is an unlikely causal factor (e. g., Richards et al. 1993, Bradford et al. 1994a, Vertucci and Corn 1996).

Natural Population Fluctuations

Population sizes of amphibians, like those of many other taxa, may fluctuate widely due to natural causes. Drought, predation, and other natural factors may even cause local extinctions, necessitating recolonization from other sites. Some declines and disappearances of amphibian populations in areas little affected by humans may be natural occurrences from which the populations may eventually recover, provided that source populations exist elsewhere in the general area (Blaustein et al. 1994b; Pechmann and Wilbur 1994). Natural processes may account for declines and extinctions in the Atlantic forest of Brazil (severe frost and drought, Heyer et al. 1988; Weygoldt 1989), and the loss of some montane populations of the northern leopard frog in Colorado (drought and demographic stochasticity, Corn and Fogleman 1984). Natural fluctuations may also interact with human impacts, resulting in losses from which recovery is unlikely.

Pechmann et al. (1991) used 12 years of census data for amphibians breeding at a pond in South Carolina to illustrate how extreme natural fluctuations may be, and how difficult it can be to distinguish them from declines due to human activities (Figure C). Even "long-term" ecological studies rarely capture the full range of variability in population sizes (Blaustein et al. 1994b; Pechmann and Wilbur 1994). Formulating "null models" of the expected distribution of trends in amphibian populations around the world, against which recent losses may be compared, is a challenging task (see Pechmann 2003 for a review). For example, populations in which juvenile recruitment is more variable than adult survival may decrease more often than they increase, because the average increase is larger than the average decrease (Alford and Richards 1999). The expectation for these populations is that more than half will exhibit a decline over any given time interval even if there is no true overall trend.

(I)

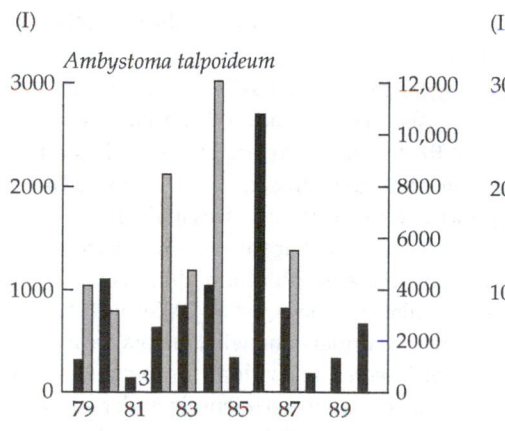

(II)

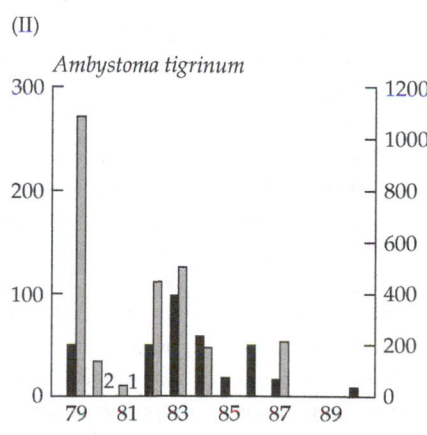

Figure C Natural fluctuations in population size of two species of salamanders in a temporary pond, Rainbow Bay, South Carolina from 1979 to 1990. Numbers of both breeding females (black bars, left axis) and metamorphosing juveniles (gray bars, right axis) vary greatly annually, and some species disappear from and reappear in the system. Numbers in the figure refer to extremely low counts of 3 or fewer individuals. Such species may be poor choices as indicators for the larger system since their natural population fluctuations are so large. (Modified from Pechmann et al. 1991.)

Climate Change

Changes in amphibian population sizes and distributions are often related to changes in environmental temperatures and precipitation (e.g., Bannikov 1948; Bragg 1960; Semlitsch et al. 1996). This variation has traditionally been viewed as natural (Pechmann and Wilbur 1994). The effects of anthropogenic greenhouse gas increases on global climates suggest alternative interpretations in some cases, however.

Some of these alternatives are mediated by the possible effects of greenhouse warming on the El Niño/Southern Oscillation (ENSO), a cyclical warming (El Niño) and cooling (La Niña) of the eastern tropical Pacific. Kiesecker et al. (2001a) found that during El Niño events, precipitation in the Oregon Cascade Mountains was low, resulting in low water levels at *Bufo boreas* breeding sites. The shallow water afforded toad eggs little protection from UV-B radiation, which increased their vulnerability to a pathogenic fungus (*Saprolegnia ferax*), causing high mortality. Kiesecker et al. (2001a) hypothesized that if global warming is increasing the intensity and frequency of El Niño conditions, this may result in amphibian population declines via scenarios analogous to that elucidated by their research with western toads. The effects of global warming on El Niño conditions remain unresolved, however (Cane 2005; Cobb et al. 2003, see Chapter 10).

The disappearance of the golden toad (*Bufo periglenes*) and other frogs from the Monteverde Cloud Forest Preserve in Costa Rica in the late 1980s was associated with an El Niño event, as were subsequent cyclical reductions of some remaining amphibian populations (Pounds et al. 1999). During El Niños, the height at which clouds form increases at Monteverde, reducing the deposition of mist and cloud water that is critical to the cloud forest during the dry season. Pounds et al. (1999) suggested that greenhouse warming has exacerbated this El Niño effect (see Case Study 10.3), citing a global climate model simulation that predicted warming will increase cloud heights at Monteverde during the dry season (Still et al. 1999). This model represents only a crude proxy, because its spatial resolution is too coarse to explicitly model cloud formation on a particular mountain (Lawton et al. 2001; Still et al. 1999). Regional atmospheric simulations and satellite imagery suggest that lowland deforestation upwind also may have increased cloud base heights at Monteverde (Lawton et al. 2001).

Atelopus ignescens and several other *Atelopus* species are thought to have disappeared from the Andes of Ecuador during 1987–1988, which included the most extreme combination of dry and warm conditions in 90 years (Ron et al. 2003). Temperatures in this region increased 2°C over the last century, probably largely due to greenhouse warming (Ron et al. 2003). Other studies also have detected associations between enigmatic declines and disappearances of amphibian populations and temperature and precipitation anomalies (Laurance 1996; Alexander and Eischeid 2001; Burrowes et al. 2004). In these cases the anomalies were within the range of natural variation, thus it is unlikely that they were the direct cause of the amphibian losses, although they may have been a contributing factor. Global warming has been associated with earlier breeding of some amphibian species (e.g., Beebee 1995; Gibbs and Breisch 2001). Although these phenological changes could potentially result in demographic or distributional changes, there is no evidence that they have to date, except when earlier breeding is associated with dry conditions as for western toads in Oregon (Blaustein et al. 2003b, see also Corn 2003; Kiesecker et al. 2001a).

Subtle Habitat Changes

Subtle habitat changes are an understudied potential cause of enigmatic losses of amphibians. For example, fire suppression in Lassen Volcanic National Park, California, has allowed encroachment of trees and shrubs in and around open meadow ponds, streams, and marshes, rendering these sites unsuitable for *Rana cascadae* breeding (Fellers and Drost 1993). Pond canopy closure is known to have local effects on amphibian biodiversity elsewhere (Halverson et al. 2003; Skelly et al. 1999). Construction of dams in the Sierra Nevada has altered temperatures and hydrological regimes downstream, making the habitat unacceptable for *Rana boylii* breeding (Jennings 1996).

Conclusions

The current thinking of the majority of researchers is that there are many interacting causes for enigmatic amphibian losses. Great challenges face those studying declining amphibian populations in scaling individual effects to the population level. Further challenges include directly testing hypotheses formulated from correlative studies with well-designed field experiments to elucidate those mechanisms that are driving the perceived patterns of decline. Synergistic studies can start by dealing with individual phenomena such as the effects of introduced species, and from this foundation scale up to include multiple factors.

Meanwhile, is there nothing we can do to combat these declines? Not at all. A good example is the rapid rebound of *Rana muscosa* populations following trout removal in the high Sierra Nevada (Vredenburg 2004; see Figure B). Management changes are most likely to be put into effect when direct evidence of the phenomenon has been demonstrated, as has been the case in large national parks in California where fish removal is being actively pursued. Even though fish do not explain all of the declines detected, removal of fishes nonetheless has a salutary effect.

A greater challenge is the infectious disease problem, especially chytridiomycosis. Whereas adult *Rana muscosa* with chytridiomycosis in the laboratory invariably die, field studies have shown that some infected adults can survive the summer (at a minimum) under certain circumstances in the field (Briggs et al. 2005). Models indicate that survival of at least some fraction of infected post-metamorphic individuals is crucial to the persistence of populations (Briggs et al. 2005), and studies in progress should help us understand what factors are associated with survival of infected individuals in this and other species. Achieving a balance between scientific understanding of problems and conservation action is a continuous challenge.

CASE STUDY 3.2

Hope for a Hotspot
Preventing an Extinction Crisis in Madagascar

Carly Vynne, University of Washington and Frank Hawkins, Conservation International

Madagascar as a Biodiversity Hotspot

The island nation of Madagascar is world-renowned for its high levels of endemism and biodiversity richness. Despite its proximity to the African mainland, Madagascar does not share any of the typical African animal groups such as elephants, antelopes, or monkeys. Its 160 million year separation has instead resulted in a unique assemblage of plants and animals. Madagascar is characterized by tropical rainforest in the east, dry deciduous forests in the west, and a spiny desert in the far south.

The Madagascar and Indian Ocean Islands Hotspot, which includes the neighboring islands of the Mascarenes, Comoros, and Seychelles, has an impressive 24 endemic families, thus making it one of the highest-priority biodiversity hotspots on earth. At the species level, 93% of mammals, 58% of birds, 96% of reptiles, and 99% of the amphibians are endemic (Figure A). All 651 terrestrial snails and 90% of the estimated 13,000–15,000 plant species are endemic (Mittermeier et al. 2004). In the last 15 years, thousands of new plant species have been identified, as well as many new animal species (22 mammal, 40 fish, 40 butterfly, 150 reptile and amphibian, and 800 ant species). Certainly, many new species remain to be discovered and described.

Hotspot under Siege

Madagascar stands out as one of the highest global priorities for conservation not only because of its high biodiversity values, but also because of the grave threats facing the country. It is estimated that 83% of the primary forest cover of Madagascar has already been lost. The percentage is even higher for the hotspot's smaller islands, and only 10% of the original 600,461 square kilometers of natural habitat remain throughout the entire hotspot. Should current rates of forest destruction continue, all of Madagascar's forests would be lost within 40 years.

The current human population of between 17 and 18 million is growing at about 2.8% per year, doubling every 20 to 25 years. This population is mostly rural, and extremely poor; the per capita annual income in 2002 was U.S.$268. Currently about 2.4 million people live within 10 km of protected areas.

Due to its geographic isolation, humans only arrived on Madagascar about 2000 years ago. The Malagasy people, who are of both African and Asian descent, brought with them agricultural methods such as slash and burn agriculture and extensive cattle grazing on pasture that needs to be burned regularly. These land uses have been particularly detrimental because of infertile soils and low-life resistance of native vegetation. *Tavy*, the traditional slash and burn method of clearing rainforest for rain-fed rice production, continues to be a primary threat today, with around 1%–2% of forest being lost annually to this cause.

The disappearance of forest cover has reduced soil fertility and increased erosion, and poses a dire threat to biodiversity. Estimates show that Madagascar loses the equivalent of 5% to 15% of its gross domestic product annually through declines in soil productivity, loss of forests, and damage to infrastructure linked to forest destruction. The high central plateau of Madagascar has been essentially deforested, and soils washed into the sea are visible from space, giving Madagascar the appearance of having a "red ring" around the island.

Subsistence fuel wood collection, charcoal production, and hunting, also affect Madagascar's forests. Wildlife trade has a serious impact on some endemic amphibians, reptiles, and succulent plants. The proliferation of invasive species affects some ecosystems and species diversity, particularly in aquatic areas. For example, *Bedotia tricolor*, a native endemic fish, is

Figure A   Madagascar is home to an enormous number of endemic species, including this panther chameleon (*Furcifer pardalis*). (Photograph by F. Hawkins.)

now listed as Critically Endangered due in part to competition from an introduced predator, the spotted snakehead fish (Loiselle et al. 2004).

Impacts of commercial timber exploitation are greater than in some parts of the tropics, as trees are generally smaller and compete poorly with introduced species. Regeneration of primary rain forest after logging and slash-and-burn consists mostly of introduced species. Presence of invasive species is not a temporary state; rather, invasive species dramatically alter the trajectory of forest succession (Brown and Gurevitch 2004).

With such human pressures, the resultant threats to biodiversity are not surprising. Currently, 283 birds, mammals, and amphibians are globally threatened with extinction. Of these, all but two are endemic. To help thwart a major extinction spasm from one of the most biologically rich places on Earth, a substantial effort in conservation is currently underway.

Promise for Protected Area Expansion

At the 2003 World Parks Congress held in Durban, South Africa, the President of Madagascar, Marc Ravalomanana, announced the Durban Vision—the Government of Madagascar's commitment to increase protected areas. In his words, "We can no longer afford to let one single hectare of forest go up in smoke or let our many lakes, marshes and wetlands dry up, nor can we inconsiderately exhaust our marine resources. I would like to inform you of our resolve to bring the protected areas from 1.7 million hectares to 6 million hectares over the next five years. This expansion will take place through strengthening of the present national network and implementation of new mechanisms for establishment of new conservation areas."

Prior to this announcement, only 2.4% of the hotspot is designated under the most highly protected categories I–IV of IUCN protected areas; overall only 3% of the land area has any protection. The President of Madagascar pledged to seek $50 million from the international community to make the protected area expansion a reality, and within six months $18 million had been pledged to the trust fund.

Maximizing Protected Area Design for Biodiversity Conservation

To ensure that investments in Madagascar's protected areas provide the highest possible coverage of species diversity, many conservation organizations and independent researchers are working to assemble scientific data that will highlight areas most urgently in need of protection.

New databases make systematic planning for a network of protected areas possible. First, geographically associated, digital GIS files of Madagascar's protected areas were developed in preparation for the World Parks Congress. From these maps it is possible to identify "gaps" where important areas remain without park protection (see Essay 14.2). Second, distribution maps now exist for many of Madagascar's threatened species. A third important dataset is a forest cover change map. A map of deforestation between 1950 and 2000 is shown in Figure B. From this dataset and map, planners may see areas that are undergoing rapid habitat conversion and areas where suitable habitat patches remain for protection in reserves. In addition it can be used to evaluate specific threats to species that occupy small ranges or limited habitats.

The Durban Vision has spurred conservation biologists and planners to take a countrywide view and use emerging information both to develop tools that can help guide decision mak-

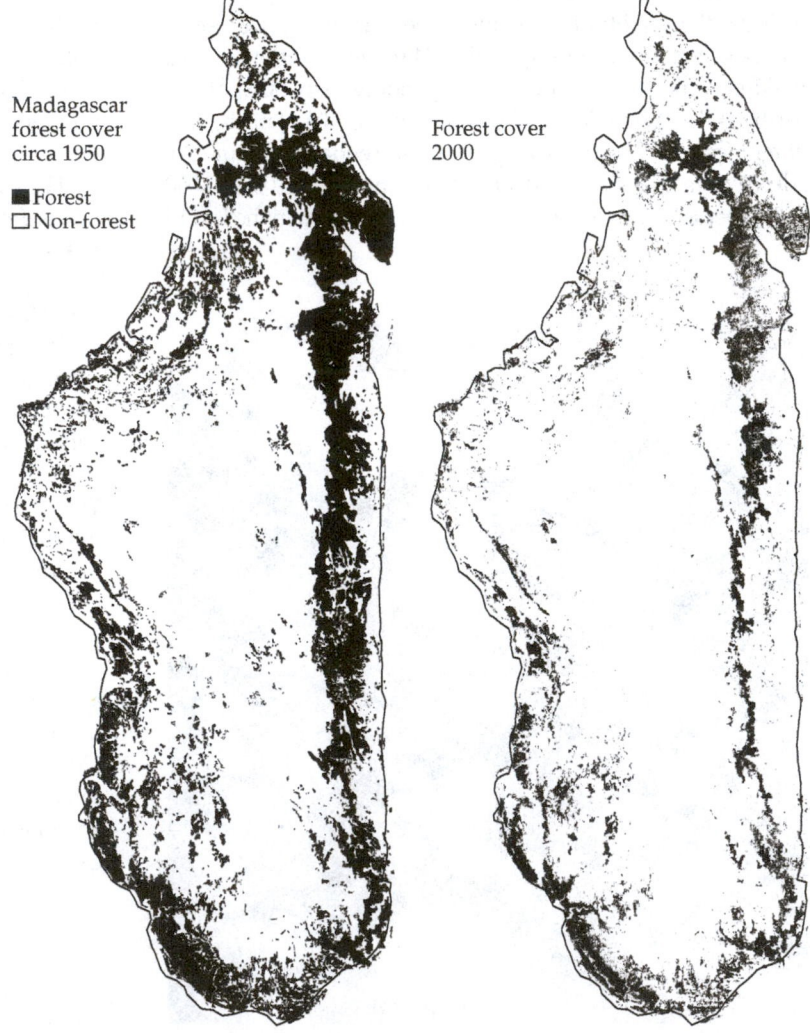

Madagascar forest cover circa 1950

■ Forest
□ Non-forest

Forest cover 2000

Figure B   Loss of forest cover on Madagascar between 1950 and 2000. (Cartography by M. Denil © Conservation International.)

ing and to define key biodiversity areas. Not only will these processes be used to optimize expansion of the protected area network, but the information can help inform assignment of management categories so that the highest priority areas are afforded the greatest protection.

From Planning to Park Establishment: The Case of Masoala

Even with the emerging tools now available for reserve selection, seldom are protected areas implemented based on scientific design principles. A notable exception to this is Masoala National Park, established in Madagascar in 1997. The establishment of this park created Madagascar's largest protected area. The planning process was based on biological and socioeconomic principles of conservation design. There is much to learn from the Masoala design phase to ensure creation of successful reserves in the future.

The Masoala Peninsula of northeast Madagascar is one of the largest forest blocks in the country. Beginning in 1990, efforts were underway to establish an integrated conservation and development project (ICDP), as well as a protected area, on the peninsula. As local use of the ICDP was part of the strategy, an important component of park design was to demarcate core forest areas for biodiversity conservation. Thus, design criteria emphasized that the park be of sufficient size to buffer against disturbances and protect its resources. The second component of the design aimed to enable socioeconomic sustainability and prevent conflicts with local communities (Kremen et al. 1999).

To design the park, maps of different forest types were compiled. Next, a biological inventory was completed, with surveys for birds, primates, small mammals, butterflies, and cicindelid and scarab beetles along five environmental gradients. These data were used to ensure as many species as possible were represented in the core area of the park. To incorporate socioeconomic data, all villages and settlements on the peninsula were mapped and focus-group interviews were held to obtain socioeconomic and agricultural data. Kremen and colleagues subsequently conducted an economic use analysis to determine where to zone for selective logging. Within the use zone, a forest inventory assessed timber resources. Finally, a threats analysis identified and ranked the major threats to biodiversity and identified areas likely to be lost without increased levels of protection.

The full range of elevation gradients present on the peninsula were required if the park was to meet the objective of protecting all known biodiversity. The peripheral, threatened zone in the park's eastern limit had a high edge-to-area ratio and would have been extremely difficult to protect. Thus, this area was left outside of the park to serve as an ecological buffer and an economic support zone for local people. The economic analyses suggested that sustainable forest management in this zone would provide better returns to local communities than slash-and-burn practices.

The recommended design led to the establishment of the present day Masoala National Park, which is 2300 km² and is surrounded by a 1000 km² buffer of community managed forest (Kremen et al. 2000). Slash-and-burn farming for subsistence rice production is the principal threat to unprotected forests. Thus, the park management is working with local communities to create economic incentives for maintaining the boundary forest. Continued assistance to the local communities will be required over a long time period.

That Masoala exists as a national park today should be credited not only to sound design, but also to persistent and persuasive pressure from NGOs. As described by Kremen et al. (2000), "Several timber companies were prospecting for concessions on the Masoala Peninsula during the time that the National Park was being established, and the government nearly abandoned the park project in favor of a logging company. The conservation and diplomatic community played a large role in persuading the government to reject the logging companies' proposals, using both political and economic arguments. Without this pressure, the Masoala Peninsula, one of Madagascar's most important reservoirs of biodiversity, would perhaps have become a forestry concession instead of a national park."

Continuing the Legacy: Sustainability and Financing of a New Reserve System

While local incentives are important to protecting resources, incentives at national and global scales also are essential to the success of conservation. To assess benefits and costs, Kremen et al. (2000) compared estimates of benefits from the Masoala ICDP at local, national, and global scales. The results from their opportunity–cost scenario showed how sustainable harvest versus large-scale timber clearing would benefit local communities, the national government, and the global community. These results have strong implications for long term sustainability of not only the Masoala site, but for protected areas throughout Madagascar.

Kremen et al. (2000) found that, at the national level, the highest short-term return would come from large-scale industrial forestry concessions. The scenario is different at the local and global scales, however. Under the assumption that the forestry concessions would be foreign-owned, the local communities would benefit more from long-term sustainable harvest than from commercial timber harvest. At the global scale, the greatest benefit is also from protection of forests. Loss of the forest would cost the international community an estimated U.S. $68 to U.S. $645 million, based solely on the area's contribution to reducing greenhouse gases. Thus, both the local and global communities benefit from forest protection, whereas, at the national scale, the greatest economic payoff would be to sell off the forest for timber concessions. This so-called "split-incentive" situation suggests that continued pressure on national government through financial and political incentives may be critical to long-term park protection.

Work is underway to estimate the costs of managing the entire protected area network (national parks and conservation

sites). Between 1997 and 2003, U.S.$50 million, mostly from international donors, was spent on Madagascar's protected area system. However, benefits at the national level exceeded this figure by 25%, through water delivery, erosion avoidance, and ecotourism revenue (Carret and Loyer 2003). This study changed the government view on the economic value of protected areas, and it was decided that a trust fund was needed to ensure that the planned reserve network would receive the required operational and management funds. To plan for this trust fund and determine the core costs for the forthcoming expanded network, efforts are underway to model expected costs.

The cost model being developed is innovative in that it takes into account priorities for management by considering overall biodiversity values as well as threat levels facing the reserve. Specifically, management costs allotted to a protected area depend on (1) the importance of the area for biodiversity, (2) the strategic function of each area (i.e., potential for developing research, ecotourism, or educational opportunities), (3) the size of the area, and (4) the threat level. Table A shows the number of park wardens required based on these parameters. The model incorporates supervisory, administrative, and equipment costs based on the number of wardens. This adaptive and predictive cost tool allows a cost estimate for each new site proposed to the system. It provides the global costs of managing the entire reserve system as well as cost by site over years. This tool will not only provide the trust fund and donors with a transparent analysis of anticipated costs of managing the network, but will serve as the basis for a cost accounting and monitoring system for the national governing agency. These estimates also help conservationists and the Madagascar government work together to leverage the required funds.

Protected Areas: Providing Promise for Madagascar

There are many reasons to believe that a strengthened and expanded protected area network and expanded conservation measures will make a difference in stopping extinction on Madagascar. Compared with an average annual rate of loss during the 1990s of 0.9% per year, deforestation from within protected areas was only 0.04% to 0.02%. Species also appear to be faring better in protected areas, as reef fish biomass is greater within the relatively new Masoala National Park

(about 810g/m^2) than outside (525 g/m^2). The annual burning of the Alaotra marshes declined from 32% to 2 % from 2000 to 2002. At the landscape scale, the Zahamena-Mantadia priority conservation corridor experienced one-third the forest loss (2.2%) than the neighboring area (6.7%). Furthermore, Madagascar has invested heavily in its National Environmental Action Plan over the last 10 years. A new Malagasy Parks Service has been established that is helping to professionalize the service and implement management plans.

From Priorities to Projects

Within the planning process for increasing Madagascar's protected area network, there are continued efforts to identify the gaps in protected areas. Once the gaps are identified and new areas for protection proposed, projects will be initiated to help conserve biodiversity values as intended. The site-conservation process requires three elements: an international donor interested in investing in biodiversity, a "conservation broker" that has contact with both the donor and local groups working in the area (usually an international NGO or similar body), and a self-motivated and self-sustaining, field-based NGO network that manages biodiversity based on a general recognition of the economic and environmental benefits that this brings. Implementation of conservation programs is increasingly done at the landscape or biodiversity scale. Rarely can site conservation be achieved by focusing solely on the site of interest; competing interests and threats are too strong for narrow conservation solutions.

The Menabe region of western Madagascar provides an example in which many conservation interest groups are coming together to realize on-the-ground conservation projects that are desperately needed. The Menabe is a genuine hotspot within a hotspot and one of the most important conservation priorities in Madagascar and Africa. The forest is one of the few sites on Earth where baobab trees grow in dense colonies. Indeed, Grandidier's baobab (*Adansonia grandidieri*), one of the largest and most beautiful, is probably functionally extinct outside this area. Menabe's rich biodiversity is under enormous human pressures that have led to a 25% loss in area over the last 40 years. This rate of loss has accelerated in the last 10 years.

Illegal slash-and-burn agriculture to cultivate maize has been the most important cause of forest loss in recent years. The soils of the region are sandy, dry, and poor, and once the organic matter in standing vegetation has been used up, crop yields diminish exponentially and the farmers are forced to move on. The second key threat is logging. While the direct impacts of tree felling seem not to have a great short-term impact on most forest biodiversity, additional hunting and increased risk of fire post-logging often means that wood-cutting ultimately leads to destruction of the forest. Therefore, long-term conservation is difficult to reconcile with logging with-

TABLE A *Amount of Management Allotted to a Protected Area Based on Threat, Biodiversity Values, and Size*

Attributes	Protected area Size in hectares	Hectares/warden
High threat with exceptional biodiversity	5000–20,000	1500
	20,000–60,000	3000
High threat with high biodiversity	5000–10,000	4000
	More than 10,000	6000
Low threat with higher exceptional biodiversity	Independent of size	7000

out significant investment in monitoring, oversight, and enforcement, all of which have been absent up to now in the weak policy and government context.

To implement conservation goals in this region, the Regional Development Committee, helped by a range of national and international NGOs, is working to engage local communities and regional authorities in conservation, and support their capacity to manage the biodiversity resources in the region. Additionally, they work to help develop economic conditions that support conservation. Improved local capacity and a more favorable environment for sustainable economic activities will address systemic conditions that enable these threats to persist. Current conditions and expected change due to conservation projects being implemented in the region are listed in Table B.

More favorable conditions will promote access to viable livelihood options compatible with the long-term conservation of the corridor's biodiversity. This will help to address proximate threats that cause loss of high-priority habitats and will improve community, local, and national government capacity to monitor, manage, and control forest use in the following ways:

- Dramatically increasing and widening the economic benefits from biodiversity-friendly activities such as ecotourism

- Making sustainable agricultural practices more attractive and economically viable than slash-and-burn practices

- Providing long-term sources of wood for charcoal, building materials, and firewood through community management and private sector involvement

- Creating a commercially- and ecologically-viable timber production system

Meeting these objectives is only possible because of revived interest of local NGOs in collaborative processes to address biodiversity concerns in the region. Fanamby and Durrell Wildlife Conservation Trust are two NGOs that have been leading this effort. They work with the National Waters and Forests Authority, the Center for Professional Forestry Training, the German Primate Center, ANGAP (National Protected Area Management Agency), local community groups and other local NGOs, and mayors. This work is chiefly aimed to build local capacity to design and implement forest management plans and to enforce regulations. Durrell Wildlife Conservation Trust also monitors the conditions of wildlife and runs an education and awareness campaign to decrease bushmeat hunting, wildlife trade, and illegal logging. Activities of the collaboration include (1) increasing participation in rural resource management planning, (2) identifying the highest biodiversity priority areas, (3) fostering broad support for conservation enforcement, and (4) promoting the development of private sector forest plantations to meet local demand for wood, charcoal and firewood.

A collaborative approach has much to offer. For example, in one area the local population has long been hostile to conventional development action and the area was considered a lost cause. However, recent negotiations with village elders for a traditional Malagasy law, or *dina*, has resulted in a substantial reduction of forest cutting and an openness to new income-generating activities. The key to project sustainability is developing the capacity of regional and local stakeholders to benefit from the sound management of environmental resources, especially biodiversity. While some investment in biodiversity conservation will always be required, the overall strategy is to develop the technical capacity of the region so that stakeholders in Menabe are independently capable of accessing investment in biodiversity conservation. The economic opportunities offered by endemic biodiversity, especially in the context of ecotourism, are likely to contribute substantially over the long term to the very depressed economy of the region. Current agricultural and logging processes are totally unsustainable, but opportunities exist to move these sectors away from resource mining and into genuinely sustainable practices.

One important aspect of sustainability is social sustainability—the capacity of the region to absorb the changes that are necessary for the development envisaged and to establish the skills and institutional capacity and mechanisms to sustain the changes. The current constraint is the combination of very poor management oversight and a very depressed local economy that has little or no scope or motivation for investment. With this program, the partners hope to overcome both of these con-

TABLE B *Current Problems Facing the Menabe Corridor and Improvements Expected from Conservation Collaboration*

Current systemic conditions	Expected change
Low local institutional (community, government, and NGO) management and technical capacity	Improved local and regional capacity for natural resource management
Inadequate protection of priority biodiversity areas within the corridor	Targeted investment in biodiversity conservation in most important areas
Government ownership of natural resources leading to land-grabs in forested areas	Transfer to local communities of management rights and economic benefits derived from sustainable natural resources
Unregulated forestry sector, with local forestry administration often heavily involved	Improved compliance with national forestry legislation
Inadequate local benefit from biodiversity conservation and low community support for conservation	Increased local knowledge and participation of the local population conservation in sustainable livelihood options resulting in support for biodiversity

straints and encourage those initiatives that are already clearly manifest in the region.

The Malagasy as Stewards

"Tsy misy ala, tsy misy rano, tsy misy vary"

The Malagasy recognize that the fate of the forest is their own. As the above saying goes, without forest, there is no water, there is no rice. Many efforts underway are channeling international interest in biodiversity conservation to the local communities who are neighbors to critical habitats. For example, half of all entrance fees to parks now go back to the local communities. Conserving the country's unique heritage into the future will certainly pose many challenges, but the actions taken by the country's leaders are providing great hope for the future of this hotspot. While all acknowledge the importance of Madagascar for global biodiversity, there have been many who have questioned whether saving the hotspot would be worthwhile, or even feasible. Fortunately, Madagascar has recognized its own position as unique home to biodiversity, and is making optimists out of many who once doubted that anything could be done to save the country from an extinction crisis.

CASE STUDY 3.3

Key International and U.S. Laws Governing Management and Conservation of Biodiversity

Daniel J. Rohlf, Lewis and Clark Law School

A primary focus of conservation biology is on understanding how one species—*Homo sapiens*—has affected life forms and ecosystems across the planet. If conservation biologists wish to build knowledge that is not only relevant to making choices about the future of Earth's biological resources, but is indeed influential in shaping this future, it is imperative that they develop at least a good working knowledge of social decision-making processes and regulatory systems in addition to honing their scientific expertise. Here, I summarize some of the major International and U.S. laws that influence biodiversity conservation. See Box 3.3 for a short description of other national biodiversity laws.

International Legal Regimes that Affect Biodiversity

International law is unique in that, unlike individual nations' systems of governance, there is no global sovereign with recognized authority to impose rules for conduct. This means that much of international law actually consists of agreements by countries that choose to cooperate with one another toward a specific goal; these agreements typically take the form of treaties, conventions, and voluntary participation in international organizations. The following discussion summarizes some of the agreements most important to managing and protecting biodiversity.

The Convention on International Trade in Endangered Species of Wild Fauna and Flora

The Convention on International Trade in Endangered Species of Wild Fauna and Flora (CITES)—first opened for signature in 1973—attempts to regulate international trade in species whose existence is or may be imperiled as a result of commerce and other trafficking. The agreement regulates only international trade in designated species and their parts; it does not attempt to govern habitat management or other species management decisions made by individual countries.

CITES' trade restrictions apply only to species (and parts or products made from a particular species) that member countries vote to add to one of the convention's three appendices. Species that appear in Appendix I—the most imperiled classification—are subject to the most stringent trade restrictions, including a ban on international trade for "primarily commercial purposes." Facing lesser threats, Appendix II species may move in international commerce, but like those species in Appendix I, any shipment must be accompanied by an export permit. CITES does not set up any sort of international authority to implement and enforce the convention; accordingly, each individual country bears responsibility for making necessary findings and issuing required permits. The agency that serves as the "Scientific Authority" for the government of an exporting country must certify that the transaction "will not be detrimental to the survival of that species" in the wild. However, nothing in CITES or its implementing documents defines what constitutes a "detrimental" impact to Appendix I and II species. As a result, each country's Scientific Authority must formulate and apply its own meaning of this term.

Finally, any CITES member country may unilaterally add a species to Appendix III. Appendix III species are subject to restrictions only when the trade originates from the country that listed the species; this provision thus allows individual countries to regulate international commerce in a species that is in trouble only within that country.

Convention on Biological Diversity

The Convention on Biological Diversity (CBD) was opened for signature in 1992 at the Earth Summit in Rio de Janeiro, and now includes 188 countries as parties (but not the United States). The agreement is the centerpiece of the international community's efforts to craft a comprehensive approach to conserving biological diversity. Administratively, a Secretariat headquartered in Montreal serves as the permanent body for day-to-day business, and a Conference of the Parties takes place every two years.

The CBD requires countries to develop national strategies "for the conservation and sustainable use of biological diversity." In meeting this goal, parties pledge to carry out a litany of conservation actions, including creating or maintaining a system of protected areas, making efforts to restore degraded ecosystems and recover threatened species, eradicating nonnative species, pursuing programs for ex situ conservation, and using knowledge derived from "traditional lifestyles" of indigenous communities to foster biodiversity protection as well as sustainable use of biological resources. However, the CBD spells out these obligations only in very general terms, leaving the task of deciding precisely the sort and extent of biodiversity protections and programs to adopt to individual countries with help from advisory committees. The agreement also specifies that developed countries have a responsibility to provide "new and additional financial resources" to their less-developed counterparts; and in a telling nod to the link between economic status and ability to pursue conservation objectives, it notes that developing countries' adherence to their commitments under the convention will depend largely on the extent to which developed countries follow through with transfers of funds and technology to their less well-off neighbors.

The Convention on Biological Diversity also attempts to tackle the sensitive issues of access to and use of biological and genetic resources. It holds that each country has a right to control access to its biological resources, and that use of such resources requires prior informed consent of the nation of origin; a country's consent can be contingent on reaching "mutually agreed terms," (i.e., payment). The agreement also provides that developed states have a responsibility to share technologies and biotechnologies with countries that are not as well off; concern about how this provision affects intellectual property rights has proven to be one of the most significant obstacles to U.S. ratification of the convention.

International Habitat and Ecosystem Protections

There are a number of international efforts to identify and protect either ecosystems in general or specific types of ecosystems. The following paragraphs summarize three of the most prominent.

The United Nations' Educational, Social and Cultural Organization's (UNESCO) Man and the Biosphere Programme provides an institutional umbrella for a series of international

Biosphere Reserves. UN member states, acting on a purely voluntary basis, nominate for inclusion within the system areas within their jurisdictions that already receive protections under domestic legislation. Nominated areas must meet a list of UNESCO criteria, including a requirement that a proposed reserve "contribute to the conservation of landscapes, ecosystems, species and genetic variation," and encompass "a mosaic of ecological systems representative of major biogeographic regions." While formal designation as a Biosphere Reserve brings a measure of international recognition, there are no international standards for management of these reserves; national laws supply the sole directives for protecting biodiversity and managing human activities in these areas. More than 400 Biosphere Reserves have been designated in 94 countries.

The Convention on Wetlands, also known as the Ramsar Convention after the city in Iran where it was originally signed in 1971, sets forth a process for identifying important wetlands analogous to that of UNESCO's program for Biosphere Reserves. The 138 countries that have signed the convention pledge to designate at least one wetland within their borders for inclusion of the convention's "List of Wetlands of International Importance," as well as to include wetland protections in their national land use planning. Again, national and local laws provide the exclusive management standards for wetlands included on the list, which now includes over 1300 wetland areas covering over 100 million hectares.

Lastly, 13 marine or large freshwater areas are the focus of coordinated management and protection efforts under the United Nation Environment Program's Regional Seas initiative. Working with UNEP, neighboring countries develop an "Action Plan" for coordinated action to conserve and manage a regional sea. These plans typically set out broad "framework" strategies that rely on more specific international conventions and protocols among the countries surrounding a regional sea to initiate steps toward implementing the Action Plans' broad outlines. Such agreements also usually rely on national laws to actually carry out identified actions. While some Regional Seas programs, such as that covering the Mediterranean, have developed a number of conventions and protocols that countries have integrated into their national laws, other programs exist mainly on paper.

Federal Laws in the United States that Affect Biodiversity

The federal government in the United States has regulated various elements of biodiversity for over a century, but many of the most prominent laws date back only to the 1970s. The U.S. has a number of important statutes that deal with biodiversity, a fact which itself provides insights into this complex arena: On one hand, the many laws dealing with biological resources serve as evidence of the country's recognition of the importance of biodiversity and the United States' leadership role in crafting laws to manage and protect it; on the other hand, this multitude of not-always-consistent laws and policies illustrate

the federal government's fragmented approach to the natural world.

Endangered Species Act

Since 1973, the Endangered Species Act (ESA) has served as the United States' principal wildlife conservation law, and its high-profile controversies coupled with a public fascination with endangered species have made this statute the country's (and one of the world's) best-known biodiversity protection schemes. It provides a number of substantive protections for species classified as either endangered or threatened, including mechanisms for conserving not only individual organisms, but their habitat as well. Two federal agencies, the U.S. Fish and Wildlife Service and National Oceanic and Atmospheric Association (NOAA) Fisheries (collectively "the Services"), are responsible for day-to-day implementation of the law.

Particularly in more recent years, determining what groupings of organisms are eligible for protection under the ESA has generated substantial controversy. The law defines "species" eligible for protection to include, in addition to full species, "subspecies of fish or wildlife or plants, and any distinct population segment of any species of vertebrate fish or wildlife which interbreeds when mature." It is important to note that in formulating its legal definition of species, U.S. Congress drew a distinction between vertebrates, plants, and all other species. Populations of vertebrates can thus be listed in some places even if they are abundant in other parts of their ranges; insects, plants, and other invertebrates must face wholesale elimination at the species or subspecies level before they can receive legal protections under the ESA. However, protracted and sometimes rancorous debate has marked efforts to determine precisely the meaning of a "distinct population segment" of vertebrate species; a policy adopted by the implementing agencies in 1996 considers two primary factors—whether a given population is (1) "discrete" from other populations of the same species, and (2) "significant" to the species as a whole.

Rather than providing biologically specific definitions of "threatened" and "endangered," to guide the Services in making listing decisions for eligible species, the ESA contains a list of factors that the agencies must consider in making listing (as well as delisting) calls. This includes a litany of biological threats such as habitat destruction, disease, and over-harvest, as well as a catchall category that encompasses any "other natural or manmade factors." Significantly, the statute also specifies that a species can receive the law's protections as a result of "the inadequacy of existing regulatory mechanisms" to ensure the species' future. After listing a species, the appropriate service must also designate its "critical habitat," which the law defines as the habitat essential for the species' recovery. Unlike when making listing decisions, an agency may consider economic and other nonbiological factors in designating critical habitat.

The ESA contains two key protections for species listed as threatened or endangered. First, actions carried out or authorized by federal agencies cannot result in impacts that would "jeopardize the continued existence" of listed species or "destroy or adversely modify" designated critical habitat. The Services have generally interpreted these two standards to prohibit similar levels of impact; an agency action must put in doubt the future of the entire listed species to run afoul of these restrictions. Second, the law prohibits anyone—private landowners included—from "taking," (i.e., killing or injuring) a listed species. Regulations define the term "take" broadly to include habitat impacts that result in death or injury to members of a listed species. This provision of the law has generated considerable opposition because it effectively gives the federal government wide authority over land-use decisions across the country, a power traditionally exercised by state and local governments. However, two factors have helped to keep controversies over the ban on take below the boiling point. First, Congress amended the law in 1982 to add the ESA's so-called Habitat Conservation Plan (HCP) provisions, which allow the Services to grant a permit that allows some take of listed species in exchange for an agreement on the landowner's part to adopt at least some land-use limitations for the benefit of these species. Additionally, the federal government has (unofficially) adopted a rather relaxed stance on enforcing the ESA's take prohibition.

Finally, the ESA requires the Services to formulate "recovery plans" for listed species. These plans spell out specific measures to improve the status of the species they cover, and must include "objective, measurable criteria" that, when achieved, will trigger a process to delist the species as "recovered." Agency statistics reveal that while the populations of many species listed as threatened or endangered have stabilized, few have actually recovered. The fact that there is no clear legal mechanism requiring implementation of recovery plans probably plays a role in the ESA's mixed record of conservation successes.

National Forest Management Act

Enacted by Congress in 1976 after a series of controversies over logging practices on federal land, the National Forest Management Act (NFMA) sets management standards for the 192 million acre national forest system. This vast area encompasses some of the most valuable—as well as most intact—habitat remaining in the contiguous states. These lands also harbor substantial economic and recreational resources as well, making the trade-offs involved in national forest management among the most difficult and controversial of those facing federal land managers. To a large extent, the NFMA set up the U.S. Forest Service to face endless conflict, both internally as well as with outside interests, by making "multiple use" the touchstone of forest management. The statute encompasses this "do it all" approach by setting out only very general substantive standards—essentially giving agency officials a vaguely-worded set of marching orders to please all constituencies by both producing and protecting resources.

The NFMA's provisions addressing biodiversity reflect the law's broad and imprecise mandates. Each national forest must prepare a land-management plan setting forth the overall blueprint for forest management, but the plans' specific provisions for biodiversity are largely influenced by interpretations of vague statutory requirements. The law directs the Forest Service to "provide for diversity of plant and animal communities based on the suitability and capability of the specific land area in order to meet overall multiple use objectives (set forth in an individual national forest's management plan)." The statute contains no definition of "diversity" or other key terms in this mandate, however, thus leaving open to the agency's own interpretation the outlines of the Forest Service's duties in this area.

For nearly a quarter century, regulations interpreting the statute's diversity mandate called on the Forest Service to "preserve and enhance the diversity of plant and animal communities, including endemic and desirable naturalized plant and animal species, so that it is at least as great as that which would be expected in a natural forest." The regulations specified that forest managers must "maintain viable populations of existing native and desired nonnative vertebrate species" found in each national forest. A "viable population" was defined as "one which has the estimated numbers and distribution of reproductive individuals to ensure its existence is well distributed in the planning area." The regulations further mandated that "habitat must be provided to support, at least, a minimum number of reproductive individuals and that habitat must be well distributed so that those individuals can interact with others in the planning area." To assist agency officials in carrying out these duties, the regulations also prescribed designation and monitoring of "management indicator species," species whose population trends the Forest Service believed would indicate the ecological effects of its management activities.

The Forest Service's interpretation of its duties under the ESA grew murky in 2000. Shortly before leaving office, the Clinton administration adopted sweeping changes to the NFMA regulations that established "ecological sustainability" as the management aim for biodiversity on national forests; this term was defined as "the maintenance or restoration of the composition, structure, and processes of ecosystems including the diversity of plant and animal communities and the productive capacity of ecological systems." However, the regulations retained a complimentary focus on species; managers were required to provide a "high likelihood" of maintaining viable populations of native and desired nonnative species, with a "viable" species defined as "consisting of self-sustaining and interacting populations that are well-distributed through the species' range."

Within weeks after they were adopted, however, the incoming Bush administration put the new NFMA regulations on hold and later published its own proposal to replace them. This interpretation of the law's diversity mandate also emphasized ecosystem-level management, but proposed to give Forest Service officials significant latitude to depart from ecologically sustainable practices if necessary to meet multiple-use goals. Additionally, in the most significant change from previous interpretations of the NFMA's diversity provision, the new proposal dropped the requirement to maintain viable populations of forest species. Instead, the proposal called for management that, "to the extent feasible, should foster the maintenance or restoration of biological diversity in the plan area, at ecosystem and species levels, within the range of biological diversity characteristic of native ecosystems within the larger landscape in which the plan area is embedded." Whatever the shape of the new regulations, their changing and widely varied forms in recent years illustrate the broad and ill-defined nature of the NFMA's underlying legal mandate for managing biodiversity.

The Magnusun Act

The Magnusun Fishery Conservation and Management Act of 1976 (amended substantially in 1996) controls marine resources, particularly commercially valuable fish species, within the United States' Exclusive Economic Zone, the area within 200 nautical miles of the country's coastline. The statute's dual—and often conflicting—purposes are to promote a domestic fishing industry while establishing a federal program for "conservation and management" of the country's fishery resources. Eight regional fishery management councils, composed of both fishing industry officials and other interests, develop and implement "Fishery Management Plans" for each "stock" of commercially valuable fish; these plans set an allowable catch and other needed "conservation and management" requirements designed to result in "optimum yield" of a fishery.

The Magnusun Act has not prevented the collapse or near-collapse of many of the nation's fisheries. Consequently, the amended law sets out several requirements applicable to stocks that are "over-fished," which it defines as a rate of fishing mortality that jeopardizes a fishery's capacity to produce its maximum sustained yield on a "continuing basis." Within one year of a determination by NOAA Fisheries or one of the councils that a fishery is over-fished, or is likely to be over-fished within two years, the relevant council must revise its management plan to "rebuild" the stock, though this process may take up to a decade to accommodate the "needs of fishing communities."

Revisions to the Magnusun Act in 1996 added tentative steps toward protecting marine habitat and ecosystems. The regional councils must identify "essential fish habitat," defined as "those waters and substrate necessary to fish for spawning, breeding, feeding or growth to maturity." NOAA Fisheries and the Department of Commerce must ensure that any of their "relevant programs" further the "conservation and enhancement" of this habitat. Additionally, the law sets up a process whereby other federal agencies whose actions "may adversely affect" essential fish habitat must consult with NOAA Fisheries, which in turn receives comments on the proposed action

from the relevant council. If these fisheries experts determine that an adverse impact to identified habitat is in fact likely, NOAA Fisheries provides the federal agency with recommendations for measures that would "conserve" the affected essential fish habitat. The federal agency must issue a written response that either details how the recommendations were implemented or explains why they were not.

Marine Mammal Protection Act

The Marine Mammal Protection Act (MMPA) takes a somewhat similar management approach as that of the Magnusun Act. The MMPA targets for conservation "stocks" of marine mammals, which the statute defines as "a group of marine mammals of the same species or smaller taxa in a common spatial arrangement, that interbreed when mature." NOAA Fisheries, the agency primarily responsible for implementing the act, must seek to maintain these stocks at their "optimum sustainable populations." In addition to enforcing a general ban on intentionally killing or injuring marine mammals, the agency must take steps to reduce the incidental impacts of fishing operations on marine mammal stocks suffering levels of mortality that exceed their "potential biological removal levels." The latter is calculated according to a detailed statutory formula that provides the number of animals that can be "removed" from the population by human-caused mortality. If this level is exceeded by incidental mortalities from commercial fishing operations, NOAA Fisheries must develop "take reduction plans" that place mandatory conditions on these fisheries to reduce marine mammal deaths.

The MMPA obliquely encompasses ecosystem conservation in its conservation scheme. The law emphasizes that "stocks should not be permitted to diminish beyond the point at which they cease to be a significant functioning element of the ecosystem of which they are a part," and thus includes ecosystem health as a factor in determining stocks' optimum sustainable populations. The statute also specifies that "efforts should be made to protect essential habitat" of marine mammals, but establishes no substantive standards or binding requirements to actually carry out such actions

Conclusion

Future human efforts to better manage and protect the planet's biological riches will depend on innovations in both the sciences and the world of policies and regulation. A broad array of international and domestic regulatory regimes already play a significant role in shaping the status of biodiversity, but thoughtful input from the scientific community could make their implementation and evolution vastly more effective. Such contributions, however, can only come from conservation biologists and other scientists that have developed a sound working knowledge of these laws and policies. Better integration of science into law and policy through the work of new generations of professionals is one of the keys to sustaining biodiversity.

Summary

1. Human impacts are a pervasive facet of life on Earth. Increasing numbers of humans and increasing levels of consumption by humans create conditions that endanger the existence of many species and ecosystems: Habitat degradation and loss, habitat fragmentation, overexploitation, spread of invasive species, pollution, and global climate change.

2. Many threats are synergistic, where the total impact of two or more threats is greater than what you would expect from their independent impacts. Also, many threats intensify as they progress. Further, populations often suffer from genetic and demographic problems once they are pushed to low density or small sizes. All these factors make it more difficult to design effective conservation strategies.

3. Extinction can be either global or local, where a species is lost from all of the Earth or from only one site or region, respectively. In addition, ecological extinction can occur when a population is reduced to such a low density that although it is present, it no longer interacts with other species in the community to any significant extent. All these forms of extinction can compromise community functioning, perhaps leading to further losses.

4. Paleontological evidence reveals a long history of human-caused extinctions, although many may have resulted from an interaction between human hunting or species introductions and climate or habitat changes. Extinctions can cause cascade effects, where a first extinction indirectly causes secondary extinctions of species strongly interacting with the first. Extinction of dominant species, keystone species, or ecosystem engineers, as well as the introduction of such species, all can cause cascade effects.

5. The IUCN Red List of Threatened Species is the most comprehensive source of information on patterns of global endangerment. Overall, 41% of all evaluated species are threatened. Only a few taxa have been comprehensively surveyed (mammals, birds, amphibians, and gymnosperms), and all show high fractions of threatened species. Habitat loss and degradation is the leading cause of species endangerment.

6. Roughly one third of the species in the United States are threatened, the majority by habitat loss and degradation. Freshwater mussels are particularly threatened, as are other freshwater taxa, showing that human impacts have been particularly large on aquatic systems.

7. Attributes of species that confer rarity or specialization seem to increase vulnerability to extinction among many creatures, on land and at sea. Species with low reproductive rates are vulnerable due to the rate of extreme changes, and are particularly vulnerable to overexploitation. Island endemics and species with small population sizes are highly vulnerable to perturbations. Island endemics are particularly vulnerable to introduced species, especially when those species are from guilds not represented on the island.

8. Economic and social changes over the next century will have a large impact on the degree to which biodiversity is compromised over the next century. The extremely large number of people in abject poverty compels us to actions that will help these populations, but this need not be at the expense of biodiversity. Changes in consumptive habits among people in China and India will have enormous impacts on global biodiversity.

9. International agreements, laws and regulations are critical tools for slowing biodiversity losses. These policy instruments range from very comprehensive and powerful to more limited in their power and use. At the 2002 Johannesburg World Summit on Sustainable Development, 190 countries committed to reduce biodiversity losses significantly by 2010, which has become a rallying point for biodiversity conservation during this first decade of the 21st century.

10. Strategies to reduce biodiversity losses focus on prioritizing places to work, understanding the causes of declines, and creating strategies that will be effective in reducing threats. Although efforts focused on ecosystems and landscapes are more likely to be effective in conserving multiple layers of biodiversity, strategic approaches often act on the site or species level.

Please refer to the website www.sinauer.com/groom for Suggested Readings, Web links, additional questions, and supplementary resources.

Questions for Discussion

1. How have anthropogenic threats to biodiversity changed over human history? Which do you expect to be the most important threats in 2050?

2. Despite considerable efforts, we know very little about the status of most of the world's species. How can we improve our knowledge on species endangerment? What kinds of information might be most helpful to garner, and why?

3. How can we use information on endangerment to motivate more conservation-oriented policies? What kinds of information are particularly persuasive to you? How might they serve to persuade national or local government officials?

4. Are any species "expendable"? Given the limited resources we have for conservation, how should we prioritize conservation efforts among species? Should we give up on those most likely to go extinct?

5. In two or three decades' time, your children or other youngsters may ask you a question along the lines of, "When the biodiversity crisis became apparent in its full scope during the 1990s, what did you do about it?" What will your answer be?

4

Conservation Values and Ethics

J. Baird Callicott*

It is inconceivable to me that an ethical relation to land can exist without love, respect, and admiration for land, and a high regard for its value. By value, I of course mean something far broader than mere economic value; I mean value in the philosophical sense.

Aldo Leopold, 1949

The Value of Biodiversity

Conservation biologists often treat the value of biodiversity as a given. To many laypeople, however, the value of biodiversity may not be so obvious. Because conservation efforts require broad public support, the conservation biologist should be able to articulate fully the value of biodiversity. Why should we care about—that is, value—biodiversity?

Environmental philosophers customarily divide value into two main types, expressed by alternative pairs of terms: **instrumental** or **utilitarian** as opposed to **intrinsic** or **inherent**. Instrumental or utilitarian value is the value that something has as a means to another's ends. Intrinsic or inherent value is the value that something has as an end in itself. The intrinsic value of human beings is rarely contested. The intrinsic value of nonhuman natural entities and nature as a whole has been the subject of much controversy. Perhaps because the suggestion that nonhuman natural entities and nature as a whole may also have intrinsic value is so new and controversial, some prominent conservationists (e.g., Myers 1983) have preferred to provide a purely utilitarian rationale for conserving biodiversity. The view that biodiversity has value only as a means to human ends is called **anthropocentric** (human-centered). On the other hand, the view that biodiversity is valuable simply because it exists, independently of its use to human beings, is called **biocentric** or **ecocentric**.

Instrumental value

The anthropocentric instrumental (or utilitarian) value of biodiversity may be unequivocally divided into three basic categories—goods, services, and information.

*With contributions by Michael Gillespie, University of Washington, Bothell.

TABLE 4.1   *Four Categories of the Instrumental Value of Biodiversity*

Category	Examples
Goods	Food, fuel, fiber, medicine
Services	Pollination, recycling, nitrogen fixation, homeostatic regulation
Information	Genetic engineering, applied biology, pure science
Psycho-spiritual	Aesthetic beauty, religious awe, scientific knowledge

The psycho-spiritual value of biodiversity may be accounted a fourth kind of anthropocentric utilitarian value, as we have done here, or it may be lumped with intrinsic value (Table 4.1).

GOODS   Human beings eat, heat with, build with, and otherwise consume many other living beings. But our use of nature is surprisingly limited: While 12,000 plants have been identified as food sources, the vast majority of people survive on 6–8 cultivated species, of which two grains (rice, wheat) are of critical importance. Only a small fraction of all life-forms have been investigated for their utility as food, fuel, fiber, and other commodities. Many potential food plants and animals may await discovery. And many of these may be grown on a horticultural or agricultural scale, as well as harvested in the wild, adding variety at least to the human diet, and possibly even saving us from starvation if conventional crops fail due to incurable plant diseases or uncontrollable pests (Vietmeyer 1986a,b). Fast-growing trees—useful for fuelwood or making charcoal, or for pulp or timber—may still be undiscovered in tropical forests. New organic pesticides may be manufactured from yet-to-be screened or discovered plants (Plotkin 1988). The medicinal potential of hitherto undiscovered and unassayed plants and animals seems to be the most popular and persuasive rationale of this type for preserving biodiversity. Vincristine, extracted from the Madagascar periwinkle, is the drug of choice for the treatment of childhood leukemia (Farnsworth 1988). Discovered in the late 1950s, it is the most often cited example of a recent and dramatic cure for cancer manufactured from a species found in a place where the native biota is now threatened with wholesale destruction. Doubtless many other hitherto unscreened, perhaps even undiscovered, species might turn out to have equally important medical uses—if we can save them.

The degree to which conservationists rely on the possibility that potential medicines may be lost if we do not reduce the rate of species extinction is revealing. It reflects the reverence and esteem with which pharmaceuticals are held in contemporary Western culture—a culture, it would seem, of hypochondriacs. Spare no expense or inconvenience to save them, if unexplored ecosystems may harbor undiscovered cures for our diseases! According to Meadows (1990), "Some ecologists are so tired of this line of reasoning that they refer wearily to the 'Madagascar periwinkle argument.'. . . [Those] ecologists hate the argument because it is both arrogant and trivial. It assumes that the Earth's millions of species are here to serve the economic purposes of just one species. And even if you buy that idea, it misses the larger and more valuable ways that nature serves us."

SERVICES   Often overlooked by people who identify themselves first and foremost as "consumers" are the varied services performed by other species working diligently in the complexly orchestrated economy of nature (Meadows 1990; Daily 1997; Heal 2001). Green plants replenish the atmosphere with oxygen and remove carbon dioxide. Certain kinds of insects, birds, and bats pollinate flowering plants, including many agricultural species, and are being lost at a frightening rate (Buchmann and Nabhan 1996). Fungi and microbes in the soil decompose dead organic material and play a key role in recycling plant nutrients. Rhizobial bacteria turn atmospheric nitrogen into usable nitrate fertilizer for plants. If the Gaia hypothesis (Lovelock 1988) is correct, Earth's temperature and the salinity of its oceans are organically regulated. The human economy is a subsystem of the economy of nature and would abruptly collapse if any of these and other major service sectors of the larger natural economy were to be disrupted. The scale of the impact of the human economy on the economy of nature has now become pervasive. According to Vitousek et al. (1997), "Many ecosystems are dominated directly by humanity, and no ecosystem on Earth's surface is free of pervasive human influence."

INFORMATION   Each species' genome is a storehouse of information. Desirable characteristics encoded in isolatable genes are transferable, by means of gene-splicing, to enhance or create edible or medical resources. Genetic information, in other words, is a potential economic good. Such information also has another utility, more difficult to express. Meadows (1990), however, captures it nicely:

Biodiversity contains the accumulated wisdom of nature and the key to its future. If you ever wanted to destroy a society, you would burn its libraries and kill its intellectuals. You would destroy its knowledge. Nature's knowledge is contained in the DNA within living cells. The variety of genetic information is the driving engine of evolution, the immune system for life, the source of adaptability.

Approximately 1.75 million species have been formally named and described (see Chapter 2). The most conservative estimates suggest that between five and ten million species exist on Earth, which means that only

20%–40%, at most, are known to science (Gaston 1991). Based on more liberal estimates of the total—30 million or more—the number known to science could represent less than 5% (Erwin 1988). Many species may become extinct in the coming quarter-century, before they can even be scientifically named and described.

The vast majority of these threatened species are not vascular plants or vertebrate animals; they are insects (Wilson 1985). The reason that Erwin (1988) suspects that there may be so many species of invertebrates is that so many may be endemic or host-specific. Most of these unknown insects at risk of extinction would probably prove to be useless as human food or medicine—either as whole organisms, as sources of chemical extracts, or as sources of gene fragments—nor would many likely play a vital role in the functioning of regional ecosystems (Ehrenfeld 1988). Though it may be difficult to so callously view such a tragedy, we may account their loss, nevertheless, in purely utilitarian terms—as a significant loss of a potential nonmaterial human good, namely pure human knowledge of the biota.

PSYCHO-SPIRITUAL RESOURCES Aldo Leopold (1953) hoped that, through science, people would acquire "a refined taste in natural objects." A beetle, however tiny and ordinary as beetles go, is as potentially beautiful as any work of fine art. And natural variety—a rich and diverse biota—is something that Soulé (1985) thinks nearly everyone prefers to monotony. Wilson (1984) finds a special wonder, awe, and mystery in nature—which he calls "biophilia," and which for him seems almost to lie at the foundations of a religion of natural history. To be moved by the beauty of organisms and whole, healthy ecosystems, to experience a sense of wonder and awe in the face of nature's inexhaustible marvels is to become a better person, according to Norton (1987).

From the point of view of the value of information—genetic and otherwise—the mindless destruction of biodiversity is like book burning. From the point of view of natural aesthetics and religion it is like vandalizing an art gallery or desecrating a church. There has been little doubt expressed that the value of pure scientific knowledge is anthropocentric. And the aesthetic and spiritual value of nature is often understood to be a pretentious kind of utilitarian value. Ehrenfeld (1976) thinks that aesthetic and spiritual rationales for the conservation of biodiversity are "still rooted in the homocentric, humanistic worldview that is responsible for bringing the natural world, including us, to its present condition." Nevertheless, the beauty and sanctity of nature has sometimes been referred to as an intrinsic, not an instrumental value. According to Sagoff (1980), for example, "We enjoy an object because it is valuable; we do not value it merely because we enjoy it. . . . Esthetic experience is a perception, as it were, of a certain kind of worth."

Intrinsic value

Unlike instrumental value, intrinsic value is not divisible into various categories. Discussion of intrinsic value has focused on two other issues: the sorts of things that may possess intrinsic value, and whether intrinsic value exists objectively or is subjectively conferred.

In response to mounting concern about human destruction of nonhuman life, some contemporary philosophers have broken with Western religious and philosophical tradition and attributed intrinsic value, by whatever name, to the following: robustly conscious animals (Regan 1983); sentient animals (Warnock 1971); all living things (Taylor 1986); species (Callicott 1986; Rolston 1988; Johnson 1991); biotic communities (Callicott 1989); ecosystems (Rolston 1988; Johnson 1991) and evolutionary processes (Rolston 1988). Leopold (1949, 1953) attributed "value in the philosophical sense"—by which he could only mean what philosophers call "intrinsic value"—to "land," defined as "all of the things on, over, or in the earth" (Callicott 1987a). Soulé (1985) categorically asserts that "biotic diversity has intrinsic value"; and Ehrenfeld (1988) categorically asserts "value is an intrinsic part of diversity."

Biocentric environmental philosophers who claim that intrinsic value exists objectively in human beings and other organisms reason as follows. In contrast to a machine, such as a car or a vacuum cleaner, an organism is self-organizing and self-directed (Fox 1990). A car is manufactured, in other words; it does not grow up, orchestrated by its own DNA. And a car's purposes—to transport people and to confer status on its owner—are imposed on it from a source outside itself. Machines do not have their own goals or purposes, as organisms do—neither consciously chosen goals nor genetically determined goals. What are an organism's self-set goals? They may be many and complex. For human beings they may include everything from winning an Olympic gold medal to watching as much television as possible. All organisms, however, strive (usually unconsciously and in an evolutionary sense) to achieve certain basic predetermined goals—to grow, to reach maturity, to reproduce (Taylor 1986).

Thus, interests may be intelligibly attributed to organisms, but not to machines. Having ample sunlight, water, and rich soil is in an oak tree's interest, though the oak tree may not be actively interested in these things, just as eating fresh vegetables may be in a child's interest, though the child may be actively interested only in junk food. One may counter that, by parity of reasoning, getting regular oil changes is in a car's interest, but because a car has no purpose of its own, being well-maintained is not in its own interest, but in the interest of its user, whose purposes it serves exclusively. Another way of saying that ever striving and often thriving organisms have interests is to say that they have a good of their own. But *good* is just an

older, simpler word meaning pretty much the same thing as *value*. Hence to acknowledge that organisms have interests—have goods of their own—is to acknowledge that they have what philosophers call intrinsic value.

One problem with objective intrinsic value, thus understood, is that it seems limited to individual organisms. Conservationists, however, are not professionally concerned with the welfare of individuals, but with the preservation of species, the health and integrity of ecosystems, and evolutionary and ecological processes (Soulé 1985). Johnson (1991) attributes interests to species, based on the controversial claim by Ghiselin (1974) and Hull (1976, 1978) that species are best conceived not as taxa, but as individuals protracted in space and time. He also attributes interests to ecosystems, claiming that they are superorganisms. But such a conception of ecosystems has fallen so far out of fashion in ecology that Johnson can cite no contemporary ecologists to support his claim. Rolston (1988) prudently avoids basing his attribution of intrinsic value to species and ecosystems on either obsolete or suspect scientific ideas. He argues that because a basic evolutionary "goal" of an organism is to reproduce, its species is therefore one of its primary goods. Species evolved, however, not in isolation, but in a matrix of ecosystems. Therefore, Rolston (1988) concludes, ecosystems also have intrinsic value. Evolutionary processes—going all the way back to the Big Bang—produced ("projected") beings with goods of their own; thus they too, Rolston (1988) reasons, have intrinsic value. The breadth of diversity existed before we did and enabled our own evolution, and thus, although we are relative newcomers, if we can value any of that diversity now, it must all have intrinsic value since it preceded us.

More strictly observing the classic distinction between objective "facts" and subjective values, some environmental philosophers argue that all value, including intrinsic value, is subjectively conferred (Callicott 1986; Elliot 1992): no conscious subjects, no value. *Value*, they think, is (or should be) first a verb, and only derivatively a noun. On this view, something has instrumental value if it is valued for its utility, while something has intrinsic value if it is valued (transitive verb) for its own sake. Subjects—at least those capable of the intentional conscious act of valuing—value themselves intrinsically. But some conscious subjects, among them human beings, are quite capable of intrinsically valuing other human beings, other living things, species, ecosystems, and ecological and evolutionary processes. Human beings are literally kin to all other species and are members of biotic as well as human communities (Leopold 1949). Thus, according to Callicott (1992), other species and the hierarchically ordered biotic communities and ecosystems to which we human beings belong are—no less

than our fellow human beings and our hierarchically ordered human communities—the sorts of entities that we should value intrinsically.

Intrinsic and instrumental value are not mutually exclusive; many things may be valued both for their utility and for themselves. We may value a forest both for its wood and for its inherent value. Similarly, intrinsically valuing biodiversity does not preclude appreciating the various ways in which it is instrumentally valuable.

By claiming that biodiversity has intrinsic value (or is intrinsically valuable), Norton (1991) argues that some environmental philosophers and conservation biologists have actually done more harm than good for the cause of conservation. Why? Because the intrinsic value issue divides conservationists into two mutually suspicious factions—anthropocentrists versus bio- and ecocentrists. The latter dismiss the former as "shallow resourcists"; and the former think that the latter have gone off the deep end (Norton 1991). If biodiversity is valuable because it ensures the continuation of ecological services, represents a pool of potential resources, satisfies us aesthetically, inspires us religiously, and makes better people out of us, the practical upshot is the same as attributing intrinsic value to it: We should conserve it. Instrumentally valuing biodiversity and intrinsically valuing it "converge" on identical conservation policies, in Norton's (1991) view (Figure 4.1); thus, we don't really need to appeal to the intrinsic value of biodiversity to ground conservation policy. Hence, Norton (1991) argues, the controversial and divisive proposition that biodiversity has intrinsic value should be abandoned. A wide and long anthropocentrism, he thinks, is an adequate value package for conservation biology.

Attributing intrinsic value to biodiversity, however, makes a practical difference in one fundamental way that Norton seems not to have considered. If biodiversity's intrinsic value were as widely recognized as is the intrinsic value of human beings, would it make much difference? All forms of natural resource exploitation that might put it at risk would not be absolutely prohibited, as intrinsic value can be overridden. After all, recognizing the intrinsic value of human beings does not absolutely prohibit putting people at risk when the benefits to the general welfare (or "aggregate utility") of doing so are sufficiently great. For example, soldiers

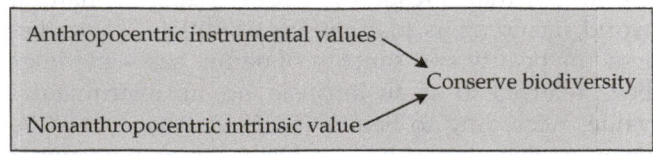

Figure 4.1   Norton's convergence hypothesis.

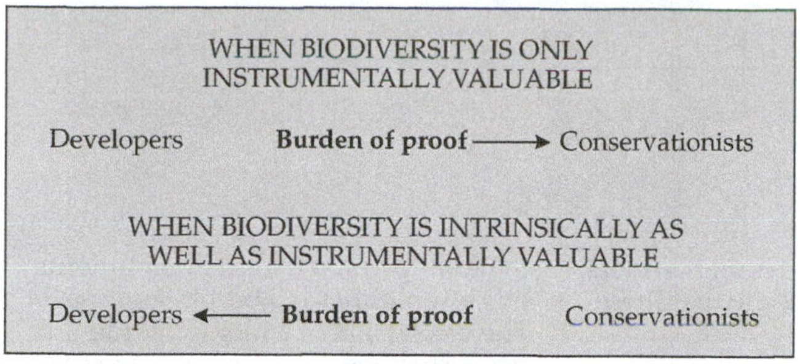

WHEN BIODIVERSITY IS ONLY INSTRUMENTALLY VALUABLE

Developers   **Burden of proof ⟶** Conservationists

WHEN BIODIVERSITY IS INTRINSICALLY AS WELL AS INSTRUMENTALLY VALUABLE

Developers **⟵ Burden of proof**   Conservationists

Figure 4.2   Burden of proof according to instrumental and intrinsic value systems.

from the United States and other nations have been sent into combat and killed or wounded (as are citizens of the countries where they fight), not only to protect themselves and their fellow citizens from imminent annihilation, but also to achieve geopolitical goals.

Rather, if the intrinsic value of biodiversity were widely recognized, then sufficient justification must be offered for putting it at risk—just as we demand sufficient justification for putting soldiers at risk by sending them to war. The practical difference that attributing intrinsic value to biodiversity makes is to shift the burden of proof from conservationists who are trying to protect it to those whose actions might jeopardize it (Figure 4.2). Fox (1993) puts this point clearly and forcefully:

> Recognizing the intrinsic value of the nonhuman world has a dramatic effect upon the framework of environmental debate and decision-making. If the nonhuman world is only considered to be instrumentally valuable then people are permitted to use and otherwise interfere with any aspect of it for whatever reasons they wish. If anyone objects to such interference then, within this framework of reference, the onus is clearly on the person who objects to justify why it is more useful to humans to leave that aspect of the world alone. If, however, the nonhuman world is considered to be intrinsically valuable then the onus shifts to the person who wants to interfere with it to justify why they should be allowed to do so.

Monetizing the Value of Biodiversity

Monetizing the value of biodiversity, where the instrumental value to human beings is expressed in monetary terms, is a technical task for economists. Here, only the basic ways of putting a dollar value on biodiversity and the philosophical issues raised by the prospect of doing so can be discussed. A fuller discussion of economic "valuation" is provided in Chapter 5. It might seem that only the instrumental value of biodiversity is subject to expression in monetary terms. Some environmental economists, accordingly, explicitly endorse a strict anthropocentrism (Randall 1986). However, as we shall see,

even the intrinsic value of biodiversity can be taken into account in an economic assessment of conservation goals.

Some endangered species have a market price: notoriously, elephants for their tusks; rhinoceroses for their horns; baleen whales for their meat, bone, and oil; and tigers for their pelts and bones. In some cases—the blue and sperm whales, for example—their monetary value is the only reason they are threatened with extinction. In other cases—the tiger and the mountain gorilla, for example—habitat destruction is also a factor in their endangerment. Myers (1981), however, suggests that taking advantage of their monetary value may be the key to conserving many species. Holmes Rolston III provides an alternative perspective in Essay 4.1.

According to modern economic theory, what is necessary for transforming a species' market price from a conservation liability into a conservation asset is to take it out of a condition that economists call a "commons" and "enclose" it. *Enclosing* here does not mean literally building a fence around a species population; it means, rather, assigning rights to cull it. A wild species that has a market value is subject to over-harvesting when property rights to it cannot be legitimately asserted and enforced. This leads to the **tragedy of the commons** (Hardin 1968). If a resource can be owned (either privately or publicly) and property rights to it can be enforced, then the species will be conserved, so the theory goes, because the owner will not be tempted to "kill the goose that lays the golden egg." Or will he or she? Other factors, such as species' reproductive rates and growth rates in relationship to interest rates, discount rates, and so on, confound this simple picture. As Haneman (1988) points out, "The interest rate level, the nature of the net benefit function and its movement over time, and the dynamics of the resource's natural growth process combine to determine the optimal intertemporal path of exploitation . . . Other things being equal, the higher the interest rate at which future consequences are discounted, the more it is optimal to deplete the resource now."

The blue whale is a case in point. The International Whaling Commission effectively protects whale populations, despite occasional poaching, by allotting species harvest quotas to whaling nations and, since 1982, imposing a moratorium on commercial whaling (Forcan 1979; Birnie 1985). Clark (1973) concludes, however, that because they reproduce so slowly, it would be more profitable to hunt blue whales to complete extinction and invest the proceeds in some other industry than to wait for the species population to recover and harvest blue whales at sustainable levels indefinitely. Clark does not recommend this course of action. On the contrary, his point is that market forces alone cannot always be made to further conservation goals.

ESSAY 4.1

Our Duties to Endangered Species

Holmes Rolston III, *Colorado State University*

■ Few persons doubt that we have obligations *concerning* endangered species, because persons are helped or hurt by the condition of their environment, which includes a wealth of wild species, currently under alarming threat of extinction. Whether humans have duties directly *to* endangered species is a deeper question, important in both ethics and conservation biology, in both practice and theory. Many believe that we do. The U.N. *World Charter for Nature* states, "Every form of life is unique, warranting respect regardless of its worth to man." The *Convention on Biological Diversity* affirms "the intrinsic value of biological diversity." Both documents are signed by well over a hundred nations. A rationale that centers on species' worth to persons is anthropocentric; a rationale that includes their intrinsic and ecosystem values is naturalistic.

Many endangered species have no resource value, nor are they particularly important for the usual humanistic reasons: scientific study, recreation, ecosystem stability, and so on. Is there any reason to save such "worthless" species? A well-developed environmental ethics argues that species are good in their own right, whether or not they are "good for" anything. The duties-to-persons-only line of argument leaves deeper reasons untouched; such justification is not fully moral and is fundamentally exploitive and self-serving on the part of humans, even if subtly so. Ethics has never been very convincing when pleaded as enlightened self-interest (that one ought always to do what is in one's intelligent self-interest).

An account of duties to species makes claims at two levels: one is about facts (a scientific issue, about species); the other is about values (an ethical issue, involving duties). Sometimes, species can seem simply made up, since taxonomists regularly revise species designations and routinely put after a species the name of the "author" who, they say, "erected" the taxon. If a species is only a category or class, boundary lines may be arbitrarily drawn, and the species is nothing more than a convenient grouping of its members, an artifact of taxonomists. No one pro-

poses duties to genera, families, orders, or phyla; biologists concede that these do not exist in nature.

On a more realistic account, a biological species is a living historical form, propagated in individual organisms, that flows dynamically over generations. A species is a coherent, ongoing, dynamic lineage expressed in organisms, encoded in gene flow. In this sense, species are objectively there—found, not made, by taxonomists. Species are real historical entities, interbreeding populations. By contrast, families, orders, and genera are not levels at which biological reproduction takes place. Far from being arbitrary, species are the real survival units.

This claim—that there are specific forms of life historically maintained over time—does not seem fictional, but rather is as certain as anything else we believe about the empirical world, even though at times scientists revise the theories and taxa with which they map these forms. Species are not so much like lines of latitude and longitude as like mountains and rivers, phenomena objectively there to be mapped. The edges of such natural kinds will sometimes be fuzzy, and to some extent discretionary (see Chapter 2). One species will slide into another over evolutionary time. But it does not follow from the fact that speciation is sometimes in progress that species are merely made up, rather than found as evolutionary lines.

At the level of values and duties, an environmental ethics finds that such species are good kinds, and that humans ought not, without overriding justification, to cause their extinction. A consideration of species offers a biologically based counterexample to the focus on individuals—typically sentient and usually persons—so characteristic of Western ethics. In an evolutionary ecosystem, it is not mere individuality that counts. The individual represents, or re-presents anew, a species in each subsequent generation. It is a token of an entity, and the entity is more important than the token. Though species are not moral agents, a biological identity—a kind of value—is here defended. The dignity resides in the

dynamic form; the individual inherits this, exemplifies it, and passes it on. The possession of a biological identity reasserted genetically over time is as characteristic of the species as of the individual. Respecting that identity generates duties to species.

The species is a bigger event than the individual, although species are always exemplified in individuals. Biological conservation goes on at this level too, and, really, this level is the more appropriate one for moral concern, a more comprehensive survival unit than the organism. When an individual dies, another one replaces it. Tracking its environment over time, the species is conserved and modified. With extinction, this stops. Extinction shuts down the generative processes in a kind of superkilling. It kills forms (species) beyond individuals. It kills collectively, not just distributively. To kill a particular plant is to stop a life of a few years or decades, while other lives of such kind continue unabated; to eliminate a particular species is to shut down a story of many millennia, and leave no future possibilities.

Because a species lacks moral agency, reflective self-awareness, sentience, or organic individuality, some hold that species-level processes cannot count morally. But each ongoing species represents a form of life, and these forms are, on the whole, good kinds. Such speciation has achieved all the planetary richness of life. All ethicists say that in *Homo sapiens* one species has appeared that not only exists but ought to exist. A naturalistic ethic refuses to say this exclusively of one late-coming, highly developed form, but extends this duty more broadly to the other species—though not with equal intensity over them all, in view of varied levels of development.

The wrong that humans are doing, or allowing to happen through carelessness, is stopping the historical gene flow in which the vitality of life lies. A shutdown of the life stream is the most destructive event possible. Humans ought not to play the role of murderers. The duty to species can be overridden, for example, with pests or disease organisms. But a *prima facie* duty stands

nevertheless. What is wrong with human-caused extinction is not just the loss of human resources, but the loss of biotic sources. The question is not: What is this rare plant or animal good for? But: What good is here? Not: Is this species good for my kind, *Homo sapiens*? But: Is *Rhododendron chapmanii* a good of its kind, a good kind? To care about a plant or animal species is to be quite nonanthropocentric and objective about botanical and zoological processes that take place independently of human preferences.

Increasingly, we humans have a vital role in whether these stories continue. The duties that such power generates no longer attach simply to individuals or persons, but are emerging duties to specific forms of life. The species line is the more fundamental living system, the whole, of which individual organisms are the essential parts. The species too has its integrity, its individuality, and it is more important to protect this than to protect individual integrity. The appropriate survival unit is the appropriate level of moral concern.

A species is what it is, inseparable from the environmental niche into which it fits. Particular species may not be essential in the sense that the ecosystem can survive the loss of individual species without adverse effect. But habitats are essential to species, and an endangered species typically means an endangered habitat. Integrity of the species fits into integrity of the ecosystem. Endangered species conservation must be ecosystem-oriented. It is not preservation of *species* that we wish, but the preservation of *species in the system*. It is not merely *what* they are, but *where* they are that we must value correctly.

It might seem that for humans to terminate species now and again is quite natural. Species go extinct all the time. But there are important theoretical and practical differences between natural and anthropogenic extinctions. In natural extinction, a species dies when it has become unfit in its habitat, and other species appear in its place. Such extinction is normal turnover. Though harmful to a species, extinction in nature is seldom an evil in the sys-

tem. It is rather the key to tomorrow. The species is employed in, but abandoned to, the larger historical evolution of life. By contrast, artificial extinction shuts down tomorrow because it shuts down speciation. One opens doors, the other closes them. Humans generate and regenerate nothing; they only dead-end these lines. Relevant differences make the two as morally distinct as death by natural causes is from murder.

On the scale of evolutionary time, humans appear late and suddenly. Even more lately and suddenly they increase the extinction rate dramatically. What is offensive in such conduct is not merely senseless loss of resources, but the maelstrom of killing and insensitivity to forms of life. What is required is not prudence, but principled responsibility to the biospheric earth. Only the human species contains moral agents, but conscience ought not be used to exempt every other form of life from consideration, with the resulting paradox that the sole moral species acts only in its collective self-interest toward all the rest. ■

The idea of conserving economically exploitable threatened species by enclosing them and harvesting them in a sustainable manner may work well enough in conserving species with relatively high reproductive and growth rates (such as ungulates), but may not work at all well in conserving species that have relatively low reproductive and growth rates (such as whales, although perhaps this is not true of some species, such as elephants. Because males keep growing throughout their lives, some have argued that the best way to grow ivory is to allow male elephants to die a natural death). Hence, enlisting the market in the cause of conservation must be done very carefully on a case-by-case basis.

Potential goods—new foods, fuels, medicines, and the like—have no market price, obviously, because they remain unknown or undeveloped. To destroy species willy-nilly, however, before they can be discovered and examined for their resource potential is to eliminate the chance that a desirable commodity will become available in the future. Hence, biodiversity may be assigned an "option price" defined as "the amount people would be willing to pay in advance to guarantee an option for future use" (Raven et al. 1992). The option price of any given undiscovered or unassayed species may be very small because the chance that a given species may prove to be useful is also probably very small (Ehrenfeld 1988).

But added together, the option price of the million or more species currently threatened with wholesale extinction might be quite formidable.

The market confers a dollar value on biodiversity in other ways than the price of the actual and potential goods that nature affords. People pay fees to visit national parks, for example, and to hike in wilderness areas. Such fees—no less than the price of vincristine or of wildebeest steaks—express the value of a bit of biodiversity in money. But often, because user fees are usually low, the true monetary value of the psycho-spiritual "resource" is under-expressed by those fees alone. Subsidies provided from local, state, and federal tax revenues might also be factored in when assessing the monetary value of a psycho-spiritual resource. The money people spend—for such things as gasoline, food, lodging, and camping equipment—to get to a particular spot and visit it may be credited to the resource by employing the "travel cost method" (Peterson and Randall 1984; see also Chapter 5). "Contingent valuation," in which people are polled and asked what they would be willing to pay for the opportunity to enjoy a certain experience—say to hear wolves howling in Yellowstone National Park in the United States—is also used to calculate the dollar value of psycho-spiritual resources (Peterson and Randall 1984).

Economists also recognize—and of course attempt to monetize—the "existence value" of biodiversity (Randall 1988). Some people take a modicum of satisfaction in just knowing that biodiversity is being protected even if they have no intention of consuming exotic meats or personally enjoying a wilderness experience. Existence value has a price; one way to ascertain it would be to calculate the amount of money sedentary people contribute to conservation organizations, such as The Nature Conservancy or the Rainforest Action Network. Further, economists now also recognize "bequest value"—the amount people would be willing to pay to assure that future generations of *Homo sapiens* will inherit a biologically diverse world (Raven et al. 1992).

Accurately monetizing the value of the often free or under-priced recreational, aesthetic, intellectual, and spiritual utility of nature is more often attempted than monetizing the value of the services provided to the human economy by the economy of nature. Meadows (1990) hints at one way of monetizing natural services: "How would you like the job," she asks, "of pollinating trillions of apple blossoms some sunny afternoon in May? It's conceivable, maybe, that you could invent a machine to do it, but inconceivable that the machine could work as elegantly and cheaply as the honey bee, much less make honey on the side." The value of nature's service economy could be monetized by calculating the cost of replacing natural services with artificial ones. Put in terms of scarcity and option, what would be the cost of employing human labor or machines to pollinate plants, if—because of present economic practices, such as excessive use of insecticides—in the future pollinating organisms were to become alarmingly scarce? A more sophisticated version of this approach along with other methods was employed by Costanza et al. (1997) to calculate the astounding value of $33 trillion per year for 17 ecological services provided by the biosphere. By comparison, the aggregated value of the gross national product of all the world's human economies is $18 trillion. As Costanza et al. (1997) point out, monetizing the value of ecosystem services is fraught with multiple uncertainties; and so their figure, large as it is, is likely to be an underestimate (Heal 2001; Howarth and Farber 2002).

Ehrenfeld (1988) notes however, that, just as many species have little potential value as goods, many species are likely to have little importance in the service sector of the economy of nature: "The species whose members are the fewest in number, the rarest, the most narrowly distributed—in short, the ones most likely to become extinct—are obviously the ones least likely to be missed by the biosphere. Many of these species were never common or ecologically influential; by no stretch of the imagination can we make them out to be vital cogs in the ecological machine."

Some philosophers and conservation biologists strenuously object to the penchant of economists for reducing all value to monetary terms (Ehrenfeld 1988; Sagoff 1988). As the enormously influential philosopher Immanuel Kant (1959) remarked, some things have a price, others have a dignity. And, as a familiar matter of fact, we have attempted to exclude certain things from the market that we believe have a dignity—things, in other words, to which we attribute intrinsic value. Indeed, one possible motive for claiming that biodiversity has intrinsic value (or is intrinsically valuable) is to exclude it from economic valuation, and thus to put it beyond the vagaries of the market. We have, for example, attempted to take human beings off the market by outlawing slavery, and attempted to take sex off the market by outlawing prostitution. Why not take intrinsically valuable biodiversity off the market by outlawing environmentally destructive human activities?

Sagoff (1988) argues that we have two parallel and mutually incommensurable systems for determining the value of things: the market and its surrogates, on the one hand, and the ballot box, on the other. As private individuals most of us would refuse to sell our parents, spouses, or children—at any price. And as citizens united into polities, we may refuse to trade diminished biodiversity for any "benefit" projected in a cost–benefit analysis. Indeed, the United States Endangered Species Act of 1973 is a splendid example of a political decision to take biodiversity off the market.

Economists counter that we must often make hard choices between such things as the need to bring arable land into production and saving the habitat of endangered species (Randall 1986; Costanza et al. 1997). While we may like to believe, piously and innocently, that intrinsically valuable people are literally priceless, the value of a human life is not uncommonly monetized. The dollar value of a human life, for example, might be reflected by the amount that an automobile insurance company pays a beneficiary when a customer kills another person in an accident, or by the maximum amount that an industry is willing to pay (or is required by law to pay) to protect the health and safety of its employees. Similarly, recognizing the intrinsic value of biodiversity does not imply that it cannot be priced. The only way we can make informed choices is to express the entire spectrum of natural values from "goods" and "services" to "existence" in comparable terms: dollars.

The Endangered Species Act was amended in 1978 to create a high-level interagency committee, the so-called "God Squad," which could allow a project that put a listed species in jeopardy of extinction to go forward if its economic benefits were deemed sufficiently great. This legislation affirms that we do indeed have two incommensurable systems of determining value—one economic and the other political. It also affirms the original

political decision to exempt biodiversity from being routinely monetized and traded off for greater economic benefits, while acknowledging that politically and economically determined values often clash in the real world. And it provides for when the opportunity cost of conserving biodiversity exceeds an unspecified threshold, the God Squad can allow economic considerations to override the general will of the citizens of the United States, which is democratically expressed through their Congressional representatives, so that the nation's extant native species be conserved, period.

Bishop (1978) formalizes the reasoning behind the God Squad amendment to the U.S. Endangered Species Act. He advocates the safe minimum standard (SMS) approach, an alternative to the practice of aggregating everything from the market price to the shadow price of biodiversity, plugging it into a cost–benefit analysis (CBA), and choosing the economically most efficient course of action (see Chapter 5). Instead, the SMS assumes that biodiversity has incalculable value and should be conserved unless the cost of doing so is prohibitively high (Figure 4.3). As Randall (1988) explains,

> Whereas the . . . [CBA] approach starts each case with a clean slate and painstakingly builds from the ground up a body of evidence about the benefits and costs of preservation, the SMS approach starts with a presumption that the maintenance of the SMS for any species is a positive good. The empirical economic question is, "Can we afford it?" Or, more technically, "How high are the opportunity costs of satisfying the SMS?" The SMS decision rule is to maintain the SMS unless the opportunity costs of doing so are intolerably high. In other words, the SMS approach asks, how much will we lose in other domains of human concern by achieving the safe minimum standard of biodiversity? The burden of proof is assigned to the case against maintaining the SMS.

As noted earlier in this chapter, the practical effect of recognizing the intrinsic value of something is not to make it inviolable, but to shift the burden of proof, the onus of justification, onto those whose actions would adversely affect it. Because the safe minimum standard approach to monetizing the value of biodiversity shifts the burden of proof from conservationists to users, it tacitly acknowledges, and incorporates into economic appraisal, biodiversity's intrinsic value.

Conservation Ethics

According to Leopold (1949), ethics, biologically understood, constitutes "a limitation on freedom of action." Ethics, in other words, constrains self-serving behavior in deference to some other good (Table 4.2).

Anthropocentrism

In the Western religious and philosophical tradition, only human beings are worthy of ethical consideration. All other things are regarded as mere means to human ends. Indeed, anthropocentrism seems to be set out in no uncertain terms at the beginning of the Bible. Man alone is created in the image of God, is given dominion over Earth and all the other creatures and, finally, is commanded to subdue the whole creation. White (1967) claimed that because Jews and Christians believed for many centuries that it was not only their God-given right but their positive religious duty to dominate all other forms of life, science and an eventually aggressive, environmentally destructive technology developed uniquely in Western civilization.

As Norton (1991) has shown, an effective conservation ethic can be constructed on the basis of traditional Western anthropocentrism, whether grounded in traditional biblical ideas or in the belief typical of traditional Western philosophy that "man" is uniquely rational and thus uniquely valuable intrinsically. Ecology has revealed a world that is far more systemically integrated than the biblical authors could have imagined, and subduing nature has untoward ecological consequences. An anthropocentric conservation ethic would require individuals, corporations, and other interest groups to fairly consider how their actions that directly affect the natural environment indirectly affect other human beings. Logging tropical forests, for example, may make fine hardwoods available to wealthy consumers, turn a handsome profit for timber companies, employ workers, and earn foreign exchange for debt-ridden countries. But it may also deprive indigenous peoples of their homes and traditional means of subsistence, and people everywhere of undiscovered resources, valuable ecosystem services, aesthetic experience, and scientific knowledge. And unchecked, logging may leave future generations of human beings a depauperate world (what is called intergenerational inequity). Thus, environmentally destructive types of resource development may be judged unethical without any fundamental change in the framework of traditional Western moral thought.

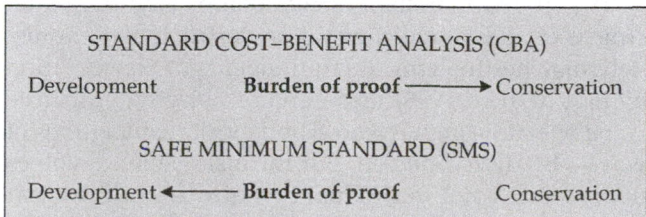

Figure 4.3 Burdens of proof according to the standard CBA and the SMS approaches.

TABLE 4.2   *A Comparison of Western Environmental Ethics*

Value	Anthropocentrism	Judeo-Christian stewardship ethic	Biocentrism	Ecocentrism
Intrinsic value	Human beings	Species/creation as a whole	Individual organisms	Species, ecosystems, biosphere
The value of nature	Instrumental	Holistic–intrinsic	Individualistic–intrinsic	Holistic–intrinsic
"Man's" place in nature	Lord and master	Caretaker	One among equals	Plain member and citizen

With biodiversity so greatly threatened, we frequently frame the problem as one of satisfying human needs and wants at the expense of other species. But we should bear in mind that human wants often greatly exceed human needs and that all human beings are not equally empowered to satisfy their wants and needs; some are able to do so more readily than others. Figueroa and Mills (2001) point out "People of color, the poor, and under-represented groups, such as indigenous tribes and nations are faced with a disproportionate amount of environmental burdens." A branch of anthropocentric environmental ethics is "environmental justice," which calls for environmental burdens and benefits to be fairly shared. According to Figueroa and Mills (2001) environmental justice is not only about fair distribution, but also fair participation "in the environmental movement and in other arenas that bear on how environmental benefits and burdens are assigned."

Further, environmental justice requires taking account of the effects of producers and consumers in industrial economies on such spatially distant places as Alaskan watersheds, the Everglades, and the Great Barrier Reef (Harte 1993). Global climate change, caused mainly by fossil fuel consumption in industrialized societies affects all ecosystems, some more destructively than others. Oceanic islands and other low-lying areas may, for example, eventually be inundated by rising sea levels. Is it just for the people of Micronesia or Bangladesh to bear the burden of the "progress" and "development" of people living in Northern Europe, North America, and Eastern Asia? Is it just for those who may suffer the most from global warming to have the least influence on international negotiations to mitigate climate change? Is a highly developed nation obligated, by considerations of environmental justice, to help less-developed nations cope with the environmental problems to which it disproportionately contributed in the course of its development?

The Judeo–Christian stewardship conservation ethic

Stung by the allegation that the Judeo–Christian worldview was ultimately responsible for bringing about the contemporary environmental crisis, some environmentally concerned Christians and Jews challenged White's (1967) interpretation of biblical environmental attitudes and values (Barr 1972). After all, God pronounced everything that He created to be "good," during the five days before He created human beings. Thus, God appears to have conferred intrinsic value on every kind of creature, not just on humanity. Indeed, the text suggests that God intended His creation to be replete and teeming with life:

> And God said, Let the waters bring forth abundantly the moving creature that hath life, and fowl that may fly above the earth in the open firmament of heaven. And God created great whales, and every living creature that moveth, which the waters brought forth abundantly, after their kind, and every winged fowl after his kind: and God saw that it was good. And God blessed them, saying, Be fruitful and multiply, and fill the waters in the seas, and let fowl multiply in the earth (Genesis 1: 20–22).

Further, "dominion" is an ambiguous notion. Just what does it mean for "man" to have dominion over nature? White (1967) argues that in the past, at least, Jews and Christians took it to mean that people should exercise a despotic reign over nature. Later in Genesis, however, God put Adam (who may represent all human beings) in the Garden of Eden (which may represent all of nature) "to dress it and to keep it" (Genesis 2:15). Our "dominion," this suggests, should be that of a responsible caretaker—a *steward*—rather than a tyrant. But what about "man" alone being created in the image of God? That could be taken to confer unique responsibilities on human beings. As God cares for humanity, so we who are created in the image of God must care for Earth.

The Judeo–Christian stewardship environmental ethic is especially elegant and powerful. It also exquisitely matches the ethical requirements of conservation biology (Baker 1996) and confers objective, intrinsic value on nature in the clearest and most unambiguous of ways—by divine decree. But intrinsic value devolves upon species, not individual organisms. For it is clear that during His several acts of creation God is creating species, "kinds," not individual animals and plants— whales, in other words, not specifically the one that

swallowed Jonah or the one named Moby Dick. Thus, it is species, not individual organisms, that God pronounces good. Hence, human beings may freely use other living beings as long as we do not endanger their species—as long, in other words, as we do not compromise the diversity of the creation. As Ehrenfeld (1988) points out, the Judeo–Christian stewardship environmental ethic makes human beings directly accountable to God for conserving biodiversity: "Diversity is God's property, and we, who bear the relationship to it of strangers and sojourners, have no right to destroy it." More discussion of the possibility of environmental ethics grounded in Judaism and Christianity may be found in *Christianity and Ecology*, edited by Hessel and Radford Reuther (2000); *Judaism and Ecology*, edited by Tirosh-Samuelson (2002); and *Judaism and Environmental Ethics*, edited by Yaffe (2001).

Traditional non-Western environmental ethics

Christianity is a world religion, but so are Islam and Buddhism. Other major religious traditions, such as Hinduism and Confucianism, while more regionally restricted, nevertheless claim millions of devotees. Ordinary people are powerfully motivated to do things that can be justified in terms of their religious beliefs. Therefore, to distill environmental ethics from the world's living religions is extremely important for global conservation. The well-documented effort of Jewish and Christian conservationists to formulate the Judeo–Christian stewardship environmental ethic in biblical terms suggests an important new line of inquiry: How can effective conservation ethics be formulated in terms of other sacred texts? Callicott (1994) offers a preliminary comprehensive survey. Beginning in 1998 a three-year series of thirteen conferences was held at the Harvard Center for the Study of World Religions on the theme of "Religions of the World and Ecology." The proceedings of those conferences are published in a Harvard University Press book series of the same title, edited by Tucker and Grim. To provide even a synopsis of these works would be impossible here. However, a few sketches of traditional non-Western religious conservation ethics may be suggestive.

Muslims believe that Islam was founded, in the seventh century C.E., by Allah (God) communicating to humanity through the Arabian prophet, Mohammed, who regarded himself to be in the same prophetic tradition as Moses and Jesus. Therefore, because the Hebrew Bible and the New Testament are earlier divine revelations underlying distinctly Muslim belief, the basic Islamic worldview has much in common with the basic Judeo–Christian worldview. In particular, Islam teaches that human beings have a privileged place in nature, and going further in this regard than Judaism and Christianity, that indeed all other natural beings were created to serve humanity. Hence, there has been a strong tendency among Muslims to take a purely instrumental approach to the human–nature relationship.

Islam does not distinguish between religious and secular law. Hence, new conservation regulations in Islamic states must be grounded in the Koran, Mohammed's book of divine revelations. In the early 1980s, a group of Saudi scholars scoured the Koran for environmentally relevant passages and drafted *The Islamic Principles for the Conservation of the Natural Environment*. While reaffirming "a relationship of utilization, development, and subjugation for man's benefit and the fulfillment of his interests," this landmark document also clearly articulates an Islamic version of stewardship: "He [man] is only a manger of the earth and not a proprietor, a beneficiary not a disposer or ordainer" (Kadr et al. 1983). The Saudi scholars also emphasize a just distribution of "natural resources," not only among members of the present generation, but among members of future generations. And as Norton (1991) has argued, conservation goals are well served when future human beings are accorded a moral status equal to that of those currently living. Saudi scholars have even found passages in the Koran that are vaguely ecological. For example, God "produced therein all kinds of things in due balance" (Kadr et al. 1983). More discussion of an Islamic environmental ethic may be found in *Islam and Ecology*, edited by Richard Foltz, Frederick Denny, and Azizan Baharudin (2003).

Ralph Waldo Emerson and Henry David Thoreau, thinkers at the fountainhead of North American conservation philosophy (discussed in Chapter 1), were influenced by the subtle philosophical doctrines of Hinduism, a major religion in India. Hindu thought also inspired Naess's (1989) contemporary "Deep Ecology" conservation philosophy. Hindus believe that at the core of all phenomena there is one and only one Reality or Being (Figure 4.4). God, in other words, is not a supreme Being among other lesser and subordinate beings, as in the Judeo–Christian–Islamic tradition. Rather, all beings are a manifestation of the one essential Being—called *Brahman*. And all plurality, all difference, is illusory or at best only apparent.

Such a view would not seem to be a promising point of departure for the conservation of biological diversity because the actual existence of diversity, biological or otherwise, seems to be denied. Yet in the Hindu concept of *Brahman*, Naess (1989) finds an analogue to the way ecological relationships unite organisms into a systemic whole. However that may be, Hinduism unambiguously invites human beings to identify with other forms of life, for all life-forms share the same essence. Believing that one's own inner self, *atman*, is identical, as an expression of *Brahman*, with the selves of all other creatures leads to compassion for them. The suffering of one life-

Figure 4.4   Hindu woman praying in the Ganges River. (Photograph © Mark Downey/Painet Inc.)

form is the suffering of all others; to harm other beings is to harm oneself. As a matter of fact, this way of thinking has inspired and helped motivate one of the most persistent and successful conservation movements in the world, the Chipko movement, which has managed to rescue many of India's Himalayan forests from commercial exploitation (Guha 1989; Shiva 1989). More discussion of a Hindu environmental ethic may be found in *Hinduism and Ecology* (Chappel and Tucker 2000).

Jainism is a religion of relatively few adherents, but a religion of great influence in India. Jains believe that every living thing is inhabited by an immaterial soul, no less pure and immortal than the human soul. Bad deeds in past lives, however, have crusted these souls over with *karma*-matter. *Ahimsa* (noninjury of all living things) and asceticism (eschewing all forms of physical pleasure) are parallel paths that will eventually free the soul from future rebirth in the material realm. Hence, Jains take great care to avoid harming other forms of life and to resist the fleeting pleasure of material consumption. Extreme practitioners refuse to eat anything but leftover food prepared for others, and carefully strain their water to avoid ingesting any water-borne organisms—not for the sake of their own health, but to avoid inadvertently killing other living beings. Less extreme practitioners are strict vegetarians and own few material possessions. The Jains are bidding for global leadership in environmental ethics. Their low-on-the-food-chain and low-level-of-

consumption lifestyle is held up as a model of ecological right livelihood (Chappel 1986). And the author of the *Jain Declaration on Nature* claims that the central Jain moral precept of *ahimsa* "is nothing but environmentalism" (Singhvi n.d.). A deeper discussion of a Jain environmental ethic may be found in *Jainism and Ecology* (Chappel 2002).

Though now virtually extinct in its native India, Buddhism has flourished for many hundreds of years elsewhere in Asia. Its founder, Sidhartha Gautama, first followed the path of meditation to experience the oneness of *Atman–Brahman*, and then the path of extreme asceticism in order to free his soul from his body—all to little effect. Then he realized that his frustration, including his spiritual frustration, was the result of desire. Not by obtaining what one desires—which only leads one to desire something more—but by stilling desire itself can one achieve enlightenment and liberation. Further, desire distorts one's perceptions, exaggerating the importance of some things and diminishing the importance of others. When one overcomes desire, one can appreciate each thing for what it is.

When the Buddha realized all this, he was filled with a sense of joy, and he radiated loving-kindness toward the world around him. He shared his enlightenment with others, and formulated a code of moral conduct for his followers. Many Buddhists believe that all living beings are in the same predicament: We are driven by desire to a life of continuous frustration, and all can be liberated if all can attain enlightenment. Thus Buddhists can regard other living beings as companions on the path to Buddhahood and *nirvana* (Figure 4.5).

Buddhists, no less than Jains and Christians, are assuming a leadership role in the global conservation movement. Perhaps most notably, the Dalai Lama of Tibet is the foremost conservationist among world-renowned religious leaders. In 1985, the Buddhist Perception of Nature Project was launched to extract and collate the many environmentally relevant passages from Buddhist scriptures and secondary literature. Thus, the relevance of Buddhism to contemporary conservation concerns could be demonstrated and the level of conservation consciousness and conscience in Buddhist monasteries, schools, colleges, and other institutions could be raised (Davies 1987). Bodhi (1987) provides a succinct summary of Buddhist environmental ethics: "With its philosophic insight into the interconnectedness and thoroughgoing interdependence of all conditioned things, with its thesis that happiness is to be found through the restraint of desire, with its goal of enlightenment through renunciation and contemplation and its ethic of non-injury and boundless loving-kindness for all beings, Buddhism provides all the essential elements for a relationship to the natural world characterized by respect, care, and compassion." More discussion of the

Figure 4.5   Roots of an enormous Bodhi tree (*Ficus religiosa*) have grown around this head of a Buddha at Wat Maha That. Ayuthaya, Thailand. Buddha attained enlightenment under a bodhi tree. (Photograph © Davis Crossland/Alamy.)

possibility of a Buddhist environmental ethic may be found in *Buddhism and Ecology* (Tucker and Duncan Ryuken Williams 1997).

One-fourth of the world's population is Chinese. Fortunately, traditional Chinese thought provides excellent conceptual resources for a conservation ethic. The Chinese word *tao* means *way* or *road*. The Taoists believe that there is a *Tao*, a Way, of nature. That is, natural processes occur not only in an orderly but also in a harmonious fashion. Human beings can discern the *Tao*, the natural well-orchestrated flow of things. And human activities can be either well adapted to the *Tao*, or they can buck it. In the former case, human goals are accomplished with ease and grace and without disturbing the natural environment; but in the latter they are accomplished, if at all, with difficulty and at the price of considerable disruption of neighboring social and natural systems. Capital-intensive Western technology—such as nuclear power plants and industrial agriculture—and other forms of development that have increased in China in recent years are very "unTaoist" in sprit and motif. More discussion of the possibility of a Taoist environmental ethic may be found in *Daoism and Ecology* (Girardot et al. 2001).

Contemporary conservationists may find in Taoism an ancient analogue of today's countermovement toward appropriate technology and sustainable development (Sylvan and Bennet 1988). The great Mississippi Valley flood of 1993 is a case in point. The river system was not managed in accordance with the *Tao*. Thus, levees and flood walls only exacerbated the big flood when it finally came. Better to have located cities and towns outside the flood plain and allow the mighty Mississippi River occasionally to overflow. The rich alluvial soils in the river's flood plains could be farmed in drier years, but no permanent structures should be located there. That way, the floodwaters could periodically spread over the land, enriching the soil and replenishing wetlands for wildlife, and the human dwellings on higher ground could remain safe and secure. Restoring the natural flood regime to the Mississippi would have other significant benefits. It would slow the erosion of coastal south Louisiana and reduce the hypoxic zone that is spreading from the mouth of the river in the Gulf of Mexico (Mitsch et al. 2001).

In another example, the world's largest dam, the Three Gorges Dam, is currently being completed on the Yangtze River in China. Long debated on a variety of grounds, the Three Gorges Dam will flood a 600-km reservoir, inundating over 1300 archeological sites and the habitat of numerous endangered species. Although hydropower is far cleaner than the coal power that has been the dominant source of energy in China, many warned that the benefits of the Dam would be outweighed by the damage of submerging so large an area behind the Dam. Perhaps the officers of the U.S. Corps of Engineers or the engineers of the Three Gorges Dam should study Taoism. We might wish that the proponents of the Three Gorges Dam had abandoned newfangled Maoism for old-fashioned Taoism before trying to contain, rather than cooperate with, the Yangtze River.

The other ancient Chinese religious worldview is Confucianism. To most people, Asian and Western alike, Confucianism connotes conservatism, adherence to custom and social forms, filial piety, and resignation to feudal inequality. Hence, it seems to hold little promise as an intellectual soil in which to cultivate a conservation ethic. Ames (1992), however, contradicts the received view: "There is a common ground shared by the teachings of classical Confucianism and Taoism. . . . Both express a 'this-worldly' concern for the concrete details of immediate experience rather than . . . grand abstractions and ideals. Both acknowledge the uniqueness, importance, and primacy of particular persons and their contributions to the world, while at the same time expressing the ecological interrelatedness and interdependence of this person with his context."

From a Confucian point of view, a person is not a separate immortal soul temporarily residing in a physical

body; a person is rather the unique center of a network of relationships. Because his or her identity is constituted by these relationships, the destruction of one's social and environmental context is equivalent to self-destruction. Biocide, in other words, is tantamount to suicide.

In the West, because individuals are not ordinarily conceived to be robustly related to and dependent upon their context—either for their existence or for their very identity—it is possible to imagine that they can remain themselves and be "better off" at the expense of both their social and natural environments. But from a Confucian point of view, it is impossible to abstract persons from their contexts. Thus, if *context* is expanded from its classical social to its current environmental connotation, Confucianism offers a very firm foundation upon which to build a contemporary Chinese conservation ethic. More discussion of the possibility of a Confucian environmental ethic may be found in *Confucianism and Ecology* (Tucker and Berthrong 1998).

Even if we can attribute the spectacularly non-Taoist behavior in contemporary China, such as the construction of the Three Gorges Dam on the Yangtze River, to outside influences, a long history of environmental degradation in China suggests that there may be a discrepancy between belief and behavior. Such a discrepancy was noted by Tuan (1968) and Paper (2001) who questioned the efficacy of religious worldviews to motivate actions. One might reply that in contemporary China, at least, religious belief has been undermined by half a century of anti-religion Communist-Party rule. But Williams (2003) indicates that in medieval China—long before the advent of communism and other colonizing Western ideologies, such as mechanistic materialism—there was extensive deforestation. Anderson (1996), however, points out that the practical demands of any people's economic needs and wants may conflict with their religious ideas. Thus in premodern China the outcome of this tension between economic and ethical imperatives may be traced on the landscape where one can find evidence of forest conservation as well as evidence of deforestation. Anderson (1996) also suggests that much can be learned from a study of the way a community actually practices their land-oriented religious and folk beliefs. Such studies reveal the complex ways that moral fallibility and emotional conflicts also complicate the intellectually idealized relationship between belief and behavior, and often lead to sustainable—albeit imperfectly so—interactions between people and their natural environments.

In any case, Peterson (2003) notes that the turn of the millennium has been characterized by an explosion of interest in, and writing about, the actual and potential ways that spiritual and religious traditions can support nature conservation. In addition to the way religious tenets and spiritual beliefs might be adapted to jibe with ecology and conservation biology, Gardner (2002) also notes the potential for social cooperation and political alliance among religious institutions and conservation organizations. Religions not only develop environmental ethics grounded in their foundational worldviews, but historically they have also motivated people to act on their beliefs and values—however imperfectly—in ways seemingly more powerful than do science and philosophy (Oelschlaeger 1994). Further, religions touch the lives of more people than do science and philosophy, and religious institutions control enormous physical and financial resources. Thus despite some competition and conflicts in values and motivations, the rewards of social cooperation and political alliance between conservation organizations and religious institutions represent a tremendous opportunity for the cause of nature conservation. According to Gardner (2002), "With great effort, the two communities could indeed bring about a historic reconciliation and generate the societal energy needed to sustain the planet and its people."

The tenets and conservation implications of these various non-Western religions are summarized in Table 4.3, and in Essay 4.2 by Susan Bratton, which explores the role of religion in conservation further.

Biocentrism

Although it is vital to base conservation ethics in religious worldviews in order to appeal to the broadest spectrum of laypeople, it is more problematic for conservation biologists to do so in their professional capacity. Various religiously based conservation ethics may not be mutually consistent. A more universal conservation ethic that transcends cultural as well as political boundaries is necessary if the conservation policies informed by that ethic are to be coherent.

Before the advent of environmental ethics, moral philosophers in the Western tradition granted moral standing to human beings alone, not by appeal to a mystical property, such as the image of God, but by appeal to an observable trait, such as rationality or linguistic ability. Because only people, they argued, can reason or speak, only people are worthy of ethical treatment. In the eighteenth century, Kant (1959) argued that human beings are intrinsically valuable ends because we are rational, while animals (and other forms of life) are only instrumentally valuable means because they are not. Contemporary environmental philosophers have attempted to construct a nonanthropocentric environmental ethic without appeal to mystical religious concepts, such as God, the Tao, or the universal Buddha-nature. And some have done so by arguing that reason and linguistic ability are inappropriate qualifications for moral standing, and that other observable traits are more appropriate.

TABLE 4.3 *Types of Traditional Non-Western Conservation Ethics*

Characteristic	Islam	Hinduism	Jainism	Buddhism	Taoism	Confucianism
Source of value in nature	External; *Allah* (God)	Internal; *Atman-Brah-man*	Internal; soul (*jiva*)	Internal; Buddha-nature	Emergent; the *Tao* (Way)	Emergent; relational
Human attitude toward nature	Respect for creation is respect for Creator	Identification; self-realization	*Ahimsa* (noninjury)	Loving-kindness; solidarity	Harmony; cooperation	Interrelated; interdependent
Conservation practicum	Conserve resources for future generations	Conserve trees and other beings that manifest *Atman-Brahman*	Low on the food chain; low level of consumption	Still desires; reduce consumption; contemplate nature	Adapt human economy to nature's economy	Conserve nature to preserve human society

ESSAY 4.2

Monks, Temples, and Trees
The Spirit of Diversity

Susan P. Bratton, *Baylor University*

◼ A Buddhist monk bends over and carefully waters a small seedling in the temple garden. Others of its kind are nearby. Older, taller trees shade the sanctuary paths with their fan-shaped leaves, and produce a crop of edible nuts each year. The monk looks at the little ginkgo and reflects that he never has seen one growing on its own in the surrounding mountains. Only in the temple gardens and their environs has the ginkgo survived, at least in his region of China.

From a venerable lineage, datable to the lower Jurassic, *Ginkgo biloba* is the only known remaining species of an entire division of vascular plants, the Ginkgophyta. Often called a "living fossil," the modern shade tree is little different from the ginkgos of the early Cretaceous period. *Ginkgo* is also a taxon that may or may not exist in the wild. One of the largest "seminatural" populations, at Tian Mu Shan, is near the Kaishan temple, and thus may have been under partial human protection, if not management, for centuries. Over the last several thousand years, Buddhist monks have probably slowly replaced the ginkgo's natural dispersal agents, such as leopard cats (*Felis bengalensis*) and helped preserve the species for posterity (del Tredici et al. 1992).

Our contemporary technocratic and scientifically oriented society often mistakenly considers religion to be either uninterested or uninformed when it comes to protection and management of the natural world. We also assume that if religion is interested, it is the more "primitive" religions and those that practice magic that attempt to relate to or manipulate wild nature, while the great religions of the world—particularly the "peoples of the book," Judaism, Christianity, and Islam—are too theological and otherworldly to concern themselves with the various small pieces that make up the cosmos. The truth is, religious values have often helped to protect natural diversity, and religion remains one of the most important wellsprings of human concern for other species. E. O. Wilson has suggested that science alone cannot protect biodiversity; other cultural values must be called on as well.

Science attempts to understand the world through objective comparison. The various elements in the environment become "other," or differentiated from the scientist, who makes a conscious effort to distance herself from the phenomena she is observing. Religion, in contrast, establishes relationship or identification with the "other." The shaman becomes an intermediary with nature and links the village with the surrounding forests and their creatures; the Buddhist monk works in the temple garden and increases his spiritual understanding of the cosmos as a whole; the Hebrew psalmist sees the glory of God in the diversity of the wild and praises divine wisdom for placing the stork in the cedars and for maintaining both birds and forests with water gushing from mountain springs. Religion has a freedom of symbolic and aesthetic expression inappropriate to science. Religion can speak with nature, science can only speak about it.

Religion forwards the preservation of natural diversity in several different ways. The first is by providing ethical and social models for living respectfully with nature. For most cultures, religion is a primary means of defining right and wrong. The Koyukon of Alaska, for example, do not separate the natural and the spiritual world, and explain the spiritual power residing in nature through Distant Time stories about the evolution of the cosmos. Since nature has spiritual power, it commands respect and is included in the religious code of morality and etiquette. The Koyukon avoid waste in food harvest and take only what they can use from their fragile far-northern lands. They do not kill female waterfowl preparing to nest, nor do they take young animals. They fear retribution in the form of bad luck if they violate taboos or are disrespectful of the animals they hunt, so their husbandry of natural resources is tightly tied to an animist worldview (Nelson 1983). Other religions with very different notions of the otherworldly

may have rather similar rules. For example, the Hebrew scriptures, with their one transcendent God, forbid removal of a mother bird from her nest.

Secondly, religion often provides direct protection for wild and cultivated plants and animals. Many cultures have holy places, including mountains, that humans may approach only for religious purposes, if at all. Rivers or forests may be sacred environs, where wildlife and vegetation are not to be disturbed. Sites are sometimes set aside specifically to protect taxa that have medicinal value or are utilized in religious ritual. Taboos or special religious significance can prevent the killing of individual wildlife species. Buddhism, one of the most abstract and philosophical of all religions, has protected numerous organisms, from ginkgos to cranes to monkeys, resident on the grounds of its temples. Some early Christian monks would not allow the native oak forests to be cleared from around their monasteries. St. Francis of Assisi instructed his followers to leave the borders of a cultivated garden unweeded to provide space for wildflowers, so that the blossoms, in their beauty, could praise the creator God. Even our contemporary wilderness areas in the United States are, among other purposes, supposed to preserve and protect "spiritual values."

Lastly, religion ties the nonhuman residents of the cosmos to the divine or to the overall meaning of human existence. This gives the biota a value that science alone cannot provide. The saffron-robed initiate caring for the temple landscape sees each individual creature as beautiful in itself and beautiful in its interrelationship with its neighbors. The trees, the small clump of flowers, the rock and the sand, become more than xylem and chloroplasts, or feldspar and quartz. For the dedicated practitioner, the sanctity of the environment is an inspiration and a blessing. The spiritual realization of the Buddhist, in turn, blesses the environment (14th Dalai Lama 1992). In early and medieval Christianity, where love and compassion were key values and holiness was fervently pursued, the monks and desert ascetics often cared for wildlife, healing animals with injuries and even rescuing them from hunters. The early Christians thought animals could recognize the pure of heart, and that even wild lions and wolves would show affection for the great saints.

The religious myths and stories that teach us about the importance of other species are often so basic that we, in our human-dominated, industrial world, miss the critical message. Take, for example, the tale of Noah's ark. Noah did not save the animals just to be nice. Noah saved the animals because humans need the animals—all the animals, not just the domestic and the edible. Also, in the Genesis original, it is God who instructs Noah to build the ark. The great God of Israel wanted the animals rescued, and put Noah to a great deal of trouble during a very damp climatic period to accomplish this. God had created the animals in wondrous diversity and in marvelous order, and had blessed them as both good and beautiful well before the Garden of Eden was an official mailing address. When the animals march onto the ark according to their kinds, it is divine organization that is being honored, and when Noah saves them all, not just a few, it is the glory of divine handiwork that is being preserved (Bratton 1993). Modern conservation biology can perhaps take a lesson from this. ■

Singer (1975) and Regan (1983) have done much to raise awareness of ethical issues about how human beings treat nonhuman beings. They exposed classic Western anthropocentric ethics to the following dilemma: If the qualification for ethical standing—or "criterion for moral considerability" as it is more technically called —is pitched high enough to exclude nonhuman beings, then it will exclude as well those human beings who also fail to measure up. Human infants, the severely retarded, and the profoundly senile are not rational. If, following Kant, we make rationality the criterion of moral considerability, then these human "marginal cases" may be treated just as we treat nonhuman beings who fail to meet it. They may become, for example, unwilling subjects of painful medical tests and experiments; they may be hunted for sport; or they may be made into dog food. No one would want that to happen. To avoid those repugnant consequences—and to do so in a logically consistent manner—Singer (1975) and Regan (1983) argue that we must lower the criterion for moral considerability. But if it is pitched low enough to include the human marginal cases, then it will also include a number of nonhuman animals. Singer (1975) follows Kant's eighteenth century contemporary, Jeremy Bentham, and argues that **sentience**, the capacity to experience pleasure and pain, ought to be the criterion for ethical standing.

Goodpaster (1978) first took the step from animal liberation to biocentric (literally "life-centered") environmental ethics. From a biological point of view, sentience, he argued, evolved not as an end in itself, but as a means to animals' survival. Hence if there is something morally relevant about sentience, how much more morally relevant is that which sentience evolved to serve—namely, life. Moreover, all living things, as explained earlier in this chapter, have a good of their own and therefore have interests. That fact too, according to Goodpaster (1978), ought to entitle all living things to ethical standing.

Delineating a more extreme view, Taylor (1986) argues that all living things are of equal "inherent worth" (Figure 4.6). Apart from the ethically problematic and practically impossible task of according equal moral consideration to each and every living thing, Taylor's pure and extreme biocentrism has little relevance to conservation biology—which, once more, is not concerned with the fate of individual organisms but of species, ecosystems, and evolutionary processes.

Nonetheless, conservation biologists should recognize the importance of this biocentric or "reverence-for-life" (as

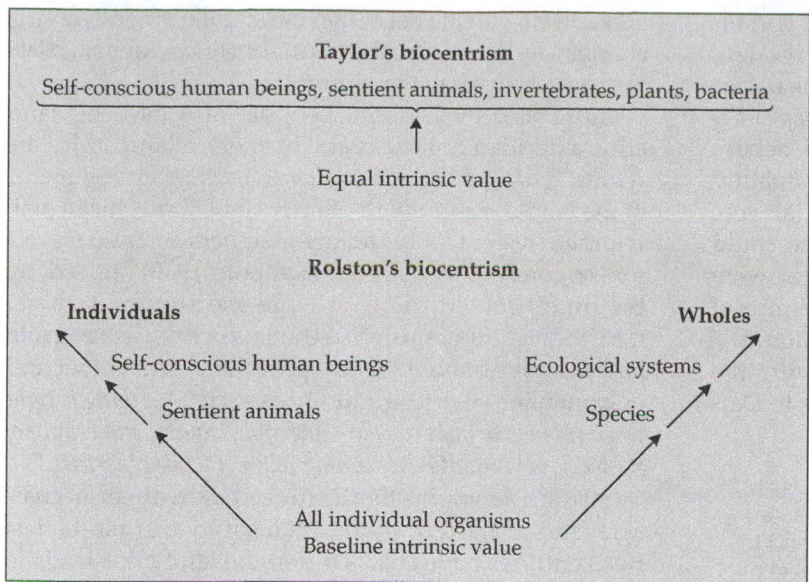

Figure 4.6 Taylor's biocentrism, in which all individual organisms have equal intrinsic value, and Rolston's, in which the baseline intrinsic value at the level of individual organisms is augmented by sentience and self-consciousness; that is, organisms incur increasing intrinsic value for sentience and self-consciousness. Rolston also provides a parallel valuation scheme for "wholes": species and ecosystems.

it is popularly known) ethical orientation. It captures and attempts to explicate an attitude that many people seem to hold deeply—and one that resonates with some of the religious and spiritual traditions sketched previously: The presence of life on Earth is a wonder, and a natural and appropriate focus for nonanthropocentric concern. Rolston (1994) argues for the derivative moral considerability of species, ecosystems, and evolutionary processes on essentially biocentric grounds. Biotic communities and ecosystems are the habitats of intrinsically valuable organisms, which play their stochastic role in the evolutionary trajectory of their species. How can we give them due moral consideration without also considering the habitats necessary for their well-being? Based on biocentric accounts of the intrinsic value of individual organisms Rolston (1988) thus provides a value "dividend," as we might think of it, for species, ecosystems, and evolutionary processes. And he argues that we therefore have a moral duty to preserve them as well.

As modified by Rolston (1988, 1994) biocentrism may address the primary concerns of conservation biologists and hence may represent a viable conservation ethic. As to the practicability of biocentrism, Rolston (1988) agrees with Taylor that all living things have intrinsic value (or inherent worth) and thus should enjoy moral standing. But he does not agree that all living things are of equal intrinsic value. To the baseline intrinsic value that all organisms possess by virtue of having interests and a good

of their own, Rolston adds a value "bonus," as we might think of it, for being sentient; and he adds an additional value bonus for being rational and self-conscious. Hence, sentient animals have more intrinsic value than insentient plants, and human beings have more intrinsic value than sentient animals (see Figure 4.6). Rolston's biocentrism thus better accords with our intuitive sense of a value hierarchy than does Taylor's, because in Rolston's version, the life of a human being is more valuable than that of a white-tailed deer, and that of a deer more valuable than that of a jack pine.

Conservation biologists should bear in mind that public discussions, debates, and ultimately policy decisions are shaped in part by biocentric concerns (Dizard 1994). It is not uncommon for those engaged in environmental management to be confronted by citizens who are moving beyond narrow anthropocentric values—concerned about the welfare of individual geese or deer or orcas—but who are opposed to culling overgrown species populations in the interest of ecosystem health or the maintenance of biodiversity. How to build up concern for the preservation of species or ecosystems and of biodiversity based on increasingly popular concern for individual organisms represents a distinctly philosophical and rhetorical challenge for conservation biologists.

Ecocentrism

For sound philosophical as well as temperamental reasons, those conservation biologists with nonanthropocentric sympathies have gravitated to the Aldo Leopold land ethic in their search for a fitting conservation ethic. Leopold was himself a conservation biologist; indeed he was, perhaps, the prototype of the breed (Meine 1992). Further, the Leopold land ethic is not based on religious beliefs, nor is it an extension of the ethical paradigm of classic Western moral philosophy; it is grounded, rather, in evolutionary and ecological biology. Hence, all nonanthropocentric conservation biologists, irrespective of religious or cultural background, will find the Leopold land ethic intellectually congenial.

In *The Descent of Man*, Darwin tackled the problem of the evolutionary origins and development of ethics. How could "limitations on freedom of action" possibly have arisen, through natural selection, given the universal "struggle for existence" (Darwin 1904)? In a nutshell, Darwin answered as follows: Social organization enhances the survival and reproductive efficiency of many

kinds of organisms. Among mammals, parental and filial affections, having spilled over to other close kin, bound individuals into small social units such as packs, troops, and bands. When one mammal—*Homo sapiens*—acquired the capacity for reflection and speech, behaviors that are conducive to social integrity and stability were dubbed "good" and those that are antisocial were dubbed "bad." Or, as Darwin wrote, "No tribe could hold together if murder, robbery, treachery, etc., were common; consequently such crimes within the limits of the same tribe, 'are branded with everlasting infamy.'" Once originated, ethics developed apace with the growth and development of society. According to Darwin (from the 1904 edition),

> As man advances in civilization, and small tribes are united into larger communities, the simplest reason would tell each individual that he ought to extend his social instincts and sympathies to all the members of the same nation though personally unknown to him. This point being once reached, there is only an artificial barrier to prevent his sympathies extending to the men of all nations and races.

Here, at the beginning of the twenty-first century, we have finally reached the point that Darwin could only envision in the middle of the nineteenth: a universal ethic of human rights. Meanwhile, ecology discovered (actually rediscovered, because many tribal peoples seem to have represented their natural environments in analogous terms) that human beings are members not only of various human communities—from the familial clan to the family of man—but are also members of a "biotic community" as well.

From Darwin we may learn that "All ethics so far evolved rest upon a single premise: that the individual is a member of a community of interdependent parts" (Darwin 1904); and from Leopold (1949) we may learn that ecology now "simply enlarges the boundaries of the community to include soils, waters, plants, and animals, or collectively: the land." If, whenever a new community came to be recognized in the past, "the simplest reason would tell each individual that he ought to extend his social instincts and sympathies," Leopold argues, then the same "simplest reason" ought to kick in again, now that ecology informs us that we are members of a biotic community.

Though altogether forgotten in Western moral philosophy over the last 200 years, human ethics has always had a strong holistic aspect. That is, human beings have felt that they had duties and obligations to their communities as such, as well as to individual members of those communities. About this Darwin (1904) was emphatic: "actions are regarded by savages, and were probably so regarded by primeval man, as good or bad, solely as they obviously affect the welfare of the tribe, not that of the species, nor that of an individual member of the tribe. This conclusion agrees well with the belief that the so-called moral sense is aboriginally derived from the social instincts, for both relate at first exclusively to the community."

Influenced by Darwin, Leopold also gave his land ethic a decided holistic cast: "In short, a land ethic," he writes, "changes the role of *Homo sapiens* from conqueror of the land community to plain member and citizen of it. It implies respect for his fellow-members and also respect for the community as such" (Leopold 1949). Indeed, by the time Leopold came to write the summary moral maxim, or "golden rule" of the land ethic, he seems to have forgotten about "fellow-members" altogether and only mentions the "community as such": "*A thing is right when it tends to preserve the integrity, stability, and beauty of the biotic community. It is wrong when it tends otherwise.*"

Staunch apologists for the rugged individualism characteristic of Western moral philosophy during the last two centuries have charged that the land ethic leads to "environmental fascism"—the subordination of the rights of individuals, including human individuals, to the good of the whole (Regan 1983; Aiken 1984). They have a point where nonhuman animals are concerned. The land ethic would permit—nay, even require—killing animals, such as feral goats or rabbits, that pose a threat to populations of endangered floral species or to the general health and integrity of biotic communities. But Leopold, following Darwin, represented the land ethic to be an ethical "accretion"—that is, an addition to, not a substitute for, our long-standing human-to-human ethics.

That human beings have recently become members of national and international communities does not mean that we are no longer members of more ancient and more narrowly circumscribed social groups, such as extended families, or that we are relieved of all the moral duties and responsibilities that attend our active family, clan, and civic affiliations (Figure 4.7). Similarly, because we now realize that we are also members of a biotic community does not entail that we are relieved of all the moral duties and responsibilities that attend our membership in the full spectrum of human communities.

This defense of the Leopold land ethic against the charge that it is a case of environmental fascism leads to the charge that it is a "paper tiger," an ecocentric environmental ethic without "teeth" (Nelson 1996). For if we must fully acknowledge all our ancient and modern human duties and obligations as well as our more recently discovered environmental ones, how can we ever justify sacrificing human interests to conserve nonhuman species and ecosystems?

Fortunately, all human–environment conflicts are not life and death issues. We rarely face a choice between killing human beings and conserving biodiversity. Rather, most choices are between human lifestyles and biodiversity. For example, Japanese and other con-

Community membership in the Leopold land ethic

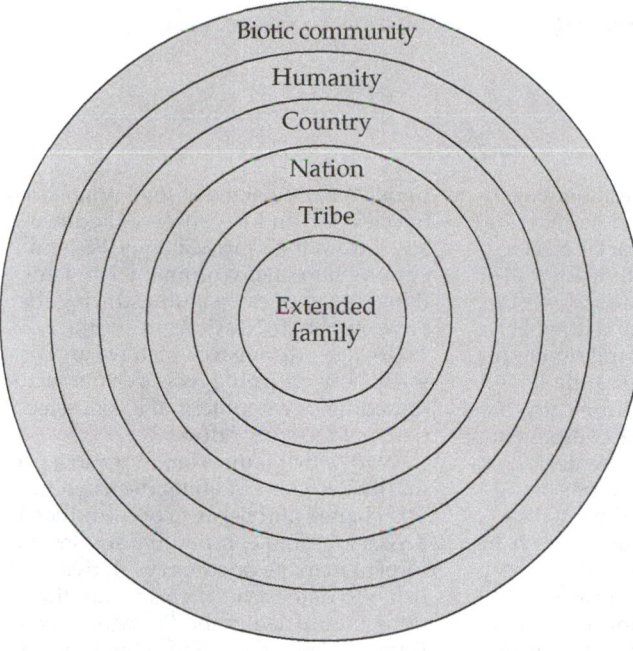

Figure 4.7   The various communities to which human beings belong, and how these communities are hierarchically ordered in the Leopold land ethic. The smallest and most intimate community is the family; the largest is the multispecies biotic community. In general, duties and obligations related to the communities at, or closer to, the center historically have taken precedence over those at, or closer to, the perimeter. But we must also consider the gravity, or weight, of duties and obligations related to these communities, as well as their proximity, when they come into conflict with one another.

sumers of whale meat are not asked to lay down their lives to save the whales, only to change their dietary preferences. To save forests we do not have to commit suicide; we can save them by using less lumber and paper, by recycling what cellulose we must extract, and by replanting forests over degraded lands. All human interests are not equal. We should be prepared to override less important or short-term human interests for the sake of the vital interests of other forms of life, and for ecological health and integrity. The ways in which the values correlative to various communities come into conflict and get resolved is explored in Case Study 4.1 by Yih-Ren Lin and Ken-Cheng Li. A discussion of how similar values have percolated through U.S. federal agencies follows in Essay 4.3 by Edwin Pister.

Leopold penned the land ethic at mid-century. Ecological science then represented nature as tending toward a static equilibrium, and portrayed disturbance and perturbation, especially those caused by *Homo sapiens*, to be abnormal and destructive (Odum 1953). In light of recent doubts about the very existence of "biotic communities"

that persist as such through time (Brubaker 1988), in view of the shift in contemporary ecology to a more dynamic paradigm (Botkin 1990), and in recognition of the incorporation of natural disturbance to patch- and landscape-scale ecological dynamics (Pickett and White 1985) we might wonder if the Leopold land ethic has become obsolete. Has the paradigm shift (discussed in Chapter 1) from "the balance of nature" to the "flux of nature" in ecology invalidated the land ethic? No, but recent developments in ecology may require revising the land ethic.

Leopold was aware of, and sensitive to, natural change. He knew that conservation must aim at a moving target. How can we conserve a biota that is dynamic, ever changing, when the very words conserve and preserve—especially when linked to integrity and stability—connote stasis? The key to solving that conundrum is the concept of temporal and spatial scale. A review of Leopold's essay "The Land Ethic" reveals that he had the key, though he may not have been aware of just how multiscalar change in nature actually is.

In "The Land Ethic" Leopold (1949) writes, "Evolutionary changes . . . are usually slow and local. Man's invention of tools has enabled him to make changes of unprecedented violence, rapidity, and scope." As noted, Leopold was keenly aware that nature is dynamic, but, under the sway of mid-century equilibrium ecology, he conceived of natural change primarily in evolutionary, not in ecological, terms. Nevertheless, scale is equally relevant when ecological change is added to evolutionary change, that is, when normal climatic oscillations and patch dynamics are added to normal rates of extinction, hybridization, and speciation.

Homo sapiens is a part of nature, "a plain member and citizen" of the "land-community" as Leopold (1949) put it. Hence, anthropogenic (human-caused) changes imposed on nature are no less natural than any other. But, because *Homo sapiens* is a moral species, capable of ethical deliberation and conscientious choice, and evolutionary kinship and biotic community membership add a land ethic to our familiar social ethics, anthropogenic changes may be land-ethically evaluated. But by what norm? The norm of appropriate scale.

Temporal and spatial scale in combination are key to the evaluation of direct human ecological impact. Long before *Homo sapiens* evolved, violent disturbances regularly occurred in nature (Pickett and White 1985), and they still occur, quite independently of human agency. Volcanoes bury the biota of whole mountains with lava and ash. Tornadoes rip through forests, leveling trees. Hurricanes erode beaches. Wild fires sweep through forests and savannas. Rivers drown floodplains. Droughts dry up lakes and streams. Why, therefore, are analogous anthropogenic disturbances—clear cuts, beach developments, hydroelectric impoundments, and the like—envi-

ESSAY 4.3

The Importance of Value Systems in Management
Considerations in Desert Fish Management

Edwin P. Pister, *Bishop, California*

■ To one who started his career in 1953, managing game fish populations for the California Department of Fish and Game in the state's vast and heavily tourist-impacted eastern Sierra and desert regions, it has been encouraging to note how, during the ensuing 45 years, the priorities of society have grown more sophisticated in matters relating to conservation. Unfortunately, programs of resource management agencies have been slow in recognizing and responding to the public will.

During the two decades following World War II, most government fisheries management efforts in the United States were directed toward satisfying the desires of a nation freed from wartime constraints and eager to explore outdoor recreation. This was the era of huge production trout hatcheries and reservoir management programs. Agency leadership assumed a complacent attitude that such programs would satisfy society's needs forever. All one had to do to accomplish this was build more fish hatcheries, and introduce into the reservoirs any alien game fish (regardless of the source or ecological consequence) that showed promise of improving angling for a while. Into this management scenario entered a cadre of fisheries biologists/managers, emerging in vast numbers from burgeoning fish and wildlife schools and eager to apply their newly learned technologies to satisfying angler demand and, where this proved to be inadequate, to top off the harvest by calling for yet another hatchery truck. The term "biodiversity," yet unborn, lay dormant in the womb of a society unaware of adverse changes in fish communities that, even at this early date, were beginning to occur throughout the American West. In the context of this essay, the terms "manage" and "conserve" are essentially synonymous. What we are attempting to do now is manage for the conservation of biodiversity.

As one might expect, considering the basic need of fish for water, changes were first noted in the desert areas of North America, where negative biological impacts of water extraction and diversion by humans were soon recog-

nized (Miller 1961). Government agencies in the Southwest (and nationwide) soon found themselves faced with responsibilities for which they were ill-prepared: management and preservation of a native fish fauna with which they were almost totally unfamiliar (or even unaware), known essentially only to academic researchers. Ironically, most agency knowledge of this component of the fauna derived from inventories conducted following chemical poisoning projects designed to eradicate native fishes. Up to that point, nongame fishes had been viewed primarily in the very negative context of being unwanted competitors with economically important, introduced game species (Pister 1991).

Agencies were caught off-guard from two major perspectives: (1) their knowledge of the biology of many of these species and related ecological interactions was totally inadequate to assure the continued existence of an intact fauna, and (2) few individuals within these agencies were of a philosophical bent that encouraged enthusiasm for nongame management. Those that were, were constantly plagued by a question posed by their peers and society: "What good are they?"—a question that, unfortunately, remains with us even today. The infancy of nongame species management is reflected in the fact that when Robert Rush Miller and I wrote a paper in 1971 on the management of the Owens pupfish, it was the first paper ever published in the *Transactions of the American Fisheries Society* relating to the management of a nongame, or commercially unimportant, species (Miller and Pister 1971).

It was the certain knowledge that government inertia in this respect would persist for at least a decade that caused a group of concerned scientists in 1969 to form the Desert Fishes Council, essentially to "hold the fort" until such time as fully funded management programs for native fishes could be implemented (Pister 1991). At this writing, over 25 years later, we still await full implementation, as both state and federal agencies struggle with perpetually underfunded programs in an often futile effort to fulfill their obligations

under the provisions of the Endangered Species Act, and to conserve biodiversity within their jurisdictions. Federal suppression of environmental and endangered species programs during the environmental "Dark Ages" of the 1980s, a syndrome to which we are sadly now returning, was a major factor impeding development of a concerted national recovery effort.

Native fish faunas are in trouble primarily because of habitat destruction and change and because of introduced predacious and competitive species. Efforts to manage or conserve native fishes to date have taken several directions, chronicled in the *Proceedings of the Desert Fishes Council* and other publications. They include establishment of small refuges, free from alien species and designed to emulate as closely as possible the evolutionary habitat of the species or species complex, restoration of damaged habitat, artificial rearing facilities such as the Dexter National Fish Hatchery, operated by the U.S. Fish and Wildlife Service near Roswell, New Mexico, and acquisition and protection of major habitat areas, utilizing an ecosystem integrity approach. In this latter instance managers must often learn to live with the existence of alien fishes and inexorable change due to increasing societal demands for water. Recent research on habitat preferences of desert fishes allows this information to be incorporated into operational plans of water development projects, thus assisting in management and recovery efforts.

The long-term importance of the refuge management approach cannot be overemphasized. If a North American desert fish species is not currently listed as endangered, it will not be long before it reaches that point. From all indications, urban development in the desert will continue indefinitely into the future, and each time a new dwelling is connected to a domestic water supply, it either directly or indirectly impacts an aquatic habitat. Very few desert aquatic habitats today even approach pristine, and the situation continues to deteriorate. We must live with the possibility that much of our desert aquatic fauna may eventually exist only in artificial

refuges, or in greatly altered native habitats. The ethics and evolutionary practicality of such a scenario provide much discussion for biologists and philosophers as we enter the next century.

Recovery efforts to date have primarily been holding actions, and very little "recovery" per se has been effected. This rather depressing situation will in all likelihood continue for as long as ever-increasing numbers of people demand an ever-increasing standard of living. We can only hope, before it is too late, that the values inherent in maintaining natural biodiversity will become sufficiently clear, and society will be willing to make the minor sacrifices necessary to retain it. Indiana University's Lynton Caldwell put it this way (Miller 1988):

The environmental crisis is an outward manifestation of a crisis of mind and spirit. There could be no greater misconception of its meaning than to believe it to be concerned only with endangered wildlife, human-made ugliness, and pollution. These are part of it, but more importantly, *the crisis is concerned with the kind of creatures we are and what we must become in order to survive* (emphasis added). ■

---

ronmentally unethical? As such, they are not; it is a question of scale. In general, frequent, intense disturbances, such as tornadoes, occur at small, widely distributed spatial scales; spatially broadcast disturbances, such as droughts, occur infrequently. And most disturbances at whatever level of intensity and scale are stochastic (random) and chaotic (unpredictable). The problem with anthropogenic perturbations—such as industrial forestry and agriculture, exurban development, drift net fishing, and such—is that they are far more frequent, widespread, and regularly occurring than are nonanthropogenic perturbations; they are well out of the normal range of spatial and temporal scales experienced by ecosystems over evolutionary time.

Pickett and Ostfeld (1995)—proponents of the natural disturbance/patch dynamics paradigm in ecology—agree that appropriate scale is the operative norm for ethically appraising anthropogenic ecological perturbation. They note that,

> ... the flux of nature is a dangerous metaphor. The metaphor and the underlying ecological paradigm may suggest to the thoughtless and greedy that because flux is a fundamental part of the natural world, any human-caused flux is justifiable. Such an inference is wrong because the flux in the natural world has severe limits. ... Two characteristics of human-induced flux would suggest that it would be excessive: fast rate and large spatial extent.

Among the abnormally frequent and widespread anthropogenic perturbations that Leopold himself censures in "The Land Ethic" are the continent-wide elimination of large predators from biotic communities, the ubiquitous substitution of domestic species for wild ones, the ecological homogenization of the planet resulting from the anthropogenic "world-wide pooling of faunas and floras," and the ubiquitous "polluting of waters or obstructing them with dams."

The summary moral maxim of the land ethic, then, must be updated in light of developments in ecology over the past quarter-century. Leopold acknowledged the existence and land-ethical significance of natural environmental change, but seems to have thought of it primarily on a very slow evolutionary time scale. But even so, he thereby incorporates the concept of inherent environmental change and the crucial norm of scale into the land ethic. In light of more recent developments in ecology, we can add norms of scale to the land ethic for both climatic and ecological dynamics in land-ethically evaluating anthropogenic changes in nature. Although one hesitates to edit Leopold's elegant prose, we attempt to formulate a dynamized summary moral maxim for the land ethic with the following:

> A thing is right when it tends to disturb the biotic community only at normal spatial and temporal scales. It is wrong when it tends otherwise.

CASE STUDY 4.1

Cypress Forest Conservation on Taiwan
A Question of Value

Yih-Ren Lin and Ken-Cheng Li, Providence University, Taiwan

Taiwan, with a total area of 36,000 km² lies in the Pacific Ocean, 160 km from Mainland China, and is bisected by the tropic of Cancer. Two-thirds of the island is ruggedly mountainous and lushly vegetated. The highest peak, Yushan, reaches 3952 m and is also the highest mountain in East Asia.

Most of the human population of 22–23 million live in the lowlands on the fertile western coastal plain. How best to conserve Taiwan's most magnificent natural treasure, its remnant cypress forests, has become a major political controversy.

The cypress trees on Taiwan, *Chamaecyparis* spp., belong to the same family, *Cupressaceae*, as those in the southeastern swamps of the United States (*Cupressus* spp). But they grow at elevations between 1800 m and 2500 m, and in size (they can grow to be 60 m tall and 20 m in girth), habitat, ecology, commercial value, and charisma they more resemble the California redwoods. There are only six species of *Chamaecyparis* in the world, two of which, *C. formosensis* and *C. obtusa* var. *formosana* are endemic to Taiwan, found growing in both pure and mixed stands (Figure A). In addition to their magnificent beauty, Taiwan's cypress forests are of great ecological importance. They provide habitat for other rare and endemic species. And they grow on steep slopes in misty mountains, which receive about 400 cm of rain per year, where hundreds of streams originate, and the geomorphology is highly unstable. Thus they also help to modulate the montane hydrology, hold the soil, and prevent floods and landslides.

Figure A   A single cypress tree. (Photograph courtesy of S. W. Chung, Taiwan Forestry Research Institute.)

Due to the excellent quality of Taiwanese cypress timber and their great size, the trees have an extremely high commercial value. In 1912, the Japanese, who were then the colonial masters of Taiwan, began extensive logging in three major forests—Alishan, Taipingshan, and Bashenshang—the most accessible sites for transport. In 1945, rule of Taiwan shifted from the Japanese to the Chinese. After being expelled from the mainland by Mao Tse-Tung and the Chinese Communist Party, Chiang Kai-Sheck and the Nationalist Party (KMT) retreated to Taiwan in 1949. The KMT government disregarded the advice of some scientists who insisted on the ecological and cultural importance of preserving the cypress forests, and, for purely economic reasons, intensified the logging activities begun by the Japanese.

In 1959, the KMT government announced a far-reaching policy, "The Principles of Forest Management in Taiwan." According to this plan, the old-growth cypress forests were to be completely clear-cut and artificially replanted over 80 years—with the exception of some isolated patches that were to be preserved for research and recreation. Also in 1959, the Council of Veterans (COV)—a cabinet-level department of the government—was established to look out for the well-being of the veterans of Chiang Kai-Sheck's war with Mao Tse-Tung. These men had left their families and villages behind on the mainland, to which they could never return. Thus they had no social and economic infrastructure in Taiwan to support them. In addition to that overseen by the Bureau of Forestry, a large portion of the cypress logging was turned over to the Forest De-

velopment Division of the COV to provide jobs for the veterans. Thus began the period of the most extensive logging of cypress trees in Taiwan's history. For nearly 25 years, approximately 70% of the total revenue of Taiwan's timber industry came from logging the cypress trees. Taiwanese environmental non-governmental organizations (NGOs) estimate that almost 90% of the country's ancient cypress forests have been destroyed.

In addition to providing the veterans with jobs and benefits, the ostensible national purpose of logging the cypress trees was to enhance economic growth and accelerate the modernization of Taiwan. However, it achieved this result at the cost of an ecological tragedy. After such ruthless logging, there followed nature's revenge. With the forest cover removed, the rain ran unimpeded off the mountains and into the densely settled valleys. In 1959 Taiwan suffered the infamous Ba-Chi flood, in which more than a thousand people died or disappeared. In 1960, the Ba-Ih flood took the lives of 210 people. In 1963, 361 people lost their lives in the flooding caused by Typhoon Gloria. These statistics grimly testify to the cost in human life of discounting a suite of nature's free ecosystem services, not to mention the collateral economic damage inflicted by these floods. Ironically, by now the threat of such flooding seems to have become an accepted part of Taiwanese daily life, especially during the rainy season.

Nearly a century of extensive logging sanctioned by several governments has left only two large remnants of cypress forest on Taiwan—the Chi-Lan (dominated by *C. obtusa* var. *formosana*) and the Sho-Ku-Luan (dominated by *C. formosensis*). As one of the six so-called "Asian Tigers, " Taiwan's economy shifted from one based principally on agriculture and natural resource exploitation to one based principally on manufacturing during the last quarter of the twentieth century. With this shift and with the democratization of the political landscape and the growth of ecological literacy, a "Logging Natural Forest Ban" was declared in 1991 due to the effects of environmentalists. However, many in the forestry industry still sought means to get at the cypress timber because of its great economic value. This began a struggle over Taiwan's cypress forests that continues today.

In 1994 the COV proposed to continue the practice of "salvaging" cypress trees that had been killed (e.g., blown over by high winds, or killed by lightening strikes or by disease), and artificially replanting young trees in the openings. They argue that doing so is necessary to regenerate the cypress forests. Under this ruse they were allowed to not only remove (and sell) the

dead cypress trees, but also to clear the subcanopies and under-stories of the forest, composed mostly of broadleaved trees and shrubs. Of course, as result of this silvicultural "treatment," the ecology of the cypress forest on some 800 ha was drastically changed.

In 1998 the COV proposed a five-year plan to remove dead trees in most of the remaining cypress forests, with the rationale that if they do not, the living forests will no longer be able to reproduce themselves and will eventually die out. The COV argument—that the cypress forest cannot survive without human help—is hardly credible; indeed it is preposterous. Environmentalists counter that for the past million years there has been no such silviculture, yet the cypress forests flourished on Taiwan's mountains. Considering the great economic value of salvaged cypress timber, one suspects that the COV is motivated less by the well-being of the cypress forests than by their own prosperity. In addition to clearing the subcanopies and understories, heavy salvage equipment damages living trees and the delicate soil structure of the cypress forests, not to mention endangering the safety of people living downstream. Moreover, the media have uncovered quite a few violations of the post-1991 forest regulations, including the poaching of living cypress trees.

This proposal provoked a number of environmental NGOs to band together and launch the "Rescue Chamaecyparis Forest Movement" in 1998. Their goal was to force the COV to stop their cypress-salvage operation and to revoke altogether their right to manage the forests. After arguing with pro-logging foresters in the press and mounting all sorts of petition drives, the environmental NGO consortium had not achieved this goal by year's end. Therefore, they intensified their efforts. First, they held rallies attracting thousands of people (Figure B) and continued to lobby Congress. In February, 1999, the KMT government finally agreed to stop the COV's cypress-salvage operation.

Undaunted, the COV and forestry industry in Taiwan continued to press their cypress-salvage scheme. With the preservation of the cypress forests again in doubt, environmentalists countered with a plan to create a new national park to protect the cypress forests permanently. On Christmas Day, 1999, the environmentalists held a rally called "Protecting the Millennium Christmas Trees, Launching the Chi-Lan Chamaecyparis National Park." This rally happened during Taiwan's presidential election campaign. Seeking the votes of environmentalists, Democratic Progressive Party (DPP) candidate Chen Shui-Bien responded by promising to establish the new national park if he were elected.

If created, this new national park, however, would have to be carved out of the traditional territory of the Atayal indigenous people. Taiwan's indigenous peoples are an ethnically distinct minority that were driven into the highlands when Chinese people began to immigrate to the island in the seventeenth century. The new national park proposed by the Rescue Chamaecyparis Forest Movement therefore introduced a new interest group and set of values into the cypress forest conservation debate. Taiwan already had six national parks comprising 8% of the island's total area. When they were created, indigenous peoples were often ignored and dispossessed. Understandably, they therefore tend to be opposed to any addition to the national park system. On the other hand, the indigenous peoples are not natural allies of the COV—which also opposes creating the new national park—because their traditional subsistence uses of the highlands do not include logging, nor is logging consistent with those uses. After a dialogue with some Atayal leaders, environmental NGOs decided to include local indigenous people in the design of the new national park, and to establish a new "co-management" protocol in the new national park that would involve indigenous people on the park's board of directors and employ indigenous people as park rangers, guides, naturalists, and in other park jobs. So conceived, the proposed new national park would ensure the preservation of the precious old-growth cypress forests, as well as revitalize indigenous culture and enhance the livelihood of indigenous people. It would put traditional ecological wisdom to work for both the forest and its native people.

To everyone's surprise, including his own, Chen Shui-Bien won the presidency in 2000. The NGOs began to pressure him to make good on his promise to create a new national park. In September of 2000, the new government did propose to fulfill his promise, but in a strange way. A new national park of about 27,000 ha was to be created, but mostly outside the main cypress forests targeted by the COV. Further, the COV would be allowed to retain its right to

Figure B Protest in favor of establishing a national park in Taiwan to protect the cypress forests. Photograph courtesy of Y. - R. Lin.)

manage those forests under the guidance of an experimental scheme called the "Sustainable Ecosystem Management of the Chi-Lan Mountain Project," which would comprise an area of 121,000 ha. This document compromise did not at all address the concerns of the Ataya. Thus the proposed "compromise" enraged both the environmentalists and indigenous groups, who refused to tolerate the finagling of the new DPP government. The compromise was technically designed to honor a political promise, but in reality was designed in such a way as to skirt its main purpose.

In October of 2000, the environmental NGOs and some indigenous groups appealed to the Ministry of the Interior—which oversees the national parks—to respect their now-allied concerns. In response, a "Maqaw National Park Advisory Committee" was formed. Over the course of eight rounds of meetings, the committee discussed issues related to the concept of co-management, its means of implementation, and the boundaries of the new national park that will both preserve the cypress forests and substantially benefit local indigenous people. This committee soon became an important forum for dialogue between conservationists, indigenous people, scientists, and the state. However, it has not succeeded in eliminating skepticism and distrust among the opposing groups.

The fate of the cypress forests very much depends on the resolution of the conflicting values of environmentalists, indigenous people, and the state. Environmentalists value the cypress forests both intrinsically and for their higher-end utility—their psycho-spiritual benefits. Indigenous people also value the forests intrinsically, but they value them too as a source of subsistence goods and ecosystem services. The state's values are more complex and internally inconsistent. The COV,

as one element of the state, values the cypress forests purely as a timber resource. To be sure, the COV and their foresters acknowledge the need for conserving cypress trees, but their conservation paradigm is that of sustained yield, under an intensive techno-scientific management regime. Other elements of the state value cypress forests for their ecosystem services both in regard to present and future generations of Taiwanese.

How these values will be reconciled or which combination will eventually prevail is not clear. What is clear is that the tacit recognition of the intrinsic value of the cypress forests has shifted the burden of proof from preservationists of all stripes to the COV and its forest management allies. The COV must now justify its exploitative use of the cypress forests under the guise of "salvage" for ecosystem health and assisted regeneration. They cannot simply assume that they have a prima facie right to take what they wish and leave the onus of proving the greater instrumental value of preserving the forests to the preservationists. Progress toward resolution will not come easily because each faction believes that its values are non-negotiable. The COV remains a powerful symbol of state concern to many of the descendents of the people who came to Taiwan from China with Chiang Kai-Sheck. To conservationists the cypress forests are a symbol of Taiwan's intrinsically valuable magnificent natural heritage. To indigenous people the forests are their storied residences.

Creating a new park will be a long-term process that demands patience, intelligence, courage, vision, and most importantly, trust between all interest groups. These elements of dialogue are rarely found in a young democracy like Taiwan's. However, despite their differences, a common starting point shared by all parties in the debate is the conservation of the remaining cypress forests for future generations of Taiwanese.

Summary

1. Conservation biology, a crisis discipline, is driven by the value of biodiversity. But why should people value biodiversity? Philosophers have distinguished two basic types of value, instrumental and intrinsic. Biodiversity is instrumentally valuable for the goods (e.g., actual and potential food, medicine, fiber, and fuel), services (e.g., pollination, nutrient recycling, oxygen production), information (e.g., practical scientific knowledge, a genetic library), and psycho-spiritual satisfaction (e.g., natural beauty, religious awe, pure scientific knowledge) that it provides for intrinsically valuable human beings. Biodiversity may also be intrinsically valuable—valuable, that is, as an end in itself, as well as a means to human well-being. Like ourselves, other forms of life are self-organizing beings with goods of their own. And we human beings are capable of valuing other beings, other species, biotic communities, and ecosystems for their own sakes, as well as for what they do for us.

2. In order to compare its value with the value of other things, economists have attempted to monetize both the instrumental and intrinsic value of biodiversity. Philosophers have also based conservation ethics on the value of biodiversity. If biodiversity is only instrumentally valuable to human beings, its destruction by one party in pursuit of personal gain may be harmful to another—in which case, the destruction of biodiversity may be immoral. If biodiversity also has intrinsic value its destruction may be doubly immoral.

3. The Bible recognizes the intrinsic value of other species (God declared them to be "good"). Accordingly, contemporary Jewish and Christian theologians have formulated a Judeo–Christian stewardship conservation ethic. Many other world religions are also developing distinct conservation ethics based on their scriptures and traditions.

4. The Aldo Leopold land ethic is not based on any religion, but on contemporary evolutionary and eco-

logical biology. From an evolutionary perspective, human beings are kin to all other forms of life, and from an ecological perspective, human beings are "plain members and citizens" of the "biotic community." According to Leopold, these general scientific facts generate ethical obligations to our "fellow voyagers in the odyssey of evolution," to "fellow-members of the biotic community," and to that "community as such." Though ecology now acknowledges the normalcy of change and disturbance in nature, the Leopold land ethic, appropriately revised in light of these recent developments in science, remains the guiding environmental ethic for conservation biology.

> Please refer to the website www.sinauer.com/groom for Suggested Readings, Web links, additional questions, and supplementary resources.

Questions for Discussion

1. Should conservation biologists explain the value of biodiversity to the general public in purely instrumental (or utilitarian) terms or should they also offer reasons for thinking that biodiversity has intrinsic (or inherent) value?

2. How should a conservation biologist who is trying to save a small endangered plant species, such as Furbish's lousewort, respond to the question, "What good is it?"

3. How does the understanding of human nature and the place of human beings as explained in Genesis (in the Bible) compare with the understanding of human nature and the place of human beings in nature forthcoming from science?

4. Suppose a population of weedy sentient animals, say feral goats, is threatening a plant species endemic to an island with extinction. What ethical concerns should a conservation biologist take into account before proposing a course of action?

5. If, in Rolston's version of biocentrism the life of a white-tailed deer is more intrinsically valuable than that of a jack pine, would it also follow that the life of a gray squirrel is more intrinsically valuable than the life of a thousand-year-old redwood tree? Is the life of a human being more intrinsically valuable than that of a thousand-year-old redwood tree? Why?

5

Ecological Economics and Nature Conservation

Gareth Edwards-Jones

You can be as rich as you like—but still the loveliest things in the world aren't to be bought.

Enid Blyton, 1968

Why Do We Need Ecological Economics?

As an academic discipline, economics is older than ecology, with a history going back at least as far as Adam Smith's *The Wealth of Nations* published in 1776. This predates the major works of Charles Darwin (1859) by 83 years, and the establishment in 1913 of the world's first ecological society, the British Ecological Society, by 137 years. Despite the absence of a formal discipline called "ecology," early economics was concerned with several aspects of the environment, such as the use of land, timber, and fish. This was because the supply of these natural resources was crucial to the development of industry and society in Western Europe and North America during the eighteenth and nineteenth centuries. It was only toward the end of the twentieth century, when industrialization in the West was complete, that economics was linked with ecology and applied to issues like nature conservation and sustainable development.

A major driver in bringing ecology and economics together was the realization that neither ecology nor any other natural science, could on its own, prevent environmental degradation, nor could economics provide complete guidance for resource management without a natural science foundation. This is because habitat and species loss, like most other environmental problems, are caused by people, so it seems only sensible that finding solutions to these problems will also require an understanding of people and the factors that affect their behavior. This is where economics comes in, and along with other social sciences like sociology, anthropology, and psychology, it can provide valuable insights into conservation problems. This chapter seeks to explain the basics of economics for conservationists, with the hope of shining some light on the way economists think, and give some idea of how economics can potenitally help solve some aspects of conservation problems. The chapter concludes with case studies that show both the interactions of economics with

conservation and the possibility of using economic methods to help make conservation decisions.

Let us begin by considering a hypothetical debate—that of building a new road to bypass a town of historical and cultural interest. Often on such occasions, politicians seem in favor of building new roads, as do many local residents. On the other hand, environmentalists often seem to be against them. Making a logical and defensible decision in such a situation is difficult, as each group identifies different advantages and disadvantages (costs and benefits) arising from the project. Benefits arise for some local people from a reduction in town center traffic, and these manifest themselves as reduced noise, reduced air pollution, and a safer environment. Other locals though, see only costs arising from the project, and they believe that the reduction in town center traffic will adversely affect their businesses. The politicians suggest there will be economic benefits on a larger scale, as communications between different regions will be improved, while environmentalists believe that new road construction should be kept to a minimum as it fragments habitats and may adversely affect rare or critical species. The ultimate problem is that the project will affect different groups of people in very different ways and thus, it is difficult to make a rational judgment of the project's costs and benefits.

One approach to solving such problems is to express the costs and benefits of such a project on a similar scale. Then it would be simple to decide whether or not the benefits of the project were greater than the costs. Economists claim to have a solution to this challenge—convert all costs and benefits into a monetary scale. For an economist this solution makes sense as the approach fits neoclassical economic theory, which suggests that money can act, albeit imperfectly, as a measure of the extent to which the well-being (utility) of individuals are affected.

Utility is an important concept in economics, and as stated above it is a measure of "well-being." An increase in an individual's utility is felt to be a good thing, although the exact cause of any increase may be due to many different events such as a pay raise, the purchase of a new DVD, an improvement in health, or an improvement in local biodiversity. The exact causes of increases (or decreases) in utility vary between people, and between times in an individual's life; in other words, what makes me a bit happier now may not be anything like what you need to make yourself a bit happier. For this reason utility remains a useful concept but is nearly impossible to measure.

Despite the problems in quantifying utility there is little doubt that utility is a real phenomenon. For example, consider your situation when faced with a choice between two goods—let's say money and a cola drink. Standard economic theory suggests that individuals can (implicitly or explicitly) identify a satisfactory trade-off between the quantities they want of these two goods—that is, they are somehow "balancing" their utility. For example, when you are thirsty you may choose to buy a drink of cola for $1; you have traded off a quantity of one good, money, for another good, cola. If you walked away from such a deal feeling content with the trade, then you are said to be "indifferent" to trading $1 for a drink of cola (no net change in utility). However, had the shop offered you cola at a price of $20, you may not have chosen to buy it, as you weren't prepared to trade such a large amount of money for a drink of cola. Indeed had you been forced to pay such a large amount for a drink, you may have felt angry at paying so much. In such a situation economists would say you had lost utility as a result of the trade, as evidenced by your dissatisfaction. Conversely of course, the shop may have sold cola at the low price of 1¢, in which case you would feel very good about the trade, feeling you had gained from the deal (i.e., you had achieved a net gain in utility). This type of argument can be applied to a whole range of goods including environmental goods, and is fundamental to many economic analyses.

In our road construction example, each individual will view the decision from the point of view of their own utility. Economic theory suggests that in such a situation most people would seek to balance (or improve) their utility, and as part of this "balancing," most people would be willing to accept as compensation for a loss, some quantity of a "substitute good" that provides an equal amount (or more) utility to them than the utility they would lose from the development of the road. Generally when we talk about compensation we tend to think of monetary compensation—and this is certainly the way many of the world's legal systems work. But all of the above arguments also apply to environmental goods; therefore, in theory, if increased pollution from car exhaust fumes was identified as a major problem arising from the new road, then there is some amount of money that could exactly substitute for a given decline in air quality, leaving individuals no worse off after the road was built than before (i.e., no net loss of utility). The exact amount of money needed to achieve this may vary among individuals and situations according to a whole range of factors, such as level of pollution, the age of a person, and their health, wealth, and attitudes toward the environment (from a theoretical view point this variation does not matter, as economics tends to work on the level of society, and thus averages out extreme values). What is crucial to understanding the rationale of economics is the firmly held belief of traditional economics that trading environmental goods for money to maintain utility is theoretically possible, regardless of how desirable you personally may or may not feel it to be.

If we accept these arguments about utility and the substitutability of goods, then comparisons between the

effects of different actions become very straightforward. All we need to do is to measure the overall level of utility before the project, then estimate utility after the project, and if overall utility has increased, the project should go ahead. As stated previously, measuring utility is near impossible, but as we know that people will substitute the utility provided by one good for utility provided by another, then the simpler option is to undertake the assessment of the project in monetary terms—assuming that this is a good proxy for utility. Such a process is well known to economists and is termed **cost–benefit analysis** (**CBA**) (Hanley and Spash 1993). There are many technical and philosophical difficulties associated with converting all the costs (losses of utility) and benefits

(gains in utility) of any project or policy into monetary terms, especially when many of them relate to environmental goods. For these reasons, there have been many objections to the cost–benefit framework on the grounds that it does not represent the full complexity of the natural and social environment. As a result, several other means of making decisions about the environment have been devised, some of which are more frequently used than others. For further details see the final section of this chapter and the essays written by leading environmental economists Herman Daly (Essay 5.1) and Robert Costanza (Essay 5.2) (also see Edwards-Jones et al. 2000).

Because CBA is the dominant method used in public planning decisions and in ecological economic analyses,

ESSAY 5.1

Steady-State Economics

Herman E. Daly, *University of Maryland*

■ Any system in steady state is characterized by balanced, opposing forces or fluxes. This does not imply stagnation; indeed steady-state systems internally may be highly dynamic. For example, as long as the environment does not change in a major way, the species composition of an old-growth forest will remain in steady state, neither losing nor gaining species overall, although considerable turnover of species may be occurring locally, and seasonal changes will occur. Similarly, an economic steady-state system may be highly dynamic, but, on average, the many flows and exchanges are balanced.

The worldview underlying standard economics is that the economy is a system isolated from the natural world, a circular flow of exchange value just between businesses and households. In such an isolated system neither matter nor energy enters or exits, and the system has no relation with its environment; for all practical purposes there is no environment. Thus, standard economics ignores the origin of resources or the fates of wastes; they are "external" to the economic system. While this vision may be useful for analyzing exchange between producers and consumers, and related questions of price and income determination, it is useless for studying the relation of the economy to the environment. It is as if a biologist's vision of an animal contained a circulatory system, but no digestive tract. The animal would be an isolated system, completely independent of its

environment. If it could move it would be a perpetual motion machine.

Whatever flows through a system, entering as input and exiting as output, is called "throughput." Just as an organism maintains its physical structure by a metabolic flow and is connected to the environment at both ends of its digestive tract, so too an economy requires a throughput, which must to some degree both deplete and pollute the environment. As long as the scale of the human economy was very small relative to ecosystems, one could ignore throughput since no apparent sacrifice was involved in increasing it. The economy has now grown to a scale such that this is no longer reasonable.

Standard economics has also failed to make the elementary distinction between **growth** (physical increase in size resulting from accretion or assimilation of materials; a *quantitative* change), and **development** (realization of potentialities, evolution to a fuller, better or different state; a *qualitative* change). Quantitative and qualitative changes follow different laws. It is clearly possible to have growth without development or to have development without growth.

The usual worldview, the one that supports most economic analysis today, is that the economy is not a subsystem of any larger environment, and is unconstrained in its growth by anything. Nature may be finite, but it is just one sector of the economy, for which other sectors can substitute,

without limiting overall growth in any important way. If the economy is seen as an isolated system then there is no environment to constrain its continual growth. But if we see the economy as one subsystem of a larger, but finite and nongrowing Earth, then obviously its growth is limited. The economy may continue to develop qualitatively without growing quantitatively, just as the planet Earth does, but it cannot continue to grow; beyond some point it must approximate a steady state in its physical dimensions.

The worldview from which steady-state economics (Daly 1992) emerges is that the economy, in its physical dimensions, is an open subsystem of a finite, nongrowing, and materially closed total system—the biosphere. An "open" system is one with a "digestive tract," that takes matter and energy from the environment in low entropy form (raw materials), and returns it to the environment in high entropy form (waste). A "closed" system is one in which only energy flows through, while matter circulates within the system. A **steady-state economy** is an open system whose throughput remains constant at a level that neither depletes the environment beyond its regenerative capacity, nor pollutes it beyond its absorptive capacity. A result of steady-state economics is sustainable development, or development without growth—a physically steady-state economy that may continue to develop greater capacity to satisfy human wants by increasing the efficiency of resource

use, but not by increasing the resource throughput.

Economic growth is further limited by the complementary relation between man-made and natural capital. If the two forms of capital were good substitutes, then natural capital could be totally replaced by man-made capital. But, in fact, man-made capital loses its value without an appropriate complement of natural capital. What good are fishing boats without populations of fish, or saw mills without forests? And even if we could convert the whole ocean into a fishpond we would still need the natural capital of solar energy, photosynthetic organisms, nutrient recyclers, and so forth. The standard economists' emphasis on substitution while ignoring complementarity in analyzing technical relations among factors of production seems a reflection of their preference for competition (substitution) over cooperation (complementarity) in social relations.

In an empty world, increasing throughput implies no sacrifice of ecosystem services, but in a full world it does. The ultimate cost of increasing throughput is loss of ecosystem services. Throughput begins with depletion of natural stock and ends with pollution, both of which are costs in a full world. Therefore, it makes sense to minimize throughput for any given level of stock. If we recognize that the economy grows by converting ever more of the ecosystem (natural capital) into economy (man-made capital), then we see that the benefit of that expansion is the extra services from man-made capital and the cost is the loss of service from reduced natural capital.

The efficiency with which we use the world to satisfy our wants depends on the amount of service we get per unit of man-made capital, and the amount of service we sacrifice per unit of natural capital lost as a result of its conversion into man-made capital. This overall **ecological-economic efficiency** is the ratio of man-made capital services gained (MMK) to natural capital services sacrificed (NK), or MMK/NK.

In an empty world there is no noticeable sacrifice of NK services required by increases in MMK, so the denominator is irrelevant. In a full world any increase in MMK comes at a noticeable reduction in NK and its services.

The steady-state economic view recognizes that economic systems are not isolated from the natural world, but are fully dependant on ecosystems for the goods and services they provide. Inasmuch as the overall size of the natural world cannot increase (and in fact decreases steadily at the hands of humankind), our economic systems cannot continually increase; they must operate as a steady-state system, one that does not quantitatively grow without bounds, but that can qualitatively develop. The internal workings of the economic machine must fully account for the raw natural materials consumed and the resultant wastes eliminated. The ultimate accounting rules for a realistic economy are the first and second laws of thermodynamics, which are inviolable. ■

this chapter will focus on CBA, discussing the technicalities of how to do a CBA, and highlighting some of the difficulties in applying this technique to environmental issues. Further, we have found that a focus on one framework is most effective in allowing biologists with limited economic background to become acquainted with many key economic concepts in a reasonably structured way, and thus understand how they work (and don't work). The decision to follow this focus is not meant to suggest that CBA is the best, or only, economic method that can be applied to conservation problems. Many alternate approaches to CBA involve extensive negotiations, more detailed analysis, and often, different philosophical frameworks. Such alternatives ultimately may end up resolved in political or legal processes. CBA has the benefit of being an efficient decision-making tool that is widely used and understood by a broad array of actors. Ecological economic approaches to CBA allow greater scope for nonmarket values to play a role in critical conservation decisions.

Cost–Benefit Analysis and its Application to Conservation

Cost–benefit analysis (CBA) aims to compare the monetary costs and the monetary benefits of any project or policy to determine whether the benefits outweigh the costs. The assumption is that if benefits are greater than costs then the project should proceed, but if the costs outweigh the benefits then it should not proceed. CBA has been widely applied in practice and endorsed in the context of both public and private decision-making. Gramlich (1990) cites Benjamin Franklin and Abraham Lincoln as early proponents of CBA. More recently, the U.S. Presidential Executive Order 12291 (1981) made the application of CBA a necessary requirement for all new policies in the U.S. In the U.K., the government has endorsed the inclusion of environmental appraisal methodologies into formal procedures governing both policy and projects.

The underlying justification for CBA lies within the theory of welfare economics. It is founded on a directly utilitarian approach to decision-making that should, under certain circumstances, enable government to select those projects that maximize social welfare (that is, do what is best for the whole of society, even if some groups suffer some costs). The practice of CBA is therefore often considered to have the "hypothetical consent of the citizenry," since the citizens of any democratic country elect governments precisely on the basis that those governments will work to raise the people's welfare (Leonard and Zeckhauser 1986). (In fact, although the citizenry may "consent" to the general principle of CBA in decision-making, this does not mean that they will endorse its application in every particular case). In practice there

ESSAY 5.2

Valuation Of Ecological Systems

Robert Costanza, *University of Vermont*

The issue of valuation is inseparable from the choices and decisions we have to make about ecological systems. Some argue that valuation of ecosystems is either impossible or unwise, that we cannot place a value on such "intangibles" as human life, environmental aesthetics, or long-term ecological benefits. But, in fact, we do so every day. When we set construction standards for highways, bridges, and the like, we value human life—acknowledged or not—because spending more money on construction would save lives. Another frequent argument is that we should protect ecosystems for purely moral or aesthetic reasons, and we do not need valuations of ecosystems for this purpose. But there are equally compelling moral arguments that may be in direct conflict with the moral argument to protect ecosystems—for example, the moral argument that no one should go hungry. All we have done is to translate the valuation and decision problem into a new set of dimensions and a new language of discourse, one that, in my view, makes the valuation and choice problem more difficult and less explicit.

So while ecosystem valuation is certainly difficult, one choice that we do *not* have is whether or not to do it. Rather, the decisions we make as a society about ecosystems *imply* valuations. We can choose to make these valuations explicit or not; we can undertake them using the best available ecological science and understanding or not; we can do them with an explicit acknowledgment of the huge uncertainties involved or not; but as long as we are forced to make choices we are doing valuation. The valuations are simply the relative weights we give to the various aspects of the decision problem.

I believe that society can make better choices about ecosystems if the valuation issue is made as explicit as possible. This means taking advantage of the best information we can muster and making uncertainties about valuations explicit too. It also means developing new and better ways to make good decisions in the face of these uncertainties. Ultimately, it means being explicit about our goals as a society, both in the short term and long term. Society has begun to do this with the recent growing consensus that sustainability is the appropriate long run, global goal (World Commission on Environment and Development 1987).

Sustainability and the Valuation of Natural Capital

Sustainability has been variously construed, but one useful definition is the amount of consumption that can be sustained indefinitely without degrading capital stocks—including "natural capital" stocks (Pearce and Turner 1989; Costanza and Daly 1992). Since "capital" is traditionally defined as a produced (manufactured) means of production, the term "natural capital" needs explanation. It is based on a more functional definition of capital as "a stock that yields a flow of valuable goods or services into the future." What is functionally important is the relation of a stock yielding a flow— whether the stock is manufactured or natural is in this view a distinction between kinds of capital, and not a defining characteristic of capital itself. For example, a stock or population of trees or fish provides a flow or annual yield of new trees or fish, a flow that can be sustainable year after year. The sustainable flow is "natural income," the stock that yields the sustainable flow is "natural capital." Natural capital may also provide services like recycling waste materials or water catchment and erosion control, which are also counted as natural income. Since the flow of services from ecosystems requires that they function as whole systems, the structure and diversity of the ecosystem is a critical component in natural capital.

To achieve sustainability, we must incorporate natural capital, and the ecosystem goods and services that it provides, into our economic and social accounting and our systems of social choice. In estimating these values we must consider how much of our ecological life support systems we can afford to lose. To what extent can we substitute manufactured capital for natural capital, and how much of our natural capital is irreplaceable? For example, could we replace the radiation screening services of the ozone layer if it were destroyed, or the pollination services of honeybees if they were eliminated?

Because natural capital is not captured in existing markets, special methods must be used to estimate its value. These range from attempts to mimic market behavior using surveys and questionnaires to elicit the preferences of current resource users (i.e., willingness to pay), to methods based on energy analysis of flows in natural ecosystems, which do not depend on current human preferences at all (Farber and Costanza 1987; Costanza et al. 1989). Because of the inherent difficulties and uncertainties in determining these values, we are better off with a pluralistic approach that acknowledges and utilizes these different, independent approaches.

The point that must be stressed is that the economic value of ecosystems is connected to their physical, chemical, and biological role in the long-term, global system—whether the present generation of individuals fully recognizes that role or not. If it is accepted that each species, no matter how seemingly uninteresting or lacking in immediate utility, has a role in natural ecosystems (which *do* provide many direct benefits to humans), it is possible to shift the focus away from our imperfect short-term perceptions and toward the goal of developing more accurate values for long-term ecosystem services. Ultimately, this will involve the collaborative construction of dynamic, evolutionary models of linked ecological economic systems that adequately address long-term responses and uncertainties.

Toward an Ecological Economics

Valuation of natural capital will always involve large uncertainties. How do we make good decisions in the face of this uncertainty? Current systems of direct environmental regulation attempt to directly command specific behavior from the top down in an effort to control pollution. They are not very efficient at managing environmental resources for sustainability, particularly in the face of

uncertainty about long-term values and impacts. They are inherently reactive rather than proactive. They induce legal confrontation, obfuscation, and government intrusion into business in a way that reduces efficiency. Rather than encouraging long-range technical and social innovation, they tend to suppress it. They do not mesh well with the market signals that firms and individuals use to make decisions, and do not effectively translate long-term global goals into short-term local incentives.

We need to explore promising alternatives to our current command and control of environmental management systems, and we need to modify existing government agencies and other institutions accordingly. The enormous uncertainty about local and transnational environmental impacts needs to be incorporated into decision-making. We also need to better understand the sociological, cultural, and political criteria for acceptance or rejection of policy instruments.

One example of an innovative policy instrument currently being studied is a flexible environmental assurance bonding system designed to incorporate environmental criteria and uncertainty into the market system, and to induce positive environmental technological innovation (Costanza and Perrings 1990). In addition to direct charges for known environmental damages, a company would be required to post an assurance bond equal to the current best estimate of the largest potential future environmental damages; the money would be kept in interest-bearing escrow accounts. The bond (plus a portion of the interest) would be returned if the firm could show that the suspected damages had not occurred or would not occur. If damages occurred, the bond would be used to rehabilitate or repair the environment and to compensate injured parties. Thus, the burden of proof would be shifted from the public to the resource-user, and a strong economic incentive would be provided to determine the true costs of environmentally innovative activities and to develop cost-effective pollution control technologies.

Ecological economic thinking leads one to conclude that instead of being mesmerized into inaction by scientific uncertainty over our future, we should acknowledge uncertainty as a fundamental part of the system. We must develop better methods to model and value ecological goods and services, and devise policies to translate those values into appropriate incentives. If we continue to segregate ecology and economics we are courting disaster. ∎

---

are five main stages within CBA: project definition, classification of impacts, conversion into monetary terms, project assessment, and sensitivity analysis. Each of these is discussed in more detail in the following sections.

Project definition

The first step of CBA is to define precisely the project being assessed and where the boundary of enquiry will lie. This is largely an arbitrary process and relates to the question being asked, and the person or organization asking the question. In a small business situation a farmer may ask what the costs and benefits would be of employing a new staff member. Here the boundary of enquiry is quite small and may only include the financial outflow related to paying the new staff member and the improvements in productivity that such a staff member would make to the business. The boundary may be different in a big business situation, such as when a petroleum company considers building a new oil-rig. Here the company would still be interested in the financial return on its investment over time, but may need to have a wide boundary of enquiry that could include the local costs of the construction, the costs of transporting oil to a processing facility, and the future global demand for oil. However, the company itself may not be interested in considering the value of any biodiversity lost as a result of the development, or the future costs of global warming, which is exacerbated through the burning of fossil fuel. Similarly the company may not be too interested in whether the project would affect local people in terms of pollution, increased job opportunities, and elevated property prices.

This highlights an important point; private business is primarily concerned with financial returns on investments and will set a relevant project boundary, but government should be concerned about both financial matters and the general well-being of society. Therefore, government-sponsored projects should set a wide project boundary that explicitly includes biodiversity, clean up costs, and long-term issues of sustainability. This theoretical division between the boundaries of government and business CBA becomes blurred when government wants private businesses to achieve certain things on its behalf—for example, power generation, airport developments, or water management—and there is potential for governments to reduce the boundaries of enquiry to exclude some of the environmental and social factors we may expect them to include. When pressed on these issues politicians often argue that such decisions are made in the "national interest." Indeed some may argue that the boundaries between business and government is perpetually blurred as governments in Western market-led countries want to see more businesses of every kind, regardless of their exact aims, and often do not want to put too many constraints on their development. In these situations, legal environmental standards and planning law become important baselines, which in essence are stopping business from causing too many "costs," while not requiring them to undertake individual CBAs for all of their activities.

We will discuss these issues again later, but it is critical to recognize that defining the boundaries of a CBA is a key issue, with much in common with setting the agenda at a meeting—if it's not on the agenda, it doesn't get discussed. It is also important to note that the wider

the boundaries the more extensive and expensive the study will be, and that the results of the CBA are only relevant within the boundaries set at the outset.

Classification of impacts

After definition, the impacts arising from the project must be identified. For the oil-rig project the impacts may include:

- Cost of materials such as steel and concrete
- Effects on local employment levels, including direct employment (initially by the construction company and then by the rig's operatives), and indirect employment generated by increased spending in the local economy
- Impacts on the environment, from the construction process itself, through the operational life of the rig, and to the disposal of the rig

All the impacts to be included in the CBA must be quantified and then tabulated in terms of when they arise. If the construction project is estimated to take 18 months for completion, then there will be an associated estimate for the input of, for instance, engineers' labor hours for each of these months. The same procedure is applied to each impact for each discrete time interval. This is important for the process of discounting, as explained later in this chapter

Having identified the types of impacts that will arise from the project, it is necessary to estimate their magnitude and duration. One difficult issue in determining the magnitude of any impact is to decide exactly which changes are explicitly related to the project, and which would happen anyway for other reasons. Consider an oil-rig project situated in Scotland, and the hypothetical example that the oil-rig will reduce the breeding habitat of a fictional bird—the Tartan Finch—by 30%. Biologists working for a bird conservation group estimate that such a loss of habitat would have a serious impact on population viability of the Tartan Finch, such that the probability of extinction in the next 100 years would increase from the pre-project level of 10% to 30% after the project was complete (Figure 5.1). However, ecologists working for the oil company dispute this figure and suggest that the oil-rig will not have any significant impact on the long-term viability of the Tartan Finch. The company's ecologists argue that conservation scientists have failed to note that the Tartan Finch is migratory, spending its winters in the Sahel of Western Africa, and that the greatest threat to populations of the Tartan Finch arises from degradation of their traditional over-wintering grounds by African farmers. Should this degradation continue, populations would fall by 70% within 10 years; when this is taken into account, the probability of extinction in the next 100 years would be 80% if the oil-rig project did not go ahead and only 81% if it did go ahead. This, of course, assumes that the oil-rig and the African environment can be considered independent influences on the birds' survival. The company therefore argues that the impact of the project itself on the Tartan Finch is negligible. This hypothetical example serves to reiterate the importance of being able to separate the impacts of the project per se, from general background changes. CBA should only consider those changes directly attributable to the project; any inability to identify these will affect calculations of the project's real worth, and may lead to the wrong decision being made.

Conversion into monetary terms

To allow the relatively simple mechanics of CBA to work it is necessary to represent all the project's impacts on a single scale. Money is usually chosen as that scale. CBA practitioners generally state that one reason for choosing money is its convenience. This is because until recently

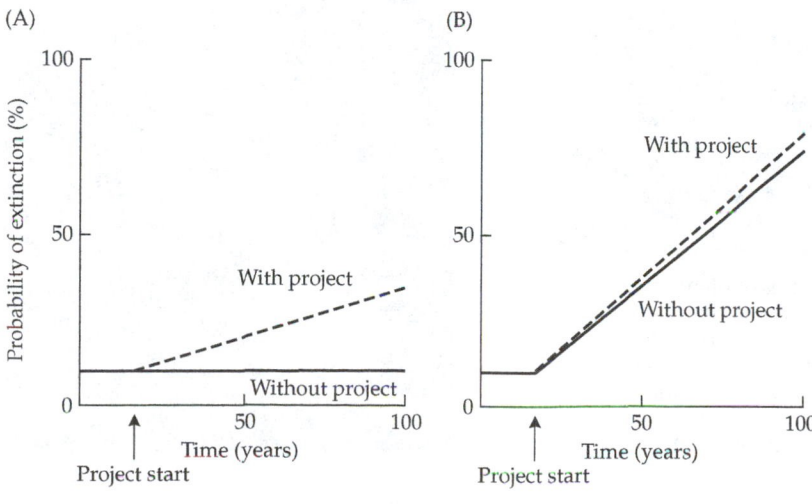

Figure 5.1 Two views of the impacts of a hypothetical project on the probability of extinction over 100 years for a fictitious bird, the Tartan Finch. (A) The estimates of a bird conservation group that considers only the direct impact of the project on the breeding habitat, and (B) the estimates of the company scientists that includes impacts in the migratory overwintering grounds. If a CBA was being done as part of the project's appraisal it would be necessary to ascertain which of these predictions is correct, and then to estimate the difference between the "with" and "without" project projections, and to place a financial value on this difference.

the majority of CBAs were only concerned with items that could be bought and sold with money, such as steel, concrete, or labor (so-called market goods). However, as environmental awareness has grown, so has the desire to include so-called nonmarket goods (many social and environment goods, such as clean air, or the persistence of an endangered species, that cannot be bought in a normal market) within CBA. While at one level it seems more sensible to include environmental issues in CBA than to exclude them, the requirement to assign monetary values to environmental goods causes philosophical problems.

Regardless of the philosophical debate surrounding valuation of the environment, at a pragmatic level the advantage of including the environment in CBA is that it should force decision-makers (both in government and business) to explicitly consider the environment in their decision-making and project analyses. If environmental issues were not included in the CBA then they may either be left off the decision-making agenda altogether, or their value could be trivialized. Such trivialization may happen quite naturally. For example, if the benefits of a project can be expressed in terms of millions of dollars (or thousands of jobs) and the cost is the loss of a few Northern Spotted Owls (*Strix occidentalis caurina*; see Case Study 17.4), then these very tangible benefits may make it difficult for many decision-makers to really consider the value of the owls.

A major problem with including the environment in CBA concerns valuing environmental goods and converting these values into monetary terms. This has been an area of great academic activity in recent years and while some convergence of methodology has been achieved, there is still much ongoing technical and philosophical debate. Before proceeding with a discussion of CBA it is worth considering the types of values the environment has and how these can be assigned monetary values.

What Kind of Values Does Biodiversity Have?

Environmental goods clearly affect human welfare in a wide variety of ways, and it is important to determine the different categories of value that the environment brings. A taxonomy of the economic values associated with the environment has been proposed by a number of environmental economists (Weisbrod 1964; Krutilla 1967). Although there is not yet complete agreement on this distinction, it is widely accepted that environmental values can be split into two broad categories: **use values** and **nonuse values**.

Use values are associated with the values that come as a result of contact with, or use of, natural resources. Use values are of three types. **Direct use values** are associated with the direct use of resources, and these can be consumptive, as in the extraction of timber from forests or fish from the sea, or nonconsumptive, such as the appreciation of a landscape (Figure 5.2). **Indirect use values** arise when the environment performs some service for

(A)

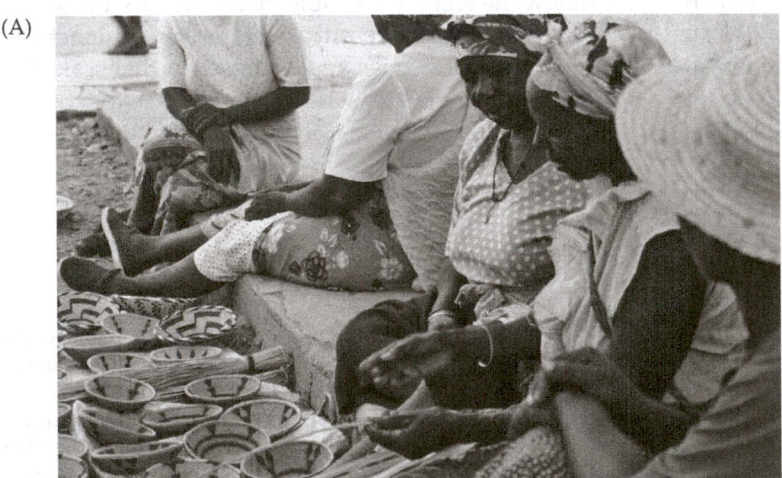

(B)

Figure 5.2 Examples of local people making direct uses of plants and animals in their environment. (A) A group of women in an arid region of southeastern Zimbabwe who make crafts and furniture from palm leaves. These will either be used in the home or sold in local markets. (B) Lody, a villager collecting honey from a wild nest in the forest near Bevoahazo, Madagascar. Her family will eat the honey as part of their normal diet. (A, photograph by N. Hockley; B, photograph by G. Edwards-Jones.)

humankind simply through its normal functioning. For example, forests sequester carbon, thereby partly mitigating climate change, trees can stabilize soils, and wetlands can purify water. Most indirect use values are identical to what we call **ecosystem services** (see Chapters 2 and 4). In addition to these values environmental economists have identified an **option value**, which is defined as the value placed on environmental assets by those people who want to secure the use of the good in the future.

Nonuse values correspond to those benefits that do not imply contact between the consumers and the good. People do not need to use the good in any way, either directly or indirectly, to derive value from it. Such values include **existence values**, which are values that derive simply from the knowledge that a particular good exists. To test this idea, ask yourself how you would feel if you picked up a newspaper tomorrow and discovered that the mountain gorilla had become extinct. If such news would make you feel a little sad, then you had clearly lost utility. But to have lost utility in this way, you must have had some utility to start with. In other words, without realizing it, you have positive utility for the mere existence of the mountain gorilla, and probably also for many other species and natural features too.

Exactly how the utility you hold varies between different species is known only to you. Do we hold the same level of existence value for all species, or is the existence of some species more valued than others? If so, which are most and least valued and why? Do you think you hold equal existence values for mountain gorilla, white-tailed deer, cod, and the smallpox virus? Can we hold existence values for extinct species? That is, can we gain some value from knowing that dinosaurs did exist, even though they are extinct? As not all species on Earth are described yet, do undescribed species have any value to us? How does this value change as we learn more about them? Can we have negative existence values (i.e., for pests and disease vectors)? Many of these questions remain open for debate, but there is little doubt that we do hold real existence values for many species. In a similar way that we hold existence values, some economists suggest we hold **bequest values**, which are values that we receive from knowing that we will hand something on to future generations.

The sum of the components of use and nonuse values is called the **total economic value** (**TEV**) and represents the entire value of a resource. Figure 5.3 lists the values usually attributed to wetland ecosystems. To explore TEV in more detail, consider the value of a forest from the viewpoint of different people, each typical of their profession. Consider an economist working for a logging firm. She would consider the forest's value to be constituted largely from consumptive, direct use values, like timber. Conversely, if we viewed the same forest from

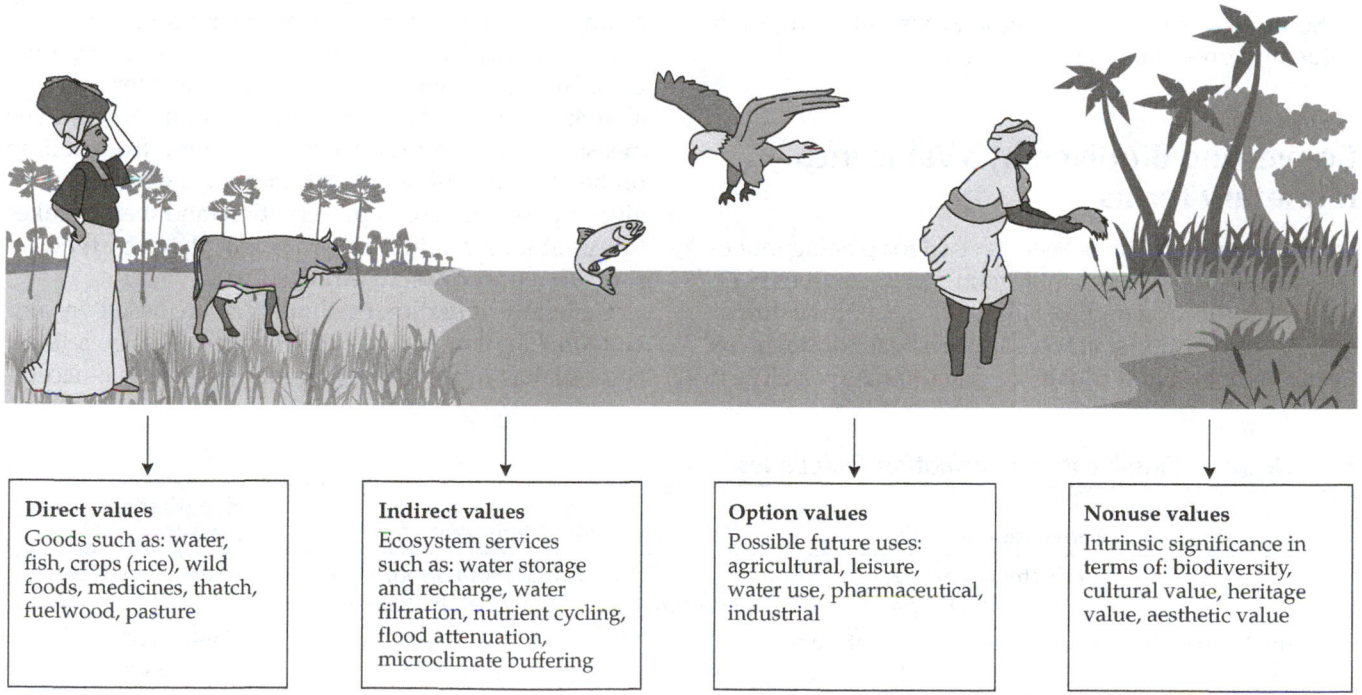

Direct values	**Indirect values**	**Option values**	**Nonuse values**
Goods such as: water, fish, crops (rice), wild foods, medicines, thatch, fuelwood, pasture	Ecosystem services such as: water storage and recharge, water filtration, nutrient cycling, flood attenuation, microclimate buffering	Possible future uses: agricultural, leisure, water use, pharmaceutical, industrial	Intrinsic significance in terms of: biodiversity, cultural value, heritage value, aesthetic value

Figure 5.3   Total economic value of wetlands is composed of direct, indirect, option, and nonuse values. (Modified from Emerton 1999.)

the point of view of a landscape artist, we may find that she considered the primary value of the forest to be aesthetic (direct use value, nonconsumptive), while a teenager may consider the forest's value to be mainly recreational—a place to mountain bike, hike, and have fun. Each of these perceptions of the forest's values are true—it does have a timber value, an aesthetic value, and a recreational value—but none of these alone adequately represent its full value (i.e., its total economic value). This TEV is the sum of all these individual values, and its calculation requires each individual value to be recognized, and then quantified in monetary terms.

It could be argued that the reason so much habitat has been lost over the last few centuries in North America and Europe is because the decision-makers at the time either did not recognize all the individual values associated with such habitats, or did not ascribe a particularly high quantitative value to their continued existence, relative to other values. So in the early twentieth century, farmers in the Midwest of the U.S. valued the agricultural potential of natural prairie much higher than any other uses; during the 1950s foresters in the U.K. did not value the biodiversity of upland peat bogs when they planted new plantations of nonnative trees for timber production; and currently marine fishermen do not value the natural state of benthic habitats, which they disturb when they trawl for fish. To make more rational decisions about the future of our natural resources, that is, to encourage conservation, we must consider all values that constitute the TEV in a quantitative manner.

Some of the economic methods for obtaining such values are presented in the next section.

Converting Biodiversity's Value into Monetary Terms

Several techniques have been devised for placing monetary values on environmental goods. Munasinghe (1993) classified these economic methods into three broad groups: **conventional market approaches**, **implicit market approaches**, and **constructed market approaches**. It is not possible to discuss the details of these methods here, but there are many references on the topic (e.g., Mitchell and Carson 1989, Arrow et al. 1993, and Edwards-Jones et al. 2000). A brief outline of the three major groups is given in the following sections and in Table 5.1.

Conventional market approaches

The principle behind conventional market approaches is to establish a link between an environmental good and some other good that already has a market value.

The **substitute**, or **alternative cost approach**, uses the cost of available substitutes for the particular environmental good to give value to that good. For example, the value of firewood collected from an Indonesian forest and used in home cooking could be valued by comparing it to the price of buying firewood in a local market. Similarly, the recreational value of clean fresh water in lakes and rivers may be estimated from the costs of building and running private swimming pools.

The **production function approach** is slightly different. Here the environment is considered as a factor of production. Changes in environmental quality then lead to changes in productivity and production costs. Thus the value of an environmental change, such as an increase in soil fertility would be the value of *additional* crop resulting from the environmental improvement, as compared to the level of production without such an improvement.

The **opportunity cost approach** uses market prices to estimate the value derived from using a resource in a particular way by examining the values of alternative uses. Thus the benefits of preserving forests for a national park rather than harvesting them for timber would be measured by using the foregone income from selling timber. In brief, this approach measures what has to be given up for the sake of preservation (and thereby takes this as at least a minimum value for the preserved resource if it continues to be preserved).

Generally valuation methods that are based on conventional market approaches are deemed to be reliable because they relate to existing markets, and it is accept-

TABLE 5.1 *Classification of Valuation Techniques*

	Conventional market	Implicit market	Hypothetical market
Actual behavior	Productivity change, opportunity cost, dose–response, preventative expenditures	Travel cost method, wage differential, hedonic pricing	Artificial market
Potential behavior	Shadow projects, substitute costs		Contingent valuation
Other	Cost effectiveness		

Source: Modifed from Munasinghe 1993.

ed that the further the good is from the market place, the less reliable is the estimate of its value.

Implicit market techniques

These approaches assume that the behavior of people reflects their implicit valuations of the environment. This may be through wages accepted to work in locations with different levels of environmental quality, prices or rents paid for residential properties that have particular levels of environmental amenities or the actual expenditure made on reaching certain recreational sites of high environmental quality. The two main methods are **hedonic pricing (HP)** and **travel cost methodology**.

HEDONIC PRICING METHODS (HP)   Hedonic pricing assumes that the total value of any good is a function of the set of characteristics associated with that type of good. Thus any particular make of car has a set of associated speed, fuel-efficiency, color, and other characteristics, and the final value that consumers are prepared to pay for a car is related to how it is evaluated across these characteristics. If the level of any one characteristic, say fuel efficiency, is increased while holding all other characteristics constant, it should be possible to observe how the price of the car changes. This change will give an estimate of the value of that particular characteristic to the purchaser.

Applications of hedonic pricing to environmental goods have focused on residential property prices (considering either rentals or purchases) to see how they vary with associated environmental attributes. The technique requires analysts to collect a large quantity of data relating to all the factors that influence the value of a property, such as number of rooms, central heating, garage space, proximity to parkland, and so on. These data are collected for a large number of houses, and statistical regression techniques are then used to isolate the variations in purchase or rental price that come about as a result of changes in environmental factors, while holding standardized values for all other characteristics constant. The proportion of the price differential between two otherwise identical houses accounted for by the change in the environmental quality characteristic reveals an individual purchaser's valuation of the importance of environmental quality.

For example, Geoghegan (2002) considered the impact of different types of "open space" on property prices in Howard County, Maryland, U.S. The open spaces were classified as either "permanent" or "developable." Permanent open spaces were areas of amenity land and farmland that for some reason could not be developed in the future. Developable open space on the other hand was available for future building development. The results showed that houses near permanent open spaces were on average three times the value of

similar properties near developable open space, thereby showing the value people place on the landscape and amenity characteristics of the Maryland countryside. In a similar study in Windhoek townships of Namibia, Humavindu and Stage (2003) showed that properties close to a garbage dump had reduced prices, while those close to a dam recreation area showed increased prices. A pleasant view of the recreation area has a greater influence on property prices than does access to a dump.

Although it has been widely used, hedonic pricing has several limits in its application. It requires huge amounts of data to undertake the required statistical analysis, and the reliability of data may be a shortcoming of this method. House prices, for example, are often distorted and owners of houses frequently sell or rent at lower prices than the maximum offer received; therefore, the observed price may not correspond to the marginal willingness to pay (WTP). Finally, this method does not capture nonuse values and does not take into account the effect on prices of individual's expectations on the future quality of landscape (Ablest et al. 1985).

TRAVEL COST METHOD (TCM)   First proposed in 1947 by Harold Hotelling, the travel cost method centers on the expenditures incurred by households or individuals to reach recreational sites. It uses these expenditures as a means of measuring willingness to pay for the recreational activity (Trice and Wood 1958; Clawson 1959). The sum of the cost of traveling (including the opportunity cost of time) and any entrance fee gives a proxy for market prices in estimating demand for the recreational opportunity provided by the site under investigation. By observing these costs and the number of trips that take place at each of the range of prices, it is possible to derive a demand curve for the particular site being visited.

TCM is considered one of the most effective approaches in valuing recreation services (Ward and Loomis 1986; Smith 1989; Bockstael et al. 1991), although several potential problems can be encountered with this method. These include difficulties in determining the costs of time when people are engaged in travel and recreational activities, and how to deal with multi-purpose visits—specifically, what component of overall cost should be allocated to one particular site when people make trips to visit several sites.

Hypothetical market

On some occasions it is difficult to gain a suitable valuation from either a conventional or implicit market technique, for example, when considering the existence value of a species. There simply is no market analogy for this type of good, yet not to attach a value to such existence values could have severe impacts on estimation of TEV or subsequent conservation decisions (Case Study 5.1). To

gain an economic value for such goods it is necessary to construct a hypothetical market for that good. In other words we simply "pretend" there was a market for the good and then use the values obtained in this hypothetical market as if they were real. The technique most widely used when doing this is the **contingent valuation method** (**CVM**). At its heart this is a simple technique that works by asking people (usually via some form of questionnaire) to state what their **willingness to pay** (**WTP**) would be to preserve some named environmental good. Such a question may be in the form: "Imagine that the only way to save the mountain gorilla from extinction was to establish a trust fund which would be used to finance the gorilla's conservation. Would you be willing to contribute to such a fund? Yes or No." If the respondent replied, "Yes," the next question may be: "If you were asked to make a one time contribution to the fund, how much would you be willing to contribute?"

A CVM survey would normally ask a large sample of people to answer such a question. Technically speaking such questionnaires simulate a hypothetical (contingent) market for the good in question—here, the continued existence of the mountain gorilla. In effect, individuals are asked to reveal their indifference between sums of money and the supply of the good in question. This is the same issue as discussed earlier concerning the balancing of utility. If the survey was done correctly and the sample was representative of the relevant human population, it would be possible to estimate the mean (or median) WTP to save the mountain gorilla. Mean bids can then be summed by the relevant human population to give a final value for the good in question. So if such a survey was conducted in Canada, we may be able to state: The mean WTP of the Canadian public to conserve the mountain gorilla is $20, and as the human population of Canada is 32 million, the total existence value of Canadians for the mountain gorilla is $640 million (20 × 32 million). While this sounds relatively straightforward, there are some inherent complications of such calculations; these are discussed in Box 5.1,

which highlights a study by Gunawardena et al. (1999) aimed at assessing the total economic value of the Sinharaja Rain Forest reserve in Sri Lanka.

Examples of monetary values placed on biodiversity

DIRECT USE VALUES   Humankind makes direct use of numerous species for a wide variety of purposes. These include food, clothing, shelter, pharmaceuticals, and ornamental and aesthetic uses, both in developed and developing countries. For example, it has been estimated that only nine plant species—wheat, rice, maize, barley, sorghum, potato, yam, sugarcane, and soybean—provide 75% of the energy humankind derives from plants (Fowler and Mooney 1990). In addition to using these species as food items, plants and plant derivatives are used in construction, engineering, power generation, and for clothing. However, not all species that are useful to humans are traded in markets. Particularly in developing countries, many species are harvested and used directly without any recourse to a market. For example, the annual consumption of fuelwood and crayfish are estimated to have an equivalent monetary value of $38 and $12 to average households in Madagascar, a significant contribution given that average annual income is little more than $200 (Kramer et al. 1994). Plants harvested for medicinal use can have a significant value, even as high as $3,327 per ha (Balick and Mendelsohn 1992).

Some species provide tangible benefits to humans in other ways besides being the source of food and pharmaceuticals. These include being the quarry of recreational hunting and fishing trips, and the focal interest of many ecotourist activities. Some estimates of the value individuals place on engaging in these activities are shown in Table 5.2.

INDIRECT USE VALUES OR ECOSYSTEM SERVICES   Ecosystem services are the services provided by ecosystems that provide some benefit to humankind. One such service is

TABLE 5.2   *WTP for Wildlife Related Activities*

Good	Value (U.S.$)	Valuation technique	Source
Recreational value of Khechiopalri Lake, India	0.88 for locals, $7.19 for international tourists	CVM[a]	Maharma et al. 2000
Visits by mountain bikers to New Mexico	150/trip	TCM[b]	Hesseln 2003
Visits by hikers to New Mexico	130/trip	TCM	Hesseln 2003
Recreational fishing in Brazilian Pantanal	540–869/trip	TCM	Shrestha 2002
Tourist visits to Komodo National Park, Indonesia	11.70	CVM	Walpole et al. 2000

[a]Contingent valuation methodology.

[b]Travel cost methodology.

BOX 5.1 Estimating the Total Economic Value of Sinharaja Rain Forest, Sri Lanka

Sinharaja forest has been a World Heritage Site since 1990, although prior to that date it had undergone alternating periods of logging and protection. Sinharaja is now one of the least-disturbed forests in the wet zone of Sri Lanka. Gunawardena et al. (1999) attempted to estimate the TEV for rainforest, and as part of this sought the WTP to preserve the forest in its current state for posterity from four different groups of people. These groups were:

- Villagers who lived in the periphery of the forest (within 3 km of the reserve)
- Rural people who lived at a distance from the forest
- Urban people living in Colombo, Sri Lanka's capital city, a long distance from the forest
- Urban people living in Edinburgh, the capital of Scotland, U.K., many thousands of kilometers from the forest

These sample populations differ in their use of the forest. Local villagers are very dependant on the forest and use it for life-sustaining activities like hunting and collecting food, while the other groups make minimal direct use of the forest for any purpose. The four populations also showed decreasing household income in the order: U.K., urban Sri Lanka, rural Sri

Lanka, local villagers. Thus at the extremes, the village respondents were the poorest group and made use of the forest for food and building materials, and those from the U.K. sample were the richest and most had never heard of Sinharaja, although they may hold existence values for rainforests generally.

Despite these differences between samples, the CVM used the same basic questionnaire structure to assess the mean WTP of each population. A summary of the results is shown in Table A, and not surprisingly, the wealthier the respondents were, the more they actually bid for forest preservation. However, if we look beyond the absolute monetary amounts and concentrate on the proportion of respondents' income that these bids comprise, we can see the village respondents bid a far greater proportion of their income for use and bequest values than other populations. This highlights a major problem in estimating global values for natural resources—poor people will inevitably bid less in absolute terms for any given resource than people from richer countries. But this does not necessarily mean they value it any less—they simply have fewer resources with which to bid. Dealing with issues like this remains quite a challenge for environmental economists.

TABLE A *Mean WTP Bid (U.S.$) and Percentage of Income for Four Samples Surveyed in a Contingent Valuation for Protecting Sinharaja Forest Reserve in Sri Lanka*

	Value type					
	Use value		Bequest value		Existence value	
Sample	WTP	% income	WTP	% income	WTP	% income
Sri Lankan village	1.02	0.52	0.84	0.42	0.41	0.18
Rural Sri Lanka	1.27	0.21	1.26	0.16	1.73	0.26
Urban Sri Lanka	6.04	0.33	3.65	0.23	3.39	0.22
United Kingdom	53.38	0.14	47.43	0.13	47.06	0.12

Note: Village sample size is 211, rural sample size is 220, urban sample size is 223, and United Kingdom sample size is 204.
Source: After Gunawardena 1997.

the decomposition of the large quantity of organic wastes produced by humans and their enterprises, which weigh about 38 billion tons worldwide (Pimental 1998). This decomposition process, which is an ecosystem function, actually provides more than one service to humankind. It removes the vast volume of waste, and serves to recycle the waste and ensure that important nutrients re-enter managed and natural ecosystems, and thereby enabling continued plant growth. Also, the decomposition of these organic wastes reduces the risk of them causing environmental pollution or disease.

Monetary values have been attached to 17 classes of ecosystem services, ranging from climate regulation to recreation. Costanza et al. (1997) estimated that the total value of ecosystem services is U.S.$16–54 trillion ($10^{12}$) per year, with an average of U.S.$33 trillion per year. The ecosystems that provide the greatest value per hectare

are estuaries, wetlands, swamps, and floodplains. The value of these ecosystems is largely related to their role in nutrient recycling and waste treatment.

Ecosystem services are rarely included in traditional calculations of value. Yet the non-market values of watershed protection, non-timber forest products, carbon sequestration, and recreation can exceed the marketed values for timber, fuelwood, and grazing (Millennium Ecosystem Assessment 2005; Figure 5.4, Case Study 5.2).

EXISTENCE VALUES In addition to their direct and indirect values to humans, many species may also have an existence value. This is the value that humans derive from simply knowing that a certain environmental good exists, regardless of whether or not that individual will ever use the good directly or indirectly. Such values are difficult to measure and impossible to validate. They are

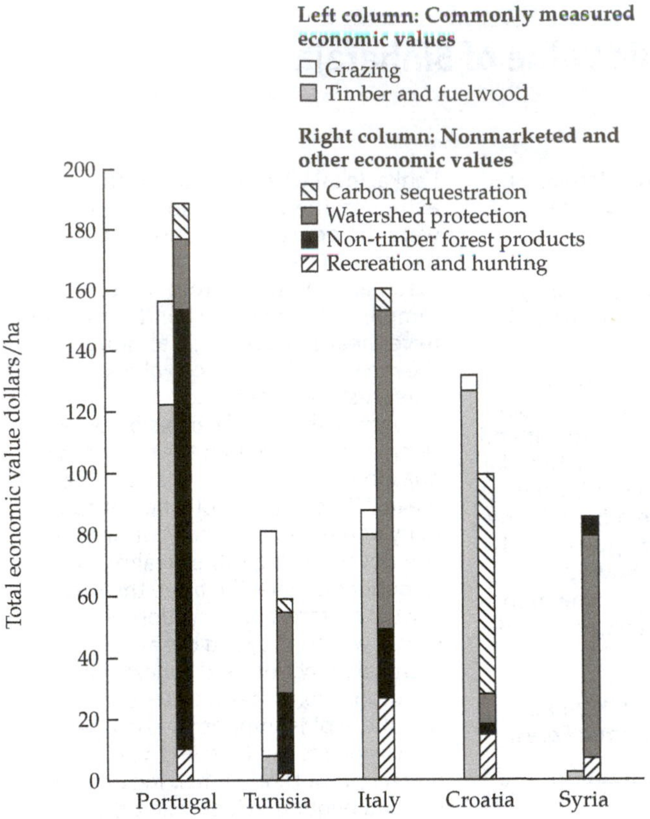
Left column: Commonly measured economic values
Grazing
Timber and fuelwood
Right column: Nonmarketed and other economic values
Carbon sequestration
Watershed protection
Non-timber forest products
Recreation and hunting
Total economic value dollars/ha
0
20
40
60
80
100
120
140
160
180
200
Portugal
Tunisia
Italy
Croatia
Syria

Figure 5.4 Nonmarket values of forest can be substantial, often nearly as high (Croatia, Tunisia), or higher than commercial values of forests (Portugal, Italy, and Syria). (Modified from the Millennium Ecosystem Assessment 2005.)

TABLE 5.3 ***Examples of WTP Bids for Threatened and Endangered Species***

	Willingness to pay		
Species	**Low value**	**High value**	**Average of all studies**
Northern Spotted Owl*	58	124	92
Pacific salmon/ Steelhead*	41	115	83
Grizzly bear*			60
Red-Cockaded Woodpecker*	13	20	17
Sea otter*			38
Gray whale*	22	43	34
Bald Eagle*	20	43	31
Bighorn sheep*	16	39	28
Atlantic salmon*	9	10	10
Striped shiner*			8
Bald Eagle†	233	333	283
Humpback whale†			227
Gray wolf†	21	155	88
Cutthroat trout†	17	22	20

Note: All values in 2004 U.S.$.
Source: Adapted from Loomis and White 1996.
*Studies reporting annual WTP.
†Studies reporting lump sum WTP.

accessible only through the use of CVM, and are the source of much debate. Table 5.3 provides some examples of existence values for a range of species.

Controversy surrounding the use of CVM

Despite many studies highlighting methodological and philosophical problems with the CVM approach, it remains popular with environmental economists. This popularity continues because:

1. It is theoretically sound (although CVM may sound rather odd at first, it does in fact have a solid theoretical basis).
2. Compared with other techniques, it does not require a huge amount of data and it relies on relatively simple estimation techniques.
3. It is applicable to all goods and services (including use and nonuse values), and is the only technique available for the evaluation of nonuse values, which makes it particularly significant both economically and philosophically.

However, from a wider perspective, and despite its relative popularity among environmental economists, the technique remains controversial for several reasons: First, it is by definition based on a hypothetical market, and it is difficult, if not impossible, to validate values obtained in a CVM against any data available in "real markets." This raises question marks over the values its produces. A number of potential problems associated with the basic technique have been identified. Many of these are concerned with the form of the questionnaire such as, the exact nature of the WTP question, the nature of the payment vehicle (how people are asked to pay, e.g., in a tax, a trust fund, or an entrance fee), and how much information people should be given about the good in question prior to asking the WTP question. Over the years these issues have been considered by a range of international experts in the field, and they have provided guidance on how to undertake a CVM in the most appropriate manner (e.g., see Mitchell and Carson 1989, Arrow et al. 1993). Second, CVM assumes that the public can make rational value judgments. What if the public is wrong? In these politically correct times we do not often discuss the possibility that the public's opinion may be wrong. Who

should we believe in issues of conservation, highly trained scientists or the general public? Survey-based methods almost assume implicitly that the public has good information on the topic of the survey and is able to make intelligent decisions. Is this assumption safe?

One attempt to test for possible differences between experts and the public is reported in Edwards-Jones et al. (1995). They undertook ecological surveys of three mountain sites and two coastal sites in Scotland, and recorded the presence and abundance of species of birds, vascular plants, and ground beetles (Coleoptera: Carabidae). They then considered the Shannon-Weiner diversity of the sites and the rarity and distribution of the recorded species. These data where then passed to nine independent professional ecologists who were asked to rank the sites in order of importance. At the same time a CVM was undertaken to ascertain the WTP of the public visiting these sites to preserve them into the future. The results, shown in Table 5.4, indicate that the experts and the public arrived at entirely different relative rankings of the sites. This is because when the experts evaluated the sites they considered characteristics such as species diversity and rarity, whereas the public evaluations where dominated by landscape, peacefulness, and ease of access. So even though both groups evaluated the same areas of land, they did so from entirely different bases. This sort of difficulty becomes really important if CVM were to be used to aid land-use decisions. Now that you know all this, if you had to pick only one upland site to conserve, whom would you believe—the experts or the public?

Third, related to the previous point, is a set of issues surrounding the underlying philosophy of the technique and respondents' reactions to it. For example, is it philosophically reasonable to expect people to be able to convert existence values into monetary terms? Do existence values constitute real values that people hold consistently over long periods of time, or are they simply formed in response to the CVM survey? Is it morally acceptable to even try to place monetary values on species? How do WTP bids vary among cultures, and how should we view the WTP bids of people in developing countries, which are inevitably smaller than those obtained in more developed countries? (Essay 5.3 by Mark Sagoff offers some answers to these questions.)

Project Assessment in CBA

By the time the project assessment phase of a CBA has been reached, the analytical boundary of the analysis has been drawn, the impacts to be considered have been identified and converted into monetary terms, and the timing of their occurrence has been estimated. So it should now be possible to develop a table that shows the impacts of the project, their monetary value, and the time over which they will occur. A simplified example for a hypothetical factory development is shown in Table 5.5, which we'll use to help make decisions about the project. However, a few key questions arise such as, from whose perspective should we view the costs and benefits, and over what time scale should we consider these costs and benefits?

Table 5.5 shows that the costs to the developer over the first 10 years of the project are $40,000), while the developer's benefits are also $40,000. So if the developer was thinking of only operating the factory for 10 years,

TABLE 5.4 *Rankings of Sites by Ecological "Experts" and the Public[a]*

Site	Mean WTP ($)	Ranking based on WTP	Ranking by experts
Upland			
Bonaly	17.15	1	2
Glencourse	14.88	2	3
The Howe	13.00	3	1
Coastal			
Yellowcraigs	17.86	1	2
North Berwick	17.65	2	1

Source: Modified from Edwards-Jones et al. 1995.
[a]Rankings inferred from WTP bids to protect the sites in perpetuity.

TABLE 5.5 *A Simplified Representation of the Costs and Benefits ($ × 1000) Over Time Associated with Construction of a Hypothetical Factory*

Item	Time Scale		
	10 years	20 years	100 years
Developer costs	40	70	70
Developer benefits	40	90	90
Total environmental costs	27	47	207
Benefit to local economy	30	60	10
Financial B[a] – C[b]	0	20	20
Economic B – C	3	13	–197
Total B – C	3	33	–177

Note: Environmental costs and benefits and the wider economic costs and benefits are often called social cost benefits, and the CBA that considers these values is termed an economic CBA. The CBA pertaining only to the developer is termed a financial CBA. There is also such a thing as a social CBA, which considers the distributional aspects of the costs and benefits—who the actual winners and losers are.

[a]B, benefits.
[b]C, costs.

ESSAY 5.3

A Non-Economic View of the Value of Biodiversity

Mark Sagoff, *University of Maryland*

■ Both environmental economics and conservation biology are normative sciences in that they presuppose and often advocate a view of the goal or purpose of environmental policy. "Advocacy for the preservation of biodiversity is part of the scientific practice of conservation biology," according to Barry and Oeschlaeger (1996). Conservation biologists believe that the natural diversity of organisms, the history of evolution, and the complexity of ecosystems are goods in themselves; these goods have intrinsic value and should be protected and preserved (Soulé 1985).

Environmental economics, as a normative discipline, posits the welfare of individuals—"utility" or "well-being"—as the goal of environmental and social policy, at least when equity issues are not pressing. The disciplines of conservation biology and environmental economics assert and advocate different and perhaps opposing goals or principles for environmental policy. One urges the conservation of nature, the other the maximization of utility.

To understand the policy goal economists advocate, one must grapple with the concept of welfare or utility, explain how it is measured, and determine why society should try to maximize it. Utility may refer to judgments people make—or things they want to possess or to happen—because they think they will benefit as a result. Economists argue plausibly that the individual in general is the best judge of what may make him or her better off. Thus welfare may relate to beliefs people entertain about what is good for them or what benefits them.

When people debate public policies and choices, however, they rarely justify their positions in terms of what they think benefits them as individuals. On the contrary, they adopt the role of citizens in presenting their views of the public interest, collective obligations, and community objectives. Thus, when an individual takes a position about a public choice and defends it on the merits, he or she supports a judgment others may agree with or oppose on the basis of reasons they regard as arguments. The reasons may have little to do with what the individual believes benefits her or him.

For example, when a person takes a position about foreign policy—about intervention in a foreign land, for example—he or she offers reasons and arguments. To the economist, however, a person (other than an economist) neither entertains nor expresses objective opinions about public policy. To the economist, a citizen is a location at which welfare or utility is found; a citizen is a source of subjective views or beliefs about what benefits her or him; he or she is an authority only on his or her own well-being. Alan Randall has wrote that in the view of the economist, "what the individual wants is [in his estimation] good for the individual" (Peterson and Randall 1984). Thus, beliefs are reduced to benefits; what the individual thinks is good for any reason is assimilated to what the individual thinks improves his or her well-being.

The contingent valuation method (CVM), asks people to state what they would be willing to pay (WTP, or "willingness to pay") to preserve some named environmental good, such as the preservation of the mountain gorilla. Oddly, economists interpret WTP to preserve a mountain gorilla as if it were WTP for some gain in utility or benefit. To ask respondents to state their WTP for a particular *policy* or *outcome*, for example, the protection of a species, however, is not to ask them to state their WTP for a benefit or gain in *utility* or *well-being*. Economists interpret the stated WTP for the policy as if it were WTP for a benefit the respondent expects that policy to afford her or him. Yet a person who believes that society ought to protect a mountain gorilla, may have no expectation at all that he or she will benefit as a result (Sen 1977). Indeed, as Tietenberg (2000) observes, people who do not expect to benefit in any way from an environmental good may still be committed to its preservation. He also notes, "people reveal strong support for environmental resources even when those resources provide no direct or even indirect benefit" (Tietenberg 1994).

Empirical research shows that responses to CV questionnaires reflect moral commitments rather than concerns about personal welfare. In one example, a careful study showed that ethical considerations dominate eco-nomic ones in responses to CV surveys. "Our results provide an assessment of the frequency and seriousness of these considerations in our sample: they are frequent and they are significant determinants of WTP responses" (Schkade and Payne 1994). In another study, researchers found that existence value "is almost entirely driven by ethical considerations precisely because it is disinterested value" (Barbier et al. 1995).

Why should society or social policy seek to increase or maximize utility or welfare, as economists suppose, rather than seek to protect nature, as conservation biologists argue? One might answer that utility or welfare, as economists understand it, has something to do with human happiness and contentment, which are certainly worth pursuing. Thus, one could suppose that economic approaches have a basis in the ethical theory of Utilitarianism—the view that happiness or pleasure matters most. In fact, no connection whatsoever exists between concepts of welfare or utility, as economists use them, and the concepts of happiness or satisfaction that are found in Utilitarian philosophy. The concepts of welfare or utility refer to the satisfaction of preferences, weighed by WTP, taken as they come, bound by indifference. We have already seen that people base their preferences, particularly about public policy, on principled concerns rather than their own happiness. The connection between utility (preference-satisfaction) and happiness is doubtful at best.

The concept of utility or welfare, in any event, is measured by WTP not by happiness. Judge Richard Posner (1981) makes the point that the "most important thing to bear in mind about the concept of value [in the economist's sense] is that it is based on what people are willing to pay for something rather than the happiness they would derive from having it." If economic value is a function of WTP for something rather than the happiness the person would derive from having it, it is unsurprising that those willing to pay the most for goods derive the most economic value from them. The term "economic value" simply coincides with WTP and has no connection to anything else and thus lacks value, meaning, or importance.

Indeed, the relation between welfare in the economic sense and happiness in the Utilitarian sense rests on little more than the ambiguity of the term "satisfy" or "satisfaction." The word "satisfaction" as in the phrase "the satisfaction of preference" may mean either of two different things. It may mean that a preference or desire is met or fulfilled; in this sense terms, conditions, equations, predictions, etc. may be "satisfied." Second, it may mean that the person who has the preference is content or happy. The satisfaction of preference in the first sense does not clearly relate to satisfaction in the second sense, that is, pleasure or contentment.

Many people report they become *less* happy as their income and purchasing power increases (Lane, 1991; Samuelson 1997; Argyle 1999). Studies relating wealth to perceived happiness find that "rising prosperity in the USA since 1957 has been accompanied by a falling level of satisfaction. Studies of satisfaction and changing economic conditions have found overall no stable relationship at all." (Argyle 1987) One major survey states, "None of the respondents believed that money is a major source of happiness."(Diener et al. 1986 and Diener et al. 1993). That money (and thus preference-satisfaction) does not bring happiness after basic needs are met may be one of the best established findings of social science research (Lane 1993). Thus, the thesis that preference satisfaction promotes welfare appears either to be trivially true (if "welfare" is defined as preference satisfaction) or empirically false (if "welfare" is defined as perceived happiness). Happiness seems to depend on the things money cannot buy, e.g., love, friendship, and faith, not on the extent of one's possessions or one one's income and thus ability to satisfy preferences.

If preferences are mental states, they cannot be observed. If they are inferred or "constructed" from behavior, they're also indeterminate, since there are many ways to interpret a person's actions as enacting a choice, depending on the opportunities or alternatives the observer assumes define the context. For example, the act of purchasing Girl Scout cookies could "reveal" a preference for eating cookies, supporting scouting, not turning away the neighbor's daughter, feeling good about doing the right thing, avoiding shame, or any of a thousand other possibilities. Choice appears to be no more observable than preference because its

description presupposes one of many possible ways of framing the situation and determining the available options. If cost–benefit analysis takes preferences as its data, then it has no data. Preferences and choices are constructed together by the beholder—who must size up the situation, construct the choice the individual makes, and then attribute to the individual the preferences that already are implicit in that way of describing his or her behavior.

Finally, few if any data indicate maximum WTP—the concept that coincides with economic value—for any ordinary good. When you run out of toothpaste, get a flat tire, or have to buy the next gallon of milk or carton of eggs, you are unlikely to know or even have an idea about the *maximum* you are willing to pay for it. Instead, you check the advertisements to find the *minimum* you have to pay for it. It is not clear how economists can estimate maximum WTP when all they can observe are competitive market prices. Competition drives price down to producer cost, not up to consumer benefit. For example, you might be willing to pay a fortune for a life-saving antibiotic, but competition by generics may make the price you actually pay negligible. The competitive market prices one pays are no indication of maximum WTP—since they usually represents bargains, not benefits. Indeed, maximum WTP, which economists use as a measure of utility, cannot be measured, because all the data hover around market prices.

These and many other arguments suggest that the goal environmental economists advocate—the satisfaction of preferences or utility—has no connection with any conception of the good, any conception of value, but itself. Having a preference gives the individual a reason to try to satisfy it—and he or she should be free to try to do so through economic and political institutions, such as markets and electoral politics that are as open, fair, and convenient as possible. Absent any relationship to happiness or well-being in any substantive sense, however, preference-satisfaction, and with it utility and welfare, are not plausible goals for social policy.

Some economists argue that citizens want to have their preferences satisfied; therefore they implicitly or hypothetically assent to cost–benefit approaches to social policy that attempt to maximize the satisfaction of preferences insofar as the resource base allows. There are at least two problems for this view. First, to override in the name of

efficiency, scientific management, or centralized planning what citizens themselves arrange through their own behavior in markets and political processes, economists would have to know more than the individuals themselves about what they want and how to arrange events to provide it. This places enormous information costs on economists—even greater, perhaps, then individuals face themselves.

Second, in a democracy, the problem is not implicit or hypothetical consent, which every theory may claim, but actual consent, which is given either through voluntary exchanges in the market or through political deliberation in legislatures ending in a vote. Legislatures are constantly seized with moral questions—not economic trade-offs—about such issues as capital punishment, abortion, aid to the poor, education, health policy, environmental policy, and so on. Democratic councils discriminate on the merits between reasonable and unreasonable preferences, basic needs and frivolous desires, and goals that express or contradict our best aspirations as a people. As a result, Congress has explicitly, repeatedly, and properly rejected a cost–benefit test in all major environmental legislation. As Cropper and Oates (1992) observe, "The cornerstones of federal environmental policy in the United States," such as the Clean Air and Clean Water Acts, "explicitly prohibited the weighing of benefits against costs in the setting of environmental standards."

Environmental policy essentially turns on questions of justice, on aesthetic judgments, and on spiritual beliefs. Questions of justice concern pollution, that is, the use by one person of the person and property of another as a catchment for waste without his or her consent. Aesthetic judgments arise over the beauty and magnificence of the natural world as compared to the rather less impressive, but sometimes more utilitarian, houses, factories, mines, and roads we put in its place. Spiritual beliefs inform our wonder and awe at creation or evolution. It is clear that conservation biologists have contributed immensely to our appreciation and understanding of the moral, aesthetic, and spiritual judgments that form the basis of environmental valuation in meaningful sense. Although economic science has been more influential, its conflation of value with preference defeats whatever contribution it may make to environmental policy. ■

his costs would match his benefits and his profits would appear to be zero. In this situation it may not be rational for a developer to proceed with the factory. However, if the developer planned to operate the factory for 20 years, by the end of this time the profits (benefits minus costs) would be $20,000. It is very hard to plan an industrial process much beyond 20–30 years, so the developer is uncertain about the profits from year 20 onward, and may only plan a 20-year active life for the factory.

While a developer may feel confident that benefits would exceed costs over a 20-year time scale, and would therefore want to proceed with the project, an environmentalist may take a different viewpoint. He may note that the cost of habitat loss over 20 years is $40,000, and this is on top of soil erosion in the early construction years costing $7,000. From his point of view, the factory would only bring costs, and hence it may not be rational for him to support its development.

The state governor, on the other hand may have a different viewpoint. She may be concerned about employment in the region, and the factory will offer jobs and create additional economic effects (multiplier benefits) to the value of $3,000 a year. Thus the economic benefits (i.e., financial benefits to others apart from the factory's developers), would be substantial, and over a 20-year period would outweigh the environmental costs. Therefore, as shown in Table 5.5, over a 20-year time scale the financial CBA would be positive (i.e., the developer would make money) and the economic CBA would be positive (i.e., society would benefit more from job creation than it would suffer from environmental loss). In this situation it may seem rational to proceed with the factory developments as the benefits clearly exceed costs over the planned life of the project.

While some CBAs are carried out in the manner described above, such a procedure is flawed in several respects. First, just because the project is only planned to operate for 20 years does not mean that the environmental costs will only last 20 years. The habitats lost during the factory development are lost forever. Indeed if we extend the CBA to 100 years we see that the costs of the lost habitat are far greater than the financial and economic benefits provided by the factory in its 20-year life (see Table 5.5).

Not only was the initial analysis flawed in its choice of time scale, but also it has not considered the changing value of costs and benefits over time. It may seem unusual to suggest that the value of a benefit may vary with time but consider the following question: Which would you rather receive, $100 now or $100 in a year's time? Most people say they would take the money now. So then we can pose a second question: Which would you rather receive, $100 now or $1000 in year's time? Many people may now opt to receive more money in a year's time. But if you keep repeating this question with ever closer gaps between the offered present and future payments, eventually you get to a point where people are indifferent about receiving money now and in the future. This general preference to receive good things sooner rather than later (and conversely put off bad things to the future) is termed time preference, and time preference and the productivity of capital are reasons why many economists suggest that all future costs and benefits should be converted to a **present value** (**PV**); that is, the value of some future sum of money should be worked out in today's terms. This process is based on a process of **discounting**, and is a vitally important concept in economics, as described in Box 5.2.

Discounting a future benefit to a present value has several interesting effects (Table 5.6). Today's value of $10,000 received in 10 years' time is $6,139 at a discount rate of 5% (this is equivalent to people being indifferent to a choice of $100 today and $105 in a year's time), but is only $3,855 at a 10% discount rate. So we can quickly conclude that the choice of discount rate is crucial in any analysis. Also, the value of $10,000 dollars received in 100 year's time is very small in today's terms at both discount rates.

This issue is important for environmentalists, as the environment should offer benefits into perpetuity (e.g., indirect services like nutrient cycling in a wetland will last as long as the wetland does, and existence values from large vertebrates will last as long as those species remain extant). However when these future benefits are discounted they appear to be negligibly small in today's terms. For this reason when long-term environmental costs are included in a CBA they can easily be outweighed by benefits that occur earlier in the project's life.

TABLE 5.6   *The Present Value (After Discounting) of $10,000 over 10, 20, 50, and 100 Years at Three Different Discount Rates*

Discount rate	Time (years)				
	0	10	20	50	100
0	10,000	10,000	10,000	10,000	10,000
5	10,000	6,139	3,768	872	76
10	10,000	3,855	1,486	85	0.73

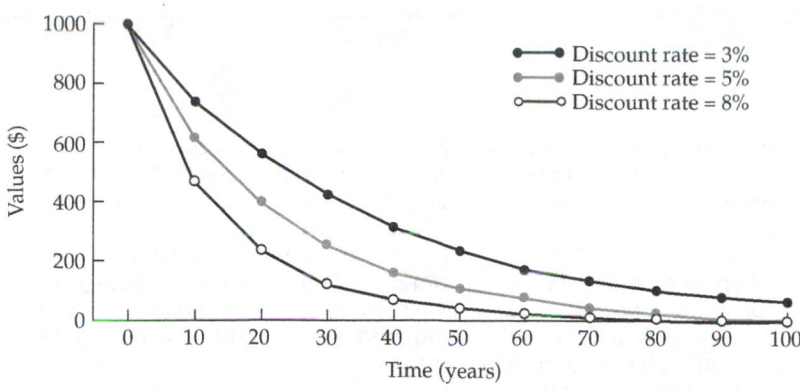

Figure 5.5 Example of how the net present value of a sum of $1000 varies over 100 years at three different discount rates. Note that the higher the discount rate, the higher the value in the near future, but over the longer term, present value is low for all discount rates shown.

Also note that the larger the discount rate, the smaller future benefits are (Figure 5.5). For this reason many environmentalists call for small discount rates to be used in CBAs to permit the long-term benefits of the environment to have influence in the analysis. However, the use of small discount rates (or even a discount rate of 0) is usually strongly opposed by financial institutions, as they are not felt to be realistic. Some recognition of the impact that different discount rates have is evident in the U.K. where government projects are planned using different discount rates according to project type: 8% is applied to publicly financed roads, 6% to management decisions, and 3% to land acquisitions for new tree planting carried out by the Forestry Commission (HMSO 1991). In these situations, different discount rates are used to reflect the flow of benefits arising from specific projects over time, so in forest projects it is important to have low discount rates to reflect the long-term nature of the investment. There is no requirement for most business ventures to consider such long-term benefits, and hence they usually adopt larger discount rates. The fact that some parts of government use different discount rates in different situations shows that the public servants actually understand the importance of discounting, and that they are concerned with making investments for reasons other than short-term returns.

Net present value (NPV)

The impact of discounting within our example CBA is to suggest that the value of a future benefit (say a $2,000 profit in year 20) may not actually be equivalent in today's terms to a $2,000 profit in year 4. To examine the impact of the timing of costs and benefits on the overall project, it is common to calculate the **net present value** (**NPV**) of all costs and benefits in a project.

Calculating NPV requires all monetary values to be discounted at an agreed rate so that they are all expressed in present value (PV)

terms. The sum of the discounted benefits minus the sum of the discounted costs is equal to the net present value (NPV) of the project. The NPV is the keystone of the CBA appraisal methodology. If the NPV is positive, then the project represents an efficient shift in resource allocation according to the estimations and calculations of the CBA, thereby increasing social welfare. More simply, a positive NPV justifies the project, and a large NPV is better than a small one.

The results of estimating NPV for our hypothetical factory project (Table 5.7) shows that after 20 years this is positive at all discount rates—that is, the project is viable over 20 years, even though environmental damage is being caused. If the factory ran for only 10 years then it would not be viable either at the 5% or 10% discount levels. This is due to the fact that it would not have been running long enough to cover the development costs experienced in the early years of the project. However, over longer time spans the project shows a negative

TABLE 5.7 *Present Value of Costs, Benefits, and Net Present Value of a Hypothetical Factory*

Time	Discount rate	PV[a] costs	PV benefits	NPV[b]
10	0%	67.00	70.00	3.00
20	0%	117.00	150.00	33.00
100	0%	277.00	100.00	−177.00
10	5%	55.25	52.62	−2.63
20	5%	78.95	90.54	11.59
100	5%	93.71	90.54	−3.17
10	10%	47.06	40.75	−6.31
20	10%	58.91	59.70	0.79
100	10%	61.88	59.70	−2.18

Note: Value numbers represent U.S.$ × 1000; values are calculated at three different discount rates and over different time scales.

[a]Present value.

[b]Net present value.

BOX 5.2 Discounting

Why do we discount? This is a difficult question for many people to grasp. In essence we discount to take into account humankind's natural greed and short-term perspective. Humans want instant gratification—whether it be money, capital goods, or food. Similarly we tend to put off bad things, both in time and space; for example, we often delay going to the doctor to receive painful treatment, and given a choice we tend not to want to live next to smelly, polluting industrial sites. This perfectly understandable desire to have good things sooner rather than later while seeking to delay bad things may work for us in our daily lives—but it makes formal analysis of future costs (bad things) and benefits (good things) rather complicated. Discounting seeks to remove these complications—although discounting itself may seem complicated. The fundamental supposition of discounting is that in the estimation of the value of a project, future costs and benefits count for less than present ones. Two immediate qualifications must be made here:

First, discounting is entirely separate from the process of adjusting for inflation. All calculations of discounting take place in *real terms*—that is—keeping the "purchasing power" of money values constant across time by deflating by a chosen rate. Even if the rate of inflation was zero, the arguments for discounting still stand.

Second, discounting applies to the values attributed to future *consumption*, not to future *utility*. This is an extremely important distinction. The utility, or welfare, of all individuals in all time periods is considered as equally valuable; what is discounted is the value of consumption in different time periods, because the relationship between consumption and utility is generally thought to be time-dependent, for reasons given in the following sections.

With the above qualifications in mind, there are two basic reasons for discounting: time preference and the productivity of capital.

Time Preference

Time preference describes the preference of an individual for *when* he or she receives a benefit (or has to bear a cost). Is $1000 received today considered as valuable as $1000 received this time next year, or in ten year's time? If the individual's assessment of the value of the benefit does change depending on when it is received, that individual is said to have a "time preference." What justifies having a positive (i.e., sooner rather than later) time preference? The justification is due to two factors: "pure" time preference and opportunity costs.

Pure Time Preference

Pure time preference relates essentially to the factors of impatience and uncertainty when considering any future good. Individuals are often myopic in their behavior, acting for the short term because the future somehow seems less important than the here and now. This is basically impatience; as an extreme, yet unfortunately commonly experienced example for many millions of people today, someone who is starving will have an extremely high time preference for immediate consumption for food and bear little regard to the future implications of that consumption—because if he doesn't eat now, he won't have a future. The desire for goods now is also related to uncertainty about the future; future goods may prove to be less valuable in the future than they are now because of a change in circumstances (of course, the reverse might also be true). This generally leads the future to be discounted because of uncertainty about future benefits, whereas we are much more certain of how a good will impact on our immediate welfare.

Opportunity Costs

The second reason for a positive time preference rate is **opportunity costs**, that is, the opportunities for generating further benefits that come from having a good now rather than in the future. In the simple case of $1000 now or in a year's time, with a prevailing bank interest rate of 10%, you can choose to either spend immediately or deposit the money in the bank. If you choose the latter option, you will have $1100 at the end of the first year. If you keep the money on deposit for a second year, you will have $1210, the year after $1331, and so on. So if you were offered $1000 as a gift, you should opt to take it now, rather than in a year's time, as by not having your money you are missing out on the chance of earning more money through investing the money.

Productivity of Capital

The productivity of capital argument for discounting is basically the flip side of the opportunity cost argument for having a positive time preference. That is, the productivity of capital accounts for the fact that there are opportunities within the economy to earn interest on savings, because the process of saving (with that money reinvested by banks in productive projects) leads to increased overall output in the future. If the economy was completely static, that is, not growing at all, there would essentially be no rationale for discounting on the basis of future opportunity costs, because money invested now would yield no surplus returns. Even in this situation though, the apparently basic human characteristic of pure time preference would still justify discounting at some level.

Calculating Present Values

When we undertake discounting, we calculate the present value of a future sum of money. The present value of some future value (a cost or benefit) is given by: $PV = V_0 / (1 + d)^t$, where PV = present value, V_0 = value of the sum at time 0, d = discount rate (expressed as a proportion), and t = year for which the PV is required. For example, suppose we wanted to estimate the present value of $1000 in taxes from an ecotourism lodge, which will be paid to the government once the lodge turns a profit in 3 year's time. Then, assuming the discount rate is 5% the calculation is: $PV = 1000 / (1.05)^3$, which gives: $1000 / 1.157 = 864. So the value of $1000 made in 3 year's time is $864 in today's money (assuming a 5% discount rate). You do not need to calculate these figures yourself all the time, as many spreadsheets have discounting functions.

Effects of Varying the Discount Rate

The effect of discounting values is that costs and benefits that occur in the future are worth less in *financial terms* than *identical* costs and benefits arising in the present. This tends to support decisions in favor of those projects that have greater short-term benefits, and those projects in which costs arise in the medium to long-term. Any positive discount rate leads asymptotically to a fixed net present value, as values accruing several decades hence become negligible. It is this aspect of assigning a fixed value to assets that may provide a constant benefit stream indefinitely into the future that makes many people intuitively suspicious of its justification.

At the heart of most of these criticisms is the issue of whether the discount rate should be adapted in some way in order to bring about a social improvement when applied within a CBA, or whether discounting should operate on market or near market determined rates, and other ethical or political criteria should be adopted separately as constraints on development. It should be noted that changing the discount rate—in particular, using a lower rate as many environmentalists advocate—may in itself have very adverse environmental consequences. This is for three reasons:

1. Low rates encourage wide investment in projects due to the cheaply available capital; this accelerates the rates of depletion of resources overall.

2. High rates concomitantly reduce this rate of growth, tending toward less development overall and a greater stock of resources inherited by the next generation.

3. Primary industries (agriculture, timber, fishing) tend to have relatively low rates of return compared to more high-tech industries. A high discount rate tends to make these types of projects look less attractive to investors, and this may have subsequent negative effects on rural communities. Conversely, a low discount rate may encourage investment in these types of projects, which in some areas may have negative impacts on the environment.

NPV for all discount rates. This is due to the continuing environmental costs and the absence of profit from the factory's output in years 20–100.

So answering the apparently simple question, Is the project viable? is actually a rather difficult proposition. From the perspective of the private investor it would be worth developing the factory and running it for 20 years, whatever the discount rate. Similarly from the state governor's perspective such a proposition would be viable, as the benefits to the local economy would be greater than the environmental losses. However, once the time scale of analysis is expanded, this latter argument is reversed, and based on these results it would not be sensible for the state governor to support the project. In reality though, the state governor may still decide to support the project, regardless of the CBA results, if she placed significantly greater weight on short-term economic gains than on long-term environmental losses.

Sensitivity analysis

The final output of a CBA is quantitative and relatively simple to understand. For example, it could be stated simply: The benefit/cost ratio for this project is 3:1 (i.e., benefits are 3 times greater than the costs). Such simple statements do not really communicate the complexity of the thinking and the uncertainty surrounding many of the figures used in the calculations inherent in a CBA. Sensitivity analyses allow analysts to consider how sensitive the final output of the CBA may be to variations in specific inputs. In essence this means changing one of the key inputs in a systematic way and rerunning the CBA, as we did in Table 5.7 for the dif-

ferent discount rates. By repeating this process for a range of inputs it is possible to understand how robust the final result is to uncertainties in input values. In other words, it lets analysts answer the question: What if my estimate of this input to the CBA is wrong—will it affect the final result?

While it would theoretically be possible to consider the effect of varying any input to the CBA, usually sensitivity analyses are conducted on inputs in one or more of the following major classes of uncertainty:

- The quantity and quality of physical inputs and outputs to and from the project
- The respective financial values of these inputs and outputs
- The value of changes in environmental quality
- The appropriate discount rate

It is wise to consider the influence upon NPV of changing each of these estimations. From these calculations, it should be possible to identify the estimates to which NPV is relatively sensitive. For example, although the estimated NPVs of two projects might be identical at, say, $10 million, one might be significantly more sensitive to errors in estimation. This provides important additional information for decision-makers, and such analyses allow the targeting of research expenditure into the better estimation of certain critical costs and benefits.

Objections to CBA

Although CBA is widely practiced, it has also been strongly criticized from perspectives outside economics,

particularly from the environmental and social spheres. Problems with the technique include the following:

1. There is uncertainty associated with the estimation of the physical costs and benefits associated with a project. This arises because of the difficulty in accurately defining the project and its building and maintenance requirements ahead of time. It remains common for large projects to actually cost far more than the original budget suggested— witness the frequent problems with large developments like the Olympic Games and military contracts. Such miscalculations are a problem because if actual costs increase relative to budgeted costs and the benefits stay the same, the cost/benefit ratio inevitably decreases. Thus, critics worry that CBA may encourage development at the expense of conservation, unless special care is taken to account for such uncertainties in costs and benefits.

2. There is also uncertainty associated with the estimation of the monetary valuations of environmental impacts, and potential unease in using such figures in decision-making. In the case of threats to rare species, for example, there may be controversy associated with the valuation figure for their protection, including variability around the mean, sample size, and the soundness of the methods used in its estimation. These issues may be exacerbated if a large number of people dispute the underlying philosophy of such calculations.

3. There may be issues associated with the distribution of costs and benefits among the population. Standard CBA practice does not distinguish between the sections of society that receive benefits and those that endure costs as a result of a project. At best this may seem unfair, at worst downright discriminatory, especially if those suffering the costs all live in the same geographical area or belong to the same social or ethnic group. For example, the benefits of an oil development by a French oil company in an African country may accrue to the French investors and the governing elite of the country, while the costs may fall on the poorer people living adjacent to the development. This sort of issue can be addressed by widening the boundary of a cost–benefit analysis to include distributional issues within the analysis, and on some occasions this is done (e.g., Ninan and Lakshmikanthamma 2001; Kumar 2002).

   Similarly, there may be concerns about unequal distributions of costs and benefits to humans versus all other species and ecosystems. Problems in achieving equity among humans may be even more pronounced when considering ethical considerations among species. Some economists have advocated even larger breaks with traditional economic practice on these grounds (see Essay 5.3). However, environmental economists attempt to address these issues by monetizing the effects on nonhumans as thoroughly as possible, and bringing these values into the cost–benefit analysis.

4. Although discounting is a commonly accepted practice it still causes philosophical and practical problems. The practical problems relate to which discount rate to use when, while the philosophical problems relate to whether or not we should use a discount rate at all. As we saw above, any positive discount rate will reduce the significance of future costs and benefits, a fact that is often felt to downplay the importance of long-term environmental protection.

5. Standard CBA does not take any particular note of irreversible actions in development decisions. So the fact that a project may drive a species extinct does not carry any particular weight within a standard CBA. Rather such analyses simply calculate whether the benefits of the project outweigh the costs of causing that extinction, and if they do, the implied attitude is: that it does not matter that some creature has gone extinct. Such a blunt mechanical approach may be difficult for many people to accept.

6. While in some ways CBA is a blunt, mechanical approach, it is also open to subtle manipulation by the people carrying it out, for example by varying the project boundary and the discount rate. It is therefore susceptible to cynical manipulation by unscrupulous analysts, and in this way institutions undertaking the CBA can seek to gain a result that best meets it needs. The best way to avoid this is to ask for all the assumptions inherent in a CBA to be made available for public scrutiny—and to then get some public-minded people who understand CBA to view these data.

7. Finally, and related to some of the above criticisms, there is no explicit sustainability criterion in cost–benefit analysis. Because CBA is generally applied to individual projects, the methodology is not designed to ensure overall sustainability of the impacts of a series of projects or developments, or the size of the economy as a whole (note: this criticism is certainly not exclusive to CBA).

How can we best use CBA in conservation?

It is possible to make CBA more sensitive to environmental concerns through lowering discount rates, taking a longer time perspective, and perhaps even giving

greater weighting to environmental and social objectives than financial ones. But despite these improvements, there is an assumption that lies at the heart of CBA that many conservationists find difficult to accept—the assumption of substitutability. Substitutability implies that more of one good can make up for loss of another. So under this assumption more money can in some way substitute for loss of a wetland or even of a species. Environmentalists point out that this cannot be true in all cases. When a species is extinct it is gone forever; similarly when the ozone layer is gone, no amount of money can replace it in the short term (see Case Study 5.1 for a discussion about whether or not the nonsubstitutability of human life may be a challenge to biodiversity conservation).

Many of the people who recognize the problems of nonsubstitutability can be termed ecological economists. Generally ecological economics recognizes the importance of economics in our society but seeks to broaden the philosophical and methodological base of traditional economics. As such ecological economists seek to explicitly incorporate an ethical and environmental dimension into decision-making. This is a lot harder to do than it sounds, but one simple way of achieving this is to adopt decision-making mechanisms that focus on environmental issues and run parallel to traditional economic ones.

Alternative Decision-Making Methods for the Environment

There are several broad classes of methods for making decisions about environmental issues, and many different specific techniques are used within each class of method. Generally each broad class of method is best suited to aid a particular type of decision, be that a project or policy related decision. Several decision-making methods are very briefly described here.

Environmental impact assessment

Environmental impact assessment (EIA) is the process of identifying the potential impacts of a project or policy before it goes ahead. The final output of an EIA is a document called the environmental statement (ES). ESs are used to help decide whether or not the environmental impacts of a project are acceptable, and can be considered in conjuction with other project-related information when deciding whether or not a project should go ahead. The impacts considered in an EIA can include such things as the effect of a project on habitats, species, landscapes, economies, and health. Completing an EIA often requires a combination of fieldwork, modeling, and expert opinion. Depending on the context, ESs either can be used to inform the CBA, or can be considered independently of a CBA.

EIAs have been legally required for certain projects in the U.S. since 1969, and in many other countries since the 1970s, and have been applied to a wide range of projects from building houses to oil-rigs. It would be possible to stipulate that any project should have to be acceptable both financially, as determined by CBA, and environmentally as determined by EIA before it could be sanctioned. However, there are three main problems with such an approach. First many EIAs are done to a poor standard and rarely live up to their potential. Second, few national planning systems are strong enough to enforce the submission of adequate EIA and CBA for every project of concern. Finally, even when an EIA is done well, there remains a tendency among decision makers at all levels to weight financial gains over environmental losses. For more information see Petts (1999a, 1999b) and Modak and Biswas (1999).

Risk assessment and management

Formally, risk is defined as the probability of a given hazard causing a given consequence, multiplied by the magnitude (severity) of that consequence. This may be represented algebraically as: $R = P \times M$, where R = risk, P = probability of a given hazard causing a given consequence, and M = magnitude of harm that the given consequence would cause.

A hazard is something, or a set of circumstances, that has the potential to cause harm, and a consequence is the outcome of concern. To understand this in a conservation context, consider the case of a proposal to build a wind farm as an alternative energy source. The wind turbines can create a hazard for birds and bats, and the consequence could be a decrease in bird or bat populations as a result of increased mortality when encountering the wind farm. The probability that the wind farm would harm bird or bat populations would depend on the location of the wind farm in relation to major flyways of migratory birds, or major nesting or roosting areas (Kerlinger 2000). Thus, the risk to bird and bat populations can be minimized by careful selection of locations for the wind farm, or additional design elements that are repellent to wildlife such as strobes (Pimental et al. 2002).

This type of analysis can be applied to a range of issues relating to the environment such as the regulation of pesticides and other chemicals, the planning and design of urban areas, the allocation of effort in fisheries, the creation of protected areas, the design and management of restored areas, and the actions within species conservation programs. While the mechanisms for formal **risk assessment** itself are well known, there are several difficult issues with its implementation. One relates to what constitutes an acceptable level of risk. For example, if after undertaking a risk assessment for a new pesticide, it was shown that if used legally it had a one in ten million chance of causing egg shell thinning in a named species

of bird, would that be acceptable? Or would we want the probability to be one in one hundred million?

A second issue relates to communicating risk to the general public. No action we undertake carries zero risk. In terms of causing harm to human health, three of the most risky things humans can do are to ride in motor vehicles, smoke cigarettes, and drink alcohol. The risk of harm arising to humans from these activities is normally far, far greater than the risk of harm to their health from a pesticide. Despite scientists regularly stating what the risks of actions like driving a car actually are, we do not see much evidence of the public changing their car-driving behavior. Partly this is because of the risk:benefit ratio. While driving a car is a very risky activity, the benefits of doing so are perceived to be huge, and for this reason we continue driving cars. However, many other activities that are undertaken in society do not bring many perceived benefits to certain individuals, so why should these individuals live with the risk when the risks outweigh the benefits? This is too complex a subject to deal with in more detail here, but there is a long literature on public attitudes to risk. It is important to note that risk assessment is a well established process that is frequently used as part of official regulatory work in our governments. For more information see Regan et al. (2003) and Matsuda (2003).

THE PRECAUTIONARY PRINCIPLE   The precautionary principle is an ethical principle that states that if the consequences of an action are unknown but are judged by some scientists to have a high probability of being negative, then from an ethical point of view, it is better not to carry out the action rather than risk the uncertain, but possibly negative consequences (see Chapter 1).

New chemicals and new technologies are constantly being used in the Western world, and while there is active risk assessment of some types of chemicals, generally the assumption is that the use of new technologies will be permitted unless it can be shown that there are negative consequences associated with their use. The problem with this type of approach is that experience shows us that not all possible negative consequences are known at the time a technology is licensed. Unfortunately there are cases where these negative consequences only become apparent over time. For example, the effect of the pesticide DDT on the eggshell thickness of birds of prey only became apparent after many years of using DDT. The precautionary principle suggests that we don't need to know all possible negative consequences of a new technology or action, but we should be honest about our ignorance and not proceed until we are more certain of the facts.

Although the precautionary principle does appear in several laws around the world, there is much concern about its implementation. Some people worry that should the precautionary principle be implemented and upheld in law, proponents of new technologies would need to prove that no negative consequences would arise from use of the technology. While such a philosophy has much to commend it, we must also realize that enacting it wholeheartedly may act as a brake on innovation and technological advances, while simultaneously tying up a lot of resources in testing and debating new technologies prior to their introduction. This is on top of the philosophical difficulties of proving a negative.

The precautionary principle does not necessarily have to be applied in all cases, and some definitions suggest that it should only be applied where there are threats of serious or irreversible damage from the introduction of new technology. This would remove the need to undertake large scale and expensive testing on all products. Hence, given our long and largely benign experience of plastic chairs, we would expect there to be no serious threat of them causing cancer and they could be marketed without being tested.

In the same vein, many industries may try to make similar claims for their products, and thereby avoid the need for testing. Because of this strategic behavior by industry, the current debate revolves around what exactly constitutes a "serious threat." While the example of the plastic chair is a facetious, though relatively simple one, there are heated debates on-going around the globe about threats posed by climate change, pesticides, genetically modified organisms, and other technologies, and whether or not the precautionary principle should apply to developments in these fields. Generally speaking, because our ability to define the risks posed to natural communities and individual species is drastically less than our experience in evaluating health risks from plastics, our abilities to apply the precautionary principle in conservation contexts is much reduced. Much political and legal wrangling remains to be done before the precautionary principle can be fully adopted and implemented. For more information see O'Riordan and Cameron (1994), Freestone and Hey (1996), and Brauer (2003).

MULTI-CRITERIA ANALYSIS (MCA)   **Multi-criteria analysis** is not a stand alone decision-making method; rather, it provides a formal structure for integrating the results from all other approaches to help decision-makers chose a plan or project that best fits with their own priorities and objectives. In a nutshell, an MCA evaluates a set of plans or activities against a set of criteria to determine how best to achieve pre-determined objectives. It assesses the impact of each plan (or the activities that will compromise a plan) on each criteria. Then, through a formal programming procedure it compares these individual criterion impacts to evaluate and select the plan or combination of activities that most successfully meets the objectives.

Multi-criteria methods were pioneered in the 1970s following the development of single-objective programming techniques such as **linear programming (LP)**. LP is a method of mathematical analysis that calculates the optimal solution to a planning problem in which there is a single objective, and several constraints on action. For example, a farmer may want to maximize his profits from farming, but he must work within constraints on the amount of land he can farm, the amount of water available for irrigation, the amount of money he can borrow, and so on. Given these constraints and the objective of the decision-maker, LP calculates the most profitable combination of crops to plant.

A multi-objective analysis extends the problem-solving approach by introducing additional objectives. A farmer may want not only to generate profits, but also reduce his risk of crop failure, and create a landscape that provides corridors for native species between fields. The type of problem modeled by this kind of multi-objective appraisal is generally closer to the real priorities of a decision-maker than a single objective approach. In fact, there are probably very few situations in which a decision can realistically be based on only a single objective.

While it might be possible to provide a financial valuation for some of the objectives in an MCA, it may be more sensible to measure each one by specific and appropriate criteria—for example, profits in dollars, landscape by acreage of tree cover and length of hedgerows, and so on. This introduces multiple-criteria into the analysis. These criteria measure objective achievement on a range of scales from simple quantities to purpose-built indices. This freedom in data gathering is a major advantage of MCA techniques, as it allows natural scientists to express their data in a consistent and formal way without necessarily using a monetary scale (Figure 5.6).

While MCA has many advantages, there are also some disadvantages. First, it is a very technical process that requires a good understanding of the mathematics involved. Second, many scientists are attracted to it because it avoids the need to explicitly value the environment in monetary terms; unfortunately, the avoidance of the monetary scale does not mean you can avoid the underlying issues of "evaluating" data, priorities, and objectives. This is because many of the philosophical problems inherent in converting environmental criteria to a monetary scale are relevant for all scales of measurement—even if that scale seems to be a relatively simple one, like 1 to 10. For more information see Zeleny (1976) and Dreschler (2004).

Adding nonsubstitutability to CBA

Finally, we can explore means to insert the concept of nonsubstitutabilty into CBA by specifying that certain attributes of the environment are inviolate and should never be affected by any project or policy. This is the basis of the concept of **critical natural capital**, where certain parts of the environment are deemed to be "critical," and should not be damaged, or substituted for by other goods. This is a good theoretical idea, but in reality it is very difficult to define which parts of the environment are critical. The ozone layer and atmosphere seem to be critical to all life on Earth, yet there remains great political debate about whether certain countries such as the U.S. should reduce emissions of atmospheric pollutants at a cost to their traditional industries.

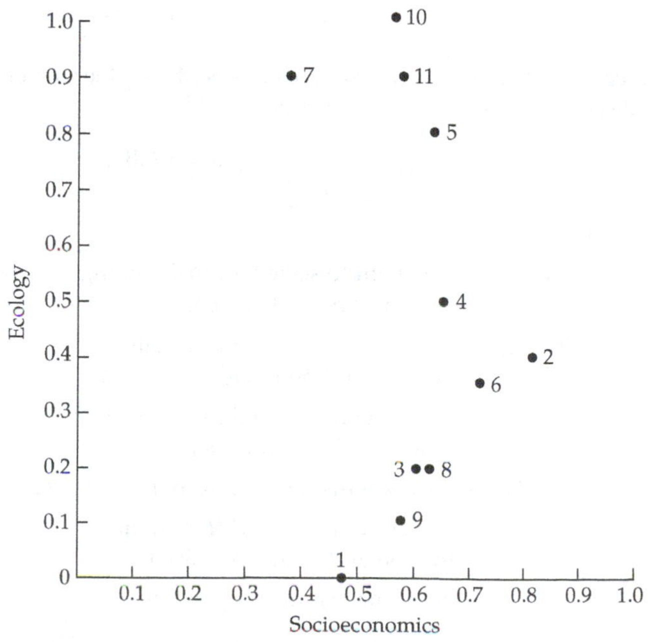

Figure 5.6 Application of multi-criteria analysis to the problem of diffuse pollution from agricultural land to an inland lake in Scotland, U.K. The figure shows the ranking of eleven alternative management systems to control eutrophication in terms of their ecological and socioeconomic impacts. The maximum score any management alternative can achieve is 1. So assuming that we are equally concerned about maximizing ecological gain and minimizing socioeconomic losses, the best option is the one that is closest to the top right-hand corner of the graph (i.e., closest to achieving a maximum score of 1 on both scales). This example shows the simple case of considering only a set of management alternatives against only criteria (i.e., the two axes, sometimes called dimensions). More complex analyses can consider many more criteria, and their results can be expressed in multidimensional space. The particular technique used here is called composite programming, but there are many different sorts of MCA technique that can be applied to different types of problems. (Modified from Bailey et al. 1995.)

The argument is also difficult at a biological level. Are all species critical? No conservationist would sanction species extinction, but we know that extinction of species does not necessarily damage human welfare or economic output. What, for example, have been the impacts of extinction of species like the Dodo, Passenger Pigeon or thylacine (the Tasmanian wolf)? With hindsight one could suggest that as their extinction has not caused a collapse of biological or social systems, these species were not critical. We may then extrapolate the argument and suggest that very few species are actually critical to maintaining ecological integrity and human welfare. This type of argument carries some power with certain sections of society, but it would seem to be counterproductive for conservationists to support this type of reasoning, as it will almost inevitably lead to the sanctioning of further extinctions. Whatever your viewpoint, it is true to say that there is tremendous uncertainty regarding the costs of species loss. For this reason, we may not be able to reasonably argue that it is "safe" to lose any species. However, with continuing economic development we are sure to lose species, and of course have already lost many. This is a problem, as making species inviolate, while appealing from ethical and precautionary grounds, may be impractical in resolving conservation disputes.

A less controversial approach may be to argue that certain nature reserves should be designated as critical, and thereby protected from damage. However, even this simple strategy has two major problems. First, very few reserves worldwide have sufficient legal protection to render them exempt from damage—that is—they are not legally defined as critical. Second, many reserves suffer from diffuse impacts. This means that actions outside the reserve are causing damage within the reserve. Examples of these include disruption of hydrological regimes by upstream land use change or coastal developments, and the deposition of atmospheric pollutants like nitrogen and sulfur. So if reserves were to be categorized as critical, then arguably these diffuse impacts should be stopped, and this immediately brings us into conflict with the financial interests of the damaging industries.

For these reasons the search for a theoretically and practically acceptable means of including environment and conservation into decision-making continues. While new methodological developments would be welcome, much depends on changing the attitudes of the public, private business, and government.

CASE STUDY 5.1

The Costs of Biodiversity Conservation
The Case of Uganda

Peter C. Howard and Gareth Edwards-Jones, University of Wales, Bangor

---

Much of the discussion so far has centered on identifying and estimating the values of biodiversity. Economics, however, is concerned with the balance of costs and benefits. This case study is concerned with considering the costs associated with conserving biodiversity. In particular, there is the question of whether or not the benefits of biodiversity conservation in many developing countries outweigh its costs. To explore these questions some original data collected by Howard (1995), and subsequently reworked with the author, will be presented.

The aim of the study was to quantify the costs and benefits associated with protected areas in Uganda to estimate their total economic value (TEV). This was done by valuing the direct and indirect use and option values of Uganda's protected areas. It also examined the costs involved in maintaining them, both in terms of direct management costs, and the opportunity costs of protecting the land and resources, rather than using them in some other way such as agriculture or livestock husbandry. The net benefit to society (NB) of Uganda's protected areas is then derived as the sum of all benefits less the sum of all costs. This can be summarized as:

$$NB_{PAs} = GB_{DU/M} + GB_{DU/NM} + GB_{IU} + GB_{OE} - C_M - C_{LO} - C_D$$

where:

NB_{PAs} = Net benefit to society of maintaining Uganda's protected areas

$GB_{DU/M}$ = Gross benefits derived from direct use of marketed products

$GB_{DU/NM}$ = Gross benefits derived from direct use of nonmarketed products

GB_{IU} = Gross benefits derived from indirect uses

GB_{OE} = Gross benefits derived from nonuse (option and existence value)

C_M = Cost of management operations

C_{LO} = Costs of protection, in terms
        of lost of opportunities for alternative
        development

C_D = Cost to local people in terms of
        damage to crops and livestock

Background on Uganda and Its Protected Areas

Uganda is located on the equator in east-central Africa, covering an area of about 236,000 km² on the central African plateau. The country supports exceptional biological diversity since it lies in a zone of overlap between biotas typical of the east African savannas and those of the west African rainforests. The concentration of this biological wealth in such a small area offers outstanding opportunities to achieve global biodiversity conservation objectives in a cost-effective manner.

Despite these opportunities for biodiversity conservation, several social issues have complicated their achievement. Uganda became politically independent in 1962. After a relatively peaceful few years, an economic crisis in Uganda led to a military coup in 1971 in which Idi Amin Dada came to power. Amin, who ran a brutal regime, declared himself president for life, but was forced from office in 1979, and Milton Obote was elected president in 1980. Opposition guerrillas were largely responsible for Obote's removal from the presidency in 1986. During the 1990s, the government promulgated nonparty presidential and legislative elections. While these are genuine elections, all candidates belong to the same party. The country has had some political stability in recent years after President Yoweri Musaweni was elected president in 1996 and re-elected in 2001.

Conservation efforts are challenged by the dense and growing human population. In 2003 the human population of Uganda was estimated to be 25.6 million, up from 4.8 million in 1950. Current models predict a population of 53.1 million by 2025. Uganda is already Africa's fourth most densely populated country, after Rwanda, Burundi (its near neighbors), and Nigeria, so the prospects of the predicted continued growth are truly frightening. Uganda is predominantly an agricultural country with 83% of the labor force engaged in subsistence farming, largely on very small farms of less than 2 ha in size. Agriculture accounts for about 60% of the gross domestic product (GDP), and more than 90% of exports, with coffee being the most im-

portant of these. There are few proven mineral reserves and little prospects of industrial development over the short and medium term (Hamilton 1994).

The bulk of Uganda's protected areas were established during the first decades of the twentieth century, at a time when the human population was about one-fifth its present size. Currently, 32,440 km², or 13.7% of the total area of Uganda is designated as a national park, or as a wildlife or forest reserve (Figure A). All three categories are "protected" inasmuch as they support predominantly unmodified natural systems, and are managed to ensure long-term protection and maintenance of biological diversity, largely to the exclusion of human settlement, cultivation, and domestic livestock.

Estimating the Benefits of Protected Areas

Data on tourism, the sale of forest products, and the revenue from hunting, game cropping, and other animal products were obtained from the accounts sections of the relevant agencies for the period of study (1989–94). Data on the amounts and types of goods collected from national parks by local people were obtained from undertaking a survey of local people in three different study areas.

Only two estimates were made of indirect ecosystem values: watershed protection and carbon sequestration. An estimate of

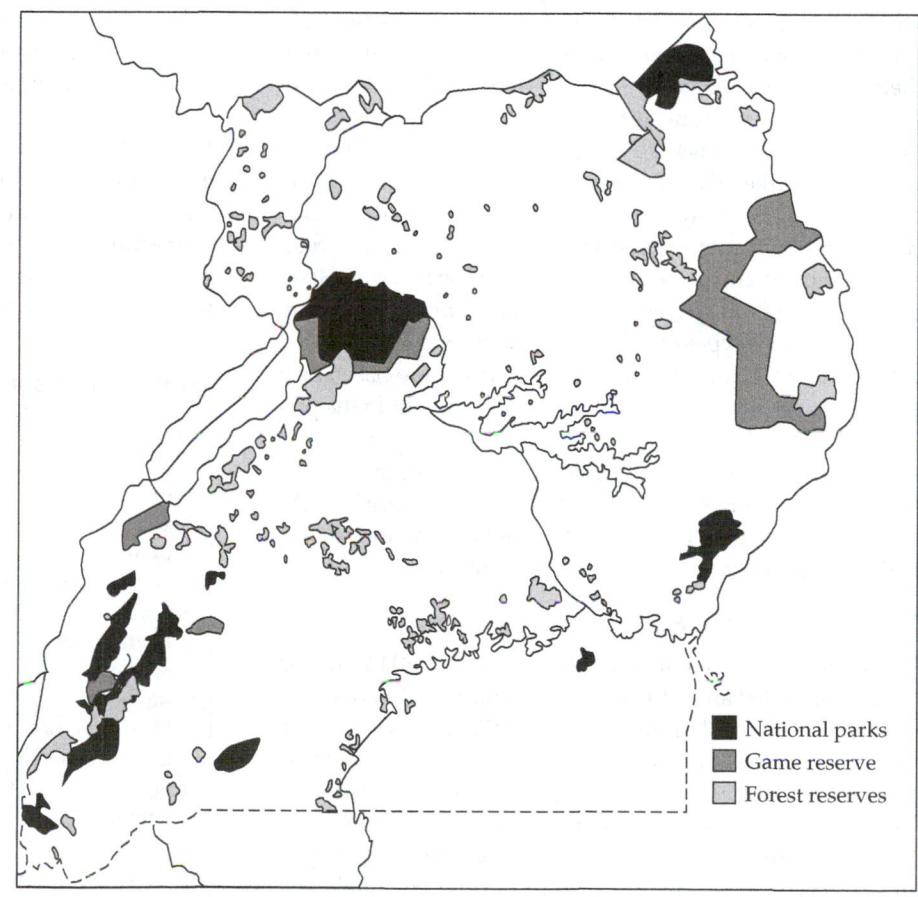

■	National parks
▨	Game reserve
▢	Forest reserves

Figure A   The protected area network in Uganda in the mid-1990s. (From Howard 1995.)

watershed protection in Uganda's protected catchments was represented by the rather arbitrary figures of 70% of the annual fish catch from Lakes Edward and George, and 20% of the annual catches from Lakes Kyoga and Albert, and the Albert Nile. Carbon sequestration was valued by considering the amount of carbon held in the different types of vegetation found in Uganda. Using aboveground biomass estimates produced by Uganda's National Biomass Study (Velle and Drichi 1992) and standard values of the carbon content of different vegetation types, it was possible to estimate that at present values, the carbon locked up by the protected areas was worth U.S.$245 million.

Option and existence values are undoubtedly the least-tangible benefits associated with Uganda's protected areas, despite their likely importance. Given the enormous challenges and controversial nature of estimating existence values, only option values of biodiversity were estimated, and these were estimated for the pharmaceutical value of natural vegetation and genetic material for coffee breeding.

Estimating the Costs of Protected Areas

The costs of managing Uganda's protected areas were estimated in terms of actual costs incurred in each year for a five-year review period (1989–94), as recorded by the authorities and donors concerned. There are also opportunity costs associated with protected areas, namely the difference between the financial return from their current use and the financial returns from their most profitable potential use. For example, if a piece of land in the protected area was producing no financial benefits now, but could, if converted to agriculture, produce a return of $100 per year, then the opportunity cost of keeping that land in the protected area is $100. Following this logic, the opportunity costs of the entire protected area system was estimated by comparing their current financial returns with those that could have been obtained from conversion to relevant agriculture uses. Information on the types of cultivation and livestock production that could occur, and their financial returns were based on a survey of agricultural systems in villages adjacent to the protected areas.

Finally, a village survey provided data on losses suffered by local people living close to protected areas in terms of the amount of crops and livestock lost to wildlife, and the amount of time spent on protecting crops and livestock.

The Cost–Benefit Ratio

Uganda earned gross annual revenues of U.S.$11.7 million from its protected areas, 91% of which came from international donors. A net annual profit of U.S.$8.6 million was generated, after U.S. $3.1 million of internal costs were paid by Uganda. Thus from a purely financial perspective the protected areas had a benefit:cost ratio of 3.7:1 (Table A). However, the results of the analysis change dramatically when the boundary of analysis is expanded to incorporate a wider economic perspective (Table B). Despite annual benefits totaling U.S.$67.5

TABLE A *Annual Financial Return to the Ugandan Government and Its Agencies from Protected Areas in Millions of U.S. Dollars**

Item	Annual flow (million U.S.$)
Benefits	
Total revenues	1.007
Donor contributions	10.677
Subtotal	11.684
Costs	
Government capital development expenditure	1.108
Government recurrent expenditure	1.169
Ugandan National park revenue reinvested	0.851
Subtotal	3.128
Net financial benefit to Uganda	8.556

*Based on 1993–94 data.

million, a net annual cost of U.S$189.2 million was incurred, largely because of the very high opportunity cost of land and the heavy losses suffered by local communities to their crops and livestock. The benefit:cost ratio then becomes 0.36 ($67.5 million:$189.2 million).

The results of this study suggest that it may be in the Ugandan Government's short-term financial interests to maintain its current protected areas, but from a wider economic perspective there are obviously serious questions over their viability. If the analysis were correct, then from an international biodiversity conservation perspective, this would be clear cause for concern. However, before getting too concerned, it is important to consider the weaknesses in the data set. There are undoubtedly some errors and omissions in the CBA estimate, and we may suspect that the opportunity costs of conservation may be over estimated, while the indirect value of ecosystems may be undervalued. Further it is important to remember that existence values have been omitted entirely, and should we as-

TABLE B *Economic Analysis of Uganda's Protected Areas in Millions of U.S. Dollars**

Item	Annual flow (million U.S.$)
Benefits	
Total revenues	1
Community-use	33
Watershed benefit to fisheries	13.8
Carbon sequestration	17.4
Biodiversity option value	2.3
Subtotal	67.5
Costs	
Financial cost to Uganda	3.1
Crop and stock losses	75.5
Opportunity cost of land	110.6
Subtotal	189.2
Benefit – costs	–121.7
Benefits/costs	0.36

*Based on 1989–94 survey data.

TABLE C *Mortality and Food Consumption of the 22 Most Developed Countries versus Uganda and Kenya*

Country	Measure					
	Mortality rate under 5 years of age*	Underweight children under 5 years of age (%)	Adult mortality rate under 40 years of age (%)	Life expectancy at birth (years)	Daily per capita supply of calories	Daily per capita supply (g) of protein
Most developed countries	35	<1	8	70.1	2858	74
Uganda	141	26	44	40.5	2249	53
Kenya	90	23	15	53.8	1980	52

Note: All data are for 1995.

Source: Human Nations Development Programme 1998.

*Child mortality rate under 5 years of age is given per 1000 live births.

sume that the mountain gorilla has a similar existence value to the grizzly bear, then the aggregated existence values of the population of North America and Europe would go a long way to redressing the difference in costs and benefits.

Although there may be argument over the exact figures obtained for Uganda, it is interesting to note that Norton-Griffiths and Southey (1995) reported similarly high costs associated with protected areas in Kenya, and in light of this, it is worthwhile considering some of the ethical issues associated with the results from this type of analysis.

For example, consider that malnutrition and related health problems are much more common in developing countries than developed ones (Table C). It could be argued that if more land were available for cultivation in developing countries, more food could be produced, and malnutrition and related illnesses would become less prevalent (although it should be noted that a recent survey showed the agricultural potential of most protected areas to be low [Gorenflo and Brandon 2005], and in fact that their unsuitability to agriculture may explain why these areas were still in a relatively pristine state).

Many of the benefits of biodiversity conservation flow to people in developed countries (e.g., existence value, carbon sequestration, and the majority of tourism revenue), while many of the costs fall on local people (e.g., loss of land for agriculture, and arguably the associated malnutrition and premature death accompanying food shortages). Would we in Europe and North America be willing to suffer the same balance of costs and benefits as those reported for Uganda and Kenya? To investigate the ethical boundaries of this issue it is interesting to personalize the debate to the following situation.

Gareth Edwards-Jones (GEJ) lives on the edge of Snowdonia, a mountainous National Park in North Wales, U.K. One of the local rarities is the Snowdon lily (*Lloydia serotina*), a small white flower found on the upper cliff faces of Mount Snowdon and other high peaks in the area. GEJ also has a six-year old son and three-year old daughter). Imagine the hypothetical situation where my local government currently spends $1 million on pediatric medicine and $50,000 on conservation of the Snowdon lily and the surrounding ecosystem. In the days before a local election, a candidate for an extreme environmental

group knocks on GEJ's house door and tells of his plans for public expenditure should he be elected. One part of his budget proposal is to reduce spending on pediatric medicine by 80% and shift the $800,000 saved from this toward conservation of the Snowdon lily. The candidate then asks for GEJ's vote for him in the forthcoming election.

GEJ: "I would not vote for such a plan. Such a severe reduction in the pediatric health budget would significantly decrease the health provision available to my children. As a parent I would not trade any risk to the health of my children for any gains in biodiversity conservation. While such a statement may seem odd to people who do not have children, I do not believe I am unusual in taking this stance, and many parents would agree with me."

But if the analyses of the costs of conservation in developing countries are correct, then this is exactly what we are asking many other people to do. Through supporting the maintenance of protected areas for biodiversity we may be preventing local people from cultivating land and growing food that could keep them and their families alive (Figure B). Is this a defensible ethical stance?

Figure B Should this Ugandan national park be used for food production? (Photograph by G. Edwards-Jones.)

CASE STUDY 5.2

Costs and Benefits of Restoration of Post-Industrial Landscapes

Gareth Edwards-Jones

The United Kingdom (U.K.) is a densely populated country where land is scarce and the amount of land affected by quarrying and mineral workings continues to expand. Restoration of post-industrial land has received great interest over the past decade, as the law now requires developers to restore land previously exploited by them. While restoration has a real cost, usually at least $100,000/ha, if left unrestored quarry wastes may also pose an economic cost. They pose a hazard to safety and may also reduce the aesthetic beauty of the landscape (i.e., diminish nonconsumptive use values). Finally, because people find quarries and their wastes ugly, unrestored they may also have a real economic impact as they may reduce the desirability of the area for new investment and development.

While quarries are clearly degraded environments, some may have developed some biodiversity value as nest sites for certain birds and homes for unique plant communities, such as scree specialists. Restoring the quarries may remove such biodiversity interest where it occurs, but it should also reduce the safety hazard and turn a negative element of the landscape into something more positive. Furthermore well-designed restorations may also provide opportunities for recreation and/or biodiversity gains. Because of these benefits, it is generally felt to be a good thing to restore old quarries and mineral wastes. However, it remains unclear what the endpoint or goal of restoration should be (see Chapter 15), and who should decide on the goals of a restoration project. Quarries are usually privately owned, but the benefits of their restoration will accrue largely to the local community. There is potential for conflict in aims between the desires of the quarry owners, who may bear the brunt of any restoration costs, and local people who will suffer the costs of an unrestored quarry, and similarly benefit from any restoration.

The Quarry

This case study, based on work by Sharpe (2000), is concerned with Penrhyn quarry in North Wales, one of the largest slate quarries in Europe. The quarry currently covers 1400 hectares, and over the last 100 years has transformed the landscape, as it has literally destroyed one side of a mountain, while simultaneously creating large areas of spoil (waste slate and other materials) which together cover an area of 2.9 square kilometers of the quarry. Very little vegetation grows on these spoil heaps as they lack any soil or water-holding capacity, hence significant restoration activity is necessary to enable vegetation establishment. Despite the environmental impact of the quarry it is important to say it has been the source of significant economic

wealth over the last century and provides many jobs in the local area.

The Valuation Study

A contingent valuation methodology (CVM) study of the general public was undertaken to consider their willingness to pay (WTP) for four restoration options on the slate quarry. This was part of a larger project that sought to explore the following questions:

- Does the public express a preference for restored sites over derelict wasteland?
- What are the nonmarket benefits of different restoration options?
- Would feasible restoration options be economically viable?

Four general approaches to restoration for the quarry were developed, which were characterized as biodiversity, landscape, heritage and recreation, and unrestored slate quarry. These options included a varying range of access, wildlife, aesthetic quality, heritage, and recreation amenities.

- The biodiversity option involves restoring the quarry to native woodland composed of silver birch (*Betula pendula*), sessile oak (*Quercus petraea*), and rowan (*Sorbus aucuparia*). Such a woodland would take a long time to establish, but would create a diverse plant and wildlife habitat; however this option would offer limited recreational opportunities.

- The landscape option establishes woodland in key areas of the quarry to reduce visual impact of waste heaps. It includes creation of a mixed broadleaf and conifer woodland. Although designed to be aesthetically pleasing, it would not support such a diverse range of wildlife or look as natural compared to the conservation option. It could be created relatively quickly and there would be some opportunities for recreational activities.

- The heritage and recreation option has little biodiversity value, but concentrates on conserving historic features, particularly buildings, as well as quarry faces. It includes a wide range of potential recreational activities from specialist tourism to low-key recreation, like walking. Additional attractions and facilities include a visitor center, a picnic site, toilets, and a car park.

- The unrestored slate quarry option involves leaving the slate quarry as it currently stands, allowing natural regeneration. Sheep grazing will continue and heaps of waste materials will remain as a reminder of industrial heritage. Fewer recreational activities would be possible due to limited access imposed for safety reasons.

The CVM questionnaire consisted of three parts. Part 1 requested information regarding respondents' residence, gender, occupation, and recreational and environmental interests. Part 2 introduced the four management options and sought respondents' willingness to pay, while Part 3 sought socio-economic data on respondents' income and level of education. In Part 2, photomontages of the differing possible future management options were displayed to respondents (Figure A). For each scenario, respondents were asked if they would visit the site if it were restored for that type of management, and were asked about the likely frequency and nature of their future visits. Respondents were then asked if they would like to make a one-time contribution to a trust fund that would be responsible for managing the restoration project. Respondents were told that contributions would not be used to restore a scenario they did not prefer and no further payments would be required.

Over 200 people were surveyed and their willingness to pay for the four options fell in the order biodiversity ($54.83), landscape ($53.10), heritage and recreation ($49.44), and unrestored quarry ($21.67). To estimate aggregate benefits, these mean individual bids were multiplied by the population of the region. (Table A).

Cost–Benefit Analysis

The aggregate benefits were combined with appropriate costs for each management option in a cost–benefit analysis. The costs included labor, fencing to exclude grazing, tree planting, and subsequent management. Benefits were restricted to the direct benefits provided by the site to the public, and ignored multipliers such as employment salaries, increased property values in the vicinity of the site, direct income from visitors, and resultant impacts on the local town's economy. This results in

(I)

(II)

(III)

(IV)

Figure A The four restoration options for Penrhyn slate quarry in North Wales, U.K. (I) Biodiversity conservation, (II) landscape, (III) heritage and recreation, and (IV) unrestored. (Images by Sharpe 2000.)

TABLE A *Total Aggregate Benefits for Each Management Option*

Management option	Mean WTP (U.S.$)	Total aggregate benefits (U.S.$)
Biodiversity	54.83	7,086,261
Landscape	53.10	6,862,712
Heritage and recreation	49.44	6,289,173
Unrestored	21.67	2,800,371

Source: Sharpe 2000.

an underestimation of market benefits. Despite these reductions in benefits, the biodiversity, landscape, and unrestored quarry management options were all shown to have positive NPV's and are therefore economically viable. In the example shown in Table B, landscape was the most cost-effective scheme, while heritage and recreation was not viable, largely due to the capital costs of building an expensive heritage center. However, the ranking of the options varied according to how the benefits

TABLE B *Comparison of Estimated Total Costs, Benefits, and NPV across Each Quarry Restoration Scenario*

Management option	Total estimated costs	Total estimated benefits	Difference (B – C)	NPV
Biodiversity	2,520,417	7,086,261	4,565,844	2,062,558
Landscape	1,795,159	6,862,712	5,067,553	2,914,596
Heritage and recreation	6,000,877	6,289,173	388,295	–611,016
Unrestored	21,576	2,800,371	2,778,795	2,623,645

Note: Discounted NPV at 6% for the individual duration of the project. Assumed total aggregate benefits were distributed evenly through all years and benefits were invested each year to maintain their real value. All values are expressed in 2004 U.S.$.
Source: Sharpe 2000.

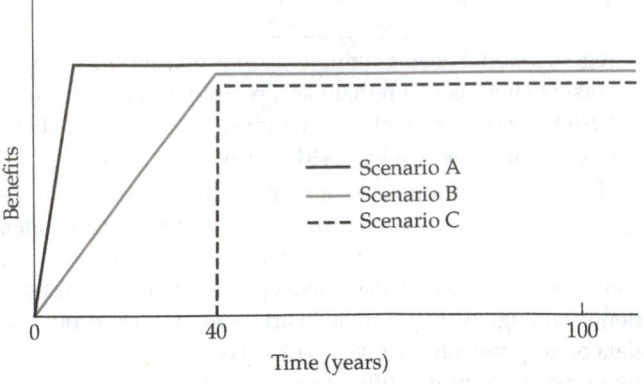

Figure B Three different potential benefit curves for a restoration project involving the growth of trees. The trees are planted at time 0 and will reach full height at year 40. The benefits relate to the biodiversity, landscape, and recreational restoration options. Scenario A assumes that all the benefits arise very soon after planting and are constant for the next 99 years. Scenario B assumes that the benefits increase linearly as the trees grow and are constant from year 40 to year 100. Scenario C assumes there are no benefits until the trees are fully mature, and then they stay constant until year 100. Which benefit curve should you choose when developing a cost/benefit analysis?

were assumed to flow over the lifetime of the project. In this example the benefits were assumed to flow equally over the timescale to the project (i.e., equal benefits in each year). However, it could be argued that the value derived from the biodiversity and landscape options does not reach a maximum until the trees are mature. In other words benefits would not be constant over time. If the assumption underlying the CBA is changed so that no benefits accrue until year 30, the assumed date of tree maturity, then the ranking of preferred projects is altered and the biodiversity option becomes the preferred option (Figure B). Thus, in this case, the result of the CBA rests on a philosophical debate of how the benefits derived from landscape and biodiversity vary over time—an interesting and subtle debate.

Summary

1. Conservation requires more than scientific information and analysis. In practice, successful conservation requires an understanding of people and the way they think and behave, while at the policy level it is important that decision-makers make the right strategic decision for the right reason. The discipline of economics can help with both of these issues. At first sight this may seem strange, but fundamentally economics is concerned with the distribution of scarce resources, and many of the methods developed in economics can be applied to conservation problems. For many people, money is a scarce resource, and so their behavior is affected by their de-

sire to get more. Economics can help understand what drives their behavior and offers some insight into how this behavior can be modified. Biodiversity also may be a scarce resource, and economics can offer insights into how governments and other decision-makers should manage biodiversity to achieve maximum benefit for all.

2. This chapter describes some of the basic issues central to economics and shows how they can be applied to conservation problems. Several decision-making frameworks are discussed, including environmental impact assessment, risk assessment, multi-criteria decision-making, and the precautionary principle; but most of the chapter follows the steps of the most com-

mon decision-making framework in economics—cost–benefit analysis (CBA). CBA requires all costs and benefits of a project to be expressed in monetary terms, and some of the techniques for obtaining monetary values for environmental goods are discussed. However, not all techniques can be used in all situations. Biodiversity has different sorts of value to humankind, including direct use value (the value we get from using part of the environment), indirect use value (the value provided by ecosystems without us being really aware of it, e.g., the ozone layer) and existence value (the value we all have from knowing that a certain species or environmental good exists, regardless of whether or not we will ever use it). Each of these values can be estimated in monetary terms so long as the appropriate technique is used.

3. Having identified the individual monetary costs and benefits of a project, including the environmental costs and benefits, it is important to consider how these may change over time. Understanding this requires an understanding of discounting. This is a fundamental, but sometimes controversial, principle in economics. Discounting tries to capture the future value of something in today's terms. For example, how can I work out the value of a $100 that I may receive in 10 years time, in today's terms? Or using a conservation example, how can I calculate the value of a forest every year over each of the next 100 years and express this value in today's terms? Without careful consideration about the value of the environment at some time in the future, it is very easy to reach the wrong decision about the conservation decisions we make today.

4. A final complication in undertaking cost–benefit analyses relates to the distribution of costs and benefits within society. In other words, who benefits and who loses from our actions? For example, is it fair for local people in a developing country to suffer the costs of establishing a protected area near their village, while most of the financial benefits of doing this flow to the government and tourism companies? Is it fair for some countries to keep on polluting the atmosphere with greenhouse gases, while many other countries seek to reduce their emissions? Economics can offer some insight onto how to deal with these so-called distributional effects, but other disciplines such as ethics and political science also offer valuable insights. Considering these sorts of issues shows an important overlap between the application of economics to conservation, and a consideration of the ethics of any decision.

Please refer to the website www.sinauer.com/groom for Suggested Readings, Web links, additional questions, and supplementary resources.

Questions for Discussion

1. Imagine you had it in your power to authorize a relatively cheap control mechanism that would drive the malaria-carrying mosquitoes of the genus *Anopheles* extinct? Given the massive costs that malaria brings around the world in terms of human suffering and financial losses in productivity, but would you authorize the extinction?

2. Do a small questionnaire in class and find out your existence values for a range of species such as Bengal tiger, polar bear, alligators, honeybee, a venomous snake, a ground beetle and the HIV virus. If they differ, why, and what might this mean for conservation?

3. What is your personal time preference? Would you rather have a $100 now or in 10 year's time? How may your time preference change over your lifetime as you go through different life stages or have different degrees of financial security and responsibility (e.g., being a student, raising a child, getting old, having a job or not, owning a home or not). How do you think the time preference of a poor person in a developing country may differ from that of a person in a developed country?

4. Do you think that politicians in your city or region give biodiversity enough weight in their decision-making? Do you think they would make better or worse decisions if they were forced by law to obtain monetary valuations of relevant environmental goods (including biodiversity) before making any decisions?

5. How can the rich and developed nations of the world encourage poorer, less-developed nations to conserve their biodiversity? Indeed should the rich nations try to do this at all, or should we leave each country to make up their own minds about the importance of conservation?

UNIT II

Focus on Primary Threats to Biodiversity

6

Habitat Degradation and Loss

Martha J. Groom and Carly H. Vynne

Don't it always seem to go that you don't know what you've got 'til it's gone?
They paved paradise and put up a parking lot.

Joni Mitchell, 1970

Habitat degradation and loss is the most serious threat to biodiversity largely because it is the most sweeping in its effects. Habitat conversion or loss corresponds to a complete change in community and ecosystem states, and thus causes losses at all levels of biodiversity. Population and species extinction caused by habitat degradation not only represents losses at those levels of biodiversity, but also loss at the genetic level. Habitat degradation that eliminates or reduces some populations simplifies and disrupts communities, and in extreme cases reduces or eliminates key ecosystem processes.

The majority of Earth's land surface (83%) has been transformed by human activities to a greater or lesser extent (Sanderson et al. 2002; see Figure 1.4). About 60% of Earth's ecosystems are considered degraded or unsustainably used (Millennium Ecosystem Assessment 2005a,b). Temperate grasslands, savannas, and shrublands have lost the greatest proportion of their area (more than 80% of their original extent), followed by Mediterranean habitats (72% degraded) (Figure 6.1). Losses in aquatic systems are also substantial. More than 20% of coral reefs have been destroyed and an additional 20% degraded, and 35% of assessed mangrove ecosystems have been destroyed in the past two decades (Millennium Ecosystem Assessment 2005b). Three to six times more water is now stored behind dams and in reservoirs than flows in all the world's rivers (Millennium Ecosystem Assessment 2005b). Clearly, habitat transformations have been considerable, and future trends suggest continuing, or even accelerating rates of change in most biomes (see Figure 6.1).

Habitat degradation is the primary cause of extinction and endangerment globally (see Figure 3.8) and in most nations (e.g., the U.S.; see Table 3.4). Species are two to four times more likely to be threatened by habitat degradation than by any other type of threat; this should not be surprising given the broad array of impacts caused by habitat transformations, and of factors that comprise habitat degradation. Further, despite notable successes (see Chapter 14), habitat degradation even threatens our

Figure 6.1 Habitat transformation has been more extensive in some biomes prior to 1950, and losses between 1950 and 1990 have particularly affected tropical areas. Trends in habitat conversion suggest that 50% or more of the original extent of most biomes will have been greatly reduced in value for biodiversity by 2050. (Modified from Millennium Ecosystem Assessment 2005b.)

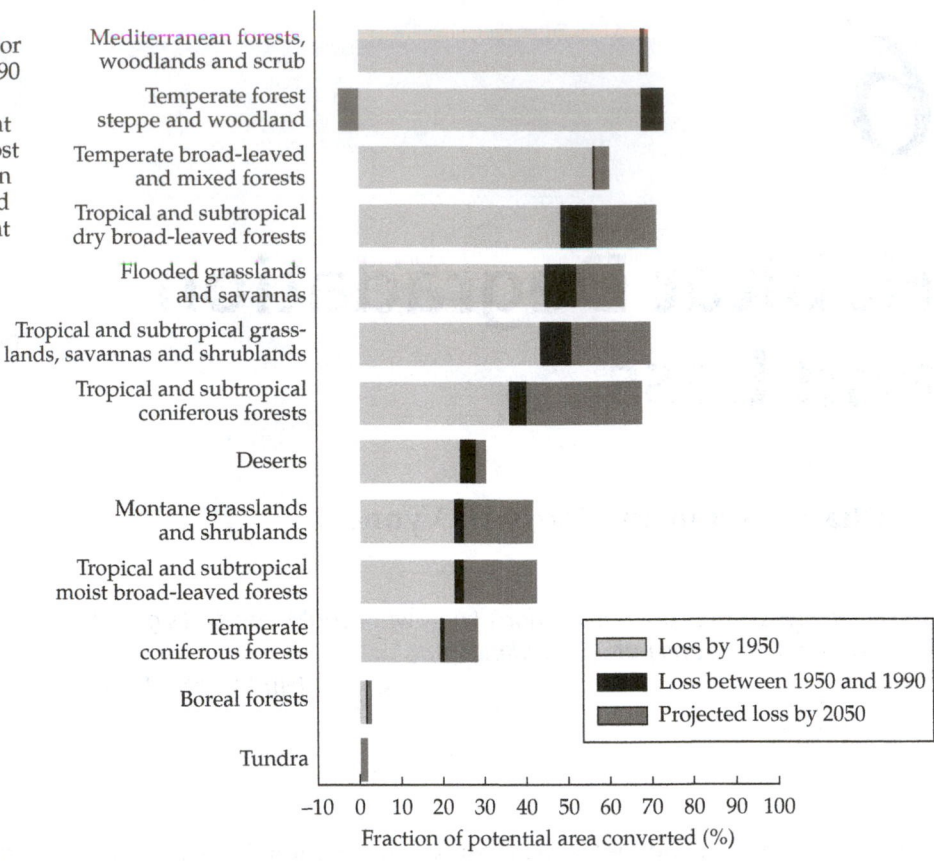

protected areas through increased isolation (DeFries et al. 2005) and loss of areas within reserves (Kinnaird et al. 2003; Mayaux et al. 2005).

A detailed knowledge of the range of impacts caused by different human activities, and their distribution on local through global scales can help us address the threat of habitat degradation. With this information, we can identify priorities for action, develop strategies to counteract the problems caused by habitat degradation, and slow or reverse its spread.

What Constitutes Habitat Degradation and When Is Habitat "Lost"?

Habitat degradation and loss are caused by a wide variety of human activities, including agriculture, mining, forestry, fisheries, aquaculture, groundwater extraction, fires, infrastructure development, dams, urbanization, industry, pollution, and changes in community and ecosystem structure wrought by changes in the abundance of native and invasive species (Table 6.1). Pollution is not always considered a cause of habitat degradation, but in most instances, the presence of pollutants in the form of light, noise, or toxic chemicals that contami-

nate water, land, and air degrades the quality of the environment for many, and at the extreme, for all, species.

Impacts of habitat change range from light and temporary (e.g., small fires), to devastating and permanent (e.g., urbanization). **Habitat degradation** generally refers to impacts that affect many but not all species, and that may be temporary (although many impacts are persistent at low-to medium levels of intensity). In contrast, **habitat loss** usually refers to impacts so severe that all, or nearly all, species are adversely affected, or to cases where the time span needed for recovery is extremely long. Here, we will use *habitat loss* to refer to extreme changes in habitats that make them unable to support more than a fraction of their original functions and species, and the broader term *habitat degradation* to refer to all other cases. **Habitat transformation** and **habitat conversion** are two additional terms often used interchangeably with degradation and loss, and we will use them when referring to processes of change (e.g., conversion from forest to cropland). Both can constitute either habitat degradation or loss depending on their severity, although habitat conversion especially is often meant to imply habitat loss.

Habitat degradation need not result in total loss of habitat to all species, but instead may translate into a re-

TABLE 6.1 *Forms of Habitat Degradation*

Agriculture	**Infrastructure development**
Crops	**Industry**
Shifting agriculture	**Human settlement**
Small-holder farming	Small-scale
Agro-industry farming	Villages to towns
Wood plantations	Suburban-urban sprawl
Small-scale	Urban areas
Large-scale	**Land, air, water transportation**
Nontimber plantations	**Dams**
Small-scale	**Tourism/recreation**
Large-scale	**Telecommunications and power lines**
Livestock	
Nomadic	**Pollution**
Small-holder	**Atmospheric pollution**
Agro-industry	Global warming/oceanic warming
Abandonment post-agricultural use	Acid precipitation
Marine aquaculture	Ozone hole effects
Freshwater aquaculture	Smog
	Land pollution
Extraction	Agricultural chemicals
Mining	Household chemicals
Small-scale	Commercial/industrial chemicals
Gold-mining	Light pollution
Strip-mining	**Water pollution**
Fisheries	Agricultural chemicals
Subsistence	Household chemicals
Artisinal/small-scale	Commercial/industrial chemicals
Large-scale/industrial	Thermal pollution
Wood	Oil slicks
Small-scale subsistence	Sediment
Selective logging	Sewage (nitrogen and phosphorus)
Clear-cutting	Agricultural fertilizers
Nonwoody vegetation collection	Solid waste
Coral removal	Noise pollution
Groundwater extraction	
	Biotic changes
Fires	Invasive nonnative species
War/civil unrest	Change in native species dynamics

Source: Modified from IUCN Red List threat categories, IUCN 2004.

duction in the capacity of an ecosystem to support some subset of species (Figure 6.2). To be considered degraded, changes in the environment must significantly reduce or eliminate some populations from the site. For example, food supply, shelter, and appropriate abiotic conditions may be less available. Pollution may reduce the survival or reproductive ability of individuals, though not necessarily making it impossible for all individuals to survive. Biotic interactions among species also may be degraded, but not to the point where further degradation via cascade effects is likely (see Chapter 3). Less severe forms of habitat degradation generally have less severe impacts

Figure 6.2 Effect of increasing land use intensity on the fraction of species in vertebrate taxa and plants that remain compared with the inferred population sizes from the pre-colonial period in southern Africa. Protected areas are expected to represent the most natural condition, and average population sizes under each type of use are scaled in relation to this category, which is designated as 100%. Light-use areas are those where use is sustainable (e.g., light grazing). Degraded areas are those where use exceeds productive capacity of the land. Cultivated areas are both croplands and planted pastures. Plantations are nonnative monocultures. Urban are built-up areas and high impact mining regions. Taxa respond differently to this gradient of use, but all decline as use becomes more intense. Light-use and urban areas have increased numbers of ponds relative to natural conditions, and therefore support larger amphibian populations. (Modified from Scholes and Biggs 2004.)

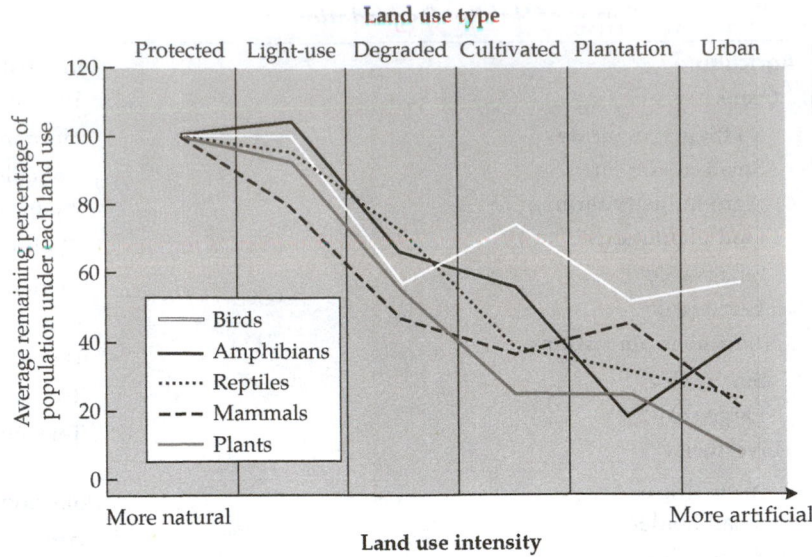

Patterns of Habitat Transformation on Land and In the Sea

For most of human history, we have strongly modified the places we inhabit. In addition to the indirect effects of human-caused extinctions on community composition and ecosystem characters described in Chapter 3, prehistoric humans directly modified ecosystems, and in turn influenced the nature of vegetative and associated animal communities. Transformation of forest to cropland can be seen as early as 4000–7000 years before present (B.P.) in many sites in Asia, with evidence of permanent clearings showing up by 1000–2000 years B.P., and permanent transformation of grasslands to croplands can be seen even earlier in the Middle East and Africa.

Although humans have transformed their environments for thousands of years, the pace of change has accelerated sharply in the last century, and particularly since 1960, as the world population more than doubled since then. In the U.S., only 42% of native vegetation remains, with less than 25% remaining in the midwest (Stein et al. 2000) (Figure 6.3). Certain ecosystems have been virtually eliminated, including native grasslands and vernal pools of California, floodplain wetlands and

that can, in time, be reversed to some degree (e.g., native ecosystems can recover following forestry and agriculture, depending on the intensity and duration of those land use activities). However, as the nature of degradation of habitats grows in spatial extent and severity, the ability for recovery is diminished, and we begin to shift from habitat degradation to habitat loss.

streams in Mississippi, seagrass meadows in many coastal areas, and old-growth forests in the lower 48 states (Noss and Peters 1995; see Figure 6.3). Florida, California, and Hawaii have the greatest number of threatened ecosystems (Noss and Peters 1995). Worldwide, habitat degradation has been particularly extensive in Europe, South and East Asia, New Zealand, and coastal regions adjacent to large cities.

Evidence of human influence is found even in relatively remote areas. The Amazon Basin once held large human populations who practiced intensive agriculture on its floodplains, and perhaps to a lesser extent in the uplands (Denevan 1992; Cleary 2001). When Europeans colonized South America they introduced diseases that decimated native Amazonian populations, leaving the region sparsely populated until recent times, during which cultivated areas regrew into tropical forests. Even where vegetation was not radically changed, prior human use influenced species composition and more subtle components of ecosystem character (e.g., cultivated "garden" areas of forest that yielded greater quantities of key plant species, or attracted game to enhance hunting success; Denevan et al. 1984; Rival 1998). Indeed, nearly all habitats on Earth have experienced substantive influence by humans (Redford and Feinsinger 2001).

Habitat degradation and loss is traced using vegetation analysis via sediment cores coupled with climate data to predict prior vegetation states, and through historical accounts, maps, photos and, most recently, satellite imagery for more direct data on habitat change. Case Study 6.1 by David Foster describes the many contributions historical ecology can make to conservation. Satellite images from across the globe have made it possible

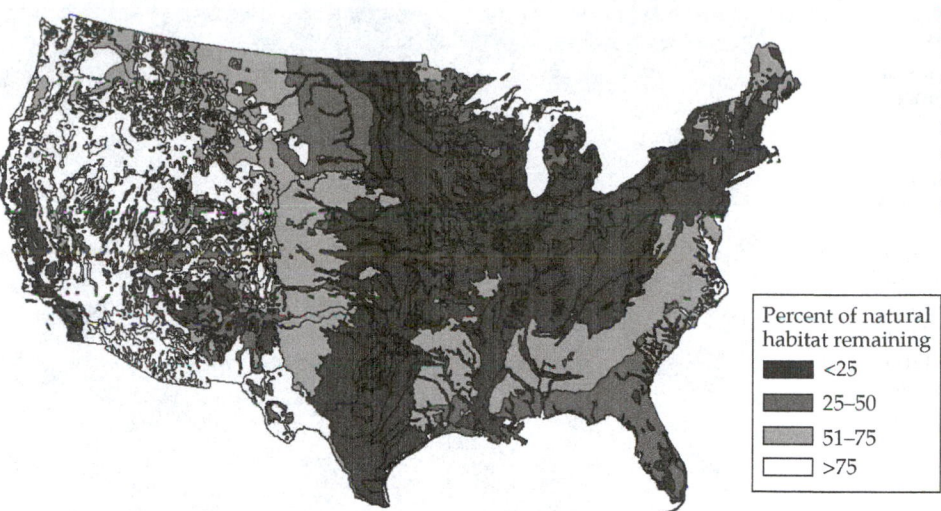

Figure 6.3 Natural vegetation has been degraded extensively in the U.S. A comparison of land cover shown in 1992 satellite images with potential natural vegetation based on regional climate and soils indicates substantive change in 58% of the natural vegetation. (Modified from Stein et al. 2000.)

Percent of natural habitat remaining

- ■ <25
- ■ 25–50
- ■ 51–75
- □ >75

to track changes in detail over the past 30 years, although the cost and accuracy of estimates based on these methods varies (e.g., UN Environment Programme 2003, 2004). Nonetheless, particularly for forest systems, satellite images are a powerful tool that provide an efficient means to gather data on habitat change.

Forest systems and deforestation patterns

Over the past three centuries, about half of the world's forest cover has been removed to make way for croplands, pastures, and settlements. In 25 countries, no forest remains, and in an additional 29, more than 90% of forests have been lost (Millennium Ecosystem Assessment 2005b). Europe has the least forest cover of any continent, most of which was lost to agriculture before 1700. Eastern North America's forests were almost completely removed by 1850, and in recent decades have begun regenerating, as have those of Russia.

Worldwide, between 1990 and 2000, the proportion of land area covered by forest decreased from 30.4% to 29.7% (from 38.79 million km² to 37.85 million km²; UN Environment Programme 2002). This translates to an annual loss in forest area equivalent in size to a country such as Egypt or Colombia. Forest losses during the 1990s were greatest in Africa, which lost 8% of its forested area during this single decade. More than 60% of annual forest losses worldwide have occurred in the tropics, with a further 2.3 million ha of tropical forest sufficiently degraded to be detectable via satellite imagery (Achard et al. 2002). A typical example is shown in paired satellite images of the Iguazú and Paraná river region of South America, where deforestation occurred over 30 years, particularly in Paraguay following creation of a dam for irrigation and new access roads into the forest (Figure 6.4). Between 1990 and 1997, tropical

deforestation rates were greatest in central Sumatra, Madagascar, and in Acre, Brazil (Achard et al. 2002).

Forests provide many ecosystem services including regulating water supplies by mediating flow rates, controlling erosion, and affecting climate through gas exchange. Over two-thirds of the world's people depend on water flows through forests for their water supply. In some cases, cities and states have acted to protect forested watersheds to maintain their water supplies (e.g., New York City invested in habitat protections in the Catskills; Seattle in the Cedar River watershed). Retaining forest cover may be equally important to human settlements to prevent erosional damages. Extreme erosion and flooding caused by deforestation may have contributed to the collapse of earlier cultures by decreasing agricultural productivity (e.g., the large settlement at Cahokia in the Mississippi River floodplain east of what is now St. Louis; Woods 2004).

Heavy deforestation in a watershed increases runoff and can cause flash flooding of cities and other settlements. The Panama Canal watershed suffered high rates of deforestation during the 1950s–1970s, and although rates have slowed, continued clearing worries government officials that there may be insufficient water coming into the canal zone to move ships through the locks. To prevent this, Panamanian officials are now increasing their vigilance over forest clearing, and promoting reforestation within the watershed.

Deforestation can also cause changes in local, regional, and global climate (see Chapter 10). For example, in the mid 1980s, ecosystem scientists calculated that nearly 50% of the rain that falls over the Amazon Basin is returned to the atmosphere via evapotranspiration by forests (Salati and Vose 1984). Deforestation results in decreased evapotransporation, which, if occurring on a

Figure 6.4 Land cover change in the past 30 years has been dramatic in the Iguazú, the Paraná River region where the boundaries of Paraguay, Brazil, and Argentina meet. Since the first image was captured via satellite in 1973 (A), the Itaipú Dam, the world's largest hydroelectric project, was built, inundating a huge area (B). Extensive agricultural development was promoted in Paraguay, beginning just before the 1970s (note road leading from the Ciudad del Este in (A) and the "herring-bone" pattern of initial settlement and forest clearing), and has resulted in nearly total conversion of forest to soybeans fields. The large reserves in Argentina and Brazil surrounding the Iguazú Falls protect the last remnants of these forests. (Satellite images courtesy of NASA. From United Nations Environment Programme 2004.)

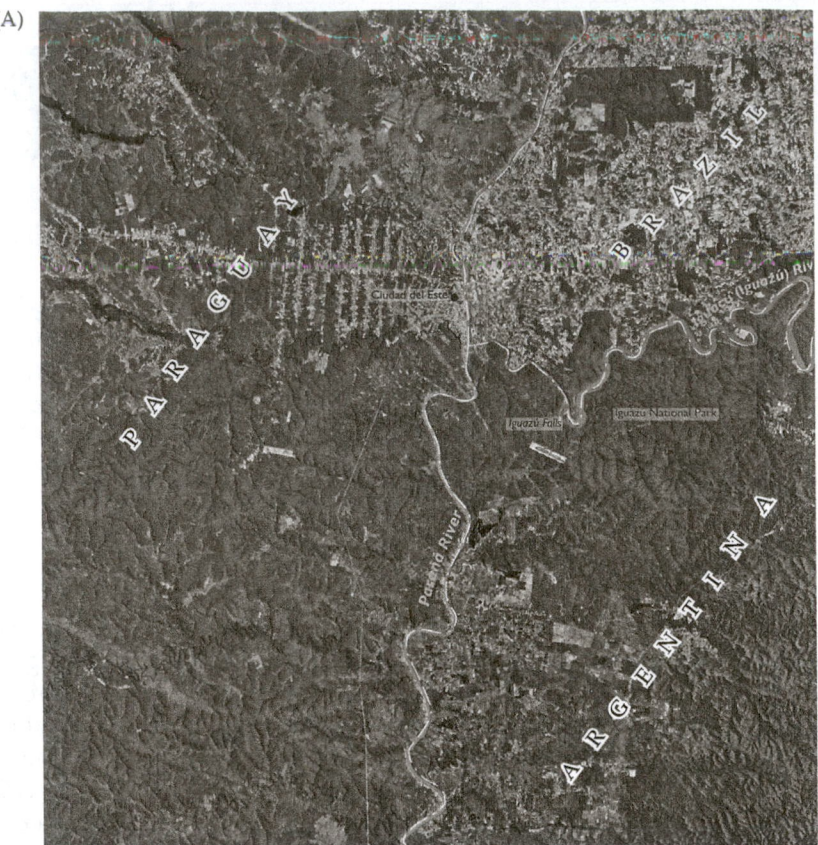

large enough scale can cause regional drying and may change temperatures (Betts 2004). Widespread clearing in temperate forest in the last few centuries appears to have contributed to a cooling of global climate, and reforestation in eastern North America and Russia may thus contribute to warming (Betts 2004). Deforestation in the tropics is expected to increase global warming, as well as regional warming (Houghton et al. 2000), and also to create conditions that favor the spread of fires as the regional climates become drier (Nepstad et al. 1999), which is likely to further exacerbate climate warming.

The Amazon Basin is the world's largest expanse of tropical forest, and despite efforts to reduce deforestation, rates of forest clearing remain high. As forests are generally burned after they are cut, and fires set to control invasive species in pasture and croplands within the Amazon, the Brazilian Amazon has shifted from being a carbon sink to a major source of CO_2 emissions, with over 200 tons of carbon emitted to the atmosphere each year (Houghton et al. 2000).

Although satellite imagery provides an efficient means to track gross-scale change, particularly in tracking recent deforestation, it underestimates the effects of forest degradation overall (Nepstad et al. 1999). For example, in Brazil's Amazon, the extent of intact forest is reduced by logging damage, fragmentation effects (see Chapter 7), and increases in fire frequency and extent due to escape of intentionally set fires. Fires are exacerbated by the

drying effects of deforestation and increased frequencies of El Niño events caused by global climate change (Nepstad et al. 1999, 2001). Fires can cause a variety of damage, including agricultural losses and respiratory illness, and also contribute to CO_2 emissions, which during El Niño years may cost up to 9% of the region's GDP (de Mendonça et al. 2004).

Loss of and damage to grassland, savanna, and shrubland habitats

Grasslands, savannas, and shrublands cover 52.5 million km^2 or 40.5% of Earth's surface, and are found predominately in dry subtropical and dry-humid temperate zones (White et al. 2000). These ecosystems are naturally dominated by herbaceous and shrub vegetation, and are maintained by drought, fire, freezing, or grazing by wild ungulates and thus do not grow into forests or dense woodlands. As such, they often provide ideal conditions for agriculture or livestock, and thus have been transformed to a great extent worldwide (see Figure 6.1), although their extent in some areas has been increased by conversion of forest to pasturelands.

Temperate grasslands have been the most heavily converted to agriculture, and secondarily to urban environments. The prairie ecosystems of North America provide one of the best examples: nearly 97% of tallgrass prairies, and over 60% of mixedgrass and shortgrass prairies have been converted since the mid 1800s (White et al. 2000) (Figure 6.5). Flooded and tropical-subtropical grasslands increasingly have been replaced by agricultural lands in the last few decades, with over 20% converted to other uses by the mid 1990s (White et al. 2000). Both urbanization and permanent cultivation represent transformations that result in net habitat loss, while less-intensive agriculture and agricultural mosaics provide areas where some other species and attributes of original ecosystems may persist.

In general, livestock have more often been raised within grassland, savanna and shrubland ecosystems than in converted forest, in large part because these systems already support large populations of wild ungulates and are replete with forage for domestic livestock. However, although domestic livestock are often added to these ecosystems with little other modification, they are generally raised to reach high biomass, and extensive fencing leads to sedentary habits that increase soil erosion, change vegetation, and in extreme cases, eliminate most vegetative cover (Evans 1998; Frank et al. 1998).

Desertification in the Middle East and Asia and northern and eastern Africa has occurred following a long history of occupation, and more recently as a result of poor management of land. Through a combination of climate change and human action, dry forests were replaced by grasslands, and fragile woodlands and marshlands were transformed into scrublands and deserts (Lamb 1977; Grove and Rackham 2001; Nyssen et al. 2004). Initially, throughout these regions, small human settlements began to grow cereals in fertile soils. Gradually, people felled neighboring forests for additional cropland, wood, and fuel, and larger civilizations arose as agricultural expansion began in these cleared lands. Over thousands of years, moist conditions prevailing during the rise of the Roman Empire reversed, and coupled with increasing water use for irrigation and direct support of large, human civilizations changed the character of the land from a mosaic of woodlands and grasslands to much less productive shrub or desert landscapes (Lamb 1977; Lamb et al. 1991). As local conditions became drier, forests and woodlands could not recover, facilitating a permanent change to ecosystems adapted to arid conditions. Similar trends have occurred and accelerated greatly in modern times via the introduction of intensive

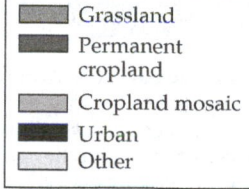

Figure 6.5 The central tallgrass prairie ecosystems of North America have been almost entirely replaced by permanent cropland, urban areas, and cropland mosaics. (Modified from White et al. 2000.)

grazing in semiarid grasslands (Millenium Ecosystem Assessment 2005c).

Degradation of freshwater systems

With agricultural and urban expansion, freshwater systems are degraded through water diversions, dams, and extensive wetland losses. Regulation of river flows and impoundments created for agricultural and city use has become the norm. Presently, only 2% of U.S. rivers run unimpeded, and less than one-third of rivers worldwide run without regulation (Abramovitz 1996). Water development contributes to endangerment for 91% and 99% of U.S. federally-listed fishes and mussels, respectively (Stein et al. 2000). Agriculture and land clearing have increased sedimentation in wetlands and streams, which in turn leads to an overall loss of these aquatic habitats, as well as a shift in stream character from faster flows and deeper waters to still, shallower waters that favor a different assemblage of species (Poff 2002).

Efforts to vastly increase agricultural productivity in arid zones has led to massive water withdrawals from lake systems as well. In one of the most extreme cases, the Aral Sea, Kazakhstan, once the fourth largest lake in the world, has been reduced by 60% in extent over just two decades (Figure 6.6). The diversion of water from the Aral Sea has become one of the most dramatic natural management disasters. Although irrigation allowed an economically valuable cotton industry to develop, the environmental and biodiversity costs have been extreme. The lake rapidly shrank as water was diverted from both it and the two major rivers that feed it, the Amu Darya and the Syr Darya. Neither river reaches the Aral Sea today. The lake bed became exposed, and windblown dust has carried toxic, pesticide-laden fumes across the region. The shrinking of the Aral Sea has precipitated extinction of native fishes and other aquatic species, resulting in the loss of major fisheries and thus reducing the food security of people living around the lake. It has also created a regionally drier local climate, decreased water quality generally, causing salinization of local water sources in particular, and it has caused an increase in disease incidence in local populations (Micklin 1988; UN Environment Programme 2002). The southern sections of the Aral Sea are expected to disappear within 15 years if current trends continue.

There is not a complete assessment of the degree to which the world's wetland habitats have been lost or degraded, but it appears that in developed countries losses have been extensive. Wetland losses in the U.S. reached 53% by 1980, mostly due to agriculture (Figure 6.7), and have continued wherever urban populations are growing primarily due to urbanization and infrastructure development. In Europe, losses are likely to have been even heavier, ranging from 60%–70% (Revenga et al. 2000). In developing countries, wetland impacts have often come as these habitats are reduced or eliminated by dams for irrigated agriculture, as was the case for the Aral Sea. In another example, in the 1970s, the Waza-Logone flood-

(A)

(B)

Figure 6.6   The Aral Sea shrank to 40% of its original volume within a decade following its use for irrigation. It also became so salty that its commercial fisheries entirely collapsed by 1984. Further water withdrawals broke the sea into separate areas, and it is expected that the southern Aral Sea will disappear entirely within 15 years. (Satellite images courtesy of NASA. From United Nations Environment Programme 2004.)

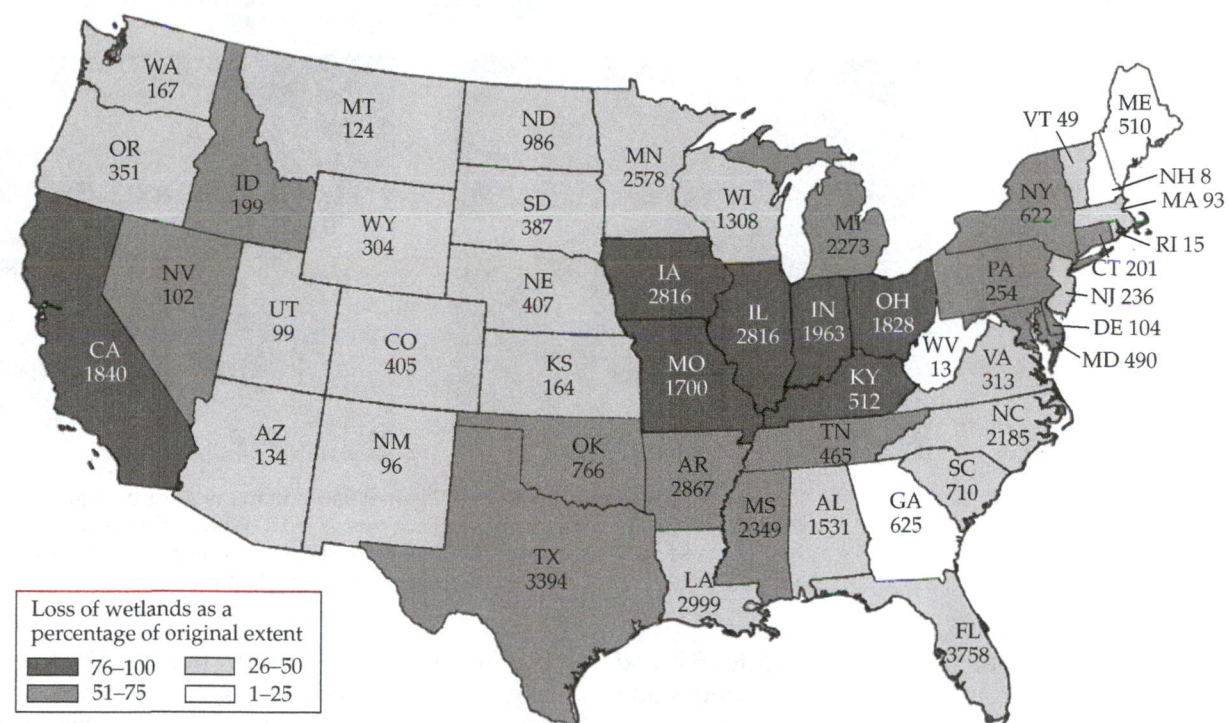

Figure 6.7 Historical loss of wetlands by State in the U.S., 1780s–1980s. Numbers in each state represent wetland loss in thousands of ha. (Modified from Revenga et al. 2000.)

plain in Cameroon was dammed to provide irrigation for rice production. The floodplain had been nourished by annual floods, and supported abundant mammal, bird and fish populations, as well as over 100,000 people. After the dam was built, the ecosystem essentially collapsed, creating a disaster for biodiversity and the local people. A rehabilitation project is now underway to restore the floodplain (Ngantou and Braund 1999).

Degradation of marine ecosystems

Roughly 60% of the world's human population lives within 100 km of a coast, and as a result approximately 20% of ecosystems adjacent to the oceans have been highly modified (lost to urban development or intensive agriculture) (Burke et al. 2000). Further, many populations are heavily reliant on marine resources for food and income. Thus, human impacts on marine systems, particularly coastal ones, have been large. However, we are only just beginning to understand marine habitat degradation. Many of the gravest impacts on marine, estuarine, and freshwater systems are caused by toxic chemicals, solid waste, and nitrogen enrichment. Because of their general importance, these are discussed separately in the next sections.

Coastal estuaries and other wetlands have been destroyed through filling, draining, dredging, shoreline stabilization, and conversion for aquaculture (Burke et al. 2000). As urban populations grow, coastal environments shrink rapidly as they make way for ports, homes, and stabilization efforts. Major rivers such as the Nile, the Colorado, and the Ganges are so heavily used for irrigation that the vast deltas that once channeled their flows into the ocean have become minor trickles. Sediments are tied up behind dams and impoundments, which deprives estuaries of nutrients and thereby interrupts these food webs. Further, the lack of sediment flows themselves has caused recession of coastlines once maintained by these flows (Nilsson et al. 2005). Loss of freshwater influence can increase salinity in estuaries, forcing major ecosystem change, which in turn can affect productivity and fisheries. Water divisions from the Ganges have even changed local climate, making it more extreme and drier (Adel 2002). The myriad ecological and economic consequences of these upstream influences are only recently being appreciated.

Mangroves are subtropical-tropical forests growing at the interface of the land and the sea. Mangrove forests are often extensive, and provide critical functions, such

(A)   (B)

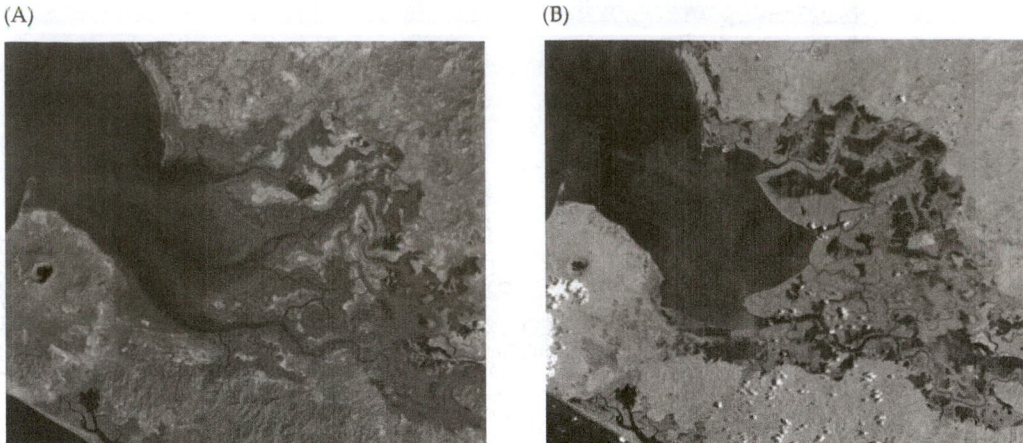

Figure 6.8 Mangrove forests have been replaced in many areas by aquaculture. This pair of satellite images shows mangrove forests in the Gulf of Fonseca, Honduras in 1987 (A), and their loss to roughly 775 km^2 of shrimp ponds in 2000 (B). (Images courtesy of NASA. From United Nations Environment Programme 2004.)

as serving as nursery and breeding grounds for fish and invertebrates, supporting human use of fisheries and wood, and buffering against coastal erosion. Many mangrove ecosystems are being transformed for aquaculture of shrimp, shellfish, and finfish (Figure 6.8), and have long been used as a source of wood, but increasingly are being logged for commercial timber (Alongi 2002). Mangrove forests and other coastal habitats are also cleared for coastal development to support tourism and spreading urban populations.

Other coastal habitats have been devastated by development, including the peat swamps of Vietnam and Cambodia (which are almost entirely destroyed) and seagrass beds worldwide, which have suffered from nutrient and sediment loading (Phillips 1984; Burke et al. 2000). Although seagrass systems are typically low in species richness, they are extremely important nursery habitats for juvenile fish and for invertebrates (Phillips 1984), and are widely distributed throughout temperate and tropical regions. Losses of seagrass beds can be extreme, with more than 90% degraded or lost in some coastal systems. Not only do these losses affect associated animal populations and their fisheries, but the stabilizing effects of seagrass root masses is lost, resulting in significant erosion of soft-sediment habitats.

Coral reef ecosystems have suffered from degradation due to pollution and sedimentation, direct exploitation of corals, destructive fishing at reefs, and more recently, disease outbreaks and bleaching associated with climate change. Reefs have become popular recreation sites for scuba divers, who can cause damage to fragile coral structures. Coral reefs are more threatened where coastal human populations are larger due to degradation from sedimentation and pollution, as well as from recreation. In some places, corals have also become damaged by de-

bris from fisheries, or directly damaged by trawls. Globally, 27% of coral reefs are highly threatened due to combinations of all these factors, and 31% experience moderate levels of disturbance (Bryant et al. 1998).

Marine trawling has led to extensive damage of marine benthic habitats (Watling and Norse 1998). Indeed, effects of trawling can be compared to forest clearing in both the breadth of effects, and the radical transformation that it causes in some habitats (Table 6.2). Les Watling describes some of these impacts in greater detail in Essay 6.1.

Human Activities That Cause Habitat Degradation

Agricultural activities (including crop and livestock farming, timber plantations, and aquaculture), extraction activities (mining, fisheries, logging, and harvesting), and development (human settlements, industry, and associated infrastructure) are the three main proximate causes of habitat degradation and loss. Not surprisingly, these are also the main causes of endangerment for birds (Birdlife International 2004; Figure 6.9) and amphibians (Stuart et al. 2004) due to habitat change. War and other violent conflict can include among their tragedies extensive habitat degradation. In addition, habitats are degraded via invasion of nonnative species (e.g., European and African grasses, and other invasive species now dominate vast expanses of rangeland in the Western U.S. to the exclusion of some native species; see Chapter 9). For ease in interpretation, we will discuss the impacts of pollution in the next section, as these include both direct transformation of habitats and more subtle degradation of the quality of an area for species.

TABLE 6.2   *A Comparison of the Impacts of Forest Clear-Cutting and Trawling of the Seabed*

Impact	Clear-cutting	Bottom trawling
Effects on substratum	Exposes soils to erosion and compresses them	Overturns, moves, and buries boulders and cobbles, homogenizes sediments, eliminates existing microtopography, leaves long-lasting grooves
Effects on roots of infauna	Stimulates, then eliminates saprotrophs that decay roots	Crushes and buries some infauna; exposes others, thus stimulating scavenger populations
Effects on emergent biogenic structures and structure formers	Removes or burns snags, down logs, and most structure-forming species aboveground	Removes, damages, or displaces most structure-forming species above sediment-water interface
Effects on associated species	Eliminates most late-successional species and encourages pioneer species	Eliminates most late-successional species and encourages pioneer species
Effects on biogeochemistry	Releases large pulse of carbon to atmosphere by removing and oxidizing accumulated organic material; eliminates nitrogen fixation by arboreal lichens	Releases large pulse of carbon to water column and atmosphere by removing and oxidizing accumulated organic material; increases oxygen demand
Recovery to original structure	Decades to centuries	Years to centuries
Typical return time	40–200 years	40 days–10 years
Area covered per year globally	~0.1 million km^2 (net forest and woodland loss)	~14.8 million km^2
Latitudinal range	Subpolar to tropical	Subpolar to tropical
Ownership of areas where it occurs	Private and public	Public
Published scientific studies	Many	Few
Public consciousness	Substantial	Very little
Legal status	Activity increasingly modified to lessen impacts or not allowed in favor of alternative logging methods and preservation	Activity not allowed in few areas

Source: Modified from Watling and Norse 1998.

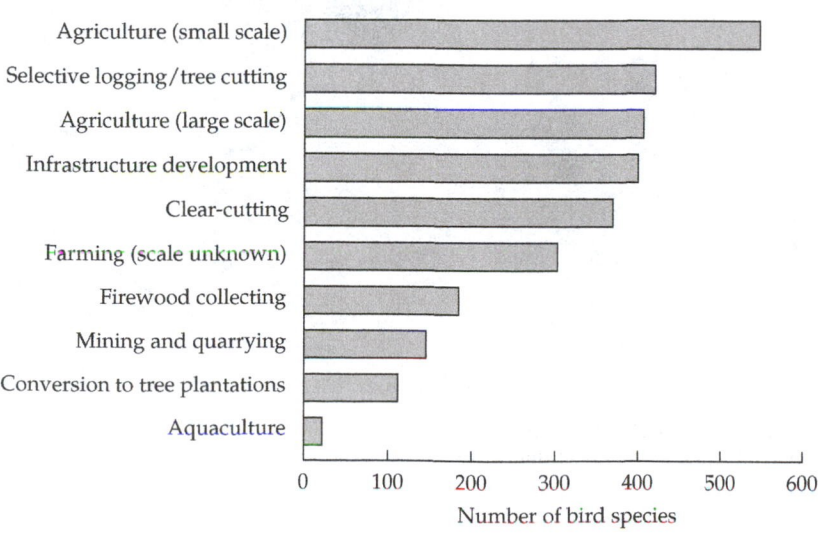

Figure 6.9   Bird species that are threatened by habitat degradation and loss are primarily affected by forms of agriculture, logging, and other tree-clearing, and by infrastructure development. (Modified from Birdlife International 2004.)

Agriculture

Historically, agricultural expansion has been the primary cause of ecosystem change, with a great acceleration of impacts seen from technological advances such as the invention of the plow and with economic expansions stimulated by commerce (Henley 2002). With the arrival of the industrial era, farming became mechanized, and global markets were linked by swifter commerce. These changes stimulated even more rapid conversion of habitats to agricultural use. More recently, the green revolution expanded the intensity with which crops are grown. Impoundments for irrigation allow rapid agricultural conversion. Overall, agricultural land use grew from around 265 million ha in 1700 to 1.2 billion ha in 1950, and now stands at over 1.5 billion ha (Wood et al. 2000). Annually, agricultural lands are expanded by about 0.3%—an area equivalent to the country of Greece or Nicaragua (Wood et al. 2000).

Arable lands are almost fully exploited worldwide, with over 98% transformed (Sanderson et al. 2002), representing an almost complete loss of the communities and ecosystems that once occupied those areas (Figure 6.10). For example, over 97% of the former tallgrass prairies in North America have been farmed for more than a century (see Figure 6.6). Expanding agricultural frontiers have converted 50% of woodland and grassland habitats in Brazil's Cerrado region to soybean fields and other cash crops, and rates of conversion continue to be high (Klink and Machado 2005). Over 70% of the land areas of Europe and South Asia are under cultivation (Wood et al. 2000).

Most croplands are used to grow annual crops, such as wheat, rice, and soybeans, and a smaller fraction (9%) are used for permanent crops such as coffee, tea, sugarcane, and fruits and nuts (Wood et al. 2000). Crop-based agriculture predominates in India (94%), South Asia (91%), and Southeast Asia (84%, where a third of this is permanent cropland). Roughly 5% of crop-based agriculture is irrigated, which increases the total environmental costs of agriculture as explained earlier.

The most intense forms of agriculture with vast, irrigated monocultures that are heavily treated with pesticides, herbicides, and fertilizers can be considered mostly "lost" to biodiversity. These practices are particularly prevalent in the U.S., but are increasingly used in other regions (such as in Brazil's Cerrado). Indeed, the disruption of most ecosystem processes is such that farming practices cause results in losses at every level of biodiversity. However, less-intensive forms of agriculture may be more compatible with biodiversity conservation—although these are not without impacts. Even shifting cultivation, which is more benign than most other forms of farming, has been shown to lead to long-term degradation of tree species diversity in sites in Borneo (Lawrence 2004).

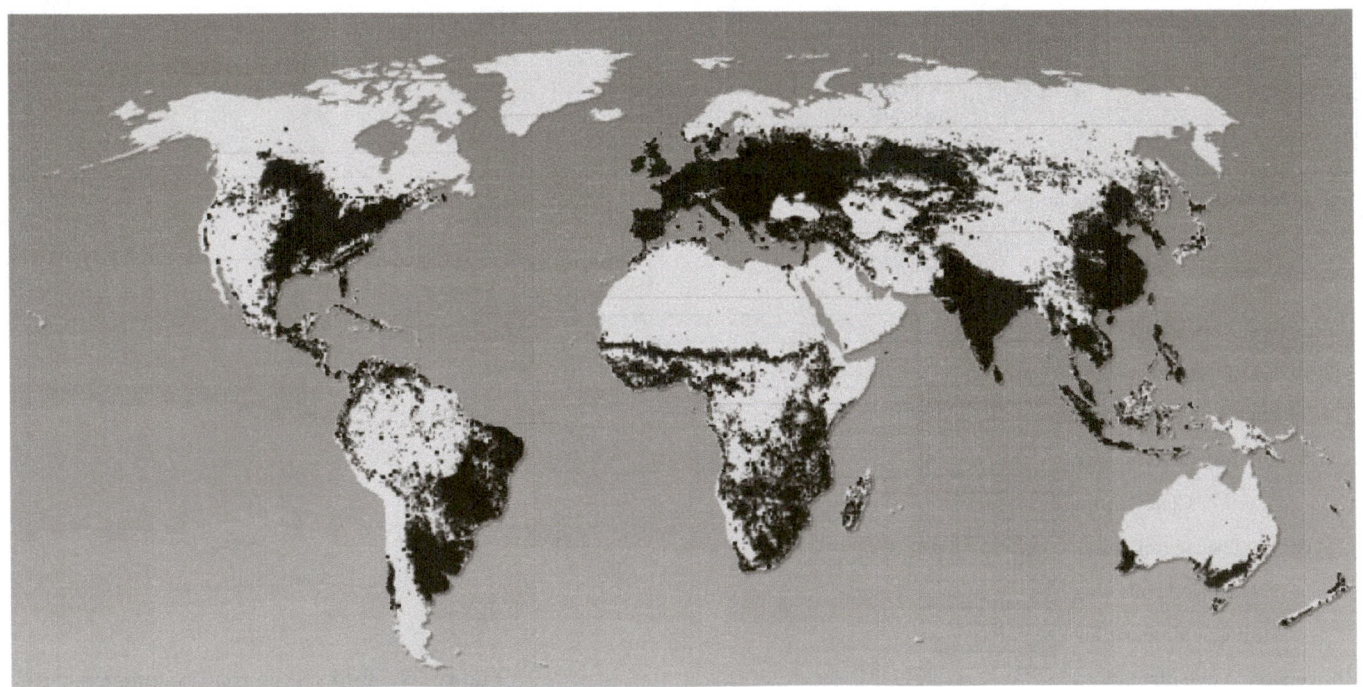

Figure 6.10   Distribution of cultivated systems (dark areas) worldwide (defined as areas in which at least 30% of the landscape is cultivated). (Modified from Millennium Ecosystem Assessment 2005b.)

However, efforts that enhance yields and the stability of production, and prevent expansion of agricultural lands may be developed to be compatible with biodiversity conservation. Polyculture systems usually offer more opportunity for wild species to be maintained within these systems. Many species can be maintained in an agricultural matrix, particularly if agricultural practices are more aligned with biodiversity conservation (e.g., organic production methods, low water use). Jeff McNeeley and Sara Scherr discuss promising agricultural practices in Case Study 6.2.

Highly modified pasturelands, which have been overgrazed for long periods, represent 6%–8% of current land cover worldwide (Sanderson et al. 2002). Overgrazing has damaged an additional 12%–14% of the world's pastureland. In total over 680 million ha are considered degraded to the point that they are less able to support livestock, and at worst, desertified (Millenium Ecoystem Assessment 2005c; Table 6.3). In comparison, agricultural mismanagement accounts for degradation of over 550 million ha across the world (see Table 6.3), about the same amount that is degraded by deforestation. About 70% of agricultural lands are used as pasture, with pasture the predominant use (>75%) in Australia and New Zealand, sub-Saharan and North Africa, East Asia, and Latin America, and the Caribbean (Wood et al. 2000). Latin America supports the largest proportion of the world's beef cattle production (26%), much of this on marginal lands cut from tropical dry and moist forests.

A large fraction of cereal production, particularly of maize, is used for livestock feed (76% in developed countries and 56% in developing countries). Increasingly, soybean meal that is left over after extracting oil is used for livestock feed as well. Livestock that are fed grains have a far larger ecological footprint in terms of land and water resources than those raised in pastures.

However, livestock that are an integrated part of farms primarily eat what would have been wasted. Nonetheless, the great increase in demand for meat and milk has a substantially higher ecological cost than demand for most crops.

Although we know that at the extremes agroecosystems greatly degrade natural habitats, we also know there are many instances in which most ecosystem processes are retained and numerous species are able to coexist. Unfortunately, little is known about the true status of wild biodiversity within agricultural systems.

Extractive activities

Extractive activities, particularly some forms of mining and quarrying, have devastating impacts on native habitat, and can create soil disturbances so profound that it takes affected areas centuries to recover. Nearly 60% of IUCN Red List plant species are threatened by mining or logging (IUCN 2004). Exploitation of oil and gas resources, and particularly the refining of petroleum products leads to wide-scale habitat degradation. These impacts are particularly severe in coastal environments. Mining and quarrying are linked strongly to urban development, and has increased as these areas have expanded.

Logging can consist of clear-cuts that radically change forested habitats to extremely selective logging of one or a few species at low densities. However, from the perspective of biodiversity, even low-intensity forestry can cause local extinctions due to the disturbances caused in the logging process and the need to create access routes (e.g., Johns 1985; Thiollay 1992, 1997). Commercial logging may cause smaller changes in geochemical processes than conversion to open pasture or other agricultural uses, but differences from primary forests are still detectable, and may be important in global greenhouse

TABLE 6.3 *Extent of Land Degradation by Several Causes*

Degradation extent	Cause
680	Overgrazing: Losses are most severe in Africa and Asia where pasturelands have become overgrazed. About 20% of the world's pastures are degraded.
580	Deforestation: Large-scale logging and clearance for agriculture and urban use. 220 million ha of tropical forest were destroyed between 1975–1990, mainly for pasture and cropland.
550	Agricultural mismanagement: Soil erosion has reached 25,000 million tonnes annually. Salinization and waterlogging affect about 40 million ha annually.
137	Fuelwood consumption: Woodfuel is the primary source of energy in many developing countries, and represents a substantial form of degradation to forests and woodlands.
19.5	Urbanization and industry: Urban growth, road construction, mining, and industry create complete habitat loss. Useful agricultural land is often lost as well.

Note: Degradation extent is given in millions of hectares.
Source: UN Environmental Programme 2002.

ESSAY 6.1

Scraping Bottom
The Impact of Fishing on Seafloor Habitats

Les Watling, *University of Maine*

■ Most of the species found in seafood markets, from fish to shrimp, crabs, and lobsters, live in or on the bottom of the sea. Some are caught using traps or "pots," some are taken using long lines of hooks, but the vast majority are taken with gear that is hauled across the bottom. The term "mobile fishing gear" is often used to describe the trawls and dredges involved in this fishery.

Throughout nearly all the continental shelves of the world (those areas adjacent to continents that are up to 200 m deep), mobile fishing gear is used to catch bottom-dwelling species. In colder waters, most of this effort is directed toward fish such as cod and haddock, and flatfishes such as halibut and flounders, but there is also a small shrimp fishery, especially in the northern hemisphere. In warm temperate to tropical waters, most of the fishery is for shrimps and prawns, but there are also areas, such as the northwestern Australia shelf, where fish are the target species. Watling and Norse (1998) estimated that about half of the area of the world's continental shelves are subject to trawling each year. This estimate had a few flaws; for example, the number of trawling vessels was underestimated by about 25%, which would increase the estimated area, and the fact that some areas are fished multiple numbers of times was not accounted for, which would decrease the estimated area. In the end, however, their estimate was probably not far off the mark.

The primary gear types that are used are trawls and dredges. Trawls are usually of two types, the otter trawl and the beam trawl. Otter trawls are the largest, although often not the heaviest. An otter trawl consists of a large conical net held open by floats on the top rope, chain and weights on the bottom rope, and two large, slightly curved, usually steel, plates called "doors" that are pulled through the water by the ship. The doors are connected to the net by a series of long cables. Since about the mid-1980s, the bottom rope has been equipped with large rubber disks so that the net can be pulled over bottoms with large (0.5 m diameter) boulders without rocks becoming part of the catch. Beam trawls also have a long, conical net, but in their case the net is attached to two D-shaped frames separated by a long (up to 12 to 16 m) steel beam. The bottom front portion of the net is replaced by rows of chains that are designed to stir fish from the sediments. These chains can penetrate sandy bottoms up to about 8 cm. Dredges are smaller than trawls, but much heavier. They usually have an iron bar at the front, and an iron frame to which a chain bag is attached. In some areas, the front iron bar may also be equipped with long teeth designed to dig into the sediment. A variation on this dredge design is the hydraulic dredge, which is used in shallow sandy sediments. At the front of the dredge are nozzles that are connected by hose to a water pump on the ship. Water is forced from these nozzles under high pressure into the sediment, thus dislodging large clams and other target species.

The seafloor is not a flat, featureless, muddy plain. In some areas, especially at high northern and southern latitudes where there was glacial activity in the past, the bottom of the ocean consists of gravel and boulders, sometimes with a covering of fine mud. The boulders in these areas can be up to 1 m in diameter. Wherever there is hard substrate like this, there will be high numbers of sponges, corals, and various colonial species, all of which provide living space for other, smaller, species. In other areas, such as off the east coast of the U.S., the bottom is made up almost entirely of sand, not unlike what is seen on the beaches from Cape Cod to Florida. In these areas, sand may be moved by waves and currents into ripples that are 10–20 cm apart and 5–10 cm high. Several species, such as hermit crabs and worms, along with juvenile fish, use ripples as shelter from predators, and find food in the depressions between them. Most of the seafloor, especially in deeper water, however, consists of fine, soft, mud, but these areas are not truly featureless. There are a large number of small invertebrate species that construct burrows and tubes in the mud so that an undisturbed bottom often looks like it is full of holes. These structures are used for feeding, either in an effort to obtain food particles buried deep in the sediment, or to allow the animal to raise its feeding structures higher off the bottom.

Mobile fishing gear impacts bottom habitat in several ways. In gravel and boulder areas, most of the species that are standing erect from the bottom are removed by the gear, either because they are directly crushed, tangled in the net, or crushed by large stones that are rolled by the foot rope chains and disks. Removal of these larger species (the incidental nontarget species are referred to as "bycatch") means that their associated species are also lost, so the overall effect is to lower the biodiversity of these species-rich areas. Fishing with trawls in these areas has been likened to cutting a whole forest to get the deer. In mud bottoms, the upper 5 cm or so of the sediment may be stirred upward into the water by the turbulence caused by the passage of the foot rope and the net, or the weight of the foot rope gear can compact the sediment. Either way, burrows constructed in the mud are filled in, tubes made of muddy substrate are broken, and the animals living in them are forced to expend energy to reconstruct their dwelling places. Some lose their burrows, and are left exposed on the sediment surface (Figure A).

Were it not for the fact that fishing the bottom with mobile gear occurs over such vast areas, and at almost any time of the year, these impacts might be minimal. Unfortunately, the scale of the activity is such that in some areas of the world, such as the North Sea, the Gulf of Maine, the coast of California, or the northwest shelf of Australia, there are almost no undisturbed areas left. In the Gulf of Maine, for example, there are small pockets of the seafloor that have not been trawled recently, and there are even a few small areas we discovered in 2003 that appear to never have been trawled. But, on a worldwide basis, we do not know what the percentage of

(I)   (II)

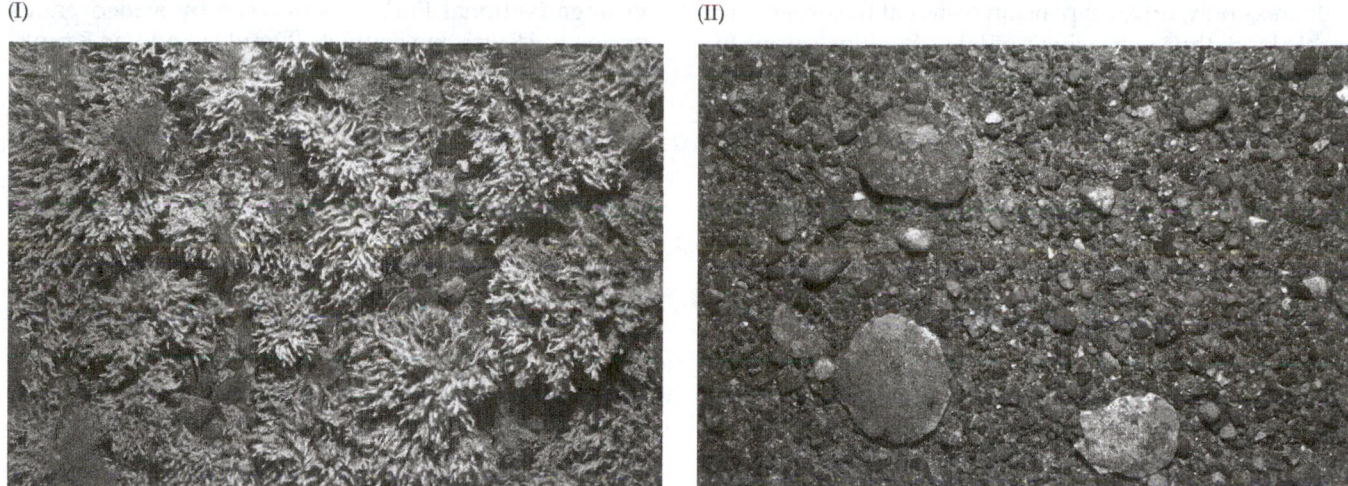

Figure A   Marine benthic habitats are typically complex (I), but can be converted to barren grounds by heavy trawls (II). (Photographs by the U. S. Geological Survey, courtesy of Page Valentine.)

"never disturbed" bottom might be. Very likely it is a small portion of the total shelf area.

In recent years, as continental shelf fisheries have become depleted, bottom fishing has moved offshore into deeper waters and onto isolated seamounts. This action has to be viewed as a true ecological tragedy. The deep sea (fishing occurs at depths up to 1800 m) is a place where there is naturally little or no physical habitat disturbance, where the environment might be stable for hundreds or thousands of years, and where species diversity is very high. Fishing in these areas creates distur-

bance for which no species is prepared, so diversity loss is highly probable. In the South Pacific, thousands of miles from land, fishing on seamounts for orange roughy (*Hoplostethus atlanticus*) results in bycatch of deep sea corals and other large suspension-feeding invertebrates that can be measured by the ton. Most of these species are hundreds of years old, and almost none have been described by scientists, so we have no idea what we are losing.

At present, trawling is rarely regulated by national governments; however, there are a few exceptions. For example, the Norwegian government

banned trawling in all areas where deep water coral reefs are known to occur. There are some no-trawling areas on the northwestern Australian shelf and within the waters of some states in the U.S. But the overwhelming proportion of the world's continental shelves and slopes are still open to fishing with mobile gear. We know that, as a result, considerable amounts of habitat of all kinds has been disturbed, modified in ways that we are only beginning to understand, but we do not know yet what the consequences are for most of the species living in those disturbed areas. ■

---

emissions or in long-term productivity of such land uses. The impacts of forestry are discussed in much greater detail in Chapter 8, and an assessment of the sustainability of selective logging systems in tropical forests is given in Case Study 8.3.

As mentioned previously, many fisheries' practices are more like mining or logging in their impacts on marine systems. Gear impacts on ocean bottoms (see Essay 6.1), as well as destructive use of explosives or cyanide at coral reefs have devastating impacts on local biodiversity and will require long periods for recovery.

Urbanization and infrastructure development

Cities and other heavily built environments represent the most extreme transformation of habitats, to what might most easily be termed "lost" rather than degraded. Approximately 3% of Earth's surface has been converted to built, urban environments. Not only do urban

environments obliterate original land cover, but they have enormous influence on nearby aquatic systems, and because they cannot supply all of their own resource requirements, they have large effects on ecosystems far away that provide cities with water, food, and other materials. Presently, about half of the world's population lives in cities, and urban populations are growing by about 2% per year (UN Environment Programme 2002). Thus, although the total area covered by urban environments may appear small, the ecological footprint of cities is many times as great. For example, Vancouver, British Columbia has a footprint equivalent to 174 times its area, and London a footprint 125 times its area—which is equivalent to all the productive land in the United Kingdom to support 12% of the population of the country. Cities in the Third World often have an order of magnitude smaller requirements, but still larger than their area (UN Environment Programme 2002).

Increasingly, urban expansion comes at the expense of arable land, thus increasing potential food insecurity. In China, 5 million ha of farmlands were lost to urban growth between 1987 and 1992 alone (UN Environment Programme 2002). Waste generated by urban populations is a major source of environmental degradation both through pollution and conversion of natural habitats to landfills.

Damage to coastal ecosystems is highly correlated with urban density (Burke et al. 2000). Similarly, urban environments have high impacts on freshwater systems through draining and filling of wetlands, water use from lakes and streams, and pollution. Urban environments have a high percentage of paved or covered surfaces that are impervious to water, and thus runoff from cities is rapid. Further, cities require a great deal of water for the maintenance of their populations, and withdraw considerable fractions of underground aquifers and river flows, at times capturing flows hundreds of km from their centers.

Urbanization has tremendous social costs as well. In the developing world, one out of every four city households lives in poverty, and a third of cities have no solid waste collection for low- to mid-income people, or sewage treatment. Certainly, addressing means of making cities more sustainable and liveable is a necessity on many fronts (see Case Study 18.2).

War and violent conflict

Wars and civil unrest can have devastating consequences on natural habitats. Extensive use of defoliants (e.g., Agent Orange) by U.S. military forces in the Vietnam War left toxins throughout an enormous area, which are known to have created human health problems and presumably caused problems for wildlife as well. Political instability in Africa has exacted an enormous human toll, and created vast populations of refugees that have had a detrimental impact on land. Displaced people need to find food and wood for fuel, and thus will hunt animals (bushmeat; see Chapter 8) and cut down trees. In the Democratic Republic of Congo, civil wars have disrupted logging concessions initially leading to a decrease in deforestation rates, but any gains were generally erased by forest degradation for fuelwood collection (Draulen and Van Krunksalen 2002). War in Afghanistan directly damaged ecosystems, and also left the country with no resource management for over two decades, which resulted in extensive harm to woodlands and wild animal populations (Zahler 2003).

At times, political conflicts can have the beneficial effect for biodiversity by halting or greatly slowing development. For example, biodiversity has flourished in the demilitarized zone between North and South Korea (Kim 1997). However, conflicts have led to a lapse in enforcement of protected area boundaries. For example,

Virunga National Park was overrun by armed gangs during the Rwandan conflicts (Draulen and Van Krunksalen 2002). However, the efforts of many dedicated Congolese conservation biologists and general respect given to some high-profile species allowed populations of the mountain gorilla (*Gorilla gorilla beringei*), elephant (*Loxodonta africana africana*), okapi (*Okapi johnstoni*), and northern white rhino (*Ceratotherium simumcottoni*) to survive the conflicts essentially unscathed, even though some populations declined (Hart and Hart 2003). Westing (1992) suggested that protected areas should be demilitarized to prevent such occurrences. Certainly, confronting the problems of political instability must number among critical initiatives for conservationists.

Pollution as a Form of Habitat Degradation

Environmental pollutants are so pervasive in our world today that few if any corners of Earth remain untouched by human-synthesized chemicals. While direct habitat destruction is usually the most obvious form of habitat degradation, pollutants accumulated in otherwise intact habitats are likely the most pervasive and widespread environmental problems today. Pollutants may affect biological resources directly by altering the chemical balance of water or soils, lead to mortality in wildlife that accidentally ingest alien objects, or affect ecosystem functioning. Common forms of pollution include light pollution, direct waste disposal, and release of synthetic chemicals into the environment. In some cases, clean-up may require active removal of the contamination and also restoration efforts. When pollution is less severe or a habitat is minimally altered, a habitat may recover simply by removing the source of pollution.

Light pollution

The invention and rapid proliferation of electric lights over much of Earth's surface has transformed nighttime for many animals. Sources of light pollution include "sky glow" produced by cities, illumination of buildings, lighting on vessels and offshore oil rigs, and car and street lights. Ecological light pollution alters the natural light regime in terrestrial and aquatic systems and interrupts evolved patterns of behavior—cuing diurnal individuals to be active at night, and suppressing activity of nocturnal species.

Changes in light levels may most profoundly affect nocturnal animals, which have a broad range of anatomical adaptations to see at night. For frogs, a sudden increase in illumination can reduce their visual capacity for minutes to hours (Buchanan 1993). Nocturnally migrating birds become disoriented and "trapped" in lit areas, often leading to exhaustion or mortality due to collision

with structures. Beachfront lighting deters sea turtles from coming onto the beach to nest and disorients hatchlings (Peters and Verhoeven 1994). Increased light levels may also affect predator–prey relationships, as small rodents, some lagomorphs, marsupials, and snakes forage less at high-illumination levels. Patterns of interspecific competition can alter if species differ in their activity levels in lit areas.

Artificial light levels may also impact communication. For example, coyote packs howl during new moon when light levels are lowest. This communication is thought to be essential for assembling the pack before hunts and for establishing territories. Sky glow could increase ambient light enough to eliminate this behavior (Longcore and Rich 2004).

Air pollution and acid rain

More toxic chemicals are released into the air than into any other medium (Bryner 1993). Air pollution may take a variety of forms: smoke produced from combustion, fumes from vapors during chemical reactions, dust resulting from industrial and agricultural processes, or mist released in spraying or by the effect of sunlight on automobile exhaust. Various air pollutants can have detrimental effects on human health as well as environmental resources. The major sources of air pollution are transportation, fuel combustion from power plants and space heating of buildings, industrial processes, solid waste disposal, and burning of forests and agricultural lands. Smog caused by traffic and industrial factories in cities may travel hundreds of miles to an otherwise "pristine" area.

The U.S. National Park Service is concerned with air pollution, most of which comes from outside the Parks. Pollution can injure various species of trees and other plants, acidify streams and lakes, and leach nutrients from soils. For example, ozone pollution is damaging over 30 plant species in the Great Smoky Mountains National Park in Tennessee and North Carolina, as well as forested habitats in many areas downwind from industrialized areas worldwide. Jeanne Panek explains how ozone damages trees in Essay 6.2.

The ability to appreciate scenic vistas is highly dependent on good visibility, which is often compromised due to poor air quality. The economic value of visibility in U.S. national parks was estimated based on surveys showing the public is willing to pay, on average, U.S.$46.31 to U.S.$76.06 per household per year to improve visibility or to prevent it from being degraded in National Parks in the Southwest, the Southeast, and California. When these values are multiplied by the number of households in the country or even in a particular region, these dollar amounts grow to the billions (National Park Conservation Association 2005). If America's national parks become known for their haze, visitation might drop off, threatening the economic resource base that supports their management (Figure 6.11).

Acid deposition (or, more commonly, acid rain) has caused extensive degradation of freshwater and some forested habitats in many countries. In the eastern U.S. and Canada, acid deposition is 30–40 times higher than it was in the 1980s. The primary source is sulfur oxides from power plants burning fossil fuels. Gases transformed in the atmosphere into sulfuric and nitric acids usually remain there for weeks and may travel hundreds of miles before settling on Earth as dry or rained particulates. Within the U.S., acid rain poses a serious threat to watersheds, lakes, and streams in the Northeast, forests and coastal plains of the mid-Atlantic, and forests in northern Florida (Bryner 1993). Acid rain is also believed to harm red spruce (*Picea rubens*) and other tree species in the eastern and southeastern U.S. Rainfall in the Great Smoky Mountains National Park is now five to ten times more acidic than normal rainwater on average, which may pose particular risks to the region's diverse amphibian populations (see Case Study 3.1).

(A)    (B)    (C)

Figure 6.11   Air pollution in Grand Canyon National Park, Arizona, as shown in a fixed camera at the Desert View Watchtower on three different days; visibility is (A) 303 km, (B) 156 km, and (C) zero. Visibility can change daily due to pollution from copper smelters and power plants in the region and from smog from Los Angeles and Las Vegas. (Photographs courtesy of U.S. National Park Service.)

ESSAY 6.2

Forest Ozone Injury
How Physiology and Climate Play a Role

Jeanne Panek, Panek and Associates, *Berkeley, CA*

■ One of the most widespread air pollution problems worldwide is the increase of ground-level ozone concentrations. Ozone (O_3) is toxic to plants even at moderate concentrations, and damages crops, forests, and other vegetation throughout the industrialized world. Rising global CO_2 concentrations are getting a great deal of attention, but ozone concentrations are actually increasing more quickly (Hough and Derwent 1990; Finlayson-Pitts and Pitts 1999). By 2100, 50% of forests worldwide are expected to be exposed to elevated O_3 concentrations (Fowler et al. 1999), as well as a changing climate (International Govememental Panel on Climate Change 2004). Forest researchers, therefore, are focusing efforts to understand how ozone, climate, and forest physiology interact. A question of major importance is, with ambient ozone concentrations rising, how much will ozone injury to forests increase? Getting at the answer is not straightforward.

To cause injury, ozone needs to get *into* the plant; therefore plant physiology plays an important role. Ozone enters plants through tiny pores (stomata) on the leaves, which open and close in response to climatic factors such as soil water availability, humidity, temperature, and light. Once inside, ozone is like a bull in a china shop—it breaks open cells and wreaks havoc on photosynthesis. Ozone *uptake* by plants is therefore the key to ozone injury to plants, and stomata are the most important factor controlling ozone uptake.

In the U.S. and in Europe, current ozone exposure standards designed to protect crops and forests are not based on ozone uptake, but instead on ambient ozone concentrations *outside* the plant. This is because it is difficult to measure ozone uptake whereas ambient ozone concentrations are easy to measure. Furthermore, ozone standards are based on ozone concentrations summed over the "growing" season, the four months from June to September. But in parts of the U.S., forests have a longer, or different, growing season. Thus the ozone standard does not reflect well the ozone that the forests are "seeing" in those regions (Panek et al. 2002). In Europe at least, regulatory agencies are working to change the ozone standard to rely on estimates of ozone uptake by vegetation, reflecting the current understanding of the role of physiology in ozone injury.

How Much Can Physiology Matter?

Climate affects forest physiology, which then affects ozone uptake. Eastern forests, with hot, humid summers ideal for maximum physiological activity, take up most of their ozone in the summer. Here, ozone uptake correlates well with ambient ozone concentrations outside the leaves. Western forests are different. In the Mediterranean climate of California and in the Pacific Northwest there is little rainfall during the summer months, the soils slowly dry out, and by August there is little water in the soil or the air. Late summer and early fall is a time of dormancy for many plant species in these areas. However winters are moist and relatively warm, unlike winters in the eastern U.S., so many plant species are active during this time.

Ponderosa pine (*Pinus ponderosa*) is one of those plant species—it is a widespread and important forest species, ecologically and economically, found west of the Rocky Mountains. It is also extremely susceptible to ozone pollution (Miller et al. 1963). Late spring and early summer is a time when soils are full of rain- or snowmelt-water, the air is warm and moist, and thus the time when ponderosa pines are at peak activity, absorbing their greatest amounts of CO_2, water, and ozone (Panek and Ustin 2004). While frozen eastern forests are dormant, western forests are physiologically active and have been shown to take up to 15% of their ozone during the December–March winter months (Panek and Ustin 2004).

One of the most interesting features to the pattern of ozone uptake in a Mediterranean climate like California is the large difference between ozone concentrations in the air and forest ozone uptake. During the hot and sunny summer months, ambient ozone concentrations rise to their annual maxima. However, because water availability in the soils and the air decreases over summer and forests become increasingly dormant, the forests take up only a very small fraction of the ambient ozone during the time when ozone concentrations are highest (Figure A).

How Will Drought-Adapted Forests Respond to Increased Precipitation?

Predicted climate change will cause rainfall patterns to alter in many regions of the U.S. Within the vast range of ponderosa pine, some regions will see increased precipitation, some will see less. An important question is, how will species respond to changes in water availability? Clearly, less precipitation and snowmelt during the summer will lead to increased dormancy. But can drought-adapted species respond to increased summer precipitation by becoming physiologically more active? That is, is the summertime dormancy in Mediterranean-climate forests "hardwired" or is it plastic? A watering experiment was performed on ponderosa pine saplings during the driest part of summer 1998 to address this question. Unwatered trees exhibited a typical drought dormancy pattern, whereas watered trees remained at peak physiological activity throughout the drought. Daily ozone uptake in watered trees was 40% greater by the end of the experiment (Panek and Goldstein 2001). Thus, ponderosa pine can change its behavior to adapt to increased summer rainfall; unfortunately at the same time this leaves it vulnerable to increased ozone uptake.

Ozone Uptake at the Continental Scale: The European Approach

European regulatory agencies have collaborated with many of the world experts on ozone effects over a period of years to evaluate the European ozone standard, which, like the U.S. standard, is based on ozone concentra-

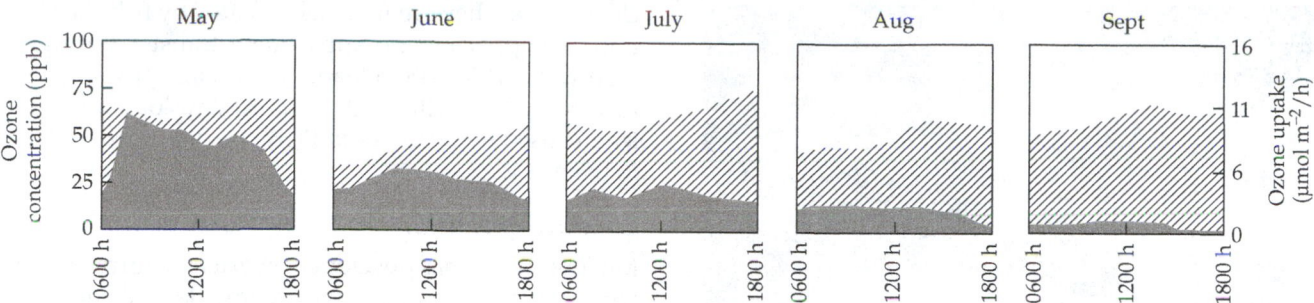

Figure A Ozone concentration (hatched area) plotted with ozone uptake (solid areas) for Yosemite National Park, CA, showing increasing decoupling of concentration and uptake over the season. Individual panels present 1 day of measurements taken every 2 hours from 0600 to 1800 hours PST in each month during the growing season. (Modified from Panek and Ustin 2004.)

tions. They have rejected the concentration-based standard in favor of an uptake-based approach for most vegetation types. This is a major step forward in protecting crops and forests from ozone injury. This effort is ongoing in Europe as a part of the United Nations Economic Commission for Europe Convention on Long-Range Trans-boundary Air Pollution.

This new approach requires modeling forest physiology at a large scale. A group of European scientists (Emberson et al. 2000a,b; Simpson et al. 2001) developed an approach that explicitly incorporates ozone uptake at the regional scale across Europe. Their approach makes use of three existing pan-European data sets: measured and estimated ozone concentrations at a 50 × 50 km spatial resolution (Jonson et al. 1999); climate data; and land cover and soil-type maps. A large climatic gradient exists in Europe, from Scandinavia to the Mediterranean, even larger than seen in the continental U.S. However, their model describes uptake across this broad region accurately for a number of different vegetation types. This approach is part of the developing ozone standard in Europe, an effort that is leading the way worldwide to a new understanding of ozone uptake in the context of climate change.

With ozone concentrations rising as quickly as they are, ozone injury to crops and forests is likely to increase worldwide. As a result of changing climate, drought will increase in some regions of the world, and forests and non-irrigated crops in these regions might receive some protection from ozone injury because of their physiological dormancy during drought. However, in dry regions that are predicted to become wetter because of climate change, vegetation is likely to become more susceptible to ozone injury. Thus, plant physiology plays an important role in the terrestrial response to concurrent changes in climate and ozone. ■

Air pollution is recognized by governments as a threat to human health and environmental security. For example in 1990, the U.S. enacted major amendments to the Clean Air Act in an effort to curb acid rain, urban air pollution, and toxic air emissions. The law sets standards for alternative clean fuels, promotes the use of low-sulfur coal and natural gas, seeks to reduce energy waste through conservation, and encourages the use of market-based principles such as performance-based standards and emission banking and trading. While much remains to be done to bring air-pollution levels to a safe and desirable level, the Act has been successful in improving air quality for human health, as well as for reducing acid rain and ozone levels.

Solid waste and plastics

Improper disposal of waste is another cause of habitat degradation and is particularly problematic in the world's oceans. Plastics in the form of industrial pellets as well as manufactured items have increased in the ocean and along coasts in the last 50 years (Page et al. 2004) and pollution of our oceans with these and other nondegradable waste products is a leading cause of mortality for many marine organisms. Plastic bags floating at the water's surface look like jellyfish and thus are often ingested by sea turtles. Entanglement and subsequent drowning is a major cause of mortality for marine mammals, sea turtles, and seabirds that must return to the surface of the ocean to breathe (Burger and Gochfeld 2002; Mascarenhas et al. 2004; Page et al. 2004). Nearly 1500 fur seals die annually from entanglement in fisheries debris in Australia alone (Page et al. 2004) (Figure 6.12).

Ingestion of plastics is a major cause of mortality and injury of seabirds. Plastics may cause intestinal blockage, or decrease feeding efficiency. Albatrosses, which may accidentally swallow plastics mistaken for surface-swimming squid, may be unable to regurgitate food, thus leading to an inability to successfully rear chicks. Plastics have been found in the guts of the majority of seabirds examined (Burger and Gochfeld 2002; Derraik 2002). Across all taxa of marine organisms, 267 species (including 86% of all sea turtle species and 43% of all marine

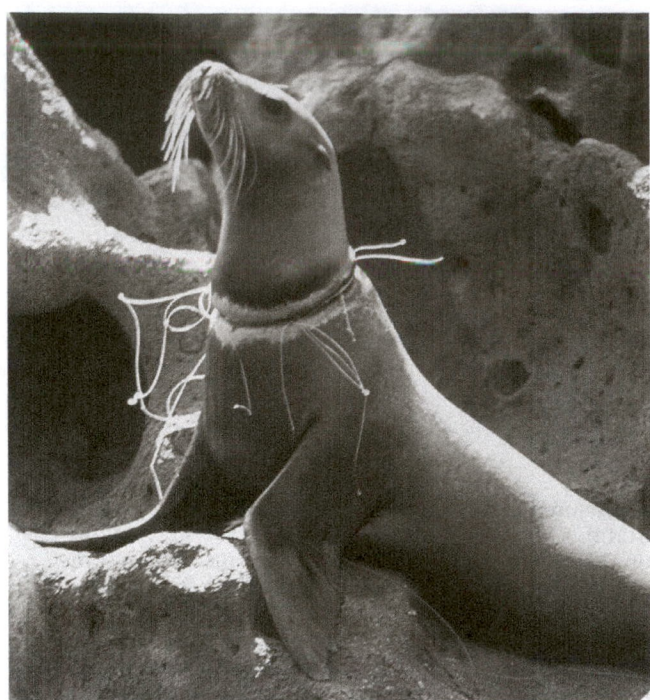

Figure 6.12 Entanglement in plastic debris can cause deep wounds and restrict feeding, as seen here for this California sea lion (*Zalophus californianus*). (Photograph © Hal Beral/ Visuals Unlimited.).

mammal species) have been documented to be affected by plastic debris (Laist 1997).

Plastics pose a particular problem because they do not biodegrade as most garbage does, but instead "photodegrade." This means that sunlight breaks them into smaller and smaller pieces, but they do not ever entirely disappear. Furthermore, these plastics particles are concentrated in areas of the ocean called subtropical gyres, which are created by mountainous flows of air moving from the tropics toward the polar regions. High-pressure systems depress the ocean surface, and the rotating air mass drives a slow surface current that moves with the air in a clockwise spiral. Winds in the high-pressure region are normally light or calm, thus floating debris do not get mixed into the water column.

Within the gyres is a great variety of filter-feeding organisms that feed on the ever-renewed crop of phytoplankton. While entanglement and ingestion are clearly problems for larger marine organisms, the worst threat from the ubiquitous plastic pollution may be the plastic polymers that act as sponges for DDT, PCBs, and other oily pollutants. Plastic resin pellets have been shown to concentrate poisons to levels as high as a million times their concentrations in the water. Organisms ingesting these products become concentrators and dispersers of

chemicals as they are themselves eaten by fish. Concentration of poisons and subsequent transfer through a food web, which often leads to humans, poses serious risks to both wildlife and humans. Physiological problems caused by increases in PCBs are discussed in detail in Essay 3.1.

Chemical pollution

Modern society has developed an extensive array of synthetic chemicals, many of which provide convenience in our lives, increase food production, and help prevent disease. Increasingly, we discover that the chemicals used to produce common items such as clothing, computers, and food containers are contaminating the environment and having adverse effects. Pollutants may be harmful to wildlife, humans, and entire ecosystems due to their physical aspect, toxicity, or by alteration of habitat due to overload. Toxic chemicals have been documented in all systems, from tropical forests, to pelagic habitats, to arctic tundras (Brown 2003).

The field of environmental toxicity is devoted to understanding the relationship between toxins, our environment, and their effects on biotic systems. Scientists are becoming increasingly aware of the impacts and ecological risks due to persistent and chronic exposure to sublethal levels of toxins. The case of the 1989 oil spill from the *Exxon Valdez* in Prince William Sound, Alaska, not only caused immediate mortality, but chronic exposure to sublethal levels of oil continues to adversely affect recovery of wildlife populations. The *Valdez* spill released some 42 million liters of crude oil, contaminating at least 1990 km of pristine shoreline in Prince William Sound, the Kenai Peninsula, Kodiak Archipelago, and the Alaska Peninsula (Figure 6.13).

Oiling of fur or feathers in marine mammals and birds, which require routine contact with the sea surface, can lead to death resulting from hypothermia, smothering, drowning, and ingestion of toxic hydrocarbons. The immediate effect of the *Valdez* oil spill was disastrous and in the days following the spill between 1000 and 2800 sea otters and 250,000 seabirds died as a direct result of oiling. A further 302 harbor seals died in the days following the spill, likely from inhalation of toxic fumes leading to brain lesions, stress, and disorientation (Peterson et al. 2003).

The magnitude of the spill prompted long-term ecological research on recovery of the system and some surprising results have ensued. Rates of dispersion and degradation of oil diminished through time, due in part to "ecosystem sequestration" whereby most oil remaining after 1992 (an estimated 806,000 kilograms) was sequestered in environments where degradation was suppressed by physical barriers to disturbance, oxygenation, and photolysis. The result is that a survey occurring 10

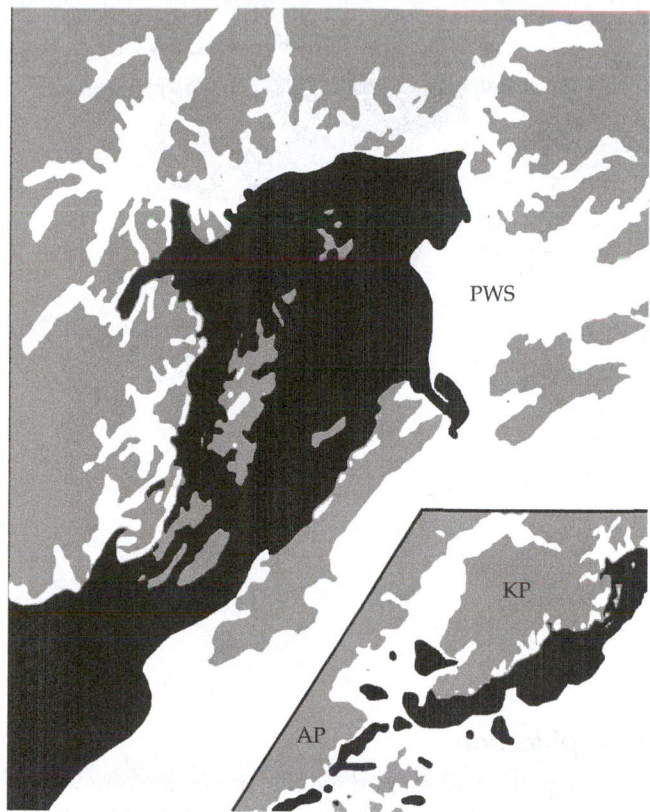

Figure 6.13 The *Exxon Valdez* oil spill contaminated over 1900 km of shoreline in Alaska, from its origin in Prince William Sound (PWS), along the Kenai Peninsula (KP) to the Alaska Peninsula (AP) and Kodiak Island (KI). (Reprinted with permission from Peterson et al. 2003 © 2003 AAAS.)

years later revealed 55,600 kg of little-weathered *Exxon Valdez* oil in intertidal subsurface sediments. Thus, sediment-affiliated wildlife populations such as fish, sea otters, and seaducks have been subjected to chronic exposure. These sublethal exposures have lead to increased mortality in these populations due to compromised health, growth, or reproduction (Peterson et al. 2003).

Acute chemical exposures often cause direct mortality, or impairment of reproductive capacity. While acute events such as oil spills have severe visible impacts, daily release and subsequent accumulation of chemicals in our environment are threatening wildlife, people, and entire ecosystems around the globe. Toxic chemicals can alter sexual, neurological, and behavioral development, impair reproduction, and undermine immune systems (Brown 2003). Many animals are known to pass chemicals to their offspring through the placenta, via suckling, or in their eggs. Research in endocrinology on how chemicals disrupt or inhibit animals' reproductive systems is unraveling some of the mysteries of how chemicals are affecting fitness in a variety of taxonomic groups. Lou Guillette discusses the problems posed by endocrine disrupters in Essay 6.3.

Beyond effects on individuals, pollution has had substantial impacts on habitat. For example, chemical pollution is responsible for 12% of degraded soils. Once these soils are contaminated beyond a certain level they become dangerous for agriculture or human habitation, as well as to wildlife populations (UN Environment Programme 2002). Efforts to reduce and mitigate contamination at large scales are therefore needed.

Habitat degradation due to excessive nitrogen inputs

Not all forms of pollution are derived from noxious or toxic chemicals. Human activities have had profound impacts on the global biogeochemical cycles of carbon, nitrogen, and phosphorous, and humans have approximately doubled the rate of nitrogen input into the terrestrial nitrogen cycle (Vitousek et al. 1997). A major source of nitrogen is in fertilizers applied to croplands in the form of animal manures and synthetic fertilizers, of which there are greater than 80 million metric tonnes produced annually. To date, major agricultural areas in Europe, North America, India, and China have experienced the most extreme increases in nitrogen over the past 50 years (Figure 6.14).

The deposition of anthropogenically fixed nitrogen from the atmosphere onto plant and soil surfaces is one of the most important factors currently causing global-scale changes to terrestrial ecosystems (Throop and Lerdau 2004). Nutrient supplies have generally positive relationships with growth of terrestrial, marine, and freshwater plants. Great increases in nutrient supplies, however, can lead to highly undesirable changes in ecosystem structure and function (Smith et al. 1999).

Because nitrogen is important in controlling biological processes from organismal physiology to ecosystem nutrient cycling, changes in nitrogen availability may have substantial consequences at every level of biodiversity (Vitousek et al. 1997). Increased nitrogen deposition may act either directly on biological processes (e.g., altered nutrient cycling) or indirectly (e.g., causing shifts in herbivory that result from altered host plant biology) (Throop and Lerdau 2004). Recently, nitrogen pollution has been found to be associated with the formation of harmful algal blooms, in which species exude toxic compounds that cause wildlife or human health problems (Galloway et al. 2003).

Individual interactions between herbivores and plants may translate to changes in population, community, and ecosystem processes. Nitrogen deposition may influence the composition of herbivore communities directly by altering the plant host community. Large-scale changes in herbivory caused by nitrogen deposition can

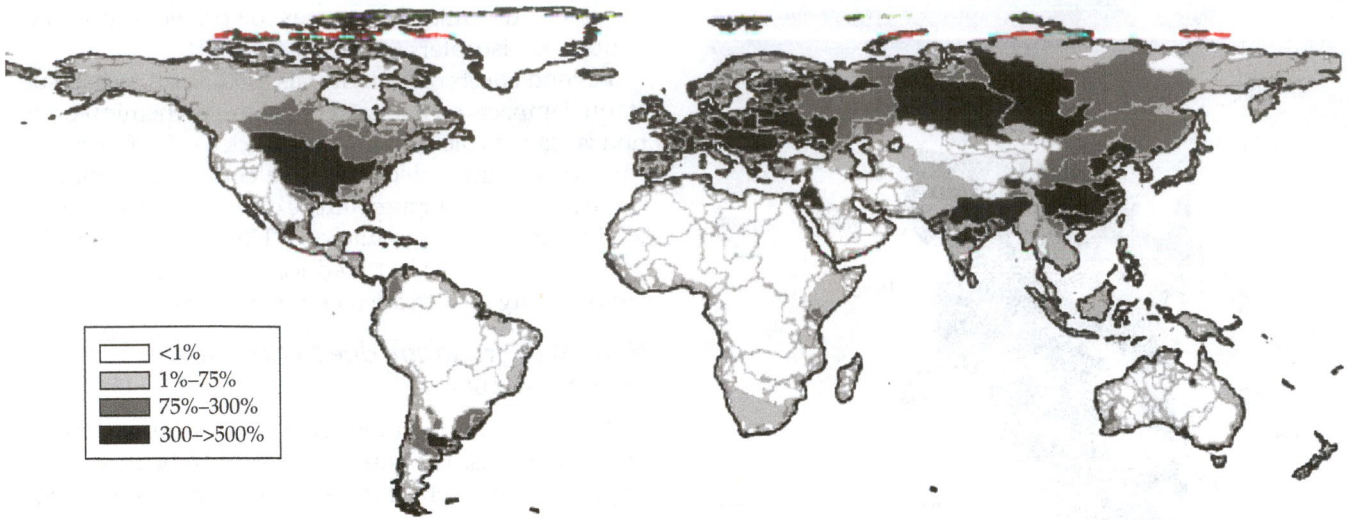

Figure 6.14 The amount of nitrogen deposited in the world's major watersheds has increased tremendously from predisturbance levels to the mid-1990s. (Modified from Millennium Ecosystem Assessment 2005b.)

Legend:
- <1%
- 1%–75%
- 75%–300%
- 300–>500%

affect ecosystem productivity and carbon storage, as well as the dynamics of the cycling of nitrogen and other elements. Such changes can alter ecosystem processes by changing patterns of litterfall, for example. Thus, changes in herbivory as a result of increased nitrogen deposition could have large economic consequences by impacting the growth and mortality patterns of commercially important species (Throop and Lerdau 2004), as well as creating cascade effects that may be of lasting harm to some components of biodiversity.

Not all ecosystems respond to nitrogen deposition similarly. Aquatic systems have faced the greatest challenges from excess nitrogen, in part because such systems are typically nitrogen-limited and because wetlands, lakes, and estuaries are the natural repositories of waters and sediments draining from a watershed. Response of a given habitat depends on factors such as successional state, ecosystem type, nitrogen demand or retention capacity, land-use history, soil type, topography, climate, and the rate, timing, and type of nitrogen deposition (Matson et al. 2002). In temperate, unmanaged forests where many of the species are adapted to nutrient-poor conditions, for example, additional nitrogen inputs are expected to alter species composition and diversity. Experimental nitrogen enrichment in a North American grassland increased total insect abundance but decreased plant and insect species richness (Figure 6.15) (Haddad et al. 2000). Nitrogen-deficient areas of the tropical and subtropical oceans are acutely vulnerable to nitrogen pollution, as an increased rate of decomposition depletes water column oxygen concentrations (Beman et al. 2005).

Eutrophication

In the aquatic and marine realms, the most common effects of increased nutrient loads are rapid increases in the abundance of algae and aquatic plants, and reduced water clarity. **Eutrophication**, the transformation of an ecosystem from low nutrient levels to high nutrient-enrichment, is a widespread outcome of excessive anthropogenic inputs of nitrogen. Although stimulating photosynthetic organisms increases productivity initially, it can degrade water resources rapidly through a chain of effects that permeate the ecosystem. Zooplankton proliferate as

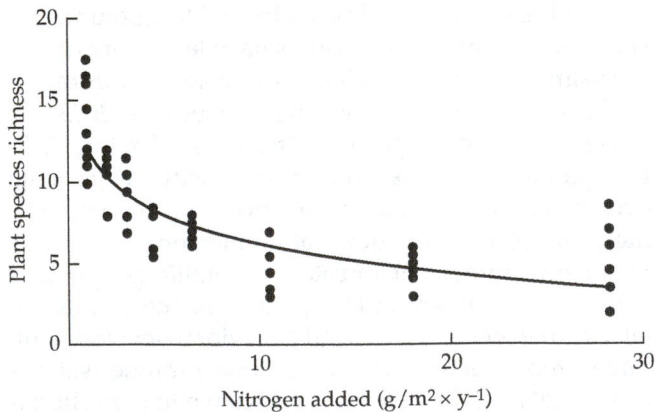

Figure 6.15 Plant species richness declines along a nitrogen gradient in a prairie ecosystem at Cedar Creek, Minnesota. (x-axis is the rate of nitrogen addition plus 1). Each point represents the average of data collected in 1995 and 1996. (Modified from Haddad et al. 2000.)

ESSAY 6.3

Endocrine Disrupting Contaminants, Conservation, and the Future

Louis J. Guillette, Jr., *University of Florida*

■ Evidence is mounting that environmental contaminants may be having strong effects on wildlife, and possibly on human populations. As a result, some wildlife populations may be experiencing severe declines or extinction. Others may show no obvious decline but individuals may exhibit various symptoms of stress resulting in reduced fecundity and offspring survival, and increased susceptibility to disease. Examples include declines of fishes of various species in the United States, Europe, and Asia that exhibit ovotestes, a pathological condition in which both male and female cellular features are found in the gonad (Jobling et al. 1998). The mechanisms underlying these and other endocrine changes are poorly understood, yet these declines are clearly the result of complex phenomena involving human influences, including wide-scale environmental pollution.

In 1991, a small group of scientists from various disciplines met and generated a consensus statement regarding impacts of environmental pollutants: "We are certain of the following: A large number of man-made chemicals that have been released into the environment, as well as a few natural ones, have the potential to disrupt the endocrine system of animals, including humans" (Colborn and Clement 1992). Since that declaration, a wide array of chemical contaminants—termed endocrine disrupting contaminants, or EDCs—has been documented to disrupt normal endocrine function. EDCs disrupt the normal cell-to-cell signaling required for development, growth, and reproduction, among other activities, in vertebrates and invertebrates alike (Guillette and Crain 2000).

Chemicals with the potential to disrupt the endocrine system function through a variety of mechanisms. Chemicals with endocrine-disruptive activity include chemical components of various pesticides; chemical stabilizing products called phthalates and bisphenol A, found in some pesticides, plastics, health care products, cosmetics, perfumes as well as other personal care products; flame retardants such as polychlorinated biphenyls (PCBs) and polybrominated diphenyl ethers (PBDEs); combustion products such as dioxin (TCDD); surfactants such as nonylphenol and octylphenol, found in some soaps and commercial products; nitrate derived from fertilizers and other sources; and heavy metals such as cadmium and arsenic (Colborn et al. 1993; Rooney and Guillette 2000; Markey et al. 2002; Guillette and Edwards 2005). Reports of new chemicals with endocrine disruptive activity are published almost weekly (see http://www.ourstolenfuture.org).

Although a debate continues concerning whether the health and development of human populations are at risk from normal background exposure to EDCs, clear evidence exists that some human populations have been affected by high exposures in a manner similar to that seen in wildlife or laboratory animals (Guo et al. 1993; Guillette et al. 1998; Sharpe and Irvine 2004). In contrast to human populations, there is strong consensus that wildlife has been affected by the endocrine-disruptive actions of various environmental contaminants (for representative reviews, see Tyler et al. 1998; Crain et al. 2000; Oberdorster and Cheek 2000; Guillette and Gunderson 2001). Alterations have been described in fishes, amphibians, reptiles, birds, and mammals.

One well-documented case is that of American alligators (*Alligator mississippiensis*) living in contaminated lakes in Florida (Guillette et al. 2000). Contaminants can affect organisms in two fundamentally different ways. First, embryonic exposure can cause organizational abnormalities by altering the chemical signals required for normal embryonic development (Guillette et al. 1995). Organizational modifications are those that permanently change the structure and functioning of a tissue or organ system, thus modifying the phenotype of the exposed individual. Many of the abnormalities reported in the contaminant-exposed alligator populations appear to be organizational, as they are present at birth and apparently continue throughout life. These alterations include altered ovarian follicle formation, altered gonadal steroid hormone (androgens and estrogens) synthesis, small penis size, and altered enzyme activity in the liver. A number of contaminants have been identified in alligator eggs and serum from those populations with individuals displaying these alterations. These contaminants include various pesticides and their metabolites and nitrates. Several of these contaminants exhibit an affinity for the alligator estrogen (ER) and/or progesterone (PR) receptors (Vonier et al. 1996; Guillette et al. 2002). These receptors are transcription factors that play a central role in the control of gene expression. Recent studies have shown that various contaminants are capable of sex reversing (male to female) reptilian embryos at concentrations as low as 100 parts per thousand (ppt), 10 to 100 times lower than concentrations measured in the eggs themselves (Crain et al. 1997; Matter et al. 1998; Willingham and Crews 1999; Willingham et al. 2000).

Second, endocrine-disrupting contaminants also can induce activational abnormalities whereby a contaminant can induce or suppress normal endocrine responses (Guillette et al. 1995). Juvenile alligators exhibit alterations of the reproductive and endocrine systems years after hatching that become more pronounced if these animals remain in the contaminated lakes (Guillette et al. 2000). In males these changes include reductions in androgenic steroid hormones, such as testosterone and dihydrotesterone, elevations in plasma estradiol-17β (the principle estrogenic steroid hormone in female alligators that usually is found at almost undetectable levels in the plasma of normal males), and altered thyroxine concentrations (the principle thyroid hormone associated with the control of metabolism). Altered endocrine parameters in juvenile alligators occur in various Florida wetlands not associated with significant pesticide spills or point-source contamination, such as Lake Okeechobee, Lake Griffin, and the northern Everglades (Guillette et al. 1999; Gunderson et al. 2004). These alterations in the reproductive and endocrine systems of hatchling and juvenile alligators are found in the

same populations exhibiting high embryonic and neonatal mortality. Interestingly, these observations are not limited to alligators, as mosquitofish (*Gambusia holbrooki*) living in the same lakes show many of the same alterations, including altered steroid hormone profiles, small gonopodia, reduced sperm counts, and altered reproductive behavior in males (Toft et al. 2003; Toft and Guillette 2005).

Over the last decade, the conclusions on the stability and health of various wildlife populations have been influenced by the contaminant exposure history of the animals under study. These studies, some of which have been discussed above, document the sublethal effects of contaminants on the cellular signaling mechanisms of animals (McLachlan 2001). A growing scientific literature suggests that researchers examining the biology of any and all species, irrespective of ecosystem type or location should investigate the contaminant history of the ecosystem they work in. That is, all biologists studying natural systems, or even laboratory-based systems, must now consider contaminant exposure as a confounding and important variable in their studies. Ubiquitous global contamination now has the potential to alter the normal functioning of the endocrine, nervous, and immune systems of all organisms with resulting changes in gene expression and phenotypes. Contaminants act as one major factor by which the environment alters the expression of the genotype and thus the phenotype (Fox 1995; Iguchi et al. 2001). Phenotypic changes are associated with altered reproductive potential, developmental pathways, behavior, and survival. Conservation biology, with its goal of preserving and restoring populations and ecosystems, must actively embrace this new world view, and incorporate an understanding of the multiple mechanisms by which contaminants alter the biology of organisms, populations, and ecosystems. To do less would be to exclude a major factor influencing the potential success of such programs.

The scientific community must recognize that laboratory or even microcosm studies do not represent the temporal and spatial complexity of contaminant exposure seen by free-ranging organisms. Further, it has become obvious that future studies must examine the organizational as well as activational roles of pollutants—snapshots of individuals at a single point in time are not going to provide the insight needed to access the health of the populations under study or allow us to develop effective plans for the conservation of healthy ecosystems. Modern conservation biology must embrace our growing understanding of the complexity, variation of exposure, and impacts from genomic to ecosystem levels of organization, associated with contaminant exposure. To do less would guarantee that we continue to underestimate the impact of chemical pollution on the conservation of the world's ecosystems and the species they comprise. ◼

---

they eat the bounty of new growth in the phytoplankton, and when they die, their bodies sink to lake or ocean bottoms where they are decomposed by bacteria and other organisms. Although fish and other consumers benefit temporarily from the spikes in phytoplankton and zooplankton populations, the increased action of decomposers removes oxygen from the water at such a great rate that it can create anoxic conditions that may result in large-scale fish kills. Ultimately, the ecosystem may undergo a radical change in species composition.

Nitrogen-based agricultural fertilizers, whose use is predicted to double or triple over the next 50 years, are the primary source of nitrogen pollution worldwide. A global expansion of agriculture is inarguably affecting marine habitats around the globe (Beman et al. 2005). In the Mississippi River Basin, which has had two centuries of land use, fertilizer use from farming has had a more significant effect on water quality than has land drainage or the conversion of native vegetation to cropland and grazing pastures (Turner and Rabalais 2003). The northern portion of the Gulf of Mexico ecosystem, which contains almost half of the nation's coastal wetlands and supports commercial and recreational fisheries that generate U.S.$2.8 billion annually, has undergone profound changes. Due to eutrophication caused by excess nitrogen discharged from the watershed, large areas of hypoxic bottom water, known as the Dead Zone, are a re-

current feature in the Gulf during the summer. Hypoxic waters can stress and suffocate marine organisms, therefore affecting commercial harvests and the health of impacted ecosystems, yet these are now widespread (Figure 6.16).

Large portions of tropical and subtropical marine regions may be similarly vulnerable to nitrogen in agricultural runoff. Rapidly developing agricultural areas in the tropical Americas, West Africa, South Asia, and Southeast Asia will experience large increases in fertilizer use over the next few decades, thus potentially affecting marine ecosystem processes in adjacent, large marine habitats (Beman et al. 2005).

Eutrophication can be reversed when the supply of excess nitrogen is removed. A widely recognized success story is that of Lake Washington, located in Washington State. This lake had grown eutrophic due to sewage inputs, and shortly after these inputs were shifted and treated, the lake recovered (Edmondson 1991). Untreated sewage is discharged directly into freshwater and marine habitats worldwide. For example, 90% of wastewater is untreated before being discharged into the Caribbean (UN Environment Programme 2002). Certainly, many eutrophic conditions could be ameliorated by sewage treatment.

While removing excess nitrogen and phosphorus from sewage and detergents has been shown to be effective,

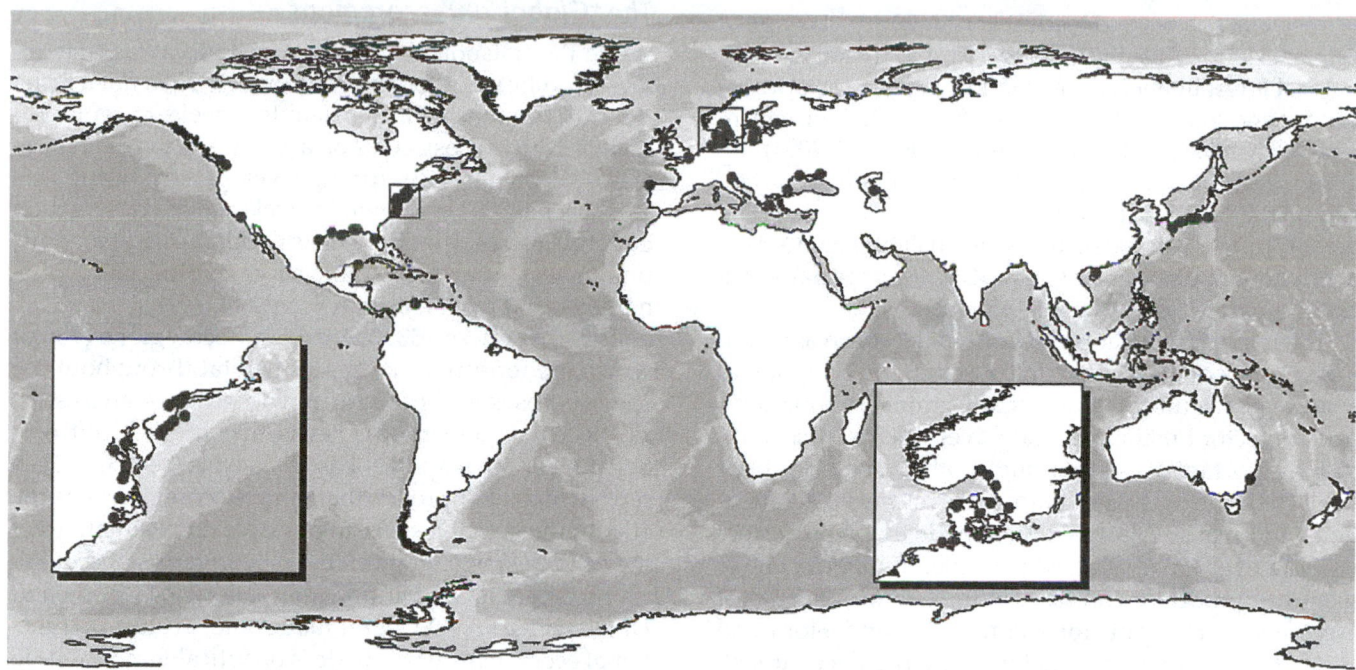

Figure 6.16 World distribution of hypoxic or "dead zones" is correlated to position of watersheds that have greatly enhanced nitrogen inputs, and to major cities. (Modified from Burke et al. 2000.)

curbing nitrogen pollution from agriculture has been more challenging. Presumably, reductions in fertilizer use might lead to similar improvements. The Yaqui Valley of Mexico is one of the world's most productive agricultural regions, particularly for wheat. However, fertilizer use in this watershed had become excessive (>300% increase over background nitrogen inputs, see Figure 6.14), leading to eutrophication in the Gulf of California. Experiments showed, however, that by altering the timing and amount of fertilizer input, nitrogen outputs could be reduced without affecting crop yield or quality (Matson et al. 1990). Such reductions in nitrogen benefit not only associated aquatic systems, but also reduce greenhouse gas emissions and lower costs to farmers. Thus, detailed studies might help guide areas where restriction of nutrient inputs into water bodies could lead to successful eutrophication management, as well as other gains that enhance agricultural sustainability.

Protecting What's Left: Approaches to Global Habitat Conservation

Given the degree to which much of Earth's surface has been degraded and the limits to our ability, funding, and time for action, conservationists prioritize areas that are most in need of conservation and restoration. Globally oriented conservation groups such as Conservation In-

ternational (CI), The Nature Conservancy (TNC), the Wildlife Conservation Society (WCS), and World Wildlife Fund (WWF), base their conservation strategies on the identification of critical places that demand their efforts. While their targets and tactics often differ, the general goal of each of the conservation organizations is to maintain intact ecosystems and thwart human-caused species extinctions. To approach this goal, these organizations prioritize resources based on the biological importance or representation of an area, threat or vulnerability facing the region, financial costs of conservation, and likelihood of success. In prioritizing how to allocate scarce resources, conservation managers often face a paradox in determining how to conserve the most threatened resources, which are often the most costly to protect, while maximizing investment in conservation. We discuss next a variety of approaches that these and other groups are taking to identify and conserve critical and endangered habitats around the world.

Biodiversity hotspots

The hotspots concept was first articulated by Norman Myers to inform terrestrial conservation priorities based on principles of irreplaceability and threat. Myers' original ten tropical forest hotspots have been expanded upon by Russell Mittermeier and colleagues at CI. The hotspots are now represented by 34 regions of exceptional endemism and habitat loss. Each of these areas holds at least 1500 en-

demic plant species and has lost at least 70% of its original habitat extent. While the majority (22 of 34) of hotspots are tropical forest biomes, the expanded analysis includes savannas and grasslands, temperate forest, Mediterranean-type ecosystems, and deserts (Mittermeier et al. 2004). Together, these 34 areas occupy 15.7% of Earth's land surface and hold 77% of Earth's terrestrial vertebrate species (Figure 6.17). Over 300,000 plant species, half of which are endemic to just one hotspot, and 42% of terrestrial vertebrates are restricted to these 34 regions.

The impact of the hotspots concept has been astounding in terms of investment in conservation, adoption as a strategy for multilateral organizations, foundations, private sector businesses, and governments. Also, critically important areas that might otherwise have been neglected due to their high threat levels received attention and great strides have been made that many critics would never have believed possible. In Liberia, one of the most important countries in the Guinean Forests of West Africa Hotspot, for example, CI and Flora and Fauna International have maintained a presence through periods of severe instability and violence. In large part due to their efforts, the Liberian Senate in 2003 enacted legislation to expand their protected area system and make conservation a priority for their reconstruction.

The "Global 200" ecoregions

While the hotspots approach prioritizes areas that are most significant in terms of biodiversity, it is not a *representative* approach. One system that seeks to represent each of Earth's ecosystems on a global scale is the "Global 200" Ecoregions approach, developed by WWF–U.S. The "Global 200" ecoregions are selected as conservation targets based on species richness, endemism, taxonomic uniqueness, unusual ecological or evolutionary phenomena, and global rarity (Olson et al. 2001).

Together, these 233 "Global 200" ecoregions (136 of which are terrestrial) represent habitat throughout the special ecosystems of Earth and account for areas such as the Arctic and tundra areas that are left out of a hotspots or other species-based analysis (see Plate 3). A representative approach that targets ecoregions has the advantage in that it seeks to preserve representatives of the full range of biological diversity in terms of ecological processes and evolutionary phenomena (Olson and Dinerstein 1998). On a pragmatic note, protecting functional ecosystems worldwide is of critical importance to local human populations, both for current prospects and the potential for future evolutionary change (Kareiva and Marvier 2003). Further, species richness is not a surrogate for evolutionary uniqueness or evolutionary

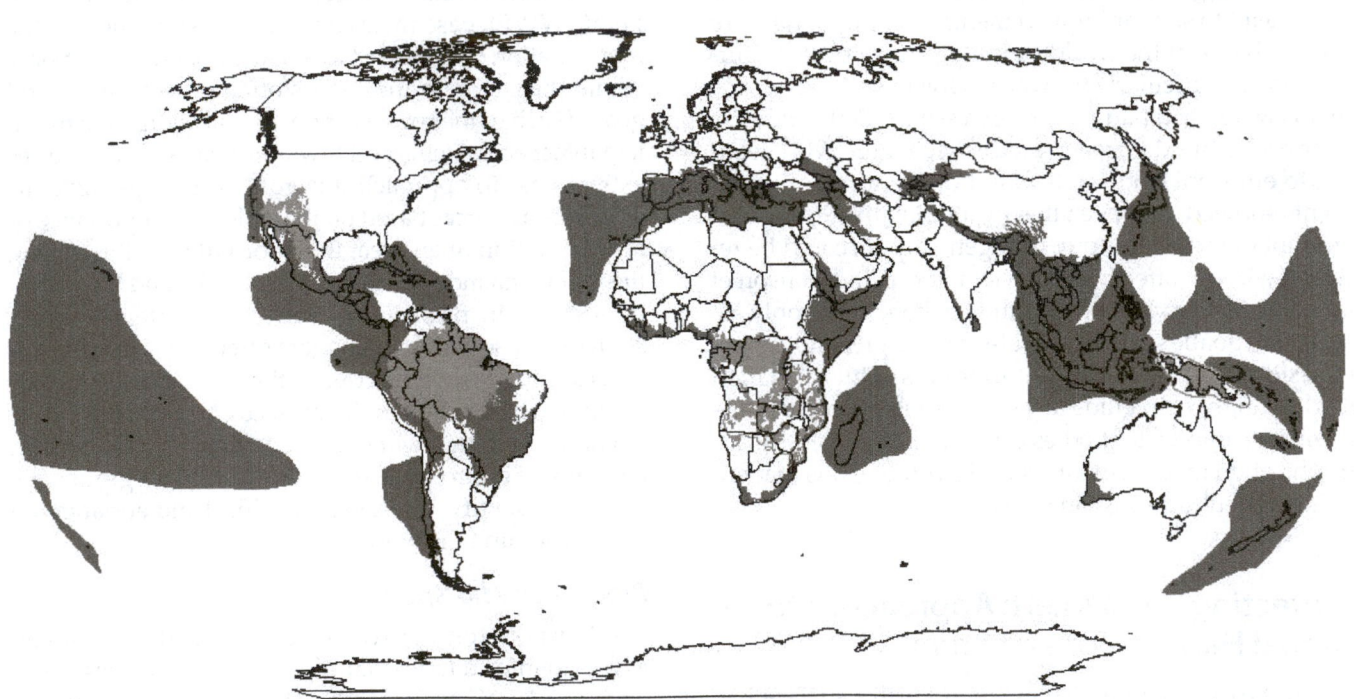

Figure 6.17 Biodiversity hotspots are the most immediately threatened regions with the greatest richness of plant species on Earth. Hotspots (dark gray) are priority regions for conservation efforts. Major tropical wilderness areas (light gray) are also a conservation priority by CI. (Courtesy of Conservation International.)

age—in fact, speciose groups tend to be evolutionarily younger. Thus, conserving only the species-rich areas misses an important part of our evolutionary legacy (Kareiva and Marvier 2003), which can be captured through a more representational approach. Another benefit of setting priorities at the scale of ecoregions is that these areas approximate the dynamic arena within which ecological processes most strongly interact (Orians 1993).

Crisis ecoregions

Recently, in conjunction with WWF, The Nature Conservancy (TNC) completed a strategic analysis of the status of the world's ecoregions. Hoekstra et al. (2005) calculated a conservation risk index (CRI) of threat as the ratio of the percent of habitat converted to human uses to the percent protected by biome (Figure 6.18) and by ecoregion. They then classified three levels of crisis for ecoregions comparable to the IUCN Red List cate-

gories: Critically Endangered (>50% conversion and a CRI >25; 64 ecoregions), Endangered (>40% conversion and a CRI >10; 80 ecoregions) and Vulnerable (>20% conversion and a CRI >2; 161 ecoregions) (Figure 6.19). Temperate grassland, savannas, and scrublands, and Mediterranean ecoregions scored particularly high on this index, and are strongly represented in the set of crisis ecoregions.

There is broad overlap among ecoregions prioritized according to threat and those highlighted for their biological distinctiveness in the Global 200 (44% overlap) and for high species-richness in the hotspots (50% overlap). Yet this analysis highlights many critical regions not identified in the hotspots or Global 200 approaches, and thus is an important complement to defining where conservation priorities should lie. Importantly, the crisis ecoregions approach highlights whole ecoregions at grave risk of elimination in the near term at a somewhat smaller spatial scale.

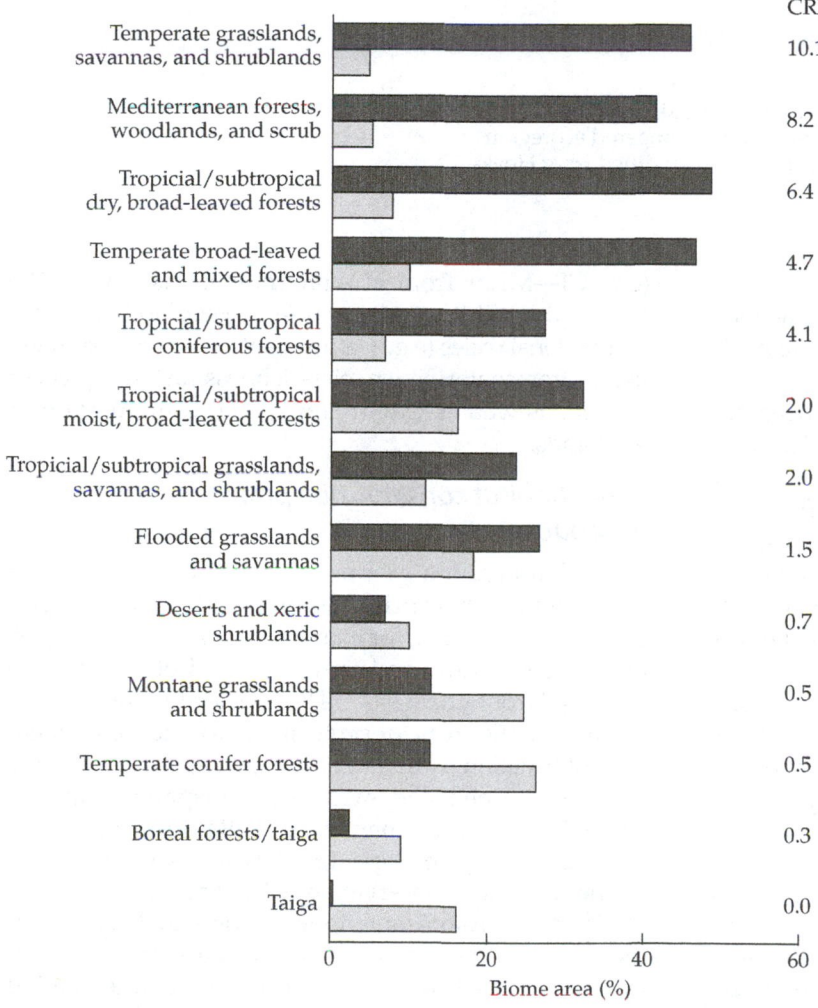

CRI

Biome	CRI
Temperate grasslands, savannas, and shrublands	10.1
Mediterranean forests, woodlands, and scrub	8.2
Tropicial/subtropical dry, broad-leaved forests	6.4
Temperate broad-leaved and mixed forests	4.7
Tropicial/subtropical coniferous forests	4.1
Tropicial/subtropical moist, broad-leaved forests	2.0
Tropicial/subtropical grasslands, savannas, and shrublands	2.0
Flooded grasslands and savannas	1.5
Deserts and xeric shrublands	0.7
Montane grasslands and shrublands	0.5
Temperate conifer forests	0.5
Boreal forests/taiga	0.3
Taiga	0.0

Biome area (%)

Figure 6.18 Protected areas provide better coverage of some biomes (e.g., temperate conifer forests) than of others (Mediterranean habitats), particularly relative to the proportion of total habitat area has been converted to human uses. The Conservation Risk Index (CRI; dark gray) is the ratio of the percent habitat converted to percent habitat protected (light gray). (Modified from Hoekstra et al. 2005 © Blackwell Publishing.)

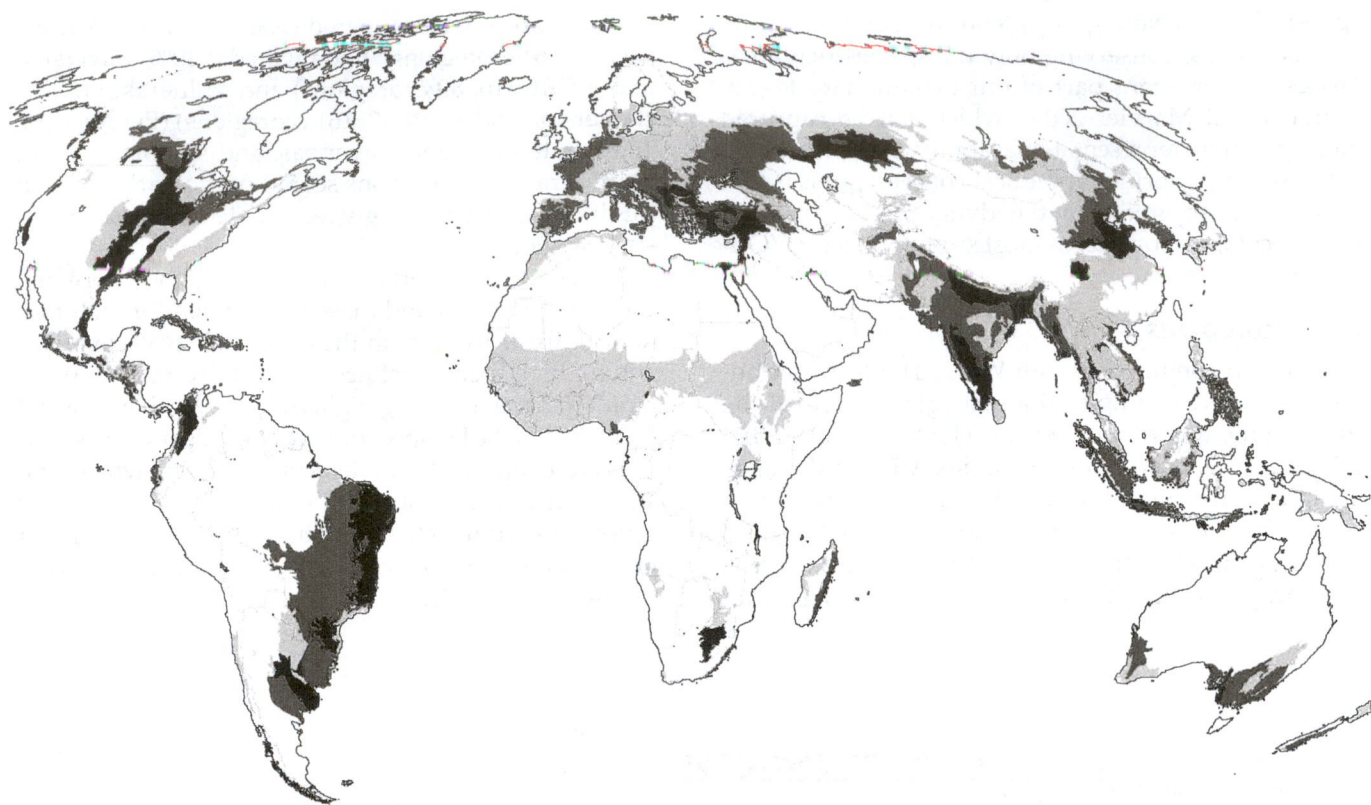

Figure 6.19   Distribution of the 305 crisis ecoregions that are at-risk of elimination due to habitat conversion. Vulnerable ecoregions are in light gray, endangered ecoregions in medium gray, and critically endangered ecoregions in black. (Modified from Hoekstra et al. 2005 © Blackwell Publishing.)

Wilderness protection

Another approach used to identify and prioritize actions for key areas has been to focus on areas where the habitat is largely in a natural, intact state. Wilderness areas are Earth's last frontiers where habitat has not been destroyed, impacts of humans are less noticeable, and threats are less imminent. Wilderness areas are of critical importance for maintaining intact faunal and floral assemblages, ecosystem services, and traditional lifestyles of tribal peoples (Mittermeier et al. 2001). It is here that conservation may have the greatest opportunity for success and where investments may protect the largest amount of habitat.

To complement the hotspots approach, CI also prioritized wilderness areas in tropical regions. Of 37 wilderness areas identified based on intactness, size, and low human population density, a few qualify as high-biodiversity wilderness areas, with levels of species-richness and endemism paralleling that of the richest hotspots (Mittermeier et al. 2001; see Figure 6.17). In contrast, WCS has drawn attention to the "last of the wild"—the 10% of each biome that is least affected by human development (Sanderson et al. 2002). Approaches have been undertaken to identify remaining wilderness areas at the global

(e.g., CI—Major Tropical Wilderness Areas, WCS—The Last of the Wild), continental (e.g., Wildlands Project), and regional scales (e.g., California Wilderness Coalition), using representation, species-richness and uniqueness criteria, as well as extent and spatial configuration of wildlands.

Other habitat conservation priority-setting approaches

Many conservation groups establish conservation targets and base their priorities on specific habitat types. Global Forest Watch, for example, uses satellite images, mapping software, and on-the-ground observation to identify "frontier forests." They seek to identify minimum dynamic areas for protection, provide comprehensive information on threatening activities as or before the activities happen, and encourage transparency and accountability in forest management (Redford et al. 2003). The National Speleological Society focuses specifically on the study and conservation of karst and cave habitats. Under the Ramsar Convention, the only global environmental convention to target a particular ecosystem, wetlands of international importance are identified and, if nominated, receive special resources for their conserva-

tion (see Chapter 14). The U.S. has a suite of environmental laws dedicated specifically to the conservation of free-flowing rivers.

The majority of priority-setting approaches focus on terrestrial, and to a lesser extent freshwater systems. Regional priority-setting for the marine realm has been more challenging due to a lack of data on species occurrences, as well as the status of marine ecosystems. One of the first global, fully data-driven analyses for the marine realm applied a hotspots-like approach for tropical coral reefs. In this analysis, geographic range maps of 3235 reef fish, corals, snails, and lobsters were used to identify the richest centers of endemism for coral reefs (Roberts et al. 2002). Eighteen areas were identified that together cover 0.028% of the ocean's area (35% of coral reefs) and include about 65% of range-restricted species from the four taxa considered in analysis. A threats analysis was then conducted to prioritize the areas based on conservation urgency.

Conservation of Habitats: The How

While each of the above-mentioned approaches highlights *where* to implement conservation, it does not prescribe *how* it should be done. To achieve conservation goals, efforts need to address the specific activities that cause these destructive impacts. A number of social and economic factors are the underlying driving force of land use change decisions, which ultimately influence habitat degradation (Figure 6.20). As economic circumstances favor different land uses, these become predominant. Social and political actions will circumscribe land-use decisions. The success of conservation programs to reduce habitat degradation will be most influenced by our understanding of what factors drive practices that degrade ecosystems, and the ability of human societies to embrace such efforts. Many organizations and government agencies have developed a large range of approaches to conserve habitats. In all cases, the

solution for how conservation should be done depends on the scale of the conservation target.

Approaches to conserving wilderness include many strategies. For example, some efforts focus in the policy arena to establish national parks, or to fund conservation through the purchase of development rights or "debt-for-nature swaps." Social interventions, such as working with indigenous groups to help them maintain their legal rights to lands, or preventing the development of roads or other permanent infrastructure in these areas are also widely used conservation strategies. International conservation groups have been successful in pioneering debt-for-nature swaps where countries agree to set aside wilderness in exchange for debt relief. In 2004, for example, CI, TNC, and WWF brokered a deal whereby the U.S. government forgave a grant that in turn is allowing Colombia to apply U.S.$10 million toward conserving 11 million acres of its tropical forests. Debt-for-nature approaches are described in greater detail by Neal Hockley in Essay 6.4. Other financial incentives such as purchasing the rights to timber concessions have helped thwart development of some of the world's wilderness areas, at least for the medium term. Comparison of the economic return of degrading versus sustainable uses has sometimes convinced governments to promote more ecologically sustainable practices, and to protect habitats (Figure 6.21).

Domestically within the U.S., groups such as the Wilderness Society and California Wilderness Coalition (CWC) have played major roles in getting wilderness legally defined and protected, and then expanding the legal wilderness system and ensuring protection of its values. The 1964 Wilderness Act legally mandated the preservation of substantial areas "where Earth and its community of life are untrammeled by man, where people may visit but not remain." Regional or local groups such as CWC work to designate new areas for inclusion in the wilderness system and thus remove the threat of development of these areas. These groups also strive to

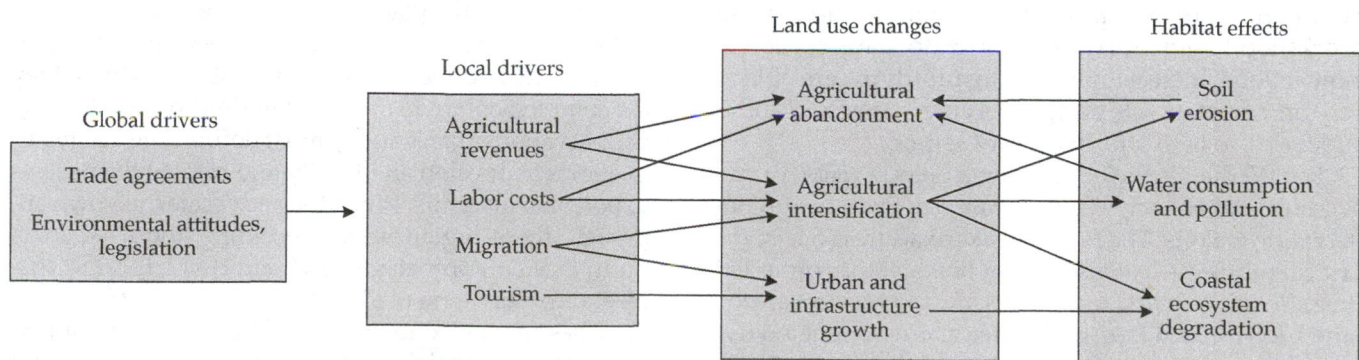

Figure 6.20 An example of how habitat degradation caused by land use change is based on economic and social drivers of land use decisions. (Modified from Millennium Ecosystem Assessment 2005a.)

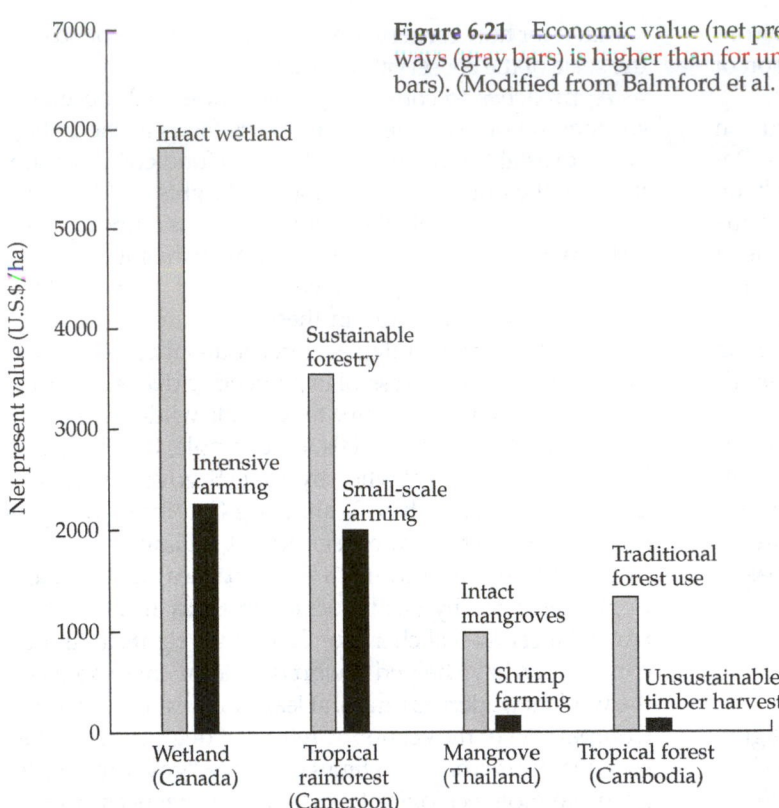

Figure 6.21 Economic value (net present value) from managing ecosystems in sustainable ways (gray bars) is higher than for unsustainable uses of that same ecosystem type (black bars). (Modified from Balmford et al. 2002.)

preserve as-of-yet undesignated but intact lands from threats that including commercial logging, mining, off-road vehicle abuses, energy extraction, and unwise development. They also work to prevent any weakening or abuse of existing policies.

One approach used by several international conservation NGOs (e.g., CI, WWF, and TNC) employs expert-based conservation priority-setting processes to guide conservation efforts on the ground (e.g., Groves 2003). These processes have served to further define geographic priorities as well as to identify critical actions. By bringing together taxonomic experts, conservation professionals, and governments to agree upon and set a common vision for priorities in conservation, these workshop-based processes have led to many conservation actions. Increasingly, the organizations are able to rely on new data sets on species, ecoregions, and social processes, to help inform priority-setting.

The Wildlands Project is using a unique vision of "re-wilding" habitat across North America to motivate conservation actions. The Project seeks to address one of the critical problems in conservation biology: that protected areas (in this case in Canada, the U.S., and Mexico) are too small, too isolated from each other, and represent too few types of ecosystems to sustain native wildlife (Locke 2000). The Wildlands Project works to reconnect the continent's "megalinkages" from the Yukon to Mexico by working through local networks. Within each of the linkages, re-

gional systems of protected areas connected by mosaics of public and private lands that provide safe passageways for wildlife are identified in conservation plans. These plans are then used by conservation groups, state and federal agencies, local governments, land trusts, private landowners, and others to make more informed decisions on the ground in their efforts to protect wildlife and wildlands.

Reaching the goal of "re-wilding" North America necessarily involves the participation of private landowners, who are encouraged to participate in voluntary actions to protect landscape linkages. Ambitious visions are turned into prescriptions through meetings with stakeholders and landscape mapping. This incredibly ambitious project is underway and has made great progress through regional networks. The Wildlands Project thus provides an excellent example of how mapping habitat loss, reaching scientific consensus on the conservation outcomes, and then local activities to protect, restore, and reconnect ensure actions are targeted to meet the conservation objectives (Locke 2000).

International conservation organizations currently employ a sophisticated combination of tactics to meet their conservation goals. A brief review of the website of any conservation organization reveals programs addressing economics and conservation, the illegal wildlife trade, healthy people and ecosystems, businesses and biodiversity, ecotourism, environmental policy, conservation enterprises, and climate change. Developing in-country expertise to implement these programs is critical to the organizations' mission success and each of the groups strives to develop technical capacity in resource conservation in the places where they work.

While the approaches used to define where conservation should be done and how to go about it differ, they are complementary to one another (Redford et al. 2003). Conservation organizations must define their strategies in efforts to develop an identity and market themselves to potential investors. While they advocate their own approach, these organizations recognize the need for a multi-faceted approach as each effort is critical to successful conservation of global biodiversity. Different approaches are needed for the very different regions around the world; recognizing opportunities for success and building on lessons learned will ensure that each of the approaches contributes as much as possible to helping to conserve and protect Earth's remaining natural places.

ESSAY 6.4

Debt-for-Nature Swaps

Neal Hockley, *University of Wales, Bangor*

■ Since 1987 more than U.S.$1 billion has been channeled into nature conservation in developing countries, through a mechanism known as "debt-for-nature swaps" (DFNS). These swaps grew out of two issues that came to the fore in the 1980s—the developing world debt crisis, and increasing concern about the rapid loss of natural ecosystems, particularly tropical forests. Interest in DFNS has waxed and waned several times over the last two decades, but they remain a significant route for funding conservation in some poor countries.

Throughout the 1960s, governments of developed nations made bilateral loans to less-developed countries (LDCs), many of them in the form of "concessional" loans, at below-market interest rates. In the 1970s, commercial banks started to see these LDCs as potential clients and after the first oil crisis of 1973 the banks had large amounts of oil producers' capital, which they needed to invest. They lent it at attractive, but variable, interest rates to LDCs, who borrowed heavily on the basis of the high prices their export commodities had been fetching, low interest rates, and a favorable exchange rate against the U.S. dollar. However, in 1981, with developed countries in recession following the second oil shock of 1979, interest rates increased as the West tried to combat inflation. Commodity prices tumbled and the dollar rose in value. Soon poor countries struggled to pay their debts, and in 1982 Mexico became the first to announce it could not make its payments. Forty two countries followed, and the "debt crisis" had begun, affecting both "commercial" debt owed to the banks and the bilateral debt owed to the rich countries' governments.

The banks viewed the debts owed by LDCs as unlikely to be repaid, and a secondary market in "bad debt" began, where debt was bought and sold at prices discounted from its "face value," according to how likely it seemed that the country would ever pay it back. Banks and LDCs sought ways to reduce the stock of bad debts, but they were constrained in their ability to do so and reduced LDC debt by only U.S.$38.9 billion (a fraction of the total) between 1982 and 1989 (Minzi 1993). In addition, the International Monetary Fund (IMF) offered new loans to LDCs to assist them through the crisis, conditional on the LDCs submitting to economic restructuring programs intended to make their economies more efficient.

The debt crisis and IMF restructuring programs forced LDCs to try to increase their export earnings. Because most relied on primary industries—agriculture, forestry and mining—for the bulk of their exports, the debt crisis increased pressure on the natural environment (Moran 1992). Debt was therefore seen as a cause of deforestation (Kahn and McDonald 1995) and the twin crises of debt and environmental destruction became linked in many people's minds.

Although the direct causal link between debt and environmental destruction has been disputed (Gullison and Losos 1993), the debt crisis undoubtedly exacerbates poverty in LDCs. The poorer a country is, the less able it is to fund nature conservation and the less willing it may be to resist pressures from its own people for agricultural land or jobs in mining and logging industries.

The Debt-for-Nature Swaps Idea

In 1984 Thomas Lovejoy of the World Wildlife Fund (WWF) suggested that the crises of debt and deforestation could both be tackled if, "debtor nations willing to protect natural resources could be made eligible for discounts or credits against their debts" (Lovejoy 1984). Three years later, the first "debt-for-nature swap" was concluded by Conservation International (CI), the Government of Bolivia, and Citicorp bank.

A debt-for-nature swap is an agreement whereby debt in "hard" currency (e.g., U.S. dollars) is forgiven (cancelled) by the owner of the debt, in exchange for commitments to support conservation on the part of the borrower (usually an LDC government). These may be financial commitments (usually paid in the local currency) to fund conservation activities, or they may be actions that the debtor government promises to undertake, such as the designation of protected areas.

The debtor government aims to reduce its foreign currency liability and also benefits by funding environmental projects in its own country that might not otherwise be funded. In the case of bilateral debt, the creditor is a rich country, which makes the best of a bad situation by canceling debt that would not otherwise be repaid, in exchange for environmental action in developing countries that addresses the concerns of the rich country's own citizens. In the case of commercial bank debt, the bank either sells the debt at the market price to a conservation organization, or it donates or discounts the debt below the market price in return for favorable publicity or tax credits. The conservation organization aims to persuade the debtor country to conserve nature and tries to maximize the amount and effectiveness of conservation spending in that country. Local conservation organizations usually receive the local currency funds and carry out projects on the ground.

There have been many DFNSs since 1987, and more than U.S.$1 billion of funds has been allocated to conservation through this mechanism (Paddack 2003).

Evaluating the Debt-for-Nature Concept

Interest in DFNS has varied dramatically over the last 20 years and the most important factor in determining their importance has been the presence of other schemes for reducing LDC debt, against which DFNS must compete. DFNS have only ever accounted for a small part of the debt burden, and this fact, combined with the uncertain nature of the links between debt reduction and deforestation, means that DFNS should be evaluated primarily on their ability to fund conservation programs in LDCs.

Evidence suggests that the potential for conservation organizations to greatly multiply conservation funds through DFNS is limited. Any apparently attractive conversions may offer a large risk, due to lost control over funds. DFNS are most likely to be successful in countries with stable governments and that have a genuine interest in nature conservation. However, it is

likely that in these countries, other sources for funding of conservation will exist and may avoid the high transaction costs of DFNS.

Perhaps the most important role of DFNS, as originally suggested by Lovejoy (1984), could be to even out the distribution of costs and benefits of conservation, which at present are shared unequally between LDCs who bear the costs, and rich countries, whose citizens gain benefits (Balmford and Whitten 2003). Where nature conservation is not in the best interests of poor countries, rich countries will have to pay them to do it and there is evidence to suggest that "generous" bilateral DFNS have occurred (e.g., Torras 2003). However, distributional inequalities of conservation costs and benefits also exist within LDCs, with the rural poor bearing disproportionate costs of forest preservation (e.g., Ferraro 2002). Any conservation program in poor countries will have to address this issue, yet so far, most attempts to compensate for the local costs of conservation have been indirect and often unsuccessful (Ferraro 2001). DFNS, like all conservation programs in poor countries, will be increasingly judged on their contribution to poverty alleviation and must demonstrate that nature conservation can help the poor and not just the rich. ■

CASE STUDY 6.1

The Importance of Land Use History to Conservation Biology

David R. Foster, Harvard Forest, Harvard University

"Yes, but what is its history?"

Answering this single question in regards to a species, habitat, or landscape often provides a wealth of ecological insight as well as essential tools for conservation biologists as they seek to restore, maintain, or manage conditions into the future. Today it is easy to compile thousands of instantaneous field measurements on dataloggers, couple these with results from controlled experiments, and then feed them all into sophisticated models and statistical analyses. Consequently, there is a strong tendency to concentrate on the present without reflecting on the past. And yet, studies from forests to grasslands, lakes to streams, and from wilderness areas to reservoirs have shown that every species, habitat, and landscape is strongly conditioned by its history. Each is on a trajectory of change determined by processes and events occurring years to centuries ago, but that are invisible to even the most precise sensors. Therefore, it is always worthwhile to pause, reflect, and dig a bit (literally and figuratively) to interpret the past as we face modern conservation challenges. A historical perspective is especially useful to conservation biologists, for we seek first to establish an ecological understanding of the factors that contribute to the abundance and distribution of organisms, assemblages, and habitats and then apply that understanding toward managing for particular goals. Effort devoted to historical reconstructions can often save us time, money, and grief as we move forward.

One impact of historical studies on the fields of ecology and conservation biology has been to increase appreciation of the importance of land-use activity in shaping modern conditions. In diverse ecosystems, from the Amazon to the north slope of Alaska, we have found evidence for the near ubiquity of human activity, albeit often from prehistoric times. Ranging from forest-clearing, agriculture, and landscape management with fire, to hunting and plant collection, humans have exerted direct and indirect impacts on species abundances and ecosystem structure and function. Often the ecological consequences of these impacts persist for centuries, even beyond the society's duration. Clearly, we have come to be attuned to the history of anthropogenic as well as natural processes. Fortunately, we can employ the same tools and approaches toward reconstructing the past regardless of the relative importance of either process.

Why should ecologists investigate history? Reconstruction of a species, site, or landscape's history may yield the following ecological insights:

- *The range of variability of species abundance, changes in species assemblages, or shifts in environmental conditions and landscape patterns.* How do we evaluate whether the change in a population is worrisome or not without information on its past dynamics and the processes that have shaped these? How do we interpret the current climatic or landscape setting without an appreciation for how these have varied in the past? A long temporal perspective provides critical reference for interpreting the significance and relative magnitude for recent changes and dynamics.

- *The role of past natural and human disturbances in controlling modern ecological patterns and conditioning future changes.* In many ecosystems disturbances are a regular, though often infrequent process that is responsible for maintaining critical conditions or suites of species. In some situations extreme events such as a catastrophic

windstorm, a pest or pathogen outbreak, or ancient land use impacts have selectively favored or removed particular species or altered soil conditions or stream chemistry in lasting ways. For example, an area that was once plowed may look superficially the same as one that never had been, but the species composition and relative abundance may differ even a hundred years post-disturbance (Figure A). Though the evidence for such events may be hidden by successional changes and the resilience of ecosystems, their impacts may be enduring.

• *The future trajectories of the ecosystem under study.* All species, communities, and landscapes are changing. Frequently, it is impossible to measure the rate and direction of that change through short-term studies or observations. Nonetheless, projecting or anticipating future changes is a critical component of most studies in conservation biology. By lengthening the record of observations from a human observational scale to the appropriate ecological scale, historical studies can reveal the nature of these trajectories and enable more reasonable projections of future conditions.

These critical ecological insights allow conservation biologists to move from interpretation to management discussions with land managers, the public, and various constituencies. Here history often helps us to identify:

• *Feasible targets for conservation.* Knowing the history of a site often facilitates the identification of a range of possible future conditions or outcomes that may be achievable through management. Management for the restoration, introduction, or maintenance of particular conditions are often disappointing precisely because the initial goals were unreasonable, outside the range of past or expected conditions, or based on a poor understanding of the inherent trajectory of change in the system. By framing goals against an understanding of past conditions rather than subjective expectations, we have a stronger likelihood of success.

• *Effective tools and approaches to management.* Without a good understanding of the factors and processes responsible for developing modern conditions, managers may use the wrong tools in an attempt to achieve

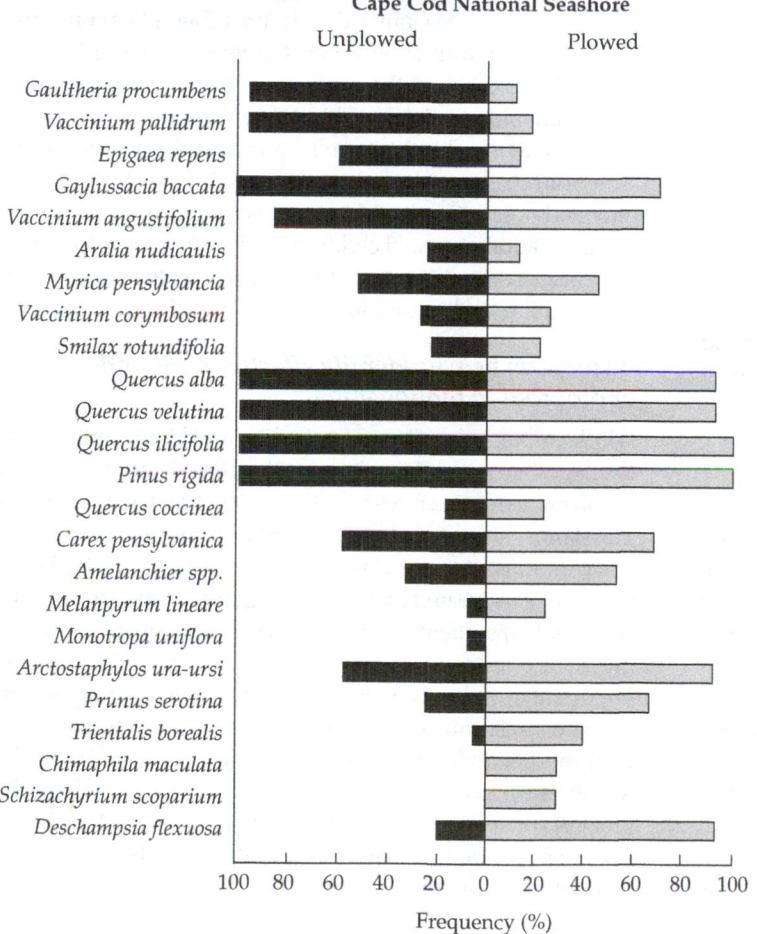

Cape Cod National Seashore

Unplowed   Plowed

Figure A   The understory flora of woodlands that had once been cultivated for agriculture (plowed) differs from that found in sites that had always been woodlots (unplowed) in the Cape Cod National Seashore, MA. In this region, as in much of New England, lands were cleared of trees to make way for agriculture, plowed, and beginning in the nineteenth century, abandoned and allowed to reforest. Both plowed and unplowed sites have been forested and managed since the late nineteenth century, yet a century after their use, the flora clearly differs. There are three groups of species represented, (1) species (especially ericaceous shrubs) that are more abundant in former woodlots that were eliminated by agriculture and have failed to reestablish, (2) weedy species that established well in agricultral sites and persisted there, but that are incapable of establishing widely in intact forest areas, and (3) species that are excellent dispersers and establish readily in any site. (Modified from Foster et al. 2003.)

desired goals. A great example of this is early management of many national parks in which fire was interpreted as having a negative influence on natural ecosystems and was contained as much as possible. Subsequent study, often initiated by undesirable changes in park conditions, indicated that fire, ignited by lightning or people, was often a critical ecosystem process. Policy changes to allow for natural or prescribed fires were an appropriate response to these historical insights.

- *A compelling historical narrative to accompany ecological interpretations and conservation management plans.* A historical narrative is often extremely effective in engaging and educating managers and the public when depicting the ecological status of an uncommon species or unusual and important habitat. People enjoy knowing the history of the landscapes that they live in or care for, and they frequently recognize that this history provides critical background for management. While scientists may like to think of management as an objective process grounded solely in data, subjective impressions are critical to the development of policy and management. Historical research yields both data and rich narrative that are mutually supportive.

Examples of History Providing Insights into Conservation

There are two examples from New England that describe how learning from ecological history informs conservation decisions. The first example discusses the unrealized goal of restoring Atlantic salmon in the Connecticut River. Unfortunately this is a painful tale of the pitfalls of ignoring history in framing conservation management. The second example concerns efforts to maintain and restore uncommon grassland plant and animal species in Massachusetts. This is a more upbeat story of learning and adapting management based on history.

History can help us identify feasible targets for conservation

For the past three decades state agencies, private groups, industry, and many thousands of citizens and school children have joined a national effort to restore the Atlantic salmon to the Connecticut River, New England's largest watercourse. Based on the belief that massive runs of salmon had been eliminated by industrial pollution and dams in the nineteenth century, an effort exceeding U.S.$100 million has been directed toward cleaning the river, facilitating the passage of anadromous fish, and stocking salmon by the millions to small tributaries. Living in the river's watershed where I read news articles on this effort, I was perplexed by the low number of returning fish, only a handful after 30 years of effort. Researching the federal program I was surprised to discover that it had never investigated the historic or prehistoric dynamics of salmon. In fact, the entire program was based on anecdotes of people pulling bucketfuls of fish from the river. Further reading revealed that Cathy Carlson, a doctoral student at the University of Massachusetts, had comprehensively investigated all of the watershed's archaeological sites and could find no evidence for either the remains or the use of salmon by Native Americans.

Carlson provided a biologically reasonable interpretation of salmon's history that contains a sobering message for conservationists. The Connecticut River lies just beyond the southern range limit for Atlantic salmon and did so for most of the last 5000 years. However, early Colonial times coincided with the Little Ice Age, a period of globally cool climate and sea temperatures when the salmon's range extended slightly further south. At this point modest numbers of fish entered the Connecticut. These historical runs and the great abundance of many other anadramous fishes, especially American shad, led to the grossly exaggerated anecdotes. Although nineteenth century industrialization may have contributed to the decline of salmon, it coincided with warming climates, which continue today.

This historical perspective suggests that restoration of Atlantic Salmon to the Connecticut is infeasible, a fact supported by last year's return of only 45 fish from more than 10 million released. Nonetheless, the program continues. However, there is a bright side to this story because this effort yielded immense benefits. The river is now swimmable, nearly 200,000 American shad migrate upstream annually, the regional economy and population benefit from river recreation, and an entire generation of school children in the four state watersheds have been exposed to important lessons in biology, geography, conservation, and restoration.

History can help us identify effective tools and approaches to management

A little-appreciated aspect of New England biodiversity is that the largest group of declining and threatened species does not live in our extensive forests, but occupies open habitats such as grasslands, shrublands, and heathlands. Birds, small mammals, and insects (especially lepidopterans) thrive in the diverse habitats created by varying heights and species of grasses, forbs, and scattered shrubs. These habitats are disappearing due to development and forest succession. Many of the most important open land habitats are now protected by state agencies or conservation organizations, especially sandplain grasslands and heathlands in the coastal landscape of Cape Cod, Martha's Vineyard, Block Island, and Nantucket where the greatest diversity of uncommon species occur. As tall shrubs and trees invade these sites conservation efforts increasingly focus on restoring these habitats.

Early restoration attempts were based on an interpretation that grasslands and heathlands were natural communities largely created and maintained by fire, a natural process and one manipulated by prehistoric humans. This line of thinking, which paralleled national recognition of fire's importance, led to programs of prescribed burning by conservation organizations and agencies. However, results were frequently disappointing. What was wrong?

New historical studies yielded ecological interpretations that are gradually broadening the management toolbox that conservationists employ on these areas. Drawing from diverse studies using paleoecology, historical maps and documents, interpretation of soil morphology, and compelling sets of paired historical photographs, we now conclude that the vast majority of these habitats are of cultural origin. However, it was early European land use activity, not Native American practices, that created these habitats from a forested landscape. Essentially every important sandplain grassland and heathland bears the subtle marks of ancient agriculture, especially grazing by sheep and cattle. It is easy to forget, with the current extent of forest, that through the mid-nineteenth century New England was a deforested agricultural landscape (Figure B). Over the past two centuries most farms and farmland have been abandoned and forests have expanded regionally through a natural process.

The sandplain grasslands and heathlands are remnants of old pastures that now appear perfectly natural under all but the most intense scrutiny or through the lens of historical insight. From this perspective it is clear that fire was not a primary mechanism for the development and formation of these habitats; they were produced by removal of the trees and intensive grazing. However, fire was one of many processes, including mowing, that colonial farmers used to keep them open and that added structural and compositional diversity to these habitats. Based on these new insights, conservation groups and land managers are now shifting toward employing a much more diversified and historically accurate set of tools to create, restore, and manage these areas. Today this conservation toolbox includes grazing, mowing, brush cutting, soil scarification, and fire. Under this diversified management regime it is possible to maintain a wide array of habitats that can accommodate the diverse needs of many important and unusual species.

A Balanced Approach That Raises New Questions

In the end, are historical studies and an understanding of past land use a panacea that will transform ecological studies and their application in conservation biology? Obviously not. The reconstruction of past environments and landscapes is only one tool among many that help interpret modern landscapes and manage them effectively. But when combined with intense studies of the species and habitats that we work on, history provides a rich source of important information.

And like any other piece of scientific information it inevitably introduces new complexities, questions, and conundrums into our interpretations and into the ways in which they may be used by different constituencies. For example, because all landscapes have changed in the past and there are no absolute baseline conditions, which of all the historical and future conditions should we choose for our conservation goal? Furthermore, because change is natural and humans have affected so many landscapes over time, why should we worry about the changes wrought by human activity we are witnessing today? Finally, how do we value "nature" when it displays abundant evidence of cultural activity—should we attempt to manage for cultural as well as natural conditions? The extent of change and the cultural origins of the sandplain grasslands described above raise many of these questions.

Obviously, these and many other issues are part of what makes all of our management efforts so challenging, interesting, and complex. Injecting some historical perspective inevitably expands these interpretations and discussions.

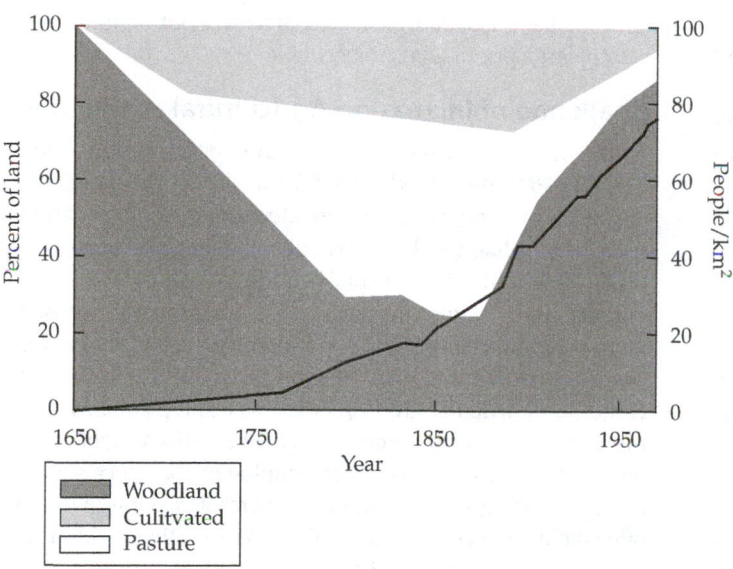

Figure B   Beginning in the mid-1600s, the landscapes of New England were largely converted to pasture and croplands, but with the industrial revolution were abandoned and reforested until the present day. Population density (black line) increased over the same period, as people moved to cities from the countryside. (Modified from Foster 2002.)

CASE STUDY 6.2

Promoting Wildlife Conservation in Agricultural Landscapes

Jeffrey A. McNeely, IUCN-The World Conservation Union and Sara J. Scherr, Ecoagriculture Partners

Conventional wisdom holds that crop and livestock production are largely incompatible with wildlife conservation. Thus, policies to protect wildlife typically rely on land use segregation, establishing protected areas from which agriculture is excluded (at least legally). Farmers are seen as sources of problems by those promoting this view of wildlife conservation. However, new approaches indicate that farming systems can make important contributions to biodiversity conservation. These contributions can be enhanced by technical, institutional, and policy research.

Challenge of Conserving Biodiversity in Agricultural Landscapes

Evidence from satellite images shows that agricultural production systems dominate global land use. Over 12% of the world's land area exists in landscapes where greater than 60% are croplands or planted pastures. Croplands account for at least 30% of land use in another 20% of the world's land area (Wood et al. 2000). Overall, more than 80% of terrestrial area is affected by agriculture if you count land that is being fallowed as part of the farming cycle, tree crops, ranching and pastoral systems, and community forestry. About a third of global agricultural extent is in high-intensity production systems that use high levels of agrochemicals for continuous cropping, and often reshape land and waterways. The rest of the agricultural extent is under extensive farming systems that use far fewer inputs, but require relatively large expanses of land to produce lower crop and livestock yields.

Agriculture as it is often practiced—in both intensive and extensive systems—is one of the chief threats to wild species and their habitats. Nearly half of all temperate broad-leaved forest and tropical and subtropical dry forest, and a third of temperate grass and shrubland have been lost as wildlife habitat through conversion to agricultural use; conversion rates are especially high in Asia and Europe. Over half of the most species-rich areas contain large human populations whose livelihoods depend on farming, forestry, herding or fisheries. Many of these human populations are plagued by chronic poverty and hunger. Unfortunately, in such areas, biodiversity is often depleted in ways that threaten productivity and sustainability, as well as species conservation.

Agriculture also has large impacts on freshwater systems. Irrigation is practiced on over 250 million ha, and over 70% of the freshwater resources withdrawn for human use (89% in some developing countries) are used for irrigating farms, thus modifying natural hydrological systems. Over half of the world's wetlands—among the planet's most valuable wildlife—have been converted to agriculture. Critical watersheds are home to large and growing human populations that must be used for food and fiber production as well as for terrestrial and aquatic wildlife.

Finally, agriculture mismanagement has led to decline in the sustainability of many areas. For example, significant soil degradation has occurred on 16% of all crop, pasture and forestland worldwide, and half of all land within the agricultural extent, thereby affecting the diversity of soil microorganisms. Excessive use and poor management of crop nutrients, pesticides, and penned livestock wastes are a major cause of habitat pollution that can kill wildlife directly or impair reproduction.

Such evidence suggests a need to redouble efforts to establish protected areas "off limits" to agriculture. But this is not sufficient. Of over 100,000 major sites already devoted to conserving wild biodiversity, agricultural use can occur on up to 30% of the land area within a majority of these sites. Most of the rest are islands within a "sea" of agriculture. Some ecologists calculate that even if the existing protected areas remain as wildlife habitat, 30%–50% of their species may still be lost because such isolated protected areas do not contain sufficiently large populations to be viable.

Challenge of Increasing Agricultural Production

Some experts suggest that 30 years from now, we will need at least 50%–60% more food to feed the world's population and to enable at least some degree of greater affluence. Yet providing more food is likely to be a particular challenge because of the decline in global environmental conditions. Loss of biodiversity, deforestation, water shortages, desertification, soil erosion, global climate change, and various other factors make it increasingly difficult to increase food productivity. Soil erosion could be a particular limiting factor, as lost top soil has led to the abandonment of some 1,000,000 km^2 of croplands each decade (Myers 1999). The biotechnologies that once seemed most promising are coming under increasing public scrutiny, often failing to meet the most critical needs for the world's poor.

Irrigation historically has been an effective means of increasing productivity, and the area of irrigated croplands increased by an average of 2.8% per year between 1950 and 1980. Subsequent advances have been much smaller, falling to only 1.2% per year, and some experts expect future expansion to slowly come to a halt. On a per capita basis, irrigated lands have shrunk by 6% between 1978 and 1990 and are expected to

contract by a further 12% per capita by 2010. Even worse, nearly 10% of irrigated lands have become so salinized as to reduce crop yields (Myers 1999). Agricultural land is increasingly becoming a limited resource, and per capita arable land has declined by an average of 1.9% per year since 1984; some agricultural lands are so abused that they are losing much of their agricultural value, while other areas are falling victim to urban sprawl and the spread of industry into farming areas.

Reconciling Agricultural Production and Biodiversity Conservation: Ecoagriculture

Can ways be found to reduce, or even reverse, the impacts of agriculture on wild biodiversity? Given present agricultural technologies and policies, most farmers can increase biodiversity significantly only by reducing production and livelihood security. Initiatives to promote more ecologically sensitive farming systems (called "sustainable," "regenerative," or "organic" agriculture) are expanding, often with positive impacts on wild biodiversity, but they focus mainly on preserving "useful" wild species, such as pollinators or beneficial soil microfauna.

Farmers rely on many levels of biodiversity in their livelihood, drawing on genetic resources to maintain productivity in their crops, using pollinators (usually from the wild), and drawing on water from watersheds whose productivity is maintained through functional ecosystems. But how can farming communities be convinced to conserve biodiversity as part of their daily work? In fact, quite a lot of effort is going into such conservation, in virtually all parts of the world, often led by agricultural communities themselves to protect biodiversity and ecosystem services important to their own livelihoods and culture (Figure A).

Figure A Shade and mulch trees enhance sustainability of this tea tree estate in Sri Lanka, and may provide habitat for some wild species. (Photograph courtesy of Ecoagriculture Partners, by L. Simmons.)

In such places, a new framework for land and resource management is needed: "**ecoagriculture**," that is, landscapes that sustain rural livelihoods through productive and sustainable agriculture (crops, livestock, forests, fishes) and *also* protect biodiversity and ecosystem functions effectively at a landscape scale (McNeely and Scherr 2003). Ecoagriculture adapts the concept of "ecosystem management" to production landscapes. Enhancing rural livelihoods through more productive and profitable farming systems becomes a core strategy for both agricultural development and conservation of biodiversity. The ecoagriculture framework is needed wherever demands for food, ecosystem services, and rural livelihoods converge, especially in rural farming communities living in or around protected areas and other habitats of high biodiversity value or endangered species, in critical watersheds serving human and wildlife populations, in biologically degraded landscapes where ecosystem services are essential for sustainable agriculture and local livelihoods need urgent rehabilitation, or in urban agricultural regions where production–ecosystem–livelihood interactions are especially intense.

Strategies and Synergies of Ecoagriculture

Ecoagriculture encompasses two sets of strategies with diverse elements combined to meet location-specific challenges for land and resource management at the landscape scale (Table A). The first set seeks to increase wildlife habitat in non-farmed patches in agricultural landscapes, creating mosaics of wild and cultivated land uses.

One approach that has been taken is to establish habitat networks and corridors in "in-between" spaces that have ecological synergies with farming (such as hedgerows or windbreaks). For example, farmers in 19 communities in Costa Rican cloud forest created 150 ha of windbreaks by planting a mix of indigenous and exotic tree species. The windbreaks increased the herd-carrying capacity of the pastures and raised coffee and milk yields, and also serve as important biological corridors connecting remnant forest patches.

Another approach has been to create new protected areas that also directly benefit local farming communities (by increasing the flow of wild or cultivated products, enhancing locally valued environmental services, or increasing agricultural sustainability). For example, in Australia, farmers have cooperated to rehabilitate degraded agricultural areas by fencing off key areas for revegetation, and then used them to introduce threatened and endangered native marsupial species.

A third approach has been to raise the productivity of existing farmland to prevent or reverse conversion of natural areas, along with explicit measures to protect or restore the biodiversity value of uncultivated lands. For example, dairy

TABLE A *The Differences in Agricultural Systems*

Percent landscape in agricultural use	Low-intensity production	Moderate-intensity production	High-intensity production
Low amount (<30%)	Shifting cultivation	Mountain rainfed crop-forest systems	Irrigated oasis agriculture
Moderate amount (30%–70%)	Pastoral herding systems	Agropastoral systems; short-fallow rainfed crop; plantations	Permanent crop system in lower/mixed potential areas
High amount (>70%)	Low-input dryland crop-scape in agricultural use livestock system	Agroforests; continuous cropping	Asian "rice bowls," U.S. "bread-baskets"

farmers in the threatened Atlantic Forest of Brazil, who benefited from improved pastures and livestock breeds, were willing to work with conservation organizations to convert marginal pastures back to native forest.

A second set of ecoagriculture strategies enhance the habitat quality of productive farmland in ways that also increase yields or income, or reduce costs. This can be done by reducing agricultural pollution through new methods of nutrient and pest management, and farm and waterway filters. For example, in Yunnan Province in southern China, farmers reduced the need for pesticides by using more diverse rice varieties to control rice blast disease. The mixture increased rice yields by 89%, as the rice blast declined by 94%. In Africa, fallow plantings of "fertilizer shrubs" not only restore soil organic matter and replace costly fertilizer, but also provide improved habitat for wildlife.

Another approach is to modify the management of soil, water, and natural vegetation to enhance habitat quality, or the management of wildlife or domestic livestock for more compatible resource use. For example, in California, rice farmers are disposing of waste straw and controlling pests and diseases cost-effectively by winter-flooding fallow fields, which provides high-quality wetland habitats for migratory birds. In east Africa, herders have learned to use and manage rangeland resources in ways that minimize competition with wild ungulates.

Modifying the mix and configuration of agricultural species to mimic the structure and function of natural vegetation, often with greater emphasis on perennial species that have economic as well as environmental value is another approach. For example, in Central America, scattered trees provide shade to cattle, as well as timber, firewood, and fence posts to farmers. They also retain rich communities of plants that would otherwise not be present in the agricultural landscape, providing food for migratory birds, such as Three-wattled Bellbirds (*Procnias tricarunculata*), Resplendent Quetzals (*Pharomachrus mocinno*), and Keel-billed Toucans (*Ramphastos sulfuratus*), as well as to bats and other animals that live on or near the farms.

Research Directions

While these examples illustrate the potential for dramatic benefits from ecoagriculture, in most landscapes, locally-known options are limited and often the costs of innovation discour-

age action. Major improvements can be anticipated from more effective cross-site information exchange, and from adopting "best practices" for agriculture and conservation management, improving these through trial and error to design landscapes that address both local livelihood and conservation objectives. Traditional knowledge, particularly from more extensive and biodiverse land-use systems, as well as experimentation by diverse local stakeholders are rich sources of innovation.

The next generation of innovations beyond these, particularly in more intensively managed landscapes, will depend on more systematic research. Indeed, ecoagriculture on a globally meaningful scale is feasible now in large part because of our greater capacity to find synergies through scientific management. Advances in conservation biology, agricultural ecology, plant breeding, ecosystem monitoring systems, and modelling are revolutionizing our ability to understand and manipulate wildlife–habitat–agriculture interactions, to the benefit of both people and the rest of nature. However, the necessary work has barely begun. Our basic understanding of patterns of biodiversity in production landscapes is limited—only 6% of published conservation biology studies were implemented in agricultural landscapes. Where agricultural research encompasses biodiversity, it typically focuses on impacts at the plot or farm level, rather than at the landscape/habitat scale. Major new scientific initiatives are required, using methods and tools from various disciplines. Research is critically needed on institutional innovations that enable more effective, lower-cost processes of engaging stakeholders in landscape-scale planning, coordination and assessment, as well as policies to support ecoagriculture; examples include:

- Veterinary research to develop a livestock vaccine against rinderpest, a viral disease, has not only protected domestic cattle in east Africa, but also protected millions of wild buffalo, eland, kudu, wildebeest, giraffes and warthogs that share rangelands and reserves, and that are also susceptible to the disease. New park zoning and use regulations, as well as communications systems with local herders, are being developed for more successful co-management.

- Crop breeders in the U.S. are developing native perennial grains (such as bundleflower [*Desmanthus*

illinoensis], wild rye [*Leymus racemosus*], eastern gama-grass [*Tripsacum dactyloides*, a relative of *Zea mays*], and Maximilian sunflower [*Helianthus maximiliani*]) that can be grown more sustainably with much less environmental damage in dryland farming regions. The systems are not yet economically competitive, but yields have reached 70% of annual wheat varieties. Production costs are lower, and habitat value for wildlife is many times higher than in conventional wheat fields.

- In the humid tropics, research has demonstrated the benefits of both sustainability of production and biodiversity conservation of farming systems that "mimic" the structure of the natural forest ecosystems. Millions of ha of multi-strata "agroforests" in Indonesia produce commercial rubber, fruits, spices and timber, often in a mosaic with rice fields and rice fallows. The number of wild plant and animal species in these agroforests are often nearly as high as in natural forests. Maintaining these systems involves policy reforms to strengthen farmers' tenure claims, and "level the playing field" with subsidized rice production.

- In Central America, researchers are developing systems of shade-grown coffee with domesticated native shade tree species, that maintain coffee yields while also diversifying income sources and conserving wild biodiversity. Farmer adoption of these systems has been promoted through changes in public coffee policy to favor shade systems, technical assistance, and in some cases price premiums in international markets for certified "biodiversity-friendly" coffee.

- In South Africa, research has shown that both the population of helmeted guinea fowl (*Nuzmida meleagris*) and overall avian diversity declined with increasingly intensive agriculture and disappearance of edge habitat and associated, habitat mosaic typical of traditional agriculture. On the other hand, traditional agriculture in the form of contouring (plowing in parallel contours to the topography to reduce erosion and maximize water retention) in a pesticide-free environment resulted in extensive edge habitat that provided food and cover for birds, leading to an increase in overall bird diversity (Ratcliffe and Crow 2001).

To have meaningful impact on biodiversity conservation at global or regional scales, ecoagriculture must be broadly promoted. The vision of ecoagriculture is that agricultural communities in ecologically important and sensitive areas around the world will actively manage their landscapes as *ecoagriculture*—enhancing rural livelihoods and sustainable production of crops, livestock, fish and forests, while also conserving biodiversity, watersheds and ecosystem services.

Summary

1. Continued habitat degradation is the largest threat to biodiversity. Nearly 60% of Earth's ecosystems are degraded or used unsustainably. Temperate grasslands, savannas and shrublands, and Mediterranean habitats have lost the most area to human activities. Habitat degradation is the primary cause of extinction and endangerment globally. In the U.S. only 42% of native vegetation remains.

2. Habitat degradation and loss are caused by agriculture, mining, forestry, fisheries, aquaculture, groundwater extraction, fires, infrastructure development, dams, urbanization, industry, pollution, and changes in community and ecosystem structure wrought by changes in the abundance of native and invasive species. Impacts of habitat change range from light and temporary (e.g., small fires), to devastating and permanent (e.g., urbanization). The term habitat loss is reserved for impacts so severe that all or nearly all species are affected, and for which recovery times are long.

3. Humans have transformed habitats throughout their history, primarily for agriculture. Even seemingly pristine areas have been influenced by human activities at some point in the past. Forests are strongly degraded, with global annual losses in forest area equivalent to a country the size of Egypt or Colombia. Consequences of deforestation include increased erosion and flooding, local and global extinctions, and climate change. Grasslands, savannas, and shrublands have been widely replaced by cropland and pasturelands. Extreme degradation has resulted in desertification of these systems. Freshwater systems have been degraded by dams, water diversions, and wetland losses. Radical transformation of freshwater systems has caused extreme environmental changes in many places, including the devastation of even large water bodies, such as the Aral Sea. Marine habitats are also strongly modified, although data for evaluation are fewer than for freshwater habitats. Estuaries are degraded by water diversion and nutrient inputs and by engineering projects to manipulate water flows and build urban environments. Coral reefs and many coastal ecosystems have been greatly damaged in the past few decades.

4. Agriculture is the primary land use that leads to habitat degradation, and intensive cropping systems

can lead to complete habitat loss. Overgrazing has resulted in more degraded land than any other mismanaged land use. Extractive activities such as mining and quarrying cause habitat loss, as do some forms of petroleum exploration and refining. Logging and trawling can strongly degrade forested and marine benthic ecosystems, respectively. Urbanization causes habitat loss, and cities have ecological footprints that can extend more than 100 times their area. Wars and other violent conflicts cause widescale habitat degradation both directly from the conflict and via the degradation caused by displaced human populations and limited ability to enforce conservation regulations.

5. Pollution is a special form of habitat degradation that can directly cause habitat loss (e.g., when soils become contaminated with toxic chemicals) and indirectly compromise the physiological functioning of individual organisms, thus degrading the value of an area. Pollution can take many forms including light pollution, air pollution, acid rain, solid and plastic waste, and toxic chemical pollution of water and soil at chronic to acute levels.

6. Nitrogen (and to a lesser extent phosphorus) pollution has grown to be a highly degrading force through its effects on biological processes from organismal physiology to ecosystem nutrient cycling. Nitrogen deposition has more than doubled through human activities, causing widespread eutrophication of aquatic systems. Eutrophication caused by untreated sewage can be reversed, but it has proved more difficult to reduce agricultural nitrogen inputs.

7. Efforts to protect remaining natural habitats have centered on setting priorities to govern where conservation efforts are focused. Among these efforts are priotization schemes that emphasize areas with high species-richness and immediate threats, representativeness of global ecoregions, and areas of remaining wilderness. Conservation practice attempts to address the underlying social, political, and economic factors that influence land-use decisions, and promote conservation of these remaining natural areas. Conservationists are embracing a diversity of approaches as no one approach will succeed alone.

Please refer to the website www.sinauer.com/groom for Suggested Readings, Web links, additional questions, and supplementary resources.

Questions for Discussion

1. Why is habitat degradation the greatest threat to biodiversity?

2. What are the primary factors leading to degradation of forested ecosystems; grasslands, savannas and shrublands; freshwater systems; and marine systems?

3. Why is pollution so large a problem in aquatic ecosystems? How can these problems be addressed (e.g., plastics, eutrophication)?

4. Some have argued that the most important factor that conservation biologists should focus on is improving agricultural yields to conserve biodiversity. Do you agree or disagree? Why?

5. How should conservation biologists seek to preserve remaining natural areas? What should their priorities be? How should they approach achieving their goals?

7

Habitat Fragmentation

Reed Noss, Blair Csuti, and Martha J. Groom

Frag·ment (frag ment) n. 1. A part broken off or detached from a whole. 2. Something incomplete; an odd bit or piece...

The American Heritage Dictionary

Destruction or alteration of natural habitat by humans is the major proximate threat to biodiversity worldwide. Most blatant is direct habitat loss, as when a forest is converted to a cattle pasture, a wetland is drained for cropland, a river is dammed to create a reservoir, or a remnant prairie is replaced by a shopping mall. Such destruction, however, rarely sweeps across the landscape uniformly as a moving front. When we view the broader landscape, as from an airplane, the most striking pattern is often scattered bits and pieces of what once was continuous natural habitat.

Habitat fragmentation has two components: (1) A reduction in the area covered by a habitat type, or natural habitat generally, in a landscape; and (2) a change in habitat configuration, with the remaining habitat apportioned into smaller and more isolated patches (Harris 1984; Wilcove et al. 1986; Saunders et al. 1991). Although only the latter component is fragmentation in the literal sense (Fahrig 2003), in the real world changes in habitat configuration almost always occur in tandem with a reduction in habitat area. There is considerable variation, however, in the pattern of fragmentation. In some cases a landscape may be more "shredded" than fragmented (see Essay 7.1 by Peter Feinsinger). In other cases, "variegation," or creation of a shifting, fuzzy-edged mosaic of continuously varying suitability, more accurately describes the landscape from the standpoint of species' habitat requirements (McIntyre and Barrett 1992; Ingham and Samways 1996; McIntyre and Hobbs 1999). The **landscape matrix** that surrounds habitat fragments may be hospitable to many native species, or at least can be used for movement among fragments (Ricketts 2001). Nevertheless, the end result of human settlement, agriculture, or intensive resource extraction in a landscape is often a patchwork of small, isolated natural areas in a sea of altered land (Figure 7.1). The forces behind such landscape conversion and their effects are graphically illustrated in Case Study 7.1 by Richard Knight et al.

In recent decades, thousands of studies worldwide have sought to illuminate the mechanisms underlying the loss of native biodiversity associated with fragmentation, predict which kinds of species are most sensitive to fragmentation, and suggest

ESSAY 7.1

Habitat "Shredding"

Peter Feinsinger, *Northern Arizona University*

■ When modern-day humans convert a landscape and reduce the original habitat to a small fraction of its former area, the term *habitat fragmentation* is most commonly employed. This label conjures up an image of circumscribed islands of natural habitat jutting from an advancing sea of agriculture or other form of land development, isolated by quite inhospitable terrain from one another and from the nearest unconverted "continent." The insularization analogy is compelling and powerful, with the result that most work on habitat remnants and nature reserves assumes, implicitly or explicitly, that these are configured as islands. For example, MacArthur and Wilson (1967) used a fragmentation example to lead off their classic treatise on island biogeography.

Even the debate surrounding very different physical layouts—corridors and networks—now treats these as means of connecting remnant habitats of greatest concern, assumed to be island-like, rather than as entities themselves of conservation interest. The island analogy also underlies models for the genetic and demographic consequences of small population size (Chapters 11 and 12). The observation that "nature is patchy" generates many potent ecological concepts, such as metapopulations, source–sink topographies, dispersal–diffusion processes, or landscape mosaics. When these concepts are used to model events on landscapes of conservation concern, again we often assume that natural habitat in such landscapes is insularized. And for many landscapes the assumption is valid (Figure A).

Other configurations are possible, however. At least in Latin America, and I suspect in other regions, it is often difficult to find landscapes that fit the image of insularization well enough for the island analogy and its corollaries to be applied with a clear conscience. Graphic examples abound, though, of **habitat shredding**. Along many advancing agricultural frontiers, and even in some stabilized agricultural landscapes, land use practices *shred* the original habitat into long narrow strips (Figure B) rather than fragment it into two-dimensional isolates. Shreds of native vegetation snake

Figure A   A fragmented landscape seen in the study plots of the Biological Dynamics for Forest Fragments project near Manaus, Brazil. (Photograph © Mark Moffett/Minden Pictures.)

Figure B   A shredded landscape caused by subsistence farming in the Ruhengeri district, Rwanda. (Photograph © Bruce Davidson/naturepl.com.)

along watercourses or ridges, survive on disputed boundaries between different landholders, persist as buffers between different crops, or serve as cheap (if rather ineffective) fences for cattle pasture. Shreds may be several meters wide or several hundreds, encompassing hectares or thousands of hectares. Typically, shreds are not isolates; they connect directly with the as-yet unconverted habitat (which may itself be a protected reserve) that persists beyond the inva-

sion front. Some shreds protrude as simple peninsulas, others link with one another in complex networks, and a few, corridor-like, run into a second large tract at the far end, although this is rare.

Does it really make a difference to the biota, or to its investigators, whether habitats are fragmented or shredded? Yes! In shredded landscapes, populations of native plants and animals do not necessarily languish in isolated, two-dimensional patches; rather, they are tentacles extending from a corpus that still resides in unconverted habitat, probing the converted countryside. The key questions are not about which populations persist despite low population size, for how long they persist, and in which size of fragments; rather, the questions are about which populations extend outward, for what distance they extend, and along which sort of shreds. Long and narrow, shreds may be "all edge." Thus, edge effects and the interaction of a shred's contents with its landscape context are of paramount importance (Forman 1995). The investigator's focus must of necessity broaden beyond the focus on contents alone that often characterizes studies of fragments.

Metapopulation models, landscape-mosaic metaphors, and other "nature is patchy" concepts do not fit the shred configuration well. Instead, shredded landscapes may demand that ecologists develop new or modified models of population dynamics, demography, dispersal, and genetics, or that they apply the principles of landscape ecology (Forman 1995). Most importantly, the conservation consequences of shredded habitats may differ significantly from those of fragmented habitats.

So, what questions might conservation ecologists ask of shredded landscapes? Here are a few; I trust that readers will be able to generate many additional ones.

1. At the population level:

- For a given plant or animal species, what demographic changes occur along a shred?

- What genetic structure characterizes the tentacular demes occupying shreds, as compared to demes in the intact habitat? Genetically speaking, are shred populations robust or decaying?

2. At the community level:

- What changes in species composition occur along a peninsular shred, from base to tip? Who drops out,

who appears, where do these changes occur, and why?

- What is the nature of species replacement (if any) along the shred? Is there a one-for-one replacement of natives by robust opportunists and exotics, or is there a monotonic shift in species richness?

- What happens to community function, structure, and dynamics? For example, do guild structure, life-form spectrum, community-level plant phenology, and internal disturbance regimes remain fairly consistent or do these attributes change with distance along the shred?

- Does the nature of species interactions change along a shred, and if so what are the consequences to the interactors? For example, might pollination of native flora depend on animal populations centered in habitat beyond the invasion front, and thus decline along the shred's length? In contrast, might seed predation levels respond most directly to exposure to edge, and thus remain quite constant throughout the shred's length?

3. At the landscape level (see also discussions in Forman 1995):

- Are edge effects constant along the length of a shred?

- How do edge effects vary with the shred's context? For example, what differences exist among the edge effects exerted by neighboring habitat consisting of unimproved cattle pasture, improved cattle pasture, crops of various kinds with various levels of treatment with fertilizer and pesticides, variously treated shrub or tree plantations, and second growth of various ages and consistencies?

- What sorts of interchange occur between the shred's contents and its context, and how do these vary with different contexts? Do animals native to the shredded habitat exploit, on a daily or seasonal basis, resources in the context? Do animals resident to the context exploit resources in the shred?

- Do shreds merely absorb exotics from nearby converted landscapes, or do shreds spray the converted landscapes with propagules of native species?

- Are shreds doomed to decay, to fade into the species composition of the

converted landscape, or might they serve as reservoirs of native species for eventual restoration of that landscape?

- How do shreds themselves affect the ecology of their context? Aside from effects on physical attributes such as hydrology and nutrient flow, how do shreds affect their agricultural context regarding weed control, arthropod herbivores, natural enemies of herbivores, or pollinators of animal-pollinated crop plants?

- Thus, what role might shreds, and various shred configurations, play in the management of converted landscapes?

In short, shredded habitats invite empirical study and modeling in their own right, as ecologically interesting and significant landscape features along agricultural frontiers in the Neotropics at least. They should arouse conservation concern, as possible refuges for native species, as corridors for, or barriers to exotics, and as potential reservoirs of native species for future restoration of their surroundings. Shreds' manifest potential for interchange with developed lands surrounding them suggests a potentially critical role in landscape management. But how much do we really know about the ecology of shreds and their role in conservation? Aside from extensive work in European landscapes summarized by Forman and Godron (1986), we know very little (which has enabled me to speculate shamelessly here). Need a research topic? Think "shreds."

This essay is not intended to disparage the "nature is patchy" perspective, an approach having tremendous explanatory and predictive power for most landscapes and a metaphor that has guided most of this author's own research career. Many converted landscapes may fit the island analogy well, and others that now display shreds may be doing so only temporarily, as a transitional phase on the way to true insularization. Nevertheless, rather than accepting without question the island analogy, conservation ecologists working at population or community levels should allow landscapes themselves to instruct them as to which metaphor—islands, shreds, or yet another analogy—best fits the current layout of remnant habitats, and then should ask the questions most appropriate to that layout. ■

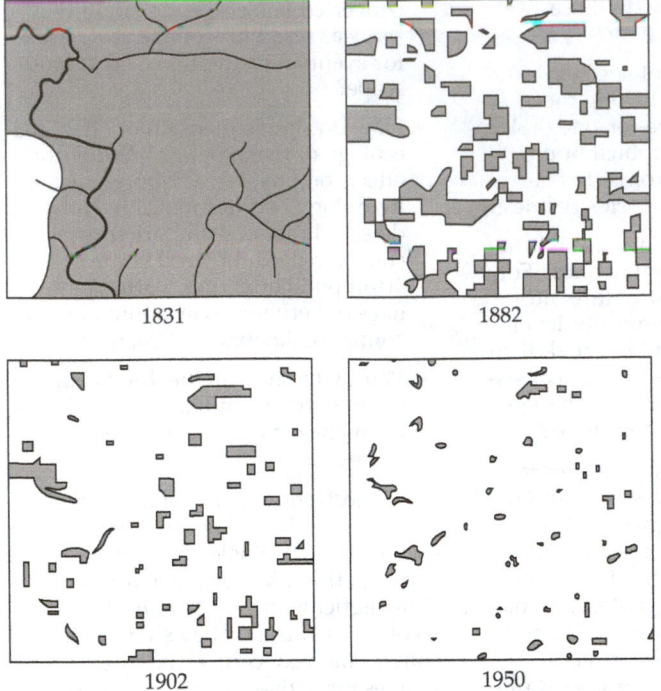

1831   1882

1902   1950

Figure 7.1 Changes in a wooded area of Cadiz Township, Green County, Wisconsin, during the period of European settlement. Shaded area represents the amount of land in forest in each year. (From Curtis 1956.)

measures to reduce or mitigate the effects of fragmentation. Such studies have documented local extinctions, shifts in composition and abundance patterns to favor weedy species, and other forms of biotic impoverishment in fragmented landscapes (Burgess and Sharpe 1981; Noss 1983; Harris 1984; Wilcox and Murphy 1985; Saunders et al. 1991; Laurance and Bierregaard 1997; Debinski and Holt 2000; Knight et al. 2000). As might be expected for a phenomenon as complex as fragmentation, however, empirical generalizations that apply across ecosystems have been hard to come by. Biological effects of fragmentation that are pronounced in some regions are apparent-

ly missing or are of minor importance in others. Despite these complications, scientists are beginning to understand how fragmentation reduces native biodiversity and what kinds of policy and management actions are prudent to apply in the absence of site-specific data.

In this chapter, we review some differences between fragmented landscapes and naturally heterogeneous landscapes; discuss species–area relationships, island effects, and the important role of the landscape matrix; identify kinds of species most likely to be sensitive to fragmentation; and summarize the biological consequences of fragmentation. We conclude with recommendations for countering fragmentation.

Fragmentation and Heterogeneity

A superficial view of fragmentation portrays a large area of homogeneous habitat being broken into pieces and embedded in nonhabitat. Thus, the historical forest cover in Figure 7.1 is shown as uniformly gray and the agricultural matrix (in the more recent scenes) as uniformly white. But the apparent homogeneity of these forest patches, as well as the matrix, is an artifact. If we zoom in and map any forest at higher resolution (Figures 7.2 and 7.3), we see that it is far from uniform. In fact, virtually all landscapes and the patches within them are mosaics at one scale or another (Essay 7.2 by Steward Pickett). Across a natural landscape, the distribution of vegetation typically reflects changes in elevation (with corresponding temperature and precipitation gradients), slope and aspect (with corresponding gradients in soil moisture and other properties), as well as differences in parent material. This heterogeneity is vividly displayed in mountainous regions, such as the Great Smoky Mountains in the U.S. (see Figure 7.2), but is also true in relatively flat landscapes such as the southeastern coastal plain of the United States. An elevation gradient of only a few meters in Florida may lead through a progression of longleaf pine (*Pinus palustris*) and turkey oak (*Quercus laevis*) on

Figure 7.2 Topographic distribution of vegetation on an idealized west-facing slope on the Great Smoky Mountains National Park. Vegetation types: BG, beech gap; CF, cove forest; F, Fraser fir; GB, grassy bald; H, hemlock; HB, heath bald; OCF, chestnut oak–chestnut; OCH, chestnut oak–chestnut heath; OH, oak–hickory; P, pine forest and heath; ROC, red oak–chestnut oak; S, spruce; SF, spruce–fir. (From Whittaker 1956.)

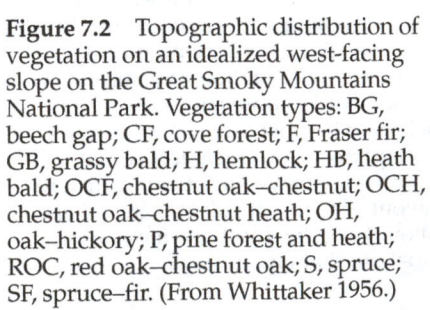

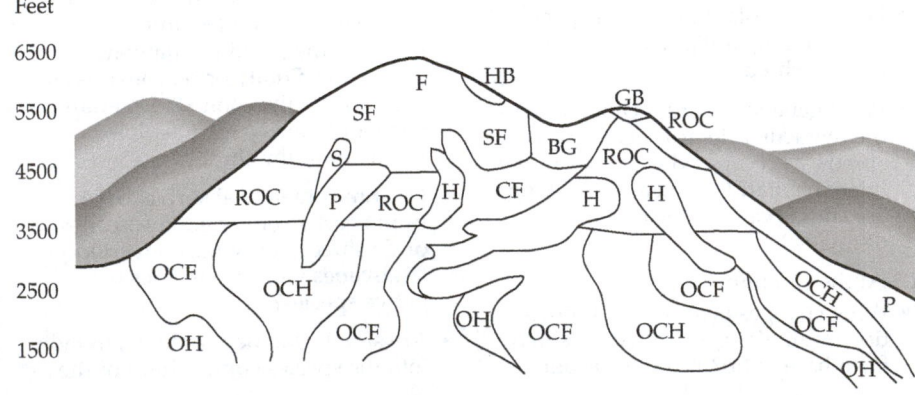

Feet

6500

5500

4500

3500

2500

1500

(A)

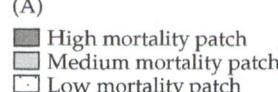

■ High mortality patch
▨ Medium mortality patch
☐ Low mortality patch

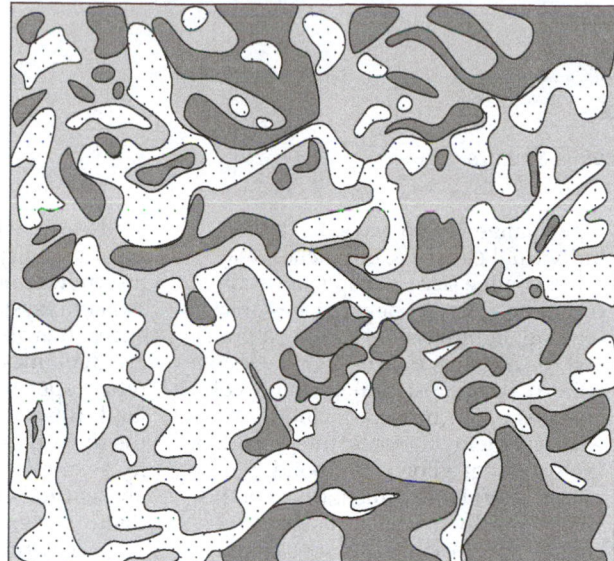

Figure 7.3   (A) Fire mortality patches from 1800 to 1900 in the Cook-Quentin study area, Willamette National Forest, Oregon. Scale is about 10 km from left to right. (B) Stand development phases in a 1 km wide section of virgin forest in Yugoslavia. Patches average about 0.5 ha in size. Phases represent stages in a continuous cycle of forest dieback and recovery. (A, from Morrison and Swanson 1990; B, from Mueller-Dombois 1987.)

(B)

☐ Rejuvenation phase

■ Stand-reestablishment or building phase

▨ Optimal phase

▨ Terminal phase

▨ Breakdown or dieback phase

▨ Regeneration phase

▨ Mixed-structure phase

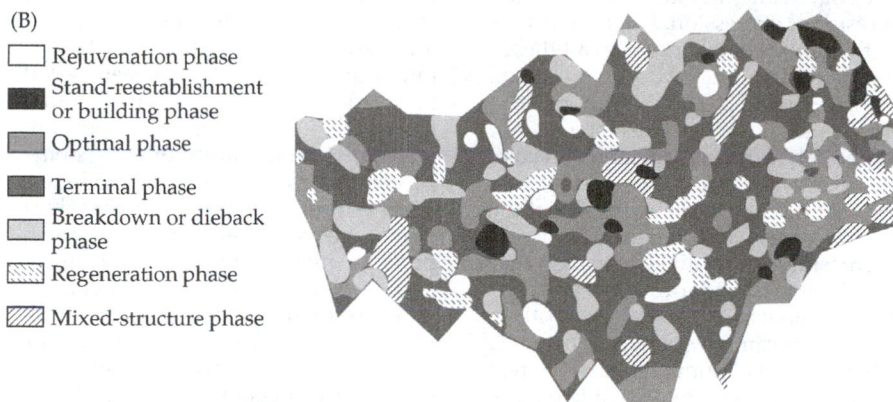

dry sandhills, down through flatwoods with longleaf pine, then slash pine (*P. elliottii*) on wetter sites, and sometimes pond pine (*P. serotina*) on the wettest sites. Slopes may have seepage bogs grading down into shrub swamps. This gradient-aligned vegetation pattern is a product of fire interacting with the slope–moisture gradient (Wolfe et al. 1988; Noss and Harris 1990).

Natural disturbances create additional heterogeneity in vegetation, beyond that generated by the physical environment. The **grain** of a landscape is often determined by the spatial scale of disturbance, that is, by the size and distribution of disturbance-generated patches. Relatively large disturbances—for instance extensive stand-replacing fires—create a coarse-grained pattern (see Figure 7.3A), whereas canopy gaps caused by death and fall of individual trees or small groups of trees create a fine-grained pattern (see Figure 7.3B). Because a given ecosystem will often be affected by many different kinds and

scales of disturbance, several grains of pattern may be overlaid on one another. Furthermore, disturbances are typically patchy in time as well as space, so that new disturbances occur in some areas while previously disturbed sites are recovering. This continuously changing pattern has been called a space–time mosaic (Watt 1947) or shifting mosaic (Bormann and Likens 1979). Ecologists generalize that natural disturbances, especially when they are of moderate intensity or intermediate frequency of occurrence, increase the diversity of habitats, microhabitats, and species in an area (see Chapter 2).

The patterns portrayed in Figures 7.2 and 7.3 are examples of natural patchiness or horizontal complexity. Every landscape and seascape is patchy, though some are more patchy than others (Forman and Godron 1986). As a consequence of patchiness, habitat quality for species varies spatially, and many species are distributed as **metapopulations**—systems of local populations

ESSAY 7.2

Mosaics and Patch Dynamics

Steward T. A. Pickett, *Institute of Ecosystems Studies*

■ Mosaics are patterns composed of smaller elements, like the tile or glasswork that reached an artistic zenith during the Byzantine era. One of the marvels of such mosaics is that in spite of being made of individual bits fixed in place by mortar, the best works seem animated and lively. It is ironic that a static entity can suggest such motion and liveliness.

Like mosaic art, most of the landscapes in which we must practice conservation are composed of smaller elements—individual forest stands, lakes, hedgerows, shrubland patches, highways, farms, or towns. Because the time scale of human observation is short relative to many landscape changes, people have often assumed mosaic landscapes to be static, with unchanging bits of nature and culture cemented into place.

Most often, landscape mosaics have been looked at only from the perspective of specific elements within them, rather than as an entire array that might interact. Focus may be on a stand of a rare plant, or the breeding ground of an unusual animal. The local spatial scale of observation is linked to a tacit assumption that the status of a particular population, community, or ecosystem could be understood by studying a particular patch in a mosaic. The conditions in adjoining or distant elements of the mosaic were ignored.

Ecologists have learned, however, that these two assumptions often do not hold in the real world. First, virtually all landscapes are in fact dynamic. Although the mosaics of art only *seem* to vibrate and shimmer, mosaic landscapes *do* in fact change. Bormann and Likens (1979) coined the phrase "shifting mosaic" to label the insight that landscapes are dynamic. Landscapes

may change in two ways. First, the individual elements, or patches, may arise, change size or shape, or disappear. For example, new patches arise by logging, lightning fire, turning a prairie sod for farming, converting a farm to a suburb, or reforestation. Examples of changes in patch shape include encroachment of a forest into a field, or the spread of a bog into a pond. Second, the structure, function, or composition of individual patches may change. For instance, the species composition, and hence rate of nutrient cycling, in an ecosystem may change as a result of succession. Patch dynamics is the term that incorporates all these fluid possibilities.

The second incorrect assumption about mosaics is that the elements act separately from one another. As ecologists began to look longer and more mechanistically at the dynamics of the specific sites they studied, it became increasingly clear that even when there are distinct boundaries between systems, organisms, materials, and other influences can flow between them. Thus, mammal populations in a forest may rely on food from outside the forest, certain insect populations may be maintained by migration, and some forest successions can be driven primarily by seed input rather than by species interactions at the site. It is safer to assume that ecological systems in a mosaic landscape are open, rather than closed and isolated.

The changes within landscapes, suggested by the metaphor of a shifting mosaic of patches, and the fluxes among patches, are crucial to conservation. Conservation strategies and tactics that ignore these two dynamic aspects of landscapes are doomed to fail. This conclusion is all the more germane in

landscapes where humans, with their great mobility, energy subsidies, and insertion of novel organisms and materials into systems, are dominant influences. It is also important to realize that the dynamics of a landscape may reflect specific human behaviors and land uses, either now or in the past. So the dynamics of landscapes are in reality a complex mixture of human effects and natural effects such as land use, disturbance, and succession.

Successful conservation requires knowing what the patches are, how they change, and how they are affected by fluxes from outside the target area (or even the region). There may be important fluxes that have been halted or reversed by human activities in the landscape. Or there may be important population, community, or ecosystem processes within the site that no longer occur naturally. Conservation involves not only choosing areas in which the processes that are responsible for the existence of a conservation target are intact, but also compensating for natural or anthropogenic processes that no longer occur. Conservation is, in a sense, active maintenance of patch dynamics.

Some of the most important insights for conservation to arise from modern ecology are the need to treat ecological systems, be they populations, communities, or ecosystems, as open to outside influences, and to manage systems to maintain the dynamics that created them. All landscapes in which we have an interest in practicing conservation should be treated as shifting and interconnected mosaics. The smaller or more delicate the target for conservation, the more critical the patch dynamic view becomes. ■

linked by occasional dispersal. Because patches of habitat suitable to a species are often spatially separated, and individual populations are often small and prone to extinction, the persistence of a metapopulation is tied to dispersal by individuals or propagules from one patch to another. If dispersal between patches becomes impossi-

ble due to distance or a lack of corridors, stepping stones, or a suitable matrix through which the species of concern will travel, the entire metapopulation eventually may go extinct. Some species are inherently poor dispersers. For example, the samango monkey (*Cercopithecus mitis labiatus*) in South Africa has a group-dynamic

social behavior that severely limits dispersal among forest fragments. In a fragmented landscape no functional metapopulation exists for this species; hence, it is at high risk of extinction except in very large blocks of forest (Lawes et al. 2000). Flightless insects and ant-dispersed herbs provide other examples of poor dispersers.

Dispersal is more likely to maintain metapopulations in naturally patchy landscapes than in formerly continuous landscapes fragmented by human activity (den Boer 1970; Hansson 1991), probably because the organisms living in patchy landscapes are accustomed to traveling across a diverse mosaic. The metapopulation model also suggests that habitat patches currently unoccupied may be critical to survival, because they represent sites for possible recolonization. Establishing small populations on vacant patches may help prevent downward spirals in metapopulations (Smith and Peacock 1990).

Recognizing that conservation biology is a value-laden science and that conservation biologists place high value on diversity, it is easy to conclude that patchiness is "good." Old-growth forests, for example, have been of interest to ecologists and conservationists in part because they are so heterogeneous. With trees of many ages, and canopies that are tall and uneven, old-growth forests have higher rates of gap formation than do younger stands (Clebsch and Busing 1989; Lorimer 1989). High levels of habitat heterogeneity, expressed horizontally and vertically, contribute to high species diversity.

But if patchiness is good, then why is fragmentation caused by humans perceived as bad? Surely fragmentation creates a patchy landscape; superficially, at least, the patterns in Figures 7.1, 7.2, and 7.3 are similar. Are conservation biologists just being misanthropic? Or are there fundamental differences between naturally patchy landscapes and fragmented landscapes? What precisely are these differences? These are not trivial questions. Answering them may allow the design of land-use plans and management practices that mimic natural processes and patterns and thereby maintain biodiversity; failing to answer them will likely lead to further biotic impoverishment.

The differences between naturally patchy and fragmented landscapes are only beginning to be explored scientifically. The nature of patchiness and the effects of increases in fragmentation in marine realms is even less studied, thus the bulk of our discussion will focus on terrestrial systems. We can hypothesize that the following distinctions are ecologically significant:

1. Fragmentation has resulted in a reduction of the extent and connectivity of habitats, and species may or may not adjust to this change in habitat availability and configuration.

2. A naturally patchy landscape has rich internal patch structure (lots of tree-fall gaps, logs, and different layers of vegetation), whereas a fragmented landscape typically has simplified patches and matrix, such as parking lots, cornfields, clear-cuts, and tree farms.

3. Largely because of the previous point, a natural landscape often has less contrast (less pronounced structural differences) between adjacent patches than does a fragmented landscape, and therefore potentially less-intense edge effects.

4. Certain features of fragmented landscapes, such as roads and various human activities, pose specific threats to population viability.

In short, fragmentation creates a landscape different from that shaped by the natural disturbances to which species have adapted over evolutionary time (Noss and Cooperrider 1994). The greater the difference, the greater the threat to species persistence. So, for example, species in landscapes where clear-cuts and regenerating forests form a dynamic mosaic show fewer harmful effects of fragmentation than species inhabiting more static landscapes with forest patches surrounded by agricultural fields or suburbs (Schmiegelow et al. 1997; Hansen and Rotella 2000; Schmiegelow and Mönkkönen 2002). We will attempt to shed light on these differences by explaining how fragmentation threatens biodiversity. We emphasize that the mechanisms underlying differences in the viability of populations in natural and fragmented landscapes are still largely inferred, not proven (Harrison and Bruna 1999; Fahrig 2003). There are many degrees and scales of fragmentation. It is a process with unpredictable thresholds, not simply an either/or condition. Spatial heterogeneity and patchy distributions of species in intact landscapes lead to complex distributions when those landscapes are fragmented, often confounding predictions of biotic responses to fragmentation (Norton et al. 1995). No two landscapes are likely to show identical trajectories of change.

The Fragmentation Process

In terrestrial ecosystems, fragmentation typically begins with gap formation or perforation of the vegetative matrix as humans colonize a landscape or begin extracting resources there. For a while, the matrix (that is, the most common habitat type) remains as natural vegetation, and species composition and abundance patterns may be little affected (Figure 7.4). But as the gaps get bigger or more numerous, they eventually become the matrix. The connectivity of the original vegetation has been broken. At this point, or even earlier in the fragmentation process, the fragmented landscape ceases to support some species.

Fragmentation requires quantitative measurement to make meaningful comparisons between study areas.

Figure 7.4 A fragmentation sequence begins with gap formation of perforation of the landscape A. Gaps become bigger or more numerous (landscape B) until the landscape matrix shifts from forest to anthropogenic habitat (landscape C). (From Wiens 1989.)

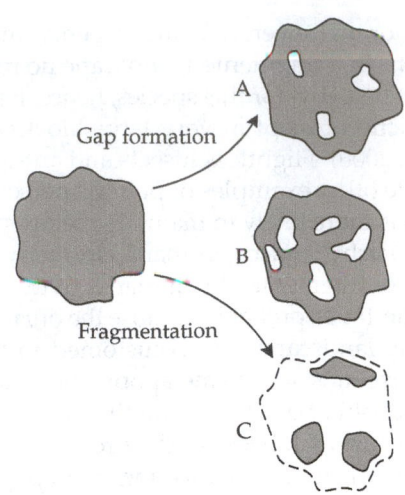

Landscapes differ in their spatial pattern, for example patch sizes and shapes, the isolation of patches from one another, and the complexity of their edges. Ecologists have suggested a number of quantitative measures of landscape pattern (O'Neill et al. 1988; Turner 1989; Noss 1990; Mladenoff et al. 1993; see Box 7.1 by David Mladenoff).

Unfortunately, many landscape metrics derived from theory (e.g., percolation theory; Gardner et al. 1987, 1992) are not always as useful for conservation planning as hoped. For example, percolation theory predicts that at 59.28% cover, the probability of finding a corridor across the landscape increases abruptly; below this value, patches are isolated (O'Neill et al. 1988). This information provides practical advice for conservation only if it predicts responses of real species. Because species vary enormously in their mobility, response to potential barriers, and habitat preferences for movement and dispersal, abstract indices of connectivity and fragmentation are meaningless until they are validated by population-

level data. Continued study of the differences in landscape patterns and species responses between near-pristine (reference) landscapes and landscapes altered to various degrees by humans is necessary to validate proposed fragmentation indicators. The point in the fragmentation process at which biological integrity declines dramatically is usually not known, as few fragmentation studies have been conducted over a long enough period of time.

BOX 7.1 Quantifying Landscape Pattern and Fragmentation

David J. Mladenoff, *University of Wisconsin-Madison*

Habitat fragmentation is an important aspect of landscape pattern, but how does one quantify fragmentation? A variety of analytical techniques exist to describe spatial patterns of landscapes (Turner and Gardner 1991). These methods are often based on using computerized geographic information systems (GIS) software that allows complex manipulations such as changing and combining maps, and analyzing their content (Burrough 1986). These methods can produce simple summaries of mapped habitat types, including patch area, number, size-class distribution, and relative abundance. Patch size-class distributions can be analyzed much the way tree size-class distributions might be compared among forest stands.

Because maps often contain too much information to grasp visually, more complex indices can be calculated from basic map information, and many have been described and proposed (Turner 1989), and evaluated for their usefulness and meaning (Riitters et al. 1995, Cain et al. 1997; Li and Wu 2004). Many have been shown to be redundant. One must also be cautious in selecting formulas that reduce great complexity into simple numerical values, as information can be lost. To select among various indices we might ask: (1) Is the scale of the map data appropriate to our question and analysis, including the spatial extent of the mapped area, the detail of the map classes, and the resolution? (2) Does the index provide information that is biologically interpretable in relation to our question? And, (3) how is the index scaled numerically, and how can we interpret the range of values it generates from one landscape to another? All of these factors will affect the

results of any index, so the map data must not only be appropriate for the question at hand, but must be consistent among areas or landscapes if comparisons are made.

A number of indices begin by using the proportion (*p*) occupied by the various map classes or habitat types (O'Neill et al. 1988). The Shannon-Wiener diversity index can be used to measure landscape diversity based on the relative abundance of map classes. This is similar to how diversity might be calculated for plant or animal communities. Thus,

$$H = \frac{-\Sigma(p_k)\ln(p_k)}{\ln(s)}$$

Here, p_k is the proportion of the landscape occupied by map class or habitat k, s is the total number of classes in the map, and ln (*s*) is the maximum diversity possible for *s* classes in the

map; dividing by this value scales the index from 0 to 1. Values approaching 1 indicate higher diversity or greater evenness of habitat classes in a landscape.

Other indices measure the complexity of habitat patch shapes on a landscape. For a given habitat patch area, a more complex shape will have a higher edge-to-interior ratio than a simpler shape, such as a circle. The amount of edge and interior can be measured directly on a map using GIS and the ratios can be compared, or the information can be used to calculate another type of index called the fractal dimension. This is based on the relation:

$$\log (A) \sim d \log (P)$$

where *A* is the area of a patch, *P* is the perimeter of the patch, and *d* is the fractal dimension (Krummel et al. 1987; Turner 1989). For all patches in a map or class, log (*P*) is regressed against log (*A*), yielding *d,* the slope of the regression line (Krummel et al. 1987). The purpose here is not whether or not the landscape pattern is fractal (i.e., exhibits self-similarity across the scales of measurement). We are instead using *d* as a relative index. In this index, values are scaled from 1 to 2, with 1 indicating a simple, circular patch, and values approaching 2 indicating complex, convoluted patch shapes.

Another useful approach examines not only the relative abundance and shape of patches, but the distribution of habitat patches to assess their isolation. One metric of this type is the proximity index (Gustafson and Parker 1992):

$$PX = 3(S_k / n_k)$$

where S_k is the area of patch *k*, and n_k is the distance to the nearest neighbor patch *k*. The summation is calculated for a focal patch and each neighboring patch within a buffer distance that is determined by the user's needs.

Another useful group of metrics measures contagion or "clumpiness" of a habitat, indicating how aggregated a particular patch type is on the landscape. A related group measures the significance of association or adjacency relationships among different habitat types. Both of these can reflect characteristics that may be important

for evaluating species use and habitat quality of a landscape.

Another quite different group of techniques, known as **spatial statistics** or **geostatistics,** are not indices but involve more complex statistical analysis of spatial patterns. For example, semivariograms are used to measure the spatial autocorrelation shown by the distribution of elements or points within a map (Burrough 1986). The book *Learning Landscape Ecology* by Gergel and Turner (2002) provides very helpful lab exercises in using some of the indices described here, other related techniques, and variograms.

Application of these tools to landscape patterns was demonstrated in a study that quantitatively compared forest patch size, shape, distribution, number, and spatial associations between an old-growth landscape and a young, managed forest (Mladenoff et al. 1993). The old-growth landscape contained a greater range of forest patch sizes. Although most patches in both landscapes were small (<5 ha), patches in the managed landscape ranged to only 200 ha whereas those in the natural landscape ranged over 1000 ha. Similarly, fractal dimension analysis showed that patches in the managed landscape were much simpler in shape than in the old-growth landscape. The natural landscape had a dominant matrix type with both larger patches and greater patch complexity, thus maximizing both forest patch interior (habitat) and interspersion (connectivity) at the same time.

Fractal dimension analysis also showed, somewhat surprisingly, that patch complexity did not increase linearly with patch size. Instead, there were three distinct peaks in patch complexity across the range of sizes of a given forest type in the natural landscape. This suggests that different processes may operate at various scales in patch creation of a particular forest type. Such a relationship did not occur in the managed landscape.

Adjacency analysis also showed that the natural landscape had certain significant associations of various forest types—for example, a positive association between upland hemlock (*Tsuga canadensis*) and lowland conifers—that did not occur in the

more fragmented managed landscape. These relationships may be critical when a given species requires different habitats in proximity to one another for different uses. It also may be important in tree species recolonization following disturbance.

The patterns we found illustrate that there is great fragmentation of a given forest type into smaller, more dispersed patches within a managed landscape. This effect is increased by the greater proportion of successional types occurring in managed areas, which means that uncommon old forest patches are further isolated. Under high forest cutting rates, these smaller, more isolated forest patches are vulnerable to high edge:interior ratios. Lack of a dominant matrix forest type with large, integrating forest patches and characteristic shape and adjacency relationships are the major changes on these managed forest landscapes.

As a result, natural primary forest landscapes have a different spatial pattern than human-disturbed or managed landscapes. Some of these patterns and their effects can be explained in relation to tree species dispersal and colonization, and habitat use by animals. Analyzing landscape pattern can be useful in assessing the relative effects of management scenarios, in monitoring change through time (White and Mladenoff 1994), and in setting objectives for restoring spatial patterns hypothesized to be important for conservation and sustainable functioning of ecosystems (Mladenoff et al. 1994). Many of these hypotheses remain to be tested, and the patterns found would benefit from testing in other regions. The challenges in applying these new and powerful techniques are in using them consistently with appropriate data, and using them in ways that are interpretable and biologically meaningful. There are many applications of landscape metrics and spatial analysis applied to habitats and ecosystems at many possible scales, from beetles to wide-ranging vertebrates (Turner et al. 2001). Papers in the journal *Landscape Ecology* are especially useful in learning more about recent applications of these techniques.

Fragmentation and edge-effect studies in landscapes that are still largely forested typically fail to find deleterious impacts (e.g., Rudnicky and Hunter 1993; Hansen and Rotella 2000), and sometimes find positive effects of fragmentation on such measures as species diversity and abundances (McGarigal and McComb 1995). Site tenacity, best documented in birds, is one of many factors that may create time lags in response to fragmentation and other disturbances. Individuals may return to a site where they have bred successfully in the past, sometimes several years after the habitat has been altered, despite having low or no reproductive success once fragmentation has taken place (Wiens 1985). Thresholds of fragmentation often will be species-specific, as species vary tremendously in their sensitivity to fragmentation (and some species thrive in fragmented landscapes). The time lag between habitat loss and eventual species loss defines a region's extinction debt (Tilman et al. 1994), which can be considerable. The extinction debt for African forest primates is estimated to consist of over 30% of the current primate fauna; that is, 30% of the primate species in Africa are already essentially committed to extinction (Cowlishaw 1999).

In fragmentation studies researchers may see the final outcome of fragmentation without observing the process. Alternatively, they may observe parts of the process, but not the long-term consequences. Many models of habitat fragmentation or population subdivision ignore details of spatial structure, assuming that all habitat patches are equivalent in size or quality, and that all local populations are equally accessible to dispersing individuals (Fahrig and Merriam 1994). For all these reasons fragmentation remains a topic of great debate and active research interest.

As we discuss the consequences of fragmentation in the remainder of this chapter, bear in mind that the process can occur at many different spatial and temporal scales and in any kind of habitat. Essentially, fragmentation is the "disruption of continuity" in pattern or processes (Lord and Norton 1990). At a broad biogeographic scale, regions that were once connected by wide expanses (many kilometers) of natural habitat may now be isolated by agriculture; the severance of such biogeographic corridors may take place over hundreds of years. The occasional interchanges that once occurred between the faunas and floras of these regions are now precluded, with unknown evolutionary consequences. What if there had been no Bering Land Bridge or Isthmus of Panama? Migration of species in response to climate change, which in the past occurred over hundreds of kilometers (Davis 1981; Clark et al. 1998), may not be possible when regions are heavily fragmented (Peters and Darling 1985; Noss 2001).

At an intermediate scale, the kind of landscape fragmentation portrayed in Figure 7.1 typically takes place over decades. Although researchers often arrive too late to observe mechanisms leading to species loss, intensive field studies of populations and communities in regions currently being fragmented may teach us a great deal.

At a finer scale, the internal fragmentation of once pristine natural areas by roads, trails, powerlines, fences, canals, vegetation removal by livestock, and other human-related activities has not been well studied, but it has potentially dramatic effects on native biodiversity and ecological processes. Lord and Norton (1990) discussed structural fragmentation of short-tussock grasslands in New Zealand by grazing and other disturbances, followed by invasion of nonnative plants, which filled most of the matrix around native plants (Figure 7.5). They concluded that ecosystem functioning is more likely to be disrupted at finer scales of fragmentation, although the organisms affected are smaller and the overall process is less noticeable to human observers.

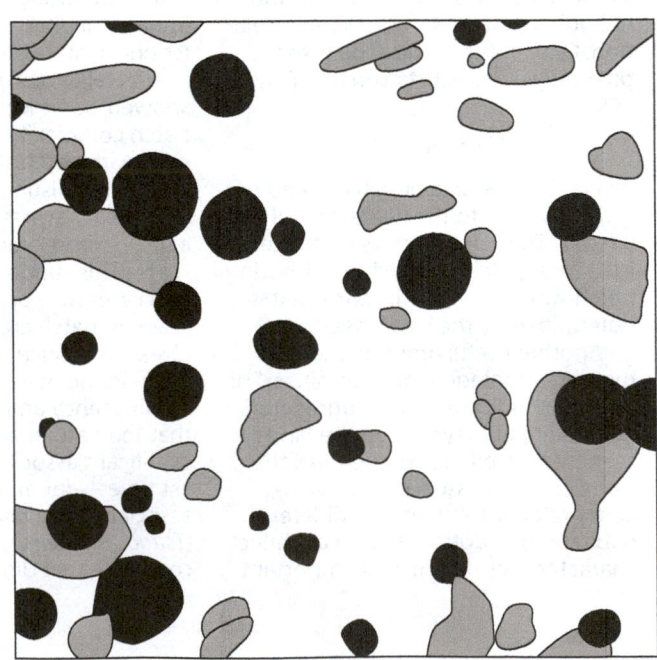

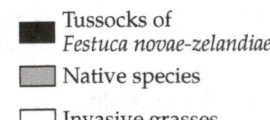

■ Tussocks of
Festuca novae-zelandiae

■ Native species

☐ Invasive grasses

Figure 7.5 This short-tussock grassland in New Zealand illustrates fine-scale fragmentation. Following grazing, invasive grasses filled the spaces normally occupied by a diverse suite of native species, thus fragmentating the grassland for plant and animal species. Area depicted is 2 m × 2 m. (Modified from Lord and Norton 1990.)

Biological Consequences of Fragmentation

Some effects of fragmentation on biodiversity have been conspicuous; others have been subtle and indirect. Some have occurred almost immediately after the initial disturbance, whereas others have developed over decades or are still unfolding. Fragmentation of some forested regions, such as the Georgia Piedmont (Turner and Ruscher 1988; Odum 1989) and Europe as a whole (Groombridge and Jenkins 2002), has been partially reversed through abandonment of agriculture and maturation of second-growth forests. Forest cover in these regions is generally increasing. However, forest fragmentation in other regions, such as the Pacific Northwest, Florida, and much of the Tropics, continues. Most deforestation in Central America has occurred since 1950 and is increasing in rate (Hartshorn 1992). Prairies, wetlands, benthic marine habitats, and many other ecosystems are still declining in area in many regions; in North America, the most critically endangered terrestrial ecosystems are grasslands, savannas, and barrens (Noss et al. 1995), whereas freshwater ecosystems have suffered the most extinctions and have the highest proportion of imperiled taxa (Stein et al. 2000). Loss of connectivity, for example as a result of fragmentation of river systems by dams, is one of the greatest threats to these systems (Pringle 2001). Fragmentation of coral reefs can affect fish and invertebrate communities in complex ways, with some effects immediately observed (e.g., species loss) and others taking longer to be seen (e.g., broad shifts in species composition, greater variability in species composition or abundance) (Ault and Johnson 1998; Nanami and Nishihira 2003).

Initial exclusion

One of the most rapid and obvious effects of fragmentation is elimination of species that occurred only in the portions of the landscape destroyed by development. Many rare species are endemics with very narrow distributions, occurring in only one or a few patches of suitable habitat. Recall the loss of up to 90 species of plants on the Centinela Ridge in Ecuador (Chapter 3) when that small patch of forest was destroyed through logging (Gentry 1986). Similarly, Cerro Tacarcuna is a mountain on the Panama-Colombia border that supports at least 71 species of angiosperms (24% of the mountain's flora) that are "extremely endemic"; that is, these species have ranges of only 5–10 km^2, and could easily be lost through fragmentation (Gentry 1986). In Colombia and Ecuador, existing national parks do not include the ranges of most of the bird species unique to those countries (Terborgh and Winter 1983). If habitat outside the parks is eliminated, these species will also be lost by ex-

clusion unless they are capable of moving rapidly to suitable habitat elsewhere. Eventually, as habitat destruction continues, suitable habitat may not be available anywhere.

Crowding effect

When an area is isolated by destruction of the surrounding natural habitat, population densities of mobile animal species may initially increase in the fragment as animals are displaced from their former homes. This packing phenomenon has been called "crowding on the ark" and has been described for tropical (Leck 1979) and temperate (Noss 1981) forest reserves, as well as for coral reefs (Nanami and Nishihira 2003). The initial increase in population densities in isolated fragments usually is followed by population collapse (Debinski and Holt 2000). The crowding effect has been convincingly demonstrated in tropical forest patches in the large-scale experiment in the Amazonian forest of Brazil (Lovejoy et al. 1986; Bierregaard et al. 1992). In this study, the capture rate of understory birds in an isolated 10-ha fragment more than doubled in the first few days following its isolation, but rapidly fell in subsequent days.

Longer-term crowding effects are likely in many cases, but not proven. The biological consequences of the crowding effect have been poorly studied. In Maine, densities of Ovenbirds (*Seiurus aurocapillus*) increased in forest fragments newly formed by logging; however, density was inversely related to productivity, as pairing success was lower in fragments than in nonfragments (Hagan et al. 1996). For these and other reasons many fragments may be population sinks, despite an abundance of birds and other organisms.

Insularization and area effects

Rapid settlement of regions such as the eastern and midwestern U.S. left behind scraps of the original vegetation as habitat fragments. The analogy with islands, though imperfect, was easy to make. Biogeographers have long known that as the area of any insular habitat declines, so does the number of species it contains. In 1855 the Swiss phytogeographer Alphonse de Candolle predicted that "the breakup of a large landmass into smaller units would necessarily lead to the extinction or local extermination of one or more species and the differential preservation of others" (cited and translated in Browne 1983). This statement may be the first written recognition of the potential negative effects of habitat fragmentation on biodiversity (Harris and Silva-Lopez 1992).

A small island or nature reserve may be smaller than the territory or home range of a single individual of some species. For example, a cougar (*Puma concolor*) is unlikely to remain long within a 100-ha park. In the Rocky Mountains of the U.S. and Canada, annual home

ranges of cougars average over 400 km² and those of grizzly bears (*Ursus arctos horribilis*) average nearly 900 km² (Noss et al. 1996). Populations of large carnivores already have become fragmented at a regional scale by habitat alteration in the Rockies (Carroll et al. 2001). Large carnivores and other wide-ranging animals are typically among the species most threatened by habitat fragmentation, in part because small areas fail to provide enough prey, but also because these animals are vulnerable to mortality due to humans and vehicles when they attempt to travel through fragmented landscapes (Harris and Gallagher 1989). In small reserves, a greater proportion of the home ranges of carnivores abut reserve boundaries, where direct conflicts with humans lead to substantial mortality (Woodroffe and Ginsberg 1998).

Other species, for reasons not entirely understood, avoid settling in small tracts of seemingly suitable habitat. Studies in the eastern U.S. confirmed that some songbird species are "area-sensitive" and usually breed only in tracts of forest many times larger than the size of their territories (Figure 7.6). Essentially the same phenomenon has been described in Japan, where several bird species are restricted to large forests, with minimum areas larger than expected by chance (Kurosawa and Askins 2003). Studies of birds on grassland remnants show that several species occur only on fragments larger than 10 or even 50 ha, even though their territories are much smaller (Samson 1983; Herkert 1994; Vickery et al. 1994). In all these cases, the probability of finding breeding pairs of area-sensitive species increases with the size of the fragment.

A common distinction in island biogeographic studies is between oceanic and land-bridge islands (MacArthur 1972). Land-bridge islands were connected to each other or to continents during the Pleistocene, when sea level was as much as 100 m lower than today. Presumably, at that time, land-bridge islands contained numbers of species similar to those in areas of equal size on the mainland. Since becoming isolated, these islands apparently lost species over time, a phenomenon called **relaxation**, with some species being more extinction-prone than others (Diamond 1972; Terborgh 1974; Faaborg 1979). However, equilibrium species richness on land-bridge islands is typically higher than on similar-sized oceanic islands that were never connected to larger land bodies (Figure 7.7; Harris 1984).

The analogy between land-bridge islands and terrestrial habitat patches isolated by development of the surrounding landscape was persuasive and spawned a series of papers proposing rules for the design of nature reserves (Terborgh 1974; Willis 1974; Diamond 1975; Wilson and Willis 1975; Diamond and May 1976). The usefulness of land-bridge island analogies for conservation has been marred by, among other problems, the weak evidence for relaxation being strongly related to island size (Abele and Connor 1979; Faeth and Connor 1979). Nevertheless, the many similarities between habitat fragments and land-bridge islands keep the analogy

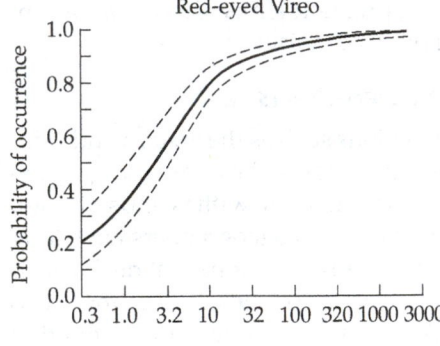

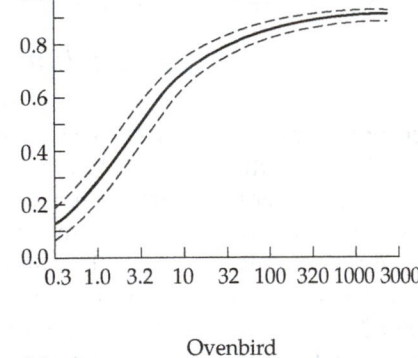

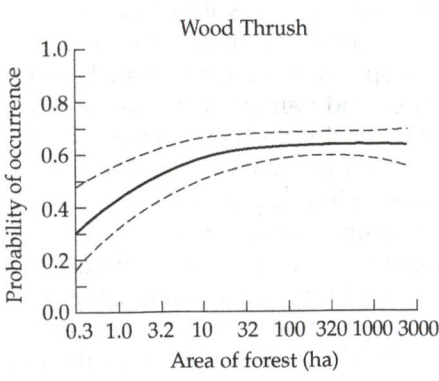

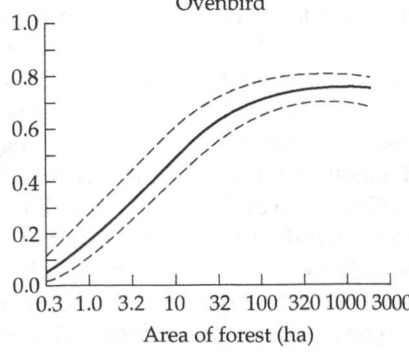

Figure 7.6 Probability of four species of common forest interior Neotropical migrant birds nesting in United States mid-Atlantic forests of various sizes, based on point counts. Dotted lines indicate 95% confidence intervals. (From Robbins et al. 1989.)

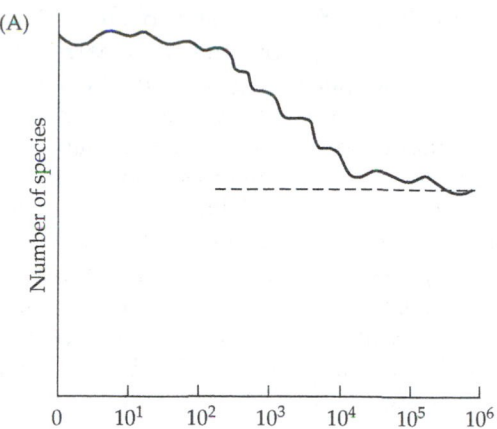

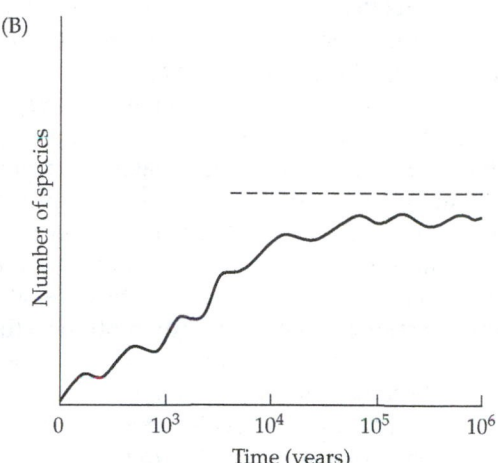

Figure 7.7 Predicted species richness over time for land-bridge islands (A) and oceanic islands (B). Land-bridge islands were once part of a large land area and contained more species than they could retain after their isolation by rising sea level; the decline in species richness after isolation is known as "relaxation." Oceanic islands, often of volcanic origin, are slowly colonized by long-distance dispersal, so species richness builds gradually to an equilibrium level (dashed line). (From Harris 1984.)

alive. For example, for national parks in western North America the number of extinctions has exceeded the number of colonizations since the parks were established, and extinction rates are inversely related to park area—both as predicted by the land-bridge island hypothesis (Newmark 1995). Few parks are large enough to maintain their historic complement of mammals. In eastern North America, for instance, no park smaller than 950 km² has sustained a population of wolves (Gurd et al. 2001). On the other hand, in the northern Canadian Rocky Mountains, an area-isolation model for parks is not useful for predicting extinctions of large carnivores under current conditions because most of the landscape matrix remains as suitable habitat. If habitat

outside parks is substantially altered, however, the parks will become more island-like and their limited area will fall below the threshold for persistence of carnivore populations (Carroll et al. 2004).

Relaxation of species richness, as predicted by the land-bridge hypothesis, has been documented in reserves that were formerly connected to extensive wild areas. A review of the fragmentation literature (Debinski and Holt 2000) found that one of the most common patterns was an initial crowding of individuals on habitat fragments, followed by a relaxation in abundances. The relaxation in abundances, over time, often is followed by a loss of species. Density-dependent models of population persistence show that fragmentation can greatly increase extinction risk (Burkey 1989). Furthermore, empirical data on extinction rates on islands and island-like habitats show that sets of smaller islands lose more species than single large islands of the same total area (Burkey 1995). In a cloud forest fragment in Colombia isolated for several decades, 40 species of birds (31% of the avifauna present in 1911) have gone extinct (Kattan et al. 1994). Fragmentation of the Atlantic forest of Brazil over 70 years led to the extinction of 28 species of birds and the endangerment of many others, altogether representing 61% of the avifauna (Ribon et al. 2003). Similarly, 50.9% of the plant species in an isolated fragment of lowland tropical rainforest, the Singapore Botanic Gardens, have gone extinct over the last century (Turner et al. 1996). Changes in the recruitment of plants in fragments can have profound impacts on community structure; for example, in central Amazonia, the regenerative plant pool in fragments is shifting rapidly toward a species-poor seedling community, with significant reductions also in the abundance of tree seedlings (Benítez-Malvido and Martínez-Ramos 2003).

The response variable in the classic equilibrium model—species richness—is a community-level property and not necessarily the most appropriate variable for conservation planning in fragmented landscapes. Subdivision or fragmentation of habitats may increase species richness, but often favors weedy species—those that thrive in areas disturbed by humans—over more sensitive ones. Many small, isolated nature reserves are quite rich in species, but nonnative species have replaced native ones (Noss 1983). In a 400-ha woodland park in metropolitan Boston, Massachusetts, the number of native species has declined by 0.36% per year over the last century, while the number of nonnative species has been increasing at a rate of 0.18% per year. The proportion of native species in the flora declined from 83% in 1894 to 74% in 1993 (Drayton and Primack 1996). Thus, species richness alone may tell us little of value for conservation planning and management. Instead, we might focus on the ratio of nonnative to native species or on population

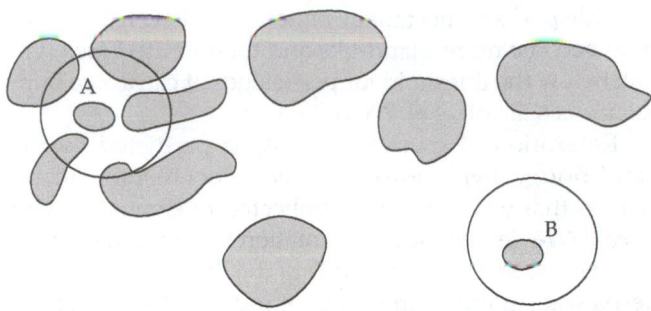

Figure 7.8 A constellation of separate habitat patches may be critical to the survival of individuals or populations. If a species requires resources in the shaded habitat patches, site A will be preferable to site B. Although no single patch is large enough by itself to support a population, the close grouping of patches in site A provides sufficient resources within the accessible part of the landscape (circle). In contrast, site B consists of one small, isolated patch and will not support a population. If human activities create impenetrable barriers to movement between the patches in site A, that site will no longer be superior to site B. (From Dunning et al. 1992.)

trends of particular species sensitive to fragmentation (Noss and Cooperrider 1994).

Isolation

Isolation of habitats and populations is an effect of fragmentation as consequential as reduction in habitat and population size. Species that are restricted to certain kinds of habitat may depend on a constellation of habitat patches in relatively close proximity, if no single patch is large enough to meet the needs of individuals or groups (Figure 7.8). As noted earlier, the viability of metapopulations may depend on movement of individuals between patches great enough to balance extirpation from local patches. Also, many animal species require a mix of different habitats with distinct resources—for example, food patches, roost sites, and breeding sites—in order to meet their life history requirements (Figure 7.9). If these critical areas become separated by barriers, populations may decline rapidly to extinction.

What constitutes a movement barrier is highly species-specific; a hedgerow that is a barrier to the movement of some species (e.g., livestock) will be a corridor to others. Unfortunately, very little information exists on the qualities of suitable dispersal habitat or on barriers for various species. Species- and habitat-specific dispersal studies are essential for gaining a better understanding of fragmentation effects. However, what we do know suggests that human-created structures and habitats—roads, urban areas, agricultural fields, clearcuts—can greatly inhibit movement and negatively affect population viability of many kinds of animals, as well as plants (especially those pollinated or dispersed by animals).

The long-term effects of dispersal barriers on population dynamics and genetic structure are becoming better documented. Still, effects of isolation are usually only inferred (i.e., a low level of genetic diversity is interpreted as resulting from isolation). For example, for a tropical tree, *Pithecellobium elegans*, three measures of genetic variation among seedlings were lowest in populations that were smaller and more isolated from other populations (Hall et al. 1996). In another tropical tree, *Elaeocarpus williamsianus*, the observed low genetic diversity appears to be the result of habitat fragmentation reducing sexual reproduction relative to vegetative reproduction, such that single clones are present in most sites (Rossetto et al. 2004). For two plant species of Swiss fen meadows, populations isolated by habitat fragmentation over the last 150 years show changes associated with genetic drift, which appears to have reduced individual fitness (Hooftman et al. 2004). For plants that are pollinated by animals, populations isolated by habitat fragmentation often show reduced reproductive success (Aizen and Feinsinger 1994a, Groom 2001).

Inbreeding depression is now well documented in wild populations of both plants and animals, and is known to increase in populations isolated by fragmentation (Keller and Waller 2002). As a rule, inbreeding depression and genetic drift increase extinction risk. Fragmentation of the native prairies of the Willamette Valley,

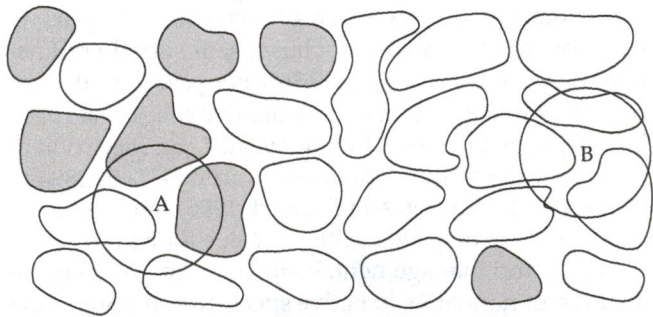

Figure 7.9 Many animals require a suite of different habitats or resources to meet life history needs. If a species requires nonsubstitutable resources found in two habitat types (shaded and open), regions of the landscape where the two habitats are in close proximity (site A) will support larger populations than regions where one habitat type is rare (site B). However, as in the example on Figure 7.8, barriers between habitat patches will destroy any advantage of site A. (From Dunning et al. 1992.)

Oregon, has led to inbreeding depression and reduced seed production in the threatened Kincaid's lupine (*Lupinus sulphureus* ssp. kincaidii), the primary larval host plant of the endangered Fender's blue butterfly (*Icaricia icarioides fenderi*) (Severns 2003). Although fragmentation may increase the among-population component of genetic diversity as isolated populations diverge over time (Simberloff and Cox 1987; Simberloff 1988), high among-population diversity is not of much value if the individual populations have greatly reduced fitness and a higher probability of extinction. The good news is that even relatively low rates of genetic interchange between populations (on the order of one successful migrant per generation) of plants and animals appear sufficient to prevent inbreeding depression and other genetic problems (see Chapter 11). Hence, measures that maintain or restore natural levels of connectivity among populations will promote persistence.

Barrier effects are both relative and cumulative, and relate to the permeability of the landscape dispersing individuals or propagules of various species. A city block can be expected to be more of a barrier (less permeable) to most forest species than a cornfield, which will be less permeable than a pine plantation. In a multiple-species context, the landscape matrix in which habitat islands are embedded is better seen as a "filter" than as a barrier to movement. The matrix will allow individuals of some species, but not of others, to pass through. Individual features, such as a river or a highway, are also filters because individuals of some species, but not others, cross them (Figure 7.10). A diversity of responses to fragmentation may be prevalent in marine systems as well, where the nature of the intervening area (soft-bottom versus hard substrate, seagrass or kelp beds versus unvegetated substrate) may influence movement of species between fragments (e.g., blue crab movement among oyster reefs is facilitated by the presence of seagrasses; Micheli and Peterson 1999).

Human-created barriers also fragment freshwater habitats. Fragmentation of river systems is widespread, as river flows are interrupted by dams or water withdrawls for irrigation. Dams, for example, block access of migratory fishes to upstream areas, and prevent recolonization of stream segments by any species whose local populations have vanished due to natural or human causes (Moyle and Leidy 1992; Dynesius and Nilsson 1992). Populations of riparian plants also can be fragmented by dams. In Sweden, adjacent impoundments in similar environmental settings develop different riparian floras because species with poor floating abilities become unevenly distributed among impoundments (Jansson et al. 2000). The flattened musk turtle (*Sternotherus depressus*) has been lost from over half of its range in the Warrior River Basin of Alabama because habitat modification of stream and river channels has fragmented suitable habitat. The remaining populations are small and

Figure 7.10 Natural or artificial barriers to movement are common in many landscapes. These barriers act as filters, since many species cannot cross large rivers or abrupt slopes, as are seen in this photo. The Columbia River between Washington and Oregon, looking like a barrier to wildlife. (Photograph © Painet, Inc.)

isolated, and therefore at high risk of extinction (Dodd 1990).

Sixty percent of the major river basins of the world are moderately to strongly affected by fragmentation, channelization, and altered flows, and all river systems with part or all of their basins in arid areas are strongly affected (Nilsson et al. 2000). Only a few large river systems in northern tundra regions are wholly free-flowing today. The need to maintain and restore hydrologic connectivity has emerged as one of the highest priorities in aquatic conservation (Rosenburg et al. 2000; Pringle 2001). Cathy Pringle describes the effects of fragmentation in freshwater systems in greater depth in Case Study 7.2.

Edge effects

Among the best documented impacts of fragmentation are edge effects (Harrison and Bruna 1999). The outer boundary of any habitat fragment is not a line, but rather a zone of influence that varies in width depending on what is measured (Murcia 1995). Edge effects have been most thoroughly studied in forests. Sunlight and wind impinge on a forest fragment from the edge, creating strong microclimatic gradients and wind turbulence. Edge zones are usually drier, less shady, and warmer than forest interiors, favoring shade-intolerant, xeric plants over mesic forest plants. In southern Wisconsin forests, edge zones of shade-intolerant vegetation may extend 10–15 m into a forest on the east, north, and south sides, and up to 30 m on the west side (Ranney et al. 1981). In Douglas fir (*Pseudotsuga menziesii*) forests of the Pacific Northwest, increased rates of blowdown, reduced humidity, and other physical edge effects may extend over 200 m into a forest (Harris 1984; Franklin and Forman 1987; Chen and Franklin 1990). These physical edge effects have been shown to increase growth rates, elevate mortality, reduce stocking density, and differentially affect regeneration of conifer species in old-growth forests up to 137 m from clear-cuts in Washington and Oregon (Chen et al. 1992). In Japan, the altered microclimate near forest edges reduces seed germination and subsequent seedling density in the understory herb Trillium camschatcense (Tomimatsu and Ohara 2004). Elevated rates of canopy and subcanopy damage, as well as proliferation of disturbance-adapted plants, occur up to 500 m from edges of tropical forest fragments in Queensland, Australia (Laurance 1991).

In some landscapes, especially in warmer climates, sealing of edges occurs through accelerated growth and increased regeneration of understory trees and shrubs. These lead to higher density and basal area of woody plants near the edge relative to the forest interior (Williams-Linera 1990). Edge sealing takes much longer in cooler environments, leading to persistently reduced

tree density and basal area near edges (Chen et al. 1992). The wide range of trends in microenvironments and vegetation and depth-of-edge influences found in various studies were reviewed by Baker and Dillon (2000).

In some cases, animals are attracted to edge, which may then function as an "ecological trap" (Gates and Gysel 1978). Many passerine birds were attracted to a field–forest edge in Michigan and nested at greater densities near the edge than in the forest interior. However, birds nesting near the edge suffered higher rates of nest predation and brood parasitism by Brown-headed Cowbirds, and as a result had greatly reduced fledging success (Gates and Gysel 1978). Roads and powerline corridors as narrow as 8 m may produce significant edge effects by attracting cowbirds and nest predators to the corridor and adjacent forest (Rich et al. 1994). Cowbird parasitism can be significant for hundreds of meters into a forest from an edge and is a major reason for the decline of forest birds in heavily fragmented landscapes (Brittingham and Temple 1983; Robinson et al. 1995). Increased rates of nest predation by jays (e.g., *Cyanocitta cristata*), crows (*Corvus brachyrhynchos*), raccoons (*Procyon lotor*), opossums (*Didelphis marsupialis*), foxes (e.g., *Vulpes fulva*), squirrels (*Sciurus carolinensis*), skunks (*Mephitis mephitis*), and other opportunistic predators may extend hundreds of meters from edges in eastern North America (Figure 7.11). Similar problems have been observed in Swedish forest fragments (Andren and Angelstam 1988). Furthermore, predation and parasitism edge effects are not limited to forests. A study of birds in tallgrass prairie fragments in Minnesota found higher rates of nest predation in small fragments, in areas close to wooded edges, and in vegetation that had

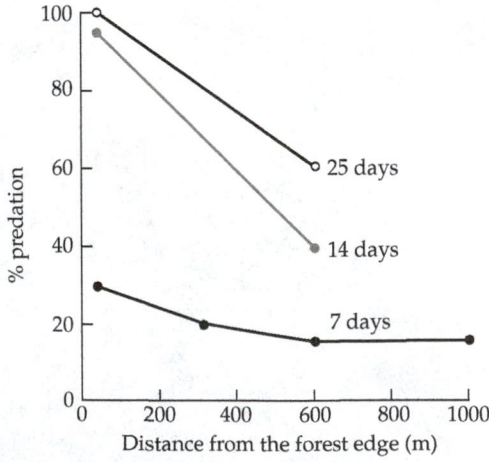

Figure 7.11   Percent of experimental nests (quail eggs) preyed upon as a function of distance from forest edge. Nests at the edge of forest are all consumed within 14 days, and even those 600 m away from the forest edge show high rates of predation after 25 days. (From Wilcove et al. 1986.)

not recently burned. Rates of brood parasitism by cowbirds were also higher near wooded edges (Johnson and Temple 1990). Fragmented prairies throughout the mid-continental U.S. show higher rates of nest predation for grassland birds in small fragments than in large fragments (Herkert et al. 2003).

Some studies have failed to confirm the ecological trap hypothesis with regard to nest predation (Ratti and Reese 1988) or have found inconsistent or conflicting evidence (Paton 1994; Murcia 1995). Part of the problem relates to inconsistencies in study design and to biases associated with the use of artificial nests (Haskell 1995; Murcia 1995). The spatial scale of investigation is also significant, with fragmentation effects on nesting success being more evident on a landscape scale than at smaller scales (Chalfoun et al. 2002; Stephens et al. 2003). True regional differences also exist. Studies from several regions show modest, if any, negative effects of forest fragmentation in cases where the landscape matrix remains dominated by forest in various stages of succession (McGarigal and McComb 1995; Tewksbury et al. 1998; Hansen and Rotella 2000; Schmiegelow and Mönkkönen 2002). In heavily fragmented landscapes dominated by disturbed lands, edge effects on birds may not be observed because the remaining patches of natural habitat are saturated with nest predators and brood parasites. Cowbirds saturate even the largest available tracts (about 2000 ha) and even areas more than 800 m from edges in southern Illinois (Robinson 1992a). In such situations, nesting success of forest birds may be so low that the entire region is a population sink. Persistence of forest interior birds in such landscapes is tenuous and may depend on immigration from landscapes with greater forest cover and better reproductive success (Temple and Cary 1988; Robinson 1992b; Donovan et al. 1995; Robinson et al. 1995).

The pervasiveness of edge effects implies that habitat patches below a certain size will lack the true interior or "core" habitat that some species require. If 600 m is determined to be the penetration distance of significant nest predation, then a circular reserve smaller than 100 ha (250 acres) will be all edge (Wilcove et al. 1986). Using a two tree-height edge width of 160 m, patches of old-growth Douglas-fir forest in the Pacific Northwest smaller than 10 ha (25 acres) are all edge; a landscape that is 50% cutover in a typical checkerboard harvest system contains no true interior forest habitat (Franklin and Forman 1987). Temple (1986) assumed a 100-m edge width for forest fragments in south-central Wisconsin. Sixteen bird species were found to be sensitive to fragmentation in this landscape, breeding less frequently or not at all in smaller sites. In a comparison of two forest fragments, one without any core habitat (due to its shape) lacked successful breeding by interior birds. The other frag-

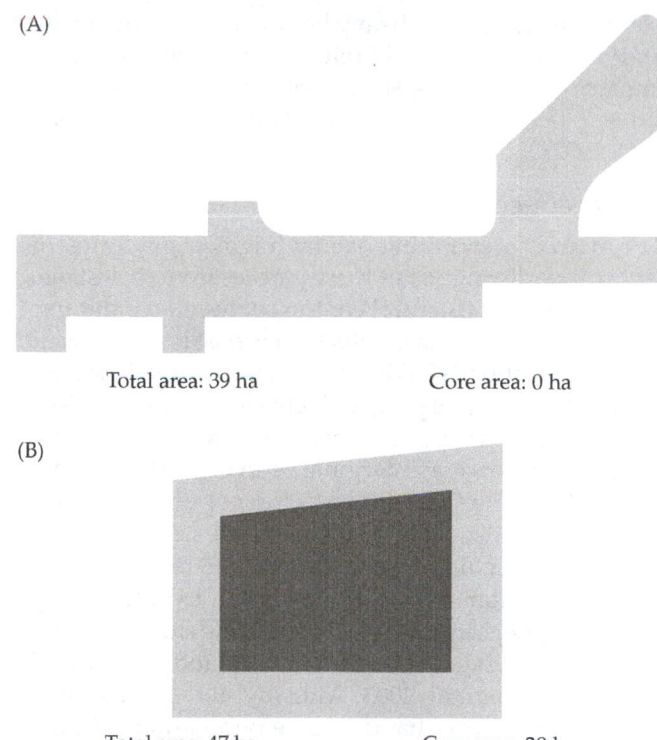

(A)

Total area: 39 ha   Core area: 0 ha

(B)

Total area: 47 ha   Core area: 20 ha

Figure 7.12 A comparison of breeding success of fragmentation-sensitive birds in two forest fragments with similar total areas but vastly different core areas (forest interior). (A) A fragment that is entirely edge habitat (light). (B) A fragment that contains 20 ha of core habitat (dark region). Of 16 species known to be sensitive to fragmentation, none bred in the fragment in (A), and 6 bred in fragment in (B). (From Temple 1986.)

ment, of similar total area but with a core area of 20 ha (50 acres), contained successful breeding pairs of 6 of the 16 fragmentation-sensitive bird species (Figure 7.12).

Human influences such as hunting commonly extend several kilometers into tropical forests from edges (Laurance 2000). Local extinction of species, such as primates, in small patches of forest with high perimeter-to-area ratios is often related to hunting pressure (González-Solís et al. 2001).

Deleterious edge effects contradict the message, long promoted by wildlife managers, that edge habitat benefits wildlife. One popular wildlife textbook urged managers to "develop as much edge as possible" because "wildlife is a product of the places where two habitats meet" (Yoakum and Dasmann 1971). It is true that most terrestrial game animals in the U.S. are edge-adapted, as are many animals characteristic of urban and intensive agricultural landscapes (Leopold 1933; Noss 1983). Forest fragmentation, combined with elimination of large carnivores, has increased deer densities so much in some regions that many species of woody and herbaceous

plants are at risk from heavy herbivory (e.g., Alverson et al. 1988; Augustine and Frelich 1998), with subsequent impacts on butterflies, shrub-nesting birds, humans sensitive to tick-borne diseases, and other species (McShea et al. 1997).

Matrix effects

The matrix surrounding habitat fragments in terrestrial landscapes distinguishes these patches from real islands. The amount of structural contrast between habitat fragments and the matrix in which they exist is one measure of fragmentation (Harris 1984); that is, as the landscape around fragments is progressively altered, the functional isolation of those fragments increases. A structurally rich matrix may serve as marginal or even highly suitable habitat for some species and buffer population fluctuations; it may also encourage dispersal among patches. Plants, in particular, are hypothesized to respond to gradients in habitat suitability more than to patches (e.g., fragments) per se, and may find suitable sites for growth in much of what is considered the landscape matrix (Jules and Shahani 2003; Murphy and Lovett-Doust 2004). On the other hand, source populations of many species may depend on large, secure patches. For example, the presence of old-growth species in a managed forest matrix may give the impression that these species are not dependent on old growth; but once old-growth source populations are eliminated, populations in the matrix may gradually or rapidly disappear (Noss and Cooperrider 1994).

For many species, as the landscape matrix departs more and more from natural habitat, isolation increases as individuals are less willing or able to travel from one patch of natural habitat to another. This process is a very common one, and occurs when the intensity of development or resource extraction increases in a landscape. Use of riparian forests by breeding birds in Idaho was influenced primarily by landscape characteristics and only secondarily by local habitat variables (Saab 1999). The landscape matrix, in particular, determined species richness and the distribution and frequency of occurrence of 32 species of breeding birds. Most important was whether the matrix was natural or agricultural habitat. Landscape-scale studies are needed that measure birth and death rates across a range of habitats, as well as dispersal among those habitats. Only then can we say with assurance which kinds of matrix are optimal for a network of natural areas.

Much of the variation in the results of edge-effect studies undoubtedly relates to differences in matrix type, management regime, and other uncontrolled factors (Murcia 1995; Hansen and Rotella 2000). Generally, the greater the structural contrast between adjacent terrestrial habitats, the more intense the edge effects. In Maryland, nest predation rates are higher in woodlots surrounded by suburbs than in woodlots surrounded by agriculture, probably because garbage and other food subsidies in suburban landscapes encourage proliferation of opportunistic **mesopredators** (small to medium-sized carnivores; Wilcove 1985). Similarly, in Ontario the diversity and abundance of songbirds in forest blocks surrounded by suburbs was much lower than in forests with few or no nearby houses, likely because populations of house cats, squirrels, and other nest predators are higher in suburbs (Friesen et al. 1995). In southern California, predation by house cats (*Felis silvestris catus*), raccoons, and other mesopredators in patches of coastal sage scrub isolated by development markedly reduced nesting success of songbirds. The problem was much less severe, however, in patches that retain populations of coyotes (*Canis latrans*), which prey on the mesopredators (Soulé et al. 1988; Crooks and Soulé 1999). Curiously, Janzen (1983, 1986) found that weedy species more readily invaded naturally disturbed sites in Costa Rican forests when the forests were surrounded by successional habitats than when surrounded by croplands and heavily grazed pastures, with invasion extending at least 5 km into a forest.

The special problem of roads

Habitat fragmentation is usually accompanied and augmented by road building. Road-building is one of the most common and insidious causes of habitat fragmentation; in many regions it surpasses all other forms of fragmentation in magnitude of effects. Yet many studies and reviews of fragmentation (e.g., Fahrig 2003) have ignored roads entirely. Trombulak and Frissell (2000) listed seven general effects of roads that are documented in the literature: mortality from road construction, mortality from collision with vehicles, modification of animal behavior, alteration of the physical environment, alteration of the chemical environment, spread of nonnative species, and increased use of areas by humans. Among the behavioral influences are the movement-barrier effects that roads create for many animals, which is a direct form of population fragmentation. Edge effects are often very prominent along roads, with documented negative consequences. For example, most Amazonian understory birds respond negatively to road edges (Laurance 2004). When the fragmentation caused by roads is considered, the average forest patch size in the U.S. is only 72 ha, only 34 ha of which could be considered core area (Heilman et al. 2002). These results are probably unrealistically conservative, as the actual density of roads in a landscape may be more than twice that shown in digital databases (Hawbaker and Radeloff 2004).

To the extent that individual animals hesitate to cross roads, roads fragment populations into smaller demo-

graphic units that are more vulnerable to extinction. A study in southeastern Ontario and Quebec found that several species of small mammals rarely ventured onto road surfaces when the road clearance (distance between road margins) exceeded 20 m (Oxley et al. 1974). In Oregon, dusky-footed woodrats (*Neotoma fuscipes*) and red-backed voles (*Clethrionomys californicus*) were trapped at all distances from an interstate highway right-of-way, but never in the right-of-way itself, suggesting that these rodents did not cross the highway (Adams and Geis 1983). In Germany, several species of carabid beetles and two species of forest rodents rarely or never crossed two-lane roads (Figure 7.13); even a narrow, unpaved forest road, closed to public traffic, served as a barrier to these species (Mader 1984). Another study found that roads and railroads inhibited normal movements of lycosid spiders and carabid beetles; although crossings were rare, longitudinal movements along these barriers were stimulated (Mader et al. 1990). Even birds may find roads to be a barrier; in Amazonia, flocks of birds avoid crossing the open 30–50 m swath created by major roads (Develey and Stouffer 2001).

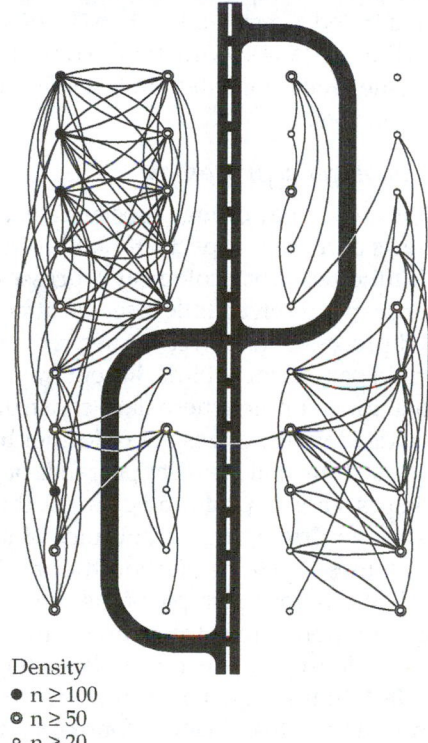

Density
● n ≥ 100
◉ n ≥ 50
○ n ≥ 20

Figure 7.13   Roads can be significant barriers to the movement of small vertebrates and invertebrates. In this example, populations of the forest-dwelling carabid beetle *Abax ater* were almost completely divided by a road and even by parking loops. Lines represent movements of marked beetles between capture and recapture points. (Modified from Mader 1984.)

Road clearances can be barriers in a wide variety of habitat types. In a study of the effects of a highway on rodents in the Mojave Desert (Garland and Bradley 1984), only one white-tailed antelope squirrel (*Ammospermophilus leucurus*), out of 612 individuals of eight rodent species captured and 387 individuals recaptured, was ever recorded as having crossed the road. The endangered Stephen's kangaroo rat (*Dipodomys stephensi*) in southern California perceives gravel roads as movement barriers, but uses dirt roads as corridors (Brock and Kelt 2004). A 9-year study in a Kansas grassland found that very few prairie voles (*Microtus ochrogaster*) and cotton rats (*Sigmodon hispidus*) ever crossed a dirt road 3 m wide (Swihart and Slade 1984). In a forested landscape in Maine, habitat use and movements of frogs and toads were affected little by roads; salamanders, however, were much less abundant near roads and seldom crossed roads (DeMaynadier and Hunter 2000). Many other studies have documented the barrier effects of roads, even for animals as large as black bears (*Ursus americanus*) (Brody and Pelton 1989). In the latter study, the frequency at which bears crossed roads of any type varied with traffic volume. An interstate highway was the most significant barrier, and bears that attempted to cross it were often killed. Roads are among the largest sources of mortality for endangered Florida panthers (Figure 7.14) and many other large mammals (Noss et al. 1996; Baker and Knight 2000), and have been implicated in widespread amphibian declines as traffic volume has increased in many regions (Fahrig et al. 1995). Land turtles and large-bodied pond turtles are also highly susceptible to road kill, such that current densities of roads in many regions may threaten their persistence (Gibbs

Figure 7.14   Florida panthers are highly endangered. Automobile collisions are the single largest source of mortality for this species, a situation that is made worse as more people move to south Florida. Efforts to reduce this mortality include posting of caution signs in areas where panthers frequently cross roads. (Photograph © Painet, Inc.)

and Shriver 2002). Disproportionate mortality of female turtles, which roam further than males in their annual migrations to nesting sites, has been shown to skew sex ratios in some species (Steen and Gibbs 2004).

The barrier effects of roads can have implications not just for short-term population viability, but also at the genetic and community levels of organization. A German study found that separation of populations of common frogs (*Rana temporaria*) by highways reduced the average heterozygosity and genetic polymorphism of local populations (Reh and Seitz 1990). A more recent German study found that populations of bank voles (*Clethrionomys glareolus*) separated by a highway were genetically distinct. No voles marked on one side of the highway were ever recaptured on the opposite side. A small country road, however, did not serve as a movement barrier (Gerlach and Musolf 2000).

Just as roads serve as barriers to the movement of some species, they serve as conduits for the invasion of others. For example, many nonnative plants, insect pests, and fungal diseases of trees are known to disperse and invade natural habitats via roads and vehicles (Schowalter 1988; Tyser and Worley 1992; Wilson et al. 1992; Lonsdale and Lane 1994; Parendes and Jones 2000). Disturbed roadsides with high light levels harbor many weeds, which disperse along the route of the road and often invade adjacent forests and other habitats. Vehicles using the road transport seeds and spores long distances, sometimes hundreds of miles.

Species invasions

The addition of species to landscapes is as important to consider in conservation biology as is species loss (see Chapter 9). Nonnative species may be second only to direct habitat destruction as a threat to imperiled species in the U.S. (Wilcove et al. 1998), and are problematic in many other regions as well. Invasive species, many of which are nonnative, often prosper as a result of habitat fragmentation because soils are disturbed and new pathways for invasion are opened up.

Roads favor species with good dispersal abilities in disturbed habitats at the expense of species with limited mobility. Although roads fragment many natural habitats and pose barriers to animal movement, they often increase connectivity of disturbed sites and promote the spread of nonnative weeds, pests, and pathogens. A tragic example is the fatal root disease fungus (*Phytophthora lateralis*) that afflicts Port Orford cedar (*Chamaecyparis lawsoniana*), a tree endemic to a small region of northwestern California and southwestern Oregon (Jules et al. 2002). This disease travels from one watershed to another along logging road systems, presumably in mud on vehicles. In grasslands, unfragmented roadless habitats can be important refugia for native species

against invasion by nonnative plants (Gelbard and Harrison 2003).

Even when there is no evidence of direct impacts of nonnative species on native species, increases in species richness at a local scale due to invasions by weedy or nonnative species are often accompanied by declines in diversity at broader scales as sensitive native species are progressively lost, even though overall species richness may remain the same or even increase. As cosmopolitan species invade more and more regions, regional biotas are homogenized and lose their distinctiveness. This process of homogenization, which can occur at several levels of biological organization (i.e., from genes to ecosystem), is a prominent form of biotic impoverishment worldwide (Olden et al. 2004; see Chapter 9).

Sometimes habitat fragments can be relatively resistant to invasion. Patches of old-growth forest in Indiana had nonnative plants along their edges, but forest interiors were virtually free of exotics (Brothers and Springarn 1992). The dense wall of vegetation that developed along the edges of fragments in this study was thought to discourage invasion by reducing interior light levels and wind speeds. In a North Carolina study, however, invasive species such as Japanese honeysuckle (*Lonicera japonica*) penetrated forests up to 60+ m from south-facing edges (Fraver 1994). With time, some nonnative species become well established in forest interiors and outcompete natives.

Effects on ecological processes

Ecological processes may change substantially as a result of edge effects and other aspects of habitat fragmentation. Until quite recently, ecological processes in fragmented landscapes received little research attention. The best-studied processes, reviewed earlier, are nest predation and brood parasitism of birds. Recent fragmentation experiments, for example where dam construction has created islands of various sizes out of former hilltops in tropical forests, show that the disappearance of area-sensitive predators leads to profound changes throughout the food web, confirming the importance of top-down regulation in these systems (Terborgh et al. 2001; see Case Study 7.3). Another example of cascading effects of forest fragmentation comes from forests in southwest Oregon, where densities of deer mice (*Peromyscus maniculatus*) are 3–4 times higher in forest fragments surrounded by clear-cuts than in intact forests, leading to increased predation on trillium (*Trillium ovatum*) seeds and reduced recruitment (Tallmon et al. 2003). Other studies of species interactions in fragmented landscapes show changes in seed dispersal and herbivory near forest edges (Murcia 1995).

Some of the strongest effects of fragmentation on ecological processes may involve the invertebrate commu-

nity. Invertebrates are critically important in decomposition, nutrient cycling, disturbance regimes, and other natural processes in ecosystems, and they appear to be quite sensitive to disruption of microclimate and other effects of fragmentation (Didham et al. 1996). Rates of defoliation during insect outbreaks may be higher along edges than in forest interiors (Kouki et al. 1999). Climatic edge effects may explain why dung and carrion beetle communities in 1-ha and 10-ha forest fragments in Brazil contain fewer species, sparser populations, and smaller beetles than do comparable areas within intact forest (Klein 1989). The drier conditions in small fragments, which are largely edge habitat, may lead to increased fatal desiccation of beetle larvae in the soil. Loss of beetles results in reduced decomposition rates of dung and probably other ripple effects throughout the ecosystem (Klein 1989). Other processes that depend on invertebrates and that are known to be disrupted by fragmentation include pollination, seed predation, and parasitism (e.g., control of phytophagous insects by parasitic Hymenoptera), all of which require more study (Didham et al. 1996).

A growing literature documents the vulnerability of animal–plant mutualisms to habitat fragmentation. Plant–pollinator interactions appear to be especially vulnerable, with spatial isolation of habitats, reduced patch size, and edge effects all known to affect pollinator populations and the plants that depend on them (Kearns et al. 1998). As fragmented plant populations become smaller, they may suffer from the **Allee effect**, where below some threshold population size they are no longer visited by pollinators (e.g., Lamont et al. 1993, Groom 1998). Allee effects can lead to population extinction, particularly in very small, isolated patches (Groom 1998; Figure 7.15). In an Argentine dry forest, most of 16 plant species examined showed fragmentation-related declines in pollination, fruit set, and seed set (Aizen and Feinsinger 1994a, 1994b). Even marine plants for which pollen is transported longer distances by water have low seed set if their populations are sufficiently isolated (Reusch 2003). Highly specialized pollination systems— for example the South African tree *Oxyanthus pyriformis*, which is pollinated by long-tongued hawkmoths—appear especially vulnerable to fragmentation. In this example, isolated populations of the tree show little or no recruitment and require hand pollination (Johnson et al. 2004).

Efforts to reduce fragmentation by enhancing habitat connectivity may benefit animal–plant mutualisms. Plant populations connected by habitat corridors in fragmented landscapes, which facilitate the movement of insect pollinators and birds that disperse seeds, show increased rates of pollination and seed dispersal compared to isolated populations (Tewksbury et al. 2002).

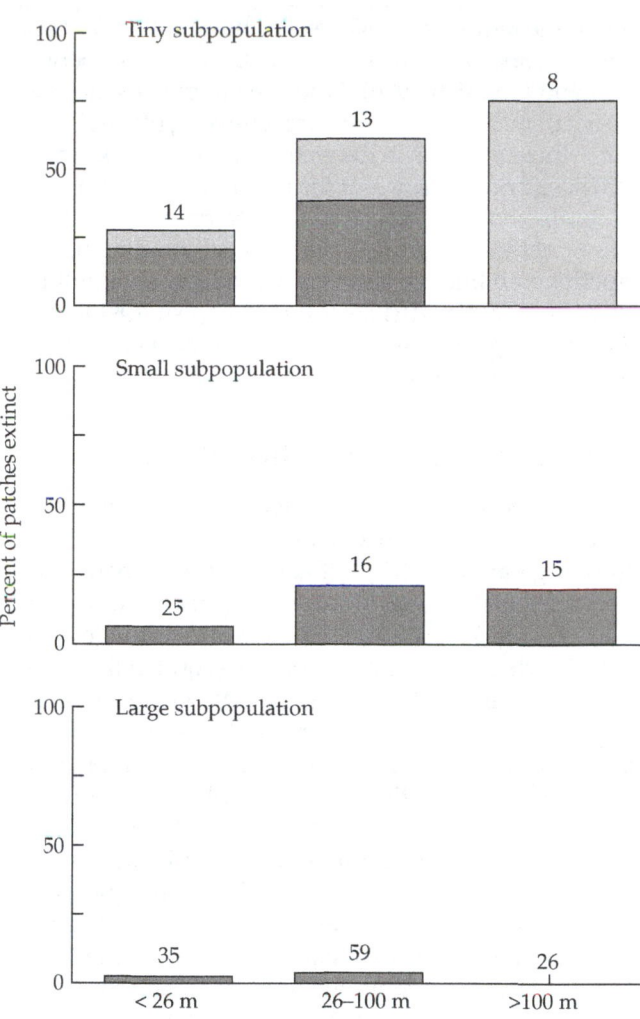

Figure 7.15 Extinction rates were highest in small, isolated patches (subpopulations) of the annual plant *Clarkia concinna* (Onagraceae). White portions of bar denote patches, in which individuals had no, or low, reproductive success, and dark gray portions denote patches that experienced a catastrophe (e.g., a flood); sample sizes are above bars. (From Groom 1998.)

The disruption of natural disturbance regimes by habitat fragmentation has been little studied, but may have striking effects on biological communities. For instance, when a landscape is fragmented, natural fires that once spread for dozens or even hundreds of kilometers from their ignition points are stopped by artificial firebreaks such as roads, farm fields, and urban areas. Lightning is unlikely to strike small habitat fragments often enough to support fire-dependent communities such as grasslands and savannas. In Florida, models of historic and current fire regimes suggest that as little as 10% increase in anthropogenic features such as roads can reduce the extent of fires by 50% (Duncan and Schmalzer 2004). Recensuses of

small prairie remnants in Wisconsin showed that 8%–60% of the plant species were lost over a 32- to 52-year period (Leach and Givinish 1996). Losses were greatest among rare, short, small-seeded, and nitrogen-fixing plants, in addition to those growing in the wettest and most productive sites. These findings support the hypotheses that passive fire suppression caused by landscape fragmentation leads to loss of native species and that plants with the poorest competitive abilities in fire-suppressed habitats will be most heavily affected (Leach and Givinish 1996). Prescribed burning and other active management are urgently needed in such cases.

Nested Species Distribution Patterns

The loss of species from fragmented landscapes may follow a predictable sequence (Patterson 1987; Blake 1991). In studying habitat patches of various sizes, a pattern of nested subsets in the distribution of species is often observed (Figure 7.16). A **nested subset** is a geographic pattern in which larger habitat patches contain the same species found in smaller patches, along with additional species found only in the larger patches. The species found only in larger areas are generally those most vulnerable to fragmentation. For example, on mountain ranges in North America's Great Basin, which are natural habitat islands, boreal mammals and birds show a nested distribution pattern that may be a consequence of selective extinction of area-dependent species on smaller islands. In this case, extinctions occurred in the same basic sequence throughout the region, despite considerable variation in extinction rates (Cutler 1991).

Distribution of bird species among woodlots in agricultural landscapes is typically nonrandom; species found in small woodlots are also found in the larger sites. In east-central Illinois, the most highly nested pattern was found for species requiring forest interior habitat for nesting and for species that migrate to the Neotropics (Blake and Karr 1984; Blake 1991). A similar nonrandom pattern, with all species occupying large patches but many species not occurring in small patches, has been documented for birds on islands in Swedish lakes, forest patches in central Spain, and in many other temperate communities (Nilsson 1986; Tellaria and Santos 1994). Studies of birds in the Wheatbelt of Western Australia documented loss of many species from small habitat remnants since isolation (Saunders 1989), a relaxation effect predictable from island biogeographic theory. Such results support previous suggestions that although a collection of small sites may harbor more species, large sites are needed to maintain populations of species sensitive to human disturbance (Diamond 1976).

Nested species distribution patterns do not always have straightforward explanations. There can be other

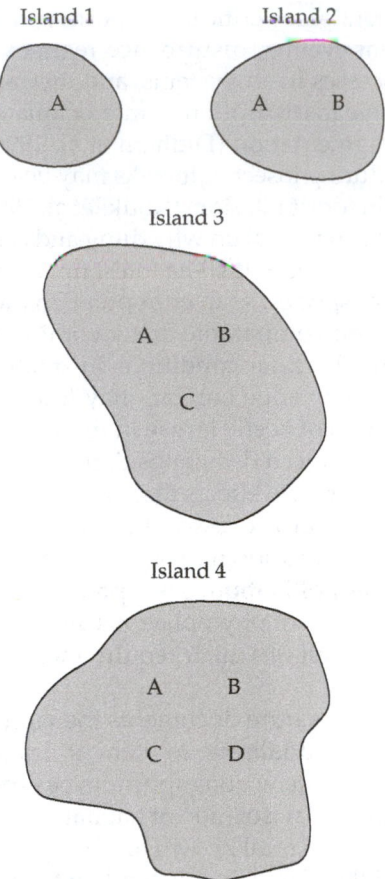

Figure 7.16 Hypothetical nested subset distribution of species on islands of different sizes. The letters A through D represent different species. Species are added in a predictable sequence with increasing island size and number of habitats. The largest island contains all four species. (From Cutler 1991.)

reasons for nested subsets besides a predictable sequence of extinctions as habitats are progressively fragmented. An examination of nestedness from the perspective of individual species showed that different kinds of factors, most related to habitat requirements, determined the distribution patterns of different species (Simberloff and Martin 1991). Most studies suggesting loss of species from small fragments have failed to test the null hypothesis that assemblages in small patches are simply "samples" from the larger regional species pool. A study of breeding birds in forest fragments in the southern taiga of Finland confirmed predictions from the null model that the location of breeding pairs varies randomly among fragments from year to year, and that the pattern of species additions with increasing sample size is similar to that expected from random sampling (Haila et al. 1993). Thus, local turnover of species in small fragments may represent simple changes in terri-

tory locations from year to year, rather than true "extinctions" and "recolonizations." Such studies suggest a need for greater experimental rigor, so that the true effects of fragmentation can be separated from random patterns and statistical artifacts.

Species Vulnerable to Fragmentation

Effects of fragmentation can be seen at several levels of biological organization, from changes in gene frequencies within populations to continent-wide changes in the distributions of species and ecosystems. At the species level, there are essentially three options for persistence in a highly fragmented landscape. First, a species might survive or even thrive in the matrix of human land use. Marine species tolerant of the more variable temperatures and salinities of human-modified estuaries predominate in fragmented estuaries (Layman et al. 2004). Many species of plants and animals worldwide thrive in human-modified habitats; although these species are typically considered "weedy" and of little conservation concern, they may comprise a substantial portion of a region's native fauna and flora.

Second, a species might survive in a fragmented landscape by maintaining viable populations within individual habitat fragments; this is an option for species with small home ranges or otherwise modest area requirements, such as many plants and invertebrates. Many of these species can meet all of their life history requirements and maintain viable populations within the boundaries of a single fragment; they might persist there indefinitely, barring major environmental change such as global warming.

A third way to survive in a fragmented landscape is to be highly mobile. A mobile species might integrate a number of habitat patches, either into individual home ranges or into an interbreeding population. The Pileated Woodpecker (*Dryocopus pileatus*) adapted to fragmented landscapes, particularly in eastern North America. Foraging individuals now travel among a number of small woodlots in landscapes that were formerly continuous forest, often using wooded fencerows as travel corridors (Whitcomb et al. 1981; Merriam 1991). White-footed mice (*Peromyscus leucopus*) and eastern chipmunks (*Tamias striatus*) maintain populations in fragmented landscapes only when dispersal between woodlots, aided by fencerow corridors (Figure 7.17), is great enough to balance local extinctions (Fahrig and Merriam 1985; Henderson et al. 1985). A species incapable of pursuing one or more of these three options is bound for eventual extinction in a fragmented landscape, unless the population is sustained by individuals continually dispersing in from source populations in less fragmented regions (e.g., Robinson et al. 1995).

What kinds of species are most vulnerable to local and regional extinction following habitat fragmentation? Consideration of life histories has produced a number of hypotheses about which life-history attributes may make species more vulnerable to fragmentation than others (Box 7.2). The proportion of species possessing such attributes may vary regionally, such that some regions have more fragmentation-sensitive taxa than others (Hansen and Rotella 2000).

For purposes of comparison, studies of fragmentation should concentrate on species that are predicted to be vulnerable due to the kinds of traits reviewed in Box 7.2, as well as on related species with similar and dissimilar life histories. Land managers might also concentrate monitoring and conservation efforts on vulnerable species. Knowing more about the autecology of such species will be fundamental to conservation success (Simberloff 1988).

Fragmentation versus Habitat Loss, and Regional Differences

Fragmentation is often confounded with other trends in landscape pattern and condition. In particular, separating fragmentation in the literal sense from overall habitat loss can be difficult. Most of the studies documenting deleterious effects of fragmentation have been carried out in highly fragmented landscapes; in less fragmented landscapes, results are mixed. For example, predation rates on birds' nests, both near edges and in forest interior, were much higher in forest remnants in an agricultural landscape than in contiguous landscapes or logged landscapes where the matrix remained as forest (Bayne and Hobson 1997). Andren (1994) hypothesized that habitat loss would be the most important process explaining species declines in landscapes with a high proportion of suitable habitat, whereas patch size and

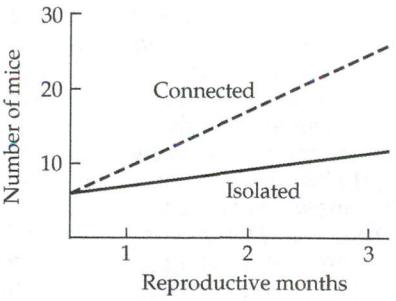

Figure 7.17 Isolated woodlots in a fragmented landscape are predicted by simulation models to have lower rates of population growth than woodlots connected by fencerow corridors. These predictions were verified by studies of white-footed mice in southern Ontario. (From Merriam 1991, based on Fahrig and Merriam 1985.)

BOX 7.2 Species Vulnerable to Fragmentation

Wide-Ranging Species

Some animals, such as large carnivores and migratory ungulates, roam a large area in the course of their daily or seasonal movements. Even rather large fragments may not provide enough area for viable populations of these species; thus, they must travel widely and will often attempt to move through heavily fragmented landscapes. In so doing, they encounter roads, people with guns, and other sources of mortality (Mladenoff et al. 1995; Noss et al. 1996; Carroll et al. 2001, 2004). As discussed earlier, (see Figure 7.9), animals of heterogeneous landscapes, such as amphibians, turtles, and other species that depend on distinct habitats for different phases of their life cycles, are vulnerable to roads and other barriers. Resplendent Quetzals (*Pharomachrus mocinno*), which require fruits from spatially separated habitats at different times of the year, cannot maintain year-round populations in small reserves; if reserves are isolated by fragmentation, Quetzals cannot migrate to track fruiting schedules and probably will go extinct in these landscapes (Wheelwright 1983; Powell and Bjork 1995).

Nonvagile Species

Species with poor dispersal abilities may not travel far from where they were born, or may be stopped by barriers as seemingly insignificant as a two-lane road or clear-cut. Many insects of old-growth forests are flightless and are poor dispersers (Moldenke and Lattin 1990). Clear-cuts are substantial barriers to carrion and dung beetles in Amazonian forests being fragmented by pasture development (Klein 1989). Perhaps surprisingly, some species of birds have very low colonizing abilities and will not cross relatively narrow areas of unsuitable habitat (Diamond 1975; Opdam et al. 1984; van Dorp and Opdam 1987). In Atlantic forest fragments in Brazil, small mammals that will move through the grassy matrix in which fragments are embedded are much more extinction-resistant than species unwilling to enter the matrix (Viveiros de Castro and Fernandez 2004). Without the occasional arrival of immigrants to provide a rescue effect and bolster genetic diversity, populations of nonvagile species may not persist long in habitat fragments.

Variation in dispersal ability has been shown to be a critical determinant of survival for mammals in fragmented tropical forests in Queensland, Australia (Laurance 1990). Animals with group social structures, such as many primates, are often poor dispersers, and hence vulnerable to fragmentation (Lawes et al. 2000). In Florida scrub, measures of patch area and isolation predicted patch occupancy by scrub lizards (*Sceloporus woodi*) very well, but not by six-lined racerunners (*Cnemidophorus sexlineatus*). Although these lizard species are ecologically similar, racerunners are much better dispersers (Hokit et al. 1999).

Limited dispersal capacity also can be a problem for plants in fragmented landscapes. When pollinators move less among fragments, plant fitness decreases, as discussed earlier. Temperate forest herbs may be restricted in distribution according to dispersal ability, with large-seeded species showing reduced occupancy of suitable patches (Ehrlen and Eriksson 2000). Even in marine environments where currents might be expected to move organisms long distances at some point in their life history, many species of invertebrates and fishes have very poor dispersal abilities, even as larvae (Ruckelshaus and Hays 1998). Further, even some species that are quite mobile may be effectively limited to fragments due to predation intensity outside of a reef, an outcrop, a seagrass bed, or kelp forest. For example, many small fishes rely on hiding places within coral reefs to avoid predation from larger fish swimming outside the reef (Shulman 1985; Posey and Ambrose 1994; Almony 2004). Thus, many marine species become effectively isolated when their preferred habitat is fragmented.

Species with Specialized Requirements

Species with specialized habitat or resource requirements are often vulnerable to extinction, especially when those resources are unpredictable in time or space. For example, Australian mammals that specialize on certain plant communities were abundant in continuous forest, less abundant in forested corridors, and generally absent from small forest fragments (Bentley et al. 2000). When resources fluctuate seasonally or annually, species dependent on those resources also fluctuate. Population variability predisposes species to extinction. The higher the level of fluctuation, the greater the chance of extinction (Karr 1982; Pimm et al. 1988), with some interesting exceptions (Pimm 1993). Drought years, for instance, often cause population crashes of wading birds, amphibians, and other species dependent on ephemeral wetlands. Reductions in fruit or mast abundance due to drought will affect frugivorous and mast-dependent animals. Habitat fragmentation makes such species vulnerable in two ways: by reducing the number of sites that contain critical resources and by isolating suitable sites and making them harder to find.

Studies of metapopulation dynamics in the Bay checkerspot butterfly (*Euphydryas editha bayensis*) suggest that local extinction is frequent on small patches of serpentine grassland, to which the species is now restricted due to fragmentation of the original, more extensive native grassland (Figure A). Persistence of the metapopulation is heavily dependent on dispersal from a source population to recolonize vacated patches. Because the species is a relatively poor disperser, stepping-stone habitat patches that reduce isolation may be important (Murphy and Weiss 1988). However, only those patches of suitable habitat closest to the large source population were found to be occupied by Bay checkerspots (Harrison et al. 1988).

Large-Patch or Interior Species

Some species occur only in large patches of forests, prairies, shrublands, or other habitats, and are absent from small patches with little or no true interior habitat. California red-backed voles (*Clethrionomys californicus*) in southwestern Oregon are restricted to remnants of old-growth forest, preferring the interiors of forest remnants to the edges and making little use of regenerating clear-cuts (Mills 1995). In a hardwood forest in northern Florida, four breeding bird species showed significantly reduced densities within 50 m of the forest edge (Noss 1991). Forest interior birds in the eastern Usambara Mountains of Tanzania are more vulnerable to extinction in fragmented landscapes than are edge species, perhaps because they avoid crossing large clearings; populations are therefore easily isolated (Newmark 1991). In central Amazonia, many forest understory birds—including army ant followers, solitary species, members of mixed-species flocks, and terrestrial species—strongly avoid forest edges (Laurance 2004). In contrast, understory hummingbirds readily use gaps, edges, and second growth and are far less vulnerable to fragmentation than insectivorous birds (Stouffer and Bierregaard 1995). Ground-foraging, insectivorous birds of tropical forests appear inordinately vulnerable to fragmentation. Amazonian forest fragments less than 100 ha in size do not maintain populations of these species even over the short term; much larger areas are probably necessary for long-term persistence (Stratford and Stouffer 1999).

Some animals are restricted to large patches of forest because their food resources are limited in small fragments. A study of Ovenbirds in Ontario found significantly lower biomass of their invertebrate prey in smaller fragments. The researchers concluded that food limitation restricted Ovenbirds from small patches (Burke and Nol 1998). Similar findings have been reported for the Eastern Yellow Robin (*Eopsaltria australis*) in Australia (Zanette et al. 2000), suggesting that the role of food supply should receive increased attention from fragmentation researchers.

Species with Low Fecundity or Recruitment

A species with low reproductive capacity cannot quickly rebuild its population after a severe reduction caused by any number of factors. As an example, Neotropical migrant birds often have low reproductive potential in comparison with permanent resident species, which perhaps is one factor responsible for their decline in fragmented forests of eastern North America (Whitcomb et al. 1981). Low fecundity and recruitment may be related to altered food web structure in fragments. For example, in central Chile, granivores (especially rodents) are more abundant in small forest fragments than in continuous forest, which results in increased consumption of large seeds, and may be altering the composition of the tree community (Donoso et al. 2003).

Species Vulnerable to Human Exploitation or Persecution

Some species are actively sought by people for food, furs, medicine, pets, or other uses, whereas other species, such as snakes and large predators, may be killed on sight. Most habitat fragments are readily accessible to humans due to high edge–interior ratios and the ubiquitous presence of roads. In traveling between habitat fragments, animals may be visible and easily killed or collected by people. The Iberian lynx (*Felis parinda*), the most endangered carnivore in Europe, is declining because fragmentation of its habitat has increased human access and has led to high levels of illegal trapping, road mortality, and hunting with dogs (Ferreras et al. 1992). As noted earlier, primates and other tropical mammals are often absent from small fragments because of their vulnerability to hunting (González-Solís et al. 2001).

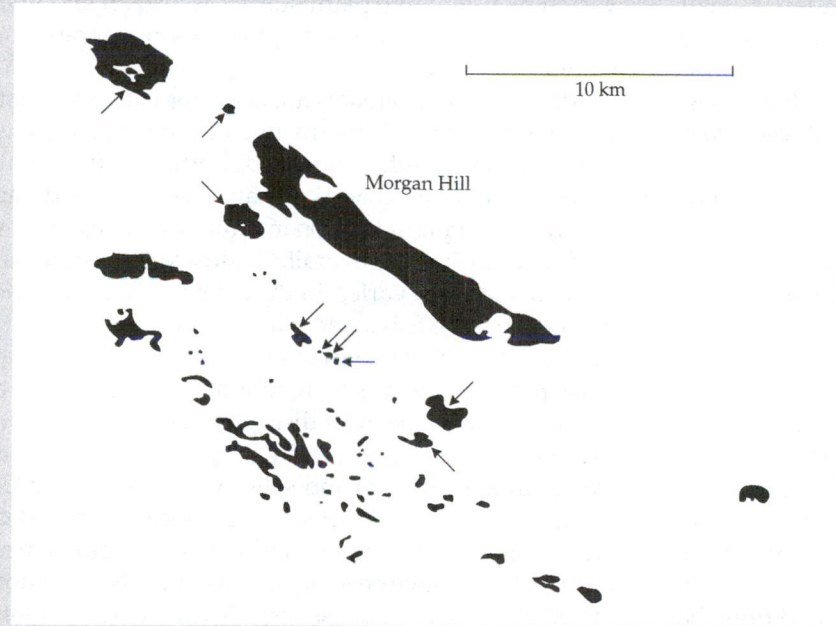

Figure A   Habitat for the Bay checkerspot butterfly (*Euphydryas editha bayensis*) metapopulation is fragmented due to natural and anthropogenic factors. The black areas represent the butterfly's serpentine grassland habitat. Of the small patches of grassland, only those closest to the large Morgan Hill population (denoted by arrows) are usually occupied, suggesting that this butterfly is a poor disperser. Extinctions in small patches are apparently common and can be reversed only when isolation is minimal. (From Harrison et al. 1988.)

isolation would become more important in landscapes with a low proportion of suitable habitat. Using computer simulations, Fahrig (1997) found that beyond 20% habitat cover, species persistence was virtually assured, regardless of habitat configuration. Fahrig (1997, 2003) went on to suggest that concern about habitat pattern (e.g., fragmentation per se) is misplaced in light of the more severe problem of habitat loss. Other studies, however, have confirmed the importance of habitat configuration and the threat posed by fragmentation to populations, especially when fragmentation is considered on a landscape (as opposed to single-site) scale and over long time periods (e.g., Brooker and Brooker 2002; Chalfoun et al. 2002; Stephens et al. 2003).

Habitat fragmentation has not been studied as thoroughly in western as in eastern North America. Are lessons learned from heavily fragmented eastern forests transferable to the West? Hansen and Urban (1992) hypothesized that bird communities from different geographic regions have different suites of life-history attributes, which will cause species to respond differently to fragmentation and other trajectories of landscape change. Based on a literature review, Hansen and Rotella (2000) identified only one bird species, the Veery (*Catharus fuscescens*), in the southern Rocky Mountains that possessed the life-history attributes that Whitcomb et al. (1981) found characteristic of fragmentation-sensitive birds of the eastern deciduous forest. Only a small percentage (7%–20%) of bird species in the Rockies were less abundant in small forest patches or near forest edges, and no species was absent from small patches. Also in contrast to eastern studies, limited studies from the Rockies that examined predation on artificial nests found that rates did not differ from forest edge to interior (Hansen and Rotella 2000).

In the Oregon Coast Range, landscape structure (i.e., the pattern of habitats across the landscape) explained less than 50% of the variation in the abundance of breeding birds, with most species more abundant in the more heavily fragmented landscapes (McGarigal and McComb 1995). Only one species in this study, the Winter Wren (*Troglodytes troglodytes*), was found sensitive to fragmentation, although sample sizes of the species that might have been most sensitive (e.g., Northern Spotted Owl [*Strix occidentalis caurina*] and Marbled Murrelet [*Brachyramphus marmoratus*]) were too small for analysis. These and some other species probably had already disappeared from the most heavily fragmented landscapes. A study in the Bitterroot Valley of Montana found patterns of nest predation opposite to those documented in the East, with higher predation rates in forested landscapes than in fragmented landscapes dominated by agriculture (Tewksbury et al. 1998). Brood parasitism by cowbirds in this study was higher in agricultural landscapes, but was more strongly related to the abundance of houses and farms than to forest cover. On the other hand, Black Grouse (*Tetrao tetrix*) and Capercaille (*T. urogallus*) in boreal forest landscapes in Finland showed reduced breeding success in areas fragmented by agriculture or clear-cutting. Nest predation by generalist predators, such as the red fox (*Vulpes vulpes*) was higher in the more fragmented landscapes and was the most likely cause of grouse nest failures (Kurki et al. 2000).

We know less about fragmentation in nonforested habitat types, but evidence of harmful effects is increasing. In North America, grassland ecosystems generally are extremely fragmented, as vast expanses have been transformed into agricultural fields. About 37% of the grassland ecoregions in the Western Hemisphere are reduced to very small and few fragments, while another 15% experience moderate fragmentation (Ricketts et al. 1999). Among bird species monitored by the Breeding Bird Survey in the U.S. and Canada, grassland species have experienced the most consistent declines as a result of changing agricultural practices that create habitat loss and a variety of fragmentation effects (Sauer et al. 1997; Robinson 1999). Strong evidence now suggests that higher rates of nest predation in fragmented grasslands are contributing to the regional declines of grassland birds in the midcontinental U.S. (Herkert et al. 2003). Invasion by nonnative plant species is also an extremely common consequence of fragmentation in grasslands (White et al. 2000).

Marine benthic habitats are also subject to fragmentation. Because most humans live in coastal areas, these ecosystems are highly modified, leading to fragmentation of mangroves, coastal wetlands, seagrass beds, and reef systems. Fragmentation in estuaries can lead to lowered species diversity overall. Censuses of 30 estuaries in the Bahamas that varied in the extent of fragmentation showed that the fish assemblages were correlated to the degree of fragmentation (Layton et al. 2004). The most diverse fish assemblages were found in unfragmented estuaries, whereas the most disrupted estuaries had lower diversity consisting primarily of species tolerant of extremes in temperature and salinity (Layton et al. 2004). Mangroves and coral reef systems appear to be particularly vulnerable to fragmentation effects. For example, coral reefs that have been degraded tend to become dominated by "weedy" coral species that are more tolerant to fragmentation (Knowlton 2001). Because many species of coral and the species associated with coral reefs do not disperse widely, it is likely that fragmentation of coral reefs will lead to highly isolated groupings of communities that can be subject to many of the detrimental effects of fragmentation discussed in this chapter.

Fragmentation effects can be quite variable in marine systems. In an experimental study, fragmentation did not affect the abundance of two common coral commensals, a shrimp (*Palaemonella* sp.) and a crab (*Trapezia cymodoce*), but death of the corals reduced abundance of these species, and also reduced species diversity (Caley et al. 2001). Many other studies of relationships between diversity and patchiness or fragmentation of coral reefs has shown that fish and invertebrate diversity is positively correlated with larger, more continuous and complex coral reef structures (Ault and Johnson 1998).

Studies in seagrass beds have shown a variety of effects, with details of each case (including the cause of fragmentation, faunal community composition, and fishing activities) influencing outcomes of each case (Bell et al. 2001; Hovel 2003). More extensive damage to seagrass beds caused by intensive propeller scarring is more likely to result in negative impacts to fauna, probably because such impacts included damage to the substrate, not only to the configuration of the seagrass beds (Bell et al. 2001).

The Problem of Climate Change

Fragmentation is a threat to biodiversity even in a relatively stable world. If we add the phenomenon of rapid climate change, then we have perhaps the most ominous of all potential threats to biodiversity (Peters and Lovejoy 1994). Species migrating in response to climate change have always had to cope with dispersal barriers such as rivers, lakes, mountain ranges, and desert basins. The additional set of barriers created by human activities will make migration all the more difficult (Peters and Darling 1985; Collingham and Huntley 2000; Noss 2001). New climates will render reserves set aside to protect certain species or communities unsuitable for them. Weeds may dominate many fragments.

Even natural rates of climate change threaten species restricted to fragments surrounded by inhospitable habitat. The increased rates of change predicted with greenhouse warming may eliminate all but the most vagile species as they fail to track shifting climatic conditions. High-elevation and high-latitude habitats may be lost entirely. Collingham and Huntley (2000) used a spatially explicit model to investigate the impacts of habitat fragmentation on the ability of a wind-dispersed tree (*Tilia cordata*) to migrate in response to changing climate. Simulated dispersal rates were reduced dramatically when habitat availability fell below 25% of landscape area. Landscapes with a "blocky" (coarse-grained) pattern had the strongest negative effect on migration, which suggests that multiple small reserves of suitable habitat, which could serve as "stepping stones" for dispersal, might be preferable to a smaller number of large reserves. Other species, more dependent on large habitat blocks and requiring intact habitat corridors to migrate, would be expected to favor a different landscape pattern. Although wide habitat corridors, stepping stones, and artificial translocations of populations northward and upslope may help some species in some areas, these solutions will not suffice for whole communities, especially if climate change is as rapid as predicted.

Conclusions and Recommendations

This chapter has reviewed evidence that fragmentation occurs at many spatial scales and may have a variety of short-term and long-term ecological effects. Some species, especially those considered weedy, benefit from fragmentation, whereas many others are at increased risk of extinction. Global biodiversity can be maintained in large part by devoting conservation resources to those species and ecosystems at greatest risk of loss—balanced by a strategy of maintaining large wild areas that are not presently at high risk but may become so in the future—and by controlling or reversing the processes that place species and ecosystems at risk. There is also a need to develop effective management strategies for landscapes that are already fragmented, including management of the internal dynamics of remnant natural areas and the external influences on those areas.

Strategies for countering fragmentation follow logically from consideration of the fragmentation process and its effects. Because fragmentation causes a reduction in the size of natural areas and isolation of remaining areas in a sea of unsuitable or less suitable habitat, corrective action should include maintenance or restoration of large, intact core areas that span substantial portions of regional landscapes (Noss and Cooperrider 1994). These core areas will contain the source populations of many fragmentation-sensitive species (Robinson et al. 1995; Carroll et al. 2001). Where circumstances prohibit establishment of truly large reserves, biodiversity can be well served by land-use practices—for example, clustered developments, reduced road building, and less intensive forestry methods—that minimize fragmentation and optimize connectivity of similar natural habitats. Connectivity is the antithesis of fragmentation. Connectivity must be defined functionally as the movement and genetic interchange of species that enhances their viability (Beier and Noss 1998) or, on a broader spatiotemporal scale, as the successful migration of floras and faunas in response to environmental change.

Large, interconnected nature reserves are only part of the solution to the fragmentation problem. Entire landscapes, including private and multiple-use public lands,

should be managed in ways that minimize destruction and isolation of natural habitats. Opportunities to reduce and reverse fragmentation abound on public lands in the U.S. Unfortunately, fragmentation continues on these lands (Noss and Cooperrider 1994; Shinneman and Baker 2000), albeit it has begun to slow very recently. Between 1972 and 1987, average forest patch size in two ranger districts of the Willamette National Forest, Oregon, decreased by 17%, the amount of forest–clear-cut edge doubled, and the amount of forest interior at least 100 m from an edge declined by 18% (Ripple et al. 1991). In the Olympic National Forest, Washington, more than 8% of the old growth in 1940 was in patches larger than 4000 ha; in 1988, only one patch larger than 4000 ha remained, and 60% of the old growth was in patches smaller than 40 ha in size. Of the remaining old growth in 1988, 41% was within 170 m of an edge (Morrison 1990). Roads and road-edge habitat cover more than one third of a 37,233-ha study area in the Black Hills National Forest of Wyoming and South Dakota (Shinneman and Baker 2000). Over 600,000 miles of roads have been built in U.S. national forests, enough to circle Earth 15 times (Coghlan and Sowa 1998). Stopping destructive management of public lands is essential to conserving biodiversity in the U.S. The situation, especially with respect to deforestation, is typically worse in developing countries that have fewer and less rigorously enforced environmental laws (e.g., Indonesia; Kinnaird et al. 2003), although at least many poor countries are unable to publicly finance road-building on the scale the U.S. does.

Some recommendations for maintaining biodiversity in fragmented landscapes and seascapes (or those in danger of being fragmented) follow from the information presented in this chapter:

1. Conduct a landscape or seascape analysis. Determine the pattern of habitats and connections at multiple spatial scales, and relate these to the needs of native species in the landscape. Where are the major, unfragmented blocks of habitat? Can natural connections between habitats be maintained or restored?

2. Evaluate the landscape or seascape of interest within a larger context. Does it form part of a critical linkage of ecosystems at a regional scale? What is the significance of this landscape to conservation goals at regional, national, and global scales?

3. Avoid any further fragmentation or isolation of natural areas. Developments, resource extraction activities, and other uses should be clustered (and minimized) so that large blocks of natural habitat

remain intact. In planning protected areas, emphasize large areas whenever possible, though smaller areas are often useful as habitat for some species and as stepping stones for animal movement.

4. Minimize edge effects around remnant natural areas. This can be done by establishing buffer zones with low-intensity human use. Be careful, however, not to produce population sinks that lure sensitive species out of protected areas and into areas where mortality is greater or reproduction is reduced.

5. While conserving large, unfragmented patches of habitat, don't "write off" the small fragments. Such areas may be the last refuges for many species in highly fragmented regions and can maintain populations of many species for decades. Moreover, these small fragments may contain the sources for recolonization of the surrounding landscape, should destructive activities cease (Turner and Corlett 1996).

6. Also, do not write off the landscape matrix as nonhabitat. There will rarely be enough area in reserves to conserve all of a region's biodiversity. In many cases—for example, on public lands and large private land holdings—opportunities exist to maintain habitat conditions in the matrix that will meet the needs of the majority of native species in the region.

7. Identify traditional wildlife migration routes and protect them. Steer human activities away from critical wildlife movement areas.

8. Maintain native vegetation along streams, fencerows, roadsides, powerline rights-of-way, and other remnant corridors in strips as wide as possible, in order to minimize edge effects and human disturbances.

9. Minimize the area and continuity of artificially disturbed habitats dominated by weedy or non-native species, such as roadsides, in order to reduce the potential for biological invasions of natural areas.

10. Small fragments often suffer from disruption of natural processes, such as fire regimes. Active management will be needed to maintain the native flora and fauna of these fragments.

11. Avoid dam construction, water diversions, and other activities that disrupt aquatic or hydrologic connectivity, and reverse these disruptions where possible.

CASE STUDY 7.1

Subdividing the West

Richard L. Knight, Colorado State University; John Mitchell, Rocky Mountain Forest and Range Experiment Station; Eric A. Odell, Colorado Division of Wildlife; Jeremy D. Maestas, Colorado State University

It's boom time in the Rockies. Beginning in the early 1990s, the Intermountain West of the United States has been experiencing a sudden increase in population growth. This region, consisting of eight states (Arizona, Colorado, Idaho, Montana, Nevada, New Mexico, Utah, and Wyoming) is not only the fastest-growing area in the U.S., its growth rate rivals that of Africa and exceeds that of Mexico. Attracted by a higher quality of life in the West compared to other regions in the U.S. where crime, traffic congestion, air pollution and cost-of-living are worse, it is projected that people will continue to move to this area for decades to come. Unlike past "booms," which were fueled by energy and mineral development, this growth is driven by expansion of the service, recreation, and information industries and is marked by the conversion of private agricultural lands to rural

subdivisions (exurban development). Most importantly, this growth may forever alter the native biodiversity of this vast region, resulting in more human-adapted species and fewer species whose evolutionary histories do not predispose them to live among elevated densities of people.

Until recently the West was principally urban, with most of its citizens living in cities; rural areas sustained low population densities on ranches and farms. Today, these once rural Western landscapes are experiencing rapid population growth, principally through people moving to small-acreage subdivisions (Figure A). During the 1980s metropolitan areas in the U.S. began growing at rates faster than nonmetropolitan areas, but in the Intermountain West rural counties grew faster than the metropolitan centers (Cromartie 1994; Theobald 1995). It is

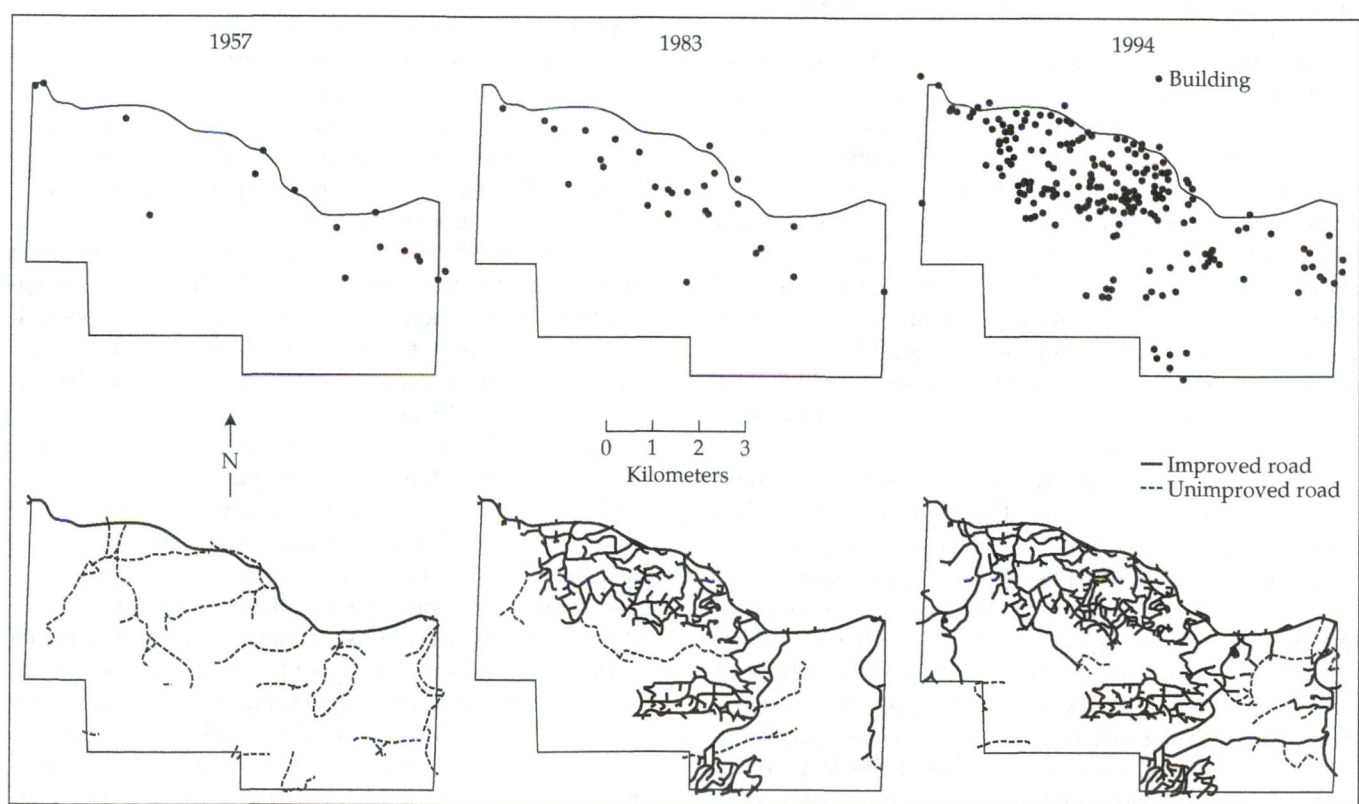

Figure A A former ranch in Colorado that has been subdivided into ranchettes. The upper maps show the increase in the number of houses over time; the lower maps show the increase in roads built to access the homes.

this rapid development of formerly rural landscapes that bodes poorly for the region's natural heritage.

In the American West, land that is not in cities and towns is likely to be farm and ranchland. Between 1992 and 1997 Colorado ranch and farmland has declined by 10,900 hectares per year, much of it converted to subdivisions and commercial development (Colorado Department of Agriculture 2000). Between 1969 and 1987 in Park County, Wyoming, 19% of ranchland was platted for subdivision, while in Teton County, Idaho the rate was 16%. In Gallatin County, Montana, the rate was 23% (Greater Yellowstone Coalition 1994).

The New West is one of ranchette developments, rural subdivisions as vast as the former ranches they now occupy. This style of development is creating a new landscape, evenly sliced and diced at 20–40 acre parcels (see Figure A). Because approximately half of the West is in public lands, private lands are equally important in supporting viable populations of biodiversity. Because private lands are often at lower elevations and contain some of the West's most important wetlands, these areas are among some of the region's most productive sites. Based largely on the patterns of early settlement history, riparian areas were homesteaded and have remained the cornerstones of much of the farm and ranchlands in the West. Today, prime development property includes wetlands, streams and rivers. Whereas historically human densities were low along riparian areas, today home sites are now crowding along these desirable areas.

What are the conservation implications of this new settlement pattern for the West? Two conspicuous changes associated with exurban development are an increase in human density and an accompanying increase in buildings, roads, and fences (Knight and Clark 1998). These changes translate to more dogs and cats, more automobiles and road-killed wildlife, an increase in wildlife nuisance problems, landscaping with non-native plants, more yard lights left on at night, more noise, and more people walking across the land. Preliminary studies indicate that this will result in an altered natural heritage, with species that thrive in association with humans becoming more pervasive and with species sensitive to humans and their activities becoming more scarce.

Two recent studies support this generalization. We examined bird and mammal communities associated with exurban development in western Colorado (Odell and Knight 2001). Generalist species, like the Black-billed Magpie, Brown-headed Cowbird, and American Robin were more numerous near ranchettes while other species, such as the Blue-gray Gnatcatcher, Orange-crowned Warbler, and Dusky Flycatcher, had reduced densities near homes and only regained higher densities with increasing distance from these homes (Figure B). Likewise, dogs and cats were more likely to be encountered near homes while coyotes and foxes were only common away from homes. The second study examined bird communities across three common land uses of the rural Intermountain West: protected areas, ranches, and ranchette developments

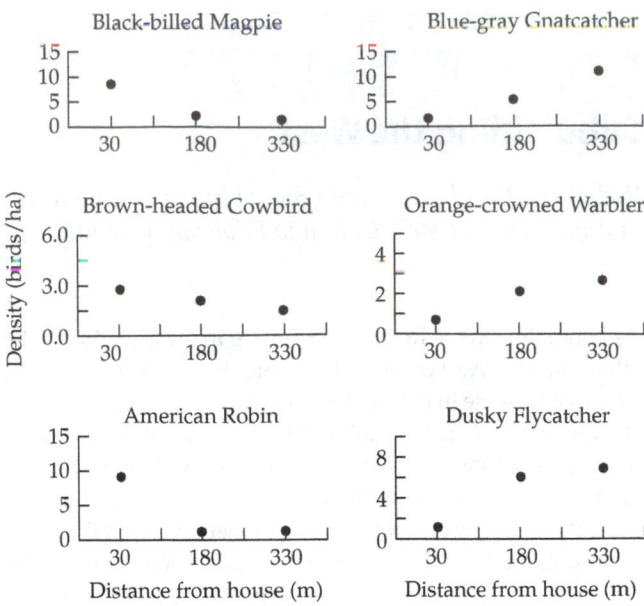

Figure B Densities of songbirds with increasing distances (30 m, 180 m, and 330 m) from rural ranchettes. Note that generalist species (left column) are more abundant close to homes while specialist species (right column) are more common farther from homes.

(Maestas et al. 2003). Preliminary evidence showed that generalist species, such as the European Starling and Black-billed Magpie are more numerous on the ranchette developments while more sensitive species, like the Vesper Sparrow, are only common on protected areas and land devoted to ranching (Figure C). This latter information is important because it provides the data to support the work of environmental organizations such as The Nature Conservancy that are working with ranchers to keep their lands out of development. This approach to conservation assumes that biodiversity is as well served on ranches as it is on protected areas—or at the least, that biodiversity is better protected on ranch lands than on rural housing developments.

Ranches adjacent to public lands are also eagerly bought and subdivided. Suddenly, national parks, forests, and other public lands are ringed by home sites, with every home accessed by individual roads. This results in a whole new suite of challenges for public land managers. Ecosystem management stresses managing landscapes based on ecological rather than administrative boundaries (see Chapter 13). When suddenly you have neighbors rimming your boundary who view fire as a legitimate threat to their homes, rather than an ecological process that needs to be restored to the land, does fire remain a management tool? When suddenly entry onto public lands is across the fence rather than through the entrance, what are the effects of ranchette owner's chainsaws, dogs, cats, weapons, garbage, and nonnative plants? When these same lands are viewed as sources for black bears, cougars, and ungulates, all of

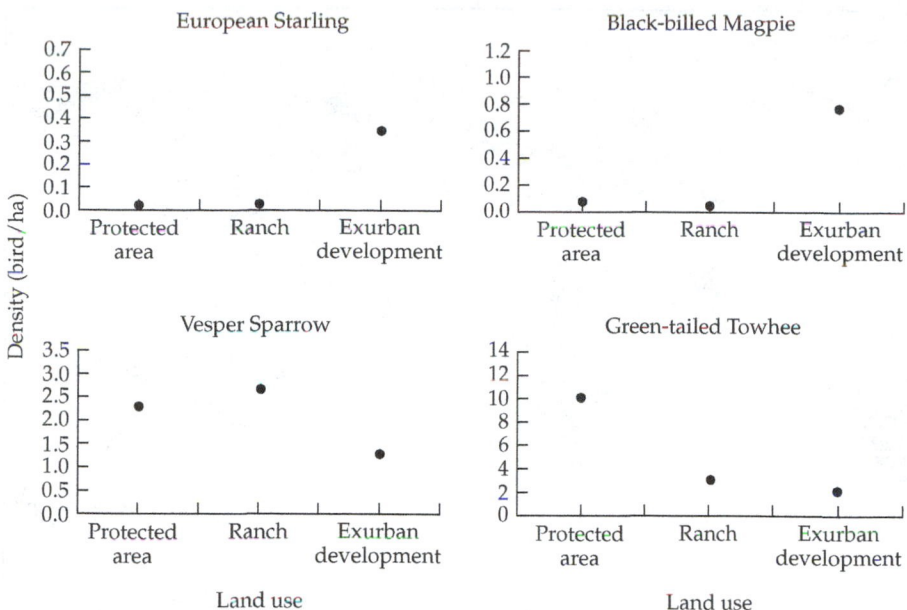

Figure C Densities of songbirds on three different land uses in the American West: protected areas, ranches, and ranchette developments. Several generalist species were more common on lands devoted to exurban development than on protected areas and ranches. Also some species, such as Vesper Sparrows, occurred in higher densities on protected areas and ranches than on ranchettes.

which now become suspect as threats or nuisances, does wildlife win?

Given the continuing exurbanization of the American West, what does the future offer for humans and the maintenance of the regional biological diversity? Will newcomers to this region come to understand how their dogs and cats, their exotic plants, their yard lights beaming through the once black night are altering the wildlife and the silence they came to the region seeking? Will they come to learn how to manage their horses to minimize overgrazing, how to place their access roads to minimize soil erosion, learn to live with rattlesnakes, black bears, and cougars? Will they come to appreciate the sublime beauty of their own

landscape so as not to build on its ridges and clifflines, or up against the stream banks? Do these ranchettes promise more than a crowded and congested Mountain West (Knight 1997)?

Because of these human settlement patterns and their associated influences, there will be a New West, one quite different from the old, and one that may very well last as long as the old. It is being made today, in the region and on the land by the region's new inhabitants. How these once-rural lands will appear and what native biological diversity they might support will depend on how these new occupants listen to the land and hear what it says, and whether or not they understand and accept what the land can and cannot do.

CASE STUDY 7.2

The Fragmentation of Aquatic Ecosystems and the Alteration of Hydrologic Connectivity

Neglected Dimensions of Conservation Ecology

Cathy Pringle, University of Georgia

As terrestrial landscapes become increasingly fragmented, so do hydrologic connections between landscape elements. Ecosystems throughout the world are threatened by habitat fragmentation caused by cumulative alterations in hydrologic connectivity (Figure A). The term *hydrologic connectivity* (Pringle 2001) is used here in an ecological sense to refer to water-mediated transfer of matter, energy, and organisms within or be-

tween elements of the hydrologic cycle. Human influences that alter this property include dams, associated flow regulation, groundwater extraction, and water diversion—all of which can result in a cascade of ecological effects in both aquatic and terrestrial ecosystems. Factors such as nutrient and toxic pollution are perpetuated by hydrologic connectivity and their effects can be exacerbated by changes in this property—such as when so

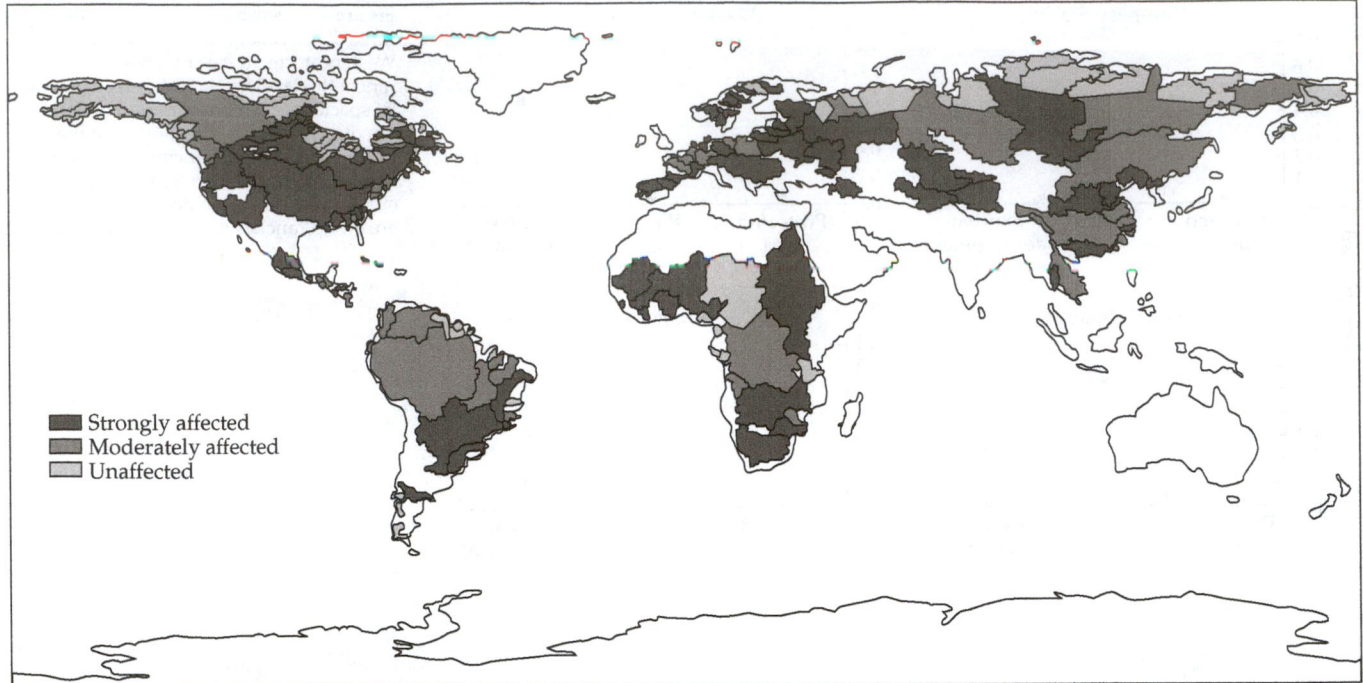

Figure A   Degree of fragmentation of major river basins of the world. Of the 227 large river basins assessed by Dynesius and Nilsson (1994) and Nilsson et al. (2000), 37% are strongly affected by fragmentation and altered flows, 23% are moderately affected, and 40% are unaffected. Strongly affected systems include those with less than one quarter of their main channel left without dams, where the largest tributary has at least one dam, as well as rivers whose annual flow patterns have changed substantially. Sub-basins of the Amazon, Orinono, and Congo are unaffected, although each of these major rivers has some sub-basins that are affected by fragmentation. The Yangtze River in China, which currently is classified as moderately affected, will become strongly affected once the Three Gorges Dam is completed.

much water is extracted from a river that there is little left for dilution of pollutants (e.g., sewage, nutrients), which become highly concentrated (Pringle 2003a,b).

Freshwater ecosystems are becoming fragmented at a rate unprecedented in geologic history, contributing to dramatic losses in global aquatic biodiversity and associated ecosystem integrity (e.g., Rosenberg et al. 2000; Pringle et al. 2000). Of the 3.2 million miles of streams in the lower 48 states, only 2% remain free-flowing and relatively undeveloped. Less than 42 free-flowing rivers of over 125 miles in length exist; the remaining 98% of U.S. streams have been fragmented by dams and water diversion projects (Benke 1990). The U.S. has lost over half the wetlands that existed at the time of the American Revolution. Accordingly, the World Wildlife Fund's Living Planet Index (which measures the average change over time in populations of almost 200 species of freshwater birds, mammals, reptiles, amphibians, and fishes) has declined by 50% globally over the 29-year period of 1970–1999 (WWF 2000). Current rates of extinction of many freshwater taxa are more than 1000 times the normal "background" rate, and as a whole in the U.S. freshwater species are more imperiled than terrestrial species (Master et al. 1998).

The fragmentation of freshwater ecosystems has not received the attention it deserves within the field of conservation biology. One reason for this is that most of the theoretical un-derpinnings of the conservation biology of fragmented landscapes were developed under a conceptual model of landscapes that were not yet entirely fragmented, and when awareness of hydrologic connectivity was in its infancy (Pringle 2001a). Only in the last decade have we begun operating on the premise that groundwaters and surface waters are interconnected as a single resource (e.g., Winter et al. 1998). While there are excellent books on the subject of habitat fragmentation, the words *stream* or *river* do not even appear in the indices of major books on this subject (e.g., see Shafer 1990, Schelhas and Greenberg 1996, Laurance and Bierregaard 1997, and Soulé and Terborgh 1999). Also, while the importance of size, shape, and configuration of reserves is much discussed, less consideration has gone to reserve location with respect to watershed boundaries, underlying regional aquifers, and atmospheric deposition patterns.

Parks and reserves are not impervious to the fragmentation of aquatic ecosystems outside of their boundaries. Expanding human populations require more water. Consequently, there is increasing pressure to dam rivers and pump aquifers within or near "protected areas" as a result of water shortages exacerbated by droughts, coupled with increased demands from municipal and agricultural growth. As just one illustration of the magnitude of outside pressures facing national wildlife refuges in

TABLE A *Examples of Upstream Biological Degradation Caused by Dams and Human Activity*

Downstream human activities	Upstream biological legacies
Urbanization; dams and impoundments; gravel mining; channelization	Genetic isolation; population-level changes: "source" of native species "sink" for nonnative species; ecosystem-level changes: primary production, nutrient cycling, decomposition

the western U.S.: In 1994, 150 out of 224 refuges reported conflicts with other water users, and only 98 reported that their existing water rights assured delivery of adequate water in an average year (Pringle 2000). Indeed, the fragmentation of aquatic ecosystems is affecting protected areas throughout the world, resulting in changes in faunal composition and food-web dynamics (Pringle 2001a,b). In the next section I highlight consequences of fragmentation of aquatic ecosystems at levels from genes to ecosystems.

Genetic and Species-Level Changes

As rivers become increasingly fragmented by dams, water extraction, or pollution, populations of aquatic organisms are subjected to reduced gene flow and loss of genetic variation. The Cherokee darter (*Etheostoma scotti*) provides one example. This small fish is endemic to portions of the Etowah River system in the Peidmont region of Georgia in the southeastern U. S. (Bauer et al. 1994). Because of its current isolated range, which is fragmented by degraded habitat (urbanization and other human land use), there is limited potential for genetic exchange among populations. For obligate riverine species with large home ranges, impoundments may similarly fragment the range, causing losses in genetic diversity and local extinctions (Table A).

The genetic and species-level effects of dams on economically important migratory fishes such as salmonids have received much attention (Pacific Rivers Council 1993). Over 100 major salmon and steelhead populations or stocks have been extirpated on the West Coast of the United States and Canada, while at least 214 more are at risk of extinction (Nehlsen et al. 1991). Less is known about the genetics and species-level effects of stream fragmentation on North American biota of less economic importance (nongame fishes, freshwater shrimps, and other invertebrates). In tropical areas such as the Amazon, fish migratory patterns are so complex (covering huge drainage areas) that the effects of stream fragmentation are unknown even for economically important fish species (Goulding et al 1996).

Population- and Community-Level Changes

North American mussel species have been extirpated or have declined, in part, because dams have resulted in the extirpation of their migratory fish host species (which host their parasitic glochidial stage). One example is the dwarf wedge mussel (*Alasmidonta heterodon*). In the past twenty years, known populations of the dwarf wedge mussel have dropped from 70 to 19 (Middleton and Liitschwager 1994). A leading theory for the cause of its demise is that the fish species that serves as its host

during the critical stage of development also may be in decline. It is suspected that the host fish is anadromous, and that dam construction has blocked its access to upstream mussel populations. The mussel is also extremely sensitive to the sedimentation, genetic population isolation, and changes in ecosystem function that results from dams and riverbank erosion, and to toxic chemicals in agricultural and industrial effluents.

Degraded downstream areas can act as population sinks for native riverine species and, alternatively, as sources of nonnative species or facultative riverine species (Pringle 1997). For example, in his studies of streams draining urban areas in the Appalachicola-Chattahoochee-Flint River Basin, DeVivo (1996) found highly variable fish faunas and atypical age structures that may be related to rich heavy metal and pesticide levels. Only those fishes most tolerant of degraded conditions (often nonnative species) had established populations. Consequently, stream reaches upstream of degraded downstream reaches are vulnerable to invasion by nonnative species common to degraded areas, which then serve as source populations of invasives (see Table A).

Ecosystem- and Landscape-Level Changes

When dominant faunal components of an ecosystem are excluded from upper portions of the watershed as a result of downstream human activities, a cascade of ecosystem-level effects may occur, particularly when the extirpated component was an important food source, predator, host species, or habitat modifier. We are just beginning to acknowledge the magnitude of ecosystem-level consequences of migratory faunal depletion caused by dams (Freeman et al. 2003). As just one example, populations of Bald Eagles and grizzly bears that depend on migrating salmon as a food source may decrease dramatically if salmon are eliminated (Spencer et al. 1991). Faunal components that are vulnerable to river fragmentation can also play key roles in ecosystem-level properties and processes such as water quality and nutrient cycling. It is well documented that anadromous fishes such as salmon can provide major input of nutrients and energy to freshwater systems when spawning adults return from the sea (Ben-David et al. 1998; Gresh et al. 2000). Consequently, when dams block salmonid migration routes, patterns of nutrient cycling in entire riverine ecosystems can be altered.

The loss of mussel species from streams where they were once diverse and abundant is yet another legacy of river fragmentation that often goes unacknowledged. Ninety percent of the world's freshwater mussel species are found in North

America, and 73% of all mussel species in the U.S. are at risk of extinction or are already extinct. The prognosis is not good: In 1990, 90% of the listed mussels were still declining, and only 3% were increasing (Master 1990). Given that mussels filter an enormous amount of water, and that they were once plentiful, consequences of their elimination likely include substantial losses in system productivity, decreased local retention of nutrients, and alterations in the structure and stability of the benthic stream environment (Strayer et al.1999).

Groundwater exploitation in stream watersheds can sever lateral connections between stream channels and adjacent springs and wetlands, resulting in landscape-level changes in the drainage network and the distribution of biota (Winter et al. 1998; Pringle and Triska 2000). The increasing exploitation of groundwater reserves for municipal, industrial, and agricultural use is having profound effects on riverine ecosystems as groundwater tables are lowered. For example, populations of the anadromous striped bass (*Morone saxatilis*) are dependent on coldwater refuges within riverine systems during hot summer periods because of their high oxygen requirement. In the southeastern U.S., spring-fed stream systems are home to healthy and productive populations of striped bass. These streams have a high thermal diversity, and the fish can actively search out and use spring-fed areas as refuges (Van Den Avyle and Evans 1990). Extensive groundwater withdrawals threaten the springs, and thus the survival of biota dependent on coldwater refuges.

An Example from Puerto Rico

Consequences of river fragmentation are illustrated in the Caribbean National Forest (CNF) of Puerto Rico. The CNF (113 km^2) is located in the northeastern corner of Puerto Rico and is drained by nine major rivers, which are characterized by a simple food chain (typical of oceanic islands) dominated by migratory freshwater shrimps and fishes that are important food sources for both aquatic and terrestrial organisms (including humans). The migrations of fishes and shrimps form a dynamic link between stream headwaters and estuaries. In the case of shrimps, newly hatched larvae migrate downstream and complete their larval stage in the estuary; upon metamorphosis, juveniles migrate upstream where they live as adults. When water is removed from rivers for human use (e.g., drinking water or irrigation), this results in direct mortality of shrimp larvae migrating to the ocean (Benstead et al. 1999). At present, all except one of the nine stream drainages within the CNF have low-head dams and associated water intakes on their main channels. The extent of water extractions is so severe that on an average day, approximately 70% of riverine water draining the forest is withdrawn into municipal water supplies before it reaches the ocean (Crook 2005). Massive larval mortality of shrimp resulting from these water withdrawals could potentially affect upstream recruitment of adults and other ecosystem processes. If migratory shrimps and fishes were to be extirpated above dams and water intakes, as has occurred above high dams without water spillways in other regions of Puerto Rico (e.g., Homquist et al. 1998), concomitant changes in ecosystem structure and function might occur (Pringle et al. 1999).

In summary, hydrologic connectivity plays an important role in maintaining the biological integrity of natural landscapes. Human activities have significantly fragmented freshwater ecosystems and altered hydrologic connectivity on regional and even global scales. It is critical that an understanding of hydrologic connectivity be incorporated into the field of conservation biology (where it has largely been ignored)—into both theoretical constructs and practical applications.

As just one example, studies of habitat fragmentation would be strengthened by considering the question: How does the size, shape, and configuration of a given forest fragment (or biological reserve), *with respect to its location in the watershed*, affect biodiversity? A major challenge is to develop a more predictive understanding of how hydrologic connectivity, and alterations of this property, influence ecological patterns on regional and global scales so that we can proactively address environmental challenges before they become crises.

CASE STUDY 7.3

Dissecting Nature
The Islands of Lago Guri

John Terborgh, Duke University

Studies of land-bridge islands—bits of a continental landmass that have been isolated by rising water—provided some of the first early warnings that habitat fragmentation results in the extinction of species (Brown 1971; Diamond 1972; Terborgh 1974). Many land-bridge islands were created at the end of the Ice Age when meltwater from continental glaciers raised sea level by roughly 120 m. Such islands (including Taiwan, Sri Lanka, and Trinidad), appear to have lost substantial numbers of species since they were separated from the mainland 10–14 thousand years ago (Terborgh and Winter 1980). The specific mechanisms that caused certain species and not others to disappear from these post-Pleistocene land-bridge islands are ob-

scured by time and remain largely unknown. But because the analogous process of habitat fragmentation is occurring all over the world, it is important that conservation scientists understand the processes that impel extinctions on land-bridge islands.

Not all land-bridge islands formed in the distant past. Humans have created thousands of them, albeit relatively small ones, by damming rivers to create artificial impoundments. For a decade my research team has been studying a set of such land-bridge islands in Lago Guri, a huge hydroelectric impoundment in east-central Venezuela. Lago Guri islands are former hilltops that were cut off from the mainland in 1986 after closing of the 160-m tall Raul Leoní dam, an event that flooded an area nearly twice the size of Rhode Island. Lago Guri contains literally hundreds of islands, ranging from tiny blips of land supporting only a single tree, to landmasses of many square kilometers.

The Islands of Lago Guri, Venezuela

For understanding the mechanisms of extinction in habitat fragments, recently created islands, such as those in Lago Guri, offer scientific advantages over habitat fragments in a mainland land-use matrix. For one, the matrix in which islands are embedded—water—is entirely uniform. For another, water makes a much better barrier to the movements of animals than a typical land-use mosaic. Many wide-ranging species, such as some raptors, frugivores, and herbivores freely move between habitats on the mainland, but rarely if ever undertake long journeys across water. Species initially present in mainland fragments can wander away, and others can invade, so it is difficult to say just what species are exerting effects within a given fragment, and how strong those effects might be. In contrast, the populations of nonflying vertebrates on islands are essentially trapped, so one can confidently say which species are present and which are absent. Islands thus afford a degree of control over community composition that could not readily be achieved on a mainland.

At the time of isolation in 1986, Lago Guri islands initially held samples of the mainland flora and fauna. These were probably not random samples, at least for animals, because some individuals undoubtedly flew or swam away when they discovered that their haunts were steadily shrinking in a rising pool of water. Species that were reluctant to leave must have found themselves increasingly crowded as the hilltops on which they were stranded continued to contract. Some species may have been induced to swim away in response to crowding and shrinkage of their habitat, whereas others may have suffered high mortality.

Howler monkeys (*Alouatta palliata*) provide a good example of the conditional behavior shown by trapped populations. A dozen years after inundation, we documented the presence of howlers on seven small islands located more than 0.5 km from a larger landmass, but failed to find them on any of more than 20 closer small islands. By inference, it appears that howlers are willing to swim up to 0.5 km to escape entrapment, but will not attempt longer crossings. Other animals show different re-

sponses to water barriers. For example, we once saw a collared peccary (*Tayassu tajacu*) swimming between islands situated nearly 2 km apart, and a red brocket deer (*Mazama americana*) swimming from the mainland to an island some 800 m offshore. It can be assumed that other species react in individualistic ways to the prospect of crossing water.

Because of the willingness of at least some large vertebrates to swim considerable distances, we have concentrated our research on islands located a kilometer or more from the mainland, and at least several hundred meters from any larger landmass. By using only islands that satisfy these criteria, we are confident that the populations we are studying really are trapped, and that undetected visits of wide-ranging species, especially large predators, are rare to nonexistent.

Our work has emphasized two size classes of islands, small (0.5–2.0 ha) and medium (4–12 ha). Two large islands (>150 ha) and two sites on the nearby mainland serve as controls. The climate alternates between wet and dry periods with an annual rainfall of around 1200 mm concentrated in the period of May to October. The prevalent habitat is semideciduous tropical dry forest.

In the early 1990s, 7–8 years after inundation, we conducted systematic inventories of the vertebrates and some invertebrates of a dozen small and medium islands (Terborgh et al. 1997a, 1997b). The results revealed consistent patterns of presence and absence of a number of key species (Table A). Small islands supported a limited fauna composed of small birds, rodents, amphibians, lizards, and invertebrates. In addition, some of these islands held howler monkeys (*Alouatta seniculus*) or tortoises (*Geochelone carbonaria*), or both species. The faunas of medium islands were somewhat larger and more variable, but always included howler monkeys, tortoises, and armadillos (*Dasypus novemcinctus*), and some also included capuchin monkeys (*Cebus olivaceus*), a larger armadillo (*Dasypus kappleri*), agouti (*Dasyprocta leporina*), paca (*Agouti paca*), and porcupine (*Coendou prehensilis*). Collared peccaries and perhaps ocelot (*Felis pardalis*) are intermittently present on the closer medium islands. The large islands and mainland control sites retained complete or nearly complete vertebrate faunas including, jaguar, puma, ocelot, tapir, deer, peccary, coati-mundi, anteaters, tayra, Harpy Eagle, Spix's Guan, Black Curassow, Red-and-green Macaw and many others.

While the question of which species are able to persist on small and medium islands is interesting in itself, it becomes more interesting when the species are assigned to functional groups, according to their ecological roles. At least seven functional groups comprise the complete fauna of the mainland. These include predators of vertebrates (cats, large raptors, large snakes), specialized predators of social insects (armadillos, tamandua, giant anteater), frugivores (bats, guans and other birds), omnivores (capuchin monkey, coati-mundi, tayra, many birds), seed predators (small and large rodents, bearded saki monkey, peccary, parrots and macaws), predators of arthropods (bats, small birds, lizards, amphibians, spiders) and herbivores (howler monkey, deer, tapir, porcupine, iguana, tor-

TABLE A *Counts of Selected Species on Lago Guri Islands and Mainland*

	Small, far				Small, near				Medium			Large	
Mammals	Col	Igu	Mie	Tri	Bum	Pal	Per	Roc	Cor	Lom	Pan	DM	TF
Dasypus spp.	0	+[a]	0	0	0	0	1	0	2	1	1	1	+
Rodent, small	2	1	0	1	4	6	6	+	2	1	7	1	1
Reptiles													
Ameiva ameiva	0	5	24	6	11	3	5	8	1	0	4	2	1
Geochelone carboneria	0	1	0	1	1	0	1	2	0	2	2	1	0
Amphibians													
Bufo marinus	2	0	1	0	1	0	2	0	0	+	0	1	0
Dendrobates leuco	10	3	4	6	11	0	0	7	1	10	3	2	0
Invertebrates													
Atta spp. (colonies)	3	5	1	0	6	1	1	2	2	2	1	0	0
Tarantula spp.	8	3	20	9	2	0	6	7	5	9	10	9	6

Note: Data normalized to 1 ha.

[a] + is recorded on landmass, but not on formal census.

toise, and many insects). Only members of the last three functional groups (seed predators, predators of arthropods, and herbivores) are able to persist on small Lago Guri islands; the remaining groups are absent or nearly so. The situation of medium islands was made more complex by the presence of armadillos, agoutis, and (sometimes) capuchin monkeys.

The Phenomenon of Hyperabundance

A striking and consistent feature of the animal populations trapped on small and medium islands was their hyperabundance (population densities higher than those measured at our control sites). In some cases the levels of hyperabundance were truly remarkable, reaching or exceeding a factor of 10 (Terborgh et al. 2001). We found this to be true for small rodents (several genera), common iguanas, howler monkeys (when present), and leaf-cutter ants (*Atta* spp., and *Acromyrmex* spp.). All these animals are consumers, either seed predators (rodents) or folivores (iguanas, howler monkeys, and leaf-cutter ants).

Consumer hyperabundance could be caused by reduced competition from other seed predators and herbivores. However, that explanation fails to account for the observations: For over a decade, rodent densities have been elevated more than 30 times higher than controls, an enormous increase that has persisted beyond what one might expect from a reduction in competition alone, and three persistent folivores (leaf-cutter ants, iguanas, and howler monkeys) are simultaneously hyperabundant, although they compete with one another. Release from competition thus fails to provide a plausible explanation for such extraordinary levels of hyperabundance.

An alternative hypothesis is that consumers increased in density in response to low or negligible predation. The islands we are studying were selected expressly because of their remoteness in the hope that all predators of vertebrates would be absent, or at most, present only as rare short-term visitors. Despite numerous visits to all the islands included in the study, we did not detect predators of vertebrates on any of them.

While the predator hypothesis is difficult to test directly, indirect evidence points to armadillos as an important control agent of leaf-cutter ants. Armadillos are found on all medium islands, where leaf-cutter ant densities are substantially lower than on small islands. To test whether armadillos were contributing to the mortality of young leaf-cutter ant colonies, Madhu Rao placed wire cages that exclude armadillos over the entrances of colonies on medium and small islands. A year later, the protected colonies on medium islands had all survived, whereas unprotected colonies suffered significant mortality. On small islands, where armadillos were not present, the cages had no effect (Rao 2000). Predation by armadillos, where present, thus seems to be important in restricting leaf-cutter ant densities. By inference, predation may be limiting the densities of other consumers on large islands and the mainland. Clearly food resources are not; otherwise consumers would not be able to increase so dramatically and persist at high densities for more than a decade.

A Top-Down Trophic Cascade?

Fragmentation of the once continuous forested habitat by inundation eliminated predators of vertebrates from all but the largest landmasses because the maintenance of a predator population requires a large prey base, which in turn requires sizeable areas of habitat. Increased population densities of consumers in the absence of predators constitutes the first step in an ecological chain reaction known as a top-down trophic cascade. The term refers to a perturbation that propagates downward through two or more trophic levels of an ecological system, resulting in alternating positive and negative impacts on successive levels. If a trophic cascade were operating on small Lago Guri islands, abnormally high levels of consumers would be expected to result in adverse effects at the producer level (vegetation).

There are two major routes by which hyperabundant consumers could affect the vegetation. One is by increasing the mortality of already established stems; the other is via reduced re-

cruitment of new stems. Our results indicate that the mortality of already established trees is substantially higher on small islands than at control sites. Moreover, the pattern of mortality is significant. Trees dying on small islands tend to belong to species preferred by leaf-cutter ants, whereas those recruiting into the adult class tend to belong to less-preferred species. Thus, high levels of herbivory appear not only to be causing accelerated mortality of canopy trees, but are also driving systematic changes in the species composition of the forest (Rao et al. 2001).

To investigate whether hyperabundant consumers reduce the recruitment of new stems, we tagged and mapped 4771 saplings at sites on 12 small, medium, and large (control) landmasses (Table B). The islands were already 11 years old in 1997 when we began to monitor saplings and it was apparent then that sapling densities on small islands were severely reduced, being only 37% of controls. Analysis of variance indicated a highly significant effect of landmass size in the stem counts, but no discernible effect of environmental variables, such as exposure to wind. As expected by the trophic cascade hypothesis, densities of small saplings were low on small islands in the presence of hyperabundant herbivores.

Five years later in 2002 when we recensused the plots, the density of small saplings on small islands had fallen still further to 25% of control levels, yet still there was no hint of a wind effect. Even more strikingly, the recruitment of new saplings was suppressed by 81% on small islands relative to controls and 67% on medium islands. High mortality of adult trees coupled with low recruitment of saplings seemed to foretell disaster for the forests of these islands. We examined this possibility by using data on the survival and recruitment of saplings and adult trees to fit a simple matrix model. The model predicts how the numbers of stems in various size classes would change over time, given the assumption that survival and recruitment rates remain constant into the future (Figure A). The results foresee dramatic declines in the numbers of small and large saplings preceding similar declines in the adult tree stands of small and medium islands. In contrast, the sapling and tree stands of the large (control) landmasses are projected to remain constant into the future.

These predictions are driven primarily by the sharply depressed rates of recruitment of small saplings on small and medium islands due to herbivory by leaf-cutter ants on seedlings and small saplings. By eliminating other possibilities through experiments and observations, including hyperabundant rodents, other arthropods, and environmental effects of prevailing winds, leaf-cutter ants emerged as the most probable agents of the decline, for they are by far the most abundant herbivorous arthropods in the understory of these islands. Our results thus support the trophic cascade hypothesis and implicate both pathways by which a top-down trophic cascade might impact vegetation (accelerated mortality of established individuals and reduced recruitment of new stems).

Looking to the Future

What will happen to the ecosystems of these islands in the future? As the trees comprising the original canopy gradually die off, some will be replaced by saplings already present in the understory, but eventually all of these will either die or become adult trees. At that point, all further recruitment will be confined to species that are resistant to leaf-cutter ants and any other adverse conditions that prevail on the islands. Observations of severely impacted islands suggest that herbivore-resistant lianas and perhaps some understory shrubs will predominate, creating a vegetation of greatly altered physical structure and species composition. Along the way, much plant and animal diversity likely will be lost. The endpoint will be an impoverished system quite unlike the biologically rich dry forest that covered the landscape before it was flooded.

What we are seeing in Lago Guri has parallels in other ecosystems, where hyperabundant herbivores have reduced species-rich natural vegetation to an odd collection of resistant plants. Long-term overgrazing of semi-arid range, for example, will convert grassland to thornscrub or shrub–steppe (Schlesinger et al. 1990). Plant species with relatively low investments in chemical and mechanical defenses grow faster and compete better than species that invest heavily in anti-herbivore defenses (Coley et al. 1985), but only flourish while herbivore numbers are held in check by predators. Intense grazing predictably reduces the ratio of palatable to unpalatable plant species, thereby increasing the resistance of the vegetation and reducing the carrying capacity for grazers. Our findings demonstrate that these same processes can operate in forests, with similar implications for reduced diversity, increased plant defenses, and lowered carrying capacity for consumers.

Are the processes unleashed on small Guri islands by the absence of predators applicable to larger spatial scales? Whether our findings from Lago Guri provide a sufficient explanation for historical species losses from larger land-bridge islands, such as Trinidad, will only be determined through further research. However, there is mounting evidence that similar, if less extreme, processes are at work in parts of the U.S., where the absence of predators (wolves, cougars, and bears) brought about directly through persecution or indirectly by fragmenting the landscape has released herbivore populations. Hyperabundant populations of white-tailed deer (East), feral hogs (South, Hawaii),

TABLE B *Number of Small and Large Saplings in Sample Plots on Lago Guri Landmasses*

	Landmass					
	Small (n^a = 6)		Medium (n = 4)		Large (n = 2)	
Sapling size	Exposed	Protected	Exposed	Protected	Exposed	Protected
Small	69	67	119	165	166	155
Large	131	155	173	193	139	101

$^a n$ = Number of saplings sampled.

burros (Southwest) and goats (California islands, Hawaii) have massively impacted natural vegetation (Alverson et al. 1988; McShea et al. 1997), and in turn affect animals dependent on those plant communities.

Today we occupy a world distorted by the widespread extirpation of large carnivores. This should be read as a wake-up call, because much of North America may already be experiencing the initial stages of a top-down trophic cascade similar to the one that is transforming the vegetation of Lago Guri islands (Pedersen and Wallis 2001; Ripple et al. 2001).

Predators are an essential component of a healthy ecological system and play a vital role in maintaining diversity over time. As top-predators are removed, negative effects can propagate throughout the community in a top-down trophic cascade (Terborgh et al. 1999). As this process runs its course, the entire ecological community will change, and much diversity will probably be lost.

For biodiversity to prosper over the long run, it is imperative to reestablish predators wherever possible. Restoring balance to native ecosystems through the reintroduction of predators like wolves and cougars will require more than just good science—it will require good dissemination of the science to make the public aware of the benefits of predators. At present, relatively few citizens understand that a superabundance of deer and other animals signals an unbalanced ecosystem that will increase the ranks of the endangered species list. The public must be educated about predators so that ordinary citizens think of them in a positive light. Now that the science points in a clear direction, building the social consensus needed to implement the science remains a substantial challenge to conservation.

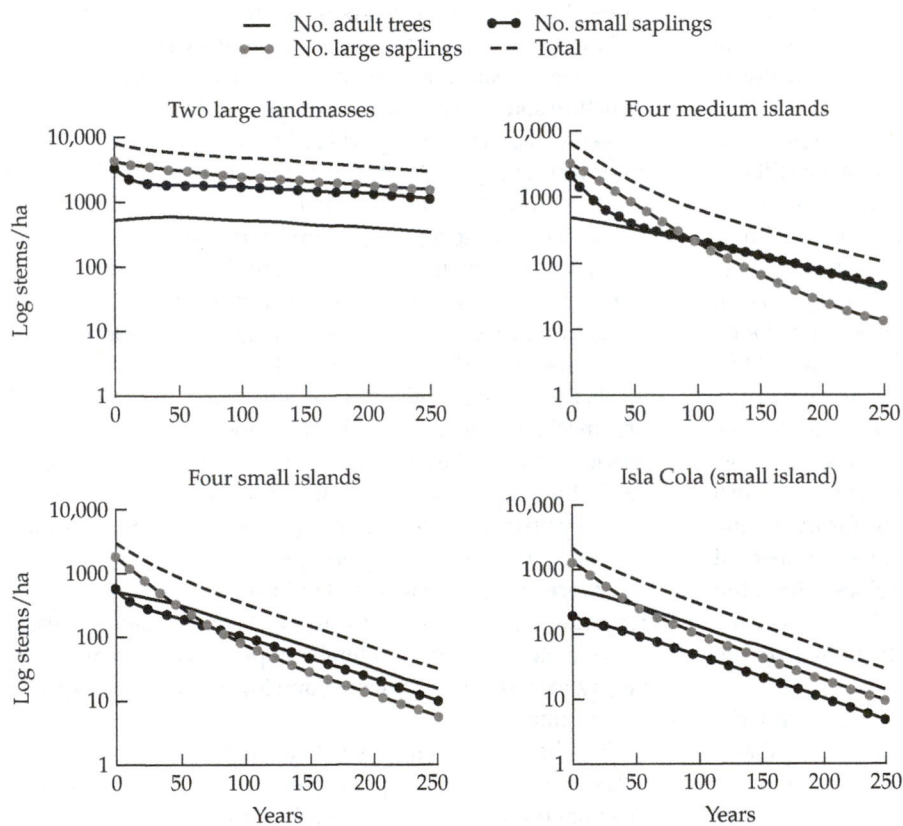

Figure A   Dramatic declines in total density of trees (as well as of adult, large sapling, and small sapling stage classes) are predicted on medium and small islands in Lago Guri given current survival and recruitement rates. In contrast, trees on large landmasses are predicted to decline only slightly.

Summary

1. Fragmentation, the loss and isolation of natural habitats, is one of the greatest threats to regional and global biodiversity. Whereas natural disturbances and other processes create heterogeneous landscapes rich in native species, human land uses often create islands of natural habitat embedded in a hostile matrix. Such fragmentation reduces or prevents normal dispersal, which is critical to long-term genetic integrity and population viability for species, and increases edge effects and other threats to population persistence.

2. Although the effects of fragmentation vary regionally, no region escapes all of the threats to biodiversity posed by fragmentation. Fragmentation acts to reduce native biodiversity through several major mech-

anisms, of which five stand out as especially noteworthy. First, because remaining fragments represent only a sample of the original habitat, many species will be eliminated by chance (initial exclusion). Second, small fragments contain fewer habitats, support smaller populations of native species that are more susceptible to extinction, and are less likely to intercept paths of dispersing individuals (island-area effect). Third, the modified landscape in which fragments exist is often inhospitable to many native species, thus preventing normal movements and dispersal and often leading to genetic problems (isolation). Fourth, climatic influences and opportunistic predators and competitors from the disturbed landscape penetrate into fragments, reducing the core area of suitable habitat (edge effects). Fifth, roads are a significant source of habitat fragmentation and have many other deleterious impacts on ecosystems, including the facilitation of nonnative species invasions. Finally, the disruption of natural disturbance regimes, animal–plant mutualisms, food web dynamics, and other processes in fragmented landscapes will lead to changes in biological communities. Generally speaking, the greater the contrast between fragments of natural habitat and the surrounding landscape matrix, the more severe the impacts.

3. Some kinds of species, in terms of life-history traits, are more vulnerable to the effects of fragmentation than others: rare species, wide-ranging species with large home ranges, species with limited powers of dispersal, species with low reproductive potential, species dependent on resources that are unpredictable in time or space or that are otherwise highly variable in population size, species inhabiting the interiors or large patches, and species exploited or persecuted by people. Some of the most ominous effects of habitat fragmentation may not become apparent for decades or centuries, for example the effects of fragmentation on ecological processes and the inability of species in isolated fragments to track changes in habitat conditions related to changing climate.

4. Some common sense strategies for countering fragmentation are: (1) do not destroy or fragment intact wildlands, waterways, or other natural areas, few of which remain in most regions; (2) minimize road construction, clearing of vegetation, development, and creation of other barriers to dispersal; (3) maintain or restore wide habitat corridors, a structurally rich landscape matrix, or other forms of functional connectivity between natural areas; (4) minimize creation of artificial dispersal corridors (such as weedy roadsides) that encourage invasions of exotic species; and (5) actively manage habitat fragments as necessary to compensate for disrupted natural processes.

Please refer to the website www.sinauer.com/groom for Suggested Readings, Web links, additional questions, and supplementary resources.

Questions for Discussion

1. Lake Gatun was formed when the Panama Canal was built in 1914, creating the 17-km^2 Barro Colorado Island. Since then, careful observations have documented a 25% decline in the island's avifauna. Speculate on the reasons for this decline.

2. The Northern Spotted Owl, an obligate resident of old-growth coniferous forests in the Pacific Northwest, is thought to be sensitive to high temperatures, it feeds on small mammals (many of them arboreal), and is subject to predation by the Great Horned Owl, a habitat generalist. Old-growth forests within the range of the Northern Spotted Owl once covered 60%–70% of the region, but have been reduced by 90%. How might different approaches to forest management affect the long-term viability of Spotted Owl populations?

3. Many species of small vertebrates and invertebrates are sensitive to the barrier effects of roads, refusing to cross even two-lane roads in some cases. Larger animals, including endangered species such as the Florida panther, are vulnerable to road kill. How might these kinds of problems be corrected?

4. Population models suggest that the grizzly bear and the wolf in the Rocky Mountains may each require many millions of acres of wild habitat for long-term viability. Yellowstone Park, for comparison, is only 2.2 million acres. Thus, habitat fragmentation at a regional scale is a threat to persistence of these species. The grizzly bear, along with the wolf, is vulnerable to human encounters, many related to livestock production or road access by poachers. Housing subdivisions are also being constructed in prime habitat. How might conservation strategy address these regional fragmentation problems for large carnivores? What kinds of information—biological and otherwise—might be used to determine necessary reserve sizes, linkage widths, and management guidelines?

5. Suppose you are the manager of a small (20-ha) forested nature reserve in the midwestern U.S. What kind of management would you consider to reduce edge effects and maintain ecological integrity in this reserve? What variables (biotic and abiotic) would you monitor to determine whether or not management is effective?

8

Overexploitation

John D. Reynolds and Carlos A. Peres

There's enough on this planet for everyone's needs, but not enough for everyone's greed.

Gandhi

Exploitation involves living off the land or seas, such that wild animals, plants, and their products are taken for purposes ranging from food to medicines, shelter, and fiber. The term **harvesting** is often used synonymously with **exploitation**, though harvesting is more appropriate for farming and aquaculture, where we reap what we sow.

In a world that seems intent on liquidating natural resources, overexploitation has become the second most important threat to the survival of the world's birds, mammals, and plants (see Figure 3.8). Many of these species are threatened by subsistence hunting in tropical regions, though others are also threatened in temperate and arctic regions by hunting, fishing, and other forms of exploitation. Exploitation is also the third most important driver of freshwater fish extinction events, behind the effects of habitat loss and introduced species (Harrison and Stiassny 1999). Thus, while problems stemming from habitat loss and degradation quite rightly receive a great deal of attention in this book, conservationists must also contend with the specter of the "empty forest" and the "empty sea."

We begin this chapter with a brief historical context of exploitation, which also provides an overview of some of the diverse reasons people have for using wild populations of plants and animals. This is followed by reviews of recent impacts of exploitation on both target and nontarget species in a variety of habitats. To understand the responses of populations, we then review the theory behind sustainability. The chapter ends with a consideration of the culture clash that exists between people who are concerned with resource management and people who worry about extinction risk.

History of, and Motivations for, Exploitation

Humans have exploited wild plants and animals since the earliest times, and most contemporary aboriginal societies remain primarily extractive in their daily quest for food, medicines, fiber, and other biotic sources of raw materials to produce a wide

range of utilitarian and ornamental artifacts. Modern hunter-gatherers in tropical ecosystems, at varying stages of transition to agriculture, still exploit a large number of plant and animal populations. By definition, these species have been able to coexist with some background level of human exploitation. However, the archaeological and paleontological evidence suggests that premodern peoples have been driving other species to extinction since long before the emergence of recorded history. Human colonization into previously unexploited islands and continents has often coincided with a rapid wave of extinction pulses resulting from overexploitation by novel consumers. Mass extinction events of large-bodied vertebrates in Europe, parts of Asia, North and South America, Madagascar, and several archipelagos have all been attributed to post-Pleistocene human overkill (Martin 1984; McKinney 1997). These are relatively well documented in the fossil and subfossil record, but many more obscure target species extirpated by human exploitation will remain undetected.

In more recent times, extinction events induced by exploitation have also been common as European settlers wielding superior technology expanded their territorial frontiers and introduced market and sport hunting. The death of the last Passenger Pigeon (*Ecopistes migratorius*) in the Cincinnati Zoo in 1914 provided a notorious example of the impact that humans can have on habitats and species. The Passenger Pigeon probably was the most numerous bird in the world, with estimates of 1–5 billion individuals (Schorger 1955). Hunting for sport and markets, combined with clearance of their nesting forests (Bucher 1992), had significant impacts on their numbers by the mid-1800s, and the last known wild birds were shot in the Great Lakes region of the U.S. at the end of the century. After that, it was simply a question of which of the birds in captivity would survive the longest, and "Martha" in Cincinnati "won."

The decimation of the vast bison herds in North America followed a similar time-line, but here hunting for meat, skins, or merely recreation, was the sole cause. In the 1850s, tens of millions of these animals roamed the Great Plains in herds exceeding those ever known for any other mega-herbivore, but by the century's close, the bison was all but extinct (Dary 1974). Following an expensive population recovery program, both the plains (*Bison bison bison*) and wood bison (*Bison bison athabascae*) are currently classified by the 2004 IUCN Red List as Lower Risk/Conservation Dependent, although the wood bison is listed as endangered by the U.S. Endangered Species Act and Appendix II of CITES.

An example that is lesser known is the extirpation of monodominant stands of Pau-Brasil trees (*Caesalpinia echinata*) from eastern Brazil, a source of red dye and hardwood for carving that gave Brazil its name. Pau-Brasil once formed dense clusters along 3,000 km of the Brazilian Atlantic forest, and the species sustained the first major trade cycle between the new Portuguese colony and European markets. It was exploited relentlessly from 1500 to 1875, when it finally became economically extinct. Since the advent of synthetic dyes, the species has been used primarily for the manufacture of high-quality violin bows, and Pau-Brasil specimens are currently largely confined to herbaria, arboretums, and a few small protected areas (Dean 1996). These examples suggest that even some of the most abundant populations can be driven to extinction in the wild by exploitation. Exploitation of both locally common and rare species thus needs to be adequately managed if populations are to remain demographically viable in the long term.

People exploit wild plants and animals for a variety of reasons, which need to be understood if management is to be effective. There may be more than food and money involved. The recreational importance of hunting and fishing in developed countries is well known. For example, hunting creates more than 700,000 jobs in the U.S. and a nationwide economic impact of about $61 billion per year, supporting nearly 1% of the entire civilian labor force in all sectors of the U.S. economy (LaBarbera 2003). Over 20 million hunters in the U.S. spend nearly half a billion days afield in pursuit of game, and fees levied to game hunters finance a vast acreage of conservation areas in North America (Warren 1997).

The importance of exploitation as a recreational activity is not restricted to wealthy countries. For example, in reef fisheries in Fiji, capture rates are highest with spears or nets, and while these techniques are used when fish are in short supply, the rest of the time people adopt a more leisurely pace with less efficient hand-lines, using the extended time for social and recreational purposes (Jennings and Polunin 1995). Cultural and religious practices are often important. For example, feeding taboos switch on or off among hunters in tropical forests according to availability of alternative prey species (Ross 1978; Hames and Vickers 1982). The fate of some endangered species is closely bound to religious practices, as in the case of the babirusa wild pig, *Babyrousa babyrousa*, in Sulawesi. This species is consumed heavily in the Christian-dominated eastern tip of Sulawesi, but rarely consumed over the Muslim-dominated remainder of the island (Clayton et al. 1997).

Impacts of Exploitation on Target Species

Many of the best-known impacts of exploitation on populations involve cases of direct targeting, whereby hunting, fishing, logging, and related activities are selective,

aimed at a particular species. In this section we present examples from major ecosystems in both temperate and tropical areas.

Tropical terrestrial ecosystems

TIMBER EXTRACTION   Tropical deforestation is driven primarily by frontier expansion of subsistence agriculture and large-scale development programs involving improved infrastructure and access. However, animal and plant population declines are typically preceded by hunting and logging activity well before the coup de grâce of complete deforestation is delivered. Approximately 5.8 million ha of tropical forests are logged each year (Food and Agriculture Organization 1999; Achard et al. 2002). Tropical forests account for about 25% of the global industrial wood production, worth U.S.$400 billion or about 2% of the global gross domestic product (World Commission of Forests and Sustainable Development 1998). Much of this logging activity opens up new frontiers to wildlife and nontimber resource exploitation, and catalyzes the transition into a landscape dominated by slash-and-burn and large-scale agriculture.

Few studies have examined the impacts of selective logging on commercially valuable timber species, and comparisons among studies are limited because they often fail to employ comparable methods that are reported adequately. The best case studies come from the most valuable timber species that have already declined in much of their natural geographic distributions. For instance, the highly selective logging of broadleaf mahogany (*Swietenia macrophylla*) is driven by the extraordinarily high value of this species on international markets. These conditions make it lucrative for loggers to pay royalty payments as well as high transportation costs of reaching remote wilderness areas. Selective logging of mahogany and other prime timber species affects the forest by creating canopy gaps and causing severe collateral damage due to construction of roads and skid trails, particularly in the case of mechanized operations.

One of the major obstacles to implementing a sustainable forestry sector in tropical countries is the lack of financial incentives for producers to limit offtakes to sustainable levels and invest in regeneration (see Essay 8.1 by Steve Ball for an example involving East African blackwood). Economic logic dictates that trees should be felled whenever their rate of timber volume increment drops below the prevailing interest rate (Pearce 1990). Postponing exploitation beyond this point would incur an opportunity cost because profits from logging could be reinvested at a higher rate elsewhere (see Chapter 5 for a full discussion of economic discounting). This is particularly the case where land tenure systems are unstable, and where there are no disincentives against mining the resource capital at one site and moving else-

where once this is depleted. This is clearly shown in a mahogany study in Bolivia where the smallest trees felled are about 40 cm in diameter, well below the legal minimum size (Gullison 1998). At this size, trees are increasing in volume at about 4% per year, whereas real mahogany price increases have averaged only 1%, so that a 40-cm mahogany tree increases in value at about 5% annually, slowing down as the tree becomes larger (Figure 8.1). In contrast, real interest rates in Bolivia in the mid-1990s averaged 17%, creating a strong economic incentive to liquidate all trees with any value regardless of resource ownership. The challenges and prospects of sustainable forestry in the Tropics are discussed in more detail in a case study by Pinard and colleagues in Case Study 8.3.

SUBSISTENCE HUNTING   Humans have been hunting wildlife in tropical forests for more than 100,000 years, but consumption has greatly increased over the last few decades. Exploitation of bushmeat (the meat from wild animals) by tropical forest dwellers has increased due to larger numbers of consumers, changes in hunting technology, scarcity of alternative sources of protein, and because it is often a preferred food. Recent estimates of annual hunting rates are 25,850 tons of wild meat in Sarawak (Bennett 2002), 73,890–181,161 tons in the Brazilian Amazon (Peres 2000a), and 1–3.7 million tons in Central Africa (Fa et al. 2001).

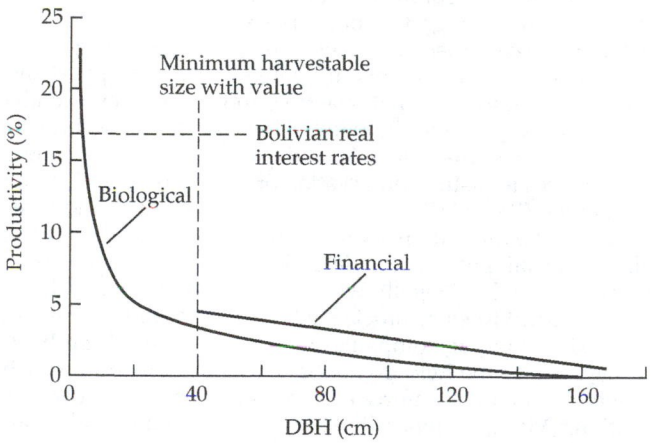

Figure 8.1   Productivity of Bolivian mahogany trees as a function of size expressed as tree diameter at breast height (DBH). Financial productivity is equivalent to size increments combined with a 1% real increase in price. Trees equal to or smaller than 40 cm in DBH have no commercial value, but at this size they increase at 5% in value per year. To maximize their profits, loggers should fell trees at a size when the financial productivity drops below the prevailing interest rate, which in Bolivia was 17% in 1989–1994. The discrepancy between the interest rates and tree growth rate provides a strong incentive to overexploit. (Modified from Gullison 1998.)

ESSAY 8.1

East African Blackwood Exploitation

Steve M. Ball, *Mpingo Conservation Project*

The Miombo woodlands, which cover a large swath of southern Africa, are a highly species-diverse ecosystem and the mammalian fauna requires large ranges. Thus, large contiguous areas of land must be protected for conservation of biodiversity to be effective. Even if the protected area system were well maintained, fragmentation would have a significant impact on the integrity of the conserved areas. Community-based conservation could fill the gap, providing buffer zones and corridors between reserves, but the communities need adequate incentive. The East African blackwood (*Dalbergia melanoxylon*) has the potential to do this.

The tree gets its name from its beautiful dark heartwood, which is inky black in the best-quality timber. The wood is used in the West to make woodwind instruments, especially clarinets and oboes, and locally it is a prime choice for wood carvers. It is the most valuable timber growing in the Miombo woodlands of southern Tanzania (Figure A). Both the species and its habitat are under threat from commercial exploitation and inward migration of people as a result of recent road improvements. The blackwood's high economic value, its status as Tanzania's national tree, and its cultural significance both in the West and in Tanzania make it an ideal flagship species to justify conservation of the habitat (Ball 2004).

Current exploitation levels are probably unsustainable, with at least 50% of trees being felled illegally without a license. Local forestry officials who depend on logging companies for transport into the field do not have sufficient resources to enforce existing regulations. Villagers generally know

Figure A A mpingo, an African blackwood tree. (Photograph courtesy of Steve M. Ball.)

when illegal logging is taking place and could apprehend suspects, but they have no incentive to do so because all hardwood timbers remain the exclusive property of the government. Intentional fires are also thought to prevent regeneration and facilitate heart-rot in adult trees. Plantations are not an option because of their high costs, uncertain return, and long rotation time (blackwood is estimated to take 70–100 years to reach timber size).

Community-based forest management is now a major theme of conservation and rural development activities in Tanzania. In largely deforested areas these can succeed by focusing on firewood and water-catchment issues, but this approach is unlikely to succeed in southeastern Tanzania where forest cover currently exceeds 70%. However new laws allow for villages to take ownership of local forest resources if they can produce a viable management plan for the forest. This includes tenure rights over even high-value timber trees, such as East African blackwood, which were previously government-reserved species. License fees for felling timber could significantly increase village income, giving local people an incentive to look after their natural resource assets. Once schemes are well established, rural communities could use the promise of future income as collateral against micro-finance loans to increase the up-front benefits.

Simple economic analysis suggests that sustainable, community-based management of East African blackwood offers a powerful argument to justify conservation of the forest on public lands and to complement the protected areas system, providing income in buffer zones and corridors between reserves. However, it will be important to work with the suppliers of musical instrument manufacturers, and avoid a significant increase to their costs. If these challenges can be met, blackwood could be used as an economic key to conserve—as managed areas—large tracts of Miombo woodland. This would not happen through a system of forest reserves, but rather with a more inclusive strategy of sustainable exploitation of the region's natural resources. For more information, see www.mpingoconservation.org.

Hunting rates are already unsustainably high across large tracts of tropical forests, averaging six times the maximum sustainable rate in Central Africa, for example (Figure 8.2). Consumption by both rural and urban communities is often at the end of supply chains that are hundreds of kilometers long, and that extend into many previously inaccessible areas (Milner-Gulland and Bennett 2003). The rapid acceleration in tropical forest defaunation due to unsus-

tainable hunting occurred initially in Asia, is now sweeping through Africa, and is likely to move to even the remotest parts of the Neotropics. This pattern reflects human demographics in different continents: There are 522 people per km^2 of remaining forest in South and Southeast Asia, 99 in Central-West Africa, and 46 in Latin America.

Subsistence game hunting affects the structure of tropical forest mammal assemblages, as revealed by vil-

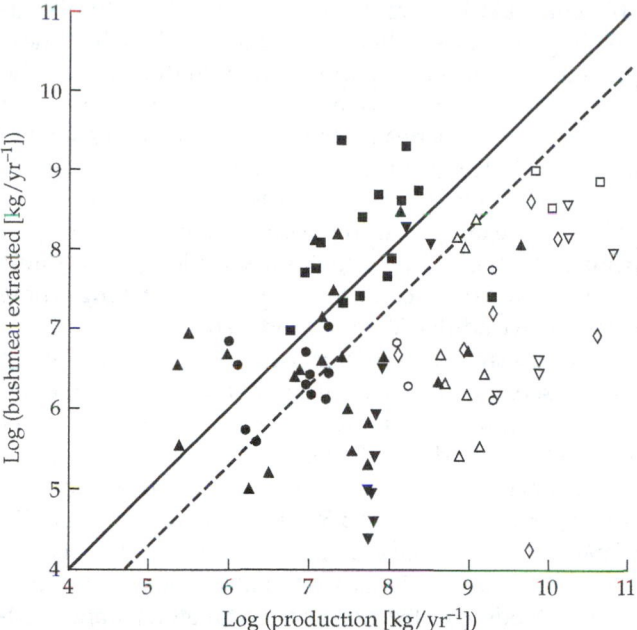

Figure 8.2 Hunting rates are unsustainably high across large tracts of tropical forests as seen in the relationship between total extraction and total production of game meat throughout the Congo and Amazon basin (solid and open symbols, respectively) by mammalian taxa. The solid line indicates where extraction equals production; the dashed line indicates exploitation levels at 20% of production, considered to be sustainable for long-lived taxa. Taxon symbols are as follows: ungulates (squares), primates (triangles), carnivores (circles), rodents (inverse triangles), and other taxa (diamonds). (Modified from Fa et al. 2002.)

lage-based kill profiles in Neotropical (Jerozolimski and Peres 2003) and African forests (Fa and Peres 2001). This can be seen in the composition of residual game stocks at forest sites subjected to varying degrees of hunting pressure where overhunting often results in faunal biomass collapses, mainly through declines and local extinctions of large-bodied mammals (Bodmer 1995; Peres 2000a).

NONTIMBER FOREST PRODUCTS   Nontimber forest products (NTFP) are biological resources other than timber that are taken from either natural or managed forests (Peters 1994). Examples of exploited products (of whole plants or plant parts) include fruits, nuts, oil seeds, latexes, resins, gums, medicinal plants, spices, dyes, ornamental plants, and raw materials and fiber such as Desmoncus climbing palms, bamboo, and rattan. Forest wildlife and bushmeat can also be considered as a prime NTFP, but for a number of reasons they will be treated as a distinct category.

The socioeconomic importance of NTFP extraction to indigenous peoples cannot be underestimated. Many ethnobotanical studies have catalogued the wide variety

of useful plants or plant parts used by different aborigine groups throughout the Tropics. For example, the Waimiri-Atroari Indians of central Brazilian Amazonia make use of 79% of the tree species occurring in a single 1-ha terra firme (upland) forest plot (Milliken et al. 1992), and out of the ≥16,000 species of angiosperms in India, 6000 are used for Ayurvedic or other traditional medicine, and over 3000 are officially recognized by the government for their medicinal uses.

Exploitation of NTFPs often involves the systematic removal of reproductive units from the population, but the level of mortality in the exploited population depends on the method of extraction and whether vital parts are removed. Traditional NTFP extractive practices, for either subsistence or commercial purposes, are often considered to comprise a desirable, low-impact economic activity in tropical forests compared to alternative forms of land use involving structural disturbance, such as selective logging and shifting agriculture (Plotkin and Famalore 1992). As such, the exploitation of NTFPs is usually assumed to be sustainable and is viewed as a promising compromise between the requirements of biodiversity conservation and those of extractive communities under varying degrees of market integration.

A recent study of Brazil nuts (*Bertholletia excelsa*) questions the standard assumptions about sustainability of NTFPs (Peres et al. 2003). Brazil nuts are the base of a major extractive industry supporting over half of the tribal and nontribal rural population in many parts of the Brazilian, Peruvian, and Bolivian Amazonia, either for their direct subsistence value or as a source of income. This wild seed crop is firmly established in domestic and export markets, has a history of over 150 years of commercial exploitation, and is one of the most valuable nontimber extractive industries in tropical forests anywhere. Brazil nuts have been widely held as a prime example of a sustainably extracted NTFP, yet the persistent collection of *B. excelsa* seeds has severely undermined the patterns of seedling recruitment of Brazil nut trees. This has drastically affected the age structure of many natural populations to the point where persistently overexploited stands will succumb to a process of senescence and demographic collapse, threatening this cornerstone of the Amazonian extractive economy (Peres et al. 2003; Figure 8.3). Nevertheless, the concept of sustainable NTFP extraction is now so deeply entrenched into national resource use policy that extractive reserves (or their functional analogues) have become one of the fastest growing categories of protected areas in tropical forests (Fagan et al. 2005). The implicit assumption is that traditional methods of NTFP exploitation have little or no impact on forest ecosystems and tend to be sustainable because they have been practiced over many generations. However, virtually any type of NTFP exploitation in tropical forests

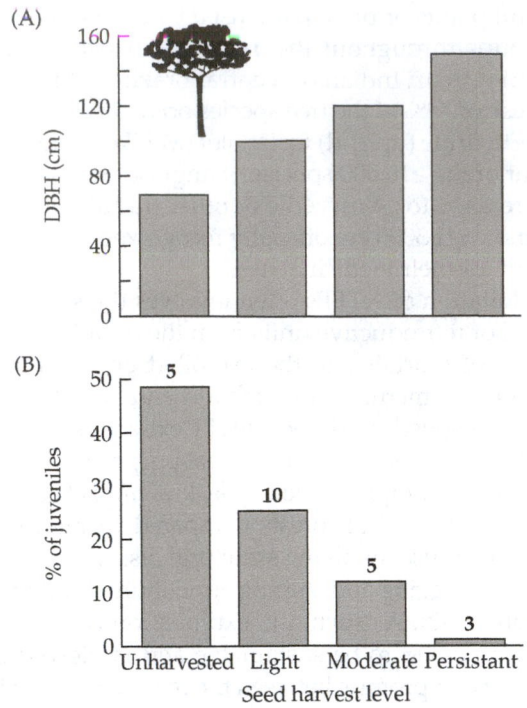

Figure 8.3 Relationships between historical levels of Brazil nut collection and mean tree size, expressed in terms of DBH, cm (A); and percentage of juvenile trees (B) in different populations. Numbers above bars indicate the number of populations studied throughout the Brazilian, Bolivian, and Peruvian Amazon. (Modified from Peres et al. 2003.)

will have an ecological impact. The exact extent and magnitude of this impact depends on the accessibility of the resource stock, the floristic composition of the forest, the nature and intensity of extraction, and the particular species or plant part under exploitation (Peres and Lake 2003).

Few studies have quantitatively assessed the demographic viability of nontimber plant products. A boom in the use of homeopathic remedies sustained by over-collection of therapeutic and aromatic plants is threatening at least 150 species of European wild plants and driving many populations to extinction (TRAFFIC 1998). Commercial exploitation of the pau-rosa or rosewood tree (*Aniba rosaeodora*), which contains linalol, a key ingredient in fine perfumes, involves a destructive technique that almost invariably kills the tree. This species has consequently been extirpated from virtually its entire geographic range in Brazilian Amazonia (Mitja and Lescure 2000). Chanel N°5® and other perfumes made with pau-rosa fragrance gained an enormous international market decades ago after being popularized by Hollywood stars like Marilyn Monroe, but the number of processing plants in Brazil fell from 103 in 1966 to fewer than 20 in 1986, due to the dwindling resource base. Yet French perfume con-

noisseurs have been reluctant to replace the natural pau-rosa fragrance by synthetic substitutes, and the last unexploited populations of pau-rosa remain threatened. The same could be argued for a number of NTFPs for which the exploitation by destructive practices involves a lethal insult to whole reproductive individuals, such as the extraction of fruits and palm-hearts in many arborescent palms. For example, in the Iquitos region of Peru, the fruits of *Mauritia flexuosa* palm trees, a long-lived forest emergent, are often collected only once by felling whole reproductive adults (Vasquez and Gentry 1998).

Enthusiasm for NTFPs in community development and conservation partly results from unrealistic studies reporting their high economic value. For example, Peters et al. (1989) reported that the net present value of fruit and latex extraction in the Rio Nanay of the Peruvian Amazon was $6,330/ha, assuming a 5% discount rate and that 25% of the crop was not taken. This is in sharp contrast with a 30-month study in Honduras that measured the local value of foods, construction and craft materials, and medicines extracted from the forest by 32 Indian households (Godoy et al. 2000). The combined value of consumption and sale of forest goods ranged from U.S.$18 to U.S.$24 per hectare per year, at the lower end of previous estimates (between U.S.$49 and U.S.$1,089 per hectare per year). NTFP extraction thus cannot be seen as a solution for rural development and in many studies the potential value of NTFPs is exaggerated by the assumption of unrealistically high discount rates, unlimited market demands, availability of transportation facilities, and absence of product substitution.

What, then, is the impact of NTFP extraction on the dynamics of natural populations? How does the impact vary with the life history of plants and animals? Are current extraction rates truly sustainable? These are all questions that could steer a future research agenda but the demographic viability of NTFP populations will depend on the species' ability to recruit new adults either continuously or in sporadic pulses while being subjected to repeated exploitation.

Temperate terrestrial ecosystems

FORESTRY   According to one assessment, only 22% of the world's original forest cover remains in large, relatively natural ecosystems, or so-called "frontier forests" (Bryant et al. 1997). These remaining frontier forests are predominantly classed as either boreal (48%) or tropical (44%), with only a small fraction (3%) remaining in the temperate zone. However, much like tropical forests, most of the frontier forests in mid to high latitude regions are also increasingly threatened by logging and agricultural clearing (World Resources Institute 1998).

The overall impact of different sources of structural disturbance generated by modern forestry in the temper-

ate and boreal zones may depend on the spatial scale and intensity of disturbance, the history of analogous forms of natural disturbance, the groups of organisms considered, and whether forest ecosystems are left to recover over sufficiently long intervals following a disturbance event such as commercial thinning or clearcutting. The patterns of landscape-scale human disturbance in these forests can be widely variable in intensity, duration, and periodicity, and are often mediated by economic incentives to cut timber from high-biomass old-growth forests, rather than natural regrowth or fast-growing tree plantations on a long rotation cycle.

The expanding frontier of commercial forestry into hitherto remote, roadless wildlands often results in high levels of forest conversion and fragmentation of remaining stands, with significant impacts to forest wildlife. A review of 50 studies in Canadian boreal forests on the effects of postlogging silviculture on vertebrate wildlife concluded that large impacts are universal when native forests are replaced by even-aged stands of rapidly-growing nonnative tree species (Thompson et al. 2003). Loss of special structures, such as snags containing nest cavities and large decaying woody debris, is particularly important in the decline of forest species depending on those structures (McComb and Lindenmeyer 1999). Unlogged boreal coniferous forests within protected areas in Finland, where population trends of land birds have been exceptionally well documented, are extremely important to the native old-growth avifauna, such as the Siberian Jay (*Perisoreus infaustus*) and the hole-nesting Three-toed Woodpecker (*Picoides tridactylus*), which have declined in areas under timber extraction (Virkkala et al. 1994). Indeed, logging is often considered to be the most important threat to species in boreal forests (Hansen et al. 1991; Imbeau et al. 2001). Some 50% of the red-listed Fennoscandian species are threatened because of forestry (Berg et al. 1994). Forests actively managed for biodiversity could support 100% of the species occurring in Washington state, whereas timber management on a 50-year rotation at the landscape level could support a maximum of 87% and a mode of 64% of the species potentially occurring in forests (Carey et al. 1996).

In summary, many of the detrimental impacts of forest management on biodiversity are associated with the large-scale structural simplification of the ecosystem at all forest levels, age-class truncation, and other consequences of intensive forest management intended to increase the yield of the desired forest component.

A reduction of this impact often involves modern principles of ecological forestry (Seymour and Hunter 1999), which may include a partial removal of the stand, often in a patch mosaic using relatively small clear-cut sizes so that each stage of forest development is represented somewhere in the landscape. As long as most stages are present at any one time, the requirements of nearly all species can be met somewhere (although see examples in Chapter 7 on the negative effects of habitat fragmentation). Furthermore, clear-cutting may only be used when and where absolutely necessary. Other exploitation methods include thinning and partial cutting, each with different effects on wildlife habitat. Landscape-level decisions to maximize the biodiversity value of the forest may include allocating different portions of the total area to successively longer rotations, ranging from 50 years for short-lived species up to 300 years for late-successional habitat.

HUNTING Large mammals, small game, and waterfowl are also major targets in temperate countries, where recreational hunting can be a popular sport across a large section of the constituents. Annual culls of large ungulates have been rather successful in North America, as shown by population trends, in both controlling populations and stocking the freezer of the average hunter. White-tailed deer (*Odocoileus virginianus*), the most common and widespread of wild ungulates in the U.S., increased from fewer than 500,000 around the turn of last century, to about 30 million today. Deer are among the most heavily hunted species in the U.S., with some 5 million killed by over 10 million hunters every year, generating about 20 billion dollars in hunting-related revenue in 2001 (U.S. Fish and Wildlife Service 2002). In Texas alone, it is estimated that the white-tailed deer population numbered more than 3.1 million in 1991 in spite of heavy hunting pressure, and approximately 474,000 animals were shot by hunters in that year. Resident birds and small to mid sized mammals frequently hunted or trapped for sport are often referred to as "small game." These may include game birds such as quail, pheasant, partridge and grouse, rodents and lagomorphs such as squirrels, rabbits, and hares, and even carnivores such as coyotes and raccoons. The effects of different hunting strategies upon populations of the most popular game bird in the U.S., the bobwhite quail, have been simulated showing that maximum yields can be sustained from annual capture rates of about 55% (Roseberry 1979). Hunting at such high rates, however, leaves little room for error in calculated bag quotas because it depresses the following spring populations by 53% below unexploited levels.

Ducks, geese, and swans are gregarious and often migratory species that also attract enormous attention from game hunters. As such, the establishment of annual waterfowl hunting regulations is a complex procedure shared by various governmental levels and private organizations, involving thousands of wildlife biologists and habitat managers under the jurisdiction of wildlife agencies. For example, retrieved duck and goose har-

vests during 2001 were 19.4 million in the U.S. and Canada, down by 6.6% on the total numbers bagged in the previous year (Martin and Padding 2002). In the U.S. alone, in 2001, this involved annual sales of 1.66 million federal duck stamps to 1.59 million hunters who collectively spent nearly 15 million hunter-days in pursuit of waterfowl. Needless to say, seasonal hunting license fees and leases of private hunting areas generate welcome cash used for both population management and for protection of habitats against other forms of land use.

The relatively orderly use of wildlife in North America is not necessarily the rule for all temperate regions. Illegal use and commercialization of wildlife continue to generate a substantial clandestine traffic—even for species that are fully protected on paper—in several mid- to high-latitude countries ranging from Chile to Russia. The problem usually lies not in the regulations, which are often already extensive and strict, but rather in lack of law enforcement that is often attributed to weak institutional capacity.

Aquatic ecosystems

MARINE ECOSYSTEMS   The impacts of marine fisheries on target species are well known. Roughly three-quarters of the world's fish stocks are considered to be fully fished or overexploited (Food and Agriculture Organization 2002). Since the 1990s global catches have leveled off for the first time in human history, despite continuing advances in capture technology (Figure 8.4). Exploitation of aquatic animals now seems to be following the pattern that occurred in many terrestrial ecosystems long ago, with reliance on hunting of wild animals being supplemented by the captive rearing of domestic stock, through aquaculture (see Figure 8.4). No one should delude themselves into thinking that the move toward fish farming

will save wild stocks, as many of the fish that are reared, such as salmon and trout, are fed with meal derived from wild fish! Indeed, stocks of many wild fishes have continued to decline. Recent surveys of 232 stocks showed a median decline of 83% over the past 25 years (Hutchings 2000, 2001; Hutchings and Reynolds 2004). A stock of Atlantic cod (*Gadus morhua*) off eastern Canada, which once seemed absolutely limitless, has undergone a decline of 99.9%, since the 1960s, leading to a designation of "endangered" under the Canadian Species at Risk Act (Committee on the Status of Endangered Wildlife in Canada 2003). This decision is not an isolated case; several other species of commercially exploited fishes have been listed recently as threatened with extinction (Musick et al. 2000; IUCN 2004). These developments show that some fisheries concerns are moving from the traditional focus on sustainability into the realm of extinction risk (Reynolds et al. 2002; Dulvy et al. 2003).

Concerns about extinction risk in the sea are most acute for those targeted species that have combinations of traits that make them most susceptible to capture, and biologically least productive (Reynolds et al. 2002; Dulvy et al. 2003). For example, the marine species that are most likely to be targeted are those that occupy shallow waters accessible to fishing gear, those that form dense shoals in predictable places, or those that are most valuable. If a species has one or more of these characteristics in combination with long generation times, the results can be hazardous to the health of the population. The Chinese bahaba (*Bahaba taipingensis*) is a fish that meets many of these criteria (Sadovy and Cheung 2003). It is a huge species of croaker (family Scienidae), which may exceed 2 m in length (Figure 8.5). It has traditionally been caught along its coastal range from Shanghai to Hong Kong. This species has been

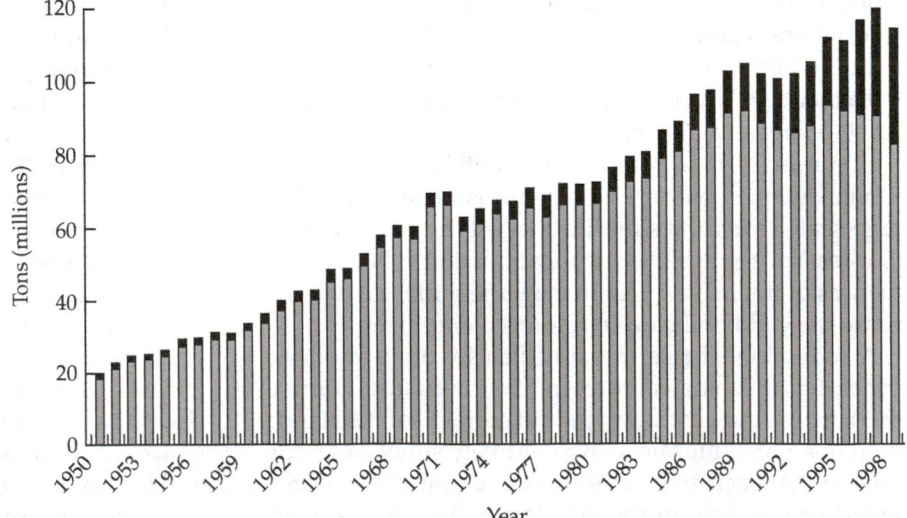

Figure 8.4   Trends in global fisheries. The gray portion of the bars indicate capture fisheries and the black portion of the bars indicate aquaculture. (From Hart and Reynolds 2002, based on data from FAO 2000; corrected for misreporting of capture fisheries of China by Watson and Pauly 2001.)

Figure 8.5 Chinese bahaba (*Bahaba taipingensis)* caught as an incidental by-catch by a trawler west of Hong Kong. (Photograph courtesy of C. Tai-sing.)

popular for the medicinal properties of its swim bladder. Its numbers have declined to 1% of its abundance in the 1960s. During this time its price skyrocketed to the point where swim bladders in 2001 were worth seven times the price of gold (Sadovy and Cheung 2003). Scientists do not have enough information to say much about its population demography, but its large size invariably implies that it will take many years to reach maturity. This biological feature, combined with such strong economic incentives for people to catch it, have conspired to push the species to the edge of extinction.

The same basic rules of vulnerability for fish species apply to other taxa. For example, species of abalone along the Pacific coast of North America often occur in shallow waters, where they are readily accessible to divers (Figure 8.6). There has been serial depletion of these species, beginning with the most valuable and then moving on to less valuable ones (Tegner et al. 1996). This is reminiscent of the pattern that occurred with the

great whales before the International Whaling Commission's moratorium on commercial whaling in 1986. The white abalone (*Haliotus sorenseni*) has been hit particularly hard. Ranging from California to Baja California, white abalone densities in some locations during the early 1970s were estimated at 1,000–5,000 per acre (Lafferty 2003). Commercial and recreational capture reduced its numbers to less than one per acre by the 1990s. In May 2001 this species became the first marine mollusc to come under the U.S. Endangered Species Act, with an estimated population in the wild of only about 3,000 individuals. Most remaining animals are restricted to deep waters beyond the reach of fishing (Lafferty et al. 2003). While the abalones' value and accessibility provided the motive and the means for overexploitation, it has also suffered from an additional problem: the Allee effect. This is the phenomenon whereby per capita fitness declines as a population becomes smaller (see Chapter 12). The ensuing feedback can cause populations to become more vulnerable as they become increasingly rare, potentially spiraling to extinction. In the case of abalone, the Allee effect arises through the need for individuals to be within 1 m of each other for a male's sperm to reach a female. Biologists have had some success with artificial fertilization in captivity, with the intention of rearing abalones for release into the wild.

FRESHWATER ECOSYSTEMS   Many freshwater taxa are subject to exploitation for food and a variety of additional products. In 2000, the global estimate for all inland capture fisheries was estimated at 8.8 million tons (FAO

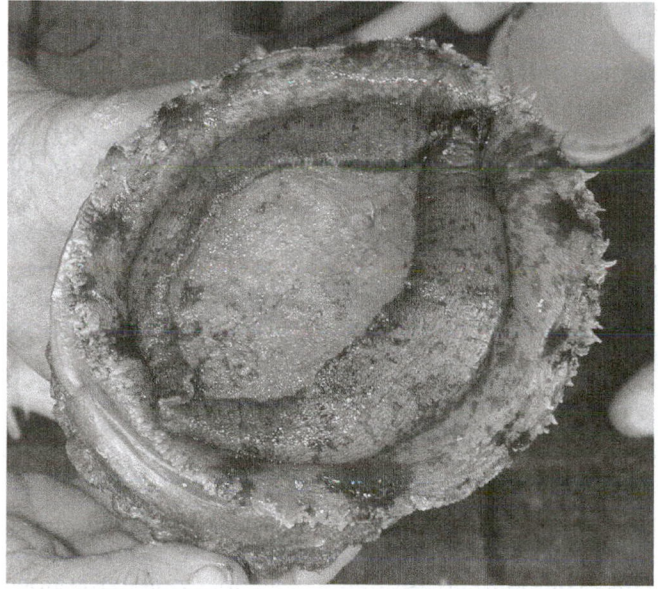

Figure 8.6 White abalone (*Haliotis sorenseni*) underside. (Photograph courtesy of K. Lafferty, USGS.)

2002). This figure is considerably less than the estimate for inland aquaculture (22.4 million tons). In many temperate countries, recreational fishing in fresh waters is a major past-time, yielding an estimated 2 million tons of fish (Cowx 2002). In 1996 it was estimated that 35 million people in the U.S. spent 514 million angler-days fishing, spending U.S. $38 billion in the process (U.S. Fish and Wildlife Service 1997). Among 22 European countries, the number of anglers was estimated at 21.3 million (Cowx 1998).

The world's salmon species illustrate the vulnerability of fish species that spend all or part of their life cycle in freshwater. Populations of four of the seven species of eastern Pacific salmon and trout in the genus *Oncorhyncus* are currently listed under the U.S. Endangered Species Act. Over-fishing has contributed to severe declines in many populations, often in combination with habitat loss through construction of dams that block their migrations, as well as degradation of spawning streams due to forest clearance and water abstraction (Lynch et al. 2002). The plight of salmon species is particularly sobering when one considers the enormous amount of attention that has been paid to these populations by scientists and conservationists, as well as their high economic and cultural significance. The most recent response of the U.S. government at the time of writing (May 2004) has been the announcement of the intention to count hatchery-released fish as "wild" (Kaiser 2004). Thus, the hundreds of millions of fish that are reared in captivity each year will elevate the population counts of many of the 27 populations of Pacific salmon and cutthroat trout that are endangered. This can lead to their removal from protection under the Endangered Species Act and presumably, the discontinuation of many habitat restoration programs. The decision ignores the fact that hatchery fish become domesticated rapidly, showing genetic and phenotypic divergence from wild stocks (e.g., Fleming et al. 1994; Heath et al. 2003). Furthermore, there is little evidence that hatcheries enhance the viability of wild stocks (Myers et al. 2004). Yet the regional director of the National Marine Fisheries Service, which has jurisdiction over salmon management, has declared that "Just as natural habitat provides a place for fish to spawn and to rear, also hatcheries can do that..." Presumably, zoos can also be considered natural habitats.

Fish are not the only taxa exploited from fresh waters. During the late 1800s and early 1900s millions of freshwater mussels (family Unionidae and Margaritiferae) were collected from freshwaters of Canada and the U.S., primarily to make buttons from their shells (Williams et al. 1993; Helfrich et al. 2003; Williams and Neves 2004). This exploitation was rarely sustainable, due to the slow generation times of these species. Mussels were saved from major impacts of exploitation by the switch to plas-

tics for manufacturing buttons during the 1940s. However, today several million kilograms of mussels are still exported annually to Asia, where small beads are made from their shells and inserted into other bivalves for the production of pearls. Unfortunately, mussels now face much more critical threats from the "usual suspects" in freshwater conservation: pollution, damming, dredging, channelization, siltation, and competition with nonnative bivalves such as the Asian clam (*Corbicula fluminea*) and the zebra mussel (*Dreissina polymorpha*) (Strayer et al. 2004). These problems have led to 72% of the 297 species that occur in the U.S. being considered endangered, threatened, or of special concern, including 21 species that are extinct (Williams and Neves 2004).

Impacts of Exploitation on Nontarget Species and Ecosystems

Most hunting, fishing, and collecting activities affect not only the species that are the primary targets, but also those that are taken accidentally or opportunistically. Furthermore, exploitation may cause physical damage to the environment, and also may have ramifications for other species through cascading interactions, phase shifts in the structure of the ecosystem, and changes in food webs. Here we discuss a few examples of how extractive activities targeted to one or a few species can drastically affect the structure and functioning of terrestrial and aquatic ecosystems.

Tropical terrestrial ecosystems

LOGGING AND FOREST FLAMMABILITY   Even logging targeted to a single timber species can puncture the forest canopy and increase the density of treefall gaps. This increase can trigger major ecological changes by increasing light and creating a warmer and drier microclimate in the understory, which thereby affects the dynamics of plant regeneration, and increases forest susceptibility to fire disturbance. In fact, even highly selective logging operations with modest levels of incidental damage to nontarget trees can generate enough structural disturbance to greatly augment understory desiccation and dry fuel loads, thereby breaching the forest flammability threshold (Holdsworth and Uhl 1999; Nepstad et al. 1999). Any source of ignition during subsequent severe droughts can initiate extensive ground fires that will dramatically reduce the functional and biodiversity value of previously unburned tropical forests (Barlow and Peres 2004). Surface wildfires that are at least partly induced by logging disturbance currently threaten millions of hectares of Amazonian, Mesoamerican, and Southeast Asian forests (Cochrane 2001; Siegert et al. 2001). Despite these undesirable direct and indirect effects, large-scale mechanized

logging operations continue unchecked in many seasonally dry tropical forest regions.

HUNTING AND LOSS OF SEED DISPERSAL SERVICES   Successful seedling recruitment in many flowering plants depends on seed dispersal services provided by large-bodied frugivores (Howe 1984). In tropical forests, the proportion of plant species with an endozoochorous dispersal mode (bearing seeds dispersed by an animal's digestive tract) is often more than 90% (Peres and Roosmalen 2002). Undispersed seeds simply fall to the ground underneath the parent's canopy and have a low survival probability (Augspurger 1984; Chapman and Chapman 1996). For example, 99.96% of *Virola surinamensis* seeds that drop under the parent are killed within 12 weeks (Howe et al. 1985). Many studies have shown lower seed mortality rates caused by fungal attack or vertebrate and invertebrate seed predators are lower at greater distances from parents. However, there have been only a few studies that have examined the effects of removing seed dispersers are lower on the demography of gut-dispersed plants. Wright et al. (2000) explored how hunting alters seed dispersal, seed predation, and seedling recruitment for two palms (*Attalea butyraceae* and *Astrocaryum standleyanum*) in Panama. They found that where hunters had not reduced mammal numbers, most seeds were dispersed away from the parent palms, but were subsequently eaten by rodents. Where hunters had reduced mammal abundance, few seeds were dispersed, but these tended to escape rodent predation. Thus, seedling density increased by 3–5-fold at heavily hunted sites compared to unhunted sites. In contrast, Asquith et al. (1999) demonstrated that recruitment of *Hymenaea courbaril* required scatterhoarding of their large seeds by agoutis (*Dasyprocta* sp.), and recruitment rates of many plant species that produce very large seeds cached by rodents is likely to be very low in heavily hunted areas.

Studies examining seedling recruitment under different levels of hunting pressure (or abundance of large-bodied seed dispersers) reveal very different outcomes. At the community level, seedling density in overhunted forests can be indistinguishable, greater, or less than that in the undisturbed forests (Dirzo and Miranda 1991; Chapman and Onderdonk 1998; Wright et al. 2000), but the consequences of increased hunting pressure to plant regeneration is likely to depend on the target species. In persistently hunted Amazonian forests, where large primates are either driven to local extinction or severely reduced in numbers, the probability of effective dispersal of large-seeded plants ingested primarily by these frugivores can decline by more than 60% compared to nonhunted forests (Peres and Roosmalen 2002). However, more conclusive evidence is required before the importance of the loss or reduction of effective animal dispersal services can be properly understood for different plant species.

Temperate ecosystems

Higher order interactions resulting from selective extinction or severe population declines of large mammals that play an important role as landscapers at the ecosystem scale have been documented in the temperate zone. Beavers (*Castor canadensis*) and their Eurasian congener (*Castor fiber*) are prime examples of ecosystem engineers, which can be defined as organisms that have the potential to dramatically alter the structure and function of ecosystems at large spatial scales. Large ponds created by the labor-intensive stream-damming activity of beaver colonies create large-scale semi-permanent flood disturbance that drastically changes the structure and patch dynamics of wetlands (Naiman et al. 1986; Wright et al. 2002). Beavers were once locally abundant in many parts of North America and Eurasia but their populations plummeted due to the pelt trade and habitat conversion in the nineteenth and early twentieth century, radically changing wetland ecosystems. No one knows the precise extent of these changes but it is clear that beaver damming activity has profound effects on the biogeochemistry of wetland systems (Naiman et al. 1994), the dynamics of shifting successional mosaics of aquatic patches (Johnston and Naiman 1990), and ultimately the population dynamics of other wetland species including waterfowl (McCall et al. 1996). Beavers are now making a gradual comeback in many parts of North America (although they were exterminated by 1700 and are yet to be reintroduced in many parts of Europe including Britain) but they often continue to be perceived as a nuisance requiring controversial measures of population control.

In temperate terrestrial ecosystems, large mammals that once had profound large-scale effects on the structure of plant communities but have been hunted to near extinction in historic times include bison, bears, and wolves. Joel Berger and colleagues (2001) demonstrated a cascade of ecological events that were triggered by the local extinction of grizzly bears (*Ursus arctos horribilis*) and wolves (*Canis lupus*) from the Yellowstone ecosystem. These include large increases in the population of a riparian-dependent herbivore, the moose (*Alces alces*), the subsequent alteration of riparian vegetation structure and density by ungulate herbivory, and the coincident reduction of Neotropical migrant birds in the affected willow communities, including riparian specialists such as Gray Catbirds (*Dumetella carolinensis*) and MacGillivray's Warblers (*Oporornis tolmiei*).

Where large predators (wolves, bears) have been removed through predator control programs, or other forms of direct persecution, ungulate populations can

greatly increase in size, with subsequent impacts on plant communities. White-tailed deer in the U.S. and reindeer in Europe were once controlled by wolves and bear, and despite annual harvests, their numbers are larger than they were when controlled by native predators. The result is that their larger populations may over-browse forests, which can cause extensive damage in some places (e.g., Väre et al. 1996 and Horsley et al. 2003). Particularly in areas where hunting is prohibited, deer populations can become so large that most tree and herb seedlings are consumed, which may lead to a change in patterns of forest succession (e.g., Horsley et al. 2003 and Pedersen and Wallis 2004), or to decline of rare species (e.g., Gregg 2004). Plant populations can be slow to recover from episodes of excessive herbivory. For example, a population of the rare orchid showy lady slipper (*Cypripedium reginae*) in West Virginia lost up to 95% of all stems during a 3-year period of excessive deer browsing, from which the population took 11–12 years to recover (Gregg 2004). Beyond effects on plants, uncontrolled herbivore populations can have indirect effects on decomposer (Wardle and Bardgett 2004) and spider communities (Miyashita et al. 2004).

Aquatic ecosystems

MARINE ECOSYSTEMS Marine fisheries are estimated to have a global by-catch of roughly 27 million tons annually, which is between one-third and one-fourth of the total marine landings (Alverson et al. 1994). This is probably a conservative estimate. Most of these by-catches were from trawl fisheries, followed by drift nets and gill nets. Shrimp trawls, with their small mesh sizes, account for roughly one-third of the total by-catch, with ratios of weight of discarded animals per weight of shrimp caught typically about 5:1. It has been estimated that shrimp trawlers in the Australian northern prawn fishery typically discard 70,000 individual organisms during each night of fishing.

Some by-catches are threatening species with extinction. All but two of the world's 21 species of albatross are considered threatened with extinction, and most of these have been severely affected by longline fisheries in the Southern Ocean (IUCN 2004). Albatrosses, as well as other seabirds such as petrels, drown when they take baited hooks as the lines are being set in fisheries aimed at species such as the Patagonian toothfish (*Dissostichus eleginoides*). While the rate of accidental capture per hook is very low, the potential risks are enormous when scaled up by the total fishing effort, with individual vessels often setting many thousands of hooks each day, and a total of over 250 million hooks being set each year since the 1990s south of 30° South (Tuck et al. 2003). Long-lived species such as seabirds have extremely low resilience against elevated adult mortality, due to their very long generation times and low rates of productivity. Recent re-assessments by the IUCN (2004) have led to six species of albatross recently being "upgraded" toward more severely threatened status, with fisheries by-catch playing a significant role in each case. These problems are not confined to the Southern Ocean. For example, it has been estimated that 10,000 Black-footed Albatrosses (*Phoebastria nigripes*) may be killed each year in the central North Pacific (Lewison and Crowder 2003). Measures adopted to reduce mortality on seabirds include bird-scaring devices, the release of baited lines from below the water surface (where albatrosses cannot reach them), use of heavy lines that sink immediately, and use of fish oil on the surface, which dissuades seabirds from landing on the water.

Sea turtles are caught incidentally by longlines and trawlers. For example, it has been estimated that 20,000 turtles have been killed each year in the Mediterranean in longlines set for swordfish. Turtle excluder devices are now in widespread use in trawl fisheries, and these reduce turtle mortality by providing an exit flap that turtles can push through, while retaining the fish. Turtles are not the only reptiles taken as by-catch. It has been estimated that 120,000 sea snakes are taken annually by prawn trawlers in the Gulf of Carpentaria, Australia.

There are mounting concerns about the impacts of by-catches on populations of many fish species as well. The best-known cases involve sharks, skates, and rays (see Case Study 8.1 by Julia Baum). These fishes are toward the seabird–turtle end of the life-history continuum. For example, the European common skate (*Raja batis*) does not reach maturity until it is 11 years old. They used to be taken by the thousands as by-catch in prawn trawls in the Irish Sea (Brander 1981), but only six individuals were captured by extensive government bottom fish surveys between 1989 and 1997 (Dulvy et al. 2000). In this same region, there is evidence for complete local extinction of one and possibly two more species.

FRESHWATER ECOSYSTEMS In fisheries, it is easy to find freshwater equivalents of the kinds of difficulties that face many marine fish species. An example of nontarget fishes is the Mekong giant catfish (*Pangasianodon gigas*) (Figure 8.7), a species that is in a similar situation to the Chinese bahaba discussed earlier. This fish is restricted to the Mekong River Basin in Thailand (where no individuals have been caught since 2001), Laos, and Cambodia. Studies by Zeb Hogan and colleagues (2004) have shown that individuals can reach 3 m in length and weigh 300 kg, and are therefore extremely valuable to fishers. Furthermore, they are migratory, thereby running a gauntlet of nets set for other species. Finally, their only spawning grounds, in whirlpools and rapids in the Chiang Khong-Chiang Saen region of the Thai–Laos border, are being

Figure 8.7   Mekong giant catfish, *Pangasianodon gigas*, from the Mekong River. (Photograph courtesy of Z. Hogan.)

dynamited as part of a plan to enhance river navigation. Conservation biologists have set up a scheme whereby fishers telephone them day or night if they catch a Mekong giant catfish, which the biologists then purchase at market price, tag, and release away from the nets. This is a stop-gap measure, being used to save a small number of fish while working on longer-term conservation measures. The prospects for this species do not look good, and its threat status has now been raised to "critically endangered" by the IUCN (2004).

Overexploitation can also have unforeseen effects on biodiversity through trophic interactions among species in an ecosystem. A freshwater example that is currently receiving a great deal of attention involves the effects of reduced numbers of Pacific salmon on productivity of streams and their riparian vegetation in northwestern North America (e.g., Cederholm et al. 1999). All native Pacific salmon species die after they spawn, and their decomposing bodies provide a large proportion of the nutrients received by streams in their range. These nutrients have been acquired when the fish were growing during the marine phase of their life cycle. Experimental studies have shown a variety of effects of nutrients from salmon carcasses. For example, carcass enrichment of artificial streams has led to an eight-fold increase in densities of macroinvertebrates, which also increased twenty-five-fold in artificially enhanced natural streams (Wipfli et al. 1998). These effects have been shown to have important feedback to productivity of the streams for salmonids themselves, because juvenile salmon prey on many of these insects (Wipfli et al. 2003). Studies have also shown higher growth of riparian trees such as white spruce, *Picea glauca* (Helfield and Naiman 2002). Indeed, there has been preliminary success in matching records of the sizes of salmon runs in southeastern Alaska from 1924 to 1994 with tree-ring growth in Pacific coastal rainforests (Drake et al. 2002). All of these examples imply that stream ecosystems, as well as associated terrestrial components, will have been affected strongly by the many extinctions and depletions of salmon populations that have occurred in recent decades.

Biological Theory of Sustainable Exploitation

The examples in the preceding sections give a flavor of the great diversity of forms of exploitation and impacts on species and ecosystems. Some of these forms of exploitation, such as fisheries, hunting, and forestry in temperate countries, have been occurring under scientifically informed management. Because this is a chapter on "overexploitation," most of the examples that we have chosen have not been very encouraging. In fact, species and ecosystems show a wide variety of responses to exploitation. In this section we review the theory that has been advanced to explain how populations and ecosystems respond to exploitation, and tie this to management options.

Biological populations are by definition renewable, but why should we ever expect plant and animal populations to be able to withstand the elevated mortality that occurs with most forms of exploitation? The key is the ability of birth or death rates to compensate when we remove individuals from the population. As populations are reduced by exploitation, there may be reduced competition for food, territories, shelter, and a lower transmission rate of diseases. This can lead to greater birth rates or enhanced survival. The tendency for such components of fitness to limit populations at high density is known as density dependence (Figure 8.8). This does not mean that populations do not experience episodes of density-independent growth, as would occur after sudden environmental changes, for example. But the tendency of populations to fluctuate around some sort of mean value, rather than growing indefinitely, can be attributed to density dependence.

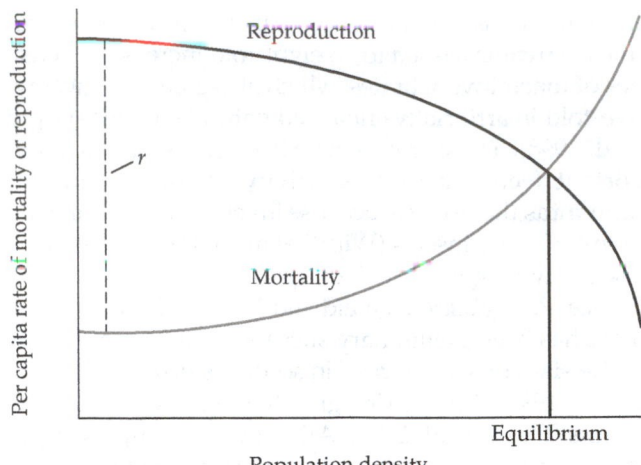

Figure 8.8 Density dependence stems from relationships between population density and per capita rates of birth and death. The difference between these at a small population size is the intrinsic rate of natural increase, r, and the density at which these rates are balanced is the equilibrium population size.

To understand sustainable and unsustainable levels of exploitation, we need to ask whether removal of individuals is equivalent to "thinning" the population, thereby allowing the survivors to grow more quickly or survive better, or is the removal occurring too rapidly for populations to compensate.

The simplest way to ask about the ability of a population to compensate for elevated mortality is to start with the logistic model, which gives the number of individuals at time t as:

$$N_t = \frac{N_{max}}{1 + \left(\dfrac{N_{max}}{N_0} - 1\right)e^{-rt}} \qquad [8.1]$$

Here N_{max} is the maximum population size (Figure 8.9A). This is often called the "carrying capacity," or equilibrium population size. None of these terms is meant to imply that populations remain stable, but they convey the idea of some sort of average size over a given period of time. N_0 is the initial number of individuals. The parameter r is the intrinsic rate of natural increase, that is, the difference between per capita birth and death rates at small population sizes where there is no density dependence. For many species we may want to substitute biomass for number of individuals, as in the case of fisheries, in which case we often see the terms B substituted for N.

Figure 8.9A shows the classical population growth rate that matches the logistic equation. This is the sort of trajectory that we would expect, for example, if we put a few *Daphnia* into a tank of water with phytoplankton as

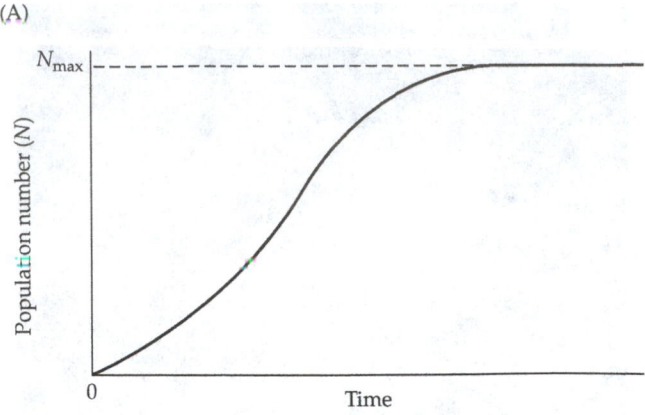

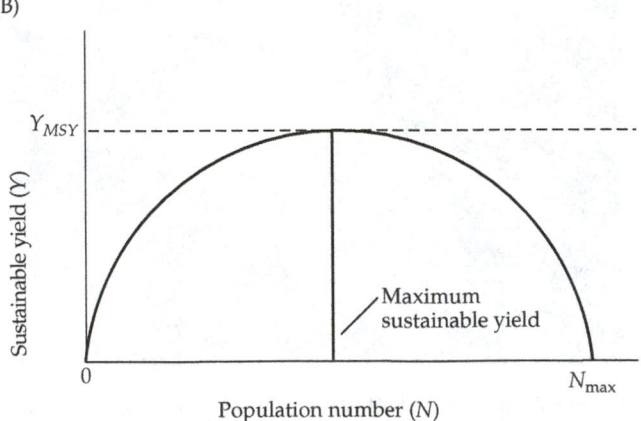

Figure 8.9 (A) Logistic population growth of a population up to a maximum population size, N_{max}. (B) Sustainable yield, Y (surplus production) against population size for the logistic case shown in (A). The maximum sustainable yield (Y_{MSY}) occurs at 50% of the maximum population size.

food. The population grows slowly at first, because there are few individuals producing offspring, but growth accelerates up to a point where the animals start running out of food, whereupon growth stops. In the real world, populations will be buffeted by changing environmental conditions, interactions with their predators, and so on. These can cause population fluctuations, cycles, or crashes. But the logistic is still a reasonable "default" option, conveying the essence of the potential for density-dependence. Specific details of the biology of species will determine whether the initial upward slope and final decline due to density dependence are steeper, shallower, or occur sooner or later than shown here.

The growth rate of this population before exploitation can be considered its surplus production, $g(N)$, and is given as:

$$g(N) = rN\left(1 - \frac{N}{N_{max}}\right) \qquad [8.2]$$

This indicates, sensibly, that as the number of individuals, N, approaches the maximum population size, N_{max}, there is no growth, and even more sensibly, that growth stops when the population size is zero.

If the population in Figure 8.9A is being exploited at a steady rate, then its rate of change per unit time, dN/dt will be the difference between its surplus production and the yield, Y, for a given level of exploitation:

$$\frac{dN}{dt} = rN\left(1 - \frac{N}{N_{max}}\right) - Y \qquad [8.3]$$

The population's surplus production is the yield that can be removed sustainably. We can plot this property against different population sizes (Figure 8.9B). This gives us the classic dome-shaped yield curve that underpins all models of exploitation. This shows that the maximum sustainable yield (Y_{MSY}) occurs at an intermediate population size, which coincides with the inflection point in the logistic curve shown in Figure 8.9A. This makes intuitive sense: We can take the most from a population when it is at the size where it can grow most quickly. The dome does not have to be symmetric; a failure of the logistic assumption can cause it to lean to the left or to the right. For example, density-dependent processes such as cannibalism, competition, predation, or disease may occur at smaller or larger population sizes than depicted in Figure 8.9A (Sutherland and Gill 2001). If the curve leans to the right, we should allow the population to remain at higher numbers.

Stability of exploitation

In theory, we have discovered how many individuals we can take from the population to maximize the yield. But in practice, we will have a very difficult time taking exactly that number. For one thing, people rarely do what they are told, and common experience suggests that we have to assume that the population will probably be exploited at a higher rate than we recommend. Even the best-controlled fisheries and hunts are subject to the vagaries of uncertainty in the population estimates, or unforeseen circumstances such as weather conditions causing populations to behave in unpredictable ways. The question is, what happens when (not if) the population is exploited at a rate that differs from the one that theory suggests should be maximally sustainable? The answer depends on how the exploitation rate is managed.

Constant quota exploitation

Our formulation implies a system of exploitation in which the number of individuals removed is independent of the population size. That is, in Equation 8.3 we have subtracted Y from the surplus yield, rather than making Y proportional to N. This case is typically known as constant

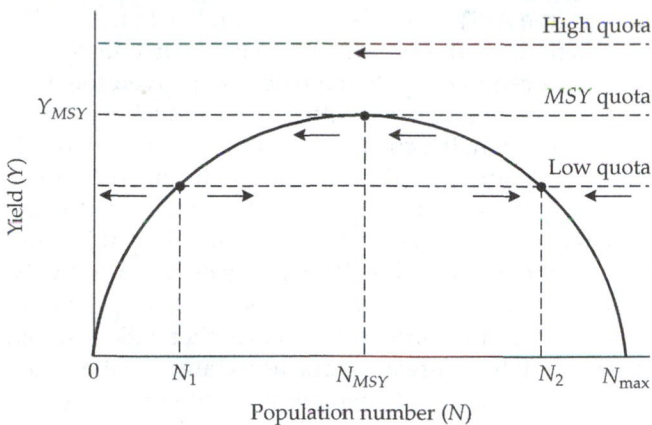

Figure 8.10 Equilibria and population stability under constant quota exploitation. The arrows indicate the directions of change in population size for each quota.

quota because the numbers removed are "constant" in the sense of being independent of the population size. Examples include fisheries that set quotas on numbers of fish that are caught, regardless of subsequent changes in the number of fish in the sea, or hunters who are given fixed limits that do not vary as the number of target animals goes up or down over time. This may sound extreme, because common sense suggests that quotas should be adjusted according to such changes in the quarry. And it is extreme. And so was the collapse of Peruvian anchovy in the 1970s, which occurred in part because people did not fully appreciate the implications of this assumption (see the discussion later in this chapter).

The constant quota situation is depicted in Figure 8.10, where each of three potential quotas runs horizontally across the graph, implying no relationship with population size. First, consider the high-quota line. Here, the yield exceeds the population's surplus production capability, perhaps because of illegal hunting activity, or because the population's resilience had been over estimated. The arrow beneath this line shows that with constant exploitation at this rate, the population will become extinct. Year after year the quota exceeds the population's ability to keep up with the elevated mortality. An alternative that seems more sensible is the MSY quota. However, this is a very risky target because it is impossible to get precise measurements of either the surplus production curve, or N. Furthermore, as we have argued, even if this were possible, there is still a good chance that we will be unable to set quotas sufficiently accurately to score a bulls eye when we shoot for the MSY point. If the population is at or smaller than N_{MSY}, the quota will exceed its surplus production and it will decline to zero. On the other hand, the N_{MSY} point will be stable if the population is initially higher, because al-

though the *MSY* quota is initially higher than surplus production, as the population declines it will benefit from reduced density dependence. Still, given the difficulty of estimating *MSY* with precision, such a quota is dangerous. Finally, consider the low quota example. If the population is initially below N_1 it will still crash. However, if it is between N_1 and N_2, then its production will exceed the quota, and it will grow to N_2. If it is initially higher than N_2, it will be driven down to N_2 by the quota exceeding its surplus production. So N_1 is an unstable equilibrium and exploitation in that region should be avoided. In contrast, as long as we are sure the population is well above N_1, exploiting it with the low quota indicated would allow it to grow to a stable equilibrium at N_2. Here, finally, we have sustainable exploitation.

Proportional (constant effort) exploitation

A much more sensible way to set exploitation targets is to tie them directly to the size of the population. In practice, this is always bound to happen to some extent, because as plants and animals become scarce, people tend to switch to alternative species or other activities, even if no one is forcing them to do so. For example, local demand for game species in tropical forests tends to increase sharply when catches in the marine fishing sector fair poorly (Brashares et al. unpublished data), and tropical forest hunters become heavily reliant on smaller-bodied game species once large-bodied species are depleted (Jerozolimski and Peres 2003). Conservationists and resource managers amplify this kind of common sense by encouraging or forcing people to reduce pressures on populations that reach low abundance. So, unlike the case of constant quota above, we have exploitation effort that is proportional to the population size.

In this scenario the size of the yield, *Y*, will be equal to the exploitation rate, *E*, multiplied by the population size, *N*:

$$Y = EN \qquad [8.4]$$

From Equations 8.2 and 8.3 we know that a steady state will occur between the yield and the surplus production when $g(N) = Y$. This means that $g(N) = EN$. Therefore,

$$rN\left(1 - \frac{N}{N_{max}}\right) = EN \qquad [8.5]$$

We can rearrange this equation to find the equilibrium population size for any rate of exploitation, as long as *E* is below *r*:

$$N = N_{max}\left(1 - \frac{E}{r}\right) \qquad [8.6]$$

These equilibria are shown in Figure 8.11. The advantage of this form of management is that as long as the ex-

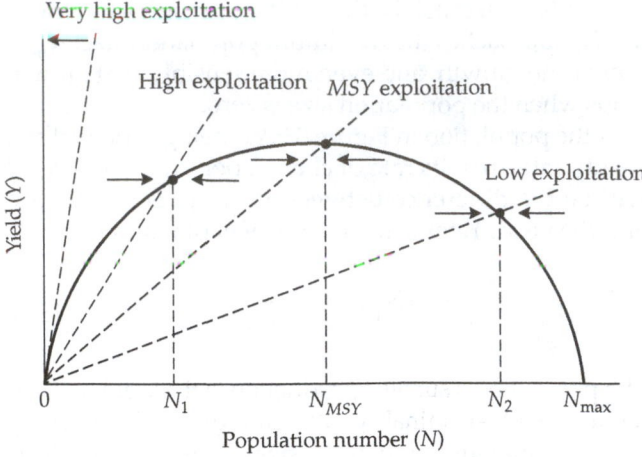

Figure 8.11 Equilibria and population stability under proportional exploitation (also called "constant effort exploitation"). Here, a constant fraction of the population is removed. The dashed lines have slopes *E* corresponding to different rates of exploitation. The arrows indicate the directions of change in population size for each scenario.

ploitation rate is below the intrinsic rate of natural increase, *r*, then all equilibria are stable. For example, whereas the high quota crashed the population under the constant quota scenario in Figure 8.10, here we find that if the population is initially below this removal rate, it will increase to N_1, and if it is above this point, it will decrease to N_1. The *MSY* is also stable, and it still occurs at half of the maximum population size.

Figure 8.12 compares the risk of extinction from exploitation based on either constant quotas or proportional rates for a study of the American marten (*Martes americana*) in southern Ontario, Canada (Fryxell et al. 2001). The marten is a member of the weasel family (Mustelidae) found primarily in coniferous forests where it preys on a wide variety of small vertebrates as well as some invertebrates. Marten are trapped for their fur and trappers are granted licenses for exclusive access to trapping grounds. Fryxell et al. (2001) use a simulation model of data from commercial trapping from 1972 to 1991 to evaluate effects of different types of harvest. This analysis showed that whereas exploitation in proportion to the population size has a negligible chance of causing local extinction, this risk becomes quite high for moderate levels of exploitation under a constant quota system that does not track the population.

Threshold exploitation

The final class of exploitation strategies involves the use of population size thresholds to determine not only the rate of exploitation but also whether exploitation should take place at all (Lande et al. 1997, 2001). All populations are subject to random (stochastic) variations, for exam-

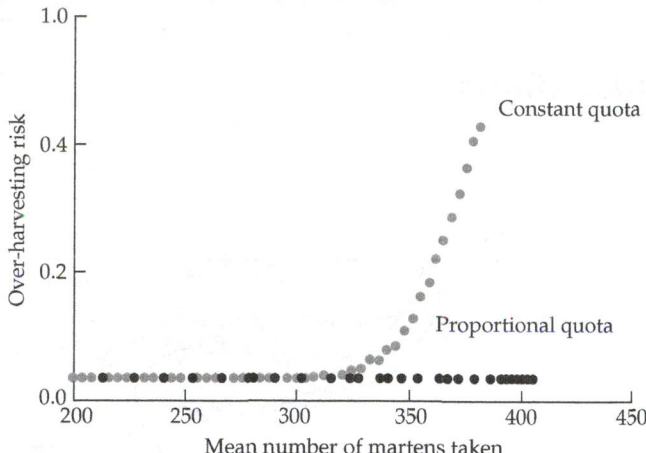

Figure 8.12   Probability of overexploitation (local extinction) in relation to mean yield of marten in commercial trapping in southern Ontario, Canada. The proportional quota line (black) refers to yields that track annual changes in the population whereas the constant quota line (gray) is based on the same absolute number of individuals being trapped each year regardless of population size.

ple, due to environmental variation. In the absence of exploitation, they will sometimes exceed their carrying capacities temporarily. We can take advantage of this by taking the entire "surplus" whenever the population is above its carrying capacity, but otherwise ceasing exploitation completely. This would maximize the cumulative yield while minimizing the chances of collapse. A more precautionary variant on this is to remove only a proportion of the surplus. This is an even safer version of the proportional exploitation method outlined above because it maintains the population near its carrying capacity. However, while such low rates of exploitation are excellent for conservation, it is difficult to convince people to accept such severe restrictions.

Bioeconomics

Understanding biological constraints is necessary to achieving a sustainable level of exploitation. But as we saw earlier in the chapter, we also need to understand the hunters' incentives and disincentives if we are to understand their impacts on the hunted, and provide management advice to mitigate such impacts. Bioeconomic models incorporate the costs and benefits of exploitation. Even in societies where exploitation is a necessity rather than a source of revenue, as in subsistence hunting, a cost–benefit framework can be very useful for understanding peoples' behavior (see Essay 8.2 for an example).

Open access and the tragedy of the commons

The **"tragedy of the commons"** (Hardin 1968) provides a powerful explanation for a lot of the damage that peo-

ple are doing to the environment and to each other. Imagine you have sole access to the trout in a lake on your property. You will probably manage these trout quite carefully, because if you take too many, you will be the one who suffers in future. But if the pond straddles the boundary with your neighbor, then each of you should only take half the number you would take if you had exclusive access. Will you both be so prudent? Unless you get along very well with each other, probably not, because each person's self-restraint can be exploited by the other one. Imagine the outcome if we scale this example up to the North Sea, bordered by many countries, each with thousands of fishers, all competing for the same fish. The sea is a "common" fishing ground, and indeed, the massive over-fishing of the past century has been tragic for both fishers and their prey.

We can incorporate the tragedy of the commons into the scheme developed in the previous sections by assuming that the benefits from exploitation are directly proportional to the yield. So the dome in Figure 8.13 represents the benefits. Assume that the costs of exploitation are proportional to the effort necessary to obtain that yield. For example, in a fishery the more days spent at sea, the greater the costs of fuel and labor. This is depicted by the straight line rising from the origin. If we ignore the costs, the maximum benefits will be found at E_{MSY}, or the effort that provides the maximum sustainable yield. However, if the goal is to maximize profits, this would lead to a lower exploitation rate corresponding to the effort at which the difference between benefits and costs is maximized, at E_P. But the most important fundamental truth of

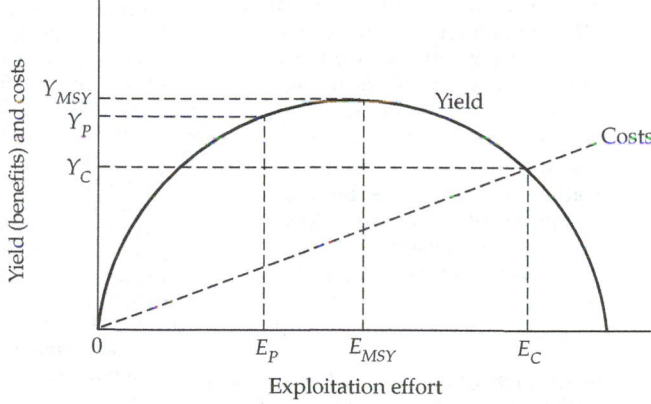

Figure 8.13   The economics of exploitation, represented by costs that are proportional to effort, and benefits proportional to yields. At E_P the profits are maximized, at E_{MSY} the maximum sustainable yield is obtained, and at E_C the costs are equal to the benefits. Note that unlike Figures 8.10 and 8.11, the x-axis here represents fishing effort (which increases from left to right, corresponding to population number decreasing from left to right).

ESSAY 8.2

Using Economic Analysis to Bolster Conservation Efforts
Marine Aquaria and Coral Reefs

Gareth-Edwards Jones, *University of Wales, Bangor*

■ Advances in technology combined with increased interest in marine systems have contributed to a large increase in American households keeping marine aquaria. As part of the large global business associated with establishing and maintaining these aquaria, approximately 350 million fish are harvested annually from the wild and sold worldwide with a value of $963 million (Young 1997). This continued wild harvest is necessary as some species of aquarium fish do not breed well in captivity. Some of the more popular aquarium species are associated with coral reefs, and 85% of aquarium fishes are caught from coral reefs around the Philippines and Indonesia. Catching these fishes can be an important source of income for some communities, but it is increasingly happening on a large scale and there are 4000 aquarium fish collectors in the Philippines alone.

A network of traders are involved in the marketing of the fish; the fishermen themselves may sell to local traders who then sell to exporters and so on. There are significant financial gains at each step in the marketing chain. For example, an orange and white striped clownfish costs 10 cents when bought from a Filipino collector, but sells for $25 or more in an American pet store (Simpson 2001). Thus, traders in developed countries typically make the largest gains, while collectors or even the first traders make pennies. Because the returns to collectors are so low, there is room for conservation alternatives to be attractive; it does not take as much return to exceed the earnings of the people supplying the fish.

The impacts of collecting these fishes from the wild are quite varied. When using traditional netting techniques for catching the fish the most direct impact is a reduction in the population density of the harvested species, as shown for the Banggai cardinalfish (*Pterapoggon kauderni*) in Sulawesi, Indonesia (Kolm and Berglund 2003) (Figure A).

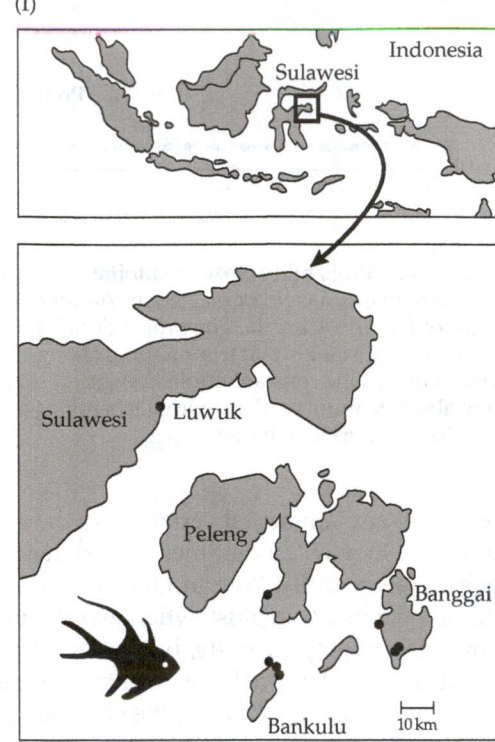

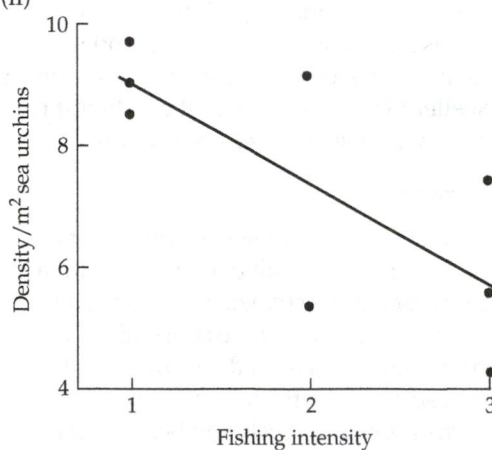

Figure A (I) Map of the Banggai Archipelago of Indonesia, showing study sites as black circles. The Banggai cardinalfish is pictured in the lower left corner. (II) Correlation between Banggai cardinalfish density and the intensity of fishing at eight separate sites in the Banggai Archipelago near Sulawesi, Indonesia. Because the cardinalfish is associated with sea urchins, their density is expressed per m^2 of sea urchins on the sea bed. Intensity of fishing was estimated from interviews with local people. Banggai cardinalfish are caught by nets only, so reductions in density are solely related to fishing pressure, not to reef destruction (Modified from Kolm and Berglund 2003.)

While traditional forms of harvesting can reduce species densities, an issue of greater concern is the use of cyanide to catch the fish. Fishermen typically crush hydrogen cyanide tablets in a squeeze bottle and squirt the resulting liquid into the crevices where the reef fish hide. The cyanide triggers asphyxiation or muscle spasm in fish, stuns them, and makes them easier to catch. These methods have been used since the 1950s and are believed to have quite a large impact on fishes and the coral reefs themselves. Experiments have tested

the response of 10 species of coral to concentrations of hydrogen cyanide lower than those typically used by fishers. Eight of the coral species died immediately, and the other two died within three months. Death occurred because the cyanide disrupted the relationship between the coral and its symbiotic zooxanthellae. Doses as low of 50 mg/L cause death of the zooxanthellae, and one spray (approximately 20 cc) can kill the corals over an area of 5 m^2 within 3–6 months. These impacts are of concern as Southeast Asia has 30% of the world's coral reefs, and these have been declining in quality due to a variety of factors, of which cyanide fishing is but one. Now only 4.7% of Philippine and 6.7% of Indonesian reefs are in perfect condition and quite naturally these are the reefs targeted by collectors. The cyanide also affects fishes themselves. Half the poisoned fish die on the reef, and 40% of those caught alive die before reaching an aquarium (Simpson 2001).

Estimates of the economics of this practice suggest that the profitability of cyanide fishing to the fishermen is about $33,000 per km^2 over 25 years (10% discount rate). The direct losses to fisheries total $40,000 per km^2 and the tourism losses can range between $3,000 and $436,000 per km^2 depending on location (Cesar 1997). The nonuse costs of losing the coral reef are likely to be larger than these direct values.

Solutions?

This is an interesting situation where poor local people are seeking to harvest a local resource to sell it to richer Westerners, but a side effect of this harvest is the loss of another valuable resource. The marketed resource is nonessential to Westerners, (i.e., we could all live without a marine aquarium if we had to), but we like the aquariums, and many conservationists may argue that through keeping aquaria, people's interests in marine conservation could be heightened. So what should we do?

The solution seems to be to permit a sustainable harvest of fishes caught in a nondestructive manner. Attempts to achieve this can be legislative; for example the application of international law like CITES (Convention on the International Trade in Endangered Wild Fauna and Flora) to the relevant fish species could stop all trade in these species. Alternatively, tighter import controls could be implemented in the destination countries (e.g., in North America and Europe) whereby only fish harvested to approved standards

would be permitted entry. While some legislative backing of this nature could be beneficial, in practice it will be very hard to police.

Alternatively, it may be possible to achieve more sustainable harvest by apportioning property rights to different sets of actors. Currently most marine fisheries are open access, and because of this are susceptible to the so-called "tragedy of the commons." In theory the solution to this is to allocate property rights to different groups of people. In the presence of property rights, the fishermen should seek to manage their resource with a long-term perspective, as opposed to the extreme short-term perspective that open access encourages (Ostrom and Schlager 1996). The allocation of property rights to fishermen is on-going around the world; for example, the Chilean government is offering the rights to harvest a benthic gastropod *Concholepas concholepa*, known locally as "loco" (Castilla and Fernandez 1998). This involves allocating certain areas of seabed to defined groups of local people from local communities who then manage them collectively.

While the theory of property rights is simple, recent work on the Chilean experience shows that the allocation of property rights is not a panacea and can lead to considerable social tension between different groups in a community. One important issue in all such efforts relates to policing the agreements, as it is difficult for hard-pressed officials to spot breaches of agreements at sea. In the absence of official policing, disagreements between groups about who can harvest where can become difficult, if not violent. These sorts of issues highlight the need for strong, functional institutions to help coordinate and manage any network of common property resources. While the need for such institutions is clear from theory and practice, their development can be a long and complex process requiring considerable effort from all involved (Ostrom 1990).

A third approach to dealing with these issues is being tested by the Marine Aquarium Council. They aim to set up a certification scheme whereby fish caught in traditional nets would be labeled in some way, which would enable consumers to buy these fish in preference to cyanide-caught individuals. If all consumers preferentially purchased certified fishes, then the advantage of using cyanide would disappear. To get such a certification scheme to

work several issues have to be followed in parallel:

- There needs to be accurate identification of non-cyanide caught fishes near the point of capture. This can be done through testing samples of fishes in export warehouses for cyanide exposure.

- Fisherman need to be informed about the certification scheme and trained to use alternative collecting techniques, such as hand nets.

- Fishes caught without cyanide should be labeled so that fish buyers can choose to support practices that preserve reefs.

- Consumers should be educated about the environmental issues associated with the use of cyanide for catching fish.

Many such certification schemes are currently under development around the globe, particularly for food and forest products, and are easily linked to other management schemes such as the allocation of property rights. In most situations certified products will cost more than non-certified products. This is almost inevitable if, as in the case of aquarium fishes, the damaging fishing techniques are used because they are easier and cheaper than less-damaging techniques. But if consumers can be educated about the certification scheme, then the hope is that they would chose to buy the more expensive, certified products rather than the cheaper alternatives. In this way the combination of an efficient market economy and an educated consumer can bring real benefits to the environment. One of the most positive examples of this comes from Home Depot® stores in the U.S., patronized by millions of people each year. Home Depot® has begun selling and promoting certified forest products, thus vastly increasing their availability and visibility, and due to their enormous market share, reducing the costs.

Unfortunately, research done to date on food products suggests that ultimately price determines consumers' purchasing patterns. Generally only wealthier and more educated consumers are prepared to pay more for certified products. Thus if economic means are to be used to help conservation, there is a need to increase the general level of education and awareness in society—a continuing challenge to all conservation biologists. ■

exploitation revealed by this diagram is that in open access operations, we can expect people to join the exploitation until their individual costs become equal to their profits, at E_C. Exploitation will proceed to this risky breakeven point because people are competing with each other.

The lesson from this analysis is that open-access exploitation leads to much greater rates of exploitation than are most profitable or safest for the long-term survival of the population. This is tragic both for the resource and for consumers, because each individual is catching less than they could if they had fewer competitors. Governments often respond by providing perverse subsidies (Myers 1998; Myers and Kent 2001), leading to lower apparent costs, and hence encouraging further overexploitation (Repetto and Gillis 1988). The capital invested in many activities such as commercial fisheries and logging operations cannot easily be converted to other uses, making it difficult for people to leave the business when times are hard. Naturally, this leads to resistance against restrictions on exploitation rates. In fact, exploitation can have a one-way ratchet effect, with governments providing aid to encourage overexploitation when populations are already low, as well as supporting investment in the activity when times are good, to increase profits. The consequent over-capitalization leads to a one-way trip toward overexploitation.

Discounting

Even when people have exclusive rights to exploit a resource, they may still be tempted to exploit it heavily now, rather than conserve it for the future. This is because we place a higher value on current than on future worth. This is due to economic discounting (see Chapter 5). Discounting can have serious implications for efforts to restrain overexploitation of renewable resources (Clark 1976). For example, suppose you could either let a forest grow for 100 years, when the larger trees will fetch twice their current value, or chop it down now. Which should you do? From a purely economic perspective the most sensible thing might be to cut it now and invest the money because the interest may accumulate faster than the growth rate of the standing timber.

This carries a rather sobering message for conservationists: Even with small discounting rates that are well within reasonable bounds for economics, it may be economically rational to exploit populations to extinction rather than leaving them to provide future returns. This may be especially true for slow-growing species such as whales and hardwood trees, which take a long time to accumulate value. This does not imply that conservationists should roll over in the face of economics, but merely that they will need to consider more than the economic investment value of renewable resources to formulate cogent argues for conservation. As many of the chapters of this book make clear, this is not difficult.

Comparison of Methods for Calculating Sustainable Yields

There are many ways of putting the theory of sustainable exploitation into practice, depending on what people need to know, and how much time and money they can spend getting the information. For example, the International Whaling Commission uses some of the most sophisticated population models in the world to calculate the impacts of hunting on whale populations. Their models need to be robust, in order to stand up to the tremendous pressures exerted by those who are "for" or "against" whaling. Yet, to be used effectively, these models often demand a lot of information about the biology of whales that is not yet available.

Methods for calculating sustainable yields have been reviewed by Hilborn and Walters (1992), Milner-Gulland and Mace (1998), Quinn and Deriso (1999), Jennings et al. (2001), among others. Here we give only the bare minimum needed to understand the logic of each method and the pros and cons of using them.

Surplus production

Surplus production models are also called surplus yield models or simply "production" models. They are most often used in fisheries. The simplest ones require very little data, thanks to an ingenious derivation by Schaefer (1954) who pointed out that if you know how yields have responded to different levels of exploitation effort over time, then you could estimate the dome-shaped yield curve shown in Figure 8.9B. Effort might be measured as number of fishing boats out each day of the year, but in non-fisheries contexts it could also be the number of days spent hunting by all hunters in a given area per season. The corresponding yields might be the thousands of tons of bluefin tuna (*Thunnus maccoyii*) caught by the fishery each year (or number of mallards, *Anas platyrhincos*, shot per year in an area).

While this is a helpful simplification, it has several pitfalls that have led to extremely dangerous outcomes when this method was first applied. First, it treats each year as an independent replicate, which it will not be, due to lags in the ability of the population to respond to different levels of mortality. For example, cod in the North Sea take about four years to reach maturity. Therefore, no matter how much you reduce density dependence by reducing the number of adults in any given year, it will take four years for the survivors to produce offspring that become old enough to be caught. So if you double the fishing effort from one year to the next, you may catch twice as many fish. But obviously this cannot be due to the survivors producing twice as many young, because in that second year, those young fish will only be one year old, and not big enough to be caught. What is really happening is that more fish are caught with a

doubling of effort simply because you are scooping more adults out of the population. You are not learning much about the ability of populations to respond to lower densities, you are simply cutting more deeply into your original stock of fish. The population is decidedly not in equilibrium with the fishery. This dangerous assumption was the downfall of the original production models (Hilborn and Walters 1992). The biggest collapse in the history of fisheries happened to the Peruvian anchovy (*Engraulis ringens*) in the early 1970s when the fishery was operating at a level that a surplus production model suggested to be safe (Boerema and Gulland 1973).

A number of much more sophisticated surplus production models have been developed since the rise and fall in popularity of the original Schaefer version (Polacheck et al. 1993; Schnute and Richards 2001). While these require more information than the elegantly simple early version, they can get around the dangerous equilibrium assumption. In fact, in fisheries they have been shown to sometimes outperform more complex models.

Yield per recruit

Yield-per-recruit models were developed originally as part of the "dynamic pool" concept in the landmark fisheries book written by Beverton and Holt (1957). The dynamic pool refers to models that keep track of separate processes that add to a population, such as recruitment and growth, or that subtract from it, such as natural mortality and fishing mortality. The basic principle can be applied to many other forms of exploitation.

Imagine a species that becomes more valuable as it ages, due to increasing size. Older fish provide more meat and older trees provide more wood. If exploitation is weak, most of the fish or trees will be big when they are taken, which is good. But if we wait too long, natural mortality will have taken its toll and there won't be as many individuals available to us. Yield-per-recruit models search for the level of mortality that maximizes the yield under this tradeoff between numbers and value. Here, a "recruit" is defined as an individual that has become big enough to be captured, not in the demographic sense of an individual that has reached maturity. Once the level of mortality has been found that maximizes the yield per recruit, we can calculate the total yield that will be obtained from a given level of mortality, if we know how many recruits are coming each year. That is how fishing quotas are set in many countries today (Jennings et al. 2001). Yield in this context is measured in biomass (number of fish caught in each age cohort multiplied by that cohort's mean weight).

This method requires information about how the value (often size) of individuals increases with age, as well as an estimate of natural mortality rates, because natural mortality determines the rate at which individuals die before we have a chance to get them. While growth data are often available, natural mortality can be difficult to estimate because scientists are usually late on the scene, with exploitation well underway by the time they start collecting data. This makes it difficult to disentangle natural from human-induced mortality. Yield-per-recruit models are the standard approach used in many temperate fisheries, but developing countries in the Tropics rarely have the resources to collect the growth and recruitment data needed to use this technique as a management tool.

Full demography

For some species there has been sufficient economic value or conservation concern to lead to the production of full-blown population models. These models combine information on vital rates such as births, juvenile survival, age at maturity, and adult survival. These parameters are often analyzed with matrix models in which populations are divided into discrete classes based on their age or stage in the life cycle (reviewed by Caswell 2001, and Lande et al. 2003). The most comprehensive models incorporate stochastic variation in parameters, and quantify uncertainty in the outputs.

Assessment of population growth rates is a common practice to deduce whether populations are declining as a result of direct exploitation, or exploitation of their habitats. For example, Lande (1988) used a model to make a preliminary determination of whether the Northern Spotted Owl (*Strix occidentalis caurina*) was in decline in the northwestern United States. This species was suffering from overexploitation of old-growth forests, prompting potential listing under the U.S. Endangered Species Act. A significant population decline would be indicated by a population growth rate, λ, that is significantly less than 1.0. Lande's model was based on estimates of clutch size, fledgling survival probability, the probability of successful dispersal, and survival of subadults and adults. The estimated population growth rate was 0.961. However, this estimate had a high degree of uncertainty, due to uncertainty in the underlying parameter estimates. When these uncertainties were scaled up through the model, the estimate of λ had a confidence interval of ± 0.0562. Therefore, λ could not be distinguished statistically from a value of 1. Further field research ensued, yielding better parameter estimates, and worse news for the owls, with significant population declines detected in 10 of 11 sites (Forsman et al. 1996).

Other uses of demographic analyses are to diagnose the reasons for population declines, and also to suggest the likelihood of success of various management procedures (Caswell 2001). For example, many populations of sea turtles suffer from egg collecting, hunting of adults, and bycatch of juveniles and adults in fisheries. Should conservationists concentrate on protecting eggs, juveniles or adults? Crouse et al. (1987) examined the effects on population growth rates of enhancing survival of each of

these and other life stages in loggerhead turtles (*Caretta caretta*). They found that protection of eggs was far less important to population demography than protection of older life stages. This and subsequent research (e.g., Crowder et al. 1994, Heppell et al. 1996) contributed strongly to changes in conservation tactics, most notably through the adoption of mandatory turtle excluder devices in trawlers in the United States. Spatial analyses also can improve understanding of what management interventions may be most effective. For example, analyses of the coincidence of turtle sightings and shrimp trawling activities allows specific recommendations for area closures that can reduce bycatch mortality by avoiding larger concentrations of sea turtles in areas of lower economic importance (McDaniel et al. 2000). Finally, continued monitoring coupled with population modeling allows evaluation of how well management efforts are doing, and suggestions for improvements (e.g., increased compliance; Lewison et al. 2003) (Figure 8.14).

The simplest structured population models often ignore density dependence. That is, the probability of surviving long enough to move from one age class (or stage) in the life cycle and eventually reproducing, is kept constant as the population changes in size. As we argued earlier, this assumption may be defendable in some situations, but because the concept of sustainability is founded on density-dependent compensation, some people would argue that it makes little sense to ignore density dependence in exploitation models. A study of the edible palm, *Euterpe edulis*, showed the value of incorporating density dependence explicitly (Freckleton et al. 2003). This edible palm is found in the Brazilian Atlantic forests. Edible "palm hearts" correspond to the apical meristem of this species. Extraction of the apical meristem kills the plant. Freckleton et al. (2003) modeled a population of edible palms divided into seven stages according to size of the plant and its reproductive state. Probabilities of surviving and growing from one stage to the next, as well as annual reproductive output, were calculated from field studies, which showed strong density dependence in these parameters. For example, seedlings (the first stage in the model), compete with each other for light and resources, and are also shaded out by high densities of adults (Silva Matos et al. 1999). Analyses showed that when the correct amount of density dependence in survival and growth from one stage to the next were incorporated, populations were predicted to be able to sustain considerably higher levels of exploitation before they become eradicated.

Adjustments based on recent results

If we can monitor either the population itself, or the numbers of individuals that are taken as a result of various management measures, then we can adjust the quotas each year to take account of new information. For example, in North America hunting licenses for waterfowl and mammals are allocated each year according to the health of the populations. So, if goose populations increase from one year to the next, the hunting season may be extended. Box 8.1 provides an example of this method applied to the setting of quotas for trapping marten in Canada.

Demographic rules of thumb

Parameter-hungry models are useless for the vast majority of the world's exploited species, because we usually lack even the most fundamental information about the biology of the organism and the activities of the people who exploit them (Johannes 1998). The cruel irony is that it is often the species that we know least about that are in most trouble from intentional or accidental exploitation. Most of the tropical vertebrates that are hunted for their meat fall into this category, as do most whales, sea turtles, sharks, and rays. Various demographic rules of thumb have been developed that seek to overcome this problem.

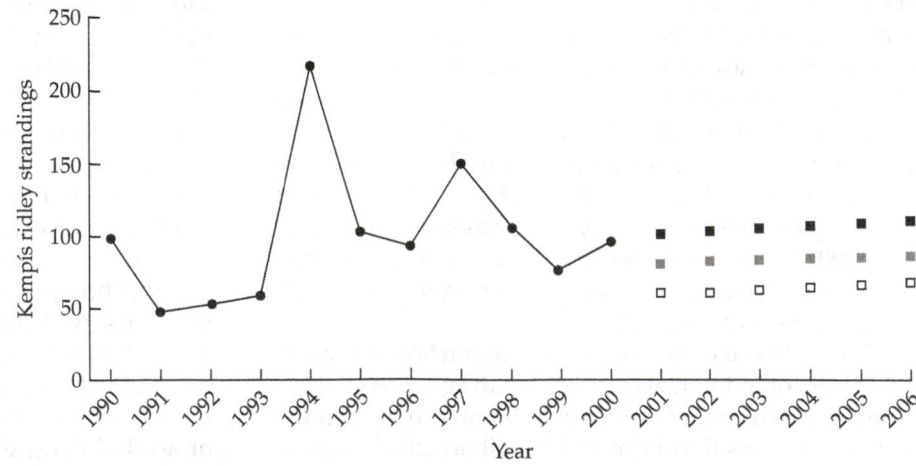

Figure 8.14 Projected annual strandings for Kemp's ridley sea turtles based on three potential levels of compliance by shrimpers in using turtle excluder devices in their nets. Black squares = compliance at 2000 levels, gray squares = 50% improved compliance over 2000 levels, and open squares = full compliance. Fluctuations in strandings reflects variation in compliance and sea turtle abundance in areas where shrimp fisheries are most active. (Modified from Lewison et al. 2003.)

BOX 8.1   Adjustments of Quotas for the Marten (*Martes americana*) According to Previous Results

John Reynolds, *Simon Fraser University*

Commercial trapping of martens in Ontario is regulated by trap-line quotas, which are issued by the Ontario Ministry of Natural Resources. Fryxell et al. (2001) analyzed data from a district in the southern part of the province where trappers were compelled to report how many animals they caught each year. Trappers had also been asked to submit carcasses voluntarily so that biologists could determine the age structure of the animals. Over the period of 1972–1991, 53% of the trapped animals were brought in for age determination. The numbers and ages of animals caught were the only information available to local managers, who used this information to adjust quotas upward or downward each year.

How successful were the local managers in guessing the right quotas according to recent trapping results? To answer that question, Fryxell et al. (2001) used an age-structured model to calculate population sizes and optimal quotas over the time period. They found that the marten population had varied three-fold over the 20 years, and that although the local managers did not have this clear information, they set quotas that matched this variation quite well (Figure A). The proportion of young animals in each year was correlated with the quotas that the managers set the following year (Figure B). This proved to be a very sensible strategy, because the proportion of young trapped was closely correlated with estimates of the annual rate of increase of the population (Figure C). Simulations by Fryxell et al. (2001) suggested that the highest average yield during this period would have been obtained if 36% of the animals had been trapped each year. This was remarkably similar to the average value of 34% that had been set by the local managers!

How did the local managers do so well? Close cooperation between managers and the trappers was essential, as were the strong links between the proportion of young animals trapped

each year, the population growth rate, and the population size. This example shows that good cooperation combined with an ability to predict popu-

lation growth rates can lead to effective conservation management, even in the absence of detailed information.

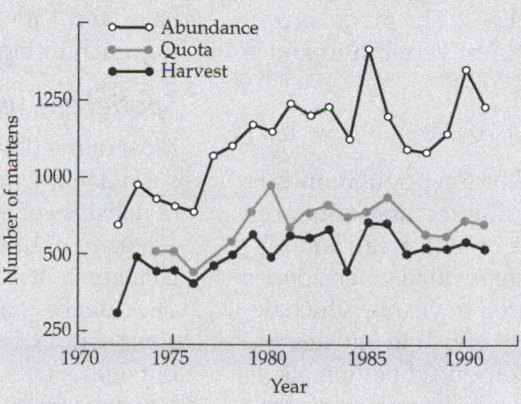

Figure A   Data from commercial trapping of marten, *Martes americana*, in a district in southern Ontario, Canada. Estimated population sizes (open circles) were mirrored by quotas (gray circles) and number of martens harvested (black circles).

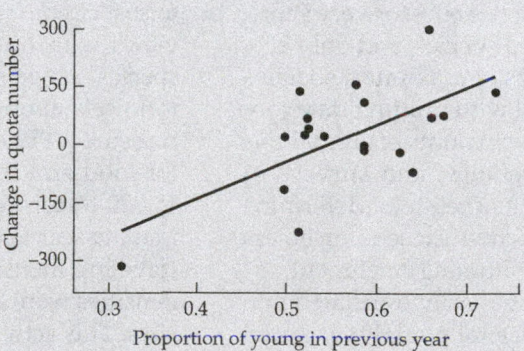

Figure B   Changes made by local managers to quotas were related to the proportion of young that were trapped in the previous year.

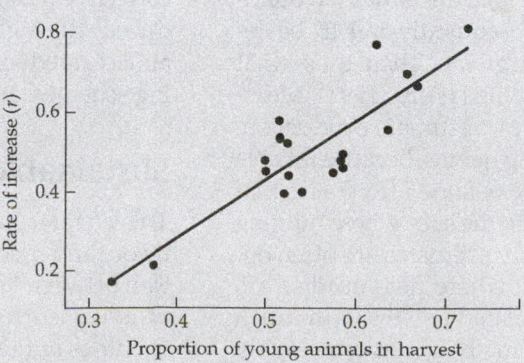

Figure C   The proportion of young trapped was a good predictor of estimates of annual population growth rates, calculated as $ln(N_t / N_s)$, where N_t is the abundance at the start of the trapping season and N_s is the number of survivors from the previous year's trapping season.

One of the best-known rules of thumb has come to be known as the Robinson and Redford model, which was developed as a quick shortcut to assess whether exploitation of tropical mammals can be defined as sustainable (Robinson and Redford 1991). This model simply calculates the total annual sustainable production (population increments through births and immigration), assuming that the maximum potential production (P_{max}) occurs at about 60% of carrying capacity (K). This figure accounts for any density-dependent effects on population growth and is intermediate between the (MSY) at 50% of K, as used in classic logistic models of population growth, and models where MSY is reached at about 70% of K (McCullough 1982). The model is possibly more realistic for tropical forest vertebrates and is expressed as:

$$P_{max} = (0.6D \times \lambda_{max}) - 0.6D \qquad [8.7]$$

In this equation, D is an equilibrium population density estimate near K (or in a nonhunted area similar in structure and composition to the hunted area); and λ_{max} is the maximum finite rate of increase for a given species from time t to time $t + 1$ (measured in years), which depends primarily on the number of breeding females and the annual number of offspring produced per female. In addition, this method calculates maximum quotas for a species by assuming that the proportion of the maximum production that can be taken is 60% for very short-lived species, 40% for short-lived species, and only 20% for long-lived species. Estimates of maximum sustainable offtakes are then compared with hunting data (obtained from household interviews, counts of animal carcasses consumed at forest dwellings, and surveys of wildlife meat sold in urban markets) to determine whether the population production exceeds or is less than the demand within a given hunting catchment.

These data, however, cannot be easily translated into actual cull rates because the collateral mortality induced by hunters may be considerable or even exceed the number of kills that are actually retrieved and consumed. In the Brazilian Amazon, for example, the number of woolly monkeys (*Lagothrix lagotricha*) that are lethally wounded by shotgun pellets, but subsequently fail to be retrieved by hunters, can be far greater than those that actually reach consumer households (Peres 1991). Moreover, many studies are likely to overestimate production, and therefore sustainable hunting quotas, because reliable population density estimates are obtained from sites that are not comparable ecologically to the sites where hunting is taking place. Moreover, density estimates are often obtained from high-density sites, where field studies are most feasible, and then extrapolated to sites with much lower densities (Peres 2000b). Finally, λ_{max} can be highly

variable, is rarely known for any given target species at the sites being evaluated, and can easily be overestimated if the only available data come from populations in captivity or long-term studies in high-quality habitats. Nevertheless, this model provides a preliminary measurement of hunting sustainability in poorly studied systems and has the advantage of simplicity and side-stepping more data-hungry approaches that could not be applied to most game hunting scenarios in tropical forests. There have been a number of applications (e.g., Fitzgibbon et al. 1995, Muchaal and Ngandjui 1999) and criticisms and extensions of the Robinson and Redford model (e.g., Slade et al. 1998, Peres 2000b, Milner-Gulland and Akcakaya 2001, Salas and Kim 2002), but it continues to be popular in tropical hunting studies.

Spatial and temporal comparisons

Most of the time, we have to settle for inferences about sustainability from simple evidence such as comparisons of densities of target species in locations that vary in the intensity of hunting, or interviews with hunters about changes in the abundance of their prey. Often these types of evidence can be combined to yield powerful conclusions about whether extractive activities are sustainable, but without being able to make precise quantifications of benchmarks such as maximum sustainable yields. For example, a study of hunting of Madagascar radiated tortoises, *Geochelone radiata*, used a combination of interviews with hunters, a comparison of changes in the species' range size over time, and comparisons of the tortoise's abundance in sites that differed in hunting pressure (O'Brien et al. 2003). The tortoises are collected for food and the pet trade. This study estimated that 45,000 tortoises are collected each year from southern Madagascar in the catchment of two cities. Hunters are traveling increasingly far to obtain their animals, and densities were highest in the most remote, unexploited sites. This activity is taking place in spite of the species being protected by law. This information, combined with documentation of range contraction, led the authors to conclude that the species is headed for extinction unless corrective measures are taken. Hunting rates could be reduced through education and awareness programs aimed at reducing demand, alternative income-generating schemes, and enhanced legal enforcement.

Sustainable Use Meets Biodiversity

This chapter has repeatedly run into contrasts between theory and practice, goals and implementation. A comparison between "north" and "south" illustrates the diversity of issues surrounding exploitation. In most temperate countries hunting, fishing, and forestry management pro-

tools have been developed through a long history of trial and error based on biological principles. In most tropical countries, exploitation regulations are typically nonexistent or unenforceable. The concepts of game wardens, fisheries officers, bag limits, no-take areas, and hunting licenses are completely unfamiliar to the vast majority of tropical subsistence hunters and fishers who often rely heavily on wild animals as a critical protein component of their diet. In many tropical countries, wildlife is an "invisible" commodity and local offtakes are often not restrained until the stock is depleted. This is reflected in the contrast between carefully regulated and unregulated systems where large numbers of hunters may operate. For example, Minnesota hunters sustainably kill over 700,000 wild white-tailed deer every year, whereas Costa Rica can hardly sustain an annual hunt of a few thousand individuals without pushing the same cervid species, albeit in a very different food environment, to local extinction.

But even well-meaning management prescriptions can be completely misguided, bringing once highly abundant target species to the brink of extinction. The 97% decline of saiga antelopes (from >1 million to <30,000) in the steppes of Russia and Kazakhstan over a 10-year period has been partly attributed to conservationists actively promoting exports of saiga horn to the Chinese traditional medicine market as a substitute for the horn of endangered rhinos (Milner-Gulland et al. 2001). Saiga antelopes were finally placed on the Red List of Threatened Species in October 2002 following this population crash.

Even within cultures, there may be a diversity of perceptions about the objectives of management. For example, in countries that can afford to engage in fisheries management, the standard goal has been to achieve maximum sustainable yields of the target species. In the past 15 years, however, collapses of fish stocks and concerns about by-catches and damage to marine habitats have led to new objectives, such as obtaining sustainable yields while minimizing impacts on ecosystems (Reynolds et al. 2002). Not everyone agrees with these objectives, especially those who perceive that their traditional livelihoods are at risk from reduced quotas. Furthermore, there is also debate within the scientific community about what these objectives mean in practice, and how to implement them. In some sectors, such as forestry, managers are used to dealing with highly polarized debates between those who seek to preserve forests and those who seek to log them. Today, fisheries managers are facing similar scrutiny from many conservationists, who are asking new questions about impacts on habitats and ecosystems. Indeed, the past decade has seen the first listing by the IUCN of commercially exploited fishes as threatened with extinction (e.g., Atlantic cod), and the first cases of exploited marine fishes and mollusks being listed under

the U.S. Endangered Species Act (e.g., smalltooth sawfish, *Pristis pectinata*, and white abalone).

In many instances failure to protect species from overexploitation has more to do with institutional short-comings than lack of scientific knowledge. European and North American fisheries scientists know a tremendous amount about the demography of the major commercially exploited fish species, yet this information has not prevented some species from reaching historically low population levels in the past decade. Of course it would help to know even more about the animals that we are trying to manage, such as impacts of low densities on reproductive behavior, trophic interactions in relation to habitat alterations, and links between environmental variability and recruitment. But translating this information into management action for the long-term benefit of fishers and fishes would still run into the age-old problem of lack of political strength in the face of criticism from those affected negatively by restrictions. In many developing countries, such problems are compounded by the dire lack of alternatives for food and employment. When combined with a lack of institutional structures to implement any policies that may be acceptable, it is easy to understand why exploitation often runs a course that is slowed only by accessibility of the resource.

As with most of the other conservation issues covered in this book, we feel that while the future holds some enormous challenges, it is by no means entirely bleak. For example, the adoption of precautionary principles in exploitation by many countries, combined with greater concern for impacts on ecosystems, is certainly a move in the right direction. We are also encouraged by the large number of imaginative programs in many countries that are aimed at giving local people incentive to make sustainable use of plant and animal populations and their habitats; Elizabeth Bennett et al. discuss some of these programs in Case Study 8.2 and Michelle Pinard et al. discuss them in Case Study 8.3 (see examples in Chapter 16, also see Case Study 6.2). Indeed, there is a growing appreciation for the enormous economic and social value of ecosystem services from wild nature that far exceed the economic benefits of conversion to agriculture or other uses (Costanza et al. 1997; Balmford et al. 2003). For example, a forest can provide far more than just timber, including flood defense, carbon storage and sequestration, and pollination of adjacent crops. Many researchers are now working to find novel ways of rewarding local people for exploiting resources sustainably in return for benefits received by others. These are exciting times for integrating science with the policy of exploitation, and it is imperative that these efforts are successful if we are to meet the challenge of sustaining wild populations and their habitats.

CASE STUDY 8.1

Overexploitation of Highly Vulnerable Species
Rational Management and Restoration of Sharks

Julia Baum, Dalhousie University

"Sharks ... may be considered as one of the richest 'strikes' in the only inexhaustible mine—the sea."

> Barrett, 1928, *Scientific Monthly*

Anthropogenic impacts on natural ecosystems have almost always begun with the overexploitation of large vertebrates. Consequent species collapses or extinctions had markedly transformed terrestrial, freshwater, and coastal ecosystems by the early twentieth century, while oceanic species were still largely protected by their distant location and large geographic ranges. Oceans, it seemed, were too vast and their creatures too plentiful and widespread to be unduly impacted by humans. But in the past few decades, humans have come to dominate these ecosystems as well. Today, industrialized fishing fleets are ubiquitous throughout the world's oceans, reaching even the remotest locations and leaving few refuges for marine species. As a result, overexploitation now threatens the future of many large oceanic vertebrates, including whales, tunas, billfishes, and sea turtles. And sharks, rather than being one of the richest "strikes" in the ocean are one of the most vulnerable (Figure A).

Sharks are among Earth's oldest inhabitants, having evolved over 400 million years ago. Along with skates and rays, sharks comprise the subclass Elasmobranchii within the class Chondrichthyes, distinguished from bony fishes by their cartilaginous skeletons. The world's 350 shark species are a diverse group, ranging in size from dwarf sharks that reach less than 20 cm to the 12 m long whale shark, occupying habitats from freshwater lakes to entire ocean basins, and occurring in arctic and tropical waters from the surface to depths of a thousand meters. Our perception of sharks as ferocious predators belies the fragility of their populations. Sharks tend to grow slowly, mature at a late age, and give birth to few pups following a long gestation period—characteristics more reminiscent of marine mammals than fishes. These traits result in low intrinsic population growth rates, leaving sharks unable to withstand heavy fishing pressure, and with little compensatory ability to recover from over fishing.

Localized shark fisheries that developed in the few decades following Barrett's description were characterized by "boom and bust" patterns. Fisheries for the soupfin shark in California and Australia, dogfish in Scotland, and porbeagle shark in the Northwest Atlantic all collapsed in under a decade.

Despite this history of population collapses and sharks' known vulnerability, their exploitation has intensified globally. Shark fisheries underwent a resurgence in the 1980s, driven both by the decline of traditional food fish species and by increased demand for shark products. Shark meat is increasingly consumed in Western societies, and the high demand for shark-fin soup in Asia has made shark fins one of the most highly valued marine products (Rose 1996). The practice of finning—removing the fins and returning the remaining carcass to the water—has now been made illegal in the Atlantic, but a worldwide ban is needed. Besides these directed fisheries, many sharks are caught incidentally, in fisheries targeting other species. In particular, pelagic longline fisheries, which target swordfish and tunas throughout temperate and tropical oceans, are the world's largest source of shark by-catch (Bonfil 1994).

Quantifying the impact of current exploitation levels on shark species has proven difficult. Sharks have typically been a low research and management priority because of their historically low economic value and because their catches are often incidental. As a result, basic population parameters such as age at maturity and longevity are unknown for many species, and shark catches have been poorly monitored. Fisheries observers usually monitor only a fraction of the overall fishing effort—perhaps 5% in domestic waters, and even less in international waters—and until recently most fisheries recorded shark by-catch in a generic "shark" category rather than identifying individual species. Thus, many shark assessments are done on aggregated species groups, obfuscating any variation in species' responses

Figure A   Scalloped hammerhead sharks (*Sphyrna lewini*) swimming at Cocos Island, Costa Rica. (Photograph © Phillip Colla www.oceanlight.com.)

to exploitation. Compounding these problems are the wide ranges of many sharks, which commonly cross international boundaries, and even entire oceans, imposing serious constraints on our ability to monitor their populations.

New research provides strong evidence that coastal and oceanic shark populations have undergone rapid, large declines in the Northwest Atlantic (Baum et al. 2003). To demonstrate this, researchers analyzed the largest dataset sampling the Northwest Atlantic pelagic ecosystem, that of the U.S. pelagic longline fishery. This type of gear consists of mainlines tens of miles long suspended horizontally in pelagic waters (upper water column) by floatlines and buoys, with baited hooks attached on branchlines at set intervals. Thus it resembles a transect through the pelagic ecosystem. The fishery, which operates from Newfoundland into the Gulf of Mexico and south to Brazil, and includes both coastal and distant offshore waters, covers the entire range of Northwest Atlantic coastal shark populations and a substantial proportion of oceanic shark populations. Because fishers were required to record shark catches from each longline set and because their fishing effort was intense, researchers could estimate changes in abundance even for rare shark species: Between 1986, when fishers first started recording some sharks to species, and 2000, the dataset comprises over 200,000 longline sets and 117 million hooks! Researchers used statistical models to standardize catch rates for variations in the gear, location, and timing of sets. Estimated mean catch rates can then be compared across years as an index of a species' relative abundance.

All shark populations examined in the study are estimated to have declined by 40%–89% since the mid-1980s (Baum et al. 2003). Researchers estimate that since 1986 hammerhead sharks have declined by 89%, great white sharks by 79%, thresher sharks by 80%, and tiger sharks by 65% (Figure B). The remaining coastal sharks (CST), examined together because of the potential for misidentification in the logbooks of these similar looking congenerics, declined by 61% just since 1992. Blue and mako sharks are estimated to have declined by 60% and 40% respectively since 1986; oceanic whitetip sharks by 70% since 1992 (see Figure B). These latter species range across the entire Atlantic. Thus, while the data do not cover their entire populations, because other longline fisheries exert intense fishing pressure across the Atlantic, it is quite plausible that the pattern found in the Northwest Atlantic is representative of the entire region. Despite the enormity of documented declines, they are likely underestimates because they do not account for changes that occurred during the first three decades of offshore shark exploitation in the Northwest Atlantic (i.e., from 1957 to 1985).

Halting shark population declines is necessary if we are to avoid species extinctions; reversing them is essential if we are to restore populations to their former levels. Restoration requires that we understand the composition and abundance of natural shark assemblages, that is, that we have some knowledge of what it is we want to restore. Jackson (2001) notes that because most ecological studies began years, and sometimes

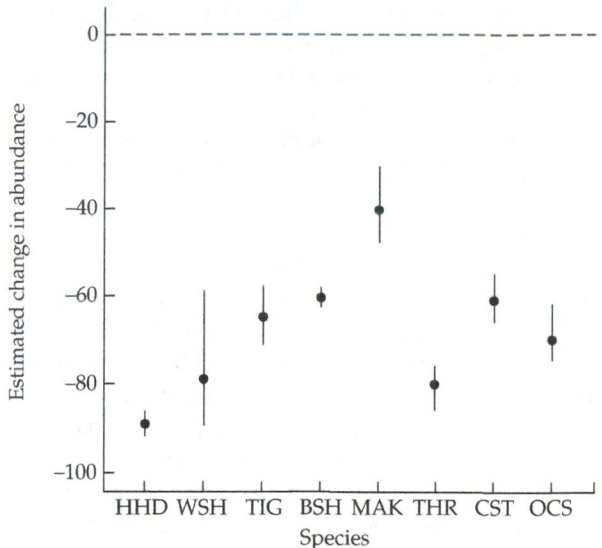

Figure B Total estimated change in relative abundance (± 95% confidence interval) for scalloped, great and smooth hammerhead (HHD), great white (WSH), tiger (TIG), blue (BSH), shortfin and longfin mako (MAK), common and bigeye thresher (THR) sharks between 1986 and 1999, and for bignose, blacktip, dusky, night, silky, and spinner (CST) and oceanic whitetip (OCS) sharks between 1992 and 1999.

centuries, after anthropogenic influences began, the former natural abundance of many large vertebrates in coastal ecosystems was fantastically greater than today. Assessments based on recent data alone obscure the original abundance of species, such that we may become complacent about species that are now rare. This lack of historical perspective on what is natural is coined the "shifting baseline" (Pauly 1995).

Estimating the baseline abundance of pelagic sharks in the Northwest Atlantic is possible because of the short history of disturbance in the open ocean compared to most other ecosystems: We need only look back to the early development of industrial offshore fisheries in the late 1950s. In the Gulf of Mexico, research surveys were conducted in the mid-1950s using pelagic longlines, to determine the potential for a commercial tuna fishery (Wathne 1959). These surveys provided some of the first biological information available on pelagic sharks, and on the region's offshore resources. Sharks were so abundant on these surveys that they damaged a high proportion of tuna on the longlines and were considered a serious problem. Researchers in the 1950s documented that between 2 and 25 oceanic whitetip sharks were usually seen around their vessel during gear retrieval (Wathne 1959). Oceanic whitetip and silky sharks were the second and fourth most commonly caught fishes overall on the pelagic longline sets, comprising almost 15% of the total catch, and 85% of all shark catches. Matching these data to comparable information from the current pelagic longline fishery reveals a very different picture. By the late 1990s these two species accounted for less than 0.5% of the total catch, and only 16% of shark catches, suggesting that substantial

changes have occurred in the Gulf of Mexico's shark assemblage in less than half a century. Recent research papers on sharks have either not mentioned the oceanic whitetip shark or have dismissed it as a rare exception in the Gulf of Mexico, with no recognition of its former prevalence in the ecosystem. In contrast, a new analysis that incorporates these historic data estimates that oceanic whitetip sharks have declined by over 99% in 50 years, suggesting that this species is ecologically extinct in the Gulf of Mexico (Baum and Myers 2004).

Beyond the risk of species extirpations in the Northwest Atlantic, estimated shark declines may indicate broad ecosystem changes. Consumers such as sharks usually exert important controls on food web structure, diversity, and ecosystem functioning. The generality of the loss of pelagic consumers in the Northwest Atlantic is particularly worrisome: Most pelagic shark populations appear to be remnants of their natural abundance, and many other apex predators, including the swordfish and marlin, are drastically reduced. Although the ecosystem impacts of overexploitation in the open ocean remain largely unexplored, any changes are likely to be massive in scale given the vastness of these ecosystems, and are likely to be difficult to reverse given the life history of sharks.

The estimated loss of pelagic sharks in the Northwest Atlantic may also reflect a common global pattern of decline in shark species. Longlines and other pelagic fisheries are pervasive in all oceans, catching many of the same shark species as in the Northwest Atlantic, turning the risk of extirpation in that region into the risk of global extinction. Apart from the species discussed here, most other sharks face similar exploitation pressures and

monitoring constraints. Around the world, most sharks are caught in multispecies fisheries, be it pelagic longlines, bottom trawls targeting other fishes, or directed shark fisheries that catch several species. This is problematic because the pursuit of productive target species will continue to drive the fishery, even after less-productive shark populations have collapsed. Finally, because of the difficulty in adequately sampling the large geographic ranges of shark populations, shark collapses and local extinctions may occur unnoticed. Local extinctions of two other elasmobranch species, the common skate (*Raja batis*) and the barndoor skate (*Dipturus laevis*), for example, were only documented after the fact (Brander 1981; Casey and Myers 1998).

Restoration of shark populations will require fundamental changes to their management. If we are to avoid extinctions of shark populations, fisheries need to be managed for the viability of the most vulnerable, rather than the most productive, species. This means that substantial reductions in fishing pressure in most fisheries that catch sharks are needed. Recovery targets that incorporate a historical perspective are needed, and management must work toward these. Effective international management plans need to be implemented and enforced to account for sharks' wide geographic ranges. Finally, as individuals we need to make responsible choices about the fish we consume by becoming informed about their status and the effect of fisheries on incidentally caught species.

Sharks have existed for over 400 million years, but have been threatened with extinction in less than 50. We may now be a critical juncture—where the decisions made today will decide the ultimate fate of these species.

CASE STUDY 8.2

The Bushmeat Crisis
Approaches for Conservation

Richard G. Ruggiero, U.S. Fish and Wildlife Service; Heather E. Eves, Natalie D. Bailey, and Andrew Tobiason, Bushmeat Crisis Task Force; Elizabeth L. Bennett, Wildlife Conservation Society

In Africa, forest is often referred to as "the bush," thus wildlife and the meat derived from it is referred to as "**bushmeat.**" Bushmeat is eaten for subsistence, and is also traded in local and regional markets to supply protein to people in cities and towns. The bushmeat trade in Africa is a multi-billion dollar a year industry involving millions of animals from cane rats to elephants (Wilkie and Carpenter 1999; Fa et al. 2002), and millions of hunters, traders, market sellers, and consumers (Figure A, Mendelson et al. 2003). Scientific studies show that the majority of this unregulated, commercial hunting for wild meat is unsustainable and threatens biodiversity conservation goals

(Robinson and Bennett 2000). Recent estimates of hunting rates in the Congo Basin show that a majority of species are hunted unsustainably (Fa et al. 2002). The expansion of bushmeat hunting poses a risk not only to the viability of wildlife species, but also to the livelihoods of the people who share the land with them (Figure B; Milner-Gulland et al. 2003), as well as to human, wildlife, and livestock health (Hardin and Auzel 2001; Wolfe et al. 2004).

Recommendations for promoting and regulating the bushmeat trade as a means to support rural livelihoods (Brown 2003) are generally not supported by current management ca-

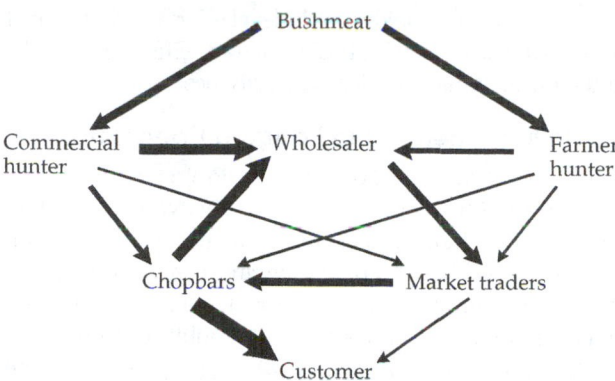

Figure A The cycle of the bushmeat trade.

pacity, policy, ecological, or economic realities (Fa et al. 2003; Wilkie et al. in press). While some suggest that a sustainable trade in wildlife for bushmeat can exist (Cowlishaw et al. 2005), these claims are generally based on species with high reproductive potential, such as rodents, in degraded habitats (Robinson and Bennett 2004). Such scenarios generally emerge after all the large-bodied mammals have already been hunted out (Bennett et al. 2002). Long-distance commercial trade in wildlife from tropical forests is almost inevitably unsustainable, and under the very limited circumstances in which it might be biologically feasible, strong management capacity must be in place. Such capacity is lacking across central Africa.

Robinson and Bennett (2000) document numerous root causes of the bushmeat crisis that must be addressed in efforts to conserve wildlife. Expanding human populations increase demand for bushmeat, while poverty and food insecurity create increased reliance on natural resources. Private companies in extractive industries usually do not engage in wildlife management plan-

ning, thereby facilitating increased access and unsustainable exploitation of wildlife resources. Governments and other land managers often have low capacity for monitoring and enforcement, or have poorly designed or implemented wildlife policies. Rural and urban human populations frequently have inaccurate perceptions of the limitlessness of wildlife and a general lack of awareness of wildlife use impacts. The poor, in particular, lack protein and income alternatives to bushmeat. These driving causes, compounded with the unsustainable commercial wildlife hunting levels impacting populations all across Africa have resulted in a crisis which threatens massive population declines; particularly vulnerable are the forest species which are an order of magnitude less productive than grassland species.

Solutions to the bushmeat crisis must simultaneously address these causes while supporting both undisturbed core areas and some regulated-use areas. Such plans may assure that wildlife and biodiversity are conserved for future generations and that the livelihoods of marginalized rural peoples are supported (Milner-Gulland et al. 2003). Protected areas are an essential component to any landscape management plan. Biodiversity conservation goals and alternatives for income must be coupled with appropriate wildlife regulations and enforcement to limit hunting to local areas and to nonendangered species with high reproductive rates (Eves and Ruggiero 2000).

Solutions include policy development, capacity building for education and enforcement personnel, development of both protein and income-generating alternatives, incorporating wildlife management planning into economic development and extractive industry activities (especially logging, mining, and road-building), and collaborative information sharing and management for development of best practices, adaptive management, and long-term monitoring. Such solutions necessitate the commitment and collaboration of government, nongovernmental organizations (NGOs), the private sector, and communities to support essential wildlife management planning activities that include governance, enforcement, and community participation to support biodiversity protection (Ruggiero 1998).

Progress in many of these areas is now being made. While the bushmeat issue was not well-known or understood before the late 1990s, it has since become a central issue for many conservation projects and international efforts (Bushmeat Crisis Task Force 2004). The Bushmeat Crisis Task Force (BCTF) was created in 1999 as the first coordinated effort by North American wildlife conservation groups and professionals from a range of disciplines to focus efforts on the growing unsustainable, illegal, and commercial trade in bushmeat in Africa and around the world. BCTF has prioritized four areas of engagement for this collaborative action: information management; engaging with key

Figure B Duikers on the back of a logging truck in Gabon. (Photograph by R. Ruggiero.)

decision-makers; formal education/training program development; and public awareness. Professionals from conservation, government, industry and development fields work with the BCTF staff and steering committee to identify and analyze available information, develop consensus on solutions, produce resources, and identify areas of collaboration potential. Major BCTF initiatives include working with key decision makers (KDM) in Africa and the U.S. toward policy development and mobilization of resources; the Bushmeat Education Resource Guide (BERG), which supports public education by zoological parks and other institutions; the Bushmeat Promise, a campaign designed to engage the public toward action; and the Bushmeat Information Management and Analysis Project (IMAP), which provides geo-referenced BCTF project and publication databases online, assembles bushmeat-relevant datasets from around the world, and gathers new data on relevant information in the Congo Basin. For more information about the bushmeat crisis and what is being done to address it, visit the Bushmeat Crisis Task Force website: www.bushmeat.org.

Since the establishment of BCTF and other bushmeat efforts, public awareness in the U.S. and Europe has been raised considerably. Only a handful of media articles were produced prior to 2000, but more than 750 were produced in the five years that followed (Bushmeat Crisis Task Force 2004). A number of awareness programs in urban and rural Africa communicate the impacts of the bushmeat trade on wildlife and human communities that enable consumers to make better-informed decisions (Obadia et al. 2002). Formal education and training tools are also developed in Africa and the U.S. to further educate key decision makers, field personnel, and the public (Bushmeat Crisis Task Force 2004; Bailey and Groff 2003).

At the international policy level, the World Conservation Union (IUCN) adopted an official bushmeat resolution in 2000, the Convention on International Trade in Endangered Species of Wild Fauna and Flora (CITES) created a bushmeat working group in 2000, the Africa Law Enforcement and Governance (AFLEG) treaty signed in 2003 by both timber producer and consumer nations identified bushmeat as a priority concern, and the Convention on Biological Diversity (CBD) established a working group to review the issues related to nontimber forest products, including bushmeat.

Further evidence of an increased focus on addressing the unsustainable bushmeat trade is the Congo Basin Forest Partnership (CBFP). Officially established at the World Summit on Sustainable Development in September 2002, CBFP is an association of 29 governments and international NGOs collaborating to promote biodiversity conservation and improved livelihoods. CBFP partner organizations and their projects employ aspects of the bushmeat trade as indicators of the success of these landscape-level management efforts. Three leading success indicators are directly related to bushmeat: wildlife management planning and enforcement, protein and income-generating alternatives, and training and awareness to enable behavior change.

While numerous projects focus on one or two focal aspects of the bushmeat trade, there are a few that stand out as examples of the multi-stakeholder, multi-level effort that is essential to address the unsustainable bushmeat trade effectively, amidst a landscape of human development activities.

Wildlife Management in Logging Concessions

One example of such collaboration is the Project for Ecosystem Management of the Peripheral Zone of the Nouabalé-Ndoki National Park (PROGEPP), a wildlife management program based in the Kabo and Pokola logging concessions in the northern Republic of Congo. This project, initiated in 1999, is a joint effort of the Government of the Republic of Congo, the *Congolaise Industrielle des Bois* (CIB) logging company, the Wildlife Conservation Society (WCS), and the communities associated with the concessions. Aside from the government, the CIB logging company is the single largest employer in Congo and their logging concessions have acted as a magnet for immigration. The population of Pokola, the largest town near the concession, has more than doubled since 1999, from 7,200 to 16,000. Many would hunt bushmeat if not controlled and if other food options were unavailable. Instead, as a result of the project, wildlife populations are thriving, including gorillas, chimpanzees, elephants, and bongo, and people are maintaining a healthy mixed, protein-rich diet. The key to this success is conducting a complex of activities that include wildlife management planning, training, monitoring, education, and enforcement while simultaneously developing alternative sources of income and protein-rich foods.

Elkan and Elkan (2002) provide an overview of PROGEPP including three key rules adopted with regard to hunting in the concession—no snare hunting, no hunting of legally protected species, and no export of bushmeat outside the immediate community—a model established earlier with local communities associated with the adjacent Nouabalé-Ndoki National Park. The rules are made known to all workers and their families, especially truck drivers who often serve as intermediaries for illegal trade, as well as communities not affiliated with CIB, and are enforced by "eco-guards" at numerous checkpoints. In addition, conservation and land-use zoning was created according to traditional community zones and natural resource use. No-hunting areas, community hunting zones, and buffer zones around the Nouabalé-Ndoki National Park were adopted and established in the logging concessions.

CIB, like any large company working in an area of limited government capacity is in a position to dramatically influence how land and resources are being managed beyond their mandated exploitation, and fulfills many public services for its employees and nearby communities. As a complement to the social, health, and education programs CIB already provides, the wildlife management plan includes company provision of affordable livestock and protein-rich food alternatives.

Since the project's inception, snare rates are significantly reduced, endangered species are found even in areas where regulated hunting is taking place, and protein alternatives are being incorporated into the culture of the logging concession (Elkan and Elkan 2002). In addition, national policies regard-

ing the provision and training of eco-guards within all logging concessions in Congo have been adopted.

Conservation is Good Business

The information emerging from PROGEPP identifies wildlife management planning and enforcement, partnership development, and provision of protein and income-generating alternatives as keys to bushmeat management success. Similarly, a project in Zambia involving an agricultural cooperative is showing signs of conservation success based on these same priority actions (Lewis 2004). The basic premise for this conservation program is that "food secure, farm-based communities with alternative sources of income to illegal use of wildlife can contribute positively to wildlife production" (Lewis 2004). The Community Markets for Conservation and Rural Livelihoods (COMACO) efforts developed by WCS in collaboration with the Zambia Wildlife Authority (ZAWA) and the Ministry of Tourism, Environment and Natural Resources (MTENR) are designed to directly address the causes that drive overexploitation of wildlife for bushmeat in this region—food insecurity, need for income, poor land-use practices, and access to markets.

The COMACO project is innovative among conservation programs as it actively engages individuals and communities in a business cooperative and has been very effective in coordinating the support of the private sector. Producer depots supply a central trading post and households and communities can belong to the cooperative as long as they make a commitment to improving land-use practices and supporting conservation. That is, families must agree to a specific land-use management plan and producer-group conservation by-laws before being accepted in the program.

The COMACO project in 2003–2004 had over 750 conservation farming groups supplying depots with rice, chickens, groundnuts, and honey. The cooperative negotiates higher prices for these products, and in turn farming groups turn in hunting materials; over 30,000 snares and nearly 500 guns were collected during one three-year period. This model is most applicable to subsistence and cash-crop farmers who have enough food but not enough income, live at a medium density, and produce conservation-friendly products (Lewis 2004).

Critical to the entire initiative is the linked focus on wildlife protection. Experience across the entire tropical globe shows that providing protein and income alternatives alone do not reduce hunting. COMACO overcomes this by linking the cooperative with increased protection of wildlife; all cooperative members are also involved in wildlife enforcement, and their doing so is a prerequisite of their eligibility for the program. Hence, they are both benefiting from improved livelihoods while also protecting their local wildlife.

While these examples are promising it is important to note that such systems are site specific, require trained and committed experts in wildlife, community development, enforcement, and education and demand long-term commitment of personnel, financial resources, monitoring, and evaluation. Application of similar programs without such resources and capacity in place are unlikely to succeed. Caution should be used in responding to plans that support broad-scale policies of regulated trade in bushmeat as they are unlikely to lead to achieving either biodiversity conservation or development goals in the long, or even in the short-term. Quite the opposite, they are more likely to result in further impoverishment of rural communities and degradation of the natural resources upon which they depend. What is required is a global commitment and response that simultaneously and sufficiently supports nations that have identified the bushmeat problem as a threat to their natural and cultural heritage so that locally-designed efforts such as those described have the opportunity to succeed. This will include sufficient resources to enable training, enforcement, development of protein and economic alternatives, landscape-level management that includes protected area as well as multiple-use zones, and awareness programs for the public.

CASE STUDY 8.3

Managing Natural Tropical Forests for Timber

Experiences, Challenges, and Opportunities

Michelle A. Pinard, University of Aberdeen; Manuel R. Guariguata, Environmental Affairs, Convention on Biological Diversity; Francis E. Putz, University of Florida; Diego Pérez-Salicrup, Universidad Nacional Autónoma de México

Applied forest ecologists, or silviculturalists, find managing the immense diversity of natural tropical forests for various goods and services a challenge, not only because of the complexity of these forests but because the requirements for sustainability are also diverse and complex. While the diversity may often seem overwhelming, natural forests in the moist tropics have many of the attributes that one might consider conducive to management. For example, they are productive in terms of the diversity of goods and services, and are known to be highly resilient; many areas that currently support old-growth tropical forest

are the result of secondary forest succession after agricultural fields were abandoned hundreds of years ago (e.g., Darien, Panama; Bush and Colinvaux 1994). Despite the opportunities represented by these attributes, examples of good management are few. There is huge variation in forestry practices across the tropics, with some countries and forest owners moving toward more sustainable practices, while others continue to convert or degrade forest, often under the pretense of timber management.

Our aim is to explore some of the variation among forest management practices and to identify what seems to be working to improve management. We also discuss what we see as the main challenges and opportunities for sustainable management of natural tropical forests. Although we focus on the ecological, economic, and social bases of sustainable forest management for timber, our motivation is principally forest conservation. It may seem strange to be trying to save tropical forests with chainsaws and bulldozers, but the majority of the tropical forests that remain are only likely to be saved if they can be managed profitably for timber. In many cases, if people cannot make money from forests, then the forests will likely be destroyed in favor of some more lucrative land use, such as agriculture.

Figure A Selective logging typical involves cutting a few valuable species, and extracting them via skidders or draft animals. (Photograph by M. Pinard).

tion (nominally, the time it takes for a tree to grow to maturity). In theory, cutting cycle lengths and harvesting intensities are determined on the basis of the rate of commercial timber productivity. Harvesting also can be done more strategically, as a silvicultural treatment to promote the regeneration of ecologically similar groups of trees. For example, light-demanding tree species are favored by intensive stand interventions, whereas if less timber is harvested per unit area, more shade-tolerant species benefit. Unfortunately, the temptation of marketable standing timber often overwhelms efforts to secure sustained yields. As a result, timber-mining operations favor both slow-growing, shade-tolerant species, and fast-growing pioneers, not the species with marketable timber. Moreover, given the general lack of reliable data on forest productivity, managers establish cutting cycles by making educated guesses at how fast trees grow.

The deleterious environmental impacts of logging can be minimized if well trained and closely supervised crews fell and extract timber following a detailed harvest plan. In forests from Borneo to Brazil, loggers participating in research projects on "reduced-impact logging" have demonstrated that good logging practices are technically fea-

Background and Context for Tropical Forest Management

Because of the high diversity of tropical trees and the concomitant low population densities of marketable species, logging in natural forests in the tropics is typically selective; a relatively small number of trees above a threshold size (the minimum felling diameter) are felled and extracted. The logs are usually skidded out of the forest using bulldozers, rubber-tired skidders, farm tractors, or draft animals (Figure A). Where machines and draft animals are not available, large logs are often sawn into thick boards that are manually carried out of the forest.

Logging impacts on the residual forest vary with harvest intensity, level of pre-harvest planning and control, soil type and terrain, and forest stature and structure, among other factors (Figure B). Due to the selective nature of harvesting, forests are repeatedly logged after intervals of a few years to a few decades. Where logging is scheduled so as to achieve a sustained yield of timber, this approach to forest management is referred to as a polycyclic silvicultural system. Where properly designed, several harvesting episodes (or cuts) are scheduled within a rota-

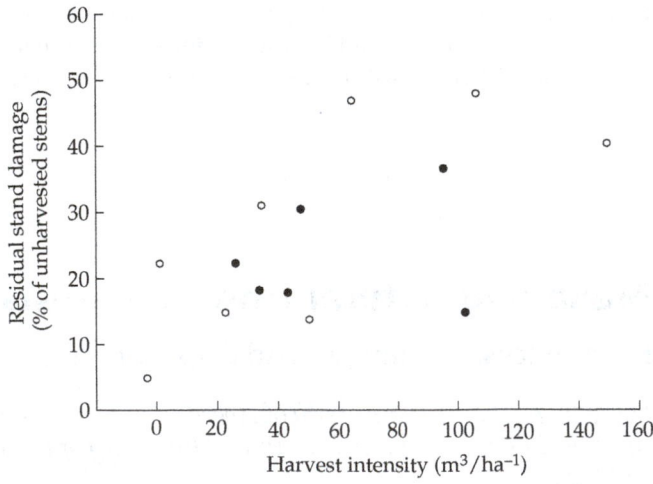

Figure B Variability in harvest intensities (m^3/ha^{-1}) and incidental damage to the residual stand (percentage of unharvested trees) across the tropics. Open symbols are areas in which harvesting was planned and controlled; closed symbols are areas in which harvesting was carried out in a conventional way (typical for the region, generally with little planning and control over operations). (Modified from Pinard et al. 2005.)

sible and that their application is financially viable (e.g., Holmes et al. 2002). Logging damage to the residual stand can be reduced by up to 50% when trees are felled in predetermined directions and logs are skidded out of the forest along pre-planned pathways (e.g., Pinard et al. 2000). Unfortunately, due to unsupportive forest policies, industry resistance to change, corruption, and the massive profits from exploitative logging, reduced-impact logging methods are seldom fully used (Putz et al. 2000).

The socioeconomic and political contexts in which forests are exploitatively logged or managed for the sustained yield of timber vary tremendously across the tropics. Most tropical forest-rich countries are poor in terms of most economic measures, many stagger under huge international debts to development banks, and some are poorly governed. The expertise needed to manage forests is often scarce even in countries with governments committed to maintaining a permanent forest estate. Furthermore, corruption can seriously impede the best-intended efforts at management. Making matters more complicated is the fact that forests may be owned by individuals, communities, corporations, or governments that grant concessions to industrial corporations and, all too often, claims of ownership overlap or are otherwise contested. Equally worrisome is the fact that although most of the large forest tracts that remain in the tropics are currently in remote locations far from markets, accessibility is rapidly increasing as major road-building efforts push back forest frontiers. It also should be noted that many forests planned for forest management are actually inhabited by people who are not involved in management planning and other decision-making processes. These social, economic, and political challenges are coupled with the biophysical challenges of managing little-known forests under what are often adverse environmental conditions.

High Financial Opportunity Costs of Maintaining Forests

Anyone promoting sustainable forest management must grapple with the high financial opportunity costs of managing forests relative to the profits from extracting all of the marketable trees and converting the forest to some more lucrative land use. By "opportunity costs," economists refer to the cost associated with giving up or postponing an opportunity, such as the chance to invest in alternative income-generating activities. This challenge is particularly powerful in many tropical countries where industrial agriculture, cattle ranching, and predatory logging pay so much better than maintaining forest under sustainable management (e.g., Pearce et al. 1999; Southgate 1998). Due to the "time value of money" (future profits and future costs are "discounted" in terms of current value, see Chapter 5), even unsustainable and eventually very costly practices can be financially very attractive. Market failures, particularly the serious undervaluing of forest goods and services, and policy failures, such as the provision of subsidies for unsustainable agricultural practices, further undermine efforts

to promote sustainable forest management (Richards 1999).

Examples of how high opportunity costs interact with policy failures to promote deforestation in the tropics could be cited from almost anywhere, but the liquidation of forests in the 1980s in Costa Rica is especially illustrative. Costa Rican forest management is conducted on a strictly private basis. There are no government concessions, nor does the government directly manage production forests. Up until recently, the opportunity costs of managing forests for sustained timber production were apparently too high to make the option attractive (Kishor and Constantino 1993). In addition, Costa Rica's relatively young, volcanically-derived soils can support pasture for many more years than those established in less fertile locations such as in eastern Amazonia. Up until 1969, when the first modern forestry law was passed in Costa Rica, forests were legally considered impediments to land conversion for pasture, agriculture, and development activities. During the 1970s and 1980s, forest conversion was fuelled by government subsidies for cattle production, coupled with excessive regulation of private landowners that caused them to lose interest in managing forests on a sustainable basis. As a result, annual deforestation rates and consequent rates of forest fragmentation were very high (Sanchez-Azofeifa et al. 2001). The driving forces for forest conversion in Malaysia and Mexico were different from those in Costa Rica, but ended up with the same massive scale of deforestation.

Almost all Malaysian forests were claimed by the government and let out to logging concessionaires, but after logging, many were converted into oil palm plantations. Societal financial benefits from logging, while a great deal less than they might have been due to corruption and mismanagement, fuelled national development. In Malaysia today, slash-and-burn agriculture is rare and many people live in cities and work in high-tech industries but the biological costs of this development were immense.

In Mexico, the extensive deforestation that occurred in the 1960s and 1970s was an indirect result of the "green revolution," a global initiative to terminate hunger. The government implemented a "national deforestation plan" in which small-scale farmers were given the right to lands they converted to agriculture. The objectives of the program were to produce more agricultural goods and to relieve political pressure from landless farmers. Of course, the cost of such program was paid by society at large, as the value of tropical forests was simply ignored (Dirzo and García 1991; Masera et al. 1997; Turner et al. 2001).

Pressures to Overexploit

As countries rich in forests develop economically, the growth of capacity in wood-processing industries often exceeds the sustainable rate of supply of raw materials. Typically, where industries initially thrive on a large supply of timber from forest conversion and first cuts in heavily stocked, old-growth forests, with time, forests available for conversion decline and

the industries need to rely on the more modest harvests from the permanent forest estate (Southgate 1998). This type of "boom and bust" cycle has occurred in many different parts of the tropics. For example, the government of Ghana is struggling to impose an annual allowable cut of only 1 million m^3, which is based on the best available growth and yield data but is far below the industry's capacity (Adam 2003). As in many other places, Ghana's timber-processing industry has great political power and is applying pressure on the government to increase the allowable annual cut beyond the maximum sustainable yield (Adam 2003). Any further increases in harvesting could only be achieved by reducing the lengths of cutting cycles or by increasing the intensity of each harvest by selecting more species or more trees of the favored species by reducing minimum felling diameters. Available data on standing volume, forest productivity, and ecological impacts suggest that these options are not ecologically viable in even the short term and are not economically viable in the long term.

Insecurity of Land Tenure

Among the problems plaguing tropical countries, property rights issues figure prominently. Because forest management is a long-term endeavor, insecurity of land tenure, or at least guaranteed access to the resources the forest provides, is a necessary prerequisite for making investments in sustainable forestry. The failure of many tropical governments to assure landowners and forest concession holders that they will continue to control their property forever, or at least receive appropriate compensation if it is taken away from them, is a powerful justification for treating forests as if their natural resources are not renewable. But while security of tenure seems like an important prerequisite for sustainable management, it is not in itself sufficient to assure that forests will be treated as if the future matters. Under some conditions, providing secure tenure can promote deforestation if forest owners use their property as collateral for bank loans to buy cattle or chainsaws.

Marketing and Managing for More Species

Although technical challenges for tropical silviculture (e.g., timber harvesting and stand tending) tend to be secondary to socioeconomic and political problems, applied forest ecologists still have important roles to play. These roles are expanding as conservation and environmental service provisions are included more explicitly in management priorities. Furthermore, as markets begin to accept formerly noncommercial timber species, studies on the population biology and physiological ecology of these species become critical.

Tropical silviculture has its roots in the management of a few high value Asian and African species such as teak (*Tectona grandis*) and the African mahoganies (*Khaya* spp., *Entan-*

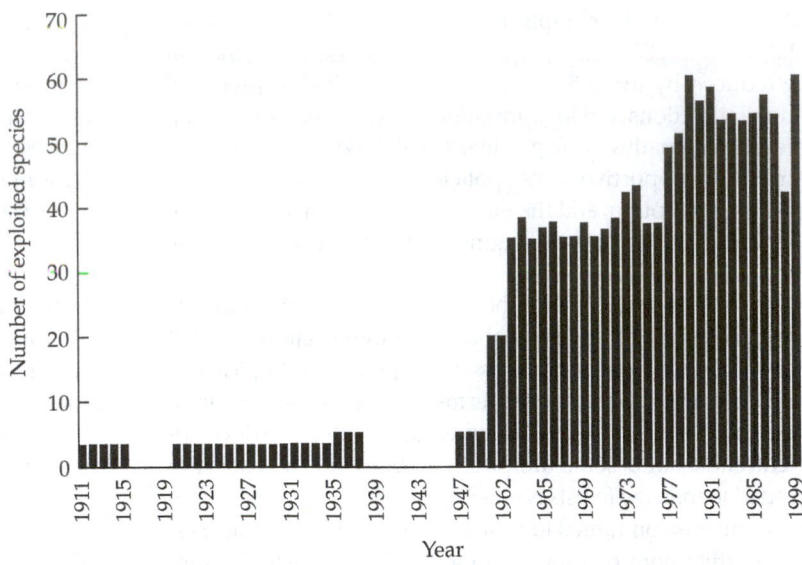

Figure C   The number of tree species exploited for timber in Ghana between 1911 and 1989. Data compiled from annual reports of the Ghana Forestry Department. For those years without bars, no data were available. (Modified from Adam 2003.)

dophragma spp.). In contrast to conditions in the first half of the twentieth century, forest managers in the moist tropics now can market dozens of species (Figure C). In Amazônas, for example, in the early 1990s, 90% of the state's timber production was based on four species, *Copaifera multijuga, Ceiba pentandra, Nacleosis caloneura,* and *Virola surinamensis.* With the exhaustion of the economically accessible stocks of these species, increased demand for sawn timber, and improvement of law enforcement, new industries have been established and the list of species used by industry has expanded considerably (e.g., to 70 species in one 40,000-ha project; Freitas 2004).

Long lists of marketable timber-producing tree species present managers with both opportunities and risks. Greater numbers of commercial species increases a manager's flexibility in selecting trees to be harvested, trees to be retained as seed sources, and future crop trees to be favored by judicious thinning around their crowns. This flexibility is important given the uncertainties in timber markets and the wide range of conditions appropriate for the regeneration and growth of valuable tree species. On the other hand, there is a risk of over-harvesting, particularly where there is little control of forest harvesting operations or where harvesting decisions are based solely on minimum felling diameters without concern for the regeneration requirements of the harvested species.

Variation among Forests

Tropical forests are justly famous for their overall species diversity, but ecologists have only recently recognized how much diversity there is *among* forests in the tropics. Similarly, forest managers are becoming increasingly aware that the silvicultural treatments they apply must be appropriate for the particular characteristics of the species and forests they are managing. For example, retaining canopy cover by carefully harvesting spa-

tially isolated trees is critical for managing forests for shade-tolerant trees that are abundant in the subcanopy and understory. The same approach to harvesting light-demanding species, in contrast, would not contribute to sustainability because these species only regenerate in large clearings. Foresters in the Dipterocarpaceae-dominated forests of southeast Asia as well as those working in the forests on the Guyana Shield of South America, where many of the commercial timber tree species are at least moderately shade-tolerant, are justifiably adamant about reducing the impacts of logging on the residual stand. Conversely, silvicultural intensification, including exposure of mineral soil in large canopy openings, is being called for in seasonal forests on the rim of the Amazon where most of the commercially important timber tree species are light-demanding (Fredericksen and Putz 2003). While many forest management practices are widely applicable, such as the need to avoid sediment loading of streams with runoff from logging roads, no single silvicultural system is appropriate for all tropical forests.

Variation among forests also means that some forests may be easier to manage than others. For example, some of greatest success in management of tropical forests has been in low diversity forests with a high density of commercial timber trees. Mangrove and swamp forests are particularly notable, although low diversity forests can also be found in uplands. For example, silviculture in oak-dominated montane forests in Costa Rica has proven to be straightforward partially because the canopy trees produce very large crops of acorns that satiate their seed predators resulting in high seedling densities in the understory (Saenz and Guariguata 2001; Guariguata and Saenz 2002). The lowland forests of Peninsular Malaysia are extremely diverse, but their management is also relatively easy because the canopy is dominated by species of Dipterocarpaceae that share silvicultural requirements. Unfortunately, the successful regeneration of an extensive area by good forest management has been erased as most of these areas have been cut to make way for oil palm plantations (Figure D).

Working With Less Valuable and Fragmented Forests

After widespread deforestation for cattle ranching, industrial farming and other agricultural uses, as well as oil palm and wood fiber plantations, the land left for natural forest management for timber is mostly remote, steep, rocky, or swampy. The conditions that render land unsuitable for more intensive uses also add

Figure D Oil palm plantations are becoming more common, replacing selective logging practices in much of Southeast Asia. (Photograph by M. Pinard).

to the challenge of forest management. Natural forest managers have to be comfortable working in remote areas and need to be very sensitive to environmental issues where slopes are steep and soils are easily damaged by heavy equipment. Additionally, the costs of applying even the simplest of stand improvement treatments, such as cutting vines on future crop trees, can become excessively expensive under these adverse conditions.

Managers should also become aware that timber-producing forests are getting smaller. In the neotropics, while the *annual* cutting area of a single logging company in the Bolivian Amazon might average 2000 ha, the *total* area of production forest of individual concessionaires or forest owners in Nicaragua, Costa Rica, and Honduras rarely exceed this same figure (Table A; one

TABLE A *A Regional (Central versus South American) Contrast in the Estimated Areas of Mixed-Species, Lowland Broad-Leaved Forest Currently Managed for Timber*

Country	Mean (range)[a]	N[b]	Observations
Honduras	1342 (196–4149)	45	Includes community and privately-owned forest in the northern lowlands
Nicaragua	2814 (5–42887)	39	Includes the entire country; management plans granted during 1999 only
Costa Rica	62 (2–481)	404	Privately-owned forests in the northern lowlands (excludes *Carapa-Pentaclethra* forests on poorly-drained soils)
Peru	17460 (2697–49620)	31	State and privately owned forest, Amazonian lowlands
Bolivia	62420 (1171–365847)	95	Concessions and community forests, Amazonian lowlands

Source: Data from the following governmental institutions: Corporación Hondureña de Desarrollo Forestal (COHDEFOR) and Instituto Nacional Forestal (INAFOR). Data also from the non-governmental organization Fundación para el Desarrollo de la Cordillera Volcánica Central (FUNDECOR), and from the bilateral/multilateral research initiative BOLFOR and Center for International Forestry Research (CIFOR).
[a]The mean (range) of the forest area is measured in hectares.
[b] Number of sites.

notable exception to this pattern is the forest community concessions in Petén, Guatemala that reach thousands of hectares [ha] [Ortiz and Ormeño. 2002]). Although generalizing how fragmentation affects the biological sustainability of timber management is currently difficult, penetration of light and wind into fragmented forests from their edges results in increased flammability, a huge concern given the frequency of human induced ignitions of wildfires in many areas of the tropics (Laurance and Cochrane 2001). Also, vines and other forest weeds proliferate on forest edges, and can greatly increase the mortality of edge-exposed trees (Laurance et al. 2001). Seed dispersal by vertebrates can also vary between fragmented and extensively forested areas (Guariguata et al. 2002).

Biological considerations aside, a potential constraint on timber production in small fragments of forest relates to landscape-level planning of management operations. Forest fragmentation may limit the viability of selective logging schemes that might work well in extensive forests because of the low probability of securing an adequate volume of a given timber species. Even well intentioned policies can exacerbate this problem. By law in Costa Rica, for example, harvesting quotas for a given timber species should not exceed 60% of the total number of harvestable individuals (Comisión Nacional de Certificación Forestal 1999). Given that most Costa Rican forests destined for timber production are only a few tens of ha in extent, harvesting quotas guided by the "60%" rule may be financially unrealistic in many forest patches due to the small absolute number of trees.

How the Challenges of Tropical Forestry Can Be Confronted

Many approaches may be helpful in reducing unsustainable forestry practices. Among these, demand from consumers, adjusted markets, policy changes, and technological advances all may play a part, often together.

Pressure from socially and environmentally concerned consumers

Frustrated by continued unsustainable forestry practices in the tropics, environmentally concerned consumers of forest products are using their market power to promote better management. Initially, boycotts of tropical forest products were staged, but they back-fired—driving down the market price of tropical timber only served to increase the rate of logging as companies strove to maintain their profits. Much more successful has been an international program of voluntary third-party certification of good forest management practices coordinated by the Forest Stewardship Council (FSC).

While every year the area of tropical forest certified as well managed according to FSC Principles (Table B) increases by about 1 million ha, global demand for certified forest products continues to exceed the supply. Unfortunately, consumer awareness of the beneficial influences of selecting certified forest products in the U.S. still lags far behind Europe. As more and more Americans move from the countryside to cities and suburban areas where they have few opportunities to learn the difference between sound and unsound forest management practices, efforts to promote forest management as a conservation strategy will become increasingly challenging. Counterbalancing this trend, country of origin labeling, organic food certification, "fair-trade" programs, and marketing of "bird-friendly, shade-grown" coffee, are all helping to make consumers in the U.S. aware of their potential power to increase social welfare and protect the environment.

Paying the true value for tropical timber

One way to reduce the opportunity costs inherent in managing forests for timber is to capture the full economic value of tropical forests. Many of the values of forests are not included in traditional market transactions for timber and even most non-timber forest resources (e.g., rattan canes, Brazil nuts, and palm hearts). Although we recognize the roles of tropical forests as storehouses of fantastic biodiversity and for their effects on local and global climates, forest owners are generally not financially compensated for these values. The costs or benefits of forests that accrue to people other than the property owners are known as externalities. Economists are working with ecologists around the world to capture these externalities in economic analyses by creating markets for sequestered carbon, transpired water, and protected biodiversity. In these efforts, Costa Rica has been a world leader.

An innovative mechanism for forest conservation, Payment for Environmental Services, was included in Costa Rica's most recent forestry law and implemented through a decentralized body (National Forestry Financing Fund [FONAFIFO]). These payments represent an attempt to financially compensate small landholders for the services their forests provide including carbon sequestration, watershed functions, maintenance of biological diversity, and protection of scenic beauty, going some way to reduce the opportunity costs of maintaining forest cover.

Markets for sequestered carbon are developing all over the world, but use of carbon credits to promote forest conservation through improved management faces a number of serious challenges. Unfortunately, reforestation of deforested areas, afforestation of naturally nonforested areas (e.g., grasslands and savannas), and, to a lesser extent, strict forest protection have all received more support from international policymakers than reduced-impact logging and other changes in forest management that promote carbon sequestration (e.g., Putz and Pinard 1993). Part of the problem is that many policymakers cannot recognize the difference between well and poorly managed forests, and doubt that anyone else can either.

Policy and legal reform

Policy and legal reform are often prerequisite to the removal of disincentives to sustainable forest management. For example, *ejidatarios* (communal owners) or indigenous communities control 80% of the forest resources of Mexico. Initially they were

TABLE B *Ten Principles of Forest Stewardship According to the Forest Stewardship Council*

Principle 1: Compliance with laws and FSC principles

Forest management shall respect all applicable laws of the country in which they occur, and international treaties and agreements to which the country is a signatory, and comply with all FSC principles and criteria.

Principle 2: Tenure and use rights and responsibilities

Long-term tenure and use rights to the land and forest resources shall be clearly defined, documented and legally established.

Principle 3: Indigenous peoples' rights

The legal and customary rights of indigenous people to own, use and manage their lands, territories, and resources shall be recognized and respected.

Principle 4: Community relations and worker's rights

Forest management operations shall maintain or enhance the long-term social and economic well-being of forest workers and local communities.

Principle 5: Benefits from the forest

Forest management operations shall encourage the efficient use of the forest's multiple products and services to ensure economic viability and a wide range of environmental and social benefits.

Principle 6: Environmental impact

Forest management shall conserve biological diversity and its associated values, water resources, soils, and unique and fragile ecosystems and landscapes, and, by so doing, maintain the ecological functions and the integrity of the forest.

Principle 7: Management plan

A management plan—appropriate to the scale and intensity of the operations—shall be written, implemented, and kept up to date. The long term objectives of management, and the means of achieving them, shall be clearly stated.

Principle 8: Monitoring and assessment

Monitoring shall be conducted—appropriate to the scale and intensity of forest management—to assess the condition of the forest, yields of forest products, chain of custody, management activities and their social and environmental impacts.

Principle 9: Maintenance of high conservation value forests

Management activities in high conservation value forests shall maintain or enhance the attributes which define such forests. Decisions regarding high conservation value forests shall always be considered in the context of a precautionary approach.

Principle 10: Plantations

Plantations shall be planned and managed in accordance with principles and criteria 1–9, and principle 10 and its criteria. While plantations can provide an array of social and economic benefits, and can contribute to satisfying the world's needs for forest products, they should complement the management of, reduce pressures on, and promote the restoration and conservation of natural forests.

Source: Modified from Forest Stewardship Council website, http://www.fscoax.org/.

obliged by law to sell their forest products to companies and thus had limited real control over the fates of their forests. This legal arrangement discouraged sustainable use and fuelled conflicts between the companies and the *ejidatorios*. The companies were not interested in sustainable management because their concessions were short term and they had no prospects of becoming forest owners. The *ejidatarios*, in turn, had little interest in the forests because they had such little control over their use. This conflict was resolved in the 1980s when a new law allowed *ejidatarios* to manage their forests directly. Today, the best examples of good forest management in Mexico come from *ejidatarios* and indigenous communities that recognize the long-term value of keeping their forests (Flachsenberg and Galletti 1998; Bray 2004). Moreover, some of the indigenous communities ac-

tively promote reforestation of lands formerly cleared for agriculture (Bocco et al. 2000).

Policy reforms are only effective in improving forest management practices when supported by the appropriate regulations and the institutions needed to implement and enforce the new policies. In Bolivia, for example, the passage of new forestry legislation in 1996 was followed by the creation of a new institution that was assigned the task of assuring compliance with the legislation. In the first year of activity, the new national forestry authority resolved millions of hectares of conflicting land claims, issued long-term cutting contracts, and increased tax revenue to the Bolivian government four-fold (Nittler and Nash 1999).

Technological strategies

Tropical silviculturalists have approached the problem of managing high diversity in tropical forests in various ways. In the 1960s, uniform silvicultural systems (e.g., shelterwood systems or patch clearcuts) were promoted as a means of domesticating the forest and increasing productivity by favoring a few light-demanding species at the expense of other more slow-growing, shade tolerant species. Polycyclic management systems (those more commonly used today) were considered less appropriate because of the high losses of volume caused by incidental damage associated with selective timber harvesting, the risk of a shift in forest composition to more shade-tolerant species, and the risk of over-harvesting if cutting cycles are too short (Dawkins and Philip 1998).

Domestication of tropical forests through intensive silvicultural interventions is generally not practiced today despite the fact that the silvicultural objectives of increasing stocking and growth rates appear to have been met in many forests treated as early as the 1950s (e.g., Alder 1993; Manokaran 1998; Osafo 1968). Two factors appear related to the lack of support for radical domestication—one is the relatively high risk of forest clearing for agriculture during the first decades in the management cycle when the forest holds little value, and the other is an assumed incompatibility between intensive silviculture and biodiversity conservation (Dawkins and Philip 1998; but see de Graaf 2000).

Typical approaches to timber management in natural tropical forests currently are conservative, aimed at reducing the impacts of interventions (Fredericksen and Putz 2003). To a great extent, reduced-impact logging has become synonymous with sustainable forest management, even though the former is only one component of the latter. Where light-demanding species are being harvested yet failing to regenerate, more intensive, locally imposed interventions are being promoted (Dickinson and Whigham 1999; Fredericksen and Mostacedo 2000; Snook 1996). For example, thinning interventions that liberate individual crop trees and maintain overall diversity in structure are sometimes recommended (see Wadsworth 1997 for review). In forests in which the commercial timber is produced by both light-demanding and shade-tolerant species, a mixed system would seem most appropriate. For example, single-tree selection might be applied where there is advance regeneration of shade-tolerant species, while in other areas large gaps might be created near potential seed trees to promote regeneration of light-demanding species (Pinard et al. 2000).

Looking Forward: Where Will the Next Decades Lead?

Whilst uncertainty in timber prices, politics, and priorities make it difficult to predict where the next decades will lead tropical forestry, a few pathways seem more likely than others. Overall, while deforestation rates are likely to remain high, forest management practices should continue to improve under the influence of the FSC's timber certification processes (see Table B). These improvements will be a consequence of more forest owners working towards certification, certified producers improving their practices through the monitoring and feedback programs built into the certification process, and in response to indirect influences of certification such as the exchange of information and experience among forest owners, managers, and regulators.

It is also clear that the focus of the timber industry will increasingly move to marginal lands where opportunity costs are low and natural forest management is relatively attractive as a land use. With widespread devolution of control of forest lands to indigenous groups and other rural communities, timber buyers will be motivated to develop socially acceptable approaches to company–community partnerships (Mayers and

Vermeulen 2002). Farm forestry will also continue to develop, contributing to a shift in milling capacity to process smaller stems. As the big profit margins associated with exploiting old growth forest give way to the more modest profits associated with sustainable harvests, smaller operators are likely to be attracted to the business. Provision of technical support to many small operators may be more difficult than to fewer large operators, but levels of motivation and innovation may also be higher. Where managers are committed to sustainable practices, more silviculture will be introduced into management. Meanwhile, integration of inventory data, harvest planning, and monitoring operations into geographic information systems (GIS) will facilitate the implementation of more intensive and directed operations.

The trends in policy towards broadening societal participation in forestry are likely to continue with more efforts to devolve forest management to rural communities. But early experience gained across a broad range of socio-political and economic conditions emphasizes the need for strategies that combine both short- and long-term objectives (du Toit et al. 2004), institutional development, conflict resolution (Castro and Nielson 2001), and a recognition that in some cases, it will take a lot of time and effort for communities of subsistence farmers to become sustainable forest managers.

As pressure to place remaining tracts of natural forest under protection, the divide between those arguing for sustainable use and those arguing for complete protection may grow wider (see Dickinson et al. 1996 and Pearce et al. 2003 for summaries and primary literature therein), one consequence being increased competition for sources of funds for development. All progress towards sustainable forest management depends on good governance. Where corruption and illegal logging continue, it will be difficult to improve forest management practices (Ravenal et al. 2004).

The challenges facing tropical forest management are many. The field of natural forest management in the tropics needs energetic and dedicated people who are willing to put up with the rigors of tropical forestry work. Input from biologists is needed to ensure that the development of good practices, and the criteria and indicators used to assess and monitor these practices, are based on relevant biological and ecological information (Putz and Viana 1996).

Summary

1. Overexploitation is ranked second only to habitat loss as a cause of extinction risk in species whose status has been assessed globally. It involves the unsustainable use of wild animals, plants, and their products for purposes such as food, medicines, shelter, and fiber. The motivations for such exploitation are as varied as the plants and animals that are taken, ranging from subsistence hunting and fishing, to recreational and economic pursuits carried out by wealthy individuals and corporations. While most environmentalists are well aware that fisheries and large-bodied birds and mammals are often overexploited, recent studies have shown that these changes can be far more dramatic than previously recognized. Fisheries can reduce the biomass of targeted and nontargetted species by 80% or more in

10–20 years, and subsistence hunters in the Neotropics can create a vacuum of large-bodied species within 10–20 miles of their villages.

2. The biological theory of sustainable exploitation is firmly rooted in the field of population ecology, which seeks to understand the responses of populations to increased mortality of individuals through density-dependent compensation. This theory has produced a range of methods for estimating sustainable limits of exploitation from simple rules of thumb based on life histories to highly sophisticated models that are appropriate for only a small fraction of cases where the necessary input data are available. This theory cannot be put into practice to make exploitation sustainable without a clear understanding of the motivations of people who use wild animals and plants, the alternatives available to them, and the effects of management recommendations on their livelihoods. The "tragedy of the commons" often looms large in overexploitation in both terrestrial and aquatic ecosystems, exacerbated by "discounting," whereby we give wild populations a higher value for what we can get from them today than in the future. Yet there is a growing realization of the enormous economic and social benefits that we receive from ecosystem goods and services. With encouragement from processes such as the World Summit on Sustainable Development in 2002 and the Millennium Ecosystem Assessment, many people are now rising to the challenge of finding innovative ways helping people to reap the rewards of sustainable exploitation.

Please refer to the website www.sinauer.com/groom for Suggested Readings, Web links, additional questions, and supplementary resources.

Questions for Discussion

1. How can rich countries help encourage poor countries to exploit their plants and animals sustainably?

2. Actions taken to protect vulnerable species may affect some people's livelihoods. Because the decision to protect biodiversity is an expression of public will, do you think that governments have an obligation to assist families and businesses whose livelihoods are harmed by a conservation plan?

3. Many cases of overexploitation involve not only people's livelihoods, but their subsistence. How can we prioritize conservation versus human subsistence? What mechanisms might we foster to reduce this conflict?

4. In what ways can models help us exploit wild populations more sustainably? What are some of the limitations of the models typically used to estimate safe harvest levels?

5. Why is overexploitation so large a problem in the marine realm, and relatively less important in most terrestrial biomes?

9

Species Invasions

Marjorie Wonham

I am earth's native: No rearranging it!

Robert Browning

A carnivorous snail exterminates over 56 endemic snail species in the southeast Pacific. A meter-long tree-climbing snake consumes endemic birds, bats, lizards, and skinks on Guam. A fast-growing leafy green seaweed carpets the eelgrass meadows of the northern Mediterranean. A 10-kg fish species begins devouring small endemic lake fishes in Africa. Aggressive ants and bees outcompete their native counterparts across the Americas. Smallpox decimates native human populations in North America, and myxomatosis decimates European rabbits in Australia. What do these high-profile species have in common? All are nonnative species that have been introduced by human activity—either intentional or unwitting—beyond their native ranges.

We live in a world of introduced species (Table 9.1). European trees (*Pinus* spp.) live in Africa, African dung beetles (family Scarabaeidae) in Australia, Australian possums (*Trichosurus vulpecula*) in New Zealand, New Zealand snails (*Potamopyrgus antipodarum*) in North America, North American beavers (*Castor canadensis*) in South America, South American trees (*Mimosa pigra*) in Asia, and Asian crabs (*Eriocheir sinensis*) in Europe. The sheep herded on Navaho and New Zealand pastures, the horses ridden by Argentine Gauchos, the cattle herded in Texas and on the African Masai Mara—these emblematic and economically important species are also introduced.

What are the conservation implications of introduced species? A general context is provided in the following section, and then four major conservation-focused questions regarding introduced species are addressed. First, what are the impacts of introduced species? How do they alter ecological communities and at what spatial and temporal scales? When are impacts likely to be minor, and when will they demand our attention? Second, what factors determine whether an invasion succeeds or fails? Are there certain traits that make a potential invader more likely to succeed, or a community more likely to be invaded? Third, how are species introduced? What human activities act as pathways for accidental and intentional invasions? And finally, how can we best control or even prevent invasions to reduce their conservation impacts?

TABLE 9.1   *Introduced Species Terminology*

Cryptogenic	Applies to species whose status—native or nonnative—cannot readily be determined; commonly used for species whose cosmopolitan distributions or unclear taxonomy make their geographic origins uncertain
Indigenous	Synonymous to native
Introduced	Refers to a species that has been released outside its native range; synonymous with nonnative, nonindigenous
Introduction	The release of escape of a nonnative species
Invasion	The establishment and spread of an introduced species
Invader	An introduced species
Native	Describes species that evolved in a region
Nonindigenous	Synonymous with introduced, nonnative
Reintroduced	Refers to intentionally released individuals of a native species that was locally endangered or extinct

Note: The terms adventive, alien, exotic, invasive, naturalized, nuisance, pest, and weedy are also used to refer to introduced species, but are either loosely defined, applied to both native and nonnative species, or are rich with non-scientific connotations. The terms introduced, nonnative, or nonindigenous are therefore more clear when referring to species outside their native ranges.

Each of these questions represents an area of active research with rapid development and turnover of ideas and principles. The science of biological invasions is, like the larger science of conservation, a relatively young field in which far more is unknown than known. Thus, we will address some of the key ideas as perceived at present, with the caveat that many may be provisional and all are worthy of investigation, testing, refinement, and even replacement by future students of conservation.

What Are the Conservation Implications of Introduced Species?

In terms of their number and ubiquity, the scale of biological invasions is staggering. On the plains and prairies of North America 11% of plants are nonnative (White et al. 2000) and in Hawaii 35% are likewise nonnative (Pimentel et al. 2000). Approximately 8000 shrub and herb species and 750 tree species have been introduced to South Africa. Over 3000 plant species have been introduced to the state of California alone. In the U.S., introduced species are ranked as one of the top threats to endangered species (Wilcove et al. 1998; Sala et al. 2000). Worldwide, 20% of endangered vertebrate species are threatened by introduced species—13% of those on mainlands and 31% on islands (Heywood 1995). Among freshwater fishes, 67% of tropical and over 50% of temperate introductions established successfully (Welcomme 1988); a review of 31 fish introductions found that 77% caused native fish populations to decline (Ross 1991). In coastal waters, nearly 300 introduced species are known in North America, over 160 in the Mediterranean, more than 140 in the North American

Great Lakes basin, more than 50 in Australian coastal waters, and more than 40 in the Black Sea. Numbers are continuing to increase: In San Francisco Bay, the estimated rate of successful new invasions has increased from one every 55 weeks from 1851 to 1960, to one every 14 weeks from 1961 to 1995 (Cohen and Carlton 1998) (Figure 9.1). This influx of nonnative species has utterly changed the biotic makeup of land, freshwater, and marine communities.

Species range expansions, like extinctions, are natural processes over evolutionary time. In these ways, biological communities develop and change slowly as continents shift and volcanic islands are colonized anew. But

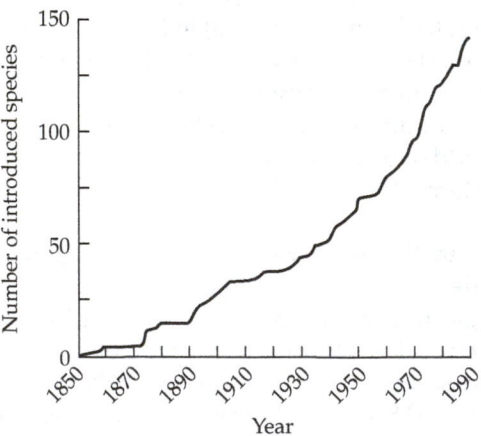

Figure 9.1 Accumulation of established estuarine and marine invaders in San Francisco Bay, California, over 140 years (records with uncertain collection dates and records resulting from specialized sampling for invaders have been excluded). (Modified from Cohen and Carlton 1998.)

today, the rate of human-caused introductions, like that of extinctions, exceeds background levels at an unprecedented scale (Office of Technology Assessment 1993; National Research Council 1995). Although only a small fraction of introduced species establish in their new environments, each one that does has some ecological impact, and a small proportion have dramatic impacts that must be addressed in conservation, economic, aesthetic, and ethical realms.

Species invasions affect economics, public health, and biodiversity. Socioeconomic effects of invaders stem from both intentionally and unintentionally introduced species. Much of the world's industrialized agriculture depends on only a few broadly distributed species: Around the world, European cows, pigs, and chickens are raised, European wheat and South American potatoes are farmed, Asian oysters are grown, and European brown trout are stocked. The estimated economic benefit of these marketable nonnative species runs to hundreds of billions of dollars per year in the U.S. alone (Pimentel et al. 2000). On the other hand are the many introduced species that become pests. Some of these are accidental arrivals with crop species; others hitch a ride on boats, airplanes, or trucks. In the U.S., costs associated with the impacts and control of the Eurasian zebra mussel (*Dreissena polymorpha*) are estimated at $1 billion over the first 10 years of its invasion, the European gypsy moth (*Lymantria dispar*) at $11 million per year, European ship worms (*Teredo navalis*) at $200 million per year, and purple loosestrife (*Lythrum salicaria*) at $45 million per year—and these are only a few of the hundreds of unwanted species introduced worldwide (Pimentel et al. 2000).

Many public health threats are caused by introduced diseases. Smallpox, avian and human malaria, and the red tides caused by toxic phytoplankton can outbreak in new areas and have devastating effects, particularly on immunologically naive populations. The global epidemic of HIV/AIDS has spread around the world from its native Africa, claiming approximately 18 million victims to date, with 5 million new infections annually. The seventh global cholera pandemic, which began in 1961 in Indonesia and which now persists in Latin America, Africa, and Asia, claims an annual global death toll of tens of thousands of victims. With today's rapid global transport of humans, crops, and livestock, the potential for disease spread is increasing. This is an immediate conservation concern in terms of species losses, and a larger-scale concern in terms of the human health and economic stability that are essential ingredients for effective conservation.

The biodiversity impacts of invaders are complex. A successfully introduced species may increase local species richness, or α-diversity, and indeed many established invaders appear not to drive native species extinct

(see Essay 9.1 by Ariel Lugo on some of the benefits of introduced species). On the other hand, many invaders certainly do cause or contribute to extirpations and extinctions. And a broader argument focused on regional, or β-diversity would note that introductions may reduce diversity by increasing similarity among regions. The freshwater fish faunas of the U.S., for example, are today 7.2% more similar than before European colonization, largely as a result of introductions (Rahel 2002). With this brief introduction to the broad scale of biological invasions and their conservation implications, we now turn to a more focused treatment of the ecological impacts of introduced species.

What Are the Impacts of Invasions?

Invasion impacts can be evaluated across ecological and evolutionary realms. First, we can consider both direct and indirect effects of a single invader. The direct effects of an invader result from its immediate interaction with another species through predation, competition, parasitism, or disease. Indirect effects follow from there, for example, through resource competition, trophic cascades, apparent competition, and habitat modification (Wootton 1994).

Second, we can consider the consequences of multiple invasions. We might expect a given species' impacts to differ in different locations and through time, as a result of differences in invader densities and behavior, differences in the invaded community, and the interaction of both. The effects of one invader may also be enhanced or reduced by those of another, and the cumulative effects of multiple invaders may differ from a prediction based on the summed impacts of each invader.

Third, we can examine impacts across a range of ecological and evolutionary scales. Most obvious, perhaps, are impacts on the abundance of particular populations. Looking more closely at populations, we may also detect changes in the behavior or genetic makeup of individuals; stepping back, we may see changes in community-scale diversity and species interactions, and in ecosystem characteristics such as habitat structure or nutrient cycling.

It is important to note that the ecological impacts of introduced species are *inherently* neither bad nor good. Such judgments can only be made in the context of an explicit value framework, such as endangered species conservation, habitat restoration, or an economic bottom line. These value frameworks, which are essential elements of conservation biology, must be thoughtfully applied to the impacts of introduced species.

Population and community impacts

Introduced species are implicated in the endangerment and extinction of species around the world. This popu-

ESSAY 9.1

Maintaining An Open Mind on Nonnative Species

Ariel E. Lugo, *Institute of Tropical Forestry, Puerto Rico*

"To consider only the invading species themselves in developing management programs or in recommending regulatory actions is tantamount to curing symptoms and not disease."

Ewel 1986

■ Several reasons are given for the success of nonnative species. They may do well because they are freed from natural enemies, competitors, and parasites. Consequently, a common strategy of control programs is to introduce organisms from the native habitats of target nonnative plants to harm them; that is, eliminate nonnatives by introducing more nonnatives! Nonnative species may also find little competition in the ecosystems they invade; empty niches may be available to exotics, causing little effect on the invaded ecosystem. The success of exotics has also been attributed to their being aggressive colonizers, fast growing, and highly fecund species. However, Ewel (1986) noted that "species invasions often reflect the conditions of the community being invaded rather than the uniquely aggressive traits of the invader."

I subscribe to Ewel's point of view. When a good match is made between the genome of any species and the environmental conditions that support its growth, the result can be explosive population increases. This is why organisms that are rare in their natural habitats may suddenly become weedy in a new situation. The water hyacinth (*Eichhornia crassipes*) is fairly inconspicuous in its natural Amazonian habitats but exhibits explosive growth in the slow moving, highly eutrophic waters of artificial canals and reservoirs in Florida. Experiments show that it cannot grow as well when subjected to oligotrophic waters or fast flow. Close observation of successful nonnative organisms usually yields similar results.

Ecological Functions and Services of Nonnative Species

I have studied monoculture plantations of nonnative trees in Puerto Rico and have compared them with native forests of similar age. My studies include 73 comparisons of structure, composition, and function of these ecosystems and I cannot find a single comparison that suggests an ecological anomaly in the forests dominated by nonnative species. These ecosystems function like native forests, with differences mostly in the magnitude of rates and state variables (i.e., biomass and other structural features). No negative effects have been detected in the water cycle, accumulation of carbon and nutrients, or in any other site condition. Claims to the contrary, such as those leveled against *Eucalyptus* trees, cannot be substantiated when evaluated critically.

I have also found that native plant species grow and develop under canopies of nonnative species established in degraded sites, including some in which succession was arrested prior to planting nonnatives. Native birds are attracted to and use the native species understory of these nonnative tree plantations. For these reasons, I (and others) have suggested that nonnative species can be important tools for land rehabilitation and restoration of biological diversity in damaged sites where natural succession is arrested. For example, nitrogen-fixing trees can increase soil nitrogen for the entire ecosystem, and promote higher productivity. One would expect that higher productivity would eventually result in a greater capacity to fix carbon, circulate nutrients, and support more species.

Will Nonnative Species Dominate the World?

The dominance of nonnative species on the landscape will be a function of the degree of human modification of the environment. In general, human activity fragments the landscape, favors establishment of nonnative species, increases environmental heterogeneity, may cause species extinctions, and may augment the total number of species on the landscape. The Pacific Islands are instructive because they represent a worst-case scenario of the effects of intensive human activity on small land areas isolated from sources of biotic replenishment. On these islands, particularly the Hawaiian chain, isolation allowed the evolution of a highly diverse and endemic suite of organisms, originally exposed to a slow rate of invasion by nonnative species.

Humans greatly accelerated that rate and in the process transformed the flora and fauna of the Pacific Islands. The process has been ongoing for some 2000 to 4000 years and the results have been staggering in terms of species extinctions and transformation of biotic composition. The number of species across all taxonomic groups has increased from about 9000 to 12,000, and many see this trend as an erosion in global biodiversity because endemic forms are lost while pantropical weeds replace them. There is certainly truth to this argument, but it is not entirely accurate because many nonnative species are neither weedy nor pantropical. Some may be rare and endangered species that find refuge in another, more favorable location and then become weedy. In Puerto Rico, for example, *Delonix regia* is a common naturalized species in danger of extinction in its native Madagascar habitat.

To better understand why the Pacific Islands appear so vulnerable to invasive species, I compared the density of species in the Caribbean Islands with those of the Hawaiian Islands. If the area and plant species density values of the Hawaiian Islands are used to estimate the numbers of plant species expected in Cuba and Puerto Rico (same latitude as Hawaii), the results underestimate the actual number of species on the two Caribbean islands. This means that when area is corrected for, the density of species in the Caribbean is much higher than in the Hawaiian Islands. Part of this higher species density in the Caribbean is caused by the much greater density of species in Cuba and Puerto Rico than in the Hawaiian Islands. This is explained in part because these islands are closer to continental landmasses, but also by the fact that the Caribbean Islands are six times older than the Hawaiian Islands. Could this mean that the Hawaiian Islands have a greater capacity to absorb additional plant species than do Caribbean Islands? The age of islands as well as their degree of isolation influence the density of species

and invasibility of their communities. In Hawaii, nonnative species are the likely invaders of plant communities because they are actively transported from a large reservoir of genetic material (the whole world) while the native species evolve slowly and are constrained by founder effects.

I would expect that in the absence of significant climate change, environments that today support high species richness will do so in the future, but the species composition may be different. And we should not forget that the forces of evolution are not suspended for nonnative species. One could argue that the enrichment of islands with nonnative genomes provides fuel for the evolutionary process and greatly increases adaptive possibilities. This is particularly important in light of human-induced changes in the atmosphere, climate, geomorphology, and other environmental conditions.

The change in species composition taking place in the world today is not a chaotic process; it is a process that is responding to fundamental changes in the conditions of the planet. Age-old

ecological constraints such as time, energetics, biotic factors, growth conditions, and opportunity are at play, regulating which species are successful and which are not in a specific location. Human activity generates the environmental change that powers the response of organisms through adaptation, evolution, or formation of new groupings of species and communities.

Management Strategies for a Changing World

I have highlighted contradictions in the way we deal with biodiversity issues in general and nonnative species in particular. Even in Hawaii, where there is great concern about the degree of nonnative species invasion and their potentially negative effects, the government actively and successfully introduces hundreds of insect species for agricultural pest control. We correctly worry about the negative effects of human activity, but forget that this activity started thousands of years ago at a time when people depended directly on the environment for survival. Even then,

nonnatives were introduced and perhaps countless numbers of species were driven to extinction. I point these things out not to excuse introductions of species or to condone driving species to extinction. But, we must maintain an open mind and analyze the issue of nonnative species introductions and management as an intrinsic and continuous process in a world where our own species is a main driver of change.

It is in our power to take actions to mitigate the negatives of our activities and to enhance the positives. Actions that may help are learning to manage and control environmental change, recognizing when conditions are obviously beyond our control, avoiding condemning species because of successional stage or ecological function, improving our capacity to manage biotic resources, concentrating human activity to allow more space for native ecosystems, and encouraging environmental heterogeneity as a mechanism to maximize biodiversity. One thing is clear: The world will continue to change and become less familiar to those that walked on it or wrote about it centuries ago. ■

---

lation-scale impact is relatively easy to document, and may extend to major changes in community structure, measured as changes in trophic dynamics or interaction strengths between species.

INTRODUCED PREDATORS   The brown tree snake, *Boiga irregularis*, is a stunning example of the population- and community-level impacts of an introduced vertebrate predator on a naive island population. The snake's native range extends from Australia through New Guinea to the Solomon Islands. Shortly after World War II, it arrived on the island of Guam where the only native snake was a tiny, wormlike insectivore. The invader gradually spread across the island but remained at low densities for two decades until the early 1960s, when it began to appear in large numbers. More and more well-fed snakes were found in chicken coops; others were carbonized on power lines, causing electrical blackouts (Figure 9.2). During this time, native birds began disappearing.

Ten species of forest birds followed a similar pattern of decline: Each disap-

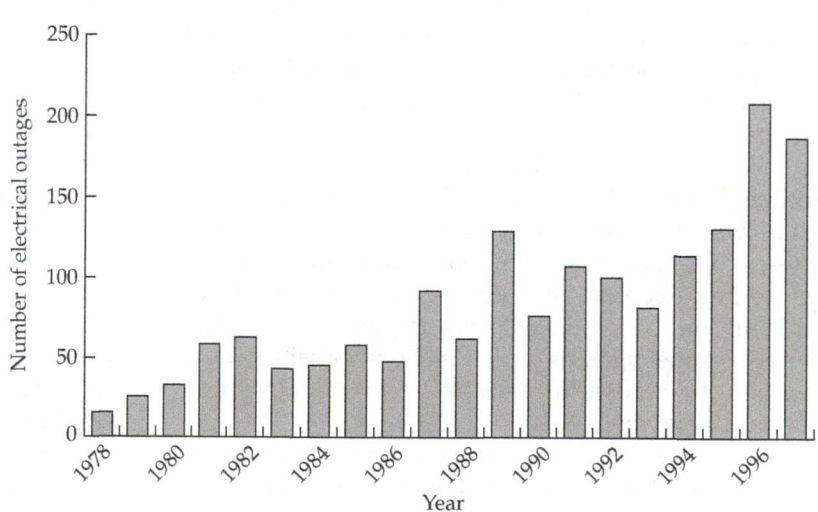

Figure 9.2   Annual electrical outages caused by the invading brown tree snake on Guam. (Modified from Fritts 2002.)

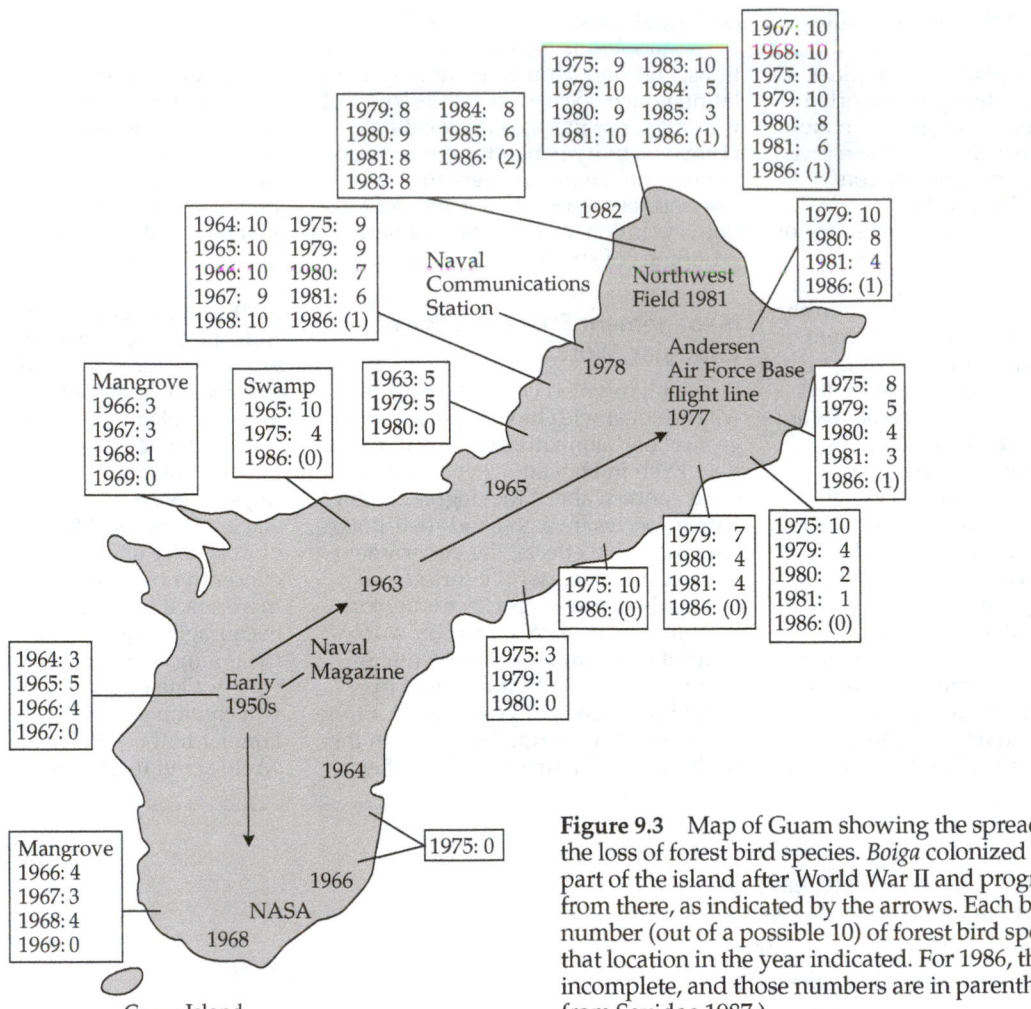

Figure 9.3 Map of Guam showing the spread of *Boiga* and the loss of forest bird species. *Boiga* colonized the southern part of the island after World War II and progressively spread from there, as indicated by the arrows. Each box lists the number (out of a possible 10) of forest bird species found at that location in the year indicated. For 1986, the surveys were incomplete, and those numbers are in parentheses. (Modified from Savidge 1987.)

peared from southern Guam by the late 1960s, and one by one were lost progressively further north (Figure 9.3). In early 1983, all ten species could still be found in one small patch of tall forest beneath a cliff line on the northern tip of the island. By 1986, they had disappeared from this area as well.

There was considerable debate about the cause of the decline in Guam's birds. Some scientists argued that pesticides were to blame, whereas others believed that avian diseases might be at fault. The native birds might, for example, be susceptible to diseases carried by introduced birds or domesticated chickens and pigeons. Introductions of competing birds could also be responsible, not to mention the familiar list of introduced predatory rats, cats, and other mammals.

Today, the cumulative evidence indicates that the nocturnal brown tree snake, which can reach abundances up to 5000/km^2, has been largely responsible for eating the birds—now totaling 15 species—to extinction (Table 9.2). Three of these species and two distinctive sub-

species were endemic to Guam. In addition to the birds, the brown tree snake has consumed native and introduced anoles, geckos, skinks, rats, and bats, radically restructuring the terrestrial food web of Guam (see Table 9.2). Although pigs and monitor lizards consume the snake, they do not seem to exert any significant population control.

Because brown tree snakes have been found in airplane wheel wells, there is evidence that they may be able to hitch a flight off Guam. In anticipation of this, other South Pacific islands have developed monitoring programs to try to prevent its further spread. Given the frequent movement of people and goods among islands, as well as nearby continental areas, it is critical that the small island developing states (SIDS) of the South Pacific act together to limit species introductions. One island's excellent program can be undermined by the failure of another island to act. Thus, SIDS have been negotiating both legally binding and voluntary programs that are proactive in approach—export inspections, prohibitions

TABLE 9.2   *Changes in Vertebrate Trophic Interactions on Guam before (1945) and after the Brown Tree Snake Invasion (1995)*

	1945	1995
Carnivores	Fairy Tern	Brown tree snake
	Mangrove Monitor*	Mangrove Monitor*
	Micronesian Kingfisher	
	Oceanic gecko*	
Insectivores	Pelagic gecko	Curious skink*
	Blue-tailed skink	Blue-tailed skink
	Mariana skink	Mourning gecko
	Spotted-belly gecko	House gecko
	Mourning gecko	Mutilating gecko*
	House gecko	Moth skink
	Mutilating gecko*	
	Moth skink	
	Rufous Fantail	
Omnivores	Black rat*	Black rat*
	Polynesian rat*	Polynesian rat*
	House mouse*	House mouse*
	Guam Rail	
	Mariana Crow	
	Bridled White-eye	
	Micronesian Honeyeater	
Herbivores	Micronesian Starling	Philippine Turtle-dove*
	Philippine Turtle-dove*	
	Mariana fruit bat	
	White-throated Ground-dove	
	Mariana Fruit-dove	

Source: After Fritts and Rodda 1998.

*Indicates historic introductions that preceded the brown tree snake.

of movement of certain dangerous taxa, and efforts to control spread via ballast water.

Other predatory invaders share a similarly infamous reputation. The giant African snail (*Achatina fulica*), introduced in Hawaii as a garden ornamental, had by 1955 become a garden and crop pest. To control it, the State Department of Agriculture introduced 15 nonnative predatory snail species over a period of six years, of which three species established (Cowie 1998). One of these, the rosy wolfsnail (*Euglandina rosea*) had a minimal impact on *A. fulica*, but contributed to the disappearance of 15 of the 20 endemic *Achatinella* snail species on the island of Oahu, resulting in the genus being listed under the United States Endangered Species Act. This story is not restricted to Hawaii: *E. rosea* has been introduced to over 20 other Pacific islands, with similarly ineffective biocontrol results and devastating extinction conse-

quences. In the Society Islands of French Polynesia, 56 of the 61 endemic partulid snails were driven to extinction, a staggering loss of biodiversity that also spelled the end of an artisanal jewelry-making economy (Coote and Loève 2003).

In aquatic systems, substantial population-level impacts of predators have been recorded, but often without the ultimate extinctions known from terrestrial and particularly island systems. Although introduced Nile perch (*Lates nilotica*) predation caused extinctions of endemic cichlid fishes in Lake Victoria (Goldschmidt 1993), recently researchers found that some of the affected populations declined but did not disappear entirely, and a few may be recovering (Witte et al. 2000).

Even familiar, backyard species can be conservation calamities. In reviewing the impacts of introduced mammals and birds, Ebenhard (1988) recorded 59 introduc-

tions of domestic cats (*Felis domesticus*). A majority of these (64%) had detrimental effects, including extinctions of native prey populations; this is twice the percentage of all other predators reviewed in this study. The story is more interesting, though, in that 35 of 49 (71%) introductions of cats to islands had detrimental effects, while only 3 of 10 (30%) introductions to mainlands, or islands once connected to mainlands, had such effects. Even the mainland 30% is high compared to the 9% average for all mammalian predators together. These impacts of cats highlight the recurring observation that islands may be more heavily affected—at least by vertebrate predators—than mainlands, and that cats themselves appear to be unusually problematic invaders.

INTRODUCED COMPETITORS The impacts of competition are often trickier to observe and measure than those of predation. Nonetheless, a number of conspicuous invaders appear to have caused large ecological changes through direct and indirect effects of competition.

One of North America's most diverse faunas is the approximately 297 species and subspecies of unionid mussels endemic to the eastern part of the continent. These hard-shelled invertebrates live buried in the soft bottoms of lakes and rivers, with just enough shell exposed to filter plankton from the water. Unionids have the dubious distinction of being one of North America's most endangered fauna: Some 40%–75% of unionid species are classified as extirpated or species of concern due to habitat loss, harvest, dam and canal construction, and introduced species. The most famous of the troublesome invaders is the Eurasian zebra mussel, which was first found in the Great Lakes in 1988 and has since spread into the St. Lawrence, Hudson, and Mississippi River drainages. The mussel is only about 2.5 cm in shell length, but it reaches densities of over 15,000/m². It differs from the unionids in two important biological aspects: Its larvae are planktonic and disperse freely, whereas unionid larvae require fish hosts for dispersal, and it has byssal threads with which it attaches to hard substrates such as docks, boat hulls, and other bivalve shells. This is the primary mechanism of its impact: Zebra mussels settle at high densities onto unionid shells, reducing the ability of the native mussels to filter water and respire, feed, or excrete wastes (Figure 9.4). In Lake Erie, Lake St. Clair, Lake Oneida, and the Detroit River, zebra mussels are estimated to have contributed to 64%–100% of local unionid extinctions, whereas habitat loss is estimated to have contributed to 2%–30% (Ricciardi 1998).

Many invaders have pronounced effects through both predation and competition. The Argentine ant *Linepithema humile*, native to South America, is a striking example: In addition to its competitive effects on native ants,

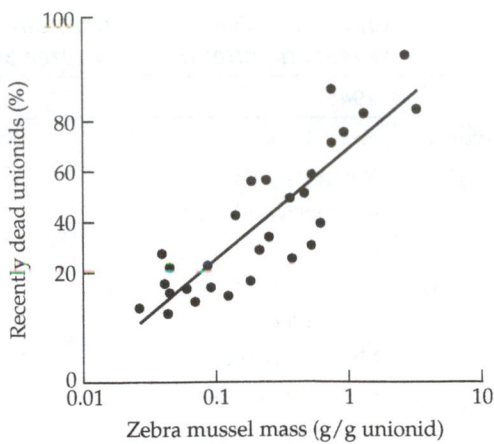

Figure 9.4 Percent of native unionid mussels recently dead as a function of the relative wet weight of introduced European zebra mussels (*Dreissena polymorpha*) attached to their shells. Note log scales. (Modified from Ricciardi 2003.)

this invader competes with other insects including honeybees, preys on herbivorous insects, and can reduce the survival and prey of birds, lizards, and small mammals (Holway et al. 2002).

In some cases, one invader may interact to increase another's effects. Prior to the 1970s, the Gulf of Maine was dominated by kelp beds (*Laminaria* spp.) and understory macroalgae. When groundfish were overharvested, their sea urchin prey (*Strongylocentrotus droebachiensis*) increased and grazed down the kelp. But when commercial harvesting reduced the urchin population in the 1990s, the kelp did not return as expected. In the interim period, two new invaders had established: an Asian macroalga (*Codium fragile* ssp. *tomentosoides*) and a European bryozoan (*Membranipora membranacea*). By combining long-term observations with local experiments, Harris and Tyrell (2001) and Levin and colleagues (2002) pieced together the interaction between these two invaders (Figure 9.5), and their impacts in altering the marine community. The bryozoan is an encrusting species that grows on the large flat kelp blades, making kelp more brittle and prone to breaking off during winter storms. Also in winter, fragments of the invading alga disperse, settle, and begin new vegetative growth. The alga can colonize new spaces much more rapidly than the kelp (which does not spread by fragmentation), and it is not a preferred food of the urchin. In this way, it has replaced the kelp beds. Because kelp forests are important nursery grounds for fish and decapods, the implications of kelp loss extend to the pelagic and benthic animal community.

Alternatively, a new invader may depress the impacts of a previous one. In the Black Sea, the ctenophore (comb jelly) *Mnemiopsis leidyi* arrived in the early 1980s and by

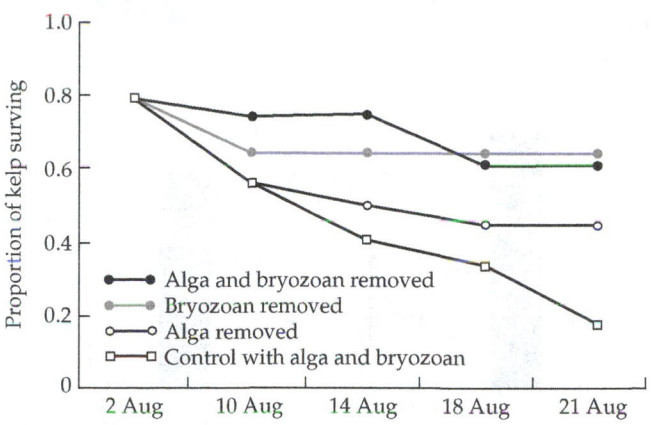

Figure 9.5 Impacts of the introduced Asian alga (*Codium fragile* ssp. *tomentosoide*) and European bryozoan (*Membranipora membranipora*) on the survival of native subtidal kelp as shown by a 1997 experiment in the Gulf of Maine. Kelp survival increased more when the bryozoan was removed than when the alga was removed. Solid circles represent sites where the alga and the bryozoan were removed. Gray circles represent removal of only the bryozoan, while open circles represent removal of only the alga. Open squares represent controls with both introduced species present. (Modified from Levin et al. 2002.)

1988 had reached a maximum biomass of over 4000 g/m^2 wet weight. Native to temperate to subtropical coastal estuaries of the Atlantic Americas, the ctenophore is a generalist planktonic carnivore that grows up to 120 mm long and can consume up to ten times its weight per day in zooplankton, including fish eggs and fish larvae. Correlated with its population boom was a decline in the abundance of zooplanktivorous anchovy, the Mediterranean horse mackerel, and sprat, and the collapse of the associated commercial fisheries (Figure 9.6). In 1997, a

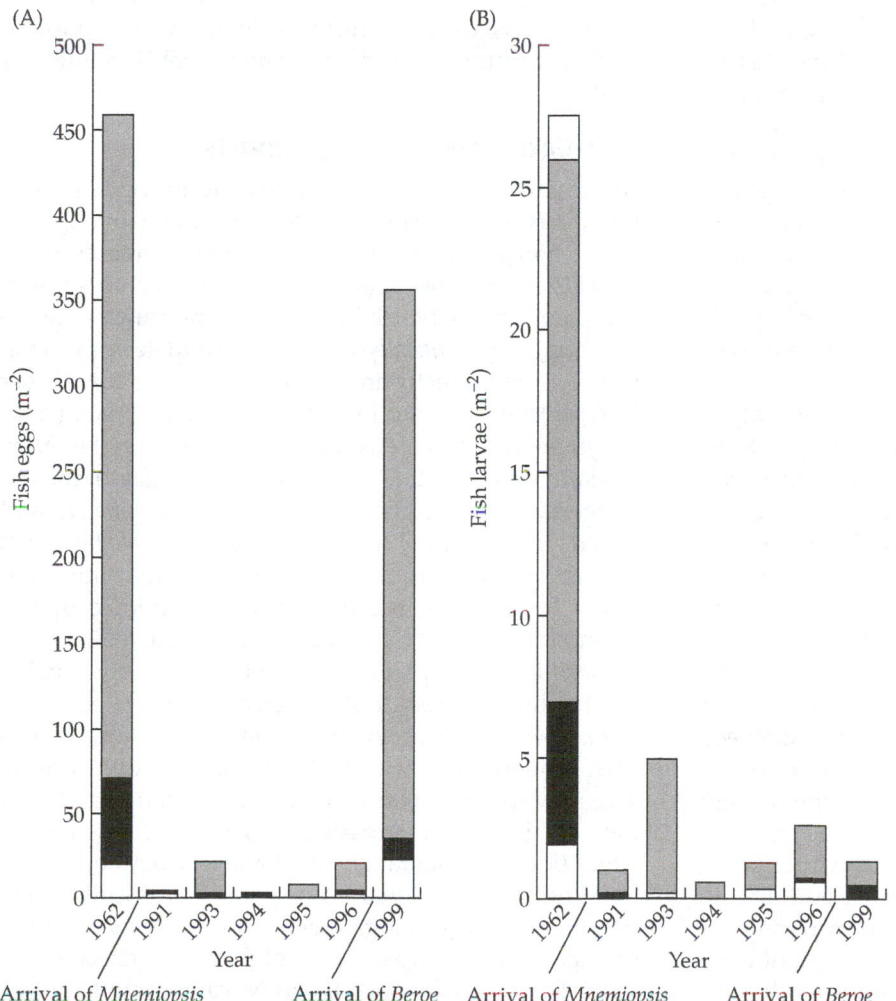

Figure 9.6 Change in the abundance of (A) fish eggs and (B) fish larvae in the northern Black Sea for anchovy (gray), Mediterranean horse mackerel (black), and other species (white) following the bloom of the Atlantic ctenophore *Mnemiopsis leidyi* in 1988, and the arrival of the Mediterranean ctenophore *Beroe ovata* in 1997. (Modified from Shiganova and Bulgakova 2000.)

second introduced ctenophore appeared: *Beroe ovata*, which is native to the Mediterranean, and is a voracious predator of other ctenophores. By 1999, *M. leidyi* had declined to 155 g/m² and the fish populations were showing early evidence of possible recovery (see Figure 9.6). Without knowing complete catch-per-unit-effort fishery data, the relative impacts of fishing pressure versus that of predation on the native fish cannot be distinguished, but correlative evidence implicates *M. leidyi* as at least partially responsible for the fish species declines, and *B. ovata* as indirectly contributing to their recoveries.

Population impacts of invaders are among the easiest to observe, but their cascading direct and indirect effects, through trophic, competitive, mutualistic, and other interactions, are wildly difficult to predict. At the same time, they teach us about the interaction linkages in ecological systems and together can inform our future choices in invasion prevention and control.

Morphological and behavioral impacts

At smaller ecological scales, invaders can cause changes in the morphology and behavior of native species. When the predatory European green crab (*Carcinus maenas*) invaded the Northwest Atlantic coastline in the early to mid-1900s, populations of a small native herbivorous intertidal snail (*Littorina obtusata*) developed a significantly thicker shell that provides some protection from predation (Vermeij 1982). In a series of laboratory experiments, Trussell and Smith (2000) found that the snails responded to the crab's chemical cues by growing thicker shells, and had a corresponding decrease in body mass. Because fecundity typically scales with body mass in these gastropods, the immediate impacts of a change in morphology may scale up to larger effects on reproductive success and population growth.

In some cases, these smaller-scale impacts propagate across multiple species. The Asian flowering plant *Impatiens glandulifera* was introduced to Europe from the Himalayas just over 100 years ago. Over 2 m high, it is the tallest annual in Europe, has a broad climate and soil type tolerance, and is spreading rapidly. The nectar of *I. glandulifera* has a sugar content comparable to that of other bumblebee-pollinated plants (48%), but its nectar production rate (0.47 mg/hour per flower) is an order of magnitude greater than that of common native species (0.01–0.04 mg/hour). Pollinating bumblebees visit the invader approximately four times more frequently than they visit the native plant with the highest nectar production, marsh woundwort (*Stachys palustris*; Chittka and Shurkens 2001). Where the invader is established, visits to the native are reduced by 50% and its seed set by 25% (Figure 9.7). By altering the foraging behavior of a native pollinator, this introduced plant indirectly affects the fecundity of native plants. Although

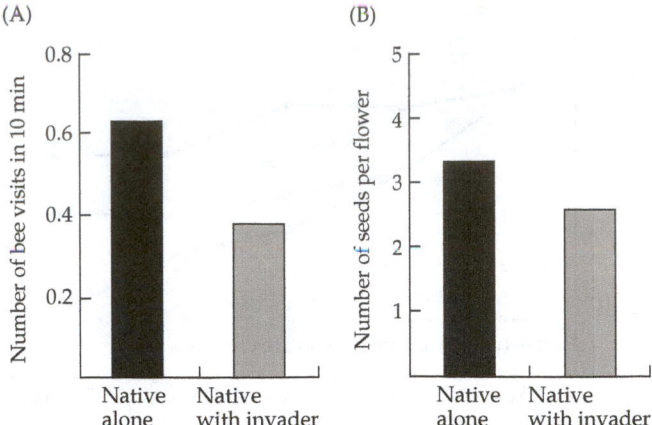

Figure 9.7 Bumblebee visits to (A) and seed set of (B) the native marsh woundwort *Stachys palustris* were significantly greater in pure stands (black bars) than in mixed patches with the introduced Asian annual *Impatiens glandulifera* (gray bars). Bars show mean values. (Modified from Chittka and Schürkens 2001.)

morphological and behavioral impacts may seem subtle, they have the potential to scale up to affect population-, community-, and ecosystem-level dynamics as well.

Genetic and evolutionary impacts

At the genetic level, hybridization and introgression by introduced species can fundamentally alter native species fitness and generate new—and sometimes aggressive—hybrids and strains. This process occurs in both animals and plants. In the North Pacific, a Mediterranean mussel (*Mytilus galloprovincialis*) appears to compete with its native congener directly and through hybridization (Geller 1999). Elsewhere, the introduced Mallard Duck (*Anas platyrhynchos*) threatens the persistence of the New Zealand Grey Duck (*A. superciliosa superciliosa*), the endangered endemic Hawaiian Duck (*A. wyvilliana*), and the endemic Florida Mottled Duck (*A. fulvigula fulvigula*) by hybridizing with all three (Rhymer and Simberloff 1996). In England, the introduced Atlantic cordgrass (*Spartina alterniflora*) hybridized with the native European *S. maritima*; the hybrids were not particularly notable until a chromosome-doubling created a new species, *S. anglica*, which was accidentally and intentionally distributed around the world for forage and erosion control. This new species has since earned a reputation as one of the world's most aggressive coastal invaders, altering water flow, sedimentation, and habitat structure.

Although evolutionary change can be difficult to document, there is growing evidence for this level of invader impact. For example, larvae of the native checkerspot butterfly (*Euphydryas editha*) in Nevada, now feed pri-

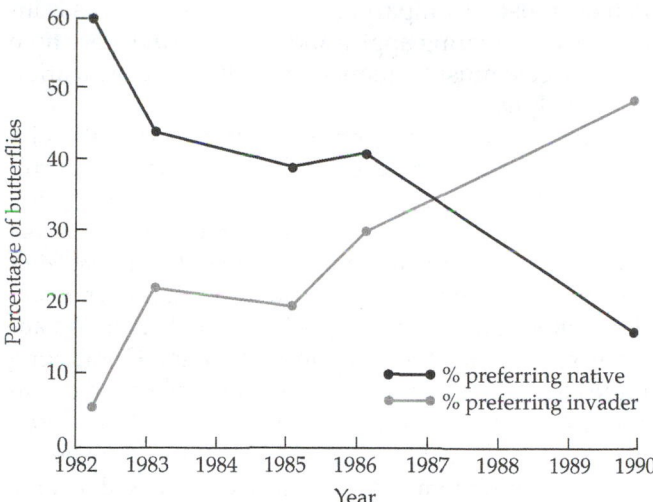

Figure 9.8 Percent of *Euphydryas editha* (a native checkerspot butterfly) preferring to oviposit on the native plant *Collinsia parviflora* (black line) versus the introduced *Plantago lanceolata* (gray line) over time in Nevada. (Modified from Singer et al. 1993.)

marily on the introduced European narrowleaf plantain (*Plantago lanceolata*), which was introduced a century ago in association with cattle ranching (Singer et al. 1993). In uninvaded areas, the butterfly prefers the native figwort (*Collinsia parviflora*), and in invaded areas it increasingly prefers the invader. Experiments with greenhouse-reared butterflies show that this preference is independent of where the animals live (Figure 9.8), strongly suggesting that the observed change is evolutionary. This dietary change increases the dependence of native butterflies on the disturbed habitats into which *P. lanceolata* is spreading.

An invading population can inflict a powerful selective pressure on its new neighbors, and likewise may be selected upon by abiotic and biotic features of its new environment. Although they are harder to document, we may expect that evolutionary impacts are the rule, rather than the exception, in many invasions. Given this general possibility, what is the role of conservation biology in protecting not only populations but also the evolutionary future of species? This is a question of growing conservation concern as a new era dawns in the use of genetically modified agricultural crops. The future impacts of gene transfer and hybridization between engineered species and their wild relatives are a ripe area for experimental and theoretical investigation.

Ecosystem impacts

Certain invaders have a disproportionate effect on ecosystem characteristics including habitat structure, disturbance regimes, and nutrient cycling. By their very nature, invaders that play essential roles in ecosystem processes may well have dramatic impacts on diversity as well.

Cordgrass salt marshes (composed primarily of *Spartina* spp.) are a major biological community type on the east coast of North America, where they are threatened by habitat loss. On the west coast, however, introduced populations of these species are flourishing as they spread across hundreds of square kilometers of tidal mudflats. One of these, *Spartina alterniflora*, was introduced inadvertently to the Pacific coast in the late 1800s, either with oyster shipments or with solid ballast emptied from ships, and was planted intentionally in some bays for duck hunting cover and erosion control. Spreading by sexual reproduction and vegetative propagation, it has covered more than 6000 ha in one estuary, Willapa Bay, alone. At shoot densities up to $600/m^2$, it enhances sediment-deposition rates and effectively raises the substrate above the water level, creating coastal salt marsh in lieu of intertidal mudflat. Densities of resident clams, amphipods, anemones, and worms are all affected; further impacts on juvenile salmon and migrating birds have yet to be determined (Daehler and Strong 1996). Its congener *S. anglica* is a similarly if not more aggressive invader both here and elsewhere around the world.

On the Atlantic coast, a major cause of habitat loss is grazing by nutria (coypu, *Myocastor coypus*, a large rodent), which was introduced from South America to develop fur farms in the 1930s. Released and escaped individuals established a feral population that spread quickly, consuming already-threatened marsh habitat and increasing erosion. Trapping, coastal habitat degradation, and an increase in the population of native alligators all contributed to the nutria decline, but in recent decades falling fur prices have led to reduced trapping pressure and the population is increasing again. The species has a worldwide distribution of wild populations that consistently damage native plants, agricultural crops, and soil drainage (Gosling et al. 1988). Introduced to the U.K. in 1929, their subsequent eradication campaign—an expensive and decades-long hunting and trapping program—eventually led to their removal from this island coastline. In South America, strikingly similar ecosystem impacts have been caused by the invasion of the North American beaver (*Castor canadensis*) (see Case Study 9.1 by Christopher Anderson).

Other invaders influence disturbance regimes. For example, the European cheatgrass *Bromus tectorum* in North American prairies, the Central American grass *Schizachyrium condensatum* in Hawaiian woodlands, the Australian shrub *Hakea serica* in the South African fynbos, and the tropical American shrub *Mimosa pigra* on Australian flood plains all increase fuel and the frequen-

cy and intensity of fires in their invaded ranges. The new fire regime may favor the invaders over native plant species, and can alter nutrient cycling by volatilizing carbon and nitrogen and reducing nitrogen availability (D'Antonio and Vitousek 1992).

Nutrient cycling can also be affected more directly. The African tree *Myrica faya* possesses nitrogen-fixing root symbionts that allow it to invade volcanic substrates in Hawaii's Volcanoes National Park. By increasing nitrogen input to the system over four-fold, it renders the soil uninhabitable by native species adapted to low-nutrient conditions (Vitousek et al. 1987). Introduced from the Azores and Canaries in the 1800s, the tree is now so abundant in the park that control efforts have been abandoned.

In Yellowstone National Park, the tiny New Zealand snail *Potamopyrgus antipodarum* was inadvertently introduced in 1994 and now reaches extraordinary densities of up to 20–500 thousand per m^2 in the streams it has invaded. Because of its high abundance, this algae and detritus eater dominates local carbon and nitrogen cycling. In a recent survey, Hall and colleagues (2003) estimated that the snail makes up the vast majority (97%) of the primary consumer biomass, and consumes most (about 75%) of the gross primary production in this system. It also produces (by excretion) some 65% of the stream's ammonium (NH_4^+), approximately 2.5 times the average contribution of total animal excretion in other freshwater systems. Like zebra mussels in other North American lakes and rivers, and the European gypsy moth in North American forests, this invader is substantially increasing nutrient cycling rates. The further effects of this ecosystem impact on other resident species are not yet known.

We have seen that invader impacts at morphological and behavioral levels may scale up to population-, community-, and ecosystem-changes. Further, impacts occurring at ecosystem levels may cascade down to affect communities and populations of native species. Given the wide-ranging and often unpredicted impacts of invaders, how can we rank invasion impacts to help prioritize our conservation strategies? How, for example, should conservation efforts be allocated among invaders with rapid impacts on a few species, compared to those with slower but larger-scale impacts on habitat and food web structure? To begin to address these questions, it is helpful to consider different approaches to measuring invader impacts.

Measuring invader impacts

Determining invasion impacts is essential to prioritizing control efforts. Given the multitude of invaders and their wide range of impacts, how can we identify those species that warrant the most attention from conserva-

tion biologists? Comparing impacts across species is indeed like comparing apples and oranges—but since time and budgets must be managed, tools for prioritization can be helpful.

A useful heuristic approach is to consider the elements of a species impact. In this sense, Parker and colleagues (1999) provided a framework for defining the impact of a species, I, as a product of the range that it occupies (R in, say, m^2), its abundance over that range (A in numbers or biomass per m^2), and its per-capita effect (E, the change caused by a single individual acting on another species or other response variable). Combining these elements gives us an expression for a species impact, $I = R \times A \times E$. The primary value of this framework is in highlighting three major ways in which a species can have a high total impact: Clearly, those with a large extent, high abundance, and strong per capita effect will have the strongest impacts. Box 9.1 by Ingrid Parker discusses this theme in more depth.

Some of the challenges in extending this framework are incorporating how E scales with A (which may be nonlinear), in totaling up all the direct and indirect impacts an invader has on multiple species or in multiple areas, and in considering how a species' impact may change over time. In applying this index, it would rapidly become clear that in general, R and A will typically be better known than E for a given invader. This is largely because R and A are easier to measure, as opposed to invader impacts.

More progress has been made in predicting invader spread, where most models draw from classic equations originally developed by Fisher (1937) and Skellam (1951) to model the spread of genes and populations. In these models, the rate of a species' population growth is a function of both its intrinsic growth rate over time (r) and its spread over space. For a single species moving in a one-dimensional habitat, its change in abundance (N) over time (t) at a point in space can be represented as:

$$\frac{\delta N}{\delta t} = rN + D\frac{\delta^2 N}{\delta x^2}$$

The lower-case Greek letter δ (delta) represents "change"; specifically, the change in N relative to one factor (time) of the two (time and space) that influence it. (If you are mathematically inclined, you may recognize δ as the symbol for a partial differential equation.) On the right-hand side of the equation, the first term is the familiar exponential population growth; it can be replaced with a logistic growth term, as appropriate. The second term represents the population's spatial movement as a diffusion coefficient, D, multiplied by a term we read as the second derivative of the population change in space—a quantity that may admittedly be difficult to visualize but is mathematically tractable. A basic

BOX 9.1 Understanding the Impacts of Nonnative Species

Ingrid M. Parker, *University of California, Santa Cruz*

It has become increasingly obvious that nonnative species are one of the primary threats to biodiversity today. Nonnative species can have a range of impacts, from suppressing native populations, to altering ecosystem properties such as nutrient cycles or disturbance regimes, to hybridizing with native species. How do we measure the impacts of introduced species? How do we compare the effects of different species? The answers to these questions have important implications for ranking invaders in the policy arena and for developing conservation priorities.

Preventing the introduction of new pest species depends on predicting which species are likely to become pests. Thousands of species have been introduced by humans to regions far from their native ranges, but only a small subset of those survive independent of human cultivation, with an even smaller set invading natural habitats beyond disturbed sites such as roadsides. Only a small subset of these invaders become aggressive weeds or pests. Mark Williamson noted that each of these steps along the invasion process reduces the number of species by about an order of magnitude; he called this "the tens rule" (Williamson and Fitter 1996). Which species make it through each of these invasion filters? It is very difficult to predict. Statistical models offer some insights into which introductions will be initially successful (National Research Council 2002), but very little quantitative research has been focused on trying to understand which traits separate high impact from low impact invasions.

One challenge to using statistical models to predict which species will have large impacts is simply the matter of defining what we are trying to predict. While it is relatively easy to determine which species achieve self-sustaining populations, people do not always agree about which ones have a significant impact. First, people with different perspectives focus on very different measures of impact. For example, a social scientist or policy-maker might use a monetary scale in quantifying how much society spends to control the common pigeon ($1.1 billion per year; Pimentel et al. 2000), or how much is lost in crop revenues to nonnative insects ($13.9 billion per year; Pimentel et al. 2000). Estimates of monetary impacts are useful in drawing attention to the general problem of biological invasions, and in directing resources toward solving particular invasion problems (for example, in 1990 the U.S. Congress passed the Non-Indigenous Aquatic Nuisance Prevention and Control Act primarily in response to the costly invasion of the zebra mussel). However, a purely financial scale is heavily weighted toward utilitarian values and may overlook species that have large negative effects on native species without greatly influencing human affairs.

A useful conceptual framework divides impact into three components: range, abundance, and local per capita effect (Parker et al. 1999). Geographic range is the most simplistic measure of impact, and it is often used when other information is not available. For example, while we have few quantitative data on the ecological interactions between *Hypericum perforatum* and native species in the western United States in the 1930's, we do know that its distribution reached more than 100,000 ha at its peak (Holloway 1957); this number is frequently used to indicate why this species was perceived as a serious weed. Geographic range is a useful measure for tracking change in impact over time, but to assume that overall impact increases precisely with area implies that ecological effects are similar everywhere.

It is naive to claim that local effects are similar everywhere, or to assume that two species that have the same range have the same impact. For one thing, abundance can vary greatly among species or among sites within a species and may determine much about the ecological impact. It makes intuitive sense that for the same invader, a population with higher density should exert a higher impact, simply because more of the available resources are being usurped by that one species. For illustration, imagine a simple case in which the limiting resource is space. A nonnative species that dominates more of the space will have a greater impact on that ecosystem in terms of reductions in the populations of native species, in terms of ecosystem function, and so on.

The third component of impact is per capita effect. Ecologists don't believe that range and abundance by themselves explain everything about impact, because all species are not equal. Some species seem to have impacts greatly out of proportion to their size or density; for example, a few introduced cats on an island can decimate a population of nesting seabirds (Keitt et al. 2002). Several "rules of thumb" have been put forward to explain why certain subsets of species seem to have large impacts.

1. Keystone predators: Top predators can have strong, top-down effects on the structure of natural communities (Paine 1969). Just as conserving populations of native top predators plays an important role in preserving natural ecosystems (see Chapter 3), adding novel top predators can have cascading effects.

2. Ecosystem engineers: Some species seem to have large impacts on other species because they change some fundamental aspect of the physical or chemical environment; these have been called "ecosystem engineers" (Jones et al. 1997). Plants such as beach grasses that stabilize dunes (Wiedemann and Pickart 1996) or nitrogen-fixing shrubs on lava flows (Vitousek and Walker 1989) qualify as ecosystem engineers.

3. Filling the empty niche: Both of the previous categories are special cases of a more general suggestion, which is that organisms that play a novel role in an ecosystem have a greater probability of caus-

Continued on next page

ing major impacts. Part of the support for this idea comes from the observation that some of the worst cases of nonnative species impacts, including extinctions of native species, come from island systems. Islands are less diverse and also tend to be missing whole functional groups of organisms. The native species are therefore poorly adapted to dealing with a nonnative species from a missing functional group, as in the case of cat predation on unsuspecting seabirds.

Beyond the effects of each individual introduction, we should try to grasp the big picture of biological invasions. There may be nonlinear threshold effects of multiple introductions, where one nonnative species facilitates the success or impact of others in an "invasional meltdown" (Simberloff and Von Holle 1999).

Over evolutionary time, only the major faunal interchanges that occurred with events such as the rise of the Panamanian Isthmus (which linked North and South America for the first time) come close to the scale of our current invasion crisis (Elton 1958). The paleontological record from these faunal interchanges provides us with the insight that we should expect massive changes in species composition, and extinctions may be a common outcome (Vermeij 1991).

prediction of this model is that over a large enough time scale, an invading species will spread with a constant speed, $c = 2\sqrt{rD}$, whether we assume exponential or logistic growth.

Perhaps surprisingly, this simple model has accurately captured the observed spread of a number of invaders, including North American muskrats in Europe and European starlings in North America. In general though, the model underestimates spread rates of many invaders, including cheatgrass, pine trees, and Argentine ants (Higgins and Richardson 1999; Suarez et al. 2001). Incorporating long-distance dispersal to represent mechanisms such as high winds, railway lines, or intercontinental shipping often generates a more accurate map of spread. More sophisticated spread models provide further insight by incorporating factors such as Allee effects, age/stage structure, interspecific interactions, spatially heterogeneous environments, and stochasticity. The insight provided by such modeling exercises into the processes driving invader spread can help inform and prioritize management.

The measurement and modeling of the third component of a species impact, its per capita effect E, is considerably more challenging. Exactly what should be measured, and how? A standard metric of species impact is its per-capita interaction strength with another species, which can be as simple as the difference in a response variable between invaded and uninvaded sites divided by the abundance of the invader. If for example we were measuring the per-capita effect of an introduced predator on its prey abundance, N, we could calculate $E = (N_I - N_U)/A$, where the subscripts refer to invaded (I) and uninvaded (U) areas and A is the invader abundance (Berlow et al. 1999). Modified versions of this index can distinguish keystone from numerically dominant invaders and account for variation in interaction strength over space or time (Berlow et al. 1999).

Because every successful invader must consume resources and in turn be consumed, every invader must have some effect on the system it invades. Studies of native species have illustrated that in many communities there are a few very strongly interacting species and the majority are comparatively weaker interactors. So it appears for invaders, too: Most introduced species have relatively smaller effects, but some have dramatic effects that can be of major conservation concern. These impacts range across ecological and evolutionary scales, and can cascade from direct to indirect effects. Invaders can also modify each other's impacts, either exacerbating or mediating them. The means of invasion impact are the familiar ecological interactions of predation, competition, mutualism, parasitism, commensalism, and amensalism. Predicting particular impacts, however, is an extraordinary challenge. This is particularly so over longer-term scales as invaders and the species with which they interact evolve. Measuring impacts is key to developing conservation strategies for responding to invasions. We can borrow from ecological metrics to help in this, but must also remember to evaluate the broader context of a species' economic and aesthetic impacts as well. Having considered the range and scale of invader impacts, we will now turn to the question of when and where invasions are most likely to occur.

What Factors Determine Whether a Nonnative Species Becomes Invasive?

Which species are the most successful invaders? Which communities are the most invasible? In answering these questions, we may be able to identify certain species, or types of species, that are high-risk invaders, and certain high-risk communities that should be prioritized for protection from, and restoration following, invasion.

A figure of the invasion process (Figure 9.9) illustrates that the final outcome of an incipient invasion depends

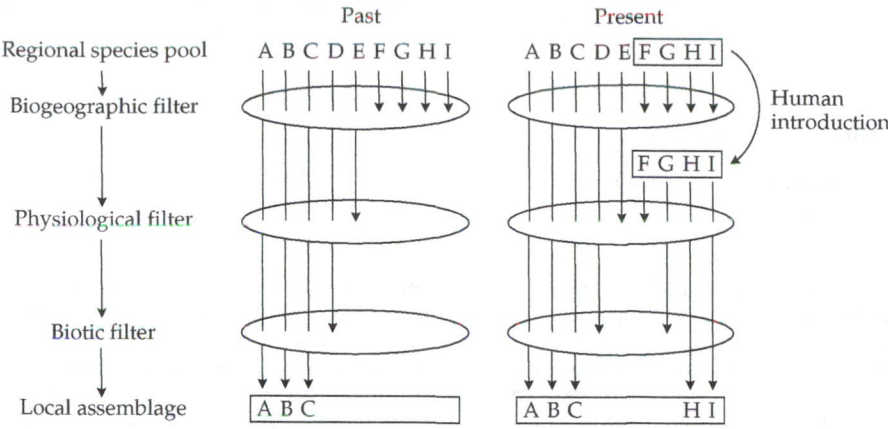

Past Present

Regional species pool A B C D E F G H I   A B C D E F G H I

Biogeographic filter

Human introductions

F G H I

Physiological filter

Biotic filter

Local assemblage A B C   A B C   H I

Figure 9.9 The series of events leading to a successful invasion can be pictured as a series of bottlenecks, where fewer individuals of each species, and fewer species overall, pass to each successive stage. The biogeographic filter represents physical barriers for some species. Human activities bring those species that were separated naturally into contact. The physiological filter represents the match between the species and existing climate, while the biotic filter represents the interactions between native and the introduced species. (Modified from Carlton 1985 and Rahel 2002.)

on a series of criteria being met. First, an invasion pathway must be in place to connect a species to a new environment outside its native range. We will examine invasion pathways in more detail in the next section; for now, we will simply note that a pathway must exist, and must deliver a sufficient propagule pressure, or number, frequency, and quality of organisms. Population growth also requires that the invader's physiological tolerances and life history match the new environment well enough for survival and reproduction. In addition, the net effects of interactions with resident species must also allow population growth. A successful invasion results when the match among invasion pathway, the species' biology, the abiotic conditions, and the biotic interactions are sufficiently good—which begs the question of what, exactly, is good enough?

Somewhat tautologically, we can say that under the right conditions any species can be a successful invader and any community can be invaded. Not every introduction succeeds, however—in fact, most do not—so clearly the "right" conditions are only infrequently met. Because each invasion involves a unique combination of the invasion pathway, the invader, the abiotic environment, and the invaded biotic community, the challenge is to tease out generalities in each of these stages that contribute to overall success.

Of course, in treating each stage separately we run the risk of missing key interactions between an invader and its pathway, or its new community. The species traits facilitating invasion are relevant only in the context of particular invasion pathways and communities, and the invasibility of a climate and community will depend on the invaders and pathways in question. For example, since the Suez Canal opened in 1869, the eastern Mediterranean has been invaded by over 200 species from the Red Sea, whereas the Red Sea has been invaded by only a handful of Mediterranean species. Does this mean the Mediterranean is particularly vulnerable to invasions? Or

that Red Sea species are particularly poor invaders? Or simply that there is a dominant direction to species transport through the canal? Distinguishing among these possibilities is clearly important for developing appropriate management actions.

Although we will discuss propagule pressure, invader characteristics, and community characteristics separately, it must be remembered that all three interact to determine the ultimate success of an invasion. These themes are discussed by Hugh MacIsaac in Case Study 9.2, which enhance the understanding of invasions in the Great Lakes of North America. Importantly, we can often learn a great deal about invasive species through international collaborations (Essay 9.2).

Propagule pressure

By definition, all invasions require a pathway that can transport species beyond their native range. Pathways can be at the global scale of international trade, the regional scale of road construction, or the local scale of planting nonnative garden shrubs and releasing pet cats. The particular invasion pathway will determine the propagule pressure, or the quantity, quality, and frequency of arriving organisms. Two of the most commonly recurring predictors of a successful invader are association with human activity, and a previous history of successful invasion (which means, among other things, a larger pool of source populations). These observations reinforce the general sense that propagule pressure is a major force in determining invasion success. Besides the obvious reason that species can't invade if they aren't delivered, the particular importance of propagule pressure may stem in large part from the dynamics of small populations.

Most invasions, whether accidental or deliberate, begin as relatively small populations. In addition to having to tolerate the abiotic and biotic features of the new environment, the invader initially faces certain problems

ESSAY 9.2

Global Exchange
Introduced Species and International Collaboration

Dianna K. Padilla, *Stony Brook University,* **Alexander Y. Karatayev,** *Stephen F. Austin State University,*
Lyubov E. Burlakova, *Stephen F. Austin State University*

■ Like most things in life, those that are closest or the most familiar to us get the most attention. This is especially true in science where we rely on our intuition, past personal experience, and literature. For most North American scientists, this means literature written in English, published in North America. Rarely is research in other parts of the world part of our scientific inquiry. This tendency creates problems for research on invasive species, especially invaders from distant countries where different languages are spoken, or where different political systems and economies result in isolation of scientific information and independent evolution of the ways in which we investigate questions as well as how we report our findings.

The zebra mussel, native to Eastern Europe, first arrived in North America in the mid-1980s. It was immediately clear that this new invader had the potential to cause enormous economic and environmental damage. North America was not the first place to be invaded by the zebra mussel, nor would it be the last. Zebra mussels had been invading across Europe from native waters flowing into the Black, Azov, and Caspian Seas for over 200 years (Karatayev et al. 2003). During this time scientists in Eastern Europe, especially in the former Soviet Union, studied many aspects of zebra mussel biology, ecology, and impacts (Karatayev et al. 1997, 1998). But, because of language and political barriers, most of this information, knowledge, and expertise was not available to North American scientists. Thus, when zebra mussels were first discovered in the Laurentian Great Lakes, North American managers and scientists scrambled, trying to figure out where they might spread, how they might be controlled, and what impacts they could have on freshwater systems. Annual meetings between Canadian and U.S. scientists and managers were held to increase information exchange, and eventually scientists from former Soviet countries were invited to attend these annual meetings. Because the cost of the trip was more than 10 years of salary for Belarus professor Alexander Karatayev, financial assistance was critical to making the meeting possible.

The Madison conference spurred interest in research and approaches that each of us might learn from the other. Although in North America we were still at the beginning of a very steep learning curve, Karatayev had studied zebra mussels for his entire career and was a leader in the long tradition of hydroecology in the former Soviet Union. Recognizing the great potential in international collaboration, he invited several North American scientists, including Dianna Padilla, to attend a conference at the Belarussian National Academy of Sciences.

The collaboration that developed among us presented many unexpected challenges and rewards. Our work has been full of the difficulties of trying to join scientific ideas and traditions that developed independently, and jargon that does not translate well. Each of us had experiences with different scientific approaches, our political and economic situations were very different, and of course we faced a language barrier.

The Internet played a crucial role in the development and continuation of our collaboration, allowing us to write a grant proposal in a short time. Our initial grant brought Karatayev and Lyubov Burlakova to Padilla's lab for two extended visits for the sole purpose of increasing information exchange. We first reviewed what was known about the impacts of zebra mussels on freshwater systems in the former Soviet Union, and tied together changes in both biotic and abiotic parts of the system as well as structural and functional changes induced by the introduction of zebra mussels (Karatayev et al. 1997). The three of us would sit around the computer terminal and go word by word, sentence by sentence, struggling over meanings, how to translate into the language of North American freshwater science, debate the data in support of conclusions, and evaluate whether the conclusions were supported by, or different from, early North American data and why. By the time *Dreissena* was found in North America, over 1000 papers on various aspects of the biology, ecology, and control of zebra mussels had been published in journals in the Soviet Union in Russian. Attempts to obtain Russian literature usually only reviewed a handful of papers that often lead to incorrect conclusions. For example, based on this limited information, several papers published in North America in the 1980s listed significant differences in the biology between European and North American *Dreissena*. However, it is now clear that there are no differences. If early collaboration had been more effective, we would have saved precious time, effort, and money, and could have made faster progress addressing key questions about this invader.

Our work has increased general awareness of research from Eastern Europe on zebra mussels. Due to their experiences, Karatayev and Burlakova have become leaders among their Eastern colleagues in writing in English, publishing in international journals, and employing experimental designs and statistical analyses that are part of Western aquatic research. Their impact on Western science thus far includes identifying the historic patterns of the spread of zebra mussels as well as providing extensive long term data on lake benthos before and after invasion, not only at the population and community levels, but also for ecosystem level parameters (Karatayev et al. 1998, 2002, 2003). Our efforts translating and providing information from the literature published in Russian have been fruitful, and more and more authors are using this information and seeking the help and cooperation of Eastern scientists.

In our efforts to cope with invasive species, we may be able to avoid reinventing the wheel, by using knowledge from elsewhere. Our experiences tell us that we can gain a great deal, both personally and scientifically, by opening the doors to international collaborations. ■

unique to small populations. These include Allee effects, demographic stochasticity, environmental stochasticity, and spatial heterogeneity. These influences on population dynamics are the same challenges faced in the recovery of small populations of endangered species (see Chapter 12). Because Allee effects and demographic stochasticity in particular are functions of population size, the initially introduced population must grow to exceed some threshold abundance to be free of them. The threshold population size will vary across species, but in general we can imagine that the higher the propagule pressure—the larger and more frequent the initial populations—the higher the chances of a successful invasion. By the same token, effective prevention of invasions may be thought of as reducing propagule pressure below a threshold level that is too low for success.

Propagule pressure is not a constant. As long as human trade and transport continue to grow and shift around the globe, invasion pathways will grow and shift as well. These pathways obviously determine propagule pressure and therefore the potential for a species to invade or a community to be invaded. Thus a widespread and successful invader could be a species with a particularly high propagule pressure, a species that is especially good at establishing and spreading once it has been delivered, or both. Similarly, a community could be invaded by many species as a result of a high incoming propagule pressure, or of its inherent invasibility. In the next sections, we consider the species and community traits that may contribute to invasion success.

Invading species characteristics

Any species could, in principle, invade successfully somewhere. But some species seem better at it than others, and it is of great interest to conservation biologists to identify which species or types of species are most likely to be successful. We might begin by identifying characteristics that we think are likely to make a species adaptable and fast-growing: for example, high fecundity (e.g., *Impatiens glandulifera*), ability to spread vegetatively (e.g., the alga *Caulerpa taxifolia*; see Case Study 9.3 by Susan Williams), parthenogenetic or hermaphroditic reproduction (e.g., the ctenophore *Mnemiopsis leidyi*), or broad physiological tolerances and diet (e.g., brown tree snake). Undoubtedly we could find an example of a successful invader for every trait we could think of. At the same time, though, there would be plenty of species with these same traits that were not invaders, and other species that invaded successfully despite their lack of these particular characteristics.

It is perhaps not surprising that a robust list of key characteristics for all invaders has proved elusive. There are, after all, likely to be important differences among taxonomic groups that make certain characteristics more relevant for some organisms than for others. Also, an invasion pathway may favor different characteristics than the establishment process, so successful invaders will be selected for different features at different stages. Although an overall list of invader characteristics may not be a realistic tool for conservation, there has been considerable progress in reviewing taxon-specific characteristics, and in isolating particular characteristics important in individual invasions.

A taxon-level perspective has been developed successfully for several plant and animal groups. For woody plants, Reichard and Hamilton (1997) classified 114 introduced species established in North America on the basis of their likelihood of spread. Using a statistical analysis, they ranked 11 traits in their ability to correctly classify these species. Five of the traits contributed significantly to the likelihood of spread (particularly, other invasions and the length of time the fruit is on the plant), and six were negatively associated (particularly evergreens and intraspecific hybrids); the ordering of traits differed when the characteristic of other successful invasions was excluded (Table 9.3).

A similar list was obtained by Rejmánek and Richardson (1996) who classified 24 species of pine tree (*Pinus* spp.) introduced worldwide for cultivation. They found that only three of ten life history characteristics contributed significantly to a *Pinus* species' likelihood of successful invasion: short juvenile period, short interval between seed crops, and small seed mass. The first two characteristics contribute to rapid population growth; the third might indicate higher numbers of seeds and broader dispersal, or faster seed germination or seedling growth.

For fishes introduced to the North American Great Lakes, Kolar and Lodge (2003) classified 45 species on the basis of 25 characteristics. They found that with only four (growth rate, temperature and salinity tolerance, and egg size), they could correctly classify 87%–94% of the fishes as successful, quickly spreading, and high impact (Table 9.4).

It is worth noting that the success of some invasions appears to result from a quirk of fate. Individuals of an invading population are a subsample of the native population, and may carry particular genetic traits or experience genetic bottleneck effects that increase their invasion ability. Such is the case for invading populations of the Argentine ant *Iridomyrmex humilis*, which in their native range devote considerable effort to intercolonial aggression and defense. Introduced populations, which are a major threat to native ants and other biota, are much more cooperative among colonies but maintain their interspecific aggression. This altered social behavior, which appears to have a genetic basis, is found only in the introduced populations.

Similarly, a species may through lucky accident shed its natural enemies in the course of invading a new location. For example, if the few individuals to start a new

TABLE 9.3   *Characteristics Asociated with Invasiveness in 114 Species of Nonnative Woody Plants in North America*

Variables	Function coefficients	
	All species	Excluding "invades elsewhere"
Invades elsewhere	−0.349	n/a
Reproduces vegetatively	−0.301	−0.348
Length of time fruit on plant	−0.281	−0.447
Flowers perfect	−0.238	—
Flowers in winter	0.260	—
Cold needed for seed germination	0.323	—
Native to temperate Asia	0.342	0.330
Leaves evergreen	0.414	0.637
Intraspecific hybrid	0.521	0.517
Native to North America	0.700	0.910
No seed pretreatment needed for germination	—	−0.332

Note: Negative values indicate characteristics that contribute significantly to spread; positive values indicate characteristics associated with *not* spreading.

Source: Reichard and Hamilton 1997.

population happen to be free of a parasite or disease, they may be able to grow and spread more rapidly. This "enemy release" hypothesis need not apply forever, as an invader is ultimately likely to acquire new parasites in the new environment, but a sufficient delay may enhance its initial prospects of success.

Because invasion success depends ultimately on the interaction between the invader and its new community, it may be important to consider both factors simultaneously in evaluating invasion success. It is a general ob-

TABLE 9.4   *Biological Characteristics Associated with Invasion Success, Rapid Spread, and Pest or Nuisance Status for Introduced Fishes in The North American Great Lakes*

Invading fish species	Biological characteristics
Successful	Faster growth, wider temperature tolerance, wider salinity tolerance
Quickly-spreading	Slower growth, wider temperature tolerance, lower survival at high temperature
Nuisance	Better survival at low temperature, wider salinity tolerance, smaller eggs

Source: After Kolar and Lodge 2003.

servation that species novelty, such as its ability to use a resource in a new way, may contribute to its success and impacts. Obviously, novelty can only be invoked in the context of the community being invaded. Such is the case, for example, for nitrogen fixation by the tree *Myrica faya* in Hawaii, and a tolerance of intertidal mudflat habitat by the Atlantic *Spartina alterniflora* in the northeast Pacific. Novelty in the context of the community may also account for the high success (and impacts) of invasions on islands. This applies both to communities invaded by a new predator (such as the species on Guam faced by the brown tree snake) and to competing species: For example, a review of plant invaders found that the percent of introduced species was 2.6 times higher on islands than on mainlands (Figure 9.10). The role of novelty is consistent with the biotic resistance hypothesis, in that an invader that uses a resource for which there is minimal competition may face less resistance than one using a resource in heavy demand. In this way, it is also consistent with the resource availability perspective on invasibility.

In other cases, novelty is not clearly of key importance. On the mudflats of northern California, for example, the western Pacific mudsnail *Batillaria attramentaria* was accidentally introduced via oyster shipments in the early 1900s. The invader is strikingly similar to a native mudsnail (*Cerithidea californica*) in size, life history, and diatom grazing rate, so it would initially have been difficult to predict which species would be the competitive dominant. Over time, it became clear that the invader was es-

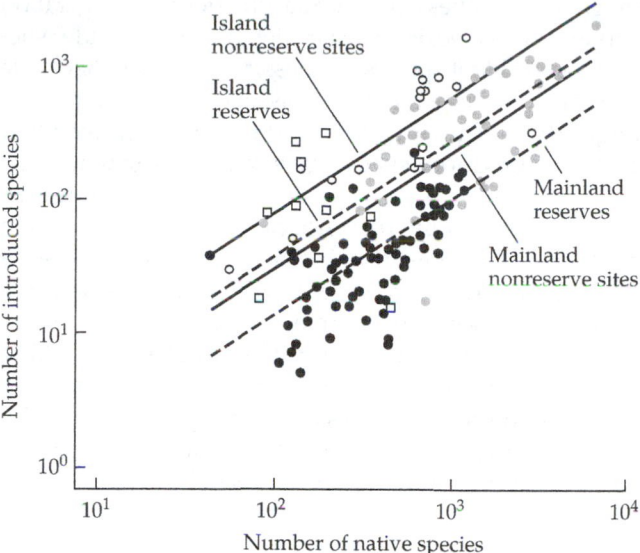

Figure 9.10 The ratio of introduced to native species for 177 sites and regions around with world, showing island nonreserve sites (open circles, solid lines) and reserves (open squares, dashed lines), and mainland nonreserve sites (gray circles, solid lines) and reserves (solid circles, dashed lines). Note log scale on both axes. (Modified from Lonsdale 1999.)

tablishing and the native was beginning to decline. Which species traits gave the invader the edge? Detailed field experiments showed that the invader had three advantages

over the native: a higher density-independent survival, a higher food conversion efficiency, and a lower susceptibility to a parasitic castrating trematode (Figure 9.11; Byers and Goldwasser 2000). By constructing a mathematical model of the interaction between the two species, it was found that the invader's survival rates were more important than its superior competitive ability or resistance to parasitism in determining its success (see Figure 9.11B). This type of detailed investigation provides insight into the specific mechanisms that drive invasion success and impacts, mechanisms that complement the broader-scale survey approach to characterizing species traits.

Invader traits that contribute to invasion success are useful in developing effective screening for proposed intentional introductions, and for prioritizing eradication efforts when unwanted new invaders are first detected. Are there similar generalities we can draw about the invasibility of different types of biotic communities?

Invaded community characteristics

Again, every biological community is evidently invadable when faced with the right introduced species. If some communities are more easily invaded than others, conservation efforts could be prioritized to protect vulnerable communities. Given a reasonable invasion pathway and propagule pressure, the invasibility of a community depends on two primary features.

First, the climate and habitat must be hospitable to potential invaders. To pick extreme examples, the Sahara

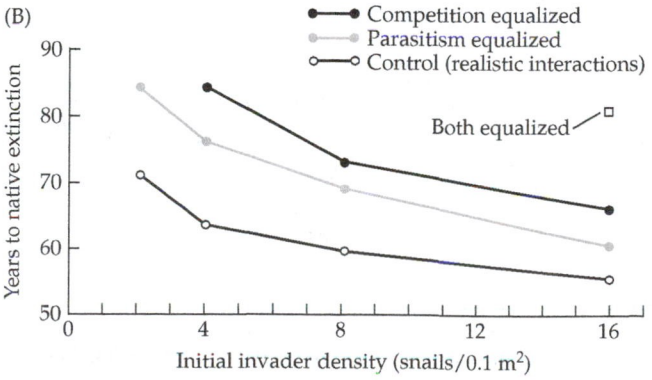

Figure 9.11 (A) Schematic of the observed interactions between the introduced mudsnail *Batillaria attramentaria* and the native *Cerithidea californica*, showing the invader's three advantages: lower mortality, stronger competition, and lower susceptibility to parasitism. Boxes show the interacting species, and arrows show the direction, type (positive or negative), and strength (indicated by arrow thickness) of impact. Dashed line represents the indirect competitive effect of the two species on each other. (B) Results of a computer experiment showing the average number of years to *C. californica*'s extinction, as a function of invader density, when the invader's advantage is modeled realistically (open circles), and when competition (black circles), parasitism (gray circles), or both (open square) are adjusted to make the species equal. The model shows that differences in both competition and parasitism contribute to the native species' mortality. (Modified from Byers and Goldwasser 2001.)

desert or Arctic ice floes may be less invadable by the majority of Earth's species simply because of their extreme abiotic environments. Second, the species richness, interaction strengths, and trophic structure of the community, must be able to accommodate new species. In his seminal work on invasions, Elton (1958) posited the **biotic resistance hypothesis** that species-rich systems were more stable and therefore less susceptible to species outbreaks and invasions. A proposed mechanism for this hypothesis is that in species-rich communities the greater overall use of resources reduces the available "niche space" for prospective invaders.

Tests of the biotic resistance hypothesis returned somewhat paradoxical results, as both positive and negative relationships have been found between invasion success and resident species richness. For example, in riverside sedge communities in California, broader-scale vegetation surveys show a positive association between invader success and resident species richness, whereas smaller-scale experiments show a negative association (Levine 2000) (Figure 9.12). This and other studies have suggested that perhaps introduced species generally do best where native species also thrive, but that in local interactions they may do worse in the presence of more competitors. On the other hand, there are numerous large- and small-scale vegetation studies showing a positive relationship between introduced and native species richness (Figure 9.13) (e.g., Macdonald et al. 1989; Keeley et al. 2003). Also, because the impacts of an invasion may include reduced or increased species richness, we clearly need further investigation to resolve the interesting relationship between invasion success, impacts, and species richness.

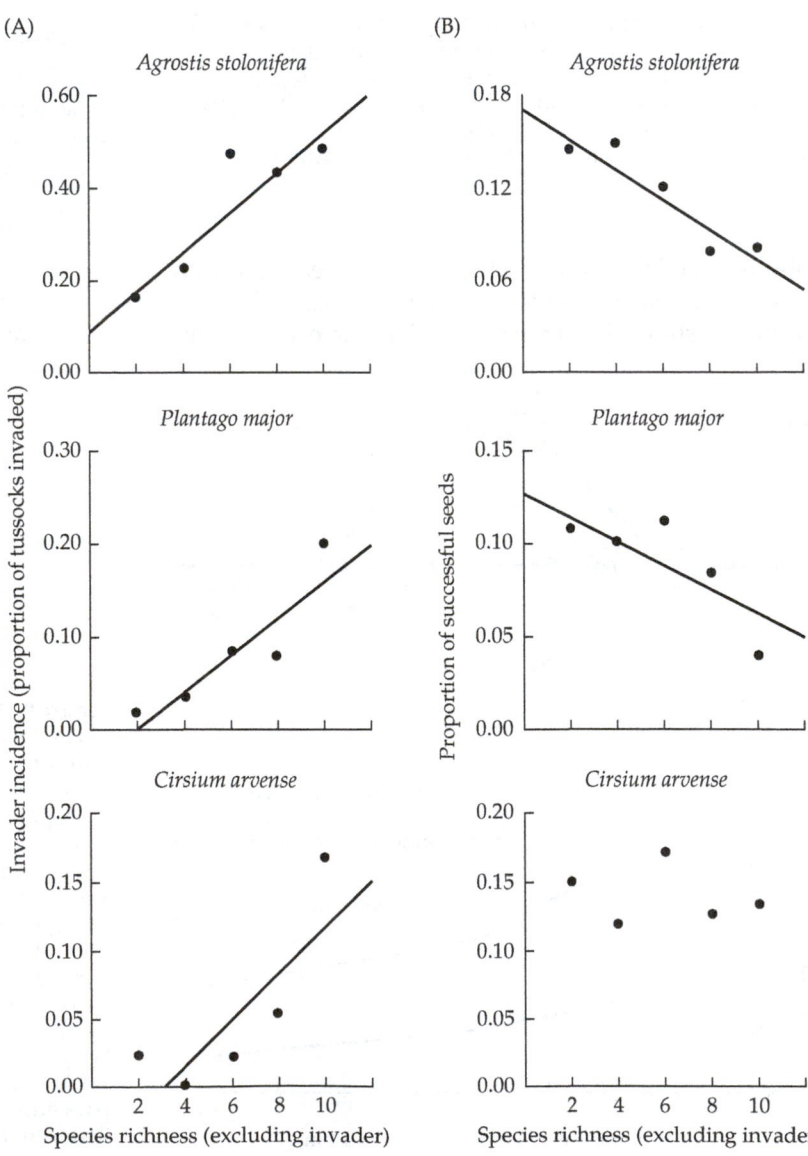

Figure 9.12 (A) In a streamside tussock plant community in northern California, the presence of the introduced Canada thistle (*Cirsium arvense*), common plantain (*Plantago major*), and creeping bent grass (*Agrostis stolonifera*) increased with increasing species richness. (B) In experimental transplants, however, the proportion of *A. stolonifera* and *P. major* (but not *C. arvense*) seeds that germinated decreased with increasing species richness. (Modified from Levine 2000.)

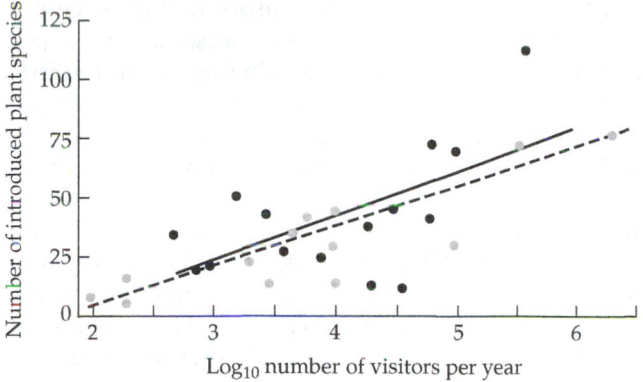

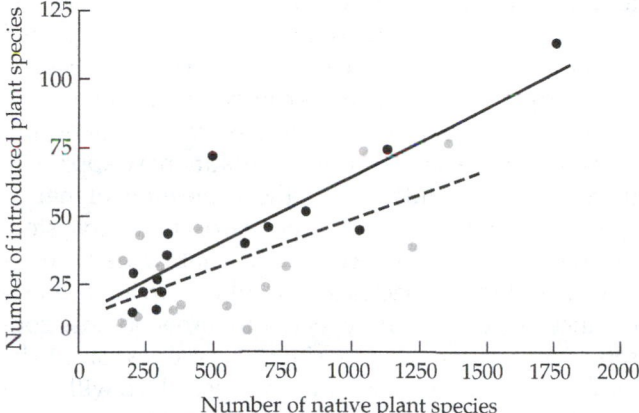

Figure 9.13 The number of introduced plants increases with (A) an increasing annual number of human visitors, and (B) the number of native plants, in terrestrial reserves in South Africa. Reserves in different biomes are shown with different symbols: solid circles (solid line regression) represent tropical dry forest, and gray circles (dashed line regression) represent evergreen forest. (Modified from Macdonald et al. 1989.)

A complementary hypothesis is that disturbance—either natural or anthropogenic—may make a community more easily invadable, particularly for plants (Lozon and MacIsaac 1997). This could be for several reasons. Disturbance may make resources available that could be exploited by any species, allowing invaders to establish a toehold. New forms of disturbance may disfavor native species that are not adapted to it, possibly favoring invaders that are. Invaders themselves can also be agents of disturbance, by altering natural disturbance regimes (e.g., fire or hydrology), by causing physical disturbance themselves (e.g., erosion or substrate replacement), and by simply consuming other species; in some cases, these alterations favor the new species and enhance its invasion success.

To fully understand the role of disturbance, it is important to distinguish between its role in delivering potential invaders and its role in facilitating the establishment of these invaders. In terrestrial reserves, for example, the number of introduced species is proportional to the number of human visitors (see Figure 9.13): Does this reflect the role of humans as an invasion pathway, or do visitors perhaps cause enough disturbance to make invasions easier? Similarly, in a review study across multiple reserves, the percent of introduced plants was 2.2 times lower than in nonreserve sites (Lonsdale 1999) (see Figure 9.10). Does this mean that creating reserves may provide some protection from invasions, or that we tend to establish reserves in areas that are already less invaded? Following establishment, the direct effects of an invader on resident species must also be clearly distinguished from those of the disturbance-mediated invasion, and the role of extrinsic disturbance must be disentangled from that of disturbance caused by the invader itself.

The biotic resistance and disturbance hypotheses are complementary if we view them from the perspective of resource availability. Davis and colleagues have proposed that a species-rich community with intense competition may be less invadable until disturbance alters the availability of resources, either by providing new resources (e.g., clearing a gap for light) or by reducing the effectiveness of competitors (e.g., through toxins), allowing an invader to establish (Davis et al. 2000) (Figure 9.14).

Empirical evidence supports this heuristic model. For example, nitrogen fixation by the introduced tree *Myrica faya* in Hawaii and by the native bush lupine *Lupinus arboreus* in coastal California has facilitated the success of other plant invaders that cannot themselves fix nitrogen. In another study, the success of an introduced Pacific

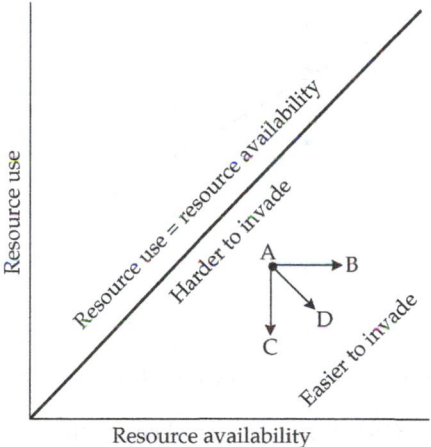

Figure 9.14 A representation of the conceptual theory that increased resource availability increases a plant community's susceptibility to invasion. The resource availability at a given time (A) could increase due to an increase in resources (B), a decline in resource use by resident species (C), or both (D). Each of these scenarios would predict an increase in the community's invasibility. (Modified from Davis et al. 2000.)

colonial ascidian (*Botrylloides diegensis*) decreased significantly with an increasing number of native Atlantic species already present (Stachowicz et al. 1999) (Figure 9.15). This appears consistent with the biotic resistance hypothesis, although the mechanism driving this result seems to be resource availability: The amount of bare space available for the invader to settle in decreased as the number of native species increased (see Figure 9.15).

Identifying general traits contributing to successful invasions is a useful exercise in understanding and predicting invasions, and in prioritizing control efforts. However, success and impacts may be difficult to predict, even for a species whose role in its native system is well understood. This is for two major reasons. First, the invading population may possess traits that make it unique in the new community, facilitating its invasion. Also, because individuals in this population are a subsample of the native population, they may carry particular genetic traits, or not carry particular parasites and diseases, that alter their potential role. Second, our understanding of a species in its native environment is determined in part by the resources and limitations the species experiences. If these differ, as might be expected in a new habitat, the species may play a different role.

Clearly, introduced species can swamp a native community and can have severe and often unpredictable impacts on native species. From a precautionary approach, therefore, it is evident that the protection of native species assemblages and their associated ecosystem processes demands protection against introduced species as well. If this approach also contributes to invasion resistance, then so much the better in terms of those conservation goals. We next consider how species come to be introduced outside their native ranges, which will motivate the subsequent sections on invasion prevention and control.

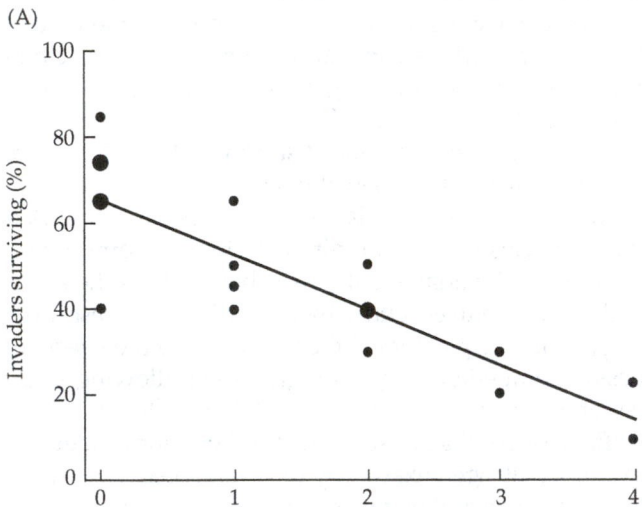

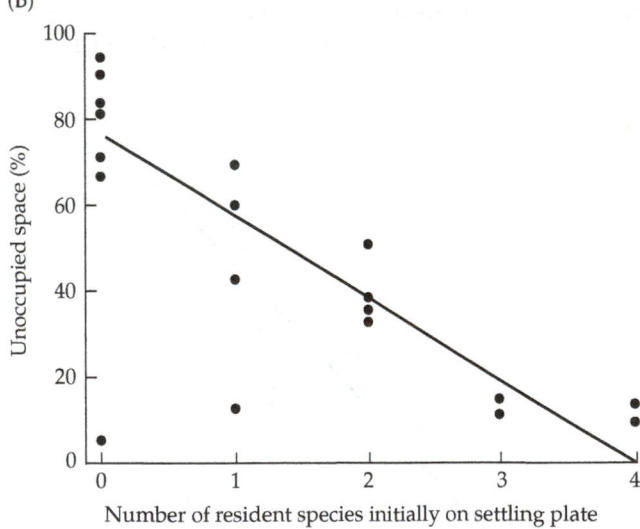

Number of resident species initially on settling plate

Figure 9.15 On subtidal settling plates colonized by 0–4 native marine invertebrate species in Long Island Sound, (A) the survival of naturally-recruiting introduced Pacific ascidians *Botrylloides diegensis*, and (B) the availability of free space (which decreased significantly with increasing number of native species). This suggests that competition for space may drive the observed relationship between invasion success and resident species richness. Larger dots indicate two data points on top of each other. (Modified from Stachowicz et al. 1999.)

How Are Species Introduced?

Through trade and travel, humans introduce species both intentionally and inadvertently. Intentional introductions began long ago, as humans carried crop and livestock species with them across Africa and Eurasia and into the Americas. Diseases and parasites—unintentional and undesirable hitchhikers—were carried along as well. Given the tremendous impacts of some invaders, and our still-developing understanding of the nature of invasive species and vulnerable communities, there is a strong motivation to prevent and control unwanted species introductions. To achieve this goal we must first understand how they occur.

Unintentional pathways

Species can hitchhike on any type of human transport from airplanes to cars to shirt buttons made of tropical nuts (Elton 1958). For example, the growth of global shipping has shaped—and continues to shape—ecological communities through species invasions.

The advent of large-scale Eurasian maritime traffic some 600 years ago ushered in a new era of dispersal and colonization. Chinese maritime trade to India and East Africa began in the early 1400s; by the end of the century,

Portuguese vessels had sailed along the coasts of Africa and India, and Spanish, British, and Dutch vessels to the Americas. In the 1700s, Russian exploration began in the North Pacific. The wooden hulls of these sailing ships carried entire communities of marine organisms, with meter-thick assemblages of crabs, barnacles, mussels, tunicates, worms, and hundreds of smaller organisms sailing across the world oceans (e.g., Carlton and Hodder 1995).

Ships at the time carried solid ballast of rock and soil that was loaded at the departure port for stability during ocean crossings, and deposited on arrival. This ballast contained scores of plant seeds and soil invertebrates, including the European earthworm *Lumbricus rubellus* that now alters soil structure and plant diversity in New Zealand, Australia, and North America. Indeed, some records of European plants in North America are known only from old piles of discharged solid ballast (Hickman 1993).

By the mid- to late 1800s, wooden sailing ships with solid ballast had been replaced by steel-hulled steam ships carrying water ballast. Water ballast is harbor water that is taken into tanks onboard a ship. The water enters through a metal grate that permits planktonic organisms, and sometimes larger crabs, shrimp, and fish, to enter freely. Any species that survive the voyage in the ballast tank may be released when the water is discharged (Figure 9.16; see Case Study 9.2). Most major invertebrate phyla, as well as algae, bacteria, and viruses, have been collected from ballast water; these are but a small sample of the thousands of species transported daily in the billions of tons of ballast water of the global commercial shipping fleet. The zebra mussel and ctenophore are both examples of high-impact species introduced via ballast water.

As global shipping evolved, new routes were constructed to shorten the travel time between major continents. The Suez Canal joined the Mediterranean and the Red Seas in 1869, the Erie and Welland Canals linked the Laurentian Great Lakes to the St. Lawrence estuary by 1829, and the Panama Canal linked the Atlantic and Pacific basins in 1914. Species have traversed all three, but primarily the Suez Canal; traversing the Panama Canal and the St. Lawrence Seaway involves sharp changes in salinity that comparatively few organisms can tolerate. Certain migratory species however, can readily make the transition: The sea lamprey (*Petromyzon marinus*) invaded through the Great Lakes from the St. Lawrence River, and has decimated native fish stocks and consistently defied containment and eradication since its arrival in the early 1800s.

Because ships were sailing the oceans long before natural historians were recording species names and distributions, we must assume that today's species include a number of unrecognized historical invaders. Indeed, many cosmopolitan marine species of uncertain origin may reflect this history of ship-mediated translocation. The refinement of genetic techniques may allow us to reconstruct the past of some of these early hitchhikers.

Figure 9.16 Ship deballasting in a harbor. (Photograph by M. Wonham.)

Today, the size and speed of ships continue to increase. Larger volumes of ballast water are being delivered in shorter transport times, increasing the quantity and quality of propagules discharged into destination ports. In addition, shifting global markets mean that the routes connecting source and destination ports continue to change, creating new combinations of prospective invaders and recipient regions. There is every reason to expect that, without regulation, the number of shipping-mediated invasions will continue to increase. We could trace a similar invasion history for the trains and train tracks and vehicles and roads that crisscross the continents and serve as pathways for plant and animal invaders on land.

Intentional pathways

Species are introduced intentionally for agriculture, aquaculture, recreation, and ornamental purposes. As European colonizers spread out across the globe, they longed for familiar species from home, and back home, a growing European market demanded the import of exotic animals from abroad. The formation of the Societé Zoologique d'Acclimatation in France, followed by the Acclimatization Society of the United Kingdom and sister societies in Russia, Australia, New Zealand, and Hawaii, led to a mass translocation of wildflowers, songbirds, and birds and mammals for food and sport (Lever 1992).

Government agencies have had a similar mandate to introduce species, particularly fishes. One of the first and most widely distributed was the German carp, which was delivered across the continent as an easy species to grow in local and private ponds. Now carp are major pests, destroying submerged macrophytes and increasing turbidity in nearshore freshwaters. Subsequent introductions of some fish are responsible for declines of some amphibian species (see Case Study 3.1).

Fishery-related invasions can also include the introduction of prey species. In the 1970s, the mysid *Mysis relicta* was introduced to Flathead Lake, Montana, to enhance the prey base for the recreational fishery on nonnative kokanee salmon (Spencer et al. 1991). As it turned out, the shrimp competed with the salmon for their prey. The resulting decline in salmon abundance was a likely cause for the reduction in their predators including Bald Eagles and grizzly bears, with a consequent loss in tourist dollars to the park. Thus, an introduction intended to increase fish abundance and tourism appears to have had the opposite effect.

Intentional introductions are commonplace for aquaculture, agriculture, erosion control, biological control, horticulture, and hunting. Unfortunately, many of these intentional introductions have resulted in unfortunate impacts to native species and communities. Our challenge is to refrain from introducing species that will become invasive (as discussed earlier), and to control the spread of the worst invaders.

How Do We Manage Species Invasions?

Invasion control

Many introduced species pose immediate threats to native species, and are currently altering biodiversity and ecosystem processes. These species are of pressing concern to managers, both within and outside reserves. At the same time, a continual flood of potential new invaders poses new threats. Control is clearly required to reduce or eliminate certain established invaders, but requires considerable effort, time, and expenditure, and may not always be effective. Unless we also focus on preventing future invasions, we will be unable to catch up with the escalating problems of invasion control.

Controlling introduced species is not a simple prospect. The removal of an unwanted species may require physical control (trapping, hunting, digging, pulling), chemical control (pesticides, herbicides, medication), or biological control (the enhancement of native or introduction of nonnative predators, parasites, or diseases). Some invaders are readily controlled by a single method, but more often several approaches are needed. When multiple invaders are established in an area, we need a more regional management perspective to address their combined control.

Species-based control

For a relatively contained invasion of a conspicuous species, physical control alone may be effective. Physical control involves trapping, digging up, and otherwise removing invaders. For example, a polychaete worm pest of abalone (*Terebrasabella heterouncinata*) was accidentally imported from its native South Africa to California in the 1980s. The worm does not directly kill its host, but by encrusting the edge of the shell it deforms the host's shell, making the abalone unmarketable. When the worm spread to native intertidal snail populations, a massive volunteer effort was initiated to reduce the host population below the level needed to sustain the worm. Some 1.6 million snails were hand picked from the rocky shore, and the worm has not been observed there since, an apparent success story in rapid-response, physical removal of an invader.

For a more widespread invasion, chemical control (e.g., with pesticides and herbicides) may be more effective (see Case Study 9.3). Yet, an obvious drawback of chemical control is the nontarget mortality suffered by other resident species, as well as the potential for such disturbances to enhance invasibility of the community.

Biological control is most highly developed in agricultural systems. Generally, biocontrol features the release of a predator, parasite, or pathogen to reduce the invader population. Early efforts prioritized the release of generalist predators, on the theory that they would persist better than specialists. Unfortunately the results can often been devastating. Nontarget biocontrol effects appear to be more the rule than the exception. Clearly, pre-release testing is essential to ensure that a prospective biocontrol species will not cause severe nontarget impacts. Judy Myers presents a discussion of the pitfalls and promise of biocontrol in terrestrial systems in Case Study 9.4.

In many cases, multiple approaches are required to target a particular invader. For the coastal Atlantic cordgrass in the northeast Pacific, physical control has included substantial digging, bagging, and removal by large groups of volunteers, and chemical control has included extensive spraying with the herbicide glyphosate (Rodeo®). Despite these efforts, the plant has continued to spread into the bay. Biocontrol with a Californian leafhopper (*Prokelesia marginata*), predicted to be the only method that will slow the spread of the plant, is underway (Grevstad et al. 2003).

An open question in conservation biology concerns the ethics of introducing a disease or similar control agent. Death caused by disease can involve sores, blindness, paralysis, or seizures; all of which are likely to be unpleasant to experience. On the other hand, if rabbits continue to propagate they will not only damage other species but also eventually die of starvation when their population exceeds carrying capacity. Current research on sterilizing pathogens may provide a solution that rests more easily with both ecosystem-scale conservation and individual-scale animal welfare. With future study, we may refine our use of other approaches to pest control, including sterile insect release, native species enhancement (integrated pest management), and changing agricultural practices.

There is tremendous effort and funding needed to develop a control program, which ultimately may or may not be effective. Ideally, each control effort for each invasive species will be tailored, through experimental and modeling research, to target the most efficient means to reduce the population and prevent unwanted impacts on nontarget species. One such research program has been effective in control of invasive rabbits in Australia (Box 9.2). Most invasions are not, of course, single events, so careful restoration plans must be developed for the removal of multiple introduced species.

In all cases, effective control of introduced species requires clear lines of authority and rapid response when an invasion is first detected. Once a species has begun to spread, eradication is vastly expensive and often impossible, and controlling it to a tolerable level becomes the goal (Figure 9.17). Island systems, which may be some of the most vulnerable to invasion, may also offer the greatest hope for restoration because they are relatively small and contained. Effective, early eradication can mean acting without complete information, and simply assuming a new invader will be a threat. This precautionary approach can prevent the larger costs of control further down the line, once a problematic invader is established. Invasion control may only make sense in partnership with prevention, as repeated control efforts could cause environmental damage themselves.

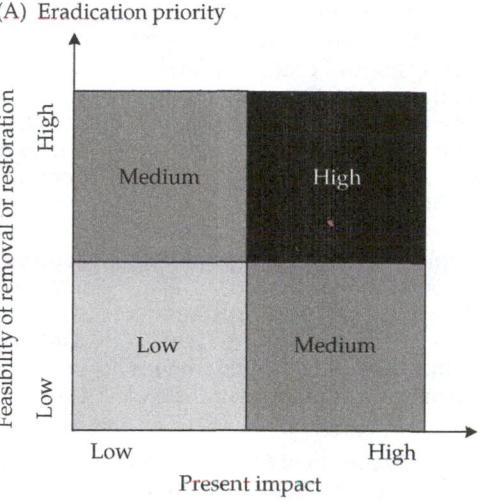

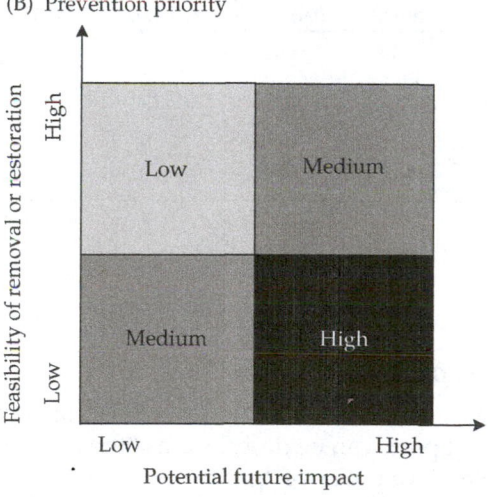

Figure 9.17 Schematic for the process faced by policymakers and managers who must decide which nonindigenous species to (A) eradicate or control, once already present in the system, or (B) keep out, if not already present in the system. (Modified from Parker et al. 1999.)

BOX 9.2 Using Models to Improve Control of Introduced Rabbits in Australia?

The European rabbit (*Oryctolagus cuniculus*) was intentionally introduced to Australia as a game animal in 1859. There they bred rapidly—a female can produce 4–5 litters of 4–6 young in one year—and reached densities up to 1500/km². By grazing on agricultural lands they competed with cattle and sheep; in other ecosystems they competed with native grazers and threatened the survival of insects, bilbies, and wombats. Soil erosion, following both grazing and warren construction, increased. Rabbits supported high populations of predatory feral foxes and cats, which also ate other prey. The total economic damage including costs of control are estimated at AUD 600 million (U.S.$457 million) per year.

Initial control efforts included shooting, trapping, netting, digging, gassing, and poisoning, but all proved insufficient. Finally, the myxoma virus was introduced in the 1930s. Myxoma is rarely lethal in its native South America where it infects a South American jungle rabbit (*Sylvilagus brasiliensis*), but it is usually fatal in European rabbits. The virus was first introduced to Australia in the 1930s and again in 1950. The second introduction was initially a stunning success, leading to 99% mortality of rabbits and an associated increase in the annual wool production equivalent to AUD 60 million (US$46 million). But only a few years later the virus appeared to be losing its effectiveness. This begged two questions: Why was the effectiveness decreasing, and would the virus continue to be useful? The first question was answered with detailed laboratory and fieldwork showing evolution in both the rabbits and the virus. The result is shown in the shift over three decades from the highly virulent strain I to the milder strains III and IV (Table A).

The second question was easier to address with a mathematical model (Dwyer et al. 1990), which we will reconstruct here. We will begin with a standard epidemiological model that tracks the number of rabbits in each of three categories: susceptible (*S*), infectious (*I*), and recovered immune (*R*) individuals (Figure A). In this convenient approach, we don't have to deal directly with the virus or the disease vectors; we can focus on the rabbits alone.

In Figure A, boxes represent the number of rabbits in each category, *S*, *I*, and *R*. The arrows represent parameters (the probabilities of birth, survival, infection, and recovery) that govern

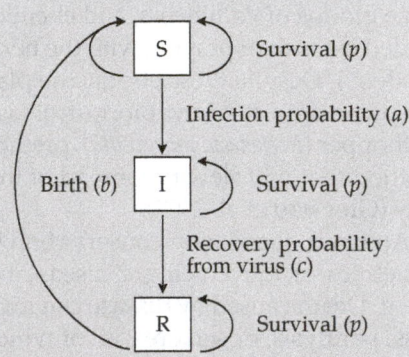

Figure A A schematic of how a disease affects a population. *S* = the number of individuals susceptible to the disease, *I* = the number of individuals that are infected, and *R* = the number of individuals that have recovered from the disease. Straight downward arrows denote the probablity that an individual will progress from one disease category to the next. Morality due to the disease is not directly shown, but can be calculated as $1 - (c + p)$. Curved arrows to the right represent the probablity of surviving and remaining in the same disease category. Curved arrows to the left represent the birth rate of new individuals (all of whom are susceptible). In this case, infected individuals do not reproduce.

TABLE A *Evolution of Myxoma Viral Strains in Australian Rabbits after the Introduction of Strain I*

Strain	Survival (days)	Mortality (%)	Proportion in 1950	Proportion in 1964–1966	Proportion in1975–1981
I	<13	>99	100	<1	2
II	14–16	95–99	0	<1	3
III	17–28	70–95	0	64	67
IV	29–50	50–70	0	34	28
V	>50	<50	0	1	0

Note: The strains differ in virulence, and over time strains III and IV become most prevalent.
Source: Modified from Dwyer et al. 1990.

Invasion prevention

The philosophy of preventing invasions from occurring in the first place can be described as "the precautionary principle," as a policy of "guilty until proven innocent," or in terms of "good lists" and "bad lists" of low- and high-risk species. Preventing invasions requires identifying and regulating invasion pathways. We have already seen the tremendous range of ways in which species are transported around the world; interrupting these pathways requires creative thinking at industrial and person-

al scales. Intentional introductions clearly require careful forethought, implementation, and assessment. In general, these processes fall under the jurisdiction of the governmental agencies responsible for the introductions.

Unintentional introductions, by contrast, occur continually through industrial, governmental, and individual pathways and are perhaps more challenging to manage. As an example, I will highlight the approaches to preventing invasions from the ballast water of commercial ships. Ballast water is a large and rather undiscrimi-

how rabbits either remain in or move from one disease category to the next. The parameters are: b, the net annual production of offspring by adults; p, the annual survival probability; a, the probability of viral transmission between an infected and a susceptible rabbit; c, the annual probability of recovery for an infected rabbit.

We can now translate the figure into equations, with one for each rabbit category. Using this more compact mathematical notation makes it easier to develop computer simulations to predict the future of the rabbit population. We will use the same notation as above, and by convention we will use t for time: This year is written simply as t and next year is $t + 1$. This notation gives us the following equations:

This kind of model is known as a system of discrete-time difference equations. A great deal of mathemati-cal theory has been developed for such models and can be used to predict the fate of the rabbit population. We can visualize these predictions using computer simulations to add together $S + I + R$ and can watch the progression of the population following the introduction of the virus (Figure B).

What insights can this model provide? Our model is a simplified version of the one developed by Dwyer et al. (1990); their full model incorporates all viral strains, the rabbit age structure (juveniles versus adults), the temporal change in virus level in an individual infectious rabbit, and the survival time of an infected rabbit. Although our model is much simpler, it can still illus-trate one of the most useful results of the original more complicated one. Namely, when natural mortality is suffi-ciently high, the virus will continue to

Figure B Simulations of the simplified rabbit myxoma model for two example rabbit populations with identical myxoma mortality. One population (gray line) has a low natural mortality rate and grows exponentially; the other (black line) has a higher natural mortality rate and the population is seen to level off.

$$\underbrace{S_{t+1}}_{\substack{\text{Susceptible rabbit} \\ \text{number next year}}} = \underbrace{\left(S_t + b\left[S_t + R_t\right]\right)}_{\substack{\text{Susceptible rabbits this year} \\ \text{plus newborns from } S \text{ and } R}} \times \underbrace{p}_{\substack{\text{Natural} \\ \text{annual} \\ \text{survival} \\ \text{probability}}} \times \underbrace{(1-a)I_t}_{\substack{\text{Probability of meeting} \\ \text{an infected rabbit} \\ \text{and } not \text{ getting the virus}}}$$

$$\underbrace{I_{t+1}}_{\substack{\text{Infected rabbit} \\ \text{number next year}}} = \underbrace{(I_t + aS_t)}_{\substack{\text{Infected rabbits this year} \\ \text{plus new infections from } S}} \times \underbrace{p}_{\substack{\text{Natural} \\ \text{annual} \\ \text{survival} \\ \text{probability}}} \times \underbrace{(1-c)}_{\substack{\text{Annual survival} \\ \text{probability of an} \\ \text{infected rabbit}}}$$

$$\underbrace{R_{t+1}}_{\substack{\text{Recovered rabbit} \\ \text{number next year}}} = \underbrace{(R + cI_t)}_{\substack{\text{Recovered rabbits this year} \\ \text{plus new recoveries from } I}} \times \underbrace{p}_{\substack{\text{Natural annual} \\ \text{survival probability}}}$$

control the rabbits, but when rabbit mortality is low, it will not. Natural mortality occurs primarily by preda-tion; when predation is high, the virus will generate enough additional mor-tality to control the population and when predation is low, further addi-tional control method (e.g., by hunt-ing) will be required to supplement the viral mortality and achieve control. This is an intuitively sensible result; the power of the model is in telling us what the threshold level of natural mortality is. The understanding and control of other introduced diseases, including HIV, malaria, and hoof-and-mouth, have been similarly enhanced by modeling.

nating pathway; thousands of freshwater and marine species are transported daily around the world. Effective prevention of this pathway must target animals, plants, and microbes, and must be applicable to a wide array of vessel types and port capacities.

Technologically, there is no shortage of proposed bal-last water treatments; heat, chlorination, ultraviolet ra-diation, filtration, and open-ocean exchange are but a few. Thus far, open-ocean exchange is the most widely adopted approach. Exchange involves releasing the orig-inal ballast in mid-ocean and refilling the tanks with ocean water for discharge on arrival. Like many pro-posed options, exchange is unlikely ever to be 100% ef-fective in removing invasion risk. A growing field of in-vestigation involves modeling and risk assessment to evaluate the costs and benefits of alternate prevention techniques for different invasion pathways.

Invasions that occur at a smaller scale—the fertilizer we dig into our gardens, the horse feed brought into na-tional forests, the shoes of racers arriving at an interna-tional competition—all may serve as pathways for non-native organisms. Although each of these may occur at a smaller spatial or temporal scale, their sum represents a massive and continual redistribution of species' propag-ules that pose an important and unsolved challenge in conservation policy and practice.

Regardless of scale, effective invasion prevention re-quires a certain amount of legislation and policy develop-ment. Species invasions are a global conservation issue, with every nation serving as both a donor and a recipient

for introductions. At the international scale, countries around the world have ratified the 1973 Code of Practices from the International Council for the Exploration of the Seas, and its subsequent updates. The 2003 version states:

> All introductions and transfers of marine organisms carry risks associated with target and non-target species (including disease agents). Once established, introduced species can spread from foci of introductions and have undesirable ecological, genetic, economic, and human health impacts. . . . This Code of Practice provides a framework to evaluate new intentional introductions, and also recommends procedures for species that are part of current commercial practices to reduce the risk of unwanted introductions, and adverse effects that can arise from species movement.

Similarly, the *Convention on Biological Diversity* states that all participating governments (>180 nations) should "as far as possible and as appropriate . . . prevent the introduction of, control or eradicate those alien species which threaten ecosystems, habitats or species . . . ."

How do these general statements of good intent play out in individual countries? In the U.S., federal legislation concerning introduced species first emerged a century ago in the form of the Lacey Act of 1900, which identifies a "dirty list" of invaders that now includes the brown tree snake, the zebra mussel, and other species that harm wildlife and are therefore prohibited from import. Subsequent federal legislation focused on the protection of agriculture rather than wildlife (Table 9.5). A more holistic approach was taken by President Carter, whose Executive Order in 1977 restricted federal agency introductions of nonnative species into "any natural ecosystem." Initially, federal agencies attempted to implement this order with a clean-list approach—that is—by specifying species that were deemed safe for import. In practice, however, the dirty-list approach of prohibiting only known pests eventually became—and remains—federal policy (Office of Technology Assessment 1993).

In marine and aquatic systems, where unintentional invasions are common, the National Aquatic Nuisance Prevention and Control Act (1990) and the National Invasive Species Act (1996) adopt a multispecies preventive approach. Motivated largely by the devastating economic and ecological impacts of the zebra mussel invasion in the Great Lakes, these laws mandated the development of a National Aquatic Nuisance Species Task Force, and the implementation of guidelines for ballast water discharge from commercial ships. The logistics of implementing these regulations are an active area of research and development, and will require monitoring and evaluation to test their effectiveness.

TABLE 9.5 *Timeline of Selected Federal U.S. Legislation Relating to Introduced Species*

1900	Lacey Act	Prohibits import of species that harm wildlife
1912	Plant Quarantine Act	Allows regulation and quarantine of nursery stock for plant diseases and insect pests
1931	Animal Damage Control Act	Allows control of any animal that damages agriculture, aquaculture, public health, or other enterprise
1939	Federal Seed Act	Regulates import and transport of seeds, especially of noxious weeds
1944	USDA Organic Act	Allows federal eradication of plant pests including noxious weeds
1957	Federal Plant Pest Act	Prohibits import and transport of plant pests (animals, fungi, other plants)
1968	Carlson-Foley Act	Allows federal government access to noxious plants in states with noxious weed acts
1974	Federal Noxious Weeds Act	Reulates the transport of plants listed as noxious
1977	President Carter's Executive Order 11987	Restricts federal agency introductions of nonnative species into "any natural ecosystem"
1990	National Aquatic Nuisance Prevention and Control Act	Establishes voluntary ballast water management guidelines
1996	National Invasive Species Act	Establishes compulsory ballast water management guidelines
1999	President Clinton's Executive Order 13112	Establishes National Invasive Species Council, with a mandate to develop a National Invasive Species Management Plan coordinating all federal on activity introduced species
2000	Plant Protection Act	Regulates plant pest transport, both accidental and intentional for biological control
2001	National Invasive Species Management Plan	Mandated by Executive Order 13112; provides a blueprint for federal prevention, management, research, and outreach on introduced species

In 1999, President Clinton's Executive Order 13112 mandated the development of a National Invasive Species Management Plan to unify federal agency policy "on those non-native species that cause or may cause significant negative impacts and do not provide an equivalent benefit to society." Among them, the multiple acts governing introduced species confer scattered and overlapping implementation and enforcement authority on a wide range of federal agencies and departments. Thus, there remains as yet no comprehensive federal legislation addressing the prevention and control of introduced species. Both in the U.S. and around the world, this remains an open area of active development of conservation policy and legislation.

CASE STUDY 9.1

Invaders in an Invasible Land

The Case of the North America Beaver (*Castor canadensis*) in the Tierra del Fuego–Cape Horn Region of South America

Christopher B. Anderson, Institute of Ecology, University of Georgia and Omora Ethnobotanical Park, Puerto Williams, Chile

The Argentine National Wildlife Commission introduced the North American beaver (*Castor canadensis*) to the Fagnano Lake area on Tierra del Fuego Island in 1946 (Lizarralde 1993). The introduction was part of a broader plan to import economically valuable pelt-bearers to create a fur industry, thus "enhancing" the ecosystem. The beaver introduction consisted of only 50 individuals, or 25 mating pairs, brought from Canada. Today, the population is estimated to be approximately 115,000 in both Argentine and Chilean territory (Lizarralde et al. 1996; Skewes et al. 1999). Over the ensuing decades the beaver spread throughout the island of Tierra del Fuego and began crossing channels and fjords to colonize adjacent islands in the southern Magellanic Archipelago. The Straits of Magellan proved to be a formidable natural barrier for many years, but has been breached recently. Reports now confirm a resident population on the Chilean mainland in the southern and western portions of the Brunswick Peninsula. No significant geographic barriers remain to prevent the beaver from expanding its range further north, colonizing the entire southern temperate forests of South America.

In its native habitat, the beaver is associated with many, often beneficial, ecosystem alterations (Naiman et al. 1988). Its wide-scale impacts on habitats and biota make it the classic example of an *ecosystem engineer*, a term coined to describe species that create, modify, and maintain habitats, thereby influencing biotic and ecological dynamics of other species and the entire system (Jones et al. 1994). In North America the beaver's ecosystem alterations include increasing retention of organic material and sediments, altering biogeochemical nutrient cycles and influencing the assemblage and dynamics of both aquatic and riparian organisms (McDowell and Naiman 1986, Naiman et al. 1986, Medin and Clary 1991; Naiman et al. 1994). These alterations can often be described as beneficial for the ecosystem. For example, Naiman et al. (1988) describe the beaver-induced increases in retention and efficiency as a factor that enhances the stream's resilience to disturbance. Furthermore, Pollock et al. (1998) and Wright et al. (2002) credit beaver-modifications with increasing diversity and richness of riparian vegetation communities.

As an introduced species in southern South America, however, we must reevaluate the beaver's ecological role and resist the urge to apply Northern Hemisphere paradigms to austral ecosystems. Unfortunately, baseline ecological studies do not exist for this region prior to the beaver's introduction. Nevertheless, we can broadly define five categories of change that the beaver has created in the Magellanic Archipelago: riparian deforestation and soil destabilization; alteration of the light regime in clearings; modification of stream and riparian habitat characteristics and biotic assemblages; expansion of wetlands and elevated water table; and sediment and organic matter accumulation that modifies biogeochemical cycles (Lizarralde and Venegas 2001).

An interesting ecological question is to think about why this area may be experiencing greater and longer-term impacts, compared to North America. Part of the answer is due to the fact that the densities described for southern South America are on the high end of the range of values reported for their native range (Gurnell 1998). Several studies suggest the carrying capacity of these ecosystems has already been reached and often exceeded (Lizarralde 1993; Skewes et al. 1999). The absence of predators and hunting on most islands is one reason that their numbers have become so high. Another factor is the fact that the forests and wetlands of Tierra del Fuego and the austral islands are ideal habitats.

In addition to a lack of mammalian predators, the tip of South America is generally species-poor with regards to most other taxa. Only ten or fewer terrestrial mammals are native to

the Magellanic Islands (Lizarralde and Escobar 2000). Likewise, only a few species of woody plants constitute the riparian vegetation assemblage. The three predominant trees species are all from one genus, *Nothofagus*. All are broad-leaved; only two species of conifer exist in southern South America, neither of which is represented in the riparian vegetation of the Tierra del Fuego–Cape Horn region.

In addition to differences in species richness, there are important contrasts in the reproductive biology and heterogeneity of this vegetation community compared to the Northern Hemisphere. In the North, beaver preferentially browse deciduous trees, often leaving behind conifers (Jenkins 1980; Naiman et al. 1988). In addition, some trees in North America, such as aspen (*Populus tremuloides*), have chemical defenses that limit beaver herbivory (Basey et al. 1988). The North American vegetation assemblage also includes several species that are adapted to wet soil conditions. In contrast the forests of southern Chile and Argentina lack such a diversity of vegetation types (Pisano 1977), meaning that there is nothing to colonize the beaver wetlands, except herbaceous plants. Further, beaver meadows provided an avenue for further invasions, facilitating an increase in the abundance of nonnative plant species.

Finally, the reproductive biology of *Nothofagus* in southern South America depends on long-lived seedlings because they rarely reproduce asexually from root suckers and seed banks are very short-lived in the cold, damp climate (Riveros et al. 1995). This reproductive strategy means that in the *Nothofagus* forests of the Tierra Del Fuego-Cape Horn region, herbivory—whether from rabbits, guanacos, or beavers—severely limits regeneration, often effectively stopping it (Armesto et al. 1995). In our own study, we observed that the creation of beaver meadows nearly eliminated *Nothofagus* seedling establishment, thus preventing forest regeneration. The cumulative effect of these differences is that the beaver's impact on riparian forests can be broader and longer lasting than expectations from the Northern Hemisphere, due to differences in community representation, a lack of herbivore defense mechanisms and distinct reproductive strategies.

As an ecosystem engineer with a relatively high reproductive rate, and generalist diet and habitat preferences, *C. canadensis* is a good candidate for a potentially invasive species. Combine those characteristics with southern South America's ecological and evolutionary singularites, compared to the Northern Hemisphere, and the product is today's broad-scale, pervasive impact that the beaver is having in austral South America (Figure A). Control efforts to manage the species were slow to initiate and often are still nonexistent. However, even with a concerted effort, mitigation of the impacts of this species will be difficult to implement. The region's isolated rugged geography would make eradication nearly impossible. Plus, in addition to the beaver, several other nonnative pelt-bearers were introduced in the 1940s and 1950s for the fur industry, including such highly invasive species as mink (*Mustela vison*) and muskrat (*Ondatra zibethica*). In fact, quite strikingly, two-thirds of all terrestrial mammals in the Tierra del Fuego-Cape Horn Region are nonnative; only 10 species are native, while 19 have been introduced in the last 100 years (Lizarralde and Escobar 2000). Several bird, fish, and invertebrate species also have successfully (often invasively) established themselves (Vila et al. 1999; Anderson and Hendrix 2002).

The very mention of Tierra del Fuego or Cape Horn conjures up images of remoteness, isolation, and severity. It even was identified as one of the 37 most pristine wilderness areas left in the world (Mittermeier et al. 2001). Yet while to many people, southern South America is *terra incognita*, an unknown land, it clearly does not escape the current global-scale ecosystem alterations caused by such phenomena as exotic species, global warming, and the ozone hole. The ecological importance of southern South America, one of the world's last remaining "pristine" sites, cannot be overlooked. However, research priorities need to account for these combined anthropogenic influences to elucidate their effects on the region's ecology. The beaver, in particular, needs special attention because as an *exotic ecosystem engineer* its impacts by definition influence the whole system.

Although southern South America remains a pristine region, it clearly does not escape the current global-scale ecosystem alterations caused by such phenomena as exotic species, global warming, and the ozone hole. If we are to retain the ecological uniqueness of this region, we will need to focus more attention on minimizing the charges brought on by these threats, particularly by this nonnative ecosystem engineer.

Figure A   Beaver strongly altered stream hydrology and riparian forest along the Ukika River, Isla Navarino, Chile. The riparian forest has been cut and flooded, leaving behind desiccated dead trunks, and numerous dams and impoundments that alter the lotic environment. (Photograph by C. Anderson.)

CASE STUDY 9.2

Tracking Aquatic Invasive Species

Hugh MacIsaac, Great Lakes Institute for Environmental Research, University of Windsor

Biological invasions by invasive species represent a serious and growing threat to global biodiversity. Human-mediated introduction of species beyond biogeographic barriers that historically constrained dispersal has allowed invasive species to establish populations far beyond their native ranges. Successful biological invasions involve a sequence of events beginning with introduction of invasive species' propagules, and terminating with the species' integration into the new community (see species H and I in Figure 9.9). Many introduced species fail to successfully establish in their new habitat (see Figure 9.9). Failure to successfully colonize a new habitat may be related to demographic considerations (e.g., too few propagules introduced, or improper age or sex ratio), or to physiological intolerance (i.e., individuals cannot tolerate physical or chemical conditions in the novel habitat; see species F in Figure 9.9). Invasions also could fail despite a large inoculum and suitable physiological conditions if colonists are heavily parasitized or preyed upon or if they are outcompeted by existing species in the new habitat (see species G in Figure 9.9). A great deal remains to be learned regarding how these different factors affect the invasion process, and careful experiments are required to separate potential causes of invasion failure.

The U.S. appears particularly vulnerable to invasive species for two reasons. First, the number of invasive species established in a country is positively correlated with its gross domestic product, and as the world's largest economy, the U.S. likely receives more invasive species than any other country. Secondly, the U.S. consists of a large number of different ecosystem types, providing a wealth of different habitats for species adapted to a broad variety of environmental conditions.

Lakes and coastal marine ecosystems are highly vulnerable to invasive species because of their close association with human development. Indeed, many of the world's great cities are located adjacent to these ecosystems, and are used extensively for transportation, recreation, and food acquisition. These activities result in unintentional or intentional introductions of species from other regions of the world to new aquatic habitats they could never reach through natural dispersal. For example, in the Great Lakes region, it has been estimated that the rate of introduction of crustacean zooplankton by humans exceeds their background natural dispersal rate by about 50,000 times (Hebert and Cristescu 2002).

The Great Lakes have been successfully invaded by at least 162 invasive species, although many more likely remain to be discovered (Figure A). These species include everything from bacteria to fishes. Notorious invaders in the Great Lakes include zebra mussels (*Dreissena polymorpha*; Figure B), sea lamprey (*Petromyzon marinus*) and common carp (*Cyprinus carpio*). These species have had diverse and often unpredictable effects on these ecosystems. Zebra mussels, for example, are considered "ecosystem engineers" because the filter-feeding activities of large populations alter physical conditions (e.g., water clarity), concentrations of key nutrients (e.g., nitrogen, phosphorus), and an array of biological conditions (e.g., species abundances of other benthic invertebrates, macrophytes, phytoplankton, zooplankton, fishes, and waterfowl).

Vectors that transport invasive species to new areas from native regions, and the pathways or routes by which they move, are topics of great interest to invasion biologists. It may be possible to reconstruct the invasion history of different habitats by reviewing the vectors and pathways associated with species that have successfully invaded. In the case of the Great Lakes, the dominant vector of invasive species introduction over the past 45 years, since the St. Lawrence Seaway opened, has been international shipping. Ships account for about 67%

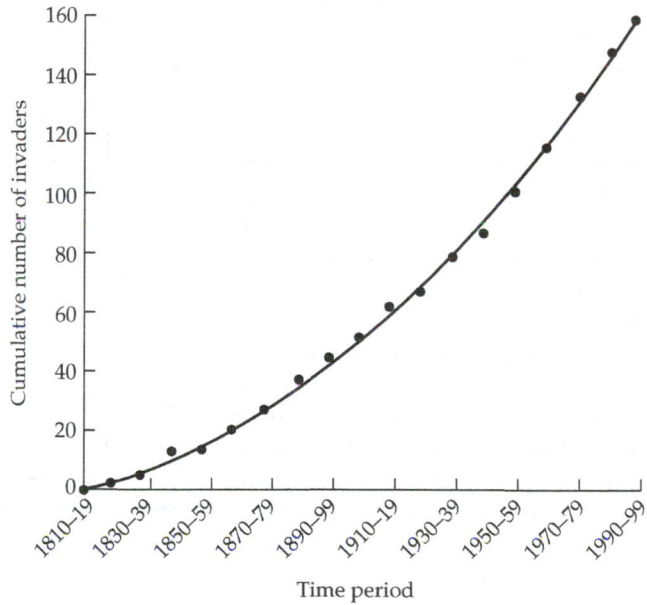

Figure A   Reported invasion rate for invasive species in the Laurentian Great Lakes. The accumulation of invaders increased during the 1990s despite a ban on discharge of foreign, fresh ballast water by transoceanic vessels in 1993. (Modified from Holeck et al. 2004.)

of invasive species that established during this period, primarily through the discharge of ballast water that was loaded at a freshwater port elsewhere in the world.

Ships were important introduction vectors throughout the 1990s, even though the U.S. Coast Guard effectively banned discharge of foreign freshwater by requiring vessels intending to dump water into the Great Lakes to exchange it for open-ocean water during transit to the Great Lakes in 1993 (see Figure A). Ballast water exchange policies should reduce invasion risk dramatically because any freshwater organisms not purged from tanks would be killed by exposure to salt water loaded into the tanks. However, ships may spread invasive species *among* the Great Lakes, as they frequently load ballast water while operating on the Great Lakes, and later discharge the mixed ballast elsewhere in the region.

Risk profiles can be developed for each of the Great Lakes based upon how many ships discharge ballast water into each lake. The more ships that discharge ballast water into a lake increases the risk of invasions. Risk profiles also can be developed to identify major donor areas contributing species to the Great Lakes. Many of the invertebrate species that established in the lakes since 1985 are native to the Black Sea region of southeastern Europe. This finding was initially surprising as fewer than 5% of inbound vessels to the Great Lakes originate from this area. However, many Black Sea species are spreading through rivers and canals throughout Europe, often bringing these species to ports from which vessels destined to the Great Lakes operate. Molecular markers including allozymes and nuclear and mitochondrial DNA can be used to compare invasive populations with populations in Europe to identify the source of North American stocks. These analyses have demonstrated that species are spreading in a "stepping stone" fashion in which major ports serve as "hubs" and international ships as "spokes."

Many other vectors contribute invasive species to the Great Lakes, notably fish stocking, aquarium releases, bait bucket fishes, and connections to other

Figure B The zebra mussel (*Dreissena polymorpha*) was first found in North America in Lake St. Clair (upstream of Lake Erie) in 1988 by undergraduate university students. The mussel has since spread to lakes, reservoirs, and rivers throughout temperate, eastern North America. (Photograph © Oxford scientific/photolibrary/ Scott Camazine.)

systems via man-made canals. Further, once species invade the Great Lakes, they may spread to inland lakes through a host of other human-mediated mechanisms. As an example, the spiny waterflea *Bythotrephes longimanus* is a large-sized zooplankton species native to Europe and Asia, with a very long caudal process (tailspine). This species was brought to the Great Lakes in 1982. By 2003, it had spread to at least 59 inland lakes in Ontario, at least 5 in Michigan, 1 in each of Wisconsin and Minnesota, and 3 reservoirs in Ohio. The species is capable of producing "resting eggs" that tolerate emersion from water. Although both natural vectors (waterbirds and humans) may move this waterflea, our research points to pleasure boaters and anglers as the most likely vectors (MacIsaac et al. 2004). Fishing line trolled through an infected lake "samples" a huge volume of water, inadvertently snagging waterfleas by their tailspine. These animals accumulate as knots on the fishing line. If the same line is subsequently used on another lake, an invasion could result if resting eggs contained within the knot detach from the line and hatch.

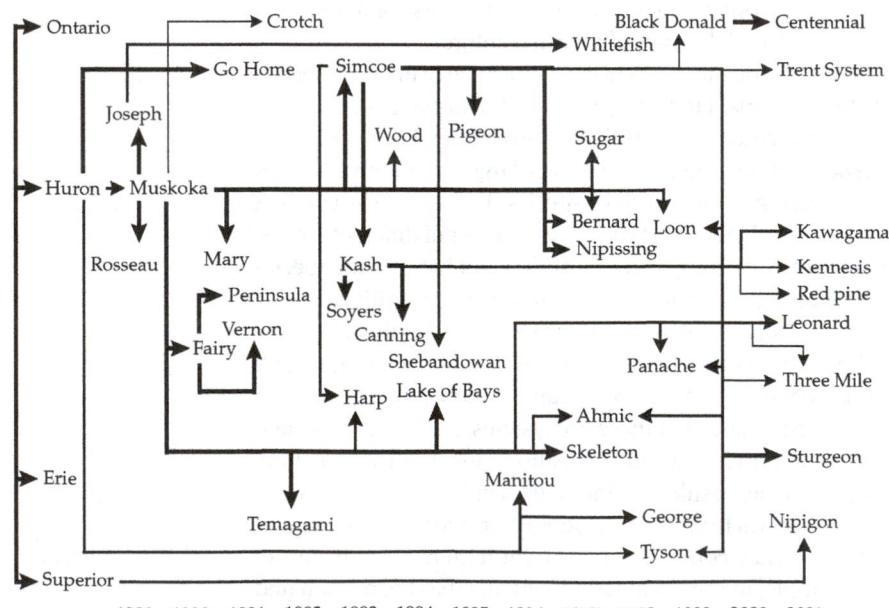

Figure C Time-series model of *Bythotrephes* spread to inland lakes in Ontario from the Great Lakes based upon movement of anglers and boaters from invaded to noninvaded lakes. Arrow thickness denotes strength of the vector from the source (left side) to the newly invaded lakes (right side). Only the dominant source lake is shown unless vector strength from two potential sources was less than 1% apart (e.g., Bernard Lake). Lake Muskoka has served as a "hub" in dispersal of the species. (Modified from MacIsaac et al. 2004 and Muirhead and MacIsaac 2005.)

By surveying recreationalists at dock and marina facilities on invaded lakes and determining which noninvaded lakes they subsequently visited, we developed a risk-based model to identify which noninvaded lakes are most vulnerable to invasion. Lakes that were noninvaded in 2000 but which had large amounts of inbound traffic (boaters and anglers) from invaded source lakes were significantly more likely to become invaded in subsequent years (2001–2003) than other noninvaded lakes that had less inbound traffic (MacIsaac et al. 2004). Even though many lakes have connections to multiple invaded lakes, in many cases a single system contributed most of the inbound traffic (Figure C). Inland lake invasions were first reported in 1989, from Lake Huron. Shortly thereafter, many other lakes were reported invaded. One lake, Muskoka, appeared to have played a pivotal role in spread of the waterflea, as it is the dominant pathway to 18 other lakes that were directly invaded, and 17 more that were colonized indirectly. Now most of the lakes near Lake Muskoka have been invaded.

However, most outbound traffic from Lakes Simcoe and Kash is directed to lakes not yet invaded. Consequently, new invasions could be prevented if public education on the importance of preventative measures was targeted to boaters and anglers who use these two lakes.

Introduction of invasive species is an unfortunate consequence of international trade and travel. Governments and international conservation organizations like the Global Invasive Species Program have become increasingly aware of the ecological, economic, and human health harm caused by invasive species, and have begun to develop strategies to reduce invasion rates. It is essential that vectors and pathways be identified and, if possible, quantified such that programs and policies can be developed to reduce future invasions. Lakes may represent model ecosystems with which to begin these efforts because they are discrete ecosystems whose links to one another can be identified and quantified.

CASE STUDY 9.3

When a Beauty Turns Beast
Caulerpa taxifolia: One of the World's Worst Invasive Species

Susan L. Williams, Bodega Marine Laboratories and University of California, Davis

Marine invasive species, along with pollution and over-fishing, are considered the principle threats to marine biodiversity and the functioning of marine ecosystems. One of the most damaging invasive species is *Caulerpa taxifolia*, a beautiful green seaweed native to tropical regions of the world (Figure A). *C. taxifolia* is prized by aquarium hobbyists for its bright color, intricate fernlike shape, and its fast growth. In its native habitat near coral reefs, it is not very common although other species of the genus *Caulerpa* are quite common and play important ecological roles (in stabilization of sediments and colonization of unvegetated bottoms after disturbance). *C. taxifolia* might be more common growing in aquaria than in nature. In 1984, a small piece of it was identified outside its native habitat, growing in the Mediterranean Sea just outside Monaco's public aquarium. Dr. Alex Meinesz, a French scientist who worked on species of *Caulerpa*, warned that it was nonnative and had a capacity to spread. His warnings went unheeded because no one expected a tropical seaweed like *C. taxifolia* to withstand cooler water in winter. The first surprise was that it not only survived, it grew rapidly. One of the hardest lessons about nonnative species is realizing how difficult it is to predict what they will do in a new environment.

Over time, in the absence of a rapid management response, *C. taxifolia* spread throughout the Mediterranean. It spread through vegetative propagation within a given area but was also dispersed over the entire geographic area, carried on boat anchors and in the nets of fishermen, which became too clogged with *Caulerpa* to effectively catch fish. Totally unexpectedly, it showed up in the colder waters of Croatia and most recently it jumped to Tunisia on the African side of the Mediterranean, and it is still spreading. *C. taxifolia* has had devastating ecological effects in many, but not all, of the areas it invaded, earning it nicknames including "The Killer Alga" and "The Devil's Alga." The seaweed assumed superstar villain status, appearing in newspaper and magazine articles, books, documentaries films, and on National Public Radio.

All this attention paid to a seaweed stemmed from the ecological and economic harm it caused in the Mediterranean. *C. taxifolia* displaced colorful native communities the way that bamboo can replace a diverse native forest in tropical locales. The displacement of native communities by *C. taxifolia* was followed by a decline in local marine tourism in the Mediterranean; SCUBA divers will not pay money to see a monotonous carpet of *C. taxifolia*. Species of Caulerpa, including *C. taxifolia*, also produce toxins that make them distasteful to animals. From this, scientists infer that the food web is being changed dramatically in places where *C. taxifolia* has displaced a diverse native community of palatable seaweeds and seagrasses.

Figure A *Caulerpa taxifolia* taking over an eelgrass bed (*Zostera capricorni*) near Sydney, Australia. First identified in the region in 2004, *Caulerpa taxifolia* covered 4–8 km² within months. (Photograph by B. Nyden.)

While too late for significant control in the Mediterranean, the awareness that *C. taxifolia* could be a severe environmental problem was raised through all the media attention. When *C. taxifolia* was discovered near San Diego in southern California in the summer of 2000, immediate action was taken. An environmental consultant had noticed a strange seaweed while doing a routine survey of an eelgrass bed. After searching the Internet and realizing she might have found *C. taxifolia*, she contacted a seaweed taxonomist at a university herbarium who verified the finding. *C. taxifolia* is not native to the west coast of North America and Hawaii is the nearest place it is found. She immediately notified local resource managers. An unofficial management team was formed, which decided to eradicate the seaweed. Within a month, chlorine bleach was being pumped under tarps placed over areas of the eelgrass infested with *C. taxifolia* to kill it. A second invasion site near Los Angeles was then identified and the bleach treatment was modified for the conditions there. Over time, more and more *C. taxifolia* was found in infested lagoons and treated, and the eradication program was cautiously declared a success after no *C. taxifolia* was found after fall 2002. However, the jury is still out because the most difficult part of an eradication program is finding and killing the last tiny piece.

The rationale for dedicating government resources to eradicate *C. taxifolia* in southern California was provided by the U.S. Noxious Weed list. *C. taxifolia* was listed as a noxious weed just a year before it was found in California, in response to a petition from many concerned scientists and environmentalists. This listing gives the U.S. Department of Agriculture (USDA) legal authority to ban noxious weeds from importation or interstate transfers. While the authority did not specifically cover eradication, the intent was clear that *C. taxifolia* was not desirable in nonnative habitats in the U.S. An interesting wrinkle in this story is that neither the USDA (federal) nor the state counterpart (California Department of Food and Agriculture) had ever worked in the marine environment or with a seaweed. The agriculture agencies were accustomed to eradicating serious weedy pests in vineyards and croplands with an arsenal of herbicides and in some cases, biological control agents. The ocean presented a very different challenge. For example, herbicides that killed terrestrial and freshwater weeds had no effect on *C. taxifolia*. The application of any chemical under water is difficult to target and chemicals dilute rapidly. Another issue was that the public is exquisitely sensitive to the fact that marine environments, even when not completely pristine, retain more of their historical biodiversity than a corn or wheat field. The public wanted to be certain that killing native marine life along with *C. taxifolia* in the short-term was necessary to preserve marine biodiversity in the long-term. And, the public wanted a method that caused the least damage to the native ecosystem and with the fewest residual effects. Public outreach had to be incorporated into the eradication program.

While the southern California *C. taxifolia* eradication program is considered a model of rapid response and government collaboration, it is also one of the first attempts to eradicate a marine invasive species. It is therefore prudent to consider how the model might be improved for the future. One area for improvement would be to initiate scientific research as part of

the eradication program. For example, the eradication program would benefit from knowing how much chlorine was actually required to kill *C. taxifolia*. Chlorine is expensive, and application of liquid chlorine near the residential areas adjacent to the invasion sites posed a hazard to humans. Without knowing what concentration of chlorine was applied, the time over which it was maintained, and the organic load of *C. taxifolia* plus the seagrass, other seaweeds, and animals to be bleached, managers will not have good information to meter the amount to use at a different invasion site. Because the tarps covering *C. taxifolia* also eliminated light required for seaweed photosynthesis and growth, perhaps opaque tarps alone would be sufficient to eradicate it, at much lower cost and less environmental impact. Another area for research remains to determine the environmental factors, such as light and temperature, that control the growth and spread of *C. taxifolia*. Understanding these factors would help pinpoint where to survey for the next invasion, thereby economizing on efforts and resources.

The *Caulerpa taxifolia* story focuses on the eradication program, but scientists and managers agree that preventing invasions from happening in the first place is the most cost effective way to protect native ecosystems from their damaging effects. Identifying the pathways for introduction and then shutting them down is critical in the management of invasive species. For *C. taxifolia*, the pathway is very clear. DNA fingerprinting identified invasive *C. taxifolia* in the Mediterranean, California, and also some sites in Australia as a strain that had been propagated in home and public aquaria. At both invasion sites in southern California, it was apparent that people discharged water from, or dumped, their home aquariums into the lagoons, providing a potential route by which *C. taxifolia* established. Pet shops and aquarium stores often sell many species of *Caulerpa*, including *C. taxifolia*. The aquarium hobby is second only to photography in the U.S. and is growing rapidly. Engaging the aquarium industry in finding environmentally-friendly alternatives to *C. taxifolia* and posting cautionary labels about the risks of dumping nonnative species will be critical in preventing new and costly invasions. Educating the public about the risks might be the best insurance against another *Caulerpa* invasion or something even worse.

CASE STUDY 9.4

Biological Control as a Conservation Tool

Judith H. Myers, University of British Colombia

Introduced species are one of the major threats to native biodiversity (Chapin et al. 2000). Introduced mammalian predators, herbivores, and plant diseases have had great impacts on native ecosystems. Introduced insects are among the most serious agricultural and forest pests (e.g., codling moth, gypsy moth, and western flower thrips, to name a few). Many plant species have also enjoyed major increases in distribution in the last 300 years, in some cases on global scales, frequently following intentional introductions by humans (Myers and Bazely 2003).

Classical biological control can reduce the impacts of introduced species and has been used particularly for the control of introduced insects and weeds (Hajek 2004). Classical biological control is the introduction of native predators, parasitoids, herbivores, or diseases of introduced species to the exotic habitat in which the target species has become established and detrimental. This approach makes ecological sense. In most cases, species in the native habitat are kept under natural control by environmental conditions and natural enemies. In the new environment they are largely free of natural enemies and enjoy an advantage over the native species. The goal of classical biological control is thus to reestablish some of the natural enemies in the exotic habitat to reduce the competitive advantage enjoyed by the introduced species (Figure A).

However classical biological control presents a paradox because it involves the introduction of more species and therefore, a further "dilution" of natural biodiversity. From a conservation perspective, successful biological control can be a valuable asset for the restoration of native habitats. On the other hand, unsuccessful biological control, in which the densities of populations of the target species remain unchanged following the addition of additional exotic species, is contrary to conservation goals. More seriously, biological control programs that involve nontarget impacts can harm native ecosystems, creating new conservation problems. An example is the ludicrous introduction of cane toads in an attempt to control beetles that attack sugar cane in places such as Australia (Low 2001). Once established, cane toads increased and spread widely to both urban and natural areas where they have had many negative interactions with native species. Another example of nontarget impacts involves one of the most successful biological control agents of exotic cactuses, the moth *Cactoblastis cactorum*. The success of this South American species in controlling cactus in Australia and South Africa preceded its introduction to the Caribbean from where it spread to Florida. Now that the moth is established in North America, there is considerable potential that it will spread into the center of cac-

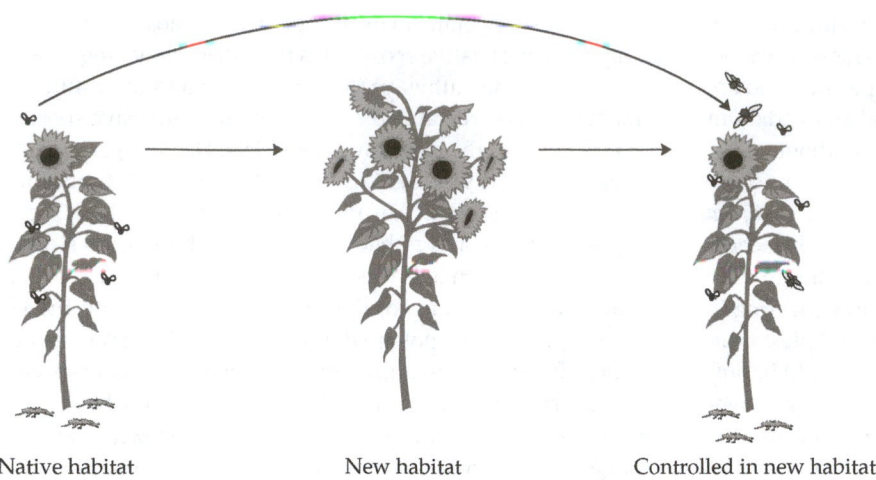

Native habitat   New habitat   Controlled in new habitat

Figure A Invasive species are often those that escape their herbivores from their native habitat in a new habitat, and grow at a much higher rate as a result. Biocontrol attempts to halt this rapid growth by introducing the herbivore(s) that kept the plant in check into its new habitat.

tus diversity in Mexico and Central America (Pemberton 1995; Myers and Bazely 2003).

Control is not always necessary for introduced species. In a number of introductions of animals, populations initially reach high densities and broad distributions but then decline. An example is the Crested Mynah (*Acridotheres cristatellus*), which was once very common in southwestern British Columbia but has now become extinct. Another is the European cranefly, *Tipula paludosa*, which was a major pest in the 1960s in the vicinity of Vancouver, B.C., but has declined to nonpest status without biological control (Myers and Iyer 1991). But many introduced species reach and maintain "outbreak" densities and dominate ecosystems. For some of these, biological control is an appropriate approach. To maximize the value of biological control as a conservation tool, the introduction of the *fewest*, *safest*, and *most effective* agents is required. Choosing these biological control agents requires the application of ecological and evolutionary theory.

How Many Agents are Necessary for Success?

With each introduction of an exotic species comes the possibility of some unpredicted, nontarget impact and the "dilution" of native biodiversity. Thus the goal must be to achieve success with the introduction of the minimum number of agents. Whether this will be one or many agents depends on how biological control works. Biological control could result from accumulated stress to target species following the introduction of a number of species of natural enemies that attack various life history stages or parts of the plant (Harris 1981). An alternate hypothesis is that only a few species of natural enemies have major impacts on target weeds—for example, the ability to kill plants. Therefore success will be achieved after the "effective" agent has been introduced. To determine if cumulative stress or a single effective species best explains successful biological

control, Denoth et al (2002) reviewed 27 successful programs for biological control of weeds that involved the release of multiple species of agents. In 60% of the programs success was attributed to a single agent, and to two agents in an additional 22% of programs. These single agents are frequently sufficient for successful biological control. However, if the target weed inhabits a broad range of habitats however, more species of agents may be required. In this case different species or biotypes of agents may be effective in different environments. An example is the biological control of leafy spurge, *Euphorbia esula*, in western North America (Anderson et al. 2000).

How Are Safe Agents Selected?

The safety of potential biological control agents is a major concern. Host specificity testing is used to avoid the introduction of species that will attack native plants or that will possibly become agricultural pests (review in Kluge 2000). Testing is usually done by caging potential agents with nontarget test plants either alone (no choice) or with the target plant species (choice). Whether females lay eggs on the nontarget plant species, and whether larvae develop on the plants is recorded. Over a 100-year history of biological control practice, 259 species of insects have been deliberately released and established for the biological control of weeds (Kluge 2000). None have become agricultural pests (McFadyen 1998). In cases in which biological control agents have moved onto nontarget plants, the shifts have all involved plants that are congeneric with the target weed (Pemberton 2000). One such shift involved the seed feeding weevil, *Rhinocyllus conicus*, introduced from Europe to North America for the control of *Carduus* thistles. At the time of the introduction *Rhinocyllus* was known to also attack *Cirsium* thistles, but the low density of native thistles was thought to make them unlikely targets (Zwölfer and Harris 1984). *Rhinocyllus* has begun to attack rare native species of thistles however, and has reduced the number of seeds in one population of *Cirsium canescens* in Nebraska by 80% (Louda 1998). This suggests that expanded ecological criteria and quantification of ecological interactions should be a greater part of the screening process for biological control agents (Louda 2000).

How Are Effective Agents Selected?

To be able to predict what species will be a successful control agent is a real challenge to the ecological and evolutionary understanding of plant–insect interactions. Coevolutionary theory predicts that host plants will evolve resistance to natural enemies. Thus one might predict that common species of

herbivores such as seed predators of thistles, will be those that have little impact on the density of their hosts and thus would be poor control agents (Myers 2001). One might also predict that insect species that are rare because they are poor competitors with other insects will have great potential for population increase if released without their competitors (Zwölfer 1973). In this case the host plants will not have evolved resistance to the rare insect herbivores and may show greater vulnerability. If this reasoning holds, species that are rare in the native habitat should be good candidates for introduction.

Another way to select agents is to look at the characteristics of previously successful agents. Of different insect orders Hemiptera, particularly scale insects, and Coleoptera, beetles, have been the most successful as biological control agents (Crawley 1990). Environmental matching and the presence of predators at release sites have been identified as important determinants of the success of biological control agents (Cullen 1995). Agents that kill plants and have prolonged attack are frequently successful. Seed predators are rarely successful control agents (Myers and Risely 2000) although they may be important in some cases for reducing the spread of nonnative species such as trees that have been introduced because of their commercial value (Hoffman and Moran 1995).

Historically, testing the efficacy of potential biological control agents has been secondary to testing their host specificity. This has led to the introduction of many species that have had little contribution to biological control success. Examples include the introduction of six species of seed predators on yellow starthistle, *Centaurea solstitiales*, in California without reduction in weed density (Pitcairn et al. 1999).

An experimental approach might help to identify successful control agents. The creation of high-density patches of weeds in the native habitat could elucidate the species of natural enemies that thrive at densities comparable to those in exotic habitats and those that have sufficiently deleterious impacts to reduce the host density. In addition to parsimoniously choosing the best and safest biological control agents, biological control of weeds involves coordinated manipulation of agent release, plant competition, and disturbance (McEvoy and Coombs 1999). It is applied ecology and conservation at its best.

Biological Control of Diffuse Knapweed: An Example of a Control-Resistant Weed

The biological control of diffuse knapweed, *Centaurea diffusa*, in western North America has been a long, slow process. Diffuse knapweed is native to Eurasia and was introduced to North America beginning in the late 1800s. In British Columbia the major introduction stemmed from contaminated alfalfa seeds imported following World War I. The control program began in the late 1960s with the introduction of two species of gall flies, *Urophora affinis* and *U. quadrifasciata*. These rapidly established and spread, and reduced seed production (Myers 1987). In the mid-1970s a root-boring beetle, *Sphenoptera jugoslavica*, was introduced from Europe. The major impact of this beetle was to

reduce the growth and seed production of attacked plants. The result of both the beetles and the gall flies was a reduction in seed production by over 95%. In an intensive study of the population ecology of knapweed, Powell (1990a,b) found that the survival of seedlings and rosettes, and the proportion of plants flowering were reduced at high plant density. This allowed populations of knapweed to compensate for insect attack that reduced seed production. In addition, the experimental removal of flowering plants increased the survival of earlier life history stages of knapweed (Myers et al. 1990). Two models (Powell 1990a; Myers and Risley 2000) predicted that knapweed densities would only decline when seed production was reduced to extremely low levels (>99% reduction). An agent capable of killing later life history stages of the knapweed was required. Altogether seven species of agents have been introduced on diffuse knapweed in British Columbia. Only recently have populations declined (Figure B) following the introduction of the last species, the weevil *Larinus minutus*.

Is the apparently successful control of knapweed an example of accumulated stress or of finally finding the most effective biocontrol agent? The only way to answer this question is by undertaking field experiments of the type carried out by McEvoy et al. (1993) with tansy ragwort, *Senecio jacobaea*. In these experiments cages were used to evaluate the impacts of agents independently. Such experiments have not yet been done for knapweed but several factors are important to the interpretation of the knapweed situation. Although larvae of *Larinus* develop by eating seeds, adult beetles feed on rosettes and bolting plants in the early spring. This damage can actually kill plants. In the last three years, areas of knapweed infestation have been very dry, and fluctuations of knapweed densities are known to be related to rainfall (Myers and Bazely 2003). Drought could have contributed to the success of *Larinus*.

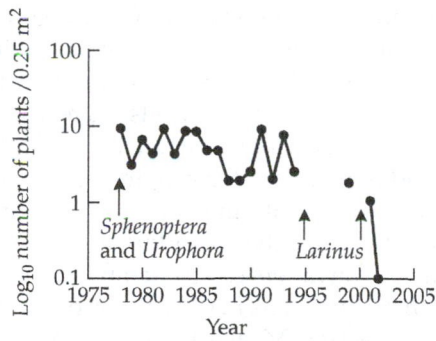

Figure B Attempts to control diffuse knapweed (*Centaurea diffusa*) where it has been a difficult invasive weed in British Columbia with various herbivores had been unsuccessful until a weevil, *Larinus minutus*, was introduced in 1995 and 2000. The control of *C. diffusa* by *L. minutus* may have been aided by a series of dry years (1987–1989; 2001–2002) that also reduced growth in the invasive weed. Unsuccessful earlier biocontrol introductions included two species of *Urophora* (gall flies), and one of *Sphenoptera* (a root-boring beetle).

The apparent successful biological control of diffuse knapweed is very exciting to those of us involved in the program for over 30 years. It is a good example of the need for basic ecological studies and the time is now right for more experimental studies on the impacts of *Larinus*, the most recently introduced agent. In addition, more information should be acquired on what plant species will replace knapweed. These are likely to be introduced grasses, but they also may include other introduced rangeland weeds. While biological control can be a tool for the conservation and restoration of ecosystems, it is only a small part of a very complex challenge facing conservationists in this world of rapid and extensive mixing of species.

Summary

1. Examining biological invasions from a conservation perspective shows that introduced species can cause striking ecological change and contribute to native species endangerment and extinction, and ecosystem alteration. The long-term impacts of these changes, in terms of ecosystem services, are difficult to predict. In terms of human needs, many important species that provide subsistence and recreational and cultural enjoyment are introduced. At the same time, introduced predators, pests, and diseases cause tremendous human, wildlife, and agricultural losses. The economic impacts of introduced species can be positive or negative, and range in scale from artisanal to industrial to public health. The successful management of biological invasions requires an understanding of their underlying biology: how invaders affect native species and alter ecological and evolutionary processes, what makes particular invasions successful, how invasions occur, and how can we control and prevent them. These questions encapsulate a rapidly evolving field of questions, debates, and preliminary answers that will demand extensive future research to fully address.

2. Every invader has some impact on the community in which it establishes. Many invaders may have minimal impacts, but some have dramatic effects. Invasion impacts extend from small-scale alterations in genetics and behavior to population-, community-, and ecosystem-level impacts. To really evaluate the impacts of invasions, we must consider both the direct and the (often extensive) indirect effects of an invader, and how its impacts may compound or mitigate those of other invaders in the same area. High-impact invaders can be plants, animals, or microbes; they can act as herbivores or predators, parasites or diseases. Measuring invasion impacts requires creative and careful study design and implementation, and can be thought of in many cases as characterizing a species' areal extent, its abundance, and its per-capita effects. In all cases, it is useful to recall that the ecological impacts of an invader cannot be considered *inherently* bad or good, as these labels apply only in the context of particular conservation or other goals.

3. Every species can be a successful invader, and every community can be successfully invaded—under the right circumstances. The first of these circumstances is the presence of an invasion pathway linking species and community and delivering sufficient propagule pressure for a population to establish. Pathways and propagule pressure are particularly important elements of invasion success when we consider how to prevent invasions from occurring.

4. We can also consider traits of species and communities that tend to increase the chances of successful invasions. For species, two of the most commonly seen traits are an association with human activity, and a large pool of potential source populations (either a broad native range or multiple past invasions). Beyond these, species characteristics associated with success are easiest to identify for particular groups, such as woody plants or fishes, or for particular cases where the presence or absence of behaviors, genetic traits, or even other species contribute significantly to success. An area of considerable current interest is applying our understanding of traits of successful invaders to the reduction of invasion risk at legislative and policy levels.

5. Communities at high risk of being invaded are those with invasion pathways delivering propagules from well-matched climates and habitats. Beyond this, it seems likely that a combination of resident species richness, resource availability, and disturbance influence the outcome of invasions. Teasing apart the importance of these factors, and their application in a conservation context, is an ongoing challenge. In contemplating the features of successful invasions, it is especially useful to distinguish between species and communities with a high invasion frequency attributable to the presence of particular invasion pathways, as opposed to traits that make these species and communities inherently good at establishing new populations or at being invaded.

6. Species are introduced both intentionally and inadvertently, and the eradication and prevention of all

invasions is probably an unrealistic and undesirable goal. However, understanding the pathways of introduction allows us to make judicious choices regarding which invasions we actively permit, and which we seek to minimize. It is clear from many case studies that eradicating, or even controlling an established invader, is a nontrivial task. Not only is it expensive, but invasion control through mechanical, chemical, and biological means can further damage an already disturbed system. Invasion control may be particularly effective when implemented early in an invasion, before a species is either widespread or a key food web component. Biological control is an especially challenging approach to invasion control, as it adds an additional level of unpredictability to the system. A precautionary approach to invasion management demands a prevention program to complement control efforts.

Please refer to the website www.sinauer.com/groom for Suggested Readings, Web links, additional questions, and supplementary resources.

Questions for Discussion

1. What do the terms *success* and *impact* mean in the context of biological invasions? Which features of a successful invasion (in terms of species, communities, or both) might you expect to be the same as, or different from, features of a high-impact invasion, and why?

2. Biological invasions are one of many factors affecting biodiversity, including direct harvest, agri/aquaculture, habitat loss, and climate change. How do these factors interact in terms of their causes, effects, and management?

3. If you were designing a reserve, how would you minimize the risk of biological invasions? How would your approach differ between a terrestrial and a marine reserve?

4. Disturbance may influence a community's invasibility, and a successful invader may cause additional disturbance. How would you test these ideas in a field study?

5. Species richness may influence a community's invasibility, and a successful invader may influence species richness. How would you use your understanding of these processes to manage a nature reserve?

10

Biological Impacts of Climate Change

Camille Parmesan and John Matthews

Ok, so you convinced me that global warming is happening, but why should my constituents care?

Anonymous U.S. Congressional Staffer, 2004

Global climate is swiftly changing, with poorly known consequences for biodiversity and human well being. In the last 90 years Earth's mean temperature rose 0.6°C, a rate of increase that has not been seen in 10,000 years. Since the mid-1990s, it has been clear that mean global temperature rose during the twentieth century, but for several more years the cause of the observed warming was much debated in scientific circles. The United Nations and the World Meteorological Organization perceived that if humans were causing a fundamental shift in global climate, the consequences could be enormous for all peoples. To facilitate reaching a global scientific consensus, the Intergovernmental Panel on Climate Change (IPCC) was formed, whose Third Assessment Report (IPCC 2001a) involved more than 1200 scientists from diverse disciplines. The consensus among the more than 200 climate experts from IPCC Working Group I was that "global rises in mean yearly temperature of the past 50 years were primarily due to global rises in anthropogenically produced greenhouse gases" (i.e., carbon dioxide, methane, nitrous oxide, and halocarbons) and that the rate of temperature change will itself accelerate over the coming century (IPCC 2001a).

As with many other conservation issues, however, the most influential debates—those in the policy and public arenas—continue well beyond the point of scientific consensus. And, indeed, consensus on the cause of global warming is only the beginning of a series of new questions. What does anthropogenic climate change signify for the future? Is climate change a reversible process, and how quickly will new weather patterns emerge? What can be done to mitigate the detrimental effects of altered climates? How will new climatic patterns influence species and ecosystems already besieged by other human-induced pressures?

Signs of climate change are already visible, although most effects have been relatively mild compared to other anthropogenic threats to wild species, particularly habitat loss and degradation. Conservation concerns about climate change are less focused on current impacts or even on impacts over the next decade but on consequences from 50 to 100 years from now. Existing climate models predict widespread

and dramatic alterations in weather patterns over coming centuries, altering all nations and virtually every ecosystem. In the long term, climate change is likely to rise to the forefront of threats to biodiversity and endangered species. The scale, duration, and probable severity of climate change separate it from other environmental issues. Indeed, national and international political responses to the challenge of climate shifts has been more comparable in scale to preventing HIV/AIDS or polio than strategies to counteract local or regional issues such as species invasion or overexploitation. Climate change is the one threat that exists everywhere, cannot be reversed by local actions, and will continue even if all nations come to agreement to tackle the problem.

In this chapter, we shall discuss some basic processes that have shaped climate in the past and how those processes have been altered during the twentieth century, the relationship between climate, wild species, and ecosystems, and the conservation implications of long-term climate shifts. Finally, since climate change has been defined as a global threat as well as a regional and local issue, we will discuss international climate change policy with special attention to the United States and Canada.

The Nature of Climate Change

Generally speaking, climate shifts have been caused by changes in the retention and distribution of solar energy across the planet. Solar radiation passes through the atmosphere as short wavelength ultraviolet (UV) waves. When UV waves hit Earth's surface, they are transformed into long wavelength infrared (IR) waves, which we perceive as heat (Figure 10.1). A remarkable aspect of Earth's biosphere is that a system has developed that has been able to maintain sufficient quantities of life-promoting liquid water for billions of years. Such a system implies a relatively stable pattern of energy retention.

The atmospheric gases that are most important in maintaining Earth's current energy balance are carbon dioxide (CO_2), methane (CH_4), and water vapor, gases that cannot be penetrated by radiation in the infrared spectrum. These gases are referred to (with more poetic license than accuracy) as "greenhouse gases" by analogy to the glass of a gardener's greenhouse. Short-wave (UV) energy is allowed to pass through our atmospheric window, but the transformed long-wave (IR) energy is prevented from dissipating back into space.

Greenhouses gases are really more like a blanket surrounding and insulating Earth's surface. The resulting greenhouse effect maintains our relatively stable surface temperatures. Natural levels of Earth's greenhouse gases maintain temperatures about 60°F above what it would be if Earth's atmosphere lacked such gases. In contrast

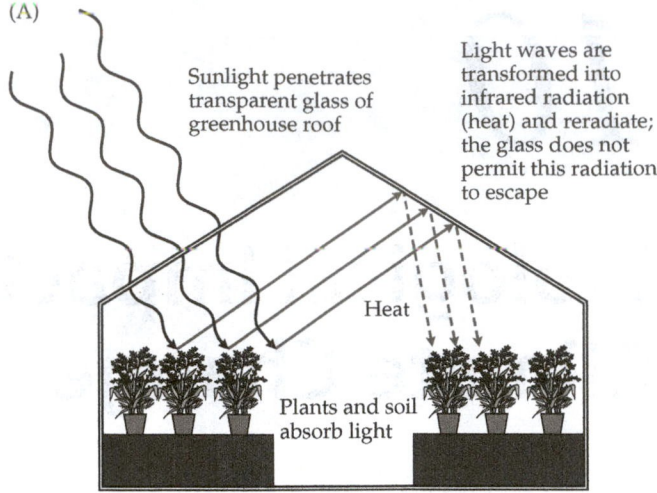

(A) Sunlight penetrates transparent glass of greenhouse roof

Light waves are transformed into infrared radiation (heat) and reradiate; the glass does not permit this radiation to escape

Heat

Plants and soil absorb light

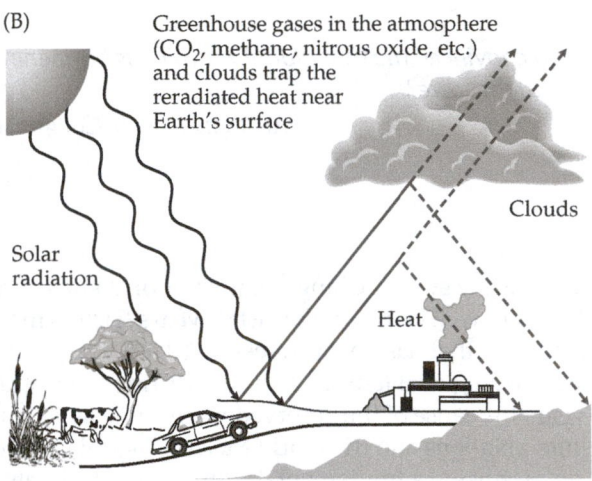

(B) Greenhouse gases in the atmosphere (CO_2, methane, nitrous oxide, etc.) and clouds trap the reradiated heat near Earth's surface

Clouds

Solar radiation

Heat

Light waves are transformed into infrared radiation (heat) reflected back to Earth by clouds and reradiated

Figure 10.1   The greenhouse effect. Earth's atmosphere already serves as a greenhouse that warms Earth's surface (A), but human influences are increasing this effect (B). About 70% of UV wavelengths pass through Earth's atmosphere, whereas about 30% are reflected. Burning of fossil fuels in industry, automobiles, and buildings, as well as deforestation give off CO_2 to the atmosphere that enhance the greenhouse effect. In addition, methane from swamps, rice paddies, and cattle enter the upper atmosphere and trap heat. (From Primack 2004.)

this natural greenhouse effect, any temperature increase attributable to human activity is referred to as the **enhanced greenhouse effect**.

Climate change through the ages

Since the industrial revolution, burning of coal, oil, and natural gas has increased levels of greenhouse gases by about 30%. Mean global temperatures increased 0.6°C

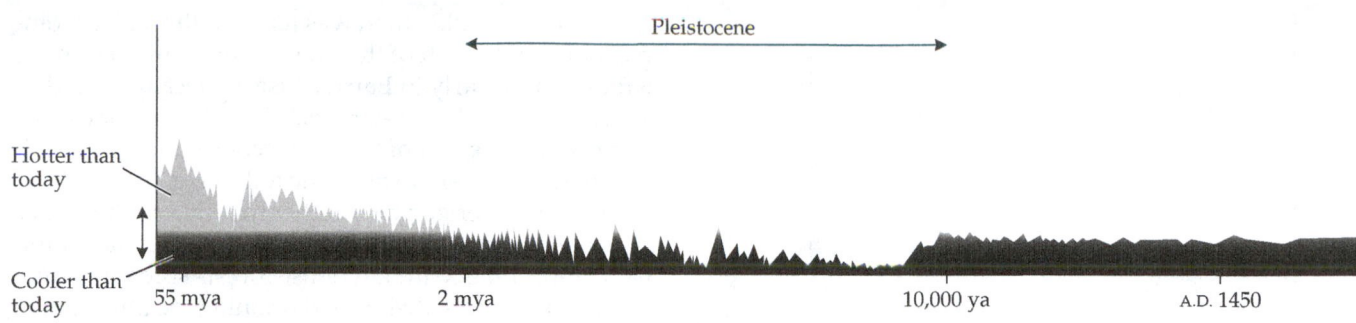

Figure 10.2 Average global temperature over the last 65 million years. Gray indicates temperatures hotter than today; black indicates temperatures cooler than today. At 65–100 million years ago (mya; the age of the dinosaurs), CO_2 was much higher than it is today and the climate was much warmer and wetter. About 13 million years ago, we start to see brief periods in which temperature and CO_2 levels were as high as they are today. The Pleistocene period began around 2 mya and ended 12,000 years ago, during which Earth cycled between glacial (frozen) and interglacial (warm) periods. Note that climate had been relatively stable over the past 10,000 years. (Modifed from *National Geographic*, May 1998.)

between 1860 (when collection of accurate temperature records began) and 1998 (IPCC 2001a). Based on what is known about the properties of these gases, it has long been clear that they would have some effect on Earth's climate system. Early in the 1990s, however, it was unclear how much of the observed temperature rise was due to the rise in greenhouse gas concentrations as compared to natural variation in climate.

Investigations of global climate change, Earth's climatic history, and the many causes of climate fluctuation have been actively pursued for over a century. Putting a 0.6°C temperature rise into appropriate context requires an analysis of climate change at several scales and an appreciation of the impermanence of "normal" climate.

Climate certainly shows natural variation, even at time scales shorter than an average human lifespan. However, short-term variation is not a reliable indicator of long-term climate trends. Indeed, there is much confusion over the contrast between weather (which refers to a specific event or a brief period, such as a thunderstorm or a spring with low rainfall) and climate (which refers to long-term averages, such as the amount of rain that can be expected to fall during an average monsoon season in Mumbai, India). Climate trends are not visible in particular weather events but can be deduced over many weather events. An especially mild winter in New England cannot be "blamed" on human-induced factors. But a trend between 1950 and 2000 for shorter and milder winters in New England could be reasonably ascribed to a climate shift.

Indeed, Earth's climate has varied significantly enough for the definition of "normal" global climate to differ sub-

stantially based on the observer's time scale. Planetary temperatures appear to have begun trending downward since the beginning of the Tertiary period, starting between 50 and 60 million years ago (mya) and falling by 10°C. Today's climate, then, is relatively cool at this level of resolution (Figure 10.2).

Finer scales make different patterns apparent. Using data from ocean sediment cores, coral reef cores, and other paleoecological data, one can see fluctuations in surface temperatures since the end of the Cretaceous. During the geologically recent Pleistocene (a period that lasted between 1.8 million to 12,000 years ago), global climate exhibited large-scale periodic fluctuations in temperature and precipitation. These fluctuations—the Milankovitch cycles—match changes in Earth's orbit and tilt relative to the sun's axis that subtly alter the amount, location, and timing of solar energy reaching Earth (Imbrie and Imbrie 1986). The strength and speed of temperature change is believed to be driven by a series of positive and negative feedbacks through the biotic and abiotic systems, with greenhouse gas concentrations being the strongest driver.

There is a strong correlation between greenhouse gas abundance and mean temperatures. For instance, cores drilled in the Arctic and Antarctic ice sheets contain tiny air bubbles embedded when the ice formed. These bubbles reveal the exact composition of the atmosphere at the time of ice deposition. Not only does this give an exact value for atmospheric carbon dioxide levels, but scientists have discovered that the ratio of two isotopes of oxygen ($^{18}O_2$ and $^{16}O_2$) can be used to accurately calculate atmospheric temperatures at the time the ice was

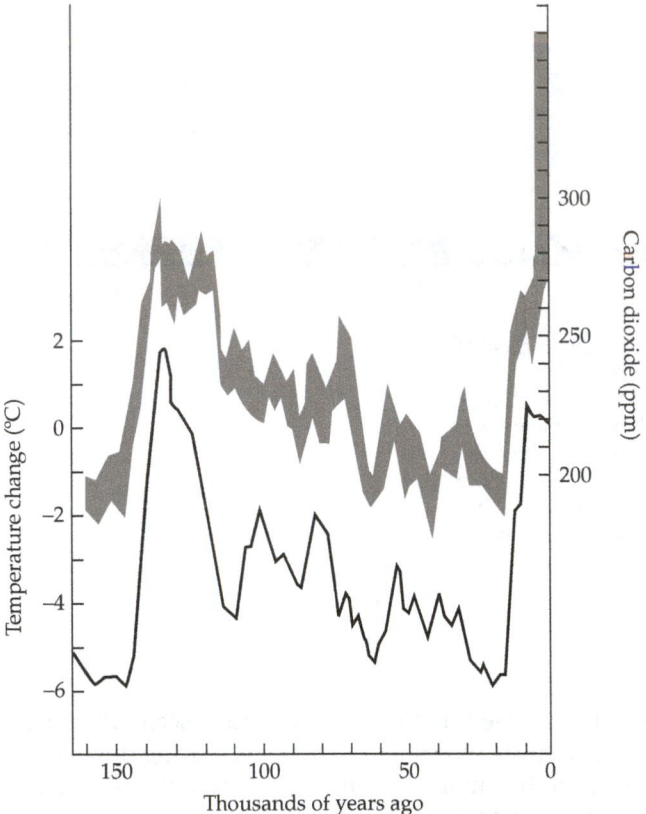

Figure 10.3 The relationship between temperature (normalized to today's average global °C), and carbon dioxide (ppm) over the past 160,000 years. Ice core samples show a clear correlation between atmospheric CO_2 concentrations (gray line) and the global temperature record (black line). Note that the current level of CO_2 (at 360 ppm) is far higher than the highest natural level. (Modified from IPCC 2001a.)

more carbon dioxide than was released through decomposition), and much of the atmospheric carbon dioxide left over from early in Earth's history became stored as what we now call "fossil fuels." After the major extinction event at the end of the Cretaceous Period, in which 76% of species went extinct, the following Tertiary Period was characterized by another massive evolutionary radiation of terrestrial plant species. Trees became dominant at this time, with tall dense forests spreading over much of the land—their wood draining the atmosphere of another large pool of carbon. Finally, at this time vast limestone deposits were formed. Carbon is a major component of limestone. Thus, over several million years, vast quantities of carbon were sucked out of the atmosphere and put into long-term storage ("sequestered") in the form of oil, coal, natural gas pockets, stands of forest, and rock formations.

The more recent temperature changes in the Pleistocene can help us understand how much of a shift in temperature is necessary to affect major physical and biological features of Earth. Pleistocene climates alternated between two extremes: glacial (cold and dry) and interglacial (warm and wet) periods. Average Pleistocene temperature cycling from the beginning of a glacial period to the end of an interglacial period operated on a scale of about 100,000 years. Peak glacial periods were about 5°C cooler than current global mean temperatures, which is substantial enough a decrease for glaciers to cover the majority of the Northern Hemisphere and for sea levels to be several hundred meters lower than current levels. Increasing global temperatures will push our current climate pattern (interglacial) to one that is warmer than any sustained period during the Pleistocene—or indeed over the past 10 million years.

Human enhancement of the greenhouse effect

The insulating properties of carbon dioxide were discovered in the mid-1800s. The greenhouse effect as a major driver of Earth's climate system was first hypothesized by Svante Arrhenius in 1896. He even predicted that the emission of carbon dioxide by combustion of coal following the advent of the Industrial Revolution (about 1750) would eventually warm the world (Weart 2003).

Our contemporary climate represents the end of an interglacial period. For unknown reasons, glacial–interglacial cycling stopped about 10,000 years ago and temperatures have remained relatively stable, varying by only 1°C (see Figure 10. 2). Over that time, atmospheric carbon levels shifted only about 10% until the Industrial Revolution. Therefore, the increase of 0.6°C since 1910 represents a large and sudden change compared to natural variations over the past 10,000 years. Further, atmospheric carbon dioxide levels rose 36% since 1910—from 280 ppm to 380 ppm—which is significantly outside the

formed. Using data from various locations (Greenland, Siberia, and Antarctica), a record of greenhouse gas composition extending back 740,000 years has been assembled, with much of this span resolvable to individual years (McManus 2004). Over this period, carbon dioxide levels and mean global temperatures are closely correlated (Figure 10.3; IPCC 2001a).

Estimates from geological evidence show that carbon dioxide in the atmosphere during the peak of the Cretaceous was much higher than now. There are two major reasons why these levels declined, one biological and one geological. The Cretaceous was a time in which terrestrial plants expanded and thrived. However, wet, anoxic conditions slowed decomposition and dead plant matter built up in large quantities. Subsequent burial and exposure to massive pressures transformed these dead organic deposits into coal and oil formations. Thus, the Cretaceous was a time in which the terrestrial plant biome was a huge carbon sink (i.e., living plants took up

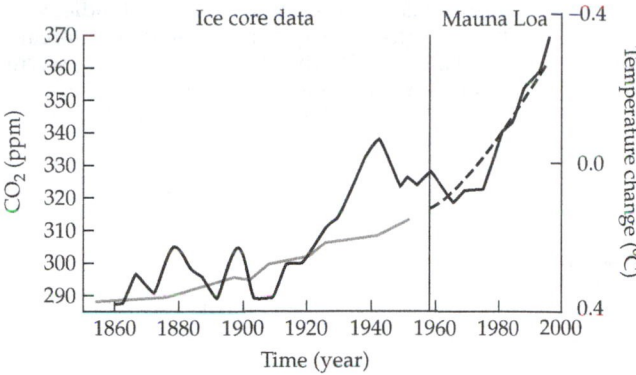

Figure 10.4   Relationship between twentieth century levels of atmospheric carbon dioxide and global temperature. The solid gray line shows CO_2 records from ice cores, the dashed black line shows annual average CO_2 records from atmospheric measurements in Hawaii, and the black line shows mean yearly global temperature trends. This steady increase of CO_2 in the atmosphere has caused greater retention of heat and a gradual warming of Earth. (Modified from IPCC 2001a.)

bounds of natural variability over the past half million years (see Figure 10.3). Like withdrawing money long stored in a huge bank vault, the Industrial Revolution has allowed these carbon deposits to once again circulate in the world's "carbon economy."

There is now a strong consensus among climate scientists that Earth's mean surface temperature has increased, and that this heating is largely due to human-induced increases in greenhouse gas concentrations (Figure 10.4; Crowley 2000; IPCC 2001a; Karl and Trenberth 2003). Most of the surface temperature increase is due to the enhanced (rather than the natural) greenhouse effect.

Mechanisms regulating the global energy budget

Humans actually produce a variety of greenhouse gases, most of which already existed in the atmosphere prior to the Industrial Revolution. Of all the greenhouse gases produced by humans, however, the greatest concern surrounds the production of CO_2. Not only is CO_2 produced in much greater quantities than any other of the greenhouse gases, but a carbon dioxide molecule remains stable in the atmosphere for over 100 years. In contrast, methane is 100 times stronger than CO_2 in its greenhouse effects, but a methane molecule is broken down in about a decade. Methane is produced in the low-oxygen conditions of rice fields, from the digestive systems of cattle and other ruminants, and as a by-product of coal mining and natural gas use. Most of the remaining greenhouse effect is from nitrous oxides and chlorofluorocarbons (CFCs) (both produced largely by industrial processes) and low-atmosphere ozone (from fossil fuel combustion).

Collectively, all of the activities and processes that increase the ability of Earth's surface and atmosphere to absorb and retain solar energy are referred to as positive radiative forcings. Direct solar radiation and greenhouse gases cause major positive radiative forcing, but others also exist. For instance, deforestation creates positive radiative forcing if it is replaced by grasslands (such as pastures following a clear-cut). Trees cool their locality by high evapotranspiration rates (like an evaporative cooler used in commercial greenhouses). A pasture at the same site has much lower rates of evapotranspiration and will be hotter than a forest on average, everything else being equal.

In contrast, negative radiative forcings tend to cool Earth's surface and lower atmosphere. For instance, sulfur dioxide (SO_2) from car exhaust and industrial processes such as coal-powered electrical plants form aerosols (small particles suspended in the atmosphere) that reflect solar energy back into space and hence cool Earth's surface air temperatures.

Mediating positive and negative radiative forcings is the relative reflectance, or albedo of Earth's surface and clouds. Different surfaces have different albedos. Snow reflects UV and cools surface air temperatures, whereas bare dirt converts UV to heat and warms Earth's surface. Taken together, the interaction of positive and negative radiative forcings and albedo is quite complex. Because SO_2 also causes acid rain, there have been recent attempts in industrialized nations to reduce SO_2 pollution. Because SO_2 remains in the atmosphere for only a week, reduction of SO_2 emissions will immediately lessen its impact as a negative radiative forcing.

Moreover, forcings can have either natural or anthropogenic origins. Mount Pinatubo in the Philippines had a major eruption in 1991 that increased global aerosol concentrations. The ash output contained 25–30 million tons of sulfur dioxide that entered the atmosphere during a very short period and which rapidly dispersed across the globe. Mount Pinatubo temporarily reduced global mean temperatures, though within three years the particulate matter had largely left the atmosphere and global temperatures rebounded to their former levels (Yang and Schlesinger 2002).

The concept of radiative forcing is used to classify and track the cycling of elements and compounds relevant to climate change through meteorological, geological, hydrological, and biological systems. Figure 10.5 shows the latest consensus on the relative contributions of known radiative forces on twentieth century temperature trends.

Much attention has been focused on the carbon cycle because carbon is such an important component in carbon dioxide and methane. The complete carbon cycle consists of carbon "sinks" (processes that remove carbon from atmospheric circulation), and carbon "sources" (processes

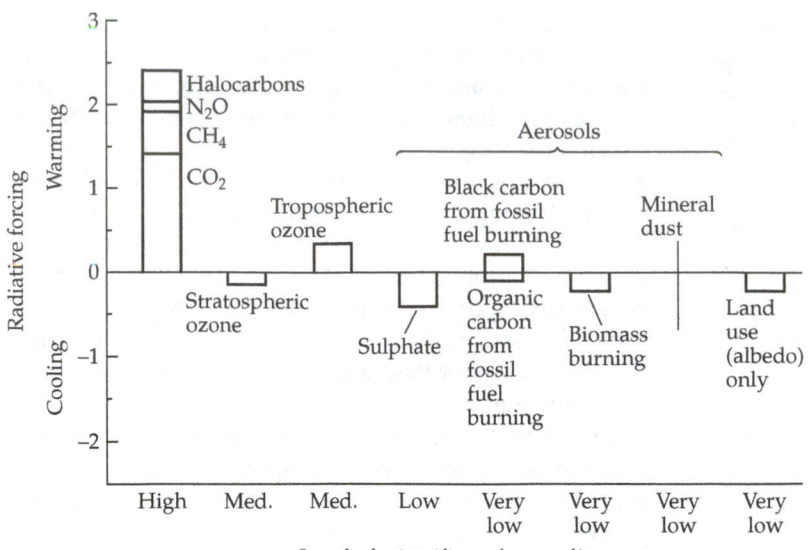

Figure 10.5 Estimates of strengths of radiative forces on global energy budget. Radiative forcing is in watts/m². Positive forces (above zero) warm Earth's surface; negative forces (below zero) cool Earth's surface. (Modified from IPCC 2001a.)

that release stored carbon into the atmosphere). There are numerous sources and sinks for carbon, encompassing both biotic and abiotic systems. While the carbon fluctuations documented by ice cores record the net effect of changes in the relative balance between sources and sinks, these records do not by themselves describe which aspects of the carbon cycle changed. One of the goals of research into the processes behind the carbon cycle therefore is to determine the relative contribution of natural and anthropogenic sources to recent changes in the concentration of greenhouse gases. Current carbon dioxide levels, for instance, could reflect an increase in the output of carbon sources (e.g., due to elevated industrial activity) or because carbon sinks have decreased their input (e.g., due to decreases in the growth rate of tropical rainforests). Such knowledge is vital to predicting greenhouse gas concentrations in the future.

Research into the cycling of other elements and molecules also attempts to untangle the synergies of various systems. For example, water vapor is the most abundant and important of all of the natural greenhouse gases, but its effects are not included in the enhanced greenhouse effect since humans do not produce water vapor in a manner comparable to the other greenhouse gases. Water vapor concentrations are probably increasing, however, as a side effect of rising global temperatures. A warmer atmosphere is able to hold more water vapor, and higher ambient temperatures increase oceanic evaporation and vegetation transpiration rates. The implications of these changes over long time scales are uncertain.

Perhaps the most frightening prospect in this regard is a so-called runaway climate shift, leading to a sudden and large increase or decrease in global temperature. Several researchers modeling periods of climate change in the past suggest that elemental cycles normally oper-

ate within tolerances that allow minor climate variation but that resist significant shifts. However, if some tolerances are exceeded, several radiative forcings may reinforce one another and push global climate through a period of rapid change before reaching a new plateau. For example, if greenhouse gases reached a certain threshold level, then the resulting increased air temperatures may trigger large-scale melting of polar land ice. This would expose dark surface rocks that have been covered for a very long time. These new dark surfaces would absorb more solar energy than the former ice cover, which would spawn additional, rapid large-scale melting.

Current and Future Climate Change

To predict future climate, scientists have developed and tested **global climate models** (**GCMs**) based on the processes by which atmospheric greenhouse gases affect global climate. Though there are some differences among models used by, say, the National Center for Atmospheric Research (NCAR) in the U.S. and by the Hadley Center in the U.K., there are also many broad agreements. We review these briefly here.

Temperature and precipitation changes

All GCMs indicate that greenhouse-gas-driven warming should be greatest at the poles and weakest in the tropics, and this is indeed the pattern seen in recent analyses of global temperature change over the twentieth century (IPCC 2001a). In some parts of Alaska, Canada, and Siberia, mean annual temperatures have increased by 2–4°C since 1900, much more than most areas of the lower 48 states of the U.S. (Oechel et al. 1993; IPCC 2001a). One of the striking trends at mid- and high latitudes has been toward fewer days and nights below freezing in wintertime and more ex-

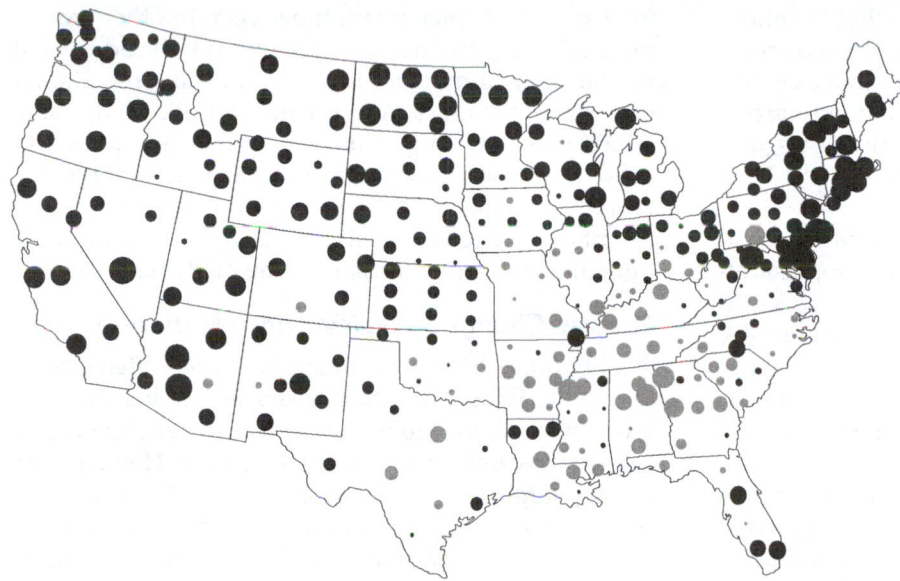

Figure 10.6 Temperature trends in the lower United States from 1901 to 1998. Black indicates areas that are warming and gray indicates areas that are cooling. The size of the dot indicates the magnitude of temperature change, with the largest dots representing a change of 3°C per 100 years. (From National Climatic Data Center/NESDIS/NOAA.)

treme heat days in summer. These events may in some cases simply reflect the same level of variation but around a new, higher mean, but there are indications that yearly variance in temperatures has also increased (Karl et al. 2000; Karl and Trenberth 2003).

In the U.S., warming averaged about 0.7°C (1°F) during the twentieth century. New regional models and analyses also conclude that the North American warming trend can largely be attributed to a rise in greenhouse gases (Karoly et al. 2003; Stott 2003; Zwiers and Zhang 2003). Still, there is variation across the U.S., with some areas even cooling (Figure 10.6).

It is becoming increasingly apparent that aspects other than temperature are also changing (reviewed by Meehl et al. 2000a,b; Easterling et al. 2000a,b). Precipitation has

increased globally, and precipitation events have been increasing in intensity. Consequently, major floods have been occurring more frequently and with greater severity (Karl et al. 1996; Karl and Knight 1998; Groisman et al. 1999; IPCC 2001a). In the U.S., total precipitation has increased by 5%–10% over the twentieth century, but there is regional variation. Winter precipitation (snow) has increased in the Great Lakes region, while other U.S. regions are becoming drier (e.g., eastern Montana; Groisman et al. 2001) (Figure 10. 7).

Making predictions for coming decades has proven difficult. Even if all human emission stopped immediately, enough greenhouse gases have been added to the atmosphere to sustain temperature increases throughout the twenty-first century. In reality, the emission of all

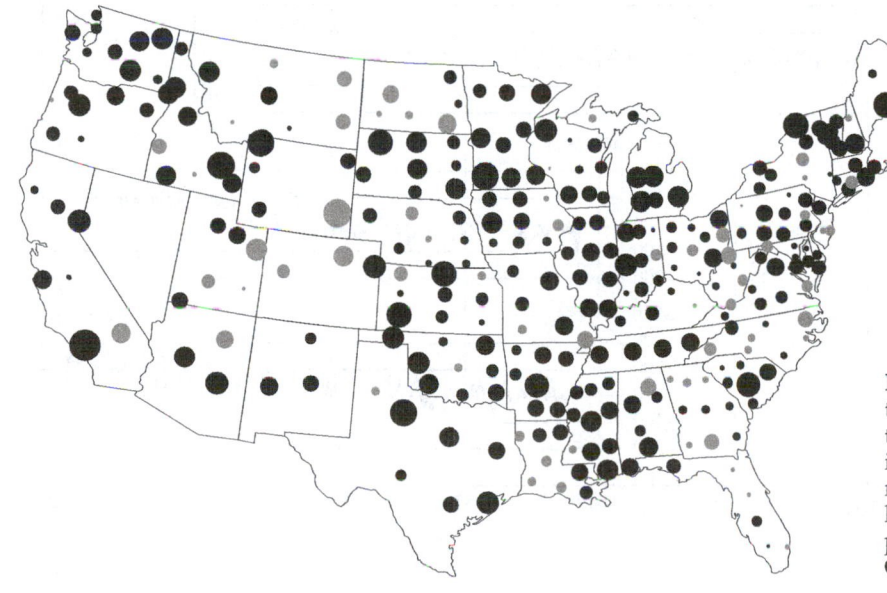

Figure 10.7 Precipitation trends from 1901 to 1998. Black indicates areas where precipitation is increasing and gray indicates where it is decreasing. The size of the dot indicates the magnitude of precipitation changes, with the largest dots indicating a 40% change in the precipitation over 100 years. (From National Climatic Data Center/NESDIS/NOAA.)

greenhouse gases (except CFCs and methyl bromide) has continued to rise since 1990, leading to predictions by the IPCC of a quickening in the rate of climate change due to the enhanced greenhouse effect. Indeed, carbon levels now increase about 10% each twenty years, and between 1976 and 2000 global mean temperatures have risen at a rate of 3°C per century.

The need for accurate climate predictions has challenged climate scientists to create increasingly sophisticated computer models to simulate atmosphere dynamics. To develop predictions that would be useful to policymakers, the IPCC (2001a,b) devised four basic "scenarios" to understand warming trends over the twenty-first century. These scenarios are based on different assumptions about global development, population growth, technological improvements in energy conservation and creation, and approaches to the management of greenhouse gas emissions. These scenarios also differ in their target carbon dioxide stabilization level between 450 ppm and 1000 ppm (remember that 1990 levels were 350 ppm). These are further subdivided into a total of 40 scenarios, which are the basis for specific climate models.

Using 1990 as a baseline, these models predict an increase in global temperature between 1.4°C and 5.8°C by 2100. Other trends are also seen across models. One of the ironies of climate projection to 2100 is that while precipitation is expected to continue to increase, higher air temperatures will lead to more rapid evaporation. Moreover, the increase in variance of precipitation events will change the patterns of water availability (seasonality and frequency), which may be more important to humans and to wildlife and ecosystems than a change in total amount of precipitation per year. In other words, more rain may fall in a given locality, but the additional rainfall may or may not result in substantially altered ecosystem dynamics. As a further complication, a sustained rise in mean global temperatures is expected to cause changes in global air and ocean circulation patterns that, in turn, will affect regional climates to varying extents (IPCC 2001a,b). All of these processes are very difficult to model, particularly as an evolving system.

Oceans: Change in sea level and circulation

The boundary between land and water has often moved over time. Twenty thousand years ago for instance, sea levels were about 120 m lower than current levels and glaciers covered much of the Northern Hemisphere. Clearly large shoreline movements are likely to have profound impacts on marine and terrestrial species and ecosystems near coastlines. On shorter time scales, however, variation has usually been small.

Measurements of sea level for the years before the nineteenth century come from geological analyses of rock formations and fossil coral reefs. Global sea level variations have averaged between 0.1 and 0.2 mm annually over the past 3000 years. Since the nineteenth century, shoreline data come from direct measurement of ocean level by tide gauges. These records document that ocean levels have risen over the past 200 years, and that the twentieth century showed a greater increase than the nineteenth (IPCC 2001a) (Figure 10.8). The rate of increase over the twentieth century has averaged between 1 and 2 mm annually, ten times the estimates for the past 3000 years (IPCC 2001a). Several recent studies have linked the

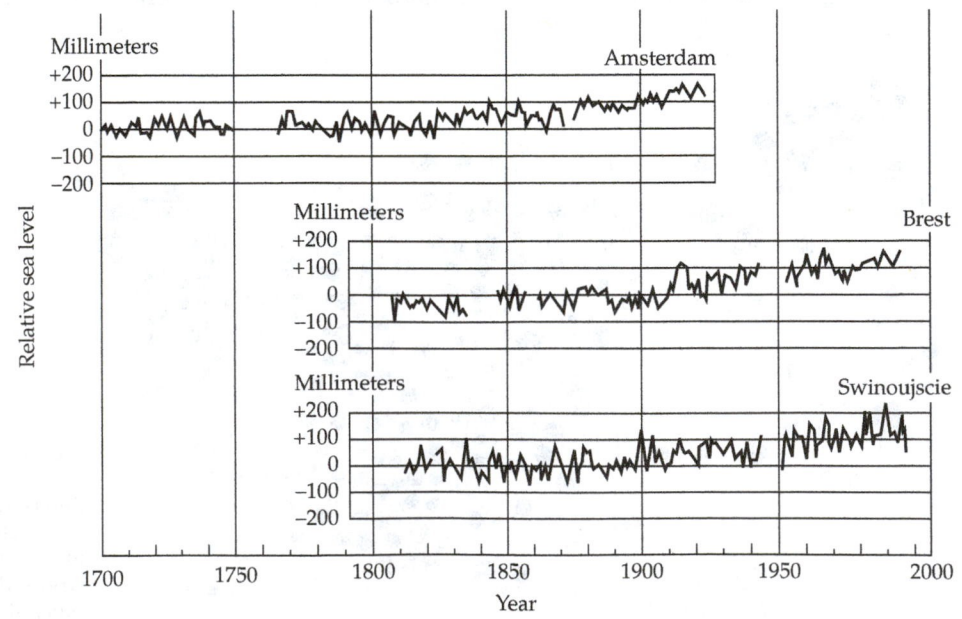

Figure 10.8   Sea level rise over the past 300 years in three European cities. A limited number of sites in Europe have nearly continuous records of sea level spanning 300 years; these records show that the greatest rise in sea level took place over the twentieth century. Records shown from Amsterdam, The Netherlands, Brest, France, and Swinoujscie, Poland, as well as other sites, confirm the accelerated rise in sea level over the twentieth century as compared to the nineteenth. (From IPCC 2001c.)

observed increase in ocean temperatures and in sea level to rises in atmospheric greenhouse gases (Barnett et al. 2001, IPCC 2001a; Levitus et al. 2001; Reichert et al. 2002; Gillett et al. 2003).

Sea level rise is due both to thermal expansion and to an increase in total water as land ice melts. Thermal expansion is simply the increase in volume of any liquid as it gets warmer. Given the very large total volume of Earth's oceans, a small temperature change can have an important effect. Observations over the twentieth century suggest that thermal expansion is responsible for an average of 1 mm annually (IPCC 2001a). Because liquid water has a much higher heat capacity than the atmosphere, there is a much longer lag time between adding heat to the ocean and observing a temperature rise than there is for warming the atmosphere. If atmospheric carbon dioxide were fixed at current levels, atmosphere temperatures would continue to rise for another 100 years and lagging thermal expansion would cause sea levels to rise for many centuries.

More directly, seas rise from an increase in water. The largest quantity of non-ocean water is locked up in massive polar ice sheets. In geological terms, sudden shifts in sea level have occurred before, as when sea levels rose at an average annual rate of 10 mm per year between 15,000 and 6,000 years ago as glaciers melted and retreated poleward. Melting of the polar ice caps has contributed between 0.2 and 0.4 mm annually to sea level increases over the twentieth century (IPCC 2001a). Sea level rise is predicted to accelerate from the 20–30 cm over the twentieth century and to 35–90 cm over the twenty-first century. (IPCC 2001a). The effects of rising sea levels on ecosystems will be substantial. Coastal areas are expected to suffer increased erosion rates with more severe storm surges associated with higher sea levels.

Major ocean circulation systems are already showing signs of being affected by the rise in atmospheric temperatures. The cyclical shift in Pacific Ocean currents, "El Niño," results in massive warming in the mid-Pacific and a warm tongue of water creeping up the west coast of North America and pushing away the normal cold current that flows from Alaska. An El Niño year is characterized by dryer, warmer conditions in the Pacific Northwest (Oregon and Washington) and much wetter conditions throughout much of California and the Southwest. Major El Niño events, such as the ones in 1982–1983 and 1997–1998, cause substantial flooding. El Niño events have increased in frequency and intensity, and some models predict that the "normal" state by 2050 may resemble "El Niño-like" conditions (Meehl et al. 2000b).

SNOW, ICE, AND HYDROLOGICAL CHANGES   Dramatic declines in long-term snow and ice cover already have occurred over the past century (IPCC 2001a,b). There has been a widespread retreat of mountain glaciers in North America, South America, Europe, Africa, New Zealand, and central Asia during the twentieth century. Since 1850, the glaciers of the Swiss Alps have lost about 30%–40% of their surface area and about half of their volume. The permanently snow-capped Mount Kenya and Mount Kilimanjaro in Africa have lost over 60% of their glacial volumes in the last century. In Glacier National Park, glaciers have declined by 70% and are expected to disappear completely by 2020 (Hall and Fagre 2003). It is estimated that about 30% of the projected change in sea level by 2100 will likely come from melting land ice (IPCC 2001a). The shrinking of glaciers will also have significant socioeconomic impacts. Many areas depend on permanent snowpack to keep their reservoirs filled through the summer. Rapidly melting glaciers increase rock and ice falls, which lead to high rates of erosion. Moreover, the loss of scenic beauty will likely affect areas heavily dependent on tourism (Figure 10.9).

Flood events result in more rain disappearing as runoff rather than soaking into the ground. As a result of predicted increased flooding in many regions, soil moisture may decline and underground aquifers many have longer recharge periods. Clearly, freshwater hydrology could alter substantially, but variably, across the globe.

Climate scientists have been modeling the combination of these effects on relatively large scales. For instance, most rain in western California occurs in the winter, and summer river flow is largely derived from the gradual melting of the Sierra Nevada alpine snowpack. But winter snowpack is changing, with trends toward lighter snowpack and earlier melt date at lower elevations and heavier snowpack at higher elevations (Johnson et al. 1999). Current models suggest that winter

Figure 10.9   Worldwide, glaciers are retreating. Here date signs show the retreat of the Exit Glacier in the Kenai Peninsula, Alaska. (Photograph © Y. Momatiuk/Photo Researchers, Inc.)

alpine temperatures will increase enough to lower total snow accumulation in mountain regions of the U.S. (U.S. National Assessment Report 2001). The increased precipitation will instead fall as rain, causing western slope streams and rivers to have stronger winter flows relative to spring and reducing summertime flow substantially.

Recent studies of nonregulated streams in the U.S. showed the initial signs of these changes: Peak spring flow had advanced by about two to three weeks since the 1950s (Lins and Slack 1999; Groisman et al. 2001). In mountainous regions, this shift has been related to earlier melting of winter snowpack (Dettinger and Cayan 1995; Johnson et al.1999). Shifts in hydrological trends are likely to alter local freshwater and terrestrial ecosystems and necessitate major adjustments in the current regime of human water use.

In some cases, managed freshwater systems may be buffered against climate change in regions with strongly regulated flow regimes via canalization of streams and building dams. However, these engineering solutions come at a cost to native biodiversity. For example, dams along the Columbia River in the northwestern U.S. have been shown to impede upstream migration of adult salmon as well as downstream migration of their offspring. Dams are considered a primary reason why Pacific salmon have failed to recover even after fishing was greatly reduced. At a time when conservation biologists are arguing for dam removal, the consequences of climate change for water availability may bring about greater pressure to build dams to maintain water for urban and agricultural areas.

Predicted Biological Impacts

Concern about the conservation implications of climate change stems from a substantial body of literature that affirms the importance of climate in shaping natural systems. Over a century of ecological research demonstrates the crucial role of climate in determining the systematic geographical patterns in the distribution of major biomes or vegetation communities (Woodward 1993). In turn, the distributions of the animals associated with and dependent on these vegetation communities are also a function of climate (Andrewartha and Birch 1954).

Biologists use multiple approaches to form projections of ecological changes expected to accompany various types of climatic change. Predictions of how ecological systems and species may "behave" in response to climate change come from snapshot analyses of current relationships between climate contours and species' distributions (their "**climate envelope**"), from manipulative laboratory studies on plant and animal physiologies with respect to temperature and precipitation tolerances, and from analyses of the fossil record. Most studies seek

to predict how particular species will react to climate change, and increasingly, conservationists are seeking to predict how interacting webs of species might change (for an in-depth discussion of these complexities see Case Study 10.1 by Lisa Crozier).

During the great Pleistocene glacial–interglacial cycles, the world experienced much more dramatic climatic changes than were seen during the twentieth century (detailed previously). Massive range shifts and some extinctions occurred (Huntley 1991; Davis and Zabinski 1992; Coope 1995). Over time, ecosystems adjusted to climate changes and new types of communities emerged across the landscape.

Using fossil evidence to predict ecosystem responses to current warming trends, one might first assume that ecosystems will move to new locations that are relatively similar to their current abiotic limits given new climatic constraints. Thus, a marine kelp forest might gradually shift its location to a cooler (and deeper) zone, or a large coastal salt marsh might move inland and higher in elevation as sea levels rise. Such simple movement of ecosystems may be unlikely in most cases, however. Studies of individual members of complex ecosystems show that each species tends to have different environmental needs and tolerances. Some members of the ecosystem may move and others may have no clear path for retreat. Grasses in some alpine meadows may be unable to find patches of appropriate soil at higher elevations, while the insect and mammal herbivores that feed on those grasses may be able to move to mountains with higher meadows. Paleoecologists have described just such individualistic responses to past major climate changes. Many communities existed in the past but do not occur today (Davis and Zabinski 1992). These "non-analog" communities may have been the result of different response times of component species to a changing climate, or may have been the result of non-analog climates (Pielou 1992).

Thus, an alternative prediction to simple ecosystem movement is that the denizens of diverse ecosystems and communities will have divergent responses to climate shifts. The net effect of these responses may be the eventual creation of new community types as species abundances change, new species invade, and some species go extinct. With current climate change, some existing ecosystems may disappear and new ones may arise in their stead. Unfortunately, the complexity of these effects makes them very difficult to anticipate or model with any precision.

Responses to extreme weather

What *is* clear from basic ecological and physiological research is that natural systems are strongly influenced by extremes of weather and climate (reviewed by Parmesan et al. 2000 and Easterling et al. 2000 a,b). One of the very

first such studies dates back to the nineteenth century when Bumpus (1899) documented that a severe winter storm over Lake Michigan disproportionately killed off both the largest and the smallest sparrows, thereby generating strong natural selection on body size.

Many biological processes undergo sudden shifts at particular thresholds for temperature or precipitation (Precht et al. 1973; Weiser 1973; Hoffman and Parsons 1997). Tolerances to frost and to low levels of precipitation often determine plant and animal range boundaries (Andrewartha and Birch 1954; Woodward 1987). For example, trees only occur where annual precipitation is above 300 mm. Further, tropical trees are killed by low temperatures ranging from 0°C to 10°C, whereas temperate broad-leaved deciduous trees can survive temperatures to –40°C, and many boreal species appear to be able to survive any extreme low (Woodward 1987). An extended drought in New Mexico in the 1950s caused the boundary between pine and pinyon–juniper forest to shift by 2 km, where it remains today (Allen and Breshears 1998). Animals exhibit similar climate restrictions. For instance, the nine-banded armadillo (*Dasypus novemcinctus*) requires more than 38 cm annual precipitation and is further restricted to latitudes with fewer than 20–24 days below freezing throughout winter and to fewer than nine consecutive days below freezing (Taulman and Robbins 1996).

Single extreme temperature and precipitation events can have profound and long-lasting effects on physical characteristics. Drought years in the Galápagos, induced by El Niño, cause evolution of larger beak size in Darwin's Finches (*Geospiza fortis*), while extremely wet years drive evolution of small beak (and body) size (Boag and Grant 1984). Perhaps the most extreme example is that in many reptiles, an individual's sex is determined by the maximum temperature experienced during a critical phase of embryonic development (Bull 1980; Janzen 1994). Experiments with map turtles (*Graptemys* spp.) showed that maximum incubation temperatures below 28°C produce only males, whereas maximum temperatures surpassing 30°C produce only females (Bull and Vogt 1979). Even under the current climate, single-sex nests are common. Thus even small changes in temperature extremes could result in single-sex populations, with obvious conservation implications.

Insect species are well known for their boom and bust cycles, with much of that variation having been linked to climate. Single drought years have been shown to affect individual fitness and population dynamics of many insects, causing drastic crashes in some species (Singer and Ehrlich 1979; Ehrlich et al. 1980; Hawkins and Holyoak 1998), while leading to population booms in others (Mattson and Haack 1987).

Many studies have related El Niño events to changes in marine biotic systems (Roemmich and McGowan

1995; Sagarin et al. 1999). Particularly striking were widespread massive coral bleaching events that followed the 1982–1983 and the 1997–1998 intense El Niños (Coffroth et al.1990; Glynn 1990; Hoegh-Guldberg 1999). Host–parasite interactions including infectious diseases are influenced by climatic factors (such as El Niño) in terrestrial and marine communities, and these links are likely to be altered with a changing climate (Harvell et al. 2002). Finally, ecosystem structure and function are affected by disturbance events, many of which are associated with tornadoes, floods, and tropical storms (Pickett and White 1985; Walker and Willig 1999).

In summary, even if the statistical distribution of extreme events remains the same, the absolute number of days that lie outside of the physiological tolerance of a particular species may increase (Figure 10.10; Schärr et al. 2004). Further, this increase in extremes is most likely

(A)

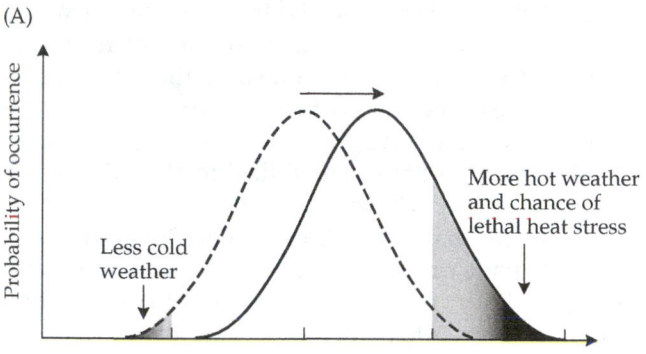

(B)

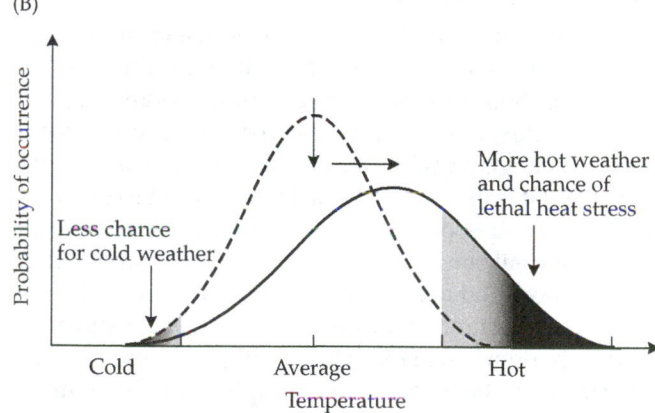

Figure 10.10 Increases in mean temperature and temperature variability will affect the probability that an individual experiences extreme events. Areas in gray represent temperatures at which an individual suffers heat stress or cold stress; areas in black represent lethal hot or cold temperatures. (A) An increase in mean temperature, leading to more record hot weather, will lead to greater heat stress and lethal heat events, and less cold stress. (B) An increase in both the mean temperature and variance, leading to much greater heat stress and lethal hot weather events, and less cold stress. (Modified from IPCC 2001c.)

to affect populations and individuals at the edge of the species range where individuals are often living at the limit of their species' physiological tolerances and thus are susceptible to stressful, harmful, or lethal weather events (Hoffman and Parsons 1997). The predicted increases in extreme temperature and precipitation events are expected to lead to physical and behavioral changes in a few species, to dramatic changes in the distributions of many species, and to population and even species extinction in the worst cases (Parmesan et al. 2000).

Observed Biological Impacts of Climate Change

Twentieth century warming was but a fraction of what flora and fauna dealt with during the last glacial–interglacial cycle. In contrast, however, humans have had enormous negative impacts on biodiversity through massive habitat destruction and fragmentation, loading of natural systems with nutrients and toxins from cars and industry, widespread overexploitation of populations, and the introduction of nonnative species over the past century. Increasingly we are observing changes in wild systems that have been linked to climatic trends. Though most effects have been relatively minor and benign, it is clear that the continuation of observed trends is likely to have profound impacts on wild species and consequently on the ways in which the conservation community strives to preserve biodiversity.

Detection and attribution

Given the many ways humans have been altering species' distributions and influencing local population abundances for thousands of years, and most strongly over the past 100 years, how is it possible to detect an influence of climate change on wildlife? Given that a response is detected, what is its relative impact with respect to other anthropogenic stressors? If the impacts of climate change are relatively weak and buried within a general framework of human-mediated habitat deterioration, is climate change, in itself, important? Addressing these questions in a scientifically rigorous manner is difficult, but is essential for determining the conservation implications of climate change.

The first question has been addressed in several recent reviews and analyses, which show that twentieth century climate change has had a wide range of consequences and has impacted many diverse taxa in disparate geographic regions (Easterling et al. 2000b; Hughes 2000; IPCC 2001b; Peñuelas and Filella 2001; Walther et al. 2002; Parmesan and Yohe 2003; Root et al. 2003). However, this seemingly simple statement glosses over the enormous difficulties biologists have had in tackling the question of climate change impacts. A basic problem is that studies used to assess climate change impacts are, of necessity, correlational rather than experimental. Therefore, interpretation of biological changes requires less direct, more inferential methods of scientific inquiry (e.g., Sagarin and Micheli 2001). Further, detecting significant trends in long-term datasets is particularly difficult when data are often patchy in quantity and quality, and when natural yearly fluctuations are typically noisy. Once change has been detected, attributing the cause of the biological change to climate change requires consideration of multiple nonclimatic factors that may confound effects of climate.

It is no surprise, then, that *detection* and *attribution* are prime issues for scientific assessment teams such as the IPCC. The search for such a climate "signal'" in natural systems has been a challenge for climate change biologists. Some studies have attempted to conduct research that filters out effects of confounding forces, such as habitat degradation, by utilizing data from relatively pristine areas. Even so, detecting long-term trends in say, population abundances, has been difficult unless the data have been consistently collected over several decades and at regional or continental scales.

Using various methods to deal with these difficulties, responses to climate change have been detected in individual studies conducted from the population to the community level, as well as on ecosystem processes. The following is a selection of such studies that exemplify the links between biotic changes and climate change since the mid-1800s.

Evolutionary and morphological changes

A factor often downplayed in discussions of responses to global warming is the propensity for evolution of a population in situ to the selective forces brought about by climate change. A single unambiguous study in the U.S. documented rapid evolution of an introduced species to fit a gradient of climate conditions in the region where it was introduced (Huey et al. 2000). In its native Europe, the fly *Drosophila subobscura* has longer wings in northern countries than in southern countries. Just after being introduced to the western U.S., the flies all had the same wing length. Twenty years later, however, the flies had mimicked their European ancestors by evolving to fit the temperature gradient from south to north: wing lengths became shorter in southern California, gradually increasing in length north to Oregon. This demonstrates that wild species can respond rapidly to local climate conditions.

However, introduced species probably comprise a somewhat artificial situation and this result may tell us little about observed or potential evolutionary responses of native U.S. species to climate change. A major eco-

logical theory postulates that the tremendous success of some invasive species can be attributed to their release from predators or coevolved parasites, diseases, and pathogens (Williamson 1996; Keane and Crawley 2002), which sufficiently decrease selective pressures to allow the organism to lose many of its adaptations for deterring or coping with enemy attack. Maintenance of defense mechanisms is thought to constrain an organism in other ways. For instance, the cost of defense might result in reduced growth or reproduction, or an inability to respond to selection for coping with climatic extremes. Freedom from the need for defense might allow the evolution of novel traits, such as those associated with climate adaptation. Thus, invasive species might be better able to evolve adaptations to cope with a changing climate than native species, which operate under more adaptive constraints.

It is inherently difficult to address the question of how common it is for a population to respond to climate change through genetic evolution. Very few studies have documented an evolutionary response, but there are few systems in which there is the potential for such documentation because the question requires historical data on frequencies of genetically based traits to compare with modern frequencies. It's not surprising, then, that several of the handful of such studies are on the model genetic system—fruit flies in the genus *Drosophila*.

Geneticists in the 1950s noticed that certain types of chromosomal inversions were associated with heat tolerance. These "hot'" genotypes were seen to increase in frequency during a season, as temperatures rose from early spring through late summer, and to increase in frequency in southern populations as compared to northern. Increases in frequencies of warm-adapted genotypes and decreases in cold-adapted genotypes have been observed in wild populations of *D. subobscura* fruit flies in Spain between studies conducted in 1976 compared to 1991 (Rodríguez-Trelles et al. 1996; Rodríguez-Trelles and Rodríguez 1998; Rodríguez-Trelles et al. 1998). An analogous study on *D. robusta* in the northeast U.S. that used genetic data going back to 1946 also found increases in warm genotypes and decreases in "cool" genotypes in all populations sampled (Levitan 2003). This change was so great that populations in New York in 2002 were converging on genotype frequencies found in Missouri in 1946.

A study in Britain indicated evolution at the northern range edge of a grasshopper toward a greater frequency of long-winged individuals, which have better dispersal capabilities (Thomas et al. 2001). Thus, this grasshopper was able to colonize northward faster then was expected from average dispersal abilities within old populations. This is an intriguing finding, in that it suggests that species may be able to shift their ranges relatively quickly in response to changing climate.

Studies of small mammals in the southwestern U.S. have provided excellent documentation of physical changes in response to both historical and current climate change. Small mammals have exhibited relatively slight range shifts since the last glacial maximum compared to insects and plants. This might give the impression that these animals are little affected by climatic regime. Quite the contrary is evident from both paleoecological studies and modern studies. A study in New Mexico on white-throated wood rats (*Neotoma albigula*) showed that these small mammals responded to warmer winters and hotter summers over eight years (1989–1998) by getting 16% smaller (Smith and Betancourt 1998). This study was able to pinpoint the 2–3°C rise in temperature as the driver of the size change, showing that changes in precipitation were not correlated with size changes. However, the detailed reasons for this response are still unknown. For instance, it is not known whether the size changes are *plastic* (i.e., changes within an individual within its lifetime) or *genetic* (i.e., changes in genotypes from one generation to the next). Neither is it known *why* size changes with temperature. Being smaller may reduce overheating in the summer, warmer winters may increase survival of the smallest individuals, or size may reflect a response to changes in food availability.

A conservation application suggested by these studies might be the deliberate transfer of southern individuals into more northerly populations in the hopes of facilitating local evolutionary response to rapid climate warming. This might be particularly important for populations of endangered species that are geographically isolated from other populations and unlikely to receive natural gene flow from populations that carry other climate-adapted genotypes. However, such translocation will change the evolutionary trajectories of such populations by altering their genetic composition, which could have negative consequences, and could alter the nature of interactions of the target species with other community members, and for these reasons, conservationists are also wary of such steps. Certainly, the potential benefits and dangers of interventions to help species cope with climate change will need to be considered carefully.

Phenological shifts

Humans have long been interested in the events marking the beginning of spring, such as blooming of the first spring flower, leafing out of popular trees, and nest-building by birds. The timing, or phenology, of these events is commonly caused by seasonal temperature changes (e.g., the number of days above a certain °C), photoperiod (the amount of daylight, which varies predictably over the course of a year), lunar tides, seasonal weather (e.g., floods or monsoons), and the phenologies of other organisms (e.g., the arrival of a migratory prey

species). Phenologies can often be very precise in their timing and coordination, as with the breeding of corals or the hatching of sea turtles. Not all of these events would be affected by climate changes, of course. But many species' phenologies are driven directly or indirectly by temperature cues.

Studies in North America and Europe provide strong evidence that global climate change has already caused changes in the timing of biological events. Brown et al. (1999) studied Mexican Jays, *Aphelocoma ultramarina*, in the Chiricuahua Mountains of southern Arizona and showed that between 1971 and 1998 the breeding season of the study birds advanced by 10 days on average. The laying date of first clutches was significantly correlated with April monthly minimum temperatures, which increased by about 2.5°C. Dunn and Winkler (1999) used more than 3400 nest records on the timing of the initiation of breeding in Tree Swallows, *Tachycineta bicolor*, throughout their range in the contiguous 48 states and Canada. During 24 years with adequate data (1959–1991), the authors were able to show that the timing of laying was significantly correlated with the mean May temperature and that the average date of laying advanced by 9 days.

Bradley and colleagues (1999) were able to take advantage of observations made by Aldo Leopold at a Wisconsin farm in the 1930s and 1940s on the timing of spring events for birds and native flowers. Comparing these data to their own surveys in the 1980s and 1990s enabled them to look for long-term trends over the 61-year period. For example, Northern Cardinals (*Cardinalis cardinalis*) were heard singing 22 days earlier, and butterfly milkweed (*Asclepias tuberosa*) was blooming 18 days earlier. Of 55 species studied, 19 (35%) showed advancement of spring events. On average, spring events occurred 7.6 days earlier by the 1990s compared to 61 years before, coinciding with March temperatures being 2.8°C warmer.

One other long-term (100-year) study focused on frogs in Ithaca, New York. Gibbs and Breisch (2001) compared recent records (1990–1999) with a turn of the century study (1900–1912). They found that in four of six species, males begin courtship calling 10–13 days earlier than they did in the early 1900s (the other two species showed no change). Maximum temperature in the study area has increased 1.0°C–2.3°C during five of the eight months critical for the frog reproductive cycle. Other studies of frogs have shown that reproduction is closely linked to both nighttime and daytime temperatures (Beebee 1995).

In the U.K. Crick et al. (1997), analyzing more than 74,000 nest records from 65 bird species between 1971 and 1995, found that the mean laying dates of first clutches for 20 species had advanced by, on average, 8.8 days, similar to the 10 and 9 days found in the U.S. studies. Butterflies in Europe show similar correlations of dates of first appearance and spring temperatures, as

well as a recent advancement of first appearance in 26 of 35 species total (Roy and Sparks 2000). In Great Britain, the red admiral (*Vanessa atalanta*) is appearing more than a month earlier than it did 20 years ago. Amphibian breeding has advanced by 1–3 weeks per decade in England (Beebee 1995). Throughout Europe, trees are leafing out earlier, shrubs and herbs are flowering earlier, and fall colors are coming later, leading to an overall lengthening of the vegetative growing season by nearly 11 days since 1960 (Menzel and Fabian 1999; Menzel 2000).

Some species may also experience phenology conflict, which can occur when breeding or growth events have different clock references. In one European bird species—the Pied Flycatcher (*Ficedula hypoleuca*)—egg-laying is determined by photoperiod but the departure for southern migration is set by temperature, squeezing just-hatched chicks between both events and forcing them to develop quickly enough to be able to fly long-distances effectively (Both and Visser 2001).

Abundance changes and community reassembly

For many species, climate has been shown to have a direct impact on population size. Sillett et al. (2000) studied the Black-throated Blue Warbler (*Dendroica caerulescens*), a species that migrates between temperate North America (the breeding range) and Central America and the Caribbean (the overwintering range). Over a ten-year period the authors tracked a correlation between population dynamics and El Niño-La Niña years. The cycle, known as the El Niño Southern Oscillation (ENSO), is a powerful but predictable weather pattern that influences precipitation and temperature during a multimonth period over much of the Western Hemisphere. El Niño years tended to reduce available food during winter; La Niña years had abundant food and much higher survival rates for overwintering birds. Fecundity and survivorship of hatchlings of birds in the breeding range was also influenced, with much lower fledgling mass during El Niño years relative to La Niña years. Again, this variance was associated with changes in food abundance based on rainfall levels. The study implies that long-distance migrants and species that depend on large-scale dispersal might be just as at risk of negative climate effects as are sedentary species.

Changes in the structure of local plant and animal communities indicate that at some study sites the more warm-adapted species are flourishing while more cold-adapted species are declining. This has been particularly well-documented in the Californian coastal waters of the Pacific, which have experienced a 60-year period of significant warming in nearshore sea temperatures (Figure 10.11; Barry et al. 1995; Holbrook et al. 1997; Sagarin et al. 1999).

A marine sanctuary in Monterey Bay has provided a valuable, relatively undisturbed environment to look for responses to this rise in ocean temperatures. In addition,

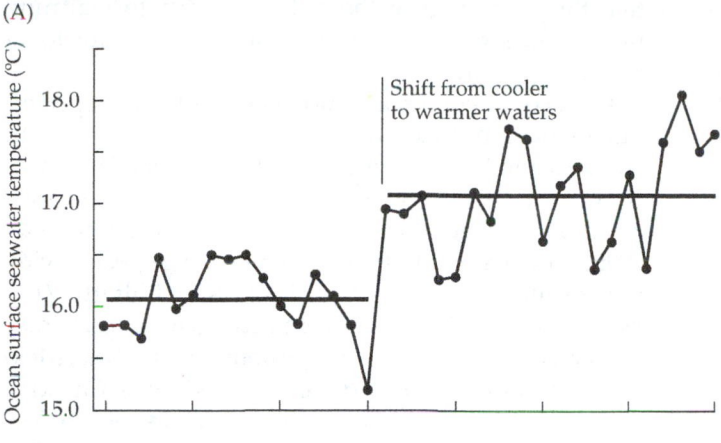

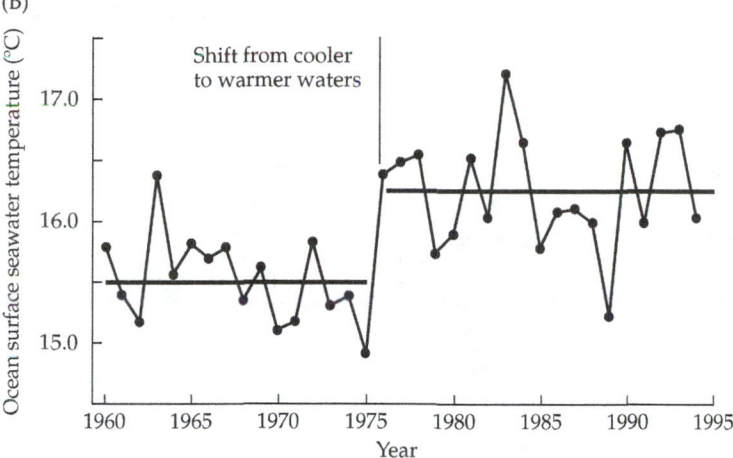

Figure 10.11 Mean annual ocean surface temperatures at (A) Scripps Pier and (B) Santa Barbara, California, from 1960 to 1995. Horizontal lines indicate long-term mean temperatures for the periods of 1960–1975 and 1976–1995. Leaders indicate a regime shift from cooler to warmer waters during 1976–1977. (Modified from Holbrook et al. 1997.)

Monterey Bay is an interesting example because it is located in a small region of overlap between northern species (that extend their ranges all the way to Alaska), and southern species (whose ranges extend down to Mexico). Fixed plots were established in 1931, which researchers in the 1990s were able to locate and resurvey. They found that, compared to the earlier survey, abundances of nearly all southern species had increased significantly while abundances of nearly all northern species decreased (Barry et al. 1995; Sagarin et al. 1999). Thus the dominance relationships in this intertidal community have shifted markedly in response to water temperature change. Holbrook et al. (1997) found similar shifts over the past 25 years in kelp forest fish communities off southern California (Figure 10.12). Southern species have greatly increased their proportionate dominance of community composition at the expense of more northern species. Farther offshore, Roemmich and McGowan (1995) showed that over 43 years (1951–1993) of this temperature change, the population abundance of plankton species also was reduced greatly. Taken as a whole, this body of work in the California Current and associated intertidal areas has demonstrated clear effects of warming sea temperatures on vertebrate and invertebrate communities.

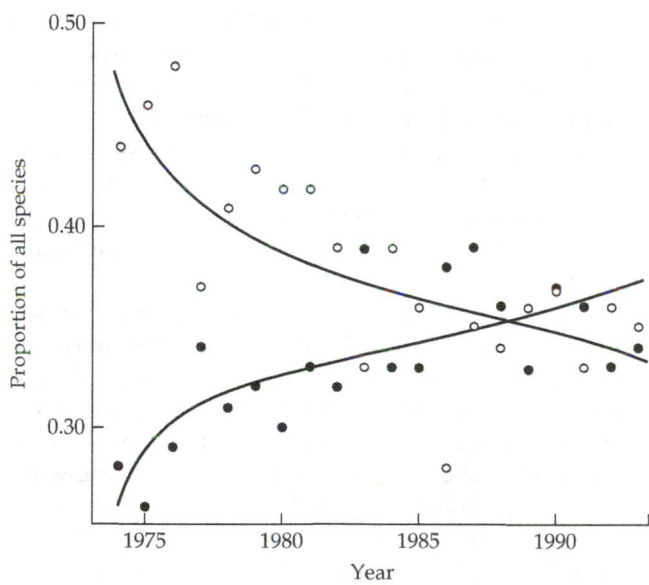

Figure 10.12 Proportions of the species of kelp forest fish present that were northern (open circles) or southern (closed circles) during censuses off of southern California from 1974 to 1995. (Modified from Holbrook et al. 1997.)

Some plant communities in North America have shown increase in woody species relative to grasses and herbs. In experimental manipulations, multiple individual factors, (increased temperature, increased water availability, and increased CO_2) have all resulted in increases in woody species in plant communities (Harte and Shaw 1995; Chapin et al. 1995). Large-scale trends in this direction appear to be occurring in diverse landscapes, from Alaskan tundra (Chapin et al. 1995; Sturm et al. 2001) to the desert Southwest (Turner 1990; McPherson and Wright 1990; Brown et al. 1997). Though overgrazing is partially responsible in given locales, experimental results suggest that the general phenomenon also stems from the joint effects of increases in all three of these factors (temperature, water, carbon dioxide) in these regions (Archer 1995).

Range shifts

Range shifts are changes in the geographic distribution of a species. By analyzing preserved remains of plants, insects, mammals, and other organisms that were deposited during the most recent glacial and interglacial cycles, scientists have been able to track where different species lived at times when global temperatures were either much warmer or much cooler than today's climate. The range of most species was several hundred km closer to the equator or several hundred meters lower in elevation during times when Earth was in a glacial period—that is, 4°C –5°C cooler than it is today (Cox and Moore 2000).

A study of the 59 breeding bird species in Great Britain showed both expansions and contractions of the different northern range boundaries, but northward movements were of a greater magnitude than southward movements, with a mean northward shift of 18.9 km over a 20-year period (Thomas and Lennon 1999). For a few well-documented species, it has been shown that the northern U.K. boundaries have tracked winter temperatures for over 130 years (Williamson 1975). Further, higher reproductive success has been linked to warmer springtime temperatures (Visser et al. 1998). Physiological studies indicate that the northern boundaries of North American songbirds may generally be limited by winter nighttime temperatures (Root 1988a,b).

In Canada, the red fox (*Vulpes vulpes*) has expanded northward over the past 70 years while the arctic fox (*Alopex lagopus*) has contracted toward the Arctic Ocean (Hersteinsson and MacDonald 1992). The timings of the boundary changes have tracked warming phases. Occasional accidental transplants of the arctic fox southward from its range limit had succeeded, provided that the red fox, which is competitively dominant was locally absent. However, prior to recent climatic warming, multiple accidental transplants of the red fox north of its range limit had failed. The red fox has physical attributes that make it less adapted to cold conditions than the arctic fox (e.g., longer ears and limbs). From this it has been inferred that the expansion of the red fox is due to warming trends, causing the competitively inferior arctic fox to retreat northward.

On a continental scale, movements of entire species' ranges have been found in butterflies in both North America and Europe, where two-thirds of the 58 species studied have shifted their ranges northward by as much as 100 km per decade (Parmesan 1996; Parmesan et al. 1999). The first study to document such a species-wide range shift was with Edith's checkerspot butterfly (*Euphydryas editha*). Population extinctions were four times higher along the southern range boundary (in Baja, Mexico) than along the northern range boundary (in Canada), and nearly three times higher at lower elevations (below 8000 feet) than at higher elevations (from 8000 to 12,500 feet) (Figure 10.13; Parmesan 1996). In concert with glob-

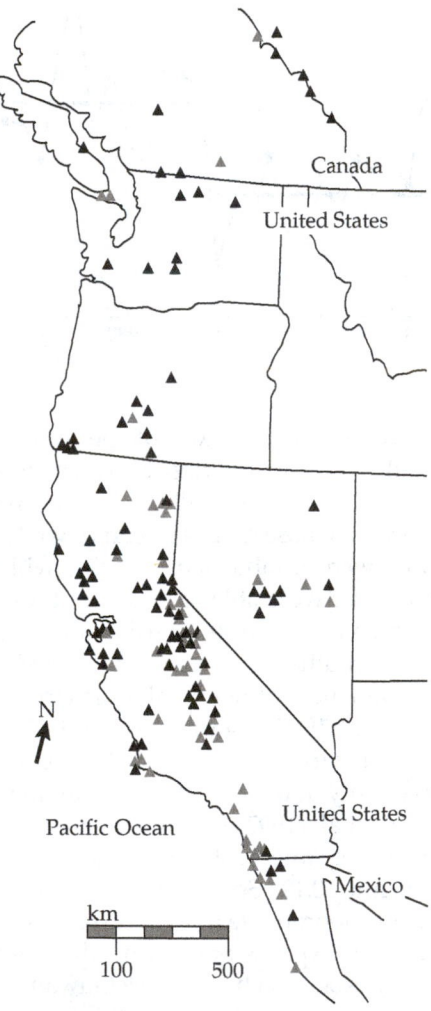

Figure 10.13   Patterns of population extinctions of *Euphydryas editha* from 1860 to 1996. Each triangle represents a single population. Historical records are from 1860 to 1983. Black represents populations still present during the 1994–1996 census. Gray represents populations recorded as extinct during the 1994–1996 census. (From Parmesan 1996.)

al warming predictions, this extinction process had effectively shifted the range of *E. editha* northward 92 km, and upward 124 m in elevation since the beginning of the century (Parmesan 1996). This closely matched the observed warming trend over the same region, in which the mean yearly temperature isotherms had shifted 105 km northward and 105 km upward (Karl et al. 1996). No other factor could explain the observed range shift.

Butterflies have, in fact, provided some of the best studies of climate constraints and climate change impacts. Seventy years of published studies document the limiting effects of temperature on butterfly population dynamics, particularly at northern range edges (Pollard 1988; Dennis 1993; Parmesan 2003). The northern boundaries of many European butterflies are correlated with summertime temperature isotherms, and recent expansion matches predictions from models based solely on temperature constraints (Thomas 1993; Warren et al. 2001). Populations toward the northern boundary become increasingly confined to the warmest microclimates (e.g., short turf and south-facing hills) (Thomas 1993; Warren et al. 2001). Transplants beyond the northern boundary have failed to sustain breeding colonies, even when the habitat appeared suitable (Ford 1945).

The sachem skipper butterfly (*Atalopedes campestris*) has expanded from California to Washington State (420 miles) in just 35 years (Crozier 2003a,b; Figure 10.14). During a single year—the warmest on record (1998)—it moved 75 miles northward. Laboratory and field manipulations showed that individuals are killed by a single, short exposure to extreme low temperatures (–10°C), or repeated exposures to –4°C, indicating that the northern range limit is dictated by winter cold extremes (Crozier 2003a). As climate continues to reduce the number and severity of winter extreme cold events, the northward spread of this species, as well as others with similar limitations, should continue.

Montane studies have generally been more scarce, but the few that exist show a general movement of species upward in elevation. Lowland birds have begun breeding on mountain slopes in Costa Rica (Pounds et al. 1999; see Case Study 10.2 by Karen Masters, Alan Pounds, and Michael Fogdon), alpine flora have expanded toward the summits in Switzerland (Grabherr et al. 1994), and Edith's checkerspot butterfly has shifted upward by 105 m in the Sierra Nevada of California (Parmesan 1996). Globally, several studies show poleward and upward movement of treeline in certain locales (Kullman 2001; Moiseev and Shiyatov 2003). Within North America, general upward movement of treeline has occurred in the Canadian Rockies as temperatures rose by 1.5°C (Luckman and Kavanagh 2000). However, the precise locations of treelines result from a complex response to temperature, precipitation, fire regimes, and outbreaks of pathogens and herbivores. This whole complex of factors determines the local elevations and lati-

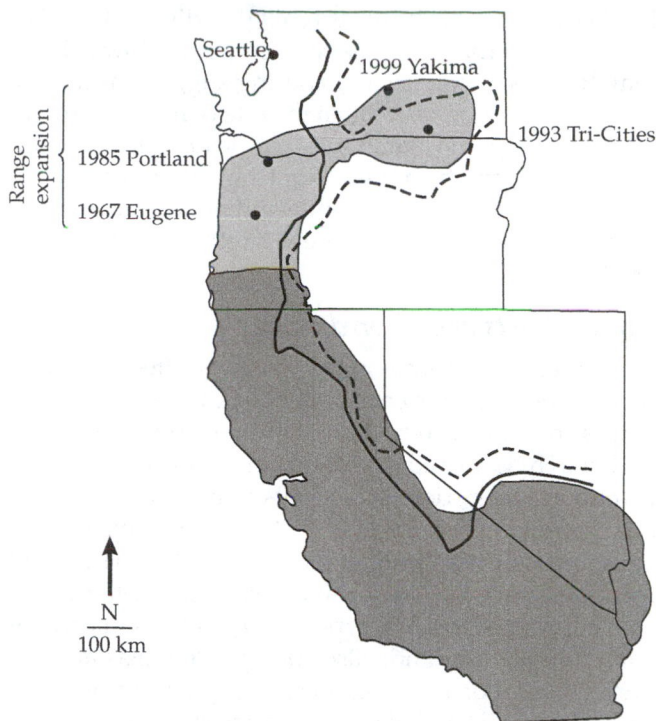

Figure 10.14 The overwintering range of the sachem skipper butterfly (*Atalopedes campestris*) includes Washington, Oregon, California, and Nevada. The historical range is shaded dark, while range expansion over the past three decades is shown in lighter shading. Colonization dates of *A. campestris* in four cities in Oregon and Washington are shown. Contour lines represent the boundary of January average minimum temperature of 0°C (solid) and –4°C (dashed). (From Crozier 2003b.)

tudes of treeline (Luckman and Kavanagh 2000; Grace et al. 2002). Further, mechanistically, patterns of variability in temperature and precipitation are considered to be far more important than simple averages (Swetnam and Betancourt 1998).

Sea level rise

Sea level rise has caused some contractions in the distributions of coastal species. The pine forest barrens in the Florida Keys have been steadily pushed out of the lowest-lying areas because of increased ground water salinity caused by sea level rise (Ross et al. 1994). They are now restricted to the hillier areas, which has resulted in habitat loss for species that depend on the pine barrens.

Since the 1940s, one of the salt marshes of Barn Island, Rhode Island, has undergone large changes in vegetation community composition, with increases in cover of low marsh species such as *Spartina alterniflora*, forbs, and graminoids, at the expense of the high marsh communities (dominated by *S. patens* and *Juncus gerardii*) (Warren and Niering 1993). At Barn Island, response to rising sea

level is not a simple landward shift of individual species, but a shift in the entire marsh community toward low marsh species. Warren and Niering suggested that sea level rise has been faster than the rate of new accumulation of appropriate marsh substrate, leading to less habitat for high marsh species. Sam Pearsall of The Nature Conservancy discusses potential mitigation measures to cope with sea level rise in eastern North Carolina in Case Study 10.3.

Direct effects of carbon dioxide

Some greenhouse gases may have biotic effects in themselves. The carbon from atmospheric CO_2 is essential for plants in many ways: Carbon is used to make sugars and starch, is a fundamental component of secondary compounds (e.g., defensive agents against being eaten or attacked by disease), and is the building block of plant structures. If the amount of carbon available to the plant increases, such as through increased atmospheric CO_2 concentrations, then the types and quantities of carbon-containing compounds also change (Peñuelas and Estiarte 1998). For these reasons, increased atmospheric carbon dioxide has been shown to act as a plant "fertilizer." In experiments that increase concentrations of CO_2, plants respond with increased growth. A review of experimental studies found that atmospheric concentrations of 650 ppm stimulated photosynthesis by 60%, the growth of young trees by 73%, and wood growth per unit of leaf area by 27% (Keeling and Whorf 1999).

Another major effect is that increasing atmospheric CO_2 has been shown to increase the carbon–nitrogen (C:N) ratio in plants (reviewed by Cotrufo et al. 1998; Bezemer and Jones 1998). Insects that feed on plant tissues are very sensitive to the C:N ratio, and may alter their diet preferences, growth rates and fecundity in response to changes in this ratio. Bezemer and Jones (1998) reviewed experimental studies that compared insect performance when fed host plants grown at ambient CO_2 levels (around 350 ppm) versus plants grown at high CO_2 levels (650–700 ppm). The study used data published for 43 insect species (from aphids to beetles to flies) and 42 plant species (from small herbs to trees). As expected, percent nitrogen was lower and carbon-containing compounds were higher (storage, structure, and defensive compounds) for plants raised in high CO_2. The surprising finding was how idiosyncratic insect responses were to the changes in their host plants. Leaf chewers (such as the lepidoptera, butterflies and moths) tended to eat more, take longer to develop, be smaller, and were more likely to die when fed high CO_2 host plants. This is expected because nitrogen is a limiting factor for their development and hence they do worse on the high CO_2/low nitrogen plants, resulting in much lower population sizes. The high CO_2 plants also had much higher concentrations of phenolics and terpenes—secondary compounds that act as a defense against lepidoptera. What is more surprising is that other insect groups, such as the phloem feeders (including aphids), actually had *faster* developmental time and attained *higher* population numbers on high CO_2 plants.

Alteration in the amount or timing of floral resources are likely to have large influences on the fitness of nectaring adults, such as bees, wasps, butterflies, and moths. This would be particularly true for insects that are tightly coevolved with a specific plant species. In a study by Rusterholz and Erhardt (1998), doubled CO_2 levels had dramatic effects on floral resources commonly used by butterflies in Switzerland. They exposed five species of common flowering herbs and shrubs to ambient or doubled CO_2. For those species that showed a significant response, high CO_2 led to earlier flowering and more flowers being produced. High CO_2 flowers also contained significantly less nectar, generally higher sugar concentrations, lower total amino acids, and altered amino acid compositions. Butterflies have strong species-specific preferences for nectar quantity and composition. The concern for conservation is that recovery by endangered butterfly species may be affected by these subtle changes in floral resources in the wild.

Ecosystem process changes

Are we seeing changes in the flows of energy and nutrients in the biosphere? This process is often referred to as "ecosystem functioning," and is difficult to assess even in the absence of environmental change. Rather than tackle that question directly, it is easier to detect changes in specific components of ecosystem flows and processes. As discussed earlier the length of the growing season is increasing (Myneni et al. 1997; Zhou et al. 2001; Lucht et al. 2002). Because plants are a major intermediary for carbon flow through ecosystems, this lengthening of the active period has altered the annual cycle of carbon dioxide levels in the atmosphere (Keeling et al. 1996).

The Alaskan tundra has historically been a CO_2 sink, but that may be changing. Historically, when tundra plants die, they quickly become part of a frozen layer of organic matter. Even in midsummer, only the very upper surface of this layer defrosts. When this dead matter is frozen, it is in storage. The soil organisms that break it down are able to function only if it gets above freezing, and the warmer it is, the faster they work. The Alaskan tundra has already experienced much stronger warming trends than the rest of the U.S. At some point, as deeper layers remain above freezing, the rate of decomposition of dead matter will exceed the rate of plant growth, and the tundra will turn from a net sink to a net source of CO_2.

This switch from net sink to net source is already occurring in some tundra areas. Oechel et al. (1993) found

that at Toolik Lake and Prudhoe Bay, Alaska, during 1983–1987 and 1990, the tundra was acting as a source of CO_2 to the atmosphere. This was associated with long-term atmospheric and soil temperature increases, soil drying, and increased depth to water table. By 1998, moderate acclimation resulted in a return to the tundra acting as a sink in summer (Oechel et al. 2000). However, warmer winters continue to cause an overall net loss of carbon.

Lengthening of the growing season is expected to generally increase carbon uptake simply by allowing more time for biomass accumulation. Indeed, net primary productivity (NPP) has increased both across North America (Hicke et al. 2002) and globally (Nemani et al. 2003). However, recent increased carbon storage in the eastern U.S. stems largely from regrowth of trees on previously logged mature forest and abandoned agricultural land and not from regional climate change (Fan et al. 1998; Schimel et al. 2000). In contrast, the switch across boreal forests to becoming carbon sources is believed to stem directly and indirectly from regional climate change. The warming and drying trends in the boreal regions of North American have been linked to reduced tree growth due to water stress (Barber et al. 2000), increased pest outbreaks, and increased incidences of wildfire (Kurz and Apps 1999).

These trends are not likely to be stable, as the underlying processes are dynamic and, to some extent, unpredictable. Once the regrowth occurring in the eastern U.S. becomes mature forest, the rate of carbon uptake will be substantially reduced. Likewise, fertilization effects of increased CO_2 and nitrogen are short term, having a positive effect only until some other resource becomes limiting to plant growth. As discussed earlier, climate changes are causing major shifts in community structure, and different vegetation communities have very different carbon cycling and storage properties (open tundra differs from spruce forest, for example). Therefore, current patterns of carbon uptake and emission from terrestrial vegetation are likely to change considerably over the next few decades.

The global picture: A synthesis of biological impacts

Several recent reviews and global analyses across individual studies provide convincing evidence that twentieth century climate change has already affected Earth's biota (Hughes 2000; IPCC 2001b; Walther et al. 2002; Parmesan and Yohe 2003; Root et al. 2003). This conclusion is particularly compelling because the patterns of biological change are similar regardless of the taxonomic group or geographic region.

However, a skeptic might be wary of extrapolating from those individual studies to global biodiversity impacts. The publication process has been accused of "cherry picking" studies that show an impact. This effect is known as a "positive publishing bias," and stems from a fundamental human trait to be more interested in reporting that something is happening than not. Consider a researcher returning to the same site every year and recording that numbers of "wattling widgets" is exactly the same now as it was 30 years ago. This type of information might go into a yearly field station report, but is unlikely to become accessible to the scientific community at large through publication in peer-reviewed literature.

A meta-study by Parmesan and Yohe (2003) sought to address this problem. This synthesis focused on multi-species studies; hence, species that have not responded to recent climate change were documented along with those exhibiting responses, allowing for an estimate of overall impact of climate change to be made. In all, this study combined data from more than 1700 species across the globe. Interestingly, the skeptics were partly right—about half of the species in this study were stable, showing no response.

On the other hand, the changes that were observed were not random, but were systematically in the direction expected from regional changes in climate. Over the past 20–140 years, an alarming 50% of all species studied exhibited significant responses to regional warming trends by showing earlier phenology (timing of breeding or emerging events), a shift in their distributions toward the poles and higher elevations, or both. These responses have been occurring in diverse ecosystems (from temperate terrestrial grasslands to marine intertidal communities to tropical cloud forest), and in many types of organisms (e.g., birds, butterflies, sea urchins, trees, and mountain flowers) (Table 10.1; Parmesan and Yohe 2003).

Important diagnostic patterns, specific to climate change impacts, helped provide a "fingerprint" of global climate change as the driver of the observed changes in natural systems. Species with more than 70 years of annual data showed signs of tracking decadal temperature swings, such as shifting southward during cool periods and northward during warm periods. At sites where northern and southern communities overlap, the northerly (cold-adapted) species declined while the southerly (warm-adapted) species increased. Such diagnostic "sign-switching" responses were observed in 294 species, spread across the globe and ranging from oceanic fish to tropical birds to European butterflies (Parmesan and Yohe 2003).

One of the most important aspects of warming trends may occur at the community level. Climate change appears to affect each species differently (hence the variation in response in Parmesan and Yohe's [2003]; see Table 10.1). Thus, individualistic changes in distribution, abundance, and phenology are likely to alter interspecies

TABLE 10.1   *A Summary of Observed Changes from Studies of 944 Species*

Climate change prediction	Changed as predicted (%)	Changed opposite to prediction (%)	Statistical likelihood of obtaining pattern by chance
Earlier timing of spring events	87	13	Very unlikely
Extensions of poleward or upper species' range boundaries	81	19	Very unlikely
Community (abundance) predictions: Cold-adapted species declining and warm-adapted species increasing	85	15	Very unlikely

Source: Data from Parmesan and Yohe 2003.

relationships such as competition, predation, and parasitism. Particular effects will be very difficult to predict, but community reassembly (potentially disassembly) and disruptions of species interactions could prove to be the most widespread and worrisome biotic impacts of climate change (see Case Study 10.1).

Conservation Implications of Climate Change

Earth's ecosystems face a scale of climate-related change that has not been seen for many thousands of years and that will continue for centuries. The best available climate models suggest that most of Earth's terrestrial zones and all of its aquatic systems will be altered as a result. Many studies also suggest that some impacts can already be seen. The logical extrapolation from these observed impacts is that there will be large changes in communities and increased numbers of species' extinctions.

Extinctions

Extinctions have been occurring at a higher rate in recent centuries for a variety of anthropogenic causes, but climate change is likely to exacerbate population declines sufficiently for some species to completely disappear. To date, there have been only two extinctions directly attributable to climate change, the golden toad and harlequin frog in Costa Rica (see Case Study 10.2). Conservation biologists, however, are also concerned with the loss of a distinct subspecies or with loss of all individuals of one species in a given geographic area (e.g., cougars in the southern U.S. Rocky Mountains). Drastic regional declines of this type have already been directly attributed to climate change.

The abundance of zooplankton (microscopic animals and immature stages of many species) has declined by 80% off the California coast. This decline has been related to the gradual warming of sea surface temperatures (Roemmich and McGowan 1995). Zooplankton are a major food source for oceanic wildlife, and the decline of this food supply has been harmful to many birds, fishes, and mammals. The Sooty Shearwater (*Puffinus griseus*) and Cassin's Auklet (*Ptychoraamphus aleuticus*) (which prey directly on zooplankton, or on species that eat zooplankton) have both declined substantially off the central California coast since 1987 (Veit et al. 1996; Oedekoven et al. 2001). However, the Common Murre (*Uria aalge*) remained stable at the same sites (Oedekoven et al. 2001). Though all three species were affected by declines and changes in food resources, their responses were individualistic and apparently related to differences in life histories, foraging behavior, and habitat preferences (Oedekoven et al. 2001).

Clearly, ability to physically track a moving climate zone will be important for species' survival. A very sedentary lifestyle does not bode well for long-term persistence. Unfortunately, many endangered species tend to have just such traits: they have strong local adaptations and habitat specificity, are restricted to small areas, and have limited dispersal ability. An initial "first pass" study as to how such species will fare by 2050 under different climate change scenarios showed that species with low dispersal had nearly twice the risk of extinction of species with high dispersal for any given climate scenario (Thomas et al. 2004).

Scientists use the current distribution of a species to construct a "climate envelope" of threshold values for temperature and precipitation variables within which the species is able to survive and reproduce. These climate envelope models can then be used to estimate what geographic area will be climatically suitable under a variety of climate change scenarios. Figure 10.15 shows the results of one such model for a rare South African flower (Parmesan 2005; Midgley et al. 2002).

The Thomas et al. (2004) study formed its extinction risk estimates by synthesizing results of many different climate envelope models applied to different data sets from around the world. The results provided a rough estimate of extinction risks by 2050. Even if one assumes perfect dispersal capabilities, an absolute reduction in climatically suitable space suggests that about 9%–13% of species could go extinct even with minimal warming

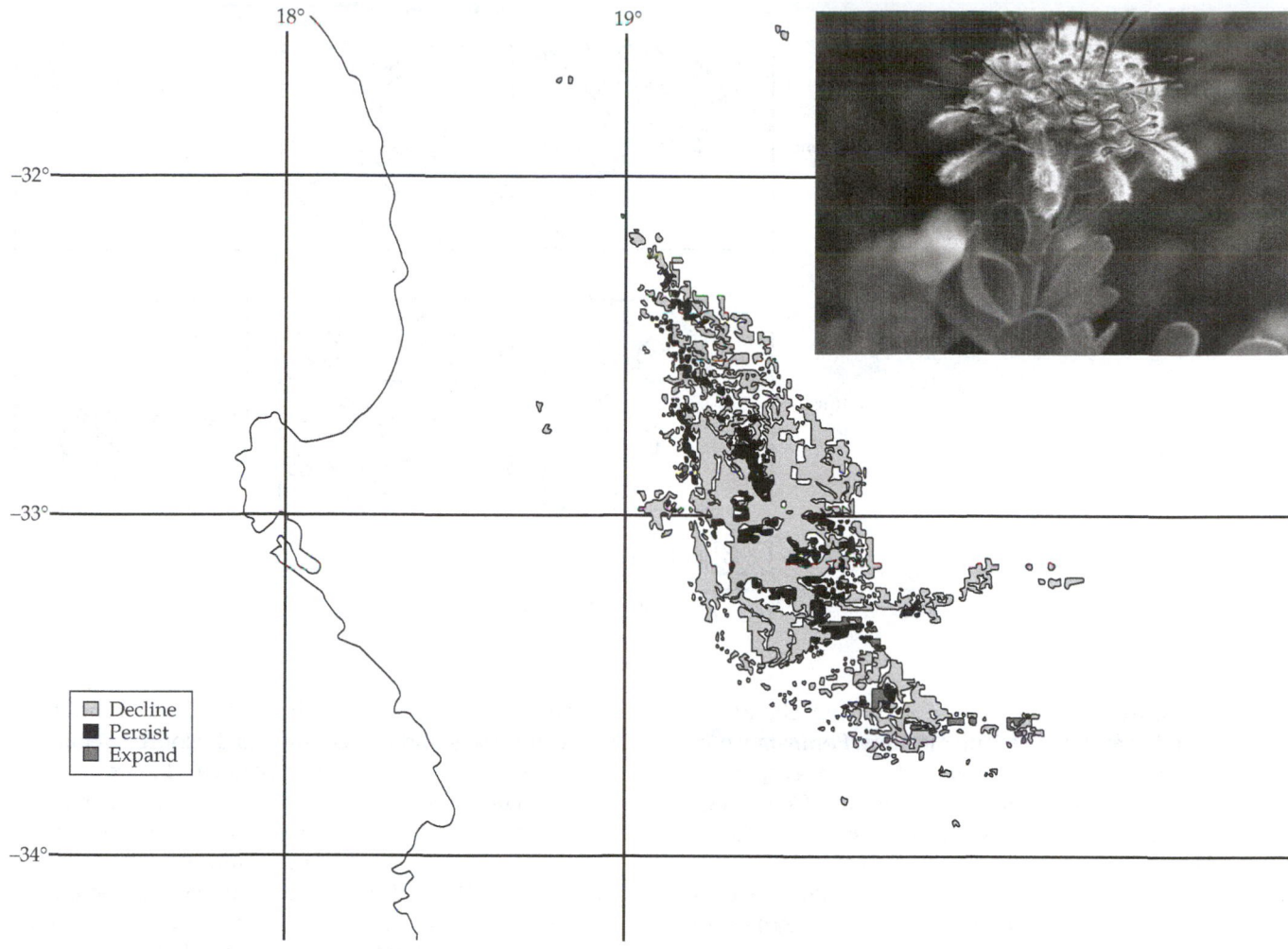

Figure 10.15   Range shift model of the plant *Vexatorella amoena*, a member of the Proteaceae family whose range is found in the mountains above Cape Town, South Africa. Black denotes range retained in 2050, light gray denotes range projected to be lost in 2050, and dark gray denotes areas in which newly suitable climatic conditions appear for the species in 2050. Gridlines mark latitude and longitude. (Modified from Midgley and Millar 2005.)

of another 1.2°C (Thomas et al. 2004). Under maximum projected warming (>2°C), extinction estimates varied from 21% to 32%.

The worrying aspect of these estimates is that they are based strictly on climate considerations. However, today's world is fundamentally different than it was during past large climatic fluctuations. Humans dominate the globe. Land has been converted to cities, agricultural fields, or ranching, reducing the quantity and quality of land available for wildlife. The small plots of good habitat that remain are surrounded by wasteland—areas in which most species cannot survive. Species may no longer have the option of simply moving into better environments if the place they are living becomes intolerably hot or dry. Jen-

nifer Gill discusses such subtleties, and describes approaches to predicting changes to coastal habitats and mitigation planning for the conservation of coastal migrant birds in Case Study 10.4.

Current protection of endangered species relies heavily on reserve systems. Often these are a fraction of the species' former range. The concern is that many of these reserves may not remain suitable for the targeted species with another 100 years of climate change. Figure 10.16 shows hypothetical scenarios of a species' climate envelope shifting in and out of existing reserves in the future. Thus species today are much more likely to go extinct than during the Pleistocene if global warming trends continue.

Figure 10.16 Changing relationships between reserve boundaries and a species' range as climate changes. (A) Decreasing range of the species within a reserve and its absence where it was formerly present; (B) decreasing range of the species within a reserve and its presence where it was formerly absent; (C) change in the location of the species' range. (Modified from Hannah et al. 2005.)

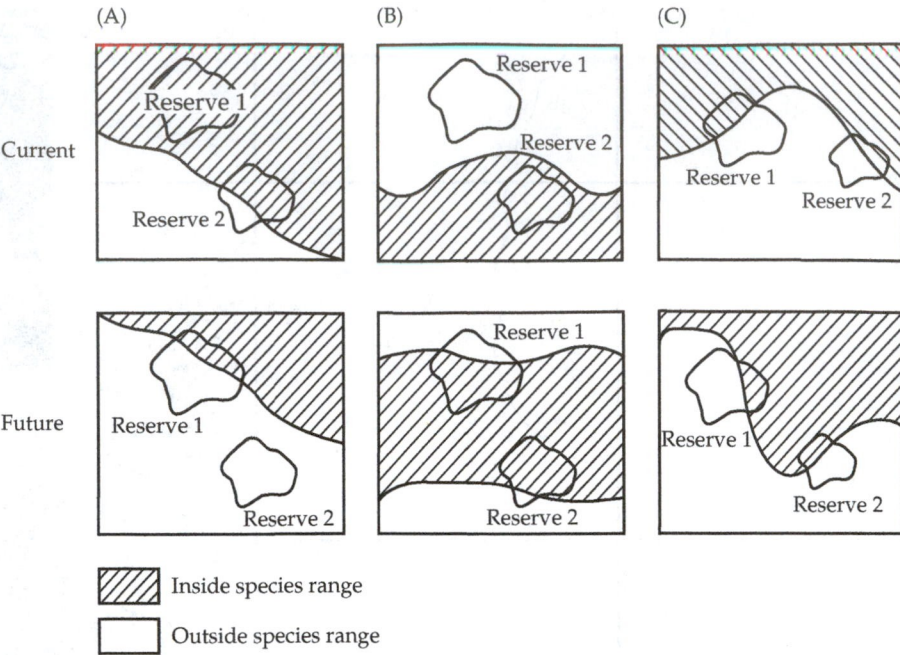

The promises and dangers of a shifting climate are hard to predict for any particular locality. For instance, climate models give conflicting predictions on temperature and precipitation changes for the U.S. Gulf Coast States (Twilley et al. 2001; U.S. National Assessment Report 2001). The only change we can be sure of is that the Gulf Coast is likely to be heavily affected by elevated sea levels.

Though exact prediction may be difficult, (if not impossible), it's clear that the conservation implications are serious. Policymakers, conservation biologists, and resource managers in each state will have various needs and responses to local climate change. Although unique local responses will have to be formulated to address specific localities and conditions, there are some general guidelines that may help in conceptualizing broad strategies.

Responses to climate change by resource managers

Most resource managers and conservation planners focus on local-scale projects such as a particular reserve, habitat, population, or community. Much of their work involves projecting population trends of endangered or threatened species and developing (and then implementing) strategies to stabilize these populations (see Chapter 12). Climate shifts present a new challenge for resource managers because most population models have assumed climate (but not weather) to be an ecosystem constant. With a stable environment, our standard approaches to conservation (e.g., the delineation of fixed areas, protection of rare species, habitat restoration and creation) may be sufficient to maintain biodiversity in

most areas. But climate change could reduce the utility of many of these standard techniques. Resource managers clearly need accurate predictions of local climate impacts. Unfortunately, the same climate models that are reasonably predictive and useful for global planning are poorly resolved at regional levels and are almost nonexistent at local levels. Worse, finer-scaled models may be a long time in coming, and the credibility of such high-resolution forecasting will take yet more time. In the meantime, the climate is already shifting at the local scale.

Fortunately, steps can be taken to minimize the negative impacts of future climate change. Rather than expecting climate scientists to present a specific set of predictions to create a targeted set of responses, local-scale resource managers could develop programs that emphasize flexibility and multiple contingencies. Such an approach fits well with the philosophy of adaptive management that became popular in the early 1990s (see Chapter 13). Some examples of climate-aware adaptive approaches include:

1. The reassessment of species and habitats in regard to their relative vulnerabilities to climate change: An evaluation of the vulnerability of an endangered species, for instance, may differ under scenarios that assume a stable versus a changing climate. Attention and resources may need to be reallocated based on such a susceptibility analysis.

2. The design of new reserves to allow for shifts in the distributions of target species within reserves:

Such shifts may be horizontal (allowing species threatened by rising sea levels, for instance, to move farther inland) or vertical (allowing species to move upward in elevation). An awareness of the need for escape corridors is leading some resource managers to place more value on selecting reserves with great topographic and elevational diversity.

3. The promotion of native habitat corridors between reserves: Corridors along fence lines, ditches, streams, and other minimally used land can aid the redistribution of species between reserves.

4. The creation of dynamic rather than static habitat conservation plans: Management plans based on empirical approaches and regular observation and reassessment are likely to be more useful than detailed long-term scenario modeling given the poor resolution of climate models at local scales. Localized studies of rainfall and temperature trends, regional sea level changes, or shifts in the presence and persistence of vernal ponds are studies that could have important implications at a local level. Moreover, traditionally defined researchers in universities and other large institutions may not be aware of local-scale projects that would be amenable to research designed to address management plans. Local managers may be the best personnel to define, plan, and carry out projects to distinguish between alternate management approaches.

5. The alleviation of the effects of nonclimate stressors: Current climate change is occurring in the context of other anthropogenic stressors (e.g., habitat destruction, industrial contaminants, and invasive species), so the fate of one population will hinge on the net effect of all stressors. In some cases it may be easiest to reduce the overall stress on a species by mitigating some of the nonclimate stressors. For example, if both climate change and an invasive species threaten a native population, the most cost-effective approach may be to focus attention on reducing the incursions of the invasives.

6. The generalization of regional or global climate impact predictions to a particular system: Many estuaries, for instance, will face a rapid loss of shallow submerged habitats and low-lying terrestrial habitats with rising sea levels. Increases in the frequency and severity of coastal storms could also cause increased erosion along shorelines and the loss of rookery islands. Salt-intolerant ecosystems such as bottomland hardwoods may face unprece-dented intrusions of brackish waters. More detailed modeling may predict the relative amount or frequency of coastal flooding. On the other hand, current models already make clear that such flooding is going to occur worldwide. A prudent approach therefore, would be to begin preparing for elevated sea levels sooner rather than later.

Specific and detailed local-scale models are vital for resource managers to develop research-based management plans. But in lieu of such models, the above strategies are necessary compromises. Indeed, many resource managers are employing multiple approaches simultaneously. For instance, The Nature Conservancy has been attempting to prepare for climate change in the Albemarle River estuary in North Carolina and Virginia by purchasing "escape zones" upstream along major waterways (see Case Study 10.3). This strategy is designed to give threatened ecosystems buffer zones, thereby allowing plant and animal species to move inland as sea levels rise. Such an approach gives these sensitive communities time and space to adjust to their new environmental conditions.

Climate change and conservation policy

The global scale of anthropogenic climate change threatens to exceed any other conservation problem. Land-use changes and habitat fragmentation are devastating, but they operate at smaller scales—from local to regional. The logging of a Costa Rican forest does not obviously alter pelagic bird abundance in British Columbia. The potential extinction of Siberian tigers has no clear connection to subsistence farmers in Central African Republic. In contrast, weather patterns and anthropogenic greenhouse gases do not discriminate national boundaries or regional biomes. Warming trends promise to transform our Earth into an unfamiliar place. Citizens and policymakers worldwide must understand that *global* climate change is a *local* issue everywhere.

The pervasiveness of climate change elevated climate-change science in the early 1980s from the province of meteorologists to an issue of great concern to scientists in many disciplines and to policymakers in many nations. Few other conservation issues have achieved such a cohesive international reaction. The creation of treaties such as the Kyoto Protocol and the formation of international climate change advisory groups represent macro-level responses to global warming trends.

Policy implies planning by governments ranging from cities to nations, but policy can also refer to activity by nongovernmental organizations such as conservation and environmental groups, many of which are directly involved in large-scale resource management. Corporations also generate and are guided by policies, and many corpora-

tions are involved in the "carbon economy." Each type of institution has had a strong influence on developing approaches to confront climate change at a global level.

Responses to climate change at national and international levels

As a general rule, resource managers at local levels focus on coping with proximal environmental issues. Almost by definition they address the "little" (rather than "big") picture. Yet, in the case of climate change, large-scale problem-solving strategies are going to be needed to mitigate the impending crisis. There is no quick fix for greenhouse-gas-driven climate change, but solutions for alleviating the worst-case scenario are not a great mystery: Existing carbon sources must be reduced or eliminated, and carbon sinks should be increased wherever possible. The result will be a slower rate of climate change over the coming centuries, which will provide ecosystems (and humans) more time to adjust.

Reduction of greenhouse gas emissions by the largest emitters would be useful in slowing the accumulation of greenhouse gases. The largest emitter of greenhouse gases is the U.S., with about 25% of global emissions created by <5% of the world's population. This is not only considerably larger than other nations in an absolute sense, but only Singapore and Qatar (both of which are very small countries) have higher per capita rates of emissions than the U.S.

However, the action of any single nation—even the U.S.—would eventually be swamped by the general increase in the rate of emission by other nations, particularly China. Further, because carbon and other greenhouse gases are emitted primarily from automobiles and industrial activity, climate change is also an economic issue. Without coordinating actions by other countries, a drop in U.S. emissions would be made up rapidly by increases in greenhouse emissions from developing countries if they follow the same industrialization pathway as the U.S. Thus, the most useful means of lowering the rate of greenhouse gas emissions is to coordinate the actions of many countries, implementing policy reforms at national and international levels simultaneously.

The role of government in climate change policy

Climate change policy differs a great deal from other environmental issues because of its global scale and because of its implications for wide-reaching economic adjustments. Current negotiations have used the agreements formed as a result of the Vienna Convention for the Protection of the Ozone Layer as a model. The most relevant of these agreements is the Montreal Protocol. The Montreal Protocol sought to limit the production of industrial chemicals that destroy stratospheric ozone and thus allow more biologically dangerous ultraviolet radiation to reach Earth's sur-

face. The Montreal Protocol has been widely judged a success. Most nations have ratified the treaty, which called for the step-wise elimination of the target chemicals to reduce economic burdens on the industries having to change production methods. The atmospheric concentrations of these chemicals have been greatly reduced, and positive effects in the stratosphere are already apparent. While previous international agreements provide good models, the scope and goals of these earlier treaties were smaller than the issues surrounding climate shifts.

Several differences are notable in contrasting the Montreal Protocol and potential climate agreements. First, the Montreal Protocol focused on families of chemicals that could be replaced relatively easily, but greenhouse gases are produced through processes that permeate our civilization. Reducing emissions of some greenhouse gases such as methane and carbon dioxide will lead to major economic adjustments. In fact, the implications of reducing greenhouse gas emissions on a large scale may be more similar to health care initiatives designed to eliminate smallpox or AIDS worldwide than to the Montreal Protocol. The economic impact of the latter represented little more than tinkering. In contrast, the global strategies to reduce widespread diseases have resulted in the formation of new kinds of global institutions, social structures, and cultural movements that had not previously existed.

Second, the evidence for anthropogenic warming trends is much more complex and relatively less direct than is evidence for anthropogenic damage to the ozone layer. In fact, climate change science could be one of the most sophisticated and multidisciplinary scientific endeavors in human history. Moreover, the records of changes in climate and biological systems are small relative to the impacts predicted for the next few centuries. In other words, policy responses to climate change must be largely based on limited trend data. Definitive proof of catastrophic climate change impacts will not exist until mitigation itself may be too late. Therefore, policymakers are faced with a simple choice: preempt further climate change based on limited information or assume that strong actions can be deferred until later. A brief history of the political process from the time of initial concern to present is described in Table 10.2.

Major themes in climate change negotiations

Policy debate about climate change made a dramatic shift during the 1990s. In the late 1980s, the scientific basis for ascribing warming trends to human activities still was being formed. Just 10 years later, global leaders were actively working on means of reducing emissions and assessing credit or blame for relative changes in carbon sinks and sources. Many states in the U.S. have already begun programs to curb greenhouse gas emissions (Fig-

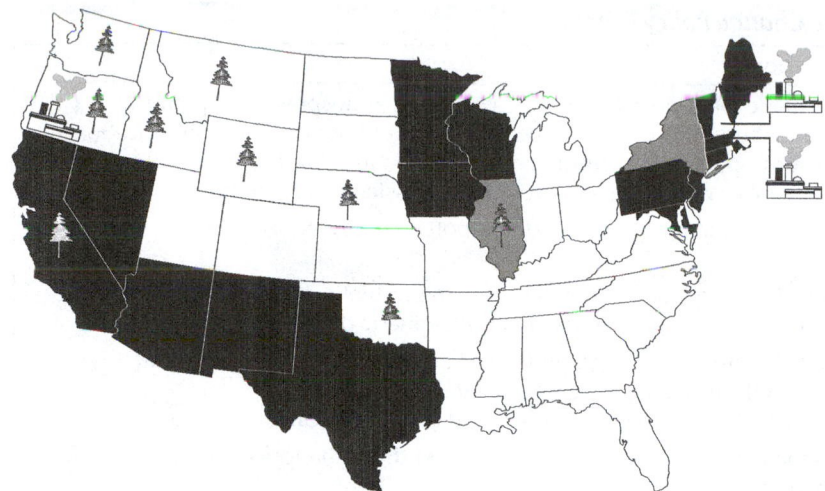

Figure 10.17 Map of states that have programs in place as of September 2003 to reduce sources of greenhouse gases. California is the only state that regulates emissions of greenhouse gases from vehicles. Black indicates states with mandatory energy diversification rules, gray indicates states with voluntary goals toward energy diversification. Power plant icons represent states with mandatory programs to reduce CO_2 emissions from utilities, trees represent states with voluntary forestry or agricultural programs that promote carbon sequestration. (Modified from American Legislative Exchange Council 2002.)

ure 10.17). Climate change is no longer questioned as a process in itself.

The Kyoto Protocol is an outgrowth of the UNFCCC (United Nations Framework Convention on Climate Change) (see Table 10.2) designed to provide specific targets for reductions, reporting of emissions levels, and creation of enforcement mechanisms to ensure that greenhouse gas emissions targets were met. More than 141 nations have ratified the Kyoto Protocol as of February 16, 2005, when the treaty went in force. Despite having broad international support, the Kyoto Protocol was crippled by the withdrawal of support from the U.S. and Russian governments, both of which are actively debating measures that may limit their economies. However, Russian President Vladimir Putin ultimately decided to support the Kyoto Protocol in 2004. It remains unclear how successful the Kyoto Protocol will be without the cooperation of the U.S., the world's largest economy. Still, many nations, particularly in Western Europe, promoted the accord as a vehicle to curbing global greenhouse gas emissions and began the implementation process even without a binding international treaty.

If nothing else, the treaty is the culmination of a broad consensus that developed among scientists and global policymakers that humans are driving changes in the atmosphere that are leading to warming trends and that action must be taken to forestall or slow these trends to avoid social and environmental disasters. However, consensus about how to implement such changes has been even harder to achieve. Several notable themes have played important roles in global talks:

1. Different rules for industrialized versus developing countries: Early in the UNFCCC negotiation process a clear distinction was made between long-industrialized countries (primarily Europe and other Western nations) and industrializing nations (such as China and India). The former group, also called Annex I countries, are responsible for most of the greenhouse gases currently in the atmosphere; they also tend to be richer than the second group and should be able to bring more technological and economic resources to bear on reducing emissions. The second group of signatories—Annex II nations—are in the process of industrializing; they are likely to contribute more to emissions levels in coming decades. Their economies tend to be less diverse and flexible, often due to their historical relationship with former colonial powers. These nations are given a longer period of time to implement the Kyoto Protocol than Annex I nations. The distinction between "developed" and "developing" countries is likely to remain important in negotiations over the next few decades.

2. Multiple mechanisms for reducing carbon emissions: The Kyoto Protocol is unclear about how emissions would be reduced, although the schedule for emissions reductions is much clearer and more defined (see Table 10.2). Several reduction mechanisms have been proposed. For example, developed nations could replace the use of fossil fuels such as coal and oil with renewable energy sources, such as solar energy, biomethane generators, wind- and water-driven turbines, or nuclear power. Also, both developed and developing nations could reforest areas that had been deforested (thereby creating carbon sinks) and reduce rates of deforestation (which is considered a carbon source).

3. New technologies for old industries: Technological solutions to shifting national economies away

TABLE 10.2 *A Chronology of Major Climate Change Policy Events*

Date	Meeting or event	Results and conclusions
1896	First attribution of the connection between atmospheric carbon and climate	Svante Arrhenius made the connection between CO_2 and atmospheric temperature and speculated that burning fossil fuels such as coal could increase the concentration of carbon in the atmosphere in the future and lead to an increase in global temperatures. His research was widely disregarded by other scientists at the time.
1979	First World Climate Conference	Human-induced climate change is identified as a potential threat.
1980	Montreal Protocol	World leaders meet to sign an agreement designed to gradually phase out the production and use of chemicals that destroy atmospheric ozone.
1988	Formation of IPCC	The United Nations Environment Program (UNEP) and the World Meteorological Organization (WMO) create the Intergovernmental Panel on Climate Change (IPCC) to coordinate research and analysis of climate change.
1990	First IPCC Report	The IPCC states global climate is clearly changing, and these changes are probably a result of human activity.
1992	Rio Convention	The United Nations Conference on the Environment and Development (also known as the Earth Summit or Rio Convention) convenes in Rio de Janeiro, Brazil. A total of 154 nations sign the United Nations Framework Convention on Climate Change (UNFCCC), which asks signatories to reduce greenhouse gas emissions and creates a feedback mechanism for future talks. Developed nations (also known as Annex 1 countries) will be asked to make larger reductions than developing nations.
1995	Second IPCC Report	The second report states that "the balance of evidence" leads the authors to conclude that there was a "discernible human influence on the global climate system." New evidence suggests that climate change processes are more serious than they were described in 1990, and the first attempt at reducing human impacts seems weak. Many policymakers and scientists believe the UNFCCC needs stronger teeth.
1995	COP1: The Berlin Mandate	At the first Council of the Parties (COPs) meeting, the Berlin Mandate is signed, which formally recognizes the ineffectiveness of UNFCCC calls for voluntary greenhouse gas reductions. A committee drafts protocols to design other strategies for the 1997 COPs meeting.

from carbon-based systems are likely to increase in importance. One example is the development of cost-efficient hydrogen fuel cells. These fuel cells emit only water vapor but have proven difficult to produce cheaply and on a small scale. Another is the development of low- and no-emission coal-fired electrical generating plants, which would operate by pumping their carbon emissions into long-term storage ("sequestration") below ground or in the deep oceans. The widespread implementation of either strategy is probably a decade or more away.

4. Fostering incentives to meet targets: Many have advocated making use of market-based tools, including the buying and selling of emissions and reduction "credits." These credits could be exchanged between developed nations and between developed and developing nations. In anticipation of global implementation of the Kyoto Protocol, the Chicago Carbon Exchange is already up and running.

Beyond these international efforts, many corporations are making changes on their own. In the petroleum industry, major voluntary shifts in corporate strategy have been taking place, reflecting the influence of public opinion and the future profitability of alternative energy sources. British Petroleum (BP), one of the world's largest corporate oil producers, recently adopted a widespread marketing campaign stating that BP stood for "Beyond Petroleum" (Wee 2002). Shell Oil Company has shifted research efforts into solar energy technology and

TABLE 10.2 *Continued*

Date	Meeting or event	Results and conclusions
1997	COP3: The Kyoto Protocol	Delegates in Kyoto, Japan, agree that Rio Convention targets are insufficient and emissions should be reduced more quickly. Global emission levels should be 5% less than 1990 levels by 2012, Annex 1 nations are asked to make much more substantial reductions to meet the global figure. The U.S. agrees to a 7% reduction and Canada to 6%; the European Union level is an 8% reduction. Some nations go much farther. Germany promises to reduce emissions by 25% and the United Kingdom by 15%. About 160 nations sign the accord, which must then be ratified or acceded to in each country. The treaty does not become activated until the 1990 emissions levels of ratifying countries totals at least 55% of 1990 levels. At this point, a stepwise series of greenhouse gas emissions kicks in. Greenhouse gas reduction mechanisms will be detailed at COP6 in 2000.
2000	COP6: The Hague, Netherlands	The U.S. under George W. Bush (elected in 2000) and Canada under Jean Chrétien want larger amounts of carbon sink credit for forest growth (thereby allowing for higher net carbon emissions) than other signatories will allow. The meeting fails to negotiate an agreement on mechanisms and breaks up. A second meeting held shortly afterwards again dissolves without an agreement. Since the U.S. is the largest emitter of greenhouse gases worldwide, many observers feel that the Kyoto Protocol will have little impact without U.S. support. The Kyoto Protocol is widely pronounced to be dead.
2001	COP7: Bonn, Germany	Some 180 countries constituting all of the Kyoto Protocol signatories except the U.S. and Australia (but now including Canada) approve the mechanism framework for implementing the accord. Supporters of the Kyoto Protocol focus on pressuring other large emitters (e.g., Russia and Japan) to ratify the treaty.
2001	Third IPCC Report	More sophisticated modeling leads IPCC authors to write "globally averaged mean surface temperature is projected to increase by 1.4° to 5.8°C over the period 1990 to 2100." The third report also compiles substantial evidence of biotic effects.
2002	Rio + 10: Johanesburg, South Africa	The U.N. World Summit on Sustainable Development follows up on issues raised by the Rio de Janiero summit in 1992 (hence the conference's alternate name: Rio + 10), with special attention to finding means to create climate-friendly development.
2004	Moscow	Russia's president Vladimir Putin ratifies the Kyoto Protocol, which immediately activates provisions of the treaty.
2007	Fourth IPCC Report	Work has begun already to prepare the next scheduled IPCC report, due in 2007.

development, and in 2004 Shell's CEO spoke out on the need for a major shift away from carbon-based fuels to avert the most severe climate change scenarios (Adam 2004). Indeed, alternative energy sources, such as nuclear, solar, and geothermal, have been an important area for research for many oil companies since the OPEC oil crisis of the 1970s, though these shifts are only small components of the portfolio for most of the large petrochemical producers and refiners even today. Nonetheless, voluntary shifts in corporate policies (see Figure 10.16) will continue to play an important role in climate outcomes and global climate policy.

The future of climate change policy

Regardless of what policy actions are taken, Earth's ecosystems have already made a commitment to climate change. In other words, even if all greenhouse gas emissions ended immediately, the rise in atmospheric CO_2 that has already taken place will continue to warm Earth for another century (Figure 10.18). The 1990s will probably be remembered as a decade when scientists and policymakers in most nations moved from simply expressing concern about climate change to beginning to commit to action. The relevance of these actions to conservation biology is in many cases indirect: Most policymakers are acting out of concern for the effects of climate shifts on humans rather than ecosystems. On the other hand, the UN-FCCC accord may be an example of an instance in which actions taken for the good of human societies are also good for ecosystems. Certainly the foresight that the UN-FCCC and the Kyoto Protocol reflect bodes well for continued co-existence of humans and natural biodiversity.

Figure 10.18 Projected rise in global mean temperature over the next 100 years depicts Earth's "commitment to climate change." This figure is a generic illustration for CO_2 stabilization at any concentration between 450 and 1000 ppm, and therefore has no units on the response axis. Responses to stabilization in this range show broadly similar time courses, but the impacts become progressively larger at higher concentrations of CO_2. After CO_2 emissions are reduced and atmospheric concentrations stabilize, surface air temperature continues to rise by a few tenths of a degree per century for a century or more. Thermal expansion of the ocean should continue long after CO_2 emissions have been reduced, while melting of ice sheets should contribute to sea level rise for many centuries. (Modified from IPCC 2001c.)

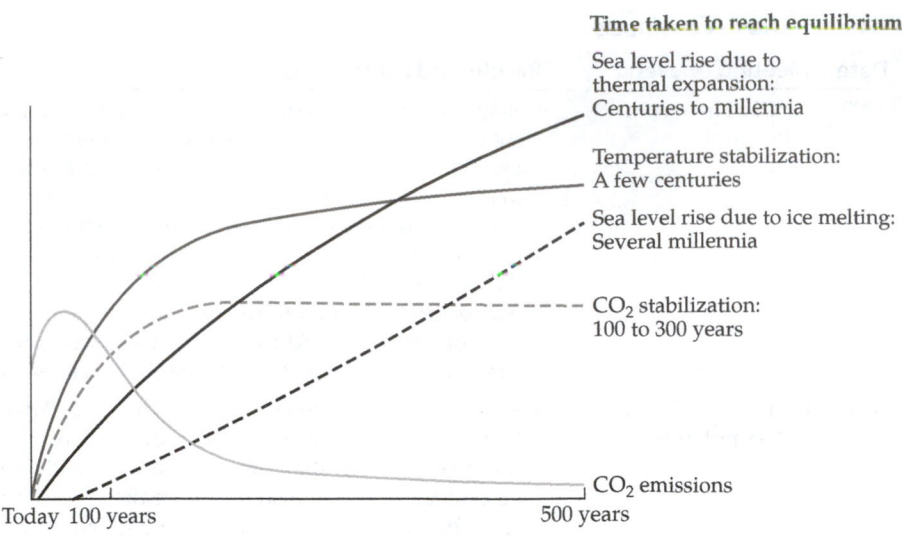

CASE STUDY 10.1

Challenges to Predicting Indirect Effects of Climate Change

Lisa Crozier, University of Chicago

This chapter demonstrates that climate change has widespread, profound, and complex effects. Here, I explore the theme of complexity. We can currently only glimpse at the depth of ecological and evolutionary understanding that is necessary to predict the biological changes to come. We know that the delicate web of ecological interactions as they exist now will be disrupted by rapid climate change, but predicting where strands will break and where new strands will be spun is extremely difficult. Each species will experience both direct effects of climate change (e.g., temperature or desiccation stress), and indirect effects that are mediated through some secondary physical or biological process (e.g., a shift in ocean currents or a change in a predator–prey interaction). Our current experimental approaches are adequate for predicting direct effects, but indirect and synergistic effects are much less tractable. These secondary effects are therefore poorly understood, and so are not included in most predictions despite evidence that they can overwhelm direct effects. Much work lies ahead to unravel the many ties between species and their environments and to clarify how behavioral or physiological processes may change in response to a changing environment.

Consider a forest ecosystem. Tree growth and survival are very sensitive to temperature, precipitation, and potentially

CO_2 concentration (i.e., direct effects). However, they are also very sensitive to disturbance regimes and herbivory (i.e., indirect effects). Climate strongly influences the frequency, intensity, and nature of major disturbances that in turn shape both species composition and ecosystem function. A change in disturbance regimes can transform communities much faster than changing temperatures alone. Major disturbances hasten the death of increasingly maladapted genotypes and open up large areas to colonization by different species. Herbivores large and small can also tilt the competitive balance between plants, and herbivores will respond directly to climate change as well as to changes in plant condition and predation rates. How will all of these dynamics interact over the next century?

Changing Disturbance Regimes Will Change Communities

Disturbance regimes such as fire and insect outbreaks are now recognized as an essential characteristic of many communities. Different species are adapted to different fire regimes. Spruce (*Picea* sp.) and juniper (*Juniperus* sp.), for example, cannot tolerate frequent fires, whereas lodgepole pine (*Pinus contorta*) depends upon fire to replace itself. The frequency, seasonality, severity, and type of fire (e.g., crown or surface) determine

postfire ecosystem structure and function (Flannigan et al. 2000). These qualities in turn depend on existing forest characteristics and climate. Fires will likely increase in severity and frequency over most of North America due to rising temperatures and more variable precipitation (Flannigan et al. 2000). This increase is expected even in some regions where average precipitation goes up, because wet years foster rapid growth and produce extra fuel that burns more intensely during dry years. Fire-tolerant species will be favored in these regions. These species in turn tend to promote fires (e.g., by having more flammable wood). Concern is growing that fires may establish a positive feedback loop and contribute to global warming by releasing stored carbon, changing reflectance properties and reducing evapotranspiration rates (Dale et al. 2001). So warming begets fire, which begets fire-loving trees that beget more fires, which beget more climate warming, and the cycle repeats itself.

Insect and pathogen outbreaks also shape plant communities. Insects and pathogens are actually the dominant force of disturbance in forests in the United States, as measured by area affected (20.4 million ha/year) and economic cost (U.S.$1.5 billion/year) (Dale et al. 2001). Outbreaks predispose some forests to major fires by increasing fuel load, providing fallen trees that act as "fire ladders" which convert surface fires to stand-killing crown fires, and providing smoldering material that eventually spreads when weather conditions are suitable (Logan and Powell 2001). Outbreaks also often occur when trees are drought stressed (Mattson and Haack 1987), which tends to occur in hot, dry years when fires are more likely. Tree-ring chronologies show that insect outbreaks have been a regular feature of forest life for thousands, and probably millions of years (Swetnam and Lynch 1993). Like fire, these so-called "normative outbreaks" are necessary for the regeneration of some forests (Mattson 1996; Logan and Powell 2001). Nonetheless, severe outbreaks can cause essentially permanent changes in vegetation, especially when conditions for regrowth are somewhat less favorable than average. For example, several large outbreaks of the autumnal moth, *Epirrata autumnata* in northern Europe transformed patches of dense birch forests into tundra (Kallio and Lehtonen 1973). Climatic factors probably contributed to both the outbreaks and poor regrowth. Reindeer and vole herbivory further hindered regeneration. An even greater concern is that as insects shift ranges they may colonize new hosts that cannot tolerate large-scale infestations.

Outbreaks will probably increase in many areas where they are now infrequent. Warmer winters increase outbreak probability in species such as the autumnal moth (Virtanen et al. 1998) and southern pine beetle (*Dendroctonus frontalis*) (Ungerer et al.1999) by increasing overwinter survivorship. Outbreaks in other species, such as spruce beetle (*D. rufipennis*) and mountain pine beetle (*D. ponderosae*), depend on the amount of time it takes for larvae to complete their life cycle, and this time period depends on temperatures during the growing season (Hansen et al. 2001; Logan and Powell 2001; Regniere and Nealis 2002).

As temperatures rise, these species will invade higher latitudes and elevations. Other species, such as gypsy moth (*Lymantria dispar*) and spruce budworm (*Choristoneura occidentalis*) will respond more to changes in precipitation (Williams and Liebhold 1995), so it is difficult to predict whether outbreaks will increase or decrease because of uncertainty in climate forecasts.

Because most species have survived warmer climates in the past, theoretically they should not be threatened by outbreaks of native insect species. However, some defoliating insects are introduced (such as the gypsy moth), so climate change will expose naive hosts to attack. Furthermore, if trees are weakened by other stresses or have greatly reduced habitat, they may be much more vulnerable. For example, whitebark pine (*Pinus albicaulis*) may be seriously threatened in the Rocky Mountains by climate-induced shifts in the range of the mountain pine beetle (Logan and Powell 2001). Whitebark pine has been decimated by an introduced blister rust throughout much of their historical range, so the remaining healthy populations are important for the species' survival. Mountain pine beetle will likely invade whitebark pine habitat in response to a 2°C rise in temperature. Whitebark pine is a very long-lived species, with some individuals reaching 1500 years old. Its life history strategy makes it unlikely that it could recover quickly from defoliating outbreaks. Decline of whitebark pine would affect the entire ecosystem because it provides essential food resources for many species. Its pine nuts are especially high in fat, making them the preferred food source of the mutualist bird, Clark's Nutcracker (*Nucifraga columbiana*), red squirrels, and bears. Grizzly and black bears raid squirrel pine nut caches in preparation for hibernation, when accumulating fat reserves is especially important (Mattson et al. 2001). These long-lived hardy trees also provide ecosystem services such as snow stabilization, which helps store water for the thirsty west.

Trophic Interactions Askew

Many of the predictions above are based on the direct effects of temperature on insect survivorship or growth rates. However, insect responses may be tempered by interactions with other species, such as host plants or predators. For example, it is important for many caterpillars to emerge from winter dormancy at the same time as their host plants. If caterpillars emerge from winter dormancy too early, they may starve before their host plants leaf out. On the other hand, if caterpillars emerge too late, they miss the more nutritious and palatable young foliage. However, each species has a unique set of cues that trigger seasonal behaviors, so synchrony between host plant and herbivore may not be maintained during climate change. For instance, egg hatch in the winter moth, *Operophera brumata*, depends on cumulative temperatures above a certain threshold, and therefore responds directly to warmer springs by hatching earlier. One of its host plants, Sitka spruce (*Picea sitchensis*), on the other hand, depends on a chilling period in addition to accumulating spring temperatures to trigger budburst. When warmer winters reduce chilling, the spring temperature

threshold rises, delaying budburst. So the asynchrony between caterpillar emergence and spruce budburst is expected to grow in response to climate change (Dewar and Watt 1992). Another host plant for this species, red oak (*Quercus rubra*), does not have a chilling requirement so its budburst has advanced in recent years—but the response has been slower than that of the winter moth (Visser and Holleman 2001). The suitability of various hosts may therefore change with climate change. However, in most cases we do not know exactly what triggers each species, so many pair-wise interactions are hard to predict (Harrington et al. 1999).

Predators of insects will also vary in their responses to climate change and their ability to track changing phenology of their prey. Parasitoids are often limited more by cool temperatures than are their prey (e.g., Randall 1982, Davis et al. 1998, and Virtanen and Neuvonen 1999), so warming will increase their relative impact. However, birds seem to be less flexible in advancing their migrating and breeding behaviors to keep up with earlier caterpillar emergence (Van Noordwijk et al. 1995; Visser and Holleman 2001; Strode 2003; Visser and Rienks 2003). Migrating birds face an even greater challenge in trying to maximize food availability at multiple stops along their migration route, because warming is occurring much more rapidly at higher latitudes (Strode 2003). Determining the limiting factors for each of these species individually and then in concert with changes in other species is a daunting task, but is essential for predicting how they will respond in years to come.

Comparing Direct and Indirect Effects in Boreal Forests

With all species responding independently to climate change and to each other, how can we hope to predict the net effects? It is extremely difficult and few authors have tried it. A comprehensive understanding of community interactions is needed, and few ecosystems have been studied thoroughly enough to achieve this level of understanding. Nonetheless, to demonstrate the potential importance of indirect effects, Niemela et al. (2001) attempted such an analysis of boreal forests (Figure A). They predicted that in some cases the effects of herbivores and secondary changes in soil quality will outweigh the direct effects of climate change on vegetation. For example, in southern Finland,

Norway, and Sweden the direct impacts of climate change are likely to be minimal because the dominant trees are relatively insensitive to changes in winter temperature, and little change in summer temperature or precipitation is predicted for this region. Insect and mammalian herbivores however, are extremely sensitive to winter temperature and precipitation. Snow depth has a strong impact on foraging success of moose (*Alces alces*) and deer, and consequently their winter survivorship (Ayres and Lombardero 2000; Niemela et al. 2001). Because browsing mammals tend to prefer broad-leaved species, their success is likely to convert a neutral tree response to a negative one. These authors predict that pine sawflies (*Nediprion sertifer*) will have a greater impact on pine, while moose and roe deer (*Capreolus capreolus*) inhibit the growth of broad-leaved trees. In sum, increased herbivory on competitors will lead to an expansion of spruce.

Farther north, the direct effects of climate change should improve growth of spruce, pine, and birch. The expected rise in snow will be deleterious for moose and reindeer, but rising winter temperatures should be advantageous for the autumnal moth and the pine sawfly. Because these moths limit birch and

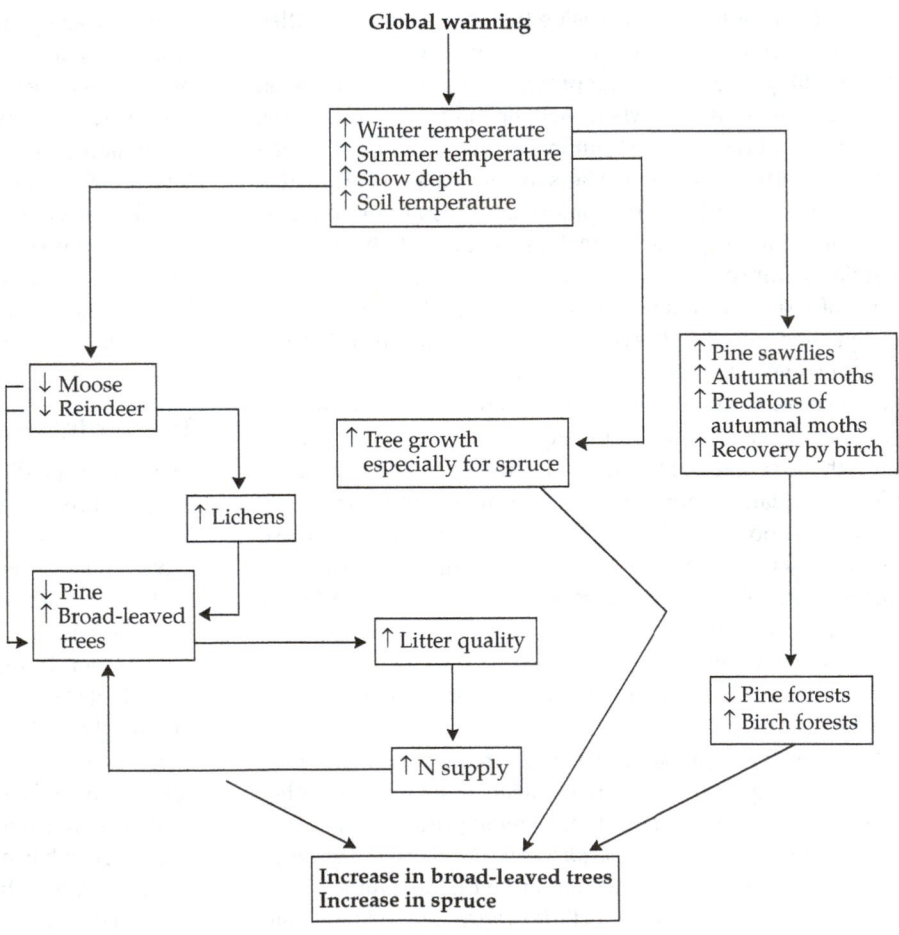

Figure A Effects of climatic warming on the biotic interactions likely to influence forest composition in northern Finland, Norway, and Sweden. (Modified from Niemela et al. 2001.)

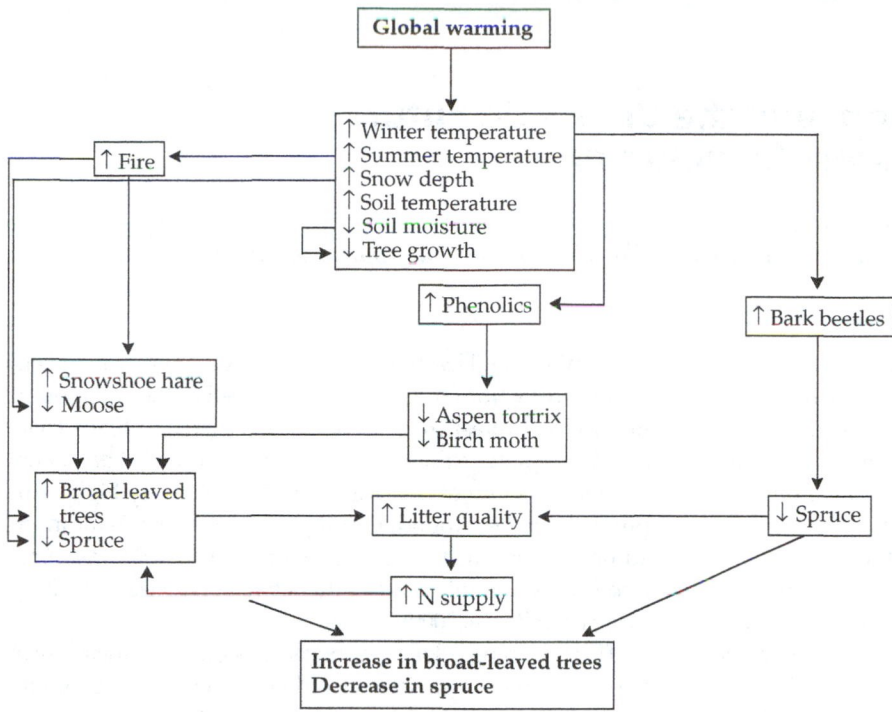

Figure B   Effects of climatic warming on the biotic interactions likely to influence forest composition in interior Alaska. (Modified from Niemela et al. 2001.)

pine, respectively, they would be expected to counteract benefits for both birch and pine. However, increasing ant predators of the autumnal moth and improved growing conditions for birch may compensate for insect damage in birch, but not in pine. Overall these authors predicted that both spruce and broad-leaved species will increase in the north. In Alaska, on the other hand, fire and bark beetles are expected to shift the advantage to broad-leaved species (Figure B).

Clearly, incorporating many species interactions into our predictions radically complicates forecasts. All of the pair-wise dynamics mentioned above are well documented, but extrapolation to the whole ecosystem is still somewhat speculative. Experiments at the ecosystem scale are very difficult, so these predictions require careful extrapolation from small-scale experiments, and specially designed factorial experiments. Furthermore, predictions are even more difficult because most of the traits of interest could change over time.

Evolutionary Potential and Constraints

A crucial assumption underlying the examples discussed above is that preferences and tolerances are constant. However, most of these traits can change, as demonstrated by artificial selection experiments in the lab, geographic variation in the field, and changes in natural populations over time. Plastic behaviors and the potential for genetic change in response to climate change are hardest to predict because they require extensive experimentation and some inherently unpredictable events.

Many of the behaviors that determine phenology are under genetic control, and therefore would require evolutionary change to maintain synchrony between trophic levels. Evolution in these traits can happen very quickly, as demonstrated by the recent change in day length necessary to induce dormancy in the pitcher plant mosquito (*Wyeomyia smithii*) (Bradshaw and Holzapfel 2001). This mosquito has adapted over the past 30 years to a longer growing season by requiring a shorter day to trigger winter hibernation. Cues for hibernation are finely tuned along latitudinal gradients in many species, so these traits are probably very labile (Bradshaw 1976; Scriber 1994; Chown et al. 2002). However, this flexibility may be the exception rather than the rule. Constraints on evolution can be subtle, reflecting the underlying genetic architecture of phenotypic variation. In a spectacular study of genetic constraints on adaptation to global warming, Etterson (2001) demonstrated that evolution of drought tolerance in prairie plants in Minnesota may be slowed by links between genes. Genes associated with traits that would be advantageous in a more arid environment are tied to other genes under different (sometimes opposing) selection pressures. Large-scale transplant experiments with complete genetic crosses like this are extremely difficult, but they reveal fundamental constraints many species may face.

Additional constraints have been imposed by humans. Rapid evolution requires genetic variation for the relevant trait, which is diminished in very small populations or populations that have been selected by humans for particular traits. For example, stress tolerance is frequently associated with slower growth rates, so forestry practices that select for faster-growing trees have reduced natural variation that would have facilitated adaptation to climate change. Highly fragmented and reduced habitat also lowers the chance that individuals will find suitable areas for colonization outside their current ranges, which will be necessary as suitable climates move to higher latitudes. Thus although ecological and evolutionary responses may have allowed species to survive previous climate changes, resilience to future changes is much less certain. Like most of the other anthropogenic impacts discussed in this book, disrupting Earth's climate is a massive and irreversible experiment. The results will not be known for decades or centuries, so we must rely on our predictions to decide how we should act now. Large pieces of the puzzle are still missing, but we can greatly improve our understanding by incorporating indirect effects into more ecological and evolutionary studies.

CASE STUDY 10.2

Climate Change, Extinction, and the Uncertain Future of a Neotropical Cloud Forest Community

Karen L. Masters, Council on International Educational Exchange, J. Alan Pounds and Michael P. L. Fogden, Golden Toad Laboratory for Conservation, Monteverde Cloud Forest Preserve and Tropical Science Center

Steep mountainsides in northern Costa Rica rise into the clouds and spread along a northwest–southeast axis, which bisects the land into two climatically distinctive regions. The mountaintops, obscured from sight by billows of blowing mist, are carpeted by verdant cloud forest. Just below the bank of clouds, on the Pacific slope, abundant sunshine pierces the blowing mist and ignites rainbows that arc overhead, recalling the legend of the pot of gold at the rainbow's end. The world-renowned Monteverde Cloud Forest is indisputably wealthy in biodiversity: Monteverde's bank of biological richness includes some 3000 species of vascular plants (Haber 2000), more than 160 species of reptiles and amphibians (Pounds 2000), and 425 resident or migratory bird species (Young and McDonald 2000).

Straddling the continental divide along the Tilarán mountain range in northern Costa Rica, Monteverde is bathed in clouds nearly year round because of two distinctive oceanic weather systems, Caribbean and Pacific, which maintain the cool, lush conditions on the mountaintop (Figure A; Clark et al. 2000). A nearly perpetual supply of nutrient-rich precipitation supports record loads and diversity of epiphytes (plants that live on top of others), including orchids, ferns, and mosses, giving the vegetation its characteristic look and feel. Fre-

quent rainfall and leaching produce nutrient-poor soils and encourage the epiphytic growth form as well as the stilt roots of many cloud forest trees, beneath which golden toads (*Bufo periglenes*), now extinct, once gathered to breed. The almost continuous input of moisture results in relatively high plant productivity year round, yielding reliable resources for myriad organisms, such as Resplendant Quetzals (*Pharomachrus moccino*), which migrate locally between patches of fruiting trees (Powell et al. 2000).

Their dazzling colors and courtship, along with their care of young, nestled in cavities of old snags, engage a booming tourist industry and fascinate field biologists. However, when decades-long research on Monteverde bird communities revealed that the Keel-billed Toucan (*Rhamphastos sulfuratus*), a known predator of hole-nesting birds and a resident of lower elevations, had colonized and begun breeding in cloud forest habitat, ornithologists questioned why this was happening. The Keel-billed Toucan was not unique: Many other cloud forest intolerant bird species were showing similar upslope shifts in breeding ranges. Simultaneously, herpetologists were becoming increasingly concerned by downward trends in populations of lizards, and were alarmed by the outright disappearances of once-abundant frog and toad species, including two—the golden toad (*Bufo periglenes*) and Monteverde harlequin frog (*Atelopus* sp.)—that were endemic to Monteverde (Pounds 2000; Pounds and Puschendorf 2004). Sharing notes and concerns, biologists began to explore possible explanations for the emerging patterns.

Habitat loss was quickly ruled out because forest regeneration was common in the Monteverde area. The spread of pathogens, while perhaps affecting specific taxa, seemed an unlikely universal explanation, given the range of species involved. Researchers began to investigate local climate and the impacts of large-scale events and processes such as El Niño and global warming. Evidence in favor of a shifting local climate, brought about by global warming but punctuated by El Niño events, is mounting; climate shifts now figure prominently as the ultimate cause of the biological changes. Indeed, it is increasingly clear that the demise of the golden toad and that of the Monteverde harlequin frog represent the first, modern-day extinctions linked to global warming.

Monteverde's climate has changed profoundly since 1972 when John H. Campbell, an original Quaker settler of the zone,

Figure A Cloud cover over mountainous regions of Costa Rica are necessary for development of cloud forest ecosystems. (Photograph courtesy of Andrew Sinauer.)

began collecting daily weather data (Pounds et al. 1999). The frequency of days with little or no measurable precipitation during the windy-misty and dry seasons (December to May) has increased. Moreover, while the total precipitation during this period has not changed, precipitation is now concentrated on fewer days, yielding longer stretches of mist-free or mist-reduced days (Pounds et al. 1999). Whereas sequences of three or more dry days were rare in the 1970s, it is now common to have stretches of five or more days with dry or near-dry conditions.

According to the lifting cloud base hypothesis, the altitudes at which clouds form as air masses ascend the mountain slopes has increased, reducing moisture inputs to the cloud forest (Pounds et al. 1999). In a warming world, both the rate of evaporation and the water-holding capacity of air increase, thereby enhancing the water vapor available for cloud formation (IPCC 2001a). However, the altitudes at which clouds form depend on local conditions of relative humidity and thus on temperature (Still et al. 1999).

The frequency of cloud cover has also increased. Clouds reduce incoming solar radiation while having an insulating effect at night. An increase in nighttime temperatures and a decrease in daytime temperatures have reduced the diurnal temperature range, the difference between the daily minimum and the daily maximum (Pounds et al. 1999). All of the above parameters are linked to large-scale climatic conditions, including global warming and El Niño events. Over the past 30 years, global sea surface and air temperatures have risen sharply (IPCC 2001a). Together with the superimposed warm episodes of El Niño, the trend has led to more frequent and longer dry spells at Monteverde (Pounds et al. 1999).

Diverse vertebrate taxa, including amphibians, reptiles, and birds, have variously responded to climate change at Monteverde (Pounds et al. 1999). The most famous cases involve the extinctions of two local endemics, the golden toad and the Monteverde harlequin frog. In the same period, local populations of 20 other amphibian species disappeared. While a few have recolonized, post-crash censuses show that populations have fluctuated in response to climate but have remained below pre-crash densities. In the case of reptiles, local populations of the cloud forest anole (*Norops tropidolepis*) and the montane anole (*N. altae*), both endemic to Costa Rica, have declined since the late 1990s, when amphibian populations first fell dramatically (Pounds et al. 1999). In contrast, the gray lichen anole (*N. intermedius*), previously limited to middle elevations, has now colonized upper regions, and the ground anole (*N. humilis*), a primarily lowland species, has increased in abundance.

Upslope colonization is better documented for bird communities, which have been extensively studied since the late 1970s. Numerous cloud forest intolerant species have extended their ranges upslope, while most cloud forest species have not altered their distribution (Pounds et al. 1999). The net colonization rate (the number of species moving up minus the number moving down) is positive in some years and negative in others, as it fluctuates with climate. Nevertheless, the average colonization rate is positive, indicating an overall pattern of upslope range extension. Compositional changes in the bird community are pervasive and striking; colonizing species represent broad taxonomic and ecological ranges from both the Atlantic and Pacific slopes (Pounds et al. 1999).

The taxonomic diversity of affected species and the variety of reactions suggests multiple proximate triggers, all linked by sensitivity to climate change. Amphibian declines worldwide have directed attention to possible taxon-specific explanations that may not easily apply to birds and reptiles at Monteverde, but may be ultimately linked to climate. For instance, the chytrid fungus *Batrachochytrium dendrobatidis*, which attacks the skin of amphibians and may under some circumstances cause mortality, is a suspected culprit in the extinction of amphibian populations (Berger et al. 2004). Pounds and Puschendorf (2004) hypothesize that increasing cloud cover may prevent chytrid-infected frogs from raising their body temperatures sufficiently to kill the fungus, thus facilitating its persistence and transmission. This hypothesis, like others that propose climate-mediated disruptions in host–parasite interactions, has not been tested. Nevertheless, it highlights the importance of species interactions and the complexity of biological response to climate change (Pounds 2001). Clearly, investigations into the proximate mechanisms of Monteverde extinctions and population declines are of pressing significance.

The observed patterns signal ongoing or imminent change in both the biotic and abiotic elements of the Monteverde cloud forest. Although Neotropical cloud forest plants adapted to the climate oscillations of the Pleistocene, the current temperature increase may be ten times faster that that during the transition from the Pleistocene to modern times (Bush et al. 2003). Increased rates of extinction are a likely consequence. The Monteverde case suggests that range-restricted or dispersal-limited species, or those adapted to narrow environmental conditions or interacting with such species will suffer most. Climatic specificity for narrow altitudinal ranges often characterizes tropical montane organisms, especially cloud forest species.

Cloud forest epiphytes may be especially vulnerable (Figure B). Many are climate-sensitive (e.g., Gentry and Dodson 1987; Benzing 1995, 1998; Atwood 2000), having adapted to particular conditions within the high microclimatic heterogeneity that characterizes cloud forests (Lawton and Dryer 1980; Clark and Nadkarni 2000; Haber 2000; Haber et al. 2000; Kappelle and Brown 2001). In the Monteverde area alone, there are over 500 species of orchids, many of which inhabit narrow altitudinal belts according to their specific temperature and moisture requirements (Atwood 2000; Hammel et al. 2003). Critical to ecosystem functioning of cloud forests, epiphytes are both "nutrient scavengers," taking up compounds in mist that would otherwise blow by the canopy, and keystone species, by providing nutrition to diverse organisms throughout the year (Coxson and Nadkarni 1995). Thus, losses of epiphytes may have consequences that reverberate through all trophic levels.

Figure B   An epiphyte such as this orchid *Encyclia ionophlebia*, is an important component of cloud forest ecosystems. (Photograph © Oxford Scientific/photolibrary/Michael Fogden.)

Species that successfully adapt to ongoing change at Monteverde or that colonize the area will create non-analog communities. The composition of these communities will depend on the climatic tolerances of species and their ability to adapt to novel interactions. Rates of community transformation may exceed those at which interacting species can adapt to one another, leading to disruptions, for instance, in natural host–parasite population cycles. Whether and how native species will cope with simultaneous changes in climate and community composition is unclear. For instance, the Resplendent Quetzal, whose altitudinal distribution rarely includes the comparatively low elevations of Monteverde (Stiles et al. 1989), is already declining (Fogden, unpublished data). Whether it can survive as communities continue to change is doubtful. What seems likely is that the singular character and quality of the Monteverde cloud forest, assembled of unique species that have evolved over the millennia in response to one and other and to the physical environment, will be forever transformed in a matter of decades.

CASE STUDY 10.3

Adapting Coastal Lowlands to Rising Seas

Sam Pearsall, The Nature Conservancy and Benjamin Poulter, Duke University

In the Albemarle Region of northeast North Carolina (Figure A), the land is subsiding and the sea is rising. Global warming from post-industrial modification of the atmosphere is causing the sea to rise, partly as the result of ice melting, but mainly due to the thermal expansion of seawater. In the Albemarle Region, the current, combined rate of relative sea level rise is 43 mm (2 inches) in 10 years (Permanent Service for Mean Sea Level 2004), and when the sea rises an inch, it extends inland many feet. Forty inches of sea level rise will inundate about a million acres of low-lying lands (see Figure A). At the present rate, this will take 200 years. If the rate triples, it will take about 65 years, less than a lifetime.

Rising seas produce adverse environmental effects locally as the result of inundation, salt poisoning, and erosion. If the Outer Banks are breached by hurricanes that are stronger and more numerous, then the local effects could be severe. The effects of rising seas also may be felt far inland of the inundation zone. For example, salt-spray, storm surges, and salt water infiltration up rivers and ditches may double the area affected by rising seas. In particular, regions that feature soils with high organic content such as the peat-based soils of much of the Albemarle Region are at greater risk because organic soils experience rapid decomposition by sulfate-reducing bacteria in the presence of salt (Hackney and Yelverton 1990). Thus, the large island of dry ground shown northeast of Lake Mattamuskeet in Figure A will not last very long as it is an area of pocosin (sclerophyllous shrub swamp) on a peat dome. The Albemarle Region is highly favored for the conservation of coastal ecosystems. The area's rich mosaic of forests, dunes, wetlands, rivers, and sounds provides an extraordinarily productive natural system. The Albemarle Sound is part of the largest closed lagoon in the world, and part of the second largest and healthiest estuary in the eastern United States. The Roanoke River floodplain contains vast areas of cypress–tupelo swamp forests and the largest and least disturbed bottomland hardwood forest ecosystems remaining in the mid-Atlantic region. Natural fires, floods, and storms are so dominant in this region that the landscape changes very quickly. Rivers routinely change their courses and emerge from their banks. The Outer Banks have been described as a "river of sand" flowing south along the

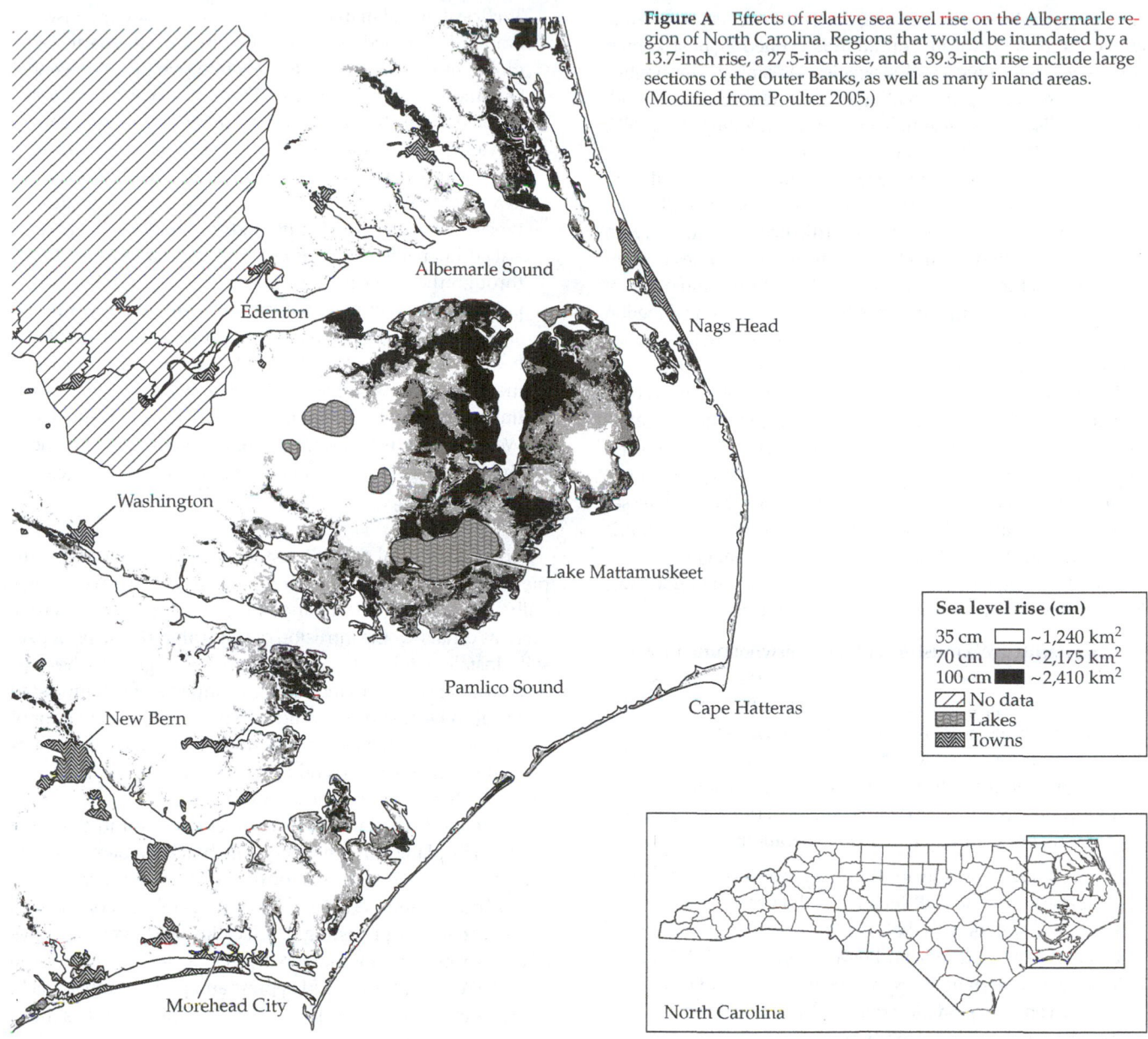

Figure A Effects of relative sea level rise on the Albermarle region of North Carolina. Regions that would be inundated by a 13.7-inch rise, a 27.5-inch rise, and a 39.3-inch rise include large sections of the Outer Banks, as well as many inland areas. (Modified from Poulter 2005.)

Sea level rise (cm)

35 cm	☐	~1,240 km^2
70 cm	▨	~2,175 km^2
100 cm	■	~2,410 km^2
▨	No data	
▨	Lakes	
▨	Towns	

North Carolina

continental shelf (Kaufman and Pilkey 1983). Xeric environments of sand dunes and ridges share ecotones with the hydric environments of sounds and pocosins. Several national wildlife refuges are located in the region, as are conservation lands established by The Nature Conservancy, the NC Coastal Land Trust, the NC Wildlife Resources Commission, and other private and public agencies. Up to half of the lands at risk from rising seas are conservation lands, established and maintained for the public benefit.

The managers of conservation lands are mainly uncertain about what actions they could take in the face of global climate change (Lavendel 2003). No one really knows how much or how fast the sea will rise. The responses of public officials (e.g., the U.S. Army Corps of Engineers) are unpredictable, but they tend toward addressing the symptoms of rising seas as temporary, local disasters. The frequency and strength of storms, impacts on precipitation (amount and distribution), and even the responses of plants and animals are not known, and may not be knowable. New species will invade, native species will be stressed, and sometimes native species themselves will become invasive. The only thing that can be predicted with certainty is that the landscapes and seascapes of the Albemarle Region soon will change

dramatically. An engineering or deterministic response—one where the desired future condition is posited against known stresses to determine ideal strategies—is simply impossible. Adaptive management—treating management as a real world, real-time, disciplined, scientific experiment (Holling 1978, 1982; Holling and Meffe 1996) is the only viable alternative.

Ecological stresses can be grouped into three general categories. Some stresses are relatively brief in duration or affect relatively small areas. These stresses (disturbances) modify an ecosystem locally without changing its normal ranges of composition, structure, or function. Other stresses maintain pressure on an ecosystem so that composition, structure, and function change gradually and adaptively. Finally, some stresses are severe enough to cause the ecosystem to change substantially and suddenly, nearly always toward simplification of composition, structure, and function. This kind of change is called a "catastrophe" (Poston and Stewart 1978; Scheffer et al. 2001). Ecological managers must conserve normal disturbance regimes (even as they change) and conserve adaptability to chronic stresses (even as they change). The most important strategy, therefore, is to prevent catastrophic (sudden, simplifying) changes.

The challenge we face in the Albemarle Region of northeast North Carolina can be summarized as follows:

Uncertainty: We must accept the overwhelming probability that the sea will continue to rise at an increasing rate; that temperatures will rise; that frequency and severity of storms will probably increase; and the amount, frequency, and duration of precipitation will change unpredictably. Ecosystems and organisms will respond unpredictably in both ecological (within generation) and evolutionary (between generations) time scales.

Desired outcome: Ideally, the area in conservation in the Albemarle Region will include both current conserved lands and waters with their future ecosystems, including estuaries and bottoms, and newly conserved lands and waters that seem likely, by virtue of spatial and ecological adjacency, to provide refuge or transitional habitat for ecosystems that are currently conserved.

Strategy: We cannot describe confidently the future ecosystems or, given the high probability of both invasion and adaptation, even the future organisms of the Albemarle Region. Even the ecological near future is virtually unknowable. However, we can take steps now—plausible steps based on literature review, modeling, and expert opinion—to reduce the likelihood of catastrophic transformations. We must formulate testable hypotheses about the best ways to mitigate catastrophic change. We must implement these hypotheses in many places, and we must closely monitor each replicate so that we can detect local failures. We must adapt our management as necessary to minimize the likelihood of catastrophic transformation across the region.

Tactics: Several management actions are under consideration for immediate application in the Albemarle Region. These tactics group loosely into two groups—conservation and restoration. The conservation tactics involve acquisition and preservation of additional conservation lands, especially inland and upland of existing conservation lands to facilitate the movement of species away from rising seas. Where we suspect vegetation types may move inland or upland, we should actively resist additional fragmentation of the landscape, especially through the construction of new drainage ditches and paved roads; and we must resist the tremendous pressure to armor shores (e.g., with rip-rap, sea walls, bulkheads). An armored shore is only a temporary impediment to the rising sea, while it is a near-permanent impediment to the inland migration of coastal wetlands. We should acquire easements that prevent armoring the shore, especially where it is not already developed (e.g., see Titus 1998).

There are several possible restoration tactics that can be implemented. As the sea rises, drainage ditches that were originally dug to provide farmland and prevent mosquitoes now serve as canals for the intrusion of salt water into areas of peat soils. Installing tide gates and selectively plugging ditches represents an approach to reducing the impacts of salt intrusion and peat soil reduction over much of the non-littoral interior of the region. These ditches are not widely recognized as avenues for salt water intrusion, and they are managed by powerful drainage district associations (Figure B).

Shore lands that are likely to be submerged in the short term can be planted immediately with brackish marsh species, such as brackish needle rush or bald cypress (Figure C). The native bald cypress (*Taxodium distichum*) is tolerant of brackish water, capable of persisting for decades, and even centuries after its roots are submerged by the waters of Albemarle Sound. We should plant bald cypress and perhaps other brackish-tolerant woody species, wherever the land has been cleared. We may be able to subsidize these plantings by offering carbon sequestration credits or timber investment opportunities. We should actively manage standing forests for the restoration of their missing bald cypress component.

The native beds of submerged aquatic vegetation in Albemarle Sound have been severely reduced over the years by human actions (trawling, dredging, sedimentation). We should begin restoration of these beds immediately.

Lastly, as Albemarle Sound inevitably grows more salty and more physically active as the sea rises and the Outer Banks are breached, we should establish and seed marl reefs to provide habitat for native oysters (Figure D) to maintain water quality in the sounds and to buffer the fragile, newly submerged shore from physical erosion.

Figure B Thick peat deposits on the Albemarle Peninsula are vulnerable to salt water intrusion due to sea level rise. Numerous ditches throughout the region can bring rising sea water further inland, where the thickest peat deposits can be destroyed. (Modified from Poulter 2005.)

Peat thickness (ft)
- 1–2
- 4–6
- 8–10
- 12–14

Albemarle Sound

Aligator River

Pamlico Sound

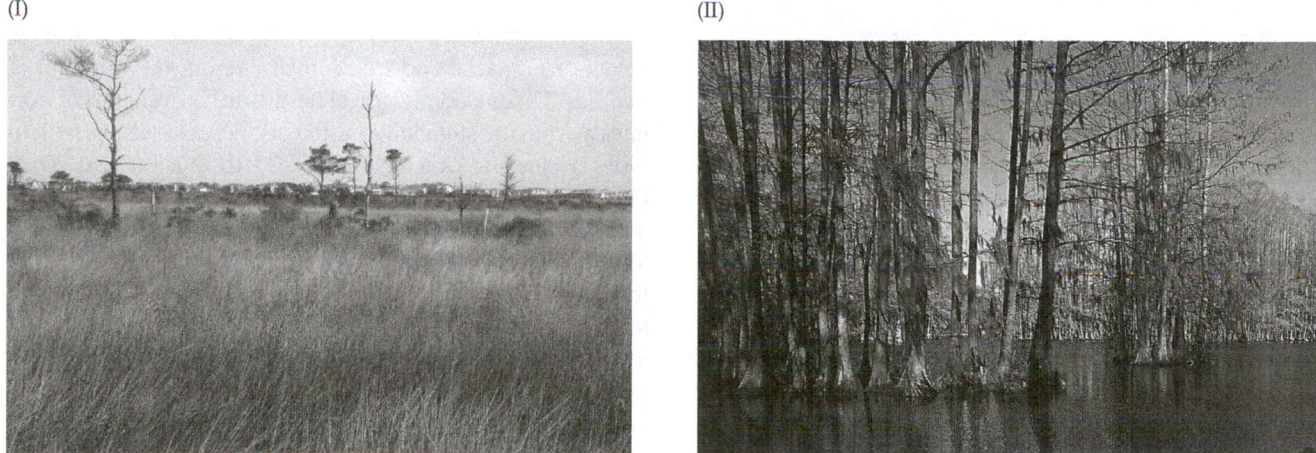

Figure C Examples of vegetation adapted to brackish and submerged conditions that can be planted in anticipation of sea level rises. (I) Brackish needle-rush marsh invading dry land due to current sea level rise. (II) A bald cypress swamp (*Taxodium distichum*) in North Carolina. (I, photograph by Sam Pearsall; II, photograph © Gary Retherford/Photo Researchers, Inc.)

Conclusions

In the Albemarle Region, the proportion of lands likely to be affected by rising seas is very high—among the highest in the world. And the proportion of those endangered lands in public ownership and conservation management is also very high. Local ecosystems and species are highly adaptable, and although the nature of the adaptations are not predictable, this is a region where manipulating the placement of species possess-

ing adaptations to rising sea levels have a good chance of avoiding catastrophic transformations. We have a list of practical management actions to try, so that adaptations over an ecosystem level have time to happen. We have professional conservation managers in place, and we have the instrument of adaptive management to stave off failure and increase the probability of success. We must begin implementation immediately, and we must continue as long as the sea continues to rise.

(I)

(II)

Figure D   (I) Building oyster reefs with marl. (II) Building oyster reefs with empty oyster shells. (I, photograph by Ashley Harraman Burgin; II, photograph courtesy of NOAA.)

CASE STUDY 10.4

Climate Change and Coastal Migrant Birds

Jennifer Gill, University of East Anglia, Norwich, U.K.

Coastlines around the world are in a time of rapid change as a consequence of rises in global sea levels combined with extensive human development and exploitation. Coastal zones also harbor some of the habitats with greatest biological diversity and economic importance. The implications of these changes are thus potentially severe, and our ability to manage coastlines to minimize these impacts will rest on our ability to predict the consequences of coastal change, in order to guide effective management policies.

Among the groups most reliant upon coastal zones are migratory birds that breed in Arctic and subarctic regions, but depend upon highly productive temperate coastal mudflats to survive the winter and fuel their journeys to and from the Arctic. Climate change may deal a double blow to these species, as both Arctic regions and temperate coastal zones are likely to change substantially in response to global warming and sea level rise (Watkinson et al. 2004). Arctic zones are extremely vulnerable to

increases in glacial and icecap melt altering the duration of freeze and thaw periods, and to northward movements of treelines as climatic suitability for forests increases at higher latitudes (Weller and Lange 1999; IPCC 2001b; Bigelow et al. 2003). Declines in the area, and potentially also the suitability, of arctic and subarctic habitats are likely to have serious consequences for the millions of migratory birds that breed in these areas. At the opposite end of the migratory range, coastal mudflats in temperate and tropical zones are threatened both by a reduction in area, as hard coastal defenses prevent the natural inland migration of coastal habitats, and by the extensive human developments that frequently replace these key habitats, together with human exploitation of the resources within them.

Figure A describes some of the main ways in which climate-induced changes to breeding or wintering habitat may influence population size in coastal migrant birds, through direct or indirect impacts on demographic processes. Large-scale

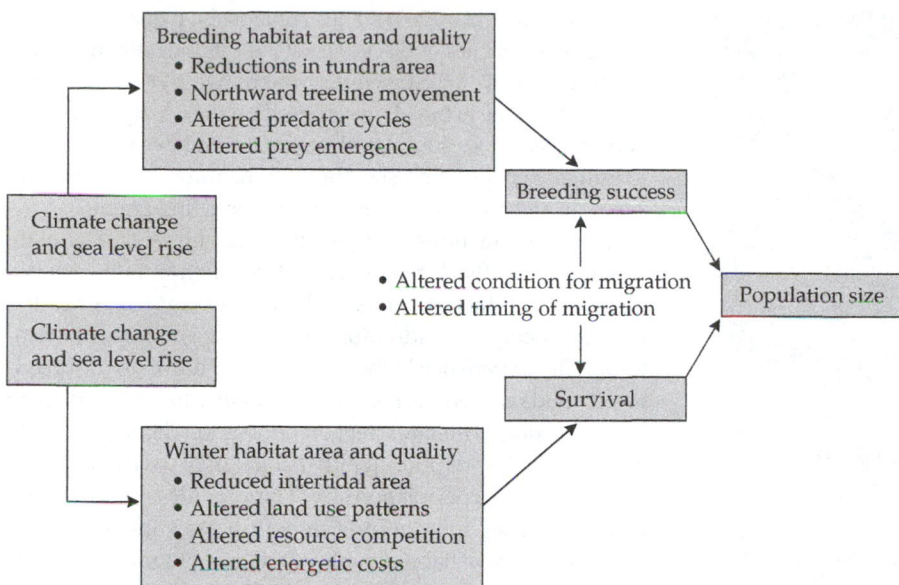

Figure A   Examples of the mechanisms by which climate change and sea level rise are likely to influence population size in migrant birds that breed in the Arctic and overwinter in temperate coastal zones.

changes to habitat structure clearly have the potential to influence population size dramatically in many of these species, through processes such as limiting the availability of breeding locations or foraging sites and hence increasing competition for the remaining resources. Models of vegetation change suggest that many arctic breeding bird species may experience losses of over 50% of current breeding habitat (Zöckler and Lysenko 2000). However, in some cases the less obvious effects (e.g., changes in predation levels or energetic costs) may also be important. Many high-arctic migrants face huge energetic costs in traveling to and from the breeding grounds each year, in some cases requiring dramatic annual physiological changes (Piersma 1998; Battley et al. 2000). This energetic expenditure needs to be fuelled by the continued availability of food resources along a network of wintering and passage sites. Similarly, many predator species at high latitudes undergo population cycles that result in cyclic levels of predation on migrant birds (e.g., Bety et al. 2002). Changes in predator abundance or the timing of the breeding season could potentially lead to a decoupling of these cycles, with unknown consequences for either party. Ultimately the influence of any of these changes on population size will depend on the density-dependent nature of each of the processes, which is frequently the most difficult aspect to quantify (Sutherland 1996).

While the loss of tundra habitat is very likely to have serious negative consequences for breeding birds, impacts of climate change at passage and wintering sites at lower latitudes are less easy to predict. In many areas, sea level rise and the presence of hard sea defenses may result in reductions in intertidal areas through coastal squeeze, with likely serious impacts on the

species present. However, in response to these processes, wetland restoration is taking place in some coastal areas and, in the U.K., managed realignment of sea defenses is being undertaken at a growing number of sites (Perrow and Davy 2002; Pethick 2002). These habitat creation initiatives could therefore result in net benefits for some species (e.g. Evans et al. 1998). However, the loss of vulnerable low-lying coastal wetlands adjacent to intertidal areas may be more difficult to mitigate. Species dependent on these habitats, or on the juxtaposition of intertidal and coastal freshwater habitats, may be the ones most likely to suffer if these habitats cannot be protected or recreated.

Environmental changes at one end of the migratory range can also have consequences elsewhere in the range. For example, seasonal declines in productivity mean that individual migrants are under strong selective pressure to return to the breeding grounds as soon as possible (Kokko 1999; Drent et al. 2003). The timing and abundance of resources on spring passage sites prior to departure for the Arctic can therefore be critical in facilitating these rapid departures (e.g., Prop et al. 2003 and Battley et al. 2004). The productivity and abundance of many intertidal invertebrates that form a key part of these resources can be strongly temperature-dependent (Beukema et al. 1996; Honkoop and Beukema 1997). Hence, changes in prey abundance through habitat change and temperature shifts are likely to influence the timing of arrival of migrants in the Arctic and the condition that they arrive in. The short window of time in which prey are available in the Arctic also magnifies the importance of the breeding period coinciding with peaks in prey abundance. Any change in spring conditions on passage sites or in the Arctic could potentially result in a mismatch between arrival time and the availability of breeding sites and prey supplies. The impact of any such mismatches will depend on the cues used by migrants to time their departures and arrivals, and their ability to adapt their migratory schedule to changing environmental conditions.

Predicting the impact of climate change and sea level rise on migratory birds therefore requires an understanding of the processes that influence population size in both the breeding and the wintering grounds, and any interactions between them. In particular, the density-dependent processes that regulate population size need to be quantified over these very large spatial scales. An example of the interactive nature of population size, distribution and demography is provided by the Black-tailed Godwit, *Limosa limosa islandica*, a migratory

Figure B Many shorebird species are particularly vulnerable to sea level changes due to climate change, such as this Black-tailed Godwit (*Limosa limosa islandica*). (Photograph © Bernard Castelein/naturepl.com.)

shorebird that breeds in Iceland and winters in temperate western Europe (Figure B). This species provides an ideal opportunity for studies of large-scale population processes because recent rapid population increases have resulted in a change in spatial distribution across U.K. wintering sites that follows a clear buffer effect; in the 1970s the population was originally concentrated in good-quality winter sites in the south of England, where rates of prey consumption are high, but since then it has expanded into poorer-quality sites in the east of England where prey intake rates are lower (Gill et al. 2001).

The establishment of a network of volunteer observers right across the range of the Icelandic Black-tailed Godwit has resulted in thousands of records of individually color-marked birds over several years. This has allowed direct comparisons of survival rates of birds wintering in traditionally and recently occupied wintering sites. The between-site differences in prey consumption rates are mirrored by differences in annual survival; birds wintering in the high-quality, traditionally occupied sites have significantly higher survival rates than those in the recently occupied, poorer-quality sites (Gill et al. 2001).

The question of whether the quality of winter habitat might affect breeding performance was then assessed by measuring the timing of arrival of individual birds in Iceland. These data demonstrated that individuals from higher-quality winter locations arrived significantly earlier than those from poorer-quali-

ty locations (Gill et al. 2001). Thus winter habitat quality not only affects survival rates in Black-tailed Godwits, but also influences timing of arrival on the breeding grounds, which is likely to confer an advantage in breeding success. Population expansion into poorer-quality locations therefore appears to have the potential to influence population size through the impact on per capita survival, and on delayed arrival on the breeding grounds.

These interactions suggest that the impact of climate change on the Black-tailed Godwit population will vary depending on which habitats and locations are subject to the greatest change. As individuals inhabiting good-quality wintering sites experience higher survival and return to the breeding grounds earlier, the loss of good-quality habitat is likely to have a disproportionate effect on population demography. By contrast, the loss of poor-quality habitat may have a relatively minor impact on population size. This provides scope for both targeted conservation efforts to both protect higher-quality sites and improve habitat quality as a form of mitigation.

While reducing the impacts of climate change in Arctic zones will require global policy responses to reduce carbon emissions, the impacts on temperate coastal zones can also be addressed to a large extent through appropriate management to reduce the human pressures on these valuable habitats. In theory, such management approaches are simple: prevent any developments that threaten the integrity of the habitat, restrict all human exploitation of the resources within the habitat, and begin a program of removing sea defenses to allow the habitats to move inland in response to sea level rise. In practice, of course, there are many constraints on these policy options and coastal managers must make key decisions about which areas of land need to be afforded different types of protection. What is required of conservation biologists therefore is to inform these decision-making processes, ideally through the development of predictive models to assess the consequences of different coastal management options. Such models require integrated, interdisciplinary approaches to science to effectively link the climatological, oceanographic, geomorphological, and ecological processes that need to be incorporated. This is a tall order, but one that research groups, such as the Tyndall Centre for Climate Change Research, are beginning to tackle. The science relating to each of these processes has been developed largely in isolation, and each has significant uncertainty yet to be unraveled. Nevertheless, decisions will be made whether or not conservation biologists are prepared to attempt to influence these processes.

Coastal zone managers across the globe are being faced with serious decisions regarding future management strategies in response to climate change and sea level rise. The challenge for conservation biologists is both to improve our understanding of the processes influencing population regulation in many more species, and to better direct our findings toward policymakers so that effective conservation forms a cornerstone of these decision-making process.

Summary

1. The greenhouse effect is nothing new. The presence of greenhouse gases, such as carbon dioxide, in Earth's atmosphere allow heat to be retained at the surface, making the world about 60°F warmer than it would be without the greenhouse effect. Our use of fossils fuels (oil, coal, and natural gas) is releasing large amounts of carbon that have been stored deep underground for many millions of years. Levels of carbon dioxide in the atmosphere are 30% higher than before the Industrial Age, and this percentage increases every year. This has enhanced the greenhouse effect, causing the surface of Earth to gradually become warmer. In 2001, the Intergovernmental Panel on Climate Change concluded that the global rise in average yearly temperature over the past 50 years was due primarily to increased concentrations of anthropogenic greenhouse gases. Because carbon dioxide is very stable, and combustion of fossil fuels continues, global warming trends are expected to continue, with large impacts on local and regional climate patterns. Current and future human-induced changes in the global climate will directly affect regional conditions, such as geographic patterns of temperature and precipitation.

2. Biologists and paleontologists have spent the past 150 years documenting the crucial role of climate in determining the geographical distribution of species and ecological communities. Climate variability and change can affect plants and animals in a number of ways, including their distributions, population sizes, and even physical structure, metabolism, and behavior. The timing of important ecological events, including the flowering of plants and the breeding times of animals also shifts in conjunction with climate change.

3. Recent analyses at both the global level and at the scale of the U.S. have estimated that more than half of all wild species have shown a response to twentieth century climate change. For many species, biologists understand the underlying mechanisms driving a species' sensitivity to particular temperature or precipitation thresholds.

   Geographic ranges for some plants and animals have shifted northward and upward in elevation, and in some cases, contracted. Range shifts have been observed in organisms as diverse as birds, mammals, intertidal invertebrates, and plants. Such major shifts in species' locations alter species composition within communities, and thus species interactions. In particular, such shifts in composition are likely to alter important competitive and preda-

tor–prey relationships, which can reduce local or regional biodiversity.

4. Ecosystem processes such as carbon cycling and storage have been altered by climate change. The lengthening of the growing season has altered the annual cycle of carbon dioxide (CO_2) levels in the atmosphere, because plants are a major intermediary for carbon flow through ecosystems, with changes in carbon storage and release patterns. Climate change has the potential to degrade ecosystem functions vital to global health.

   The findings that climate change is affecting biological systems are consistent across different geographic scales and a variety of species. Even against a background of apparently dominating forces such as direct human-driven habitat destruction and alteration, a climate "fingerprint" is discernible in natural systems.

5. Future biological consequences of anthropogenic climate change are more likely to include range contractions than range shifts. During historic glacial cycles, range shifts of hundreds to thousands of miles were common, and species extinction was rare. However, achieving such massive relocation is much less likely across the human-dominated, artificially fragmented landscapes of today. Species that are not adapted to urban and agricultural environments are likely to be confined to smaller total geographic areas as climate causes them to contract from their southern and lower boundaries. Already rare or endangered species, or those living only on high mountaintops, are believed to have the highest risk of extinction.

6. Reducing the adverse effects of climate change on ecosystems can be facilitated through a broad range of strategies, including adaptive management, promotion of transitional habitat in nonpreserved areas, and the alleviation of nonclimate stressors. The protection of transitional habitat that links natural areas might assist in enabling species migration in response to climate change. Meanwhile, promoting dynamic design and management plans for nature reserves may enable managers to facilitate the adjustment of wild species to changing climate conditions (e.g., through active relocation programs). Also, because climate change may be particularly dangerous to natural systems when superimposed on already existing stressors, alleviation of the stress due to these other anthropogenic factors may help reduce their combined effects with climate change.

7. The international community has responded to the threats of climate change by developing agreements

for reductions in greenhouse gas emissions. The most prominent of these is the Kyoto Protocol, which is likely to become international law by the end of 2005. Within the U.S., individual cities and states have fostered incentives for "green" energy programs (e.g., wind and solar), with the result that local electricity is often from a mix of traditional fossil-fuel burning plants and emission-free sources. Policies committed to reducing greenhouse gas emissions cannot stop climate change, but they can have a large influence on minimizing the magnitude and rate of change over the next 100 years. Keeping climate change to a minimum is vital to allow both wild species and human managers to adapt to a dynamically changing environment.

> Please refer to the website www.sinauer.com/groom for Suggested Readings, Web links, additional questions, and supplementary resources.

Questions for Discussion

1. What trends have been documented in twentieth century global and U.S. climate?

2. Why are certain gases called "greenhouse" gases and where do they come from? Which gas is the main one being focused on in policy and why is it increasing?

3. What are the main lines of evidence that wild species, in general, have been affected by twentieth century global climate change? How do we know these changes aren't just natural, or due to the many other things humans are doing?

4. How does twentieth century climate change compare to past major (natural) climate changes? What is different about the modern world that makes climate change more likely to cause harm now (and hence be a conservation problem) than it did in the geological past (e.g., over the several hundred thousands of years of glacial–interglacial cycles)?

5. What recommendations would you give to conservation groups to help them reduce loss of biodiversity under climate change? What would you do if you were the manager of a reserve? What would you consider in designing and buying new reserves to make the species they are protecting more resilient to climate change?

11

Conservation Genetics

The Use and Importance of Genetic Information

Kim T. Scribner, Gary K. Meffe, and Martha J. Groom

Wild species must have available a pool of genetic diversity if they are to survive environmental pressures exceeding the limits of developmental plasticity. If this is not the case, extinction would appear inevitable.

O. H. Frankel, 1983

Contemporary extinction rates are as high as any that have ever occurred on Earth. This loss of species is accompanied by a more subtle, but no less important loss of genetic diversity. When a population or species disappears, all of the genetic information carried by that population or species is lost. When a contiguous population is fragmented through habitat destruction, and many small, isolated populations result, levels of genetic diversity within each will decay over time at rates far greater than would be observed in a large and randomly mating population. In the words of Thomas Foose (1983), "Gene pools are becoming diminished and fragmented into gene puddles."

Why should we be concerned with genetic diversity when more severe threats, including habitat destruction, global warming, overexploitation, and the spread of exotic species threaten wholesale destruction of entire ecological systems? Certainly, the problems caused by massive destruction of tropical rainforests, for example, do not lend themselves to genetic solutions. Species in those systems will go extinct through loss of habitat or other sources of direct mortality, and high levels of genetic diversity will not increase the probability of population persistence in the face of such catastrophic threats. However, there are several reasons to believe that genetic principles and empirical data can be the basis of an important area of conservation biology, and can provide fundamentally important information on which conservation decisions are made.

First, the rate of evolutionary change in a population is proportional to the amount of genetic diversity available. This is the **Fundamental Theorem of Natural Selection** (Fisher 1930). Genetic variation provides the raw material for future adaptation and is the basis for a species' evolutionary flexibility and responsiveness to environmental change. Genetic variation must be preserved for both the short- and long-term survival of populations and species (Allendorf and Leary 1986).

Second, diversity as measured at the level of genes or quantitative genetic traits represents the primary level of biodiversity. Genes regulate all biological processes on the planet. Every biochemical product, every growth pattern, every instinctive behavior, every color morph is encoded in a genetic "library" of unimaginable global extent. Wilson (1985) has calculated that the billion bits of genetic information carried in the DNA of a single house mouse, if translated into equivalent English text, would fill nearly all 15 editions of the *Encyclopedia Britannica* printed since 1768. Loss of such diversity will likely decrease the ability of organisms to cope with environmental challenges, and will also discard biological information potentially useful to humans, such as crop genetic diversity or valuable biochemicals. In essence, we are losing the "blueprints" of life.

Third, there are many conservation challenges that benefit from the guidance and direction that genetic data, collected and interpreted within the constructs of sound population genetic theory, can furnish. Unique representatives of evolutionary lineages can be identified through phylogenetic analyses, which may add a weight to the priority placed on conserving particular species. Evolutionary or anthropogenic forces that affect levels of genetic variation can be monitored, and future trends predicted. Genetic markers and well-established methods of statistical inference can be used to identify forces, both natural and human-induced, that are responsible for the loss of genetic diversity. Further, genes provide an increasingly popular and effective means for understanding important aspects of the behavior, movements, and population dynamics of creatures of conservation concern.

The basic issue that links genetics to conservation is that small and isolated populations, whether in the wild or in captivity, tend to lose genetic variation over time through slow erosive processes, as well as faster deleterious effects of inbreeding (Amos and Balmford 2001).

Loss of adaptive variation may well increase the probability of population extinction or reduce opportunities for future adaptive evolutionary change. Thus, genetic analysis can help us understand the nature and trajectory of genetic threats to population viability.

Increasingly, the power of conservation genetics comes in the form of new tools for understanding populations. Genetic analyses can allow us to monitor dispersal of individuals through a landscape, estimate population sizes, or trace genetic changes over individual lineages. Moreover, the precision of population monitoring and assessment is being increased through noninvasive techniques that allow the analysis of DNA from hair, feathers, scat, or other materials that can be gathered without ever touching an animal. Molecular technologies and population genetic theory can compliment ecological studies of population demography and behavior (see Chapter 12) to help us make credible predictions of population trends, and manage populations in the face of threats to species survival and future evolutionary potential. Thus, conservation genetics has two fundamental aims: to help maintain natural patterns of genetic diversity at many levels, and thus preserve options for future evolution, and to provide tools for population monitoring and assessment that can be used for conservation planning.

In this chapter we provide an overview of molecular and population genetic principles and how this theory and technology can provide an understanding of natural, altered, or captive biological systems. Because many readers may not have a background in genetics, we begin with an introduction to the field. Many key concepts will be defined as we progress through the chapter, but many are integral to our understanding at many stages. We will then provide a broad overview of contemporary efforts and controversies in the field of conservation genetics (Table 11.1; Frankham et al. 2002). Readers can find in-depth presentations of topics referenced in a diverse and extensive peer-reviewed litera-

TABLE 11.1 *Broadly Defined Genetic Issues in Conservation Biology*

1. Deleterious effects of inbreeding on reproduction and survival (inbreeding depression)
2. Loss of genetic diversity and ability to evolve in response to environmental change
3. Fragmentation of populations and reduction in gene flow
4. Random processes (genetic drift) overriding natural selection as the main evolutionary force
5. Accumulation and loss (purging) of deleterious mutations
6. Genetic adaptation to captivity and its adverse effects on reintroduction success
7. Resolving taxonomic uncertainties
8. Defining management units within species
9. Use of genetic analyses in forensics
10. Use of molecular genetic analyses to understand aspects of species biology
11. Deleterious effects of fitness that sometimes occur as a result of out-crossing (outbreeding depression)

Source: After Frankham et al. 2002.

ture and in excellent books covering theory and background (e.g., Frankham et al. 2002 and Avise 2004), methodology (e.g., Ferraris and Palumbi 1994, Schierwater et al. 1994, Smith and Wayne 1996, and Baker 2000) and empirical applications (e.g., Loeschcke et al. 1994 and Avise and Hamrick 1996).

Genetic Variation: What Is It and Why Is It Important?

A species' pool of genetic diversity exists at three fundamental levels: genetic variation within individuals (heterozygosity), genetic differences among individuals within a population, and genetic differences among populations (Figure 11.1). Each level in this hierarchy acts as a reservoir of genetic variation of potential importance to conservation, which can be studied to provide insight for conservation practice. We describe how genetic variation is measured in Box 11.1.

Variation within individuals

Genes that code for heritable variation are located on chromosomes (large molecules mainly composed of deoxyribonucleic acid, or DNA) in the nuclei of cells (see Figure 11.1). Chromosomes consist of long sequences of nucleotides, some of which code for molecules that create the structure and the physiological functions of an organism. A gene represents a specific segment of DNA of a specific chromosome pair that (1) codes for the primary structure of proteins (structural genes), (2) codes for the formation of ribonucleic acid (RNA), or (3) regulates the location and time of gene expression. Chromosomal DNA in the nucleus and DNA found in organelles such as the mitochondria (mitochondrial DNA, or mtDNA) or chloroplasts of plants represent a diverse storehouse of information that can be studied to address various conservation questions.

The majority of genetic material does not appear to code for any product, and thus, while heritable, it is not subject to natural selection. This type of genetic material is called **neutral genetic variation**, and is the type of material primarily assayed using molecular genetic techniques. Depending on the research question, it may be important to analyze genes that represent **adaptive variation** (i.e., genes under selection), while for other questions neutral genetic variation provides the best marker of evolutionary change. In general, we are most concerned about preservation of adaptive variation, but unfortunately, we rarely know what part of the genome codes for adaptive traits. In some cases, we can examine portions of DNA that we know code for critical enzymes or other proteins, and at other times we know that a portion of the DNA has no adaptive variation (for example, tandem repeat sequences; see Box 11.1). Because neutral genetic markers may change at a different rate, or in different directions than adaptive ones, we have to interpret results based on neutral genetic variation with extreme caution (McKay and Latta 2002). It is always important to examine the limits to what we can understand depending on the type of genetic material we are using.

With the exception of identical twins and clones, every individual of a species is genetically unique. However, genes do not determine every aspect of an organism. A

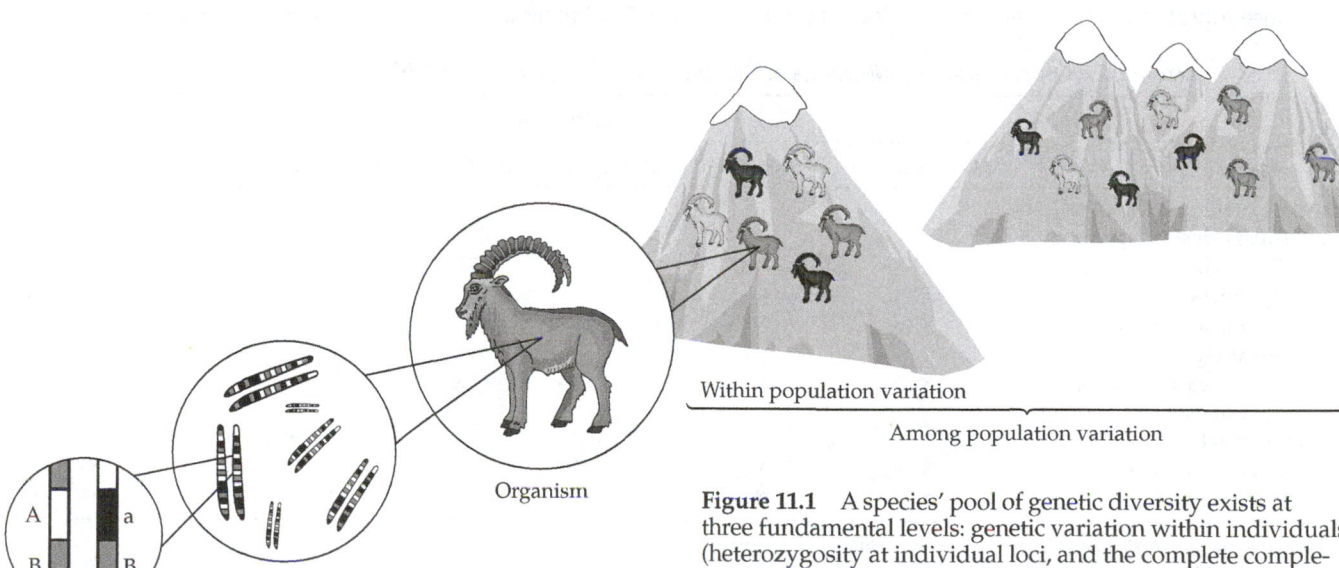

Within population variation

Among population variation

Organism

Chromosomes

A a
B B

Individual loci

Within individual variation

Figure 11.1 A species' pool of genetic diversity exists at three fundamental levels: genetic variation within individuals (heterozygosity at individual loci, and the complete complement of genetic information in the chromosomes), genetic differences among individuals within a population (different alleles at the same loci among individuals), and genetic differences among populations (dependent on the distribution of alleles among populations, and population size).

BOX 11.1 Measures of Genetic Diversity

Kim T. Scribner, Michigan State University

Genetic Variation as Assayed Using Molecular Markers

As recently as the 1950s only a very few species possessed genetic variation whose pattern of inheritance could be followed (e.g., color phases of snow geese [*Anser caerulescens*] or banding patterns of king snakes [*Lampropeltis getulus*]). Subsequently, with the advent of biochemical and molecular genetic technology, an ever-increasing array of methods are currently available whereby individual genotypes can be defined and levels of genetic variability can be quantified empirically.

Choice of a genetic marker depends on the desired application and on the level of variation required (Table A). Levels of variation are a function of mutation rates for the segment of DNA, which is a function of the mechanism of mutation (i.e., single base pair changes versus changes in length due to insertions or deletions). Rates of mutation vary across different regions of the genome. Some regions are highly conserved. When mutations occur in these regions they affect fitness and are selected against. Regions that do not code for functional gene products can be highly variable because mutations in these regions make little or no difference to the fitness of an organism.

Different research questions require different genetic markers (see Table A). If one wishes to address questions on the level of species or genera, it would be prudent to choose a marker that is more highly conserved. Markers based on segments of DNA that accumulate mutations more slowly will be more amenable for comparative studies of distantly related taxa (e.g., species, genera) because gene regions that evolve more rapidly (i.e., acquire mutations at a faster rate) are more predisposed to recurrent mutations at the same location resulting in **homoplasy** (two individuals having the same genetic type, independent of common ancestry). For analyses of genetic parentage, individual identification (forensics), or estimation of pair-wise relatedness, markers with high levels of polymorphism are required to maximize the potential to resolve variation on an individual-to-individual basis. Levels of genetic variation resolved with allozymes are typically moderate to low because not all amino acid changes reflected in different protein sequences are detectable. Higher levels of genetic variation are typically resolved using variable number of tandem repeat (VNTR) minisatellite and microsatellite loci (see Table A). Mutations at minisatellite or microsatellite loci (regions of DNA characterized by tandemly repeated stretches of simple sequence repeats such as $[GA]_n$) occur at a much higher rate than do single base pair substitutions. Variation can also be resolved at the level of individual base pairs.

Choice of a marker should also be consistent with limitations associated with collection and preservation of biological material from plants or animals. Protein allozymes typically require that the animal be sacrificed because many different tissue sources are needed. Obviously, this is a problem with rare or threatened species. In addition, samples for protein allozymes must be immediately frozen, which may not be possible in many field situations.

Markers that are resolved using the **polymerase chain reaction** (**PCR**) are preferred. Using this approach, one can amplify many copies of a specific gene region. In this way very small amounts of material (e.g., fish scales, feathers, mammalian hairs) can be used as sources of DNA. Today, use of PCR is widespread in conservation genetic research.

Choice of genetic marker can also be made based on the mode of inheri-

TABLE A *Research or Conservation Applications of Genetic Markers Currently Available*

Level of analysis or research question	Technique[a]						
	Morphology	Karyology	Allozymes	mtDNA	VNTR	AFLP or RAPD	SNP
Ecosystem/species/ subspecies/metapopulation							
Phylogenetic distinction	XXX	XXX	X	XXX	X	X	X
Hybridization	XXX	XXX	XX	XX	XX	X	XX
Phylogenetic history	XXX	XXX	X	XXX	XX	X	X
Population							
Genetic variability and population structuring	X	X	XX	XXX	XXX	XX	XXX
Individual							
Breeding system	X	X	XX	X	XXXX	X	XX
Forensics	X	X	XX	XX	XXXX	X	XX

Note: The number of X's indicates the relative applicability of the technique.

Source: After Mace et al. 1996.

[a]Morphology could include any aspect of phenotype that can be quantified such as color, shape, size, etc. Karyology refers to chromosomal polymorphisms; allozymes refer to protein electrophoretically distinguishable polymorphisms; mtDNA, maternally inherited mitochondrial DNA; VNTR, variable number of tandem repeat loci including minisatellites and microsatellites; AFLP, amplification fragment length polymorphisms; RAPD's, randomly amplified polymorphic DNA; SNP, single nucleotide polymorphisms.

tance. Many genetic markers are inherited in a Mendelian fashion. Other markers are transmitted from a single sex. For example, mitochondrial DNA (mtDNA) is an extra-chromosomal genome in the cell mitochondria outside of the nucleus, and is inherited from mother to offspring with no paternal contribution, which greatly facilitates studies of female gene flow or **philopatry**. There are other sex-linked markers (e.g., on the Y chromosomes or W chromosomes of heterogametic males and females in mammals and birds, respectively). Markers that have different inheritance patterns can be particularly useful. For example, if females are more philopatric to natal areas than are males, female-mediated gene flow can be traced particularly well via comparison of mtDNA (Scribner et al. 2001; Blundell et al. 2002).

Quantitative Genetic Traits

By definition, evolution proceeds by changes in gene frequency. Allele frequencies and levels of genetic variation as assayed using molecular markers respond in a predictable manner to population perturbations and to evolutionary forces of mutation, migration, selection, and genetic drift. However, most traits show continuous variation (e.g., height and weight) that typically exhibits a continuous distribution across a range of values. Our ability to predict changes becomes more difficult in the case of **polygenic** (quantitative) traits. However, natural selection acts on phenotypes, which result from complex interactions of heritable genetic and environmental sources of variation. To make predictions regarding changes in levels of variation for quantitative traits, we need to look at phenotypic values and to summarize the variation in terms of means, variances, and covariances in trait values across individuals.

Quantitative genetics has often been viewed as useful primarily for achieving specific phenotypic goals from the selective breeding of domestic animals and plants. However, quantitative genetic methods have been used extensively to understand genetic processes underlying adaptive evolution (Lynch and Walsh 1998). Analyses of phenotypic resemblance of individuals of known relationship permit inferences to be made about the inheritance and evolution of characters that are

controlled by variation in genes and effects across genes. Specific breeding designs and statistical techniques can be used to estimate genetic parameters that determine a population's response to selection. Experimental approaches are particularly useful for investigating patterns of adaptive evolution and comparing results in populations that exist in different environmental or geographic settings.

Variation in phenotypic values (P) for quantitative traits such as body size, color, or fecundity can be apportioned into genotypic effects (G) that define the proportion of variation attributed to sets of genes that affect a trait, and to environmental deviation (E) representing all nongenetic factors that influence the trait such as nutritional effects. The relationship can be written as ($P = G + E$). The genetic component (V_G) of phenotypic variance (V_P) can be partitioned further as $V_G = V_A + V_D + V_I$, where V_A is the additive variance or variance in breeding values, V_D is the dominance variance, and V_I is the variance attributed to the interaction among factors if loci don't contribute additively. The degree of resemblance between relatives is determined by **heritability** in the narrow sense as $h^2 = V_A/V_P$.

Although both molecular (single locus) and polygenic quantitative genetic traits can be used to measure genetic variation, several features distinguish them (Moran 2002). First, traits that contribute to fitness typically represent the cumulative effects of the actions of many loci. Second, phenotypic expression of polygenic traits is dependent on nongenetic (i.e., environmental) variation. For example, size and fecundity in many organisms is tied to nutritional regimes. Third, even in the absence of environmental variation, gene expression for polygenic traits may have many components including additive (independent effects of different loci), dominance (interaction of alleles within loci) and **epistatic** (interaction among loci) components of variance. In addition, genes often affect multiple traits (**pleiotropy**) and these effects may be antagonistic or in opposing directions.

A major difference between molecular and quantitative genetic variation is that many molecular polymorphisms routinely assayed for natural, or captive or domestic populations, are selectively neutral or nearly so (i.e., they

don't influence fitness). As such, the major forces affecting levels and partitioning of genetic variation within and among populations are genetic drift and migration. On the other hand, most polygenic traits are subject to selection. As we will see, the rate of evolution will vary depending on the evolutionary factor involved. Natural selection can very rapidly and effectively alter the phenotypic and genotypic composition of a population. In contrast, rates of change for neutral genetic markers are occurring in a time-dependent process that is influenced predictably on the basis of population size through genetic drift.

Molecular and quantitative genetic traits differ in the rate of generation of new variation via mutation. Mutation rates may be 10^2 to 10^4 higher for polygenic traits than for single-locus traits. This higher rate of generation of new variation potentially contributes to differences in the propensity for loss of variation over time.

Population levels of genetic variation as typically assayed for molecular and quantitative traits may or may not change in a similar fashion to natural or anthropogenic changes. Heterozygosity (H) is a measure of within-population genetic variance for molecular traits and h^2, or the proportion of total phenotypic variance attributed to additive genetic variance, is the quantitative genetic measure of population variance. For characters effectively neutral with respect to selection, h^2 and H are predicted to respond in a similar manner to population **bottlenecks**, or low population size. However, similarity in response may break down when genes underlying quantitative traits exhibit dominance or epistasis, or the genes are under selection (Hard 1995).

Neutral genetic markers and phenotypic traits are largely under the influence of different evolutionary forces. Molecular measures of population subdivision have been shown to provide concordant, although consistently lower, estimates of the degree of population subdivision than have estimates based on heritable quantitative traits (Lynch et al. 1999), suggesting that selection and adaptation to local environmental regimes is important. For quantitative genetic traits, the same genotype will likely show variation in phenotype in different environments (phenotypic plasticity).

particular physical or behavioral character (the phenotype, or expressed aspect of a trait, such as eye color) may be due entirely to genotype (genetic constitution), to environment (such as nutritional regime), or more likely, to a combination of both (such as skin color as an interaction between racial background [genetics] and recent exposure to sunlight [environment]). Thus, both genes and environment contribute and interact together to create a rich diversity of forms and functions among individuals of all species. But only the genetic components of phenotypes are heritable. Organisms cannot pass on aspects of phenotypes that result from environmental influences. An analytic method, quantitative genetics, is helpful for discerning which aspects of phenotypes are determined by genes. Quantitative genetics is described in more detail in Box 11.1.

The ultimate source of genetic variation is mutation. Mutations occur when DNA is changed in such a way as to alter the genetic message it carries. A mutation can range in extent from a change in one base in the nucleotide sequence, through the deletion or duplication of a group of nucleotides, to large-scale changes, such as the deletion, duplication, or translocation of large portions of chromosomes or the loss or duplication of entire chromosomes. Mutations are either neutral (meaning they have no effect), or they alter the expression of genes, and thus cause changes in an organism's form or function. Mutations are thus the ultimate basis of genetic variation among individuals.

Variation within individuals also is produced each generation by recombination (mixing of chromosomal material) during production of gametes in sexual reproduction. At any particular gene locus there are two alleles, or copies of the gene, inherited from the two parents. At the *population* level, a given locus may be said to be either monomorphic (both copies of the allele are always the same; there is no variation at that locus) or polymorphic (there are two or more types of alleles possible at that locus). Within any individual, a polymorphic locus may be either homozygous (two copies of the same allele) or heterozygous (two different alleles). The overall level of heterozygosity—the proportion of gene loci in an individual that contains alternative forms of alleles—is one measure of genetic variation within individuals.

What is the value of focusing on genetic variation within individuals? First, heritable genetic variation is the basis for evolutionary change and is essential if natural selection is to operate. The individual is the level upon which natural selection acts. Second, the individual is the level at which genetic problems such as **inbreeding** occur. Inbreeding results from matings between related individuals, where progeny have increased likelihood of inheriting alleles that are identical by descent from a common ancestor. Third, knowledge of individual genotypes may be important in some captive breeding programs, when mating schemes that attempt to maximize genetic diversity of offspring and avoid inbreeding are developed. Finally, genetic variation is always measured in individuals, and can only be estimated for collections of individuals, such as populations, through statistical summaries of measures from individuals.

Rarely do we direct our conservation efforts solely toward individuals, however. More commonly, populations or species are the units of concern. Consequently, we must begin by understanding genetic variation among individuals, and then move on to understand genetic variation of groups both within and among populations.

Variation among individuals

Variation among individuals—or population-level variation—consists of the types of alleles present and their relative frequencies across all members of a population considered together (the **gene pool**) (see Figure 11.1). Consider, for example, the gene locus that encodes for an enzyme called phosphoglucomutase (PGM) in the club moss *Lycopodium lucidulum*, a primitive plant (Table 11.2). This locus has three possible alleles in this species, designated *a*, *b*, and *c*. In four populations in New York and Connecticut there are major differences in the presence and frequencies of these alleles. One Connecticut population has alleles *b* and *c*, another has only *b*; one New York population has only *a* and *b*, and the other has only *b*. Each population may thus be described by the types (*a*, *b*, *c*) and frequencies of occurrence of these alleles. When more gene loci are sampled, a more detailed portrait of genetic differences among individuals sampled in different populations may be assembled (see Table 11.2).

When five loci are considered, it is clear that some alleles are more rare among populations (e.g., PGM *c* and G6PD-1 *b* are each found in only a single population), and that some populations have greater allelic diversity than others (e.g., Woodbridge, CT versus Litchfield, CT). Within-population diversity is often summarized as the mean individual heterozygosity (H_p)—that is, the mean of each individual's multi-locus heterozygosity. As you can see, Woodbridge has the highest H_p (with only one monomorphic locus), while Litchfield has the lowest H_p (with only one polymorphic locus). Part of the challenge facing conservation geneticists is to understand what forces contribute to low diversity (as in the Litchfield, CT population), and when it may be important to preserve or increase genetic diversity within a population over time.

Gene frequencies within a population generally change over time due to mutation, natural selection, random processes such as **genetic drift** (discussed later), unequal family sizes or number of matings, changes (especially reductions) in population size, or immigration from or emigration to other populations (called **gene**

TABLE 11.2 *Gene Frequencies at Five Polymorphic Loci in the Club Moss* Lycopodium lucidulum

Locus	Allele	Woodridge, CT	Litchfield, CT	Binghamton, NY	New Lebanon, NY
PGM					
	a	0.00	0.00	0.50	0.00
	b	0.86	1.00	0.50	1.00
	c	0.14	0.00	0.00	0.00
PGI-2					
	a	0.68	1.00	1.00	0.75
	b	0.32	0.00	0.00	0.25
G6PD-1					
	a	0.93	1.00	0.82	0.91
	b	0.07	0.00	0.18	0.09
G6PD-2					
	a	1.00	1.00	0.50	1.00
	b	0.00	0.00	0.50	0.00
LGGP-1					
	a	0.50	0.50	1.00	1.00
	b	0.50	0.50	0.00	0.00

Source: Levin and Crepet 1973.

flow). It is these changes in gene frequencies, and especially loss of alleles, that are often of concern to conservation biologists.

Many studies have documented correlations between ecological characteristics and levels of genetic variation. Nevo et al. (1984), and Hamrick et al. (1979, 1981) extensively surveyed the animal and plant literature, respectively, and found different levels of genetic variation within taxa depending on certain details of their biology. Numerically abundant and widespread species typically have higher levels of genetic variability than do local endemic species or those with limited distributions. Species that are highly specialized, highly territorial, or of large body size typically have lower genetic diversity (Table 11.3). One pervasive feature among plant and animal species is the correlation between within-population levels of genetic diversity (H_P) and population size (or traits correlated with population size). Population genetic theory predicts that loss of genetic variation within populations should occur at a slower rate with increasing numbers of individuals transmitting genes from one generation to another.

Variation among populations

Species rarely exist as single, randomly interbreeding, or **panmictic** populations spread across large areas. Instead, genetic differences typically exist among populations that are found in different places across a landscape. These geographic genetic differences are a critical component of overall genetic diversity. To understand among-population variation, consider again the four *Lycopodium* populations from Table 11.2. Levels of within-population diversity (H_P) were quite variable, yet the total variation for these four populations combined is higher than the diversity within any one population. Recalling that some alleles are only found in a single population, it is easy to understand why the combination of all the populations leads to higher total genetic diversity (H_T). Thus, genetic diversity among a set of populations consists of within-population diversity and among-population divergence (genetic differences among populations sampled from different geographic locations). A simple genetic model of this diversity is:

$$H_T = H_P + D_{PT}$$

where H_T = total genetic variation (heterozygosity), H_P = average diversity within populations, and D_{PT} = average divergence among populations surveyed (Nei 1973, 1975). Divergence may arise among populations both from random processes (founder effects, genetic drift, episodic drastic reductions in population number [**demographic bottlenecks**]), and from local selection. We will discuss these mechanisms later.

The critical point is that a species' total genetic variation may be quantified and apportioned into component parts: within- versus among-population diversity. Accordingly, researchers can determine how variation is spatially distributed and define areas of conservation interest. For example, Stangel et al. (1992) found that the

TABLE 11.3   *Ecological and Life History Correlates with Heterozygosity*

Occupancy of different life zones

  Cosmopolitan and temperate + tropical > tropical > temperate > arctic

Degree of endemism

  Species with broad geographic distribution > endemic species

General habitat requirements

  Overground > arboreal or aquatic > underground

Degree of specialization

  Generalists > specialists

Climatic conditions

  Species inhabiting ecological extremes > regions of broader climatic variation

Degree of territoriality

  Nonterritorial > territorial

Body size

  Small > medium > large > very large

Note: Organisms sharing a given life history trait to the left of the > symbol tend to have greater heterozygosity than organisms with a different life history to the right of the > symbol.
Source: After Nevo et al. 1984.

endangered Red-cockaded Woodpecker (*Picoides borealis*) in the southeastern United States had an overall mean heterozygosity level of 0.078 (or 7.8%), which is within the typical range of values for most bird species. Of the total genetic variation measured, 14% consisted of among-population differentiation (D_{PT}), and 86% was mean genetic diversity within populations (H_p). The among-population component for these woodpeckers was higher than that of other bird species examined, whose vagility typically results in high genetic exchange and thus little local genetic differentiation. These woodpeckers are more site-specific than many birds, and consequently local populations tend to diverge genetically. Conservation programs for this species should therefore protect both components of genetic diversity to retain the maximum amount of variation and maintain a natural population genetic structure. An overview of levels of total heterozygosty (H_T) and the proportion of that heterozygosity attributable to among-population divergence (D_{PT}) in natural populations is offered in Table 11.4. Note that the more vagile groups (birds, insects) have the lowest levels of among-population differentiation.

Often, ecological characteristics of a species influence how variation is organized within versus among populations. For example, consider two large ungulate species, the white-tailed deer (*Odocoileus virginianus*) and the Alpine ibex (*Capra ibex*). Ibex live exclusively on the tops of high alpine mountains and do not frequent habitats at lower elevation. Population sizes typically are small and

there is limited gene flow among populations (i.e., they are largely isolated on different mountains). One would expect that due to comparatively small population sizes and lack of immigration, a large proportion of the total genetic variation would be partitioned among ibex populations. Recent analyses (Maudet et al. 2002) revealed

TABLE 11.4   *Mean Total Heterozygosity (H_T) and Proportion Due to Among-Population Differentiation (D_{PT}) in Several Major Taxonomic Groups*

Taxon	H_T	Number of species	D_{PT}	Number of species
Vertebrates				
Fishes	5.1%	195	0.135	79
Amphibians	10.9%	116	0.315	33
Reptiles	7.8%	85	0.258	22
Mammals	6.7%	172	0.242	57
Birds	6.8%	80	0.076	16
Invertebrates				
Insects	13.7%	170	0.097	46
Crustaceans	5.2%	80	0.169	19
Molluscs	14.5%	105	0.263	44
Others	16.0%	15	0.060	5

Source: Ward et al. 1992.

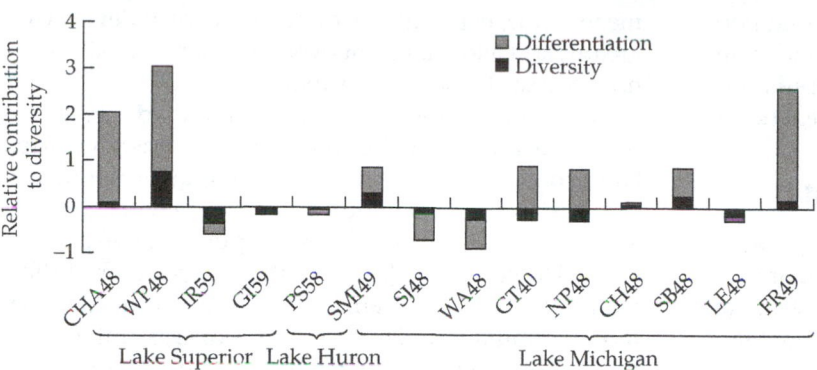

Figure 11.2 Relative contribution of past populations of lake trout (*Salvelinus namaycush*) to gene diversity in the population as a whole, H_T. Values above the horizontal line indicate populations that contribute more than average to diversity; values below the line indicate populations that contribute less than average to genetic diversity. Contributions are divided in two components; one indicates contribution to total diversity of a population due to its differentiation from other populations (in gray), and the other contributions due to their own level of diversity (in black). (Modified from Guinand et al. 2003.)

that ibex populations are indeed highly differentiated genetically and have low levels of genetic variation (heterozygosity) within each population. In contrast, white-tailed deer typically have large population sizes, are highly mobile, and are widely and often relatively uniformly distributed over large geographic areas. Compared to ibex, white-tailed deer have a greater proportion of total genetic variance contained within rather than between populations (e.g., Scribner et al. 1997).

Often, demographic declines in abundance and extirpation of populations can have disproportionately large effects on total levels of genetic diversity. For example, many lake trout (*Salvelinus namaycush*) populations in the Great Lakes were extirpated due to overexploitation and predation by nonnative sea lamprey (*Petromyzon marinus*). Estimates of genetic variation were lower in contemporary populations than in their historical counterparts (Guinand et al. 2003). Analyses revealed heterogeneous contributions of populations to total gene diversity (Figure 11.2).

One frequent scenario occurs when managers wish to know whether current mortality levels that are due to humans can be sustained while maintaining the local population at healthy levels. As just demonstrated, genetic data can help to determine whether there are genetic differences among populations. Genetic data are also used as one indicator of reproductive isolation (i.e., if populations differ significantly in the frequency of alleles, then the populations are likely, although not certainly, to be reproductively isolated). If the genetic composition of populations differ substantially, it may be wise to manage populations separately. If, however, no significant differences are found, interpreting results can be problematic. This is because statistical analyses investigating population structure usually use the null hypothesis that populations are panmictic (that is, non-isolated genetically), so in the instance when statistical power is low (when sample sizes are low), it would be possible to find no difference when in fact populations

have little gene flow. This problem of low power is common to many analyses in conservation.

Another issue concerns the type of genetic marker used. Molecular markers are usually indicative of neutral genetic variation. These neutral molecular markers are used as a proxy for adaptive variation for the simple reason that we often have not identified the genes that are important in creating structure, function, or performance. Many have found this to be a serious limitation to understanding the significance of population-level genetic variation (Amos and Balmford 2001; McKay and Latta 2002). Because neutral variation may be either more or less divergent among populations than adaptive variation (Reed and Frankham 2001), often it is difficult to evaluate whether a population is losing critical adaptive variation when analyses are based on neutral markers.

Variation at the level of metapopulations

Populations are seldom independent, but rather are interconnected to various degrees by movements and gene flow. Each population may thus be considered as belonging to a larger metapopulation (Hanski and Gilpin 1997) (see Chapters 7 and 12 for discussion of ecological aspects of metapopulation structure). In each generation, there is some probability that an individual population will go extinct. Recolonization can occur via migration from other populations. Because populations can become extinct and be recolonized, the time periods over which metapopulations may persist can greatly exceed those for individual populations.

Metapopulation structure can have important effects on levels of genetic diversity within and among populations of a metapopulation. In general, a metapopulation usually loses genetic variation (alleles and heterozygosity) at an accelerated rate relative to a single large population (Hedrick and Gilpin 1997; Whitlock and Barton 1997). Alleles are lost via extinction, initially only those possessed by founding individuals are represented in

recolonized populations, and thus diversity within populations is often low. Depending on how often individuals move among populations, the separate populations may be relatively distinct or relatively similar genetically.

Why Is Genetic Diversity Important?

The amount of adaptive variation in a population should be related to the health of a population, or to its ability to withstand stresses and challenges. Frankham (1996) summarized information from a large number of studies in a meta-analysis and identified a number of general relationships that had long been predicted to be correlated with levels of genetic variability (Table 11.5). Frankham found that genetic variation within species was positively correlated with population size, species distribution, and habitable area, and negatively correlated across species with body size. Thus, species with more diversity have larger populations and occupy more area. While this might indicate that greater diversity is beneficial for populations, it may only indicate that smaller, more geographically restricted populations have less variation, which may or may not have any impact on the viability of those smaller populations.

While it seems logical that adaptive genetic diversity should be important, evidence from nature has yielded contradictory evidence about the importance of heterozygosity. There is no absolute value of heterozygosity (H_p) that indicates a population's health (Amos and Harwood 1998). Although high heterozygosity may be correlated with individual **fitness**, not all species or populations have high levels of within-population heterozygosity (H_p) and there is no standard, "acceptable" level. Typical heterozygosity levels vary greatly among taxonomic groups (see Table 11.4). Measured values based on protein allozymes from natural populations ranged from 0 to over 0.3 (0 to >30%). Mean measured values are about 0.14 for plants, 0.10 for invertebrates, and 0.05 for vertebrates, with a great deal of variance around each

mean. Thus, expectations of heterozygosity levels vary greatly, and a low level, or even no heterozygosity, does not necessarily indicate an anomaly.

For example, Sherwin et al. (1991) studied a small, endangered, and isolated population of the eastern barred bandicoot (*Perameles gunnii*), a marsupial, in southeastern Australia. No genetic variation was detected at 27 loci, perhaps as a result of small population size and isolation. However, two large "control" populations in Tasmania also had no variation at those loci, indicating that recent population size alone cannot account for the paucity of genetic diversity in this species. Too often, control populations for comparison are lacking. Many species in nature have little detectable genetic variation (Nevo 1984; Merola 1994), yet they seem to do well. Some populations of these species clearly were numerically less abundant in historical times, whereas others were not. Part of the confusion may stem from the difficulty of measuring adaptive variation, rather than neutral variation. It is possible that if it were more feasible to measure changes in the variation that most influences individual fitness, low heterozygosity for adaptive traits might be correlated with lower success, but this analysis is not possible to perform at present.

While the absolute value of heterozygosity may not provide a clear indicator of risk, *loss* of heterozygosity in a population may indicate problems. For example, many species seem to experience lower reproductive success after inbreeding, which lowers their heterozygosity, although this is by no means universal (see review by Hedrick and Kalinowski 2000). Populations that have recently lost heterozygosity through inbreeding may be less able to handle novel challenges in their environment. For example, Frankham et al. (1999) found that more heterozygous *Drosophila* were better able to adapt to the unusual situation of increases in salinity in their environment.

Among-population divergence may play a critical role in local fitness and population survival. First, popu-

TABLE 11.5 *General Correlates of Genetic Variation among Population*

1. Genetic variation within species will be positively correlated with population size.
2. Genetic variation will be positively correlated with habitat area.
3. Genetic variation will be greater in species with wider ranges.
4. Genetic variation in animals will be negatively correlated with body size.
5. Genetic variation will be negatively correlated with rate of chromosomal evolution.
6. Genetic variation will be positively correlated with population size across species.
7. Genetic variation will be lower in vertebrates than in invertebrates or plants.
8. Genetic variation should be lower in island populations than mainland populations.
9. Genetic variation will be lower in endangered species than nonendangered species.

Source: After Frankham 1996.

lations may be locally adapted due to long-term interaction of their genetic systems with the biotic and abiotic environments of the region, and selection for that region. Local adaptation is common and readily seen in many plant and animal populations, and can lead to speciation. Second, **coadapted gene complexes** may arise in local populations (discussed in Essay 11.1 by Alan Templeton). These are gene combinations with a long history together in a population, and are thought to work particularly well in combination. For both of these reasons, among-population divergence should be considered for its potential importance as a component of overall genetic diversity, contributing to local uniqueness and evolutionary adaptability.

Forces that Affect Genetic Variation within Populations

Population geneticists have long studied the ways in which genetic variation is augmented and lost through the forces of evolution. Human impacts can alter evolutionary forces by increasing the influence of one or more forces relative to others. To predict the effects of human intervention on genetic diversity, and to identify means to ameliorate negative consequences in conservation management, we first need to understand how evolutionary forces affect genetic diversity. Once we have an understanding of evolutionary effects on genetic variation, we can examine how changes in these forces could accelerate loss of genetic diversity. In this section, we will explain how genetic diversity is changed through mutation, genetic drift, gene flow, mate selection, and natural selection. We will begin by defining in more detail a concept that is integral to understanding all of these natural processes— genetically effective population size, N_e.

The genetically effective population size (N_e)

When making projections of likely changes in allele frequency, or of generational changes in levels of genetic variation in populations, it is not the absolute number of individuals, or census breeding population size (N), that is relevant to the distribution and abundance of genetic variation, but the **genetically effective population size** (N_e). Not every individual has an equal probability of contributing genes to the next generation. An idealized population is typically conceived as a large, randomly mating population with non-overlapping generations, a 1:1 breeding sex ratio, with roughly equal family sizes, and under no selection. Real populations never meet these criteria. Accordingly, population genetic models allow corrections to be made in the sizes of real popula-

ESSAY 11.1

Coadaptation, Local Adaptation, and Outbreeding Depression

Alan R. Templeton, *Washington University*

■ When individuals from animal or plant species from different geographical areas are brought together (as in captive breeding or translocation programs), they often hybridize with one another. An outbreeding depression occurs when these hybrids, or the offspring of the hybrids, have reduced fertility and survival. There are two main causes of outbreeding depression: coadaptation and local adaptation. Coadaptation results when a local population evolves a gene pool that is internally balanced with respect to reproductive fitness. For example, the process of gamete formation normally requires a matched set of chromosomes (with the exception of the sex chromosomes). However, if the two sets of chromosomes differ in number or structure, fertility problems are possible. This type of coadaptation is illustrated by a reduction in fertility during captive breeding of the owl monkey, *Aotus trivirgatus* (de Boer 1982). It turned out that animals from different geographical populations had been mixed and that these local populations had different chromosome types. By pairing animals with similar chromosome types, successful reproduction was restored.

The second major cause of outbreeding depression is local adaptation. The geographical range of many species encompasses a variety of environmental conditions. Under such conditions, local populations often adapt to their regional environment, particularly if dispersal among populations is limited. Hybridization between individuals from different local populations can lead to individuals that are not adapted to any or to the wrong local environment. For example, ibex (*Capra ibex*) became extinct locally in the Tatra Mountains of Czechoslovakia through overhunting. Ibex were then successfully transplanted from nearby Austria, which had a similar environment. Some years later, a decision was made to augment the Tatra ibex population by importing animals from Turkey and the Sinai, areas with a much warmer and drier climate. The introduced animals readily interbred with the Tatra herd, but the resulting hybrids rutted in the early fall instead of the winter (as the native Tatra and Austrian ibex did), and the resulting kids of the hybrids were born in February, the coldest month of the year. As a consequence, the entire population went extinct (Greig 1979).

The risk of local adaptation is inherent to all translocation programs, which by definition involve the movement of organisms from one location to another. If the translocation program also involves moving organisms from more than one location to a common place,

the risk of coadaptation is also present. Both local adaptation and coadaptation were factors in the reintroduction of collared lizards (*Crotaphytus collaris*) in the Missouri Ozarks (Templeton, 1996).

Most collared lizards live in the deserts of the American Southwest. The Ozark populations represent a geographic and ecological extreme for the species. Although there are no true deserts in the Ozarks, the collared lizards can live on glades—open, rocky habitats characterized by a hot, dry microclimate. One of the few trees that can invade these hot, dry glades is the juniper *Juniperus virginiana*. Prior to European settlement, the Ozarks were subjected to frequent fires that prevented junipers from becoming established on the glades and the forest was maintained as open woodland. The suppression of fires over the last 50 to 150 years in various parts of the Ozarks has transformed the open woodland into a forest with a thick, woody understory and has allowed many former glades to become shaded over by junipers, leading to the extirpation of many glade inhabitants (Ladd 1991).

A glade restoration program was initiated in the early 1980s to counteract the destruction of these unique communities. Glade habitat was reopened by cutting trees and burning the glades (although not initially the surrounding forests). The glade flora returned remarkably well after a few years of burning. Collared lizards, however, will not disperse even through a few hundred meters of the unburned forest. Consequently, once extirpated from a glade, collared lizards were incapable of recolonizing a restored glade. Therefore, collared lizards had to be actively reintroduced as part of a general glade community restoration program. An examination of past reintroduction programs indicates that the probability of a self-sustaining population increases with increasing geographical proximity of the site of origin of the propagules to the site of release (Greig 1979; Griffith et al. 1989). Such a strategy minimizes the risk of an incompatible local adaptation. Local adaptation was a real danger for the Ozark collared lizards, which inhabit an ecological extreme for the species as a whole and which display many unique adaptive traits not found in their southwestern relatives (Templeton 1996). Therefore, all lizards obtained for the reintroduction program were captured from other Ozark glades whose habitats had not yet

degenerated to the point of causing local extinction.

Although this strategy minimized the risk of local adaptation, it accentuated the risk of coadaptation. The remaining glade populations were small in size, so only a handful of lizards could be collected from any single glade without endangering the natural glade population. This meant that releases into restored glades would have to involve lizards caught from several different natural glade populations, thereby creating the potential for outbreeding depression via coadaptation. Molecular genetic surveys revealed that the different natural glade populations are genetically distinct from one another with little or no genetic contact, making the possibility of coadaptation likely. However, these same studies revealed that there was no geographical pattern to genetic distances between glade populations in the northeastern Ozarks, implying that their fragmentation was recent and that glade populations had been extensively interconnected genetically prior to fire suppression (Hutchison and Templeton 1999). This in turn implies that there had been little time for coadapted gene complexes to have evolved. With these contradictory indicators of risk for coadaptation, it was decided to go ahead with a mixed release program.

The decision to accept the risk of coadaptation was based on an evolutionary considerations. Natural populations are the products of evolution, and as long as they retain genetic variation, they will continue to evolve. Reintroduced populations are no exception. Hence, evolutionary change, and the factors that promote it or retard it, must be incorporated into strategic decision-making. In general, when compromises must be made or contradictory evidence exists, genetic diversity and the potential for evolutionary change should have priority over other considerations. We are not wise enough to anticipate all the challenges a reintroduced population will face. If natural or released populations have the potential for evolutionary change, they can evolve solutions to problems such as outbreeding depression or environmental change; without it, their adaptive flexibility is lost. We should never forget that Earth's biodiversity is the product of past evolution and is not, nor ever has been, static. Hence, conservation programs should try to conserve processes (such as evolution) that affect living organisms and

ecosystems rather than conserve the current status quo of the living world.

Mixing lizards from different glades would maximize their genetic variation, thereby aiding them in overcoming any outbreeding depression through evolutionary change, just as captive populations have been able to overcome inbreeding depressions through evolutionary change (Templeton and Read 1984). To see if this was the case, the populations have been monitored after their initial release. In terms of the primary goal (to establish self-sustaining, genetically variable populations), the program has been successful: 67% of 9 introduced populations have persisted since release (from 7 to 18 years ago) versus a 19% success rate of translocation programs involving reptiles and amphibians in general (Dodd and Seigel 1991).

Three of the introduced populations were in an area that has been managed with frequent prescribed forest fires since 1994. In the 10 years between the first release in 1984 and the initiation of prescribed burning of both the forest and the glades, not one lizard in these three translocated populations had dispersed to another glade, some of which were separated by only 50 meters of forest (Templeton et al. 2001). In the first eight years after forest fire management, the lizards colonized over 40 new glades, dispersed up to 5 km through burned forest, established extensive gene flow between glade populations, and increased their total population size about twenty-fold. This dramatic response to fire indicates that the observed fragmentation in much of the Ozarks is indeed a consequence of recent fire suppression and that there had been no significant coadaptation. These important conclusions could not have been made if the lizard populations had not been carefully monitored after their introduction.

As shown by the initial decision in 1984 to perform mixed releases, management decisions often have to be made with incomplete or contradictory scientific knowledge. Conservation programs should be implemented in a manner that both aids in the management of biodiversity as well as in finding answers to questions about biodiversity that can be generalized to other conservation needs. In this regard, reintroduction programs should be regarded and designed as both management programs and scientific experiments simultaneously. ∎

tions to more accurately reflect the factors affecting levels of genetic diversity. Such a correction is the genetically effective population size (N_e), which is typically smaller, and often much smaller, than the census size (N). Ratios of effective population size to adult census population size are affected by fluctuations in population size, variance in family size, taxonomic group, and unequal sex ratios (Frankham 1995).

Effective population size is defined as the size of an idealized population that would have the same amount of inbreeding, loss of heterozygosity, or of random gene frequency drift, as the population under consideration (Kimura and Crow 1963). Mathematically, if demographic information is available, N_e can be estimated empirically (Box 11.2). Because N_e is nearly always significantly smaller than the census population size, conservationists have worked to estimate N_e and use this value in planning. Two examples of how N_e is used in conservation planning are discussed in Case Study 11.1 by Fred Allendorf and colleagues.

Mutation

Mutations are the ultimate source of new genetic variation. They occur during errors in replication of a nucleotide sequence or other alteration of the genome. The rate of mutation to new alleles (μ) is generally quite low for a specific locus (typically 10^{-4} to 10^{-6} per locus per generation depending on the segment of DNA), so the net effect of mutation on a single locus is usually weighed over evolutionary time scales rather than on a generation-by-generation basis.

Most mutations are selectively neutral. That is, they have no net effect on the fitness of an individual. However, most mutations that affect fitness are mildly or strongly deleterious. The effects are most often seen when expressed in a homozygous state, depending on the degree of dominance (i.e., the degree to which one or both alleles are expressed in heterozygous genotypes). Even if mutations occur at very low rates, given the large number of genes in most species, it is believed that mildly deleterious mutations arise at a rate perhaps as frequently as one per gamete per generation. In higher organisms, approximately 100 deleterious alleles may be present in each individual.

Mutations are important in a conservation sense because the probability of extinction can be affected by the accumulation of deleterious mutations. Even with very low mutation rates, if the population size is small and remains so for long periods, mutations can accumulate and eventually can become high enough in frequency to be expressed. The probability of **fixation** of a deleterious mutation (where all individuals in a population are homozygous for that mutation) decreases in large populations (i.e., populations with effective sizes greater than a few hundred individuals). Large populations with high reproductive rates can be expected to be more resistant to accumulations of deleterious mutations than populations of small size.

In many species, only a fraction of offspring survive from birth to reproduce. With low probability of individual survival, a mutant gene newly arisen in one parent will, with high probability, be eliminated. However, in small populations, mutations have a higher probability of being passed on to offspring because inbreeding is more common. Mutations may accumulate to fixation if the strength of selection (S) in eliminating the genotype possessing this allele is lower than the probability of persistence by chance alone (i.e., $S < 1/2N_e$). A mutation that is recessive can become fixed through inbreeding because of the greater probability that individuals will share these mutations, and thus pass them on to their offspring.

As deleterious mutations accumulate, there is a gradual decline in the mean viability of individuals, which can lead to more rapid declines in population size. This precipitates a downward spiral, where mutations continue to accumulate and the population size becomes smaller and smaller, which progressively increases the probability of fixation of future mutations. This phenomenon has been termed **mutational meltdown** (Lynch 1995a).

There are numerous implications of mutational accumulation for conservation and management of natural or captive populations. Current management policies that provide formal protection to species only after they have dwindled to $N = 100$ to 1000 are inadequate if mutational meltdown is possible. The results of Lynch (1995a,b) and others indicate that the accumulation of mutations can pose a substantial threat to the survival of small populations. For "closed" populations that are maintained at levels below 100 individuals, a substantial load of deleterious mutations can be expected to develop within a few dozen generations. Because much of the reduction in fitness will be due to fixation of mutations, it may be essentially irreversible if the population remains closed to immigration from outside sources. Species with low birth rates such as many birds and mammals may be particularly susceptible.

However, a test of this hypothesis using *Drosophila* did not show as severe an effect, although it is unclear why (Gilligan et al. 1997). Some have suggested that several counteracting forces may protect populations from mutational meltdown, such as active preference for mates with fewer mutations, sperm competition, or higher fecundity of individuals with fewer mutations (Jennions et al. 2001; Tregenza and Wedell 2000).

Genetic drift

Genetic drift is the random fluctuation of gene frequencies over time due to chance alone. Because each adult

BOX 11.2 Estimation of Effective Population Size

Gary K. Meffe, University of Florida

Effective population size (N_e) is affected by a number of parameters, several of which we examine here. N_e is affected by sex ratio of the breeders in the following way:

$$N_e = (4N_m \times N_f) / (N_m + N_f)$$

where N_m and N_f are the numbers of successfully breeding males and females, respectively. For example, a census population of 500 would have an N_e of 500 (at least with respect to sex ratio) if they all bred and there was a 1:1 sex ratio: $N_e = (4 \times 250 \times 250)/(250 + 250) = 500$. However, if 450 females bred with 50 males, then $N_e = (4 \times 50 \times 450)/(50 + 450) = 180$; the genetically effective population size in this case is only 36% of the census size due to few males participating in breeding. This makes intuitive sense, because only those breeding are assured of passing their genes on to the next generation.

Similarly, if not every member of a population is breeding (for example, if a proportion are immature), then N_e should be lower than the censused population. For example, if there were only 100 breeding females and 50 breeding males, and the remaining 350 individuals were immature, N_e would be about 133.

Large variation in progeny number among breeding adult males and among females can also reduce effective population size, as the genes of those parents with very large families are more represented in the next generation, and those families that have proportionally few or no progeny are poorly represented. Suppose the mean number of surviving progeny is k and the variance in number of progeny surviving from parent to parent is V_k. Then it can be shown that:

$$N_e = k(Nk - 1)/[V_k + k(k - 1)]$$

Thus, the larger the variance in progeny number (V_k), the lower N_e. Obviously, breeding systems, differences in numbers of progeny, and population structure are important concerns in effective population size and conservation of genetic diversity (Chesser et al. 1993, 1996).

Population size fluctuations tend to reduce N_e, and large population fluctuations greatly reduce N_e, because every time a population numerically declines to a small size, it experiences a demographic bottleneck. To describe the effect of bottlenecks on N_e, the harmonic mean of population sizes in each generation can be calculated:

$$1/N_e = 1/t \, (1/N_1 + 1/N_2 + \cdots + 1/N_t)$$

where t = time in generations. Generations with small numbers are weighted more heavily because they cause an immediate loss of genetic variation available to subsequent generations. Even if the population grows in subsequent generations, all of those individuals will be descended from the smaller number living during the demographic bottleneck. Some new variation may be introduced due to mutation, but the rate of mutation is very small relative to the loss of unique genes with the death of many individuals in a population during a bottleneck. A single crash in population size can produce a large reduction in N_e. Because populations may have varied greatly in size over time, current population size is not a predictor of the amount of current or historical genetic variation.

Similarly, when a population is founded, the diversity of the population in subsequent generations will be a function of the initial number of founders, and the genetic diversity among them. If the founders are not representative of the parent population, or if only a few founders are involved, then the newly established population is a biased representation of the larger gene pool from which it came, and may have lower overall genetic diversity. When the number of founders is very small, N_e will be small for many generations. This phenomenon is called the **founder effect**.

---

passes on only one copy (allele) of each of their genes to each offspring, and variation in the number of offspring produced is generally high, then by chance alone, some alleles may not be passed on and represented in the next generation. Drift leads to a loss of variation more quickly, and thus is of greatest conservation concern in small populations (Figure 11.3).

In a randomly mating population, we assume that the gametes forming zygotes in the next generation are a random sample of $2N$ gametes drawn from the previous generation, where N is the size of the adult breeding population. In a finite population there are $2N$ possible outcomes, each of which has some probability of occurring. The expectation is that the allele frequency in the next or t generations in the future, $E(p_t)$, will equal that observed presently. However, the smaller the population size contributing genes to the next generation, the larger the random drift in allele frequency is likely to be.

The ultimate outcome of genetic drift within any small population that is closed to immigration will be that all but one allele will be lost and one allele will reach a frequency of 100% (i.e., all populations will drift to fixation of one allele or the other). The probability that an allele will eventually become fixed in a population equals the initial frequency of that allele (p_0). For example, if a population was sampled at an initial time period using a single locus with two alleles (A and a), and the frequency of the A allele was found to be 0.8 we would expect the population to eventually become fixed for the A allele with probability 0.8. The probability for fixation of the alternate a allele would be $1 - p_0 = 0.2$. Thus, the alleles have a greater chance of being eliminated by drift, particularly in small populations.

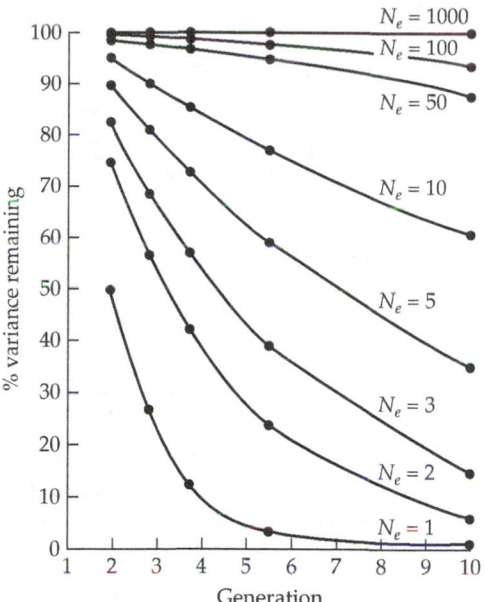

Figure 11.3 Average percentage of genetic variance remaining over 10 generations in a theoretical, idealized population at various genetically effective population sizes (N_e). Variation is lost randomly through genetic drift.

One of the earliest experiments demonstrating the effect of drift was conducted by Buri (1956) on the fruitfly *Drosophila melanogaster* using eye color polymorphisms. In flies homozygous for the scarlet locus (*st/st*), different genotypes at the *bw* locus result in three distinguishable phenotypes, all of which have the same fitness:

Genotype of fly	*st/s* *bw/bw*	*st/s* *bw/bw*^{75}	*st/st* *bw*^{75}/*bw*^{75}
Phenotype of fly	White eye	Light orange eye	Reddish orange eye

Buri started with 107 separate populations, each with 8 females and 8 males, all heterozygotes (*bw/bw*^{75}). In

each generation in each population, Buri counted the number of *bw* and *bw*^{75} alleles, and randomly selected 8 males and 8 females to start the next generation. He computed the allelic frequencies (p = frequency [*bw*] and q = frequency [*bw*^{75}]) for each generation until one or the other allele was lost. Quickly, the 107 populations went to fixation, with roughly half being homozygous for the *bw* allele and the remainder for the *bw*^{75} allele.

The direction of genetic drift is random. Populations are equally likely to exhibit an increase or decrease in the frequency of an allele from the original value of $p_0 = 0.5$. We do not know with certainty what the actual frequency will be in any future generation. However, if we know the population size of each generation, we can estimate the expected variance, and then can with some confidence establish expectations for the magnitude of change.

When the population size is small, the loss of diversity due to drift can be dramatic (see Figure 11.3). Expected rates of loss of heterozygosity due to drift can also be predicted on the basis of population size (Figure 11.4). The expected proportion of original heterozygosity remaining after one generation of drift in a randomly mating population is $1 - 1/2N_e$. If population size remains constant over many generations the heterozygosity after t generations (H_T) relative to the original level of heterozygosity (H_0) can be estimated as:

$$H_T = (1 - 1/2N_e)^t H_0$$

While this loss in heterozygosity is of concern, it is important to recognize that the rate of loss is usually slower than the time frame in which conservation actions can occur. For example, a mammal with a generation length of 10 years reduced to a N_e of 50 in 1900 would still have 90% of its heterozygosity in 2000 (Amos and Balmford 2001) (see Figure 11.3). Further, we should also recognize that adaptive genetic variation is usually subject to natural selection, and thus may be more easily retained, while genes that are neutral are lost more quickly (McKay and Latta 2002). Therefore, loss of the most im-

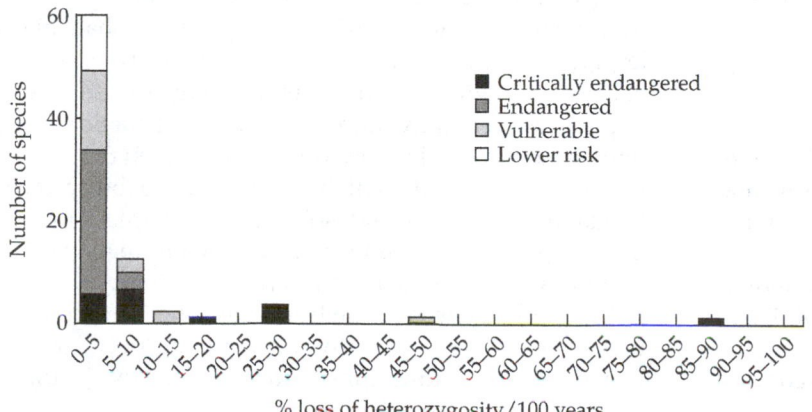

Figure 11.4 Frequency distribution of predicted percent loss of heterozygosity over 100 years for 80 mammal species, according to conservation status (based on Hilton-Taylor 2000). Predictions were based on the assumption that species declined instantly to its lowest level (usually current population size), and remained there for a century. (Modified from Amos and Balmford 2001.)

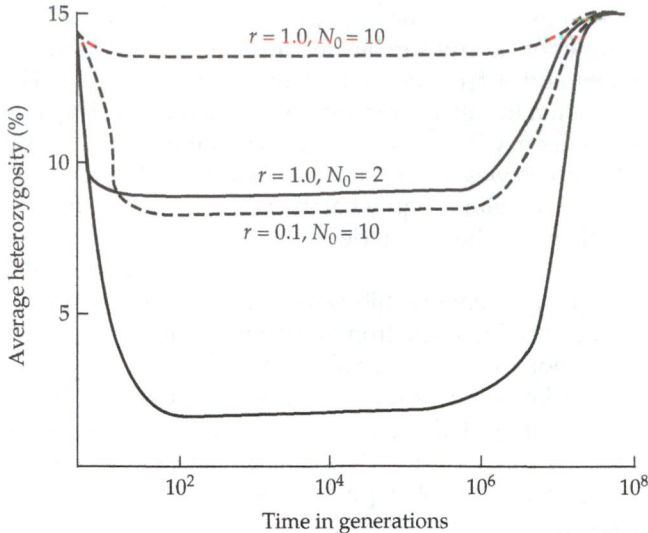

Figure 11.5 After a bottleneck, genetic variation (as measured by average heterozygosity) very slowly recovers. Recovery is quickest when populations have a high growth rate ($r = 1.0$), and when the bottleneck is less severe (founding number $N_0 = 10$ or greater). (Modified from Nei 1975.)

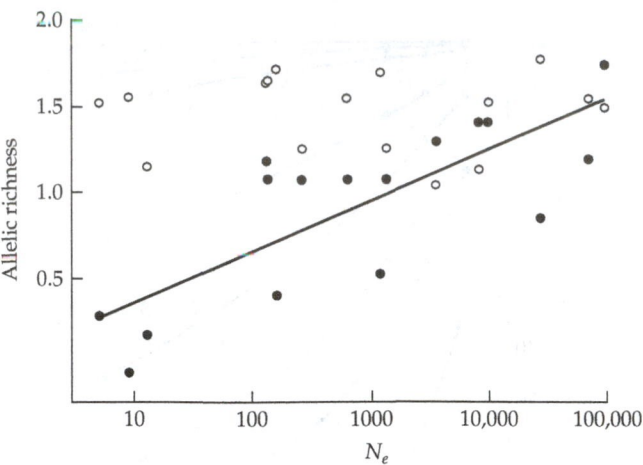

Figure 11.6 Rare alleles are lost from small, isolated populations of an endangered daisy (*Rutidosis leptorrhynchoides*) in Australia. (Modified from Young et al. 1999.)

portant variation is likely to be slower than our overall estimates or measure of heterozygosity based on molecular or biochemical markers would indicate.

When a demographic bottleneck occurs, the magnitude of the genetic loss depends not only on the size of the bottleneck but also on the growth rate of the population afterward. While a single-generation reduction to an N_e of 10–50 may not severely reduce genetic diversity, a prolonged period at small size (with significant genetic drift each generation), can have large effects (Figure 11.5).

Rare alleles are usually lost more quickly from populations (Leberg 1992). We know that rare alleles contribute little to overall genetic variation, but they may be important to a population during infrequent or periodic events such as unusual temperatures or exposure to new parasites or pathogens, and can offer unique responses to future evolutionary challenges. The expected number of alleles, E_n, remaining at a locus each generation is estimated as:

$$m - \sum_{1}^{j} \left(1 - p_j\right)^{2N_e}$$

where m = the original number of alleles, and p_j is the frequency of the jth allele (from Denniston 1978). Rare alleles are lost rapidly from small populations, even when much of overall genetic diversity (H_t) is retained. Data from an endangered daisy species in Australia demonstrates the losses of rare alleles in nature in Figure 11.6.

The effects of genetic drift have important implications for species or populations of conservation concern.

If a primary goal is to maintain genetic diversity in a population (i.e., heterozygosity and alleles), large population size is important. In a population closed to immigration, recovery of population levels of genetic variation is an extremely long process (see Figure 11.5). The only mechanism to recover variability in a closed population is mutation. Thus, gene flow from migration, either via natural or assisted means, likely will be an increasingly important conservation tool as fragmentation of habitat increases. Small populations over a prolonged period are thus to be avoided in conservation programs whenever possible, except when such species are the principle focus of concern, such as a unique, isolated endemic species. However, these principles are particularly relevant to captive breeding programs.

Gene flow

Gene flow is defined as the movement of genes from one population to another. The exchange of individuals (and their genes) among populations tends to homogenize the genetic composition of populations. Populations that may have held unique variation will come to share alleles, given enough gene flow between them. Thus, while genetic drift will tend to create differences in separate populations (enhance among-population genetic variation), gene flow will reduce differences. Given that natural populations will with increasing probability face habitat fragmentation and isolation (see Chapter 7), documenting degree of connectivity among populations has become a central issue in conservation biology.

Measuring gene flow can be problematic. Direct measures of movements of individuals among populations may differ from indirect measures of gene flow (Slatkin

1985) because those individuals who moved may not in fact successfully breed in the new population. Direct observations of dispersal or migration thus provide no information about gene flow per se. Direct observations also chronicle the extent of movements only over the period of observation but provide no information regarding historical levels of dispersal. Genetic similarities among populations could be due to recent common ancestry as well as gene flow (Slatkin 1991). Further, genetic measures of gene flow report the cumulative effects of contemporary and past events.

Several methods are available to empirically estimate gene flow (Waser and Strobeck 1998). Using assignment tests (Paetkau et al. 1995; Pritchard et al. 2000; Wilson and Rannala 2003) based on estimates of population allele and expected genotype frequencies, it is possible to probabilistically estimate whether individuals originated from the population from which it was observed or from another population. Coalescence-based analyses (e.g., Beerli and Felsenstein 2001) provide simultaneous estimates of evolutionary effective population size and rates of migration.

For many populations of conservation or management concern, present levels of gene flow are of interest. In general, genetic data are used to estimate the number of migrants received in a population per generation (m). One common rule of thumb holds that if the product of effective population size and migration rate ($N_e m$) is greater than 1 (one migrant per generation), this rate of gene flow into a population is sufficient to minimize the loss of alleles and heterozygosity while allowing for divergence in allele frequencies (and adaptive variation) among populations. This one-migrant-per-generation rule may be a desirable minimum. However, other genetic and non-genetic factors that influence desired levels of connectivity should also be closely reviewed when making conservation decisions (Mills and Allendorf 1996).

Inbreeding depression

Inbreeding, or mating between close relatives, is a potentially serious problem whose probability of occurrence increases in small, isolated populations, even if mating occurs at random. Inbreeding does not change allele frequencies, but rather increases the frequency of homozygous genotypes. Because close relatives are more likely to share alleles that are identical by descent, their progeny are more likely to be given the same alleles by both parents.

Inbreeding can lead to decreased fitness (**inbreeding depression**) either by the expression of deleterious recessive alleles in a homozygous state or the loss of heterozygosity. It is generally believed that some degree of dominance will prevent the expression of deleterious alleles except in a homozygous recessive condition. In

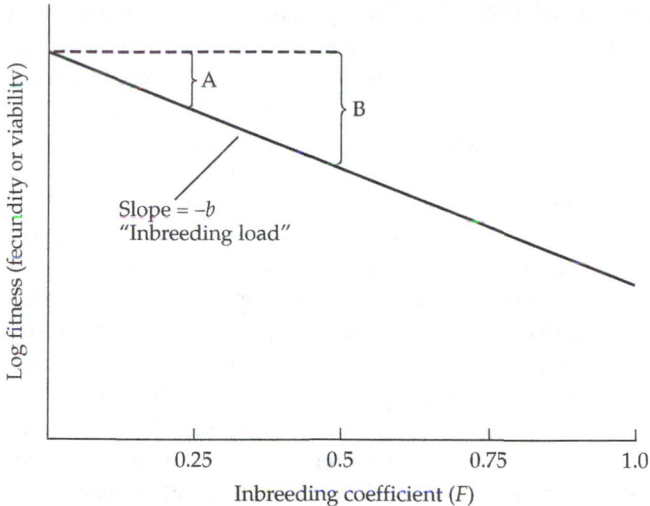

Figure 11.7   Inbreeding depression is the difference in fitness between outbred individuals (dotted line, $F = 0$) and inbred ones (solid line, $F > 0$). If inbred individuals were as fit as outbred individuals, their fitness would lie along the dotted line, but when inbreeding depression occurs, their fitness lies along a solid line of slope, $-b$, where the slope is the "inbreeding load." Individual A has a lower inbreeding load than individual B, which is more inbred. (Modified from Keller and Waller 2002.)

higher vertebrates, individuals may carry, on average, 100 deleterious or slightly deleterious alleles. If expressed in a homozygous condition, this could cause a reduction in fitness. Often, inbreeding depression is measured by comparing the fitness of progeny (based on relative fecundity and/or viability) produced from outcrossed matings with those from inbred crosses over a range of inbreeding values (Figure 11.7). Inbreeding has long been known to result in detrimental effects in captive (Ralls and Ballou 1983) and natural populations (Keller and Waller 2002).

Lerner (1954) wrote a compendium of observations on cultivated plants and domesticated animals regarding relationships among heterozygosity, growth rate, and other measures of performance and development. Lerner's experiments contrasted inbred, predominantly homozygous lines with highly heterozygous progeny from crosses between inbred lines. Lerner found that highly heterozygous individuals and lines generally exhibited traits that breeders strived to "fix" in their strains, such as superior growth rates, larger size, and more "buffered" developmental processes that led to lower degrees of asymmetry (a measure of developmental stability). Inbreeding depression may contribute to declines in metabolic efficiency, growth rate, reproductive physiology, and disease resistance seen in many populations that recently experienced greater levels of inbreeding (Gilpin

and Soulé 1986). For example, data from domesticated animals indicate that an increase in the inbreeding coefficient (ΔF) of 10% will result in a 5%–10% decline in individual reproductive traits such as clutch size or survival rates. In aggregate, total reproductive attributes may decline by 25% (Frankel and Soulé 1981).

Data on inbreeding depression in the wild are difficult to compile, as the level of inbreeding is not easily known. However, recently some robust studies have shown severe effects of inbreeding in wild populations. Inbreeding increased extinction risk in Glanville fritillary butterfly (*Melitaea cinia*) populations, confirmed through both observational (and experimental studies (Saccheri et al. 1998; Nieminen et al. 2001). Other studies have shown that inbreeding depression may play a strong role when populations are subject to additional stresses. In experimental studies, inbred and outbred land snails (*Arianta arbustorum*) (Chen 1993) and white-footed mice (*Peromyscus leucopus*) (Jimenez et al. 1994; Figure 11.8) were released into the field and survivorship was followed. In both cases, inbred individuals had significantly lower survival. Importantly, mice that remained in the laboratory showed more equal survival between inbred and outbred individuals. Keller et al. (1994) studied a natural population of Song Sparrows (*Melospiza melodia*) where inbreeding coefficients were known through **pedigree analysis**. The population crashed during severe winter weather and outbred individuals survived at a significantly higher rate than inbred individuals.

Many species are known to avoid inbreeding in the wild (Pusey and Wolf 1996). In some species nearly all young males disperse (e.g., spotted hyenas; Smale et al. 1997) which is believed to represent an evolved behavior to avoid deleterious effects of inbreeding. Such studies give further evidence that inbreeding depression is real and important. Inbreeding also may have a threshold relationship with extinction, wherein a population persists quite well until a particular level of inbreeding is reached, and then suddenly declines to extinction (Frankham 1995).

Not all inbreeding is cause for alarm. Some natural populations have apparently experienced low levels of inbreeding for many generations with no ill effects (Thornhill 1993). In these cases, it is thought that slow inbreeding has given selection an opportunity to purge the population of deleterious recessive alleles. However, two recent reviews—one by Ballou (1997) of 25 captive mammal studies, and the other by Byers and Waller (1999) of 53 plant studies—showed that inbreeding led to purging only in some cases and only in some populations, and then to only a limited extent. Thus, although purging is a possible benefit to inbreeding, it seems a highly uncertain one. Further, rapid inbreeding may compromise long-term population viability, especially if a population has little history of prior inbreeding. Inbreeding depression therefore may be more prevalent in a species or population with historically large population sizes that now occurs in small populations.

One way to avoid inbreeding is to purposely outbreed. The opposite of inbreeding depression is outbreeding enhancement, which in the domestic animal or plant literature is often referred to as **heterosis** or **hybrid vigor**. Individuals from different populations are not likely to be homozygous for the same recessive alleles. Outbreeding can lead to masking of different deleterious recessive alleles present in different populations and an increase in heterozygotes. Thus, outbreeding of an inbred population should result in increased population mean fitness. If such a simplistic perspective were indeed true, one universal conservation prescription would be to advocate mating individuals from different populations. Indeed, a widely held view is that outbreeding among individuals from very different genetic backgrounds has a high likelihood of increasing fitness out to some threshold level (Figure 11.9). However, we need to consider outbreeding in the context of the entire genome because declines in fitness can occur due to outbreeding.

Outbreeding depression

The phenomenon of **outbreeding depression** can be expressed in several ways. Under one scenario, declines in fitness of hybrids or outcrossed genotypes can occur due to "genetic swamping" of locally adaptive genes through gene flow or directed matings from another population that evolved under different ecological conditions. We can consider two genotypes *AA* and *BB* that evolved in environments 1 and 2, respectively. *AA* has higher fitness in environment 1 than the *BB* or hybrid (*AB*) genotype. Conversely, genotype *BB* has the highest fitness in environment 2. Hybrid genotype *AB* is not well adapted to either environment. The presence of inferior hybrid genotypes as a consequence of gene flow will result in decreased population fitness.

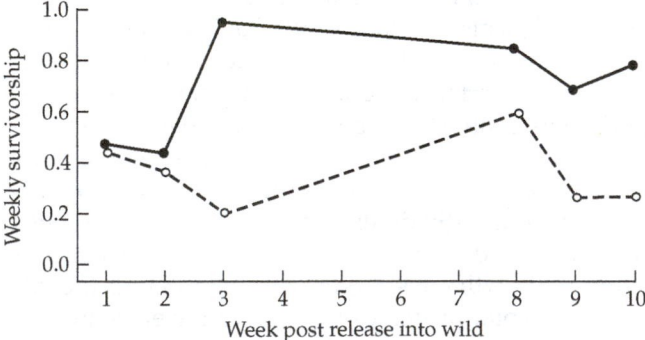

Figure 11.8 Inbred white-footed mice, (*Peromyscus leucopus*) (open circles), had lower survivorship than outbred individuals (solid circles) after release into the wild. Wild-caught mice were used to found an inbred and outbred line, and descendants of these mice were released back into the wild and followed for 10 weeks. (Modified from Jimenez et al. 1994.)

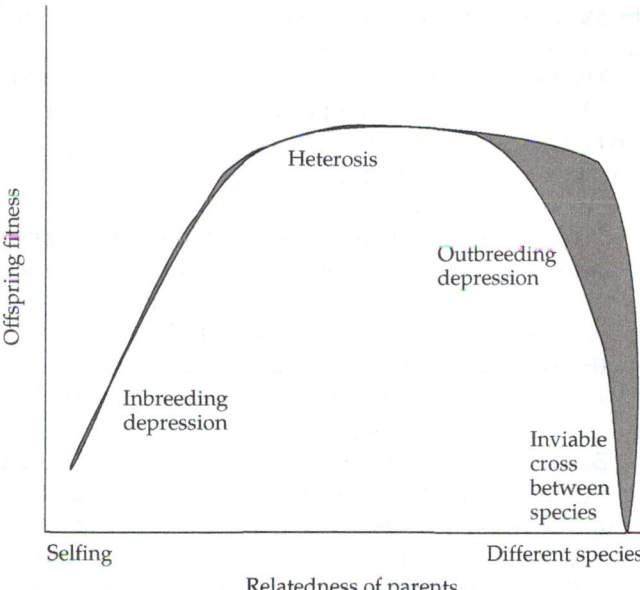

Figure 11.9 Offspring fitness is influenced by the degree of relatedness of parents. Closely related parents produce inbred young that are less fit than those of unrelated parents of the same species, leading to "inbreeding depression." When parents are unrelated, fitness rises yielding hybrid vigor or "heterosis." As parents are more distantly related, some decline in fitness may occur (outbreeding depression) and usually at some point, offspring from crosses between species are far less fit, even to the point of inviability.

The second way in which outbreeding depression can occur is by the breakdown of physiological or biochemical compatibilities between genes that have evolved in different populations. Interactions among alleles at several loci (**epistasis**) collectively affect fitness. Organisms have evolved in the context of specific environments and have evolved suites of genotypes across many genetic loci that are coadapted to each environment (coadapted gene complexes). If new alleles are introduced via gene flow into the genetic background of the resident population, a loss in fitness is expected through physiological or biochemical incompatibilities introduced through disruption of these coadapted gene complexes (see Essay 11.1).

If habitats differ and if there are population differences in heritable traits that confer fitness advantages in each different habitat type, the ability of colonists to survive may be compromised (i.e., there is increasing likelihood of lower intrinsic rates of population increase following colonization). One interesting example of the importance of habitat heterogeneity and colonization potential was described for Great Tits (*Parus caeruleus*) by Dias et al. (1996). These authors found that in patchy landscapes, life history traits were adapted to local evergreen or deciduous habitats. High rates of gene flow from high quality deciduous habitats (sources) into poorer quality evergreen (sink) habitats prevented local adaptation.

Gilk et al. (2004) examined the effects of hybridization between pink salmon (*Oncorhynchus gorbuscha*) from widely separated (1000 km) populations on fitness-related traits including homing ability, marine survival, and distribution of family size. Hybridization-reduced survival decreased return rates for F_1 and F_2 adults (a proxy measure of marine survival). These authors argue compellingly that outbreeding of geographically separated populations can erode fitness and should be recognized as a potential negative consequence of management practices.

Inbreeding and outbreeding depression can occur simultaneously in a population. Fluctuations in population size and gene flow (either natural or directed) of maladaptive alleles can result in inbreeding or outbreeding depression respectively in natural populations, potentially reducing the population fitness. Ultimately, in the design of breeding strategies, one must weigh the relative effects of potential past inbreeding in the population relative to the effects of oubreeding on locally adaptive genotypic combinations.

Natural selection

Natural selection is defined as the differential survival and reproduction of different genotypes in the population, or the differential success of genotypes passing gametes to other generations. Genotypes may have different fitness overall, or at different stages in the life cycle. Selection can be decomposed into components, including viability selection (differential survivorship to adulthood), sexual selection (differential mating success of different individuals), and fertility selection (differential production of offspring).

For a single locus with two alleles, the effect of natural selection can be derived with knowledge of the relative fitness of each genotype. Under directional selection, the fitness (w) of one homozygous genotype exceeds all others ($w_{AA} < w_{Aa} < w_{aa}$ or $w_{AA} > w_{Aa} > w_{aa}$), and the frequency of a deleterious allele steadily decreases to 0. If the heterozygote is inferior, then the fitness of the heterozygote is lower than either homozygous genotype ($w_{Aa} < w_{AA}, w_{aa}$), and the frequency of one allele (either A or a) steadily decreases to 0. Which allele is lost (A or a) depends on the initial allele frequency (p_0), with the initially more common allele becoming dominant. Finally, with **overdominance** or heterozygote superiority ($w_{Aa} > w_{AA}, w_{aa}$), both alleles are maintained in the population. Overdominance represents the only fitness relationship that will maintain both alleles at a single diallelic locus over time. Thus, the effect of natural selection is to reduce genetic variation by eliminating deleterious and rare alleles if consistent directional selection or homozygote advantage are operating, while the effect will be to maintain variation when overdominance is operating.

Sexual selection, or differential mating success, can influence gene frequencies in natural or captive popula-

tions due to competition for mates (typically male–male competition) or to mate choice (typically by females). Under male–male competition, genes that confer greater ability for males to successfully compete for females increase in the population. Examples of traits that confer competitive advantages include body size, coloration, or ornamentation. Darwin argued that mate choice is expected to occur in the sex that invests more energy in reproduction, often the female. Because the female invests more energy in reproductive activities, it would be advantageous to her if she could choose a male that would maximize the fitness of her offspring. The result of sexual selection is a change in gene frequency between the pool of potential mates and the pool of successfully mating adults. Sexual selection can also reduce effective population sizes.

Natural selection maximizes reproductive success in wild populations by optimizing reproductive success and survivorship. Selection has the potential to eliminate deleterious alleles from a population. However, the ability of selection to purge maladapted genotypes (and deleterious alleles) can be compromised in small populations. If the intensity of selection (S) is much greater than $1/2N_e$, selection can be effective at eliminating deleterious alleles. However, in situations where $S \leq 1/2N_e$, genetic drift becomes more important than selection, and mildly deleterious alleles can build up by chance due to the small breeding pool of individuals.

Using Conservation Genetics to Inform Management

If evolution is the unifying feature and driving force of natural systems, then the primary goal of genetic management in nature should be to allow continued evolutionary change in the populations and species of concern. By definition, evolution is dynamic, and change is expected. Further, ecological systems are dynamic, and generally are not at equilibrium (Chapter 1). The best way to "manage" such dynamic, changing systems is to permit and allow for change: a conservationist rather than a preservationist approach. Conservation guided by an evolutionary perspective seeks conditions that will allow populations to continually adapt to inevitably changing conditions. In this section, we present a variety of means for applying genetic techniques to better understand the status and appropriate actions for recovery of species.

Time scales of concern

Genetic conservation actions should be compatible with conservation goals that are developed on three time scales: maintenance of viable populations in the short term (extinction avoidance), maintenance of the ability to continue adaptive evolutionary change, and maintenance of the capacity for continued speciation.

The first level of concern, avoidance of population extinction, has a time scale of days to decades and is the first and most obvious goal of conservation. If this goal is not met, then further goals are automatically denied. Local extirpation and recolonization cycles can be a normal part of metapopulation dynamics. The type of population extinction to be avoided, however, is that which is not part of the natural system dynamics, does not have a recolonization source, and is caused, directly or indirectly, by human action. To the extent the loss of genetic diversity may contribute to such extinction, it becomes a primary interest of conservationists.

Because environmental change is being accelerated by humans, genetic management must also maintain the ability of populations and species to genetically adapt, or evolve in the face of rapid or extensive change. Limiting gene flow in small populations can result in reduction of genetic diversity due to inbreeding or genetic drift, and is poor long-term management. This concern has a time scale of decades to millennia.

Finally, speciation is the creative part of biodiversity, just as extinction serves to reduce genetic and species diversity. The potential for continued speciation must be maintained, especially now that extinction rates are so exceedingly high. To consider only short-term, preservationist goals is to ignore larger negative consequences of human development in Earth's history. Retention of the ability to continue to speciate is an ultimate goal of conservation, although the prolonged time scale involved is difficult to appreciate.

Identifying and Prioritizing Groups for Conservation

Given the importance of genetic variation to short-term fitness, continued adaptation, and the speciation process, a difficult and practical question confronts the resource manager. What are the units of genetic conservation? What, in fact, should we conserve? Conservation initiatives cannot save every population, every morphological variant, every unique allele. Given limited resources, managers increasingly pose the question "How do we determine and define the biologically significant units within a species that are worthy of attention?"

To conserve biological diversity we need to identify suites of characters that may be used to define a group as different or unique, and establish criteria for levels of rarity or abundance that constitute a threat to future viability. New questions arise as we consider how we identify the degree of distinction among, and status of, different species or populations. What characters should

be used to determine the degree of uniqueness and how should variation be quantified? What criteria should be used to prioritize conservation choices among candidate species? What level of "distinctness" is deserving of protection? When should governments intervene to conserve populations or species? Should conservation actions begin when a species as a whole is threatened, or when individual populations, or ecotypes are threatened? John Avise addresses some of these challenges in Essay 11.2.

In trying to answer these questions, the conservation geneticist attempts to use genetic information to provide clues about the uniqueness or evolutionary role of different taxa. However, this is extremely difficult. There are no universally accepted criteria for purposes of character selection, evaluation of degree of differentiation, classification of organisms, or evaluation of uniqueness. In his book *Genetics and the Origins of Species* (1937), Dobzhansky wrote "Biological classification is simultaneously a man-made system of pigeon-holes devised for the pragmatic purpose of recording observations in a convenient manner and an acknowledgement of the fact of organic discontinuity." Hendry et al. (2000) wrote "A frequent outcome when studying complex biological systems is that data collected to test a specific hypothesis can be interpreted several ways, and interpretations can be influenced by the paradigm through which a given scientist views the world." The primary limitation of conventional taxonomy is that all taxa placed at the same Linnaean "rank" (i.e., species, genus, family) are not equivalent in age, diversity, disparity or other consistent property of their biology or evolutionary history.

The significance of genetic variation as one of several currencies for biodiversity evaluation is recognized widely. An assumption often made is that protection of diversity at or above the species level will protect the genetic and evolutionary potential at the population level. However, misalignments between recognized taxonomy and apportionment of genetic diversity might result in a failure to adequately conserve diversity. Further, current taxonomy is an imperfect guide because it does not weight species by their evolutionary distinctiveness. For example, we might place greater value on preserving a taxon that is not related closely to any extant species than on one related to otherwise widespread and common species. We need not only be able to tell various taxa apart, but also to evaluate their larger conservation importance. Better ranking can be made on evolutionary grounds using genetic information.

In the U.S., under the Endangered Species Act (ESA), a "species" is defined to include any subspecies and "any distinct population segment of any species of vertebrate or wildlife" (Sec 3[15]). Federal agencies have struggled

to interpret and apply the phrase "distinct population segment." An accepted definition identifies a population (or group of populations) as distinct (and hence a species) under the ESA if it represents an **evolutionarily significant unit** (**ESU**) of the biological species (Waples 1991). Two criteria must be met for a group to be considered an ESU. First, the group must be substantially reproductively isolated from other conspecific population units. Isolation does not have to be absolute, but should be strong enough to permit evolutionarily important differences to accrue in different populations. Second, the group must represent an important component in the evolutionary legacy of the species. The second criterion would be met if the population contributes to the ecological or evolutionary diversity of the species as a whole. The ESU concept has been widely embraced by agencies responsible for the stewardship of our natural resources, and genetic markers and theory are the primary tools used to identify ESUs as seen in the case of Pacific salmon, described by Robin Waples in Case Study 11.2. Protection of ESUs has become both a tool for endangered species conservation, and more broadly, an important means of meeting our long-term conservation goals.

One approach to more rigorously defining ESUs in a genetic sense is called a hierarchical gene diversity analysis, and it is based on the fact that genetic diversity within or between species can be apportioned hierarchically. Our task is to partition overall genetic diversity into within-population and among-population components and determine where biologically significant breaks in genetic diversity occur. At the lowest level of the hierarchy, interbreeding individuals within a population are genetically most similar. As we move through the hierarchy, greater genetic differences occur among more geographically separated or otherwise distinct populations, until there may be very large genetic differences between populations strongly isolated by physiography or geographic distance, or based on some behavioral or ecological isolating mechanism such as the timing of reproduction. The actual hierarchical partitioning of the genetic data is accomplished through an analysis called "*F*-statistics," introduced in Box 11.3. More detailed considerations of *F*-statistics are beyond the scope of this book, but can be found in population genetics texts (Crow and Kimura 1970; Hartl and Clark 1997; Hedrick 2000).

A species can thus be visualized as having a spatial genetic architecture. The species consists of a collection of populations, with a hierarchical genetic structure based on the degree of genetic similarity among them. In turn, this is a function of geography and levels of gene flow. Genetic differences among populations as measured using molecular genetic markers accrue in a time-dependent manner as a function of the length of time since populations have been reproductively isolated, on

ESSAY 11.2

A Rose Is a Rose Is a Rose

John C. Avise, *University of Georgia*

■ In the final analysis, biodiversity reflects genetic heterogeneity. Thus, the concerns of conservation biology ultimately represent concerns about the erosion and loss of genetic diversity. Genetic diversity is arranged hierarchically, from the family units, extended kinships, and geographic population structures within species, to a graded scale of genetic differences among reproductively isolated taxa that separated phylogenetically at various depths in evolutionary time. Unfortunately, traditional taxonomic characters (the visible phenotypes of organisms) are not an infallible guide to the underlying genetic subdivisions. In the last three decades, evolutionists have acquired a new set of tools that can be employed to assess genetic diversity more directly, at the level of proteins and even the genes themselves. In general, these various molecular assays can be of service to conservation biology by contributing to the characterization of genetic diversity at any level of the biological hierarchy.

What's in a name? A "Dusky Seaside Sparrow" by any other name is just as melanistic. Nonetheless, nomenclatural assignments inevitably shape our perceptions of how the biotic world is partitioned, and hence the biological units toward which conservation efforts may be directed. In the case of the Dusky Seaside Sparrow, this dark-plumaged population near Cape Canaveral, Florida was described in the late 1800s as a species (*Ammodramus nigrescens*) distinct from other Seaside Sparrows (*Ammodramus maritimus*) common along the Atlantic and Gulf coasts of North America. Although the dusky later was demoted to subspecies status, the nomenclatural legacy stemming from the original taxonomic description prompted continued special focus on this recognized form. Thus in the late 1960s, when the population crashed due to changing land-use practices, the dusky was listed formally as "endangered" by the U.S. Fish and Wildlife Service. Despite last-ditch conservation efforts, the Dusky Seaside Sparrow went extinct in 1987.

The point of relating this sad story involves an unexpected footnote. After natural death of the last known dusky (in captivity), molecular analyses of DNA isolated from its tissues revealed an exceptionally close genetic relationship to other Atlantic coast seaside sparrows, but a deep phylogenetic distinction of all Atlantic from Gulf of Mexico birds, likely due to effects of ancient (Pleistocene) population separations (Avise and Nelson 1989). Thus, the traditional taxonomy for the Seaside Sparrow complex (upon which management efforts were based) apparently had failed to capture the true genetic relationships within the group, in two respects: (1) by giving special emphasis to a presumed biotic partition that proved to be evolutionarily shallow; and (2) by failing to recognize a deeper phylogeographic subdivision between Atlantic and Gulf coast populations. This finding from molecular genetics should not be interpreted as evidence of heartlessness over loss of the dusky. All population extinctions are regrettable, particularly in this age of accelerated habitat loss. Extinction of the dusky population is to be mourned, but perhaps we can be consoled by the knowledge that it is survived by close genetic relatives elsewhere along the Atlantic coast. Furthermore, the discovery of a deep and previously unrecognized phylogenetic subdivision between Atlantic and Gulf coast forms of the seaside sparrow should be paramount in any conservation plans for the remaining populations.

Many taxonomic assignments in use today were first proposed in the last century, often from limited phenotypic information and preliminary assessments of patterns of geographic variation. How adequately these traditional assignments summarize biological diversity remains to be determined, through continued systematic reevaluations to which molecular approaches can contribute. As with the seaside sparrows, past errors of phylogenetic commission and omission both may be anticipated, at least occasionally.

Another example involves pocket gophers in the southeastern U.S. An endangered population referable to "*Geomys colonus*" in Camden County, Georgia, first described in 1898, has proven upon molecular reexamination to represent merely a local variant of the widespread *G. pinetis* (Laerm et al. 1982). In these same genetic assays (which involved comparisons of proteins, DNAs, and chromosome karyotype), a deep but previously unrecognized phylogeographic split was shown to distinguish the pocket gophers of eastern Georgia and peninsular Florida from those to the west.

Inadequate taxonomy also can kill, as exemplified by studies of the tuatara lizards of New Zealand. This complex has been treated as a single species by government and management authorities, despite the fact that molecular (and morphological) appraisals have revealed three distinct groups (Daugherty et al. 1990). Official neglect of this described taxonomic diversity may unwittingly have consigned one form of tuatara (*Sphenodon punctatus reischeki*) to extinction, whereas another form (*S. guntheri*) has survived to this point only by good fortune. As noted by Daugherty et al. (1990), "Taxonomies are not irrelevant abstractions, but the essential foundations of conservation practice."

In other cases, molecular reappraisals of endangered forms may bolster the rationale for special conservation efforts directed toward otherwise suspect taxa. For example, recent molecular appraisals of the endangered Kemp's ridley sea turtle (*Lepidochelys kempi*) showed that this "species" (which had a controversial taxonomic history) does indeed fall outside the range of genetic variability exhibited by assayed samples from its more widespread congener, *L. olivacea* (Bowen et al. 1991).

These examples illustrate but a few of the many ways that molecular genetic methods can contribute to the assessment of biodiversity, and hence to the implementation of conservation programs. Ironically, even as these exciting molecular methods for reexploring the biological world are being developed, the biota to which they might be applied are disappearing at an unprecedented rate through direct and indirect effects of the human population explosion. ■

Nearctic
Palearctic
Oceania
Neotropic
Afrotropic
Indo-Malay
Oceania
Australasia
Antarctic
Tropical and subtropical moist broad-leaved forests
Tropical and subtropical dry broad-leaved forests
Tropical and subtropical coniferous forests
Temperate broad-leaved and mixed forests
Temperate conifer forests
Boreal forests/taiga
Tropical and subtropical grasslands, savannas and shrublands
Temperate grasslands, savannas and shrublands
Flooded grasslands and savannas
Montane grasslands and shrublands
Tundra
Mediterranean forests, woodlands and scrub
Deserts and xeric shrublands
Mangroves

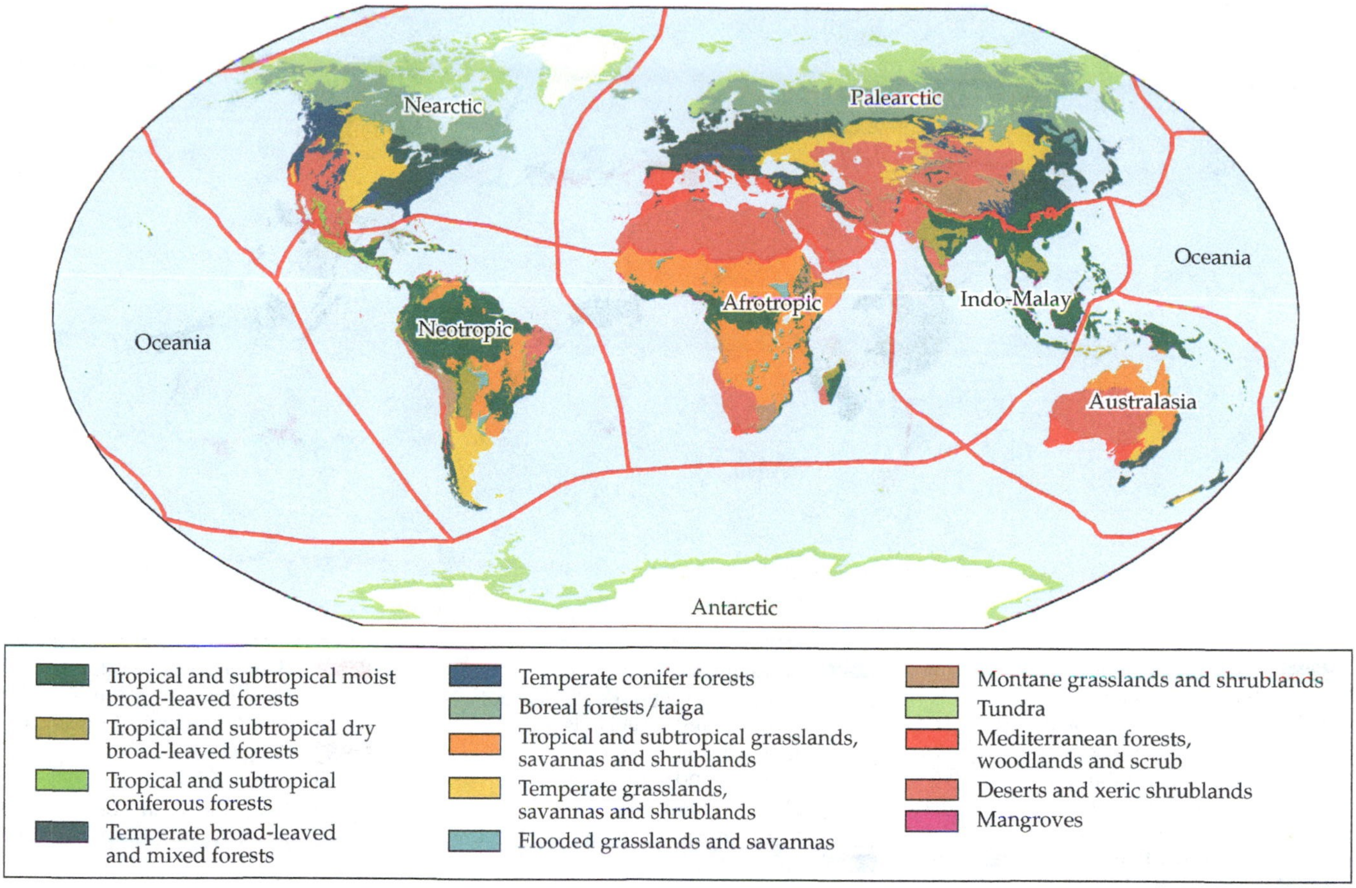

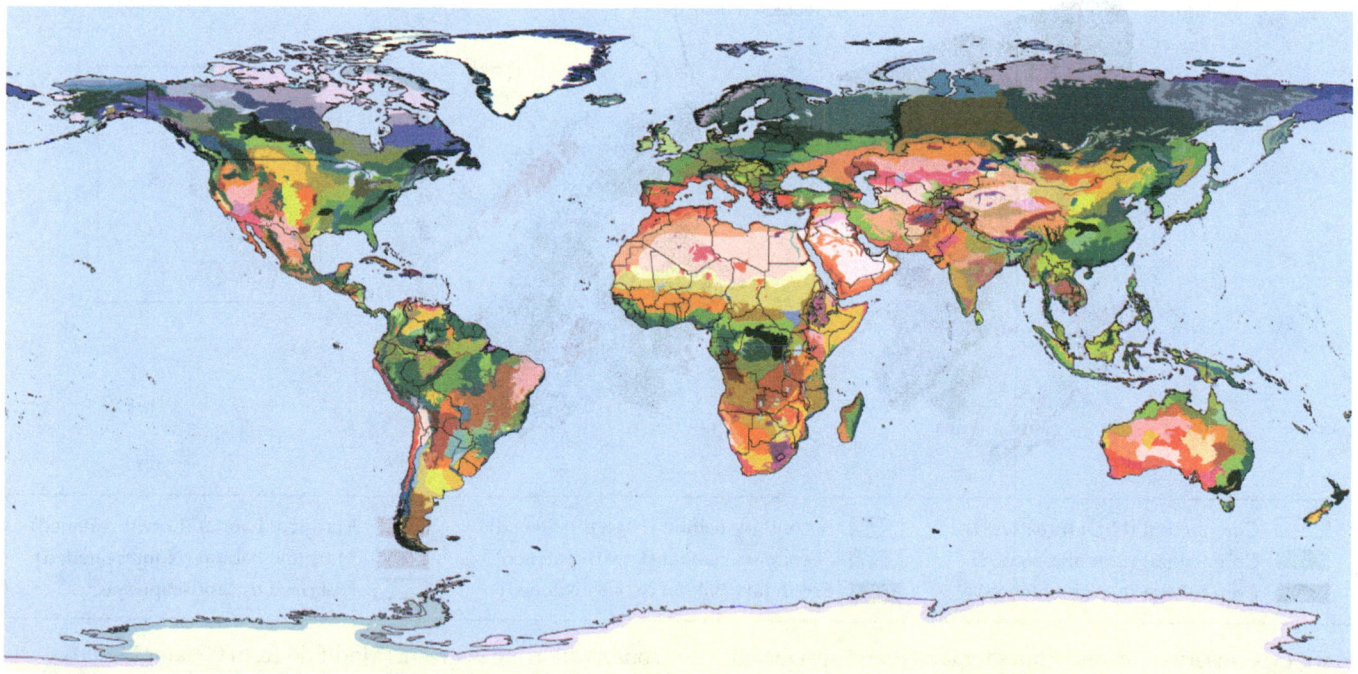

Plate 1 The eight biogeographic realms of the world, with the 14 terrestrial biomes distributed within these. (Courtesy of WWF–US, GIS map provided by J. Morrison.) [2]

Plate 2 The 867 ecoregions of the world. (Courtesy of WWF–US, GIS map provided by J. Morrison.) [2]

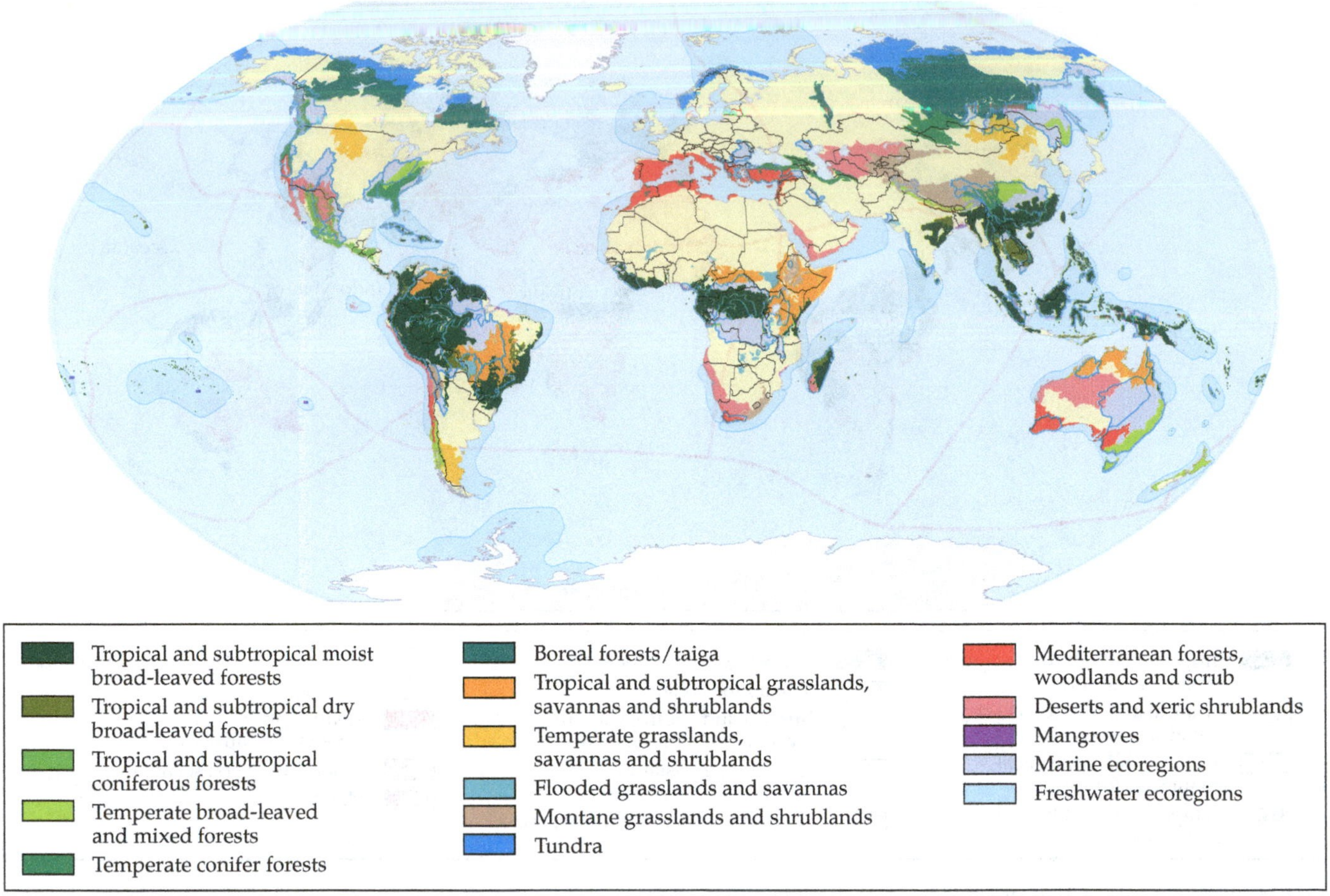

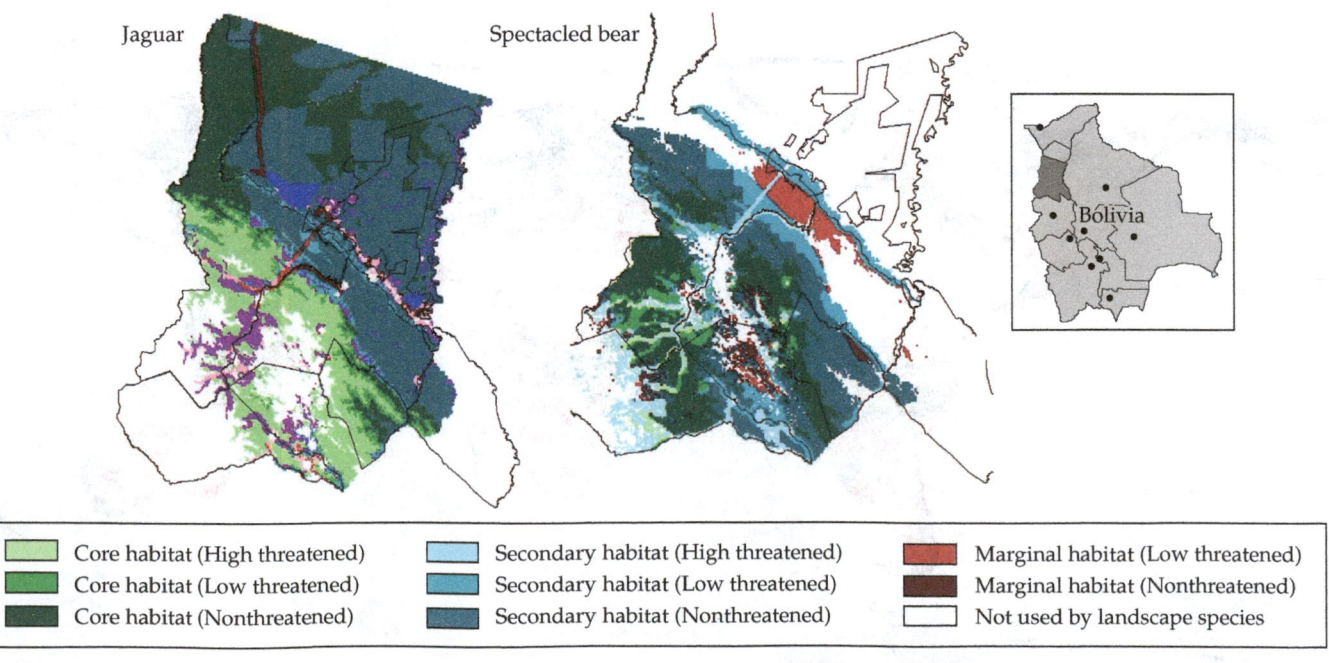

Plate 3 The Global 200 ecoregions are selected to represent the full complement of ecoregional diversity in the terrestrial, marine and freshwater realms. The most distinctive and outstanding ecoregions are selected to represent global diversity. (Courtesy of WWF-US, GIS Map provided by J. Morrison.) [6]

Legend (Plate 3):

- Tropical and subtropical moist broad-leaved forests
- Tropical and subtropical dry broad-leaved forests
- Tropical and subtropical coniferous forests
- Temperate broad-leaved and mixed forests
- Temperate conifer forests
- Boreal forests/taiga
- Tropical and subtropical grasslands, savannas and shrublands
- Temperate grasslands, savannas and shrublands
- Flooded grasslands and savannas
- Montane grasslands and shrublands
- Tundra
- Mediterranean forests, woodlands and scrub
- Deserts and xeric shrublands
- Mangroves
- Marine ecoregions
- Freshwater ecoregions

Legend (Plate 4):

- Core habitat (High threatened)
- Core habitat (Low threatened)
- Core habitat (Nonthreatened)
- Secondary habitat (High threatened)
- Secondary habitat (Low threatened)
- Secondary habitat (Nonthreatened)
- Marginal habitat (Low threatened)
- Marginal habitat (Nonthreatened)
- Not used by landscape species

Plate 4 Conservation landscapes for jaguar and spectacled bear in the Greater Madidi landscape. Lighter colors denote regions of conflict with local populations. Highest priorities for conservation are in green. (Modified from Greater Madidi Landscape Conservation Program 2004; GIS map provided by H. Gomez.) [Case Study 12.2]

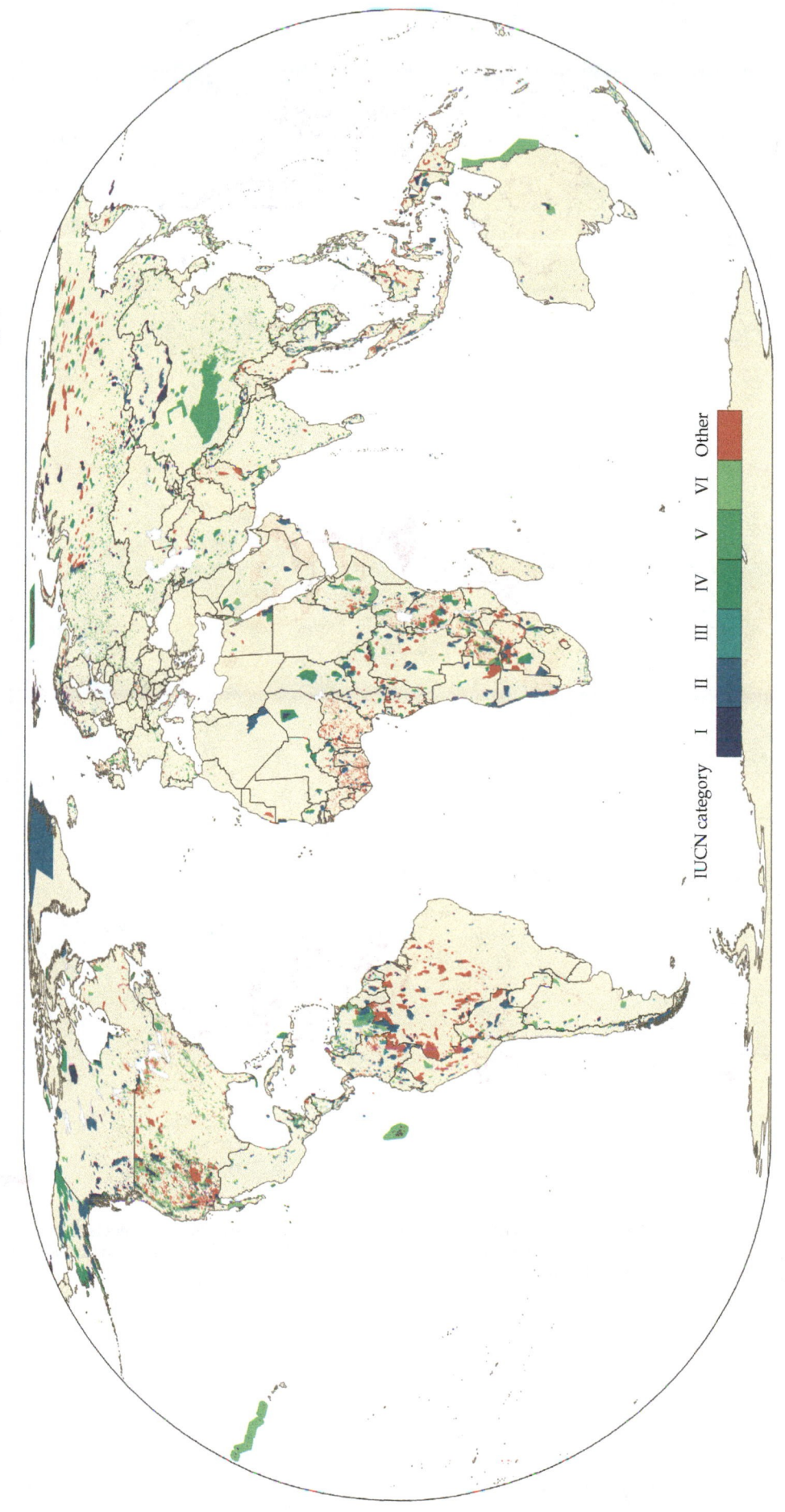

Plate 5 The world's terrestrial and marine protected areas. Shown are all protected areas in IUCN categories I-VI and those without a designation at present from the World Database on Protected Areas 2005. (Courtesy of the WDPA Consortium 2005 and the CI GIS lab, drawn by M. Denil.) [14]

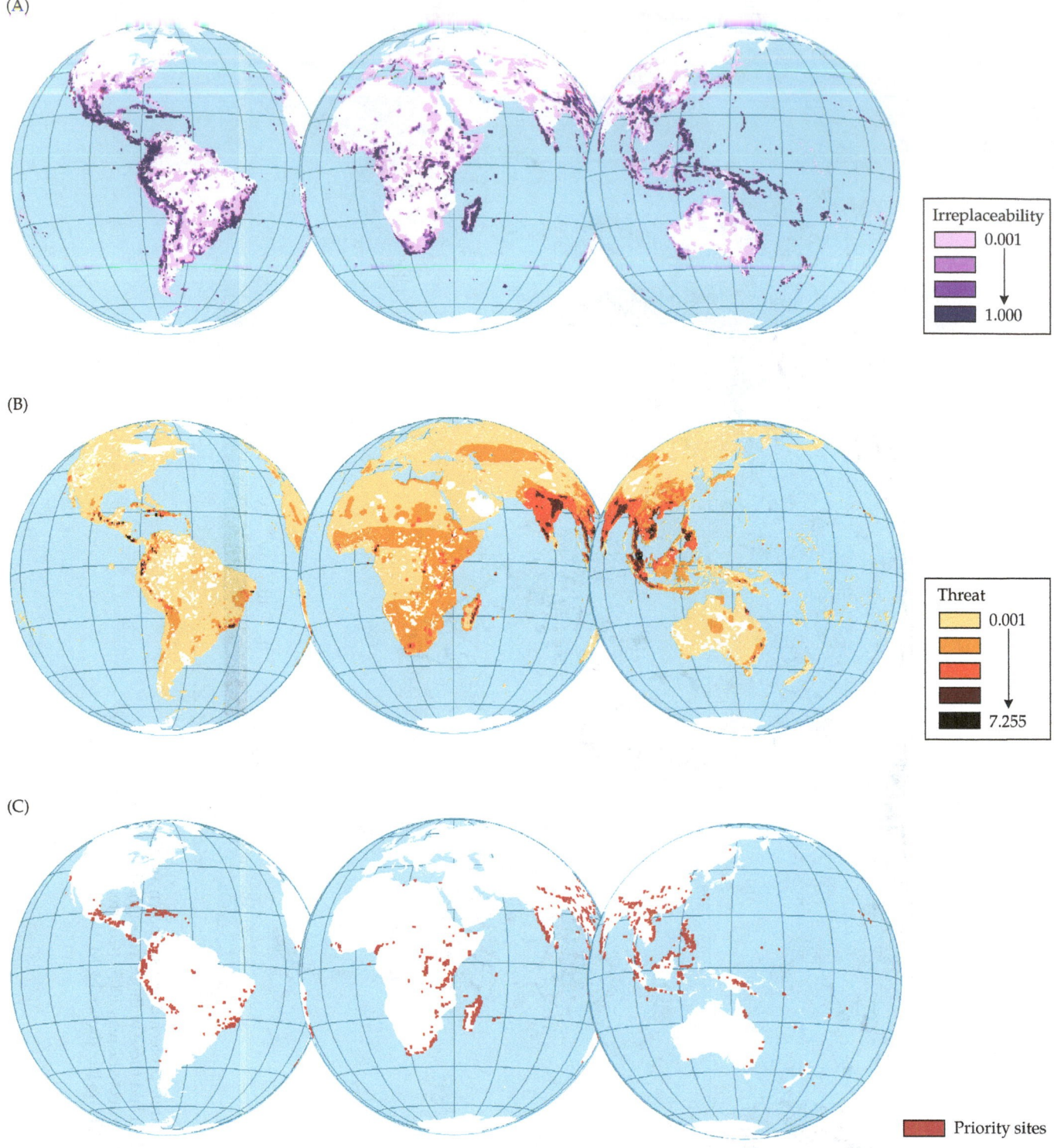

Plate 6 Global species Gap analysis. The distribution of (A) irreplaceability, (B) threat, and (C) priority sites (based on data for mammals, amphibians, turtles, and threatened birds) for the expansion of the global protected-area network (From Rodrigues et al. 2004a; ©American Institute of Biological Sciences.) [14]

BOX 11.3 Calculation of *F*-Statistics

Derrick W. Sugg, *University of Georgia, Savannah River Ecology Laboratory*

Fixation indices, or *F*-Statistics, were developed by Sewall Wright (1922, 1965, 1969, 1978) as a means to describe how genetic diversity is partitioned in a population. By partitioning genetic diversity into different components one can determine the relative amounts residing within individuals, subpopulations, and the overall population. Because adaptive evolution requires genetic variation to proceed, it is important to understand how much of the total variation is available for selection acting on individuals. More recently, conservation biologists have shown renewed interest in fixation indices because they provide a means to determine how natural populations maintain genetic variation (beneficial for developing management strategies) and to determine levels of genetic variation in threatened or captive populations (beneficial for assessing the success of management strategies).

Typically when one calculates fixation indices it is for a structured population. The classical approach is to sample individuals from different subpopulations at fairly distinct geographic locations. Such a population is said to consist of three levels of structure: individuals (*I*), subpopulations (*S*), and the total population (*T*). One calculates the average individual heterozygosity by counting the number of heterozygous individuals in a subpopulation and dividing that sum by the total number of individuals in the subpopulation. This calculation is made for every subpopulation, and the average for all subpopulations is called the average individual heterozygosity:

$$H_I = \frac{1}{k}\sum_{i=1}^{k}\frac{\#\,Heterozygotes_i}{N_i}$$

where *k* is the number of subpopulations and N_i is the number of individuals in the *i*th subpopulation. At the same time one can use those individuals to determine the frequency of the genes. The gene frequencies are used to calculate the expectations for heterozygosity in the average subpopulation $\bar{H}_S$ and the total population (H_T). The expectation for the average subpopulation is

$$\bar{H}_S = \frac{2}{k}\sum_{i=1}^{k}p_i - p_i^2$$

where p_i is the frequency of the gene in the *i*th subpopulation. The expected number of heterozygous individuals for the entire population is given by, $H_T = 2(\bar{p} - \bar{p}^2)$ where *p* is the frequency of the gene averaged over all individuals in the population without respect to the subpopulation they came from. $\bar{H}_S$ predicts the frequency of heterozygous individuals in subpopulations had they mated at random and H_T predicts the same frequency if individuals are mating at random without respect to subpopulations.

These estimates of the observed and expected frequency of heterozygous individuals can be used to calculate the fixation indices, F_{IS}, F_{IT}, and F_{ST}. Values for F_{IS} determine whether or not subpopulations have fewer or more heterozygous individuals than expected. It is calculated from:

$$F_{IS} = \frac{\bar{H}_S - H_I}{\bar{H}_S}$$

When there are fewer heterozygous individuals than expected ($\bar{H}_S > H_I$), F_{IS} will be positive. When $\bar{H}_S < H_I$ then F_{IS} will be negative. Therefore, negative values for F_{IS} indicate an excess of heterozygous individuals in subpopulations and positive values indicate the opposite condition. F_{IT} is calculated in a similar manner:

$$F_{IT} = \frac{H_T - H_I}{H_T}$$

and the interpretation of positive and negative values are the same except that they apply to the total population instead of the subpopulations. Finally, the degree of genetic differentiation among subpopulations (how unique they are) is given by:

$$F_{ST} = \frac{H_T - \bar{H}_S}{H_T}$$

which is always greater than or equal to zero. High values for F_{ST} indicate that subpopulations have very different gene frequencies, and when $F_{ST} = 1$ then subpopulations are said to be "fixed" for different genes; each subpopulation has a unique gene for each locus.

Models by Wright make simplifying assumptions including equal reproductive contributions among breeding adults and a large number of subpopulation of equal and constant size contributing dispersers to the pool of migrants. More recently, Wright's models have been recast using different methodologies or by emphasizing the importance of different evolutionary forces. Readers interested in this subject area are encouraged to read additional literature in this area including Slatkin (1991), Crow and Aoki (1984), Chesser (1991a,b), Wade and McCauley (1988), and Whitlock and McCauley (1999).

the effective size of the populations, and over longer periods of time based on the accumulation of new mutations. This process is portrayed in Figure 11.10 following Moritz (1994a,b) who considers two populations that are essentially parts of a single population. Measures of gene frequencies estimated for each population at a molecular marker such as mitochondrial DNA reveals that the two populations share the same haplotypes (A, B, and C) in nearly equal frequency. Over time, if barriers to gene flow arise and the two populations become reproductively isolated, genetic drift will result in differences in the frequencies of each haplotype to accrue. Eventually, haplotypes will be lost from one population or another. Over long periods of time new mutations

Figure 11.10 Evolution of gene frequencies within and among populations over time. Schematic representation shows that the allele frequency trajectories become increasingly variable among populations as a function of isolation time. (Modified from Moritz 1994a,b.)

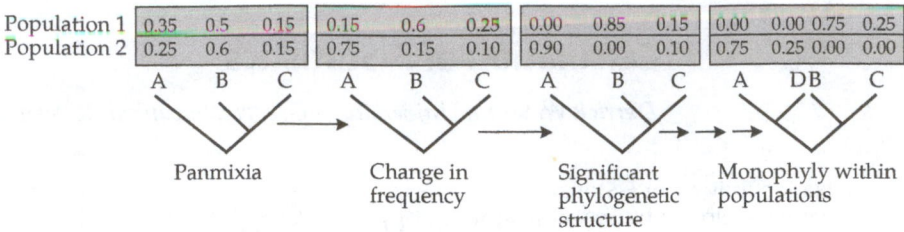

Population 1	0.35	0.5	0.15		0.15	0.6	0.25		0.00	0.85	0.15		0.00 0.00 0.75 0.25
Population 2	0.25	0.6	0.15		0.75	0.15	0.10		0.90	0.00	0.10		0.75 0.25 0.00 0.00

Panmixia — Change in frequency — Significant phylogenetic structure — Monophyly within populations

will arise in one or the other populations adding further to interpopulation divergence. Conservation geneticists have provided compelling theoretical and empirical examples showing that changes in frequency, loss of alleles or haplotypes (such as A or B), or an accumulation of new forms can be profitably used to quantify degree of distinction.

Echelle (1991) compiled hierarchical genetic data for numerous fish species of western North America. The data (Table 11.6) allowed genetic diversity to be partitioned into only three levels: within-population diversity (H_P), divergence among samples within drainages (D_{SD}) and divergence between drainages (D_{DT}). (Note that the subscripts can freely change to reflect the particular situation. There is nothing set about the subscripts used, or the levels of diversity addressed). Echelle found a great deal of variation among these species in their hierarchical patterns of genetic diversity. Some, such as *Xyrauchen texanus* (razorback sucker) and *Cyprinodon bovinus* (Leon Springs pupfish), had nearly all of their genetic variation represented as within-population heterozygosity. Others, such as *Oncorhynchus clarki henshawi* (cutthroat trout) and *Gambusia nobilis* (Pecos gambusia) had a large proportion of their diversity represented as divergence among samples

within drainages. Finally, some had appreciable variation between drainages, such as *O. clarki lewisi* (another cutthroat trout subspecies) and *C. macularius* (desert pupfish).

One prioritization strategy advocates protecting species that represent evolutionary depth (Avise 2004) by employing molecular genetic data to determine phylogeographic relationships within species and identifying the deepest evolutionary separations (Figure 11.11). Given the ability to place divergence on some relative time scale makes genetic markers advantageous for use in establishing criteria for "degree of uniqueness" for conservation purposes. For example, Avise (1992) showed that many co-distributed, but phylogenetically independent species in the southeastern U.S. exhibited remarkably similar phylogeographic patterns suggesting that prolonged periods of reproductive isolation and ge-

(A)

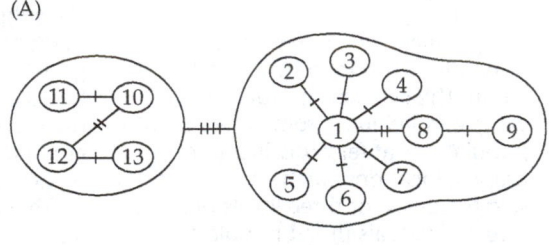

(B)

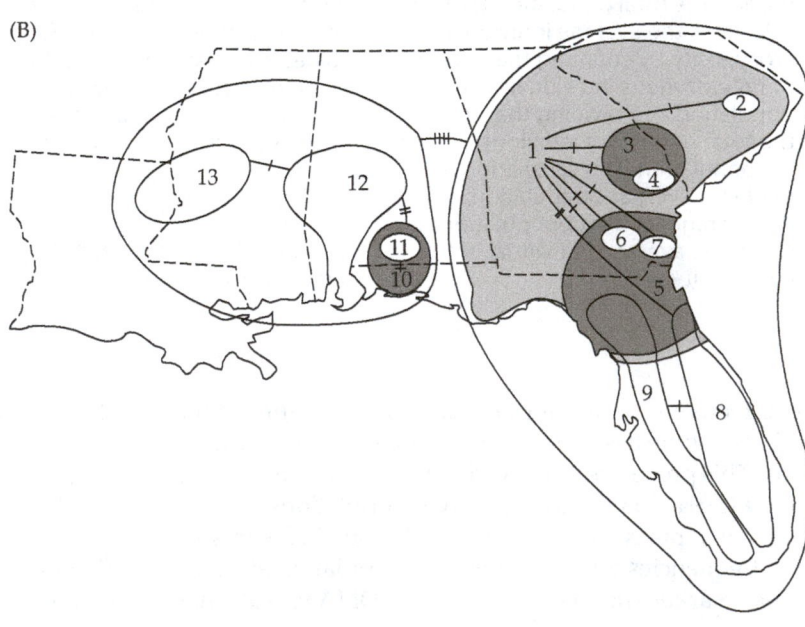

Figure 11.11 Phylogenetic relationships among mitochondrial DNA haplotypes observed in the bowfin fish (*Amia calva*) reflect deep evolutionary divisions. (A) Hand-drawn parsimony network where slashes indicate the number of base pair differences between haplotypes, which are largest between haplotypes 1–9 and 10–13. (B) The haplotype network is superimposed over a map of the southeastern United States showing the locations of origin for each haplotype. (Modified from Avise et al. 1987.)

TABLE 11.6 *Distribution of Genetic Diversity in Endangered and Threatened Fishes of Western North America*

Taxon	P/L	H_P	D_{SD}	D_{DT}
Salmonidae (trout and salmon)				
Oncorhynchus nerka	18/26	94.4	3.1	2.5
O. apache	5/35	90.5	9.5	—
O. clarki bouvieri	8/46	96.3	3.7	—
O. clarki henshawi	15/35	55.5	44.5	—
O. clarki lewisi	103/29	67.6	15.7	16.7
O. mykiss	38/16	85.0	7.7	7.3
O. gilae	4/35	86.4	13.6	—
Catostomidae (suckers)				
Catostomus discobolus yarrowi	3/45	54.8	—	45.2
C. plebeius	5/45	92.9	—	7.1
C. plebeius (second study)	4/27	11.3	—	88.8
Xyrauchen texanus	2/21	98.9	1.1	—
Cyprinodontidae (killifishes)				
Cyprinodon bovinus	5/28	98.6	1.4	—
C. elegans	7/28	89.2	10.8	—
C. macularius	3/38	70.1	—	29.9
C. pecosensis	6/28	92.3	7.7	—
C. tularosa	3/28	81.0	19.0	—
Poeciliidae (livebearers)				
Gambusia nobilis	16/24	48.4	51.6	—
Poeciliopsis o. occidentalis	10/25	59.3	40.7	—
Cichlidae (cichlids)				
Cichlosoma minckleyi	3/13	97.7	2.3	—
C. minckleyi (second study)	3/27	94.6	5.4	—
Cottidae (sculpins)				
Cottus confusus	16/33	53.9	46.1	—

Note: P/L = numbers of populations/gene loci surveyed; percentage of total genetic diversity measured is separated into heterozygosity within populations (H_P), divergence among samples within drainages (D_{SD}), and divergence among drainages (D_{DT}). Dash indicates data not measured at that level in the hierarchy.

Source: Echelle 1991.

netic divergence were common among them. Identification of areas of phylogeographic discordance thus may indicate the most distinct evolutionary lineages, which can then be used to guide prioritization strategies.

Finally, conservation geneticists have proposed a number of "rules of thumb" for prioritizing populations and taxa of evolutionary importance for conservation. Within species, habitat destruction is the leading cause of species endangerment in the U.S. (Noss et al. 1997). Populations that exist in areas threatened with development or other catastrophic events can be prioritized for

conservation using genetic criteria gained from the study of molecular genetic markers. For example, Petit et al. (1998) introduced a method whereby the relative contribution of different populations to allele diversity and to total diversity can be partitioned into two components based on diversity within the population and degree of divergence from other populations. Returning to the example used earlier from Guinand et al. (2003; see Figure 11.2), imagine that 50 years ago as lake trout spawning populations were disappearing across the Great Lakes, managers had the ability to genotype indi-

viduals from each remnant population and to quantify the relative contributions of each remnant population to the total diversity remaining. Clearly, the populations CHA48, WP48, and FR49 contributed disproportionately to the total variance of population samples (and are still in existence) at this time. Managers might have focused efforts on enhancing contributions from other populations to increase levels of allelic and total diversity. Clearly, other factors must be considered as well, but the approach portrayed offers a means of quantification.

At the level of species, several methods of quantification have also been proposed as means of prioritizing species or higher taxonomic groups for purposes of conservation. Researchers have proposed that taxa can be prioritized on the basis of phylogenetic relationships. Interrelationships among organisms can be determined using standard phylogenetic methods and methods of clustering on the basis of sharing of character states of degree of genetic similarity. Different aspects of the phylogeny have been emphasized. Inferences that can be derived from phylogenetic relationships under each of several methods are shown in Figure 11.12.

CONSERVE BASAL TAXA Basal taxa (the presumed ancestral taxa in the group under consideration) are often low in abundance and in number of species when compared with their other more recently derived groups, and frequently have extremely restricted geographic distributions. Further, their evolutionary divergence from all other organisms being considered is disproportionately large compared to other members of the group. Conservation priorities should be given to those basal taxa that contribute the most to overall phylogenetic diversity represented in the evolutionary tree.

CONSERVE SPECIES-RICH GROUPS Portions of a group of evolutionarily related species that are diverse in terms of absolute number of species should be preferentially conserved. These taxa appear to have diverged quickly, possibly indicating future potential for further adaptive radiation in response to changes in the environment.

CONSERVE SPECIES THAT ARE MOST DIFFERENT FROM ONE ANOTHER The conservation value of a taxon can be based on how different it is from other taxa by some measure of genetic distance (Crozier 1992; Crozier and Kusmierski 1994). This method can be applied to populations as well as species. Similar measures have been proposed to evaluate proportional contributions of populations or taxa to total genetic variance and allelic diversity (Petit et al. 1998).

CONSERVE THOSE TAXA THAT MAXIMIZE PHYLOGENETIC DIVERSITY **Phylogenetic diversity** (**PD**) can be derived from the branching pattern on a phylogenetic tree (a diagram of the relationship patterns among taxa). The optimal subset of organisms to conserve to maximize total diversity represented is the set that spans the greatest number of nodes on the phylogenetic tree (Faith 1992, 1994). The relative value of additional organisms can be evaluated by determining which additional organism adds the greatest number of nodes to the existing phylogeny.

All these rules of thumb share the use of phylogenetic information for prioritizing conservation efforts, although they use different criteria. There is value to such quantitative measures because criteria for conservation are based on data that are independent of societal, political, or other values. All methods have been offered as viable approaches. Clearly, knowledge of evolutionary processes and distribution of genetic variation, either adaptive vari-

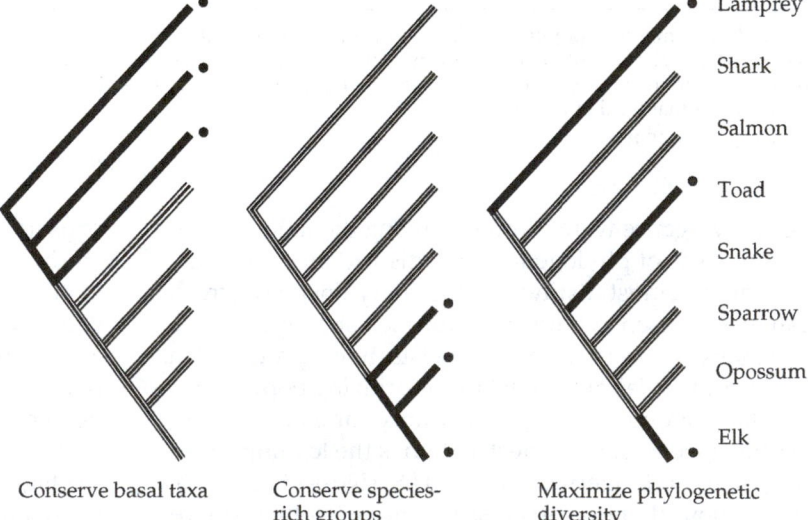

Figure 11.12 Alternative criteria for prioritizing conservation based on phylogenetic status. Examples show the use of systematic data to describe evolutionary relationships among species. Different principles can be used to offer recommendations for prioritizing species for conservation. Information on the branching order and branch length can be used similarly for different genetic variants within a species.

Conserve basal taxa

Conserve species-rich groups

Maximize phylogenetic diversity

Lamprey

Shark

Salmon

Toad

Snake

Sparrow

Opossum

Elk

ation or neutral divergence caused by isolation can provide valuable alternatives for conservation (Moritz 2002).

Genetic Information and Design and Implementation of Breeding Strategies

Early conservation biologists were alerted to the dangers of inbreeding when captive populations began to experience severe genetic defects. This led to extensive study of genetic relationships among captively bred individuals, and the creation of specialized breeding strategies.

Use of pedigrees

Pedigree analysis represents the genetic study of a multigenerational population with ancestral linkages that are known, or that can be reasonably modeled, and can be used to predict genetic changes in a population. Use of pedigrees is typically restricted to relatively small captive populations (zoos, domestic, or companion animals). Information of importance to use pedigrees for management includes the number of founders, the distribution of founder genes among living animals, the relationships among living individuals in the population, and the capacity of the population to retain genetic variation.

The concept of "genetically important individuals" has been widely used for the selective breeding of zoological and botanical collections. Genetically important individuals are those whose reproduction is most critical for the retention of genetic variability (Ballou and Lacy 1995). Implementation of breeding strategies for managed populations is based on the goals of minimizing changes in the population gene pool and retaining as much of the genetic variation of founding individuals as possible. Such intensive management is possible where pedigree information is known or can be inferred, and where selective mating is possible. One general strategy is based on maintenance of large effective population size, by maximizing the number of breeders, equalizing the number and sex ratios of breeders, decreasing the variance in reproductive success of breeding adults, and reducing fluctuations in population size over time. These principles can be more formally applied by deriving summary measures of genetic contributions within a pedigree that is either known or that can be inferred based on genetic analysis.

Estimation of degree of relatedness without knowledge of pedigree relationships

In many instances we have no, or incomplete, knowledge of genealogical relationships among individuals. We can, however, even in the absence of pedigree data, glean much information about inter-individual relationships using genotypic data obtained if many polymorphic loci are assayed (i.e., **high polymorphic information count [PIC]** or high allelic diversity and heterozygosity).

There are numerous relatedness estimators (see review in Van De Casteele et al. 2001). One measure developed by Queller and Goodnight (1989) summarizes pairwise estimates of relatedness using population allele frequencies and the number of alleles shared (0 to $2N$, where N is the number of loci assayed) by weighting relationship estimates by the likelihood they share the same alleles. For a randomly mating population, the expectation is the average level of relatedness will be 0. Individuals that share greater or fewer alleles than expected given the population allele frequency have higher estimates of pair-wise relatedness. Parent and offspring, and full siblings would on average share half of all alleles and would thus have a relatedness of 0.5.

Genetic markers were used to establish inter-individual relationships and to design a breeding program for reintroductions of a critically endangered species, the Guam Rail (*Rallus owstoni*) (Haig et al. 1990). Flightless species of birds on the island of Guam were going extinct because of introduced predators. All remaining individuals of the Guam Rail were brought into captivity. To better manage matings within the captive rail population, genetic analyses were conducted to determine genetic diversity and relatedness among founders. Simulation analyses were conducted to determine the mating strategy that would maximize retention of genetic diversity, which has proven effective.

Analyses of Parentage and Systems of Mating

Knowledge of mating systems and of the environmental, physiological, and demographic factors associated with use of different tactics is important in understanding how males and females maximize reproductive success. Mating tactics that result in polygyny (males breeding with multiple females) and polyandry (females mating with multiple males) can have marked effects on apportionment of levels of genetic variation within and among populations (Chesser 1991a,b) and effective population size (Sugg and Chesser 1993). As discussed previously, variation in reproductive success can influence effective population size. Evaluation of mating tactics and reproductive success is complex, involving ontogenetic and seasonal shifts in habitat use, movement patterns and behavioral repertoires that are inherently difficult to observe. There has been increased interest in applying mo-

lecular markers to assess biological parentage to better understand the evolutionary significance of mating systems, and to better evaluate male and female reproductive success. Studies using genetic markers have also contributed to the fundamental distinction between the social mating system (i.e., behavioral interactions typically observed in nature) and the genetic mating system (i.e., actual biological maternity and paternity). Genetically based determination of parentage has provided heretofore unobtainable information on aspects of breeding behavior and ecology needed to address important issues in conservation biology (Clemmons and Buchholz 1997; Gosling and Sutherland 2000).

Forensics and Species or Population Identification

Co-occurrence of individuals from multiple breeding populations and subspecies complicates management, especially when management objectives differ among populations. Management requires that individuals from different species or populations can be identified. However, assigning group membership on the basis of phenotype is not always possible because of environmentally induced geographic variation, and variation due to age and sex. Molecular genetic technology and statistical methods of analysis have been employed in analyses of species, populations or individual identification.

For example, efforts to restore threatened or endangered populations of Pacific salmon are often hampered by the inability to assign individuals sampled in admixtures to population of origin. Information on population assignment is of particular concern in the design of supportive breeding programs, where misidentification of adults may result in directed matings between individuals from genetically distinct populations. Production of progeny of mixed heritage results in erosion of genetic population structure and local adaptations. Olsen et al. (2000) used microsatellite loci to identify and select individuals from one spawning period for supportive breeding and prevented unintentional crosses between two genetically differentiated but co-occurring populations.

For species of conservation concern, molecular genetics has been used for forensic identification of commercial products from endangered species. Products including ivory, horn, shell, meat, feathers, dried leaves, and a host of other commercially valuable items that are derived from plant or animal materials (e.g., Baskin 1991, Milner-Gulland and Mace 1991, and Baker and Palumbi 1996). One widely acclaimed forensic case involved use of molecular markers to monitor the illegal trade in whale and dolphin products (Baker and Palumbi 1996). A global moratorium on commercial whaling was established by the International Whaling Commission (IWC) during 1985 and 1986; however whaling never actually stopped. IWC member nations were allowed to harvest small numbers of selected whale species for research purposes or for aboriginal and subsistence use. It was long suspected based on the availability of processed whale products that protected species were still being harvested and products illegally sold commercially. Baker and Palumbi (1996) used well-established statistical phylogenetic methods and known genetic relationships between protected and nonprotected species to analyze samples of whale products that were purchased in the market place (Figure 11.13). The authors were able to identify illegally harvested species and even to identify ocean of origin for some species.

Individual Identification and Estimation of Population size

Managers need information on population size, recruitment, and mortality. However, data are often not readily available. Advancements in molecular genetics have provided a means to identify individuals, facilitating mark-recapture studies (White et al. 1982). Small quantities of DNA can be routinely collected noninvasively using hairs (Woods et al. 1999; Mowat and Strobeck 2000), feathers (Pearce et al. 1997), feces (Kohn et al. 1999), and even sloughed skin (Palsboll et al. 1997; see Case Study 11.3 by Sam Wasser and Kathleen Hunt). Reviews of technology and empirical applications are provided in Cornuet et al. 1999, Mills et al. 2000, and Palsboll 1999, respectively. Often, larger sample sizes can be obtained indirectly from hair deposited on wire, feathers collected from nests, or other sources using noninvasive techniques than would be possible by handling individuals. The "mark" and "recovery" are based on observations of the same multi-locus genotype observed in samples collected on different sampling occasions.

Understanding Effects of Population Exploitation on Levels of Genetic Diversity

Many species are subjected to sport or commercial harvest. Often exploitation changes population size, sex ratio, and age structure that can have effects on genetic diversity. Scribner et al. (1991) found that male-only harvest of deer and the resulting sex ratio skew in favor of females in populations of mule deer (*Odocoileus hemionus*) contributed to observed declines in heterozygosity. Ryman et al. (1981) showed that harvest regimes that change the sex and age structure of harvested pop-

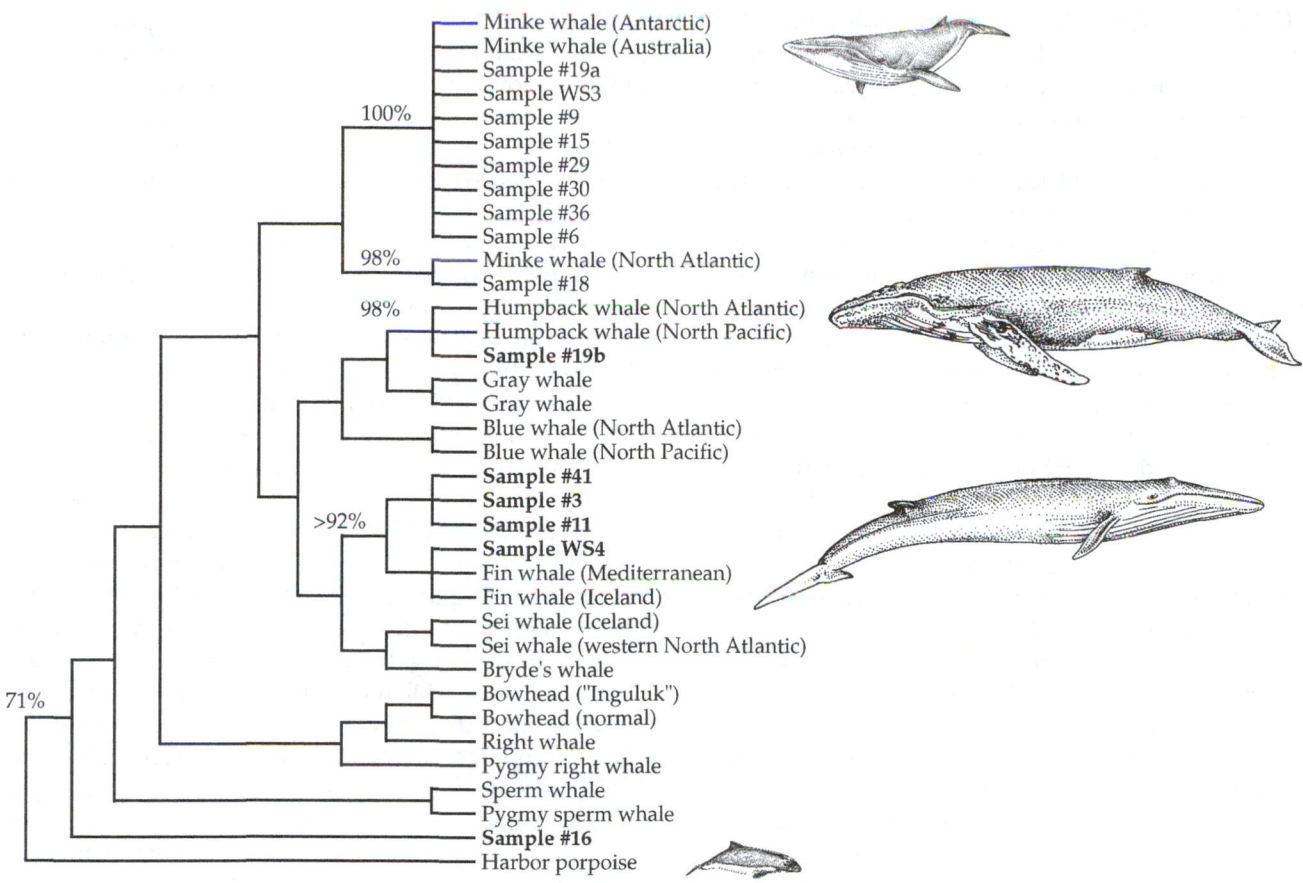

Figure 11.13 Forensic identification using mtDNA of "dolphin or minke whale meat" samples legally sold in Japanese markets. All bold faced specimens were from whale species that have not been legally harvestable since 1976, but were marketed as legal meat (dolphin or minke). (Modified from Baker and Palumbi 1996.)

ulations increased variance in reproductive success, which can greatly reduce effective population size.

One of the most widely employed uses of genetic markers has been to monitor the rates of exploitation of natural populations occurring in mixed stock marine fisheries. Pacific salmon exhibit complex patterns of spatial genetic subdivision among populations that breed in freshwater. Because of their tendency to home to locations where they were born (Quinn 1993), Pacific salmon tend to be spatially genetically structured, frequently at microgeographic scales. In the marine realm, adults undergo extensive migrations that result in the mixing of genetically differentiated populations. High rates of exploitation of commercially valuable salmonid fishes typically occur at times when members of genetically distinct populations are admixed in marine habitats, resulting in harvests of multiple populations simultaneously. Regulations are typically based on total allowable harvest without knowledge of risks to specific populations. Small populations of conservation concern are at greatest risk to overexploitation. Using statistical methods of mixed stock analysis (MSA), genetic differences among breeding populations can be used to estimate proportional contributions of populations to admixed harvests in marine habitats (Pella and Milner 1987; Shaklee et al. 1999). Similar strategies have been used to manage species of migratory waterfowl (e.g., Inman et al. 2003 and Scribner et al. 2003).

Limitations of Using Genetics in Conservation Planning

Genetic information at the population and species levels can provide important specific guidance for management and recovery programs, as well as an overall philo-

sophical perspective of the primacy of evolutionary dynamics in life's processes. The fields of evolutionary biology, and population and molecular genetics, which are the foundations of conservation genetics, are well established. Genetic technology does have limitations however, and will not alone be the savior of biodiversity. The application of genetics to conservation problems is a young science that is still developing. Many advances have been made recently that make genetic methodology more accessible to managers and conservation biologists. However, ad hoc or casual genetic analyses are not recommended. Established laboratories with experienced investigators should be consulted.

Plant or animal tissues must be obtained and properly treated. Historically, this meant that fresh or frozen samples had to be obtained from many individuals. For investigators working in remote field sites, difficulties in capture and handling methodology, and the need to return undegraded fresh or frozen samples to the laboratory precluded genetic analysis. However, with the advent of DNA technology, this can be accomplished easily and from minimal sources of material such as hairs or feathers, often noninvasively, and in a manner in which samples can be preserved for long periods at ambient (field) conditions (Palsboll 1999).

Genetics will play a pivotal role in many circumstances, such as studying small isolated populations on real or habitat islands, small numbers of charismatic vertebrates (e.g., large and rare predators or ungulates),

salmon stocks exploited by humans, and captive rearing in zoos, botanical gardens, and aquariums, and by providing critical biological data for species of conservation concern. However, a genetic approach will not save the biological diversity being lost daily in tropical forests, or the coral reefs being killed by coastal development and global climate change. Wholesale habitat destruction is a problem on a different level that cuts across genetic boundaries, and for which solutions are more economic and political than biological. We have much to learn, and we must be realistic about the limitations. Habitat availability and biological interactions and processes should be the primary focus of conservation everywhere. If the habitat is not available, and materials and energy do not flow through an ecosystem, then maintenance of genetic diversity will ultimately degrade to an exercise in ex situ care. Without suitable ecosystems and dynamic ecological processes, high levels of genetic diversity alone would not ensure long-term population viability. Woodruff (1992) stated this concisely when he said, "Genetic factors do not figure among the four major causes of extinction (the Evil Quartet; Diamond 1989): overkill, habitat destruction and fragmentation, impact of introduced species, and secondary or cascade effects." Thus, although genetic factors are major determinants of a population's long-term viability, conservationists can do more for a threatened population in the short-term by managing its habitat. Habitat management is the cheapest and most effective way of conserving genetic diversity.

CASE STUDY 11.1

Genetics And Demography of Grizzly Bear Populations

Fred W. Allendorf, University of Montana, Craig R. Miller and Lisette P. Waits, University of Idaho

Fragmentation and isolation of populations is of increasing concern in conservation. Loss of genetic variation in isolated populations of large mammals is especially serious because of their low population densities and large spatial requirements. Thus, even the largest protected reserves may be too small to maintain viable populations of large mammals.

The most useful concept to estimate the expected rate of loss of genetic variation in isolated populations is effective population size, N_e (Waples 2002). Knowledge of effective population size allows prediction of the expected time when reduced genetic variation is likely to threaten continued existence of an isolated population. In spite of universal agreement about the importance of effective population size for making

management decisions, considerable confusion persists about its estimation in natural populations.

Well-known studies with domestic animals have shown that inbreeding and the concomitant loss of genetic variation has a variety of harmful effects on development, reproduction, survival, and growth rate. Early studies with zoo populations indicated that wild species are susceptible to similar effects (Ralls et al. 1988). In the wild, where conditions tend to be harsher and more variable, the detrimental effects of inbreeding and low diversity on fitness are stronger (Keller and Waller 2002).

Several recent studies have demonstrated that these fitness reductions can increase the probability of extinction. For example, a study on adders (*Vipera berus*) showed how a long-term

decline in demographic rates and a reduction in population size to near zero were rapidly reversed by the influx of genetically diverse individuals (Madsen et al. 1999). A similar effect was observed in a Greater Prairie Chicken (*Tympanuchus cupido pinnatus*) population that had declined to very few individuals on a remnant patch of tallgrass prairie (Westemeier et al. 1998). In a metapopulation of the Glanville fritillary butterfly (*Melitaea cinxia*), subpopulations with lower heterozygosity were more likely to go extinct (Saccheri et al. 1998). It is important to recognize, however, that in all three of these cases, as in most situations, there was not a single cause of decline or extinction. Rather, extinction results from the actions and interactions of genetic, environmental, ecological, and demographic forces.

The rate of loss of genetic variation is generally measured by change in average heterozygosity per individual per locus (*H*). Heterozygosity is expected to be lost at a rate of $(1/2N)$ per generation in a theoretical "ideal" population of equal numbers of males and females that are all equally likely to contribute offspring to the next generation (Wright 1969). However, as described in the main text, a wild population of *N* individuals will lose heterozygosity much faster than $1/2N$ due to unequal sex ratios, fluctuations in population size, and nonrandom reproductive success of individuals, resulting in much greater rate of loss of heterozygosity. Effective population size (N_e) is defined as the size of the ideal population that will result in the same amount of loss in heterozygosity or change in allele frequencies as in the actual population being considered.

The number of alleles will also decline in small populations. Because distinct alleles can respond differently to evolutionary challenges, a population with high allelic diversity may be much more likely to successfully adapt than a population with low allelic diversity (Allendorf 1986). The importance of allelic diversity is perhaps most dramatically illustrated by loci associated with the immune system (MHC loci) where it is not unusual to observe more than 20 alleles at a single locus (Edwards and Hedrick 1998). Unfortunately, the concept of N_e based on the loss of heterozygosity may not be a good predictor of the loss of allelic diversity (Luikart et al. 1999). This is because rare alleles will be lost at population sizes for which there will be little loss in heterozygosity. If rare alleles compose a substantial proportion of the number of alleles in a population, their loss will result in a dramatic drop in allelic diversity. However, because rare alleles have little effect on heterozygosity, their loss will not be reflected in the decline in heterozygosity.

A variety of methods provide estimates of N_e under different violations of the assumptions of the ideal population (Waples 2002). These methods incorporate estimates of demographic parameters such as age at first reproduction, life span, variance in reproductive success, and the number of breeding individuals of each sex. Several problems make it difficult to use these estimations in wild populations. First, these formulas cannot be combined to estimate rate of loss of genetic variation in a wild population because many of the assumptions are not likely to hold. Second, many of the parameters needed to estimate N_e with these formulas are virtually impossible to estimate in wild populations. Finally, most populations do not consist of a single, randomly mating group. Existing formulas for estimating N_e have not been designed to incorporate effects of gene flow between geographically separated local populations.

The following example from an isolated grizzly bear population (*Ursus arctos*) introduces two other approaches to estimate N_e and illustrates how N_e can be used in a conservation context. In 1975, the grizzly bear was listed as threatened under the U.S. Endangered Species Act. The number of grizzly bears in the contiguous 48 states has declined from an estimated 100,000 in 1800 to less than 1000 at present (Servheen 1999). The inhabited range of the species within this area is now less than 1% of its historic range. The current verified range of the grizzly bear is approximately five million hectares in six separate subpopulations in four states (Servheen 1999). The range reduction isolated subpopulations because continuous habitat was divided and movement corridors disappeared. Population decline accelerated because these isolated subpopulations were small and subject to stochastic demographic influences.

An estimation of the rate of loss of genetic variation in grizzly bear subpopulations is needed to determine population sizes required to maintain viable subpopulations. Moreover, it is also important to determine what management actions can be taken to reduce the rate of loss of genetic variation in the remaining subpopulations. Initial minimum viable population size (MVP) guidelines for the grizzly bear were based on a comprehensive series of computer simulations of demography that did not consider genetics (Shaffer and Sampson 1985). Initial recovery targets for some of the six subpopulations were less than 100 individuals, a size that will lose genetic variation at a rate likely to decrease fitness if the subpopulations are isolated.

We developed a simulation model to estimate effective population size of grizzly bear populations (Harris and Allendorf 1989). The model is a discrete-time, stochastic computer program that follows the history and kinship of each individual (in simulated populations). Values of life history parameters used in the simulations were taken from studies of grizzly bear populations in Montana, Wyoming, and British Columbia. We compared the loss of heterozygosity per generation in model populations to that expected in ideal populations to estimate N_e (Figure A). Our results indicated that the effective population size of grizzly bears is approximately 25% of census size (Allendorf et al. 1991).

An alternative approach is to estimate N_e directly from the amount of genetic drift (as measured by changes in heterozygosity and allele frequencies) observed over time at molecular genetic markers. By definition, heterozygosity will be lost at a rate of $(1/2N_e)$ per generation. N_e can be estimated by measuring the actual loss in heterozygosity at many loci over several generations (Waples 2002). The estimated N_e can be divided by an estimate of *N* over the same interval to obtain an estimate of

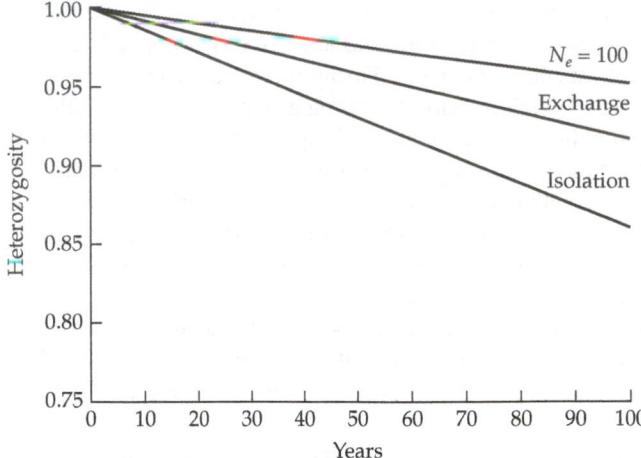

Figure A Loss of heterozygosity in populations of 100 grizzly bears. The top line is the expected rate of loss if the population behaved as an ideal population of 100 individuals. The bottom line shows the rate of loss estimated by computer simulations in an isolated population of 100 bears. The middle line shows the effect of introducing two unrelated bears every generation (10 years) into the population of 100 bears.

TABLE A *Genetic Variation in Brown Bears from North America*

Location	Allozymes		Microsatellites		mtDNA	
	H_e	A	H_e	A	h	A
Alaska/Canada	0.032	1.2	0.763	7.5	0.689	5
Kodiak Island	0.000	1.0	0.265	2.1	0.000	1
NCDE	0.014	1.1	0.702	6.8	0.611	5
YE	0.008	1.1	0.554	4.4	0.240	3

Note: NCDE, Northern Continental Divide Ecosystem; YE, Yellowstone Ecosystem; H_e, mean expected heterozygosity; A, average number of alleles observed; h, diversity. (h is computationally equivalent to H_e, but is termed diversity because mtDNA is haploid so that individuals are not heterozygous.)
Source: Data from Allendorf; Waits et al. 1998.

N_e/N. This empirical approach has the advantage of not requiring detailed information about demographic rates, but it does require having an accurate field-based estimate of N. Since the estimate of N_e will represent the harmonic mean of N_e over the interval, the estimate of N must also be the harmonic mean of the census sizes (Waples 2002).

We have used the empirical method to estimate N_e in grizzly bears from the Yellowstone ecosystem (YE) taking advantage of the historical tradition of collecting museum specimens. The Yellowstone grizzly bear population from the 1960s was sampled by extracting DNA from bone inside museum skulls. A genetic sample from the contemporary population is available from other work (Paetkau et al. 1998). We have estimated the harmonic mean of N_e at around 75 using the observed changes in allele frequencies at eight microsatellite loci (Miller and Waits 2003). Estimates of population size in Yellowstone between the 1960s and 1990s are surrounded by large uncertainty, but a summary suggests a harmonic mean N around 280 individuals. Combining these values yields an estimate of $N_e/N = 27\%$—remarkably similar to the 25% estimate obtained from the simulation approach.

Grizzly bears in the Yellowstone ecosystem have been isolated from other grizzly bear populations for at least 100 years. YE grizzly bears have substantially less genetic variation at allozyme loci, microsatellite loci, and mtDNA than their nearest neighbor population, the Northern Continental Divide ecosystem (NCDE) (Table A). Genetic analysis of museum specimens of grizzly bears suggests that most of this difference was present 100 years ago and is not due to genetic drift since the time of their isolation (Miller and Waits 2003).

The effects of inbreeding depression on survival are often measured by the mean number of "lethal equivalents" per diploid genome. A lethal equivalent is a set of deleterious alleles that would cause death if homozygous. Thus, one lethal equivalent may either be a single allele that is lethal when homozygous, two alleles each with a probability of 0.5 of causing death when homozygous, or 10 alleles each with a probability of 0.10 of causing death when homozygous. The range of lethal equivalents per individual estimated for captive mammal populations ranges from about 0 to 30. The median number of lethal equivalents for captive mammals was estimated to be 3.14 (Ralls et al. 1988). This corresponds to about a 33% reduction of juvenile survival, on average, for offspring with an inbreeding coefficient of 0.25.

Incorporation of inbreeding effects with a computer simulation model for population viability analysis (VORTEX, Lacy 1993) suggests that YE grizzly bears may be vulnerable to the genetic effects of their isolation (Allendorf and Ryman 2002). The effects of inbreeding depression on extinction were modeled in a population of approximately 280 bears (the harmonic mean of the population size for YE since the 1960s; see Figure A). The life history and demographic data for these simulations are from an analysis of grizzly bear recovery in the Rocky Mountains of the United States (Allendorf and Ryman 2002). In the absence of any inbreeding depression (zero lethal equivalents), the persistence probability is similar in the first and second hundred years of the simulations. However, even moderate inbreeding depression (2.5 lethal equivalents per genome) results in a greatly decreased probability of persistence in the second hundred years.

In 1995, a federal judge ruled that the U.S. Fish and Wildlife Service "failed to meet its obligation under the ESA to incorporate into the Grizzly Bear Recovery Plan objective, measurable criteria addressing genetic isolation." Long-term recovery of YE grizzly bears must entail reconnecting it with other grizzly bear populations to maintain genetic variation and avoid potential problems associated with inbreeding depression. How much exchange is appropriate? We extended our initial

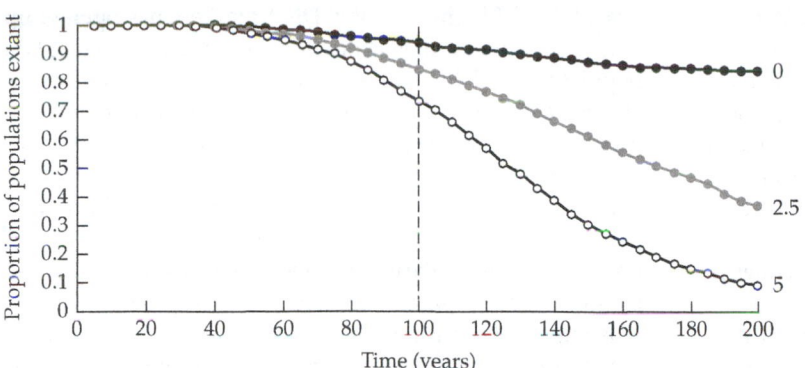

Figure B Effects of inbreeding depression on the persistence of a grizzly bear population based on a computer simulation model (VORTEX). Each point represents the proportion of 1000 simulated populations that did not go extinct during the specified time period. Simulated populations began with 50 bears and had a carrying capacity of 280. Inbreeding depression was incorporated as a different number of lethal equivalents (0, 2.5, and 5.0).

simulations to determine the amount of gene flow needed to reduce the rate of loss of genetic variation in subpopulations. The introduction of two unrelated bears each generation greatly reduced the rate of loss of genetic variation (see Figure A). This agrees with analytic results that have shown that even one migrant per generation is expected to limit genetic divergence among subpopulations (Wright 1969). This amount of gene flow is also appropriate for maintaining the historical amount of genetic divergence between YE and NCDE grizzly bears (Miller and Waits submitted). We therefore recommend that one or two effective migrants per generation from NCDE to YE is an appropriate level of gene flow. N_e is unlikely to be so small in Yellowstone grizzly bears that genetic deterioration poses an immediate threat (Figure B). Nevertheless, it must be remembered when interpreting this figure that the YE has already been isolated for 100 years. The best option available to establish gene flow is natural connectivity by restoring and protecting habitat and promoting the establishment of intervening populations. It is unlikely that the possibility of achieving natural connectivity will exist in several decades if we do not make it a priority now. However, if efforts to restore natural gene flow are not pursued or if they fail, then the artificial transplantation of a small number of bears each generation will be necessary to ensure that the Yellowstone grizzly is not threatened by genetic factors. We must also remember that although important, genetic management must not distract from addressing the most pressing threats to the Yellowstone grizzly: habitat loss and direct human mortality.

CASE STUDY 11.2

Using Genetic Analyses to Guide Management of Pacific Salmonids

Robin Waples, Northwest Fisheries Science Center, Seattle

The ecological, economic, cultural, educational, and aesthetic importance of Pacific salmon (*Oncorhynchus* spp.) to the land and peoples of the Pacific Northwest is so great that they are of high conservation value by almost any standard that could be applied. Management of these anadromous species presents a number of difficult challenges: Salmon require high quality freshwater habitat for spawning and rearing; they must undergo extensive freshwater and marine migrations within narrow time windows to successfully complete their life cycle; and they traverse multiple local, state, tribal, and international jurisdictions. As a result of numerous human and natural factors, populations of most species are in widespread decline.

Pacific salmon also hold considerable interest for evolutionary biologists (Hendry and Stearns 2004). Legendary for their long migrations and precise homing ability in which adults return to natal rivers to spawn, Pacific salmon often show strong evidence for local adaptation—genetic fine-tuning to local environmental conditions and physical features of the habitat. Nevertheless, salmon are good colonizers and have demonstrated the ability to evolve rapidly when conditions are favorable. The unusual life history traits of Pacific salmon—they all die after reproduction (like species with discrete generations) but have variable age at maturity (like species with overlapping generations)—mean that standard population genetics models have to be modified before they can be applied to these species (Waples 2004).

Pacific salmon were among the first species to be examined for genetic polymorphisms in natural populations (Utter et al.

1973), and genetic information has played an important role in conservation and management for several decades. An early application addressed one of the most pervasive problems in salmon management: Harvests typically occur at sea, where co-occurrence of fish from many discrete populations makes it difficult to target abundant (often hatchery supported) stocks while minimizing take from populations of conservation concern. This problem can be addressed by taking advantage of naturally occurring genetic markers that differ in frequency among spawning populations.

Although adult salmon in general display a remarkable ability to find and return to their natal rivers to spawn, a background level of straying, or return to another river, means that population differences will be modest for any individual genetic marker. Thus, integrating information across many gene loci is needed to have sufficient statistical power to resolve any complex mixtures. Coordination among laboratories has facilitated compilation of extensive datasets of "baseline" allele frequency data from individual spawning populations, which then can be used to analyze mixtures of unknown origin. In one recent application (Seeb et al. 2004), baseline data for over 350 chum salmon (*O. keta*) populations around the Pacific Rim were used to evaluate stock composition of high-seas harvests in the North Pacific Ocean. Evidence was found for strong seasonal changes in stock composition at fixed localities, and this information helps managers to shape fisheries to maximize harvest of healthy populations and minimize effects on at-risk populations. Results also showed that (1) the huge Bering Sea fishery for walleye pollock (*Theragra chalcogramma*) takes immature fish from throughout the Pacific Rim as bycatch, and (2) North American chum salmon migrate far to the west and are captured off Kamchatka.

Sometimes it is important to know the origin of individual fish, rather than the average composition of a mixture. Recent advances in molecular genetic and statistical methods have made it possible to "assign" individuals to their population of origin with high probability, based on their multi-locus genotype. This can be particularly important in ensuring brood stock integrity for captive propagation programs for endangered populations. In the Sacramento River, four different seasonal "runs" of Chinook salmon (*O. tshawytscha*) co-occur, with their names reflecting the season in which adults enter freshwater to begin their spawning migration. Genetic analysis of spawners used in a captive program for the endangered winter-run population indicated that some spring-run fish had inadvertently been incorporated into the brood stock, and the program was suspended. A software program (Banks 2000) was developed to allow real-time screening of maturing adults, based on genetic variation at a suite of DNA markers that can be obtained from fin clips. Because this process allows managers to exclude virtually all spring-run fish while excluding only a small fraction of fish that are actually winter run (Hedgecock et al. 2001), the captive program has been restarted.

Both of the above methods depend on mean differences among populations in genetic traits. The recent widespread availability of highly variable DNA markers has opened up new opportunities for genetic analyses that do not depend on population-level differences. These markers, through a process akin to DNA fingerprinting, allow one to identify parents of individuals. If a large sample of offspring can be assigned in this way to specific parents, one can directly evaluate how different traits affect fitness. For example, Seamons et al. (2004) used parentage analysis to examine how adult size and time of reproduction affected the number and size of offspring produced by steelhead (*O. mykiss*) spawning in a stream in western Washington. Surprisingly, large females produced only marginally more offspring than small females, and large males produced more offspring than small males in only one of four years—suggesting that the factors determining reproductive success in the wild are more complex than previously thought. Results also showed that each year only a few of the adults produced most of the offspring, a pattern that leads to substantial reduction in effective population size (N_e) compared to census size and consequently, a more rapid loss of genetic diversity. Currently, a number of studies are underway in the Pacific Northwest that will use parentage analysis to evaluate the relative reproductive success of hatchery-origin fish that spawn in the wild. Given the huge scale of hatchery operations and the depressed state of many wild populations, understanding the long-term effects of hatchery fish on diversity and fitness of wild populations is one of the most pressing conservation challenges.

These examples demonstrate the wealth of information that can be gleaned from presumably neutral genes, which serve as markers for evaluating questions of interest. But populations also differ in many morphological and life history traits that are likely to be directly related to fitness. In assigning conservation priorities, how does one decide which traits or populations are most important? Has a particular trait evolved many times within the species (hence it might be regenerated again in the future if lost) or only once (in which case populations expressing this trait might have high conservation value)? These questions can be answered by joint analysis of genetic and life history data. For example, seasonal run timing is often used by managers to help define management and conservation units for salmon. When variation in run timing of Chinook salmon from California to British Columbia was mapped onto a tree depicting genetic relationships of the same populations, two striking geographic patterns emerged (Figure A).

In coastal basins and the lower Columbia River, many distinct genetic lineages include populations with different run timing. In these areas, genetic differences between populations with different run timing but within the same river basin are small enough that they can be explained by fairly recent divergence, or more likely, low levels of ongoing gene flow (see Figure A). This result thus provides strong evidence for repeated, parallel evolution of this life history trait. In sharp contrast, a very different pattern is found in the interior Columbia and Snake River Basins, east of the ecological barrier represented

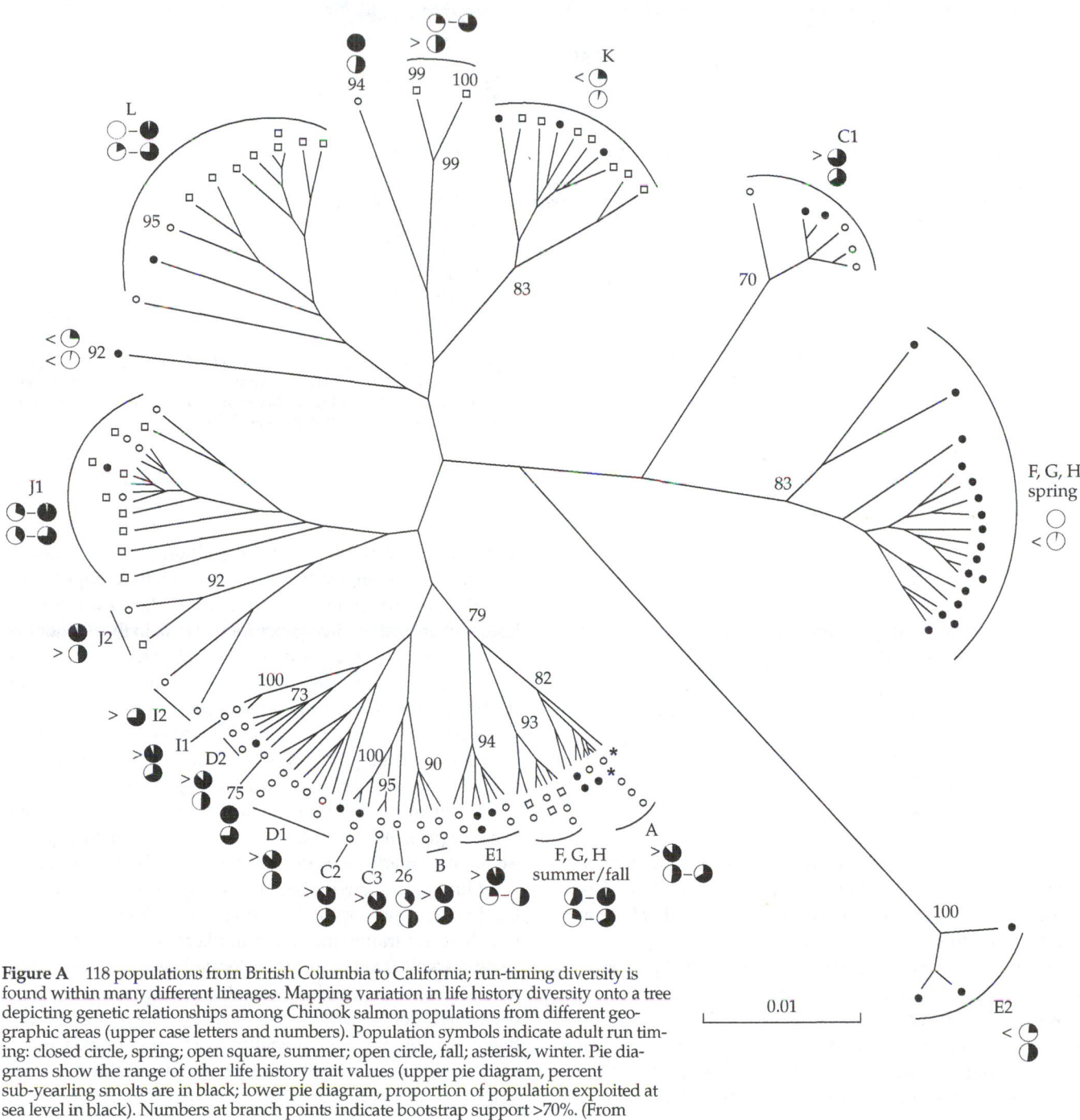

Figure A 118 populations from British Columbia to California; run-timing diversity is found within many different lineages. Mapping variation in life history diversity onto a tree depicting genetic relationships among Chinook salmon populations from different geographic areas (upper case letters and numbers). Population symbols indicate adult run timing: closed circle, spring; open square, summer; open circle, fall; asterisk, winter. Pie diagrams show the range of other life history trait values (upper pie diagram, percent sub-yearling smolts are in black; lower pie diagram, proportion of population exploited at sea level in black). Numbers at branch points indicate bootstrap support >70%. (From Waples et al. 2004.)

by the Cascade Mountains. In this region, all spring-run populations form a single evolutionary lineage that is strongly differentiated from all fall-run populations, and in fact the two life history types behave largely as separate species where they overlap in distribution (Figure B). Interior spring-run populations also have a unique suite of tightly correlated life history traits (juvenile and ocean migration patterns; run and spawn timing) that perhaps has evolved only once within the species. Because of these unique life history traits, these interior populations utilize freshwater and marine habitats in very different ways than do coastal populations, and this diversity helps maximize overall productivity and provides resilience for the

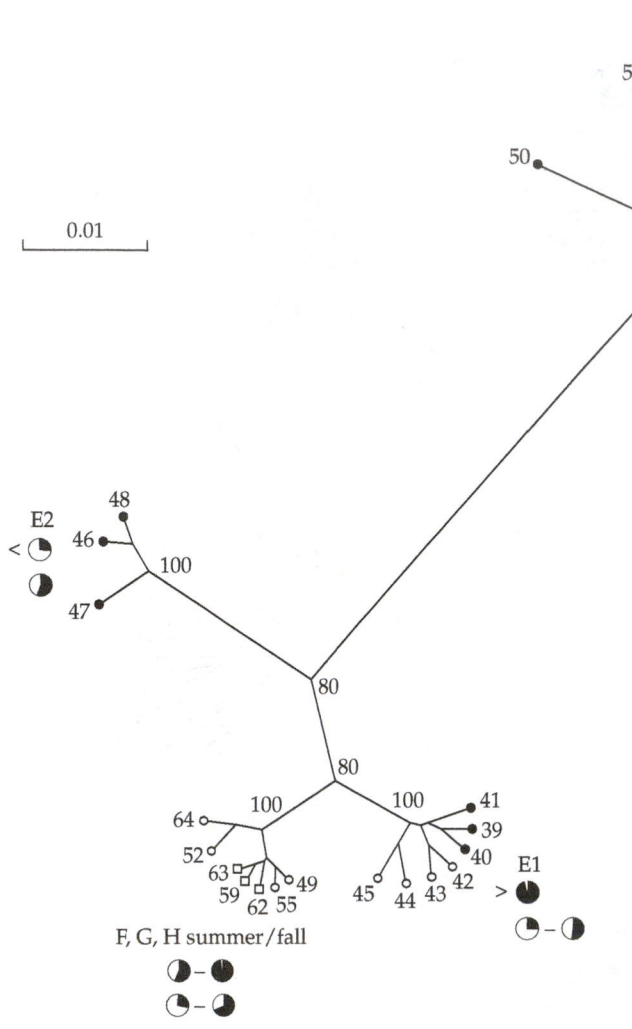

0.01

Figure B Thirty-four populations from the Columbia River; interior populations (F, G, H) form two separate genetic lineages (spring versus summer/fall). Other life history traits also separate these lineages in the same pattern. (From Waples et al. 2004.)

http://www.nwr.noaa.gov/1salmon/salmesa/fractlist.htm). In Chinook salmon, many coastal ESUs include populations with different run timing, reflecting the shallow evolutionary history of life history divergence for this trait. In the interior Columbia basin, however, spring-run and fall-run populations are in separate ESUs, consistent with data showing these life history traits have separate evolutionary origins.

Recent advances related to the human genome project have brought us to the point where we can begin in earnest the study of salmon genomics—the comprehensive study of whole sets of genes and their interactions rather than single genes or proteins. A couple of approaches in particular promise important advances in the near future, both directed toward finding sections of the genome responsible for fitness related traits. One approach combines breeding experiments with high-saturation molecular markers, in hopes of finding some markers that happen to be located near to genes coding for traits of interest, such as growth rate or disease resistance. Once identified, these markers can greatly facilitate selective breeding programs for aquaculture and agriculture. This approach also can provide information of considerable conservation value, as information about adaptive differences could be important in determining conservation priorities, in managing captive propagation and reintroduction programs, and in understanding the relationships between biodiversity and sustainability (Luikart et al. 2003). Another approach is to sift through data for large numbers of molecular markers sampled in multiple populations, looking for genes that differ among populations much more than the average neutral marker. These "outlier" genes are considered likely candidates to be under strong selection for local adaptations, and they can be targeted for more detailed analysis.

species as a whole against inevitable environmental fluctuations and climate change. For many reasons, therefore, these interior Chinook salmon populations are of considerable conservation importance.

This type of information has been instrumental in identification of formal conservation units of Pacific salmon. The past decade has witnessed considerable debate about how best to define evolutionarily significant units (ESUs) to help guide conservation efforts (reviewed by Fraser and Bernatchez 2001). Under the U.S. Endangered Species Act (ESA), populations of vertebrates (e.g., Bald Eagles and grizzly bears) can be listed if they represent "distinct population segments," or DPSs—a vague term not defined in the statute. To facilitate identification of DPSs of Pacific salmon, Waples (1995) developed an ESU concept that utilizes genetic, ecological, and life history information to define groups of populations that represent major components of diversity within the species as a whole. In the contiguous U.S., over 50 total ESUs have been identified in six salmon species, about half of which are now listed as threatened or endangered species under the ESA (for a current list see

CASE STUDY 11.3

Scat: Singing the Wildlife Conservation Blues

Samuel K. Wasser and Kathleen Hunt, Center for Conservation Biology, University of Washington

The pressures facing wildlife are often multidimensional. Landscapes vary widely over space and time and one pressure seldom occurs in isolation of others. Habitat destruction can restrict movement, limit food, create opportunities for invasive species and emergent diseases to take hold, salt the area with pollutants, and so on. This creates a dilemma. We need to be able to distinguish between these pressures to understand how to properly mitigate them. And, in many cases, we need to monitor this on a landscape scale to be sure that our results are broadly applicable; otherwise, those culpable can simply argue that the results don't apply to them. Lastly, methods used to acquire these data need to be sufficiently valid to stand up to public challenge.

Some of the most important information needed to address large scale conservation issues include temporal changes in the abundance, distribution, and physiological health of wildlife in relation to environmental pressures. Fortunately, many tools are now becoming available to enable researchers to acquire such data on a landscape scale and diverse array of species (e.g., Kohn et al. 1999, Woods et al. 1999, Turner et al. 2003, and Silver et al. 2004). This case study describes applications, of noninvasive genetic and physiological methods using fecal samples (scat), including novel ways of collecting these samples.

Feces are the most available wildlife product in the environment. Face it: Everyone poops. Feces or scat is extremely persistent in the environment. It does not soak into the ground like urine. Most importantly, it contains a treasure trove of biological information, including DNA (from the host, and also from food, parasites, and gut bacteria; Wasser et al. 1997; Garrido et al. 2000; Deagle et al.2005), hormones that reflect physiological stress (Goyman et al. 1999; Wasser et al. 2000; Young et al. 2004), as well as reproductive and nutritional health (Brown et al. 1997; Velloso et al. 1998; Rolland et al. 2005), dietary components, parasites, pathogens, immune proteins, toxins, and more (Guillette et al. 1994; Graczyk et al. 2001; Hayes et al. 2002). Recent methods using specially trained scat detection dogs have greatly improved the ability to acquire these valuable samples over large remote areas (Wasser et al. 2004a).

We present two case studies that combine fecal hormone, DNA, and detection dog methodologies to address important conservation problems. Hopefully, this overview will illustrate the enormous potential that currently exists to tackle the information acquisition challenge we are now facing to conserve our planet.

African Elephant (*Loxodonta africana*)

Poaching decreased African elephant (*Loxodonta africana*) populations from 1.3 million to 500,000 individuals between 1979 and 1987 (Said 1995). The United Nations Convention on International Trade in Endangered Species (CITES) responded by instituting an ivory ban in 1989. But demand for ivory has continued, fueled largely by markets in Japan and China. Forest elephants (*L. cyclotis*), now believed to be a different species from the savannah elephants (*L. africana*) (Roca et al. 2001; Comstock et al. 2002), have been hit especially hard because logging and associated road building has recently opened up their previously impenetrable habitats to poachers. Pressure on CITES to reopen the ivory trade has been especially intense from several southern African nations (Botswana, Namibia, South Africa, Zambia, and Zimbabwe). These countries argue that their elephant populations have become too large and damaging to their habitat and that culling—killing entire herds—is the only viable way to control them. They want to sell their ivory to help pay for these expensive culling efforts. However, many other nations, led by Kenya and India, are concerned that relaxing trade restrictions will escalate poaching rates across the African continent.

Our Center is using fecal DNA and hormone measures to address two important conservation questions pertaining to the illegal elephant ivory trade: (1) What are the long-term impacts of poaching on African elephants? More specifically, how has the selective targeting of large adult females for their tusks affected group stability and functioning? This question is being asked by Kathleen Gobush, a graduate student in our Center. (2) Can we use genetic tools to track the origin of poached ivory, helping wildlife authorities determine where poaching is most concentrated and how international trade decisions may affect illegal trade in ivory across the continent?

The measurement challenges here stemmed from the unusually large fecal mass of the African elephant. We had to show that we could reliably measure DNA and hormone concentrations in a 12 kg fecal mass, particularly since neither hormones or DNA are evenly distributed in scat. We identified and tested 16 microsatellite DNA markers for African elephants, initially using blood from a captive African elephant (Comstock et al. 2000, 2003), and later extending these techniques successfully to wild elephant feces (Wasser et al. 2004b). We now know; (1) how to process and preserve the sample in the field, (2) how important is lack of rain is as a stressor in this species (Foley et

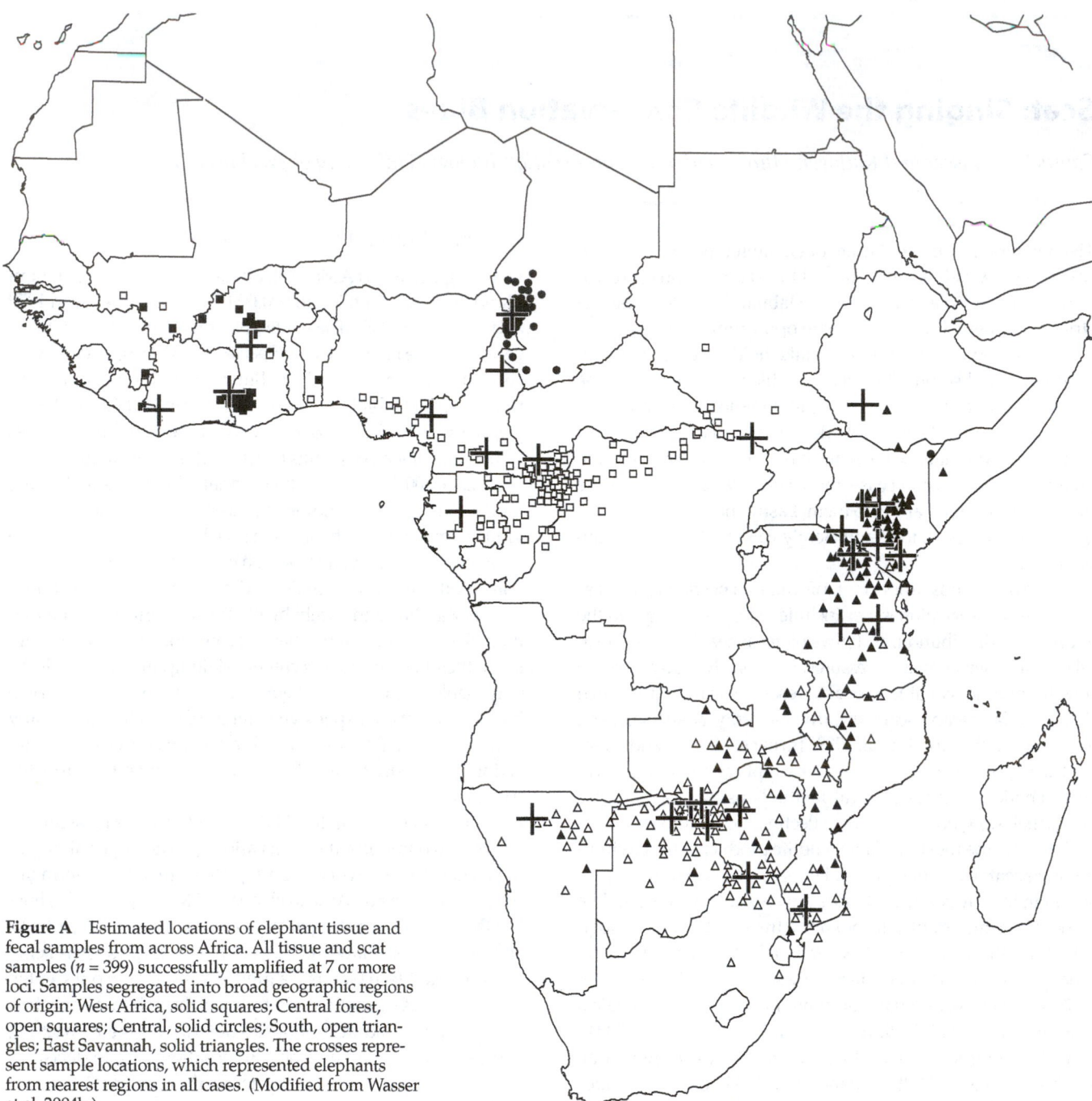

Figure A Estimated locations of elephant tissue and fecal samples from across Africa. All tissue and scat samples (*n* = 399) successfully amplified at 7 or more loci. Samples segregated into broad geographic regions of origin; West Africa, solid squares; Central forest, open squares; Central, solid circles; South, open triangles; East Savannah, solid triangles. The crosses represent sample locations, which represented elephants from nearest regions in all cases. (Modified from Wasser et al. 2004b.)

al. 2001), and (3) the need to control for rainfall effects when examining stress-related long-term impacts resulting from loss of matriarchs in these elephants across seasons and years. These results provide the bases needed for Kathleen Gobush to determine how loss of adult females affect the stress and reproduction of adult females in elephant groups.

The above genetic tools have also enabled us to develop methods to evaluate the extent and distribution of the illegal ivory trade across Africa. Fecal DNA is being used to rapidly build a continent-wide geographic-based map of elephant allele frequencies for 16 microsatellite DNA loci. Zambian and Malawian authorities were able to sample their entire countries for us in just two weeks simply by asking park rangers to collect fresh scat while out on routine patrols. We are now able to estimate the geographic origin of poached elephant tusks across Africa by matching DNA extracted from the ivory to the geographic-based allele frequencies (Wasser et al. 2004b; Figure A).

We are currently using this genetic tracking method in a collaborative investigation with INTERPOL and the Lusaka Agreement Task Force to determine the origin of the largest illegal ivory seizure in the history of the trade. In June 2002, 6.5 tons of illegal ivory were seized in Singapore, including 532 raw elephant tusks and 42,000 ivory hanko blanks (used in Asia to imprint family seals on letters). This shipment of ivory originated in Zambia, and then passed through Malawi and South Africa before being seized Singapore. We are determining the origin(s) of these ivory samples as we write this. We soon hope to use these same methods to monitor change in the movement and distribution of ivory in the major markets around the world, allowing us to track how international trade decisions influence poaching rates across the continent.

Grizzly Bears

Grizzly bears (*Ursus arctos horribilis*) are threatened throughout their range. Their numbers have been dramatically reduced by over-hunting, poaching, and habitat destruction (McLellan et al. 1999). Yet, these top-level carnivores can play an important role in the ecosystem, regulating the abundance of prey species as well as other, more prolific, sympatric carnivores such as black bears. Their large home range (350 km^2) also makes them an important umbrella species since preserving their habitat will simultaneously preserve habitat for many sympatric species. Pressure persists to continue hunting these charismatic bears, as does poaching for their body parts such as claws and gall bladders used in traditional medicine throughout Asia. This has created the need for tools to monitor change in the abundance, distribution and physiological health of this species over time in relation to environmental disturbances across large, remote areas.

One of the major problems of grizzly bear management is simply obtaining an accurate estimate of population size, since the bears are thinly distributed over huge geographic areas. Most methods of population estimation for bears rely on mark-recapture analysis, named from the original practice of actually catching and marking individuals, releasing them into the population, and then catching a second set of individuals and seeing what proportion include the originally marked individuals. In practice, mark-recapture analysis can be used in any situation in which individuals can be identified.

John Woods and colleagues (Woods et al. 1999) extended mark-recapture analysis to measure grizzly bear abundance using hair DNA from "hair snags," an apparatus designed to catch a tuft of hair from a bear. A circle of barbed wire is positioned at chest height to the bear, with a strong scent, such as aged cow's blood laced with fish oil, placed in the center of the circle. Bears are attracted to this awful stench and will crawl under the wire to investigate it, leaving a tuft of hair with its follicles on the barbs. Accurate sampling of an entire population of bears requires careful positioning of these hair snags over large remote areas and regularly re-positioning them over the sampling period. DNA extracted from the hair follicles is then used to "mark" each bear by determining its species, sex, and individual identity. Data from this fairly noninvasive method can then be used in mark-recapture analyses to estimate the population size and distribution of bears in the study area.

The hair snag technique is a powerful method that is being widely applied in grizzly bear studies (Paetkau et al. 1998; Woods et al. 1999; Mowat and Strobeck 2000). However, like any method, it has its limitations. Mark-recapture analysis requires that every animal have an equal chance of being "caught" (snagged, in this case). But with hair snags, this is generally not the case. Males venture into snags more frequently than do females, and females with cubs generally avoid hair snags (Woods et al. 1999). Researchers have developed elaborate statistical procedures to correct this problem (White and Burham 1999), although methods that provide more equal catchability are still preferable. The use of scented lures potentially causes another problem by artificially drawing bears into an area from long distances away.

Having developed methods to extract DNA from bear feces (Wasser et al.1997a), we began to pursue alternative sample collection methods using scat. However, visually searching for scat can also lead to sampling biases, since some animals may deposit their feces more conspicuously than others (e.g., males may use feces as a territorial mark). To remedy these problems, we borrowed methods for training narcotics and bomb detection dogs to instead train dogs to detect wildlife scat (Wasser et al. 2004a). These dogs proved to have remarkable detection abilities. They can locate scat at distances as far as 0.5 km away from numerous species at once. And, if well-trained, they rarely make a mistake (Smith et al. 2001; Wasser et al. 2004a).

Scat detection dogs are carefully selected for their trainability, high-energy play drive, and obsessive personality. In short, we select dogs that are crazy for a tennis ball. In training, we pair the tennis ball reward with sample detection. These dogs have their dream job—running all day in the woods looking for bear feces and being rewarded by playing with a tennis ball. These obsessive dogs will enthusiastically search all day long for samples under these circumstances. Moreover, sampling is fairly unbiased because what the dogs really want is their ball, and they know that the only way to get it is to find a sample from the target species, regardless of the age, sex, or reclusiveness of the individual that left the sample.

Starting in 1999 we applied these methods to the study of grizzly and black bears in the Yellowhead ecosystem of Alberta, Canada. Forty percent of this study area includes a portion of the protected Jasper National Park, and the remaining 60% is a multi-use area subjected to a variety of disturbances, including timber extraction, pit mining, gas and oil exploration, numerous associated roads and seismic lines, and multiple ecotourism opportunities. We wanted to examine how these pressures influenced the distribution, abundance, and physiological health of grizzly bears in the study area. Since hair snags were also being used to study this same population of

bears, we were able to compare the scat dog and hair snag methods.

Scat dogs found samples from nearly four times the number of individuals per km^2 as did the hair snags, including nearly all the individuals detected by the hair snags in the same area (Wasser et al. 2004a). We also showed that these data could be used in mark-recapture analyses to reliably estimate population size.

Reproductive and stress hormones were also extracted from these scat samples. Results were somewhat unexpected and vividly illustrate the usefulness of combining physiological with genetic measures in this way. Grizzly bears appear to be most abundant, least stressed, and have higher body fat levels and conception rates in the more disturbed northern part of the multi-use study area, outside the national park. Presumably, the 50+ years of human disturbances there created forest seral stages with a large diversity of bear foods. By contrast, the northern portion of the national the park had no grizzly bears at all, only black bears, despite what appeared to be good habitat. This may be due to the high level of ecotourism in that portion of the park, where mobs of tourists stop and exit their cars to watch the bears. Grizzly bears seem to be less tolerant of such disturbances than are black bears. While grizzly bears clearly preferred the multi-use area, this area was also found to act as a population sink, because of a relatively high frequency of bear poaching in this area. Five grizzly bears were poached in this area during our three-year study—a 5% population loss, which is unsustainable. Investigators radio-collaring bears in the study area also failed to find any female bears over 15 years of age in the multi-use area. Old female bears were only found in the national park, presumably because females in the multi-use area were poached before reaching old age. Clearly, the situation here is far more complicated than meets the eye. Noninvasive measures are greatly improving our ability to understand this complexity.

Conclusions

Noninvasive physiological and genetic methods on feces, coupled with use of scat detection dogs, can open up a whole new world of opportunities to deal with the conservation challenges currently facing our planet in a cost-effective manner. Once properly validated, considerable information can be garnered from these measures, greatly increasing the potential to address critical conservation questions.

Summary

1. Because evolution is the central concept in biology, and population genetics is a central feature of evolution, genetics plays an important role in conservation biology (see Table 11.1). Genetic variation provides the raw material from which adaptation proceeds, and is critical to continued evolutionary change. This variation occurs at three levels: within individuals, within populations, and among populations.

2. Loss of variation may have negative fitness consequences and prevent adaptive change in populations. Loss of variation occurs in small populations through founder effects, genetic drift, and inbreeding. Adaptive variation among populations can be eroded when isolated populations experience gene flow by human actions, directly via translocation or indirectly via facilitating dispersal by changes to the environment.

3. Management of genetic variation in nature should proceed with immediate (years), moderate (decades, centuries) and long-term (millennia) time scales in mind.

4. The appropriate units of conservation for a given species may be based on any reasonable biological criteria that identify separate populations or groups of higher taxonomic affiliation. Genetically, conservation units may be defined through a hierarchical gene diversity analysis, which apportions the total genetic diversity of a species into within- and among-population components, the latter in a geographically hierarchical fashion. This biogeographic perspective is important in determining natural population genetic structure and probable historical levels of gene flow among populations. This methodology also allows estimation of the relative depths of evolutionary separation of geographic groups, where older separations should receive priority conservation action.

5. Such approaches are not rote, and require analysis of individual circumstances. Cookbook prescriptions for genetic management generally should be avoided in preference to conservation based on attributes and needs of the taxa (or population) of concern. Good conservation practice necessitates fundamental knowledge of the ecology, behavior, and demographic characteristics of natural populations, and genetics can be used to gather such data.

Please refer to the website www.sinauer.com/groom for Suggested Readings, Web links, additional questions, and supplementary resources.

Questions for Discussion

1. Consider an island population of outcrossing land snails that has been naturally isolated from mainland populations for tens of thousands of years. Human activities have further isolated this population into several small units, each of which has lost most of its original heterozygosity. Should mainland individuals be introduced to the island to increase heterozygosity levels? What are the pros and cons of such a suggestion? What are some other factors to consider?

2. Conservation efforts could focus, among other things, on either uniqueness or high diversity. That is, at the genetic, or species, or habitat or even ecosystem levels, choices could be made between protecting diversity and protecting uniqueness. Where should priorities lie? Is there a ready answer to this question?

3. Genetic information can sometimes have biopolitical consequences. For example, Boone et al. (1993) found that there was no genetic basis for recognition of four subspecies of cotton mice (*Peromyscus gossypinus gossypinus*, *P. g. megacephalus*, *P. g. palmarius*, and *P. g. anastasae*) in the southeastern U.S., although there was a great deal of both within- and among-population diversity and most populations were somewhat differentiated. Their data argue that subspecific status is unwarranted, but this could also remove habitat protection for one habitat, Cumberland Island, because its protection was based on the presence of this endangered subspecies. How should such genetic information be handled with respect to the legal consequences for habitat protection?

4. Consider two populations that have experienced some level of inbreeding. It has been empirically demonstrated that average levels of inbreeding are the same in both populations. In population A, inbreeding has taken place incrementally over long periods of time. In population B, levels of relatedness among parents (and thus expected inbreeding in progeny), accrued very rapidly in only a few generations. In which population would you expect fitness to be most affected in the immediate future? Why?

5. To manage populations effectively, managers need estimates of recruitment and mortality to forecast future population trends. For many creatures that migrate from breeding to wintering areas it is difficult to determine breeding population of origin for individuals sampled on wintering areas or during migration. How might molecular genetic markers and statistical methods be used to place individuals in breeding populations?

UNIT III

Approaches to Solving Conservation Problems

12

Species and Landscape Approaches to Conservation

John B. Dunning Jr., Martha J. Groom, and H. Ronald Pulliam

It is further declared to be the policy of Congress that all Federal departments and agencies shall seek to conserve endangered species and threatened species and shall utilize their authorities in futherance of the purposes of this Act.

The U.S. Endangered Species Act of 1973

Most efforts aimed at conserving biodiversity have focused on protecting individual populations or species, although the means pursued usually involve conservation of habitat. The reasons for these species-oriented approaches are straightforward. Species and populations are the essential unit of evolution, and thus make a logical target for action. Legal mandates for preventing extinction often are explicitly centered on individual species and populations. Many nations such as Australia, India, Canada, South Africa, and Brazil, mandate protections that are modeled in part after the U.S. Endangered Species Act—including creating and maintaining a list of endangered and threatened species, identifying and protecting critical habitat that sustains those species, and developing individual recovery programs for the protection of those species. On a global scale, the IUCN's Red List of Threatened Species also focuses worldwide attention on threats at the species level.

International treaties on endangered species also tend to focus on the species level of taxonomic organization. The Convention on International Trade in Endangered Species (CITES), for instance, lists threatened species in a number of appendices to the Convention. International trade in these listed species is restricted, especially for the species that are the most critically endangered. In a few cases, entire groups of species are listed rather than individual species (for instance, bog turtles, genus *Clemmys*); this is usually done when it is difficult to distinguish endangered species within the group from less threatened forms (Tryon 1999).

An additional reason for the traditional focus on single species or populations is that of conservation education. Efforts to engage the public on conservation issues are often tied to the public's interest in certain high-profile species, especially large mammals and birds, such as the giant panda or the gray wolf (symbols of the conservation efforts of WWF and Defenders of Wildlife, respectively). For instance, education campaigns to promote the nationwide conservation programs in some Caribbean nations are oriented around the endemic species of parrots found on these

islands. Because these education efforts are species-oriented, it is natural that the conservation efforts themselves are also oriented toward protecting individual species.

A final reason for the emphasis on single species and populations in conservation is that a great deal of the theory that has formed the basis for much of conservation also emphasizes species. Measurements of species diversity and richness emphasize counts of species, making individual species the focus of attention. This approach is still a dominant driver of conservation globally, particularly as applied to efforts to conserve the greatest fraction of the world's biodiversity by protecting "hotspots" of species diversity (e.g., Rodrigues et al. 2004; see Chapters 6, and 14).

In many cases, species become the focus of conservation attention when they have been reduced to a few small, remnant populations. Therefore, to understand species-specific conservation, one must understand the ecology of small and declining populations. In this chapter, we review basic population ecology, emphasizing concepts that apply to the conservation of single populations or networks of small populations. We also discuss how natural populations are dynamic at a variety of temporal and spatial scales. Many, perhaps most, species exist in nature as networks of populations whose dynamics in space and time are interconnected by the movements of dispersing individuals. It is only by viewing populations from the perspectives of different scales and levels of interconnection that one can appreciate the enormous complexities of population abundance and distribution. Understanding this complexity is an essential prerequisite to developing a practical theory of population ecology that can aid in conserving biological diversity.

Populations and How They Change

Few topics have attracted the attention of ecologists more than fluctuations in the numbers of plants and animals through time and their variation in abundance through space. Understanding population fluctuations, and thus population conservation, requires understanding the links between demographic processes—birth, death, immigration, and emigration—and the environments in which populations exist. A thorough understanding of the mechanisms driving population change can be essential for the conservation of many species, as the effectiveness of interventions on behalf of a population or species will only be successful if our actions increase population sizes. Unfortunately, often it is not simple to identify what processes are most important in driving population change (Caughley and Gunn 1996).

Some ecologists have been particularly impressed by the relative constancy of populations, while others have been impressed with their extreme variation. The former usually postulate an "equilibrium" population size and explain the observed equilibrium by reference to density-dependent factors, which prevent populations from getting either too small or too large. On the other hand, those impressed by the magnitude of population fluctuations usually see the world as consisting of many local subpopulations, each of which has a high probability of extinction due to the unpredictable nature of factors that operate independently of population density. Given the great diversity of organisms and environments and the fact that most ecologists study only a few species in a few places, there should be little surprise in such diversity of opinions.

Organisms clearly vary with regard to both their susceptibility to the vicissitudes of nature and the duration of their life spans in relation to the frequency of natural disturbances. Organisms also vary in the extent to which they live their lives in one location or experience a wide variety of environmental conditions in different locations. These facts alone account for much of the difference among species regarding the extent to which populations fluctuate in time and space. An entire generation of rotifers or thrips may experience an unusual cold spell that reduces reproduction and increases mortality, while in the next generation conditions may be optimal for the species. Individual whales and sequoias, on the other hand, experience thousands of separate cold and warm fronts, and the whales of one generation are quite likely to experience an average environment something like that experienced by the whales of other generations. Accordingly, whales and sequoias are much less likely to fluctuate wildly in population size than are thrips, and even if they did fluctuate as much, an ecologist observing them over the course of his or her career would be less likely to record the fluctuations.

Conservation biologists can track changes in populations by using principles and techniques of population demography. Demography focuses on the intrinsic factors that contribute to a population's growth or decline, including age-dependent birth rates (natalities) and age-dependent death rates (mortalities). Rates of dispersal among subpopulations (immigration into and emigration out of habitat patches) are also components of demography. These four factors, *b*irth, *i*mmigration, *d*eath, and *e*migration, are often referred to as the **BIDE factors**, and all changes in population size can be attributed to one of these four factors. However, other population attributes, such as sex ratio, age- or stage-structure, and time of first reproduction, influence population dynamics and therefore can be considered secondary demographic factors. These additional factors are referred to as **life history characteristics** and they play an important role in long-term population trends and the evolution of population characteristics.

Figure 12.1 Florida Key deer (*Odocoileus virginianus clavium*) now encounter habitats criss-crossed with roads, which has led to high mortality rates in some populations. (Courtesy of J. Gemme.)

Studies of Florida Key deer (*Odocoileus virginianus clavium*), a small subspecies of the widespread white-tailed deer found only in the island chain at the southern end of the Florida peninsula, illustrate the importance of understanding population demography in a conservation setting (Figure 12.1). Key deer are found on about 20 islands, with about 75% of the population on only two islands: Big Pine Key and No Name Key. Most of the undeveloped land on these islands is part of a national wildlife refuge, but the deer coexist with humans (both residents and tourists) on both islands. Threats to the recovery of the Key deer population include habitat loss, mortality due to factors associated with urbanization, and deer–vehicle collisions. Lopez et al. (2003) conducted a demographic study to determine which factors appear to have the strongest effect on deer population dynamics and how deer management could be changed to increase the probability of population recovery.

The study found that annual mortality rates were different between the sexes (with adult males having particularly low survival) and that vehicle collisions were the cause of more than 50% of adult deaths. Managers of the national wildlife refuge had recognized this problem, and in 2003 were completing a project along the main highway on Big Pine Key designed to reduce deer–vehicle collisions. The project includes highway fencing, underpasses, and devices similar to cattle guards to prevent deer from entering the roadway at intersections. Lopez et al. (2003) also determined, however, that the deer populations on Big Pine Key and No Name Key were near carrying capacity, and reducing mortality due to vehicle collisions may increase both disease transmission among

members of the herd and negative human–deer interactions. The researchers concluded that reducing deer–vehicle collisions is a desirable goal, but suggest that managers consider actions to reduce population birth rates, such as contraception, as a way of compensating for the decrease in mortality associated with the highway improvement project.

Mechanisms of population regulation

Conservation biologists are interested in why some species are rare, and what keeps them so. Further, often we need to understand how mechanisms that influence population size can be manipulated to increase their size and maintain viable populations of rare, as well as more common species. Thus, conservationists are concerned with factors that might regulate population size. A population can be said to be regulated if it has the tendency to increase when rare and to decline when common. The concept of population regulation is closely tied to that of density dependence. Howard and Fiske (1911) introduced the distinction between "catastrophic mortality factors" that kill a constant proportion of a population independent of its density, and "facultative mortality factors" that kill an increasing proportion of the population as density increases. The same idea is now embodied in the distinction between density-independent and density-dependent factors affecting population growth, though density dependence can refer to birth rates as well as to death rates. **Density-independent factors** influence birth and death rates in a manner independent of population density, while the intensity of **density-dependent factors**, by definition, changes with population density (Figure 12.2).

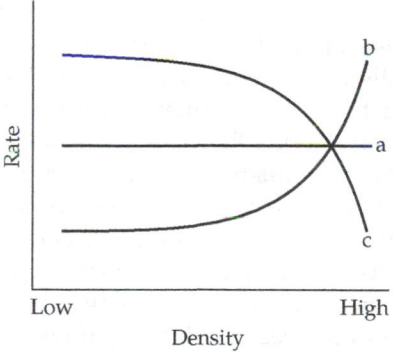

Figure 12.2 This figure demonstrates density-independent (a) and density-dependent (b and c) population responses. In the case of a density-independent response (a), density has no influence on a given parameter, such as mortality or birth rate. In the case of density-dependence, a parameter such as mortality may increase at higher densities (b), while other parameters, such as clutch size or individual growth rate, may decrease at higher densities (c). The particular shapes of the response curves, of course, would vary greatly among species and environments.

For density-dependent factors to regulate population growth, either per capita mortality must increase or per capita natality must decline as population density increases. Although there are a myriad of factors that can, in theory, contribute to density-dependent changes in mortality and natality, most of these can be grouped into several categories or mechanisms of population regulation. Among the most prominent are the following:

- Increased mortality or decreased natality due to a shortage of resources
- Increased mortality due to increased predation, parasitism, or disease
- Increased mortality or decreased natality due to increased intensity of intraspecific social interactions

Further, if individuals die before they reproduce, mortality factors may indirectly lead to fewer births as well.

Social behavior can play a direct role in regulating some animal populations, although in most cases, and perhaps all, social behavior interacts with resource shortage, disease, and predation to determine population size. In general, social behavior regulates access to resources such as food, cover, and breeding sites, and thereby affects survival and reproduction. For example, House Wrens (*Troglodytes aedon*) interfere with the breeding of many other species by puncturing the eggs in nests built within the wrens' territories. This interference may have played a major role in the near disappearance of a subspecies of the Bewick's Wren (*Thryomanes bewickii*) endemic to the eastern United States, because Bewick's Wrens appear to be particularly susceptible to this kind of reproductive interference (Kennedy and White 1996; Figure 12.3). Populations also can be regulated by intraspecific social dominance behavior or aggression.

Occasionally, density and social behavior have the opposite effect in that birth rates may increase or death rates may decrease at intermediate to high densities. In some organisms, high population densities are required to stimulate courtship and breeding activity, or to otherwise allow reproduction. This phenomenon, called the Allee effect, may affect breeding if the population drops below the required density (Allee et al. 1949). For example, pollen transfer may become difficult, if not impossible, among individuals of a rare, widely dispersed plant, resulting in failed reproduction (Bawa 1990). When habitat loss and fragmentation separate populations, entire populations may fail to reproduce (Lamont et al. 1993), and this may contribute to population extinction (Groom 1998). Colonially nesting birds may suffer poor recruitment if their densities become too low (Halliday 1980), and some fishes must nest in dense colonies to reduce predation rates on eggs and larvae (Hamilton 1971; Helfman et al. 1997). Queen conchs (*Strombus gigas*) suffer

Figure 12.3 Bewick's Wren entering a nest cavity. Bewick's Wren in the eastern United States has virtually disappeared. One cause may be social disruption by an increasing number of House Wrens, which puncture the eggs of other birds that nest within their territories. (Photograph by R. and N. Bowers.)

from Allee effects when overexploitation has driven populations to low densities (Stoner and Ray-Culp 2000; Gascoigne and Lipcius 2004; Figure 12.4). Even the spread of invading species may be slowed initially by Allee effects (e.g., the now widespread house finch [*Carpodacus mexicanus*; Veit and Lewis 1996] and zebra mussels [*Dreissena polymorpha*; Leung et al. 2004] appear to have spread slowly due to Allee effects after their introduction into New York City and the Great Lakes, respectively). Increasingly, conservation biologists are alert for the presence of Allee effects, as they can limit the ability of a small population to rebound to larger population sizes.

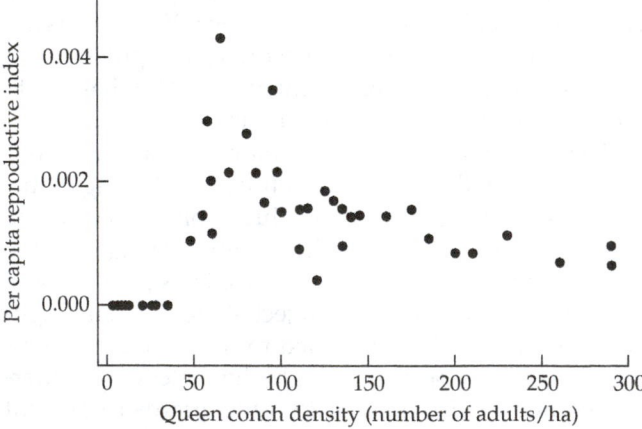

Figure 12.4 Queen conch (*Strombus gigas*) suffer from Allee effects in the Exuma Cays, Bahamas. Per capita reproductive activity is zero when adult queen conch density is below 50 individuals per hectare, but it increases once individuals congregate in greater numbers. (Modified from Stoner and Ray-Culp 2000.)

These examples emphasize the facultative nature of the factors regulating changes in population density. Other demographic factors act independently of population density; that is, the magnitude of mortality or natality does not depend on population density per se. For example, Davidson and Andrewartha (1948a,b) investigated the causes of outbreaks of *Thrips imagines*, a major pest of apples in Australia. In moist areas such as irrigated suburban residential areas, these thrips persist even during dry years, but in drier habitats their food supplies decline in midsummer, causing a catastrophic crash in their numbers. Although the crashes in thrip numbers occur following the peak of thrip populations in midsummer, there is no evidence for a density-dependent regulating mechanism, because the decline in food availability is due to an extrinsic factor, dry weather, rather than to overexploitation of the food resources by thrips.

It must be kept in mind that populations can be regulated by more than one factor, and that the simultaneous effects of several factors working in concert may be responsible for population changes. The extinction of *Trilepidea adamsii*, a New Zealand mistletoe, may have involved both density-independent and density-dependent factors (Norton 1991). The mistletoe was never widespread, so the extensive habitat destruction (a density-independent factor) that took place in New Zealand probably limited the plant to a few locations. Overcollecting by botanists and grazing by an introduced opossum are two density-dependent forces that harmed the species. Scientific collecting is density dependent because as a species becomes more rare, specimens become more valuable, and field collectors may make a special effort to find and gather the last few individuals. The added pressure of collecting, and a nonnative predator may have eliminated the few remaining *Trilepidea* populations. Our success in conserving populations thus may often depend on our ability to identify and characterize the factors that regulate populations and determine population fates (Caughley and Grimm 1996; see Chapter 3).

Special problems of very small populations

Species that have been reduced to a single or few remaining small populations present a special challenge to conservation. Of the four general causes of extinction defined by Shaffer (1981, 1987)—genetic losses, demographic variability and declines, environmental variation, and catastrophes—all can be most damaging to small populations. Genetic losses and demographic uncertainty (especially periods of decline in survival or reproduction) may not harm a moderately sized population, but could cause the extinction of very small populations, while environmental variability can harm populations of both small and larger sizes. Even species that are abundant are not immune from extinction if they are found in single populations, because natural catastrophes can wipe out entire populations if they are found only in one vulnerable location.

In Chapter 11, we discussed the problems that can be caused by loss of genetic variation, particularly adaptive variation. In extremely small populations mate choice may be very limited and individuals may be forced to breed with close relatives, which may reduce survivorship or reproductive capacity. Indeed, due to these or similar circumstances, most threatened species appear to have diminished genetic diversity compared to a nonthreatened congener (Spielman et al. 2004). One endangered species program that has dealt with this potential loss explicitly is that of the Florida panther (*Felis concolor coryi*), the southeastern subspecies of the mountain lion. By the 1980s, the panther was reduced to less than 50 individuals in the wild, spread over a large area of southern Florida. Symptoms of inbreeding, such as morphological deformities and low sperm counts, were common. The biologists responsible for recovery of the Florida panther opted to bring some western mountain lions of the geographically closest subspecies (*F. c. stanleyana*) to increase the genetic diversity of the panther population (Hedrick 1995). Although this plan ran the risk of bringing genes into the local population that were not well adapted to the southern Florida environment, it was thought that loss of genetic diversity through limited breeding options was a greater threat. However, in this case, as in many others, genetic problems developed after the population declined due to other threats. Indeed, given differences in the magnitude of their effects, we expect that habitat degradation and fragmentation, overexploitation, and conflicts with invasive species will have greater effects on most species' survival than genetic losses (Lande 1988), although managing genetic threats can still be necessary to rescue populations that are at the brink of extinction.

Another extinction threat faced by small populations is that of **demographic uncertainty**. Demographic characteristics—such as sex ratio, reproductive success, and mortality rates—can change dramatically over a short period in small populations, and these random changes can lead to rapid extinction when not buffered by large population size. One species in which the role of demographic change has been studied is *Borderea chouardii*, a small cliff-dwelling plant found only in one location in the Pyrenees Mountains of Spain (Garcia 2003). The plant is tiny, but extremely long-lived; individuals may live for 300 years. Garcia (2003) modeled the growth rate of the *Borderea* population under different ecological scenarios asking why the species was limited to a single area and to see if increased adult survivorship, reproductive success, or other demographic traits would allow the species to spread. In spite of features that suggested that the species was suffering from demographic stress (a skewed sex ratio, lack of juvenile plants, low reproductive output), Garcia's analysis suggested that the species' long life

span allowed the plant to survive. The primary limitations that restrict *Borderea* to its single location are the scattered nature of its cliff-face habitat, and an almost total lack of dispersal ability. Hand-sowing of seeds might allow the species to spread to other patches, but the primary conservation need identified by their analysis was protection of the existing adult plants and habitat.

Environmental uncertainty can cause extinction in small populations by causing a sudden increase in reproductive failure or individual mortality. One conservation project that must work around environmental uncertainty is the recovery program for the Black Stilt (*Himantopus novaezelandiae*), the rarest shorebird in the world. The stilt breeds along braided rivers in the South Island of New Zealand. Braided rivers have no set channel; instead water moves down a floodplain of pebbles and rocks, with the main channels changing seasonally as floods move the rock material to new positions (Figure 12.5). Black Stilts lay eggs directly among the pebbles in these floodplains, and regularly lose clutches to floods. Because the floods are a regular feature of the rivers, presumably the stilts could withstand the associated reproductive failures when the birds were more common. But with only a few pairs left in the world, the fate of each egg is critical. New Zealand conservationists reduce the effects of river flooding each year by finding each clutch laid by the wild stilts and replacing the eggs with ceramic duplicates. The adults are fooled by the imitation eggs and continue incubating, while their real eggs are developing in an artificial incubator. If a flood destroys the original nest, the biologists replace the clutch with another set of artificial eggs as soon as the floodwaters recede. The adults continue to incubate the ceramic dummies. When the real eggs are ready to hatch, they are transported back into the field and placed within their parents' nest. In this way, losses due to

Figure 12.5 The upper Rakaia River provides a typical example of a braided river in New Zealand. (Photograph © Bill Bachman/Photo Researchers, Inc.)

floods are reduced and the offspring are raised in the wild by their parents (Read and Merton 1991).

Conservation plans must also guard against the effects of **natural catastrophes**, environmental events that are more irregular in timing (and therefore less predictable) and larger in spatial scale than events that fall under the previous category of environmental uncertainty. For example, recovery programs that involve captive breeding must allow for the danger of catastrophic loss—for example, a single breeding facility can be hit with a disease outbreak or natural disaster (see Chapter 15). In most cases the only safeguard for these programs is to create separate facilities far enough apart that it is unlikely that the same disaster can strike all populations.

Source–sink concepts and their application to conservation

In many populations, individuals occupy habitat patches of differing quality. Individuals in highly productive habitats may be successful in producing offspring, while individuals in poor habitats may suffer poor reproductive success or survival. A population's fate as a whole may depend on whether the reproductive success of individuals in the good habitats outweighs the lack of success by individuals in the poor areas. The idea that population dynamics may depend on the relative quality of good and poor habitats is called **source and sink dynamics**, and is now recognized as an important concept in conservation biology (Pulliam 1988, 1996), with applications to grizzly bears (Doak 1995), reef fishes (e.g., Tupper and Juanes 1999), butterflies (e.g., Thomas et al. 1996) and numerous applications to birds (e.g., Brawn and Robinson 1996; Begazo and Bodmer 1998; With and King 2001).

Good habitats are called **sources**, and are defined as areas where local reproductive success is greater than local mortality. Populations in source habitats produce an excess of individuals, who must disperse outside their natal patch to find a place to settle and breed. Poor habitats, on the other hand, are areas where local productivity is less than local mortality. These areas are called **sinks** because, without immigration from other areas, populations in sink habitats inevitably spiral "down the drain" to extinction. The terms *source* and *sink* are also used to describe the populations found in these habitats: **source populations** are those found in source habitats, and **sink populations** are those found in sink habitats.

Population ecologists have discovered that many species have both source and sink populations (Pulliam 1996; Vierling 2000). The excess individuals produced by source populations can disperse to sink habitat patches and maintain the populations found in these poorer habitats. A population that consists of several subpopulations linked together by immigration and emigration is called a **metapopulation** (Figure 12.6; see Chapter 7). Metapopulation and source–sink are different but relat-

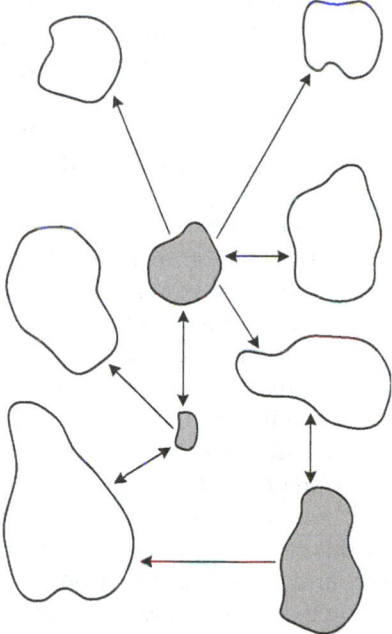

Figure 12.6 A schematic example of a metapopulation structure affected by source–sink dynamics. In this case, a few source habitats (shaded) provide excess individuals, which emigrate to and colonize sink habitats (open). Arrows indicate directions of movement of individuals. The sink habitats may be spatially larger than the sources, and may even have higher population densities and produce dispersing individuals, but their populations would go extinct were it not for the presence of the sources habitats.

ed concepts. In the original meaning of the term (Levins 1969), a metapopulation occurs when there are many populations or habitat patches of similar quality. There is some chance of extinction in all patches, but at any point in time some patches are empty and some are occupied. The metapopulation is maintained by dispersal from occupied patches and the recolonization of empty ones. In source–sink dynamics, the habitat patches are of distinctly different quality. One habitat patch (the source) is not extinction-prone, but rather continually generates a reproductive surplus. The other habitat patch (the sink) is unsuitable habitat in the sense that the population would disappear in the absence of periodic immigration from source habitats. Real populations rarely fall neatly into one category or the other and may exhibit mixed characteristics of metapopulation and source–sink dynamics (or neither).

Several results from studies of source and sink dynamics have broad implications for conservation biology. First, theoretical models of sources and sinks have shown that a small proportion of the total population may be located in the source habitats. For instance, using demographic parameters that are reasonable for many natural populations, Pulliam (1988) showed that as little as 10%

of a metapopulation may be found in source habitats and still be responsible for maintaining the 90% of the population found in sinks. Such relationships may greatly affect the ability of conservationists to identify critical habitats for endangered species. Until recently, critical habitats were defined as the places where a species was most common. Source habitats, however, are defined by demographic characteristics—habitat-specific reproductive success and survivorship—not population density. Source habitats could easily (and mistakenly) be ignored if conservationists concentrated on preserving habitat only where a species is most common, rather than where it is most productive (Van Horne 1983). For example, current direction can strongly influence whether a particular coral reef or stretch of river is a net source or sink for production of juveniles. Smaller upstream populations may serve as sources for larger downstream ones. If source habitats are not protected in a conservation plan, obviously the whole metapopulation could be threatened.

Evidence of source–sink dynamics has been found in many plant and animal species, on land as well as the sea. One recent example is the case of the Florida Key deer introduced at the beginning of this chapter. Residential and commercial development has resulted in habitat loss and fragmentation on Big Pine Key, leaving more suitable habitat on the northern end (NBPK) than on the southern end (SBPK) of the island. The deer population was about equally split between the two halves of the island, but NBPK is a source, where annual reproduction exceeds annual mortality (that is, when annual population growth rate [λ, lambda] is greater than 1; λ for NBPK = 1.02; Harveson et al. 2004). SBPK, on the other hand, is a sink where mortality is about 15% higher than recruitment from local reproduction ($\lambda = 0.87$), primarily due to collisions with automobiles (Lopez et al. 2003; Harveson et al. 2004; Figure 12.7). Importantly, 15% of the fawns born in NBPK immigrated as yearlings to SBPK while only about 5% of the yearlings from SBPK immigrated to NBPK, supporting the idea that immigration from the source helps to maintain the sink population. Harveson et al. (2004) suggest that future management should continue to try to reduce mortality factors in the sink habitat while preserving the source population in NBPK.

Source–sink dynamics have played a strong role in the recovery of the Peregrine Falcon (*Falco peregrinus*) in California. After crashing due to eggshell thinning caused by the pesticide DDT (and its metabolite DDE) in the 1950s, northern California populations of Peregrine Falcons began recovering, but the southern coastal populations did not, presumably because of higher levels of DDE that persisted there (Figure 12.8). Further a third population type, urban populations, rebounded quickly because young birds had very high survival due to the high availability of easy prey and lack of predators in urban centers (Kauffman et al. 2004). Two

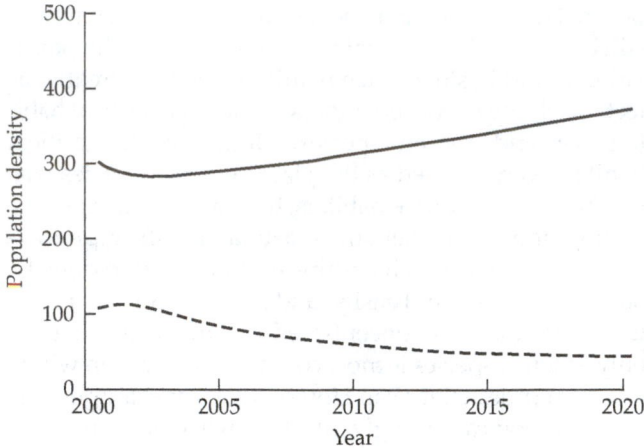

Figure 12.7 Population projection of the source population on Big Pine Key (NBPK, solid line) compared with the sink population (SBPK, dashed line). Without management to reduce mortality on SBPK, this population is at risk of extinction. (Modified from Harveson et al. 2004.)

source–sink models have been applied to this case to evaluate the effectiveness of conservation strategies for Peregrine Falcons in California. Wootton and Bell (1992) modeled this population as two subpopulations (combining coastal and urban populations in northern and southern California) linked by dispersal. Their analysis showed that the northern population could serve as a source for the smaller southern population. At the time, the regional management strategy for this species included the release of captive-reared young, and was aimed primarily at the southern population, which Wootton and Bell showed to be a sink. This led to the worry that such reintroduction efforts would be doomed to failure in the southern sink population. Recently, however, Kauffman et al. (2004) have shown that the southern coastal population most likely would have gone extinct without the addition of hundreds of young Peregrine Falcons (see Figure 12.8), because although both northern and urban populations were strong sources of young birds, very few dispersed away from these habitats. The northern population, in contrast,

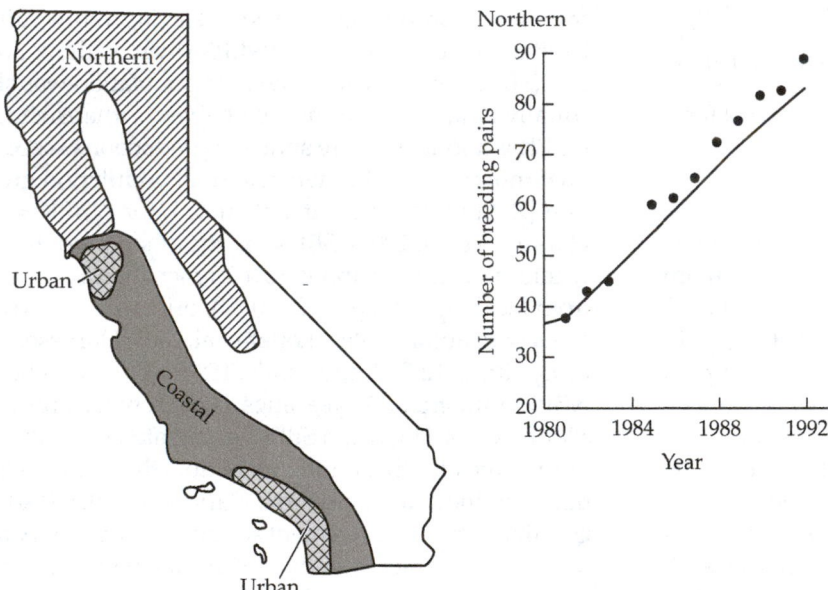

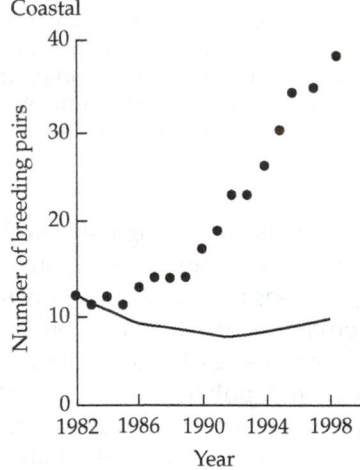

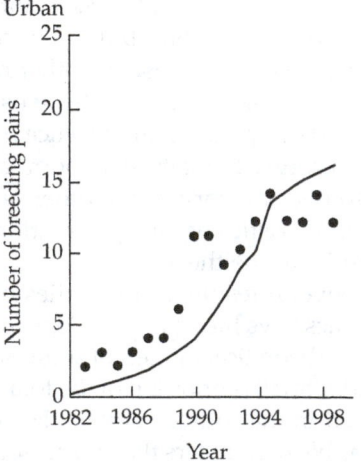

Figure 12.8 Peregrine Falcons are distributed in 3 populations in California: northern and two southern populations—coastal and urban. Populations have been supplemented with captive bred and translocated individuals since the 1980s, which has led to an augmented population size in coastal, but not northern or urban populations (black circles indicate actual population counts with additional individuals; black lines represent the size the population would be if no individuals had been added). (Modified from Kauffman et al. 2004.)

would have succeeded without intervention, as sufficient suitable habitat existed to absorb most of the population growth (see Figure 12.8). By the late 1990s, the southern population had ceased to be a sink, although many young birds moved to urban habitats where their reproductive success is higher.

In categorizing habitats as sources and sinks, care should be given to both temporal and density-dependent changes in population demography. The original source–sink concept emphasized variation in demography across space, but birth and death rates also vary over time. What may be a source one year may be a sink another year, as has been seen in butterflies (Thomas et al. 1996). Several flavors of source–sink dynamics can be distinguished dependent on the spatial and temporal fluctuations in habitat quality. All patches may be poorer in unfavorable years, with sources temporarily becoming sinks, but still being relatively better habitats than sinks. Alternatively, what is an unfavorable year in the source might be a favorable one in the sink, with source and sink temporarily reversing roles. For example, in an unusually wet year, soils may become waterlogged and unfavorable for plant germination in what is normally source habitat, while in the normally dry sink habitat there may be just enough moisture to ensure successful germination and to turn a sink into a temporary source.

Care also should be taken to distinguish absolute sinks from pseudo-sinks (Watkinson and Sutherland 1995). If organisms crowd into good habitat where survival and reproductive success are density-dependent, a habitat patch may become a pseudo-sink, where crowding is so severe that local mortality exceeds local reproduction. In an absolute sink, the removal of immigration results in local extinction, but in a pseudo-sink, the removal of immigration results in a population decrease but not in extinction. Furthermore, source–sink dynamics can be confused with what has been termed balanced dispersal by McPeek and Holt (1992). In balanced dispersal models, there are no sink habitats but patches do differ in quality and carrying capacity. Source–sink models predict a net flow of individuals from sources to sinks, whereas balanced dispersal models predict, on average, equal numbers of individuals moving in both directions. Diffendorfer (1998) reviewed studies of 28 species, mostly birds and mammals, and concluded that there are good examples of both source–sink dynamics and balanced dispersal in natural populations, but that most populations lie on a continuum, exhibiting elements of both.

Determining which habitats are sources and which are sinks requires a great deal of knowledge about the natural history of organisms. One needs to know the birth and death rates of individuals in each habitat type, details of dispersal behavior, and other aspects of the organism's life history. Rates may also vary from patch to patch; for instance, Hokit and Branch (2003) recently demonstrated how demographic rates can vary with patch size. Without such knowledge, it is impossible to design a conservation plan that considers realistic population dynamics. Critics of this approach argue that this information is often difficult to gather and thus rarely available (Wennergren et al. 1995). And, of course not all species exhibit source–sink dynamics. Yet, often conservation outcomes may depend on a clear assessment of how natural variability in habitat quality affects species, and thus, studies to obtain such basic information may be critical in the planning of management and conservation strategies.

Metapopulation concepts, threshold responses, and conservation

Levins (1969) introduced the concept of the metapopulation to describe a collection of subpopulations, each occupying a suitable patch of habitat in a landscape of otherwise unsuitable habitat. The subpopulations are linked together by the emigration and immigration of individuals between patches, allowing for local extinctions and recolonizations. Levins (1969) demonstrated that the fraction of suitable habitat patches occupied at any given time represents a balance of the rate at which subpopulations go extinct in occupied patches and the rate of colonization of empty patches (see Hanski and Gilpin 1997).

The subpopulation in each patch can fluctuate in size, and when a subpopulation is very small, local extinction can be prevented by occasional immigrants that arrive from neighboring patches. This has been termed the **rescue effect** by Brown and Kodric-Brown (1977), who argue that it is often a major factor in maintaining small populations. The rescue effect may also be important in maintaining high levels of species diversity, because poor competitors will not be excluded from patches by locally well-adapted species if the populations of the poorer competitors are maintained through immigration (Stevens 1989). The rescue effect is in part responsible for one major characteristic of metapopulations: The proportion of habitat patches that are occupied is relatively constant through time, even though populations in the individual patches may go extinct relatively frequently.

Conservation biologists use metapopulation models to describe the structure of populations that are found scattered across isolated patches and that are threatened or otherwise of management interest. In many of these metapopulation analyses, the goal is to identify particular subpopulations, habitat patches, or links between patches that are critical to maintenance of the overall metapopulation. For instance, Beier (1993) studied the cougar (*Felis concolor*) population of the Santa Ana Mountains of southern California. Beier used radiotelemetry data to show that the cougars form a series of semi-iso-

Figure 12.9   Riparian habitats provide crucial corridor elements needed to maintain a viable population of cougars in southern California. (Photograph by R. and N. Bowers.)

lated populations found mostly in small mountain ranges linked by riparian corridors (Figure 12.9). A metapopulation simulation model showed that the overall population in the region is heavily dependent on movement by individual cats through the corridors to colonize empty areas. Beier looked at the importance of specific habitat patches and corridors in maintaining the metapopulation. One corridor in the northern part of the study linked a 150 km² patch (8% of the total area) with the rest of the region. When the study was published, this corridor was slated for development by the city of Anaheim. Beier's analysis demonstrated that loss of this corridor would mean that the northern habitat patch would eventually be unoccupied due to lack of immigrants, and that the cougar population as a whole would suffer greatly. Because the metapopulation model identified the importance of this corridor, Beier proposed that the development plan be modified to leave the corridor intact.

A number of studies have shown that the local populations of a metapopulation often go extinct, but studies directly relevant to the persistence times of metapopulations per se are extremely rare. One of the first theoretical analyses to consider the persistence of metapopulations is that of Hanski et al. (1996) who estimated "minimum viable metapopulation size" (the number of subpopulations required to support metapopulation persistence).

Hanski et al.'s theoretical analyses suggest that many endangered species may be on a trajectory toward extinction due to nonviable metapopulation size, in spite of protection of current subpopulations. These concepts have been further developed to reflect metapopulations that exist on spatially realistic landscapes, as described by Ilkka Hanski in Essay 12.1.

The critical insight from these analyses is that populations that appear to be safe for many years may suddenly decline if they are subject to threshold responses. Ecologists have found evidence consistent with the theoretical concept of critical thresholds. Critical thresholds (disproportionate population declines associated with habitat loss) were found in a study of Massachusetts amphibians (Homan et al. 2004). Two study organisms, the wood frog (*Rana sylvatica*) and the spotted salamander (*Ambystoma maculatum*) require ponds for breeding and nearby upland forests in which the adults live while not breeding. Homan et al. (2004) found that pond occupancies by the two species declined drastically if the forest cover in the surrounding landscape dropped below some threshold, and that the thresholds differed depending on the overall spatial scale considered. Homan et al. (2004) concluded that the effectiveness of wetlands protection legislation, such as that found in Massachusetts, depends on the spatial scales incorporated into the laws.

Not all patchily distributed populations are metapopulations and conservation biologists should be wary of misapplications of the metapopulation concept (Harrison 1991). In discussing the regional dynamics of plants, Eriksson (2000) distinguishes metapopulations from remnant populations—systems of patchily distributed subpopulations that persist through long, unfavorable periods when local population growth rates are negative—and argues that the distribution of suitable habitat and the life history characteristics of a species interact to determine the regional dynamics of a population. According to Eriksson, short-lived species with specialized habitat requirements tend to exhibit metapopulation dynamics, but long-lived species—especially those with clonal propagation or large seed banks—are more likely to be remnant populations. The persistence of these remnant populations depends more on the life history characteristics that allow them to survive through unfavorable times than on immigration from nearby populations (the rescue effect). In fact, it is possible that remnant populations are not in need of conservation interventions, but rather their existence may stage the recovery of many other species once a stressful period (e.g., drought) has passed.

Any attempt to estimate a population's long-term health must be done in the context of social goals and political realities, and with an understanding that any

ESSAY 12.1

Metapopulations, Extinction Thresholds, and Conservation

Ilkka Hanski, *University of Helsinki*

■ Species living in fragmented landscapes have a fragmented population structure. Individuals occupying one habitat fragment constitute a local population, whereas a set of such local populations in a landscape makes up a metapopulation. Individuals interact with other individuals in the same local population. Different local populations are coupled by dispersing individuals moving from one local population to another and possibly to previously unoccupied habitat patches.

To acquire a working knowledge of metapopulation dynamics, it is helpful to become familiar with a simple model originally proposed by Richard Levins (1969) for changes in the size of a metapopulation in the course of time. Levins simplified the modeling task by assuming an infinitely large network of identical habitat patches. Furthermore, we only consider whether these patches are occupied at any point in time by the species or are empty. With these assumptions, the size of the metapopulation inhabiting the patch network can be described by the fraction of currently occupied patches, denoted by p. Changes in the value of p in the course of time are described by the following equation:

$$dp/dt = cp(1 - p) - ep$$

where c is the colonization rate and e is the extinction rate of patches.

The model assumes that the fraction of occupied patches (p) decreases because of local extinctions in the currently occupied patches (ep), and it increases due to colonization of currently unoccupied patches ($cp[1 - p]$). Each local population is assumed to have the same risk of extinction, and the existing populations, which are assumed to be similar to each other, contribute equally to the pool of dispersers. The dispersers are spread out across the entire patch network, hence they "encounter" and colonize empty patches out of all patches in proportion to how many empty patches there are ($1 - p$).

The growth of such a metapopulation is limited by the availability of empty patches for colonization. When the metapopulation is very small (i.e., p is very small), almost all patches are unoccupied and available for colonization, hence the rate of colonization is almost cp. A species can invade such a network if the colonization rate is greater than the extinction rate, in other words if $cp > ep$, or simply $c > e$. This condition defines an important concept in metapopulation dynamics, the **extinction threshold**. If the condition is not met, the entire metapopulation will go extinct. At equilibrium, the local extinctions and colonizations are equally frequent and hence the rate of change in metapopulation size equals zero ($dp/dt = 0$). We can solve the value of p at equilibrium, which is $p^* = 1 - e/c$. We thus see that, in the Levins model, if the extinction threshold $c > e$ is met, the metapopulation size at equilibrium is positive. Note also another key prediction of the theory: The species will occupy a decreasing *fraction* of the available habitat when the threshold value is approached, and at the threshold point the species is not able to maintain a positive balance between local extinctions and recolonizations, and hence it goes regionally extinct *even if some habitat still remains*.

The Levins model is immensely valuable in clarifying some of the most essential issues about the dynamics of metapopulations in fragmented landscapes. But all models have their limitations. One limitation of the Levins model is that it assumes an infinitely large network of independent patches, in which case stochasticity in local extinctions and colonizations becomes averaged out at the metapopulation level. In reality, a metapopulation may go extinct because of the inevitable stochasticity in extinctions and colonizations even if the threshold condition $c > e$ is satisfied. Extinction is more likely to occur the closer the metapopulation is to the extinction threshold.

Another shortcoming of the simple metapopulation model is that there is no description of landscape structure in the model; hence it is not really possible to analyze the consequences of habitat loss and fragmentation with this (and any comparable) model. Real metapopulations live in patch networks with a finite number of habitat patches; the habitat patches are not identical to each other (patches vary in size, quality, and degree of isolation from each other), which presumably affects the chances of dispersers finding them. These considerations have been incorporated into the spatially realistic metapopulation theory (Hanski 2001). This theory is similar to the matrix population theory for age-structured populations, though now the structure in the population is due to individuals being distributed among different kinds of habitat patches rather than among different age or size classes. The key idea is to include in models the effects of habitat patch area, quality, and connectivity (inverse of isolation) on the processes of local extinction and colonization. Generally, the extinction risk decreases with increasing patch area because large patches tend to have large populations with a small risk of extinction, and the colonization rate increases with connectivity to existing populations. The spatially realistic metapopulation theory provides a link between metapopulation ecology and landscape ecology (Hanski 2001).

The theory provides a measure to describe the capacity of an entire habitat patch network to support a metapopulation, which we denote by λ_M and call the metapopulation capacity of the fragmented landscape (Hanski and Ovaskainen 2000). Mathematically, λ_M is the leading eigenvalue of a "landscape" matrix, which is constructed with the assumptions about how habitat patch areas and connectivities influence extinctions and colonizations. In this model, λ_M can be closely approximated by the average *colonization potential* of the habitat patches, which tells us how many other occupied patches each occupied patch will give rise to during its lifetime (Hanski 1999).

The size of the metapopulation at equilibrium (p^*) is given by the following equation:

$$p_\lambda^* = 1 - e/(c\lambda_M)$$

This is similar to the equilibrium in the Levins model, with the only difference being that now metapopulation size at equilibrium depends also on the metapopulation capacity, and metapopulation size is measured somewhat differently (by a weighted average of the probabilities of different habitat patches being occupied; Hanski and Ovaskainen 2000). The threshold condition for metapopulation persistence in this spatially realistic model is given by:

$$\lambda_M > e/c$$

In words, metapopulation capacity has to exceed a threshold value, which is set by the extinction proneness (e) and colonization capacity (c) of the species, for long-term persistence. To compute λ_M for a particular landscape one needs to know the scale of connectivity, set by the dispersal range of the species, and the areas and spatial locations of the habitat patches. For a given species, the metapopulation capacity can be used to rank different fragmented landscapes in terms of their capacity to support a viable metapopulation of the species: The larger the value of λ_M, the better the landscape.

The metapopulation capacity has been employed in analysis of the extinction threshold of the Glanville fritillary butterfly (*Melitaea cinxia*) in the Åland Islands in Finland, where the butterfly occurs in a very large network of some 4000 dry meadows within an area of 50 by 70 km, with much regional variation in the average size and density of meadows. The 4000-meadow network has been divided into smaller networks that are inhabited by relatively independent butterfly metapopulations (Hanski et al. 1996a). The x-axis in Figure A shows the metapopulation capacity for 25 such networks in western Åland. The y-axis shows the size of the respective metapopulation, based on a survey of habitat patch occupancy in one year. It is apparent that the butterfly did not occur in networks with metapopulation capacity smaller than around –1.5, whereas in networks with a $\lambda_M > -1.5$, the size of the butterfly metapopulation increased as predicted by the theory. This is a clear example of an extinction threshold.

To summarize, the metapopulation theory indicates that a network of habitat patches must satisfy a certain necessary condition in terms of the number, sizes, and spatial configuration of the

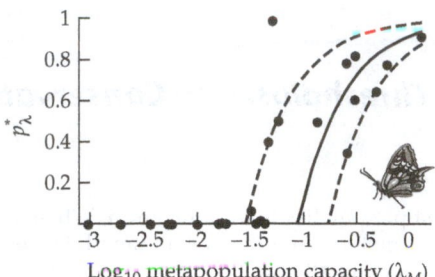

Figure A Extinction threshold in the Glanville fritillary butterfly. The x-axis gives the metapopulation capacity for each network; the y-axis shows the size of the respective metapopulation based on a survey of habitat patch occupancy in one year. (Modified from Hanski and Ovaskainen 2000.)

habitat patches to allow the long-term persistence of the species. The extinction threshold delimits networks that satisfy this condition from those networks that do not satisfy it. The threshold condition is fundamental for conservation biology and for practical conservation, though it must be realized that the condition refers to the equilibrium state in a deterministic model, which is an approximation of the corresponding stochastic model and reality.

So far we have considered the dynamics of species in static landscapes. However, changes in landscape structure, and not only in metapopulation size, are commonplace due to both natural causes and to human alteration of many landscapes and habitats. Indeed, on-going habitat loss and fragmentation is the number one threat to populations and species at all spatial scales, from local to global. How do metapopulations respond to changing landscape structure?

In the first place, we should recognize that it takes some time following a change in landscape structure, especially if the change is abrupt, before the metapopulation has reached the new equilibrium. Considering a community of species, Tilman et al. (1994) coined the term *extinction debt* to refer to situations in which, following habitat loss, the threshold condition is not met for some species, but these species have not yet gone extinct because they respond relatively slowly to environmental changes. More precisely, the extinction debt at a given point in time is the number of extant species that are predicted to go extinct (sooner or later)

because the threshold condition is not satisfied for them.

To take another example on the Glanville fritillary, habitat that is suitable for this butterfly has been lost due to changes in farming and other forms of land use, and in Åland the current area of meadows and pastures suitable for the butterfly is estimated to be roughly 20% of the habitat that existed 50 years ago (Nieminen et al. 2004). Figure B gives a detailed example covering a period of 20 years for an area of 5 by 5 km. In the map (Figure B.I), gray areas represent habitat that was suitable in 1973 but became unsuitable over the subsequent 20 years, while the black areas remained suitable. Figure B.II shows the response of the butterfly metapopulation to this loss and fragmentation of habitat, based on the predictions of a metapopulation model that was fitted to empirical data (Hanski et al.1996b). For the purpose of this example, we assume that before and after the 20-year period when habitat was lost, the amount of habitat remained constant. In this case, the metapopulation size is predicted to follow closely the equilibrium size. But this is not always the case. Figure B.III gives similar results for a hypothetical scenario of further loss of 50% of the area of each of the patches remaining in 1993. The metapopulation equilibrium (solid line) now moves all the way to extinction; in other words, the additional loss of habitat has moved the metapopulation beyond its extinction threshold. But note that now the predicted change in metapopulation size shows a long extinction lag, where it takes a long time for ultimately doomed populations to go extinct. The reason for the long extinction lag in Figure B.III is that here the new equilibrium metapopulation size is located close to the extinction threshold. Figure B.IV shows a general theoretical result of the length of the time delay in metapopulation response (vertical axis) to a change in landscape structure (Ovaskainen and Hanski 2002). The horizontal axis gives the new equilibrium size of the metapopulation following the change in landscape structure. The time delay is especially long when the new equilibrium is close to the extinction threshold, as was the case in the example shown in Figure B.III.

The observation that extinction lags are especially long when the new

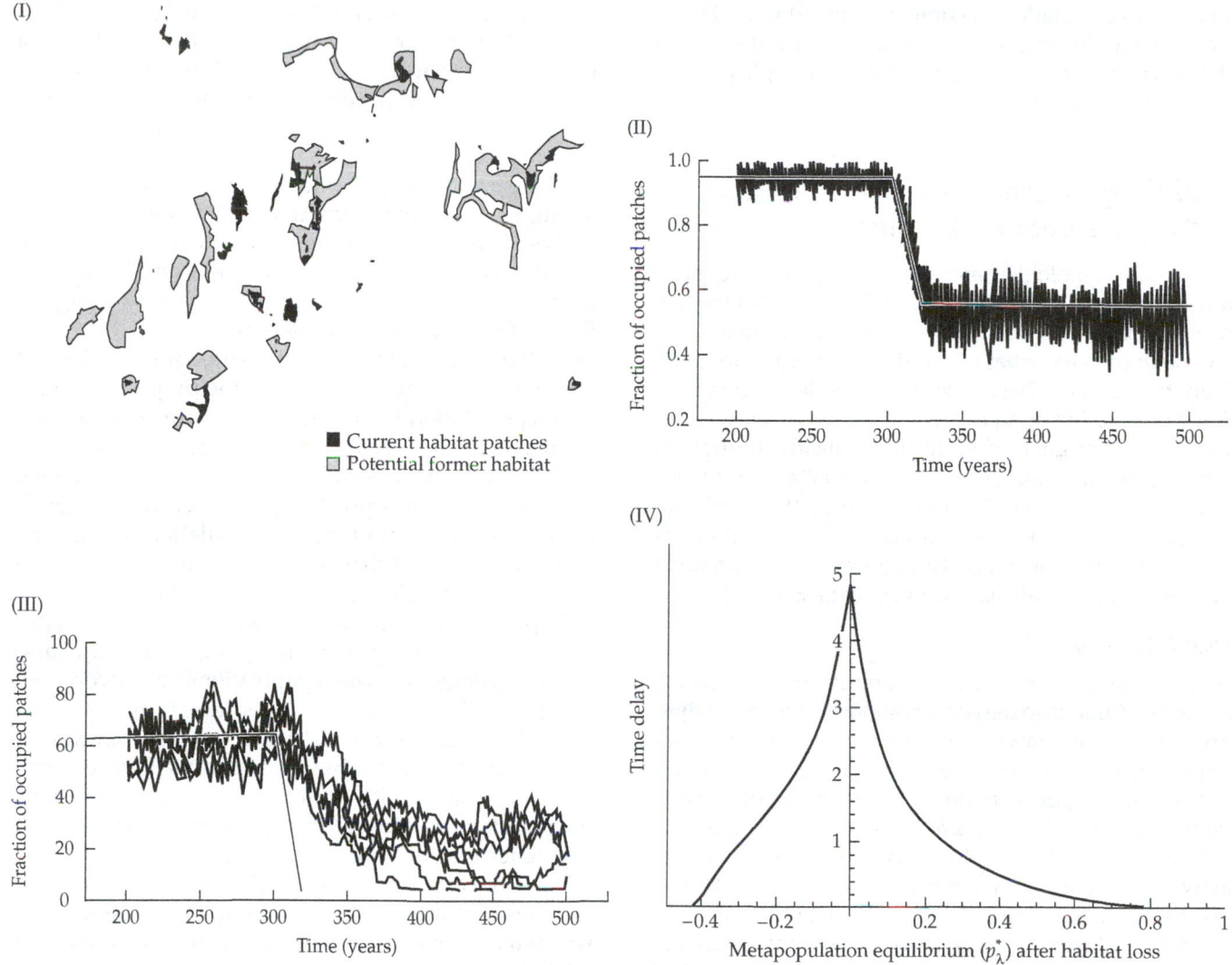

Figure B Extinction dynamics in the Glanville fritillary following habitat loss and fragmentation. (I) An example of habitat loss and fragmentation in a 5 × 5 km area during a 20-year period, 1973–1993. (II) The model-predicted response of the butterfly metapopulation to the loss and fragmentation of habitat shown in (I). The center line gives the model-predicted equilibrium metapopulation size corresponding to the prevailing structure of the landscape, while the surrounding lines show the model-predicted actual changes in metapopulation size before, during, and following the observed reduction in habitat area. (III) Similar results as in (II) for a hypothetical scenario of further loss of 50% of the area of each of the patches remaining in 1993. (IV) A general theoretical result giving the length of the extinction lag in metapopulation response (y-axis, relative time units) in relation to the new equilibrium size of the metapopulation following the change in landscape structure. (Modified from Nieminen et al. 2004; Hanski et al. 1996; Ovaskainen and Hanski 2002.)

metapopulation equilibrium is located close to the extinction threshold has very important implications for conservation. By definition, endangered species are located close to their extinction thresholds, and the theoretical result we have just learned tells us that the metapopulation response is especially long in these cases. This means that we are very likely to underestimate the level of threat to endangered species because they are only slowly responding to changing landscape structure. Their environment may have changed recently, but they have not yet had time to reach the new equilibrium corresponding to the present structure of the landscape, and the new equilibrium may be extinction. ∎

predictions are made in a context of uncertainty. Therefore, in virtually all cases conservation of multiple populations is preferable to an emphasis on a single protected population or site.

Modeling Approaches for Prediction and Conservation Planning

Conservation biologists increasingly depend on quantitative models of population dynamics to predict the fate of populations, and to test scenarios that may explain decline or indicate what kinds of interventions are more likely to succeed. These models can be developed for a single population or an interacting group of subpopulations, or species interactions (often implicitly through incorporating variation in human impacts) across a landscape scale. Here we describe some of the modeling approaches that have been used for conservation, including population viability analyses and landscape models using spatially explicit population models.

Population viability analysis

Models of factors contributing to population viability are valuable additions to the conservationist's toolbox. **Population viability analysis**, or **PVA**, examines the demographic effect of different threats or management practices on a population, or set of populations, by projecting into the future (Soulé 1987; Shaffer 1990; Beissinger and McCullough 2002). Essentially, PVA is a quantitative risk analysis aimed at refining our understanding of the factors that influence population fate. Many published PVAs combine field studies of important demographic parameters with simulation modeling of the possible effects of various threats. Generally, the object of these analyses is to generate a prediction of extinction risk. Others explore how an intervention such as periodic controlled burns or new limits to a fishery may affect the future growth of a population. Because the techniques of PVA allow us a means to assess the severity of threat of a population's extinction, or explore potential future outcomes of our conservation activities, many have called for its extensive use in endangered species management (e.g., Carroll et al. 1996; National Research Council 1996). Enthusiasm for PVAs, however, is tempered by concerns about their data needs and the dangers of misunderstanding limits to the accuracy of their predictions (Groom and Pascual 1997; Reed et al. 2002).

The primary value of a PVA is that it is a systematic way of exploring how the distribution of possible future population sizes (including extinction, or a population size of 0) varies as a consequence of different assumption and management options. A PVA can be a valuable tool for identifying weak links in our understanding of a species and may provide important insights into how different management options alter the likelihood of long-term persistence of the species. Bill Morris and Dan Doak describe the varieties of PVA in more detail and offer their practical philosophy to approaching PVA in Essay 12.2.

There has been a great deal of skepticism about PVA because of data limitations, uncertainties, and the rarity that predictions can be tested in the near term. Brook et al. (2000) conducted retrospective PVAs on 21 long-term data sets of mostly mammals and birds. By using the first half of the data to parameterize the PVA, they were then able to project the population trajectory for the next half of the data set, and compare the projections to the actual population trends. The PVAs were generally accurate, predicting declines that did occur. However, the datasets they used were all of very high quality, allowing robust estimates of population parameters for their models, and equally important, the populations did not experience any abrupt change between the first and second half of the data series (Coulson et al. 2001).

Population viability analysis is considered useful primarily for examining extinction risks, but PVA can also be used to determine when populations may become too small to fulfill their ecological roles. Increasingly, we have recognized that it is not just population and species extinction that is harmful, but the loss of the functions that the species may fulfill within a community or ecosystem. As populations become rare or sparse, they may no longer act as significant predators, pollinators, seed dispersers, or competitors, and their comparative absence can cause cascading changes among those species with whom they formerly interacted quite strongly. Thus, PVA might be harnessed to understand when a species might drop below its "ecologically functional population size," and to suggest the point where interventions are most useful to prevent these sorts of debilitating changes to ecosystems. O'Brien et al. present the concept of ecologically functional populations or EFPs, and describe an example of using a PVA to calculate critical population sizes of seed dispersers to maintain dominant tree species in Indonesia in Essay 12.3.

The value of hierarchical analysis for understanding population change

Most models of population dynamics project future population sizes based on current population size and per capita birth and death rates. Some population models and studies go further by attempting to incorporate the causal factors that determine birth and death rates. Although the latter approach is sometimes described as incorporating the mechanisms of population regulation, there are mechanisms of population regulation that operate at more than one level in a hierarchy of causation.

ESSAY 12.2

Population Viability Analysis and Conservation Decision Making

William F. Morris, *Duke University,* **and Daniel F. Doak,** *University of California, Santa Cruz**

■ Broadly defined, population viability analysis (PVA) is the use of quantitative methods to predict the likely future status of a population or collection of populations of conservation concern. PVA refers to a wide range of methods to analyze data and link the results to models of population growth and decline. In Table A, we group eight of these uses under the headings of assessment (in which our goal is simply to ask how well the population is doing) and management (in which our goal is to determine what interventions will reduce the population's likelihood of extinction).

Although the acronym PVA is commonly used as though it signified a single method or analytical tool, PVAs are in fact based upon a range of data analysis and modeling methods that vary widely both in their complexity and in the kinds and amount of data they require. At the simpler end of the spectrum lie models that predict only the total number of individuals in a single population. The data needed to fit these models to a particular population consist of either exhaustive counts or estimates of the total number of indi-

viduals in the population or in an easily recognized subset of the population (such as territory-holding males or mothers with dependent offspring). These counts are obtained from a series of (typically annual) censuses, and are referred to as **count-based PVAs**.

Count-based PVAs treat all individuals in the population as though they were identical. Yet we know that for species such as the loggerhead sea turtle (*Caretta caretta*), some individuals (e.g., mature breeders) are likely to contribute far more to future population growth than others (e.g., hatchlings, who must survive for many years to become reproductive), and that very different fractions of the population may be in categories with high versus low survival or reproduction (e.g., mature trees versus seedlings). Thus, we may be able to obtain a more accurate assessment of the viability of such a population if we account for these differences in individual contributions and for the fractions of individuals of each type currently in the population.

Species in which individuals differ substantially in age, size, developmental stage, social status, or any other attribute that affects their contributions to population growth are said to have "structured populations." Building a PVA for a structured population requires more information than can be obtained from simple counts of the total

number of individuals in the population. Specifically, we must estimate the rates of important demographic processes (survival, growth, and reproduction) separately for each type of individual in the population. Estimating these rates typically requires us to conduct a demographic study in which we mark individuals of each type and follow them for several years, each year recording whether they survived, their type (e.g., age or size), and the number of offspring they produced. Thus, PVAs that require data from a demographic study are referred to as **demographic PVAs**. The data needed to perform a structured PVA are more expensive and labor-intensive to collect, but they allow us to explore how well management techniques aimed at different types of individuals will fare. Demographic PVAs are based on models known as population projection matrices or demographic matrix models. Most published PVAs fall into this category, and many authors still use the term PVA to strictly mean this type of quantitative analysis. While demographic PVAs can yield more informative predictions for most species, this is only true when we have enough data to estimate the many parameters these models require. With less information, the simpler count-based approach can give much more reliable results, even for species with highly structured populations.

*Reprinted with modification from Chapter 1 of *Quantitative Conservation Biology: Theory and Practice of Population Viability Analysis.* W. F. Morris and D. F. Doak. 2002. Sinauer Associates Inc., Sunderland, MA.

TABLE A *Potential Uses of PVA "Products"*

Category of use	Specific use	Sources for examples
Assessment of extinction risk	Assessing the extinction risk of a single population	Shaffer 1981, Shaffer and Samson 1985, Lande 1988
	Comparing relative risks of two or more populations	Forsman et al. 1996, Menges 1990, Allendorf et al, 1997
	Analyzing and synthesizing monitoring data	Menges and Gordon 1996, Gerber et al. 1999
Guiding management	Identifying key life stages or demographic processes as management targets	Crouse et al. 1987
	Determining how large a reserve needs to be to gain a desired level of protection from extinction	Shaffer 1981, Armbruster and Lande 1993
	Determining how many individuals to release to establish a new population	Bustamante 1996, Howells and Edwards-Jones 1997, Marshall and Edwards-Jones 1998, South et al. 2000
	Setting limits on the harvest or "take" from a population that are compatible with its continued existence	Nantel et al. 1996, Ratsirarson et al. 1996, Tufto et al. 1999, Caswell et al. 1998
	Deciding how many populations are needed to protect a species from regional or global extinction	Menges 1990, Lindenmayer and Possingham 1996

At their simplest, both count-based and structured PVAs describe single populations, and they do not keep track of the actual spatial locations of individuals or of suitable habitat. Yet for species such as Furbish's lousewort (*Pedicularis furbishiae*) that exist as metapopulations, assessing the fate of a single population will tell us little about the species' vulnerability to regional or global extinction. Moreover, for organisms such as Northern Spotted Owl (*Strix occidentalis caurina*) and Leadbeater's possum (*Gymnobelideus leadbeateri*) that are threatened by habitat fragmentation, the spatial arrangement of remaining fragments, and hence the ease with which individuals can disperse between them, can strongly dictate population viability. Thus there is often a need for PVA models to explicitly include more than one local population or patch of suitable habitat. **Multi-site PVAs** include a broad range of model types. "Economy-class" multi-site models simply represent a set of isolated populations of a single species, with no movement of individuals between them, and are useful for calculating the probability that at least one population will persist for a given length of time. At the extreme opposite end of the spectrum from the simplest count-based PVAs lie so-called "spatially-explicit individually-based" (or SEIB) models, which track the actual positions of all individuals on a detailed habitat landscape (sometimes constructed using geographical information systems [GIS] databases) as they are born, move, reproduce, and die. Not surprisingly, these highly complex, multi-site PVAs have the most demanding data requirements of all PVA models. Not only does one need to estimate the contributions of each type of individual to population growth, one also needs to quantify their detailed movement behaviors, as well as the locations of different types of habitat on the landscape. Consequently, highly detailed multi-site PVAs will only be made possible by extraordinary data collection efforts.

Carrying out a PVA requires biologists to consider a wide array of issues, including whether and how to incorporate density-dependence or environmental correlations to demographic rates. An important issue common to all PVAs is how to account for the effects of observation error on measures of population viability. Whenever population size is determined by a sampling procedure (e.g., aerial surveys, mark-recapture methods, or extrapolation from quadrat samples) rather than exhaustive enumeration, vagaries of the sampling process will introduce variation into the counts. That is, the counts will only be estimates of the true population size, and some of the variation in the counts from year to year will reflect variation in the sampling process rather than biologically meaningful variation that actually influences population viability. Similarly, more detailed demographic data are obtained by sampling, and also subject to observation errors. Beyond counts or demographic information, multi-site PVAs usually require that we measure two additional factors, dispersal of individuals and spatial environmental correlations, each of which will have its own observation error. In addition to inflating estimates of variability, observation error introduces uncertainty into our estimated viability measures. While observation error can occur in all quantitative assessments, it is particularly difficult to measure dispersal, and thus multi-site PVAs are particularly prone to added uncertainty.

All conservation biologists are familiar with the following conundrum. It is difficult to make an accurate prediction of the future status of a population with only a limited amount of data. Yet it is precisely the rarity of threatened and endangered species that makes it both difficult to obtain a large quantity of population data and especially desirable to have some sort of population assessment. As a result, practitioners of population viability analysis operate in an arena in which data scarcity is the rule rather than the exception. In recognition of this reality, we argue that measures of population viability should be viewed not as ironclad predictions of population fate but as "works in progress" subject to updating as more data become available. But when data on a particular species are truly scarce, performing a PVA may do more harm than good by engendering a false sense of rigor. In such cases, basing conservation decisions on other methods makes far better sense. Although PVA methods are a potentially useful set of conservation tools, we do not see PVA as a panacea for all conservation problems.

The Importance of Keeping Models Simple

Population biologists have developed a vast array of complex and mathematically sophisticated models, many of which can be adapted to predict the likelihood of population extinction. We recommend using two simple and related rules of thumb to guide the use of population models in conservation biology.

The first rule is: Let the available data tell you which type of PVA to perform. More specifically, we should not seek to build a PVA model that is more complex than the data warrant. It is our view that when data are limited (as they almost always will be when we are dealing with the rare, seldom-studied species that are the typical concern of conservation planners) the benefits of using complex models to perform population viability analyses will often be illusory. That is, while more complex models may promise to yield more accurate estimates of population viability because they include more biological detail (such as migration among semi-isolated populations, the effects of spatial arrangement of habitat patches, and the nuances of genetic processes such as inbreeding depression, gene flow, and genetic drift), this gain in accuracy will be undermined if the use of a more complex model requires us to guess at critical components about which we have no data. Instead, we advocate choosing models and methods in PVA determined primarily by the type and quantity of data that are available, and not the desire to include all possibly important processes and rates. It is better to use a simple approach (keeping the simplifications in mind) than to construct a complex house of cards that relies on numbers with no empirical justification.

The second rule is: Make sure you know what your model is doing. This rule will be easier to follow if we also heed the first rule—that is, if we keep our models simple. This second rule is germane to the question of whether one should build one's own PVA model or use one of the software packages designed specifically for PVA that are now widely available, such as ALEX (Possingham and Davies 1995), GAPPS (Harris et al. 1986; Downer 1993), INMAT (Mills and Smouse 1994), RAMAS (Ferson 1994; Akçakaya and Ferson 1992; Akçakaya 1997), ULM (Ferrière et al. 1996), or VORTEX (Lacy et al. 1995). Careful use of these programs can certainly lead to a defensible viability assessment. However, naive users of the programs run the risk of making two errors. First, without fully understanding the underlying models used by the programs, users may build into the program incorrect assumptions

about the biology of the species or population under study, leading to incorrect estimates of population viability. For example, some of these programs incorporate density dependence in very specific ways that may not be appropriate for the organism under study (Mills et al. 1996). Comparisons involving some of these programs have been made that can help users understand the differences among the programs (see Lindenmayer et al. 1995, Mills et al. 1996, and Brook et al. 1997). Second, inexperienced users may be lured by the array of options proffered by the software into including risk factors about which no data exist, thus violating our first rule. We hold that, even if one ultimately plans to use one of the software packages to perform PVAs, it is essential that one begins by learning how to build one's own model from scratch, so as to fully understand what the packages are doing. One strength of building your own PVA model is that it can be far more flexible than any prepackaged software would allow.

Many people have criticized PVA practices, providing useful cautions against interpreting the results too literally, or ignoring the potential effects of uncertainty on model outcomes (e.g., Beissinger and Westphal 1998, Ludwig 1999). Certainly, in more than a few cases, limitations due to poor information will be so severe as to render performing a PVA unproductive at best, and detrimental at worst. After all, when data limitations are extreme, using a quantitative analysis invites us to view as mathematically rigorous a viability assessment that is really not much better than a guess. However, these criticisms do not invalidate PVA in general. It is important to remember that the alternatives to some type of quantitative analysis are unsubstantiated "expert" opinions, sweeping generalities, and politically motivated priority setting.

PVAs can be improved by following a number of common sense practices. First, population viability estimates should always be presented with confidence intervals. Extremely broad confidence intervals indicate that we do not have enough data yet to justify a formal PVA, but may also heighten awareness of how uncertain we are about effects on a population of any given action. Second, viability metrics should be used as relative, not absolute, gauges of population status, such as whether a species is in decline, or which population is at greater risk of extinction. Third, do not try to project population viability far into the future. Errors and consequently confidence intervals grow large in long-term projections. Thus, shorter-term predictions are in keeping with both the demands of statistical rigor and the realities of political and economic constraints. Fourth, consider a PVA to be a work in progress, which can be improved as more data become available. PVAs can contribute most to conservation planning when used to sharpen our understanding, and combine diverse, quantitative information into unified predictions. ■

ESSAY 12.3

Ecologically Functional Populations

Tim O'Brien, Margaret Kinnaird, *Wildlife Conservation Society*, **Martha J. Groom,** *University of Washington, Bothell and Seattle,* **and Peter Coppolillo,** *Wildlife Conservation Society*

■ One of the most insidious aspects of the current conservation crisis is not the extinction of species or populations but the catastrophic declines in wildlife abundance. These declines lead to a slow dissolution of the complex interactions among species, and between species and their environments. Whether we consider the control exerted by top carnivores on prey communities or plant–pollinator relationships, the significant reduction or loss of these interactions can weaken the fabric of natural communities by undermining the forces that structure them. Natural communities and ecosystems with diminished species interactions are mere shells of their intact forms.

If we wish to maintain this rich web of community interactions, we should work to ensure more than the simple persistence of target species. Instead, our first goal should be to conserve species populations that are large enough to fulfill their ecological roles and by extension preserve the intricacies of community interactions. We define these "large enough" populations as **ecologically functional populations (EFP)**. Critical to the definition is the concept that as the density of a species decreases, functionality also decreases. Thus as the species declines, it may cease to be an important interactor with other species, may no longer have a noticeable impact on the ecosystem, or be unable to fulfill its ecological role. A species' ecological role may be defined by the effects it has on the abundance, distribution, or behavior of other species, an abiotic component of its ecosystem, or by the unique ecological phenomena it represents.

The EFP concept first appeared in the literature in 1988 when Richard Conner argued that the U.S. forestry and wildlife agencies' directives to maintain species at their minimum viable population size (MVP) set our goals too low. Conner contended that we should strive first to protect EFPs and focus on MVPs only as a last resort. In 1992 Kent Redford addressed the issue more forcefully in his article *The Empty Forest*. Redford poignantly describes how botanically intact forests will ultimately deteriorate if their resident animal communities are depauperate and unable to adequately perform ecological functions such as pollination and seed dispersal. An empty forest may be deceptively beautiful today but will not persist into the future. In 2000, Redford and Peter Feinsinger further developed this line

of thinking into *The Half-Empty Forest*. Modeling populations of bumble bees, hummingbirds and guans, they argued that even relatively small declines in abundance of these key species can lead to extensive disruption of their otherwise intact communities. A variety of other studies concerning carnivore assemblages (Navaro et al. 2000; Berger et al. 2001), herbivorous mammals (Roldán and Simonetti 2001), canopy dwelling primates (Pacheco and Simonetti 2000), and fishes (Jackson et al. 2001) have shown that severe declines, particularly of harvested species, can cause cascading effects leading to degraded ecosystems or communities.

Why EFPs?

We believe there are three reasons why we should consider EFPs as an important approach to ecosystem conservation. The first is to ensure ecosystem health. Conserving functionally important species will prevent the erosion of biodiversity by preserving the interactions that structure natural communities. The argument assumes that significant species can be identified, that they perform essential functions in their communities (Walker 1992, 1995), and that direct or indirect interactions with the functionally significant species are essential for the persistence of other, less-significant species. This argument has been used to justify the conservation of focal species (Ehrlich and Wilson 1991; Walker 1995), keystone species (Paine 1966; Terborgh 1986) and ecosystem conservation (Schulze and Mooney 1993). Second, EFPs will generally be larger than MVPs, require larger areas for persistence, and thus lead to more effective ecosystem conservation. Although MVPs by definition should persist over a period defined as "long term," they may be of insufficient size to maintain full functionality within the ecosystem. Finally, we consider the rich web of interactions that define natural communities worthy of protection for intrinsic reasons alone (Callicott et al. 1999; Callicott and Mumford 1997; Redford and Richter 1999).

Strong Interactors

The EFP concept may be most relevant to those species that have strong interactions (MacArthur 1972; Powers et al. 1996) with a few critical components of a community rather than species with diffuse effects within their communities or slight effects. Species that have

large effects on their communities regardless of their abundance are called "strong interactors" (MacArthur 1972) or "fundamentally important species" (Hurlbert 1997). Within the broad class of strong interactors, we can include keystone species and dominant species (see Figure 3.5).

Dominant species that have large effects on other members of the community include trees of the family Dipterocarpaceae. Dipterocarps make up the majority of the canopy in many intact Bornean rainforests and their mast fruiting events affect the spatial and temporal distribution of many birds and mammals (Curran and Leighton 2000). Clearly, both keystone and dominant species are strong interactors and when selecting target species for conservation, the strength of the interaction should be the metric and the distinction between types of interactors should be unimportant.

Unambiguous identification of strong interactors is not an easy task. For example, Terborgh's (1986) argument that figs served as keystone species has received both support and challenge (Gautier-Hion and Michaloud 1989; Shanahan et al. 2001). Manipulative experiments are the traditional method of identifying strong interactors but these are difficult at large scales and often lack relevance for management. Comparative studies or natural experiments have the drawback that uncontrolled factors may creep into the comparison, leading to ambiguous interpretations. The most practical approach to defining strong interactions may be a combination of intuition based on natural history observations and a comparison of existing or readily gathered data from a variety of sites. Although unlikely to be extremely precise, this approach may allow a crude estimate of EFPs. The important point is to use the best data available and collect data if none are available.

An Example: Figs, Frugivores, and the Regeneration of Tropical Trees

Large fig trees are widely, though not universally, recognized as keystone resources for fruit-eating birds and mammals (Terborgh 1986; Gautier-Hion and Michaloud 1989). The term "fig" describes a genus in the family Moraceae that includes about 750 species with a global distribution and a variety of growth forms from climbers and free-standing trees to the unusual

hemi-epiphytic growth form of strangling figs. A recent review by Shanahan et al. (2000) lists almost 1300 vertebrate species that feed on figs—primarily parrots, pigeons, hornbills, and primates (Figure A). Kinnaird and O'Brien (2005) have shown that figs are strong interactors for primates and hornbill communities in Southeast Asia. Wide-ranging hornbills on Indonesian islands decline in number or leave study areas when fig availability is low. In addition, home range size of males delivering fruits to females and chicks during the breeding season expand and contract with the supply of figs. Fig availability also influences resource defense and grouping patterns of primates and hornbills. In contrast, primates and hornbills do not respond spatio-temporally or behaviorally to the availability of important diet trees in other families.

Hornbills and primates, in turn, are strong interactors with other tropical trees. In their search for widely dispersed figs, primates and hornbills encounter, consume, and disperse a broad range of fruits and seeds. For example, Sulawesi crested black macaques (*Macaca nigra*) and Sulawesi Red-knobbed (*Aceros cassidix*) and Taritic Hornbills (*Penelopides exarhatus*) disperse the seeds of more than 150 and 55 species of tropical trees, respectively. Throughout Asia, primates and hornbills disperse the seeds of more than 500 species of trees and lianas. This example illustrates the web of interactions that structure rainforest ecosystems; the figs are key players in the maintenance of frugivore populations that then provide extensive seed dispersal services to the forest tree community.

To determine the strength of the seed dispersal service provided by primates and hornbills and the EFP threshold for these seed dispersers, we attempted to determine the threshold at which the dispersal service was too low to maintain populations of tropical trees. We chose two tree species shared by Sulawesi and Sumatra and examined the key factors necessary for maintenance of populations: a canopy tree, *Dracontomelon dao* (Family Anacardiaceae), and a mid-canopy tree *Polyalthia glauca* (Family Annonaceae). We chose these trees because both genera are widespread in Asia, and both are important food resources. *Dracontomelon* produce large crops of fruits that are eaten by most primates, wild pigs, and elephants. *Polyalthia* produce

(I)   (II)

Figure A Large fig trees (I) serve as keystone resources for many species, including this Sulawesi Red-knobbed Hornbill (*Macaca nigra*) (II). (Photographs by M. Kinnaird.)

small crops of fruit consumed by primates, hornbills, and large fruit pigeons. Primates have no significant competitors for *Dracontomelon* since pigs and elephants wait for the fruit to fall, whereas primates and hornbills may compete for *Polyalthia*. We first conducted extensive censuses of tree populations. From these censuses we calculated the size distribution of the trees from seedling class to the largest sizes. We then estimated fruit production among reproductive-sized trees according to the size classes we calculated. Using demographic analyses, we determined the level of seed dispersal necessary to assure that at least one individual tree survived to reproductive size under the assumption that seeds must be moved away from the parent tree to germinate. We also censused the frugivore populations, collected data on the ingestion and processing rates of fruits by primates and hornbills, and measured or modeled the distribution of seed dispersal distances.

Our analysis for each tree species suggested that the proportion of germinating seeds reaching adult size was approximately 0.1% (1/1000) for *Dracontomelon* and 0.04% (4/10,000) for *Polyathia*. Therefore, we set the threshold for effective seed dispersal to occur when the density of primates and hornbills falls below the density at which 1000 seeds of *Dracontomelon* are dispersed and germinate in a square kilometer of forest, and 2500 germinating seeds of *Polyathia* are dispersed.

To determine the EFP threshold for primates and hornbills we simulated a square kilometer of habitat with observed densities and size structure of trees and density of seed dispersers (Figure B). We then calculated changes in seed production and number of seeds dispersed as disperser populations decline. We found that loss of seed dispersers elicited a four-step process in the tree population (Figure B.I). First, initial declines in seed disperser populations had little effect on seed production. Second, a threshold occurred when the number of seeds dispersed was insufficient to support survival of a single tree to the largest size class. At this point the largest trees began to disappear from the system, reducing fruit and seed production for the population. Third, continued loss of dispersers further reduced the supply of dispersed seeds accelerating the collapse of size class distribution in tree populations. Finally, a second threshold occurs where too few seeds are dispersed to assure that at least one tree survives to the reproductive size class. At this point the tree population ceases to produce seed and we see the loss of the tree from the community and a consequent reduction in tree diversity.

The simulation predicts that the shape of the response to loss of seed dispersers varies according to size distribution and seed production of tree populations. The average crop size of *Dracontomelon* on Sulawesi stayed constant (no loss of large individuals) as primate density declined by two thirds (67 to 20 individuals/km²), but then

collapsed quickly as the primate density declined further (Figure B.I). In this example, 6 macaques/km² was sufficient to assure a reproductive population of *Dracontomelon* but more than 20 macaques/km² were necessary to maintain a full representation of tree size classes. For *Polyalthia* on Sumatra, hornbills disperse 4 times as many seeds as primates. Because of smaller fruit crops and lower survival rates, *Polyalthia* were more sensitive to loss of seed dispersal services in our simulation (Figure B.II). In the example, hornbills disappeared first and then primates, resulting in a steep decline in dispersal service followed by a more gradual decline. The overall result however is similar to *Dracontomelon* with an early collapse of the population structure (beginning at 35 seed dispersers/km² for this example) followed by rapid loss of adult trees and local extinction at 4 seed dispersers/km².

In these examples, dispersers are only moving a portion of the total seed crop, which suggests an additive effect of dispersal rather than redundancy in the system. In a primate-hornbill disperser system in which only one disperser population declines (*Polyalthia*, for example) the tree may only experience numerical and structural changes with the loss of one set of dispersers. As both dispersers decline, the tree population once again collapses. The situation becomes more dire if dispersers abandon a tree species due to poor fruit production (Redford and Feinsinger 2000) and switch to other

(I)

(II)

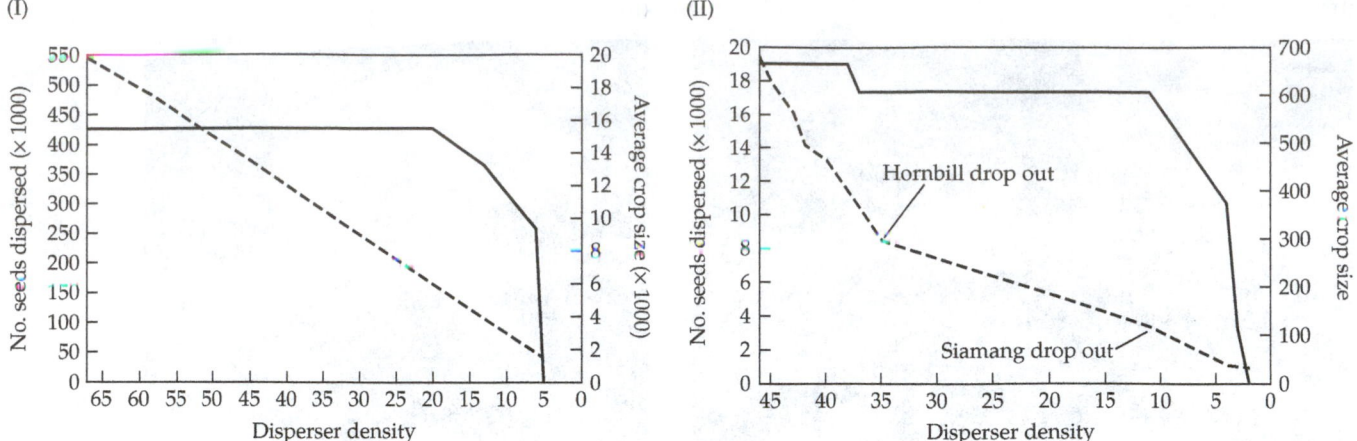

Figure B Effect of declines in disperser populations on number on seeds dispersed (dashed line) and average crop size of parent trees (solid line) for *Dracontomelon dao* (I) and *Polyalthia glauca* (II). *Dracontomelon* is dispersed by crested black macaques on Sulawesi and *Polyalthia* is dispersed by hornbills, siamang, and other primates on Sumatra. Differences in slopes reflect differences in dispersal service provided by each species or group of species.

fruit sources. In this case the extinction of the abandoned tree population may occur more quickly.

Anecdotal support for this model comes from the island of Sumba, Indonesia where the endemic Sumba Hornbill (*Aceros everetti*) has disappeared from many small forest fragments on the island. Only 30% of the hornbill-dispersed tree species remain in small patches without hornbills. The population structure in the remaining species is skewed toward old trees suggesting a lack of recruitment due to reproductive failure.

Our example illustrates that density of seed dispersers can be surprisingly small and still maintain reproductively viable populations of forest trees. To maintain a fully functioning tree population however, requires EFP densities that are 3–10 times higher. We may achieve survival of the tree population by managing for low densities of seed dispersers (an MVP strategy) but if our target is to maintain a healthy tree population *and* structural diversity of the forest, an EFP at densities high enough to assure adequate seed dispersal service is essential. ■

Thus, population regulation may be viewed most productively as a hierarchical process.

An example of this view comes from studies of population fluctuations in granivorous birds such as sparrows. Typically, these studies have looked for regulatory factors by relating sparrow birth and death rates to spatial and temporal variations in food supply. All other things being equal, the survival of sparrows during winter is highest when and where seed production has been greatest (Pulliam and Parker 1979; Pulliam and Millikan 1982). Unfortunately, all other things are rarely equal, and in the case of sparrows, not only food supply, but also habitat availability may vary dramatically from year to year. Sparrows live primarily in early successional habitats, and the availability of such habitats depends on a complex of factors, ranging from the decisions of farmers to abandon land to the rate of old-field succession. Year-to-year variation in the abundance of wintering sparrows in the southeastern U.S. may depend less on how much food is available in each patch of habitat than on how habitat availability is affected by factors such as

the influence of the global economy on the price of soybeans and the decisions made by farmers (Odum 1987).

Thus, local factors such as food supply affect sparrow populations at one level, while regional patterns of agriculture affect habitat availability for the sparrows at a very different level. To make population projections, one needs to know how many individuals there are, what habitats they occupy, and the characteristic birth and death rates for individuals in those habitats. A purely empirical model can be constructed based on population size and distribution and on habitat-specific demography. Such a model can be used to project future population sizes as long as (1) the habitat-specific birth and death rates do not change and (2) the fraction of the population in each habitat type does not change.

This type of model can be used to examine the hierarchical levels of population regulation. The model can incorporate factors such as food supply, competition, disease, and predation that influence population growth by affecting birth and death rates within the habitats where individuals occur. Such a model is called a "mechanis-

tic" model, and provides an "individual-level" explanation of the phenomenon of population growth because it emphasizes factors operating within small-scale patches. These are the factors that ecologists have emphasized in population studies (Hassell 1978; Pulliam 1983; Werner et al. 1983).

However, explanations consisting of all those factors also operate at larger spatial and temporal scales, such as the factors that determine the availability of suitable habitat (Pearson 1993). In the case of the overwintering sparrows, broad-scale, geographic factors such as land use and climate change operate at a "landscape level," beyond the habitats where the population currently resides and often over relatively long periods of time. Factors at this landscape level determine the amount and location of suitable habitat for each particular species. Figure 12.10 illustrates this hierarchical approach. A complete explanation of past trends or a projection of future population trends requires an understanding of both individual-level factors determining birth and death rates within habitats and the landscape-level factors determining regional trends in habitat availability. To understand the population dynamics of species of interest, conservationists therefore need to be concerned with environmental factors that operate at a variety of spatial scales.

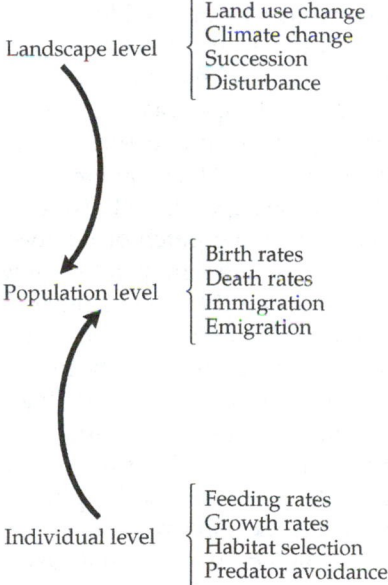

Figure 12.10 Population dynamics should be understood as resulting from a hierarchy of processes affecting populations at different levels. Landscape-level changes in the availability of habitat determine how much suitable habitat exists for a given species, and its configuration (and therefore its accessibility). The availability of suitable habitat and the behavior and physiology of individual organisms combine to influence the dynamics of populations.

One region where individual-level and landscape-level factors are both involved in a controversial conservation situation is the Greater Yellowstone area, which includes Yellowstone National Park (Keiter 1997). American bison (*Bos bison*) in Yellowstone are one of the last free-ranging bison populations in North America, and are managed by the National Park Service with limited human control. The bison spend most of the year within the national park, but in harsh winters some animals move out of the park onto private and public grazing lands to the north and east of the park's boundaries. This creates a management problem, because these lands are leased to local ranchers for cattle grazing.

A portion of Yellowstone bison carry brucellosis, an infectious disease that causes females to abort their fetuses (Keiter 1997; see Essay 3.1). Brucellosis can be an economic disaster if it becomes established in cattle herds. Infected cows lose their calves through abortion, and meat from infected cattle cannot be sold at market, as the *Brucella abortus* organism that causes the disease is also associated with a related disease in humans. Ranches that get infected with the disease face economic ruin, and if brucellosis becomes established throughout a region, expensive steps must be taken to establish that the region is "brucellosis-free" before meat or cattle can be exported. The ranching industry therefore desires to restrict the possibility of disease transmission as much as possible.

Brucellosis is transmitted between animals in spring, when uninfected animals can come in contact with infected tissues (the aborted fetuses and placentas; Keiter 1997). It is known that cattle can give the disease to bison, but it has never been demonstrated that bison can transmit the disease to cattle under the free-ranging conditions found in the Greater Yellowstone area. Nonetheless, the ranching industry has convinced the affected states that the bison must be controlled, to prevent even the possibility of disease transmission. When bison leave the park in winter, state wildlife biologists try to harass the animals back into the park. If that does not succeed, the bison are shot. All animals are killed, males as well as females, regardless of whether they are infected or not. This policy became especially controversial in the winter of 1996–1997, when over 1000 bison were killed. Combined with other mortality factors, about half of the 3000 bison in Yellowstone died that winter. Most subsequent winters have seen fewer deaths.

Management of the bison herd at Yellowstone is complicated by both individual-level and landscape-level factors. The population growth rate of the Yellowstone herd is determined by local factors such as food supplies within the park and the lack of controlling predators, both of which affect reproductive success and mortality rates. Bison leave the park in winters when local winter food is inadequate, especially during harsh weather.

Local ranchers complain that the Park Service should keep the bison in the park, but the movements outside the park are not "unnatural." Instead, the boundaries of the park were drawn without including the bison's natural winter range.

But the dynamics of the bison are also affected by landscape-scale factors such as the distribution of private and public grazing lands outside the park, the presence of cattle on some of these lands, and even winter tourism. The primary reason for the killing policy is to reduce rancher's fears of disease transmission. Yet, the fact that *Brucella* is also found in the region's 30,000 elk strongly suggests that the bison killing policy will not prevent the possibility of cattle exposure to the disease, because the elk also use the same winter range as the cattle and bison. Shooting of the elk is not politically possible, however, because elk are a game species and therefore have the protection of the regional hunting community (bison are not a legal game species). Winter tourism also affects movements of the bison. Tourism businesses outside the park rent snowmobiles to tourists to travel on packed trails inside and outside the park. Where these trails enter the park's valleys containing bison, the animals may use the trails to move out of the park (Bjornlie and Garrott 2001).

The management program for Yellowstone bison is therefore affected by factors that operate at different scales, making this a complicated situation for wildlife biologists and park policymakers. Dobson and Meagher (1996) examined the population ecology of the disease and its bison host, and concluded that eradication of the disease in wild bison is probably impossible because of the low population threshold at which the bison become infected. Thus, unless cattle are removed from some ranges in harsh winters, the potential for disease transmission will color the management situation for the foreseeable future. In 2003, a conservation organization arranged a swap of grazing leases to reduce the number of cattle grazing in winter ranges just north of the park. Because of the multiple issues that are involved, the Yellowstone bison situation provides an effective vehicle for discussing the ethics of conservation strategies (see Dunning 2000).

Landscape models for conservation

Individuals often move among suitable patches in the landscape, or in the sea. Because populations in the various patches are linked by movement of dispersing individuals, the fates of the populations are interconnected. The landscape approach recognizes the interconnectedness of populations and incorporates this concept into models and management plans.

Individuals of the same species living in relative proximity to one another may experience quite different physical and biotic environments, even to the extent that some may not be able to survive and reproduce while others do very well. At spatial scales substantially larger than what one individual encounters, the landscape experienced by a population represents a mosaic of good and bad places for the species. The growth, or lack thereof, of the population is determined not only by the quality of the individual microsites occupied, but also by the spatial and temporal distribution of suitable and unsuitable microsites or patches of habitat.

Increasingly, conservation biologists are adopting the landscape perspective when designing management plans and analyzing the environmental factors affecting species of interest (Noss 1983; Carroll et al. 2001, 2003; Sanderson et al. 2002). Conservation strategies recognize that organisms move over heterogeneous landscapes, and therefore that saving a single patch of "critical habitat" will rarely be enough to maintain a population. Researchers working with aquatic systems were among the first to realize this principle, because a river or lake cannot be considered protected if the watershed that feeds the system is not included in the management plan. Similarly, concepts such as source–sink dynamics require conservationists and land managers to adopt a landscape perspective. In many areas, this broader perspective requires planners to consider landscape patterns and land use strategies outside of the land unit (park, forest district, county) that they are working with directly (Noss 1983). Thus, even in a region as large as Yellowstone National Park, wildlife managers recognize that their management strategies must cut across artificial political and agency boundaries to establish a region-wide conservation plan (Goldstein 1992; Turner et al. 1995; Noss et al. 2002).

How is this landscape perspective developed and applied in the field? A landscape is a mosaic of habitat patches across which organisms move, settle, reproduce, and eventually die (Forman and Godron 1986; Forman 1995). The size of a landscape depends on the organism. A field or woodlot that is a single patch of relatively uniform quality for a bird or mammal may, at the same time, be a mosaic of patches of quite different quality for individual nematodes or shrubs. For any species, the landscape containing a population can, in principle, be mapped as a mosaic of patches varying from unsuitable to highly suitable. Such maps are the basic tool of the landscape ecologist. Each map is specific to the habitat requirements of one species and must be drawn at a scale appropriate to that organism. In general, the scale must be fine enough to resolve the areas occupied by individuals over significant portions of their lifetimes (Turner 1989; Turner et al. 1989).

To map the patches of suitable habitat for a particular species, one must have a set of criteria for drawing the habitat boundaries. Following Elton (1949) and Andrewartha and Birch (1984), a habitat boundary is chosen so as to circumscribe a "certain homogeneity with respect to the sort of environments it might provide for animals" (Andrewartha and Birch 1984). In managed landscapes, the drawing of habitat boundaries is usual-

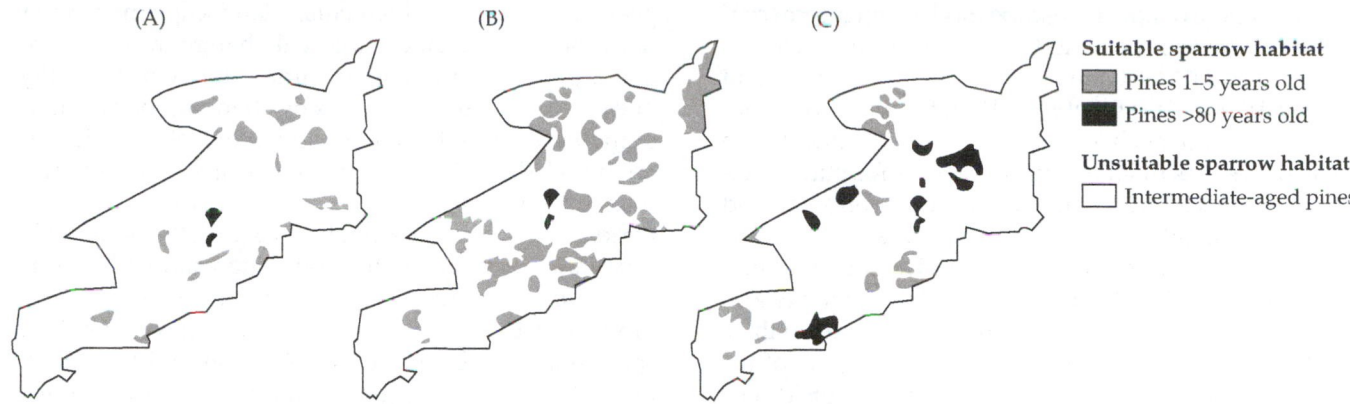

Figure 12.11   Distribution of suitable breeding habitat for Bachman's Sparrow in (A) 1970, (B) 1990, and (C) 2010 on a 5000-ha tract at the Savannah River Site, South Carolina. Bachman's Sparrow breeds both in older-growth pine forests and in young clearcuts, but not in middle-aged pine stands. The distribution of habitats in 1970 is based on the land use history for the area; the distribution in 2010 is based on a proposed management plan. Notice the island-like nature of suitable habitat patches.

ly simplified by the strong contrasts between habitats subjected to different management histories. For example, a pine plantation is clearly discernible from a neighboring field or deciduous woodlot. In landscapes less dominated by human activities, habitat boundaries are often "softer" and more arbitrary. Boundaries are typically very ambiguous or absent altogether in marine systems, where application of landscape concepts is difficult and sometimes inappropriate.

Suitable sites for a particular species are often distributed as isolated patches embedded in a matrix of unsuitable habitat. Figure 12.11 provides an example of this type of landscape, showing suitable habitat for Bachman's Sparrow (*Aimophila aestivalis*), a species of management interest to the U.S. Forest Service. This bird is found on pinelands where timber is harvested. The maps in Figure 12.11 show distributions of suitable habitat on a 5000-ha tract of timberland at the Savannah River Site, a U.S. Department of Energy facility in South Carolina. At the Savannah River Site the sparrow is found in two habitats: frequently burned old-growth pine stands, and very young pine clear-cuts. Both of these habitat types have the appropriate vegetative structure (an open understory) and are therefore suitable breeding sites for this species (Dunning and Watts 1990). Hardwood stands and pine stands between 5 and about 80 years of age are unsuitable for Bachman's Sparrows. Figure 12.11B shows the distribution of suitable sites on the study area in 1990. Figure 12.11A shows the probable distribution of suitable sites 20 years earlier (1970) based on the known land use history of the area, while Figure 12.11C shows the projected distribution of suitable sites 20 years in the future (2010) based on a proposed management plan for the site (Liu et al. 1995). The maps indicate that the locations of

suitable habitat patches change for this species on a relatively short time scale (in 20 years there will be 4–6 sparrow generations). The fact that the locations of suitable patches changes from year to year could pose a problem for the sparrow, as field studies show that isolated, short-lived suitable patches often are not colonized (Dunning et al. 1995). The sparrow's regional population may be reduced if a sizable portion of the suitable habitat is not occupied (Dunning et al. 2000).

A number of factors that influence the location of suitable habitat for Bachman's Sparrow are quite general in that they influence the location of habitat for many terrestrial species. Factors such as soil type, topography, and vegetative cover all provide information on the suitability of a site for a particular species, and all of these factors can be readily mapped. For the Bachman's Sparrow, soil type and topography influence the rate at which seedling trees grow and therefore, the ages of pine stands that have vegetation profiles suitable for the sparrow. In addition, time since disturbance, successional status, and management history may provide additional information on the suitability of a site. Knowledge of these factors improves our ability to map out suitable habitat patches for a species.

The combination of factors determining site suitability will be different for every species. Such information alone will not be enough to determine unambiguously the presence or absence of a species, but it can often be used to categorize habitats as suitable versus unsuitable, and in some cases to assign a probability of occupation. A landscape map based on the number and location of suitable sites under existing or proposed land use patterns can be an invaluable tool for species management (Zabel et al. 2003).

An example of how landscape models can inform conservation efforts is provided by the study conducted by Liu et al. (2001) of China's effort to protect the giant panda (*Ailuropoda melanoleuca*). With the support of conservation groups such as the World Wildlife Fund, China set aside a series of large nature reserves for panda protection. Reserves are located in mountainous areas, and often only a small portion of each reserve is suitable habitat for the bamboo on which the pandas depend. Local people living within the reserves benefit by an increase in ecotourism, which helps to stabilize the economy of these rural areas. Liu et al. (2001) studied the impact of the growth in the human population within and outside one major park, the Woolong Nature Reserve, which has been a flagship project for this program. Surprisingly, Liu et al. determined that the rate of habitat destruction of panda habitat inside the reserve had increased since the reserve establishment. Apparently, ecotourism had improved the local economy enough to sustain a high human population growth rate. The local population cleared land for agriculture and to provide firewood needed in the production of objects to sell to tourists. Unfortunately, the land cleared by local residents included much of the prime habitat within the reserve for pandas, and the rate of habitat destruction within the reserve was actually greater than that in the surrounding region. Liu et al. (2001) used extensive ground surveys of bamboo habitat suitability and panda observations to develop landscape maps of changing panda habitat distributions, allowing analysis of how potential changes in reserve design and management might better benefit the pandas at landscape to regional scales.

Spatially explicit population models

Metapopulation models of the sort discussed earlier are very general conceptual models that do not attempt to incorporate the complexities of real landscapes. As Levins (1966) pointed out, general models typically offer general insights, but they are neither very precise nor very realistic. One of the primary themes of both conservation biology and landscape ecology is that details, such as the geometry of habitat patches in a landscape, can influence population trends and extinction probabilities. In contrast, models that incorporate realistic details of particular species and landscapes may be better suited for managing particular species on particular landscapes, but because they are so specific, the conclusions reached from them cannot easily be generalized to other species and landscapes.

Spatially explicit population models, or **SEPM**, incorporate the actual locations of organisms and suitable patches of habitat, and explicitly consider the movement of organisms among such patches (Dunning, Stewart et al. 1995). Spatially explicit models are generally com-

posed of three major elements: a landscape map, a scenario of how the landscape will change in the future, and a population dynamics simulation. Conceptually, one can model many systems by changing one or more of these elements. Different populations of a species can be modeled by changing the landscape map; different management plans by changing the future-change scenarios, and different species by changing the natural history variables built into the population simulation. Very few research programs have actually extended their models in this fashion; most models have remained species- and landscape-specific. Case Study 12.1 by Kimberly With provides a good example of a research program using SEPMs that satisfy both local and broader conservation concerns.

The mobile animal population, or MAP, is a class of spatially explicit population simulation models (Pulliam et al. 1992; Liu 1993; Liu et al. 1995) that simulate habitat-specific demography and the dispersal behavior of organisms on computer representations of real landscapes. In MAP models the landscape is represented as a grid of cells, each of which is the size of an individual territory of the species being simulated (e.g., 2.5 ha for Bachman's Sparrow or 1000 ha for Northern Spotted Owl; see map inserts in Figure 12.12). Clusters of adjacent cells represent the size and location of forest tracts in the landscape; these tracts are assumed to be relatively uniform in terms of suitability for the species of interest. MAP models contain subroutines that specify forest management practices, succession, and in some cases, the growth rates of tree species. Thus, such a model can depict current landscape structure and project the landscape structure in the future based on a management plan specifying a harvest and replanting schedule. Other management activities such as thinning or burning stands, which might influence stand suitability, can be incorporated easily into MAP models. The most realistic MAP models are run on landscape maps generated by **geographic information systems** (**GIS**, discussed further in Chapter 14), which incorporate the actual distribution of habitat patches in a region. Whereas most MAP models run on landscapes of less than 10,000 ha, MAP models can simulate landscape change and population dynamics over 100,000 ha or more (Pulliam et al. 1995).

Analyses using MAP models have proven valuable in a variety of conservation situations. For example, BACHMAP, a MAP model developed specifically for Bachman's Sparrow, was used to determine how forest management practices may influence the population viability of the sparrow in pine forests in the southeastern U.S. One analysis using BACHMAP demonstrated that aspects of a management plan designed to improve habitat for one endangered species (the Red-cockaded Woodpecker) may cause a short-term decline in the sparrow's numbers, thus potentially creating a second species requiring manage-

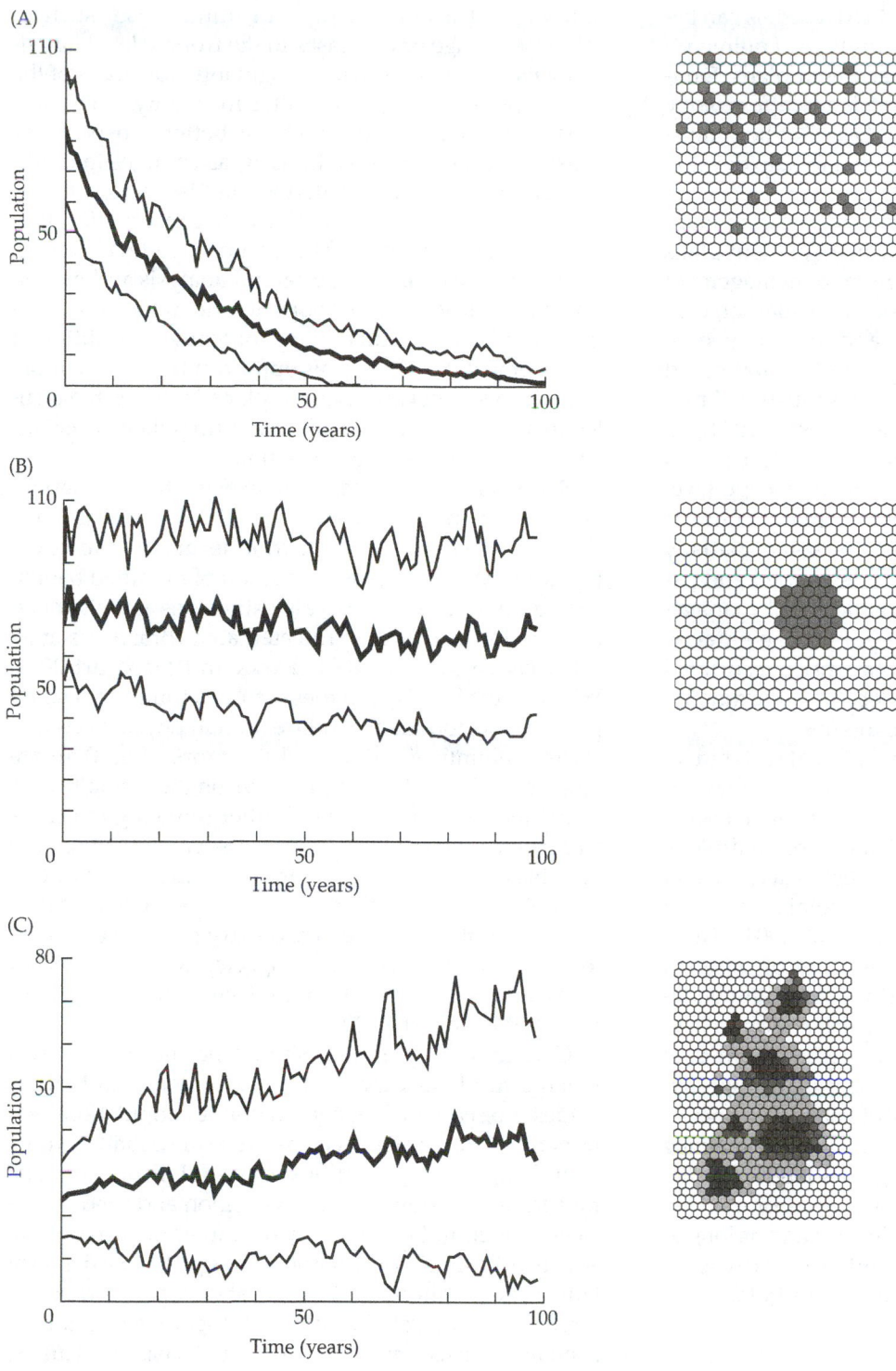

Figure 12.12 Results of a simulation model for Northern Spotted Owls that varied only the configuration of suitable and unsuitable habitat; all other population parameters were held constant. All results are based on 30 simulations; the heavy line in each graph is the mean population response, and the thin lines are one standard deviation from the mean. (A) Suitable habitat is randomly scattered. (B) Suitable habitat is arrayed in one large block. (C) Clusters of suitable habitat are surrounded by marginal habitat (buffers). (From McKelvey et al. 1993.)

ment intervention (Liu et al. 1995). MAP models can become innovative tools for avoiding this type of unintended effect during the design of management plans. Dunning et al. (2002) describe exercises in landscape ecology involving simulations with BACHMAP that help illustrate how GIS linked SEPMs can be used to help develop conservation plans.

Spatially explicit models provide conservation managers with tools for predicting how management plans and land-use changes may affect species of management concern. As their usefulness has become apparent, conservationists are advocating expanding the use of SEPMs into new areas. For example, individual-based models linked to GIS maps could enormously aid prediction of impacts of toxic spills (Carlsen et al. 2004), allowing quicker mobilization of efforts to protect critical populations. Expansion of SEPM for landscape-level restoration also could be beneficial in many contexts (Holl et al. 2003; see Chapter 15). While some spatially explicit models are sufficiently well developed and parameterized to be used in conservation and management, for most species the relevant population parameters are not known, and several years of concentrated fieldwork would be required to collect the large amount of data needed to parameterize the models.

For example, to understand how different water management regimes in the Everglades might affect fish and wading bird populations, conservationists have used a spatially-explicit model, ALFISH, that models hydrology across marsh habitat and predicts fish survival and movements based on water level, food availability, and predation levels (Gaff et al. 2000; Gaff et al. 2004). These models are used for a relative assessment of different water management scenarios, and their recommendations are considered qualitatively correct, although field validation shows that they cannot predict more than 20%–40% of the variation in fish density. This high uncertainty comes from the sensitivity of fish to the exact locations and sizes of ponds, and the difficulty of obtaining such fine scale data across the Everglades landscape. Because this model is primarily used as a decision-making tool, rather than a predictor of fish density, and as long as the qualitative depiction of the Everglades landscape is correct, such models can provide helpful insights.

Challenges and Opportunities of Conservation at the Landscape Scale

As landscapes become increasingly dominated by human-altered habitats, conservationists not only need to evaluate the viability of species across a large spatial scale, but also to project ecological, social, and economic influences that will alter how humans interact across the landscape. Tracking changes in human use (such as land-use change or decreases in the trophic level targeted by fisheries) can provide insight into how parts of the landscape or seascape will alter in quality over time, causing individuals to search for better prospects, or causing local extinctions. Equally, as changes in biodiversity affect human populations, landscape-level evaluations may point to incentives for conservation that support human welfare. Thus, landscape-level conservation may be enhanced by careful analysis and modeling that can tease apart important influences on population trends, or predict likely outcomes of different management decisions. This opportunity to create more specific, relevant conservation plans includes the challenge of integrating such analyses into policy decisions, and influencing local citizen action.

Recent analyses of mammalian extinctions in national parks in the U.S., Canada, and West Africa have shown that landscape context, in terms of land cover types, human density, or proportion of modified habitat surrounding each park explain extinctions as well or better than the size of the protected area (Brashares et al. 2001; Harcourt et al. 2001; Parks and Harcourt 2002; Wiersma et al. 2004). All these studies found that small parks are surrounded by dense human populations and lower amounts of natural habitat more often than are larger ones. Thus, human pressures on these small parks might increase edge effects, further reducing the effective size of these already small reserves, producing a "double jeopardy" for small protected areas (Harcourt et al. 2001). Understanding the landscape context of protected areas thus becomes a necessary part of creating effective plans for conserving species within protected areas, or among networks of protected areas (see Chapter 14) (Wiersma et al. 2004).

Conservation planning should be more successful when a full landscape analysis is undertaken that includes projections of changes in human populations. **Alternative-futures analysis** makes use of spatially-explicit landscape-scale projections of several distinct options for future development within a region and predicts socioeconomic and biodiversity outcomes of each option. A consortium of conservation biologists (particularly landscape architects and ecologists), city and regional planners, and local citizens work together to examine probable consequences of distinct decisions visually, through an iterative process of creating and analyzing integrative landscape maps of change and impacts (Baker et al. 2004; Figure 12.13). Specific analyses then focus on the outcomes of distinct "alternative futures"—for example, predicted trends for threatened species when channeling growth into a new urban center versus low-density sprawl in the Camp Pendelton are of California (Steinitz 1996).

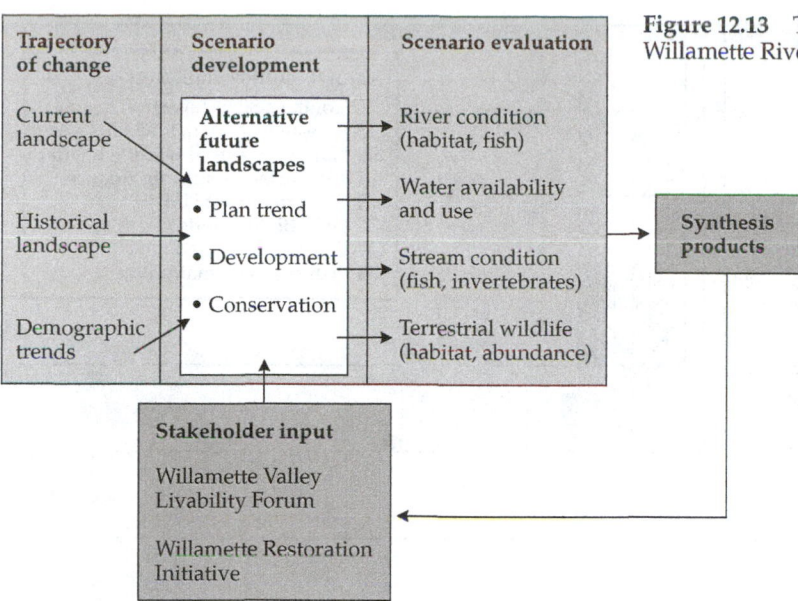

Figure 12.13   The alternative-futures analysis process for the Willamette River Basin. (Modified from Baker et al. 2004.)

Alternative-futures analysis for the Willamette River Basin, Oregon, evaluated the impacts on four resources (water availability, the Willamette River's condition, wildlife abundance and habitat, and stream community health) of three possible future scenarios: (1) projecting the trends of 1990 forward to 2050 ("Plan Trend"), (2) projecting 1990 trends, but with enhanced development ("Development"), and (3) projecting 1990 trends with enhanced, but politically acceptable, conservation and restoration goals ("Conservation") (Baker et al. 2004). Only conservation led to significant improvement over 1990 in habitat, wildlife abundance, and stream health measures (Baker et al. 2004; Schumaker et al. 2004: Figure 12.14), but this plan also led to trade-offs for agriculture. Although this analysis does not dictate the path that will be taken by policymakers, the process is likely to influence the decisions made because the consequences of decisions have been made concrete, and due to enduring working relationships among the consortium working on the analysis (Hulse et al. 2004). Importantly, the project identified several conservation and restoration opportunities throughout the Willamette River Basin that have been adopted for use in guiding salmon restoration, as well as other projects in the region.

Some other examples of landscape-scale conservation analysis and planning are derived from a **landscape species approach**. This approach selects focal species that range across large scales (such as large carnivores), and for which essential habitat can be defined, and interactions with human activities characterized. Using GIS, regions of the landscape that are required for the species, and a separate layer of human activities and political jurisdictions are mapped. These combined layers are analyzed to identify the suite of connected landscape elements of essential and high value to the species that are feasible to conserve (Sanderson et al. 2002; Figure 12.15). Probable increases in threats to the focal landscape needed to protect this species are evaluated to determine where conservation interventions might be most important. This procedure is then applied to several different kinds of species to develop a conservation strategy across the entire landscape. A further example of this approach is detailed in Case Study 12.2 by Lilian Painter, Robert Wallace, and Humberto Gomez.

Margaret Kinnaird and her colleagues used a landscape approach to evaluate where targeted conservation efforts might be most fruitful in the face of rapid deforestation inside and surrounding a large forest reserve on Sumatra (Kinnaird et al. 2003). Three species, the Sumatran tiger (*Panthera tigris sumatrae*), Sumatran rhinoceros (*Dicerorhinus sumatrensis*), and elephant (*Elephas maximus*), all used interior forest primarily based on surveys using automatic cameras. Projecting patterns of forest loss from analysis of satellite images of Bukit Barisan Selatan National Park between 1985 and 1999, Kinnaird et al (2003) determined where large blocks of forest should remain and their degree of fragmentation, and therefore what areas should be targeted most in future conservation efforts. The challenge will be in creating enduring partnerships with local communities that provide incentives for forest conservation and restoration that could save these endangered species in the face of government instability.

The complications of conserving wide-ranging species are compounded when species move across national boundaries. Conserving migratory species, such as cranes (Case Study 12.3) and penguins (Essay 12.4), may

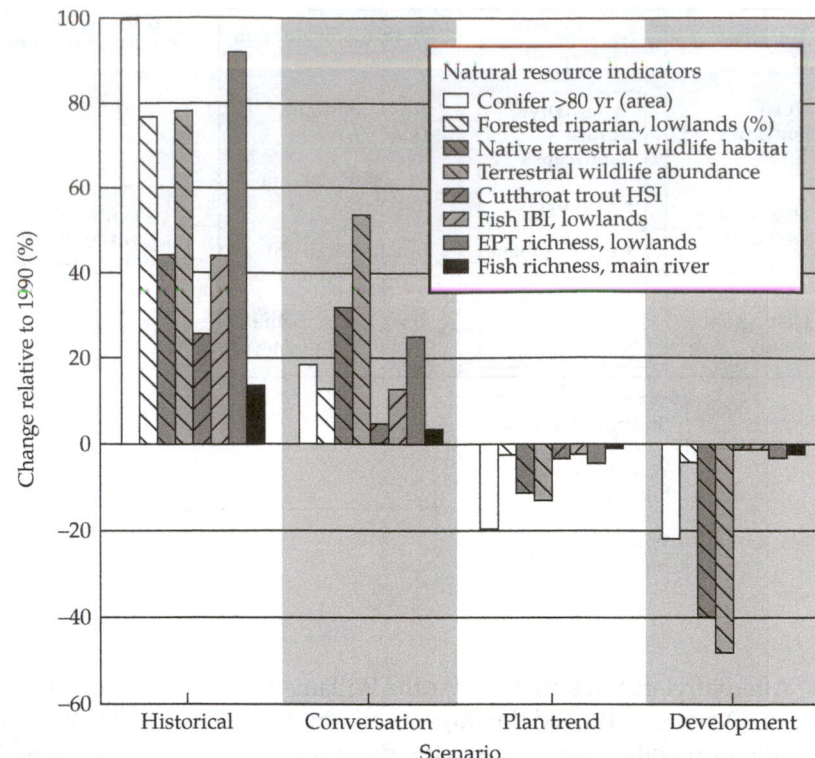

Figure 12.14 Percent change in several habitat, wildlife, and stream health indicators in the Willamette River Basin, in three alternative future scenarios, as explained in the text, relative to 1990 conditions. HSI, habitat suitability index; IBI, index of biotic integrity; EPT, Ephemoptera, Plecoptera, and Tricoptera richness in streams. (Historical conifer cover was 184% that of 1990 extent.) (Modified from Baker et al. 2004.)

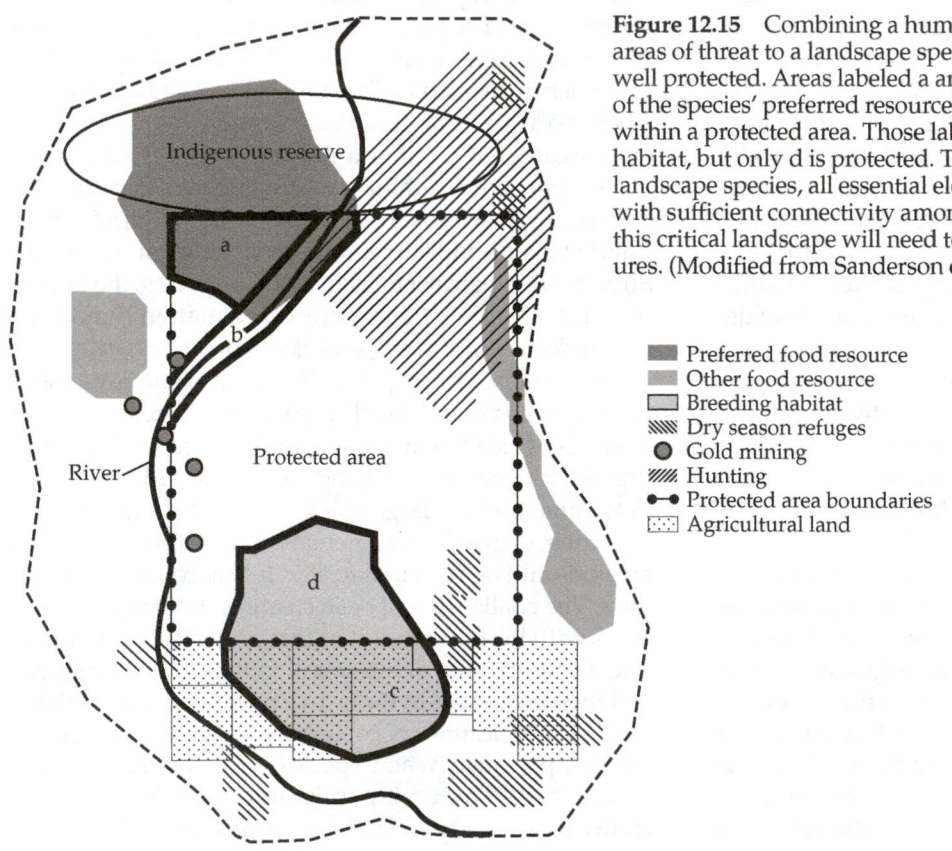

Figure 12.15 Combining a human and biological landscape reveals areas of threat to a landscape species, as well as areas that are already well protected. Areas labeled a and b represent regions where portions of the species' preferred resources and main water source (a river) lie within a protected area. Those labeled c and d are important breeding habitat, but only d is protected. To develop a full plan to conserve the landscape species, all essential elements need to be represented along with sufficient connectivity among these. Threats to the species within this critical landscape will need to be addressed through active measures. (Modified from Sanderson et al. 2002.)

create some of the greatest challenges. In these examples, landscape-level analyses of population movement, locations of population sources and sinks, and the potential for local, regional, and international partnerships are all integral parts of developing comprehensive plans for effective conservation. Both careful ecological analysis of each species and collaborative work with citizens and policymakers are essential to success in these, as in many cases (see Chapter 17). Conservation biologists will best succeed as they learn the tools of demography and landscape planning, and are able to partner with diverse stakeholders to analyze and create potential solutions.

ESSAY 12.4

Landscape-Level Conservation for the Sea

P. Dee Boersma, *Wadsworth Chair of Conservation Science, University of Washington and Scientific Fellow of the Wildlife Conservation Society*

■ Conservation of the sea, one of the least explored and least known parts of the globe, has largely been ignored because we don't live in the sea, and it is mostly is a commons owned by no nation and no person. Conservation of marine systems can't be achieved in the same way as on land because ownership, biology, and the liquid medium of life in the ocean differ so much from land. Yet, oceans cover 72% of Earth's surface, and provide a hefty proportion of our food.

Three primary boundaries are recognized in the sea, but not one of them is easily visible. The first—the continental shelves—are where most of life is concentrated. The second—the exclusive economic zones (EEZ)—include the continental shelves of countries where most fisheries occur. The third—the open ocean—is a commons where ownership is determined by who can get there and use the resources. The continental shelves and the open ocean contribute almost half of the planet's primary production and support our most important protein source, fish. Less than 10% of the ocean has zoning to govern use of ocean resources, and only 0.5% has serious protection for conservation purposes (Costanza et al. 1998). As a result of overexploitation and continued degradation of the oceans, there is widespread decline and collapse of major fish stocks. We still lack a comprehensive understanding of where, and how, humans use the ocean. Governments are subsidizing many ocean exploitation activities with tax incentives, cheap fuel, and other subsidies that foster over-use and degradation of the ocean.

Marine protection through designated marine protected areas and zoning has increased as humans seek to combat overexploitation and preserve functioning in marine ecosystems. My research focused on Magellanic Penguins (Figure A), breeding at Punta Tombo, Argentina, and their use of the marine environment near their breeding colony. By understanding their behavior at sea we can begin to predict where people and penguins may come into conflict.

Currently we are working to define ocean zoning that can place conservation-compatible uses in regions of high use by wildlife. Australia and Antarctica already use extensive marine zoning. Now, in the South Atlantic, a group of local nongovernmental organizations, with the help of the Wildlife Conservation Society, have formed a coalition interested in marine zoning intended to better protect wildlife from albatrosses to penguins to seals to whales.

Penguins, as well as most other sea life, face the challenges from human over-consumption of fish and petroleum. We fish down the food chain, taking smaller and smaller fishes, the ones that penguins like to eat. Not only are many of the fisheries unsustainable, but they depend on cheap energy to be profitable. Moreover, many fisheries use monofilament nets that catch and drown penguins and other nontarget species.

Petroleum shipping results in the occasional oil spill that is well publicized, but high petroleum use results in chronic smaller discharges of petroleum into the oceans. We estimated that a small spill of less than 10,000 liters of oil in 1991 killed 18,000 penguins along the Chubut coast of Argentina. Between 1980 and 1990 about 40,000 Magellanic Penguins were killed each year by

Figure A   Magellanic Penguins at their breeding colony at Punto Tombo, Argentina. (Photograph by P. D. Boersma.)

chronic discharge of oil along the coast of Chubut (Gandini et al 1994). As a result of these data and the determination of government and industry to reduce the impacts of chronic oil pollution in the early 1990s, shipping lanes along the coast of Chubut were moved 100 km farther offshore and the number of oiled penguins dropped. This is an example where conflicts with humans can be reduced by ocean zoning and management of the marine environment that is based on science.

Over-fishing poses another problem for Magellanic Penguins. The South Atlantic has one of the fastest-growing fisheries in the world. Squid, hake, anchovy, and shrimp are all exploited. Many of these fisheries have collapsed or have varied in harvest so much between years that management action such as closures were required to protect the stock. Penguins don't have the benefit of any management of their fishing stock and when they don't have food, they starve.

To learn where penguins were foraging, we first tried radio telemetry to track them. However, the range for radio telemetry is short, and even when we used airplanes to fly over the water to look for penguins, we had trouble finding them. We had expected to find them close to the colony. Zoo studies showed that Magellanic Penguins with full stomachs had empty stomachs in 8 hours, so we thought penguins must feed within 30 km of their nest sites. We knew penguins flew through the water at 8 km/hr so 30 km seemed like a reasonable distance for a penguin to forage and still bring food back for a chick. Yet, when we flew within about 30 km of the breeding colony we saw less than one-quarter of the population. We thought a 30 km marine reserve for penguins would protect their foraging area when they had chicks. We were wrong.

So where were the penguins? Ultimately, we used satellite tags attached to the backs of penguins to track the birds (Boersma et al. 2002). Each tag communicates the position of the bird to satellites, thus allowing the penguins to be tracked far offshore. From this technique, we found that penguins went from 400 to 600 km from their nesting site during incubation (Figure B). Even with chicks to feed, adults often went more than 100 km. Thus, a marine reserve is unlikely to protect so wide-ranging a bird (Boersma and Parrish 1999). Instead, ocean zoning may be of more help in protecting penguins

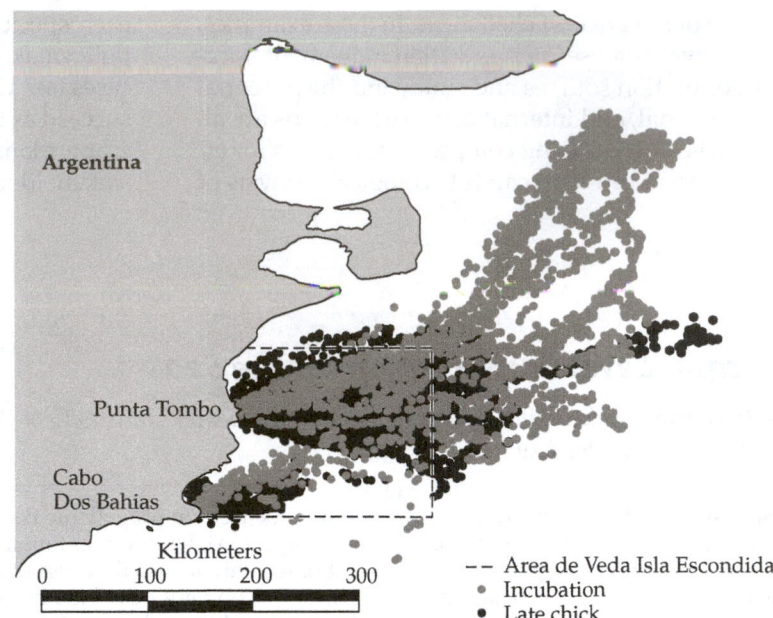

Figure B   Satellite tracks of Magellanic Penguins from two breeding colonies in Argentina (Punta Tombo and Cabo Dos Bahias) during incubation (gray dots) and when they have large chicks (black dots). The Area de Veda Isla Escondida is where large fishing boats are excluded to protect the hake spawning grounds. Note that when the penguins have large chicks they forage primarily within this management area.

and reducing conflicts between their needs and human uses of the ocean.

By suggesting changes to existing fisheries management practices, we found an opportunity to help conserve penguins. Argentina designated a zone to protect spawning hake, the "Area de Veda Isla Escondida," where only small local fishing boats are allowed to fish and large fishing boats are excluded beginning in October each year (see Figure B). Penguins often are caught in nets and drown, particularly in nets of the large fishing boats. The management area does not provide complete protection because penguins are still caught in the nets of local fishing boats and may compete with fishers for food. However, the exclusion of large boats is still beneficial to penguins. During incubation Magellanic Penguins feed inside and outside this protected area, but during the period when breeding penguins have large chicks, all of their foraging is within the fisheries management area (see Figure B). The government originally opened the "Area de Veda Isla Escondida" for all fishing boats in December. However, excluding large boats through the end of April, when penguins leave the area, would have the highest conservation value. Based on our data showing high penguin use of

this area, the government agreed to extend the closure to large boats to March starting in 1999. Following this change in practice, Magellanic Penguins fledged at heavier weights and survival of young penguins has been particularly high since 2000. We cannot be sure that the closure of this zone to large fishing boats is responsible, but it does seem likely to have been beneficial.

Knowing how penguins use the marine environment allows us to predict where protection will be needed or where problems will occur. How penguins use the marine environment depends upon the time of year. Magellanic Penguins may travel more than 2000 nautical miles from their breeding grounds in the south to their wintering grounds off Brazil. The whole South Atlantic does not need to be protected, but more fine-scale tuning can be done when we know in detail how organisms use the environment. For example, during the incubation stage, penguins at Punta Tombo generally go northeast from the colony, not south (see Figure B). Thus potential conflicts between people and penguins will be restricted to this area and other areas could be used without harm to the penguins. If people and penguins can use the marine environment in compatible ways, the future for both will be brighter. ■

CASE STUDY 12.1

Assessing Extinction Risk in Neotropical Migratory Songbirds
The Need for Landscape-Based Demographic Models

Kimberly A. With, Kansas State University

More than half of the bird species that breed in North America are not year-round residents, but are Neotropical migrants that spend the majority of the year in Central and South America. These species undertake a long, arduous journey to their breeding grounds in the north, where having survived the perils of migration, they are faced with dwindling habitat that challenges their ability to breed successfully. The widespread loss and fragmentation of breeding habitat has been implicated in the decline of many Neotropical migratory songbirds (Robbins et al. 1989a; Askins et al. 1990; Donovan and Flather 2002). Because processes affecting the population dynamics of Neotropical migrants occur at such broad scales, a landscape or regional perspective is ultimately required for the study and management of Neotropical migratory songbirds (Petit et al. 1995).

Given that songbirds must be studied at broad landscape or regional scales, how can we possibly evaluate what critical level of habitat loss will precipitate population declines, or what land-management options would minimize risk to a particular species of conservation concern? It would be difficult to execute a well-replicated field experiment to explore how different levels and rates of habitat destruction, across a gradient of fragmentation, affected the long-term persistence of various songbirds at the scale of an entire landscape. That is not to say that broad-scale field experiments on the effects of fragmentation on birds have not been performed, only that they are usually not practical and are generally done at the patch scale (experimentally creating habitat fragments of different sizes or degrees of isolation) rather than at the landscape-scale (e.g., Schmiegelow and Hannon 1999). Spatially explicit population models (SEPM), in which a simulation model of population dynamics is coupled with a landscape map, have thus become indispensable for assessing species' extinction risk in fragmented landscapes (Dunning et al. 1995; Lindenmayer et al. 1995). With a SEPM, it becomes possible to explore systematically how different components of landscape change might contribute to extinction risk, or to forecast how different species might respond to various land management scenarios (Turner et al. 1995).

SEPMs thus represent a complementary approach to broad-scale field studies and an alternative to landscape-scale experimental manipulations, which are generally not feasible or are prohibitive given the spatiotemporal scales at which Neotropical migrants operate. Computer-generated landscape scenarios afford a greater degree of control and replication than is possible in the field for exploring general questions related to how landscape change affects species' extinction risk. SEPMs are generally used in two ways. In one type of application, SEPMs are applied *strategically* to provide generalizations and insights into the relationship between the sensitivity of species to fragmentation and landscape patterns, and thus how changes in landscape structure might affect species' extinction risk. For example, this might involve evaluating the relative effects of the amount and fragmentation of habitat on extinction risk (e.g., With and King 1999). SEPMS have also been used *tactically*, as an assessment tool to evaluate the response of a target species to different land management scenarios (e.g., Liu et al. 1995).

Landscape Sources and Sinks: Linking Avian Demography to Landscape Patterns

Fragmented landscapes may function as overall population sinks for Neotropical migrants (Donovan et al. 1995). Even if sufficient breeding habitat is available, a population within a fragmented landscape may still be unable to sustain itself owing to low nesting success within habitat fragments. Birds nesting in habitat fragments are subject to edge effects—such as increased nest predation and brood parasitism—that negatively affect their reproductive success (Kremaster and Bunnell 1999). The challenge is to integrate data on the negative effects of habitat edges on nesting success across all patches in a landscape and generate a landscape-scale assessment of population viability, in terms of whether the landscape is an overall population source or sink for the species of interest.

To explore the effects of habitat fragmentation on the ability of landscapes to support viable songbird populations, With and King (2001) coupled a spatially structured avian demographic (SSAD) model with **neutral landscape models**, which are produced using a theoretical spatial distribution to create complex landscape patterns (With 1997), and can be used to generate different scenarios of habitat loss and fragmentation (Figure A). In these models, both the proportion of habitat that is left (%) and the degree to which fragments are associated spatially (H) are varied. The most fragmented habitats have low association among fragments ($H = 0$). The least fragmented habitats—where remaining habitat is grouped into one or few large blocks—have high association values ($H = 1$). Even in fairly high proportions of remaining habitat (e.g., 50%–70%), a low value of H leads to significant degrees of fragmentation (see Figure A).

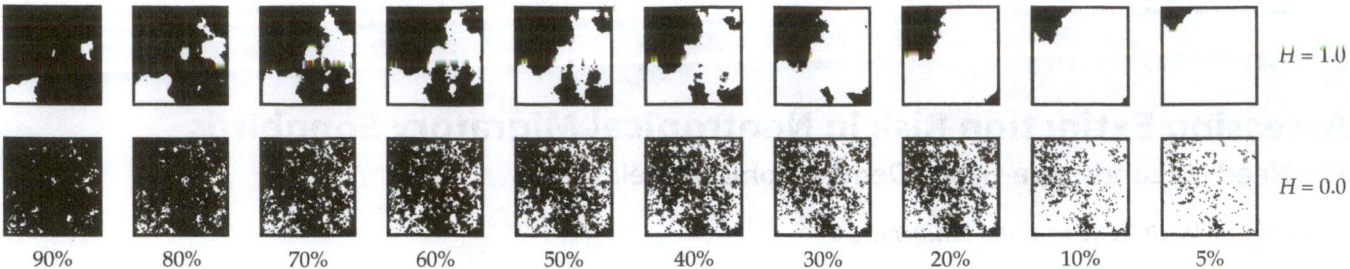

Figure A Different scenarios of landscape change created by a neutral landscape model in which both the amount (%) and degree of habitat fragmentation (*H*) were varied. *H* represents the degree of clumping in the landscape, and ranges from 0 to 1. When *H* = 1, the fragments are most closely associated spatially and form large blocks of habitat, whereas when *H* = 0, fragments are least associated, and spread widely throughout the landscape. (Modified from With 1997.)

We modeled the response of several hypothetical songbird species, created by combining two traits involving their relative sensitivity to patch area and edge effects. Area-sensitivity relates to the probability that a species will nest within patches of a given size, given that many species will only breed in habitat patches that are much larger than their individual territory requirements. Data on patch occupancy as a function of patch size are often collected in field investigations and are used to generate incidence functions that equate to the area sensitivity of the species (Robbins et al. 1989b). Species with a low probability of nesting in all but the largest patches are considered to exhibit a high degree of area-sensitivity. Edge-sensitivity is related to how reproductive success declines as a function of the amount of edge in a given patch, which is assayed by the patch edge-to-area ratio (Figure B.I). Small patches have proportionately more edge than large patches, and thus nesting success should be lowest in small patches owing to the greater potential for edge effects. Species for which nesting success is low in all but the largest patches are considered to be extremely edge-sensitive.

Nesting success is an important contributor to fecundity (*b*), a population vital rate, and is thus expected to vary as a function of landscape pattern. Most SEPMs treat fecundity as a species-specific rate, however, rather than as a landscape-specific rate. Fecundity was thus landscape-dependent in our model, and was obtained by computing the patch-specific nesting success for all territorial pairs within a given patch (see Figure B.1) across all patches on the landscape. Our model is a landscape-based demographic model, because a population vital rate (fecundity) is linked to landscape pattern. Neither adult nor juvenile survivorship (*s*) is known to be affected by fragmentation in songbirds, however, and so we used generic rates that have been reported for songbirds in the literature ($s_{adult} = 0.6$, $s_{juvenile} = 0.3$). Given information on fecundity and survivorship, it was then possible to calculate the intrinsic population growth rate (λ) for the entire landscape population, which was used to assess extinction risk. Potential sources are

(I)

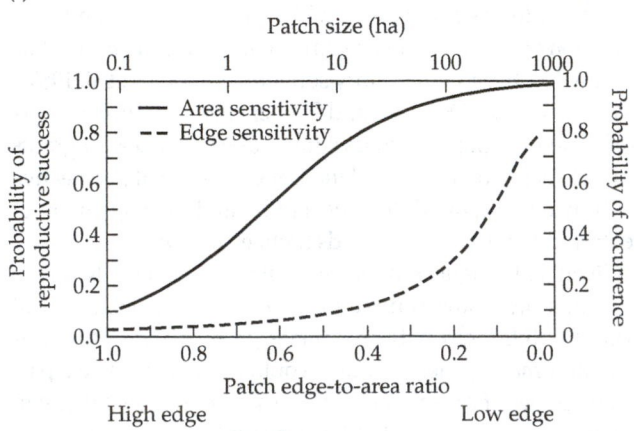

(II)

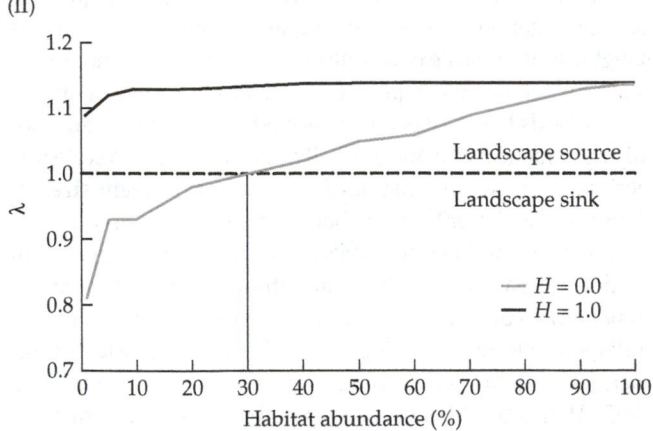

Figure B (I) Songbirds were characterized by two functions related to their area-sensitivity (probability of patch occupancy as a function of patch size) and edge-sensitivity (probability of reproductive success as a function of patch geometry). Shown here is a species with low area-sensitivity and intermediate edge-sensitivity. (II) Extinction risk for songbird populations in the two different landscape scenarios shown in (I). Landscapes function as potential population sources where population growth rates, λ, are greater than 1.0, and are at low risk of extinction. Landscapes function as population sinks where λ is less than 1.0, and are at high risk of extinction in the absence of immigration. A viability threshold thus occurs at λ = 1.0. (Modified from With and King 2001.)

those in which the landscape population is increasing ($\lambda > 1.0$), and are thus assessed as having a low extinction risk. Population sinks are those in which the landscape population is decreasing ($\lambda < 1.0$), and are thus at high extinction risk unless rescued by immigration from outlying landscapes.

The interaction between a species' sensitivity to fragmentation and landscape pattern is a complex one, but one that ultimately affects whether populations are viable beyond a critical level of habitat loss. As an example, consider a species with low area sensitivity (able to nest in small patches) and an intermediate degree of edge-sensitivity (see Figure B.I). Under a scenario of intense landscape fragmentation ($H = 0.0$; see Figure A), the population declines rapidly as a function of habitat loss (Figure B.II). This species is expected to cross the viability threshold ($\lambda = 1.0$) when 70% of the original habitat had been destroyed, at a critical level of 30% habitat. Below 30% habitat, the population is no longer viable but may nevertheless continue to persist on the landscape owing to immigration (i.e., the landscape is a population sink). Managing the landscape to preserve large patches of intact habitat ($H = 1.0$, see Figure A), however, would enable this same species to persist even at low levels of available habitat (5%), given that this species is able to nest within small habitat fragments (recall that this species has low area sensitivity). Such landscapes are thus capable of functioning as population sources for this species, because they generate a surplus of individuals relative to the amount of available habitat ($\lambda > 1.0$). In this case, landscape management might be able to mitigate extinction risk for this type of species. These results do not indicate how populations will fare in landscapes subjected to different rates of habitat loss and fragmentation, however, which is explored next.

Assessment of Extinction Risk in Dynamic Landscapes

How does assessment of extinction risk change if landscapes are undergoing chronic habitat loss and fragmentation? The previous modeling study assumed a static landscape, in which estimates of population viability (λ) were compared across a series of landscapes representing a gradient of habitat loss and fragmentation. This assumes that all landscapes lie on the same trajectory of change. Landscapes may undergo different rates of change, however, such that two seemingly similar landscapes could have arrived at their current state by different trajectories, which may have different consequences for the status and long-term viability of songbird populations in these landscapes.

By extending the SSAD model to a dynamic landscape scenario, we have explored how assessments of extinction risk are affected by the various dimensions of landscape change (Schrott et al. 2005a). Different trajectories of landscape change were created by varying both the rate of habitat destruction (0.5%, 1%, or 5%/year) and the degree of fragmentation ($H = 0.0$ versus $H = 1.0$; see Figure A). In this model, we assumed a closed population to isolate the effects of landscape change on population viability. Thus, populations can only decline under these dynamic landscape scenarios ($\lambda \leq 1.0$), and thus an analysis of viability thresholds is no longer informative. Instead, a more useful measure of population viability is the rate of change in the decline of population growth rates ($\Delta\lambda = \lambda_{t+1} - \lambda_t$), relative to the rate at which habitat is lost ($\Delta h = h_t - h_{t+1}$, where h is the percent habitat on the landscape). The point where this rate of decline begins to accelerate may indicate that the species requires immediate conservation action. We thus defined a **vulnerability threshold** as $\Delta\lambda/\Delta h < -0.01$, analogous to the most conservative IUCN criterion for species assessed as vulnerable to extinction (i.e., ≥ 1% population decline per year; Caswell 2001).

Consider once again a species with intermediate edge-sensitivity subjected to different trajectories of landscape change (Figure C). The vulnerability threshold coincides with the sharp nonlinear decline in λ as a function of percent habitat remaining on the landscape. This species crosses the vulnerability threshold at 75% habitat (i.e., 25% habitat has been destroyed) when habitat is lost gradually ($\Delta h = r = 0.5$%/year), but does not exceed the criterion for the vulnerability threshold until only 3% habitat remains (97% habitat destroyed) when the rate of loss is much more rapid (5%/year). From this, it would appear that the population is *less* vulnerable to extinction when habitat loss is rapid! This is misleading however, because it ignores the time to extinction. Given that total denudation of the landscape would occur within 20 years if

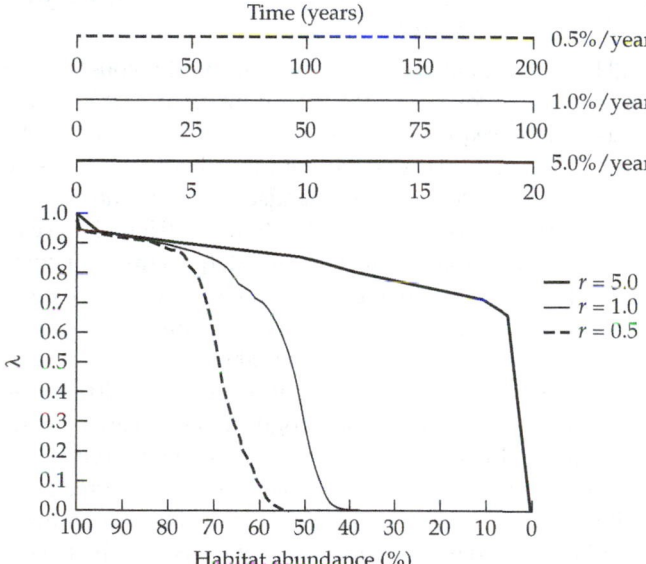

Figure C Extinction risk for migratory songbirds in dynamic landscapes undergoing chronic habitat destruction at different rates (r) and at an intermediate degree of fragmentation ($H = 0.5$). Because habitat destruction is occurring at different rates, the interpretation of the rate of population decline (λ) is time-dependent. Thus, the time scale above the graph depicts the period (number of years) over which the decline in populations occurs with respect to the rate of habitat loss (r).

habitat is lost at a rate of 5%/year, the population would go extinct within 20 years, as opposed to 55 years if habitat is lost at a rate of 0.5%/year (see Figure C).

This example illustrates that species within landscapes subjected to rapid habitat destruction may not be assessed at risk of extinction if the rate of landscape change exceeds the demographic potential of the population. In such cases, the landscape is changing faster than the population can respond; there is not sufficient time for habitat loss to reduce the population growth rate (λ) to the level of the vulnerability threshold (≥1% population decline per habitat lost per year) before the entire landscape is denuded. Some individuals persist until nearly all habitat has been destroyed, but are not reproducing successfully; they are the "living dead." Populations may thus exhibit a lagged response to habitat loss, incurring an extinction debt before they are even assessed as being at risk of extinction. This study also illustrates that populations within similar landscapes—in terms of the amount and degree of habitat fragmentation—may have very different fates depending upon how quickly those landscapes arrived at that state. For example, consider two landscapes in which half the habitat has been destroyed (50% remaining). The population may be declining if habitat loss has been recent and rapid (5%/year), or the species may have gone extinct locally if habitat loss occurred slowly over an extended period of time (0.5%/year for 100 years) (see Figure C). Thus, information on the current landscape state may not be sufficient for evaluating a species' extinction risk in the absence of information on the history of landscape change.

Demographic Limitations to Recovery through Habitat Restoration

Will habitat restoration ultimately be an effective conservation measure for recovering songbird populations that have declined as a consequence of chronic habitat loss and fragmentation? To evaluate this, we applied the dynamic SSAD model to a restoration context, in which landscapes were subjected to chronic habitat loss and fragmentation ($r = 0.5$ or 1%/year) until complete restoration (100% habitat) occurred at some point in time (t_R). The timing of restoration (t_R) was set to occur either before, at or after the population's vulnerability threshold had been exceeded (Schrott et al. 2005b).

Consider a species with moderate edge-sensitivity (Figure D). If habitat restoration occurred right at this species' vulnerability threshold, then restoration was ultimately successful in stabilizing populations if initial habitat loss had occurred gradually ($r = 0.5$%/year) and fragmentation had been minimized ($H = 1.0$). Lagged effects were evident if landscapes had previously experienced severe fragmentation ($H = 0.0$), however. In such cases the population initially appeared to remain stable after restoration for several decades, but then exhibited a rapid decline about 45 years post-restoration (see Figure D). For other species, such as highly edge-sensitive species, restoration had to be implemented some 10–15 years prior to the vulnerability

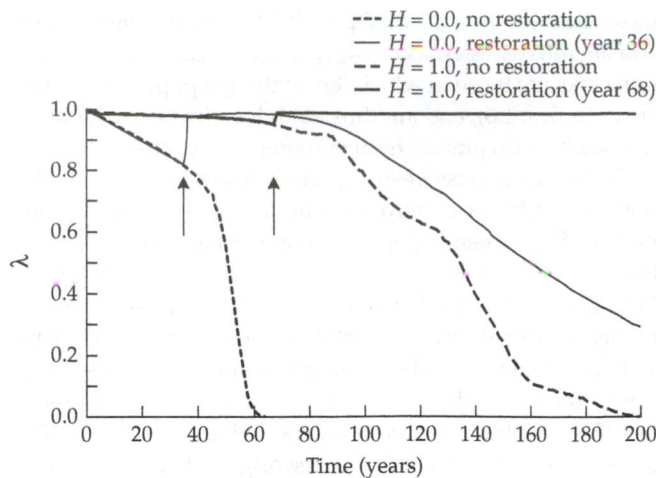

Figure D Effect of habitat restoration on the ability to recover migratory songbird populations. Habitat was initially destroyed at a rate of 1%/year, at a high ($H = 0.0$) and a low ($H = 1.0$) degree of fragmentation, until the population reached its vulnerability threshold ($\Delta\lambda/\Delta h < -0.01$). At this point (depicted by the arrows), the landscape was completely restored to its predisturbed state (100% habitat) where it remained throughout the duration of the simulation.

threshold (i.e., before the species is officially declared to be "vulnerable" to extinction) in order to stabilize population declines (not shown). Thus, conservation is more likely to be effective when measures such as habitat restoration are implemented proactively. As a final caveat, however, no single measure by itself is sufficient for evaluating population recovery. In the case of population growth rates (λ), slowing or stabilizing population growth after a period of decline will not recover initial population sizes, even with habitat restoration. There are thus demographic limitations to recovery through habitat restoration. Populations subjected to long-term habitat loss suffer an erosion of demographic potential that may compromise recovery even when the landscape is completely restored. Habitat restoration is not a panacea in conservation, but understanding when it is likely to be effective, and when it is not, is a useful application of landscape-based demographic models.

What Have We Learned from Landscape-Based Demographic Models?

The landscape-based demographic models presented here have contributed to our understanding of extinction risk in Neotropical migrants in the following ways:

1. Demonstrating how demography interacts with components of landscape change (rate, amount, fragmentation of habitat) to determine a species' susceptibility to extinction

2. Providing an integrated assessment of landscapes as overall population sources or sinks, at a scale that is commensurate to that of breeding songbirds

3. Highlighting the importance of historical effects, such

as different trajectories of landscape change; estimates of extinction risk based only on the extant landscape may be biased

4. Emphasizing the potential for lagged population responses to landscape change, in regards to both the loss as well as the restoration of habitat

5. Illustrating the potential for demographic limitations to population recovery following habitat restoration by identifying scenarios where habitat restoration may not be sufficient for recovering populations of declining songbirds

Ultimately, landscape-based demographic models can be used not just strategically, to help fill gaps in our current understanding or to generate hypotheses to be tested, but also for risk assessment, as a tactical tool in the management of a specific species in a given landscape. For example, the SSAD modeling approach discussed in this case study (With and King 2001) was originally developed and used as an assessment tool to evaluate the status of the Henslow's Sparrow (*Ammodramus*

henslowii) in a heavily disturbed landscape (King et al. 2000). The Henslow's Sparrow was found to be declining at an annual rate of 14%/year such that the landscape was a sink for this species. An increase in the landscape-wide nesting success from 39% to 58% would ultimately be required to reverse the current decline in the population. This might be achieved, for example, by minimizing disturbances that contribute to habitat fragmentation of grassland habitat favored by this species, thereby increasing patch sizes and minimizing edge effects that decrease nesting success. In general, SEPMs have proven to be valuable tools for evaluating how different scenarios of land-use change might affect extinction risk (Dunning et al. 1995). They are thus useful for identifying scenarios where intervention through land management or habitat restoration is likely to be successful in effecting a recovery, and for which species. Perhaps just as importantly, they are also useful in identifying those scenarios where such measures *cannot* recover the species, saving time and valuable resources that could be better applied elsewhere in developing alternate strategies for recovering the species.

CASE STUDY 12.2

Landscape Conservation in the Greater Madidi Landscape, Bolivia
Planning for Wildlife Across Different Scales and Jurisdictions

Lilian Painter and Robert Wallace, Wildlife Conservation Society and Greater Madidi Landscape Conservation Program, and Humberto Gomez, Greater Madidi Landscape Conservation Program

---

The Greater Madidi Landscape

The Greater Madidi Landscape (GML) is found on the eastern flanks of the tropical Andes in northwestern Bolivia and has great elevational diversity (180–6100 meters above sea level; Figure A). The GML falls within the Tropical Andes Biodiversity Hotspot (Mittermeier et al. 1998), is designated as one of the Global 200 Ecoregions (Olson and Dinerstein 1998), and is also a region of great human cultural diversity. The topographical and climatic diversity is largely responsible for its exceptional species and ecosystem richness, with plant and animal representatives of both the Andean and Amazonian ecoregions and a high degree of endemism across a variety of habitats (Gentry 1992), from high Andean puna to lowland tropical forest. Lowland and montane humid forests dominate the area but a rain shadow effect in the Tuichi and Machariapo river valleys creates a swathe of regionally important montane dry forest. Patches of regionally threatened *Polylepis* forest are found at the cloud forest-páramo interface. This area also encompasses the best-remaining example of pristine savannas in South America.

Many ecological services, including watershed protection and microclimate regulation, are provided by the forests in the GML. Furthermore, this area maintains regionally important populations of many threatened species.

In response to the strategic conservation importance of this region, Bolivia established three protected areas (see Figure A): Madidi National Park and Natural Area of Integrated Management (1,895,750 ha), Apolobamba Natural Area of Integrated Management (483,744 ha), and Pilón Lajas Biosphere Reserve and Indigenous Territory (400,000 ha).

Conservation in the GML is complicated by a great variety of territorial jurisdictions, many of which are overlapping. Within the greater landscape a human population of 36,500 people living in 173 localities is spread over two departments, ten municipalities, and several indigenous territorial demands and titled territories (see Figure A). Rural communities are represented by six campesino federations, a mining federation, and four indigenous councils. Large expanses of land outside the protected areas are found within indigenous territories,

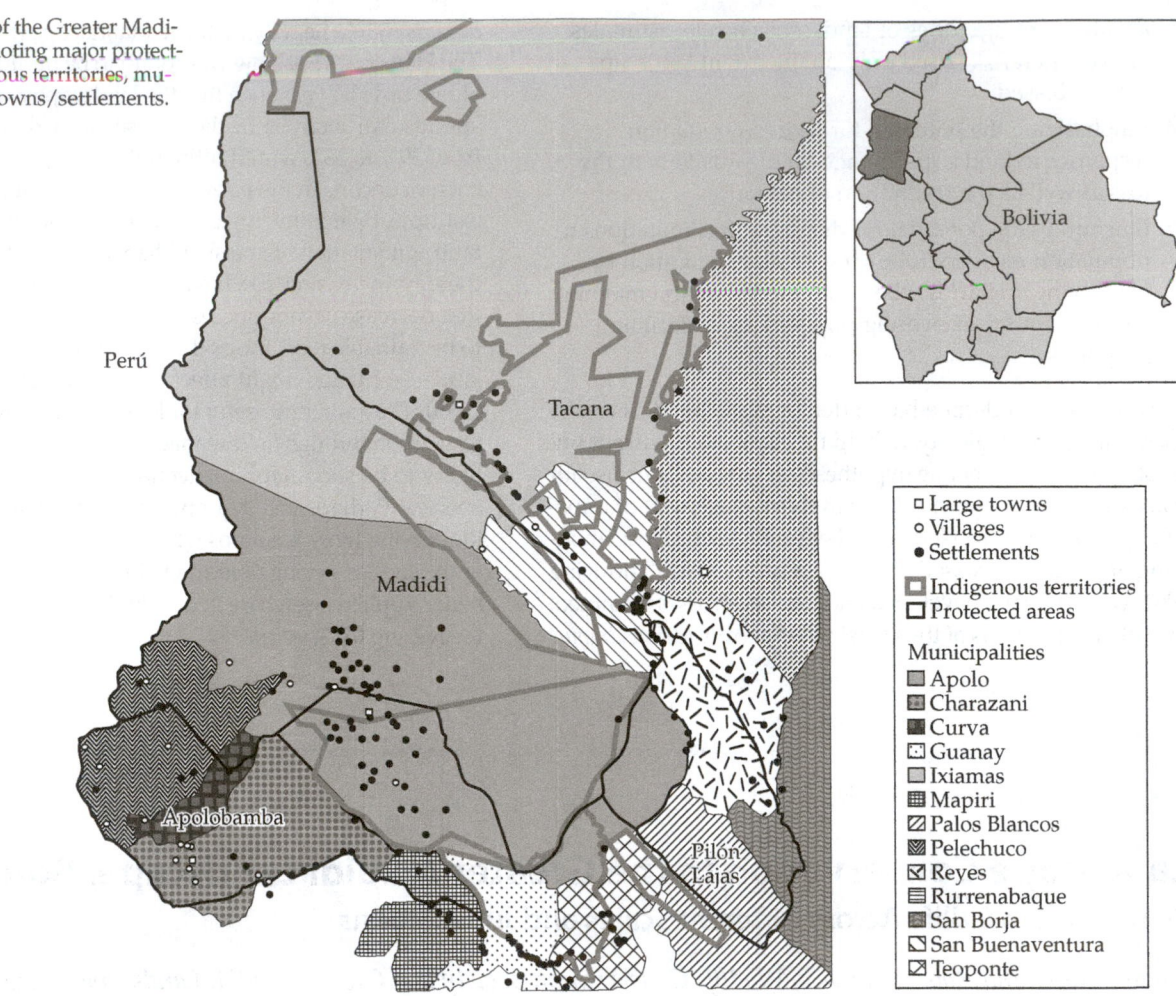

Figure A Map of the Greater Madidi landscape, denoting major protected areas, indigenous territories, municipalities, and towns/settlements.

Legend:
- □ Large towns
- ○ Villages
- ● Settlements
- ▭ Indigenous territories
- ▭ Protected areas

Municipalities
- Apolo
- Charazani
- Curva
- Guanay
- Ixiamas
- Mapiri
- Palos Blancos
- Pelechuco
- Reyes
- Rurrenabaque
- San Borja
- San Buenaventura
- Teoponte

forestry concessions, and to a lesser extent within private land. Thus, natural resource use and management in the GML is a complex institutional scenario characterized by a great diversity of local, regional, and national stakeholders whose use areas often overlap.

Landscape-Scale Conservation

Conceptually, we have followed the landscape species approach, which is based on the selection of focal species whose biological requirements in time and space make them particularly useful for identifying where different human activities may have the greatest impact on biological conservation. It represents a spatially explicit methodology for focusing conservation action through the intersection of values related to biological importance with those of human use. Landscape ecology and GIS techniques are used to map the needs of a healthy, functioning, and viable wildlife population, and at the same time map human activities and interests. The intersection of both allows the identification of areas of conflict, and hence areas where we need to focus conservation efforts while taking into account human interests (Sanderson et al. 2002; Figure B).

Landscape species are selected on the basis of area requirements, habitat heterogeneity requirements, ecological functionality, socio-cultural value, and vulnerability to threats (Coppolillo et al. 2004). These species represent a challenge for conservation because they often have the greatest spatial requirements to maintain minimum viable populations. Based on these criteria we selected a suite of species covering the elevational gradient: vicuña (*Vicugna vicugna*), Andean Condor (*Vultur gryphus*), spectacled bear (*Tremarctos ornatus*), jaguar (*Panthera onca*), and white-lipped peccary (*Tayassu pecari*).

Simultaneously, we initiated research on how human activities relate to form threats to species and biodiversity. Existing and planned human activities in the area include new and existing roads and resulting colonization and expansion of the agricultural frontier, mining and oil concessions, overgrazing, hunting, fishing, nontimber forest product extraction, and to a lesser extent, unmanaged tourism. Critically, direct threats are often accentuated by indirect threats, which include conflicting government policies and interventions, a lack of clarity in land tenure and natural resource access rights, a lack of vision for sustainable development related to

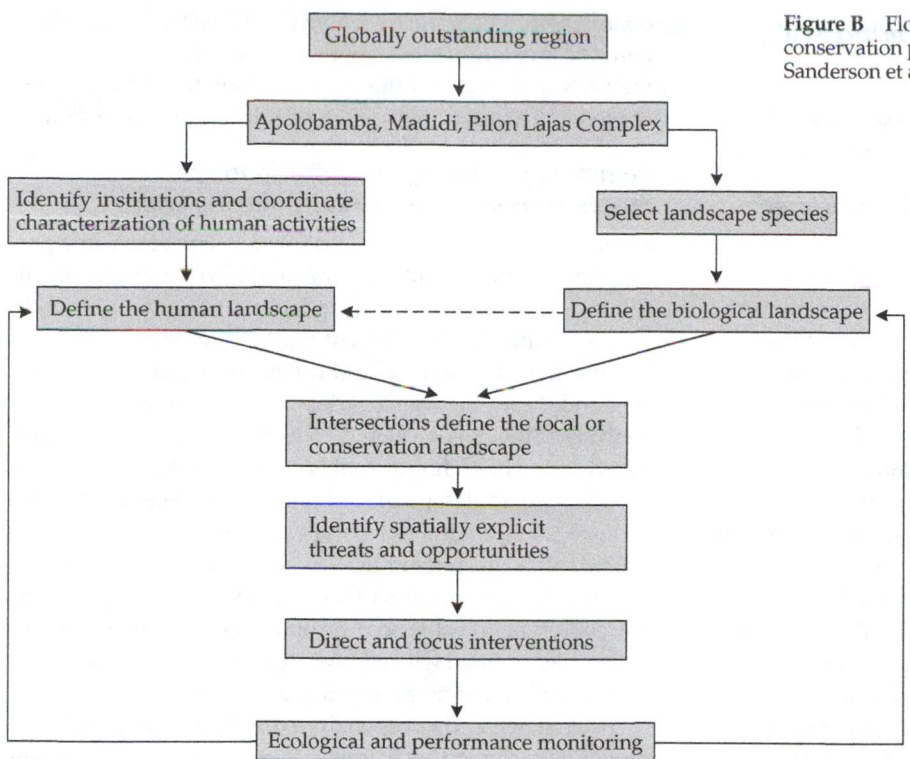

Figure B Flowchart of the landscape species approach to conservation planning, activities, and actions. (Modified from Sanderson et al. 2002.)

a poor appreciation of conservation and its benefits, a low capacity for natural resource and territorial management, a weak financial base for conservation and natural resource management programs, and information gaps and inefficient use of existing information.

A map of the key local factors involved both in conservation and in creating the most serious threats was used to describe the potential management units for landscape species. Protected areas are a logical starting point for starting to build a conservation landscape for a suite of landscape species and we made these core areas for our interventions. Other management units were identified by building out from the protected area cores, for example, indigenous territories, communal lands, and municipalities.

The Broader Socio-Political Context

The paradigm behind protected area management has changed drastically since the 1960s when local people were typically excluded from park management. Today, protected areas worldwide are increasingly run with social as well as conservation objectives in mind (see Chapter 14). The 2003 World's Park Congress (IUCN 2003) recognized that people who are struggling to meet the basic needs for survival should not be expected to make sacrifices for national and global interests. Thus, we focus on facilitating increased involvement of local stakeholders in protected area management (Jeanrenaud 2002).

In Bolivia, there has been an emphasis on promoting social and conservation objectives since the consolidation of the national park system in 1992. Primary principles guiding park

management today include respect for cultural as well as natural diversity, recognition of traditional rights over resource management, equitable distribution of benefits and participation, transference of management capacities to the local level, and integration into the regional context (SERNAP 2002). Bolivia's protected areas are large and therefore cannot be administered without the support of local populations found within and around them that have recognized traditional use rights, as well as the prohibitive costs of control and vigilance associated with patrolling extensive perimeters.

Clear definition of land tenure is an essential step for natural resource management. The Land Reform Law of 1996 in Bolivia permits the legal consolidation of indigenous collective land in the form of private and indivisible titles. This provides the necessary vehicle on which to build indigenous resource and territorial management models. Additionally, administrative decentralization, formalized in Bolivia in 1995 through the Popular Participation Law, grants municipal governments normative, operative, administrative, and technical functions over their territorial jurisdictions. However, most local governments are unable to fulfill these obligations because of technical and financial constraints, as well as an overwhelming domination of private interest groups. Participation should be a key aspect of municipal management and was a fundamental justification for government decentralization. Local people are more likely to identify and prioritize their problems accurately than central government, making resource allocation more efficient and information costs lower, while increasing the sense of ownership through local decision making.

Conservation Interventions at Differing Scales and across Jurisdictions

To be successful, conservation efforts must take place at different scales, and must link among communities and jurisdictions.

Engagement of local actors in the Madidi protected area management plan

The first five years of administration for the Madidi protected area (1995–2000) were spent establishing the park administration, and most interventions were focused on controlling illicit activities within the protected area, in particular selective logging. This led to initial conflicts with the local population because of a reduction in their access to natural resources. In addition, the territorial overlap between communities and the protected area clearly required the search for management options compatible with both the needs of the communities and the conservation objectives of the protected area.

In this context, and to ensure the social consolidation of Madidi, the development of a common vision with local stakeholders for protected area management was an essential component of the management plan process. In early 2002 we began developing the vision and strategic objectives through visits to all communities within Madidi and to those surrounding communities known or likely to have natural resource use areas within the park. During these visits the communities were informed about the protected area and the management plan process, and we learned about their community natural resource use areas, priority socioeconomic issues, and natural resource management problems. Although communities represented themselves as cohesive units they are not homogenous (Agrawal and Gibson 2001), and care was taken during the communal workshops to work separately with different user groups to obtain accurate maps of communal areas, as well as a fair prioritization of natural resource management issues.

In Bolivian protected areas, the Local Management Committee provides a forum for park authorities and local stakeholders, and for most protected areas it is the main vehicle for local participation in park management. Thus the Madidi Local Management Committee was considered a critical conduit for formalizing the vision statement, strategic objectives, and management priorities of the management plan, as well as gaining local approval of the final product. The Madidi Local Management Committee contains 23 members from local stakeholders including municipal governments, grass roots organizations, and the protected area administration.

In this light, results from the community assessments together with updated technical evaluations regarding the region's history, soil aptitude, biodiversity value, socioeconomic situation, extractive and non-extractive economic alternatives, land tenure, and park administration were presented to the Local Management Committee. Through these presentations, a revision and statement of the desired future situation for the protected area was agreed upon, including what needed to be done, where, when, and by whom to achieve the desired situation. These determinations represent a social contract between the protected area and the human population living in and around it that is essential for successful conservation action.

Community mapping and participatory zoning Processes as key conservation planning tools

Participatory Rural Appraisals allow revision of existing protected area management categories and help define zoning categories to establish permitted land uses. All communities within and immediately adjacent to Madidi were consulted regarding their vision for land use in areas under their influence and these data were overlapped with a spatially explicit analysis of the biological and conservation value of the protected area. The technical analysis considered archaeological sites, tourism routes, protection camps, biodiversity value, endemism, and importance to landscape species.

Although most communities targeted areas close to the communities for future activities, which generally possessed low conservation value because of human intervention, negotiation was required when designating areas for extensive uses such as timber and nontimber forest product extraction.

Taken together these spatially explicit analyses have led to proposals to change the boundaries of the broad management categories within the protected area; for example, the current national park category, which implies strict protection, includes the agricultural areas of seven communities and has led to inconsistencies with existing and permitted uses. Furthermore, certain areas of high biodiversity value were included within the Natural Area of Integrated Management category

Supra-communal spatial planning and zoning in the Tacana indigenous territory

The eastern border of Madidi's protected area represents both a threat and an opportunity for conservation. The San Buenaventura-Ixiamas road is a spearhead for the colonization process. Conversely, the alluvial plain between Madidi and the Beni River is an important stronghold for wildlife such as jaguar and white-lipped peccary since the majority of the area away from the colonized road is in good condition.

Less than 10% of this area is held by small landowners; the majority is held by long-term forestry concessions or by the collective indigenous territory. Given this situation we embarked on a partnership with the Tacana People's Indigenous Council (CIPTA) with two clear objectives: technical assistance for the legal consolidation of their indigenous territorial demand and a parallel participatory process to develop a natural resource management strategy for the indigenous territory once titled. In the longer term, forestry concessions will also need to be engaged in order to promote their contributions to the landscape scale management strategy.

The planning process in the indigenous territory followed a bottom-up approach, building from community Participatory Rural Appraisals up to the whole management unit level.

All 20 Tacana communities participating in the territorial demand were visited and strategic objectives, activities, as well as a zoning proposal were developed through community-level prioritization. Again, care was given to work separately with the different user groups, and during mapping of hunting areas it was sometimes necessary to work at the individual family level. Community-level zoning proposals were then combined in subregional workshops where inter-community disputes were resolved through participation of community leaders and finally these subregional proposals were combined at an Indigenous Territory scale workshop.

Community zoning proposals and maps also proved useful as a conflict resolution tool between CIPTA and other interest groups during the land titling process. Clear spatial indicators of the importance of different areas for various Tacana communities provided Tacana representatives more confidence for negotiation with neighboring properties, in particular colonist settlements. Thus a technical process of territorial planning reinforced the land titling efforts to return traditional rights over natural resources to the Tacana people.

Because management of the indigenous territory involves communities that are represented by the same indigenous organization, further advances such as development of natural resource use and access regulations have been possible. CIPTA holds the land title for the indigenous territory and therefore has some level of autonomy in developing access and harvesting rules, which makes them attractive partners to develop models of appropriate natural resource management (Barret et al. 2001; Varughese and Ostrom 2001). Natural resource regulations were initially developed at the community level, working separately with different user groups, and using the indigenous territory zoning proposal and previously defined sustainability principles broadly grouped into ecological, social, and economical sustainability considerations. Subsequently, community results were discussed in a general assembly of community authorities to develop a common regulation for the whole indigenous territory. This process requires constant communication efforts to reinforce the linkages between zoning, strategic objectives, and natural resource regulations since many of these concepts were previously unnecessary for the Tacana people.

Municipal land use planning and environmental management strategies

Although municipalities in the region have low financial resources, technical capacity, and democratic participation, they are essential for a lasting local institutional framework. Not only have they been key actors in environmental conflicts that have affected the region, but they potentially can support conservation through municipal development plans that should identify, include, and channel state support for conservation interventions identified in both protected area and indigenous territory management plans.

We have begun working with the San Buenaventura and Guanay municipalities to develop environmental management strategies and provide support for updating municipal development plans and territorial plans. These processes have involved a self-diagnostic of the main environmental problems, including broad consultation with community representatives, municipal vigilance committee representatives, and grass root organizations. Strategic objectives include promoting sustainable natural resource use activities that respond to territorial planning exercises, to build the technical and administrative capacity of the municipal unit, to build capacity in the municipal population regarding environmental issues and civic participation and legislation, and finally, to strengthen mechanisms of local participation and control of municipal management. These strategies will begin to be implemented through the municipal annual work plans and supporting institutions. Establishing coherence between municipal annual work plans and those of the regions protected areas will be an important next step.

Community-Level Interventions

The community diagnostics carried out within the Madidi protected area and the Tacana indigenous community all identified the need for support of community natural resource management projects as a priority. Further, our threats analysis identified the lack of sustainable economic activities as a critical indirect threat. Hence, our interventions at the community level have centered on two aspects: improving the sustainability of existing activities and investigating the potential for new sustainable activities that would decrease the advancement of the agricultural frontier by providing additional utilitarian values to natural habitats.

Support of existing activities includes technical assistance to indigenous communities wishing to evaluate the sustainability of, and actively manage, their subsistence hunting. The two communities that began this activity recently made preliminary management decisions regarding the reduction in harvesting of locally threatened wildlife species: marsh deer (*Blastoceros dichotomus*), lowland tapir (*Tapirus terrestris*), black spider monkey (*Ateles chamek*), and red howler monkey (*Alouatta sara*). Four adjacent communities have recently begun to monitor subsistence hunting harvests and in this way we hope to move toward a hunting management plan for the entire Tacana indigenous territory based on source–sink management models.

Support of new natural resource use activities includes technical assistance to a native bee (family Meliponidae) honey production initiative targeted at the tourist market in Rurrenabaque, commercialization of wild chocolate, handicraft production, identification of wildlife and wilderness tourism attractions, fostering planning and community decision-making processes to ensure that the new natural resource use activity is integrated to other community activities, and assisting in the organizational processes related to the functioning of these fledgling natural resource use groups.

As an associated benefit the internal reflection process required in the implementation of communal projects—for example, to define benefit distribution and responsibilities—has

strengthened community organization. These projects have also promoted a more general internal analysis of natural resource use issues within participating communities. Community natural resource use projects can provide local user groups with the technical assistance required to develop management systems for natural resource use, to embark in new economic activities, and finally, to strengthen local governance over territory and natural resources. In many cases local user groups will need to be established, comprising members from several communities, either because of biological management considerations, as in the case of wildlife, or because of the need to achieve larger volumes of production in the case of commercial activities.

Lessons from Applying the Landscape Species Approach

As we have worked to apply the landscape species approach and associated population models for the selected landscapes species in the Greater Madidi landscape, our decision to work at a landscape scale and across jurisdictional boundaries is justified, particularly for spectacled bears and jaguars. We estimate that the current intervention landscape that stretches from the Tacana Indigenous Territory in the lowlands to the Apolobamba highlands would secure around 1800 jaguars and 2000 spectacled bears (see Plate 4). According to minimum viable population estimations based on simple models of genetic variability a population size of around 2000 is required to avoid the accumulation of deleterious alleles and permit beneficial mutations (Whitlock 2000). Hence conservation interventions must respond minimally to maintaining continuity within these populations.

Landscape-scale conservation efforts require the identification of appropriate territorial units with clear administrative and legal basis. National, municipal, and departmental protected areas, indigenous territories, and other forms of private property, as well as long-term natural resource use concessions over fiscal land such as forestry concessions, all allow territorial planning, administration of resources, and implementation of management actions oriented toward conservation and local development objectives.

To build a landscape conservation plan we have embarked on a process that first develops management instruments for individual jurisdictions. To work across different scales and in overlapping jurisdictions it was necessary to build alliances, taking into account the legality and legitimacy of the different actors to play a positive role in building a solid institutional framework for conservation in the region. We began working with different stakeholders based on a variety of incentives that included resource tenure security, institutional strengthening, and direct economic returns.

The biggest challenge during this process is building participation and negotiation mechanisms between legitimate local stakeholders and different management units. These negotiation mechanisms will be the backbone of developing a landscape level of governance. Where conflicts about access to resources arise, the basis for their long-term resolution must be the existing legal framework. However, because of the weakness of the central government, respect for the rights of "others" established by law must be built at the local level in what could be described as the construction of a democratic culture.

Integrating communal planning with supra-communal management has been possible by carrying out bottom-up processes, during which community mapping has been a powerful tool to present community interests. Local stakeholders are slowly developing capabilities that may make them better stewards of local natural resources than the national government. However, if decentralization of this responsibility has a positive impact on landscape conservation it will be a result of the establishment of an appropriate control and cooperation framework, which will require integration mechanisms within and between the different planning levels. Institutional strengthening directed at improving communication and control mechanisms is therefore very important to maintain local accountability, particularly given that community interests are heard with difficulty outside local levels.

A critical next step will be influencing departmental and national-level plans, which can have huge impacts on local conservation plans through large infrastructure projects and granting of mining and hydrocarbon concessions. Nevertheless, without ignoring the importance of national and departmental interventions, we have prioritized building local constituencies for landscape conservation because of the difficulties that Bolivia's weak central government has in implementing and enforcing policy.

Although working with a wide range of actors is highly complex, we are privileged to be working in an area where integrating conservation and development is possible because there is still room for protected areas, extractive areas, and intensive use areas to coexist. The landscape species approach has allowed us to keep our focus on wildlife while working within this highly complex social and institutional challenge.

CASE STUDY 12.3

Putting the Pieces Together
Conserving Cranes and Their Habitats Around the World

Curt D. Meine, James T. Harris, Jeb Barzen, and Richard Beilfuss, International Crane Foundation

The cranes (family Gruidae) belong to an ancient family of birds, with fossil records dating back more than 50 million years. The family's 15 extant species (Figure A) are widely distributed, occurring in more than 110 countries on five continents; only South America and Antarctica lack cranes (Johnsgard 1983). Primarily birds of open wetlands, grasslands, and savannas, cranes have in some regions been able to adapt to, and even thrive within, humanized landscapes. Over the last 150 years however, cranes have had to cope with accelerated loss and degradation of habitat, overexploitation, and other new and intensified threats. As a result many crane populations, subspecies, and entire species are regarded as threatened.

Cranes have long commanded the respect and admiration of their human neighbors, a cultural value that has been critical in drawing attention to their biological plight. Even as cranes have declined in numbers, their beauty, dramatic migrations, and striking calls and behavior have inspired widespread conservation efforts. They have often served as flagship and umbrella species in broader ecosystem conservation programs. They have also provided a focal point for community-based projects that strive to meet the needs of both local wildlife and people. Since the early 1970s, a global campaign has been un-

derway to develop and coordinate conservation programs focused on cranes and the ecosystems that serve as crane habitat.

This unusual effort, centered on a single family of birds, yet international in scope and integrated in its approach, offers lessons of broad relevance to conservation biologists. In contrast to management efforts involving particular species in a particular place, crane conservation offers an example of what might be called "meta-management"—the coordination of efforts to conserve an entire group of species throughout the world.

Conservation Status of Cranes

Although cranes are subject to a wide array of direct and indirect threats, the most significant long-term conservation issues worldwide involve the loss and degradation of wetlands. This issue affects crane distribution, movement, and breeding success, and involves habitats used by both migratory and nonmigratory species throughout the year (Harris 1994; Meine and Archibald 1996). Species that use upland grasslands and savannahs have also been heavily affected by the conversion and degradation of these ecosystems.

Because of the cranes' low reproductive potential—in most species, pairs do not breed until 3–5 years of age, and raise less

Figure A The world's 15 species of cranes are here depicted in a wall poster that has also been used as an attractive educational tool. (Original artwork by David Rankin; photograph courtesy of International Crane Foundation.)

TABLE A *Conservation Status of Cranes under the IUCN Red List Criteria*

Species	Conservation status
Black-crowned Crane (*Balearica pavonina*)	Lower risk/near threatened
Grey Crowned Crane (*Balearica regulorum*)	Least concern
Blue Crane (*Anthropoides paradiseus*)	Vulnerable
Demoiselle Crane (*Anthropoides virgo*)	Least concern
Wattled Crane (*Bugeranus carunculatus*)	Vulnerable
Siberian Crane (*Grus leucogeranus*)	Critically endangered
Sandhill Crane (*Grus canadensis*)	Least concern
Sarus Crane (*Grus antigone*)	Vulnerable
Brolga (*Grus rubicundus*)	Least concern
White-naped Crane (*Grus vipio*)	Vulnerable
Hooded Crane (*Grus monacha*)	Vulnerable
Eurasian Crane (*Grus grus*)	Least concern
Whooping Crane (*Grus americana*)	Endangered
Black-necked Crane (*Grus nigricollis*)	Vulnerable
Red-crowned Crane (*Grus japonensis*)	Endangered

Note: This table includes only the conservation status of cranes at the species level. Crane conservationists are also concerned with the status of several subspecies, including the West African, Sudan, South African, and East African Crowned Cranes; Mississippi and Cuban Sandhill Cranes; and Indian and Eastern Sarus Cranes. (From Meine and Archibald 1996.)
Source: From IUCN 2004.

than one chick per year on average—increases in mortality caused by hunting, poisons, and powerline collisions can easily depress crane populations. Other important threats to cranes include overexploitation, dam construction, water diversions, urban expansion, invasive plant species, genetic and demographic problems associated with small populations, commercial trade in cranes, lack of effective environmental law enforcement, and political instability. As a result of these multiple threats, 10 of the 15 crane species are now included on the IUCN Red List of Threatened Species (IUCN 2004) (Table A). Several crane species, subspecies, and populations, including North America's Greater Sandhill Cranes and western Europe's Eurasian Cranes, are now recovering from past declines. These success stories pose new and different conservation challenges and opportunities.

Effective crane conservation depends on the ability to identify the combination of actions that are available and needed to respond to the highly varied circumstances on the ground. Such actions include stronger legal protections; use of international agreements and cooperative international programs (such as the Ramsar Convention for wetlands protection and the Convention on the Conservation of Migratory Species of Wild Animals); development of community-based conservation projects; establishment and management of protected areas; ecosystem protection and restoration; long-term monitoring and research; support for non-governmental organizations active in key crane locations; public education and professional training; and captive propagation and reintroduction. In some cases, as with the Whooping Crane, necessity has often been the mother of invention, dictating critical responses to immediate needs. In other cases, as with the endangered cranes of East Asia, conservationists have taken steps incrementally and opportunistically amid complicated sociopolitical circumstances.

Three cases from around the world illustrate how cranes have served as catalysts for innovative approaches to conservation.

Cranes and Conservation in the Amur River Basin

The Amur River along the Russia–China border is the world's eighth longest river, and the longest without a dam on its main stem. Its basin is rich in species diversity, a reflection of its unique mix of elements from the northern coniferous forests, southern deciduous forests, and Eurasian steppes. For migratory birds, the Amur Basin is an important link between arctic breeding grounds and temperate to subtropical wintering areas. It is also a center of diversity for cranes, with six species (four of which are threatened) occurring in the region (Halvorson et al. 1995).

International tensions have for decades prevented intensive development of the Amur Basin; however, in recent years development pressures have been growing. A series of dams has been proposed for the Amur River, threatening the river itself and adjacent wetlands. Rapid agricultural conversion of wetlands in the associated Sanjiang Plain in China has deprived Red-crowned and White-naped Cranes and other wetland species of critical breeding habitat. In Russia, economic uncertainty has contributed to inefficient agriculture and exploitative forestry.

Since 1980 cranes have played a key role in stimulating regional conservation initiatives. Important wetlands in Russia and China have been protected, including international reserves at Lake Khanka on the Russia–China border, and in the China-Mongolia-Russia border region. A Russian NGO, the Socio-Ecological Union (SEU), has established Muraviovka Park (Figure B), the first private conservation area created in Russia since 1917, by leasing prime crane habitat (Pryde 1999). Muraviovka is located amid farmlands; crucial community support has been fostered through an exchange program involving schoolteachers and student conservationists from the area and from the United States. Russian and American teachers are now involving Chinese colleagues from the south side of the Amur River in environmental education activities. The park has developed a model farm to demonstrate agricultural practices that safeguard soil, water, and other natural resources while enhancing production. In 2004 the park celebrated its tenth anniversary by bringing together conservationists from northeast Asia, Europe, and North America to review successes and the challenges of local, regional, and international collaboration in conserving wildlife and wetlands.

Cranes and Community-Based Conservation at Cao Hai

In the karst mountains of western Guizhou and northeastern Yunnan Province, China, lie scattered, high plateau wetlands

Figure B   Entrance to a crane refuge in the Amur Basin. (Photograph courtesy of International Crane Foundation.)

that are unique in Asia. These wetlands, dominated by slightly alkaline water, produce abundant submerged aquatic plants that in turn support a wide variety of wintering waterbirds, including the threatened Black-necked Crane and the much more common Eurasian Crane. Crane conservationists have worked at the Cao Hai Nature Reserve in western Guizhou Province since the early 1980s. Cao Hai's 20 km² wetland rests within a 98 km² watershed (Figure C). Early conservation efforts focused primarily on protecting the cranes and their wetland home. With time however, Cao Hai's wetlands diminished as the shallows were converted to agriculture.

By 1994 it was clear that the initial approach was failing to address the conservation needs of this watershed and the more

Figure C   Black-necked Cranes (*Grus nigricullis*) at Cao Hai Nature Reserve, Yunnan Province, China. (Photograph © Tim Davis/Photo Researchers, Inc.)

than 20,000 people living within it. More tangible links between conservation activities and the gripping poverty of the region were needed. The biologists and managers working at Cao Hai were ill-prepared to address these larger social issues. In response, conservationists entered into a collaboration with the Yunnan Institute of Geography and the New York-based Trickle Up Program, organizations with extensive experience in poverty alleviation. Together the partners developed a process to link poverty alleviation and wetland conservation at Cao Hai (Shouli et al. 2001).

The Trickle Up Program has worked in many countries, providing conditional $100 grants and business training to very poor people, enabling them to start their own businesses. At Cao Hai, farmers agreed to use their grants only for businesses compatible with protecting the natural resources on which cranes and people depend. The goals of the program were clear on paper. It took time however, for these goals to be shared among those involved and for attitudes to evolve—a reflection of the divergent views of what poverty alleviation and resource protection entailed. As one example, the first grants were given to relatively wealthy farmers because some people felt that people who are poor don't know how to spend money.

Eventually, poor farmers did receive grants, enabling them to reduce their demands upon local natural resources. Relations between local communities and the reserve staff, which had deteriorated to the point of physical conflict before the program began, improved dramatically. Agriculture encroachment upon the Cao Hai wetland ceased. As management within the reserve became more effective, numbers of Black-necked Cranes doubled. The success of the process developed at Cao Hai has led to its being adapted to other nature reserves in China and other parts of Asia. The basic lessons from Cao Hai have proven to be broadly applicable: Act on common interest, build trust through shaping a shared vision, and leave outcomes open.

Wattled Cranes, Wetlands, and People in the Zambezi Basin

The lower Zambezi River valley is the lifeline of Mozambique, ancient home to more than a million people and one of the most productive and biologically diverse river floodplain systems in Africa. Over the millennia, the annual spread of Zambezi floodwaters nourished the fertile floodplains and its human and non-human inhabitants. The floodplain once provided spawning grounds for fish and critical dry season grazing lands for both livestock and wildlife. The Zambezi Delta's vast, seasonally flooded grasslands were home to a rich array of wildlife, including an important population of the world's endangered

Wattled Cranes. Extensive coastal mangroves and estuaries supported a productive prawn fishery.

Since the 1960s, however, large upstream dams have disrupted the hydrological cycle of the lower Zambezi, eliminating natural flooding and greatly increasing dry-season flows (Beilfuss and Davies 1999). These changes have affected the availability of water supplies, fuel wood, building materials, and medicinal plants, while weakening the cultural relationship between local people and the river. With the loss of the annual flood, subsistence fishing, farming, livestock grazing, and the commercial prawn fishery collapsed. The dams have also altered conditions in the delta. Saltwater has intruded, the area of wetland and open water has declined, and exotic vegetation has taken over stagnant waterways. Upland plant communities have replaced wetland vegetation, while grassland fires have degraded fire-sensitive communities (Beilfuss et al. 2000). Desiccation of the floodplain has also opened the area to widespread poaching. Populations of Cape buffalo, waterbuck, reedbuck, zebra, and hippopotamus plummeted in the 1980s and 1990s. Wattled Cranes, which serve as an important flagship species for the conservation of many flood-dependent waterbird species of the Zambezi system, have ceased to breed across most of the delta (Bento 2002).

In the 1990s, conservation biologists began to collaborate closely with social scientists, dam operators, government officials, and local communities to promote sustainable management of the lower Zambezi. The immediate goal of this work has been to develop and implement a practical and equitable plan for managing water releases from the upstream Cahora Bassa Dam. An interdisciplinary team is working with a wide variety of users to assess the environmental, social, and economic benefits and drawbacks of different river management scenarios. The broader goal is to implement releases as part of a comprehensive program for poverty alleviation, natural resource conservation, and ecological restoration in the lower Zambezi. Although it is only one of many species that will benefit from these efforts, the Wattled Crane has already proven to be a valuable symbol of the potential rebirth of this endangered system.

Coordinating Crane Conservation Response

As these examples illustrate, cranes provide important opportunities to build conservation programs that combine varied goals, activities, and techniques. With limited time, money, and personnel, crane conservationists have had to develop ways to coordinate efforts effectively at local, regional, and international levels. A number of mechanisms and organizations have emerged to help integrate the various components of a balanced and comprehensive conservation program.

At international levels of the International Crane Foundation (ICF), the IUCN/SSC Crane Specialist Group, the Global Captive Crane Working Groups and additional working groups all serve to coordinate efforts and to share techniques and guidelines. National working groups, such as the U.S. Whooping Crane Recovery Team or the China Crane and Waterbird Working Group focus attention on national efforts, but also collaborate with other national groups.

Recovery teams and recovery plans

The U.S. Endangered Species Act of 1973 provided for development and implementation of recovery plans for endangered species. These plans are prepared and periodically updated by recovery teams appointed by the U.S. Secretary of the Interior. For example, the U.S. Whooping Crane Recovery Team was appointed in 1976 and the USFWS published its first Whooping Crane Recovery Plan in 1980. The Canadian Whooping Crane Recovery Team was established in 1987 to coordinate recovery activities within Canada.

Recovery activities have been closely coordinated between the two nations. In 1995, a Memorandum of Understanding on Conservation of the Whooping Crane was signed, calling for the preparation of a combined plan and the formation of a single recovery team comprising five U.S. and five Canadian members. These steps are especially important as precedents for other nations that share endangered migratory crane populations. For example, in 1995 representatives of the range nations of the rare central and western populations of the Siberian Crane met for the first time in Moscow, laying the foundation for establishing a Siberian Crane Recovery Team. In 2001 the Siberian team met in Wisconsin, and for the first time included representatives from the species' eastern flyway.

International Crane Foundation (ICF)

Since 1973 the International Crane Foundation (located in Baraboo, Wisconsin) has carried out conservation programs around the world. ICF's programs in ecosystem restoration and management, aviculture, research, education, and training have helped to strengthen the global network of crane conservationists. Its website (www.savingcranes.org) and publications provide communication links for tthe network.

ICF works actively with other local, national, and international organizations in developing community-based conservation programs that address the interrelated needs of cranes and human communities. ICF also maintains a "species bank" of threatened cranes at its headquarters, and is one of the three primary breeding facilities for the Whooping Crane. ICF has successfully bred all 15 crane species in captivity, developing new techniques that have been used in propagation of other endangered birds. ICF also provides training opportunities for biologists, managers, and educators, and supports a wide range of public education and outreach projects at its headquarters and around the world.

IUCN/SSC Crane Specialist Group

IUCN Crane Specialist Group serves as a vital hub of communications for crane researchers and conservationists worldwide. In 1906 the group published its first comprehensive action plan.

Crane working groups

Crane working groups have played a key role in supporting research, information exchange, and development of conservation programs. Crane working groups have been organized at the regional, national, and local levels. At the regional level, working groups are active in North America, Europe, and Northeast Asia. National-level working groups are best developed in Europe. In the late 1990s several local working groups in South Africa joined together under the umbrella of the South African Crane Working Group.

Global Captive Crane Working Group

The appropriate integration of ex situ conservation techniques and ecosystem restoration programs is a critical challenge for crane conservationists. Most crane species can now be reliably bred in captivity. Based on this success, the emphasis in captive programs has shifted from management of individual birds to management of healthy populations to meet conservation needs. A Global Captive Crane Working Group sets regional target populations, defines genetic and demographic objectives, allocates limited space among species, and coordinates work with field conservation projects. In addition, captive management techniques have been summarized in a crane propagation and husbandry manual (Ellis et al. 1996).

Crane workshops and meetings

Since 1975, more than 40 national, regional, international, and species-specific crane workshops and meetings have been held. These gatherings serve as important forums for information exchange, and allow scientists and conservationists from throughout the world to meet and learn from one another. Proceedings from most of the workshops have been published, making this information available to even broader audiences.

Lessons for Conservation Biologists

Each of the 15 crane species requires a different suite of conservation actions to ensure a secure future, and crane conservationists have had to integrate conservation programs under varied circumstances. Several basic guiding principles can be derived from this collective experience:

- Conservation mesures must be solidly grounded in the natural sciences, but should also involve the social sciences, humanities, law, education, economics, and other fields. Fortunately, cranes are among the best-studied groups of organisms on Earth. Effective conservation however, requires that scientific knowledge be linked with an understanding of the human dimensions of the challenge—the social forces and trends that affect crane populations and habitats. Consequently, in situ conservation programs must be broadly conceived and combine research with legal protection, habitat protection and management, education, community participation, and other components. All of these features can and

must contribute to balanced programs that sustain crane populations, crane habitats, and local human communities.

- Conservation measures should be envisioned at multiple scales of time and space. Conservation programs for cranes have spanned broad temporal and spatial scales, from highly localized and immediate efforts to save threatened habitats and populations, to longer-term programs in, for example, ecosystem restoration, watershed-scale planning, and maintenance of viable populations in captivity.

- Conservation measures should seek to harmonize species-oriented and ecosystem-oriented approaches. As well-known birds that serve as umbrella and flagship species, cranes have drawn attention to, and provided protection for, a broad array of other species as well as the processes that maintain ecosystem health. In the long run, cranes must be viewed within the larger landscapes, watersheds, and ecosystems that support them, and conservation activities must be coordinated at these scales. In particular, managers must appreciate the role of flooding, fire, vegetation change, and other processes in these dynamic systems.

- Conservation measures should take into account biological attributes and processes at all levels of the biological hierarchy. Crane conservation has required attention to problems at the genetic, individual, population, subspecies, species, and family levels. Especially in the case of the Whooping Crane and the other endangered species, these problems need to be considered simultaneously to minimize risk.

- Conservation measures should work across national, cultural, and ecological boundaries. Because most cranes are migratory, and all the species occur in more than one country, successful conservation requires clear consensus on goals and responsibilities among parties from different parts of the species' range, constant communication of reliable scientific information, and support from various governments, international institutions, and nongovernmental organizations.

- Conservation measures should seek to address local community development and conservation needs in an integrated fashion. Efforts to conserve cranes—especially the 13 species occurring in Asia and Africa—are interwoven with the challenges of local sustainable development. Wild resources of wetlands and their watersheds cannot be conserved without the active involvement and leadership of local resource users. Invariably, local people provide essential insights into the best response to the threats that cranes face.

- The relationship between in situ and ex situ conservation measures should be well defined. Captive propagation and reintroduction programs should be under-

taken only as a last resort, and not as a substitute for in situ programs. Should ex situ programs become necessary, they should be developed based on clear goals and management guidelines. Priority should be placed on maintenance and enhancement of genetic diversity within populations, on safe and effective methods for reintroduction, and on assurance of high quality care.

- Conservation biologists should be prepared for success. The work of conservation biology does not stop once a species or population has recovered, or a critical habitat has been protected. In particular, the involvement of local people in species recovery and ecosystem management programs must continue after the initial challenges are met. Failure to do so risks undermining success in the long run.

- Education should be integrated into all conservation programs. Throughout the world, crane conservation programs have taken advantage of cranes as a special vehicle for communicating basic information about wetlands and endangered species management. Ultimately the conservation of cranes requires an informed public understanding, involvement, and support for activities that sustain the ecosystems where cranes occur.

Cranes, along with much of the world's biodiversity, will face difficult circumstances in the coming decades. History provides somber lessons about the speed with which even abundant species can become threatened. History, however, also shows that recovery is possible when species are provided with the necessary environmental conditions. Although the survival and recovery of the world's cranes cannot be assured, many steps are being taken to enhance their chances. Compared to the prospects 50 years ago—when most crane species and populations were dwindling, scientific knowledge was scarce, and conservation efforts were essentially nonexistent—there is reason for cautious optimism. And in safeguarding cranes, we may ensure a more secure future for other members of the ecosystems—including people—where cranes occur.

Summary

1. An understanding of the fluctuations in numbers of natural populations and the application of this understanding to the conservation of species must be based firmly on an understanding of the factors influencing spatial and temporal variability in population demography.

2. All populations exist in heterogeneous landscapes, and different individuals experience different conditions depending on when and where they are located. Some locations (termed "sources") are highly productive and produce an excess of individuals that often populate less productive locations ("sinks"), where local mortality exceeds reproduction. In some cases, sink populations may be larger than the source populations that support them. Accordingly, great care must be exercised in the design of reserves to identify and protect source habitats.

3. Population viability depends not just on the quality of local patches of habitat, but also on the number and location of patches and the amount of movement between them. The dispersal mode of a population is a key factor in determining its viability. In many cases population dynamics must be studied at the level of many local patches of habitat, and population models must incorporate immigration and emigration explicitly.

4. Metapopulation models that consider the dynamics of many interacting subpopulations in a habitat mosaic demonstrate that the fraction of a landscape suitable for a given species and the magnitude of dispersal between suitable patches are critical components of population viability.

5. Unlike metapopulation models, spatially explicit population models consider the exact location of habitat patches and can incorporate detailed behavioral information about how dispersing individuals locate suitable habitat. Spatially explicit models can be useful tools for testing specific conservation strategies in a given region, while metapopulation models are more useful for testing general landscape influences using hypothetical populations.

6. A number of theoretical models are now available for quantitative analysis of population viability and extinction probability. These models should be viewed as useful additions to the collection of tools available for understanding the dynamics of natural populations. The more realistic models are, however, very data hungry and only as good as the natural history insights and field studies that support them.

7. Increasingly, efforts to track and predict changes across landscape scales are possible due to the use of GIS datasets coupled with spatial models. Although these models are certainly among the more data

hungry, they have the advantage of providing concrete examples that can influence planners and motivate actions where words and statistics alone may not. Where these models also include reciprocal influences of development choices on biodiversity and human welfare, the incentives to seek benefits for human populations and for wild habitats and species may be greatest.

> Please refer to the website www.sinauer.com/groom for Suggested Readings, Web links, additional questions, and supplementary resources.

Questions for Discussion

1. Traditionally, population studies have been done at a local scale on homogeneous populations. Recently there has been a shift toward landscape studies. What are some of the advantages of working at the landscape scale? Are there any disadvantages?

2. What are the advantages to conceptualizing population regulation from a hierarchical perspective? Can you think of a real population in which different factors at the local and the regional scales limit population growth? How could policies in effect at a regional scale make it difficult to manage populations by manipulating local factors?

3. This chapter emphasizes that conservationists must consider habitat quality as well as quantity when managing for threatened species. What is meant by habitat quality, and how would managers measure and improve it?

4. Could management designed to improve conditions for an endangered species have negative effects on another endangered species? Can you think of any case where this has occurred? What could a conservation biologist do to prevent this situation from occuring?

5. What is the role of mathematical models in conservation of species? How can these models best be used? Are there times when they are not appropriate, and how would you judge this?

13

Ecosystem Approaches to Conservation

Responses to a Complex World

Gary K. Meffe, Martha J. Groom, and C. Ronald Carroll

A thing is right when it tends to preserve the integrity, stability, and beauty of the biotic community. It is wrong when it tends otherwise.

Aldo Leopold, 1949

One of the challenges to conservation is the fact that ecological systems are vastly complex, and thus difficult to understand in their entirety, let alone to manage in any comprehensive sense. Protecting biological diversity involves vast areas where humans use natural resources and are an integral part of the landscape. With a world population well over 6 billion and growing, it is naive to think that we can set aside or restore enough land or seascapes to accomplish the goals of conservation biology. Thus, we need to learn to manage and protect biodiversity in a world of heavy use, and therefore to work closely with people at local, national, and international levels to achieve conservation goals.

Equally, a majority of people worldwide suffer from the deficiencies of a degraded environment, with over a third lacking such necessities as clean water and sufficient food. We recognize that improving human welfare requires ecosystem-based management to secure the ecosystem services on which human livelihoods depend; biodiversity conservation must be an integral part of that socioeconomic development. Thus, when we focus solely on either conserving biodiversity or improving human welfare, we neglect a substantive portion of the means to achieve these aims. Success in conservation and securing human livelihoods are inextricably linked.

Further, we often cannot accomplish our goals by focusing on only part of an ecosystem. Management practices that benefit a subset of species are less likely to serve all species. As we concentrate on protecting a single marsh, grassland, forest, or stream, we tend to overlook important relationships with other habitats on land, in freshwater, or the sea that could both affect the success of our protections of these single areas, as well as the health of the ecosystem as a whole.

Pursuing an "ecosystem approach" to conservation entails considering interacting human and natural systems on large spatial and temporal scales, and doing our best

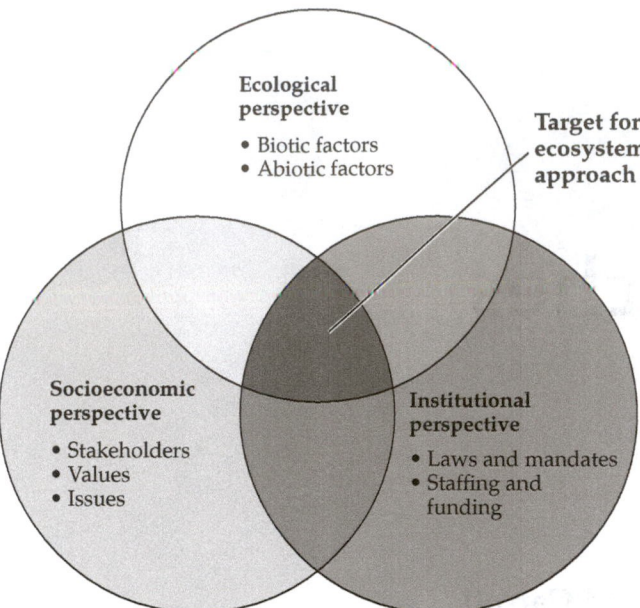

Figure 13.1 A conceptual basis for ecosystem management. Good ecosystem management lies at the intersection of ecological, socioeconomic, and institutional perspectives. (Courtesy of Dennis A. Schenborn.)

to work at such larger scales (Christensen et al. 1996). A variety of different approaches combine a focus on whole ecosystems—and on people in ecosystems—including terms such as "ecosystem management," "ecosystem-based management," "ecosystem approach," "integrated coastal and marine management," and "integrated river management." We draw from these related approaches for examples, and overall themes for conservation practice. Importantly, these approaches are complementary to many other valid approaches, including the species-based, landscape-level approaches considered in Chapter 12, and the use of protected areas (see Chapter 14), restoration and reintroduction (see Chapter 15), and efforts to develop sustainable forms of economic development (see Chapter 16). Ecosystem approaches all share the premise that conservation efforts must recognize the central role of people in creating and implementing conservation solutions.

A good way to envision an ecosystem approach to conservation is to conceive of it as three intersecting circles of focus, interest, or concern (Figure 13.1)—ecological, socioeconomic, and institutional—all of which influence how we use and interact with natural systems. A successful ecosystem approach occurs at the intersection of all three circles, where concerns for each circle have been satisfied; dwelling in any single circle, or even at the intersection of two of the three circles, indicates too narrow a perspective. For example, concern *only* for preservation of ecosystems (an ecological focus) is unre-

alistic because it ignores society's interests in those systems or institutional momentum that could have a strong bearing on how those systems are used. Likewise, there may be strong societal support for restoring a particular ecosystem, along with the ecological knowledge of the system to do it well, but if the institutions who need to do the work are not engaged, do not have the technical ability to perform the tasks, or are not appropriately funded, then the effort will fail. Sustainable management at an ecosystem level will not succeed if ecological perspectives dominate to the exclusion of human welfare, if narrow economic considerations prevent good ecological stewardship, or if institutions remain rigid and cannot adapt to incorporate alternative perspectives and ideas. The best solutions to complex natural resource management problems will occur where decisions make ecological sense, where they satisfy socioeconomic demands, and where institutions are willing to, and capable of, carrying them out.

With this conceptual model guiding us, we suggest the following as a good working definition of an ecosystem approach to conservation: *An approach to maintaining or restoring the composition, structure, and function of natural and modified ecosystems for the goal of long-term ecological and human sustainability. It is based on a collaboratively developed vision of desired future conditions that integrates ecological, socioeconomic, and institutional perspectives, applied within a geographic framework defined primarily by natural ecological boundaries.*

Ecosystem approaches to conservation may be applied across many kinds of ecological systems, ranging from natural systems such as wilderness areas, to managed systems such as national forests, rangelands, or marine fishing grounds, to highly modified areas such as agricultural, and suburban and urban landscapes. An ecosystem approach encompasses the myriad of places that together determine the health and integrity of ecological and human systems. The challenges in pursuing an ecosystem approach are many, and although the principles provide coherent, considered goals, the extent to which these goals will be reached will depend on the quality of the science associated with the plans, and the ability of stakeholders to resolve their differing perspectives amidst diverse value systems and interests. Thus, ecosystem approaches in practice are a significant scientific and social challenge.

Key Elements of an Ecosystem Approach

An ecosystem approach expands traditional natural resource management and human-land/sea relationships in three dimensions: time, space, and degree of inclusion in decision-making (Meffe et al. 2002) (Figure 13.2). First, the temporal dimension is expanded because we are concerned with the health and vitality of ecosystems into

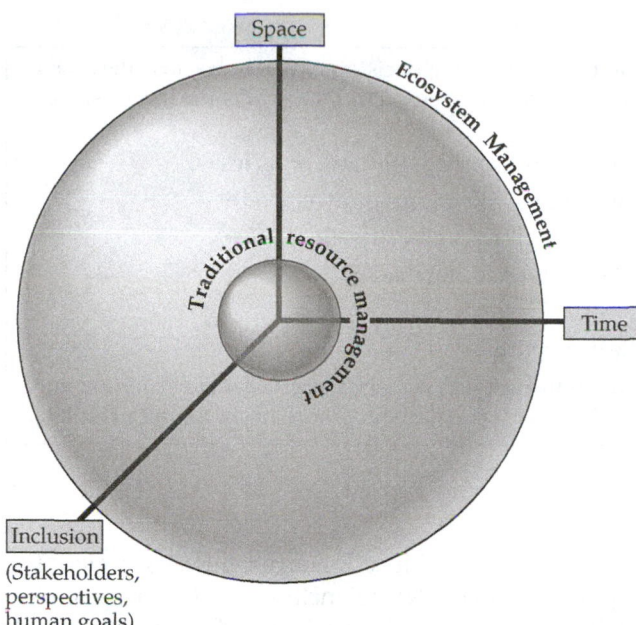

Figure 13.2 Ecosystem management as an expansion of traditional natural resource management in three dimensions. Temporal and spatial scales of concern, as well as degree of inclusion in those making decisions, are all expanded by ecosystem management.

the indefinite future rather than simply what they can do for us here and now. Our time frame should go beyond the next year or the present budget cycle, to include decades and even centuries. This does not mean that ecosystems are "frozen in time," that is, managed such that natural change is not permitted. Rather, management goals include ensuring that ecosystem dynamics occur within ranges that do not exceed the resiliency of the system (discussed later in the chapter). It makes no sense in an ecosystem perspective to hold only a short-term view when effects of human activities on ecosystems can last many decades and even centuries. Of course, in practical terms, policies work in shorter cycles of 5–20 years. The challenge is to link shorter-term actions into a long-term perspective.

Second, the spatial dimension is expanded beyond a particular location to include the larger landscape, seascape, and connections to other land or seascapes. Conservation actions rarely are sufficient within a single national park or local marine protected area because management practices in these areas have effects across boundaries, and are in turn affected by external events (Knight and Landres 1998). Scientific studies to recognize and understand the spatial connections present in nature are thus needed to more effectively manage these systems. Ideally, the spatial scale should include sufficient heterogeneity to provide resources for species dur-

ing years of resource scarcity. Beyond food, these resources could include such things as habitat cover that facilitate movement throughout the landscape, or provide necessary resources for breeding, and for the diversity necessary to maintain key interactors, such as plant pollinators or seed dispersers. In marine environments where many species have extensive potential ranges, the consideration of appropriate spatial scale must include considerations of circulation patterns and heterogeneity in coastal or benthic habitats, such as the availability of appropriate spawning grounds and areas for settlement along primary currents.

Finally, the human dimension is expanded to include a broader diversity of interests, talents, and perspectives in natural resource decision-making and conservation. Single-institution, top-down, command-and-control decisions are not tenable in a true ecosystem approach, for they usually exclude the majority of persons and interests affected by the decisions, and additionally, ignore community goals, relevant information, and talents that can contribute to problem solving. Human societies have diverse interests, perspectives, and knowledge bases, and thus will have distinct objectives for and contributions to conservation practice. Inclusive decision-making can result in more equitable solutions. Further, the prospects for finding and maintaining more successful conservation practices will be brighter when a broad community of stakeholders participate in decision-making, particularly when the needs of the natural world and future generations are well represented (Table 13.1 describes stakeholders).

Examples of ecosystem approaches

To illustrate the variety of goals and applications of ecosystem approaches to conservation, we briefly describe two of the most far-reaching and best-designed agreements that apply ecosystem-based management, the Convention on the Conservation of Antarctic Marine Living Resources and the Northwest Forest Plan. We then describe two major initiatives to integrate ecosystem approaches into conservation planning—ecosystem management as applied in the United States since the 1990s, and the ecosystem approach adopted by the IUCN to meet conservation and sustainable development goals.

CONVENTION ON THE CONSERVATION OF ANTARCTIC MARINE LIVING RESOURCES (CCAMLR)   In 1980, nations with interests in the natural resources of the Antarctic region together adopted a treaty to ensure cooperative oversight and conservation of these resources, and created the first ecosystem approach to conservation and management (Commission for the Conservation of Antarctic Marine Living Resources 2004). Taking force in 1982, the Convention on

TABLE 13.1 *What Is a Stakeholder?*

Stakeholders are people who want to or should be involved in a decision or action because they have some interest or stake in it. Their level of interest can vary from mild to intense. People can be stakeholders for a variety of reasons; they:

1. have a real or perceived interest in the resource, its use, its protection, or its users;
2. are dependent on a resource (e.g., subsistence users, sole means of livelihood);
3. have a belief that management decisions will directly or indirectly affect them;
4. are located in or near areas about which decisions are being made;
5. pay for the decision;
6. are in a position of authority to review the decisions.

Inclusion of stakeholders in decision-making helps to ensure that their concerns are met early on and that they "buy into" the decision through partial ownership, thus being more likely to support it later.

Source: Modifed from materials provided by Dennis A. Schenborn.

the Conservation of Antarctic Marine Living Resources (CCAMLR), uses a precautionary approach to preventing ecosystem degradation from exploitation of a subset of species. The Commission that executes the treaty includes representatives of all nations that use Antarctic resources and a diversity of stakeholder groups who collaboratively set fishing targets over the scale of the entire Antarctic basin (Figure 13.3). A scientific committee is charged with providing data and analysis on which the decisions of the Commission are based. A wide variety of exploited and unexploited species are monitored across the Antarctic, and monitoring data are carefully analyzed to detect trends and change management to prevent overexploitation and unintended decimation of nontarget populations. Thus, CCAMLR governs exploitation in the Antarctic via an inclusive decision-making framework among all stakeholders, basing management decisions on the analyses of the dynamics of a wide group of interacting species across the entire spatial domain of the Antarctic.

NORTHWEST FOREST PLAN (NWFP) The Northwest Forest Plan (NWFP) was created in 1994 to manage late-successional and old-growth forests within the range of the Northern Spotted Owl (*Strix occidentalis caurina*) in the northwestern U.S. After years of stormy conflict and heated court battles, the Clinton Administration successfully brokered an agreement among representatives of public agencies, timber interests, conservation groups and affected citizens to adopt the NWFP as a

mechanism for continued management of the region. The key elements of the NWFP include a system of reserves to protect these ecosystems, an Aquatic Conservation Strategy to conserve pacific salmon and provide connectivity among protected areas, management for sustainable timber extraction, employment programs for rural commu-

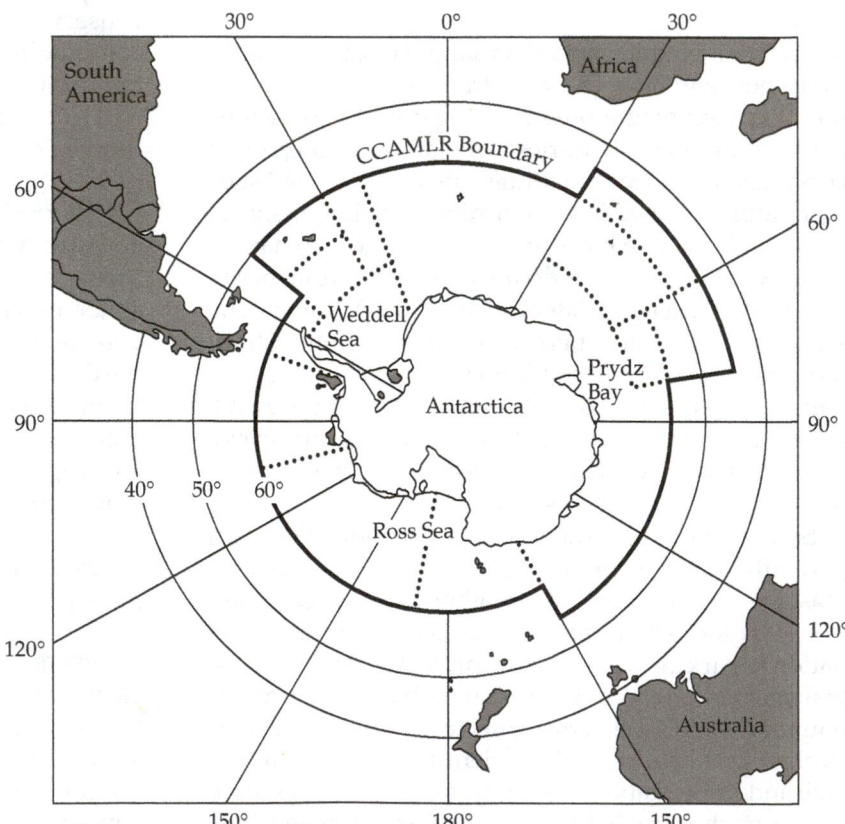

Figure 13.3 The CCAMLR manages a multi-national management area across a wide expanse surrounding Antarctica. (Modified from http://www.ccamlr.org/pu/e/conv/map2.htm.)

nities and tribes in restoration activities, and guidelines for adaptive management of the forest resources across the region (Tuchmann et al.1996). The Northwest Forest Plan incorporates conservation principles of maintaining (1) connectivity across the landscape; (2) landscape heterogeneity; (3) structural complexity; and, (4) the integrity of aquatic systems. Importantly, the plan represents an agreement among diverse stakeholders to manage across wide forest and institutional boundaries in the Pacific Northwest, based on careful analysis of the responses of a very large number of affected species to management practice.

Ecosystem management

In response to repeated conflicts in natural resource management, and an increasing recognition of the interconnectedness of ecosystems, the **ecosystem management** approach was developed in the U.S. in the 1990s (Grumbine 1994). An Interagency Ecosystem Management Task Force (1995)—a presidentially appointed group of representatives from several U.S. federal departments—characterized the ecosystem management approach as: more partnerships and greater collaboration; broader program perspective; broader resource perspective; broader geographic and temporal perspective; more dynamic planning processes; less reactive and more proactive. Their key actions for ecosystem management included:

1. Develop a shared vision of the desired ecosystem condition that takes into account existing social and economic conditions in the ecosystem, and identify ways in which all parties can contribute to, and benefit from, achieving ecosystem goals.

2. Develop coordinated approaches among federal agencies to accomplish ecosystem objectives, collaborating on a continuous basis with state, local, and tribal governments, and other stakeholders to address mutual concerns.

3. Use ecological approaches that restore or maintain the biological diversity and sustainability of the ecosystem.

4. Support actions that incorporate sustained economic, sociocultural, and community goals.

5. Respect and ensure private property rights and work cooperatively with private landowners to accomplish shared goals.

6. Recognize that ecosystems and institutions are complex, dynamic, characteristically heterogeneous over space and time, and constantly changing.

7. Use an adaptive approach to management to achieve desired goals and a new understanding of ecosystems.

8. Integrate the best science available into the decision-making process, while continuing scientific research to improve the knowledge base.

9. Establish baseline conditions for ecosystem functioning and sustainability against which change can be measured; monitor and evaluate actions to determine if goals and objectives are being achieved.

Note that these elements are a mixture of ecological, socioeconomic, and institutional considerations (see Figure 13.1), not dominated by any one concern. They recognize, for example, certain underlying ecological realities such as dynamics, complexity, and heterogeneity, while also recognizing private property rights, the need for inter-institutional cooperation, and the desire for continued economic benefits from the environment. Mike Dombeck and colleagues explore the development of ecosystem management and discuss several examples of its use in Case Study 13.1. An expanded, comprehensive treatment of ecosystem management is presented in a major, three-volume treatise that covers the subject in the U.S. from biological, social, legal, cultural, and economic dimensions (Johnson et al. 1999; Sexton et al. 1999; Szaro et al. 1999).

Generally, like many ecosystem approaches to conservation, specific cases where ecosystem management has been applied are young (10 years old or less), and thus we do not yet know much about their success. Initial efforts often focus on developing frameworks for problem definition and in gathering necessary baseline data for planning purposes (and against which to measure success, and therefore, be able to practice adaptive management). Yet, because early work involves building institutional and stakeholder participation and support, conservation efforts per se may not begin until after some of the social groundwork is laid. Nonetheless, in a survey of 105 ecosystem management projects initiated by 1995, Brush et al. (2000) found that most reported many positive outcomes of initial efforts, including improved communication and cooperation, increased public awareness of ecosystem ecology and functioning, and greater involvement of a wider variety of stakeholders. Soon, it should be feasible to assess these projects in terms of improvements in baseline data on ecological integrity and their success in meeting the needs of local communities (Brush et al. 2000).

Certainly, there is a potential for conflict between meeting the desires of stakeholders and meeting requirements for ecosystem sustainability (e.g., Lee 1993). Indeed, a major task for social research is to understand the conditions under which stakeholders might replace narrow and short-term self-interest decisions with ones that are long term, contribute to ecosystem sustainability, and are socially cooperative. There are also legal pitfalls that can inhibit good, collaborative ecosystem

management in the U.S., based on otherwise well-intended anti-trust laws (Thompson et al. 2004). Obviously such roadblocks need to be addressed if reasonable and constructive efforts are to move forward.

Using ecosystem approaches to meet the goals of the Convention on Biological Diversity

In the October 2000 Convention of the Parties to the Convention on Biological Diversity (CBD), ecosystem approaches were endorsed as the primary conceptual framework to use for achieving the goals of the CBD (Convention on Biological Diversity 2000). Accordingly, the parties drew up guidelines for increasing the use of ecosystem approaches, and subsequently the IUCN dedicated a Commission on Ecosystem Management (CEM) to provide expert guidance on the implementation of integrated management of ecosystems. The CEM defines an ecosystem approach as "a strategy for the integrated management of land, water and living resources that promotes conservation and sustainable use in an equitable way."

An ecosystem approach is advocated as a tool for developing capacity and making decisions on actions to manage resources sustainably. The CEM set forth 12 overarching principles and five points of operational guidance (Table 13.2). Similar to the experience of implementing ecosystem management in the U.S., the CEM has found that the first issue that needs to be addressed is to define the area in consideration and the key stakeholders. In most developing countries, defining the area under consideration is a bit more difficult than in the U.S. where jurisdictional boundaries are usually clear. Because in many applications in developing countries stakeholders may have very different power and capacities, the CEM further defines those who are most dependent upon resources in an area as primary stakeholders and others who are powerful, but not directly dependent (such as na-

TABLE 13.2 *Twelve Principles of the Ecosystem Approach and Five Points of Operational Guidance Adopted by the Convention on Biological Diversity*

Overarching principles

1. The objectives of management of land, water, and living resources are a matter of societal choice.

2. Management should be decentralized to the lowest appropriate level.

3. Ecosystem managers should consider the effects of their activities on adjacent and other ecosystems.

4. There is usually a need to understand and manage the ecosystem in an economic context and to:

    a. reduce market distortions that adversely affect biological diversity.

    b. align incentives to promote biodiversity conservation and sustainable use.

    c. internalize costs and benefits in the given ecosystem.

5. Conservation of ecosystem structure and function, to maintain ecosystem services, should be a priority target.

6. Ecosystems must be managed within the limits of their functioning.

7. The ecosystem approach should be undertaken at the appropriate spatial and temporal scales.

8. Recognizing the varying temporal scales and lag-effects that characterize ecosystem processes, objectives for ecosystem management should be set for the long term.

9. Management must recognize change is inevitable.

10. The ecosystem approach should seek the appropriate balance between, and integration of, conservation and use of biological diversity.

11. The ecosystem approach should consider all forms of relevant information.

12. The ecosystem approach should involve all relevant sectors of society and scientific disciplines.

Points of operational guidance

1. Focus on the functional relationships and processes within ecosystems.

2. Enhance benefit-sharing.

3. Use adaptive management practices.

4. Carry out management actions at the scale appropriate for the issue being addressed, with decentralization to lowest level, as appropriate.

5. Ensure intersectoral cooperation.

Source: Modified from IUCN 2000.

tional policymakers and nongovernmental organizations, NGOs) as secondary and tertiary stakeholders (Shepherd 2004). The ecosystem boundaries are set by the combination of ecological criteria discussed here, and also to a size appropriate to the management capacity of the stakeholders, the "counterpart range of human institutions which are likely to be able to protect, manage, and take decisions over the medium to long term within these boundaries" (Shepherd 2004). Importantly, in each case where an ecosystem approach is implemented, a local coordinating group (an "ecosystem stakeholder forum") oversees actions across the landscape, which will be a mosaic of areas that are managed by different stakeholders, at different intensities.

The central focus of planning and management then becomes the fulfillment of principles 5, 6, and 10 (see Table 13.2), where the priority target is the conservation of ecosystem structure and function, within the limits of natural ecosystem functioning, to achieve maintenance of ecosystem services through balanced conservation and sustainable use of biological diversity. Importantly, these efforts are best met by joint ventures of scientists and local inhabitants as their knowledge bases will differ, but both contributions are crucial for project success. The very process of working together builds trust in project partners, investment in project goals, and capacity among local stakeholders to manage effectively.

Development of sustainable use requires careful attention to the current structure of incentives, and their influence on human behavior. Are subsidies causing people to work natural resources unsustainably? Would positive incentives (e.g., better land rights) influence people to be better land stewards? Next, attention is given to how to connect sustainable practices to greater benefits, and crucially, to internalize both those benefits and costs of less sustainable practices to local communities (Shepherd 2004). This means that those who protect and manage resources benefit directly, and those who create environmental damages must pay for them.

Finally, all work using an ecosystem approach adheres to the tenets of adaptive management over space and time, recognizing impacts of efforts in one ecosystem on another, and developing a flexible, experimental design that will allow stakeholders to adapt to inevitable change.

The IUCN CEM has only just begun working with groups across the world using these principles to guide conservation and sustainable use.

Biophysical Ecosystems as Appropriate Management Units

An "ecosystem" is typically defined as a community of organisms interacting among themselves and with their physical environment. In this sense the ecosystem repre-

sents a higher level of organization beyond the population or the community because it also includes abiotic components. People are included in ecosystems to the extent that their activities modify the structure and processes of the biophysical ecosystem or that their lives would be affected by management decisions. "Ecosystem" takes on an expanded meaning when it is used in an ecosystem-based approach to conservation; generally, management areas associated with these approaches are large and encompass more than a single type of biophysical ecosystem. For example, if an ecosystem scientist were describing a large watershed comprising coniferous forest, lakes, and rivers, the description would be given in terms of three ecosystems with distinct properties but which were ecologically linked in many ways. Although an aquatic-minded ecologist might focus studies on ecosystem properties of the lake or river, the ecosystem scientist would also be cognizant of transport from the watershed of organic matter, water from surface runoff, and other inputs. Similarly, marine scientists would focus on long-range transport of organisms and materials in ocean currents, as well as on input from rivers and coastal ecosystems (Figure 13.4)

An example of a major ecosystem project that includes several biophysical ecosystem types is the Greater Yellowstone Ecosystem (GYE; Figure 13.5). The GYE includes as a minimum the following biophysical ecosystems: coniferous forests, grasslands, hot springs, rivers, lakes, grazing rangelands, and agricultural fields. The management logic and scientific rationale that justifies including this heterogeneous landscape in a single, huge management unit is that this network of biophysical ecosystems is linked through ecological interactions and through social institutions.

Herds of large, grazing animals—mainly bison, elk, and pronghorn—have adapted to the Yellowstone landscape by seasonally migrating to more favorable regions. One of the most significant of these migrations was the winter movement of herds down from the high plateaus to the valleys and river bottoms, regions that are now developed for agriculture and second homes. Other important migratory routes were onto national and state forest lands that surround the park. Through their need to migrate, the bison, elk, and pronghorn are dependent on access to, and utilization of, these surrounding nonparklands. Thus several biophysical ecosystems should be considered in an ecosystem approach.

In choosing the boundaries of the management area there are three biophysical considerations, which we list in descending order of importance. First, the biophysical ecosystem should be completely included within the management area. Attempting to manage only part of a watershed or half of a lake is unlikely to be successful. Second, if the area includes more than one biophysical ecosystem type, the decision to include or exclude them

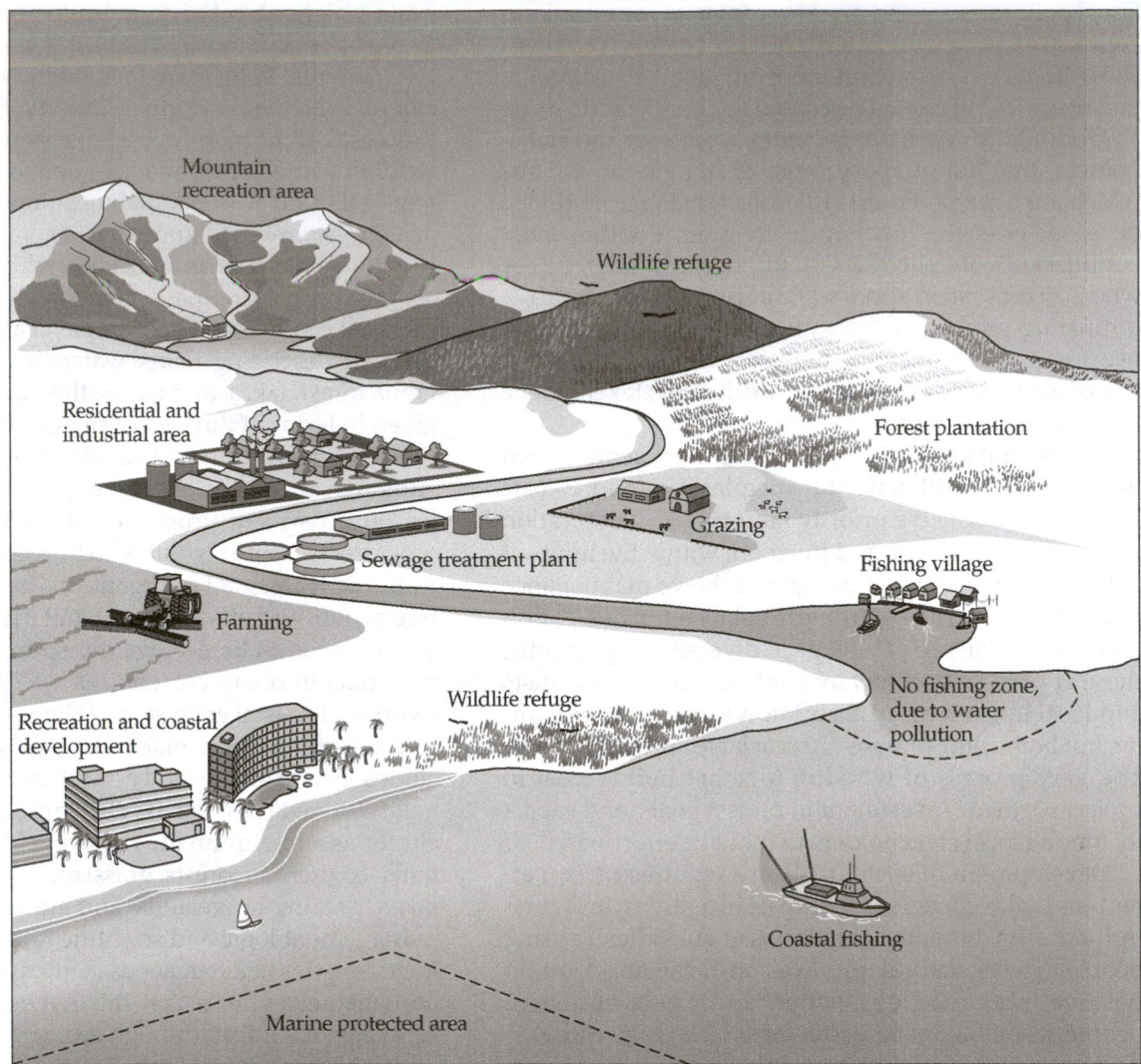

Figure 13.4 Ecosystem management attempts to involve all stakeholders that affect a large ecosystem and whom are affected by management practices within that system. In this case, a watershed is managed for many purposes, and thus a large network of stakeholders are involved in its management. Even the management of any component may be affected by the actions of other other users of the system. (Modified from Miller 1996.)

should be based on the degree to which they have functional linkages, as with lakes and watersheds or estuaries and rivers. In this case the inclusion of different ecosystems that are functionally linked is logical because management decisions that directly affect one ecosystem might indirectly affect another ecosystem. A third, less commonly used biophysical consideration is where the ecosystems are independent—that is, where they have no or only weak linkages, but where they are affected by the same stresses. Common examples of the latter case include management programs to mitigate pollutants such as acid precipitation and other industrial fallout

that can be transported long distances and across heterogeneous landscapes.

Generally, areas considered for an ecosystem approach will be large enough to encompass the resources needed by the various populations of the ecosystem. The significance of spatially explicit landscapes to ecosystem management derives from three well-established principles: (1) The spatial configuration of food sources will influence which food sources are used; (2) resources may not always be available at a particular place, but this absence may be compensated by availability at some other site; and (3) some parts of the landscape may not be used

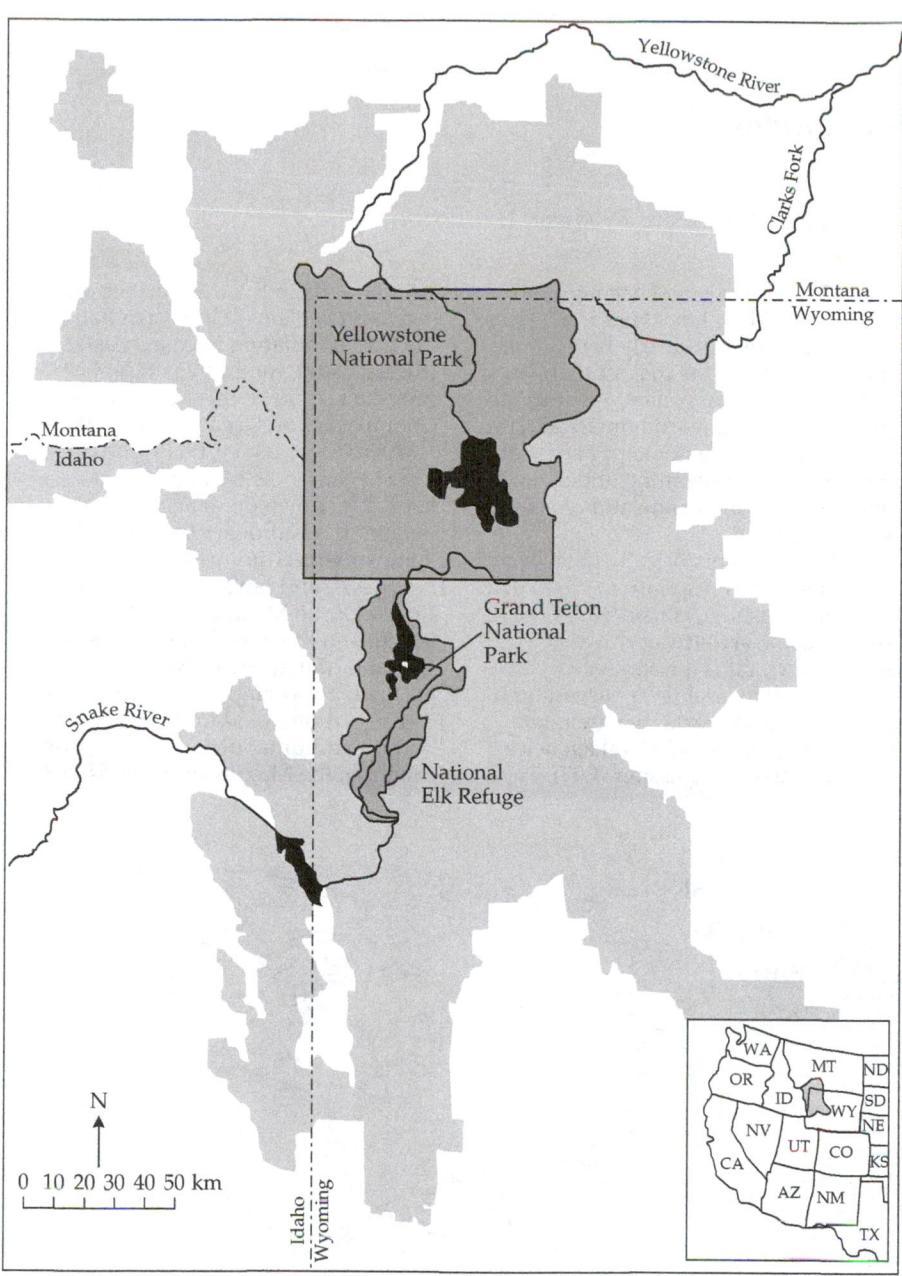

Figure 13.5 The Greater Yellowstone ecosystem. Note the relationship of Yellowstone and Grand Teton National Parks (dark shading) to the various surrounding national forests, wildlife refuges, and other wild lands (light shading). Black denotes major lakes.

because risk from predators is too great. Thus, spatial scale should encompass sufficient area and landscape heterogeneity such that populations can maintain themselves in spite of temporal variation in resource abundance. This spatial scale, what Fleming et al. (1994) refer to as the "functional landscape mosaic," largely defines the boundaries of the "ecosystem" in the context of an ecosystem approach.

In the world's oceans, the large marine ecosystem (LME) concept was developed to deal with the need to address environmental and resource management issues in the oceans on appropriate geophysical and sociopoliti-

cal scales (Sherman 1993). Because over 90% of the world's fisheries are maintained in coastal zones, the focus on the LME concept is exclusively coastal, but encompasses both oceanic regions and adjacent watersheds. Large marine ecosystems are defined as marine coastal zones that are characterized by similar bathymetry, productivity, and trophic linkages. LMEs are linked crucially to the major marine catchment basins (MCBs), or watersheds, that interact with the coastal waters, and from which much of the human impacts on these marine systems are generated (see Figure 13.4). Essay 13.1 by Karen McLeod and Heather Leslie discusses U.S. ocean policy.

ESSAY 13.1

Marine Ecosystem-Based Management
Transforming U.S. Ocean Policy

Karen L. McLeod, *Oregon State University* **and Heather Leslie,** *Princeton University*

■ Oceans and coastal areas around the world are experiencing unprecedented levels of degradation. The combined effects of human activities on land, along the coasts, and in the oceans, including agriculture, aquaculture, coastal development, fishing, ports and shipping, military activities, and tourism, are unintentionally, yet seriously harming marine ecosystems. Numerous threats associated with these activities, particularly the alteration of marine food webs, climate change, habitat damage and fragmentation, coastal erosion, invasive species, and pollution, substantially compound the damage.

United States' oceans represent the largest ocean jurisdiction in the world. They comprise an area 23% larger than the U.S. landmass, and encompass about 1.8 million km² (Figure A). Thus, U.S. ocean policies clearly have global implications. Yet the major ocean laws, policies, and management institutions in the U.S. that govern human activities in the oceans have not kept pace with the evolution of scientific understanding of the marine realm, and of ecosystems in general.

U.S. policies historically have been implemented in response to crises in a piecemeal fashion, considering only a single sector, activity, or threat at a time. The result is a fragmented patchwork of 140 federal laws pertaining to the oceans and coasts that are interpreted by dozens of federal agencies (Pew Oceans Commission 2003). Population trends in the U.S. further exacerbate the problems. More than half of the U.S. population lives in coastal counties, and another 25 million are projected to move there by 2015 (U.S. Commission on Ocean Policy 2004).

As recently as the 1970s, the oceans were regarded as vast and inexhaustible. Key federal agencies—notably the National Oceanic and Atmospheric Administration (NOAA)—and laws—such as the Coastal Zone Management Act and the Magnuson-Stevens Fishery Conservation and Management Act—that make up the U.S. ocean policy were created during this era of seemingly endless bounty and untapped resources. For example, the Magnuson-Stevens Fish-

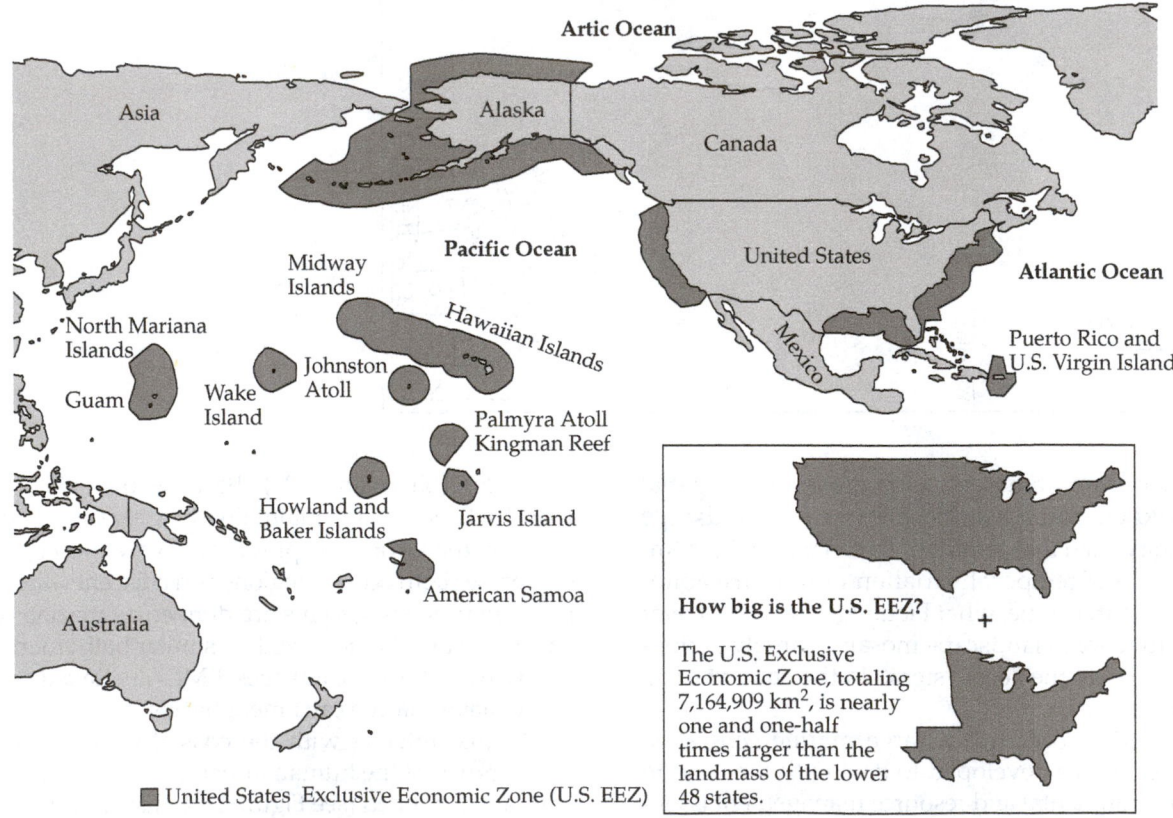

Figure A   The United States' Exclusive Economic Zone. (Modified from The Pew Oceans Commission 2003.)

ery Conservation and Management Act has numerous mandates, including the management of fishery resources for optimum yield, the development of "underutilized" fisheries, and protection of marine habitat. Also, the agency with primary responsibility for implementing the act—NOAA Fisheries—is located within the Department of Commerce. Commerce is focused on business and international trade, as well as weather prediction, marine sanctuaries, and fisheries management. The point is that the multiple, potentially conflicting objectives of U.S. coastal and ocean management institutions, as well as the governance structure itself, present challenges to ecological sustainability for policymakers, scientists, managers, and stakeholders alike.

Recently, two high-level ocean commissions, the Pew Oceans Commission (2003) and the U.S. Commission on Ocean Policy (2004), reviewed the status of U.S. ocean resources and policies for the first time since 1969. Although the final recommendations of the two commissions differed in some respects,

there was remarkable overlap in their main messages: the oceans are in trouble, there is an urgent need for action, and policies must shift toward a more comprehensive, integrated, ecosystem-based approach to managing the threats to marine ecosystems.

Ecosystem-based management (EBM) for the oceans is the integrated management of key activities that affect the marine environment (United Nations 2002; Pew Oceans Commission 2003; U.S. Commission on Ocean Policy 2004). Central to this approach are a focus on managing the cumulative effects of human activities in the context of entire ecosystems and an acknowledgement of the interrelationships among all ecosystem components. Approaches to implementing EBM vary, but share a focus on protecting ecosystem structure, functioning, and key processes. Note that we use the term "ecosystem-based" rather than "ecosystem." We do so because we believe that we can manage people's activities that influence ecosystems, but not ecosystems themselves.

The goal of EBM is to maintain or restore ecosystem integrity so that marine ecosystems can continue to provide the services that people want and need over the long term. Management of individual activities (e.g., fishing) in the context of the ecosystems in which they are embedded is necessary but not sufficient to achieve this goal. Why? Marine management to date has tended to focus on the production of a few goods (particularly fish) to the detriment of other nonconsumptive services such as the detoxification of pollutants, protection of shorelines from erosion, control of pests and pathogens, nutrient cycling, the moderation of climate and weather, and recreation, spiritual, and other nonmaterial benefits (Daily 1997). Fishing, for example, can affect marine ecosystems by removing substantial biomass, reducing the average size and age of fish, removing a large percentage of top predators, affecting populations of nontarget species via bycatch, and degrading or destroying bottom habitats through the use of harmful gear (Jennings and Kaiser 1998; Goñi 2000; Dayton et al. 2002; Figure B) These

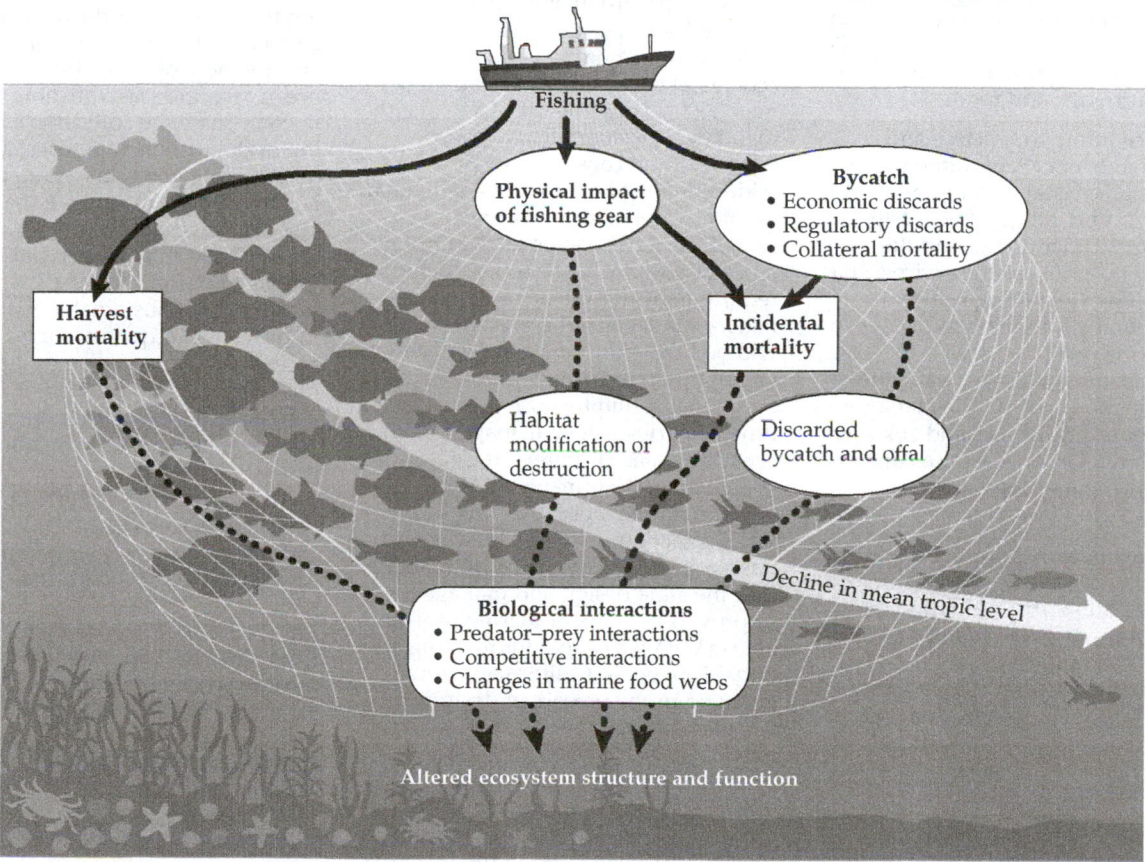

Figure B   The ecological effects of ecosystem over-fishing. (Modified from The Pew Oceans Commission 2003.)

effects can in turn influence ecological structure and functioning and reduce the overall productivity of the system. Consequently, while the incorporation of ecosystem considerations into fishery management is crucial, that alone will not result in effective management for the full range of ecosystem services that the oceans provide.

Moving toward EBM will necessitate a stepwise, multi-pronged approach—there will be no single silver bullet. The following are core components of this shift.

1. Involvement of a broad range of constituencies, stakeholders, and institutions from the outset; ecological and social systems are inextricably linked, and thus an explicit integration of these perspectives will be critical to success. The trade-offs inherent in choosing among alternative management strategies must be clearly articulated to all parties.

2. An explicit recognition of the complex linkages within and among ecosystems, including species interactions and land–sea connections; management should also strive to maintain connectivity between and within marine ecosystems by facilitating the import and export of larvae, nutrients, and food.

3. A focus on the interactive and cumulative effects of different activities (e.g., coastal development, pollution, and fishing) on ecosystem structure, functioning, and the delivery of ecosystem services; spatial zoning (including fully protected marine reserves, other types of marine protected areas, and fisheries closures) will be a valuable tool for coordinating the management of multiple uses and acknowledging the larger seascape context within which ecosystems are embedded.

4. A means of incorporating flexibility and adaptation into management plans to allow for evaluation and readjustment as new information becomes available; this should include the establishment of long-term observing, monitoring and research programs. Precaution should increase as uncertainty increases.

5. Creation of complementary and coordinated policies at global, international, national, regional, and local scales; ecosystem processes operate over a range of spatial scales, and thus there is no single appropriate scale for management.

Moving Forward with EBM in Practice

For the first time in several decades, the U.S. is primed to make major changes in ocean management and governance. These changes are of great importance to all citizens—whether or not they live near the coast—given the wide range of services provided by coastal and ocean ecosystems.

There are two areas within the U.S. where coastal and ocean stewardship is evolving in innovative and rapid ways. Related efforts internationally include Canada's Ocean Strategy, Australia's Ocean Policy, and a number of initiatives underway in both developed and developing nations under the auspices of the European Union and the multilateral Large Marine Ecosystems Program.

In 1999, California passed the Marine Life Management Act (MLMA) requiring all human activities in state waters to be sustainable, defining "sustainable" as those uses that secure the fullest range of present and long-term ecological benefits (Weber and Heneman 2000). Thus, the MLMA explicitly mandates consideration of nonconsumptive and consumptive uses, addressing the breadth of benefits that humans receive from marine ecosystems. The act also stresses the importance of sustainable fisheries, including requirements for the development of fishery management plans. These plans will be linked through a single overarching plan that must meet ecosystem and nonconsumptive use provisions; thus, fisheries will be managed in an ecosystem-based context in California's state waters (Kaufman et al. 2004).

Companion legislation, the Marine Life Protection Act (MLPA), was also passed in 1999. The MLPA mandated that the state design and manage an improved network of marine protected areas (MPAs) as a tool for meeting sustainability goals. California is now engaged in the process of designing a network of MPAs for all state waters to meet the objectives of the MLPA.

At the same time that the MLMA and MLPA processes were moving forward, the first planned network of marine protected areas in the U.S. was implemented in the Channel Islands through a distinct but related process jointly sponsored by the state and federal governments (see Case Study 14.2). Multiple stakeholders, including commercial and recreational fishermen, environmentalists, marine resource managers, scientists, educators, and local residents, participated in the design and implementation of the network through the Marine Reserves Working Group. The working group was supported by scientific and socioeconomic advisory panels and received more than 9,000 public comments over two years. The result was the establishment of an MPA network in 2003. It includes 10 fully protected marine reserves as well as two less restrictive MPAs.

In 2004, California passed the California Ocean Protection Act (COPA). This legislation is the first to move forward with some of the recommendations of the two ocean commissions, namely an explicit effort to coordinate efforts among agencies. COPA creates an Ocean Protection Council comprised of all state agencies that have responsibility for coastal and ocean management. In addition to coordinating the activities of the agencies, this council is charged with improving the effectiveness of state efforts to protect ocean resources, establishing policies to coordinate the collection and sharing of data among agencies, and recommending necessary changes to state law and policy to achieve these goals. The act also establishes a $10 million Ocean Protection Trust Fund to reduce threats to marine ecosystems, create incentives for sustainable fisheries, improve water quality, increase public access to marine resources, improve management and protection of coastal waters, and finance monitoring and other scientific data collection. If COPA proves to be effective, California could serve as a model for shifts in federal oceans governance.

Steps toward implementing marine ecosystem-based management also are being taken in the Gulf of Maine (Figure C). The Gulf of Maine Council on the Marine Environment, a decentralized network of government and nongovernmental organizations, business people, and concerned citizens from the U.S. and Canada, is working to manage the gulf as a single ecosystem (Gulf of Maine Council on the Marine Environment 2002). Since its founding in 1989, the Council has promoted an inclusive and transparent public dia-

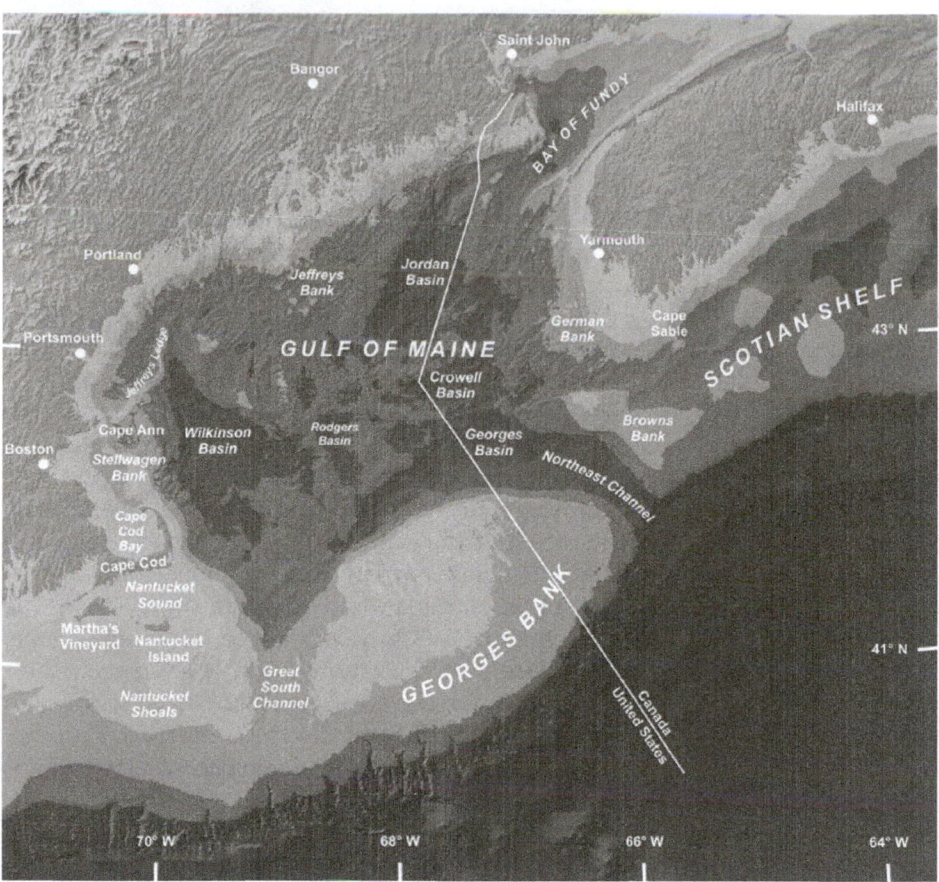

Figure C   The Gulf of Maine. (Modified from the USGS and The Woods Hole Field Center.)

logue about the Gulf and its natural resources. The Council's mission—to maintain and enhance environmental quality in the Gulf of Maine and to allow for sustainable resource use by existing and future generations—recognizes the linkages between ecological and social systems throughout the Gulf of Maine watershed.

One of the Council's guiding principles is to support ecosystem-based planning and management, which they define as "collaborative management that integrates economic and ecological values and objectives, emphasizing natural rather than political boundaries" (Gulf of Maine Council on the Marine Environment 2002). Ocean zoning, which would distribute human activities in coastal and ocean areas in a proactive, integrated, and scientifically informed manner, has been suggested as a key element of marine ecosystem-based management for the region (Doherty 2003). As noted previ-

ously, there is no one "right" scale for implementing marine ecosystem-based management: In addition to working at the Gulf-wide level on projects such as seafloor mapping and ocean zoning, the Council also supports community-based habitat restoration and resource monitoring projects (Gulf of Maine Council on the Marine Environment 2002).

The prospects for implementing ecosystem-based management off the coast of California and in the Gulf of Maine are good, but face several challenges. As noted above, marine management in the U.S. has tended to be sector-based and lacks a strong federal policy framework. In California, the state government has responded to that gap by passing COPA, which is intended to dramatically alter the marine governance and management landscape in the state. In the Gulf of Maine, diverse institutions and stakeholder groups are engaged in a long-

term effort to improve stewardship of the Gulf from the bottom up. Given the trans-boundary nature of this effort, the governance structure and management approaches that emerge will likely differ significantly from the state-driven effort in California. In both cases, scientists and other stakeholders must confront and deal with uncertainty (in the data, in the political processes, in the choices that individual households and interests groups will make, etc.), and be prepared to manage adaptively and learn from their mistakes: Information on both the ecological and social dynamics of the systems are inevitably partial, and yet these processes are moving forward. ◾

Understanding Ecosystem Dynamics and Resilience

Ecosystems are dynamic, not static, and the limits to ranges of variation in ecosystem structure and function must be identified if we are to manage in an ecosystem context. All ecosystems have a natural range of variation in structure and processes, and this range is ecosystem-dependent. For example, over the period of a month, the composition and structure of woody species in an old-growth deciduous forest will not typically change in a significant way; unless a major disturbance occurs, the forest will consist of the same trees in the same positions. However, over a period of a decade, and certainly over a century, the forest may experience great variation in composition and structure: Trees will die, others will replace them, diseases or herbivores will attack the forest, fire may break out, and so forth. There is an expected range of variation over the long term for this system.

Natural systems have expected ranges of variation, both short and long term. Many, but not all kinds of short-term variability are relatively easy to incorporate into management plans; rarer, more extreme events are more difficult to incorporate. Yet, the uncommon extreme events, the so-called episodic events such as catastrophic fires or storms, disease outbreaks, extreme floods and so forth, can unduly shape ecosystems because their effects are often large and long-term. A major and unanswered question in ecosystem management is: How are extreme but rare events incorporated into management plans?

Because ecosystems experience so much natural variation from so many sources, yet are not fundamentally altered by any but the most severe natural (and many human) disturbances, they are said to possess **resilience**. Resilience is the magnitude of disturbance that can be absorbed or accommodated by an ecosystem before its structure is fundamentally changed to a different state (Holling 1973, 1986), such as a shrubland becoming a grassland, or a grassland becoming a desert. Of course ecosystems do not have a single resiliency property, but rather any ecosystem may be resilient for some properties and vulnerable in others. For example, a system with dense vegetation may be resilient to damage caused by ice storms but, because of fuel loading, have poor resilience to fires. Similarly, coral reefs may be more resistant to damage from severe storms than to ocean warming accompanying global climate change.

Any ecosystem is expected to be resilient to disturbances and variation that are within the normal "repertoire" of what it has experienced over ecological time (i.e., millennia). Consequently, rivers are expected to be resilient to flooding, chaparral and longleaf pine–wiregrass systems to frequent fire, and high-latitude lakes to winter freezing. In fact, ecosystem composition and structure may change significantly when those disturbances do *not* occur, such as when riverine flows are stabilized by dams, fires are suppressed where they normally occur, or global warming causes lakes to remain unfrozen throughout winter. Then, the normal range of variation is not experienced and ecosystems change.

The importance of variation and disturbance is nicely illustrated in a model developed by Holling (1995), who discussed the *constructive* role that variation and disturbance play in maintaining the integrity of ecosystem function (Figure 13.6). Holling's model is, of course, a generalized abstraction of the real world, but it provides a useful construct for comparing the behavior of any particular ecosystem. In his model, terrestrial ecosystems go through four functional stages as they develop. The first stage is "exploitation," or early succession after a disturbance. Here, rapidly growing pioneer and opportunist species dominate and exploit the open space. The system then moves toward the "conservation" stage, where more mature communities develop (i.e., composed of a mature age structure of long-lived species). In this stage, which can be long-term, complex community structure develops and strong interspecific relationships, especially mutualistic ones, prevail.

As system organization and "connectance" increases, when a major biotic and abiotic disturbance, such as insect infestations, storms, fires, or pathogens occurs, the system quickly shifts into the "release" phase, whereupon the complexities, structure, and stored energies are quickly disorganized. Although release destroys structure it also creates opportunity, in the sense that the system can begin to reassemble; this is called, "creative destruction" and is an important aspect of disturbance. This rapid and chaotic release phase is followed by "reorganization," when the system begins to rebuild, sometimes in quite different ways (see Figure 13.6).

Holling's model seems to reflect reality in fire-prone forest ecosystems, but its generality has not been demonstrated. Some ecosystems, particularly riparian and coastal zones, may be better characterized by shorter, pulsing patterns produced by the daily and lunar cycle hydroperiods. In these cases, tidal flow and seasonal changes in water level periodically provide energy and material subsidies to the ecosystem (Odum et al. 1995). When appropriate, indicators of sustainability should reflect regular or periodic pulsing events and respond to conditions of too much or too little energy and material subsidy. For example, one set of indicators that might be used in evaluating an experimental pulse release of water from Lake Powell into the Grand Canyon could be based on the extent to which amounts of material accumulation in the form of sand bars and other sediment deposits approach desired levels. The important point is

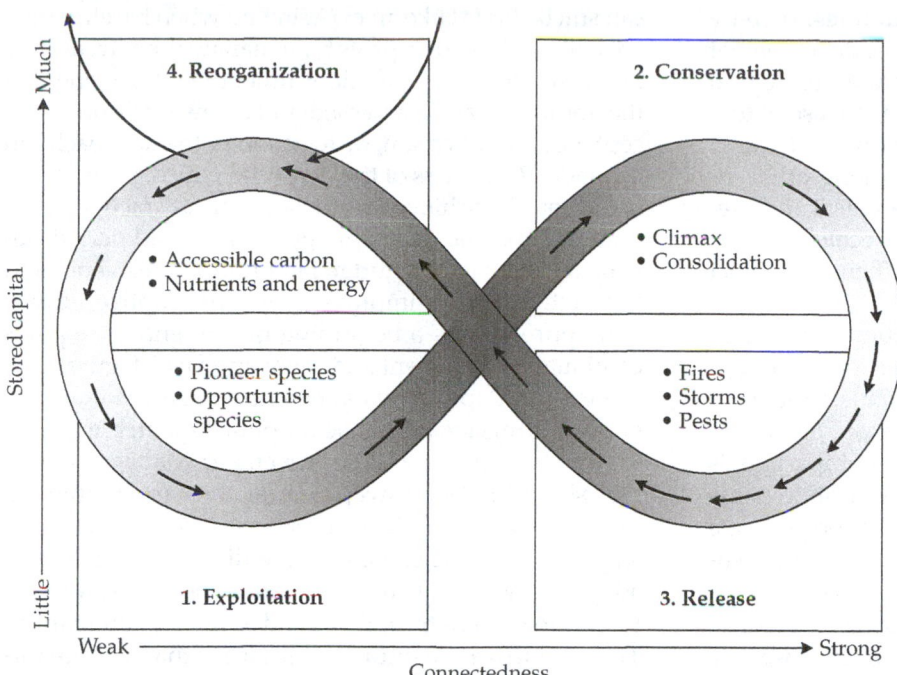

Figure 13.6 Holling's model: the functional stages of ecosystem development and the flow of events between them. (Modified from Holling 1992.)

that ecosystems undergo characteristic cycles of change. While the length of the cycles and the type of change is specific to the particular ecosystem, maintenance of change is essential to ecosystem management.

Ecosystems with keystone species and those with few compensatory responses and low redundancy present particular problems. In these systems, small changes to key players can produce very large responses. This phenomenon has been most rigorously studied in north temperate oligotrophic lakes. In these lakes, variability in predator–prey interactions can cause a "trophic cascade" of effects through the plankton and thereby influence ecosystem processes (Carpenter and Kitchell 1993).

Fire-suppression is a strong example of how changing (in this case restricting) the normal range of variation can lead to devastating consequences (Holling and Meffe 1996). Fire-suppression tends to be successful in reducing the short-term probability of fire on public lands and in suburban, fire-prone regions. But the result of constraining natural system behavior is an accumulation of fuel that eventually produces fires of much greater intensity, extent, and human cost than if fires had never been suppressed (Kilgore 1976; Christensen et al. 1989). This was strikingly demonstrated in the Yellowstone National Park fires of 1988, and subsequent fires in the West in 2002.

Whether ecosystem management goes beyond the natural ranges of variation or suppresses them, the system may change to a fundamentally different state as a result, and will not continue to function as it normally would. Consequently, identifying the driving distur-

bances and typical ranges of variation in a given ecosystem is critical to understanding the system and managing it in a sustainable way that does not alter system resilience. An example of a management plan to increase resilience of coral reefs is described by Jordan West and colleagues in Case Study 13.2.

Adaptive Management: Preparing for Change in Conservation Practice

A major consequence of the dynamic nature of ecosystems is that their management must also be dynamic— that is, flexible and responsive. Management that is approached as an experiment, and that responds in creative and innovative ways to changes in complex systems, is **adaptive management** (Holling 1978; Walters 1986; Cook et al. 2004) and is a salient feature in an ecosystem-based approach to conservation.

Traditional resource management has followed a strongly hierarchical chain-of-command model, with decisions made at the top and carried out at the bottom. This rigid, command-and-control approach to decision-making is the antithesis of an ecosystem approach (Meffe et al. 2002). There is a significant liability to such rigidly hierarchical approaches to natural resource management. Ecological systems are complex and characterized by many nonlinear processes and indirect effects. Such systems inevitably have threshold responses and produce surprises. Furthermore, to varying degrees colonization and extinction processes constantly shuffle the

biological players in most local communities, thereby changing biological interactions. Economic development in the management area also can greatly affect ecological processes through habitat fragmentation, release of toxins, spread of nonnative species, withdrawal of water, and increased frequency of wildfires, among other insults. Consequently, taking a rigid hierarchical and prescriptive approach to the management of ecological systems that are inherently dynamic and difficult to predict is a serious mismatch.

U.S. natural resource agencies have begun to recognize the need to build more flexibility into local management and to seek feedback from their management experiments. For example, forest plans developed periodically by the U.S. Forest Service set broad goals but now allow for considerable local modification of the plan by district rangers. However, conservation goals are more likely to be achieved over the long-term if they engage an adaptive management model characterized by a program of baseline and continual monitoring of indicators that measure progress toward goals, on-going analyses of policy alternatives, and an institutional capacity to change management practices when better alternatives are available and current practices are not achieving their objectives (Figure 13.7).

Adaptive management approaches all management actions as scientific experiments. In its most "pure" form, this literally involves application of treatment and control units in management whose behavior can be statistically compared. Though this is an excellent idea, in practice it is rare that actual, tightly controlled management experiments can be conducted on a large scale with true replicates. Short of that, however, management can still be *treated* like an experiment; when baseline data are collected, smaller prototype management trials may be instituted. After sufficient time, the effectiveness of the approach can be assessed and a new decision on acceptance, modification, or rejection of the approach can be made. Regardless of the degree to which a true experiment may be achieved or the level of satisfaction of the outcome, the management approach should periodically be revisited to both judge its continued relevance and to search for further improvements. An adaptive, ecosystem approach has a beginning but no end; it requires continual reassessment and innovation (see Figure 13.7).

Finally, adaptive management recognizes that mistakes or errors are not cause for punishment, but opportunities for learning and improvement. Management errors should not be "swept under the carpet," hidden from public view in fear of reprisals, for that is a good way to ensure the same errors will occur again. To be adaptive and useful, mistakes should be openly aired so that all may learn from them and avoid repeating them. This requires an institutional mindset that is more forgiving—one that acknowledges that management is a continual experiment and that mistakes will be made, and one that is assured enough in its vision and mission to learn, adapt, and modify, even if it means recognition of human flaws and limitations.

Adaptive management is an attractive concept, providing security in our planning for many stakeholders against the inherent uncertainties of ecosystem processes and the lack of our understanding of them. Widely cited as a good model for wise management, most ecosystem approaches call for its use, but could apply adaptive management more carefully or completely (Payne et al. in press). Adaptive management is employed particularly well in several cases, including conservation efforts in the Everglades, Chesapeake Bay (see Case Study 13.3 by Ron Carroll and Chrissa Carlson), and under CCAMLR and NWFP. In CCAMLR, careful monitoring of the responses of target and nontarget populations to distinct fishing regimes is undertaken over time, and in comparisons across regions in the Antarctic Ocean. When declines are detected, fishing quotas are adjusted to allow for recovery. The NWFP applies standards for pre- and postlogging disturbance monitoring and landscape-scale surveys for roughly 300 rare species, and 12 species of special concern. These data are used to gauge the effectiveness of management practices across the entire region, and future adjustments will be based on analyses of these data (Tuchmann et al. 1996). A major challenge in these plans is to devise a means to monitor a large number of species and ecosystem variables. In CCAMLR, this was achieved by simplifying the Antarctic food web, and monitoring only key species and ecosystem attributes (Kock 2001), whereas in the NWFP, an original list of 400 rare species was reduced

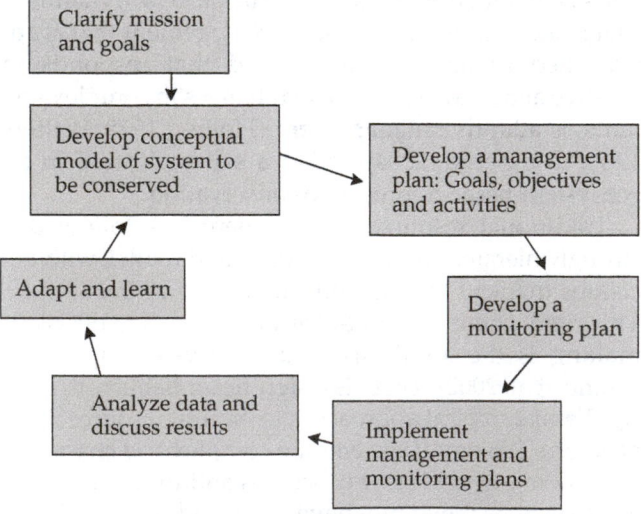

Figure 13.7 The adaptive management cycle allows for continual monitoring and refinement of management objectives. (Modified from Margoluis and Salafsky 1998.)

via careful consideration of cases where other monitoring was likely to be sufficient to indicate trends for these species as well (USFS and USBLM 2001). All of these cases represent the best of adaptive management, seeking to reduce risks to species and ecosystems by gathering exhaustive data prior to making management decisions, adjusting prior decisions in light of new information, and acting in the forefront of conservation in creating new approaches to successful management.

Analytic approaches used in adaptive management

Adaptive management approaches are now used in a wide array of complex systems, from policy analysis to industrial organization. Consequently, many analytical tools and models have been developed (e.g., mass-balance models such as Ecopath, and more individual-based models similar to the SEPMs described in Chapter 12). The majority of these use some form of simulation modeling to explore potential consequences of different management decisions. When simulation modeling was first applied to adaptive management more than twenty

years ago it was considered a major conceptual and practical breakthrough. Simulation of management decisions allows managers to substitute computer programming for actual decisions, some of which may have consequences difficult to reverse if they are be proven to be damaging—hence it is more prudent to conceptually test potential effects through models.

A suite of mass-balance models, Ecopath, Ecosim and Ecospace, have been used extensively in fisheries applications, and more recently, in marine reserve planning (Pauly et al. 2000). These models create trophically linked "pools" of resources, which may be single species, functional groups, or guilds, and project how these resources will change when subjected to different harvest regimes or other management policies (Figure 13.8). Refinements of the original static Ecopath models (Ecosim and Ecospace) have allowed exploration of environmental and social consequences of different management decisions, which often serve to reveal critical data gaps and refocus consideration of different actions (Piketch et al. 2004). Details on these models and a library of applications can be found at http://www.ecopath.org.

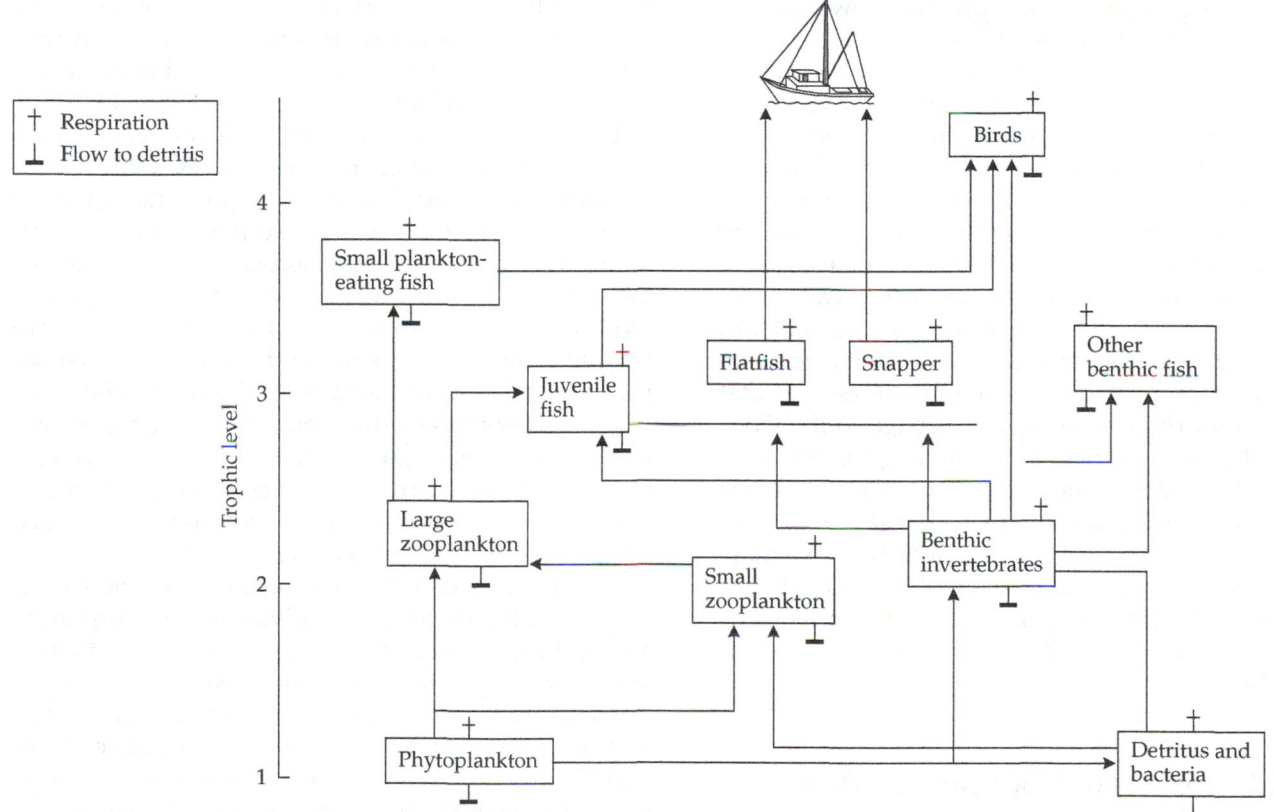

Figure 13.8   Ecopath and other mass balance models are often developed to include trophic groups of species and their connections to one another, and to human effects on them through environmental change and direct exploitation. In this model, the effects of human exploitation of two groups of species (flatfish and snapper) on the entire ecosystem can be evaluated. (Modified from National Institute of Water and Atmosphere Research 2005.)

Large-scale environmental management programs may include different biophysical ecosystems, more than one public or private institution with management responsibilities, commercial interests, advocacy groups, private citizens, and the constraints of layers of local through national law. Here, adaptive management must explore complex management options.

For example, an innovative approach for understanding how the Greater Everglades ecosystem can be maintained was developed by the U.S. State Department's Man and the Biosphere Program (Harwell et al. 1996). In this holistic approach, land use and hydrology scenarios were developed through the use of geographic information systems and simulation models that, if implemented, would provide various degrees of protection of water flow and quality for the Everglades while permitting some economic development in South Florida. These scenarios range from land use and patterns of water flow to the Everglades that mimic that of early, pre-European settlement of Florida and greatly constrain economic development, to a pattern of land use and water flow that supports rapid economic development with only minimal protection of the Everglades. Each scenario is then evaluated by teams that represent social and ecological perspectives in a process known as "scenario-consequence analysis." What has emerged from this study is a perspective that modest changes to water use in the agricultural areas around the Everglades would serve two purposes: first, to maintain appropriate water recharge to the Everglades, and second, to contribute to the sustainability of South Florida agriculture.

Perhaps even more importantly, when the different stakeholder groups were asked to make management decisions that would maximize their self interests they discovered trade-offs that they considered unacceptable. For example, when sugarcane growers maximized production with redirected water flow, they realized that their actions would cause severe damage to the Everglades but would not slow the expanding suburban development that posed the greatest threat to sugarcane lands. Scenario-consequence analysis did not indicate the best management decisions but it did create a more open forum in which stakeholders could more easily discuss goals and approaches. The case of the Everglades is developed in more detail by Lance Gunderson in Case Study 13.4.

Should Ecosystem Approaches Mimic Natural Processes?

Given that ecological systems are dynamic and disturbance driven on many different temporal and spatial scales, it seems reasonable that an ecosystem approach should incorporate natural disturbances into management regimes in an adaptive fashion. However, this is easier said than done. In practical terms it can be difficult to adequately define the historical frequency, magnitude, and extent of natural disturbances, and then more difficult to mimic them. In the absence of *precise* knowledge, *approximations* of natural disturbance regimes may be the best we can do. This is quite acceptable if we pursue such regimes in an adaptive manner with a willingness to continuously learn from our management experiments and improve our methods.

When, and in what manner, should ecological managers intervene to restore natural processes? This question assumes that the manager understands and can recognize for each of these processes their acceptable range of variation in the ecosystem and knows when these ranges have been exceeded. This level of understanding goes beyond just knowing something about historical patterns, by requiring a deeper understanding of when the extremes become so great that they will cause long-term, inappropriate changes to the ecosystem—that is, when the resiliency of the system is exceeded. For example, a forest manager may learn something about the historical frequency of fires by examining fire scars and tree rings. But it is probably more important for the manager to understand how frequently fires can occur before the ecosystem makes a fundamental shift to a different kind of ecosystem, as depicted in Holling's model (see Figure 13.6).

Historical information about the patterns of occurrence of natural disturbances such as fires, killing frosts, damaging storms, or floods helps to guide management decisions. However, we should keep in mind that mimicking historical patterns of disturbance may not produce the intended management results because contemporary ecosystems are usually embedded in landscapes that are very different from ancestral conditions. Fragmentation, isolation, and reduced area will all influence how the ecosystem responds to disturbance regimes, translocation of populations, invasion by nonnative species, or any other potentially disruptive process. Managers need to know the resiliency of their ecosystem to various kinds, intensities, and frequencies of perturbations.

To get this kind of information, managers may need to conduct experiments or analyze the results of management practices on rather large scales in an adaptive fashion. Examples include burning experimental forest blocks at different frequencies to find the threshold at which regeneration fails or invasive species increase, increasing the amplitude of water released from reservoirs to find the point at which downstream riffle habitats for fish spawning begin to degrade, or increasing stocking densities of native ungulates or cattle in desert grasslands to find the threshold at which cactus and other unpalatable plants begin to replace desert grasses.

Consider the dilemma for managers posed by the following historical analysis. The semiarid highlands where the states of New Mexico, Arizona, Colorado, and Utah meet are known as the "trans-Pecos" region. Truett (1996) asked the fundamental question, "Are current population densities of bison and elk representative of population densities in the region when indigenous cultures were dominant?" Using information such as composition of the animal remains in ancient Native American midden piles and early accounts of European explorers, he came to the conclusion that current densities are much higher than presettlement densities. Two factors seemed most likely to contribute to the higher densities found today. First, stock tanks and other impoundments for cattle have greatly increased the amount of available water in this semiarid region. Second, hunting pressure from Native Americans during the presettlement period imposed much greater mortality than current sport hunting practices.

Here then is the dilemma for managers in this region: If they take the position that a management goal is to recreate ecological systems as they existed before the landscape was modified by ranches, mining, and other development, they would have to remove artificial sources of water, reestablish migratory routes, and perhaps annually cull bison and elk. What they would achieve by this effort, of course, would not be re-creation of an ancient ecological system but rather a pale image of certain elements of ancient times. What then should the ecological manager do?

We do not suggest that the proper role of ecosystem management is to "freeze" a system in some arbitrary time period. In the real world, resource managers must balance many competing concerns. The human imprint on nature, in its myriad of forms and intensity, will always be present; this is not new. For thousands of years, humans have been major contributors to the form and function of most terrestrial and freshwater ecosystems and have been stakeholders in the benefits and costs of natural services that are supplied by those ecosystems. The oak savannas of California, oak woodlands in Europe, the open glades ("balds") on southern Appalachian mountain tops, inter-Andean valleys, African savannas, and salt marshes illustrate the range of ecosystems that have long histories of strong influence by people. The ecological manager cannot *precisely* imitate natural processes and re-create historically remote environments. But, at the other extreme, in this era of rampant population growth and development, neither can ecological managers responsibly take a laissez-faire approach to become mere spectators to the continuing erosion of biodiversity and the loss of our natural heritage.

In Box 13.1 we use fire, a common tool for terrestrial ecosystem management in the U.S., to explore how man-agers might seek a rational balance between mimicking natural processes and meeting immediate management concerns.

The Critical Role of Participatory Decision-Making Processes

In the U.S., natural resource agencies have traditionally kept management planning within the agency (Cortner and Moote 1999). When people outside the agency were consulted they usually fell into one of two classes: Scientists were contracted to answer narrowly defined technical questions, or the ranchers, miners, and loggers who held public land leases were consulted about possible economic consequences to proposed changes in policy.

Through their requirements for more frequent and timely public hearings, the U.S. National Environmental Protection Act, Endangered Species Act, and other environmental legislation opened up the process of decision-making in public resource agencies to greater public participation. Until expansion of ecosystem management and other integrated approaches, public participation in the U.S. was largely limited to commentary during public hearings. In part, this led to a widespread feeling among private landowners that environmental protection unfairly restricts private property rights, resulting in an erosion of public support for environmental protections. Because a major tenet of an ecosystem approach is that all stakeholders should participate in decision-making to engender long-term public support for conservation, where these approaches are being applied, integration of stakeholders has increased (Brush et al. 2000; Wondolleck and Yaffee 2000).

Principal stakeholders are those people whose livelihoods, residences, or financial or other interests are connected to the management area, as well as those institutions—public and private—with activities in the management area. Principal stakeholders should play major roles in the development, implementation, monitoring, and evaluation of the ecosystem management plan. Participation by minor stakeholders is best served through forums for public commentary.

A contradiction frequently exists when stakeholders are brought into decision making. Individuals often make decisions about resource use that are based on short-term and self-interest perspectives. Yet, the same stakeholders, under other circumstances, may make decisions that are long term, community based, and which favor sustainability of the resource. For participatory decision making to contribute to ecosystem sustainability, mechanisms need to be found to balance self interest with the overriding need for resource sustainability. For example, community education and local project involvement to reduce

BOX 13.1 Using Fire as a Natural Process in Ecosystem Management

C. Ronald Carroll, *University of Georgia*

Fire is both a naturally occurring agent of disturbance and a management tool used to reduce risk to property damage and to manipulate habitat pattern. Fires that are intentionally set and controlled by managers (prescribed burns) represent a common practice for terrestrial ecosystem management. Prescribed burns reduce the likelihood of catastrophic wildfires by reducing the amount of accumulated "fuel" in the form of woody debris, dry brushy undergrowth, and other highly flammable vegetation. Prescribed burning is also used to create habitat, to generate spatial heterogeneity, and to remove undesirable vegetation, among other uses. Although budget, time, and personnel limitations often constrain management practices in the "real world," it is important to reflect on what should be done in a more "ideal world." There is room for more creative approaches to the use of fire as a management tool.

Let's first explore what information would be required to use prescribed burning as a natural process and then explore when it may be appropriate to depart from using fire as a natural historical process. We would want to know, over some historic time period, the answer to several questions. How frequently, and during which parts of the annual cycle did fires occur? Were the lengths of intervals between fires normally distributed or did years with fires tend to be clustered or overly dispersed? Were the fires "hot,"—that is, did they leave cleared mineralized land and volatilize large amounts of soil nitrogen—or were the fires "cool," leaving patches of vegetation and downed logs untouched? Were the fires localized or widespread? Did the fires cover a simple or a complex topography such that postfire vegetation and animal population responses might vary? Can we burn in the historically appropriate season? Finally, we would need an answer to the critically important question: If we imposed a historically accurate fire regime, would the ecosystem respond now in the same manner

as it did in the past?

Scars of old fires, preserved in the annual growth rings of trees, allow the occurrence of fires to be dated through dendrochronology, that is, by counting annual tree rings from the present back to the location of the fire scar. Although the dating of a particular scar in a tree is straightforward, determining the temporal distribution of fire intervals, the areal extent of fires, whether the fire was "hot" or "cool," and differences in these patterns between sites requires a rigorous analytical approach. Suppose, for example, a manager wanted to know the distribution of fire intervals in two different management areas to implement a prescribed burning regime that followed the "natural" historical pattern. Assuming that the choice for the historical period of reference could be logically defended, the ecologically minded manager might reasonably ask several questions. How often did fires occur? Is the mean interval between fires an adequate estimator of fire frequency for management purposes or should variance and temporal trends of the interval lengths also be considered? Were short interval, patchy burns the norm? The manager might also want to know if the historical patterns differed between the two management areas.

The behavior of naturally occurring fires is not only a function of the interval length since the previous fire, but also influenced by atmospheric conditions such as wind and humidity. Yet, the atmospheric conditions of high wind and low humidity that enhance large hot fires are not the conditions a cautious manager would choose to conduct a prescribed burn. Consequently, prescribed fires generally result in patchy burns that may imitate some kinds of natural burns but would not, for example, imitate the effects of a hot fire that volatilizes large amounts of nitrogen, sterilizes seed beds, and removes all of the accumulated coarse woody debris.

We are left with important questions that trouble all ecological man-

agers who are attempting to recreate historical patterns of disturbance, whether by fire, flood, or any other agent. How close does the approximation need to be? What indicators should be used to monitor success or failure of the disturbance regime? Is it better to mimic the historical fire regime or to impose any fire regime that maintains the contemporary ecosystem in a desired state? Because of extensive changes in ecosystems, mimicking historic natural disturbance patterns and processes can cause an erosion of the contemporary native biodiversity of the ecosystem, enhance the establishment and spread of non-native species, or reduce the inherent fertility of the soil. Given these undesirable outcomes the manager is likely to implement a different, albeit more artificial, disturbance regime to maintain the ecosystem. This is an ecologically rational response because it recognizes that historically accurate disturbance patterns and processes might exceed the resiliency of contemporary ecosystems.

These concerns become especially important in the intensive management that is often required of small-sized protected areas. For example, if historical patterns of fire intervals and average areal extent of burns were used to establish prescribed burning regimes in small reserves, much of the reserve might be represented by a single stage in postfire recovery, thus reducing habitat heterogeneity.

Where do all these considerations leave us with regard to the development of fire management? First, historical records of fire frequency are a useful "first cut" for developing prescribed burn plans (see Lesica 1996 for a successful example of the use of fire history to establish prescribed burns). The next step would involve evaluation of any other concerns that might preclude the use of historic fire patterns. This evaluation phase will, of course, require significant involvement of stakeholders in the management area. These discussions should be framed in such a way that they can lead to tenta-

tive fire management plans that can be implemented, tested, and modified in the context of adaptive management. This is not an easy process. Because fire can affect large areas and because there is always some risk that fires will escape containment, the public is wary of the deliberate use of fire for management.

Although fire is one of the most important management tools in terrestrial ecosystems, we may soon need to find alternatives. This is well illustrated in management plans for the longleaf pine–wiregrass savanna of the U.S. southeastern coastal plain, which once extended nearly unbroken from Virginia to east Texas. Natural fires from frequent lightning strikes maintained the savanna by preventing hardwood encroachment and facilitating reproduction of pines, grasses, and forbs. Today, the savanna exists only in isolated remnant patches and must be maintained through frequent prescribed burns. Fire intervals of 2–3 years are typically needed to maintain the savanna; however, these short fire intervals make the savanna vulnerable to two important invaders. The imported red fire ant (*Solenopsis invicta*) has invaded the savanna and is imposing significant mortality on ground and shrub-nesting vertebrates. The fire ant disappears when fire intervals are greater than five years, but these long intervals encourage hardwood invasion and conversion of the savanna into thickets (Carroll and Hoffman 2000). Of even greater concern, the highly invasive cogon grass (*Imperata cylindrica*) is well adapted to frequent fires on low-fertility soils. Thus, the devil's dilemma is that we burn the savanna frequently to prevent hardwood conversion but by burning frequently we greatly increase the risk of converting the savanna to cogon grass.

Here is one promising line of research. Cogon grass has well-developed storage rhizomes. Thus, when the tillers are burned, photosynthate can be withdrawn from the rhizomes for vigorous regrowth. Wiregrass lacks storage rhizomes and recovers more slowly from fires. So, the critical question becomes: How can we shift the advantage to wire grass? Sugarcane growers may hold the answer. They manipulate the reallocation of sugar from rhizomes to tillers (the cane) prior to harvest by applying very dilute solutions of glyphosate, a chemical that at higher concentrations is an important herbicide. Dilute glyphosate can force cogon grass to translocate its stored photosynthate into tillers prior to imposing prescribed burns, thus making it vulnerable to destruction during a fire. The take home message is that we need to be thinking more creatively about the use of prescribed burns and to find ecologically and socially acceptable alternatives to fire as a primary management tool.

---

mistrust can minimize self-interest decisions and contribute to what Crance and Draper (1996) call "socially cooperative choices." An example of an applied, stakeholder-inclusion approach is offered in Box 13.2.

Effective ecosystem approaches involve a willingness to give up some degree of control. Much of the history of natural resource management is a history of control and domination: of people, of land, of resources, of other organizations. This is antithetical to good ecosystem management, wherein cooperation, consensus, and inclusion are surer roads to success. Political boundaries on a map are meaningless in an ecosystem context, but are often treated by institutions as territories to be jealously guarded (Knight and Meffe 1997). Thus, in the traditional view of natural resource management, national forests are seen as different and isolated from national parks, which have no relation to military lands, which are separate from private lands. In reality, all of these lands are interconnected, and activities on one will have consequences for others. Institutional territoriality, which acts to "control" a designated piece of a map, does nothing to integrate across the landscape in an ecosystem fashion, but merely perpetuates a closed, command-and-control mentality that is ultimately to the detriment of land health or larger human prosperity.

In addition to giving up some control, institutions pursuing effective ecosystem approaches must be willing to accept responsibilities and break down internal barriers to the flow of information. Another way of stating this is that institutions must be willing and able to learn and adapt (Lee 1993). An example of the types of changes that must be incorporated into institutional "psyches" was graphically presented by Westrum (1994) and is reviewed in Table 13.3.

An ecosystem approach requires a greater degree of partnerships among stakeholders, including international collaboration, inter-agency cooperation at all levels of government, private citizen advocacy groups, research scientists, business interests, and nongovernmental organizations. Because managers have been severely constrained by budgets that are barely adequate to support basic programmatic needs, much less allow them to respond to new innovative proposals, it is through broad partnerships that consensus on complex issues may be reached while significant new sources of financial support for management can be developed. It is worth noting that broad-based financial support has two important consequences for the manager that go beyond just increasing the size of the budget. First, risk of losing financial support is spread among the various sources of funds, thereby reducing dependency on government appropriations. Second, managers may be freer to develop innovative programs when they are less dependent on restrictive funds from state and federal sources.

BOX 13.2 Natural Community Conservation Planning

Michael O'Connell, *The Irvine Company*

Land planning and zoning are commonly employed to regulate land uses that range from single unit family dwellings on large lots to apartment houses and commercial uses such as shopping malls and industrial complexes. Although land planning as a means for influencing development has a long history, it has only recently been applied to resolve conflicts among advocates for conservation of natural areas, public agencies charged with protecting species or critical habitat, private landowners, and commercial land developers. Some individual private landowners, of course, have willingly foregone a portion of the potential property market value through conservation easements or by joining land trusts to provide long-term conservation protection for their land. But community-level conservation planning is a new approach, albeit with some significant problems to be resolved.

This process, known as natural community conservation planning, or NCCP, is an application of conservation biology principles to local land-use planning within the social, political, and legal context of land use (Reid and Murphy 1995). It emphasizes broad habitat conservation rather than single species-oriented conservation, and allows planned, controlled development while protecting natural areas. An example from Orange County and surrounding areas in southern California that has been featured in the news media is the plan to protect the coastal scrub habitat of a federally protected bird, the California Gnatcatcher (*Polioptila californica*). Among the major players in this NCCP are federal and state agencies that are charged with enforcement of the Endangered Species Act, county and municipal land planning and fire control agencies, The Irvine Company as the major landowner and developer in the region, The Nature Conservancy and various citizen groups.

The uncertainty that potential enforcement of the Endangered Species Act introduced into the development market provided a major incentive to bring pro-development players into the negotiating process. Although the process has not been easy, it was facilitated by the presence of the single major land developer, The Irvine Company, which has a history of accommodating conservation goals, such as through wetland mitigation processes. When completed, the conservation plan will set aside protected critical coastal scrub habitat and allow development to occur outside the protected area without fear of future intervention from enforcement of the Endangered Species Act.

This is an innovative approach for meeting conservation objectives in areas that are under threat of development. There are, however, three aspects that remain problematic. First, it is uncertain whether the negotiation process and successful land planning will occur when many different land developers are involved. Second, this approach may be limited to habitats that contain federal or state protected species, and may be ineffective without the associated litigation threat. Third, the conservation plan "freezes" the habitat of the species to what is circumscribed in the plan. This violates an essential feature of adaptive management, that is, the need to revise and modify protection plans. However, in regions where develop pressures are intense, such community-based conservation land use plans may be the only realistic option for protecting biodiversity.

TABLE 13.3 *A Stereotyped Presentation of Three Types of Organizations and How They Handle Information and Responsibilities*

Pathological organization	Bureaucratic organization	Generative, adaptive, or progressive organization
The organization doesn't want to learn new information	The organization may not learn new information	The organization actively seeks information
Messengers are "shot" (i.e., treated poorly)	Messengers are listened to— if they arrive	Messengers are trained
Responsibility is shirked	Responsibility is compartmentalized	Responsibility is shared
Bridging is discouraged	Bridging is allowed but neglected	Bridging is rewarded
Failure is punished or covered up	Organization is just and merciful	Failure results in learning and redirection
New ideas are crushed	New ideas present problems	New ideas are sought and welcome

Source: Modified from Westrum 1994.

Ecosystem-based conservation approaches not only recognize the importance of highly participatory decision making, they also identify people as integral parts of the ecosystem. This recognition and framing of people within ecosystems for conservation planning arose from several considerations. First, through the sheer size of the human population coupled with our material desires and needs, we have a pervasive influence on virtually all ecosystems; no culture, lifestyle, or political system is completely exempt from this influence. Second, for reasons that are ethical or religious, we have a responsibility to act as stewards to counter the potential degradative consequences that our economic lives have on natural ecosystems. Third, we cannot understand the dynamics of ecosystem functions, or the variation in ecosystem structure, without explicitly including people, in a sense as a "keystone species."

Future Directions in Ecosystem-Based Conservation

Ecosystem approaches have attracted a great deal of attention because they provide a useful framework for shaping conservation and management actions. The challenge before us is to translate these ideals into practice. This means we need to conduct a great deal of research, work intensively with affected stakeholders to identify goals and means to achieve them, and work in an adaptive fashion while recognizing many sources of uncertainty.

Scientifically, we need to accomplish a great deal, including the following:

1. Develop clear hypotheses about the nature of structuring factors in ecosystems and how management of those factors affects community structure and the outcomes for human social and economic welfare.

2. Explicitly model ecosystem-human interactions and test management with such models.

3. Execute the best in adaptive management in which we set up monitoring and management as experiments as much as possible via instituting controls, and pre- and post-management data collection.

4. Explicitly confront uncertainty through sensitivity analyses, and "what if" discussions.

These same challenges confront conservation at any scale, but are yet larger in scope and level of complexity in ecosystem approaches.

Ecosystem approaches do not replace species and landscape approaches to conservation, but rather are complementary to these. As Hilborn (2004) stated, it is not that single-species approaches fail for their science, but that they fail to recognize the central importance of people and to manage people. Much of the hope for these approaches comes from the shift in framework to placing people at the center of analysis and implementation. By working together with those most affected by ecosystem change, we surely are more likely to find common ground and agree upon collective action for long-term goals than if policies are imposed from above.

Socially, economically, and politically, we have the challenge of finding means to meet human needs and aspirations sustainably, and to come to an accord where sustainable use of ecosystem resources is a shared value. We need also to shift from short-term planning horizons that are politically feasible to inclusion of a long-term policy perspective that guides actions. We will see in the near future how these approaches are working to create and maintain local involvement, and achievement of social and conservation goals.

CASE STUDY 13.1

Ecosystem Management on "People's Land" in the United States

Mike Dombeck, University of Wisconsin, Stevens Point; Chris Wood, Trout Unlimited; Brian Kermath, University of Wisconsin, Stevens Point

Citizens of the Unites States collectively own 78 million ha of national forests managed by the U.S. Forest Service and 110 million ha of public lands managed by the Bureau of Land Management (BLM). These lands, often referred to as the "people's lands," contain remarkable and abundant cultural, archaeological, historical, and natural resources. Decisions made today about their management will determine whether or not tomorrow's generations will benefit from them.

A century ago, when the national forests were first designated, the mandate was to protect the forests, provide a sustainable supply of timber, and secure favorable conditions for water flows. This was in response to the "cut-and-run" era of

timber harvests that left the United States with 80 million acres of denuded "cutovers," mostly in the East and Midwest. Slash fires, raging flash floods, and soil erosion devastated huge portions of the landscape following the clear-cuts. Similarly, by 1900 the public lands now managed by the BLM, mostly in the arid west, suffered from the "tragedy of the commons" and were overgrazed and badly degraded.

Historically, the agencies managed these lands to meet a variety of needs. The national forests are often referred to collectively as the "land of many uses" and BLM lands, the "working lands." The BLM and the Forest Service have a similar multiple-use legal mandate, which includes commodity production and recreation and other values. However, multiple-use management is not easy and has always been subject to interpretation and controversy, as social values and needs are constantly changing. Politics, economics, the law, interest group pressures, and court decisions further complicate multiple use management decisions.

The multiple-use mandate has served the country for many decades. Extractive products including livestock, timber, fossil fuels, and minerals that flow from these lands have contributed to the nation's development, wealth, and employment base, especially in many rural areas. And, tens of millions of Americans now visit public lands for recreation, education, and inspiration.

But times are constantly changing. We can no longer view public lands as vast stretches of open space that exist to supply the nation with timber, minerals, and oil. With the U.S. population approaching 300 million, we have roughly four times as many people as a century ago, and per capita resource consumption is greater now than ever. Sprawl and developed land are growing at an increasing rate, now consuming nearly 3642 ha a day. Within 30 years, the urban footprint of the U.S. will double in size to become equal to the combined area of its national forests.

The economy has also changed. Although extractive products still flow from the public lands, recreation has become the economic mainstay. Of the $145 billion generated annually from national forests, for example, timber accounts for less than 3%, whereas recreation is responsible for more than 75% and supports nearly 3 million jobs. Economists are also beginning to assign values to goods and services that historically were not accounted for. We now know, for example, that the water flowing through our national forests is worth nearly $4 billion annually (Sedell et al. 2000), and its quality and quantity—and therefore value—is optimized in the most naturally functioning forests.

At the same time, the land's value goes way beyond its production of goods and services—values that clearly need to be considered by managers. Take endangered species for example. Of the 214 stocks of Pacific salmon at risk of extinction (Nehlsen et al. 1991), BLM and Forest Service managed lands contain habitat for 109 and 134 stocks, respectively (Williams and Williams 1997). To protect these and other species, as well as the genetic diversity that keeps populations fit and the evolutionary potential on which future biodiversity depends, we must manage our lands in the context of the larger ecosystems in which they occur.

In response to ever-changing circumstances and demands land management agencies began shifting away from the narrowly targeted management of commodities and individual species to a broader, landscape-wide, ecosystem approach. Although its antecedents are difficult to pinpoint, the Forest Service first used the expression "ecosystem management" as official policy in 1992.

The intention of ecosystem management is usually not to return the land to some pristine state, but rather to manage the land in a condition that will best meet society's broadly-defined needs for the present and into the future. The success of ecosystem management depends on the application of the best science available to assess the land's potential to meet those needs—including forest goods and services, watershed function and water supply, and biodiversity in all its forms.

Principles of Ecosystem Management

Regardless of who owns or manages the land, or how big the land unit is, there are several essential operating principles of ecosystem management (modified from USDI BLM 1994) including:

1. Maintain the ecological integrity of the management area and all affected lands.

2. Gather and use the best available scientific information in making decisions and remember that our understanding of ecosystems is greater today than when our system of public lands was first established, our understanding of ecological thresholds is limited, we cannot always predict the outcomes of our actions, and that our understanding of ecology continues to advance.

3. Involve the public in the planning process and coordinate with other public agencies (federal, state, county, municipal), nongovernmental organizations, private landowners, and commercial interests.

4. Determine desired future conditions based on historic, natural, economic, and social considerations and on society's larger responsibilities of guaranteeing resources for the future as well as providing goods and services for the present.

5. Minimize the negative impacts to the land that result from extractive activities, eliminate unnecessary activities that degrade the managed land and other lands that might be affected, and restore previously disturbed land to the extent possible. This includes reconnecting isolated and fragmented parts of the landscape with corridors, greenways, and buffers.

6. Base planning and management on distant time horizons and goals with measurable short- and long-term targets.

7. Adopt holistic approaches that cross discipline boundaries and synthesize knowledge.

8. Practice adaptive management. Be flexible and willing to adjust to new information, the ever-changing natural and human landscapes, and the ever-changing human and conservation needs.

9. Continually update and disseminate information that is useful to all stakeholders.

The first principle is key to the approach's success and it brings focus, meaning, and direction to the others. The overriding priority should be to restore and maintain biodiversity in all its incarnations—as we believe that is the best way to guarantee the land's productivity. At its root, ecosystem management seeks to safeguard long-term ecological processes in the provision of goods and services, while also broadening what constitutes "goods and services."

The approach recognizes that ecological processes and our use and abuse of the land occur across varied geographic and temporal scales and that the environmental effects of human activities often have lag periods that can be several human generations. Ecosystem management attempts to operate appropriately.

Ecosystem management also calls for land managers to manage the land in the context of larger ecosystems and distant time horizons. In the words of John Muir (1869), "When we try to pick anything out by itself, we find it hitched to everything else . . . " Because social spaces rarely coincide with natural ecological patterns, managers cannot overlook the connections between the lands under their charge and the adjoining lands.

Finally, land managers must consider that our public lands alone cannot safeguard biodiversity, meet our natural resource commodity needs, and provide the ecosystem services we depend on for the long haul. It is therefore crucial that private property owners are enlisted to help. In the end, the integrity of our protected areas will depend not only on how our public lands are managed, but also on what we're doing on the lands that surround and otherwise affect them. The outcome of well-orchestrated management should be a fine-grained mosaic of sustainable practices across the landscape.

Land Management Advisory Groups

Given the increasingly difficult challenges of resource management, land managers must find effective approaches to public land management. One way, and perhaps the only way, is through an adaptive process of intelligent shared visioning. To this end, the federal land management agencies began forming partnerships with state and local governments, private landowners, nongovernmental organizations, and other public land users to ensure local involvement in the planning and management process. Even before the term "ecosystem management" was used, public–private partnerships began taking shape in the late 1980s, especially in the West.

The western states became the focus of many of the early efforts because they contain most of the public lands and the nation's fastest-growing populations with large proportions in urban areas. Conflicts in the West were multiplying and the political climate of resource management was becoming increasingly charged. City dwellers and suburbanites generally appreciate public lands for their aesthetic, recreational, and spiritual values, whereas rural people historically have depended on these lands for their extractive potential. Of course, these partnerships are challenging to form and maintain, particularly where stakeholder interests are quite divergent.

Examples of Collaborative Management

The Henry's Fork Watershed Council was organized in 1993 to address land management issues, coordinate research, and help write policy to mange 688,000 ha in the Henry's Fork Basin of eastern Idaho and western Wyoming. The council is made up of scientists, federal, state, and local agencies, nonprofit organizations, and citizens, and functions as a collaborative forum designed to overcome previously adversarial relationships. To date, the council has completed research on the historical relationship between cutthroat and the nonnative trout species; conducted habitat and population surveys on over 483 km of stream; identified potential habitat improvement projects; launched an awareness campaign that has improved local understanding of the issues and has helped facilitate communication; developed an informative brochure and interpretive signs and posters; restored portions of the stream and its tributaries; and has banned new dams, diversions, and hydroelectric projects on 314 km of the Henry's Fork and its tributaries. More information on the council can be found at http://www.snre.umich.edu/emi/cases/henrysfork/.

The Owl Mountain Partnership (OMP) is a collaboration of multiple federal, state, county, and private ranch representatives to jointly manage 404,688 ha (originally 97,125 ha) of mixed-ownership land. The integrated partnership was formed to serve the economic, cultural, and social needs of the community while developing adaptive, long-term landscape management programs, policies, and practices that ensure ecosystem sustainability. Initially a project to solve livestock–wildlife conflicts, the partnership has protected and improved resources across the area including habitat for waterfowl, upland wildlife, big game, and fish, in addition to protecting livestock and crops from wildlife. The partnership has also produced vegetation, wildlife, and water quality databases to help assess rangeland health. Between 1993 and 2001, the OMP had completed about 120 projects. More information on the partnership can be found at http://www.northpark.org/owlmtn/.

The Coalition for Unified Recreation in the Eastern Sierra formed in 1991 to preserve the eastern Sierra's natural, cultural, and economic resources and enrich visitor and resident satisfaction in the area. If we agree that the ecosystem management approach is the best approach to deliver the best

long-term prognosis, then the greatest challenge to success may rest with overcoming the obstacles to effective communication and forming joint resolutions between diverse, and often divergent interest groups. This project has been successful in building bridges between such groups. Environmentalists have suggested that one of the project's greatest accomplishments has been in getting business and environmental interests to sit at the same table and communicate. More information on the coalition can be found at http://www.snre.umich.edu/emi/cases/cures/.

The Elkhorns Cooperative Management Area covers approximately 101,170 ha of mountainous grasslands and forests some 26 km southeast of Helena, Montana. The BLM, Forest Service, Fish and Wildlife Service, and the Montana Department of Fish, Wildlife, and Parks signed an agreement in 1992, to jointly manage the area. Although cooperation has been difficult at times, owing to "turf battles" and personal disagreements, the unit continues to operate toward effective management of the land (Thomas et al. 2002). More information can be found at http://www.snre.umich.edu/emi/cases/elkhorn/.

The Trout Creek Mountains Working Group formed in 1988 in response to overgrazing of the high desert country of southeastern Oregon with the goal of improving watershed health. In 1991 grazing on 211,650 ha of public lands faced a shutdown when the Lahontan cutthroat trout (*Oncorhynchus clarki henshawi*), a federally listed threatened species, was discovered in Willow and Whitehorse creeks. The working group, comprising ranchers, environmentalists, and state and federal agencies, began a dialogue and, using a collaborative process, searched for common goals and avoided costly litigation. Grazing, which had been nonstop for decades, was put on a "rest/rotation" program that helped woody vegetation return, native trout populations to rebound, and riparian areas and water quality to improve. Today some of the healthiest populations of Lahontan cutthroat trout are found in the area while ranchers still grazing their cattle. More information about the group can be found at http://www.sustainablenorthwest.org/founders/pdf/tcmwg.pdf.

Although we will not be able to fully assess the successfulness of ecosystem management for some time, these examples illustrate that positive outcomes are attainable if stakeholders are willing to cooperate. They also demonstrate the complex nature of assembling diverse interest groups to deliver meaningful plans. All the technical expertise in the world cannot overcome public disinterest in, or worse, distrust of the forces behind conservation and restoration.

The examples also illustrate that as a collaborative process, ecosystem management's first logistical priority must be to work with all stakeholders, including businesses, government agencies, municipalities, academia, and nongovernmental organizations to develop shared goals for healthy ecosystems. Developing "ecological sideboards" or thresholds that measure ecosystem health provides managers and citizens with performance measures. As eminent wildlife ecologist Aldo Leopold noted, "the oldest task in human history is to live on a piece of land without spoiling it."

Too often, federal agencies allow operating procedures to complicate relatively straightforward missions and, consequently, public acceptance and effectiveness. Resource management agencies cannot become complacent. Ecosystem management can and does work when resource professionals spend more time on the land with local interests, community leaders, user and conservation groups, state officials, and others, communicating ideas, educating people, and building community support for healthy, diverse, and productive ecosystems.

The challenge for the collaborative process is not to model or predict the annual flow of wood fiber, forage, or energy produced from public lands, nor is it to anticipate shifting political winds and act accordingly. The role of the resource manager is to state what is in the best long-term interest of the land, and what services the land can provide in an ecologically sustainable manner based on the best available science. Elected officials and their staffs often shape policies to fit political agendas. As controversies over the Northern Spotted Owl in the Pacific Northwest, the recent efforts to drill in the Arctic National Wildlife Refuge, and the exploitation of coal-bed methane energy resources in the Rocky Mountains demonstrate, in the end it is the citizens that own these lands who speak loudest and last. Every citizen has the ability, right, and duty, to tell elected officials and agency personnel how to manage their lands—and if that fails, to seek recourse in the court system or through the ballot box.

Management of the public lands requires keeping an eye to the future. What are the values that make our network of public lands so unique? What will we want from these lands in 50 and 100 years and beyond? If public land managers and the people they serve are not continually asking these questions, they need not worry about the answers because odds are the public land network we now enjoy will cease to exist, either through privatization, dismemberment, or misuse. This long-term perspective is often confounded by election cycles of 2–6 years that impel some political leaders to demand more from the public lands and waters than they can sustain over the long term—the values and characteristics of public lands that will provide for what Pinchot called, "the greatest good for the greatest number for the longest time."

CASE STUDY 13.2

Coral Bleaching
Managing for Resilience in a Changing World

Jordan M. West*, U.S. Environmental Protection Agency; Paul A. Marshall, Great Barrier Reef Marine Park Authority; Rodney V. Salm, The Nature Conservancy; Heidi Z. Schuttenberg*, National Oceanic and Atmospheric Administration

Over the last two decades, significant increases in the incidence and severity of temperature-related, mass coral bleaching have raised questions in research, management, and policy forums about how to respond to these events. From a management perspective, mass coral bleaching poses a unique threat to coral reefs in that its root cause—warmer temperatures—is beyond the control of local reef managers. Nevertheless, pragmatic ideas are emerging as to how to boost the long-term resilience of coral reefs in the face of anomalously high temperatures associated with climate variability and change. In particular, managers have concluded that it is now more important than ever to consider the cumulative impacts of simultaneous interacting stresses on coral reefs, and that practical options exist for maximizing coral resistance to bleaching, coral survival after bleaching, and reef recovery after bleaching-related mortalities have occurred.

Coral Bleaching as an Emergent Issue

Coral reefs are renowned for their high productivity (Moberg and Folke 1999). Although coral reefs cover less than 1% of Earth's surface, they produce complex, limestone structures that are home to one-quarter of all marine life (Spalding et al. 2001). The ability of coral reefs to support such biological diversity depends largely upon the symbiotic relationship between corals and the microscopic dinoflagellate algae (zooxanthellae) that live in their tissues. Corals are strongly dependent on their zooxanthellae, which provide the corals with up to 90% of their energy requirements (Muscatine 1990). However, stressful conditions can cause this relationship to break down, resulting in dramatic decreases in zooxanthellae densities within the coral tissues. Because zooxanthellae also provide much of the color to coral tissues, their loss leaves the tissue transparent, revealing the white coral skeleton and giving the coral the appearance of having been "bleached."

Bleaching events may occur at local scales due to a variety of stressors, including storms, disease, sedimentation, pollution, and salinity changes (Brown 1997). Mass bleaching, however, affects reefs at regional to global scales and is caused by increased sea temperatures associated with climate events such as El Niño Southern Oscillation (ENSO) cycles (Wilkinson 2002; Buddemeier et al. 2004). Temperature increases of 1°C–2°C above the long-term average maximum are all that is required to trigger mass bleaching; both the magnitude and duration of the temperature anomaly are important, with higher temperatures requiring shorter exposure times (Hoegh-Guldberg 1999; Coles and Brown 2003). While temperature is the trigger, light also interacts to influence the severity of bleaching (Jones et al. 1998), so the degree to which corals are shaded from light exposure can also be important.

Bleached corals are still living, and if stressful conditions subside soon enough, they can regenerate their zooxanthellae populations and survive the bleaching event. However, even corals that survive experience reduced growth rates (Goreau and Macfarlane 1990), decreased reproductive capacity (Ward et al. 2000), and increased susceptibility to disease (Harvell et al. 1999). If bleaching stresses are severe or persistent, corals eventually die. In extreme cases, bleaching events cause catastrophic mortality of corals; for example, the mass coral bleaching event associated with the 1997–1998 ENSO resulted in extensive bleaching of reefs around the world, with 16% mortality globally (Wilkinson 1998; Wilkinson 2000).

Resilience

While mass bleaching events occur at large scales, the bleaching response itself can be extremely patchy at local scales. Even in the severest cases, scattered colonies and patches of reef survive (Westmacott et al. 2000; West and Salm 2003) because there is considerable variation in the microenvironments of coral colonies, creating differences in exposure to heat, light or other stressors (West and Salm 2003). There are also intrinsic differences in susceptibility due to variation in the genetic compositions of corals and their zooxanthellae (Rowan et al. 1997). With an understanding of what factors generate such variability in bleaching susceptibility, it should be possible to identify which coral reefs have the greatest natural resilience to coral bleaching—and to make use of this information to design management strategies that can best protect coral reefs over the long term.

Ecosystem resilience relates to the speed of return to equilibrium after a disturbance and the magnitude of disturbance that can be absorbed by a system before it shifts from one sta-

---

* The views expressed are the authors' own and do not reflect official policies of the U.S. Environmental Protection Agency or the National Oceanic and Atmospheric Administration.

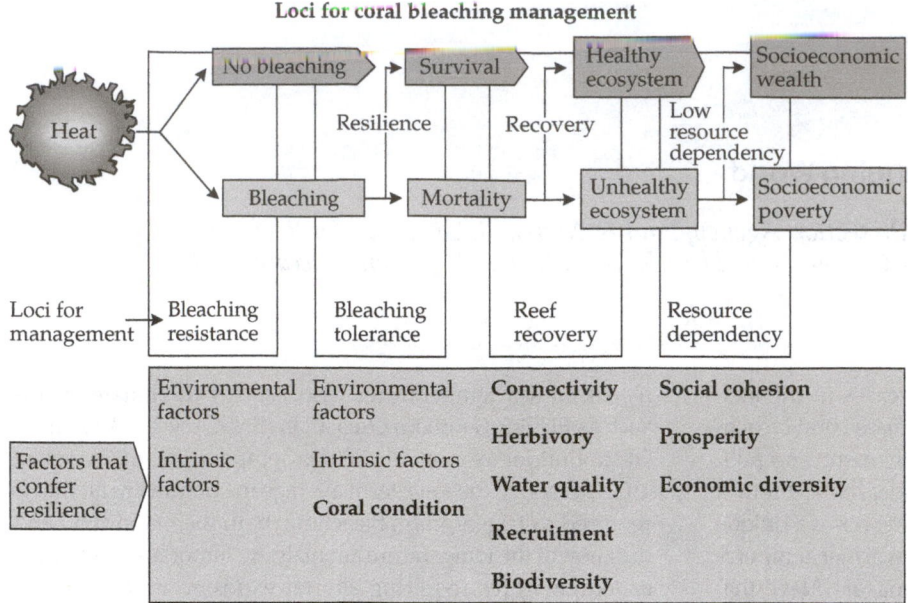

Figure A Management can aim to build resilience to mass coral bleaching in four key areas: bleaching resistance, bleaching tolerance, reef recovery, and reef resource dependency by humans. Factors that confer resilience are listed in the diagram; those shown in bold can be directly promoted through management interventions while the rest can be utilized in management interventions even though they cannot be directly controlled or enhanced. (Modified from Obura 2003.)

ble state to another, such as from a coral-dominated to an algal-dominated community (Gunderson 2000; Nyström et al. 2000). During a mass coral bleaching event, a cascading sequence of impacts to corals and coral reef ecosystems can occur, and these impacts may ultimately result in negative socioeconomic consequences for the humans who depend on reef resources. Figure A describes this sequence and identifies opportunities for management interventions to minimize impacts at each stage. First, bleaching resistance refers to whether or not corals bleach when exposed to heat stress. Second, if bleaching does occur, then bleaching tolerance determines whether corals survive or die during the bleaching event. If coral mortality does occur, then reef recovery depends on the ability of the reef system to bounce back and maintain ecosystem processes. Finally, if long-term impacts to coral reef health occur, then degree of reef resource dependency by humans determines how the ecological degradation affects the well being of societies that depend on reef resources.

In a mass bleaching context, then, resilience will depend on the ability of corals to resist or survive bleaching, or the ability of reefs to recover after mortalities have occurred. The capacity of coral reefs to resist and recover from disturbances will become increasingly important if the frequency and severity of bleaching events continue to increase as a result of climate change.

Identifying Resilient Reefs

Based on evidence from the literature and systematically compiled observations from researchers in the field, a number of environmental and biological factors that correlate with resilience to coral bleaching have been identified (West and Salm 2003). Reef scientists and managers are now exploring ways to include such considerations in coral reef management plan-

ning. The Nature Conservancy has since expanded on this information considerably to develop a "R^2-Reef Resilience Toolkit" (The Nature Conservancy and Partners 2004) that provides detailed guidance and data sources for gathering information about resilience factors and for adapting management principles accordingly. Some general categories of reef resilience factors are reviewed here:

- *Cool Water.* Some sites may have consistently cooler water due to upwelling or proximity to deep water. Local bathymetry, ocean and tidal currents, and prevailing winds may all play an important role in reducing the temperature of water bathing a reef. However, some researchers have suggested that ocean currents may not be a reliable source of long-term resilience because climate change may result in new current patterns.

- *Shading.* Some reefs may also be protected from bleaching stress by shading. The role of sunlight in the bleaching process means that corals sheltered from direct sunlight by topographic or bathymetric features are at reduced risk of bleaching. While many reef areas are unlikely to be associated with features that can provide shade, reef complexes built around steep-sided limestone or volcanic islands, such as in Palau, Indonesia, and the Philippines, may have many shaded sites.

- *Screening.* Unnatural levels of sediments and excessive phytoplankton growth from nutrient enrichment can stress and kill corals. However, naturally turbid conditions may provide a measure of protection for corals exposed to anomalously warm water. Ongoing research suggests that organic matter in turbid areas may absorb UV wavelengths with the result that corals at these sites may be less susceptible to bleaching. However, turbid

conditions are often suboptimal for coral reef development, and biological diversity may be low in these areas.

- *Resistant and Tolerant Coral Communities.* Knowledge about the composition of coral communities can also help predict sites that are more resilient to bleaching. Observations during past coral bleaching events from around the world indicate that certain types of corals are generally more resistant to bleaching than others (McClanahan et al. 2003). If a site is dominated by resistant species, then it is less likely to bleach in response to thermal stress. Similarly, certain corals appear to be able to tolerate a bleached state for an extended period and are therefore less likely to die even if they bleach. While less work has been done on bleaching tolerance, it appears that corals with a massive morphology and thick tissue, such as those from the families Poritidae, Favidae, and Mussidae, have greater tolerance to bleaching.

- *High Recovery Rates.* The ability to recover after a disturbance is another characteristic of coral reef resilience to mass bleaching. Sites that showed strong recovery from previous disturbances, such as storms, are likely to recover more quickly from bleaching events. Where recovery rates are not known, managers can infer a site's capacity for recovery by evaluating whether conditions are conducive to coral recruitment and survival.

Different reef sites may be assessed for the presence of the above resilience factors using data from past bleaching events or through surveys of relevant reef characteristics. Because there is uncertainty about the extent to which past patterns will be repeated during future mass bleaching events—and the extent to which current environmental characteristics will remain consistent during weather events—all data should be interpreted carefully. Once resilience data have been gathered, the decision tree in Figure B can be used to identify areas to target for management based on their potential resilience to sea temperature anomalies.

A site's potential resilience is only one of several important factors that should go into decisions about identifying areas for increased management. Thus, the decision tree (see Figure B) begins with the application of conven-

tional selection criteria to evaluate the eligibility of sites for protection based on social, economic, ecological, regional or pragmatic criteria (Salm et al. 2000). The criteria used should be carefully chosen so that the selection process meets the specific objectives of the planners and stakeholders of the particular reef system.

Next, any information that is available on past heat exposure and bleaching responses should be evaluated for each of the candidate sites. Sites should initially be divided into those that have been exposed to heat stress previously, and those that have not. Information about thermal stress, presented as sea surface temperature anomalies, is now readily accessible to most reef managers through the NOAA *HotSpot* program, freely available on the Internet (http://www.orbit.nesdis. noaa.gov/sod/orad/). Currently available *HotSpot* maps (50 × 50 km) enable managers to distinguish differences in exposure to thermal stress at a larger spatial scale. This can be readily supplemented with reef-scale measurements obtained from direct temperature readings with thermometers or inexpensive data loggers.

Proceeding down Figure B, we can examine possible reasons for some reefs having no recorded history of anomalously high water temperatures. If sites have low risk of exposure to high water temperatures because of their oceanography or

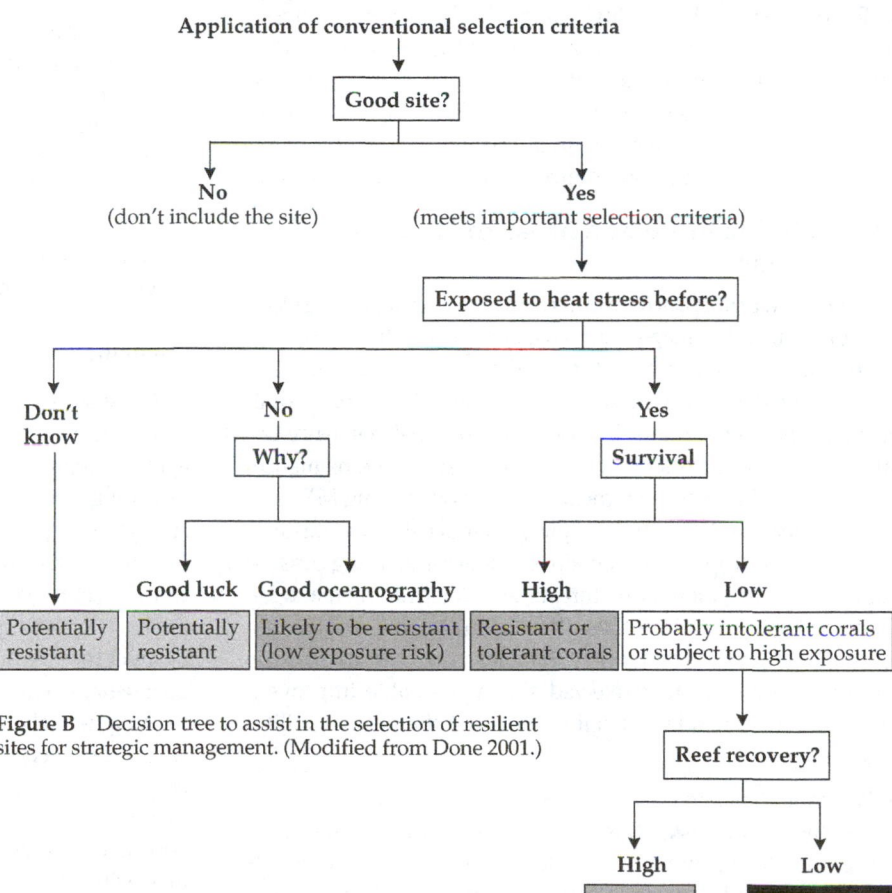

Figure B   Decision tree to assist in the selection of resilient sites for strategic management. (Modified from Done 2001.)

other physical characteristics, then they may be at low risk to bleaching in the future. In examining the reasons for low exposure, it may be useful to question whether the feature conferring resilience in the past is likely to remain unchanged in the future (e.g., shading from mountains is unlikely to change, but currents may shift under a different climate scenario). However, in many cases it may not be possible to identify the mechanisms or characteristics that have resulted in the "good luck" of a site having been spared exposure to heat stress in the past. In other cases, it may not be possible to even ascertain whether or not a site previously experienced heat stress. In both of these situations, the resilience of the reef community to bleaching cannot be assessed. Managers should consider implementing a monitoring program at these sites to document their response to any future episodes of thermal stress.

The next step is to examine the response of reefs that are known to have experienced thermal stress in the past. Reefs that have suffered only minor coral mortality during previous temperature anomalies are likely to be populated by corals that are resistant to bleaching or that have a high tolerance for bleaching. These reefs are probably sites of high resilience—unless they were only exposed to minor stress, in which case they may still not be resistant or tolerant to more extreme temperature stresses in the future.

The remaining category of sites includes those that have suffered substantial mortality as a result of exposure to stressful temperatures. The rate of recovery at these sites provides important information about their resilience. Damaged sites that show high rates of recovery are resilient. Sites with low rates of recovery are not resilient, unless the causes of slow recovery can be identified and remedied by management action.

Using Marine Protected Areas to Manage for Resilience

Traditionally, principles for Marine Protected Area (MPA) selection, design, and management have not specifically addressed the threat of mass coral bleaching (Salm and Coles 2001); however, expected increases in the extent and severity of mass coral bleaching now warrant the inclusion of additional, resilience-related criteria in MPA site selection and design. The resilience principles outlined here are meant to build on existing MPA selection criteria and design principles, not replace them. Existing MPA planning approaches remain essential for defining conservation objectives, identifying threats, and determining management strategies to address these threats. The intention of the following additional resilience principles, then, is to further increase the chance that selected sites will enable improved management of coral reefs in the context of climate variability and change:

- *Representation and Replication.* Sometimes called "spreading the risk," this principle recommends that in the uncertain context of climate change, MPA network design should aim to replicate a range of reef types and

related habitats. This reduces the risk of any one type being totally lost during a major bleaching event, hurricane, or other disturbance. This principle aims to maximize the probability that—across species and habitats—there will be sufficient survival and recovery to maintain functional coral reef ecosystems.

- *Refugia.* The refugia principle aims to take advantage of coral reef areas of natural resilience, as identified in the previous section. In the context of mass coral bleaching, refugia can serve as "seed banks" or source reefs for less resilient areas. For refugia to serve this role, they must be effectively protected from local stressors (e.g., over-fishing, sedimentation, or pollution) and thus are high priority for increased management attention.

- *Connectivity.* Connectivity strengthens coral reef resilience by promoting recovery after mass coral bleaching events and other disturbances. Implementing this principle in MPA design involves considering prevailing currents and adjacent non-reef areas. Linking MPAs along prevailing, larvae-carrying currents can replenish downstream reefs, increasing the probability of recovery at multiple coral reef sites. Adjacent non-reef areas are important to connectivity because they can become important staging areas for coral recruits as they move between reefs and into new areas.

- *Effective Management.* Coral reef ecosystems in good condition are better able to survive and recover from mass bleaching events. This principle refers to the ability to effectively manage local stressors at a potential site to optimize good coral reef condition. High coral cover, abundant fish populations, and good water quality are all elements of coral reef ecosystem health that support recovery. To implement this principle, MPA selection should give priority to sites where levels of resource use and effective management can help maintain these supportive attributes.

Once sites are selected for inclusion in an MPA network, managers must decide on the management objectives and management regime for each protected area. In the context of mass bleaching, one goal for management is to increase resilience by strengthening the factors that confer resilience. MPAs are particularly suited to providing management that protects reefs from direct threats, such as those from over-fishing and recreational overuse or misuse. While MPAs can assist in addressing indirect threats, such as land-based pollution, achieving this goal usually requires broader management activities.

A high-level objective of MPA management that aims to ensure reef resilience should be to protect fish abundance, with an emphasis given to herbivorous fishes. The role of herbivores in maintaining conditions that are conducive for coral recruitment and survival makes their protection critical for reefs subject to increasing sea temperatures. While some level of harvest may be sustainable, the importance of herbivores to future reef re-

silience means that managers should carefully manage fishing activity to ensure adequate levels of herbivory are sustained (i.e., a conservation objective), and not merely to ensure a sustainable or maximum harvest (i.e., a fisheries objectives).

Controlling recreational use of MPAs is another way managers can support the resilience of reef ecosystems. Recreational activities can result in physical damage from diving and boat anchoring, and from release of nutrients and combustion products from vessels. Where MPAs have been established to protect important bleaching refuges, even localized stresses associated with recreational activities may pose a significant threat to resilience. Options include carefully controlling snorkeling, diving, and boat usage to minimize stress to corals, especially during or following a bleaching event. While MPA managers may already have regulations and best-practice guidelines in place, measures to ensure that users avoid imposing additional stresses during periods of temperature stress may also be considered.

Broader Management Interventions to Increase Resilience

Many managers have a range of authorities and tools that can be used to protect resilient reef areas from local stressors and to increase coral reef resilience. These include fishery regulations, tourism permitting, permitting of coastal development, and watershed management. Expected increases in the extent and severity of mass coral bleaching have implications for effective application of these traditional management tools. At present, these implications are largely understood as conceptual guidelines that will benefit from refinement with additional experience and research.

Manage for the factors that confer resilience

Broader reef management efforts have a key role to play in supporting the factors that confer reef resilience. In particular, efforts to address indirect threats to reefs that can degrade coral reef condition can normally not be achieved without broader, collaborative efforts. Examples of indirect threats include degraded water quality that might result from coastal development or agricultural land use.

One of the greatest challenges to achieving such "Integrated Coastal Management" (ICM) is successful coordination among partners with a variety of interests. For example, protecting resilient areas may require that farmers change their land-use practices. While designation as an MPA can assist in focusing this type of coordination, in many cases additional mandates from high levels of government may be needed.

The importance of ICM has been recognized in law and polices for over 30 years (www.icm.gov). There are examples of successful ICM projects using an area-based approach that can serve as models for these efforts. However, implementing ICM remains challenging, and efforts to create supportive conditions—such as fostering leadership, building constituencies, developing technical capacity, and leveraging financial resources—will remain important factors for success.

Recognize the cumulative effects of multiple stressors

With the additional threat of mass coral bleaching, management of localized stressors, such as water quality, over-fishing, or recreational misuse, may need to become more conservative if levels of ecological condition and services are to be maintained. Managers need to consider how additional stress should be integrated into management regimes. For example, permits for coastal development might be revised to specify that these activities will be conducted during cooler times of year to avoid compounding heat stress with degradation of water quality. In the future, targets and expectations about fish abundance, water quality, and physical damage from recreational use should be revised to reflect the cumulative impacts of global and local stressors.

Manage adaptively

A key challenge in implementing broader measures to build reef resilience is the uncertainty in management targets. Conceptually, it is well known that maintaining high coral cover, abundant herbivore populations, and good water quality are important in promoting resilience. In implementing management, however, it would be extremely useful to know the answers to questions relating to how good water quality has to be, what herbivore abundance is sufficient and what land-use practices or fishery regulations will meet needed targets. Ongoing work with ecological models is aiming to clarify the relative importance of different management goals and to increase understanding of the management targets required.

The detailed answers to these questions will vary based on different scenarios of sea warming and differing ecological locations. The need to manage with imperfect information is likely to characterize this situation for some time to come. Given the pace of environmental change, an adaptive management approach (Gunderson 2000) is required. In this approach, different management policies are viewed as hypotheses, such that management actions become treatments in an experimental sense. This allows coral reef managers to begin acting immediately to conserve coral reef biodiversity based on what is already known about coral bleaching resilience, while also building into the process the capacity for future adjustment and refinement of management practices as new information becomes available.

Resilience Theory in Practice: The Rock Islands of Palau

Known as one of the "Seven Underwater Wonders of the World," The Rock Islands of Palau support over 425 hard and soft coral species, over 300 species of sponges, and some 1278 species of reef fishes (Spalding et al. 2001). Here, TNC is helping its partners apply reef resilience theory in the design of the first nationwide network of MPAs that take into account new resilience principles. The network will build on and strengthen Palau's existing MPAs (Figure C) and will be supported by

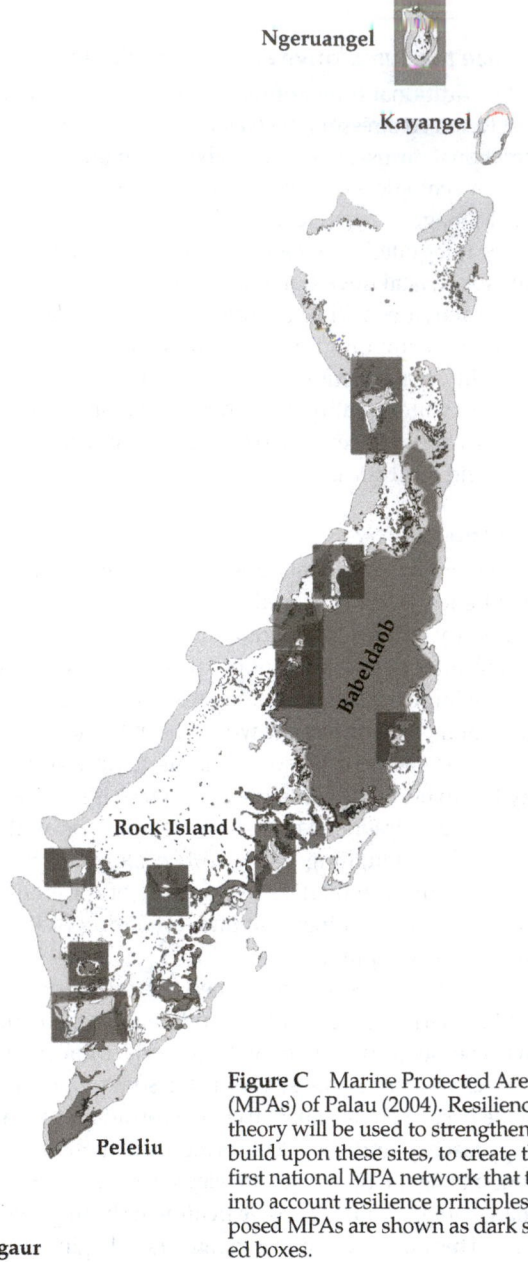

Ngeruangel

Kayangel

Babeldaob

Rock Island

Peleliu

Angaur

Figure C   Marine Protected Areas (MPAs) of Palau (2004). Resilience theory will be used to strengthen and build upon these sites, to create the first national MPA network that takes into account resilience principles. Proposed MPAs are shown as dark shaded boxes.

national protected areas network legislation that was signed into law in 2003. The objective of the project is to improve understanding of ecosystem integrity and connectivity within and among reef areas and to create a mutually replenishing network of well-managed MPAs that enhances the resilience prospects of Palau's reef ecosystem in the context of climate change.

One way that TNC and partners are further refining the resilience principles and supporting their application is through the development of "bleaching risk models," which aim to integrate data from current meters, conductivity sensors, temperature profiles, and pressure gauges with hydrodynamic models that can map water mixing and thermal stress across coral reefs. Once field-tested and refined, such tools may be generally applicable for design of future MPA networks globally, with the goal of maximizing coral reef biodiversity conservation in the face of a changing world.

CASE STUDY 13.3

Large-Scale Ecosystem Management
The Chesapeake Bay

C. Ronald Carroll and Chrissa Carlson, University of Georgia

The 11,400 km^2 Chesapeake Bay on the United States mid-Atlantic coast receives runoff from a 165,800 km^2 watershed that includes all of the state of Maryland and Washington D.C., substantial parts of Pennsylvania, Virginia, and Delaware, and that extends into New York (Figure A). Nineteen rivers drain the watershed, with three of them (the Susquehanna, Potomac, and James) supplying more than 80% of the freshwater that enters the bay. It is the largest bay and estuary ecosystem in North America. Because the bay is shallow (the main stem channel depth averaging approximately 6.5 meters), runoff

Figure A The Chesapeake Bay Watershed (light gray area) spans five states in the eastern United States.

sive. In recent years, outbreaks of highly toxic *Pfisteria* have been responsible for significant fish kills (Boesch 2001). Turbidity has increased as larger amounts of sediments are carried into the bay and as phytoplankton populations continue to expand. Nutrient loading (especially phosphorus and nitrogen) increased from sources such as fertilizers used excessively on crop fields, lawns, and golf courses, as well as from livestock manure, leaking septic systems, and overburdened sewage treatment plants. Because livestock outnumber people 11 to 1 in the watershed, manure runoff has been especially important. The runoff soup also contains toxins such as heavy metals, PCBs, and pesticides as well as endocrine disrupters and antibiotics (Chesapeake Bay Foundation 2004).

The large decline in the once abundant filter-feeding oyster populations no doubt exacerbated these problems. An important consequence of the nutrient loading has been summer blooms of phytoplankton. As these dense populations die and decompose, bacteria deplete Chesapeake Bay waters of oxygen. Oxygen concentrations below 2 mg/L, too low to sustain shellfish and finfish populations, are commonly reached by midsummer. Most of the stresses on the bay are chronic consequences of human behavior and economic development; but superimposed on these continuous stressors are important episodic weather events. For example, tropical storms Agnes (1973) and Eloise (1975) together transported 40 million tons of sediment into the estuary, the equivalent of a decade of normal sediment transport (USGS 2003). These tropical storms may have exceeded the resiliency of the bay ecosystem to sediment and nutrient loading, thereby pushing the bay into a more rapid downward spiral of degradation. Finally, adding to the long list of stressors on the bay ecosystem, as fish and shellfish populations declined, fishing efforts increased in the 1960s and 1970s in futile attempts to compensate for falling yields, pushing these populations to extremely low levels. For example, as oysters declined, harvest pressure on blue crabs increased.

Although the emphasis here is on the last 40 years, it is important to recognize that the bay began to degrade as far back as the earliest time of colonial settlement. Tobacco farming was introduced widely around the Chesapeake Bay in the 1600s. Colonial production practices for tobacco left bare soil over much of the year and fields often extended to the edges of streams and rivers. Rates of soil erosion and sediment transport into the bay must have been substantial. Sediment cores from various parts of the bay indicate that a shift in diatom species assemblages took place early in the colonial period. Pennate (elongate) diatoms that indicate low nutrient loading gave way to centric (circular) diatoms that indicate more eutrophic conditions (Cooper 1995). It is important to keep this

from the watershed contributes excessive amounts of sediment, nutrients, and toxins. The shallow depth and large watershed mean that the integrity of the bay's ecosystem is highly vulnerable to poor rural and urban land use, industrial output, sewage, and other human endeavors because the volume of the bay cannot adequately dilute the large discharge of pollutants.

Historically, Chesapeake Bay has been the home of major commercial and recreational fisheries. For recreational fishermen, the bay is famous for its striped bass, known locally as rockfish. Commercially, the bay has supplied nearly half of the nation's blue crabs and substantial amounts of oysters, food fish such as drum and mackerel, and oily fish such as menhaden used in pet and livestock feeds, as well as cosmetics, medicine, and industry. Most finfish and shellfish of commercial or recreational importance are either fully exploited (i.e., additional harvest pressure would cause population decline) or already overexploited (NOAA 2004).

For several decades, harvests of oysters and crab, and the recruitment rates for shad, menhaden and several other fishes, have been declining. Striped bass harvests declined until fishing moratoriums were imposed in the mid 1980s. The stock has since recovered in some places, suggesting that over-exploitation was an important cause of the earlier decline. Along with the decline in finfish and shellfish, other disturbing trends developed. The once-extensive beds of seagrass, an important habitat for juvenile crabs and fish, began to shrink. Noxious and even toxic algal blooms became more frequent and exten-

long legacy of pollution in mind when evaluating the success or failure of recent attempts to improve the bay. A few decades of good environmental practices are unlikely to erase several hundred years of degradation. Patience and long-term commitment are necessary.

Because the value of the Chesapeake Bay to communities in its watershed is both culturally and economically beyond measure, there has been widespread alarm to the growing evidence about the failing health of the bay. In response to the large and growing public concern that was becoming evident in the 1960s, there arose a complex mixture of government programs and nongovernmental environmental advocacy groups. These ranged from federal, state, and municipal agencies on the public side to multi-state umbrella nongovernmental organizations (NGOs) down to community-level organizations. Approximately 680 bay-focused NGOs are active in the Chesapeake Bay watershed. Such a large number of stakeholder groups would naturally represent many approaches for saving the bay, ranging from rigorous data collecting and species monitoring by federal agencies to confrontational advocacy tactics by some of the NGOs. The potential for chaos was real, but with leadership from several government agencies and the larger NGOs, a collaborative structure developed.

One of the earliest and most influential NGOs was the Chesapeake Bay Foundation (CBF). Established in the mid-1960s, CBF is a private NGO that works throughout the watershed to promote environmental education, to advocate for improved legislation, and to promote private and public sector partnerships. The CBF provides an annual assessment of the ecological health of the bay in their report, *State of the Bay*. The public sector counterpart to CBF is the Chesapeake Bay Program (CBP). Restoring and maintaining the ecological and economic health of the Bay and watershed is the mission of CBP which functions as a multi-level partnership that includes a large group of federal, state, and county agencies. In the late 1980s, CBF played a seminal role in developing the multi-state Chesapeake Agreement. All the states within the watershed were signatories to the agreement, which set target goals for the restoration of the bay; Maryland has been in the forefront of efforts to restore the bay. To fund actions called for in the Chesapeake Agreement, Maryland established the Chesapeake Bay Trust as a public–private partnership. The trust receives funds from sales of commemorative license plates, a state tax check-off, and gifts. Trust funds have been used to support private NGO and public efforts to restore the bay. One of the more systemic efforts supported in part by the trust is Maryland's participation in Tributary Strategies, an effort coordinated throughout the multi-state watershed by the Chesapeake Bay Program. In Maryland's Tributary Strategies effort, teams are assigned to each of its ten major tributary basins. The teams comprise representatives of public agencies, NGOs, and private citizens. The teams work with landowners, industry, and public agencies within their tributary basin to encourage better management practices to reduce pollutant loading of the bay. To monitor environmental changes the teams measure such key variables as total nitrogen and phosphorus concentrations, abundance of algae, total suspended solids, water clarity, summer bottom dissolved oxygen, and stream index of biotic integrity (IBI).

In parts of the upper Chesapeake Bay, environmental conditions have improved. In these areas, striped bass populations have recovered somewhat, the areal extent of submerged bay seagrasses has increased, and water quality has improved. It is difficult to identify the primary causes for these improvements. It seems clear though that watershed-wide requirements for better land management, reduced application of fertilizer on farms and lawns, improved sewage and septic systems, limitations on the use of phosphorus-based detergents, and increased use of forest as streamside buffers have all contributed to restoration of the upper bay. However, despite similar efforts in the lower bay, improvements are slow in coming. Perhaps this is not surprising when we consider that the effects of all the multiple stressors on the bay reach their maximum impact in the lower bay where pollutants accumulate.

Efforts to Restore Oysters and the Ecological Function Oysters Provide

The role of science as a foundation knowledge base for ecological restoration of the Bay is indisputable. However, science plays only a contributory role in resolving the commonly conflicting social and economic goals of ecosystem restoration. The case of oyster decline in the Chesapeake is instructive. From the beginning of human settlement around the Chesapeake, oysters have been an important resource. For the past several hundred years, the native eastern oyster (*Crassostrea virginica*) have been gathered to supply seafood markets throughout the eastern United States and Canada (Figure B.I).

Annual oyster harvests that used to be measured in millions of bushels are now measured in few hundreds of bushels (see Figure B.II). Between 1967 and 1997, oyster harvests dropped from 3922 metric tons to 786 metric tons. Since 1998, oyster harvests have risen and stabilized around 1000 metric tons. The decline was probably due to the combination of two diseases, over-harvesting in some areas, physical damage to oyster reefs, and stress from toxins and sediments. The oyster watermen were caught in an untenable economic squeeze between declining yields and decreasing prices due to growing supplies of cheaper imported oysters from Asia. Some of the watermen wanted to stock the bay with nonnative oysters thought to be more resistant to disease and to the degraded bay environment. They could point to successful introductions of the Asian Pacific oyster (*C. gigas*) and subsequent increase in yields in France, Australia, and the west coast of North America. The environmental community and NGOs favored efforts to restore the native oysters by rebuilding the oyster reefs, limiting harvests, promoting aquaculture of native oysters, and generally improving the environment of the bay.

To help resolve this conflict, the CBP requested that the Ocean Studies Board of the National Academy of Science com-

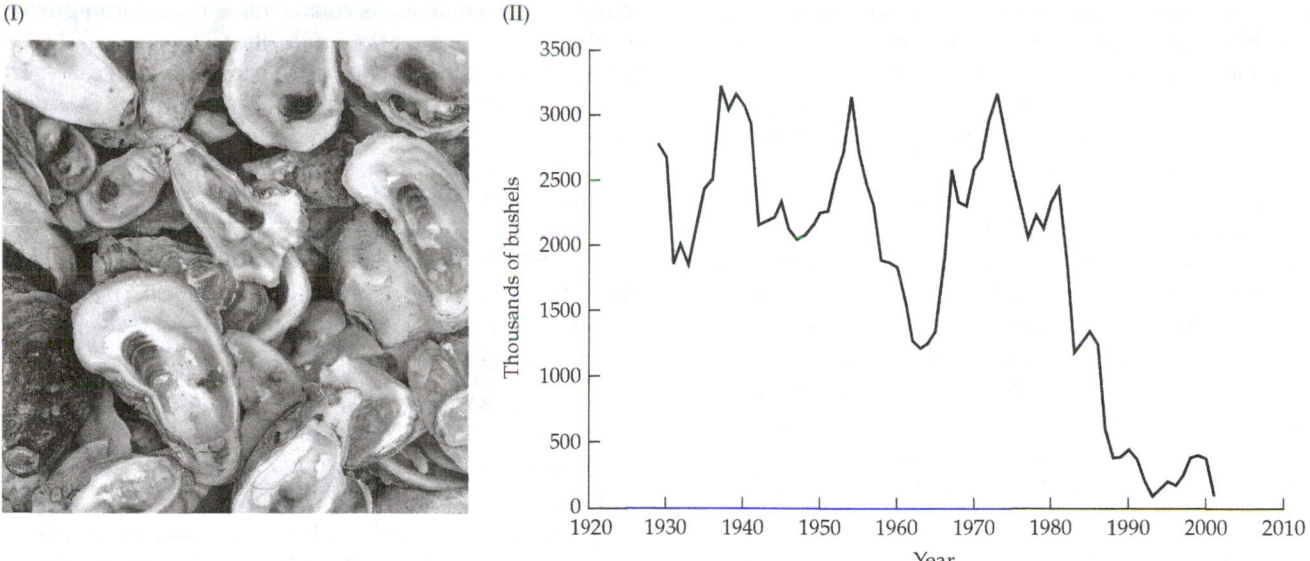

Figure B (I) The Eastern oyster (*Crassostrea virginica*) is an economically important shellfish. Creating a sustainable harvest of this species is a major focus of ecosystem management in Chesapeake Bay. (II) Quantities of native oysters gathered per year has fallen precipitously since 1970. (I, Photograph © David Sieren/Visuals Unlimited; II, modified from Maryland Department of Natural Resources 2005.)

mission a study to investigate the potential risks and benefits of introducing nonnative oysters into the Chesapeake. The report is a remarkably thorough risk assessment (Ocean Studies Board 2004). The two principal environmental risks of introducing nonnative oysters are that unintentional introductions of pathogens and other invasive species may occur along with the introduced oysters and that the nonnative oyster may displace native oysters and other native species. There is a long record of noxious species piggybacking on purposeful introductions of native species, and oysters are no exception. However, this risk may be minimized by using pure cultures of nursery-raised spat (juvenile oysters) for the introduction. The risk of spread of the nonnative oyster may be reduced by using triploid spat that develops into infertile adults, although this strategy has not been evaluated via a full risk assessment.

The Ocean Board Studies report recommends a three-pronged approach to estimate environmental risks and possible economic benefits. First, they call for more comparative ecological studies between nonnative and native oysters throughout the entire life cycle. Second, they recommend a highly controlled experimental introduction of sterile triploid populations of the Asian Suminoe oyster (*C. ariakensis*). Third, they recommend more investigation into the feasibility of using disease-resistant native oysters. Interestingly, they recognize that restoration has social and economic dimensions and they call for more involvement by social scientists.

The oyster example illustrates a common approach for addressing problems in the Chesapeake Bay. First, in the short term, look for solutions that address economic problems without making long-term solutions more problematic. Second, address the systemic causes of degradation to the bay ecosystem. Public agencies and nongovernmental environmental organizations play complementary roles in efforts to achieve these goals. It is worth noting that the process of restoring oysters to historic levels is more than species restoration or recovery of a commercial enterprise. Oysters are filter feeders and therefore circulate large volumes of water, removing suspended particles in the process. In the earlier part of the 1900s oyster populations were so large that they processed the entire volume of bay water every 3–4 days. This suggests an important environmental service for oysters. As stressors to the bay are reduced, recovering populations of native or nonnative oysters will aid the recovery process by removing nutrients and suspended sediments.

The many varied efforts to restore and protect the Chesapeake Bay ecosystem illustrate an important goal for ecosystem management in general. Given that environmental NGOs and public agencies have differing primary roles but share the broad objective of advancing good stewardship, an important goal for both parties is to develop productive relationships and, indeed, productive relationships among the many NGOs and agencies themselves. We note that productive relationships may not always be amiable. Advocacy NGOs and public agencies may at times have testy interactions and advocacy NGOs may go so far as engaging the agencies in lawsuits. The important point is that the outcomes of these interactions should contribute to better environmental stewardship. But, "better environmental stewardship" is a slippery concept, difficult to pin down. Better for whom or what? Better only in the long term? And, who or what loses? So the key first step is to find common ground by developing a shared vision of envi-

ronmental stewardship for the bay. This means coming to an agreement on the appropriate spatial and temporal scales, defining the stakeholders and their roles, taking an ecosystem approach rather than piecemeal, single-species management, and placing emphasis on identifying and reducing the key stressors on the bay ecosystem.

We can point to a number of ways in which NGOs and public agencies in the Chesapeake Bay ecosystem are attempting to reach this common ground. For example, CBF in its annual report, *State of the Bay*, uses data from various sources—including public agency scientists—to generate indices of improvement or decline. Important system-level stressors are included such as nitrogen and phosphorus concentrations and levels of dissolved oxygen. Changes in a key habitat, submerged bay seagrasses, are reported, as well as an annual assessment of how well restoration efforts are performing. Population changes of species of particular interest to the general public and agencies, such as the blue crab, oysters, and striped bass, are reported as qualitative indices. Other examples of NGOs taking more of an ecosystem approach include several that are encouraging better management practices for forestry and agriculture, and better stormwater control in urban environments. The Chesapeake Bay Alliance provides an example of working with key stakeholders. The Alliance is working with homebuilders associations to encourage better site designs to reduce surface runoff.

On the public agency side, the Chesapeake Bay Program plays a coordinating role among the public environmental agencies and encourages citizen and NGO participation. Although a shared vision for environmental stewardship of the Bay has not been explicitly articulated, the individual efforts of the public agencies and the NGO community are increasingly convergent. There remains considerable room for improved collaboration. As a recent example, the Chesapeake Fisheries Ecosystem Plan (NOAA 2004) is an excellent technical report to guide fisheries management. However, the report is heavily dominated by the viewpoint of regulatory agency scientists and academics. The viewpoints of the watermen, NGOs, or other stakeholders are scarcely mentioned.

Some Take-Home Lessons

The history of the efforts to restore the Chesapeake Bay contain important general lessons that apply to the management of many complex ecosystems.

1. A long legacy of ecosystem abuse cannot be quickly erased. Restoration efforts must be monitored but it is unrealistic to expect a steadily improving ecosystem.

2. Public and private efforts are both important, and coordinated efforts are essential. A certain amount of tension between NGOs and public agencies is to be expected and is probably healthy.

3. Defining a shared vision for restoration should be a common goal among all stakeholders.

4. A stable core source of financial support for restoration efforts that is accessible by all stakeholder groups is important.

5. Science should inform the policy process but so should social and economic considerations. Ecosystem management policy represents trade-offs among the contributions from science, economics, and social concerns. But, the various risks that derive from these trade-offs should be explicitly known and made part of the political dialogue.

CASE STUDY 13.4

The Everglades
Trials in Ecosystem Management

Lance H. Gunderson, Emory University

One of the most visible attempts at ecosystem management is occurring in the Florida Everglades. With huge and unabated population growth in the region, as well as an important and entrenched agricultural base, the challenges for good management are as difficult here as anywhere. Effective management of the Everglades requires understanding the key driving forces that historically structured the system, the many complexities and surprises associated with human manipulations, and attention to cross-scale processes that influence the natural system.

The Everglades is one of the most recognized ecosystems in the world. During the twentieth century, dramatic changes in population and human development have transformed this large, once-contiguous wetland into a partitioned, highly managed system for water control. Because water is central to both the ecological and human system, ecosystem-level management in the Everglades has been and continues to be centered on hydrologic processes. Floods, droughts, fires, storms, and other landscape-scale processes are key, self-organizing elements in the remnant wetland ecosystem. Human management

institutions have been forced to adapt to these cycles, and have undergone periodic shifts and reformation as a result of unanticipated variation or surprises in these ecosystem processes (heavy rainfall, hurricanes, droughts, eutrophication), or from the maturation of latent management deficiencies.

In the coupled human and natural system, the recurring sequence appears to be that unforeseen ecological changes are perceived as crises, followed by periods of dramatic institutional reformation. These crises provide opportunities for innovation, creativity, and renaissance. Indeed, the relatively fast dynamics of the ecosystem have produced many crises and surprises during the last century. The trials of the Everglades provide a useful case study for lessons in management at the ecosystem level, while only beginning to embrace the concepts of ecosystem management as discussed in this chapter.

Although Everglades management has focused on ecosystem-level processes (primarily the hydrologic system), until recently that focus was to control unwanted variation in the hydrology by compartmentalization, drainage, impoundment, and the use of technology for moving and cleaning water. Other pieces of an ecosystem management approach—sustainability, system resilience, cross-scale issues, and adaptive management—are only beginning to be addressed in an effort to revive and sustain ecologic values around the notion of ecosystem restoration.

The remainder of this case study is structured in three parts. The first is a brief review of the cross-scale relationships of the Everglades ecosystem, presented as a way of compressing the complexity of ecologic relationships into a framework that provides understanding. The second recounts the history of water management during this century, revealing the coupling between ecological and institutional dynamics. The lesson is that human institutions must adapt to unpredictable changes in the ecosystem. The final section recounts incipient ecosystem management activities focused on ecosystem restoration.

Understanding the Ecosystem

One of the precepts of ecosystem management is that it should be based on ecological understanding at multiple scales. That is, ecosystem management does not just refocus the scale of management pervue, like changing ocular power on a microscope, but attempts to deal with issues by understanding how ecosystems are structured and how they function across spatial and temporal scales (Figure A). Consequently, this section portrays elements of the Everglades at three snapshots in space, and addresses how structures and functions interact across scales. The three scales include the local habitat scale, the landscape scale, and the regional scale; they will be used to illustrate the structures and processes that are objects of management at each of these scale ranges, and how they interact.

Alligator hole

A characteristic habitat of the Everglades is the alligator hole (Figure B), which is surrounded by a variety of emergent

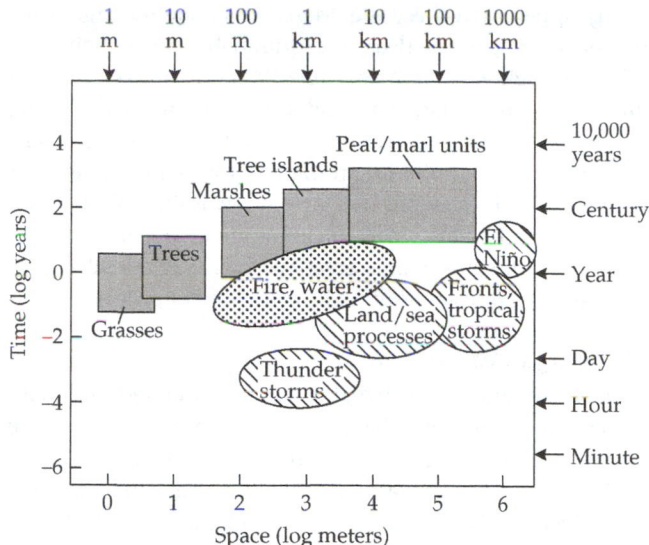

Figure A Spatial and temporal domains of critical ecosystem components in the Everglades. The squares represent the domains within a vegetation hierarchy of individual plants, associations, and landscape units. The crosshatched ellipses represent the atmospheric hierarchy, the primary sources of water input. Fires and surface water conditions are in an intermediate position, and integrate the vegetation and atmospheric hierarchies. (Modified from Light et al. 1995.)

marsh plants, including sawgrass—the most common plant in the Everglades. The area of open water is kept free of plants by the activity of American alligators (*Alligator mississippiensis*). At this scale, patterns are created by biotic processes such as competition for light and nutrients, as well as by animal activity. Many processes, including photosynthesis, respiration, decomposition, nutrient mobilization, soil accretion, or bedrock dissolution, occur at smaller scales than seen here but may be

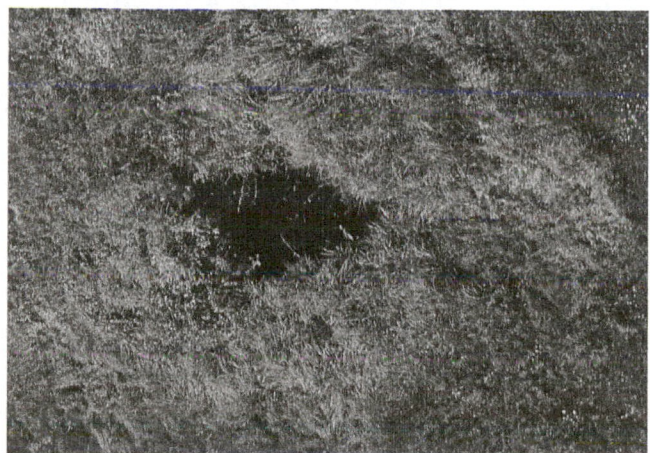

Figure B Aerial photograph (10 mile window) showing the landscape mosaic in a conserved area (Everglades National Park) at the scale of a single alligator hole. (Photograph courtesy of the South Florida Water Management District.)

partly apparent at this scale. Many small aquatic organisms (grass shrimp, water fleas, mosquitofish) spend their lives within a single alligator hole, while other organisms, including alligators, require larger spatial scales for survival. During droughts, the holes are refugia for many aquatic organisms (Craighead 1971), offering resilience for populations over broader spatial scales. The habitat window is the scale at which populations of plants, small aquatic animals, and alligators are monitored, but most management in the Everglades occurs at broader spatial scales, especially at the landscape level.

The Everglades landscape

In the Everglades ecosystem, the biota are adapted to, and interact with, a few key landscape-scale processes—surface water hydrology, fires, storms—that fluctuate on time scales of decades and cover areas up to hundreds of square kilometers. The underlying topography of both the bedrock and the soil surface is very flat, with elevation differences between the highest, driest hardwood forests and the lowest alligator holes less than one meter.

Fires have been part of the ecosystem for as long as the wetland complex has existed, and occur at multiple frequencies, with annual and decadal intervals. Severe, peat-consuming fires lower soil elevation, leave little organic matter or seeds of local vegetation, and generally result in dramatic shifts in vegetation assemblages (Craighead 1971). Less severe fires generally consume aboveground biomass, leaving viable roots and rapid regeneration of composition and structure similar to that before the fire. Fine-scale patterns of sawgrass stands and wet marshes are created by this interplay of severe fires that create the wet prairies and less severe fires and hydrology that moderate the slow spatial spread of remnant sawgrass stands.

The landscape scale contains the home ranges of a suite of small animals (fish, small mammals), but is only a part of foraging area of wading birds. The wading birds are a critical cross-scale link, as they feed over larger spatial scales, up to a regional level. Fire management (control and prescribed fires) is applied at the landscape scale, whereas water management is applied at the regional scale.

The region

The Everglades wetland is part of a larger watershed that covers most of the southern peninsula of Florida (Figure C). Three sub-basins make up the watershed: the Kissimmee River basin, Lake Okeechobee, and the Everglades. Rainfall is the primary input, with an annual mean of about 130 cm (range from 95 to 270 cm; MacVicar and Lin 1984). Most rainfall occurs during the summer months associated with convective thunderstorms, but other frequencies (3 months and 12 years) indicate cycles at longer and shorter time frames (Gunderson 1992); the longer-term cycles of flood and drought have been linked to El Niño activity (Rasmussen 1985; Ropelewski and Halpert 1987).

The annual pattern of wet and dry seasons is one of the dominant cycles organizing energy flow through the system. Water levels rise during June and remain high into early fall

due to summer rains; water levels then slowly decline through the fall and winter months. In spring, high temperatures, high evaporation, and little or no rain result in the lowest water levels of the year. This rainfall variation translates to large spatial variation in water depths. In the wet season, most of the system is inundated, with high primary and secondary aquatic productivity. As the system dries, the spatial extent of wetted area decreases and aquatic organisms move with the drying front, concentrating their numbers and providing forage for predators such as alligators and wading birds. If the drought is severe, most of the organisms die or are consumed, thereby offering little stock for recovery; if less severe, the remnant stocks allow for rapid recovery when water returns.

Development within the Everglades is scarcely 100 years old. Human population growth in the region has been dramatic, exploding from less than 30,000 at the turn of the century to just over 5 million in 1990. Development was predicated on several factors that enabled humans to control undesirable aspects of summers in southern Florida, including air conditioning, mosquito control through pesticide application, and controlling natural flooding in many areas including eastern marshlands of the Everglades.

Human land uses and geologic and hydrologic features are key structures apparent at the regional scale (see Figure C).

Figure C A satellite image of South Florida, showing the hydrologic system of the Everglades and current land uses. The Everglades was once the southern third of a hydrologic system that included the Kissimmee River Basin and Lake Okeechobee (the large lake near the center of the photograph). Land uses of the Everglades ecosystem are indicated by the Everglades Agriculture Area in the north, Water Conservation Areas (dark areas), urban development along the east coast, and the national park and preserve in the south. (Photograph courtesy of the South Florida Water Management District.)

Land in the historic Everglades (excluding the Kissimmee River Basin and Lake Okeechobee) is divided among four sectors of use: urban, agriculture, water control, and protected park. Urban systems cover 12%, and are seen as the lighter, mottled patterns along the eastern rim of the Everglades. Various forms of agricultural uses cover 27% (Everglades Agricultural Area, EAA), most of it found in the north Everglades and south of Lake Okeechobee. The central third of the historic marshes has been designated as Water Conservation Areas (the large, dark regions in Figure C). About 21% of the historic system is preserved in Everglades National Park or Preserve, at the southern terminus of the freshwater wetland. Of the original Everglades ecosystem, less than half remains in areas where conservation is the primary management objective (Gunderson and Loftus 1993; Davis et al. 1994).

The geologic features under the Everglades have structured much of the development and management history. The deepest soils—categorized as peats and mucks—are Holocene sediments, which occur in the northern Everglades and attracted agriculture. The higher, coastal ridge is where urban development occurred. These geologic features also contributed to the low nutrient or oligotrophic status of the historic Everglades system.

Development of Water Management in the Everglades

In the past century, the Everglades ecosystem has been transformed from a vast subtropical wetland into a highly managed, multiple-use system as a result of one of the largest public works projects in the world. This transformation was not a linear process, but characterized by turbulence and punctuated change. For the most part, management was driven by a series of events perceived as crises that threatened exploitation of the resources. Each crisis precipitated actions resulting in a reconfiguration and emergence of a new system.

Crises appear to arise from two causes: those created by external environmental events and those from human activities. The former are mostly exogenous to the region, and arise from larger scale processes such as hurricanes or too much or too little rainfall over the system. The other type of crises occur over a longer period of time, tend to be endogenous to the system, and reflect a chronic problem defined by slower variables, such as water quality degradation that leads to dramatic shifts in dominant taxa. Such crises derive from human development of natural resources, and are associated with agricultural activities, primarily in the form of soil loss and water pollution.

These crises have resulted in at least four major eras of water management (Light et al. 1995), based on a recurring theme of flooding impacts. The first two eras were a result of flooding from high rainfall events, the third was related to drought, and the fourth resulted from attempts to rectify latent or previously unattended problems.

The earliest settlers were intent on reclaiming land "lost" to natural flooding, in order to farm the rich muck soils. Early attempts at drainage were able to control water levels during average water conditions; the approach was to dig canals and drain the land as fast as possible. Periods of dry years allowed for agricultural expansion, until the next wet year. However, these attempts at drainage were unable to cope with the full variation in climatic regimes that would inevitably occur. Flooding crises occurred in 1903 as a result of high rains, and in 1926 and 1928 from severe hurricanes.

In 1947, the same year when Everglades National Park was established, over twice the normal amount of rain fell on south Florida, severely affecting the coastal communities and inundating some areas for several months. The acute flooding resulted in implementation of a widespread plan to avoid this type of flooding in the future. The massive control plan, developed by the U.S. Army Corps of Engineers, called for creation of specific land use areas (agriculture, water conservation, and national park) and a water management infrastructure (2240 km of canals and levees, pumping stations with 3.8 billion liters/day capacity, and requisite water regulation schedules) needed to regulate flood waters. This era of water management lasted from 1947 through the early 1970s.

The crisis that precipitated the next era of water management was a drought in 1971. Less than 82 cm fell over the system, creating the worst drought in 40 years (Blake 1980). By the early 1970s, the population of southern Florida topped the 2 million mark and sugarcane production in the EAA more than tripled following the communist takeover in Cuba and the subsequent U.S. ban on Cuban sugar imports. The low rainfall, coupled with increased urban and agricultural water demands, prompted concern for an adequate water supply. Serious problems arose in trying to retool a system designed for flood control to now meet water supply concerns. The situation was exacerbated by the difficulty inherent in decisions involving trade-offs among water use categories. The major reformation was the creation of state water management districts, which would manage the system for water supply on a watershed-wide basis.

The history of water management provides a few lessons for ecosystem management. First, there will always be inherent uncertainty in the dynamics of the system being managed; consequently, management institutions must be capable of adapting, renewal, and learning. Second, the focus of management must be on multiple scales, in both space and time. Management focused on one scale (e.g. local drainage of farmland) is likely to cause surprises and crises at larger scales. Third, a key lesson is that the success of ecosystem management is not in clear articulation of its goals, but in what kind of management framework is established to meet those goals and to seek learning while managing. Conservation, preservation of ecological structures and processes, and minimizing external threats while maximizing external benefits, are noble objectives and in one shape or form have been part of the history of conservation in the Everglades. Yet clearly management has failed in at least half a century of trying. To see some of the reasons and possible solutions, the final section discusses the most recent attempts at adaptive, ecosystem management around the goal of ecosystem restoration.

Restoring the Everglades for Conservation

Since the early 1980s the focus of management in the Everglades has been on ecosystem restoration. Water management via water control has benefited both urban and agricultural interests. However, conservation and preservation of the remnant wetland has been less successful and is now being sought through ecosystem restoration. That restoration is predicated on resolving current resource issues and conflicts, and on recreating as much of the historic attributes of the ecological system as feasible. The historic ecosystem was characterized by three attributes: It supported a suite of animals with large spatial requirements, it was oligotrophic, and it supported a spatially diverse vegetation mosaic. The specific environmental issues are categorized and discussed in the next sections.

Vegetation issues

The key vegetation issues include changes in spatial distribution due to water management, losses in native cover types due to land use conversions, changes in species composition due to nutrients, and invasion of exotic flora (Gunderson 1994). Vegetation changes have been related to modifications of the hydrologic regime. Sawgrass and tree island communities have been replaced by open-water marshes in areas of water impoundment. Where water levels have been reduced, broad-leaved hardwoods and invasive trees have replaced the grassy marsh vegetation.

Nutrient enhancements or additions change species composition of both macrophytic and periphyton assemblages (Davis 1991). The addition of phosphorus-laden water (linked to runoff from agriculture) into portions of Water Conservation Areas has resulted in a shift in dominance from sawgrass to cattail communities. Also in these areas, the periphyton (algae community) has changed to dominance by pollution-tolerant taxa.

Although about 17% of the flora is from areas outside the southeastern coastal plain, only a few have become aggressive invaders. The main invasive trees are the melaleuca (*Melaleuca quinquenervia*), Australian pine (*Casuarina* spp.), and Brazilian pepper (*Schinus terebinthifolius*). Melaleuca invades the drained marshes of the eastern glades, where water levels have been lowered and fires burn frequently (Myers 1983). Australian pine invades the southern and eastern glades, but on slightly higher sites. Brazilian pepper is found on unburned upland sites and on abandoned farm fields. All are adapted to short periods of flooding.

Animal issues

Many changes have been noted in the fauna during this century. Most notable are an increase in nonnative species, endangerment and extirpation of other species, and dramatic changes in populations of key taxa. Several hundred species of animals have naturalized along the southeastern coast of Florida; of those, a few dozen (primarily fishes) exploit altered niches within the Everglades system (Gunderson and Loftus 1993). Nonnative fishes live primarily in the canals, drainage ditches, and borrow pits throughout the region, although a few species survive in the mangrove–freshwater ecotone. Nonnative birds have naturalized in the coastal ridge area. As a rule of thumb, the abundance of introduced animals declines further from the coastal area.

At least 17 taxa have been recognized as federally endangered or threatened from southern Florida; most notable are the Florida panther (*Puma concolor coryi*), Cape Sable Seaside Sparrow (*Ammodramus maritimus mirabili*), and Wood Stork (*Mycteria americana*). Although many of these taxa have small populations and are vulnerable to extinction, none have disappeared from the wetlands of the Everglades. Extirpations have occurred in the fauna of the upland communities due to conversion for other human uses, with habitat loss cited as the primary cause.

The most dramatic changes in the fauna have been observed in nesting populations of wading birds (Frederick and Collopy 1987; Bancroft 1989; Ogden 1994). From the 1920s through the 1960s, the Everglades provided the primary nesting area for populations in the southeastern United States. Since then, only 5% to 10% of the populations that once nested continue to use the area (Ogden 1978, 1994), possibly due to a spatial decrease of early wet season habitat via development, less water flow through the system, alteration of hydrologic regimes necessary for successful feeding and breeding, and higher quality of other sites in the southeast. In spite of the loss of nesting, the Everglades still provides an important feeding area for winter and transient populations.

A number of other issues haunt the Everglades, including the presence of toxins such as mercury, which settles from the atmosphere, is trapped by organic sediments, and then is mobilized into the food webs. There is also a growing recognition of linkages between the Everglades and other areas, such as the need for freshwater flow into the estuaries of Florida Bay, and the importance of the area as an overwinter feeding stop for many birds.

Attempts at recovery

By the late 1980s, a group of scientists from South Florida Water Management District and Everglades Park began a process to synthesize existing ecological understanding and to translate that into restoration strategies. A symposium was held, followed by three years of workshops where a computer model to improve communication among scientists, engineers and resource managers was developed and tested. The synthesis emerged in a book (Davis and Ogden 1994) where, for the first time, changes to the system are described, restoration goals are defined, and restoration polices were prescribed.

This informal collaborative effort led to a series of activities to restore the Everglades ecosystem (Holling et al. 1994; Walters and Gunderson 1994), with a major focus on restoring hydrologic processes. The process of hydrologic restoration distills to a few key points. First, more water should flow through what is left of the system, into the estuaries of Florida Bay (Walters et al. 1992). Second, that water should be clean. Third, there is enough water in Lake Okeechobee, the EAA, and the urban sectors to supply restoration needs. Each of these sources has specific costs associated with improving water

quality. Finally, restoration should move forward as an adaptive process, with management actions viewed as experiments derived from working hypotheses that deal with the many uncertainties of the system. These key points laid the foundation for the restoration process that is still ongoing.

Led by the U.S. Corps of Engineers, a formal interagency process was established in the mid 1990s, to restudy the entire system and plan options for restoration. The work was largely a technical committee, but was overseen by a federal taskforce and state commission that actively engaged stakeholders. These groups were able to define restoration actions that were technically, socially, and politically feasible and recommend those alternatives to the U.S. Congress in the form of a comprehensive restoration plan.

In 2000, the U.S. Congress authorized the initial funding to implement the Comprehensive Everglades Restoration Plan. The plan calls for 68 projects estimated to take 36 years to complete and cost $7.8 billion. The plan focuses on increasing water storage, including untested technology, to make more water available to users during dry periods. The passage of this act represents a major commitment by society to ecosystem management. However, will 8 billion dollars restore degraded environmental values of this world-renowned ecosystem? It will take us decades to answer this question.

Summary

1. An ecosystem approach to management has become the centerpiece for conservation management in recent years. Although definitions of ecosystem approaches abound and serve many different interests, we believe that it consists of simultaneous attention to ecological, socioeconomic, and institutional concerns, and satisfaction of each in conservation management. If approached on large spatial and temporal scales in an adaptive manner and with inclusion of all relevant stakeholders, ecosystem-level management should result in progress toward both our conservation and human development ideals.

2. Uncertainty and risk are inherent in conservation. Understanding and dealing with the contribution from sources of uncertainty and risk, such as environmental stochasticity, non-independent effects, and cumulative effects, is essential to good conservation planning.

3. Conservation at the ecosystem level offers great promise for successful biodiversity conservation. The identification of strong functional linkages among biophysical ecosystems within the landscape provide an objective and scientifically defensible means for identifying natural boundaries.

4. The principles of an ecosystem approach articulated here offer those who seek to create and implement conservation strategies some reasonable mechanisms for establishing rational goals. The principle of long-term sustainability calls for managers to establish functional landscape mosaics that contain the necessary resources to support biodiversity and ecosystem functions. The principle of adaptive management allows the manager to make decisions in a climate of uncertainty and, through continuous monitoring and the use of indicators, to make necessary corrections. The principle of stakeholder involvement in decision-making allows managers to find an acceptable balance between exclusionary preservation at one extreme and total economic exploitation at the other extreme.

> Please refer to the website www.sinauer.com/groom for Suggested Readings, Web links, additional questions, and supplementary resources.

Questions for Discussion

1. How does an ecosystem approach differ from a species or a landscape approach to conservation?

2. What are the challenges in applying an ecosystem approach to areas of diverse ownership (e.g., lands in the western U.S. with mixed federal, state, and private ownership)? How do these compare to those of applying ecosystem approaches to marine areas that have clear and undefined ownership?

3. Imagine that you have become part of a team to manage a watershed, such as the one pictured in Figure 13.4. Develop a list of concerns you might have about this watershed, and of potential goals of stakeholders in the watershed. If your charge is to manage this watershed to allow continued beneficial use by all of the current users, what principles would you bring to bear on its management? How would you approach creating a plan that could ensure benefits to future generations (e.g., a plan that maintains an ecologically functional ecosystem)?

4. Continuing with the example based on Figure 13.4, how might you set up your efforts in an adaptive management context? What would you need to do to make adaptive management possible?

5. How might one engage stakeholders who are new to an area in ecosystem-based management?

14

Protected Areas

Goals, Limitations, and Design

Hugh P. Possingham, Kerrie A. Wilson, Sandy J. Andelman, and Carly H. Vynne

All we have to decide is what to do with the time that is given us.

J. R. R. Tolkien, 1954

The increasing pressures exerted on the environment by humans make preservation of natural areas crucial for the persistence of biological diversity (McNeely 1994; Groombridge and Jenkins 2002). Protected areas are one of the most effective tools available for conserving biodiversity. While protected areas can be degraded by external pressures, the majority of terrestrial protected areas are successful at stopping deforestation and mitigating the damaging effects of logging, hunting, fire, and grazing (Bruner et al. 2001). **Marine protected areas** (**MPAs**) often have bans on fishing or may require actions that reduce pollution. Therefore, while dedicating protected areas is only one of many actions we can take to conserve biodiversity, it can abate some of the key threats: habitat degradation, overexploitation, and to a lesser extent, pollution and nonnative species invasion. They form the foundation on which many of our conservation efforts are based.

Terrestrial and marine protected areas around the world not only assist to safeguard biodiversity, but often provide other benefits, such as protecting water supplies, providing flood protection, protecting cultural values, and sustaining the livelihoods of indigenous groups. Protected areas also provide an increasingly urbanized society with much-needed contact with nature.

Today, in regions of particularly intense human settlement, protected areas often contain the only remaining examples of particular habitat types and species populations. For some species restricted to protected areas, such as the northern hairy-nosed wombat (*Lasiorhinus krefftii*) which occurs only in Epping Forest National Park of Australia (Woolnough and Johnson 2000), avoiding extinction is entirely dependent upon the continued protection of its habitat. Other species with large area requirements, such as the elephants of east Africa, may not be afforded adequate protection by one protected area alone. Protecting these species may require a system of protected areas linked by corridors that allow movement from one area to another.

(A)

(B)

Figure 14.1 Many of the large, attractive natural reserves in the United States, such as (A) Grand Canyon and (B) Yosemite were created as geological attractions and for their aesthetic appeal. Consequently, they may not be especially effective for biodiversity conservation. (A, photograph by G. K. Meffe; B, © J. Hughes/Visuals Unlimited.)

The concept of setting aside areas for the preservation of natural values is not a recent phenomenon. Historical examples include the sacred groves of Asia and Africa and the indirect protection of biodiversity afforded by royal hunting forests (Wright and Mattson 1996; Chandrashekara and Sankar 1998). More recently we have moved to more formal establishment of protected areas. The first national protected areas were Yosemite and Yellowstone National Parks in North America, designated in 1864 and 1872 respectively, with the third being Royal National Park near Sydney, Australia, in 1879, followed by Kruger National Park in South Africa in 1892 (Spellerberg 1994). Many of the earliest national parks in the U.S. were established primarily to preserve their dramatic landscapes (Figure 14.1).

The establishment of protected areas has tended to be in response to the loss of natural areas in which people have a vested interest. Those who called for the first national parks in America had witnessed a loss of wild places important to them (Nash 1990; Pimm et al. 1995). Hunters in the U.S. were among the first to favor federal land protection as many of them came to understand the crucial role of wildlife refuges and national parks in maintaining their sport. Although hunting was off-limits in these areas, it was here that populations of game animals were sustained (Nash 1990). Contemporary "no take" MPAs play a similar role. By protecting breeding and spawning grounds, they help safeguard fish populations that are caught outside of protected areas (Gerber et al. 2003).

The establishment of protected areas is now a vital legislative component of most national and regional strategies to counter biodiversity loss. In addition, protected area establishment is a requirement for many international environmental agreements and conventions, including the Convention on Biological Diversity (CBD) (Box 14.1), the Convention on the Conservation of Migratory Species of Wild Animals, the Convention on International Trade in Endangered Species of Wild Flora and Fauna (CITES), the Convention for the Protection of the World Cultural and Natural Heritage, and the Convention on Wetlands of International Importance.

This chapter focuses on the functions, design, and limitations of protected areas and the process of reserve system planning. Throughout this chapter, we use the term **protected area** to denote any area of land or sea managed for the persistence of biodiversity and other natural processes in situ, through constraints on incompatible land uses. The term **reserve system** describes a system of protected areas.

The Current State of Protected Areas

The number of protected areas increased rapidly worldwide beginning in the early 1960s (Chape et al. 2004; Figure 14.2). Over 80% of the world's protected areas have been established since the First World Parks Congress, held in 1962. In total, there are 104,791 protected areas covering approximately 18.38 million km² on land and 1.89 million km² at sea worldwide (see Plate 5). The total

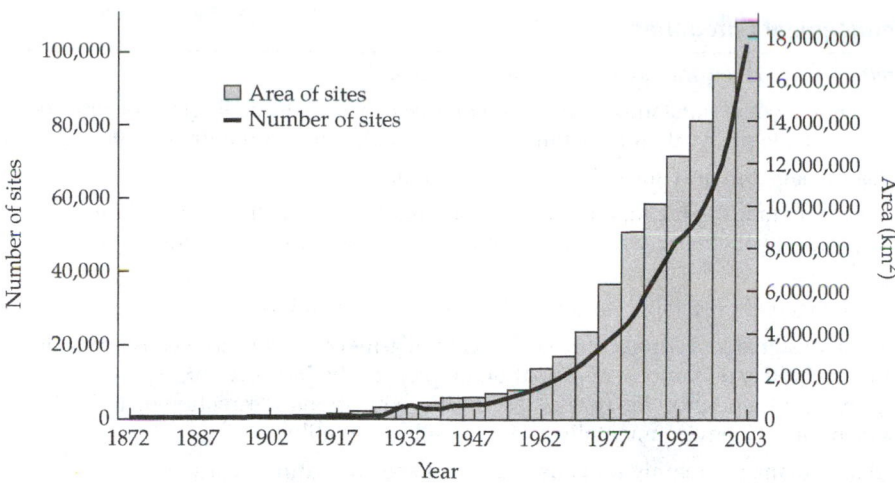

Figure 14.2   Cumulative growth in protected areas from 1872–2003. Number of sites includes both terrestrial and marine reserves, but area of sites includes only terrestrial reserves. (Modified from Chape et al. 2003.)

coverage of protected areas has more than doubled over the last decade to approximately 12.65% of Earth's land surface (Chape et al. 2003). Data on the distribution and status of protected areas are maintained in a freely accessible database by a consortium of organizations (World Database on Protected Areas Consortium 2004).

Types of Protected Areas

Protected areas fall under several different categories and each is accorded a different level of protection. Importantly, these include not only protection in strictly protected areas, but also in areas subject to a variety of man-agement arrangements that attempt to balance competing uses (land or marine) with biodiversity conservation objectives. The definition of a protected area adopted by the IUCN is, "an area of land and/or sea especially dedicated to the protection and maintenance of biological diversity, and of natural and associated cultural resources, and managed through legal or other effective means."

Protected areas vary in management intent from strict protection to sustainable extraction of natural resources. The IUCN has defined six protected area management categories, based on the primary management objective, which are summarized in Table 14.1, and reviewed next with some examples.

BOX 14.1   Convention on Biological Diversity

The importance of in situ conservation was highlighted in the United Nations Convention on Biological Diversity—a key document to arise from the Rio de Janeiro Earth Summit in 1992—where it is noted that "the fundamental requirement for the conservation of biological diversity is the in situ conservation of ecosystems and natural habitats and the maintenance and recovery of viable populations of species in their natural surroundings." With regard to in situ conservation, Article 8 of the Convention on Biological Diversity states that each contracting party must as far as possible and as appropriate:

1. Establish a system of protected areas or areas where special meas-

ures need to be taken to conserve biological diversity.

2. Develop, where necessary, guidelines for the selection, establishment and management of protected areas or areas where special measures need to be taken to conserve biological diversity.

3. Regulate or manage biological resources important for the conservation of biological diversity, whether within or outside protected areas, with a view to ensuring their conservation and sustainable use.

4. Promote the protection of ecosystems, natural habitats, and the maintenance of viable populations of species in natural surroundings.

5. Promote environmentally sound and sustainable development in areas adjacent to protected areas with a view to furthering protection of these areas.

6. Rehabilitate and restore degraded ecosystems and promote the recovery of threatened species (among other things) through the development and implementation of plans or other management strategies.

7. Endeavor to provide the conditions needed for compatibility between present uses and the conservation of biological diversity and the sustainable use of its components.

TABLE 14.1   *IUCN Protected Area Management Categories*

Category Ia	Strict nature reserve: Protected area managed mainly for scientific research.
	An area of land and/or sea possessing some outstanding or representative ecosystems, geological, or physiological features and/or species; available primarily for scientific research and/or environmental monitoring.
Category Ib	Wilderness area: Protected area managed mainly for wilderness protection.
	A large area of unmodified or slightly modified land, and/or sea, retaining its natural character and influence, without permanent or significant habitation, which is protected and managed so as to preserve its natural condition.
Category II	National park: Protected area managed mainly for ecosystem protection and recreation.
	A natural area of land and/or sea, designated to protect the ecological integrity of one or more ecosystems for present and future generations, exclude exploitation or occupation inimical to the purposes of designation of the area, and provide a foundation for spiritual, scientific, educational, recreational, and visitor opportunities, all of which must be environmentally and culturally compatible.
Category III	Natural monument: Protected area managed mainly for conservation of specific natural features.
	An area containing one or more specific natural, or natural/cultural feature that is of outstanding or unique value because of its inherent rarity, representative or aesthetic qualities, or cultural significance.
Category IV	Habitat/species management area: Protected area managed mainly for conservation through management intervention.
	An area of land and/or sea subject to active intervention for management purposes to ensure the maintenance of habitats and/or to meet the requirements of specific species.
Category V	Protected landscape/seascape: Protected area managed mainly for landscape/seascape conservation and recreation.
	An area of land (with coast and sea as appropriate) where the interaction of people and nature over time has produced an area of distinct character with significant aesthetic, ecological and/or cultural value, and often with high biological diversity. Safeguarding the integrity of this traditional interaction is vital to the protection, maintenance and evolution of such an area.
Category VI	Managed resource protected area: Protected area managed mainly for the sustainable use of natural ecosystems.
	An area containing predominantly unmodified natural systems, managed to ensure long term protection and maintenance of biological diversity, at the same time providing a sustainable flow of natural products and services to meet community needs.

Source: IUCN 1994.

Strict nature reserves and wilderness areas (Category I)

The primary purpose of Category I protected areas is to protect biodiversity and maintain evolutionary and ecosystem processes, as well as ecological services. Category I areas also are managed for scientific research and environmental monitoring. Recreation is excluded so that research and monitoring can be undertaken in minimally-disturbed sites. Selection of these areas is based not only on their representative character, but on the adequacy of their size for protecting their values of importance. The areas should be significantly free of direct human intervention, although the objectives of wilderness areas may be to allow indigenous people to maintain their lifestyle and traditional forms of ecosystem management. The establishment of Category I protected areas is difficult because they exclude mechanized forms of transportation and extractive use, and limit access.

National parks (Category II)

National parks are protected areas managed mainly for ecosystem protection and human enjoyment or recreation. Direct exploitation is excluded and parks are designated to provide for environmental preservation as well as spiritual, scientific, educational, and recreational opportunities that are environmentally and culturally compatible. Achieving this dual mandate of providing both ecosystem protection and opportunities for recreation can be complicated. For example, in the U.S. a controversial issue is whether the recreational use of snowmobiles should be allowed in Yellowstone National Park's backcountry in wintertime. It has been argued that their use compromises the Park's mandate for environmental preservation due to their associated noise and exhaust pollution.

An example of a Category II protected area is the 32,000-ha Tubbataha Reef Marine Park of the Philip-

pines. This park is comprised of two atolls and protects outstanding marine resources, which are considered to be of great importance for sustaining the region's fisheries, and also support an active recreational diving industry. Case Study 14.1 by Carlos Fernández-Delgado highlights the Doñana National Park of Spain, which is a globally important stopover for migrant bird species.

Natural monuments (Category III)

Category III protected areas are managed for the conservation of specific natural or cultural features and thus are generally more limited in size and scope than Category I or II protected areas. Natural monuments might protect natural features such as a waterfall, cave, or dune, as well as culturally significant features such as ancient archaeological sites. Natural Monuments may also protect significant biological features, such as the Giant Sequoia National Monument in the U.S.

Habitat/species management area (Category IV)

Category IV protected areas are established for conservation purposes but require management intervention to ensure their biodiversity values are sustained. Scientific research and environmental monitoring are often the primary activities undertaken in these areas. The Baiyer River Sanctuary in Papua New Guinea, for example, provides wildlife habitat and protects one of the largest populations of birds of paradise in the world in a region where much of the landscape is cultivated for coffee and tea.

Protected landscape/seascape (Category V)

Category V protected areas are designed to protect the historical interaction of people and nature. Management objectives focus on safeguarding the tradition of this interaction, which may involve protecting traditional land uses or building practices, and social and cultural values. The Mount Emei and Leshan Giant Buddha in Sichuan, China, for example, contains both natural and cultural features of significance. The area is one of the four holy lands of Chinese Buddhism. Cultural artifacts present at the site include the 71 m-high statue of Buddha, which was carved into a prominent mountain peak in the early eighth century. Some 2000 people continue to live inside the area, including monks and nuns that reside in the temples and monasteries. Protection of the site is important for a number of endemic and globally threatened species of flora and fauna.

Managed resource protected area (Category VI)

Category VI protected areas are managed to ensure long-term protection of biological diversity and allow for sustainable resource use by communities. The Ngorogoro Crater Conservation Area of northern Tanzania is an area of global significance for its geological, cultural, and nat-

ural history. The pastoral Maasai people use the area for cattle grazing: it is estimated that some 285,000 cattle graze approximately 75% of the conservation area. While the conservation area was initially developed to benefit the Maasai, it is now recognized for its conservation value. The area is one of the largest, inactive, unbroken, and unflooded calderas in the world and is home to one of Africa's largest wildlife aggregations and an isolated relict population of the black rhinoceros (*Diceros bicornis*).

Biosphere Reserves, Ramsar Wetlands, and World Heritage Sites

Protected areas of all category levels can also be classified as Biosphere Reserves, Ramsar Wetlands, and World Heritage Sites. For example, the Ngorogoro Crater Conservation Area (a Category VI protected area) and Yellowstone National Park (a Category II protected area) are both Biosphere Reserves and World Heritage Sites.

There are currently 411 biosphere reserves in 94 countries dedicated under the UNESCO Man and the Biosphere Programme (Groombridge and Jenkins 2002). Biosphere Reserves are dedicated for a variety of objectives including research, monitoring, training, and demonstration as well as conservation. Importantly, biosphere reserves in the ideal are designed to create one or two areas of low-intensity human uses surrounding a strictly protected area at the core (Figure 14.3). In reality,

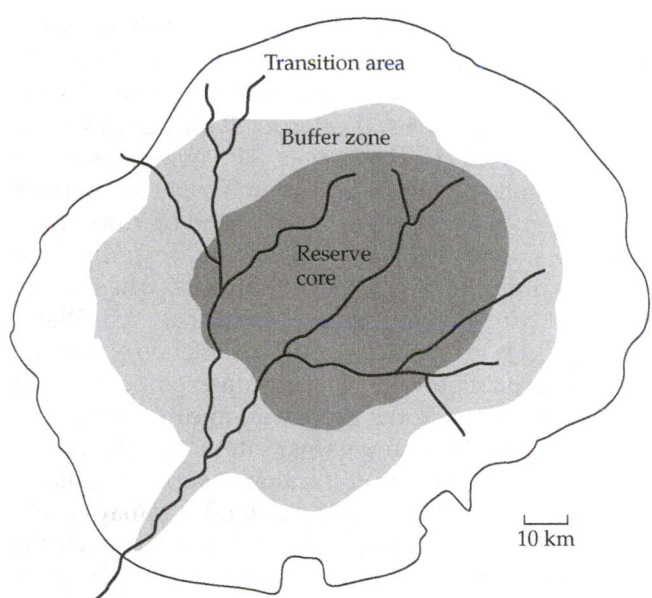

Figure 14.3   A schematic diagram of a biosphere reserve system, showing a core area that is accorded complete protection surrounded by a buffer zone where human uses are highly compatible with biodiversity conservation, which is in turn surrounded by a transition zone in which more intensive human uses are allowed. Beyond the biosphere reserve the full spectrum of human uses exist.

biosphere reserves often have a greater mix of uses, and are not as ideally shaped for biodiversity conservation (e.g., circular with a strict protected area at the core away from most human influences).

The Convention on Wetlands of International Importance was signed in 1971 in Ramsar (Iran) and provides a framework for international cooperation for the conservation of wetland habitats. The convention's mission is "the conservation and wise use of all wetlands through local, regional and national actions and international cooperation, as a contribution towards achieving sustainable development throughout the world." This is the only global environmental convention whose mission is to target particular ecosystem types. Wetlands are broadly defined by the convention to include rivers, lakes, swamps, wet grasslands, coral reefs, estuaries, deltas, tidal flats, and near-shore marine areas, as well as some human-made wetlands such as fishponds, and rice paddies. Contracting parties and member countries of the convention commit to designating eligible areas as Ramsar wetlands, promoting wise use of wetlands, and consulting with other parties about management and implementation of convention regulations. In return, management tools and technical expertise are made available through the formation of partnerships and financing. As of March 2002, 1148 Ramsar wetlands had been designated, which cover approximately 96 million ha (Groombridge and Jenkins 2002).

The Convention Concerning the Protection of the World Cultural and Natural Heritage was adopted in Paris in 1972, and provides for the designation of areas of "outstanding universal value to all the peoples of the world" as World Heritage sites. The objectives of the Convention are to encourage member parties to track and report on the state of conservation of World Heritage sites; to provide technical assistance and professional training for site preservation, and when necessary, to provide emergency assistance for World Heritage sites in immediate danger. Other objectives are to enhance public awareness, encourage participation of local populations in the preservation of their cultural and natural heritage, and garner international cooperation in conservation of cultural and natural heritage. Signatory parties that uphold the standards may benefit from international recognition and assistance. Of the 788 World Heritage sites distributed among 124 countries, 611 are recognized for cultural values, 154 for natural values, and 23 for both cultural and natural values.

Strict protection versus multiple use

The role of strictly protected (Category I–IV) versus multiple use areas (Category V–VI) in meeting biodiversity conservation goals has been hotly debated. On the one hand, strictly protected areas that exclude hunting and other extractive uses are likely to be most efficient at meeting biodiversity conservation goals. National parks and monuments, wildlife sanctuaries, and game reserves have formed the cornerstone of efforts to conserve biodiversity worldwide. However, exclusionary tactics may alienate people who benefit from extractive use of these resources, thus making multiple-use designations more amenable to achieving broad conservation goals. While each of the different types of protected area can contribute to conserving biodiversity, they vary in their contribution to biodiversity conservation relative to support of human populations (Redford and Sanderson 2000; Terborgh 2000; Adams 2004).

For example, territories established to protect the rights of traditional peoples may play a major role in conserving biodiversity in the region by prohibiting outsiders from pursuing extractive industries. However, local extinction of several species has been documented in indigenous reserves. Thus, we should recognize that indigenous people are not necessarily more likely than people from affluent, modern societies to be conservationists (Redford and Sanderson 2000). While each of the different types of protected area can contribute to a goal, consensus is emerging that specific reserves should be established for specific objectives. In the case of indigenous reserves, it may be advisable to establish a reserve that protects the rights of local peoples to their lands, but may or may not have specific biodiversity outcomes as a goal. If conservation goals are unlikely to be met in a given protected area, then alternative areas where they can be met should be identified.

There is growing recognition that conservation objectives are unlikely to be achieved by the dedication of strictly protected areas alone (Reid 1996). Thus, 23.3% of the total extent of the world's protected areas are assigned to Category VI (managed resource protected areas); protected areas managed mainly for the sustainable use of natural ecosystems (IUCN 1994; Chape et al. 2003; Figure 14.4). This category recognizes the role that protected areas play in sustaining the livelihoods of local people and therefore accommodate a degree of sustainable use as part of their management (Chape et al. 2003). Two of the world's largest protected areas are classified as Category VI: the Ar-Rub'al-Khali Wildlife Management Area in Saudi Arabia (640,000 km^2) and the Great Barrier Reef Marine Park in Australia (345,400 km^2). By comparison, less than 11% of the extent of the world's protected areas is assigned to Category Ia (strict nature reserve) and Category Ib (wilderness area) and 23.5% is assigned to Category II (national park) (IUCN 1994; Chape et al. 2003). Although there are many Category III (natural monument) protected areas, they generally cover only small geographic areas. A third of the world's protected areas have not been assigned an IUCN management category (see Figure 14.4).

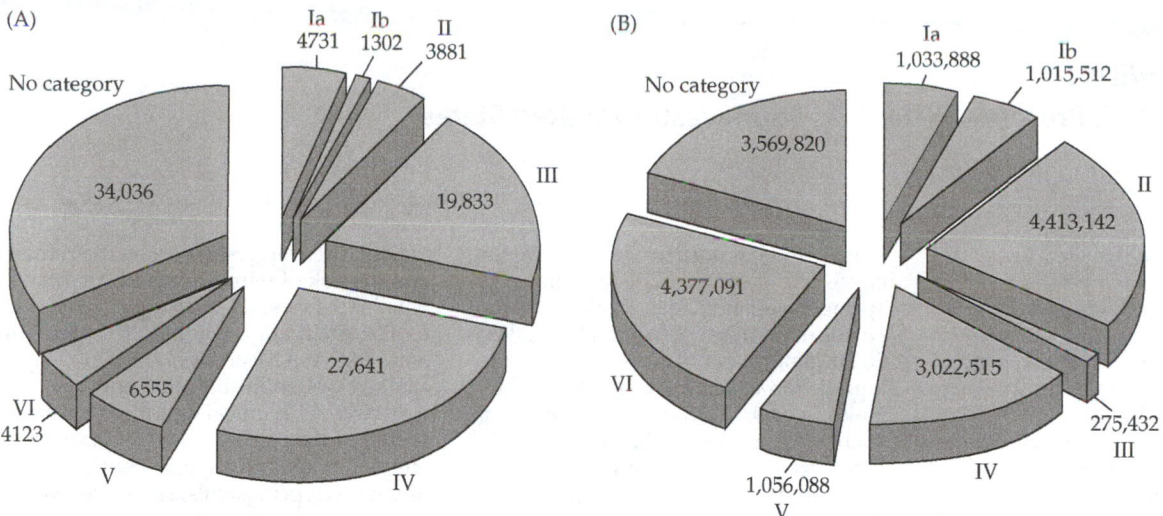

Figure 14.4 Number of protected areas (A) and area covered by protected areas (B) in km² by IUCN categories I–VI, and those unassigned to any IUCN category. (Modified from Chape et al. 2003.)

Management Effectiveness of Protected Areas

Threats to protected areas must be eliminated if the protected areas are to meet their objectives and contribute to biodiversity conservation (Essay 14.1 provides an example from the U.S. southwest). The IUCN established a framework for measuring management effectiveness, which considers: (1) issues related to design; (2) appropriateness of management; and (3) whether the objectives of protected areas are delivered. Design issues include considerations of size and shape, buffer zones, and linkage to other areas. Inappropriate design may result in protected areas that are too small to meet their conservation objectives. Without adequate or appropriate management, threats may continue in spite of legal designation of protected status (Case Study 14.1 discusses such challenges). Assessment of the level to which protected areas are meeting their stated objectives involves evaluation of both biological and social outcomes. These three criteria are being assessed for protected areas around the globe so that management can be improved and resources for protected area establishment can be mobilized.

For example, due to concern about deterioration of natural areas in Brazil, the World Wildlife Fund (WWF) and the Brazilian Environment Institute (IBAMA) evaluated 86 protected areas. Indicators were selected to measure both implementation of protected area goals, as well as vulnerability of each protected area. The study concluded that 47 of the protected areas were largely unimplemented, 32 were minimally implemented, and only seven were implemented to a reasonable degree. These results were used to lobby the Brazilian govern-

ment for increased funding and support for its reserve system. The approval of an investment scheme, whereby funds generated from protected areas are reinvested in the protected areas, resulted from this lobbying (Hockings et al. 2000).

Many protected areas lack resources, are inadequately managed, and thus do not achieve the conservation goals for which they were established. In addition, many face significant threats and challenges from inappropriate development both within and outside their boundaries (Reilly 1985; Liu et al. 2001). While legislative protection may be adequate to stop the exploitation of wildlife and habitat in some countries, in others, protected areas are vulnerable to hunting, encroachment, and timber harvesting. In some regions, protected areas are not even secure from vegetation clearing (Peres and Terborgh 1995; Menon et al. 2001).

In spite of these inadequacies, protected areas have been shown to effectively reach many of their goals (Chomitz 1996; Bruner et al. 2001). Even in cases of little or no management or infrastructure, species abundance and diversity is usually higher within protected areas than in the surrounding landscape, and many extinctions have been prevented by setting them aside. In regions of intense human impacts, protected areas are often the only remaining patches of native vegetation. In the marine realm, "there is compelling, irrefutable evidence that protecting areas from fishing leads to rapid increases in abundance, average body size, and biomass of exploited species. It also leads to increased diversity of species and recovery of habitat from fishing disturbance" (Roberts 2000).

In one study of management effectiveness, researchers used a questionnaire to collect data on land-use pressure, local conditions, and management activities of 93 pro-

ESSAY 14.1

Constant Vigilance
Maintaining a Fish Preserve in the Arid Southwestern United States

Jack E. Williams, *USDA Forest Service*

■ The desert oasis Ash Meadows has been the scene of many battles over land and water use. Endangered fishes—mostly species of pupfish—have been the focus of concern. One pivotal battle between the diminutive Devils Hole pupfish (*Cyprinodon diabolis*) and agricultural interests intent on withdrawing water from underground aquifers that feed springs in Ash Meadows was ultimately decided in the U.S. Supreme Court. On June 7, 1976, the Court upheld a lower court ruling for a permanent injunction of groundwater pumping until a safe water level for the pupfish could be established. During the court proceedings, the fate of the pupfish became the rallying cry for both conservationist and development interests (Figure A). Now, more than two decades later, and despite numerous endangered species listings and the creation of a National Wildlife Refuge at Ash Meadows, questions still remain concerning the survival of the Devils Hole pupfish and the area's many other endemic species.

Ash Meadows, located along the Nevada–California border approxi-mately 145 km northwest of Las Vegas, consists of dozens of crystal-blue springs, wetlands, and alkaline uplands surrounded by the Mojave Desert. With 26 endemic taxa of fishes, springsnails, aquatic insects, and plants within the 95-km² oasis, it is the smallest area with such a rich and specialized biota in the U.S. (Deacon and Williams 1991).

As in most preserves, protection of Ash Meadows has come in fits and starts. On January 17, 1952, President Harry S. Truman declared 16 ha around Devil's Hole a disjunct portion of Death Valley National Monument. The natural values of Devil's Hole received further protection through court rulings during the 1970s. The major breakthrough for protection of the entire Ash Meadows area, however, came in June of 1984, when 5154 ha was acquired with the help of The Nature Conservancy and designated by Congress as the Ash Meadows National Wildlife Refuge. This halted plans for a large commercial and residential development, but not before some spring systems were drained and ditched and their flows diverted to create reservoirs. Certain areas within the refuge boundary remain privately owned, while the rest are managed by the Bureau of Land Management.

Introduction of nonnative species has been a long-term problem at Ash Meadows. During the 1960s, an illegal tropical fish farm provided the source for a wide variety of exotic species that flourished in the warm spring waters. Large populations of introduced mosquitofish (*Gambusia affinis*) and sailfin mollies (*Poecilia latipinna*) persist in many areas. Largemouth bass (*Micropterus salmoides*) and channel catfish (*Ictalurus punctatus*) have been introduced into reservoirs and have invaded springpools, where they prey on the native pupfish and speckled dace. Introduced bullfrogs and crayfish also prey on native species.

Recent management efforts have focused on control of exotics and restoration of natural spring channels and wetland habitats (Stein et al. 1999, 2000). Largemouth bass and channel catfish have been removed from spring systems but still occur in reservoirs on the refuge and the risk of reinvasion is substantial. Smaller fishes such as mosquitofish and mollies are notoriously hard to control in areas like Ash Meadows, with its many shallow wetlands and interconnected waterways. Chemical treatment has been tried, but the effects of such control efforts on nontarget species, such as the tiny native springsnails, can be severe. Meanwhile, work continues to restore habitats. Channelized spring outflows have been restored and other spring flows have been diverted back into natural meandering channels.

Long-term maintenance of groundwater aquifers that feed Ash Meadows' spring systems is likely to be an even bigger challenge than control of nonnative species. Removal of groundwater from areas outside Ash Meadows may have negative consequences in the future by reducing springflows within Ash Meadows. With deep groundwater throughout much of the region flowing from northeast to southwest, water rights acquisition by the City of Las Vegas across much of southern Nevada is cause for concern.

These issues cannot be resolved easily by the National Wildlife Refuge system, which traditionally has focused on waterfowl production and hunting and is poorly equipped to deal with problems originating outside refuge boundaries. Even fishing activities on the refuge's reservoirs conflict with protection of the endemic spring-dwelling species because of the likelihood for exotic predatory fishes to reinvade spring systems. Restoration of natural springs, their outflows, and desert wetlands may conflict with desires for improved vehicle access, recreation facilities, and our tendency to intensively manage landscapes. And, as with the groundwater concerns, we are finding that ecosystem boundaries seldom conform to the administrative boundaries of the preserve.

To date, many urban and agricultural centers of the arid western U.S. have flourished with little regard for water consumption rates or effects on native biota. There are better alternatives for meeting the growing urban

Figure A Reserves protecting species such as the Devils Hole pupfish typically provoke visible responses on both sides of the development/conservation debate. (Photograph by E. P. Pister.)

needs for water than tapping our already-depleted surface and ground-waters. Professor James Deacon of the University of Nevada, Las Vegas, correctly questioned why society should spend billions of dollars on new water projects when it would be cheaper and environmentally more sound to "get serious about retrofitting Las Vegas for water efficiency [and] then get serious about converting agriculture in the Colorado River basin to water efficiency and use the savings for urban needs— in both Nevada and California."

Water use, whether surface waters on the refuge or groundwater from out-side refuge boundaries, will continue to garner political attention. How society responds to these issues may be the ultimate court case for the Devils Hole pupfish and the other endemic life forms in Ash Meadows. ■

---

tected areas in 22 countries (Bruner et al. 2001). Only 17% of the protected areas, which had a median age of 23 years, had experienced net clearing since establishment (Figure 14.5). The protected areas were more heavily impacted by hunting and logging than by clearing, but these impacts were considerably reduced inside their boundaries as compared to their surroundings. Management effectiveness was correlated with basic management activities such as enforcement, degree of boundary demarcation, direct compensation to local communities, and deterrents to illegal activities (e.g., probability that a violator was sanctioned). The density of guards was especially important: the median density of guards in the 15 most effective protected areas was more than eight times higher than in the 15 least-effective protected areas.

A recent report by World Wildlife Fund (WWF) showed that management effectiveness is correlated with IUCN category, and that World Heritage, UNESCO Man and Biosphere Reserve, and Ramsar sites are as effective as other types of protected areas at meeting conservation objectives. An analysis of threats showed that poaching, logging, encroachment, and collection of nontimber forest products, were more of a problem than all other threats combined (WWF International 2004).

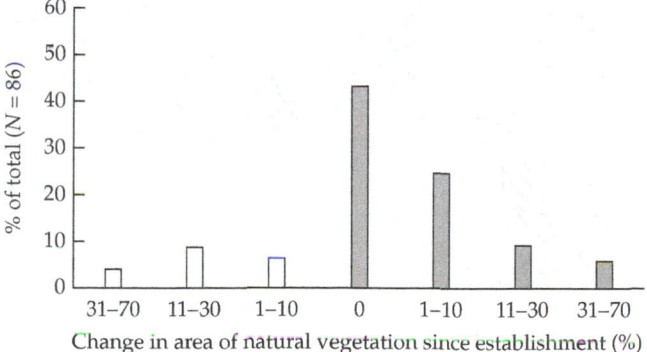

Figure 14.5 Change in the area of natural vegetation since establishment for 86 tropical protected areas. White bars denote loss of areas and gray bars denote an increase in area (due to regrowth and restoration). The majority of protected areas have either experienced no net clearing or have actually increased natural vegetative cover. Protected areas have a median age of 23 years. (Modified from Bruner et al. 2001.)

The Need for Reserve Systems

Single protected areas will rarely be of adequate size or scope to conserve a representative sample of the biodiversity of a region, therefore entire reserve systems are critical for the conservation of biodiversity. However, a reserve system need not be comprised entirely of areas with strict protection. For example, it might be comprised of a combination of strictly protected areas that are off-limits, areas that afford protection while also allowing light use, and indigenous lands where traditional harvesting practices can continue.

This is not to say that single protected areas are unimportant, even if they do not harbor great species richness. Certain areas may, for example, be low in richness but hold significant populations of unique or critically threatened species, or be the only site in the world where a species is found. The Kihansi Spray Toad *Nectophrynoides asperginnis*, for example, occurs only in the spray of a single waterfall on the escarpment of the Udzungwa Mountains, Tanzania (Poynton et al. 1998). The fate of the world's population of this toad thus depends on the protection of this single waterfall. In most cases, however, single protected areas are rarely, if ever, sufficient to represent an entire region, and thus we need to develop networks of reserve systems that can achieve our larger goals for biodiversity conservation.

The distribution of protected areas relative to habitat type is uneven, which is partly a legacy of ad hoc or politically expedient decisions about protected area establishment (Pressey 1994). Early efforts toward protected area establishment tended to focus on designation of single protected areas rather than entire reserve systems. Analyses of reserve systems at global, regional, or national scales indicate that there are gaps and biases in the representation of biodiversity (e.g., Scott et al. 2001; Andelman and Willig 2003; Rodrigues et al. 2004a). **Gap analysis** (Scott et al. 1987; Jennings 2000) is an approach used to identify "gaps," or areas of under-representation in the existing reserve system, by comparing the distribution of protected areas with the distribution of species, vegetation types, or other types of biodiversity (Essay 14.2).

In 1992, the Fourth Congress on National Parks and Protected Areas (Caracas, Venezuela) called for protection of at least 10% of each major biome by the Year 2000

ESSAY 14.2

Gap Analysis
A Spatial Tool for Conservation Planning

J. Michael Scott and Jan Schipper, *University of Idaho*

■ Gap analysis is an analytical approach to conservation assessment that provides information that can be used to "keep common species common" by identifying those animal and plant species or communities that are not adequately represented in existing conservation lands. Gap analysis can be conducted at a variety of spatial scales and for whatever suite of animal and plant communities is of biological or political interest. Information from gap analyses can be used by land managers, planners, scientists, and policymakers to provide information they need to make more informed decisions when identifying priority areas for conservation.

Gap analysis had its beginnings in Hawaii (Kepler and Scott 1985; Scott et al. 1986; Scott et al. 1993) when information on the distribution of endangered Hawaiian forest birds was compared, using a geographical information system (GIS), with the occurrence of areas dedicated to the long-term conservation of native plants and animals. Results of this comparison showed almost complete lack of overlap between area of occupancy of endangered birds and areas established to protect native species (Figure A). This information was used by agency personnel and policymakers to establish Hakalau Forest National Wildlife Refuge. The first state-wide gap analysis to assess conservation status of vertebrate species and dominant habitats was in Idaho (Caicco et al. 1995; Kiester et al. 1996). Since then, gap analyses have been completed or are underway in every state of the U.S. and in many other countries.

Gap assessments are conducted using maps of species predicted distributions (Scott et al. 2002) and mapped land cover types (Table A). Land cover maps are developed using satellite imagery (e.g., Landsat 7) as the base data in a GIS. Then, using field plots, serial photos, and other sources are used to help classify the unique spectral classes derived from the imagery into distinct vegetation classes.

Predicted species distributions are based on existing range maps,

known occurrences, and other distributional information combined with information on the habitat affinities for the species (Scott et. al. 1993; Karl et. al. 1999) Distribution maps for individual species can be overlaid using a GIS to create maps of species richness (Scott et al. 1987) for any group of species that is of political or biological interest. Additional maps of land ownership and land management are also created. These maps are then overlaid with the

maps of species distribution and mapped cover types to assess what area, and percentages of mapped occurrences of the species and cover types occurring in that area, fall within the existing conservation estate.

Gap analyses have been conducted at a diversity of spatial and thematic scales, from state-and ecoregional- to global-level assessments, and including terrestrial, freshwater, and marine components. Dozens of U.S. states have

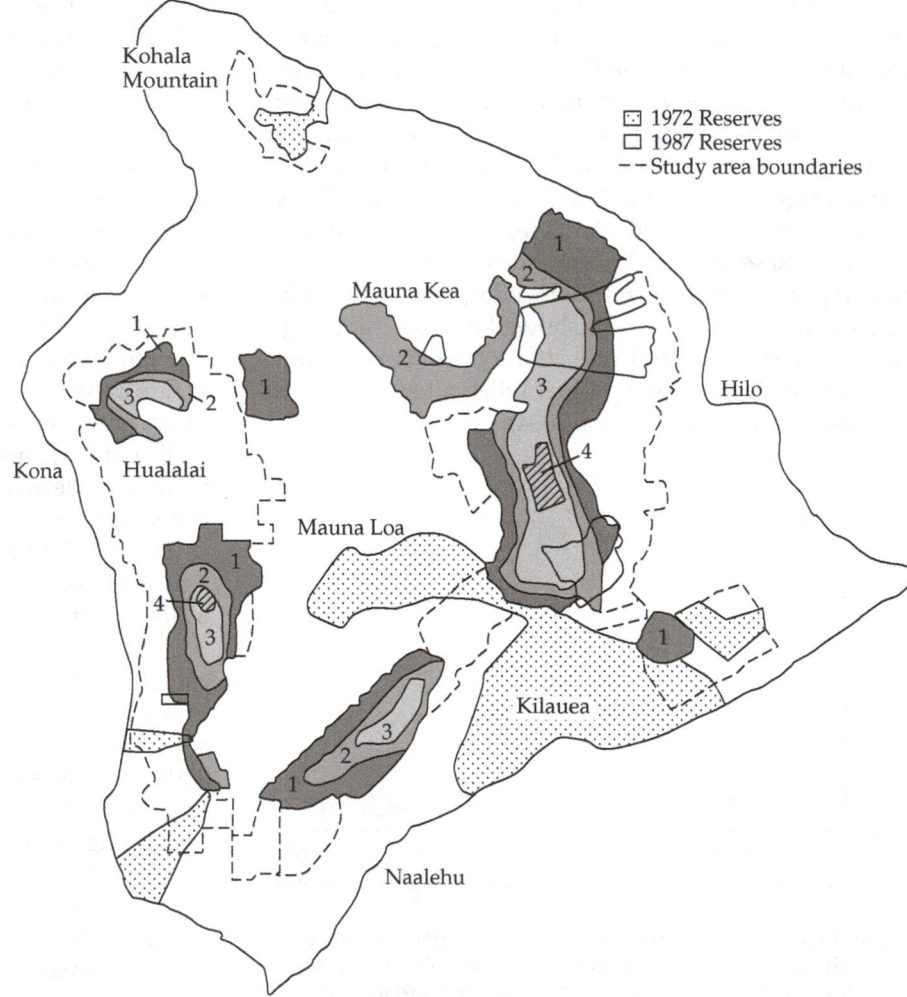

Figure A   Distribution of endangered Hawaiian finches in relation to existing nature reserves on the island of Hawaii in 1982. Numbers refer to the number of finch species occurring over a given range. Note that reserves established by 1972 had almost no overlap with any finch species. New reserves established in 1987 have improved this situation based on the results of gap analyis. (Modified from Kepler and Scott 1985.)

TABLE A *Six Fundamentals of Gap Analysis*

1. Maps existing vegetation
2. Maps predicted distribution of native species
3. Maps land ownership for public lands and land management status for all lands
4. Shows the current distribution of protected areas
5. Compares distribution of any species, group of species or vegetation types of interest with the conservation network
6. Provides an objective data set for local, regional, state, and national interests to make decisions regarding conservation of species and ecosystems

completed gap analysis, and many more are in progress (Figure B shows a flowchart of state-level gap processes). Wright et al. (2001) and Davis et al. (1995) have conducted multi-state conservation assessments using gap analysis, while Stoms et al. (1998) have conduced gap analyses within a single ecoregion. At the national scale, Scott et al. (2001) conducted a conservation assessment of the occurrence of dominant cover types within existing nature reserves and of the distribution of existing nature reserves by elevation and soil types. Crumpacker et al. (1998) assessed the occurrence of potential vegetation types on all public lands in the U.S. Additionally, there has recently been a conservation assessment of eco-

logical content and context of refuges in the U.S. Fish and Wildlife Services National Wildlife Refuge System in the coterminous U.S. (Scott and Loveland in press). Internationally, gap analyses have been conducted in both the terrestrial and marine realm, across parts of Africa, Asia, Europe, and the Middle East.

Gap provides decision-makers with information that can be used to make more-informed decisions regarding land use. The biggest single misapplication of gap is to take information from a gap analysis conducted at one scale and test its predictions or apply it at another scale. For example, a manager could be badly misled if he used predicted occurrence of species and vege-

tation types from a gap analysis conducted with a minimum mapping unit of 40 ha to make decisions within a 200 ha reserve. Gap is not a substitute for endangered and threatened species management planning or research, neither is it a thorough nationwide inventory of biological resources, many of which cannot be mapped at a continental or nationwide scale. Also related to scale, gap maps for species distribution should be done at a scale appropriate to the species or a segment of the population of the species.

Gap analysis was born of the realization that a species-by-species approach to conservation is not effective for regional planning (Scott and Csuti 1997). Single species planning has proven reactive, delayed until a species was teetering on the brink of extinction and thus requiring large sums of money and personnel to mount a species recovery effort. Gap analysis is a *proactive* effort to keep species off the endangered species list and special management lists by protecting them and their associated habitat types while they are still common. This approach avoids many of the conflicts inherent in recovering endangered species.

For further information on gap analysis we refer readers to the USGS Gap Analysis Program website (see http://www.gap.uidaho.edu/) which provides an extensive data base of literature and information on state, regional, national, and international gap analysis efforts. ■

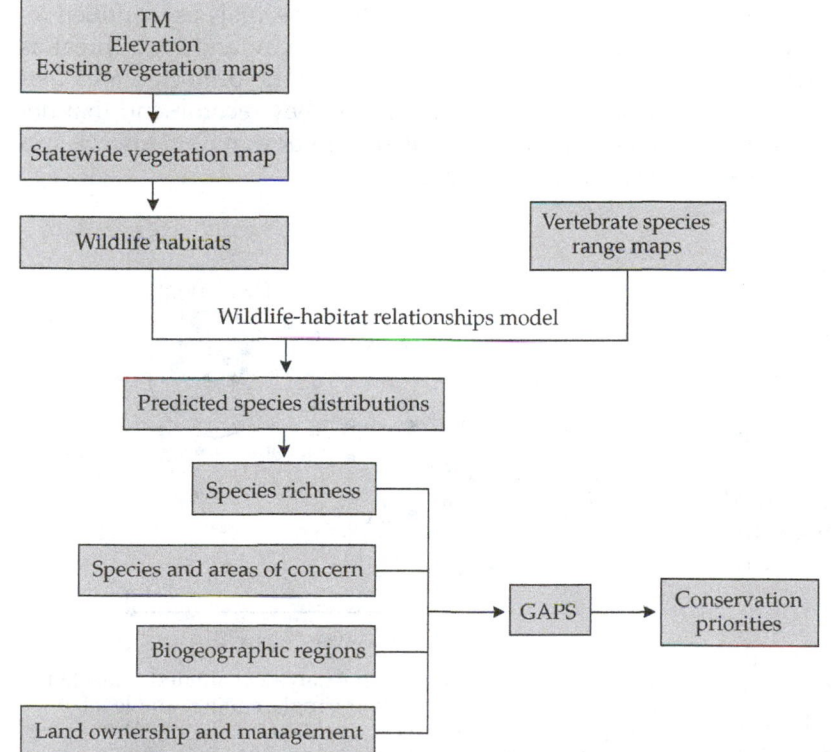

Figure B Flow chart of Gap analysis as applied to state-level analysis, but which can be adapted to other scales. Analyses of overlapping data layers allow identification of gaps between predicted and actual species occurences and protected areas, and thus provide material for deciding conservation priorities. (Courtesy of California Gap Analysis.)

Figure 14.6 Illustration of the methods used in the Global Gap Analysis. Data on species distributions were overlaid with the distribution of protected areas using GIS. Species whose distribution coincided at least partially with a protected area were considered "covered" and those whose distributions did not were considered "gap species." An overlay of the distributions of gap species produces a map that highlights regions where new protected areas are needed to protect these vertebrate gap species. (Modified from Rodrigues et al. 2003.)

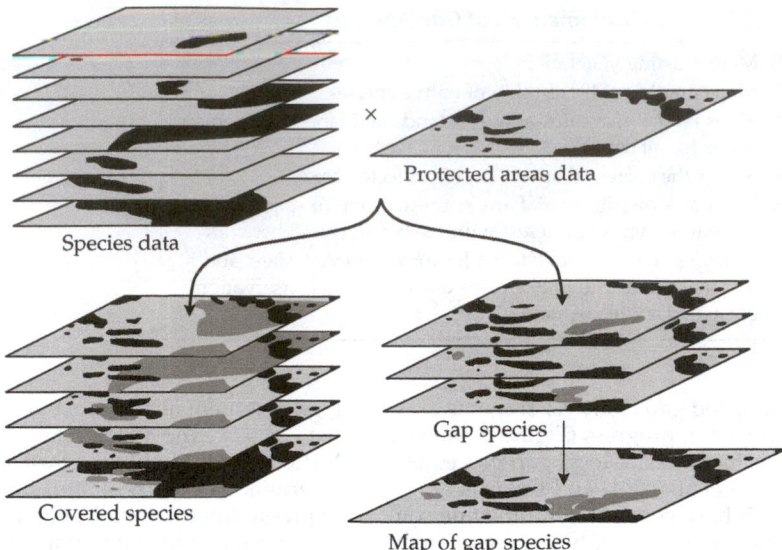

Species data

Protected areas data

Covered species

Gap species

Map of gap species

(IUCN 1993), and this has become a major national and international guideline. In 2003, the Fifth World Parks Congress (Durban, South Africa) announced that 11.5% of Earth's land surface is now under some form of protection, and that the 10% target has been reached for nine out of 14 major biomes (IUCN 2003).

Ironically, often it is the species of most conservation concern (i.e., threatened and small range species) that are most poorly represented within protected areas. Recently, the extent to which our existing protected areas include vertebrate species was assessed in the Global Gap Analysis Project (Rodrigues et al. 2004a,b; Brooks et al. 2004). The Global Gap Analysis combined data from the World Database on Protected Areas with distributional data for 11,633 species of mammals, amphibians, freshwater turtles and tortoises, and globally threatened birds. Species distribution maps were overlaid onto protected area maps using geographic information systems (GIS) to assess how well each species was represented in protected areas, and to identify *gap species*; those that are not covered in any part of their range (Figure 14.6). The analysis revealed 1424 species that are not protected in any part of their range, 804 of which threatened with extinction (20% of all threatened species analyzed). These numbers nearly double when considering the species that are represented only by very marginal overlaps with existing protected areas. Amphibians, overall, are the least well represented in protected areas, mainly due to their smaller ranges and higher levels of endemism, but also because they have received relatively less conservation attention.

The Global Gap Analysis also highlighted the skew in the distribution of protected areas, both geographically and in terms of size. For example, in the New World, the median size of strictly protected areas (Category I and II) is only 4.86 km^2 and 57% are less than 10 km^2 (Andelman and Willig 2003). In addition, the distribution of protected areas in the New World is skewed towards higher latitudes: 35% of the total area of strictly protected areas is found in Alaska (Andelman and Willig 2003). Globally, only 46% of protected areas are found in the tropics, where 76% of all species reside (Rodrigues et al. 2004a). Tropical forests, especially areas of topographic complexity, and islands were highlighted in the Global Gap Analysis as urgent priorities for the expansion of the global network of protected areas (see Plate 6). These areas are in regions long recognized to be centers of endemism that are suffering high levels of habitat destruction (Myers et al. 2000). Thus, the analysis identified regions of both high irreplaceability and high threat as priorities for the expansion of the protected area network (Figure 14.7). Further, they recommend that detailed analyses be initiated to design new reserve systems in these priority areas.

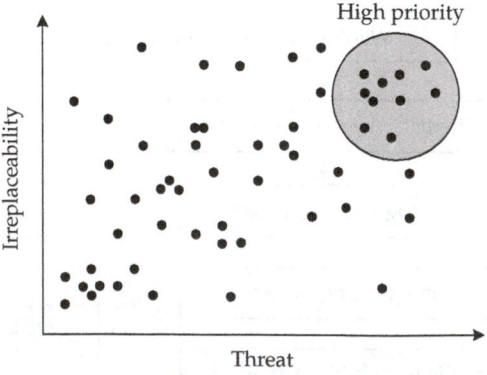

High priority

Irreplaceability

Threat

Figure 14.7 The Global Gap Analysis evaluated protected and unprotected sites for their irreplaceability and level of threat. Unprotected areas that ranked highly in both axes were highlighted as highest priority regions for the expansion of reserve systems. (Modified from Rodrigues et al. 2003.)

Creating a global, ecologically representative reserve system will require investments of U.S.$3–$11 billion per year over the next 30 years (according to estimates by James et al. 2001; Pimm et al. 2001). At the same time, globally intact ecosystems are being converted to human-dominated uses at a rate of over 1% per year (Balmford et al. 2002), creating an even more urgent need for the designation of new protected areas. Thus, there is a need to prioritize the allocation of scarce conservation resources to the expansion of existing protected areas and the creation of new ones so that the returns for biodiversity conservation are maximized.

Although protected areas comprise 12.65% of Earth's surface as of 2005, marine protected areas still make up a very small component: 0.5% of the surface area of the oceans are protected, which is equal to 1.89 million km^2 or 9.1% of the area of all protected areas (see Plate 5). The largest marine protected area is the Great Barrier Reef Marine Park in Australia (345,400 km^2), for which one-third is fully protected from extractive uses, such as fishing. Thus, there is a large need to expand our network of marine protected areas. One such effort is described by Satie Airame and Deborah Brosnan in Case Study 14.2.

Approaches to Planning Reserve Systems

The ecological purpose of a reserve system is to include and sustain representative samples of the full range of biodiversity and ecosystem processes of the region in which it lies (Margules and Pressey 2000). Some of the first reserves created primarily for biodiversity protection aimed to conserve a **flagship species**, usually a large mammal that is compelling to the public (Figure 14.8). While protected areas should represent and maintain biodiversity and separate it from the processes that threaten its persistence (Margules and Pressey 2000), historically, biodiversity considerations have played a small role in the selection of protected areas. Other factors, such as suitability for alternative land uses, availability, scenic beauty, and recreational value have played a more important role.

The trend of protecting land left over from major exploitative uses is recognized throughout the world, with examples from Australia (Pressey 1994; Pressey and Tully 1994), New Zealand (Mark 1985), Canada, the United Kingdom (Henderson 1992), the United States (Runte 1972; Shands and Healy 1977; Scott et al. 2001), Africa, and Japan (Pressey 1994). Strong evidence for the perception of protected areas as "worthless lands" in Australia comes from the revocations of protected areas after it was realized that they were preventing the exploitation of some resource of commercial value. For example, between 1939 and 1984 there were 23 revocations

of protected areas in Tasmania (Australia), the majority of which were national parks (Mercer and Peterson 1986). Some of the larger scale national park revocations were due to pressure from forestry, hydroelectric developments, and mining interests. In the 1950s there were revocations of protected areas in South Australia due to the expansion of the rural sector and even more recently to allow for mineral exploration.

Selecting protected areas in an ad hoc or opportunistic manner generally results in the conservation of economically marginal land and reserve systems that do not represent the full range of biodiversity (Pressey and Tully 1994). Representational biases in reserve systems often occur because certain environments are more politically and economically expedient to protect, leaving other areas poorly protected, regardless of their conservation value. Consequently, many reserve systems are dominated by dry, infertile, waterlogged, saline, or steep habitats.

When biodiversity conservation is the primary goal, reserve systems are often created for one of three purposes:

1. Protect particular species (e.g., threatened, flagship, or **umbrella species** [wide-ranging species whose requirements include those of many other species])

2. Preserve biodiversity, focusing on areas of high species richness (e.g., tropical rainforests or coral reefs, or areas with high endemism)

3. Preserve large and functioning ecosystems and their associated ecosystem services (e.g., catchments or watersheds)

Scoring systems based on these reserve system goals were developed in the 1980s in an attempt to provide an explicit and rational basis for reserve system design (Margules and Usher 1981; Smith and Theberge 1987; Pressey and Nicholls 1989). These systems scored or rated potential protected areas against several criteria to provide an overall indication of their conservation value. The main limitation of scoring approaches is that they: lack explicit goals, cannot deal with synergistic benefits of multiple areas, and cannot tell us when we have conserved enough. Therefore, when using scoring approaches, it is impossible to determine when full representation has been accomplished. For example, knowing that one set of species is globally imperiled, while another set of species is widespread but demonstrably secure, can be very effective in focusing our attention where it is most needed. However, this alone does not help us decide how many populations of the globally-imperiled species we should include in our reserve system. Scoring approaches generally lead to inefficient and unrepresentative reserve systems (Bedward et al. 1991; Margules et al. 1991; Pressey 1997).

(A)

(B)

(C)

Figure 14.8 Protected areas are often established primarily to conserve a single species, usually large, high-profile vertebrates, such as (A) African elephant (*Loxodonta africana*), (B) giant panda (*Ailuropoda melanoleuca*), or (C) Bengal tiger (*Panthera tigris*). (A, photograph by G. J. James/Biological Photo Service; B, by Gerry Ellis/Digital Vision; C, by Vivek Sinha/Biological Photo Service.)

The selection of protected areas using statistical analyses of multiple variables (multivariate environmental space; Belbin 1995; Faith and Walker 1996) or using gap analysis (Scott et al. 1993) also aims to identify a system of representative and complementary protected areas. Multivariate environmental space procedures select protected areas that allow the biggest incremental gain in capturing the range of environmental variation in a region. An underlying assumption of these approaches is that environmental space can serve as a surrogate for the range of biodiversity of a region.

Recent research efforts in the field of conservation planning have focused on the development of principles and tools to design efficient reserve systems that represent as much biodiversity as possible for a fixed cost. In the next sections, we review the principles behind systematic approaches for creating and designing reserve systems, referred to herein as **systematic conservation planning**. After the reserve system is created, the next set of challenges for conservation planning relate to the political and physical establishment of the reserve system.

Systematic conservation planning

Approaches to systematic conservation planning recognize that, due to constraints on the amount of land that can be set aside for biodiversity conservation, there is a need to conserve biodiversity in the most efficient manner possible (Pressey et al. 1993). Therefore, a common goal of

systematic conservation planning is to meet quantitative conservation objectives, such as conserving 15% of each habitat type within a system of complementary protected areas, as cheaply as possible. Conservation objectives (referred to by some authors as "targets") are operational definitions of a decision to reach a certain level of conservation for particular biodiversity features (Pressey et al. 2003). These objectives provide a clear purpose for conservation planning and improve the accountability and defensibility of the process.

Systematic conservation planning involves finding the best set of potential protected areas that satisfies a number of principles: **comprehensiveness**, **representativeness**, **adequacy**, **efficiency**, flexibility, risk spreading, and **irreplaceability**. In addition, principles regarding protected areas connectivity and shape are usually included. We explain and consider these concepts next.

COMPREHENSIVENESS   A comprehensive reserve system is one that contains examples of many biodiversity features, where biodiversity features might include species, habitats, or ecological processes. While ideally we would like to include a sample of every kind of biodiversity feature in our reserve system this is rarely achieved.

REPRESENTATIVENESS   Realistically, a fully protected reserve system will cover only a fraction of the landscape. Consequently, we will not be able to protect all the vari-

ety within most features, but rather our reserve system will only sample each biodiversity feature. Ideally, we would like each sample to be representative of that feature. For example, if we wish to conserve populations of a particular species, or samples of a habitat, we would prefer that the samples we choose cover the range of variation in that species and/or habitat. Despite its importance, the notion of representativeness is rarely explicitly dealt with in conservation planning. The concept of representativeness is closely related to the idea of comprehensiveness because if we define biodiversity features at a finer scale (e.g., subdivide a habitat type into several types), then comprehensively sampling the new, less coarsely defined habitat types, is equivalent to representing the coarser habitat type.

ADEQUACY   Even if we manage to dedicate enough protected areas to comprehensively protect representative samples of all biodiversity features, we cannot be sure that these features will persist in the reserve system. Ideally we would like to have a reserve system that is adequate to ensure the persistence and continued evolution of all features contained within it. This aspect of conservation planning is not well developed. Practitioners generally deal with the idea of adequacy by setting conservation objectives in the form of a target percentage of original extent or a target population size for each species as determined by population viability analysis (see Essay 12.2) or expert opinion.

EFFICIENCY   A simple way to abide by the principles of comprehensiveness, representativeness, and adequacy is to conserve everything. In practice this is impossible. The principle of efficiency is that we try to achieve our objectives for the least possible cost. Cost is often taken to be the area of land or sea protected, but may include purchase and management costs, or the costs of lost economic development. Efficiency is important, because it minimizes the possibility of constructing a reserve system that is too large and expensive to manage. An efficient reserve system is more likely to succeed in the face of competing interests. It is also a sound platform from which to expand a reserve system.

FLEXIBILITY   A flexible conservation plan is one that enables us to achieve our objectives efficiently in a number of ways. Flexibility might be important to take advantage of opportunities that arise for conservation, such as a block of land with high conservation value becoming available for purchase.

RISK SPREADING   There is a natural tension between the principle of connectivity and the idea of risk spreading. Catastrophes such as hurricanes and disease, and illegal habitat destruction can affect the conservation values of

any area, even if it is protected. Close, well connected protected areas might be a disadvantage if the arrangement increases the chance of disease spread, nonnative species invasion, or allows for disturbance events to move from one protected area to another. There are values of spreading risk that may outweigh any advantages to be gained from the elevated dispersal rates and increased recolonization potential that might be provided when protected areas are in close proximity. Risk spreading can be achieved by separating protected areas by a minimum distance.

IRREPLACEABILITY   The irreplaceability of an area reflects how important its inclusion is in the reserve system if we are to meet all our conservation objectives efficiently. A completely irreplaceable area is essential to meeting conservation objectives, whereas an area with a very low irreplaceability can be substituted by other sites. The measure of irreplaceability provides a quantitative assessment of the contribution of unselected areas for meeting conservation objectives (Pressey et al. 1993; Pressey et al. 1994; Ferrier et al. 2000). Irreplacability can be viewed in two contexts; the likelihood that an area is necessary to achieve the conservation objectives for the features it contains, or the extent to which the options for achieving conservation objectives are reduced if the area is unavailable for conservation.

If an area of high irreplaceability is not conserved, one or more features could be lost. Some areas are completely unique and vital to conservation objectives, and these must be included in the reserve system if at all possible. Irreplaceability can help determine which areas are priorities for conservation, but other constraints and considerations may mean that areas with lower irreplaceability are more suitable for conservation. For example, the vulnerability, condition, and/or cost of an area might influence its priority for protection.

CONNECTIVITY   Maintaining connections among protected areas is often an essential component of a reserve system. There are several reasons why connectivity might be important. First, species with populations distributed patchily across the landscape depend on dispersal for maintaining genetic variability; without movement between protected areas, populations may experience problems associated with inbreeding. Second, wide-ranging species, such as the African elephant (*Loxodonta africana*), may require more space than could reasonably occur in a single protected area, and their survival will depend on their ability to move between protected areas. Third, climate change is likely to alter the ranges of many species, and ensuring that the landscape mosaic between reserves of natural habitat remains hospitable will influence whether such species will persist into the future. Finally, conservation activities outside protected

areas considerably increase the opportunities for land-scape planning, which can address the sources of threats to biodiversity and improve ecological functioning. Therefore, while the primary focus of a reserve system is to secure areas most critical for the protection of biodiversity, thinking beyond the borders and planning for the landscape mosaic is also imperative to meeting the objectives of reserve systems.

Connectivity for ecosystem processes, such as water flow or fire regimes, also should be explicitly incorporated into conservation planning (Possingham et al. 2005), however there are few examples of this being achieved. For example in conserving riparian systems we might be interested in reserve systems that not only conserve a representative sample of each kind of riparian habitat, but that also ensure that each protected area is a self-contained catchment (i.e., all flows of water are contained within a given protected area). Similarly with processes such as fire, we may wish to minimize the chance that a single protected area is completely affected by a single fire. In cases in which vegetation mosaics are important for biodiversity conservation, each protected area should be of sufficient size to contain a complete range of fire-induced successional states. Using such process-driven objectives as opposed to pattern-driven objectives in systematic conservation planning is still in its infancy (Pressey et al. 2003).

PROTECTED AREA SHAPE   There are both ecological and economic reasons why protected areas should be large with low edge-to-area ratios and be well connected (as discussed previously). Long thin protected areas with high edge-to-area ratios are sensitive to edge effects caused by biotic interactions such as predation or abiotic factors such as humidity and wind (Fagan et al. 1999). For edge-sensitive species, the effectiveness of a protected area is reduced if it has a high edge-to-area ratio. Equally, from an economic perspective, boundaries should be minimized to reduce maintenance costs.

The concept of corridors between protected areas arose as a corollary of the theory of island biogeography (see Chapter 7). The definition of corridors includes the traditional "strip of land that differs from the adjacent landscape on both sides, linking two or more protected areas," and "stepping stones" of suitable habitat for various species within a matrix of less-suitable habitat that can serve as a "path" between protected areas (Earn et al. 2000). A large-scale biodiversity conservation corridor may contain several isolated or semi-isolated protected areas, each of which constitutes a "nucleus" of habitat connected through corridors (Sanderson et al. 2002). The size, shape, and management of corridors varies, but their primary purpose is to promote connectivity and allow movement of individuals between otherwise isolated patches of habitat.

Corridors and stepping stones are particularly prone to problems arising from small size and high edge-to-area ratios. Thus, a significant challenge is to identify potential corridors that are relatively robust to these problems. However, achieving this can add significantly to the costs of a reserve system.

MINIMIZING FRAGMENTATION OF RESERVE SYSTEMS   In fragmented and convoluted reserve systems, edges are long relative to area. The degree of fragmentation among protected areas in a reserve system can be measured by the boundary length of the reserve system divided by the area. The higher this ratio the more fragmented is the reserve system. One simple measure of how compact or clustered a reserve system is can be calculated from the ratio of the boundary length of the reserve system and the circumference of a circle of the same area (Possingham et al. 2000):

$$\frac{Boundary\ Length}{2\sqrt{\pi \times Area}}$$

A circle is the most compact shape possible, so this is the ratio of the boundary length to the theoretical minimum and is a dimensionless measure. Values approaching 1 resemble the shape of a circle and are relatively compact and unfragmented.

We can make a reserve system that is not fragmented or convoluted by setting an objective that takes into account the length of the reserve system boundary. By minimizing a combination of reserve system boundary length and cost, we can create efficient and compact reserve systems. As we place more emphasis on minimizing boundary length, costs may be compromised, and vice versa (Figure 14.9).

Each of these principles can help create a reserve system that will achieve broad conservation objectives of adequately representing biodiversity in viable populations. However, in practice, applying these principles often requires data we do not have, and may involve complex optimization algorithms and difficult tradeoffs.

The use of surrogates for reserve system planning

The principles of comprehensiveness, adequacy, and representativeness require that a reserve system must capture all the biodiversity of a region. However, information on biodiversity, even for well-known taxa such as birds and mammals, is incomplete. Therefore, the features used for conservation planning must act as **surrogates** for total biodiversity (Ferrier et al. 2000). Many types of surrogates have been proposed including well-known taxonomic groups, species assemblages, plant communities, and various spatial classifications of land and water, such as forest or marine benthic types (Margules and Pressey 2000; Ferrier 2002; Ferrier et al. 2002).

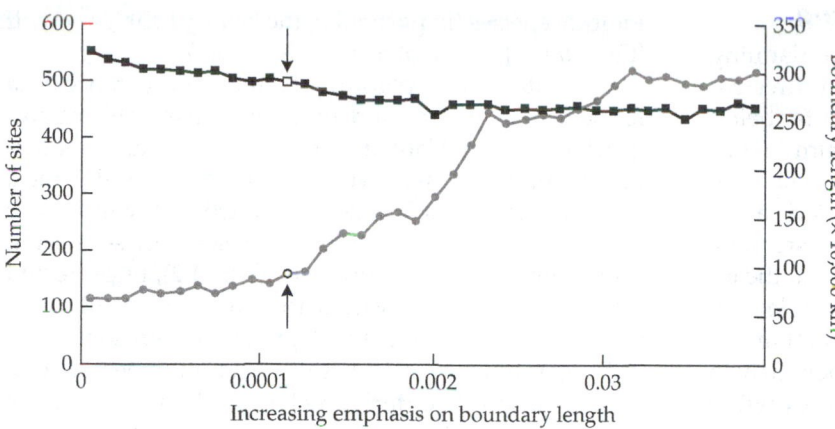

Figure 14.9 Maximizing the ecological effectiveness of a reserve system can involve analyses of the tradeoff between boundary length (which increases edge effects) and the number of areas protected. Pictured is the increase in boundary length (black line, measured in km × 10,000) and number of additional areas protected (gray line) for a hypothetical reserve system. The arrows indicate a reserve system that represents a reasonable compromise between increasing number of areas protected and boundary length: it reduces the boundary length by 6.7%, while it increases the number of areas protected by 22%.

The level of support in the literature for various surrogates has been variable (see Reyers and van Jaarsveld 2000; Beger et al. 2003; Ferrier 2002; Faith et al. 2004; Pressey 2004 for an introduction to the literature). Ideas about the properties that surrogates should have are emerging to efficiently represent other elements of biodiversity (Howard et al. 1998; Oliver et al. 1998; Su et al. 2004). However, a reserve system designed to be optimal and adequate for a single or even a set of species is not likely to satisfy the requirements of all species or biodiversity, even if species have similar spatial distributions (Andelman and Fagan 2000; Figure 14.10). Further, since it is unlikely that it will ever be possible to measure the variation of biodiversity within or between regions, the true effectiveness of biodiversity surrogates is indeterminable (Flather et al. 1997).

One conclusion from all studies testing the effectiveness of surrogates is that there will never be a perfect surrogate or suite of surrogates. The choice of surrogate will depend on both the presumed effectiveness of the surrogates available, and the amount of time, cost and effort required to develop alternative ones (Pressey and Ferrier 1995; Ferrier 1997, 2002). Conservation planning practitioners therefore should make the best use of all available environmental and biological data to inform decision-making.

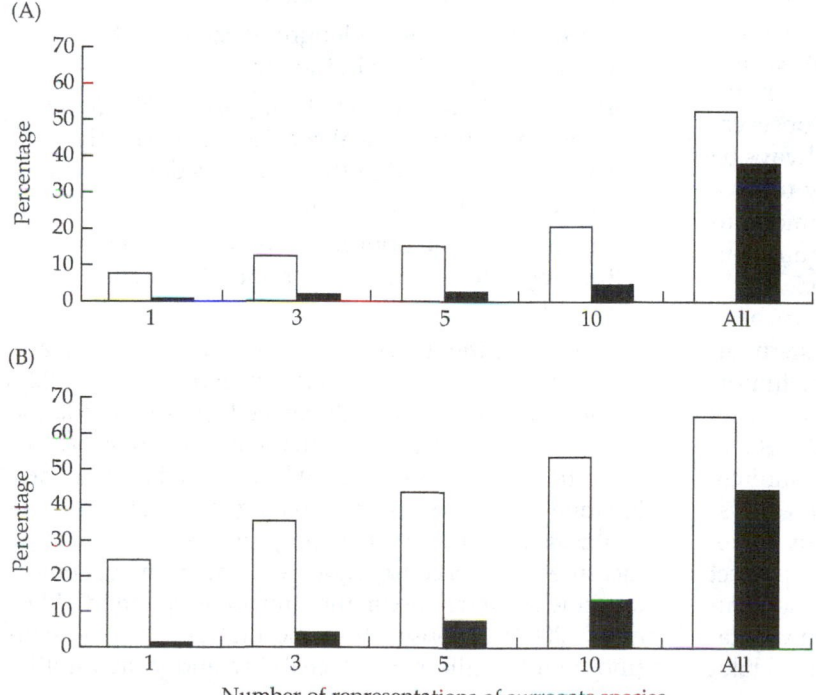

Figure 14.10 Percentage of all species (white bars) represented in a reserve system selected by conserving 1, 3, 5, 10 and all occurrences of charismatic (A) or habitat specialist (B) surrogate species using data from the Columbia Plateau region of Washington. As the number of occurrences increases, so does the number of sites needed in (and therefore the cost of) the reserve system (black bars). (Modified from Andelman and Fagan 2000.)

Tools for systematic conservation planning

Recently, an array of systematic conservation planning techniques have been developed that are goal-directed, transparent, defensible, flexible, amenable to being solved with mathematical algorithms, and aim to efficiently meet quantitative objectives within a system of representative and complementary protected areas (Margules and Pressey 2000; Pressey 2002). Systematic conservation planning is an evolving discipline at the interface of biological, mathematical, and social sciences. The process does not preclude expert judgment and it has been recognized that systematic and expert-driven approaches should be combined (Pressey and Cowling 2001; Cowling et al. 2003b). Systematic conservation planning has informed conservation in both terrestrial (e.g., Pressey 1998; Cowling and Pressey 2003) and marine realms (e.g., Ferdana 2002; Airame et al. 2003; Great Barrier Reef Marine Park Authority 2003).

Conservation planning has made extensive use of computer algorithms to aid decision making. The aim of systematic conservation planning is to select areas to be part of a reserve system by either minimizing or maximizing the value of an objective function, subject to constraints that control the choices made. These formulations are referred to as the minimum-set and the maximal coverage problems. These methods are described in Box 14.2.

The objective of the minimum-set problem is to minimize resources expended, subject to the constraint that all features meet their conservation objectives. The objective of the maximal coverage problem is to maximize protection of features subject to the constraint that the resources expended do not exceed a fixed cost. The minimum-set problem aims to achieve the conservation objectives allocated to each feature in the most efficient manner and specifies a baseline reserve system obtainable for minimal cost. From the perspective of biodiversity conservation, the largest reserve system possible will always be desirable; however, in reality the extent of any reserve system will be limited by social and economic constraints. Therefore, building a reserve system requires observing the principle of efficiency defined earlier.

For example, a systematic conservation plan has been developed for the Greater Yellowstone Ecosystem, an area with high wilderness qualities that serves as important habitat for native carnivores, including grizzly bears (*Ursos artos*) (Noss et al. 2002). Presently, 36% of the ecosystem is privately owned and 64% is publicly owned. Existing protected areas cover 27% of the ecosystem, but these protected areas are not ecologically representative. The aim of the conservation plan is to protect special elements (including threatened species and communities), represent environmental variation (in vegetation, geoclimate, and aquatic habitat), and secure habitat for focal species (in particular, the grizzly bear, gray wolf [*Canis lupus*], and wolverine [*Gulo gulo*]).

The data used in planning included occurrence data for special elements, and remotely sensed abiotic and geoclimatic data. Habitat suitability models and population viability models were constructed for the focal species to guide choices on how much and which types of habitat should be included to conserve these species. Using simulated annealing (see Box 14.2), unprotected areas within the ecosystem that are biologically irreplaceable and vulnerable to degradation were identified.

Vulnerability was assessed using an expert assessment of multiple criteria, including the proportion of each unprotected area in private versus public ownership; presence of active grazing, mining, or timber leases or potential for such activities in the near future; road density and trends; human population and housing density and trends; and, disruptive recreational uses and trends. The irreplaceability of the unprotected areas was assigned using the following criteria, which were weighted equally:

- Protects at a minimum 50% (or 100% for critically imperiled globally or imperiled globally) of the viable occurrences of: imperiled local-scale species, vulnerable and declining birds, coarse scale and regional scale aquatic fish species, and plant communities.

- Represents, at a minimum, 25% of the area of each wetland vegetation type, and at a minimum 15% of the area of each other vegetation type in the region.

- Represents, at a minimum, 15% of the area of each geoclimatic class in the region.

- Represents 20% of the length of each aquatic (stream) habitat type in the region.

- Protects habitat capable of supporting 75% of the population of each focal species that currently could be supported in the region, as determined by the species distribution models.

- Maintains viable population of focal species over time, as determined by the population viability model.

Protecting the areas of highest irreplaceability and vulnerability would expand the reserve system by 22% and increase protection of threatened species by 46% and the representation of geoclimatic classes by 49%. The protection of these areas might be achieved through designating new, or extending existing, national parks, but could also be achieved through less strict protection measures such as conservation easements, management agreements, and national monument designations. Noss et al. (2002) proposed that the highest priority areas (those with highest irreplaceability and vulnerability)

BOX 14.2 Formulation of the Conservation Planning Problem for Designing Reserve Systems

The classic systematic conservation planning problems are to either maximize or minimize the value of an objective function to obtain desired coverage in a reserve system. These are simply expressed mathematically, and then solved by one of a number of methods.

Minimum-Set Problem

The objective of the minimum-set problem is to minimize the resources expended while meeting the conservation objectives.

Let each site have a cost c_j and each feature a conservation objective r_j. The variable x_j equals 1 if site j is selected for the reserve system, otherwise it equals 0. The representation of feature i in site j is contained in a matrix with elements a_{ij}. The objective is to minimize:

$$\sum_{j \in J} c_j x_j$$

subject to the equation:

$$\sum_{j \in J} a_{ij} x_j \geq r_i$$

for every feature i.

Specific versions of this general problem include varying the costs, varying the currency of the features' representation a_{ij}, and varying the objectives. For example, to minimize the number of sites needed to represent each feature, c_j equals 1, because the sites have equal size and/or equal cost; to minimize the total area needed to represent each feature at a particular level, the cost of a site c_j is its area; and to minimize the total cost needed to represent each feature at a particular level, the cost of a site c_j is its monetary value.

We may need to deal with different sorts of objectives for different features. Where we only have presence absence data the variable a_{ij} equals 1 if feature i is in site j, otherwise it equals 0. In this case the target for that feature, r_i, equals 1, if a single representation of each feature is sought, and r_i is a number > 1, if multiple representa-

tions of each feature are sought. If we wish to conserve a fraction of the area of a particular feature and the data, a_{ij}, represents the area covered by feature i in site j, then r_i should be the amount of habitat that we wish to conserve.

Maximal Coverage Problem

The objective of the maximal coverage problem is to maximize the level of feature representation given a fixed amount of resources.

For feature i in site j the objective is to maximize:

$$\sum_{i \in J} y_i$$

subject to:

$$\sum_{j \in J} c_j x_j \leq T$$

where c_j and x_j are as previously defined. If the conservation objective allocated for feature i is achieved y_i equals 1, otherwise it equals 0. The maximum available expendable resource is T, which is in the same units as c_j.

The methods for solving systematic conservation planning problems fall into several classes: local heuristic algorithms, which select sites in a stepwise manner (Pressey et al. 1993; Pressey et al. 1994), global heuristic algorithms, which select sites in sets (e.g., simulated annealing, Ball and Possingham 2000, 2001), and optimization algorithms (Cocks and Baird 1989; Underhill 1994).

Kirkpatrick (1983) was the first to define the minimum set problem and used a heuristic method to find the solution. An iterative **heuristic algorithm** iterates through a list of sites, choosing the best site at each step according to explicit rules (Nicholls and Margules 1993). The process implicitly considers complementarity, as the contribution of unselected sites to meeting the conservation objectives is recalculated each time a site is added to the reserve system.

A greedy heuristic algorithm adds sites to the reserve system sequentially by selecting the site that adds

the most unprotected species to the set that has already been selected. Therefore, it "greedily" attempts to maximize the rate of progress toward the objective at each step. With our example from the Cape Floristic Region described in the text, this algorithm would first select site 4 from Table 14.2 because it would protect nine species, and after that select either sites 1 or 6 as each would add three additional species. Regardless of whether site 1 or 6 is selected first, both will be selected. After that site 8 would be required. Therefore, four sites are required to form the minimum set. Note that sites 1, 4, and 6 do not form a very compact set. If we were concerned about fragmentation then we may want to find a different solution. An alternative algorithm is to represent those species that occur in only a handful of sites first (that is, represent the rare species first). For example, seven species are only represented in one site each. Sites 1, 4, 6 and 8 are essential to represent all of these species. A rarity algorithm first selects any sites that are essential and then selects sites that add the most unprotected species to the reserve system. Therefore, rare species are targeted first and a complementary set is built from there. However, local heuristic algorithms that choose sites sequentially are not guaranteed to find the optimal solution; indeed they are unlikely to find optimal solutions for anything but small problems.

An alternative approach is to express the conservation planning problem in the form of an integer linear program and use mathematical programming techniques to find the optimal solution (Cocks and Baird 1989; Church et al. 1996). However, integer linear programming may fail with large datasets and have the further disadvantage of producing only a single optimal solution and therefore not allowing for flexibility as a range of solutions is not provided.

Continued on next page

Box 14.2 *continued*

Global heuristic algorithms, such as **simulated annealing** and genetic algorithms, use some of the principles of local heuristics, but can allow us to make bad decisions in the short-term to improve the chances of getting better overall solutions. While they do not ensure that the optimal solution will be found, they are the most reliable and advanced method for large problems and facilitate flexible decision making by offering many very good solutions. Practically, this is what decision-makers need.

In the context of conservation planning, simulated annealing begins by generating a completely random reserve system (Kirkpatrick et al. 1983). During each iteration, a site, which may or may not be in the reserve system, is randomly chosen. The change to the value of the reserve system by adding or removing this site is evaluated. This change is combined with a parameter referred to as the "temperature" and then compared to a uniform, random number. The site is then added to or removed from the reserve system, depending on this comparison. When the temperature is high, at the start of the process, both good and bad changes are accepted. As the temperature decreases, the chance of accepting a bad change is lessened and eventually only good changes are accepted (Ball and Poss-ingham 2000; Possingham et al. 2000). The acceptance of bad changes early in the process allows the reserve system solution to move temporarily through suboptimal solution space. The advantage of allowing bad changes as well as good is that it can avoid getting trapped in local optima and increases the number of routes by which the global minimum might be reached, improving the chance of obtaining an optimal or near-optimal reserve system.

These methods to solve systematic conservation planning problems have been used in many places throughout the world to guide the design and expansion of reserve systems.

should receive the highest levels of protection, whereas lower priority areas could accommodate greater levels of human use.

Recent rezoning of the Great Barrier Reef represents another case where systematic conservation planning has aided decision-making. The Great Barrier Reef Marine Protected Area extends about 400 km outward from the coast for a length of over 2300 km along the north east coast of Australia (Figure 14.11). It is one of the largest marine protected areas in the world and is managed by a single Federal Government Agency—The Great Barrier Reef Marine Park Authority (GBRMPA).

On July 1, 2004 the Marine Park was rezoned, becoming the largest systematically designed reserve system in the world. From a scientific perspective the first step in the rezoning was to classify the reefs and the area between reefs into 70 "bioregions." The different bioregions synthesize a large amount of expert opinion and biological data on the distribution and abundance of marine organisms. Each bioregion is sufficiently different from the other bioregions that a comprehensive reserve system needs to contain samples of every bioregion. Hence, a major goal of the rezoning was to conserve at least 20% of every bioregion in "no take" zones that prohibit extraction of marine resources for everyone, or all but traditional peoples. Further principles guiding the planning process included a minimum size for individual protected areas of about 400 km^2, replication of each bioregion in three separate protected areas, minimizing fragmentation, and ensuring geographic diversity to improve the representativeness of the reserve system.

Ultimately all of the principles, data, and socioeconomic information were combined to define various reserve system options. MARXAN, freely available conservation planning software (www.ecology.uq.edu.au/marxan.htm), was used to integrate the biological principles with data on social preferences and economic information about uses of the region to provide a range of

Figure 14.11 Great Barrier Reef is one of the most diverse areas on Earth. The Great Barrier Reef marine protected area extends over 345,400 km^2 and about one-third is newly protected from commercial fishing. (Photograph courtesy of NASA.)

plausible conservation options. These options were used by GRBMPA and the community to select a rezoning plan acceptable to the Federal Government. This is the largest scale and most complex application of systematic conservation planning principles and tools to date.

The Great Barrier Reef Marine Park is divided into eight zones, each of which allows different uses. In particular, the scientific research and marine national park zones do not allow more than traditional use of marine resources, and no extractive uses whatsoever (e.g., fishing) are allowed in preservation zones. Until 2004 less than 5% of the park was zoned no take and now about one third of the entire park does not permit commercial take. The conservation prospects for this remarkable region are certainly brightened by this outcome, while the abilities to use this productive ecosystem for a variety of human uses has been preserved.

Designing a reserve system for the Cape Floristic Region of South Africa

To understand systematic conservation planning, it is useful to overview the planning process in detail for a specific example—the Cape Floristic Region of South Africa (Cowling et al. 2003a). This area is recognized as a center of plant and animal diversity and endemism, and as a global hotspot of biodiversity (Cowling et al. 1996; Olson and Dinerstein 1998; Myers et al. 2000). The region is the world's smallest floral province, yet is home to many more indigenous plant species than any other similar-sized area, with 70% of its plant species found nowhere else on Earth. Moreover, the region is the most threatened of the world's six floral kingdoms (Rouget et al. 2003b), with 1406 threatened plant species (Cowling and Hilton-Taylor 1994). In total 30% of the region has been altered, with the major threats to biodiversity being cultivation, urban development, and the spread of non-native invasive trees (Rouget et al. 2003b). Protected areas in the region account for 22% of the land area and include both strictly and informally protected areas, however, these are biased towards upland areas (Rouget et al. 2003a). Cowling and Pressey (2003) outlined three major reasons why a systematic conservation plan was required for the region: (1) the existing reserve system did not represent biodiversity; (2) there were increasing threats to biodiversity; and (3) there was diminishing institutional capacity for conservation.

Systematically designing a reserve system for the Cape Floristic Region required several tasks and a large amount of biological and biophysical data. The first task was to delineate the geographical extent of the planning region, within which decisions would be made about the location, configuration, and management of protected areas. The planning region for the Cape study was centered on the Cape Floristic Region—an area of 87,892 km^2; the region extends approximately 60 km beyond the boundaries of the Cape Floristic Region to capture ecological processes that transcend this boundary.

The second task was to identify the elements of biodiversity requiring protection in the planning region. These elements, referred to here as features, can include species, populations, species assemblages (for example, vegetation types), land classifications (for example, environmental classes), and features that represent important natural processes. In the Cape Floristic Region, data for five different kinds of biodiversity features (Cowling et al. 2003a) were used to develop the systematic conservation plan:

1. Land classes
2. Plant species in the family Proteaceae
3. Selected lower vertebrates (i.e., fishes, amphibians, and reptiles)
4. Large- and medium-sized mammals
5. Features important for ecological and evolutionary processes (for example, edaphic interfaces and macroclimatic gradients)

The Cape Floristic Region study was one of the first systematic conservation plans to incorporate a large variety of features, in particular, **ecological process features**.

The third task was to delineate the geographical units of evaluation, referred to here as planning units. Planning units can be any discrete part of the landscape and any size or shape, including rectangular or hexagonal grids cells, ownership parcels, vegetation types, subcatchments, vegetation fragments, or logging compartments. Information on the occurrence or extent of each feature within each planning unit is required; for example, how much of a particular land class is in the unit and whether or not there is a record of a particular species. In the Cape Floristic Region study, the planning unit layer was a grid of sixteenth-degree squares (approximately 3900 ha each; Figure 14.12) (Cowling and Pressey 2003). The existing protected areas were also used as planning units because they contribute to conserving features; however, it was assumed that the existing protected areas were fixed and they were themselves not considered candidates for selection. The boundaries of ecological process features (e.g., edaphic interfaces and riverine corridors) were incorporated as process planning units. The planning unit layer for the Cape Floristic Region consisted of 3014 grid cells (74.5% of the region), 2993 process planning units (6.2% of the region), and 1032 protected areas (19.3% of the region).

The fourth task was to generate conservation objectives, which are in the same units employed to record the occurrence or extent of each feature in each planning unit. Each feature may be given its own objective to reflect its ecological requirements, natural rarity, or vulnerability to threatening processes; otherwise, a generic

Figure 14.12 Distribution of the 13 species and three habitat types in a portion of the Cape Floristic Region. (Reprinted with modification from Cowling et al. 2003b, © 2003, with permission from Elsevier.)

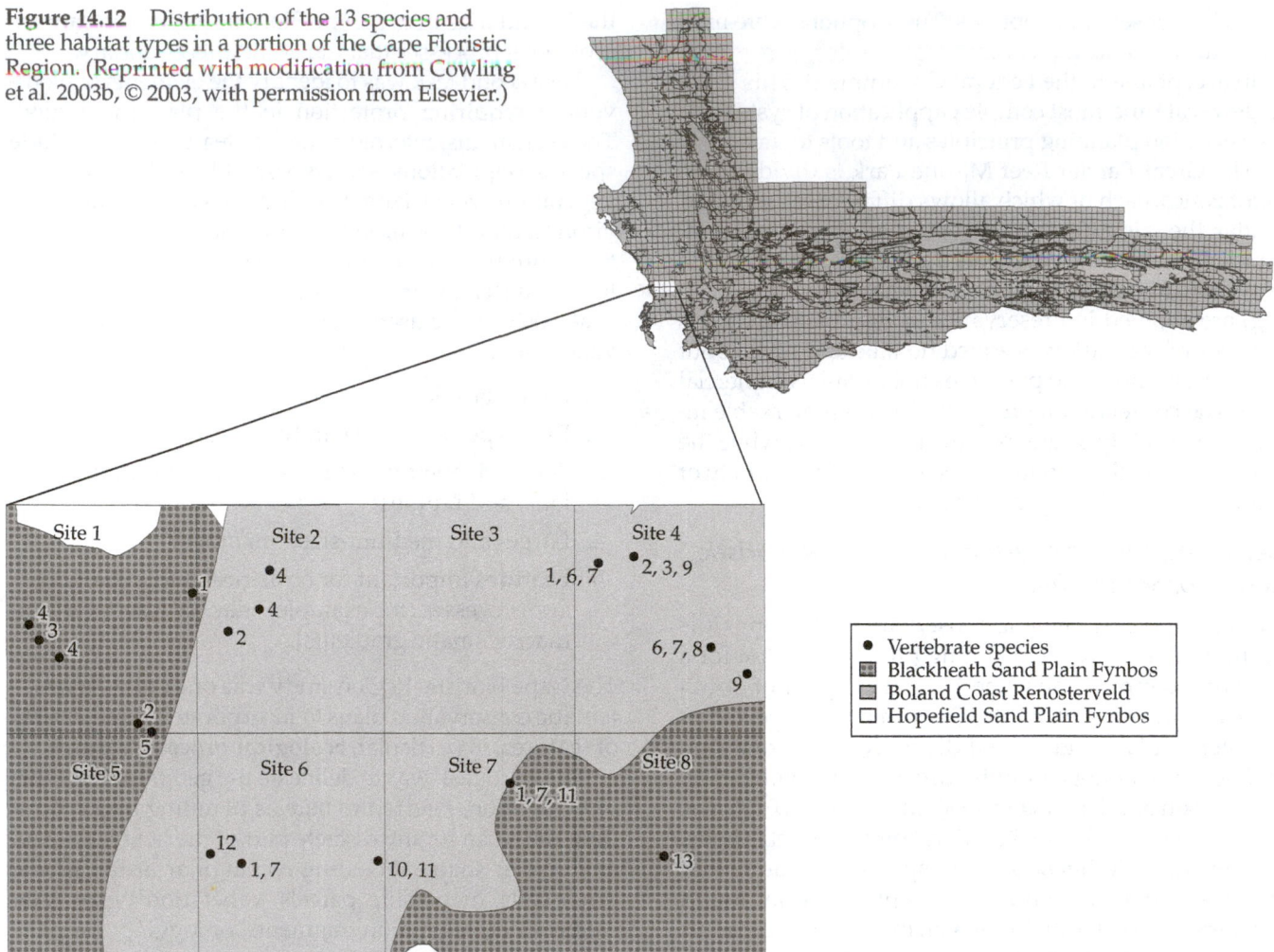

objective can be assigned to all features (e.g., conserve 20% of each vegetation type). The development of conservation objectives for the Cape Floristic Region is described in Pressey et al. (2003).

Planning in the Cape Floristic Region relied heavily upon expert knowledge to derive many of the data layers, for example the land classes, mammal distributions, and the spatial distribution of threats. This would be true in any region, even where the region is well studied and the data are abundant. Expert knowledge was also used in the final stages of deriving the conservation plan (Cowling et al. 2003b).

The conservation plan sought to achieve some minimum representation of biodiversity features for the smallest possible cost (a minimum set problem: see Box 14.2). Therefore, the objective was to minimize costs and biodiversity protection entered as a constraint. This is best illustrated by an example taken from the larger data set. Here, for a sample of thirteen species and three habitat types, the objective is to conserve at least one popula-

tion of every species and at least one representation of each habitat type. The presence or absence of each of the 16 features is known for eight sites as illustrated in Table 14.2 in a site-by-features matrix (see Figure 14.12 for an illustration of the distribution of the species and habitat types). The minimum set problem is to find the smallest number of sites that will represent every species once. A "1" denotes a presence of either a species or habitat type, and a "0" denotes the absence of either a species or habitat type. In this case, the minimum set reserve system is sites 1, 4, 6 and 8 as no other set of sites will conserve all species.

While this solution can be calculated by hand, for bigger datasets this would be impossible. For example if there are 5000 sites, the number of possible reserve systems is 2^{5000}. Therefore, methods have been devised for searching for the minimum set that represent efficient solutions to reserve system design problem (see Box 14.2). Such a computer-aided approach was used in this case.

TABLE 14.2   *An Example of a Site by Features Matrix for the Cape Floristic Region*

Species number	Feature	Sites								Species/ habitat range
		1	2	3	4	5	6	7	8	
1	Ocellated gecko (*Pachydactylus geitje*)	1	0	0	1	0	1	1	0	4
2	Sand toad (*Bufo angusticeps*)	1	1	0	1	0	0	0	0	3
3	Cape caco (*Cacosternum capense*)	1	0	0	1	0	0	0	0	2
4	Cape sand frog (*Tomopterna delalandii*)	1	1	0	0	0	0	0	0	2
5	Clawed frog (*Xenopus laevis laevis*)	1	0	0	0	0	0	0	0	1
6	Cape legless skink (*Acontias meleagris meleagris*)	0	0	0	1	0	0	0	0	1
7	Karoo girdled lizard (*Cordylus polyzonus*)	0	0	0	1	0	1	1	0	3
8	Marico gecko (*Pachydactylus mariquensis mariquensis*)	0	0	0	1	0	0	0	0	1
9	Common caco (*Cacosternum boettgeri*)	0	0	0	1	0	0	0	0	1
10	African leaf-toed gecko (*Afrogecko porphyreus*)	0	0	0	0	0	1	0	0	1
11	Cape girdled lizard (*Cordylus cordylus*)	0	0	0	0	0	1	1	0	2
12	Common egg-eater (*Dasypeltis scabra*)	0	0	0	0	0	1	0	0	1
13	Delalande's beaked blind snake (*Rhinotyphlops lalandei*)	0	0	0	0	0	0	0	1	1
	Blackheath Sand Plain Fynbos	1	1	0	1	1	0	1	1	6
	Hopefield Sand Plain Fynbos	1	1	0	0	0	0	0	0	2
	Boland Coast Renosterveld	1	1	1	1	1	1	1	1	8
Site species/habitat richness		8	5	1	9	2	6	5	3	

The final conservation plan for the Cape Floristic Region covered 52.3% (49,958 km²) of the extant habitat in the planning region (Figure 14.13). The conservation plan was developed in a series of stages and built upon existing statutory protected areas (stage 0). The first stage incorporated planning units for four spatially fixed process components (edaphic interfaces, upland-lowland interfaces, sand movement corridors, and inter-basin riverine corridors). The second stage incorporated planning units of maximum irreplaceability for achieving the conservation objectives for broad habitat units, plant species in the Proteaceae, and vertebrate species. The third stage incorporated planning units for achieving the conservation objectives for large- and medium-sized mammals. The fourth stage planning units were selected to represent macroclimatic gradients, and the fifth stage considered upland–lowland gradients. The final stage of the planning process involved selecting planning units to achieve all outstanding conservation objectives for broad habitat units, endemic plants in the family Proteaceae, and vertebrates, while minimizing the inclusion of highly vulnerable areas (Cowling et al. 2003a).

Confronting Threats in Protected Areas

As described throughout this chapter, conservation planning involves locating and designing protected areas to promote the persistence of biodiversity in situ. Therefore, protected areas must be able to mitigate at least some of the processes that threaten biodiversity. Information on threatening processes and the relative vulnerability of areas and natural features to these processes is therefore crucial for effective conservation planning.

Pressey et al. (1996) defined vulnerability as the likelihood or imminence of biodiversity loss to current or impending threatening processes, and this definition can be extended to distinguish three dimensions of vulnerability: exposure, intensity, and impact (Wilson et al. 2005). Exposure is either the probability of a threatening process affecting an area over a specified time, or the expected time until an area is affected. Areas with the same exposure to a threatening process can be affected at different levels of intensity. The intensity of a threat can take many forms, such as the volume of timber extracted per ha of a forest type or the density of an invasive plant species. Impact refers to the effects of a threatening

(A)

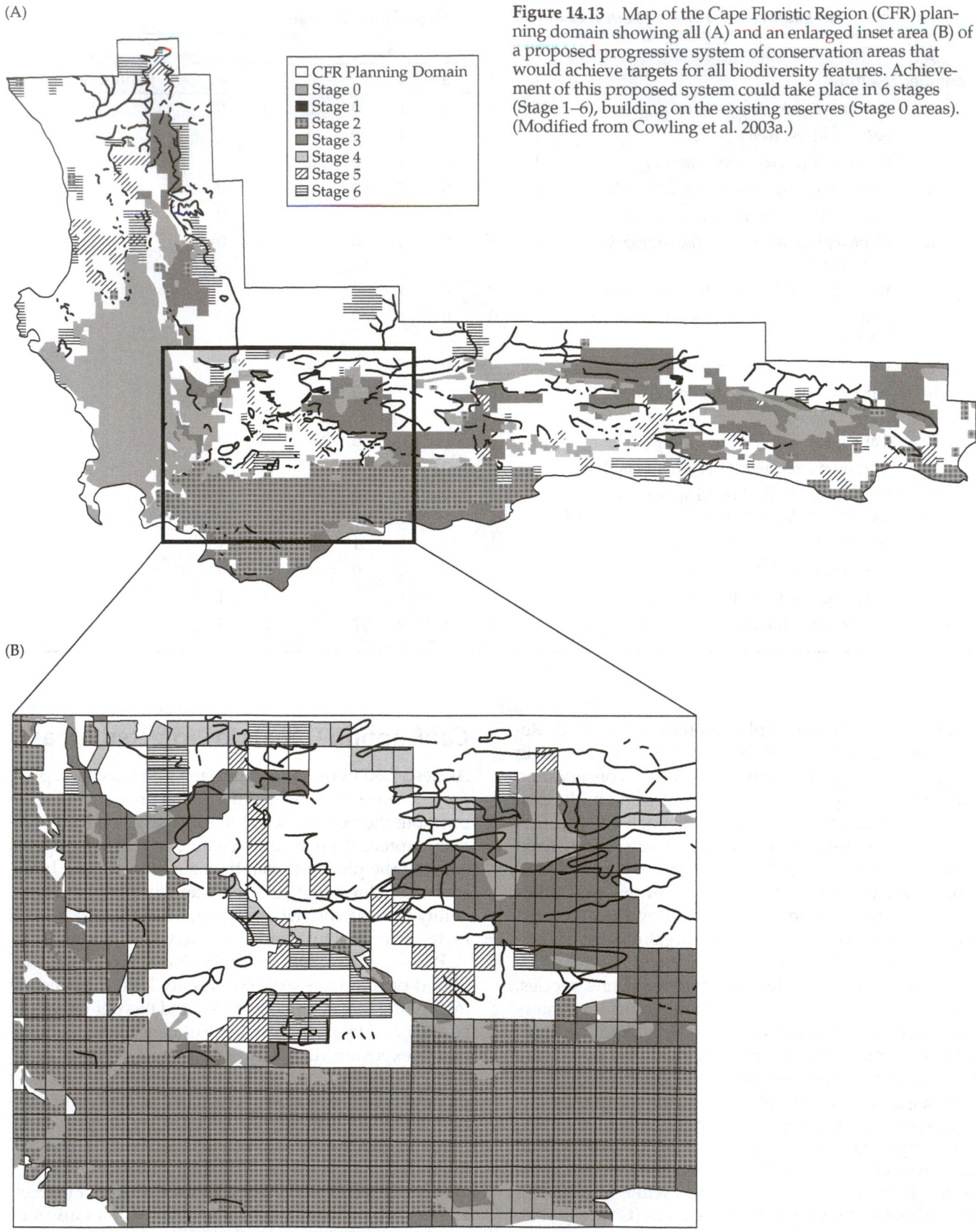

CFR Planning Domain
Stage 0
Stage 1
Stage 2
Stage 3
Stage 4
Stage 5
Stage 6

Figure 14.13 Map of the Cape Floristic Region (CFR) planning domain showing all (A) and an enlarged inset area (B) of a proposed progressive system of conservation areas that would achieve targets for all biodiversity features. Achievement of this proposed system could take place in 6 stages (Stage 1–6), building on the existing reserves (Stage 0 areas). (Modified from Cowling et al. 2003a.)

(B)

process on particular features and could indicate effects on distribution of species, their abundance, or likelihood of persistence. Areas of particular concern for conservation planners have high exposure to very intense threatening processes. Features of concern will be those occurring in such areas and experiencing strongly negative impacts.

Wilson et al. (2005) review methods that have been used to assess vulnerability and categorized them into four groups based mainly on the types of data employed. All methods estimate exposure, but some deal also with intensity and impact. The first method uses information on permitted or projected land uses. The second method identifies the extent of past impacts on features and uses these data to predict future impacts on the same features. In some circumstances, the underlying spatial (e.g., proximity to cities and roads) and environmental characteristics (e.g., soil type, slope, climate) believed to have predisposed areas to threatening processes in the past are determined, and areas that are presently unaffected and share these characteristics are then identified. The third method identifies vulnerable areas as those with high concentrations of taxa with high probabilities of extinction, and the final method is based on expert knowledge. All four methods have been employed at a variety of spatial scales and resolutions in countries with differing levels of development, even in those typically regarded as data-poor. The data underpinning many of the methods are globally available and so most methods are applicable anywhere, at least at a coarse scale.

When developing a conservation plan, vulnerable areas might be avoided so that objectives are achieved, as far as possible, in areas without liabilities for implementation and management. Considerations of defensibility, or avoiding vulnerable areas, can be especially important if resources are likely to be insufficient for effective management (Peres and Terborgh 1995). When implementation of new protected areas commences, an important consideration in scheduling their implementation will often be their relative vulnerability. The more vulnerable areas might receive higher priority, especially if there are few or no alternative areas available to protect the features they contain (Pressey and Taffs 2001; Noss et al. 2002; Lawler et al. 2003). This strategy can minimize the extent to which conservation objectives are compromised by threatening processes during the frequently protracted process of establishing protected areas on the ground.

Protected areas can be important in mitigating proximate threats arising from activities such as agriculture, logging, mining, or grazing of domestic livestock. In some cases, and depending on resources for management, protected areas can also prevent or reduce the spread of nonnative plants and animals and mitigate the adverse effects of changes to fire regimes and other natural disturbances.

However, protected areas might be ineffective in excluding invasive nonnative plants and animals or mitigating hydrological impacts from nearby developments unless complemented with intensive on-site management and changes in land-use patterns throughout the region. Given the limitations of protected areas in preventing all threats to biodiversity, conservation planning must operate as part of a broader conservation strategy involving policy, legislation, education, and economics.

Conservation objectives and persistence

Conservation plans are most often formulated to represent biodiversity pattern, where pattern might be measured, for example, by the distribution of species or habitat types. More recently, attempts have been made to improve the likelihood that biodiversity persists in protected areas by setting conservation objectives for viability (Noss et al. 2002) and natural processes (Pressey et al. 2003), and by considering spatial design (McDonnell et al. 2002).

While conservation objectives (such as conserving 15% of a remaining habitat type) are politically expedient and have generally helped to grow the global reserve system, these objectives are general and do not recognize that some regions will require significantly higher levels of protection than others. Even though the goal of protecting 10% of the terrestrial world has been surpassed, the global reserve system is far from complete in terms of conserving global biodiversity (Rodrigues et al. 2004a,b; Hoekstra et al. 2005). In addition to using blanket area objectives, we may also need species-based and ecoregion-based objectives that will better encompass the spatial distribution of biodiversity, need for protection, and other factors, such as levels of endemism and rarity.

Dynamics and uncertainty

In the context of extensive and increasing threats to biodiversity, theoretical and procedural advances in conservation planning over recent decades (such as those described in Box 14.2) have been important but insufficient to guide the investment of limited resources for conservation. A major shortcoming is that, while there is a considerable body of theory and procedures for designing reserve systems in a static world, there is limited theory for their design in a dynamic world.

Static approaches to conservation planning assume that proposed protected areas can be acquired instantly and that these areas will remain unchanged prior to protection. In the "real world," the process of identifying and implementing reserve systems is rarely instantaneous and even if an "optimal" reserve system can be identified and prioritized, budget and opportunity constraints may mean that it takes decades of negotiation and land purchases to translate a conservation plan into

a functioning reserve system (James et al. 2001; Pimm et al. 2001). In the interim, the acquisition of proposed protected areas might be hindered and the areas may undergo changes. Some biodiversity might be lost and some areas may be destroyed or degraded before they are acquired. Others may no longer be available for conservation. Changes in the political and economic climate may also constrain conservation action. These problems highlight another source of complexity in conservation planning: the existence of uncertainty.

Meir et al. (2004) have considered the consequences for conservation planning of assuming a static world and ignoring uncertainty. They found that using relatively simple rules for selecting areas to be protected, such as choosing the available area with the highest species richness or the highest richness of rare species, worked better than designing optimal reserve systems, particularly in the context of land-use change and uncertainty affecting where opportunities for conservation investments might arise. Although the performance of optimal sets (see Box 14.2) and comprehensive conservation plans will undoubtedly improve if the plans are iteratively updated, they found that, given the rates of habitat loss reported in the literature, comprehensive conservation plans would need to be updated annually to perform as well as the simple decision rules. Because information contained within conservation data bases is updated relatively slowly, and it takes a considerable amount of work to develop comprehensive conservation plans, updating these plans annually seems unrealistic. Thus, conservation resources might be better invested in determining the biodiversity value and relative importance of particular sites, rather than in developing comprehensive designs for large-scale systems of protected areas. In Case Study 14.3, Gustavo Kattan discusses the difficulties of applying conservation planning tools when biological information is very sparse, as it is in the Andes region of Colombia.

Despite the contribution of recent work that has improved reserve system planning, it is recognized that there is room for further developments. The development of theories and procedures for undertaking conservation planning in a dynamic and uncertain world is an active area of research.

Complex economic considerations

The conservation principle of efficiency incorporates some economics into conservation planning, however, there are a complex range of economic issues that require further consideration. Often these are considered through cost–benefit analyses, as described in Chapter 5. Here we explore just a few additional ways that economics could play a greater role in conservation planning, but there remain many unanswered questions and unsolved problems at the interface of biodiversity conservation and economics.

First, most of the conservation planning literature ignores the impact of protected area dedication on areas outside the reserve system. For example, in the marine realm closing areas to fishing may only serve to displace that fishing effort to other areas, making a once sustainable industry, unsustainable. In the terrestrial realm, all acquisition of property affects local property prices. Consequently, acquiring areas for protection can reduce the cost-effectiveness of future additions to a reserve system.

Second, much of the conservation planning literature focuses on the cost of acquiring a reserve system, ignoring the cost of maintaining the biodiversity values for which the protected areas were acquired. More sophisticated conservation planning theory should account for both the initial cost of purchase and the long-term maintenance of the protected area, including all the uncertainties inherent in protected area management.

Finally we need to recognize complex trade-offs between the different values society places on different types of land use. The classic conservation planning problems (see Box 14.2) of minimizing costs while reaching all conservation objectives, or maximizing biodiversity benefits for a fixed cost allow no trade-offs between biodiversity and economics. In each case one of the two currencies is set as a hard constraint and the other currency is maximized (or minimized). Trading between different societal values, however, occurs all the time. Hence, theories and tools for making such trade-offs are needed. The need becomes even more pressing when we consider other values of landscapes, such as amenities, water filtration, and soil protection. Ultimately economists can play an active role in helping determine how society would like to trade-off between these different values by incorporating nonuse values into planning analyses (see Chapter 5).

Incorporating Social and Cultural Contexts

Over the past few decades, we have greatly advanced our technical capacity to design effective reserve systems. Yet, sound systematic conservation planning cannot ensure the success of a protected area or a reserve system if it does not account for the social and cultural context of a region. Currently, between 300 and 420 million people live in a state of chronic poverty globally (Adams 2004). How we reconcile biodiversity protection goals with the needs of hundreds of millions of the world's poor requires careful consideration.

The modern model of federally-owned and managed protected areas has its roots in the nineteenth century

movement for national parks in the U.S. In the U.S., clearly delineated boundaries that demarcate wilderness areas from areas that can be logged and a management focus towards enforcing restrictions on human use or interference might be politically appropriate. The problem with applying this model without consideration of the social and cultural contexts is that communities living near newly created protected areas may be denied access to the resources that form the basis of their livelihoods. Furthermore, it may be impossible to enforce a method of border demarcation or use restrictions that are not part of the cultural context of a region. For example, in Papua New Guinea, where a complicated land tenure system controls access to resources, agreements to protect an "area" must be based on use of the resource, not a line on the ground.

While conservation planning is important, in the end, protected area boundaries will be largely influenced by social and political factors. Thus, consideration of the history of an area is not an esoteric issue but an important element for gauging its likelihood of success as a protected area. How boundaries are defined when protected areas are established and what features they include, as well as land tenure rights within and surrounding protected areas are thus issues of tremendous importance in ensuring effective conservation and management of biodiversity. Governments, nongovernmental organizations, multinational corporations, local peoples, and funding agencies may all influence success.

The case of Amboró National Park, Bolivia illustrates the problems that may arise when a protected area is established without proper consideration of, or consultation with, local communities (Moreno et al. 1998). Amboró National Park ranges from the humid zone of the Andes mountains to the dry area of the Chaco to the east. The site was first designated as a natural reserve in 1973, but this declaration did not include a management plan, nor did it restrict or change local use of the area. Resource extraction and human settlement continued. In 1984, however, the reserve was reclassified to national park status. With the elevated status came new management objectives that prohibited hunting and fishing and the extraction of timber. Forestry concessions were annulled and all existing within-park settlements were subjected to restrictions. Then, in 1991, the government of Bolivia issued a decree expanding Amboró from 180,000 to 637,000 ha, more than tripling its initial size.

The expansion of the Amboró was celebrated by conservationists as it incorporated watersheds and ecosystems that were previously unrepresented. However, the government's expansion of the protected area was performed without regard to social factors, an analysis of existing land tenure, and consultation of the communities residing in or around the park (see Case Study 12.2). Existing human settlements were incorporated by the expansion, thus implicitly contradicting national law that prohibits resource use in protected areas. The incorporated communities and people living nearby became hostile to the park for having usurped their lands and their rights to its use. It was soon recognized that the current exclusionary management of the park was unsustainable. In an attempt to reduce tension over the new park boundaries, a program was initiated to negotiate new zones for internal protection of the park, after community consultation (Moreno et al. 1998).

This program, termed the Red Line Project, involved community members clearing a 1.5 m wide path that would represent the limits of their use rights within the park. This visible boundary helped reinforce differences in uses allowed outside, rather than inside, the park. In certain areas the project moved forward and raised hope that the new line would come to constitute the new park boundaries. In other areas a stalemate occurred because park administration and local people could not agree over boundaries, as some communities claimed land deep inside the park.

In spite of Bolivian law disallowing resource use in national parks, intense use continued inside of Amboró in areas where Red Line agreement could not be reached. Deforestation resulting from slash-and-burn agriculture and small-scale timber extraction continued, along with hunting. The land tenure situation and means of initial establishment of the park made it unlikely that resource use would cease. In 1996, a decree was issued to reduce the size of Amboró to 440,000 ha. While this could be seen as a loss to conservation, the history of the Park's formation made it inevitable that its conservation goals would fail. Albeit a smaller protected area than it once was, resource extraction inside the park has been reduced and improved management is helping the park successfully meet its current biodiversity conservation goals (Moreno et al. 1998).

The case of Amboró illustrates the importance of considering social and political factors in conservation planning. While conservation objectives and reserve design goals should inform siting of protected areas, achieving a successful protected area necessarily will be tied to social and political context.

The vast majority of regions where protected area expansion is most urgent are in low-income, tropical countries that can least afford the costs of establishing and managing new reserves (Bruner et al. 2004). The majority of the benefits of protected areas are realized at the global scale, even if we include long-term local benefits. Thus, the costs of establishment of protected areas in priority regions largely should be borne by the glob-

al community. Financing on behalf of the global community may be done through multi and bilateral institutions, foundations, private corporations, and individuals. While the task of global coverage will be challenging, protected areas are highly cost-effective when it comes to protecting biodiversity. Advances in data availability coupled with advances in the science of conservation planning are allowing us unprecedented opportunity to move forward with the urgent task of conserving biodiversity.

CASE STUDY 14.1

Conservation Management of a European Natural Area
Doñana National Park, Spain

Carlos Fernández-Delgado, Universidad de Córdoba, Spain

In contrast to the Americas, Europe has been settled by technological western cultures for thousands of years, and these people have had a long time to leave their mark upon the land. Consequently, fewer natural areas remain in Europe than in many other regions, and there are fewer opportunities to manage them in a semi-pristine state. The challenges of managing the remaining natural areas are great, as the relatively small remnants exist in a matrix dominated by humanity. This phenomenon is not unique to Europe—indeed, many protected areas in densely populated regions face many challenges, and as human populations continue to grow, this phenomenon will become more common.

Doñana National Park of Spain serves a critical role in biodiversity conservation, and is embedded in a matrix of high human use. Doñana is unique in many respects; it is a major stopover point in the migration route of birds moving between Europe and Africa, it is home to some of the most endangered mammals in the world, and it contains perhaps the most significant wetland in Europe. Despite thousands of years of use, Doñana remains one of the most ecologically important sites left in the midst of lands wholly or mostly converted to human uses. Yet Doñana's ability to support biodiversity is under constant threat due to its proximity to culturally and economically critical locations in Spain. In this case study, I will explore the importance of this national park, historically and ecologically. As I discuss the external and internal challenges to the protection and management of Doñana, I will build perspective of what is necessary for protecting its future.

Location and Description of Doñana National Park

The 50,720 ha Doñana National Park, located in Andalucía on the southwestern coast of Spain (Figure A), is part of the Guadalquivir River Basin. The Guadalquivir is one of the largest rivers in the country and is the only navigable one. The area has a subhumid Mediterranean climate, influenced by the Atlantic Ocean, with alternating dry and rainy seasons (Em-

berger et al. 1976; Font 1983; Siljeström et al. 2002). Some three thousand years ago, during the pre-Roman period, a lake (Lacus Ligustinus) covered much of the area (Rodríguez and Clemente 2002a). The lake remained during Roman colonization until approximately the fourth or fifth Century A.D., after which the progressive appearance of a littoral bar (Rodríguez and Clemente 2002a) and alluvial deposits due to severe deforestation in the uplands transformed the lake into a marsh with a tidal influence, and ultimately into a marsh with a pluvial influence (Bernúes 1990; Rodríguez and Clemente 2002b). It is on this marsh that most of present-day Doñana National Park is located.

Doñana has three dominant ecosystem types: fixed dunes (or "cotos"), mobile dunes, and marshes (Montes et al. 1998). The fixed dunes are affected by the depth of the water table, the mobile dunes are driven by substratum mobility, and the marshes are created by seasonal rains (García-Novo et al. 1977). The mobile dunes, which run parallel to the coastline for 30 km, are the most important dunes on the Iberian Peninsula and some of the most extensive in Europe. They move an average of 4–6 m per year, with some sections moving up to 20 m per year (García-Novo et al. 1976). The marshes occupy some 27,000 ha, and are highly productive and seasonally variable, with winter water depths up to 1 m and little to no water during summer. The broad ecotone between the marshes and uplands has a complex vegetational structure and a high faunal diversity, including many herbivores and predators.

The Biogeographical Importance of Doñana

Doñana's geographical position, between the European and African continents and between the Atlantic Ocean and the Mediterranean Sea, results in a rich flora and fauna, with some 838 species of vascular plants, 39 species of fishes, 11 amphibians, 21 reptiles, 370 birds, and 29 mammals recorded. The park is especially critical to bird diversity, as some three-fourths of all European species are found in Doñana due to its position on the migratory routes of many species and its abundant food re-

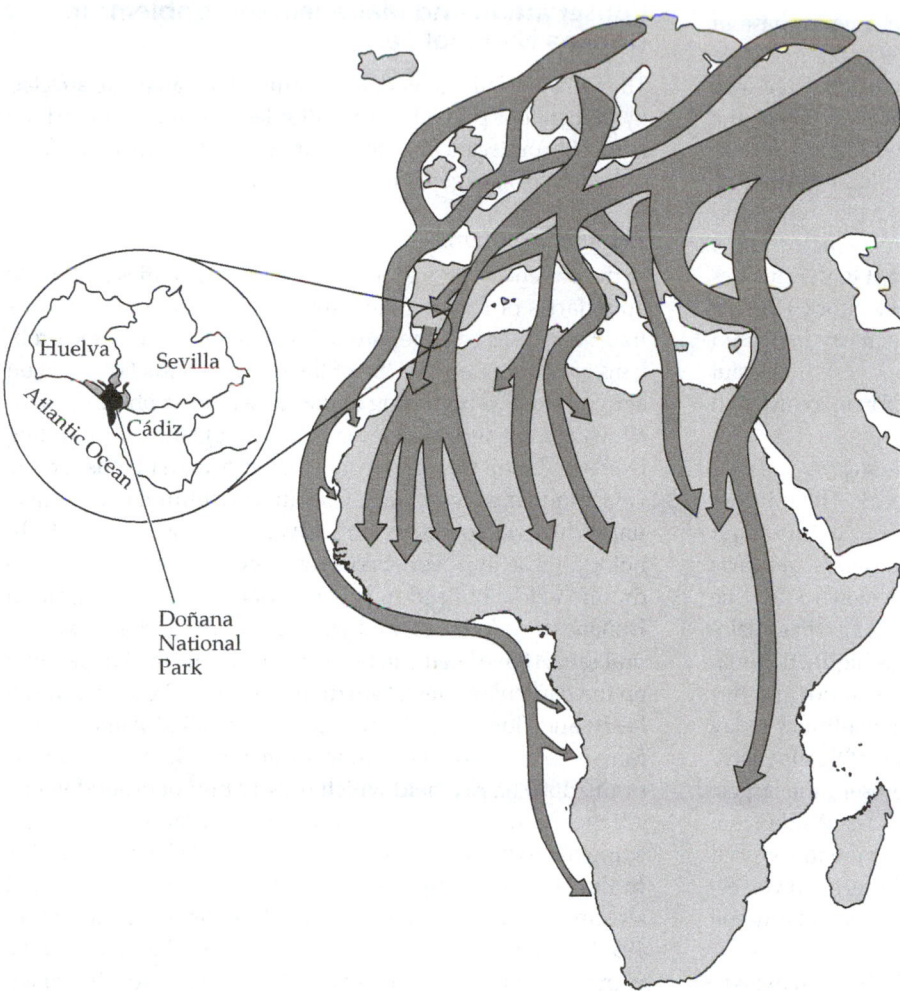

Figure A Doñana National Park is located in a critically important area of southern Europe on the Iberian Peninsula. Major bird migration pathways for the entire European and African continents pass through Doñana.

sources (see Figure A) (García-Novo et al. 1977; Amat 1980; Máñez and Garrido 2002). During spring and autumn migration, it is easy to observe more than 200 species of birds.

Many Doñana species are endemic, threatened, endangered, or otherwise of ecological interest. This includes an endemic cyprinodontid (*Aphanius baeticus*) recently described as a new pupfish species (Doadrio et al. 2002), the White-headed Duck (*Oxyura leucocephala*), European Bittern (*Botaurus stellaris*), Squacco Heron (*Ardeola ralloides*), Marbled Teal (*Marmaronetta angustirostris*), Ferruginous Duck (*Aythya nyroca*), Crested Coot (*Fulica cristata*), Andalusian Hemipode (*Turnix sylvatica*), Curlew (*Numenius arquata*), Black Tern (*Chlidonias nigra*), Ruddy Shelduck (*Tadorna ferruginea*), Slender-billed Gull (*Larus genei*), Spanish Imperial Eagle (*Aquila adalberti*), Purple Gallinule (*Porphyrio porphyrio*), the only European species of mongoose (*Herpestes ichneumon*), and the Iberian lynx (*Lynx pardinus*), considered by the IUCN (2004) as the most threatened felid in the world.

This rich natural area means that Doñana is one of the most important wetlands in Europe. This was recognized when it was declared a national park by the Spanish Government in 1969, a wetland of international importance by the Ramsar Convention, an International Biosphere Reserve in 1981, and a World Heritage Site by UNESCO in 1995.

Human History in Doñana

Human habitation and use of the Doñana area goes back 3000 years B.C. (Rivera 2002). All the great European civilizations have passed through and used this area, including Tartesics, Phoenicians, Greeks, Romans, Visigoths, and Arabs. Prior to 1262, the area was controlled by Arabs, and the marshlands were used for grazing by Arabian horses, while the surrounding hills were used for timber extraction and wax and honey harvesting from beehives. In 1262, after conquering the area, the Christian King Alfonso X established a hunting area in Doñana, beginning a hunting phase that lasted 400 years (Granados 1987). Hunting centered on wild pigs and red and fallow deer in the forest, and on waterfowl in the marshes. This activity largely protected the forest, which served as a shelter for game species. However, hunting encouraged the eradication of predator species such as foxes, wolves, birds of prey, lynx, and mountain cats. There were large rewards for each animal killed, which greatly reduced the predator populations, while the herbivore populations, freed of their enemies, increased considerably.

In 1628, the introduction of cattle began a new phase in Doñana (Granados 1987). In addition to cattle grazing, people were able to cut firewood for their personal needs. Excessive cattle populations resulted in overgrazing, reduced the Doñana cork oaks (*Quercus suber*), and increased dune formation. Growing interest in forest development led to a forestry phase. Initially, forestry was centered on cork oaks, white and black poplars, willows, ashes, and junipers (Granados 1987). Many animal species that were unable to adapt to the changes were locally extirpated, including the Black Stork (*Ciconia nigra*) and the Swan (*Cygnus* sp.). Forestry activity increased with the introduction of pines in 1737, which was so successful that they are the principal component of the forest today, with a resulting loss of oaks.

In approximately 1940, river red gum (*Eucalyptus camaldulensis*) was introduced in many areas of the park. The purpose of this introduction was to produce cellulose for pulp and paper companies. The species, a native of Australia, withdraws great amounts of soil nutrients and lowers the water table due to its fast growth and excessive transpiration; it also has allelopathic effects on other vegetation and contains highly flammable volatile oils, which can trigger forest fires in hot and dry seasons (FAO 1985). An eradication program continues today to eliminate residual specimens. Introduction of this invasive species and other agricultural activities represent the largest human intervention in Doñana in recent years. Aside from river red gum, an additional 50 plant species were introduced both accidentally and intentionally; most of these, however, became scarce because of inappropriate conditions and low soil fertility.

Modern scientific interest began in the 1950s, with expeditions led by the Spanish naturalist José A. Valverde. The area was eventually visited by renowned naturalists such as Guy Mountfort, Roger Tory Peterson, and the Nobel laureate Sir Julian Huxley, which raised its public visibility. During that time, Doñana was seriously threatened by government forestry and agricultural initiatives, and Valverde asked for help from many international organizations (IUCN, International Wildlife Research Bureau, International Council for Bird Preservation). As a result of these activities, in 1964 the World Wildlife Fund (WWF) helped a Spanish scientific group buy 6794 ha in the Doñana area, establishing the first protected area, the Doñana Biological Station. In 1969, Doñana National Park was officially created with an initial extension of 37,425 ha; this was enlarged to 50,720 ha in 1978 when the present boundaries were established. In 1989, the regional government declared 53,709 ha surrounding the park as protected area, designated the Doñana Natural Park, ushering in the present conservation phase. Together the national and natural park constitute one of the most important in the European Continent, with nearly 40,000 ha of pristine marshes, more than 5000 ha of rivers, ponds, channels, and lagoons, 7000 ha of coastal and sand dunes, 43,000 ha of coniferous forests and 24,000 ha of shrublands (Figure B).

Conservation and Management Problems in Doñana National Park

The Doñana National Park and Natural Park areas are affected by tremendous pressures from all sides. Despite the large total size of the protected area, these external and internal threats require active management.

External problems

In the present conservation phase, human activities within the boundaries of Doñana have drastically diminished, and its traditional uses have begun to be managed. Most of the problems are now external. One of the biggest conflicts centers on agriculture. The beginning of the conservation phase coincided with recent agricultural development of lands surrounding Doñana. Until that time, most of the marshlands of the Guadalquivir estuary were, from an agricultural perspective, unproductive; however, with advances in agricultural technology, these impoverished soils were improved. Large hydraulic works built to improve agriculture in areas around Doñana directly influenced water flow within the park. Natural canals have been cut, blocked, or transformed depending on the particular interest surrounding each. Presently, strawberry and flower cultivation in the so-called Almonte-Salt Marsh Sector extracts groundwater for irrigation. There is even a 1000 ha rice field which uses 10 hm^3 of ground water. All these activities are lowering the water table, thus threatening the park's vegetation (Rodríguez and Clemente 2002b). In the last 30 years groundwater extractions have reduced stream flow in the area to about 50% of historic flows (ITGE 1999). This process is altering the hydrological dynamic of the marshland, increasing the length of the dry period. Other negative impacts are the accumulation of agricultural plastic waste residues and contamination of surface waters with pesticides and excess nutrients. Some areas are still being deforested to allow for new cultivation.

On April 25, 1998 part of the dike of a tailings pond at the "Los Frailes" zinc mine, located in Aznalcóllar collapsed. The accident released about 4 million m^3 of acidic water and 2 million m^3 of mud rich in toxic metals to the Guadiamar River, the lower tributary of the Guadalquivir, whose waters flow into Doñana (Grimalt and Macpherson 1999; Borja et al. 2001). As a consequence, 67 km of the Guadiamar main channel and 4634 ha of the floodplain were polluted with heavy metals, particularly Fe (34–37%), S (35–40%), Zn (0.8%), Pb (0.8%), As (0.5%), and Cu (0.2%) (Grimalt et al. 1999). The avalanche of water and sludge was stopped just before Doñana National Park, thanks to the rapid construction of an emergency dam.

Cleaning efforts continued until the summer of 2000 and constituted an unprecedented effort that removed more than 98% of the sludge. Nevertheless, mechanical removal of the contaminants increased the negative effects of the toxic spill by causing major impacts on geomorphological characteristics of the area (Gallart et al. 1999). Most of the studies on the flora and fauna conducted after the accident showed that pollution

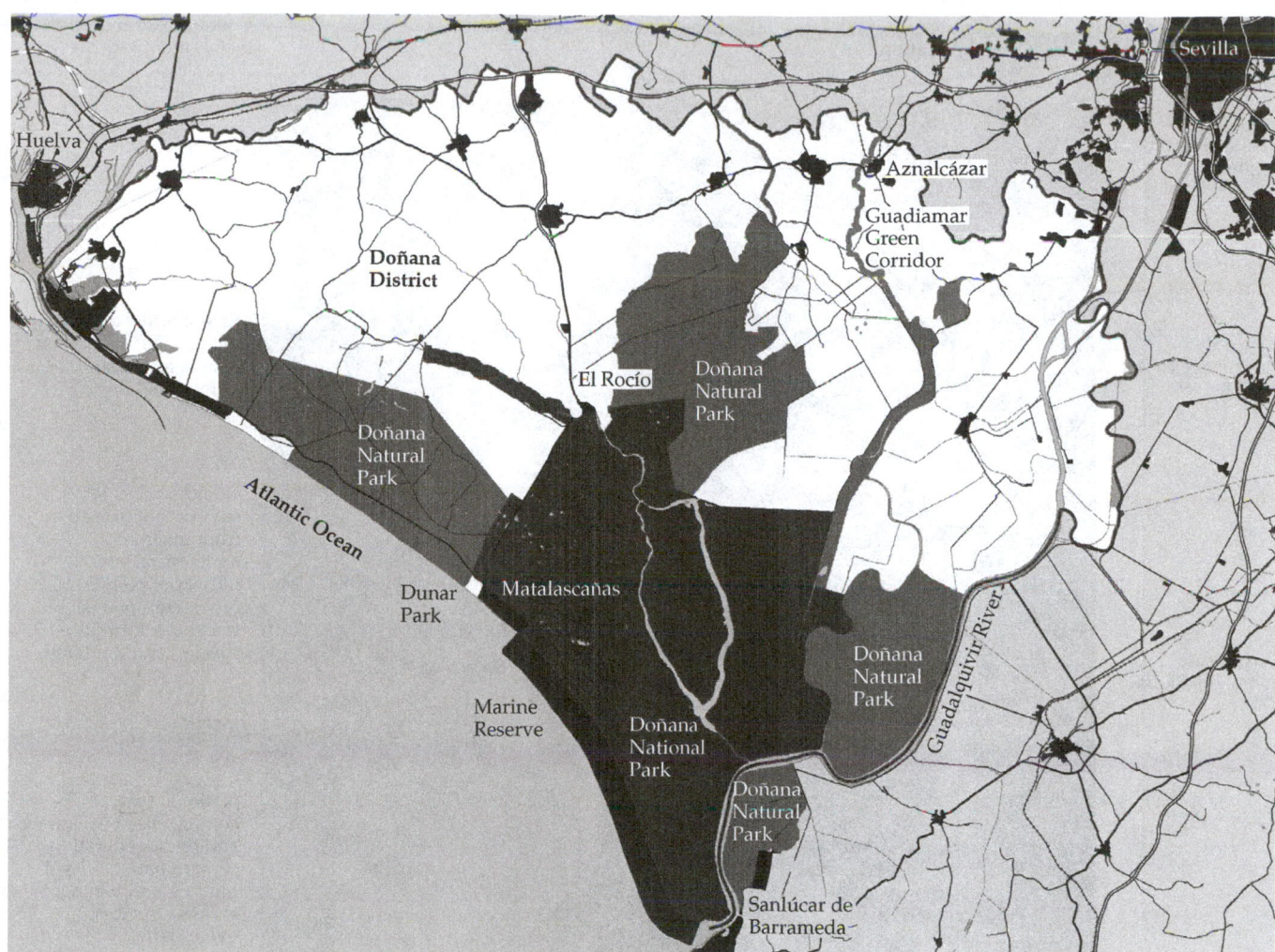

Figure B Doñana National Park is surrounded by the Natural Park, both included in the so-called Doñana District, a larger area which includes 14 villages and some 150,000 residents. Distributed in different villages and natural areas of the National and Natural parks are 8 visitor centers, 7 observatories, 2 research centers, 4 other natural areas, 6 recreation areas, and many hiking, biking, and horse trails. (Modified from Plan de Ordenación del Territorio 2002.)

did not seriously affect the trophic web in Doñana (Junta de Andalucía 2003). However, those zones of the mine devoted to processing raw materials have not been cleaned and the large areas where the old tailing deposits are located have not been restored. All of this will make the mine a long-term source of contamination, perhaps for centuries. The accident also demonstrated the vulnerability of Doñana and the problems involved with the management of a natural area if the whole ecosystem is not considered.

Aside from agricultural and heavy metal residual contamination, waters are also polluted with organic contamination coming from the neighboring villages (the majority of their waste water has not been purified) (Arambarri et al. 1996; Arribas et al. 2003). Pollutants arriving to Doñana have caused fish and bird mass mortality; 50,000 and 20,000 birds were

killed in 1973 and 1986, respectively (Castroviejo 1993, Arribas et al. 2003; Saldaña et al. 2003).

Matalascañas, an old fishing village on the beaches of Doñana, has become a major urbanized tourist area with about 80,000 people visiting each summer. Major impacts of tourism include lowering of the water table, coastal organic pollution, and straying of domestic animals into the park. Political pressures for new urbanized areas are constant. Ultimately this pressure has led to a modification of one of the main conservation programs for Doñana (Plan for Territorial Direction of Doñana Area of 1988) to allow the construction of a major urbanized area in Sanlúcar de Barrameda, close to Doñana (Figure C).

Since the 1960s there has been a great demand for new roads within Doñana to promote tourism. Thus, numerous roads have been built, which has improved transit through the

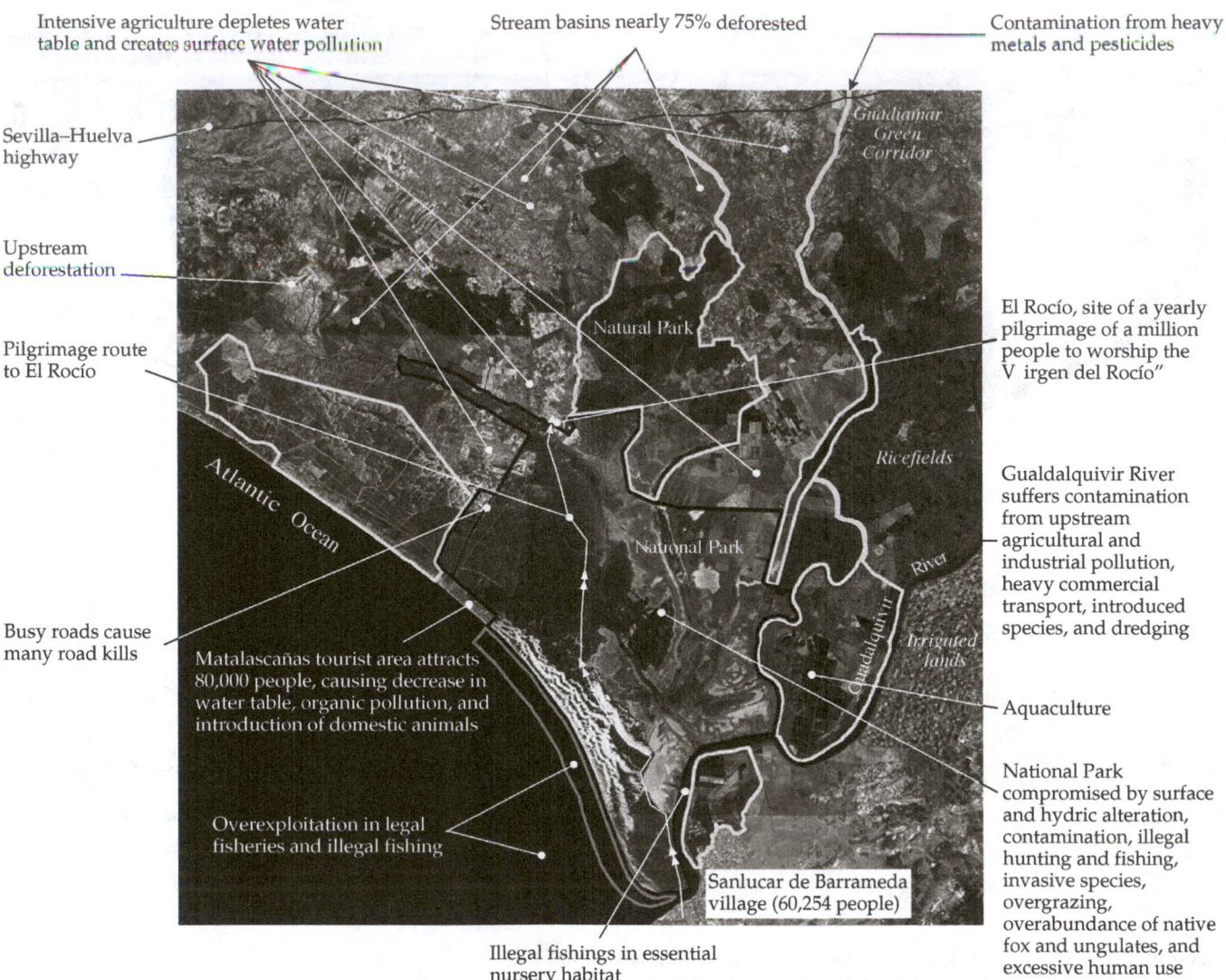

Intensive agriculture depletes water table and creates surface water pollution

Stream basins nearly 75% deforested

Contamination from heavy metals and pesticides

Sevilla–Huelva highway

Upstream deforestation

Pilgrimage route to El Rocío

Busy roads cause many road kills

Matalascañas tourist area attracts 80,000 people, causing decrease in water table, organic pollution, and introduction of domestic animals

Overexploitation in legal fisheries and illegal fishing

Illegal fishings in essential nursery habitat

Guadiamar Green Corridor

Natural Park

Atlantic Ocean

National Park

Ricefields

River

Guadalquivir

Irrigated lands

Sanlucar de Barrameda village (60,254 people)

El Rocío, site of a yearly pilgrimage of a million people to worship the Virgen del Rocío"

Gualdalquivir River suffers contamination from upstream agricultural and industrial pollution, heavy commercial transport, introduced species, and dredging

Aquaculture

National Park compromised by surface and hydric alteration, contamination, illegal hunting and fishing, invasive species, overgrazing, overabundance of native fox and ungulates, and excessive human use

Figure C   A schematic overview of the many conservation problems facing Doñana National Park. (Modified from Junta de Andalucia 1995.)

park (see Figure C). However, roads fragment habitats and are a primary source of mortality for many Doñana vertebrates. Almost 50% of all lynx mortality in the area is due to automobiles (Ferreras et al. 1992; Ramos and Soriguer 2002). In 2002, a rural road passing through one of the best habitats for Doñana's lynx was asphalted; two lynx have subsequently been killed by cars. This is significant for this species, as there are no more than 200 individuals in the wild, of which 30 are living in Doñana (Guzman et al. 2002).

Another negative impact on Doñana National Park has its origin in the cultural aspects of the area, especially in a religious tradition centered in the nearby village of El Rocío, where pilgrims have gathered annually since the sixteenth Century to celebrate the Virgin of Rocío. Initially this was a local celebration, but today it is a national event, with the number of visitors (called "romeros") increasing every year. The

pilgrimage, or "romería," is a festival lasting one week, during which more than 1 million visitors arrive in a variety of vehicles, including nearly 250,000 horses. A diversity of groups or, "brotherhoods," use traditional roads to get to the village, one of which crosses the National Park from south to north (see Figure C). More than 10,000 people and 3000 vehicles use this road twice a year. One of the biggest problems is that this activity takes place during one of the most delicate times of year in the National Park, during spring, when many birds are breeding. Elimination of this traditional activity is not possible due to traditional rights and the strong public outcry and revolt that it would engender. Nevertheless, strong regulations are increasing every year under the so called "Plan Romero."

Hunting has increased considerably in areas around the park because of the large numbers of birds that are attracted to Doñana. Nevertheless, since the Aznalcóllar accident, hunting

was forbidden in the surrounding areas and restrictions continue in some of them. Predators are also attracted to the park, and adjacent landowners set illegal traps or poison them; this is a leading cause of death for many carnivores, including the Imperial Eagle and the lynx (Estación Biológica de Doñana 1991). Illegal fishing in the lower reaches of the Guadalquivir River is damaging an important nursery area for fisheries in the Gulf of Cádiz. Considered an important biodiversity hot spot for the aquatic communities (Fernández-Delgado et al. 2000), many are particularly concerned by these activities.

Navigation in the Guadalquivir River toward the port of Seville constitutes another environmental problem due to riverbank erosion as well as continuous dredging to deepen the navigation canal. At this time, there is a large project to extract about 10 hm^3 of sediments to deepen the navigation canal to increase the numbers of vessels going to Seville. The project has been approved by the Spanish Environmental Minister (Boletín Oficial del Estado 2003) in spite of the protest of scientific and conservation groups. A second project to construct a large dam in the Guadalquivir River near Doñana, which could destroy the entire estuarine ecosystem (Fernández-Delgado 1996) has been temporarily delayed.

In addition to the increase in maritime traffic activity in such a sensitive ecological area, there is also an increased risk of introduction of species transported in ballast water or on the hulls of vessels (Carlton and Geller 1993; Ruiz et al. 2000). Several invertebrate species have already been introduced by this process (Cuesta et al. 1991, 1996; Fernández-Delgado 2003), including the Chinese Mitten Crab (*Eriocheir sinensis*) an invasive species that has severely impacted the San Francisco Estuary in California (Aquatic Nuisance Species Task Force 2002). At this time the Seville Port Authorities have not addressed this problem.

Finally, capacity of dams built in the Guadalquivir river basin reach 7000 hm^3; this could increase salinity intrusion in the estuary and decrease productivity of the area due to nutrient retention. In addition to these predictable effects is possible alteration of the trophic dynamics of the whole estuarine ecosystem.

Internal problems

Inside Doñana there are also problems. Deforestation of the stream basins surrounding the National Park has increased erosion, generating siltation and sand deposition problems in some areas. Thus, 350 ha of marshlands have been inundated with sand coming from the El Partido stream.

The red and fallow deer populations are overabundant and from time to time must be controlled artificially. Nevertheless, the greatest damage on the terrestrial vegetation is from the 5000 cows and horses living inside the national park. Many areas in Doñana are affected by this cattle pressure (Soriguer et al. 2001). Local people have traditional rights to use the pastures, and park managers have had only limited success regulating the number of cattle inside the park. Another problem is

that they are passing diseases such as tuberculosis to wild mammals (Ramos and Soriguer 2002).

Foxes and wild pigs are also very abundant and difficult to control because of their high reproductive rates; the density of foxes in Doñana is one of the highest in Europe (Rau 1987), which has fatal consequences for rabbit populations already affected by introduced viral infections (myxomatosis and the hemorrhagic viral disease). Reduced rabbit populations directly affect the Imperial Eagle and lynx populations inside the park, which rely on rabbits as primary prey (Ramos and Soriguer 2002). The eagle has decreased from 15 pairs in 1976 to nine at present (Ferrer 1993; Reserva Biológia de Doñana 2002). There are some 30 lynx living in Doñana, and only five of these are mature females (Ramos and Soriguer 2002).

Botulism, an endemic illness in Doñana affecting waterfowl, may become a major problem if water quality declines further (INITAA 1992), a likely event considering the low water quality entering the marshlands (Castells et al. 1992; Saura et al. 2001). The accumulation of heavy metals and pesticides, especially in birds of prey, threatens to reduce their reproductive success (González et al. 1984; Hernández et al. 1986, 1988; RBD 2002). Illegal hunting and fishing are also problems in spite of the large number of guards controlling the national and natural parks.

Most of the water management in Doñana has been oriented toward waterfowl; thus, many sluices were built in the past to prevent fresh water from flowing from the park into the Guadalquivir River. This activity has isolated the various water bodies of the park from the main river, preventing fishes and aquatic invertebrates from colonizing after a dry period. The number of fish species inside the national park (17) is much lower than outside (46), with some of them critically endangered.

Nonnative invasive species have markedly changed the ecology of Doñana. The most spectacular effect has been due to the intentional introduction of the red swamp crayfish (*Procambarus clarkii*) in 1974. This species is so well adapted to the area that it has not only altered trophic relationships, but has also influenced the economy of the region (Montes et al. 1993). Diversity and biomass of the macrophyte community has also been negatively affected by this species (Duarte et al. 1990; García-Murillo et al. 1993). Presently, the crayfish occupies nearly all of the Iberian Peninsula, and annual harvest in the Doñana area is 2700–4300 metric tons.

The water fern (*Azolla filiculoides*), an invasive aquatic pteridophyte, threatens the marshland ecosystem (Cobo et al. 2002). Six (35%) of the 17 fish species living in Doñana are nonnative. Carp (*Cyprinus carpio*) and goldfish (*Carassius auratus*) reduce water quality. The eastern mosquitofish (*Gambusia holbrooki*) and the mummichog (*Fundulus heteroclitus*), threaten the endemic pupfish (*Aphanius baeticus*) with near extinction (Fernández-Delgado et al. 2000) just as largemouth bass (*Micropterus salmoides*) contributed to disappearance of the stickleback (*Gasterosteus aculeatus*) (Fernández-Delgado 1987).

The most recent fish colonization is by the pumpkinseed (*Lepomis gibbosus*), which has stable populations inside the park (Fernández-Delgado et al. 2000). In 1999, the Florida turtle (*Trachemys scripta*) was detected in some water bodies of Doñana (Díaz-Paniagua et al. 2002).

Measures to Protect Doñana

At this time, four programs are being developed to protect both the National and Natural Parks, one focused on the national park (The Use and Management Plan of the Doñana National Park), two on the natural park (The Ordination Plan of Natural Resources of the Nature Park, and The Use and Management Plan of the Nature Park) and a fourth on the Doñana region (The Coordination Territorial Managing Plan of Doñana and their Surroundings).

Conservation measures within the park fall under "The Doñana National Park Management Plan" (Boletín Oficial del Estado 1991), which are currently being updated. This plan is implemented by a committee of people who represent local, regional, and national governments, universities, citizens, and conservation associations. The committee's guiding philosophy is that conservation takes priority over any other activities within the park, and that the park's natural richness depends on conservation in the surrounding areas.

The park has been divided into four zones: Zones of Special Use (173 ha), which include installations for park management and visitor information centers, Zones of Moderate Use (382 ha) intended to preserve the traditional roads that cross the park, Zones of Restricted Use (100 ha) near the information centers, where people are allowed to walk freely, and Reserve Zones (50,065 ha), the bulk of the park, with entry restricted to managers, researchers, landowners, and other authorized people.

The program is designed for management of natural resources, research, public use, compatible extractive uses, and improving the relationship between the park administration and its neighbors. Included in the natural resource management plan is an attempt to restore the park's water system to its state before the transformations adjacent to Doñana. Managers are also trying to maintain and/or recover the vegetation formations characteristic of Doñana by eliminating exotic species, controlling animal plagues and illnesses, and by preventing fires.

Both the flora and fauna of Doñana are managed toward preservation of native species, protection of threatened or endangered populations, control of overabundant species, and elimination of nonnative species where possible. Two management programs focus on the most visible and charismatic species of Doñana: the lynx and the Imperial Eagle. Both programs are oriented toward habitat improvement, which will benefit many other species as well. Additionally, densities of ungulates (red deer, fallow deer, wild pigs) and foxes are being controlled, and the most abundant introduced plants are being eradicated.

The management plan allows the park's use by local residents for traditional resource extraction, such as coal mining, beekeeping, harvest of mollusks (*Donax trunculus*) on the beach, and of pinecones in the forest, hunting and fishing in designated areas, and extensive cattle grazing. A public use program provides visitors with information about the national park and argues for the need for its conservation. Seven reception centers have been built at various points to inform visitors of Doñana's history and natural riches.

A series of outreach activities is attempting to improve the relationship between the national park and the residents of surrounding areas. These activities include environmental education programs in the neighboring villages; information points about the park have been established in several of them.

The 53,709 ha of Doñana Natural Park is managed by the regional administration through the Ordination Plan of Natural Resources of the Nature Park and The Use and Management Plan of the Nature Park in Seville. Both plans work together to manage all the activities in this buffer area compatible with the conservation of the national park. As that of the national park, the plans are run by a committee of people representing local, regional, and national governments, universities, communities, and conservation associations. The natural park is divided into three parts; Reserve Zones (10.5%) include natural areas with high natural values. These areas are exclusively devoted to conservation, research, education, or ecological restoration. Special Use Zones (74.5%) represent areas with high natural values but with some degree of human intervention. The activities here are oriented to preserve the system through the sustainable use of resources. Finally, there are the Common Regulation Zones (15.0%), modified areas of light–moderate use that have some natural interest. Here, conservation measures are oriented to the development of restoration and management programs, which try to reduce the impact of human activities.

There is a fourth plan called the Territorial Regulation Plan for the Doñana district. Its main objective is to establish the jurisdictional basis for the regulations and sustainable development of the surrounding Doñana areas (see Figure C) to secure and make compatible the preservation of the natural resources with socioeconomic progress and the increment in the quality of life of their people. The area affected includes 14 villages with about 150,000 people and has a budget of some U.S.$40 million for the next 12 years.

The Future of Doñana

Two questions remain. Will Doñana ultimately be saved from the human pressures that surround it? And what will be its conservation status when it passes to the next generation?

At the end of the 1980s, Doñana had serious conservations problems. Conflicts increased between conservation of the natural areas and increase of economic activities, especially with the expansion of strawberry cultivation, tourist areas, and ur-

banization projects. Inside the park there were also many problems with the local people relative to traditional uses of resources. Local people felt that the national park was delaying economic development of the area. This led to the President of the Regional Government to create an International Committee to evaluate the situation, identify the problems, and suggest solutions (Castells et al. 1992). The report set the basis for the development of "The Sustainable Development Plan for Doñana's Neighboring Areas," the first plan of this type in Spain. The plan foresaw an investment of some U.S.$350 million between 1992 and 2000 to develop economic activities in the surrounding villages compatible with conservation goals of the national park. The funds came from the European Union (59%) and the Regional (36%) and Central (5%) governments. Only 10% of the European funds came specifically for the plan; the rest were monies previously adjudicated to the Regional and Central administrations and that were redirected to this plan (Requejo and Belis 2003). The Doñana 21 foundation (see: http://www.donana.es/index.php) was created to develop the plan. Ten years later, this plan has been evaluated through the Doñana+10 Foundation (see: http://www.donana-mas10.com). The majority of this investment has been for water management (23%), equipment and road infrastructure (31%), agriculture (11%), and environment (18%). The rest paid for a variety of programs such as education (6%), tourism (4%), and promotion of economic activities (6%) (Belis et al. 2003). Unfortunately, even with this investment 13% of the objectives were not met, and some of them, such as the agriculture and water management programs, clearly failed. The most successful were road construction, tourist infrastructures, and enterprise opportunities.

To the advantages associated with the Sustainable Development Plan for Doñana's Neighboring Areas we have to add the improved quality of life for the people of the Doñana region in recent years. The rent per capita is now similar to the rest of the Andalucía Region. Between 1991 and 2001 unemployment decreased by 42%. The number of enterprises increased from 400 to 800 in 10 years and in the last 30 years the population has grown 30% (Junta de Andalucía 2002). This process is driven by agriculture, mainly strawberry cultivation, which now covers about 4000 ha in the lands surrounding the national park.

Because of these developments and the Aznalcóllar toxic spill, two big restoration programs are being conducted in the region: the Guadiamar Green Corridor and the Doñana 2005 Restoration Program. The former, promoted by the Regional Government, aims at restoration of the Guadiamar basin and reestablishment of an ecological corridor between the mountainous area of Sierra Morena and the littoral systems of Doñana. At the same time, the program seeks improvement of the quality of life of Guadiamar basin inhabitants by developing a socioeconomic system that is environmentally sustainable and integrated into the natural context. It had a budget of about U.S.$40 million for 1998–2001 with a research program of U.S.$7 million in the subsequent five years. The program is in line with the European Water Framework Directive because it covers the whole riverine system and strives to preserve the natural dynamics of the riverine ecosystems.

The second program, funded by Spain's Ministry for Environment, promotes the restoration of large degraded areas. It aims at the hydrologic regeneration of the watershed and river bed flowing into the marshland of the national park to recover the water supply to the marshlands, ensure quality and quantity of water, and stop wetland degradation. The project is the most important wetland restoration effort ever undertaken in Spain, both for its budget (U.S.$129 million) and for the extent of its target area.

The number of publications about Doñana have increased sharply in the last few years, and the system is better known, which is good for conservation. Several museums have been opened (Sea World Museum, Religious Museum) and new natural areas have been acquired for public use, such as the Dunar Park (see Figure B). At this time in the Doñana region there are eight visitor centers, seven observatories, two research centers, six recreational areas, six horse trails, six bike trails, seven hiking trails, and four jointly coordinated nature reserves. This infrastructure led to more than 400,000 visitors to the area in 2002. On the other hand, one the most successful aspects of the Doñana sustainable plan has been a decrease of aggressiveness toward the national park and development of a perception by the local people that Doñana represents a positive natural heritage that still has not been efficiently exploited, especially relative to tourism opportunities (Belis et al. 2003). Thus, in spite of the many challenges to Doñana, there are also many indications that it will be saved, though much work remains. A focused collaboration between central and regional governments, more active local participation in environmental decision making processes, more expert technical advice and, of course, more funding, are basic ingredients needed to preserve this important natural heritage.

CASE STUDY 14.2

The California Channel Islands Marine Reserves
Scientists Informing Policy and Management

Satie Airamé, University of California, Santa Barbara and Deborah Brosnan, Sustainable Ecosystems Institute

Marine Protected Areas (MPAs) have become an increasingly important tool for conserving coasts and oceans. MPAs are areas of the ocean where some or all activities are limited or prohibited to protect natural and cultural resources. MPAs have different shapes, sizes, and management characteristics, and have been established for different purposes. There are now well over 4200 MPAs worldwide and hundreds more are in the planning stages (Kelleher et al. 1995; Roberts et al. 2003). Their size spans over six orders of magnitude ranging from the very tiny (0.002 km^2) to very large (846 km^2; Halpern 2003).

Marine reserves (also known as "no take" zones) are a special class of MPAs in which an area of ocean is completely protected from all extractive and destructive activities. There is abundant evidence that protecting areas of the ocean in no take marine reserves leads to rapid increases in abundance, size, biomass, and diversity of animals that are fished or impacted indirectly by fishing, regardless of where in the world reserves are located. Halpern (2003) reviewed 76 studies of reserves that were protected from at least one form of fishing. Across all reserves, abundance (measured as density) approximately doubled. Biomass, or the weight of all organisms combined, increased 2.5 times in reserves as compared to fished areas. Average body size of organisms protected in marine reserves increased by approximately 30%. The increase in size contributes to greater reproductive potential (Béné and Tewfik 2003). In addition to changes in biomass, abundance, size, and reproductive potential, the number of species in each sample increased by 30%. However, marine reserves will do little to protect ocean life against external threats and influences such as climate change and pollution. Additional legislation and agreements will be needed to address these issues.

Marine reserves can be contentious because they limit consumptive activities, such as recreational or commercial fishing. The proposal of a marine reserve elicits strong emotions and responses in many people, including both those who feel that their livelihood or traditional way of life will be affected and those who believe that marine reserves are necessary to stop degradation of marine ecosystems.

Many early MPAs were created on the basis of common sense rather than on science; the underlying assumption was that some protection was better than none. In fact, many early MPAs that were designed for protection and conservation still allowed fishing and other extractive practices within their boundaries.

Scientific input is an important and necessary component of the design, implementation, and monitoring of MPAs world-wide. MPA science continues to evolve as practical experience and research provide new insights and information. For instance, scientists have made important advances in how to meet multiple goals such as combining the protection of biodiversity (which argues for few large reserves) with fisheries goals (which favors many small reserves; Hastings and Botsford 2003). Scientists also have contributed knowledge on site selection (e.g., Airamé et al. 2003; Roberts et al. 2003; Leslie et al. 2003), dispersal and genetics (e.g., Palumbi 2003; Shanks et al., 2003), and reserve size and spacing (e.g., National Research Council 2000).

The California Channel Islands Marine Reserves

In 1998, the California Fish and Game Commission, a state-appointed board with the authority to establish marine reserves, received recommendations to create no-take marine reserves around the northern Channel Islands off the coast of California (Figure A). Local environmental organizations, some fishing groups, and other concerned stakeholders—many of whom were alarmed by declines in fished species—developed a proposal for establishing marine reserves around the islands. However, because the broader community was not involved in its development, the proposal was rejected and a process was

Figure A The northern Channel Islands off the coast of California. (Photograph © Gary Crabbe/Alamy.)

established to bring together social, economic, and ecological information to support the decision about the marine reserves.

The process was developed by the California Department of Fish and Game, the state agency charged with managing fisheries and enforcing fishery regulations within California state waters (0–3 nautical miles offshore), and the Channel Islands National Marine Sanctuary, a federal agency that manages an area of 1252 square nautical miles of ocean around the northern Channel Islands of San Miguel, Santa Rosa, Santa Cruz, Anacapa, and Santa Barbara. The sanctuary was established in 1980 to protect the region from oil and gas drilling. However, commercial and recreational fishing are allowed in the sanctuary, as long as the activities comply with regulations established by state and federal fisheries agencies.

Together, the California Department of Fish and Game and the Channel Islands National Marine Sanctuary began a community-based process to consider the appropriateness of marine reserves within the Sanctuary waters. To support the process, a Marine Reserves Working Group (MRWG) was established to represent the full range of community perspectives, including commercial and recreational fishing interests, diving and other nonconsumptive interests, and the general public. The MRWG worked together toward consensus using professional facilitators, and agreed on a set of goals (Table A) for the marine reserves. The MRWG met monthly for nearly two years to receive, consider, and integrate information about marine reserves and their potential role in the management of the Channel Islands.

The Channel Islands National Marine Sanctuary Advisory Committee (SAC) determined that decisions should be based on the best available scientific and socioeconomic information. Thus, the SAC established two advisory panels to assist the MRWG. A team of economists and social scientists was appointed to gather and evaluate socioeconomic data, and a panel of 16 marine scientists—the Science Advisory Panel (SAP)—was formed, including marine ecologists, oceanographers, ichthyologists, phycologists, fishery managers, geneticists, statisticians, and modelers. The SAP was asked to perform four tasks: (1) gather and evaluate information about marine reserves and

their effects; (2) determine the status and trends of marine species and habitats in the study region; (3) develop ecological design criteria for marine reserves; and (4) generate and evaluate options for networks of marine reserves. For over two years, the SAP worked closely with the MRWG to provide them with scientific advice. The process proved to be a challenge. Not only did scientists have to cope with limited scientific information, they also had to deal with challenges and criticisms from those opposed to marine reserves (or the inclusion of certain sites recommended by the scientists as important for protection).

At the same time, the public was active and vocal with their opinions. The California Department of Fish and Game and the Channel Islands National Marine Sanctuary received close to 10,000 public comments via email, faxes, phone calls, and letters. 94% were in favor of marine reserves and 6% opposed them. Many of those opposed wanted the size of proposed reserves reduced while many in favor thought that a greater extent—at least 30%–50% of the coastal areas—should be set aside as reserves. Moreover those opposed to the establishment of marine reserves frequently questioned the quality of the science and the biases of the scientists who served on the SAP. The scientists, who worked hard to ensure the quality and objectivity of their research, were not fully prepared for the strength and vigor of public criticism of the science and the scientists themselves.

Ecological Criteria for the Channel Islands Marine Reserves

To meet goals for biodiversity conservation, marine reserves should include each distinct biogeographic region in the study area (Roberts et al. 2003). A biogeographic region is an area of animal and plant distributions having similar or shared characteristics throughout. Three main biogeographic regions were identified in the Channel Islands waters based on biological and physical differences: a region of cool water to the northwest, a region of warm water to the southeast, and a zone of mixing between the two bodies of water. The SAP recommended setting aside several reserves within each of the three distinct biogeographic regions.

TABLE A *Marine Reserves Goals for the Channel Islands National Marine Sanctuary*

Category	Goal
Ecosystem and biodiversity	To protect representative and unique marine habitats, ecological processes, and populations of interest in the Channel Islands National Marine Sanctuary
Sustainable fisheries	To achieve sustainable fisheries by integrating marine reserves into fisheries management
Socioeconomic	To maintain long-term socioeconomic viability while minimizing short-term socioeconomic losses to all users and dependent parties
Natural and cultural heritage	To maintain visitor areas and spiritual and recreational opportunities which include cultural and ecological features and their associated values
Education	To foster stewardship of the marine environment by providing educational opportunities to increase awareness and encourage responsible use of resources

Source: Marine Reserves Working Group.

Conserving biodiversity also requires protection of a range of marine habitats within each biogeographic region (Roberts et al. 2003). The scientists developed a simple multidimensional habitat classification, using depth, exposure, substrate type, dominant plant assemblages, and other features. Their goal was to ensure that a suitable amount of each habitat was protected within each biogeographic region.

To conserve biodiversity, marine reserves should protect species of special concern, including keystone species, species of economic importance, and threatened and endangered species. The MRWG identified 119 species of special concern in the Channel Islands, including plants, invertebrates, fishes, seabirds and marine mammals. For species whose distributions were known, such as marine mammals and seabirds, the scientists used data to help locate potential sites for marine reserves. Fisheries data were used in a separate economic impact analysis to evaluate potential costs of reserve designs (Leeworthy and Wiley 2000).

One of the most important questions related to reserve planning is how much area should be in reserves. Ideally, the size of a marine reserve depends on the potential dispersal distance, population growth rate, and fishing pressure on species of special concern (Roberts et al. 2001). However, this information was not known for the majority of species of interest and therefore, the SAP followed more general guidelines. For conservation of biodiversity, the benefit of a reserve increases with size (e.g., Margules et al. 1988; Dayton et al. 2000; Roberts and Hawkins 2000). Larger reserves protect more habitats and populations, providing buffers against losses from environmental fluctuations or other natural factors that may increase mortality rates or reduce population growth rates. For fisheries management goals, however, the benefit of a reserve does not increase directly with the area it occupies. The maximum benefit of fully protected reserves for fisheries, in terms of sustainability and yield, occurs when the reserve is large enough to export sufficient larvae and adults, and small enough to minimize the initial economic impact on fisheries. After reviewing of the literature, the SAP recommended a wide range of possible reserve sizes, from 30% to 50% percent of the Channel Islands National Marine Sanctuary, to achieve both conservation and fisheries goals of the MRWG.

The SAP also addressed the important question of whether the marine reserves should be concentrated in a few large areas, or more spread out among a greater number of smaller areas. In the ocean, the ecological connections between different geographic locations can be much greater than on land because animals can drift and swim through the ocean with few barriers to movement. For example, larvae may drift at sea for several months as plankton, move as juveniles into shallow areas near the shore to grow, and settle as adults in deeper waters. Consequently, scientists developed the concept of networks of marine reserves (Roberts et al. 2001). A network of marine reserves is a set of marine reserves connected by larval dispersal and juvenile or adult migration (NRC 2000). Networks include reserves that may perform different functions, such as providing nursery, spawning, and feeding areas. The conservation goals for a network of marine reserves are achieved through the combined effects of each reserve.

Applying Ecological Criteria to the Design of a Marine Reserve Network

Once they identified the ecological criteria, the scientists' next task was to apply the criteria to develop a variety of network designs for consideration by the MRWG. The scientists used a computer-based modeling tool known as MARXAN to evaluate the data and identify areas that could be effective marine reserves. To begin the process, the scientists divided the planning region, and associated data, into 1500 "planning units" of 1×1 minute (approximately 1×1 square nautical miles) following lines of longitude and latitude. The scientists entered the ecological criteria into the modeling tool. Then, the SAP used simulated annealing to explore and generate options for networks of marine reserves.

First, the program randomly generated an initial reserve system, made up of a set of planning units, including the target percentage of each habitat and feature. The program then calculated the "cost" of the reserve system based on the total size of the set of planning units. The program then randomly selected a planning unit and evaluated the impact of adding or removing it from the reserve system. This process continued a million times until a set of good solutions was produced. The ultimate goal of the process was to identify a solution that met the ecological criteria (in the form of a network of marine reserves) in the smallest area possible (to minimize the social and economic costs). At the end of the analysis, the SAP presented the MRWG with the ecological criteria used to run the model, the original data used in the model, and a portfolio of ten different options for networks of marine reserves.

The MRWG reviewed the findings of the advisory panels using an interactive Geographic Information System (GIS) tool (Killpack et al. 2000). The GIS tool included the information provided by the SAP as well as socioeconomic information about the major commercial and recreational activities in the Channel Islands. Technical facilitators assisted the MRWG with the tool at several public meetings (Killpack et al. 2000), which allowed the MRWG and members of the public to view and query the data, as well as to develop and evaluate designs for the marine reserves. If a particular design did not meet the ecological criteria, then the members of the MRWG adjusted the boundaries to satisfy conservation goals. Alternately, if a particular design had a high potential economic impact, members of the MRWG were able to adjust the boundary of the proposed reserve to limit the potential impacts.

Various designs for networks of marine reserves in the Channel Islands, developed by the MRWG, were given to the advisory panels for review. The SAP evaluated the designs based on the ecological criteria and provided suggestions for improvement to meet goals for biodiversity conservation and sustainable fisheries. The economic advisory panel conducted

economic impact analysis to determine the potential impacts to commercial and recreational activities. Using information from the advisory panels, the MRWG adjusted designs to meet the greatest number of goals. During the process, the MRWG developed and reviewed over 40 designs.

After evaluating a broad range of options, members of the MRWG were not able to reach consensus on a single design. However, in May 2001 the MRWG produced a map representing the two alternative designs and all of the information generated during deliberations of the MRWG and its advisory panels to the Sanctuary Advisory Council (SAC). The SAC reviewed and forwarded the information to the California Department of Fish and Game and the Sanctuary. In June 2001, the Department and the Sanctuary developed a compromise between the two maps, based on the information provided by the MRWG and its advisory panels.

After required analysis and public review, and the California Fish and Game Commission approved a final map, the network of state MPAs was established in April 2003. The network includes ten no take marine reserves and two conservation areas that allow limited commercial and recreational fishing (see Figure B).

The network of marine reserves includes many of the sites identified by the scientists, but the size of the network was smaller than the scientists had recommended. The SAP determined that the adopted network is likely to contribute to local conservation of biodiversity, but is unlikely to contribute substantial spillover or export to surrounding fisheries due to the small size of individual reserves. Thus, the adopted design is likely to achieve some, but not all of the goals established by the MRWG.

Challenges to MPAs in the Channel Islands

Controversy over marine reserves was intense during the planning stages and did not stop after the state MPA network was formally approved. For instance, while many recreational and commercial fishermen supported the reserves, and were even among the first to propose them, other fishermen opposed them. There was heated and often bitter public debate in meetings and in the written and electronic media. Some commercial fishermen objected to the MPAs because they felt that they might have unbearable short-term economic impacts. Recreational fishing groups felt that they were not the cause of the problem and so should not have to bear the burden of conservation.

Fishermen who objected to the reserves sought a temporary restraining order in Ventura County Superior Court to halt implementation. The group argued that the reserves would financially impact their businesses and that the California state agencies did not comply with existing regulations during the multi-year process to establish the reserves. The judge ruled that blocking the reserves was against the public's interest and that the plaintiffs were unlikely to prevail in court and so he refused to issue a temporary restraining order. On April 9, 2003 the state MPAs officially went into effect (Title 14, California Code of Regulations, Sections 27.82, 530, and 632).

Although challenging, the process of considering and then creating marine reserves as a management tool in the Channel Islands brought together science, economics, and public opinion to develop a solution to a complex management problem.

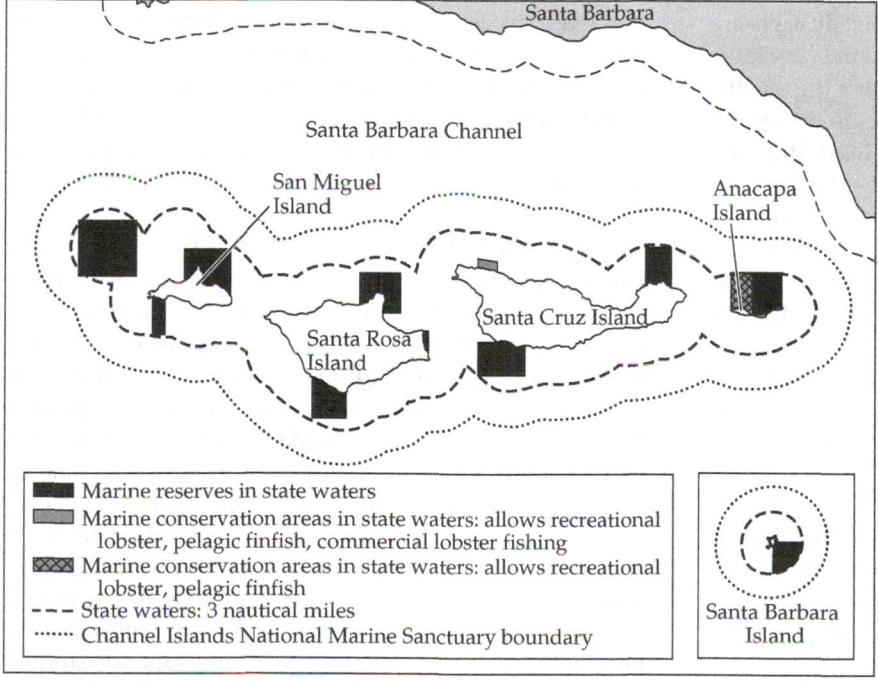

Figure B The Channel Islands network of marine protected areas. Santa Barbara Island is located about 40 km southeast of the other islands, so it is shown in the inset box. (Modified from the Channel Islands National Marine Sanctuary 2004.)

Santa Barbara

Santa Barbara Channel

San Miguel Island

Anacapa Island

Santa Cruz Island

Santa Rosa Island

- ■ Marine reserves in state waters
- ▨ Marine conservation areas in state waters: allows recreational lobster, pelagic finfish, commercial lobster fishing
- ▨ Marine conservation areas in state waters: allows recreational lobster, pelagic finfish
- - - - State waters: 3 nautical miles
- ······ Channel Islands National Marine Sanctuary boundary

Santa Barbara Island

CASE STUDY 14.3

Reconciling Theory and Practice in Designing a Regional Reserve System

A Colombian Case Study

Gustavo Kattan, Wildlife Conservation Society, Colombia Program

A regional reserve system should be representative, resilient, redundant, and restorative. This so-called "four-R framework" (Groves 2003) sounds simple, but presents formidable challenges to a practitioner involved in conservation planning. In this case study, I present the perspective of a person standing between conservation science and practice, with the additional handicap of working in a scientifically developing country. How do we reconcile the rigor of science with the reality in the field?

The Implications of the Four-R Framwork

Representative means that the system should represent the variety of biological manifestations in a region, from landscapes and communities to genes. However, seeking to attain this is not a practical objective, so The Nature Conservancy (1997) proposed an operational concept, the so-called coarse filter (ecosystems)—fine filter (species) approach. The coarse filter is an application of the umbrella principle. By conserving ecosystems (represented in areas large enough to be viable), we hope to conserve all their communities and ecological processes. However, focal species (species of special interest) may not be represented in these areas, so a special effort is made to ensure their inclusion in protected areas. Therefore, at the very least, achieving adequate representation requires knowing the distribution of native ecosystems and focal species in the region, with enough precision to be able to map them. In addition, applying the principle of representation requires setting quantitative goals: what proportion of original ecosystems do we want to preserve? In how many areas? Certainly the consequences of representing 10% of the extent of original ecosystems in one large, contiguous area or in ten small, disjunct areas differ in many ways.

The resilience of the reserves, which is a function of area, also needs to be considered. Larger reserves contain more complete sets of species and have a higher probability of maintaining their biological integrity over time.

The principle of redundancy is also relevant for deciding on number of areas. Redundant reserves—reserves that contain ecosystems and species that are already represented in the system—are an insurance policy against the possible catastrophic loss of some areas. Redundancy is particularly important for small reserves.

Finally, habitat restoration is a tool for achieving conservation goals when the area of an ecosystem needs to be increased or isolated reserves need to be connected.

The Need for Reserve System Planning in Colombia

Colombia is presently revising its national system of protected areas. The system has been in place for over 40 years, but there are gaps in ecosystem representation. For example, many parks and reserves in the Andean region were established to protect the headwaters of important river systems. Thus, parks are concentrated at the upper elevations, whereas lower slopes and inter-Andean valleys are unrepresented. This parallels the pattern of human occupation and landscape transformation, which in turn, was determined by the more benign climate, favorable topography and fertile soils of the mid-elevations. Unfortunately, these are also hot spots of biological diversity and endemism.

To facilitate the process of constructing the system of protected areas, the country was regionalized, and each region made responsible for designing its own system. This follows a political trend of decentralization, increased regional autonomy, and transfer of environmental responsibilities from the central government to regions and municipalities. Most regions are defined by jurisdictions of government agencies and socioeconomic and cultural factors, but little biology. Eventually, regional systems of protected areas will be integrated into a national system.

Developing a Reserve System in Eje Cafetero

One region currently constructing its regional system of protected areas is the "Eje Cafetero" or main coffee growing region of the country (hence "SIRAP-EC" for Sistema Regional de Areas Protegidas del Eje Cafetero). This is a region of about 30,000 km² encompassing the middle portion of the western and central ranges of the Andes and the Cauca Valley. Two features of this region present serious challenges for the planning process. First is a lack of biological knowledge. In spite of having several major urban centers with important academic institutions in or around the region, biological knowledge is insufficient for a rigorous planning process. Thus, we can refer to regional distributions of ecosystems and species only on a very general level. Second is the degree of landscape transforma-

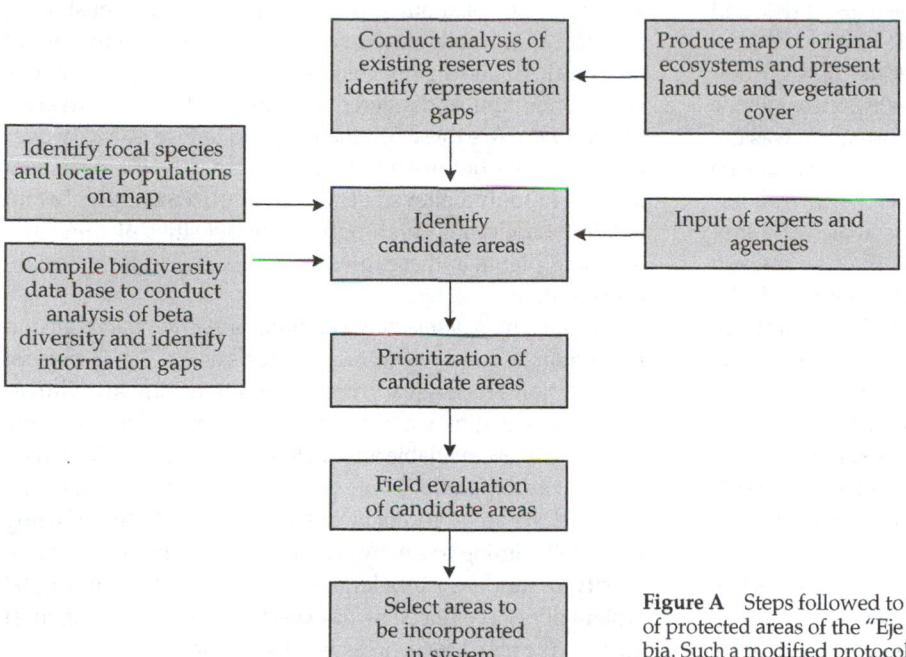

Figure A Steps followed to identify candidate areas for the regional system of protected areas of the "Eje Cafetero" or coffee growing region of Columbia. Such a modified protocol for systematic conservation planning may be useful in other regions where detailed data on biodiversity are lacking.

tion. As expected, for one of the most economically productive regions of the country (coffee is one of the pillars of Colombian economy), only a few small and isolated patches of natural ecosystems remain below 2000 m of elevation. These features are common to most of Andean Colombia, and probably other countries as well.

The planning process followed several steps (Groves et al. 2002; Figure A). We started by producing a map of original ecosystems and current land use and vegetation cover (from a variety of sources, including satellite images). Our first challenge was to define original ecosystems. Although the region was significantly transformed as recently as the first half of the twentieth century, there are no descriptions of original ecosystems and their distribution, much less of species distribution. Our only option was to use the Holdridge life-zone classification system, as a general guide to the variety of potential vegetation cover types in the region. For practical reasons, we simplified the system and divided the region into subregions and elevational zones (Kattan and Franco 2004; Kattan et al. 2004). This method probably ignores internal heterogeneity of subregions, but present knowledge does not permit a finer classification.

Using these maps, we conducted a gap analysis to evaluate the existing set of protected areas. For each subregion and elevational zone, we calculated the proportion of remnant natural vegetative cover, and the proportion currently included in protected areas (national, regional, municipal, and private reserves). Not surprisingly, we found that some ecosystems are unrepresented in the system, and that many protected areas are very small, which may reduce their long-term viability.

At the same time, we started constructing a regional biodiversity data base that has two components. The first is a data base of potential biota by subregions. This was only possible for groups such as birds and mammals, for which some general knowledge of geographic distribution was available (see Kattan et al. 2004). This information was used to conduct an analysis of potential beta diversity among subregions. The second data base is a compilation of all available locality records (in the published literature, reports, and museum specimens) of species in the region. In addition to compiling information on regional species distributions, this data base serves to identify geographic and taxonomic information gaps. This exercise revealed the second major challenge—the insufficiency of our biological knowledge of the region. At present the data base has about 30,000 species-locality records, or one per square km, which is dismally inadequate for such a biologically heterogeneous region. There is also great geographic and taxonomic bias in this knowledge. Some regions are totally unexplored, and most records that exist are for birds and plants. For example, there are on average nine records per bird species, but only one per insect species (for the few insects included). This means that for many species there is only one record for an entire region. This is hardly adequate for predicting species distributions and making sure that viable populations are properly protected.

A parallel line of work concentrated on focal species (fine filter approach). By expert consensus, two categories of focal species were made: at-risk species (i.e., under some degree of national or regional threat), and species that could be used as surrogates for conservation planning (in the sense of Lambeck

1997 and Sanderson et al. 2002). This region, however, is very rich in restricted-range species, most of which are at risk, and a first analysis produced an unmanageable number of focal species. We settled on selecting the vertebrate (fishes, frogs, birds, and mammals) and plant (timber trees) species with the highest priority as conservation targets (information was unavailable for invertebrates). Our objective was to make sure they were included in protected areas, which required information on distribution and population status, which of course we lack. One of the top priority species is the Cauca Guan (*Penelope perspicax*, Cracidae), a galliform, fruit-eating bird with a geographic distribution restricted to the middle Cauca Valley. This species has lost 95% of its habitat and is presently reduced to a few isolated populations (Renjifo 2002).

To proceed to the next step, (i.e., selecting candidate areas), we needed to set representation goals. The next major hurdle was to define quantitative goals. Some options have been suggested (Groves 2003) including: (1) species-area relationships; (2) beta diversity; (3) minimum dynamic area (Pickett and Thompson 1978). The species–area relationship dictates that 30%–40% of the area of a given ecosystem is required to preserve 80%–90% of species. The question then becomes: Do we create one reserve or several reserves adding up to this percentage? Beta diversity may be of help, as it dictates the number and location of areas necessary to represent regional species diversity. We know that each of our subregions represents different ecosystem types and species assemblages; thus, we need reserves in each subregion. We also know that there is significant beta diversity within subregions. A study of bird diversity on the western slope of the Central Cordillera revealed high beta diversity among elevational belts and among river drainages (Kattan et al. unpublished data; this study also revealed that fragmentation increases beta diversity because of differential extinction of local populations). We have no information on minimum dynamic area (patch dynamics), except that we know the larger, the better. We then decided to aim for 30%–40%, trying to distribute areas over entire subregions (covering entire altitudinal gradients whenever possible), and to make each area as large as possible.

The reality, of course, is that in our region there are not enough remnants to represent 30%–40% of original ecosystems, at least in some subregions. Some ecosystems, such as dry forest and wetlands on the valley floor, and pre-montane forest fragments in the coffee-growing elevational belt (1000 m to 1800 m), are in a critical state, with only a few small and isolated fragments remaining. So we revised our plan, and aimed to preserve 100% of remnants of critical ecosystems while identifying restoration opportunities to increase their cover.

Is it worth it to invest in the preservation of small, probably degraded fragments? The answer is a categorical yes. It is clear that many species have been lost. Two studies, based on historical data, demonstrated that forest fragmentation caused local extinction of about 30% of bird species (Kattan et al. 1994; Renjifo 1999). Habitat fragments, however, are still important repositories of the regional biodiversity, including populations of endemic and endangered species (Kattan and Alvarez-López 1996). Are these fragments going to lose integrity over time? Possibly, but pre-montane forests may be more resilient than commonly believed. This has to do with the inherent patch dynamics and high population densities of some animals and plants in these forests (Murcia and Kattan, unpublished data).

The next step is selecting candidate areas to be included in the system. A variety of criteria and tools are available to assist the selection process (e.g., reserve-selection software; Groves 2003). Our situation was simple, however, as our options were limited to a few available areas. These areas were delimited on a map in an experts' workshop, following a simple algorithm: select all the areas you believe have any possibility of being protected, aiming to satisfy criteria of representation, beta diversity, redundancy, and large size (we called this our digital system of reserve selection, as experts pointed their fingers at areas in the map). Still, areas had to be prioritized, for which we used the irreplaceability and vulnerability concepts as described in this chapter (Groves 2003). Areas were ranked by experts according to these two variables taking into account predefined criteria, and the highest priority was assigned to areas scoring in the top 50% for both variables. The second priority was given to areas with high irreplaceability and low vulnerability, third priority to areas with high vulnerability and low irreplaceability, and the lowest priority to areas scoring low on both variables. We are still working to refine our proposed reserve system to allow viable populations of focal species to persist.

We approached the design of the SIRAP-EC with a scientific frame of mind, but faced two major limitations: lack of sufficient biological knowledge for a rigorous design, and a highly fragmented region, which severely limited design options. Under these circumstances, our version of the four-R framework (coarse filter approach) can be summed up in the following words: complement, expand and connect areas as much as possible. That is, (1) include complementary areas to represent all regional ecosystems and beta diversity, even if it means representing a small percentage of the original ecosystems; (2) identify opportunities to expand protected areas, either by adding habitat blocks where available, or by restoring habitat; and (3) connect habitat fragments as much as possible, either by habitat restoration or by identifying productive matrices that foster connectivity (Durán and Kattan 2005). And our version of a fine filter approach is to try to include all known populations of at-risk species in protected areas.

Summary

1. Setting aside protected areas are one of the most effective tools available for conserving biodiversity and can also provide other benefits such as protecting water supplies and cultural values, and sustaining the livelihoods of indigenous groups. Protected areas range from those that are strictly protected, and in which extractive uses are excluded, to multiple-use areas in which the sustainable extraction of natural resources is allowed. While the coverage of protected areas has more than doubled in the past decade, analyses of reserve systems at global, regional, or national scales indicate that there are gaps and biases in the representation of biodiversity. Creating a global, ecologically representative reserve system that is well-managed will require substantial financial investment. Thus, there is a need to prioritize the allocation of scarce conservation resources to the expansion of existing protected areas so that returns for biodiversity conservation are maximized.

2. Recent research efforts in the field of conservation planning have focused on the development of principles and tools to design efficient reserve systems that represent as much biodiversity as possible at a fixed cost. Systematic conservation planning has informed reserve system design in both terrestrial and marine realms. For example, conservation planning principles and tools informed the conservation plan recently developed for the Cape Floristic Region of South Africa and were the basis of the recent rezoning of the world's largest marine park—the Great Barrier Reef Marine Park of Australia.

3. This chapter focuses on the functions, design, and limitations of protected areas and the processes of conservation planning. By considering all the complexities of conservation planning, we can see that the science of reserve system planning is still in its infancy. How to address the limitations of data and incorporate ecological, social and political processes are active areas of research.

> Please refer to the website www.sinauer.com/groom for Suggested Readings, Web links, additional questions, and supplementary resources.

Questions for Discussion

1. What is the irreplaceability of a protected area? Explain your answer using an example of both a completely replaceable and completely irreplaceable protected area. What is adequacy in the context of conservation planning for reserve systems and how could we measure and plan for adequate reserve systems?

2. Are all protected areas managed in the same way? Discuss the IUCN protected area management categories and provide explanations and illustrations of each category.

3. Why wouldn't we add any area we can cheaply acquire to an existing reserve system as fast as possible?

4. In what way does the phrase "the whole is more than the sum of the parts" apply to a systematically designed reserve system?

5. If we conserved 15% of every biome in the world would that be enough to ensure the long-term persistence of global biodiversity?

15

Restoration of Damaged Ecosystems and Endangered Populations

Peggy L. Fiedler and Martha J. Groom

Re·store (v.): 1. To bring back into existence or use; reestablish; 2. To bring back to an original condition; 3. To make almost as good as new.

American Heritage Dictionary, 2000

Given the extensive damage already caused by human activities, many scientists and practitioners turn to restoration as the primary means to conserve biodiversity. Yet, attempts to restore or repair damaged ecosystems or endangered species are fraught with difficulties, and meaningful execution requires some measure of tenacity, clairvoyance, and dumb luck. As restoration science and practice matures, we are increasingly able to make substantial improvements to damaged sites or to recover endangered species, but only very rarely can we say restoration has fully reinstated an ecosystem to its condition prior to degradation.

Our goal in this chapter is to present an overview of the field of ecological restoration and species reintroduction, and provide some insight into the primary theories guiding their practice. In illustrating both theory and practice, we refer to a suite of case studies that provide details of particular projects, for in restoration or reintroduction, more than in many aspects of conservation, the "devil is in the details." We first describe the field of restoration ecology, and then move on to describe animal reintroduction. Many principles in restoration ecology apply also to reintroduction efforts, and we highlight only differences between these practices.

Ecological Restoration

In many ways, restoration began as a legitimate subdiscipline of ecosystem ecology when Aldo Leopold initiated plant community restoration at the University of Wisconsin Arboretum during the 1930s. Leopold and his colleagues restored approximately 120 ha of forest and mixed-grass prairie, primarily through manipulating ecosystem processes first, and vegetative structure second. At that time, although there were no clearly recognized theoretical constructs guiding restoration efforts,

Leopold's work was founded in careful observation of natural and modified plant communities, and guided by strong ecological intuition (Meine 1991).

Today restoration ecology draws on all major disciplines and subdisciplines of the natural sciences, including ecosystem and landscape ecology, geomorphology, hydrology, soil science, geochemistry, animal behavior, theoretical ecology, population biology, invasion biology, and evolutionary ecology. Ecological restoration projects encompass a wide variety of scales, goals, and participants, ranging from the planting of native species by community volunteers in a neighborhood park to the restoration of tens of thousands of acres of wetland ecosystems coordinated by an amalgam of private, non-profit, and government agencies.

Most broadly, ecological restoration is "the process of intentionally altering a site to establish a defined, indigenous, historic ecosystem. The goal of this process is to emulate the structure, function, diversity and dynamics of the specified ecosystem" (Society for Ecological Restoration [SER] 1991). The renowned British ecologist, A. D. Bradshaw conceptualized restoration as an effort to move a degraded ecosystem or community toward the greater structural and functional complexity that characterize intact ecosystems. Several possible outcomes may result from restoration efforts (Figure 15.1). Outcomes are determined both by what can be achieved given the circumstances affecting a particular site, and by deliberate choices made about what is desired, be it rehabilitation of some structural or functional aspect of a community, replacement of one community by another, or restoration of both function and structure of the "original," historic state.

Restoration is an iterative process that includes:

1. Examining preexisting, historic, and current reference conditions prior to designing the restoration plan

2. Developing a restoration design or plan

3. Obtaining the necessary permits, where relevant, to perform the work

4. Implementing the design, which can include modifications to soil, hydrology, and plant and animal communities as appropriate

5. Monitoring of the restored site

Increasingly, restoration projects apply a landscape approach to restoration, primarily because restoration projects are de facto embedded within a larger ecological matrix (Bell et al. 1997). External inputs and influences are real, and easily can derail even the best-planned restoration efforts. Further, restoration can be active, passive, or both, depending upon whether translocation or installation of species is required, or if the principles of plant and animal succession are relied upon, or if some combination of active planting and passive recruitment is designed into the project.

Any land-based change, especially those undertaken with public monies for the public good, demands a well thought-out, scientifically defensible approach, even if the details are untried or novel in a restoration context. Perhaps what distinguishes restoration most from other environmental sciences is the acknowledgement of the need for, and the willingness to risk direct (and at times substantial) manipulation of natural systems to achieve restoration goals.

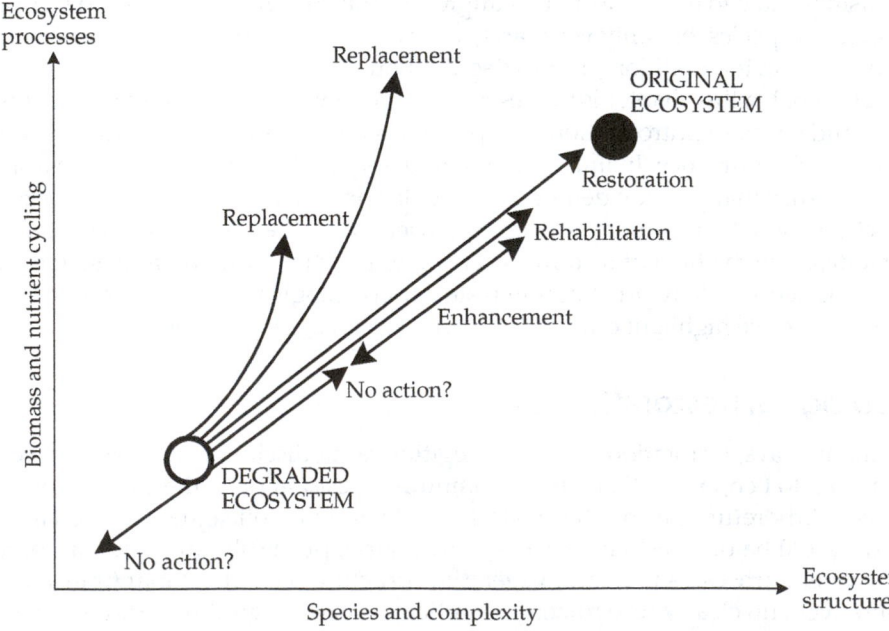

Figure 15.1 The trajectory of a restoration project may be viewed in terms of ecosystem structure (species richness and complexity of interactions) and processes (e.g., primary productivity, nutrient cycling). A reduction in both dimensions occurs upon degradation; the restoration process is an attempt to direct the system back toward the original state. Complete restoration would involve return to that state; a partial, but significant return would result in rehabilitation, and a modest improvement would constitute enhancement. Improvement along another trajectory (generally moving to recover a greater fraction of ecosystem processes typical of the original ecosystem) results in replacement. If left alone, a site can improve or further degrade. (Modified from Bradshaw 1984.)

Ecological restorationists are involved in a variety of activities, including restoration, enhancement, reclamation, re-creation, rehabilitation, remediation, augmentation, and translocation. Often, the term **ecological restoration** is used to refer to all these activities, and we entitled this chapter in this spirit. Yet, it is important to understand the differences between these actions, as they yield separate outcomes. For example, **rehabilitation** is quite similar to restoration, but instead of returning the region to a model condition, the intent is to improve it from some degraded state (see Figure 15.1). The objectives of **reclamation** work, often undertaken on lands that have been mined, frequently prioritize stabilization of the terrain, removal of pollutants, assurance of public safety, aesthetic improvement, revegetation, or some combination of these activities. Alternatively, **re-creation** returns a habitat to a particular historic condition, but there is no implication that the historic condition is an accurate analog to undisturbed regional reference ecosystems; re-creation could return a site to any chosen historic condition (see Figure 15.1). Similarly, **replacement** specifies a community type to be created on a site, but implies that this community was not present on the site previous to human disturbance, but rather is a choice made to accomplish some conservation objective. Often restoration work includes ecological **enhancement** or **augmentation**, but these methods aim only to add or increase one or a few ecosystem functions, not return all ecosystem functions to a specified level of function. Each type of restoration activity reflects a different set of priorities and opportunities.

While restoration efforts generally aim to rehabilitate or restore plant communities, animal communities rarely are directly addressed. Animal "restoration" usually is undertaken only for species that are highly endangered, typically through intensive **ex situ breeding reintroduction** and **translocation** programs. Full restoration of all trophic levels in an ecosystem has not been attempted anywhere, although very often the motivation for ecosystem restoration is the conservation of one or more animal species (e.g., restoring prairie and host plants for Fender's blue butterfly [*Icaricia icarioides fenderi*]; Schultz 2001). Another such example is riverine restoration to enhance the survival of salmon stocks along the Pacific coast of North America, where improvements to water quality, increases in longitudinal connectivity by the removal of dams, or elimination of nonnative predators or pests are activities implemented under the rubric of restoration. Most of this chapter, therefore, will focus on the restoration of ecosystems, referring more directly to plants than to animals. However, we also will consider animal reintroduction programs in which substantial manipulation of animal populations are undertaken as one large, special case of restoration efforts.

Role of restoration ecology in conservation

Restoration ecology is a young and controversial discipline. Efforts to restore ecosystem functioning to severely degraded areas are seen by some as inspiring and important, while others view these efforts as overreaching, expensive, and inefficient uses of resources that could be directed to preservation of intact natural areas. Despite this controversy, restoration ecology is evolving into a key discipline for environmental protection, particularly as applied to the restoration of severely degraded sites, as well as a means to enhance management of relatively undisturbed natural areas. Some of our most far-reaching and effective federal laws, such as the Clean Water Act of 1972, require restoration as a form of mitigation for unavoidable impacts as a result of a particular project (see Case Studies 15.1 by Joy Zelder and 15.2 by Lyndon Lee, for examples). Phenomena such as the large scale wildfires in the western U.S. have fed an increasing awareness of the need for active management to restore ecological processes to those natural areas that are considered to be relatively undisturbed.

Furthermore, restoration ecology offers the opportunity to conduct experiments that provide insights into important basic biological questions, such as community assembly dynamics, secondary succession, fire cycles, the role of keystone species, and the nature of invasibility of ecosystems (Ehrenfeld and Toth 1997; Palmer et al. 1997; Holl et al. 2003). Insights from this type of research can be invaluable for the management of natural areas, addressing issues such as controlling nonnative species invasions, reintroducing keystone species such as large predators, and restoring natural disturbance regimes, such as fire cycles or flooding.

A serious concern regarding ecological restoration is its effect on the regulation of ecosystem conversion for human purposes (Katz 1992). Some see environmental regulation as providing federal, state, or local mandates to restore ecosystem functions. The concern is that ecosystems (or populations of sensitive species), even if they are located in areas valuable for other purposes, will be viewed as more expendable because they can be destroyed and "replaced" by restoration efforts elsewhere, believing we have the knowledge and technology to do so. Restoration ecologists and conservation biologists agree that this is a dangerous perspective (Richter 1997; Young 2000). However, pro-development forces increasingly do view ecological restoration as an alternative to in situ conservation. Similarly, ex situ breeding programs and reintroductions may make it appear that we can maintain plant or animal diversity in botanical gardens or zoos, thereby undermining efforts to conserve species in the wild. Therefore, we stress that ecological restoration and conservation are complementary parts of an overall ecosystem protection and man-

agement strategy (Guerrant et al. 2004). Critically, ecological restoration should never be seen as a substitute for protection of intact ecosystems, but rather should be pursued to augment such conservation efforts, or as the primary effort to rehabilitate already degraded ecosystems (Young 2000). Ecological restoration is, therefore, both a means to an end, and an end in and of itself—depending on the project purpose. Restoration can at best reverse, and at worst abate, the current degradation of Earth's fragile ecosystems.

Steps in designing and implementing ecological restorations

Conducting ecological restoration requires a multidisciplinary, iterative approach. Restoration goals and design should be reviewed and revised as data on site conditions are collected, community concerns are addressed, and as the constructed restoration evolves (Figure 15.2). Necessary steps in conducting a restoration include site assessment, goal setting, restoration design, implementation, monitoring, and adaptive management. Multiple stakeholders typically are involved at all stages of the restoration effort, a process that likely gains the broadest possible support.

SITE ASSESSMENT The first step in exploring a site's restoration potential is to assess what is there, what was there, and what could be there. To assess what is currently present at the restoration site, scientists consult existing documents, such as plant or animal surveys, geologic and soil maps, weather data, environmental reports, published literature, and student research projects, among other types of information. Sometimes the most useful information is not published in the scientific literature but instead is found in private files of local scientists, county offices, or the lay public. Thus, a thorough description of current conditions at the site is the essential starting point, in addition to a clear under-

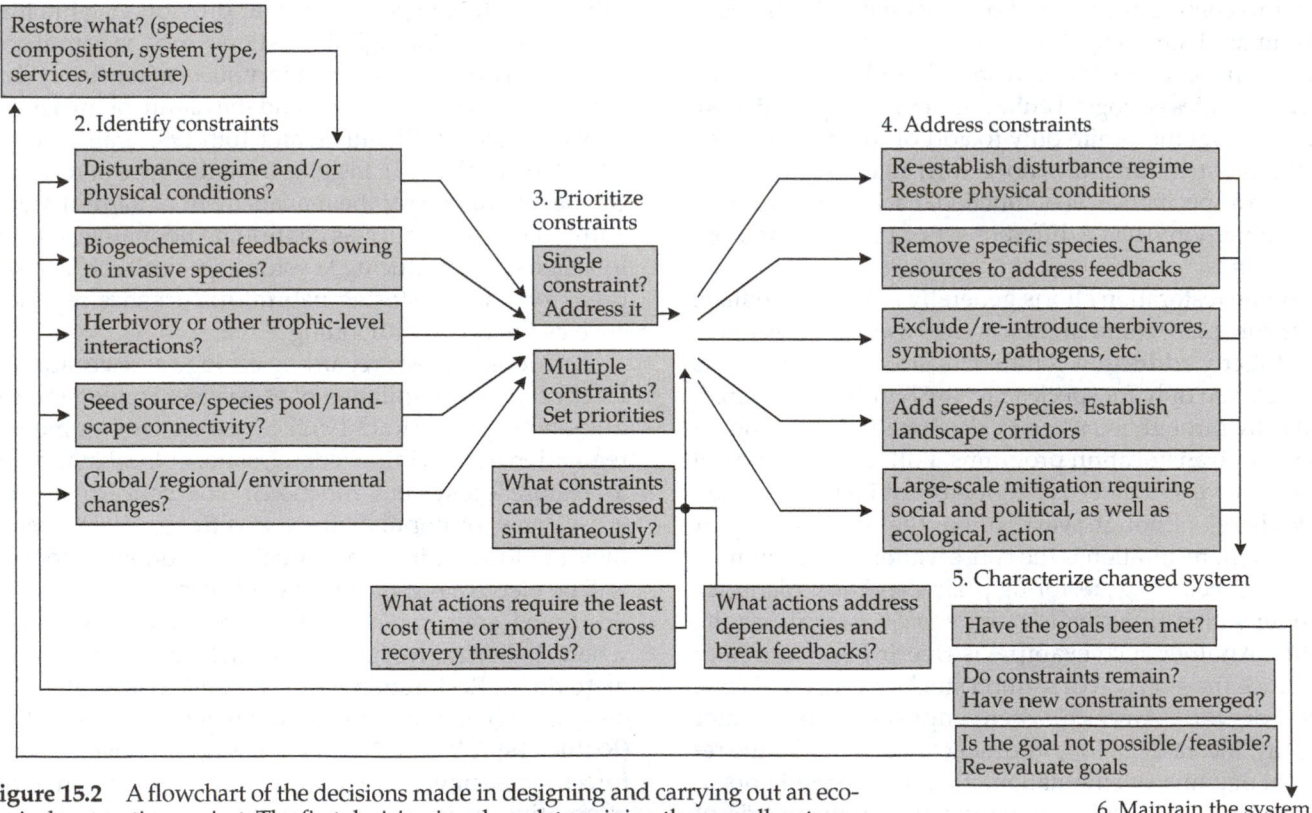

Figure 15.2 A flowchart of the decisions made in designing and carrying out an ecological restoration project. The first decision involves determining the overall restoration goal (1), which is critical to all the decisions that follow. To be as successful as possible, the next effort involves identifying the constraints that could limit achieving that goal (2), and to devise a strategy to address these constraints (3 and 4). Once the restoration is undertaken, a review of progress, and additional efforts to remove ecological barriers to the restoration goal are made (as needed). (5) If these cannot be overcome, reevaluation of the restoration goals may be necessary. Once the restoration goal is met, many sites need some additional maintenance (6). (Modified from Suding et al. 2004.)

standing of ecological processes in the surrounding landscape that strongly affect site conditions (Bell et al. 1997; Holl et al. 2003).

Critically, legal restraints on the property, such as land ownership, property easements, deed restrictions, and local zoning must be defined. Legal constraints set very real boundaries on where restoration can occur and what kind of restoration can be done.

Finally, to assess what the proposed restoration site was like in the past, the environmental history of the site should be reconstructed (see Case Study 6.1). Diverse types of data, including pollen cores, historical accounts by early explorers, survey records, and old photographs and maps can be pieced together to reconstruct site history. Yet, learning about a site's history and understanding current conditions does not provide a complete picture of what *could* exist at the restoration site, even in the rare cases where abundant historical data are available.

Data also should be collected from similar ecosystems in the region with a variety of disturbance histories, comprising a body of data typically known as "regional reference" (Egan and Howell 2001). Relatively undisturbed reference sites help restorationists understand how similar ecosystems have evolved in the absence of substantial human disturbance, suggesting possible restoration targets for what could exist at the site. Conversely, severely degraded reference sites help restorationists know what is likely to occur if ecosystem processes such as nonnative species invasion, or cessation or suppression of natural disturbance cycles, are allowed to continue unabated at the restoration site. Selection and interpretation of reference sites will influence the design of the project, so this needs to be approached carefully (White and Walker 1997).

SETTING GOALS Ecological restoration is an inherently subjective process. It is seldom if ever possible to construct an ecosystem that existed before significant human disturbance. Therefore, socioeconomic values must play a role in setting the goals of restoration projects. These goals will differ depending on the cultural values, objectives, and context of the participants. For example, a wetland ecosystem restored by a coalition of duck hunters will probably seek to maximize the ecosystem functions that provide the best duck habitat, while a wetland restored by the U.S. National Park Service is more likely to consider a broad variety of ecosystem functions, such as long-term water storage, biogeochemical processing, and sediment storage, as well as habitat for wildlife.

Constraints to restoration must be considered carefully in the goal-setting process. Existing infrastructure, available resources, lack of knowledge about how the ecosystem functions, limited funding, and conflicting

stakeholder interests all may limit restoration potential. An obvious example is the reintroduction of top predators, such as wolves or grizzly bears, to a landscape from which they have been exterminated. Their presence may be perceived as dangerous by citizens living in that landscape, particularly to those citizens who own livestock. This has been the case in efforts to reintroduce the Harpy Eagle (*Harpia harpyja*) to Panama, which is described by Ursula Valdezand and colleagues in Essay 15.1.

DESIGN Restoration design requires a multidisciplinary approach. The consideration of all levels of ecosystem organization, from genes to ecosystem processes, requires practitioners to draw from genetics, ecology, hydrology, geology, botany, and other fields. Depending on available resources, site constraints, and goals, restorationists may target one or more of the multiple organizational levels that make up an ecosystem, including genes, populations, communities, and landscape-level processes. In targeting specific components of an ecosystem for restoration, it is important to consider the ecological roles of each component, as well as traits they posses that may enhance their success at a restoration site (Pywell et al. 2003). This is because it is possible (even likely) to ignore or compromise the functioning of one component, while increasing the functioning of another (Bradshaw 1987). Thus, those involved in the design must work cooperatively as well as iteratively to ensure that the design works as an integrated whole.

In some cases, however, restoration that results in an increase in ecosystem functioning at multiple trophic levels may involve relatively simple steps; ecosystem functioning may increase substantially across very large areas as a result. For example, ecosystem functioning in Lake Washington, an 87 km^2 lake near Seattle, Washington, was increased substantially by diverting raw and treated sewage flows away from the lake. Water quality improved and heavy algal blooms declined (Edmondson 1991). Biologists and their collaborators have facilitated the expansion of tropical dry forest across the Guanacaste area of northwestern Costa Rica by restoring the landscape-level disturbance regime (i.e., reestablishment of natural wildfire frequency) and facilitating seed dispersal (Jansen 2004; Figure 15.3). An overview of tropical forest restoration efforts is given by Patricia Townsend in Case Study 15.3.

Although we know relatively little about certain ecosystem components such as soil bacteria and other microorganisms, these components still may be restored in some cases. For example, topsoil may be removed from a site and sequestered while the site is graded or otherwise modified. The original soil may then be used to inoculate the restored site with the appropriate soil flora and fauna (Rokich 2003). In a vernal pool restora-

(A)

(B)

Figure 15.3 Restoration of tropical dry forest from degraded pasture in the Guanacaste region of Costa Rica via fire suppression. (A) Site in 1980 (before restoration) had been a pasture for 200 years, and is dominated by nonnative grasses that thrive on fires set every 1–3 years to maintain an open pasture. Forest seen to the left and in the background is old secondary succession oak (*Quercus oleoides*), with greater than 100 native species. (B) Same view as in (A) in 2000 following 17 years of elimination of anthropogenic fire. The canopy of the oak forest is still visible as the horizon, but the young forest has grown in thickly with wind-dispersed *Rehdera trinervis* (family Verbenaceae), intermixed with another 70 woody species. Such rapid natural forest invasion of pasture in the absence of fire is characteristic of tens of thousands of hectares in the Guanacaste Conservation area. (Photographs courtesy of D. Janzen.)

tion project in Santa Barbara, California, Ferren and his colleagues collected the upper layer of soil from established vernal pools and used this inoculum for restoration of re-created vernal pools (Ferren et al. 1998). Only pools that had received inoculum from extant natural pools rapidly took on characteristics of these natural pools. Uninoculated pools remained species poor and were more prone to invasion by nonnative grasses (Ferren et al. 1998).

Finally, before a project can be implemented, design documents must be generated to guide the construction. These documents should include construction specifications detailing site grading, sediment and erosion control plans, planting plans, irrigation layout, and monitoring points and transects.

IMPLEMENTATION Depending on the project goals, constraints, costs, and design, restoration projects are implemented in a wide variety of ways. Restoration work can range, from a 62,000 kg, fully loaded scraper moving many cubic meters of soil to achieve the design grade, to

a single individual sowing seeds of native prairie plants along a road verge. Thus, implementing a restoration project requires an understanding of such practical topics as construction machinery, surveying, erosion control techniques, plant propagation, irrigation materials, and animal behavior to complement the ecological theory that informs the design stage.

When possible, local community members should be included in implementation. Involving volunteers in steps such as weeding and planting can be cost effective if the scale of the restoration project is small and it builds a commitment to, and ownership of, the restoration effort on the part of local residents. Ultimately, this commitment may be critical to maintenance, monitoring, and management of the restoration site over the long term. Training and supervision of volunteers can prevent their involvement from compromising the quality of the restoration (Geist and Galatowitsch 1999).

Incorporating an experimental design in the implementation of a restoration can provide useful information for the scientific community and for the future of the restored ecosystem (Figure 15.4). For example, Rokich (2003) experimented to find the best soil handling practices to promote regrowth of native species following sand mining in a semiarid shrubland in Australia. A summary of these efforts is described by Deanna Rokich in Essay 15.2.

Maintaining control groups and applying treatments to portions of the restoration site allow for adaptive management and for statistical analysis of restoration outcomes. For example, at Stanford University in California, the Center for Conservation Biology (CCB) teamed with restoration consultants to restore the breeding habitat for the California tiger salamander (*Ambystoma californiense*). Prior to the breeding habitat restoration, the CCB research staff studied the tiger salamander for more than ten years, including: (1) monitoring daily, seasonal, and annual patterns of tiger salamander migration; (2) identifying, minimizing, and

ESSAY 15.1

The Harpy Eagle Conservation Program
Research, Conservation, and Community-Based Education to Save the National Bird of Panama

Ursula **Valdez,** *University of Washington* **and Angel Muela and Marta Curti,** *The Peregrine Fund*

The Harpy Eagle (*Harpy harpyja*), one of the most powerful birds of prey in the world, is a top predator that lives in Neotropical rainforests. It eats monkeys and sloths, and occasionally other birds and small deer. Habitat loss and direct persecution have resulted in dramatic Harpy Eagle declines. Consequently, Harpy Eagle populations, formerly distributed throughout tropical rainforests from southern Mexico to northern Argentina, have been drastically reduced in Central America and are declining in South America, with wild Harpy Eagle populations currently extant only in Panama.

Since 1995, The Peregrine Fund (TPF) has been working to conserve Harpy Eagles in Central America. After preliminary studies of their natural history, TPF biologists conducted successful experimental releases of captive-bred individuals in the lowland forests of Panama. Unfortunately, these initial releases were thwarted by the shooting of two eagles in the forests surrounding the release site, prompting TPF biologists to recapture the surviving eagles for their own safety. Although a disappointing and frustrating experience, these events compelled TPF biologists to evaluate the strategies originally used and to design an integrated conservation program. Today, the Harpy Eagle Conservation Program (HECP) involves not only research and captive propagation, but also incorporates an equally important public education component.

Biologists and educators now work hand in hand in the conservation of Harpy Eagles and their habitats in Panama by using the species as a "flagship" for conservation. The preservation of this bird of prey is expected to create an umbrella of protection for other animal and plant species that live in the same large tracts of forest that Harpy Eagles do. Current goals of TPF's Harpy Eagle Conservation Program are to: (1) identify factors that limit wild populations of Harpy Eagles by conducting research on the breeding biology and habitat use of this raptor; (2) improve captive breeding and

Figure A   A newly released Harpy Eagle surveys the forest. (Photograph courtesy of A. Muela.)

releases, adapting techniques previously successful in recovering Peregrine Falcons (*Falco peregrinus*) and other birds of prey in North America, and; (3) minimize persecution of Harpy Eagles by raising awareness of their important ecological roles.

One of the key aspects of this conservation program is participation at all levels by Panamanians. In Darien Province, TPF has formed a strong partnership with indigenous communities in areas where healthy populations of wild Harpy Eagles still exist. A cooperative agreement with the Embera-Wounaan people allows for ecological research in their territories and provides biologists and local community members the opportunity to exchange traditional and scientific knowledge. A team composed of program biologists and indigenous "parabiologists" is monitoring at least 22 Harpy Eagle nests in this region. The parabiologists share their knowledge of the forest, wildlife, and plants, while learning rap-

tor and habitat evaluation techniques and the use of modern instruments necessary to measure habitat variables. Additionally, the parabiologists, who are formally employed by the program and also trained in conservation education, serve as the critical link between villagers and program educators. In many cases they conduct the educational activities themselves and in their own language, which is a more effective way to reach target audiences.

Beginning in 2001, TPF began an ongoing, systematic environmental education program aimed at raising awareness through public campaigns and talks in schools and community centers. The Harpy Eagle was designated the National Bird of Panama in 2002. Along with this official status, Panamanian lawmakers passed a protective law to conserve and prohibit shooting of this rare species. Also at this time, the first Harpy Eagle chicks hatched at TPF's breeding facilities in Panama. Working with the media, this news reached even the most remote places in the country. In a relatively short time, the results were measurable: Panamanians developed pride for and an interest in the conservation of their national bird.

People living in rural communities surrounding areas where young eagles will be released are a principal target audience of the education program. Program educators and biologists visit these communities regularly to talk with community leaders and together plan and conduct educational and conservation activities. Adults, children, and school teachers actively participate in talks and campaigns. As a result, many have developed their own local conservation education initiatives. At the same time, public and private schools in Panama City are helping spread the conservation message, not only by inviting program educators to give presentations at their institutions, but also by working on their own conservation education projects. In cooperation with the Panamanian Ministry of Education, TPF has developed a supplement to the national environmental

education curriculum on raptors, Harpy Eagles, and the functioning of ecosystems. This supplement will be made available to all schools in the country.

As research continues in the field, important information is also being obtained during the captive breeding and release process. By early 2005, 28 Harpy Eagle chicks had been successfully captive-bred in Panama. These chicks are kept in captivity until they are approximately 6 months of age. Now that public opinion in local release sites had grown favorable to Harpy Eagles, releases restarted in 2002 in the Panama Canal Watershed. Young

captive-bred eaglets are radio-tracked on a daily basis, allowing researchers to learn about their movements, ecology, and behavior, while at the same time providing the birds with needed care while they learn how to hunt. When young Harpy Eagles no longer need human care they are recaptured and relocated permanently in remote, suitable habitat. In cooperation with the Panamanian government, Ministry of Education, local and international volunteers, and private NGOs, TPF biologists and educators are working diligently to keep the birds and their habitats safe. Recently, release efforts as

well as conservation education have been extended to Belize where the Belize Zoo and the government are the main partners in this endeavor. In future years, the Harpy Eagle program might also extend its efforts into other countries of Central America.

Although Harpy Eagle conservation is not guaranteed success in Central America, the creation of a program that integrates research, breeding and release, and community education, is an essential step to reach this long-term conservation goal. ■

ESSAY 15.2

Achieving Success in Mine Reclamation
An Example from Semiarid Lands in Western Australia

Deanna Rokich, *Botanic Gardens and Parks Authority, Western Australia*

■ Energy production and resource use in Western Australia involves the mining of minerals in arid and semiarid landscapes. The ability to restore mine sites following such disturbances is of utmost practical importance. Rapid plant recovery is important to minimize soil erosion due to wind and water movement. Unfortunately, given the harsh climatic conditions in much of Australia, particularly at a mining site, plant establishment tends to be slow and reliant on a narrow window of opportunity. Hence, it is both difficult and costly to restore previously mined sites. Further, while the focus of previous mine reclamations has been on restoration of the biological components of the system, soil and hydrologic components generally have been neglected. Growth and long-term establishment of the native flora rely on restoration (or remediation) of soil characteristics and hydrologic processes to some targeted predisturbance, reference condition.

As part of standard sand mine rehabilitation practices, mining companies typically strip the topsoil and overburden from the mine site, to a depth of 0.3–1.0 m. Stripped topsoils are stockpiled for up to 3 years, and post-restoration, spread across a restoration site at depths up to 1 m. Commonly, these topsoil handling operations are undertaken at various times of the year, without consideration of their seasonal effects on the soil seed bank. After the

stored topsoil is replaced, the restoration site is typically mulched and irrigated. The results of such postmined procedures are a limited species richness and abundance.

In the region of Perth, Western Australia, Rocla Quarry Products operates a sand extraction facility that impacts *Banksia* woodland communities in the path of the mine. Rocla is committed to restoring postmined sites to an ecosystem closely resembling the premined ecosystems dominated by a species complement typical of undisturbed *Banksia* woodlands. To accomplish this, Rocla teamed with the Science Division of Kings Park and Botanic Garden to investigate how best to restore the distinctive *Banksia* woodlands of the Swan River coastal plain.

At the commencement of the restoration work at the rehabilited postmined sites, the diversity and sustainability of the postmined vegetation were extremely limited. It was initially hypothesized that reasons for the depauperate communities may relate to handling of the topsoil of the original community, the soil storage methods employed, potential changes in the soil environment as a result of the manipulation, or microclimatic changes in the soil environment associated with sand mining. To understand the requirements for successful *Banksia* woodland restoration, Kings Park scientists

focused on two key areas of inquiry—seedling recruitment and plant survival, and individual species' responses to the postmined soil environment. These autecological studies yielded essential information for developing appropriate ecological restoration practices for the *Banksia* woodlands.

Following mining, the first step in mine reclamation is to evaluate the state of the site. This is because during mining, dramatic changes occur in the soil environment. Kings Park scientists gathered soil and plant data at the Rocla mine site near Perth to begin their evaluation, including the seed ecology and regenerative potential of the soil seed bank, detailed methods of topsoil handling and storage, seed pretreatments, mulching practices, the effect of soil stabilizers, autecology of several of the dominant local weeds, irrigation and soil ripping practices, and plant-soil-water relations.

The key result from the Kings Park research team was the development of sound principles for the handling and storage of topsoil for the restoration of *Banksia* woodland ecosystems. Sand mining restoration practices now involve stripping and spreading topsoil to only 10 cm in depth, stripping and spreading fresh and dry topsoil, and cessation of the common practice of mulching. Thus, the Kings Park study provided ecological information relat-

ing to the dynamics and regenerative potential of the soil seed banks of *Banksia* woodland species, seedling recruitment, and survival patterns in *Banksia* woodland, disturbed and post-mined sites, and factors that reflect seedling recruitment and survival opportunities following ecosystem restoration. Work currently is under-way to restore degraded *Banksia* wood-land sites throughout the Perth metro-politan area based upon the findings of this study on topsoil handling at the Rocla mine site.

Therefore, although focused on the restoration of *Banksia* woodlands in a mining context, Kings Park scientists provided information of practical benefit to other urban bushland areas in West-ern Australia where ecological restora-tion and management is a priority. ■

Figure 15.4 A prairie restoration experiment at the Curtis Prairie, WI, ca. 1985. The highest-quality prairies have been planted using such labor-intensive horticultural methods, and consequently are small in scale. (Photograph courtesy of the University of Wisconsin Arboretum.)

mitigating threats to the salamander (i.e., constructing an underground tunnel under a heavily traveled road along the migration path); and (3) developing a long-term con-servation plan. Then, after designing, planning, and con-structing restored ponds, staff scientists evaluated the work critically, including monitoring a variety of wet-land ecosystem functions (i.e., seed recruitment, sala-mander use, timing and duration of inundation) to doc-ument the performance of the restored wetlands (Stanford Restoration Ecology Group 2003). Unless proj-ects are in collaboration with such research institutes, rarely are there the funds necessary to do such detailed experimental analyses. In short, using an experimental approach typically requires more resources and time than a regulatory approach.

Often during restoration implementation, discoveries are made about site constraints that could not have been predicted in the design. Thus, although restorationists attempt to adhere to design specifications, changes to the design occur frequently in the field. For example, changes to a grading plan may need to be made to avoid

an undocumented archaeological site. These uncertain-ties demonstrate the need for qualified scientists to be on-site during all construction and implementation ac-tivities. Finally, at the conclusion of the construction phase, "as-built conditions" should be documented. Im-portantly, as-built maps depict the final grade (eleva-tions) and design features, and not just those called for in the design. As-built documents usually include maps of plantings or other major aspects of the project (e.g., in-stallation of an irrigation system).

MONITORING AND ADAPTIVE MANAGEMENT   Ecosystem restor-ations are long-term propositions. This is because after the implementation phase, long-term monitoring is vital to guide the adaptive management of the restored site. To provide useful information, monitoring should follow a sampling design that allows for statistical analysis (Mich-ener 1997). Monitoring can reveal diverse and unexpect-ed factors that may compromise ecosystem functioning at the restoration site, or conversely, advantageous but un-expected outcomes. Simply taking an inventory of the number and types of plant species is not particularly in-formative unless the monitoring plan is designed to assess possible causes of population trends and potential re-sponses to undesirable trends.

One of the most common problems in ecological restor-ation is the natural recruitment of disturbance-adapted, invasive species at the expense of the indigenous flora and fauna (D'Antonio and Meyerson 2002). These prob-lems can be addressed by various management tech-niques, ranging from handweeding, mowing, and burn-ing to control invasives, to replanting, fertilization and addition of soil amendments, and irrigation to quickly es-tablish native plants and the fauna that depend on them for habitat (Figure 15.5). In addition, at times, insufficient amelioration of site conditions or faulty plant materials may lead to the wholesale failure of planted nursery stock to establish. Unfortunately, often the monitoring of restoration projects is completed only to comply with var-ious regulatory permit conditions. Compliance monitor-ing typically yields different information than scientific monitoring.

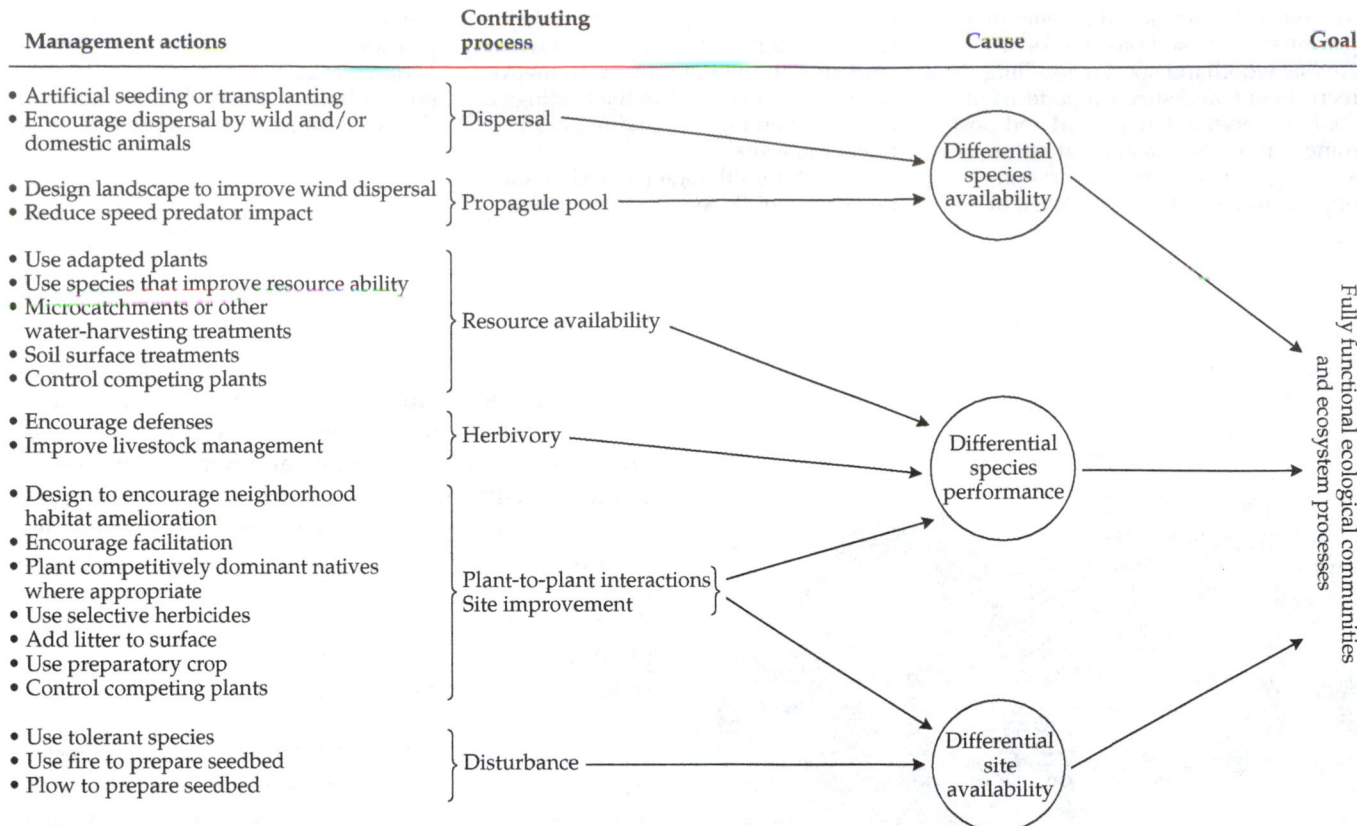

Figure 15.5 Schematic of how recovery of a fully functional and self-maintaining ecosystem might be approached by focusing on contributing processes and causes of change. (Modified from Whisenant 1999.)

Success in the establishment of animal populations in a restored area often is assessed at the monitoring stage. Surveys can reveal how animal species composition has changed in the restored area, and these data can be compared to data from "control" areas. Behavioral studies also can evaluate whether the use of restored habitats by a target species is similar to that in relatively undamaged, natural habitats.

Restoration challenges

Ecological restoration is fraught with challenge. Bradshaw and Chadwick (1980) wrote that restoration represented the ultimate test of ecology, for we need to understand ecological systems very well to restore them properly. Beyond the limits of our ecological knowledge, we are often challenged to work within a particular scale, which usually does not match the scale of ecological processes or the catchments of human impacts. Finally, we are beset with a number of difficulties in implementation, discussed next.

LACK OF KNOWLEDGE It is often difficult to design restoration projects that target multiple trophic levels of ecosystems because basic knowledge of many ecosystem components is absent. For example, much more information is available on birds and mammals than on bacterial communities across the globe, yet the success of a project may be dependent on appropriate soil bacteria being present (Wilson 2000). By and large, the extent of our knowledge of biological groups and ecosystems reflects what researchers have found to be charismatic and intellectually compelling, not necessarily what is most important to ecosystem functioning. We are fortunate that it is sometimes possible to restore elements of an ecosystem, albeit inadvertently, without thoroughly understanding them. Ignorance of these components, however, can seriously compromise restoration efforts. Joy Zedler describes one such case where a wetland restoration project failed to achieve its goal of providing Light-footed Clapper Rail (*Rallus longirostis levipes*) habitat because of problems with soil structure, and bacterial and insect communities (see Case Study 15.1).

Often we are able to identify our knowledge gaps, and then fill them in, by working backwards from our goals to identify primary causes of change that are needed to allow us to meet that goal (see Figure 15.5). Once we have identified these primary causes of change, we can identify contributing processes that can be affected

by management actions to affect the desired change. For example, if we discover that herbivory is limiting reestablishment of native species in a grassland, we can work to exclude livestock and wild ungulates from the restoration site, at least until individual plants become well established. Or if we find that soils are relatively nutrient poor, we might try planting nitrogen fixing species to facilitate improvements to the soil and enhance success of other species, thereby improving both species performance and site quality (see Figure 15.5). This kind of approach at least identifies promising avenues for our efforts, and we eventually learn from the successes and failures what kinds of management actions are most likely to work at sites with different types of limiting processes.

We know, too, that animals often play key roles in structuring ecosystems. However, the majority of restoration efforts are focused (sometimes exclusively) on plant communities. Typically, plants are cheaper and simpler to cultivate and establish than animals. Particularly for small restoration efforts constrained by limited funds and short time frames, removing nonnative plant species and planting natives is a more attainable goal than the reintroduction of a keystone animal taxon. Further, providing suitable habitat for animals through the establishment of native vegetation often is the most that can be afforded to restore animal populations, even those of conservation concern. In many cases, a "bottom-up" approach may be the most effective—that is, once essential ecosystem components, such as soil structure and geochemistry, hydrological functions, and vegetative structure are restored, animal communities may assemble themselves. However, suitable habitat does not by itself guarantee the presence and viability of specific animal populations (as illustrated in Case Studies 15.4 and 15.5). A schematic of the various influences determining both population viability and community composition is given in Figure 15.6.

Population genetics plays an important role in restoration. Attention to the genetic makeup of restored populations is necessary because populations that are not genetically diverse and sufficiently adapted to local conditions may fail to establish successfully (Gray 2002). Some restorationists neglect the importance of local adaptation by obtaining propagules from any convenient commercial source, even when this means planting seeds in Massachusetts that were harvested in Minnesota (Franson 1995). Others insist that populations on or adjacent to the restoration site are sufficiently diverse that supplementation by other populations in the region is unnecessary and even unwise. A balanced approach between these two extreme views may be what is needed in most cases.

Extreme habitats such as mine spoils may support one or a few genotypes that have evolved to tolerate the stresses imposed by such environments. Alternatively, these habitats may require a broader range of genetic

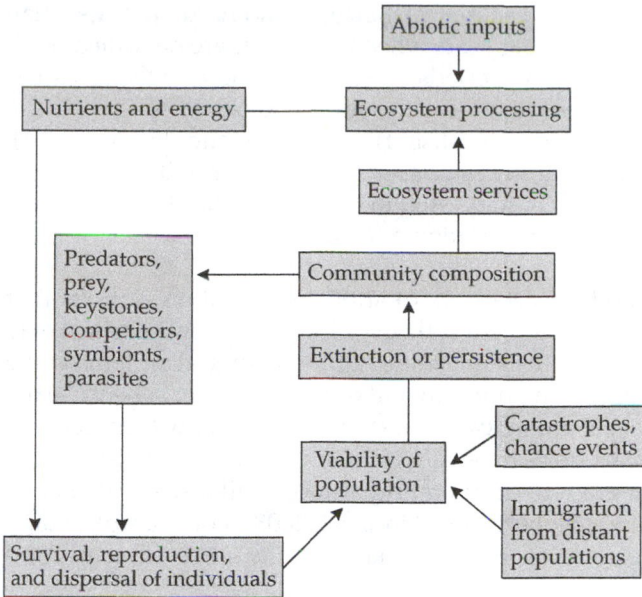

Figure 15.6   Schematic of interactions among individual, population, community, and ecosystem level properties that influence community composition in a restoration project. Individual fitness, as well as dispersal and chance events, affect population viability, which in turn will determine community composition. Ecosystem processes are governed in large part by community composition, and in turn, ecosystem and community properties modify community composition through their effects on individual fitness. Thus, to create and maintain diverse communities it is necessary to attend to factors on all these levels. (Modified from Marzluff and Ewing 2001.)

variation than is present in the local populations. Moderately disturbed habitats can be expected to support a variety of local genotypes. Restoration projects, however, cannot avoid the introduction of new genes or genotypes, and potentially the loss of locally adapted genotypes, a process that is an inevitable result of project implementation (Schultz 2001). In short, a restoration effort may threaten the genetic integrity of nearby local populations, especially in very large-scale restoration projects when many hundreds of thousands of individual "new" plants are installed. Ultimately, large and severe disturbances that can occur in a large restoration project call for a mix of genotypes from both local and non-local sources.

Collection of plant material should be done carefully (Guerrant et al. 2004). Collection of seed may under-represent genetic variation if there is not sufficient temporal and spatial sampling. Strong selection during propagation for characteristics such as rapid growth may result in linkage to traits such as short life span. This, or lack of cold hardiness or disease resistance, is typical of native cultivars sometimes used in reclamation. Founder effects are real and underappreciated problems in small isolated restoration projects, particularly those with a species-rich planting scheme. Loss of local alleles conferring, for

example, disease resistance or cold hardiness, also may occur during years when these traits are not called upon (e.g., during periods of warm weather or if the disease is not present, ultimately, populations will suffer because the alleles are absent) (Guerrant and Fiedler 2004). Adopting a holistic approach to community reestablishment, as is suggested in Figure 5.6, should result in the strongest restoration outcomes.

SCALE ISSUES IN RESTORATION   Restoration projects that focus on a small scale (e.g., less than a few hectares) may succeed in establishing native ecological systems in the short term, but may fail in the longer term because the larger ecological context required to allow these restoration efforts to be self-sustaining is either not present, too degraded, or operating at too small a scale (Bell et al. 1997; Moberg and Rönnbäck 2003). For example, many restoration projects take place in urban or semi-rural

areas where restoring natural disturbance or hydrological regimes may be extremely difficult, if not prohibited by laws. Habitat connectivity, essential for gene flow and recolonization of a restoration site in the event of local extirpations, may be possible over short distances, but impossible over critical lengths or corridors without huge sums of money and a reorientation of public policy (Simberloff et al. 1999). In increasingly more landscapes, such as the Los Angeles Basin, metropolitan Las Vegas, Manhattan, or the south shore of Lake Michigan, irreversible civil commitments to an urban lifestyle preclude biologically meaningful large-scale restoration.

When ecosystem degradation has been extremely intense or of great spatial extent, restoration can be particularly difficult to achieve (Whisenant 1999; Moberg and Rönnbäck 2003). In these cases, substantial remediation of the abiotic environment may be necessary before successful ecosystem restoration is possible (Figure 15.7;

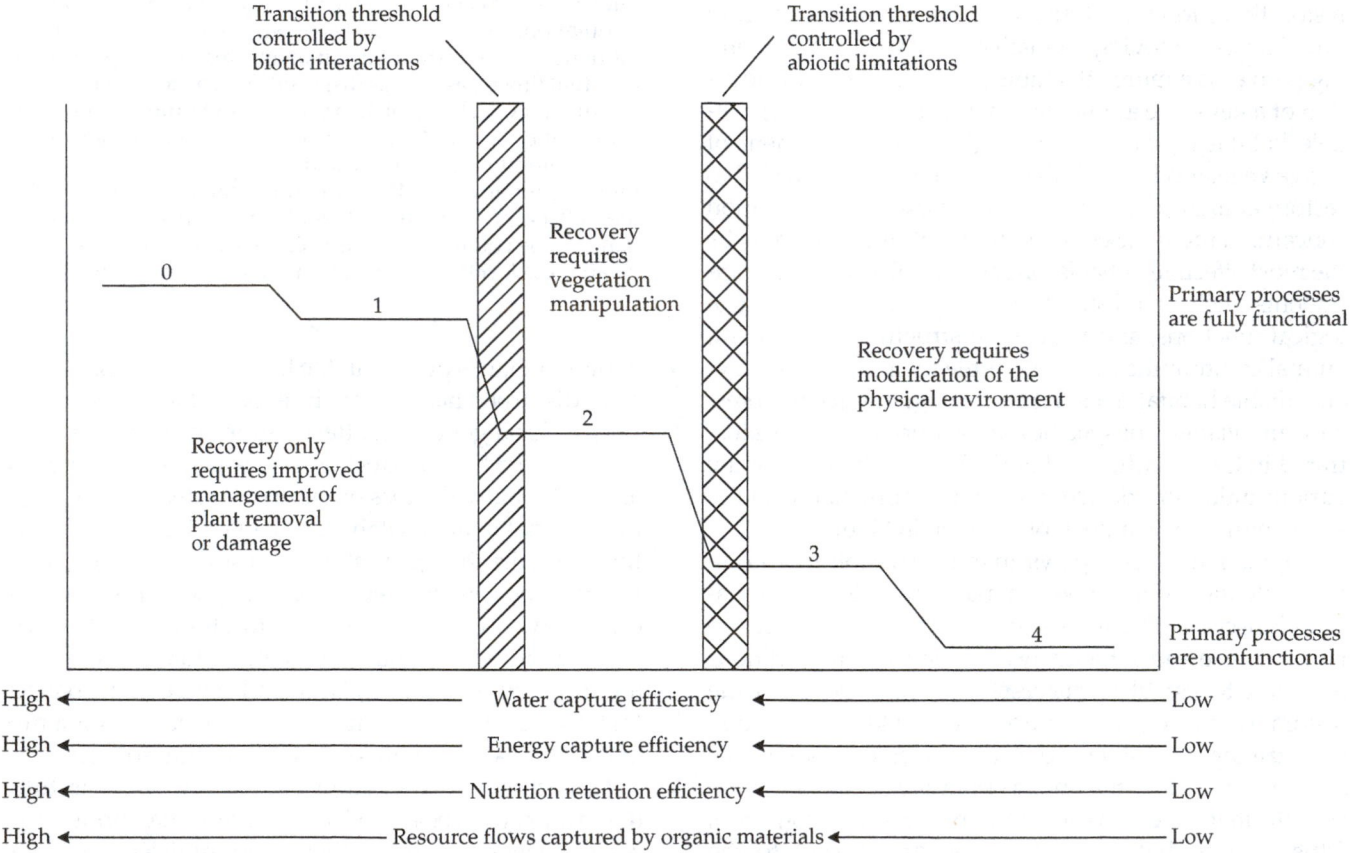

Figure 15.7 Degradation at a site brings a highly functional ecosystem (at 0) through states that are progressively less functional (1–4). Ecosystems commonly pass through state transitions, during which recovery requires greater levels of intervention. In a slightly degraded site, recovery requires minimal intervention, but after the threshold is crossed (between 1 and 2), manipulations of vegetative structure and composition are necessary. If degradation is more severe, the second threshold is crossed (between 2 and 3) in which physical interventions (e.g., soil improvements) are required to restore the site. (Modified from Whisenant 1999.)

Whisenant 1999). More optimistically, natural regeneration might be enhanced through restoration to overcome some of these limitations. For example, although widespread regrowth of forests following agricultural abandonment has taken place over much of the northeastern U.S., many plant species cannot recolonize because they are in tiny remnants of native forest that are too distant from the regenerating forest (Velland 2003). In such cases, active reintroduction of typical forest species might recover the character, structure, and functions of these forests that may not take place without such intervention.

Unfortunately, providing habitat connectivity often is beyond the scope of small-scale restoration projects. Frequently, inadequate funding and lack of political support constrain efforts to restore landscape-level processes and connectivity, even if the site potential would allow it to occur. In these cases, a long-term effort by community stakeholders (e.g., local universities, regulators, interest groups) could be made to restore such features as opportunities arise. Unless landscape-level processes are restored, ecological restorations within the larger landscape will require constant attention, funding, and maintenance. We may do well to recall that ecological systems cannot usually be recovered fully to their predisturbance state (see Figure 15.1), but it may be possible to bring them back from an alternatively stable state that is characterized by low species diversity and poor ecosystem functioning (Suding et al. 2004).

IMPLEMENTATION IN PRACTICE   Many restoration projects do not use a multidisciplinary approach. Instead, they focus more narrowly on a single taxon (e.g., an endangered species) or a single ecosystem function (e.g., short- or long-term water storage). In some cases, restoration efforts focus on a single factor to meet environmental regulatory requirements, or because the resources and expertise needed for an ecosystem approach is lacking. Although such projects may be overly simplistic, even these small univariate projects can constitute an improvement over prior degraded conditions. More and more restoration practitioners are taking advantage of recent advances in the natural sciences that can inform their work (Palmer et al. 1997; Holl et al. 2003; Suding et al. 2004). Restorations not only make use of scientific knowledge, but add to it as well; when resources are available to support an experimental approach and to allow long-term data collection, ecological restoration offers unique opportunities to add to our scientific knowledge while increasing ecosystem functioning.

Animal Reintroduction

Most restoration of animal populations has taken either the form of habitat restoration with no direct management of animal populations, or efforts that are more manipulative involving ex situ breeding programs with the eventual aim of reintroducing individuals of an extirpated threatened species to their native range, or the translocation of individuals from one area to another. Thus, animal restoration tends to parallel plant community restoration by using techniques that bring animals back into a site, rather than efforts that foster or enhance a preexisting population.

Because reintroduction programs have as their primary goal the eventual reestablishment of a viable, free-ranging population, species thus targeted have tended to be large, charismatic vertebrates—those that have captured the imaginations of a broad array of people. Importantly, these species usually play large roles in their native ecosystems: They are top predators (e.g., gray wolf [*Canis lupus*]), scavengers (e.g., California Condor [*Gymnogyps californianus*]) principle herbivores (e.g., American bison [*Bos bison*]), or modifiers of habitat (e.g., beavers [*Castor canadensis*]). Often, reintroductions provide incentive for broader conservation efforts (IUCN 1995).

The Reintroduction Specialist Group of the World Conservation Union's (IUCN) Species Survival Commission, a group charged with monitoring and planning reintroduction actions for some of the world's most threatened species, created guidelines for the practice of reintroduction. The aim of these guidelines is to ensure that "reintroductions are both justifiable and likely to succeed, and that the conservation world can learn from each initiative, whether successful or not" (IUCN 1995). Examples of well-justified reintroduction objectives include enhancing the long-term survival of a species, reestablishing a keystone species, and providing long-term economic benefits to the local and/or national economy. The following are IUCN (1995) guidelines designed to maximize success:

Step 1   Conduct a feasibility study, including assessment of the biology of the species, availability of individuals of the same taxonomic status for reintroduction, and whether other species have taken up the ecological role of the species that has been extirpated from the wild.

Step 2   Select and evaluate sites within the historic range of the species, ensuring that suitable habitat is available that is not subject to the same threats that caused the original demise of the population, and that has long-term protected status.

Step 3   Identify and evaluate suitability of stock to be reintroduced, including genetic factors.

Step 4   Evaluate social, political, and economic conditions at the reintroduction site to ensure that long-term financial and political support will be available. Efforts to minimize conflicts between

human populations and the reintroduced species should be evaluated in advance.

Step 5 Plan a properly financed reintroduction with approval by all stakeholders, and in coordination with management agencies. Design pre- and post-release monitoring to make the reintroduction a carefully designed experiment, with the capability to test methodology, thereby allowing improvements for future releases.

Step 6 Post-release monitoring should be done using an adaptive model, ensuring that necessary intervention can be carried out.

Steps recommended for reintroductions require careful review and analysis of whether the conditions that led to extirpation of the population have improved sufficiently to allow a population to thrive now, as well as an evaluation of the long-term potential for the species to be conserved. This differs somewhat from many habitat restoration projects, particularly wetland mitigation sites, which often are selected more based on convenience rather than long-term sustainability.

Steps taken for animal reintroductions are generally similar to those for restoration activities, but usually necessitate more detailed analysis of the behavior, ecology, and genetics of the individuals that will be reintroduced. High juvenile mortality, loss of rare alleles and genetic diversity, reproductive dysfunction, and other problems associated with extreme inbreeding in captive-bred species (e.g., Ralls et al. 1979; Fuerst and Maruyama 1986) led to detailed pedigree analysis and genetic management of captive breeding efforts (e.g., Foose and Ballou 1988; Ralls et al. 1988). Similarly, genetic problems that can arise with translocation led to greater oversight of the genetics involved in these activities (Griffith et al. 1989). Genetic considerations in captive breeding, reintroduction, and translocation are discussed in greater detail in Box 15.1.

Captive breeding can be extremely challenging, and often necessitates careful analysis of the behavior of a species to encourage mating and successful rearing of young. Reintroduction can be still more complex as individuals that have been raised in captivity may require tutoring to be able to survive in the wild. Detailed accounts of reintroduction programs for the black-footed ferret (*Mustel nigripes*) and the Puerto Rican Parrot (*Amazona vittata*) are described in Case Studies 15.4 by Dean Biggins and colleagues, and 15.5 by Jaime Collazo and colleagues. Both of these species require training to become adept at foraging and evading predators in the wild. Further, public education about reintroduction efforts, particularly regarding large predators, is increasingly critical to success, as often humans may deliberately or mistakenly kill these species if they feel them to be a threat (see Essay 15.1). Each of these examples highlight key steps needed to maximize success, as well as the learning process that takes place as conservationists work to reintroduce critically endangered species.

Restoration in Marine Environments

Marine restoration ecology has received far less attention than its terrestrial and freshwater counterparts. However, restoration activities are widespread, particularly where restoration enhances commercially important fisheries (Petersen et al. 2003). For example, reforestation projects to restore mangrove ecosystems from areas cleared for shrimp ponds and other development have been advocated to restore productive fisheries and increase fuelwood to local communities (Kaly and Jones 1998). Some of these efforts are informed by a great deal of ecological study (e.g., French McCay et al. 2003), whereas many are still in the trial-and-error stage (Moburg and Rönnbäck 2003). Following an enormously destructive typhoon season in 1970 that took over 300,000 lives in Bangledesh, a large-scale mangrove afforestation program began in the hopes of better protecting fragile coastal communities. However, because these were established on mudflats that had not previously supported these ecosystems, and powerful storms are an annual event, the success of these ventures has been relatively poor (Moburg and Rönnbäck 2003). Unfortunately, in this case, the well-intentioned restoration efforts resulted in the degradation of intertidal mudflats (which are critical areas for seabird foraging, and a wealth of other species).

In many coral reef restoration projects physical structures are placed on the ocean bottom to mimic reefs. These can be concrete structures designed with structural heterogeneity similar to that found in natural reefs or large debris that required disposal (e.g., airplanes, ships, oil rigs). Most projects are considered complete at that point, with the hope that nearby healthy reef communities will provide sufficient propagules for a reef community to assemble at the site (Moburg and Rönnbäck 2003). Another approach to reef restoration includes translocation of healthy coral fragments to a damaged reef in hopes that this will speed up recovery. Corals grow extremely slowly, and without intervention reef recovery, coral ecosystem recovery can take decades, perhaps a century (Pearson 1981). These methods are very labor-intensive, and thus far, not easily conducted on a large scale.

The enormous catastrophe caused by the tsunamis of December 26, 2004 have prompted many to look more closely at the roles of mangroves and coral reefs in protecting inland areas, and provided a new motivation for restoration of these areas. Further, the economic importance of these ecosystems for fisheries and tourism has long created an incentive for their restoration (which

BOX 15.1 Genetic Considerations in Reintroduction

Kim T. Scribner, *Michigan State University*

Supplementing populations by releasing captive-bred animals and cultivated plants into natural environments, or translocating individuals from one wild population into another, is an increasingly common conservation practice. The goal of supplementation is to increase the size of existing natural populations when levels of natural recruitment are not sufficient to ensure long-term population viability.

However, several issues should be considered from a genetics perspective when supplementation or reintroduction is considered. First, consider the case of supplementing a population with captive-bred individuals. Genetic changes may occur in captivity or cultivation due to intentional or inadvertent selection. For example, manipulation of reproductive and survival rates during captive breeding (e.g., higher nutritional regime and concomitantly higher fecundities and survivorship) will likely have an impact on the genetic characteristics of these captive-bred or cultivated populations. These changes may have an appreciable negative effect on performance under natural conditions, which could undermine the effectiveness of the supplementation strategy. Further, when genetic characteristics of a captive population differ from those of the wild populations and these individuals interbreed, undesirable characteristics from the captive population may be transferred to the natural populations. At some point, a natural population may be considered compromised, as the original wild genes will have been replaced by those from the captive population.

Next, consider the case of supplementation by translocation. When individuals are transferred between wild populations, and if the wild populations have significantly different genetic characteristics, interbreeding may cause the disruption of coadapted gene complexes or other forms of local adaptation in the target population. Outbreeding depression can result from matings between supplemental individuals (either wild-caught or captive-bred) and those in the target wild populations if the degree of genetic divergence is too great. Supplementa-

tion thus becomes less desirable when the level of genetic divergence between populations is high and the probability of introgression with native populations (mating between introduced and native individuals) is high. Further, we should recognize that supplementation undertaken in a conservation context typically would only increase effective population size (N_e) to a small degree, and only in certain situations (Ryman and Laikre 1991; Figure A). For example, in hatcheries, the gametes of multiple individuals are often pooled (e.g., sperm of multiple males mixed with eggs of multiple females). This practice results in a lower N_e because few individuals are mating relative to a typical wild population. In other words, mixing of gametes during the fertilization

process can lead to extreme variance in reproductive success, with only a tiny fraction of the population effectively contributing genes to offspring each year.

When considering the reintroduction of individuals to a location where the species has been extirpated, conservation geneticists are generally concerned most with maintaining the diversity of the founder population, and in the rate of growth of the reintroduced population. Generally, introductions of large numbers of individuals are preferred because larger founder populations promote the preservation of higher levels of genetic diversity. Scribner (1993) and Maudet et al. (2002), in studies of reintroduced Alpine ibex (*Capra ibex*) in Europe, found that estimates of

Continued on next page

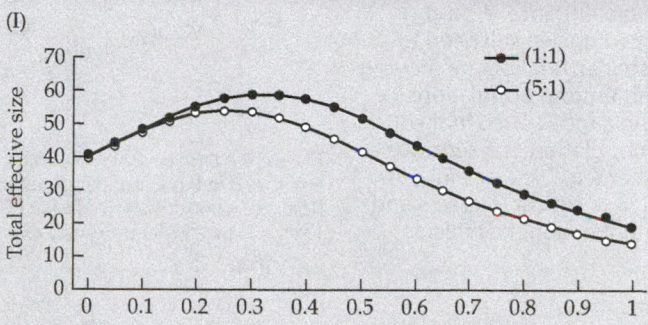

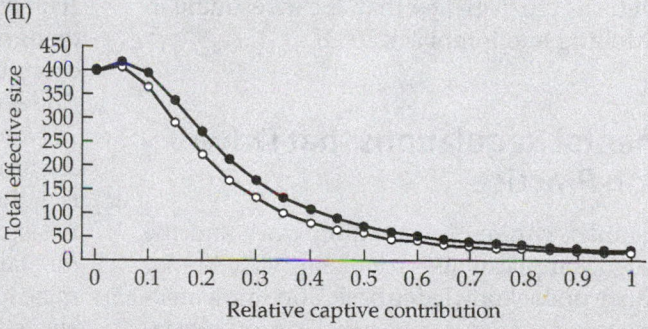

Figure A   Sex ratios used in captive breeding and the relative contribution of captive-bred individuals to wild populations influences N_e. (I) Wild population size = 40; (II) Wild population size = 400. Additions of captive bred individuals enhances N_e to a point, and then decreases N_e (rapidly in the case of large populations as in II). N_e is affected by the manner in which gametes are mixed during fertilization, as in fish hatcheries. When the sperm of multiple males are mixed with eggs (e.g., 5 males per female, 5:1), the effective population size can be considerably lower than if the eggs of a single female were combined with the sperm of a single male (1:1). The effect is more pronounced for small (I) rather than large (II) wild populations. (Modified from Bartron and Scribner, unpublished data.)

Box 15.1 *continued*

population heterozygosity were larger for populations with larger initial founding population size and in populations that exhibited high growth rates following introductions. Several releases may improve the long-term probability of maintaining levels of genetic variation. Stocking adults with equitable sex ratios will maximize N_e for the numbers introduced. If large numbers of individuals are not available for any single introduction event, stocking from multiple source populations or over multiple years can dramatically increase genetic diversity and minimize genetic drift. Bodkin et al. (1999), in studies of reintroduced sea otters (*Enhydra lutra*), found that high levels of haplotype diversity (a maternally inherited mitochondrial DNA genotype) were correlated with the larger founder populations that grew quickly in size (i.e., with a population bottleneck that was short in duration, and not too severe; Figure B).

Although certain translocation techniques can enhance N_e, and reduce inbreeding, we still need to remember that any deliberate moving of individuals among populations, or reintroducing captive-bred individuals into a wild population, is a sensitive and difficult enterprise. Certainly, in this context, it is critical to assess and monitor potential genetic effects.

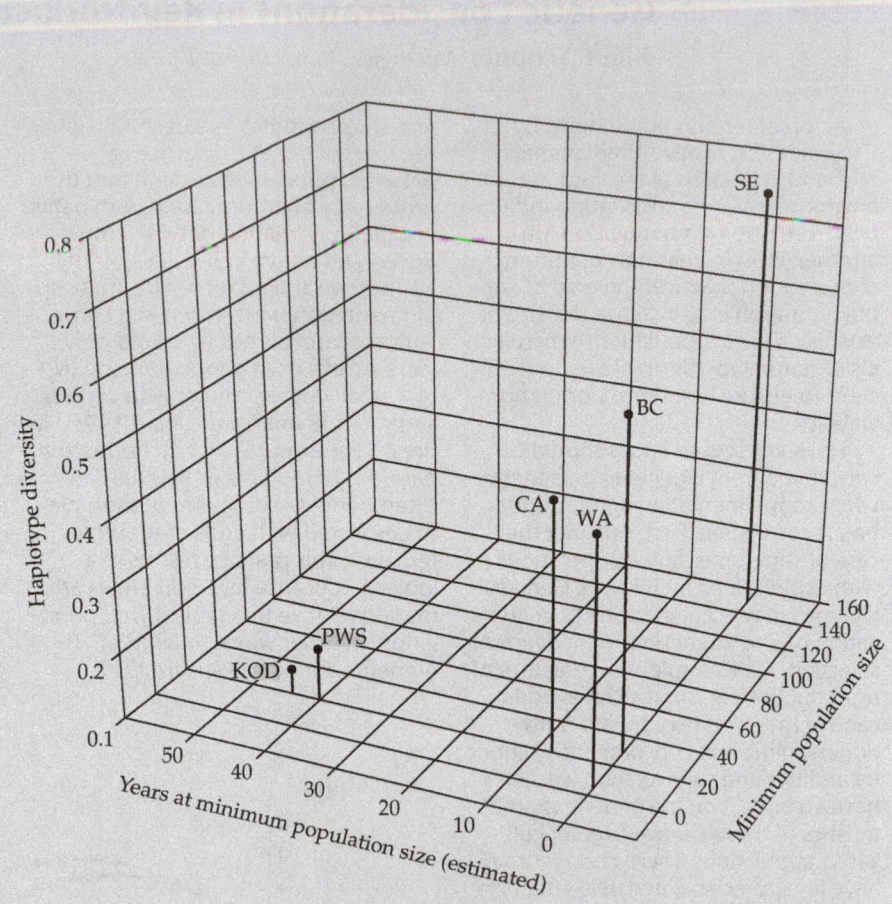

Figure B Diversity of mtDNA haplotypes is highest in reintroduced populations of sea otters that had large founder population sizes, and that grew rapidly after reintroduction. Locations: SE = southeastern Alaska; BC = British Columbia; WA = Washington, CA = California, PWS = Prince William Sound, KOD = Kodiak Island. (Modified from Bodkin et al. 1998.)

often yields returns much greater than the investment in restoration; Moburg and Rönnbäck 2003).

Environmental Regulations that Drive Restoration Practice

Given the complex nature of restoration work and the many associated complications—conflicting stakeholder interests, infancy of the knowledge base, and implementation challenges—it is not surprising that any one obstacle can be insurmountable. Furthermore, the financial burden of restoration is significant, with the total cost of a typical restoration project in the U.S. mounting to as much as U.S.$3.00 per square foot, or over U.S.$130,000 per acre. Such expense renders even small-scale ecosystem restoration projects problematic, often leading to compromises as elements of a restoration design are eliminated or project goals redefined. The reality is that restoration projects ex-

ecuted on an ecologically meaningful scale require hundreds of thousands of dollars (or more) to design, permit, execute, and monitor, and the exorbitant price tag excludes many from pursuing them.

Nonetheless, ecological restoration is an active field of scientific inquiry and applied science, so much so, in fact, that restoration to mitigate for landscape alteration is a legal mandate in national and international environmental regulations. These regulations not only determine how restoration can be accomplished, but encourage federal and state government funding for the work. Still other regulations necessitate ecological restoration projects as ends in themselves.

Regulations in the United States

The U.S. has a large infrastructure of regulations, policies, and laws that support environmental protection, management, and restoration. Origins of government-mandated environmental protection trace to the 1930s' "Dust

Bowl," a period of time when agricultural systems primarily in the Great Plains were completely devastated. This period of agricultural, environmental, and social ruin ultimately led to the creation of what is now known as the Natural Resources Conservation Service (NRCS). The mandate for this U.S. government agency is to develop and disseminate comprehensive information on management (and conservation) techniques for water and soil (Handbook of Ecological Restoration Volume 2, 2002).

Federal mandates for more broadly based ecological restoration made significant advances beginning in 1969, when the U.S. Congress passed the National Environmental Policy Act (NEPA). With NEPA's passage, the federal government took a proactive stand by establishing standards to protect or enhance our environment, its natural resources, and consequently human health. The U.S. Environmental Protection Agency (EPA) was subsequently created by then President Richard Nixon to coordinate and implement environmental regulations.

Wetland restoration occurs in large part because of the current "no net loss of area and/or function" policy of the U.S. Clean Water Act of 1972, as amended. Other ecological restoration programs and projects are required by laws developed to encourage the federal government to fund restoration of particular ecosystem components, particularly those with commercial value. These laws include the Federal Aid in Wildlife Restoration Act of 1937 to restore game for hunting, the Federal Aid in Fish Restoration Act of 1950 for sport fishing, and the National Forest Management Practices Act of 1976 for timber and associated wildlife (Berger 1991). However, restoration efforts that focus on commercially valuable ecosystem components frequently do not consider broader ecosystem functioning, and often augment commercially valuable populations at the expense of other ecosystem components.

CLEAN WATER ACT (CWA), 1972   The objective of the Clean Water Act is "to restore and maintain the chemical, physical and biological integrity of the Nation's waters." It is the foundation for surface water quality protection throughout the U.S., although it does not specifically protect ground water. The aim of the CWA is to reduce pollutant discharges and manage polluted runoff to the Nation's waters, so as to protect the propagation of fish, shellfish, and wildlife in and on America's waterways.

In the first two decades after the CWA was passed, its implementation focused predominantly on the chemical aspects of water quality, regulating discharges from point-source polluters like sewage and industrial facilities. However, since the early 1990s, considerably more attention has been given to examining the physical and biological integrity of the waters as well as examining sources of nonpoint-source pollution such as stormwa-

ter runoff from roads. The shift to a watershed approach in regulating pollutants and their sources emphasizes both protecting healthy waters and restoring impaired waters (EPA 2003b). In 1990, a policy of no net loss of area and/or functions of waters, including wetlands was formalized in a Memorandum of Agreement between the EPA and the U.S. Army Corps of Engineers (Berger 1991). While this policy has not achieved its stated goal of eliminating short-term wetlands losses (National Wetland Assessment 1997), it has led to more aggressive efforts to reduce them.

A large proportion of wetlands ecosystem restoration efforts in the U.S. is the result of the mitigation requirements of the CWA. Specifically, the CWA requires that, if there are significant unavoidable impacts to jurisdictional waters of the U.S., those responsible for the impacts must mitigate for waters/wetland loss by restoring, if possible, the same class of waters/wetlands as on the project site. Mitigation proposed to compensate for unavoidable impacts must be approved by the relevant federal, state, and local regulatory agencies prior to commencement of the project.

ENDANGERED SPECIES ACT (ESA), 1973   The objective of the Endangered Species Act is "to provide a means whereby the ecosystems upon which endangered and threatened species depend may be conserved and to provide a program for the conservation of such endangered and threatened species...." The ESA specifically prohibits any action that could harm, harass, or further endanger federally designated endangered or threatened species, or their associated critical habitat.

Within the ESA, Section 9 prohibits the destruction of endangered species, while Section 10 lays out the circumstances under which a "take" (e.g., harassing, harming, killing, collecting) by private entities is permitted. A take is permitted only if the action will not endanger the species, and the responsible party is taking all possible steps to mitigate the impact. The ESA allows the taking of plants, but not animals, on nonfederal lands, but if an endangered animal is present, endangered plants typically are protected along with the animal species. Federally protected plant populations on private land cannot be harmed in violation of state law, however.

In instances where projects are expected to impact endangered species, those responsible are required to minimize their impacts. Regulators from the U.S. Fish and Wildlife Service, the National Marine Fisheries Service, and state agencies may require restoration as mitigation for species whose habitat will be damaged by a proposed project. Restoration may be required at the same site or another location. With respect to the protection of plant species, in some instances, individuals of federally protected species are translocated to new locations and the project is allowed to proceed.

BOX 15.2 Other Pertinent U.S. Legislation Requiring Restoration Activities

Coastal Wetlands Planning Protection and Restoration Act of 1990
The Coastal Wetlands Planning Protection and Restoration Act of 1990 authorizes the U.S. Fish and Wildlife Service to make matching National Coastal Wetlands Conservation grants to coastal states for acquiring, managing, restoring, or enhancing wetlands.

Estuaries and Clean Water Act of 2000   The Estuaries and Clean Water Act of 2000 established an Estuary

Habitat Restoration Council responsible for developing a National Habitat Restoration Strategy. The Council also is responsible for reviewing and establishing funding priorities among restoration projects.

Federal Aid in Wildlife Restoration Act of 1937   Also known as the Pittman-Robertson Act, this provides federal aid to states for wildlife restoration work. Funds from an excise tax on sporting arms and ammunition are

provided to states on a matching basis for land acquisition, development, and management projects.

North American Wetlands Conservation Act of 1989   Signed by President G. H. W. Bush, the North American Wetlands Conservation Act of 1989 encouraged partnerships among federal agencies and others to protect, restore, enhance, and manage wetlands, and other habitats for migratory birds, fish and wildlife.

SURFACE MINING CONTROL AND RECLAMATION ACT (SMCRA), 1977   The objectives of the SMCRA are to prevent the adverse environmental effects of surface mining (particularly of coal) and restore lands that have been disturbed by mining. Regulatory authority lies either within individual states or within the Office of Surface Mining Reclamation and Enforcement. The SMCRA also established the Abandoned Mine Land program to reclaim mining-related degraded land (Perrow and Davy 2002).

Under the SMRCA, a mining company must follow strict guidelines to obtain a permit for coal mining. Permit applications must document pre-mining environmental conditions, detail mining activities, and specify proposed reclamation and monitoring activities. Post-mining, the company is required to begin reclamation via seeding the area, followed by proven reclamation activities over a five- to ten-year period. Mining companies must restore the site including preexisting landforms and land uses, and restoration of vegetation and water resources. If the responsible company fails to properly restore the mined lands, regulatory agencies will not release the final bond (Perrow and Davy 2002).

While this legislation has driven the reclamation of tens of thousands of acres of land and hundreds of kilometers of streams (Wali et al. 2002), there are shortcomings. Examples include water-quality impairment as a result of ten to hundreds of years of acid mine drainage, lack of reclamation in abandoned pre-1977 mine sites, and lack of regulation of other types of mining (e.g., silver, gold, lead, uranium and copper) (Perrow and Davy 2002).

Coal mining is highly disruptive to ecosystems (e.g., destroys fish and wildlife habitat, causes erosion and landslides, pollutes water). All coal mining activities must be permitted by the appropriate regulatory agency and companies are required to restore mined lands to either the former or better use. This entails returning the site to its pre-project topographic conditions; controlling erosion, and air and water pollution; minimizing disturbance to the hydrologic regime; removing, retaining, and redistributing the pre-project soil; and reestablishing vegetation. Some additional legislation that requires restoration is described in Box 15.2.

International regulations

Legislation mandating and promoting ecological restoration varies significantly from country to country. Many countries, however, share common elements in their restoration-related legislation. For example, mining regulations require reclamation of degraded lands in China (Wong and Bradshaw 2002), India (Dhar and Chakraborty 2002), and Canada (Wali et al. 2002). The 1972 United Nations Conference on Human Environment held in Stockholm is considered by most conservation biologists as the landmark international event that first brought attention to the decline of ecosystem health worldwide, and addressed what could be done to deal with further declines. Since that time, an increasing number of international conferences, summits, and conventions have been held that specifically address conservation issues. Many of these have focused on issues relating to ecological restoration.

Most recently, international efforts to protect the world's biodiversity culminated in the internationally legally binding treaty that originated at the Earth Summit in Rio de Janeiro in 1992. The resulting international document is known as the *Convention on Biological Diversity*. Article 8 (f) of the Convention specifically addresses restoration activities, promoting in situ conservation by requiring signatories of the convention to restore degraded ecosystems and threatened species, with particular emphasis on forests, inland waters, and marine ecosystems, including coral reefs.

Concluding Thoughts

The science of restoration ecology is on a steep trajectory. From the perspective of some who have been engaged in the field over the past 30 years, it is clear we have come a long way. Even critics of restoration agree. But just how far has this field progressed? Looking back, it is clear that efforts to restore ecosystems in the 1970s were based not on good science or bad science, but on no science at all. At best, many projects were dubbed "science lite." In the 1980s, many ecologists disparaged ecological restoration as "glorified gardening," and indeed, many restoration efforts (then and now) involve simply the replanting of native nursery stock. In the 1990s, working in interdisciplinary teams with diverse backgrounds and objectives, restoration ecology was more "ad hoc" or "compromise" ecology than a science based primarily on sound ecological principles. In this new millennium, however, we've seen a half decade of experimentation and adaptive management of restoration projects. We've also seen new college programs developed—programs that are devoted to the training of ecosystem restorationists (unlike the last cohort, who were in college when few if any programs for restoration ecology existed). Ultimately, while there is still plenty of basic science to come by to advance the field, restoration ecology poses some of the most important and exciting questions to conservation biologists today and in the decades ahead.

Increasingly, restoration may become a vital component of conservation practice, as we seek to improve degraded habitats (Wilson 1992; Hobbs and Harris 2001). Coupling restoration efforts to larger conservation efforts may offer many opportunities. For example, although single-species restoration and reintroduction efforts often involve some ecosystem restoration, rarely are they explicitly tied together (see Case Study 15.6). Increasing opportunities for larger scale restoration could be of enormous conservation importance (Simberloff et al. 1999; Young 2000; Holl et al. 2003). Young (2000) argues that restoration offers a long-term perspective that is helpful to conservation, which typically is more reactive to crises on shorter time scales. Because restoration is itself a long-term process, it forces us to consider how we might influence biodiversity conservation over longer time scales. For example, restoration can be integrated into management of highly fragmented areas using landscape-scale analysis; land uses that are most harmful and recoverable could become priority targets for restoration, while land uses that have beneficial or neutral aspects could be promoted (Young 2000). Restorationists might help communities adapt to global climate change through a careful consideration of which plant genetic backgrounds may best foster adaptation to new environmental conditions (Hufford and Mazer 2003; Rice and Emery 2003). This might be achieved by selecting seed sources that are appropriate to probable future states, and that also include representatives of a variety of specific microclimates that could be experienced at the site (Rice and Emery 2003).

One of the greatest needs for future work in restoration as a conservation tool is to increase our capacity to learn from restoration efforts, via comparative analyses and experimental approaches (Holl et al. 2003). As we can evaluate which factors are responsible for restoration successes and failures, we will be best able to create expanded opportunities to enhance species and ecosystem recovery through restoration.

CASE STUDY 15.1

Restoring the Nation's Wetlands
Why, Where, and How?

Joy B. Zedler, University of Wisconsin

Over the past two centuries, 53% of the wetland area in the contiguous United States has been destroyed, mostly through drainage for agriculture (Dahl 1990). That is an average rate of about 24 ha per hour, for a total loss of over 47 million ha in 200 years. The lower 48 states have only about 40 million ha of wetlands left. No wonder conservation leaders have developed a policy of "no net loss of wetland acreage and function" (The Conservation Foundation 1988). No wonder the National Research Council (1992) has called for restoration of 4 million ha by 2010. This essay considers why, where, and how we should go about meeting this goal.

Why wetlands should be restored relates to their many landscape-level functions. Wetlands act like "sponges"; they provide flood protection by reducing flood peaks and reducing shoreline erosion. To downstream water users, they are the "kidneys" of the landscape, because they filter sediments, nutrients, and contaminants from inflowing waters, thereby improving water quality. To local and migratory animals, they are "supermarkets" that provide a wide variety of foods. Wetlands produce timber, waterfowl, and other products of economic value. In addition, they are aesthetically pleasing, which is part of the reason that over 160 million Americans spend U.S.$14.3

billion each year observing, photographing, and enjoying nature (Duda 1991). Resource agencies and managers agree that wetlands perform critical functions that benefit humankind.

This does not mean that every wetland undergoing human-induced changes should be restored to some earlier condition as compensation for wetland losses elsewhere in the region (i.e., within the mitigation process; National Research Council 2001). In New Jersey, several mitigation projects aim to convert wetlands dominated by the nonnative invasive Phragmites (*Phragmites australis*) to more saline marshes dominated by the native species smooth cord grass (*Spartina alterniflora*). Sediments are removed to lower the topography and incise tidal channels, but regrowth of *P. australis* is a continual threat. However, *P. australis* may function similarly to other native grasses for several aspects of ecosystem functioning (Findlay et al. 1999). The mandate to mitigate damages to the nation's remaining wetlands would be better met by restoring sites that no longer function as wetlands, rather than remodeling one type of wetland in hopes of providing a different type with more or different functions.

Where wetlands should be restored is a more controversial issue. Perhaps the greatest future threat to wetlands lies with coastal wetlands. Only 7% of the nation's remaining wetlands occur along the coast, toward which the bulk of the human population is moving, with increased pressure to develop wetlands. Coastal wetlands also are threatened by rising sea level due to global warming. A 3°C increase in temperature by 2100 is predicted to raise sea level 1 m and to eliminate 65% of the coastal marshes of the contiguous U.S. (Park et al. 1989). Thus, coastal wetlands may have the greatest need for restoration.

Wetland restoration is not always feasible, but it can take place where wetland topography and hydrology are not terribly degraded and where the economic tradeoff is not unreasonable. Prime candidates are wetlands that have been ditched or tiled for agriculture but are only marginally productive (National Research Council 1992).

The most difficult question is *how* wetlands can be restored on a large scale. In some places it is a relatively simple matter of recreating the hydrology that allowed natural wetlands to develop. In Florida's Everglades, managers are returning the Kissimmee River to its historic meandering channel by undoing the 90-km straight channel that was cut in the 1960s and that eliminated 18,200 ha of river floodplain wetlands and reduced waterfowl populations by 90% (Koebel 1995). Forcing the river back into its natural channel will rejuvenate much of the river's floodplain (Toth et al. 1995). This restoration project has high potential for success because the channel morphology still exists, and native plants and animals persist nearby.

In other places, restoration is more difficult. In the upper Midwest, tiles can be plugged and drainage ditches can be filled, allowing wetland characteristics to develop without great cost; however, invasive plants (e.g., reed canarygrass *Phalaris arundinacea*) dominate newly exposed soil. Seeds of native plants can be sown to reduce weed invasions, but a dense canopy is needed to deter germination of *P. arundinacea* seeds (Lindig-Cisneros and Zedler 2002a). Further, until the demand for such restoration work increases, the price of native plant seeds will remain high.

Although invasive species threats are somewhat less in the most saline wetlands, salt marsh restoration still faces multiple challenges, especially in urban areas. In the San Diego region, over 85% of the naturally occurring salt marsh is developed, watersheds are greatly modified, streams are dammed, and degraded water flows into each coastal wetland. Sites are criss-crossed with roads, power lines, and sewer lines, and they are surrounded by buildings and roads, with no buffer between the wetland and heavily-used urban land. Many of southern California's coastal wetland species are considered threatened with extinction; for example, the list of sensitive species at Tijuana Estuary includes one plant, seven invertebrates, two reptiles, and 14 birds, and the region's salt marshes continue to experience extirpations (e.g., of short-lived plant species) when estuary mouths close to tidal flushing, even when the loss of tidal flow is temporary (Zedler 2001).

Considerable attention has been given to restoring nesting habitat for one bird on the endangered species list, the Light-footed Clapper Rail (*Rallus longirostris levipes*). This bird is a year-round resident of southern California salt marshes. Two salt marshes were created at San Diego Bay expressly for Clapper Rails as mitigation for damages to natural wetlands caused by highway widening, a new freeway interchange, and a new flood control channel. The first marsh (1984) was a 12-acre excavation (from dredge spoil) with eight islands and channels. In 1990, an additional 17-acre site was excavated nearby. To date, rails have not nested at either site. Several inadequacies of the constructed salt marshes were revealed in comparisons with natural reference marshes. The constructed marshes had less-abundant epibenthic invertebrates (Scatolini and Zedler 1996), shorter vegetation (Zedler 1993), and lower concentrations of soil organic matter and soil nitrogen (Langis et al. 1991). From the standpoint of the Clapper Rail, the short stature of the plants may be the biggest problem—when the tide rises, the plant canopy is fully submerged, leaving no cover for rails, their nests or chicks.

A chain of events explains the short plant canopies at these constructed marshes. Sandy sediments are "leaky," so nutrients do not accumulate in the soil or belowground tissues (Boyer et al. 2000). Nitrogen limits plant growth, especially height. Low organic matter concentrations further limit nitrogen fixation rates and perhaps the invertebrates that help recycle nutrients. Finally, the short vegetation appears to be inadequate for use by beetles (*Coleomegilla fuscilabris*) that consume scale insects (*Haliaspis spartina*), which are native herbivores on the cordgrass vegetation (Boyer and Zedler 1996). Scale insect outbreaks further impair cordgrass growth. Adding nitrogen to the soil improves plant growth while fertilizer is being added (1–4 years), but plants resume their short stature the year after nitrogen addition ceases.

From the problems that have plagued restoration attempts in San Diego Bay, I conclude that we are not yet able to recreate

self-sustaining salt marshes or to reestablish self-sustaining populations of endangered salt marsh birds. Endangered species may well be the most difficult components to restore to wetland ecosystems; because of their high habitat specificity, they are the first to decline when sites are modified and perhaps the last to return when artificial habitats are created for them.

Multiple examples of trial-and-error approaches to wetland restoration have not provided the community structure or ecosystem functions desired by managers or promised by mitigators (National Research Council 2001). The time is ripe to begin *testing* alternative restoration actions so that we can *explain* why some are effective and others are not. Preliminary results from Tijuana Estuary indicate that "adaptive restoration" can both identify the most effective actions and provide cause-and-effect explanations. Adaptive restoration is the design of the restoration project as an experiment that tests one or more alternative restoration actions, with follow-up research to understand outcomes (Zedler 2001). When restoration projects are very large (e.g., 200 ha are planned for Tijuana Estuary), the restoration efforts can be phased, and results of early modules used to inform later modules. The first restoration module at Tijuana Estuary tested the need to plant all eight members of the marsh–plain community, as well as the benefits of planting assemblages with up to six species (Figure A). Only three of the eight species recruited readily; the rest clearly need to be planted (Lindig-Cisneros and Zedler 2002b). One aggressive native, Virginia glasswort (*Salicornia virginica*), reduces recruitment and growth of others and, hence, is best left to colonize restoration sites on its own. Species-rich plantings had greater biomass, especially belowground, accumulated more nitrogen, and produced more complex canopies, documenting the importance of higher diversity to ecosystem functioning (Zedler et al. in 2001).

The second restoration module made use of findings from the first (e.g., *S. virginica* was not planted; species-rich assemblages were introduced) and tested the need to incise tidal creek networks into the otherwise flat marsh plain to enhance ecosystem functioning (Figure B). Among the response variables being evaluated are the diets and growth of fishes that move onto the marsh surface. Additional experiments were superimposed to test planting densities for California cord grass (*Spartina foliosa*) and marsh–plain assemblages, as well as seedling establishment with and without nurse plants. Future research will indicate the benefits of adding tidal creek networks relative to the extra cost of site preparation, so that future modules can be designed to meet restoration goals cost effectively.

Adaptive restoration is an appealing approach because restoration can proceed without waiting for the results of long-term research. If there is a question about how to proceed and alternative suggestions for restoration actions, the alternatives can become experimental treatments that are implemented in replicate. With a set of alternative actions and replicates of each, any differences in outcomes that withstand statistical analysis can be attributed to the restoration action. Experiments can be simple (e.g., three areas with tidal creeks, three without) or complicated (plantings with different numbers of species, different combinations of species, different planting densities). The more complex experiments potentially yield more information, but require more replicate plots and more area (e.g., a fully crossed experiment employing three levels of planting density, three assemblages, three soil amendments, and five-fold replication would require 135 plots). Plots need to be sized appropriately for the organisms involved; hence, 2×2-m plots suffice for studies of plant canopies, whereas 1.2-hectare plots are more appropriate for assessing bird use.

Figure A   Representation of salt marsh restoration at Tijuana Estuary's "Tidal Linkage," California. Example of how a bare salt-marsh plain can be experimentally vegetated. Each square is a 2×2 meter plot, planted with 6, 3, 1, or 0 species of halophyte seedlings to test the effect of species richness on canopy development, recruitment, and the accumulation of biomass and nitrogen. Numbers represent the number of species planted except for single-species plantings, which are labeled with the initials of the species: Lc = *Limonium californicum*, Bm = *Batis maritima*, Sv = *Salicornia virginica*, Jc = *Jaumea carnosa*, Sb = *Salicornia bigelovii*, Tc = *Triglochin concinna*, Fs = *Frankenia salina*, Se = *Suaeda esteroa*. Each planted plot had 90 seedlings. Researchers learned that plantings were essential for five of the eight species (Lc, Bm, Jc, Tc, and Fs) and for Sb and Se if no parent plants were in the vicinity. In addition, it was clear that species-rich plots developed more complex canopies and higher biomass and nitrogen crops. Thus, diversity accelerated development. (From Keer and Zedler 2002; Lindig-Cisneros and Zedler 2002b; and Callaway et al. 2003.)

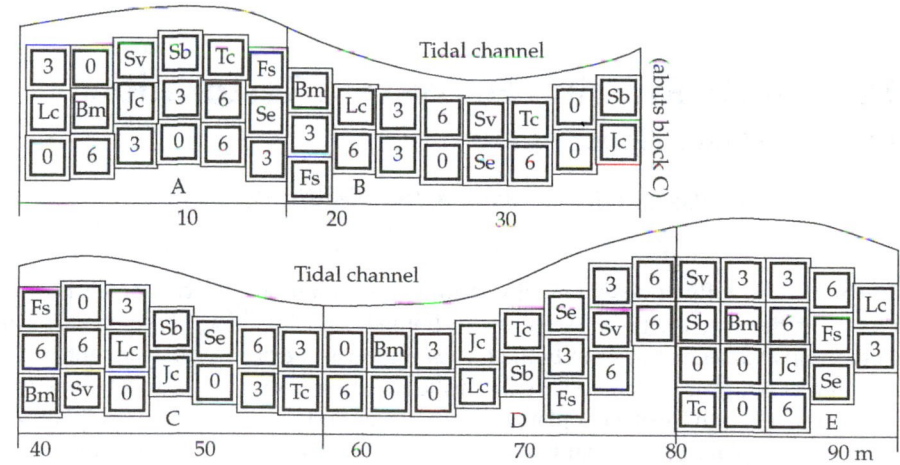

Figure B An 8 ha adaptive restoration project at the "Friendship Marsh," Tijuana Estuary, California. This ongoing experiment shows how a large, 8 ha salt marsh restoration can be implemented. The site consists of six, approximately 1 ha "cells," that either have or lack a tidal creek network. Additional experiments tested amending soil with kelp compost and alternative spacing of transplants of both *Spartina foliosa* near the tidal channel and five halophytes on the marsh plain. Researchers at the University of Wisconsin expect most attributes of community structure and ecosystem function to be affected by the presence of tidal creeks; hence, studies of algal mats, vascular plants, invertebrates, and fish are underway.

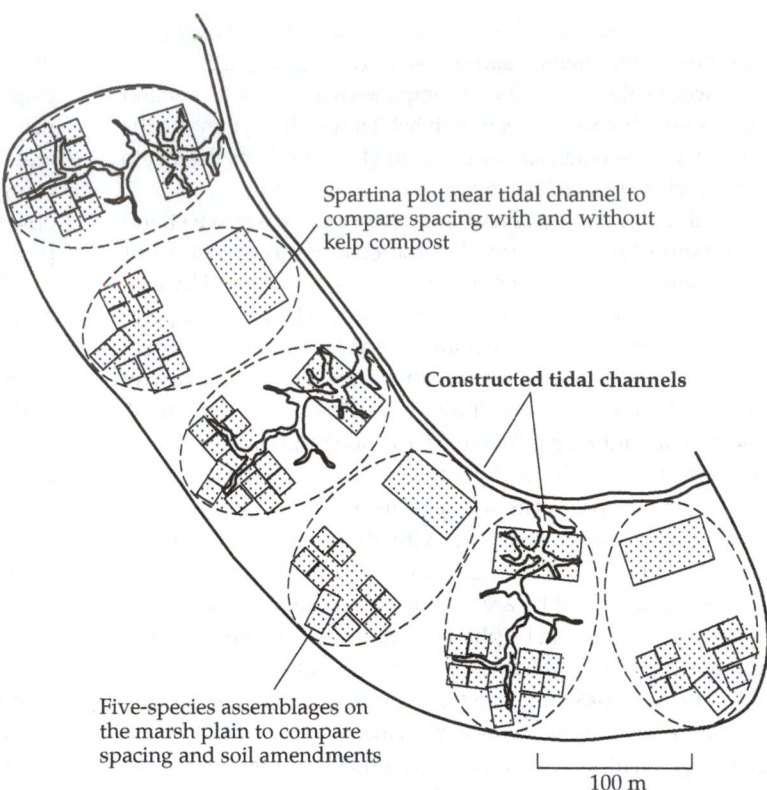

Spartina plot near tidal channel to compare spacing with and without kelp compost

Constructed tidal channels

Five-species assemblages on the marsh plain to compare spacing and soil amendments

100 m

In summary, we know *why* wetlands need to be restored: to restore native plant communities and hydrologic, water quality, and ecosystem functions. We have some ideas *where* the greatest gains can be made in the shortest period of time: marginally productive agricultural lands. And we know *how to begin* to restore wetlands that are not too damaged in regions where biodiversity is not too depleted: restore the hydrology, provide suitable substrate, transplant native vegetation, control invasive species, and hope that native animal populations can expand into the new habitats. But some wetlands are not restorable, and results for alternative approaches cannot be predicted, especially for sensitive species (such as endangered birds) in highly disturbed sites and in highly modified landscapes where source populations are depleted (Zedler 2000). Few, if any, restorationists can guarantee that desired targets can be met within a specific time frame.

In my opinion, the greatest gains in both theory and practice can be made by designing restoration projects as experiments that test alternative restoration actions. The time has come for adaptive restoration to replace trial and error approaches. My "field of dreams" is: If you build it *as an experiment*, they (students and other researchers) will come *to study and understand the effectiveness of alternative restoration approaches*. If we don't restore wetlands adaptively, we will remain unable to predict, with sufficient accuracy, the trajectories and outcomes of restoration.

CASE STUDY 15.2

Temperate Riverine Ecosystem Restoration
The North Creek Floodplain

Lyndon C. Lee, Blasland, Bouck and Lee, Inc., Seattle, WA

The North Creek Riverine Floodplain Ecosystem Restoration Project at the University of Washington, Bothell/Cascadia Community College campus in Bothell, Washington is one of the largest floodplain restorations in the Pacific Northwest. Importantly, the North Creek floodplain restoration was designed to be an exemplary project that would yield multiple educational benefits for regional K–12, undergraduate and graduate students, as well as community outreach opportunities. The core of this floodplain restoration project involved relocating a large perennial stream, North Creek, from its pre-project diked and straightened channel, to a newly constructed, meandering channel, thereby reconnecting the stream with its floodplain. The floodplain ecosystem was restored from its pre-project ditched and drained agricultural pasturelands to a mosaic of native floodplain forest, scrub–shrub and persistent emergent wetland plant communities.

Restoration of the North Creek Floodplain might never have taken place without federal, state and local environmental regulations that required mitigation for damages to wetlands caused by the proposed development, and if the development was not to create a campus for higher education. The site included a large number of wetlands and the stream was used by federally protected salmon species. Development of the campus would destroy 2.5 hectares of wetlands. This necessitated mitigation under federal, state, and local environmental regulations of at least 5 ha of wetlands. In keeping with its goals of being environmentally proactive, the State committed to a much larger project—to restore ecosystem functioning to 23.5 ha of lower North Creek by reconnecting it to portions of its historical floodplain through the creation of a new primary and secondary creek channel.

Site Description and Land Use History

The 23.5 ha North Creek Floodplain Ecosystem Restoration site is at the terminus of the 20-km long North Creek channel, which has a watershed area of nearly 7280 ha. North Creek flows through the campus of University of Washington, Bothell/Cascadia Community College before joining the Sammamish River on the south side of the campus.

Before European settlement, the lowland portion of the campus consisted of a complex array of channels, small backwater lakes, and alluvial depressions that were part of a junction environment where North Creek flowed into what is now the Sammamish River. Hydrologic characteristics of the site once were controlled by the interaction of runoff from the North Creek watershed and natural fluctuations in water levels of Lake Washington. When lake levels were highest, the restoration site was a shallow part of a very large lake that included contemporary Lake Sammamish and Lake Washington. However, in its recent geologic history, the site supported a complex marsh-forested wetland environment, and floodplain soils formed as a complex mixture of organic deposits.

Historically, the floodplain was dominated by several tree species, including western red cedar (*Thuja plicata*), black cottonwood (*Populus trichocarpa*), Douglas-fir (*Pseudotsuga menziesii*), big-leaf maple (*Acer macrophyllum*), and red alder (*Alnus rubra*). By 1895, the site had been logged for the first time, and by 1916, large portions of the North Creek channel system had been straightened and confined within artificial levees, and the restoration site was used for grazing and farming (Figure A.I). Extensive drainage systems (e.g., ditches and tile drains) were constructed in attempts to manage both groundwater and surface waters. In addition, construction of the Hiram Chittendom locks between 1913 and 1916 resulted in the lowering of Lake Washington by approximately 3.5 m.

(I)

(II)

Figure A   (I) The site prior to restoration in 1997. The farmhouse overlooks a pasture dominated by invasive, nonnative species. The stream channel was straightened to serve as a log flume when the watershed was logged, and lies at the base of a hill. (II) The North Creek Floodplain Restoration site in August, 2004. Notice the meandering channel for the creek, and the numerous small ponds and tree mounds introduced into the wetland. The site is bordered by major highways and the new campus. (I, photograph courtesy of BBL, Inc. II, photograph © George White Location Photography, www.gwlocphoto.com.)

This steepened significantly the longitudinal gradient in lower reach of North Creek.

Urbanization in the upper portions of North Creek during the last 60 to 70 years has greatly increased runoff and peak flow rates. This hydrological modification has had direct impacts on water quality, and on food, cover, and breeding habitat for salmon and other aquatic species in the North Creek ecosystem. In addition, a suite of aggressive, nonnative herbaceous weeds has dominated the site such that the structure of the pre-project plant communities did not provide the variety of habitats found in native forested wetlands. Thus, the ecosystem functions performed by existing wetlands prior to the implementation of the North Creek Floodplain Ecosystem Restoration Project were at

TABLE A *Channel and Wetland Design Targets for the North Creek Floodplain Restoration*

1. Maximize stream channel length while creating pools and riffles typical of floodplain streams to enhance stream functions and habitat for endangered salmonids.
2. Maximize contact time between water and wetlands by creating depressions for both short- and long-term water storage to enhance floodplain ecosystem processes.
3. Place large woody debris in-channel to guide channel alignment and morphology, to mimic natural features of streams, and to prevent flooding of the University of Washington, Bothell/Cascadia Community College campus and nearby highways.
4. Provide for increased peak flows as a result of urbanization of the North Creek watershed by creating a secondary stream channel.
5. Allow lateral channel movement within design parameters, as would be typical of natural streams.
6. Provide visual access from both the campus and highway corridors.

a level well below that of the ecosystem functions of reference wetlands under relatively undisturbed conditions.

Choosing the Goals for Restoration

Landscape-level modifications of the North Creek watershed and the greater Lake Washington Basin eliminated the possibility of returning the site to its original, historical condition. Thus, in 1977, the restoration team designed a floodplain ecosystem that would function at the highest level possible given the current hydrologic conditions, the urban environment in which the restoration project was located, and the physical constraints of the site (Table A). The final restoration design was based on historical information, reference data from similar stream ecosystems in the Puget Sound lowlands, the physical and biological characteristics of the existing North Creek, and the best professional judgment of the scientists and engineers involved in the floodplain restoration design. Overriding design objectives included reconnecting North Creek to portions of its historic floodplain, restoring natural stream morphology to the channel, reintroducing large wood (woody debris) to the channel and the floodplain ecosystem, restoring native plant communities to the floodplain, and providing improved habitat for aquatic and terrestrial wildlife species (see Table A; Figure A.11).

The North Creek Floodplain Ecosystem Restoration is based on a hydrogeomorphic (HGM) approach that emphasizes restoration of wetland function (Brinson 1993). The HGM approach first classifies wetlands based on position in the landscape, dominant source(s) of water, and flow and fluctuation of the water in and through the wetland. Wetland ecosystem functions then are identified for each wetland class and subclass, and data are collected from reference sites to provide an indication of the range of ecosystem functions for each wetland subclass.

Restoration Design and Implementation

Riverine ecosystem restoration usually requires a phased approach. One cannot fully restore any highly degraded forested ecosystem in a year or two. The North Creek Floodplain Ecosystem Restoration therefore requires near-term (1–5 years) and long-term (5–50 years) monitoring and management to achieve the project's targets set by the restoration scientists. Controlling weeds on the site is essential to the project's success, requiring an ongoing, integrated, and adaptive weed management strategy.

Floodplain Grading, Stream Channel, and Addition of Woody Debris

Grading of the site occurred in two distinct phases. The first involved stripping one foot of topsoil and weed material to eliminate a particularly pernicious invasive species, reed canary-grass (*Phalaris arundinacea*). The entire site was then mass graded to produce a very low-gradient floodplain surface that sloped downgradient from the stream's entrance to the restoration site. After grading of the floodplain, the new primary and secondary stream channels were excavated and the old North Creek channel was left intact to carry existing stream flow while the newly graded restoration site stabilized. Engineered log jams and additional woody debris were installed within the stream channel and across the floodplain.

A second stage of grading involved fine-scale work, including placement of streambed gravels, and refinements to the channel geometry. Further, complex floodplain microtopography characteristic of reference wetlands was added by creating microdepressions and adding stumps and floodplain log jams that are normally created by treefalls and floods (Figure B). These physical features increase wetland ecosystem processes by forming a natural means of bank stabilization, energy dissipation, flow diversion, scour pools, short- and long-term water storage, complex water balance characteristics, and sediment retention. Following fine grading, the channel banks were planted and a temporary irrigation system was installed.

After completion of this large first phase of riverine restoration, the stream was diverted incrementally from the old channel to the new one over two full growing seasons. Once accomplished, the old channel was backfilled and graded, and the interface between the recently restored floodplain and the old channel was graded and planted.

Plant Communities

Reference conditions within the Puget Sound Lowlands were used to design plant community types, patch sizes, and native plant species composition, abundance, and spacing. Reference sites were characterized by a mosaic of forested or scrub–shrub plant communities dominated by native species. Mature native Puget Sound lowland forests typically are closed-canopy, structurally complex, uneven-aged coniferous forests domi-

Figure B Microdepressions (small ponds such as the one in the foreground) and woody debris (seen in the background to the left) were added throughout the restoration site to provide habitat heterogeneity that mimics the features typically found in more mature sites that are created by treefalls and floods. Different plant species were planted within and around microdepressions or on small mounds to enhance fine-scale habitat differences that are critical to increasing plant and animal diversity on the site. (Photograph courtesy of BBL, Inc.)

nated by red cedar, western hemlock (*Tsuga heterophylla*) and Douglas-fir. The forests' understories contain a diverse mix of herbs, ferns, and bryophytes. Scrub–shrub communities support primarily deciduous shrubs and small trees such as red osier dogwood (*Cornus stolonifera*), hardhack (*Spiraea douglasii*) and several species of willows (*Salix* spp.), with a species-rich herbaceous layer.

The planting plan for the restoration site consisted of a mosaic of approximately 20 plant community types grouped into several broad categories: (1) scrub–shrub and emergent graminoids along the primary and secondary stream channel banks; (2) riparian forest; (3) scrub-shrub floodplain; (4) forested floodplain (deciduous); (5) forested floodplain (coniferous); (6) forested peatland; and (7) evergreen upland forest. Plant community types were positioned on the restored floodplain landscape according to site water balance, frequency of flooding, proximity to the channel, substrate condition, and overall aesthetics.

Development of native plant communities on the restoration site is a long-term project target because species composition, cover, height, densities, age distribution, and spacing at the reference sites reflect long-term ecosystem dynamics. At the outset, the intent of the restoration project was to install plant community compositions and densities of the woody species characteristic of self-sustaining, mature communities. Thus these plant communities were "jump started" by planting fast-growing, early seral stage plants in addition to the more slowly growing plants. However, many of the late seral stage species typical of mature reference sites are sun-intolerant species that require shade for survival and reproduction. Therefore, initial planting focused on establishing native species that grow rapidly on an open site, rapidly creating shade and conditions suit-

able for plants installed during later phases of the project. Plantings will be necessary in later years, and will focus on the replacement of plants lost to mortality as well as the introduction of shade-tolerant natives.

Weed Control

The North Creek Floodplain Restoration site was managed for many years as pasture for domestic livestock and was dominated by nonnative species, in particular, reed canarygrass and large patches of Himalayan blackberry (*Rubus discolor*). Our weed management policy relies on an integrated and adaptive management strategy. Such an approach usually is more successful than a single control strategy such as the sole reliance on chemical herbicides. Our integrated weed control strategy began with mowing the entire site. Then areas to be graded were either stripped to a depth of one foot to remove the weed cover, or buried under two or more feet of fill to prevent regrowth of the existing weeds. Following mass and fine grading, the site was planted with a cover crop to discourage reinvasion by weeds. Prior to planting, areas were disked and plowed to deter the establishment of newly germinated weed seeds. Ongoing weed control relies on shading from the planted nursery stock, hand or small mechanical weeding, and spot application of chemical herbicides (i.e., EPA-approved aquatic herbicides). Because weeds are more easily controlled when they first invade a site, successful weed management requires continual monitoring and repeated intervention to catch and control invasions and regrowth in the early stages.

Restoration Development Over Time

Restorations of natural ecosystems, particularly forested ones, take a long time. Native forests of the Puget Sound require hundreds of years to reach maturity. It will take another 30–40 years before the North Creek Floodplain Restoration Site really begins to function fully as a self-sustaining forested ecosystem. Therefore, while the restored riverine floodplain ecosystem is maturing, it will be necessary to actively manage the site, by, for example, controlling weeds and replacing dead nursery stock. Management requirements at such sites usually are highest during the first three to five years after planting, and then gradually taper off.

Ultimately, the North Creek Floodplain Restoration Project has demonstrated that large-scale riverine floodplain restoration is possible, and that it can be successful. It also has demonstrated that continuity in the project team, from design and environmental permitting, to implementation, to coordination with public stakeholders, and finally, to monitoring and maintenance, is crucial to a project's success. Without continuity in the project team, important project details are neglected, the design is poorly or wrongly implemented, costs soar, and the stakeholders lose ownership. North Creek has shown us what restoration ecology, dedicated scientists, and visionary thinking can accomplish.

CASE STUDY 15.3

From Kenya to Costa Rica
Solutions for Restoring Tropical Forests

Patricia A. Townsend, University of Washington

Wangari Maathai won the 2004 Nobel Peace Prize in part because of her actions in Kenya from 1960 to 1990, a period during which one-fifth of the world's tropical forests were cleared (WRI 1996). Maathai (2004a) realized the connection between deforestation and poverty and started the Green Belt Movement of planting trees, which has spread to other countries in East and Central Africa. Unfortunately, tropical forests continue to be destroyed at a rate of 0.5% per year (Achard et al. 2002). Because tropical forests lie almost entirely within developing countries with limited resources, economic incentives and aid for restoring forests are usually necessary.

Many tropical countries have ecotourism businesses dependent on charismatic species highly susceptible to habitat loss and fragmentation. Reforestation may be essential to the viability of these species, as well as to maintaining high diversity overall, and has the potential to aid in conservation and economic development where ecotourism is a major component of the economy.

Natural regeneration or methods that assist natural regeneration are the most economical means for reforestation. Seed sources—such as remnant trees or nearby forest patches—and a means of dispersal into degraded areas are needed to facilitate natural regeneration. In contrast to temperate systems, the seeds of most tropical plants are animal dispersed (50%–100%; Howe and Smallwood 1982). However, many forest-dwelling fruit-eaters—especially birds and most mammals—have little incentive to move through deforested areas where fruit abundance is low and the risk of being eaten is high (Janzen 1990). Few mammal- and bird-dispersed seeds are thus brought into these areas even when forest is nearby (Duncan and Chapman 2002). In contrast, bats will cross open areas and disperse seeds during flight, which can lead to a dominance of bat-dispersed plants in deforested areas.

Because natural dispersal is often lacking, restorationists have worked to introduce seeds by a variety of means. Horses were used to disperse large-seeded species into Costa Rica's Guanacaste National Park, which, combined with fire suppression, has led to rapid regrowth of target tree species (Janzen 2002). In degraded pastures in Costa Rica, several researchers have used artificial perches, which are effective in enhancing seed dispersal by birds but unfortunately do not increase the number of native tree seedlings beneath the perches (Holl 1998; Shiels and Walker 2003).

Tropical deforested areas present extremely challenging environments for germination and early seedling growth for most tree species. Microclimates are typically hot and dry once trees have been cleared, and soils are often poor, compacted, or eroded (Brown and Lugo 1994). To make matters worse, tropical pastures are typically dominated by nonnative grasses that limit tree seedling establishment and growth through competition (Uhl et al. 1988; Sarmiento 1997; Holl et al. 2000). Experimental work has shown that enhancing microclimate and reducing grass competition through the use of shade cloth (Hooper et al. 2002) or addition of woody debris can enhance the germination, growth, and survival of tree seedlings. In a Costa Rican pasture, naturally occurring fallen logs had trees and shrubs growing on and adjacent to them at densities eight times higher than in surrounding pasture due to the microsite variation and weed suppression they provided (Slocum 2000).

Planting of tree seedlings is necessary in many areas because of a lack of seed dispersal. Maathai's Green Belt Movement successfully planted 30 million trees in Kenya by involving thousands of community groups, primarily women, who were paid a small amount for each seedling they grew. In many parts of the world, reforestation projects have primarily been motivated by the economic incentive of growing timber, and secondarily with concern to restore ecosystem function, provide habitat for particular species, or generally increase biodiversity. Unfortunately, this may lead to a less-diversified forest than would result from natural regeneration (Box A by Carolina Murcia).

In Costa Rica, Carpenter et al. (2004) found that the species planted had a profound effect on forest restoration, especially in badly degraded areas. In their experiment, they interplanted *Terminalia amazonia*, a native economically valuable timber tree, with different species of legumes. When planted with the native tree *Inga edulis*, *T. amazonia* grew fastest, which was possibly due to "nurse tree effects" of partial shading and/or increasing nitrogen availability with symbiotic nitrogen fixing bacteria. After 12 years, trees and soil were both successfully restored—*T. amazonia* became the dominant tree and *I. edulis* filled the understory—providing habitat for a diversity of wildlife (Figure A).

Native trees are being planted in the Monteverde region of Costa Rica to expand and connect existing forest fragments. Many fruit-eating bird species, such as the Three-wattled Bellbird (*Procnias tricarunculata*), are altitudinal migrants that are highly dependent on the few remaining forest fragments at mid elevations. Restorationists planted a variety of species, in-

BOX A   Tropical Montane Forest Restoration: Do Tree Plantations Help or Hinder?

Carolina Murcia, *Wildlife Conservation Society*

Andean forests have been significantly reduced over centuries of human occupation, but most drastically in the twentieth century. Habitat loss and fragmentation are responsible for the local extinction of populations of many Andean species (Kattan et al. 1994). Our challenge now is to restore these forests to prevent further species loss and maintain viable populations. Once the sources of disturbance or habitat loss have been removed or neutralized, there are two basic approaches to recover the vegetation: allow natural regeneration to take place, or intervene by planting trees. Both approaches depend on natural succession, but differ in the initial conditions. The question is: What is the best strategy to restore the composition, structure, and function of the original ecosystems? This work evaluates whether the end points of using these two strategies are equivalent or if different initial conditions can gener-

ate a completely different system than the one intended (e.g., Robinson and Dickerson 1987).

To test the relative effectiveness of these two strategies, I compared 35 year-old forests that resulted from natural regeneration to adjacent forests initiated from Andean alder (*Alnus acuminata*) tree plantations. In the Central Andes of Colombia, Andean alder has been planted both for commercial purposes and to recover forests in montane watersheds. Andean alder has several characteristics that make it appealing for reforestation: it is native to the Andes, it grows quickly (up to 8 m in 2 years; Carlson 1975) and it can grow in very poor soils due to its association with nitrogen-fixing actinomycetes.

The Ucumarí Regional Park (4240 ha) protects the Otún River watershed, and is located on the western slope of the Central Andes of Colombia spanning an elevation gradient between

1700 and 2400 m. Until the mid 1960s, the upper part of the park (2200–2400 m) was a cattle ranch with pastures in the valley bottom and the lower slopes. The upper portions of the mountains and the steepest slopes were never clear-cut, but the trees with the best wood were selectively logged from those remnant forests. In the 1950s, the local government acquired the land and planted some of the pastures on the slopes with Andean alder, while others were left to regenerate naturally. The result is a mosaic of intermixed patches of alder plantations and natural regeneration that lie in the lower portions of the slopes, and in close proximity to a selectively logged mature forest. At this site, in 1994, I set up five permanent plots (50 × 25 m) each in naturally regenerated forest and alder plantations, and re-measured all individuals 5 years later.

The resulting forests looked quite different: the plantations had an alder-dominated canopy and no subcanopy, with openings between 10–15 m in height, while naturally regenerated sites had a continuous vertical foliage distribution (Figure A). Although structurally quite similar (in terms of density and basal diameter profile), they contain different species. Of the 178 species found in the plots, only 23 were shared by the two forest types (Murcia 1997). Further, naturally regenerated sites were much more species-rich than alder plantation sites (on average 63.4 versus 43.2 species per 0.125 ha plot). This difference was due mostly to the plantation's nearly mono-specific canopy and its lack of subcanopy. These differences in species composition also were associated with a higher spatial turnover in species among the regeneration plots than among the plantations (Murcia 1997).

Differences in species composition are likely to cause differences in several ecological processes, especially those that involve plant–animal inter-

(I)

(II)

Figure A   Images of the forest interior in alder plantation (I) and naturally regenerated forests (II) at Ucumari Regional Park, Colombia. Notice the open middle layer where alder trunks are visible in (I), and the greater complexity in (II). (Photographs by C. Murcia.)

Continued on next page

Box A *continued*

actions. One such process is seed rain. Alder is a wind-dispersed tree and, because few species shared the canopy with alder in the plantations, there are very few resources available to frugivores and seed dispersers. Over one year, seed rain in alder plantations was far greater than in naturally regenerated sites (4,304 seeds per m² for alder versus 1,175 seeds per m² for natural regeneration); however, 85% of the seeds in the plantations were from alder. Once alder seeds were removed from analyses, the only difference was that the total number of species in the regeneration was twice that found in the plantations. In addition, the seed rain in the regeneration was dominated by animal-dispersed seeds, whereas the plantation was equally split between wind- and animal-dispersed seeds. Finally, the temporal and spatial patterns of seed rain were more homogeneous in the plantations.

These results suggest that planting alders does have a significant impact on the resulting forest. Alder seems to have created conditions that favored the establishment of some plant species but not others, and that, in turn, may be affecting processes, such as seed, rain that will affect the future composition of the forest, creating a feedback cycle. The data available so far from Ucumarí, suggests that the two forest types are on diverging successional trajectories.

By planting trees to restore forests, initial conditions are defined that are likely to have long term consequences for the ecosystems to be restored. The effects are likely to be stronger in tropical and highly diverse ecosystems where the pool of potential colonizing species is larger than in the temperate zone. Other studies have found that restoring forested ecosystems with plantations resulted in forests with lower species richness than natural regeneration (Struhsaker et al. 1989; Chapman and Chapman 1996; Fimbel and Fimbel 1996; Cavelier and Tobler 1998). In all these cases, however, appropriate seed sources were available in the landscape. When that is not the case, or the soil is severely degraded, or there are no effective seed dispersers, planting trees may be the best option to recover a forested ecosystem of any sort (Parrotta 1993).

cluding many of the Lauraceae (avocado) family, which produce fruit preferred by bellbirds. Regeneration hopefully will be facilitated further as bellbirds and other fruit-eating species have an incentive to travel to the newly reforested areas where they can disperse seeds of other native forest plants.

Many tropical tree species can regenerate vegetatively from cut branches, which may be a more economical alternative to planting tree seedlings. This technique has been used by farmers throughout southern Mexico and Central America to create "living fences." The fence-post species are able to tolerate the harsh conditions found in pastures and grow and produce fruit much faster than a tree seedling. Zahawi (2005) has experimented with 2–3 m-tall living fence posts as a restoration strategy in abandoned pastures in Pico Bonito National Park, Honduras. The living fence posts are planted in clumps to create an almost instant island of trees. Once the living fence posts begin to resemble trees, birds use them to perch and also defecate seeds, and the shade improves microclimate for seed ger-

(I)   (II)   (III)

Figure A   Rapid reforestation of a degraded pasture site in southern Costa Rica (I) followed planting of a native timber species, *Terminalia amazonia*, and legumes, such as *Inga edulis*. The site two years (II) and eight years (III) after planting. *I. edulis* was severly pruned in 1999 to reduce competition. (Photographs courtesy of L. Carpenter.)

mination and growth. The hope is that this will lead to the establishment of a growing island of trees that will eventually merge with other expanding forest patches.

Regardless of the approach taken, the restoration of tropical forest may be critical to global environmental stability by supporting biodiversity, regulating climate, and stabilizing soil (Janzen 2002; Martinez-Garza and Howe 2003). Reforestation can lead to improved water and soil quality (Brown and Lugo 1994), and the carbon sequestered by the forests may help offset CO_2 emissions (Silver et al. 2000). Maathai (2004b) remarked after receiving her prize that the Nobel committee recognized the need "....to encourage community efforts to restore the earth at a time when we face the ecological crises of deforestation, desertification, water scarcity and a lack of biological diversity." These efforts need to be made worldwide and especially in tropical forests.

CASE STUDY 15.4

Restoration of an Endangered Species
The Black-footed Ferret

Dean E. Biggins, U.S. Geological Survey; Brian J. Miller, Denver Zoological Foundation; Tim W. Clark, Yale School of Forestry and Environmental Studies; and Richard P. Reading, Denver Zoological Foundation

Restoring populations of critically endangered species can be challenging. We illustrate this in a review of the biological and social challenges faced by the black-footed ferret captive breeding and reintroduction program (described in greater detail in Clark 1989; Miller et al. 1996). The primary value of our account lies with analyzing these challenges and how well the recovery process worked to address them.

North America's black-footed ferret (*Mustela nigripes*, Figure A) is a 600–1400 g mustelid (weasel family) whose closest relatives are the European (*M. putorius*) and Siberian (*M. eversmannii*) polecats. The recent historic range of the black-footed ferret coincided with the ranges of black-tailed (*Cynomys ludovicianus*), Gunnison's (*C. gunnisoni*) and white-tailed (*C. leucurus*) prairie dogs, which collectively occupied about 100 million acres of intermountain and prairie grasslands in the North American West. Ferrets are dependent on prairie dogs as their primary prey, and also use their burrows for shelter. Despite this large range, ferrets declined precipitously by the 1960s. Quickly listed as endangered under the Endangered Species Act of 1973, ferrets seemingly disappeared by the time the U.S. Fish and Wildlife Service approved the first recovery plan in 1978.

Black-footed ferrets were rediscovered near Meeteetse, Wyoming in 1981. During the next several years, field research enhanced our understanding of ferret behavior and ecology (Biggins et al. 1986; Clark 1989), building on data gathered during studies conducted in South Dakota from 1964–1974. By May 1985 a plan was formulated to begin captive propagation, starting with six ferrets to be captured from Meeteetse. In July 1985, biologists noted a severe decline in prairie dogs, accompanied by a decline in ferrets. Plague (caused by the bacterium *Yersinia pestis*) was discovered in the prairie dogs, stimulating a massive campaign to halt its spread. The operation concentrated on killing fleas, the plague vector, by dusting 80,000 burrows with the insecticide carbaryl. In October 1985 six ferrets were captured for the breeding program, but all died from canine distemper when two carried the disease from the field into the holding facility, exposing the other four. Six additional ferrets subsequently were captured. The low point for ferrets was reached in winter of 1985–1986, with four known survivors at Meeteetse plus the six captives. No kits were born in captivity in 1986, but two litters were produced at Meeteetse. A decision was made to take the few remaining wild ferrets from the wild, bringing the captive population to 18 by early

Figure A The black-footed ferret (*Mustela nigripes*) is highly endangered, due largely to habitat loss and intentional extermination of its primary prey, various prairie dog species. (Photograph courtesy of D. E. Biggins.)

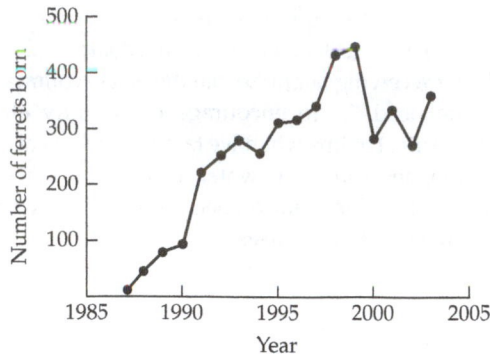

Figure B Black-footed ferret kits born in the captive population.

1987, and placing the future of the species in the captive breeding program.

After a tenuous start, the captive population expanded and has been maintained at 200–300 adults. Coordinating captive propagation is the responsibility of the Species Survival Plan (SSP) working committee (under the auspices of the American Zoo and Aquarium Association), with participants from each ferret propagation facility. The primary objective of the SSP is to produce kits for reintroduction as efficiently as possible (Figure B), and to serve as the ultimate hedge against extinction. Allocating ferrets for reintroduction is the responsibility of FWS.

By 1991, sufficient ferret kits were being produced to start reintroduction. The Meeteetse prairie dogs continued to decline, eliminating reintroduction possibilities there. An alternate site near Medicine Bow, Wyoming received ferrets from 1991–1994. New sites have been added periodically since the first releases in Wyoming, with releases continuing for multiple years (Figure C), and populations augmented by wild-born kits.

Biological Challenges

CAUSES OF DECLINE A critical component of any recovery program is identifying the cause of decline. A major problem for ferrets was the destruction of prairie dogs, which constitute about 90% of the ferrets' diet and dig the burrows that serve as ferret shelters. During 1900–1960, prairie dogs were reduced to about 2% of their former geographic area, largely due to agricultural development and extensive poisoning campaigns, wherein prairie dogs are considered pests (Figure D). Sylvatic plague, probably introduced to North America in about 1900, also caused massive die-offs of prairie dogs (Barnes 1982).

The prairie dog ecosystem supports many species, and the most tightly linked is the black-footed ferret. Fragmentation of ferret habitat (prairie dog colonies) eliminated or reduced ferret populations, with remaining populations more susceptible to extinction by disease, genetic problems, demographic events, or natural catastrophes. Sources of immigration were undoubtedly destroyed, and recolonization and genetic ex-

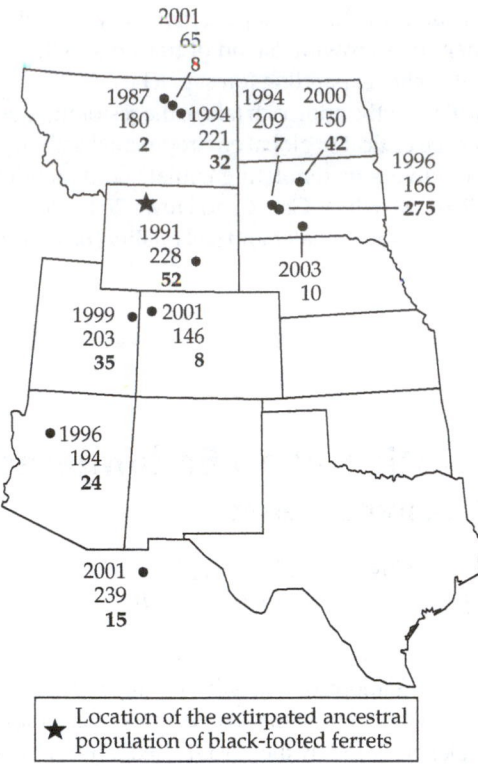

Figure C The 12 states where black-footed ferrets were historically present and sites where captive-born ferrets have been released. Numbers designate year of first release, total numbers of ferrets released (all years), and minimum population estimates (bold type) for Fall of 2003.

change hampered. Under these circumstances, risk of extinction may rise disproportionately more rapidly than habitat destruction, and reduction of one species may produce secondary extinctions (Wilcox and Murphy 1985).

In addition to habitat loss (e.g., prairie dog decline) the black-footed ferret is very sensitive to canine distemper. In the 1970s some black-footed ferrets died after vaccination with a modified live virus used safely to immunize other ferret species (Carpenter et al. 1976). Erickson's (1973) prediction that "the hazards of exposure of the highly sensitive black-footed ferret to canine distemper virus may be substantial" was realized with the apparent involvement of that disease in ferret decline at Meeteetse (Forrest et al. 1988). More recently, ferrets living in outdoor pens have died from plague (Biggins and Godbey 2003), showing unexpectedly high sensitivity to that disease (many other carnivores display resistance to plague). Plague was widespread at Meeteetse in 1985, and likely contributed to the demise of the ferrets in the region.

In summary, ferrets were first reduced in population size by massive reduction in the extent of their primary prey, and then reduced still further by disease (Biggins and Schroeder 1988). The twin challenges of protecting habitat and avoiding diseases are daunting. A proposed habitat program involves legal

(I)

(II)

Figure D (I) The black-tailed prairie dog (*Cynomys ludovicianus*) is one of the species critical to ferret survival. (II) Prairie dogs historically have been actively exterminated by ranching and farming interests, an activity that continues today. (Photographs courtesy of D. E. Biggins.)

protection of prairie dogs in areas where ferrets should otherwise be able to thrive, protected areas, public education, and converting federal subsidies for poisoning prairie dogs into positive incentives for ranchers who manage for both wildlife and livestock (Miller et al. 1994a). At present researchers of the Wyoming State Veterinary Laboratory are working on a distemper vaccine for ferrets, and plague research has begun to address several questions.

CAPTIVE BREEDING Biological challenges in captive breeding include increasing efficiency, rearing kits to maximize survival after release, protection from disease, genetic management, and proper husbandry (e.g., diet, photoperiod, stress). An unresolved dilemma concerns maximizing production and genetic management. Males that breed best are most likely to become over-represented genetically. Balancing their genetic influence requires reducing their reproduction, but that likely reduces kit production. Also, the program never completed a genetic analysis of founders, resulting in a studbook with a weak foundation (Miller et al. 1996).

In captivity, about two-thirds of prime females produce kits. Although this rate is not radically different from other captive mustelids (e.g., mink, *M. vison*), it seems lower than that for free-ranging animals at Meeteetse. High neonatal mortality decreases production; causes could include crowding, excessive human disturbance, or inbreeding. Rigorous research should address these and other questions. Such studies may pose risks to individual animals and was discouraged when ferret numbers were low, but a different philosophy should prevail now.

Ferrets required captive breeding, which successfully prevented extinction and produces animals for reintroduction. However, captive propagation is not a panacea and should be employed cautiously, as it is expensive, risky, and must be tied closely to reintroduction needs.

REINTRODUCTION To increase chances of survival after release, we tested techniques with Siberian polecats (Figure E) in 1989 and 1990 (Miller et al. 1993; Miller et al. 1996), and recommended a field test on black-footed ferrets released in 1991 (Miller et al. 1996). Pre-release conditioning was not used in 1991, but was attempted in Wyoming in 1992. Ferrets reared in outdoor pens survived significantly better after release than cage-raised counterparts and demonstrated different behaviors (Biggins et al. 1999). The replicated experiment produced similar results in Montana and South Dakota (Biggins et al. 1998), and direct translocations have proven even more effective. Several release strategies (e.g., holding animals in cages for acclimation, varying degrees of provisioning) have been used in different years and sites. Predation, primarily by coyotes (*Canis latrans*), has caused the most deaths of radio-tagged animals. In 1995, ferret deaths seemed lower in Montana when coyotes were aerially hunted and electric fencing excluded them from reintroduction sites. Coyotes have been killed at other release sites. Each coyote management and release method has proponents, but lack of sound testing has provided ambiguous data that fuel continued debate (Breck et al. 2004). Debates and delays notwithstanding, reintroductions have become more efficient.

Additional information on diseases, however is disconcerting: (1) Canine distemper is ubiquitous in other carnivores at all reintroduction sites; (2) plague is present at reintroduction sites in all states except South Dakota; and (3) reintroduced ferrets in Shirley Basin, Wyoming declined as plague increased. Also, the ultimate effect of inbreeding remains unknown. Failure to fully understand the potentially interactive causes of the ferrets' decline, ineffectiveness at reestablishing habitat (or even halting losses), and the looming diseases questions emphasize the uncertain future of reintroductions. Ferret reintro-

(I)

(II)

Figure E   Black-footed ferrets were so rare that techniques were often worked out on "surrogate" species, such as Siberian polecats (*Mustela eversmannii*). Young polecats reared and conditioned under varying procedures were tested using predator models such as (I) stuffed owls, and (II) a "robo-badger," a remote-controlled, stuffed, and motorized badger. (Photographs courtesy of D. E. Biggins.)

ductions should remain experimental for now, designed to maximize detecting factors that limit future recovery.

Several interrelated questions exist that have no easy answers, even with relevant data. For example, is it reasonable to extend the Noah's Ark model (Hutchins and Conway 1995) of zoo production and release of animals to "repeated floods" (i.e., if we cannot achieve viable, self-sustaining ferret populations, should we artificially maintain nonviable populations through periodic restocking)? What is the smallest population worthy of sustaining and at what level of perpetual restocking?

SOCIAL CHALLENGES   The social challenges of black-footed ferret recovery are equally daunting. Indeed, the attitudes and values people hold toward wildlife are the root of ferret decline. Attitudinal research suggests that the recovery program must address antagonism of ranchers, develop support among undecided or uninformed individuals, and maintain support of conservationists (Reading and Kellert 1993). Cutlip and Center (1964) suggested pressure, purchase, and persuasion to change attitudes. A simple education program in Montana that merely provided information was unsuccessful at garnering local support and alienated some conservationists (Reading 1993). Prairie dog poisoning sponsored by governmental entities (see Figure D.II) reinforces attitudes that these species have little value, complicating public relations (Miller et al. 1996).

Program participants likewise have varying perceptions. Most programs involve multiple groups, each with different values, ideologies, and definitions of the problems, goals, and methods of operation. Although diverse views can provide

multiple ideas and approaches to strengthen a program, they seemed to instigate conflict and power struggles that deflected attention from ferret recovery (Miller et al. 1996). Each agency strove to appease its primary constituents, but constituencies differ. Who should control the ferrets' destiny: local groups, states, or the country as a whole (in part, a recurring question of interpreting the U.S. Constitution)? The majority of ranchers at reintroduction sites do not favor prairie dogs or ferrets. This puts state wildlife agencies in a difficult position, as they are primarily funded by sportsmen whose access to ranch lands often depends on good relationships between the agency and landowners. Federal agencies, mostly funded through general tax funds, try to represent everyone—an impossible task.

In making crucial decisions, the pace seemed sluggish at times, perhaps partly caused by fear of failure and the notion that failure from "natural causes" (no action) is preferable to failure from bad decisions. Risk aversion is understandable given the castigation that agencies receive from almost every conceivable angle. The ferret program is replete with examples, including criticism from: (1) the Sierra Club Defense Fund, who challenged the nonessential, experimental designation for providing too little protection; (2) the Wyoming Farm Bureau who complained that ferrets received too much protection and ranchers too little; and (3) the American Humane Society which was angered over the release of adult ferrets rather than kits.

USE OF SCIENCE AND PROGRAM GOALS   Early in the Meeteetse program everyone agreed to: (1) maximize learning about the

biology of this little-understood animal; (2) do everything possible to conserve this only known population; and (3) plan for species recovery. Each of these tenets conflicted in some way with others. Goal (2) was adopted as the first priority at Meeteetse. To some degree, the research implied by goal (1) and the removal of animals for translocations or captive breeding implied in goal (3) conflict with goal (2). The situation was further confused by realization that research manipulations could place some animals at risk, but the results could advance recovery.

During early reintroductions, the primary objective was to achieve a pre-determined survival rate for a specified period (e.g., 20% for 30 days); a result was weakened intensity of monitoring designed to maximize learning. Again, anything that *might* have increased the impact on released animals was prohibited, including some research and monitoring methods. The recovery plan implied providing the highest priority to learning during early reintroductions; we argue that learning was given lower priority from the beginning.

The general tendency was for short-term, geographically localized goals to take precedence over longer-term recovery goals. Intensive study and monitoring provide localized benefits (e.g., radiotelemetry enabled rescue of some animals), but the focus was to provide information to help future efforts, often at another site or state. The conflict has been mis-stated as research versus operational ferret recovery. Everyone in the program is working toward recovery, but perceptions differ on which process will achieve recovery quickest. Demographic goals are appropriate, especially in the long-term (e.g., the Recovery Plan), but learning should not always be subordinate to them (Clark 1996).

An issue related to goal setting is interpreting success and failure. Emphasizing ferret numbers conserved or established can lead to "success" with little learning, which increases the chances for long-term, large scale failure. Conversely, a "failure" could teach us something so dramatic that it becomes far more valuable than a localized "success," such as short-term survival goals. Success should be redefined (aided by carefully prioritized goals) to avoid stifling risk-taking, innovative thinking, and experimentation. The utility of success/failure characterizations is related mostly to public relations (your supporters may become detractors with too many "failures" and too few "successes") and legal issues (a demographic "success" is prescribed for down-listing the ferret from endangered to threatened).

FUNDING   Funding inconsistencies continue to impede the captive breeding program. Five zoos produce approximately half the surplus ferrets using their own funding. The other half are produced at a facility in Wyoming, supported over the years by Federal and state funds. Greatly diminished funding jeopardized the program in 1995. Private funding (PIC Technologies) rescued the effort temporarily, but uncertainty remains. Numbers of captive ferrets should not fluctuate with available money, so stability is essential.

PLANNING AND IMPEMENTATION   Planning and implementation determines how well skills, funds, resources, organizations, and people are coordinated to meet recovery challenges. Common problems occur across species, including slow decision making, decision making without outside expertise, decision making that consolidates control at the expense of scientific and management priorities, reward of organizational loyalty instead of creativity, faulty information flow (or even blockage), failure to develop objectives that accurately evaluate program success, deviation from plans during implementation, and rigid bureaucratic hierarchy that impedes effective action. Plans can be altered substantially during implementation by incompetent execution, deliberate delay when people oppose the plan, or yielding to local political pressure. The ferret program has been criticized for all of these (Miller et al. 1996; Clark 1997).

One way to avoid such problems (or at least decrease their intensity) is by forming balanced recovery teams to make recommendations. A recovery plan can outline goals, but only teams—communicating freely and efficiently—can respond to rapidly changing events (Miller et al. 1994b). Information flow within and between teams and working groups improved substantially when the FWS began periodic conference calls. Team members should be selected to achieve a balance of expertise and not simply represent agencies. It makes no more sense for a team of upper-level managers to issue judgment on highly technical plans than it does for specialized scientists to try to generate and coordinate funding for the program. In addition, when advisors come from a high organizational level their recommendations may reflect politics more than biology (Miller et al. 1996). Because even good plans and devoted teams can outlive their usefulness, periodic outside review of the entire program is crucial. The American Zoo Association and the World Conservation Union (IUCN) have conducted several such reviews of the black-footed ferret program.

How far should we go to save the black-footed ferret? It is the only North American ferret; there is nothing else like it on this continent, and it occupies a unique niche in a declining biotic assemblage, the prairie dog ecosystem. Saving this "flagship" species will help assure a place for many of its associates. To eliminate it will, in the words of Rolston (see Essay 4.1), "shut down a story of many millennia, and leave no future possibilities." The challenges facing the ferret are not trivial, but we fervently hope that the task can be accomplished.

In the Eye of the Hurricane
Efforts to Save the Puerto Rican Parrot

Jaime A. Collazo, North Carolina State University and U.S. Geological Survey; Martha J. Groom, University of Washington, Bothell; Susan M. Haig, U.S. Geological Survey; Thomas H. White, U.S. Fish and Wildlife Service; Britta D. Muizniecks, U.S. Fish and Wildlife Service

The Puerto Rican Parrot (*Amazona vittata*) is one of the most endangered birds in the world (Figure A). Once abundant (perhaps 100,000 individuals) and distributed across Puerto Rico, the species collapsed to a mere 200 individuals, all confined to the rainforest of the Luquillo Mountains of eastern Puerto Rico by the 1940s (Snyder et al. 1987). Range contraction and concurrent population decline were associated with widespread deforestation in the late nineteenth and early twentieth centuries. At the peak of deforestation in 1930, only about 10% of the island was covered by forest (Wadsworth 1950). The vulnerability of the Puerto Rican Parrot to forest loss is three-fold: (1) it is dependent on fruit resources from forest trees and vines; (2) it relies on large-diameter trees for nesting cavities; and (3) it suffers from heavy predation by Red-tailed Hawks (*Buteo jamaicensis*) when flying in the open.

Although remaining forests are generally well-protected, and many areas are regrowing, Puerto Rican Parrots have continued to decline due to a large number of challenges to the remnant wild population; the population of Puerto Rican Parrots decreased from 200 individuals in the mid-1950s to 14–17 individuals during the early 1970s (Figure B; Rodríguez-Vidal 1959; Snyder et al. 1987). Current threats to the species are posed by Pearly-eyed Thrashers (*Margarops fuscatus*), which compete for nest cavities and prey on the parrot's eggs and chicks, competition for tree cavities with European and Africanized honeybees (*Apis mellifera*), predation of fledglings and adults by Red-tailed Hawks, egg predation by introduced rats (*Rattus rattus* and *R. norvegicus*), parasitism by warble flies (*Philornis pici*), and the impact of hurricanes (Snyder et al. 1987; USFWS 1999). Heavy precipitation also decreases breeding success, as wet cavities are unsuitable for nesting and excessive humidity contributes to chick respiratory diseases. Management efforts in the Luquillo Mountains (e.g., nest guarding, predator control, nurturing displaced chicks from the wild) primarily target these adverse impacts.

Efforts to prevent the species' extinction began with its listing as an endangered species in 1967. Initial recovery efforts focused on enhancing breeding productivity by protecting chicks from nest predation and ectoparasites (Snyder et al. 1987). However, soon it was obvious that more intervention was needed to prevent extinction of the Puerto Rican Parrot.

Meeting a Formidable Conservation Challenge: Captive Breeding, Release, and Reintroduction Efforts

The Luquillo Aviary, a captive-rearing facility was established in 1973 to rear and foster chicks into wild nests to increase breeding productivity. Given the regularity of hurricane disturbance, a second aviary (José L. Vivaldi Aviary) was created in 1993 in the limestone lowlands of north-central Puerto Rico to safeguard the population (Lacy et al. 1989). Captive-breeding efforts are aided by captive flocks of Hispaniolan Parrots (*A. ventralis*) in both aviaries, serving as incubation surrogates and foster parents for Puerto Rican Parrots (USFWS 1999). Currently there are 165 Puerto Rican Parrots housed among both aviaries.

Despite additions of fostered chicks into wild nests, population growth remained sluggish. Partly this has

Figure A   A Puerto Rican Parrot eating a palm fruit. (Photograph © Mark L. Stafford, www.parrotsinternational.org.)

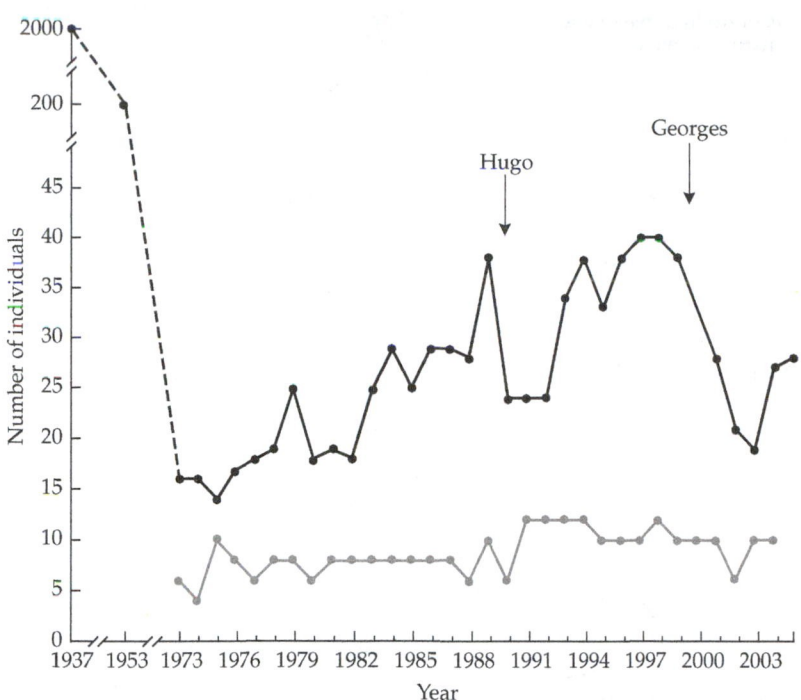

Figure B Population trajectory of Puerto Rican Parrots since recovery efforts have been underway. Black circles denote the total number of individuals in the wild population, gray circles denote numbers of individuals breeding each year. Arrows show the impacts of Hurricane Hugo and Hurricane Georges.

been due to the very low numbers of breeding pairs able to raise an additional chick and to lack of synchronicity between wild and aviary breeding cycles (Collazo et al. 2000). Thus, a release program was initiated in 2000 to bolster the wild population. We released 34 captive-reared parrots in the Caribbean National Forest (within the Luquillo Mountains) between 2000 and 2002. The first-year survival estimate was low (41%; White et al. 2005), although similar to that for parrots born in the wild. Red-tailed Hawk predation accounted for the majority of deaths (54%; Wiley et al. 2004). We are encouraged by the fact that a bird released in 2002 was recruited into the breeding population in 2004. It is disconcerting, however, that the wild population remains tiny—only 28 individuals were counted in 2004.

The prognosis for recovery remains bleak. Despite 35 years of intense recovery efforts, the wild population has never exceeded 50 individuals. Population numbers peaked at 47 in 1989, only to be reduced to 23 in the aftermath of Hurricane Hugo (Meyers et al. 1996; see Figure B). A recent population viability analysis predicted that extinction is likely within 40 years (Muiznieks 2003). This pessimistic prognosis derives from abysmal first-year survival rates seen since 1989 (ca. 40% versus 67% observed prior to 1989). The predicament is compounded by limited options available to increase breeding productivity in the wild population. It is unclear why, but the number of breeding pairs has never exceeded six during the history of the recovery program (USFWS 1999), and fledging

rates have never exceeded 1.88 per year. We estimated that breeding productivity had to be greatly increased over current levels to offset juvenile mortality, either by eight or more nests producing the current average numbers of chicks (1.56 chicks per nest attempt) or all six breeding pairs producing 2.5 or more chicks per nest attempt. Given the history of breeding productivity, neither of these scenarios seems likely. Perhaps the best chance for recovery would come from establishing a second wild population (Figure C).

The newest goal of the Puerto Rican Parrot recovery program is to establish a second wild population in the Rio Abajo Commonwealth Forest of the island's karst region beginning in 2006 (USFWS 1999; see Figure C). This is where parrots last occurred outside the Luquillo Mountains in the 1930s (USFWS 1999). The karst region may provide better habitat for a second wild population because these forests are drier, contain many food plant species required by parrots, and harbor a lower density of Red-tailed Hawks than the Luquillo Mountains (Cardona et al. 1986; Rivera-Milán 1995; Collazo and Groom 2000). While tree-nesting opportunities are not as readily available (Cardona et al. 1985), crevices in the limestone landscape and artificial nest structures could offer suitable alternatives (USFWS 1999). The forest reserve has the added advantage of harboring the José L. Vivaldi Aviary. Captive birds could serve as a "surrogate" wild population, providing a focal location where released birds could converge daily while they adjust to wild conditions.

The aviaries also provide a unique opportunity to manage genetic diversity of the species. A general goal for the maintenance of genetic diversity is retention of 90% original heterozygosity for 200 years (Ballou and Foose 1996). We found through molecular genetic and pedigree analysis that the population of wild and captive Puerto Rican Parrots has lost 7% of its heterozygosity since active management began, comparable to the rate of loss of heterozygosity in captive Guam Rails (*Rallus owstoni*), a species exhibiting serious reproductive and survival problems (Haig and Ballou 1995). Fortunately, fertility and hatching rates in Puerto Rican Parrots show no negative effects of inbreeding (Daniels et al. 2004). To prevent genetic problems that could arise from inbreeding in the future, we recommended that both captive flocks are managed in an integrated fashion following a comprehensive genetic management plan that pairs specific individuals based on mean kinship and founder contributions (see Figure C).

Recently, we developed release protocols using Hispaniolan Parrots in the Dominican Republic (Collazo et al. 2003), and subsequently, refined these protocols with the releases of Puerto Rican Parrots into the Caribbean National Forest over 2000–2002 (White et al. 2005). We now have a detailed release

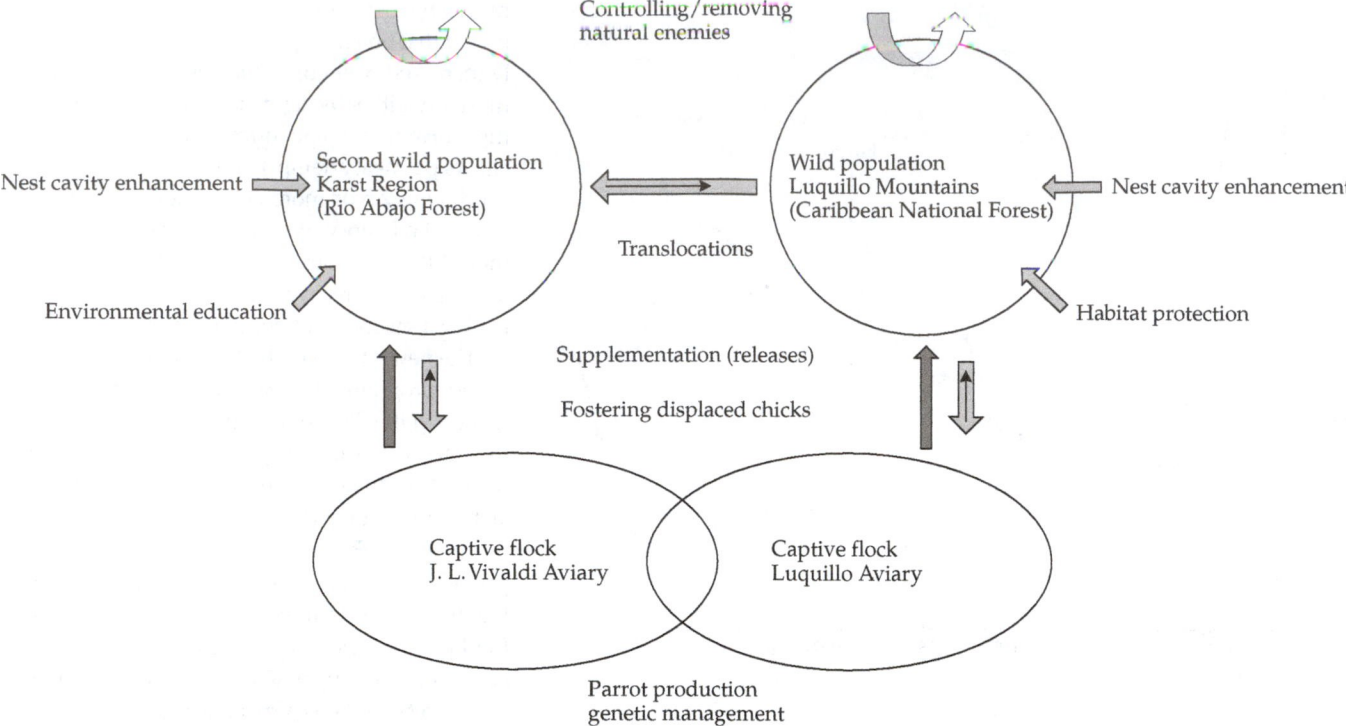

Figure C   Proposed integrated Puerto Rican Parrot recovery program for the Luquillo Mountains and Karst Region.

program for the karst; 20 birds will be released in each of three years to increase the likelihood that a "core of survivors" is established. Our fervent hope is that this program will at last allow recovery of wild Puerto Rican Parrot populations.

Funding

Puerto Rican Parrot recovery work is very expensive. On average, the U.S. Fish and Wildlife Service (USFWS) has spent U.S.$1 million per year since 1990. Over the same period, the U.S. Forest Service (USFS) has assumed the costs of maintaining the infrastructure for field operations (e.g., trails, nests, roads). The Puerto Rico Department of Natural and Environmental Resources spends about U.S.$200,000 annually on the operation of the J. L. Vivaldi Aviary, and will spend nearly U.S.$800,000 over 5 years on the reintroduction project in the Rio Abajo Forest. To reduce public costs of the recovery program we have formed partnerships with private groups to support construction of a new aviary at a lower elevation in the Caribbean National Forest, estimated at U.S.$1.5 million dollars, and land acquisition and protection in the karst region.

At current population levels, the death of a single bird constitutes a great financial as well as biological loss. The median cost of producing a fledgling parrot (in the field or in an aviary) between 1997 and 2001 was U.S.$22,105 (Engeman et al. 2003). Therefore, management practices that maximize production are justified not only biologically, but also economically. A cost–benefit analysis showed that the costs of managing multiple

predator threats were more than offset by saving a single parrot in the wild every 4 to 11 years (Engeman et al. 2003).

Social Challenges

An ever-increasing human population that continues to encroach on the natural resources of the island is by far the greatest challenge facing the Puerto Rican Parrot. Puerto Rico is an urban island. Its current population density is 425 persons per km^2, concentrated primarily in the narrow, coastal flatlands. Some trends are positive, however. Forest cover has increased to 41% (Helmer et al. 2002) as society has shifted from an agricultural- to an industrial-based economy starting in the 1950s. Protected forests in the Luquillo Mountains have increased from 5,000 ha, at the turn of the century, to over 11,000 ha at present.

Attitudes toward conservation in the island are ill defined due to the lack of education on the values and functions of natural systems. However, when some emblematic places, such as the Luquillo Mountains or the Rio Abajo forest have been threatened by development, many have responded to prevent these last large forest areas from being destroyed. Likewise, as the plight of the Puerto Rican Parrot has been widely publicized, support for its protection has increased. We hope that public support for its reintroduction in the karst region will grow and facilitate future habitat protection as the parrot population grows, and its range expands.It is more likely, however, that the strongest societal support for habitat protection will be indirect: The karst region contains the two largest aquifers

of the island, which represent a vast and clean source of water for human consumption.

The recovery of the parrot faces other serious problems. Prominent among these is the public's favorable attitude toward "exotic" pets. Puerto Rico ranks among the top locations in the U.S. supporting the import of nonnative avian species (Camacho et al. 1999), including psittacines. The potential for hybridization and competition increases as released nonnative parrots continue to expand their range. In the same vein, there is a danger that the same attitudes could encourage capture of Puerto Rican Parrots for pets or for trade, either from Luquillo or the karst region. An effective and long-term environmental education campaign, coupled with enticing arrangements for co-management of lands, will be essential components of rein-troduction efforts as the recovery program moves from the safer confines of the Caribbean National Forest to the largely privately owned forested lands in the karst region.

The recovery of the Puerto Rican Parrot is going to require continued intensive management efforts and a commitment of considerable financial resources in the near term. It is a multifaceted process that will require concerted efforts to control sources of mortality, (particularly for juvenile birds), captive propagation, habitat protection, and environmental education. Most other *Amazona* species in the West Indies are likewise threatened with extinction (Wiley et al. 2004). Therefore, lessons gained in the process of saving the Puerto Rican Parrot may be applicable to other species, amplifying the benefits of its conservation.

Summary

1. Ecological restoration is a new discipline in ecology and ecosystem studies. More than in many aspects of conservation biology, restoration and reintroduction efforts are rooted in practice, as exact site conditions, including legal requirements and the will of stakeholders, often dictate the nature of a restoration project. Nonetheless, ecological restoration relies on a wealth of ecological understanding at the genetic, population, community, ecosystem and landscape levels.

2. Ecological restoration requires teamwork. It is multidisciplinary, and works best as a bottom-up approach. Disciplines commonly contributing to restoration efforts include hydrology, geology, soil science, and animal behavior. Restoration and reintroduction require tenacity, long-term planning, creativity, and usually, lots of money.

3. To be successful, goals need to be articulated clearly at the outset of a project, and expectations tailored to not just what is wanted at the site, but what is possible. Inclusion of all stakeholders is essential to project success. Many projects require at least tolerance of neighboring communities, if not active participation in forwarding project objectives. Many projects take unexpected turns, and it is usually best for restorationists to use their creativity and allow for changes in the project to make it better.

4. Regional references are crucial in the design of ecosystem restoration. References provide better templates for "restoration" and not "replacement." Further, strong experimental design can enhance our learning from restoration efforts, as can meta-analyses of large numbers of restoration projects. We are rapidly approaching a time in which careful analyses of the trajectories of projects begun over the past few decades is yielding a great deal of information that can guide current restoration practices.

5. Animal restoration projects tend to emphasize restoration of vegetative structure. Many also involve captive breeding and reintroduction or translocation, but such efforts are only worth pursuing if the primary threats to the species are much reduced, or if habitat has been restored. Captive breeding and reintroduction are often extremely difficult, and requires a great deal of behavioral insight.

6. Marine restoration projects are not widely discussed, but a great deal of restoration efforts have been undertaken to enhance commercial fisheries.

7. In the U.S., a suite of environmental regulations, particularly the Clean Water Act (CWA), dictate a number of requirements that must be fulfilled in restoration efforts. The vast majority of restoration efforts in the U.S. are initiated due to requirements from the CWA, and to a lesser degree due to legislation, to aid game or commercial species.

8. Most restorationists find rewards in the tangible gains in improving degraded lands, or reintroducing an extirpated species. Nontraditional roles for scientists abound in the field of restoration.

Please refer to the website www.sinauer.com/groom for Suggested Readings, Web links, additional questions, and supplementary resources.

Questions for Discussion

1. Why are the distinctions between rehabilitation, reclamation, and restoration important?

2. Is restoration ethical? Why are conservation biologists concerned that restoration may be used to undermine conservation agendas? How can restoration practice be best crafted to enhance conservation?

3. What is the role of reference sites in ecological restoration? What criteria are important in selecting reference sites, and how might the selections influence a restoration project?

4. What factors limit the success of animal reintroductions? Both of the projects described in the reintroduction case studies (black-footed ferrets and Puerto Rican Parrot) are very expensive. How would you justify the expense of these projects?

5. How important is the Clean Water Act to the practice of ecological restoration in the U.S.? Why is wetland restoration so important? Do you think the provisions in the CWA for mitigating damage to wetlands make it easier to damage wetlands in the first place?

16

Sustainable Development

C. Ronald Carroll and Martha J. Groom

A sustainable society is one that ensures the health and vitality of human life and culture and of nature's capital, for present and future generations. Such a society acts to stop the activities that serve to destroy human life and culture and nature's capital, and to encourage those activities that serve to conserve what exists, restore what has been damaged, and prevent future harm.

Stephen Viederman, 1992

What Is Sustainable Development?

In the 1980s, the concept of sustainable development emerged as the means by which biodiversity and natural ecosystems would be saved while enabling humanity to continue to prosper. The concept was first promoted by the *World Conservation Strategy* (IUCN/UNEP/WWF 1980), a global conservation blueprint that grew from the United Nations Conference on the Human Environment, held in Stockholm in 1972. This report was followed by *Our Common Future*, also called the Brundtland Commission Report (World Commission on Environment and Development 1987), followed by further elaborations made for and after the next two major United Nations conferences in Rio de Janiero in 1992 and in Johannesburg in 2002. The Johannesburg Declaration on Sustainable Development and Johannesburg Plan of Implementation, like prior agreements, have been adopted by many governments and global institutions as a guide to environmentally compatible development. All of these documents promote sustainable development as a means of balancing the demands of nature and people, and to promote the welfare of both.

But what exactly is sustainable development? Does it mean the same thing to everyone? Does it hold the answer to the many conservation problems faced today and in the future? These and many other questions surround sustainable development, and, despite three decades of discussion and practice, disagreement about what sustainable development can and does achieve continues.

Sustainable development has been defined in many ways. The focus on what is to be sustained and for what purpose varies considerably from case to case, as does the means by which development will proceed (Table 16.1). Most definitions have been largely anthropocentric, focusing on human aspirations and well-being, with the natural environment providing the means by which this was to be accomplished (Robin-

TABLE 16.1 *A List of Common Concepts Included in Definitions of Sustainable Development*

What is to be sustained	What is to be developed
Nature	**People**
Biodiversity	Child survival
Ecosystems	Life expectancy
Natural habitats	Health
Evolutionary potential	Food security
Life-support	Education
Ecosystem services	Equity
Natural resources	Equal opportunity
Climate	
Ecosystem productivity	**Economy**
Culture	Living wage
	Wealth
Human communities	Productive sectors
Groups	Consumption
Places	
	Society
	Stablized population
	Housing
	Institutional capacity
	Social capital
	States
	Regions

Source: Modified from Parris and Kates 2003.

son 1993). For example, the concept of development used in the *World Conservation Strategy* emphasized that we should "satisfy human needs and improve the quality of human life," while the UN's Division for Sustainable Development defines it as "development that meets the needs of the present without compromising the ability of future generations to meet their own needs."

Particularly as it was first envisioned, sustainable development is almost exclusively an anthropocentric concept, and moreover one that promotes continued and even expanded economic prosperity. Such definitions of sustainable development are utilitarian—they perceive the environment as merely the means to an end (securing human welfare into the future), rather than having inherent good apart from human gain. The definitions do not promote biological sustainability for its own sake. While the need for improving human welfare is paramount, we hope that emerging views of sustainable development also embrace the goals of sustaining biodiversity and healthy natural ecosystems.

The Johannesburg Plan of Implementation (United Nations 2002) calls for change in consumption and produc-

tion patterns that are unsustainable, echoing sentiments voiced by ecologists, ecological economists, and conservationists for many years. The Ecological Society of America produced a "Sustainable Biosphere Initiative" in which a research agenda was proposed to help move the world toward sustainability (Lubchenco et al. 1991). In it, they defined sustainability as "management practices that will not degrade the exploited system or any adjacent system." They also recognized that a prerequisite to this vision is "consumption standards that are within the bounds of ecological possibility and to which all can aspire." We would add to this the need for humans to develop systems that allow us to live on the income from nature's capital alone, rather than on the capital itself.

Sustainable "growth" is not equivalent to sustainable "development"

There is a most critical distinction to be made between sustainable *growth* and sustainable *development*. Growth is a *quantitative* increase in the size of a system; development is a *qualitative* change in its complexity and configuration (see Essay 5.1). An economic, social, political, or biophysical system can develop without growing, and thus can be sustainable. It can also grow in size without developing or maturing; this is *not* sustainable development. Sustainable *growth* is a self-contradictory term—an oxymoron. Continued, indefinite growth on this planet or any subset of the planet is a physical impossibility. Eventually, limits of some type (space, food, waste disposal, energy) must be reached; the point at which that will happen is the only aspect open to debate. Sustainable *development*, however, focuses our attention differently. Can we make qualitative changes in complexity and configuration within existing human systems that do not place increasing *quantitative* demands on natural systems, and are in fact compatible with their continued existence? We must simultaneously address two fundamental questions, "How can we improve the quality of life for human societies through qualitative changes to our economy in ways that support healthy natural environments?" And, "How can we begin the move away from the current agenda of nonsustainable economic growth?"

If we are to attain such goals, we must first understand the patterns of human behavior and desires that brought us to this crisis in the first place (Viederman 1993). Perhaps the most important pattern is that we, especially in highly industrialized nations, treat the biosphere as if its only function is to service economic growth. We fail to recognize that there are limits to how much human waste can be assimilated by the biosphere. We have in many places exceeded the amount of pollutants that river and coastal ecosystems can tolerate before severe degradation occurs. We have exceeded the finite limits to the amount of greenhouse gases that can be put

into the atmosphere before significant global climate change occurs. Indeed, all ecosystems have limits to their resiliency. Further, there are limits to the rate at which natural environments can provide economic resources.

Too often we look for technical "fixes" rather than recognizing that we might be approaching natural limits. The examples are abundant. When fisheries decline, new, more efficient fishery gear is adopted to harvest increasingly rare and more valuable fishes. As aquifers are depleted for irrigation, we simply delay the inevitable by adopting more efficient irrigation. As petroleum sources are exhausted we adopt more efficient drilling technology to extract oil from the small remaining pockets. We do not question our acceptance of technology as the answer to all problems, despite a multitude of examples of today's problems being yesterday's solutions (Tenner 1996). This "techno-arrogance" (Meffe 1992) results in a disdain for the "natural" and a love of the "technical."

We have not distinguished between growth and development, perhaps due to our belief in technology as savior. Likewise, we fail to recognize that growth will not automatically lead to equity and justice within and among nations; that is, an "economic trickle down" effect is unlikely and unfair, and simply an excuse for a few to amass personal wealth at the expense of many. Markets are very efficient in allocating goods and services; but issues of fairness, inter-generational equity, and biotic limits to growth are typically excluded from market analyses (see Chapter 5). In large part, we may need to address the "growth myth" before we can truly fashion sustainable forms of development. As David Korten (1991) observed:

> The first step toward a sustainable human future must be to break the grip that the growth myth retains on our thinking and institutions. Growth-centered development is itself inherently unsustainable. Sustainability does not depend on ending human progress, only on abandoning the myth that erroneously equates such progress with growth.

How are sustainable development projects structured?

Sustainable development efforts have many additional names—including **integrated conservation and development projects (ICDP)**, **community-based conservation (CBC)**, community adaptation and sustainable livelihoods (CASL) projects, sustainable use, compatible use, and sustainable practices. In addition, many more sustainable alternatives to traditional practice are the focus of sustainable development projects: sustainable agriculture and sustainable forestry (see Case Studies 6.2 and 8.3), as well as sustainable fisheries and sustainable aquaculture, among others. Projects may be implemented by governments or international development agencies, or spring up from the "grassroots" of local commu-

nity organizers. The Case Studies at the end of this chapter (16.1–16.5) all describe efforts to create sustainable enterprises that promote conservation.

Sustainable development projects typically use a number of economic tools to promote sustainability, including incentives such as certification, subsidies and grants, or job creation in sustainable enterprises, and sometimes penalties, such as fines or fees for continuing unsustainable practices (Sutherland 2000; Daily and Ellison 2002). In most cases, efforts are made to exchange practices that degrade habitat and overexploit resources with those that should have lesser impacts, if not serve as truly sustainable enterprises. At times, a new, hopefully sustainable, economic activity is introduced to enhance the livelihoods of local community members. Legislation to protect key resources, and restrict or prevent damaging practices is another key tool in sustainable development. Also, legal and other policy mechanisms to increase equity, benefit sharing, and otherwise improve social conditions among local communities are integral aspects of many programs (Sutherland 2000).

Importantly, sustainable development projects often are part of a larger strategy for landscape- or ecosystem-scale conservation. For example, ICDP are usually conceptualized as including protected areas and regions of sustainable enterprise (Figure 16.1). Increasingly, there is concern that protected areas will be insufficient to conserve biodiversity (see Chapters 13 and 14), and thus efforts to identify and encourage sustainable uses of natural areas and wild species has become a part of an overall strategy for conservation.

Equally, biodiversity conservation projects are often justified as a part of larger national or international strategies to increase economic resilience, reduce disease

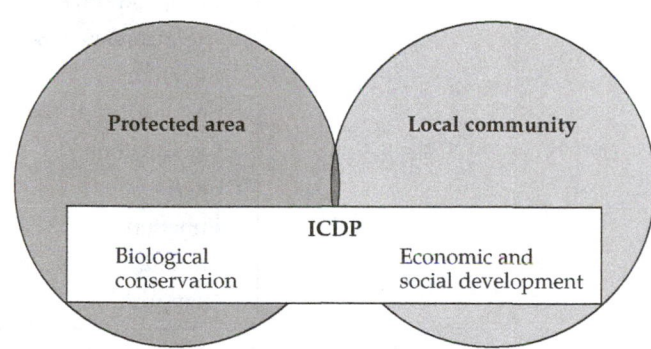

Figure 16.1   Integrated conservation and development projects (ICDPs) are meant to help conserve biodiversity, particularly in protected areas, by enhancing the benefits adjacent local communities derive from biodiversity. Project efforts target enhancing biological conservation and economic and social development that is compatible with conservation in local communities. (Modified from Alpert 1996.)

susceptibility, or decrease poverty. Here, conservation is given a higher priority because of the benefits it brings to a region, country, or global society. For example, the value of intact forests for enhancing agricultural productivity (e.g., Ricketts et al. 2004), providing clean water, or offsetting greenhouse gases (e.g., Kremen et al. 2000) may be sufficient to influence an international organization or government to encourage forest protection, or to institute reforestation programs to regain such values. Marine Protected Areas are similarly encouraged for the benefits they will supply to commercial fishing, particularly the prospect that they will enhance long-term sustainability of those fisheries.

How Successful Are Sustainable Development Projects at Conserving Biodiversity?

Sustainable development will be compatible with sustaining biodiversity depending on the intensity of use and the component of biodiversity considered. Redford and Richter (1999) observed that genetic-, population-, and species-level diversity may be particularly vulnera-

ble to increasingly intensive uses as extensive human modification typically greatly simplifies natural areas, decreasing both the number of species present at a site, and often also the genetic diversity of those species that are present (e.g., in agricultural systems). Ecosystem processes and structure, in contrast, tend to be at least partially conserved until environments are radically transformed (e.g., in urban environments). Using Noss' (1990) taxonomy of the components of biodiversity (see Essay 2.1), Redford and Richter (1999) characterized a number of different uses in terms of their ability to conserve these components (Table 16.2). The more intensive uses are largely unable to conserve most components of biodiversity.

Perhaps consideration of distinct development options side-by-side with various criteria (see Table 16.1) may best enable decision-making over the directions to take. For example, development in the Pantanal region of Brazil could take many directions from the current practices, which are dominated by ecotourism to protected areas and commercial fishing (Table 16.3; Redford and Richter 1999). The most extreme change under discussion is a project called "Hidrovia," which would involve channelizing the Paraguay river to enhance barge transport of

TABLE 16.2 *Effects of Human Alteration of Riverine[a] and Forested[b] Systems on the Components and Attributes of Biodiversity*

Biodiversity component and attribute	Human alteration			
	Built	**Cultivated**	**Managed**	**Natural**
Community/ecosystem				
Function	NC	PC	CC	CC
Structure	NC	PC	CC	CC
Composition	NC	NC	PC	CC
Population/species				
Function	NC	NC	PC	CC
Structure	NC	NC	PC	CC
Composition	NC	NC	PC	CC
Genetic				
Function	NC	NC	PC	CC
Structure	NC	NC	PC	CC
Composition	NC	NC	PC	CC

Note: NC, not conserved; PC, partially conserved; CC, completely conserved.

Source: Modified from Redford and Richter 1999.

[a]For riverine systems, *built* means heavy dam alteration and channelization; *cultivated* means heavy dam alteration and flood disturbance but with original channel; *managed* means water diversion and natural channel; *natural* means a free-flowing and natural channel.

[b]For forested systems, *built* means converted to pasture; *cultivated* means fire suppression and heavy management of natural forest; *managed* means selective logging and hunting; *natural* means large expanses of natural forest.

TABLE 16.3 *Current and Potential Resource Uses of the Pantanal, Brazil, and Their Effects on the Components and Attributes of Biodiversity*

Biodiversity component and attribute	Pantanal management options				
	Channelize (Hidrovia)	Massive landscape conversion (e.g., agriculture)	Increase fishing and hunting pressure	Current conditions (ecotourism and commercial fishing)	Eliminate commercial fishing
Community/ecosystem					
Function	NC	PC	CC	CC	CC
Structure	NC	PC	CC	CC	CC
Composition	NC	PC	CC	CC	CC
Population/species					
Function	NC	NC	PC	CC	CC
Structure	NC	NC	PC	PC	CC
Composition	NC	NC	PC	PC	CC
Genetic					
Function	NC	NC	PC	PC	CC
Structure	NC	NC	PC	PC	CC
Composition	NC	NC	PC	PC	CC

Note: NC, not conserved; PC, partially conserved; CC, completely conserved.
Source: Modified from Redford and Richter 1999.

goods from the interior of Brazil to international ports, and the creation of dams for agricultural development. These kinds of projects have been favored by international development agencies for the economic growth they will bring to Brazil and neighboring countries. However, such extreme alteration of this major river system would forever destroy the Pantanal ecosystem as it exists today, and extensive agricultural development would also greatly change its character and reduce its functionality. And certainly any development that increases human population sizes in the region is likely to result in increased hunting and fishing pressures, species introductions, and increased degradation from agricultural and ranching activities (Redford and Richter 1999).

More generally, there appears to be a strong tradeoff between the degree to which components of biodiversity can be conserved in a sustainable development proj-

ect, and the number of people whose livelihoods can be supported (Figure 16.2). While these tradeoffs in scale and sustainability are genuine, of course there are exceptions, as well as gradients within this schematic. At each scale of human population density, practices can be more or less sustainable. For example, urban environments are viewed as incompatible with biodiversity conservation, yet the ways in which urban environments are managed has enormous impact on surrounding areas used by that urban center (e.g., watersheds, agricultural lands). Thus, sustainable development will in part result from the creation of more sustainable urban environments (see Case Study 18.2).

Reports on the success of sustainable development initiatives are varied, but many point to the problems associated with slips in scale (see Figure 16.2). A program that was sustainable with a relatively small population or

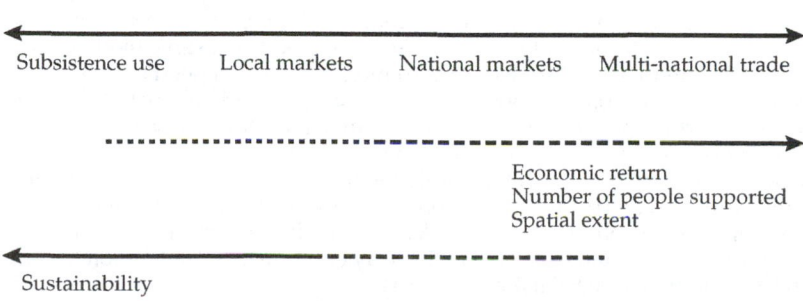

Figure 16.2 As the scale of use goes from subsistence use to increasingly larger markets, the economic returns, number of people who can be supported, and spatial scale all increase. However, the sustainability of these efforts for conserving biodiversity may generally decrease. How do we reconcile the need to provide means for sustainable livelihoods to all people with sustainability for biodiversity?

level of resource exploitation becomes unsustainable as more people move into the area, or as profitability increases and adds to incentives to overexploit and/or to use protected areas for production (e.g., coffee farmers in Indonesia convert more land from protected areas as coffee prices increase: [O'Brien and Kinnaird 2003]; expansion of aquaculture of fish-eating species leads to a net loss for wild fish species [Naylor et al. 2000]). Thus, many worry that inevitably, all sustainable development projects will grow to a size at which they begin to fail in their conservation mission. However, others argue that appropriate social supports can counteract this tendency.

A central assumption in many sustainable development projects is that when local communities receive some direct benefit from an enterprise that depends on managed or wild natural areas, they will have the incentive to conserve those areas, and will do so. This assumption, or hypothesis about human behavior, will only be met if several related conditions are met (Figure 16.3; Salafsky et al. 2001). One is that the communities must be able to reduce threats to biodiversity themselves—that is, they must have some control over the area, and can enforce policies to reduce threats through their own activities. Further, new sustainable practices either need to be more profitable than traditional ones, or perceived to add other benefits to compensate for not being as profitable to each household.

Salafsky et al. (2001) analyzed 39 efforts at community-based conservation in Asia and the Pacific to test these hypotheses, and to better understand what conditions led to coupled conservation and development. Their results showed that sustainable enterprise strategies advanced conservation goals at some, but not all sites. Many of the projects were not extremely profitable, and might require subsidies to work, if they can succeed at all when more lucrative alternatives are prevalent. Although few were profitable, successful projects were associated strongly with non-cash benefits, particularly the development of high community confidence and close relationships with the project staff. This led Salafsky et al. (2001) to conclude that education and development of trust among partners in a CBC project is an alternate path to conservation success in these ventures (see Figure 16.3, path E). Local ownership and management of the sustainable enterprise also was associated with stronger conservation outcomes. Finally, conservation threats posed by people or factors outside the community were more effectively countered than those within the community. This may be because it is more difficult to counter community members, as this increases conflict within the community. From their analyses, Salafsky et al. (2001) were able to develop a table of indicators for when a community-based, sustainable enterprise project would and would not be likely to foster conservation goals (Table 16.4).

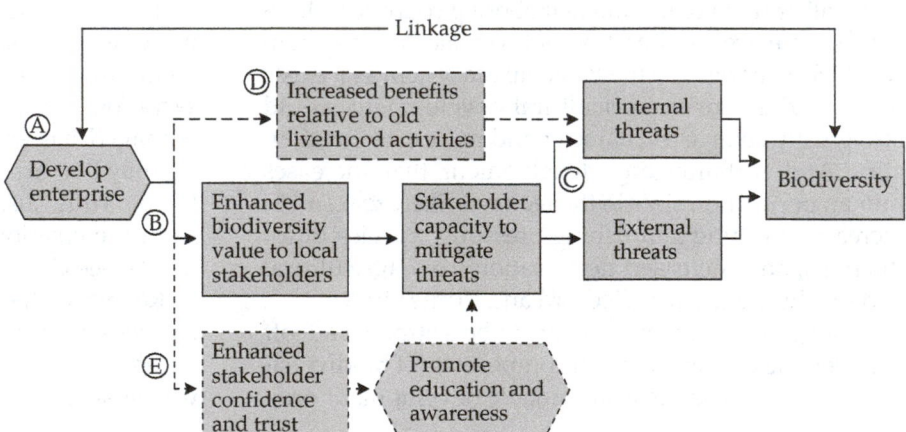

Figure 16.3 A schematic of the logic behind community-based, enterprise-based conservation projects that seek to enhance benefits from developing an enterprise with the expectation that the local community will be more motivated to conserve biodiversity that supports that enterprise. To be effective, the enterprise must depend on biodiversity to succeed or the enterprise will not enhance the value of biodiversity to local people (path A to B). Further, people in the community must have the ability to control threats to biodiversity to be effective in conserving it (path B to C), assuming that community members can control internal threats arising from their community. Alternately, the benefits from the enterprise may be more than unsustainable alternatives and thus provide a motive for conservation (path A to D). Finally, projects may enhance community understanding and trust, and in this way enhance their motivation to conserve (path E). (Modified from Salafsky et al. 2001).

TABLE 16.4   *Conditions at a Site Influence the Feasibility of a Community-Based Sustainable Enterprise Strategy*

Factor	Conditions at site			
	Unfavorable	**Not ideal**	**More favorable**	**Ideal**
Enterprise				
Potential profitability	< Variable costs	< Fixed costs	> Fixed costs	Costs + profit
Market demand	None	Low	High	Medium
Infrastructure	Poor	Marginal	Okay	Good
Local enterprise skills	None	Few	Some	Lots
Complexity	Extreme	High	Medium	Low
Linkage	None	Low	Medium	High
Benefits				
Cash benefits	None	Few	High	Moderate
Noncash benefits	None	Few	Some	High
Timing	Long wait	Unknown	Short	Immediate
Distribution	Very wide	Elites only	Limited	Targeted
Stakeholder				
Stakeholder group	Not present	Very new	Present	Established
Leadership	None	Weak	Strong	Balanced
Resource access	None	Ill-defined	Some	Full
Enforceability	None	Limited	Some	Strong
Stakeholder	Low	Minimal	Moderate	Complete
Conflict	Constant	Frequent	Occasional	Rare
Threat source	All internal	Most internal	Most external	All external
Other				
Chaos	Constant	Frequent	Some	Unlikely
Project alliance	Unwieldy	None	Strategic	Experienced
Implication	Forget it	Think hard	Maybe	Go for it

Source: Salafsky et al. 2001.

Mexico has roughly 300–480 community forest enterprises (CFEs), which are community-owned and managed forest lands for producing commercial timber, and other saleable goods. These CFEs are de facto examples of community-based, sustainable development projects, and early studies of these projects suggest that many are successful in improving economic and social benefits and their just distribution, and in creating sustainable forest management practices (Bray et al. 2003). Further, some CFEs have created protected forest areas, and many embrace environmental standards that not only promote sustainable forestry, but biodiversity conservation as well. Certainly these, and other examples discussed in Case Study 8.3 provide evidence of the potential for sustainable development in tropical forests.

Clearly, the success of sustainable development projects in biodiversity conservation will be contingent on a wide variety of conditions. New analyses of multiple projects provide some of the best prospects for deepening our understanding of the circumstances that will facilitate conservation and those that will preclude it. Despite the number of projects that have been established, and the great interest in these projects, it is only now feasible to begin learning significant lessons from them.

How Can We Best Promote Sustainability?

Given that clear guidance for how and when to pursue sustainable development projects to achieve conservation is still forthcoming, how can we best proceed? To attain sustainability for the good of humanity and the natural world, a group of conservation and development practitioners met in 1990 and drafted a set of *ecological principles of sustainability* to promote sustainability (reported by Viederman 1993):

1. Nature should be understood to be an irreplaceable source of knowledge, from which we can learn potential solutions to some of our problems.

2. We should understand that issues of environmental deterioration and human oppression and violence are linked in analysis and action. Gender and racial oppression, and efforts to dominate nature have a common root.

3. Humility must guide our actions; good stewardship begins with restraint.

4. We must appreciate the importance of "proper scale." Place and locality are the foundation for all durable economies, and must be the starting point of action to deal with our problems. Solutions are local and scale-dependent.

5. Sufficiency must replace economic efficiency. Earth's resources are finite, and this fact must be accepted in order for humanity to adopt limits. Living within our needs on a planetary scale does not mean a life of sacrifice, but of greater fulfillment. We must distinguish between "needs" and "wants."

6. Community is essential for survival. The "global community" should reflect and encourage diversity while being interdependent.

7. Biological and cultural diversity must be preserved, defended, and encouraged.

Will we adopt these principles of sustainable development? There are some reasons for guarded optimism. Major fisheries are managed through guidance from empirically rich ecological models. Marine protected areas that replenish fish stocks are becoming widespread. Sources of water supply, such as headwaters of rivers and aquifers, are better protected, surface waters are generally cleaner and more often recycled after human uses, and aquifers more frequently recharged. A global dialogue is leading to innovative ways to reduce greenhouse gases and increase the use of renewable sources of energy. Ensuring environment sustainability is one of the eight Millenium Development Goals agreed upon by the United Nations.

Some of the case studies at the end of this chapter demonstrate various aspects of and opportunities for sustainable development, and the many problems encountered in implementing projects aimed at simultaneously improving human economies and conserving species and ecosystems. These and similar approaches around the globe seem to offer some promise for conservation of biodiversity and natural systems, but not everyone is impressed by their potential. Many skeptics feel that sustainable development is a camouflaged, "politically correct" approach to further and perpetual economic growth and continued, but less overt, environmental destruction under the guise of conservation. For example Clark (1995) wrote that:

> "Sustainable development of the WECD [World Commission on Environment and Development] variety means business-as-usual. This satisfies business interests, government officialdom around the world, powerful international institutions such as the World Bank, the International Monetary Fund, UN Food and Agricultural Organization, and national elites whose international assets are relatively safe from the erratic fluctuations of their own economies. . . [The World Bank] contends that growth is the key to eradicating poverty and protecting the environment."

Some argue that there has never been a case of successful sustainability over the long term; the resource has inevitably been overexploited. This is particularly apparent in commercial fishing, in which decline of commercial stocks has been extensive and rapid over the past 50 years (see Chapter 8). As mentioned earlier, there are some signs of change, suggesting that at least some fisheries may become sustainable. But, it has required fisheries to turn away from maximizing yields and instead look at yields as a kind of biological interest that is earned from the capital represented by protected stock populations.

Ludwig et al. (1993) offer several reasons for why sustainability has been difficult to achieve. First, wealth or the prospect of wealth generates political and social power that is used to promote unlimited exploitation of resources. But there are also major problems created by lack of knowledge and the complexity of the resource. Scientific understanding and consensus is hampered by the lack of controls and replicates, so that each new problem involves learning about a new system. The complexity of the underlying biological and physical systems precludes a reductionist approach to management. Optimum levels of exploitation must be determined by trial and error. Large levels of natural variability mask the effects of overexploitation. Initial overexploitation is not detectable until it is severe and often irreversible.

Beyond the scientific constraints to managing resources sustainably, good communication between participants—natural scientists who can provide technical guidance, decision-makers who put projects in motion, and local stakeholders who are the principal actors and affected parties in any project—is also essential (Cash et al. 2003). To design and adaptively manage a sustainable development project typically requires some translation of ecological studies or even scientists' best guesses into direct applications appropriate to the exact context of the project. If issues of immediate importance to a sustainable development project are not addressed by the research of scientists, project needs will go unmet (see Chapter 17). At present, many sustainable development projects suffer from poor translation or communication.

Further, organizations and governments more often seek to maximize economic growth rather than to conserve for the future, or to contribute to creating more equitable distribution of resources. Often natural resources are seen as opportunities for rapid economic gains, and "politicians and governments tend to ally themselves with special interest groups in order to facilitate the exploitation." (Ludwig et al. 1993).

We do not offer these counter-perspectives to discourage a drive toward sustainable development or to denigrate the concept. Indeed, a truly sustainable global society, combined with a stable and secure global human population, seems the only real hope for avoiding massive biological extinctions and ecosystem collapses, and continued or worsening human suffering. Clearly, sustainable development is essential to human welfare, and the need for extensive and immediate effort to change the living conditions for the poorest people of the world is paramount. Yet, we caution that blind acceptance of the sustainable development approach can be dangerous to conservation interests because there are so many potential problems and pitfalls, and because the concept can be abused to the benefit of further growth. Additionally, enduring solutions to abject poverty can only come through sustainable forms of development, although it is certainly arguable that what will be sustainable for human societies may be well below the point that is sustainable for a majority of species. Healthy skepticism, combined with honest attempts to balance the needs of natural systems against actual, long-term human needs, seems the most sensible approach.

For those who claim that sustainability is impossible, we must ask, "What are the alternatives?" Most often, increased regulatory control over resources is given as the alternative. Certainly, legal proscriptions are necessary in some cases; for example, closing a fishery when its collapse seems apparent, to protect endangered species or protection of a critical environment, such as a national park. However, increasing top-down regulatory protection of the environment may often be counterproductive. Excluding local people from traditional resources by erecting park boundaries may lead to hostile acts such as poaching or arson that undermine the rationale for the protection. More generally, such top-down approaches serve to further alienate people from nature and make environmental protection more vulnerable to politics. We argue that it is important to find the right mix of regulatory protection and sustainability initiatives. Neither alone is likely to both protect the environment and improve quality of life.

Sustainable development as a goal certainly is possible, and may be the best hope for conservation of global biodiversity. However, three things must change radically if this goal is to be achieved. First, the value systems that lie at the core of the human fabric and drive our collective behaviors need to change drastically. In wealthy, developed countries, long-term global sustainability must replace short-term personal gain as the primary human motivation. Unless people value the environmental services provided by natural ecosystems and understand the relationship between ecosystem protection and their own long-term economic security, they will not reduce excessive resource exploitation and appreciate their natural heritage. Second, the growth-oriented economic systems that drive human existence must be replaced by steady-state economic systems that accept natural limits to our artificial economies (Czech 2000; see Essay 5.1). Finally, human population growth must slow, stop, and eventually reverse; toward this goal, sustainable development projects must contain internal incentives to limit population growth.

CASE STUDY 16.1

Ecotourism and Biodiversity Conservation

Eileen Gutiérrez, Conservation International

Human society has engaged in tourism-related activities throughout its history. Tourism encompasses a range of travel-related activities from bungee jumping on the Zambezi, to attending a business conference in the Caribbean, to hiking in a national park. Tourism is one of the largest industries in the world, involving over 700 million travelers and generating one out of every 12.3 jobs in the world in 2003 (World Travel and Tourism Council 2004), with continued growth expected. This tremendous growth, however, contributes to environmental degradation, especially in high-traffic locations in the Caribbean, Mexico, Brazil, the Mediterranean, Thailand, and China. International tourism also creates more opportunities for the introduction of new invasive species and diseases. In the past three decades, a growing trend in ecotourism and sustainable tourism development has defined an opportunity for tourism to contribute to biodiversi-

Figure A Resort towns often grow rapidly, and encroach on nearby natural areas. Here in Soliman Plage, a resort city in Tunisia, one can see the spread of the urban environment into wetlands and other natural habitats, as well as extensive coastal modification. (Photograph © Georg Gerster/ Photo Researchers, Inc.)

ty conservation efforts if it is carefully planned, implemented, and monitored.

The Evolution of Sustainable Tourism and Ecotourism Concepts

Population growth and socioeconomic trends, such as the post-World War II "baby boom" generation, increases in average family incomes, more leisure time for aging populations, and significant transportation and communication technology advances allowed tourism to grow rapidly from the early 1950s to the early 1980s. Large-scale "all-inclusive" resorts, replicating the seaside resorts of the coasts of England, France, and the United States, began to develop in the Mediterranean, the Caribbean, and Pacific Islands. Many of these developments caused extensive harm to the surrounding landscapes and marine biodiversity, which were a primary attraction to the area (Figure A).

Along with mass tourism to beautiful beaches, market demand for nature-based tourism grew. Places like Nepal saw visitation grow from less than 10,000 before 1965, to more than 240,000 in 1987. Trekkers in such high numbers damaged the fragile high-mountain vegetation and created a major solid waste management issue (Honey 1999). Today some 40% of all adults in the U.S. have visited a national park. They go to experience nature (92%), for the educational benefit (90%), to experience culture and history (89%), and to spend time with family (89%) (Travel Industry Association of America 2004).

Tourism has a cycle of development and related impacts—from initial phases to maturity to ultimate decline of a favored destination, and the negative social, cultural, economic, and environmental impacts associated with the later phases of this cycle (Gartner 1996).

Social Impacts

The presence of tourists and the change associated with tourism can cause disruption of social structures within a host community. Disparities in wealth between the tourists and hosts can promote crime; petty theft and child prostitution are problems clearly linked to tourism in many regions. Control of land and competition for resources by tourism interests, whether managed by the government or private sector, can cause social disruptions often linked to negative economic impacts. In much of the eastern Caribbean, large resorts have often attempted to fence off beaches and marine areas from local residents, causing conflicts over both recreational and fishing resources. The Maasai peoples of East Africa continue to have conflicts with park management over their rights to conduct traditional cattle migrations through many tourist destinations, such as Amboseli and Serengeti National Parks in Kenya and Tanzania.

Cultural Impacts

Tourism can often lead to negative cultural impacts. Host community members may alter their traditions (e.g., traditional clothing, food choices, cultural activities) to accommodate or emulate tourists. Younger generations tend to be most influenced. What were once revered cultural icons and representations, such as dances or songs may be produced in mass quantities and sold cheaply at tourist markets, reducing their value both as a cultural traditions and as tourist products (Figure B).

Economic Impacts

Although not as well documented as social and cultural impacts, negative economic impacts can be extensive. Destinations can become overly dependent on tourism, leaving local

Figure B Ecotourism can degrade cultural traditions by motivating people to compromise their practices or dignity for tourist dollars. Here a young boy seeks coins for exhibiting a monkey dressed up and trained to walk like a human in a public square. (Photograph © Pete Oxford/naturepl.com.)

economies extremely vulnerable to external factors such as war, natural disasters, and shifts in market demand. In rural, and in many highly biodiverse areas, local communities are often unable to fully benefit from tourism development. With limited access to financial resources, education, and training, residents may not have job skills that transfer to tourism, so tourism investments or profits may be used to purchase goods and supplies or pay salaries outside of the local economy. This phenomenon hurts the ability of the local economy and local community to develop and keep pace with tourism development. Simultaneously, tourism demand for local goods and real estate can often drive up prices and raise property taxes, particularly in areas where availability of those goods and land is limited. Often tourism development projects are supported by governments through major public sector investments in roads, electricity, water, and communication systems, and even park management, which often outweigh tax revenues or local employment benefits.

Environmental Impacts

Creation of infrastructure to support tourism (e.g., hotels) is where the most significant impacts on fragile habitats and wildlife may occur. This can lead to modification of ecosystems—especially in sensitive habitats such as mangrove forests and wetlands. Road-building to create access to natural areas opens pristine areas to even more development and deforestation. Major seasonal migrations (e.g., residents inundating nature reserves in China during major holidays, or religious pil-

grimages in Sri Lanka to national parks; see also Case Study 14.1) can overtax sewage and waste management systems, spike demand for scarce resources, such as water and gas, leading to air, ground water, and environmental pollution. Deleterious effects of tourism on animals in National Parks have been noticed in many places. Cheetah reproduction levels are depressed in heavily visited African game parks, and mountain gorillas in Uganda have died as result of exposure to the common cold brought by tourists (Goodwin et al. 1998).

Ecotourism and Sustainable Tourism Development

To address the wide range of social, cultural, economic, and environmental impacts of tourism in natural areas, efforts to foster "sustainable tourism" were initiated. Sustainable tourism focuses on improved planning and implementation to increase local community participation and to reduce overcrowding and environmental degradation (Hawkins et al. 1980). The overarching goal is to ensure the value of a region is preserved for future generations. **Ecotourism**, defined broadly as "responsible travel to natural areas that conserves the environment and improves the welfare of local peoples" (Lindberg and Hawkins 1993), became a focus among organizations and practitioners looking to integrate conservation and development efforts. Ecotourism embraces the following specific principles, which distinguish it from the wider concept of sustainable tourism:

- Contributes actively to the conservation of natural and cultural heritage
- Includes local and indigenous communities in its planning, development, and operation, contributing to their well-being
- Interprets the natural and cultural heritage of the destination to visitor(s)
- Lends itself better to independent travelers, as well as to organized tours for small size groups

The terms "sustainable tourism," "nature-based tourism" and "ecotourism" are used interchangeably. We use ecotourism to mean a form of nature-based tourism that adheres to a set of standards that help ensure environmental protection, cultural integrity, and community benefit. The ecotourism industry promotes standards through voluntary mechanisms such as certification, which audits businesses for performance in the categories just described, as well as in use of environmentally-sustainable practices (e.g., minimizing energy use), making direct contributions to conservation of local natural areas, and responsible marketing.

Examples of certification efforts include global programs such as Green Globe 21, the National Ecotourism Accreditation Program in Australia, Costa Rica's Certificate of Sustainable Tourism, and Smart Voyager in the Galápagos. Most programs, however, face challenges in determining adequate indicators of sustainable practices, as well as high costs associated with training auditors and auditing business, and the

general lack of consumer awareness of these programs that could create actual demand for certified products.

Linking Ecotourism and Biodiversity Conservation

The potential of ecotourism to contribute to biodiversity conservation is widely accepted. Globally, governments, NGOs, civil society, and private sector organizations have been working to leverage tourism for biodiversity conservation and community benefits. Parks and other protected area management organizations have been particularly interested in how tourism can cover budget needs as well as provide economic alternatives to local communities. Ecotourism can support conservation efforts in five main areas (Brandon 1996), which are reviewed next.

Financing for biodiversity conservation

Park entrance fees and payments from concessionaires can fund park ranger salaries, vehicles, and equipment to ensure monitoring of protected areas. For example, Wanglang Nature Reserve in Sichuan China covers some 30% of its operating budgets from tourism revenues. All visitors are capped to 30,000 annually to avoid over-stressing the park. Governments can also set aside revenues from tourism business taxes or airport taxes for management of protected areas, and special use fees charged to visitors can provide funding for conservation biology research. For example, popular parks such as Kruger National Park, part of the South Africa National Parks (SANP) systems, can cover all operating budgets with revenues from tourism. In 2003, SANP earned U.S.$2.4 million from visitor operations after subtracting costs for managing the parks. In some countries, revenue from highly popular parks can subsidize the operational costs of less-used but biologically important protected areas. In Ecuador, for example, revenue from the Galápagos Islands is a major source of operational funds for other lesser-known protected areas.

Alternative livelihoods for local people

Many areas of high biodiversity coincide with some of the most populated areas on the planet. Often these areas experience the highest levels of poverty with many people earning less than U.S.$1 or $2 a day. More often than not, conservation efforts have to address human welfare issues to be successful. Where viable, tourism-related employment can provide local people with an economic alternative to more destructive practices than agriculture, poaching for bush meat, or dynamite fishing for aquarium species.

One example of a project that is working to achieve good returns for the local people and promote conservation is the Chalalán Ecolodge, a joint ecotourism initiative of the rainforest community of San José de Uchupiamonas and Conservation International (CI) in Bolivia. Created in 1995 by San José villagers with a grant from the Inter-American Development Bank, the ecolodge concept was developed to provide much-needed alternatives to logging and poaching to this isolated community. CI assisted with development and design of their lodge and trained villagers in marketing and management, house keeping, food preparation, and tour guiding skills. Approximately 74 families receive regular direct economic benefits from employment and management of the ecolodge, 14 of whom are former hunters-turned-tour guides. In 2001, the lodge made its first profits enabling the community to improve primary school education.

Constituency and stewardship for biodiversity

By its very reliance on natural attractions as a resource, ecotourism can help enhance stewardship of biodiversity. Tourism stakeholders in local communities have organized to become activists against threats to natural areas. Often, however, environmental education is needed for local communities engaged in tourism to create an effective constituency for biodiversity conservation.

One promising example is the Katmandu Environmental Education Project (KEEP), which teaches guides, lodge keepers and trekkers how to minimize the impact of heavy foot traffic on very fragile montane ecosystems. The program promotes the basics of low-impact tourism in rural areas: alternatives to wood burning for cooking and heating (to slow deforestation), proper sanitation, community garbage clean-ups, and tree planting.

Providing an economic justification for protecting areas

In much of the world, tourism's demand for open space and recreational areas creates an economic justification for placing areas under protection. For example, to address economic decline, unemployment, and poverty in central Ghana, the Government of Ghana worked to conserve the area's cultural and natural resources and develop them into tourism destinations. As part of these efforts, the government created a 360 km² national park and with aid from CI, developed a major nature-based attraction—the Kakum Canopy Walkway in an adjacent secondary forest (Figure C). Today the walkway receives over 90,000 visitors per year, provides employment for 5000 people, and over U.S.$25 million in investment in ancillary services in the region.

Creating an impetus for private conservation

Customer satisfaction is often reliant on scenic, pristine natural surroundings and opportunities for wildlife viewing. This demand can encourage private investors to conserve areas around their operations to safeguard their ability to satisfy tourists. For example, Rainforest Expeditions, a tour operator based in Peru, started their ventures in the Amazon in 1989 with the Tambopata Research Center, a biological field station where tourists could experience pristine rainforest alongside working scientists. Realizing an opportunity to host larger numbers of tourists, Rainforest Expeditions entered into a joint venture with the local Ese'eja community to build Posada Amazonas, a

Figure C The Kakum Canopy Walkway has become a very popular ecotourist destination in Ghana. (Photograph © Oxford Scientific/photolibrary/Ariadne Van Zandbergen.)

24-room Ecolodge. The unique partnership splits management evenly between the company and the locals, with most staff positions and 60% of profits going to the community. Together, they also created bylaws that resulted in a 2000 ha reserve around the lodge as part of the efforts to manage the area and to safeguard nesting areas of the endangered Harpy Eagle (*Harpy harpyja*) and Scarlet Macaw (*Ara macao*).

Planning Ecotourism Development for Biodiversity Conservation

Successfully leveraging of tourism for conservation efforts requires complex planning, implementation, and management. For example, the journey of the tourist from home to a destination would have started with using a communication network for travel reservations, transportation via airports and roads paid for with public funding, accommodations and meals provided by the private sector, trained guides, hotel and restaurant staff, parks and protected areas managed to ensure possibilities of wildlife viewing, and a host community open to receiving tourists.

Ecotourism planning as it occurs today is typically not comprehensive enough to address the complexities of sustainable or ecotourism development. Often it is led by private sector interests soliciting local government support. The focus is on minimum investment and quick returns. Unfortunately, basic feasibility studies of market demand, and site and infrastructure needs neglect environmental, ecological, and socioeconomic considerations. In many parks and protected areas, investments often prioritize attracting tourists rather than effectively managing visitation for biodiversity conservation. Planning for ecotourism as part of a biodiversity conservation strategy requires a comprehensive approach, which involves the following three key issues:

1. **Stakeholder participation**   A major factor in planning for tourism development is the active support of the local community. In addition, private sector and nongovernment interests will play a critical role. Using multi-stakeholder processes at the outset of planning for tourism is necessary to address the complexities of shaping tourism to meet both biodiversity and development needs. For example, involving community members will help to address concerns and identify solutions to issues such as land tenure, natural resource management, and livelihood relationships to biodiversity, poverty, and indigenous people's rights.

2. **Government leadership**   Government can provide the policy framework and regulations necessary to ensure well-planned tourism development. For example, developing proper land use and zoning for critical ecosystems, habitats, and species or instituting and enforcing building codes and ordinances that may address siting and design issues, waste and sewage, water and energy use, and safety. Environmental impact assessments (EIAs) are often a legal requirement for tourism development; however, the capacity to monitor the quality of EIAs is limited. Government can also direct tourism tax revenue streams toward biodiversity conservation efforts and can provide planning guidance and resources for implementation.

3. **Assessments**   Planning for ecotourism that actively contributes to biodiversity conservation and benefits local people will necessarily rely on an informed process. Assessments needed to inform the process should include the following:

   • **Attractions inventory**   Attractions are the magnets that draw visitors to the destination. Potential attractions should be assessed for their draw, role in biodiversity conservation, current uses and their compatibility of sustainable tourism, and the ability to control impacts.

   • **Site and infrastructure analysis**   Accessibility, communications, transportation, and waste management are among the services essential to tourism, all of

which must be assessed for how they may support tourism and be affected by it.

- **Market demand**   Tourism trends and visitor profiles determine potential market demand in a given destination. Potential demand needs to be assessed, and a strategy for meeting and managing this demand should be developed.

- **Supply and competitiveness**   Other destinations, whether regionally or on the other side of the globe often offer similar products and services, and are in direct competition for the same travel markets. An assessment of the competition in the region can help a given destination assess how to attract visitors in light of this competition.

- **Available capacity**   Tourism is a people-oriented business and depends on quality service from trained managers and employees. Understanding the potential human resource base of a destination is critical to determining to what degree a community can participate in sustainable tourism development.

- **Socioeconomic linkages to biodiversity conservation**   It is important to assess the potential costs and benefits of tourism in terms of the social, cultural, and economic dynamics that influence tourism development and its potential to benefit biodiversity conservation. Here the process should gauge the community's attitudes and expectations, and the socioeconomic situation in relation to biodiversity conservation, as well as tourism as an economic alternative and their potential to address conservation issues.

- **Tourism impacts**   Estimates of the potential impacts, both negative and positive, of tourism on biodiversity and the environment are needed. Potential impacts on species, functioning ecosystems, and identified important biodiversity areas, as well as effects on water and energy sources and waste management systems need to be assessed.

Synthesizing the information from a tourism assessment is often an iterative process. Assessment results and recommendations may be reviewed through workshops and other forums by obtaining feedback from stakeholders and revising recommendations as necessary. Assessments can then inform the larger preparation of tourism-related planning such as:

- Park or protected area management plans
- Strategic tourism development plans
- Community development plans
- Tourism enterprise development plans

Ultimately, incorporating considerations from larger planning efforts into actual tourism enterprise design and implementation plans is critical. This is only possible with sufficient resources (expertise, funding, and an overarching philosophy of sutainability). In addition, management and monitoring will need to be adaptive, with constant review and adjustment if the long-term success of a project is to be assured.

Putting Ecotourism and Conservation in Perspective

The extent to which ecotourism is being developed sustainably and how much it is contributing to conservation goals is an open question. Certainly, only a small fraction of tourism businesses world wide are operating under sustainable or ecotourism principles. However, the number of tourism projects being developed to contribute to conservation grows every year as more governments, NGOs, communities, and private sector actors incorporate sustainability and ecotourism principles into their planning.

There are hundreds of case studies and publications based on secondary research that provide evidence that tourism is making positive contributions to biodiversity conservation efforts today. Although extremely valuable, much of this research has been neither scientific nor systematic. Without comparative analyses, gauging the net contribution of tourism to biodiversity conservation will be difficult. Much of the challenge lies in the capacity and resources available to efficiently monitor tourism's actual impact on biodiversity conservation efforts.

The potential for ecotourism as a tool for sustainable development is yet to be fully explored and understood. Clearly, it is one of several viable options for linking community benefits to biodiversity conservation and providing financing and justification for protecting Earth's biodiversity. However, rarely can it offer a whole solution, as there is a danger of incurring net negative impacts on surrounding environment and host communities and limits to the number of people it can sustainably support. When considering tourism as part of a biodiversity conservation strategy, it is important to recognize that it is a complex multifaceted and multi-sectoral endeavor, requiring careful planning, management, and monitoring in order to be successful.

CASE STUDY 16.2

The Chocó-Andean Corridor
Sustaining Livelihoods and Protecting Biodiversity

Rebeca Justicia, Maquipucuna Foundation, Ecuador, and C. Ronald Carroll, University of Georgia

When people believe that tropical forests are an impediment to their livelihoods or are only a source of short-term returns, the destruction of tropical biodiversity is an inevitable consequence. Therefore, creating economic alternatives that do not degrade forest ecosystems, have long-term sustainability, and, ideally, even find value in preserving the forest is a necessary first condition for conserving biodiversity. Helping communities to identify and implement these sustainable alternatives is the second necessary condition. There are also critically important ecological conditions that must be met. Degraded habitats may need to be restored and locally extinct species may need to be reintroduced. And above all, local communities and national agencies must share this common path to conservation and sustainable development. It was in this spirit that the Chocó-Andean Corridor was initiated.

The Chocó-Andean Corridor is located in northwestern Ecuador and connects two regions of global biodiversity significance, the lowland Chocó and tropical Andean forests. The spatial scale of the project is large: approximately 1,000,000 ha, reaching from sea-level mangroves through rainforest to paramo grasslands at 4915 m, at the crest of the western Andes Mountains (Figure A). The Corridor initiative was initiated in 1992 by the Ecuadorian nongovernmental environmental organization, Maquipucuna Foundation (MF) in collaboration with research scientists from the University of Georgia's Institute of Ecology. The conservation goal was to create continuous habitat corridors from the Andean summit to the sea, thereby allowing animals and plants the pathways to migrate as climate and resources change. The social goal was to help local communities meet their economic and social objectives in ways that protect the environment.

Although the scale is daunting, the corridor is feasible because major protected areas already occur throughout the corridor. The Cotocachi-Cayapas Reserve (202,053 ha), the Los Cedros Reserve (6000 ha), the Maquipucuna Reserve (6000 ha), and the Cayapas-Mataje Reserve (51,694 ha) are some of the more notable protected areas. However, outside the protected areas, the landscape is dotted with small and medium communities, small farms, and large oil palm and heart of palm plantations. From the beginning, environmental education, community engagement, and a commitment to help develop sustainable livelihoods were key components of the project. Many external and Ecuadorian contributors have assisted the project. In particular, the Butler Foundation, the GEF/World Bank, the Program for

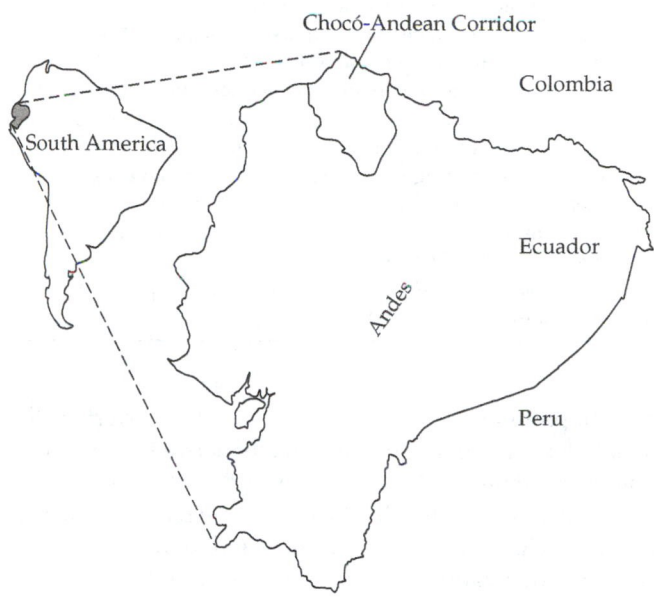

Figure A The Chocó-Andean Corridor encompasses four unique phytogeographical regions: the northwestern coast (0–900 m), the northwestern wet pre-montane forest belt or southern end of the Chocó (900–2000 m), the northwestern montane belt (2000–3000 m), and Paramo (>3000 m). (Modified from Borchsenius 1997.)

Andean Native Forests (PROBONA), and the Atlanta Botanical Garden have recognized the value of the project and have provided key support from the beginning. The Corridor initiative has been adopted by other organizations like the British Rainforest Concern as their working guideline.

The mission of the Corridor initiative is to improve long-term security of the region's vulnerable biodiversity while helping local communities achieve healthy, equitable economies. MF is pursuing this mission through 12 interrelated approaches (Table A).

Notable Progress

The Corridor Project has made significant progress over the 12 years since it began in 1992.

Ecotourism

Ecotourism is one of MF's major initiatives to fund protection of the Maquipucuna Reserve, to create sustainable employ-

TABLE A *Elements of the Chocó-Andes Corridor Project*

1. Conservation and development are community-based.
2. Reconnection of fragmented forest habitats occurs through strategic land purchases, reforestation, and encouragement of agroforestry on private lands.
3. Production and market development for high value "bird-friendly" shade-grown organic coffee and cacao is facilitated.
4. Production and training for use of native bamboo as a substitute for forest timber.
5. Community-based tissue culture of orchids and bromeliads for sale and for reintroduction into the wild.
6. Investigation of the role of human and animal diseases as affected by land use practices and long-term climate change.
7. Patterns of atmospheric greenhouse gas sequestration and emission as a function of land use are being investigated for forest, pasture, and bamboo stands.
8. Increasing gender equity in economic activities is emphasized.
9. Development and implementation of environmental education in local and regional schools and communities is strongly supported.
10. Ecotourism training is being strengthened as a means for improving environmental protection and community development.
11. Collaborative networks are being strengthened with NGOs as well as public institutions at all levels.
12. The project is contributing to improved public policy decisions for land use and environmental protection.

ment opportunities, and to reduce the grinding poverty in the region. Ecotourism has become an alternative to timber cutting, charcoal production, and cattle ranching benefitting over 120 families; generally it has helped people recognize that intact forests have economic value now and for their children's children. An organic farm has been established that produces vegetables, fruit, eggs, poultry, pork, and fish for the Maquipucuna ecotourism lodge in the corridor. The farm is also used as a demonstration site for local farmers. Composted manure is used to grow aquatic plants that support fish aquaculture and also to fertilize seedlings of native trees that will be added to coffee plantations. Maquipucuna Foundation won two prestigious international awards for the contribution of their ecotourism progress to conservation and community development (Ecotourism Showcase 2000 and the Skal 2003).

Reducing deforestation in the upper corridor

Neighboring private cooperatives and communities that control over 12,300 ha have reduced their rates of deforestation greatly. The Santa Lucia cooperative no longer clears new forest. The 53 families living in the Yunguilla cooperative used to clear an average of about 1 ha per year of new forest. Now only five families continue to clear forest and these families clear only about 0.25 ha per family per year. Over 500,000 native trees and bamboo have been planted on 2000 ha of farmland.

There are several reasons for the decline in deforestation. First, the success of the ecotourism program demonstrated that tropical forests had inherent economic value. Second, profitable alternatives to timber cutting were developed. Large diameter native bamboo grows well in pastureland and MF promotes it as an alternative to timber for building construction and furni-

ture (Figure B). Maquipucuna Foundation established a community-run center in the nearby town of Nanegal to train local people in the production and uses of bamboo. Third, the importance of natural forest as watershed protection and the significance of biodiversity as natural heritage are part of a regional environmental education initiative.

Sustainable livelihoods

In addition to the employment and revenue generated by ecotourism and bamboo production, other significant economic development initiatives are shade-grown organic coffee and cacao, and production of orchids and bromeliads.

Approximately 160 families have formed the Chocó Andes Coffee Alliance and produce shade-grown coffee without the use of pesticides or synthetic fertilizer. By developing good production practices, high-quality coffee grades, and direct market linkages, the farmers receive premium prices, nearly four times what they had been getting before the Alliance was formed. Maquipucuna Foundation purchases coffee from the Alliance and markets it directly under a master market brand, Choco-Andes, which is used for all Corridor products. Studies are underway to identify additional species of native trees that, when added into shade-grown organic coffee plantations will make the system more "bird friendly" (Carlo et al. 2004).

Maquipucuna Foundation has also partnered with the small-farm Association of Cocoa growers of North Esmeraldas in the lower part of the corridor. Native *Theobroma cacao* is grown on these farms along with "improved" varieties of cacao. Cacao from the native trees produces an exceptionally high-quality aromatic bean. Thus, the farmers can market a premier product along with a higher volume of lesser-grade chocolate.

A farm was purchased and developed into a research and training center for shade-grown organic coffee production. Experiments are underway with coffee polyculture with plantain and citrus, alder and other native trees as shade cover, and soil amendments for nematode and erosion control. Also, simulated drought research is underway to investigate how climate stress will affect pests that attack coffee when it is planted either alone, in combination with citrus and plantain, or under native tree shade. Preliminary results are highly promising. Alder enriches soil nitrogen and provides shade to young coffee plants within two years. The addition of stalk residue from local sugarcane processing reduces root-knot nematodes and increases beneficial predatory nematodes in the soil beneath coffee plants, even under simulated drought conditions. Native forest trees are also added to the polyculture. A similar research and training farm is envisioned for the cacao project.

(I)

(II)

Figure B *Guadua angustifolia* is a native bamboo species in the Chocó-Andean Corridor. It is a noninvasive and very versatile non-timber forest product (I). The Chocó-Andean training center, built entirely with *Guadua*, illustrates the great potential of bamboo for construction (II). (Photographs by W. Richerson.)

The Corridor is a major center of orchid and bromeliad diversity. In the 6000 ha Maquipucuna Reserve alone, 329 orchid species have been described, or 10% of the orchid species of the entire country of Ecuador. A tissue culture laboratory for orchids and bromeliads has been established in the town of Marianitas where several species of orchids and bromeliads are being grown for research and for sale, to reduce pressure on wild orchid populations from collectors. Orchids in tissue culture are currently sold locally to tourists and in the U.S., at the Atlanta Botanical Garden and the University of Georgia. Research is underway on reintroduction strategies to augment decimated populations of wild orchids.

Gender equity

Through their involvement in the conservation projects, women's roles have evolved from passive to active actors of development. A women's cooperative was established in the nearby town of Marianitas to produce jewelry and crafts from tagua nuts (vegetable ivory) and chonta palm. A similar cooperative exists in Yunguilla to produce jams from native fruits, cheese, and craft paper from native plant fibers. Local women are active leaders in the coffee and cacao alliances. In Marianitas a facility was constructed to help women cooperate in day care and thereby free up more time for economic activities. In this way, social capital among women has increased. Women now hold many of the elected public offices in the region.

Carbon sequestration

The amount of carbon from atmospheric greenhouse gases that is either sequestered or emitted from tropical landscapes is a significant contributor to the global atmospheric carbon budget. In the Maquipucuna Reserve, carbon stocks and emissions have been measured in the standing biomass and soil for native bamboo, pastureland, and young and old-growth forest. By protecting 22,400 ha of forest from projected deforestation, MF has avoided the emissions of approximately 3.36 million tons of carbon. Planting of 500,000 native trees and bamboo has been completed on 2000 ha of small coffee and cacao farms and reforestation of additional 2000 ha is planned. Every ha reforested is estimated to sequester between 150 and 300 tons of carbon over a 30-year period. In MF's "Grow your home" initiative families are encouraged to grow native bamboo for home construction (see Figure B). The carbon sequestered as a result could be significant at a landscape scale. University of Georgia studies have shown carbon stocks above- and below-ground for bamboo range from 80 to 240 tons of carbon per ha at normal stand densities—much higher than for pasture.

The amount of carbon emissions that are reduced through these projects is quantified through a formal process defined by the Kyoto Protocol. These emission reductions, as "certified emission reductions—CERs," will then be sold and the proceeds used to expand reforestation and promote shade-grown coffee and cacao farms. The first sale of carbon emis-

sion credits from the project is scheduled to take place in January 2006.

Emerging Challenges and Lingering Concerns

Large, complex problems seldom have simple solutions and the problems raised by the simultaneous needs for biodiversity protection and economic development in poor countries are among the most intractable. As a case in point, the communities of Santa Lucia and Yunguilla near the Maquipucuna reserve have increasingly focused on community-based ecotourism. From a simple perspective of biodiversity protection this has had some clear benefits. As explained previously, rates of deforestation have dropped sharply and people now see more economic rewards in preserving nature. However, ecotourism introduces a new economic vulnerability to these communities because they are now more closely tied to external economic decisions that are out of their control. After the events of 9/11, tourism revenue in Ecuador declined about 25% and is only now returning to previous levels. While revenue has declined, the operating costs of tourism have grown rapidly; about a four-fold increase in the past four years after Ecuador adopted the U.S. dollar as the national currency. In large part this is because transportation fuel costs, a major expense in tourism, have greatly increased.

Should net revenue from ecotourism continue to decline, the people of Santa Lucia or Yunguilla may return to clearing forest for charcoal production and cattle pasture, and it may be more difficult to encourage alternative economies in the future. For these communities, it is important to help them create more diversity in their economic lives. Yunguilla has diversified into marketing locally produced cheese, organic jams, natural fiber paper, and bromeliad plants, but the sale of these products is still closely tied to tourism.

Social capital formation is another concern. In its absence, communities are unlikely to make much economic progress and are more likely to be exploited as labor pools. Yunguilla, Santa Lucia, and, to a lesser extent, the other nearby communities of Nanegal and Palmitopamba have significant social capital in the sense that economic decisions can be made through consensus within the community and can be successfully implemented. In contrast, the town of Marianitas has divisive factions that arise from different religious affiliations, place of origin, and personal experiences. It has been difficult for the town to reach consensus on any issue and thereby to achieve much economic progress. Unfortunately, socially heterogeneous towns like Marianitas are more often the norm, rather than the exception.

We have not attempted significant communication with the oil palm industry. However, dumping of waste from oil palm processing is a major source of river contamination that may have significantly impacted fisheries in the river and estuaries. Though environmental education and a search for alternative uses of waste material from oil palm processing, we hope to find incentives for keeping the material out of the rivers.

There is a misguided belief within development agencies that research is no longer important, that we know the answers and only need to find better ways to implement them. Unfortunately this perspective has been fostered within the research community itself with social scientists disparaging the need for more ecological or agronomic research and the natural scientists dismissing the importance of social science research. It is critically important that we realize the depth and breadth of our ignorance and revitalize our efforts to build a deep and broad base of knowledge. The world is full of surprises; new knowledge is always emerging and we need to be ready to evaluate and make use of it. Here are just a few illustrative examples.

1. In our coffee ecosystem, the addition of sugar-cane residue with its high content of labile sugars increased the population of beneficial nematodes and decreased the population of pest nematodes during years of normal rainfall. However, during drought years preliminary evidence suggested that the reverse took place: beneficial nematodes decreased and pest nematodes increased. Clearly, we need a better understanding of pest control under variable climates and this will become increasingly important if broad climate changes emerge.

   It is widely held that tropical pastures will quickly revert to forest if fire and grazing are removed and if forest sources of seed are nearby. A common pasture grass used in the Andean slope lands, known as *pasto miel* (*Setaria* sp.), does not revert to forest even when grazing is absent for as long as 15 years and forest surrounds the pasture. This grass supports huge populations of bacteria on labile sugars exuded from the grass roots. The bacteria immobilize nitrogen and normally this nitrogen would be unavailable to plants. However, large populations of predatory nematodes feed on these bacteria within the root zone and release ammonia nitrogen as waste, which is quickly assimilated by the grass. Between the grass clumps, bacteria immobilize nitrogen but nematodes are scarce. Thus a kind of nitrogen account is maintained within the root zone of the grass but a nitrogen deficit occurs between the grass clumps. This nitrogen deficit strongly inhibits forest succession. With this new information we were able to develop new methods for converting *pasto miel* back to native forest or to agroforestry.

2. Shade-grown organic coffee is being widely promoted as an environmentally-friendly economic strategic. While we share the opinion that shade-grown organic coffee has much promise, we believe that there are significant ecological risks that need to be assessed and mitigated. In particular, the use of shade-grown coffee plantations by forest birds represents an important resource for forest birds and these plantations, and, if

strategically located, can help reduce the effects of forest fragmentation. However, most shade-grown organic coffee is produced on family farms totalling less than 3 ha. On these farms it is a common practice to allow chickens to forage freely among the coffee plants. On the positive side, the chickens eat many insects that could damage the coffee. However, these free-ranging chickens have a history of exposure to major avian diseases that could affect wild bird populations. Bringing forest birds that have little evolutionary exposure to avian diseases of poultry into contact with chickens could represent a major unintended risk.

3. Mosquito vectors of human diseases that use water-filled containers, both natural (e.g., tree holes) and anthropogenic (e.g., discarded tires) are also being investigated. Water-filled bamboo (untreated) segments serve as a surrogate for natural container habitats such as tree holes and bromeliad leaf axils. This is part of a larger study to investigate the migration from lower elevations of human disease vector mosquitoes that breed in container habitats. Mosquito vectors of dengue fever breed in container habitats and as the landscape is deforested the warming temperature may permit their migration from the lowlands into the highlands, which are normally too cool for them. Amphibians (mainly frogs) are declining worldwide and especially in tropical highlands, and fungal diseases are thought to be a major causative factor. One consequence of the decline of frogs is that their tadpole larvae as major potential competitors and predators of mosquito larvae also decline. For this reason we have established small terrestrial and arboreal water containers to investigate the interaction of frog larvae with mosquito larvae.

Important Lessons Learned

Over the years, we have made mistakes; however, we have learned important lessons from these mistakes. Because our "lessons learned" may be useful to other conservation and sustainable development organizations we present a few of the more important ones.

1. *Don't try to initiate a project in a community unless you are invited.* It may be slow going and frustrating but gaining trust within a community is essential. We have found that having a local person assigned specifically to maintaining good relations with communities is important.

2. *Listen to what the community has to say if you want them to listen to you.* Often, local knowledge within the community can help improve your project. But you are also bringing specialized knowledge of value to the community. Developing a dialogue to share knowledge is critical.

3. *Measure your progress.* In the effort to carry out projects on limited resources it is tempting to forego monitoring progress. You may be convinced that a project to reduce deforestation is working, but it will be more convincing to others if you can actually prove it. Aim for reports in manuscript format suitable for publishing in peer-reviewed journals. Develop measurable objectives that can be used to show the progress you are making towards your goals.

4. *The immeasurable is also important.* A family that has benefited from the project and is now able to buy shoes for their children and pay for medical care is more than a data point. As lives become enriched, as women become part of the community leadership, and as forests and biodiversity are protected fundamental progress is being made. The richness of these changes can be captured much better in stories and personal accounts than in data tables.

5. *Make long-term core funding a major priority.* This may sound obvious, but for small organizations in particular, the constant need to fund the basic metabolism of the organization means that goals become unduly shaped by what funds are available. The same goes for multilateral and bilateral projects.

6. *It is not realistic to expect sustainability after the typical 3- or 5-year project.* The key to success for the Chocó-Andean Corridor has been the long-term and place-based institutional commitment of MF and its partners, which spans beyond short-lived funding cycles. The long-term, place-based commitments make institutions and projects truly accountable.

7. *Be wary of entering agreements with much larger organizations, especially if you are at odds with their priorities and ethics.* Of course, it is naive to imagine that significant sources of funding are always pure and reflect your ideals, but it is useful to raise questions before entering into such agreements. Is it worth it? Will it affect your relations with communities? Will the agreement take you away from your primary objectives? And even, are you being used to "green" the image of an otherwise tarnished organization? It may be tempting to sign a lucrative contract but as one Ecuadorian expressed it, "It is dangerous to dance with the devil."

8. *Be conservative when setting objectives for ecoregional conservation.* Finding the right balance between working intensively at the local community scale and extensively at ecoregional scales is difficult but important. When working at large spatial scales, do not create undue expectations within communities. When working at the community level do not expect results to spread quickly or easily to other communities. Ecoregional and community-based strategies are not the same.

9. *Communities with a strong social fabric make the best partnerships with NGOs.* The best partners are communities that are characterized by responsible leadership, a capacity to establish dialogue among stakeholders, abil-

ity to set long-term goals, and community pride. In these community partnerships goals are more likely to be set jointly by the NGO and the community, and training and education is better disseminated and lasting. These characteristics, sometimes called social capital, are often not apparent, and emerge only after considerable investment of time and funds in a community.

10. *Economic incentives should be realistic and significant.* If the objective is to help communities find alternative livelihoods that reduce the rate of deforestation, the community must want the alternative. However, communities often have unrealistic expectations about alternative livelihoods and little experience with developing business plans and market outlets. Before encouraging communities to take the risk of a new economic endeavor, a thorough analysis of the need and potential for community training must be done and business plans, including marketing, need to be developed and discussed within the community. For example, we invested a lot of time in helping one community produce paper from natural fibers only to find out that the market for this

paper was weak and the return for the effort to make the paper was small. Shade-grown organic coffee had a much larger market demand and higher return but also required strong organization among growers, sophisticated business plans, a major effort to maintain high quality, initial capital investments, and market development, including brand recognition.

11. *Anticipate external threats.* External threats may come from the social or natural environment. Here we wish to emphasize external threats that come from illegal actions. Land grabs by well-funded and politically connected organizations are becoming increasingly common. When projects are successful and land values begin to rise they are at their greatest risk from these groups. It is important to work with local communities to ensure that they have strong land tenure rights. It also important for those NGOs that own land to review their land tenure rights.

CASE STUDY 16.3

Certification and the Collins–Almanor Forest

Fred Euphrat, Forest, Soil and Water, Inc., Healdsburg, California

---

As a consulting forester, I was not ready for my meeting with the forest managers of the Collins Pine Company's Collins-Almanor Forest (CAF). "We want to take care of the bugs and the lichens," said forester Barry Ford, and his words were echoed by staff and management. Their goal was to develop a forest plan that provided for wood, water, wildlife, soils, insects, fungi, and people's needs. Collins, a family operation with three large forest holdings in the U.S., wanted its plan to demonstrate productive, sustainable forest management to the state of California, to guarantee good economic returns to the owners of Collins Pine, and to serve as a blueprint for succeeding generations of managers. Their goal for a long-term management plan would document their program, as well as meet the needs for environmental certification and also reduce paperwork with the State of California.

In its goals and situation Collins Pine is unique, although the lessons it can teach us are generic. Owned jointly by the Methodist Church and the Collins family, the CAF covers over 38,000 ha of productive timberland in northeastern California. Its mill is the largest employer in the town of Chester, on the shores of Lake Almanor. Collins produces about 30 million board feet from its managed lands each year, all of which is certified as sustainably produced by Scientific Certification

Systems (SCS), a member of the international Forest Stewardship Council. Collins' goals are to maintain the flow of logs through the mill, to maintain jobs in Chester, and to take care of the land and its ecosystems; that is, the corporation seeks ecological, economic, and social sustainability.

Selective cutting began on the CAF in the 1940s based on a European model of individual tree selection. Relatively gentle slopes, relatively fast growth, and high-value wood on its lands helped to make this a feasible option. Collins cuts white fir (*Abies concolor*), sugar pine (*Pinus lambertiana*), ponderosa pine (*Pinus ponderosa*), incense cedar (*Calocedrus decurrens*), and Douglas-fir (*Pseudotsuga menziesii*). Permanent inventory plots allow its foresters to observe changes in stand composition and measure growth and tree regeneration. The corporation's sustained yield planning provides a check on the system of selective forestry, requiring the development of cutting and forest management models to maintain a viable forest and wood-production stream.

Certification was an important step on the road to public acknowledgment of Collins' sustainable practices. Among many other parameters, SCS evaluated growth versus harvest on Collins'entire landholdings to ensure that cutting did not exceed long-term forest production. The certifiers evaluated

the harvesting and road systems to ensure that no resource damage was occurring at unreasonable levels—reasonability being based on the local and regional experience of the evaluators. They looked at management of soils, of downed wood, and of standing snags for nutrient cycling and wildlife habitat. They also evaluated Collins' role in the Almanor area and the Chester community to see whether the company was a good employer and a good local steward of the landscape.

The feeling of working on Collins' land is unique. First, the land is dispersed over 200,000 ha among the holdings of other timber owners, community and power reservoirs, and National Park, Forest Service, and other public lands. There are no gates and there are no "Keep Out" signs. The surrounding communities actively use the Collins land for fishing, hunting, and camping; there are several active grazing leases as well. There are moments when one feels flung back a century in time to see open meadows within forests of large trees, cattle wandering in open range, and trust and respect among neighbors. The country is wild, with salmon, black bears (*Ursus americanus*), cougars (*Puma concolor*), Goshawks (*Accipiter gentilis*), and Bald Eagles (*Haliaeetus leucocephalus*) as the "charismatic megafauna," as well as a wide variety of "noncharismatic microfauna," such as Barry's lichens.

Can certification offer enough guidance and incentive to provide for the noncharismatics, the "boring" species that are, in fact, the building blocks of the forest ecosystem? The answer is probably yes, but maybe not in the most simple and literal of applications.

What is Collins doing differently from other companies? How does its land and forest look? Collins' land contains tall

Figure B   Lake Almanor and Mount Lassen as seen from the Collins' pine area. Such views demonstrate the potential for creating recreational activities in this area while still enjoying the economic benefits of timber harvest. (Photograph by F. Euphrat.)

pine and fir forests (Figure A). Where it occurs, the understory is manzanita (*Arctostaphylos*), ceanothus (*Ceanothus*), madrone (*Arbutus*) and tan oak (*Lithocarpu*). There are no clear-cuts, though in some areas the landscape opens up with rocky or young volcanic soils, and you can see vistas of Lake Almanor and Mount Lassen (Figure B). The forests have a park-like feel: you can see underneath the tall overstory, due to shading and brush control, and, if you look carefully, find the stumps from the previous harvest. The whole area is selectively logged about once every 15 years, so each year foresters create logging plans for about 6% of the land.

The Collins' land is interspersed with meadows, lakes, and creeks (Figure C). Salmon, steelhead, and trout use the streams for spawning, rearing, and migration. The meadows, which are also grazed, are popular sites for fishing and camping. Collins is concerned about the quality of streams within the meadows, and has an active program of riparian fencing. Riparian areas contain willows (*Salix*), cottonwoods (*Populus*), and alders (*Alnus*). The land has been managed—logged, mined, grazed and used for recreation—since before 1900, and has been under Collins' selective logging regime since 1946.

Under Collins' selective harvesting plan, individual large trees or small groups of trees up to one-tenth hectare are cut. Their foresters practice "biomassing"—whole-tree harvesting for wood chips—which reduces dense regrowth by removing stems smaller than 20 cm in diameter. Large trees are general-

Figure A   An example of the managed and harvested Collins' pine forest in northern California. (Photograph by F. Euphrat.)

Figure C Streams within the Collins' pine forest run clear and are little affected by the timber operations, as evidenced by their healthy populations of salmon and trout. (Photograph by F. Euphrat.)

ly cut with intervening spacing of 5–12 m leaving good growing room between trees following harvest. Collins' foresters manage individual trees for future growth and adjust spacing to maximize that growth, rather than harvesting all the best trees at once.

Stream zones are carefully managed, too, with harvests removing less than 30% of trees in logically constructed zones larger than required by California law. Roads and erosion processes have been evaluated in selected watersheds to determine ways to reduce the total sediment load in the streams. The feeling at Collins is that if sediment load is a problem in streams, it is up to the corporation to do its part in reducing that load, even if its share, compared with natural and other ownerships' erosion, is small.

Harvesting at Collins is conducted with standard logging practices for the region: road-based harvesting using bulldozers and skidders. The biomassing is done with a whole-tree harvester that snips tree trunks at the ground, working in tandem with a chipper. The roads in use were built in the 1950s and earlier. Where problem erosion exists on Collins land, it can generally be traced to old roads or skid trails, which would not be built today. Future harvesting may incorporate cable systems for the steepest slopes, and erosion reduction will require rehabilitation and/or closing of some roads.

In 1998, the CAF completed its sustained yield plan (SYP), a long-term management plan approved by the California State Department of Forestry. The SYP assures the State that this holding will always meet the public goal of maximum sustained productivity across the property's ecological units and timber types. The SYP models forest growth, timber harvest, and relevant changes to habitats over the coming century.

One element that is difficult to work with in the Collins' landholdings is fire. While fire is an important part of the ecosystem, the interspersed landholdings and the danger of an escaped control burn make fire, at this time, a difficult management tool. An escaped fire could create such a burden

through lawsuits or outright destruction of the forest that it would destroy the corporation. In this case, the managers must weigh the correct ecological path against the necessary economic path. Biomassing has been used in place of fire to reduce competition and undergrowth, and to create forests that are productive as well as aesthetically pleasing.

Certification and the SYP process provide guidance via feedback between the review team, which brings region-wide expertise to the evaluation, and the land managers. The team may require inventory information not presently available on trees and non-tree resources, such as soil, water, wildlife, brush species, burn history, or snag density and use. The need for this information for continued certification stimulates the corporation to work with the experts to develop effective data bases to describe the land, its productivity, and its uses.

When landowners implement monitoring, they can see for themselves and demonstrate to others the results of their improved management. With a monitoring program in place, forest managers can begin the process of adaptive management: inventory, implement management strategy, monitor, change strategy, monitor, and repeat.

Certification also increases the value of Collins' products due to their differentiation in the marketplace. Collins markets "Collinswood," certified lumber, siding, and particleboard. In addition, Collins holds an annual contest for furniture design using sustainable forest products. Collins has worked hard to inform the consumer of the quality of its products, both in terms of the quality of the board and the quality of the forest.

Certification, under the Collins' approach, increases the value of the forest land itself. A well-managed landscape is worth more, in terms of the land's productive capacity and reduced restoration needs, than other, less well-managed lands. Certification is stewardship, locally defined for incorporation into the broader marketplace, it is is a way of giving ecological goals a market value and is a way to gain value for both trees and lichens. No wonder Barry wanted lichen-oriented forestry!

CASE STUDY 16.4

Community Empowerment and Food Security

Lessons From Zimbabwe's Communal Management Programme for Indigenous Resources (CAMPFIRE)

B. Ikubolajeh Logan, Pennsylvania State University

Zimbabwe's Communal Management Programme for Indigenous Resources (CAMPFIRE) was established in 1989 under the combined initiative of the country's Department of National Parks and Wildlife Management, the World Wildlife Fund, the Zimbabwe Trust, and the University of Zimbabwe Center for Applied Social Science (Campfire Collaborative Group 1991; Murphree 1997). The initial funding for the program was secured from the United States Agency for International Development (USAID).

CAMPFIRE's main objective is to reduce rural poverty by convincing local communities in game-rich, but agriculturally unproductive areas, that wildlife is an economic asset instead of an impediment to agricultural production. In this context, arable agriculture, cattle rearing, and wildlife management are viewed as economic alternatives vying for the use of scarce land and water resources. For the community, participation in CAMPFIRE entails reduced arable agricultural production (to conserve resources and reduce tensions between humans and wildlife) and a prohibition on killing wildlife (to conserve biodiversity). In return, the community is supposed to receive the lion's share of the revenues from controlled safari hunting (economic incentive) and the meat from the kills (nutritional and food security incentive). The community is allowed to decide whether its revenues must be shared either on a per household basis or used for community-wide projects, like schools.

Within this broad framework, CAMPFIRE is constructed around a number of supporting principles: wildlife is an agricultural resource, and game management is a form of agricultural production; scarce land and water resources must be allocated to the most beneficial use, whether arable agriculture, game management, or cattle rearing; communities must be encouraged to assess the best use of their resources instead of opting automatically against game management; and finally, this resource management framework is expected to obtain a number of synergies, including rural poverty alleviation, reduced tensions between communities and wildlife, and conservation of scarce water, land, and vegetation resources in marginal areas. CAMPFIRE, therefore, offers communities a win-win framework in which wildlife ceases to be a barrier to poverty alleviation and actually becomes the source of economic empowerment.

CAMPFIRE is viewed by many as a prototype for rural development in southern Africa because of its perceived success in balancing its numerous goals and appeasing some contradictory stakeholder interests (Metcalf 1994; Dix 1996; Hill 1996; Wunder 1997). Unfortunately there are a number of problems associated with the export of the program to other countries in Africa. First, despite the claims of some of its proponents it is not clear that CAMPFIRE achieves all its goals, especially in the areas of community empowerment and poverty alleviation. Second, the program's transferability to other countries seems to be based on assumptions that a program developed in Africa by Africans is automatically replicable elsewhere on the continent. Certainly two African countries with similar resources may have to adopt idiosyncratic management options to address their different political ecologies, and may differ in overarching objectives for wildlife management programs (Neumann 1997). Furthermore, implementation of any program like CAMPFIRE would take place in the present globalization paradigm, which favors grass-roots development, local resource autonomy, and community empowerment over centralized state control in resource adjudication (Logan 2004). Donor support for all forms of rural development in Africa presently hinges on programs that reflect the potential to merge economic empowerment, local resource autonomy and ecological awareness under the aegis of sustainable development (Barrett 1996; World Bank 1996). The following discussion explores CAMPFIRE in these general contexts. The program's democratic attributes are discussed in terms of their impacts on local empowerment, and its sustainable development attributes are discussed in terms of their impacts on poverty alleviation and food security.

Since the 1990s, international NGOs have bypassed African states and taken aid funding directly to local communities to empower them via local-level control and administration of natural resources. This "government out" philosophy is born out of fatigue and frustration over three decades of state-led aid distribution during which rural livelihoods have actually deteriorated. The new orientation, which is part of the broader globalization paradigm, views international NGOs to be more transparent (less corrupt) than the African state. Compared to the African state, these organizations are believed also to have greater management capacity, be more answerable and responsive to donor concerns, and, therefore, more likely to implement community empowerment projects that are consistent with the security interests and humanitarian priorities of donors.

Unfortunately, as presently designed and implemented, international NGO-run programs raise some conceptual questions that undermine their superiority to the African state for achieving community empowerment. When international NGOs run programs, the role of the community in setting conservation and development agendas and implementing them is often weak. For example, often the imperatives of poverty alleviation are determined from above, and not linked to local priorities, and thus less successful in the long-run.

CAMPFIRE and Community Empowerment

The architects of CAMPFIRE were sensitive to its community empowerment impacts. Because the program's initial funding was from USAID and one of the implementing partners was an international NGO, there was significant international oversight to guide the program's implementation toward the new globalization paradigm. Aside from these external impetuses, the Zimbabwean state had to be sensitive to the political ecology of resource distribution because the program came on the heels of the liberation struggle, which had been waged primarily over resource autonomy and economic empowerment (see Hassler 1995; Murphree 1995).

To achieve rural empowerment, the program supported decision-making at the smallest geographical and social scales, which then informs decision making at higher levels. Although in theory this administrative structure lends itself quite well to community empowerment, its actual effectiveness has been bedeviled by lack of commitment to the concept in the program's day-to-day implementation.

There are a number of reasons why implementation has not lived up to expectations. First, wildlife management decisions may not be dictated by local interests. In one example, a village wished to get rid of a bull elephant that was causing a lot of damage to community property. The CAMPFIRE agency representative promised to have the elephant killed by the next safari hunter. Unfortunately, when the hunter arrived, he refused to cooperate because he felt the trophy too small to compensate for his expenses (Kleitz 1998). As CAMPFIRE participants, the villagers could not kill the elephant themselves and it continued to wreak havoc in the community (in one instance, razing cornfields and a hut in this researcher's presence). This example reflects a common criticism of CAMPFIRE, that implementation is often driven by control from top-down administrators who fail to understand or accept community empowerment.

A second problem surrounding CAMPFIRE's pursuit of community empowerment pertains to the very definition of community. The program's guidelines require that a community should be a self-defined entity, based on local perceptions of kinship and lineage. In principle, this is a laudable attempt at local empowerment. Yet in actual implementation this goal has been subverted by the further requirements that a community should be geographically contiguous and socially homogenous. As Logan and Moseley (2002) observe:

> A community typically possesses kinship boundaries that are semi-permeable and in constant flux; kinship

ties are often geographically and socially untidy, and may cover stakeholders in the village and relatives in cities and even overseas. Even adjacent communities may differ significantly by ethnicity, gender, income, and religion; furthermore, close kinship may not reflect proximity. Defining communities simply on the basis of geographic continuity may be administratively convenient but leads to weak or contentious decision-making at the local level. By artificially defining communities, top-level administrators weaken real community empowerment and reinforce the perception that strong central administration is necessary.

A third problem is the additional definitional requirement that a community should comprise 100 socially cohesive households. In reality, small communities can be just as socially fractured as large communities and vice versa. In Zimbabwe, this problem is further complicated by rural migration. Experience has shown that after a community is accepted into the CAMPFIRE program, its population tends to increase quite significantly from new migrants (Dzingirai 1994; Metcalf 1994). Consequently, a community of 100 households can double in size only a few years after it becomes a CAMPFIRE member.

The problems surrounding the definition and constitution of a community are further exacerbated by the definition of a "producer" community, which is the community in which a safari kill takes place. The tensions between Bulilima and Tsholotsho, two game-rich communities, reflect this problem. In the rainy season, migrating elephants come through Bulilima and cause a lot of crop and other forms of damage. By the dry season, the herd has moved to Tsholotsho. Because safari hunting is a dry-season sport, the kills are often made in Tsholotsho, which technically becomes the "producer" community. Bulilima has asserted repeatedly that it is the producer community. As there is no framework to address the problem of migrating wildlife, the CAMPFIRE agency devised an informal formula, which awards Bulilima about 40% and Tsholotsho about 60% of the meat from the kill. Neither community is satisfied with this arrangement, especially as it does not deal with the division of the revenues from the hunt, too much of which the communities complain remains in the hands of the CAMPFIRE agency. Thus, in the area of meat distribution, the two communities are adversaries; but in the area of revenue distribution, the two are allied against the CAMPFIRE agency, in their search for greater transparency.

Community reactions to the livelihood stressors imposed on them by the program are predictable. At first, they request assistance for property damage caused by wildlife. Because the agency rarely has sufficient funds to cover applications for compensation, communities take matters into their own hands by expanding their agricultural activities and engaging in more frequent kills of small and medium sized game, while leaving the big game for the safari. These adaptive activities are clearly counterproductive in terms of the program's goals. They reestablish tensions between humans and wildlife (which the program is designed to eradicate) and rather than obtain

empowerment, communities are transformed to poachers of their own resources.

In summary, even though in design CAMPFIRE is sensitive to the issue of local empowerment, its actual implementation framework undermines some of the fundamental requirements of this goal. As presently implemented, authority over some of the most critical resource management decisions resides with the CAMPFIRE agency and not local communities. CAMPFIRE's implementation is often governed by practice designed to foster safari hunting first and community empowerment and poverty alleviation second (Thomas 1995).

CAMPFIRE, Poverty Alleviation, and Food Security

One of the stated goals of CAMPFIRE is to redress poverty among rural communities located in Zimbabwe's marginal environments. Toward this end, the program provides that the producer community should receive 80% of the revenues from the safari and all of the meat from the kill. In this regard, the program possesses an important sustainable development dimension. Unfortunately, as in the case of local empowerment, this goal is subverted by some of the activities that surround the program's implementation.

Proponents of the program point to income and food receipts by local communities in the northwest as evidence of the program's success and applaud the fact that wildlife populations have increased without sacrificing community welfare. As indicated in one USAID report, "firsthand experience of the economic benefits derived from sustainable wildlife management has led to improved supervision of communally owned natural resources." (USAID 1997). USAID also claims that participation in the program of over 200,000 households can be viewed as evidence that communities appreciate the welfare-enhancing benefits that come with wildlife management.

An important direct benefit of CAMPFIRE is the revenue it brings to communities in the form of cash income, which is used for education, health, and transportation infrastructure, and as subsidies to household income. Logan and Moseley (2002) point out that the projects bring in other community benefits such as shareholder dividends, employment of the local labor force in cooperative activities, and food security through the provision of meat from organized culling of wildlife. More indirect benefits include management training and local responsibility for local resources in the form of natural resource cooperatives.

There is no doubt that CAMPFIRE brings some benefits to participating communities, but USAID estimates that communities receive only 52% of the revenues from a kill, up to 25% of which is retained at the district level (USAID 1997). By comparison, the same USAID study indicates that 48% of the revenues are shared between the CAMPFIRE agency and the safari company. Thus, on a per-capita basis, households receive very little actual income from the program. One study (Kleitz 1998) indicates that even in Guruve, considered to be one of the more successful CAMPFIRE districts, 71% of residents claim that they have never received direct monetary income from the program.

A second problem with the poverty alleviation impacts of the program relates to increased tensions between humans and wildlife. Normally, rural communities control wildlife populations through routine kills for meat and other purposes. When a community becomes a CAMPFIRE member and is banned from this activity, there is a consequent increase in wildlife populations, which impacts food security directly through crop damage, reduced meat consumption and, therefore, protein intake. Some communities are reported to lose a quarter of their crops and access to animal protein after they become program participants (Kleitz 1998). Although communities are supposed to be compensated for crop losses, this rarely happens because the CAMPFIRE agency does not have sufficient resources to meet this obligation (Dzingirai 1995). Increased wildlife populations, especially of elephants, also place stress on the environment (for example, through biomass destruction and hard-panning of land) and further reduce the productivity of marginal lands. Even though some of the larger communities like Tsholotsho have been able to use CAMPFIRE funds to build community projects like schools, it is not clear that these benefits compensate for food insecurity, which is often exacerbated by droughts.

Problems of this nature suggest that broad community benefits are necessary but not sufficient for achieving sustainable rural development. A proper assessment of the benefits of CAMPFIRE participation must involve comparisons with other options, especially in terms of opportunity costs. The question that arises is whether participation in CAMPFIRE provides communities with the best livelihood from their scarce resources. This may be analyzed by comparing participation in CAMPFIRE with non-participation or the "no action" option (maintaining the status quo) in terms of household and community livelihoods (Table A).

A community's livelihood system outside the program, the "no-action" option, would involve subsistent arable production supplemented by traditional game culling. It is also reasonable to expect that the no-action option would involve less crop damage, and greater meat consumption (higher protein intake) resulting from fewer restrictions on local hunting of destructive animals. This suggests that the no-action option may be associated with a higher degree of overall food production, when compared with the CAMPFIRE option, at least, in the short-term. The no-action option may, therefore, be preferable to communities because it does not aggravate food insecurity.

A detailed comparison of the food security benefits of no action and program options has been conducted by Logan and Moseley (2002). This study estimates that during the 1990s, CAMPFIRE households received an average of U.S$4.00 of direct income from program participation. If this amount was used only for food, it could have provided only about 5 days supply of grain staple (maize) for an average rural family of 5.6 persons. In comparison, agriculture under the no-action option could provide a minimum of 2 months supply of grain to the

same family. In essence, in the short term, CAMPFIRE participation worsens a household's food security by reducing food availability (Earl and Moseley 1996; Logan and Moseley 2002).

It is difficult to compare the short-term food advantages of the no-action option with the more long-term, community-wide benefits of CAMPFIRE. Improvements in education, health infrastructure, and managerial training are desirable for rural development. Unfortunately, the costs of these outcomes are being gained largely by the present generation while the benefits will be borne primarily by the future generation, creating a serious question of intergenerational equity that mystifies direct comparisons. One may conclude, therefore, that CAMPFIRE may have a broader array of benefits than the no-action option but that the opportunity costs of program participation may be high, especially in terms of short-term food security.

Conclusions

Zimbabwe's CAMPFIRE program is designed to empower local communities by putting them in charge of their resources, and providing them with the option of moving away from subsistence agriculture to wildlife management if this is likely to reduce poverty and improve food security. Proponents of the program, including its original financial sponsor (USAID) applaud the program's success and view it as a potential model for other game-rich southern African countries. There is, however a disconnect between the program's design and its implementation. The design includes well-meaning local empowerment and sustainable development objectives consistent with the new globalization paradigm used by donors to give financial support to poverty-alleviation programs in Africa. The program's implementation, on the other hand, conceals flaws that limit its usefulness both as a tool for poverty alleviation within Zimbabwe and as a template for resource management in southern Africa.

Program proponents point to its success in the area of wildlife preservation as evidenced by a substantial increase in

TABLE A *Generalized Comparison of the No Action and CAMPFIRE Options*

Outcome	No action option	CAMPFIRE option
Short-term benefits		
Crop yield	Higher	Lower
Available animal meat	Higher	Lower
Cash revenues	Lower	Higher
Community projects	Lower	Higher
Wildlife population	Lower	Higher
Short-term costs		
Crop damage	Higher	Lower
Long-term benefits		
Training and managerial skills	Lower	Higher
Education and health	Lower	Higher

wildlife populations in the northwest of the country in the past two decades. They also point to community benefits, especially in terms of projects like schools and clinics. The local empowerment and food security shortcomings of the program undermine the success of these other impacts, however. Overall community welfare is likely to be better in the long-term in communities that have been able to build community-wide projects from which future generations may derive benefits.

CAMPFIRE represents an important case study of sustainable development. It does a better job of preserving wildlife than meeting local economic needs and, for this reason, the future of CAMPFIRE is problematic. Communities must have a stronger role in decision-making and in controlling resources that derive from the program. To achieve this, the definition of community needs to reflect social cohesion more than simple geography; program administration should take a more facilitative role and less command and control; and, conservation ecology should inform CAMPFIRE at all levels about the impact on wildlife.

CASE STUDY 16.5

Sea Turtle Conservation and the Yolngu People of North East Arnhem Land, Australia

Rod Kennett, Charles Darwin University and Ilse Kiessling, Djalalingba Yunupingu, Djawa Yunupingu, Botha Wunungmurra, N. Munungurritj, and Raymattja Marika, Dhimurru Land Management Aboriginal Corporation

Sea turtles have occupied the world's oceans for some 100 million years. During their long evolutionary history they have witnessed dramatic upheavals and rearrangements of the continents and oceans, and major shifts in Earth's climate. These global events undoubtedly had profound impacts on sea turtles—on their nesting and feeding grounds and their migration routes—and we know from the fossil record that not all sea turtle species survived. Despite their long tenure on Earth,

Figure A The Northeast Arnhem Region of Australia, where Aboriginal people are working for the biological and cultural conservation of sea turtles.

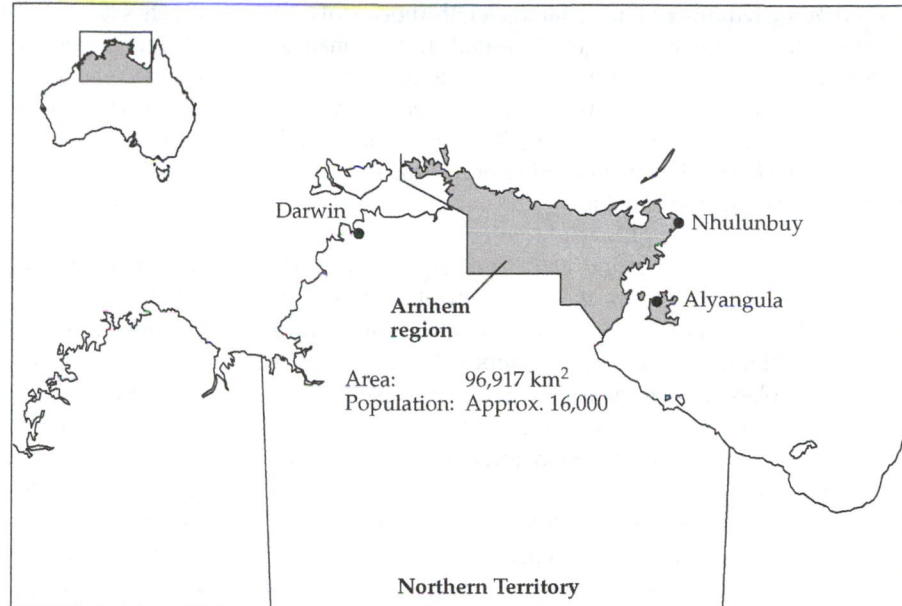

Darwin

Nhulunbuy

Arnhem region

Alyangula

Area: 96,917 km²
Population: Approx. 16,000

Northern Territory

however, the last few hundred years has seen a dramatic decline in sea turtle numbers, and the last seven extant species of sea turtle are in danger of extinction. The destruction of foraging and nesting habitats, the intentional and unintentional slaughter of turtles in fisheries, and the slow degradation of the world's oceans with pollutants have taken their toll.

People have exploited sea turtles for thousands of years and it is likely that most of the subsistence turtle fisheries around the world have been managed sustainably for much of that time. But human populations have exploded, and in many areas, the exploitation of sea turtles far exceeds that of even a hundred years ago. Sea turtles often migrate long distances between nesting beaches and home foraging grounds, so overexploitation or habitat loss in one region can have distant impacts. Many sea turtle populations (or stocks) regularly traverse domestic, national and regional political boundaries; hence to protect a turtle population over its entire range, managers must be able to identify and work with people from other regions or nations with whom they share turtles.

Recent advances in the understanding of turtle ecology and migration behavior have come from the application of contemporary scientific methods such as tagging programs, genetic analyses, and satellite telemetry. However, scientific interest and research alone cannot guarantee the long-term survival of sea turtles. Sea turtles are also of enormous cultural, spiritual and economic significance to many indigenous people around the world, many of whom retain ancestral rights, obligations, and laws that govern their use of turtles. Successful conservation programs therefore require partnerships between indigenous and nonindigenous scientists and managers.

It is within this context of declining sea turtle populations and the need to manage turtles across borders and seas in cross-cultural partnerships that indigenous turtle fishers and

managers, including the Yolngu people of northern Australia, have begun to reevaluate the ways they "look after" turtles.

Background

The Yolngu are traditional owners of northeast Arnhem Land in the Northern Territory of Australia (Figure A). Arnhem Land is a vast wet-dry tropical area of Australia—some 97,000 km² of mainland, and 6000 km² of offshore islands—that is owned and managed by Aboriginal people. Arnhem Land lies along the western edge of the Gulf of Carpentaria, a large, shallow body of water that supports a diversity of flora and fauna, much of which is endemic and significant to conservation. Unlike many other parts of Australia, Arnhem Land and the Gulf of Carpentaria region remain largely unchanged by European settlement and are sparsely populated, mostly by Aboriginal people.

Non-Aboriginal scientists believe that Aboriginal people have occupied northern Australia for somewhere between 40,000 and 100,000 years. According to Yolngu, they have occupied the land since the creation beings first formed the landscape and bestowed law and knowledge on the ancestors of today's traditional owners. Regardless of whose interpretation of the past prevails, Aboriginal people have inhabited Australia for a long time and possess a rich culture of law, ceremony, oral history, and detailed traditional ecological knowledge.

This close bond between people and country remains especially strong for the Yolngu. Today Yolngu maintain a lifestyle that accepts and embraces much non-Aboriginal (or balanda) culture, but in which Yolngu cultural practice remains paramount. Many live in remote coastal settlements where hunting and gathering native foods, including sea turtles and their eggs, remains an important part of life.

The Yolngu's strong sense of identity and culture, and determination to manage their own affairs, is reflected in the for-

mation of the Dhimurru Land Management Aboriginal Corporation. Dhimurru was originally established to manage recreation areas created around the township of Nhulunbuy for the predominantly non-Aboriginal workforce of a large, open-cut bauxite mine. More recently it has expanded its role to include the facilitation of land and resource management by traditional owners of the region.

Today, Dhimurru employs eight Yolngu rangers and two non-Aboriginal administrative staff. The Corporation works closely with rangers from the government conservation agency—the Northern Territory Parks and Wildlife Service (NTPWS). Dhimurru embraces a dual approach to conservation activities, emphasizing the value of traditional ecological knowledge and practice, while recognizing that non-Aboriginal science can inform traditional owners in developing contemporary resource management strategies. The sea turtle (or miyapunu) project is regarded as a flagship project for the organization.

The miyapunu project commenced in early 1995 in response to concerns from senior Yolngu about apparent declines in sea turtle numbers. One of the suspected causes for decline at local beaches was over-harvesting. As Djalalingba Yunupingu, senior cultural advisor, explains "In the Northern Territory, where we live, I think we catch too many turtles. We dig up too many turtle nests and we don't let them hatch and let the baby turtles go back to the sea and grow and become adult turtles." However, other factors, such as mortality as fisheries by-catch, entanglement in discarded fishing nets, increasing subsistence and commercial harvests in the Indo-Pacific region, and habitat loss in other parts of the turtles' range, are also likely to be impacting turtle populations in the Arnhem Land and Gulf of Carpentaria region.

Aims

The principal aim of the miyapunu project is to build capacity in indigenous communities to engage in informed decision making about sustainable use and management of sea turtles. The project involves a number of components, the objectives of which have evolved with time and experience.

Originally the project was focused at a local scale. What was the size and composition of the Yolngu harvest of turtles and eggs? Where were the major nesting and feeding grounds in the region and what species of turtles nested or foraged there? What other factors were affecting turtles? What laws or traditional knowledge governed where and how Yolngu hunted? As the project progressed and relationships and information exchange between Yolngu and balanda participants developed, it became clear to both parties that the project both required and was capable of addressing broader issues.

Yolgnu found that some of the green (*Chelonia mydas*) and loggerhead (*Caretta caretta*) turtles they captured had previously been tagged by researchers, often thousands of km away in Australia. Pursuing this, Yolngu visited the Mon Repos Sea Turtle Research Centre in Bundaberg, Queensland where they saw a loggerhead (Garun) nesting for the first time. As Djalalingba Yunupingu explains ". . . Garun do not come up on our beaches to nest. . . they come up to nest and lay eggs in Bundaberg, Queensland or in Western Australia. . . the reason why I came here was to see a Garun, because back home we wondered where did Garun lay its eggs and you people thought that Garun lays its eggs in the water. Nothing, it just travels around the feeding grounds and travels up here to lay its eggs and the eggs hatch here. . . " Garun also told Yolngu another important story—that to care for turtles in their waters, Yolngu must work with people in the other places where their turtles travel and so the revised project aims included tracking turtle migrations beyond Yolngu waters.

Discarded fishing nets and other marine debris have long been polluting the beaches of northern Australia. The potential implications of this debris for sea turtle populations became apparent early in the miyapunu project when during 1996, 55 turtles (half of them dead) were found stranded on approximately 20 km of beach in northeast Arnhem Land (Figure B). Such strandings continued over subsequent years and gaining an understanding of the sources and levels of impact of marine debris (and rescuing turtles) rapidly became another major component of the project.

Significance of Conservation

Six of the world's seven sea turtle species occur in Arnhem Land and Gulf of Carpentaria region, and five are known to nest there. All are listed under national and international conservation agreements and all are of enormous cultural significance to Yolngu, who know them as Dhalwaptu (Green turtle), Garriwa (Flatback turtle; *Natator depressa*), Garun (Loggerhead turtle), Guwarrtji (Hawksbill turtle; *Eretmochelys imbricate*), Muduthu (Olive Ridley turtle; *Lepidochelys olivacea*) and Wurrumbili (Leatherback turtle; *Dermochelys coriacea*).

While Australia is regarded as a stronghold for sea turtles, there are worrying signs related to their decline. For example,

Figure B   Although Dhimurru rangers attempt to find and free turtles entangled in discarded foreign fishing nets, often they are already dead when found. (Photograph by R. Kennett.)

loggerhead turtles, a species known to migrate between the Arnhem Land coast and southern Queensland, have declined by 85% over the last 15 years (Limpus and Reimer 1994). In addition, many Australian nesting turtles are known to spend at least part of their lives in Papua New Guinea or southeast Asia where they are vulnerable to commercial and subsistence harvests (Limpus and Miller 1993). The levels of harvest in Indonesia (as many as 30,000 green turtles per year in Bali alone) are unlikely to be sustainable and so may be affecting northern Australian populations. At present, there are insufficient data to make robust estimates of sea turtle numbers across much of northern Australia and hence to determine the extent of any declines.

The miyapunu project addresses national and international objectives for community-level engagement of indigenous people in the development of conservation and sustainable management strategies for sea turtles. The need to develop programs that build on traditional knowledge and practice is specifically identified in both the *IUCN Marine Turtle Conservation Strategy* (IUCN 1995) and the *Recovery Plan for Marine Turtles in Australia* (Environment Australia 2003). More broadly, Article 10 of the international *Convention on Biological Diversity* requires signatories to "protect and encourage customary use of biological resources in accordance with traditional cultural practices that are compatible with conservation or sustainable use requirements."

Major Activities and Outcomes

The miyapunu project has focused on a suite of monitoring projects, all undertaken to allow more informed management of sea turtles, and to better understand threats to their survival. Primary areas of activity include monitoring of hunting and analysis to determine sustainable hunting levels, tagging, and monitoring of long-range movements of turtles to identify other communities with which to work on sea turtle conservation, and monitoring of stranding of turtles due to entanglement in marine debris to increase awareness of this growing threat.

Sustainable harvest

To manage any resource sustainably, we need reliable data on hunting rates and demography of the hunted populations. These data are gathered through a partnership of Dhimurru rangers and community-based hunters to stockpile shells of captured turtles, and then take data from them. Data from the turtle shells include the size and species of turtle, as well as method of capture (Figure C). Turtles caught by harpoon were

Figure C   A data sheet used in the miyapunu project. This easily completed sheet provides critical information on sea turtle harvest practices and population characteristics. (Courtesy of the Dhimurru Land Management Aboriginal Corporation).

distinguished by the presence of one or two puncture wounds in the shell, while unmarked shells indicated that turtles were caught by hand, usually while nesting. The gender of the captured turtles was usually obtained by interviews with hunters while shells were being measured. The harvest data collected yielded some useful insights into the demography of the turtle harvest. Consistent with information on indigenous hunting elsewhere in Australia, green turtles are the favored target and females are taken more frequently than males. The type of turtle habitat closest to a community influenced the capture method and size range of turtles. When a community was near a foraging ground, a wider size range of turtles was taken and nesting turtles were rarely included in the harvest. Conversely, when communities were closest to a nesting or mating area, the harvest was comprised of mostly large mature turtles, and nesting turtles were often taken. These are important considerations in constructing sustainable use strategies for turtle hunting.

Probably the clearest message from the harvest-monitoring program was that indigenous involvement in the sustainable management of sea turtles was about much more than "getting Aboriginal people to count dead turtles." Collecting such data raises questions of who owns the data and what use it will be put to. Unless indigenous people are involved in planning and implementing research activities and interpreting results, in developing and producing culturally appropriate communication tools, and in formulating and assessing management options, the collection of harvest statistics is of limited value. Similarly, the mechanism for collecting harvest statistics also needs careful planning in collaboration with the community involved. Only by working through the entire process with the community does the community come to have ownership in the full process, increased capacity to gather and work with these data alone, and access to detailed information for community decisions on how to manage local hunts.

A key factor in regulating turtle harvest is the strength of customary law in an area. Traditionally, people would always seek permission to hunt on another's land before collecting eggs or turtles. Nesting beach surveys by Dhimurru rangers and discussions with indigenous elders during exchange visits (see the next section) showed that today, in some areas, customary law is weaker and for many reasons traditional owners are unable to enforce these traditional rules. Without these rules, the harvest of eggs and turtles can become unsustainable in a local area causing grave concerns to Yolngu and fueling concerns about the sustainability of Aboriginal harvests in general. Fences and gates, which allow traditional owners to regulate access, are a contemporary solution to the problem and arguably provision of funds to communities to establish and maintain such measures would contribute to turtle conservation. This approach has also enabled Dhimurru to minimize the impact of recreational vehicle activity on nesting beaches.

Tagging and Satellite Telemetry

Tagging turtles is a standard method in turtle research and long-term data sets provide invaluable information on a range of demographic parameters including growth rates, population size, and reproductive output. Metal flipper tags are easily recognized, and display a contact address, thus tag returns provide valuable data on migration destinations. Recaptures of tagged turtles play an important role in establishing links between communities. For example, Yolngu hunters recently captured a turtle originally tagged while nesting in north Western Australia by a research team that included local Aboriginal people, the Bardi. As a result, Yolngu and Bardi have organized exchange visits to talk about turtle conservation.

Not surprisingly turtle tagging has been part of the miyapunu project. Djawa Yunupingu describes the excitement of tagging the first turtle ". . . we saw a Garriwa come up the beach. . . when it had finished laying its eggs, covered the eggs, and was making its way back to the sea, it was then that we flipped it over on its back and tagged both its front flippers. It was the first turtle to be tagged by Dhimurru and all of us are very proud of it too. . . "

However, tag returns are typically low, and determining migration patterns requires long-term (a decade or more), high intensity (thousands of turtles) tagging programs. Satellite telemetry—while expensive—provides data sooner and has great potential to evoke community interest and discussion. In 1998 and 1999, a total of 20 satellite transmitters were deployed on post-nesting green turtles at Djulpan in northeast Arnhem Land. Long-term tagging studies in Australia document that green turtles travel up to 2600 km to their home foraging ground, and that turtles from the same rookery disperse to foraging grounds that may be thousands of km apart (Limpus et al. 1992). Given this wide dispersal over long distances, our initial expectations were the Djulpan turtles would travel as far afield as Western Australia and Queensland, and even Southeast Asia or Papua New Guinea, and that different turtles would migrate to widely separate locations. Surprisingly, all the Djulpan turtles traveled southward, remaining within the Gulf of Carpentaria and taking up residence on the extensive seagrass beds and reefs in the southwest corner of the gulf (Kennett et al. 2004) (Figure D). These were important findings for the miyapunu project, as it indicated that the Indonesian turtle harvest may not have a significant impact on the northeast Arnhem Land turtles and that the most important people to talk with about joint management of turtles were the Aboriginal people of Groote Eylandt and the southern Gulf of Carpentaria.

With some assistance from World Wide Fund for Nature Australia (WWF), the turtle tracking study was followed by exchange visits to Groote Eylandt, Bentinck Island, and the Sir Edward Pellew Islands where Yolngu talked with local people about turtle management issues and the need for regional cooperation. During these visits Yolngu worked with WWF and local indigenous communities to deploy a further five satellite transmitters (four on Groote Eylandt and one on Sir Edward Pellew Islands) on post-nesting green turtles. Like their counterparts from Djulpan beach, these turtles also remained in the southern region of the Gulf of Carpentaria.

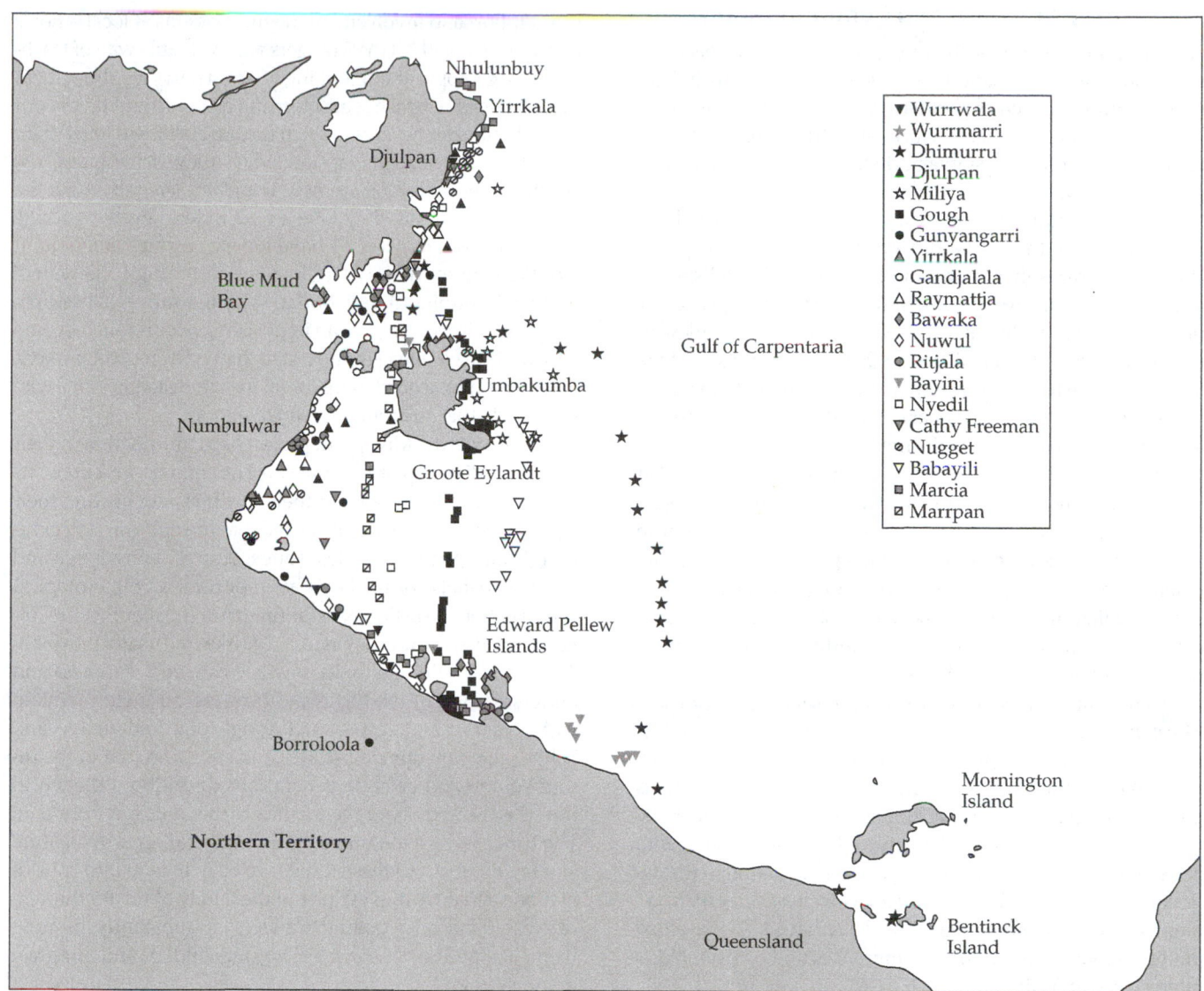

Figure D   Migration routes of 20 post-nesting green turtles from a nesting beach (Djulpan) in northeast Arnhem Land to foraging grounds in the southern Gulf of Carpentaria. Symbols and names denote individual turtles. (Map drawn by R. Kennett.)

Marine Debris and Stranding

Using helicopters to monitor otherwise inaccessible areas of coastline, Dhimurru have found more than 200 sea turtles of at least four species entangled in derelict fishing nets on a relatively small stretch of beach in northeast Arnhem Land since 1996 (Roeger 2002). Most of these turtles have been found between May and June when marine debris accumulation onshore tends to be far greater than at other times of year.

Working in collaboration with the Northern Territory government fisheries agency and WWF, Dhimurru found that trawl and drift nets of foreign manufacture, principally Taiwanese, Indonesian, and Japanese nets, are responsible for most turtle entanglements in northeast Arnhem Land (Kiessling 2003). Ongoing work under the miyapunu project to document accumulation rates of debris and numbers of turtle

strandings is one of the few sources of long-term data of this kind in northern Australia.

Results from this work have had a significant influence on raising awareness both nationally and internationally as to the extent and scale of marine debris impacts in northern Australia, and have demonstrated the critical need for international cooperation to prevent marine debris at its source. Significantly, they have also contributed to the Australian Government listing of marine debris as a "threatening process" under national environmental legislation (the Environment Protection and Biodiversity Conservation Act 1999), and the alignment of several government and research agency's work plans to address threats arising from marine debris. Dhimurru's role in the marine debris study was recognized with a prestigious national Banksia Environmental Award in 2001.

Community Networks and Communication

A significant outcome of the miyapunu project has been the networking and information exchanges between many indigenous communities across northern Australia. A major impetus was the satellite telemetry work—the tracks of the turtles formed links between people and showed Yolngu who they shared turtles with.

As Djawa Yunupingu explains, "Western scientists have taught us many things we did not know about miyapunu. What we learnt is that miyapunu travel these long distances and we were really amazed about this. We want to go to other places now—like Galiwink'u and Milingimbi—and work with people there to put transmitters on their turtles and see where they go. Maybe they (miyapunu) might go north to Indonesia?"

In some ways the maps generated by the telemetry work are a contemporary analogue of the tracks of ancestral beings from the Dreamtime. In their journeys, these ancestral beings traveled the seascapes and landscapes that now belong to many different clans and now each clan holds their own part of the complete story of the journey. Ceremonies where people gather together to tell these stories have long been occasions to discuss other issues including sharing and management of resources. So too the recent stories from the satellite tracked turtles provide a meeting point for people to discuss shared resources.

An information video *Nhaltjan Nguli Miwatj Yolngu Djaka Miyapunuwu: Sea Turtle Conservation and the Yolngu People of East Arnhem Land* has provided a valuable and culturally accessible means of sharing information with both indigenous and nonindigenous communities across northern Australia. Yolngu have also recognized the importance of engaging with government and scientific audiences, participating in the national Marine Turtle Recovery Team as well as other management workshops and similar fora. They have also presented papers at major international conferences and contributed to refereed scientific literature.

Despite their national and international success, Yolngu have not lost sight of the importance of maintaining community support and involvement. Regular lessons at local schools by Dhimurru and NTPWS rangers are a valuable way of teaching young people about looking after turtles: "we show them maps of nesting beaches and feeding grounds and talk about the right time for hunters to catch turtles. I tell them to tell their parents to not take miyapunu when they drive along the beach. Better to teach them now than later. We talk about the breeding times, that they take a long time to grow to adults and come back to nest. We have kinship through miyapunu too. We sing to them where they swim through the water" (Djawa Yunupingu). As Botha Wunungmurra, Dhimurru ranger, confirms. "Yes, it is an important job for us all. Not only for Dhimurru or the Dhimurru staff, or the Dhimurru workers. It involves all the outstations, or all the interested persons who want to do this miyapunu research with us."

Yolngu see the Miyapunu project as a continuation of their long custodianship of Australia and its animals, and know its importance extends beyond their lifetimes and beyond their region. A strong message has been that the cultural conservation (hunting rights, stories, songs, dances, knowledge and spiritual beliefs) of turtles is as important as the biological conservation of turtles. N. Munungurritj (former Senior Cultural Advisor) and Raymattja Marika (educator) explain "what we would like to do now is create an awareness and understanding of the turtle and how we can help maintain and preserve and respect it through our knowledge and beliefs, through our clan stories and songs, so that in the future Yolngu and balanda generations throughout the world can be educated about the life of a turtle in its environment. The turtle takes a long time to become a full grown creature and lay its eggs. All these things we have to take into consideration so that we don't wipe out the family of turtles through careless fishing by fishing trawlers or constantly hunting them. So that there is some left for our children and our great grandchildren."

Summary

1. Sustainable development emerged as an organizing concept for improving human welfare and maintaining biological diversity in the 1980s. Successive United Nations conferences have served as impetus to develop frameworks for sustainable development, and to assess progress toward goals. Early conceptualizations emphasized promoting human welfare exclusively, and only recently have biodiversity goals been more explicitly included as important in their own right. A growing recognition that unsustainable use dramatically undermines natural capital has motivated international communities to push for larger reforms to end unsustainable practices.

2. It is important to distinguish between the concepts of sustainable *growth* and sustainable *development*. Sustainable growth is an oxymoron, and is incompatible with conservation goals as growth implies increasing demands on natural systems. In contrast, sustainable development may be consistent with conservation in so far as qualitative changes we

make in our activities may serve both to improve human welfare and sustain biodiversity. It is critical that this distinction is understood, and that societies embrace development, but not growth goals.

3. Sustainable development projects are implemented by governments or international NGOs, or may spring up from "grassroots" of local communities. Economic tools are often used to promote practices that are sustainable and discourage those that are unsustainable. Policy mechanisms and social pressure are also used to promote the goals of a sustainable development project. Increasingly, sustainable development projects have become one component of a larger strategy for conservation in a region.

4. The record shows mixed success in the application of sustainable development strategies. Generally, more projects succeed in improving human welfare over the short term than they do in securing biodiversity protection. Therefore, the availability of ecosystem services over the long-term is likely to decline. Sustainable development projects are more difficult to sustain over larger scales and with a greater diversity of participants. Sustainable development is more likely to achieve both biodiversity and human development goals when the linkages between human enterprises and biodiversity are strong, and local communities have the capacity to enforce sustainable use themselves, and to mitigate external threats to biodiversity. As illustrated in the case studies, careful collaboration among communities and facilitators of sustainable development (NGOs, governments) is critical to positive outcomes.

5. Skeptics of sustainable development point out that in all of human history, it is most common for us to overexploit our resources. However, we may not have many good alternatives to working to enhance the sustainability of our activities. To the extent possible, sustainable development requires some revolutions in thought to be successful in the long term. Thus, the focus for development should be on sufficiency, equality, and restraint. Often working with our hopes bears fruit, as seen in many of the case studies in this chapter. Time will tell how well we

are able to create sustainable futures for humanity and biodiversity.

> Please refer to the website www.sinauer.com/groom for Suggested Readings, Web links, additional questions, and supplementary resources.

Questions for Discussion

1. Keeping in mind the difference between "growth" and "development," do you think global economies need to expand in order to maintain a state of sustainable development? If they continually expand, can there *be* sustainable development?

2. Many claims of sustainability are, under the surface, false. What criteria might you use to judge sustainability?

3. Consider this scenario: A major U.S. building-supply franchaise contracts with a South American consortium to provide mahogany doors in a sustainable fashion. Mahogany trees are carefully managed in cooperatives so that relatively few are harvested annually, at or below the replacement rate, keeping the resource sustainable. The world market is then limited to the doors that can be made from this amount of mahogany and no more, keeping both price and demand high. Is this scenario in fact sustainable? How might a "cheater" external to the system prosper and thereby destroy the process?

4. Discuss some of the potential opportunities for and problems with possible ecotourism ventures near you. What issues would need to be addressed to ensure that ecotourism would be sustainable and not damaging to natural areas?

5. One of the keys to long-term sustainability is a stable or declining human population. However, declining populations do not simply change in numbers, but also in age structure and geographic distribution. What implications might these changes have for sustainable development? Think about issues such as taxes, trade, leisure time, health, resource demand, and tourism.

17

The Integration of Conservation Science and Policy

The Pursuit of Knowledge Meets the Use of Knowledge

Deborah M. Brosnan and Martha J. Groom

The most effective way forward lies in establishing a social contract between science and society. To increase the reach and relevance of science, scientists have an obliga-tion to direct their energies toward providing society with knowledge that will enable a transition to sustainability.

Jane Lubchenco, 2002

When conservation biology emerged as a new and energetic discipline in the 1980s, a priority task was to convince scientists and those funding scientific research that losses of biodiversity constituted a global crisis, and that it was both necessary and legitimate for scientists to address this crisis in their professional work. This is no longer the case. Our challenge today is not about convincing our colleagues that there is a biodiversity crisis; rather it is about making science an integral part of how we solve that crisis at local, state, federal, and international levels.

To do this, conservationists must actively work to make conservation science a central component of environmental decisions, policies, and laws (Lubchenco 1998). Scientists, lawyers, and policymakers all agree that we need to be more deliberate in our efforts to integrate science into environmental policy; to date there exists no clear framework for doing so, and there is only limited formal training for professionals. This is a challenge, but it also creates new opportunities for conservation scientists to explore and craft new roles for themselves that combine the activities of science with the rewards of making a difference for conservation in the world (Lubchenco 1998; Clark 2001). Similarly, it presents new opportunities for economists, planners, and other social scientists to engage natural scientists in collaborative efforts toward the same ends. The focus of this chapter is on the natural scientist who strives to become better integrated into policy processes, but we hope that social scientists and others who work to foster civic engagement will also learn how to create more effective cross-disciplinary partnerships that will help achieve conservation goals.

To be effective in applying conservation science to policy and retain scientific cred-ibility in the process, a conservation scientist has to be familiar with the policy terrain

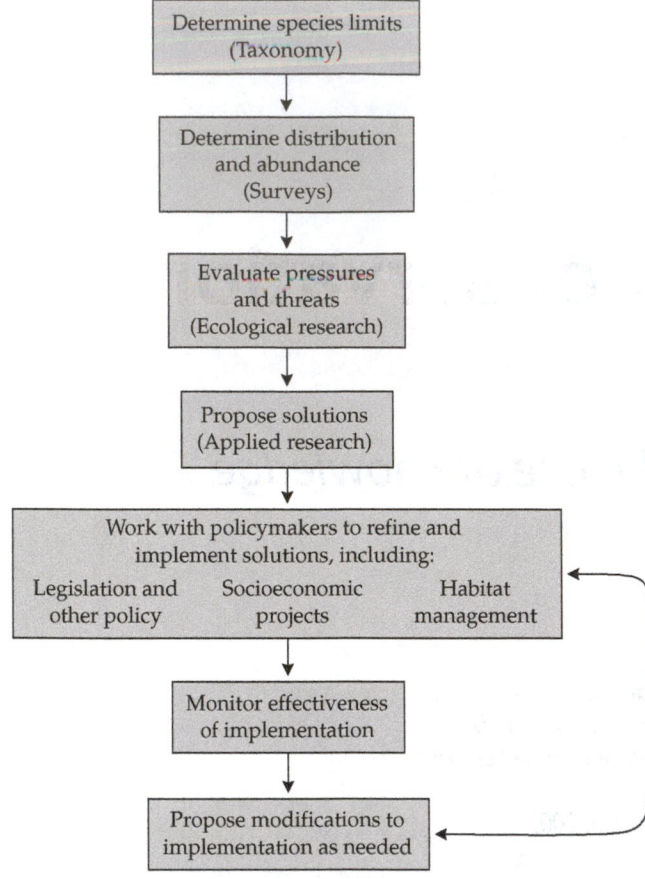

Figure 17.1 A conservation scientist can contribute at many points in the process of species conservation, but to be fully integrated into the policy process, they must work with policymakers to select and implement solutions. (Modified from Fuller et al. 2003.)

in advance. Knowledge of conservation science is necessary, but not sufficient, for successful conservation. For example, in species conservation, there are many steps in the process of formulating a solution that benefit from scientific input (Figure 17.1). However, the most critical step, the development and implementation of policies and plans for conservation, is in the hands of others, unless scientists work closely with those decision-makers (see Figure 17.1). Policies (laws, regulations, and management guidelines) that mediate the conflicting demands of society determine the outcome of conservation decisions, so a conservationist needs to know how to participate in policy processes effectively. Value systems are filters through which people interpret and apply science in policy, and are not random, unfathomable factors that influence choices (Clark 2001). An understanding of social interactions is therefore also a necessary part of being an effective conservation practitioner. Finally, a conservation scientist needs to be able to analyze problems and propose solutions in context, and in collabora-

tion with a suite of people who often have different training and perspectives (Clark 2001).

To facilitate the effective integration of conservation science into conservation action, this chapter provides an understanding of what constitutes conservation policies, which political groups make policies, and how conservation scientists can affect both through their activities and their jobs. The chapter explores the attributes of science that enable it to contribute uniquely to conservation policy and those that make the interface challenging. The case studies are designed to show the different ways people are working to translate conservation science and make it available to policymakers, or are shepherding applicable science from initial experimental research to final policy decisions.

The Need for Translational Scientists at the Interface of Science and Policy

It is impossible to integrate conservation science into environmental policy if no one but the scientist understands the science. This is one of the greatest challenges in applying conservation science to decisions. Policy and decision makers are rarely natural scientists. But they are usually faced with volumes of scientific reports, hundreds of peer-reviewed papers full of diverging opinions, statistical tables, charts, and graphs, and many of the scientific nuances are opaque. One of the greatest needs in conservation practice is a cadre of "**translational scientists**" to aid in conservation policy (Brosnan 2000). A translational scientist is a conservation scientist (from either a natural or social science background) who can translate scientific language into a policy framework so that decision-makers can use the science appropriately (Figure 17.2). A translational scientist can also translate policy questions into scientific ones so that scientists can address them (see Figure 17.2).

A second key role for the translational scientist is to articulate the implications of scientific information for biodiversity and conservation policies. Surprisingly, these implications are often not recognized by either natural scientists or policymakers because they lack the expertise and experience in each other's disciplines. The scientific and policy skill set required for this role is considerable. A translational scientist must have natural and social scientific expertise and credibility, knowledge of what constitutes basic policy or environmental law, and an understanding of how policy decisions are made. Currently there is no job announcement for a translational scientist, and the role is not confined to any one institution. But it is a core activity for any conservationist who wants to make science a part of conservation policy.

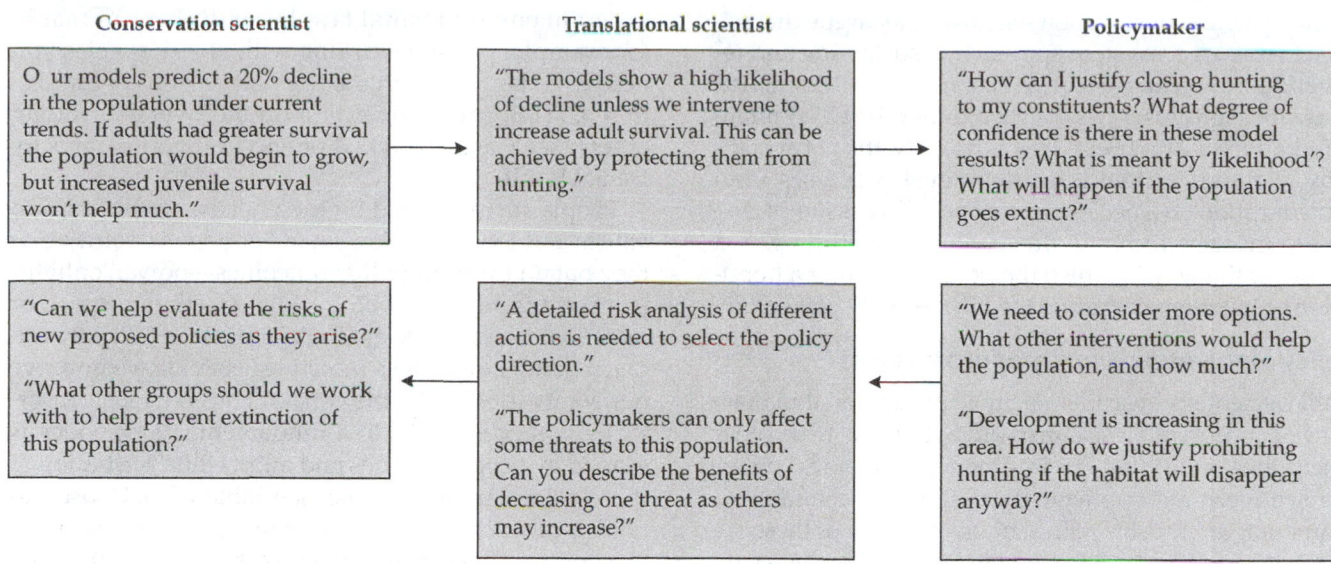

Conservation scientist

"O ur models predict a 20% decline in the population under current trends. If adults had greater survival the population would begin to grow, but increased juvenile survival won't help much."

"Can we help evaluate the risks of new proposed policies as they arise?"

"What other groups should we work with to help prevent extinction of this population?"

Translational scientist

"The models show a high likelihood of decline unless we intervene to increase adult survival. This can be achieved by protecting them from hunting."

"A detailed risk analysis of different actions is needed to select the policy direction."

"The policymakers can only affect some threats to this population. Can you describe the benefits of decreasing one threat as others may increase?"

Policymaker

"How can I justify closing hunting to my constituents? What degree of confidence is there in these model results? What is meant by 'likelihood'? What will happen if the population goes extinct?"

"We need to consider more options. What other interventions would help the population, and how much?"

"Development is increasing in this area. How do we justify prohibiting hunting if the habitat will disappear anyway?"

Figure 17.2   A translational scientist can work at the interface between conservation scientists and policymakers. A translational scientist can make the results of the scientist more accessible and useful to the policymaker, and the needs and questions of the policymaker can be made more clear to the scientist.

Fortunately, many conservation scientists are choosing to engage in a social contract to become more active in local or international environmental decisions (Lubchenco 1998). They are learning on the job that applying conservation science to environmental decisions bears no relationship to an academic discussion. This is the "real thing" and people's beliefs, careers, election prospects, way of life, income, relationships, property—indeed just about everything they value—is often at stake. People protect what they care about, and that extends to their values and way of life as much as it does to the environment. Thus, to be effective at integrating science into conservation practice, one must recognize that environmental policy decisions will always reflect political, social, and economic compromises, and that this affects how scientific knowledge is interpreted.

The Interface between the Pursuit and Use of Scientific Knowledge in Conservation

There are several key features common to all conservation science at the interface of science and policy. In an academic setting, scientists agree that the pursuit of knowledge by the scientific method is the primary goal of science. The validity of the question, objective methods, analysis, and results are central to advancing science. Scientific disagreement and debate is fostered as a way of getting closer to the truth. Two studies that reach diametrically opposite conclusions on the same question can be published in a peer-reviewed journal because each represents a valid scientific effort.

However, in the policy arena the conclusions and interpretation of science are critical. Scientific disagreements are not understood or even seen as helpful by decision-makers. They are frequently either exploited by one side or used to discredit science. For instance, in the fall of 2003, United States Senator James Inhofe (a Republican from Oklahoma) led the opposition to legislation introducing regulations to combat global warming. He went so far as to declare global warming a hoax, a viewpoint he reached from the dissenting opinions among scientists who differ in their interpretation of data and their views on the rate and amount of global temperature rise. Senator Inhofe favored the scientific opinion that suggests a minimal and slow rise, and ignored the nuances of the scientific debate and the preponderance of evidence that has convinced the international community of the reality of anthropogenic climate change (see Chapter 10). Thus, a translational scientist in the policy arena must help policy analysts sift through issues of incomplete data and conflicting results in ways that enable a robust understanding of the science, and that help individuals and groups make consequential environmental decisions, not inhibit them.

Policy-relevant science is different from ecologically relevant science

Laws and other policies often determine what scientific information can be used in conservation decisions. For

instance, while conservation scientists may argue that science favors an ecosystem approach to sustaining and recovering at-risk species, this message can be dismissed by political groups if it is couched under the U.S. Endangered Species Act (ESA). This is because the act specifically excludes certain ecosystem-level processes from consideration in species protection. This issue is discussed more by Deborah Brosnan in Case Study 17.1, a review of the way in which the policy on listing a population of killer whales under the ESA recently played out.

Policy involves a diversity of professions

Unlike scientists in academia, most professional interactions of conservation scientists are with those from other disciplines, including environmental attorneys, politicians, government policymakers, the public, and environmental or industry representatives. Many of these are highly trained and skilled in their profession. Most are present to represent and defend their own interests and needs, or those of a client.

Knowing others' concerns, constraints, and opportunities

The concerns, interests, and constraints of those who make policy are the filters through which they will interpret science. Thus, it is important to understand those filters, and also to become familiar with existing and relevant environmental policies because they directly affect how conservation decisions can be influenced. This does not mean that you have to become an expert in environmental law, but it does mean that if, for example, you are working with at-risk species you need to be familiar with the relevant sections of the U.S. ESA, CITES, and Convention on Biodiversity, among others (see Case Study 3.3 on environmental laws by Daniel Rohlf).

People are motivated by their needs, incentives, and values and tend to seek a set of "goods" or values that they obtain through policy outcomes—power, enlightenment, wealth, well-being, skill, affection, respect, or rectitude (Lasswell 1971; Clark 2001; Table 17.1). By understanding how values motivate behavior, we improve our ability to see fully the context within which conservation policy is made. In a fundamental way, policy is shaped by peoples' values and affects the distribution of those values among people (see Table 17.1). Conservationists can be best served by learning how people act to promote and balance these values through a policy context (Clark 2001).

Constraints and opportunities for conservationists

Conservationists have their own suite of opportunities and constraints. For instance, only organizations with a legislative or regulatory function (e.g., congress or federal natural resource agencies) can make environmental policies that govern the activities of others. If you work for a university or conservation group, you can work to make scientific information the foundation for policy but you cannot make the management or policy decisions. On the

TABLE 17.1 *Values, Bases of Influence, and Power that Affect Policy Outcomes and Practices*

Value	Doctrine	Example	Professionals	Institution
Power	Political doctrine	Victory or defeat in fights or elections	Presidents, politicians, leaders	Government, law, political parties
Enlightenment	Standards of disclosure	Scientific discovery, news	Scientists, reporters, editors	Scientific establishment, universities, mass media
Wealth	Economic doctrine	Income, ownership, transfer	Financiers, bankers, business people	Stock market, banks, businesses
Well-being	Nutrition, hygiene, exercise	Medical care, protectors	Doctors, nurses, health care providers	Hospitals, recreational facilities
Skill	Professional standards	Instruction, demonstration of proficiency	Teachers, crafts-people	Vocational schools, professional schools
Affection	Code of friendship	Expression of intimacy, friendship, loyalty	Friends, intimates, family, community	Familiar and friendship circles
Respect	Code of honor	Honor, discrimination, exclusion	Social elites, societies, leaders	Social classes and castes
Rectitude	Moral code	Acceptance in religious or ethical groups	Ethicists, religious leaders	Ethical and religious associations

Source: Modified from Lasswell and Kaplan 1950; Lasswell and McDougal 1992.

other hand conservation scientists who work in regulatory agencies can make specific management decisions, but in the U.S. they are prohibited by law (under The Hatch Act) to lobby the government to take specific conservation actions, whereas scientists in other institutions (e.g., conservation groups or academia) can. To some extent, management decisions made by government scientists are constrained by policies of the government in power.

Necessary roles as experts and advocates for science

While most people agree that conservation science is necessary for conservation, not everyone agrees on when and how it should play a role. In the policy arena, a conservation scientist must be a strong and tireless advocate for the scientific method. This is especially true in a crisis, when there is a tendency to first turn to science for an immediate solution and then ignore and disparage it when it fails to provide the requisite answer. Conservation scientists must equally protect against the misuse of science when battle lines have been drawn in an environmental dispute and the science is being used by one side as justification for their actions.

Conservation science can influence conservation policy decision

Conservation policy, design, and implementation decisions are made daily. Often there is a specified time frame within which a decision will be reached. Conservation scientists can ensure that science influences the decision, and their information can help parties identify the range of acceptable decisions. But a decision will be made regardless of the quality and quantity of available information, and if scientists exclude themselves from the debate, conservation policies will be made without them. In Case Study 17.2, Prithiviraj Fernando provides an example of how scientific information can expand the range of potential solutions to conserving elephants through reducing human–elephant conflict on the island of Sri Lanka. Although Fernando and other scientists have identified potential solutions, the decision of whether their advice is taken will depend on those who make land-use decisions, and how effectively these scientists can work with them.

Scientists are not value free

All scientists have values. As with all humans, we are influenced in everything that we do by our larger worldview. Scientists make every effort to remove their personal biases from research and interpretation. However, it is important to recognize how and when value systems influence the practice of science, and in some cases to acknowledge them publicly. Having values does not prevent a scientist from acting impartially, particularly when scientific integrity is one of those values.

Policymakers and the public respond to scientific information differently

Conservation scientists will evaluate a scientific study based on its questions, methods, and conclusions. The public—including policymakers—uses a wider set of evaluation criteria when deciding whether or not to accept scientific information. For example, Cash et al. (2002, 2003) conducted a five-year study on the influence of major scientific assessments on policy. Their results underscore how important it can be to acknowledge this broader set of criteria. In efforts ranging from global climate change to agriculture science, scientific assessments are likely to influence policy only if citizens, multiple stakeholders, and policymakers perceive them as having three main attributes: salience, credibility, and legitimacy (Cash et al. 2002). Essentially, each of several questions stemming from these attributes must be answered in the affirmative before scientific assessments are considered useful:

- Salience: Is the scientific assessment relevant to me because it informs the choices I make?
- Credibility: Does the study meet scientific and technical standards?
- Legitimacy: Is the assessment respectful and unbiased in addressing values, concerns, and questions of myself and others?

Cash et al. (2002) concluded that few major scientific assessments have influenced policy and this is usually because they fail to address issues of salience and legitimacy in addition to scientific credibility. Two of the most important scientific assessments, the Intergovernmental Panel on Climate Change (IPCC) and the Millennium Ecosystem Assessment (MA), however, have attained all three attributes through a conscious process of consulting and refining efforts on behalf of international decision-making bodies. Already, the recommendations of the IPCC panels are highly regarded and worked into international agreements to curb greenhouse gas emissions. The MA was only published in March–September 2005, but it is expected that the recommendations and guidance in these reports will similarly be better integrated for having been viewed as legitimate during their development, and because many of the issues addressed were at the request of international policymakers.

As more conservation scientists are learning about policy and engaging with a diverse mix of decision-makers, and as more humanists and social scientists are learning about science and engaging in work with scientists, conservation science gains a stronger and more effective voice in environmental policy. The key message is that you can make a critical difference to conservation policy when you actively engage in translational activities among the scientists and policymakers.

What Is Conservation Policy?

Conservation policy is concerned with how we regulate human activities that affect the environment, the continued existence of biodiversity in all its forms, and the support of human welfare. People achieve this through environmental laws, regulations, policies, and agreement. Conservation scientists tend to use the terms *policy* and *policymaker* under this broad definition and make little distinction between environmental law, an agency regulation, or a conservation organization's policy. However, the opportunities and mechanisms for influencing environmental decisions vary tremendously depending on the type of policy and the arena (e.g., nonprofit, government, or academia).

Environmental law refers to any and all regulatory strategies adopted for the protection of the environment. Laws are enacted at the local, state, federal, or international levels. Environmental law includes both legislation (i.e., statutes enacted by the U.S. Congress or a state legislature, such as the National Forest Management Act), and regulations. Regulations are the rules of implementation produced by government agencies to operationalize the laws. Environmental laws also include international agreements, such as the Migratory Bird Treaty Act and the Law of the Sea.

Only governments have the power to enact or change environmental laws. But the implementation of a law can be changed through agency actions, most often through a formal process known as rulemaking, or less formal changes in environmental policy statements or guidance documents. Changes in policy statements or guidance documents are simple processes that can be made for a variety of reasons, including on the basis of an agency attorney's opinion on the law. By contrast, rulemaking is a relatively formal process in which agencies develop a proposed regulation. In the U.S., the draft regulation is published for public comment in the Federal Register. Before adopting the regulation, agencies must respond to these comments and show how they have addressed them. Conservation scientists may comment on proposed rules and frequently do, but when they respond in this format their comments are treated as part of the general public's comments and opinions and not given extra weight as "expert scientific input." In contrast to rulemaking, changes to policy statements and guidance documents do not have to be published in the Federal Register and may require no public input.

Environmental policy is a broad category and includes all the ways that society tries to address environmental problems, including laws and regulations. It includes other policies that agencies adopt to enforce laws and regulations, such as incentives for landowners to place part of their land into conservation, or education programs for citizens. It includes policies and actions that are adopted by NGOs. The Nature Conservancy, for instance, has adopted an ecoregional approach to its land conservation. Local communities may adopt environmental policies around land use. Professional scientific societies such as the American Association for the Advancement of Science (AAAS), The Wildlife Society, or the Society for Conservation Biology (SCB), may develop **policy statements** on many conservation topics including scientific review, global warming, and the conduct of conservation scientists. The Ecological Society of America (ESA) develops **position papers** on many issues concerning the environment and advocates for these policies (e.g., the proposed priorities for environmental research; Ecological Society of America 1999). The AAAS issues **policy briefs**, which provide summaries of important science and technology.

Conservation policy covers a wide spectrum of activities and decisions. By contrast, management refers to how policies are implemented. In a few situations conservation scientists, particularly those in government agencies, may find themselves in the roles of both policymaker and manager when, for instance, they participate in making policy decisions and then must implement them.

An ideal in conservation science is the objective and impartial pursuit of knowledge using the scientific method. While science can provide knowledge about the world in which we live, and can help us to evaluate the consequences of our actions, it cannot provide the moral justification for environmental decisions. In contrast, policy incorporates both objective knowledge and values to arrive at compromises. In practice, the lines between conservation science and policy are blurred, and what constitutes one versus the other often can be a matter of opinion—a hotly disputed opinion.

This is illustrated by the experiences of a team of environmental attorneys including Daniel Rohlf, who represented environmental clients on a case involving endangered salmon species in the Pacific Northwest of U.S. (Figure 17.3). Environmentalists argued that dams would drive the species extinct, and industrial groups argued that dam operations were too protective of salmon. After developing a computer model to help predict salmon population trends, the National Oceanic and Atmospheric Administration (NOAA) calculated that their plans for operating dams on the Columbia River would allow salmon populations to increase to levels that would avoid jeopardizing the salmon's chances of survival. The agency classified salmon as not being in jeopardy when the models predicted that they had a 50% (or greater) chance of increasing to full recovery levels. Rohlf and his team argued that NOAA constituted its current standard for jeopardy only after it had run the models (in other words so that it could support its decision on how to operate the dams), and that giving salmon a 50% chance for

Figure 17.3 Many Pacific salmonids are threatened with extinction. A great deal of controversy surrounds how the magnitude of threats is determined and how these threats should be addressed. (Photograph © Jeff Foott/naturepl.com.)

cause the judge ruled that this dispute was entirely one of science—that is, both decisions were *scientific* ones, and consequently did not belong in a court of law.

Who Makes Conservation Policy? Breaking the Science–Policy Barrier

Conservation policy is rarely made by scientists. Policy decisions often are made by elected officials (or political appointees) whose profession is to make decisions based on their perceptions of the views of people who elect them (Blockstein 2002) (Figure 17.4). However, even when a decision is presented as science based, or "based on the best science available" it does not mean that scientists either participated in or actually made the decision.

An exception to this rule can often be found in regulatory natural resource agencies where scientists participate in management decisions and often act as the bridge between science and policy. For instance, in the U.S. Fish and Wildlife Service—which enforces conservation laws, in addition to carrying out research—conservation scientists may be asked to make judgments on whether an action will jeopardize an endangered species. On this basis they may have to then decide whether to issue or deny an environmental permit (e.g., a fishing permit or a river-dredging permit). But for the most part agencies and other institutions have deliberate barriers to maintain a distance between scientists and policy making. The U.S. Forest Service has two separate branches. One is the research arm

recovery still constituted too great of a risk. The court dismissed these claims arguing that the jeopardy standard was a *scientific* standard, and that any changes the agency made to the standard simply represented incorporating scientific knowledge based on the ways that expert scientists reach agreements on science. The court also ruled that the agency had, in accordance with its responsibility, made a *scientific* decision on the likelihood of recovery by giving salmon a 50% chance of increasing to the recovery level. The plaintiffs' appeal was thrown out of court be-

"Sir, would you take this latest warning of ecological disaster and pooh-pooh it for me?"

Figure 17.4 Policy decisions often are dominated by business interests. Businesses worldwide have a poor reputation regarding their concern for environmental issues. However, many business interests are now "greening," that is, becoming more environmentally astute—usually because this makes economic as well as ecological sense. Conservationists are beginning to make better connections to businesses, and to engage their interest in supporting policies that aid biodiversity conservationists. (Drawing by Dana Fradon; © 1992 The New Yorker Magazine, Inc.)

and it is staffed by research scientists who work at one of several forest research labs throughout the country. The other arm of the Forest Service is concerned with enforcing policy and management decisions. The rationale for these approaches is that this protects science and scientists from external influences that might bias their results and judgment. But the barrier is a double-edged sword. The consequences are that scientists are frequently absent when the actual decisions are being made, and thus, often the most relevant experts are not consulted. As a result many conservation scientists express frustration with decisions that apparently are not based on science. For example, a group of external expert scientists chosen to review the U.S. Forest Service plans for the Tongass National Forest in Alaska complained bitterly when the plan was released, saying that the Forest Service had ignored the science and advice of the panel of scientists (Brosnan 2000).

International bodies also maintain boundaries between science and policy. The International Whaling Commission (IWC) is composed of 52 member nations established to provide the proper conservation of whale stocks and make possible the orderly development of the whaling industry. The IWC consists of a decision-making commission (which sets whaling policy and quotas) and a scientific committee. The scientific committee conducts research, analyzes data, publishes reports, and provides information on the status of whale populations throughout the world. The committee provides their information and annual reports to the commissioners. But the commissioners and science committee do not come together to articulate and debate the science, and reach some agreement on the conclusions. Indeed the two groups meet separately; the science committee meets before the commissioners. When the commissioners meet to make policies and agree on quotas, it is in a highly contentious political forum and the focus is on making deals and reaching compromises. In that forum the scientific basis for catch quotas is hotly disputed and argued by the commissioners.

At the national level, when the legislature is enacting or changing environmental laws, scientists are usually invited to testify and may be consulted by the legislative staff. But there is no burden on the lawmakers to have their laws peer reviewed by scientists for scientific quality or appropriateness. The burden is on conservation scientists to monitor such laws and policies to ensure that they reflect scientific thinking and information.

Conservation policy—even that which claims to be science-based policy—invariably reflects compromise among the social, political, economic issues. This has been true across public lands in the U.S., most of which are explicitly managed for multiple uses, and which cover extensive areas in the West (Figure 17.5). For instance, while the U.S. Forest Service must use science to shape its policies, its actions must also consistently conform to the mandate it was given by Congress. In this case the U.S. Forest Service must make science-based environmental decisions that sustain the health, diversity, and productivity of the 191 million acres of national forests and grasslands to provide multiple benefits to the country from traditional commodities such as timber, range, forage, minerals, and to opportunities for recreation. Through the land and resource management planning process, this agency must address the sustainability of ecosystems by restoring and maintaining species diversity and ecological productivity to provide for recreation, range, water, timber, fish, and wildlife. Meeting all these goals at the same time is a tall order! Clearly decision-makers often have to evaluate whether, by acting to protect an endangered species, they will violate their other mandates such as providing timber. Consequently, agencies and other groups tend to focus on those principals of conservation science and results that best fit their framework, and they may tend to ignore other frameworks that conservation scientists feel are essential for conservation.

There is undoubtedly an argument to be made that science is objective, and to maintain that objectivity scientists should not be in the fray when economics, politics, and social factors are being debated. At the same time if scientists are not present, then who is maintaining the integrity of the science, and ensuring its appropriate interpretation in the final outcome? While there is an almost universal call for conservation scientists to become more active in policy, there is a divergence of opinion on what that actually means. Many see science in an advisory role. Policymakers rarely, if ever, call for conservation scientists to be leaders in the field of conservation decisions out of the belief that this will limit the full exercise of democracy (Huffman 1991). Policymakers seek the best solution for many, often divergent interest groups, and this may mean selecting an option that is not the best scientifically. While a fair number of conservation scientists may agree that solutions should be selected democratically, they may counter that integral participation by scientists is more likely to lead to better outcomes because they can provide clarification of consequences and possible solutions (Mills and Clark 2001).

The separation of science and values is an ongoing challenge for everyone. Unless, as a conservation biologist, you are content to leave the final decisions to others, you have to be prepared to advocate for the science and for the presence of scientists when decisions are made. Nonscientists such as attorneys are very willing to make conservation policy decisions and exercise their own judgment and influence based on whatever information is available (Loevinger 1992; Hildreth and Jarman 2001).

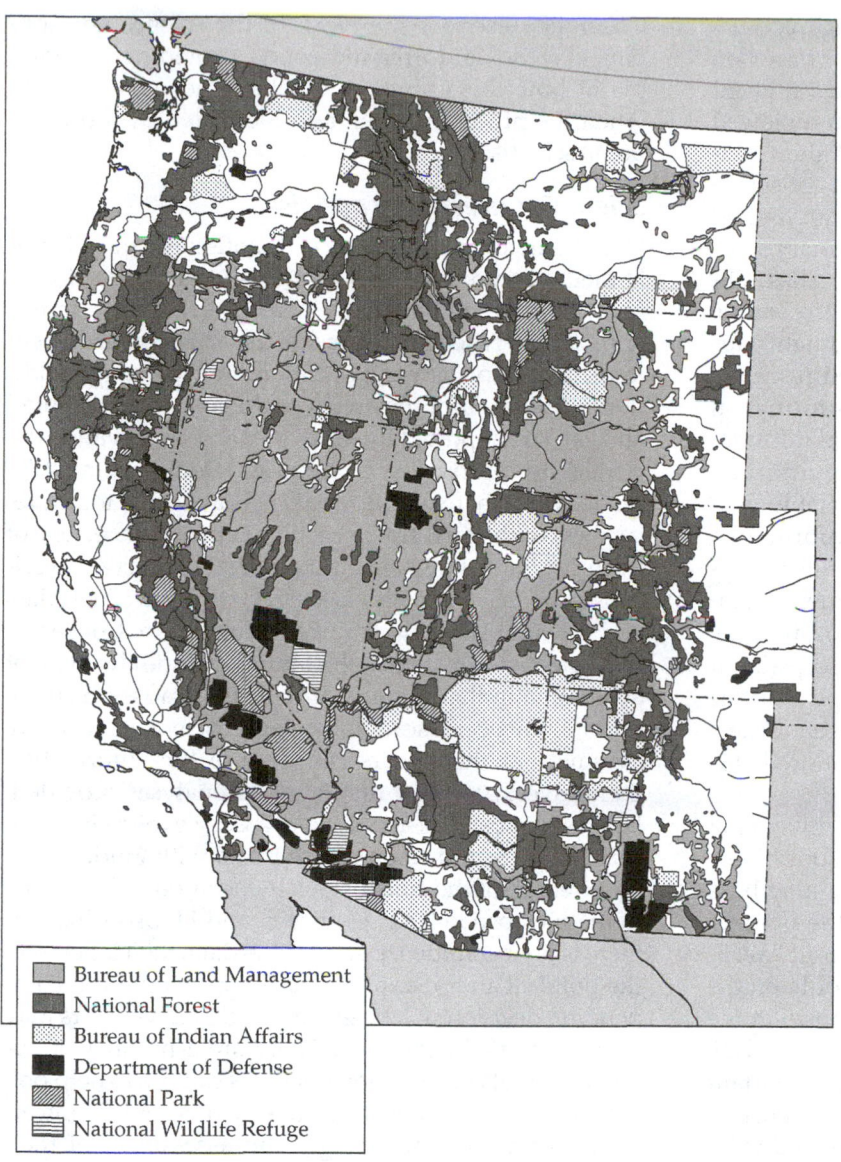

Figure 17.5 Land ownership in the western U.S. is largely federal, and thus conservation decisions must be made on behalf of the U.S. at large. (Data from the National Geographic Society.)

Legend:
- Bureau of Land Management
- National Forest
- Bureau of Indian Affairs
- Department of Defense
- National Park
- National Wildlife Refuge

Scientists, too, need to be willing to contribute their judgment to conservation policy decisions even when more research is needed (Hildreth and Jarman 2001).

The best available science: Quality and quantity of scientific information for conservation decisions

Many conservation policies and laws demand that decisions be made on the "best available science" (Brosnan 2000). But the term is not accompanied by any guidance on its definition or standards, or on how it should be used in decision-making (Doremus 1997; Smallwood et al. 1999; Bisbal 2002). On top of that, there is no guarantee that the best available science will be sufficient to inform the decisions, or that the science will meet standards set by scientific peer review. Unfortunately, this situation is a common occurrence in conservation decisions. Understanding what constitutes the best science available is important not only because the law demands it, but because conservation decisions have to be made now. There are no opportunities to send incomplete or preliminary studies back to researchers to do them over, or to do them differently.

Decision-makers, and more recently conservation scientists, have been struggling with how to bring the best science to bear on important conservation policies. The issue has become so critical that the U.S. National Academy of Sciences is now working on a study to provide advice and guidance on defining the best available science for fisheries management. The courts have also weighed in on this, often issuing conflicting rulings on what constitutes adequate or best available science (Bisbal 2002).

The scientific method consists of generating hypotheses, gathering data, and drawing conclusions based on data that may falsify the null hypothesis as an argument of proof. Scientific information that has been reviewed and accepted by peers is considered "better" than non-reviewed reports (the "gray literature"). But conservation decisions frequently have to be made with limited peer-reviewed science, often with only preliminary studies and in-the-field observations as support. These observations—while not strictly following the scientific method—can provide useful information, highlight gaps in knowledge, and identify areas for needed research. Bisbal (2002) calls these other sources of information "suggestive information" and "supplemental information." Suggestive information consists of empirical information, detailed observations, or reasonable estimates that provide valuable management information as well as partial relief to knowledge deficiencies. Supplementary information captures all other sources of information, including traditional ecological knowledge, which does not fit standard scientific knowledge (i.e., it is not based on hypothesis testing). This type of information can be especially useful when science is scarce and local knowledge may be all the information that is available.

Decision-makers, government and organizational staff, and conservation scientists are often responsible for determining what science is available. They may base this on what they find or know to exist in the peer-reviewed literature or reports. They may also limit "available" science to studies that are carried out only on the species or local habitat of interest. For instance, in the Columbia River Basin of the Pacific Northwest, U.S. management of endangered salmon is an important issue. There is a tendency to consider only the information that has been generated in the river basin and to ignore studies from elsewhere in the region or from other similar systems worldwide (Bisbal 2002; see Essay 9.2). This ignores a large amount of available information that may be relevant. Thus, it is important that in determining what is "available," those responsible identify the boundaries that they themselves have placed on available information. It can also be helpful to articulate what information is missing (Bisbal 2002).

What constitute the best science? In academia best science is that which is published in peer-reviewed journals. But peer review is not a panacea, and as Brosnan (2000) argues, traditional peer review is often inadequate for conservation decisions. Relying on a large pool of leading non-advocate scientific advisors or reviewers who collectively serve on a formal panel is one of the most effective means of articulating and using the best available science (Meffe et al. 1998; Brosnan 2000; Courtney and Hudson 2001; Bisbal 2002). Essay 17.1 by Steven Courtney describes the work of the science advisory panel for the California redwoods, providing an example of how this approach can be effective, especially when the process is open and facilitated by a translational scientist.

Scientific uncertainty and risks

Doubt is ubiquitous in science. Scientists use the scientific method to reduce uncertainty and get closer to the truth, but we accept that we will never have perfect knowledge. When scientists feel that there is insufficient information to answer a question, they carry out additional studies until they feel conclusions are warranted and can be supported by the evidence. Conservation decisions do not have such luxuries. Decisions must be made with whatever information is available today, regardless of the quantity or quality of that science. One of the most important contributions that conservation scientists can make to policy is to evaluate the available data and articulate the uncertainty around them. Conservation scientists are often reluctant to include this type of evaluation in their papers or reports because it is not a normal practice in academic science. Scientists correctly assume that their peers can deduce this information themselves, based on the data and analyses provided. However, most decision-makers are not experts in scientific analyses, and so presenting data with guidance on the strength of the conclusions is enormously beneficial to them (Brosnan 2000; Courtney and Hudson 2001). If there is poor confidence around the data, then inherently the political and scientific risks associated with the impending decision will be greater, and policymakers may need to be more cautious. There are a number of approaches that conservation scientists can use to help policymakers understand and cope with scientific uncertainty. They can verbally give an assessment of their confidence surrounding the data and conclusions. But there are also more structured approaches to articulating uncertainty and reducing risk, particularly forms of decision analysis (described by Lynn Maguire in Essay 17.2).

All conservation efforts directed at complex ecosystems have many sources of uncertainty including:

- The natural variation and inherent stochasticity of ecological systems (process uncertainty)
- Inaccurate measurement of the state of ecological systems (observation uncertainty)
- Abstract and simplified models used to predict the response of managed systems to management actions (model uncertainty)
- Fundamental misunderstanding of variables and the functional form of the model (model error)
- Uncertainty arising from the interpretation of incomplete data (subjective uncertainty)

ESSAY 17.1

Conservation Science and Policy in the Real World
The Headwaters Agreement

Steven Courtney, *Sustainable Ecosystems Institute*

During the mid-1990s no forest conservation and management issue was more contentious than the battle over the Headwaters and Pacific Lumber Company lands—80,938 ha of old-growth redwood forests in coastal northern California (Figure A). These forests were the subject of lawsuits and the stage for acts of civil disobedience. Ultimately, through an amalgam of science and politics, an agreement was reached that protected the majority of old-growth forest and allowed the landowners to realize some value from their property. Scientists played a key role in the conservation and management of these forests.

The stakes were high on all sides. One owner, Pacific Lumber Company (PalCo), was unique in still having large areas of uncut old-growth redwoods; all other comparable companies had long ago harvested such big trees. PalCo was owned and operated by the Murphy family since 1904. The family operated as a logging company but had uniquely maintained large stands of old-growth forests in their forest practices. The company had strong connections in the community, and in the nearby company town of Scotia, 1100 residents worked for PalCo. Generations of families worked for PalCo, and were proud of their relationship with the forests and the company. In the 1980s, the company changed hands in a surprise take-over by the Maxxam Corporation that quickly changed the established forest practices. The appraised value of PalCo's old-growth forests as timber was reckoned to be in billions of dollars, and under new direction the company began a program of harvest. However, many environmentalists were opposed to any loss of large redwoods, and set out on a campaign of protests, tree sit-ins, and litigation.

Because environmental laws require that companies comply with all environmental policies and have their timber harvest plans approved, both the federal and state governments became involved. The regulating agencies halted any logging that would lead to loss of endangered species. PalCo counter-sued the government, demanding compensation for lost property values, as a potential "takings" under the fifth amendment of the United States Constitution.

At this point, PalCo and the government agencies agreed to try to resolve the issue by developing a compromise plan that would attempt to protect endangered species while allowing some timber extraction (mostly on second-growth areas). This agreement would take the form of a Habitat Conservation Plan (HCP) allowable under section 10 of the Endangered Species Act (ESA). Essentially an HCP is a plan that allows private landowners to alter or destroy habitat that is home to an endangered species in return for minimizing and mitigating the impact to the species. Thus, the intent is a short-term harm for a longer-term benefit. An HCP is voluntary and is initiated by the landowner, who must then agree to the environmental terms of the agency. If an agreement (HCP) can be reached, then in return the landowner gets certainty that they will be allowed to develop or alter their lands in accordance with the plan, and will not be subject to new environmental restrictions.

Many species of wildlife occur within PalCo's old-growth forests, including three threatened species listed under the ESA—the Northern Spotted Owl (*Strix occidentalis caurina*), Coho salmon (*Oncorhynchus kisutch*), and Marbled Murrelet (*Brachyramphus marmoratus*). Agreement was reached relatively easily on protection measures for the Northern Spotted Owl. In this area of its range, the owl appears to survive well in mixed habitats, including second-growth forests (Franklin et al. 2000) and so there appeared to be many conservation options. Coho salmon protection measures were more complex, and were embedded in an overall aquatics and watershed strategy. However, agreement was elusive for the Marbled Murrelet, and this

Figure A   Intact redwood forests are now rare, and highly valued by many people. Thus, any proposal to log in redwood forests engenders passionate responses. (Photograph © Royalty-Free/Corbis.)

species rapidly became the focus of intense and acrimonious debate.

The Marbled Murrelet has a unique life history. It is a seabird, related to puffins and murres, and feeds on small fishes in the ocean. However, it nests inland, primarily high on the limbs of large trees. The bird flies as much as 80 km from the coast to a suitable nest site. Its plumage is dark brown, making it cryptic in the forest canopy. Adding to the difficulties of finding the bird in the forest is its habit of flying to and from the nest at high speed (over 100 km/hr), primarily at or before dawn. Not surprisingly, the species was little known for many years.

In 1992, the Marbled Murrelet was listed as threatened under the ESA, primarily on the basis of loss of habitat to timber harvest. However, the ecology of the species remained poorly understood. There was little information on habitat associations (e.g., selection of nest sites), demographic parameters (such as reproductive and survival rates), or population trends. Little had improved by 1997 when PalCo and the state and federal agencies were thrust into negotiating an HCP over PalCo's approximately 80,938 ha of redwood and Douglas-fir forests.

Initially, the lack of information on the murrelet was a major obstacle to developing a plan. Agencies such as the U.S. Fish and Wildlife Service (USFWS) sought to maximize protection, while PalCo's prime objective was to maximize value from the land. Essentially the parties were negotiating in different "currencies"—conservation and dollars—without having basic information on, for instance, how many murrelets were present, and where in the forests they were nesting. Negotiations were deteriorating and the parties recognized the need for outside help in developing a common set of data. Consultants and others, particularly U.S. Forest Service scientists, developed sophisticated GIS maps of forest type, and collated all sightings of murrelets in the area. The parties approached the Sustainable Ecosystems Institute (SEI) to facilitate the negotiation on issues of science.

SEI convened a panel of independent experts who, over the course of several years, guided the parties to an agreed consensus on the scientific facts. An important part of the panel selection was that the parties had faith in the expertise, credibility, and willingness of each of the scientists on the panels to be objective. This was irrespective of whether they considered them to be more or less "environmentalist" in their personal values. A panel approach also allowed for broad representation of skills and expertise. The seven scientists on the panel included some murrelet biologists, but also mathematicians and forest ecologists from academia and the U.S. Forest Service. SEI insisted that the science available to the panel and the scientific deliberations be made publicly available on its website. Initially, this was a challenge for the company because, as with all for-profit corporations, knowledge and internal issues are considered privileged information. Their release has the potential to negatively influence competitiveness and profits. Nevertheless, SEI insisted on this openness and the company agreed. This step ultimately provided greater confidence in the science and outcome.

The Marbled Murrelet Science Review Panel (MMSRP) met regularly to review scientific information and to listen to scientific input from scientists who were working on the species and its habitat. Ultimately the MMSRP were the arbiters on issues under debate or contention. SEI's role was to ensure that science issues were addressed appropriately, but also to protect the independence of the panel, and to ensure that the panel was not asked to make policy or management decisions. When issues became highly contentious or the divide between science and policy was opaque, then pressure on the panel to provide more policy guidance or make more management statements increased. The parties sometimes became frustrated at the MMSRP for not providing clearer policy guidelines, but this was outside of their role.

Under the ESA, regulatory agencies such as the USFWS must evaluate projects such as HCPs in terms of "take" and "jeopardy." Take is essentially a quantification of harm done to a species—numbers of birds killed or hectares of habitat lost. Jeopardy, on the other hand, measures such harm in terms of extinction risks of a population or species. Such a change in risk can be measured at a local level, or in terms of damage to the species as a whole. For the PalCo HCP, take was very hard to assess, given that the distribution of breeding murrelets in the forest was unknown. It was not easy to determine the effects of harvest in any particular area. Given these difficulties, it was then impossible to estimate jeopardy.

The MMSRP advised that the parties develop a program of research. Over three years the parties conducted extensive surveys for murrelets in the forest, which resulted in a map of murrelet occurrence. In essence this served to provide an estimate of take. The MMSRP also commissioned a population viability analysis (PVA), which helped to establish the sensitivity of murrelet populations to different factors, in terms of change in extinction risk (analogous to jeopardy).

The population modeling effort was of little direct use in estimating change in extinction risks as a result of proposed management because the uncertainties regarding murrelet ecology made the errors in prediction very large. However, the models were crucial in identifying which factors had the potentially largest effects on murrelet populations, by means of a sensitivity analysis. Although some scientists and politicians argue that models are a waste of resources, in this situation they were essential to the decision-makers in providing good information on which factors could be most important, and then define areas where caution is particularly warranted. Thus, the MMSRP advised a conservative approach, which PalCo and the USFWS and other regulatory agencies adopted. Further, all parties agreed that a program of monitoring to determine HCP effectiveness and further research to resolve outstanding issue was warranted.

Political support for the Headwaters agreement and HCP ultimately led to important conservation measures. As part of the deal the government paid $480 million to purchase old-growth forests. The largest block of old-growth redwoods (the Headwaters stand itself—2025 ha in total) was purchased from PalCo, together with two smaller stands, and these are now permanently protected. Lands were placed in protection under the jurisdiction of the Bureau of Land Management. Most of the remaining old-growth areas on PalCo's ownership are protected for the duration of the HCP, which is 50 years. Logging on the remaining lands is concentrated in younger forests, with added protections around streams and owl nests.

A key question is, Did the involvement of scientists materially contribute to the quality of the final HCP? Using the objective techniques developed in a national evaluation of HCPs (Kareiva et al. 1999), Cody and colleagues (1999)

assessed the quality of the science and the effectiveness of its use in this plan by comparing early and late drafts of the PalCo HCP, which represented the effort before and after the MMSRP involvement, and other recently negotiated murrelet HCPs. The final PalCo HCP was rated as excellent and outperformed other plans, most notably the early draft, which was rated inadequate. Other murrelet HCPs that did not have as integral science input were rated as significantly lacking data or analysis to reach conclusions (Cody et al. 1999). Government agencies also found the panel process so valuable that they enshrined it in the final HCP—the MMSRP meets annually to evaluate the scientific results of the ongoing research and advise on future scientific projects and directions.

Thus, this effort demonstrated that direct involvement of credible scientists can help increase the quality of management plans, and resolve deadlocks between environmentalists and business interests. Without the independent, impartial advice of the panel, there would have been little common ground on which to negotiate. All of the participants knew that their arguments were fairly evaluated, and even if they lost a particular point, this was on scientific grounds alone. The presence of a translational scientist was key to facilitating the communication of science and maintaining a buffer between science and policy, particularly when there was pressure on scientists. This same panel process has since been applied in other resource management disputes, and has been particularly useful in resolving seemingly intractable problems (Courtney and Hudson 2001). ■

- Uncertainty arising from changes in social values or restoration policy, including uncertainty about the location, size, and timing of future stressors to the system (predictive uncertainty)

Conservation scientists can provide formal measures of uncertainty. Moss and Schneider, two scientists at Stanford University, proposed a structured approach to articulating uncertainty tied to statistical likelihood of given observations (Moss and Schneider 2000). For instance, they suggest that scientists report high confidence in their conclusions when statistical analyses reflect high likelihoods (e.g., "very likely" when the probability falls between 90%–99%), but report low confidence when likelihoods are lower than 33% (Figure 17.6). Conservation issues such as global climate change, an issue loaded with uncertainty—can benefit from better discussion of the certainty around the data. Indeed, one of the working groups of the Intergovernmental Panel on Climate Change (IPCC)—which focuses on the atmospheric and oceanographic sciences behind climate change—adopted the Stanford scientists' method of stating uncertainty in its third assessment report.

When data are lacking, Bayesian statistics can be used to generate an initial assessment that can be altered as more data become available (Wade 2000; Ludwig et al. 2001; Ellison 2004). Fisheries management has increasingly relied on Bayesian methods to improve decision-making by allowing more explicit evaluation of uncertainty and its effects on predictions (Harwood and Stokes 2003). Policy analysts often use Bayesian updating as well in their research reports.

Dealing with Uncertainty and Risk through Adaptive Management

As described in Chapter 13, adaptive management was designed to allow resource managers to act in the face of acknowledged uncertainty, designing management actions to reduce uncertainty over time while permitting change in response to surprising outcomes. Instead of making static, "precise" predictions in advance, adaptive management highlights a range of possible outcomes. It treats management as an element of the learning process rather than as an independent step that follows learning, and to be most effective, explicit experimental design to test management outcomes must be built into the project. Bayesian methods are often used to design adaptive management protocols as they are useful for predicting consequences and uncertainty of different decisions, as well as comparing costs associated with those decisions (e.g., evaluating the uncertainty associated with decisions that have different costs; Wade 2000).

Under the adaptive management paradigm, decisions are provisional and contingent upon how the system responds to management action. Adaptive management also is intended to increase the ability of managers to respond to new information. One of the most critical areas in which conservation scientists can improve conservation practice is by helping to design and carry out adaptive management programs.

Beyond adaptive management of individual conservation projects, we can also think about hedging uncertainties in conservation via a conservation portfolio approach. Here, a diversity of conservation approaches are applied, rather than relying on a single approach to

Likelihood
{
Virtually certain (probability >99%)
Very likely (probability ≥90%, but ≤99%)
Likely (probability >66%, but <90%)
Medium likelihood (probability >33%, but ≤66%)
Low confidence (probability <33%)
}

Figure 17.6 Proposed ranges of confidence in scientific results, based on likelihood probabilities. (Modified from Moss and Schneider 2000.)

ESSAY 17.2

Collaborating for Conservation
Using Decision Analysis to Manage "Facts" and "Values" in Conservation Disputes

Lynn A. Maguire, *Nicholas School of the Environment and Earth Sciences, Duke University*

■ Most conservation projects require the concerted action of many participants who may not share the same goals or have the same priorities. Finding a set of conservation activities that everyone can support is a daunting task. Fortunately, there are some structured tools that can help disputing parties develop a common agenda for action. These tools blend concepts from interest-based negotiation (Fisher et al. 1991) with those from decision analysis (Keeney 1992; Clemen and Reilly 2001) to analyze environmental conflicts and promote consensus solutions.

A common feature of disputes, environmental or otherwise, is that the parties focus on actions they want to undertake or prohibit: No clear-cutting! Drill for oil in the Arctic National Wildlife Refuge! They forget to talk about their motivations for advocating these actions—*Why* don't they want any clear-cutting? *Why* should we drill for oil in Arctic National Wildlife Refuge? The answers to these questions reveal the "interests" that underlie their "positions." Clear articulation of these underlying interests provides the raw material for creative solutions to disputes, solutions that may be quite different from the range of actions originally advocated by any of the parties (Fisher et al. 1991). Discovering that an environmental group opposes clear-cutting because of its anticipated effects on black bear habitat opens up opportunities to find other ways to enhance bear habitat; discovering that the underlying interest is reducing stream sedimentation might send the search for solutions in a different direction.

Constructing an "objectives hierarchy" (Keeney 1992; Clemen 2001) representing the important underlying interests of all the parties to a dispute can help steer the conversation toward solutions that satisfy, at least to some extent, the important interests of all the parties. Table A is an objectives hierarchy that represents a portion of the underlying interests of participants who collaborated to plan forest management in the Cedar Creek area of the Sumter National Forest in western

South Carolina (Maguire 1999). The fundamental interest of all parties was the same: to manage this part of the Sumter to meet social goals. As usual, the trouble was that different participants had quite different ideas about which social goals deserved to be included.

An illustration of the second level of the hierarchy is expanded in Table A, which articulates the goals for timber production under the social goal "provide variety of goods/services." Not all parties endorsed all of these goals, and different parties definitely had different priorities when it came to making tradeoffs in reaching these goals. Notice that this hierarchy is a composite in that it incorporates the important goals of all the participants. This approach fosters a collaborative approach to a common problem by showing everyone that a successful solution must address all of the important goals to some extent. In contrast, constructing individual objectives hierarchies that show the underlying interests of just one party at a time would only emphasize their differences.

Each successively lower level on the objectives hierarchy (further to the right in Table A) reveals more detail about the next level up, providing an answer to the question: What aspects of this objective are important? Looking from the bottom up, each higher level in the objectives hierarchy provides an answer to the question: Why is that objective important? To construct a complete objectives hierarchy, a facilitator or analyst can ask participants to propose some fundamental goals and then use these questions to move up and down in the hierarchy, adding details that will help open doors to creative solutions.

The bottom level on an objectives hierarchy (see Table A) consists of attributes (or criteria) that could be used to measure how well a particular objective is being met. Calculating board feet of pine sawtimber is one way to measure how well the objective of producing a variety of wood products is being met.

TABLE A *Partial Fundamental Objectives Hierarchy for the Cedar Creek Area of the Sumter National Forest, SC*

Objective
Manage Cedar Creek area of Sumter National Forest to meet social goals
1. Provide variety of goods/services
a. Timber
i. Sustained yield
1. Indefinitely into future
2. Predictable number of acres each year
3. Variety of mature sawtimber
• hardwood
• poplar
• pine
2. Board feet of pine sawtimber
3. Protect/restore ecosystem
4. Preserve quality of life
5. Educate public
6. Promote economic well-being
7. Preserve options for future generations
8. Encourage research

Source: Modified from Maguire 1999.

A composite objectives hierarchy provides a good starting point for developing a suite of actions that address the major goals of all parties. This is the time to elicit participants' ideas about what actions should be taken or avoided and organize those into a "means/ends network" (Keeney 1992). Figure A shows part of a means/ends network for the Sumter National Forest example. The left-hand side of the diagram is the "top" of the means/ends network; it consists of the "ends" toward which the "means," or actions, are being addressed. Each successive level in the means/ends network (further to the right in Figure A) gives more detail about actions that might be taken to accomplish the desired goals. Lower levels in the network provide answers to the question: How would you accomplish that goal? And higher levels in the network provide answers to the question: Why do you want to take that action? The lowest (far right) elements in the means/ends network are specific activities, from which a combination of compatible activities could be chosen as one of the management alternatives to be evaluated by the participants.

A means/ends network serves as a working hypothesis about how actions taken are translated into impacts on desired goals. Just as participants may disagree on goals and priorities when it comes to assembling an objectives hierarchy, they may have different mental models for how ecological and social systems work. A means/ends network, or a set of alternative means/ends networks, provides a way for participants in a dispute to articulate their views and to begin assembling information to help clarify which models are most plausible. Unlike objectives hierarchies, where the participants themselves are the voices of authority on their own goals, means/ends networks are hypotheses about how the world works and are subject to modification in light of new information. Joint fact-finding is a hallmark of interest-based negotiation, in contrast to the advocacy science pursued by opposing experts in courtroom testimony (Fisher et al. 1991). Jointly sponsored studies and consultation with experts acceptable to all parties can help zero in on a means/ends network that represents a scientific consensus. Short of full consensus, using means/ends networks to display the conflicting views of different parties at

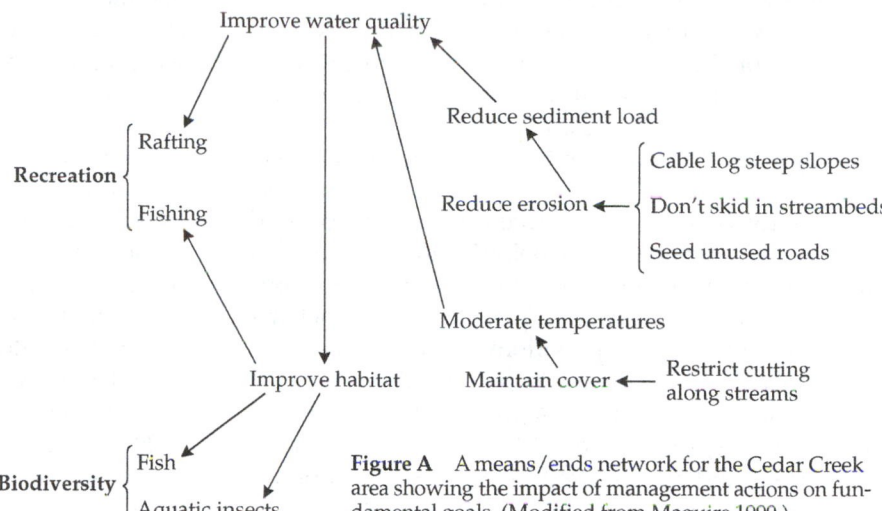

Means/Ends Network

Figure A   A means/ends network for the Cedar Creek area showing the impact of management actions on fundamental goals. (Modified from Maguire 1999.)

least clarifies the sources of their disagreements.

The combination of objectives hierarchies and means/ends networks forges the link between social values and scientific facts that are so necessary to finding solutions to conservation problems that are both acceptable to the disputing parties and technically sound. The bottom level of the objectives hierarchy, the measurable attributes, guide the collection of information needed to evaluate alternative actions in light of social goals. In the case of the Sumter National Forest, the availability of a spatial database made it possible to consider the ecological and financial consequences of alternate schemes for allocating forest land among competing uses. This database helped guide negotiations about what uses would be permitted in specific terrains (Maguire 1999).

After a particular action has been selected and implemented, follow-up monitoring of measurable attributes provides feedback on how well desired objectives are being met over time, thus improving the scientific basis for similar decisions in the future. This monitoring information also serves as a check on the scientific hypotheses incorporated in the means/ends network, so that participants can improve their representations of how the world works and resolve technical disagreements as new information becomes available.

Too often attempts to find consensus solutions to conservation disputes fall

victim to a jumble of arguments that mix disagreements about facts with disagreements about values. The decision analytic tools of objectives hierarchies and means/ends networks can help disentangle these disagreements into their component parts and provide a structure for moving toward consensus.

Of course, these tools are no panacea; there are still many ways to go astray. One common failing is to leave out the views of some decision-makers or affected parties when assembling the composite objectives hierarchy and the means/ends networks. Another way the benefits of decision analysis can fail to be realized is reluctance of some parties to articulate important objectives, such as political aspirations or financial constraints, due to embarrassment or due to the belief that disclosure will compromise their negotiating positions. Internal disagreements within the constituencies represented in a consensus process can also disrupt their effective participation.

That said, wider use of objectives hierarchies and means/ends networks in seeking consensus on conservation problems, by the participants themselves or with the help of trained facilitators and analysts, offers a route to solutions that are responsive to the many goals of the disputing parties and well-grounded in ecological and social science. ■

achieve conservation ends. This approach is described in detail in Essay 17.3 by Holly Doremus.

Science and advocacy conservation scientists have strong personal opinions on how we should conserve species and habitats. But how far should scientists go when speaking out? Where is the line between advocacy for science and advocacy for the environment? Conservation scientists and professional societies have debated this issue for years. All agree that the scientific code of ethics requires scientists to be objective in their methods and nonpartisan in their presentation of science. But should they also then refrain from advocating actions that stem from their interpretations of that science?

For instance, let's pose a hypothetical but realistic scenario. Assume that there is a proposal to log a mature second-growth grove of trees. Many people like to hike in this grove and are opposed to logging. They oppose this particular logging effort on the basis that it will harm the habitat and a threatened species that lives there. Moreover, they know that forests are in decline globally and here is one opportunity to halt their destruction. However, the scientific data do not conclusively show that any threatened species will be harmed by logging this grove. The conservation scientist with the data is also opposed to logging beautiful groves of mature second growth forests. What does the conservation scientist do? They are there to represent the science but they also have opinions and concerns for this forest, and for forests worldwide. Public pressure and their own desires point toward conserving the forest, but the scientific arguments around the threatened species may not support them. Should the scientist present their results? If they do the forest may be logged and conservationists may denounce them. Should they present their data but then provide reams of information on the precautionary principle to persuade decision-makers to take a broader view of the decision? Should they present their data, and also present their personal views distinguishing between when they are speaking as a scientist or an advocate? Should they advocate for a particular management plan or conservation option?

The central issue here is whether scientists should express values by advocating for a particular decision in a debate over conservation policies and management. This particularly concerns the situation when scientists are being relied upon for neutral, credible, scientific information. Mills (2000) identifies two types of "position advocacy" that conservation scientists may accidentally stumble into or deliberately employ to advance their personal agenda. The first is when a conservation scientist advocates for a position while pretending that the position is scientifically supported when it is actually a personal value statement. If science has standing in the debate then they may do this to gain support for their position.

This is simply unethical. Indeed Mills argues that it is as unethical to present personal values as science as it is to misrepresent experimental data in a scientific paper. The second type can create confusion. Here the conservation scientist carefully clarifies that they are expressing a personal value rather than a scientific one. Mills argues that the audience will be confused about whether the scientist is representing science or themselves, particularly if they use scientific jargon when expressing personal opinions. Although the intent is honorable in the second type, sometimes the outcome of both types of advocacy can be the same: The scientist's views become suspect. Mills argues that an attempt by scientists to simultaneously be a provider of scientific information and a position advocate is an inherent conflict of interest.

However, what about the situation when the conservation scientist's own research uncovers what they believe is a critical conservation problem that needs to be brought to the attention of managers, policymakers, or government? The global climate change debate is just such an issue. The science itself is so complex that it is mostly understood only by scientists. It is scientists who are leading the efforts to make the results known to the public and have governments take action. Is this science or advocacy?

Some global climate change scientists are courted to speak out by environmental groups who are concerned that global warming is real and want to take steps to mitigate it. Other scientists, who take a more conservative approach to the data, are courted by industry groups who feel that human-induced global climate is simply not happening (Anderson 1992). By hiring or supporting "friendly scientists" each side is ensuring that its message gets out there and that it is seen to be supported by science. In turn scientists have an audience for their science and interpretations. The upside is that the science is being heard; the downside is that the public perceives a war of words between scientists on one or other "side." Nevertheless, some scientists do believe that in these situations researchers must become advocates to get an important societal message that stems from their scientific work brought before the public. As long as the message is based on the scientific method, many feel comfortable working at the interface of science and advocacy.

Barry Noon and Dennis Murphy describe the scientific and policy process that has surrounded conservation of Northern Spotted Owls in the Pacific Northwest of the U.S. in Case Study 17.3. As biologists working both within and outside of federal agencies, they have worked through the pressures and difficulties of defining and defending their scientific work, and have helped in the complex process of applying scientific insight to long-term conservation plans for Northern Spotted Owls on federal and private lands.

ESSAY 17.3

A Policy Portfolio Approach to Biodiversity Protection on Private Lands

Holly Doremus, *School of Law, University of California at Davis*

■ Because so much of the land and other resources on which biota depend are privately owned, effective conservation policy must include strategies for biodiversity protection on private lands. Although discussions often focus entirely on top-down regulation or voluntary incentives, in fact a broad spectrum of strategies is available to both government and nongovernmental organizations. A portfolio approach employs a mix of strategies to achieve conservation objectives. Because it relies on a diversity of approaches it may be more successful than any single option. While this essay focuses on private lands, the portfolio concept is readily adaptable to management of public lands or the marine environment. The following are some of the strategies that can become part of a policy portfolio.

Educational programs

Education promotes conservation action without compelling it. Educational strategies can be highly general, like the establishment of an environmental education office within the Environmental Protection Agency in 1990. Or they can be quite specific, like programs in recovery plans for endangered species that encourage concern for the species. Education can focus on increasing the ability of those willing to engage in conservation to do so, as the California Legacy Project does by offering technical assistance to landowners interested in conservation planning.

Government acquisition of land or resource rights

A second approach is the purchase of property rights for conservation. Government or nongovernmental organizations can acquire full land ownership or carefully tailored conservation easements that protect target species or ecosystems while allowing compatible land uses. Water rights can be purchased to increase stream flows, or permits that authorize the emission of air pollutants can be bought and retired. Rights can be acquired permanently or temporarily. Short-term agreements, which may be easier to negotiate, can

protect resources while permanent solutions are sought (Neuman and Chapman 1999; Zellmer and Johnson 2002).

Direct Incentives for private conservation action

A third major strategy is to use incentives to encourage private conservation action. Economically rational landowners typically conserve less than the socially optimal amount because they bear all costs of conservation but are not able to capture all benefits. Incentives can correct that imbalance, matching the landowner's interests with the community's. Incentives may be positive (payments for desirable action) or negative (taxes imposed on undesirable action). Cash payments, tax credits, debt forgiveness, technical assistance, regulatory assurances, and even insurance against financial loss have all been used effectively to encourage conservation. Incentives need not have monetary value; they need only motivate behavior. Under the right circumstances, awards that signal community recognition can be effective, low-cost incentives (Uphoff and Langholz 1998; Defenders of Wildlife 2002).

Market creation and improvement

A fourth general strategy is to create or improve markets for conservation. Like incentives, market creation or improvement can help resource owners realize the conservation costs and benefits of their actions. Regulatory requirements can create new markets; mitigation requirements under the Clean Water Act have catalyzed development of a wetlands restoration industry (Gardner 1996; see Chapter 15). Creative use of markets can encourage conservation by providing a partial substitute for lost income from consumptive resource uses. Ecotourism, for example, has been touted as a means to allow communities to profit from biodiversity without destroying it.

Information can contribute to market-based approaches by making consumers aware of the conservation impacts of their choices. The Monterey Bay Aquarium's simple but effective "Seafood Watch" program of green, yellow, and red lists, for example, helps

consumers adjust their seafood purchasing decisions. Certification programs, such as the Forest Stewardship Council's standards for sustainable forestry, attempt to convey more complex messages through a simple signal.

Regulatory prohibitions and requirements

Government can rely on regulatory mandates or prohibitions enforced by sanctions. Regulations can be fine-tuned to balance conservation with other interests. Rather than flatly prohibit an activity, they may simply limit the manner in which it is carried out or require the production or disclosure of information.

Regulation is typically thought of as a "top-down" approach, but it is increasingly common for regulations to be developed by negotiation, with a single regulated party or an entire regulated community (Harter 2000; Coglianese 1997). An example from the biodiversity context is the approval of Habitat Conservation Plans (discussed in Essay 17.1), the details of which are worked out in negotiations between the permit applicant and the USFWS.

Evaluating Policy Options

When evaluating conservation policy, conservation biologists tend to focus exclusively on maximizing conservation results. Economists focus instead on maximizing societal welfare. Both views are too narrow. When deciding how to address a particular conservation problem, policymakers should evaluate a range of alternatives for feasibility, effectiveness, fairness, and implications for the future.

Feasibility

Feasibility asks whether the strategy can be adopted and implemented. No matter how theoretically appealing a strategy, if political resistance prevents its adoption it will not provide conservation benefits. Taxes, for example, can efficiently address many environmental problems (Oates and Baumol 1975), but resistance to novel taxes in the United States limits their usefulness. Political resistance can also limit implementation. Many observers have concluded

that USFWS resists listing species that lack public appeal under the Endangered Species Act (ESA), particularly if protection would conflict with economic activities (Yaffee 1982; Ando 1999). Finally, informational constraints can limit the feasibility of conservation strategies. Because the value of biodiversity conservation is notoriously difficult to quantify (National Research Council 1999), conservation strategies requiring precise measurement of biodiversity values are generally impractical.

Effectiveness

Policymakers must also ask whether conservation strategies will produce conservation benefits. Lack of basic biological information is often an important limitation on policy design. In the absence of reliable information about the target species or ecosystem, it is difficult to identify the best areas for acquisition or regulatory targets. Other kinds of information can also be limiting. Unless violations can be detected and sanctioned, policies will not produce their expected benefits. The effectiveness of the ESA is famously limited by the ability of landowners to "shoot, shovel, and shut up," surreptitiously killing endangered species (Polasky and Doremus 1998). Even when violations are detected, community hostility to regulation can interfere with enforcement. The ability of landowners to foresee and evade future regulation, as by destroying habitat while an ESA listing works its way through the pipeline or by preventing habitat regeneration, must also be considered (Mann and Plummer 1995; Lueck and Michael 2003).

Fairness

Conservation strategies must also be evaluated for their fairness. Costs can be imposed fairly either on those responsible for the problem or on those who benefit from the solution. Many of the benefits of biodiversity conservation are widely shared, which argues for wide spreading of the costs. But it is equally true that activities that destroy ecosystems and harm species impose costs on the wider community. This view argues for imposition of conservation costs far more narrowly on those whose actions threaten biodiversity. Choosing where to impose costs, therefore, requires deciding whether private development rights or public rights to biodiversity should take precedence. Biodiversity policy decisions are highly controversial because the context, historic background, and societal values often allow a strong case to be made on both sides.

Effects on the future

The final factor that must be considered in evaluating conservation choices is their effect on the future. Biodiversity conservation is necessarily a long-term endeavor, undertaken in the hope that future generations can enjoy the option and existence values of biodiversity, the experience of nature, and the benefit of ecosystem services. Ideally, conservation measures should be flexible enough to accommodate unexpected developments and take account of new scientific discoveries, yet permanent enough to withstand pressures for short-term economic or political gains.

The Benefits of a Portfolio of Conservation Policies

A mix of strategies can best address the divergent goals typical of conservation policy decisions, profit from synergies among various strategies, reduce the pervasive uncertainty that makes conservation choices so difficult, and decrease the risk of conservation failure.

Addressing multiple goals

Biodiversity protection is never the only goal of conservation policy. Other goals may include maintaining ecosystem services, protecting species with special cultural value, preserving open space for aesthetic value or recreational use, and sustaining the biotic distinctiveness of different regions. Different goals are best served by different policy measures. Land acquisition will generally be required for goals dependent upon public access, such as providing for recreation. But conservation easements may work just as well at lower cost for habitat protection. Where it is important to avoid threshold effects, incentives and market-based approaches will be riskier than regulation. If protecting ecosystem services whose value can be quantified is the goal, market measures that force producers and consumers to take those values into account may be most useful.

Synergistic effects

Combining conservation strategies can create synergistic advantages. For example, where resources or threats can cross property boundaries, land acquisition may need to be combined with regulations or incentives. Regulation of polluting industries or incentives to reduce pollution, can help ensure the long-term viability of reserves. By the same token, incentives can increase the effectiveness of regulation where affirmative management measures are needed. The Red-cockaded Woodpecker (Picoides borealis) nests in old trees in fire-dependent pine forests of the southeastern U.S., three-quarters of which is privately owned (Bean 1998). Unless the hardwood understory is thinned or burned, the forests become too dense for the woodpeckers. Regulations forcing landowners to undertake expensive management would be difficult to enforce. Combining regulations forbidding destruction of existing nest trees with incentives to encourage understory control is likely to be more feasible, effective, and fair.

Capacity-building education can be combined effectively with incentives. In Mongolia, an NGO known as Irbis teaches herders to make handicrafts, then buys those handicrafts for sale in the global market. That educational program is tied to incentives for conservation through agreements with herding families not to kill snow leopards or the leopards' prey. If no animals are killed, Irbis pays all area families a bonus on their handicrafts. If anyone in the area violates the agreement, however, no one gets the bonus.

Reducing uncertainty and risk

Just as an investment portfolio can reduce risks of financial loss, a portfolio of conservation policies can reduce the risk of biodiversity loss. Poor knowledge of species' needs or unforeseen ecological events can undermine the value of land acquisition or regulation. That risk can be reduced by carefully selecting a portfolio of land parcels and combining them with a suite of regulation, incentives, and education. Ideally, some conservation policy decisions should be treated as experiments, setting up controlled trials of different strategies in places where the biological and social conditions are sufficiently similar to permit meaningful comparison of the outcomes. This is the underlying philosophy of adaptive management (see Chapter 13). But there are a number of barriers to a deliberately experimental approach to management (Uphoff and Langholz 1998; Doremus 2001). First, if failure can bring irreversible consequences, reluctance to experiment is understandable. Second, policy experiments may be susceptible to manipulation. Third, the administrative costs of setting up policy experiments may be unacceptably high. Finally, there may be few areas large

enough, and biologically and culturally homogenous enough, to make suitable experimental subjects.

The biological effectiveness of conservation policies should be, but too often is not, systematically monitored. Creative studies that compare different strategies should also be encouraged. By comparing the status of endangered plant species in different U.S. states, which differed in the extent of regulatory protection for plants, for example, Rachlinski (1998) found a positive correlation between species status and regulatory protection. Of course, this one study is not definitive. Others are needed to inform future choices by helping us understand which strategies succeed, under which conditions, and at what cost. ■

Being a Conservation Scientist in the Real World

It is unthinkable for a conservation scientist to approach a project or experiment without having considered the scientific framework, the specific questions, assumptions, methodologies, options, and analyses. This structured approach allows us to be better scientists. Yet many will stumble blindly into politically charged conservation disputes without considering the real problems that need to be solved: what scientific information is needed, and how to deliver that information to the audience.

Clark (1997, 2001) recommends that conservation professionals follow a structured process for policy involvement that includes five tasks:

1. Identify and clarify goals of those involved or affected by problem and its solution.

2. Describe the trends and history of the problem with respect to the goals, including the strategies participants have been using to affect various outcomes.

3. Analyze the conditions under which these trends have taken place.

4. Project possible future trends in the problem.

5. Invent alternatives to solve the problem and to realize their goals, evaluate how well each alternative might work, and choose one (or a combination) to implement.

Critical in this formulation is the awareness of how people will be seeking to achieve their goals or values, as well as differences in their capability in achieving their goals, and how these affect their interactions. Often scientific information can aid the process by helping to recognize trends and identify their causes, and to make projections into the future. The tasks of identifying and clarifying goals and finding solutions may be most effective when scientists work closely with those formulating policy decisions, and when there is clear communication among all parties.

Working with the policy process

To work effectively in a policy process, scientists need to become adept at summarizing critical information, and developing visual aids or other material to communicate key points (Blockstein 2002). Although most scientists interact with policy processes as advisors, understanding many aspects of the framing of the process, and the role science or scientists may play can greatly improve how much science is ultimately integrated into the process:

Know what the issues are and how science affects them. Often the problem is couched in policy terms, in a social framework, or in a dispute (e.g., one group wants to ban fishing in a bay, and another wants to allow it). Evaluate the issues and concerns, frame them scientifically, and then see what science can offer.

Know who your audience is. Surprisingly, many scientists do not consider this and often believe that just showing up as a scientist is sufficient. Knowing who you are addressing is important for how you structure your message. For instance, are you participating in a peer review panel, providing expert testimony in court or before congress, or speaking at an environmental rally or on TV? Who is your audience, how familiar are they with science, and how diverse are their opinions? How you present information can determine how it will be interpreted, and affect the decisions made.

Know why you are there. For instance, have you been invited to present the most current scientific information on a topic because your science and expertise is relevant, or because you share the values of the group and are seen as someone who will advance an agenda? Are you there because you represent an agency or an organization or yourself?

Be clear on your responsibilities and expectations. Are you responsible for the contents of a report? Are you providing scientific data, interpreting results, or giving opinions? Are you a science advisor? Will you be making the decision or will that be left to others? Are you expected to persuade others of the importance of scientific evidence? Do you expect your audience to support your statements, or do you expect a hostile reception? Do others expect you to support their stance on a conservation issue?

Disclose any potential conflict of interests upfront. Many lobbyists, politicians, and advocacy groups love to use terms like *biased scientists*, and *junk science* to discredit scientists who they feel disagree with them. It makes for great media fodder. There is little you can do about it.

However, do not provide unnecessary ammunition, and if you have a potential or apparent conflict of interest disclose it upfront. For instance, you may be on a committee evaluating the effects of particular types of fishing gear on the marine ecosystem where a decision will result in either allowing or banning the gear. At one time in your career you may have had direct or indirect contacts with one of the groups who will be affected by the outcome. The connection may be a distant one; perhaps one of the affected groups supported your institution, or perhaps you published a paper with one of their scientists. These relationships do not mean that you will not be objective, impartial, and fair, and act out of scientific integrity. But in today's political climate, full disclosure is important to avoid perceptions of bias and conflict of interest. Most committees have conflict of interest disclosure forms; if they do not you can and should request an opportunity for all involved to address any potential conflict of interest.

Distinguish between speaking as an expert scientist and giving your own opinion. There will be many situations where you have a personal opinion that is not supported by data. It is your right to have these opinions, but where there is likely to be confusion between your comments based on scientific evidence versus those based on your own values, identify the source of your comments.

Contributing to conservation policy as a conservation scientist

There are many ways to engage in conservation science and policy, from working in natural resources agencies, to contributing time and expertise to environmental nonprofit organizations. Some conservation scientists prefer to ensure that their results are accessible and understood by decision-makers, others prefer to give advice through science panels, and still others become active in lawsuits and give expert testimony. The common theme is that these individuals are interacting "as conservation scientists" and thus bringing their knowledge and expertise to the issues, in addition to their global citizenship. The

challenge for a conservation scientist is often in combining both science and advocacy.

Conservation science spans a continuum of activities (Table 17.2), but not all of them incorporate policy. Carrying out research and publishing it in peer-reviewed journals is an essential conservation science activity, but it is not equivalent to making science count in conservation policy. Every year hundreds of scientific papers that specifically address conservation questions are published in peer-reviewed journals. Most of them are ignored by the decision-makers who could benefit from them, but who are unaware of or do not understand them. Publications make a difference when either the scientist recognizes the implications of their work and communicates it effectively, or when those who commissioned the work or understand its relevance champion the results. For instance, the scientific conclusions of the Pew Oceans Commission and the U.S. Commission on the Oceans have been championed by those who commissioned them and consequently the results and messages have received much attention. But they will need ongoing advocates to ensure that the scientific recommendations are implemented. Frequently a conservation scientist's job will determine where along the spectrum of activities they focus their efforts. A university professor must focus on research and teaching, but their commitment to conservation may lead them to spend personal time working locally or nationally to protect habitats, and to participate in scientific assessment committees.

The role that will be right for each person involved in conservation practice will depend on one's own values, the skills one enjoys cultivating, and where one remains motivated to be persistent through all the ups and downs of conservation practice. There is a broad array of professional roles for the conservation scientists, as well as many activities and arenas for engagement in the policy process (Table 17.3). The urgency of the challenges before us guarantees that finding a way to be policy-relevant in our professional pursuits will make a lasting contribution to the future of Earth.

TABLE 17.2 *Activities of Conservation Science*

Research	Carrying out conservation-relevant research; this can include policy-driven research and analysis
Publication	Publishing results in natural and social scientific journals
Policy-relevant outreach	Making others aware of the science and results
Interpreting science	Translating science and explaining its implications
Integrating science in political and social arenas	Science advising, review; advising or serving on decision-making committees
Public collaboration	Listening to community needs, working with local groups toward conservation goals

TABLE 17.3 *Examples of Professional Roles that Combine Conservation Science and Policy*

Roles	Application
Professional jobs	
Academia	Teaching and research and opportunities for broader outreach
Nongovernmental organizations	Science organizations, conservation organizations, zoos and aquaria
Government agencies	Forestry, wildlife, fisheries, and other conservation and natural resources departments worldwide at the national, subnational, or local level
Other government roles	Congressional staff (natural resources expert to a senator), committee staff (e.g., staff on the Committee for Environment and Public Works)
International government bodies and organizations	United Nations, World Bank
Private sector	Consulting company, natural resources corporation
Foundations	Program officer for an organization that funds conservation
Arenas for engagement	
Academic activities	Conservation science research, teaching
Political process	Integration of science into laws and policies at government level
Professional societies	Developing guidance for policies, conduct for scientists, means to enhance integration of science and policy
Working with advocacy groups	Conservation groups, trade and industry groups
Media	Effective communication about scientific issues, public outreach and engagement
Legal process	Expert testimony, work to interpret rules and laws
Advisory and review panels	National Academy of Sciences, National Research Council, AAAS panels
Foundations	Collaboration on conservation projects

Activities that can be undertaken in all roles include policy-relevant research; providing policy-relevant information; interpretation of scientific results; convening and managing science advising and review panels; translation and facilitation of science-policy interactions; advising and counsel (including peer review); congressional testimony; expert testimony (legal)

CASE STUDY 17.1

Should the Southern Resident Population of Orcas Be Listed as Threatened or Endangered?

A Scientific, Legal, or Policy Decision?

Deborah M. Brosnan, Sustainable Ecosystems Institute

Orcas, or killer whales (*Orcinus orca*), are among the most recognized and loved marine mammals (Figure A). Orcas occur in most of the world's oceans, but are most abundant in coastal habitats and in high latitudes. Orcas live in matriarchal units that congregate with other units into pods. These pods in turn congregate into populations that have distinct dialects of clicks and whistles. In the Pacific Northwest of the United States, orcas are an ancient cultural symbol for many native tribes. More recently they have become an important source of revenue as people flock to the coast to whale watch. Today, the "southern resident" population is rapidly becoming the center of a major conservation controversy over whether this population should be protected under the Endangered Species Act (ESA), which would trigger actions to prevent extinction and recover the population. As with many conservation decisions, the outcome, and thus the fate of the southern resident orcas rests on a complex mix of science, policy, and interpretation of the law.

Cetacean biologists classify the orcas of the eastern North Pacific Ocean into three reproductively isolated forms: resi-

Figure A Killer whales or orca (*Orcinus orca*) are found throughout the world's oceans, and are particularly iconic to people of the Pacific Northwest of North America. (Photograph © Image State/Alamy.)

dents, transients, and offshore. The three forms differ in morphology, ecology, behavior, and genetic characteristics (Baird 2001). Residents possess a rounded fin, congregate in large pods, and prey primarily on fish. Transient forms have a more erect and pointed dorsal fin, congregate in smaller pods, and feed primarily on other marine mammals. They travel great distances in pursuit of prey and although they move through the habitats of resident populations they do not intermingle with resident or offshore orcas. Offshore orcas are difficult to study, and thus are poorly known. They appear to range from Mexico to Alaska in both coastal and offshore waters, and include fish in their diet. Genetic data suggest that they are reproductively isolated from residents and transients, and that they are more closely related to the southern residents than to the northern residents (Hoelzel et al. 1998).

The southern resident population has been well studied. All individuals are photo-identified, and historical information is available for many that dates back as far as 1960 (e.g., Hammond et al. 1990). The southern residents consist of three pods, and from spring to fall they inhabit the inland waterways of Puget Sound, the Strait of Juan de Fuca, and Georgia Strait. In 2001 there were 78 southern resident whales. Population analyses indicate that southern residents once numbered between 140 and 200, but while there is scientific consensus that the population was larger than it is today, data are scarce, making it difficult to estimate past abundances or carrying capacity of this population reliably. Capture records document a population of more than 100 individuals in the mid-1960s. Major population declines occurred since that time (Figure B). During the early 1960s and 1970s, at least 68 (southern and northern resident) whales were removed or killed during capture for pub-

lic display in aquariums and zoos, leading to a decline of 31% in the southern residents. A second decline of 11% occurred during 1980–1984 as a result of the shortage of reproductive females. Both declines were followed by periods of population growth. Most recently, between 1996 and 2001 the population declined by 20% from 97 whales to 79 whales, although they have increased slightly during 2002–2004 (Krahn et al. 2004; see Figure B).

It is unclear what has caused the recent decline. A decrease in prey availability is one possible candidate. Salmon make up the majority of orcas' diet (Heimlich-Boran 1986; Ford et al. 1998) and they may specifically target Chinook (Ford et al. 1998). However, fall, winter, and spring salmon runs are at historically low levels (e.g., Weitcamp et al. 1995; Johnson et al. 1997; Yoshiyama et al. 1998), leading to prey deprivation during some months (Plater 2001). Pollution is another possible factor contributing to orca declines. These whales carry a high organochlorine burden that currently exceeds levels shown to have serious, sublethal effects in other marine mammal species (Ross et al. 2000; Ylitalo et al. 2001; see Essay 3.1). Organochlorines are stored in fat and it has been suggested that during times of nutritional stress organochlorines could be mobilized into the blood stream along with stored lipids from blubber. Once in the blood stream they may cause immune suppression or have other detrimental effects to the health. Observed mortality patterns suggest that this may indeed be occurring (Krahn et al. 2002). Increased disturbance from whale watch boats and ship traffic has also been suggested as a contributing factor (Krahn et al. 2002).

Alarmed by the rapid and continuing decline of the population, in 2001 the Center for Biological Diversity and ten other groups petitioned the National Marine Fisheries Service (NMFS) to protect and recover the orca population by listing it

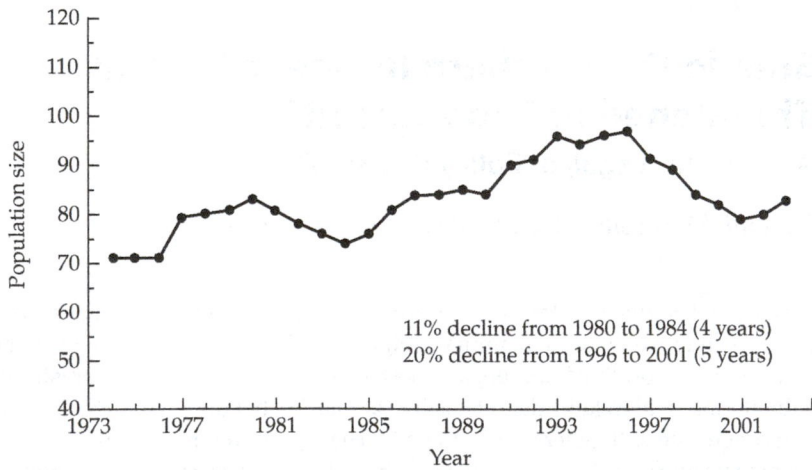

11% decline from 1980 to 1984 (4 years)
20% decline from 1996 to 2001 (5 years)

Figure B The population size of the southern resident orca experienced a 20% decline between 1996 and 2001, prompting a petition to list the population as a threatened or endangered "distinct population segment" under the Endangered Species Act. The population is recovering from a 31% decline in the 1960s and 1970s when animals were removed for aquariums and zoo displays worldwide.

as threatened or endangered under the ESA. Under the ESA, vertebrate species can be protected as full species, subspecies, or a distinct population segment (DPS). A DPS for purposes of ESA protections must be discrete and significant. Discrete means separated from other populations by physical, physiological, ecological, or behavioral factors, or separated by an international political boundary that has different protection, regulatory, or conservation measures. A population is significant if it occurs in an unusual ecological setting, if its loss would leave a significant gap in the species' range, if there is evidence that this population is the only surviving occurrence in its historic range, or that the population differs genetically in a marked way from other populations.

When the petition to list the orcas as endangered was received, NMFS concluded that there was sufficient scientific evidence to consider listing the population. NMFS then assembled a biological review team consisting of 11 of their agency's scientists with expertise in conservation biology, genetics, risk assessment, risk modeling, toxicology, and contaminants. This team produced a comprehensive scientific status review and forwarded it to the NMFS decision-makers. The scientific report reached a number of conclusions including:

1. The southern resident orcas are discrete from other orca populations. Both microsatellite DNA and mitochondrial DNA data clearly show that southern resident orcas belong to an independent population. Genetic results are supported by differences in summer ranges.

2. The factors that threaten the southern residents are likely to continue into the future. Reduced quality and quality of prey is expected to continue to affect them. The population may be at risk for chronic, serious, sublethal effects from organochlorine concentrations. They are also at some risk from oil spills and whale watching.

3. Using population viability analysis (PVA) the scientists on the biological review team calculated an extinction probability of 12%–30% in 100 years based on the average mortality rate from 1992 to 2001. The risk of extinction decreased to 1%–5% in 100 years when they averaged the mortality rate from 1974 to 2001.

4. The scientific review team unanimously agreed that the current taxonomy (which recognizes only one species of orca worldwide) is outdated and probably incorrect. Genetic and ecological differences exist to warrant splitting the species into two or more species or subspecies. Moreover, ecological, morphological, and genetic evidence is sufficient to suggest that transients and residents belong to two separate taxa. The science team pointed out that the current classification—although likely inadequate—lingers because formal taxonomic changes are slow to occur and lag behind current knowledge.

5. However, the scientists concluded that based on the one-species classification currently in effect in the published literature, the southern resident orcas did not constitute a DPS of the currently recognized global species taxon. Recognizing the inadequacy of the taxonomy, the team attempted to identify possible taxonomic breakdowns for the species and the DPS to which the southern residents belonged, but they could not reach agreement among themselves on the preferred or most likely alternative classification.

Although there is no agreed upon standard for endangered status, a scientific group convened by NMFS (Angliss and Lodge 2002) recommended that a 1% probability of extinction in 100 years could meet the requirements for listing a DPS as endangered. Based on these standards and the conclusions of the status report, southern resident Orcas exceed the threshold for extinction, and thus should be listed. Moreover, they also fall under the critically endangered status criterion of the IUCN Red List. However if the southern resident Orcas are not judged to be a DPS, and the species as a whole is evaluated, then the 1% criterion is not met and they should not be listed.

It is worth reflecting on the scientific information before the NMFS decision makers. The scientific evidence that the population is declining is undisputed. Equally, all the scientists agree that there is a substantial risk of extinction. They believe that the current taxonomy is probably wrong, but disagree on what might be the correct classification. How should the agency evaluate this information? For instance, should they make their decision based on the declining population trends, the risks to the population from ongoing threats, or the population's status as a DPS? How should the agency deal with taxonomic uncertainty and disagreement among its scientists? Is this issue even important to consider in the conservation of this population?

On completion of the biological status review, NMFS decision-makers concluded the southern resident population "faces a relatively high risk of extinction." They further concluded that the population—although discrete—is not a significant population segment based on its classification as a single species. Thus it failed to meet the standard of significance that is required by NMFS' own DPS policy, and consequently failed to meet the standard required by law for listing a distinct population segment. Therefore NMFS decided not to list the population as threatened or endangered and opted not to take action to prevent its extinction or enable its recovery.

Shortly afterward, the Center for Biological Diversity (CBD) sued NMFS (Center for Biological Diversity et al. versus Robert Lohn et al. 2001). In their lawsuit, the CBD claimed that NMFS could not use the significance criteria when evaluating a DPS for listing (indeed they argued that discrete and significant are surely synonymous), that NMFS had failed to protect the only resident orca population indigenous to the contiguous U.S., and that NMFS relied on a definition of the orca species that falls below the best available scientific information standard.

At first blush, this seems like a straightforward scientific issue for the courts to decide. But it is instructive to examine how the scientific and policy arguments played out in the courtroom. To a conservation biologist, it seems appropriate that if a decision is being made that determines the survival of a population or species and that decision is going to be made on the best available science, then all the relevant scientific evidence should be evaluated first. However, in a legal setting this is not what happens. The first consideration of the court in these situations is procedural. In the case of the southern resident orcas, the court first looked at this suit under the Administrative Procedure Act (APA) standard. This standard means that a court will only interfere with an agency action (including an environmental decision) if the decision is arbitrary, capricious, an abuse of discretion, or otherwise not in accordance with the law. This standard presumes that the agency decision is a valid one and it affirms that decision if it is reasonable. Moreover, in the case of scientific information, when there are competing scientific data or expert opinions a court must always defer to the agency's technical expertise even if "a court might find contrary views more persuasive" (Natural Resources Defense Council 1989). Thus, scientific evidence and policies as interpreted by the law are highly deferential to the regulating agencies, and they place a high burden of proof on external scientists and others who may disagree with agency findings and conclusions.

In evaluating this case, the court ruled that NMFS had been reasonable in requiring a "significance" as well as "discrete" criterion for listing a DPS. Moreover the court considered that NMFS, in developing its DPS policy, was acting within its authority and following the intent of U.S. Congress who passed the statute. Thus, it upheld NMFS policy on DPS and concluded that claims of the Center for Biological Diversity were not supported. However, the court ruled that the agency had ignored its own experts who concluded that the global taxonomy is inaccurate. The court decided that the best available science, as stated by the agency biologists, demonstrates that transients and residents should be considered separate subspecies. Thus, the court ruled that NMFS decision-makers likely made an erroneous evaluation of the significance factor. The court did not order NMFS to reverse its decision and to list the southern residents as had been requested by the petitioners. But the agency was ordered to reevaluate its listing decision in accordance with the finding and standards of the ruling. Here the judge gave NMFS some additional instructions on what he considered to be important scientific evidence. He asked them to determine whether the population's social structure, behavior, rituals, language, and knowledge should be considered in its evaluation of significance. NMFS was given 12 months to issue a new listing decision.

In reviewing this decision there are a number of elements common to many conservation decisions. First, scientists played key roles in research, analysis, and synthesis (both for government and for conservation groups). However, actual decisions were made by lawyers, judges, agency managers, and policymakers, and most of the decisions were based on existing laws and policies, with science playing a supporting role. This scenario plays out hundreds of times a year in conservation disputes around the world.

Second, only very specific aspects of the science were considered relevant. What constituted relevant science was determined by the law and policy, and not by conservation scientists. In this case, the NMFS biological review team considered only population trends and taxonomy because this is required by the ESA. Other factors including the role of the species (e.g., a top predator or possible keystone species) in the biological community, or the effects of its extinction on the dynamics of the biological community were deliberately excluded because they fall outside the ESA statute.

Third, different scientists reached different conclusions on extinction risks. Each set of conclusions was equally valid based on the data used in the analyses. The scientists who worked with the Center for Biological Diversity carried out a population viability analysis (PVA) based on different parameters (including using different age distribution, survival rates, carrying capacity, and model iterations) from those applied by NMFS, and consequently produced results that reflected higher extinction risk than those of the agency scientists (Taylor and Plater 2001). Such differences in opinion are not unusual in science; indeed these differences are considered healthy and key to scientific progress. However, in the courtroom one result will usually win out, and deference will be given to agency results and conclusions provided that they are reasonable. Moreover, even if the court feels that other scientific evidence and opinions are more persuasive, they must favor the agency science provided it is reasonable.

In the case of the orcas it was the fact that agency scientists articulated their uncertainty with current taxonomy that played a large part in the court's decision to order a reevaluation of the listing decision. Had agency scientists simply based their conclusion on current peer-reviewed published literature (a reasonable action) and other scientists argued its inadequacy, agency science would have been given preferential treatment.

In December 2004, NMFS recommended listing the southern resident population. The new status review concluded that the southern resident orca are probably a DPS of an as yet unclassified subspecies of orca that includes all resident populations in the North Pacific (Krahn et al. 2004). Further, this DPS appears to be facing continued threats, and the biological review team is concerned about its viability, and therefore recommends its listing under ESA as "threatened" (Krahn et al. 2004). Thus, although the legal process at times seemed to move away from the greatest safety for the species, the development of the policy over a longer time appears far more promising. Regardless of the ultimate decision, this case illustrates the complex role that scientific information plays in conservation policy.

CASE STUDY 17.2

Elephant Conservation in Sri Lanka
Integrating Scientific Information to Guide Policy

Prithiviraj Fernando, Center for Conservation and Research, Sri Lanka

On the island of Sri Lanka scientists are trying to make information from their studies of elephant biology available to policymakers, managers, and the community. Changing existing policies to make use of new science is always challenging, and the conservation scientists must be both advocates for science and messengers to the larger community. In many cases, creating productive partnerships with community groups most affected by any conservation plan is critical. In the case of Asian elephants (Figure A), no meaningful conservation can take place unless surrounding human communities are supportive.

The Asian elephant's (*Elephas maximus*) range has decreased to approximately 15% of its extent of 2–3 millennia ago (Olivier 1978; Sukumar 1989), and it is listed as endangered on the IUCN Red List (IUCN 2004). Habitat loss from conversion of natural landscapes to human-dominated ones that exclude elephants is the underlying cause. But proximal threats include capture for domestication, hunting for ivory, and conflict with people.

Elephants are "edge species" preferring the ecotone between forest and disturbed habitat, as long as there is sufficient freshwater. Asian elephants prefer to feed on pioneer species that colonize gaps in the vegetation and are common in secondary forests. Sri Lanka has no perennial, natural, lentic systems. Consequently, during prehistoric times, water resources available for elephants were limited to the rivers that radiate out from the central highlands. Prior to the fifth century B.C., the total elephant population in Sri Lanka was likely less than 5000 individuals, as Australoid hunter-gatherers did not disturb primary forests (Fernando 2000). However, Sri Lanka was colonized by people of Indian origin in the fifth century B.C., and the subsequent rise of their culture had a huge impact on land use patterns and elephants. Countless numbers of freshwater reservoirs were constructed by damming and diverting rivers, and primary forest in the dry zone was converted to irrigated agriculture. Elephants were used extensively in cultural events, wars, and as work animals (Jayewardene 1994). Thus, agricultural land use excluded elephants from the centers of civilization, and large-scale domestication depleted elephant populations in surrounding areas. Consequently, this ancient civilization drastically reduced the numbers of free-ranging elephants to about 2000 animals, mostly restricted to refugia (Fernando 2000).

After the decline of this civilization in the fifteenth century A.D., secondary forests regrew and thousands of freshwater bodies dotted the landscape. Sri Lanka's dry zone was trans-

Figure A   An Asian elephant (*Elephas maximus*). (Photograph by P. Fernando.)

formed into enriched elephant habitat, which set the stage for a massive resurgence of elephant populations. Elephant numbers in the dry zone shot up to about 8000 animals (Fernando 2000). Concurrently, the country came under colonial rule, and the wet zone was intensively settled and cleared for cash crops. Elephants were hunted and practically eliminated from this area. During the colonial era, the dry zone was thinly populated by people practicing shifting cultivation. Slash-and-burn agriculture creates and maintains secondary growth, and together with the innumerable fresh water bodies, the dry zone of the country thus remained optimal elephant habitat for over 500 years.

Sri Lanka's post-colonial drive to develop the dry zone led to large areas of elephant habitat being destroyed for development by irrigated agriculture. However, in contrast to the ancient civilization and colonial era, the present-day approach toward elephants is more conservation minded. The environmental atti-

tudes of the twentieth century, as well as awareness of the threatened status of Asian elephants, led to a national desire to conserve the elephants that inhabited rapidly expanding agriculture areas. Because agricultural development and elephant presence are mutually incompatible (Figure B), mitigating the ensuing human–elephant conflict became the main focus of elephant conservation and management.

Challenges Arising from Early Conservation Efforts

Conservation policy focused on using and expanding existing protected areas to prevent total elimination of elephant habitat, and to reduce conflict with farmers. Elephants were translocated from areas slated for agriculture into protected areas. Further, an attempt was made to confine elephants to protected areas by constructing electric fences on the boundaries. But translocation did not eliminate elephants from human habitats, and electric fencing did not restrict them to protected areas, and so this strategy failed to eliminate human–elephant conflict (Fernando 1993; Rudran et al. 1993; Jayewardene 1994, 1996). Worse, the translocation effort, while well intentioned, did not consider one key point: Translocation of elephants into protected areas endangers those already there (Fernando 1997). Elephant populations had inhabited these "protected areas" for centuries prior to their designation, and these populations were already at well-established carrying capacity. Introducing new animals creates competition for resources, which can destabilize elephant population structure.

Elephants in Sri Lanka do not migrate and have well-defined home ranges that are around 100 km^2 in extent (Fernando 1998; Weerakoon 1999; Fernando and Lande 2000). Genetic studies have shown that these patterns are not of recent origin, but have been preserved for thousands of years (Fernando et al. 2000). Elephants that range entirely inside protected areas have an excellent conservation future as they do not come into conflict with humans, and hence should be considered key populations for conservation of the species in Sri Lanka. However, translocating large numbers of elephants into protected areas, where resources are limited, endangers these key populations that otherwise had good conservation prospects (Fernando 1997).

Compounding the detriment of translocations into protection areas, management of these sites was decreasing the effectiveness of elephant conservation. Most areas designated as protected areas within elephant range were part of the slash-and-burn agricultural cycle. Slash-and-burn agriculture is generally viewed as a destructive method at odds with conservation (Abeywickrama et al. 1991). However, in the lowland dry zone of Sri Lanka, traditional slash-and-burn agriculture is an intermediate disturbance regime. In its traditional practice, a plot is cultivated for two to three successive years, and then left fallow for five to ten years. Studies of elephant habitat use have shown that traditional slash-and-burn cultivation leads to a mosaic of successional stages that is ideal for elephants,

Figure B In an effort to keep elephants out of crop fields, farmers use tactics such as this nail trap, which is intended for the elephant to step on and become lame. (Photograph © Vivek Menon/naturepl.com.)

supporting high elephant densities. However, agricultural methods such as irrigation, mechanized plowing, and use of fertilizer and herbicides enables permanent intensive cultivation of these areas. This manner of land use creates bare land that elephants cannot use. Unfortunately, as modern agriculture made large areas entirely unsuitable for elephants, the protected areas meant to support them declined in quality because people practicing slash-and-burn agriculture were excluded from newly designated protected areas. The "hands off" management regime resulted in a secondary climax of mature scrub habitat where elephants can find little fodder (Mueller-Dombois 1971). Thus, the protected areas meant to support and protect additional elephants displaced by agricultural development began changing into suboptimal elephant habitat.

Currently, more than two-thirds of elephant habitat in Sri Lanka is outside protected areas. With a human population growth rate of over 750 people per day in Sri Lanka, there can be little doubt that more and more elephant range will be developed for human use. Given the need to accommodate elephants displaced by development and the importance of not jeopardizing existing elephant populations, or other species in protected areas, an alternative approach to conservation and management is essential for the long-term survival of elephants in Sri Lanka.

Directions for the Future

The conservation and management of elephants face challenges that are scientific, practical, and cultural. All of these will need to be integrated into an overall strategy. From a purely management point of view, the ideal approach would be to eliminate elephants that are in areas developed for human use by culling or capture for domestication, and to manage the habitat in selected areas to increase their carrying capacities, so as to maintain a viable population size. Even though the con-

tinued killing of elephants by farmers in poorly planned development and settlement projects in elephant habitat amounts to culling, for socio-cultural and political reasons, officially culling elephants as a management policy is unacceptable in Sri Lanka. Similarly, given the environmental attitudes and the endangered status of Asian elephants, large-scale capture for domestication is also unacceptable. Therefore, a way of accommodating elephants displaced by development has to be devised.

Because the elephant is an edge species, habitat management for them is not difficult. Elephants can survive and be managed at high densities provided there is sufficient availability of fodder and water. Based on their research on elephant habits and needs, conservation scientists have suggested that traditional slash-and-burn agriculture be used to create better elephant habitat, and be incorporated into a holistic management approach. This approach envisions management of some present day elephant habitat outside protected areas as "elephant ranges," by regulating the practice of slash-and-burn agriculture. But because of elephant feeding preferences and the need for pioneer species, regulation needs to ensure that a suitable cycling regime for the creation of an optimal combination of successional stages of vegetation is adhered to, and inappropriate methods of cultivation are prevented. This form of management is more likely to be suitable for habitat contiguous with protected areas and large areas of land outside protected areas. Such a management regime is more in line with current approaches to conservation, as it takes into account human requirements and is conducted with the participation of local communities.

Thus, a successful elephant management strategy for the future should consist of land-use planning that recognizes and delineates the landscape into three zones: "human habitat," from which elephants are excluded; "managed elephant range," where habitat management is conducted through appropriate land-use management practices, enabling high elephant densities; and "protected areas," where no specific habitat management for elephants is conducted, but where elephants would exist at lower densities. Managed elephant range and protected areas would be contiguous and elephant barriers would be constructed on the ecological boundaries between "human habitat" and "elephant ranges" would help keep elephants out of high production agricultural areas.

Translocation of elephants from areas opened up for development into areas managed for them would enable the mitigation of human–elephant conflict, and would not jeopardize the future of elephants in protected areas; hence it would address the main shortcomings of the current strategy. In contrast to the present practice of translocation into protected areas, translocation into areas managed for elephants would be effective, as the managed habitat would support a larger number of elephants than were present prior to the translocation. However, the conservation scientists recommend that such management interventions be carried out with pre- and post-translocation monitoring of translocated animals as well as those already in the managed areas. This will enable detection of any unforseen detrimental effects and their appropriate management.

Supporting activities will also need to play an important role in such a management strategy. Managing "elephant ranges" by allowing regulated slash-and-burn agriculture will cause some level of crop depredation by elephants. Even if barriers are sited solely on ecological boundaries occasional breaches will still occur. Therefore activities such as compensation or insurance schemes may need to be streamlined to make them more efficient and acceptable to farmers. Community organization for crop protection can be developed and instituted. Although rural residents are conservation-minded generally, and over 78% support elephant conservation, many express their concern about the costs they bear from elephant damage to their crops (Bandara and Tisdell 2003). Economic compensation would improve local support for elephant conservation measures. A contingent valuation analysis showed that over 93% of urban residents surveyed in Colombo, Sri Lanka's capital, are willing to contribute to elephant conservation efforts, suggesting that a broad national strategy could be successful (Bandara and Tisdell 2004).

Additional forms of income-generating activity related to elephants such as community-based ecotourism can provide local communities with economic incentives that can offset losses from elephants. Elephant populations not subject to off-take by poaching or human–elephant conflict can increase at rates up to 4%–5% annually and can soon exceed what restricted areas can support (Fayrer-Hosken et al. 1997). Therefore, concurrent with habitat management, close monitoring of the health and demography of elephants as well as the effects of elephants on the environment needs to be carried out. To be successful, new efforts must convince local residents that elephant populations can coexist without significant costs to their households.

Making the Science Count in the Adoption of New Strategies

Conservation scientists have been active in bringing their research forward and proposing science-based strategies for elephant conservation. But as with many policymakers, the response to new ideas has been slow and the scientists continually confront the same challenges that all conservation scientists face the world over—integrating scientific information to a system where the decisions are based on social, personal, economic, and cultural factors. It takes time to get the information out to all the players; and it takes effort to present it in useful ways and to get science adopted into policies and on-the-ground action. Elephant management in Sri Lanka continues to be focused mainly on limiting elephants to protected areas by translocation from outside areas. Currently, the conservation scientists are continuing their dialogue with the authorities. But now they are beginning a program to provide wider exposure of the new science and conservation proposals

through the popular press and the media. In addition to working with the Department of Wildlife Conservation, they are also working with different ministries and departments that have an impact on land use management, such as the Forest Department and the irrigation and development authorities.

Through working together with the authorities and the people residing in areas with elephants, they hope to develop a management and conservation plan for nonconservation area elephant range that is practical and will benefit both people and elephants.

CASE STUDY 17.3

Management of Spotted Owls
The Interaction of Science, Policy, Politics, and Litigation

Barry R. Noon, Colorado State University and Dennis D. Murphy, University of Nevada, Reno

Application of the theoretical and applied principles of population and community ecology in conservation planning is an important step in realizing species protection on the ground. But, political pressures, legal proceedings, and policy decisions can dictate the success or failure of a species management plan. The challenge of conserving the Spotted Owl (*Strix occidentalis*) has epitomized the high-visibility struggle that can ensue when protective legislation restricts uses of natural resources.

The process of developing a scientifically sound conservation strategy that is defensible to political attacks, and is likely to be adopted and implemented, can be the most difficult and least discussed aspect of conservation biology. Many strategies, even those built on a firm foundation of defensible science, will fail if the biologists involved are inept at defending their plan against inevitable criticisms, or are unable to convince decision-makers of the true costs to society of failing to implement conservation actions.

The Northern Spotted Owl

Perhaps more than any other threatened or endangered species, the Northern Spotted Owl (Figure A) has epitomized the struggle between groups representing disparate value systems in a land of limited resources and unlimited demands. The debate has been oversimplified as a choice between employment and economic vitality on the one hand, and the survival of species and ecosystems on the other. This dichotomy provoked lawsuits and intense public and scientific disagreement; and, with rapidly diminishing options in the 1990s, it led the United States Congress and then President Clinton to judge the Spotted Owl–timber harvest situation a conflict requiring resolution at the highest political levels.

A 1989 directive by Congress to four U.S. federal agencies (Fish and Wildlife Service, Forest Service, National Park Service, and Bureau of Land Management) convened the Interagency Spotted Owl Scientific Committee (ISC) to "develop a scientifically credible conservation strategy for the Northern Spotted Owl." As members of the ISC, we struggled to devel-

Figure A   The Northern Spotted Owl (*Strix occidentalis caurina*) is a focal point of conflicts between endangered species preservation and short-term economic interests. (Photograph by D. Johnson.)

op a scientific protocol using the rigor of strong inference (that is, hypothesis testing, discussed later), which would allow consideration of both biological and nonbiological factors in the development of a Habitat Conservation Plan. The strategy developed by the ISC and presented to Congress (Thomas et al. 1990) was foundational to the development of the Northwest Forest Plan (FEMAT 1993) and the process, logic, and rationale employed was the basis for all subsequent conservation planning proposals for the subspecies.

Conservation planning for the Northern Spotted Owl has had a long and complex history that reads like the plot of a po-

litical novel. Protagonists, antagonists, confrontations, disputes, secret memos, political pressure, litigation, media distortion, and personal attacks have been everyday players and events. Set against this background of political and legal turmoil, our challenge was to bring to the forefront all information pertinent to the preservation of the species and to provide a defensible conservation plan that appropriately considered scientific uncertainty and competing value systems.

The California Spotted Owl

Scientific and management interest in the California Spotted Owl was stimulated by release of the ISC report (Thomas et al. 1990), and by concern that this subspecies was about to be petitioned for listing under the Endangered Species Act. In response, the U.S. Forest Service, the California Department of Forestry, and other state and federal agencies established a scientific team (Verner et al. 1992) to evaluate the status of the California subspecies and, if needed, to recommend changes in land-management practices. The expectation was that a proactive management response would preclude the need to list the California Spotted Owl at a later date.

The California Spotted Owl is beginning to accumulate the contentious history of legal, political, and scientific debate that characterizes its northern counterpart. The debate focuses on the management of 11 national forests that extend the length of California, with topics that are strikingly similar to those in the Pacific Northwest, focusing on tradeoffs between the harvest of large trees and political pressures to increase short-term exploitation.

A petition to list the California Spotted Owl as a threatened species under the ESA was filed in 2000. In 2003, the U.S. Fish and Wildlife Service determined that a listing was not warranted. This decision was justified, in large part, because of a new, range-wide management plan adopted by the U.S. Forest Service in 2002 that conferred significant protections to California Spotted Owl habitat. Importantly, the 2002 management plan was rescinded in 2004, and replaced with a new plan that offers fewer protections and allows considerably higher levels of timber harvest.

Contrasting Conservation Strategies

The ultimate measure of the success of a conservation strategy is long-term persistence of populations in the natural habitats that support them. Nonetheless, the immediate targets of conservation planning are not normally populations of specific sizes, but explicit management guidelines, habitat reserves, or other set-aside lands that are designed to assure species persistence. Conservation plans can differ widely in their design and subsequent implementation. Two extremes are represented by (1) plans that designate fixed, or static reserve boundaries—spatially referenced distributions of habitat blocks designed to support locally stable populations given the condition of facilitated dispersal among blocks—the Northern Spotted Owl strategy; and (2) plans that do not specify fixed reserve boundaries but attempt to retained suitable habitats widely distributed across the landscape—the California Spotted Owl strategy.

The reserve strategy is most appropriate for species that occur in largely stable (or climax) communities that are in decline due to habitat loss and fragmentation. For such species, population declines must be arrested by stabilizing both the amount and distribution of suitable habitat (see discussion in Lande 1987, and Lamberson et al. 1992, 1994). In contrast, dynamic management plans may be appropriate for species that do not demonstrate immediate and significant local population declines, but are exposed to ongoing, landscape-wide, degradation of their habitat.

In those distinct planning exercises, two important features proved central to producing conservation science that was credible, defensible, and repeatable (Murphy and Noon 1991). The first was the rigorous application of the scientific method, both in the process of gathering and analyzing data, and in communicating results to land use planners. The second was the development of clear operational definitions for the crucial terminology that recurs in legislation, management standards, and guidelines. Good science is rendered worthless when delivered to meet vague goals that are phrased as abstract biological concepts.

Creating a Scientifically Rigorous Conservation Plan

Scientific rigor in conservation planning is often difficult to achieve because most ecosystems are not amenable to experimental manipulations. The targets for conservation are often large, mobile species that exhibit complex behaviors and can be widely distributed across highly diverse landscapes that have been logged, grazed, cultivated, drained, roaded, and beset by introduced species. This study context is often highly opaque to systematic, rigorous experimental design. As a result, most management decisions are made on the basis of incomplete information drawn from disparate sources. Nevertheless, all management plans have properties that can be stated as falsifiable hypotheses and subject to testing with data (Murphy and Noon 1992).

The conservation planning process is dependent on spatially explicit ecological information—particularly the distributions of populations, habitats, and resources. That information is best portrayed as independent map layers, including such information as political boundaries and ownership, topographic features, vegetation types, distribution of roads, and elevation demarcations. When overlaid to form a composite map, these layers collectively provide an initial outline of habitat "polygons" that are candidates for inclusion in a reserve network (see Chapter 14). This preliminary reserve design map has many spatial properties that can be expressed as falsifiable hypotheses and subject to testing. Successful reserve designs are tested with available information and adjusted as necessary. The final product, a conservation strategy, is then portrayed as a map of reserve boundaries or special management zones.

The basic tests of hypotheses and emerging guidelines that dictate the number, size, shape, and spacing of reserve areas

for the static reserve design are also applicable to dynamic management designs. Implementation of a management-based conservation plan, however, can be considerably more complex. To be successful, managers must be able to project the shifting configuration of a defacto reserve system that will result from expected management actions. For Spotted Owls, this ongoing process requires the use of dynamic maps integrated with vegetation succession models to forecast where, and at what future time, degraded habitat becomes suitable. The ultimate task in dynamic planning is to schedule contemporary management actions to assure that essential habitat components will be present at all future times within the planning horizon. This challenging process should also be carried out within a hypothesis-testing framework.

Hypotheses Tested and Reserve Design Principles Invoked

To determine whether the Northern and California Spotted Owls were threatened subspecies and in jeopardy due to logging practices, we tested the following null hypotheses (Murphy and Noon 1992):

1. The population is stable or growing (i.e., the finite rate of population change (λ) of owls is ≥ 1.0).

2. Spotted Owls do not differentiate among habitats on the basis of forest age or structure.

3. No decline has occurred in the areal extent of habitat types selected by Spotted Owls for foraging, roosting, or nesting.

Analyses for the Northern Spotted Owl

For the Northern Spotted Owl, the first null hypothesis was rejected based on the observation that populations were declining substantially (λ was significantly less than 1.0 at two long-term study sites; Murphy and Noon 1992). Subsequent tests of this hypothesis, based on additional study sites and additional years of data (USDI 1992), indicated that populations of resident, territorial females declined significantly, at an estimated rate of 7.5% per year, during the 1985–1991 period. No studies found areas of stable or increasing populations.

The majority of Northern Spotted Owl habitat studies supported rejection of the second null hypothesis and provided evidence that Northern Spotted Owls select old-growth forests, or forests that retained the characteristics of old forests (Murphy and Noon 1992). The exception to this pattern was in coastal redwood forests of northern California (<7% of the owl's range) where owls are also found in younger forests that retain some residual old-growth components. Since the ISC report, numerous studies have confirmed that Northern and California Spotted Owls prefer old-growth habitat, providing additional falsification of hypothesis 2 (FEMAT 1993; Noon and McKelvey 1996; Blakesley et al. in press).

The rejection of hypothesis 2 led to the test of hypothesis 3. Based on data from national forest lands in Oregon and Wash-

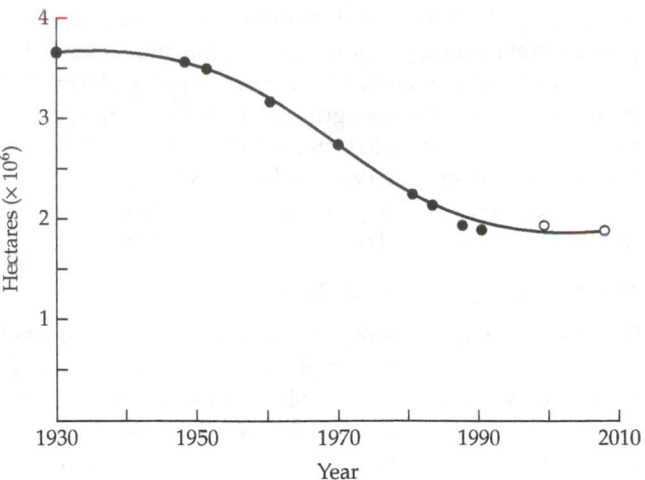

Figure B Estimated trend in the areal extent of suitable Northern Spotted Owl habitat in U.S. National Forest lands in Oregon and Washington from 1930 to 2010. Estimates beyond 1990 (open circles) are projections based on the national forest plans.

ington, the ISC found significant declines in the extent of owl habitat, a trend that was projected to continue into the future (Figure B). Additional analyses provided evidence of significant habitat declines in California (McKelvey and Johnston 1992), including regionally specific estimates of habitat loss that were provided in the species' Draft Recovery Plan (USDI 1992).

Subsequent development of the conservation strategy was based, in large part, on the results of map-based tests of five basic principles of reserve design, stated as falsifiable hypotheses (Murphy and Noon 1992):

1. Species that are well distributed across their ranges are less prone to extinction than species confined to small portions of their ranges.

2. Large blocks of habitat containing many individuals of a given species are more likely to sustain that species than are small blocks of habitat with only a few individuals.

3. Habitat patches (blocks) in close proximity are preferable to widely dispersed habitat patches.

4. Contiguous, unfragmented blocks of habitat are superior to highly fragmented blocks of habitat.

5. Habitat between protected areas is more easily traversed by dispersing individuals the more closely it resembles suitable habitat for the species in question.

Particularly relevant to a territorial species with obligate juvenile dispersal, such as the Spotted Owl, was the prediction from theoretical models of sharp thresholds for species extinction (Lande 1987; Lamberson et al. 1992, 1994; see also Chapter 12). One threat arises when the amount of suitable habitat is reduced to such a small fraction of the landscape that the difficulty of individuals finding territories compromises population persistence. Another occurs if population density is so low

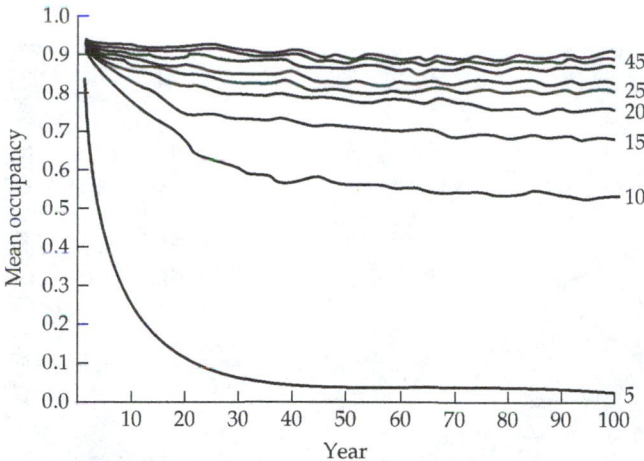

Figure C Some of the predictions for Spotted Owl persistence upon which management is based come from computer modeling. Shown here are predications of mean occupancy rates of suitable sites by pairs of Spotted Owls over time for various sizes of clusters (patches of suitable habitat sites), assuming that 60% of the sites within a cluster are suitable. Numbers on the right are total sites per cluster. These results predict that clusters with more individual sites will support more owls.

that the probability of finding a mate drops below that required to maintain a stable population.

One area of scientific uncertainty relevant to Northern Spotted Owl reserve design was the size and spacing of reserve areas. Existing biogeographic principles were helpful, but too broad for specific application to the Spotted Owl challenge. To address this uncertainty, we used computer simulation models, premised on Lande (1987), structured and parameterized in terms of the life history of the Northern Spotted Owl. The ISC determined the goal for conservation to be a 95% certainty of range-wide persistence for 100 years. Given estimates of the current amount of habitat, and its ability to regrow within 100 years, model results suggested that a minimum habitat size for locally stable populations would be a network of blocks, each capable of supporting at least 20 pairs of birds (Figure C; Thomas et al. 1990; Lamberson et al. 1992, 1994).

Analyses for the California Spotted Owl

Initial demographic studies allowed preliminary estimates of population growth rate (λ) for two study populations in the northern Sierra Nevada. Although power was low, neither were declining significantly (λ was not significantly different from 0; Noon et al. 1992). Owls were found to select stands of large old trees with closed canopies at both the landscape and home-range scale for both Sierra Nevada populations (Gutiérrez et al. 1992) particularly in nest and roost stands. Thus, hypothesis 2 was rejected.

Consistent rejection of hypothesis 2 led to tests of hypothesis 3. Forests in the Sierra Nevada have been markedly affected by human activities within the last 150 years (McKelvey and Johnston 1992). A combination of logging and natural at-

trition of old-growth forest had led to a decline in the number of large, old trees (particularly pines), had broken up the patchy mosaic of the natural forest, and encouraged development of dense understory conifer regeneration. The result was a landscape-wide loss of old-growth forest elements (e.g., large standing live and dead trees, and large downed logs) that are strongly associated with the habitat use patterns of California Spotted Owls.

Based on then current U.S. Forest Service land management plans, loss of old-growth forest elements was projected to continue, resulting in forests susceptible to fire disturbance and nearly devoid of large old trees. Given these projections, Verner et al. (1992) proposed interim (5–10 year) guidelines that both reduced allowable harvest levels and restricted silvicultural activities in habitats selected by Spotted Owls. These restrictions, invoked at a landscape scale, would serve to retain the large tree components in harvested stands in order to accelerate the rate at which these degraded stands would become suitable habitat in the future. The locations of suitable blocks of habitat would shift dynamically across the landscape, but with reduced harvest levels, an adequate amount and distribution of suitable habitat would always be available.

Why Two Different Conservation Strategies?

At both landscape- and home range-scales, the Northern and California Spotted Owls select habitats that retain old-growth forest characteristics. Consequently, timber harvest of old-growth forests, or their components, threatens both subspecies' long-term persistence. Given the wide acceptance by the scientific community of the ISC reserve design for the northern subspecies, why wasn't a similar strategy adopted for the California Spotted Owl?

First, during the past 70 years the number and distribution of Northern Spotted Owls may have been reduced by as much as 50% from pre-twentieth century levels (Thomas et al. 1990). No evidence of similar declines in number or distribution exists for California Spotted Owls, despite the fact that forests in the Sierra Nevada have been logged for the past 100 years. Currently, Spotted Owls in the Sierra Nevada are widely distributed throughout the conifer zone.

Second, clear-cutting is the primary silvicultural method in the Pacific Northwest, west of the Cascade crest, particularly over the last 50 years. Thus, habitat within the range of the Northern Spotted Owl is either undisturbed and suitable, or cut within the last 50 years and unsuitable. The result has been an island-like distribution of suitable habitat. The ISC opted for a Northern Spotted Owl strategy that clearly differentiated habitat reserves from areas where logging could occur. In contrast, selective tree harvest has been the predominant harvest method over most of the range of the California Spotted Owl (Figure D). Current data indicate that some levels of selective harvest does not render habitat unsuitable for owls, at least over the long term. As a result, the current distribution of Spotted Owls in the Sierra Nevada is relatively continuous in both

(I)

(II)

Figure D   Forest management in the Pacific Northwest typically involves clear-cutting (I), while in most of California, forests are selectively harvested (II). Selective harvest in the case pictured involved thinning an 80-year old stand of Douglas-fir to 50 trees/acre at age 40, and planting the understory with hemlock. (I, photograph by B. R. Noon; II, photograph by J. Tappenier.)

coniferous and adjacent foothill riparian–hardwood forest. Imposing a static reserve design here would leave many owls outside of reserves and vulnerable to habitat loss.

Third, fire is not a major threat to forests west of the Cascade crest in Washington or Oregon (Agee 1993). Despite the fact that fire spreads contagiously, even large contiguous blocks of old-growth forest within the region would face little risk of catastrophic loss. In contrast, mixed conifer forests where most California Spotted Owls occur are drier, and given a history of fire exclusion, very prone to catastrophic fires. A habitat reserve strategy there could deal with the uncertainties of logging, but not fire.

Collectively, the above considerations led Verner et al. (1992) to propose an interim landscape-level conservation strategy that would retain the old-growth forest components that are apparently needed by California Spotted Owls for roosting and nesting.

Toward Multi-Species Planning

In their report, the ISC emphasized that conservation of Spotted Owls was not the only, or even the primary, issue affecting forest management in the Pacific Northwest. Rather, it was the conservation and sustainable use of late-seral forests. Specifically, the ISC noted that at least 34 other species appeared to be declining due to their association with late-seral forests, and that a broader planning effort was needed to conserve these additional species. As a consequence of concerns for other species and numerous legal and political battles that followed release of the ISC report, President Clinton convened the Forest Summit in Portland, Oregon, in 1993. The summit led to the designation of another science team, dubbed the Forest Ecosystem Management Assessment Team (FEMAT). Analyses conducted by FEMAT (1993) expanded beyond Spotted Owls and

took an ecosystem approach that eventually addressed more than 1000 plant and animal species. Specific attention was given to the Marbled Murrelet (*Brachyramphus marmoratus*), which had been declared threatened under the ESA during the intervening period, and to numerous runs of salmon considered to be at risk.

The FEMAT delivered ten management options to the President. Key to discrimination among the options were viability analyses of Spotted Owl populations under the different alternatives, based on spatially explicit population models coupled with projected landscape changes (McKelvey et al. 1993; Noon and McKelvey 1996). The President selected option nine, implementing it as the Northwest Forest Plan, which designated late-successional forest reserves (LSRs), steamside buffer zones (riparian reserves), adaptive management areas, and a forest matrix to be available for timber harvest (Thomas 2004). In addition to LSRs, key watersheds were protected to provide landscape connectivity for many wildlife species and to meet the needs of anadromous fish. Option nine, however, was not judged to offer the greatest likelihood of long-term persistence for Spotted Owls; instead it represented a compromise.

Unfortunately, whether full implementation of the Northwest Forest Plan can provide for long-term viability for owls under projected levels of timber harvest has not been tested, and data do not exist to evaluate the functionality of the designated habitat reserves. Although the plan included nearly 90% of the available suitable habitat within the range of the Northern Spotted Owl, and reduces annual timber harvest to barely one-tenth of its peak in the 1980s, close monitoring of several populations suggests that the owl continues to decline (Burnham et al. 1996; Anthony et al. 2004). One hypothesis for continuing declines is that owl populations are experiencing transitions to new, lower equilibriums, even where habitat loss has

been arrested. But new threats have arisen, including invasion of Barred Owls into Northern Spotted Owl habitats, where they may compete for the same resources; political pressure to increase levels of timber harvest; and changes in the laws that govern management of national forest lands. For the past 15 years, Spotted Owl researchers have used common methods of data collection and analysis allowing for comparisons across populations; nonetheless relationships between environmental causes and population trend effects remain uncertain since the field experiments necessary to explain the declines in owl populations have not been carried out (Noon and Franklin 2002).

Far-reaching planning efforts, continuing demographic research, and several management plans have been proposed for the California Spotted Owl since the interim strategy proposed by Verner et al. (1992). These efforts culminated in 2002 with the Sierra Nevada Framework, a U.S. Forest Service management plan that emphasized the conservation of all species associated with late-seral forests in 11 national forests ranging from northeastern to southern California.

However, the Framework was short-lived and in fact, was never implemented before the Bush administration rescinded it in January 2004. In part, the U.S. Forest Service claimed that new information had become available that required the Framework decision to be revisited. Specific mention was made to a comprehensive analysis of demographic data collected from 1990 to 2000 from five long-term studies (Franklin et al. 2004). Even though the authors concluded that "all the demographic evidence available—such as estimated vital rates, rates of population change, and the differences between paired studies—suggest substantial caution in owl conservation and management efforts" (Franklin et al. 2004), the U.S. Forest Service concluded that the trend data were too uncertain to justify the constraints on timber harvest required by the Framework plan. A new plan was issued by the U.S. Forest Service in 2004 that significantly increased allowable levels of timber harvest, including the harvest of trees up to 30 inches in diameter.

Issues that Arise after the Conservation Plans Are Put Forward

Confronting scientific uncertainty

Contentious debate surrounding the value of threatened and endangered species like the Spotted Owl inevitably arises if their conservation is accompanied by significant economic costs. As a result, conservation biologists and their colleagues in forestry, range sciences, and wildlife biology have been swept into public debates that take them from the status of sequestered experts to that of key players in development of public policy. Often scientists are required to defend the merits of their science against sometimes savage criticism.

Scientists are trained to treat facts with doubt and to question their validity, a circumstance that lawyers use to advantage. Uncertainty is inherent in the scientific process because the goal of science is to reduce levels of uncertainty by subjecting alternative hypotheses to rigorous tests. Thus, scientists do not construct conclusions from data; rather, they construct hypotheses that are tested with further data. They cannot prove the truth of an assertion; rather, they fail to disprove that assertion, and thus support it (Murphy and Noon 1991).

Special interest groups employ lawyers and consultants to seek flaws and weaknesses in scientific analyses and data. They exaggerate and misconstrue the inherent, inevitable uncertainty that accompanies the best scientific efforts. In those cases in which no obvious flaws exist, critics will note how little scientists actually know. Worse still, they may use disinformation and distortion in an attempt to discredit the scientist. At times, critical data that could significantly contribute to problem definition and resolution may be purposely excluded as lawyers manipulate the litigation process.

A disproportionate amount of criticism of the ISC strategy, both in industry press releases and during litigation, was directed toward the computer simulation models and the inferences drawn from them. Models are ready candidates for commentary because any model simple enough to be operational is necessarily too simple to be completely realistic. Like all simulation models, those used in the conservation assessments of both owl subspecies were characterized by abstractions and simplifying assumptions. And like all models, they were open to criticism if one demands (unrealistically) that a model be a complete representation of the real world.

The motivation to discredit both conservation strategies rested on the simple fact that owl protection meant reducing allowable tree harvest. From the timber industry's perspective, access to large-diameter, economically valuable trees on public lands would simply be too restricted. This stipulation, however, was not a consequence of the model results, but was dictated by the habitat associations of the Spotted Owl.

Burden of proof in conservation debates

The allocation of burden of proof can often determine the results of decision-making. Some entity must assume the responsibility for providing sufficient information to compel a decision-maker to adopt a solution. In the Northern Spotted Owl litigation, the strategy of the timber industry, and to some extent the federal agency lawyers, was to put the burden on the scientists to prove an adverse effect of timber harvest on Spotted Owl persistence. In the absence of compelling information and arguments, the lawyers argued the status quo (high levels of harvest in late-seral stage forests) should continue. Failure to make a decision to change management practices for Spotted Owls was a de facto decision to continue current practices.

Federal environmental laws do not require judges to be scientific experts. Rather, the law requires public agencies to fully disclose all pertinent information, and to openly consider competing interpretations of this information. Despite attacks on the credibility and objectivity of the Spotted Owl scientists, these courtroom tactics failed because the judges ruled that ex-

isting environmental laws require a full disclosure and analysis of existing data. The analyses provided by the ISC and other scientists provided convincing evidence of risk to the species, thus mandating conservation action. Defensible science and open debate prevailed in the courtroom, and eventually led to more responsible decision-making.

Ethics and science

Most scientists involved in conservation biology are motivated by a strong sense of responsibility to natural resources and to future generations. Lawyers attempt to label such scientists as "advocates" who are inherently biased, and they refuse to recognize that one can support a position in the absence of bias; bias does not necessarily follow from advocacy. Science is not value-free, nor should it be. "Resource stewardship" is not a buzz phrase but a meaningful expression of responsibility to future generations. Conservation scientists recognize that meeting this responsibility will often come at the expense of maximizing short-term economic gain.

Some scientists involved in the Spotted Owl debates chose not to participate in the normative process to render data scientifically credible (e.g., peer review and publication). Instead, they exploited the uncertainty inherent in the scientific process to justify maintaining the status quo or to obscure reasonable hypotheses. They were often able to stir up doubt, not because a hypothesis was unreasonable, but simply because irrefutable proof was a standard that could not be met.

Improving the Role of Science in Conservation Policy

The courts have assumed an increasing role in rendering land management decisions based on procedural aspects of law, as well as deciding substantive issues that should be discussed and resolved in other arenas. Because of society's continuing failure to acknowledge that hard choices must be made, and then to move forward and make them, we have lacked an adequate forum and process for environmental problem resolution.

We need to develop alternative strategies for problem resolution, and scientists should be key contributors to this process. The forum for decision-making must be expanded to include all affected parties, representing a diversity of perspectives. Given such a forum, behavior must be governed by a set of rigidly enforced ground rules, including: (1) participating parties must treat one another with professional respect; (2) the strength of any argument put on the table should be a function of the information content of the argument; (3) no pertinent data may be withheld or suppressed; and (4) the reliability of the data should be judged by the degree to which they have been exposed to the scientific process of peer criticism and repeated attempts at falsification (see Anderson et al. 1999).

Such a forum for problem resolution would be a significant step toward solving emerging crises in land use and natural resource management. Once solutions were offered, the final responsibility would be to conduct risk assessments to accompany each of the alternative management plans. Thus, the decision-makers would be the final arbiters of which conservation plans were implemented and would be obligated to make known the risks, to both present and future generations, associated with the decisions they made.

Ultimately, the decisions we make as a society regarding management of declining resources come back to a fundamental question: Does the value gained from the continued existence of a species equal the cost incurred to assure its persistence? How we respond to this question will determine the fate of many species, including the Spotted Owl.

Summary

1. Conservation actions are often determined in policy decisions that necessarily weigh many factors, of which scientific information and recommendations form only one part. Traditionally, scientists have kept apart from policy, understanding little of the policy process and communicating primarily with other scientists. Given the needs for rapid and effective action today, scientists need to learn how to engage in the policy process by becoming effective communicators and partners in developing policy. This will require learning about the policy process, listening to what policymakers require for making decisions, and developing effective means of conveying the lessons from their research.

2. We need more translational scientists who can work at the interface between conservation science and conservation policy. Such individuals can work to convey the nuances of scientific understandings to policymakers and the concerns and needs of policymakers to scientists. Effective policies may better emerge from teams of scientists, translational scientists, social scientists, stakeholder and public representatives, and policymakers working together.

3. Conservation policy spans a broad range including laws, regulations, policy statements, and agreements. Policies seek to guide actions by governments, private businesses and citizens, often balancing conflicting needs and will of distinct groups. Conservation scientists need to understand the con-

text in which conservation policies are made, especially constraints and opportunities that may affect decisions. Further, conservation scientists need to recognize their own constraints and opportunities, and learn to work with these.

4. Conservationists are not value free, but they need to serve as consistent advocates for "good science" if they are not to be dismissed for being biased. The best use of scientific information requires an understanding of how we work to form our best understanding of issues, and advocates the strengths and weaknesses of scientific conclusions based on their merits alone—regardless of how these conclusions align or do not align with our values. Without these efforts, conservation science is not credible.

5. Scientific information is most likely to be used in conservation decisions if it meets the criteria of being salient to the issues considered by a group, credible, and legitimate in its application to these issues. Many scientific assessments have failed to influence public policy substantively because often the information is not seen as legitimate or salient to the needs of the policy maker. Two major international scientific efforts, the Intergovernmental Panel on Climate Change (IPCC) and the Millennium Ecosystem Assessment (MA), however, have attained all three attributes through a conscious process of consulting and refining efforts on behalf of international decision-making bodies. Accordingly, the conclusions and recommendations of these groups carry weight with decision makers.

6. Becoming an active participant in policy processes will require conservation scientists to learn a number of engagement and communication skills. Many opportunities for contribution to conservation policy exist in distinct arenas and activities. The role that will be right for each person involved in conservation practice will depend on one's own values, the skills one enjoys cultivating, and where one remains motivated to be persistent through all the ups and downs of conservation practice.

> Please refer to the website www.sinauer.com/groom for Suggested Readings, Web links, additional questions, and supplementary resources.

Questions for Discussion

1. In what ways could your training in conservation biology—including in this text—be organized to better increase your abilities to contribute to public policy now and in the future?

2. How can we balance the need for objective conservation science and subjective judgment in the policy process? Many of the examples in this chapter concern issues prominent in the U.S. Does your view of this balance change when you consider conservation in an area where the majority of people live in abject poverty?

3. How did the scientists in each of the case studies contribute to formulating conservation policies? What factors most contributed to successful outcomes in terms of biodiversity conservation, and what detracted most from successful outcomes?

4. Consider a prominent conservation problem—for example, the decision whether to open up the Arctic National Wildlife Refuge to oil exploration. How have conservation scientists contributed to this decision making process? What factors are considered in the decision and how could conservationists work to achieve an outcome that conserves biodiversity?

5. How can conservation scientists better work with policy makers to achieve more effective conservation policies?

18

Meeting Conservation Challenges in the Twenty-First Century

Martha J. Groom, C. Ronald Carroll, and Gary K. Meffe

Never doubt that a small group of thoughtful, committed citizens can change the world. Indeed, it's the only thing that ever has.

Margaret Mead

In this book, we have traced the patterns of biodiversity and many details governing how it is threatened by human activities. We have also examined strategies to prioritize, protect, and restore biodiversity while focusing on genes, populations, species, landscapes, or ecosystems, and explored how to better integrate policy processes and activities that can achieve biodiversity conservation. Our goal in this final chapter is to look more squarely at the challenges of meeting the needs of humans and of biodiversity over the next century, particularly given many uncertainties about our future, and to consider a number of potential approaches to integrate these aims in more detail.

Countering the Impacts on Biodiversity from Poverty in Many Countries and Over-Consumption in a Few Countries

Addressing the tragedy of abject poverty is both a moral imperative and a practical concern in conserving biodiversity. Global consumption of energy, food, timber, land for built environments, and water resources (particularly for agriculture) strongly increased as the world population grew from 3 to over 6 billion people (World Wildlife Fund 2004). Patterns of exploitation in marine fisheries show even more extreme trends. The primary driver of these increases has been increased consumption and intensified use of global resources in industrialized countries. The gap between rich and poor has become extreme, and this has led to political and economic inequalities that are resistant to change (Sachs 2005). Wide-scale resource depletion disproportionately affects developing countries, and inhibits far too many people from attaining an acceptable standard of living (Figure 18.1).

Figure 18.1 This shantytown on the edge of Rio de Janeiro, Brazil has no running water, or waste disposal services, conditions typically facing most of the world's poor. (Photograph courtesy of UPI/Bettmann.)

When conditions for life are so desperate, the ability of people to conserve biodiversity is compromised (to say the least), and the loss of biodiversity and ecosystem services reinforces the cycle. If such poverty increases as the world population grows by another 3 billion in the next 50 years, not only will human suffering be extreme, the effects on biodiversity will be devastating.

Many priority areas for conservation, such as the biodiversity hotspots and Global 200 ecoregions, overlap areas of high human population density and growth rate (UN Environment Programme 2002). Thus, if current trends continue, greatly increased habitat degradation (see Figure 6.1) and species extinctions (up to 35% of all species in the Afrotropical, 20% in the Indo-Malay, 15% in the Neotropical realms, and 8%–10% in the rest of the world) are expected by 2050 (Millennium Ecosystem Assessment 2005a,b). Unfortunately, increasing pressures from growing human populations have begun to negatively affect protected areas. Nearly 70% of buffers surrounding 198 strictly protected areas (IUCN Category I and II) in tropical forests have lost cover in the last few decades, and 25% of these protected areas lost cover within their boundaries (Mayaux et al. 2005; see also Essay 3.3, Figure B). These losses of biodiversity, even in our most protected areas, have garnered a sobering global report card on the status of ecosystem services (Millennium Ecosystem Assessment 2005b) (Table 18.1). Only four of 24 ecosystem services have improved over the last 50 years, and most of the rest have declined sharply. Thus, not only has biodiversity been degraded, but so have the prospects to end poverty.

Yet we do have reasons for optimism. Efforts to fight poverty worldwide have had some success, as in most regions (except parts of Africa) the number of people living on less than U.S.$1.00 per day has declined in the past 20 years (Sachs 2005). World population growth rates have declined, with a decrease in fertility now seen in all areas of the world (Figure 18.2). This is good news,

Figure 18.2 Fertility rates are declining worldwide, and are projected to be just 2.5 children per woman by 2025. (Modified from UN 2004.)

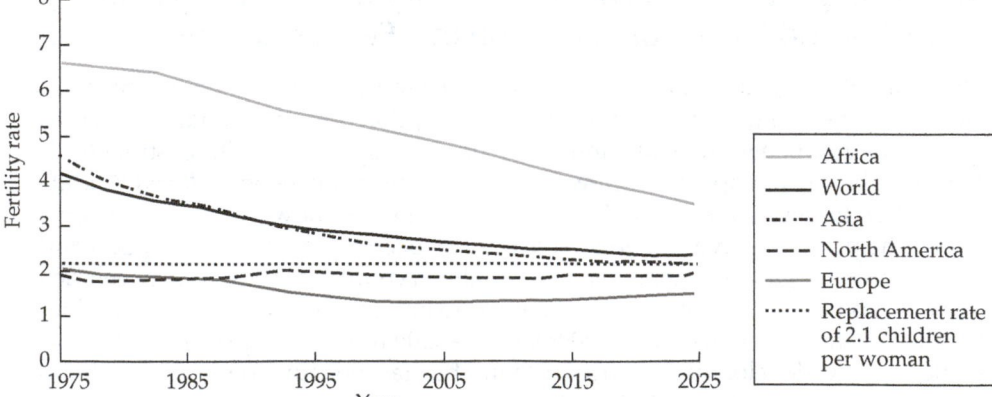

and will help the capacity of all countries to live within their means. Continued efforts to reduce family sizes, to educate and empower women, and to eliminate poverty are critical to continuing these trends.

In the past 20 years, we have greatly increased the number and size of protected areas. Now more than 12% of the terrestrial surface of Earth has some level of protected status. We are creating marine reserves, and while the total area covered lags well behind terrestrial coverage, a great deal of effort is going into the design and evaluation of these new marine reserves. There has been a tremendous increase in our capacity to analyze problems at local, na-

tional, and global scales, with increased cooperation among parties working to conserve biodiversity. The need to work with people, instead of against them, has become a major part of most conservation planning and implementation, and a broader diversity of perspectives and expertise is brought in to aid these processes than ever before. We are learning how to improve degraded habitats through restoration efforts, and to enhance plant and animal populations within restored sites. And we are learning how to create contexts for sustainable use of wild nature.

There remains much to be done. Most centrally, we need to decrease per capita levels of consumption and

TABLE 18.1 *Global Status and Trends of Ecosystem Services*

Ecosystem Service	Subcategory	Status	Notes
Provisioning services			
Food	Crops	↑	Substantial production increase
	Livestock	↑	Substantial production increase
	Capture fisheries	↓	Declining production due to over-harvest
Fiber	Timber	+/−	Forest loss in many regions, growth in some
	Cotton, hemp, silk	+/−	Declining production in some fibers, growth in others
	Wood fuel	↓	Declining production
Genetic resources	—	↓	Loss through extinction of populations and species and of crop varieties
Biochemicals	Natural medicines, pharmaceuticals	↓	Loss through extinction, over-harvest
Water	Fresh water	↓	Unsustainable use for drinking, industry, and irrigation; hydropower unchanged, but dams increasing ability to use this power source
Regulating services			
Air purification	—	↓	Decline in ability of atmosphere to cleanse itself
Climate regulation	Global	↓/+	Increases in carbon sequestration since 1950; yet net decline in forest cover
	Local, regional	↓	Preponderance of negative impacts
Water regulation	Flood control, etc.	—	Varies depending on location and forces of ecosystem change
Erosion control	—	↓	Increased soil degradation
Water purification, waste treatment	—	↓	Declining water quality
Disease regulation	—	+/−	Varies depending on ecosystem change
Pest regulation	—	↓	Natural control degraded through pesticide use
Pollination	—	↓	Apparent global decline in pollinator populations
Natural hazard regulation	—	↓	Loss of natural buffers (e.g., mangroves, wetlands)
Cultural services			
Spiritual and religious values	—	↓	Rapid decline in sacred groves, species
Aesthetic values	—	↓	Decline in quality and quantity of natural land
Recreation and ecotourism	—	+/−	More areas accessible, but many degraded

Note: Most ecosystem services have declined since 1950. Only food production has increased substantially, but at the cost of many other essential ecosystem services.

Source: Modified from Millenium Ecosystem Assessment 2005a.

exploitation of wild nature by the developed world while improving conditions for the world's poor. Achieving this will require mobilization of considerable resources and wisdom to act for the long-term good. Substantive societal change will be needed to accept the shifts in lifestyle necessary in the developed world and among the roughly one billion aspiring consumers of China, India, and other nations experiencing swift economic transitions (Myers and Kent 2004). Similarly, we need to act as one world to create conditions that favor economic growth for the world's poor without sacrificing biodiversity. The increasing interconnection of human societies in the twenty-first century creates greater challenges for biodiversity conservation, as it also increases our capacity for collective action.

The challenges ahead require a cadre of dedicated professionals and citizens to help solve the dual crises of poverty in many countries and over-consumption in a few, and reduce their impacts on biodiversity. David Ehrenfeld provides some advice on how to prepare for and respond to the constraints of our time in Essay 18.1.

Working with Uncertainty in Ecological, Social, Economic, and Political Systems

Nothing is certain but change. Yet uncertainty poses challenges that greatly hamper our ability to manage our use of resources and to conserve biodiversity. The most important of these are: the propensity for complex systems to behave in confusing and nonlinear ways and the possibility that our actions will push ecosystems into alternate states from which they may recover only slowly, or not at all.

Uncertainty exists due to the complexity and dynamic nature of ecological systems, and our own limited understanding of mechanisms driving processes at all levels of biodiversity. We now understand many phenomena that were surprises to us just 30 years ago, such as the evolution of antibiotic resistance in disease organisms, cascade effects in communities from population extinctions or introductions (see Table 3.2), and the adverse effects of many manufactured chemicals on endocrine function, particularly for species at higher trophic levels (see Essays 3.1 and 6.4). We have now seen threshold changes in population sizes due to exploitation, and expect other threshold phenomena to cause diverse ecosystem changes as climates change over this century. Certainly, this century will reveal innumerable shifts and changes we could not have predicted in advance. Our challenge is to plan for these ecological uncertainties.

In addition to these forms of biological uncertainty, there are considerable social, economic, and political uncertainties that affect conservation practice. For example,

market fluctuations will influence citizen, business, and policy choices. An unexpected crash in prices can undermine a sustainable use project, pushing people to switch practices to make a living. Governments can buffer these fluctuations to some extent via price supports. Conservation groups have attempted to do this for various forms of sustainable coffee production (shade-grown, fair-trade, organic) by offering growers prices that support decent living conditions, regardless of the market price, and also promote sustainable growing practices (TransFair U.S.A. 2004). These supports can make a large difference: farmers certified by fair trade and eco-friendly labeling in Nicaragua are less vulnerable to price fluctuations (Bacon 2005). Nonetheless, uncertainty over economic costs and benefits may prevent many actors from joining into conservation enterprises. Further, these kinds of cooperatives currently have a small market share, and thus cannot cover all the expenses of a farmer's family (Bacon 2005). Market fluctuations also place the poor at greater risk of food shortages, and thus may lead to uncontrolled or illegal bushmeat hunting, plant collection, and small-scale agriculture beyond the bounds of what is safe for long-term conservation. Finding and implementing site-appropriate agricultural methods that increase food security, and linking these to sustainable trade mechanisms is thus an important goal for both biodiversity and human welfare.

Political uncertainty has been driven by election cycles where distinct political regimes follow distinct conservation policies. In recent years, we have seen a shift in U.S. domestic policy that systematically undercuts our long history of environmental legislation through changes in administrative procedures and riders on unrelated new legislation. And, we have seen a remarkable and demoralizing decline in budgets for biodiversity protection. A challenge for conservation is to develop flexible frameworks that can sustain biodiversity targets in the face of such a large shift in political will.

To improve our ability to work with the inevitable uncertainties that surround conservation practice, we need to increase our capacity to monitor and understand trends. This will involve both developing improved metrics to monitor trends and inform decision-making, and focusing our research to better understand these sources of uncertainty and provide information needed by decision-makers and stakeholders.

Indicators needed to describe trends and guide policy

As the complexity of conservation and social problems has grown, so has the call for indicators of status and change that might better reflect the erosion of biodiversity and our capacity to sustain human populations and improve human welfare. The dominant indicators gov-

ESSAY 18.1

Conservation Biology in the Twenty-First Century

David Ehrenfeld, *Rutgers University*

■ The first decades of the twenty-first century are a critical time for students of conservation biology. These are the years in which your careers will prosper or stagnate, the years when conservation biology will either grow to meet the enormous challenges that confront it, or fail. Time spent now thinking about the probable condition of the world in which you will work is as important a part of your career preparation as is learning about the determination of minimum viable population size or techniques for the assessment of habitat quality.

In a static world, there would be no need to follow this advice. If next year is going to be the same as the last, why bother to prepare to do anything differently from the way we have been doing it? But conservation biology is grounded in the perception of a rapidly changing world. Unlike the majority of economists who act as if there is no environmental change that cannot be reversed by market forces, we know that human actions are causing momentous changes—some irreversible—in fundamental global conditions and processes, changes that we cannot control and often cannot even understand. Conservation biologists recognize change and should not be taken by surprise when it occurs.

Many of the most important changes that are bound to affect us in the coming years will not be biological or physical—something quite different from the rapid fragmentation of the Amazonian rainforest or the sighting, for the first time in human history, of open water at the North Pole. The equally momentous changes I am referring to—and what I will discuss in this essay—will be social, political, and economic in nature. Whether you accept my predictions is not important; what matters is that you understand the changes I suggest so that you can think about the future in a creative and open-minded way.

It does not take a crystal ball to predict that the most striking difference between the 1980s, when conservation biology became a recognized discipline, and the first ten to twenty years of the twenty-first century will be the depletion of a variety of critical resources—

especially oil—that have been an accepted part of our world since the end of the Second World War (Campbell 1997; Duncan and Youngquist 1999; Youngquist 1999). The decline is well underway, and affects us strongly through its effects on at least four different kinds of resources that are important to conservation biology: money for salaries and research, social stability, general environmental awareness, and specialized knowledge of plants and animals.

Money

Conservation biology has never been a wealthy field like computer science or genetic engineering. We are not accustomed to having luxurious buildings and sumptuous grants showered upon us, nor are the "help wanted" pages of journals like *Science* and *Nature* filled with dozens of advertisements in which academia and industry compete to attract recent graduates or senior conservation biologists. Nevertheless, we have benefitted from the wealth that has been spent on the sciences in the past 50 years, and we will be seriously affected when it dries up. We have piggybacked conservation research onto grants for nonconservation projects; we have built conservation and restoration into the research agendas of established ecological granting panels and field stations; we have diverted government and university resources into conservation biology and restoration programs; we have received support from private foundations and conservation organizations.

How much of this support is likely to continue? In a world that is essentially broke, a world whose wealth and whose enormously complex administrative, corporate, and financial structures depend on readily available energy and other resources that are nearly exhausted, the flow of money that has sustained nonindustrial, nonmilitary "frills" such as conservation biology is being slowly turned off at the tap (Tainter 1988). When most of the grants and the fellowships disappear, when the field stations are closed, when the foundations see the inflation-

adjusted market value of their corporate stock portfolios cut in half, and then cut in half again, what is going to happen to conservation biology?

Social Stability

The constant shifting of jobs from industrial nations to those with the lowest pay scales, the downsizing that comes with (and without) corporate mergers, the widening of the gap between rich and poor, the damaging of communities as mall-based superstores put local shops out of business, the perverting of the best parts of our culture by a mass media concerned only with commercial profits, and the resulting violence and social instability are all bound to affect conservation biologists. We cannot stand above these processes; they will plague us where we work and where we live (van Creveld 1991; Goldsmith 1994; Rifkin 1995; Sale 1995; Ehrenfeld 2000). Will it be possible to remain a conservation biologist in a society in angry turmoil? I believe there may be a way.

Environmental Awareness

Conservation biology is ethically driven—a scientific response to a societally perceived crisis. The public is aware that many species are threatened with extinction and many ecosystems are being degraded, and most people *feel* a sense of impending or actual loss and are willing to devote energy and resources to reduce the threat. Ultimately, conservation biology depends on the deep public feeling of environmental loss, and this feeling only comes to those who have tasted enough of nature to miss the savor when it is gone.

The years since the 1980s have seen a dramatic reduction in many people's day-to-day contact with nature, and a corresponding increase in the proportion of time spent in the human-created, electronic world of television, email, the internet, and other substitutes for direct experience of the sights, smells, touches, tastes, and sounds of the natural world from which we evolved (McKibben 1992). This reduced experience of nature in daily life is particularly noticeable in the young and

this does not bode well for conservation biology. As Gary Nabhan and Stephen Trimble (1994) put it: "We are concerned about how few children now grow up incorporating plants, animals, and places into their sense of *home*." How can a public that does not know nature be expected to be passionately aroused—or even nostalgically uneasy—about its demise? When the crunch comes, how will you gain support for your work from a public that has only experienced "nature" while on vacation at a Disney theme park?

Specialized Knowledge

A preoccupation with the specific characterizes all conservation biology. Most degraded places still have some kinds of plants and some kinds of animals living there. If the preservation of nature simply meant the maintenance of any sort of generic life form—grass, weeds, trees, insects, vertebrates, without regard to species or to ecosystem associations—there would be no need for conservation biologists. Our purpose is to make sure that particular species in particular ecosystems continue to exist in the place where they evolved. But to do that, we have to know how to tell one species from another, and this is becoming a problem.

Until now, conservation has been able to depend on a rich resource of plant and animal taxonomists, both professional and amateur, to identify species and define taxonomic relationships. During the late 1960s and 1970s, however, it became evident that biology departments would no longer support taxonomy and natural history: molecular biology occupied the entire stage. More than thirty years later, clas-

sical taxonomists are an aging lot. There are few graduate students who call themselves taxonomists or systematists, and natural history has all but vanished from the academic scene (Ehrenfeld 1993). True, molecular systematics is thriving in some places, but most of its practitioners are lab-based rather than field-based—the demise of classical taxonomy and natural history has left an enormous void. Today, there are many genera, even families, of organisms that have only one or no practicing taxonomists familiar with them. The situation is worsening, and is especially troublesome for tropical species (Parnell 1993). Will you still be able to be a conservation biologist if neither you nor anyone else knows exactly what it is you are trying to save?

A Strategy

The first decades of the twenty-first century are certain to be unsettled and often dangerous for the majority of people in both the industrialized and less-industrialized countries. There will be no magic formula to guarantee that a conservation biologist can work effectively under these conditions, although some individuals will be lucky. But I believe that there is a strategy that can help you cope with the resource problems I have described—a strategy that may, at the very least, tip the odds in your favor.

This strategy has six elements:

1. Minimize the cost and logistic complexity of your research.
2. Design your research to be flexible, so that your methods and even some of your objectives can be mod-

ified as changing circumstances warrant.

3. Take every opportunity that you can find to learn, in depth, the taxonomy of the groups you are studying and the natural history of all parts of your ecosystem.

4. Have a practical trade, skill, or alternative occupation that you can resort to if conservation biology cannot support you on a full-time basis. There are trades that are always in demand, regardless of circumstances. Pick one.

5. Whenever possible, design your research to include the participation and wisdom of the local community (Castilla and Fernandez 1998). Make it your goal that local people understand, approve of, participate in, and benefit from your work. Make a special effort to involve local schools and schoolchildren; work with the teachers and budget a modest amount of money to help them incorporate your project into their curricula.

6. Before the project ends, develop a mechanism to monitor the system and continue local participation after you are gone.

If you are not satisfied with this strategy, by all means invent one of your own. A diversity of strategies is bound to be more successful than a single one. What matters most is that the future finds you prepared to do your work, fully able to experience the joy and challenge of being a conservation biologist, regardless of the changing times. ■

---

erning most national and international governance have been economic, and to a lesser extent, social indicators of change. Ultimately, our interest is that biodiversity and human societies are sustained at a decent quality over the long-term, in an evolving environmental and human context. Thus, if we are to gauge how well we are doing, and also to identify details about our actions that may undermine sustainability, we need better metrics.

The Environmental Sustainability Index (ESI) is one such composite index that tracks the relative success of countries in 21 environmental indicators derived from 76 variables that indicate: (1) the health of environmental systems; (2) degree of stress on ecosystems; (3) vulnerability of human populations to environmental change; (4) the in-

stitutional and social capacity in a country to cope with environmental challenges; and (5) the degree to which the country participates in global efforts to reduce environmental damages. The ESI was designed to reflect each of these distinct areas because sustainability is seen as arising not only from wise management of natural resources and conservation of biodiversity, but also from healthy social and economic systems (Table 18.2) (Esty et al. 2005). Although focused more broadly on environment than on biodiversity, roughly 30% of the indicators are strongly related to biodiversity conservation, habitat degradation, or natural resource management and most of the balance to sustainability of human societies and their capacity to govern well.

TABLE 18.2   *The Logic behind the Selection of Indicators, Grouped into the Five Components of the Environmental Sustainability Index (ESI)*

Component	Logic: "A country is more likely to be environmentally sustainable…"
Environmental systems	…to the extent that its vital environmental systems are maintained at healthy levels, and to the extent to which levels are improving rather than deteriorating.
Reducing environmental stresses	…if the levels of anthropogenic stress are low enough to engender no demonstrable harm to its environmental systems.
Reducing human vulnerability	…to the extent that people and social systems are not vulnerable to environmental disturbances that affect basic human well-being.
Social and institutional capacity	…to the extent that it has in place institutions and underlying social patterns of skills, attitudes, and networks that foster effective responses to environmental challenges.
Global stewardship	…if it cooperates with other countries to manage common environmental problems, and if it reduces negative transboundary environmental impacts on other countries to levels that cause no serious harm.

Source: Modified from Esty et al. 2005.

The composite scores allow a relative ESI ranking of 146 countries (Figure 18.3). The countries with the highest rankings are all those with substantial natural resources, but low population density and high capacity for governance. The lowest-ranking countries face numerous challenges, both due to human mismanagement of natural resources and difficult environmental conditions (Esty et al. 2005). No country is above or below average in all indicators, and thus the ESI may reveal best practices among some countries, and target priority areas for improvement in every country.

A more policy-process oriented index such as the ESI might be usefully explored for causal drivers of changes captured by metrics focused solely on the status of biodiversity, such as the Living Planet Index (LPI; see Figure 3.13), change in habitat cover or proportion degraded (see Chapter 6), or the ecological footprint (see Figure 1.3). We may make most progress in defining priorities and challenges by exploring causal models of the interactions among the biodiversity, environmental, social, and economic dimensions of these, and other metrics (Balmford et al. 2005; Dobson 2005). We also need to develop a broader set of indices that are more representative of the full spectrum of biodiversity than those based primarily on terrestrial vertebrates or forest cover (Dobson 2005). Importantly, we need a suite of monitoring tools that can inform decision-making at each necessary level, from changes in populations to ecosystems, from single site to global scales. James Karr discusses these issues in greater detail in Essay 18.2.

Research approaches needed to inform decision-making

Beyond the need to find appropriate indicators, conservation biologists need to conduct more research to understand the "subtle" effects of human influence on biodiversity, where subtle includes a variety of often inconspicuous or unexpected interactions of humans with ecosystems, including indirect effects and lagged effects (McDonnell and Pickett 1993).

Equally important, new conservation science is needed to help define the responses of species, ecosystems, and other biological conservation targets to threats and to potential interventions clearly enough to guide management actions. Repeatedly, managers have noted that while much of conservation science is useful for alerting us to potential and actual problems, often we lack results precise enough to help differentiate which actions are necessary and most likely to succeed (e.g., Stinchcombe et al. 2002).

Often our science does not address the most pressing problems, in part because scientists have not been in sufficient collaboration with policymakers, stakeholders, and citizens to understand these critical questions (Robertson and Hull 2003; see Chapter 17). Thus, many have called for conservation scientists to answer the call for public ecology (where scientists work in collaboration with experts in diverse fields and with a wide range of stakeholders; Robertson and Hull 2003) and to engage in the social contract to contribute to society by working to solve our most pressing problems (Lubchenco 1998). One emerging need is for more collaborative research, with questions and the search for answers driven by consensus among a diverse group of researchers, policymakers, and citizens. As we have emphasized from the beginning of this text, conservation biology is inherently a multidisciplinary field. Yet, we need to become more comfortable and active in engaging in interdisciplinary teams if we are to make rapid progress in finding workable solutions to the biodiversity crisis.

A second emerging need is for conservation research to become more sophisticated in its ability to provide use-

(A)

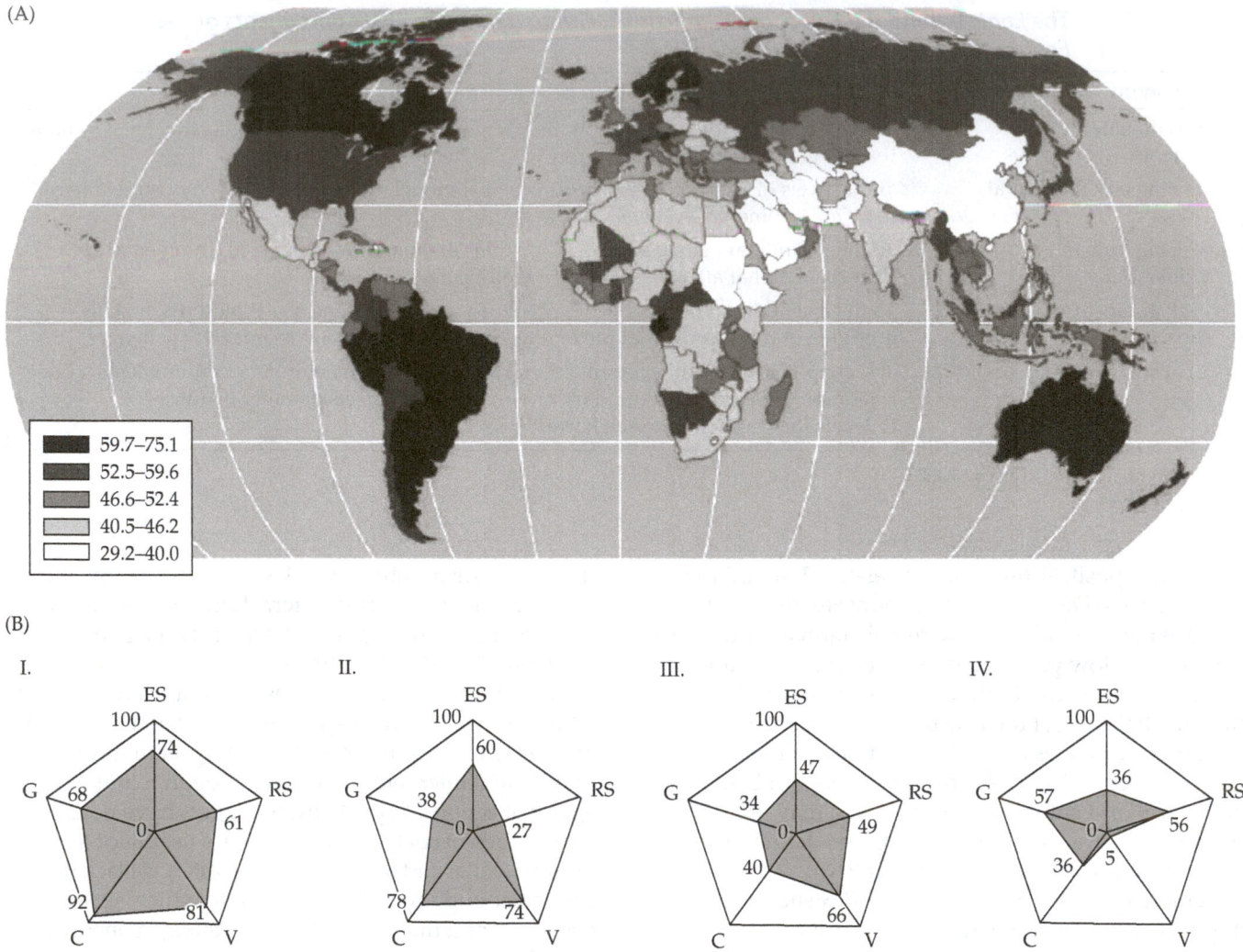

(B)

Figure 18.3 (A)The environmental sustainability index (ESI) calculates the performance of 146 countries on 21 indicators to yield an additive score. Countries with strong governance and relatively low population density rank highest, while countries with significant environmental degradation and limited capacity to prevent further degradation rank lowest by this index. (B) Summary scores for the five environmental indicators used in the ESI for I. Finland (ranked 1), II. U.S. (ranked 45), III. Macedonia (ranked 90), and IV. Ethiopia (ranked 135 of 144 countries). The scores are grouped into five primary indicators of the country's capacity to manage resources sustainably: ES = status of environmental systems, RS = effectiveness at reducing environmental stresses, V = effectiveness at reducing human vulnerability, C = social and institutional capacity for change, and G = engagement in global efforts to increase sustainability. (Modified from Esty et al. 2005.)

ful feedback to key questions for management—that is, we need to pursue "science in support of decision making" (Poff et al. 2003). An interesting example is the comparative studies of marine protected areas and traditional fisheries protection strategies carried out by Baskett et al. (2005). Through a combination of empirical studies and models they demonstrated that large no-take marine protected areas are particularly effective in maintaining large size classes of harvested fish, especially under conditions of environmental uncertainty. Such research helps guide the design of marine protected areas. Another example is the development of new theory surrounding the management of evolutionary processes. By focusing on expanding the potential to retain adaptive generation, we may be able to protect crops and endangered species in the face of rapid changes in environment wrought by an-

ESSAY 18.2

Indicators
It Matters What We Measure

James R. Karr, *University of Washington*

■ Humans measure things to understand the past, to document the present, and sometimes to predict the future. Many people measure things because they give pleasure—batting averages of baseball players, for instance—and they measure other things because they are assumed to reflect individual or collective well-being—a person's cholesterol level or annual income or a community's crime or unemployment rate. In short, we humans measure what matters to us, and, conversely, we come to value the things we measure.

For the past century, economic growth has been the overarching priority and dominant idea of humanity (McNeill 2000), and so the dominant measures, or indicators, for monitoring societal well-being have been measures of the economy. Closely watched economic indicators like the Dow Jones Industrial Average or Gross National Product (GNP) dominate the news. But the focus on these economic indicators has masked social, moral, and ecological ills that also matter to societal well-being (Davidson 2000; Manno 2000). Although people may be aware of such ills, the relentless dominance of economic indicators allows policymakers to make decisions that ignore those ills.

As a result, humanity now faces a paradox. The economic indicators we watch most closely suggest that all is well, but less-watched indicators show that the state of the biosphere is worsening. Human prosperity for thousands of years has depended directly on what societies take from Earth's natural systems. As human economic activity has intensified, parts of the natural system have eroded, progressively impoverishing the biosphere and threatening the very foundations of society. Such biotic impoverishment now takes many forms, including degraded soils, declining biodiversity, failed fisheries, changing climate, rising asthma rates, food insecurity, mounting numbers of refugees fleeing ecological degradation, and the cumulative impacts of many simultaneous and mutually reinforcing events (Chu and Karr 2001).

To reverse the erosion of living systems, society needs a new generation of indicators that no longer disguises the state of economic, social, or ecological parameters that matter to human well-being. If we watch such new-generation indicators as closely as we watch the Dow, perhaps we will again come to value the state of all Earth's living and nonliving systems and thus reverse the harm.

Better Economic Indicators

As a sign of societal well-being, Gross Domestic Product (GDP) is as flawed as it is influential, and it illustrates the problem with conventional economic measures (Cobb et al. 1995). Rising GDP, typically viewed as a sign of prosperity (i.e., total human welfare), measures economic throughput—the amount of money that passes through the economy. However, blind to "good" or "bad," GDP does not account for social or environmental costs, such as pollutants, resource depletion, cancer, crime, or auto accidents; perversely, it counts these costs as benefits when money changes hands. The state of Earth's ecosystems is left completely out of GDP accounting, as are nonmonetary contributions to human fulfillment, such as health, education, freedom, security, and peace. When society pays attention only to GDP and its companion economic metrics—the Dow, the index of leading economic indicators, the consumer price index—the resulting balance sheet is incomplete, and biotic impoverishment worsens.

Better economic indicators would refocus attention on the connections between humans and their environments. They would take into account the fact that the planet itself changes over time without expanding; likewise, development of the human economy must reflect increasing efficiency instead of ever-expanding throughput (Daly 1991).

Efforts to improve econometrics include the genuine progress indicator (GPI) and the index of sustainable eco-nomic welfare (ISEW). ISEW, for example, adjusts GNP for negative impacts on the natural systems it regards as "natural capital," wealth disparities across classes, the effects of pollutants, and other long-term social and environmental damage (Costanza et al. 1997b). A number of other efforts aim to clarify the relationship between economic and ecological systems (Figure A). Some researchers, for example, catalog the flow of goods and services from natural systems to human society (Daily 1997); they may even calculate their economic value (Costanza et al. 1997a; Pimentel et al. 1997). Others convert the rate of human consumption to land–area equivalents, estimating the size of humanity's ecological footprint (Wackernagel and Rees 1996). Still others express the influence of humans by determining the proportion of Earth's annual production consumed by humans (Pimm 2001). None of these approaches, however, explicitly measure the condition of natural biological capital—human and nonhuman—that is the sustaining wealth of the world.

Beyond Economic Indicators

Society needs indicators of the world's biological capital and how depletion of this capital is affecting human societies. Efforts to produce social indicators measuring the condition of living human systems are gaining ground. Concern about our inability to monitor public human services the way we monitor financial markets stimulated the state of Connecticut to develop an annual social index (Miringoff and Miringoff 1999; Stille 2002). The social index of leading indicators combines 16 measures of social health, including child poverty, teenage suicide rates, average weekly wages, income inequality, homicide rates, health insurance coverage, and alcohol-related traffic deaths. When examined on a national scale and aggregated at state levels, three indicators in particular—child poverty, high school completion, and health insurance—were indicators of overall social health (Stille 2002). The

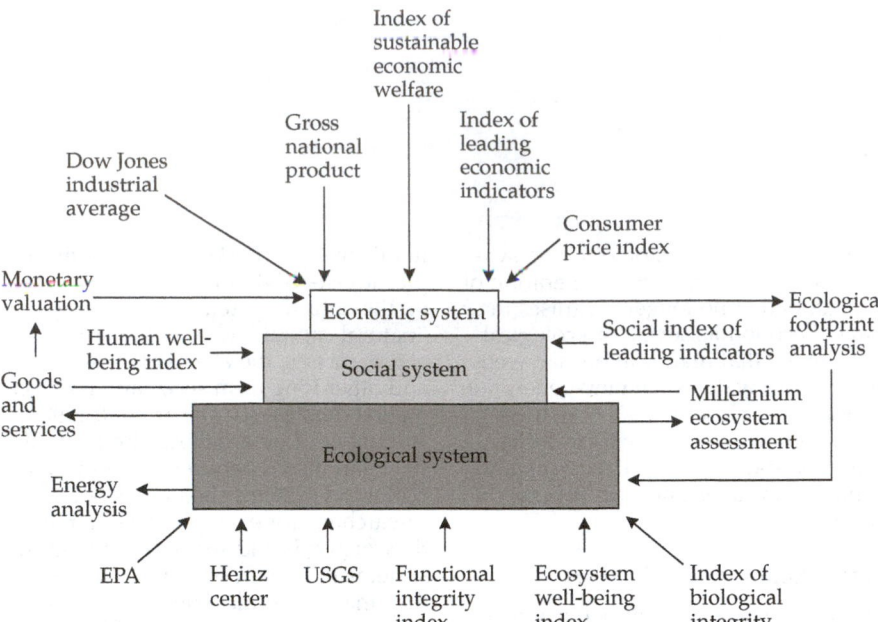

Figure A Selected indicators for ecological, social, and economic systems, including some that measure their interactions.

Miringoff study showed that although GDP continued to grow over 30 years, Americans' social health simultaneously plummeted, as problems like child poverty, average wages, teen suicide rates, income inequality, and health insurance coverage all worsened (Miringoff and Miringoff 1999). Infant mortality rate, another important indicator of societal well-being, increased in the U.S. in 2002 for the first time since 1958; 41 countries had lower infant mortality rates than the U.S. An infant born in Beijing or Havana, for example, was less likely to die in the first year of life than one born in the U.S.

Beginning in the 1970s, at least eight European nations formalized "national social health reports"; other nations including Canada, Cyprus, Hungary, Turkey, and Australia joined these countries in the 1980s and 1990s. In 2002 the ruler of Bhutan mandated production of a "gross national happiness" report. As of early 2005, the U.S. had not joined the nearly 20 countries explicitly mandating systematic evaluation of social well-being, despite, or perhaps because of, the very different picture from the one based on GDP that such an evaluation paints (Figure B).

Development of indicators to improve our knowledge of the condition of Earth's nonhuman living systems is also under way. Perhaps the farthest-reaching project of this kind is the Millennium Ecosystem Assessment, an international undertaking intended to supply decision-makers with scientific information on the consequences of ecosystem change for human well-being and on the available options for mitigating undesired change (Millennium Ecosystem Assessment 2005a). Although much work has been done to develop the assessment's concepts and approaches, few data are yet available to guide decision-makers.

Several U.S. efforts aim to understand the state of the nation's ecosystems, biological resources, or the current state of the environment at the national level. The Heinz Center (2002) identifies more than 100 specific indicators of the physical, chemical, and biological condition of the nation's ecosystems and human uses of those systems, provides data on current conditions and past trends, and highlights significant gaps in our ability to adequately describe key characteristics of these systems. In the 1990s, the U.S. Geological Survey (Mac et al. 1998) and the fledgling U.S Department of Interior's National Biological Service (before it was moved to the Geological Survey; LaRoe et al. 1995) were directed to improve information about the nation's living resources. The Geological Survey reported on factors affecting biological resources and resource trends. The National Biological Service report focused on selected species and ecosystems. Neither agency explicitly focused on developing a coherent set of indicators, and neither seems to have had much influence on natural resource or environmental policy to date.

As of early 2005, the only biological indicator project (*specifically* biological) to make it into the policy arena was the index of biotic integrity (IBI). Begun in the midwestern U.S. soon after passage of the 1972 Amendments to the Water Pollution Control Act (a.k.a. the Clean Water Act), it was the first effort to measure directly and in biological terms the unraveling of living aquatic systems caused by human activities (Karr and Dudley 1981; Karr and Chu 1999; Knopman and Smith 1993). The goal was to shift water resource agencies' focus away from chemical indicators and engineering solutions. The index, analogous to multimetric economic indexes, combines biological measures into a single index of biotic condition. In the 30 years or so since its development, this index has been used worldwide to assess the biological condition of water bodies, diagnose the cause(s) of degradation, and track the success of conservation and restoration

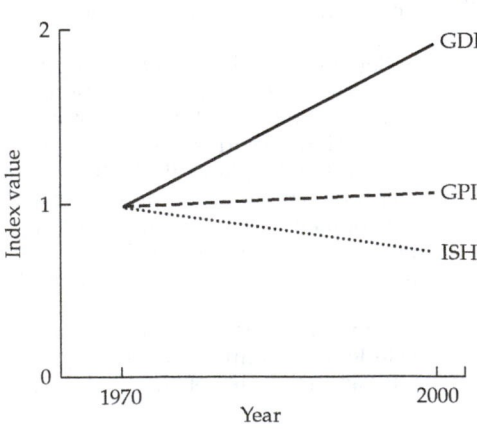

Figure B Differing views of progress based on three measures of well-being indexed to 1.0 in 1970. GDP = gross domestic product; GPI = genuine progress indicator; and ISH = index of social health.

programs (e.g., Karr and Rossano 2001; Niemi and McDonald 2004; U.S. Environmental Protection Agency 2005).

Better biological and social indicators are long overdue. By leading us to value mainly the things money can buy, narrow economic indicators have implicitly endorsed unsustainable lifestyles and blinkered us to the effects of our actions on noneconomic entities.

Until we have comprehensive social and biological measurements of well-being, we will not fully perceive the erosion of the baselayers of Earth's support systems, and our policymakers will lack the crucial foundation for informed decision-making. If we do not develop, use, and watch broad indicators of well-being, we will not fully understand the status of and trends in

living systems, how human actions influence those trends, or how we can avoid activities that threaten the well being of life on Earth. If, on the contrary, we couple improved biological indicators with carefully defined social indicators and better economic indicators, we may improve the state of the biosphere as well as our own lives. ■

---

thropogenic climate change or species introductions (Rice and Emery 2003). All conservation research can be more usefully applied if brought to the standards of "strong inference" (Platt 1964; Caughley and Gunn 1996). By placing both biological and social research in hypothesis-testing frameworks, we will be able to contribute to finding workable solutions to our conservation dilemmas.

Conservation practice necessarily must be tailored to distinct situations, yet we also need to search for commonalities and contingent guiding principles that can help us prioritize and plan. Poff et al. (2003) suggest several means for accomplishing this including: (1) conducting large-scale experiments in the context of adaptive management; (2) comparative analyses of case studies to help us understand which ecosystem responses to threats or to conservation interventions are general, and which are context specific; and (3) engaging stakeholders in cooperative ventures of observation and experimentation that engage public support of learning through conservation practice.

Experiments conducted at an appropriate scale are irreplaceable learning tools. For example, by collaborating with resource managers controlling water flow of the Colorado River through the Grand Canyon, freshwater ecologists and hydrologists were able to design a large-scale experimental water release to evaluate hypotheses concerning sediment availability and how habitat forms within the river ecosystem that now guides management of the river's flow (Rubin et al. 2002 described in Poff et al. 2003). Smaller-scale experiments also can serve to illuminate potential changes. For example, experimental climate warming can be simulated to observe effects on plant and insect communities. In one such experiment, simulated warming caused a rapid decrease in species-richness in a mountain meadow and in shrubland habitats in the Tibetan Plateau, although simulated grazing decreased this effect somewhat (Klein et al. 2004).

How do we overcome uncertainties that hinder decision-making?

Deliberative, information-gathering, and decision-making tools all can aid efforts to guide conservation (Millennium Ecosystem Assessment 2005a). Community fo-

rums, consensus conferences, and other mechanisms can help groups deliberate about how to solve key problems and build stakeholder investment (see Essay 17.2). Information that can guide deliberation can be brought in to the analysis by conservation scientists, and can be generated through collaborative efforts to define problems and assess public opinion or capacities for particular conservation actions. Many widely used decision-making methods are well suited to helping sort through uncertainties, or can be made to support better outcomes if modified to incorporate ecological economic principles (see Chapter 5). These practices help reveal uncertainties, allow stakeholders to discuss them, and place them in a context that facilitates decision-making.

One lesson is not to ignore local or indigenous wisdom and decision-making processes. Biodiversity loss and degradation has increased uncertainty for human societies, leading to negative consequences for health, food security, social relations, and material well-being. Human communities worldwide are less buffered against natural disasters, particularly flooding, as natural habitats such as mangroves and coral reefs have been degraded or replaced by drastically simpler ones (Millennium Ecosystem Assessment 2005b). This is widely understood by people experiencing these disasters, and provides a powerful motivation in favor of greater habitat conservation. Further, people in rural communities analyzed as part of the subregional assessments of the Millenium Ecosystem Assessment (MA) "cherish and promote ecosystem variability and diversity as a risk management strategy" against uncertainty and unforeseen food shortages or natural disasters and have become skeptical of solutions that limit diversity of "species, food, and landscapes [that] serve as 'savings banks' rural communities use to cope with change and ensure sustainable livelihoods" (Millennium Ecosystem Assessment 2005b). Unfortunately, many rural communities are either pushed toward less sustainable practices by well-meaning development efforts, or compromised by other activities that undermine ecosystem services in their vicinity. Governmental decisions to create dams or promote particular methods of crop production and markets may well undermine long-term security of local

communities. Decision-making partnerships between communities and policymakers in which all parties can learn from one another would likely result in better solutions.

In its analysis of the process and outcomes of a wide variety of decisions, the MA emphasized that the use of decision-making methods that adopt a pluralistic perspective are particularly helpful as they allow exploration among differing viewpoints, and can be conducted at any spatial scale (Millennium Ecosystem Assessment 2005a). Further, they noted that better outcomes for ecosystem services, and thus indirectly for biodiversity more broadly, resulted from decision-making processes that:

- Use the best available information, including considerations of the value of both marketed and non-marketed ecosystem services
- Ensure that the basis and process for decision-making is open and transparent
- Ensure the effective and informed participation of important stakeholders
- Recognize that not all values at stake can be quantified, and thus quantification can provide a false objectivity in decision-making processes that have significant subjective elements
- Strive for efficiency, but not at the expense of effectiveness
- Consider equity and vulnerability in terms of the distribution of costs and benefits
- Ensure accountability and provide for regular monitoring and evaluation
- Consider cumulative and cross-scale effects and, in particular, assess trade-offs across different ecosystem services

Responding to intensification of threats

One additional way to handle uncertainty in conservation planning is to explicitly prepare for intensification of threats and synergistic effects among them. Essentially this means broadening our sense of uncertainty even farther, which is often accomplished by creating plausible scenarios of change, and evaluating what kinds of conservation actions may be needed in light of these scenarios. The need for such precautionary thinking is obvious particularly when considering the wild card of climate change. Historical analysis may be an effective tool to inform conservation planning in the face of climate change (Gillson and Willis 2004; see Case Study 6.1). Such studies may help us predict at what climate thresholds changes in ecosystem composition may occur. Conservation strategies are just beginning to incorporate models of potential ecosystem change to inform planning (see Case Studies 10.3 and 13.3).

The 2050 scenarios developed by the MA provide a portrait of how human systems and biodiversity may be affected by plausible societal priorities and decision-making mechanisms (Figure 18.4). The scenarios are described as follows in the MA:

- *Order from Strength.* This scenario represents a regionalized and fragmented world that is concerned with security and protection, emphasizes primarily regional markets, pays little attention to public goods, and takes a reactive approach to ecosystem problems.
- *Global Orchestration.* This scenario depicts a globally connected society that focuses on global trade and economic liberalization and takes a reactive approach to ecosystem problems but that also takes strong steps to reduce poverty and inequality and to invest in public goods such as infrastructure and education.
- *Adapting Mosaic.* In this scenario, regional watershed-scale ecosystems are the focus of political and economic activity. Local institutions are strengthened and local ecosystem management strategies are common; societies develop a strongly proactive approach to the management of ecosystems.
- *TechnoGarden.* This scenario depicts a globally connected world relying strongly on environmentally sound technology, using highly managed, often engineered, ecosystems to deliver ecosystem services, and taking a proactive approach to the management of ecosystems.

All the scenarios except Order from Strength are predicted to improve human welfare. All predict losses of biodiversity in terms of land degradation and species extinction, although these losses are of the lowest magnitude with the Adapting Mosaic and TechnoGarden scenarios, and greatest in the Order from Strength scenario (see Figure 18.4). The scenarios each have elements that are quite plausible, and all but Order from Strength have outcomes that would be desirable. Poverty and food security are most dramatically improved in the Global Orchestration, followed by the TechnoGarden scenarios. Biodiversity and ecosystem services are most protected by the Adapting Mosaic and TechnoGarden scenarios, but cultural services and social freedoms are generally reduced under the TechnoGarden scenario. These scenarios are newly published as of this writing, but we expect that their implications will be discussed widely, and efforts will be made to push global coordination and local conservation planning along the most productive lines indicated from the scenario analysis. Although the 2050 scenarios are not specific as to particular actions and outcomes beyond these aggregate measures, similar scenarios could be developed at

Projected change in human welfare

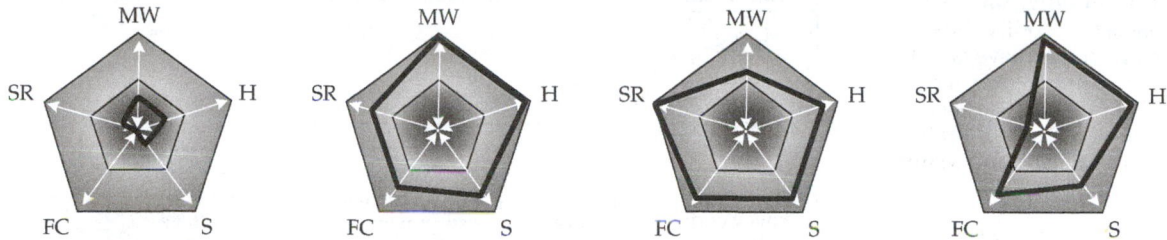

Projected losses of biodiversity

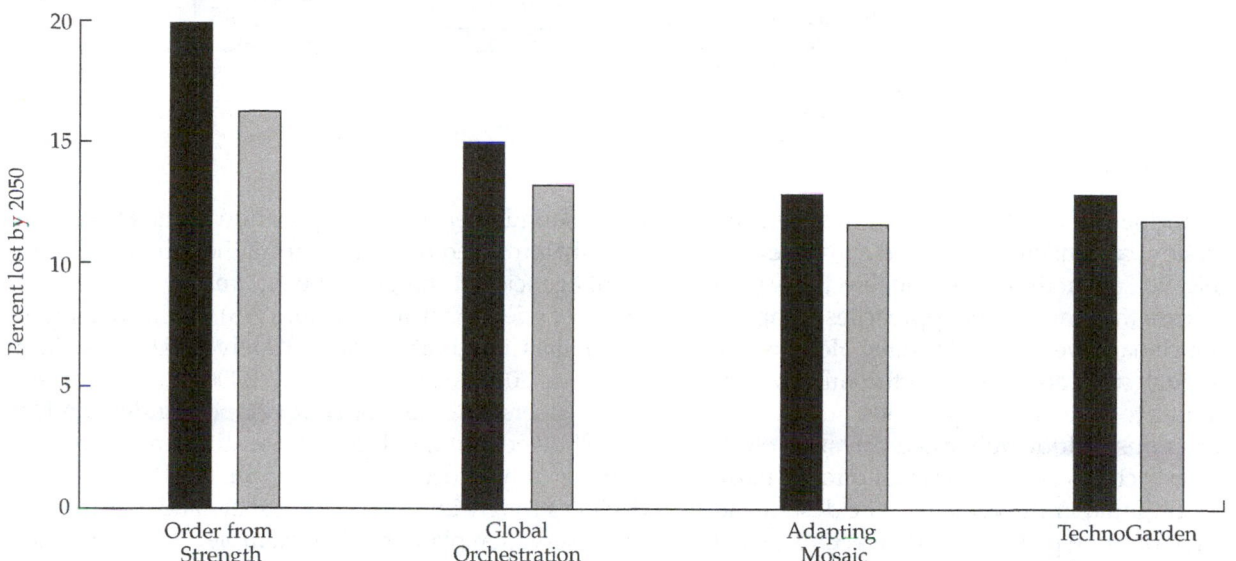

Figure 18.4 Three of four plausible scenarios of future trends in management of environmental resources developed in the Millennium Ecosystem Assessment are predicted to lead to improvements in human well-being, and relatively lower losses in biodiversity in terms of habitat degradation (black bars) or species extinctions (gray bars) by 2050. The path most similar to our current one, unfortunately, is the Order from Strength scenario in which both biodiversity and human welfare fare most poorly. Percent losses in biodiversity are relative to 1970 levels. Five axes of human welfare are represented in each pentagram figure: MW = material welfare, H = health, S = security, FC = freedom and choice, and SR = social relations. In each pentagon, the inner white pentagon represents conditions in 2000. Movement toward the center represents a worsening of human conditions, while movement toward the outer edge represents improvement by 2050. (Modified from Millennium Ecosystem Assessment 2005b.)

local and regional scales. Such engagement with plausible alternate futures can help spur discussion, and consideration of the complex linkages that drive change for biodiversity and human welfare.

Are Conservation Efforts Succeeding and How Can We Improve?

The array of approaches already used to promote conservation fall into five general categories (Figure 18.5):

1. Direct protection
2. Species and ecosystem management and restoration
3. Policy, laws, advocacy, and enforcement
4. Education and outreach
5. Economic and social incentives

These approaches seek to reduce direct and indirect threats affecting species, habitats, and ecosystem processes, with the ultimate goal of enhancing conservation of biodiversity in specific places on Earth, and each can be expanded

Figure 18.5 Biodiversity conservation benefits from a diversity of approaches that act to modify direct and
indirect threats, and achieve target
conditions. Most interventions act on
threats, while a few can also act directly on targets (such as restoration).
(Modified from Salafsky and Margulois 1999.)

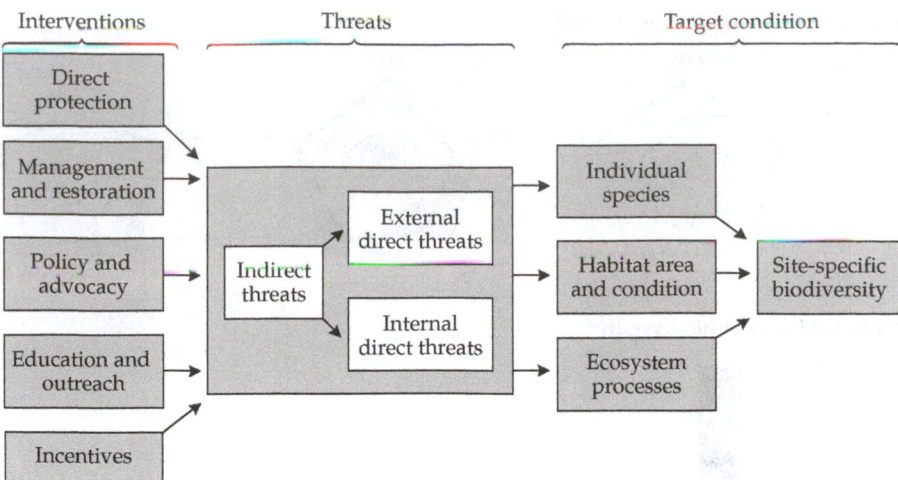

and refined to further conservation. To improve, we need
evaluations of how current efforts are working and extensive communication about these results to direct new projects. How well are these and other approaches being applied? How much have we achieved to date? How can we
improve? Although some answers are forthcoming, we are
still just beginning to answer such questions.

Direct protections include public and private protected areas, and restrictions on activities that directly harm
biodiversity. Continual improvement in the design and
implementation of these protections is an active area of
practice-based research. Management and restoration efforts seek to prevent further degradation by active management in conservation areas overlapping zones of
human use. Many of these approaches are new, and we
are just beginning to learn about how well these work.
Certainly, policy instruments have been critical in conservation, and will continue to be so. One of the greatest
lessons from the shifts seen as new governments come
into power and shift conservation agendas is that policies require continual advocacy to be effective. We are
also still in the beginning stages of creating effective education for conservation at many levels, as well as both
positive and negative incentives for conservation, which
we will cover in greater depth later in this chapter. Here
we summarize existing assessments of conservation
practice using a specific example (implementation of the
U.S. Endangered Species Act) and then move on to discuss the problems caused by limited funding, and how
we might be able to improve assessments of conservation projects in the future, and hopefully secure sufficient
funding for them to succeed.

Enhancing conservation under the
U.S. Endangered Species Act

The U.S. Endangered Species Act (ESA) is a powerful
conservation law that was developed and is applied

using sound science. The application of the ESA is continually improved by reviews from the primary government agencies in charge of the Act and by independent
teams of scientists (Carroll et al. 1996). Detailed analyses
of Habitat Conservation Plans (Kareiva et al. 1998; Harding et al. 2001; Watchman et al. 20001) and Recovery
Plans (Boersma et al. 2001) developed under the U.S.
ESA showed that the depth of scientific understanding
of threatening processes and their relationship to species
decline strongly influenced the quality and effectiveness
of conservation planning. For example, although threats
to species are generally known, specific information
about the nature of the threats, such as where and how
frequently they occurred, and how severely they affected the species when they did occur, are not known
(Lawler et al. 2002). This lack of understanding results in
less effective conservation plans through both poorer design and implementation. Unfortunately, our knowledge
gaps concerning many species are large, and only dedicated research to fill those gaps will help improve most
plans. However, these studies did find instances in
which plans were insufficiently precautionary. Thus, at
times, a shift to greater precaution would be a substantial improvement toward effective conservation under
the ESA.

Perhaps the greatest area for improvement is in how
plans are executed. Implementation of many programs
is incomplete, or in some cases even nonexistent. Until
recently, most threatened and endangered species in the
U.S. lacked their legally mandated recovery plans, and
in many cases the plans were not based on clear scientific principles (Schemske et al. 1994; Tear et al. 1995), were
not or only partially implemented (Lundquist et al. 2002;
Figure 18.6), or had targets that were too low (Tear et al.
1993). As of June 20, 2005, 80% of the 518 federally listed
animals and 82% of 746 listed plants have prepared recovery plans (U.S. Fish and Wildlife Service 2005). This

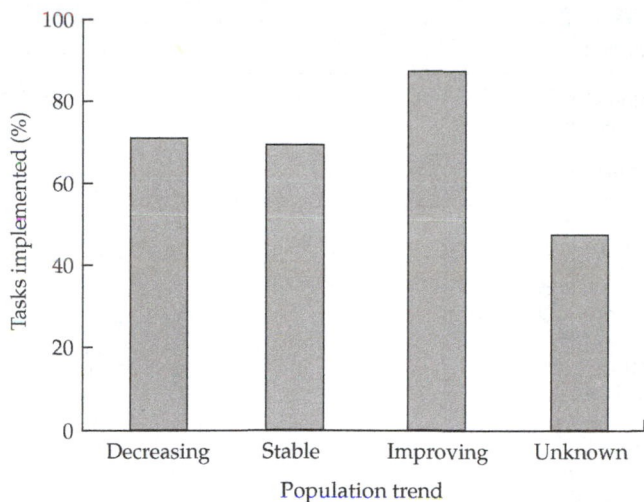

Figure 18.6 Population trends for U.S. threatened and endangered species were more often improving when their recovery plans had been fully implemented. (Modified from Lundquist et al. 2002.)

is the result of a great deal of effort to improve recovery planning. Unfortunately, many plans do not specify recovery tasks that directly address threats (Schultz and Gerber 2002), often because the nature of these threats is so poorly understood (Lawler et al. 2002).

Recent evaluations of species recovery under the ESA have revealed many patterns that can help improve its implementation. Listed species that are recovering or have recovered were more likely to have a dedicated (single-species) recovery plan, and have a greater fraction of recommended recovery tasks implemented, while species without recovery plans or with few implemented tasks were more likely to be declining (Boersma et al. 2001; Abbitt and Scott 2001; Taylor et al. 2005). Multi-species recovery plans should have advantages, such as covering larger spatial scales, and species facing similar threats, but can be weakened if they use poor species-specific understanding of biology and threats, or are not designed with an adaptive management framework (Clark and Harvey 2002). Recovered species seem to have had threats that are more easily addressed (direct sources of mortality, such as persecution or a single pollutant), while those facing water diversions, nonnative predators, and development (none of which are as tractable) tend to be declining (Abbitt and Scott 2001). The fraction of former habitat that was occupied at the time of listing was also much higher for recovered than declining species (Abbitt and Scott 2001). Finally, the longer a species has been listed and harm to individuals has been regulated under the ESA the more likely the species was to have recovered (Taylor et al. 2005).

These studies, and others, have made several recommendations for improving threatened and endangered species protection under ESA. First, planning would be improved through using conceptual frameworks that focus on threat reduction and make this a primary focus of recovery efforts, creating adaptive management plans that are adequately monitoring species status, recovery tasks, and threat levels, and finally, making use of current, quantitative data on population trends (Clark et al. 2002; Stem et al. 2005). Other recommendations concern the details of how plans are developed and implemented (Clark et al. 2002). The U.S. Fish and Wildlife Service has been very responsive to these recommendations (Crouse et al. 2002), as well as to earlier ones made with respect to Habitat Conservation Plans (Kareiva et al. 1998). Certainly, stronger partnerships between conservation scientists and natural resource managers will improve these plans (Meffe et al. 1998; Boersma and DeWeerdt 2001; Crouse et al. 2002; see Chapter 17). It is important to keep in mind that the ESA is by far the most powerful piece of legislation for protecting imperiled species in the U.S. (Carroll et al. 1996). We must continue to use the best science to improve the ESA and resist the ongoing and pervasive political efforts to weaken this remarkable conservation law.

Limited implementation and funding of conservation programs globally

In the recent revisions of the Red List and the Global Amphibian Assessment, conservationists evaluated how frequently recommended conservation programs are in place for birds and amphibians. Only 5% of conservation plans for birds are fully implemented, although an additional 62% are partially implemented. Of these, 6% have been strongly beneficial, while an additional 35% have shown some benefit. Unfortunately, no improvement has been seen in 35% of the cases, despite partial implementation of recovery plans. Generally, most plans that have been fully implemented are showing benefits, but all those cases in which no benefits have been observed are ones in which the conservation plans are only partially implemented (Baillie et al. 2004). Similarly, although about 65% of threatened amphibian species have at least part of their global population occurring in a protected area, very few other conservation measures are in place, despite recognition that they would help species recovery (e.g., corridors, habitat restoration).

Planning and implementation often are dependent on adequate funding of these activities. Listed species in the U.S. were more likely to be recovering if plans were well funded (Miller et al. 2002), because funding relates to the number of recovery tasks that can be implemented. Further, conservation costs generally are not met in any region of the world (Figure 18.7). Half of all the monies

Figure 18.7 The costs of existing conservation programs are generally not met in most areas of the world. (Modified from Balmford et al. 2003.)

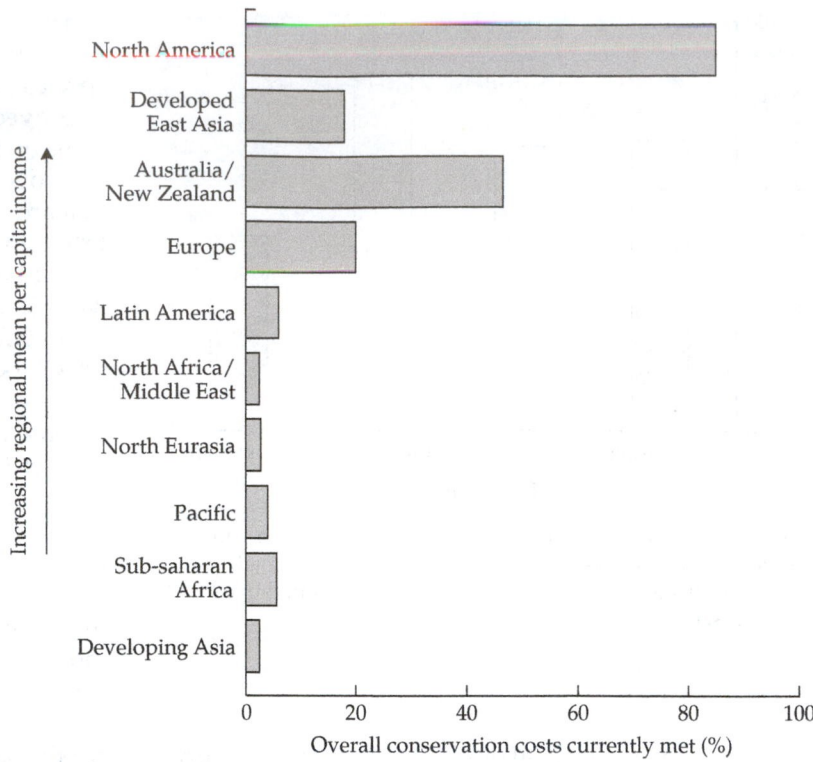

spent on conservation are spent in the U.S. (Balmford et al. 2003), and after this high, conservation funding drops off to less than 20% of what is needed on most continents. Convincing decision-makers that conservation efforts should receive higher fractions of public funds, and securing private sector investment in conservation are high priorities for the future.

Will we meet the 2010 goals of the Convention for Biological Diversity?

The 2010 goals of the Convention for Biological Diversity (CBD) were developed to motivate action to slow biodiversity loss. Representatives of 190 countries committed to working to meet these goals met at the 2002 Johannesburg World Summit on Sustainable Development (United Nations 2002) (Box 18.1). Prospects for meeting these goals are decidedly mixed, yet some are attainable depending on our collective actions in the near term. The degree to which we meet the 2010 goals—and subsequent ones—will depend on how well we are able to adapt to the uncertainties we face, learn from our efforts, and engage society in substantive biodiversity conservation.

Improving assessment of conservation efforts

Unfortunately, we know less than we would like to about the success of specific conservation programs for several reasons. First, conservation action takes place in a shifting background of further degradation. Conserva-

tion efforts are countered, at times overwhelmed, by larger trends and changes. Thus, threatened species may not recover rapidly, despite active intervention, because gains are countered by losses in other areas (e.g., captive breeding and reintroduction may bolster a population, but unless habitat loss and exploitation are also halted, these efforts may not succeed). Second, conservation programs are not always designed for efficient evaluation. Thus, conservationists have worked to create mechanisms and methods for evaluating and learning from conservation projects (e.g., Margulois and Salafsky 1998; Woodhill 2000). Doria Gordon and colleagues describe some of the sophisticated protocols developed by The Nature Conservancy in Case Study 18.1. Finally, the need to secure funding often forces conservation groups working for governments or NGOs to emphasize their successes, often in descriptive terms, and downplay or overlook their failures. While this tendency is understandable, it is unfortunate because it obscures lessons we desperately need to develop successful programs. Fortunately, most conservation organizations have been creating and using assessment strategies, and communicating about these lessons with others. A formal collaboration among conservation groups has added greatly to these efforts (Conservation Monitoring Partnership 2003).

Monitoring and evaluation (M&E) approaches generally undertake a series of steps, guided by a conceptual framework and a more detailed evaluation framework

BOX 18.1 Prospects for Achieving the 2010 Goals of the Convention for Biological Diversity

The Convention for Biological Diversity (CBD) set a suite of goals for conservation that all signatory nations must work toward over the first decade of the twenty-first century. Although a decade is a short period of time to affect large changes, the goals have helped countries set priorities and have guided international collaboration. In the table below, each goal and specific targets of each goal are listed, as are the prospects for progress by 2010 as evaluated by the Millennium Ecosystem Assessment team.

TABLE A *The 2010 Goals and Targets of the CBD and Expected Progress by 2010*

Goals and targets	Prospects for progress by 2010
Goal 1 Promote the conservation of the biological diversity of ecosystems, habitats, and biomes	
Target 1.1 At least 10% of each of the world's ecological regions effectively conserved. Target 1.2 Areas of particular importance to biodiversity protected.	Good prospects for most terrestrial regions. Major challenge to achieve for marine regions. Difficult to provide adequate protection of inland water systems.
Goal 2 Promote the conservation of species diversity	
Target 2.1 Restore, maintain, or reduce the decline of populations of species of selected taxonomic groups.	Many species will continue to decline in abundance and distribution, but restoration and maintenance of priority species possible.
Target 2.2 Status of threatened species improved.	More species will become threatened, but species-based actions will improve status of some.
Goal 3 Promote the conservation of genetic diversity	
Target 3.1 Genetic diversity of crops, livestock, and harvested species of trees, fish, and wildlife and other valuable species conserved, and associated indigenous and local knowledge maintained.	Good prospects for ex situ conservation. Overall, agricultural systems likely to continue to be simplified. Significant losses of genetic diversity in fish likely. Genetic resources in situ and traditional knowledge will be protected through some projects, but likely to decline overall.
Goal 4 Promote sustainable use and consumption	
Target 4.1 Biodiversity-based products derived from sources that are sustainably managed, and production areas managed consistent with the conservation of biodiversity.	Progress expected for some components of biodiversity. Sustainable use unlikely to be a large share of total products and production areas.
Target 4.2 Unsustainable consumption of biological resources or the impact on biodiversity reduced.	Unsustainable consumption likely to increase.
Target 4.3 No species of wild flora or fauna endangered by international trade.	Progress possible; for example, through implementation of the Convention on International Trade in Endangered Species of Wild Fauna and Flora.
Goal 5 Pressures from habitat loss, land use change and degradation, and unsustainable water use reduced	
Target 5.1 Rate of loss and degradation of natural habitats decreased.	Unlikely to reduce overall pressures in the most biodiversity-sensitive regions. However, proactive protection of some of the most important sites is possible.
Goal 6 Control threats from invasive alien species	
Target 6.1 Pathways for major potential alien invasive species controlled.	Pressure is likely to increase (from greater transport, trade, and tourism). Measures to address major pathways could be put in place (especially if globally coordinated).
Target 6.2 Management plans in place for major alien species that threaten ecosystems, habitats, or species.	Management plans could be developed.

Continued on next page

Box 18.1 *continued*

TABLE A *Continued*

Goals and targets	Prospects for progress by 2010
Goal 7 Address challenges to biodiversity from climate change and pollution	
Target 7.1 Maintain and enhance resilience of the components of biodiversity to adapt to climate change.	Pressures from both climate change and pollution, especially nitrogen deposition, will increase. These increases can be mitigated under treaties to reduce climate change and through agricultural, trade, and energy policies. Mitigation measures include carbon sequestration and use of wetlands to sequester or dentrify reactive nitrogen.
Target 7.2 Reduce pollution and its impacts on biodiversity.	Proactive measures to reduce impacts on biodiversity possible, but challenging given other pressures.
Goal 8 Maintain capacity of ecosystems to deliver goods and services and support livelihoods	
Target 8.1 Capacity of ecosystems to deliver goods and services maintained.	Given expected increases in drivers, can probably be achieved only on a selective basis by 2010.
Target 8.2 Biological resources that support sustainable livelihoods, local food security, and health care (especially for poor people) are maintained.	
Goal 9 Maintain sociocultural diversity of indigenous and local communities	
Target 9.1 Protect traditional knowledge, innovations, and practices.	It is possible to take measures to protect traditional knowledge and rights, but continued long-term decline in traditional knowledge likely.
Target 9.2 Protect the rights of indigenous and local communities over their traditional knowledge, innovations, and practices, including their rights to benefit sharing.	
Goal 10 Ensure the fair and equitable sharing of benefits arising out of the use of genetic resources	
Target 10.1 All transfers of genetic resources are in line with the CBD, the International Treaty on Plant Genetic Resources for Food and Agriculture, and other applicable agreements.	Progress is possible. More equitable outcomes will be achieved if these resources are coordinated globally.
Target 10.2 Benefits arising from the commercial and other utilization of genetic resources shared with the countries providing such resources.	
Goal 11 Parties have improved financial, human, scientific, technical, and technological capacity to implement the Convention	
Target 11.1 New and additional financial resources are transferred to participants in developing-countries to allow for the effective implementation of their commitments under the Convention, in accordance with Article 20.	Progress is possible. This outcome is more likely to be under globally, rather than locally, coordinated efforts.
Target 11.2 Technology is transferred to participants in developing countries to allow for the effective implementation of their commitments under the Convention, in accordance with Article 20.	

Source: Millennium Ecosystem Assessment 2005b.

(Stem et al. 2005). A conceptual framework defines expected cause-and-effect relationships between drivers of change and the status of biodiversity targets, while an evaluation framework specifies the expected results of conservation actions, and how those will affect biodiversity targets. A critical component in assessment is to focus not only on biological (and social or economic) targets that a conservation program is designed for, but also on social and economic drivers of biodiversity decline (Salafksy and Margulois 1999; Balmford et al. 2005; Stem et al. 2005). The net result is to place conservation projects in a full adaptive management context, with the expectation that both quantitative and qualitative assessments will be used as is appropriate to evaluate the influence of interventions on biodiversity targets, and on drivers of change (Figure 18.8).

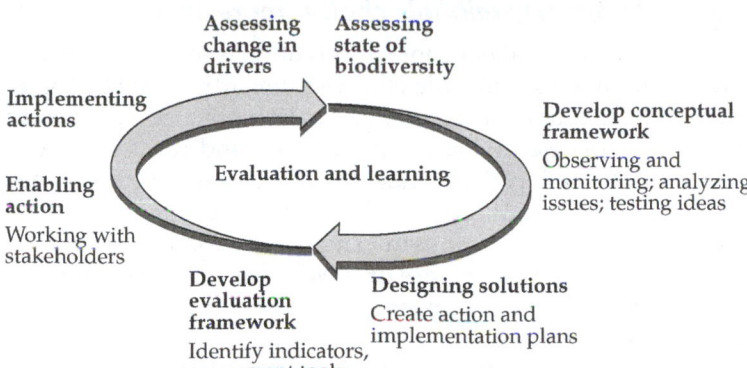

Figure 18.8 Biodiversity conservation projects should be designed within an adaptive framework. Assessment, analysis, and design of conservation projects should target both biodiversity targets and drivers of biodiversity loss. Conservation plans and actions, where possible, should target both improvements in the status of biodiversity targets, and a lessening of underlying drivers. Throughout the process, there are many opportunities for evaluation and learning, which will improve overall project success.

Influencing People's Habits: Reducing Destructive Impacts

Clearly, success in conservation in part requires convincing people to reduce their destructive habits, and offering them some palatable alternatives. But how do we get most people to change? Although many citizens are concerned about the environment, and care about biodiversity in a broad sense, few seem motivated to make substantial lifestyle changes (Figure 18.9).

Certainly, an essential component of conservation is public education. In the United States' K–12 school sys-

Figure 18.9 Most people—the general public, politicians, and conservationists alike—have a habit of denying the ills of our future, and refuse the sacrifices that will be required. (Calvin and Hobbes © 1992 Watterson. Dist. by Universal Press Syndicate. Reprinted with permission. All rights reserved.)

tem, environmental education is at best considered an enrichment to the curriculum. Recent accountability measures ("No Child Left Behind") have emphasized student success on test scores rather than innovative forms of learning that might involve the natural environment. In spite of these restrictive efforts many teachers encourage environmental education. Susan Jacobson discusses elements of public conservation education, and provides some case study examples of successful programs in the U.S. in Essay 18.3.

Universities have a special role in conservation education both in helping develop an informed and motivated citizenry, and in developing specialists in many disciplines who might contribute to future solutions. The Talloires Declaration is a recent movement to engage university communities in promoting sustainable practices both through teaching, outreach to local, regional, and global communities, and by example of institutional reform to uphold sustainability standards (Table 18.3). Service learning, where classes engage their local communities to identify environmental concerns and learn problem-solving skills, are becoming widespread.

Among the new roles universities might take on is enhancing conservation awareness among groups of students and professionals not traditionally allied with biodiversity conservation. Partnerships with MBA, Law, Public Health, civil engineering, and communications programs could all lead to changes in professional practice. At present, few professional programs provide any conservation education. It is relatively simple to encourage the use of examples from conservation in other disciplines (e.g., hedonic valuation of biodiversity in economics courses), and to offer occasional seminars focused on how conservation can be bought into disciplines not traditionally associated with conservation. A stronger approach is to engage students and faculty from various disciplines on large conservation projects. In this way, dialogue across disciplines is more focused and the values brought by different disciplines are appreciated.

Making sustainable choices more attractive

A major challenge for conservation is to find the means to make sustainable choices more attractive to people. Certainly, personal change can be motivated by a greater awareness of conservation issues and can result from public education, but it is also necessary to create other attractions. While not all conservation alternatives appear to be a "win-win" situation, we can certainly increase the prevalence of those cases that do simultaneously serve the needs of conservation and human development.

For example, in many developed countries, development is driven by demand and profit. As long as housing subdivisions are more attractive to people than compact, low-impact alternatives, then sprawling housing developments will remain profitable and will be pursued unless state or local governments regulate their expansion more aggressively. This may result from low-impact alternatives being more poorly known, or from their failure to meet key consumer demands in their design. Analysis of what factors influence such choices can help to create more effective low-impact housing designs, which could go a long way toward reducing negative impacts from development. For example, sprawl increases community tax burdens due to the higher costs of utilities and services (Nelson and Dorfman 2000). This provides an empirically supported argument that green space and lower taxes are compatible. However, such planning should also be done in a broad context, in conjunction with land use zoning and broader development trajectories, or success will be very limited.

Another tool that may help change people's attitudes and habits is conservation advertising. Advertising is a multi-billion dollar business in most developed countries, and is mainly designed to foster consumption of products. If advertising success can be garnered for countering that demand, and fostering conservation and sustainable habits, we may make some real advances in reaching the public. For example, WWF-UK and the UN Environment Programme (UNEP) have been creating effective posters that encourage the public to become involved in conservation—from helping reduce carbon emissions, protecting wildlife, reducing fisheries impacts, living more sustainably, and appreciating the ecosystem services of wild nature.

There are many arenas where conservation advertising might help push people to practice more responsible consumption patterns. Simple attractive displays for sustainable products might draw customers to them in large supermarkets. For example, photos of migratory birds could be posted above the shade-grown coffee and organic bananas to encourage a

---

TABLE 18.3 *The Talloires Declaration 10-Point Action Plan*

1. Increase awareness of environmentally sustainable development
2. Create an institutional culture of sustainability
3. Educate for environmentally responsible citizenship
4. Foster environmental literacy for all
5. Practice institutional ecology
6. Involve all stakeholders
7. Collaborate for interdisciplinary approaches
8. Enhance capacity of primary and secondary schools
9. Broaden service and outreach nationally and internationally
10. Maintain the movement

Source: University Leaders for a Sustainable Future 1994.

ESSAY 18.3

The Importance of Public Education for Biological Conservation

Susan K. Jacobson, *University of Florida*

Think of a challenging conservation problem you have encountered—protecting rare birds, cleaning up wetlands, or sustainably managing forests. Inevitably, people are part of the problem and education will be part of the solution. Effective education and outreach are essential for promoting conservation policy, changing people's behaviors, garnering funds, and recruiting volunteers. The fate of our ecosystems and the plants, animals, and people that depend on them lies in part with successful education for a variety of audiences—from children to adults in settings from schools to farms and forests.

The Challenge

Public support is often necessary to the success of environmental management efforts. Researchers could spend years designing plans or studying biological processes, but conservation goals may not be achieved without adequate public support. In such cases, failure to accurately assess and target public knowledge and opinion can result in failed conservation initiatives.

The public is exposed to conservation issues through print media, radio, television, and the Internet, through interpersonal activities such as demonstrations and workshops, and through the formal school system and informal clubs and other groups. Although public opinion polls show most people are interested in the environment, public knowledge about conservation is minimal. Concern for wildlife is largely confined to attractive and emotionally appealing species. In one poll, 89% of the people surveyed agreed that endangered Bald Eagles (*Haliaeetus leucocephalus*) should be protected, but only 24% believed the endangered Kauai wolf spider (*Adelocosa anops*) deserved protection; most of the public viewed invertebrates with indifference or aversion (Kellert 1996).

Few K–12 school curricula offer classes in biological conservation or environmental management. Environmental topics are only sometimes infused into traditional courses. Links between ecosystem services and human health and livelihoods are seldom made. Although many innovative extracurricular materials have been developed, teachers may not find the time or motivation to teach them. Yet, the need for conservation education continues to increase as problems become more complex. From cumulative impacts on wetlands restoration to climate change and declining biodiversity, a knowledgeable public is needed to effectively address conservation issues (Figure A).

Designing Conservation Education

Education is an essential management tool that recognizes the central role of people in all conservation efforts. Indeed, although a conservation goal may be focused on a biological problem, effective conservation strategies must incorporate education and communication programs designed to affect people's awareness, knowledge, attitudes, and behaviors toward nature, land management, and development.

The goals of conservation education are many. Researchers (e.g., Knudson et al. 1995, Jacobson 1999; Jacobson et al. 2005) have found that appropriate education programs can help:

- Increase public knowledge and consequent support for the development of appropriate environmental policies

- Foster a conservation ethic that enables responsible natural resource stewardship

- Alter patterns of natural resource consumption

- Enhance technical capabilities of natural resource managers

- Incorporate resource management concerns into private sector and government policymaking processes

To succeed, educators must respond to their audiences' existing interests and behaviors. Some programs target a broad audience with a public awareness campaign, such as providing information about recycling to all homeowners in Colorado, or introducing wetlands ecology themes into a curriculum. Other programs target groups practicing specific behaviors the organization wishes to change, such as providing demonstrations for xeroscaping homeowners' gardens, or modeling proper

Figure A Educating children regarding nature and conservation is an easy and rewarding activity. The natural curiosity and open minds of children make them wonderful recipients for conservation education. Here a member of the Savannah River Ecology Laboratory Outreach Program discusses snakes and their ecological roles with an eager group of elementary school students. (Photograph courtesy of D. Scott.)

activities for boating safely with marine mammals. Other programs teach problem-solving skills or issue analysis to youth or adults involved in land management or community development.

Systematic planning enables conservation educators to identify their immediate and long-term goals and to design activities to target people's knowledge, attitudes, and behaviors. Involving stakeholders in an assessment of educational needs helps determine appropriate educational interventions to solve conservation problems (Jacobson 1999). Once specific objectives for an education program are delineated, alternative methods and approaches for each audience can be considered. The examples that follow highlight some of the methods that have been successfully integrated with conservation programs to achieve specific results.

Conservation Education Examples

The activities of a conservation education program that dramatically reversed the decline of seabirds populations, such as Razorbills (*Alca torda*), Common Murres (*Uria aalge*) and Atlantic Puffins (*Fratercula arctica*), along the north shore of the Gulf of St. Lawrence in Canada included: residential youth programs on seabird ecology and wildlife law held at an island sanctuary; local student instructors with environmental training; conservation clubs; school seabird curriculum materials and theater productions; ornithological and pedagogical training for 50 local volunteers and staff; seabird posters, calendars and other publications; radio, film, and television specials on seabirds; and study tours for leaders of national and regional conservation groups.

The Quebec-Labrador Foundation, in collaboration with the Canadian Wildlife Service, developed the conservation education program as part of their seabird management plan (Blanchard 1995). Their education program resulted in decreased human predation on seabirds and eggs, which is illegal, as well as positively changing conservation knowledge and behaviors of local residents. Initial interviews with residents of communities where bird population declines were most severe revealed a lack of public knowledge concerning wildlife regulations and a high occurrence of illegal hunting. Ten years after implementing the education program, significant increases in seabird populations were recorded. A concomitant follow-up survey of com-

munity members revealed improvement in residents' attitudes concerning seabirds, and a decrease in perceived need to hunt and consume them.

This program demonstrates many aspects of successful conservation education. The audience participated in all aspects of program development. Experiential and interdisciplinary approaches were employed, from hands-on monitoring of seabirds to community theater productions starring puffin-costumed children. A comprehensive system was designed to monitor and evaluate program success—in this case through censuses of the biological resources and before/after surveys of the knowledge, attitudes and behaviors of the targeted audiences.

Understanding knowledge, attitudes, and behaviors of target audiences is essential for effective resource management and the first step in designing an education program. A land management program of the U.S. Department of Defense illustrates the importance of assessing and involving key audiences in the process. The Natural Resources Division staff at Eglin Air Force Base in the Florida Panhandle realized the need for public support as they adopted an ecosystem management plan in the 1990s. The new plan called for increased prescribed burns on up to 30,000 ha per year to restore the native longleaf pine (*Pinus palustris*) forest to their 193,000 ha forested base. Longleaf pine forest was once the most widespread ecosystem in the southeastern U.S., but now occupies only 3% of its former range.

A survey of the public in surrounding counties, however, revealed that only 12% of the residents realized that fire was a natural and beneficial process in the longleaf pine ecosystem (Jacobson and Marynowski 1997). Imagine the public indignation if large fires and the accompanying smoke disturbed their communities unannounced. Fortunately, Eglin resource managers incorporated a comprehensive public education program into their management plan. Educational materials—mass media, publications, and public events—significantly increased audience awareness of the benefits of fire. Continuity and repetition of conservation information has continued to increase public knowledge and support for appropriate forest management at Eglin as the ecosystem plan is fully implemented.

Holistic approaches to conservation education such as these are less com-

mon within the formal school system. However, because most environmental attitudes are formed during childhood, school children are an important target for conservation educators. Conservation education in schools varies among countries in content, scope, and disciplinary base, and often is overlooked or ignored relative to traditional subjects. Extracurricular materials and innovative curricular supplements often provide the main exposure for students to environmental conservation knowledge.

A challenge for conservation educators working in the schools is finding instructional strategies that can enable students not only to learn about local and global conservation issues, but also to learn how to act in response to environmental problems. To be effective, conservation education cannot be confined to the boundaries of a single classroom, but rather must operate within the context of the environment it seeks to conserve.

The Global Rivers Environmental Education Network (GREEN) exemplifies an ideal approach—hands-on, participatory, and interdisciplinary—to involve children in the complexities of environmental problem-solving. The GREEN program, developed by the University of Michigan (Stapp et al. 1995), is a water quality monitoring program that has been adopted in more than 60 countries.

GREEN began in a Michigan biology class on the banks of the Huron River. Students became alarmed about the water quality when Huron High School students contracted hepatitis A after falling off their windsurfers into the river. The students' concern and subsequent testing of the polluted water by the teacher prompted a University of Michigan class to develop a water quality monitoring program appropriate for the secondary school students. The program included instructional materials, such as maps of the local watershed, a manual outlining standards for performing nine water quality tests, material on monitoring water for macroinvertebrates, a slide-tape presentation, and a set of water quality testing kits.

At the Huron River, students found high fecal coliform counts. Student actions resulted in the discovery of faulty underground storm drains as the pollution source, and led to subsequent correction by the city. Enthusiasm for the program from this initial school resulted in the expansion of GREEN nationally and globally. Now, an international communication network

allows students to share their experiences of watershed quality and diversity across geographical and cultural boundaries. International workshops held in 18 countries allowed educators to exchange ideas on watershed programs appropriate for different geographic areas around the world. An interactive website provides a data base of locally generated water quality data and leads users through a step-by-step process of improving water quality within their watershed. The site serves as a source for exchange of ideas on river conservation and restoration for schools in every state in the U.S.

Future Needs for Conservation Education

Despite success stories such as these, conservation education around the world still lacks funding, resources, and support. Widespread methods for reaching adult and youth audiences are still lacking and conservation education has not been institutionalized into the formal educational system in most countries. Surveys of teenagers reveal that mass media, not schools, were students' most important source of information on environmental issues. Yet conservation educators typically rely on publications and curriculum supplements, and biologists seldom embrace the media or engage in public dialogue.

Conservation education is needed at many levels and in many forms. New approaches by conservation educators to incorporate mass media tools and innovative marketing techniques would strengthen public awareness of the need for conservation. As educators make better use of new technologies to reach larger audiences, efforts must be made to reach people's hearts as well as their minds. Methods to attract new audiences and engage emotions, such as using the arts to promote conservation, deserve greater attention (Jacobson et al. 2005). More widespread programs incorporating participatory, experiential techniques into project activities will foster pro-conservation attitudes and behaviors. The Convention on Biological Diversity and the Millennium Ecosystem Assessment recognize that biological diversity is about more than plants and animals and their ecosystems—it is about people and our need for food, security, medicines, fresh air and water, shelter, and a clean and healthy environment in which to live (Millenium Ecosystem Assessment 2003). Conservation education can help make these connections. Only people's knowledge, values, motivations, and consequent action will dictate whether we preserve our planet's wild species and the ecosystems that support us all. ■

---

positive association between those products and protecting wild nature. In other cases, targeted advertising may be most effective. For example, TRAFFIC, a consortium of the WWF and the IUCN which works to reduce trade in endangered species, researched how traditional medicines are used in San Francisco and then developed an ad campaign to promote preservation of tigers and rhinos in a community partnership. Both types of endangered species were sold as traditional medicines throughout the city prior to the campaign, but afterward had virtually disappeared from the market (Fox 2005). This campaign may have worked in part because the social pressures to conform to community standards were high (Ottman 1997).

Encouraging conservation through incentives

We still need to better understand the motivations for human behavior and develop both positive and negative incentives that could help influence people to act for the common good, and to conserve biodiversity. Many promising approaches exist in shifting economic incentives for conservation, as well as motivating behavior through appeal to moral/religious beliefs and the social sanctions and approval that come from acting in accordance with these beliefs.

Although we typically consider only the economic costs of conservation in terms of opportunity costs, land prices, or salaries for enforcement and management, perhaps we should begin making it clear how substantial the economic benefits of conservation are. Costanza et al. (1997) calculated as a low estimate that ecosystem services globally provide an average of U.S.$33 trillion per year, combined with a worldwide GNP of U.S.$18 trillion. A more recent analysis calculated the costs and benefits of biodiversity protection via reserve systems (Balmford et al. 2002). The total economic value lost by habitat conversion annually was computed to be on the order of U.S.$250 billion in the first year, and for every year into the future (unless there is substantial recovery or restoration). The costs of expanding the world network of protected areas to include 15% of all regions, and a network of marine reserves that covers 30% of the total area of the marine realm (roughly 100 times current levels of protection) would cost about U.S.$45 billion per year, and would protect ecosystem services valued at over U.S.$4500 billion (Balmford et al. 2002). This is a ratio of roughly 100:1; a beneficial return on a conservation investment. Balmford et al. (2002) emphasize that a putative expansion of protected areas need not come at the expense of development for the world's poor. Instead, they point to the fact that the very erosion of biological resources for short-term gains is largely responsible for current desperate poverty, and that true development is impossible without the preservation of substantive portions of biodiversity. These assertions are reinforced by the assessments undertaken by the MA (see Figure 18.4). Instead they suggest that a "judicious combination of sustainable use, conservation, and, where necessary, compensation for resulting opportunity costs...makes overwhelming economic as well as moral sense."

Following in this spirit, many advocate formalizing conservation incentives to round out a range of conservation approaches that includes protected areas, regulations, and conservation management. Conservation incentives could take many forms. Landowners could be subsidized for maintaining natural habitat on their lands, as they are in Costa Rica, or for restoring habitat, as has taken place in Brazil's Atlantic Forest or Australia's Murray-Darling Basin (Ellison and Daily 2003).

Some incentives come in the form of offsets for damages in one arena by conservation in another (Daily and Ellison 2002). Carbon credits in the form of investments in reforestation and maintenance of intact forests are an attractive option for many utilities to meet requirements to reduce net emissions (Swingland 2003). Similarly, land banks allow developers to offset damage to wetlands or habitat of endangered species by pooling their purchasing power to buy and protect larger blocks of land. Still another approach seeks to create sustainable livelihoods for people living in poverty by allowing them to earn stewardship income to guard or monitor a protected area or to subsidize an income-generating practice provided use remains below a sustainable level. While all of these approaches have positive outcomes, conservation biologists worry that unless carefully overseen, these practices will serve to justify destruction of higher-quality habitats for the preservation or restoration of inferior ones. In contrast, where the lands that may be destroyed are of marginal quality, such schemes may indeed represent net conservation gains.

Finally, many religious institutions are expanding their activities in conservation. For example, the Episcopal Ecological Network has become active in promoting biodiversity conservation and sustainable practices among congregants. Rather than having the media serve as the link between conservationists and the public, increasingly conservationists are making direct connections with religious groups. Religious organizations have the advantage of bringing together people who are exploring their moral values, and thus can be open to conservation considerations although they often lack a knowledge base about these issues (Nadkarni 2004).

Fostering sustainable use

Sustainability that allows for conservation of biodiversity must involve increases in sustainable practices in the largest sectors of human impacts—agriculture (croplands and pasturelands), fisheries, forestry, and urban systems. For example, urban ecological footprints are always much larger than their physical area, but increasingly people are working to change this (McGranahan and Satterthwaite 2003). Nicholas You and Charles Wambua present several examples of efforts to enhance the sustainability of urban environments in Case Study

18.2. Without breakthroughs in promoting sustainable practices in all these areas, we will continue to see substantial erosion of biodiversity on land and at sea.

Sustainable agricultural practices include avoiding expansion of cultivation into natural ecosystems through increases in yields, increased nitrogen use efficiency to reduce nitrogen deposition to the aquatic and atmospheric systems, and improved soil quality to maintain yields over the long term (Cassman et al. 2003; see Case Study 6.2). Recently, attention to cultural practices for reclaiming water in arid landscapes has fostered a renewal of more sustainable, and cost-effective means for irrigating crops than extensive dam and water diversion projects (Pearce 2005). We need to focus more research on potential mechanisms for enhancing biodiversity-friendly agriculture and means of increasing higher yields on less land, all the while maintaining good economic returns (Green et al. 2005). Gretchen Daily describes such efforts to enhance biodiversity conservation within human-dominated landscapes in Essay 18.4. Exchange of information on practices that improve yields and serve to increase the hospitality of the agricultural matrix among local farmers can be an invaluable tool for enhancing sustainability (DeVore 2003). Often successful approaches are developed only after years of observation, trial, error, and learning—and may be highly site-specific in detail.

Similarly, we greatly need new solutions to reverse overexploitation of marine fisheries. As most stocks are over-fished, an emphasis has to be placed on recovery. More successful efforts to reduce over-harvest have tended to be structured such that the incentives for individual fishing operators are consistent with conservation (Hilborn et al. 2005). Many suggest aquaculture as a primary means to avoid depleting wild fish, but ironically, at fish farms, predaceous fish, and even some herbivorous ones, are fed wild-caught fish. Often a greater biomass of wild fish are fed to farm-raised fish than are produced for human consumption, and only in the case of some farmed herbivorous fishes does aquaculture not create a net drain on wild fish stocks (Goldburg and Naylor 2005). Several groups now publish guides to sustainable fisheries for use in purchasing seafood. These guides highlight stocks to avoid (because they are over-harvested) and stocks that are recommended that are relatively healthy.

Sustainable practices are better promoted when individuals and institutions serve to enhance communication and translation of information among all interested parties (Cash et al. 2003). Further, good governance is critical to ensure that sustainable practices are a priority. For example, in Germany every government document is printed on 100% post consumer recycled paper. In contrast, the U.S. government is the greatest consumer of

ESSAY 18.4

Countryside Biogeography

Gretchen C. Daily, *Stanford University*

■ The future of biodiversity, and the benefits it supplies society, will be dictated largely by what happens in human-dominated landscapes. Reserve networks alone are unlikely to protect more than a tiny fraction of Earth's biodiversity over the long-run. The areas involved are (and are likely to remain) too small, too isolated, and too susceptible to change to protect more (Daily 2003; Rosenzweig 2003). At the same time, the leading proximate drivers of biodiversity loss—agricultural, pastoral, and silvicultural activities (Heywood 1995; Sala et al. 2000)—are projected to expand greatly over coming decades (e.g., Tilman et al. 2001). The threat embodied in this expansion could be mitigated, in part, through efforts to conserve, in human-dominated "countryside," many species whose native habitats are rapidly disappearing (Pain et al. 1997; McNeely and Scherr 2003).

Countryside refers to the growing fraction of Earth's land surface whose ecosystem qualities are strongly influenced by humanity, but which is not urbanized (Daily 2001). It includes active and fallow agricultural plots, gardens and pasture, plantation or managed forest, and remnants of native vegetation in landscapes otherwise devoted primarily to human activities. The countryside thus offers a broad arena where the functioning and fate of biodiversity can be studied under alternative futures.

Countryside is critical not only to the future of biodiversity, but also to the future supply of ecosystem services; the stream of societal benefits derived from ecosystems and their biodiversity (Daily 1997; Millennium Ecosystem Assessment 2003). These include pollination of crops, renewal of soil fertility, purification of water, and stabilization of climate. Many of these services are supplied on local and regional scales, and their delivery hinges on the capacity of countryside species and ecosystems to generate them in the midst of human activity. Other services such as carbon sequestration for climate stability, which are important globally, also hinge on countryside because the global reserve system will never be large enough to ensure an adequate supply of benefits.

Thus, both the success of conservation efforts and many aspects of human well-being are intimately linked to the management of human-dominated countryside. Yet, surprisingly little is known about the relative conservation value—for biodiversity and/or ecosystem services—of alternative agricultural production regimes and landscape configurations. Initial studies and conservation efforts understandably focused on natural ecosystems and, more recently, on their remaining fragments. "Island biogeography," a paradigm assuming islands of natural habitat surrounded by an inhospitable sea of agriculture and other development, informed much of the early work. Now, however, the "sea" itself has become a focus of study, for two basic reasons: (1) the fraction of the biosphere free from human influence is tiny and rapidly shrinking; and (2) many countryside habitats are not as inhospitable as was thought (Daily 2001; Perfecto 2003).

Countryside biogeography is a new conceptual framework for exploring spatial patterns of biodiversity in rural and other human-dominated landscapes (Daily et al. 2001). There is an urgent need worldwide for basic understanding of countryside biogeography to inform conservation investments (Figure A). The lack of understanding is evident even in the European Union, where extensive human-dominated countryside was created long ago, where its associated biodiversity has been the subject of detailed inquiry, where farmland is the land cover upon which many threatened species depend most (e.g., Tucker 1997), and where roughly 20% of farmland is presently under environmentally sensitive management (e.g., Pienkowski 1998; Kleijn and Sutherland 2003). In The Netherlands, for instance, over 20 years of biodiversity management schemes on farmland have yielded little perceptible benefit. The diversity of plants and abundance of target bird species is no higher on fields under management agreements than on those under conventional management (Kleijn et al. 2001). Although farmers abided by their management agreements, conservation goals were not achieved because of poorly understood constraints on the conservation and restoration of biodiversity at both landscape and local scales (such as atmospheric deposition of nitrogenous and sulphuric compounds, dispersal and seedbank dynamics of plants, and possible decoupling of nesting cues used by birds) (Bakker and Berendse 1999; Kleijn et al. 2001).

Recent work shows that countryside has potentially high biodiversity conservation value. For instance, in largely deforested, low-intensity farming landscapes of southern Costa Rica, 50% or more of the species in major vertebrate, arthropod, and plant groups occur commonly in deforested habitats. Adding to those the species that occur in small (0.1–100 ha) forested habitats, it appears that 75% or more of the native biota survives in this densely populated, heavily farmed region of the country (Hughes et al. 2002; Pereira et al. 2004; Daily et al. 2003; Horner-Devine et al. 2003; Ricketts et al. 2001; Mayfield and Daily 2005).

Whether the composition and configuration of such landscapes can sustain rich, native biotas over the long

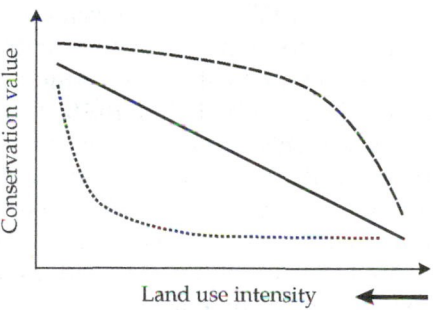

Figure A Some hypothetical relationships between conservation value and land use intensity in the countryside of a region. This is a cartoon; the metrics on both axes are actually complex and multifaceted. If the lower curve applies, the shift in land use intensity indicated by the black arrow would have little impact on conservation value. Conversely, if the upper curve applies, a small reduction in land use intensity would yield great conservation benefits.

TABLE A *Key Questions for Forecasting the Future of Biodiversity and Ecosystem Services*

1. Which species traits are advantageous in the face of major habitat alterations, and why? How will the current extinction episode shape future diversity and evolution?
2. What roles can countryside biotas play in supplying ecosystem services, such as pollination, pest control, water purification, flood control, climate stabilization, and preservation of options for the future?
3. Can the application of existing tools (ecological theory, satellite imagery, GIS, interdisciplinary models for forecasting land-use change) enable rapid and useful prediction of the persistence and functioning of countryside biotas?
4. What practical measures can be taken to enhance the capacity of countryside habitats to sustain biodiversity and ecosystem services? What are the tradeoffs for society of alternative land-use regimes?

term is an open question (e.g., Laurance and Bierregaard 1997; Donald et al. 2001), the answer to which requires both comparative studies in countrysides of contrasting ages and histories, as well as mechanistic studies that examine habitat use, movement, and population dynamics of strategically selected study organisms (e.g., Graham 2001). As work in Kenya and elsewhere has shown, the path to extinction can be many decades or centuries long (Brooks et al. 1999), cautioning against interpreting presence of biodiversity as persistence. Simple models relating biodiversity and land use suggest that significant reductions in the countryside bird fauna of southern Costa Rica may be underway (Hughes et al. 2002), as key habitat features are removed through agricultural intensification.

Three broad questions confront us as wilderness fades away. First, assuming human impacts intensify along the lines currently projected, what sorts of species and ecosystems will persist over the coming decades and centuries? Second, what sorts do we want—and how do we decide? And third, how, and to what extent, can we achieve our desires?

Inspiring progress on all three questions has been made over the past two decades. A variety of new approaches and perspectives is emerging, oriented around building a science of forecasting change in biodiversity and ecosystem services (Table A). Many efforts aim further to integrate ecology with anthropology, climatology, economics, epidemiology, history, law and other fields to characterize the consequences to society of ecosystem change and possible responses to it.

Much work and little time remain to help steer a course that will at least protect the most vital of Earth's life support services. A three-pronged scientific effort is underway to serve this end. It involves the daunting task of building scenarios of ecosystem change, with emphasis on understanding the functional roles of species and ecosystems (Balvanera et al. 2001; Chapin et al. 2000), especially in countryside. To preserve options, it also involves developing more comprehensive strategies to conserve native biotas and their services on human-dominated, as well as remaining "natural" lands. Finally, there are interdisciplinary efforts underway to characterize the biophysical and socioeconomic tradeoffs associated with alternative regimes of land management—and to devise and implement innovative conservation finance schemes that align short-term economic interests with conservation (Daily and Ellison 2002; Millennium Ecosystem Assessment 2003). For conservation to have enduring success, it must become economically attractive and commonplace in countryside. ∎

---

paper goods in the world. Efforts to enhance the sustainable use of water resources relies on a combination of good science, effective communication, and sound governance (IUCN 2000; Figure 18.10). Certainly, if governments lead in adopting sustainable practices, this will create momentum to rapidly progress toward sustainable habits in the citizenry as well. Further, if environmental stewardship is made an important issue by government, citizens may hold their elected officials more accountable for their records of stewardship. Thus, a candidate who can make the case for cost-saving and environmentally sound practices of government might well claim to be the better leader.

One innovative program to enhance knowledge and motivation of political and business leaders to initiate protections for biodiversity is the Environmental Science

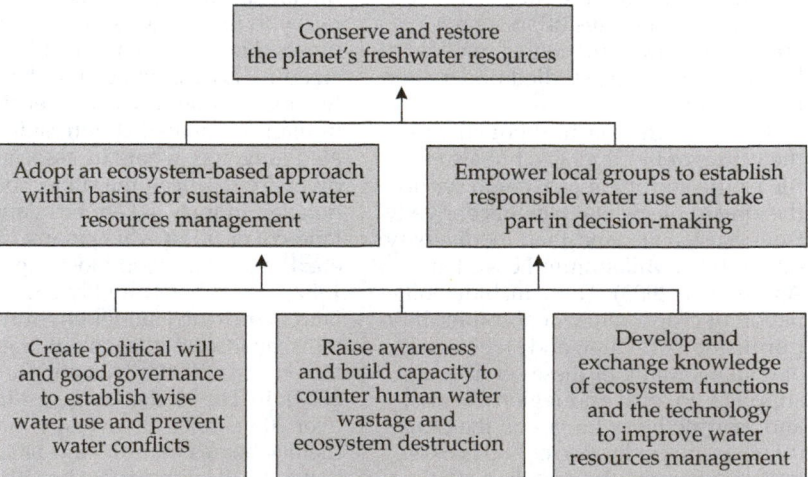

Figure 18.10 Successful conservation increasingly depends on a combination of good governance and scientifically-based management strategies that are widely communicated, targeting improvements both in the status of biodiversity and the capacity of local populations to govern sustainably. This example describes a vision for protecting the world's freshwater resources. (Modified from IUCN 2000.)

and Policy Program at the Organization of Tropical Studies (OTS). This field-based course brings congressional staff, corporate executives, and civil servants on a trip through Costa Rica to learn about tropical biodiversity and the numerous challenges facing wild species and people in developing countries. The course has been extremely successful in engaging political and business leaders in international conservation.

Our Decisions Will Determine the Fate of Biodiversity

The decisions we make over the coming decade will strongly influence how much degradation of biodiversity occurs throughout this century (Figure 18.11). We need to promote radical changes in practices to reduce population growth and consumption, and to conserve biodiversity and improve human welfare throughout the globe. Everyone can assist in the large-scale societal transitions that must be made. Indeed, without the efforts of all communities toward this end, we will end up in a more impoverished world. We advocate that every person commit their efforts—both great and small—to creating a legacy of collective action to conserve biodiversity of which we can be proud.

Not everyone using this text will choose a career in conservation biology; however, no matter your career choice you can significantly help to preserve environmental quality. You can commit to using the most sustainable practices you can—even incremental improvements have an impact. You can become a community activist, and help build coalitions to work together for sustainability. Most humans have a tendency to think locally and in immediate time frames; that is, they discount things that are far away in time or space. This is understandable; immediate needs such as food, shelter, and comfort dominate our senses, and they must be satisfied before we can contemplate larger temporal and spatial scales. However, we also have the ability as a species to think abstractly and understand patterns and trends over space and through time. To the extent possible, we each have a responsibility to think and act to promote conservation and sustainability locally, and globally.

People can make profound changes in their own lives. For example, a long-time poacher of snapping turtles (*Chelydra serpentina*) on the Flint River in Georgia noticed nodules on the inside of the shells of large turtles. When he cut them open he began finding musket balls and arrowheads, proving they had been alive at least 300 years. The idea that these turtles were ancient affected him so profoundly that he is now a proponent of turtle conservation. Finding the points of resonance that enhance the meaning of conservation for individuals will help us succeed in conservation.

Much will depend on the capacity of our leaders to accept the imperatives for conservation and sustainable development. Our hope is that decision-makers will heed the messages coming from all quarters that show our needs for change, and act now.

Good conservation for both humanity and nature comes down to nothing less than a revolution in human thought, on par with the social changes wrought by the Industrial Revolution or struggles for racial equality. Such revolutions are not easy, quick, or painless, but they are necessary and inevitable. Human ethics and value systems are certain to evolve and become more sophisticated over time. Our values must evolve in the direction of sustainability and cooperation with nature, rather than conquest and destruction of it, if humankind is to continue to experience the rich rewards of living on a biologically complex and unique planet. Over 3 billion years of the history of life, and the future of conscious human thought and achievement, are at stake.

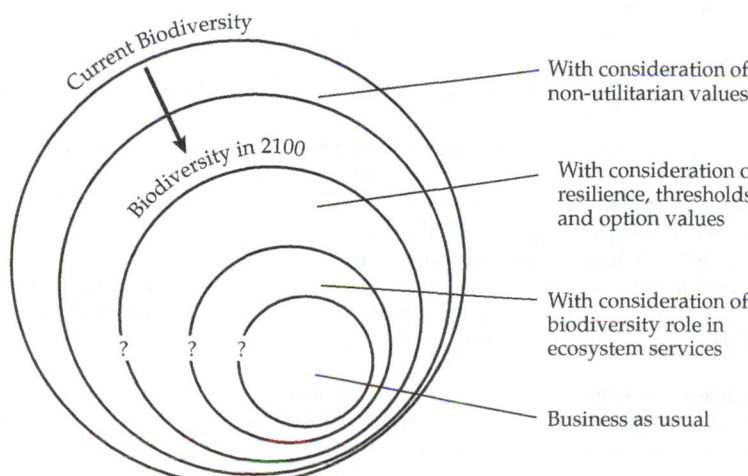

Figure 18.11 Our decisions over the next decade will commit the world to a path where a decrease in biodiversity could be drastic or moderate. Note that this diagram is conceptual, and the relative sizes of the circles do not represent any estimate of change. How will we make the best choices? (Modified from Millennium Ecosystem Assessment 2005b.)

CASE STUDY 18.1

The Nature Conservancy's Approach to Measuring Biodiversity Status and the Effectiveness of Conservation Strategies

Doria R. Gordon, Jeffrey D. Parrish, Daniel W. Salzer, Timothy H. Tear, and Beatriz Pace-Aldana,
The Nature Conservancy

The Nature Conservancy (TNC) is an international biodiversity conservation organization with the mission of preserving the plants, animals, and natural communities that represent the diversity of life on Earth by protecting the lands and waters they need to survive. To achieve this mission, TNC implements a framework called Conservation by Design (The Nature Conservancy 2000), which aims to conserve functional conservation areas within and across ecoregions (Bailey 1989; Olson et al. 2001). Once conservation areas (the "ecoregional portfolio"; Table A) are established for action (Groves 2003), we design conservation strategies using a specified planning methodology, implement those strategies, and subsequently measure our progress toward conservation goals.

Conservation areas may be aggregated into a single planning effort because of their geographic proximity and ecological similarities. Such is the case in the Lake Wales Ridge (LWR) Landscape Conservation Project in Florida. The LWR is an archipelago of inland sand ridges and dune systems that were deposited over a million years ago and remained as refuges during high sea levels of the Pleistocene and earlier (Weekly and Menges 2003). This landscape (roughly 243,000 ha) has specific soil and hydrological characteristics that contain 37 conservation areas that have been identified in the ecoregional portfolio (Figure A).

Soils are excessively drained sands, resulting in drought-adaptation in many of the species despite the 125–140 cm of annual rainfall (Myers 1990). Evergreen hardwood shrubs dominate this community, sometimes with a sparse overstory of pines. The groundcover includes exposed soil, and a diversity of patchy herbaceous species. Structure of the habitat is fire-maintained, as thunderstorms are prevalent across peninsular Florida during the summer. Historically, as sea levels fluctuated, these habitats became more or less connected with surrounding habitat, repeatedly joining and isolating populations of plants and animals. As a result, these areas now harbor the highest number of endemic plant species of any communi-

TABLE A *The Nature Conservancy Terminology to Describe Components of the Conservation Planning Process*

Term	Definition
Ecoregional portfolio	The areas of biodiversity significance identified in an ecoregional assessment that can conserve representative occurrences of biological diversity targeted to meet conservation goals.
Conservation area	An area identified in the portfolio and defined by features such as vegetation, geology, elevation, landform, ownership, or other features that is the focus of strategies designed to conserve many conservation targets. Conservation areas are designed to maintain the targets and their supporting ecological processes within their natural ranges of variability. Conservation areas range along a continuum of complexity and scale, from landscapes that seek to conserve a large number of conservation targets and multiple scales, to small sites that seek to conserve a limited number of targets.
Conservation targets	Populations of imperiled species, natural communities, and ecosystems identified through the conservation planning process as priorities for maintenance of long-term persistence within a defined area.
Conservation strategy	Actions designed to achieve a specific objective or outcome that abates a threat or enhances the ecological integrity of a conservation target.
Ecological integrity	Maintenance of viability for a species target, or maintenance of processes, composition, structure, and function within the natural range of variation for a natural community or system-level target.
Key ecological attributes	The most critical components of biological composition, structure, interactions and processes, environmental regimes, and landscape configuration that sustain a target's ecological integrity.
Attribute categories	Size: the population size needed for viability or area needed to support the natural processes maintaining a community or ecological system; condition: the demography, structure, species composition, and vigor of the conservation target; landscape context: distribution, connectedness, proximity to other habitats, and land use of adjacent areas necessary to maintain viable populations, or to sustain natural community or system-level conservation targets.
Indicators	Measurable entities that are used to assess the status and trend of a key ecological attribute or other factor. A good indicator meets the criteria of being measurable, precise, consistent, relevant, and sensitive.

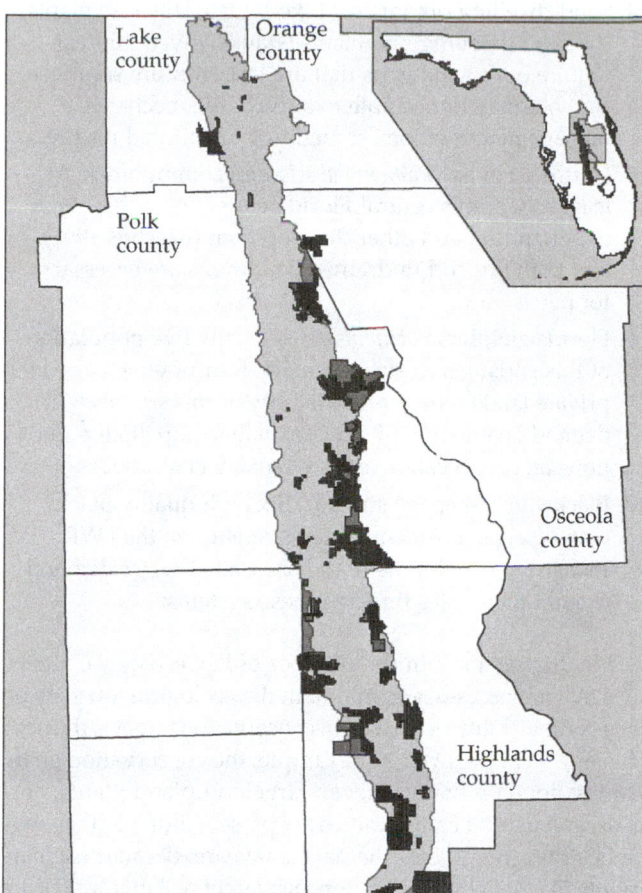

Figure A The Lake Wales Ridge Conservation Area, Florida. The entire Ridge is highlighted in gray, while the portfolio of sites included in the project are black, current conservation ownership in dark gray, and Florida Forever projects proposed for purchase by the State of Florida in 2004 in light gray. Data on managed area and Florida Forever project boundaries provided by the Florida Natural Areas Inventory. (Modified from The Nature Conservancy 2002.)

ty in Florida. Unfortunately, conversion of scrub for agriculture and development has reduced this habitat by over 85% (Weekly and Menges 2003). As a result 22 plants, three reptiles, and one bird species are now listed as federally threatened or endangered. Not surprisingly, the Ridge has long been identified as a critical landscape for conservation in Florida by TNC.

The design of effective conservation strategies (see Table A), and measurement of the change in biodiversity status as a result of those actions, requires a rigorous assessment of both the ecological integrity of the conservation areas and threats to them. Such an assessment ensures that conservation actions are prioritized according to the greatest challenges facing biodiversity and that their effectiveness can be tracked over time. Although TNC has had methods for conducting such assessments for many years (Baumgartner 2001), we needed to develop a more scientifically, temporally, and spatially consistent method for measuring ecological integrity (see Table A) and the extent and severity of threats to biodiversity of concern.

In 2002, an effort was initiated within TNC to improve methods for assessing the conservation status of focal biodiversity: the imperiled species, functional communities, and natural systems (collectively called "conservation targets"; Noss 1996) (see Table A) across the ecoregion. We also wanted to better understand whether conservation actions were effectively abating threats to those conservation targets. Measuring ecological integrity and threat status of biodiversity is an attempt to more directly assess conservation outcomes. The indirect measures of number of acres in conservation status and numbers of dollars raised for conservation action traditionally tracked by the organization are not as accurate (Salafsky et al. 2002; Christensen 2003).

The improved process defines a framework for developing conservation plans that identify efficient strategies designed to meet explicit conservation objectives, and whose outcome can be directly measured in terms of the species, communities, and natural systems identified for conservation (The Nature Conservancy 2003c; Parrish et al. 2003). The steps include:

1. Identifying the planning area of concern and a limited number of focal conservation targets. Focal conservation targets are a small subset of the biodiversity that are selected to comprehensively represent the full range of biodiversity that occurs within a given area.

2. Identifying and assessing the status of key ecological attributes (see Table A) for each focal conservation target that drive its composition, structure, and ecological function. This and can be used to help assess target ecological integrity. A conservation target is said to have ecological integrity when its dominant ecological components (e.g., composition, structure, function, ecological processes, biotic interactions) are within their natural ranges of variation such that it can withstand or recover from most natural or anthropogenic disturbances (Parrish et al 2003).

3. Identifying and rating the status of measurable indicators of the target's key ecological attributes and describing their natural range of variation.

4. Identifying and prioritizing the direct threats to the focal conservation targets and identifying which attributes of integrity are impacted.

5. Linking targets, threats, and other factors in a chain-of-causation and/or conceptual model to assist in conceptualizing and prioritizing actions.

6. Developing measurable, outcome-oriented objectives for improving the ecological integrity of a conservation target or abating critical threats to that target.

7. Developing strategies, with time lines, resources needed, and actions that should be taken to accomplish the objectives.

8. Developing and implementing a monitoring program for the identified indicators to measure changes in eco-

logical integrity and threat status and whether proposed objectives were achieved.

9. Modifying strategies and objectives as appropriate to ensure that targets progress toward improved or maintained ecological integrity.

10. Sharing information and lessons learned across the broader conservation community.

These steps are not conducted linearly, but require iterative efforts until an initial plan is developed. Further, this process is intended to involve people from outside of TNC, both to incorporate all data and expertise available, and to develop partnerships and commitment to plan implementation with input from many organizations and individuals.

The first step is the identification of focal conservation targets that represent the terrestrial, freshwater, and marine (as applicable) biodiversity across a range of geographic scales in the conservation area or project (Poiani et al. 2000). Given our limited knowledge of the ecology of most species and communities, it is difficult to identify a small number of targets that reflect the status of all the species present. Appropriate choices are species that play keystone roles, require a range of habitats over a large area, or respond rapidly to environmental changes that will eventually impact many other species. However, the "coarse filter" approach (Noss and Cooperrider 1994) suggests that ecological communities or larger systems that support numerous species should be considered first and then supplemented with species or communities likely to require more focused attention for conservation (Parrish et al. 2003).

Seven targets for the Ridge were selected to represent the ecosystem and species for conservation planning efforts:

1. Xeric upland matrix and embedded wetlands. The matrix target consists of the xeric upland "islands," their associated seasonal ponds, and the surrounding lowlands, all of which have experienced substantial losses in area (Weekly and Menges 2003). To maintain both species composition and structure, these areas require fire (Florida Natural Areas Inventory and Department of Natural Resources 1990).

2. Florida Scrub-jays (*Aphelocoma coerulescens*). The Florida Scrub-jay is federally listed as threatened, and the state population has declined by almost 50% over the past 100 years (Fitzpatrick et al. 1991). Scrub-jays inhabit rosemary-oak scrub and scrubby flatwoods (types of the xeric upland matrix) where fire has limited pine invasion and shrub heights. The Scrub-jay was made a target separate from the xeric upland matrix because of its endangerment and resulting urgency for fire management to create suitable habitat.

3. Rare upland plants of concern. Five rare upland species are of particular concern because of their restricted ranges, few protected sites, and fire suppression of much of the remaining habitat.

4. Sand-dwelling organisms. Like some of the rare plants, the sand-dwelling organisms (skinks, invertebrates) require open sand gaps that are lost with fire suppression but may not be able to survive the mechanical management methods sometimes substituted for fire.

5. Cutthroat grass (*Panicum abscissum*) communities. At least 80% of this central Florida endemic has been lost to agriculture and other development (Bacchus 1991), and both fire and undrained conditions are necessary for persistence.

6. Florida ziziphus (*Ziziphus celata*). Only five populations of this endangered shrub remain, four of which are on private lands. Long-term viability of this species will depend on creation of new, sexually compatible populations on conservation areas (Weekley et al. 2002).

7. Blackwater/seepage streams. Six high-quality blackwater/seepage stream systems remain on the LWR. Increasing development and associated water demand around those sites threaten these systems.

Identifying a minimum number of key ecological attributes that are necessary to maintain the ecological integrity or long-term viability of each focal conservation target is the next challenge. On the LWR, for example, the conservation goals require not only that the seven targeted upland plants, animals, and natural communities are present, but that they will be healthy over at least the next 100 years. Because each attribute that defines "health" for each target will involve one or more indicators, and our ability to monitor indicators in any given conservation project is typically low, keeping the numbers of attributes low is critical. One attribute for each of three categories of ecological attributes that influence biodiversity—size, condition, and landscape context (see Table A)—may be sufficient for estimating the status of the conservation target. Often ecological models are useful for identifying the most critical attributes for status and integrity assessment. On the LWR, the interaction between fire frequency and soil drainage may determine the plant community that develops on these xeric sites. For example, key attributes for the xeric upland matrix and embedded wetlands target include fire regime, vegetation structure, contiguity between wetlands and adjacent uplands, and contiguity between wetlands less than 500 m apart.

Key ecological attributes may be impossible to measure directly, requiring identification of one or more indicators that can be feasibly and effectively measured. Again, only a limited number of indicators should be selected. If another organization or agency is already monitoring several relevant variables, indicators would ideally be selected from that group (e.g., water flows and turbidity, soil or water chemistry, breeding pairs). Indicators need to be measurable, precise, sensitive, relevant, and cost-effective (Noss 1990, Margoluis and Salafsky 1998). Ideally, they will provide early warning of changes in the attribute that would impact target viability. Indicators are

TABLE B *The Nature Conservancy's Definitions of Indicator Ratings*

Indicator rating	Definition
Very good	The indicator is functioning within an ecologically desirable status, requiring little human intervention for maintenance within the natural range of variation (i.e., is as close to "natural" as possible and has little chance of being degraded by some random event).
Good	The indicator is functioning within its range of acceptable variation, although it may require some human intervention for maintenance (e.g., prescribed fire).
Fair	The indicator lies outside of its range of acceptable variation and requires human intervention for maintenance. If unchecked, the target will be vulnerable to serious degradation.
Poor	Allowing the indicator to remain in this condition for an extended period will make restoration or prevention of extirpation of the target practically impossible (e.g., too complicated, costly, and/or uncertain to reverse the alteration).

Note: Indicators must be in the good or very good category for a target to be considered viable over the long-term. However, shorter-term conservation objectives may be to maintain a specific indicator above the poor rating.
Source: Parrish et al. 2003.

likely to be refined over time as our knowledge of the target and its measurement improves.

Ultimately our goal at conservation areas is to conserve the focal conservation targets. A target can be considered "conserved" when each of its key ecological attributes are within their natural or acceptable range of variation (Parrish et al. 2003). For example, the different scrub systems that comprise a component of the upland xeric matrix on the LWR vary in the proportion of naturally occurring bare soil cover. Those openings in the shrubby canopy are critical for Florida Scrub-jay acorn burial (DeGrange et al. 1989), recruitment of herbaceous species (Young and Menges 1999), and habitat for a variety of invertebrates (Deyrup and Eisner 1993). Prior to the shifts in management and composition caused by recent human activities, the natural bare soil cover depended only on the soils, vegetation composition, local disturbance (e.g., tree falls, animal burrows), and fire history. Literature review and expert consultation suggest that bare soil might naturally range between 16% and 50% cover (The Nature Conservancy 2002). Percent cover above or below that range is considered outside of the natural range of variation for bare soil, requiring restoration before scrub can support viable populations of its constituent species.

The concept of natural, or acceptable ranges of variation underpins TNC's conservation area planning methodology and rating of ecological integrity, which uses quantitative, continuous data to define categorical "indicator ratings" of status as "very good," "good," "fair," and "poor" (Table B). The status of a key ecological attribute is considered good or very good if its indicators tell us that the attribute is within its natural (or acceptable) range of variation, and fair or poor if the attributes fall outside that minimum desired range. This ecological concept grounds the rating of ecological integrity status in the ecological principle of ranges of variation (Landres et al. 1999), providing greater rigor in our measurement, and ensures consistency in interpretation of rating scores across conservation areas and within a conservation area over time (Parrish et al. 2003).

Because we generally have insufficient data to develop these ranges, indicator ratings are considered hypotheses that

will also be refined. However, making these ratings quantitative allows evaluation of status of the indicator over time. If the ranges defining the rating are modified as our understanding increases, earlier data may be reevaluated. Table C shows the ratings for some indicators of key attributes for the xeric upland matrix and Florida Scrub-jay focal targets across the LWR landscape. The ratings were developed from published literature (e.g., Fitzpatrick et al. 1991; Young and Menges 1999) and expert input. Once the ratings have been developed, both current condition (bold in Table C) and desired future condition (italics in Table C) can be determined. The latter define the quantitative viability objectives for which strategies and actions are developed. For example, an objective for the xeric upland matrix target is that by 2013, over 80% of the area remaining in this community type (including the embedded wetlands) has a fire return interval of between 5 and 15 years (first row of Table C).

TNC specifies that objectives identify quantitative, clear, desired conditions and time frames that are feasible, appropriate, and based on the best scientific information available (The Nature Conservancy 2003b). These objectives then form the basis for strategic actions designed to move the indicator and target condition from the current status to the desired status. When the indicator rating is already good or very good, an objective is only necessary if the rating is likely to decrease without conservation action. Strategies are developed to accomplish the objective within the specified time frame and therefore should focus on maximizing effectiveness and feasibility. The actions necessary to accomplish the strategies need similar specification.

For the Florida Scrub-jay, an LWR project focal target—recruitment within sites—is considered a critical attribute of viability. The indicator used is the mean number of juveniles per family per site in July. The objective specifies that by 2013, there will be an average of 1.5 juveniles per family within the majority of Scrub-jay sites over at least 3 years of any 5-year period (see Table C). The primary strategy necessary involves the fire regime: both the area burned and fire frequency should be sufficient to maintain productive local populations.

TABLE C *Partial Viability Assessment for the Xeric Upland Matrix and Embedded Wetlands and Florida Scrub-Jay Targets*

Conservation target and category	Key ecological attribute	Indicator	Indicator ratings				Date of current rating[a]	Date of desired rating[b]
			Poor	Fair	Good	Very Good		
Xeric upland matrix								
Landscape context	Fire regime sandhill	Fire frequency[c]	<60%	**60–80%**	*>80%*	n/a	2003	2013
Condition	Community structure; sandhill	Cover of shrub midstory (1–3 m) (%)	>75%	51–75%	**26–50%**	≤25%	2003	2015
Condition	Community structure; scrub	Oak composition of shrubs within Scrub-jay habitat (%)	<20%	20–35%	**36–50%**	>50%	2003	2015
Condition	Community structure; sandhill	Cover native herbaceous layer (%)	0–4%	**5–10%**	*11–30%*	>30%	2003	2015
Condition	Community structure; scrub	Cover bare soil (%)	0–9% or >80%	**10–15% 51–80%**	16–25%	26–50%	2003	2015
Condition	Community structure; scrub	Mean shrub height	<0.5 m or >4 m	**3–4 m or 0.5–1 m**	2–3 m	*1–2 m*	2003	2015
Size	Area	Area in protected status[d]	≥400 ha	**≥200 ha**	≥80 ha	*≥40 ha*	2003	2008
Florida Scrub-jay								
Landscape context	Connectivity	Distance between scrub habitat (>1 km forest)	≥12 km	***>4 and <12 km***	2–4 km	<2 km	2003	2008
Condition	Recruitment	Mean number of juveniles and/or breeding group per site in July[e]	<0.5	0.5–1	**1–1.5**	>1.5	2002	2013
Size	Population size	Number of contiguous territories per site in natural habitat	**<15**	*15–20*	20–29	≥30	2003	2017

Note: Bold-faced text = current condition; Italic text = desired future condition; Roman text = other possible conditions.

Source: The Nature Conservancy 2003a; modified from the Lake Wales Ridge Conservation Area Plan.

[a]Ratings were done in February, with the exception of Florida Scrub-jay, which was done in July.

[b]By end of calendar year given

[c]Area burned within 5–15 years

[d]All natural uplands as of 2003

[e]Within a majority of sites over a 3–5 year period

However, measured changes in the status of the indicators for ecological integrity may not sensitively foreshadow conditions that are developing slowly, but that have critical impacts. Levels of a pathogen, for example, may slowly accumulate before large population declines are observed. Conservation activities should anticipate threats to a target's ecological integrity before irreversible declines occur. As a result, we should also consider threats to integrity and the development of more responsive, "leading" threat indicators that should supplement "lagging" ecological integrity indicators.

The process for developing threat indicators differs from that for ecological integrity. Threats to the focal conservation targets are identified and prioritized in step four of the conservation planning process previously outlined. If a particular threat, such as sedimentation in a riparian zone, has several sources, the relative contribution of those sources needs to be identified. While runoff from roads and trails may contribute to sedimentation in a particular area, runoff from agricultural fields may be the source of 95% of the sediment. Any effort to reduce sedimentation should therefore focus on changing agricultural practices rather than on road stabilization. Thus, the threats assessment is designed to identify where conservation efforts will be most cost-effective (The Nature Conservancy 2003c).

The most critical threats to Florida Scrub-jays on the LWR are altered habitat composition and structure, and habitat fragmentation (The Nature Conservancy 2002). The main source of the former threat is fire suppression. Habitat fragmentation has more sources: commercial and residential development is the primary cause, with the related development of roads and utility corridors and fire suppression considered secondary. Conversion of habitat for agricultural purposes is a tertiary cause. Thus, to reduce threats to Scrub-jays as well as other focal conservation targets, strategies must focus on slowing development and restoring the natural fire regime to the scrub habitat.

But what is the relationship between threats and the key ecological attributes of a target, and how do we know what

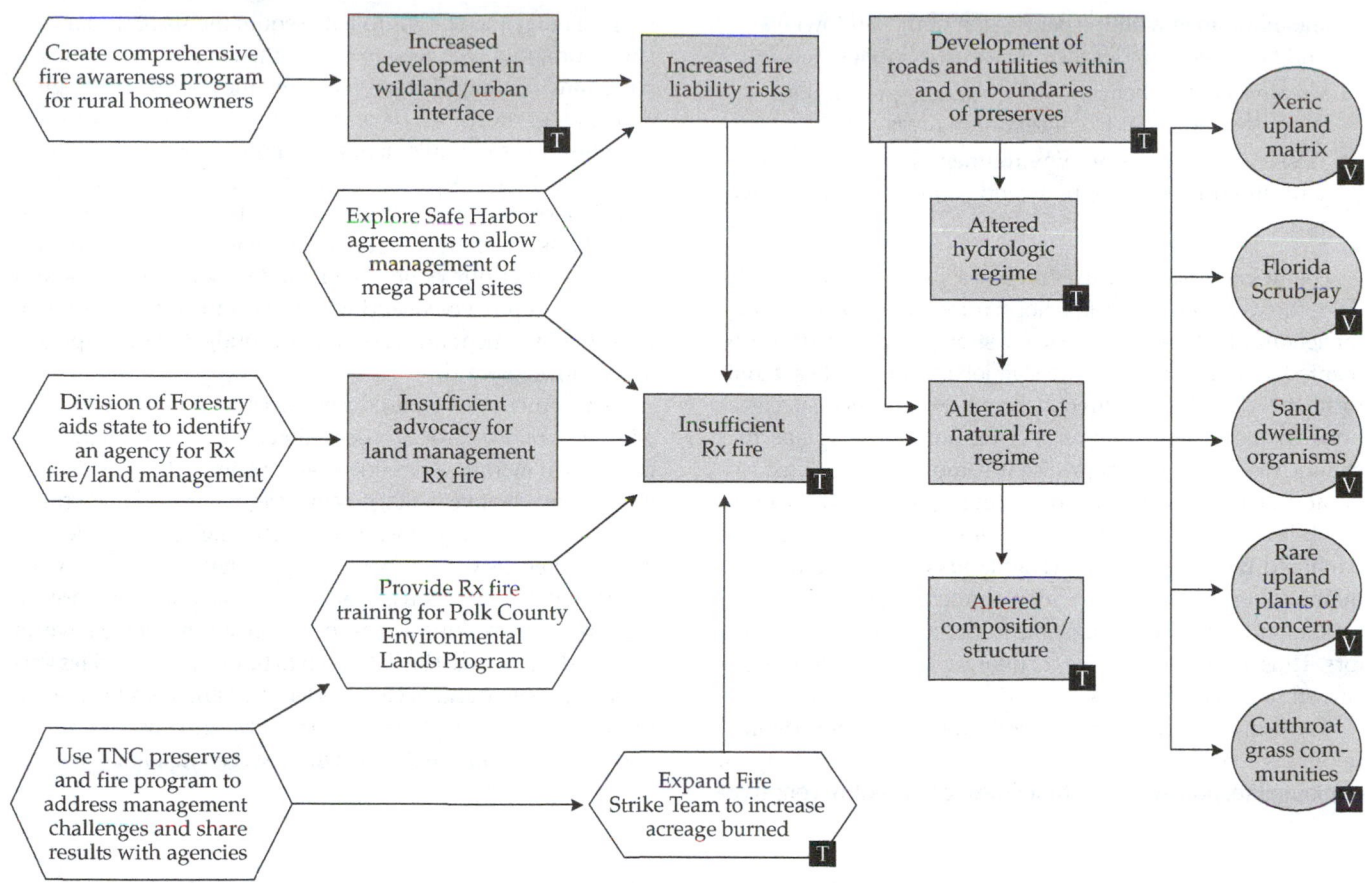

Figure B   Partial situation or chain of causation diagram for the threat of altered fire regime that impacts five of the focal conservation targets of the Lake Wales Ridge Conservation Area. Arrows link factors (rectangles) affecting the targets (circles) and potential strategies (hexagons) for meeting objectives. Monitoring should address both whether the actions necessary to accomplish the strategies have been implemented and whether the objectives are being met. (Rx fire = prescribed fire; V = viability objective; T = threat abatement objective) (Modified from The Nature Conservancy 2002.)

should be the focus of our conservation actions? Development of a situation or chain-of-causation diagram (Figure B) can help to link the most critical threats and other factors affecting the integrity of focal conservation targets to their sources (Margoluis and Salafsky 1998). These diagrams, which articulate the direct and indirect influences on integrity, help identify where conservation action may best be applied for long-term, effective conservation results. For the LWR targets, the ultimate objectives are to ensure prescribed burning to mimic the natural fire regime and maintain or restore natural hydrological regimes in conservation areas.

This assessment leads to development of quantitative threat abatement objectives addressing specific factors in the situation diagram (see Figure B). Once both viability objectives associated with specific indicators and threat abatement objectives have been defined, strategies are then developed. Selection of the most effective and efficient strategies is based on evaluation of the benefits, feasibility, and cost of each (The Nature Conservancy 2003b). This assessment is designed to identify those strategies that produce high benefits at reason-

able cost, but acknowledges that threats to conservation targets may require that expensive strategies be included. Once the primary strategies are determined, the necessary actions associated with each strategy, time frame, and responsibility may be identified.

To attain the fire-return-interval objective stated in Table C (by 2013, over 80% of the xeric upland area remaining has a fire return interval of between 5 and 15 years), , the LWR planning team identified four strategies with associated actions that were implemented in 2004 (The Nature Conservancy 2002):

1. Expand the interagency "Fire Strike Team" developed by TNC to increase the acreage burned on property of all the agencies to average 800 ha/year.

2. Explore establishing Safe Harbor agreements to allow management of multi-parcel sites. Safe Harbor is a federal program of voluntary agreements with nonfederal landowners to carry out specified conservation measures to benefit a listed species. TNC is exploring whether conservation area land managers may be permitted to burn

the inholdings with the permission of the landowners and be covered under the Endangered Species Act.

3. Create a comprehensive fire awareness program in properties adjacent to conservation areas.
4. Assist the Polk County Environmental Lands Program with fire training, planning, and conducting prescribed fire.

Following these steps, the practical aspects of work plan development remain. Specific actions, roles, timing, and funding for actions identified to achieve the strategies need to be documented and incorporated into staff job responsibilities. Monitoring needs will stem directly from the quantified specifications in the objectives, but specific monitoring plans that include the measurement precision, sampling design, frequency, and analysis will have to be developed. For the Florida Scrub-jay and its habitat, TNC is working with scientists at Archbold Biological Station (Highlands Co., FL), to develop methodology and a Ridge-wide volunteer program to collect the data necessary to evaluate the status of the viability indicators. Thus, implementing the monitoring may also require specific strategies and actions.

The work plan will contain both the action items defined under each strategy and the associated monitoring. Budgets, personnel responsible, and time lines for the entire conserva-tion area may then be compiled. Requirements of the conservation program will need to be weighed against program capacity and funding. Thus, even when the complex ecological and strategic decisions have been made, acquisition of sufficient resources to implement the conservation plan will limit effective action. However, we believe that this framework results in a compelling case for conservation action, with both better basis and better accountability than our past efforts. We anticipate that our donors and funders will recognize and value this improvement and will support the incorporation of these measurement processes into the budget of every project as we move forward.

Development of focused objectives and monitoring programs can only improve conservation if we learn from the data collected to modify and refine strategies and actions. This step, closing the adaptive management cycle, is the final and critical component of TNC's conservation planning process. Effective and comprehensive conservation approaches will only be widely adopted if information is shared among all public and private organizations working across the landscape and if we have an educated electorate. The data need to be evaluated and lessons learned (positive and negative) identified and communicated. Only then will we be able to evaluate if we are conserving what we say we are and, if not, find a more successful path.

CASE STUDY 18.2

Sustainable Urbanization and Biodiversity

Nicholas You and Charles Wambua, Best Practices and Local Leadership Programme, UN-HABITAT

Most of the literature on biodiversity focuses on the effects of human activity on the biosphere, linking issues of global warming, deforestation, and overexploitation of plant and animal life. The proposed remedies are well known. They point to the need for major changes in public policy to reduce emissions of greenhouse gases and for protection of endangered species. This case study looks at sustainability from a different perspective. We focus on a recent historical phenomenon—the fact that over half of the world's population now lives in cities—and argue that this unprecedented and ongoing transformation of human society places the challenge of sustainability increasingly in the urban arena. To illustrate our point, we describe some of the environmental problems associated with urbanization and some promising approaches that are being taken by communities around the world. Since 1997, over 2000 peer-reviewed good and best practices for sustaining urban communities from 150 countries have been documented and can be found on the Internet at: www.bestprac-tices.org; all of the examples discussed here can be found in the data base on this website.

Cities and Ecology

Historically, cities were established near the richest agricultural land and fresh water resources to satisfy the needs of sustenance, trade, and production. The ecological impact of cities on nature and life-support systems was relatively minor until the advent of rapid urbanization and industrialization, which began in Europe in the eighteenth century and continues today in developing countries.

However, it is the combined impact of urbanization and the relatively recent phenomenon of globalization that poses one of the greatest threats to biodiversity and sustainable development (Wakely and You 2001). In the developed world, urbanization no longer implies massive shifts of people from rural to urban areas. It is the "suburban revolution" whereby urban sprawl and the extensive use of the automobile are consuming

large areas of land at lower densities (see Case Study 7.1). In developing countries, where many cities are doubling their population every decade, rapid urbanization and globalization is resulting in widespread poverty and social marginalization, with up to 70% of the population living in slums and low-income settlements.

In both cases, urbanization and globalization lead to irreversible changes in the way we produce and consume (Wakely and You 2001). These changes result in more intensive use of land, including forests and watersheds; greater energy consumption for transport, heating and cooling; more intensive use of fresh water resources; and qualitative and quantitative changes in solid and liquid waste production, the disposal of which often returns to nature in the form of pollution (UN-ESCAP 2005). All of these changes directly affect biodiversity.

Ecological Footprints and Changing Consumption Patterns

An emerging trend is the growing concern and awareness of the ecological footprint of cities and the necessity to change consumption patterns (WWF 2004). Several consumer associations in Western Europe have partnered with businesses and local authorities to reduce wasteful patterns of consumption and, together with a growing consumer preference for safer and organic foods, to raise awareness of "food miles" and the ecological footprint of cities. A food mile is the distance food travels from where it is grown or raised to where it is ultimately purchased by the consumer or end user. Using fresh produce as an example, carrots grown in the San Joaquin Valley in California and transported to supermarkets in Des Moines, Iowa will travel approximately 1400 miles (2250 km). In the U.S. food travels an average of 1300 miles (2100 km) before it reaches the table (Schafer 1992). Food waste is another consumption problem that can be addressed. For example, in the UK, people routinely throw away up to 30% of edible food.

In Vienna, Austria, one of the most important tasks in its sustainability drive has been to convince its citizens to shed the "throwaway" habit (Table A). Viennese residents are encouraged to separate waste at the source, and biodegradable waste is collected separately to produce organic compost at the city-owned composting plant. Compost is distributed free of charge and a very successful do-it-yourself gardening project has been established. The long-term goal is to make it possible for citizens of Vienna to buy organic produce from local gardens thus closing the waste management and nutrient cycle.

Although urban planners and economists are often hostile to urban agriculture and tend to think of it as an unproductive use of urban land, hundreds of thousands of allotments are being cultivated on the outskirts of cities throughout Europe. There are health risks associated with urban agriculture when it involves the use of untreated sewage, which can contain heavy metals and other contaminants. Yet, food insecurity is common among the urban poor, even in the U.S., and agricultural production on small lots in urban and suburban environments can provide significant amounts of food for urban families (Sommers and Smit 1994; Brown et al. 2002). The Green-Up Program in the Bronx promotes the use of vacant lots between houses; many of these had been littered with garbage and taken over by substance abusers and drug pushers. Community groups manage these lots and cultivate vegetables with the application of compost originating from the New York Botanical Gardens, thus providing a productive pastime for residents and contributing to the greening and safety of neighborhoods. Provision of land for urban agriculture has become a planning option for the cities of New York and Detroit where thousands of acres of land have been allocated to unemployed workers for growing food. Community Supported Agriculture (CSA) and local farmers' markets are also becoming popular in several European and North American communities. With CSAs, residents pay an advance fee entitling them to fresh produce of their choice. The scheme also encourages recycling of organic waste and organic production.

Urban agriculture not only reduces food miles and the energy required to transport food, it can also provide jobs and income opportunities. In the case of Rosario, Argentina, a city of 906,000 people, the monetary crises of 2001 and subsequent recession led to widespread poverty. The Urban Agriculture Programme (UAP) was initiated in response to the needs of an urban population, the majority of whom overnight fell under the poverty line (U.S.$90.00 per month). Over 790 community gardens were established leading to improvements in nutrition, quality of life, and urban greening. Currently, more than 10,000 families are involved in the production of organic vegetables, which are consumed by 40,000 people. The value of the program both in terms of quality-of-life and in providing low-income families, especially women, with a source of income, has led the city to adopt a set of ordinances and by-laws to institutionalize urban agriculture and to provide the urban poor with security of tenure for engaging in local food production.

The recycling of wastes such as paper, metals, plastics, and glass is particularly effective when nearby industries use the waste as raw material for production. This type of recycling is widely practiced in many developed countries where waste recycling and reuse is not only a form of environmental conser-

TABLE A *Percent Urban Waste Composition in Different Parts of the World*

Product	United States	Europe	Middle East	Asia	Africa
Vegetable	28.6	21.7	50.0	75.0	85.4
Paper	43.8	31.7	16.0	2.0	6.9
Metals	9.1	5.8	5.0	0.1	3.1
Glass	9.0	8.0	2.0	0.2	0.7
Textiles	2.7	—	3.0	3.0	1.2
Plastic	3.0	3.6	1.0	1.0	1.4
Other	3.7	29.2	23.0	18.7	1.3

Source: From UN Economic and Social Commmission for Asia and the Pacific, www.unescap.org/stat/envstat/stwes-2waste.02.pdf.

vation, but a source of income for the urban poor. In Cotonou, Benin, one of the poorest countries in the world, a local waste-management program promotes waste reuse and recycling of household waste while improving revenue-generating activities. The local community pays monthly nominal fees for collection of household waste. The initiative has provided a source of gainful employment for youth and has cleaned up the mountains of trash that used to blight the landscape. Rubber, metals, plastics, and glass are converted into products for daily household use, such as kerosene lamps, pots, pans, containers, and sandals which are sold on the local market. The Benin project lays great emphasis on providing protective gear to all the youth involved: rubber and leather gloves, boots, and overalls.

The Need for More Compact and Complete Communities

The root cause of excess carbon emisisons and pollution of cities lies in land-use patterns that oblige people to use their cars to go to different places to shop, to work, and to live. This is often a result of what is called the "competing jurisdictions syndrome" where adjacent municipalities and/or counties compete with each other for shopping malls, entertainment centers, and suburban development. In the case of Greater Vancouver Livable Region project in Canada, 21 municipalities adopted a common strategic plan that is designed to protect and preserve green space, create more complete and compact communities, and offer transport alternatives.

These initiatives have enhanced the region's social, economic, and environmental health. Implementation of the plan is integrated into the budgets of the member municipalities through their official community plans. As a result of this initiative, the protected green space has increased by approximately 60,000 ha since 1991. Air-quality improvements have been significant as a result of reductions in emissions from industry and vehicles. Vancouver demonstrates that sustainable development is a planning concept that benefits development and the environment.

Changing Production Patterns through Eco-Procurement

Cities purchase a wide variety of goods and services for their own offices and operations. They also have a say in how a myriad of public institutions are run including schools, hospitals, transport and port authorities, and utilities. In Vienna, Austria, the "Oekokauf Wien" project was launched in 1999 to restructure purchasing and procurement processes in the City Administration, in Viennese hospitals and in public enterprises with a view to promote ecological sustainability. Every year, the City of Vienna purchases goods and services in excess of U.S.$3.5 billion, ranging from textiles, detergents, vehicles and spare vehicle parts, office equipment, furniture, and supplies to construction contracts and janitors' services. "Oekokauf Wien" has compiled ecological criteria for almost the entire range of products, materials and services required to run the city. Simple and rapid methods were developed for assessing products and services from an ecological point of view, without compromising on economics, quality, or usability. The resulting checklists, guidelines, and specifications not only support the purchasing of environmentally friendly products and services but also serve as a stimulus to consider the necessity of certain products and their environmentally friendly alternatives. The impact of this coordinated procurement policy has made the entire public procurement process more efficient and cost effective—it also influences the supply chain and the environmental awareness and social responsibility of business.

Poverty and Sustainable Development

Most threats to biodiversity are linked to over-consumption and the wasteful habits of the more developed nations and wealthy societies. Yet, in many developing countries, poverty is not only generating social and economic problems, it also results in the destruction of ecosystems. In Nairobi, Kenya, home to both authors of this article, it is estimated that 70% of the urban population lives below the poverty line (U.S.$1.00 per day; UN-HABITAT 2003). The majority of the poor live in slums where inadequate sanitation is a direct source of pollution. As most families can barely afford one square meal a day, fuel for cooking is major expense and they rely on wood or charcoal rather than gas, kerosene, or electricity. Charcoal production and consumption has become a major cause of deforestation. Furthermore, those living in semi-urban areas rely heavily on local food production. This can be particularly damaging in more densely populated semi-urban areas, where landless people tend to exploit scarce natural resources in an effort to earn a living.

This was the case in Malindi, Kenya where subsistence farmers living on the outskirts of the town petitioned the government to exploit the adjacent forest, which is renowned as a tourist attraction. The Arabuko-Sokoke Forest is home to six endangered species, and is ranked as the second-most important forest in Africa for avian biodiversity. It is the largest remaining piece of the coastal forest that once stretched from Southern Somalia to Mozambique. A survey conducted in the early 1990s showed that 96% of households living near the forest wanted it cleared, because they needed more land. They were also losing half their crops to forest animals. As a result, political pressure to clear the forest was gaining momentum. In 1993, with an initial grant of U.S.$50,000 from UNDP, the East African Natural History Society and the National Museums of Kenya initiated the Kipepeo (Butterfly) Project which combined the goals of poverty reduction and environmental conservation. The project was developed to train farmers to breed butterfly pupae from the forest for the global market.

The project targeted those farmers living on the fringes of the forest who were the most affected by the current situation. Meetings were organized to explain the project and participating farmers were trained in butterfly breeding. By 1996, eight self-

TABLE B *1994–2000 Earnings from the Sale of Butterfly Pupae Reared in the Kipepeo Project*

	1994	1995	1996	1997	1998	1999	2000	Totals
Number of pupae reared	10,262	12,593	18,807	21,823	21,390	54,939	56,023	195,837
Export earnings (U.S.$)[a]	15,888	18,286	27,163	41,378	39,397	105,289	103,659	351,060
Community earnings (Ksh)[b]	263,828	329,905	538,216	780,480	882,371	2,726,928	2,806,415	8,328,143
Number of ecotourists	200	843	1341	1581	1146	2233	2784	10,128
Ecotourism earnings	11,390	61,829	114,275	155,100	135,790	206,670	306,865	991,919

Source: http://www.kipepeo.org.

[a]Export earnings are really about 15% lower owing to pupal losses in transit.

[b]Community earnings are roughly 40% of export earnings each year.

help groups were created to manage the project locally. Each group had two members who were licensed to gather butterfly pupae in the forests. These groups also worked as local development committees to address other issues such as water supply, health, and sanitation. Overall, results have been very positive, both in terms of building local commitment to protect the forest and in creating income-earning opportunities for local people. By 1999, some 55,000 pupae were exported, earning U.S.$100,000 a year, of which U.S.$40,000 went directly to farmers. The average farmer now earns U.S.$75.00 a year from butterfly sales, more than double their total per capita cash income in 1993 (Table B). A local survey of farmers showed a change in attitudes, with 84% advocating preservation of the forest.

Sustainable Use of Water Resources

The world is facing a major water crisis, not only in places where water has always been scarce, but also in many countries where water is abundant. This crisis is due to a combination of factors that include rising levels of demand, shortage of nearby sources of water, wastewater treatment, and flood control and drainage.

As cities grow, nearby water sources often become inadequate, and more distant sources must be found. In the 1950s, cities such Barcelona and Beijing sourced their water within a 50-km radius of the city. By 2000, both cities were obtaining their water from over 400 km away (Barcelona Field Studies Centre 2005). By 2025 these same cities may have to resort to importing water from more than 1600 km away. While a large part of the increase in demand for water in these cities is due to rapid population growth, the increase is also due to lifestyle changes and increases in per capita consumption. The rapid spread of the use of household appliances accounts for a large part of the increase in consumption, not to mention swimming pools, car washing, and lawn watering. Added to this, in many cities, losses from leaking pipes can total 50% of the water supply. This is the case in many cities in developing countries, where water supply to slums and low-income areas is inadequate and irregular.

Water Demand Management

Efforts to supply water to meet increasing demand have well-known effects on biodiversity. Dams, reservoirs, and the diversion of water from rivers and lakes can have devastating effects on the natural habitats of aquatic life. Effective water demand management is therefore a key factor in sustainable water use. In Fukuoka, Japan, a prolonged drought in 1978 forced the city government to curb water supply for 287 days in a year. Without abundant water resources and subject to serious periodic droughts, Fukuoka faced an enormous challenge in securing a stable water supply to serve a population of over 1.3 million. To respond to this alarming situation, Fukuoka city, in partnership with the citizens and the private sector, launched various initiatives promoting a "Water Conservation Conscious City" in 1979.

Several innovations and measures were implemented. Water-saving apparatuses were introduced and, currently, approximately 96% of users have water flow-reducing devices installed in their faucets. Water savings have been realized with an average family saving up to 1000 liters per month. The Fukuoka City Water Bureau has been addressing water leakage by replacing old pipes with new ones, and through such efforts, Fukuoka has the lowest water leakage rate of Japan (2.7% in 2001). The city is also actively promoting reuse of treated wastewater. Using the Wide-Area Circulation System and Individual Circulation System for large buildings, used and treated water is used to flush toilets and for watering plants. The amount of water conserved amounts to about 7000 cubic meters a day.

Water, unlike food or fuel, can be treated and reused. Increases in water use pose the problem of wastewater treatment. While many cities in the developed world have installed wastewater treatment plants and facilities at great expense, relatively few cities in the developing world possess such systems. Untreated wastewater eventually ends up polluting rivers, lakes, and groundwater, further exacerbating the shortage of water for human consumption. However, even relatively sophisticated treatment methods cannot remove all traces of chemical or biological substances, such as hormones that remain in treated wastewater (Kolpin et al. 2002).

Some Lessons Learned and Pointers for the Future

By bringing people closer together urbanization can have many potential benefits such as better access to education and

health care. It can also facilitate social integration and civic engagement. Similarly, globalization tends to increase trade and commerce, creating new wealth and economic opportunities. But urbanization and globalization also have negative consequences. As cities increase in size, population, and percentage of the world's population, they tend to bring with them more congestion, generate more waste and pollution, and place more taxing demands on land and natural resources. Globalization has thus far resulted in an increase in the gap between "haves" and "have-nots" within and between countries thereby exacerbating poverty, social exclusion, and environmental degradation.

This case study highlights examples of what some communities and cities are doing to mitigate these negative effects. They point to several lessons learned.

First, the sustainable use of biodiversity is greatly facilitated when initiatives make use of existing knowledge and local practices and involve benefit sharing. Farmers in Kenya who were involved in the Kipepeo "Butterfly" project were empowered to pursue new income-generating opportunities without disrupting their principle agricultural activity. The forest is protected, retaining its attraction to tourists. This is quite different from what is happening in many parts of the world, where changes in economic activity and production technology result in radical disruptions of people's livelihoods and to the environment.

Second, the conservation of biodiversity and the protection of ecosystems requires the engagement of people and their communities in decision-making. While governments endeavor to address environmental issues such as air and water quality, successful solutions require a high degree of awareness on behalf of all citizens of the consequences of their own behavior including their daily consumption patterns. Costly infrastructure—water, waste, and sewage treatment plants, and new roads to ease congestion and air pollution—has its own negative impact on ecosystems. Communities will forever play "catch up," unless people and communities agree to tackle root causes. The examples of Vienna and Rosario show that the reuse and recycling of waste can bring environmental as well as economic benefits to the community. The city of Fukuoka provides a compelling example of how an effective water demand management system depends on the combined efforts of individual consumers, water companies, and the city administration. One of the key issues remains, however: the realization of more compact and complete communities to prevent urban sprawl and single-purpose suburban development. While individual municipalities can address this issue through more rational land use planning, the example of the Greater Vancouver Livable Region shows that long-lasting solutions ultimately require adjacent municipalities and counties to work together, not to compete with one another to achieve common goals.

Summary

1. At the beginning of the twenty-first century, it is clear that we simultaneously need to find ways to improve human welfare and conserve biodiversity. There are practical reasons for pursuing these aims together. Human welfare is dependent on ecosystem services, and there is now widespread evidence that our degradation of the environment has undercut our present and future ability to support human populations. Growing human populations and consumption patterns are the root cause of biodiversity loss, and without stabilizing these we will vastly erode the biological richness of Earth. Moreover, both aims are compelling moral imperatives. We cannot continue "business as usual" but must strive to do the utmost for both people and wild nature.

2. To successfully conserve biodiversity, we need to prepare for the considerable uncertainties we face due to our limited knowledge of ecological systems, and the inevitable truth that they change in rather unpredictable ways. Further, global climate change is a wild card that may cause massive ecological change, the results of which we can only predict to a limited extent now. Social and economic uncertainties, such as shifts in markets or governments in power, must also be addressed to succeed in conservation. Synthetic indicators of trends that combine metrics of social, economic, and ecological change may help us discern connections among these variables. Such indices may also be useful policy tools to motivate decision-makers to make changes that increase sustainable practices.

3. Research that allows a mechanistic understanding of the responses of species, ecosystems, and other biological conservation targets to threats and to potential conservation interventions is needed to help guide management actions. Such research may be most successful when undertaken with stakeholders, rather than in isolation. Careful use of experiments, including those presented by adaptive management efforts, case study analyses, and strategic modeling can help us refine our scientific understandings for the good of conservation. Using decision-making tools that build stakeholder involvement may be of tremendous assistance in overcoming some of the social consequences of uncertainty. This will be particularly true if scientists can make the nature of uncertainties understood and work collaboratively with stakeholders to develop solutions that can accomodate these uncertainties.

4. Scenario-building can be a means to help stakeholders understand the potential consequences of their decisions. For example, the Millennium Ecosystem Assessment produced four plausible alternative scenarios of how human welfare and biodiversity would change by 2050 depending on the type of decision-making that is undertaken. In their analyses, a scenario most similar to our current path leads to the worst outcomes. The other three scenarios all lead to better outcomes, but each leads to different trade-offs. Aspects that enhance their outcomes include an emphasis on coordination of efforts to enhance sustainability at either a global or watershed-scale, or on technological solutions that improve our abilities to increase ecosystem supports for humans. The hope is that these, and other, scenarios can help decision-makers to examine what efforts should be made within and among nations.

5. Conservation efforts fall into five general categories, all of which can be highly effective: (1) direct protection; (2) species and ecosystem management and restoration; (3) policy, laws, advocacy, and enforcement; (4) education and outreach; and (5) economic and social incentives. To maximize our successes, we need to be able to assess which aspects of conservation projects are working best, and understand the reasons behind failures in our conservation efforts. Thus, we will be best able to learn from conservation projects that are set up in an adaptive management framework. Assessments of conservation projects are relatively new, but they have already guided improvements in several programs, including the implementation of the ESA. Overall, one of the primary limitations for most conservation efforts is inadequate funding. Continuing to find means to fund conservation will be a struggle, but may succeed most in cases where conservation and the needs for sustainable development overlap. Recently, conservation groups have developed sophisticated assessment tools, and have also come together to employ the best standards of assessment, and conservation practice more generally.

6. It is imperative in the coming decades to reduce the destructive impacts of over-consumption, as well as overexploitation of land and wild species. This may be achieved through a combination of improved education and outreach to diverse citizen groups and economic and social incentives. These efforts are very new, and it is difficult to judge how well they will work, but all are promising. However, none will work without simultaneously increasing the capacity for sustainable use through improving our understanding of where and how conservation-compatible uses can be pursued.

7. The decisions we make over the coming decade will strongly influence how much degradation of biodiversity occurs throughout this century. We need to promote radical changes in practices to reduce population growth and consumption, and to conserve biodiversity and improve human welfare throughout the globe. Everyone can assist in the large-scale societal transitions that must be made. Indeed, without the efforts of all communities toward this end, we will end up in a more impoverished world. We advocate that every person commit their efforts—both great and small—to creating a legacy of collective action to conserve biodiversity of which we can be proud.

Please refer to the website www.sinauer.com/groom for Suggested Readings, Web links, additional questions, and supplementary resources.

Questions for Discussion

1. Uncertainty is always with us when we think about causes and solutions to significant environmental problems. How would you respond to the following assertion? There is so much scientific uncertainty about the causes and effects of climate change that we should just wait until we better understand the problem. To take action now might cause undue economic hardship.

2. The ESA is meant to give equal protection to large charismatic species such as grizzly bears (*Ursus arctos horribilis*) and small obscure ones such as the northern beach tiger beetle (*Cicindela dorsalis dorsalis*). What argument would you make in response to the question, "What good is a beetle in comparison with economic development"?

3. The population growth rate of nonindustrialized nations is slowing. One consequence is that these populations are also becoming relatively older. What implications for conservation do you think this demographic shift might have in the future?

4. Conservation biologists have been remiss by failing to link human health to healthy ecosystems. Consider an ecosystem of any size in your region and ask how might protecting or restoring that ecosystem benefit human health?

5. What actions can you take to help conserve biodiversity? What can you do today? What can you do over the next several months? How might a program for your personal efforts to enhance biodiversity conservation and promote human welfare take shape over your lifetime?

Glossary

A

adaptation Process of genetic change within a population due to natural selection, whereby the average state of a character becomes better suited to some feature of the environment.

adaptive management The practice of revisiting management decisions and revising them in the light of new information.

adequacy A single reserve that is large enough to fulfill its conservation functions, or in a reserve system, a sufficient number or area of reserved sites to achieve conservation objectives.

agroecosystem Land used for crops, pasture, and livestock; the adjacent uncultivated land that supports other vegetation and wildlife; and the associated atmosphere, the underlying soils, groundwater, and drainage networks.

Allee effect The phenomenon where at the point when population density is too low for individuals to find mates, reproductive success sharply declines.

allele One of a pair of genes at a particular genetic locus. Different alleles are usually named for having a specific phenotypic affect (e.g., wild type versus mutant, albino versus normal, fine-spot versus large-spot).

allopatric Describes two or more populations or species that occur in geographically separate areas. *See also* **sympatric**.

allozyme One of several possible forms of an enzyme that is the product of a particular allele at a given gene locus.

alpha- or α-richness The number of species occurring within a given habitat.

alternative-futures analysis A form of conservation planning where stakeholders and conservation biologists work together to propose and examine the consequences of distinct possible future states for an area, and use these projections as a basis for land use planning.

anthropocentrism Any human-oriented perspective of the environment, but usually used to emphasize a distinction between humans and nonhumans. For example, assessing a tropical forest in terms of its potential timber value is an exclusively anthropocentric perspective.

anthropogenic climate change The change in global, regional, and local climate as a result of human activities. Often referred to only as climate change or global warming.

area/perimeter ratio The ratio of internal area to edge habitat of a region. The area/perimeter ratio is an indication of the amount of interior habitat with respect to edge habitat, and may indicate potential success of a reserve in protecting interior species.

augmentation Restoration undertaken to expand a site in area or quality.

B

background extinction rate Historical rates of extinction due to environmental causes not influenced by human activities, such as the rate of species going extinct because of long-term climate change.

Bayesian statistics A branch of modern statistics that bases statistical inferences and decisions on a combination of information derived from observation or experiment and from prior knowledge or expert judgment. Contrast this approach with classical statistics, which regards only the data from observations or experiments as useful for estimation and inference.

beta or β-richness The change or turnover of species from one habitat to another.

bequest value This is the value that people place on goods that they may wish to save for future generations, usually similar to **existence value**.

biocentrism A perception of the world that values the existence and diversity of all biological species, as opposed to a human-centered perspective (**anthropocentrism**).

biodiversity The variety of living organisms considered at all levels of organization, including the genetic, species, and higher taxonomic levels; and the variety of habitats and ecosystems, as well as the processes occuring therein.

biological control The use of a species to consume or otherwise control the population of a pest or invasive species.

biological integrity The presence of a biota with all the parts (genes, species) created and sustained by natural evolutionary and biogeographic processes in a given region.

biome A large, regional ecological unit, usually defined by some dominant vegetative pattern, such as the coniferous forest biome.

Biosphere Reserve A concept of reserve design in which a large tract of natural area is set aside, containing an inviolate core area for ecosystem protection, a surrounding buffer zone in which nondestructive human activities are permitted, and a transition zone in which human activities of greater impact are permitted. Three goals of a biosphere reserve are conservation, training (education), and sustainable human development compatible with conservation.

buffer zone An area in a reserve surrounding the central core zone, in which nondestructive human activities such as ecotourism, traditional (low-intensity) agricul-

ture, or extraction of renewable natural products, are permitted.

bushmeat Meat from animals, including edible invertebrates, that is harvested in the wild.

C

cascade effect The phenomenon where the extinction or change in abundance of a species causes changes in the abundance or extinction of many other species, which in turn causes such changes in still more species.

cladistics A system of classification based on historical (chronological) sequences of divergence from a common ancestor.

cladogram A diagram of cladistic relationships. An estimate or hypothesis of true genealogical relationships among species or other groupings.

coadapted gene complex A concept in which particular genes from multiple loci have coevolved to collectively enhance fitness under a given set of environmental conditions.

commons Originally referred to lands in medieval Europe that were owned by townships rather than by private individuals. Now used to include any exploitable resource that is not privately owned. Sometimes applied to so-called "open resources" that are neither privately owned nor regulated by a country or agency.

community-based conservation (CBC) A conservation approach that is rooted in local community development, is highly participatory, and seeks to make both conservation and development goals primary.

comprehensiveness In a reserve system context, a site selection process that includes many or most biodiversity features (e.g., species, ecological communities).

conservation biology An integrative approach to the protection and management of biodiversity that uses appropriate principles and experiences from basic biological fields such as genetics and ecology; from natural resource management fields such as fisheries and wildlife; from social sciences such as anthropology, sociology, philosophy, and economics; as well as other fields such as the creative arts and communications.

constructed market approach Also known as hypothetical market approach. Used in **contingent valuation** to express nonmarket values. For example, by determining a representative population's **willingness to pay** to save a natural area or species a hypothetical market value can be assigned.

contingent valuation method (CVM) Uses questionnaires and other inquiry methods to find nonmarket or indirect market values. Often expressed as a question, "How much would you be willing to pay to save the Artic National Wildlife Refuge?" *See also* **willingness to pay**.

conventional market approach Assigns a market value to environmental goods or services by comparing them to similar goods and services that have known market values. The general methods include **opportunity costs**, **production-function**, and **substitute/alternative cost**.

cost–benefit analysis (CBA) A method of evaluating projects by assessing all project costs and benefits, usually in monetary units, over the lifetime of the project. *See also* **discounting** and **net present value**.

critical habitat According to U.S. Federal law, the ecosystems upon which endangered and threatened species depend.

critical natural capital A precautionary approach that argues that certain environmental goods, services, and species are too important to be traded for gains in other forms of capital (e.g., financial, human-generated goods, human capital). Critical natural capital is said to be non-substitutable. An example might be the last remaining population of an endangered species.

D

demand The aggregate desire for economic goods and services. The quantity of a good or service that consumers are willing to purchase at different prices. Demand involves the relationship between quantity and price.

deme A randomly interbreeding (panmictic) local population.

demographic bottleneck A significant, usually temporary, reduction in genetically effective population size, either from a population "crash" or a colonization event by a few founders.

demographic PVAs *See* **population viability analysis**.

demographic uncertainty Chance populational events, such as sex ratios or the act of finding a mate that influence survival in small populations.

density-dependent factors Life history or population parameters that are a function of population density.

density-independent factors Life history or population parameters that are independent of population density.

development Formation of human, social, financial or man-made capital; need not be equivalent to economic growth. *See also* **steady state economics**.

direct use value The value we receive by using some part of the environment. For example, logging is a direct use value of a forest.

discounting The technique of placing a value on a project or resource at different times in the future. A high discount rate assumes that current values will decline rapidly in the future.

dominance The condition when an allele exerts its full phenotypic effect despite the presence of a different allele at the same locus. For example, if allele *A* is dominant over *a*, then genotypes *AA* and *Aa* will have the same phenotype.

E

ecocentric value The value of ecological entities irrespective of their usefulness to humans.

ecological-economic efficiency This is the ratio of man made capital services gained to natural capital services lost as a result. Clear-cutting a watershed results in poor efficiency. Sustainable logging would have higher efficiency.

ecological footprint The impact a human community or nation has in terms of the amount of land or sea needed to produce the resources consumed.

ecological indicator A characteristic of an ecosystem that is related to, or derived from, a measure of biotic or abiotic variable that can provide quantitative information on ecological structure and function. An indicator can contribute to a measure of integrity and sustainability.

ecological integrity A living system exhibits integrity if, when subjected to disturbance, it sustains and organizes self-correcting ability to recover the state that is normal for that system. *See also* **biological integrity.**

ecological release Habitat expansion or density increase of a species when one or more competing species are not present.

ecological restoration The process of using ecological principles and experience to return a degraded ecological system to a more ecologically functional state. The goal of this process is to emulate the structure, function, diversity, and dynamics of the specified ecosystem.

ecologically functional population A population that is of sufficient size to fulfill its ecological roles.

ecoregion A relatively large area of land or water that contains a geographically distinct assemblage of natural communities.

ecosystem management An approach to maintaining or restoring the composition, structure, and function of natural and modified ecosystems for the goal of long-term sustainability. It is based on a collaboratively developed vision of desired future conditions that integrates ecological, socioeconomic, and institutional perspectives, applied within a geographic framework defined primarily by natural ecological boundaries.

ecosystem service The conditions and processes of natural ecosystems and species that provide some, usually utilitarian, human value. As examples, riparian swamps protect downstream property from flood damage or wild bees pollinate certain food crops. Ecosystem services are examples of **indirect use values.**

ecotourism Nature-based tourism; sometimes visiting new cultures is included in the definition.

edge effect The negative influence of a habitat edge on interior conditions of a habitat, or on species that use interior habitat. Also, the effect of adjoining habitat types on populations in the edge ecotone, often resulting in more species in the edge than in either habitat alone.

efficiency Achieving conservation objectives with the least possible cost. In a reserve system this often equates to minimizing the number of sites that need protection.

efficient allocation An economic term that refers to the market's ability to match resources with material ends. The apportionment of resources to the production of different goods and services.

endangered species Species threatened with extinction by anthropogenic or natural changes in their environment. Requirements for declaring a species globally Endangered are described in the IUCN Red List. Countries often also have legal requirements for designating a species as endangered, such as in the U.S. Endangered Species Act. According to U.S. Federal law, an endangered species is a species in imminent danger of extinction throughout all or a significant portion of its range.

endemic Any localized process or pattern, but usually applied to a highly localized or restrictive geographic distribution of a species.

enhancement Management technique (seeding, transplantation, fencing, watershed manipulations, etc.) that attempts to restore to predisturbance conditions those areas only partially disturbed by human influence.

environmental impact assessment (EIA) An analysis of the total beneficial and negative impacts of a project on the environment.

environmental modification Modification of the phenotype as a result of environmental influences on the genotype.

environmental uncertainty Unpredictable sources of density-independent mortality, such as an early snowstorm, that jeopardize the survival of a small population by pushing it below its minimum viable population size.

epistasis Interactions among alleles at two or more loci that affect the state of a single trait, where the combined effects differ from the sum of the individual locus effects.

equilibrium A state reached when a population's birth and immigration rates are equal to its mortality and emigration rates. Also applied to species changes in a community or to any other ecological process in which the rate of increase equals the rate of decrease, resulting in a steady state.

eutrophication Naturally, the slow aging process during which a lake, estuary, or bay evolves into a bog or marsh and eventually disappears. During the later stages of eutrophication the water body is choked by abundant plant life due to higher levels of nutritive compounds such as nitrogen and phosphorus. Human activities can accelerate the process, leading to rapid algal growth, and later **hypoxia.**

evolutionarily significant unit (ESU) Partially genetically differentiated populations that represent a significant component of the evolutionary legacy of the species and that are considered to require management as separate units.

Evolutionary-Ecological Land Ethic A philosophical approach to conservation derived from the evolutionary and ecological perspective, first advanced by Aldo Leopold. Nature is seen not as a collection of independent parts, to be used as needed, but as an integrated system of interdependent processes and components, in which the disruption of some components may greatly affect others. This ethic is the philosophical foundation for modern conservation biology.

existence value The value, usually nonmarket, that is assigned by society to a place or species simply because its existence is considered important. For example, the Arctic Wildlife Refuge has a significant existence value for many people even though they may never visit it.

exploitation The consumptive use of any natural resource.

ex situ conservation Conservation efforts that take place in zoos, aquaria, greenhouses, or in other facilities.

Usually involves storing and rearing individuals or genetic material for future reintroduction. Contrast with **in situ conservation**.

externality A cost, usually in terms of environmental degradation, that results from an economic transaction but which is not included as a debit against economic returns.

extinction threshold A population size or density below which a population becomes at risk of immediate extinction. Used to describe the phenomenon when risk of extinction rises sharply with only a small change in population size or number of habitat patches occupied.

F

feedback Refers to a system whose output modifies input to the system. Prices play this role in market systems.

fitness The relative contribution of an individual's genotype to the next generation in the context of the population's gene pool. Relative reproductive success.

fixation All individuals in a population are identically homozygous for a locus (e.g., all A_1A_1).

founder effect The principle that the founders of a new population carry only a random fraction of the genetic diversity found in the larger, parent population. Change in the genetic composition of a population due to founding or origin from a small number of individuals.

fragmentation *See* **habitat fragmentation**.

Fundamental Theorem of Natural Selection The basic theorem of population genetics, which states that the rate of evolutionary change in a population is proportional to the amount of genetic diversity (specifically, additive genetic variance) available in the population.

G

gamma- or γ-richness The number of species found within a large region, which typically includes several habitats.

GAP analysis The use of various remote sensing data sets to build overlaid sets of maps of various parameters (e.g., vegetation, soils, protected areas, species distributions) to identify spatial gaps in species protection and management programs.

gene flow The uni- or bidirectional exchange of genes between populations due to migration of individuals and subsequent successful reproduction in the new population.

gene locus The site on a chromosome occupied by a specific gene. *See also* **locus**.

gene pool Group of interbreeding adults and progeny that collectively represent the bank of genetic material available in the population for future adaptation.

genetic drift Random changes in population allele frequency and levels of genetic diversity due to the finite number of individuals contributing genes to the next generation.

genetically effective population size (N_e) The number of individuals that would result in the same level of inbreeding, or decrease in genetic diversity through drift, if the population behaved in the manner of an idealized and randomly mating population. The functional size of a population, in a genetic sense, based on numbers of actual breeding individuals and the distribution of offspring among families. N_e is typically smaller than the census size of the population.

genotype The entire genetic constitution of an organism, or the genetic composition at a specific gene locus or set of loci.

geographic information system (GIS) A computerized system of organizing and analyzing any spatial array of data and information.

geographic variation Change in a species' trait over distance or among different distinct populations. Measurable character divergence among geographically distinct populations that are often, though not necessarily, the result of local selection.

geostatistics A form of spatial statistics originally created for analyzing irregular geological features. Now used in ecology to create probability maps from a limited set of location-specific data points.

grain Refers to how organisms perceive spatial heterogeneity. The more time that an organism spends in any particular kind of environment, the more coarsely grained that environment is said to be.

greenhouse gas A gas, such as carbon dioxide or methane, which contributes to potential climate change.

growth In conventional economics, any rate of increase in value that exceeds the rate of inflation.

H

habitat degradation The decrease in the quality of an area due to human activities.

habitat fragmentation The disruption of extensive habitats into isolated and small patches; or the result of development in a large area where habitat is now fragmented into separate units; often applied to forested habitats that have been fragmented by agricultural development or logging.

habitat loss The conversion or transformation of a natural area into a wholly human occupied area of little or no use to wild species.

habitat shredding A form of habitat fragmentation in which the fragments often remain as strips or "shreds" in ravines and other inaccessible areas.

Hardy-Weinberg equilibrium The stability of gene frequencies expected in a sexual, diploid population when a number of assumptions are met, including random mating, a large population, and no migration, mutation or selection.

harvesting Collecting any natural resource, usually biological, for **exploitation**.

hedonic pricing (HP) Recognizes that goods may be described by an array of objectively measurable characteristics. For example, the price of a house may be made up of the prices ascribed to the number and size of rooms, the size of the garden, the quality of local services and the nature of the surrounding environment. *See* **implicit market approach**.

heritability The proportion of observed variation in phenotype that can be attributed to differences in genotype.

heterosis Hybrid vigor or superior performance of hybrid genotypes, usually based on comparison to parental genotypes. Heterosis often results from the masking of deleterious alleles that occur in a homozygous state in high frequency in populations. Sometimes called **hybrid vigor**.

heterozygosity A measure of the genetic diversity in a population, as measured by the number of heterozygous loci across individuals.

heterozygous The condition in which an individual has two different alleles at a given gene locus (e.g., *Aa*). Heterozygous individuals can produce two different gametes.

heuristic algorithm A set of ordered steps for solving a problem whose general purpose is not to find a optimal solution, but an approximate solution where the time or resources to find a perfect solution are not practical.

hierarchical analysis of genetic diversity An approach to defining population genetic structure for a species in nature that defines components of total genetic diversity in a spatially hierarchical fashion. Genetic diversity can be apportioned in a hierarchical fashion (e.g., alleles within individuals, individuals within populations, populations within regions) to accomplish this goal.

homoplasy Possession by two or more species (or groups) of a similar or identical trait that has not been derived by both species (or groups) from their common ancestor. Homoplasy could result from convergence, parallel evolution, or character reversal.

homozygous Individual having two copies of the same allele at a locus (*AA* or *aa*). The individual is genetically invariant because it can produce only one kind of gamete.

hotspot A geographic location characterized by unusually high species richness, often of endemic species.

hybrid vigor *See also* **heterosis**.

hypoxia/hypoxic waters Waters with dissolved oxygen concentrations of less than 2 ppm, the level generally accepted as the minimum required for most marine life to survive and reproduce

I

I = PAT An equation describing the total impact of humans on natural systems as a function of population size (*P*), level of affluence (*A*), and technological sophistication employed (*T*).

implicit market approach Assigns market values to goods and activities that include the perceived value of the environment. For example, one might accept less wages to wait tables in a national park restaurant because the surrounding environment has perceived value. Two common methods are **travel cost** and **hedonic pricing**.

inbreeding The mating of individuals who are more closely related than by chance alone, such as relatives. Also the correlation of genes within individuals relative to that expected if individuals had mated at random.

inbreeding depression A reduction in fitness and vigor of individuals as a result of increased homozygosity

through inbreeding in a normally outbreeding population.

indicator species or taxon A species or higher taxonomic group used as a gauge for the condition of a particular habitat, community, or ecosystem. A characteristic, or surrogate species for a community or ecosystem.

indirect use value The value, either market or nonmarket, that is provided by an environmental service. *See also* **option value**.

inherent value *See* **intrinsic value**.

in situ conservation Conservation efforts that take place in the wild. Contrast with **ex situ conservation**.

instrumental value The worth of an entity as judged by its utility or usefulness to humans.

integrated conservation and development project (ICDP) A term applied to large conservation projects that have the dual goal of improving people's livelihoods while protecting biodiversity and ecosystem functions.

intermediate disturbance hypothesis An hypothesis (with good empirical support) that posits that maximum species richness in many systems occurs at an intermediate level (of intensity or frequency, or both) of natural disturbance.

intrinsic value The worth of an entity independent from external circumstances or its value to humans; value judged on inherent qualities of an entity rather than value to other entities.

invasive species A nonnative species that spreads rapidly and outcompetes, preys on and otherwise reduces or eliminates populations of native species.

irreplaceability A conservation element that is nonsubstitutable, that is, an element that must be protected if conservation goals are to be met.

K

keystone mutualist species Keystone species that perform a mutualistic function, such as plant species that are broadly used as pollen sources.

keystone species A species whose impacts on its community or ecosystem are large, and much larger than would be expected from its abundance.

L

land-bridge island Areas that are presently island habitats, but were formerly connected to the mainland during periods of lower ocean levels. Land-bridge islands tend to lose species over time in a process called "relaxation."

landscape matrix The intervening area among a set of habitat fragments. Also the spatial array of habitats across a landscape.

landscape species approach A conservation approach developed by the Wildlife Conservation Society that uses wide-ranging focal species to design conservation efforts over a large landscape with both conservation and human uses.

linear programming (LP) A mathematical approach that finds the optimal solution, e.g., maximum net profit, by changing the values of the various constraints. For example, the optimal critical habitat for a species might be determined by varying the constraints of reserve

size, migration corridors, food abundance, and density of predators to determine which combination of real constraints provides the critical habitat with the highest probability of survival for the species in question.

locus Physical location occupied by a gene on a chromosome. Used interchangeably with gene. *See* **gene locus**.

M

marine protected areas (MPAs) A marine site in which some or all human uses are prohibited.

mass extinction The extinction of large numbers of taxa during a relatively brief geologic time frame, such as the extinction of dinosaurs at the end of the Cretaceous Period.

mate choice Form of sexual selection where one sex (usually females) determines whether mating occurs or not, often on the basis of phenotypic or behavioral characteristics of the opposite sex.

maximum sustained yield (MSY) The largest harvest level of a renewable resource that can be sustained over a period of many generations. Harvest of a natural population at the population size representing the maximum rate of recruitment into the population, based on a logistic growth curve.

mesopredators Refers to medium-sized predators, such as raccoons and foxes, that often increase in abundance when larger predators are eliminated.

metapopulation A network of semi-isolated populations with some level of regular or intermittent migration and gene flow among them, in which individual populations may go extinct but can then be recolonized from other populations.

minimum dynamic area The smallest area necessary for a reserve to have a complete, natural disturbance regime in which discrete habitat patches may be colonized from other patches within the reserve.

minimum viable population (MVP) The smallest isolated population that has a specified statistical chance of remaining extant for a specified period of time in the face of foreseeable demographic, genetic, and environmental stochasticities, plus natural catastrophes.

monetizing The process of placing monetary value on typically non-monetary goods and processes such as biological material or ecological processes. The process of converting values to economic units.

monomorphic Presence of only one allele at a locus. Lack of genetic diversity. Contrast wth polymorphic.

morphospecies Species that are distinguished on the basis of appearance; often used in field studies when taxonomies and keys are incomplete; a morphospecies is not a formal taxonomic entity.

multi-criteria analysis Similar to **linear programming** but more than one criterion is optimized. For example, one might want to jointly optimize the combination of tourist access and habitat quality for a species that has value for ecotourism.

multiple gene traits Those traits whose phenotype is affected by many genes. Also called polygenic traits. Often form continuous distributions (e.g., length, weight, condition factor).

multiple use concept Refers to the simultaneous and compatible use of public land and water resources by different interest groups. For example, U.S. public law requires that national forests be open to recreational use, timber extraction, mining or other concessions, and biodiversity protection. In reality, the activities of the various interest groups generally conflict, and are often incompatible with biodiversity protection.

mutation A spontaneous change in the genotype of an organism at the genetic, chromosomal, or genomic level, usually caused by errors in replication. "Mutation" usually refers to alterations to new allelic forms, and represents new material for evolutionary change, although most mutations are either mildly or strongly deleterious.

mutational meltdown The decline in reproductive rate and downward spiral towards extinction due to chance fixation of new mildly deleterious mutations in small populations.

mutualism An interspecific relationship in which both organisms benefit; frequently a relationship of complete dependence. Examples include flower pollination and parasite cleaning.

N

natural catastrophe A major environmental cause of mortality, such as a volcanic eruption, that can affect the probability of survival for both large and small populations.

natural selection A process by which differential reproductive success of individuals in a population results from differences in one or more hereditary characteristics. Natural selection is a function of genetically based variation in a trait, fitness differences (differential reproductive success) among individuals possessing different forms of that trait, and inheritance of that trait by offspring.

nested subset A pattern of species biogeographic distribution in which larger habitats contain the same subset of species in smaller habitats, plus new species found only in the larger habitat. Common species are found in all habitat sizes, but some species are found only in progressively larger habitats.

net present value (NPV) This is the cumulative value of a project when all costs and benefits are **discounted** over the lifetime of the project. By comparing alternative projects, the one with the highest NPV has the highest net benefits.

net primary productivity (NPP) Net production of plant biomass, and the basis of all food webs.

neutral genetic variation Genetic variation (alleles or genotypes) that is not or appears not to be subject to natural selection.

nonequilibrium A condition in which the rate of increase does not equal the rate of decrease. In nonequilibrial population growth, environmental stochasticity disrupts the equilibrium.

nonuse value *See* **existence value**.

nutrient pollution Contamination of water resources by excessive inputs of nutrients. In surface waters, excess algal production is a major concern.

O

opportunity cost The value forgone when not taking a particular option. When choosing one option precludes other options, the value assigned to those other options is the opportunity cost.

opportunity cost approach Uses **conventional markets** to assign forgone value when one action is taken over another. For example, protecting trees in a reserve may have many benefits but it also has an opportunity cost in the form of forgone timber sales.

option value An economic term that refers to assigning a value to some resource whose consumption is deferred to the future; the value placed on environmental assets by people who want the use of it in the future. Yet to be discovered medicinal plants means that rainforests have an option value.

outbreeding depression Reduction in reproductive fitness due to crossing of individuals from two genetically differentiated populations.

overexploitation The consumptive use of a natural resource beyond its capacity to replenish what has been taken.

overdominance The condition in which a heterozygote at a given locus has higher fitness than either homozygote. Also called heterozygote superiority.

P

panmictic Exhibiting random breeding among individuals of a population.

paradigm An established pattern of thinking. Often applied to a dominant ecological or evolutionary viewpoint; e.g., during earlier decades, the dominant paradigm in ecology held that communities were shaped by equilibrial processes.

patch dynamics A conceptual approach to ecosystem and habitat analysis that emphasizes dynamics of heterogeneity within a system. Diverse patches of habitat created by natural disturbance regimes are seen as critical to maintenance of diversity.

pedigree Multigenerational chart of parent–offspring relationships.

pedigree analysis Use of pedigrees in conservation planning, particularly for captive propagation, to avoid inbreeding and maximize genetic diversity.

phenotype The physical expression (outward appearance) of a trait of an organism, which may be due to genetics, environment, or an interaction of the two.

philopatry Individuals returning to the same area to breed.

phylogenetic diversity (PD) The evolutionary relatedness of the species present in an area.

phylogeny The evolutionary or cladistic relationships among species or higher taxa; the relatedness through descent of any taxon; e.g., the phylogeny of birds leads back to certain lines of dinosaurs.

phylogeographic Evolutionary relationships among species populations based on geographic relationships and historical gene flow patterns.

phylogeography Study of the geographical distribution of genealogical lineages, especially within species.

plasticity The condition of genetically based, environmentally induced variation in characteristics of an organism.

pleiotropy One gene affects two or more traits.

pluralism A school of thought which holds that species concepts should vary with the taxon under consideration. Many different species definitions would be employed.

point richness The number of species found at a single point in space.

policy briefs Policy briefs are concise analyses designed to give legislators and other decision-makers key features of a particular policy. Sometimes briefs are given to the news media.

policy statements Provides an administration's or an NGO's position on a particular policy.

polymerase chain reaction (PCR) A process in which a particular DNA segment from a mixture of DNA chains is rapidly replicated, producing a large, readily analyzed sample of a piece of DNA; the process is sometimes called DNA amplification.

polymorphic More than one allele at a genetic locus. Contrast with **monomorphic**.

polyploidy Possessing more than two complete sets of chromosomes.

population viability analysis (PVA) A quantitative analysis of the many environmental and demographic factors that affect survival of a population, usually applied to small populations at risk of extinction.

position paper An in-depth analysis to support **policy briefs** and **policy statements**.

precautionary principle The principle that when information about potential risks is incomplete, decisions about the future policies should be based on a preference for avoiding unnecessary environmental or health risks.

prescribed burn The process of burning an area of land to suppress or kill certain plant species and to favor the growth of others, resulting in a desired plant community.

present value (PV) The value of a project at any point in the future determined by subtracting costs from benefits and multiplying by the discount rate. *See also* **net present value.**

production function approach Uses **conventional markets** to assign a value to some change in environmental conditions that affect productivity. For example, the value of reforesting a watershed could be expressed in the market value from increases in crop production that results from a more abundant water supply.

protected area A site where human uses are restricted or prohibited and where conservation of biodiversity is a primary goal.

Q

quantitative genetics The study of phenotypic traits that are influenced by multiple genetic and environmental factors (polygenic traits).

R

reclamation A revegetation or land management goal that includes a lower diversity of species and may include substitutions by introduced species.

re-creation The act of entirely reconstructing a site denuded of its terrestrial and/or aquatic systems. This commonly occurs on surface mined lands and in brownfields (severely damaged urban and industrial lands). Also termed creation, but creation implies transforming a site to a completely different ecosystem than had previously existed on the site.

rehabilitation Creation of an alternative ecosystem following a disturbance, different from the original and having utilitarian rather than conservation values. To repair damaged or blocked ecosystem functions, with the primary goal of raising ecosystem productivity for the benefit of local people.

reintroduction An attempt to establish a species in an area which was once part of its historical range, but from which it has been extirpated or become extinct.

regional scale The largest scale of corridors in which major swaths of habitat connect regional networks of reserves.

relaxation The loss of species on land-bridge islands following separation from the mainland or the loss of species during any process of habitat fragmentation and isolation.

remediation Removal of toxicants from a contaminated environment using chemical, physical, or biological means.

remote sensing Any technique for analyzing landscape patterns and trends using low altitude aerial photography or satellite imagery. Any environmental measurement that is done at a distance.

representativeness In a reserve system, the quality of a set of sites that together include all or most biodiversity elements (e.g., species or ecological communities).

rescue effect The recolonization of a habitat when a subpopulation of a metapopulation has gone locally extinct.

reserve *See* **protected area**.

reserve system A set of protected areas in a region, usually coordinated to conserve biodiversity of that region to some extent.

Resource Conservation Ethic A philosophical approach to conservation derived from the views of forester Gifford Pinchot, based on the utilitarian philosophy of John Stuart Mill. Nature is seen as a collection of natural resources, to be used for "the greatest good of the greatest number for the longest time."

restoration or **restoration ecology** *See* **ecological restoration**.

risk assessment The process of identifying, evaluating and managing the risks that may arise from a given action or activity; e.g., the risk assessment of releasing a species as part of a biological control or a species reintroduction program.

Romantic-Transcendental Conservation Ethic A philosophical approach to conservation derived from the writings of Emerson, Thoreau, and Muir, in which nature is seen in a quasi-religious sense, and as having uses other than human economic gain. This ethic strives to preserve nature in a wild and pristine state.

S

scale The magnitude of a region or process. Refers to both spatial size—for example, a relatively small-scale patch or a relatively large-scale landscape; and temporal rate—for example, relatively rapid ecological succession or relatively slow evolutionary speciation.

secondary effects When the loss of population change in a species affects other species, often through their trophic interactions.

secondary extinctions Loss of a species as a direct or indirect result of the loss of another species.

selection Differential survival (viability selection) and reproduction (fertility selection) by different genotypes in the population or the differential success of genotypes passing gametes to other generations (sexual selection). *See also* **natural selection**.

sentient Capable of feeling or perception. Refers to a state of self-awareness among organisms, usually applied only to vertebrates.

simulated annealing A common method used in **systematic conservation planning** where a search for potential sites to include in a reserve system can include both optimal (good) and nonoptimal (bad) choices initially to achieve a better final result.

single gene traits Traits whose phenotype is largely controlled by a single gene. Usually falls into discrete or qualitative categories such as normal versus albino coat coloration.

sink habitat A habitat in which local mortality exceeds local reproductive success for a given species.

sink population A population in a low-quality habitat in which the birth rate is generally lower than the death rate and population density is maintained by immigrants from source populations.

SLOSS An acronym for "single large or several small," reflecting a debate that raged for several years asking whether, all else being equal, it was better to have one large reserve or several small reserves of the same total size.

source and sink dynamics Spatial linkage of population dynamics such that high-quality habitats (sources) provide excess individuals that maintain population density, through migration, in low-quality habitats (sinks).

source habitat A habitat in which local reproductive success exceeds local mortality for a given species.

source population A population in a high-quality habitat in which the birth rate greatly exceeds the death rate and the excess individuals leave as emigrants.

spatial statistics A broad class of statistical models that are applied to data that have an explicit spatial distribution. *See also* **geostatistics** as a special subset.

spatially-explicit population model A population model, especially a simulation model, that takes space, differences in habitat quality, and inter-habitat movement into consideration.

speciation Any of the processes by which new species form.

species diversity Usually synonymous with "species richness," but may also include the proportional distribution of species.

species invasion The rapid spread or population growth of a nonnative species into new areas.

species richness The number of species in a region, site, or sample. *See also* α-, β- and γ-**richness**.

stewardship Management of natural resources that conserves them for future generations. Usually used to distinguish from short-term, utilitarian management objectives.

steady-state economy An economy that has a constant stock of people and artifacts maintained at some desired sufficient levels by low rates of maintenance (or throughput). A steady state economy does not assume that continued economic growth, as measured by indicators like Gross Domestic Product (GDP), is prerequisite for sustainable development.

stochastic Random; specifically refers to any random process, such as mortality due to weather extremes.

substitute/alternative cost approach Estimates the market value of an environmental good by comparing it to a logical substitute that has an established market value. For example, in this approach the value of hunting a deer would be approximated by the value of an equivalent amount of purchased meat.

succession The natural, sequential change of species composition of a community in a given area.

supplementation Addition of individuals to an existing population of conspecifics. Sometimes called reinforcement.

supply The aggregate amount of goods or services available to satisfy economic needs or wants. The quantity of a good or service which producers are willing to sell at different prices. Supply involves the relationship between quantity and price.

surrogates One or more species selected as focal species for conservation planning with the intention that other species will also be conserved by the same efforts. Surrogate species are often ones with large-ranges, or sensitive to disturbance, or which have similar ecology to target species.

sustainable development In general, the attempts to meet economic objectives in ways that do not degrade the underlying environmental support system. Note that there is considerable debate over the meaning of this term. In Chapter 18, we define it as "human activities conducted in a manner that respects the intrinsic value of the natural world, the role of the natural world in human well-being, and the need for humans to live on the income from nature's capital rather than the capital itself."

synergism An interaction that has more than additive effects; for example, when the joint toxicity of two compounds is greater than their combined, independent toxicities.

systematic conservation planning The deliberate planning process of selecting sites to include in a reserve system using explicit criteria.

T

threatened species Used internationally to refer to any species listed as Critically Endangered, Endangered, or Vulnerable according to the IUCN Red List of Threatened Species definitions. According to U.S. Federal law, a threatened species is a species that is likely to become endangered in the foreseeable future.

time preference The general desire to have goods and services sooner rather than later. The use of **discounting** allows future goods and services to be valued in **present value** thereby accounting for the bias of time preference.

total economic value (TEV) The sum of all use and nonuse values including both market and nonmarket.

tragedy of the commons An idea (set forth primarily by Garrett Hardin) that unregulated use of a common, public resource for private, personal gain will result in overexploitation and destruction of the resource.

translational scientist A natural or social scientist who can translate scientific language into a policy framework to enable decision-makers to use science effectively, and who can translate policy concerns into scientific questions so that scientists can address them.

translocation Management technique often used in mitigation for endangered species protection whereby an individual, population, or species is removed from its habitat to be established in another area of similar or identical habitat.

travel cost methodology Commonly used to evaluate recreational or conservation areas. Estimates how much people are willing to pay in travel and opportunity costs to visit a particular site, or to see a particular phenomenon, e.g., a wild Bald Eagle. It is part of the **implicit market approach**.

U

use value All values except **existence value**.

utilitarian value *See* **instrumental value**.

utilitarian view A philosophical term applied to any activity that produces a product useful to humans, typically in some economic sense. Also used to describe a system of values which is measured by its contribution to human well-being, usually in terms of health and economic standard of living.

utility The difficult to measure sense of well-being. Economists generally use monetary terms as surrogate measures of utility. Also the "want-satisfying" power of goods; personal satisfaction received through an economic gain.

V

vicariance The process of a continuously distributed biota becoming separated by an intervening geographic event (such as mountain uplift or river flow), or extinction of intervening populations, resulting in subsequent independent histories of the fragmented biotas, and possible speciation events.

vulnerability threshold A population size or density below which a population becomes vulnerable to extinction. Used particularly to describe the phenomenon when risk of extinction rises sharply with only a small change in population size or number of habitat patches occupied.

W

watershed The land area that drains into a stream; the watershed for a major river may encompass a number of smaller watersheds that ultimately combine at a common point.

wetlands Lands whose saturation with water is the dominant factor determining the nature of soil development and the types of plant and animal communities that live in the soil and on its surface (e.g., Mangrove forests).

willingness to pay (WTP) Used in **contingent valuation** to estimate the value of a nonmarket good. WTP would generally be determined through questionnaires distributed to a representative population asking something like, "How much would you be willing to pay to protect a certain reserve or species?"

Z

zoning An important component of reserve design that controls human activities within and adjacent to conservation reserves, so that reserve function may be protected while some human activities, including those supplying some economic benefit, may take place.

Bibliography

Abbitt, R. J. F. and J. M. Scott. 2001. Examining differences between recovered and declining endangered species. *Conserv. Biol.*15:1274–1284. [18]

Abele, L. G. and E. F. Connor. 1979. Application of island biogeographic theory to refuge design: Making the right decision for the wrong reasons. In R. M. Linn (ed.), *Proceedings of the First Conference on Scientific Research in the National Parks*, Vol. I., pp. 89–94. USDI National Park Service, Washington, D.C. [7]

Abeywickrama, B. A., M. F. Baldwin, M. A. B. Jansen, et al. 1991. *Natural Resources of Sri Lanka*. Natural Resources, Energy and Science Authority of Sri Lanka. [17]

Abramovitz, J. N. 1996. *Imperiled Waters, Impoverished Future: The Decline of Freshwater Ecosystems*. Worldwatch Institute,Washington, D.C. [6]

Abugov, R. 1982. Species diversity and phasing of disturbance. *Ecology* 63:289–293. [2]

Achard, F., H. D. Eva, H. J. Stibig, P. Mayaux, J. Gallego, T. Richards, and J. P. Malingreau. 2002. Determination of deforestation rates of the world's humid tropical forests. *Science* 297:999–1002. [6, 8]

Adam, D. "Oil Chief: My Fears for the Planet; Shell Boss's 'Confession' Shocks Industry." *The Guardian* (London), 17 June 2004. [10]

Adam, K. A. 2003. Tree Selection in Selecting Logging: Ecological and Silvicultural Considerations for Natural Forest Management in Ghana. Ph.D. dissertation, University of Aberdeen, Aberdeen, UK. [8]

Adams, L. W. and A. D. Geis. 1983. Effects of roads on small mammals. *J. Appl. Ecol.* 20:403–415. [7]

Adams, M. J. 2000. Pond permanence and the effects of exotic vertebrates on anurans. *Ecol. Appl.* 10:559–568. [3]

Adel, M. M. 2002. Man-made climatic changes in the Ganges basin. *Int. J. Climatol.* 22: 993–1016. [6]

Agee, J. K. 1993. *Fire Ecology of Pacific Northwest Forests*. Island Press, Washington, D.C. [17]

Agrawal, A. 2002. A southern perspective on curbing global climate change. In S. H. Schneider, A. Rosencranz, and J. O. Niles *Climate Change Policy: A Survey*, pp. 375–391. Island Press, Washington, D.C. [4]

Agrawal, A. and C. C. Gibson. 2001. The role of community in natural resource conservation. In A. Agrawal and C. C Gibson (eds.), *Communities and the Environment: Ethnicity, Gender and the State in Community Based Conservation*, pp. 1–31.

Rutgers University Press, New Brunswick, N.J. [12]

Aiken, W. 1984. Ethical issues in agriculture. In T. Regan (ed.), *Earthbound: New Introductory Essays in Environmental Ethics*, pp. 247–288. Random House, New York. [4]

Airamé S. J., K. D. Dugan, H. Lafferty, D.A. Leslie, D. McArdle, and R. Warner. 2003. Applying ecological criteria to marine reserve design: a case study from the California Channel Islands. *Ecol. Appl.* 13:S170–S184. [14]

Aizen, M. A. and P. Feinsinger. 1994a. Forest fragmentation, pollination, and plant reproduction in a chaco dry forest, Argentina. *Ecology* 75:330–351. [7]

Aizen, M. A. and P. Feinsinger. 1994b. Habitat fragmentation, native insect pollination, and feral honey bees in Argentine "Chaco Serrano." *Ecol. Appl.* 4:378–392. [7]

Akçakaya, H. R. 1997. *RAMAS/Metapop: Viability analysis for stage-structured metapopulations. Version 2.* Applied Biomathematics: Setauket, NY [12]

Akçakaya, H. R. and S. Ferson. 1992. *RAMAS/Space: Spatially-structured population models for conservation biology.* Applied Biomathematics: Setauket, NY. [12]

Alder, D. 1993. *Growth and Yield Research in Bobiri Forest Reserve.* United Kingdom Overseas Development Agency Consultancy Report No. 14, Oxford. [8]

Alexander, M. A. and J. K. Eischeid. 2001. Climate variability in regions of amphibian declines. *Conserv. Biol.* 15:930–942. [3]

Alford, R. A. and S. J. Richards. 1997. Lack of evidence for epidemic disease as an agent in the catastrophic decline of Australian rain forest frogs. *Conserv. Biol.* 11:1026–1029. [3]

Alford, R. A. and S. J. Richards. 1999. Global amphibian declines: a problem in applied ecology. *Ann. Rev. Ecol. Syst.* 30:133–165. [3]

Alford, R. A., P. M. Dixon, and J. H. K. Pechmann. 2001. Ecology: Global amphibian population declines. *Nature* 412:499–500. [3]

Allee, W. C., A. E. Emerson, O. Park, T. Park, and K. P. Schmidt. 1949. *Principles of Animal Ecology.* Saunders, Philadelphia. [12]

Allen, C. D. and D. D. Breshears 1998. Drought-induced shift of a forest-woodland ecotone: Rapid landscape response to climate variation. *Proc. Nat. Acad. Sci. USA* 95:14839–14842. [10]

Allendorf, F. W. 1986. Genetic drift and the loss of alleles versus heterozygosity. *Zoo Biol.* 5:181–190. [11]

Allendorf, F. W. and N. Ryman. 2002. The role of genetics in population viability analysis. In S. R. Beissinger and D. R. McCullough (eds.), *Population Viability Analysis*, pp. 50–85. University of Chicago Press, Chicago. [11]

Allendorf, F. W. and R. F. Leary. 1986. Heterozygosity and fitness in natural populations of animals. In M. E. Soulé (ed.), *Conservation Biology: The Science of Scarcity and Diversity*, pp. 57–76. Sinauer Associates, Sunderland, MA. [11]

Allendorf, F. W., R. B. Harris, and L. H. Metzgar. 1991. Estimation of effective population size of grizzly bears by computer simulation. In E. C. Dudley, (ed.), *The Unity of Evolutionary Biology*, pp. 650–654. Proceeding of the Fourth International Congress of Systematic and Evolutionary Biology. Dioscorides Press, Portland, OR. [11]

Allendorf, F. W., D. Bayles, D. L. Bottom, K. P. Currens, C. A. Frissell, D. Hankin, J. A. Lichatowich, W. Nehlsen, P. C. Trotter, and T. H. Williams. 1997. Prioritizing Pacific salmon stocks for conservation. *Conserv. Biol.* 11:140–152. [12]

Alongi, D. M. 2002. Present and future state of the world's mangrove forests. *Environ. Conserv.* 29:331–349. [6]

Alpert, P. 1996. Integrated conservation and development projects: Examples from Africa. *BioScience* 46:845–855. [16]

Alverson, D., M. Freeberg, J. Pope, and S. Murawski. 1994. *A Global Assessment of Fisheries By-catch and Discards.* FAO, Rome. [8]

Alverson, W. S., D. M. Waller, and S. L. Solheim. 1988. Forests too deer: Edge effects in northern Wisconsin. *Conserv. Biol.* 2:348–358. [7]

Amat, J. A. 1980. Biología y ecología de la comunidad de patos del Parque Nacional de Doñana. Ph. D. dissertation, University of Sevilla. Sevilla. [14]

American Legislative Exchange Council. 2002. State Responses to Kyoto Climate Change Protocol Act. www.alec.org. [10]

Ames, R.T. 1992. Taoist ethics. In L. Becker (ed.), *Encyclopedia of Ethics*, pp. 1126–1230. Garland Press, New York. [4]

Amos, W. and A. Balmford. 2001. When does conservation matter? *Heredity* 87:257–265. [11]

Amos, W. and J. Harwood. 1998. Factors affecting levels of genetic diversity in natural populations. *Philos. T. Roy. Soc. B.* 353:177–186. [11]

Andelman, S. J. and M. R. Willig. 2003. Present patterns and future prospects for bio-

diversity in the Western Hemisphere. *Ecol. Lett.* 6:818–824. [14]

Andelman, S. J. and W. F. Fagan. 2000. Umbrellas and flagships: Efficient conservation surrogates or expensive mistakes? *Proc. Nat. Acad. Sci. USA* 97:5954–5959. [14]

Andersen, A. N. 1983. Species diversity and temporal distribution of ants in the semiarid mallee region of northwestern Victoria. *Aust. J. Ecol.* 8:127–137. [2]

Andersen, E. 1999. Seed dispersal by monkeys and the fate of dispersed seeds in a Peruvian rainforest. *Biotropica* 31:145–158. [3]

Andersen, R. A. 1992. Diversity of eukaryotic algae. *Biodivers. Conserv.* 1:267–292. [2]

Anderson, C. 1992. How much green in the greenhouse? *Nature* 356:369. [17]

Anderson, C. B. and P. F. Hendrix. 2002. Hallazgo de *Eisniella tetraedra* (Savigny 1826) (*Annelida: Oligochaeta*) en Isla Navarino, Chile. *Anales del Instituto de la Patagonia, Serie de Ciencias Naturales* 30:143–146.[9]

Anderson, D. R., K. B. Burnham, R. J. Gutierrez, E. D. Forsman, R. G. Anthony, G. C. White, and T. M. Shenk. 1999. A protocol for conflict resolution in analyzing empirical data related to natural resource controversies. *Wildlife Soc. B.* 27:1050–1058. [17]

Anderson, E. N. 1996. *Ecologies of the Heart: Emotion, Belief, and the Environment*. Oxford University Press, New York. [4]

Anderson, G., E. S. Delfosse, N. Spencer, C. Prosser, and R. Richard. 2000. Biological control of leafy spurge: An emerging success story. In N. R. Spencer (ed.), *Proceedings X International Symposium on Biological Control of Weeds*, pp. 15–25. Montana State University, Bozeman. [9]

Anderson, R. M. and R. M. May. 1986. The invasion, persistence and spread of infectious diseases within animal and plant communities. *Philos. T. Roy. Soc. B.* 314:533–570. [3]

Anderson, R. M. and R. M. May. 1991. *Infectious Diseases of Humans: Dynamics and Control*. Oxford University Press, Oxford. [3]

Anderson, R. M., C. A. Donnelly, N. M. Ferguson, M. E. J. Woolhouse, C. J. Watt, H. J. Udy, S. MaWhinney, S. P. Dunstan, T. R. E. Southwood, J. W. Wilesmith, J. B. M. Ryan, L. J. Hoinville, J. E. Hillerton, A. R. Austin, and G. A. H. Wells. 1996. Transmission dynamics and epidemiology of BSE in British cattle. *Nature* 382:779–788. [3]

Ando, A. W. 1999. Waiting to be protected under the Endangered Species Act: The political economy of regulatory delay. *J. L. & Econ.* 4:29–60. [17]

Andren, H. 1994. Effects of habitat fragmentation on birds and mammals in landscapes with different proportions of suitable habitat: a review. *Oikos* 71:355–366. [7]

Andren, H. and P. Angelstam. 1988. Elevated predation rates as an edge effect in habitat islands: Experimental evidence. *Ecology* 69:544–547. [7]

Andrewartha, H. G. and L. C. Birch. 1954. *The Distribution and Abundance of Animals*. University of Chicago Press, Chicago.[10]

Andrewartha, H. G. and L. C. Birch. 1984. *The Ecological Web*. University of Chicago Press, Chicago. [12]

Angliss, R. P. and K. L. Lodge. 2002. *Alaska Marine Mammal Stock Assessments*. U.S. Dept. Commer., NOAA Tech. Memo. NMFS-AFSC-133. [17]

Anthony R. G. and 28 coauthors. 2004. Status and Trends in Demography of Northern Spotted Owls, 1985–2003. Draft Report to the Interagency Regional Monitoring Program, USDA Forest Service, Portland, OR. [17]

Antonovics, J., A. D. Bradshaw, and R. G. Turner. 1971. Heavy metal tolerance in plants. *Adv. Ecol. Res.* 7:1–85. [2]

Aquatic Nuisance Species Task Force. 2002. A Draft National Management Plan for the Genus *Eriochheir*. Prepared by the Chinese mitten crab Control Committee. California Department of Fish and Game, Central Valley Bay, Delta Branch at http://www.delta.dfg.ca.gov/mittencrab/. [14]

Arambarri, P., F. Cabrera, and R. González-Quesada. 1996. Quality of the surface waters entering the Doñana National Park (southwest Spain). *Sci. Total Environ.* 191:185–196. [14]

Archer, S. 1995. Mechanisms of shrubland expansion: Land use, Climate or CO_2? *Climatic Change* 29:91–99. [10]

Argyle, M. 1987. *The Psychology of Happiness*. Routledge, New York. [5]

Argyle, M. 1999. Causes and correlates of happiness. In D. Kahneman, E. Diener, and N. Schwarz (eds.), *Well-Being: The Foundations of Hedonic Psychology*, pp. 353–373. Russell Sage Foundation, New York. [5]

Armbruster, P. and R. Lande. 1993. A population viability analysis for African elephant (*Loxodonta africana*): How big should reserves be? *Conserv. Biol.* 7:602–610. [12]

Armesto, J. J., C. Villagrán, and M. Kalin-Arroyo. 1995. *Ecología de los Bosque Nativos de Chile*. Universidad de Chile, Santiago, Chile. [9]

Arrhenius, O. 1921. Species and area. *J. Ecol.* 9:95–99. [2]

Arribas, C., P. Guarnizo, D. García de Jalón, C. Granado-Lorencio, and C. Fernández-Delgado. 2003. Fauna Piscícola de la cuenca del río Guadiamar: estado de Conservación, problemática y directrices de restauración. In *Ciencia y Restauración del Río Guadiamar. PICOVER 1998–2003*, pp. 438–445. Consejería de Medio Ambiente. Junta de Andalucía. [14]

Arrow, K., R. Solow, P. R. Portney, E. E. Leamer, R. Radner, and E. H. Schuman. 1993. *Report of the NOAA Panel on Contingent Valuation*. Report to the General Counsel of the U.S. National Oceanic and Atmospheric Administration, Resources for the Future, Washington, D.C. [5]

Ascanio, G. H. Adler, T. D. Lambert, and L. Balbas. 2001. Ecological meltdown in predator-free forest fragments. *Science* 294:1923–1926. [3]

Askins, R. A., J. F. Lynch, and R. Greenberg. 1990. Population declines in migratory birds in eastern North America. *Curr. Ornithol.* 7:1–57. [12]

Asquith N. M., J. Terborgh, A. E. Arnold, C. M. Riveros. 1999. The fruits the agouti ate: *Hymenaea courbaril* seed fate when its disperser is absent. *J. Trop. Ecol.* 15:299–235. [8]

Athanasiou, T and P. Baer. 2002. *Dead Heat: Global Justice and Global Warming*. Seven Stories Press, New York. [4]

Atwood, J. T. 2000. Orchids. In N. M. Nadkarni and N. T. Wheelwright (eds.), *Monteverde: Ecology and Conservation of a Tropical Cloud Forest*, p. 74. Oxford University Press, New York. [10]

Augspurger, C. K. 1984. Seedling survival of tropical tree species: Interactions of dispersal distance, light-gaps, and pathogens. *Ecology* 65:1705–1712. [8]

Augustine, D. J. and L. E. Frelich. 1998. Effects of white-tailed deer on populations of an understory forb in fragmented deciduous forests. *Conserv. Biol.* 12:995–1004. [7]

Ault, T. R. and C. R. Johnson. 1998. Spatially and temporally predictable fish communities on coral reefs. *Ecol. Monogr.* 68:25–50. [7]

Australian Government Department of the Environment and Heritage. Threatened Species and Ecological Communities. 2005. www.deh.gov.au/biodiversity/threatened/index.html. [3]

Avise, J. C. 1992. Molecular population structure and the biogeographic history of a regional fauna: A case history with lessons for conservation biology. *Oikos* 63:62–76. [11]

Avise, J. C. 2004. *Molecular Markers, Natural History and Evolution*, 2nd Edition. Sinauer Associates, Sunderland, MA. [11]

Avise, J. C. and J. Hamrick. 1996. *Conservation Genetics: Case Studies from Nature*. Chapman & Hall, New York. [11]

Avise, J. C. and W. S. Nelson. 1989. Molecular genetic relationships of the extinct dusky seaside sparrow. *Science* 243:646–648. [11]

Avise, J. C., J. Arnold, R. M. Ball Jr., E. Bermingham, T. Lamb, J. E. Neigel, C. A. Reeb, and N. C. Saunders. 1987. Intraspecific phylogeography: The mitochondrial DNA bridge between population genetics and systematics. *Ann. Rev. Ecol. Syst.* 18:489–522. [11]

Ayres, M. P. and M. J. Lombardero. 2000. Assessing the consequences of global change for forest disturbance from herbivores and pathogens. *Sci. Total Environ.* 262:263–286. [10]

Bacchus, S. T. 1991. *Identification and review procedure for cutthroat seeps*. Central Florida Regional Planning Council. Bartow, FL. [18]

Bacon, C. 2005. Confronting the coffee crisis: Can Fair Trade, organic, and specialty

coffees reduce small-scale farmer vulnerability in northern Nicaragua? *World Dev.* 33:497–511. [18]

Bailey, A. M., J. McGregor, B. B. Davies, G. Edwards-Jones, B. Dent, and R. Fawcett. 1995. *Development and validation of a framework for evaluating the trade-offs between the benefits and disbenefits of agriculture.* Report to the Ministry of Agriculture, Fisheries and Food, U.K. [5]

Bailey, N. D. and A. Groff. 2003. *Bushmeat Education Resource Guide.* Silver Spring, Maryland: Bushmeat Crisis Task Force. Accessed 15 April 2005: www.bushmeat.org/BERG. [8]

Bailey, R. G. 1989. *Ecoregions of the Continents.* U.S. Department of Agriculture Forest Service, Washington D.C. Map 1:30,000,000. [18]

Baillie, J. E. M., C. Hilton-Taylor, and S. N. Stuart. 2004. *2004 IUCN Red List of Threatened Species.* IUCN, Gland, Switzerland. [3, 18]

Baird, R. W. 2001. Status of killer whales, *Orcinus orca,* in Canada. *Can. Field Nat.* 115:676–701. [17]

Baker, A. J. (ed.). 2000. *Molecular Methods in Ecology.* Blackwell Science, London. [11]

Baker, B. 1996. A reverent approach to the natural world. *BioScience* 46:475–478. [4]

Baker, C. S. and S. R. Palumbi. 1996. Population structure, molecular systematics, and forensic identification of whales and dolphins. In J. C. Avise and J. Hamrick (eds.), *Conservation Genetics: Case Studies from Nature,* pp. 10–49. Chapman and Hall, New York. [11]

Baker, J. P., D. W. Hulse, S. V. Gregory, D. White, J. Van Sickle, P. A. Berger, D. Dole, and N. H. Schumaker. 2004. Alternative futures for the Willamette River Basin, Oregon. *Ecol. Appl.* 14:313–324. [12]

Baker, W. L. and G. K. Dillon. 2000. Plant and vegetation responses to edges in the Southern Rocky Mountains. In R. L. Knight, F. W. Smith, S. W. Buskirk, W. H. Romme, and W. L. Baker, (eds.), *Forest Fragmentation in the Southern Rocky Mountains,* pp. 221–245. University Press of Colorado, Boulder. [7]

Baker, W. L. and R. L. Knight. 2000. Roads and forest fragmentation in the southern Rocky Mountains. In R. L. Knight, F. W. Smith, S. W. Buskirk, W. H. Romme, and W. L. Baker, (eds.), *Forest Fragmentation in the Southern Rocky Mountains.* University Press of Colorado, Boulder. [7]

Bakker, J. P. and F. Berendse. 1999. Constraints in the restoration of ecological diversity in grassland and heathland communities. *Trends Ecol. Evol.* 14:63–68. [18]

Balick, M. J. and R. Mendelsohn. 1992. Assessing the economic value of traditional medicines from tropical rain forests. *Conserv. Biol.* 6:128–130. [5]

Ball, I. R. and H. P. Possingham. 2000. Marxan (v 1.8.6): Marine reserve design using spatially explicit annealing. User Manual. [14]

Ball, I. R. and H. P. Possingham. 2001. The design of marine protected areas: adapting terrestrial techniques. In F. Ghassemi,

R. Whetton, R. Little, and M. Littleboy, (eds.), *Volume 2 of Proceedings of the International Congress on Modelling and Simulation,* pp. 769–774. The Australian National University, Canberra, Australia. [14]

Ball, S. M. J. 2004. Stocks and exploitation of East African blackwood: A flagship species for Tanzania's Miombo woodlands? *Oryx* 38:266–272. [8]

Ballou, J. 1997. Ancestral inbreeding only minimally affects inbreeding depression in mammalian populations. *J. Heredity* 88:169–178. [11]

Ballou, J. and R. Lacy. 1995. Identifying genetically important individuals for management of genetic variation in pedigreed populations. In J. D. Ballou, M. E. Gilpin, and T. J. Foose (eds.), *Population Management for Survival and Recovery,* pp.76–111. Columbia University Press, New York. [11]

Ballou, J. D. and T. J. Foose. 1996. Demographic and genetic management of captive populations. In D. G. Kleiman, M. E. Allen, K. V. Thompson, and S. Lumpkin, (eds.), *Wild Mammals in Captivity: Principles and Techniques,* pp. 263–283. University of Chicago Press, Chicago, IL. [15]

Balmford, A. 1996. Extinction filters and current resilience: The significance of past selection pressures for conservation biology. *Trends Ecol. Evol.* 11:193–196. [3]

Balmford, A, and 27 others. 2005. The Convention on Biological Diversity's 2010 target. *Science* 307:212–213. [18]

Balmford, A. and T. Whitten. 2003. Who should pay for tropical conservation, and how could the costs be met? *Oryx* 37(2):238–250. [6]

Balmford, A., K. J. Gaston, S. Blyth, A. James, and V. Kapos. 2003. Global variation in terrestrial conservation costs, conservation benefits, and unmet conservation needs. *Proceedings of the National Academy of Sciences U.S.A.* 100:1046–1050. [18]

Balmford, A., A. Bruner, P. Cooper, R. Costanza, S. Farber, R. E. Green, M. Jenkins, P. Jefferiss, V. Jessamy, J. Madden, K. Munro, N. Myers, S. Naeem, J. Paavola, M. Rayment, S. Rosendo, J. Roughgarden, K. Trumper, and R. K. Turner. 2002. Why conserving wild nature makes economic sense. *Science* 297:950–953. [6, 8, 14, 18]

Balvanera, P., G. C. Daily, P. R. Ehrlich, T. Ricketts, S. A. Bailey, S. Kark, C. Kremen, and H. Pereira. 2001. Conserving biodiversity and ecosystem services. *Science* 291:2047. [18]

Bancroft, G. T. 1989. Status and conservation of wading birds in the Everglades. *Am. Birds* 43:1258–1265. [13]

Bandara, R. and C. Tisdell. 2003. Comparison of rural and urban attitudes to the conservation of Asian elephants in Sri Lanka: Empirical evidence. *Biol. Conserv.* 110:327–342. [17]

Bandara, R and C. Tisdell. 2004. The net benefit of saving the Asian elephant: A policy and contingent valuation study. *Ecol. Econ.* 48:93–107. [17]

Banks, M. A. and W. Eichert. 2000. WHICHRUN: A computer program for population assignment of individuals

based on multilocus genotype data. *J. Heredity* 91:87–89. [11]

Bannikov, A. G. 1948. On fluctuations in abundance of anuran amphibians. *Dokl. Akad. Nauk. SSSR* 61:131–134. [3]

Barber, V. A., G. P. Juday, and B. P. Finney. 2000. Reduced growth of Alaskan white spruce in the twentieth century from temperature-induced drought stress. *Nature* 405:668–673. [10]

Barbier et al. 1995. Economic value of biodiversity. In V. H. Heywood et al. (eds.), *Global Biodiversity Assessment.* Cambridge University Press, New York. [5]

Barcelona Field Studies Centre. 2005. Water Supply in the Barcelona Region. http://geographyfieldwork.com/barcelonawatersupply.htm [18]

Bargagli, R. 2000. Trace metals in Antarctica related to climate change and increasing human impact. *Rev. Environ. Contam. T.* 166:129–173. [3]

Barlow, J. and C. A. Peres. 2004. Ecological responses to El Niño-induced surface fires in central Amazonia: Management implications for flammable tropical forests. *Phil. Trans. R. Soc. Lond. B.* 359:367–380. [8]

Barnes, A. M. 1982. Surveillance and control of bubonic plague in the United States. *Symp. Zool. Soc. Lond.* 50:237–270. [15]

Barnett, T. P., D. W. Pierce, and R. Schnur. 2001. Detection of anthropogenic climate change in the world's oceans. *Science* 292:270–274. [10]

Barnosky, A. D., P. L. Koch, R. S. Feranec, S. L. Wing, and A. B. Shabel. 2004. Assessing the causes of Late Pleistocene extinctions on the continents. *Science* 306:70–75. [3]

Barr, J. 1972. Man and nature: The ecological controversy and the Old Testament. *Bull. John Rylands Library* 55:9–32. [4]

Barrett, C. 1996. Fairness, stewardship and sustainable development. *Ecol. Econ.* 19:11–17. [16]

Barrett, C. B., K. Brandon, C. Gibson, and H. Gjertsen. 2001. Conserving tropical biodiversity amid weak institutions. *BioScience* 51:497–502. [12]

Barrett, O. W. 1928. Shark fishing in the West Indies. *Scientific Monthly* 27:125–133. [8]

Barry, D. and M. Oelschlaeger. 1996. A science for survival: values and conservation biology. *Conserv. Biol.* 10:905–911. [1, 5]

Barry, J. P., C. H. Baxter, R. D. Sagarin, and S. E. Gilman. 1995. Climate-related long-term faunal changes in a California rocky intertidal community. *Science* 267:672–675. [10]

Basey, J. M., S. H. Jenkins, and P. E. Busher. 1988. Optimal central-place foraging by beavers: Tree-size selection in relation to defensive chemicals of quaking aspen. *Oecologia* 76:278–282. [9]

Baskett, M. L., S. A. Levin, S. D. Gaines, and J. Dushoff. 2005. Marine reserve design and the evolution of size at maturation in harvested fish. *Ecol. Appl.* 15:882–901. [18]

Baskin, Y. 1991. Archeaologist lends a technique to rhino protection. *BioScience* 41:532–534. [11]

Battley, P. F., T. Piersma, D. I. Rogers, A. Dekinga, B. Spaans, and J. A. van Gils. 2004. Do body condition and plumage during fuelling predict northwards departure dates of Great Knots *Calidris tenuirostris* from north-west Australia? *Ibis* 146:46–60. [10]

Battley, P. F., T. Piersma, M. W. Dietz, S. X. Tang, A. Dekinga, and K. Hulsman. 2000. Empirical evidence for differential organ reductions during trans-oceanic bird flight. *Proc. R. Soc Lon. Ser. B* 267:191–195. [10]

Bauer, B. H., D. A. Etneir, and N. M. Burkehead. 1995. *Etheostoma (Ulocentra) scotti* (Osteichthyes: Percidae), a new darter from the Etowah River system in Georgia. *Bull. Ala. Mus. Nat. Hist.* 17:1–16. [7]

Baum, J. K. and R. A. Myers. 2004. Shifting baselines and decline of pelagic sharks in the Gulf of Mexico. *Ecology Letters* 7:135–145. [8]

Baum, J. K., R. A. Myers, D. G. Kehler, B. Worm, S. J. Harley, and P. A. Doherty. 2003. Collapse and conservation of shark populations in the Northwest Atlantic. *Science* 299:389–392. [8]

Baumgartner, J. V. 2001. The Nature Conservancy's five-S decision-making framework for site-based conservation and relationship to the proposed ecological risk management framework. In R. G. Stahl, R. A Bachman, A. L Barton, J. R. Clark, P. L. deFur, S. J. Ells, C. A. Pittinger, M. W. Slimak, and R. S. Wentsel, (eds.), *Risk Management: Ecological Risk-Based Decision-Making*, pp. 163–173. SETAC Press, FL. [18]

Bawa, K. S. 1990. Plant–pollinator interactions in tropical rain forests. *Ann. Rev. Ecol. Syst.* 21:399–422. [12]

Bayne, E. M. and K. A. Hobson. 1997. Comparing the effects of landscape fragmentation by forestry and agriculture on predation of artificial nests. *Conserv. Biol.* 11:1418–1429. [7]

BBC 2002. Disposable Planet? http://news.bbc.co.uk/hi/english/static/in_depth/world/2002/disposable_planet/ [18]

Bean, M. J. 1998. The Endangered Species Act and private land: four lessons learned from the past quarter century. *Envtl. L. Rep.* 28:10710. [17]

Beaver, R. A. 1979. Host specificity of temperate and tropical animals. *Nature* 281:139–141. [2]

Bedward, M., R. L. Pressey, and A. O. Nicholls. 1991. Scores and score classes for evaluation criteria: a comparison based on the cost of reserving all natural features. *Biol. Conserv.* 56:281–294. [14]

Beebee, T. J. C. 1995. Amphibian breeding and climate. *Nature* 374:219–220. [3, 10]

Beebee, T. J. C., R. J. Flower, A. C. Stevenson, S. T. Patrick, P. G. Appleby, C. Fletcher, C. Marsh, J. Natkanski, B. Rippey, and R. W. Battarbee. 1990. Decline of the natterjack toad *Bufo clamita* in Britain: Palaeoecological, documentary and experimental evidence for breeding site acidification. *Biol. Conserv.* 53:1–20. [3]

Beerli, P. and J. Felsenstein. 2001. Maximum likelihood estimation of a migration matrix and effective population sizes in subpopulations by using a coalescent approach. *Proc. Nat. Acad. Sci. USA* 98:4563–4568. [11]

Begazo, A. J. and R. E. Bodmer. 1998. Use and conservation of Cracidae (Aves: Galliformes) in the Peruvian Amazon. *Oryx* 32:301–309. [12]

Beger, M., G. P. Jones, and P. L. Munday. 2002. Conservation of coral reef biodiversity: a comparison of reserve selection procedures for corals and fishes. *Biol. Conserv.* 111:53–62. [14]

Beier, P. 1993. Determining minimum habitat areas and habitat corridors for cougars. *Conserv. Biol.* 7:94–108. [12]

Beier, P. and R. F. Noss. 1998. Do habitat corridors provide connectivity? *Conserv. Biol.* 12:1241–1252. [7]

Beilfuss, R. D. and B. R. Davies. 1999. Prescribed flooding and wetland rehabilitation in the Zambezi Delta, Mozambique. In W. Streever (ed.), *An International Perspective on Wetland Rehabilitation*, pp.143–158. Kluwer Academic Publishers, Amsterdam. [12]

Beilfuss, R. D., P. Dutton, and D. Moore. 2000. Land cover and land use changes in the Zambezi Delta. In J. Timberlake (ed.), *Biodiversity of the Zambezi Basin Wetlands, Volume III. Land Use Change and Human Impacts*, pp. 31–106. IUCN Regional Office for Southern Africa, Harare, Zimbabwe. [12]

Beissinger, S. R. and D. R. McCullough. 2002. *Population Viability Analysis*. University Chicago Press, Chicago. [12]

Beissinger, S. R. and M. I. Westphal. 1998. On the use of demographic models of population viability in endangered species management. *J. Wildl. Mgmt.* 62:821–841. [12]

Belbin, L. 1995. A multivariate approach to the selection of biological reserves. *Biodivers. Conserv* 4:951–963. [14]

Belden, L. K. and A. R. Blaustein. 2002. Population differences in sensitivity to UV-B radiation for larval long-toed salamanders. *Ecology* 83:1586–1590. [3]

Belis, J., J. Requejo, and E. Herrera. 2003. *Evaluación final del Plan de Desarrollo Sostenible de Doñana 1993–2000*. Doñana 21 Foundation (www.donana.es). [14]

Bell, B. D., S. Carver, N. J. Mitchell, and S. Pledger. 2004. The recent decline of a New Zealand endemic: How and why did populations of Archey's frog *Leiopelma archeyi* crash over 1996–2001? *Biol. Conserv.* 120:189–199. [3]

Bell, S. S., M. S. Fonseca, and L. B. Motten. 1997. Linking restoration and landscape ecology. *Restor. Ecol.* 5: 318–323. [15]

Bell, S. S., R. A. Brooks, B. D. Robbins, M. S. Fonseca, and M. O. Hall. 2001. Faunal response to fragmentation in seagrass habitats: Implications for seagrass conservation. *Conserv. Biol.* 100:115–123. [7]

Beman, J. M., K. R. Arrigo, and P. A. Matson. 2005. Agricultural runoff fuels large phytoplankton blooms in vulnerable areas of the ocean. *Nature* 434:211–214. [6]

Ben-David, M., T. A. Hanley, and D. M. Schell. 1998. Fertilization of terrestrial vegetation by spawning pacific salmon: the role of flooding and predator activity. *Oikos* 83: 47–55. [7]

Béné, C. and A. Tewfik. 2003. Biological evaluation of a marine protected area: Evidence of crowing effect on a protected population of queen conch in the Caribbean. *Marine Ecology – Pubblicazioni della Stazione Zoologica di Napoli I* 24:45–58. [14]

Benítez-Malvido, J. and M. Martínez-Ramos. 2003. Impact of forest fragmentation on understory plant species richness in Amazonia. *Conserv. Biol.* 17:389–400. [7]

Benke, A. C. 1990. A perspective on America's vanishing streams. *J. N. Am. Benthol. Soc.* 9:77–88. [7]

Benndorf, J., W. Böing, J. Koop, and I. Neubauer. 2002. Top-down control of phytoplankton: the role of time scale, lake depth and trophic state. *Freshwater Biology* 47:2282–2295. [3]

Bennett, E. L. 2002. Is there a link between wild meat and food security? *Conserv. Biol.* 16:590–592. [8]

Bennett, E., H. Eves, J. Robinson, and D. Wilkie. 2002. Why is eating bushmeat a biodiversity crisis? *Conservation in Practice* 3:28–29. [8]

Benstead, J. P., J. G. March, C. M. Pringle, and F. N. Scatena. 1999. Effects of a low-head dam and water extraction on migratory tropical stream biota. *Ecol. Appl.* 9:656–668. [7]

Bentley, J. M., C. P. Catterall, and G. C. Smith. 2000. Effects of fragmentation of Araucarian vine forest on small mammal communities. *Conserv. Biol.* 14:1075–1087. [7]

Bento, C. M. 2002. The status and prospects of Wattled Cranes *Grus carunculatus* in the Marromeu Complex of the Zambezi Delta. M.S. Thesis, University of Cape Town, Cape Town, South Africa. [12]

Benzing, D. H. 1995. Vascular epiphytes. In M. D. Lowman and N. M. Nadkarni (eds.), *Forest Canopies*. pp. 225–254. Academic Press, New York. [10]

Benzing, D. H. 1998. Vulnerabilities of tropical forests to climate change: the significance of resident epiphytes. *Climatic Change* 39:519–540. [10]

Berg A., B. Ehnstrom, L. Gustafsson, T. Hallingback, M. Jonsell, and J. Weslien. 1994. Threatened plant, animal, and fungus species in Swedish forests: Distribution and habitat associations. *Conservation Biology* 8:718–731. [8]

Berg, R. Y. 1975. Myrmecochorus plants in Australia and their dispersal by ants. *Aust. J. Bot.* 23:475–508. [2]

Berger, J. J. 1991. The federal mandate to restore: laws and policies on environmental restoration. *The Environmental Professional* 13: 195–206. [15]

Berger, J., P. B. Stacey, L. Bellis, and M. P. Johnson. 2001. A mammalian predator-prey imbalance: Grizzly bear and wolf extinction affect avian Neotropical migrants. *Ecol. Appl.* 11:947–960. [8, 12]

Berger, L., R. Speare, H. B. Hines, G. Marantelli, A. D. Hyatt, K. R. McDonald, L. F. Skerratt, V. Olsen, J. M. Clarke, G. Gillespie, M. Mahony, N. Sheppard, C. Williams, and M. J. Tyler. 2004. Effect of season and temperature on mortality in amphibians due to chytridiomycosis. *Austr. Vet. J.* 82:434–439. [3, 10]

Berger, L., R. Speare, P. Daszak, D. E. Green, A. A. Cunningham, C. L. Goggin, R. Slocombe, M. A. Ragan, A. D. Hyatt, K. R. McDonald, H. B. Hines, K. R. Lips, G. Marantelli, and H. Parkes. 1998. Chytridiomycosis causes amphibian mortality associated with population declines in the rain forests of Australia and Central America. *Proc. Nat. Acad. Sci. USA* 95:9031–9036. [3]

Berlow, E. L., S. A. Navarrete, C. J. Briggs, M. E. Power, and B. A. Menge. 1999. Quantifying variation in the strengths of species interactions. *Ecology* 80:2206–2224. [3, 9]

Bernúes, M. 1990. Limnología de los ecosistemas acuáticos del Parque Nacional de Doñana. Ph. D. dissertation. Autonomous University of Madrid. Madrid. [14]

Berrill, M., S. Bertram, A. Wilson, S. Louis, D. Brigham, and C. Stromberg. 1993. Lethal and sublethal impacts of pyrethroid insecticides on amphibian embryos and tadpoles. *Environ. Toxicol. Chem.* 12:525–539. [3]

Best Practices and Local Leadership Program (BLP). 2005. Best Practices Database. http://www.bestpractices.org. [18]

Bettenay, E. 1984. Origin and nature of the sandplains. In Pate, J. S. and J. S. Beard (eds.), *Kwongan: Plant Life of the Sandplain*, pp. 51–68. University of Western Australia Press, Nedlands. [2]

Betts, R. A. 2004. Global vegetation and climate: self-beneficial effects, climate forcings and climate feedbacks. *J. Phys. IV France* 121:37–60. [6]

Bety, J., G. Gauthier, E. Korpimaki, and J. F. Giroux. 2002. Shared predators and indirect trophic interactions: Lemming cycles and arctic-nesting geese. *J. Anim. Ecol.* 71:88–98. [10]

Beukema, J. J., K. Essink, and H. Michaelis. 1996. The geographic scale of synchronized fluctuation patterns in zoobenthos populations as a key to underlying factors: Climatic or man-induced. *ICES J. Mar. Sci.* 53:964–971. [10]

Beverton, R. J. H. and S. J. Holt. 1957. *On the Dynamics of Exploited Fish Populations*. Fish. Invest. Ser.II, Vol.XIX, H.M.S.O., London. [8]

Bezemer, T. M. and T. H. Jones. 1998. Plant-insect herbivore interactions in elevated atmospheric CO_2: Quantitative analyses and guild effects. *Oikos* 82: 212–222. [10]

Bidleman, T. F. 1999. Atmospheric transport and air-surface exchange of pesticides. *Water Air Soil Poll.* 115:115–166. [3]

Biek, R., W. C. Funk, B. A. Maxell, and L. S. Mills. 2002. What is missing in amphibian decline research: Insights from ecological sensitivity analysis. *Conserv. Biol.* 16:728–734. [3]

Bierregaard, R. O., T. E. Lovejoy, V. Kapos, A. A. dos Santos, and R. W. Hutchings. 1992. The biological dynamics of tropical rainforest fragments. *BioScience* 42:859–866. [7]

Bigelow N. H., and 26 others. 2003. Climate change and arctic ecosystems: 1. Vegetation changes north of 55 degrees N between the last glacial maximum, mid-Holocene, and present. *J. Geophys. Res.-Atmos.* 108:(D19)8170. [10]

Biggins, D. E. and J. L. Godbey. 2003. Challenges to reestablishment of free-ranging populations of black-footed ferrets. *CR Biol.* 326:S104–S111. [15]

Biggins, D. E. and M. H. Schroeder. 1988. Historical and present status of the black-footed ferret. In *Proceedings of the Eighth Great Plains Wildlife Damage Control Workshop*, pp. 93–97. General Technical Report RM-154, USDA Forest Service, Fort Collins, Colorado. [15]

Biggins, D. E., A. Vargas, J. L. Godbey, and S. H. Anderson. 1999. Influence of pre-release experience on reintroduced black-footed ferrets (*Mustela nigripes*). *Biol. Conserv.* 89:121–129. [15]

Biggins, D. E., M. H. Schroeder, S. C. Forrest, and L. Richardson. 1986. Activity of radio-tagged black-footed ferrets. *Great Basin Nat. Mem.* 8:135–140. [15]

Biggins, D. E., J. L. Godbey, L. H. Hanebury, B. Luce, P. E. Marinari, M. R. Matchett, and A. Vargas. 1998. Survival of reintroduced black-footed ferrets. *J. Wildl. Mgmt.* 62:643–653. [15]

Birdlife International. 2004. *State of the World's Birds 2004: Indicators for our changing world*. Birdlife International, Cambridge. [3, 6]

Birnie, P. (ed.). 1985. *International Regulation of Whaling: From the Conservation of Whaling to the Conservation of Whales and Regulation of Whale Watching*. Oceana Publications, New York. [4]

Bisbal. G. A. 2002. The best available science for the management of anadromous salmonids in the Columbia River Basin. *Can. J. Fish. Aquat. Sci* 59:1952–1959. [17]

Bishop, R. 1978. Endangered species and uncertainty: The economics of a safe minimum standard. *Am. J. Agric. Econ.* 60:10–18. [4]

Bjornlie, D. D., and R. A. Garrott. 2001. Effects of winter road grooming on bison in Yellowstone National Park. *J. Wildl. Mgmt.* 65:560–572. [12]

Blake, J. G. 1991. Nested subsets and the distribution of birds on isolated woodlots. *Conserv. Biol.* 5:58–66. [7]

Blake, J. G. and J. R. Karr. 1984. Species composition of bird communities and the conservation benefit of large versus small forests. *Biol. Conserv.* 30:173–187. [7]

Blake, N. M. 1980. *Land Into Water—Water Into Land: A History of Water Management in Florida (Everglades)*. University Presses of Florida, Gainesville, FL. [13]

Blakesley, J. A., B. R. Noon, and D. R. Anderson. 2005. Site occupancy, apparent survival and reproduction in California Spotted Owls in relation to forest stand characteristics. *J. Wildl. Mgmt.* In press. [17]

Blakesley, J. A., B. R. Noon, and D. W. H. Shaw. 2001. Demography of the California spotted owl in northeastern California. *Condor* 103:667–677. [17]

Blanchard, K. A. 1995. Reversing population declines in seabirds on the north shore of the Gulf of St. Lawrence, Canada. In S.K. Jacobson (ed.), *Conserving Wildlife: International Education and Communication Approaches*, pp. 51–63. Columbia University Press, NY. [18]

Blaustein, A. R. 1994. Chicken Little or Nero's fiddle? A perspective on declining amphibian populations. *Herpetologica* 50:85–97. [3]

Blaustein, A. R., and J. M. Kiesecker. 2002. Complexity in conservation: lessons from the global decline of amphibian populations. *Ecol. Lett.* 5:597–608. [3]

Blaustein, A. R., D. B. Wake, and W. P. Sousa. 1994b. Amphibian declines: judging stability, persistence, and susceptibility of populations to local and global extinctions. *Conserv. Biol.* 8:60–71. [3]

Blaustein, A. R., J. M. Romansic, J. M. Kiesecker, and A. C. Hatch. 2003a. Ultraviolet radiation, toxic chemicals and amphibian population declines. *Divers. Distrib.* 9:123–140. [3]

Blaustein, A. R., P. D. Hoffman, D. G. Hokit, J. M. Kiesecker, S. C. Walls, and J. B. Hays. 1994a. UV repair and resistance to solar UV-B in amphibian eggs: A link to population declines? *Proc. Nat. Acad. Sci. USA* 91:1791–1795. [3]

Blaustein, A. R., T. L. Root, J. M. Kiesecker, L. K. Belden, D. H. Olson, and D. M. Green. 2003b. Amphibian breeding and climate change: Reply to Corn. *Conserv. Biol.* 17:626–627. [3]

Blaustein, A. R., B. Han, B. Fasy, J. Romansic, E. A. Scheessele, R. G. Anthony, A. Marco, D. P. Chivers, L. K. Belden, J. M. Kiesecker, T. Garcia, M. Lizana, and L. B. Kats. 2004. Variable breeding phenology affects the exposure of amphibian embryos to ultraviolet radiation and optical characteristics of natural waters protect amphibians from UV-B in the U.S. Pacific Northwest: Comment. *Ecology* 85:1747–1754. [3]

Blockstein, D. E. 2002. How to lose your political virginity while keeping your scientific credibility. *BioScience* 52:91–96. [17]

Blundell, G. M., M. Ben-David, P. Groves, R. T. Bowyer, and E. Geffen. 2002. Characteristics of sex-biased dispersal and gene flow in coastal river otters: Implications for natural recolonization of extirpated populations. *Mol. Ecol.* 11:289–303. [11]

Boag, P. T. and P. R. Grant. 1984. The classical case of character release: Darwin's finches (*Geospiza*) on Isla Daphne Major, Galápagos. *Biol. J. Linn. Soc.* 22:243–287. [10]

Bocco, G., A. Velázquez, and A. Torres. 2000. Ciencia, comunidades indígenas y manejo de recursos naturales, un caso de investigación participativa en México. *Interciencia* 25:64–70. [8]

Bockstael, N. E., K. E. McConnell, and I. E. Strand Jr. 1991. Recreation. In J. Braden

and C. D. Kolstad (eds.), *Measuring the Demand for Environmental Quality*. North Holland, Amsterdam. [5]

Bodhi, B. 1997. Forward. In *Buddhist Perspectives on the Ecocrisis*, pp. i–x. Buddhist Publication Society, Kandy, Sri Lanka. [4]

Bodkin, J. L., B. E. Ballachey, M. A. Cronin, and K. T. Scribner. 1999. Population demographics and genetic diversity: Case histories from remnant and reestablished populations of sea otters (*Enhydra lutris*). *Conserv. Biol.* 13:1378–1385. [11]

Bodmer, R. E. 1995. Managing Amazonian wildlife: Biological correlates of game choice by detribalized hunters. *Ecol. Appl.* 5:872–877.[8]

Boerema, L. K. and J. A. Gulland. 1973. Stock assessment of Peruvian anchovy (*Engraulis ringens*) and management of fishery. *Journal of the Fisheries Research Board of Canada* 30:2226–2235. [8]

Boersma, D., J. Ogden, G. Branch, R. Bustamante, C. Campagna, G. Harris, and E. K. Pikitch. 2004. 6: Lines on the Water: Ocean-use planning in large marine ecosystems. In L. K. Glover and S. A. Earle (eds.), *Defying Ocean's End: An Agenda for Action*, pp. 125–138. Island Press, Washington, D.C. [12]

Boersma, P. D. 1998. Plight of the penguins. *Wildlife Conserv.* 101:20–27. [12]

Boersma, P. D. and J. Parrish. 1999. Limiting abuse: Marine protected areas, a limited solution. *Ecol. Econ.* 31:287–304. [12]

Boersma, P. D. and S. DeWeerdt. 2001. Tapping the ivory tower: how academic-agency partnerships can advance conservation. *Conservation in Practice* 2:28–32. [18]

Boersma, P. D., D. L. Stokes, and I. Strange. 2002. Applying ecology to conservation: Tracking breeding penguins at New Island South Reserve, Falkland Islands. *Aquat. Conserv.* 12:1–11. [12]

Boersma, P. D., P. Kareiva, W. F. Fagan, J. A. Clark, and J. M. Hoekstra. 2001. How good are endangered species recovery plans? *BioScience* 51:643–650. [18]

Boesch, D. E. 2001. *Pfiesteria* and the Chesapeake Bay. *BioScience* 51: 827. [13]

Boletín Oficial del Estado 1991. Real Decreto 16 Diciembre 1991 N 1771/1991. Ministerio de Agricultura, Pesca y Alimentación. Parques Nacionales. Plan Rector de Uso y Gestión del de Doñana. Boletín Oficial del Estado N 301 del 17 de Diciembre: 8401–8411. [14]

Boletín Oficial del Estado 2003. RESOLUCIÓN de 26 de septiembre de 2003, de la Secretaría General de Medio Ambiente, por la que se formula declaración de impacto ambiental sobre el proyecto «actuaciones de mejora en accesos marítimos al puerto de Sevilla», de la Autoridad Portuaria de Sevilla. Boletín Oficial del Estado N 236 del 2 de Octubre: 35938–35953. [14]

Bolton, J. J. 1994. Global seaweed diversity: patterns and anomalies. *Bot. Mar.* 37:241–245. [2]

Bond, W. 1983. On alpha-diversity and the richness of the Cape flora: A study in southern Cape fynbos. In Kruger, F. J., D.

T. Mitchell, and J. U. M. Jarvis (eds.), *Mediterranean-Type Ecosystems: the Role of Nutrients*, pp. 337–356. Spinger-Verlag, Berlin. [2]

Bonfil, R. 1994. Overview of world elasmobranch fisheries. FAO Fisheries Technical Paper 341. FAO, Rome. [8]

Borja, F. J., A. López-Geta, M. Martín-Machuca, R. Mantecón, C. Mediavilla, P. del Olmo M. Palancar and R. Vives. 2001. Marco geográfico, geológico e hidrológico regional de la cuenca del Guadiamar. *Boletín Geológico Minero*. Special Issue. 13–34. [14]

Bormann, F. H. and G. E. Likens. 1979. *Pattern and Process in a Forested Ecosystem*. Springer-Verlag, New York. [7]

Bosch, J., I. Martinez-Solano, and M. Garcia-Paris. 2001. Evidence of a chytrid fungus infection involved in the decline of the common midwife toad (*Alytes obstetricans*) in protected areas of central Spain. *Biol. Conserv.* 97:331–337. [3]

Both, C. and M. E. Visser. (2001) Adjustment to climate change is constrained by arrival date in a long-distance migrant bird. *Nature* 411:268–98. [10]

Botkin, D. B. 1990. *Discordant harmonies: A New Ecology for the Twenty-first Century*. Oxford University Press, New York. [1, 4]

Bowen, B. W., A. B. Meylan, and J. C. Avise. 1991. Evolutionary distinctiveness of the endangered Kemp's ridley sea turtle. *Nature* 352:709–711. [11]

Boyer, K. E. and J. B. Zedler. 1996. Damage to cordgrass by scale insects in a constructed salt marsh: effects of nitrogen additions. *Estuaries* 19: 1–12. [15]

Boyer, K. E., J. C. Callaway, and J. B. Zedler. 2000. Evaluating the progress of restored cordgrass (*Spartina foliosa*) marshes: Belowground biomass and tissue N. *Estuaries* 23:711–721. [15]

Bradford, D. F. 1989. Allotopic distribution of native frogs and introduced fishes in high Sierra Nevada lakes of California: Implication of the negative effect of fish introductions. *Copeia* 1989:775–778. [3]

Bradford, D. F. 1991. Mass mortality and extinction in a high-elevation population of *Rana muscosa*. *J. Herpetol.* 25:174–177. [3]

Bradford, D. F., D. M. Graber, and F. Tabatabai. 1994b. Population declines of the native frog, *Rana muscosa*, in Sequoia and Kings Canyon National Parks, California. *Southwest. Nat.* 39:323–327. [3]

Bradford, D. F., F. Tabatabai, and D. M. Graber. 1993. Isolation of remaining populations of the native frog, *Rana muscosa*, by introduced fishes in Sequoia and Kings Canyon National Parks, California. *Conserv. Biol.* 7:882–888. [3]

Bradford, D. F., M. S. Gordon, D. F. Johnson, R. D. Andrews, and W. B. Jennings. 1994a. Acidic deposition as an unlikely cause for amphibian population declines in the Sierra Nevada, California. *Biol. Conserv.* 69:155–161. [3]

Bradford, D. F., S. D. Cooper, T. M. Jenkins, K. Kratz, O. Sarnelle, and A. D. Brown. 1998. Influences of natural acidity and introduced fish on faunal assemblages in

California alpine lakes. *Can. J. Fish. Aquat. Sci.* 55:2478–2491. [3]

Bradley, G. A., P. C. Rosen, M. J. Sredl, T. R. Jones, and J. E. Longcore. 2002. Chytridiomycosis in native Arizona frogs. *J. Wildlife Dis.* 38:206–212. [3]

Bradley, N. L., A. C. Leopold, J. Ross, and W. Huffaker. 1999. Phenological changes reflect climate change in Wisconsin. *Proc. Nat. Acad. Sci. USA* 96:9701–9704. [10]

Bradshaw, A. D. 1987. The reclamation of derelict land and the ecology of ecosystems. In W. R. Jordan, M. E. Gilpin, and J. D. Aber (eds.), *Restoration Ecology*, pp. 53–74. Cambridge University Press, Cambridge, MA. [15]

Bradshaw, A. D. 2002. The Background: Introduction and Philosophy. In M. R. Perrow and A. J. Davy (eds.), *Handbook of Ecological Restoration, Volume I: Principles of Restoration*. Cambridge University Press, Cambridge. [15]

Bradshaw, A. D. and M. J. Chadwick. 1980. *The Restoration of Land: The ecology and reclamation of derelict and degraded land*. Blackwell Publishing, Oxford. [15]

Bradshaw, W. 1976. Geography of photoperiodic response in diapausing mosquito. *Nature* 262:384–386. [10]

Bradshaw, W. and C. Holzapfel. 2001. Genetic shift in photoperiodic response correlated with global warming. *Proc. Nat. Acad. Sci. USA* 98:14509–14511. [10]

Bragg, A. N. 1960. Population fluctuation in the amphibian fauna of Cleveland County, Oklahoma during the past twenty-five years. *Southwest. Nat.* 5:165–169. [3]

Brana, F., L. Frechilla, and G. Orizaola. 1996. Effect of introduced fish on amphibian assemblages in mountain lakes of northern Spain. *Herpetol. J.* 6:145–148. [3]

Brander, K. 1981. Disappearance of common skate *Raja batis* from Irish Sea. *Nature* 290:48–49. [8]

Brandon, K. 1996. *Ecotourism and Conservation: A Review of Key Issues*. World Bank, Environment Dept., Washington, D.C. 1996. Biodiv REF G155 .A1 B74 1996. [16]

Brashares, J. S., P. Arcese, and M. K. Sam. 2001. Human demography and reserve size predict wildlife extinction in West Africa. *Proc. R. Soc. Lon. Ser. B* 268:2473–2478. [12]

Brashares, J., P. Arcese, M. K. Sam, P. B. Coppolillo, A. R. E. Sinclair, and A. Balmford. 2004. Bushmeat hunting, wildlife declines, and fish supply in West Africa. *Science* 306:1180–1183. [8]

Bratton, S. P., 1993. *Christianity, Wilderness and Wildlife: The Original Desert Solitaire*. University of Scranton Press, Scranton, PA. [4]

Brauer, I. 2003. Money as an indicator: To make use of economic evaluation for biodiversity conservation. *Agricult. Ecosys. Environ.* 98:483–491. [5]

Brawn, J. D. and S. K. Robinson. 1996. Source-sink population dynamics may complicate the interpretation of long-term census data. *Ecology* 77:3–12. [12]

Bray, D. B. 2004. Community forestry as a strategy for sustainable management: Perspectives from Quintana Roo, Mexi-

co. In D. Zarin, F. E. Putz, J. Alavalapati, and M. Schmink, (eds), *Working Forests in the Tropics*, pp. 221–237. Columbia University Press, New York. [8]

Bray, D. B., L. Merino-Pérez, P. Negreros-Castillo, G. Segura-Warnholtz, J-M. Torres-Rojo, and H. F. M. Vester. 2003. Mexico's community-managed forests as a global model for sustainable landscapes. *Conserv. Biol.* 17:672–677. [16]

Breck, S., D. Biggins, T. Livieri, R. Matchett, and V. Kopcso. 2004. Does predator management enhance survival of black-footed ferrets? In *Proceedings of the Symposium on the Status of the Black-footed Ferret and its Habitat*, pp. 28–29. U.S. Geological Survey, Fort Collins, CO. [15]

Brennan, S. and J. Withgott 2005. *Environment: the Science behind the Stories*. Pearson/Benjamin Cummings, New York. [3]

Briggs, C. J., V. T. Vredenburg, R. A. Knapp, and L. J. Rachowicz. 2005. Investigating the population-level effects of chytridiomycosis: an emerging infectious disease of amphibians. *Ecology*. In press. [3]

Briggs, J. C. 1996. *Global Biogeography*. Elsevier, Amsterdam. [2]

Brittingham, M. C. and S. A. Temple. 1983. Have cowbirds caused forest songbirds to decline? *BioScience* 33:31–35. [7]

Brock, R. E. and D. A. Kelt. 2004. Influence of roads on the endangered Stephens' kangaroo rat (*Dipodomys stephensi*): Are dirt and gravel roads different? *Biol. Coserv.* 118:633–640. [7]

Brody, A. J. and M. P. Pelton. 1989. Effects of roads on black bear movements in western North Carolina. *Wildl. Soc. Bull.* 17:5–10. [7]

Brook, B. W., L. Lim, R. Harden, and R. Frankham. 1997. Does population viability analysis software predict the behaviour of real populations? A retrospective study on the Lord Howe Island Woodhen *Ticholimnas sylvestris* (Sclater). *Biolog. Conserv.* 82:119–128. [12]

Brook, B. W., J. J. O'Grady, A. P. Chapman, M. A. Burgman, H. R. Akçakaya, and R. Frankham. 2000. Predictive accuracy of population viability analysis in conservation biology. *Nature* 404:385–387. [12]

Brooker, L. and M. Brooker. 2002. Dispersal and population dynamics of the blue-breasted fairy wren, *Malurus pulcherrimus*, in fragmented habitat in the Western Australian wheatbelt. *Wildlife Res.* 29:225–233. [7]

Brooks, R. 1988. *The Net Economic Value of Deer Hunting in Montana*. Montana Department of Fish, Wildlife, and Parks, Helena, MT. [5]

Brooks, T. M., S. L. Pimm, and J. O. Oyugi. 1999. Time lag between deforestation and bird extinction in tropical forest fragments. *Conserv. Biol.* 13:1140–1150. [18]

Brooks, T. M., M. I. Bakarr, T. Boucher, G. A. B. Da Fonseca, C. Hilton-Taylor, J. M. Hoekstra, T. Moritz, S. Olivier, J. Parrish, R. L. Pressey, A. S. L Rodrigues, W. Sechrest, A. Stattersfield, W. Strahm, and S. N. Stuart. 2004. Coverage provided by

the global protected-area system: Is it enough? *BioScience* 54:1081–1091. [14]

Brosnan, D. M. 2000. Can peer review help resolve natural resource conflicts? *Issues Sci. Technol.* XVI:32–36. [17]

Brothers, T. S. and A. Springarn. 1992. Forest fragmentation and alien plant invasion of central Indiana old-growth forests. *Conserv. Biol.* 6:91–100. [7]

Brower, L. P. 1995. Understanding and misunderstanding the migration of the monarch butterfly (Nymphalidae) in North America: 1857–1995. *J. Lepidopterists' Soc.* 49:304–385. [3]

Brower, L. P., and S. B. Malcolm. 1991. Animal migrations: Endangered phenomena. *Am. Zool.* 31:265–276. [3]

Brower, L. P., G. Castilleja, A. Peralta, J. Lopez-Garcia, L. Bojorquez-Tapia, S. Diaz, D. Melgarejo, and M. Missrie. 2002. Quantitative Changes in Forest Quality in a Principal Overwintering Area for Monarch Butterfly in Mexico 1971–1999. *Conserv. Biol.* 16:346–359. [3]

Brown, B. E. 1997. Coral bleaching: causes and consequences. *Coral Reefs* 16:S129–S138. [13]

Brown, D. 2003. *Bushmeat and poverty alleviation: implications for development policy*. ODI Wildlife Policy Briefing Number 2, November. [8]

Brown, J. H. 1971. Mammals on mountaintops: Non-equilibrium insular biogeography. *Am. Nat.* 105: 467–478. [7]

Brown, J. H. and A. Kodric-Brown. 1977. Turnover rates in insular biogeography: Effect of immigration on extinction. *Ecology* 58:445–449. [12]

Brown, J. H., T. J. Valone, and C. G. Curtin. 1997. Reorganization of an arid ecosystem in response to recent climate change. *Proc. Nat. Acad. Sci. USA* 94:9729–9733. [10]

Brown, J. L., S. H. Li, and N. Bhagabati. 1999. Long-term trend toward earlier breeding in an American bird: A response to global warming? *Proc. Nat. Acad. Sci. USA* 96:5565–5569. [10]

Brown, J. L., S. K. Wasser, D. E. Wildt, L. F. Graham, and S. L. Monfort. 1997. Fecal steroid analysis for monitoring ovarian and testicular function in diverse wild carnivore, primate and ungulate species. *Int. J. Mamm. Biol.* (Suppl. 2):27–31. [11]

Brown, K. A. and J. Gurevitch. 2004. Long-term impacts of logging on forest diversity in Madagascar. *Proc. Nat. Acad. Sci. USA* 101:6045–6049. [3]

Brown, K. H., M. Bailkey, A. Meares-Cohen, J. Nasr, J. Smit, and T. Buchanan. 2002. Urban Agriculture and Community Food Security in the United States: Farming from the City Center to the Urban Fringe. http://www.foodsecurity.org/urbanag.html#intro. [18]

Brown, S. and A. E. Lugo. 1994. Rehabilitation of tropical lands: a key to sustaining development. *Restor. Ecol.* 2:97–111. [15]

Brown, V. 2003. *Causes for Concern: Chemicals and Wildlife*. World Wildlife Fund, London. [6]

Brubaker, L. B. 1988. Vegetation history and anticipating future vegetation change. In

J. K. Agee and D. R. Johnson (eds.), *Ecosystem Management for Parks and Wilderness*, pp. 42–58. University of Washington Press, Seattle. [4]

Bruner, A. G., R. E. Gullison, and A. Balmford. 2004. Financial costs and shortfalls of managing and expanding protected-area systems in developing countries. *BioScience* 54:1119–1126. [14]

Bruner, A. G., R. E. Gullison, R. E. Rice, and G. A. B. de Fonseca. 2001. Effectiveness of Parks in Protecting Tropical Biodiversity. *Science* 291:125–128. [14]

Brush, M. T., A. S. Hance, K. S. Judd, and E. A. Rettenmaier. 2000. *Recent Trends in Ecosystem Management*. School of Natural Resources, University of Michigan. [13]

Bryant, D., D. Nielsen, and L. Tangley. 1997. *The Last Frontier Forests: Ecosystems and Economies on the Edge*. World Resources Institute, Washington, D.C. [8]

Bryant, D., L. Burke, J. McManus, and M. Spalding. 1998. *Reefs at Risk: A Map-based Indicator of Threats to the World's Coral Reefs*. World Resources Institute, Washington, D.C. [6]

Bryner, G. C. 1993. *Blue Skies, Green Politics: The Clean Air Act of 1990*. Congressional Quarterly Press, Washington, D.C. [6]

Buchanan, B. 1993. Effects of enhanced lighting on the behavior of nocturnal frogs. *Anim. Behav.* 45:893–899. [6]

Bucher, E. H. 1992. The causes of extinction of the passenger pigeon. *Curr. Ornith.* 9:1–36. [8]

Buchmann, S. L. and G. P. Nabhan. 1996. *The Forgotten Pollinators*. Island Press, Washington, D.C. [4]

Buddemeier, R. W., J. A. Kleypas, and R. Aronson. 2004. *Coral Reefs and Global Climate Change: Potential Contributions of Climate Change to Stresses on Coral Reef Ecosystems*. Prepared for the Pew Center on Global Climate Change. [13]

Bull, J. J. 1980. Sex determination in reptiles. *Q. Rev. Biol.* 55:3–21. [10]

Bull, J. J. and R. C. Vogt. 1979. Temperature-dependent sex determination in turtles. *Science* 206:1186–1188. [10]

Bumpus, H. C. 1899. The elimination of the unfit as illustrated by the introduced sparrow, *Passer domesticus*. *Biol. Lec. Mar. Biol. Woods Hole* 11:209–266. [10]

Bunt, J. S. 1975. Primary productivity of marine ecosystems. In H. Leith and R. H. Whittaker (eds.), *Primary Productivity of the Biosphere*, pp. 169–215. Springer-Verlag, New York. [2]

Burger, J. and M. Gochfeld. 2002. Effects of Chemicals and Pollution on Seabirds. In E. A. Schrieber and J. Burger, (eds.), *Biology of Marine Birds*, pp 485–525. CRC Press. [6]

Burgess, R. L. and D. M. Sharpe (eds.). 1981. *Forest Island Dynamics in Man-Dominated Landscapes*. Springer-Verlag, New York. [7]

Burgman, M. A. 2002. Are listed threatened plant species actually at risk? *Aust. J. Bot.* 50:1–13. [3]

Buri, P. 1956. Gene frequencies in small populations of mutant *Drosophila*. *Evolution*. 10:367–402. [11]

Burke, D. M. and E. Nol. 1998. Influence of food abundance, nest-site habitat, and forest fragmentation on breeding Ovenbirds. *Auk* 115:96–104. [7]

Burke, L. Y. Kura, K. Kassem, C. Revenga, M. Spalding and D. McAllister. 2000. *Pilot Assessment of Global Ecosystems: Coastal Ecosystems*. World Resources Institute,Washington, D.C. [6]

Burkey, T. V. 1995. Extinction rates in archipelagoes: Implications for populations in fragmented habitats. *Conserv. Biol.* 9:527–541. [7]

Burnham, K. P., D. R. Anderson, and G. C. White. 1996. Meta-analysis of vital rates of the northern Spotted Owl. *Stud. Avian Biol.* 17:92–101. [17]

Burrough, P. A. 1986. *Principles Of Geographic Information Systems For Land Resources Assessment*. Oxford University Press, Oxford. [7]

Burrowes, P. A., R. L. Joglar, and D. E. Green. 2004. Potential causes for amphibian declines in Puerto Rico. *Herpetologica* 60:141–154. [3]

Bush, M. B. and P. A. Colinvaux. 1994. Tropical forest disturbance: paleoecological records from Darien, Panama. *Ecology* 75:1761–1768. [8]

Bush, M. B., M. R. Silman, and D. H. Urrego. 2003. 48,000 years of climate and forest change in a biodiversity hot spot. *Science* 303:827–829. [10]

Bushmeat Crisis Task Force 2004. *BCTF Phase I Report*. Washington D.C.: Bushmeat Crisis Task Force. Accessed 15 April 2005: www.bushmeat.org/cd. [8]

Bustamante, J. 1996. Population viability analysis of captive and released bearded vulture populations. *Conserv. Biol.* 10:822–831. [12]

Butchart, S. H. M., A. J. Stattersfield, L. A. Bennun, S. M. Shutes, H. R. Akçakaya, J. E. M. Baillie, S. N. Stuart, C. Hilton-Taylor, and G. M. Mace. 2004. Measuring global trends in the status of biodiversity: Red List Indices for birds. *PloS Biol.* 2:2294–2304. [3]

Buzas, M. A. and T. G. Gibson. 1969. Species diversity: Benthic foraminifera in Western North Atlantic. *Science* 163:72–75. [2]

Byers, D. L. and D. M. Waller. 1999. Do plant populations purge their genetic load? Effects of population size and mating history on inbreeding depression. *Ann. Rev. Ecol. Syst.* 30:479–513. [11]

Byers, J. E. and L. Goldwasser. 2001. Exposing the mechanism and timing of impact of non-indigenous species on native species. *Ecology* 82:1330–1343. [9]

Caicco, S. L., J. M. Scott, B. Butterfield, and B. Csuti. 1995. A gap analysis of the management status of the vegetation of Idaho (USA). *Conserv Biol.* 9:498–511. [14]

Cain, D. H, K. Riitters, and K. Orvis. 1997. A multi-scale analysis of landscape statistics. *Landscape Ecol.* 12:199–212. [7]

Caley, M. J., K. A. Buckley, and G. P. Jones. 2001. Separating ecological effects of habitat fragmentation, degradation, and loss on coral commensals. *Ecology* 82 (12): 3435–3448. [7]

California State Coastal Conservancy. 1986. *Del Sol Enhancement*. Staff Recommendation, 17 April 1986, File No. 86–020. [15]

Callaway, J. C., G. Sullivan, and J. B. Zedler. 2003. Species-rich plantings increase biomass and nitrogen accumulation in a wetland restoration experiment. *Ecol. Appl.* 13:1626–1639. [15]

Callicott, J. B. 1986. On the intrinsic value of nonhuman species. In B. G. Norton (ed.), *The Preservation of Species: The Value of Biological Diversity*, pp. 138–172. Princeton University Press, Princeton. [4]

Callicott, J. B. 1987a. The philosophical value of wildlife. In D. Decker and G. Goff (eds.), *Valuing Wildlife: Economic and Social Perspectives*, pp. 214–221. Westview Press, Boulder, CO. [4]

Callicott, J. B. 1987b. Conceptual foundations of the land ethic. In J. B. Callicott (ed.), *Companion to A Sand County Almanac. Interpretive and Critical Essays*, pp. 186–217. University of Wisconsin Press, Madison. [4]

Callicott, J. B. 1989. *In Defense of the Land Ethic: Essays in Environmental Philosophy*. State University of New York Press, Albany. [4]

Callicott, J. B. 1990. Whither conservation ethics? *Conserv. Biol.* 4:15–20. [1]

Callicott, J. B. 1990. Standards of conservation: then and now. *Conserv. Biol.* 4:229–232. [13]

Callicott, J. B. 1992. Can a theory of moral sentiments support a genuinely normative environmental ethic? *Inquiry* 35:183–198. [4]

Callicott, J. B. 1994. *Earth's Insights: A Multicultural Survey of Ecological Wisdom*. University California Press, Berkeley. [4]

Callicott, J. B. and K. Mumford. 1997. Ecological sustainability as a conservation concept. *Conserv. Biol.* 11:32–40. [12]

Callicott, J. B., J. B. Crowder, and K. Mumford. 1999. Current normative concepts in conservation. *Conserv. Biol.* 13:22–35. [12]

Camacho Rodriguez, M. J., L. Chabert Llompart, and M. López Flores. 1999. *Guia para Identificacion de las Aves Exoticas Establecidas en Puerto Rico*. Departamento de Recursos Naturales de Puerto Rico. [15]

Campbell, C. J. 1997. *The Coming Oil Crisis*. Multi-Science and Petroconsultants S. A., Brentwood, Essex, UK. [18]

Campbell, G. L., A. A. Marfin, R. S. Lanciotti, and D. J. Gubler. 2002. West Nile virus. *Lancet Infectious Diseases* 2:519–529. [3]

Campfire Collaborative Group. 1991. The CAMPFIRE Program in Zimbabwe: Information for Visitors, Journalists and Prospective Researchers. Zimbabwe Department of National Parks and Wildlife Management: Harare. [16]

Cane, M. A. 2005. The evolution of El Niño, past and future. *Earth Planet. Sc. Lett.* 230:227–240. [3]

Cannon, J. R., J. M. Dietz, and L. A. Dietz. 1996. Training conservation biologists in human interaction skills. *Conserv. Biol.* 10:1277–1282. [1]

Cardona, J. E., M. Rivera, M. Vazquez-Otero, and C. R. Laboy. 1985. Nesting cavities in Rio Abajo Forest. Puerto Rico Department of Natural Resources, Final Project Report W-10-2. [15]

Cardona, J. E., M. Rivera, M. Vázquez-Otero, and C. R. Laboy. 1986. Availability of food resources for the Puerto Rican Parrot and the Puerto Rican Plain Pigeon in Río Abajo Forest. Puerto Rico Department of Natural Resources, Draft Report Project No. W-10(ES–1). [15]

Carey, A. B., C. Elliott, B. R. Lippke, J. Sessions, C. J. Chambers, C. D. Oliver, J. F. Franklin, and M. J. Raphael. 1996. *A Pragmatic, Ecological Approach to Small-Landscape Management*. Washington Forest Landscape Management Project Final Report. Wash. For. Landscape Manage. Proj. Rep. No. 2, Wash. Dep. Nat. Resour., Olympia. [8]

Carey, C. 1993. Hypothesis concerning the causes of the disappearance of boreal toads from the mountains of Colorado. *Conserv. Biol.* 7:355–362. [3]

Carey, C., D. F. Bradford, J. L. Brunner, J. P. Collins, E. W. Davidson, J. E. Longcore, M. Ouellet, A. Pessier, and D. M. Schock. 2003. Biotic factors in amphibian population declines. In G. Linder, S. Krest, and D. Sparling (eds.), *Amphibian Decline: An Integrated Analysis of Multiple Stressor Effects*, pp. 153–208. Society for Environmental Toxicology and Chemistry Press, Pensacola, FL. [3]

Carlo, T. A., J. A. Collazo, and M. J. Groom. 2004. Influence of fruit diversity and abundance on bird use of two shaded coffee plantations. *Biotropica* 36:602–614. [16]

Carlsen, T. M., J. D. Coty, and J. R. Kercher. 2004. The spatial extent of contaminants and the landscape scale: An analysis of the wildlife, conservation biology, and population modeling literature. *Environ. Toxicol. Chem.* 23:798–811. [12]

Carlson, P. J. 1975. Effects of soil drainage on early growth and soil nitrogen accretion of *Alnus jorulensis* in the Andean Highlands of Colombia. M.Sc. Thesis. University of Illinois, Urbana-Champaign, IL. [15]

Carlton, J. T. 1985. Transoceanic and interoceanic dispersal of coastal marine organisms: The biology of ballast water. *Oceanogr. Mar. Biol. Ann. Rev.* 23:13–371. [9]

Carlton, J. T. and J. B. Geller. 1993. Ecological roulette: The global transport of non-indigenous marine organisms. *Science* 261: 78–82. [14]

Carlton, J. T. and J. Hodder. 1995. Biogeography of coastal marine organisms: Experimental studies on a replica of a sixteenth century sailing vessel. *Mar. Biol.* 121:721–730. [9]

Carpenter, F. L., J. D. Nichols, R. T. Pratt, and K. C. Young. 2004. Methods of facilitating reforestation of tropical degraded land with the native timber tree, *Terminalia amazonia. Forest Ecol. and Manag.* 202:281–291. [15]

Carpenter, J. W., M. J. G. Appel, R. C. Erickson, and M. N. Novilla. 1976. Fatal vaccine-induced canine distemper virus infection in black-footed ferrets. *J. Am. Vet. Med Assoc.* 169:961–964. [15]

Carpenter, S. R. and J. F. Kitchell (eds.). 1993. *The Trophic Cascade in Lakes.* Cambridge University Press, Cambridge. [13]

Carpenter, S. R., J. F. Kitchell, and J. R. Hodgson. 1985. Cascading trophic interactions and lake productivity. *BioScience* 35:634–639. [3]

Carpenter, S. R., J. J. Cole, J. R. Hodgson, J. F. Kitchell, M. L. Pace, D. Bade, K. L. Cottingham, T. E. Essington, J. N. Houser, and D. E. Schindler. 2001. Trophic cascades, nutrients, and lake productivity: whole-lake experiments. *Ecological Monographs* 71:163–186. [3]

Carroll, C. R. 1990. The interface between natural areas and agroecosystems. In C. R. Carroll, J. H. Vandermeer, and P. M. Rosset (eds.), *Agroecology.* Biological Resource Management Series. McGraw-Hill Company, New York. [1]

Carroll, C. R. and C. A. Hoffman 2000. *The Pervasive Ecological Effects of Invasive Species: Exotic and Native Fire Ants.* In D. C. Coleman and P. Hendrix (eds), *Invertebrates as Webmasters in Ecosystems,* pp. 221–232. CABI Publ, Div. Of CAB International, Oxom, UK. [13]

Carroll, C., R. F. Noss, and P. C. Paquet. 2001. Carnivores as focal species for conservation planning in the Rocky Mountain region. *Ecol. Appl.* 11:961–980. [7, 12]

Carroll, C., M. K. Phillips, N. H. Schumaker, and D. W. Smith. 2003. Impacts of landscape change on wolf restoration success: Planning a reintroduction program based on static and dynamic spatial models. *Conserv. Biol.* 17:536–548. [12]

Carroll, C., R. F. Noss, P. C. Paquet, and N. H. Schumaker. 2004. Extinction debt of protected areas in developing landscapes. *Conserv. Biol.* 18:1110–1120. [7]

Carroll, C. R., C. Augspurger, A. Dobson, J. Franklin, G. Orians, W. V. Reid, C. R. Tracy, D. Wilcove, and J. Wilson. 1996. Science and reauthorization of the Endangered Species Act: Report of the Ecological Society of America. *Ecology* 60:1–11. [18]

Carroll, C. R., C. Augspurger, A. Dobson, J. Franklin, G. Orians, W. V. Reid, C. R. Tracy, D. Wilcove, and J. Wilson. 1996. Strengthening the use of science in achieving the goals of the Endangered Species Act: an assessment by the Ecological Society of America. *Ecol. Appl.* 6:1–11. [12]

Casey, J. M. and R. A. Myers. 1998. Near extinction of a large, widely distributed fish. *Science.* 281:690–692. [8]

Cash, D., W. Clark, F. Alcock, N. Dickson, N. Eckley, and J. Jager. 2002. *Salience, Credibility, Legitimacy, and Boundaries: Linking Research, Scientific Assessment and Decision-Making.* John F. Kennedy School of Government, Harvard University, Faculty Working Papers Series, RWP02–046. [17]

Cash, D. W., W. C. Clark, F. Alcock., N. M. Dickson, N. Eckley, D. H. Guston, J. Jäger, and R. B. Mitchell. 2003. Knowledge systems for sustainable devlopment. *P. Natl. Acad. Sci. USA* 100:8086–8091. [17, 18]

Cassman, K. G., A. Dobermann, D. T. Walters, and H. Yang. 2003. Meeting cereal demand while protecting natural resources and improving environmental quality. *Annu. Rev. Env. Resour.* 28:315–58. [18]

Castells, M., J. Cruz, E. Custodio, F. García-Novo, J. P. Gaudemar, J. L. González Valvé, V. Granados, A. Magraner, C. Román, M. Smart, and E. Van der Maarel. 1992. Dictámen sobre estrategias para el desarrollo socioeconómico sostenible del entorno de Doñana. Comisión Internacional de Expertos sobre el Desarrollo del Entorno de Doñana. Junta de Andalucía. Sevilla. [14]

Castilla, J. C. and M. Fernandez. 1998. Small-scale benthic fisheries in Chile: On co-management and sustainable use of benthic invertebrates. *Ecol. Appl.* 8:S124–S132. [8, 18]

Castro, A. P. and E. Nielsen. 2001. Indigenous people and co-management: Implications for conflict management. *Environ. Sci. Policy* 4:229–239. [8]

Castroviejo, J. 1993. Memoria. Mapa del Parque Nacional de Doñana. Consejo Superior de Investigaciones Científicas. Agencia de Medio Ambiente, Junta de Andalucía. [14]

Caswell, H. 2001. *Matrix Population Models,* 2nd Edition. Sinauer Associates, Sunderland, MA. [8, 12]

Caswell, H., S. Brault, A. Read, and T. Smith. 1998. Harbor porpoise and fisheries: An uncertainty analysis of incidental mortality. *Ecol. Appl.* 8:1226–1238. [12]

Caughley, G. and A. Gunn. 1996. *Conservation Biology in Theory and Practice.* Blackwell Scientific, Cambridge, MA. [3, 12, 18]

Cavelier, J. and A. Tobler. 1998. The effect of abandoned plantations of *Pinus patula* and *Cupressus lusitanica* on soils and regeneration of a tropical montane rain forest in Colombia. *Biodivers. and Conserv.* 7: 335–347. [15]

Cayetano, P. 2003. Indigenous land rights, development and environment: a Garifuna perspective. Paper submitted at Indigenous Rights in the Commonwealth Caribbean and Americas Regional Expert Meeting Amerindian Peoples' Association (APA), Guyana, June 2003. National Garifuna Council (NGC), Belize. [16]

Cederholm, C. J., M. D. Kunze, T. Murota, and A. Sibatani. 1999. Pacific salmon carcasses: Essential contributions of nutrients and energy for aquatic and terrestrial ecosystems. *Fisheries* 24:6–15. [8]

Cesar, H. 1997. *Economic Analysis of Indonesian Coral Reefs.* Environment Department, World Bank, Washington, D.C. [8]

Chalfoun, A. D., F. R. Thompson III, and M. J. Ratnaswamy. 2002. Nest predators and fragmentation: A review and meta-analysis. *Conserv. Biol.* 16:306–318. [7]

Chandrashekara, U. M. and S. Sankar. 1998. Ecology and management of sacred groves in Kerala, India. *Forest Ecol. Manag.* 112:165–177. [14]

Chape, S., S. Blyth, L. Fish, P. Fox, and M. Spalding. 2003. 2003 *United Nations List of Protected Areas. IUCN and UNEP-WCMC,* Gland, Switzerland and Cambridge, UK. [14]

Chapin, F. S. III, G. R. Shaver, A. E. Giblin, K. G. Nadelhoffer, and J. A. Laundre. 1995. Response of arctic tundra to experimental and observed changes in climate. *Ecology* 76:694–711. [10]

Chapin, F. S. III, B. H. Walker, R. J. Hobbs, D. U. Hooper, J. H. Lawton, O. E. Sala, and D. Tilman. 1997. Biotic control over the functioning of ecosystems. *Science* 277:500–504. [2]

Chapin, F. S., E. S. Zavaleta, V. T. Eviner, R. L. Naylor, P. M. Vitousek, H. Reynolds, D. U. Hooper, S. Lavorel, O. E. Sala, S. Hobbie et. al. 2000. *Nature* 405:234–242. [18]

Chapman, C.A. and D. A. Onderdonk. 1998. Forests without primates: Primate/plant codependency. *Am. J. Primatol.* 45:127–141. [3, 8]

Chapman, C. A. and L. J. Chapman. 1996. Exotic tree plantations and the regeneration of natural forests in Kibale National Park, Uganda. *Biol. Conserv.* 76: 253–257. [15]

Chapman, C. A. and L. J. Chapman. 1996. Frugivory and the fate of dispersed and non-dispersed seeds in six African tree species. *J. Trop. Ecol.* 12:491–504. [8]

Chappel, C. 2002. *Jainism and Ecology.* Harvard University Press, Cambridge, MA. [4]

Chappel, C. and M. E. Tucker. 2000. *Hinduism and Ecology.* Harvard University Press, Cambridge, MA. [4]

Chappel, C. K. 1986. Contemporary Jaina and Hindu responses to the environment. In R. C. Foltz (ed.), *Worldviews, Religion, and the Environment,* pp. 113–119. Wadsworth, Toronto. [4]

Chen, J. and J. F. Franklin. 1990. Microclimatic pattern and basic biological responses at the clearcut edges of old-growth Douglas fir stands. *Northwest Environ. J.* 6:424–425. [7]

Chen, J., J. F. Franklin, and T. A. Spies. 1992. Vegetation responses to edge environments in old-growth Douglas fir forests. *Ecol. Applic.* 2:387–396. [7]

Chen, X. 1993. Comparison of inbreeding and outbreeding in hermaphroditic *Arianta arbustorum* (L.) (land snail). *Heredity* 71:456–461. [11]

Chesapeake Bay Foundation. 2004. *Manure's Impact on Rivers, Streams and the Chesapeake Bay.* A Report by the Chesapeake Bay Foundation (CBF), July 28, 2004. [13]

Chesser, R. K. 1991a. Gene diversity and female philopatry. *Genetics.* 127:437–447. [11]

Chesser, R. K. 1991b. Influence of gene flow and breeding tactics on gene diversity within populations. *Genetics* 129:573–583. [11]

Chesser, R. K., and R. J. Baker. 1996. Effective sizes and dynamics of uniparentally and

biparentally inherited genes. *Genetics* 144:1225–1235. [11]

Chesser, R. K., O. E. Rhodes, Jr., and M. H. Smith. 1996. Gene conservation. In O. E. Rhodes, Jr., R. K. Chesser, and M. H. Smith (eds.), *Population Dynamics in Ecological Space and Time*, pp. 237–252. University of Chicago Press, Chicago. [11]

Chesser, R. K., O. E. Rhodes, Jr., D. W. Sugg, and A. Schnabel. 1993. Effective sizes for subdivided populations. *Genetics* 135:1221–1232. [11]

Chittka, L. and S. Schürkens. 2001. Successful invasion of a floral market. *Nature* 411:653–653. [9]

Chomitz, K. M. and K. Kumari. 1998. The domestic benefits of tropical forests: a critical review. *World Bank Research Observer* 13:13–35. [14]

Chown, S. L., A. Addo-Bediako, and K. J. Gaston. 2002. Physiological variation in insects: Large-scale patterns and their implications. *Comp. Biochem. Phys. B* 131:587–602. [10]

Christensen, J. 2003. Auditing conservation in an age of accountability. *Conserv. Biol. in Practice* 4:12–19. [18]

Christensen, N. L., A. Bartuska, J. Brown, S. Carpenter, C. D'Antonio, R. Francis, J. Franklin, J. MacMahon, R. Noss, D. Parsons, C. Peterson, M. Turner, and R. Woodmansee. 1996. The report of the Ecological Society of America Committee on the scientific basis for ecosystem management. *Ecol. Appl.* 6:665–691. [13]

Chu, E. W. and J. R. Karr. 2001. Environmental impact, Concept, and Measurement of. In S. A. Levin (ed.), *Encyclopedia of Biodiversity*, Vol. 2, pp. 557–577. Academic Press, Orlando, FL.

Church, R. L., D. M. Stoms, and F. W. Davis. 1996. Reserve selection as a maximal covering location problem. *Biol. Conserv.* 76:105–112. [14]

Clark, C. W. 1973. Profit maximization and the extinction of animal species. *J. Pol. Econ.* 81:950–961. [4]

Clark, C.W. 1976. *Mathematical Bioeconomics: The Optimal Management Of Renewable Resources.* Wiley, New York. [8]

Clark, J. A. and E. Harvey. 2002. Assessing multi-species recovery plans under the Endangered Species Act. *Ecol. Appl.* 12:655–662. [18]

Clark, J. A., J. M. Hoekstra, P. D. Boersma, and P. Kareiva. 2002. Improving U.S. Endangered Species Act Recovery Plans: Key findings and recommendations of the SEB Recovery Plan Project. *Conserv. Biol.* 16:1510–1519. [18]

Clark, J. G. 1995. Economic development vs. sustainable societies: Reflections on the players in a crucial contest. *Ann. Rev. Ecol. Syst.* 26:225–248. [16]

Clark, J. S., C. Fastie, G. Hurtt, S. T. Jackson, C. Johnson, G. A. King, M. Lewis, J. Lynch, S. Pacala, C. Prentice, E. W. Schupp, T. Webb III, and P. Wyckoff. 1998. Reid's paradox of rapid plant migration. *BioScience* 48:13–24. [7]

Clark, K. L. and N. M. Nadkarni. 2000. Microclimate variability. In N. M. Nadkarni and N. T. Wheelwright (eds.), *Mon-*

teverde: Ecology and Conservation of a Tropical Cloud Forest*, pp. 33–34. Oxford University Press, New York. [10]

Clark, K. L., R. O. Lawton, and P. R. Butler. 2000. The physical environment. In N. M. Nadkarni and N. T. Wheelwright (eds.), *Monteverde: Ecology and Conservation of a Tropical Cloud Forest*, pp. 15–38. Oxford University Press, New York. [10]

Clark, T. W. 1996. Learning as a strategy for improving endangered species conservation. *Endangered Species Update* 13 (1 and 2):5–6 and 22–24. [15]

Clark, T. W. 1997. *Averting Extinction: Restructuring the Endangered Species Recovery Process.* Yale University Press, New Haven, CT. [15]

Clark, T. W. 1997. Conservation biologists in the policy process: Learning how to be practical and effective. In G. K. Meffe and C. R. Carroll (eds.), *Principles of Conservation Biology*, 2nd Edition., pp. 575–597. Sinauer Associates, Sunderland, MA. [17]

Clark, T. W. 1989. Conservation biology of the black-footed ferret, *Mustela nigripes.* Wildlife Preservation Trust Special Report no. 3, Philadelphia, Pennsylvania. [15]

Clark, T. W. 2001. Developing policy-oriented curricula in conservation biology: Professional and leadership education for the public interest. *Conserv. Biol.* 15:31–39. [17]

Clarke, A. and C. M. Harris. 2003. Polar marine ecosystems: Major threats and future change. *Environ. Conserv.* 30:1–25. [3]

Clausen, J., D. D. Keck, and W. M. Heisey. 1940. *Experimental Studies on the Nature of Species. I. Effects of Varied Environments on Western North American Plants.* Publication 520, Carnegie Institution of Washington, Washington, D.C. [2]

Clawson, M. 1959. *Methods of Measuring the Demand and Value of Outdoors Recreation.* Resources for the Future, Washington D.C. [5]

Clayton, L., M. Keeling, and E. J. Milner-Gulland. 1997. Bringing home the bacon: A spatial model of wild pig harvesting in Sulawesi, Indonesia. *Ecol. Appl.* 7:642–652. [8]

Cleary, R. 2001. Toward an environmental history of the Amazon: from prehistory to the nineteenth century. *Lat. Am. Res. Rev.* 36:65–96. [6]

Clebsch, E. E. C. and R. T. Busing. 1989. Secondary succession, gap dynamics, and community structure in a southern Appalachian cove forest. *Ecology* 70:728–735. [7]

Clemen, R. T. and T. Reilly. 2001. *Making Hard Decisions with Decision Tools*, 2nd Edition. Duxbury Press, Pacific Grove, CA. [17]

Clemmons, J. R. and R. Buchholz. 1997. *Behavioral Approaches to Conservation in the Wild.* Cambridge University Press, New York. [11]

Cobb, C., T. Halstead, and J. Rowe. 1995. If the GDP is up, why is America down? *Atlantic Monthly*, October 1995. [18]

Cobb, K. M., C. D. Charles, H. Cheng, and R. L. Edwards. 2003. El Niño /Southern Oscillation and tropical Pacific climate during the last millennium. *Nature* 424:271–276. [3]

Cobo, D., E. Sánchez, and P. García-Murillo. 2002. Flora y Vegetación. In Canseco Editores, pp. 109–174. *Parque Nacional de Doñana.* Madrid. [14]

Cochrane, M. A. 2001. In the line of fire: Understanding the impacts of tropical forest fires. *Environment* 43:28–38. [8]

Cocks, K. D. and I. A. Baird. 1989. Using mathematical programming to address the multiple reserve selection problem: and example from the Eyre Peninsula, South Australia. *Biol. Conserv.* 49:113–130. [14]

Cody, M. L. 1975. Towards a theory of continental species diversity: bird distributions on mediterranean habitat gradients. In M. L. Cody and J. M. Diamond (eds.), *Ecology and Evolution of Communities*, pp. 214–257. Harvard University Press, Cambridge. [2]

Cody, M. L. 1986. Structural niches in plant communities. In J. Diamond and T. J. Case (eds.), *Community Ecology*, pp. 381–405. Harper & Row, New York. [2]

Cody, M. L. 1993. Bird diversity within and between habitats in Australia. In R. Ricklefs and D. Schluter (eds.), *Species Diversity In Ecological Communities: Historical And Geographical Perspectives*, pp. 147–158. University of Chicago Press, Chicago. [2]

Cody, M., S. Courtney, and D. Bigger 1999. *Murrelets in the mist: Conservation biology and the "Headwaters" HCP.* Symposium on Science and Policy, Society for Conservation Biology Annual Meeting University of Maryland, Maryland USA. [17]

Coffroth, M. A., H. R. Lasker, and J. K. Oliver. 1990. Coral mortality outside of the eastern pacific during 1982–1983: Relationship to el Niño. In P. W. Glynn (ed.), *Global Ecological Consequences of the 1982–1983 El Niño-Southern Oscillation*, pp. 141–182. Elsevier, Amsterdam. [10]

Coghlan, G. and R. Sowa. 1998. *National Forest Road System And Uses.* USDA Forest Service, Engineering Staff, Washington, D.C. [7]

Coglianese, C. 1997. Assessing consensus: The promise and performance of negotiated rulemaking. *Duke L.J.* 46:1255–1350. [17]

Cohen, A. N. and Carlton J. T. 1998. Accelerating invasion rate in a highly invaded estuary. *Science* 279:55–558. [9]

Colborn, T. and C. Clement. 1992. *Chemically-Induced Alterations In Sexual and Functional Development: The Wildlife/Human Connection.* Princeton Sci. Publ. Co. Inc., Princeton. [6]

Colborn, T., F. S. vom Saal, and A. M. Soto. 1993. Developmental effects of endocrine-disrupting chemicals in wildlife and humans. *Environ. Health Persp.* 101:378–384. [6]

Coles, S. L. and B. E. Brown. 2003. Coral bleaching-capacity for acclimatization

and adaptation. *Adv. Mar. Biol.* 46:183–223. [13]

Coley, P. D., J. P. Bryant, and F. S. Chapin III. 1985. Resource availability and plant anti-herbivore defense. *Science* 230:895–899. [7]

Collazo, J. A. and M. J. Groom. 2000. Avian conservation in north-central forested habitats in Puerto Rico. Final report, Puerto Rico Dept. of Natural and Environmental Resources, Federal Aid Project W–20, San Juan, PR. [15]

Collazo, J. A., F. J. Vilella, T. H. White, and S. A. Guerrero. 2000. Survival, use of habitat, and movements of captive-reared Hispaniolan Parrots released in historical, occupied habitat: Implications for the recovery of the Puerto Rican Parrot. Final Report. North Carolina Cooperative Fish and Wildlife Research Unit. Raleigh, NC. [15]

Collazo, J. A., T. H. White Jr., F. J. Vilella, and S. A. Guerrero. 2003. Survival of captive-reared Hispaniolan parrots released in Parque Nacional del Este, Dominican Republic. *Condor* 105:198–207. [15]

Collingham, Y. C. and B. Huntley. 2000. Impacts of habitat fragmentation and patch size upon migration rates. *Ecol. Appl.* 10:131–144. [7]

Colorado Department of Agriculture. 2000. *Tracking Agricultural Land Conversion in Colorado.* Colorado Department of Agriculture, Denver. [7]

Colwell, R. K. and G. C. Hurtt. 1994. Nonbiological gradients in species richness and a spurious Rapoport effect. *Am. Nat.* 144:570–595. [2]

Colwell, R. K. and J. A. Coddington, 1994. Estimating terrestrial biodiversity through extrapolation. *Phil. Trans. Roy. Soc. London B* 345:101–118. [2]

Comisión Nacional de Certificación Forestal. 1999. Estándares y procedimientos para el manejo sostenible y la certificación forestal en Costa Rica. Programa de las Naciones Unidas para el Desarrollo Humano Sostenible y Unidad de Manejo de Bosques Naturales, CATIE, Costa Rica. [8]

Commission for the Conservation of Antarctic Marine Living Resources. 2004. Basic Documents, December 2004. Convention for the Conservation of Antarctic Marine Living Resources, North Hobart, Tasmania. [13]

Comstock, K. E., E. A. Ostrander, and S. K. Wasser. 2003. Amplifying nuclear and mitochondrial DNA from African elephant ivory: A tool for monitoring the ivory trade. *Conserv. Biol.* 17:1–4. [11]

Comstock, K. E., S. K. Wasser, and E. A. Ostrander. 2000. Polymorphic microsatellite DNA loci identified in the African elephant (*Loxodona africana*). *Mol. Ecol.* 9:1004–1009. [11]

Comstock, K. E., N. Georgiadis, J. Pecon-Slattery, A. L. Rocca, S. J. O'Brien, and S. K Wasser. 2002. Patterns of molecular genetic variation among African elephant populations. *Mol. Ecol.* 11:2489–2498. [11]

Connell, J. H. 1975. Some mechanisms producing structure in natural communities.

In Cody, M. L. and J. M. Diamond (eds.), *Ecology and Evolution of Communities*, pp. 460–490. Belknap Press, Cambridge, MA. [2]

Connell, J. H. 1978. Diversity in tropical rains forests and coral reefs. *Science* 199:1302–1310. [2]

Connell, J. H. 1983. On the prevalence and relative importance of interspecific competition: Evidence from field experiments. *Am. Nat.* 122:661–696. [2]

Connell, J. H. and E. Orias. 1964. The ecological regulation of species diversity. *Am. Nat.* 111:1119–1144. [2]

Conner, R. N. 1988. Wildlife populations: Minimally viable or ecologically functional? *Wildlife Soc. B.* 16:80–84. [12]

Conservation Monitoring Partnership. 2003. Open standards for the practice of conservation. Conservation Monitoring Partnership, Washington, D.C. [18]

Convention on Biological Diversity. 2000. Decisions adopted by the Conference of the Parties to the Convention on Biological Diversity at its 5th meeting, Nairobi. [13]

Cook, W. M., D. G. Casagrande, D. Hope, P. M. Groffman, and S. L. Collins. 2004. Learning to roll with the punches: Adaptive experimentation in human-dominated systems. *Front. Ecol. Environ.* 2:467–474. [13]

Cooney, R. 2004. *The Precautionary Principle in Biodiversity Conservation and Natural Resource Management: An Issues Paper for Policy-makers, Researchers and Practitioners.* IUCN, Gland, Switzerland and Cambridge. [1]

Coope, G. R. 1995. Insect faunas in ice age environments: Why so little extinction? In J. H. Lawton and R. M. May (eds.), *Extinction Rates*, pp. 55–74. Oxford University Press, Oxford. [10]

Cooper, S. R. 1995. Chesapeake Bay watershed historical land use: impacts on water quality and diatom communities. *Ecol. Appl.* 703–723. [13]

Coote T. and É. Loève. 2003. From 61 species to five: Endemic tree snails of the Society Islands fall prey to an ill-judged biological control programme. *Oryx* 37:91–96. [9]

Coppolillo, P., H. Gomez, F. Maisels, and R. Wallace. 2004. Selection criteria for suites of landscape species as a basis for site-based conservation. *Biol. Conserv.* 115:419–430. [12]

Corn, P. S. 1998. Effects of ultraviolet radiation on boreal toads in Colorado. *Ecol. Appl.* 8:18–26. [3]

Corn, P. S. 2003. Amphibian breeding and climate change: Importance of snow in the mountains. *Conserv. Biol.* 17:622–625. [3]

Corn, P. S. and E. Muths. 2002. Variable breeding phenology affects the exposure of amphibian embryos to ultraviolet radiation. *Ecology* 83:2958–2963. [3]

Corn, P. S. and E. Muths. 2004. Variable breeding phenology affects the exposure of amphibian embryos to ultraviolet radiation: Reply. *Ecology* 85:1759–1763. [3]

Corn, P. S. and F. A. Vertucci. 1992. Descriptive risk assessment of the effects of

acidic deposition on Rocky Mountain amphibians. *J. Herpetol.* 26:361–369. [3]

Corn, P. S., and J. C. Fogleman. 1984. Extinction of montane populations of the northern leopard frog (*Rana pipiens*) in Colorado. *J. Herpetol.* 18:147–152. [3]

Cornelius, C., H. Cofré, and P. A. Marquet. 2000. Effects of habitat fragmentation on bird species in a relict temperate forest in semiarid Chile. *Conserv. Biol.* 14:534–543. [7]

Cornell, H. V. 1985. Species assemblages of cynipid gall wasps are not saturated. *Am. Nat.* 126:565–569. [2]

Cornell, H. V. 1999. Unsaturation and regional influences on species richness in ecological communities: A review of the evidence. *Ecoscience* 6:303–315. [2]

Cornuet, J. M., S. Piry, G. Luikart, A. Estoup, M. Solignac. 1999. Comparison of methods employing multilocus genotypes to select or exclude populations as origins of individuals. *Genetics* 153:1989–2000. [11]

Cortner, H. J. and M. A. Moote. 1999. *The Politics of Ecosystem Management.* Island Press, Washington, D.C. [13]

COSEWIC. 2003. *Cosewic Assessment and Update Status Report on the Atlantic cod Gadus morhua, Newfoundland and Labrador Population, Laurentian North Population, Maritimes Population, Arctic population, in Canada.* Committee on the Status of Endangered Wildlife in Canada. Ottawa, Canada. [8]

Costanza, R. and C. Perrings. 1990. A flexible assurance bonding system for improved environmental management. *Ecol. Econ.* 2:57–76. [5]

Costanza, R. and H. E. Daly. 1992. Natural capital and sustainable development *Conserv. Biol.* 6:37–46. [5]

Costanza, R., S. C. Farber, and J. Maxwell. 1989. The valuation and management of wetland ecosystems. *Ecol. Econ.* 1:335–362. [5]

Costanza, R., J. Cumberland, H. Daly, R. Goodland, and R. Norgaard. 1997b. *An Introduction to Ecological Economics.* St. Lucie Press, Boca Raton, FL. [18]

Costanza, R., R. d'Arge, R. de Groot, S. Farber, M. Grasso, B. Hannon, K. Limburg, S. Naeem, R. V. O'Neill, J. Paruelo, R. G. Raskin, P. Sutton, M. van den Belt. 1997. The value of the world's ecosystem services and natural capital. *Nature* 387:253–260. [2, 4, 5, 8, 18]

Costanza, R., F. Andrade, P. Antunes, J. van den Belt, D. Boersma, D. Boesch, F. Catarino, S. Hanna, K. Lomburg, B. Low, M. Molitor, J. Pereira, S. Rayner, R. Santos, J. Wilson, and M. Young. 1998. Principles for sustainable governance of the oceans. *Science* 281:198–199. [12]

Cotrufo, M. F., P. Ineson, and A. Scott. 1998. Elevated CO_2 reduces the nitrogen concentration of plant tissues. *Glob. Change Biol.* 4:43–54. [10]

Coughenour, M. B. and F. J. Singer. 1996. Elk population processes in Yellowstone National Park under the policy of natural regulation. *Ecol. Appl.* 6:573–593. [12]

Coulson, T., G. M. Mace, E. Hudson, and H. Possingham. 2001. The use and abuse of population viability analysis. *Trends Ecol. Evol.* 16:219–221. [12]

Courtney S. P. and W. Hudson 2001 *Reducing Risk and Uncertainty*. Sustainable Ecosystems Institute, Portland, OR. [17]

Cowie, R. H. 1998. Patterns of introduction of non-indigenous non-marine snails and slugs in the Hawaiian Islands. *Biodivers. Conserv.* 7:349–368. [9]

Cowling, R. M. (ed.). 1992. *The Ecology of Fynbos*. Oxford University Press, Cape Town. [2]

Cowling, R. M. and C. Hilton-Taylor. 1994. Plant diversity and endemism in southern Africa: an overview. In B. J. Huntley, (ed.), *Botanical Diversity in Southern Africa*, pp. 31–52. National Botanical Institute, Kirstenbosch. [14]

Cowling, R. M. and R. L. Pressey. 2003. Introduction to systematic conservation planning in the Cape Floristic Region. *Biol. Conserv.* 112:1–13. [14]

Cowling, R. M., P. M. Holmes, and A. G. Rebelo. 1992. Plant diversity and endemism. In R. M. Cowling (ed.), *The Ecology of Fynbos*, pp. 623–2112. Oxford University Press, Cape Town. [2]

Cowling, R. M., R. L. Pressey, M. Rouget, and A. T. Lombard. 2003a. A conservation plan for a global biodiversity hotspot–the Cape Floristic Region, South Africa. *Biol. Conserv.* 112:191–216. [14]

Cowling, R. M., P. W. Rundel, B. B. Lamont, M. K. Arroyo, and M. Arianoutsou. 1996. Plant diversity in Mediterranean-climate regions. *Trends Ecol. Evol.* 11:362–366. [14]

Cowling, R. M., R. L. Pressey, R. Sims-Castley, E. Baard, C. J. Burgers, A. le Roux, and G. Palmer. 2003b. The expert or the algorithm?—comparison of priority conservation areas in the Cape Floristic Region identified by park managers and reserve selection software. *Biol. Conserv.* 112:147–167. [14]

Cowlishaw, G. 1999. Predicting the pattern of decline of African primate diversity: An extinction debt from historical deforestation. *Conserv. Biol.* 13:1183–1193. [7]

Cowlishaw, G., S. Mendelson, and J. M. Rowcliffe. 2005. Structure and operation of a bushmeat commodity chain in western Ghana. *Conserv. Biol.* 19:139–149. [8]

Cowx, I. G. 1998. *Stocking and Introduction of Fish*. Fishing News Books, Oxford. [8]

Cowx, I. G. 2002. Recreational fishing. In P. J. B. Hart and J. D. Reynolds (eds.), *Handbook of Fish Biology and Fisheries: Volume 2, Fisheries*, pp. 367–390. Blackwell Publishing, Oxford. [8]

Cox, C. B. and P. D. Moore. 2000. *Biogeography: An Ecological and Evolutionary Approach*, 6th Edition. Blackwell Science, Malden, MA. [10]

Coxson, D. S. and N. M. Nadkarni. 1995. Ecological roles of epiphytes in nutrient cycles of forest ecosystems. In M. D. Lowman and N. M. Nadkarni (eds.), *Forest Canopies* pp. 495–546. Academic Press, New York. [10]

Craighead, F. C. 1971. *The Trees of South Florida*. University of Miami Press, Coral Gables, FL. [13]

Crain, D. A., A. A. Rooney, E. F. Orlando, and L. J. Guillette, Jr. 2000. Endocrine-disrupting contaminants and hormone dynamics: lessons from wildlife. In L. J. Guillette, Jr. and D. A. Crain (eds.), *Endocrine Disrupting Contaminants: An Evolutionary Perspective*, pp. 1–21. Francis and Taylor Inc., Philadelphia. [6]

Crain, D. A., L. J., Guillette Jr., A. A. Rooney, and D. B. Pickford. 1997. Alteration in steroidogenesis in alligators (*Alligator mississippiensis*) exposed naturally and experimentally to environmental contaminants. *Environ. Health Persp.* 105:528–533. [6]

Crance, C. and D. Draper. 1996. Socially cooperative choices: an approach to achieving resource sustainability in the coastal zone. *Environ. Manage.* 20:175–184. [13]

Crawley, M. 1990. Plant life-history and the success of weed biological control projects. In E. Delfosse (ed.), *Proceedings VII International Symposium on Biological Control of Weeds*, pp. 17–26. Inst. Sper. Patol. Veg., Rome. [9]

Creel, S., J. E. Fox, A. Hardy, J. Sands, R. Garrott, and R. Peterson. 2002. Snowmobile activity and glucocorticoid stress responses in wolves and elk. *Conserv. Biol.* 16:809–814. [11]

Crick, H. Q. P., C. Dudley, D. E. Glue, and D. L. Thompson. 1997. U.K. birds are laying eggs earlier. *Nature* 388:526. [10]

Cromartie, J. 1994. *Recent Demographic And Economic Changes in the West*. Economic Research Service, U. S. Department of Agriculture, Washington, D.C. [7]

Crooks, K. R. and M. E. Soulé. 1999. Mesopredator release and avifaunal extinctions in a fragmented system. *Nature* 400:563–566. [3, 7]

Cropper, M. L. and W. E. Oates. 1992. Environmental economics: A survey. *J. Econ. Lit.* 30:675–740. [5]

Crouse, D. T., L. B. Crowder, and H. Caswell. 1987. A stage-based population model for loggerhead sea turtles and implications for conservation. *Ecology* 68:1412–1423. [8, 12]

Crouse, D. T., L. A. Mehrhoff, M. J. Parkin, D. R. Elam, and L. Y. Chen. 2002. Endangered species recovery and the SCB study: A U.S. Fish and Wildlife Service perspective. *Ecol. Appl.* 12:719–723. [18]

Crow, J. F. and K. Aoki. 1984. Group selection for a polygenic trait: Estimating the degree of population subdivision. *Proc. Nat. Acad. Sci. USA* 81:6073–6077. [11]

Crow, J. F. and M. Kimura. 1970. *An Introduction to Population Genetic Theory*. Harper & Row, New York. [11]

Crowder, L. B., D. T. Crouse, S. S. Heppell, and T. H. Martin. 1994. Predicting the impact of turtle excluder devices on loggerhead sea turtle populations. *Ecol. Appl.* 4:437–445. [8]

Crowley, T. J. 2000. Causes of climate change over the past 1,000 years. *Science* 289:270–277. [10]

Crozier, L. 2003a. Winter warming facilitates range expansion: cold tolerance of the butterfly *Atalopedes campestris*. *Oecologia* 135:648–656. [10]

Crozier, L. 2003b. Winter warming facilitates butterfly range expansion: Population dynamics and larval survivorship at the northern range edge. *Ecology* 85:231–241. [10]

Crozier, R. H. 1992. Genetic diversity and the agony of choice. *Biol. Conserv.* 61:11–15. [11]

Crozier, R. H. and R. M. Kusmierski. 1993. Genetic distances and the setting of conservation priorities. In V. Loeschcke, J. Tomiuk, and S. K. Jain (eds.), *Conservation Genetics*, pp. 227–237. Birkhauser-Verlag, Berlin. [11]

Crump, M. L. and R. H. Kaplan. 1979. Clutch energy partitioning of tropical tree frogs (Hylidae). *Copeia* 1979:626–635. [3]

Crump, M. L., F. R. Hensley, and K. L. Clark. 1992. Apparent decline of the golden toad: Underground or extinct? *Copeia* 1992:413–420. [3]

Crumpacker, D. W., S. W. Hodge, D. Friedley, and W. P. Greg Jr. 1988. A preliminary assessment of the status of major terrestrial and wetland ecosystems of federal and Indian lands in the United States. *Conserv. Biol.* 2:103–115. [14]

Cuesta, J. A., J. E. García-Raso, and J. L. González Gordilla. 1991. Primera cita de *Rhitropanopeus harrisii* (Gould, 1841) (Crustacea, Decapoda, Brachyura, Xanthidae) en la Península Ibérica. *Bol. Inst. Esp. Oceanog.* 7(2):149–153. [14]

Cuesta, J. A., L. Serrano, M. R. Bravo, and J. Toja. 1996. Four new crustaceans in the Guadalquivir river estuary (SW Spain), including an introduced species. *Limnética* 12(1):41–45. [14]

Cullen, J. 1995. Predicting effectiveness: Fact and fantasy. In E. Delfosse and R. Scott, (eds.), *Proceedings VIII International Symposium on Biological Control of Weeds*, pp. 103–109. DSIR/CSIRO, Melbourne. [9]

Curran, L. M. and M. Leighton. 2000. Vertebrate responses to spatio-temporal variation in seed production of mast-fruiting Dipterocarpaceae. *Ecol. Monogr.* 70:101–128. [12]

Currie, D. J. and V. Paquin. 1987. Large-scale biogeographical patterns of species richness of trees. *Nature* 329:326–327. [2]

Curtis, J. T. 1956. The modification of mid-latitude grasslands and forests by man. In W. L. Thomas (ed.), *Man's Role in Changing the Face of the Earth*, pp. 721–736. University of Chicago Press, Chicago. [7]

Cutler, A. 1991. Nested faunas and extinction in fragmented habitats. *Conserv. Biol.* 5:496–505. [7]

Cutlip, S. M. and A. H. Center. 1964. *Effective Public Relations*. 3rd Edition. Prentice Hall, Englewood Cliffs, N.J. [15]

Czech, B. 2000. Economic growth as the limiting factor for wildlife conservation. *Wildlife Soc. Bull.* 28:4–15. [16]

Czech, B. and P. R. Kaufman. 2001. *The Endangered Species Act: History, Conservation*

Biology, and Public Policy. Johns Hopkins University Press, Baltimore. [3]

Czech, B., P. R. Kaufman, and P. K. Devers. 2000. Economic associations among causes of species endangerment in the United States. *BioScience* 50:593–601. [3]

Czechura, G. V. and G. J. Ingram. 1990. *Taudactylus diurnus* and the case of the disappearing frogs. *Mem. Queensl. Mus.* 29:361–365. [3]

D'Antonio, C. and L. A. Meyerson. 2002. Exotic plant species as problems and solutions in ecological restoration: a synthesis. *Restoration Ecology* 10:703–713. [15]

D'Antonio, C. M. and P. Vitousek. 1992. Biological invasions by exotic grasses, the grass/fire cycle, and global change. *Ann. Rev. Ecol. Syst.* 23:63–87. [9]

Daehler, C. C. and D. R. Strong. 1996. Status, prediction, and prevention of introduced cordgrass *Spartina* spp. invasions in Pacific estuaries, U.S.A. *Biological Conservation* 78:57–58. [9]

Dahl, T. E. 1990. *Wetland losses in the United States 1780's to 1980's.* U.S. Department of Interior, Fish and Wildlife Service, Washington, D.C. [15]

Daily, G. C. (ed.). 1997. *Ecosystem Services: Their Nature and Value.* Island Press, Washington, D.C. [1, 4, 18]

Daily, G. C. (ed.). 1997. *Nature's Services: Societal Dependence on Natural Ecosystems.* Island Press, Washington D.C. [13]

Daily, G. C. 2001. Ecological forecasts. *Nature* 411:245. [18]

Daily, G. C. 2003. Time to rethink conservation strategy. *Science* 300:1508–1509. [18]

Daily, G. C. and K. Ellison. 2002. *The New Economy of Nature: The Quest to Make Conservation Profitable.* Island Press, Washington, D.C. [16, 18]

Daily, G. C. and P. R. Ehrlich. 1996. Nocturnality and species survival. *P. Natl. Acad. Sci. USA* 93:11709–11712. [18]

Daily, G. C., P. R. Ehrlich, and G. A. Sánchez-Azofeifa. 2001. Countryside biogeography: Utilization of human-dominated habitats by the avifauna of southern Costa Rica. *Ecol. Appl.* 11:1–3. [18]

Daily, G. C., G. Ceballos, J. Pacheco, G. Suzán, and A. Sánchez-Azofeifa. 2003. Countryside biogeography of Neotropical mammals: Conservation opportunities in agricultural landscapes of Costa Rica. *Conserv. Biol.* 17:1814–1826. [18]

Dale, V. H., L. A. Joyce, S. McNulty, R. P. Neilson, M. P. Ayres, M. D. Flannigan, P. J. Hanson, L. C. Irland, A. E. Lugo, C. J. Peterson, D. Simberloff, F. J. Swanson, B. J. Stocks, and B. M. Wotton. 2001. Climate change and forest disturbances. *BioScience* 51:723–734. [10]

Daly, H. 1991. *Steady-State Economics.* 2nd Edition. Island Press, Washington, D.C. [5]

Daly, H. E. 1991. Boundless bull. *Orion* Summer: 59–61. [18]

Daniels, S., S. M. Haig, and J. A. Collazo. 2001. Preliminary Pedigree Analyses for the Puerto Rican Parrot. Report to the U.S. Fish and Wildlife Service, Atlanta, GA. [15]

Dansereau, P. 1957. Description and recording of vegetation on a structural basis. *Ecology* 32:172–229. [2]

Darwin, C. 1859. *On the Origin of Species.* Watts, London (reprint 1959). [5]

Darwin, C. 1904. *The Descent of Man and Selection in Relation to Sex.* J. A. Hill and Company, New York. [4]

Dary, D. A. 1974. *The Buffalo Book: The Full Saga of the American Animal.* Sage Books, Chicago. [8]

Dasmann, R. F. 1959. *Environmental Conservation.* John Wiley & Sons, Inc., New York. [1]

Daszak, P., A. A. Cunningham, and A. D. Hyatt. 2003. Infectious disease and amphibian population declines. *Divers. Distrib.* 9:141–150. [3]

Daszak, P., A. Strieby, A. A. Cunningham, J. E. Longcore, C. C. Brown, and D. Porter. 2004. Experimental evidence that the bullfrog (*Rana catesbeiana*) is a potential carrier of chytridiomycosis, an emerging fungal disease of amphibians. *Herpetol. J.* 14:201–207. [3]

Daugherty, C. H., A. Cree, J. M. Hay, and M. B. Thompson. 1990. Neglected taxonomy and continuing extinctions of tuatara (*Sphenodon*). *Nature* 347:177–179. [11]

Davidson, C. 2004. Declining downwind: Amphibian population declines in California and historical pesticide use. *Ecol. Appl.* 14:1892–1902. [3]

Davidson, C., H. B. Shaffer, and M. R. Jennings. 2001. Declines of the California red-legged frog: Climate, UV-B, habitat, and pesticides hypotheses. *Ecol. Appl.* 11:464–479. [3]

Davidson, C., H. B. Shaffer, and M. R. Jennings. 2002. Spatial tests of the pesticide drift, habitat destruction, UV-B, and climate-change hypotheses for California amphibian declines. *Conserv. Biol.* 16:1588–1601. [3]

Davidson, E. A. 2000. *You Can't Eat GNP: Economics as if Ecology Mattered.* Perseus, Cambridge, MA. [18]

Davidson, E. W., M. Parris, J. P. Collins, J. E. Longcore, A. P. Pessier, and J. Brunner. 2003. Pathogenicity and transmission of chytridiomycosis in tiger salamanders (*Ambystoma tigrinum*). *Copeia* 2003:601–607. [3]

Davidson, J. and H. G. Andrewartha. 1948a. Annual trends in a natural population of *Thrips imaginis* (Thysanoptera). *J. Anim. Ecol.* 17:193–199. [12]

Davidson, J. and H. G. Andrewartha. 1948b. The influence of rainfall, evaporation and atmospheric temperature on fluctations in the size of a natural population of *Thrips imaginis* (Thysanoptera). *J. Anim. Ecol.* 17:200–222. [12]

Davies, S. 1987. *Tree of Life: Buddhism and Protection of Nature.* Buddhist Perception of Nature Project, Hong Kong. [4]

Davis, A. J., J. H. Lawton, B. Shorrocks, and L. S. Jenkinson. 1998. Individualistic species responses invalidate simple physiological models of community dynamics under global environmental change. *J. Anim. Ecol.* 67:600–612. [10]

Davis, F. W., P. A. Stine, D. M. Stoms, M. Borchert, and D. Hollander 1995. Gap analysis of the actual vegetation of California: the southwestern region. *Madrona* 42:40–78. [14]

Davis, M. A., J. P. Grime, and K. Thompson. 2000. Fluctuating resources in plant communities: A general theory of invasibility. *J. Ecol.* 88:528–534. [9]

Davis, M. B. 1981. Quaternary history and the stability of forest communities. In D. C. West, H. H. Shugart, and D. B. Botkin (eds.), *Forest Succession: Concepts and Application*, pp. 132–153. Springer-Verlag, New York. [7]

Davis, M. B. and C. Zabinski. 1992. Changes in geographical range resulting from greenhouse warming: Effects on biodiversity in forests. In R. L. Peters and T. E. Lovejoy (eds.), *Global Warming and Biological Diversity*, pp. 297–308. Yale University Press, New Haven. [10]

Davis, S. M. 1991. Phosphorus inputs and vegetation sensitivity in an oligotrophic Everglades ecosystem. South Florida Water Management District. [13]

Davis, S. M. and J. C. Ogden (eds.). 1994. *Everglades, The Ecosystem and Its Restoration.* St. Lucie Press, Delray, FL. [13]

Davis, S. M., L. H. Gunderson, W. Park, J. Richardson, and J. Mattson. 1994. Landscape dimension, composition and function in a changing Everglades ecosystem. In S. M. Davis and J. C. Ogden (eds.), *Everglades, The Ecosystem and Its Restoration.* St. Lucie Press, Delray, FL. [13]

Dawkins, H. C. and M. S. Philip. 1998. *Tropical Moist Forest Silviculture and Management. A History of Success and Failure.* CAB International, Oxon, UK. [8]

Dayton, P. K., S. Thrush, and F. C. Coleman. 2002. *Ecological Effects of Fishing in Marine Ecosystems of the United States.* Pew Oceans Commission, Arlington, VA. [13]

Dayton, P. K., E. Sala, M. J. Tegner, and S. Thrush. 2000. Marine reserves: parks, baselines, and fishery enhancement. *B. Mar. Sci.* 66:617–634. [14]

Dayton, P. K., M. J. Tegner, P. B. Edwards, and K. L. Riser. 1998. Sliding baselines, ghosts, and reduced expectations in kelp forest communities. *Ecol. Appl.* 8:309–322. [3]

De Graaf, N. R. 2000. Reduced impact logging as part of the domestication of neotropical rainforest. *Int. Forest. Rev.* 2:40–44. [8]

de Mendonça, M. J. C., M. D. V. Diaz, D. Nepstad, R. S. da Motta, A. Alencar, J. C. Gomes, and R. A. Ortiz. 2004. The economic cost of the use of fire in the Amazon. *Ecol. Econ.* 49: 89–105. [6]

Deacon, J. E., and C. D. Williams. 1991. Ash Meadows and the legacy of the Devils Hole pupfish. In W. L. Minckley and J. E. Deacon (eds.), *Battle against Extinction: Native Fish Management in the American West*, pp. 69–91. University of Arizona Press, Tucson. [13, 14]

Deagle, B. E., D. J. Tollet, S. N. Jarman, M. A. Hindell, A. W. Trites, and N. J. Gales. 2005. Molecular scatology as a tool to study diet: Analysis of prey DNA in

scats from captive Steller sea lions. *Mol. Ecol.* 14:1831–1842. [11]

Dean, W. 1996. *A Ferro e Fogo,* 2nd Edition. Companhia das Letras, Rio de Janeiro, Brazil. [8]

Debinski, D. M. and R. H. Holt. 2000. A survey and overview of habitat fragmentation experiments. *Conserv. Biol.* 14:342–355. [7]

Defenders of Wildlife. 2002. *Conservation in America: State Government Incentives for Habitat Conservation.* Defenders of Wildlife, Washington, D.C. [17]

DeFries, R., A. Hansen, A. C. Newton, and M. C. Hansen. 2005. Increasing isolation of protected areas in tropical forests over the past twenty years. *Ecol. Appl.* 15:19–26. [6]

DeGrange, A. R., J. W. Fitzpatrick, J. N. Layne, and G. E. Woolfenden. 1989. Acorn harvesting by Florida scrub jays. *Ecology* 70:348–356. [18]

del Tredici, P., H. Ling, and G. Yang. 1992. The *Gingkos* of Tian Mu Shan. *Conserv. Biol.* 6:202–209. [4]

Delaplane, K. S. and D. F. Mayer 2000. *Crop Pollination by Bees.* CABI Publishing, New York. [2]

DeMaynadier, P. G. and M. L. Hunter. 2000. Road effects on amphibian movements in a forested landscape. *Nat. Areas J.* 20:56–65. [7]

den Boer, P. J. 1970. On the significance of dispersal power for populations of carabid beetles (Coleoptera, Carabidae). *Oecologia* 4:1–28. [7]

Denevan, W. M. 1992. The pristine myth: The landscape of the Americas in 1492. *Ann. Assoc. Am. Geogr.* 82:369–385. [6]

Denevan, W. M., J. M. Treacy, J. B. Alcorn, C. Padoch, J. Denslow, and S. F. Paitan. 1984. Indigenous agroforestry in the Peruvian Amazon: Bora management of swidden fallows. *Interciencia* 9:346–357. [6]

Dennis, R. L. H. 1993. *Butterflies and Climate Change.* Manchester University Press, Manchester, U.K. [10]

Denniston, C. 1978. Small population size and genetic diversity: Implications for endangered species. In S. A. Temple (ed.), *Endangered Birds: Management Techniques for Preserving Threatened Species,* pp. 281–289. University of Wisconsin Press, Madison. [11]

Denoth, M., L. Frid, and J. H. Myers. 2002. Multiple agents in biological control: Improving the odds? *Biol. Control* 24:20–30. [9]

Derraik, J. G. B. 2002. The pollution of the marine environment by plastic debris: a review. *Mar. Pollut. Bull.* 44:842–852. [6]

Dettinger, M. D. and D. R. Cayan. 1995. Large-scale atmospheric forcing of recent trends to-wards early snowmelt runoff in California. *J. Climate.* 8:606–623. [10]

Develey, P. F. and P. C. Stouffer. 2001. Effects of roads on movements of understory birds in mixed-species flocks in Central Amazonian Brazil. *Conserv. Biol.* 15:1416–1422. [7]

DeVivo, J. C. 1996. *Fish assemblages as indicators of water-quality within the Appalachico-*

la-Chattahoochee-Flint (ACF) River Basin. Master's thesis, University of Georgia, Athens. [7]

DeVore, B. 2003. Creating habitat on farms: the land stewardship project and monitoring on agricultural land. *Conservation in Practice* 4:28–36. [18]

Dewar, R. C. and A. D. Watt. 1992. Predicted changes in the synchrony of larval emergence and budburst under climatic warming. *Oecologia* 89:557–559. [10]

Deyrup, M. and T. Eisner. 1993. Last stand in the sand. *Nat. Hist.* 102(12):42–47. [18]

Dhar, B. B. and M. K. Chakraborty. 2002. Restoration Policy and Infrastructure: India. In Perrow, M. R. and A. J. Davy (eds.), *Handbook of Ecological Restoration, Volume 2: Restoration in Practice,* pp. 78–88. Cambridge University Press, Cambridge. [15]

Diamond, J. 1989. Overview of recent extinctions. In M. Pearl and D. Western (eds.), *Conservation for the Twenty-first Century,* pp. 37–41. Oxford University Press, New York. [11]

Diamond, J. 1992. *The Third Chimpanzee: The Evolution and Future of the Human Animal.* Harper Perennial, New York. [1]

Diamond, J. M. 1972. Biogeographic kinetics: Estimation of relaxation times for avifaunas of southwest Pacific islands. *Proc. Nat. Acad. Sci. USA* 69:3199–3203. [7]

Diamond, J. M. 1975. The island dilemma: Lessons of modern biogeographic studies for the design of natural preserves. *Biol. Conserv.* 7:129–146. [7]

Diamond, J. M. 1976. Island biogeography and conservation: Strategy and limitations. *Science* 193:1027–1029. [7]

Diamond, J. M. 1989. The present, past and future of human-caused extinctions. *Philos. T. Roy. Soc. B.* 325:469–477. [3]

Diamond, J. M. and R. M. May. 1976. Island biogeography and the design of natural reserves. In R. M. May (ed.), *Theoretical Ecology: Principles and Applications,* pp. 163–186. W. B. Saunders, Philadelphia. [7]

Diamond, S. A., G. S. Peterson, J. E. Tietge, and G. T. Ankley. 2002. Assessment of the risk of solar ultraviolet radiation to amphibians. III. Predictions of impacts in selected northern midwestern wetlands. *Environ. Sci. Technol.* 36:2866–2874. [3]

Dias, P. C., G. R. Verheyen, and M. Raymond. 1996. Source-sink populations in Mediterranean blue tits: Evidence using single-locus minisatellite probes. *J. Evolution Biol.* 9:965–978. [11]

Diaz, S. and M. Cabido. 2001. Vive la difference: Plant functional diversity matters to ecosystem processes. *Trends Ecol. Evol.* 16:646–655. [2]

Diaz, S., A. J. Symstad, F. S. Chapin, D. A. Wardle, and L. F. Huenneke. 2003. Functional diversity revealed by removal experiments. *Trends Ecol. Evol.* 18:140–146. [2]

Díaz-Paniagua, C., A. Marco, A. C. Andreu, C. Sánchez, L. Peña, M. Acosta and I. Molina. 2002. *Trachemys scripta en Doñana.* Report for the Spanish Herpetological Association. Estación Biológica de

Doñana-Parque Nacional de Doñana. [14]

Dickinson, M. B. and M. F. Whigham. 1999. Regeneration of mahogany (*Swietenia macrophylla*) in the Yucatan. *Int. Forest. Rev* 1:35–39. [8]

Dickinson, M. B., J. Dickinson, and F. E. Putz. 1996. Natural forest management as a conservation tool in the tropics: Divergent views on possibilities and alternatives. *Commonw. Forest. Rev.* 75:309–315. [8]

Didham, R. K., J. Ghazoul, N. E. Stork, and A. J. Davis. 1996. Insects in fragmented forests: A functional approach. *Trends Ecol. Evol.* 11:255–260. [7]

Diener, E., J. Horwitz, and R. A. Emmons. 1985. Happiness of the very wealthy. *Soc. Indic. Res.* 16:263–274. [5]

Diener, E., E. Sandvik, L. Seidlitz, and M. Diener. 1993. The relationship between income and subjective well-being: relative or absolute? *Soc. Indic. Res.* 28:195–223. [5]

Diffendorfer, J. E. 1998. Testing models of source-sink dynamics and balanced dispersal. *Oikos* 81:417–433. [12]

Dirzo, R and A. Miranda. 1991. Altered patterns of herbivory and diversity in the forest understory: A case study of the possible consequences of contemporary defaunation. In P. W. Price, T. M. Lewinsohn, G. W. Fernandes, and W. W. Benson (eds.), *Plant-animal Interactions: Evolutionary Ecology in Tropical and Temperate Regions,* pp 273–287. Wiley, New York. [3, 8]

Dirzo, R. and C. García. 1991. Rates of deforestation in Los Tuxtlas, a neotropical area in southeast Mexico. *Conserv. Biol.* 6:84–90. [8]

Dix, A. 1996. *CAMPFIRE: An Annotated Bibliography (1985-1996).* CASS, University of Zimbabwe: Harare. [16]

Dizard, J. 1994. *Going Wild: Hunting, Animal Rights, and the Contested Meaning of Nature.* University of Massachusetts Press, Amherst. [4]

Doadrio, I., J. A. Carmona, and C. Fernández-Delgado. 2002. Morphometric study of the Iberian *Aphanius* (Actinopterygii, Cyprinodontiformes), with description of a new species. *Folia Zoologica* 1: 51–69. [14]

Doak, D. 1995. Source–sink models and the problem of habitat degradation: General models and applications to the Yellowstone grizzly. *Conserv. Biol.* 9:1370–1379. [12]

Dobson A. 2005. Monitoring global rates of biodiversity change: challenges that arise in meeting the Convention on Biological Diversity (CBD) 2010 goals. *Phil Trans. R. Soc. B.* 360:229–241. [18]

Dobson, A., and M. Meagher. 1996. The population dynamics of brucellosis in the Yellowstone National Park. *Ecology* 77:1026–1036. [12]

Dobson, A. P. 1988. Restoring island ecosystems: The potential of parasites to control introduced mammals. *Conserv. Biol.* 2:31–39. [3]

Dobson, A. P. 2004. Population dynamics of pathogens with multiple hosts. *Am. Nat.* 164:S64–S78. [3]

Dobson, A. P. and P. J. Hudson. 1994. The interaction between the parasites and predators of Red Grouse *Lagopus lagopus scoticus*. *Ibis* 137:S87–S96. [3]

Dobson, A. P., J. P. Rodriguez, W. M. Roberts and D. S. Wilcove. 1997. Geographic distribution of endangered species in the United States. *Science* 275:550–553.

Dobzanski, T. 1937. *Genetics and the Origin of Species*. Columbia University Press, New York. [11]

Dodd, C. K. 1990. Effects of habitat fragmentation on a stream-dwelling species, the flattened musk turtle *Sternotherus depressus*. *Biol. Conserv.* 54:33–45. [7]

Dodson, C. H. and A. H. Gentry. 1991. Biological extinction in Western Ecuador. *Ann. Mo. Bot. Gard.* 78:273–295. [3]

Doherty, P. 2003. *Ocean Zoning: Perspectives on a New Vision for the Scotian Shelf and Gulf of Maine*. Ecology Action Centre, Halifax, Nova Scotia. [13]

Donald, P. F., R. E. Green, and M. F. Heath. 2001. Agricultural intensification and the collapse of Europe's farmland bird populations. *P. Roy. Soc. Lond. B* 268:25–29. [18]

Done, T. 2001. Scientific principles for establishing MPAs to alleviate coral bleaching and promote recovery. In S. V. Salm and S. L. Coles (eds.), *Coral Bleaching and Marine Protected Areas. Proceedings of the Workshop on Mitigating Coral Bleaching Impact Through MPA Design, Bishop Museum*, pp. 53–59. Asia Pacific Coastal Marine Program Report No. 102. The Nature Conservancy, Honolulu. [13]

Donoso, D. S., A. A. Grez, and J. A. Simonetti. 2003. Effects of forest fragmentation on the granivory of differently sized trees. *Biol. Conserv.*115:63–70. [7]

Donovan, T. M. and C. H. Flather. 2002. Relationships among North American songbird trends, habitat fragmentation and landscape occupancy. *Ecology* 12:364–374. [12]

Donovan, T. M., F. R. Thompson III, J. Faaborg, and J. R. Probst. 1995. Reproductive success of migratory birds in habitat sources and sinks. *Conserv. Biol.* 9:1380–1395. [7]

Donovan, T. M., R. H. Lamberson, A. Kimber, F. R. Thompson III, and J. Faaborg. 1995. Modeling the effects of habitat fragmentation on source and sink demography of Neotropical migrant birds. *Conserv. Biol.* 9:1396–1407. [12]

Doremus, H. 1997. Listing decisions under the Endangered Species Act: Why better science isn't always better policy. *Wash. U. L.Q.* 75:1029–1153. [17]

Doremus, H. 2001. Adaptive management, the Endangered Species Act, and the institutional challenges of "new age" environmental protection. *Washburn L.J.* 41:50–89. [17]

Downer, R. 1993. *GAPPS II User Manual. Version 1.3*. Applied Biomathematics: Setauket, New York. [12]

Drake, D. C., R. J. Naiman, and M. Helfield. 2002. Reconstructing salmon abundance in rivers: An initial dendrocrinological evaluation. *Ecology* 83:2971–2977. [8]

Draulen, D. and E. Van Krunksalen. 2002. The impact of war on forest areas in the Democratic Republic of Congo. *Oryx* 36:35–40. [6]

Drayton, B. and R. B. Primack. 1996. Plant species lost in an isolated conservation area in Metropolitan Boston from 1894 to 1993. *Conserv. Biol.* 10:30–39. [7]

Drechsler, M. 2004. Model-based conservation decision aiding in the presence of goal conflicts and uncertainty. *Biodivers. Conserv.* 13:141–164. [5]

Drent R., C. Both, M. Green, J. Madsen, and T. Piersma. 2003. Pay-offs and penalties of competing migratory schedules. *Oikos* 103:274–292. [10]

Drost, C. A. and G. M. Fellers. 1996. Collapse of a regional frog fauna in the Yosemite area of the California Sierra Nevada, USA. *Conserv. Biol.* 10:414–425. [3]

du Toit, J. T., B. H. Walker, and B. M. Campbell. 2004. Conserving tropical nature: Current challenges for ecologists. *Trends Ecol. Evol.* 19:12–17. [8]

Duarte, C., C. Montes, S. Agustí, P. Martino, M. Bernúes, and J. Kalff. 1990. Biomasa de marófitos acuáticos en la marisma del Parque Nacional de Doñana (SW España): importancia y factores ambientales que controlan su distribución. *Limnetica*, 6:1–12. [14]

Duda, M. D. 1991. A bridge to the future: The wildlife diversity funding initiative. A needs assessment for the Fish and Wildlife Conservation Act. Western Association of Fish and Wildlife Agencies. [15]

Dulvy, N. K., Y. Sadovy, and J. D. Reynolds. 2003. Extinction vulnerability in marine populations. *Fish and Fisheries* 4:25–64. [8]

Dulvy, N. K., J. D. Metcalfe, J. Glanville, M. G. Pawson, and J. D. Reynolds. 2000. Fishery stability, local extinctions, and shifts in community structure in skates. *Conserv. Biol.* 14:283–293. [3]

Duncan, B. W. and P. A. Schmalzer. 2004. Anthropogenic influences on potential fire spread in a pyrogenic ecosystem of Florida, USA. *Landscape Ecol.* 19:153–165. [7]

Duncan, J. R. and J. L. Lockwood. 2001. Extinction in a field of bullets: A search for causes in the decline of the world's freshwater fishes. *Biol. Conserv.* 102:97–105. [3]

Duncan, R. C. and W. Youngquist. 1999. Encircling the peak of world oil production. *Natural Resources Research* 8:219–232. [18]

Duncan, R. S. and C. A. Chapman. 2002. Limitation of animal seed dispersal for enhancing forest succession on degraded lands. In J. Levey, W. R. Silva and M. Galetti (eds.), *Seed Dispersal and Frugivory: Ecology, Evolution and Conservation*, pp 437–450. DCABI, New York, NY. [15]

Dunn, P. O. and D. W. Winkler. 1999. Climate change has affected the breeding date of tree swallows throughout North America. *P. Roy. Soc. Lon. B Bio.* 266:2487–2490. [10]

Dunning, J. B. 2000. They shoot bison, don't they? Discussing ethics in conservation courses. *North American Colleges and Teachers of Agriculture Journal* 44:40–44. [12]

Dunning, J. B. and B. D. Watts. 1990. Regional differences in habitat occupancy by Bachman's Sparrow. *Auk* 107:463–472. [12]

Dunning, J. B., D. J. Stewart, and J. Liu. 2002. Individual-based modeling: The Bachman's Sparrow. In S. E. Gergel and M. G. Turner (eds.), *Learning Landscape Ecology: A Practical Guide to Concepts and Techniques*, pp. 228–248. Springer, New York. [12]

Dunning, J. B., B. J. Danielson, and H. R. Pulliam. 1992. Ecological processes that affect populations in complex landscapes. *Oikos* 65:169–175. [7]

Dunning, J. B., R. Borgella, K. Clements, and G. K. Meffe. 1995a. Patch isolation, corridor effects, and colonization by a resident sparrow in a managed pine woodland. *Conserv. Biol.* 9:542–550. [12]

Dunning, J. B., B. J. Danielson, B. D. Watts, J. Liu, and D. G. Krementz. 2000. Studying wildlife at local and landscape scales: Bachman's Sparrows at the Savannah River Site. *Stud. Avian Biol.* 21:75–80. [12]

Dunning, J. B., Jr., D. J. Stewart, B. J. Danielson, B. R. Noon, T. L. Root, R. H. Lamberson, and E. E. Stevens. 1995b. Spatially explicit population models: Current forms and future uses. *Ecol. Appl.* 5:3–11. [12]

Durán, S. M. and G. H. Kattan. 2005 A test of the utility of exotic tree plantations for understory birds and food resources in the Colombian Andes. *Biotropica* 37:129–135. [14]

Dwyer, G., S. A. Levin, and L. Buttel. 1990. A simulation model of the population dynamics and evolution of myxomatosis. *Ecol. Monogr.* 60:423–447. [9]

Dynesius, M. and C. Nilsson. 1994. Fragmentation and flow regulation of river systems in the northern third of the world. *Science* 266:753–762. [7]

Dynesius, M. and R. Jansson. 2000. Evolutionary consequences of changes in species' geographical distributions driven by Milankovitch climatic oscillations. *Proc. Nat. Acad. Sci. USA* 97:9115–9120. [2]

Dzingirai, V. 1994. CAMPFIRE politics and emmigration in Binga: co-management as a solution to the immigration problem. Paper presented at the Conference on Lessons of the Natural Resource Management Project, Harare. [16]

Dzingirai, V. 1995. Take back your CAMP-FIRE. A study of local perceptions to electric fencing in the Framework of Binga's CAMPFIRE Programme. CASS, University of Zimbabwe, Harare. [16]

Earl, J. and W. Moseley. 1996. *RiskMap Report: Zimbabwe*. London: Save the Children Fund, U.K. [16]

Earn, D. J. D., S. A. Levin, and P. Rohani. 2000. Coherence and conservation. *Science* 290: 1360–1364. [10]

Easterling, D. R., G. A. Meehl, C. Parmesan, S. A. Chagnon, T. R. Karl, and L. O. Mearns. 2000b. Climate extremes: Observations, modeling, and impacts. *Science* 289:2068–2074. [10]

Easterling, D. R., J. L. Evans, P. Y. Groisman, T. R. Karl, K. E. Kunkel, and P. Ambenje. 2000a. Observed variability and trends in extreme climate events: A brief review. *B. Am. Meteorol. Soc.* 81:417–425. [10]

Ebenhard, T. 1988. Introduced birds and mammals and their ecological effects. *Swedish Wildlife Research Viltrevy* 13:1–107. [9]

Echelle, A. A. 1991. Conservation genetics and genic diversity in freshwater fishes of western North America. In W. L. Minckley and J. E. Deacon (eds.), *Battle Against Extinction: Native Fish Management in the American West*, pp. 141–153. University of Arizona Press, Tucson. [11]

Ecological Society of America. 1999. *Ecological Principles and Guidelines for Managing the Use of Land*. Ecological Society of America's Committee on Land Use, Ecological Society of America, Washington, D.C. [17]

Edmondson, W. T. 1991. *Uses of Ecology: Lake Washington and Beyond*. University of Washington Press, Seattle. [6, 15]

Edwards, S. V. and P. W. Hedrick. 1998. Evolution and ecology of MHC molecules: From genomics to sexual selection. *Trends Ecol. Evol.* 3:305–311. [11]

Edwards-Jones, G., B. B. Davies, and S. Hussein. 2000. *Ecological Economics: An Introduction*. Blackwell Scientific, Oxford. [5]

Edwards-Jones, G., E. S. Edwards-Jones, and K. Mitchell. 1995. A comparison of contingent valuation methodology and ecological assessment as techniques for incorporating ecological goods in to land-use decisions. *J. Environ. Planning Manage.* 38:215–230. [5]

Eeley, H. A. C. and M. J. Lawes. 1999. In J. G. Feagle, C. Janson, and K. E. Reed (eds.), *Primate Communities*, pp. 191–219. Cambridge University Press, Cambridge. [2]

Ehrenfeld, D. 1993. Forgetting. In *Beginning Again: People and Nature in the New Millennium*, pp. 65–72. Oxford University Press, NY. [18]

Ehrenfeld, D. 2000. War and peace and conservation biology. *Conserv. Biol.* 14:105–112. [18]

Ehrenfeld, D. W. 1970. *Biological Conservation*. Holt, Rinehart and Winston, Inc., New York. [1]

Ehrenfeld, D. W. 1976. The conservation of non-resources. *Am. Sci.* 64:660–668. [4]

Ehrenfeld, D. W. 1988. Why put a value on biodiversity? In E. O. Wilson (ed.), *Biodiversity*, pp. 212–216. National Academy Press, Washington, D.C. [4]

Ehrenfeld, J. G. and L. A. Toth. 1997. Restoration ecology and the ecosystem perspective. *Restor. Ecol.* 5:307–317. [15]

Ehrlen, J. and O. Eriksson. 2000. Dispersal limitation and patch occupancy in forest herbs. *Ecology* 81:1667–1674. [7]

Ehrlich, P. R. and E. O. Wilson. 1991. Biodiversity studies: Science and policy. *Science* 253:258–262. [12]

Ehrlich, P. R., D. D. Murphy, M. C. Singer, C. B. Sherwood, R. R. White, and I. L. Brown. 1980. Extinction, reduction, stability and increase: The responses of checkerspot butterfly (*Euphydryas editha*) populations to the California drought. *Oecologia* 46:101–105. [10]

Elkan, P. and S. Elkan. 2002. Engaging the Private Sector: A case study of the WCS-CIB-Republic of Congo project to reduce commercial bushmeat hunting, trading and consumption inside a logging concession. *Communiqué* November: 40–42. [8]

Elliot, R. 1992. Intrinsic value, environmental obligation and naturalness. *Monist* 75:138–160. [4]

Ellis, A. J. 1975. Geothermal systems and power development. *Am. Sci.* 63:510–521. [13]

Ellis, D. H., G. F. Gee, and C. M. Mirande. 1996. *Cranes: Their Biology, Husbandry, and Conservation*, 2nd Edition. Hancock. [12]

Ellison, A. M. 2004. Bayesian inference in ecology. *Ecol. Lett.* 7:509–520. [17]

Ellison, K. and G. C. Daily. 2003. Making conservation profitable. *Conservation in Practice* 4:12–19. [18]

Elton, C. S. 1949. Population interspersion: An essay on animal community patterns. *J. Ecol.* 37:1–23. [12]

Elton, C. S. 1958. *The Ecology of Invasions by Animals and Plants*. Methuen, London. [9]

Emberger, L., H. Gaussen, M. Kass, and A. de Phillips. 1976. Carte Bioclimatique de la zone Mediterranenne. UNESCO-FAO. Paris-Rome. [14]

Emberson, L. D., G. Wieser, and M. R. Ashmore. 2000a. Modelling of stomatal conductance and ozone flux of Norway spruce: comparison with field data. *Environ. Pollut.* 109:393–402. [6]

Emberson, L. D., M. R. Ashmore, H. M. Cambridge, D. Simpson and J.-P. Tuovinen. 2000b. Modelling stomatal ozone flux across Europe. *Environ. Pollut.* 109:403–413. [6]

Engeman, R. M., S. A. Shwiff, F. Cano, and B. Constantin. 2003. An economic assessment of the potential for predator management to benefit Puerto Rican parrots. *Ecol. Econ.* 46:283–292. [15]

Environment Australia. 2003. Recovery plan for marine turtles in Australia. Prepared by the Marine Species Section, Approvals and Wildlife Division, Environment Australia in consultation with the Marine Turtle Recovery Team, Commonwealth of Australia, July, 2003. [16]

Environmental Protection Agency. 2003b. *Introduction to the Clean Water Act*. http://www.epa.gov/watertrain/cwa/ [15]

Erickson, R. C. 1973. Some black-footed ferret research needs. In R. L. Linder and C. N. Hillman (eds.), *Proceedings of the Black-footed Ferret and Prairie Dog Workshop*, pp. 153–164. South Dakota State University, Brookings, South Dakota. [15]

Eriksson, O. 2000. Functional roles of remnant plant populations in communities and ecosystems. *Global Ecol. Biogeogr.* 9:443–449. [12]

Erwin, T. 1982. Tropical forests: Their richness in Coleoptera and other arthropod species. *Coleopts. Bull.* 36:74–82. [2]

Erwin, T. L. 1888. The tropical forest canopy: The heart of biotic diversity. In E. O. Wilson (ed.), *Biodiversity*, pp. 123–129. National Academy Press, Washington, D.C. [4]

Erwin, T. L. 1991. How many species are there? Revisited. *Conserv. Biol.* 5:330–333. [2]

Estación Biológica de Doñana. 1991. Radio-rastreo y mortalidad del águila imperial y el lince en el Parque Nacional de Doñana y su entorno. Report. [14]

Estes, J. A., D. O. Duggins, and G. B. Rathbun. 1989. The ecology of extinctions in kelp forest communities. *Conserv. Biol.* 3:251–264. [3]

Esty, D. C., M. Levy, T. Srebotnjak, and A. de Sherbinin. 2005. *2005 Environmental Sustainability Index: Benchmarking National Environmental Stewardship*. Yale Center for Environmental Law & Policy, New Haven, CT. [18]

Etterson, J. R. and R. G. Shaw. 2001. Constraint to adaptive evolution in response to global warming. *Science* 294:151–154. [10]

Evans, P. R., R. M. Ward, M. Bone, and M. Leakey. 1998. Creation of temperate-climate intertidal mudflats: Factors affecting colonization and use by benthic invertebrates and their bird predators. *Marine Pollution Bulletin* 37:535–545. [10]

Evans, R. 1998. The erosional impacts of grazing animals. *Prog. Phys. Geog.* 22:251–268. [6]

Eves, H. E. and R. G. Ruggiero. 2000. Socioeconomics and sustainability of hunting in the forests of northern Congo. In J. G. Robinson and E. Bennett (eds.) *Hunting for Sustainability in Tropical Forests*, pp. 421–448. Columbia University Press. [8]

Ewel, J. J. 1986. Invasibility: Lessons from south Florida. In H.A. Mooney and J.A. Drake (eds), *Ecology of Biological Invasions of North America and Hawaii*, pp. 214–230. Springer Verlag, New York. [9]

Fa, J. E., and C. A. Peres. 2001. Game vertebrate extraction in African and Neotropical forests: An Intercontinental Comparison. In J. D. Reynolds, G. M. Mace, K. H. Redford, and J. G. Robinson (eds.), *Conservation of Exploited Species*, pp. 203–241. Cambridge University Press, Cambridge. [8]

Fa, J. E, C. A. Peres, and J. Meeuwig. 2001. Bushmeat exploitation in tropical forests: An intercontinental comparison. *Conserv. Biol.* 16:232–237. [8]

Fa., J., D. Currie, and J. Meeuwig. 2003. Bushmeat and food security in the Congo Basin: linkages between wildlife and people's future. *Environ. Conserv.* 30:71–78. [8]

Faaborg, J. 1979. Qualitative patterns of avian extinction on Neotropical land-bridge is-

lands: Lessons for conservation. *J. Appl. Ecol.* 16:99–107. [7]

Faeth, S. H. and E. F. Connor. 1979. Supersaturated and relaxing island faunas: A critique of the species–age relationship. *J. Biogeogr.* 6:311–316. [7]

Fagan, C., C. A. Peres, and J. Terborgh. 2005. Tropical forests: A protected area strategy for the 21st Century. In press. In W. F. Laurance and C.A. Peres (eds.), *Emerging Threats to Tropical Forests.* University of Chicago Press, Chicago. [8]

Fagan, W. F., R. S. Cantrell, and C. Cosner. 1999. How habitat edges change species interactions. *The Am. Nat.* 153:165–182. [14]

Fahrig, L. 1996. Fragmentation and corridors: The misuse of theory in Conservation Biology. *Suppl. Bull. Ecol. Soc. Am.* 77(3):134. [7]

Fahrig, L. 1997. Relative effects of habitat loss and fragmentation on population extinction. *J. Wildl. Mgmt.* 61:603–610. [7]

Fahrig, L. 2003. Effects of habitat fragmentation on biodiversity. *Ann. Rev. Ecol. Evol. Syst.* 34:487–515. [7]

Fahrig, L. and G. Merriam. 1985. Habitat patch connectivity and population survival. *Ecology* 66:1762–1768. [7]

Fahrig, L. and G. Merriam. 1994. Conservation of fragmented populations. *Conserv. Biol.* 8:50–59. [7]

Fahrig, L., J. H. Pedlar, S. E. Pope, P. D. Taylor, and J. F. Wegner. 1995. Effects of road traffic on amphibian density. *Biol. Conserv.* 73:177–182. [7]

Faith, D. P. 1992. Conservation evaluation and phylogenetic diversity. *Biol. Conserv.* 61:1–10. [11]

Faith, D. P. 1994. Phylogenetic pattern and the quantification of organismal biodiversity. *Philos. T. Roy. Soc. B.* 345:45–58. [11]

Faith, D. P. and P. A. Walker. 1994. Diversity: a software package for sampling phylogenetic and environmental diversity. Reference and User's Guide Vol. 2.1. CSIRO Division of Wildlife and Ecology, Canberra. [14]

Faith, D. P. and P. A. Walker. 1996. Environmental diversity: on the best possible use of surrogate data for assessing the relative biodiversity of sets of areas. *Biodivers. Conserv.* 5:399–415. [14]

Faith, D. P., S. Ferrier, and P. A. Walker. 2004. The ED strategy: how species-level surrogates indicate general biodiversity patterns through an 'environmental diversity' perspective. *J. Biogeogr.* 31:1207–1217. [14]

Fan, S., M. Gloor, J. Mahlman, S. Pacala, J. Sarmiento, T. Takahashi, and P. Tans. 1998. A large terrestrial carbon sink in North America implied by atmospheric and oceanic carbon dioxide data and models. *Science* 282:442–446. [10]

FAO. 1985. The ecological effects of Eucalyptus. FAO Forestry Paper 59. Rome. [14]

FAO. 2002. *The State of World Fisheries and Aquaculture.* FAO Information Division. (http://www.fao.org/docrep/005/y7300e/y7300e00.htm) [8]

FAO. 1999. *State of the World's Forests.* Food and Agriculture Organization of the United Nations, Rome. [8]

Farber, S. and R. Costanza. 1987. The economic value of wetlands systems. *J. Environ. Manage.* 24:41–51. [5]

Farnsworth, N. R. 1988. Screening plants for new medicines. In E. O. Wilson (ed.), *Biodiversity,* pp. 83–97. National Academy Press, Washington, D.C. [4]

Fayer-Hosken, R. A., P. Brooks, H. J. Bertschinger, J. F. Kirkpatrick, J. W. Turner, and I. K. M. Liu. 1997. Management of African elephant populations by immunocontraception. *Wildlife Soc. B.* 25:18–21. [17]

Fellers, G. M. and C. A. Drost. 1993. Disappearance of the Cascades frog *Rana cascadae* at the southern end of its range, California, USA. *Biol. Conserv.* 65:177–181. [3]

Fellers, G. M., D. E. Green, and J. E. Longcore. 2001. Oral chytridiomycosis in the mountain yellow-legged frog (*Rana muscosa*). *Copeia* 2001:945–953. [3]

Fellers, G. M., L. L. McConnell, D. Pratt, and S. Datta. 2004. Pesticides in mountain yellow-legged frogs (*Rana muscosa*) from the Sierra Nevada Mountains of California, USA. *Environ. Toxicol. Chem.* 23:2170–2177. [3]

Felsenstein, J. 2004. PHYLIP (Phylogenetic Inference Program) version 3.6. Distributed by the author. Department of Genomic Sciences, University of Washington, Seattle. [2]

FEMAT (Forest Ecosystem Management Assessment Team).1993. *Forest Ecosystem Management: An Ecological, Economic, and Social Assessment: Report of the Forest Ecosystem Management Assessment Team.* U.S. Government Printing Office, Washington, D.C. [17]

FEPCB. 2004. *Fisheries Ecosystem Planning for Chesapeake Bay.* NOAA Chesapeake Bay Office. http://noaa.chesapeakebay.net/Fish/FEP_DRAFT.pdf. [13]

Ferdana, Z. 2002. Approaches to integrating a marine GIS into The Nature Conservancy's ecoregional planning process. In J. Breman, (ed.), *Marine Geography: GIS for Oceans and Seas,* pp. 151–158. ESRI, Redlands, WA. [14]

Fernández-Delgado, C. 1987. Ictiofauna del estuario del Guadalquivir: su distribución y biología de las especies sedentarias. Ph. D. dissertation. University of Córdoba. Córdoba. [14]

Fernández-Delgado, C. 1996. La construcción de la presa de cierre en el Bajo Guadalquivir: una nueva amenaza para Doñana y su entorno. Quercus (127). [14]

Fernández-Delgado, C. 2003. *Introducción de especies exóticas a través del agua de lastre de los barcos. Aplicación al caso del Guadalquivir.* Consejería de Medio Ambiente. Junta de Andalucía. Final Report. [14]

Fernández-Delgado, C., P. Drake, A. M. Arias y D. García 2000. *Peces de Doñana y su entorno.* Colección Técnica, Organismo Autónomo Parques Nacionales, Madrid. [14]

Fernando, A. B. 1993. Recent elephant conservation in Sri Lanka: A tragic story. *Gajah* 10:19–25. [17]

Fernando, P. 1997. Keeping jumbo afloat: Is translocation an answer to the human elephant conflict? *Sri Lanka Nature* 1:4–12. [17]

Fernando, P. 1998. Genetics, ecology, and conservation of the Asian elephant. Ph.D. dissertation, University of Oregon, Eugene. [17]

Fernando, P. 2000. Elephants in Sri Lanka: Past, present, and future. *Loris* 22:38–44. [17]

Fernando, P. and R. Lande. 2000. Molecular genetic and behavioral analyses of social organization in the Asian elephant. *Behavioral Ecology and Sociobiology* 48:84–91. [17]

Fernando, P., M. E. Pfrender, S. Encalada, and R. Lande. 2000. Mitochondrial DNA variation, phylogeography, and population structure of the Asian elephant. *Heredity* 84:362–372. [17]

Ferrari, M. J. and R. A. Garrott. 2002. Bison and elk: Brucellosis seroprevalence on a shared winter range. *J. Wildl. Mgmt.* 66:1246–1254. [12]

Ferraris, J. D. and S. R. Palumbi (eds.). 1996. *Molecular Zoology.* John Wiley & Sons, Inc., New York. [11]

Ferraro, P. J. 2001. Global habitat protection: Limitations of development interventions and a role for conservation performance payments. *Conserv. Biol.* 15(4):990–1000. [6]

Ferraro, P. J. 2002. The local costs of establishing protected areas in low-income nations: Ranomafana National Park, Madagascar. *Ecol. Econ.* 43:261–275. [6]

Ferren, W. R. Jr. and D. A. Pritchett. 1988. *Enhancement, restoration, and creation of vernal pools at Del Sol Open Space and Vernal Pool Reserve, Santa Barbara County, California. Environmental Report No. 13.* The Herbarium, Department of Biological Sciences, University of California, Santa Barbara. [15]

Ferren, W. R. Jr. and E. M. Gevirtz. 1990. Restoration and Creation of Vernal Pools: Cookbook Recipes or Complex Science? In D. H. Ikeda and R. A. Schlising (eds.), *Vernal Pools—Their Habitat and Biology,* pp. 147–148. Studies from the Herbarium No. 8, California State University, Chico. [15]

Ferren, W. R. Jr., D. M. Hubbard, S. Wiseman, A. K. Parikh, and N. Gale. 1998. Review of ten years of vernal pool restoration and creation in Santa Barbara, California. In C. W. Witham, E. T. Bauder, D. Belk, W. R. Ferren Jr., and R. Ornduff (eds.), *Ecology, Conservation, and Management of Vernal Pool Ecosystems—Proceeding from a 1996 Conference.* California Native Plant Society, Sacramento, CA. [15]

Ferreras, P., J. J. Aldama, J. F. Beltrán, and M. Delibes. 1992. Rates and causes of mortality in a fragmented population of Iberian lynx *Felis pardina* (Temminck). *Biol. Conserv.* 61:197–202. [7, 14]

Ferrier, S. 1997. Biodiversity data for reserve selection: making best use of incomplete information. In J. J. Pigram and R. C.

Sundell, (eds.), *National Parks and Protected Areas: Selection, Delimitation, and Management.* Center for Water Policy Research, Armidale, Australia. [14]

Ferrier, S. 2002. Mapping spatial pattern in biodiversity for regional conservation planning: Where to from here? *Syst. Biol.* 51:331–363. [14]

Ferrier, S., G. Watson, J. Pearce, and M. Drielsma. 2002. Extended statistical approaches to modelling spatial pattern in biodiversity in northeast New South Wales. I. Species-level modelling. *Biodivers. Conserv.* 11:2275–2307. [14]

Ferrier, S., R. L. Pressey, and T. W. Barrett. 2000. A new predictor of the irreplaceability of areas for achieving a conservation goal, its application to real-world planning, and a research agenda for further refinement. *Biol. Conserv.* 93:30–325. [14]

Ferrière R., F. Sarrazin, S. Legendre, and J. P. Baron. 1996. Matrix population models applied to viability analysis and conservation: Theory and practice using the ULM software. *Acta Oecologia* 17:629–656. [12]

Ferson, S. 1994. *RAMAS/stage: Generalized stage-based modeling for population dynamics.* Applied Biomathematics, Setauket, NY. [12]

Figueroa, R. and C. Mills. 2001. Environmental justice. In D. Jamieson (ed.), *A Companion to Environmental Philosophy,* pp. 426–438. Blackwell, Oxford. [4]

Fimbel, R. A. and C. C. Fimbel. 1996. The role of exotic conifer plantations in rehabilitating degraded tropical forest lands: A case study from the Kibale Forest in Uganda. *Forest Ecol. Manag.* 81: 215–226. [15]

Findlay, S., K. Kuehn, P. Grofman, and S. Dye. 1999. Effects of invasive marsh plants on nutrient sequestration and mineralization. Estuarine Research Federation Biennial Conference Abstracts. Estuarine Research Federation, Port Republic, Maryland. [15]

Finlayson-Pitts, B. J. and J. N. J. Pitts. 1999. *Chemistry of the Upper and Lower Atmosphere.* Academic Press, San Diego, CA. [6]

Fischer, M. and J. Stöcklin. 1997. Local extinctions of plants in remnants of extensively used calcareous grasslands 1950–1985. *Conserv. Biol.* 11:727–737. [7]

Fisher, R. A. 1930. *The Genetical Theory of Natural Selection.* The Clarendon Press, Oxford. [11]

Fisher, R. A. 1937. The wave of advance of advantageous genes. *Ann. Eugenic.* 7:353–369. [9]

Fisher, R., W. Ury, and B. Patton. 1991. *Getting to Yes,* 2nd Edition. Penguin Books, New York. [17]

Fitzgibbon C. D., H. Mogaka, and J. H. Fanshawe. 1995. Subsistence hunting in Arabuko-Sokoke forest, Kenya, and its effects on mammal populations. *Conserv. Biol.* 9:1116–1126. [8]

Fitzpatrick, J. W., G. E. Woolfenden, and M. T. Kopeny. 1991. *Ecology and Development-related Habitat Requirements of the*

Florida Scrub Jay (*Aphelocoma coerulescens coerulescens*). Nongame Wildlife Program Technical Report No. 8, Office of Env. Serv., Florida Game and Freshwater Fish Commission, Tallahassee, FL. [18]

Fitzpatrick, J. W., M. Lammertink, M. D. Luneau, Jr., T. W. Gallagher, B. R. Harrison, G. M. Sparling, K. V. Rosenberg, R. W. Rohrbaugh, E. C. H. Swarthout, P. H. Wrege, S. Barker Swarthout, M. S. Dantzker, R. A. Charif, T. R. Barksdale, J. V. Remsen, Jr., S. D. Simon, and D. Zollner. 2005. Ivory-billed Woodpecker (*Campephilus principalis*) Persists in Continental North America. *Science* 308:1460–1462. [3]

Flachsenberg, H. and H. Galletti. 1998. Forest management in Quintana Roo, Mexico. In R. Primack, D. Bray, H. Galletti, and I. Ponciano (eds.), *Timber, Tourists, and Temples, Conservation and Development in the Maya Forest of Belize, Guatemala and Mexico,* pp. 33–34. Island Press, Covelo, CA. [8]

Flader, S. L. and J. B. Callicott. 1991. *The River of the Mother of God and Other Essays by Aldo Leopold.* University of Wisconsin Press, Madison. [7]

Flannigan, M. D., B. J. Stocks, and B. M. Wotton. 2000. Climate change and forest fires. *Sci. Total Environ.* 262:221–229. [10]

Flather, C. H., K. R. Wilson, D. J. Dean, and W. C. McComb. 1997. Identifying gaps in conservation networks: of indicators and uncertainty in geographic-based analyses. *Ecol. Appl.* 7:531–542. [14]

Flecker, A. S. 1996. Ecosystem engineering by a dominant detritivore in a diverse tropical stream. *Ecology* 77:1845–1854. [3]

Fleming, D. M., W. F. Wolf, and D. L. DeAngelis. 1994. Importance of landscape heterogeneity to Wood Storks in the Florida Everglades. *Environ. Manage.* 18:743–757. [13]

Fleming, I. A., B. Johnsson, and M. R. Gross. 1994. Phenotypic of sea-ranched, farmed, and wild salmon. *Can. J.Fish. Aquat. Sci.* 51:2808–2824. [8]

Florida Natural Areas Inventory and Department of Natural Resources. 1990. Guide to the natural communities of Florida. Tallahassee, FL. http://fnai.org/PDF/Natural_Communities_Guide.pdf. [18]

Foley, C. A. H., S. Papageorge, and S. K. Wasser. 2001. Non-invasive stress and reproductive measures of social and ecological pressures in free-ranging African elephants (*Loxodonta africana*). *Conserv. Biol.* 15:1134–1142. [11]

Foltz, R., F. Denny, and A. Baharudin. 2003. *Islam and Ecology.* Harvard University Press, Cambridge, MA. [4]

Font, I. 1983. Climatología de España y Portugal. Instituto Nacional de Meteorología. Madrid. [14]

Foose, T. J. 1983. The relevance of captive populations to the conservation of biotic diversity. In C. M. Schonewald-Cox, S. M. Chambers, B. MacBryde, and L. Thomas (eds.), *Genetics and Conservation: A Reference for Managing Wild Animal and*

Plant Populations, pp. 374–401. Benjamin/Cummings, Menlo Park, CA. [11]

Foose, T. J. and J. D. Ballou. 1988. Population management: theory and practice. *Int. Zoo. Yrbk.* 27:26–41. [15]

Forcan, P. 1979. A world order for whales. In T. Wilkes (ed.), *Project Interspeak,* pp. 77–82. Graphic Arts Center, Portland. [4]

Ford, E. B. 1945. *Butterflies.* Collins, London. [10]

Ford, J. K. B., G. M. Ellis, and K. C. Balcomb. 2000. *Killer Whales.* UBC Press, Vancouver. [3]

Ford, J. K. B., G. M. Ellis, L. G. Barrett-Lennard, A. B. Morton, R. S. Palm, and K. C. Balcomb. 1998. Dietary specialization in two sympatric populations of killer whales (*Orcinus orca*) in coastal British Columbia and adjacent waters. *Can. J. Zool.* 76:1456–1471. [17]

Foreman, D, J. Davis, D. John, R. Noss, and M. Soulé. 1992. The Wildlands Project Mission Statement. *Wild Earth* (Special Issue):3–4. [4]

Forman, R. T. T. 1995. *Land Mosaics: The Ecology of Landscapes and Regions.* Cambridge University Press, Cambridge. [12]

Forman, R. T. T. and M. Godron. 1986. *Landscape Ecology.* John Wiley & Sons, New York. [7]

Forrest, S. C., D. E. Biggins, L. Richardson, T. W. Clark, T. M. Campbell III, K. A. Fagerstone, and E. T. Thorne. 1988. Population attributes for the black-footed ferret (*Mustela nigripes*) at Meeteetse, Wyoming, 1981–1985. *J. Mammal.* 69:261–273. [15]

Forsman, E. D., S. DeStefano, M. G. Raphael, and R. J. Guitiérrez. 1996. Demography of the Northern Spotted Owl. *Stud. Avian Biol.* 17:1–122. [8, 12]

Foster, D. R. 2002. Thoreau's country: a historical-ecological perspective on conservation in the New England landscape. *Journal of Biogeography* 29:1537–1555. [6]

Foster, D., F. Swanson, J. Aber, I. Burke, N. Brokaw, D. Tilman, and A. Knapp. 2003.The importance of land use legacies to ecology and conservation. *BioScience* 53:77–88. [6]

Fourteenth Dalai Lama (Tenzin Gyatso). 1992. A Tibetan Buddhist perspective on spirit in nature. In S. C. Rockefeller and J. C. Elder (eds.), *Spirit and Nature: Why the Environment is a Religious Issue,* pp. 109–122. Beacon Press, Boston. [4]

Fowler, C. and C. Mooney. 1990. *The Threatened Gene: Food, Politics, and the Loss of Genetic Diversity.* The Lutterworth Press, Cambridge. [5]

Fowler, D., J. N. Cape, M. Coyle, C. Flechard, J. Kuylenstierna, K. Hicks, D. Derwent, C. Hough, A. D. and R. G. Derwent 1990. Changes in the global concentration of tropospheric ozone due to human activities. *Nature* 344:645–648. [6]

Fox, D. 2005. Healing powers. *Conservation in Practice* 6:28–34. [18]

Fox, G. A. 1995. Tinkering with the tinkerer: Pollution verses evolution. *Environ. Health Persp.* 103 Suppl. 4: 93–100. [6]

Fox, W. 1990. *Toward a Transpersonal Ecology: Developing New Directions for Environmentalism.* Shambala, Boston. [4]

Fox, W. 1993. What does the recognition of intrinsic value entail? *Trumpeter* 10:101. [4]

Francis, R. C. 1997. Sustainable use of salmon: its effect on biodiversity and ecosystem function. In C. H. Freese (ed.), *Harvesting Wild Species: Implications for biodiversity conservation,* pp. 626–670. Johns Hopkins University Press, Baltimore, Maryland. [3]

Frank, D. A., S. J. McNaughton, and B. F. Tracy. 1998. The ecology of the Earth's grazing ecosystems. *BioScience* 48:513–521. [6]

Frankel, O. H. 1983. The place of management in conservation. In C. M. Schonewald-Cox, S. M. Chambers, B. MacBryde, and L. Thomas (eds.), *Genetics and Conservation: A Reference for Managing Wild Animal and Plant Populations,* pp. 1–14. The Benjamin/Cummings, Menlo Park, CA. [11]

Frankel, O. H. and M. E. Soulé. 1981. *Conservation and Evolution.* Cambridge University Press, Cambridge. [1, 11]

Frankham, R. 1995. Inbreeding and extinction: A threshold effect. *Conserv. Biol.* 9:792–799. [11]

Frankham, R. 1996. Relationship of genetic variation to population size in wildlife. *Conserv. Biol.* 10:1500–1508. [11]

Frankham, R., J. D. Ballou, and D. A. Briscoe. 2002. *Introduction to Conservation Genetics.* Cambridge University Press, Cambridge, U.K. [11]

Frankham, R., K. Lees, M. Montgomery, P. R. England, E. H. Lowe, and D. A. Briscoe. 1999. Do population size bottlenecks reduce evolutionary potential? *Anim. Conserv.* 2:255–260. [11]

Franklin, A. B., R. J. Gutierrez, J. D. Nichols, M. E. Seamans, G. C. White, G. S. Zimmerman, J. E. Hines, T. E. Munton, W. S. LaHaye, J. A. Blakesley, G. N. Steger, B. R. Noon, D. W. H. Shaw, J. J. Keane, T. L. McDonald, and S. Britting. 2004. Population dynamics of the California Spotted Owl (*Strix occidentalis occidentalis*): A meta-analysis. *Ornithol. Monogr.* 54. [17]

Franklin, J. F. and R. T. T. Forman. 1987. Creating landscape patterns by forest cutting: Ecological consequences and principles. *Landscape Ecol.* 1:5–18. [7]

Franklin, J. F., T. A. Spies, R. Van Pelt, et al. 2002. Disturbances and structural development of natural forest ecosystems with silvicultural implications, using Douglas-fir forests as an example. *Forest Ecol. Manag.* 155:399–423. [17]

Franson, R. L. 1995. *What a restorationist needs to know about plant population genetics.* Abstracts, Society for Ecological Restoration Conference, Seattle, WA. [15]

Fraser, D. J. and L. Bernatchez. 2001. Adaptive evolutionary conservation: Towards a unified concept for defining conservation units. *Mol. Ecol.* 10:2741–2752. [11]

Fraver, S. 1994. Vegetation responses along edge-to-interior gradients in the mixed hardwood forests of the Roanoke River Basin, North Carolina. *Conserv. Biol.* 8:822–832. [7]

Freckleton, R. P., D. M. Silva Matos, M. L. A. Bovi, and A. R. Watkinson. 2003. Predicting the impacts of harvesting using structured population models: The importance of density-dependence and timing of harvest for a tropical palm tree. *J. Appl. Ecol.* 40:846–858. [8]

Frederick, P. C. and M. W. Collopy. 1989. Nesting success of five Ciconiiform species in relation to water conditions in the Florida Everglades. *Auk* 106:625–634. [13]

Fredericksen, T. S. and B. Mostacedo. 2000. Regeneration of timber species following selection logging in a Bolivian tropical dry forest. *Forest Ecology and Management* 131:47–55. [8]

Fredericksen, T. S. and F. E. Putz. 2003. Silvicultural intensification for tropical forest conservation. *Biodivers. Conserv.* 12:1445–1453. [8]

Free, J. B. 1993. *Insect Pollination of Crops.* Academic Press, San Diego. [2]

Freedman, J. 1978. *Happy People: What Happiness Is, Who Has It, and Why.* Harcourt Brace Jovanovich, New York. [5]

Freeman, S. 2002. *Biological Science.* Prentice Hall, Upper Saddle River, N.J. [2]

Freeman, M, C. M. Pringle, E. Greathouse, and B. Freeman. 2003. Ecosystem-level consequences of migratory faunal depletion caused by dams. In K. E. Limburg and J. R. Waldman, (eds), *Biodiversity and Conservation of the World's Shads. Am. Fish. Soc. Symp.* 35:255–266. [7]

Freese, C. H. (ed.). 1997. *Harvesting Wild Species: Implications for Biodiversity.* Johns Hopkins University Press, Baltimore. [8]

Freestone, D. and E. Hey. 1996. *The Precautionary Principle and International Law: The Challenge of implementation.* Kluwer Law International, The Hague Netherlands. [5]

Freitas, J. V. 2004. Alternative Approaches to Tree Selection for Felling and Retention in the Management of Natural Forests in the Brazilian Amazon. Ph.D. dissertation, University of Aberdeen, Aberdeen, UK. [8]

French McCay, D. P., C. H. Peterson, J. T. DeAlteris, and J. Catena. 2003. Restoration that targets function as opposed to structure: the feasibility of replacing lost bivalve production and filtration. *Mar. Ecol. Prog. Ser.* 264:197–212. [15]

Friesen, L. E., P. F. J. Eagles, and R. J. MacKay. 1995. Effects of residential development on forest-dwelling neotropical migrant songbirds. *Conserv. Biol.* 9:1408–1414. [7]

Fritts, T. H. 2002. Economic costs of electrical system instability and power outages caused by snakes on the island of Guam. *Int. Biodeter. Biodegrad.* 49:93–100. [9]

Fritts, T. H. and D. Leasman-Tanner. 2001. The brown tree snake on Guam: How the arrival of one invasive species damaged the ecology, commerce, electrical systems, and human health on Gaum. A comprehensive information source. http://www.fort.usgs.gov/resources/education/bts/bts_home.asp. [9]

Fritts, T. H. and G. H. Rodda. 1998. The role of introduced species in the degradation of island ecosystems: A case history of Guam. *Ann. Rev. Ecol. Syst.* 29:113–140. [9]

Fryxell, J., J. B. Falls, E. A. Falls, R. J. Brooks, L. Dix, and M. Strickland. 2001. Harvest dynamics of mustelid carnivores in Ontario, Canada. *Wildlife Biol.* 7:151–159. [8]

Fuerst, P. A and T. Maruyama. 1986. Considerations on the conservation of alleles and genic heterozygosity in small managed populations. *Zoo Biol.* 5:171–180. [15]

Fuller, R. A, P. J. K. McGowan, J. P. Carroll, R. W. R. J. Dekker, and P. J. Garson. 2003. What does IUCN species action planning contribute to the conservation process? *Biol. Conserv.* 112:343–349. [17]

Funk, W. C. and W. W. Dunlap. 1999. Colonization of high-elevation lakes by long-toed salamanders (*Ambystoma macrodactylum*) after the extinction of introduced trout populations. *Can. J. Zool.* 77:1759–1767. [3]

Furhman, J. A. and L. Campbell. 1998. Marine ecology: Microbial diversity. *Nature* 393:410–411. [2]

Gaff, H., D. L. DeAngelis, L. J. Gross, R. Salinas, and M. Shorrosh. 2000. A dynamic landscape model for fish in the Everglades and its application to restoration. *Ecol. Model.* 127:33–52. [12]

Gaff, H., J. Chick, J. Trexler, D. DeAngelis, L. Gross, and R. Salinas. 2004. Evaluation of and insights from ALFISH: A spatially explicit, landscape-level simulation of fish populations in the Everglades. *Hydrobiologia* 520:73–87. [12]

Gaffney, J. F. 1996. Ecophysiological and technological factors influencing the management of cogon grass (*Imperata cylindrica*). Ph.D. Dissertation. Agronomy Department, University of Florida, Gainesville, Florida. [13]

Galindo, C. and J. Honey-Roses. 2004. *La tala Ilegal y su impacto en la Reserva de la Biosfera Mariposa Monarca.* WWF-Programa, Mexico City. http://www.wwf.org.mx/wwfmex/descargas/010604_Informe_Tala_Reserva.doc (accessed June 1, 2004). [3]

Galloway, J. N., J. D. Aber, J. W. Erisman, S. P. Seitzinger, R. W. Howarth, E. B. Cowling, and B. J. Cosby. 2003. The nitrogen cascade. *BioScience* 53:341–356. [6]

Gandini, P., P. D. Boersma, E. Frere, M. Gandini, T. Holik, and V. Lichtschein. 1994. Magellanic penguins (*Spheniscus magellanicus*) are affected by chronic petroleum pollution along the coast of Chubut, Argentina. *Auk* 111:20–27. [12]

Ganzhorn, J. U., J. Fietz, E. Rakotovao, D. Schwab, and D. Zinner. 1999. Lemurs and the regeneration of dry deciduous forest in Madagascar. *Conserv. Biol.* 13:794–804. [3]

Garcia, M. B. 2003. Demographic viability of a relict population of the critically en-

dangered plant *Borderea chouardii*. *Conserv. Biol.* 17:1672–1680. [12]

García-Murillo, P., M. Bernúes, and C. Montes. 1993. Los macrófitos acuáticos del Parque nacional de Doñana (SW España). Aspectos florísticos. *Actas VI Congreso Español de Limnología*. [14]

García-Novo, F.; J. Merino Ortega, L. Ramírez, L. Rodenas, F. Sancho, A. Torres, F. González- Bernaldez, F. Díaz, C. Allier, V. Bresset, and A. Lacoste. 1977. Doñana. Prospección e inventario de sus ecosistemas. ICONA. Monografía N 18. Publicaciones del Ministerio de Agricultura, Pesca y Alimentación. Secretaría General Técnica. Madrid. [14]

García-Novo, F.; L. Ramírez, and A. Martínez. 1976. El sistema de dunas de Doñana. Naturalia Hispanica N 5. Publicaciones del Ministerio de Agricultura, Pesca y Alimentación. Secretaría General Técnica. Madrid. [14]

Gardner, G. I. 2002. *Invoking the Spirit: Religion and Spirituality in the Quest for a Sustainable World*. Worldwatch Institute, New York. [4]

Gardner, R. C. 1996. Banking on entrepreneurs: Wetlands, mitigation banking, and takings. *Iowa L. Rev.* 81:527–588. [17]

Gardner, R. H., B. T. Milne, M. G. Turner, and R. V. O'Neill. 1987. Neutral models for the analysis of broad-scale landscape pattern. *Landscape Ecol.* 1:19–28. [7]

Gardner, R. H., V. H. Dale, R. V. O'Neill, and M. G. Turner. 1992. A percolation model of ecological flows. In A. J. Hansen and F. diCastri, (eds.), *Landscape Boundaries: Consequences for Biotic Diversity and Ecological Flows*, pp. 259–269. Springer-Verlag, New York. [7]

Garland, T. and W. G. Bradley. 1984. Effects of a highway on Mojave Desert rodent populations. *Am. Midl. Nat.* 111:47–56. [7]

Garrido, J. M., N. Cortabaria, J. A. Oguiza, and R. A. Juste. 2000. Use of a PCR method on fecal samples for diagnosis of sheep paratuberculosis. *Vet. Microbiol.* 77:379–386. [11]

Garrity, D. P. 2004. Agroforestry and the achievement of the Millennium Development Goals. *Agroforest. Syst.* 62:5–17. [1]

Gartner, W. C. 1996. *Tourism Development: Principles, Processes, and Policies*. John Wiley & Sons, Inc., Hoboken, N.J. [16]

Gascoigne, J. and R. N. Lipcius. 2004. Conserving populations at low abundance: Delayed functional maturity and Allee effects in reproductive behaviour of the queen conch. *Strombus gigas*. *Mar. Ecol. Prog. Ser.* 284:185–194. [12]

Gaston, K. J. 1991. The magnitude of global insect species richness. *Conserv. Biol.* 5:283–296. [4]

Gaston, K. J. 1994. *Rarity*. Chapman & Hall, London. [3]

Gaston, K. J. 2000. Global Patterns in Biodiversity. *Nature* 405:220–227. [2]

Gaston, K. J. and J. I. Spicer 2004. *Biodiversity: An Introduction*, 2nd Edition. Blackwell, Oxford. [2]

Gates, J. E. and L. W. Gysel. 1978. Avian nest dispersion and fledgling success in field-forest ecotones. *Ecology* 59:871–883. [7]

Gautier-Hion, A. and G. Michaloud. 1989. Are figs always keystone resources for tropical frugivorous vertebrates? A test in Gabon. *Ecology* 70:1826–1833. [12]

Gelbard, J. L. and S. Harrison. 2003. Roadless habitats as refuges for native grasslands: interactions with soil, aspect, and grazing. *Ecol. Appl.* 13:404–415. [7]

Geller, J. B. 1999. Decline of a native mussel masked by sibling species invasion. *Conserv. Biol.* 13:661–664. [9]

Gentry, A. H. 1986. Endemism in tropical versus temperate plant communities. In M. E. Soulé (ed.), *Conservation Biology: The Science of Scarcity and Diversity*, pp. 153–181. Sinauer Associates, Sunderland, MA. [7]

Gentry, A. H. 1988. Changes in plant community diversity and floristic composition of environmental and geographical gradients. *Annals Missouri Botanical Garden* 75:1–34. [2]

Gentry, A. H. 1992. Tropical forest biodiversity: Distributional patterns and their conservational significance. *Oikos* 63:19–28. [12]

Gentry, A. H. and C. H. Dodson. 1987. Diversity and biogeography of Neotropical vascular epiphytes. *Annals of the Missouri Botanical Garden* 74: 205–233. [10]

Geoghegan, J. 2002. The value of open spaces in residential land use. *Land Use Policy* 19:91–98. [5]

Gerber, L. R., D. P. DeMaster, and P. M. Karieva. 1999. Gray whales and the value of monitoring data in implementing the U.S. Endangered Species Act. *Conserv. Biol.* 13:1215–1219. [12]

Gerber, L. R., L. Botsford, A. Hastings, H. Possingham, S. Gaines, S. Palumbi, and S. Andelman. 2003. Population models for reserve design: A retrospective and prospective synthesis. *Ecol. Appl* 13:S47–S64. [14]

Gergel, S. E. and M. G. Turner (eds.). 2002. *Learning Landscape Ecology*. Springer-Verlag. New York, NY. [7]

Gerlach, G. and K. Musolf. 2000. Fragmentation of landscapes as a cause for genetic subdivision in bank voles. *Conserv. Biol.* 14:1066–1074. [7]

Ghiselin, M. T., 1974. A radical solution to the species problem. *Syst. Zool.* 23:536–44. [4]

Gibbons, J. W., D. E. Scott, T. J. Ryan, K. A. Buhlmann, T. D. Tuberville, B. S. Metts, J. L. Greene, T. Mills, Y. Leiden, S. Poppy, and C. T. Winne. 2000. The global decline of reptiles, deja vu amphibians. *BioScience* 50:653–666. [3]

Gibbs, J. P. and A. R. Breisch. 2001. Climate warming and calling phenology of frogs near Ithaca, New York, 1900–1999. *Conserv. Biol.* 15:1175–1178. [3, 10]

Gibbs, J. P. and W. G. Shriver. 2002. Estimating the effects of road mortality on turtle populations. *Conserv. Biol.* 16:1647–1652. [7]

Gilbertson, M. K., G. D. Haffner, K. G. Drouillard, A. Albert, and B. Dixon. 2003. Immunosuppression in the northern leopard frog (*Rana pipiens*) induced by pesticide exposure. *Environ. Toxicol. Chem.* 22:101–110. [3]

Gilk, S. E., I. A. Wang, C. L. Hoover, W. W. Smoker, S. G. Taylor, A. K. Gray, and A. J. Gharrett. 2004. Outbreeding depression in hybrids between spatially separated pink salmon, *Oncorhynchus gorbuscha*, populations: Marine survival, homing ability, and variability in family size. *Environ. Biol. Fish.* 69:287–297. [11]

Gill, J. A., K. Norris, P. M. Potts, T. G. Gunnarsson, P. W. Atkinson, and W. J. Sutherland. 2001. The buffer effect and large-scale population regulation in migratory birds. *Nature* 412:436–438. [10]

Gillespie, G. R. 2001. The role of introduced trout in the decline of the spotted tree frog (*Litoria spenceri*) in south-eastern Australia. *Conserv. Biol.* 100:187–198. [3]

Gillett, N. P., F. W. Zwiers, A. J. Weaver, and P. A. Stott. 2003. Detection of human influence on sea-level pressure. *Nature* 422:292–294. [10]

Gilligan, D. M., L. M. Woodworth, M. E. Montgomery, D. A. Briscoe, and R. Frankham. 1997. Is mutation accumulation a threat to the survival of endangered populations? *Conserv. Biol.* 11:1235–1241. [11]

Gillson, L. and K. J. Willis. 2004. 'As Earth's testimonies tell:' Wilderness conservation in a changing world. *Ecol. Lett.* 7:990–998. [18]

Gilpin, M. E. and M. E. Soulé. 1986. Minimum viable populations: Processes of species extinction. In M. E. Soule (ed.), *Conservation Biology: The Science of Scarcity and Diversity*, pp. 19–34. Sinauer Associates, Sunderland, MA. [11, 12]

Ginoga, K., O. Cacho, E. Erwidodo, M. Lugina, and D. Djaenudin. 2002. Economic performance of common agroforestry systems in Southern Sumatra: Implications for carbon sequestration services. Working paper CC03. ACIAR project ASEM 1999/093. (http://www.une.edu.au/febl/Econ/carbon/) [1]

Girardot, N. J., X. Liu, and J. Miller. 2001. *Daoism and Ecology*. Harvard University Press, Cambridge, MA. [4]

Glynn, P. W. 1990 Coral mortality and disturbance to the corals reefs in the tropical Eastern Pacific. In P. W. Glynn (ed.), *Global Ecological Consequences of the 1982–1983 El Niño-Southern Oscillation*, pp. 55–126. Elsevier, Amsterdam. [10]

Godoy, R., D. Wilkie, H. Overman, A. Cubas, G. Cubas, J. Demmer, K. Mcsweeney, and N. Brokaw. 2000. Valuation of consumption and sale of forest goods from a Central American rain forest. *Nature* 406:62–63. [8]

Goldburg, R. and R. Naylor. 2005. Future seascapes, fishing, and fish farming. *Front. Ecol. Environ.* 3:21–28. [18]

Goldschmidt, T., F. Witte, and J. Wanink. 1993. Cascading effects of the introduced Nile perch on the detritivorous/phytoplanktivorous species in the sublittoral areas of Lake Victoria. *Conserv. Biol.* 7:686–700. [9]

Goldsmith, Sir James. 1994. *The Trap*. Carroll & Graf Publishers, Inc., NY. [18]

Goldstein, B. 1992. The struggle over ecosystem management at Yellowstone. *BioScience* 42:83–187. [12]

Goñi, R. 2000. Fisheries effects on ecosystems. In C. R. C. Sheppard (ed.), *Seas at the Millennium: An Environmental Evaluation*, pp.117–133. Pergamon, Amsterdam. [13]

González, M. J., L. M. Hernández, M. C. Rico, and G. Baluja. 1984. Residues of organochlorine pesticides, polychlorinated biphenzis and heavy metals in the eggs of predatory birds from Doñana National Park (Spain), 1980–1983. *J. Environ. Sci. Health*, B19 (8,9): 759–772. [14]

González-Solís, J., J. C. Guix, E. Mateos, and L. Llorens. 2001. Population density of primates in a large fragment of the Brazilian rainforest. *Biodivers. Conserv.* 10:1267–1282. [7]

Goodpaster, K. E. 1978. On being morally considerable. *J. Philos.* 75:308–325. [4]

Goodwin, H., I. Kent, K. Parker, and M. Walpole. 1998. Tourism, conservation and sustainable development: Case Studies from Asia and Africa. International Institute for Environment and Development. *Wildlife and Development Series, No. 12.* [16]

Goreau, T. J. and A. H. Macfarlane. 1990. Reduced growth rate of *Montastrea annularis* following the 1987–1988 coral bleaching event. *Coral Reefs* 8:211–215. [13]

Gorenflo, L. J. and K. Brandon. 2005. Agricultural capacity and conservation in high biodiversity forest ecosystems. *Ambio.* 34:199–204. [5]

Gosling, L. M., S. J. Baker, and C. N. Clarke. 1988. An attempt to remove coypus (*Myocastor coypus*) from a wetland habitat in East Anglia. *J. Appl. Ecol.* 25:49–62. [9]

Gosling, M. and W. J. Sutherland. 2000. *Behavior and Conservation.* Cambridge University Press, New York. [11]

Gotelli, N. J. 2004. A taxonomic wish-list for community ecology. *Philos. T. Roy. Soc. B* 359(1444):585–597 [2]

Gould, S. J. 1989. *Wonderful Life: The Burgess Shale and the Nature of History.* W. W. Norton, New York. [2]

Goulding, M. N., J. H. Smith, and D. J. Mahar. 1996. *Floods of Fortune: Ecology and Economy along the Amazon.* Columbia University Press, New York. [7]

Goymann, W., E. Mostl, T. Van't Hof, M. L. East, and H. Hofer. 1999. Non-invasive fecal monitoring of glucocorticoids in spotted hyenas, *Crocuta crocuta. Gen. Comp. Endocr.* 14:340–348. [11]

Grabherr, G., M. Gottfried, and H. Pauli, 1994: Climate effects on mountain plants. *Nature* 369:448. [10]

Grace, J., B. Frank, and N. Laszlo. 2002. Impacts of climate change on the tree line. *Ann. Bot.* (London) 90: 537–544. [10]

Graczyk, T. K., A. J. DaSilva, M. R. Cranfield, J. B. Nizeyu, G R Kalima, and N. J. Pieniazek. 2001. *Cryptosporidium parvum* phenotype 2 infections in free-ranging mountain gorillas (*Gorilla gorilla beringei*) of the Bwindi Impenetrable National Park, Uganda. *Parasitol. Res.* 87:368–370. [11]

Gradwhol J. and R. Greenberg 1988. *Saving the Tropical Forests.* Island Press, Washington, D.C. [4]

Graham, C. H. 2001. Factors influencing movement patterns of keel-billed toucans in a fragmented tropical landscape in southern Mexico. *Conserv. Biol.* 15:1789–1798. [18]

Gramlich, E. M. 1990. *A Guide to Benefit-Cost Analysis.* Prentice-Hall, Englewood Cliffs, N.J. [5]

Granados, M. 1987. Transformaciones históricas de los ecosistemas del Parque Nacional de Doñana. Ph. D. dissertation. University of Sevilla. Sevilla. [14]

Grassle, J. F. 1989. Species diversity in deep-sea communities. *Trends Ecol. Syst.* 4:12–15. [2]

Grassle, J. F. 1991. Deep-sea benthic biodiversity. *BioScience* 41:464–469.[2]

Gray, A. 2002. The evolutionary context: a species perspective. In M. R. Perrow, and A. J. Davy (eds.), *Handbook of Ecological Restoration: Volume I, Principles of Restoration.* Cambridge University Press, Cambridge. [15]

Great Barrier Reef Marine Park Authority. 2003. Representative Areas in the Marine Park. www.gbrmpa.gov.au/corp_site/key_issues/conservation/rep_areas/. Accessed 10 December 2004. [14]

Greater Yellowstone Ecosystem. 1994. The natural and scenic values of rural private lands. In *Sustaining Greater Yellowstone, A Blueprint for the Future*, pp. 12–13. Greater Yellowstone Coalition, Bozeman, MT. [7]

Green, D. E., and C. K. Sherman. 2001. Diagnostic histological findings in Yosemite toads (*Bufo canorus*) from a die-off in the 1970s. *J. Herpetol.* 35:92–103. [3]

Green, R. E., S. J. Cornell, J. P. W. Scharlemann, and A. Balmford. 2005. Farming and the fate of wild nature. *Science* 307:550–555. [18]

Greenslade, P. J. M. and P. Greenslade. 1984. Soil surface insects of the Australian arid zone. In H. G. Cogger and E. E. Cameron (eds.), *Arid Australia*, pp. 153–176. Australian Museum, Sydney. [2]

Greenstone, M. H. 1984. Determinants of web spider species diversity: Vegetation structural diversity vs. prey availability. *Oecologia* 62:299–304. [2]

Gregg K. B. 2004. Recovery of showy lady's slippers (*Cypripedium reginae* Walter) from moderate and severe herbivory by white-tailed deer (*Odocoileus virginianus* Zimmerman). *Nat. Area. J.* 24: 232–241. [8]

Greig, J. C. 1979. Principles of genetic conservation in relation to wildlife management in Southern Africa. *S. Afr. Tydskr. Naturnav.* 9:57–78. [11]

Grenfell, B. T. and A. P. Dobson. 1995. *Ecology of Infectious Diseases in Natural Populations.* Cambridge University Press, Cambridge. [3]

Gresh, T., J. Lichatowich, and P. Schoonmaker. 2000. An estimation of historic and current levels of salmon production in the northeast Pacific ecosystem. *Fisheries* 25: 15–21. [7]

Grevstad, F. S., D. R. Sg, D. Garcia-Rossi, R. W. Switzer, and M. S. Wecker. 2003. Biological control of *Sparlina alterniflora* in Willapa Bay, Washington using the planthopper *Prokelisia marginata*: Agent specificity and early results. *Biological Control* 27:32–42. [9]

Griffith, B., J. M. Scott, J. W. Carpenter, and C. Reed. 1989. Translocation as a species conservation tool: status and strategy. *Science* 245:477–480. [11, 15]

Griffiths, D. 1997. Local and regional species richness in North American lacustrine fish. *J. Anim. Ecol.* 66:49–56. [2]

Grimalt, J. O. and E. Macpherson, (eds) 1999. The environmental impact of the mine tailing accident in Aznalcóllar (Southwest Spain). *The Science of the Total Environment.* Special Issue, 242: 1–337. [14]

Grinnell, J., and T. I. Storer 1924. *Animal Life in the Yosemite.* University of California Press, Berkely. [3]

Groisman, P. Y., R. W. Knight, T. R. Karl 2001. Heavy precipitation and high streamflow in the contiguous united states: Trends in the twentieth century. *B. Am. Meteorol. Soc.* 82:219–246. [10]

Groisman, P. Y., T. R. Karl, D. R. Easterling, R. W. Knight, P. F. Jamason, K. J. Hennessy, R. Suppiah, C. M. Page, J. Wibig, K. Fortuniak, V. N. Razuvaev, A. Douglas, E. Førland, and P. Zhai. 1999. Changes in the probability of heavy precipitation: Important indicators of climatic change. *Climatic Change* 42:243–283. [10]

Groom, M. J. 1998. Allee effects limit population viability of an annual plant. *Am. Nat.* 151:487–496. [7, 12]

Groom, M. J. 2001. Consequences of subpopulation isolation for pollination, herbivory, and population growth in *Clarkia concinna concinna* (Onagraceae). *Biol. Conserv.* 100:55–63. [7]

Groom, M. J. and M. A. Pascual. 1997. The analysis of population persistence: an outlook on the practice of viability analysis. In P. L. Fiedler and P. M. Kareiva (eds.), *Conservation Biology for the Coming Decade*, pp. 3–33. Chapman and Hall, New York. [12]

Groombridge, B. and M. D. Jenkins. 2002. Global biodiversity: responding to the change. In B. Groombridge, and M. D. Jenkins, (eds.), *World Atlas of Biodiversity: Earth's Living Resources in the 21st Century*, pp. 195–223. University of California Press, Berkeley, CA. [7, 14]

Grove, A. T. and O. Rackham. 2001. *The Nature of Mediterranean Europe: An Ecological History.* Yale University Press, New Haven, CT. [6]

Groves, C. R. 2003. *Drafting a Conservation Blueprint: A Practitioner's Guide to Planning for Biodiversity.* Island Press, Washington, D. C. [14, 18]

Groves, C. R., D. B. Jensen, L. L. Valutis, K. H. Redford, M. L. Shaffer, J. M. Scott, J. V. Baumgartner, J. V. Higgins, M. W. Beck, and M. G. Anderson. 2002. Planning for biodiversity conservation:

Putting conservation science into practice. *BioScience* 52:499–512. [2, 14]

Grumbine, R. E. 1992. *Ghost Bears: Exploring the Biodiversity Crisis*. Island Press, Washington, D.C. [1]

Grumbine, R. E. 1994. What is ecosystem management? *Conserv. Biol.* 8:27–38. [13]

Guariguata, M. R. and G. P. Saenz. 2002. Post-logging acorn production and oak regeneration in a tropical montane forest Costa Rica. *Forest Ecol. Manag.* 167:285–293. [8]

Guariguata, M. R., H. Arias-Le Claire, and G. Jones. 2002. Tree seed fate in a logged and fragmented forest landscape, Northeastern Costa Rica. *Biotropica* 34:405–415. [8]

Guerrant, E. O. Jr. and P. L. Fiedler. 2004. Accounting for sample decline during ex situ storage and reintroduction. In Guerrant et al., *Ex Situ Plant Conservation. Supporting Species in the Wild*, pp. 365–386. Island Press, Washington D.C. [15]

Guerrant, E. O. Jr., K. Havens, and M. Maunder. 2004. *Ex Situ Conservation. Supporting Species Survival in the Wild*. Island Press, Washington, D.C. [15]

Guha, R. 1989a. *The Unquiet Woods: Ecological Change and Peasant Resistance in the Himalaya*. University of California Press, Berkeley. [4]

Guha, R. 1989b. Radical American environmentalism: A Third World critique. *Envir. Ethics* 11:71-83. [4]

Guillette, E. A., M. M. Meza, M. G. Aquilar, A. D. Soto, and I. E. Garcia. 1998. An anthropological approach to the evaluation of children exposed to pesticides in Mexico. *Environ. Health Persp.* 106:347–353. [6]

Guillette, L. J Jr, T. S. Gross, G. R. Masson, J. M. Matter, H. F. Percival, and A. R. Woodward. 1994. Developmental abnormalities of the gonad and abnormal sex hormone concentrations in juvenile alligators from contaminated and control lakes in Florida. *Environ. Health Perspect.* 102:680–688. [11]

Guillette, L. J., Jr. 2000. Contaminant-induced endocrine disruption in wildlife. 2000. *Growth Horm. IGF Res.* Suppl. B:S45–S50. [6]

Guillette, L. J., Jr. and D. A. Crain. 2000. *Endocrine Disrupting Contaminants: An Evolutionary Perspective*. Taylor and Francis, Inc., Philadelphia. [6]

Guillette, L. J., Jr. and M. P. Gunderson. 2001. Alterations in the development of the reproductive and endocrine systems of wildlife exposed to endocrine disrupting contaminants. *Reproduction* 122:857–864. [6]

Guillette, L. J., Jr. and T. M. Edwards. 2005. Is nitrate an ecologically relevant endocrine disruptor in vertebrates? *Integr. Comp. Biol.* 45:19–27. [6]

Guillette, L. J., Jr., A. R. Woodward, D. A. Crain, D. B. Pickford, A. A. Rooney, and H. F. Percival. 1999. Plasma steroid concentrations and male phallus size in juvenile alligators from seven Florida lakes. *Gen. Comp. Endocr.* 116:356–372. [6]

Guillette, L. J., Jr., D. A. Crain, A. A. Rooney, and D. B. Pickford. 1995. Organization versus activation: The role of endocrine-disrupting contaminants (EDCs) during embryonic development in wildlife. *Environ. Health Persp.* 103 Suppl. 7:157–164. [6]

Guillette, L. J., Jr., D. A. Crain, M. Gunderson, S. Kools, M. R. Milnes, E. F. Orlando, A. A. Rooney, and A. R. Woodward. 2000. Alligators and endocrine disrupting contaminants: A current perspective. *Am. Zool.* 40:438–452. [6]

Guillette, L. J., Jr., P. M. Vonier, and J. A. McLachlan. 2002. Affinity of the alligator estrogen receptor for serum pesticide contaminants. *Toxicology* 181–182:151–154. [6]

Guinand, B., K. T. Scribner, K. S. Page, and M. K. Burnham-Curtis. 2003. Genetic variation over space and time: Analyses of extinct and remnant lake trout populations in the upper Great Lakes. *Proc. R. Soc Lon. Ser. B* 270:425–434. [11]

Gulf of Maine Council on the Marine Environment. 2002. *Action Plan 2001–2006*. http://www.gulfofmaine.org. [13]

Gullison R. E. 1998. Will bigleaf mahogany be conserved through sustainable use? In E. J. Milner-Gulland and R. Mace (eds.), *Conservation of Biological Resources*, pp. 193–205. Blackwell Science, Oxford. [8]

Gullison, R. E. and E. C. Losos. 1993. The role of foreign debt in deforestation in Latin-America. *Conserv. Biol.* 7(1):140–147. [6]

Gunawardena, U. A. D. P. 1997. *Economic evaluation of conservation benefits: A case study of Sinharaja Rain Forest Reserve in Sri Lanka*. Unpublished Ph.D. thesis, University of Edinburgh, U.K. [5]

Gunawardena, U. A. D. P., G. Edwards-Jones, M. J. McGregor, and P. Abeygunawardena. 1999. A contingent valuation approach for a tropical rain forest: A case-study of Sinharaja rain forest reserve in Sri Lanka. In C. S. Roper and A. Park (eds.), *The Living Forest: Non-Market Benefits of Forestry*, pp. 275–285. The Stationery Office, London. [5]

Gunderson, L. H. 1992. Spatial and temporal hierarchies in the Everglades ecosystem. Ph.D. dissertation, University of Florida, Gainesville. [13]

Gunderson, L. H. 1994. Vegetation of the Everglades: Composition and determinants. In S. M. Davis and J. C. Ogden (eds.), *Everglades, The Ecosystem and Its Restoration*. St. Lucie Press, Delray Florida. [13]

Gunderson, L. H. 2000. Resilience in theory and practice. *Ann. Rev. Ecol. Syst.* 31:425–439. [13]

Gunderson, L. H. and W. F. Loftus. 1993. The Everglades. In W. H. Martin, S. G. Boyce, and A. C. Echternacht (eds.), *Biodiversity of the Southeastern United States*. John Wiley and Sons, New York. [13]

Gunderson, M. P., D. S. Bermudez, T. A. Bryan, S. Degala, T. M. Edwards, S. A. E. Kools, M. R. Milnes, A. Woodward, and L. J. Guillette, Jr. 2004. Variations in sex steroids and phallus size in juvenile American alligators (*Alligator mississippiensis*) from 3 sites within the Kissimmee-Everglades drainage in Florida (USA). *Chemosphere* 56:335–345. [6]

Guo, Y. L., T. J. Lai, S. H. Ju, Y. C. Chen, and C. C. Hsu. 1993. Sexual developments and biological findings in Yucheng children. *Chemosphere* 14:235–238. [6]

Gurd, D. B., T. D. Nudds, and D. H. Rivard. 2001. Conservation of mammals in eastern North American wildlife reserves: How small is too small? *Conserv. Biol.* 15:1355–1363. [7]

Gurnell, A. M. 1998. The hydrogeomorphological effects of beaver dam-building activity. *Prog. Phys. Geogr.* 22:167–189. [9]

Gustafson, E. J. and G. R. Parker. 1992. Relationships between landcover proportion and indices of landscape spatial pattern. *Landscape Ecol.* 7:101–110. [7]

Guthrie, R. D. 2003. Rapid body size decline in Alaskan Pleistocene horses before extinction. *Nature* 426:169–171. [3]

Gutierrez, R. J. and eight others. 1992. In *The California Spotted Owl: A Technical Assessment of its Current Status*, pp. 79–147. General Technical Report PSW-GTR-133. Albany, California, Pacific Southwest Research Station, U.S. Forest Service, U.S. Department of Agriculture. [17]

Guzman, J. N., F. J. García, and G. Garrote. 2002. *Censo diagnóstico de las poblaciones de Lince Ibérico (Lynx pardinus) en España, 200-2002*. Dirección General de Conservación de la Naturaleza. Ministerio de Medio Ambiente. Madrid. [14]

Haber, W. A. 2000. Plants and vegetation. In N. M. Nadkarni and N. T. Wheelwright (eds.), *Monteverde: Ecology and Conservation of a Tropical Cloud Forest*, pp. 39–94. Oxford University Press, New York. [10]

Haber, W. A., W. Zuchowski, and E. Bello. 2000. *An Introduction to Cloud Forest Trees Monteverde, Costa Rica*, 2nd Edition. Mountain Gem Publications, Monteverde, Costa Rica. [10]

Hackney, C. T. and G. F. Yelverton. 1990. Effects of human activities and sea level rise on wetland ecosystems in the Cape Fear River Estuary, North Carolina, USA. In D. F. Whigham, J. Kuet, and R. E. Good (eds.), *Wetland Ecology and Management: Case Studies*, pp. 55–61. Kluwer Academic Publishers, Dordrect, The Netherlands. [10]

Haddad, N. M., J. Haarstad, and D. Tilman. 2000. The effects of long-term nitrogen loading on grassland insect communities. *Oecologia* 124:73–84. [6]

Hagan, J. M., W. M. Vander Haegen, and P. S. McKinley. 1996. The early development of forest fragmentation effects on birds. *Conserv. Biol.* 10:188–202. [7]

Haig., J., D. Ballou, and S. R. Derrickson. 1990. Management options for preserving genetic diversity: Reintroduction of Guam rails to the wild. *Conserv. Biol.* 4:290–300. [11]

Haig, S. M. and J. D. Ballou. 1995. Genetic diversity among two avian species formerly endemic to Guam. *Auk* 112:445–455. [15]

Haila, Y., I. K. Hanski, and S. Raivio. 1993. Turnover of breeding birds in small forest fragments: The "sampling" colonization hypothesis corroborated. *Ecology* 74:714–725. [7]

Hajek, A. 2004. *Natural Enemies: An Introduction to Biological Control.* Cambridge University Press, Cambridge. [9]

Hall, M. P. and D. B. Fagre. 2003. Modeled climate-induced glacier change in Glacier National Park, 1850–2100. *BioScience* 53:131–140. [10]

Hall, P., S. Walker, and K. Bawa. 1996. Effect of forest fragmentation on genetic diversity and mating system in a tropical tree, *Pithecellobium elegans. Conserv. Biol.* 10:757–768. [7]

Hall, R., J. L. Tank, and M. F. Dybdahl. 2003. Exotic snails dominate nitrogen and carbon cycling in a highly productive stream. *Front. Ecol. Environ.* 1:407–411. [9]

Halle, F. R., A. A. Oldemann, and P. B. Tomlinson. 1978. *Tropical trees and forests: An architectural analysis.* Springer-Verlag, Berlin. [2]

Halliday, T. R. 1980. The extinction of the Passenger Pigeon, *Ectopistes migratorius*, and its relevance to contemporary conservation. *Biol. Conserv.* 17:157–162. [12]

Halpern B. S. 2003. The impact of marine reserves: Do reserves work and does reserve size matter? *Ecol. Appl.* 13:S117–S137. [14]

Halverson, M. A., D. K. Skelly, J. M. Kiesecker, and L. K. Freidenburg. 2003. Forest mediated light regime linked to amphibian distribution and performance. *Oecologia* 134:360–364. [3]

Halvorson, C. H., J. T. Harris, and S. M. Smirenski (eds.). 1995. *Cranes and Storks of the Amur River: The Proceedings of the International Workshop.* Arts Literature Publishers, Moscow. [12]

Hamann, A., and E. Curio. 1999. Interactions among frugivores and fleshy fruit trees in a Philippine submontane rainforest. *Conserv. Biol.* 13:766–773. [3]

Hames, R. B., and W. T. Vickers. 1982. Optimal diet breadth theory as a model to explain variability in Amazonian hunting. *Am. Ethnol.* 9:358–378. [8]

Hamilton, A. C. 1994. *Deforestation in Uganda.* Oxford University Press, Nairobi. [5]

Hamilton, J., A. Juvik, and F. N. Scatena (eds.). *Tropical Montane Cloud Forests.* Springer Verlag, New York, NY [15]

Hamilton, W. D. 1971. Geometry for the selfish herd. *J. Theor. Biol.* 31:295–311. [12]

Hammel, B. E., M. H. Grayum, C. Herrera, and N. Zamora (eds.). 2003. *Manual de Plantas de Costa Rica, Vol. III Monocotiledóneas (Orchidaceae-Zingiberaceae).* Missouri Botanical Garden Press, St. Louis. [10]

Hammond, P. S., S. A. Mizroch, and G. P. Donovan (eds.). 1990. *Individual Recognition Of Cetaceans: Use Of Photo-Identification and Other Techniques to Estimate Population Parameters: Incorporating the Proceedings of the Symposium and Workshop on Individual Recognition and the Estimation of Cetacean Population Parameters.* Report of the International Whaling Commission (Spec. Issue 12). International Whaling Commission, Cambridge, U.K. [17]

Hamrick, J. L., J. B. Mitton, and Y. B. Linhart. 1981. Levels of genetic variation in trees: Influence of life history characteristics. In M. T. Conkle (ed.), *Isozymes of North American Forest Trees and Forest Insects*, pp. 35–41. University of California, Berkley. [11]

Hamrick, J. L., Y. B. Linhart, and J. B. Mitton. 1979. Relationships between life history characteristics and electrophoretically detectable genetic variation in plants. *Ann. Rev. Ecol. Syst.* 10:173–200. [11]

Haneman, W. M. 1988. Economics and the preservation of biodiversity. In E. O. Wilson (ed.), *Biodiversity*, pp. 193–199. National Academy Press, Washington, D.C. [4]

Hanley, N. and C. Spash. 1993. *Cost-Benefit Analysis and the Environment.* Edward Elgar, Aldershot, U.K. [5]

Hanselmann, R., A. Rodriguez, M. Lampo, L. Fajardo-Ramos, A. A. Aguirre, A. M. Kilpatrick, J. P. Rodriguez, and P. Daszak. 2004. Presence of an emerging pathogen of amphibians in introduced bullfrogs *Rana catesbeiana* in Venezuela. *Biol. Conserv.* 120:115–119. [3]

Hansen, A. J. and D. L. Urban. 1992. Avian response to landscape pattern: The role of species life histories. *Landscape Ecol.* 7:163–180. [7]

Hansen, A. J. and J. J. Rotella. 2000. Bird responses to forest fragmentation. In R. L. Knight, F. W. Smith, S. W. Buskirk, W. H. Romme, and W. L. Baker, (eds.), *Forest Fragmentation in the Southern Rocky Mountains*, pp. 201–219. University Press of Colorado, Boulder. [7]

Hansen, A. J., T. A. Spies, F. J. Swanson, and J. L. Ohmann. 1991. Conserving biodiversity in managed forests: Lessons from natural forests. *BioScience* 41:382–392. [8]

Hansen, M., B. Benz, and D. Turner. 2001. Temperature-based model for predicting univoltine brood proportions in spruce beetle (Coleoptera: Scolytidae). *Can. Entomol.* 133:827–841. [10]

Hanski, I. 1990. Density dependence, regulation and variability in animal populations. *Philos. T. Roy. Soc. B.* 330:141–150. [3]

Hanski, I. 1998. Metapopulation dynamics. *Nature* 396:41–49. [3]

Hanski, I. 1999. Habitat connectivity, habitat continuity, and metapopulations in dynamic landscapes. *Oikos* 87:209–219. [12]

Hanski, I. 2001. Spatially realistic theory of metapopulation ecology. *Naturwissenschaften* 88:372–381. [12]

Hanski, I. and M. Gilpin. 1997. *Metapopulation Biology: Ecology, Genetics and Evolution.* Academic Press, New York. [7, 11, 12]

Hanski, I. and O. Ovaskainen. 2000. The metapopulation capacity of a fragmented landscape. *Nature* 404:755–758. [12]

Hanski, I. and O. Ovaskainen. 2002. Extinction debt at extinction threshold. *Conserv. Biol.* 16:666–673. [3]

Hanski, I., A. Moilanen, and M. Gyllenberg. 1996b. Minimum viable metapopulation size. *Am. Nat.* 147:527–541. [12]

Hanski, I., A. Moilanen, T. Pakkala, and M. Kuussaari. 1996a. The quantitative incidence function model and persistence of an endangered butterfly metapopulation. *Conserv. Biol.* 10:578–590. [12]

Hansson, L. 1991. Dispersal and connectivity in metapopulations. In M. E. Gilpin and I. Hanski (eds.), *Metapopulaton Dynamics: Empirical and Theoretical Investigations*, pp. 89–103. Linnaean Society of London and Academic Press, London. [7]

Harcourt, A. H., S. A. Parks, and R. Woodroffe. 2001. Human density as an influence on species/area relationships: Double jeopardy for small African reserves? *Biodivers. Conserv.* 10:1011–1026. [12]

Hard, J. J. 1995. A quantitative genetic perspective on the conservation of intraspecific diversity. *Am. Fish. Soc. Symp.* 17:304–326. [11]

Hardin, G. 1968. The tragedy of the commons. *Science* 162:1243–1248. [4, 8]

Hardin, R. and P. Auzel. 2001. Wildlife utilization and the emergence of viral diseases. In M. I. Bakarr, G. A. B. de Fonseca, R. Mittermeier, A. B. Rylands, and K. Walker Painemilla (eds.). *Hunting and Bushmeat Utilization in the African Rain Forest: Perspectives toward a blueprint for conservation action*, pp. 85–92. Washington DC: Conservation International-Center for Applied Biodiversity Sciences. [8]

Harding, E. K., E. E. Crone, B. D. Elderd, J. M. Hoekstra, A. J. McKerrow, J. D. Perrine, J. Regetz, L. J. Rissler, A. G. Stanley, E. L. Walters, and the Habitat Conservation Plan Working Group of the National Center for Ecological Analysis and Synthesis. 2001. The scientific foundations of habitat conservation plans: a quantitative assessment. *Conserv. Biol.* 15:488–500. [18]

Harmon, D. 1996. Losing species, losing languages: connections between biological and linguistic diversity. *Southwest Journal of Linguistics* 15:89–108. [2]

Harrington, R., I. Woiwod, and T. Sparks. 1999. Climate change and trophic interactions. *Trends Ecol. Evol.* 14:146–150. [10]

Harris, J. T. 1994. Cranes, people, and nature: Preserving the balance. In H. Higuchi and J. Minten (eds.), *The Future of Cranes and Wetlands*, pp. 1–14. Proceedings of the International Symposium, Wild Bird Society of Japan, Tokyo. [12]

Harris, L. D. 1984. *The Fragmented Forest: Island Biogeography Theory and the Preservation of Biotic Diversity.* University of Chicago Press, Chicago. [7]

Harris, L. D. and G. Silva-Lopez. 1992. Forest fragmentation and the conservation of biological diversity. In P. L. Fiedler and S. K. Jain (eds.), *Conservation Biology: The Theory and Practice of Nature Conservation, Preservation, and Management*, pp. 197–237. Chapman and Hall, New York. [7]

Harris, L. D. and P. B. Gallagher. 1989. New initiatives for wildlife conservation: The need for movement corridors. In G.

MacKintosh (ed.), *Preserving Communities and Corridors*, pp. 11–34. Defenders of Wildlife, Washington, D.C. [7]

Harris, L. G. and M. C. Tyrell. 2001. Changing community states in the Gulf of Maine: Synergism between invaders, overfishing and climate change. *Biol. Invasions* 3:9–21. [9]

Harris, P. 1981. Stress as a strategy in the biological control of weeds. In G. Papavizas (ed.), Beltsville Symposia in Agricultural Research. 5. *Biological Control in Crop Production*, pp. 333–340. Allanheld, Osmun, Totowa, Beltsville, N.J.[9]

Harris, R. B. and F. W. Allendorf. 1989. Genetically effective population size of large mammals: An assessment of estimators. *Conserv. Biol.* 3:181–191. [11]

Harris, R. B., L. H. Metzgar and C. D. Bevins. 1986. *GAPPS: Generalized Animal Population Projection System, Version 3.0.* Montana Cooperative Wildlife Research Unit, University of Montana, Missoula. [12]

Harris, S. C., T. H. Martin, and K. W. Cummins. 1995. A model for aquatic invertebrate response to Kissimmee River restoration. *Restoration Ecology* 3: 181–194. [15]

Harrison, I. J. and M. L. J. Stiassny. 1999. The quiet crisis: A preliminary listing of the freshwater fishes of the world that are extinct or "missing in action." In R. D. E. MacPhee (ed.), *Extinctions in Near Times: Causes, Contexts, and Consequences*, pp. 271–332. Kluwer Academic/Plenum Publishers. [8]

Harrison, S. 1991. Local extinction in a metapopulation context: An empirical evaluation. *Biol. J. Linn. Soc.* 42:73–88. [12]

Harrison, S. 1994. Metapopulations and Conservation. In P. J. Edwards, R. M. May, and N. R. Webb (eds.), *Large-Scale Ecology and Conservation Biology*, pp. 111–128. Blackwell Science, Oxford, U.K. [7]

Harrison, S. 1997. How natural habitat patchiness affects the distribution of diversity in Californian serpentine chaparral. *Ecology* 78:1898–1906. [2]

Harrison, S. 1999. Local and regional diversity in a patchy landscape: Native, alien and endemic herbs on serpentine. *Ecology* 80:70–80. [2]

Harrison, S. and E. Bruna. 1999. Habitat fragmentation and large-scale conservation: What do we know for sure? *Ecography* 22:225–232. [7]

Harrison, S., B. D. Inouye, and H. D. Safford, 2003. Ecological heterogeneity in the effects of grazing and fire on grassland diversity. *Conserv. Biol.* 17:837–845. [2]

Harrison, S., D. D. Murphy and P. R. Ehrlich. 1988. Distribution of the Bay checkerspot butterfly, *Euphydryas editha bayensis*: Evidence for a metapopulation model. *Am. Nat.* 132:360–382. [7]

Hart, J. and T. Hart. 2003. Rules of engagement for conservation. *Conservation in Practice* 4:14–22. [6]

Hart, P. J. B. and J. D. Reynolds (eds.). 2002. *Handbook of Fish Biology and Fisheries*, Vol. 2: *Fisheries*. Blackwell Publishing, Oxford. [8]

Harte, J. 1993. *The Green Fuse: An Ecological Odyssey*. University of California Press, Berkeley. [4]

Harte, J. and E. Hoffman. 1989. Possible effects of acidic deposition on a Rocky Mountain population of the tiger salamander *Ambystoma tigrinum*. *Conserv. Biol.* 3:149–158. [3]

Harte, J. and R. Shaw. 1995. Shifting dominance within a montane vegetation community: Results of a climate-warming experiment. *Science* 267: 876–880. [10]

Harter, P. J., 2000. Assessing the assessors: The actual performance of negotiated rulemaking. *NYU. Envtl. L.J.* 9:32–59. [17]

Hartl, D. L. and A. G. Clark. 1997. *Principles of Population Genetics*. Sinauer Associates, Sunderland, MA. [11]

Hartshorn, G. S. 1992. Forest loss and future options in Central America. In J. M. Hagan and D. W. Johnston (eds.), *Ecology and Conservation of Neotropical Migrant Landbirds*, pp. 13–19. Smithsonian Institution Press, Washington, D.C. [7]

Harvell, C. D., C. E. Mitchell, J. R. Ward, S. Altizer, A. P. Dobson, R. S. Ostfeld, and M. D. Samuel. 2002. Climate warming and disease risk for terrestrial and marine biota. *Science* 296:2158–2162. [10]

Harvell, C. D., K. Kim, J. M. Burkholder, R. R. Colwell, P. R. Epstein, D. J. Grimes, E. E. Hofmann, E. K. Lipp, A. D. M. E. Osterhaus, R. M. Overstreet, J. W. Porter, G. W. Smith, and G. R. Vasta. 1999. Review: Marine ecology—emerging marine diseases—climate links and anthropogenic factors. *Science* 285:1505–1510. [3, 13]

Harveson, P. M., R. R. Lopez, N. J. Silvy, and P. A. Frank. 2004. Source-sink dynamics of Florida Key deer on Big Pine Key, Florida. *J. Wildl. Mgmt.* 68:909–915. [12]

Harwell, M. A., J. F. Long, A. M. Bartuska, J. H. Gentile, C. C. Harwell, V. Meyers, and J. C. Ogden. 1996. Ecosystem management to achieve ecological sustainability: The case of South Florida. *Environ. Manage.* 20:497–521. [13]

Harwood, J. and K. Stokes. 2003. Coping with uncertainty in ecological advice: Lessons from fisheries. *Trends Ecol. Evol.* 18:617–622. [17]

Haskell, D. G. 1995. A reevaluation of the effects of forest fragmentation on rates of bird-nest predation. *Conserv. Biol.* 9:1316–1318. [7]

Hassell, M. P. 1978. The dynamics of arthropod predator-prey systems. *Monogr. Popul. Biol.*, Number 13. Princeton University Press, Princeton. [12]

Hastings, A. and L. W. Botsford. 2003. Comparing designs of marine reserves for fisheries and for biodiversity. *Ecol. Appl.* 13:S65–S70. [14]

Hawbaker, T. J. and V. C. Radeloff. 2004. Roads and landscape pattern in northern Wisconsin based on a comparison of four road data sets. *Conserv. Biol.* 18:1233–1244. [7]

Hawken, P., A. B. Lovins, and L. H. Lovins. 1999. *Natural Capitalism: Creating the Next Industrial Revolution*. Little, Brown and Company, Boston. [1]

Hawkins, B. A., and M. Holyoak.1998. Transcontinental crashes of insect populations? *Am. Nat.* 152:480–484. [10]

Hawkins, D., E. Shafer, and J. Rovelstad. 1980. *Tourism Planning and Development Issues.* George Washington University, Washington D.C. [16]

Hawksworth, D. L. 1991. The fungal dimension of biodiversity: Magnitude, significance and conservation. *Mycol. Res.* 95:641–655. [2]

Hayes, M. P. and M. R. Jennings. 1986. Decline of ranid frog species in western North America: Are bullfrogs (*Rana catesbeiana*) responsible? *J. Herpetology* 20:490–509. [3]

Hayes, T. B., A. Collins, M. Lee, M. Mendoza, N. Noriega, A. A. Stuart, and A. Vonk. 2002. Hermaphroditic, demasculinized frogs after exposure to the herbicide atrazine at low ecologically relevant doses. *Proc. Nat. Acad. Sci. USA* 99:5476–5480. [3, 11]

Hayes, T., K. Haston, M. Tsui, A. Hoang, C. Haeffele, and A. Vonk. 2003. Atrazine-induced hermaphroditism at 0.1 ppb in American leopard frogs (*Rana pipiens*): Laboratory and field evidence. *Environ. Health Persp.* 111:568–575. [3]

Heal, G. 2001. *Nature and the Marketplace: Capturing the Value of Ecosystem Services.* Island Press. Washington, D.C. [4]

Heath D. D, J. W. Heath, C. A. Bryden, R. M. Johnson, and C. W. Fox. 2003. Rapid evolution of egg size in captive salmon. *Science* 299:1738–1740. [8]

Hebert, P. D. N. and M. E. A. Cristescu. 2002. Genetic perspectives on invasions: The case of the Cladocera. *Can. J. Fish. Aquat. Sci.* 59:1229–1234. [9]

Hecht, S. and A. Cockburn. 1989. *The Fate of the Forest, Developers, Destroyers, and Defenders of the Amazon*. Verso, New York. [4]

Hedgecock, D., M. A. Banks, V. K. Rashbrook, C. A. Dean, S. M. Blankenship. 2001. Applications of population genetics to conservation of Chinook salmon diversity in the Central Valley. In R. L. Brown (ed.), *Fish Bulletin 179: Contributions to the Biology of Central Valley Salmonids*, pp. 45–69. California Department of Fish and Game, Sacramento. [11]

Hedrick, P. W. 1995. Gene flow and genetic restoration: The Florida panther as a case study. *Conserv. Biol.* 9:996–1007. [12]

Hedrick, P. W. 2000. *Genetics of Populations*, 2nd Edition. Jones and Bartlett. [11]

Hedrick, P. W. and M. E. Gilpin. 1997. Genetic effective size of a metapopulation. In I. A. Hanski and M. E. Gilpin (eds.), *Metapopulation Biology: Ecology, Genetics, and Evolution*, pp. 166–181. Academic Press, New York. [11]

Hedrick, P. W. and S. Kalinowski. 2000. Inbreeding depression and conservation biology. *Ann. Rev. Ecol. Syst.* 31:139–162. [11]

Heemsbergen, D. A., M. P. Berg, M. Loreau, J. R. van Haj, J. H. Faber, and H. A. Verhoef. 2004. Biodiversity effects on soil processes explained by interspecific

functional dissimilarity. *Science* 306:1019–1020. [2]

Heilman, G. E., Jr., J. R. Strittholt, N. C. Slosser, and D. A. Dellasala. 2002. Forest fragmentation of the conterminous United States: Assessing forest intactness through road density and spatial characteristics. *BioScience* 52:411–422. [7]

Heimlich-Boran, J. R. 1988. Behavioral ecology of killer whales (*Orcinus orca*) in the Pacific Northwest. *Can. J. Zool.* 66:565–578. [17]

Helfield, J. M. and R. J. Naiman. 2002. Salmon and alder as nitrogen sources to riparian forests in a boreal Alaskan watershed. *Oecologia* 133:573–582. [8]

Helfman, G. S., B. B. Collette, and D. E. Facey. 1997. *The Diversity of Fishes.* Blackwell Science, Malden MA. [12]

Helfrich, L. A., R. J. Neves, and H. Chapman. 2003. Sustaining America's Aquatic Biodiversity: Freshwater Mussel Biodiversity and Conservation. Virginia Cooperative Extension Publication Number 420–523. http://www.ext.vt.edu/pubs/fisheries/420–523/420–523.html. [8]

Helmer, E. H., O. Ramos, T. del M. López, M. Quiñonez, and W. Díaz. 2002. Mapping the forest type and land cover of Puerto Rico, a component of the Caribbean biodiversity hotspot. *Caribb. J. Sci.* 38:165–183. [15]

Henderson, M. T., G. Merriam, and J. Wegner. 1985. Patchy environments and species survival: Chipmunks in an agricultural mosaic. *Biol. Conserv.* 31:95–105. [7]

Henderson, N. 1992. Wilderness and the nature conservation ideal: Britain, Canada and the United States contrasted. *Ambio* 21:394–399. [14]

Hendry, A. P. and S. C. Stearns (eds.). 2004. *Evolution Illuminated: Salmon and Their Relatives.* Oxford University Press, Oxford. [11]

Hendry, A. P., S. M. Vamosi, S. J. Latham, J. C. Heilbuth, and T. Day. 2000. Questioning species realities. *Conserv. Genet.* 1:67–76. [11]

Henley, D. 2002. Population, economy, and environment in island southeast Asia: An historical view with special reference to northern Sulawesi. *Singapore J. Trop. Geo.* 23:167–206. [6]

Heppell, S. S., L. B. Crowder, and D. T. Crouse. 1996. Models to evaluate headstarting as a management tool for long-lived turtles. *Ecological Applications* 6:556–565. [8]

Her Majesty's Stationary Office. 1991. *Economic Appraisal in Central Government: A Technical Guide for Government Departments.* HMSO, London. [5]

Herkert, J. R. 1994. The effects of habitat fragmentation on midwestern grassland bird communities. *Ecol. Appl.* 4:461–471. [7]

Herkert, J. R., D. L. Reinking, D. A. Wiedenfeld, M. Winter, J. L. Zimmerman, W. E. Jensen, E. J. Finck, R. R. Koford, D. H. Wolfe, S. K. Sherrod, M. A. Jenkins, J. Faaborg, and S. K. Robinson. 2003. Effects of prairie fragmentation on the nest success of breeding birds in the midcon-tinental United States. *Conserv. Biol.* 17:587–594. [7]

Hernández, L. M., M. C. Rico; M. J. González, M. A. Hernán, and M. A. Fernández. 1986. Presence and time trends of organochlorine pollutants and heavy metals in eggs of predatory birds of Spain. *J. Field Ornithol.* 57: 270–282. [14]

Hernández, L. M., M. J. González, M. C. Rico, M. A. Fernández, and A. Aranda. 1988. Organochlorine and heavy metal residues in falconiforme and ciconiforme eggs (Spain). *B. Environ. Contam. Tox.,* 40: 86–91. [14]

Hero, J. M. and G. R. Gillespie. 1997. Epidemic disease and amphibian declines in Australia. *Conserv. Biol.* 11:1023–1025. [3]

Hersteinsson, P. and D. W. Macdonald. 1992. Interspecific competition and the geographical distribution of red and arctic foxes *Vulpes vulpes* and *Alopex lagopus*. *Oikos* 64:505–515. [10]

Hessel, D. and R. R. Reuther. 2000. *Christianity and Ecology.* Harvard University Press, Cambridge, MA. [4]

Hesseln H., J. B. Loomis, A. Gonzalaz-Caban, and S. Alexander. 2003. Wildfire effects on hiking and biking demand in New Mexico: A travel cost study. *Journal of Environmental Management* 69:(4) 359–368. [5]

Heyer, W. R., A. S. Rand, C. A. G. da Cruz, and O. L. Peixoto. 1988. Decimations, extinctions, and colonizations of frog populations in southeast Brazil and their evolutionary implications. *Biotropica* 20:230–235. [3]

Heywood, V. H. (ed.). 1995. *Global Biodiversity Assessment.* United Nations Environment Programme. Cambridge University Press, Cambridge. [9, 18]

Hicke, J. A., G. P. Asner, J. T. Randerson, C. Tucker, S. Los, R. Birdsey, J. C. Jenkins, C. Field, and E. Holland. 2002. Satellite-derived increases in net primary productivity across North America, 1982–1998. *Geophys. Res. Lett.* 29: 69-1–69-4. [10]

Hickman, J. C. 1993. *The Jepson Manual: Higher Plants of California.* University of California Press, Berkeley. [9]

Higgins, S. I. and D. M. Richardson. 1999. Predicting plant migration rates in a changing world: The role of long-distance dispersal. *Am. Nat.* 153:464–475. [9]

Hilborn, R. 2004. Ecosystem-based fisheries management: the carrot of the stick? *Marine Ecology Progress Series* 274: 275–278. [13]

Hilborn, R. and C. J. Walters. 1992. Quantitative Fisheries Stock Assessment: Choice, Dynamics, and Uncertainty. Chapman and Hall, New York. [8]

Hilborn, R., J. M. Orensanz, and A. M. Parma. 2005. Institutions, incentives and the future of fisheries. *Philos. T. Roy. Soc. B.* 360:47–57. [18]

Hildreth, R. G. and M. C. Jarman. 2001. The use of science in marine resource management: Can we reconcile the paradigms of science, law, and policy? *Oceans 2001MTS IEEE Proc. Mar. Technol. Soc.* 3:1428–1435. [17]

Hill, A. M. and D. M. Lodge. 1999. Replacement of resident crayfishes by an exotic crayfish: the roles of competition and predation. *Ecological Applications* 9:678–690 [3]

Hill, K. 1996. Zimbabwe's wildlife utilization programs: Grassroots democracy or an extension of state power? *African Studies Review* 39:103–123. [16]

Hobbs, R. J. and J. A. Harris. 2001. Restoration ecology: repairing earth's ecosystems in the new millennium. *Restor. Ecol.* 9: 239–246 [15]

Hockings, M., S. Stolton, and N. Dudley. 2000. *Evaluating Effectiveness: A Framework for Assessing the Management of Protected Areas.* IUCN, Gland, Switzerland and Cambridge, U.K. [14]

Hoegh-Guldberg, O. 1999. Coral bleaching, climate change, and the future of the world's coral reefs. *Mar. Freshwater Res.* 50:839–866. [10, 13]

Hoekstra, J. M., T. M. Boucher, T. H. Ricketts, and C. Roberts. 2005. Confronting a biome crisis: Global disparities of habitat loss and protection. *Ecol. Lett.* 8:23–29. [6]

Hoelzel, A. R., M. Dahlheim, and S. J. Stern. 1998. Low genetic variation among killer whales (*Orcinus orca*) in the eastern North Pacific and genetic differentiation between foraging specialists. *J. Hered.* 89:121–128. [17]

Hoffman, A. A. and P. A. Parsons. 1997. Extreme Environmental Change and Evolution. Cambridge University Press, Cambridge. [10]

Hoffman, J., and V. Moran. 1995. Biological control of *Sesbania punicea* with *Trichapion lativentre*: Diminshed seed production reduces seeding but not the density of a perennial weed. In E. Delfosse and R. Scott (eds), *Proceedings of the VIII International Symposium on Biological Control of Weeds*, p. 203. DSIR/CSIRO, Melbourne. [9]

Hogan, Z. S, P. B. Moyle, B. May, J. Vander Zanden, and I. G. Baird. 2004. The imperiled giants of the Mekong. *Am. Sci.* 92:228–237. [8]

Hokit, D. G. and L. C. Branch. 2003. Associations between patch area and vital rates: Consequences for local and regional populations. *Ecol. Appl.* 13:1060–1068. [12]

Hokit, D. G., B. M. Stith, and L. C. Branch. 1999. Effects of landscape structure in Florida scrub: A population perspective. *Ecol. Appl.* 9:124–134. [7]

Holbrook, S. J., A. J. Brooks, and R. J. Schmitt. 2002 Variation in structural attributes of patch-forming corals and in patterns of abundance of associated fishes. *Marine and Freshwater Research* 53:1045–1053. [2]

Holbrook, S. J., R. J. Schmitt, and J. S. Stephens, Jr. 1997. Changes in an assemblage of temperate reef fishes associated with a climatic shift. *Ecol. Appl.* 7:1299–1310. [10]

Holdridge, L. R. 1967. *Life Zone Ecology.* Tropical Science Center, San Jose, Costa Rica. [2]

Holdsworth, A. R. and C. Uhl. 1997. Fire in Amazonian selectively logged rain forest and the potential for fire reduction. *Ecol. Appl.* 7:713–725. [8]

Holeck, K. T., E. L. Mills, H. J. MacIsaac, M. Dochoda, R. I. Colautti, and A. Ricciardi. 2004. Bridging troubled waters: Understanding links between biological invasions, transoceanic shipping, and other entry vectors in the Laurentian Great Lakes. *Bioscience* 54:919–929. [9]

Holl, K. D. 1998. Do bird perching structures elevate seed rain and seedling establishment in abandoned tropical pasture? *Restor. Ecol.* 6:253–261. [15]

Holl, K. D., E. E. Crone, and C. B. Schultz. 2003. Landscape restoration: moving from generalities to methodologies. *BioScience* 53: 491–502. [12, 15]

Holl, K. D., M. E. Loik, E. H. V. Lin, and I. A. Samuels. 2000. Tropical montane forest restoration in Costa Rica: Overcoming barriers to dispersal and establishment. *Restor. Ecol.* 8:339–349. [15]

Holling, C. S. 1973. Resilience and stability of ecological systems. *Ann. Rev. Ecol. Syst.* 4:1–23. [13]

Holling, C. S. 1978. *Adaptive Environmental Assessment and Management*. John Wiley and Sons, New York. [13]

Holling, C. S. 1978. Overview and conclusions: The approach. In C. S. Holling (ed.), *Adaptive Environmental Assessment and Management*, pp. 1–142. Wiley Intersience, Chichester, NY. [10]

Holling, C. S. 1982. Science for public policy: Highlights of adaptive environmental assessment and management. In W. T. Mason, Jr. and S. Iker (eds.), *Research on Fish and Wildlife Habitat*, pp. 78–91. U.S. Environmental Protection Agency, Washington, D.C. [10]

Holling, C. S. 1986. Resilience of terrestrial ecosystems: local surprise and global change. In W. C. Clark and R. E. Munn (eds.), *Sustainable Development of the Biosphere*, pp. 292–317. Cambridge University Press, Cambridge. [13]

Holling, C. S. 1995. What barriers? What bridges? In L. H. Gunderson, C. S. Holling, and S. S. Light (eds.), *Barriers and Bridges to the Renewal of Ecosystems*, pp. 3–34. Columbia University Press, New York. [13]

Holling, C. S. and G. K. Meffe. 1996. Command and control and the pathology of natural resource management. *Conserv. Biol.* 10:328–337. [10, 13]

Holling, C. S., C. J. Walters, S. M. Davis, J. C. Ogden, and L. H. Gunderson. 1994. The structure and dynamics of the Everglades system: Guidelines for ecosystem restoration. In S. Davis and J. Ogden (ed.), *Everglades, The Ecosystem and its Restoration*. St. Lucie Press, Delray, FL. [13]

Holloway, J. K. 1957. Weed control by an insect. *Scientific American* 187:56–62. [9]

Holmes, T. P., G. M. Blate, J. C. Zweede, R. Periera, P. Barretto, F. Boltz, and R. Bauch. 2002. Financial and ecological indicators of reduced impact logging per-

formance in the eastern Amazon. *Forest Ecol. Manag.* 163:93–110. [8]

Holmquist, J. G., J. M. Schmidt-Gengenback, and B. B. Yoshioka. 1998. High dams and marine-freshwater linkages: Effects on native and introduced fauna in the Caribbean. *Conserv. Biol.* 12:621–630. [7]

Holway, D. A., L. Lach, A. V. Suarez, N. D. Tsutsui, and T. J. Case. 2002. The causes and consequences of ant invasions. *Ann. Rev. Ecol. Syst.* 33:181–233. [9]

Homan, R. N., B. S. Windmiller, and J. M. Reed. 2004. Critical thresholds associated with habitat loss for two vernal pool-breeding amphibians. *Ecol. Appl.* 14:1547–1553. [12]

Honey, M. 1999. *Ecotourism and Sustainable Development: Who owns paradise?* Island Press, Carado, CA. [16]

Honkoop, P. J. C and J. J. Beukema. 1997. Loss of body mass in winter of three intertidal bivalve species: An experimental and observational study of the interacting effects between water temperature, feeding time and feeding behaviour. *J. Exp. Mar. Biol. Ecol.* 212:277–297. [10]

Hooftman, D. A. P., R. C. Billeter, B. Schmid, and M. Diemer. 2004. Genetic effects of habitat fragmentation on common species of Swiss fen meadows. *Conserv. Biol.* 18:1043–1051. [7]

Hooper, E., R. Condit, and P. Legendre. 2002. Responses of 20 native tree species to reforestation strategies for abandoned farmland in Panama. *Ecol. Appl.* 12:1626–1641. [15]

Horner-Devine, M. C., G. C. Daily, P. R. Ehrlich, and C. L. Boggs. 2003. Countryside biogeography of tropical butterflies. *Conserv. Biol.* 17:168–177. [18]

Horsley S. B., S. L. Stout, and D. S. de Calesta. 2003. White-tailed deer impact on the vegetation dynamics of a northern hardwood forest. *Ecol. Appl.* 13:98–118. [8]

Hough, A. D. and R. G. Derwent. 1990. Changes in the global concentration of tropospheric ozone due to human activities. *Nature* 344:645–648. [6]

Houghton, R., D. Skole, and C. Nobre. 2000. Annual fluxes of carbon from deforestation and regrowth in the Brazilian Amazon. *Nature* 403: 301–304. [6]

Houlahan, J. E., C. S. Findlay, B. R. Schmidt, A. H. Meyer, and S. L. Kuzmin. 2000. Quantitative evidence for global amphibian population declines. *Nature* 404:752–755. [3]

Hovel, K. A. 2003. Habitat fragmentation in marine landscapes: Relative effects of habitat cover and configuration on juvenile crab survival in California and North Carolina seagrass beds. *Biol. Conserv.* 110:401–412. [7]

Howard, L. O. and W. F. Fiske. 1911. The importation into the United States of the parasites of the gypsy moth and the brown-tailed moth. *B. U.S. Bur. Entomol.*, Number 91. [12]

Howard, P. C. 1995. *The Economics of Protected Areas in Uganda: Costs, Benefits and Policy Issues.* Unpublished M.S. thesis, University of Edinburgh, U.K. [5]

Howard, P. C., P. Viskanic, T. R. B. Davenport, F. W. Kigenyi, M. Baltzer, C. J. Dickinson, J. S. Lwanga, R. A. Matthews, and A. Balmford. 1998. Complementarity and the use of indicator groups for reserve selection in Uganda. *Nature* 394:472–475. [14]

Howarth, R. B. and S. Farber. 2002. Accounting for the value of ecosystem services. *Ecol. Econ.* 41:421–429. [4]

Howe H. F. 1984. Implications of seed dispersal by animals for tropical reserve management. *Biol. Conserv.* 30:264–281. [8]

Howe, H. F. and J. Smallwood. 1982. Ecology of seed dispersal. *Ann. Rev. of Ecol. Syst.* 13:201–228. [15]

Howe H. F., E. W. Schupp, and L. C. Westley. 1985. Early consequences of seed dispersal for a neotropical tree (*Virola surinamensis*). *Ecology* 66:781–791. [8]

Howells, O. and G. Edwards-Jones. 1997. A feasibility study of reintroducing wild boar *Sus scrofa* to Scotland: Are existing woodlands large enough to support minimum viable populations? *Biolog. Conserv.* 81:77–89. [12]

Hudson, P. J., A. P. Dobson, and D. Newborn. 1992. Do parasites make prey vulnerable to predation? Red grouse and parasites. *J. Anim. Ecol.* 61:681–692. [3]

Huey, R. B., G. W. Gilchrist, and M. Carlsen. 2000. Rapid evolution of a latitudinal cline in body size in an introduced fly. *Science* 287:308–309. [10]

Huffman, J. L. 1991. Truth, purpose, and public policy: Science and democracy in the search for safety. *Envtl. L.* 21:1091–1107. [17]

Hufford, K. M. and S. J. Mazer. 2003. Plant ecotypes: genetic differentiation in the age of ecological restoration. *Trends Ecol. Evol.* 18:147–55. [15]

Hughes, J. B., G. C Daily, and P. R. Ehrlich. 2002. Agricultural policy can help preserve tropical forest birds in countryside habitats. *Ecol. Lett.* 5:121–129. [18]

Hughes, L. 2000. Biological consequences of global warming: Is the signal already apparent? *Trends Ecol. Evol.* 15:56–61. [10]

Hulbert, S. H. 1997. Functional importance vs. keystoneness: Reformulating some questions in theoretical biocenology. *Australian Journal of Ecology* 22:369–382. [12]

Hull, D. L. 1976. Are species really individuals? *Syst. Zool.* 25:174–91. [4]

Hull, D. L. 1978. A matter of individuality. *Philos. Sci.* 45:335–360. [4]

Hulse, D. W., A. Branscomb, and S. G. Payne. 2004. Envisioning alternatives: Using citizen guidance to map future land and water use. *Ecol. Appl.* 14:325–341. [12]

Humavindu, M. N. and J. Stage. 2003. Hedonic pricing in Windhoek townships. *Environ. Devel. Econ.* 8:391–404. [5]

Hunsaker, C. T., B. B. Boroski, and G. N. Steger. 2002. Relations between canopy cover and the occurrence and productivity in California Spotted Owls. In J. M. Scott, P. J. Heglund, M. L. Morrison, J. B. Haufler, M. G. Raphael, W. A. Wall, and F. B. Samson, (eds.), *Predicting Species Oc-*

currences: Issues of Accuracy and Scale, pp. 697–700. Island Press, Washington, D.C. [17]

Hunt, K .E. and S. K. Wasser. 2003. Effect of long-term preservation methods on fecal glucocorticoid concentrations of grizzly bear and African elephant. *Physiol. Biochem. Zool.* 76:918–928. [11]

Huntley, B. 1991. How plants respond to climate change: migration rates, individualism and the consequences for the plant communities. *J. Bot.* 67:15–22. [10]

Huston, M. A. 1980. Soil nutrients and tree species richness in Costa Rican forests. *J. Biogeogr.* 7:147–157. [2]

Huston, M. A. 1994. *Biological Diversity: The Coexistence of Species on Changing Landscapes*. Cambridge University Press, Cambridge. [2]

Hutchings, J. A. 2000. Collapse and recovery of marine fishes. *Nature* 406: 882–5. [8]

Hutchings, J. A. 2001. Conservation biology of marine fishes: Perceptions and caveats regarding assignment of extinction risk. *Can. J. Fish. Aquat. Sci.* 58:108–21. [8]

Hutchings, J. A. and J. D. Reynolds. 2004. Marine fish population collapses: Consequences for recovery and extinction risk. *BioScience* 54:297–309. [8]

Hutchins, M. and W. Conway. 1995. Beyond Noah's Ark: the evolving role of modern zoological parks and aquariums in field conservation. *Int. Zoo. Yrbk.* 34:117–130. [15]

Hutchison, D. W. and A. R. Templeton, 1999. Correlation of pairwise genetic and geographic distance measures: Inferring the relative influences of gene flow and drift on the distribution of genetic variability. *Evolution* 53:1898–1914.

Iguchi, T., H. Watanabe, and Y. Katsu. 2001. Developmental effects of estrogenic agents on mice, fish and frogs: A mini-review. *Horm. Behav.* 40:248–251. [6]

Imbeau, L., M. Monkkonen, and A. Desrochers. 2001. Long term effects of forestry on birds of the eastern Canadian boreal forests: A comparison with Fennoscandia. *Conserv. Biol.* 15:1151–1162. [8]

Imbrie, J. and K. P. Imbrie. 1986. *Ice Ages: Solving the mystery*. Harvard University Press, Cambridge, MA. [10]

Imhoff, M. L., L. Bounoua, T. Ricketts, C. Loucks, R. Harriss, and W. T. Lawrence. 2004. Global patterns in human consumption of net primary production. *Nature* 429:810–873. [1]

INBIO (Instituto Nacional de Biodiversidad Memoria Annual). 1995. Santo Domingo de Heredia, Costa Rica [2]

Ingham, D. S. and M. J. Samways. 1996. Application of fragmentation and variegation models to epigaeic invertebrates in South Africa. *Conserv. Biol.* 10:1353–1358. [7]

Ingram, G. J. and K. R. McDonald. 1993. An update on the decline of Queensland's frogs. In D. Lunney, and D. Ayers (eds.) *Herpetology in Australia: A Diverse Discipline*, pp. 297–303. Royal Zoological Soci-

ety of New South Wales, Mosman, New South Wales, Australia. [3]

INITAA. 1992. Estudio sobre las mortalidades estivales de aves acuáticas del Parque Nacional de Doñana. Informe preliminar. Junio 1992. Centro de Investigación y Tecnología del INITAA. Ministerio de Agricultura, Pesca y Alimentación. Report. [14]

Inman, R., K. T. Scribner, J. Warrillow, H. H. Prince, and D. Luukkonen. 2003. Canada goose harvest derivation from genetic analysis of tail feathers. *Wildlife Soc. B.* 31:1126–1131. [11]

Interagency Ecosystem Management Task Force. 1995. *The Ecosystem Approach: Healthy Ecosystems and Sustainable Economies.* Vol. I: Overview; Vol. II: Implementation Issues; Vol. III: Case Studies. National Technical Information Service, U.S. Department of Commerce, Springfield, VA. [13]

Intergovernmental Panel on Climate Change (IPCC). 2004. *Climate Change 2001: Synthesis Report. A Contribution of Working Groups I, II and III to the Third Assessment Report of the Intergovernmental Panel on Climate Change.* R. T. Watson and the Core Writing Team (eds.). Cambridge University Press, UK. [6]

International Institute for Environment and Development (IIED) in association with the CAMPFIRE Collaborative Group. [16]

IPCC. 2001a. *Climate Change 2001: The Scientific Basis.* Intergovernmental Panel on Climate Change Third Assessment Report. Cambridge University Press, Cambridge and New York. [10]

IPCC. 2001b. *Climate Change 2001: Impacts, Adaptations, and Vulnerability.* Intergovernmental Panel on Climate Change Third Assessment Report. Cambridge University Press, Cambridge, U.K. [10]

IPCC. 2001c. *Climate Change 2001: Synthesis Report.* Intergovernmental Panel on Climate Change Third Assessment Report. Cambridge University Press, Cambridge. [10]

Isaac, N. J. B. and G. Cowlishaw. 2004. How species respond to multiple extinction threats. *Proc. R. Soc. Lon. Ser. B* 271:1135–1141. [3]

ITGE. 1999. *Modelo regional de flujo subterráneo del sistema acuífero Almonte-Marismas y su entorno.* Departamento de Ingeniería del Terreno y Cartográfica. E.T.S. de Ingenieros de Caminos, Canales y Puertos. Universidad Politécnica de Cataluña. Barcelona. Inédito. [14]

IUCN, Conservation International, and NatureServe. 2004. Global Amphibian Assessment. www.globalamphibian.org. [3]

IUCN. 1994. *Guidelines for Protected Areas Management Categories.* Page 261. IUCN, Cambridge, UK and Gland, Switzerland. [14]

IUCN. 1995. A global strategy for the conservation of marine turtles. IUCN Species Survival Commission, Marine Turtle Specialist Group, Gland, Switzerland. [16]

IUCN. 1995. IUCN/SSC guidelines for re-introduction. IUCN Species Survival Commission, Re-introduction Specialist Group. Gland, Switzerland. [15]

IUCN. 2000. *Vision for Water and Nature: A World Strategy for Conservation and Sustainable Management of Water Resources for the 21st Century.* IUCN, Gland, Switzerland. [18]

IUCN. 2001. *IUCN Red List Categories and Criteria: Version 3.1.* IUCN Species Survival Commission. IUCN, Gland, Switzerland. [3]

IUCN. 2003. *Recommendation 29: Poverty and Protected Areas.* World Parks Congress, Durban, South Africa. [12, 14]

IUCN. 2004. *2004 IUCN Red List of Threatened Species.* www.redlist.org. [2, 3, 6, 8, 12, 14, 17]

IUCN/UNEP/WWF. 1980. World Conservation Strategy: Living Resource Conservation for Sustainable Development. Gland, Switzerland. [16]

IUCN/UNEP/WWF. 1991. *Caring for the Earth: A Strategy for Sustainable Living.* Gland, Switzerland. [16]

Jablonski, D. 1991. Extinctions—a paleontological perspective. *Science* 253:754–757. [3]

Jablonski, D. 1995. Extinctions in the fossil record. In J. H. Lawton and R. M. May (eds.), *Extinction Rates* pp. 25–44. Oxford University Press, Oxford. [3]

Jackson, J. B. C. 2001. What was natural in the coastal oceans? *Proc. Nat. Acad. Sci. USA* 98:5411–5418. [8]

Jackson, J. B. C., M. X. Kirby, W. H. Berger, K. A. Bjorndal, L. W. Botsford, B. J. Bourque, R. H. Bradbury, R. Cooke, J. Erlandson, J. A, Estes, T. P. Hughes, S. Kidwell, C. B. Lange, H. S. Lenihan, J. M. Pandolfi, C. H. Peterson, R. S. Steneck, M. J. Tegner, and R. R. Warner. 2001. Historical overfishing and the recent collapse of coastal ecosystems. *Science* 239:629–637. [12]

Jackson, W. 1980. *New Roots for Agriculture.* University of Nebraska Press, Lincoln. [4]

Jackson, W. 1987. *Alters of Unhewn Stone: Science and the Earth.* North Point Press, San Francisco. [4]

Jacobson, S. K. (ed.). 1995. *Conserving Wildlife: International Education and Communication Approaches* Columbia University Press, New York. [18]

Jacobson, S. K. 1990. Graduate education in conservation biology. *Conserv. Biol.* 4:431–440. [1]

Jacobson, S. K. 1999. *Communication Skills for Conservation Professionals.* Island Press. Washington, D.C. [18]

Jacobson, S. K. and S. B. Marynowski. 1997. Public attitudes and knowledge about ecosystem management on Department of Defense Lands in Florida. *Conserv. Biol.* 11:770–781. [18]

Jacobson, S. K., M. D. McDuff, and M. C. Monroe. 2005. *Conservation Education and Outreach Techniques.* Oxford University Press, Oxford. [18]

James, A., K. Gaston, and A. Balmford. 2001. Can we afford to conserve biodiversity? *BioScience* 51:43–52. [14]

Jamieson, D. 2001. Climate change and gobal environmental justice. In C. A. Miller and P. N. Edwards (eds.), *Changing the Atmosphere: Expert Knowledge and Environmental Governance*, pp. 288–307. MIT Press, Cambridge. [4]

Jansson, R., C. Nilsson, and R. Renöfält. 2000. Fragmentation of riparian floras in rivers with multiple dams. *Ecology* 81:988–903. [7]

Janzen, D. H. 1967. Why mountain passes are higher in the tropics. *Am. Nat.* 101:233–249. [2]

Janzen, D. H. 1986. The eternal external threat. In M. E. Soulé (ed.), *Conservation Biology: The Science of Scarcity and Diversity*, pp. 286–303. Sinauer Associates, Sunderland, MA. [3]

Janzen, D. H. 1990. An abandoned field is not a treefall gap. *Vida Silvestre Neotropical* 2:64–67. [15]

Janzen, D. H. 2002. Tropical dry forest: Area de Conservación Guanacaste, northwestern Costa Rica. In M. R. Perrow and A. J. Davy (eds.), *Handbook of Ecological Restoration, Volume 2: Restoration in Practice*. Cambridge University Press, Cambridge, UK. [15]

Janzen, F. J. 1994. Climate change and temperature-dependent sex determination in reptiles *Proc. Nat. Acad. Sci. USA* 91:7487–7490. [10]

Jayewardene, J. 1994. *The Elephant in Sri Lanka*. Distributed by The Wildlife Heritage Trust of Sri Lanka, Colombo. [17]

Jayewardene, J. 1996. Elephant management and conservation in the Mahaweli project areas. *Gajah* 11:6–15. [17]

Jeanrenaud, S. 2002. *People-Oriented Approaches to Global Conservation: Is the Leopard Changing its Spots?* International Institute for Environment and Development, London. [12]

Jenkins, S. H. 1980. A size-distance relation in food selection by beavers. *Ecology* 61:740–746. [9]

Jennings, M. D. 2000. Gap analysis: concepts, methods, and recent results. *Landscape Ecol.* 15:5–20. [14]

Jennings, M. R. 1996. Status of amphibians. In *Sierra Nevada Ecosystem Project: Final Report to Congress: Assessments and Scientific Basis for Management Options*, pp. 921–944. Centers for Water and Wildland Resources, Davis, CA. [3]

Jennings, S. and M. J. Kaiser. 1998. The effects of fishing on marine ecosystems. *Adv. Mar. Biol.* 34:203–314. [13]

Jennings, S., and N.V.C. Polunin. 1995. Comparative size and composition of yield from six Fijian reef fisheries. *J. Fish Biol.* 46:28–46. [8]

Jennings, S., J. D. Reynolds, and S. C. Mills. 1998. Life history correlates of responses to fisheries exploitation. *Proc. R. Soc. Lon. Ser. B* 265:333–339. [3]

Jennings, S., M. J. Kaiser, and J. D. Reynolds. 2001. *Marine Fisheries Ecology*. Blackwell Science, Oxford. [8]

Jennions, M. D., A. P. Moller, and M. Petrie. 2001. Sexually selected traits and adult survival: A meta-analysis. *Q. Rev. Biol.* 76:3–36. [11]

Jerozolimski, A. and C. A. Peres. 2003. Bringing home the biggest bacon: A cross-site analysis of the structure of hunter-kill profiles in Neotropical forests. *Biol. Conserv.* 111:1–11. [8]

Jewell, E. J. (ed.). 2001. *The New Oxford American Dictionary*. Oxford University Press, New York, NY. [15]

Jiminez, J. A., K. A. Hughes, G. Alaks, L. Graham, and R. C. Lacy. 1994. An experimental study of inbreeding depression in a natural habitat. *Science* 266:271–273. [11]

Jobling, S., M. Nolan, C. R. Tyler, G. Brighty, and J. P. Sumpter. 1998. Widespread sexual disruption in wild fish. *Environ. Sci. Technol.* 32:2498–2506. [6]

Joglar, R. L. and P. A. Burrowes. 1996. Declining amphibian populations in Puerto Rico. In R. Powell and R. W. Henderson (eds.), *Contributions to West Indian Herpetology: A Tribute to Albert Schwartz*, pp. 371–380. Society for the Study of Amphibians and Reptiles, Ithaca, NY. [3]

Johannes, R. E. 1998. The case for data-less marine resource management: examples from tropical nearshore finfisheries. *Trends Ecol. Evol.* 13:243–246. [8]

Johns, A. D. 1985. Selective logging and wildlife conservation in tropical rain forests: Problems and recommendations. *Biological Conservation* 31:355–375. [6]

Johnsgard, P. A. 1983. *Cranes of the World*. Indiana University Press, Bloomington. [12]

Johnson, C. N. 2002. Determinants of loss of mammal species during the Late Quaternary "megafauna" extinctions: Life history and ecology, but not body size. *Proc. R. Soc. Lon. Ser. B* 269: 2221–2227. [3]

Johnson, E. A. and S. L. Gutsell. 1994. Fire-frequency models, methods and interpretations. In M. Begon and A.H. Fitter (eds.), *Advances in Ecological Research, Vol. 25*, pp. 239–287. Academic Press, New York. [13]

Johnson, L. E. 1991. *A Morally Deep World: An Essay on Moral Significance and Environmental Ethics*. Cambridge University Press, Cambridge. [4]

Johnson, N. C., A. J. Malk, R. C. Szaro, and W. T. Sexton (eds.). 1999. *Ecological Stewardship: A Common Reference for Ecosystem Management, Volume I*. Elsevier Science, Ltd., Oxford, U.K. [13]

Johnson, O. W., W. S. Grant, R. G. Kope, K. Neely, F. W. Waknitz, and R. S. Waples. 1997. *Status review of chum salmon from Washington, Oregon and California*. NOAA Technical Memorandum NMFS-NWFSC-32. U. S. Department of Commerce. [17]

Johnson, R. G. and S. A. Temple. 1990. Nest predation and brood parasitism of tallgrass prairie birds. *J. Wildl. Mgmt.* 54:106–111. [7]

Johnson, S. D., P. R. Neal, C. I. Peter, and T. J. Edwards. 2004. Fruiting failure and limited recruitment in remnant populations of the hawkmoth-pollinated tree *Oxyanthus pyriformis* subsp. *pyriformis* (Rubiaceae). *Biol. Conserv.* 120:31–39. [7]

Johnson, T., J. Dozier, and J. Michaelsen. 1999. Climate change and Sierra Nevada snowpack. *IAHS Publ.* 256:63–70. [10]

Johnston, C. A. and R. J. Naiman. 1990. Aquatic patch creation in relation to beaver population trends. *Ecology* 71:1617–1621 [8]

Jones, C. G., J. H. Lawton, and M. Schachak. 1994. Organisms as ecosystem engineers. *Oikos* 69:373–386. [9]

Jones, C. G., J. H. Lawton, and M. Shachak. 1997. Positive and negative effects of organisms as physical ecosystem engineers. *Ecology* 78:1946–1957. [9]

Jones, H. L. and J. Diamond. 1976. Short-term-base studies of turnover in breeding bird populations on the California Channel Islands. *Condor* 78:526–549. [3]

Jones R. J., O. Hoegh-Guldberg, A. W. D. Larkum, and U. Schreiber. 1998. Temperature-induced bleaching of corals begins with impairment of the CO_2 fixation mechanism in zooxanthellae. *Plant Cell Environ.* 21:1219–1230. [13]

Jonson, J. E., L. Tarrason, and J. Sundet. 1999. Calculation of ozone and other pollutants for the summer, 1996. *Environmental Management and Health* 10:245–257. [6]

Jules, E. S. and P. Shahani. 2003. A broader ecological context to habitat fragmentation: Why matrix habitat is more important than we thought. *J. Veg. Sci.* 14:459–464. [7]

Jules, E. S., M. J. Kauffman, W. D. Ritts, and A. L. Carroll. 2002. Spread of an invasive pathogen over a variable landscape: A nonnative root rot on Port Orford cedar. *Ecology* 83:3167–3181. [7]

Junta de Andalucía. 2002. *Plan de Ordenación del Territorio. Ambito de Doñana*. Dirección General de Ordenación del Territorio y Urbanismo. Consejería de Obras Públicas y Transportes. Junta de Andalucía. Final Draft. February 2002. [14]

Junta de Andalucía. 2003. *Ciencia y Restauración del Río Guadiamar. PICOVER 1998–2003*. Consejería de Medio Ambiente. Junta de Andalucía. [14]

Kadr, A. B. A. B., A. L. T. E. S. A. Sabbagh, M. A. S. A. Glenid, M.Y. S. Izzidien. 1983. *Islamic Principles for the Conservation of the Natural Environment*. International Union for the Conservation of Nature and Natural Resources, Gland, Switzerland. [4]

Kahn, J. R. and J. A. McDonald. 1995. Third-World Debt and Tropical Deforestation. *Ecol. Econ.* 12(2):107–123. [6]

Kaiser, J. 2004. Plan to count hatchery salmon criticized. *Science* 304:807. [8]

Kallio, P. and J. Lehtonen. 1973. Birch forest damage caused by *Oporinia autumnata* in 1965–66 in Utsjoki, N. Finland. *Rep. Kevo Subarctic.* 10:55–69. [10]

Kaly, U. L. and G. P. Jones. 1998. Mangrove restoration: a potential tool for coastal management in tropical developing countries. *Ambio* 27:656–61. [15]

Kant, I. 1959. *Foundations of the Metaphysics of Morals*. Library of Liberal Arts, New York. [4]

Kappelle, M. and A. D. Brown (eds.). 2001. *Bosques Nublados del Neotrópico*. Instituto Nacional de Biodiversidad, Santo Domingo de Heredia, Costa Rica. [10]

Karatayev A. Y, L. E. Burlakova, and D. K. Padilla. 2002. Impacts of zebra mussels on aquatic communities and their role as ecosystem engineers. In E. Leppäkoski, S. Gollasch, and S. Olenin (eds), *Invasive Aquatic Species of Europe: Distribution, Impacts, and Management*, pp. 433–446. Kluwer Academic Publishers, Dordrecht, The Netherlands. [9]

Karatayev A. Y., L. E. Burlakova, D. K. Padilla, and L. E. Johnson. 2003. Patterns of spread of the zebra mussel (*Dreissena polymorpha* [Pallas]): The continuing invasion of Belarussian lakes. *Biol. Invasions* 5:213–221. [9]

Karatayev, A. Y., L. E. Burlakova, and D. K. Padilla. 1997. The effects of *Dreissena polymorpha* (Pallas) invasion on aquatic communities in Eastern Europe. *J. Shellfish Res.*16:187–203. [9]

Karatayev, A. Y., L. E. Burlakova, and D. K. Padilla. 1998. Physical factors that limit the distribution and abundance of *Dreissena polymorpha* (Pall.). *J. Shellfish Res.* 17:1219–1235. [9]

Kareiva P., S. Andelman, D. Doak, B. Elderd, M. Groom. J. Hoekstra, L. Hood, F. Janes. J. Lamoureux, G. le Buhn, C Mc Culloch, J. Regetz, L. Savage, M. Ruckelshaus, D. D. Kelly, W. Wilbur, K. Zamudio and the NCEAS HCP working group. 1999 *Using Science in Habitat Conservation Plans*. National Center for Ecological Analysis and Synthesis, Santa Barbara. [17]

Kareiva, P. and M. Marvier. 2003. Conserving biodiversity coldspots: recent calls to direct conservation funding to the world's biodiversity hotspots may be bad investment advice. *Am. Sci.* 91: 344–351. [6]

Kareiva, P., S. Andelman, D. Doak, B. Elderd, M. Groom, J. Hoekstra, L. Hood, F. James, J. Lamoreux, G. LeBuhn, C. McCulloch, J. Regetz, L. Savage, M. Ruckelshaus, D. Skelly, H. Wilbur, K. Zamudio and the Habitat Conservation Plan Working Group of the National Center for Ecological Analysis and Synthesis (NCEAS). 1998. *Using Science in Habitat Conservation Plans*. Report. American Institute of Biological Sciences, Washington, D.C., and the NCEAS, Santa Barbara, California. Available from www.nceas.ucsb.edu/projects/hcp. [18]

Karl, J. W., N. M. Wright, P. J. Heglund, and J. M. Scott. 1999. Obtaining environmental measures to facilitate vertebrate habitat modeling. *Wildlife Soc. B.* 27:357–365. [14]

Karl, T. R. and K. E. Trenberth. 2003. Modern global climate change. *Science* 302:1719–1723. [10]

Karl, T. R. and R. W. Knight. 1998. Secular trends of precipitation amount, frequency, and intensity in the United States. *B. Am. Meteorol. Soc.* 79:232–241. [10]

Karl, T. R., R. W. Knight, and B. Baker. 2000. The record-breaking global temperatures of 1997 and 1998: Evidence for an increase in global warming? *Geophys. Res. Lett.* 27:719–722. [10]

Karl, T. R., R. W. Knight, D. R. Easterling, and R. G. Quayle. 1996. Indices of climate change for the United States. *B. Am. Meteorol. Soc.*77:279–292. [10]

Karoly, D. J., K. Braganza, P. A. Stott, J. M. Arblaster, G. A. Meehl, A. J. Broccoli, and K. W. Dixon. 2003. Detection of a human influence on North American climate. *Science* 302:1200–1203. [10]

Karr, J. R. 1982a. Avian extinction on Barro Colorado Island, Panama: A reassessment. *Am. Nat.* 119:220–239. [7]

Karr, J. R. 1982b. Population variability and extinction in the avifauna of a tropical land bridge island. *Ecology* 63:1975–1978. [7]

Karr, J. R. 1991. Biological integrity: A long-neglected aspect of water resource management. *Ecol. Appl.* 1:66–84. [18]

Karr, J. R. and D. R. Dudley. 1981. Ecological perspective on water quality goals. *Environ. Manag.* 5:55–68. [18]

Karr, J. R. and E. M. Rossano. 2001. Applying public health lessons to protect river health. *Ecol. Civil Eng.* 4:3–18. [18]

Karr, J. R. and E. W. Chu. 1999. *Restoring Life In Running Waters: Better Biological Monitoring*. Island Press, Washington, D.C. [18]

Kattan, G. H. and H. Alvarez-López. 1996. Preservation and management of biodiversity in fragmented landscapes in the Colombian Andes. In J. Schelhas and R. Greenberg (eds.), *Forest Patches in Tropical Landscapes*, pp 3–18. Island Press, Washington, D.C. [14]

Kattan, G. H. and P. Franco. 2004. Bird diversity along elevational gradients in the Andes of Colombia: area and mass effects. *Global Ecol. Biogeogr.* 13:451–458. [14]

Kattan, G. H., H. Alvarez-López, and M. Giraldo. 1994. Forest fragmentation and bird extinctions: San Antonio eighty years later. *Conserv. Biol.* 8:138–146. [7, 14, 15]

Kattan, G. H., P. Franco, V. Rojas, and G. Morales. 2004. Biological diversification in a complex region: a spatial analysis of faunistic diversity and biogeography of the Andes of Colombia. *J. Biogeogr.* 31:1829–1839. [14]

Katz, E. 1991. The big lie: human restoration of nature. *Philosphy and Technology* 12: 231–41. [15]

Kaufman, L., B. Heneman, J. T. Barnes, and R. Fugita. 2004. Transition from low to high data richness: An experiment in ecosystem-based fishery management from California. *B. Mar. Sci.* 74:693–708. [13]

Kauffman, M. J., J. F. Pollock, and B. Walton. 2004. Spatial structure, dispersal, and management of a recovering raptor population. *Am. Nat.* 164:582–597. [12]

Kaufman, W. and O. Pilkey. 1983. *The Beaches are Moving: The Drowning of America's Shoreline*. Duke University Press, Durham. [10]

Keane, R. N. and M. J. Crawley 2002. Exotic plant invasions and the enemy release hypothesis. *Trends Ecol. Evol.* 17:164–170. [10]

Kearns, C. A., D. W. Inouye, and N. M. Waser. 1998. Endangered mutualisms: The Conservation of plant-pollinator interactions. *Ann. Rev. Ecol. Syst.* 29: 83–112. [7]

Keeley, J. E., D. Lubin, and C. J. Fotheringham. 2003. Fire and grazing on impacts on plant diversity and alien plant invasions in the Southern Sierra Nevada. *Ecol. Appl.* 13:1355–1374. [9]

Keeling, C. D. and T. P. Whorf. 2004. Atmospheric CO_2 records from sites in the SIO air sampling network. In *Trends: A Compendium of Data on Global Change*. Carbon Dioxide Information Analysis Center, Oak Ridge National Laboratory, Oak Ridge, TN. [10]

Keeling, C. D., J. F. S. Chin, and T. P. Whorf. 1996. Increased activity of northern vegetation inferred from atmospheric carbon dioxide measurements. *Nature* 382:146–149. [10]

Keeney, R. L. 1992. *Value-focused Thinking*. Harvard University Press, Cambridge, MA. [17]

Keer, G. H. and J. B. Zedler. 2002. Salt marsh architecture differs with the number and composition of species. *Ecol. Appl.* 12:456–473. [15]

Keiter, R. B. 1997. Greater Yellowstone's bison: Unraveling of an early American wildlife conservation achievement. *J. Wildl. Mgmt.* 61:1–11. [12]

Keitt, B. S., C. Wilcox, B. R. Tershy, D. A. Croll, and C. J. Donlan. 2002. The effect of feral cats on the population viability of black-vented shearwaters (*Puffinus opisthomelas*) on Natividad Island, Mexico. *Anim. Conserv.* 5:217–223. [9]

Kelleher, G., C. Bleakley, and S. Wells, (eds.) 1995. *A Global Representative System of Marine Protected Areas. Volume 1*. Great Barrier Reef Marine Authority, World Bank and World Conservation Union (IUCN). Environment Department, World Bank Washington, D.C. [14]

Keller, L. F. and D. M. Waller. 2002. Inbreeding effects in wild populations. *Trends Ecol. Evol.* 17:230–241. [7, 11]

Keller, L. F., P. Arcese, J. N. M. Smith, W. M. Hochachka, and S. C. Stearns. 1994. Selection against inbred son sparrows during a natural population bottleneck. *Nature* 372:356–357. [11]

Kellert, S. R. 1996. *The Value of Life: Biological Diversity and Human Society*. Island Press, Washington, D.C. [18]

Kennedy, E. D. and D. W. White. 1996. Interference competition from House Wrens as a factor in the decline of Bewick's Wrens. *Conserv. Biol.* 10:281–284. [12]

Kennett, R., N. Munungurritj, and D.Yunupingu. 2004. Migration patterns of marine turtles in the Gulf of Carpentaria, northern Australia: implications for Aboriginal management. *Wildlife Res.* 31:241–248. [16]

Kepler, C. B. and J. M. Scott. 1985. Conservation of island ecosystems. In P. O. Moors, (ed.), *Conservation of Island Birds*, pp

255–271. Int. Counc. Bird Preservation Tech. Publ. 3. [14]

Kerlinger, P. 2000. *Avian Mortality and Communication Towers: A Review of Recent Literature, Research, and Methodology.* U.S. Fish and Wildlife Service, Office of Migratory Bird Management, Washington, D.C. [5]

Kevan, P. G., E. A. Clark, and V. G. Thomas. 1990. Insect pollinators and sustainable agriculture. *Am. J. Alternative Agr.* 5:12–22. [2]

Kiesecker, J. M. and A. R. Blaustein. 1995. Synergism between UV-B radiation and a pathogen magnifies amphibian embryo mortality in nature. *Proc. Nat. Acad. Sci. USA* 92:11049–11052. [3]

Kiesecker, J. M., A. R. Blaustein, and C. L. Miller. 2001b. Potential mechanisms underlying the displacement of native red-legged frogs by introduced bullfrogs. *Ecology* 82:1964–1970. [3]

Kiesecker, J. M., A. R. Blaustein, and L. K. Belden. 2001a. Complex causes of amphibian population declines. *Nature* 410:681–684. [3]

Kiessling, I. 2003. Finding solutions to derelict fishing gear and other marine debris in Northern Australia. Report to the National Oceans Office, Department of Environment and Heritage, Dhimurru Land Management Aboriginal Corporation and World Wide Fund for Nature Australia. Key Centre for Tropical Wildlife Management, Charles Darwin University, Northern Territory, Australia. [16]

Kiester, R., J. M. Scott, B. C. Csuti, R. Noss, B. Butterfield, and D. White. 1996. Conservation prioritization using gap data. *Conserv. Biol.* 10: 1332–1342. [14]

Killpack, D., B. Waltenberger, and C. Fowler. 2000. Using the Channel Islands spatial support and analysis tool to support group-based decision making for marine reserves. NOAA Coastal Services Center, Charleston, SC. <http://www.csc.noaa.gov/pagis/html/web_cissat_paper.pdf> [14]

Kim, K. 1997. Preserving biodiversity in Korea's demilitarized zone. *Science* 278:242. [6]

Kimura, M. and J. F. Crow. 1963. The measurement of effective population number. *Evolution* 17:279–288. [11]

King, A. W., L. K. Mann, W. W. Hargrove, T. L. Ashwood, and V. H, Dale. 2000. Assessing The Persistence of an Avian Population in a Managed Landscape: A Case Study with Henslow's Sparrow at Fort Knox, Kentucky. ORNL/TM–13734. Oak Ridge National Laboratory, Oak Ridge, Tennessee. [12]

King, C. 1935. *Mountaineering in the Sierra Nevada.* W.W. Norton, New York. [13]

Kinnaird, M. F. and T. G. O'Brien. 2005. Fast foods of the forest: The influence of figs on primates and hornbills across Wallace's Line. In L. Dew and J. Boubli (eds.), *Fruits and Frugivores: The Search for Strong Interactors,* pp. 154–188. Springer-Verlag: Netherlands. [12]

Kinnaird, M. F., E. W. Sanderson, T. H. O'Brien, H. T. Wibisono, and G. Woolmer. 2003. Deforestation trends in a tropical landscape and implications for endangered large mammals. *Conserv. Biol.* 17:245–257. [7, 12]

Kinne, O. (ed.). 1971. *Marine Ecology: A Comprehensive, Integrated Treatise on Life in Oceans and Coastal Waters.* Vol. I. Wiley Interscience, London. [2]

Kirchner, J. W. and A. Weil. 2000. Delayed biological recovery from extinctions throughout the fossil record. *Nature* 404:177–180. [2]

Kirkpatrick, J. B. 1983. An iterative method for establishing priorities for the selection of nature reserves: an example from Tasmania. *Biol. Conserv.* 25:127–134. [14]

Kirkpatrick, S., C. D. Gelatt, and M. P. Vecchi. 1983. Optimisation by simulated annealing. *Science* 220:671–680. [14]

Kishor, N. M. and L. F. Constantino. 1993. Forest management and competing land uses: an economic analysis from Costa Rica. LATEN Dissemination Note 7. The World Bank, Washington, D. C., USA. [8]

Kleijn, D. and W. J. Sutherland. 2003. How effective are European agri-environment schemes in conserving and promoting biodiversity? *J. Appl. Ecol.* 40:947–969. [18]

Kleijn, D., F. Berendse, R. Smit, and N. Gilissen. 2001. Agri-environment schemes do not effectively protect biodiversity in Dutch agricultural landscapes. *Nature* 413:723–725. [18]

Klein, A. M., I. Steffan-Dewenter, and T. Tscharntke. 2003a. Fruit set of highland coffee increases with the diversity of pollinating bees. *Proc. R. Soc. Lond. B Biol. Sci.* 270:955–961. [2]

Klein, A. M., I. Steffan-Dewenter, and T. Tscharntke. 2003b. Pollination of *Coffea canephora* in relation to local and regional agroforestry management. *J. Appl. Ecol.* 40:837–845. [2]

Klein, B. C. 1989. Effects of forest fragmentation on dung and carrion beetle communities in central Amazonia. *Ecology* 70:1715–1725. [7]

Klein, J. A., J. Harte, and X-Q Zhao. 2004. Experimental warming causes large and rapid species loss, dampened by simulated grazing, on the Tibetan Plateau. *Ecol. Lett.* 7:1170–1179. [18]

Kleitz, G. 1998. Les elephants, les paysans et les planificateurs font la course: conservation de la grande faune africaine dans la valley du Zambezie. *Table Ronde Dynamiques Sociales et Environement.* September 1998. Bordeaux: UMR CNRS–ORSTOM. [16]

Klink, C. A. and R. B. Machado. 2005. Conservation of the Brazilian Cerrado. *Conserv. Biol.* 19:707–713. [6]

Knapp, A. K., J. M. Blair, J. M. Briggs, S. L. Collins, D. C. Hartnett, L. C. Johnson, and E. G. Towne. 1999. The keystone role of bison in North American tallgrass prarie. *BioScience* 49:39–50. [3]

Knapp, R. A. 1996. Non-native trout in the natural lakes of the Sierra Nevada: An analysis of their distribution and impacts on native aquatic biota. In *Sierra Nevada Ecosystem Project, Final Report to Congress,* pp. 363–390. Center for Water and Wildland Resources, Davis, CA. [3]

Knapp, R. A. and K. R. Matthews. 2000. Non-native fish introductions and the decline of the mountain yellow-legged frog from within protected areas. *Conserv. Biol.* 14:428–438. [3]

Knapp, R. A., K. R. Matthews, and O. Sarnelle. 2001. Resistance and resilience of alpine lake fauna to fish introductions. *Ecol. Monogr.* 71:401–421. [3]

Knick, S. T. and J. T. Rotenberry. 1995. Landscape characteristics of fragmented shrubsteppe habitats and breeding passerine birds. *Conserv. Biol.* 9:1059–1071. [7]

Knight, R. L. 1997. A field report from the new American West. In C. Meine (ed.), *Wallace Stenger and the Continental Vision,* pp. 181–200. Island Press, Washington, D.C. [7]

Knight, R. L. and G. K. Meffe. 1997. Ecosystem management: agency liberation from command and control. *Wildlife Soc. B.* 25:676–678. [13]

Knight, R. L. and P. B. Landres (eds.). 1998. *Stewardship Across Boundaries.* Island Press, Washington, D.C. [13]

Knight, R. L. and T. W. Clark. 1998. Boundaries between public and private lands: Defining obstacles, finding solutions. In R. L. Knight and P. B. Landres (eds.), *Stewardship Across Boundaries,* pp. 175–191. Island Press, Washington, D.C. [7]

Knight, R. L., G. W. Wallace, and W. E. Riebsame. 1995. Ranching the view: Subdivisions versus agriculture. *Conserv. Biol.* 9:459–461. [7]

Knight, R. L., F. W. Smith, S. W. Buskirk, W. H. Romme, and W. L. Baker (eds.). 2000. *Forest fragmentation in the Southern Rocky Mountains.* University Press of Colorado, Boulder. [7]

Knopman, D. S. and R. A. Smith. 1993. Twenty years of the Clean Water Act. *Environment* 35:16–20, 34–41. [18]

Knowlton, N. 2001. The future of coral reefs. *P. Natl. Acad. Sci. USA* 98:5419–5425. [7]

Koblentz-Mishke, O. J., V. V. Volkovinsky, and J. G. Kabanova. 1970. Plankton primary productivity of the world ocean. In W. S. Wooster (ed.), *Scientific Exploration of the South Pacific,* pp. 183–193. National Academy of Sciences Press, Washington, D.C. [2]

Kock, K-H. 2001. The direct influence of fishing and fishery-related activities on non-target species in the Southern Ocean with particular emphasis on longline fishing and its impact on albatrosses and petrels—a review. *Reviews in Fish Biology and Fisheries* 11: 31–56. [13]

Koebel, J. W., Jr. 1995. An historical perspective on the Kissimmee River restoration project. *Restoration Ecology* 3:149–159. [15]

Kohn, M. H., E. C. York, D. A. Kamraadt, G. Haught, R. M. Sauvajot, and R. K. Wayne. 1999. Estimating population size

by genotyping faeces. *Proc. R. Soc. Lond.* B 266:657–663. [11]

Kokko, H. 1999. Competition for early arrival in migratory birds. *J. Anim. Ecol.* 68:940–950. [10]

Kolar, C. S. and D. M. Lodge. 2002. Ecological predictions and risk assessment for alien fishes in North America. *Science.* 298:1233–1236. [9]

Kolm, N. and A. Berglund. 2003. Wild populations of a reef fish suffer from the "nondestuctive" aquarium trade fishery *Conserv. Biol.* 17:910–914. [8]

Kolpin, D. W., et al. 2002. Pharmaceuticals, Hormones, and Other Organic Wastewater Contaminants in U.S. Streams, 1999–2000: A National Reconnaissance. http://pubs.acs.org/hotartcl/est/es011055j_rev.html [18]

Kondolf, G. M., J. W. Webb, and T. Felando. 1988. Basic hydrologic studies for assessing impacts of flow diversions on riparian vegetation: examples from streams of the eastern Sierra Nevada, California, USA. *Environ. Manage.* 11(6). [13]

Köppen, W. 1884. Die wärmezonen der erde, nach dauer der heissen, gemässigten und kalten zeit, und nach der wirkung der wärme auf die organische welt betrachtet. *Meterologische Zeitschrift* 1:215–226. [2]

Korten, D. C. 1991. Sustainable development. *World Policy J.* 9:157–190. [16]

Kouki, J., D. G. McCullough, and L. D. Marshall. 1997. Effects of forest stand and edge characteristics on the vulnerability of jack pine stands to jack pine budworm (*Choristoneura pinus pinus*) damage. *Can. J. Forest Res.* 27:1765–1772. [7]

Krahn, M. M., M. J. Ford, W. F. Perrin, P. R. Wade, R. P. Angliss, M. B. Hanson, B. L. Taylor, G. M. Ylitalo, M. E. Dahlheim, J. E. Stein, and R. S. Waples. 2004. *2004 Status Review of Southern Resident Killer Whales (Orcinus orca) under the Endangered Species Act.* U.S. Dept. Commer., NOAA Tech. Memo. NMFSNWFSC-62, 73. [17]

Krahn, M. M., P. R. Wide, S. T. Kalinowski, M. E. Dahlheim, B. L. Taylor, M. B. Hanson, G. M. Ylitalo, R. P. Angliss, J. E. Steinn, and R. S. Waples. 2002 *Status Review of Southern Resident Killer Whales (Orcinus orca) under the Endangered Species Act.* NOAA Technical Memorandum NMFS-NWFSC-54.[17]

Kramer R., M. Munasinghe, N. Sharma, E. Mercer, and P. Shyamsundar. 1994. Valuing a protected topical forest: A case study in Madagascar. In M. Munasinghe and J. McNeely (eds.), *Linking conservation and sustainable development*, pp. 191–203. Protected Area Economics and Policy World Bank and World Conservation Union (IUCN), Washington D.C. [5]

Kremaster, L. L. and F. L. Bunnell 1999. Edge effects: Theory, evidence and implications to management of western North American forests. In J. A. Rochelle, L. A. Lehmann, and J. Wisniewski (eds.), *Forest Fragmentation: Wildlife and Management Implications*, pp. 117–153. Brill, Leiden, The Netherlands. [12]

Kremen, C. 2004. Pollination services and community composition: Does it depend on diversity, abundance, biomass, or species traits? In B. M. Freitas, and J. O. P. Pereira, (eds.), *Solitary Bees: Conservation, Rearing and Management For Pollintion*, pp.115–124. University Dederal do Ceara, Ceara. [2]

Kremen, C. 2005. Managing for ecosystem services: What do we need to know about their ecology? *Ecol. Lett.* 8:468–479. [2]

Kremen, C., J. O. Niles, M. G. Dalton, G. C. Daily, P. R. Ehrlich, J. P. Fay, D. Grewal, and R. P. Guillery. 2000. Economic incentives for rain forest conservation across scales. *Science* 288:1828–1832. [3, 16]

Kremen, C., N. M. Williams, and R. W. Thorp. 2002b. Crop pollination from native bees at risk from agricultural intensification. *Proc. Nat. Acad. Sci. USA* 99:16812–16816. [2]

Kremen, C., N. M. Williams, R. L. Bugg, J. P. Fay, and R. W. Thorp. 2004. The area requirements of an ecosystem service: crop pollination by native bee communities in California. *Ecol. Lett.* 7:1109–1119. [2]

Kremen, C., R. L. Bugg, N. Nicola, S. A. Smith, R. W. Thorp, and N. M. Williams. 2002a. Native bees, native plants and crop pollination in California. *Fremontia*: 41–49. [2]

Kremen, C., V. Razafimahatratra, R. P. Guillery, J. Rakotamalala, A. Weiss, and J. S. Ratsisompatrarivo. 1999. Designing the Masoala National Park in Madagascar based on biological and socioeconomic data. *Conserv. Biol.* 13:1055–1068. [3]

Kruger, F. J. 1981. Seasonal growth and flowering rhythms: South African heathlands. In R. L. Specht (ed.), *Heathlands and Related Shrublands*, pp. 1–4. Ecosystems of the World, Vol. 9B. Elsevier, Amsterdam. [2]

Kruger, F. J. and H. C. Taylor. 1979. Plant species diversity in Cape Fynbos: Gamma and delta diversity. *Vegetatio* 41:85–93. [2]

Krummel, J. R., R. H. Gardner, G. Sugihara, R. V. O'Neill, and P. R. Coleman. 1987. Landscape pattern in a disturbed environment. *Oikos* 48:321–324. [7]

Krutilla, J. V. 1967. Conservation reconsidered. *Amer. Econ. Rev.* 57:777–786. [5]

Kuhn, T. S. 1972. *The Structure of Scientific Revolutions*, 2nd Edition. University of Chicago Press, Chicago. [1]

Kullman, L. 2001. Twentieth century climate warming and tree-limit rise in the southern Scandes of Sweden. *Ambio* 30:72–80. [10]

Kumar, S. 2002. Does "participation" in common pool resource management help the poor? A social cost-benefit analysis of joint forest management in Jharkhand, India. *World Devel.* 30:763–782. [5]

Kunin, W. E. and K. J. Gaston (eds.). 1997. *The Biology of Rarity.* Chapman and Hall, London. [3]

Kupferberg, S. J. 1997. Bullfrog (*Rana catesbeiana*) invasion of a California river: The role of larval competition. *Ecology* 78:1736–1751. [3]

Kurki, S., A. Nikula, P. Helle, and H. Lindén. 2000. Landscape fragmentation and forest composition effects on grouse breeding success in boreal forests. *Ecology* 81:1985–1997. [7]

Kurosawa, R. and R. A. Askins. 2003. Effects of habitat fragmentation on birds in deciduous forests in Japan. *Conserv. Biol.* 17:695–707. [7]

Kurz, W. A. and M. J. Apps. 1999. A 70-year retrospective analysis of carbon fluxes in the Canadian forest sector. *Ecol. Appl.* 9:526–547. [10]

LaBarbera, M. 2003. *Economic Importance of Hunting in America.* International Association of Fish and Wildlife Agencies, Washington, D.C. [8]

Lacy, R. C. 1993. VORTEX: A computer simulation model for population viability analysis. *Wildlife Res.* 20:45–65. [11]

Lacy, R. C., K. A. Hughes, and P. S. Miller. 1995. *VORTEX Users Manual, Version 7.* IUCN/SSC Conservation Breeding Specialist Group, Apple Valley, Michigan. [12]

Lacy, R. C., N. R. Flesness, and U. S. Seal. 1989. Puerto Rican Parrot population viability analysis. Captive Breeding Specialist Group, Apple Valley, MN. [15]

Ladd, D. 1991. Reexaminations of the role of fire in Missouri oak woodlands. In G. V. Burger, J. E. Ebinger, and G. S. Wilhelm (eds.). Oak Woods Management Workshop. Eastern Illinois University, IL. [11]

Laerm, J., J. C. Avise, J. C. Patton, and R. A. Lansman. 1982. Genetic determination of the status of an endangered species of pocket gopher in Georgia. *J. Wildl. Mgmt.* 46:513–518. [11]

Lafferty, K. D. 2003. White abalone restoration. United States Geological Survey. (http://www.werc.usgs.gov/coastal/abalone.html) [8]

Lafferty, K. D., M. D. Behrens, G. E. Davis, P. L. Haaker, D. J. Kushner, D. V. Richards, I. K. Taniguchi, and M. J. Tegner 2003. Habitat of endangered white abalone, I. *Biol. Conserv.* 116:191–194. [8]

Laist, D. W. 1997. Impacts of marine debris: entanglement of marine life in marine debris including a comprehensive list of species with entanglement and ingestion records. In J. M. Coe and D. B. Rogers, (eds.), *Marine Debris—Sources, Impacts and Solutions*, pp 99–139. Springer-Verlag, New York. [6]

Lamb, H. F, F. Damblon, and R. W. Maxted. 1991. Human impact on the vegetation of the Middle Atlas, Morocco, during the last 5000 years. *J. Biogeogr.* 18: 519–532. [6]

Lamb, H. H. 1977. *Climate: Past, Present, and Future.* Methuen, London. [6]

Lambeck, R. J. 1997. Focal species: a multispecies umbrella for nature conservation. *Conserv. Biol.* 11:849–856. [14]

Lamberson, R. H., B. R. Noon, C. Voss, and K. S. McKelvey. 1994. Reserve design for territorial species: The effects of patch size and spacing on the viability of the

Spotted Owl. *Conserv. Biol.* 8:185–195. [17]

Lamberson, R. H., R. McKelvey, B. R. Noon, and C. Voss. 1992. A dynamic analysis of northern Spotted Owl viability in a fragmented forest landscape. *Conserv. Biol.* 6:505–512. [17]

Lamont, B. B., A. J. M. Hopkins, and R. J. Hnatiuk. 1984. The Flora—Composition, Diversity and Origins. In J. S. Pate and J. S. Berd (eds.), *Kwongan: Plant Life of the Sandplain*, pp. 27–50. University of Western Australia Press, Nedlands. [2]

Lamont, B. B., P. G. L. Klinkhamer, and E. T. F. Witkowski. 1993. Population fragmentation may reduce fertility to zero in *Banksia goodii*: A demonstration of the Allee effect. *Oecologia* 94:446–450. [7, 12]

Lande, R. 1987. Extinction thresholds in demographic models of territorial populations. *Am. Nat.* 130:624–635. [17]

Lande, R. 1988. Genetics and demography in biological conservation. *Science* 241:1455–1460. [12]

Lande, R. 1996. Statistics and partitioning of species diversity, and similarity among multiple communities. *Oikos* 76:5–13. [2]

Lande, R., B.-E. Saether, and S. Engen. 1997. Threshold harvesting for sustainability of fluctuating resources. *Ecology* 78:1341–1350. [8]

Lande, R., B.-E. Saether, and S. Engen. 2001. Sustainable exploitation of fluctuating populations. In J. D. Reynolds, G. M. Mace, K. H. Redford, and J. G. Robinson (eds.), *Conservation of Exploited Species*, pp. 67–86. Cambridge University Press, Cambridge. [8]

Lande, R., S. Engen, and B.-E. Saether. 2003. *Stochastic Populations Dynamics in Ecology and Conservation*. Oxford University Press, Oxford. [8]

Landres, P. B., P. Morgan, and F. J. Swanson. 1999. Overview of the use of natural variability concepts in managing ecological systems. *Ecol. Appl.* 9:1179–1188. [18]

Lane, R. E. 1991. *The Market Experience*. Cambridge University Press, New York. [5]

Lane, R. E. 1993. Does money buy happiness? *Public Interest* 113:56–65. [5]

Langis, R., M. Zalejko, and J. B. Zedler. 1991. Nitrogen assessments in a constructed and a natural salt marsh of San Diego Bay, California. *Ecological Applications* 1:40–51. [15]

LaRoe, E. T., G. S. Farris, C. E. Puckett, P. D. Doran, and M. J. Mac, (eds.). 1995. *Our Living Resources: A report to the Nation on the Distribution, Abundance, and Health of U.S. Plants, Animals, and Ecosystems*. U.S. Department of the Interior, National Biological Service, Washington, D.C. [18]

Larson, S., C. J. Casson, and S. K. Wasser. 2003. Noninvasive reproductive steroid hormone estimates from fecal samples of captive female sea otters (*Enhydra lutris*). *Gen. Comp. Endocr.* 134:18–25. [11]

Lasswell, H. D. 1971. *A Pre-view to Policy Sciences*. Elsevier Press, New York. [17]

Laurance, S. G. 2004. Responses of understory rain forest birds to road edges in central Amazonia. *Ecol. Appl.* 14:1344–1357. [7]

Laurance, W. F. 1990. Comparative responses of five arboreal marsupials to tropical forest fragmentation. *J. Mammal.* 71:641–653. [7]

Laurance, W. F. 1991. Edge effects in tropical forest fragments: Application of a model for the design of nature reserves. *Biol. Conserv.* 57:205–219. [7]

Laurance, W. F. 1996. Catastrophic declines of Australian rainforest frogs: Is unusual weather responsible? *Biol. Conserv.* 77:203–212. [3]

Laurance, W. F. 2000. Do edge effects occur over large spatial scales? *Trends Ecol. Evol.* 15:134–135. [7]

Laurance, W. F. and M. A. Cochrane. 2001. Synergistic effects in fragmented landscapes. *Conserv. Biol.* 15:1488–1489. [8]

Laurance, W. F. and R. O. Bierregaard, Jr. (eds.). 1997. *Tropical Forest Remnants:Ecology, Management, and Conservation of Fragmented Communities*. University of Chicago Press. Chicago. [7, 18]

Laurance, W. F., D. R. Pérez-Salicrup, P. Delamônica, P. M. Fearnside, S. Agra, A. Jerozolinski, L. Pohl, and T. E. Lovejoy. 2001. Rain forest fragmentation and the structure of Amazonian liana communities. *Ecology* 82:105–116. [8]

Laurance, W. F., K. R. McDonald, and R. Speare. 1996. Epidemic disease and the catastrophic decline of Australian rain forest frogs. *Conserv. Biol.* 10:406–413. [3]

Laurance, W. F., K. R. McDonald, and R. Speare. 1997. In defense of the epidemic disease hypothesis. *Conserv. Biol.* 11:1030–1034. [3]

Lavendel, B. 2003. Ecological restoration in the face of global climate change: Obstacles and initiatives. *Ecol. Rest.* 21:199–203. [10]

Lawes, M. J., P. E. Mealin, and S. E. Piper. 2000. Patch occupancy and potential metapopulation dynamics of three forest mammals in fragmented afromontane forest in South Africa. *Conserv. Biol.* 14:1088–1098. [7]

Lawler, J. J., D. White, and L. L. Master. 2003. Integrating representation and vulnerability: Two approaches for prioritising areas for conservation. *Ecol. Appl.* 13:1762–1772. [14]

Lawler, J. J., S. P. Campbell, A. D. Guerry, M. B. Kolozsvary, R. J. O'Connor, and L. C. N. Seward. 2002. The scope and treatment of threats in endangered species recovery plans. *Ecol. Appl.* 12:663–667. [18]

Lawrence, D. 2004. Erosion of tree diversity during 200 years of shifting cultivation in Bornean rain forest. *Ecol. Appl.* 14:1855–1869. [6]

Lawton, R. O. and V. Dryer. 1980. The vegetation of the Monteverde Cloud Forest Preserve. *Brenesia* 18:101–116. [10]

Lawton, R. O., U. S. Nair, R. A. Pielke, and R. M. Welch. 2001. Climatic impact of tropical lowland deforestation on nearby montane cloud forests. *Science* 294:584–587. [3]

Layman, C. A., D. A. Arrington, R. B. Langerhans, and B. R. Silliman. 2004. Degree of fragmentation affects fish assemblage structure in Andros Island (Bahamas) estuaries. *Caribbean J. Sci.* 40: 232–244. [7]

Lazzaro, X., M. Bouvy, R. A. Ribeiro, V. S. Oliviera, L. T. Sales, A. R. M. Vasconcelos, and M. R. Mata. 2003. Do fish regulate phytoplankton in shallow eutrophic northeast Brazilian reservoirs? *Freshwater Biology* 48:649–668. [3]

Leach, M. K. and T. J. Givnish. 1996. Ecological determinants of species loss in remnant prairies. *Science* 273:1555–1558. [7]

Leberg, P. 1992. Effects of population bottlenecks on genetic diversity as measured by allozymes electrophoresis. *Evolution* 46:477–494. [11]

Leck, C. F. 1979. Avian extinctions in an isolated tropical wet-forest preserve, Ecuador. *Auk* 96:343–352. [7]

Lecointre, G., H. LeGuyader, and D. Visset. 2001. *La Classification Phylogénétique du Vivant*. Belin, Paris. [2]

Lee, K. N. 1993. *Compass and Gyroscope: Integrating Science and Politics for the Environment*. Island Press, Washington, D.C. [13]

Leeworthy, V. R. and P. C. Wiley. 2002. Socioeconomic impact analysis of marine reserve alternatives for the Channel Islands National Marine Sanctuary. U.S. Department of Commerce.National Oceanic and Atmospheric Administration. National Ocean Service. Special Projects.Silver Spring, Maryland. [14]

Leigh, E. G., Jr., S. J. Wright, E. A. Herre, and F. E. Putz. 1993. The decline of tree diversity on newly isolated tropical islands: A test of a null hypothesis and the implications. *Evol. Ecol.* 7:76–102. [3, 7]

LeNoir, J. S., L. L. McConnell, G. M. Fellers, T. M. Cahill, and J. N. Seiber. 1999. Summertime transport of current-use pesticides from California's Central Valley to the Sierra Nevada Mountain Range, USA. *Environ. Toxicol. Chem.* 18:2715–2722. [3]

Leonard, H. and R. Zeckhauser. 1986. Cost–benefit analysis applied to risks: Its philosophy and legitimacy. In D. McLean (ed.), *Values at Risk*, pp. 31–48. Rowman and Littlefield, Totowa, N.J. [5]

Leopold, A. 1933. *Game Management*. Charles Scribners Sons, New York. [7]

Leopold, A. 1949. *A Sand County Almanac and Sketches Here and There*. Oxford University Press, New York. [1, 4]

Leopold, A. 1953. *Round River: From the Journals of Aldo Leopold*. Oxford University Press, New York. [4]

Lerner, I. M. 1954. *Genetic Homeostasis*. Oliver & Boyd, Edinburgh, U.K. [11]

Lesica, P. 1996. Using fire history models to estimate proportions of old growth forest in northwest Montana, USA. *Biol. Conserv.* 77:33–39. [13]

Leslie, H., M. Ruckelshaus, I. R. Ball, S. Andleman, and H. P. Possingham 2003. Using siting algorithms in the design of marine reserve networks. *Ecol. Appl.* 13:S170–S198. [14]

Letourneau, D. K., L. A. Dyer, and G. C. Vega. 2004. Indirect effects of a top predator on a rain forest understory plant community. *Ecology* 85:2144–2152. [3]

Leung, B, J. M. Drake, and D. M. Lodge. 2004. Predicting invasions: Propagule pressure and the gravity of Allee effects. *Ecology* 85:1651–1660. [12]

Lever, C. 1992. *They Dined on Eland: Story of the Acclimatisation Societies.* Quiller Press Limited, Hindringham, U.K. [9]

Levin, D. A. and Crepet, W. L. 1973. Genetic variation in *Lycopodium lucidulum*: A phylogenetic relic. *Evolution* 27:622–632. [11]

Levin, P. S., J. A. Coyer, R. Petrik, and T. P. Good. 2002. Community-wide effects of nonindigenous species on temperate rocky reefs. *Ecology* 83:3182–3193. [9]

Levine, J. M. 2000. Species diversity and biological invasions: Relating local process to community pattern. *Science* 288:852–854. [9]

Levins, R. 1966. Strategy of model building in population biology. *Am. Sci.* 54:421–431. [12]

Levins, R. 1969. Some demographic and genetic consequences of environmental heterogeneity for biological control. *B. Entomol. Soc. Am.* 15:237–240. [12]

Levitan, M. 2003. Climatic factors and increased frequencies of "southern" chromosome forms in natural populations of *Drosophia robusta. Evol. Ecol. Res.* 5:597–604. [10]

Levitus, S., J. I. Antonov, J. Wang, T. L. Delworth, K. W. Dixon, and A. J. Broccoli. 2001. Anthropogenic warming of Earth's climate system. *Science* 292:267–270. [10]

Lewis, D. 2004. Snares vs. Hoes: Why Food Security is Fundamental to Wildlife Conservation. Presentation to the Africa Biodiversity Collaborative Group Food Security and Wildlife Conservation in Africa Meeting, 29 October 2004, Washington, DC. Accessed 14 April 2005: http://www.frameweb.org/ev.php?ID=10410_201&ID2=DO_TOPIC [8]

Lewison, R. L. and L. B. Crowder. 2003. Estimating fishery bycatch and effects on a vulnerable seabird population. *Ecol. Appl.* 13:743–753. [8]

Lewison, R. L., L. B. Crowder, and D. J. Shaver. 2003. The impact of turtle excluder devices and fisheries closures on loggerhead and Kemp's ridley strandings in the western Gulf of Mexico. *Conserv. Biol.* 17:1089–1097. [8]

Li, H. and J. Wu. 2004. Use and misuse of landscape indices. 2004. *Landscape Ecol.* 19:389–399. [7]

Li, Y. and D. S. Wilcove. 2005. Threats to vertebrate species in China and the United States. *BioScience* 55:147–153. [3]

Licht, L. E. 2003. Shedding light on ultraviolet radiation and amphibian embryos. *BioScience* 53:551–561. [3]

Light, S. S., L. H. Gunderson, and C. S. Holling. 1995. The Everglades: Evolution of management in a turbulent environment. In L. H. Gunderson, C. S. Holling, and S. S. Light (eds.), *Barriers and Bridges to the Renewal of Ecosystems and Institutions.* Columbia University Press, New York. [13]

Limpus C. J., J. D. Miller, C. J. Parmenter, D. Reimer, N. McLachlan, R. Webb. 1992.

Migration of green (*Chelonia mydas*) and loggerhead turtles (*Caretta caretta*) turtles to and from Australian rookeries. *Wildlife Res.* 19:347–358. [16]

Limpus, C and D. Reimer. 1994. The loggerhead turtle, *Caretta caretta*, in Queensland: a population in decline. In *Proceedings of the Australian Marine Turtle Workshop*, pp 39–59. Sea World Nara Resort, Gold Coast, 1990. [16]

Limpus, C. J and J. Miller. 1993. Family Cheloniidae. In C. J. Glasby, G. J. B. Ross, and P. L. Beesley (eds.), *Fauna of Australia. Vol 2A Amphibia and Reptilia*, pp. 133–138. Australian Government Publishing Service, Canberra, Australia. [16]

Lindberg, K. and D. Hawkins. 1993. *Ecotourism: a Guide for Planners and Managers (Vol. 1).* The International Ecotourism Society, North Bennington, VT. [16]

Lindenmayer, D. B. and H. P. Possingham. 1996. Ranking conservation and timber management options for Leadbeater's possum in Southeastern Austalia using population viability analysis. *Conserv. Biol.* 10:235–251. [12]

Lindenmayer, D. B., M. A. Burgman, H. R. Akçkaya, R. C. Lacy, and H. P. Possingham. 1995. A review of the generic computer programs ALEX, RAMAS/space and VORTEX for modeling the viability of wildlife metapopulations. *Ecol. Model.* 82:161–174. [12]

Linder, G., S. K. Krest, and D. W. Sparling (eds.). 2003. *Amphibian Decline: An Integrated Analysis of Multiple Stressor Effects.* Society of Environmental Toxicology and Chemistry (SETAC), Pensacola, FL. [3]

Lindig-Cisneros, R. A. 2001. Interactions among species richness, canopy structure, and seedling recruitment. Ph.D. Dissertation, University of Wisconsin-Madison. [15]

Lindig-Cisneros, R. A. and J. B. Zedler. 2002a. Relationships between canopy complexity and germination microsites for *Phalaris arundinacea L. Oecologia* 133:159–167. [15]

Lindig-Cisneros, R. A. and J. B. Zedler. 2002b. Halophyte recruitment in a salt marsh restoration site. *Estuaries* 25:1174–1183. [15]

Linke-Gamenick, I., V. E. Forbes, and R. M. Sibly. 1999. Density-dependent effects of a toxicant on life-history traits and population dynamics of a capitellid polychaete. *Mar. Ecol. Progr. Ser.* 184:139–148. [3]

Lins, H. F. and J. R. Slack. 1999. Streamflow trends in the United States. *Geophys. Res. Lett.* 26:227–230. [10]

Lippincott, C. L. 1997. Ecological consequences of *Imperata cylindrica* (cogon grass): Invasion in Florida Sandhill. Ph.D. Dissertation, University of Florida, Gainesville, Florida. [13]

Lips, K. R. 1998. Decline of a tropical montane amphibian fauna. *Conserv. Biol.* 12:106–117. [3]

Lips, K. R. 1999. Mass mortality and population declines of anurans at an upland site in western Panama. *Conserv. Biol.* 13:117–125. [3]

Lips, K. R., D. E. Green, and R. Papendick. 2003a. Chytridiomycosis in wild frogs from southern Costa Rica. *Journal of Herpetology* 37:215–218. [3]

Lips, K. R., J. D. Reeve, and L. R. Witters. 2003b. Ecological traits predicting amphibian population declines in Central America. *Conserv. Biol.* 17:1078–1088. [3]

Liu, J. 1993. ECOLECON: An ECOLogical-ECONomic model for species conservation in complex forest landscapes. *Ecol. Model.* 70:63–87. [12]

Liu, J., J. B. Dunning, Jr., and H. R. Pulliam. 1995. Potential effects of a forest management plan on Bachman's Sparrows (*Aimophila aestivalis*): Linking a spatially explicit model with GIS. *Conserv. Biol.* 9:62–75. [12]

Liu, J., M. Linderman, Z. Ouyang, L. An, J. Yang, and H. Zhang. 2001. Ecological degradation in protected areas: the case of Wolong Nature Reserve for giant pandas. *Science* 292:98–101. [12, 14]

Lizarralde, M. S. 1993. Current status of the introduced beaver (*Castor canadensis*) population in Tierra del Fuego, Argentina. *Ambio* 22:351–358. [9]

Lizarralde, M. S. and C. Venegas. 2001. El castor: un ingeniero exótico en las tierras más australes del planeta. In R. Primack, R. Rozzi, P. Feinsinger, R. Dirzo, and F. Massardo (eds.), *Fundamentos de Conservación Biológica: Perspectivas Latinoamericanas*, pp. 231–232. Fondo de Cultura Económica, Ciudad de México. [9]

Lizarralde, M. S., and J. Escobar. 2000. Mamíferos exóticos en la Tierra del Fuego. *Ciencia Hoy* 10:52–63. [9]

Lizarralde, M. S., G. DeFerrari, J. Escobar, and S. Álvarez. 1996. Estado de la población de *Castor canadensis* introducida en Tierra del Fuego y su estudio cromosómico. PID-BID 50/92, Dirección General de Recursos Naturales de la Provincia de Tierra del Fuego, Antártida e Islas del Atlántico Sur, Ushuaia, Argentina. [9]

Locke, H. 2000. The Wildlands Project: A Balanced Approach to Sharing North America. *Wild Earth Spring* 2000:7–10. [6]

Loeschcke, V., J. Tomiuk, and S. J. Jain. 1994. *Conservation Genetics.* Birkauser Verlag, Basel, Switzerland. [11]

Loevinger, L. 1992. Science and legal rules of evidence: A review of Galileo's Revenge: Junk science in the courtroom. *Jurimetrics* 32:487–502. [17]

Logan, B. I. 2004. Ideology and the political ecology of resource management: From sustainable development to environmental security in Africa. In W. Moseley and B. I. Logan (eds.), *African Environment and Development*, pp. 17–41. Ashgate, London. [16]

Logan, B. I. and W. Moseley. 2002. The political ecology of poverty-alleviation in Zimbabwe's CAMPFIRE program. *Geoforum* 33: 1–14. [16]

Logan, J. and J. Powell. 2001. Ghost forests, global warming, and the mountain pine beetle. *Am. Entomol.* 47:160–173. [10]

LoGuidice, K., R. S. Ostfeld, K.A. Schmidt, and F. Keesing. 2003. The ecology of in-

fectious disease: Effects of host diversity and community composition on Lyme disease risk. *Proc. Nat. Acad. Sci. USA* 100:567–571. [3]

Loh, J., R. E. Green, T. Ricketts, J. Lamoreux, M. Jenkins, V. Kapos, and J. Randers. 2005. The living planet index: Using species population time series to track trends in biodiversity. *Philos. T. Roy. Soc. B.* 360:289–295. [3]

Loiselle, P. and participants of the CBSG/ANGAP CAMP "Faune de Madagascar" workshop 2004. *Bedotia tricolor.* In IUCN, *2004 IUCN Red List of Threatened Species.* www.redlist.org. [3]

Long, L. E., L. S. Saylor, and M. E. Soule. 1995. A pH/UV-B synergism in amphibians. *Conserv. Biol.* 9:1301–1303. [3]

Longcore, J. E., A. P. Pessier, and D. K. Nichols. 1999. *Batrachochytrium dendrobatidis* gen et sp nov, a chytrid pathogenic to amphibians. *Mycologia* 91:219–227. [3]

Longcore, T. and C. Rich. 2004. Ecological light pollution. *Front. Ecol. Environ.* 2:191–198. [6]

Longhurst, A. 1998. *Ecological Geography of the Sea.* Academic Press. [2]

Lonsdale, W. M. 1999. Concepts and synthesis: Global patterns of plant invasions, and the concept of invasibility. *Ecology* 80:1522–1536. [9]

Lonsdale, W. M. and A. M. Lane. 1994. Tourist vehicles as vectors of weed seeds in Kakuda National Park, northern Australia. *Biol. Conserv.* 69:277–283. [7]

Loomis, J. B. and D. S. White. 1996. Economic benefits of rare and endangered species: Summary and meta-analysis. *Ecolog. Econ.* 18:197–206. [5]

Lopez, R. R., M. E. P. Vieira, N. J. Silvy, P. A. Frank, S. W. Whisenant, D. A. Jones. 2003. Survival, mortality, and life expectancy of Florida Key deer. *J. Wildl. Mgmt.* 67:34–45. [12]

Lord, J. M. and D. A. Norton. 1990. Scale and the spatial concept of fragmentation. *Conserv. Biol.* 4:197–202. [7]

Loreau, M., S. Naeem, P. Inchausti, J. Bengtsson, J. Grime, A. Hector, D.U. Hooper, M. A. Huston, D. Raffaelli, B. Schmid, D. Tilman, and D. A. Wardle. 2001. Biodiversity and Ecosystem Functioning: Current knowledge and future challenges. *Science* 294:804–808. [2]

Lorimer, C. G. 1989. Relative effects of small and large disturbances on temperate forest structure. *Ecology* 70:565–567. [7]

Louda, S. 1998. Population growth of *Rhinocyllus conicus* (Coleoptera: Curculionidae) on two species of native thistles in prairie. *Environ. Entomol.* 27:834–841. [9]

Louda, S. 2000. *Rhinocyllus conicus:* Insights to improve predictability and minimize risk of biological control of weeds. In N. Spencer (ed.), *Proceedings X International Symposium Biological Control of Weeds,* pp. 187–194. Montana State University, Bozeman. [9]

Lovejoy, T. E. 1984. Aid Debtors Nations' Ecology. *New York Times,* Oct. 4, 1984. p. 31. [6]

Lovejoy, T. E. and ten others. 1986. Edge and other effects of isolation on Amazon forest fragments. In M. E. Soulé (ed.), *Conservation Biology: The Science of Scarcity and Diversity,* pp. 257–285. Sinauer Associates, Sunderland, MA. [7]

Lovelock, J. 1988. *The Ages of Gaia: A Biography of Our Living Earth.* W. W. Norton, New York. [4]

Low, T. 2001. *Feral Future: The Untold Story of Australia's Exotic Invaders.* University of Chicago Press, Chicago. [9]

Lozon, J. D. and H. J. MacIsaac. 1997. Biological invasions: Are they dependent on disturbance? *Environ. Res.* 5:131–144. [9]

Lubchenco, J. 1998. Entering the century of the environment: a new social contract for science. *Science* 279:491–497. [17, 18]

Lubchenco, J. et al. 1991. The sustainable biosphere initiative: An ecological research agenda. *Ecology* 72:371–412. [16]

Lucht, W., I. C. Prentice, R. B. Myneni, S. Sitch, P. Friedlingstein, W. Cramer, P. Bousquet, W. Buermann, and B. Smith. 2002. Climatic control of the high-latitude vegetation greening trend and Pinatubo effect. *Science* 296:1687–1689. [10]

Luck, G. W., and G. C. Daily. 2003. Tropical countryside bird assemblages: Richness, composition, and foraging differ by landscape context. *Ecol. Appl.* 13:235–247. [18]

Luck, G. W., G. C. Daily, and P. R. Ehrlich. 2003. Population diversity and ecosystem services. *Trends Ecol. Evol.* 18:331–336. [18]

Luckman B. and T. Kavanagh. 2000. Impact of climate fluctuations on mountain environments in the Canadian Rockies. *Ambio* 29:371–380. [10]

Ludwig, D. 1999. Is it meaningful to estimate a probability of extinction? *Ecology* 80:298–310. [12]

Ludwig, D., M. Mangel, and B. Haddad. 2001. Ecology, conservation and public policy. *Ann. Rev. Ecol. Syst.* 32:481–517. [17]

Ludwig, D., R. Hilborn, and C. Walters. 1993. Uncertainty, resource exploitation, and conservation: Lessons from history. *Science* 260:17, 36. [16]

Lueck, D. and J. A. Michael. 2003. Preemptive habitat destruction under the Endangered Species Act. *J. L. & Econ.* 46:27–60. [17]

Luikart, G., J. M. Cornuet, and F. W. Allendorf. 1999. Temporal changes in allele frequencies provide estimates of population bottleneck size. *Conserv. Biol.* 13:523–530. [11]

Luikart, G., P. R. England, D. Tallmon, S. Jordan, and P. Taberlet. 2003. The power and promise of population genomics: From genotyping to genome typing. *Nat. Rev. Genet.* 4:981–994. [11]

Lundquist, C. J., J. M. Diehl, E. Harvey, and L. W. Botsford. 2002. Factors affecting implementation of recovery plans. *Ecol. Appl.* 12:713–718. [18]

Lynch, K. D., M. L. Jones, and W. W. Taylor (eds.). 2002. *Sustaining North American Salmon: Perspectives Across Regions and Disciplines.* American Fisheries Society, Bethesda, Maryland. [8]

Lynch, M. and B. Walsh. 1998. *Genetics and Analysis of Quantitative Traits.* Sinauer Associates, Sunderland, MA. [11]

Lynch, M., J. Conery, and R. Burger. 1995a. Mutational meltdown in sexual populations. *Evolution* 49:1067–1080. [11]

Lynch, M., J. Conery, and R. Burger. 1995b. Mutation accumulation and the extinction of small populations. *Am. Nat.* 146:489–518. [11]

Lynch, M., M. Perender, K. Spitze, N. Lehman, J. Hicks, D. Allen, L. Latta, M. Ottene, F. Bouge, and J. Colbourne. 1999. The quantitative and molecular genetic architecture of a subdivided species. *Evolution* 53:100–110. [11]

Maathai, W. 2004a. *The Green Belt Movement: Sharing the Approach and the Experience.* Lantern Books. New York, NY. [15]

Maathai, W. 2004b. Trees for democracy. New York Times Op-ed. December 10. Section A, p. 41. [15]

Mac, M. J., P. A. Opler, C. E. Puckett Haecker, and P. D. Doran. 1998. *Status and Trends of the Nation's Biological Resources.* (2 vols.) U.S. Department of the Interior, U.S. Geological Survey, Reston, VA. [18]

MacArthur, R. H. 1964. Environmental factors affecting bird species diversity. *Am. Nat.* 98:387–397. [2]

MacArthur, R. H. 1972. *Geographical Ecology: Patterns in the Distribution of Species.* Princeton University Press, Princeton, NJ. [7]

MacArthur, R. H. 1972. *Geographical Ecology.* Harper & Row, New York. [2]

MacArthur, R. H. 1972. Strong or weak interactions? *Transactions of the Connecticut Academy of Arts and Sciences* 44: 177–188. [12]

MacArthur, R. H. and E. O. Wilson 1967. *The Theory of Island Biogeography.* Princeton University Press, Princeton, NJ. [2, 3, 7]

MacArthur, R. H. and J. MacArthur. 1961. On bird species diversity. *Ecology* 42:594–598. [2]

Macdonald, I. A. W., L. L. Loope, M. B. Usher, and O. Hamann. 1989. Wildlife conservation and the invasion of nature reserves by introduced species: A global perspective. In J. A. Drake, H. A. Mooney, F. di Castri, F. J. Kruger, M. Rejmánek, and M. Williamson (eds.), *Biological Invasions: A Global Perspective,* pp. 215–255 . Scientific Committee on the Protection of the Environment 37. John Wiley & Sons Ltd., Hoboken, N.J. [9]

MacIsaac, H. J., J. Borbely, J. Muirhead, and P. Graniero. 2004. Backcasting and forecasting biological invasion of inland lakes. *Ecol. Appl.* 14:773–783. [9]

MacVicar, T. K. and S. S. T. Lin. 1984. Historical rainfall activity in central and southern Florida: Average, return period estimates and selected extremes. In P. J. Gleason (ed.), *Environments of South Florida: Present and Past II.* Miami Geological Society, Coral Gables, FL. [13]

Maddison, D. R. and K.-S. Schulz (ed.) 2004. *The Tree of Life Web Project.* Internet ad-

dress: http://tolweb.org Accessed on 2 January 2005. [2]

Mader, H. J. 1984. Animal habitat isolation by roads and agricultural fields. *Biol. Conserv.* 29:81–96. [7]

Mader, H. J., C. Schell, and P. Kornacker. 1990. Linear barriers to movements in the landscape. *Biol. Conserv.* 54:209–222. [7]

Maestas, J. D., R. L. Knight, and W. C. Gilgert. 2003. Biodiversity across a rural land-use gradient. *Conserv. Biol.* 17(5):1425.

Madsen, T., R. Shine, M. Olsson, and H. Wittzell. 1999. Conservation biology: Restoration of an inbred adder population. *Nature* 402:34–35. [11]

Maguire, L. A. 1999. Social perspectives. In M. L. Hunter Jr. (ed.), *Maintaining Biodiversity in Forest Ecosystems*, pp. 639–665. Cambridge University Press, Cambridge. [17]

Mann, C. C. and M. L. Plummer. 1995. *Noah's Choice: The Future of Endangered Species.* Alfred A. Knopf, New York. [17]

Manne, L. L., T. M. Brooks, and S. L. Pimm. 1999. Relative risk of extinction of passerine birds on continents and islands. *Nature* 399:258–261. [3]

Manno, J. P. 2000. *Privileged Goods: Commoditization and Its Impact on Environment and Society.* Lewis, Boca Raton, FL. [18]

Manokaran, N. 1998. Effect, 34 years later of selective logging in the lowland dipterocarp forest at Pasoh, Peninsular Malaysia, and implications on present-day logging in the hill forests. In S. S. Lee, D. Y. May, I. D. Gauld, and J. Bishop (eds.), *Conservation, Management and Development of Forest Resources*, pp. 41-60. Proceedings of the Malaysia-United Kingdom Programme Workshop 21–24 October 1996, Kuala Lumpur, Malaysia. Forest Research Institute Malaysia, Kepong, Malaysia. [8]

Marco, A., C. Quilchano, and A. R. Blaustein. 1999. Sensitivity to nitrate and nitrite in pond-breeding amphibians from the Pacific northwest, USA. *Environ. Toxicol. Chem.* 18:2836–2839. [3]

Margoluis, R. and N. Salafsky. 1998. *Measures of Success: Designing, Managing, and Monitoring Conservation and Development Projects.* Island Press, Washington, D.C. [13, 18]

Margules, C. R. and M. B. Usher. 1981. Criteria used in assessing wildlife conservation potential: a review. *Biol. Conserv.* 21:79–109. [14]

Margules, C. R. and R. L. Pressey 2000. Systematic conservation planning. *Nature* 405:243–253. [2, 14]

Margules, C. R., A. O. Nicholls, and R. L. Pressey 1998. Selecting networks of reserves to maximize biological diversity. *Biol. Conserv.* 43:663–676. [14]

Margules, C. R., R. L. Pressey, and A. O. Nicholls. 1991. Selecting nature reserves. In C. R. Margules and M. P. Austin, (eds.), *Nature Conservation: Cost Effective Biological Surveys and Data Analysis*, pp 90–97. CSIRO, Canberra. [14]

Margulis, L. and K. V. Schwartz 1998. *Five Kingdoms: An Illustrated Guide to the Phyla of Life on Earth*, 3rd Edition. W. H. Freeman & Company, New York. [2]

Mark, A. F. 1985. The botanical component of conservation in New Zealand. *New Zeal. J. Bot.* 23:789–810. [14]

Markey, C. M., B. S. Rubin, A. M. Soto, and C. Sonnenschein. 2002. Endocrine disruptors: From wingspread to environmental developmental biology. *J. Steroid Biochem.* 83:235–244. [6]

Marquis, R. and E. Braker. 1994. Plant-herbivore interactions: Diversity, specificity, and impact. In L. A. McDade, K. S. Bawa, H. A. Hespenheide, and G. S. Hartshorn (eds.), *La Selva: Ecology and Natural History of a Neotropical Rainforest*, pp. 261–281. University of Chicago Press, Chicago. [2]

Marshall, K. and G. Edwards-Jones. 1998. Reintroducing Capercaillie (*Tetrao urogallus*) into southern Scotland: Identification of minimum viable populations at potential release sites. *Biodivers. Conserv.* 7:275–296. [12]

Martin E. M. and P. I. Padding. 2002. *Preliminary Estimates of Waterfowl Harvest and Hunter Activity in the United States During the 2001 Hunting Season.* U.S. Fish and Wildlife Service Division of Migratory Bird Management, Laurel, Maryland. [8]

Martin, P. R. and J. K. McKay 2004. Latitudinal variation in the divergence of populations and the potential for future speciation. *Evolution* 58:938–945. [2]

Martin, P. S. 1984. Prehistoric overkill: The global model. In P. S. Martin and R. G. Klein, *Quaternary extinctions: a prehistoric revolution*, pp. 354–403. University of Arizona Press, Tucson. [8]

Martin, P. S. 1990. 40,000 years of extinctions on the planet of doom. *Global and Planetary Change* 82:187–201. [3]

Martinez-Garza, C. and H. F. Howe. 2003. Restoring tropical diversity: beating the time tax on species loss. *J. Appl. Ecol.* 40:423–429. [15]

Martinez-Solano, I., J. Bosch, and M. Garcia-Paris. 2003b. Demographic trends and community stability in a montane amphibian assemblage. *Conserv. Biol.* 17:238–244. [3]

Martinez-Solano, I., L. J. Barbadillo, and M. Lapena. 2003a. Effect of introduced fish on amphibian species richness and densities at a montane assemblage in the Sierra de Neila, Spain. *Herpetol. J.* 13:167–173. [3]

Mary, F. and G. Michon. 1987. When agroforestry drives back natural forests: A socio-economic analysis of a rice-agroforest system in Sumatra. *Agroforest. Syst.* 5:27–55. [1]

Maryland Department of Natural Resources. 2005. http://mddnr.chesapeakebay.net/mdcomfish/oyster/totoymcfquery.cfm?value=oyster. [13]

Mascarenhas, R., R. Santos, and D. Zeppelini. 2004. Plastic debris ingestion by sea turtle in Paraiba, Brazil. *Marine Pollut. Bull.* 49:354–355. [6]

Masera, O., M. J. Ordóñez, and R. Dirzo. 1997. Carbon emissions from Mexican forests: Current situation and long-term scenarios. *Climatic Change* 35:265–295. [8]

Master, L. 1990. The imperiled status of North American aquatic animals. *Biodiversity Network News* (The Nature Conservancy) 3:1–8. [7]

Master, L., S. R. Flack, and B. A. Stein. 1998. *Rivers of Life: critical watersheds for protecting freshwater biodiversity.* The Nature Conservancy, NatureServe, Arlington, VA. [7]

Matson, P., K. A. Lohse, and S. J. Hall. 2002. The globalization of nitrogen deposition: consequences for terrestrial ecosystems. *Ambio* 31:113–119. [6]

Matson, P., R. Naylor, and I. Ortiz-Monasterio. 1990. Integration of environmental, agronomic, and economic aspects of fertilizer management. *Science* 280:112–115. [6]

Matsuda, H. 2003. Challenges posed by the precautionary principle and accountability in ecological risk assessment *Environmetrics* 14:245–254. [5]

Matter, J. M., D. A. Crain, C. Sills-McMurry, D. B. Pickford, T. R. Rainwater, K. D. Reynolds, A. A. Rooney, R. L. Dickerson, and L. J. Guillette, Jr. 1998. Effects of endocrine-disrupting contaminants in reptiles: Alligators. In W. Suk, (ed.), *Principles and Processes for Evaluating Endocrine Disruption in Wildlife*, pp. 267–289. SETAC Press, Pensacola, FL. [6]

Matthews, K. R., K. L. Pope, H. K. Preisler, and R. A. Knapp. 2001. Effects of nonnative trout on Pacific treefrogs (*Hyla regilla*) in the Sierra Nevada. *Copeia* 2001:1130–1137. [3]

Mattson, W. 1996. Escalating anthropogenic stresses on forest ecosystems: forcing benign plant-insect interactions into new interaction trajectories. In E. Korpilahti, H. Mikkela, and T. Salonen (eds.), *Caring for the Forest: Research in a Changing World*, pp. 338–342. IUFRO World Congress Organizing Committee, Finland. [10]

Mattson, W. and R. Haack. 1987. The role of drought stress in provoking outbreaks of phytophagous insects. In P. Barbosa and J. Schultz (eds.), *Insect Outbreaks*, pp. 365–407. Academic Press, San Diego. [10]

Mattson, W . J. and R. A. Haack. 1987. The role of drought in outbreaks of plant-eating insects. *Bioscience* 37:110–118. [10]

Maudet, C., C. Miller, B. Bassano, C. Breitenmoser-Wursten, D. Gauthier, R. Obexer-Ruff, J. Michallet, P. Taberlet, and G. Luikart. 2002. Microsatellite DNA and recent statistical methods in wildlife conservation management: Applications in alpine ibex (*Capra ibex ibex*). *Mol. Ecol.* 11:421–436. [11]

May, R. M. 1988. How many species are there on earth? *Science* 241:1441–1449. [2]

May, R. M., J. H. Lawton, and N. E. Stark. 1995. Assessing Extinction Rates. In J.H. Lawton and R.M. May (eds.), *Extinction Rates*, pp. 1–24. Oxford University Press, Oxford. [3]

Mayaux, P., P. Holmgren, F. Achard, H. Eva, H-J Stibig, and A. Branthomme. 2005. Tropical forest cover change in the 1990s and options for future monitoring. *Phil. Trans. R. Soc. B.* 360:373–384. [6, 18]

Mayers, J. and S. Vermeulen. 2002. *Company-Community Forestry Partnerships: From Raw Deals to Mutual Gains?* International Institute for Environment and Development, London. [8]

Mayfield, M. and G. C. Daily. 2005. Country-side biogeography of Neotropical herbaceous and shrubby plants. *Ecol. Appl.* 15:423–439. [18]

McCall, T. C., T. P. Hodgman, D. R. Diefenbach, and R.B Owen. 1996. Beaver populations and their relation to wetland habitat and breeding waterfowl in Maine. *Wetlands* 16:163–172. [8]

McClanahan, T., T. Done, and N. C. Polunin. 2003. Resiliency of coral reefs. In L. H. Gunderson and L. Pritchard (eds.), *Resilience and the Behavior of Large Scale Systems*. Island Press, Washington D.C. [13]

McComb, W. and D. Lindenmeyer. 1999. Dying, dead and down trees. In M. L. Hunter Jr. (ed.), *Maintaining Biodiversity in Forest Ecosystems*, pp. 335–372. Cambridge University Press, Cambridge. [8]

McCorquodale, S. M. and R. F. DiGiacomo. 1985. The role of wild North American ungulates in the epidemiology of bovine brucellosis: A review. *J. Wildlife Dis.* 21:351–357. [3]

McCoy, E. D. 1994. "Amphibian decline": A scientific dilemma in more ways than one. *Herpetologica* 50:98–103. [3]

McCullough, D. R. 1982. The theory and management of *Odoicoleus* populations. In C. M. Wemmer (ed.), *Biology and Management of the Cervidae*, pp. 535–549. Smithsonian Institution, Washington, D.C. [8]

McDaniel, C. J., L. B. Crowder, and J. A. Priddy. 2000. Spatial dynamics of sea turtle abundance and shrimping intensity in the U.S. Gulf of Mexico. *Conserv. Ecol.* 4:5. [8]

McDonald, K. R. 1990. Rheobatrachus Liem and Taudactylus Straughan and Lee (Anura: Leptodactylidae) in Eungella National Park, Queensland: distribution and decline. *T. Roy. Soc. South Aust.* 114:187–194. [3]

McDonnell, M. J. and S. T. Pickett. 1993. *Humans as Components of Ecosystems: The Ecology of Subtle Human Effects and Populated Areas*. Springer-Verlag, New York. [18]

McDowell, D. M. and R. J. Naiman. 1986. Structure and function of a benthic invertebrate stream community as influenced by beaver (*Castor canadensis*). *Oecologia* 68:481–489. [9]

McEvoy, P. and E. Coombs. 1999. Biological control of plant invaders: Regional patterns, field experiments, and structured population models. *Ecol. Appl.* 9:387–401. [9]

McEvoy, P., N. Rudd, C. Cox, and M. Huso. 1993. Disturbance, competition and herbivory effects on ragwort, *Senecio ja-*cobaea populations. *Ecol. Monogr.* 63:55–75. [9]

McFadyen, R. E. 1998. Biological control of weeds. *Annu. Rev. Entomol.* 43:369–393. [9]

McGarigal, K. and W. C. McComb. 1995. Relationships between landscape structure and breeding birds in the Oregon Coast Range. *Ecol. Monogr.* 65:235–260. [7]

McGranahan, G. and D. Satterthwaite. 2003. Urban centers: an assessment of sustainability. *Annu. Rev. Env. Resour.* 28:243–274. [18]

McIntyre, S. and G. W. Barrett. 1992. Habitat variegation, an alternative to fragmentation. *Conserv. Biol.* 6:146–147. [7]

McIntyre, S. and R. Hobbs. 1999. A framework for conceptualizing human effects on landscapes and its relevance to management and research models. *Conserv. Biol.* 13:1282–1292. [7]

McKay, J. K. and R. G. Latta. 2002. Adaptive population divergence: Markers, QTL and traits. *Trends Ecol. Evol.* 17:285–291. [11]

McKee, J. K., P. W. Sciulli, C. D. Fooce, and T. A. Waite. 2003. Forecasting global biodiversity threats associated with human population growth. *Biol. Conserv.* 115:161–164. [3]

McKelvey, K. S. and J. D. Johnson 1992. In J. Verner *The California Spotted Owl: A Technical Assessment of its Current Status*, pp. 225–246. General Technical Report PSW-GTR-133. Pacific Southwest Research Station, U.S. Forest Service, U.S. Department of Agriculture, Albany, California. [17]

McKelvey, K., B. R. Noon, and R. H. Lamberson. 1993. Conservation planning for species occupying fragmented landscapes: The case of the northern spotted owl. In P. M. Kareiva, J. G. Kingsolver, and R. B. Huey (eds.), *Biotic Interactions and Global Change*, pp. 424–450. Sinauer Associates, Sunderland, MA. [7, 12, 17]

McKibben, B. 1989. *The End of Nature*. Anchor Books, New York. [1]

McKibben, B. 1992. *The Age of Missing Information*. Random House, NY. [18]

McKinney, M. L. 1997. Extinction vulnerability and selectivity: Combining ecological and paleontological views. *Ann. Rev. Ecol. Syst.* 28:495–516. [3, 8]

McLachlan, J. A. 2001. Environmental Signaling: What embryos and evolution teach us about endocrine disrupting chemicals. *Endocr. Rev.* 22:319–341. [6]

McLellan, B. N., F. W. Hovey, R. D. Mace, J. G. Woods, D. W. Carney, M. L. Gibeau, W. L. Wakkinen, and W. F. Kasworm. 1999. Rates and causes of grizzly bear mortality in the interior mountains of British Columbia, Alberta, Montana, Washington, and Idaho. *J. Wildl. Mgmt.* 63:911–920. [11]

McManus, J. F. 2004. Paleoclimate: A great grand-daddy of ice cores. *Nature* 429:611–612.

McNeely, J. A. 1994. Protected areas for the 21st Century: working to provide benefits for society. *Biodivers. Conserv.* 3:3–20. [14]

McNeely, J. A. and S. J. Scherr. 2003. *Ecoagriculture: Strategies to Feed the World and Save Wild Biodiversity*. Island Press, Washington, D.C. [6, 18]

McNeill, J. R. 2000. *Something New Under the Sun: An Environmental History of the Twentieth-Century World*. W. W. Norton, NY. [18]

McPeek, M. A. and R. D. Holt. 1992. The evolution of dispersal in spatially and temporally varying environments. *Am. Nat.* 134:1010–1027. [12]

McPherson, G. R. and H. A. Wright.1990. Establishment of *Juniperus pinchotii* in western Texas, USA: Environmental effects. *J. Arid Environ.* 19:283–288. [10]

McShea, W. J., H. B. Underwood, and J. H. Rappole. 1997. *The Science of Overabundance: Deer Ecology And Population Management*. Smithsonian Institution Press, Washington, D.C. [7]

Meadows, D. H. 1990. Biodiversity: The key to saving life on earth. *Land Stewardship Letter* (Summer):4–5. [4]

Meagher, M. and M. E. Meyer. 1994. On the origin of brucellosis in bison of Yellowstone National Park: A review. *Conserv. Biol.* 8:645–653. [3]

Medin, D. E. and W. P. Clary. 1991. Small mammals of a beaver pond ecosystem and adjacent riparian habitat in Idaho. INT-445, USDA Forest Service Intermountain Research Station. [9]

Medley, K. E. 1993. Primate Conservation along the Tana River, Kenya: An examination of the forest habitat. *Conserv. Biol.* 7:109–121. [7]

Meehl. G. A., T. Karl, D. R. Easterling, S. Changnon, R. Pielke Jr., D. Changnon, J. Evans, P. Y. Groisman, T. R. Knutson, K. E. Kunkel, L. O. Mearns, C. Parmesan, R. Pulwarty, T. Root, R. T. Sylves, P. Whetton, and F. Zwiers. 2000. An introduction to trends in extreme weather and climate events: observations, socioeconomic impacts, terrestrial ecological impacts, and model projections. *B. Am. Meteorol. Soc.* 81:413–416. [10]

Meffe, G. K. 1992. Techno-arrogance and halfway technologies: Salmon hatcheries on the Pacific coast of North America. *Conserv. Biol.* 6:350–354. [16]

Meffe, G. K. 1993. Sustainability, Natural Law, and the "Real World." *The George Wright Forum* 10(4):48–52. [1]

Meffe, G. K. 1996. Conservation science: A creative tension. *Oryx* 30:226–228. [1]

Meffe, G. K., L. A. Nielsen, R. L. Knight, and D. A. Schenborn. 2002. *Ecosystem Management: Adaptive, Community-Based Conservation*. Island Press, Washington, D.C. [13]

Meffe, G. K., P. D. Boersma, D. D. Murphy, B. R. Noon, H. R. Pulliam, M. E. Soulé, and D. M. Waller. 1998. Independent scientific review in natural resource management. *Conserv. Biol.* 12: 268–270. [17, 18]

Meine, C. 1991. *Aldo Leopold: His Life and Work*. University of Wisconsin Press, Madison. [15]

Meine, C. 1992. Conservation biology and sustainable societies. In M. Oelschlaeger (ed.), *After Earth Day: Continuing the Con-*

servation Effort, pp. 37–65. University of North Texas Press, Denton. [4]

Meine, C. M. and G. W. Archibald. 1996. *The Cranes: Status survey and Conservation Action Plan*. IUCN, Gland, Switzerland. [12]

Meir, E., S. Andelman, and H. P. Possingham. 2004. Does conservation planning matter in a dynamic and uncertain world. *Ecol. Lett.* 7:615–622. [14]

Mendelson, S., G. Cowlishaw, and J. M. Rowcliffe. 2003. Anatomy of a bushmeat commodity chain in Takoradi, Ghana. *Journal of Peasant Studies* 31:73–100. [8]

Menges, E. 1990. Population viability analysis for an endangered plant. *Conserv. Biol.* 4:52–62. [12]

Menges, E. and D. R. Gordon. 1996. Three levels of monitoring intensity for rare plant species. *Natural Areas Journal* 16:227–237. [12]

Menon, S., R. G. Pontius, J. Rose, M. I. Khan, and K. S. Bawa. 2001. Identifying conservation-priority areas in the tropics: a land-use change modeling approach. *Conserv. Biol.* 15:502–512. [14]

Menzel, A. 2000. Trends in phenological phases in Europe between 1951 and 1996. *Intl. J. Biometeorology* 44:76–81. [10]

Menzel, A. and P. Fabian. 1999. Growing season extended in Europe. *Nature* 397:659. [10]

Mercer, D. and J. Peterson. 1986. The revocation of national parks and equivalent reserves in Tasmania. *Search* 17:134–140. [14]

Merola, M. 1994. A reassessment of homozygosity and the case for inbreeding depression in the cheetah, *Acionyx jubatus*: Implications for conservation. *Conserv. Biol.* 8:961–971. [11]

Merriam, G. 1991. Corridors and connectivity: Animal populations in heterogeneous environments. In D. A. Saunders and R. J. Hobbs (eds.), *Nature Conservation 2: The Role of Corridors*, pp. 133–142. Surrey Beatty, Chipping Norton, Australia. [7]

Metcalf, S. 1994. The Zimbabwe communal areas management programme for indigenous resources CAMPFIRE. In Western, D. and R. Wright (eds.), *Natural Connections: Perspectives in Community-Based Conservation*. Washington D.C.: Island Press. [16]

Meyer, M. E. and M. Meagher. 1995. Brucellosis in free-ranging bison (*Bison bison*) in Yellowstone, Grand Teton, and Wood Buffalo National Parks: A review. *J. Wildlife Dis.* 31:579–598. [3]

Meyers, J. M., W. J. Arendt, and G. D. Lindsey. 1996. Survival of radio-collared nestling Puerto Rican Parrots. *Wilson Bull.* 108:159–163. [15]

Micheli, F. and C. H. Peterson. 1999. Estuarine vegetated habitats as corridors for animal movement. *Conserv. Biol.* 13:869–881. [7]

Michener, W. K. 1997. Quantitatively evaluating restoration experiments: research design, statistical analysis, and data management considerations *Restor. Ecol.* 5:324–337. [15]

Micklin, P. P. 1988. Dessication of the Aral Sea: a water management disaster in the Soviet Union. *Science* 241:1170–1176. [6]

Middleton, E. M., J. R. Herman, E. A. Celarier, J. W. Wilkinson, C. Carey, and R. J. Rusin. 2001. Evaluating ultraviolet radiation exposure with satellite data at sites of amphibian declines in Central and South America. *Conserv. Biol.* 15:914–929. [3]

Middleton, S. and D. Liittschwager. 1994. *Witness: Endangered Species of North America*. Chronicle Books, San Francisco. [7]

Midgley, G. F. and D. Millar. 2005. Modeling species range shifts in two biodiversity hotspots. In T. Lovejoy and L. Hannah (eds.), *Climate Change and Biodiversity*, pp. 229–231. Yale Univerity Press, New Haven, CT. [10]

Midgley, G. F., L. Hannah, D. Millar, M. C. Rutherford, and L. W. Powrie. 2002. Assessing the vulnerability of species richness to anthropogenic climate change in a biodiversity hotspot. *Global Ecol. Biogeogr.* 11:445–451. [10]

Mikkelsen, P. M. and J. Cracraft. 2001 Marine biodiversity and the need for systematic inventories. *B. Mar. Sci.* 69:525–534. [2]

Milewski, A. V. and W. J. Bond. 1982. Convergence of myrmecochory in Mediterranean Australia and South Africa. In B. C.Buckley (ed.), *Ant-Plant Interactions in Australia*, pp. 89–98. Junk, The Hague. [2]

Millennium Ecosystem Assessment, 2005a. *Ecosystems and Human Well-being: Synthesis Report*. Island Press, Washington, D.C. [1, 3, 6, 18]

Millennium Ecosystem Assessment, 2005c. *Ecosystems and Human Well-being: Desertification Synthesis*. World Resources Institute, Washington, DC. [6]

Millennium Ecosystem Assessment. 2003. *Ecosystems and Human Well-Being: A Framework for Assessment*. Island Press, Covelo, CA. [2, 18]

Millennium Ecosystem Assessment. 2005b. *Ecosystems and Human Well-being: Biodiversity Synthesis*. World Resources Institute, Washington, D.C. [6, 18]

Miller, B., G. Ceballos, and R. Reading. 1994a. Prairie dogs, poison, and biotic diversity. *Conserv. Biol.* 8:677–681. [15]

Miller, B., R. Reading, and S. Forrest. 1996. *Prairie Night: Black-footed Ferrets and Recovery of Endangered Species*. Smithsonian Institution Press, Washington, D.C. [15]

Miller, B., D. Biggins, L. Hanebury, and A. Vargas. 1993. Reintroduction of the black-footed ferret. In P. J. S. Olney, G. M. Mace, and A. T. C. Feister (eds.), *Creative Conservation: Interactive Management of Wild and Captive Animals*, pp. 455–463. Chapman and Hall, London. [15]

Miller, B., R. Reading, C. Conway, J. A. Jackson, M. Hutchins, N. Snyder, S. Forrest, J. Frazier, and S. Derrickson. 1994b. Improving endangered species programs: Avoiding organizational pitfalls, tapping the resources, and adding accountability. *Environ. Manage.* 18:637–645. [15]

Miller, C. R. and L. P. Waits. 2003. The history of effective population size and genetic diversity in the Yellowstone grizzly (*Ursus arctos*): Implications for conservation. *Proc. Nat. Acad. Sci. USA* 100:4334–4339. [11]

Miller, G. H., J. W. Magee, B. J. Johnson, M. L. Fogel, N. A. Spooner, M. T. McCulloch, and L. K. Ayliffe. 1999. Pleistocene extinction of *Genyornis newtoni*: Human impact on Australian megafauna. *Science* 283:205–208. [3]

Miller, J. K., J. M. Scott, C. R. Miller, and L. P. Waits. 2002. The Endangered Species Act: dollars and sense? *BioScience* 52:163–168. [18]

Miller, K. R. 1996. *Balancing the Scales: Guidelines for Increasing Biodiversity's Chances through Bioregional Management.* World Resources Institute, Washington, D.C. [13]

Miller, P. R., J. R. Parameter, Jr., O. C. Taylor, and E. A. Cardiff. 1963. Ozone injury in the foliage of *Pinus ponderosa*. *Phytopathology* 52:1072–1076. [6]

Milliken, W., R. P. Miller, S. R. Pollard, and E. V. Wandelli. 1992. *Ethnobotany of the Waimiri-Atroari Indians of Brazil*. Royal Botanic Gardens, Kew. [8]

Mills, L. C. 1995. Edge effects and isolation: Red-backed voles on forest remnants. *Conserv. Biol.* 9:395–403. [7]

Mills, L. S. and P. E. Smouse. 1994. Demographic consequences of inbreeding in remnant populations. *Am. Nat.* 144:412–431. [12]

Mills, L. S. and R. W. Allendorf. 1996. The one-migrant-per-generation rule in conservation. *Conserv. Biol.* 10:1509–1518. [11]

Mills, L. S., J. J. Citta, K. P. Lair, M. K. Schwartz, and D. A. Tallmon. 2000. Estimating animal abundance using noninvasive DNA sampling: promises and pitfalls. *Ecol. Appl.* 10:283–294. [11]

Mills, L. S., S. G. Hayes, C. Baldwin, M. J. Wisdom, J. Citta, D. J. Mattson, and K. Murphy. 1996. Factors leading to different viability predictions for a grizzly bear data set. *Conserv. Biol.* 10:863–873. [12]

Mills, T. J. 2000. Position advocacy by scientists risks science credibility and may be unethical. *Northwest Sci.* 74:165–168. [17]

Mills, T. J. and R. N. Clark. 2001. Roles of research scientists in natural resource decision-making. *Forest Ecol. Manag.* 53:189–198. [17]

Millspaugh, J. J., R. J. Woods, K. E. Hunt, K. J. Raedeke, G. C. Brundige, B. E. Washburn, and S. K. Wasser. 2001. Fecal glucocorticoid assays and the physiological stress response in elk. *Wildlife Soc. B.* 29:899–907. [11]

Milner-Gulland, E. J. and E. L. Bennett. 2003. Wild meat: The bigger picture. *Trends Ecol. Evol.* 18:351–357. [8]

Milner-Gulland, E. J. and H. R. Akcakaya. 2001. Sustainability indices for exploited populations. *Trends Ecol. Evol.* 16:686–692. [8]

Milner-Gulland, E. J. and R. Mace. 1991. The impact of the ivory trade on the African elephant *Loxodonta africana* population as

assessed by data form the trade. *Biol. Conserv.* 55:215–229. [11]

Milner-Gulland, E. J. and R. Mace. 1998. *Conservation of Biological Resources.* Blackwell Science Ltd., Oxford. [8]

Milner-Gulland, E. J., E. L. Bennett, K. Abernethy, M. Bakarr, R. Bodmer, J. Brashares, G. Cowlishaw, P. Elkan, H. Eves, J. Fa, C. Peres, C. Roberts, J. Robinson, M. Rowcliffe, and D. Wilkie. 2003. Wild meat: the bigger picture. *Trends Ecol. Evol.* 18:351–357. [8]

Milner-Gulland, E. J., M. V. Kholodova, A. Bekenov, O. M. Bukreeva, I. A. Grachev, L. Amgalan, and A. A. Lushchekina. 2001. Dramatic declines in saiga antelope populations. *Oryx* 35:340–345. [8]

Ministry of Finance and Economic Planning. 1994. *The 1991 population and housing census, Uganda.* National Summary. Entebbe, Statistics Department. [5]

Minzi, M. L. 1993. The Pied Piper of Debt-for-Nature Swaps. *University of Pennsylvania Journal of International Business Law* 14(1):37–62. [6]

Miringoff, M. L. and M.-L. Miringoff. 1999. *The Social Health of the Nation: How is America Really Doing?.* Oxford University Press, NY. [18]

Mitchell, C. E. and A. G. Power. 2003. Release of invasive plants from fungal and viral pathogens. *Nature* 421: 625–627. [3]

Mitchell, R. C. and R. T. Carson. 1989. *Using Surveys to Value Public Goods: The Contingent Value Method.* Resources for the Future, Washington, D.C. [5]

Mitja, D. and J.-P. Lescure. 2000. Madeira para perfume: Qual será o destino do pau-rosa? In *A Floresta em Jogo: o Extrativismo na Amazônia Central.* Editora UNESP: Imprensa Oficial do Estado, São Paulo, Brazil. [8]

Mitsch, W. J., J. W. Daly, J. G. Wendell, P. M. Groffman, D. L. Hey, G. W. Randall, and N. Wang. 2001. Reducing nutrient loading from the Mississippi River basin to the Gulf of Mexico. *BioScience* 51:373–388. [4]

Mittermeier, R. A., N. Myers, P. Robles Gil, and C. G. Mittermeier (eds.). 1999. *Hotspots: Earth's Biologically Richest and Most Endangered Terrestrial Ecoregions.* CEMEX, Mexico. [12]

Mittermeier, R. A., C. G. Mittermeier, P. R. Gil, J. Pilgrim, G. Fonseca, T. Brooks, and W. R. Konstant. 2001. *Wilderness: Earth's Last Great Wild Places.* CEMEX, Mexico City. [6]

Mittermeier, R. A., P. Robles-Gil, M. Hoffmann, J. D. Pilgrim, T. M. Brooks, C. G. Mittermeier, J. L. Lamoreux, and G. Fonseca 2004. *Hotspots Revisited: Earth's Biologically Richest and Most Endangered Ecoregions.* CEMEX, Mexico. [6]

Miyashita, T., M. Takada, and A. Shimazaki. 2004. Indirect effects of herbivory by deer reduce abundance and species richness of web spiders. *Ecoscience* 11:74–79. [8]

Mladenoff, D. J., M. A. White, J. Pastor, and T. R. Crow. 1993. Comparing spatial pattern in unaltered old-growth and disturbed forest landscapes for. *Ecol. Appl.* 3:293–305. [7]

Mladenoff, D. J., M. A. White, T. R. Crow, and J. Pastor. 1994. Applying principles of landscape design and management to integrate old-growth forest enhancement and commodity use. *Conserv. Biol.* 8:752–762. [7]

Mladenoff, D. J., T. A. Sickley, R. G. Haight, and A. P. Wydeven. 1995. A regional landscape analysis and prediction of favorable gray wolf habitat in the northern Great Lakes region. *Conserv. Biol.* 9:279–294. [7]

Moberg, F. and C. Folke. 1999. Ecological goods and services of coral reef ecosystems. *Ecol. Econ.* 29:215–233. [13]

Moberg, F. and P. Rönnbäck. 2003. Ecosystem services of the tropical seascape: interactions, substitutions and restoration. *Ocean. Coast. Manage.* 46:27–46. [15]

Modak, P. and A. K. Biswas. 1999. *Conducting environmental impact assessment in developing countries.* United Nations University Press, Tokyo. [5]

Moiseev P. A. and S. G. Shiyatov. 2003. The use of old landscape photographs for studying vegetation dynamics at the treeline ecotone in the Ural Highlands, Russia. In L. Nagy, G. Grabherr, C. Körner, and D. B. A. Thompson (eds.), *Alpine Biodiversity in Europe.* Springer-Verlag, Heidelberg-Berlin. [10]

Moldenke, A. R. and J. D. Lattin. 1990. Dispersal characteristics of old-growth soil arthropods: The potential for loss of diversity and biological function. *Northwest Environ. J.* 6:408–409. [7]

Monfort, S. L., S. K. Wasser, K. L. Mashburn, M. Burke, B. A. Brewer, and S. R. Creel. 1997. Steroid metabolism and validation of noninvasive endocrine monitoring in the African wild dog (*Lycaon pictus*). *Zoo Biol.* 6:533–548. [11]

Montes, C., F. Borja, M. Bravo, and J. M. Mporeira. 1998. *Reconocimiento biofísico de espacios naturales protegidos. Doñana: una aproximación cosistémica.* Consejería de medio Ambiente. Junta de Andalucía. [14]

Montes, C., M. A. Bravo, A. Baltanás, P. J. Gutiérrez, G. Sancho, A. M. Marcos, J. R. Jordá, C. M. Duarte, P. Alcorlo, O. García, M. E. González, and M. Otero. 1993. Bases ecológicas para la gestión integral del cangrejo rojo de la marisma (*Procambarus clarkii*) en el Parque Nacional de Doñana. Convenio Instituto para la Conservación de la Naturaleza (ICONA)—Universidad Autónoma de Madrid. II vols. Final Report. [14]

Moore, D. 1997. Clear waters and muddied histories: Environmental history and the politics of community in Zimbabwe's eastern highlands. Paper presented at the Conference on Representing Communities: Histories and Politics of Community–Based Resource Management. Helen, Georgia, USA, June. [16]

Moran, K. 1992. Debt-for-nature swaps: A response to Debt and Deforestation in Developing Countries? In T. E. Downing, S. B. Hecht, T. E. Pearson, and C. Garcia-Downing (eds.), *Development or Destruction—The conversion of Tropical Forest to Pasture in Latin America,* pp. 305–316. Westview Press: Boulder, CO. [6]

Moran, P. 2002. Current conservation genetics: Building an ecological approach to the synthesis of molecular and quantitative genetic methods. *Ecol. Freshwater Fish.* 11:30–55. [11]

Morehouse, E. A., T. Y. James, A. R. D. Ganley, R. Vilgalys, L. Berger, P. J. Murphy, and J. E. Longcore. 2003. Multilocus sequence typing suggests the chytrid pathogen of amphibians is a recently emerged clone. *Mol. Ecol.* 12:395–403. [3]

Moreno, A., R. Margoluis, and K. Brandon. 1998. Bolivia: Amboró National Park. In K. Brandon, K. H. Redford, and S. E. Sanderson (eds.), *Parks in Peril: People, Politics and Protected Areas,* pp. 323–352. The Nature Conservancy and Island Press, Washington, D.C. [14]

Moritz, C. 1994a. Defining evolutionary significant units for conservation. *Ann. Rev. Ecol. Syst.* 9:373–375. [11]

Moritz, C. 1994b. Applications of mitochondrial DNA analysis in conservation: A critical review. *Mol. Ecol.* 3:401–411. [11]

Moritz, C. 2002. Strategies to protect biological diversity and the evolutionary processes that sustain it. *Syst. Biol.* 51:238–254. [11]

Morrison, P. H. 1990. *Ancient Forests in the Olympic National Forest: Analysis From a Historical and Landscape Perspective.* The Wilderness Society, Washington, D.C. [7]

Morrison, P. H. and F. J. Swanson. 1990. *Fire History and Pattern in a Cascade Range Landscape.* PNW-GTR-254. USDA Forest Service, Portland, OR. [7]

Moses, n. d. Genesis. In: King James, et al. (trs.), *The Holy Bible,* pp. 5–50. World Publishing Company, Cleveland, OH. [4]

Moss, R. H. and S. H. Schneider. 2000. Uncertainties. In R. Pachauri, T. Taniguchi, and K. Tanaka (eds.), *Guidance Papers on the Cross Cutting Issues of the Third Assessment Report of the IPCC,* pp. 31–51. World Meteorological Organization, Geneva. [17]

Mountfort, G. 1958. *Portrait of a Wilderness. The Story of the Coto Doñana Expeditions.* Hutchinson & Co. (Publishers) Ltd. Great Britain. [14]

Mowat, G. and C. Strobeck. 2000. Estimating population size of grizzly bears using hair capture, DNA profiling, and mark-recapture analysis. *J. Wildl. Mgmt.* 64:183–193. [11]

Moyle, P. B. and R. A. Leidy. 1992. Loss of biodiversity in aquatic ecosystems: Evidence from fish faunas. In P. L. Fiedler and S. K. Jain (eds.), *Conservation Biology: The Theory and Practice of Nature Conservation, Preservation, and Management,* pp. 127–169. Chapman and Hall, New York. [7]

Muchaal, P. K. and G. Ngandjui. 1999. Impact of village hunting on wildlife populations in the western Dia Reserve, Cameroon. *Conserv. Biol.* 13:385–396. [8]

Mueller-Dombois, D. 1971. Crown distortion and elephant distribution in the woody

vegetations of Ruhuna National Park, Ceylon. *Ecology* 53:208–226. [17]

Muir, J. 1869. *The Unpublished Journals of John Muir*. [13]

Muirhead, J. R. and H. J. MacIsaac. 2005. Development of inland lakes as hubs in an invasion network. *J. Appl. Ecol.* 42:80–90. [9]

Muiznieks, B. D. 2003. *Population viability analysis of the Puerto Rican Parrot: An evaluation of its current status and prognosis for recovery*. M.S. Thesis, North Carolina State University. [15]

Mulder, M. B. and P. Coppolillo. 2005. *Conservation: Linking Ecology, Economics and Culture*. Princeton University Press, Princeton, N.J. [2]

Munasinghe, M. 1993. Environmental economics and biodiversity management in developing countries. *Ambio* 22:126–135. [5]

Murcia, C. 1995. Edge effects in fragmented forests: Implications for Conservation. *Trends Ecol. Evol.* 10:58–62. [7]

Murcia, C. 1997. Evaluation of Andean alder as a catalyst for the recovery of tropical cloud forests in Colombia. *Forest Ecol. Manag.* 99:163–170. [15]

Murombedzi, J. 1992. Decentralization or recentralization? Implementing CAMP-FIRE in the Omay Communal Lands of the Nyaminyami District. CASS, University of Zimbabwe: Harare. [16]

Murphy, D. D. and S. B. Weiss. 1988. Ecological studies and the conservation of the Bay checkerspot butterfly, *Euphydryas editha bayensis. Biol. Conserv.* 46:183–200. [7]

Murphy, H. T. and J. Lovett-Doust. 2004. Context and connectivity in plant metapopulations and landscape mosaics: Does the matrix matter? *Oikos* 105:3–14. [7]

Murphree, M. 1995. The Lesson from Mahenye: Rural poverty, democracy and wildlife conservation. *Wildlife and Development Series, No. 1*. Published by the International Institute for Environment and Development (IIED) in association with the CAMPFIRE Collaborative Group. http://wildnetafrica.co.za/bushcraft/arcles/document_campfire1.html [16]

Murphree, M. 1997. Congruent objectives, competing interests and strategic compromise. Paper presented at the Conference on Representing Communities: Histories and Politics of Community-Based Resource Management. Helen, Georgia, June. [16]

Murphy, D. D. and B. R. Noon. 1991. Coping with uncertainty in wildlife biology. *J. Wildl. Mgmt.* 55:773–782. [17]

Murphy, D. D. and B. R. Noon. 1992. Integrating scientific methods with habitat conservation planning: Reserve design for the northern Spotted Owl. *Ecol. Appl.* 2:3–17. [17]

Murphy, M. A., L. P. Waits, K. C. Kendall, S. K. Wasser, J. A. Higbee, and R. Bogden. 2002. An evaluation of long-term preservation methods for brown bear (*Ursus*

arctos) faecal DNA samples. *Conserv. Genet.* 3:435–440. [11]

Muscatine, L. 1990. The role of symbiotic algae in carbon and energy flux in reef corals. *Coral Reefs* 25:1–29. [13]

Musick, J. A., M. M. Harbin, S. A. Berkeley, G. H. Burgess, A. M. Eklund, L. Findley, R. G. Gilmore, J. T. Golden, D. S. Ha, G. R. Huntsman, J. C. McGovern, S. J. Parker, S. G. Poss, E. Sala, T. W. Schmidt, G. R. Sedberry, H. Weeks, and S. G. Wright. 2000. Marine, estuarine, and diadromous fish stocks at risk of extinction in North America (exclusive of Pacific salmonids). *Fisheries* 6:6–30. [8]

Muths, E., P. S. Corn, A. P. Pessier, and D. E. Green. 2003. Evidence for disease-related amphibian decline in Colorado. *Biol. Conserv.* 110:357–365. [3]

Myers, J. H. 1987. Population outbreaks of introduced insects: Lessons from the biological control of weeds. In P. Barbosa and J. Schultz (eds.), *Insect Outbreaks*, pp.173–193. Academic Press, New York. [9]

Myers, J. H. 2001. Predicting the outcome of biological control. In C. Fox, D. Roff, and D. Fairbairn (eds.), *Evolutionary Ecology: Concepts and Case Studies*, pp. 361–370. Oxford University Press, Oxford. [9]

Myers, J. H. and C. Risley. 2000. Why reduced seed production is not necessarily translated into successful biological weed control. In N. R. Spencer (ed.), *Proceedings X International Symposium on Biological Control of Weeds*, pp. 569–580. Montana State University, Bozeman. [9]

Myers, J. H. and D. R. Bazely. 2003. *Ecology and Control of Introduced Plants*. Cambridge University Press, Cambridge. [9]

Myers, J. H., and R. Iyer. 1981. Phenotypic and genetic characteristics of the European cranefly following its introduction and spread in western North America. *J. Anim. Ecol.* 50:519–531. [9]

Myers, J., C. Risley, and R. Eng. 1990. The ability of plants to compensate for insect attack: Why biological control of weeds with insects is so difficult. In E. Delfosse (ed.), *VII. International Symposium on Biological Control of Weeds*, pp. 67–73. Patol. Veg., Rome, Italy. [9]

Myers, N. 1981. A farewell to Africa. *Internat. Wildl.* 11:36, 40, 44, 46. [4]

Myers, N. 1983. *A Wealth of Wild Species*. Westview Press, Boulder, CO. [4]

Myers, N. 1987. The extinction spasm impending: Synergisms at work. *Conserv. Biol.* 1:14–21. [1, 3]

Myers, N. 1998. Lifting the veil on perverse subsidies. *Nature* 392:327–328.

Myers, N. 1999. Pushed to the Edge. *Natural History* 108: 20–22. [6]

Myers, N. and A. Knoll, (eds.). 2001. The Future Course of Evolution. Special Issue of *P. Natl. Acad. Sci.* [18]

Myers, N. and A. Knoll. 2001. The biotic crisis and the future of evolution. *Proc. Nat. Acad. Sci. USA* 98:5389–5392. [3, 18]

Myers, N. and J. Kent. 2001. *Perverse Subsidies: How Misused Tax Dollars Harm the Environment and the Economy*. Island Press, Washington, D.C. [8]

Myers, N. and J. Kent. 2004. *The New Consumers: The Influence of Affluence on the Environment*. Island Press, Washington, D.C. [18]

Myers, N., R. A. Mittermeier, C. G. Mittermeier, G. A. B. Fonseca, and J. Kent. 2000. Biodiversity hotspots for conservation priorities. *Nature* 403:853–858. [2, 14]

Myers, R. 1983. Site susceptibility to invasion by the exotic tree *Melaleuca quinquenervia* in south Florida. *J. Appl. Ecol.* 20:645–658. [13]

Myers, R. A., S. A. Levin, R. Lande, F. C. James, W. W. Murdoch, and R. T. Paine. 2004. Hatcheries and endangered salmon. *Science* 303:1980. [8]

Myers, R. L. 1990. Scrub and high pine. In R. L. Myers and J. J. Ewel (eds.), *Ecosystems of Florida*, pp. 150–193. University of Central Florida Press, Orlando, FL. [18]

Myneni, R. B., C. D. Keeling, C. J. Tucker, G. Asrar, and R. R. Nemani. 1997. Increased plant growth in the northern high latitudes from 1981 to 1991. *Nature* 386:698–702. [10]

Nabhan, G. and S. Trimble. 1994. *The Geography of Childhood: Why Children Need Wild Places*. Beacon Press, Boston. [18]

Nadkarni, N. M. 2004. Not preaching to the choir: communicating the importance of forest conservation to non-traditional audiences. *Conserv. Biol.*18:606–609. [18]

Naeem, S. and S. Li 1997. Biodiversity enhances ecosystem reliability. *Nature* 390:507–509. [2]

Naess, A. 1989. *Ecology, Community, and Lifestyle*. Cambridge University Press, Cambridge. [4]

Naiman, R. J., C. A. Johnston, and J. C. Kelley. 1988. Alteration of North American streams by beaver: The structure and dynamics of streams are changing as beaver recolonize their historic habitat. *BioScience* 38:753–762. [9]

Naiman, R. J., J. M. Melillo, and J. E. Hobbie. 1986. Ecosystem alteration of boreal forest streams by beaver (*Castor canadensis*). *Ecology* 67:1254–1269. [3, 8, 9]

Naiman, R. J., G. Pinay, C. A. Johnston, and J. Pastor. 1994. Beaver influences on the long-term biogeochemical characteristics of boreal forest drainage networks. *Ecology* 75:905–921. [9]

Nanami, A. and M. Nishihira. 2003. Population dynamics and spatial distribution of coral reef fishes: Comparison between continuous and isolated habitats. *Environ. Biol. Fishes* 68:101–112. [7]

Nantel, P., D. Gagnon, and A. Nault. 1996. Population viability analysis of American Ginseng and Wild Leek harvested in stochastic environments. *Conserv. Biol.* 10:608–621. [12]

Nash, R. F. 1990. *American Environmentalism: Readings in Conservation History*. McGraw-Hill Publishing Company, New York. [14]

National Institute of Water and Atmospheric Research. 2005. Fisheries and Aquaculture Update. http://www.niwa.co.nz/ncfa/fau/2002-04/seafood#seafood_large.jpg [13]

National Park Conservation Association. 2005. Clean Air Campaign. www.npca.org/across_the_nation/visitor_experience/code_red/additional_impacts.asp. [6]

National Research Council. 1992. *Restoration of Aquatic Ecosystems: Science, Technology, and Public Policy.* National Academy Press, Washington, D.C. [15]

National Research Council. 1996. *Science and the Endangered Species Act.* U.S. Government Printing Office, Washington, D.C. [12]

National Research Council. 1996. *Stemming the Tide: Controlling Introductions of Nonindigenous Species by Ships' Ballast Water.* National Academy of Sciences, Washington, D.C. [9]

National Research Council. 1999. *Perspectives on Biodiversity: Valuing Its Role in an Ever-changing World.* National Academy Press, Washington, D.C. [17]

National Research Council. 2000. *Ecological Indicators for the Nation.* National Academy Press, Washington, D.C. [2, 14]

National Research Council. 2001. *Compensating for Wetland Losses under the Clean Water Act.* National Academy Press. Washington, D.C. [15]

National Research Council. 2002. *Predicting Invasions of Nonindigenous Plants and Plant Pests.* National Academy Press, Washington, D.C. [9]

NatureServe. 2005. http://www.natureserve.org. [3]

Naumann, M. 1994. A water-use budget for the Caribbean National Forest of Puerto Rico. Special Report, USDA Forest Service. [7]

Navaro, A. J., J. C. Funes, and R. S. Walker. 2000. Ecological extinction of native prey of a carnivore assemblage in Argentine Patagonia. *Biolo. Conserv.* 92:25–33. [12]

Naylor, R. L., R. J. Goldburg, J. H. Primavera, N. Kautsky, M. C. M. Beveridge, J. Clay, C. Folke, J. Lubchenco, H. Mooney, and M. Troell. 2000. Effect of aquaculture on world fish supplies. *Nature* 406:1017–1024. [16]

Nehlsen, W., J. E Williams, and J. A. Lichatowich. 1991. Pacific salmon at the crossroads: Stocks at risk from California, Oregon, Idaho, and Washington. *Fisheries* 16:4–21. [7, 13]

Nei, M. 1973. Analysis of gene diversity in subdivided populations. *Proc. Nat. Acad. Sci. USA* 70:3321–3323. [11]

Nei, M. 1975. *Molecular Population Genetics and Evolution.* North-Holland Publishing Company, Amsterdam. [11]

Nelson, M. P. 1996. Holists and fascists and paper tigers. *Ethics and the Environment* 1: 91–102. [4]

Nelson, N. and J. H. Dorfman. 2000. *Cost of Community Service Studies for Habersham and Oconee Counties, Georgia.* The University of Georgia, Center for Agribusiness and Economic Development, Center Report CR-00-5. February 11, 2000. [18]

Nelson, R. K. 1983. *Make Prayers to the Raven: A Koyukon View of the Northern Forest.* University of Chicago Press, Chicago. [4]

Nemani, R. R., C. D. Keeling, H. Hashimoto, W. M. Jolly, S. C. Piper, C. J. Tucker, R. B. Myneni, and S. W. Running. 2003. Climate-driven increases in global terrestrial net primary production from 1982–1999. *Science* 300:1560–1563. [10]

Nepstad, D., A. Moreira, and A. Alencar. 1999. Large-scale impoverishment of Amazonian forests by logging and fire. *Nature* 398: 505–508. [6]

Nepstad, D.C., G. Carvalho, A. C. Barros, A. Alencar, J. P. Capobianco, J. Bishop, P. Moutinho, P. Lefebvre, U. L. Silva, and E. Prins. 2001. Road paving, fire regime feedbacks, and the future of Amazon forests. *Forest Ecol. Manag.* 154:395–407. [6]

Nepstad, D. C., A. Verissimo, A. Alencar, C. Nobre, E. Lima, P. Lefebvre, P. Schlesinger, C. Potter, P. Moutinho, E. Mendoza, M. Cochrane, and V. Brooks. 1999. Large-scale impoverishment of Amazonian forests by logging and fire. *Nature* 398:505–508. [8]

Neuman, J. C. and C. Chapman. 1999. Wading into the water market: The first 5 years of the Oregon Water Trust. *J. Envtl. L. & Litig.* 14:135–184. [17]

Neumann, R. 1997. Model, panacea, or exception? Contextualizing CAMPFIRE and related programs in Africa. Paper presented at the Conference on Representing Communities: Histories and Politics of Community-Based Resource Management. Helen, Georgia, USA. [16]

Nevo, E., A. Bieles, and R. Ben-Shlomo. 1984. The evolutionary significance of genetic diversity: Ecological, demographic, and life history correlates. In G. S. Mani (ed.), *Evolutionary Dynamics of Genetic Diversity*, pp. 12–213. Springer-Verlag, Berlin, Germany. [11]

Newmark, W. D. 1987. Animal species vanishing from U. S. parks. *Internat. Wildl.* 17:1–25. [2]

Newmark, W. D. 1991. Tropical forest fragmentation and the local extinction of understory birds in the eastern Usambara Mountains, Tanzania. *Conserv. Biol.* 5:67–78. [7]

Newmark, W. D. 1995. Extinction of mammal populations in western North American national parks. *Conserv. Biol.* 9:512–526. [7]

Ngantou, D. and R. Braund. 1999. Waza-Lagobe: Restoring the good life. *World Conservation* 30:19–20.

Nicholls, A. O. and C. R. Margules. 1993. An upgraded reserve selection algorithm. *Biol. Conserv.* 64:165–169. [14]

Nicholls, K. H. 1999. Evidence for a trophic cascade effect on north-shore western Lake Erie phytoplankton prior to the zebra mussel invasion *Journal of Great Lakes Research* 25:942–949. [3]

Niemela, P., F. S. Chapin, K. Danell, and J. P. Bryant. 2001. Herbivory-mediated responses of selected boreal forests to climatic change. *Climatic Change* 48:427–440. [10]

Niemi, G. J. and M. E. McDonald. 2004. Application of ecological indicators. *Annu. Rev. Ecol. Evol. S.* 35:89–111. [18]

Nieminen, M., M. Siljander, and I. Hanski. 2004. Structure and dynamics of *Melitaea cinxia* metapopulations. In P. R. Ehrlich and I. Hanski. (eds.), *On the Wings of Checkerspots: A Model System for Population Biology*, pp. 63–91. Oxford University Press, Oxford. [12]

Nieminen, M., M. C. Singer, W. Fortelius, K. Schoeps, and I. Hanski. 2001. Experimental confirmation that inbreeding depression increases extinction risk in butterfly populations. *Am. Nat.* 157:237–244. [11]

Nilsson, C., C. A. Reidy, M. Dynesius, and C. Revenga. 2005. Fragmentation and flow regulation of the world's large river systems. *Science* 308:405–408. [6]

Nilsson, C., M. Svedmark, P. Hansson, S. Xiong, and K. Berggren. 2000. River fragmentation and flow regulation analysis. In C. Revenga, J. Brunner, N. Henninger, K. Kassem, and R. Payne (eds.), *Pilot Analysis of Global Ecosystems: Freshwater Systems*. World Resources Institute, Washington, D.C. [7]

Nilsson, S. G. 1986. Are bird communities in small biotope patches random samples from communities in large patches? *Biol. Conserv.* 38:179–204. [7]

Ninan, K. N. and S. Lakshmikanthamma. 2001. Social cost-benefit analysis of a watershed development project in Karnataka, India. *Ambio* 30:157–161. [5]

Nittler, J. B. and D. W. Nash. 1999. The certification model for forestry in Bolivia. *J. Forest* 97:32–36. [8]

Noon, B. R. and A. B. Franklin. 2002. Scientific research and the Spotted Owl (*Strix occidentalis*): Opportunities for major contributions to avian population ecology. *Auk* 119:311–320. [17]

Noon, B. R. and K. S. McKelvey. 1996. Management of the Spotted Owl: A case history in conservation biology. *Ann. Rev. Ecol. Syst.* 27:135–162. [17]

Noon, B. R., K. S. McKelvey, D. W. Lutz, W. S. LaHaye, R. J. Gutierrez, and C. A. Moen. 1992. Estimates of demographic parameters and rates of population change. In J. Verner et al., (eds.), *The California Spotted Owl: A Technical Assessment of its Current Status*, pp. 175–186. General Technical Report PSW-GTR-133. Pacific Southwest Research Station, U.S. Forest Service, U.S. Department of Agriculture, Albany, California. [17]

Norton, B. G. 1987. *Why Preserve Natural Variety?* Princeton University Press, Princeton. [4]

Norton, B. G. 1991. *Toward Unity Among Environmentalists.* Oxford University Press, New York. [4]

Norton, D. A. 1991. *Trilepidea adamsii*: An obituary for a species. *Conserv. Biol.* 5:52–57. [12]

Norton, D. A., R. J. Hobbs, and L. Atkins. 1995. Fragmentation, disturbance, and plant distribution: Mistletoes in woodland remnants in the Western Australian Wheatbelt. *Conserv. Biol.* 9:426–438. [7]

Norton-Griffiths, M. and C. Southey. 1995. The opportunity costs of biodiversity

conservation in Kenya. *Ecol. Econ.* 12:125–139. [5]

Noss, R. F. 1981. The birds of Sugarcreek, an Ohio nature reserve. *Ohio J. Sci.* 81:29–40. [7]

Noss, R. F. 1983. A regional landscape approach to maintain diversity. *BioScience* 33:700–706. [7]

Noss, R. F. 1990. Indicators for monitoring biodiversity: A hierarchical approach. *Conserv. Biol.* 4:355–364. [2, 18]

Noss, R. F. 1991a. Effects of edge and internal patchiness on avian habitat use in an old-growth Florida hammock. *Natural Areas J.* 11:34–47. [7]

Noss, R. F. 1991b. Landscape connectivity: Different functions at different scales. In W. E. Hudson (ed.), *Landscape Linkages and Biodiversity*, pp. 27–39. Island Press, Washington, D.C. [7]

Noss, R. F. 1996. Ecosystems as conservation targets. *Trends Ecol. Evol.* 11:351. [18]

Noss, R. F. 2001. Beyond Kyoto: Forest management in a time of rapid climate change. *Conserv. Biol.* 15:578–590. [7]

Noss, R. F. and A. Y. Cooperrider. 1994. *Saving Nature's legacy: Protecting and Restoring Biodiversity*. Defenders of Wildlife and Island Press, Washington, D.C. [7, 18]

Noss, R. F. and L. D. Harris. 1986. Nodes, networks, and MUMs: Preserving diversity at all scales. *Environ. Mgmt.* 10:299–309. [7]

Noss, R. F. and R. L. Peters. 1995. *Endangered Ecosystems: A Status Report on America's Vanishing Habitat and Wildlife*. Defenders of Wildife, Washington, D.C. [6]

Noss, R. F., E. T. LaRoe, and J. M. Scott. 1995. *Endangered Ecosystems of the United States: A Preliminary Assessment of Loss and Degradation*. Biological Report 28. USDI National Biological Service, Washington, D.C. [7]

Noss, R. F., M. A. O'Connell, and D. D. Murphy. 1997. *The Science of Conservation Planning*. Island Press, Washington, D.C. [11]

Noss, R. F., C. Carroll, K. Vance-Borland, and G. Wuerthner. 2002. A multicriteria assessment of the irreplaceability and vulnerability of sites in the Greater Yellowstone Ecosystem. *Conserv. Biol.* 16:895–908. [12, 14]

Noss, R. F., H. B. Quigley, M. G. Hornocker, T. Merrill, and P. C. Paquet. 1996. Conservation. Biology and carnivore conservation. *Conserv. Biol.* 10:949–963. [7]

Nyssen, J., J. Poesen, J. Moeyersons, J. Deckers, M. Haile, and A. Land. 2004. Human impact on the environment in the Ethiopian and Eritrean highlands—a state of the art. *Earth-Science Reviews* 64: 273–320. [6]

Nyström, M., C. Folke, and F. Moberg. 2000. Coral-reef disturbance and resilience in a human-dominated environment. *Trends Ecol. Evol.* 15:413–417. [13]

Nyström, P., O. Svensson, B. Lardner, C. Brönmark, and W. Granéli. 2001. The influence of multiple introduced predators on a littoral pond community. *Ecology* 82: 1023–1039 [3]

Ocean Studies Board. 2004. *Nonnative Oysters in the Chesapeake Bay*. Committee on Nonnative Oysters in the Chesapeake Bay, National Research Council of the Nation Academies. The National Academies Press, Washington, D.C. [13]

O'Brien, S., E. R. Emahalala, V. Beard, R. M. Rakotondrainy, A. Reid, V. Rahariosa, and T. Coulson. 2003. Decline of the Madagascar radiated tortoise *Geochelone radiata* due to overexploitation. *Oryx* 37:338–343. [8]

O'Brien, T. G. and M. F. Kinnaird. 2003. Caffeine and conservation. *Science* 300:587. [16]

O'Neill, R. V. and eleven others. 1988. Indices of landscape pattern. *Landscape Ecol.* 1:153–162. [7]

O'Riordan, T. and J. Cameron. 1994. *Interpreting the Precautionary Principle*. Earthscan Publications, London. [5]

Oates, W. E. and W. J. Baumol. 1975. The instruments for environmental policy. In E. Mills (ed.), *Economic Analysis of Environmental Problems*, pp. 95–128. Columbia University Press, New York. [17]

Obadia, J., M. Caro, and N. Anderson. 2002. Adapting communications to a dynamic cultural landscape: Recommendations for the development and implementation of a Bushmeat Crisis Task Force public awareness campaign in Central Africa. College Park, Maryland: University of Maryland. Available online: http://bushmeat.org/pdf/PS_PACAfrica.pdf. [8]

Oberdorster, E. and A. O. Cheek. 2000. Gender benders at the beach: Endocrine disruption in marine and estuarine organisms. *Environ. Toxicol. Chem.* 20:23–36. [6]

Obura, D. 2003. Rapid management response to bleaching. Coral Reefs, Climate, and Coral Bleaching. A workshop hosted by the U.S. Department of Commerce (NOAA), U.S. Environmental Protection Agency (EPA), and U.S. Department of the Interior (DOI). 18–20 June, 2003 Hawaii. [13]

Odell, E. and R. L. Knight. 2001. Songbird and medium-sized mammal communities associated with exurban development in Pitkin County, Colorado. *Conserv. Biol.* 15:1143–1150. [7]

Odum, E. P. 1953. *Fundamentals of Ecology*. W. B. Saunders, Philadelphia. [4]

Odum, E. P. 1987. The Georgia landscape: A changing resource: Final report of the Kellogg Physical Resources Task Force. Institute of Ecology, University of Georgia, Athens. [12]

Odum, E. P. 1989. *Ecology and Our Endangered Life-Support Systems*. Sinauer Associates, Sunderland, MA. [1]

Odum, E. P. 1989. Input management of production systems. *Science* 243:177–182. [7]

Oechel, W. C., S. J. Hastings, G. L. Vourlitis, and M. Jenkins. 1993. Recent change of arctic tundra ecosystems from net carbon dioxide sink to a source. *Nature* 361:520–523. [10]

Oechel, W. C., G. L. Vourlitis, S. J. Hastings, R. C. Zulueta, L. Hinzman, and D. Kane.

2000. Acclimation of ecosystem CO_2 exchange in the Alaska Arctic in response to decadal warming. *Nature* 406:978–981. [10]

Oedekoven, C. S., D. G. Ainley, and L. B. Spear. 2001. Variable responses of seabirds to change in marine climate: California Current, 1985–1994. *Mar. Ecol.–Prog. Ser.* 212:265–281. [10]

Oelschlaeger, M. 1994. *Caring for Creation: An Ecumenical Approach to the Environmental Crisis*. Yale University Press, New Haven, CT. [4]

Office of Technology Assessment. 1993. *Harmful nonindigenous species in the United States*. U.S. Congress, Washington, D.C. [9]

Ogden, J. C. 1978. Recent population trends of colonial wading birds on the Atlantic and Gulf Coastal plains. In A. Sprunt, J. C. Ogden and S. Winckler (eds.), *Wading Birds. Research Report no. 7*, pp. 137–153. National Audubon Society, New York. [13]

Ogden, J. C. 1994. A comparison of wading bird nesting colony dynamics as an indication of ecosystem conditions in the southern Everglades. In S. M. Davis and J. C. Ogden (eds.), *Everglades, The Ecosystem and its Restoration*. St. Lucie Press, Delray, FL. [13]

Olden, J. D., N. L. Poff, M. R. Douglas, M. E. Douglas, and K. D. Fausch. 2004. Ecological and evolutionary consequences of biotic homogenization. *Trends Ecol. Evol.* 19:18–24. [7]

Olff, H. and M. E. Ritchie. 1999. Effects of herbivores on grassland plant diversity. *Trends Ecol. Evol.* 13:261–265. [2]

Oliver, I., A. J. Beattie, and A. York. 1998. Spatial fidelity of plant, vertebrate, and invertebrate assemblages in multiple-use forest in Eastern Australia. *Conserv. Biol.* 12:822–825. [14]

Olivier, R. 1978. Distribution and status of the Asian elephant. *Oryx* 14:379–424. [17]

Olsen, G. J. and C. R. Woese. 1996. Lessons from an Archaeal genome: What are we learning from *Methanococcus jannaschii*? *Trends in Genetics* 12:377–379. [2]

Olsen, J. B., P. Bentzen, M. A. Banks, J. B. Shaklee, and S. Young. 2000. Microsatellites reveal population identity of individual pink salmon to allow supportive breeding of a population at risk of extinction. *T. Am. Fish. Soc.* 129:232–242. [11]

Olsen, T. M., D. M. Lodge, G. M. Capelli, and R. J. Houlihan. 1991. Mechanism of impact of an introduced crayfish (*Orconectes rusticus*) on littoral congeners, snails, and macrophytes. *Canadian Journal of Fisheries and Aquatic Sciences* 48:1853–1861. [3]

Olson, D. M. and E. Dinerstein. 1998. The Global 200: a representation approach to conserving the Earth's most biologically valuable ecoregions. *Conserv. Biol.* 12:502–515. [6, 12, 14]

Olson, D. M., E. Dinerstein, E. D. Wikramanayake, N. D. Burgess, G. V. N. Powell, E. C. Underwood, J. A. D'Amico, I. Itoua, H. E. Strand, J. C. Morrison, C. J.

Loucks, T. F. Allnutt, T. H. Ricketts, Y. Kura, J. F. Wettengel, P. Hedao, and K. R. Kassem. 2001. Terrestrial ecoregions of the world: A new map of life on Earth. *BioScience* 51:933–938. [2, 6, 18]

Opdam, P., D. van Dorp, and C. J. F. ter Braak. 1984. The effect of isolation on the number of woodland birds in small woods in the Netherlands. *J. Biogeogr.* 11:473–478. [7]

Orians, G. H. 1993. Endangered at what level? *Ecol. Appl.* 3:206–208. [6]

Orians, G. H. and W. E. Kunin. 1991. Ecological uniqueness and loss of species. In G. H. Orians, G. M. Brown, W. E. Kunin, and J. E. Swierzbinski (eds.). *The Preservation and Valuation of Biological Resources*, pp. 146–184. University of Washington Press, Seattle. [2]

Ortiz, S., F. Carrera, and L. M. Ormeño. 2002. Comercialización de productos maderables en concesiones forestales comunitarias en Petén, Guatemala. Serie Técnica, Informe Técnico No. 326. Colección Manejo Diversificado de Bosques Naturales. CATIE, Turrrialba, Costa Rica. [8]

Osafo, E. D. 1968. *The application of the Tropical Shelterwood System (TSS) in Coupe 2 of Bobiri Research Centre*. FPRI Ghana, Technical Note No. 3. [8]

Ostfeld, R. S. and F. Keesing. 2000. Biodiversity and disease risk: The case of Lyme disease. *Conserv. Biol.* 14:722–728. [3]

Ostfeld, R. S., C. G. Jones, and J. O. Wolff. 1996. Of mice and mast. *BioScience* 46: 323–330. [3]

Ostrom, E. 1990. *Governing the Commons: The Evolution of Institutions for Collective Action*. University Press, Cambridge. [8]

Ostrom, E. and E. Schlager. 1996. The formation of property rights. In S. Hanna, C. Folke, and K. Maler (eds.), *Rights to Nature: Ecological, Economic, Cultural and Political Principals of Institutions for the Environment*, pp. 127–157. Island Press, Washington, D.C. [8]

Ottman, J. A. 1997. *Green Marketing: Opportunity for Innovation*. NTC Publishing Group, Lincolnwood, IL. [18]

Ovaskainen, O. and I. Hanski. 2002. Transient dynamics in metapopulation response to perturbation. *Theor. Popul. Biol.* 61:285–295. [12]

Owens, I. P. F. and P. M. Bennett. 2000. Ecological basis of extinction risk in birds: Habitat loss versus human persecution and introduced predators. *Proc. Nat. Acad. Sci. USA* 97:12144–12148. [3]

Oxley, D. J., M. B. Fenton, and G. R. Carmody. 1974. The effects of roads on populations of small mammals. *J. Appl. Ecol.* 11:51–59. [7]

Pacheco, J., G. Ceballos, G. C. Daily, P. R. Ehrlich, G. Suzán, B. Rodríguez-H., and E. Marcé. Diversidad, historia natural y conservación de los mamíferos de la región de San Vito de Coto Brus, Costa Rica. *Revista de Biología Tropical*. In press. [18]

Pacheco, L. F. and J. A. Simonetti. 2000. Genetic structure of a Mimosoid tree de-

prived of its seed disperser, the spider monkey. *Conserv. Biol.* 14:1766–1775. [12]

Pacific Rivers Council. 1993. *The Decline of Coho Salmon and the Need for Protection under the Endangered Species Act*. Report of the Pacific Rivers Council, Eugene, OR. [7]

Packer, M. J., R. D. Holt, P. J. Hudson, K. D. Latherly, and A. P. Dodson. 2003. Keeping the herds healthy and alert: Implications of predator control for infectious disease. *Ecol. Lett.* 6:797–802. [3]

Paddack, J.-P. 2003. *Mobilizing Funding For Biodiversity Conservation: A User-Friendly Training Guide*. Available from: www.worldwildlife.org/conservationfinance. [6]

Paetkau, D. 2003. An empirical exploration of data quality in DNA-based population inventories. *Mol. Ecol.* 12:1375–1387. [11]

Paetkau, D., C. Strobeck, and I. Stirling. 1995. Microsatellite analysis of population structure in Canadian polar bears. *Mol. Ecol.* 4:347–354. [11]

Paetkau, D., L. P. Waits, P. L. Clarkson, L. Craighead, E. Vyse, R. Ward, and C. Strobeck. 1998. Variation in genetic diversity across the range of North American brown bears. *Conserv. Biol.* 12:418–429. [11]

Page, B., J. McKenzie, R. McIntosh, A. Baylis, A. Morrissey, N. Calvert, T. Haase, M. Berris, D. Dowie, P. D. Shaughnessy, and S. D. Goldsworthy. 2004. Entanglement of Australian sea lions and New Zealand fur seals in lost fishing gear and other marine debris before and after Government and industry attempts to reduce the problem. *Marine Pollut. Bull.* 49:33–42. [6]

Pain, D. J., D. Hill, and D. I. McCracken. 1997. Impact of agricultural intensification on bird distributions in Britain 1970–1990. *Agr. Ecosyst. Environ.* 64:19–32. [18]

Paine, R. T. 1966. Food web complexity and species diversity. *Am. Nat.* 103:91–93. [12]

Paine, R. T. 1969. A note on trophic complexity and community stability. *Am. Nat.* 110:91–93. [3]

Paine, R. T. 1969. The *Pisaster-Tegula* interaction: Prey patches, predator food preference, and intertidal community structure. *Ecology* 50:950–961. [9]

Paine, R. T. 1974. Intertidal community structure: Experimental studies on the relationship between a dominant competitor and its principal predator. *Oecologia* 15:93–120. [2]

Palen, W. J., D. E. Schindler, M. J. Adams, C. A. Pearl, R. B. Bury, and S. A. Diamonds. 2002. Optical characteristics of natural waters protect amphibians from UV-B in the U.S. Pacific Northwest. *Ecology* 83:2951–2957. [3]

Palen, W. J., D. E. Schindler, M. J. Adams, C. A. Pearl, R. B. Bury, and S. A. Diamond. 2004. Optical characteristics of natural waters protect amphibians from UV-B in the U.S. Pacific Northwest: reply. *Ecology* 85:1754–1759. [3]

Palmer, M. A., R. F. Ambrose, and N. L. Poff. 1997. Ecological theory and community

restoration ecology. *Restoration Ecology* 5:291–300. [15]

Palomares, F., P. Gaona, P. Ferreras, and M. Delibes. 1995. Positive effects on game species of top predators by controlling smaller predator populations: an example with lynx, mongooses, and rabbits. *Conserv. Biol.* 9:295–305. [3]

Palsboll, P. J. 1999. Genetic tagging: Contemporary molecular ecology. *Biol. J. Linn. Soc.* 68:3–22. [11]

Palsboll, P. J., J. Allen, M. Berube, P. J. Clapham, T. P. Feddeersen, P. L Hammond, R. R. Hudson, H. Jogensen, S. Katona, A. H. Larsen, F. Larsen, J. Lien, D. K. Mattila, J. Sigurjonsson, R. Sears, T. Smith, R. Sponer, P. Stevick, and N. Oien. 1997. Genetic tagging of humpback whales. *Nature* 388:767–769. [11]

Palumbi, S. P. 2003. Population genetics, demographic connectivity and the design of marine reserves. *Ecol. Appl.* 13:S138–S145. [14]

Panek, J. A. and A. H. Goldstein. 2001. Response of stomatal conductance to drought in ponderosa pine: implications for carbon and ozone uptake. *Tree Physiol.* 21:335–342. [6]

Panek, J. A. and S. Ustin. 2004. Ozone uptake in relation to water availability in ponderosa pine forests: Measurements, modeling, and remote-sensing. Report to National Park Service, Air Resources Division, Denver, CO. [6]

Panek, J. A., M. Kurpius, and A. H. Goldstein. 2002. An evaluation of ozone exposure metrics for a ponderosa pine ecosystem. *Environ. Pollut.* 117:93–100. [6]

Paper, J. 2001. Chinese religion, Daoism and deep ecology. In D. Barnhill and R. Gottlieb (eds.), *Deep Ecology and World Religions: New Essays on Sacred Ground*, pp. 107–126. State University of New York Press, Albany. [4]

Parendes, L. A. and J. A. Jones. 2000. Role of light availability and dispersal in exotic plant invasion along roads and streams in the H. J. Andrews Experimental Forest, Oregon. *Conserv. Biol.* 14:64–75. [7]

Park, R., M. Trehan, P. Mausel, and R. Howe. 1989. The effects of sea level rise on U.S. coastal wetlands. In J.B. Smith and D.A. Tirpak (eds.), *The Potential Effects of Global Climate Change on the United States: Appendix B: Sea Level Rise*, pp. 1–55. U. S. Environmental Protection Agency, Washington, D.C. [15]

Parker, F. D., S. W. T. Batra, and V. J. Tepedino. 1987. New pollinators for our crops. *Agr. Zool. Rev.* 2:279–304. [2]

Parker, I. M., D. Simberloff, W. M. Lonsdale, K. Goodell, M. Wonham, P. M. Kareiva, M. H. Williamson, B. Von Holle, P. B. Moyle, J. E. Byers, and L. Goldwasser. 1999. Impact: Toward a framework for understanding the ecological effect of invaders. *Biol. Invasions* 1:3–19. [9]

Parks, S. A. and A. H. Harcourt. 2002. Reserve size, local human density, and mammalian extinctions in U.S. protected areas. *Conserv. Biol.* 16:800–808. [12]

Parmesan, C. 1996. Climate and species' range. *Nature* 382:765–766. [10]

Parmesan, C. 2003. Butterflies as bioindicators for climate change effects. In C. L. Boggs, W. B. Watt, and P. R. Ehrlich (eds.), *Butterflies: Ecology and Evolution Taking Flight*, pp. 541–560. University of Chicago Press, Chicago. [10]

Parmesan, C. 2005. Range and abundance changes. In T. Lovejoy and L. Hannah (eds.), *Climate Change and Biodiversity*, pp. 41–55. Yale University Press, New Haven. [10]

Parmesan, C. and G. Yohe 2003. A globally coherent fingerprint of climate change impacts across natural systems. *Nature* 421:37–42. [10]

Parmesan, C., T. L. Root, and M. R. Willig. 2000. Impacts of extreme weather and climate on terrestrial biota. *B. Am. Meteorol. Soc.* 81:443–450. [10]

Parmesan, C., N. Ryrholm, C. Stefanescu, J. K. Hill, C. D. Thomas, H. Descimon, B. Huntley, L. Kaila, J. Kullberg, T. Tammaru, W. J. Tennent, J. A. Thomas, and M. Warren 1999. Poleward shifts in geographical ranges of butterfly species associated with regional warming. *Nature* 399:579–583. [10]

Parnell, J. 1993. Plant taxonomic research, with special reference to the tropics: problems and potential solutions. *Conserv. Biol.* 7(4):809–814. [18]

Parra-Olea, G., M. Garcia-Paris, and D. B. Wake. 1999. Status of some populations of Mexican salamanders (Amphibia: Plethodontidae). *Rev. Biol. Trop.* 47:217–223. [3]

Parris, M. J. and D. R. Baud. 2004. Interactive effects of a heavy metal and chytridiomycosis on Gray Treefrog larvae (*Hyla chrysoscelis*). *Copeia* 2004:344–350. [3]

Parris, M. J. and J. G. Beaudoin. 2004. Chytridiomycosis impacts predator-prey interactions in larval amphibian communities. *Oecologia* 140:626–632. [3]

Parris, T. W. and R. W. Kates. 2003. Characterizing and measuring sustainable development. *Ann. Rev. Env. Resour.* 28:559–586. [16]

Parrish, J. D., D. P. Braun, and R. S. Unnasch. 2003. Are we conserving what we say we are? Measuring ecological integrity within protected areas. *Bioscience* 53:851–860. [18]

Parrotta, J. A. 1993. Secondary forest regeneration on degraded tropical lands: The role of plantations as "foster ecosystems." In H. Lieth and M. Lohman (eds.), *Restoration of Tropical Forest Ecosystems*, pp. 63–73. Kluwer Academic Publishers, Amsterdam, The Netherlands. [15]

Parsons, T. R. 1996. Taking stock of fisheries management. *Fisheries Oceanography* 5:224–226. [3]

Paton, P. W. C. 1994. The effect of edge on avian nest success: How strong is the evidence? *Conserv. Biol.* 8:17–26. [7]

Patterson, B. D. 1987. The principle of nested subsets and its implications for biological conservation. *Conserv. Biol.* 1:323–334. [7]

Pauly, D. 1995. Anecdotes and the shifting baseline syndrome of fisheries. *Trends. Ecol. Evol.* 10:430. [8]

Pauly, D. and V. Christensen. 1995. Primary production necessary to sustain global fisheries. *Nature* 374:255–247. [1, 3]

Pauly, D., V. Christensen, and C. Walters. 2000. Ecopath, Ecosim, and Ecospace as tools for evaluating ecosystem impact of fisheries. *ICES J. Mar. Sci.* 57:697–706. [13]

Pauly, D., V. Christensen, J. Dalsgaard, R. Froese, and F. Torres Jr. 1998. Fishing down marine food webs. *Science* 279:860–863. [3]

Payne, J., E. Holmes, and A. Edwards. 2005. Is there ecology in ecosystem management? Report of the Working Group on Ecosystem Management. Cumulative Risk Initiative, Northwest Fisheries Science Center, Seattle, WA. [13]

Pearce, D. W. and R. K. Turner. 1989. *Economics of Natural Resources and the Environment*. Wheatsheaf, Brighton. [5]

Pearce, D., F. E. Putz, and J. Vanclay. 2003. Sustainable forestry: Panacea or pipedream? *Forest Ecol. Manag.* 172:229–247. [8]

Pearce, D. 2005. Pipe dreams. *Conservation in Practice* 6:20–27. [18]

Pearce, J., R. L. Fields, and K. T. Scribner. 1997. Nest materials as a source of genetic data for avian behavioral studies. *J. Field Ornithol.* 68:471–481. [11]

Pearce, P. 1990. *Introduction to Forestry Economics*. University of British Columbia Press, Vancouver. [8]

Pearson, D. L. and S. A. Juliano 1993. Evidence for the influence of historical processes in co-occurrence and diversity of tiger beetle species. In R. E. Ricklefs and D. Schluter (eds.), *Species Diversity in Ecological Communities*, pp. 194–202. University of Chicago Press, Chicago. [2]

Pearson, R. G. 1981. Recovery and recolonization of coral reefs. *Marine Ecology Progress Series* 4: 105–22. [15]

Pearson, S. M. 1993. The spatial extent and relative influence of landscape-level factors on wintering bird populations. *Landscape Ecol.* 8:3–18. [12]

Pechmann, J. H. K. 2003. Natural population fluctuations and human influences: Null models and interactions. In R. D. Semlitsch (ed.), *Amphibian Conservation*, pp. 85–93. Smithsonian Institution Press, Washington, D.C. [3]

Pechmann, J. H. K. and H. M. Wilbur. 1994. Putting declining amphibian populations in perspective: Natural fluctuations and human impacts. *Herpetologica* 50:65–84. [3]

Pechmann, J. H. K., D. E. Scott, R. D. Semlitsch, J. P. Caldwell, L. J. Vitt, and J. W. Gibbons. 1991. Declining amphibian populations: The problem of separating human impacts from natural fluctuations. *Science* 253:892–895. [3]

Pedersen, B. S. and A. M. Wallis. 2004. Effects of white-tailed deer herbivory on forest gap dynamics in a wildlife preserve, Pennsylvania, USA. *Nat. Area. J.* 24:82–94. [7, 8]

Pella, J. J. and G. B. Milner. 1987. Use of genetic marks in stock composition analysis. In N. Ryman and F. Utter (eds.), *Population Genetics and Fishery Management*, pp. 247–276. Washington Sea Grant, Seattle. [11]

Pemberton, R. W. 1995. *Cactoblastis cactorum* (Lepidoptera: Pyralidae) in the United States: An immigrant biological control agent or an introduction of the nursery industry? *Am. Entomol.* 41:230–232. [9]

Pemberton, R. W. 2000. Predictable risk to native plants in weed biological control. *Oecologia* 125:489–494. [9]

Peñuelas, J. and I. Filella. 2001. Phenology: Responses to a changing world. *Science* 294:793. [10]

Peñuelas, J. and M. Estiarte 1998. Can elevated CO_2 affect secondary metabolism and ecosystem function? *Trends Ecol. Evol.* 13:20–24. [10]

Pereira, H. M., G. C. Daily, and J. Roughgarden. 2004. A framework for assessing the relative vulnerability of species to land-use change. *Ecol. Appl.* 14:730–742. [18]

Peres, C. A. 2000a. Effects of subsistence hunting on vertebrate community structure in Amazonian forests. *Conserv. Biol.* 14:240–253. [8]

Peres, C. A. 2000b. Evaluating the impact and sustainability of subsistence hunting at multiple Amazonian forest sites. In J. G. Robinson and E. L. Bennett (eds.), *Hunting for Sustainability in Tropical Forests*, pp. 31–57. Columbia University Press, New York. [8]

Peres, C. A. and I. R. Lake. 2003. Extent of nontimber resource extraction in tropical forests: Accessibility to game vertebrates by hunters in the Amazon basin. *Conserv. Biol.* 17:1–17. [8]

Peres, C. A. and J. W. Terborgh. 1995. Amazonian nature reserves: an analysis of the defensibility status of existing conservation units and design criteria for the future. *Conserv. Biol.* 9:34–46. [14]

Peres, C. A. and M. van Roosmalen. 2002. Patterns of primate frugivory in Amazonia and the Guianan shield: Implications to the demography of large-seeded plants in overhunted tropical forests. In D. Levey, W. Silva and M. Galetti (eds.), *Seed Dispersal and Frugivory: Ecology, Evolution and Conservation*, pp. 407–423. CABI International, Oxford. [8]

Peres, C. A., C. Baider, P. A. Zuidema, L. H. O. Wadt, K. A. Kainer, D. A. P. Gomes-Silva, R. P. Salomão, L. L. Simões, E. R. N. Franciosi, F. Cornejo Valverde, R. Gribel, G. H. Shepard Jr., M. Kanashiro, P. Coventry, D. W. Yu, A. R. Watkinson, and R. P. Freckleton. 2003. Demographic threats to the sustainability of Brazil nut exploitation. *Science* 302:2112–2114. [8]

Perfecto, I., A. Mas, T. Dietsch, and J. Vandermeer. 2003. Conservation of biodiversity in coffee agroecosystems: a tri-taxa comparison in southern Mexico. *Biodivers. Conserv.* 12:1239–1252. [18]

Perrow, M. R. and A. J. Davy (eds.). 2002. *Handbook of Ecological Restoration*. Cambridge University Press, Cambridge. [10]

Perrow, M. R. and A. J. Davy (eds.). 2002a. *Handbook of Ecological Restoration. Volume 1. Principles of Restoration.* Cambridge University Press, Cambridge. [15]

Perrow, M. R. and A. J. Davy (eds.). 2002b. *Handbook of Ecological Restoration. Volume 2. Restoration in Practice.* Cambridge University Press, Cambridge. [15]

Pessier, A. P., D. K. Nichols, J. E. Longcore, and M. S. Fuller. 1999. Cutaneous chytridiomycosis in poison dart frogs (*Dendrobates* spp.) and White's tree frogs (*Litoria caerulea*). *J. Vet. Diagn. Invest.* 11:194–199. [3]

Peters, A. and K. J. F. Verhoeven. 1994. Impact of artificial lighting on the seaward orientation of hatchling loggerhead turtles. *J. Herpetol.* 28:112–114. [6]

Peters, C. M. 1994. *Sustainable Harvest of Nontimber Plant Resources in Tropical Moist Forest: An Ecological Primer.* Biodiversity Support Program, Washington, D.C. [8]

Peters, C. M., A. H. Gentry, and R. O. Mendelsohn. 1989. Valuation of an Amazonian Rainforest. *Nature* 339:655–656. [4, 8]

Peters, R. L. and J. D. S. Darling. 1985. The greenhouse effect and nature reserves. *BioScience* 35:707–717. [7]

Peters, R. L. and T. E. Lovejoy (eds.). 1992. *Global Warming and Biological Diversity.* Yale University Press, New Haven, CT. [7]

Peterson, A. L. 2003. Review of D. Barnhill and R. Gottlieb (eds.). Deep Ecology and World Religions: New Essays on Sacred Ground. *Environ. Ethics* 25:215–219. [4]

Peterson, C. H. and R. N. Lipcius. 2003. Conceptual progress towards predicting quantitative ecosystem benefits of ecological restorations. *Mar. Ecol. Prog. Ser.* 264:297–307. [15]

Peterson, C. H., J. H. Grabowski, and S. P. Powers. 2003. Estimated enhancement of fish production resulting from restoring oyster reef habitat: quantitative valuation. *Mar. Ecol. Prog. Ser.* 264:249–264. [15]

Peterson, C. H., S. D. Rice, J. W. Short, D. Esler, J. L. Bodkin, B. E. Ballachey, and D. B. Irons. 2003. Long-Term Ecosystem Response to the Exxon Valdez Oil Spill. *Science* 302:2082–2086. [6]

Peterson, G. L. and A. Randall, (eds.). 1984. *The Valuation of Wildland Benefits.* Westview Press, Boulder, CO. [4, 5]

Pethick, J. 2002. Estuarine and tidal wetland restoration in the United Kingdom: Policy versus practice. *Rest. Ecol.* 10:431–437. [10]

Petit, L. J., D. R. Petit, and T. E. Martin. 1995. Landscape-level management of migratory birds: looking past the trees to see the forest. *Wildlife Soc. B.* 23:420–429. [12]

Petit, R. J., A. E. Mousadik, and O. Pons. 1998. Identifying populations for conservation on the basis of genetic markers. *Conserv. Biol.* 12:844–855. [11]

Petraitis, P. S., R. E. Latham, and R. A. Niesenbaum. 1989. The maintenance of species diversity by disturbance. *Quart. Rev. Biol.* 64:418–464. [1]

Petranka, J. W. 1998. *Salamanders of the United States and Canada.* Smithsonian Institution Press, Washington, D.C. [3]

Petts, J. (ed.). 1999a. *Handbook of Environmental Impact Assessment.* Vol. 1, Environmental impact assessment: Process, methods and potential. Blackwell Science, Oxford. [5]

Petts, J. (ed.). 1999b. *Handbook of Environmental Impact Assessment.* Vol. 2, Environmental impact assessment in practice: Impact and limitations. Blackwell Science, Oxford. [5]

Pew Oceans Commission. 2003. *America's Living Oceans: Charting a Course for Sea Change.* Pew Oceans Commission, Arlington, VA. [13]

Phillips, R. C. 1984. The ecology of eelgrass meadows in the Pacific Northwest: A community profile. U.S. Fish and Wildlife Service. FWS/OBS-84/24. pp.85. [6]

Pianka, E. R. 1986. *Ecology and Natural History of Desert Lizards.* Princeton University Press, Princeton. [2]

Pickett, S. T. A. and J. N. Thompson. 1978. Patch dynamics and the design of nature reserves. *Biol. Conserv.* 13:27–37. [14]

Pickett, S. T. A. and P. S. White. 1985. *The Ecology of Natural Disturbance and Patch Dynamics.* Academic Press, San Diego. [4, 10]

Pickett, S. T. A. and R. S. Ostfeld. 1995. The shifting paradigm in ecology. In R. L. Knight and S. F. Bates (eds.), *A New Century for Natural Resources Management,* pp. 261–278. Island Press, Washington, D.C. [1, 4]

Pickett, S. T. A., V. T. Parker, and P. L. Fiedler. 1992. The new paradigm in ecology: Implications for conservation biology above the species level. In P. L. Fiedler and S. K. Jain (eds.), *Conservation Biology: The Theory and Practice of Nature Conservation Preservation and Management,* pp. 65–88. Chapman and Hall, New York. [1]

Pielou, E. C. 1992. *After the Ice Age: The Return of Life to Glaciated North America.* University of Chicago Press, Chicago. [10]

Pienkowski, M. W., (ed.). 1998. Forum: biodiversity and high-nature-value farming systems. *J. Appl. Ecol.* 35:948–990. [18]

Piersma, T. 1998. Phenotypic flexibility during migration: optimization of organ size contingent on the risks and rewards of fueling and flight? *J. Avian Biol.* 29:511–520. [10]

Piketch, E. K., C. Santora, E. A. Babcock, A. Bakum, R. Bonfil, D. O. Conover, P. Dayton, P. Doukakis, D. Fluharty, B. Heneman, E. D. Houde, J. Link, P. A. Livingston, M. Mangel, M. K. McAllister, J. Pope, and K. J. Sainsbury. 2004. Ecosystem-based fisheries management. *Science* 305:346–347. [13]

Pimental, D. 1998. Economic benefits of natural biota. *Ecol. Econ.* 25:45–47. [5]

Pimentel, D., L. Lach, R. Zuniga, and D. Morrison. 2000. Environmental and economic costs of nonindigenous species in the United States. *BioScience* 50:53–65. [9]

Pimentel, D., C. Wilson, C. McCullum, R. Huang, P. Dwen, J. Flack, Q. Tran, T. Saltman, and B. Cliff. 1997. Economic and environmental benefits of biodiversity. *BioScience* 47:747–757. [18]

Pimental, D., M. Herz, M. Glickstein, M. Zimmerman, R. Allen, K. Becker, J. Evans, B. Hussain, R. Sarsfeld, A. Grosfeld, and T. Seidel. 2002. Renewable energy: Current and potential issues. *BioScience,* 52:1111–1120. [5]

Pimm, S. L. 1991. *The Balance of Nature?* University of Chicago Press, Chicago. [7]

Pimm, S. L. 1993. Life on an intermittent edge. *Trends Ecol. Evol.* 8:45–46. [7]

Pimm, S. L. 2001. *The World According to Pimm: A Scientist Audits the Earth.* McGraw-Hill, NY. [18]

Pimm, S. L., H. L. Jones, and J. Diamond. 1988. On the risk of extinction. *Am. Nat.* 132:757–785. [7]

Pimm, S. L., G. J. Russell, J. L. Gittleman, and T. M. Brooks. 1995. The future of biodiversity. *Science* 269:347–351. [14]

Pimm, S. L., M. Ayres, A. Balmford, G. Branch, K. Brandon, T. Brooks, R. Bustamante, R. Costanza, R. Cowling, L. M. Curran, A. Dobson, S. Farber, G. A. B. da Fonseca, C. Gascon, R. Kitching, J. McNeely, T. Lovejoy, R. A. Mittermeier, N. Myers, J. A. Patz, B. Raffle, D. Rapport, P. Raven, C. Roberts, J. P. Rodríguez, A. B. Rylands, C. Tucker, C. Safina, C. Samper, M. L. J. Stiassny, J. Supriatna, D. H. Wall, and D. Wilcove. 2001. Can we defy nature's end? *Science* 293:2207–2208. [14]

Pinard, M. A. 2005. Changes in plant communities associated with timber management in natural forests in the moist tropics. In D. Burslem, M. Pinard, and S. Hartley, (eds.), *Biotic Interactions in the Tropics.* Cambridge University Press, London, UK. [8]

Pinard, M. A., F. E. Putz, and J. Tay. 2000. Lessons learned from the implementation of reduced-impact logging in hilly terrain in Sabah, Malaysia. *Int. For. Rev.* 2:33–39. [8]

Pinchot, G. 1947. *Breaking New Ground.* Harcourt, Brace and Co., New York. [1, 4]

Piotrowski, J. S., S. L. Annis, and J. E. Longcore. 2004. Physiology of *Batrachochytrium dendrobatidis,* a chytrid pathogen of amphibians. *Mycologia* 96:9–15. [3]

Pisano, E. 1977. Fitogeografía de Fuego-Patagonia Chilena. I. Comunidades vegetales entre las latitudes 52° y 56°S. *Anales del Instituto de la Patagonia* 8:121–250. [9]

Pister, E. P. 1991. The Desert Fishes Council: Catalyst for change. In W. L. Minckley and J. E. Deacon (eds.), *Battle Against Extinction: Native Fish Management in the American West,* pp. 55–68. University of Arizona Press, Tucson. [1]

Pitcairn, M., D. Woods, D. Joley, C. Turner, and J. Balciunas. 1999. Population buildup and combined impact of introduced insects of yellow starthistle, *Centaurea solstitialis,* in California. In N. R. Spencer (ed.), *Proceedings X International Symposium on Biological Control of Weeds,*

pp. 747–751. Montana State University, Bozeman. [9]

Plater, B. 2001. *Petition to List the Southern Resident killer whale (Orcinus orca) as an Endangered Species under the Endangered Species Act.* The Center for Biological Diversity, Berkeley. [17]

Platt, J. R. 1964. Strong inference. *Science* 146:347–353. [18]

Plotkin, M. and L. Famalore. 1992. *Sustainable Harvest and Marketing of Rain Forest Products.* Island Press, Washington, D.C. [8]

Plotkin, M. J. 1988. The outlook for new agricultural and industrial products from the tropics. In E. O. Wilson (ed.), *Biodiversity,* pp. 106–116. National Academy Press, Washington, D.C. [4]

Plowright, W. 1982. The effects of rinderpest and rinderpest control on wildlife in Africa. *Symposia of the Zoological Society of London.* 50:1–28. [3]

Poff, N. L. 2002. Ecological response to and management of increased flooding caused by climate change. *Phil. Trans. Roy. Soc. London-A* 360: 1497–1510. [6]

Poff, N. L., J. D. Allan, M. A. Palmer, D. D. Hart, B. D. Richter, A. H. Arthington, K. H. Rogers, J. L. Meyer, and J. A. Stanford. 2003. River flows and water wards: emerging science for environmental decision making. *Frontiers in Ecology and the Environment* 1:298–306. [18]

Poiani, K. A., B. D. Richter, M. G. Anderson, and H. E. Richter. 2000. Biodiversity conservation at multiple scales: Functional sites, landscapes, and networks. *BioScience* 50:133–146. [18]

Poinar, G. O. 1983. *The Natural History of Nematodes.* Prentice-Hall, Englewood Cliffs, N.J. [2]

Polacheck, T., R. Hilborn, and A. E. Punt, 1993. Fitting surplus production models: Comparing methods and measuring uncertainty. *Can. J. Fish. Aquat. Sci.* 50:2597–2607. [8]

Polasky, S. and H. Doremus. 1998. When the truth hurts: Endangered species policy on private land with imperfect information. *J. Environ. Econ. Manage.* 35:22–47. [17]

Pollard, E. 1988. Temperature, rainfall, and butterfly numbers. *J. Appl. Ecol.* 25:819–898. [10]

Pollock, M. M., R. J. Naiman, and T. A. Hanley. 1998. Plant species richness in riparian wetlands—a test of biodiversity theory. *Ecology* 79:94–105. [9]

Pope, S. E., L. Fahrig, and G. Merriam. 2000. Landscape complementation and metapopulation effects on leopard frog populations. *Ecology* 81:2498–2508. [7]

Posey, M. H. and W. G. Ambrose. 1994. Effects of proximity to an off-shore hard-bottom reef on infaunal abundances. *Mar. Biol.* 118:745–753. [7]

Posner, R. 1980. *The Economics of Justice.* Harvard University Press Cambridge, MA. [5]

Possingham, H. P. and I. Davies. 1995. ALEX: A model for the viability analysis of spatially structured populations. *Biol. Conserv.* 73:143–150. [12]

Possingham, H. P., J. Franklin, K. Wilson, and T. J. Regan. 2005. The roles of spatial heterogeneity and ecological processes in conservation planning. In G. M. Lovett, C. G. Jones, M. G. Turner, and K. C. Weathers, (eds.), *Ecosystem Function in Heterogeneous Landscapes.* Springer-Verlag, New York. [14]

Possingham, H. P., S. J. Andelman, M. A. Burgman, R. A. Medellín, L. L. Master, and D. A. Keith. 2002. Limits to the use of threatened species lists. *Trends Ecol. Evol.* 17:503–507. [3]

Possingham, H., I. Ball, and S. Andelman. 2000. Mathematical methods for identifying representative reserve networks. In S. Ferson and M. Burgman, (eds.), *Quantitative Methods for Conservation Biology,* pp. 291–305. Springer-Verlag, New York. [14]

Postel, S. L., G. C. Daily, and P. R. Ehrlich. 1996. Human appropriation of renewable freshwater. *Science* 271:785–788. [1, 3]

Poston, T. and I. Stewart. 1978. *Catastrophe Theory and Its Applications.* Pitman, San Francisco. [10]

Poulter, B. 2005. Interactions between landscape disturbance and gradual environmental change: Plant community migration in response to fire and sea level rise. PhD Thesis. Duke University.

Pounds, J. A. 2000. Amphibians and reptiles. In N. M. Nadkarni and N. T. Wheelwright (eds.), *Monteverde: Ecology and Conservation of a Tropical Cloud Forest,* pp. 149–177. Oxford University Press, New York. [3, 10]

Pounds, J. A. 2001. Climate and amphibian declines. *Nature* 410:639–640. [10]

Pounds, J. A. and M. L. Crump. 1994. Amphibian declines and climate disturbance: The case of the golden toad and the harlequin frog. *Conserv. Biol.* 8:72–85. [3]

Pounds, J. A. and R. Puschendorf. 2004. Clouded futures. *Nature* 427:107–109. [10]

Pounds, J. A., M. P. L. Fogden, and J. H. Campbell, 1999: Biological response to climate change on a tropical mountain. *Nature* 398:611–615. [10]

Pounds, J. A., M. P. L. Fogden, J. M. Savage, and G. C. Gorman. 1997. Tests of null models for amphibian declines on a tropical mountain. *Conserv. Biol.* 11:1307–1322. [3]

Powell, G. V. N. and R. Bjork. 1995. Implications of intratropical migration on reserve design: A case study using *Pharomachrus mocinno. Conserv. Biol.* 9:354–362. [7]

Powell, G. V. N., R. D. Bjork, S. Barrios, and V. Espinoza. 2000. Elevational migrations and habitat linkages: Using the resplendent quetzal as an indicator for evaluating the design of the Monteverde Reserve Complex. In N. M. Nadkarni and N. T. Wheelwright (eds.), *Monteverde: Ecology and Conservation of a Tropical Cloud Forest,* pp. 439–442. Oxford University Press, New York. [10]

Powell, R. 1990a. The functional forms of density-dependent birth and death rates in diffuse knapweed (*Centaurea diffusa*) explain why it has not been controlled by *Urophora affinis, U. quadrifasciata* and *Sphenoptera jugoslavica.* In E. Delfosse (ed.), *Proceedings VII International Symposium on Biological Control of Weeds,* pp. 195–202. Inst. Sper. Patol. Veg., Rome. [9]

Powell, R. 1990b. The role of spatial pattern in the population biology of *Centaurea diffusa. J. Ecol.* 78:374–388. [9]

Power, M. E., D. Tilman, J. A. Estes, B. A. Menge, W. J. Bond, L. S. Mills, G. Daily, J. C. Castilla, J. Lubchenco, and R. T. Paine. 1996. Challenges in the quest for keystones. *BioScience* 46:609–620. [3, 12]

Poynton, J. C., K. M. Howell, B. T. Clarke, and J. C. Lovett. 1998. A critically endangered new species of *Nectophrynoides* (Anura: Bufonidae) from the Kihansi Gorge, Udzungwa Mountains, Tanzania. *African J. Herpetology* 47:59–67. [3]

Precht, H., J. Christophersen, H. Hensel, and W. Larcher. 1973. *Temperature and Life.* Springer-Verlag, New York. [10]

Pressey, R. and S. Ferrier. 1995. Types, limitations and uses of geographic data in conservation planning. *The Globe* 41:45–52. [14]

Pressey, R. L. 1994. *Ad hoc* reservations: forward or backward steps in developing representative reserve systems? *Conserv. Biol.* 8:662–668. [14]

Pressey, R. L. 1997. Priority conservation areas: towards an operational definition for regional assessments. In J. J. Pigram and R. C. Sundell, (eds.), *National Parks and Protected Areas: Selection, Delimitation, and Management,* pp. 337–357. Centre for Water Policy Research, Armidale, New South Wales. [14]

Pressey, R. L. 1998. Algorithms, politics and timber: an example of the role of science in a public, political negotiation process over new conservation areas in production forests. In R. T. Wills and R. J. Hobbs, (eds.), *Ecology for Everyone: Communicating Ecology to Scientists, the Public and the Politicians,* pp. 73–87. Chipping Norton, Surrey Beatty and Sons, NSW. [14]

Pressey, R. L. 2002. The first reserve selection algorithm—a retrospective on Jamie Kirkpatrick's 1983 paper. *Prog. Phys. Geog.* 26:434–441. [14]

Pressey, R. L. 2004. Conservation planning and biodiversity: Assembling the best data for the job. *Conserv. Biol.* 18:1677–1681. [14]

Pressey, R. L. and A. O. Nicholls. 1989. Efficiency in conservation planning: scoring versus iterative approaches. *Biol. Conserv.* 50:199–218. [14]

Pressey, R. L. and K. H. Taffs. 2001. Scheduling conservation action in production landscapes: priority areas in western New South Wales defined by irreplaceability and vulnerability to vegetation loss. *Biol. Conserv.* 100:355–376. [14]

Pressey, R. L., and R. M. Cowling. 2001. Reserve selection algorithms and the real world. *Conserv. Biol.* 15:275–277. [14]

Pressey, R. L. and S. L. Tully. 1994. The cost of *ad hoc* reservation: a case study in western New South Wales. *Aust. J. Ecol.* 19:375–384. [14]

Pressey, R. L., I. R. Johnson, and P. D. Wilson. 1994. Shades of irreplaceability: towards a measure of the contribution of sites to a reservation goal. *Biodivers. Conserv.* 3:242–262. [14]

Pressey, R. L., R. M. Cowling, and M. Rouget. 2003. Formulating conservation targets for biodiversity pattern and process in the Cape Floristic Region, South Africa. *Biol. Conserv.* 112:99–127. [14]

Pressey, R. L., C. J. Humphries, C. R. Margules, R. I. Vane-Wright, and P. H. Williams. 1993. Beyond opportunism: key principles for systematic reserve selection. *Trends in Ecol. Evol.* 8:124–128. [14]

Pressey, R. L., S. Ferrier, T. C. Hager, C. A. Woods, S. L. Tully, and K. M. Weinman. 1996. How well protected are the forests of north eastern New South Wales? Analyses of forest environments in relation to formal protection measures, land tenure and vulnerability to clearing. *Forest Ecol. Manag.* 85:311–333. [14]

Price, P. W. 1980. *Evolutionary Biology of Parasites*. Princeton University Press, Princeton. [3]

Primack, R. B. 2004. *Essentials of Conservation Biology*, 3rd Edition. Sinauer Associates, Inc. Sunderland, MA. [10]

Pringle, C. M. 1997. Exploring how disturbance is transmitted upstream: Going against the flow. *J. N. Am. Benthol. Soc.* 16:425–438. [7]

Pringle, C. M. 2000. Threats to U.S. Public lands from cumulative hydrologic alterations outside of their boundaries. *Ecol. Appl.* 10: 971–989. [7]

Pringle, C. M. 2001. Hydrologic connectivity and the management of biological reserves: A global perspective. *Ecol. Appl.* 11:981–998. [7]

Pringle, C. M. 2001b. Hydrologic connectivity: A call for greater emphasis in wilderness management. *International Journal of Wilderness* 7:21–26. [7]

Pringle, C. M. 2003a. What is hydrologic connectivity and why is it ecologically important? *Hydrological Processes* 17:2685–2689. [7]

Pringle, C. M. 2003b. Interacting effects of altered hydrology and contaminant transport: Emerging ecological patterns of global concern. In M. Holland, E. Blood, and L. Shaffer (eds.), *Achieving sustainable freshwater systems: A web of connections*, pp. 85–107. Island Press, Covalo, CA. [7]

Pringle, C. M. and F. J. Triska. 2000. Emergent biological patterns and surface-subsurface interactions at landscape scales. In J. B. Jones and P. J. Mulholland (eds.), *Stream and Gound Waters*, pp. 167–193. Academic Press, NY. [7]

Pringle, C. M., M. C. Freeman, and B. J. Freeman. 2000. Regional effects of hydrologic alterations on riverine macrobiota in the New World: Tropical-temperate comparisons. *BioScience* 50: 807–823. [7]

Pringle, C. M., N. H. Hemphill, W. McDowell, A. Bednarek, and J. March. 1999. Linking species and ecosystems: Different biotic assemblages cause interstream differences in organic matter. *Ecology* 80: 1860–1872. [7]

Pritchard, J. K., M. Stephens, and P. Donnelly. 2000. Inference of population structure using multilocus genotype data. *Genetics* 155:945–959. [11]

Prop J., J. M. Black, and P. Shimmings. 2003. Travel schedules to the high arctic: Barnacle geese trade-off the timing of migration with accumulation of fat deposits. *Oikos* 103:403–414. [10]

Prusiner, S. B. 1994. Prion diseases of humans and animals. *J. Roy. Coll. Phys. Lond.* 28 Suppl.: 1–30. [3]

Pryde, P. R. 1999. The privatization of nature conservation in Russia. *Post-Sov. Geogr. Econ.* 40:383–393. [12]

Pulliam, H. R. 1983. Ecological community theory and the coexistence of sparrows. *Ecology* 64:45–52. [12]

Pulliam, H. R. 1988. Sources, sinks, and population regulation. *Am. Nat.* 132:652–661. [7, 12]

Pulliam, H. R. 1996. Sources and sinks: Empirical evidence and population consequences. In O. E. Rhodes, R. K. Chesser, and M. H. Smith (eds.), *Population Dynamics in Ecological Space and Time*, pp. 45–69. University of Chicago Press, Chicago. [12]

Pulliam, H. R. and G. C. Millikan. 1982. Social organization in the nonreproductive season. *Avian Biol.* 6:169–193. [12]

Pulliam, H. R. and T. A. Parker. 1979. Population regulation of sparrows. *Fortschr. Zool.* 25:137–147. [12]

Pulliam, H. R., J. B. Dunning Jr., and J. Liu. 1992. Population dynamics in a complex landscape: A case study. *Ecol. Appl.* 2:165–177. [12]

Pulliam, H. R., J. Liu, J. B. Dunning, D. J. Stewart, and T. D. Bishop. 1995. Modelling animal populations in changing landscapes. *Ibis* 137:S120–S126. [12]

Purvis, A., K. E. Jones, and G. M. Mace. 2000. Extinction. *Bioessays* 22:1122–1133. [3]

Pusey, A. and M. Wolf. 1996. Inbreeding avoidance in animals. *Trends Ecol. Evol.* 11:201–206. [11]

Putz, F. E. and M. Pinard. 1993. Reducing the impacts of logging as a carbon-offset method. *Conserv. Biol.* 7:755–757. [8]

Putz, F. E. and V. Viana. 1996. Biological challenges for certification of tropical timber. *Biotropica* 28:323–330. [8]

Putz, F. E., D. P. Dykstra, and R. Heinrich. 2000. Why poor logging practices persist in the tropics. *Conserv. Biol.* 14:951–956. [8]

Pywell, R. F., J. M. Bullock, D. B. Roy, L. Warman, K. J. Walker, and P. Rothery. 2003. Plant traits as predictors of ecological performance in restoration. *J. Appl. Ecol.* 40:65–77. [15]

Queller, D. C. and K. F. Goodnight. 1989. Estimating relatedness using genetic markers. *Evolution* 43:258–275. [11]

Quinn, J. F. and G. R. Robinson 1987. The effects of experimental subdivision on flowering plant diversity in a California annual grassland. *J. Ecol.* 75:837–856. [2]

Quinn, J. F. and S. Harrison 1988. Effects of habitat fragmentation and isolation on species richness: Evidence from biogeographic patterns. *Oecologia* 75:132–140. [2]

Quinn, T. J., II and R. B. Deriso. 1999. *Quantitative Fish Dynamics*. Oxford University Press, New York. [8]

Quinn, T. P. 1993. A review of homing and straying of wild and hatchery-produced salmon. *Fish. Res.* 18:29–44. [11]

Quintana-Ascencio, P. F. and E. S. Menges. 1996. Inferring metapopulation dynamics from patch-level incidence of Florida scrub plants. *Conserv. Biol.* 10:1210–1219. [7]

Rabinowitz, D., S. Cairns, and T. Dillon. 1986. Seven forms of rarity and their frequency in the flora of the British Isles. In M. E. Soulé (ed.), *Conservation Biology: The Science of Scarcity and Diversity*, pp. 182–204. Sinauer Associates, Sunderland, MA. [3]

Rachlinski, J. J. 1998. Protecting endangered species without regulating private landowners: the case of endangered plants. *Cornell J.L. & Pub. Pol.* 8:1–36. [17]

Rachowicz, L. J. and V. T. Vredenburg. 2004. Transmission of *Batrachochytrium dendrobatidis* within and between amphibian life stages. *Dis. Aquat. Organ.* 61:75–83. [3]

Rahel, F. J. 2002. Homogenization of freshwater faunas. *Ann. Rev. Ecol. Syst.* 33: 291–315. [9]

Ralls, K and J. Ballou. 1983. Extinction: Lessons from zoos. In C. M. Shonewald-Cox, S. M. Chambers, B. MacBryde, and L. Thomas, (eds.), *Genetic and Conservation: A Reference for Managing Wild Animal and Plant Populations*, pp. 164–184. Benjamin/Cummings, Menlo Park, CA. [11]

Ralls, K. and J. D. Ballou. 2004. Genetic status and management of California Condors. *Condor* 106:215–228. [12]

Ralls, K., J. D. Ballou, and A. Templeton. 1988. Estimates of lethal equivalents and the cost of inbreeding in mammals. *Conserv. Biol.* 2:185–193. [11, 15]

Ralls, K., K. Brugger, and J. Ballou. 1979. Inbreeding and juvenile mortality in small populations of ungulates. *Science* 206:1101–1103. [15]

Ramos, B. and R. C. Soriguer. 2002. Mamíferos. In *Parque Nacional de Doñana*, pp. 315–338. Canseco Editores. Madrid. [14]

Randall, A. 1986. Human preferences, economics, and the preservation of species. In B. G. Norton (ed.), *The Preservation of Species: The Value of Biological Diversity*, pp. 79–109. Princeton University Press, Princeton. [4]

Randall, A. 1988. What mainstream economists have to say about biodiversity. In E. O. Wilson (ed.), *Biodiversity*, pp. 217–223. National Academy Press, Washington, D.C. [4]

Randall, M. G. M. 1982. The dynamics of an insect population throughout its altitudinal distribution: *Coleophora alticolella* (Lepidoptera) in northern England. *J. Anim. Ecol.* 51:993–1016. [10]

Ranney, J. W., M. C. Bruner, and J. B. Levenson. 1981. The importance of edge in the structure and dynamics of forest islands. In R. L. Burgess and D. M. Sharpe (eds.), *Forest Island Dynamics in Man-Dominated Landscapes*, pp. 67–95. Springer-Verlag, New York. [7]

Rao, M. 2000. Variation in leaf-cutter ant (*Atta* sp.) densities in forest isolates: The potential role of predation. *J. Trop. Ecol.* 16:209–225. [7]

Rao, M., J. Terborgh, and P. Nuñez. 2001. Increased herbivory in forest isolates: Implications for plant community structure and composition. *Conserv. Biol.* 15: 624–633. [7]

Rasmussen, E. M. 1985. El Niño and variations in climate. *Am. Sci.* 73:168–177. [13]

Ratcliffe, C. S. and T. M. Crow. 2001. The effects of agriculture and the availability of edge habitat on populations of helmeted guinea fowl *Numida meleagris* and on the diversity and composition of associated bird assemblages in Quazulu-Natal Province, South Africa. *Biodivers. Conserv.* 10:2109–2127. [6]

Ratsirarson, J., J. A. Silander, and A. F. Richard. 1996. Conservation and management of a threatened Madagascar palm species, *Neodysis decaryi* Jumelle. *Conserv. Biol.* 10:40–52. [12]

Ratti, J. T. and K. P. Reese. 1988. Preliminary test of the ecological trap hypothesis. *J. Wildl. Mgmt.* 52:484–491. [7]

Rau, J. R. 1987. Ecología del zorro (*Vulpes vulpes*) en la Reserva Biológica de Doñana. Ph. D. dissertation. University of Sevilla. Sevilla. [14]

Raunkaier, C. 1934. *The Life Forms of Plants and Statistical Plant Geography*. Clarendon Press, Oxford. [2]

Raup, D. M. 1991. *Extinction: Bad Genes or Bad Luck?* Norton, New York. [3]

Raup, D. M. 1994. The role of extinction in evolution. *Proc. Nat. Acad. Sci. U.S.A.* 91:6758–6763. [3]

Raven, P., R. Norgaard, C. Padoch, T. Panayotou, A. Randall, M. Robinson, and J. Rodman. 1992. *Conserving Biodiversity: A Research Agenda for Development Agencies*. National Academy Press, Washington, D.C. [4]

Ravenal, R. M., I. M. E. Granoff, and C. A. Magee (eds.), 2004. *Illegal Logging in the Tropics: Strategies for Cutting Crime*. Haworth Press, Binghamton, NY. [8]

Read, C. and D. Merton. 1991. Behavioural manipulation of endangered New Zealand birds as an aid towards species recovery. In B. D. Bell, R. O. Cossee, J. E. C. Flux, B. D. Heather, R. A. Hitchmough, C. J. R. Robertson, and M. J. Williams (eds.), *Acta XX Congressus Internationalis Ornithologici*. 4:2514–2522. [12]

Reading, C. J. 1998. The effect of winter temperatures on the timing of breeding activity in the common toad *Bufo bufo*. *Oecologia* 117:469–475. [3]

Reading, R. P. 1993. *Towards an endangered species reintroduction paradigm: A case study of the black-footed ferret*. Ph.D. dissertation, Yale University, New Haven CT. [15]

Reading, R. P. and S. R. Kellert 1993. Attitudes of Montanans towards a proposed black-footed ferret (*Mustela nigripes*) reintroduction, with special reference to ranchers. *Conserv. Biol.* 7:569–580. [15]

Redford, K. H. 1992. The empty forest. *BioScience* 42:412–422. [1, 3, 12]

Redford, K. H. and B. D. Richter. 1999. Conservation of biodiversity in a world of use. *Conserv. Biol.* 13:1246–1256. [12, 16]

Redford, K. H. and M. A. Sanjayan. 2003. Retiring Cassandra. *Conserv. Biol.* 17:1463–1464. [1]

Redford, K. H. and P. Feinsinger. 2001. The half empty forest: Sustainable use and the ecology of interactions. In J. D. Reynolds, G. M. Mace, K. H. Redford, and J. G. Robinson (eds.), *Conservation of Exploited Species*, pp. 370–399. Cambridge University Press, Cambridge. [3, 6, 12]

Redford, K. H. and S. E. Sanderson. 2000. Extracting humans from nature. *Conserv. Biol.* 14:1362–1364. [14]

Redford, K. H., P. Coppolillo, E. W. Sanderson, G. A. B. da Fonseca, E. Dinerstein, C. Groves, G. Mace, S. Maginnis, R. A. Mittermeier, R. Noss, D. Olson, J. G. Robinson, A. Vedder, and M. Wright. 2003. Mapping the conservation landscape. *Conserv. Biol.* 17:116–131. [6]

Reed, D. H. and R. Frankham. 2001. Correlation between fitness and genetic diversity. *Conserv. Biol.* 17:230–237. [11]

Reed, J. M., L. S. Mills, J. B. Dunning, E. S. Menges, K. S. McKelvey, R. Frye, S. R. Beissinger, M-C. Anstett, and P. Miller. 2002. Emerging issues in population viability analysis. *Conserv. Biol.* 16:7–19. [12]

Regan, H. M., H. R. Akcakaya, S. Ferson, K. V. Root, S. Carroll, and L. R. Ginzburg. 2003. Treatments of uncertainty and variability in ecological risk assessment of single-species populations. *Hum. Ecol. Risk Assess.* 9:889–906. [5]

Regan, T. 1983. *The Case for Animal Rights*. The University of California Press, Berkeley. [4]

Regniere, J. and V. Nealis. 2002. Modelling seasonality of gypsy moth, *Lymantria dispar* (Lepidoptera: Lymantriidae), to evaluate probability of its persistence in novel environments. *Can. Entomol.* 134:805–824. [10]

Reh, W. and A. Seitz. 1990. The influence of land use on the genetic structure of populations of the common frog Rana temporaria. *Biol. Conserv.* 54:239–249. [7]

Reichard, S. H. and C. W. Hamilton. 1997. Predicting invasions of woody plants introduced into North America. *Conserv. Biol.* 11:193–203. [9]

Reichert, B. K., R. Schnur, and L. Bengtsson. 2002. Global ocean warming tied to anthropogenic forcing. *Geophys. Res. Lett.* 10:20-1–20-4. [10]

Reid, T. S. and D. D. Murphy. 1995. Providing a regional context for local conservation action. *BioScience* S84–S90. [13]

Reid, W. V. 1996. Beyond protected areas: Changing perceptions of ecological management objectives. In R. C. Szaro, and D. W. Johnston, (eds.), *Biodiversity in Managed Landscapes: Theory and Practice*, pp. 442–453. Oxford University Press, New York. [14]

Reilly, W. K. 1985. National parks for a new generation. *Environment* 27:19–20; 40–45. [14]

Rejmánek, M. and D. M. Richardson. 1996. What attributes make some plant species more invasive? *Ecology* 77:1655–1661. [9]

Relyea, R. A. and N. Mills. 2001. Predator-induced stress makes the pesticide carbaryl more deadly to gray treefrog tadpoles (*Hyla versicolor*). *Proc. Nat. Acad. Sci. USA* 98:2491–2496. [3]

Renjifo, L. M. 1999. Composition changes in a sub-Andean avifauna after long-term forest fragmentation. *Conserv. Biol.* 13:1124–1139. [14]

Renjifo, L. M. 2002. *Penelope perspicax*. In L. M. Renjifo, A. M. Franco, J. D. Amaya, G. H. Kattan, and B. López (eds.), *Libro rojo de Aves de Colombia*, pp. 124–130. Instituto de Investigación de Recursos Biológicos Alexander von Humboldt, Bogotá, Colombia. [14]

Repetto, R. and M. Gillis (eds.). 1988. *Public Policies and the Misuse of Forest Resources*. Cambridge University Press, Cambridge. [8]

Requejo, J. and J. Belis. 2003. La experiencia Doñana. La experiencia de la planificación del Desarrollo Sostenible. Jornadas Internacionales sobre Desarrollo Sostenible. Almonte 5–8 november 2002. *Sostenible* 3:145–157. Special Issue. [14]

Reserva Biológica de Doñana (RBD) 2002. Evolución de la población de águila imperial ibérica (*Aquila adalberti*) en Doñana. Periodo 1988–2001 in *www-rbd.ebd.es/segu/aves/aguilimp/evo8800.htm*. [14]

Retallick, R. W. R., H. McCallum, and R. Speare. 2004. Endemic infection of the amphibian chytrid fungus in a frog community post-decline. *PLoS Biol.* 2:e351. [3]

Reusch, T. B. H. 2003. Floral neighbourhoods in the sea: How floral density, opportunity for outcrossing and population fragmentation affect seed set in *Zostera marina*. *J. Ecol.* 91:610–615. [7]

Revenga, C., J. Brunner, N. Henninger, K. Kassen, and R. Payne. 2000. *Pilot Assessment of Global Ecosystems: Freshwater Systems*. World Resources Institute: Washington, D.C. [6]

Rex, M. A. 1973. Deep-sea species diversity: Decreased gastropod diversity at abyssal depths. *Science* 181:1051–1053. [2]

Rex, M. A. 1983. Geographical patterns of species diversity in the deep-sea benthos. In Rowe, G. T. (ed.), *The Sea*, Vol. 8, pp. 453–472. Wiley, New York. [2]

Reyers, B. and A. S. van Jaarsveld. 2000. Assessment techniques for biodiversity surrogates. *S. Afr. J. Sci.* 96:406–408. [14]

Reynolds, J. D., N. K. Dulvy, and C. R. Roberts. 2002. Exploitation and other threats to fish conservation. In P. J. B.

Hart and J. D. Reynolds (eds.), *Handbook of Fish Biology and Fisheries: Vol. 2, Fisheries*, pp. 319–341. Blackwell Publishing, Oxford. [8]

Reznick, D. N. and H. Bryga. 1987. Life-history evolution in guppies (*Poecilia reticulata*): 1. Phenotypic and genetic changes in an introduction experiment. *Evolution* 41:1370–1385. [2]

Reznick, D. N., H. Bryga, and J. A. Endler. 1990. Experimentally induced life-history evolution in a natural population. *Nature* 346:357–359. [2]

Rhymer, J. M. and D. Simberloff. 1996. Extinction by hybridization and introgression. *Ann. Rev. Ecol. Syst.* 27:83–109. [9]

Ribon, R., J. E. Simon, and H. Theodoro de Mattos. 2003. Bird extinctions in Atlantic forest fragments of the Viçosa region, southeastern Brazil. *Conserv. Biol.* 17:1827–1839. [7]

Ricciardi, A. 2003. Predicting the impacts of an introduced species from its invasion history: An empirical approach applied to zebra mussel invasions. *Freshwater Biol.* 48:972–981. [9]

Ricciardi, A., R. J. Neves, and J. B. Rasmussen. 1998. Impending extinctions of North American freshwater mussels (Unionoida) following the zebra mussel (*Dreissena polymorpha*) invasion. *J. Anim. Ecol.* 67:613–619. [9]

Rice, B. and M. Westoby. 1983. Species richness at tenth-hectare scale in Australian vegetation compared to other continents. *Vegetatio* 52:129–140. [2]

Rice, K. J. and N. C. Emery. 2003. Managing microevolution: restoration in the face of global change. *Front. Ecol. Environ.* 1:469–478. [15, 18]

Rich, A. C., D. S. Dobkin, and L. J. Niles. 1994. Defining forest fragmentation by corridor width: The influence of narrow forest-dividing corridors on forest-nesting birds in southern New Jersey. *Conserv. Biol.* 8:1109–1121. [7]

Richards, M. 1999. 'Internalising the externalities' of tropical forestry: A review of innovative financing and incentive mechanisms. European Union Tropical Forestry Paper 1. Overseas Development Institute, London. [8]

Richards, S. J., K. R. McDonald, and R. A. Alford. 1993. Declines in populations of Austalia's endemic tropical rainforest frogs. *Pacific. Conserv. Biol.* 1:66–77. [3]

Richardson, D. M., R. M. Cowling, B. B. Lamont. 1996. Non-linearities, synergisms and plant extinctions in South African fynbos and Australian kwongan. *Biodivers. Conserv.* 5:1035–1046. [3]

Ricketts, T. H. 2001. The matrix matters: Effective isolation in fragmented landscapes. *Am. Nat.* 158:87–99. [7]

Ricketts, T. H. 2004. Tropical forest fragments enhance pollinator activity in nearby coffee crops. *Conserv. Biol.* 18:1–10. [2]

Ricketts, T. H., G. C. Daily, P. R. Ehrlich, and J. P. Fay. 2001. Countryside biogeography of moths in a fragmented landscape: biodiversity in native and agricultural habitats. *Conserv. Biol.* 15:378–388. [18]

Ricketts, T. H., G. C. Daily, P. R. Ehrlich, and C. D. Michener. 2004. Economic value of tropical forest to coffee production. *P. Nat. Acad. Sci. USA.* 101:12579–12582. [2, 16]

Ricketts, T. H., E. Dinerstein, D. M. Olson, C. J. Loucks, W. Eichbaum, D. DellaSala, K. Kavanagh, P. Hedao, P. T. Hurley, K. M. Carney, R. Abell, and S. Walters (eds.). 1999. *Terrestrial Ecoregions of North America: A Conservation Assessment.* Island Press, Washington, D.C. [7]

Ricklefs, R. E. 1987. Community diversity: Relative roles of local and regional processes. *Science* 235:167–171. [2]

Rifkin, J. 1995. *The End of Work: The Decline of the Global Labor Force and the Dawn of the Post-Market Era.* G. P. Putnam's Sons, NY. [18]

Riiters, K. H., R. V. O'Neill, C. T. Hunsaker, J . D. Wickham, D. H. Yankee, S. P. Timmins, K. B. Jones, and B. L. Jackson. 1995. A factor analysis of landscape pattern and structure metrics. *Landscape Ecol.* 10:23–40. [7]

Ripple, W. J., G. A. Bradshaw, and T. A. Spies. 1991a. Measuring forest landscape patterns in the Cascade Range of Oregon, USA. *Biol. Conserv.* 57:73–88. [7]

Ripple, W. J., E. J. Larsen, R. A. Renkin, and D. W. Smith. 2001. Trophic cascades among wolves, elk, and aspen on Yellowstone National Park's northern range. *Biol. Conserv.* 102:227–234. [7]

Rival, L. 1998. Domestication as a historical and symbolic process: wild gardens and cultivated forest in the Ecuadorian Amazon. In W. Balée (ed.), *Advances in Historical Ecology*, pp. 232–250. Columbia University Press, New York. [6]

Rivera, M. L. 2002. Arqueología e historia in *Parque Nacional de Doñana.* Canseco Editores. Madrid, pp. 339–352. [14]

Riveros, M., M. A. Parades, M. T. Rosas, E. Cardenas, J. J. Armesto, M. T. K. Arroyo, and B. Palma. 1995. Reproductive biology in species of the genus *Nothofagus. Environ. Exp. Bot.* 35:519–524. [9]

Robbins, C. S., D. K. Dawson, and B. A. Dowell. 1989 (1989b). Habitat area requirements of breeding forest birds of the Middle Atlantic states. *Wildl. Monogr.* 103:1–34. [7, 12]

Robbins, C. S., J. R. Sauer, R. Greenberg, and S. Droege. 1989a. Population declines in North American birds that migrate to the Neotropics. *Proc. Nat. Acad. Sci. USA.* 86:7658–7662. [12]

Robinson, S. K. 1992a. *Effects of Forest Fragmentation on Migrant Songbirds in the Shawnee National Forest.* Report to Illinois Department of Energy and Natural Resources. Illinois Natural History Survey, Champaign, IL. [7]

Robinson, S. K. 1992b. Population dynamics of breeding Neotropical migrants in a fragmented Illinois landscape. In J. M. Hagan and D. W. Johnston (eds.), *Ecology and Conservation of Neotropical Migrant Landbirds*, pp. 408–418. Smithsonian Institution Press, Washington, D.C. [7]

Robinson, S. K. 1999. The most threatened birds of the continental North America.

In Ricketts, T. H., E. Dinerstein, D. M. Olson, C. J. Loucks, W. Eichbaum, D. DellaSala, K. Kavanagh, P. Hedao, P. T. Hurley, K. M. Carney, R. Abell, and S. Walters (eds.), *Terrestrial Ecoregions of North America: A Conservation Assessment,* pp. 69–70. Island Press, Washington, D.C. [7]

Robinson, S. K., F. R. Thompson III, T. M. Donovan, D. R. Whitehead, and J. Faaborg. 1995. Regional forest fragmentation and the nesting success of migratory birds. *Science* 267:1987–1990. [7]

Roberts, C. M. 2000. Selecting marine reserve locations: Optimality versus opportunism. *Bulletin of Marine Science* 66: 581–592 [14]

Roberts, C. M. 2003. Our shifting perspectives on the oceans. *Oryx* 37:166–177. [3]

Roberts, C. M. and J. P. Hawkins 2000. *Fully Protected Marine Reserves: A Guide.* WWF Endangered Seas Campaign, Washington D.C. USA, and Environment Department, University of York, UK. [14]

Roberts, C. M., C. J. McClean, J. E. N. Veron, J. P. Hawkins, G. R. Allen, D. E. McAllister, C. G. Mittermeier, F. W. Schueler, M. Spalding, F. Wells, C. Vynne, and T. B. Warner. 2002. Marine biodiversity hotspots and conservation prioroties for coral reefs. *Science* 295:1280–1284. [2]

Roberts, C., B. Halpern, S. R. Palumbi, and R. R. Warner. 2001. Designing marine reserve networks: why small isolated protected areas are not enough. *Conserv. in Practice* 2:10–17. [14]

Roberts, C., C. McLean, J. E. N. Veron, J. P. Hawkins, G. R. Allen, D. E. McAllister, C. G. Mittermeier, F. W. Schueller, M. Spalding, F. Wells, and C. Vynne. 2002. Marine biodiversity hotspots and conservation priorities for tropical reefs. *Science* 295:1280–1284. [6]

Roberts, C., G. Branch, R. H. Bustanamte, J. C. Castilla, J. Dugan, B. S. Halpern, K D. Lafferty, H. Leslie, J. Lubchenco, D. McArdle, M. Ruckelshaus, and R. R. Waraner. 2003. Application of ecological criteria in selecting marine reserves and developing reserve networks. *Ecol. Appl.* 13:S125–S228. [14]

Robertson, D. P. and R. B. Hull. 2003. Public ecology: and environmental science and policy for global society. *Environ. Sci. Policy* 6:399–410. [18]

Robinson, J. G. 1993. The limits to caring: Sustainable living and the loss of biodiversity. *Conserv. Biol.* 7:20–28. [16]

Robinson, J. G. and E. L. Bennett (eds.). 2000. *Hunting for Sustainability in Tropical Forests.* Columbia University Press, New York. [8]

Robinson, J. G. and K. H. Redford. (eds.). 1991. *Neotropical Wildlife Use and Conservation.* University of Chicago Press, Chicago. [8]

Robinson, J. R. and E. L. Bennett. 2004. Having your wildlife and eating it too: An analysis of hunting sustainability across tropical ecosystems. *Animal Conservation* 7:397–408. [8]

Robinson, J. V. and J. E. Dickerson Jr. 1987. Does invasion sequence affect community structure? *Ecology* 68:587–595. [15]

Roca, A. L., N. Georgiadis, J. Pecon-Slattery, and S. J. O'Brien. 2001. Genetic evidence for two species of elephant in Africa. *Science* 293:1473–1477. [11]

Rodrigues, A. S. L., S. J. Andelman, M. I. Bakarr, L. Boitani, T. M. Brooks, R. M. Cowling, L. D. C. Fishpool, G. A. B. Fonseca, K. J. Gaston, M. Hoffman, J. Long, P. A. Marquet, J. D. Pilgrim, R. L. Pressey, J. Schipper, W. Sechrest, S. N. Stuart, L. G. Underhill, R. W. Waller, M. E. J. Watts, Y. Xie. 2003. Global Gap Analysis: Towards a representative network of protected areas. Advances in Applied Biodiversity Science 5. Conservation International, Washington, D.C. [14]

Rodrigues, A. S. L., H. R. Akcakaya, S.. J. Andelman, M. I. Bakarr, L. Boitani, T. M. Brooks, J. S Chanson, L. D. C. Fishpool, G. A. B. Da Fonseca, K. J. Gaston, M. Hoffmann, P. A. Marquet, J. D.Pilgrim, R. L. Pressey, J. Schipper, W., Sechrest, S. N. Stuart, L. G. Underhill, R. W. Waller, M. E. J. Watts, and X. Yan, 2004a. Global gap analysis: Priority regions for expanding the global protected-area network. *BioScience* 54:1092–1100. [14]

Rodrigues, A. S. L., S. J. Andelman, M. I. Bakarr, L. Boitani, T. M. Brooks, R. M. Cowling, L. D. C. Fishpool, G. A. B. da Fonseca, K. J. Gaston, M. Hoffmann, J. S. Long, P. A. Marquet, J. D. Pilgrim, R. L. Pressey, J. Schipper, W. Sechrest, S. N. Stuart, L. G. Underhill, R. W. Waller, M. E. J. Watts, and X. Yan. 2004. Effectiveness of the global protected area network in representing species diversity. *Nature* 428:640–643. [12]

Rodrigues, A. S. L., S. J. Andelman, M. I. Bakarr, L. Boitani, T. M. Brooks, R. M. Cowling, L. D. C. Fishpool, G. A. B. da Fonseca, K. J. Gaston, M. Hoffmann, J. S. Long, P. A. Marquet, J. D. Pilgrim, R. L. Pressey, J. Schipper, W. Sechrest, S. N. Stuart, L. G. Underhill, R. W. Waller, M. E. J. Watts, and X. Yan. 2004b. Effectiveness of the global protected area network in representing species diversity. *Nature* 428:640–643. [14]

Rodríguez, A. and L. Clemente. 2002a. Geomorfología in *Parque Nacional de Doñana*. Canseco Editores. Madrid, pp. 19–42. [14]

Rodríguez, A. and L. Clemente. 2002b. Hidrología superficial in *Parque Nacional de Doñana*. Canseco Editores. Madrid, pp. 57–68. [14]

Rodríguez-Trelles, F. and M. A. Rodríguez. 1998. Rapid micro-evolution and loss of chromosomal diversity in *Drosophila* in response to climate warming. *Evol. Ecol.* 12:829–838. [10]

Rodríguez-Trelles, F., G. Álvarez, and C. Zapata. 1996. Time-series analysis of seasonal changes of the O inversion polymorphism of *Drosophila subobscura*. *Genetics* 142:179–187. [10]

Rodríguez-Trelles, F., M. A. Rodríguez, and S. M. Scheiner. 1998. Tracking the genetic

effects of global warming: *Drosophila* and other model systems. *Conserv. Ecol.* [online] 2:2. http://www.consecol.org/vol2/iss2/art2/.[10]

Rodriguez-Vidal, J. A. 1959. Puerto Rican parrot (*Amazona vittata vittata*) study. *Monog. Dept. Agric. & Commerce, P.R.* 1:1–15. [15]

Roeger, S. 2002. Entanglement of marine turtles in netting: Northeast Arnhem Land, Northern Territory, Australia. Report to Alcan Gove Pty Ltd., World Wide Fund for Nature Australia, Humane Society International, Northern Land Council. Dhimurru Land Management Aboriginal Corporation, Northern Territory. [16]

Roemmich, D. and J. McGowan. 1995. Climatic warming and the decline of zooplankton in the California Current. *Science* 267:1324–1326. [10]

Rojstaczer, S., S. M. Sterling, and N. J. Moore. 2001. Human appropriation of photosynthesis products. *Science* 294:2549–2552. [1]

Roldán, A. I. and J. A. Simonetti. 2001. Plant-mammal interactions in tropical Bolivian forests with different hunting pressures. *Conserv. Biol.* 15:617–623. [12]

Rolland, R. M., K. E. Hunt, S. D. Kraus, and S. K. Wasser. 2005. Assessing reproductive status of right whales (*Eubalaena glacialis*) using fecal hormone metabolites. *Gen. Comp. Endocr.* In press. [11]

Rollins-Smith, L. A., L. K. Reinert, V. Miera, and J. M. Conlon. 2002a. Antimicrobial peptide defenses of the Tarahumara frog, *Rana tarahumarae*. *Biochem. Bioph. Res. Co.* 297:361–367. [3]

Rollins-Smith, L. A., J. K. Doersam, J. E. Longcore, S. K. Taylor, J. C. Shamblin, C. Carey, and M. A. Zasloff. 2002b. Antimicrobial peptide defenses against pathogens associated with global amphibian declines. *Dev. Comp. Immunol.* 26:63–72. [3]

Rolston, H. 1988. *Environmental Ethics: Duties to and Values in the Natural World*. Temple University Press, Philadelphia. [4]

Rolston, H. 1994. *Conserving Natural Value*. Columbia University Press, New York. [4]

Romain-Bondi, K. A., R. B. Wielgus, L. Waits, W. F. Kasworm, M. Austin, and W. Wakkinen. 2004. Density and population size estimates for North Cascade grizzly bears using DNA hair-sampling techniques. *Biol. Conserv.* 117:417–428. [11]

Ron, S. R., W. E. Duellman, L. A. Coloma, and M. R. Bustamante. 2003. Population decline of the Jambato Toad *Atelopus ignescens* (Anura: Bufonidae) in the Andes of Ecuador. *J. Herpetol.* 37:116–126. [3]

Rooney, A. A. and L. J. Guillette, Jr. 2000. Contaminant interactions with steroid receptors: Evidence for receptor binding. In L. J. Guillette, Jr. and D. A. Crain (eds.), *Endocrine Disrupting Contaminants: An Evolutionary Perspective*, pp. 82–125. Francis and Taylor Inc., Philadelphia. [6]

Root, T. L. 1988a. Environmental factors associated with avian distributional limits. *J. Biogeogr.* 15:489–505. [10]

Root, T. L. 1988b. Energy constraints on avian distributions and abundances. *Ecology* 69:330–339. [10]

Root, T. L., J. T. Price, K. R. Hall, S. H. Schneider, C. Rosenzweig, and J. A. Pounds. 2003. Fingerprints of global warming on wild animals and plants. *Nature* 421:57–60. [10]

Ropelewski, C. F. and M. S. Halpert. 1987. Global and regional scale precipitation patterns associated with the El Niño/Southern Oscillation. *Monthly Weather Rev.* 115:1606–1626. [13]

Rose, D. 1996. *An Overview of World Trade in Sharks and Other Cartilaginous Fishes*. TRAFFIC International, Cambridge. [8]

Roseberry, J. L. 1979. Bobwhite population responses to exploitation: Real and simulated. *J. Wildl. Mgmt.* 43:285–305. [8]

Rosenberg, D. M., P. McCully, and C. M. Pringle. 2000. Global-scale environmental effects of hydrological alterations: Introduction. *BioScience* 50:746–751. [7]

Rosenheim, J. A., L. R. Wilhoit, and C. A. Armer. 1993. Influence of intraguild predation among generalist insect predators on the suppression of an herbivore population. *Oecologica* 96:439–449. [3]

Rosenzweig, M. L. 1971. The paradox of enrichment: Destabilization of exploitation ecosystems in ecological time. *Science* 171:385–387. [2]

Rosenzweig, M. L. 1995. *Species Diversity in Space and Time*. Cambridge University Press, Cambridge. [2, 3]

Rosenzweig, M. L. 2003. *Win-win ecology: how the Earth's species can survive in the midst of human enterprise*. Oxford University Press, Oxford. [18]

Ross, E. B. 1978. Food taboos, diet, and hunting strategy: The adaptation to animals in Amazon cultural ecology. *Curr. Anthropol.* 19:1–36. [8]

Ross, M. S., J. J. O'Brien, L. Da Silveira, and L. Sternberg. 1994. Sea-level rise and the reduction in pine forests in the Florida Keys. *Ecol. Appl.* 4:144–156. [10]

Ross, P. S. 2000. Marine mammals as sentinels in ecological risk assessment. *HERA* 6:29–46. [3]

Ross, P. S. and L. S. Birnbam. 2003. Integrated human and ecological risk assessment: A case study of persistent organic pollutants (POPs) in humans and wildlife. *Hum. Ecol. Risk Assess.* 9:303–324. [3]

Ross, P. S., G. M. Ellis, M. G. Ikonomou, L. G. Barrett-Lennard, and R. F. Addison. 2000. High PCB concentrations in free-ranging Pacific killer whales, *Orcinus orca*: effects of age, sex and dietary preference. *Mar. Pollut. Bull.* 40:504–515. [3, 17]

Ross, P. S., R. L. De Swart, R. F. Addison, H. Van Loveren, J. G. Vos, and A. D. M. E. Osterhaus. 1996. Contaminant-induced immunotoxicity in harbour seals: Wildlife at risk? *Toxicology* 112:157–169. [3]

Ross, S. T. 1991. Mechanisms structuring stream fish assemblages: Are there lessons from introduced species. *Environ. Biol. Fish.* 30:359–368. [9]

Rossetto, M., C. L. Gross, R. Jones, and J. Hunter. 2004. The impact of clonality on an endangered tree (*Elaeocarpus williamsianus*) in a fragmented rainforest. *Biol. Conserv.* 117:33–39. [7]

Roubik, D. W. 1995. *Pollination of Cultivated Plants in the Tropics*. Food and Agriculture Organization, Rome. [2]

Rouget, M., D. M. Richardson, and R. M. Cowling. 2003a. The current configuration of protected areas in the Cape Floristic Region, South Africa—reservation bias and representation of biodiversity patterns and processes. *Biol. Conserv.* 112:129–145. [14]

Rouget, M., D. M. Richardson, R. M. Cowling, J. W. Lloyd, and A. T. Lombard. 2003b. Current patterns of habitat transformation and future threats to biodiversity in terrestrial ecosystems of the Cape Floristic Region, South Africa. *Biol. Conserv.* 112:63–85. [14]

Rowan, R., N. Knowlton, A. Baker, and J. Jara. 1997. Landscape ecology of algal symbionts creates variation in episodes of coral bleaching. *Nature* 388:265–269. [13]

Roy, D. B. and T. H. Sparks. 2000. Phenology of British butterflies and climate change. *Glob. Change Biol.* 6:407–416. [10]

Ruckelshaus, M. R. and C. A. Hays. 1998. Conservation and management of species in the sea. In P. L. Fiedler and P. M. Kareiva, (eds.), *Conservation Biology for the Coming Decade*, pp. 112–156. Chapman and Hall, New York. [7]

Rudnicky, T. C. and M. L. Hunter. 1993. Avian nest predation in clearcuts, forests, and edges in a forest-dominated landscape. *J. Wildl. Mgmt.* 57:358–364. [7]

Rudran, R., J. Jayewardene, and W. A. Jayasinghe. 1993. Need for an integrated approach to elephant conservation in Sri Lanka. In J. C. Daniel and H. Datye (eds.), *A Week with Elephants: Proceedings of the International Seminar on Asian Elephants*, pp. 197–216. Bombay Natural History Society, Oxford University Press, Bombay. [17]

Ruggiero, R. G. 1998. The Nouabalé-Ndoki Project: Development of a Practical Conservation Model in Central Africa. In H. E. Eves, R. Hardin, and S. Rupp (eds). *Resource Use in the Trinational Sangha River Region of Equatorial Africa: Histories, Knowledge Forms and Institutions*, pp. 176–188. Yale School of Forestry and Environmental Studies Bulletin Series Number 102. New Haven, CT. http://www.yale.edu/sangha/PDF_FILES/ENGLISH_.PDF/SEC._3/RUGGIERO.PDF [8]

Ruiz, G. M., T. K. Rawlings, F. C. Dobbs, L. A. Drake, A. Huq, and R. R. Colwell. 2000. Global spread of microorganisms by ships: Ballast water discharged from vessels harbours a cocktail of potential pathogens. *Nature* 408:49–50. [14]

Runte, A. 1972. Yellowstone: It's useless, so why not a park? *National Parks and Conservation Magazine* March:4–7. [14]

Russell, H. S. 1976. *A Long Deep Furrow*. University Press of New England, Hanover, NH. [1]

Rusterholz, H. P. and A. Erhardt. 1998. Effects of elevated CO_2 on flowering phenology and nectar production of nectar plants important for butterflies of calcareous grasslands. *Oecologia* 113:341–349. [10]

Ryman, N. and L. Laikre. 1991. Effects of supportive breeding on the genetically effective population size. *Conserv. Biol.* 5:325–329. [11]

Ryman, N., R. Baccus, C. Reuterwall, and M. H. Smith. 1981. Effective population size, generation interval, and potential loss of genetic variability in game species under different hunting regimes. *Oikos* 36:257–266. [11]

Saab, V. 1999. Importance of spatial scale to habitat use by breeding birds in riparian forests: A heirarchical analysis. *Ecol. App.* 9(1):135–151. [7]

Saccheri, I., M. Kuussaari, M. Kankare, P. Vikman, W. Fortelius, and I. Hanski. 1998. Inbreeding and extinction in a butterfly metapopulation. *Nature* 392:491–494. [11]

Sachs, J. 2005. *The End of Poverty: Economic Possibilities for Our Time*. Penguin Group, East Rutherford, N.J. [3, 18]

Sadovy, Y. and W. L. Cheung. 2003. Near extinction of a highly fecund fish: The one that nearly got away. *Fish and Fisheries* 4:86–99. [8]

Saenz, G. P. and M. R. Guariguata. 2001. Demographic response of tree juveniles to reduced-logging in a Costa Rican montane forest. *Forest Ecol. Manag.* 140:75–84. [8]

Sagarin, R. D. and F. Micheli. 2001. Climate change in nontraditional data sets. *Science* 294:811 [10]

Sagarin, R. D., J. P. Barry, S. E. Gilman, and C. H. Baxter. 1999. Climate-related change in an intertidal community over short and long time scales. *Ecol. Monogr.* 69:465–490. [10]

Sagoff, M. 1980. On the preservation of species. *Columbia J. Environ. Law* 7:33–76. [4]

Sagoff, M. 1988. *The Economy of the Earth: Philosophy, Law, and the Environment*. Cambridge University Press, Cambridge. [4]

Said, M. Y. 1995. *African Elephant Database*. IUCN, Gland, Switzerland. [11]

Sala, O. E., F. S. Chapin III, J. J. Armesto, R. Berlow, J. Bloomfield, J. R. Dirzo, E. Huber-Sanwald, L. F. Huenneke, R. B. Jackson, A. Kinzig, R. Leemans, D. Lodge, H. A. Mooney, M. Oesterheld, N. L. Poff, M. T. Sykes, B. H. Walker, M. Walker, and D. H. Wall. 2000. Global biodiversity scenarios for the year 2100. *Science* 287:1770–1774. [9, 18]

Salafsky, N. and R. Margoluis. 1999. Threat reduction assessment: a practical and cost-effective approach to evaluating conservation and development projects. *Conserv. Biol.* 13:830–841. [16, 18]

Salafsky, N., R. Margoluis, K. H. Redford, and J. G. Robinson. 2002. Improving the practice of conservation: A conceptual framework and research agenda for conservation science. *Conserv. Biol.* 16:1469–1479. [18]

Salafsky, N., H. Cauley, G. Balachander, B. Cordes, J. Parks, C. Margoluis, S. Bhatt, C. Encarnacion, D. Russell, and R. Margoluis. 2001. A systematic test of an enterprise strategy for community-based biodiversity conservation. *Conserv. Biol.* 15:1585–1595. [16]

Salas L. A. and J. B. Kim. 2002. Spatial factors and stochasticity in the evaluation of sustainable hunting of tapirs. *Conserv. Biol.* 16:86–96. [8]

Salati, E. and P. B. Vose. 1984. Amazon Basin: A system in equilibrium. *Science* 225:129–138. [6]

Saldaña, T., P. Guarnizo, C. Arribas, C. G. Utrilla, D. García-González, D. Fletcher, J. A. Carmona, and C. Fernández-Delgado. 2003. Efecto del vertido tóxico de las minas de Aznalcóllar sobre la fauna piscícola del río Guadiamar in *Ciencia y Restauración del Río Guadiamar. PICOVER 1998–2003*. Consejería de Medio Ambiente. Junta de Andalucía, pp. 170–181. [14]

Sale, K. 1995. *Rebels Against the Future: The Luddites and Their War on the Industrial Revolution: Lessons for the Computer Age*. Addison-Wesley Publishing Co., Reading, MA. [18]

Salm, R. V. and S. L. Coles. 2001. *Coral Bleaching and Marine Protected Areas: Proceedings of the Workshop on Mitigating Coral Bleaching impact through MPA Design*. Bishop Museum, Honolulu, 29–31 May, 2001. Asia Pacific Coastal Marine Program Report No. 0102, The Nature Conservancy, Honolulu, Hawaii. http://www.conserveonline.org [13]

Salm, R. V., J. Clark, and E. Siirila. 2000. *Marine and Coastal Protected Areas: A Guideline for Planners and Managers*. IUCN, Washington D.C. [13]

Salm, R.V., T. Done, and E. McLeod. In Press. Marine Protected Area (MPA) planning in a changing climate. In *Coral Reefs and Climate Change: Scientific and Management Implications*, American Geophysical Union. [13]

Samson, F. B. 1983. Minimum viable populations—A review. *Nat. Areas J.* 3(3):15–23. [7]

Samuelson, R. 1997. *The Good Life and Its Discontents: The American Dream in the Age of Entitlement*. Knopf, New York. [5]

Sanchez-Azofeifa, G. A., R. C. Harriss, and D. L. Skole. 2001. Deforestation in Costa Rica: A quantitative analysis using remote sensing imagery. *Biotropica* 33:378–384. [8]

Sanders, H. L. 1968. Marine benthic diversity: A comparative study. *Am. Nat.* 102:243–282. [2]

Sanders, H. L. and R. R. Hessler. 1969. Ecology of the deep-sea benthos. *Science* 163:1419–1424. [2]

Sanderson, E. W., K. H. Redford, A. Vedder, P. B. Coppolillo, and S. E. Ward. 2002. A conceptual model for conservation planning based on landscape species require-

ments. *Landscape Urban Plan.* 58:41–56. [12, 14]

Sanderson, E., M. Jaiteh, M. A. Levy, K. H. Redford, A. V. Wannebo, and G. Woolmer. 2002. The human footprint and the last of the wild. *BioScience* 52:891–904. [1, 3, 6]

Sarmiento, F. O. 1997. Arrested succession in pastures hinders regeneration of tropical Andean forest and shreds mountain landscapes. *Environ. Conserv.* 24:14–23. [15]

Sauer, J. R., J. E. Hines, G. Gough, I. Thomas, and B. G. Perterjohn. 1997. *The North American Breeding Bird Survey Results and Analysis.* Patuxent Wildlife Research Center, Laurel, MD. [7]

Saunders, D. A. 1989. Changes in the avifauna of a region, district, and remnant as a result of fragmentation of native vegetation: The Wheatbelt of Western Australia. A case study. *Biol. Conserv.* 50:99–135. [7]

Saunders, D. A., R. J. Hobbs, and C. R. Margules. 1991. Biological consequences of ecosystem fragmentation: A review. *Conserv. Biol.* 5:18–32. [7]

Saunders, D. A., R. J. Hobbs, and P. R. Ehrlich. 1993. Reconstruction of fragmented ecosystems: global and regional perspectives. Surrey Beatty & Sons, Chipping Norton, New South Wales. [18]

Saura Martínez, J., B. Bayán, J. Casas, A. Ruiz de Larramendi, y C. Urdiales. 2001. *Documento Marco para el Desarrollo del Proyecto Doñana 2005. Regeneración hídrica de las cuencas y cauces vertientes a las marismas del parque nacional de Doñana.* Ministerio de Medio Ambiente. Madrid. [14]

Savidge, J. A. 1987. Extinction of an island avifauna by an introduced snake. *Ecology* 68:660–668. [9]

Scatolini, S. R., and J. B. Zedler. 1996. Epibenthic invertebrates of natural and constructed marshes of San Diego Bay. *Wetlands* 16:24–37. [15]

Schaefer, M. B. 1954. Some aspects of the dynamics of populations important to the management of commercial marine fisheries. *Bull. I-ATTC* 1:25–56.

Schafer, M. 1992. Sustainable Agriculture and Sustainable Diet in the Ecocity, Urban Ecology (USA). http://www.urbanecology.org.au/ecocity2/agriculture.html [18]

Schärr, C., P. L. Vidale, D. Lüthi, C. Frei, C. Häberli, M. A. Liniger, and C. Appenzeller. 2004. The role of increasing temperature variability in European summer heatwaves. *Nature* 427:322–326. [10]

Scheffer, M., S. Carpenter, J. A. Foley, C. Foulke, and B. Walker. 2001. Catastrophic shifts in ecosystems. *Nature* 413:591–596. [10]

Schelhas, J. and R. Greenberg. 1996. *Forest Patches in Tropical Landscapes.* Island Press, Washington, D.C. [7]

Schemske, D. W., B. C. Husband, M. H. Ruckelshaus, C. Goodwillie, I. Parker, and J. G. Bishop. 1994. Evaluating approaches to the conservation of rare and endangered plants. *Ecology* 75:584–606. [18]

Schierwater, B., B. Streit, G. P. Wagner, and R. Desalle. (eds.). 1994. *Molecular Ecology and Evolution: Approaches and Applications.* Birkhauser, Basel, Switzerland. [11]

Schimel, D., J. Melillo, H. Tian, A. D. McGuire, D. Kicklighter, T. Kittel, N Rosenbloom, S. Running, P. Thornton, D. Ojima, W. Parton, R. Kelly, M. Sykes, R. Neilson, and B. Rizzo. 2000. Contribution of increasing CO_2 and climate to carbon storage by ecosystems in the United States. *Science* 287:2004–2006. [10]

Schindler, D. E., M. D. Scheuerell, J. W. Moore, S. M. Gende, T. B. Francis, and W. J. Palen. 2003. Pacific salmon and the ecology of coastal ecosystems. *Frontiers in Ecology and the Environment* 1:31–37. [3]

Schindler, D. W., P. J. Curtis, B. R. Parker, and M. P. Stainton. 1996. Consequences of climate warming and lake acidification for UV-B penetration in North American boreal lakes. *Nature* 379:705–708. [3]

Schlesinger, W. H., J. F. Reynolds, G. L. Cunningham, L. F. Huenneke, W. M. Jarrell, R. A. Virginia, and W. G. Whitford. 1990. Biological feedbacks in global desertification. *Science* 247:1043–1048. [7]

Schmiegelow, F. K. A. and M. Mönkkönen. 2002. Habitat loss and fragmentation in dynamic landscapes: Avian perspectives from the boreal forest. *Ecol. Appl.* 12:375–389. [7]

Schmiegelow, F. K. A. and S. J. Hannon. 1999. Forest-level effects of management on Boreal songbirds: The Calling Lake fragmentation studies. In J. A. Rochelle, L. A. Lehmann, and J. Wisniewski (eds.), *Forest Fragmentation: Wildlife and Management Implications*, pp. 201–221. Brill, Leiden, The Netherlands. [12]

Schmiegelow, F. K. A., C. S. Machtans, and S. J. Hannon. 1997. Are boreal birds resilient to forest fragmentation: An experimental study of short-term community responses. *Ecology* 78:1914–1932. [7]

Schneider, S. H. 1989. *Global Warming: Are We entering the Greenhouse Century.* Sierra Club Books, San Francisco, CA. [4]

Schnute, J. T. and L. J. Richards. 2001. Use and abuse of fishery models. *Canadian Journal of Fisheries and Aquatic Sciences* 58:10–17. [8]

Schoener, T. W. 1983. Field experiments on interspecific competition. *Am. Nat.* 122:240–285. [2]

Scholes, R. J. and R. Biggs (eds.). 2004. Ecosystem Services in Southern Africa: A Regional Assessment. The Regional-Scale Component of the Southern African Millennium Ecosystem Assessment. CSIR, Pretoria, South Africa. [6]

Schonewald-Cox, C. M., S. M. Chambers, B. MacBryde, and L. Thomas (eds.). 1983. *Genetics and Conservation: A Reference for Managing Wild Animal and Plant Populations.* Benjamin/Cummings, Menlo Park, CA. [1]

Schorger, A. W. 1955. *The Passenger Pigeon, its Natural History and Extinction.* University of Wisconsin Press, Madison. [8]

Schowalter, T. D. 1988. Forest pest management: a synopsis. *Northwest Environ. J.* 4:313–318. [7]

Schrott, G. R., K. A. With, and A. W. King. 2005a. On the importance of landscape history for assessing extinction risk. *Ecol. Appl.* 15:493–506. [12]

Schrott, G. R., K. A. With, and A. W. King. 2005b. Demographic limitations on the ability of habitat restoration to rescue declining populations. *Conserv. Biol.* 19: In Press. [12]

Schuldt, J. A. and A. E. Hershey. 1995. Effect of salmon carcass decomposition on Lake Superior tributary streams. *J. N. Am. Benthol. Soc.* 14:259–268. [7]

Schultz, C. B. 2001. Restoring resources for an endangered butterfly. *J. Appl. Ecol.* 38: 1007–1019. [15]

Schultz, C. B. and L. R. Gerber. 2002. Are recovery plans improving with practice? *Ecol. Appl.* 12:641–647.[18]

Schulze, E. D. and H. A. Mooney. 1993. *Biodiversity and Ecosystem Function.* Springer Verlag: New York. [12]

Schumaker, N. H., T. Ernst, D. White, J. Baker, and P. Haggerty. 2004. Projecting wildlife responses to alternative future landscapes in Oregon's Willamette Basin. *Ecol. Appl.* 14:381–400. [12]

Schwartz M. W., C. A. Brigham, J. D. Hoeksema, K. G. Lyons, M. H. Mills, and P. J. van Mantgem. 2000. Linking biodiversity to ecosystem function: implications for conservation ecology. *Oecologia* 122:297–305. [2]

Schwarzenberger, F., E. Mostl, E. Bamberg, J. Pammer, and O. Schmehlik. 1991. Concentrations of progestagens and oestrogens of pregnant Lipizzan, trotter and thoroughbred mares. *J. Reprod. Fert.* 44 (Suppl):489–499. [11]

Scott, J. M. and B. Csuti. 1997. Gap Analysis for biodiversity survey and maintenance II. In M. L. Reaka-Kudla, D. E. Wilson, and E. O. Wilson (eds.), *Biodiversity: Getting the Job Done*, pp. 321–340. National Academy Press, Washington D.C. [14]

Scott, J. M. and T. W. Loveland. Ecological content and context of national wildlife refuges. In Press. [14]

Scott, J. M., B. Csuti, J. Jacobi, and J. E. Estes. 1987. Species richness: A geographic approach to protecting future biological diversity. *BioScience* 37:783–788. [14]

Scott, J. M., J. B. Haufler, P. J. Heglund, and M. L. Morrison. 2002. *Predicting Species Occurrences: Issues of Accuracy and Scale.* Island Press, Washington, D.C. [2]

Scott, J. M., S. Mountainspring, F. L. Ramsey, and C. B. Kepler. 1986. Forest bird communities of the Hawaiian Islands: Their dynamics, ecology, and conservation. *Studies in Avian Biology* No. 9. 431. [14]

Scott, J. M., F. W. Davis, R. G. Mcghie, R. G. Wright, C. Groves, and J. Estes 2001. Nature reserves do they capture the full range of America's biological diversity? *Ecol. Appl.* 11:999–1007. [14]

Scott, J. M., F. Davis, B. Csuti, R. Noss, B. Butterfield, C. Groves, H. Anderson, S. Caicco, F. Derchia, T. C. Edwards, J. Ulliman, and R. G. Wright. 1993. Gap analy-

sis: A geographic approach to protection of biological diversity. *Wildl. Monogr.* 123:1–41. [2, 14]

Scriber, J. M. 1994. Climatic legacies and sex chromosomes: Latitudinal patterns of voltinism, diapause, size, and host-plant selection in two species of swallowtail butterflies at their hybrid zone. In H. V. Danks (ed.), *Insect Life-Cycle Polymorphism.* Kluwer Academic Publishers, Dordrect, The Netherlands. [10]

Scribner, K. T. 1993. Conservation genetics of managed ungulate populations. *Acta. Theolo.* 38, Suppl. 2:89–101. [11]

Scribner, K. T., R. K. Chesser, and M. H. Smith. 1997. Defining genetically contiguous sub-populations of white-tailed deer: Spatial and temporal permeability of microgeographic structure. *J. Mammal.* 78:744–755. [11]

Scribner, K. T., M. H. Smith, R. A. Garrott, and L. H. Carpenter. 1991. Temporal, spatial, and age-specific changes in genotypic composition of mule deer. *J. Mammal.* 72:126–137. [11]

Scribner, K. T., R. Fields, S. Talbot, J. Pearce, and M. Petersen. 2001. Sex-biased dispersal in threatened Spectacled eiders: Evaluation using molecular markers with contrasting modes of inheritance. *Evolution* 55:2105–2115. [11]

Scribner, K. T., J. A. Warrillow, J. O. Leafloor, H. H. Prince, R. L. Inman, D. R. Luukkonen, and C. S. Flegel. 2003. Genetic methods for determining racial composition of Canada goose harvests. *J. Wildl. Mgmt.* 67:123–136. [11]

Seamons, T. R., P. Bentzen, and T. P. Quinn. 2004. The effects of adult length and arrival date on individual reproductive success in wild steelhead trout (*Oncorhynchus mykiss*). *Can. J. Fish. Aquat. Sci.* 61:193–204. [11]

Sedell, J., M. Sharpe, D. D. Apple, M. Copenhagen, and M. Furniss. 2000. *Water and the Forest Service.* USDA Forest Service Publication FS-660. Washington, D.C. [13]

Seeb, L. W., P. A. Crane, C. M. Kondzela, R. L. Wilmot, S. Urawa, N. V. Varnavskaya, and J. E. Seeb. 2004. Migration of Pacific Rim chum salmon on the high seas: Inferences from genetic data. In A. J. Gharrett et al. (eds.), *Genetics of Subpolar Fish and Invertebrates*, pp. 21–36. Kluwer, London. [11]

Semlitsch, R. D., D. E. Scott, J. H. K. Pechmann, and J. W. Gibbons. 1996. Structure and dynamics of an amphibian community: Evidence from a 16-year study of a natural pond. In M. L. Cody and J. A. Smallwood (eds.), *Long-Term Studies of Vertebrate Communities*, pp. 217–248. Academic Press, San Diego. [3]

Sepkoski, J. J. Jr. and D. M. Raup. 1986. Periodicity in marine extinction events. In D. K. Elliott. (ed.), *Dynamics of Extinction*, pp. 3–36. John Wiley & Sons, New York. [2]

SERNAP. 2002. *Políticas para el Sistema Nacional de Áreas Protegidas.* La Paz, Bolivia. [12]

Servheen, C. 1999 Status and management of the grizzly bear in the lower 48 United States. In C. Servheen, S. Herrero, and B. Peyton (eds.), *Bears: Status Survey and Conservation Action Plan*, pp. 50–54. IUCN, Cambridge, U.K. [11]

Severns, P. 2003. Inbreeding and small population size reduce seed set in a threatened and fragmented plant species, *Lupinus sulphureus* ssp. *kincaidii* (Fabaceae). *Biol. Conserv.* 110:221–229. [7]

Sexton, W. T., A. J. Malk, R. C. Szaro, and N. C. Johnson (eds.). 1999. *Ecological Stewardship: A Common Reference for Ecosystem Management, Volume III.* Elsevier Science, Ltd., Oxford, U.K. [13]

Seymour, R. S. and M. L. Hunter, Jr. 1999. Principles of ecological forestry. In M. L. Hunter Jr. (ed.), *Maintaining Biodiversity in Forest Ecosystems*, pp. 335–372. Cambridge University Press, Cambridge. [8]

Shafer, C. L. 1990. *Nature Reserves.* Smithsonian Institution Press, Washington D.C. [7]

Shaffer, M. L. 1981. Minimum population sizes for species conservation. *BioScience* 31:131–134. [7, 12]

Shaffer, M. L. 1987. Minimum viable populations: Coping with uncertainty. In M. E. Soulé (ed.), *Viable Populations for Conservation*, pp. 69–86. Cambridge University Press, Cambridge. [12]

Shaffer, M. L. 1990. Population viability analysis. *Conserv. Biol.* 4:39–40. [12]

Shaffer, M. L. and F. B. Samson. 1985. Population size and extinction: A note on determining critical population sizes. *Am. Nat.* 125:144–152. [11, 12]

Shaklee, J. B., T. D. Beacham, L. Seeb, and B. A. White. 1999. Managing fisheries using genetic data: Case studies from four species of Pacific salmon. *Fish. Res.* 43:45–78. [11]

Shanahan, M., S. So, S. G. Compton, and R. Corlett. 2001. Fig-eating by vertebrate frugivores: A global review. *Biological Review* 76:529–572. [12]

Shands, W. E. and R. G. Healy 1977. *The Lands Nobody Wanted.* The Conservation Foundation, Washington, D.C. [14]

Shanks A.L., B.A. Grantham, and M. H. Carr. 2003. Propagule dispersal distance and the size and spacing of marine reserves. *Geol. Appl.* 13: 5159 – 5169. [14]

Sharpe, N. L. 2000. *Estimating benefits for slate quarry restoration projects using contingent valuation and cost benefit analysis.* Unpublished M.S. thesis, University of Wales, Bangor. [5]

Sharpe, R. M. and D. S. Irvine. 2004. How strong is the evidence of a link between environmental chemicals and adverse effects on human reproductive health. *Brit. Med. J.* 328:447–451. [6]

Shepherd, G. 2004. *The Ecosystem Approach: Five Steps to Implementation.* IUCN, Gland, Switzerland. [13]

Sherman, C. K. and M. L. Morton. 1993. Population declines of Yosemite toads in the eastern Sierra Nevada of California. *J. Herpetol.* 27:186–198. [3]

Sherman, K. 1993. Large marine ecosystems as global units for marine resources management—An ecological perspective. In K. Sherman, L. M. Alexander, and B. D. Gold (eds). *Large Marine Ecosystems. Stress, Mitigation and Sustainability.* American Association for the Advancement of Science: Washington. [13]

Sherwin, W. B., N. D. Murray, J. A. M. Graves, and P. R. Brown. 1991. Measurement of genetic variation in endangered populations: Bandicoots (Marsupialia: Peramelidae) as an example. *Conserv. Biol.* 5:103–108. [11]

Shiels, A. B. and L. R. Walker. 2003. Bird perches increase forest seeds on Puerto Rican landslides. *Restor. Ecol.* 11:457–465. [15]

Shiganova, T. A. and Y. V. Bulgakova. 2000. Effect of gelatinous plankton on the black and Azov Sea fish and their food resources. *ICES J. Mar. Sci.* 57:641–648. [9]

Shinneman, D. J. and W. L. Baker. 2000. Impact of logging and roads on a Black Hills ponderosa pine forest landscape. In R. L. Knight, F. W. Smith, S. W. Buskirk, W. H. Romme, and W. L. Baker (eds.), *Forest Fragmentation in the Southern Rocky Mountains*, pp. 311–335. University Press of Colorado, Boulder. [7]

Shiva, V. 1989. *Staying Alive: Women, Ecology and Development.* Zed Books, London. [4]

Shmida, A. and M. V. Wilson. 1985. Biological determinants of species diversity. *J. Biogeogr.* 12:1–20. [2]

Shouli, H., J. Harris, W. Wanying, and Y. Yongqing (eds.). 2001. Community-Based Conservation and Development: Strategies and Practice at Cao Hai. Guizhou Nationalities Publishing House, Guizhou, People's Republic of China. [12]

Shrader-Frechette, K. 2002. *Environmental Justice: Creating Equality, Reclaiming Democracy.* Oxford University Press, New York. [4]

Shrestha, R. K., A. F. Seidl, and A. S. Moraes. 2002. Value of recreational fishing in the Brazilian Pantanal: A travel cost analysis using count data models. *Ecol. Econ.* 42:(1–2) 289–299. [5]

Shue, H. 1999. Global environment and international inequality. *International Affairs* 75, Reprinted in D. Schmidtz, and E. Willott (eds.) 2002. *Environmental Ethics: What Really Matters, What Really Works.* Oxford University Press, New York. [4]

Shulman, M. J. 1985. Recruitment of coral reef fishes: Effects of distribution of predators and shelter. *Ecology* 66:1056–1066. [7]

Siegert, F., G. Ruecker, A. Hinrichs, and A. A. Hoffmann. 2001. Increased damage from fires in logged forests during droughts caused by El Niño. *Nature* 414:437–440. [8]

Signor, P. W. 1990. The geological history of diversity. *Annu. Rev. Ecol. Syst.* 21:509–539. [2]

Siguer, R.C., A. Rodríguez, and L. Domínguez. 2001. *Análisis de la incidencia de los grandes herbívoros en la marisma y la vera del Parque Nacional de Doñana.* Ministerio de Medio Ambiente. Madrid. [14]

Siljeström, P., L. Clemente, and A. Rodríguez. 2002. Clima in *Parque Nacional de Doñana*. Canseco Editores. Madrid. pp. 43–56. [14]

Sillett, T. S., R. T. Holmes, and T. W. Sherry. 2000. Impacts of a global climate cycle on population dynamics of a migratory songbird. *Science* 288:2040–2042. [10]

Silva Matos, D. M., R. P. Freckleton, and A. R. Watkinson. 1999. The role of density-dependence in the population dynamics of a tropical palm. *Ecology* 80:2635–2650. [8]

Silver, S. C., L. E. T Ostro, L. K. Marsh, L. Maffei, A. J. Noss, M. J. Kelly, R. B. Wallace, H. Gomez, and G. Ayala. 2004. The use of camera traps for estimating jaguar *Panthera onca* abundance and density using capture/recapture analysis. *Oryx* 38:1–7. [11]

Silver, W. L., R. Ostertag, and A. E. Lugo. 2000. The potential for carbon sequestration through reforestation or abandoned tropical agricultural pasture lands. *Restor. Ecol.* 8: 394–407. [15]

Simberloff, D. 1988. The contribution of population and community biology to conservation science. *Annu. Rev. Ecol. Syst.* 19:473–511. [7]

Simberloff, D. and B. Von Holle. 1999. Positive interactions of nonindigenous species: Invasional meltdown? *Biol. Invasions* 1:21–32. [9]

Simberloff, D. and J. Cox. 1987. Consequences and costs of conservation corridors. *Conserv. Biol.* 1:63–71. [7]

Simberloff, D. and J. L. Martin. 1991. Nestedness of insular avifaunas: Simple summary statistics masking species patterns. *Ornis Fennica* 68:178–192. [7]

Simberloff, D., D. Doak, M. Groom, S. Trombulak, A. Dobson, S. Gatewood, M.E. Soulé, M. Gilpin, C. Martinez del Rio, and L. Mills. 1999. Regional and continental restoration. In M. Soule and J. Terborgh (eds.), *Continental Conservation: Scientific Foundations of Regional Conservation Networks*, pp. 65–98. Island Press, Washington, D.C. [15]

Simon, N. 1962. *Between the Sunlight and the Thunder. The Wildlife of Kenya*. Collins. London. [3]

Simpson, D., J.-P. Tuovinen, L. Emberson, and M. R. Ashmore. 2001. Characteristics of an ozone deposition module II: Sensitivity analysis. *Water, Air, and Soil Pollution* 143:123–137. [6]

Simpson, S. 2001. Fishy business. *Sci. Am.* 285:82–89. [8]

Singer, M. C. and P. R. Ehrlich. 1979. Population dynamics of the checkerspot butterfly *Euphydryas editha*. *Forts. Zool.* 25:53–60. [10]

Singer, M. C., C. D. Thomas, and C. Parmesan. 1993. Rapid human-induced evolution of insect-host associations. *Nature* 366:681–683. [9]

Singer, P. 1975. *Animal Liberation: A New Ethics for Our Treatment of Animals*. The New York Review, New York. [4]

Singhvi, L. M. n. d. *The Jain Declaration on Nature*. Federation of Jain Associations of North America, Cincinnati. [4]

Skellam, J. G. 1951. Random disperal in theoretical populations. *Biometrika* 38:196–218. [9]

Skelly, D. K., E. E. Werner, and S. A. Cortwright. 1999. Long-term distributional dynamics of a Michigan amphibian assemblage. *Ecology* 80:2326–2337. [3]

Skewes, O., F. González, L. Rubilar, and M. Quezada. 1999. *Investigación, Aprovechamiento y Control de Castor, Islas Tierra del Fuego y Navarino*. Instituto Forestal-Universidad de Concepción, Punta Arenas. [9]

Slade N. A., R. Gomulkiewicz, and H. M. Alexander. 1998. Alternatives to Robinson and Redford's method of assessing overharvest from incomplete demographic data. *Conserv. Biol.* 12:148–155. [8]

Slatkin, M. 1985. Gene flow in natural populations. *Ann. Rev. Ecol. Syst.* 16:393–430. [11]

Slatkin, M. 1991. Inbreeding coefficients and coalescence times. *Genet. Res.* 58:167–175. [11]

Slocum, M. G. 2000. Logs and fern patches as recruitment sites in a tropical pasture. *Restor. Ecol.* 8:408–413. [15]

Smale, L., L. G. Frank, and K. E. Holekamp. 1997. Ontogeny of dominance in free-living spotted hyenas: Juvenile rank relations with adult females and immigrant males. *Anim. Behav.* 46:467–477. [11]

Smallwood, K. S., J. Beya, and M. L. Morrison 1999. Using the best scientific data for endangered species conservation. *Environ. Manage.* 24:421–435. [17]

Smith, A. T. and M. M. Peacock. 1990. Conspecific attraction and the determination of metapopulation colonization rates. *Conserv. Biol.* 4:320–327. [7]

Smith, D. A., K. Ralls, B. Davenport, N. Adams, and J. E. Maldonado. 2001. Canine assistants for conservation. *Science* 291:435. [11]

Smith, F. A. and J. L. Betancourt. 1998. Response of bushy-tailed woodrats (*Neotoma cinerea*) to late Quaternary climatic change in the Colorado Plateau. *Quaternary Res.* 50:1–11. [10]

Smith, P. G. R., and J. B. Theberge. 1987. Evaluating natural areas using multiple criteria: theory and practice. *Environmental Management* 11:447–460. [14]

Smith, T. B. and R. K. Wayne (eds.). 1996. *Molecular Genetic Approaches in Conservation*. Oxford University Press, Oxford. [11]

Smith, V. H., G. D. Tilman, and J. C. Nekola. 1999. Eutrophication: impacts of excess nutrient inputs on freshwater, marine, and terrestrial ecosystems. *Environ. Pollut.* 100:179–196. [6]

Smith, V. K. 1989. Taking stock of progress with travel cost recreation demand methods: Theory and implementation. *Marine Resource Econ.* 6:279–310. [5]

Snook, L. K. 1996. Catastrophic disturbance, logging and the ecology of mahogany (*Swietenia macrophylla* King): Grounds for listing a major tropical timber species in CITES. *Botanical Journal of the Linnean Society* 122:35–46. [8]

Snyder, N. F. R., J. W. Wiley, and C. B. Kepler. 1987. *The parrots of Luquillo: natural history and conservation of the Puerto Rican parrot*. West. Found. Vert. Zool., Los Angeles. [15]

Society for Ecological Restoration (SER). 1991. Program and abstracts, 3rd Annual Conference, Orlando, Florida 18–23 May, 1991. [15]

Sommers, P and J. Smit. 1994. CFP Report 9 – Promoting Urban Agriculture: A Strategy Framework for Planners in North America, Europe and Asia. http://web.idrc.ca/es/ev–2124–201–1–DO_TOPIC.html [18]

Soulé, M. E. (ed.). 1987. *Viable Populations for Conservation*. Cambridge University Press, New York. [12]

Soulé, M. E. 1985. What is conservation biology? *BioScience* 35:727–734. [1, 4]

Soulé, M. E. 1986. Conservation biology and the "real world." In M. E. Soulé (ed.), *Conservation Biology: The Science of Scarcity and Diversity*, pp. 1–12. Sinauer Associates, Sunderland, MA. [1]

Soulé, M. E. and B. A. Wilcox. 1980. *Conservation Biology. An Evolutionary-Ecological Perspective*. Sinauer Associates, Sunderland, MA. [1]

Soulé, M. E. and J. Terborgh (eds.). 1999. *Continental Conservation: Scientific Foundations for Regional Reserve Networks*. Island Press, Washington DC. [7]

Soulé, M. E., D. T. Bolger, A. C. Alberts, J. Wright, M. Sorice, and S. Hill. 1988. Reconstructed dynamics of rapid extinctions of chaparral-requiring birds in urban habitat islands. *Conserv. Biol.* 2:75–92. [7]

South, A., S. Rushtan, and D. McDonald. 2000. Simulating the proposed reintroduction of European beaver (*Castor fiber*) to Scotland. *Biol. Conserv.* 93:103–116. [12]

Southgate, D. 1998. *Tropical Forest Conservation: An Economic Assessment of the Alternatives in Latin America*. Oxford University Press, NY. [8]

Spalding, M. D., C. Ravilious, and E. P. Green. 2001. *World Atlas of Coral Reefs*. University of California Press, Berkeley. [13]

Sparling, D. W., C. A. Bishop, and G. Linder (eds.). 2000. Ecotoxicology of amphibians and reptiles. Society of Environmental Toxicology and Chemistry (SETAC), Pensacola, FL. [3]

Sparling, D. W., G. M. Fellers, and L. L. McConnell. 2001. Pesticides and amphibian population declines in California, USA. *Environ. Toxicol. Chem.* 20:1591–1595. [3]

Specht, R. L. and E. J. Moll. 1983. Mediterranean-type heathlands and sclerophyllous shrublands of the world: An overview. In F. J. Kruger, D. T. Mitchell, and J. U. M. Jarvis (eds.), *Mediterranean-Type Ecosystems: The Role of Nutrients*, Ecological Studies No. 43, pp. 41–65. Springer-Verlag, Berlin. [2]

Spellerberg, I. F. 1994. *Evaluation and Assessment for Conservation: Ecological Guidelines for Determining Priorities for Nature Con-*

servation. Chapman and Hall, London. [14]

Spencer, C. N., B. R. McClelland, and J. A. Stanford. 1991. Shrimp stocking, salmon collapse, and eagle displacement: Cascading interactions in the food web of a large aquatic ecosystem. *BioScience* 41:14–21. [7, 9]

Speth, J. G. 2003. Two perspectives on globalization and the environment. In J. G. Speth (ed.), *Worlds Apart: Globalization and the Environment*, pp. 1–18. Island Press, Washington, D.C. [4]

Spielman, D., B. W. Brook, and R. Frankham. 2004. Most species are not driven to extinction before genetic factors impact them. *Proc. Nat. Acad. Sci. USA*. 101:15261–15264. [12]

Sprent, J. F. A. 1992. Parasites lost? *Int. J. Parasitol*. 22:139–151. [3]

Stachowicz, J. J., R. B. Whitlatch, and R. W. Osman. 1999. Species diversity and invasion resistance in a marine ecosystem. *Science* 286:1577–1579. [9]

Stanford Restoration Ecology Group. 2003. Seasonal Wetlands Restoration Project, Center for Conservation Biology, Stanford University, California. http://www.stanford.edu/~paulojco/fo othills/wetlands.html [15]

Stangel, P. W., M. R. Lennartz, and M. H. Smith. 1992. Genetic variation and population structure of red-cockaded woodpeckers. *Conserv. Biol*. 6:283–292. [11]

Stapp, W. B., M. M. Cromwell, and A. Walls. 1995. The global rivers environmental education network. In S. K. Jacobson (ed.), *Conserving Wildlife: International Education and Communication Approaches*, pp. 177–197. Columbia University Press, NY. [18]

State Coastal Conservancy. 1986. Staff Recommendation 17 April 1986, Del Sol Vernal Pools Enhancement, File No. 86-020. [15]

Steadman, D. W. 1995. Prehistoric extinctions of Pacific Island birds: Biodiversity meets zooarchaeology. *Science* 267:1123–1131. [3]

Steadman, D. W. and P. S. Martin. 2003. The late Quaternary extinction and future resurrection of birds on Pacific Islands. *Earth-Sci. Rev*. 61:133–147. [3]

Steeger, T., and J. E. Tietge. 2003. *White Paper on Potential Developmental Effects of Atrazine on Amphibians*. Office of Prevention, Pesticides, and Toxic Substances, U.S. Environmental Protection Agency, Washington, D.C. [3]

Steen, D. A. and J. P. Gibbs. 2004. Effects of roads on the structure of freshwater turtle populations. *Conserv. Biol*. 18:1143–1148. [7]

Steffan-Dewenter, I. and T. Tscharntke. 2001. Succession of bee communities on fallows. *Ecography* 24:83–93. [2]

Steffan-Dewenter, I., and A. Kuhn. 2003. Honeybee foraging in differentially structured landscapes. *Proc. R. Soc. Lond. B. Biol. Sci*. 270:569–575. [2]

Steffen, W. and P. Tyson (eds.). 2001. *Global Change and the Earth System: A Planet Under Pressure*. International Geosphere Program, Stockhom. [1]

Stehli, F. G., R. G. Douglas, and N. D. Newell. 1969. Generation and maintenance of gradients in taxonomic diversity. *Science* 164:947–949. [2]

Stein, B. A., L. L. Master and L. E. Morse. 2002. Taxonomic bias and vulnerable species. *Science* 297:1807–1807. [3]

Stein, B. A., L. S. Kutner, and J. S. Adams (eds.). 2000. *Precious Heritage: The Status of Biodiversity in the United States*. Oxford University Press, New York. [2, 3, 6, 7]

Stein, J. R., J. E. Heinrich, B. M. Hobbs, and D. St. George. 1999. Southern Nevada eco-region report. *Proc. Desert Fishes Council* 30:54–57. [14]

Stein, J. R., J. E. Heinrich, B. M. Hobbs, and J. C. Sjoberg. 2000. Native fish and amphibian management in southern Nevada. *Proc. Desert Fishes Council* 31:7–9. [14]

Steinitz, C. 1996. *Biodiversity and landscape planning: Alternative futures for the region of Camp Pendelton, California*. Harvard University Graduate School of Design, [12]

Stem, C., R. Margulois, N. Salafsky, and M. Brown. 2005. Monitoring and evaluation in conservation: a review of trends and approaches. *Conserv. Biol*. 19:295–309. [18]

Stephens, S. E., D. N. Koons, J. J. Rotella, and D. W. Willey. 2003. Effects of habitat fragmentation on avian nesting success: A review of the evidence at multiple spatial scales. *Biol. Conserv*. 115:101–110. [7]

Stevens, G. C. 1989. The latitudinal gradient in geographical range: How so many species coexist in the tropics. *Am. Nat*. 133:240–256. [12]

Stewart, M. M. 1995. Climate driven population fluctuations in rain-forest frogs. *J. Herpetol*. 29:437–446. [3]

Stiles, F. G., A. F. Skutch, and D. Gardner. 1989. *A Guide to the Birds of Costa Rica*. Cornell University Press, Ithaca. [10]

Still, C. J., P. N. Foster, and S. H. Schneider. 1999. Simulating the effects of climate change on tropical montane cloud forests. *Nature* 398:608–610. [3, 10]

Stille, A. 2002. With the Index of Leading Economic Indicators, a Social Report Card. *New York Times*, April 27, Section B, page 11. [18]

Stinchcombe, J., L. C. Moyle, B. R. Hudgens, P. L. Bloch, S. Chinnadurai, and W. F. Morris. 2002. The influence of the academic conservation biology literature on endangered species recovery planning. *Conserv. Ecol*. 6(2): Art. No. 15. [18]

Stohlgren, T. J., L. D. Schell, and B. Vanden Heuvel. 1999. How grazing and soil quality affect native and exotic plant diversity in Rocky Mountain grasslands. *Ecol. Appl*. 9:45–64. [2]

Stolarski, R., R. Bojkov, L. Bishop, C. Zerefos, J. Staehelin, and J. Zawodny. 1992. Measured trends in stratospheric ozone. *Science* 256:342–349. [3]

Stoms, D. M., F. W. Davis, K. L. Driese, K. M. Cassiday, and P. M. Murray 1998 Gap analysis of the vegetation of the inter-mountain semi-arid desert ecoregion. *Great Basin Nat*. 58:199–216. [14]

Stoner, A. W., and M. Ray-Culp. 2000. Evidence for Allee effects in an over-harvested marine gastropod: density-dependent mating and egg production. *Mar. Ecol. Prog. Ser*. 202:297–302. [12]

Stott, P. A. 2003. Attribution of regional-scale temperature changes to anthropogenic and natural causes. *Geophys. Res. Lett*. 30:1728. [10]

Stouffer, P. C. and R. O. Bierregaard. 1995. Effects of forest fragmentation on understory hummingbirds in Amazonian Brazil. *Conserv. Biol*. 9:1085–1094. [7]

Stratford, J. A. and P. C. Stouffer. 1999. Local extinctions of terrestrial insectivorous birds in a fragmented landscape near Manaus, Brazil. *Conserv. Biol*. 13:1416–1423. [7]

Strayer, D. L., N. F. Caraco, J. J. Cole, S. Findlay, and M. L. Pace. 1999. Transformation of freshwater ecosystems by bivalves. *BioScience* 49:19–27. [7]

Strayer, D. L., J. A. Downing, W. R. Haag, T. L. King, J. B. Layzer, T. J. Newton, and S. J. Nichols. 2004. Changing perspectives on pearly mussels, North America's most imperiled animals. *BioScience* 54:429–439. [8]

Strode, P. K. 2003. Implications of climate change for North American wood warblers (Parulidae). *Glob. Change Biol*. 9:1137–1144. [10]

Struhsaker, T. T., J. M. Kasenene, J. C. Gaither Jr., N. Larsen, S. Musango, and R. Bancroft. 1989. Tree mortality in the Kibale forest, Uganda: a case study of dieback in a tropical rain forest adjacent to exotic conifer plantations. *Forest Ecol. Manag*. 29:165–185. [15]

Stuart, S. N., J. S. Chanson, N. A. Cox, B. E. Young, A. S. L. Rodrigues, D. L. Fischman, and R. W. Waller. 2004. Status and trends of amphibian declines and extinctions worldwide. *Science* 306:1783–1786. [3, 6]

Sturm, M., C. Racine, and K. Tape. 2001. Increasing shrub abundance in the Arctic. *Nature* 411:546–547. [10]

Su, C. J., D. M. Debinski, M. E. Jakubauskas, and K. Kindscher. 2004. Beyond species richness: community similarity as a measure of cross-taxon congruence for coarse-filter conservation. *Conserv. Biol*. 18:167–173. [14]

Suarez, A. V., D. A. Holway, and T. J. Case. 2001. Patterns of spread in biological invasions dominated by long-distance jump dispersal: Insights from Argentine ants. *Proc. Nat. Acad. Sci. USA*. 98:1095–1100. [9]

Suding, K. N., K. L. Gross, and G. R. Houseman. 2004. Alternative states and positive feedbacks in restoration ecology. *Trends Ecol. Evol*. 19:46–53. [15]

Sugg, D. W. and R. K. Chesser. 1993. Effective population sizes with multiple paternity. *Genetics* 137:1147–1155. [11]

Sukumar, R. 1989. *The Asian elephant: Ecology and management*. Cambridge University Press, Cambridge. [17]

Sutherland, W. J. 1996. *From Individual Behaviour to Population Ecology*. Oxford University Press, Oxford. [10]

Sutherland, W. J. 2000. *The Conservation Handbook: Research, Management and Policy*. Blackwell Science, Ltd: Oxford. [16]

Sutherland, W. J. and J. A. Gill. 2001. The role of behaviour in studying sustainable exploitation. In J. D. Reynolds, G. M. Mace, K. H. Redford, and J. G. Robinson (eds.), *Conservation of Exploited Species*, pp. 259–280. Cambridge University Press, Cambridge. [8]

Swetnam, T. and A. M. Lynch. 1993. Multi-century, regional-scale patterns of western spruce budworm outbreaks. *Ecol. Monogr.* 63:399–424. [10]

Swetnam, T. W. and J. L. Betancourt. 1998. Mesoscale disturbance and ecological response to decadal climatic variability in the American Southwest. *J. Climate* 11:3128–3147. [10]

Swihart, R. K. and N. A. Slade. 1984. Road crossing in *Sigmodon hispidus* and *Microtus ochrogaster*. *J. Mammal.* 65:357–360. [7]

Swingland, I. R. 2003. *Capturing Carbon and Conserving Biodiversity*. EarthScan Publications Ltd., London. [18]

Sylvan, R. and D. H. Bennet. 1988. Taoism and deep ecology. *The Ecologist* 18:148–158. [4]

Szaro, R. C., N. C. Johnson, W. T. Sexton, and A. J. Malk (eds.). 1999. *Ecological Stewardship: A Common Reference for Ecosystem Management, Volume II*. Elsevier Science, Ltd., Oxford, U.K. [13]

Tainter, J. A. 1988. *The Collapse of Complex Societies*. Cambridge University Press, Cambridge, UK. [18]

Talbot, L. M. and M. H. Talbot. 1963. The wildebeest in western Maasailand. *Wildl. Monographs*. 12:1–122. [3]

Tallmon, D. A., E. S. Jules, N. J. Radke, and L. S. Mills. 2003. Of mice and men and trillium: Cascading effects of forest fragmentation. *Ecol. Appl.* 13:1193–1203. [7]

Taulman, J. F. and L. W. Robbins. 1996. Recent range expansion and distributional limits of the nine-banded armadillo (*Dasypus novemcinctus*) in the United States. *J. Biogeogr.* 23:635–648. [10]

Taylor, M. and B. Plater. 2001. *Population Viability Analysis for the Southern Resident Population of the Killer Whale (Orcinus orca)*. The Center for Biological Diversity, Tucson. [17]

Taylor, M. J. F., K. F. Suckling, and J. J. Rachlinski. 2005. The effectiveness of the Endangered Species Act: A quantitative analysis. *BioScience* 55:360–367. [18]

Taylor, P. W. 1986. *Respect for Nature. A Theory of Environmental Ethics*. Princeton University Press, Princeton. [4]

Taylor, S. K., E. S. Williams, and K. W. Mills. 1999. Effects of malathion on disease susceptibility in Woodhouse's toads. *J. Wildlife Dis.* 35:536–541. [3]

Tear, T. H., J. M. Scott, P. H. Hayward, and B. Griffith. 1993. Status and prospects for success of the Endangered Species Act: a look at recovery plans. *Science* 262:976–977. [18]

Tear, T. H., J. M. Scott, P. H. Hayward, and B. Griffith. 1995. Recovery plans and the Endangered Species Act: are criticisms supported by data? *Conserv. Biol.* 9:182–192. [18]

Tegner, M. J., L. V. Basch, and P. K. Dayton. 1996. Near extinction of an exploited marine invertebrate. *Trends Ecol. Evol.* 11:278–280. [8]

Tellaria, J. L. and T. Santos. 1995. Effects of forest fragmentation on a guild of wintering passerines: The role of habitat selection. *Conserv. Biol.* 71:61–67. [7]

Temple, S. A. 1986. Predicting impacts of habitat fragmentation on forest birds: A comparison of two models. In J. Verner, M. L. Morrison, and C. J. Ralph (eds.), *Wildlife 2000: Modeling Habitat Relationships of Terrestrial Vertebrates*, pp. 301–304. University of Wisconsin Press, Madison. [7]

Temple, S. A. and J. R. Cary. 1988. Modeling dynamics of habitat-interior bird populations in fragmented landscapes. *Conserv. Biol.* 2:340–347. [7]

Templeton, A. R., 1996. Translocation in conservation. In R. C. Szaro and D. W. Johnston (eds.). *Biodiversity in Managed Landscapes: Theory and Practice*, pp. 315–325. Oxford University Press. Oxford. [11]

Templeton, A. R., R. J. Robertson, J. Brisson, and J. Strasburg. 2001. Disrupting evolutionary processes: The effect of habitat fragmentation on collared lizards in the Missouri Ozarks. *Proc. Natl. Acad. Sci. U. S. A.* 98:5426–5432. [11]

Tenner, E. 1996. *Why Things Bite Back: Technology and the Revenge of Unintended Consequences*. Knopf, New York. [16]

Terborgh, J. 1974. Preservation of natural diversity: The problem of extinction prone species. *BioScience* 24:715–22. [3, 7]

Terborgh, J. 1986. Keystone plant resources in the tropical forest. In M. Soulé (ed). *Conservation Biology: Science of Scarcity and Diversity*, pp. 330–344. Sinauer Associates, Sunderland, MA. [12]

Terborgh, J. 2000. The fate of tropical forests: a matter of stewardship. *Conserv. Biol.* 14: 1358–1361. [14]

Terborgh, J. and B. Winter. 1980. Some causes of extinction. In M. E. Soulé and B. A. Wilcox (eds), *Conservation Biology: An Evolutionary-Ecological Perspective*, pp. 119–133. Sinauer, Associates, Sunderland, MA. [7]

Terborgh, J. and B. Winter. 1983. A method for siting parks and reserves with special reference to Columbia and Ecuador. *Biol. Conserv.* 27:45–58. [7]

Terborgh, J. and S. J. Wright. 1994. Effects of mammalian herbivores on plant recruitment in two neotropical forests. *Ecology* 75:1829–1833. [3]

Terborgh, J., L. Lopez, and J. Tello. 1997a. Bird communities in transition: The Lago Guri islands. *Ecology* 78:1494–1501. [7]

Terborgh, J., L. Lopez, J. Tello, D. Yu, and A. R. Bruni. 1997b. Transitory states in relaxing land bridge islands. In W. F. Laurance and R. O. Bierregaard, Jr. (eds), *Tropical Forest Remnants: Ecology, Management, and Conservation of Fragmented Communities*, pp. 256–274. University of Chicago Press, Chicago. [7]

Terborgh, J., J. Estes, P. Paquet, K. Ralls, D. Boyd-Heger, B. Miller, and R. Noss. 1999. The role of top carnivores in regulating terrestrial ecosystems. In M. E. Soulé and J. Terborgh (eds.), *Continental Conservation: Scientific Foundations Of Regional Reserve Networks*, pp. 39–64. Island Press, Washington, D.C. [3, 7]

Terborgh, J., L. Lopez, P. Nuñez V., M. Rao, G. Shahabuddin, G. Orihuela, M. Riveros, R. Ascanio, G. H. Adler, T. D. Lambert, and L. Balbas. 2001. Ecological meltdown in predator-free forest fragments. *Science* 294:1923–1926. [7]

Tewksbury, J. J., S. J. Hejl, and T. E. Martin. 1998. Breeding productivity does not decline with increasing fragmentation in a western landscape. *Ecology* 79:2890–2903.

Tewksbury, J. J., D. J. Levey, N. M. Haddad, S. Sargent, J. L. Orrock, A. Weldon, B. J. Danielson, J. Brinkerhoff, E. I. Damschen, and P. Townsend. 2002. Corridors affect plants, animals, and their interactions in fragmented landscapes. *Proc. Nat. Acad. Sci. USA.* 99:12923–12926. [7]

The Conservation Foundation. 1988. Protecting America's Wetlands: An action agenda. The Conservation Foundation. Washington, D.C. [15]

The Heinz Center (The H. John Heinz III Center for Science, Economics and the Environment). 2002. *The State of The Nation's Ecosystems: Measuring the Lands, Waters, and Living Resources of the United States*. Cambridge University Press, NY. [18]

The Nature Conservancy. 1997. *Designing a Geography of Hope: Guidelines for Ecoregion-Based Conservation in The Nature Conservancy*. The Nature Conservancy, Washington, D.C. [14]

The Nature Conservancy. 2000. *Conservation by Design—A framework for mission success*. Arlington, VA. [18]

The Nature Conservancy. 2002. *Lake Wales Ridge Conservation Project Planning Workbook*. Lake Wales, FL. http://www.conserveonline.org/2003/07/s/ConPrjMgmt_v4 [18]

The Nature Conservancy. 2003a. *Assessment of Target Viability Worksheet: Conservation Project Management Workbook Versions 3 (CAP) and 4*. Arlington, VA. http://www.conserveonline.org/2003/10/v/TV_Guide_Version_July_03. [18]

The Nature Conservancy. 2003b Guidelines for Designing and Selecting Conservation Strategies. Arlington, VA. http://www.conserveonline.org/2003/06/b/Strategy_Guidelines_ver6_03. [18]

The Nature Conservancy. 2003c. *The Enhanced 5-S Project Management Process*. Arlington, VA. http://www.conserveonline.org/2003/10/v/Enhanced_5-S_Project_Mngmt_Process_07-28-03. [18]

The Nature Conservancy and Partners. 2004. *R²– Reef Resilience: Building Resilience into Coral Reef Conservation; Additional Tools for Managers, Volume 2.0*. CD ROM

Toolkit. The Nature Conservancy, Marine Initiative. [13]

The Wildlands Project. 2005. http://www.twp.org/cms/page1090.cfm. [6]

Theobald, D. M. 1995. Landscape morphology and effects of mountain development in Colorado: A multi-scale analysis. Ph.D. dissertation, Univ. of Colorado, Boulder. [7]

Thiollay, J. M. 1992. Influence of selective logging on bird species-diversity in a Guianan rain-forest. *Conserv. Biol.* 6:47–63. [6]

Thiollay, J. M. 1997. Disturbance, selective logging and bird diversity: A Neotropical forest study. *Biodiversity and Conservation* 6:1155–1173. [6]

Thomas, C. D. 1993. Holocene climate changes and warm man-made refugia may explain why a sixth of British butterflies possess unnatural early-successional habitats. *Ecography* 16:278–284. [10]

Thomas, C. D., and J. J. Lennon. 1999. Birds extend their ranges northwards. *Nature* 399:213. [10]

Thomas, C. D., M. C. Singer, and D. A. Boughton. 1996. Catastrophic extinction of population sources in a butterfly metapopulation. *Am. Nat.* 148:957–975. [12]

Thomas, C. D., E. J. Bodsworth, R. J. Wilson, A. D. Simmons, Z. G. Davies, M. Musche, and L. Conradt. 2001. Ecological and evolutionary processes at expanding range margins. *Nature* 411:577–581. [10]

Thomas, C. D., A. Cameron, R. E. Green, M. Bakkenes, L. J. Beaumont, Y. C. Collingham, B. F. N. Erasmus, et al. 2004. Extinction risk from climate change. *Nature* 427:145–148. [10]

Thomas, J., J. M. Wondolleck, and S. L. Yaffee. 2002. *Elkhorn Mountains Cooperative Management Area: Ecosystem Management Initiative*. University of Michigan, Ann Arbor. [13]

Thomas, J. W. 2004. Sustainability of the Northwest Forest Plan: Still to be tested. In K. Arabas and J. Bowersox (eds.), *Forest Futures: Science, Policy, and Policy for the Next Century*, pp. 3–22. Rowman and Littlefield Publishers, Lanham, MD. [17]

Thomas, J. W., E. D. Forsman, J. B. Lint, E. C. Meslow, B. R. Noon, and J. Verner. 1990. *A conservation strategy for the Northern Spotted Owl: Interagency Committee to Address the Conservation of the Northern Spotted Owl*. U. S. Department of Agriculture, Portland, OR. [17]

Thomas, S. 1995. The legacy of dualism in decision-making within CAMPFIRE. Wildlife and Development Series, No. 4. Published by the International Institute for Environment and Development (IIED) in association with the CAMPFIRE Collaborative Group. http://wild-netafrica.co.za/bushcraft/arcles/document_campfire4.html. [16]

Thompson, I. D., J. A. Baker, and M. Ter-Mikaelian. 2003. A review of the long-term effects of post harvest silviculture on vertebrate wildlife, and predictive models, with an emphasis on boreal forests in Ontario, Canada. *For. Ecol. Manag.* 177:441–469. [8]

Thompson, J. R., M. D. Anderson, and K. N. Johnson. 2004. Ecosystem management across ownerships: The potential for collision with antitrust laws. *Conserv. Biol.* 18:1475–1481. [13]

Thornhill, N. W. (ed.). 1993. *The Natural History of Inbreeding and Outbreeding: Theoretical and Empirical Perspectives*. [11]

Thorson, G. 1957. Bottom communities (sublittoral and shallow shelf). In H. Ladd (ed.), *Treatise of Marine Ecology and Paleoecology*. Geological Society of America Memoir 67. [2]

Throop, H. L. and M. T. Lerdau. 2004. Effects of nitrogen deposition on insect herbivory: Implications for community and ecosystem processes. *Ecosystems* 7:109–133. [6]

Tilman, D. 1982. *Resource Competition and Community Structure*. Princeton University Press, Princeton. [2]

Tilman, D. 1985. The resource ratio hypothesis of succession. *Am. Nat.* 125:827–852. [2]

Tilman, D. 1996. Biodiversity: population versus ecosystem stability. *Ecology* 77:350–363. [2]

Tilman, D. and J. A. Downing. 1994. Biodiversity and stability in grasslands. *Nature* 367:363–365. [2]

Tilman, D., R. M. May, C. L. Lehman, and M. A. Nowak. 1994. Habitat destruction and the extinction debt. *Nature* 371:65–66. [7, 12]

Tilman, D., J. Fargione, B. Wolff, C. D'Antonio, A. Dobson, R. Howarth, D. Schindler, W. H. Schlesinger, D. Simberloff, and D. Swackhamer. 2001. Forecasting agriculturally driven global environmental change. *Science* 292:281–284. [18]

Timmermann, A., J. Oberhuber, A. Bacher, M. Esch, M. Latif, and E. Roeckner. 1999. Increased El Niño frequency in a climate model forced by future greenhouse warming. *Nature* 398:694–697. [3]

Tirosh-Samuelson, H. 2002. *Judaism and Ecology*. Harvard University Press, Cambridge, MA. [4]

Titus, J. G. 1998. Rising Seas, coastal erosion, and the takings clause: How to save wetlands and beaches without hurting property rights. *Maryland Law Review* 57:1279–1399. [10]

Toft, C. A. 1991. An ecological perspective: The population and community consequences of parasitism. In C. A. Toft, A. Aeschilmann, and L. Bolis, (eds.), *Parasite-Host Associations: Coexistence or Conflict?*, pp. 319–343. Oxford University Press, Oxford. [3]

Toft, G. and L. J. Guillette, Jr. 2005. Decreased sperm count and sexual behavior in mosquitofish (*Gambusia holbrooki*) exposed to water from a pesticide contaminated lake. *Ecotox. Environ. Safe.* 60:15–20. [6]

Toft, G., T. Edwards, E. Baatrup, and L. J. Guillette, Jr. 2003. Disturbed sexual characteristics in male mosquitofish (*Gambusia holbrooki*) from a lake contaminated with endocrine disrupters. *Environ. Health Persp.* 111:695–701. [6]

Tomimatsu, H. and M. Ohara. 2004. Edge effects on recruitment of *Trillium camschatcense* in small forest fragments. *Biol. Conserv.* 117:509–519. [7]

Torchin, M. E., A. M. Kuris, and K. D. Lafferty. 2002. Parasites of the green crab in its native and introduced range. *Parasitology* 124:S137–S151. [3]

Torchin, M. E., K. D. Lafferty, A. P. Dobson, A. P. Mckenzie, and A. M. Kuris. 2003. Introduced species and their missing parasites. *Nature* 421:628–630. [3]

Torras, M. 2003. An ecological footprint approach to external debt relief. *World Dev.* 31(12):2161–2171. [6]

Torsvik, V., L. Øvreås, and T. F. Thingstad. 2002. Prokaryotic diversity—magnitude, dynamics and controlling factors. *Science* 296:1064–1066. [2]

Toth, L. A. 1995. Conceptual evaluation of factors potentially affecting restoration of habitat structure within the channelized Kissimmee River ecosystem. *Restoration Ecology* 3: 160–180. [15]

TRAFFIC 1998. *Europe's Medicinal and Aromatic Plants: Their Use, Trade and Conservation*. TRAFFIC International, Cambridge. [8]

TransFair USA. 2004. Fair trade certification overview. www.transfairusa.org/content/certification/overview.php. 2005. [18]

Travel Industry Association of America. 2004. *The National Parks Traveler*. 2004 Edition. [16]

Travis, J. 1994. Calibrating our expectations in studying amphibian populations. *Herpetologica* 50:104–108. [3]

Tregenza, T. and N. Wedell. 2000. Genetic compatibility, mate choice and patterns of parentage: Invited review. *Mol. Ecol.* 9:1013–1027. [11]

Trexler, J. C. 1995. Restoration of the Kissimmee River: A conceptual model of past and present fish communities and its consequences for evaluating restoration success. *Restoration Ecology* 3:195–210. [15]

Trice, A. H. and S. E. Wood. 1958. Measurement of recreation benefits. *Land Economics* 66:589–597. [5]

Trombulak, S. C. and C. A. Frissell. 2000. Review of ecological effects of roads on terrestrial and aquatic communities. *Conserv. Biol.* 14:18–30. [7]

Truett, J. 1996. Bison and elk in the American southwest: in search of the pristine. *Environ. Manage.* 20:195–206. [13]

Trüper, H. G. 1992. Prokaryotes: An overview with respect to biodiversity and environmental importance. *Biodivers. Conserv.* 1:227–236. [2]

Trussell, G. C., and L. D. Smith. 2000. Induced defenses in response to an invading crab predator: An explanation of historical and geographic phenotypic change *Proc. Nat. Acad. Sci. USA* 97:2123–2127. [9]

Tryon, B. W. 1999. Bog turtles, southern style. *Endangered Species Bulletin* 24(3):12–13. [12]

Trzcinski, M. K., L. Fahrig, and G. Merriam. 1999. Independent effects of forest cover and fragmentation on the distribution of forest breeding birds. *Ecol. Appl.* 9:586–593. [7]

Tuan, Y-F. 1968. Discrepancies between environmental attitude and behavior: Examples from Europe and China. *Can. Geogr.* 12:176–191. [4]

Tuchmann, E. T., K. P. Connaughton, L. E. Freedman, and C. B. Moriwaki. 1996. *The Northwest Forest Plan: A report to the President and Congress*. U.S. Department of Agriculture, Forest Service, Pacific Northwest Research Station, Portland, OR. [13]

Tuck, G. N., T. Polacheck, and C. M. Bulman. 2003. Spatio-temporal trends of longline fishing effort in the Southern Ocean and implications for seabird bycatch. *Biol. Conserv.* 114:1–27. [8]

Tucker, G. M. 1997. Priorities for bird conservation in Europe: the importance of the farmed landscape. In D. J. Pain and M.W. Pienkowski, (eds.), *Farming and Birds in Europe: the Common Agricultural Policy and its Implications for Bird Conservation*, pp. 79–116. Academic Press, San Diego, California, U.S. [18]

Tucker, M. E. and D. R. Williams. 1997. *Buddhism and Ecology*. Harvard University Press, Cambridge, MA. [4]

Tucker, M. E. and J. Berthrong. 1998. *Confucianism and Ecology*. Harvard University Press, Cambridge, MA. [4]

Tufto, J., B.-E. Saether, S. Engen, J. E. Swenson, and F. Sandegren. 1999. Harvesting strategies for conserving minimum viable populations based on World Conservation Union criteria: Brown bears in Norway. *Proc. Royal. Soc. Lon. Ser. B.* 266:961–967. [12]

Tupper, M. and F. Juanes. 1999. Effects of a marine reserve on recruitment of grunts (Pisces: Haemulidae) at Barbados, West Indies. *Environ. Biol. Fish.* 55:53–63. [12]

Turner, B. L. II, S. Cortina Villar, D. Foster, J. Geoghegan, E. Keys, P. Klepeis, D. Lawrence, P. Macario Mendoza, S. Manson, Y. Ogneva-Himmelberger, A. B. Plotkin, D. R. Pérez-Salicrup, R. Roy Chowdhury, B. Savitsky, L. Schneider, B. Schmook, C. Vance. 2001. Deforestation in the southern Yucatán Peninsular region: An integrative approach. *Forest Ecol. Manag.* 154:353–370. [8]

Turner, I. M. and R. T. Corlett. 1996. The Conservation value of small, isolated fragments of lowland tropical rain forest. *Trends Ecol. Evol.* 11:330–333. [7]

Turner, I. M., K. S. Chua, J. S. Y. Ong, B. C. Soong, and H. T. W. Tan. 1996. A century of plant species loss from an isolated fragment of lowland tropical rain forest. *Conserv. Biol.* 10:1229–1244. [7]

Turner, J. W., R. Nemeth, and C. Rogers. 2003. Measurement of fecal glucocorticoids in parrotfishes to assess stress. *Gen. Comp. Endocr.* 133:341–352. [11]

Turner, M. G. 1989. Landscape ecology: The effect of pattern on process. *Annu. Rev. Ecol. Syst.* 20:171–197. [7, 12]

Turner, M. G. and C. L. Ruscher. 1988. Changes in landscape patterns in Georgia, USA. *Landscape Ecol.* 1:241–251. [7]

Turner, M. G. and R. H. Gardner (eds.). 1991. *Quantitative Methods in Landscape Ecology: The Analysis and Interpretation of Landscape Heterogeneity*. Springer-Verlag, New York. [7]

Turner, M. G., R. H. Gardner, and R. V. O'Neill. 2001. *Landscape Ecology in Theory and Practice*. Springer-Verlag, New York. [7]

Turner, M. G., V. H. Dale, and R. H. Gardner. 1989. Predicting across scales: Theory development and testing. *Landscape Ecol.* 3:245–252. [12]

Turner, M. G., G. J. Arthaud, R. T. Engstrom, S. J. Hejl, J. Liu, S. Loeb, and K. S. McKelvey. 1995. Usefulness of spatially explicit population models in land management. *Ecol. Appl.* 5:12–16. [12]

Turner, R. E. and N. N. Rabalais. 2003. Linking landscape and water quality in the Mississippi River Basin for 200 years. *BioScience* 53:563–571. [6]

Turner, R. M. 1990. Long-term vegetation change at a fully protected Sonoran desert site. *Ecology* 71:464–477. [10]

Twilley, R. R., E. J. Barron, H. L. Gholz, M. A. Harwell, R. L. Miller, D. J. Reed, J. B. Rose, E. H. Siemann, R. G. Wetzel, and R. J. Zimmerman. 2001. *Confronting Climate Change in the Gulf Coast Region: Prospects for Sustaining Our Ecological Heritage*. Union of Concerned Scientists and The Ecological Society of America. [10]

Tyler, C. R., S. Jobling, and J. P. Sumpter. 1998. Endocrine disruption in wildlife: A critical review of the evidence. *Crit. Rev. Toxicol.* 28:319–361. [6]

Tyler, M. J. 1991. Declining amphibian populations: A global phenomenon? An Australian perspective. *Alytes* 9:43–50. [3]

Tyler, T., W. J. Liss, L. M. Ganio, G. L. Larson, R. Hoffman, E. Deimling, and G. Lomnicky. 1998. Interaction between introduced trout and larval salamanders (*Ambystoma macrodactylum*) in high-elevation lakes. *Conserv. Biol.* 12:94–105. [3]

Tyser, R. W. and C. A. Worley. 1992. Alien flora in grasslands adjacent to road and trail corridors in Glacier National Park, Montana (USA). *Conserv. Biol.* 6:253–262. [7]

Uhl, C., R. Buschbacher, and E. A. S. Serrao. 1988. Abandoned pastures in eastern Amazonia. I. Patterns of plant succession. *J. Ecol.* 76:663–681. [15]

UN Environmental Programme. 2002. *Global Environmental Outlook 3: Past, Present and Future Perspectives*. Earthscan Publications Ltd., London. [6, 18]

UN Environmental Programme. 2003. *Selected Satellite Images of Our Changing Environment*. Division of Early Warning and Assessment, United National Environmental Programme, Nairobi, Kenya. [6]

UN Environmental Programme. 2004. *Analyzing Environmental Trends Using Satellite Data : Selected Cases*. Division of Early Warning and Assessment, United National Environmental Programme, Nairobi, Kenya. [6]

UN Millennium Project. 2005. *Investing in Development: A Practical Plan to Achieve the Millennium Development Goals*. New York. [3]

Underhill, L. G. 1994. Optimal and sub-optimal reserve selection algorithms. *Biol. Conserv.* 70:85–87. [14]

UNESCAP. 2005. *Solid Waste*. http://www.unescap.org/stat/envstat/stwes-2waste.02.pdf. [18]

Ungerer, M. J. A., P. Matthew, and M. J. Lombardero. 1999. Climate and the northern distribution limits of *Dendroctonus frontalis* Zimmermann (Coleoptera: Scolytidae). *J. Biogeogr.* 26:1133–1145. [10]

UN–HABITAT. 1999. *Managing water for African cities*. City Reports and Strategy Background Papers, Vol. 2. [18]

UN–HABITAT. 2003. The challenge of slums. Global Report on Human Settlements, 2003. [8]

United Nations Development Programme. 1998. *Human Development Report*. Oxford University Press, Oxford. [5]

United Nations. 2002. CBD goals declaration at World Sustainability Summit, Johannesburg, South Africa. [18]

United Nations. 2002. Plan of Implementation of the World Summit on Sustainable Development. In *Report of the World Summit on Sustainable Development, Johannesburg, South Africa, 26 August–4 September 2002*, pp. 6–72. A/CONF.199/20. United Nations, New York. [16]

United Nations. 2004. *United Nations World Population Division*. Global Data. United Nations, Geneva. [1, 18]

United States Environmental Protection Agency. 2005. Use of biological information to better define designated aquatic life uses in state and tribal water quality standards: Tiered aquatic life uses. EPA-822-R-05-001, Office of Water, United States Environmental Protection Agency, Washington D.C. [18]

United States Fish and Wildlife Service. 2005. Summary of Listed Species. ecos.fws.gov/tess_public/TESSBoxscore. [3]

United States National Assessment Report 2001. Climate Change Impacts on the United States: The Potential Consequences of Climate Variability and Change. Cambridge University Press, Cambridge. [10]

Uphoff, N. and J. Langholz. 1998. Incentives for avoiding the tragedy of the commons. *Environ. Conserv.* 25:251–261. [17]

USAID. 1997. CAMPFIRE: U.S. Agency for International Development Fact Sheet. http://www.info.usaid.gov/press/releases/9700101.html. [16]

U.S. Commission on Ocean Policy. 2004. *An Ocean Blueprint for the 21st Century: Final Report of the U.S. Commission on Ocean Policy*, Washington, D.C. [13]

USDA Forest Service. 1982a. *Golden Trout Wilderness Management Plan, Inyo and Sequoia National Forests.* USDA Forest Service, San Francisco. [13]

USDA Forest Service. 1982b. *Golden Trout Habitat and Wilderness Restoration on the Kern Plateau, Inyo National Forest.* USDA Forest Service, San Francisco. [13]

USDI BLM. 1994. *Ecosystem Management in the BLM: From Concept to Commitment.* United States Department of the Interior, Bureau of Land Management, Washington, D.C. [13]

USDI (U. S. Department of Interior, Fish and Wildlife Service).1992. *Final Draft Recovery Plan for the Northern Spotted Owl.* U. S. Fish and Wildlife Service, Portland, Oregon. [17]

U.S. Department of Energy. 2002. Energy Information Administration, International Energy Annual. Washington, D.C. [1]

U.S. Environmental Protection Agency. 2005. Use of biological information to tier designated aquatic life uses in state and tribal water quality standards. Office of Water, U.S. Environmental Protection Agency, Washington, D.C. [18]

U.S. Fish and Wildlife Service. 1997. *1996 National Survey of Fishing, Hunting, and Wildlife-Associated Recreation.* U.S. Department of the Interior, Fish and Wildlife Service and U.S. Department of Commerce, U.S. Census Bureau. [8]

U.S. Fish and Wildlife Service. 1999. Technical/Agency Draft Revised Recovery Plan for the Puerto Rican Parrot (*Amazona vittata*). Atlanta, Georgia. [15]

U.S. Fish and Wildlife Service. 2002. *2001 National Survey of Fishing, Hunting, and Wildlife-Associated Recreation.* U.S. Department of the Interior, Fish and Wildlife Service and U.S. Department of Commerce, U.S. Census Bureau. [8]

U.S. Fish and Wildlife Service. 2005. *Summary of Listed Species and Recovery Plans.* http://ecos.fws.gov/tess_public/TESS-Boxscore. [18]

U.S. Forest Service and U.S. Bureau of Land Management. 2001. Record of Decision and standards and guidelines for amendments to the survey and manage, protection buffer, and other mitigating measures standards and guidelines. USFS and BLM, Portland, OR. [13]

USGS 2003. *A Summary Report of Sediment Processes in Chesapeake Bay and Watershed.* U.S. Geological Survey, Water-Resources Investigations Report 03–4123. [13]

Utter, F. M., F. W. Allendorf, and O. Hodgins. 1973. Genetic variability and relationships in Pacific salmon and related trout based on protein variations. *Syst. Zool.* 22:257–270. [11]

van Creveld, M. 1991. *The Transformation of War.* The Free Press, New York. [18]

Van De Casteele, T., P. Galbusera, and E. Matthysen. 2001. A comparison of microsatellite-based pair-wise relatedness estimators. *Mol. Ecol.* 10:1539–1549. [11]

Van Den Avyle, M. J. and J. W. Evans. 1990. Temperature selection by striped bass in a Gulf of Mexico coastal river system. *N. Am. J. Fish. Mgmt.* 10:58–66. [7]

Van Dorp, D. and P. F. M. Opdam. 1987. Effects of patch size, isolation and regional abundance on forest bird communities. *Landscape Ecol.* 1:59–73. [7]

Van Horne, B. 1983. Density as a misleading indicator of habitat quality. *J. Wildl. Mgmt.* 47:893–901. [12]

Van Noordwijk, A. J., R. H. McCleery, and C. M. Perrins. 1995. Selection for the timing of great tit breeding in relation to caterpillar growth and temperature. *J. Anim. Ecol.* 64:451–458. [10]

Van Riper, C., S. G. Van Riper, M. L. Goff, and M. Laird. 1986. The epizootiology and ecological significance of malaria in Hawaiian land birds. *Ecol. Monogr.* 56:327–344. [3]

Vanni, M. J., C. Luecke, J. F. Kitchell, Y. Allen, J. Temte, and J. J. Magnuson. 1990. Effects on lower trophic levels of massive fish mortality. *Nature* 344:333–335. [3]

Väre H., R. Ohtonen, and K. Mikkola. 1996. The effect and extent of heavy grazing by reindeer in oligotrophic pine heaths in northeastern Fennoscandia. *Ecography* 19:245–53. [8]

Varughese, G. and E. Ostrom. 2001. The contested role of heterogeneity in collective action: Some evidence from community forestry in Nepal. *World Dev.* 29:747–765. [12]

Vasquez, R. and A. H. Gentry. 1989. Use and misuse of forest-harvested fruits in the Iquitos area. *Conserv. Biol.* 3:350–361. [8]

Veit, R. R. and M. A. Lewis. 1996. Dispersal, population growth, and the Allee effect: Dynamics of the house finch invasion of eastern North America. *Am. Nat.* 148:255–274. [12]

Veit R. R., P. Pyle, and J. A. McGowan. 1996. Ocean warming and long-term change in pelagic bird abundance within the California Current system. *Mar. Ecol. Prog. Ser.* 139:11–18. [10]

Velland, M. 2003. Habitat loss inhibits recovery of plant diversity as forests regrow. *Ecology* 84:1158–1164. [15]

Velle, K. and P. Drichi. 1992. *National Biomass Study, Phase I Technical Report.* Forest Department, Kampala, Uganda. [5]

Velloso, A. L., S. K. Wasser, S. L. Monfort, and J. M. Dietz. 1998. Longitudinal fecal steroid excretion in the maned wolves (*Chrysocyon brachyurus*). *Gen. Comp. Endocr.* 112:96–107. [11]

Vermeij, G. J. 1982. Phenotypic evolution in a poorly dispersing snail after arrival of a predator. *Nature* 299:349–350. [9]

Vermeij, G. J. 1991. When biotas meet: Understanding biotic interchange. *Science* 253:1099–1104. [9]

Verner, J., K. S. McKelvey, B. R. Noon, R. J. Gutierrez, G. I. Gould Jr., and T. W. Beck. 1992. *The California Spotted Owl: A Technical Assessment of its Current Status.* General Technical Report PSW-GTR-133. Albany, California, Pacific Southwest Research Station, U.S. Forest Service, U.S. Department of Agriculture. [17]

Vertucci, F. A. and P. S. Corn. 1996. Evaluation of episodic acidification and amphibian declines in the Rocky Mountains. *Ecol. Appl.* 6:449–457. [3]

Vickerman, K. 1992. The diversity and ecological significance of Protozoa. *Biodivers. Conserv.* 1:334–341. [2]

Vickery, P. D., M. L. Hunter, and S. M. Melvin. 1994. Effects of habitat area on the distribution of grassland birds in Maine. *Conserv. Biol.* 10:1087–1097. [7]

Viederman, S. 1992. *Public Policy: Challenge to Ecological Economics.* Unpublished manuscript. [16]

Viederman, S. 1993. Sustainable development—what is it and how do we get there? *Curr. Hist.* 92:180–185. [16]

Vierling, K. T. 2000. Source and sink habitats of Red-winged Blackbirds in a rural/suburban landscape. *Ecol. Appl.* 10:1211–1218. [12]

Vietmeyer, N. 1986a. Exotic edibles are altering America's diet and agriculture. *Smithsonian* 16(9):34–43. [4]

Vietmeyer, N. 1986b. Lesser-known plants of potential use in agriculture and forestry. *Science* 232:1379–1384. [4]

Vila, I., L. S. Fuentes, and M. Saavedra. 1999. Ictiofauna en los sistemas límnicos de la Isla Grande, Tierra del Fuego, Chile. *Rev. Chil. Hist. Nat.* 72:273–284. [9]

Villard, M. -A., P. R. Martin, and C. G. Drummond. 1993. Habitat fragmentation and pairing success in the Ovenbird (*Seiurus aurocapillus*). *Auk* 110:759–768. [7]

Villard, M.-A., M. K. Trzcinski, and G. Merriam. 1999. Fragmentation effects on forest birds: Relative influence of woodland cover and configuration on landscape occupancy. *Conserv. Biol.* 13:774–783. [7]

Virkkala, R., A. Rajasarkka, R. A. Vaisanen, M. Vickholm, and E. Virolainen. 1994. The significance of protected areas for the land birds of southern Finland. *Conserv. Biol.* 8:532–544. [8]

Virtanen, T. and S. Neuvonen. 1999. Performance of moth larvae on birch in relation to altitude, climate, host quality and parasitoids. *Oecologia* 120:92–101. [10]

Virtanen, T., S. Neuvonen, and A. Nikula. 1998. Modelling topoclimatic patterns of egg mortality of *Epirrita autumnata* (Lepidoptera: Geometridae) with a Geographical Information System: Predictions for current climate and warmer climate scenarios. *J. Appl. Ecol.* 35:311–322. [10]

Visser, M. and F. Rienks. 2003. Shifting links: Climate change disrupts food chains. *Levende Natuur* 104:110–113. [10]

Visser, M. and L. Holleman. 2001. Warmer springs disrupt the synchrony of oak and winter moth phenology. *Proc. R. Soc Lon. Ser. B* 268:289–294. [10]

Visser, M. E., A. J. van Noordwijk, J. M. Tinbergen, and C. M. Lessells. 1998. Warmer springs lead to mistimed reproduction in great tits (*Parus major*). *Proc. R. Soc Lon. Ser. B* 265:1867–1870. [10]

Vitousek, P. M. and L. R. Walker. 1989. Biological invasions by *Myrica faya* in Hawaii: Olant demography, nitrogen fixation, ecosystem effects. *Ecol. Monogr.* 59:247–265. [9]

Vitousek, P. M., H. A. Mooney, J. Lubchenco, and J. M. Melillo. 1997. Human domina-

tion of Earth's ecosystems. *Science* 277:494–499. [1]

Vitousek, P. M., L. L. Loope, and C. P. Stone. 1987. Introduced species in Hawaii: Biological effects and opportunities for ecological research. *Trends Ecol. Evol.* 2:24–227. [9]

Vitousek, P. M., P. Ehrlich, A. Ehrlich, and P. M. Matson. 1986. Human appropriation of the products of photosynthesis. *Bio-Science* 36:368–373. [1, 3]

Vitousek, P., H. Mooney, J. Lubchenko, and J. Melillo. 1997. Human domination of Earth's ecosystems. *Science* 277:494. [4]

Vitousek, P., J. Aber, R. Howarth, G. Likens, P. Matson, D. Schindler, W. Schlesinger, and D. Tilman. 1997. Human alteration of the global nitrogen cycle: sources and consequences. *Ecol. Appl.* 73:737–750. [6]

Viveiros de Castro, E. B. and F. A. S. Fernandez. 2004. Determinants of differential extinction vulnerabilities of small mammals in Atlantic forest fragments in Brazil. *Biol. Conserv.* 119:73–80. [7]

von Droste, B. 1988. The role of biosphere reserves at a time of increasing globalization. In V. Martin (ed.), *For the Conservation of the Earth*, pp. 89–93. Fulcrum, Golden, CO. [4]

von Humboldt, A. 1806. *The Physiognomy of Plants*. English Edition. H. G. Bohn London, 1849. [2]

Vonesh, J. R. and O. De la Cruz. 2002. Complex life cycles and density dependence: Assessing the contribution of egg mortality to amphibian declines. *Oecologia* 133:325–333. [3]

Vonier, P. M., D. A. Crain, J. A. McLachlan, L. J. Guillette, Jr., and S. F. Arnold. 1996. Interaction of environmental chemicals with the estrogen and progesterone receptors from the oviduct of the American alligator. *Environ. Health Persp.* 104:1318–1322. [6]

Vredenburg, V. T. 2004. Reversing introduced species effects: Experimental removal of introduced fish leads to rapid recovery of a declining frog. *Proc. Nat. Acad. Sci. USA* 101:7646–7650. [3]

Wackernagel, M. and W. Rees. 1996. *Our Ecological Footprint: Reducing Human Impact on the Earth*. New Society Publishers, Philadelphia. [3, 18]

Wackernagel, M. et al. 2002. Tracking the ecological overshoot of the human economy. *Proc. Nat. Acad. Sci. USA* 99:9266–9271. [1]

Wade, M. and D. E. McCauley. 1988. Extinction and recolonization: Their effects on the genetic differentiation of local populations. *Evolution* 42:995–1005. [11]

Wade, P. R. 2000. Bayesian methods for conservation biology. *Conserv. Biol.* 14:1308–1316. [17]

Wadsworth, F. H. 1950. Notes on the climax forest of Puerto Rico and their destruction and conservation prior to 1900. *Caribbean Forester* 11:38–47. [15]

Wadsworth, F. H. 1997. Secondary forests and their culture. In *Forest Production for Tropical Americ*, pp. 101–154. United

States Department of Agriculture Forest Service, Agriculture Handbook 710. [8]

Waits, L. P., S. L. Talbot, R. H. Ward, and G. F. Shields. 1998. Mitochondrial DNA phylogeography of the North American brown bear and implications for conservation. *Conserv. Biol.* 12:408–417. [11]

Wake, D. B. and H. J. Morowitz. 1991. Declining amphibian populations: A global phenomenon? *Alytes* 9:33–42. [3]

Wakely, P. and N. You. 2001. *Implementing the Habitat Agenda. In Search of Urban Sustainability*. Development Planning Unit, University College of London: London. [18]

Wali, M. K., N. M. Safaya, and F. Evrendilek. 2002. The Americas: with special reference to the United States of America. In M. R. Perrow and A. J. Davy, (eds), *Handbook of Ecological Restoration, Volume II: Restoration in Practice*, pp. 2–31. Cambridge University Press, Cambridge. [15]

Walker, B. 1995. Conserving biological diversity through ecosystem resilience. *Conserv. Biol.* 9:747–752. [12]

Walker, B. H. 1992. Biodiversity and ecological redundancy. *Conserv. Biol.* 6:18–23. [12]

Walker, L. R. and M. R. Willig. 1999. An introduction to terrestrial disturbance. In L. R. Walker (ed.), *Ecosystems of Disturbed Ground*. Ecosystems of the World Series. Elsevier Science, Amsterdam. [10]

Wallace, A. R. 1863. On the physical geology of the Malay archipelago. *J. R. Geogr. Soc.* 33:217–34. [4]

Walpole M. J., H. J. Goodwin, and K. G. R. Ward. 2000. Pricing policy for tourism in protected areas: Lessons from Komodo National Park, Indonesia. *Conser. Biol.* 15:218–227. [5]

Walters, C. J. 1986. *Adaptive Management of Renewable Resources*. Macmillan, New York. [13]

Walters, C. J. and L. H. Gunderson. 1994. Screening water policy alternatives for ecological restoration in the Everglades. In S. M. Davis and J. C. Ogden (eds.), *Everglades, The Ecosystem and its Restoration*. St. Lucie Press, Delray Florida. [13]

Walters, C. J., L. H. Gunderson, and C. S. Holling. 1992. Experimental policies for water management in the Everglades. *Ecol. Appl.* 2:189–202. [13]

Walther, G.-R., E. Post, A. Menzel, P. Convey, C. Parmesan, F. Bairlen, T. Beebee, J. M. Fromont, and O. Hoegh-Guldberg. 2002. Ecological responses to recent climate change. *Nature* 416:389–395. [10]

Waples, R. S. 1991. *Definition Of "Species" Under The Endangered Species Act: Application to Pacific Salmon*. NOAA Tech. Memorandum NMFS F/NWC–194. National Marine Fisheries Service, Seattle. [11]

Waples, R. S. 1995. Evolutionarily significant units and the conservation of biological diversity under the Endangered Species Act. In J. L. Nielsen (ed.), *Evolution and the Aquatic Ecosystem: Defining Unique Units in Population Conservation*, pp 8–27. American Fisheries Society, Bethesda, MD. [11]

Waples, R. S. 2002. Definition and estimation of effective population size in the conservation of endangered species. In S. R. Beissinger and D. R. McCullough (eds.), *Population Viability Analysis*, pp. 147–168. University of Chicago Press, Chicago. [11]

Waples, R. S. 2004. Salmonid insights into effective population size. In A. P. Hendry and S. C. Stearns (eds.), *Evolution Illuminated: Salmon and Their Relatives*, pp. 295–314. Oxford University Press, Oxford. [11]

Waples, R. S., D. J. Teel, J. Myers, and A. Marshall. 2004. Life history divergence in Chinook salmon: Historic contingency and parallel evolution. *Evolution* 58:386–403. [11]

Ward, S., P. Harrison, and O. Hoegh-Guldberg. 2000. Coral bleaching reduces reproduction of scleractinian corals and increases susceptibility to future stress. In M. K. Moosa, S. Soemodihardjo, A. Soegiarto, K. Romimohtarto, A. Nontji, Soekarno, and Suharsono (eds.), *Proceedings of the Ninth International Coral Reef Symposium*, pp 1123–1128. Bali. [13]

Ward, W. A. and J. B. Loomis 1986. The travel cost demand model as an environmental policy assessment tool: A review of the literature. *Western J. Agr. Econ.* 11:164–178. [5]

Wardell, A. G., P. Bernhardt, R. Bitner, A. Burquez, S. Buchmann, J. Cane, P. A. Cox, V. Dalton, P. Feinsinger, M. Ingram, D. Inouye, C. E. Jones, K. Kennedy, P. Kevan, H. Koopowitz, R. Medellin, S. MedellinMorales, G. P. Nabhan, B. Pavlik, V. Tepedino, P. Torchio, and S. Walker. 1998. The potential consequences of pollinator declines on the conservation of biodiversity and stability of food crop yields. *Conserv. Biol.* 12:8–17. [2]

Wardle D. A. and R. D. Bardgett. 2004. Human-induced changes in large herbivorous mammal density: The consequences for decomposers. *Front. Ecol. Environ.* 2:145–153. [8]

Warnock, G. J., 1971. *The Object of Morality*. Methuen, London. [4]

Warren, L. S. 1997. *The Hunter's Game: Poachers and Conservationists in Twentieth-Century America*. Yale University Press, New Haven, CT. [8]

Warren, M. S., J. K. Hill, J. A. Thomas, J. Asher, R. Fox, B. Huntley, D. B. Roy, M. G. Telfer, S. Jeffcoate, S. G. Willis, J. N. Greatorex-Davies, D. Moss, and C. D. Thomas. 2001. Climate versus habitat change: opposing forces underlie rapid changes to the distribution and abundance of British butterflies. *Nature* 414:65–69. [10]

Warren, R. S. and W. A. Niering. 1993. Vegetation change on a northeast tidal marsh: Interaction of sea-level rise and marsh accretion. *Ecology* 74:96–103. [10]

Waser, P. M. and C. Strobeck. 1998. Genetic signatures of interpopulation dispersal. *Trends Ecol. Evol.* 13:43–44. [11]

Wasser, S. K. 1996. Reproductive control in wild baboons measured by fecal steroids. *Biol. Reprod.* 55:393–399. [11]

Wasser, S. K., S. Papageorge, C. Foley, and J. L. Brown. 1996. Excretory fate of estradiol and progesterone in the African elephant (*Loxodonta africana*) and patterns of fecal steroid concentrations throughout the estrous cycle. *Gen. Comp. Endocr.* 102:255–262. [11]

Wasser, S. K., C. S. Houston, G. M. Koehler, G. G. Cadd, and S. R. Fain. 1997. Techniques for application of fecal DNA studies of Ursids. *Mol. Ecol.* 6:1091–1097. [11]

Wasser, S. K., A. M. Shedlock, K. Comstock, E. A. Ostrander, B. Mutayoba, and M. Stephens. 2004. Assigning African Elephant DNA to Geographic Region of Origin: Applications to the Ivory Trade. *Proc. Nat. Acad. Sci. USA* 101:14847–14852. [11]

Wasser, S. K., B. Davenport, E. R. Ramage, K. E. Hunt, M. Parker, C. Clarke, and G. Stenhouse. 2004. Scat detection dogs in wildlife research and management: Applications to grizzly and black bears in the Yellowhead Ecosystem, Alberta, Canada. *Can. J. Zool.* 82:475–492. [11]

Wasser, S. K., K. E. Hunt, J. L. Brown, K. Cooper, C. M. Crockett, U. Bechert, J. J. Millspaugh, S. Larson, and S. L. Monfort. 2000. A generalized fecal glucocorticoid assay for use in a diverse array of non-domestic mammalian and avian species. *Gen. Comp. Endocr.* 120:260–275. [11]

Watchman, L., M. Groom, and J. Perrine. 2001. Science and uncertainty in habitat conservation planning. *Am. Sci.* 89:351–360. [18]

Wathne, F. 1959. Summary report of exploratory long-line fishing for tuna in Gulf of Mexico and Caribbean Sea, 1954–1957. *Commer. Fish. Rev.* 21:1–26. [8]

Watkinson, A. R. and W. J. Sutherland. 1995. Sources, sinks, and pseudo-sinks. *J. Anim. Ecol.* 64:126–130. [12]

Watkinson, A. R., J. A. Gill, and M. Hulme. 2005. Flying in the face of climate change: a review of climate change, past, present and future. *Ibis.* In press. [10]

Watling, L. and E. A. Norse. 1998. Disturbance of the seabed by mobile fishing gear: a comparison to forest clearcutting. *Conserv. Biol.* 12:1180–1197. [6]

Watt, A. S. 1947. Pattern and process in the plant community. *J. Ecol.* 35:12–22. [7]

WDPA Consortium 2004. World Database on Protected Areas. World Conservation Union (IUCN) and UNEP-World Conservation Monitoring Centre (UNEP–WCMC). [14]

Weart, S. R. 2003. *The Discovery of Global Warming.* Harvard University Press, Cambridge, MA. [10]

Weber, M. L. and B. Heneman. 2000. *Guide to California's Marine Life Management Act.* Common Knowledge Press, Bolinas, California. [13]

Wee, H. 2002. "Can oil giants and green energy mix?" *Business Week Online,* 25 September, 2002. [10]

Weekley, C. W. and E. S. Menges. 2003. Species and vegetation responses to prescribed fire in a long-unburned, endem-ic-rich Lake Wales Ridge scrub. *J. Torrey Bot. Soc.* 130:265–282. [18]

Weekley, C., T. L. Kubisiak, and T. Race. 2002. Genetic impoverishment and cross-incompatibility in remnant genotypes of *Ziziphus celata* (Rhamnaceae), a rare shrub endemic to the Lake Wales Ridge, Florida. *Biodivers. Conserv.* 11:2027–2046. [18]

Weerakoon, D. K. 1999. *Ecology and Ranging Behaviour of Wild Elephants and Human Elephant Conflict in Northwestern Region.* Unpublished Report. Department of Wildlife Conservation, Sri Lanka. [17]

Weisbrod, B. 1964. Collective consumption services of individual consumption goods. *Quart. J. Econ.* 78:471–77. [5]

Weiser, W. (ed.). 1973. *Effects of Temperature on Ectothermic Organisms.* Springer-Verlag, New York. [10]

Weitkamp, L. A., T. C. Wainwright, G. J. Bryant, G. B. Milner, D. J. Teel, R. G. Kope, and R. S. Waples. 1995. *Status Review of Coho Salmon from Washington, Oregon, and California.* U.S. Dept. Commer., NOAA Tech. Memo. NMFS-NWFSC-24. [17]

Welcomme, R. L., 1988. International introductions of inland aquatic species. Food and Agriculture Organization (FAO) *Fisheries Technical Papers* 294. FAO, Rome. [9]

Weldon, C., L. H. du Preez, A. D. Hyatt, R. Muller, and R. Speare. 2004. Origin of the amphibian chytrid fungus. *Emerg. Infec. Dis.* 10:2100–2105. [3]

Weller, G. and M. Lange. 1999. *Impacts of Global Climate Change in the Arctic Regions.* In International Arctic Science Committee, Center for Global Change and Arctic System Research, pp. 1–59. University of Alaska, Fairbanks. [10]

Weller, M. W. 1995. Use of two waterbed guilds as evaluation tools for the Kissimmee River restoration. *Restor. Ecol.* 3: 211–224. [15]

Wennergren, U., M. Ruckelshaus, and P. Kareiva. 1995. The promise and limitations of spatial models in conservation biology. *Oikos* 74:349–356. [12]

Werner, E. E., G. G. Mittlebach, D. J. Hall, and J. F. Gilliam. 1983. Experimental tests of optimal habitat use in fish: The role of relative habitat profitability. *Ecology* 64:1525–1539. [12]

West, J. M. and R. V. Salm. 2003. Resistance and resilience to coral bleaching: Implications for coral reef conservation and management. *Conserv. Biol.* 17:956–967. [13]

Westaway, D., G. A. Carlson, and S. B. Prusiner. 1995. On safari with PrP: Prion diseases of animals. *Trends Microbiol.* 3:141–147. [3]

Westemeier, R. L., J. D. Brawn, S. A. Simpson, T. L. Esker, R. W. Jansen, J. W. Walk, E. L. Kershner, J. L. Bouzat, and K. N. Paige. 1998. Tracking the long-term decline and recovery of an isolated population. *Science* 282:1695–1698. [11]

Westing, A. 1992. Protected natural areas and the military. *Environ. Conserv.* 19:343–348. [6]

Westmacott, S., K. Teleki, S. Wells, and J. M. West. 2000. *Management of Bleached and Severely Damaged Coral Reefs.* IUCN, Washington, D.C. [13]

Westoby, M., K. French, L. Hughes, B. Rice, and L. Rodgerson. 1991. Why do more plant species use ants for dispersal on infertile compared with fertile soils? *Aust. J. Ecol.* 16:445–455. [2]

Westphal, C., I. Steffan-Dewenter, and T. Tscharntke. 2003. Mass flowering crops enhance pollinator densities at a landscape scale. *Ecol. Lett.* 6:961–965. [2]

Westrum, R. 1994. An organizational perspective: designing recovery teams from the inside out. In T. W. Clark, R. P. Reading, and A. L. Clarke (eds.), *Endangered Species Recovery. Finding the Lessons, Improving the Process,* pp. 327-349. Island Press, Washington, D.C. [13]

Weygoldt, P. 1989. Changes in the composition of mountain stream frog communities in the Atlantic Mountains of Brazil: Frogs as indicators of environmental deteriorations? *Stud. Neotrop. Fauna E.* 243:249–255. [3]

Wheelwright, N. T. 1983. Fruits and the ecology of resplendent quetzals. *Auk* 100: 286–301. [7]

Whisenant, S. G. 1999. *Repairing Damaged Wildlands: A Process-oriented, Landscape-scale Approach.* Cambridge University Press, Cambridge. [15]

Whitcomb, R. F., C. S. Robbins, J. F. Lynch, B. L. Whitcomb, M. K. Klimkiewicz, and D. Bystrak. 1981. Effects of forest fragmentation on avifauna of the eastern deciduous forest. In R. L. Burgess and D. M. Sharpe (eds.), *Forest Island Dynamics in Man-Dominated Landscapes,* pp. 125–213. Springer-Verlag, New York. [7]

White G. C. and K. P. Burnham. 1999. Program MARK: Survival estimation from populations of marked animals. *Bird Study* (Suppl.) 46:120–138. [11]

White, G. C., D. R. Anderson, K. P. Burnham, and D. L. Otis. 1982. *Capture-Recapture and Removal Methods for Sampling Closed Populations.* Los Alamos National Laboratory, LA 8787-NERP, Los Alamos, New Mexico. [11]

White, L. W. 1967. The historical roots of our ecologic crisis. *Science* 155:1203–1207. [4]

White, M. A. and D. J. Mladenoff. 1994. Old-growth forest landscape transitions from pre-European settlement to present. *Landscape Ecol.* 9:191–205. [7]

White, P. S. and J. L. Walker. 1997. Approximating nature's variation: selecting and using reference information in restoration ecology. *Restor. Ecol.* 5:338–349. [15]

White, R. P., S. Murray, and M. Rohweder. 2000. *Pilot Assessment of Global Ecosystems: Grassland Ecosystems.* World Resources Institute: Washington, D.C. [6, 7, 9]

White, T. H., J. A. Collazo, and F. J. Vilella. 2005. Survival of Captive-reared Puerto Rican Parrots released in the Caribbean National Forest. *Condor* 107:424–432. [15]

Whiteman, H. H. and S. A. Wissinger. 2005. Multiple hypotheses for population fluctuations: The importance of long-term

data sets for amphibian conservation. In M. L. Lanoo (ed.), *Status and Conservation of U.S. Amphibians*. University of California Press, Berkeley. In press. [3]

Whitlock, M. C. 2000. Fixation of new alleles and extinction of small populations: Drift load, beneficial alleles, and sexual selection. *Evolution* 54:1855–1861. [12]

Whitlock, M. C. and D. E. McCauley. 1999. Indirect measures of gene flow and migration: *Heredity* 82:117–125. [11]

Whitlock, M. C. and N. H. Barton. 1997. The effective size of a subdivided population. *Genetics* 146:427–441. [11]

Whittaker, R. H. 1956. Vegetation of the Great Smoky Mountains. *Ecol. Monogr.* 26:1–80. [7]

Whittaker, R. H. 1960. Vegetation of the Siskiyou Mountains, Oregon and California. *Ecol. Monogr.* 30:279–338. [2]

Whittaker, R. H. 1970. *Communities and Ecosystems*. Macmillan, New York. [2]

Wiedemann, A. M. and A. Pickart. 1996. The *Ammophila* problem on the northwest coast of North America. *Landscape Urban Plan.* 34:287–299. [9]

Wiens, J. A. 1977. On competition and variable environments. *Am. Sci.* 65:590–597. [2]

Wiens, J. A. 1985. Habitat selection in variable environments: shrub-steppe birds. In M. L. Cody (ed.), *Habitat Selection in Birds*, pp. 227–251. Academic Press, San Diego.

Wiens, J. A. 1989. *The Ecology of Bird Communities*. Vol. 2. Processes and Variations. Cambridge University Press, New York. [7]

Wiersma Y. F., T. D. Nudds, and D. H. Rivard. 2004. Models to distinguish effects of landscape patterns and human population pressures associated with species loss in Canadian national parks. *Landscape Ecol.* 19:773–786. [12]

Wilcove, D. S., C. H. McLellan, and A. P. Dobson. 1986. Habitat fragmentation in the temperate zone. In M. E. Soulé (ed.), *Conservation Biology: The Science of Scarcity and Diversity*, pp. 237–256. Sinauer Associates, Sunderland MA.

Wilcove, D. S., D. Rothstein, J. Dubow, A. Phillips, and E. Losos. 1998. Quantifying threats to imperiled species in the United States. *BioScience* 48:607–615. [3, 7, 9]

Wilcox, B. A. and D. D. Murphy. 1985. Conservation strategy: The effects of fragmentation on extinction. *Am. Nat.* 125:879–887. [7, 15]

Wiley, J. W., R. S. Gnam, S. E. Koenigh, A. Dornelly, X. Galvez, P. E. Bradley, T. White, M. Zamore, P. Reillo, and D. Anthony. 2004. Status and conservation of the family Psittacidae in the West Indies. *J. Caribbean Ornith.* (Special Issue Honoring Nedra Klein):94–154. [15]

Wilkie, D. S. and J. F. Carpenter. 1999. Bushmeat hunting in the Congo Basin: An assessment of impacts and options for mitigation. *Biodivers. Conserv.* 8:927–955. [8]

Wilkie, D. S. and J. T. Finn. 1990. Slash-burn cultivation and mammal abundance in the Ituri Forest, Zaïre. *Biotropica* 22:90–99. [18]

Wilkie, D. S., E. L. Bennett, M. Starkey, K. Abernethy, R. Fotso, F. Maisels, and P. Elkan. In Press. If trade in bushmeat is legalized, can the laws be enforced and wildlife survive in Central Africa?: Evidence from Gabon. *Journal of International Wildlife Law and Policy*. [8]

Wilkinson, C. F. 1992. *Crossing the Next Meridian: Land, Water, and the Future of the West*. Island Press, Washington, D.C. and Covelo, CA. [13]

Wilkinson, C. 1998. *Status of Coral Reefs of the World: 1998*. Australian Institute of Marine Science, Queensland, Australia. [13]

Wilkinson, C. 2000. *Status of Coral Reefs of the World: 2000*. Australian Institute of Marine Science, Queensland, Australia. [13]

Wilkinson, C. 2002. *Status of Coral Reefs of the World: 2002*. Australian Institute of Marine Science, Townsville, Australia. [13]

Williams, D. W. and A. M. Liebhold. 1995. Herbivorous insects and global change: Potential changes in the spatial distribution of forest defoliator outbreaks. *J. Biogeogr.* 22:665–671. [10]

Williams, J. D. and R. J. Neves. 2004. Freshwater mussels: A neglected and declining aquatic resource. http://biology.usgs.gov/s+t/frame/f076.htm#15584. [8]

Williams, J. D., M. L. Warren Jr., K. S. Cummings, J. L. Harris, and R. J. Neves. 1993. Conservation status of freshwater mussels of the United States and Canada. *Fisheries* 18:6–22. [8]

Williams, J. E. and C. D. Williams. 1996. An ecosystem based approach to management of salmon and steelhead habitat. In D. J. Stouder, P. A. Bisson, and R. J. Naiman (eds.), *Pacific Salmon and Their Ecosystems: Status and Future Options*, pp. 541–556. Chapman and Hall, New York. [13]

Williams, J. E. and J. N. Rinne. 1992. Biodiversity management on multiple-use federal lands: An opportunity whose time has come. *Fisheries* 17(3):4–5. [13]

Williams, M. 2003. *Deforesting the Earth: From Prehistory to Global Crisis*. University of Chicago Press, Chicago. [4]

Williams, S. E., and J.-M. Hero. 1998. Rainforest frogs of the Australian wet tropics: Guild classification and the ecological similarity of declining species. *Proc. R. Soc. Lon. Ser. B* 265:597–602. [3]

Williams-Linera, G. 1990. Vegetation structure and environmental conditions of forest edges in Panama. *J. Ecol.* 78:356–373.

Williamson, K. 1975. Birds and climate change. *Bird Study* 22:143–164. [10]

Williamson, M. 1996. *Biological Invasions*. Chapman & Hall, London. [10]

Williamson, M. and A. Fitter. 1996. The varying success of invaders. *Ecology* 77:1661–1666. [9]

Willig, M. R., D. M. Kaufman, and R. D. Stevens. 2003. Latitudinal gradients of biodiversity: Pattern, process, scale and synthesis. *ARES* 34:273–309. [2]

Willingham, E. and D. Crews. 1999. Sex reversal effects of environmentally relevant xenobiotic concentrations on the red-eared slider turtle, a species with temperature-dependent sex determination. *Gen. Comp. Endocr.* 113:429–435. [6]

Willingham, E., T. Rhen, J. T. Sakata, and D. Crew. 2000. Embryonic treatment with xenobiotics disrupts steroid hormone profiles in hatchling red-eared slider turtles (*Trachemys scripta elegans*). *Environ. Health Persp.* 108:329–332. [6]

Willis, C. K., R. M. Cowling, and A. J. Lombard. 1996. Patterns of endemism in the limestone flora of South African lowland fynbos. *Biodivers. Conserv.* 5:55–74. [2]

Willis, E. O. 1974. Populations and local extinctions of birds on Barro Colorado Island, Panama. *Ecol. Monogr.* 44:153–169. [7]

Willson, M. F., S. M. Gende, and B. H. Marston. 1998. Fishes and the forest. *BioScience* 48:455–462. [3]

Wilson, E. O. 1984. *Biophilia*. Harvard University Press, Cambridge, MA. [1, 4]

Wilson, E. O. 1985. The biological diversity crisis. *BioScience* 35:700–706. [4]

Wilson, E. O. 1989. The current state of biological diversity. In E. O. Wilson (ed.), *Biodiversity*, pp. 3–18. National Academy Press, Washington, D.C. [1]

Wilson, E. O. 1992. *The Diversity of Life*. Belknap Press, Cambridge, MA. [2, 3, 15]

Wilson, E. O. 2000. On the future of conservation biology. *Conserv. Biol.* 14:1–3. [15]

Wilson, E.O. 2002. *The Future of Life*. Knopf Press, New York. [1]

Wilson, E. O. and E. O. Willis. 1975. Applied biogeography. In M. L. Cody and J. M. Diamond (eds.), *Ecology and Evolution of Communities*, pp. 522–534. Belknap Press of Harvard University Press, Cambridge, MA. [7]

Wilson, G. A. and B. Rannala. 2003. Bayesian inference of recent migration rates using multilocus genotypes. *Genetics* 163:1177–1191. [11]

Wilson, J. B., G. L. Rapson, M. T. Sykes, A. J. Watkins, and P. A. Williams. 1992. Distributions and climatic correlations of some exotic species along roadsides in South Island, New Zealand. *J. Biogeogr.* 19:83–193. [7]

Wilson, K. A., R. L. Pressey, A. N. Newton, M. A. Burgman, H. P. Possingham, and C. J. Weston. 2005. Measuring and incorporating vulnerability into conservation planning. *Environ. Manage.* 35:527–543. [14]

Winter, T. C., J. W. Harvey, O. Lehn Franke, and W. M. Alley. 1998. *Ground water and surface water: A single resource*. US Geological Survey Circular 1139. [7]

Wipfli, M. S., J. Hudson, and J. Caouette. 1998. Influence of salmon carcasses on stream productivity: Response of biofilm and benthic macroinvertebrates in southeastern Alaska, USA. *Can. J. Fish. Aquat. Sci.* 55:1503–1511. [8]

Wipfli, M. S., J. Hudson, J. Caouette, and D. T. Chaloner. 2003. Marine subsidies in freshwater ecosystems: Salmon carcasses increase the growth rates of stream-resident salmonids. *T. Am. Fish. Soc.* 132:371–381. [8]

Wissinger, S. A. and H. H. Whiteman. 1992. Fluctuation in a Rocky Mountain population of salamanders: Anthropogenic acidification or natural variation. *J. Herpetol.* 26:377–391. [3]

With, K. A. 1997. The application of neutral landscape models in conservation biology. *Conserv. Biol.* 11:1069–1080. [12]

With, K. A. and A. W. King. 1999. Extinction thresholds for species in fractal landscapes. *Conserv. Biol.* 13:314–326. [12]

With, K. A. and A. W. King. 2001. Analysis of landscape sources and sinks: the effect of spatial pattern on avian demography. *Biol. Conserv.* 100:75–88. [12]

Witte, F., B. S. Msuku, J. H. Wanink, O. Seehausen, E. F. B. Katunzi, P. C. Goudswaard, and T. Goldschmidt. 2000. Recovery of cichlid species in Lake Victoria: An examination of factors leading to differential extinction. *Reviews in Fish Biology and Fisheries* 10:233–241. [9]

Woese, C. R., O. Kandler, and M. L. Wheelis. 1990. Towards a natural system of organisms: Proposal for the domains Archaea, Bacteria, and Eucarya. *Proc. Nat. Acad. Sci. USA* 87:4576–4579. [2]

Wolfe, N. D., W. M. Switzer, J. K. Carr, V. B. Bhullar, V. Shanmugam, U. Tamoufe, A. T. Prosser, J. N. Torimiro, A. Wright, E. Mpoudi-Ngole, F. E. McCutchan, D. L. Birx, T. M. Folks, D. S. Burke, and W. Heneine. 2004. Naturally acquired simian retrovirus infections in central African hunters. *The Lancet* 363:932–937. [8]

Wolfe, S. H., J. A. Reidenauer, and D. B. Means. 1988. *An Ecological Characterization of the Florida Panhandle.* Biological Report 88(12). U.S. Fish and Wildlife Service, National Wetlands Research Center, Slidell, LA. [7]

Wondolleck, J. M. and S. L. Yaffee. 2000. *Making Collaboration Work: Lessons from Innovation in Natural Resource Management.* Island Press, Washington, D.C. [13]

Wong, M. H. and A. D. Bradshaw. 2002. Restoration Policy and Infrastructure: China: Progress in the Reclamation of Degraded Land. In M. R. Perrow and A. J. Davy (eds.), *Handbook of Ecological Restoration, Volume 2: Restoration in Practice*, pp. 89–98. Cambridge University Press, Cambridge. [15]

Wood, S., K. Sebastian, and S. J. Scherr. 2000. *Pilot Assessment of Global Ecosystems: Agroecosystems.* International Food Policy Research Institute and World Resources Institute, Washington, D.C. [6]

Woodhill, J. 2000. *Planning, Monitoring and Evaluating Programmes and Projects: An Introduction to Key Concepts, Approaches, and Terms.* IUCN, Gland, Switzerland.

Woodruff, D. S. 1992. *Biodiversity: Conservation and Genetics.* Proceedings of the Second Princess Chulabhorn Science Congress, Bangkok. [11]

Woodroffe, R. and J. R. Ginsberg. 1998. Edge effects and the extinction of populations inside protected areas. *Science* 280:2126–2128. [7]

Woods, J. G., D. Paetkau, D. Lewis, B. N. McLellan, M. Proctor, and C. Strobeck. 1999. Genetic tagging of free-ranging black and brown bears. *Wildlife Soc. B.* 27:616–627. [11]

Woods, W. I. 2004. Population nucleation, intensive agriculture, and environmental degradation: The Cahokia example. *Agriculture and Human Values* 21:255–261. [6]

Woodward, F. I. 1987. *Climate and plant distribution.* Cambridge University Press, Cambridge. [10]

Woolnough, A. P. and C. N. Johnson. 2000. Assessment of the potential for competition between two sympatric herbivores—the northern hairy-nosed wombat, *Lasiorhinus krefftii*, and the eastern grey kangaroo, *Macropus giganteus.* *Wildlife Res.* 27:301–308. [14]

Wootton, J. T. 1994. The nature and consequences of indirect effects in ecological communities. *Ann. Rev. Ecol. Syst.* 25:443–466. [9]

Wootton, J. T. and D. A. Bell. 1992. A metapopulation model of peregrine falcon in California: Visibility and management strategies. *Ecol. Appl.* 2:307–321. [12]

World Bank 1996. *Toward environmentally sustainable development in sub-saharan Africa: A World Bank Agenda.* The World Bank, Washington D.C. [16]

World Bank. 2005. Water Resources Management. http://lnweb18.worldbank.org/ESSD/ardext.nsf/18ByDocName/WaterResourcesManagement. [18]

World Commission of Forests and Sustainable Development. 1998. Final Report on Forest Capital. Cambridge University Press, Cambridge. [8]

World Commission on Environment and Development. 1987. *Our Common Future: Report of the World Commission on Environment and Development.* Oxford University Press, Oxford. [5, 16]

World Conservation and Monitoring Centre. 1992. *Global Biodiversity: State of the Earth's Living Resources.* Chapman & Hall, London. [2]

World Population Prospects. 2002. *The 2002 Revision Global Database.* UN World Population Division. [1]

World Resources Institute. 1996. *World Resources 1996–1997: The Urban Environment.* Oxford University Press, Oxford. [15]

World Resources Institute 1998. *Fragmenting Forests: The Loss of Large Frontier Forests.* World Resources Institute, Washington, D.C. [8]

World Resources Institute. 2003. *Ecosystems and Human Well-being: A Framework for Assessment.* Island Press, Washington, D.C. [18]

World Tourism Organization. 2004. *Tourism 2020 Vision* [16]

World Travel and Tourism Council. 2004. http://www.wttc.org/2004tsa/PDF/Executive%20Summary.pdf [16]

World Wildlife Fund for Nature. 2004. *Living Planet Report 2004.* World Wildlife Fund for Nature, Gland, Switzerland. www.unep.org. [1, 18]

Worster, D. 1979. *Nature's Economy. The Roots of Ecology.* Anchor Books, New York. [4]

Wright, A. H., and A. A. Wright 1949. *Handbook of Frogs and Toads of the United States And Canada.* Comstock Publishing Associates, Ithaca, NY. [3]

Wright, J. P., C. G. Jones, and A. S. Flecker. 2002. An ecosystem engineer, the beaver, increases species richness at the landscape scale. *Oecologia* 132:96–101. [8, 9]

Wright, R. G. and D. J. Mattson. 1996. The origin and purpose of national parks and protected areas. In R. G. Wright, (ed.), *National Parks and Protected Areas: Their Role in Environmental Protection*, pp. 3–14. Blackwell Science, Cambridge, MA. [14]

Wright, R. G., J. M Scott, S. Mann, and M. Murray 2001. Identifying unprotected and potentially at risk plant communities in the western United States. *Biol. Conserv.* 98:97–106. [14]

Wright, S. 1922. Coefficients of inbreeding and relationship. *Am. Nat.* 56:330–338. [11]

Wright, S. 1965. The interpretation of population genetic structure by *F*-statistics with special regard to systems of mating. *Evolution* 19:395–420. [11]

Wright, S. 1969. *Evolution and the Genetics of Populations.* Vol. 2. *The Theory of Gene Frequencies.* University of Chicago Press. [11]

Wright, S. 1978. *Evolution and the Genetics of Populations*: Vol. 4. *Variability within and among populations.* University of Chicago Press. [11]

Wright, S. J., H. Zeballos, I. Dominguez, M. M. Gallardo, M. C. Moreno, and R. Ibáñez. 2000. Poachers alter mammal abundance, seed dispersal and seed predation in a Neotropical forest. *Conserv. Biol.* 14:227–239. [3, 8]

WSSD. 2002. *World Summit on Sustainable Development (WSSD) Plan of Implementation.* Johannesburg, South Africa. [13]

Wunder, M. 1997. Of elephants and men: crop destruction, CAMPFIRE, and wildlife management in the Zambezi Valley, Zimbabwe. Ph.D. dissertation, University of Michigan, Ann Arbor. [16]

Yaffe, M. 2001. *Judaism and Environmental Ethics.* Lexington Books, Lanham, MD. [4]

Yaffee, S. L. 1982. *Prohibitive Policy: Implementing the Federal Endangered Species Act.* MIT Press, Cambridge, MA. [17]

Yan, N. D., W. Keller, N. M. Scully, D. R. S. Lean, and P. J. Dillon. 1996. Increased UV-B penetration in a lake owing to drought-induced acidification. *Nature* 381:141–143. [3]

Yang, F. L. and M. E. Schlesinger. 2002. On the surface and atmospheric changes following the 1991 Pinatubo volcanic eruption: A GCM study. *J. Geophys. Res.-Atmos.* 107 (D8):4073 [10]

Ylitalo, G. M., C. O. Matkin, J. Buzitis, M. M. Krahn, L. L. Jones, T. Rowles, and J. E. Stein. 2001. Influence of life-history parameters on organochlorine concentrations in free-ranging killer whales (*Orcinus orca*) from Prince William Sound, AK. *Sci. Total Environ.* 281:183–203. [17]

Yoakum, J. and W. P. Dasmann. 1971. Habitat manipulation practices. In R. H. Giles, (ed.), *Wildlife Management Techniques*, pp. 173–231. The Wildlife Society, Washington, D.C. [7]

Yoshiyama, R. M., F. W. Fisher, and P. B. Moyle. 1998. Historical abundance and decline of Chinook salmon in the Central Valley region of California. *N. Am. J. Fish. Manage.* 18:487–521. [17]

Young, B. E. and D. B. McDonald. 2000. Birds. In N. M. Nadkarni and N. T. Wheelwright (eds.), *Monteverde: Ecology and Conservation of a Tropical Cloud Forest*, pp. 179–222. Oxford University Press, New York. [10]

Young, B. E., S. N. Stuart, J. S. Chanson, N. A. Cox, and T. M. Boucher. 2004. *Disappearing Jewels: The Status of New World Amphibians*. NatureServe, Arlington, Virginia. [3]

Young, B. E., K. R. Lips, J. K. Reaser, R. Ibanez, A. W. Salas, J. R. Cedeno, L. A. Coloma, S. Ron, E. La Marca, J. R. Meyer, A. Munoz, F. Bolanos, G. Chaves, and D. Romo. 2001. Population declines and priorities for amphibian conservation in Latin America. *Conserv. Biol.* 15:1213–1223. [3]

Young, C. C. and E. S. Menges. 1999. Postfire gap-phase regeneration in scrubby flatwoods on the Lake Wales Ridge. *Florida Scientist* 62:1–12. [18]

Young, K. M., S. L. Walker, C. Lanthier, W. T. Waddell, S. L. Monfort, and J. L. Brown. 2004. Noninvasive monitoring of adrenocortical activity in carnivores by faecal glucocorticoid analyses. *Gen. Comp. Endocr.* 137:148–165. [11]

Young, L. G. L. 1997. Sustainability issues in the trade for wild and cultured aquarium species. In B. Faust and J. Peters (eds), *Marketing and Shipping Live Aquatic Products*, pp. 145–151. Northeastern Regional Agricultural Engineering Service Cooperative Extension, Ithaca, New York. [8]

Young, T. P. 2000. Restoration ecology and conservation biology. *Biol. Conserv.* 92:73–83. [15]

Youngquist, W. 1999. The post-petroleum paradigm—and population. *Popul. Environ.* 20:297–31. [18]

Zabel, C. J., J. R. Dunk, H. B. Stauffer, L. M. Roberts, B. S. Mulder, and A. Wright. 2003. Northern Spotted Owl habitat models for research and management application in California (USA). *Ecol. Appl.* 13:1027–1040. [12]

Zahawi, R. A. 2005. Establishment and growth of living fence species: an overlooked tool for the restoration of degraded areas in the tropics. *Restor. Ecol.* 13:1–11. [15]

Zahler, P. 2003. Top-down meets bottom-up: conservation in a post-conflict world. *Conservation in Practice* 4:23–29. [6]

Zanette, L., P. Doyle, and S. M. Trémont. 2000. Food shortage in small fragments: evidence from an area-sensitive passerine. *Ecology* 81:1564–1666. [7]

Zedler, J. B. 1993. Canopy architecture of natural and planted cordgrass marshes: selecting habitat evaluation criteria. *Ecological Applications*. 3: 123–138. [15]

Zedler, J. B. 2000. Progress in wetland restoration ecology. *Trends in Ecology and Evolution* 15:402–407. [15]

Zedler, J. B., (ed.). 2001. *Handbook for Restoring Tidal Wetlands*. Marine Science Series, CRC Press LLC, Boca Raton. Florida. [15]

Zedler, J. B. and J. C. Callaway. 2000. Evaluating the progress of engineered tidal wetlands. *Ecological Engineering* 15:211–225. [15]

Zedler, J. B., J. C. Callaway, and G. Sullivan. 2001. Declining biodiversity: Why species matter and how their functions might be restored. *BioScience* 51:1005–1017. [15]

Zedler, P. H. 1987. The Ecology of Southern California Vernal Pools: A Community Profile. *U.S. Fish and Wildlife Service Biological Report 85.* [15]

Zeleny, M. 1976. *Multiple Decision Decision-Making.* Springer-Verlag, Berlin. [5]

Zellmer, S. B. and S. A. Johnson. 2002. Biodiversity in and around McElligot's pool. *Idaho L. Rev.* 38:473–504. [17]

Zhou, L., C. J. Tucker, R. K. Kaufmann, D. Slayback, N. V. Shabanov, and R. B. Myneni. 2001. Variation in northern vegetation activity inferred from satellite data of vegetation index during 1981–1999. *J. Geophys. Res.* 106:20069–20083. [10]

Zöckler, C. and I. Lysenko. 2000. *Waterbirds on the edge: First circumpolar assessment of climate change impact on arctic breeding waterbirds.* World Conservation Press, Cambridge. [10]

Zwiers, F. W. and X. Zhang. 2003. Towards regional-scale climate change detection. *J. Climate* 16:793–797. [10]

Zwölfer, H. 1973. Competition and coexistence in phytophagous insects attacking the heads of *Carduus nutans* L. In *Proceedings II International Symposium Biological Control of Weeds*, pp. 74–80. Rome, Italy. [9]

Zwölfer, H. and P. Harris. 1984. Biology and host specificity of *Rhinocyllus conicus* (Frol.) (Col. Curculionidae), a successful agent for biocontrol of the thistle, *Carduus nutans* L. *Z. Angew. Entomol.* 97:36–62. [9]

Index